CONWAY'S

ALL THE WORLD'S FIGHTING SHIPS 1947-1995

USS *Constellation* in 1995. Since 1947, the fleet aircraft carriers of the United States Navy have been the epitome of naval power, and of global power projection. RAN

CONWAY'S

ALL THE WORLD'S FIGHTING SHIPS 1947-1995

CONWAY MARITIME PRESS

Editorial Director
ROBERT GARDINER
Editor
STEPHEN CHUMBLEY
Contributors
PRZEMYSŁAW BUDZBON
STEPHEN CHUMBLEY
NORMAN FRIEDMAN
ERIC GROVE
ANDRZEJ M JASKULA
JOHN JORDAN
GERHARD KOOP
HUGH LYON
ANTHONY PRESTON
ROBERT L SCHEINA
ADAM ŚMIGIELSKI
IAN STURTON
MAREK TWARDOWSKI
KARL-ERIC WESTERLUND
NB Contributors credited on contents page, with their initials following their respective sections. 1982 contributor given first, 1995 contributor second
Line drawings
By Przemysław Budzbon (with the assistance of Marek Twardowski), John Jordan and Ian Sturton
Book design
By Roger Lightfoot
Index
By Geoffrey Meadon

This single volume revised edition
first published in 1995 by Conway Maritime Press
An imprint of Brassey's (UK) Ltd
33 John Street, London WC1N 2AT

British Library Cataloguing in Publication Data

Conway's All the World's Fighting Ships.–
1947–1995.- 2Rev. ed
I. Gardiner, Robert
623.8250904

ISBN 0 85177 605 1

Typeset by Dorwyn Ltd, Rowlands Castle, Hants
Printed in Great Britain by The Bath Press, Avon

Contents

Foreword

This volume updates, extends and replaces the two-part *Conway's All the World's Fighting Ships 1947–1982.* The foreword to the first edition explained that the particular aim of the book was to provide a coherent overview of the postwar naval revolution, in both its technological and political aspects. It pointed out that despite the suspicions and secrecy of the Cold War era, some information was then becoming available – certainly enough to make sense of the vitally important fifteen years after 1945, and sufficient to suggest the rationale for developments during the following two decades. What made *Fighting Ships 1947–1982* different from the contemporary yearbooks, which can only provide a snapshot of the existing situation, was its intention to describe the whole period and to suggest the main driving forces behind the navies and their ships.

Since that was written, the Cold War has ended, forcing many navies to reassess their principal roles, while the disintegration of the Soviet Union has allowed a flood of new information to reach the West. The political geography has altered beyond recognition, but if anything this has enhanced the need for a work like *Fighting Ships* which can now look at the whole postwar era in the light of significantly more and better data.

So great have been the changes since 1982 that a simple appendix volume was out of the question, and so for this edition every entry from Albania to Zimbabwe has been thoroughly scrutinised and, where necessary, revised. The old Cold War division between East and West has been abandoned, reflecting the new world order, and countries are simply listed alphabetically. However, this volume covers the years from 1947 to the present, so has to account for political changes to the boundaries, status or even the names of many countries, without duplication but with the maximum ease of reference. In general, the running order and titles attempt to reflect the political reality for the majority of the period: for example, reunified countries like Germany and Vietnam have separate entries up to the point of amalgamation, and the dominant state thereafter; where countries have recently gained independence from larger entities – the successor states to the Soviet Union being the obvious case – the new navies are listed after that of the original power. In terms of the origins of particular warship classes, this seems most logical, but readers unsure of the location of any navy should consult the Contents page, where all are listed alphabetically with appropriate cross-references.

Wherever possible, each section has been revised and augmented by the original contributor to the first edition, but the circumstances of individuals as well as nations have altered in the intervening years, and some were not available. This has made many countries collaborative ventures, which makes the task of justly crediting the work somewhat complicated – especially as some countries were wrongly attributed in the first edition. For this volume, each country listed on the Contents page is followed by two sets of initials, denoting the original contributor(s) and then the person who carried out the revision. They can be identified by reference to the Contributor's list opposite the Contents page.

SCOPE

In terms of coverage, this volume follows the pattern of its predecessors, concentrating specifically on *fighting ships*. Ideally, fleet support and replenishment units should have been covered in deference to their enhanced postwar importance, but shortage of space made this impossible. Therefore, only a few auxiliaries with specific combat value have been included, and non-naval maritime forces (coast guard, army, police or fishery protection services) have been omitted unless they enjoy a close relationship (like the US Coast Guard does) with the national navy. Otherwise, the coverage is both broad and deep, extending down to minor patrol and landing craft in even the smallest navies.

ORGANISATION

This will be familiar to readers of previous volumes, and follows a standard order for each country: a general introduction, followed by a statement of fleet strength in 1947, then the post-1947 classes in type and chronological order. However, since the war the traditional typology has become confused, so this volume has adopted six basic type-divisions – Major Surface Ships (aircraft carriers, cruisers, destroyers and frigates), Submarines, Amphibious Warfare Vessels, Small Surface Combatants (corvettes to patrol craft), Mine Warfare Vessels and Miscellaneous. The ships of each navy are presented in this order, but within the main divisions classes are grouped in a manner approximating to the traditional hierarchy (carriers before cruisers, cruisers before destroyers, etc) with the oldest classes first. The names of the classes in the headings to the tables are followed by a description *in italics*, but these are not necessarily official descriptions of ship-types.

The data given in each table follows the style of the earlier volumes, with the addition of an entry for Sensors. This gives the designations of the main radar and sonar sets, and in some cases, fire control systems.

The remaining data is largely self-explanatory, except for the following points:

1. Where two tonnages and speeds are quoted for submarines, the first applies to the surface condition and the second as submerged.

2. Gun armament is given in the traditional form, namely: number of barrels – size of gun/calibre (number of mountings × number of barrels per mount). For example, '6–3in/70 (3×2)' means six 70-calibre 3in guns disposed in three twin mountings. For missiles, where the number of independent systems is the significant factor, it is the number of launchers which is quoted: for example, '2 Sea Sparrow SAM (2×8)' means that each of the two launchers has eight tubes, but there are not sixteen individual systems.

In order not to duplicate information, the basic details of all purchased or transferred warships are given under the country of origin, and this also applies to warships built for export where the design originated with the navy of the shipbuilding country concerned. Only variations from the standard type are noted in the data tables for each recipient navy. For this reason, such classes are referred to in terms of the original design – for example, as 'ex-US *Adjutant* class minesweepers' where transferred, or as 'US *Adjutant* type minesweepers' when built specifically for export. Similarly, for all transferred classes the date of acquisition is quoted in the building table, the original construction dates being available under the country of origin.

ILLUSTRATIONS

Classes of frigate-size and upward are usually represented by both a photograph and a line drawing; wherever possible these are complementary, representing either sister ships, or the same vessel before and after alteration. The emphasis on appearance changes is carried into the captions for both types of illustration, which are dated wherever and as accurately as possible.

The line drawings, which were specially commissioned for this book, are mostly reproduced to 1/1250 scale. A few very large ships are reproduced to 1/1750 scale to allow them to fit horizontally onto a page, and a few small ships are reproduced at 1/750 scale, so that a reasonable degree of detail could be included. All exceptions to the standard 1/1250 scale are clearly marked.

ACKNOWLEDGEMENTS

For help with photographs we are indebted to the following individuals and organisations; Dr S Cioglia, Norman Friedman, Ross Gillett, Pierre Hervieux, Vic Jeffery, Gerhard Koop, MG Photographic, MoD, Jacques Navarret, Len Pierce (Vosper-Thornycroft), Antony Preston, Robert Scheina, Ian Sturton, Clive Taylor, Leo van Ginderen and Karl-Erik Westerlund.

NOTE

As always we welcome 'alterations and additions' to information in this book. All correspondence on this subject should be directed to Conway Maritime Press, 33 John St, London WC1N 2AT.

Robert Gardiner

Abbreviations

The following glossary covers most of the abbreviations and technical terms used in this book, but it does not include abbreviations in common use nor designations for machinery, equipment and electronics or companies usually known by their initials. A separate list of the now widely-used US ship designations follows the glossary. See also the Introductions to France, Italy, Japan, The Netherlands, Soviet Union/Russia, United Kingdom and USA for lists of electronics and specific codenames.

AA(W), anti-aircraft (warfare)
ABC, atomic, bacteriological and chemical (warfare). See also NBC
AC, alternating current
AD, aircraft direction (British); air defence
(Weapon) Alfa, US automatic ASW rocket
'A's & 'A's, alterations and additions (British)
ASCAC, ASW Classification and Analysis Centre (US)
ASM, air-to-surface missile
ASMD, anti-ship missile defence (US)
ASROC, Anti-Submarine Rocket (US)
ASR, air-sea rescue; Admiralty Standard Range (diesels)
ASW, anti-submarine warfare; anti-submarine weapon
AUWE, Admiralty Underwater Weapons Establishment
aux, auxiliary
Avgas, aviation fuel
AWACS, Airborne Warning and Control Systems

BATRAL, *bâtiments de transport legers* (French LSMs)
BDC, *bâtiments de débarquement de chars* (French LSTs)
bhp, brake horse power
(B) PDMS, (basic) point defence missile system
BU, broken up
BuShips, Bureau of Ships (US)

c, circa
cal, calibre
CIC, combat information centre
CIWS, close-in weapon system
CMS, coastal minesweeper (British)
CNO, chief of naval operations
Co, company
CODAG, combined diesel and gas
CODOG, combined diesel or gas
COGAG, combined gas and gas
COGOG, combined gas or gas
comm, commissioned
comp, completed
COSAG, combined steam and gas
CP, controllable pitch (propellers)
CTOL, conventional take-off and landing
cwt, hundredweight
cyl, cylinder(s)

DASH, Drone Anti-Submarine Helicopter
DC, depth charge; direct current
DCT, depth charge thrower
DD, drydock
DELEX, Destroyer Life Extension Programme (Canadian)
D/F, direction-finder (-finding)
DBM, German Navy League
DP, dual purpose
DTCN, *Direction Technique des Constructions Navales* (French)
DUKW, amphibious wheeled vehicle
DYd, dockyard

ECM, electronic countermeasures
EDA, *engins de débarquement ateliers* (French landing craft)
EDATS, extra deep armed team (mine) sweeping
EDIC, *engins de débarquement infanterie et chars* (French LCTs)
EEC, European Economic Community
EEZ, exclusive economic zone
ESM, electronic support measures
ext, extreme

FAA, Fleet Air Arm (British)
FCS, fire control system
fd, flight deck
FFDS, folding fin discarding sabot
FFO, furnace fuel oil
FIS, Frontier Inspection Service (German)
FRAM, Fleet Rehabilitation And Modernisation (US)
FPB, fast patrol boat (British)
ft, foot; feet
fwd, forward
FY, fiscal year (US)

gal, gallon(s)
GB, Great Britain
GIUK, Greenland-Iceland-United Kingdom
GMSA, German Mine Sweeping Administration
GP (MG), general purpose (machine gun)
GRP, glass reinforced plastic
GUPPY, Greater Underwater Propulsive Power (US)

HA, high angle
HDML, harbour defence motor launch (British)
H/F, high frequency
HF/DF, high frequency direction-finding
HP, high pressure
hp, horsepower
HS, harbour service
HUK, hunter-killer group (US)

I, Island
IFF, Identification Friend or Foe
ihp, indicated horsepower
IMS, inshore minesweeper (British)
in, inch(es)
Ind, industry
IoW, Isle of Wight

JMSDF, Japanese Maritime Self Defence Force

K-gun, type of (US) DCT
km, kilometre(s)
kt(s), knot(s)
kW, kilowatt(s)

LAMPS, Light Airborne Multi-Purpose System (US)
LP, low pressure
LRMP, long-range maritime patrol (aircraft)
LRO, Long-Range Objectives group (US)
LSU, Labour Service Unit (Germany)

m, metre(s)
max, maximum
MCM(V), mine countermeasures (vessel)
MDAP, Mutual Defense Assistance Program (US)
M/F, medium frequency
MG, machine gun
MGB, motor gunboat (British)
MIRV, multiple independently targetted re-entry vehicle
Mk, Mark
ML, motor launch (British)
mm, millimetre(s)
Mod, model
Mousetrap, ASW rocket (US)
MSA, Maritime Safety Agency (Japan)
MSBS, French designation for SLBM
MSDF, see JMSDF
MSTS, Maritime Sea Transportation Service (US)
MTB, motor torpedo-boat (British)

NATO, North Atlantic Treaty Organisation
NBC(D), nuclear, bateriological and chemical (defence)
NECPA, National Emergency Command Post Afloat (US)
nm, nautical mile(s)
no, number
NSRDC, Naval Ship Research and Development Centre (US)
NTDS, Naval Tactical Data System (US)
NYd, Navy Yard (US; actually known as Naval Shipyards from 1950s)

oa, overall
OPV, offshore patrol vessel (British)

PDMS, point defence missile system
pdr, pounder
PIRAZ, Positive Identification and Radar Advisory Zone (USA)
pp, between perpendiculars
psi, pounds per square inch
PUFFS, passive underwater fire control feasibility study (US submarine sonar system)
PVDS, propelled variable-depth sonar

QF, quick-firing (British)

RA, rear admiral
RAF, Royal Air Force
RAM, Rolling Airframe Missile (US)
RAN, Royal Australian Navy
RARDE, Royal Artillery Research and Development Establishment
RAT, Rocket Assisted Torpedo (US)
RC, radio control
RCN, Royal Canadian Navy
R&D, research and development
RF, rapid fire (US)
R(F)L, rocket (flare) launcher
RN, Royal Navy
RPC, remote power control
rpb, rounds per barrel
rpg, rounds per gun
rpm, revolutions per minute

SAM, surface-to-air missile
SAN, South African Navy
SATCOM, satellite communication
SATNAV, satellite navigation (US)
SB, shipbuilders; shipbuilding
SCOT, satellite communication terminal (British)
SDML, seaward defence motor launch (British)
shp, shaft horsepower
SLBM, submarine launched ballistic missile
SLEP, Service Life Extension Programme (US)
SNLE, French term for SSBN
SOSUS, Sound Surveillance System (US)
SSM, surface-to-surface missile
STAAG, Stabilised Tachymetric Anti-Aircraft Gun (British)
std, standard
STOVL, short take-off, vertical landing
STWS, Ship's Torpedo Weapon System (British)
SYd, shipyard
SUBROC, submarine fired ASW rocket (US)

t, ton(s)
TACAN, Tactical Air Navigation aid (US)
TACTASS, Tactical Tower Array Surveillance System (US)
TCD, *transports de chalands de débarquement* (French LSDs)
TFCC, Tactical Flag Command Centre (US)
TRIP, Tartar Reliability Improvement Programme (US)
TS, training ship
TT, torpedo tube(s)

UHF, ultra high frequency
UNREP, underway replenishment (US)
US(N), United States (Navy)

VA, vice admiral
VCNS, Vice Chief of the Naval Staff
VDS, variable depth sonar
VLF, very low frequency
VLS, vertical launch system
VSTOL, vertical/short take-off and landing
VTE, vertical triple expansion
VTOL, vertical take-off and landing
VSRAD, Very Short Range Air Defence System (Dutch)

WEU, Western European Union
Wks, Works
wl, waterline
W/T, wireless telegraphy (obsolete British term for radio)
Y-ARD, Yarrow-Admiralty Research and Development organisation
Y-gun, type of DCT (US)

3-D, three-dimensional

US SHIP DESIGNATIONS

These consist of letter codes representing basic ship types with additional letters (usually added on the end) to further define the sub-types. The main types are listed (together with untypical designations) followed by the most common suffixes. A few of the rarer designations will be found under the relevant classes in the US section.

A. Ship types

AE, ammunition ship
AFS, combat stores ship
AGC, later LCC (*qv*)
AGMR, major communications relay ship
AGSS, auxiliary submarine
AKA, later LKA (*qv*)
AO, fleet oiler
AOE, fast combat support ship
AMS, later MSC (*qv*)
APA, later LPA (*qv*)

BYMS, YMS (*qv*) built for Britain

CA, heavy cruiser
CG, missile cruiser (ex-frigate)
CL, light cruiser
CPIC, coastal patrol and interdiction craft
CSGN, strike cruiser
CV, aircraft carrier
CVA, attack aircraft carrier
CVS, ASW support carrier
CVV, medium aircraft carrier

DD, fleet destroyer
DDE, fleet destroyer converted to escort destroyer
DE, destroyer escort
DL, destroyer leader (frigate)
DSRV, deep submergence rescue vehicle

FF, frigate

LCA, landing craft, assault
LCAC, landing craft, air cushion
LCC, amphibious command ship
LCM, landing craft, mechanised
LCP(L), landing craft, personnel (large)
LCT, landing craft, tank
LCU, landing craft, utility
LCVP, landing craft, vehicle and personnel
LFS, fire support ship
LHA, amphibious assault ship
LHD, amphibious assault ship, dock
LKA, amphibious cargo ship
LPA, amphibious personnel transport
LPD, amphibious transport, dock
LPH, amphibious helicopter carrier
LSD, landing ship, dock

MCM, mine countermeasures vessel
MHC, coastal minehunter
MSB, minesweeping boat
MSC, coastal minesweeper
MSD, drone minesweeper
MSI, inshore minesweeper
MSL, minesweeping launch
MSO, ocean minesweeper

PC, patrol craft
PCH, hydrofoil patrol craft
PF, patrol frigate
PG, patrol gunboat
PHM, missile firing hydrofoil
PT, torpedo boat
PTF, fast patrol boat

SC, submarine chaser
SCS, sea control ship
SS, submarine
SST, training submarine
VSS, VSTOL support ship
WHEC, Coast Guard high endurance cutter
WMEC, Coast Guard medium endurance cutter

YMS, later MSC (*qv*)

B. Principal suffixes

G, guided missile firing
N, nuclear powered
B, ballistic missile firing
K, ASW hunter-killer
R, radar-picket
X, experimental, or characteristics undecided

Albania

Albania, uniquely among the smaller nations, freed herself from Second World War Axis occupation using her own military forces. An underground army was set up and commanded by the Communist Party, which seized power in 1944 after the country was liberated. With the government formed by Enver Hoxha, a totalitarian communist state was set up and plans for rapid industrial development were drawn up, despite lack of capital, know-how and material wealth. The scale of industry proved to be inadequate to supply any kind of fighting craft and the status of the Albanian naval forces has reflected changes in the political sympathies of the regime since the end of the war. Until 1948 Albania was a satellite of Yugoslavia with a customs as well as a monetary union. The Albanian naval forces, part of the 'People's Army' at that time, were supplied by the Soviets with some minor craft while Yugoslavia helped to restore two old minesweeping tenders. After the expulsion of Yugoslavia from the Cominform and the overpowering of the pro-Tito group by the Stalinist faction in Albania, the country became a close Soviet ally despite territorial isolation and had to rely on the Soviet Union for both technical advice and capital loans. Albania was valuable to the Soviet Union as she provided her with Mediterranean bases for the Soviet Navy. After joining the Warsaw Pact in 1955 Albania allowed the Soviet Union to begin construction of base facilities at Sazan (Saseno) Island in the Gulf of Valona. In exchange, cancellation of the large Albanian debt was announced by the Soviet Union in 1957 and large amounts of credit made available for the modernisation and expansion of the armed forces. During 1956–61 the Albanian Navy, which was separated from the Army at that time, purchased two 'Whiskey' class submarines, six MTBs of the P 4 type, two minesweepers of the T 43 type, six inshore minesweepers of the T 301 class, four 'Kronstadt' class patrol craft, eleven minesweeping boats and a number of auxiliaries.

The new Soviet credit dried up by the late 1950s which, together with Khrushchev's dislike of the strict Stalinist course maintained by the Albanian government, inclined Tirana to flirt with Red China. When a rift between the two communist giants became evident, public disapproval of the Chinese viewpoint was disguised by the Soviet Communist Party as an attack on Albania. This resulted in diplomatic relations with the Soviet Union being severed in December 1961. Albania effectively withdrew from the Warsaw Pact and deprived the Soviet Union of her base at Sazan Island, seizing two 'Whiskey' class submarines there. The Soviet-made ships were commissioned and maintained with Chinese assistance, giving the Albanians an effective task force of three submarines and six MTBs; however lack of spares soon decreased their efficiency.

Transfer of Chinese craft was initiated in 1965 with six MTBs of the P 4 type and provided Albania with an effective squadron of fast attack craft which by the late 1970s consisted of four missile craft of the 'Hoku' class, thirty-two torpedo hydrofoils of the 'Huchwan' class and six patrol craft of the 'Shanghai II' class. This force of comparatively new craft – despite their somewhat obsolescent design – and favourable strategic location of bases gave the Albanians the ability to control traffic through the 50-mile wide Strait of Otranto, although the strained relations between Albania and China disclosed in the late 1970s may have affected the supply of spares for the Chinese-made craft.

Despite sparing use of vessels and production of rudimentary spare parts, the fleet could not be maintained for long without foreign assistance. As a result since the early 1980s some vessels have either been scrapped or placed in reserve, while others have been cannibalised for spares. In 1993 most of the vessels were in poor condition. The number of operational units continues to decline due to advanced age and lack of spare parts. The submarines are non-operational, and only one 'T 43' class and two 'T 301' class minesweepers have been seen at sea. Following standard Eastern Bloc practice, Albanian warships' recognition numbers were periodically changed.

The condition of her naval vessels reflects the economic state of Albania. After 46 years of strict Stalinist economic management, the country had the lowest GNP in Europe. This, and phantasmagorian plans such as building nuclear shelters instead of houses created fertile ground for the anti-communist ideas that came from other Warsaw Pact countries. The Communist Party was faced by growing internal opposition, which succeeded in gaining over 30 per cent of parliamentary seats in the 1991 election. However, the new government is troubled by inexperience and internal rivalry which does not help the economy.

In 1993 the Albanian Navy was reported to have a total of 3,300 men (probably since reduced), mostly ratings on three years' conscription. The main bases are in Durazzo, Valona, Sazan Island (sixteen underground berths, defended by 122mm and 88mm batteries), Sarande, Shingjin and Paşsa Liman. Coastal defence is controlled by the army (except for 400 naval personnel) which deploys three artillery battalions near important bases, while pillboxes are located along the entire coast – in the most vital areas they are spaced at 100–150m intervals. A number of small patrol vessels are deployed on lakes on the Yugoslav and Greek borders.

FLEET STRENGTH 1947

The 1947 status of the Albanian naval forces cannot be described precisely but according to some sources the following craft were in service:

Two Italian-built motor boats of the *Tirane* class although they may be former Italian MAS-boats transferred from the USSR.

Two ex-German *MFP*-craft (220t, built *c*1943, stricken early 1970s).

Two minesweeping tenders (ex-Yugoslav *Marjan* and *Mosor*, 125t, built 1931, recovered after the war, stricken 1967).

SUBMARINES

Ex-Soviet 'WHISKEY' class *submarines*

Two transferred from the USSR in 1960 while an additional pair was seized in 1961 upon the withdrawal of the Soviets from their base at Sazan Island. Three were operational (numbers *512, 514* and *516*), while the remaining boat was used for harbour training and as a charging station. She was stricken in 1976 and cannibalised for spare parts. In the 1980s one operational boat was deactivated to become a stationary training boat, while two others (bearing numbers *422* and *442* in 1990 and *552* and *523* by 1993) were extant 1995. Both probably unable to dive. Based at Sazan.

SMALL SURFACE COMBATANTS

Ex-US 'VOSPER' type *fast attack craft (torpedo)*

Seven (Lend-Lease supplies) transferred from the USSR in 1950. Stricken mid-1960s.

Ex-Soviet 'P 4' type *fast attack craft (torpedo)*

Class:
221–226, 316–321

Six transferred 1956 and augmented 1965 by six others of Chinese construction. The latter were different in having additional twin 12.7mm MG mount fitted and lacking radar which was to be installed later. All stricken from active list by 1987 but are still reported to have limited operational status in 1995.

Ex-Chinese 'HUCHWAN' class *fast attack craft (torpedo) hydrofoils*

Class:
121–129, 320–325, 326–327 (stricken 1989), *427–432, 433* (stricken 1989), *520–527*

All were built in Shanghai and transferred in 1968 (six), 1969 (fifteen), 1970 (two), 1971 (seven) and 1974 (two). Since the 1970s these vessels have provided the main operational strength of the Albanian navy; however their condition has deteriorated due to lack of maintenance and spares. At least eight were deleted 1988–93 and cannibalised for spares. Another escaped to Italy in May 1991.

Ex-Chinese 'HOKU' class *fast attack craft (missile)*

Four of this class transferred from China 1976–77. Stricken 1982–84.

Ex-Soviet 'KRONSTADT' class *patrol craft*

Class:
191, 192, 502, 504, numbers changed later to *150* (stricken 1983), *151* (stricken 1979), *350, 351*

Four craft transferred 1958. Two were sent for ASW updating to the USSR shortly before the 1961 break and were claimed back by the Albanian government in mid-1965. Two boats extant 1995.

Ex-Chinese 'SHANGHAI II' class *patrol craft*

Class:
101–106, renumbered since 1982 to *150–152, 650–652*

Four transferred from China in 1974 and two more in 1975. Probably non-operational 1995.

MINE WARFARE VESSELS

Ex-Soviet 'T 43' type *ocean minesweepers*

Class:
152, 342, (renumbered mid 1980s to *360, 460*)

Two of the short hull series were acquired August 1960. One deleted 1991, other extant 1995 but non-operational.

Ex-Soviet 'T 301' class *inshore minesweepers*

Three pairs transferred 1957, 1959 and 1960 respectively. Reportedly fitted with navigational radar in early 1980s. Numbers of five of these boats became known in the 1970s: *153, 154, 343–345. 343* and one other were deleted 1979. Remaining four had numbers *361, 362, 461* and *462* in 1991. Two in reserve 1991, two others extant 1995 with limited operational status.

Ex-Soviet 'K M 4' type *minesweeping boats*

Three transferred after 1945. All stricken 1967.

Ex-Soviet 'PO 2' type *minesweeping boats*

Eleven bought in the USSR during 1957–60. Used for general purposes; all stricken by 1988–89.

Algeria

The Algerian armed forces (including a fourth service, the National Gendarmerie) are formidable and well-equipped, much military effort having been devoted to the 150,000-man army, the fourth strongest in Africa, with a reputation and traditions based on winning the prolonged guerrilla war leading to independence in 1962. Algeria's merchant marine, boosted like much else by the country's recent oil wealth, expanded from seventy-five ships (239,815grt) in 1974 to 130 vessels (1,218,621grt) in 1980.

The navy, which began with a pair of Egyptian (ex-US) coastal minesweepers, has been gradually expanded and modernised in the past 25 years, personnel being increased from 3000 to 6000, with about 9000 reservists, and it is now probably stronger and certainly more efficient than the Libyan Navy. Previously a force of large patrol and fast attack craft, it was reinforced in 1980 by the first of two Soviet missile corvettes and a 'Koni' class frigate, giving some cruising capability. From January 1982 a Soviet 'Romeo' class submarine gave training to Algerian crews at Mers-el-Kebir, and in 1987–88 two 'Kilo' class submarines were acquired from the Soviet Union to replace the two 'Romeos' acquired earlier.

Despite receiving further Soviet equipment, and regular visits from the Soviet fleet, Algeria did not provide the USSR with base facilities. Since the early 1980s, Algeria has started to build its own small surface vessels, but lacks the funds to buy the larger replacement vessels it urgently requires. The main bases along the 800-mile coast are Mers-el-Kebir, Algiers, Arzew, Philippeville, La Senia and Annaba (Bone).

Ex-Soviet 'KONI' class *frigates*

No	Name	Builder	Launched	Comp	Fate
901	MURAT REIS	Zelenodolsk SY	*c*1979	1980	Extant 1995
902	RAIS KELLICEN	Zelenodolsk SY			Extant 1995
903	RAIS KORFU	Zelenodolsk SY			Extant 1995

901 was delivered from Black Sea 20 December 1980 and 902 was transferred 4 March 1982. 903 followed in October 1984.

Soviet 'ROMEO' class *submarines*

Particulars: See Soviet 'Romeo' class

One transferred January 1982, the second in February 1983 on five-year lease. Replaced in active service by 'Kilo' class boats in 1989 but remained in Algeria. Both extant 1995 as battery-charging and pierside training hulks.

Soviet 'KILO' class *submarines*

Particulars: See Soviet 'Kilo' class

Two units acquired 15 September 1987 and January 1988. Position for shoulder-launched Strela SAM (eight missiles) at aft end of sail. Both extant 1995

Ex-Soviet 'NANUCHKA II' class *corvettes*

No	Name	Builder	Launched	Fate
801	RAS HAMIDOU	Petrovskiy SY, Leningrad	*c*1979	Extant 1995
802	SALAH REIS	Petrovskiy SY, Leningrad	*c*1979	Extant 1995
803	RAIS ALI	Petrovskiy SY, Leningrad		Extant 1995

801 delivered 4 July 1980, 802 received 9 February 1981. Reported that more (perhaps three) are to follow.

KEBIR class *patrol craft*

Particulars: See Oman *Al Bushra* class

Brooke Marine 37.5m design. They are armed with 1–76mm OTO Melara Compact and 1–30mm. Since 1983 eleven more have been built. They are used by the Coast Guard.

Kebir class FAC 1982 — Brooke Marine

Ex-Soviet 'OSA I' and 'OSA II' *fast attack craft (missile)*

Three 'Osa I' class boats transferred in 1967 and numbered *641* to *643*; nine 'Osa II' boats transferred (four in 1976–77, two in 1978, two in 1979, one in 1981) have been numbered *644* to *651*. One damaged and one (*643*) destroyed by a missile explosion in 1981. Remainder extant 1995.

Ex-Soviet 'KOMAR' class *fast attack craft (missile)*

Six transferred in 1966 and numbered *671* to *676*. All now stricken.

Ex-Soviet 'P 6' class *fast attack craft (torpedo)*

Twelve transferred 1963–68. Two discarded in 1975, two transferred to the Coast Guard, two used for training and two are hulks. Remainder all stricken by 1984.

Rais Kellich 26.5.95 — L & L van Ginderen

Ex-Soviet 'SO 1' class *large patrol craft*

Six boats transferred October 1965–October 1967 and numbered *P 651* to *P 656*. Three had two 533mm TT taken from the 'P 6' FAC. All discarded by mid-1980s.

Ex-Chinese HAINAN class *patrol craft*

Seven transferred from 1990. Used by Coast Guard. All extant 1995.

Ex-US BYMS type *coastal minesweepers*

Class (former name):
Djebel Aures (ex-*Tor*), *Sidi Fradj* (ex-*Darfur*)

Acquired from Egypt in July 1962. *Djebel Aures* wrecked off Algiers in April 1963, and *Sidi Fradj* discarded *c*1971.

Ex-Soviet 'T 43' class *ocean minesweepers*

Class:
M 521, M 522

Launched *c*1950–56. Both transferred 1968 and both discarded 1984–85.

MISCELLANEOUS

Ex-Soviet 'Polnochny' class landing ship, transferred in August 1976 and numbered 555.

Ex-Soviet 'Zhuk' class coastal patrol boat received from the Black Sea 22 May 1981.

Ex-Soviet 'Poluchat' class coastal patrol craft, used for torpedo recovery.

Sixteen coastal patrol craft built in Italy (1976–78) for the Algerian Coast Guard.

Two Brooke Marine 93m logistic landing ships *Kalaat Beni Hammed* and *Kalaat Beni Rachid*, launched 1984 (see Oman *Nasr al Bahr* class for details). Both extant 1995.

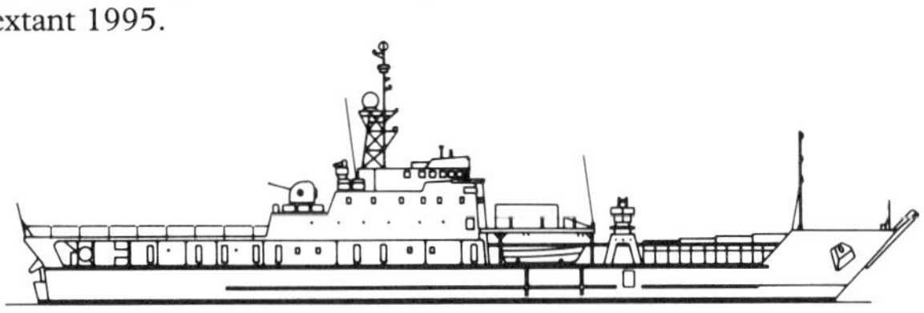

Kalaat Beni Hammad 1984

Angola

Angola's internal problems since her civil war-wracked independence from Portugal in November 1975 made excursions by her armed forces unlikely. Support had been given to the Ovambo guerrillas in Namibia (South-West Africa), but the ruling MPLA alone had neither the strength nor the ability to resist South African reprisals in the frontier area, or wipe out UNITA's increasing inland territory, and the 17,000–20,000 Cuban troops were stationed well to the north. From 1976 to the late 1980s, when a 20-year Treaty of Friendship and Co-operation was signed with the USSR, Angola's capital city and port, Luanda, became the Soviet Navy's third most visited Atlantic port for all the years since 1953, behind only Guinea and Cuba.

The very young Angolan Navy, commanded by the Deputy Minister of Defence, is and will almost certainly remain, a small ship patrol force for a 1000-mile coastline. Most of its boats were transferred by Portugal at independence, and there are also ex-Soviet craft delivered since February 1977. Maintenance, which must present considerable problems for the former, depended on Soviet and Cuban technicians, and training for the 1000 sailors was by advisers from those countries, with a Nigerian team; some officers went to the USSR. The Eastern bloc gave relatively little economic assistance, and sometimes even harmed the economy – Soviet trawlers and factory ships have fished the local waters bare. The continuing civil war with the UNITA guerrillas has left Angola in serious economic difficulties and it cannot afford the vessels it needs to patrol its long coastline.

Ex-Soviet 'SHERSHEN' class *fast attack craft (torpedo)*

Four transferred, one on 16 December 1977, two on 16 December 1978 and one in February 1979.

Ex-Soviet 'OSA II' class *fast attack craft (missile)*

Six transferred 1982–83. One lost 1988, remainder extant 1995.

Ex-Portuguese ARGOS class *patrol vessels*

Seven units transferred, four in 1975 and three in 1976. Five were renamed *P 361*, *P 362*, *P 375*, *P 379* and *P 1130*. Ex-*Argos* and ex-*Dragao* believed to be non-operational, having been transferred for spares. All discarded by 1992.

MISCELLANEOUS

The following minor fighting and amphibious force vessels have been acquired since 1975: from Portugal the former coastal patrol boats *Venus* (*Jupiter* class), *Espiga*, *Formalhaut*, *Pollux*, *Altair* and *Rigel* (all *Bellatrix* class), and the LCT *Alfance*.

From the USSR two 'Poluchat' class and at least one 'Zhuk' class coastal patrol craft (in February 1977), and three 'Polnochny' class landing ships (delivered 16 December 1977, 6 December 1978 and 1 February 1979).

There were also up to nine ex-Portuguese LDM 400s and five ex-Soviet 'T 4' LCMs. The latter distributed Soviet arms to the MPLA along the coast during the critical fighting with UNITA and South Africa in 1975–76. It was reported that three, perhaps Soviet-crewed, shelled the port of Moçâmedes for two days (10–11 February 1976) before several hundred Cubans landed in or near the port.

All ex-Portuguese vessels have now been discarded.

Argentina

The position of the Argentine Navy *vis-à-vis* its continental rivals was adversely affected by the political and naval changes wrought by the Second World War. From the late 1940s until the acquisition of the carrier *Independencia*, the Argentine Navy remained at numerical parity with the Brazilian Navy, and was only slightly superior to that of Chile. The acquisition of *Independencia*, followed by the *25 de Mayo*, gave the Navy a strike capability that neither of its neighbours could match, and thus returned superiority to the Argentine fleet.

In the decades following the Second World War, the Argentine Navy was very active. In 1955, after a number of failures, the navy deposed Juan Peron. On 16 September 1955, key sea and shore commands were seized, principally by using fictitious orders, although as anticipated, numerous naval shore facilities could not be protected and fell to Peron's forces. Rio Santiago, home of the Naval Academy, had to be evacuated, as well as the submarine base at Mar del Plata. The navy then demonstrated its muscle. Naval aircraft bombed the ground forces, which were trying to attack Puerto Belgrano, the key naval base, and fleet units bombarded an oil storage area near the recently evacuated submarine base. The cruiser *9 de Julio* and two destroyers bombarded the area for four hours. Three frigates then landed 350 marines and sailors. The success of this attack proved the navy's ability to bombard the capital. In the afternoon of 19 September, as the fleet appeared off Buenos Aires, Juan Peron resigned. This popularly supported victory strengthened the navy's position nationally.

In recent decades, the Argentine Navy has been combating 'fish piracy' within its 200-mile limits. In 1966 the destroyer *Santa Cruz* fired on and holed a Russian trawler, which had refused to be escorted into Mar del Plata. A decade later, on 21 September 1977, the destroyers *Py*, *Rosales* and *Segui* trapped four more trawlers, two Soviet and two Bulgarian; one poacher was killed by naval gunfire and three Argentine sailors drowned while attempting a boarding in rough seas. In the mid-1970s the Argentine Navy, along with sister services, fought against terrorists. The August 1975 bombing of *Santisima Trinidad* received international headlines.

In the spring of 1982, Argentina fought and lost the Falklands War over the Malvinas (Falklands). There can be little doubt that the initial decision to invade was the result of diplomatic events that took place between January and the end of March. As with most countries, Argentina was unprepared for war. The navy had no warning. The submarine force, the one potential wild card, was in a state of transition. Neither Type 209 was one hundred per cent worked up and only one of two old boats was operational. A few new naval attack aircraft were just becoming operational with almost a dozen more to arrive in the late spring.

Almost the entire navy was used to support Operation 'Rosario' – the marine landings in the Falklands and Georgias on 2–3 April, having put to sea on 28 March ostensibly for exercises with the Uruguayan Navy. Following the invasion, the fleet withdrew to the principal Argentine naval base, Puerto Belgrano. Towards the end of April, the fleet took up positions just outside the British 200-mile exclusion zone. Almost due north of the islands were the carrier *25 de Mayo*, and most of the fleet, including the two Type 42s. To the SW of the islands was the cruiser *General Belgrano* and two destroyers supported by an oiler. Late on 1 May, the carrier aircraft located a British task force NE of the islands. The Argentine northern force attempted to close but were unable to do so. The Argentine northern force was now being shadowed by two Sea Harriers. Being aware that the United States was providing intelligence to the British, the Argentine fleet commander decided to withdraw to protected waters along the coast. Shortly after the Argentine force becan to retire, the *General Belgrano* was torpedoed and sunk. The fleet did not make another appearance, the aircraft from the carrier being deployed from southern Argentina.

In recent decades Argentina, like Brazil and Peru, has turned to bi-national shipbuilding agreements with developed European shipbuilders, which specify that part or all of the combatants are to be built or assembled in Argentina. For those ships built in Argentina, the European builders are required to supply some components, technical aid and training, co-ordination and advice. The first bi-national agreement was with Howaldtswerke Deutsche Werft for two Type 209 submarines; these boats joined the fleet in the early 1970s. The second agreement was with Vickers for the construction of two Type 42 destroyers: *Hercules* commissioned on 12 July 1976 and *Santisima Trinidad* joined the fleet in 1981. The third was with Thyssen Nordseewerke for six Type TR1700 submarines, and the fourth was with Thyssen Rheinstahl-Blohm & Voss for four Type MEKO 360 frigates and six Type MEKO 140 corvettes. These bi-national agreements marked the return of 'obtaining the best that money can buy' which was the policy of Latin American navies before the Second World War.

In 1982 for an active fleet of fifty-four ships the personnel of the Argentine Navy numbered 31,000, including 2900 officers. In 1995 the reduced personnel strength was 14,000, including 2500 officers. Volunteers are supplemented by 12-month conscripts. There is also a Marine Corps of 6000 men. The main naval bases are Buenos Aires, Rio Santiago (La Plata), Mar de Plata (for submarines) and Puerto Belgrano, with a smaller facility at Ushaia on Tierra del Fuego. The main naval building yard is Astilleros y Fábricas Navales del Estado (AFNE) at Rio Santiago, but the navy also holds a majority shareholding in Darsena Norte (Tandanou) and Darsena Este.

FLEET STRENGTH 1947

BATTLESHIPS

Name	Launched	Disp (normal)	Fate
Libertad class			
INDEPENDENCIA	26.2.1891	2300t	Discarded 1951
Rivadavia class			
MORENO	23.9.11	27,700t	Sold 8.2.56
RIVADAVIA	26.8.11	27,700t	Sold 8.2.56

Moreno and *Rivadavia* were inactive in the latter years of their careers.

CRUISERS

Name	Launched	Disp (normal)	Fate
LA ARGENTINA	16.4.37	6500t	Discarded 1974
Garibaldi class			
PUEYRREDON	25.7.1898	6800t	Discarded 2.8.54
Veinticinco de Mayo class			
ALMIRANTE BROWN	28.9.29	6800t	Discarded 27.6.61
VEINTICINCO DE MAYO	11.8.29	6800t	Discarded 24.3.60

La Argentina
Modified British *Arethusa* class. Designed as a training cruiser, *La Argentina* made numerous foreign cruises, taking Isabella Peron on her European tour in 1947, visiting Spain and Italy. In 1950, her aircraft catapult was removed and her 4–4in (4×1) AA guns were replaced by 12–40mm (4×2, 1×4) AA. This, plus the addition of radar and other electronics, increased full load displacement from 7500t in 1939 to over 8500t by 1974. In 1951 she was reclassified from a school ship to a cruiser, resuming her training role in 1961. She was disarmed and decommissioned in 1970.

Garibaldi class
In her closing days, the ageing armoured cruiser *Pueyrredon* was used as a coast defence ship and for training. She was converted to oil, had her secondary battery removed, and was fitted with a tripod mast and fire control director; she was renovated at Puerto Belgrano, Argentina. She was decommissioned on 4 January 1955; her sisters *Garibaldi*, *San Marti* and *General Belgrano* were discarded 20 March 1934, 18 December 1935 and 8 May 1947 respectively.

Veinticinco de Mayo class
Modified Italian *Trento* class ships, these cruisers were often criticised in British publications as being generally unsatisfactory; however, they were popular ships in the Argentine Navy. Their stacks were raised shortly before the Second World War, altering their appearance. Both cruisers took part in the 1948 Antarctic operation, and in 1959 they were placed in reserve.

DESTROYERS

Name	Launched	Disp (normal)	Fate
La Plata class			
CORDOBA	1911	890t	Discarded 10.1.56
LA PLATA	1911	890t	Discarded 10.1.57
Jujuy class			
CATAMARCA	1911	1010t	Discarded 10.1.56
JUJUY	15.4.12	1010t	Discarded 10.1.56
Churucca class			
CERVANTES	Jun 1925	1522t	Discarded 24.6.61
JUAN DE GARAY	2.11.25	1522t	Discarded 25.3.60
Mendoza class			
LA RIOJA	26.1.29	1570t	Discarded 30.4.62
MENDOZA	13.7.28	1570t	Discarded 30.4.62
TUCUMAN	10.10.28	1570t	Discarded 20.4.62
Buenos Aires class			
BUENOS AIRES	21.9.37	1375t	Discarded 1971
ENTRE RIOS	21.9.37	1375t	Discarded 1973
MISIONES	23.9.37	1375t	Discarded 3.5.71
SAN JUAN	24.6.37	1375t	Discarded 1973
SAN LUIS	23.8.37	1375t	Discarded 3.5.71
SANTA CRUZ	3.11.37	1375t	Discarded 1973

Churucca class
Built in Spain to the First World War British Admiralty Design Flotilla Leaders, this was a highly successful series. Following the Second World War, 1–4.7in and 1–3in gun were replaced by 6–40mm AA: two pairs were located between the funnels and one pair aft the second funnel. They were reclassified as *torpederos* (destroyer escorts) in 1952. In May 1961 *Cervantes* was placed in reserve. *Juan de Garay* was used for training 1952–59.

Mendoza class
Modified *Bruce* class. Reclassified as *torpederos* (destroyer escorts) in 1952, *La Rioja* and *Tucuman* were disarmed and laid up in that same year, although they continued to show up in naval annuals as being active. *Mendoza* was decommissioned in 1961.

Buenos Aires class
Modified *Greyhound* class. By the close of their careers, armament varied considerably throughout the class. *Buenos Aires* in 1969 was representative: 3–4.7in (3×1), 4–40mm (4×1) AA, 2 Hedgehog, 2 ASW torpedo launchers, 4 K-guns, 1 DC rack. *Buenos Aires* was disarmed in 1970, *Entre Rios* in 1972, *Misiones* in 1972, *San Juan* in 1972, *San Luis* in 1970 and *Santa Cruz* in 1971.

SUBMARINES

Name	Launched	Disp	Fate
Santa Fe class			
SALTA	17.1.32	775t/920t	Discarded 5.4.61
SANTA FE	28.7.31	775t/920t	Discarded 23.4.59
SANTIAGO DE ESTERO	28.3.32	775t/920t	Discarded 23.4.59

Modified *Cavallini* class. *Salta* made her 1000th dive on 3 August 1960. *Santiago de Estero* made a record dive for the class of 374ft (114m) in the South Atlantic.

Other vessels on strength in 1947 included the training ship *Presidente Sarmiento* (launched 31 August 1897, 2733t normal, decommissioned 1961); the *Rosario* class (river gunboats) *Parana* (28 April 1908, 1050t, sold 14 September 1956) and *Rosario* (27 July 1908, 1050t, sold 23 July 1959); the *Bouchard* class (minesweepers) *Bouchard, Drummond, Fournier, Granville, Parker, Py, Robinson, Seaver* and *Murature* (launched 1943–44, 900t, in service 1982). 'King' class *Presidente Sarmiento* currently lies at Buenos Aires as a museum. The *Bouchard* class were the first modern warship class constructed in Argentina. One unit, *Fournier*, struck an unchartered rock at the entrance of the San Gabriel Channel and sank on 22 September 1949; *Parker* was ceded to Paraguay in 1964 and *Py* and *Bouchard* in 1968; the rest were discarded 1962–67.

MAJOR SURFACE SHIPS

Independencia 1963

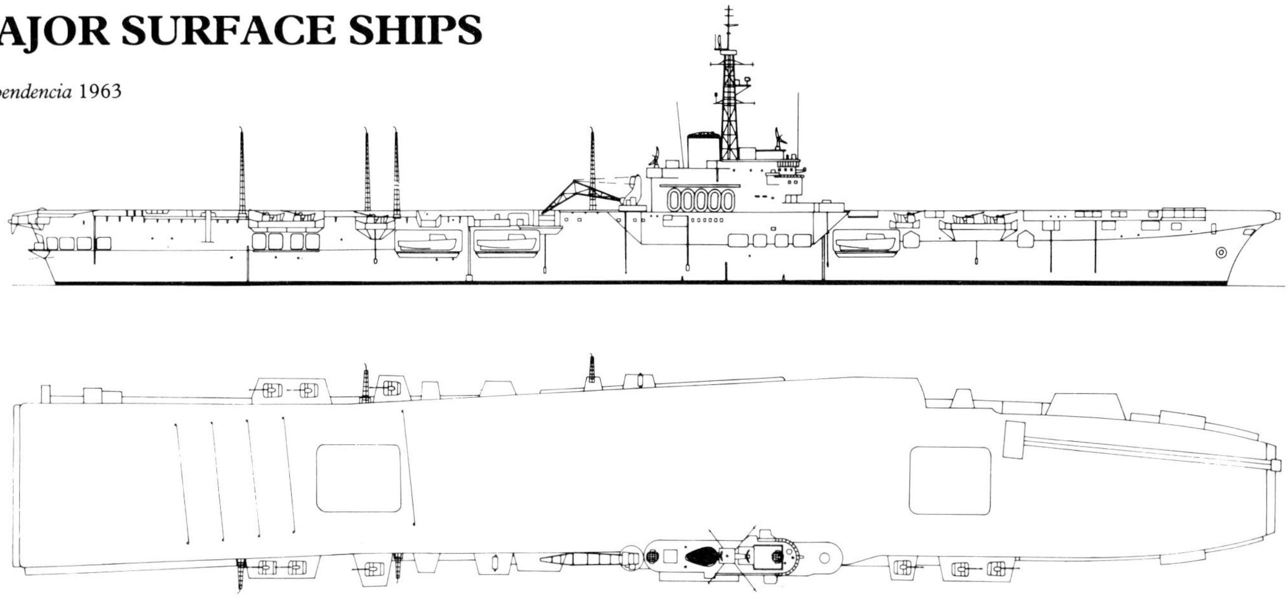

Independencia, launched 20 May 1944, was purchased from Great Britain in June 1958, in part with funds from the sale of the battleships *Moreno* and *Rivadavia* and of the armoured cruiser *Pueyrredon*. The carrier had been modernised in 1952–53 at which time she received a lattice mast and an enlarged bridge. Three years later she underwent another major refit, being fitted with a 5° angled deck. She was insulated for tropical service and partially air-conditioned. On 6 August 1958 she was named *Independencia*; on 4 November 1958 the Argentine flag was raised in Portsmouth; on 8 June 1959 the first Argentine naval aircraft operated from her decks; and on 8 July 1959 she commissioned. As transferred, she carried 12–40mm AA; however, this was reduced to 8–40mm shortly after going into service. In August 1963 the first jet aircraft, an F-9F Panther, operated from *Independencia*. She was capable of carrying twenty-four aircraft. Attack aircraft available in limited numbers for carrier operations included F-4U Corsairs, F9F Panthers, T-28 Trojans and TF-9J Cougars. At the end of her service she was carrying six S-2A Trackers and fourteen T-28s. In 1970 she was disarmed and placed in reserve.

Ex-British COLOSSUS class *aircraft carrier*

No	Name	Builder	Acquired	Fate
V 1	INDEPENDENCIA (ex-*Warrior*)	Harland & Wolff	Jun 1958	Sold 17.3.71

Independencia c1965 Robert L Scheina

Veinticinco de Mayo 1982 L & L van Ginderen

Veinticinco de Mayo 1974

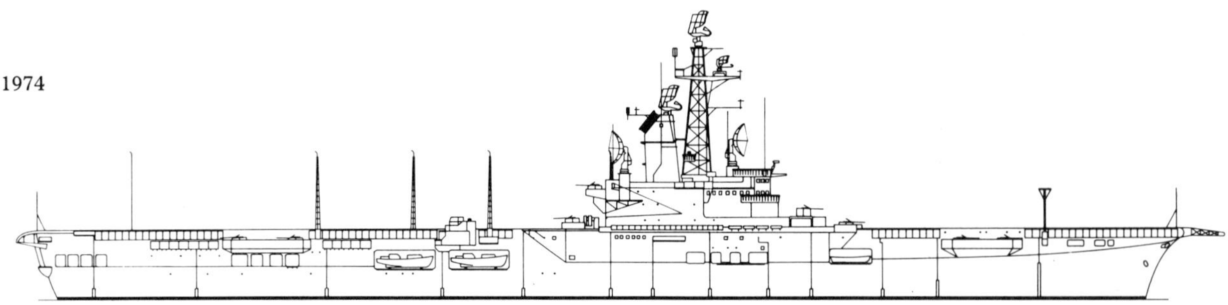

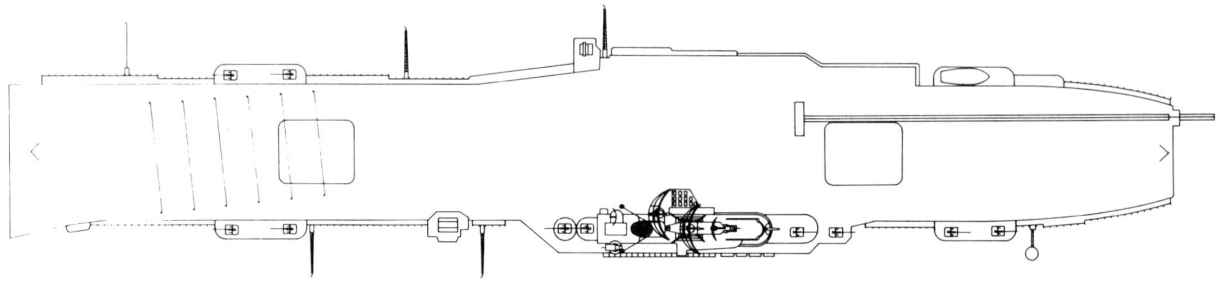

Ex-British modified COLOSSUS class *aircraft carrier*

Displacement: 15,892t standard; 19,896t full load
Dimensions: 630ft pp, 693ft 2in oa, 690ft fd × 80ft, 121ft fd × 25ft
192.0m, 211.1m, 210.3m × 24.4m, 36.9m × 7.6m
Machinery: 2-shaft Parsons geared turbines, 4 Admiralty 3-drum boilers, 42,000shp = 24kts. Fuel 3200t
Armament: 12–40mm (12×1), 22 aircraft
Complement: 1250

No	Name	Builder	Acquired	Fate
V 2	VEINTICINCO DE MAYO (ex-*Karel Doorman*, ex-*Venerable*)	Cammell Laird	8.10.68	Extant 1995

An extensive boiler room fire caused this carrier to be prematurely decommissioned from Dutch service, but she was extensively rebuilt at Wilton Fijenoord Shipyard and purchased by Argentina. Her boilers were replaced with ones from an uncompleted near-sister, *Leviathan*. She commissioned in the Argentine Navy on 8 October 1968 and embarked her first aircraft on 3 September 1969.

Veinticinco de Mayo is air-conditioned. Her hangar deck is 455ft × 52ft × 17ft 6in (*138.7m × 15.9m × 5.3m*). Two 40mm guns have been removed. In 1972, Argentina acquired sixteen A-4Q Skyhawks, which have been the backbone of the carrier's air complement, which also includes four S-2As and two SH-34J helicopters. In 1979, her flight deck was rebuilt and the area increased. In 1982 the carrier was used in the Falklands conflict. A typical aircraft complement during the months from March to May was eight Skyhawks, four Trackers, and a few general purpose helicopters. The Super Étendard attack aircraft, five of which had arrived in Argentina in the autumn of 1981, could not operate from the carrier. *Veinticinco de Mayo* was not yet fitted with the essential electronics to provide the inertial guidance data to the aircraft. Following the war, the carrier was modified so that the could handle the Super Étendards. Electronics consisted of Dutch LW-02, SGR-105 (DA-05), SGR-103 (ZW-01), TACAN with the British CAAIS data automation system.

British combat data system CAAIS was replaced in 1983 by Dutch SEWACO type to allow easy cooperation between the carrier and German built destroyers of the *Almirante Brown* class (MEKO 360). At this time one 40mm AA gun was removed. The ship for all practical purposes ceased to be operational around 1985. The post Falklands War trials showed that she was not compatible with Super Étendard fighters and could operate only the A-4Q Skyhawk aircraft.

Engineering problems left her inoperative from June 1986. The ship was planned for a major two-year refit starting 1988. It was reported that the steam plant was to be replaced by all-diesel plant or even a CODOG arrangement. Italian Fincantieri even won the contract for technical assistance to AFNE dockyard, however all these plans were suspended and in October 1991 it was announced that the ship would probably be sold in the near future. The last electronic fit of the carrier consisted of Dutch radars LW 08, DA 08, LW 02, ZW 01, also Racal Decca 1226, VI/SGR 109, as well as carrier landing radar Selenia MM/SPN 720 and Tacan. The carrier remains partly dismantled at Rio Santiago. See under Netherlands for further details.

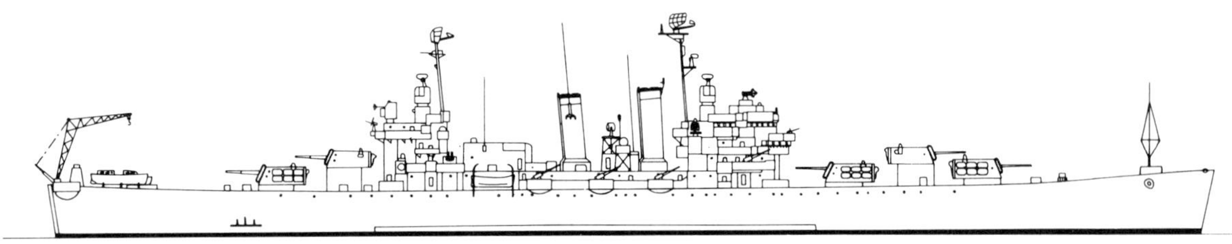

General Belgrano 1982

Ex-US BROOKLYN class *cruisers*

No	Name	Builder	Acquired	Fate
C 4	GENERAL BELGRANO (ex-*Diecisiete de Octobre*, ex-*Phoenix*)	New York SB	12.4.51	Sunk 2.5.82
C 5	NUEVE DE JULIO (ex-*Boise*)	Newport News	12.4.51	Decommissioned 1978, BU 1983

General Belgrano was launched 12 March 1938, purchased by Argentina 12 April 1951, and refitted at the Philadelphia Navy Yard. The superstructure was reduced, bulges were added, the beam was increased and mainmast derricks and catapults were removed. The hangar in *General Belgrano*'s hull could accommodate two helicopters with spare parts; the nearly flat counter and high freeboard aft provided an excellent helicopter platform. She commissioned 5 December 1951, although numerous erroneous dates are cited. She was fitted with a Dutch search radar in the early 1960s, and in 1967 two Seacat SAM launchers were installed aft of the bridge. In 1980 she still mounted her full armament (including 20mm guns). She was retained for her amphibious support role and used as a gunnery training ship. During the Falklands conflict, *General Belgrano* and two escorting *Allen M Sumner* class Exocet-armed destroyers were stationed to the southwest of the Falklands (Malvinas). She patrolled on a 200-mile line between Burwood Bank, S of the Falklands, and Isla de los Estados, E of Tierra del Fuego. On 2 May at 1600 hrs local time she was torpedoed at a range of about 1400 yards by HMS *Conqueror*. At the time she was torpedoed, *General Belgrano* was approximately 220 miles from the Falklands sailing at 10kts toward the mainland. Two Mk8** torpedoes hit, the first under the aft 5in gun director and the second just forward of A turret. The first hit tore the emergency generators from their foundations. The ship was without power. *General Belgrano* sank at 1645 hrs; 321 lives were lost (95 per cent killed by the torpedo hits) with the cruiser and 880 rescued after 24 hours in life rafts.

Nueve de Julio was launched 3 December 1936, and was commissioned 19 December 1951. In the early 1960s, she, too, was fitted with a Dutch surface search radar. Decommissioned and stripped in 1978, she served as parts reservoir for her sister-ship and was scrapped in 1983.

Nueve de Julio (as USS *Boise*) 26.10.45 L & L van Ginderen

Ex-US FLETCHER class *destroyers*

No	Name	Builder	Acquired	Fate
D 24	ALMIRANTE STORNI (ex-*Cowell*)	Bethlehem, San Pedro	17.8.71	Decommissioned 1982
D 20	BROWN (ex-*Heerman*)	Bethlehem, San Francisco	See notes	Decommissioned 1982
D 23	ALMIRANTE DOMECQ GARCIA (ex-*Braine*)	Bath Iron Wks	17.8.71	Expended as target 7.10.83
D 21	ESPORA (ex-*Dortch*)	Federal, Kearny	See notes	Decommissioned 1977
D 22	ROSALES (ex-*Stembel*)	Bath Iron Wks	See notes	Decommissioned 1982

Brown and *Espora* commissioned on 10 August 1961 and *Rosales* on 3 August 1961. *Almirante Storni* fired across the bows of the British scientific survey ship *Shackleton* on 2 April 1976 S of the Falklands.

Ex-US ALLEN M SUMNER class *destroyers*

No	Name	Builder	Acquired	Fate
D 26	HIPOLTO BOUCHARD (ex-*Borie*)	Federal, Kearny	1.7.72	Deleted 1984
D 29	PIEDRA BUENA (ex-*Collett*)	Bath Iron Wks	Apr 1974	Deleted 1984
D 25	SEGUI (ex-*Hawk*)	Federal, Kearny	1.7.72	Deleted 1983
D 31	ESPORA (ex-*Mansfield*)	Bath Iron Wks	Jun 1974	BU 1978

Bouchard was fitted with VDS, helicopter facilities and a hangar prior to transfer. *Piedra Buena* was originally activated for spares, but she was activated in 1977; the destroyer *Mansfield* was also originally transferred for spares, although for a while it was considered commissioning her as *Espora*. Four MM38 Exocet (4×1) SSMs added to operational units in 1977–78. *Segui* did not receive FRAM II modernisation as other ships, although both have since had their VDS removed.

During the Falklands conflict *Bouchard* and *Piedra Buena* were used to screen the cruiser *General Belgrano*, and delivered an unsuccessful depth charge attack following the torpedoing of that ship. *Bouchard* was probably hit by HMS *Conqueror*'s third Mk8** torpedo, which failed to explode, on the far side of the *Belgrano*. *Segui* operated with the carrier during the conflict.

Ex-US GEARING (FRAM II) class *destroyer*

No	Name	Builder	Acquired	Fate
D 27	COMODORO PY (ex-*Perkins*)	Consolidated, Orange	15.1.73	Deleted 1984

Four MM38 Exocet SSMs added 1977–78. Has SPS-40 and SPS-10 radar and SWS-23 sonar. During the Falklands conflict, she was part of the force working with the carrier. See under United States for further details.

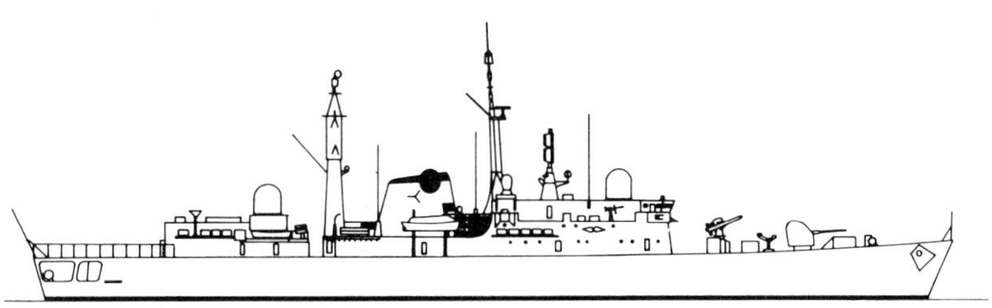

Hercules 1979

British TYPE 42 (SHEFFIELD class) *destroyers*

No	Name	Builder	Laid down	Launched	Comp	Fate
D 1	HERCULES	Vickers, Barrow	16.6.71	24.10.72	10.5.76	Extant 1995
D 2	SANTISIMA TRINIDAD	AFNE, Rio Santiago	11.10.71	9.11.74	July 1981	Extant 1995

On 18 May 1970 the Argentine Government announced the signing of a contract for two ships, one to be built in Argentina and the other in Great Britain. *Santisima Trinidad* required a long time to build owing to delays caused by the technology transfer. She was sabotaged by terrorists on 22 August 1975, causing moderate damage. *Hercules* was fitted with two MM38 Exocet above the hangar in 1979 and *Santisima Trinidad* commissioned with this system.

Both ships have since been fitted with four MM38, two either side of the funnel, rather than on top of the hangar. The Argentine Type 42s were part of the invasion force that captured the Falklands on 2 April. The 150 commandos that were landed at night on 1 April went ashore by helicopter from *Santisima Trinidad*. She was the flagship for the invasion force. During late April and early May, *Santisima Trinidad* provided a screen for the carrier. During this period, *Hercules* provided a screen for a destroyer force operating independently from the carrier forces.

Electronic fit differs from their British sisters and includes combat data system ADAWS-4, radars Type 965P, Type 1006, Type 992Q, Type 909, sonar Type 184M and Type 162M. These ships operate small French SA 319B Alouette III helicopters instead of the standard British Lynx. Pennant numbers are displayed without the D prefix. Both were laid up for some years during 1983–86 and reportedly offered for sale to Turkey in 1986 due to the inability to obtain spares from the UK. After some trials in 1987 both were put back into service in 1988. See under United Kingdom for further particulars.

Santisima Trinidad 29.6.81 L & L van Ginderen

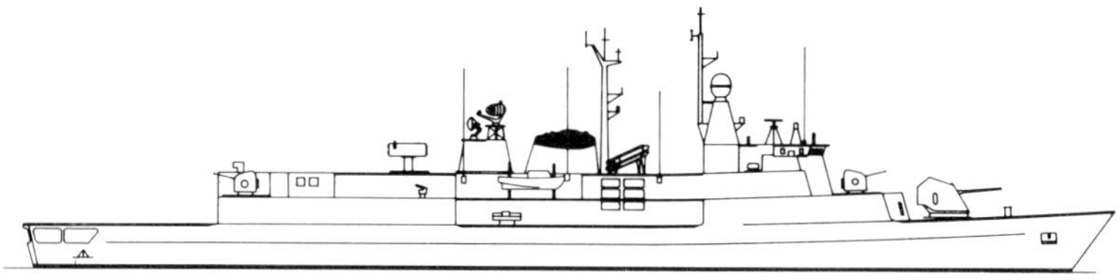

MEKO 360 H2 design

Initially to have been a class of six, four of which were to be built in Argentina, instead the Argentines decided to build four MEKO 360s and six MEKO 140s. Reloads are carried for the Aspide SAM magazine. The 40mm are controlled by two HSA LIROD optronic/radar directors. These ships were built on the modularised systems concept. Under construction during the conflict in the South Atlantic, the British-built gas turbines were embargoed for a while.

These ships are well liked in the Argentine Navy and their high operational availability was shown by *Almirante Brown* taking part in Allied Gulf operations in late 1990. Their ASW capability is at present limited as the original order for Lynx helicopters was cancelled in 1982 and the planned purchase of Italian AB 212 machines was not funded. These ships carry two small SA 319B Alouette helicopters, but plans foresee the acquisition of US Kaman SH-2 machines. The MEKO concept is well suited for rapid change of armament set or modernisation and refit. A very similar ship named *Aradu* serves in the Nigerian Navy. Smaller derivatives called MEKO 200 gained much popularity.

MEKO 360 H2 class *frigates*

Displacement: 2900t standard; 3360t full load
Dimensions: 390ft 5in pp, 413ft oa × 49ft 3in × 19ft max
125.9m, 119.0m × 15.0m × 5.8m
Machinery: 2-shaft COGOG: 2 Olympus TM 3B gas turbines, 51,800shp = 30.5kts; 2 Tyne RM-IC gas turbines, 10,200shp = 18kts. Range 4500nm at 18kts
Armament: 8 MM40 Exocet SSM missiles (4×2), 1 Albatros-Aspide SAM (1×8; 24 missiles), 1–127mm OTO Melara, 8–40mm Breda (4×2), 6–324mm ASW TT (2×3; 18 torpedoes), 2 Sea Lynx helicopters
Sensors: Combat data system SEWACO; radar HSA DA 08A, HSA ZW 06, Decca 1226, Signaal STIR, Signaal WM 25; sonar Krupp Atlas 80, DSQS 21BZ; countermeasures CSEE Dagaie, Breda SCLAR, Graseby G1738; ESM/ECM Sphinx/Scimitar
Complement: 200

No	Name	Builder	Laid down	Launched	Comp	Fate
D 10	ALMIRANTE BROWN	Blohm & Voss, Hamburg	8.9.80	28.3.81	9.2.83	Extant 1995
D 11	LA ARGENTINA	Blohm & Voss, Hamburg	31.8.81	25.9.81	19.7.83	Extant 1995
D 5	HEROINA	Blohm & Voss, Hamburg	24.8.81	17.2.82	31.10.83	Extant 1995
D 13	SARANDI	Blohm & Voss, Hamburg	9.3.82	31.8.82	16.1.84	Extant 1995

Heroina 26.10.85 L & L van Ginderen

Launched 22 August 1942. Armed in Argentina with 3–105mm (3×1), 4–47mm (4×1) AA, DC racks. Transferred to *Guardacostas* 1961 and renamed *Juan Batista Azopardo*.

Ex-British 'RIVER' class *frigate*

No	Name	Builder	Acquired	Fate
P 31	HERCULES (ex-*Asheville*, ex-*Nadir*, ex-*Adur*)	Canadian Vickers	18.2.48	Sold 1969

Ex-US TACOMA class *patrol frigates*

Class (former name, comm, fate):
Heroina (ex-*Reading*, 8.2.47, sold 5.8.64), *Sarandi* (ex-*Uniontown*, ex-*Chattanooga*, 18.2.48, sold 29.6.67), *Santisima Trinidad* (ex-*Caicos*, ex-*Hannam*, ?, sold 1971)

All units armed in Argentina with 2–105mm (2×1), 4–47mm (4×1) AA, DC racks. Pennant nos P 32–34 respectively. *Heroina* was converted 1961 to a meteorological vessel; *Santisima Trinidad* (name often incorrectly cited as *Trinidad*) was converted to a hydrographic ship in 1962 and renamed *Comodoro Lasserre*.

Ex-British modified 'FLOWER' class *frigate*

Class (no, acquired):
Republica (P 35, 28.12.48).

British corvette *Smilax* launched on 24 December 1942. Armed in Argentina with 1–105mm, 4–20mm (4×1), DC racks. See under Great Britain in the 1922–1946 volume for further details.

Drummond and *Guerrico* Apr 1980 Robert L Schiena

French A 69 type *corvettes*

Armament: 4 MM38 Exocet SSM (4×1), 1–100mm, 1–40mm, 2–20mm (2×1), 6–324mm ASW TT (2×3)
Sensors: Radar DRBV-51A, DRBC-32C, Decca 1226; sonar Diodon
Other particulars: As French *D'Estienne D'Orves* class

No	Name	Builder	Laid down	Launched	Comp	Fate
F 701	DRUMMOND (ex-*Good Hope*)	Arsenal de Lorient	12.3.76	5.3.77	Oct 78	Extant 1995
F 702	GUERRICO (ex-*Transvaal*)	Arsenal de Lorient	1.10.76	13.9.77	Oct 78	Extant 1995
F 703	GRANVILLE	Arsenal de Lorient	1.12.78	28.6.80	22.6.81	Extant 1995

Good Hope and *Transvaal* were originally built for South Africa, but delivery was embargoed. They were purchased by Argentina on 25 September 1978. For further details see under France. All three units were deployed extensively during the Falklands conflict. *Guerrico* supported the South Georgia operation on 3 April 1982. She was hit by an anti-tank rocket and heavy small arms fire. The hit was just below the Exocet launcher and the damage was modest. She spent three days in the yard being repaired. The other two ships supported the main invasion on 2 April. All three ships operated as independent force to the north of the islands during late March and the first two days of April. She retired to protected waters along with the rest of the fleet on 2 April.

Drummond and *Guerrico*, the ex-South African units, were originally ordered for the French Navy as *Lieutenant de Vaisseau le Henaff* and *Commandant l'Herminier*. Currently all three are armed with twin Breda 40mm AA guns aft replacing the old Bofors guns. Armament augmented by two single Oerlikon 20mm AA guns. Fitted with fin stabilisers. Fitted with CSEE Panda Mk2 optical director. Armament includes four MM 38 Exocet SSM. Original pennant number P 1-P 3 changed recently to F 31-F 33 (letters are not displayed).

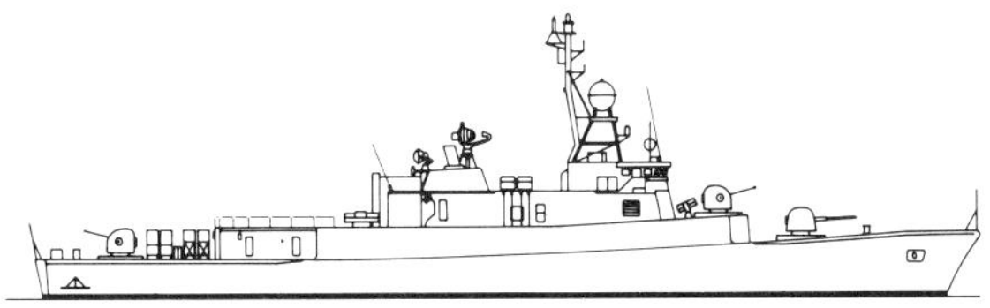

Espora as built

MEKO 140 class *corvettes*

Displacement: 1560t standard; 1790t full load
Dimensions: 299ft 3in oa, 283ft 6in pp × 40ft × 10ft 11in
91.2mm, 86.4m × 11.1m × 3.33m (hull)
Machinery: 2 shafts, 2 SEMT-Pielstick 16 PC2-5V400 diesels, 22,600bhp = 27kts. Range 4000nm at 18kts
Armament: 4 MM38 Exocet SSM (4×4), 1–76mm OTO Melara Compact, 4–40mm (2×2), 6–324mm ASW TT(3×2), 1 Sea Lynx helicopter
Sensors: Radar Decca 1226, DA-05/2, WM 28, LIROD gfes; sonar Krupp A50 80
Complement: 93 (11 officers)

No	Name	Builder	Laid down	Launched	Comp	Fate
F 4	ESPORA	AFNE, Rio Santiago	3.4.81	23.1.82	5.7.85	Extant 1995
F 5	ROSALES	AFNE, Rio Santiago	1.7.81	4.3.83	14.11.86	Extant 1995
F 6	SPIRO	AFNE, Rio Santiago	1.4.82	24.6.83	24.11.87	Extant 1995
F 7	PARKER	AFNE, Rio Santiago	2.8.82	31.3.84	17.4.90	Extant 1995
F 8	ROBINSON	AFNE, Rio Santiago	6.6.83	15.2.85		For sale incomplete 1992
F 9	GOMER ROCA	AFNE, Rio Santiago	1.12.83	14.11.86		For sale incomplete 1992

The design, based on the Portuguese *Joao Coutinho*, was ordered on 1 August 1979. Designed by Blohm & Voss, using the modular concept, these ships allow much flexibility in armament fit according to need. Whole class was built in Argentina with German assistance. First three completed without helicopter hangar. *Parker* fitted with telescoping hangar which is to be fitted to the sisters. All fitted with SEWACO combat data system as well as CSEE Dagaie chaff launchers and ESM/ECM Racal RQN-3B/TQN-2X. Armament augmented by two 12.7mm MGs. It is intended to fit whole class with eight MM40 Exocet SSM instead of four

MM38. Capable of operating Lynx helicopter, but equipped with SA 319B Alouette III. *Parker* was flooded in 1986 while fitting out which delayed completion. Fitted with stabilisers. These successful ships are used mainly for offshore patrol and fishery protection duties. *Spiro* and *Rosales* took part in Allied operations in the Gulf in 1990–91. Original planned pennant numbers F 4–F 9 changed to F 10–F 15 and in 1988 to the present F 41–F 46. Last two ships, still uncompleted due to lack of funds, are being offered for sale and recently Venezuela had shown interest in acquiring them.

Parker 24.5.95 L & L van Ginderen

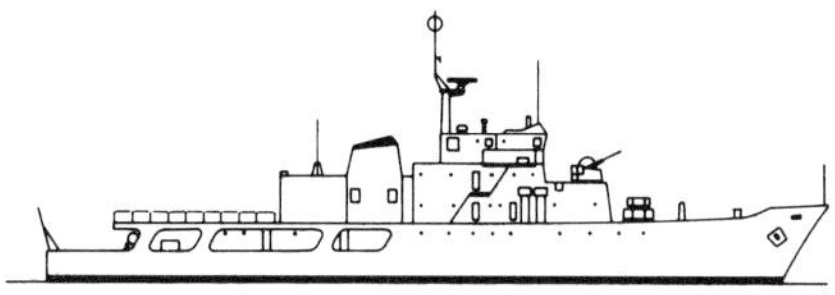

Mantilla as built

Ordered in 1979 from Bazan, Spain. Specially fitted for patrol duties in harsh conditions prevailing around southern Argentinian coasts. Currently operated by Coast Guard. Endurance 20 days; carry 144 rounds for 40mm gun. Originally equipped with Alouette III helicopters on loan from the navy. There are plans to limit the Coast Guard activities to operations inside the 12 mile territorial seas, which may cause some ships to be transferred to the navy.

Spanish HALCON type *corvettes*

Displacement: 767t standard; 910t full load
Dimensions: 220 ft × 34ft 6in × 13ft 7in
67.0m × 10.5m × 4.2m
Machinery: 2 shafts, 2 Bazan-MTU 16V 956 TB91 diesels, 9000bhp = 21kts. Range 5000nm at 18kts
Armament: 1–40mm Breda, 2–12.7mm (2×1), 1 HB 350B Esquilo helicopter
Sensors: Radar Decca 1226
Complement: 33 (10 officers + 4 cadets)

No	Name	Builder	Laid down	Launched	Comp	Fate
GC 24	MANTILLA	Bazan, Ferrol	16.2.81	29.6.81	20.12.82	Extant 1995
GC 25	AZOPARDO	Bazan, Ferrol	1.4.81	14.10.81	28.4.83	Extant 1995
GC 26	THOMSON	Bazan, Ferrol	Feb 1981	7.12.81	20.6.83	Extant 1995
GC 27	PREFECTO PIQUE	Bazan, Ferrol	Sep 1981	24.2.82	20.7.83	Extant 1995
GC 28	PREFECTO DERBES	Bazan, Ferrol	Nov 1981	16.6.82	20.11.83	Extant 1995

SUBMARINES

Santa Fe (i) was refitted at Mare Island Navy Yard, and commissioned 21 August 1960; she was decommissioned in 1971. *Santa Fe* (ii) received a GUPPY II refit in 1948–50. The *Santa Fe* (ii) was employed in the initial invasion of the Falklands. On 2 April 1982 a small force of frogmen swam from the submerged submarine off Port Stanley and secured the landing site for the 600-plus Argentine Marine Corps invasion force. On 25 April *Santa Fe* was attacked by three British helicopters while on the surface after she had dropped supplies to the Argentine garrison of S Georgia. The submarine was hit by a Wasp AS 12 missile; she managed to reach King Edward Cove where she settled on the bottom. One petty officer was severely wounded during the attack and another killed by a British sentry after the crew had been taken prisoners. The hulk has since been moved out of the anchorage.

Santiago del Estero (i) was refitted at Mare Island Navy Yard and commissioned 21 August 1960. A modernised fin was added in the United States in 1966; she was decommissioned in 1971. *Santiago del Estero* (ii) received a GUPPY IA refit in 1961. She was not operational during the Falklands conflict. Initially the submarine remained at Mar del Plata, the primary submarine base. During April she was moved to Puerto Belgrano in an attempt to keep the Royal Navy guessing.

Ex-US BALAO class *submarines*

No	Name	Builder	Acquired	Fate
S 11	SANTA FE (i) (ex-*Lamprey*)	Manitowoc	27.7.60	BU for spares 1971
S 21	SANTA FE (ii) (ex-*Catfish*)	Electric Boat	7.1.71	Beached 25.4.82, later sunk
S 12	SANTIAGO DEL ESTERO (i) (ex-*Macabi*)	Manitowoc	11.8.60	BU for spares 1972
S 22	SANTIAGO DEL ESTERO (ii) (ex-*Chivo*)	Electric Boat	Jan 1971	Deleted 1983

Santiago del Estero outboard of the *Almirante Domecq Garcia* (*Comodoro Py* astern) 1982 Robert L Schiena

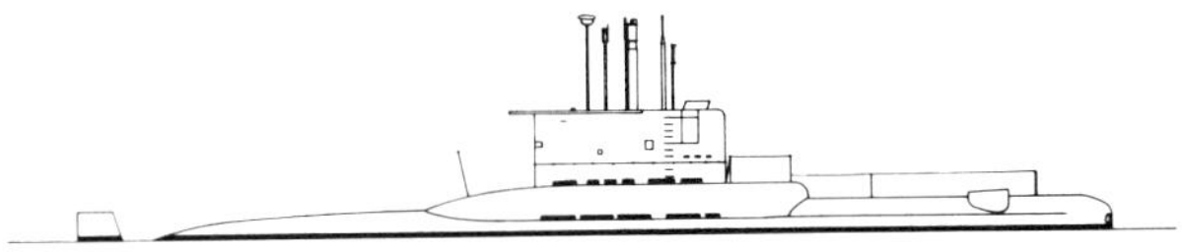

Salta as completed
NB 1/750 scale

West German Type 209 *submarines*

Displacement: 1248t surfaced; 1440t submerged
Dimensions: 183ft 4in × 20ft 7in × 18ft
55.9m × 6.3m × 5.5m
Machinery: As Greek Type 209
Armament: As Greek Type 209
Sensors: Fire control HSA M8, radar Thomson-CSF Calypso II, sonar Krupp Atlas CSU 3, Thomson-Sintra DUUX 2C and DUUG 1D
Complement: 32 (5 officers)

No	Name	Builder	Laid down	Launched	Comp	Fate
S31	SALTA	Howaldtswerke, Kiel/AFNE, Rio Santiago	30.4.70	9.11.72	7.3.74	Extant 1995
S32	SAN LUIS	Howaldtswerke, Kiel/AFNE, Rio Santiago	1.10.70	3.4.73	24.5.74	Extant 1995

These two units were built in sections in West Germany and assembled in Argentina. Both units deployed for the Falklands conflict. *Salta* returned to base suffering from some mechanical problem. *San Luis* was stationed in a 'free fire' area north of the Falklands. She attacked British forces on two occasions. On the first, on about 5 May, she fired torpedoes at a British task force; none hit. *San Luis* was pinned down for about 20 hours by British forces before escaping. On 10 May *San Luis* attacked another target without result.

Salta in refit from late 1980s and in 1991 *San Luis* also had begun refit at Domecq Garcia Shipyard. Progress is very slow due to lack of funds. Both ships may be sold and modernisation completed for new owner.

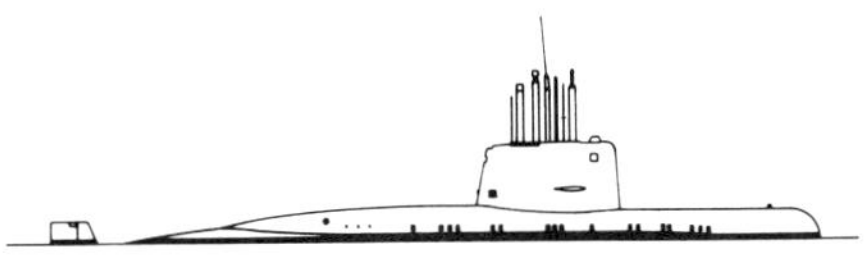

Santa Cruz 1984

SANTA CRUZ class (TR 1700) *submarines*

Displacement: 2116t surfaced; 2264t submerged
Dimensions: 216ft 6in × 24ft × 21ft 4in
66.0m × 7.3m × 6.5m
Machinery: 1 shaft diesel-electric: 4 MTU 16V 652 diesels plus 1 Siemens electric motor and 4 alternators, 6720bhp (6600/4400kW)=15kts/25kts. Range 12,000nm at 8kts surfaced, 20nm at 25kts dived, 460nm at 6kts dived
Armament: 6–533mm TT (bow, 22 torpedoes)
Sensors: Radar Thomson-CSF Calypso; sonar Atlas Elektronic CSU 3/4, Thomson Sintra DUUX-5
Complement: 26 (plus 6 spare berths)

No	Name	Builder	Laid down	Launched	Comp	Fate
S 41	SANTA CRUZ	Thyssen Nordseewerke, Emden	6.12.80	28.9.82	18.10.84	Extant 1992
S 42	SAN JUAN	Thyssen Nordseewerke, Emden	18.3.82	20.6.83	19.11.85	Extant 1992
S 43	SANTA FE	Astilleros Domecq Garcia	4.10.83			Suspended
S 44	SANTIAGO DEL ESTERO	Astilleros Domecq Garcia	5.8.85			Suspended

Contracted on 30 November 1977. Initially Argentina was to acquire four Type 1700s and two TR 1400s (a smaller version), but in 1982 the order was changed to six of the larger type. First two German-built boats completed and delivered in 1984–85, while other two being built in Argentina were suspended pending funding and in 1986 permission was given by German partner to sell all these submarines to a third country, but so far buyer not found. These are very advanced and fast diesel-electric submarines. Diving depth is 270m and pressure hull is 48m long. Fitted with torpedo auto-reload system allowing the reloading of the torpedo tubes in 50 seconds. Endurance is 30 days.

AMPHIBIOUS WARFARE VESSELS

Modified US DE SOTO COUNTY type *tank landing ship*

Displacement: 4164t normal; 7100t full load
Dimensions: 445ft × 62ft × 17ft 6in
135.6m × 18.9m × 5.3m
Machinery: 2 shafts; 6 diesels, 13,700bhp = 16kts
Armament: 12–40/60mm (3×4)
Sensors: Plessey AWS-1 radar
Complement: 124

No	Name	Builder	Launched	Fate
Q 42	CABO SAN ANTONIO	AFNE, Rio Santiago	1968	Extant 1995

Very similar to US *De Soto County* LSTs; commissioned 2 November 1978. Can carry up to twenty-three medium tanks and 700 men. She carried the reinforced 2nd Marine Battalion to the Falklands in April 1982 and laid mines off Port Stanley.

Ex-US LST 511 class *tank landing ships*

Class:
Cabo San Bartolome (ex-*BDT 2*), *Cabo San Francisco de Paula* (ex-*BDT 3*), *Cabo San Gonzalo* (ex-*BDT 4*), *BDT 5*, *Cabo San Isidro* (ex-*BDT 6*), *Dona Elora* (ex-*BDT 7*), *Dona Stella* (ex-*BDT 8*), *Dona Dorothea* (ex-*BDT 9*), *BDT 10*, *Dona Irma* (ex-*BDT 11*), *BDT 12*, *BDT 13*, *Cabo San Vicente* (ex-*BDT 14*).

Fourteen ex-US LSTs, built by Puget Sound DD in 1944, were transferred in the early 1950s. *BDT 5*, *8*, *9* and *12* were stricken 1958–60; *BDT 2*, *7*, *10*, *11* and *13* were broken up 1964; *BDT 1* was deleted 1971, *BDT 6* in 1978, *BDT 4* and *14* in 1980, and *BDT 2* in 1981.

OTHER AMPHIBIOUS SHIPS
There were also two ex-US LSMs, *BDM 1* and *BDM 2* acquired after the war. *BDM 2* became the mine warfare support ship *Corrientes* in 1969, and *BDM 1* was stricken 1971. A number of troop transports have also been employed since 1947: *Bahia Aguirre* and *Bahia Buen Suceso* (3100t, built in Canada, 1950) were still in service 1982 the latter landing the Argentine scrap metal party on S Georgia; *La Pataia*, *Le Maire* and *Les Eclaireurs* (3825t, built in Italy 1951): *La Pataia* BU 1986, *Le Maire* BU 1964, *Les Eclaireurs* accommodation vessel 1972.

Ex-US ASHLAND class *dock landing ship*

No	Name	Builder	Acquired	Fate
Q 43	CANDIDO DE LASALA (ex-*Gunston Hall*)	Moore, Oakland	24.4.70	Deleted 1982

War-built LSD armed with 12–40mm (2×4, 2×2). Doubled, when required, as a depot ship for small craft.

MINOR LANDING CRAFT
In 1995 these comprised:
Four ex-US LCM(6) type, numbered *EDM1–4*, transferred 1971.
Ex-US LCVP type, variously reported as eight and fourteen strong, numbered between *EDVP 1* and *37*; the remainder of thirty-seven craft transferred between 1946 and 1970.
There are also eight Argentine-built LCVPs, constructed since 1971.
Fifteen US LCILs were transferred after the war, and numbered *BD 1* to *15*. These were withdrawn from service between 1958 and 1971.

SMALL SURFACE COMBATANTS

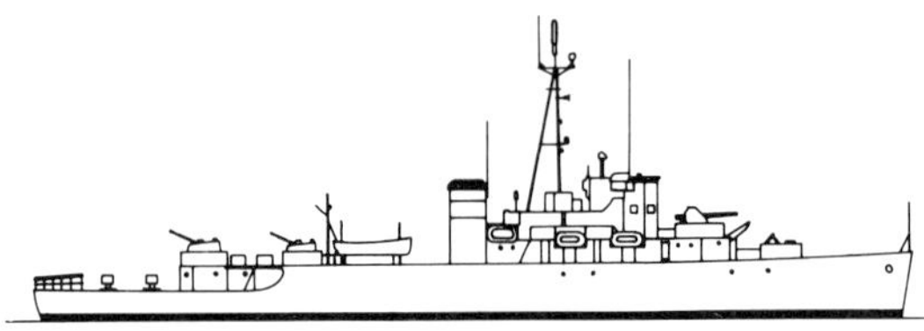

Azopardo 1966

AZOPARDO class *patrol escorts*

Displacement: 1220t standard; 1400t full load
Dimensions: 278ft 10in pp × 31ft 6in × 13ft 2in max
85.0m × 9.6m × 4.0m
Machinery: 2-shaft Parsons geared turbines, 2 watertube boilers, 5000shp = 20kts. Range 2300nm
Armament: 1–4.1in, 2–40mm (2×1) AA, 1 Hedgehog, 4 DC launchers
Complement: 170

Class (launched, comm):
Azopardo (11.12.53, Jan 1955), *Piedra Buena* (17.12.54, 16.12.58).
Both built by AFNE, Rio Santiago. Pennant nos P 35 and P 36 respectively. Originally these units were to be *King* class patrol vessels, but they were redesigned to be a larger, follow-on type. They were discarded in 1973.

US 78ft HIGGINS type *torpedo boats*

Displacement: 36t standard; 50t full load
Dimensions: 78ft oa × 21ft × 5ft
23.8m × 6.4m × 1.5m
Machinery: 3-shaft Packard petrol engines, 4500bhp = 42kts
Armament: 2–40mm (2×1), 4 torpedoes
Complement: 10 (est)

Class:
Alakush (P 82), *Towwora* (P 84); the last of LT 1–9
Commissioned in 1949, and built specifically for Argentina. Five were disposed of in 1963 and two in 1966; the final pair were deleted 1985.

West German TNC 45 type *fast attack craft*

Displacement: 240t standard; 265t full load
Dimensions: 138ft 9in pp, 147ft 4in oa × 24ft 3in × 7ft 6in
42.3m, 44.9m × 7.4m × 2.3m
Machinery: 4-shafts, 4 MTU MD 872 diesels, 14,000bhp = 37kts. Range 640nm/1700nm at 36kts/16kts
Armament: 1–76mm OTO Melara Compact, 2–40mm (2×1), 2–533mm TT
Sensors: Radar Decca 101, HSA M22
Complement: 42

Class (no, launched, comm):
Indomita (ELPR 2, 8.4.74, 12.12.74), *Intrepida* (ELPR 1, 12.12.74, 20.7.74).
Both built by Lürssen, Vegesack, ordered in 1970. Two further projected units were cancelled. The TT fire German wire-guided T4 torpedoes. Both extant 1995.

Israeli DABUR type *patrol craft*

Class (no):
Baradero (P 61), *Barranqueras* (P 62), *Corinda* (P 63), *Concepcion del Uruguay* (P 64).
Built by IAI and delivered in 1978. Armed with 2–20mm (2×1) and 4–12.7mm MG (2×2). All extant 1995.

OTHER PATROL VESSELS
The Argentine Navy also operates a number of armed tugs in the patrol role. Two 1235t ex-US *Achomawi* class ocean tugs, *Comandante General Irigoyen* (ex-*Cahuilla*, transferred 1961) and *Francisco de Churruca* (ex-*Luiseno*, purchased 1971) are armed with 4–40mm (1×2, 2×1). Three 570t ex-US *Sotoyomo* class tugs, *Yamana* (ex-*Maricopa*, transferred 1947), *Alferez Sobral* (ex-*Catawha*, transferred 1972) and *Comodoro Somellera* (ex-*Salish*, transferred 1972), are armed with 1–40mm and 2–20mm (2×1). A sister, *Daiquita* was sold in 1979; *Sobral* badly damaged by Sea Skua missiles fired from British Lynx helicopters during the Falklands conflict. Twenty patrol craft of Z-28 class ordered in 1978 at Blohm & Voss, Germany and delivered in 1979–80: *Mar del Plata, Martin Garcia, Rio Lujan, Rio Uruguay, Rio Paraguay, Rio Parana, Rio Plata, La Plata, Buenos Aires, Cabo Corrientes, Quequen, Bahia Blanca, Ingenierio White, Golfo San Matias, Madryn, Rio Desado, Ushuaia, Canal de Beagle, Islas Malvinas* and *Rio Iguaza* (GC 64–83); some took part in the Falklands conflict and GC 83 was lost and GC 82 captured by British forces and renamed *Tiger Bay* (81t full load, 27.6m, 2100bhp = 22kts, 1200nm at 12kts, 1–20mm, 2–12.7mm).

MINE WARFARE VESSELS

Ex-British 'TON' class *minesweepers*

Class (former name, recomm):
Chaco (ex-*Rennington*, Dec 1969), *Chubut* (ex-*Santon*, 19.7.68), *Formosa* (ex-*Ilmington*, 19.7.68), *Neuquen* (ex-*Hicleton*, 13.9.68), *Rio Negro* (ex-*Tarlton*, 20.6.68), *Tierra del Fuego* (ex-*Bevington*, 23.12.68).
Pennant Nos M 5, M 3, M 6, M 1, M 2 and M 4 respectively. *Chaco* and *Formosa* have been converted to minehunters. All units still in service 1995.

Australia

Before the Second World War the Royal Australian Navy had been the largest of the Commonwealth navies, but difficulties in supplying material from Great Britain after 1941 slowed down expansion by comparison with Canada. In 1946, however, there was still an impressive line-up of cruisers, destroyers and escorts.

Postwar building policy tended to follow RN lines at first, with copies of the latest destroyer designs being built locally, but in 1962, for the first time, guided missile destroyers were bought from the United States, marking a growing reliance on American technology rather than European, to avoid the shortage of spares which had plagued the RAN in 1942–45. There was still close liaison with the RN on design work, however, first on the Type 21 frigate programme and then for the DDL programme during 1966–73, but for political reasons the options were not taken up and once again US ships were purchased.

Although the former insistence on building hulls in Australia has proved impossible to sustain, every effort is made to generate work in Australia, either by developing systems or by offset agreements to make components. In conjunction with the RN an anti-submarine missile-system, Ikara, was developed in the late 1950s and early 1960s. When European and American sonars proved unable to cope with the temperature range and sound-propagation found in the southwest Pacific, the RAN went ahead with its own sonar, Mulloka.

In recent years the RAN has concentrated its efforts on providing forces capable of backing up US forces in the Pacific rather than trying to specialise in ASW or minesweeping. With an immense coastline to watch it has built a number of small patrol craft, but as tension with Indonesia has subsided, several of the older craft were transferred both to that country and to the newly-independent Papua New Guinea. At present the main commitment is to the replacing of the heterogeneous force of submarines and escorts with new construction. As always, local industrial participation is fostered; two FFG–7 type frigates have been built locally, the Swedish submarines are being built in a new yard, and a German-designed frigate is being built entirely in Australia. A GRP minehunter design is also going ahead.

The loss of fixed-wing carrier aviation when HMAS *Melbourne* was scrapped has not been made good by shore-based aircraft. Despite Air Force claims to be able to offer shore-based air cover, the RAN and the Army were increasingly aware that the new system was unworkable. The Army felt the lack of aircraft very keenly and in the early 1990s there were moves to build up to three Training and Helicopter Support Ships (THSS). This plan was dropped because of cost, and two tank landing ships were bought from the US Navy for conversion instead. The RAN is already looking ahead to modernise and replace the surface fleet.

FLEET STRENGTH 1947

CRUISERS

Name	Launched	Disp (deep load)	Fate
Kent class			
AUSTRALIA	17.3.27	14,910t	BU 1955
London class			
SHROPSHIRE	5.7.28	14,580t	BU 1955
Amphion class			
HOBART (ex-*Apollo*) (C 02)	9.10.34	9420t	BU 1962

Kent class
Australia served as the RAN flagship for several years. Following repairs to the kamikaze damage in January 1945 she had 'X' turret removed and had a modern RN-pattern director control tower on the bridge, fitted with Type 284 radar. A high-angle DCT with Type 285 radar was fitted on the centreline, forward of the mainmast. She was unusual in that her 4in guns were mounted one deck lower.

London class
Shropshire lost her 20mm guns, but was otherwise unaltered after 1945. She differed from *Australia* in having two high-angle DCTs fitted, one on the bridge and one abaft the funnels. She was laid up in 1948.

Amphion class
Hobart completed battle-damage repairs at the end of 1944, but was laid up postwar. She received an extensive refit at the Newcastle State DYd in 1953–56 for service as a training ship, but was then laid up once more. For this role she was given a lattice foremast, but had all TT and secondary guns removed.

DESTROYERS

Name	Fate	Name	Fate
'Tribal' class: launched 1940–44, 2710t deep load			
ARUNTA (D 130)	foundered 13.2.69	WARRAMUNGA (D 123)	BU 1963
BATAAN	BU 1958		
Quilliam class: launched 1941–42, 2480t deep load			
QUADRANT	Type 15 (see notes)	QUIBERON	Type 15
QUALITY	BU 1958	QUICKMATCH	Type 15
QUEENBOROUGH	Type 15		

'Tribal' class
In 1952–53 *Arunta* was modernised for ASW work, followed by *Warramunga* in 1953–54. This involved stepping a lattice mast and extending the after deckhouse from abaft the TT to the former 'Y' mounting. The original 4.7in guns were left in position forward as well as the original fire control, but a twin 4in Mk 16 AA gun, a twin Mk 5 and two Mk 7 Bofors guns were added on the deckhouse, with a Squid triple ASW mortar on the quarterdeck. *Bataan* was scheduled to be similarly modernised, but the cost of the other two ships was so high that her refit was cancelled in 1957. *Arunta* foundered in tow en route to Japan for scrapping.

Quilliam class
The transfer from the RN was made permament in June 1950 and the vessels were then scheduled to undergo conversion to ASW frigates along the lines of the Type 15 frigates in the RN (see Major Surface Ships). As with the destroyers, however, costs rose beyond the original £A400,000 planned and in 1957 *Quality*'s conversion was cancelled and she was put on the disposal list.

FRIGATES

Name	Fate	Name	Fate
Grimsby class: launched 1936 and 1940, 1510t deep load			
SWAN (F 74)	Sold 5.6.64	WARREGO (F 312)	BU 1966
'River' class: launched 1943–45, 2200t deep load			
BARCOO (F 174)	BU 1972	GASCOYNE (F 354)	BU 1972
BARWON (F402)	BU 1962	HAWKESBURY (F363)	BU 1962
BURDEKIN (F 402)	BU 1962	LACHLAN (F 364)	To New Zealand 1949
DIAMANTINA (F 377)	Preserved 1980	MACQUARIE (ex-*Culgoa*) (F 532)	BU 1962
Modified 'River' class: launched 1944, 2187t deep load			
CONDAMINE (F 698)	BU 1962	MURCHISON (F442)	BU 1962
CULGOA (ex-*Macquarie*) (F 408)	BU 1972	SHOALHAVEN (F 535)	BU 1962

Grimsby class
Swan was used for training cadets from 1955 and *Warrego* was employed as a survey ship. Both these former sloops had their armament reduced to 2–4in AA (1×2) and 1–40mm Bofors guns forward, and extra deckhouses were added aft.

'River' class
Originally similar to RN and RCN 'River' class but with different armament. *Barcoo, Hawkesbury* and *Macquarie* had the twin 4in Mk16 in 'A' position, with two Squid DC mortars in 'B' position and a single 40mm Bofors gun on the Squid handing room (*Barcoo* and *Hawkesbury* only). From the early 1950s *Barcoo* was used on survey duties; she was replaced in 1959 by *Diamantina* and *Gascoyne*. *Diamantina* paid off 1979 and preserved at Brisbane.

Modified 'River' class
The armament was modified to approximate to RN 'Bay' class AA frigates, with a twin 4in Mk 16 (2×2) forward and aft, 3–40mm Bofors (3×1) and two twin 20mm guns. To limit topweight the guns were mounted in 'A' and 'Y' positions. All four had lattice masts by 1955.

MINESWEEPERS

Name	Fate	Name	Fate
Bathurst class: launched 1940–43, about 790t			
ARARAT (M 34)	BU 1961	HORSHAM (M 235)	BU 1956
BENALLA (M 323)	BU 1958	INVERELL (M 233)	To New Zealand 1952
BOWEN (M 285)	BU 1956	JUNEE (M 362)	Sold Jul 1958
BUNBURY (M 241)	BU 1962	KAPUNDA (M 218)	BU 1962
BUNDABERG (M 231)	BU 1962	KATOOMBA	BU 1957
CASTLEMAINE (M 244)	HS *c*1956 Preserved 1974	KIAMA (M 353)	To New Zealand 1952
COLAC (M 05)	HS 1962	LATROBE (M 234)	BU 1956
COOTAMUNDRA (M 186)	For disposal 1973	LITHGOW (M 206)	BU 1956
COWRA (M 351)	BU 1962	MILDURA (M 207)	BU 1965
DELORAINE (M 252)	BU 1956	PARKES (M 361)	BU 1957
DUBBO (M 251)	BU 1958	ROCKHAMPTON (M 203)	BU 1962
ECHUCA (M 252)	To New Zealand 1952	SHEPPARTON (M 246)	BU 1958
FREMANTLE (M 246)	BU 1962	STAWELL (M 348)	To New Zealand 1952
GLADSTONE (M 324)	Sold 16.6.56	STRAHAN (M 363)	BU 1962
GLENELG (M 236)	BU 1957	TOWNSVILLE (M 205)	Sold 8.8.56
GYMPIE (M 238)	BU 1962	WAGGA (M 183)	Sold Mar 1962

Bathurst class
Based on the *Bangor* class, but in fact a new design, sometimes described as 'corvette-type' fleet sweepers. Although obsolete many continued to serve as sea-going and static training ships. *Echuca, Inverall, Kiama* and *Stawell* were transferred to the RNZN in 1952 and retained their names. *Castlemaine* presented to the Maritime Trust of Australia in 1973 and preserved as a museum ship at Williamstown.

TANK LANDING SHIPS

Name	Fate	Name	Fate
Ex-British LST (3) type: launched 1944–45, 3065t full load			
LABUAN (ex-*LST 3501*)	BU 1956	*LST 3008*	Sold 4.6.50
LAE (ex-*LST 3035*)	Sold 9.11.55	*LST 3014*	Sold 4.6.50
TARAKAN (ex-*LST 3017*)	Sold 12.3.54	*LST 3022*	Sold 4.6.50

Six vessels transferred after brief service in the Royal Navy; commissioned into the RAN on 1 July 1946. They served principally as transport and supply vessels.

MOTOR LAUNCHES
Of the thirty-five Fairmile 'B' type MLs that were built in Australia during the war, thirty-three survived, but they were rapidly decommissioned and sold at the end of hostilities.

Of the HDML type, nine were built in Australia, two in Britain and sixteen in the USA, and all survived. In 1950 *HDML 1323, 1326, 1328* and *1329* were lent to the RN for use in Hong Kong, and on their return to Australia in 1958 were transferred to the Philippines. By 1952 only *HDML 1321, 1322, 1324, 1325* and *1327* remained in the RAN, and they were rated as seaward defence motor launches (SDMLs) and from 1959 as seaward defence boats. *SDML 1322* was stricken in 1953, *SDB 1327* in 1966, *SDB 1321* in 1972, *SDB 1324* (as TS *Nepean*) stricken 1982 and *SDB 1325* deleted *c*1992.

MAJOR SURFACE SHIPS

HMS *Terrible*, the only aircraft carrier built in one of HM dockyards, was laid down on 10 April 1943, launched on 30 April 1944 and completed for the RAN on 5 February 1949. She saw considerable active service in the Korean War alongside her British sisters, operating Sea Furies and Fireflies (normal complement twelve of each, plus two Sea Otter amphibians). A major modernisation was cancelled in 1954, and from May 1953 she took on flying training duties. In 1962–63 she was relegated to a fast military transport, and her defensive armament was reduced to 4–40mm. In this role she carried vehicles stowed on deck and the hangar was used for troop accommodation, and from 1969 six LCPUs were carried under davits. She was put on the disposal list in 1973 and sold for scrapping on 30 October 1975.

NOTE
The British *Colossus* class light fleet carrier *Vengeance* was loaned to Australia during 1952–55. Taken over by the RAN on 13 November 1952, the ship saw active service until defence cuts in late 1954 reduced the ship to a training role. She was intended as a stop-gap until the commissioning of *Melbourne* and was returned to the RN in June 1955.

SYDNEY *aircraft carrier*

Displacement: 14,000t standard; 17,780t deep load
Dimensions: 630ft pp, 695ft oa × 80ft × 23ft mean deep load
192.0m, 211.8m × 24.4m × 7.0m
Machinery: 2-shaft Parsons geared turbines, 4 Admiralty 3-drum boilers, 40,000shp = 25kts. Oil 3200t
Armament: 24–40mm Bofors, 26 aircraft
Complement: 1300

No	Name	Builder	Acquired	Fate
R 17	SYDNEY (ex-*Terrible*)	Devonport DYd	1948	BU 1976

Sydney 1974, transporting Army equipment to Vietnam RAN

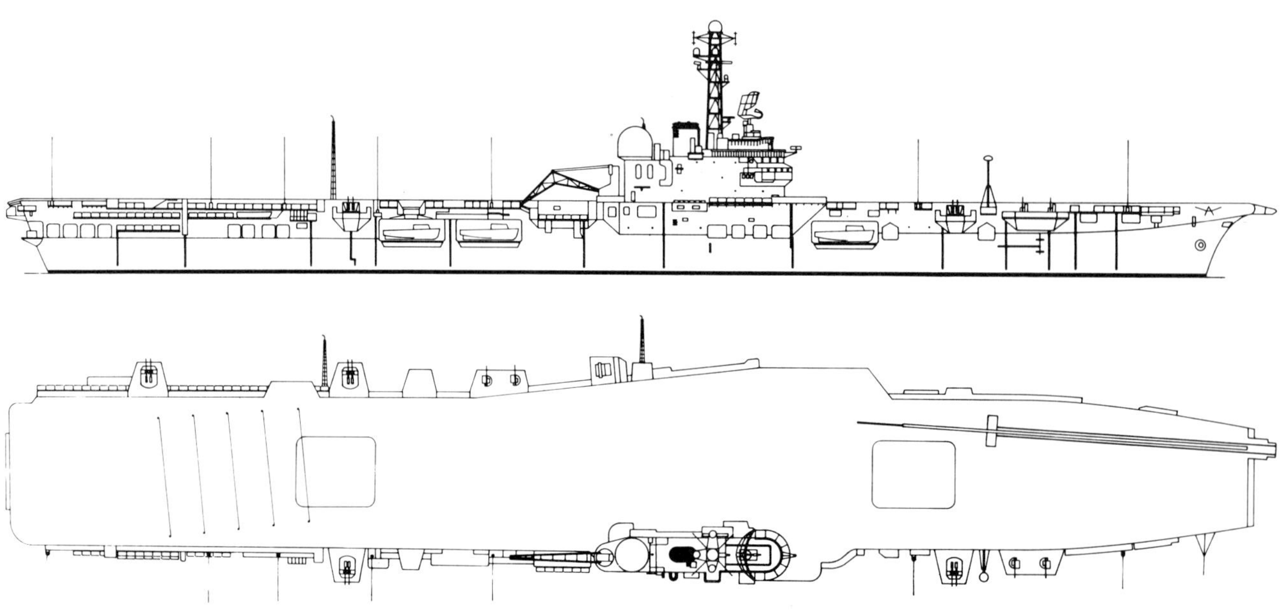

Melbourne 1980

Work was resumed on this vessel in 1949 after her purchase by the Australian government, and the opportunity was taken to incorporate as many new ideas as possible, including a steam catapult, a 6° angled deck and mirror landing sights. However, apart from a lattice mast and whip aerials she looked similar to the original *Majestics*.

A greatly enhanced radar suite was incorporated, including no fewer than three Type 277Q height-finding sets. The armament was reduced slightly to 25–40mm. She recommissioned at the yard of her builders on 28 October 1955 and embarked an air group of eight Sea Venom fighters, twelve Gannet anti-submarine aircraft (replaced in the early 1960s by Wessex helicopters) and two Sycamore helicopters; from 1963–67 this was reduced to four Sea Venoms, six Gannets and ten Wessexes.

After several years' services as Flagship of the RAN she was taken in hand for modification to allow her to operate A-4 Skyhawk fighter-bombers (December 1967–November 1968). This involved new radar and communications and strengthening of the decks, lifts, catapult and arrester gear. This time she received a mixture of Dutch and US radars, although the British Type 293 surface search set and Type 978 navigation sets were retained. The most obvious external changes were the addition of the Signaal LW-02 air surveillance scanner over the bridge, the dome covering the SPN-35 carrier-controlled approach scanner abaft the funnel and a TACAN aerial and various ECM pods on the lattice mast. In addition the funnel was fitted with an inconspicuous cap to divert smoke from the radar arrays.

MELBOURNE *aircraft carrier*

Displacement: 16,000t standard; 19,966t full load
Dimensions: 701ft 6in oa, 650ft wl × 80ft 3in × 25ft 6in
213.8m, 198.1m × 24.5m × 7.8m
Machinery: 2-shaft geared steam turbines, 4 Admiralty 3-drum boilers, 42,000shp = 23kts. Range 6,200nm/12,000nm at 23kts/14kts
Armour: Mantlets over torpedo warheads
Armament: 12–40mm/60 Bofors Mk 5 and Mk 7 (4×2, 4×1), 24 aircraft
Sensors: Radar LW-02, 293, 978, SPN-35. See notes
Complement: 1417 (inc 347 air group) as flagship

No	Name	Yard	Acquired	Fate
R 21	MELBOURNE (ex-*Majestic*)	Vickers-Armstrong, Barrow	1949	BU 1985

Melbourne 1979 RAN

The air group now consisted of four A-4G Skyhawks and six S-2E Tracker ASW aircraft and ten Wessexes, but from 1972 onwards she carried eight Skyhawks, six Trackers and ten Wessexes. Sea King Mk 50 ASW helicopters replaced the Wessexes and some of the Trackers (the composition of her air group changed to meet requirements).

In 1971 the catapult was rebuilt, with a bridle-catcher extension at the forward end of the flight deck, and the deck was strengthened once more. She was refitted again in November 1972–July 1973, and after a further refit in 1976 it was announced that the ship would serve until 1985. In the meantime *Melbourne* paid off into 'Contingent Reserve' on 30 June 1982; she was sold to Chinese breakers and left Sydney on 27 April 1985 for Shanghai.

'MELBOURNE Replacement' *aircraft carrier*

In 1980, the Australian Minister of Defence announced that a replacement for as HMAS *Melbourne* would be ordered in 1981. HMS *Invincible* was offered in 1982 at a very attractive price. The offer was accepted and the name *Australia* was allocated, but the Falklands crisis put an end to negotiations. VSEL then completed a detailed design study for a CVL/CVS to be built in Australia, but this project was dropped because of RAN budget cuts.

Modified 'BATTLE' type *destroyers*

Two ships were ordered for the RAN to a slightly enlarged (Batch 3) version of the Royal Navy's 1943 'Battle' design (eight more for the RN were cancelled in 1945). They were able to accommodate an extra 40mm STAAG mounting amidships, but had the Mk 6 fire control system of the original 1942 'Battle' design. *Tobruk*'s 4.5in guns were imported, but *Anzac*'s were built under licence in Australia; unlike their RN sisters, they had twin Mk 6 mountings as fitted to the *Darings*. In 1966 *Anzac* was converted to a training ship, with 'B' gun replaced by a classroom. Both ships had capped funnels, unlike the RN 'Battles'.

Displacement: 2400t normal; 3500t deep load
Dimensions: 355ft pp, 379ft oa × 41ft × 13ft 6in
108.2m, 115.5m × 12.5m × 4.1m
Machinery: As British 'Battle' class
Armament: 4–4.5in/45 Mk 6 (2×2), 6–40mm/STAAG Mk 2 (3×2), 6–40mm/60 Mk 7 (6×1), 10–21in TT (2×5), 1 Squid ASW mortar
Sensors: Radar 275, 262, 293; sonar 170, 174
Complement: 332

No	Name	Builder	Laid down	Launched	Comp	Fate
D 37	TOBRUK	Cockatoo I DYd	5.8.46	20.12.47	8.5.50	BU 1972
D 59	ANZAC	Williamstown DYd	23.9.46	20.8.48	14.3.51	BU 1975

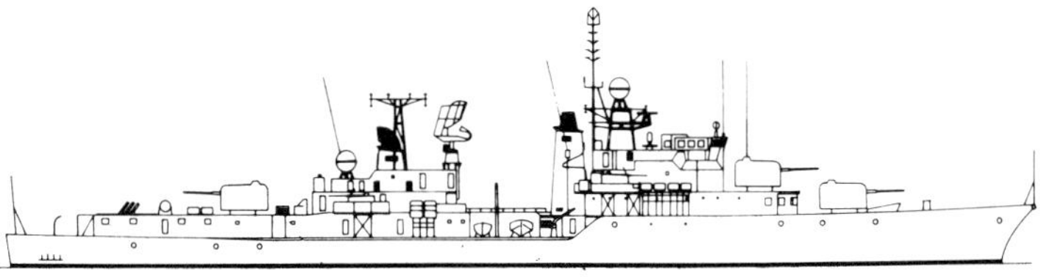

Vendetta 1977

Four modified *Daring* class destroyers were ordered in December 1946, but work did not start until 1949, with material imported from Great Britain. In general the original layout was followed, but the after bank of TT was suppressed and 'X' turret was moved forward to allow a Limbo mortar to be mounted on the centreline. *Waterhen* was cancelled in 1954. As completed, *Voyager* was very similar to the British *Daring*, but *Vampire* and *Vendetta* were given two CRBF directors on the centreline forward and aft of the funnel and two twin 40mm Mk 5 Bofors mountings sided.

Voyager was sunk on 10 February 1964 in collision with the carrier *Melbourne*, and HMS *Duchess* was transferred from the RN on loan; she was bought outright in 1972 and paid off in 1977. In 1969–73 *Vampire* and *Vendetta* underwent complete rebuilding, with Dutch Signaal WM22 fire control forward and aft, an enclosed bridge and an early-warning LW-02 radar array on a stump mast forward of the second funnel. The original lattice foremast was removed, so that a new capped funnel was now visible. *Duchess* was not refitted to the same extent, but had a new after funnel and a large deckhouse aft for classrooms in place of the twin 4.5in mounting and Squid.

Duchess and *Vendetta* were put on the disposal list in 1979, but in 1980 *Vampire* joined a new training squadron, she was paid off in August 1985. *Vendetta* was not stricken until January 1987.

VOYAGER class *destroyers*

Displacement: 2800t normal; 3600t deep load
Armament: 6–4.5in/45 Mk 6 (3×2), 4–40mm STAAG Mk 2 (2×2), 2–40mm Mk 5 (1×2), (*Vampire* and *Vendetta*, 4–40mm Mk 5 (2×2), 2–40mm Mk 7 (2×1)), 5–21in TT, 1 Limbo Mk 10 ASW mortar
Sensors: Radar 275, 262, 293; sonar 170, 174/177
Complement: 320
Other particulars: As British *Daring* class

No	Name	Builder	Laid down	Launched	Comp	Fate
D04	VOYAGER	Cockatoo I DYd	10.10.40	1.5.52	12.2.57	Sunk 10.2.64
D 08	VENDETTA	Williamstown DYd	4.7.49	3.5.54	26.11.58	BU 1987
D 11	VAMPIRE	Cockatoo I DYd	1.7.52	27.10.56	23.6.59	Museum ship 1985
—	WATERHEN	Williamstown DYd	Dec 52	—	—	Cancelled 1954
D 154	DUCHESS	Thornycroft		Acquired 1964		BU 1980

General Purpose Escort

Displacement: 1800t deep load
Dimensions: 330ft pp × 34ft × 11ft
101.5m × 10.4m × 3.5m
Machinery: 2-shaft CODOG = 37kts. Range 5200nm at 12kts, 3700nm at 18kts.
Armament: 1 Seacat SAM (1×3), 1–5in/54 Mk 45, 2–20mm (2×1), 6–324mm TT (2×3)
Sensors: Radar LW-02, M-22; sonar Type 177 or Mulloka
Complement: ?

In the early 1960s the RAN had a requirement for a small and comparatively cheap general-purpose escort (GPE) to perform a range of peacetime and wartime missions. Armament was comparatively light but balanced between anti-surface warfare (ASuW), air defence and anti-submarine warfare (ASW). The project was dropped in favour of the Attack class patrol craft and the DDLs. Names were not allocated, although the official sketch design and artist's impression show pennant number 007. Numbers are not known but at least eight is a likely total.

Vendetta as completed — Courtesy Ross Gillett

Brisbane 19.11.90 — RAN

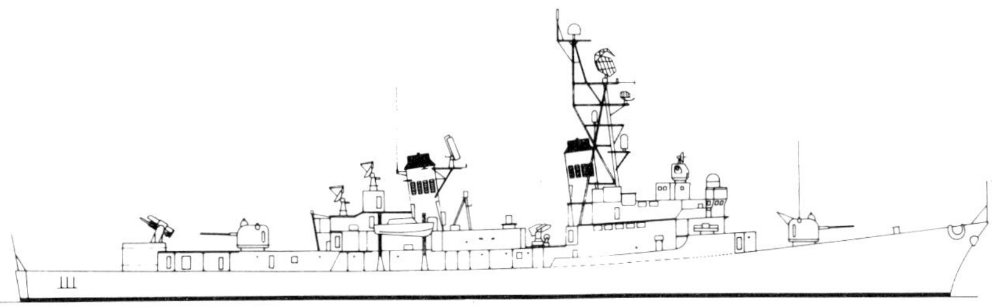

Brisbane 1978

PERTH class *guided missile destroyers*

Two *Charles F Adams* type guided missile destroyers (DDGs) were ordered from the United States in January 1962 and a third was ordered exactly a year later. They were virtually identical to the original class apart from having Ikara in place of ASROC; this necessitated building a deckhouse between the funnels housing the Ikara magazine, with two single-arm launchers port and starboard at the forward end. Between 1974 and 1979 all three were modernised, with Standard SM-1 missiles (forty carried) in place of Tartar and gun mountings converted to Mod 10. All three underwent further modernisation in 1985–1991, with upgraded electronics, etc. Ships sent to the Gulf in 1990–91 received two Mk 15 Phalanx CIWS 'Gatlings'. In 1993 the near-sister USS *Goldsborough* was bought for spares.

Armament: 2–5in/54 Mk 42 (2×1), 1 Tartar RIM24 (later Standard SM1A RIM66B) SAM system, 6–12.75in (324mm) Mk 32 ASW TT (2×3), 2 Ikara ASW systems (2×1)
Sensors: Radar SPS-40, SPS-10, SPS-52B, SPG-51C, SPG53A, Type 978; sonar SQS-23f
Other particulars: As US *Charles F Adams* class

No	Name	Builder	Laid down	Launched	Comp	Fate
D 38	PERTH (ex-*DDG 25*)	Defoe SB	21.9.62	26.963	17.7.65	Extant 1995
D 39	HOBART (ex-*DDG 26*)	Defoe SB	26.10.62	9.1.64	18.12.65	Extant 1995
D 41	BRISBANE (ex-*DDG 27*)	Defoe SB	15.2.65	5.5.66	16.12.67	Extant 1995

Official artist's impression of 1966 destroyer design — Courtesy Ross Gillett

1966 Light Destroyer Project

Displacement: 4200t normal; about 5000t deep load
Dimensions: 425ft wl × 48ft × *c*22ft
129.5m × 14.6m × 6.7m
Machinery: 2-shaft COGOG: 2 Rolls-Royce Olympus plus 2 Tyne, 50,000shp = 30kts. Range 6000nm at 18kts
Armament: 1–5in/54 Mk 42 DP, 4–35mm Emerlec-30 (2×2), 6–12.75in (324mm) Mk 32 TT (2×3), 1 Standard SM-1A RIM-66A SAM system, 6 SSMs, 2 helicopters
Complement: 210

In 1966 the RAN and RN began a joint study of a 'light destroyer' (DDL) with good air defence and a limited ASW/surface role. The joint study broke up because the RAN had no intention of departing from its policy of buying US weaponry, but the resulting ship strongly resembled the Royal Navy's contemporary Type 42 DDG. The machinery was the same, and the dimensions very similar (slightly longer and beamier), but a double hanger for two Lynx helicopters was provided and six SSMs. It was also hoped to fit the new Mulloka sonar, designed for Australian requirements. The extra beam would have given a rather better fuel supply, and the ships had a designed endurance of 6000nm at 18kts.

The detailed design work was entrusted to Y-ARD Australia in 1970 and it was hoped to lay the first keel in June 1975, followed by the others at two-yearly intervals. All three would have been built at Williamstown DYd. The first was to be ready in 1980 and the last in 1984, at an estimated cost of $A118m (£59.2m) at 1972 prices; this figure included design and support costs as well as construction. The ships would have looked very similar to the *Sheffield* class, but with a single-arm Tartar launcher at the forward end of the hangar, a single missile-tracker radar abaft the funnel and a single plated mast.

The new Labour Government which came to power in 1973 deferred a decision on the DDLs and cancelled them outright in 1974, choosing American DDGs instead.

QUADRANT class *frigates*

The wartime transfer of the five ships of the *Quilliam* class was made permanent in June 1950 and work began on conversion to the same standard as the RN's Type 15 frigates. The work was done by Williamstown DYd (*Quadrant* April 1950–July 1953), and Cookatoo Island (*Queenborough* May 1950–December 1954, *Quickmatch* April 1951–September 1955, and *Quiberon* November 1950–December 1957), but *Quality* was not converted and was scrapped in 1958. They differed little from the British ships apart from having a round-fronted bridge with the twin 40mm Bofors ahead of it. *Quadrant* had two Squid ASW mortars (2×3), whereas the other three were fitted with Mk 10 Limbo (2×3).

Particulars: As British *Rapid* class

No	Name	Builder	Laid down	Launched	Comp	Fate
F 01	QUADRANT	Hawthorn Leslie	24.9.40	28.2.42	26.11.42	BU 1963
F 02	QUEENBOROUGH	Swan Hunter	6.11.40	16.1.42	10.12.42	BU 1975
F 03	QUIBERON	White	14.10.40	31.1.42	22.7.42	BU 1972
F 04	QUICKMATCH	White	6.2.41	11.4.42	30.9.42	BU 1972

AUSTRALIA

In August 1950 plans for six Type 12 frigates were announced, but subsequently the number was cut to four; No 5 and No 6 were later reordered to a modified design. They incorporated all the improvements of the RN *Rothesay* class, and in addition had MRS3 fire control and Dutch Signaal LW–02 air warning radar. The second pair had the forecastle deck extended aft to the stern, providing additional accommodation. *Stuart* was the first to be fitted with the Ikara ASW system, which replaced one of the Limbo Mk 10 mortars aft, and *Derwent* was the first RAN ship to receive the Seacat GWS22 missile system in lieu of the twin 40mm Bofors. All were subsequently fitted with both systems; about twenty-four of each missile are carried.

Yarra ran trials with the new Mulloka sonar in 1975–76 in place of her Type 177M, and then paid off for a half-life modernisation in which the Mk 10 mortar was removed and replaced with a helicopter flight deck. She was also due to receive SPS–55 in lieu of Type 293 radar and Signaal M22 fire control for the Seacat missiles. The other three were given a more extensive modernisation, with similarly improved electronics and sonar, but with 6–12.75in (324mm) Mk 32 ASW TT replacing the Mk 10 mortar, and improvements to the boilers and to general standards of habitability. *Parramatta*'s refit extended from June 1977 to August 1981, *Stuart* was taken in hand in March 1979 (completion mid-1983), and work on *Derwent* begain in July 1981, *Yarra* paid off in January 1985 and *Parramatta* and *Stuart* followed. *Derwent* was scuttled as an artificial reef off Rottnest Island on 21 December 1994.

YARRA class *frigates*

Particulars: (As built) as British *Whitby* and *Rothesay* classes

No	Name	Builder	Laid down	Launched	Comp	Fate
F 45	YARRA	Williamstown DYd	9.4.57	30.9.58	27.7.61	BU 1985
F 46	PARRAMATTA	Cockatoo I DYd	3.1.57	31.1.59	4.7.61	BU 1991
F 48	STUART	Cockatoo I DYd	20.3.59	8.4.61	27.6.63	BU 1991
F 49	DERWENT	Williamstown DYd	16.6.59	17.6.61	30.4.64	Scuttled 21.12.94
—	No 5	Williamstown DYd	—	—	—	Cancelled 1956
—	No 6	Cockatoo I DYd	—	—	—	Cancelled 1956

Derwent as completed RAN

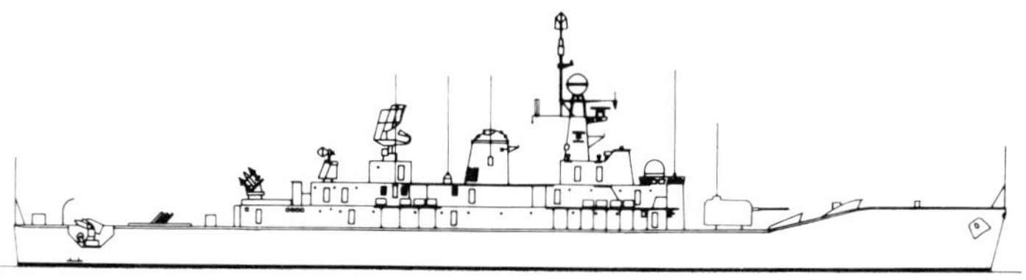

Swan as completed

The two cancelled *Yarra* class were re-ordered nine years later to a modified design which incorporated many features of the RN *Leander* class. The forecastle deck was extended right aft and a new continuous two-decked superstructure and streamlined funnel replaced the former light superstructure. The air warning radar antenna was mounted on a new stump mainmast, but no provision was made for a helicopter hangar.

Both ships received a limited half-life modernisation in 1984–85, with Mulloka sonar, Mk 32 tubes in place of the Mk 10 mortar, etc. To be taken out of service 1996–98.

SWAN class *frigates*

Armament: 2–4.5in/45 Mk 6 (1×2), 1 Ikara ASW system, 1 Seacat M22 SAM (1×4), 1 Limbo Mk 10 ASW mortar
Sensors: Radar LW-02, M22, Type 293; sonar 162, 170, 177M, 185
Complement: 247
Other particulars: As British *Rothesay* class

No	Name	Builder	Laid down	Launched	Comp	Fate
F 50	SWAN	Williamstown DYd	16.8.65	16.12.67	20.1.70	Extant 1995
F 53	TORRENS	Cockatoo I DYd	12.8.65	28.9.68	19.1.71	Extant 1995

(For illustration see US *Oliver Hazard Perry* class)

Following the cancellation of the DDL project in 1974, the Department of Defence ordered two of the new *Oliver Hazard Perry* (PF 109, later FFG 7) class frigates. Orders were placed in February 1976 for FFG 17–18 out of the USN's programme, and FFG 35 was ordered in November 1977. The growing need of replacement for the *Daring* class led to an order for FFG 48 in April 1980 and in September the same year the Defence Minister announced that the replacement for the six 'River' class DEs (the *Yarra* and *Swan* classes) would also be members of this class. Two were ordered from Williamstown DYd (later Amecon) in 1983, but Nos 7–10 were never built. *Melbourne* and *Newcastle* were modified during construction with a bigger flight deck and extended stern to accommodate the S-70B-2 Seahawk helicopter. The others were modified after completion. *Newcastle* was trials ship for the Kariwara sonar in 1995.

ADELAIDE class *guided missile frigates*

Particulars: As US *Oliver Hazard Perry* class

No	Name	Builder	Laid down	Launched	Comp	Fate
F 01	ADELAIDE (ex-*FFG-17*)	Todd, Seattle	29.7.77	21.6.78	15.11.80	Extant 1995
F 02	CANBERRA (ex-*FFG-18*)	Todd, Seattle	1.3.78	1.12.78	21.3.80	Extant 1995
F 03	SYDNEY (ex-*FFG-35*)	Todd, Seattle	21.1.80	26.9.80	29.1.83	Extant 1995
F 04	DARWIN (ex-*FFG-48*)	Todd, Seattle	3.7.81	26.3.82	21.7.84	Extant 1995
F 05	MELBOURNE	Williamstown/ Amecon	12.7.85	5.5.89	15.2.92	Extant 1995
F 06	NEWCASTLE	Williamstown/ Amecon	21.7.89	21.2.92	11.12.93	Extant 1995
—	No 7	—	—	—	—	Cancelled
—	No 8	—	—	—	—	Cancelled
—	No 9	—	—	—	—	Cancelled
—	No 10	—	—	—	—	Cancelled

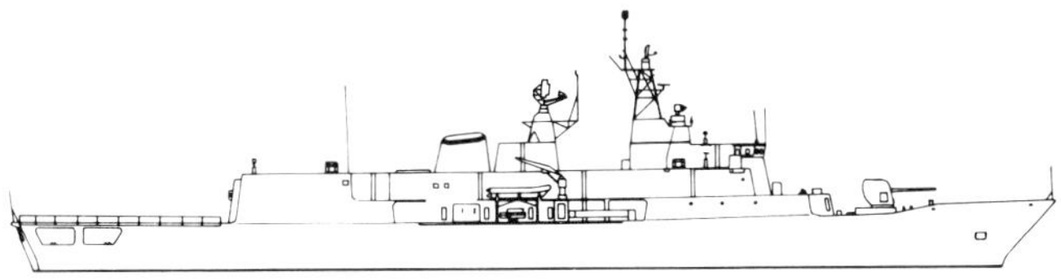

Anzac class

A contract was signed in November 1989 between Blohm + Voss and Australian Marine Engineering Consolidated (Amecon) for a variant of the MEKO 200 design. Eight were for the RAN and two for the RNZN, but an option for two more RNZN ships was not taken up. First steel was cut on 27 March 1992 and the first of class started trials early in 1995; thereafter one hull will be delivered each year. The MEKO concept permits modules to be built over a wide area, even in New Zealand, to meet local requirements for work-sharing.

The design is similar to MEKO 200s built for other navies, but differs mainly in having single-shaft machinery. Initial choice for the gun-calibre was to be 76mm but Army pressure secured a move to 127mm, and the design permits upgrading of the armament. Options include Harpoon missiles, CIWS, and additional vertical launch missile silo, triple Mk 32 ASW TT (2 × 3) and Kariwara towed sonar array.

ANZAC class *frigates*

Displacement: 2900t standard; 3200t deep load
Dimensions: 357ft 7in wl, 387ft 1in oa × 47ft 1in × 14ft 4in
109.0m, 118.0m × 14.4m × 4.4m
Machinery: Single-shaft CODOG; 1 LM 2500 gas turbine, 30,000shp, 2 MTU 12V1163 TB83 diesels, 8840bhp = 31.75kts. Fuel 300t. Range 900nm/4100nm at 31.75/18kts
Armament: 1–5in/54 MK 45, 2–12.7mm MGs, 1 Evolved Sea Sparrow Mk 41 VL SAM system
Sensors: Radar SPS-49(V), Sea Giraffe 150HC, 9LV Mk3; sonar Spherion-B
Helicopter: 1 S-70B-2
Complement: 163

No	Name	Builder	Laid down	Launched	Comp	Fate
150	ANZAC	Transfield Shipbuilding	27.3.92	19.3.94	Mar 1996	Trials 1995
151	ARUNTA (ex-*Arrente*)	Transfield Shipbuilding	22.7.95	—	—	Building
152	WARUMUNGU	Transfield Shipbuilding	—	—	—	Ordered
153	STUART	Transfield Shipbuilding	—	—	—	Ordered
154	PARRAMATTA	Transfield Shipbuilding	—	—	—	Ordered
155	BALLARAT	Transfield Shipbuilding	—	—	—	Ordered
156	TOOWOMBA	Transfield Shipbuilding	—	—	—	Ordered
157	PERTH	Transfield Shipbuilding	—	—	—	Ordered

SUBMARINES

Virtually identical to the British *Oberon* design, built 1964–78, all six were refitted with USN-pattern periscopes and weapons, Sub-Harpoon missiles and Mk 48 torpedoes. The principal elements are a Singer/Librascope SFCS Mk 1 digital fire control system and a Krupp Atlas CSU3–41 active/passive sonar (active transducer in fin, passive array in the new bow sonardome). The refit programme ran from October 1977 to October 1985. The short after TT were removed.

They serve with the 1st Submarine Squadron, four at HMAS *Platypus* in Sydney, and two at HMAS *Stirling* in Western Australia. The three survivors will be taken out of service in 1996–98.

OXLEY class *patrol submarines*

Sensors: Radar Type 1006; sonar Types 187C, 197, 2007, Micropuffs
Other particulars: As British *Oberon* class

No	Name	Builder	Laid down	Launched	Comp	Fate
S 27	OXLEY	Scotts	2.7.64	24.9.65	27.3.67	BU 1992
S 59	OTWAY	Scotts	29.6.65	29.11.66	22.4.68	HS 1995
S 70	OVENS	Scotts	17.6.66	5.12.67	18.4.69	Out of service 1995
S 60	ONSLOW	Scotts	22.5.67	29.8.68	22.12.69	Extant 1995
S 61	ORION	Scott-Lithgow	6.10.72	16.9.74	15.6.77	Extant 1995
S 62	OTAMA	Scott-Lithgow	28.5.73	2.12.75	*27.4.78*	Extant 1995

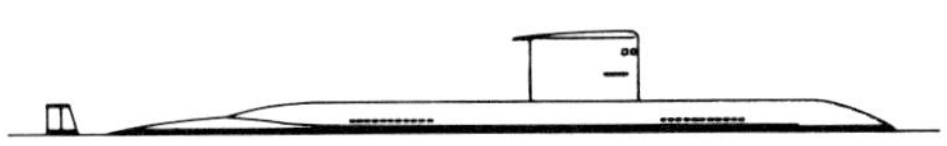

Collins class design

After a lengthy evaluation of foreign designs the Swedish Kockums Type 471 design was selected as a replacement for the *Oberon* type in the mid-1990s. Under an agreement signed in June 1987 a contract was signed between Kockums AB and the Australian Submarine Corporation (ASC) to build six boats in Adelaide, South Australia. An option for two was not taken up.

Armament and fire control are US-pattern, and the sonar is French. The option of retrofitting a Stirling air-independent propulsion system is being considered, and a shore-based test rig has been set up.

The RAN plans to base four at HMAS *Stirling* in Western Australia, and the remaining pair at HMAS *Platypus*. *Farncomb* and sisters to be commissioned in 1997–99.

COLLINS class *patrol submarines*

Displacement: 2500t/3298t
Dimensions: 254ft oa × 25ft 7in × 23ft
77.5m × *7.8m × 7m*
Machinery: 1-shaft diesel-electric, 3 Hedemora VB 210 18-cyl diesels, Jeumont-Schneider motor, 3.5mW = 10kts snorting/20kts submerged. Endurance 9000nm/10kts
Armament: 6–21in TT, 23 Mk 48 Mod 4 torpedoes and Sub-Harpoon missiles or mines
Sensors: Radar 1007; sonar Scylla (bow and flank array) and Karriwarra towed array
Complement: 42 officers and ratings + 5 trainees

No	Name	Builder	Laid down	Launched	Comp	Fate
S 71	COLLINS	ASC	14.2.90	28.8.93	Nov 1995	Extant 1995
S 72	FARNCOMB	ASC	1.3.91	Dec 1995	—	Fitting out
S 73	WALLER	ASC	19.3.92	—	—	Building
S 74	DECHAINEUX	ASC	4.3.93	—	—	Building
S 75	SHEEAN	ASC	17.2.94	—	—	Building
S 76	RANKIN	ASC	Mar 1995	—	—	Building

Collins 1994
RAN

AMPHIBIOUS WARFARE VESSELS

TOBRUK *tank landing ship*

Displacement: 3400t normal; 5800t full load
Dimensions: 417ft × 60ft × 16ft
127.1m × 18.3m × 4.9m
Machinery: 2 shafts, 2 Mirrlees Blackstone KDM8 diesels, 9600bhp = 17kts
Armanent: 2–40mm
Sensors: Radar KH 1006, RM 916
Complement: 70 (plus 350–500 troops)

No	Name	Builder	Launched	Fate
L 50	TOBRUK	Carrington Slipways	1.3.80	Laid up 1995

A modified version of the British *Sir Bedivere* class LSLs, but rated as an amphibious heavy lift ship, *Tobruk* was laid down on 7 February 1978 and commissioned on 23 April 1981. She can operate Wessex helicopters, can carry pontoons and has a powerful heavy lift system. Two LCVPs are carried in davits. Laid up 1995.

BALIKPAPAN class *heavy landing craft*

Displacement: 310t light; 503t full load
Dimensions: 146ft × 33ft × 6ft 6in
44.5m × 10.1m × 2.0m
Machinery: 2 shafts, 2 GM diesels, 675bhp = 10kts. Range 3000nm at 10kts
Armament: 2 MGs
Sensors: Radar RM 916
Complement: 16

Class:
Balikpapan (L 126), *Brunei* (L 127), *Labuan* (L 128), *Tarakan* (L 129), *Wewak* (L 130), *Salamaua* (L 131), *Buna* (L 132), *Betano* (L 133)

All were built by Walkers, Maryborough in 1971–72. They are rated as heavy landing craft (LCH) and can carry three tanks. They were originally built for the Army. *Salamaua* and *Buna* were transferred to Papua New Guinea in 1974, and five are used for subsidiary duties, with one in reserve, but all six are earmarked for Navy/Army service in time of emergency. Replacement design sought 1995.

KANIMBLA class *helicopter support ships*

Displacements: 8450t full load
Dimensions: 562ft × 69ft × 6ft
171.3m × 21.18m × 1.80m
Machinery: 2-shaft Alco 1B-251 diesels, 16,500bhp = 20kts. Range 14,250nm at 14kts
Armament: 6-12.7mm MG (6×1), 1 Mk 15 Phalanx CIWS, 4 S-70 Blackhawk assault helicopters
Compliment: 170 + 450 troops

Class (no, former name):
Kanimbla (L 51, ex-*Saginaw*), *Manoora* (L 52, ex-*Fairfax County*)

When plans for up to three Training and Helicopter Support Ships (THSSs) were dropped in 1993 the RAN started negotiations to acquire two *Newport* (LST-1179) class tank landing ships. They will replace *Tobruk* and the training ship *Jervis Bay*. Tenders are being sought for conversion, consisting of the fitting of a hangar to accommodate four army helicopters, removal of the over-the-bow ramp and addition of 230t of aviation fuel stowage. *Kanimbla* was commissioned on 29 August 1994 and *Manoora* on 25 November 1994.

SMALL SURFACE COMBATANTS

ATTACK class *patrol craft*

Displacement: 100t normal; 125t deep load
Dimensions: 107ft 6in oa × 20ft × 7ft 3in
32.8m × 6.1m × 2.1m
Armament: 1–40mm/60 Mk 7, 2-12.7mm MG (2×1)
Sensors: Radar RM 916
Complement: 19

Class:
Acute (P 81), *Adroit* (P 82), *Advance* (P 83), *Aitape* (P 84), *Samarai* (P 85), *Archer*(P 86), *Ardent* (P 87), *Arrow* (P 88), *Assail* (P 89), *Attack* (P 90), *Aware* (P91), *Ladava* (P 92), *Lae* (P 93), *Madang* (P 94), *Bandolier* (P 95), *Barbette* (P 97), *Barricade* (P 98), *Bombard* (P 99), *Buccaneer* (P 100), *Bayonet* (P 101)

Twenty 33m patrol boats were ordered in November 1965 for coastal patrol work; five – *Aitape*, *Samarai*, *Ladava*, *Lae* and *Madang* – were to be employed in New Guinea, the others in Australian home waters. P 81–82, P 85, P 89–92, P 94, P 98, P 100 were built by Evans Deakin, the rest by Walker's, Maryborough. In 1973–74 *Bandolier* and *Archer* were transferred to Indonesia and renamed *Sibarau* and *Silinan*, and in 1975 the five boats in New Guinea waters were transferred to the newly formed Papua New Guinea Defence Force. *Arrow* sank at her moorings in Darwin on 25 December 1974 during Hurricane 'Tracy', but *Attack*, although damaged, proved capable of repair. *Barricade* was sold to Indonesia in 1982, followed by *Acute* and *Bombard* in 1983, and *Assail*, *Attack* and *Barbette* in 1985. *Buccaneer* was for disposal in 1984 and was sunk as target in October 1988; *Advance* was preserved by the National Maritime Museum. *Bayonet* was for disposal in 1988.

Tobruk as completed RAN

FREMANTLE class *patrol craft*

Displacement: 240t normal; *c*275t deep load
Dimensions: 137ft 10in oa × 23ft 5in × 6ft 11in
42.0m × 7.1m × 1.9m
Machinery: 3 shafts, 2 MTU 16V538 TB 91 diesels, plus Dorman cruising diesel, 6000bhp = 30kts. Range 1450nm/4800nm at 28kts/8kts
Sensors: Radar Decca 1226
Complement: 22

Class:
Fremantle (P 203), *Warrnambool* (P 204), *Townsville* (P 205), *Wollongong* (P206), *Launceston* (P 207), *Whyalla* (P 208), *Ipswich* (P 209), *Cessnock* (P 210), *Bendigo* (P 211), *Gawler* (P 212), *Geraldton* (P 213), *Dubbo* (P 214), *Geelong* (P 215), *Gladstone* (P 216), *Bunbury* (P217), *Ballarat* (P 218), *Mildura* (P 219), *Armidale* (P 220), *Bundaberg* (P 221), *Pirie* (P 222)

Replacements for the *Attack* class, built to Brooke Marine's PCF–420 design. The lead ship, *Fremantle*, was laid down at Lowestoft in 1977 and commissioned in March 1979; the remainder were built by North Queensland Engineers and Agents at Cairns, and all were in service by the end of 1984.

Bunbury as completed RAN

Offshore Patrol Combatant Project

Displacement: 1250t full load
Dimensions: 248ft × 39ft × 13ft
75.5m × 12.0m × 3.9m
Machinery: 2-shaft diesels = 25kts
Armament: 1-76mm, 1 light gun, 2 MG, 1 light helicopter
Compliment: 64

The RAN planned to build a new class of Offshore Patrol Combatants (OPCs) at the beginning of the next century, to serve as offshore patrol vessels in peacetime or as warships in low-intensity conflicts. Under a government initiative the builders, Transfield Shipbuilding, were encouraged to negotiate with the Royal Malaysian Navy to see if its New Generation Patrol Vessel requirement could be combined with the OPC to produce a joint design. If the project goes ahead, the OPEC timetable will be brought forward; if not, the *Fremantle* class will be given a life-extension overhaul.

MINE WARFARE VESSELS

Ex-British 'TON' class *coastal minesweepers*

Particulars: As British 'Ton' class

Class (former name, fate):
Curlew (ex-*Chediston*, for disposal 1990), *Gull* (ex-*Swanston*, sold 1976), *Hawk* (ex-*Somerleyton*, sold 1976), *Ibis* (ex-*Singleton*, sold 1984), *Snipe* (ex-*Alcaston*, scrapped 1988), *Teal* (ex-*Jackton*, sold 1978)

Six CMSs bought from the Royal Navy in 1961 and modified to suit RAN requirements, with air-conditioning, re-engining with Deltics and fitting with stabilisers. *Curlew* and *Snipe* were converted to minehunters in 1967–70. *Snipe* was scrapped in 1988 and *Curlew* was paid off for disposal on 23 April 1990. *Ibis* paid off for disposal on 7 May 1984, and was sold to Australian commercial interests.

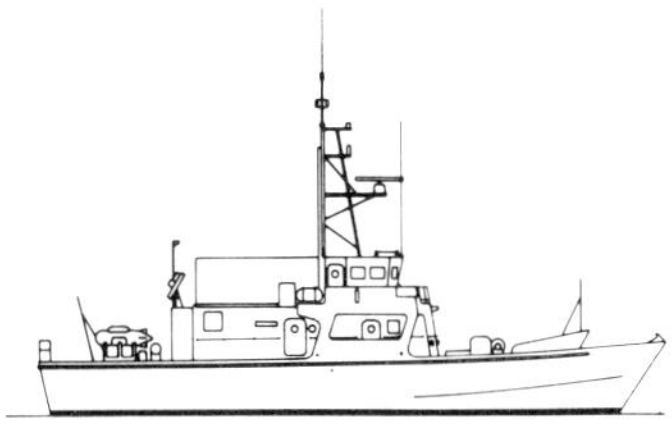

Rushcutter 1986

'BAY' class *inshore minehunters*

Displacement:	170t standard
Dimensions:	101ft 9in oa × 29ft 6in × 6ft 7in *31.0m × 9.0m × 2.0m*
Machinery:	2-shaft SACM-Poyard 520-V8-S2 diesels driving Schottel active rudders, 650hp = 10kts. Range 1200nm
Armament:	4–12.7mm MGs
Sensor;	Radar 1006; sonar DSQS-11H

Class:
Rushcutter (M 80) extant 1995, *Shoalwater* (M 81) extant 1995.

Tenders for two prototype glass-reinforced plastic (GRP) catamaran minehunters were sought in April 1981 and orders were placed with Carrington Slipways, Cairns, in January 1983. It was hoped to order four more following trials and evaluation but the hull proved too small for coastal work and severe problems were encountered with the sonar, so the project was stopped. In 1991 Thomson Sintra leased its Ibis V system for trials in *Shoalwater*, and early in 1992 Atlas Elektronik installed their new DSQS-11M in *Rushcutter* for competitive trials. The German sonar passed acceptance and was retrofitted to *Shoalwater*.

Auxiliary minesweepers
A force of Craft of Opportunity (COOP) minesweepers was created from 1986. Four fishing vessels, *Brolga*, *Kooraga*, *Salvatore V* and *Wave Rider* became Small Auxiliary Minesweepers or MAS(S), and the former tugs *Bandicoot* and *Wallaroo* (both built in 1982) became Large Auxiliary Minesweepers or MSA(L). A new MSA(S), *Carol S*, was leased for a year in 1993 to replace *Salvatore V* when her charter expired.

HUON class *coastal minehunters*

Displacement:	720t deep load
Dimensions:	172ft 3in × 32ft 6in × 9ft 9in *52.5m × 9.9m × 3m*
Machinery:	1-shaft GMT 230 diesel, 2200bhp = 12kts (4kts sweeping) Range 600nm at 12 kts
Armament:	1-40mm/60
Sensors:	Radar Type 1006, sonar Type 2093

Class (no):
Huon (M 82), *Hawkesbury* (M 83), *Norman* (M 84), *Gascoyne* (M 85), *Diamantina* (M 86), *Yarra* (M 87)

After evaluation of a number of foreign designs a modified *Gaeta* design from Intermarine was selected in 1994, to meet the coastal minehunter (MHC) requirement. All six to be built by Australian Defence Industries (ADI) at Newcastle, NSW, to be in service in 1998–2002. GEC-Marconi Nautis-M combat system, Sutec Double Eagle remote control mine disposal vehicle.

Austria

Austria was divided into four occupied zones after the Second World War; Vienna – situated in the Soviet zone – being divided into four sectors as was Berlin. Through the untiring efforts of the Austrian politicians an international treaty was signed on 15 May 1955, giving Austria self-control, a democratic constitution and perpetual neutrality. The treaty forbids her from owning nuclear weapons, submarines, assault craft, manned torpedoes and sea mines.

In accordance with Austria's strict neutrality and budget limitations, the Austrian Second Republic's riverine forces were small, consisting only of the equipment of the Army Engineering Corps. In 1968 a patrol boat detachment was formed consisting of the tiny 40ft boat *Oberst Brecht* (A 601) and an old self-propelled pontoon nicknamed 'flat-iron'. At the end of the 1960s the conservative government planned to build ten armed patrol boats nicknamed 'Prader's Panzerkreuzer' after the then Minister of Defence. Of this group only *Niederösterreich* was built. Apart from her there were ten US-built M 3 launches (normally unarmed, four delivered 1965 and six in 1976), which were discarded in 1984. These were replaced by ten German-built 'M-boot 80' class launches. All extant 1995.

NIEDERÖSTERREICH *patrol craft*

Displacement:	73t full load
Dimensions:	97ft 4in × 17ft 5in × 3ft 7in *29.67m × 5.41m × 1.1m*
Machinery:	2 shafts, 2 MWM V16 diesels, 1620bhp = 22kts. Oil 9.3t. Range 1030nm
Armament:	1–20mm AA, 2–12.7mm MG (2×1), 2–7.62mm MG 42 (2×1), 1–84mm Carl Gustav M2 AT launcher
Complement:	12

No	Name	Builder	Launched	Fate
A 604	NIEDERÖSTERREICH	Korneuberg	26.7.69	Extant 1995

This multipurpose boat was designed by Mayerform S/A, an Austrian design bureau based in Geneva, Switzerland. She can be quickly changed into three different configurations: E meaning river patrol boat, K meaning rescue boat, R meaning dispatch boat.

Based at the harbour of the 'Pionierkaserne Wien Kuchelau' and maintained by the 27 Danube Flotilla personnel of the Engineering Corps ('Heerespioniere'); she is under command of 1st Panzer Brigade should she be needed for action.

Bahamas

The Bahamas archipelago of 700 islands (40 inhabited), which became independent within the Commonwealth in July 1973, has no navy as such; patrol boats are operated by the Marine Police and should stronger defences be called for the country would turn to Britain. The Bahamas serves as a base for a US naval underwater research and development station and there are also US Navy and Coast Guard units on the islands. The patrol boats are based at Bay Shore Marina, Nassau. On 11 May 1980 the patrol boat *Flamingo*, sister ship to *Marlin*, was hit by Cuban MiG-21 aircraft and sunk; Cuba subsequently paid out £2.5m in reparations. Naval personnel in 1994 amounted to 850.

Three Protector class patrol craft *Yellow Elder*, *Port Nelson* and *Samana* (P 03–05), built by Fairey Marine, delivered in 1986 (180t full load; 33m × 6.7m × 2.1m, 4398bhp, 30kts, 1–20mm, 3 MG) and all extant 1995.

Two Vosper Thornycroft 103ft type patrol craft *Marlin* (P 01) and *Flamingo* (P 02) delivered in 1978 (109t full load; 31.4m × 6.0m × 1.7m; 25kts, 1–20mm, 2 MG); *Flamingo* sunk in 1980.

Six ex-USCG *Cape* class built 1953–59 and transferred in 1989 as *Fenrick Sturrup* (ex-*Shoalwater*), *David Tucker* (ex-*Upright*), *Austin Smith* (ex-*Current*), *Edward Williams* (ex-*York*), *San Salvador II* (ex-*Fox*), *Fort Fincastle* (ex-*Morgan*) (P 6–11), all extant 1995.

Seven Keith Nelson type patrol boats *Acklins*, *Andros*, *Eleuthera*, *San Salvador*, *Abaco*, *Exuma*, *Inagua* (P 21–27), delivered by Vosper Thornycroft in 1971 (P 21–24) and 1977 (P 25–27): (37t full load, 18.3m, 20kts, 3 MG) P 21 burned out in 1980 while P 24 deleted in 1982. P 22, P 23, P 25–27 extant 1995.

Landing craft *Fort Charlotte* (ex-YFU 97, ex-LCU 1611) transferred in 1991 (339t full load; 11kts, 2 MG), extant 1995.

Marlin 1978 C & S Taylor

Bahrain

Bahrain consists of a group of low-lying islands situated about twenty miles off the coast of Saudi Arabia. The largest of these is Bahrain, on which the capital, Manama, is situated at the northern end. The population (1991 estimate) is 516,000. A friendship treaty was signed between Britain and Bahrain after independence in 1971.

By 1992 naval personnel had increased to 650 officers and ratings. There are also about 250 men serving in the Coast Guard patrol force. During the 1980s Bahraini naval forces were increased significantly. New acquisitions include two fast helicopter-capable missile armed fast corvettes of the Lürssen 'FPB 62' type, commissioned in 1987–88, and four fast attack craft of the Lürssen 'TNC 45' type which entered service in pairs in 1984 and 1986–89. These acquisitions significantly increased the capabilities of the navy beyond its former role as a small coastal patrol force. This build-up can be seen as a success story for Lürssen, as all ships and craft of any fighting value were designed and built by them. During the early 1990s there were reports of the proposed procurement of up to three air cushion minehunters, but nothing has come of this so far. Bahrain continues to play host to Western naval forces in the Gulf. It was also a sanctuary for surviving Kuwaiti warships which fled that country during the Iraqi invasion.

Lürssen 'FPB 62' type *missile corvettes*

Displacement: 536t half load; 632t full load
Dimensions: 206ft 9in oa × 30ft 6in × 9ft 5in
63.0m × 9.3m × 2.9m
Machinery: 4 shafts, 4 MTU 20V538 TB92 diesels, 18,740bhp = 34.5kts. Range 4000nm/16kts
Armament: 4 MM40 Exocet SSM (2×2), 1–76mm/62 OTO Melara, 2–40mm/70 Breda (1×2), 2–20mm/93 Oerlikon GAM-BO1 (2×1), CSEE Dagaie chaff RL, 1 helicopter
Sensors: Radar PEAB Sea Giraffe, Racal-Decca 1226, PEAB 9LV331; ESM/ECM Racal Decca Cutlass/Cygnus
Complement: 43 (7 officers)

No	Name	Builder	Commissioned	Fate
50	AL MANAMA	Lürssen, Vegesack	14.12.87	Extant 1995
51	AL MUHARRAQ	Lürssen, Vegesack	3.2.88	Extant 1995

These are small, versatile missile corvettes capable of carrying a helicopter, ordered in February 1984. They were launched in 1986–87. The big after superstructure, topped with a landing platform, includes a helicopter hangar incorporating a lift. The Aerospatiale SA 365F Dauphin 2 helicopter is armed with AS 15TT SSM (all weather missiles tracked by radar rather than by a tail flare). Fitted with CSEE Panda Mk2 optical director. Magazine capacity is as follows: 400 rounds 76mm, 4400 rounds 40mm and 5000 rounds 20mm ammunition and eight AS 15TT SSM for helicopter launch. These corvettes can cruise at a continuous speed of 32kts with 16,000bhp engine MCR. They are much more capable than the standard fast attack craft and can serve as leaders for FAC squadrons providing over-horizon targeting for anti-ship missiles, using the embarked helicopter. Very similar ships serve in the navy of the United Arab Emirates.

Lürssen 'FPB 38' type *fast attack craft (gun)*

Displacement: 188t half load; 205t full load
Dimensions: 126ft 8in × 23ft × 7ft 3in
38.6m × 7.0m × 2.2m
Machinery: 2 shafts, 2 MTU 20V 539 TB91 diesels, 9500bhp = 34kts. Range 550nm/1100nm at 31.5kts/16kts
Armament: 2–40mm/70 Breda-Bofors Compatto (1×2), 2–20mm GAM-601 Oerlikon (2×1), mines, Dagaie chaff RL, 2–50mm RFLon 40mm, 2–3 pdr saluting guns
Sensors: Radar CSEE Lynx, Philips 9LV100G, Philips 9GR600, Racal Decca 1226
Complement: 27

No	Name	Builder	Launched	Fate
10	AL RIFFA	Lürssen, Vegesack	Apr 1981	Extant 1995
11	HAWAR	Lürssen, Vegesack	Jul 1981	Extant 1995

Two ordered 1979. Some sources incorrectly suggested three ships were ordered. Originally planned to have 1–40mm/70 Bofors. Commissioned August and November 1981 respectively.

Al Manama 1987

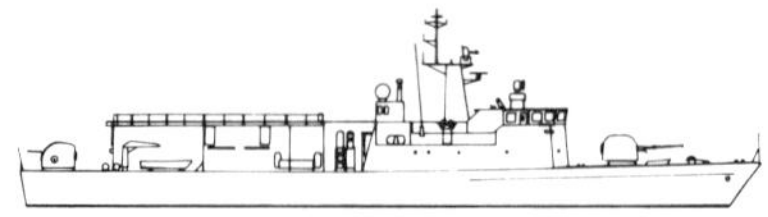

Al Riffa as completed
NB 1/750 scale

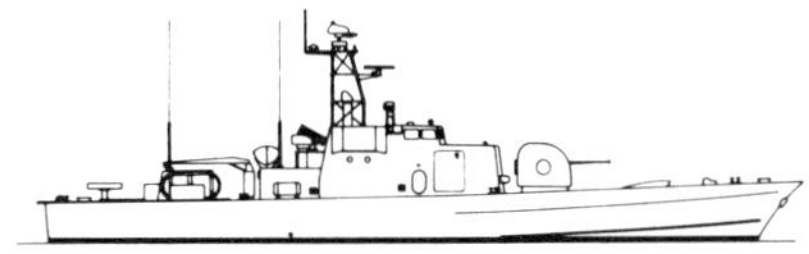

Lürssen 'TNC 45' type *fast attack craft (missile)*

Displacement: 228t standard 259t full load (270t, second pair)
Dimensions: 147ft 4in oa (153ft 10in, second pair) × 23ft (24ft, second pair) × 8ft 2in (9ft 10in, second pair)
44.9m (46.9m) × 7.0m (7.3m) × 2.5m (3.0m)
Machinery: 4 shafts, 4 MTU 16V538 TB92 diesels, 12,820bhp = 40kts.Range 1500nm/15kts, 500nm/38.5kts
Armament: 4 MM40 Exocet SSM (2×2), 1–76mm/62 OTO Melara, 2–40mm/70 Breda (1×2), 3–7.62mm MG
Sensors: Radar PEAB LV223, Racal-Decca 1226, ECM Racal Cygnus (only second pair); ESM Decca RDL2ABC, Dagaie chaff RL
Complement: 36 (6 officers) or 43 (second pair)

No	Name	Builder	Commissioned	Fate
20	AHMAD EL FATEH	Lürssen, Vegesack	5.2.84	Extant 1995
21	AL JABERI	Lürssen, Vegesack	3.5.84	Extant 1995
22	ABDUL RAHMAN AL FADEL	Lürssen, Vegesack	10.9.86	Extant 1995
23	AL TAMEELAH (ex-*Sabah*)	Lürssen, Vegesack	25.3.89	Extant 1995

First pair ordered in Germany in 1979; second pair in May 1985 to the yard's modified standard design. Rumours of a third pair have not been confirmed. Launched in 1983–84 and 1986–88 respectively. Modified boats serve in the UAE navy. Fitted with CSEE Panda optical director. Carry 250 rounds 76mm, 1800 rounds 40mm and 6000 rounds 7.62mm. Can be fitted for minelaying. Some reports give machinery rating at 15,600bhp and maximum speed as 41kts.

Al Tameelah 1989
NB 1/750 scale

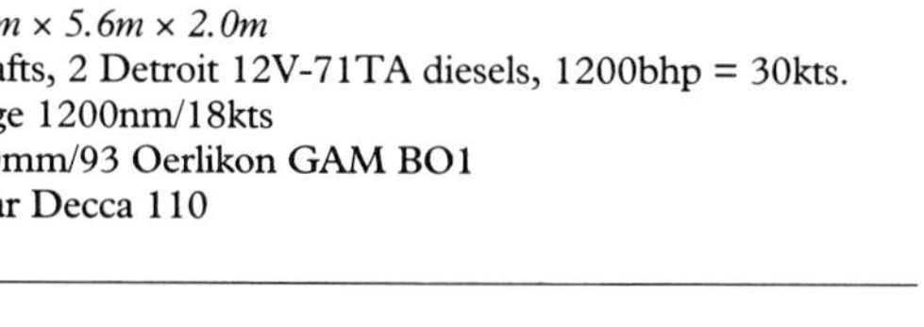

'SWIFT FPB 20' type *fast attack craft (gun)*

Displacement: 33t full load
Dimensions: 63ft oa × 18ft 4in × 6ft 7in
19.2m × 5.6m × 2.0m
Machinery: 2 shafts, 2 Detroit 12V-71TA diesels, 1200bhp = 30kts. Range 1200nm/18kts
Armament: 1–20mm/93 Oerlikon GAM BO1
Sensors: Radar Decca 110
Complement: ?

Class:
Al Jarim (No 30, launched 1982), *Al Jasrah* (No 31, 1982)

Two fast attack craft of aluminium construction built by Swiftships, Morgan City, USA, and commissioned in February 1982. Extant 1995.

Utility landing craft

Displacement: 420t full load
Dimensions: 129ft 11in oa × 36ft 1in × 5ft 11in
39.6m × 11.0m × 1.8m
Machinery: 2 shafts, 2 Detroit 16V-71N diesels, 1800bhp = 13kts. Range 1500nm/10kts
Armament: 2–12.7mm MG
Sensors: Radar Racal Decca
Complement: 8 (2 officers)

Class:
Ajirah (No 41), *Mashtan* (No 42), *Rubodh* (No 43), *Suwad* (No 44)
Landing craft utility (LCU) built by Swiftships, Morgan City, USA, can carry up to 100t of fuel and 88t of water. Fitted with bow ramp and 15t crane. Sister ships serve in the Venezuelan Navy.

COASTAL PATROL CRAFT
All these units are part of the Coast Guard under Ministry of Interior control.
Three Fairey Marine 'Tracker' class (26t full load, 64ft, 28kts, 1–20mm) *Deraa 1, Deraa 2, Deraa 3*: *Deraa 1* bought 1974, in service since 1975. Two others bought 1980. *Deraa 2* stricken 1990, remainder extant 1995.
Two Fairey Marine 'Spear' class (10t full load, 29ft, 26kts, 2 MG, 3 men): *Saham* (4) and *Khataf* (5) bought 1974, in service since 1975.
Two Vosper type (32t, 56ft, 29kts): *Howar* (3) and *Roubodh*, Singapore-built 1977.
Three small Vosper type (6.3t, 36ft, 27kts): *Al Bayneh, Junnan* and *Quaimas*, Singapore-built 1977. All three deleted 1992.
One 50ft Cheverton type (9t, 22kts): *Mashtan* (6) UK-built 1976. Deleted 1993.
Three 27ft Cheverton type (3.3t, 15kts): *Noon* (15), *Askar* (16), and *Suwad* (17) built 1977. All deleted 1993.
Wasp type GRP patrol boat (103t full load, 98ft 5in oa, 25kts, 1–30mm, 2–7.62mm MG): *Al Muharraq*; built Souters, Cowes; in service 1985. Extant 1995.
Four Halmatic class (31.5t full load, 66ft oa, 25kts, 2–7.62mm MG): *Deraa 2, 6, 7, 8*, all in service 1995.
Two Wasp 20m class (36t full load, 65ft 6in, 24.5kts): *Deraa 4* and *5*; built Souters, Cowes; GRP hull, in service 1983. Extant 1995.
Three Wasp 11m class (7.25t, 36ft, 24kts): *Saham 1, 2, 3*, built by Souter, Cowes; in service 1983. All deleted 1993.
Six Halmatic class boats (17t, 47ft 3in, 27kts): *Saif 5, 6, 7, 8, 9, 10*; GRP hull, in service 1990–91. Extant 1995.
Four Fairey 'Sword' class patrol boats (15t, 45ft oa, 28kts): *Saif 1, 2, 3, 4*; purchased 1980. All extant 1995.

AMPHIBIOUS WARFARE VESSELS
One 'Loadmaster' class landing craft, *Safra* (7) UK-built 1976 and commissioned December 1977.
One utility hovercraft built by Tropimere, UK, 1977.
One Fairey Marine landing craft (150t, 74ft oa): *Safra 2* (40) commissioned 1980, on loan to the navy. Extant 1995.
One Halmatic support craft (165t, 85ft, 13kts): *Safra 3*, commissioned 1992. Extant 1995.

Bangladesh

The People's Republic of Bangladesh was proclaimed in what had been East Pakistan in April 1971. This was followed first by civil war and then in December by the Third Indo-Pakistan War, which resulted in the complete independence of Bangladesh. Although four Soviet 'Vanya' class minesweepers operated during 1972–74 in Bangladesh coastal waters to clear Chittagong of mines and wrecks these were not incorporated in the Bangladesh Navy, but returned to the USSR. The navy came into being with its own river craft in 1972, and was formed around four old British frigates and a number of transferred gunboats. There is also a force of river gunboats. Volunteer personnel numbered approximately 7600 in 1993. Their bases are Chittagong (HQ), Dhaka, Khulna, Juldia and Kaptai. Bangladesh lacks the money to acquire new ships except for river gunboats. Some vessels were damaged in the April 1991 monsoon.

Ex-British SALISBURY class *frigate*

No	Name	Builder	Acquired	Fate
F 16	UMAR FAROOQ (ex-*Llandaff*)	Hawthorn Leslie	10.12.76	Extant 1995

Proposed acquisition of second unit, HMS *Lincoln*, did not take place.

Ali Haider May 1990 L & L van Ginderen

Ex-British LEOPARD class *frigates*

No	Name	Builder	Acquired	Fate
F 17	ALI HAIDER (ex-*Jaguar*)	Denny	6.7.78	Extant 1995
F 15	ABU BAKR (ex-*Lynx*)	John Brown	12.3.82	Extant 1995

The RN AA Type 41 frigate HMS *Jaguar* was transferred to Bangladesh for £2m. Refitted by Vosper Thornycroft Aug–Oct 1978. The ASW weapons and equipment were removed while she was still in the RN.

Ex-Chinese 'JIANGHU II' class *frigate*

No	Name	Builder	Acquired	Fate
F 18	OSMAN (ex-556)	Jiangnan, China	8.11.89	Extant 1995

Transfer of second unit proposed but not carried out.

Ex-Chinese 'HOKU' class *fast attack craft (missile)*

Class:
Duranta, Durdan, Durbar, Duruedya
First pair transferred on 6 April 1983 and second pair in October 1983. All extant 1995.

Ex-Chinese 'P 4' class *torpedo-boats*

Four transferred 1983. All discarded 1989.

Ex-Chinese 'HUANGFENG' class *fast attack craft (missile)*

Class:
Durdharsha, Durdanta, Durnibar, Durdanda
Four transferred 10 November 1988. All extant 1995.

Ex-Chinese 'HUCHUAN' class *torpedo-boats*

Four transferred in 1988 to replace 'P 4' type boats. All extant 1995.

Ex-Yugoslav KRALJEVICA class *large patrol craft*

Class (no, former name):
Karnaphuli (P 301, ex-*PBR 502*), *Tista* (P 302, ex-*PBR 505*)
Type 510 units sold to Bangladesh on 6 June 1975. Discarded by 1988.

Tista 1976 L & L van Ginderen

Ex-Indian AKSHAY class *large patrol craft*

Class (no, former name):
Padma (P 201, ex-*Akshay*), *Surma* (P 202, ex-*Ajay*)
Transferred 12 April 1973 and 26 July 1974 respectively. Both re-armed in late 1980s with 8–20mm (2 × 4).

Ex-Pakistani *large patrol craft*

PNS *Jessore* was salvaged after the 1971 War of Independence and repaired at Khulna SY, recommissioned 23 November 1978 and renamed *Bishkali*.

SHAMJALA class *large patrol craft*

Shamjala, first of a class of six built by Sumidagawa of Japan, was launched 2 July 1981 and commissioned in 1982. She displaces 160t and measures 98ft 4in (30m).

Ex-Chinese 'SHANGHAI II' class *large patrol craft*

Class:
Shaheed Daulat, Shaheed Farid, Shaheed Mohibullah, Shaheed Akhtaruddin, Towheed, Towfiq, Tamjeed, Tanveen, P 102, P 103
P 102 and P 103 transferred 1974 and discarded in the early 1980s. Of the remaining eight, four were transferred 6 July 1980 and four in May 1982. All extant 1995.

Ex-Chinese 'HAINAN' class *large patrol craft*

Class (no):
Durjoy (P 811), *Nirbhoy* (P 812)
Transferred September 1982 and December 1985 respectively. Both extant 1995.

Ex-British RIVER class *minesweepers*

Class (former name):
Saikat (ex-*Carron*), *Shaibal* (ex-*Helford*), *Shapli* (ex-*Waveney*), *Surovi* (ex-*Dovey*)
Transferred 1994 for use as patrol boats. All extant 1995.

Ex-British ISLAND class *offshore patrol boat*

Class (former name):
Shaheed Raihal Amin (ex-*Jersey*)
Transferred 28 January 1994. Extant 1995.

PABNA class *river patrol boats*

Class (no, commissioned):
Pabna (P 101, 12 June 1972), *Noakhali* (P 102, 8 July 1972), *Patuakhali* (P 103, 7 November 1974), *Bogra* (P 104, 15 July 1977), *Rangamati* (P 105, 11 February 1977)
Five were the first warships to be built in Bangladesh (by DEW Narayangon of Dhaka) between 1972–77 and are based on river steamers. They are 75ft (*22.9m*) long, have 2 Cummins diesels giving 10.8kts (range 700nm at 8kts), displace 69.5t, a crew of 33 and are armed with 1–40mm. *Noakhali*, *Pabna* and *Patuakh-ali* have the gun aft but *Bogra* and *Rangamati* have it mounted forward.

AMPHIBIOUS WARFARE VESSELS
Four ex-Chinese 'Yuchan' class landing craft transferred 1986.
Three 83t landing craft built in Bangladesh in late 1980s.
Six ex-US LCM 6 transferred 1992
All extant 1995.

Barbados

Barbados operates a coastal patrol force; a 124-man volunteer coast guard (created early 1973) employs seven patrol boats (the 37.5m *Trident* in service November 1981, one 30t, British-built 'Guardian' class boat and three 11.5t Lewis type boats also British-built, plus two converted shrimp boats). As an independent state within the Commonwealth since 1966 Barbados would rely on Britain for military support. The Coast Guard, under a Royal Navy Commander, has its HQ at St Ann's Fort, St Michael and a new base (HMBS *Willoughby Fort*) at Bridgetown, the island capital.

Trident as completed — Brooke Marine

Belgium

At the end of the Second World War the Belgian Navy was left with only two vessels of any consequence, both of which had been recovered from the Germans. Neither was in particularly good mechanical condition; *Artevelde* was incapable of proceeding to sea under her own power, and *Jan Breydel*, despite her extensive conversion while in German hands, was thirty years old. An ex-US Navy patrol frigate was added in 1947, but only with the transfer of six ex-British ocean minesweepers of the *Algerine* class between 1949 and 1953 did the postwar Belgian Navy acquire a meaningful operational capability. Membership of NATO was soon to bring an expanded role, and a corresponding increase in the size of Belgium's naval forces. Operating in conjunction with the fleets of other NATO powers of northwest Europe, these forces were to be responsible for the security of maritime communications in the North Sea, the English Channel and the Western Approaches. A large programme of mine countermeasures vessels was financed by the United States under the Mutual Defense Assistance Program (MDAP). By 1960 no fewer than forty-seven such vessels were in service with the Belgian Navy, served by two converted logistics support ships. At about the same time, the four remaining *Algerine* type ocean minesweepers – which now included two ex-Canadian units recently transferred – were allocated to more general patrol duties.

The fleet structure built up by 1960 remained basically unaltered during the next two decades, and was reinforced by the replacement of the logistics support vessels by two purpose-built units completed 1966–67, and the replacement of the *Algerines* by the small frigates of the E 71 type, sophisticated modern vessels which represented a considerable achievement by the small Belgian shipbuilding industry.

By the late 1970s the wooden minesweepers built during the 1950s were in urgent need of replacement. The larger MSOs of the US *Agile* type had all undergone conversion as minehunters during the mid-1970s, but many of the smaller MSCs of the *Adjutant* type had been transferred to other countries and the remainder were being steadily discarded. France and the Netherlands, which were in a similar position and had similar operational requirements, joined with Belgium in the 'Triparite Minehunter Project'. The 10-ship *Aster* class experienced major initial delays as a result of problems in forming a suitable shipbuilding consortium. This programme has now been completed, but a shortage of funds has effectively precluded a projected update programme for the frigates. A collaborative venture with the Netherlands to provide a new class of sea-going minesweeper to replace the surviving units of the US *Agile* and *Adjutant* types appeared to be a victim of European defence restructuring in the wake of the collapse of the Warsaw Pact, but has now been revived as a national programme.

BASES
A major new operational base was completed at Zeebrugge in 1985, and this now accommodates the frigates, most of the MCM vessels and the two MCM logistics ships, together with all reserve units. Other MCM vessels are based at Ostend, the former main base, and at Antwerp.

FLEET STRENGTH 1947

FRIGATES

Name	Launched	Disp (standard)	Fate
JAN BREYDEL (ex-*Barbara*, ex-*Zinnia*)	12.8.15	1200t	BU 1952
ARTEVELDE (ex-*K 4*, ex-*Lorelei*, ex-*Artevelde*)	28.8.40	1640t	BU 1954
LUITENANT TER ZEE VICTOR BILLET (ex-*PF 57*, *Sheboygan*)	31.7.43	1430t	Stricken 1957

Jan Breydel (formerly *Zinnia*) was an ex-British 'Flower' class sloop employed as a fishery protection vessel until captured by the Germans in 1940. She was subsequently rebuilt at the Cockerill yard, Antwerp, and emerged, named *Barbara*, with a typical German profile and an armament of 3–105mm (3 × 1), 8–37mm (4 × 2) and 12–20mm (6 × 2) guns.

Artevelde which, originally, was to have replaced her in the fishery protection role, was seized while on the slipway at the Cockerill yard in 1940, and was subsequently fitted out in similar fashion to *Zinnia* by Wilton-Fijenoord. She had an armament of 3–105mm (3 × 1), 4–37mm (2 × 2) and 10–20mm (2 × 4, 1 × 2).

Neither ship was in very good mechanical condition when they reverted to Belgian control at the end of the War. *Artevelde*, which could not even procede to sea under her own power, never had her machinery repaired, and became a stationary training ship. Both vessels were stricken in the early 1950s after some modifications to their armament.

Victor Billet was an ex-US Navy patrol frigate of the *Tacoma* class acquired in March 1947. Until 1950 she operated as part of the International North Atlantic Weather Organisation. She was subsequently transferred to fishery protection duties and became a stationary training ship in the mid-1950s.

Luitenant Ter Zee Victor Billet 1949 — L & L van Ginderen

Wandelaar 13.7.94 — L & L van Ginderen

MAJOR SURFACE SHIPS

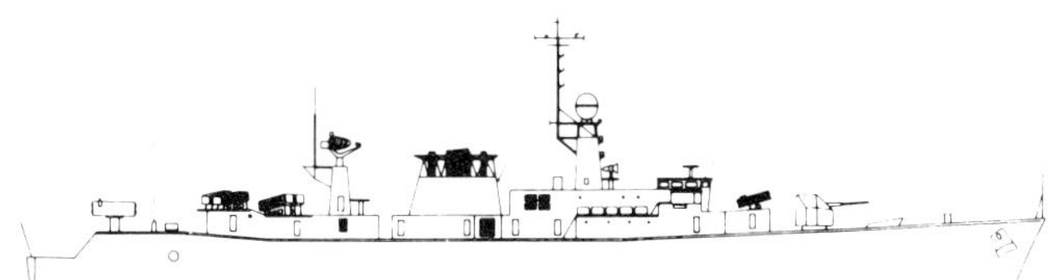

Wielingen as completed

WIELINGEN (E 71) class *frigates*

Displacement: 1880t light; 2283t full load
Dimensions: 338ft pp, 349ft oa × 40ft × 17ft
103.0m, 106.4m × 12.3m × 5.3m
Machinery: 2 shaft CODOG: 1 Olympus TM3B gas turbine, 28,000shp = 25kts, plus 2 Cockerill CO240 V12 diesels, 6000bhp = 20kts. Range 4500nm at 18kts
Armament: 4 MM38 Exocet SSM (4×1), 1 NATO Sea Sparrow SAM (1×8, 8 missiles), 1–100mm, 1–375mm Bofors ASW RL, 2 catapults for L5 torpedoes, 2 Corvus chaff RL
Sensors: Radar DA-05, WM-25; sonar SQS-505A
Complement: 160

No	Name	Builder	Laid down	Launched	Comp	Fate
F 910	WIELINGEN	Boëlwerf, Temse	5.3.74	30.3.76	20.1.78	Extant 1995
F 911	WESTDIEP	Cockerill, Hoboken	2.9.74	8.12.75	20.1.78	Extant 1995
F 912	WANDELAAR	Boëlwerf, Temse	5.3.75	21.6.77	27.10.78	Extant 1995
F 913	WESTHINDER	Cockerill, Hoboken	8.12.75	28.1.77	27.10.78	Stricken 1993

Detailed studies for these vessels were begun in 1969. Despite the limited size and capabilities of the Belgian shipbuilding industry, the government made procurement conditional on Belgian participation in their design and construction. Costs were to be kept to a minimum by providing only those features essential to the escort mission in the North Sea/English Channel area. Only weapons systems in service with other NATO navies or under development at that time were considered, and particular stress was laid on seaworthiness, watertight integrity and automation.

The design was heavily influenced by developments in both the French and the Royal Netherlands navies. The major weapons systems, with the exception of NATO Sea Sparrow, are of French origin, while the Netherlands provided the air search and fire control radars, together with the SEWACO IV tactical data system. Under the terms of the contract the Belgian shipyards were responsible for the hull and machinery, whereas the installation and testing of the weapons systems was the responsibility of the Belgian Navy.

The action information centre is between decks, and the ships can be divided into two independent gas-tight citadels for NBC close-down. All machinery can be remotely controlled from a central control station at main deck level. Plans to fit a Goalkeeper CIWS and new ESM equipment have now been shelved, but Sea Sparrow has been upgraded, and they have now received the latest variant of the L5 torpedo.

Westhinder was retired following a grounding incident in 1993; two of the remaining three ships will be retained in operational service, with the third in refit or reserve.

BELGIUM

MINE WARFARE VESSELS

Ex-British ALGERINE class *ocean minesweepers*

Displacement: 1040t standard; 1335t full load
Dimensions: 213ft pp, 225ft oa × 36ft × 10ft 6in
65.0m, 68.6m × 10.8m × 3.2m
Machinery: 2-shaft geared steam turbines (triple expansion in ex-Canadian vessels), 2000shp = 16kts. Range 4000nm at 10kts
Armament: 1–102mm (4in), 4–40mm (4×1) (except ex-Canadian vessels), 8–20mm (4×2); 4 DCT
Complement: 101

No	Name	Builder	Acquired	Fate
Ex-British				
M 900	ADRIEN DE GERLACHE (ex-*Liberty*)	Harland & Wolff	29.11.49	Stricken 1969
M 901	GEORGES LECOINTE (i, ex-*Cadmus*)	Harland & Wolff	Jan 1950	Stricken 1959
M 902	JAN VAN HAVERBEKE (ex-*Ready*)	Harland & Wolff	4.7.51	Stricken 1960
M 903	A F DUFOUR (i, ex-*Fancy*)	Blyth SB & DD Co	9.8.51	Hulked 1959
M 904	DE BROUWER (ex-*Spanker*)	Harland & Wolff	25.2.53	Stricken 1966
M 905	DE MOOR (ex-*Rosario*)	Harland & Wolff	15.1.53	Stricken 1969
Ex-Canadian				
F 901	GEORGES LECOINTE (ii, ex-*Wallaceburg*)	Port Arthur SB, Ontario	31.7.59	Stricken 1969
F 903	A F DUFOUR (ii, ex-*Winnipeg*)	Port Arthur SB, Ontario	7.8.59	Stricken 1966

These ships were formerly ocean minesweepers of the *Algerine* class, launched in 1942–44. The six ex-Royal Navy vessels were transferred during 1949–53. Their size made them well suited to general patrol duties in addition to their minesweeping role. In the mid-1950s *De Brouwer* and *De Moor* were tropicalised for service in African waters and by the late 1950s all six ships had been reclassified *drageurs-escorteurs* (minesweeping escorts). Three of the class paid off 1959–60, and a fourth, *Adrien de Gerlache* became a logistics vessel (*bâtiment logistique*). Her armament was reduced to two 40mm and she was given the number A 954. At the same time two ex-Canadian vessels of the same class were acquired, and took over the names *Lecointe* and *Dufour.* The Canadian ships differed from the ex-RN units in having reciprocating engines; they were fitted with 20mm twin AA guns, but these were quickly replaced by 40mm singles. Following the transfer of the Canadian vessels the class were redesignated *escorteurs côtiers* (coastal escorts) and were given 'F' pennant numbers (*De Brouwer* became F 904 and *De Moor* F 905). *A F Dufour* and *De Brouwer* paid off in 1966, and the remaining pair had the single 102mm replaced by a single 40mm before they followed in 1969.

US AGILE type *ocean minesweepers*

Displacement: 720t standard; 780t full load
Dimensions: 165ft pp, 172ft oa × 36ft × 14ft
50.3m, 52.4m × 11.0m × 4.2m
Machinery: 2-shafts, 2 General Motors 8–268A diesels, 1600bhp = 14kts. Range 3000nm at 10kts
Armament: 1–40mm
Sensors: Sonar AN/UQS–1
Complement: 72

No	Name	Builder	Acquired	Fate
M 902	VAN HAVERBEKE (ex-*MSO 522*)	Peterson, Wisconsin	9.12.60	Extant 1995
M 903	A F DUFOUR (iii, ex-*Lagen*, ex-*MSO 498*)	Bellingham, Washington	14.4.66	Extant 1995
M 904	DE BROUWER (ii, ex-*Nansen*, ex-*MSO 499*)	Bellingham, Washington	14.4.66	Stricken 1993
M 906	BREYDEL (ex-*AM 504*)	Tacoma, Washington	15.2.56	Stricken 1993
M 907	ARTEVELDE (ex-*AM 503*)	Tacoma, Washington	15.12.55	Stricken 1985
M 908	GEORGES TRUFFAUT (ex-*AM 515*)	Tampa SB	12.10.56	Extant 1995
M 909	FRANCOIS BOVESSE (ex-*AM 516*)	Tampa SB	25.1.57	In reserve

These wooden-hulled ocean minesweepers of the *Agile* type were built in the USA and transferred under MDAP shortly after completion. M 903, M 904 and M 907 were launched in 1954, M 906 and M 908 in 1955, M 909 in 1956, and M 902 (*Van Haverbeke*) in 1959; she was transferred in 1960 and took the name and number of a unit of the *Gerlache* class which had just been stricken.

In 1966 two further units, originally launched in 1954, were acquired from Norway; formerly named *Lagen* and *Nansen*, they were renamed *A F Dufour* (M 903) and *De Brouwer* (M 904) respectively, thereby replacing two vessels of the *Gerlache* class which were retired in the same year. In 1972 *Artevelde* was converted to a diving vessel. Conventional sweep gear was removed and in 1974 a Siebe Gorman decompression chamber was fitted. The remaining six units were converted to minehunters 1973–76; a General Electric SQQ-14 sonar was fitted and the single 40mm removed and replaced by a 12.7mm twin mounting.

De Brouwer and *Breydel* were retired in 1993. Of the remaining four ships, *Van Haverbeke* is used as a base ship for divers and *Bovesse* has been placed in reserve. It is still hoped to replace this class with a new type of coastal minesweeper (see below).

Breydel Dec 1973 — C & S Taylor

US ADJUTANT type *coastal minesweepers*

Displacement: 330t standard; 390t full load
Dimensions: 138ft pp, 144ft oa × 27ft × 8ft
42.0m, 44.0m × 8.3m × 2.6m
Machinery: 2-shafts, 2 General Motors 8–268A diesels, 880bhp = 13.5kts. Range 2700nm at 10.5kts
Armament: 1–40mm
Sensors: Radar Decca 1229; sonar AN/UQS-1
Complement: 36

Class:
Henry B Nevins, NY – *Lier* (M 912), *Sint Niklaas* (M 918), *Diksmuide* (M 920)
Quincy Adams, Mass – *Diest* (M 910), *Maaseick* (M 913)
Hodgdon Bros, Maine – *Eeklo* (M 911), *Roeselaere* (M 914), *Arlon* (M 915), *Bastogne* (M 916), *Charleroi* (M 917), *Hervé* (M 921), *Malmédy* (M 922), *Verviers* (M 934), *Veurne* (M 935)
Consolidated SB, NY – *Sint Truiden* (M 919), *Blankenberge* (M 923), *Laroche* (M 924)
Hiltebrant DD, NY – *De Panne* (M 925)
Boël & Zonen, Temse – *Mechelen* (M 926), *Spa* (M 927), *Stavelot* (M 928), *Heist* (M 929)
Béliard, Ostend – *Rochefort* (M 930), *Knokke* (M 931), *Nieuwport* (M 932), *Koksijde* (M 933)

All twenty-six of these wooden-hulled coastal minesweepers of the *Adjutant* type were financed under MDAP. The first sixteen were built in the USA and transferred 1953–55. Eight more were completed by Belgian yards 1954–55, and a further two US-built vessels were added in 1956. The Belgian-built vessels were fitted out with machinery and armament from the USA.

In 1966 M 914–16 were transferred to Norway in return for two *Agile* type MSOs. At the same time *Mechelen* became an ocean research ship (A 962). Fifteen further units were returned to the USA in 1969 and were subsequently transferred to other navies: M 910–13, M 917, M 918, M 920 and M 925 to Taiwan, and M 919, M 921–24 to Greece. Those vessels which remained in service with the Belgian Navy had the original 20mm twin mounting replaced by a 40mm single at about this time. *Veurne* and *Verviers* were converted to minehunters with a Plessey Type 193 sonar and Voith-Schneider propellers in 1969 and 1972 respectively. The former was subsequently used for survey work. In 1977 *Spa* became a munitions ship (A 963) and *Heist* a degaussing ship (A 964). *Knokke* was stricken in 1976 and transferred to sea scouts at Ghent in 1978. The remaining ships were retained either in active service or in reserve into the late 1980s to provide a conventional minesweeping capability. They will eventually be replaced by a new class of 610t minesweepers developed jointly with the Royal Netherlands Navy. The two minehunter conversions, *Verviers* and *Veurne*, were stricken in 1988 and 1987 respectively. *Heist* was restored as a minesweeper in 1985, having served since 1978 as a degaussing tender (A 964). Her sister *Spa* (A 963, ex-M 927), continues to serve as a missile/munitions transport for frigates. *Stavelot* was stricken in 1987. *Kiksijde* and *Nieuwport* were stricken in 1991; *Heist* and *Rochefort* in 1992.

Verviers Dec 1973 C & S Taylor

HERSTAL class *inshore minesweepers*

Displacement: 160t standard; 190t full load
Dimensions: 107ft pp, 113ft oa × 22ft × 7ft
32.5m, 34.5m × 6.7m × 2.1m
Machinery: 2-shafts, 2 Fiat-Mercedes Benz MB 820 diesels, 1260bhp = 15kts. Range 2300nm at 10kts
Armament: 2–20mm (1×2)
Sensors: Radar Decca 1229
Complement: 17

Class:
Temse (M 470), *Hasselt* (M 471), *Kortrijk* (M 472), *Lokeren* (M 473), *Turnhout* (M 474), *Tongeren* (M 475), *Merksem* (M 476), *Oudenaerde* (M 477), *Herstal* (M 478), *Huy* (M 479), *Seraing* (M 480), *Tournai* (M 481), *Vise* (M 482), *Ougrée* (M 473), *Dinant* (M 484), *Andenne* (M 485).

These inshore minesweepers were derived from the British 'Ham' class. All were built in Belgium by Mercantile Marineyard at Kruibeke, eight (M478–85) under MDAP and the remainder under Belgian Navy estimates. Four were launched in 1956, eight in 1957 and the last four in 1958. Designed to sweep the Schelde River, they were fitted for acoustic, magnetic and mechanical sweeping to a depth of 4.5–10m.

In the late 1960s the 20mm twin mounting began to be replaced by a new 12.7mm twin mounting. *Temse* and *Tournai* were transferred via the USA to South Korea in 1970. At the same time, *Herstal* was transferred to survey duties. In the late 1970s *Hasselt, Kortrijk* and *Huy* were modified as ready duty ships; all sweep gear was removed and replaced by a deckhouse, pollution clean-up gear and special cranes mounted on the stern. *Hasselt* was broken up in 1986, *Lokeren* paid off in 1987, *Kortrijk* and *Oudenaerde* in 1989, *Huy* and *Seraing* in 1990, and *Andenne, Turnhout, Tongeren, Herstal* and *Vise* in 1991. *Merksem, Ougrée* and *Dinant* in 1992.

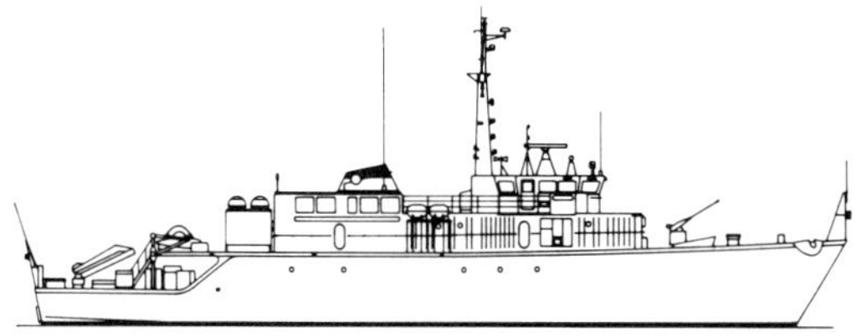

Fuschia 1989
NB 1/750 scale

ASTER ('Tripartite') class *minehunters*

Displacement: 511t standard; 595t full load
Dimensions: 169ft oa × 29ft × 8ft
51.6m × 9m × 2.5m
Machinery: 1-shaft, 1 Brons-Werkspoor A-RUB 215X V12 diesel, 1900bhp= 15kts; twin active rudders (7kts) and bow thruster. Range 3000nm at 12kts
Armament: 1–20mm, 2 PAP 104
Sensors: Radar Decca 1229; sonar DUBM 21B
Complement: 34–46

No	Name	Builder	Launched	Fate
M 915	ASTER	Béliard, Ostend	6.6.85	Extant 1995
M 916	BELLIS	Béliard, Ostend	14.2.86	Extant 1995
M 917	CROCUS	Béliard, Ostend	6.8.86	Ammunition ship 1995
M 918	DIANTHUS	Béliard, Ostend	26.2.87	Stricken 1993
M 919	FUCHSIA	Béliard, Ostend	23.9.87	Stricken 1993
M 920	IRIS	Béliard, Ostend	21.4.87	Stricken 1993
M 921	LOBELIA	Béliard, Ostend	6.1.88	Extant 1995
M 922	MYOSOTIS	Béliard, Ostend	4.8.88	Extant 1995
M 923	NARCIS	Béliard, Ostend	30.3.90	Extant 1995
M 924	PRIMULA	Béliard, Ostend	17.12.90	Extant 1995

These ten vessels were the Belgian contribution to the 'Tripartite' minehunter programme (for technical details see under France). Construction of the class was delayed by problems in forming the Polyship consortium which was to have been responsible for the programme. The consortium was eventually dissolved, and the ships were reordered from Béliard in February 1981; the hulls were built at Ostend in a purpose-built 'ship factory', then towed to Béliard Mercantile at Rupelmonde for fitting-out. The ship's company varies from twenty-three to forty-six according to task; six divers are carried when minehunting. A portable divers' decompression van can be carried on 01 deck abaft the forecastle break. M 917 was fitted with two single 12.7mm MG for deployment to the Middle East in 1987–88, M 918, M 920 and M 922 deployed to the Gulf in 1991 as Belgium's contribution to the allied MCM force.

Three of the class paid off in 1993, and *Crocus* was modified as an ammunition transfer vessel in the same year.

New minesweeper class

Displacement: 620t full load
Dimensions: 157ft 6in wl, 172ft oa × 34ft × 10ft 3in
48m, 52.4m × 10.4m × 3.1m
Machinery Range 3000 nm at 12kts
Armament: 1–20mm
Complement: 25–30

No	Name	Builder	Launched	Fate
4 ordered		Béliard Polyship NV		

The memorandum for a joint Belgium/Netherlands minesweeper project was signed in April 1989, and in November 1990 a design contract was awarded to Van der Giessen de Noord Marinebouw in conjunction with Béliard Polyship NV; initial plans were completed in 1992. Originally sixteen hulls were to be built for the Netherlands Navy and ten for Belgium, but in the changing political climate these figures were subsequently reduced to eight and six respectively. In January 1993 the Netherlands pulled out of the project due to budgetary constraints, and Belgium decided to continue alone, authorizing four hulls in July 1993. The first is due to be completed in 1999.

The requirement is for a ship capable of sweeping bottom-laid mines which have sunk so far into the sea bed as to be undetectable by mine hunters. The magnetic sweep gear, 'Sterne M', is being developed by Thomson Sintra, and comprises six bodies, towed in array, each carrying two coils. These generate a simulated ship's signature. Proven acoustic and mechanical sweep devices will also be installed.

MISCELLANEOUS

KAMINA *support ship*

Displacement: 3900t standard; 5750t full load
Dimensions: 345ft pp, 374ft oa × 48ft × 18ft
105.0m, 114.0m × 14.6m × 5.5m
Machinery: 1 shaft, 1 Blohm & Voss diesel, 3600bhp = 15kts. Range 10,000nm at 15kts
Armament: 2–76mm (2×1), 4–40mm (4×1), 5–20mm (5×1)
Complement: 175

No	Name	Builder	Acquired	Fate
AP 907	KAMINA (ex-*Royal Harold*, (ex-*Hermann von Wissmann*)	Cockerill, Hoboken	1950	Stricken 1967

One of two mercantile vessels ordered from the Cockerill yard for Poland and laid down prior to the Second World War, this ship was launched on 26 December 1940 and completed as the depot ship *Hermann von Wissmann* and was employed as a support ship for U-boats in Norway. She was seized by the British in 1945, and was transferred to the Belgian Navy in 1950. She was employed as a troop transport throughout the 1950s; her original armament was replaced and there was accommodation for 800 troops. In 1962 she was reclassified as a logistics support and training vessel, and was renumbered A 957. Cable reels replaced the after gun mountings, the armament being reduced to 1–76mm, 1–40mm and 2–20mm.

GODETIA *support ship*

Displacement: 1700t light; 2500t full load
Dimensions: 288ft wl, 301ft oa × 46ft × 11ft 6in
87.9m, 91.8m × 14m × 3.5m
Machinery: 2 shafts, 4 ACEC-MAN diesels, 5400bhp = 19kts. Range 8700nm at 12kts
Armament: 4–40mm (2×2)
Complement: 95

No	Name	Builder	Launched	Fate
A 960	GODETIA	Boëlwerf, Temse	7.12.65	Extant 1995

Designed, like her half-sister *Zinnia*, primarily as a command and logistics support ship for mine countermeasures vessels, *Godétia* also doubles as a royal yacht. The ship can be protected against radioactive fallout by closed-circuit ventilation, and has space for a laboratory and accommodation for oceanographic research personnel. She has passive tank stabilisation and controllable-pitch propellers.

A central hold is served by a single crane for the transfer of supplies to smaller vessels. As completed, *Godétia* had 40mm twin mountings fore and aft, but in the late 1960s the after mounting was removed to make room for two reels for minesweeping cables. At the same time a landing platform for a light liaison helicopter was fitted aft as a continuation of the first superstructure deck. In a second major refit 1979–80 the foremost cable reel was replaced by a new deckhouse, and a 40mm single and four 12.7mm twin mountings were fitted in place of the original armament.

Godetia 11.8.81 L & L van Ginderen

Zinnia 1975

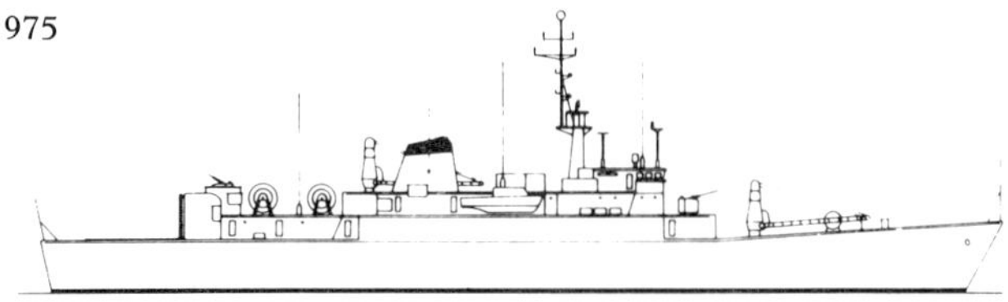

ZINNIA *support ship*

Displacement: 1705t light; 2685t full load
Dimensions: 309ft wl, 326ft oa × 46ft × 12ft
94.2m, 99.5m × 14m × 3.6m
Machinery: 1 shaft, 2 Cockerill-Ougrée V12 TR 240 CO diesels, 5000bhp = 18kts. Range 14000nm at 12.5kts
Armament: 3–40mm (3×1), Alouette III helicopter
Complement: 123

No	Name	Builder	Launched	Fate
A 961	ZINNIA	Cockerill, Hoboken	6.5.67	Extant 1995

Developed from *Godétia*, *Zinnia* incorporated a number of modifications. The bridge structure was set farther aft, allowing a second supply hold and crane to be accommodated in the bow section. She was also given a telescoping hangar and a landing platform on the stern, enabling her to operate an Alouette II (now III) liaison helicopter. Two cable reels were fitted on the after deckhouse from completion. Fin stabilisers were adopted in preference to the liquid anti-roll system installed in her half-sister.

Godétia and *Zinnia* replaced the war-built logistics vessels *Adrien de Gerlache* and *Kamina*, which were both retired from service in the late 1960s. *Zinnia* was deployed to the Gulf in 1987–88 in support of Belgian MCM vessels, and again in 1991 when she was fitted with chaff launchers. Both vessels are based at Zeebrugge. In peacetime both ships are employed in the fishery protection role.

Belize

Belize (British Honduras until June 1973) became independent on 21 September 1981 after protracted negotiations; the 202,000 inhabitants were, and still are, wary of the Guatemalan threat to peace and security. Guatemala has consistently claimed a large area of Belize as her own territory. Security against an incursion by Guatamala remains upheld by a British military presence as well as by several states in the area, namely Jamaica, Barbados and Guyana.

Belize operates two 22kt patrol boats *Belize* and *Belmopan* (15t, Brooke Marine type, 3 MG, built in 1972) along her 174-mile coastline (she has proven offshore oil reserves) and has a total of three merchant ships. Personnel amounts to fifty volunteers. A British frigate or destroyer was stationed off Belize and acted as Caribbean guard ship; she was backed up by a support ship. Two more patrol boats of Souter Wasp type were commissioned in 1984, *Dangriga* and *Toledo* (P 01–02), (36.3t full load, 20.0m, 18kts, 1–12.7mm, 2 MG). *Dangriga* was deleted in 1993.

Benin (ex-Dahomey)

In early 1978 the People's Republic of Benin (Dahomey until 1975) began forming a naval force at Cotonou; its army personnel now totals about 200. The Soviet Union delivered two 'P 4' class torpedo boats and two 'Zhuk' class coastal patrol craft in January 1979 with another two in September 1980. The 'P 4s' were discarded in 1990 and the 'Zhuks' in 1992. To patrol the country's 75-mile coastline they have been replaced by the French-built PR-362T patrol boat *Patriote*: 70t, 125ft (*38m*), 36kts, 1–40mm, 2–12.7mm MG (1×2). She was damaged on her outward journey in 1988 but was repaired and returned to service in 1991. Extant 1995.

Bolivia

An internally strife-torn country which has experienced numerous *coups*, Bolivia has less cause to fear external than internal aggression because of her mountainous and landlocked geographical location. Bolivia has been in dispute with Chile since the early 1960s over the Rio Lanca and outlets to the Pacific. The *Fuerza Naval Boliviana* with a total of 2600 men in 1982, increased to 5000 in 1992, operates riverine (12,000 miles navigable) and lake patrol craft from bases at Tiquina, Puerto Suárez, Riberalta, Trinidad, Guaramerin and Puerto Villarroel. A vice-admiral commands the five naval districts and most training is done in Argentina.

One transport ship *Libertador Bolivar* (9000grt) is operated. Acquired from Venezuela in 1977 and employed commercially.

Nine river transport boats. Extant 1995.

Up to thirty river patrol boats in operation, including about twenty-four on Lake Titicaca, and two ex-US PBR Mk 2 boats.

Brazil

Out of the Second World War, the Brazilian Navy acquired the long-desired parity with that of Argentina. Brazil supported the Allied war against the Axis almost from the beginning: Allied planes patrolled the South Atlantic from Brazilian airfields; convoys were screened by Brazilian warships; and 25,000 Brazilian troops fought in Italy. It is not, therefore, surprising that, beginning in 1944, the United States began to transfer destroyer escorts to Brazil under Lend-Lease. Five years earlier, a proposed transfer of destroyers from the United States had created a furore in Argentina: the United States was accused of upsetting the balance of power and so the arrangement was cancelled. By the late 1940s, the theory of parity with Argentina had been established.

The Brazilian Navy has been the Latin American trend-setter in warship acquisitions. This navy was the first to obtain the submarine, the dreadnought, and the aircraft carrier. In 1956 Brazil purchased a *Colossus* class carrier, HMS *Vengeance*, from the Admiralty for $9m, and the ship, renamed *Minas Gerais*, was extensively modernised. The acquisition of the carrier caused a major inter-service struggle within Brazil. At issue was the control of the aircraft to be flown from *Minas Gerais*. From the completion of her modernisation until 1963, the carrier was planeless and neither President Jusecelino Kubitschek nor his two successors were able to resolve the problem. In 1963 the Navy purchased six T-28 trainers and four helicopters, but the Air Force, which was responsible for all military aircraft, refused to let the planes be assigned to the carrier. Ultimately, numerous high-ranking Air Force and Navy officials resigned. The final decision was that fixed-wing aircraft operating from the carrier would be flown by the Air Force, and rotary-wing aircraft by the Navy. A naval air arm was formed on 26 January 1965. The significance of this decision was that the Navy would not control attack aircraft and that the carrier would be exclusively an ASW platform. Thus, once again, the Argentine Navy, with *Independencia* and, later, *25 de Mayo* would be the most powerful in the South Atlantic.

Brazil has taken the greatest advantage of bi-national shipbuilding agreements of any country in Latin America. Since independence in the early nineteenth century, Brazil's struggle to develop a shipbuilding industry has been tenacious. One-third of the Brazilian fleet which participated in the War of the Triple Alliance (1864–70) was of indigenous construction, but from 1890 until the 1950s, Brazil struggled to re-establish her shipbuilding industry. In December 1958, a major breakthrough occurred: Ishikawajima-Harima Heavy Industries of Japan founded a new shipyard in Brazil. By 1974 the labourers and skilled workmen were Brazilian, with a few Japanese filling technical and supervisory positions. In that year the yard employed 3250 persons, operated the two largest drydocks in Latin America, and had already constructed 426,906grt. This, plus other maritime industries attracted into Brazil, provided the necessary industrial base for the bi-national agreement to construct frigates. In 1970 Brazil and the British company Vosper Thornycroft agreed to construct six Mk 10 frigates, four of which were to be built in England and two in Brazil. In mid-1981 Brazil ordered a modified Mark 10 frigate to be built in Brazil. The frigate is classified as a training ship. The newly established shipbuilding industry, plus the bi-national construction agreement, have thus allowed Brazil to begin to meet much of its own shipbuilding needs. In 1994 64,700 officers and men (12-month conscripts, including 14,600 marines) served with the Brazilian Navy; main bases are at Rio de Janeiro, Aratu, Belem, Natal, Ladario (river flotilla) and Sao Pedro (naval air arm) covering a 4655-mile coastline and the world's longest river system. The active fleet totalled ninety-seven warships.

FLEET STRENGTH 1947

BATTLESHIPS

Name	Launched	Disp (normal)	Fate
Minas Gerais class			
MINAS GERAIS	10.9.08	19,200t	Discarded 1953
SAO PAULO	19.4.09	19,200t	Discarded 1951

Minas Gerais was inactive in the latter years of her career. *Sao Paulo* had not been modernised in the 1930s as her sister had been, and she was no more than a stationary guard ship from that time until her sale.

CRUISER

Name	Launched	Disp (normal)	Fate
RIO GRANDE DO SUL	20.04.09	3000t	Discarded 1948

DESTROYERS

Name	Launched	Disp (normal)	Fate
Marcilio Dias class			
GREENHALGH	8.7.41	1500t	Discarded 1966
MARCILIO DIAS	20.7.40	1500t	Discarded 1966
MARIZ E BARROS	28.12.40	1500t	Discarded 1972
Acre class			
ACRE	30.5.45	1340t	Discarded 1974
AJURICABA	30.5.45	1340t	Discarded 1964
AMAZONAS	29.11.43	1340t	Discarded 1973
APA	30.5.45	1340t	Discarded 1964
ARAGUARY	14.7.46	1340t	Discarded 1974
ARAGUAYA	29.11.43	1340t	Discarded 1974

Mariz e Barros was fitted with a Seacat SAM system in 1966 as a test platform; this sytem was transferred to *Mato Grosso*, an *Allen M Sumner* class unit, in the mid-1970s.

Mariz e Barros with Seacat SAM aft 1968 USN

BRAZIL

Cananela 1952
NB 1/750 scale

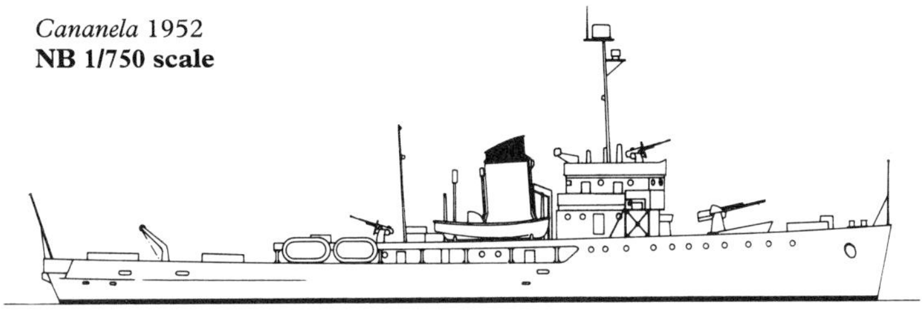

MINOR WARSHIPS

Name	Launched	Disp (normal)	Fate
Henrique Dias class trawlers			
BARRETO DE MENEZES (ex-*Paru*)	Feb 45	813t	Discarded 1960
FELIPE CAMARAO (ex-*Papaterra*)	Jul 42	813t	Discarded 1960
FERNANDES VIEIRA (ex-*Parati*)	11.6.42	813t	Discarded 1953
HENRIQUE DIAS (ex-*Pargo*)	26.8.42	813t	Discarded 1960
MATHIAS DE ALBUQUERQUE (ex-*Pampano*)	11.6.42	813t	Discarded 1952
VIDAL DE NEGREIROS (ex-*Pelegrime*)	1942	813t	Discarded 1951
Rio Pardo submarine chaser			
RIO PARDO	29.11.43	132t	Discarded 1974
Carioca class minelayers			
CABEDELO	16.9.39	552t	Discarded 1960
CAMOCIM	28.10.39	552t	Discarded 1960
CANANELA	22.10.38	552t	Discarded 1960
CARAVELAS	14.9.39	552t	Discarded 1960
CARIOCA	22.10.38	552t	Discarded 1960
Ex-US PC class submarine chasers			
GOIANA (ex-*PC 554*)	2.5.42	280t	Discarded 1952
GRAJAU (ex-*PC 1236*)	17.11.43	280t	Discarded 1960
GRAUNA (ex-*PC 561*)	6.12.43	280t	Discarded 1960
GUAIBA (ex-*PC 604*)	14.6.43	280t	Discarded 1952
GUAJARA (ex-*PC 607*)	22.10.43	280t	Discarded 1960
GUAPORE (ex-*PC 554*)	26.9.42	280t	Discarded 1958
GUAUPA (ex-*PC 605*)	14.6.43	280t	Discarded 1952
GURUPI (ex-*PC 547*)	26.9.42	280t	Discarded 1960
Ex-US SC 497 class submarine chasers			
JACUI (ex-*SC 1288*)	19.5.43	95t	Discarded 1951
JAGUARAO (ex-*SC 765*)	16.2.43	95t	Discarded 1951
JAGUARIBE (ex-*SC 767*)	16.2.43	95t	Discarded 1951
JAVARI (ex-*SC 763*)	7.12.42	95t	Discarded 1951
JUNDIAI (ex-*SC 1289*)	26.4.43	95t	Discarded 1951
JURUA (ex-*SC 764*)	31.12.44	95t	Discarded 1951
JURUENA (ex-*SC 766*)	31.12.42	95t	Discarded 1951
JUTAI (ex-*SC 762*)	31.12.42	95t	Discarded 1951

SUBMARINES

Name	Launched	Disp (normal)	Fate
Humaita class			
HUMAITA	11.6.27	1390t/1884t	Discarded 1951
Tupi class			
TAMOIO	14.2.37	620t/853t	Discarded 1960
TIMBIRA	30.12.36	620t/853t	Discarded 1960
TUPI	28.11.36	620t/853t	Discarded 1960

DESTROYER ESCORTS

Name	Launched	Disp (normal)	Fate
Ex-US Cannon class (DET type)			
BABITONGA	8.7.43	1240t	Discarded 1964
(ex-*Alger*)			
BAEPENDI	25.5.43	1240t	Discarded 1974
(ex-*Cannon*)	22.8.43	1240t	Discarded 1975
BAURU			
(ex-*Reybold*)			
BEBERIBE	5.9.43	1240t	Discarded 1968
(ex-*Herzog*)	8.7.43	1240t	Discarded 1975
BENEVENTE			
(ex-*Christopher*)	8.8.43	1240t	Discarded 1964
BERTIOGA			
(ex-*Pennewill*)	8.8.43	1240t	Discarded 1975
BOCAINA			
(ex-*Marts*)	5.9.43	1240t	Discarded 1974
BRACUI (ex-*McAnn*)			

MISCELLANEOUS

Name	Launched	Disp (normal)	Fate
River gunboats			
PARNAIBA	2.9.37	595t	Extant 1995
PARAGUACU (ex-*Victoria*, (ex-*Espiritu Santo*)	22.12.38	430t	Discarded 1972
PERNAMBUCO	1910	470t	Discarded 1948
Training ship			
ALMIRANTE SALDANHA	12.1933	3325t	Discarded 1992

Almirante Saldanha initially was a four-masted schooner. In 1961 her masts were removed and she was refitted as an oceanographic research ship.

MAJOR SURFACE SHIPS

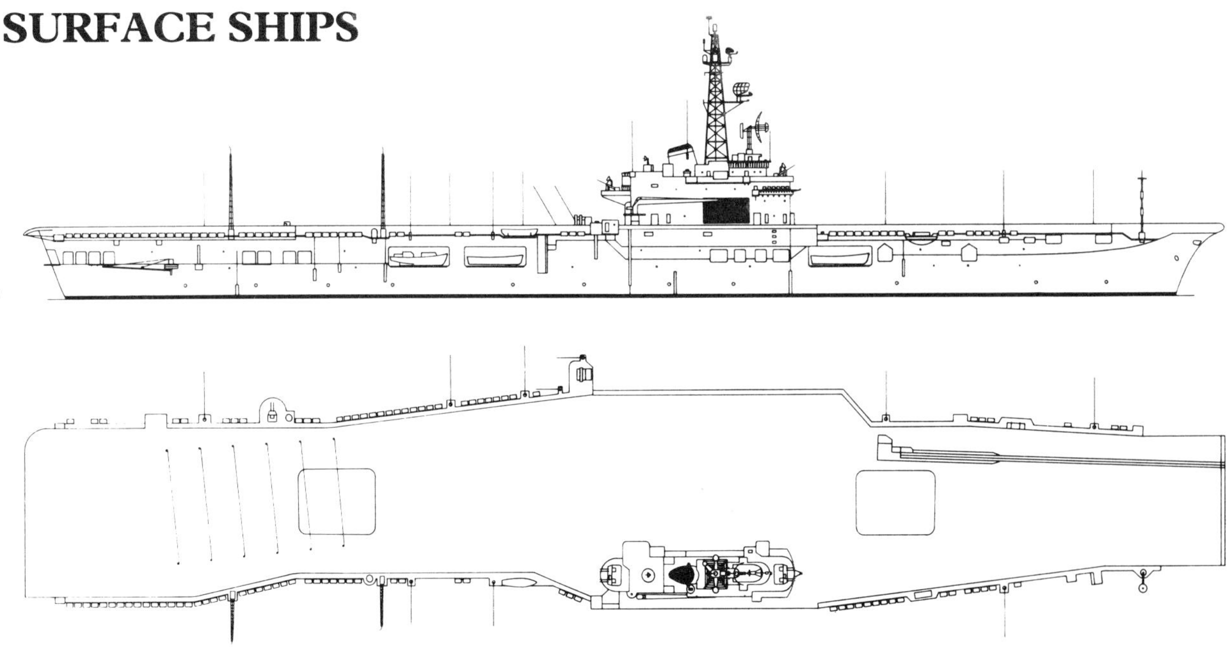

Minas Gerais 1973

Ex-British COLOSSUS class *aircraft carrier*

No	Name	Builder	Acquired	Fate
A 11	MINAS GERAIS (ex-*Vengeance*)	Swan Hunter, Wallsend	14.12.56	Extant 1995

Minas Gerais c1968 — Brazilian Navy

Laid down 16 November 1942 and launched 23 February 1944, *Minas Gerais* was purchased by Brazil for $9m and reconstructed for $27m. She was reconstructed at Verolme Dock, Rotterdam from June 1957 to December 1960. She received an angled deck, steam catapult, mirror-sight deck landing system, new fire control equipment and radar. During reconstruction a 2500kW alternating current system was installed; current is supplied by four turbo-generators and one diesel generator. Hangar dimensions are 445ft × 52ft × 17ft 6in high (*135.64m × 15.85m × 5.33m*). The flight deck and landing gear were reinforced to take aircraft up to 20,000lb. *Minas Gerais* was commissioned 6 December 1960 at Rotterdam and arrived in Rio de Janeiro 13 January 1961. A dispute between the Navy and Air Force delayed deployment until the mid 1960s. The compromise reached was that the Air Force would fly the fixed wing aircraft and the Navy would fly the helicopters. *Minas Gerais* underwent a major refit 1976–80. She is used primarily for ASW. Currently armed with 10–40mm (2×4, 1×2) with SPG-34 radar on the quad mounts; two Mk 63 and one Mk 51 fire control systems. SPS-40B has replaced SPS-12 and the rest of the radar fit comprises SPS-8B, SPS-4 and Raytheon 1402. Aircraft total about fifteen: up to seven S2E Trackers, and eight SH-3D Sea Kings (although four Sea Kings, two Jet Rangers and two Ecureuils is more common). Currently the electronic fit is as follows: radar SPS-40B, ZW 06, two SPG-34, Scanter Mil., Ferranti combat data system compatible with CAAIS-fitted ships. Aircraft capacity: fixed wing aircraft – six Grumman S-2G Trackers; helicopters – four to six Agusta SH-3A Sea King, two Aerospatiale UH-13 Ecureuil II, three Aerospatiale UH-14 Super Puma. Completed refit in 1991. The Brazilian Navy still plans construction of a new 35,000–40,000t carrier capable of 28kts which will accommodate 30–40 'modern' aircraft.

Ex-US BROOKLYN class *cruiser*

No	Name	Builder	Acquired	Fate
C 11	BARROSO (ex-*Philadelphia*)	Philadelphia N Yd	Jan 1951	BU 1974

Barroso date unknown — Robert L Scheina

Purchased by Brazil in January 1951, she commissioned 21 August 1951. Bulges were added and the armoured conning tower was removed late during the Second World War to help improve stability. The mainmast derricks and catapults were removed prior to transfer. The hangar in the hull could accommodate two helicopters with spare parts. The nearly flat counter and high freeboard aft provided an excellent helicopter platform. Her 8–5in were in single mounts and AA weapons comprised 28–40mm and 20–20mm. *Barroso* was decommissioned in 1973 and scrapped.

Ex-US ST LOUIS class *cruiser*

No	Name	Builder	Acquired	Fate
C 12	TAMANDARE (ex-*St Louis*)	Newport News	29.1.51	Deleted 1973

Purchased by Brazil 29 January 1951 and commissioned 6 February 1952. She underwent the same wartime alterations as *Barroso*. She differed substantially from the five ex-US *Brooklyn* class ships in Latin America. Her superstructure was placed closer together; her 8–5in guns were paired in gunhouses on high bases, and she had a small tripod abaft the aft funnel. Light AA comprised 28–40mm, 8–20mm. The remarks for *Barroso* concerning helicopter operations apply to *Tamandare*. *Tamandare* was sold for breaking up in 1980 and foundered off South Africa on 24 August 1980 while under tow to Taiwan.

Although sister-ships, these units possessed many differences, explicable by the fact that they were completed over a two-year period. *Para* and *Paraiba*, as early *Fletchers*, had rounded bridges whereas later units had the angular fronts. Armament varied as follows: *Pernambuco* 4–5in (4×1), 6–3in (3×2), Hedgehog, DC racks; all others 5–5in (5×1), 10–40mm (4×2, 1×2), Hedgehog, DC racks.

Ex-US FLETCHER class *destroyers*

No	Name	Builder	Acquired	Fate
D 33	MARANHAO (ex-*Shields*)	Puget Sound N Yd	1.7.72	Discarded 1990
D 27	PARA (ex-*Guest*)	Boston N Yd	5.6.59	Discarded 1978
D 28	PARAIBA (ex-*Bennett*)	Boston N Yd	15.12.59	Discarded 1978
D 29	PARANA (ex-*Franks*)	Bethlehem, Staten I	20.7.61	Discarded 1982
D 30	PERNAMBUCO (ex-*Knapp*)	Seattle-Tacoma	20.7.61	Discarded 1982
D 31	PIAUI (ex-*Lewis Hancock*)	Federal, Kearny	2.8.67	Discarded 1989
D 32	SANTA CATARINA (ex-*Irwin*)	Bethlehem, San Pedro	10.5.68	Discarded 1978

All except *Mato Grosso* had received FRAM II modernisation. *Mato Grosso* had 1 Seacat SAM (1×4) with M20 optical director. *Sergipe* and *Espirito Santo* have VDS as well as standard SQS-29/32 sonar.

Ex-US ALLEN M SUMNER class *destroyers*

No	Name	Builder	Acquired	Fate
D 34	MATO GROSSO (ex-*Compton*)	Federal, Kearny	27.9.72	Discarded 1990
D 35	SERGIPE (ex-*James C Owens*)	Bethlehem, San Pedro	16.7.73	Extant 1995
D 36	ALAGAOS (ex-*Buck*)	Bethlehem, San Francisco	16.7.73	Extant 1995
D 37	RIO GRANDE DO NORTE (ex-*Strong*)	Bethlehem, San Pedro	29.10.73	Extant 1995
D 38	ESPIRITO SANTO (ex-*Lowry*)	Bethlehem, San Pedro	29.10.73	Extant 1995

FRAM I conversions. Transfer of one extra ship, *McKean* (DD 784) is under consideration.

Ex-US GEARING class *destroyers*

No	Name	Builder	Acquired	Fate
D 25	MARCILIO DIAZ (ex-*Henry W Tucker*)	Consolidated, Orange	3.12.73	Extant 1995
D 26	MARIZ E BARROS (ex-*Brinkley Bass*)	Consolidated, Orange	3.12.73	Extant 1995

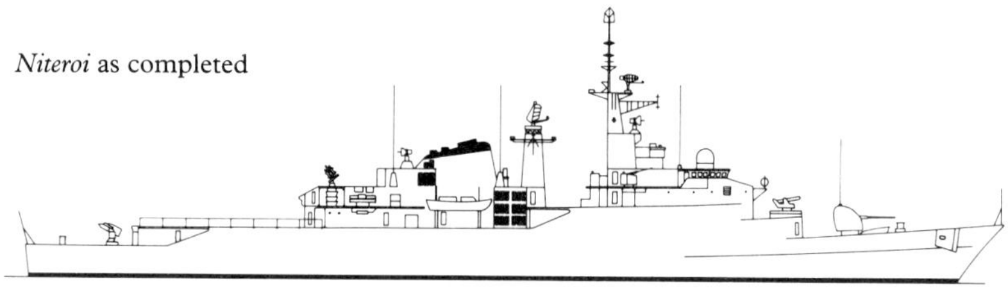
Niteroi as completed

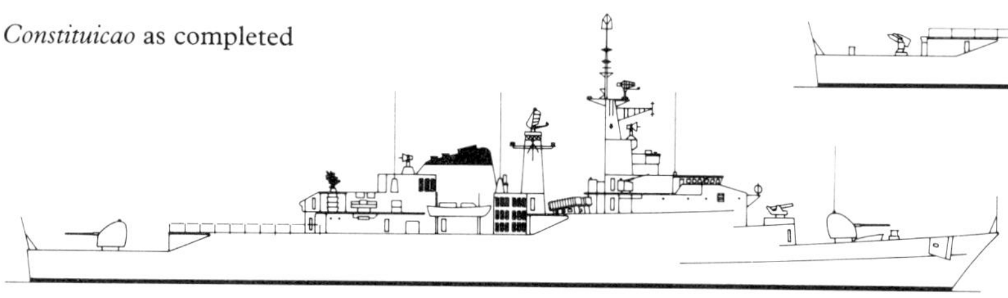
Constituicao as completed

NITEROI class *frigates*

Displacement: 3200t standard; 3800t full load
Dimensions: 400ft wl, 424ft oa × 44ft 2in × 18ft 2in max
121.9m, 129.2m × 13.5m × 5.6m
Machinery: 2-shaft CODAG: 2 Rolls-Royce Olympus gas turbines, plus 4 MTU diesels, 56,000shp + 18,000bhp = 30kts (22kts diesels). Endurance 45 days
Armament: 2 Seacat SAM (2×3), 1 Ikara ASW missile, 1–4.5in, 2–40mm (2×1), 1–375mm Bofors ASW RL (1×2), 6–324mm Mk 32 ASW TT (2×3), 1 DC rack, 1 Lynx helicopter (except *Constituição* and *Liberal* which have 1 extra 4.5in gun aft and 4 MM38 Exocet SSM instead of Ikara)
Sensors: Radar AWS-2, ZW-06, RTN-10X fire control; sonar EDO 610, EDO 700E VDS (not in *Constituição* and *Liberal*)
Complement: 200

No	Name	Builder	Laid down	Launched	Comp	Fate
F 42	CONSTITUICAO	Vosper Thornycroft	13.3.74	15.4.76	31.3.78	Extant 1995
F 41	DEFENSORA	Vosper Thornycroft	14.12.72	27.3.75	5.3.77	Extant 1995
F 44	INDEPENDENCIA	Arsenal de Marinho, Rio de Janeiro	11.6.72	2.9.74	3.9.79	Extant 1995
F 43	LIBERAL	Vosper Thornycroft	2.5.75	7.2.77	19.11.78	Extant 1995
F 40	NITEROI	Vosper Thornycroft	8.6.72	8.2.74	20.11.76	Extant 1995
F 45	UNIAO	Arsenal de Marinho, Rio de Janeiro	11.6.72	14.3.75	17.9.80	Extant 1995

Designed by Vosper Thornycroft and designated Mk 10, this handsome class is unique to the Brazilian Navy. The contract between Vosper Thornycroft Ltd, Portsmouth, and the Brazilian government was announced on 29 September 1970 for a value of £100m. Under the terms, four ships were to be built in Great Britain and two in Brazil, with materials, equipment, and leadyard services supplied by Vosper Thornycroft. These ships are comparable in size and complexity to the British Type 42, yet the Mk 10 requires a complement of 100 less men. Those ships built in Brazil required much more time to complete.

They are divided into two subtypes, for ASW and general purpose duties, the latter exchanging Ikara system (called Branik in Brazil) and VDS for SSMs and an extra gun. They are all fitted with CAAIS action data automation system and Decca ECM gear.

In June 1981 the Brazilian Navy ordered a training ship based on the design of the Mk 10 *Niteroi* from the Arsenal de Marinho, Rio de Janeiro. She is is fact considerably different from her near sisters. Her statistics are: 2333t light, 3345t full load; 430ft 9in oa × 44ft 4in × 13ft 9in max (*131.3m × 13.5 × 4.2m*), 2 shafts, 2 diesels = 19kts; armed with 1–76mm OTO Melara Compact; complement of 150 plus cadets. She was laid down 18 September 1981, launched on 23 September 1983 and named *Brazil* (U27), commissioned in 1986. The 76mm gun was replaced by 2–40mm Bofors.

Defensora 1977 C & S Taylor

Brazil 23.8.93 L & L van Ginderen

Ex-US GARCIA class *frigates*

No	Name	Builder	Launched	Recommissioned
D 27	PARA (ex-*Albert David*)	Lockheed SB & Constr. Co.	19.12.64	18.9.89
D 28	PARAIBA (ex-*Davidson*)	Avondale	2.10.64	25.7.89
D 29	PARANA (ex-*Sample*)	Lockheed SB & Constr. Co.	28.4.64	24.8.89
D 30	PERNAMBUCO (ex-*Bradley*)	Bethlehem, San Francisco	26.3.64	25.9.89

All four ships leased for five years in 1989, and arrived in Brazil on 13 December 1989. All four have the enlarged hangar. *Para* and *Parana* had their SQR 15 towed array sonar removed prior to the transfer. Armed with two obsolete US 127mm/38 WW II vintage guns. *Para* and *Parana* have an automatic ASROC reload system. For further details see under USA. All extant 1995. The Brazilian Navy was also planning to acquire up to four ex-US *Charles F. Adams* guided missile destroyers but this was cancelled in 1993.

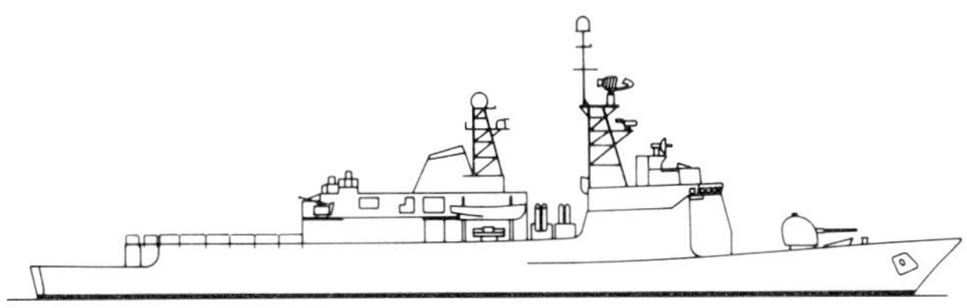

Inhauma class design

INHAUMA class *frigates*

Displacement: 1670t standard; 1970t full load
Dimensions: 314ft 3in oa, 295ft 3in pp × 37ft 5in × 12ft 1in (18ft sonar)
95.77m, 90m × 11.4m × 3.7m (5.5m)
Machinery: 2 shafts, CP propellers, CODOG, 1 GE LM2500 gas turbine, 23,000shp = 27kts, 2 MTU 16V 396 TB94 diesels, 7800bhp. Range 4000nm at 15kts
Armament: 4 MM40 Exocet SSM, 1–115mm Vickers Mk 8, 2–40mm Bofors (2×1), 6–324mm Mk 32 ASW TT (2×3), 1 Westland Lynx helicopter
Sensors: Radar Plessey AWS 4, Kelvin Hughes Type 1007, fire control Selenia Orion RTN 10X; sonar Krupp Atlas ASO 4 Mod 2
Complement: 162 (19 officers)

No	Name	Builder	Laid down	Launched	Comp	Fate
V 30	INHAUMA	Arsenal de Marinha, Rio de Janeiro	23.9.83	13.12.86	12.12.89	Extant 1995
V 31	JACEGUAY	Arsenal de Marinha, Rio de Janeiro	15.10.84	8.6.87	2.4.91	Extant 1995
V 32	JULIO DE NORONHA	Verolme/Arsenal de Marinha	8.12.86	15.12.89	7.10.92	Extant 1995
V 33	FRONTIN	Verolme/Arsenal de Marinha	14.5.87	6.2.92	Mar 1994	Extant 1995

These small frigates were designed by the Brazilian Navy with the assistance of the German firm Marine Technik. First pair ordered on 15 February 1982 and the second on 9 January 1986. Originally it was planned to build twelve such ships and in 1986 the number was raised to sixteen units. Presently only four ships are to be completed, while the Brazilian Navy is looking for a new design of cheaper corvette. V 32 and V 33 were transferred for completion to the Arsenal de Marinha upon the bankruptcy of Verolme in 1991. Fitted with Ferranti CAAIS 450 combat data system and Saab EOS-400 fire control system. Armament includes two Plessey Shield chaff launchers as well as complete ESM/ECM outfit. These ships proved somewhat top-heavy and the planned Vulcan Phalanx CIWS will not be installed on the stern, instead a single SAM launcher may be fitted in future. Plans to arm the later ships and re-arm the first four with Brazilian designed missiles (Barracuda, Avibras SSAI-N and FILA 20) was shelved. These new ships are a welcome addition to the Navy, especially as they are fitted with a landing deck and hangar large enough for the British Lynx helicopter.

An improved *Inhauma* class frigate to be named *Barroso* will have its hull 4.5m longer and have higher freeboard for better seakeeping. The ship is to receive an improved electronics suite and there will also be some changes in armament.

Inhauma 17.7.94
L & L van Ginderen

BRAZIL

Four US *Knox* class frigates were approved by the US Congress for transfer to the Brazilian Navy in 1994–95. Ships selected were listed as *Hepburn* (FF 1055), *Patterson* (FF 1061), *Francis Hammond* (FF 1067) and *Downes* (FF 1070).

Ex-US KNOX class *frigates*

SUBMARINES

Both were refitted at Philadelphia N Yd before transfer, initially as a five-year loan under MDAP. After return to the USN *Humaita* was sunk as a target.

Ex-US GATO class *submarines*

No	Name	Builder	Acquired	Fate
S 14	HUMAITA (ex-*Muskallunge*)	Electric Boat	Jan 1957	Returned 1969
S 15	RIACHVELO (ex-*Paddle*)	Electric Boat	Jan 1957	BU 1968

Both units were refitted at the Pearl Harbor N Yd and transferred in September 1963. They were overhauled in the early 1970s, at which time they received modern fins. *Rio Grande do Sul* was decommissioned in 1972 and sold for scrap June 1975. *Bahia* was sold to the Brazilian Museum of Naval Technology, Santos.

Ex-US BALAO class *submarines*

No	Name	Builder	Acquired	Fate
S 11	RIO GRANDE DO SUL (ex-*Sand Lance*)	Portsmouth N Yd	Sep 1963	BU 1975
S 12	BAHIA (ex-*Plaice*)	Portsmouth N Yd	Sep 1963	Museum ship 1972

All modernised 1948–50 under GUPPY II programme (*Odax* had already received GUPPY I). *Rio Grande* and *Rio de Janeiro* were decommissioned in 1978.

Ex-US GUPPY II type *submarines*

No	Name	Builder	Acquired	Fate
S 10	GUANA BARA (ex-*Dogfish*)	Electric Boat	28.7.72	Discarded 1983
S 11	RIO GRANDE DO SUL (ex-*Grampus*)	Boston N Yd	13.5.72	BU 1981
S 12	BAHIA (ex-*Sea Leopard*)	Portsmouth N Yd	27.3.73	Discarded 1993
S 13	RIO DE JANEIRO (ex-*Odax*)	Portsmouth N Yd	8.7.72	BU 1981
S 14	CEARA (ex-*Amberjack*)	Boston N Yd	17.10.73	Discarded 1987, BU 1990

The first two units were ordered in 1969 and the third in 1972. Completion of *Toneleros* was delayed by a serious fire. She was towed to Chatham Dockyard where the centre 60ft section was replaced and she was rewired. This fire caused the recabling of all *Oberon* units then under construction. They carry 18 torpedoes, British Mk 8 and US Mk 37s.

British OBERON class *submarines*

No	Name	Builder	Laid down	Launched	Comp	Fate
S 20	HUMAITA	Vickers, Barrow	3.11.70	5.10.71	18.6.73	Extant 1995
S 21	TONELEROS	Vickers, Barrow	18.11.71	22.11.72	8.9.78	Extant 1995
S 22	RIACHUELO	Vickers, Barrow	26.5.73	6.9.75	12.3.77	Extant 1995

Both underwent GUPPY III modernisations in 1948 followed by GUPPY III reconstruction in 1961–62.

Early in 1983 Brazil decided in favour of the West German Type 209 design for three new submarines to be built in Rio de Janeiro, to replace the early GUPPY types.

Ex-US GUPPY III type *submarines*

No	Name	Builder	Acquired	Fate
S 15	GOIAS (ex-*Trumpetfish*)	Cramp	15.10.73	Discarded 1992
S 16	AMAZONAS (ex-*Greenfish*)	Electric Boat	19.12.73	Museum 1992

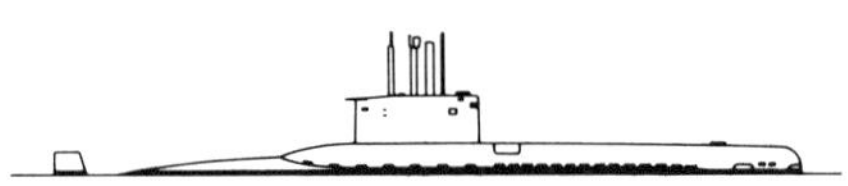

Tupi 1989

First boat of the Type 1400 (enlarged version of the Type 209) ordered in 1984 from German yard. Originally five more (two of them to the improved design) were to be ordered from Arsenal de Marinha for a total of six boats in service by the end of the 1990s. However only three will be built in Brazil. Reportedly the cancelled fourth Brazilian-built boat was to be named *Tocantins*. Diving depth is 250m, endurance 50 days, and the ships are fitted with Ferranti KAFS-A10 action data automation, Sperry Mk 29 Mod 2 (SINS) inertial navigation and two Kollmorgen Mod 76 periscopes. There are plans to order an enlarged and improved Brazilian-designed submarine of the S-NAC-1 class displacing 2420t. The last of the original series, *Tapajos*, may be cancelled in favour of the new design, the first of which can be laid down in 1996.

TUPI class (209 Type 1400) *patrol submarines*

Displacement: 1260t surfaced; 1440t submerged
Dimensions: 200ft 9in × 20ft 6in × 18ft
61.2m × 6.25m × 5.5m
Machinery: 1 shaft, diesel-electric, 4 MTU gensets 1760kW, 1 Siemens main electric motor 4600shp = 21.5kts submerged, 11kts surfaced/snorkelling. Range 400nm/50nm/8200nm at 4kts/16kts (submerged)/8kts surfaced, fuel 118t
Armament: 8–533 TT (bow, 16 Mk 24 Tigerfish Mk 1 torpedoes)
Sensors: Radar Thomson-CSF Calypso III; sonar Krupp Atlas CSU 83-1; ESM Thomson-CSF DR-4000
Complement: 30–33

No	Name	Builder	Laid down	Launched	Comp	Fate
S 30	TUPI	Howaldtswerke-Deutsche Werft, Kiel	8.3.85	28.4.87	6.5.89	Extant 1995
S 31	TAMOIO	Arsenal de Marinha, Rio de Janeiro	15.7.86	18.11.93	1994	Extant 1995
S 32	TIMBIRA	Arsenal de Marinha, Rio de Janeiro	15.9.87	1995	1996	Building
S 33	TAPAJOS	Arsenal de Marinha, Rio de Janeiro	1993	1997	1999	Building

The single submarine is planned as an intermediate stage between the diesel-electric submarines of the *Tupi* class and the planned nuclear submarines of indigenous design. Despite her recent change of name (originally she was named *Tonconatins*), the status of this project is not clear due to financial problems. Requirements are for a diving depth of 300m and an endurance of 60 days. She is also to be able to lay up to 32 mines.

TAMANDARE (Improved *Tupi*) *submarine*

Displacement: 1850t surfaced; 2425t submerged
Dimensions: 219ft × 26ft 3in × 18ft
67.0m × 8m × 5.5m
Machinery: 1 shaft diesel-electric; 4 MTU diesels plus 1 Siemens main electric motor and 4 Siemens alternators, 2400bhp (3400kW)=11kts/22kts. Range 11,000nm at 8kts surfaced, 400nm at 4kts dived
Armament: 8–533mm TT (bow, 16 torpodoes)
Sensors: Radar Thomson CSF Calypso III; sonar Atlas Elektronic CSU 83/1
Complement: 39 (8 officers)

No	Name	Builder	Laid down	Launched	Comp
S 34	TAMANDARE	Arsenal de Marinha, Rio de Janeiro	1996	1998	2000

Despite the delays in the diesel submarine programme, the plans for the building of SSN of the S-NAC-2 class have very high priority. These submarines are to displace about 2800t and their power plant is to develop 50,000kW for a speed over 25kts. A prototype nuclear reactor is being built in Sao Paulo and an uranium enrichment plant was opened in 1988. However it would take probably many years to develop and build a successful SSN, even with the possible help of ex-Soviet scientists.

S-NAC-2 CLASS NUCLEAR SUBMARINES

AMPHIBIOUS WARFARE VESSELS

Ex-US DE SOTO COUNTY class *tank landing ship*

No	Name	Builder	Acquired	Fate
G 26	DUQUE DE CAXAIS (ex-*Grant County*)	Avondale	15.1.73	Extant 1995

Launched 12 October 1956. Stülcken heavy lifting gear fitted.

Ex-US LST 511 class *tank landing ship*

No	Name	Builder	Acquired	Fate
G 28	GARCIA D'AVILA (ex-*Outagami County*)	Bethlehem, Hingham	1.12.73	Deleted 1989

Originally transferred on loan 21 May 1971.

Ex-US THOMASTON class *dock landing ships*

No	Name	Builder	Recommissioned
G 30	CEARA	Ingalls, Pascagoula	28.11.89
G 31	RIO DE JANEIRO	Ingalls, Pascagoula	21.11.90

Originally it was planned to build a 4500t LST, however instead two LSIs of the US *Thomaston* class were acquired. G 30 is ex-*Hermitage* (LSD 34) and G 31 is ex-*Alamo* (LSD 33).

MINOR LANDING CRAFT
Brazil also operates the following:
Four US *LCU 1610* type utility landing craft, named *Guarapari* (L10), *Timbau* (L 11), *Comboriu* (L 12), *Tramandai* (L 13), built at Rio de Janeiro 1974–78.
Twenty-eight Japanese-built LCVPs, 1959–60.
Thirty Brazilian-built 'EDVPs' (LCVPs), 1971.
Five EDVM 25 (LCM 8) type built at Rio de Janeiro 1992–93.
Six EDVM 301-306 of LCM 6 type.

SMALL SURFACE COMBATANTS

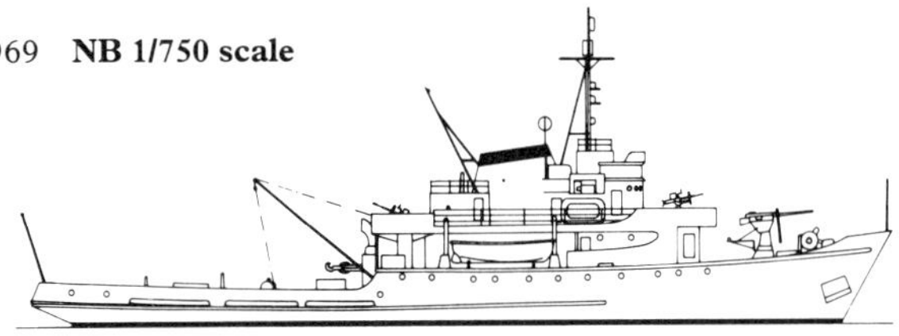

Angostura 1969 **NB 1/750 scale**

IMPERIAL MARINHEIRO class *tug/patrol ships*

Displacement: 911t standard, 960t full load
Dimensions: 183ft 7in pp × 30ft 6in × 11ft 7in max
56.0m × 9.3m × 3.60m
Machinery: 2 shafts, 2 Sulzer diesels 2160bhp = 16kts. 135t fuel
Armament: 1–3in, 4–20mm (4×1)
Complement: 60

Class (commissioned):
Angostura (1955), *Baiana* (1955), *Cabacia* (1954), *Forte de Coimbra* (1954), *Iguatemi* (1954), *Imperial Marinheiro* (1954), *Ipiranga* (1954), *Maerim* (1955), *Purus* (1955), *Solimoes* (1955). Built by Smit, Netherlands.

Fleet tugs classified by Brazil as patrol ships, *Imperial Marinheiro* serves as a submarine tender and all are equipped for fire fighting. Pennant nos between V 15 and V 24. Extant 1995.

The Brazilian Navy also used three ex-US *Sotoyomo* class tugs in a patrol role (*Tritao* R 21, *Tridente* R 22, *Triunfo* R 23). They were purchased in 1947 and now discarded.

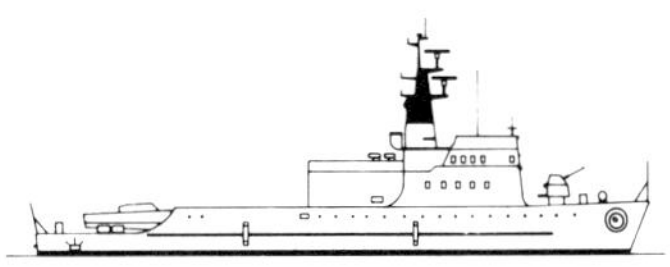

Raposo Tavares 1973

PEDRO TEIXEIRA class *river patrol ships*

Displacement: 700t standard
Dimensions: 203ft 5in pp × 30ft 9in × 6ft 1in max
62.0m × 9.4m × 1.85m
Machinery: 2 shafts, 2 MAN diesels, 1920bhp = 16kts. Range 6800nm at 10kts
Armament: 1–40mm, 6–12.7mm, 2–81mm mortars
Complement: 80 (6 officers)

Class (no, launched, commissioned):
Pedro Teixeira (P 20, 14.10.70, 17.12.73), *Raposo Tavares* (P 21, 11.6.72, 17.12.73).

Both units constructed by the Arsenal de Marinha, Rio de Janeiro. The design is indigenous. These ships have a helicopter platform and a hangar and carry one LCVP. Both extant 1995. In Amazon flotilla.

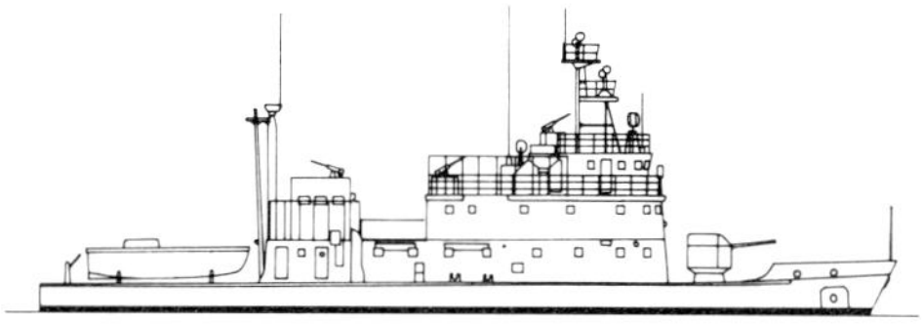

Roroima as completed
NB 1/750 scale

RORAIMA class *river patrol ships*

Displacement: 340t standard; 365t full load
Dimensions: 147ft 7in pp × 27ft 6in × 4ft 3in max
46.3m × 8.4m × 1.3m
Machinery: 2 shafts, 2 MAN diesels, 1814bhp = 14.5kts
Armament: 1–40mm, 6–12.7mm MG, 2–81mm mortars
Complement: 54 (9 officers)

Class (no, launched, commissioned):
Amapa (P 32, 9.3.73, Jan 1976), *Rondonia* (P 31, 10.1.73, 1974), *Roraima* (P 30, 9.11.72, 21.2.75).

Constructed by Maclaren, Niteroi, these ships are of indigenous design. All in service 1995, in Amazon flotilla. Two more hulls completed as hospital ships *Oswaldo Cruz* and *Carlos Chagas* (1984).

PORTO ESPERANCA *river patrol ship*

Displacement: 270t standard; 380t full load
Dimensions: 161ft 9in oa, 149ft 7in pp × 27ft 9in × 4ft 7in
49.3m oa, 45.6m pp × 8.45m × 1.4m
Machinery: 2 shafts, 2 diesels = 12kts.
Armament: 2–40mm (1×2), 6–12.7mm MG, 2–81mm mortars
Complement: 62 (8 officers)

Class (commissioned): *Porto Esperanca* (1991)
A new river patrol ship laid down in 1985 at Arsenal de Marinha, Rio de Janeiro and launched in 1989. Two planned sisters to be built in private shipyards cancelled due to financial problems. Has a helicopter deck and carries two LCVP.

Parati 1972
NB 1/750 scale

PIRATINI class *large patrol craft*

Displacement: 105t standard
Dimensions: 95ft 2in pp × 19ft × 6ft 5in max
29.0m × 5.80m × 2.0m
Machinery: 2 shafts, 4 diesels, 1100bhp = 17kts
Armament: 1–12.7mm MG, 1–81mm mortar
Complement: 15

Class (no, commissioned):
Pampeiro (P 12, May 1971), *Parati* (P 13, Jul 1971), *Penedo* (P 14, Sep 1971), *Piraja* (P 11, Mar 1971), *Piratini* (P 10, Nov 1970), *Poti* (P 15, Oct 1971).

All units were constructed by Arsenal de Marinha, Rio de Janeiro, under offshore agreement with the United States. Based on the US Coast Guard 95ft WPB type. All extant 1992.

GRAUNA class *large patrol craft*

Displacement: 410t full load
Dimensions: 152ft 6in × 24ft 7in × 7ft 6in
46.5m × 7.5m × 2.3m
Machinery: 2 shafts, 2 MTU 16V 396 TB94 diesels, 5800bhp = 22kts. Range 2000nm at 12kts
Armament: 1–40mm Bofors, 2–20mm Oerlikon
Complement: 25 (4 officers)

Class (commissioned):
Grauna P 40 (Apr 1992), *Goiana* P 41 (Nov 1992), *Grajau* P 42 (1993), *Guaiba* P 43 (1993), *Guajara* P 44, *Guapore* P 45, *Gurupa* P 46, *Gurupi* P 47.

Being built by Estaleiro Maua, Niteroi, according to the Vosper QAF design similar to Bangladesh *Meghna* class. First two ordered in 1987 and second two in 1990. Second batch of four boats is still to be ordered. The names of these four probably assigned to patrol craft ordered in Germany (see below).

German Type 151 *large patrol craft*

A class of four large patrol craft was ordered in 1994 from the Peene-Werft GmbH, Wolgast yard (formerly VEB Peene-Werft in the GDR). These ships (hull nos 151.03 to 151.06) are being built on the uncompleted hulls of Type 151 FACs originally laid down for the former East Germany and Poland. Three such ships, completely rebuilt as patrol craft with reduced speed, serve with the German Frontier Guard as the *Neusterlitz* class. It is possible that the Brazilian vessels will receive the names originally planned for the four unbuilt patrol craft of the *Grauna* class.

PIRAJU class *submarine chasers*

Displacement: 130t
Dimensions: 128ft × 19ft 6in × 6ft
39m × 6.0m × 1.8m
Machinery: 1 shaft, diesel, 890bhp = 20kts.
Armament: 1–40mm, 3–20mm (3×1), 30 DC
Complement: 15

Class (no):
Piraju (P 1), *Pirambu* (P 2), *Piranha* (P 3), *Piraque* (P 4), *Pirapia* (P 5), *Pirauna* (P 6).

All built, of wood, at Rio de Janeiro 1947–48. *Pirauna* was stricken 1960, *Pirambu* and *Pirapia* in 1964, *Piraju* and *Piranha* in 1971 and *Piraque* in 1973.

There are also some other smaller patrol boats:

Four *Tracker II* class boats *P 8002–8005* (ex-*L 01–04*) delivered 1988–89 by Ebin So., Porto Alegre (31t, 20.9m, 27kts, 2–12.7mm MG).

Ten *Swift Mk II* type boats *R 61–64, R 71–76* delivered in 1972 and 1982 (23t, 15.7m, 22kts, 1 MG).

MINE WARFARE VESSELS

West German SCHUTZE class *minesweepers*

Displacement: 230t standard; 280t full load
Dimensions: 154ft 9in oa × 23ft 6in × 6ft 9in
47.2m × 7.2m × 2.1m
Machinery: 2 shafts, 4 Maybach diesels, 4500bhp = 24kts. Range 710nm at 20kts
Armament: 1–40mm Bofors
Complement: 40 (4 officers)

Class (no, ordered, commissioned):
Abrolhos (M 19, Nov 1973, 16.4.75), *Albardao* (M 20, Nov 1973, 21.7.75), *Anhatomirim* (M 16, Apr 1969, 30.11.71), *Aracatuba* (M 18, Apr 1969, 13.12.72), *Aratu* (M 15, Apr 1969, 5 Apr 1971), *Atalaia* (M 17, Apr 1969, 13.12.72).

All units were built by Abeking and Rasmussen, Germany. They are wooden hulled and all are in service 1995.

Ex-US YMS type *minesweepers*

Class (former name, transfer date):
Javari (ex-*Cardinal*, 15.8.1960), *Jurua* (ex-*Jackdaw*, 15.8.1960), *Juruena* (ex-*Grackle*, Apr 1963), *Jutai* (ex-*Egret*, 15.8.1960).

These units were reclassified as coastal minesweepers (old) in February 1955. Pennant nos M 11–14. All discarded in 1969, except *Juruena*, 1974.

Brunei

Brunei became a British protectorate in 1888. In 1957 she became part of the Federation of Malaya and in 1963 joined the Federation of Malaysia. This oil and gas-rich Sultanate, a little larger than Washington DC, is self-governing internally but relies on Britain to assist with her defence. Independence came on 1 January 1984. The navy is run by the volunteer Royal Brunei Malay Regiment and Britain provides its commander and many of the officers. It is divided into the First Flotilla for offshore patrols and the Special Boat Squadron for policing the rivers. A powerful force of missile and gun armed patrol boats is based at Muara, and unlike many small navies is not confined to fishery protection and anti-smuggling work.

PAHLAWAN *corvette*

Particulars: As Danish *Söloven* class

The gas turbine powered *Pahlawan* was built by Vosper at Portsmouth during 1965–67 (launched 5 December 1966, commissioned 19 October 1967) to the same design as the Danish *Söloven* class. She had a wooden hull and aluminium alloy superstructure, and was originally armed with 1–40mm and 2–20mm (1 × 2). *Pahlawan* was refitted by Vosper Thornycroft, Singapore in May 1972 with 8 SS–12 SSM (4 × 2). She had Decca TM 616 radar. *Pahlawan* was discarded in 1977.

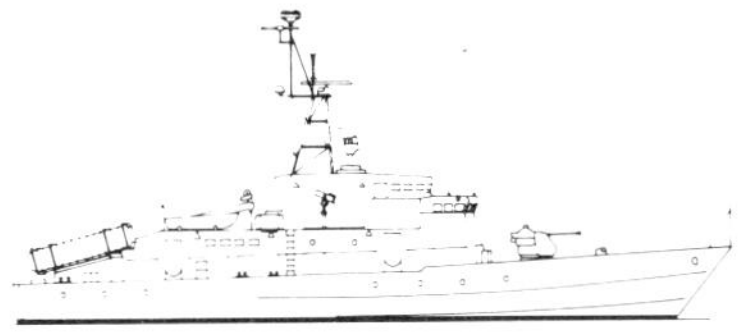

Waspada 1979
NB 1/750 scale

WASPADA class *fast attack craft (missile)*

Displacement: 150t full load
Dimensions: 110ft, 121ft oa × 23ft 6in × 6ft
33.5m, 36.9m × 7.2m × 1.8m
Machinery: 2 shafts, 2 MTU 20V538 TB91 diesels, 9000bhp = 32kts. Fuel 16t. Range 1200nm at 14kts
Armament: 2 MM38 Exocet SSM (2×1), 2–30mm Oerlikon GCM-Bol (1×2), 2–7.62mm MG, 2–2in MoD(N) R(F)L
Sensors: Radar Decca TM 1229
Complement: 24

Class (no, launched):
Waspada (P 02, 3.8.77), *Pejuang* (P 03, March 1978), *Seteria* (P 04, 22.6.78)

These boats were built for Brunei by Vosper, Singapore during 1976–79. They have welded steel hulls and aluminium superstructure identical to the Venezualan *Constitucion* class but have more powerful MTU diesels and a different armament controlled by Sperry 1412A digital computer and the Sea Archer missile system. *Waspada* (in service July 1978) differs from *Pejuang* and *Seteria* (both in service 1979) by having an enclosed bridge and training facilities. Proposed modernisation, including lengthening by 2m, was cancelled in the mid-1980. All extant 1995.

Bulgaria

During the closing stages of the Second World War Bulgaria fell into the Soviet sphere of influence and the Bulgarian Army came under Soviet command. The Soviet troops' presence and a strong pro-Russian sentiment in the country were the major factors which enabled the Communists to seize power more easily than in other countries of the recently emerged Soviet bloc. After the peace treaty with the Allies was signed and the new constitution had come into force, the Soviet troops left Bulgaria at the close of 1947 and remodelling of the country on Soviet lines began. Nationalisation of private industry was started immediately, collectivisation of peasant holdings was pursued while industrialisation became one of the principal aims of economic policy regardless of the lack of raw materials and of technically educated manpower for heavy industry.

By 1950 the strict Stalinist course was adopted throughout and in May that year a new army command was formed with a Soviet officer of Bulgarian origin taking the post of Minister of National Defence and C-in-C of the armed forces. With the help of Soviet advisers the armed forces were equipped and reorganised on the Soviet model. By 1953 they numbered about 220,000 despite the peace treaty limitations. The treaty of 10 February 1947 ordered a considerable reduction of the Bulgarian armed forces as compared with their pre-war status; maximum permitted strength included an army of 55,000, an air force of up to ninety aircraft (with no bombers) while the tonnage limitation for the navy was set at 7250t with personnel not exceeding 3500. Despite the spectacular build-up of the army, the naval forces were not so successful as the Soviet Black Sea Fleet had few resources to spare for her ally at that time. However, because of evident neglect of the Bulgarian Navy, a *Novik* class destroyer (ordered for the Imperial Russian Navy), three war-built *M* class coastal submarines and four ex-German M-boats of 1935 type were transferred from the Soviet Union to form the nucleus of the 'People's Navy'.

After Stalin's death the country gained a considerable margin of freedom. In 1955 Bulgaria joined the Warsaw Pact and the armed forces were reduced by 45,000 while supplies of modern equipment began to arrive. During the late 1950s one war-built *Ognevoi* class destroyer as well as a number of modern vessels – two 'Whiskey' class submarines, two 'Riga' class frigates, eight 'P 4' type MTBs, two 'Kronstadt' class patrol craft – were purchased in the Soviet Union. In December 1959 command of the armed forces was taken over by a Bulgarian officer and Soviet advisers were withdrawn.

There were a few additions to the Bulgarian Navy during the 1960s, but the present strength was largely built up under three successive Five Year Plans since 1970. Naval power currently encompasses short-range landing and minesweeping capabilities, backed by some ASW capacity and striking forces of FACs and naval aviation. This expansion programme, backed by construction of new auxiliaries (some built domestically), was completed by the end of 1985 and resulted in a balanced coastal navy with some, albeit limited, projection capability.

The modernisation programme commenced in the later 1980s with the transfer from the USSR of two 'Romeo' class submarines, one 'Poti' class corvette and two 'Polnocny' class landing ships, apparently to replace those acquired in the early 1970s. Possible further transfers were to include more landing ships as well as ASWs and FACs, but these plans were forestalled by the deteriorating Bulgarian economy, particularly since the mid-1980s. A reduction of 12 per cent in the military budget was announced in 1989, including the scrapping of five warships (probably corvettes and some landing craft). Following the overthrow of President Zhivkov's regime in 1990, free elections, and the break-up of the Warsaw Pact, plans for further military expenditure were shelved because of the still critical economic situation. Bulgaria now follows a policy of seeking membership of NATO and associate membership of the European Community. As the country's economy struggles to improve, the navy is being further reduced. It is planned that by the year 2000 the Bulgarian Navy will comprise only three submarines, FACs, small patrol and mine warfare craft and helicopters.

The Bulgarian navy consists of the Black Sea Fleet and the Danube Flotilla. The fleet is organised into one division each of submarines, ASWs, FAC(M)s, FAC(T)s and one brigade each of landing ships and minesweepers. In 1992, the navy had a total of 8800 men, of whom 2100 were afloat, 2200 in coastal defence (twenty batteries with over one hundred 130mm and 6in guns as well as three battalions of six truck-mounted SSC-2b ('Samlet') missile launchers), and 200 in naval aviation. By 1994, the total had been reduced to 5,400 men. Personnel are largely ratings on three years' national service (eighteen months since 1991). The fleet headquarters and main base is at Varna, and other bases are at Burgas and Sozopol. The Danube River Flotilla, which is headquartered at Vidin, and has bases at Atiya and Balchik, operates ex-Soviet 'PO 2' class boats and two fast patrol boats. The naval aviation division with stations at Varna and Burgas, has Mi-14 'Haze' and twelve Mi-2 'Hoplite' and Mi-4 'Hound' helicopters.

FLEET STRENGTH 1947

TORPEDO BOATS

Name	Launched	Disp	Fate
Drski class			
DRSKI	1907	98t	Preserved 1955
KHRABRY	1907	98t	Deleted 1954
STROGI	1907	98t	Deleted 1954

Used for escort service during the Second World War. Seized by the Soviets in September 1944 but returned in April 1945. *Drski* was landed in 1955 and exhibited at Varna.

MTBs

Name	Launched	Disp	Fate
MA *German S 2 class*			
F 1–F 4	1939	47.8t	Stricken after 1975
Ex-Netherlands T 52 class			
I–III	*c*1941	30t	Deleted 1960

German S 2 class
Two transferred from Germany in 1939, third in 1941, while fourth not delivered; seized by the Soviets in Septembered 1944 and commissioned as *TKA 958–960*. Returned in April 1945, they were recommissioned in Bulgaria and served until 1953, when they were converted to avisos.

Ex-Netherlands T 52 class
Transferred from Germany to Bulgaria 1942; seized by USSR in September 1944 and commissioned as *TKA 961–964*. Two returned in April 1945 to Bulgaria, whilst others retained by USSR and substituted by two Soviet TM-200 boats. Non-operational after the war due to faulty engines.

PATROL CRAFT

Name	Launched	Disp	Fate
Ex-US SC 1916 type			
BELOMOREC (ex-*C 27*)	1917	77t	Stricken 1952
CHERNOMOREC (ex-*C 80*)	1917	77t	Stricken 1952
Ex-Soviet Artillerist class			
281, 282	1940s	240t	Stricken *c*1965
Ex-Soviet MO-4 class			
I–IV	1938	56t	Deleted 1950s
Ex-Soviet TK class			
I–VI	1940s	50t	Deleted 1950s
Ex-Soviet OD-200 class			
I–VI	1943	43.4t	Deleted 1950s

Ex-US SC 1916 type
Seized by USSR in September 1944 and commissioned as the guard boats *SKA 757–758*. Returned in April 1945.
Ex-Soviet Artillerist, MO-4, TK and OD-200 classes
Transferred from USSR in 1947.

MAJOR SURFACE SHIPS

Ex-Soviet NOVIK class *destroyer*

No	Name	Builder	Launched	Fate
	ZHELEZNYAKOV (ex-*Petrovskij*, ex-*Korfu*)	Russud SY, Nikolayev	10.10.17	BU 1956

Commissioned into the Bulgarian Navy under her Soviet name from 18 December 1947 to 15 September 1949.

Ex-Soviet OGNEVOI class *destroyer*

No	Name	Builder	Launched	Fate
	GEORGI DYMITROV (ex-*Ozornoi*)	61 Kommunar Yd	1941	BU 1967 (?)

Transferred 1956; modernisation planned in 1961 abandoned and ship deleted 1963.

Ex-Soviet 'RIGA' class *frigates*

No	Name	Builder	Acquired	Fate
11	DRUZKI (ex-*Kobchik*)	USSR	1957–58	Deleted 1993
12	SMELY	USSR	1957–58	Deleted 1990
13	BODRY	USSR	Nov 1985	Deleted 1990

Built Nikolayev 1955. *Druzki* and *Smely* were refitted extensively during 1980–81. Two *Mesar* class fast tankers were built in Bulgaria during 1979–87 to support Mediterranean deployment of these ships. *Druzki* is the first of the class to have missile launchers installed and armament which differs from the standard outfit, namely 1 SA-N-5 SAM *Strela* (1×4) (16 reloads), 3-100mm (3×1), 2–30mm (1×2), 2–23mm (1×2), 2 or 3–21in TTs, 2 RBU-2500 ASW RL (2×16). *Smeli* and *Bodry* broken up 1992.

Ex-Soviet 'KONI I' class *frigate*

No	Name	Builder	Acquired	Fate
111	SMELI (ex-*Delfin*)	USSR	Feb 1990	Extant 1995

Transferred 1990 from Soviet Navy, having been used on the Black Sea since 1976 to train foreign crews for the exported 'Koni' class frigates.

SUBMARINES

Ex-Soviet 'M' class *submarines*

Class:
M 1–M 3
Three units of the Series XII bis transferred in the early 1950s; commissioned as *M 1–3*. Two units exchanged for two 'Whiskey' class boats in 1958; remaining unit stricken in 1967 and probably returned to the USSR.

Ex-Soviet 'WHISKEY' class *submarines*

Class:
Pobeda, Slava
Transferred 1958; given Bulgarian names. Stricken late 1970s.

Ex-Soviet 'ROMEO' class *submarines*

Class (no):
Pobeda (81), *Slava* (82), *Nadezha* (83), *Slava* (84)
Series production in the Soviet Union during 1958–61. *Pobeda* and *Slava* transferred to Bulgaria 1972–73 to replace 'Whiskey' class units, the names of which they took. Since the mid 1980s both were used for harbour training duties. The next pair were transferred in 1985 and 1986, and one took the name of *Slava* (82), which did not however pay off until 1990. Although both of these were extant in 1995, their operational status is doubtful.

AMPHIBIOUS WARFARE VESSELS

D 3 type *landing craft*

Displacement: 420t
Dimensions: 164ft × 20ft 4in × 6ft 6¾in
50.0m × 6.2m × 2.0m
Speed: 10kts
Armament: 1–37mm

Class:
436 +8
Nine were built during 1954 in Bulgaria to a design based on that of German war-built *MFP*s. Some of them were unarmed. Five were stricken 1983–84, the four boats extant in 1990 were numbered *712–715*, but in poor condition and practically non-operational and were deleted in 1991.

Ex-Soviet 'VYDRA' class *landing craft*

Class:
420–429 +13
These utility landing craft of 425t standard displacement, consisted of a first batch of ten transferred to Bulgaria 1970. A further batch was built in Bulgaria 1974–78. In 1985 renumbered *601–613* and *701–712*. *701–702* stricken 1987 and replaced by two 'Polnocny' class (see below). *703–707* and *712* converted to minelaying 1992. *708–711* deleted 1993, remainder 1995.

Ex-Soviet 'POLNOCNY A' class *landing ships medium*

Class:
Ivan Zagubanski (701), *Antares* (702)
'Project 770', built for the Soviet Navy 1965 by Polnocna shipyard, Gdansk, Poland. Transferred 1986–87 as replacements for two *Vydra* class; *Ivan Zagubanski* renamed *Sirius* in 1992. Plans to convert both ships into minelayers abandoned. Both extant 1995.

SMALL SURFACE COMBATANTS

Ex-Soviet 'POTI' class *corvettes*

Class (no):
Bditelni (45, then 42), *Bezstrashni* (43), *Khrabri* (44), *Letyashti* (41), *Naporisti* (46), *Strogi* (42).
The first three were transferred in December 1975 as replacements for the 'Kronstadt' class patrol craft. The fourth was added at the end of 1986 and two others followed in 1990. *Strogi* and *Naporisti* deleted 1993, others extant 1995.

Ex-Soviet 'PAUK' class *corvettes*

Class (no):
Reshitelni (13), *Bodri* (14)
Transferred September 1989 and December 1990 respectively. Both extant 1995.

Ex-Soviet 'TM-200' class *fast attach craft (torpedo)*

Two transferred 1950 to replace the two ex-Netherlands T 52 class boats retained by the Soviet Navy.

Ex-Soviet 'P 4' type *fast attack craft (torpedo)*

Eight boats of this type were transferred from the USSR in 1956. Four were deleted *c*1975 and the remaining boats were all stricken by 1980.

Ex-Soviet 'OSA I & II' class *fast attack craft (missile)*

Classes:
'Osa I': *Burya* (103), *Taifun* (112); 'Osa II': *Grom* (104), *Smerch* (113), *Svetkavitsa* (111), *Uragan* (102).
The 'Osa I' class boats were transferred 1971, followed by 'Osa II' in 1978, 1982 and 1984 respectively; all extant 1995.

Ex-Soviet 'SHERSHEN' class *fast attack craft (torpedo)*

Class:
25–30
Transferred from the USSR in the early 1970s, and renumbered since 1985 as *104–106* and *114–116*. Two deleted 1990, and the remaining four followed in 1992.

Ex-Soviet 'TARANTUL II' class *fast attack craft (missile)*

Class (no):
Mulniya (101)
Transferred December 1989 and extant 1995. Planned second unit not taken over because of financial problems.

Ex-Soviet 'KRONSTADT' class *large patrol craft*

Class:
94, *95*
Transferred in 1957 and commissioned as *94* and *95* (later changed to *33* and *34*). Stricken in 1975 and replaced by 'Poti' class corvettes to which their numbers were assigned.

Ex-Soviet 'SO I' type *submarine chasers*

Class:
41–46
Transferred in 1963–64 and commissioned as *41* to *46*. *41–44* stricken 1988, *45–46* stricken 1989.

Ex-Soviet 'ZHUK' class *coastal patrol craft*

Class:
511–513, *515*, *521–523*, *531–533*
Small coastal boats of 45t, transferred from USSR 1980–81, used also on Danube.

Ex-Soviet '1124' class *armoured river motor gunboats*

A number transferred from the USSR in the 1950s for the Bulgarian Danube Flotilla. Deleted after the Flotilla was disbanded in 1959.

MINE WARFARE VESSELS

Ex-German M-boat 1935 type *ocean minesweepers*

Four assigned to USSR after the war, were transferred to Bulgaria in the 1950s. Stricken *c*1962.

Ex-Soviet 'T 43' type *ocean minesweepers*

Class:
17, *18*, *21*, *22* (later *47–49*)
Four short-hull type transferred from USSR 1953; one cannibalised for spares. *48* converted to research vessel and renamed *N I Vaptzarov*. Last one stricken 1991.

Ex-Soviet 'T 301' class *inshore minesweepers*

Four were purchased from the USSR in 1955. Three were stricken in 1974, one in 1975.

Ex-Soviet 'VANYA' class *inshore minesweepers*

Class:
Dimiter Paskalev (36), *Dobrotich* (33), *Ekstati Viniarov* (35, later 34), *Iskar* (31), *Kiril Minkov* (34), *Zibar* (32)
Two pairs transferred in 1970 and 1971 respectively and a further pair in 1985. *Dimiter Paskalev* and *Kiril Minkov* deleted in 1992, remainder extant 1995.

Ex-Soviet 'SONYA' class *coastal minesweepers*

Class (no):
Briz (61), *Priboj* (63), *Shkval* (62), *Shtorm* (64)
Wooden hull boats with GRP sheath, 450t full load, transferred 1981–84; extant 1995.

Ex-Soviet 'YEVGENYA' class *inshore minesweepers*

Class:
371–374
GRP hulled boats of 90t full load, transferred from USSR 1977. Renumbered *51–54*; later *65–68*.

Ex-Soviet 'PO 2' type *minesweeping boats*

Built in Bulgaria under Soviet licence, twenty-four completed 1957–60, and relegated to Danube patrol duties 1976–82. Six deleted by 1985. Numbered *301–306*, *401–406* and six others in *450* and *600* series; Deleted since 1980s, including three in 1991 and eight in 1992. In 1995 only *57*, *58*, *218* and *219* were extant.

Soviet 'OLYA' class *inshore minesweepers*

Class:
51–55
Built in Bulgaria under licence between 1988 and 1992. All extant 1995.

Cambodia (Kampuchea)

Cambodia (now known as Kampuchea) became a French protectorate in 1867 and a French colony in 1887. She became an independent kingdom within the French Union in 1949 and became truly independent in 1954. The Marine Royal Khmer was set up on 20 April 1954 with French assistance. The Vietcong made use of eastern Cambodia and Prince Norodom Sihanouk tried to steer a neutral course. This meant that the navy received very few ships from abroad and the country lacked the money or inclination to purchase any.

There were border incidents in the late 1960s with Thailand and South Vietnam and in May 1970 the Americans and South Vietnamese invaded East Cambodia to eradicate the Vietcong. Sihanouk was overthrown and a pro-Western military government took over, renaming the navy, Marine National Khmer. American aid was lavished on the armed forces, including the navy, but the indigenous Khmer Rouge pushed them back and besieged the capital, Phnom Penh, from 1973 until, the river lifeline cut, it surrendered in April 1975. At that juncture the navy totalled 11,000 men including 4000 marines, all 18-month conscripts.

Most of the major warships fled to Thailand or the Philippines, two MTBs were sunk during the American operations to recapture the freighter *Mayaguez*, and most of the remaining ships were sunk or discarded during the Khmer Rouge dictatorship or the subsequent Vietnamese invasion. The continuing three-way civil war has retarded the country's recovery and the navy relies on vessels transferred from the USSR in the mid-1980s.

Ex-French EDIC type *landing craft*

The French LST 606, built by Franco-Belge in 1968–69, was transferred to Cambodia on completion in August 1969, and renumbered *T 916*. She was discarded in 1979.

Ex-US *landing craft*

Two ex-USN LCT 6, transferred to France in 1951 for service in Indo-China, were transferred by the French to Cambodia in 1956–57. They were deleted in *c*1973–75. Two ex-USN LCU 501 Type were transferred first to France then to Cambodia at about the same time. Two more were transferred on 31 May 1962, and two more in November 1972. Four were deleted in 1979. An ex-USN LCU 1466 Type was also transferred in October 1969. *T 919* (ex-USN LCU 1577) was sunk by a mine on 5 May 1970. The remainder were either repossessed by the USN or deleted between 1973–75. In the early 1970s twenty ex-USN LCM 6 and twenty-three ex-USN ATC (converted LCM 6 or 8) were also transferred and suffered the same fate.

Ex-US PC class *large patrol craft*

Two ex-USN PCs, transferred to the French Navy in 1951 for use in Indo-China, were transferred to the Cambodian Navy in 1955–56. They were *E 311* (ex-*Flamberge*, ex-USS PC 1086), and *E 312* (ex-*L'Inconstant*, ex-USS PC 1171). *E 311* fled to Thailand on 16 May 1975 and has been scrapped. *E 312* fled to the Philippines on 21 May 1975, and was incorporated into the Philippines Navy in December 1976 as *Negros Oriental*.

Ex-Yugoslav Type 108 *fast attack craft (torpedo)*

Two were presented to Cambodia in 1965, and renumbered *VR 1* and *VR 2*. Both sunk by US aircraft on 13 May 1975 during the *Mayaguez* incident.

Ex-US LSIL class *landing craft (support gunboats)*

P 111 (ex-USS *LSIL 9039*, ex-*LSIL 875*) and P 112 (ex-*Medecin Capitaine Le Gall*, ex-USS *LSIL ?*) had been USN rearmed large infantry landing craft transferred to the French Navy in 1951 for service in Indo-China. The French transferred them to Cambodia in 1957. Both escaped to the Philippines on 17 April 1975, and P 111 was incorporated in the Philippines Navy on 17 November 1977 as *Marinduque*.

Ex-US *patrol gunboat*

US *PGM 70* was transferred to Cambodia under MDAP in 1964. She was deleted during 1973–75.

Ex-Soviet 'TURYA' class *patrol craft*

Three delivered 1984–85 without torpedo tubes. Extant 1995.

Ex-Soviet 'STENKA' class *patrol craft*

Four delivered 1985–87. Two cannibalised for spares, remainder extant 1995.

Ex-Soviet 'T 4' class *landing craft*

Two transferred 1984–85. Both extant 1995.

Ex-British *harbour defence motor launches*

Three RN HDMLs, among twenty-six transferred to the Free French Navy in 1943, were transferred by the French to Cambodia. *VP 762* (ex-*VP 42*, ex-*HDML 1457*) was transferred in 1956; *VP 748* (ex-*HDML 1223*) and *VP 749* (ex-*HDML 1229*) followed a year later. *VP 212* (ex-*748*) was the only one to survive the Khmer Rouge takeover. Stricken by 1995.

Three ex-Chinese 'Yulin' type coastal patrol boats were transferred to Cambodia in January 1968 and renumbered *VP 1–3*. No longer in service in 1995.

Two ex-USN AVR type, seventeen ex-USN 'Swift' class PCF and sixty-five ex-USN PBR Mk 1 and Mk 2 river patrol craft were transferred to Cambodia in 1972–74. Several were sunk in 1973–75 and others reclaimed by the USN. Only two PBR reported extant 1995.

Cameroon

The United Republic of Cameroun is the result of a successful 1961 merger of the former French Cameroun and the southern part of the British Cameroons, two territories with different colonial pasts.

The French trained and equipped armed forces are small, and intended mainly for internal security among the 12.3 million inhabitants. France guarantees external security by 1974 treaty. The 600-man navy, under French command until 1973, operates a coastal patrol force of fast attack craft and patrol craft from its HQ in Douala, where an American company built a new $9 million naval base that became operational in 1982. Other ports along the 250-mile coastline are Victoria, Man O'War Bay (base for the 100 Cameroun Marines), Kribi, Tiko Idabato, Garoua and Lobé. The republic's merchant marine expanded from eighteen ships in 1974 to forty-four by 1980 which was the year the French firm of Cofras Marine delivered a study on how Cameroun is to protect her 200-mile economic zone.

Plans to enlarge the navy by acquiring three 55m and one 15m ex-US patrol craft foundered due to lack of funds.

Ex-Chinese 'SHANGHAI II' class *fast attack craft (gun)*

Class (no):
Cap Cameroun (101), *Man O'War Bay* (102)

Transferred in July 1976. Seventy Cameroun sailors trained in China before their delivery. Repaired in the Douala floating dock during 1980. Both extant 1995.

L'AUDACIEUX *large patrol craft*

Particulars: As Ivory Coast *Le Valeureux*

P 48 type boat ordered from SNCF in September 1974, laid down 10 February 1975, launched 31 October 1975 and commissioned 11 May 1976. She was hulked in 1991. The P 48S class patrol craft *Bakassi* was built by SNCF in 1981: 308t full load, 26kts, 8 MM40 Exocet SSM (2×4), 2–40mm (2×1).

MISCELLANEOUS

The French seaward patrol craft *VC 6* (P 756) and *VC 8* (P 758) acquired in 1964 and ?1969 respectively, and named *Vigilant* and *Audacieux*; both discarded in 1975. The smaller French seaward patrol craft *VP 747* (ex-British *HDML 1423*) transferred in 1961 and named *Patrie du Cameroun*. Replaced in 1963 by *VP 768* (ex-British *HDML 1228*) which was assigned the same name and discarded *c*1972.

There are three French-built small coastal patrol craft and six landing craft, built in the Ivory Coast and Gabon, as well as nine smaller assault boats.

Canada

In 1947 the Royal Canadian Navy was the most sophisticated of the British Commonwealth navies, with an aircraft carrier and two cruisers as well as a large and seasoned force of destroyers, frigates and fleet minesweepers. In addition to providing 40 per cent of the escort forces engaged in the Battle of the Atlantic, the Canadians had also built a very large number of warships, ranging from 'Tribal' class destroyers to *Algerine* class minesweepers.

During the postwar years the RCN was forced to surrender much of its broader capability in order to retain high capability in the most essential roles. The cruisers proved expensive to run and demanded skilled manpower, so they were not replaced. For a while fixed-wing aviation was maintained with the single carrier *Magnificent*, but her successor was reduced to purely anti-submarine duties. As with most small navies, the cost of maintaining even that level of sophistication proved too expensive, and *Bonaventure* was not replaced.

Although the RCN remained strongly influenced by Royal Navy ideas on ship design, the experience of the Second World War dictated a growing reliance on the United States for electronics. Thus British-pattern equipment was very soon replaced by off-the-shelf purchases or Canadian-made variants.

The RCN has concentrated its efforts on anti-submarine warfare, and made no effort to retain fixed-wing aviation when *Bonaventure* had to be scrapped. It was the first navy to explore the possibilities of using large helicopters from frigates and destroyers, operating S-55s from the modernised 'River' class in the 1950s and the S-61 Sea King from the *St Laurent* type DDHs. From this concept stemmed the first all-Canadian design, the very large and capable DDH-280 or *Iroquois* class helicopter-carrying destroyers. These ships have a number of unusual features, including provision for two Sea Kings with their associated Canadian-designed Bear Trap recovery system, and are among the best ASW ships afloat, but they were expensive. A Parliamentary inquiry concluded that the programme had run out of control and the decision was taken that no more major warships be designed 'in house'.

The result of this was a protracted search to find a design to suit the Canadian Patrol Frigate (CPF) requirement, to replace the DDEs. In 1983 a contract was awarded for six 'City' class, and in 1987 a second batch of six was ordered. In the meantime a DDE life-extension programme was started, and in 1986 a contract was awarded to modernise the 'Tribals' and to convert them from ASW ships to anti-air warfare (AAW) ships. Only two of the original DEs will be retained when the 'City' class are all in service.

Submarines were operated after the Second World War, primarily to provide training for ASW forces. Two ex-USN submarines were returned, and to replace the RN flotilla based at Halifax, three boats were ordered from British yards in 1962. These were modernised in the 1980s, but will need to be replaced by the mid-1990s.

In a surprise move the Canadian Government announced in 1985 that it would create a force of up to twelve nuclear attack submarines (SSNs). The rationale was the need to exercise sovereignty over the Arctic, reflecting friction with the United States, but the huge infrastructure costs do not appear to have been correctly estimated. Although the US was hostile to the project, the British and French fought to get their designs accepted, only to see the whole programme dropped, almost as suddenly as it had been created.

At the time of (1995) writing the issue of the *Oberon* replacement has not been settled, but the four ex-RN *Upholder* class are under negotiation. In 1991, after another lengthy debate, orders were placed for a new class of combined MCM craft and training tenders.

The title 'Royal Canadian Navy' was dropped in 1968, when a unified defence force was created by the Canadian Forces Reorganization Act, but ships retain the prefix HMCS. The long-standing bases at Halifax, Nova Scotia, and Esquimalt, British Columbia, are still active. Two destroyer squadrons and two of the submarines serve on the Atlantic coast, while the two destroyer squadrons in the Pacific are used principally for training.

'Tribal' class

When the war ended, *Nootka*, *Micmac*, *Cayuga* and *Athabaskan* were still fitting out so they incorporated several changes dictated by wartime experience. *Nootka* and *Micmac* retained 4.7in guns, but omitted the DCT; in place of the pompom aft, *Nootka* had two twin 40mm Mk 5 side by side. *Cayuga* and *Athabaskan* received four twin 4in/45 Mk 16 and a Mk 6 director, with twin 40mm Mk 5 as in *Nootka*. Between 1946 and 1951 all seven underwent piecemeal modernisation, with 4.7in replaced by twin 4in, some with USN-pattern Mk 63 radar-assisted fire control. Squid DC mortars were added on the quarterdeck, and single and twin 40mm Bofors replaced the light AA.

In 1951 the class was taken in hand for full modernisation, the last ship being recommissioned in 1955. Starting with *Micmac*, the armament and fire control were modernised, with two RN-pattern twin 4in Mk 16 forward, a USN-pattern twin 3in/50 in 'X' position and two Squid triple DC mortars on the quarterdeck. A short aluminium lattice mast replaced the previous one, and the funnels were fitted with prominent caps. *Cayuga* and *Athabaskan* retained their Mk 6 director, but the rest had a modified version of the USN Mk 63 system, with radar trackers on each 4in gunshield, and were given a stripped Mk III(W) director for surface fire. *Cayuga* and *Athabaskan* had a fibre-glass shield fitted to the 3in guns. *Haida* was preserved as a museum at Toronto, where she remains today, the last 'Tribal'.

As modernised the ships' particulars were:

Displacement: 2200t normal; 2500t deep load
Armament: 4–4in/45 Mk 16 (2x2), 2-3in/50 Mk 22 (1x2), 4-40mm/60 Boffins (4x1) (*Cayuga*, *Athabaskan* Mk 7), 4–21in TT, 2 Squid ASW mortars
Sensors: Radar 275 (2 ships), SPS-10, SPS-6, 293, 262; sonar 170, 174
Complement: 240

'Cr' and Valentine classes

Crescent and *Algonquin* were given Type 15 conversions to frigates, and re-classified as 'destroyer escorts' in 1952 (see under Major Surface Ships). *Sioux* and *Crusader* were similarly re-rated, although they remained unconverted; some partial updating of their ASW equipment was the only modernisation carried out.

FLEET STRENGTH 1947

AIRCRAFT CARRIER

Name	Launched	Disp (deep load)	Fate
Colossus class			
WARRIOR (ex-*Brave*)	20.5.44	18,300t	Returned to RN 1948

On her completion in March 1946, *Warrior* was transferred from the RN, and was returned when *Magnificent* was ready in 1948.

CRUISERS

Name	Launched	Disp (deep load)	Fate
Fiji class			
UGANDA	7.8.41	11,090t	BU 1961
Swiftsure class			
ONTARIO (ex-*Minotaur*)	29.7.43	11,480t	BU 1960

Apart from being 'winterised' with added insulation and steam heating for weapons and deck-fixtures, and the addition of USN-pattern search radar, *Uganda* remained largely unaltered. She was not renamed *Quebec* until 14 January 1952 and was employed in training at Halifax.

Ontario differed from her sister *Swiftsure* in having Mk 6 fire control, like *Superb*. As *Quebec* (ex-*Uganda*) she received USN-pattern search radar and was 'winterised'. She spent most of her time as a training ship at Esquimalt, BC.

DESTROYERS

Name	Fate	Name	Fate
'Tribal' class: launched 1941–46, 2745t deep load			
HAIDA	Preserved 1964	CAYUGA	BU 1964
HURON	BU 1965	MICMAC	BU 1964
IROQUOIS	BU 1966	NOOTKA	BU 1964
ATHABASKAN	BU 1969		
'Cr' class: launched 1944, 2535t deep load			
CRESCENT	Type 15 (see notes)	CRUSADER	BU 1965
Valentine class: launched 1943, 2545t deep load			
ALGONQUIN (ex-*Valentine*, ex-*Kempenfelt*)	Type 15	SIOUX (ex-*Vixen*)	BU 1965

FRIGATES

Name	Fate	Name	Fate
'River' class: launched 1942–44, 2180t deep load			
NEW GLASGOW	Sold 1967	LA HULLOISE	BU 1966
NEW WATERFORD	BU 1968	LANARK	BU 1966
OUTREMONT	BU 1966	LAUZON (ex-*Glace Bay*)	Sold 1964
ST CATHERINE'S	Weather ship 1950	PENETANG (ex-*Rouyn*)	To Norway 1956
SWANSEA	BU 1968	PRESTONIAN (ex-*Beauharnois*)	To Norway 1956
ANTIGONISH	Sold 1967	STE THERESE	Sold 1966
BEACON HILL	Sold 1968	STETTLER	Sold 1967
BUCKINGHAM (ex-*Royal Mount*)	BU 1966	STONETOWN	Weather ship 1950
CAP DE LA MADELEINE	BU 1966	SUSSEXVALE (ex-*Valdorian*)	BU 1967
FORT ERIE (ex-*La Tuque*)	BU 1966	TORONTO (ex-*Gifford*)	To Norway 1956
INCHARRAN	Sold 1966, BU 1970	VICTORIAVILLE	Stricken 1974
JONQUIERE	Sold 1967		

'River' class

Most of the later vessels differed from the original 'River' class design in having twin 4in Mk 16 AA guns in 'B' position. In 1951–53 *Prestonian* underwent complete rebuilding to enhance her ASW capabilities. This involved extending the forecastle deck right aft, raising the funnel and providing a much larger and higher bridge. The original generating machinery (three steam and one diesel generator) was changed to two steam and two diesel generators, with much greater capacity, and a more powerful sonar was installed. The conversion was successful, and twenty more ships were converted by the end of 1958 (all except *St Catherine's* and *Stonetown*, which became weather ships). As reconstructed the particulars were:

Displacement: 1570t standard; 2360t full load
Dimensions: 310ft 4in oa × 36ft 8in × 16ft full load (aft)
91.8m × 11.2m × 4.9m
Machinery: 2-shaft VTE, 2 Admiralty 3-drum boilers, 5500ihp = 20kts. Oil 720t
Armament: 2–4in Mk 16 AA (1 × 2), 2–40mm/60 Mk 5 AA (1 × 2), 4–40mm/60 MK 9 (4 ×1), 2 Squid ASW mortars
Sensors: USN-pattern SPQ-2, TDY-1 ECM gear
Complement: 140

CANADA

In 1956 *Buckingham* was fitted with a helicopter landing deck aft to test the feasibility of operating a large Whirlwind HO4S ASW helicopter; these trials were successful and paved the way for the even bigger SH-3A Sea King in later ships. Foreign transfers comprised *Penetang*, *Toronto* and *Prestonian* (Norwegian *Draug*, *Garm* and *Troll*, 1956); *Victoriaville* became the diving tender *Granby* in 1966.

MINESWEEPERS

Name	Fate	Name	Fate
Bangor class			
BROCKVILLE	BU 1961	GODERICH	Sold 1959
DIGBY	BU 7	KENORA	To Turkey 1957
GRANBY	BU 1961	KENTVILLE	To Turkey 1957
LACHINE	To RCMP 1950	MAHONE	To Turkey 1958
MELVILLE	To RCMP 1950	MALPEQUE	Sold 1959
NORANDA	To RCMP 1950	MEDICINE HAT	To Turkey 29.11.57
TRANSCONA	To RCMP 1950	MILLTOWN	Sold 1959
TROIS RIVIERES	To RCMP 1950	MINAS	BU 1959
TRURO	To RCMP 1950	NIPIGON	To Turkey 29.11.57
BLAIRMORE	To Turkey 1958	PORT HOPE	Sold 1959
CHIGNECTO	BU 1957	RED DEER	Sold 1959
FORT WILLIAM	To Turkey 1957	SARNIA	To Turkey 1958
GANANOQUE	Sold 1959	SWIFT CURRENT	To Turkey 1958
GEORGIAN	Sold 1959	WESTMOUNT	To Turkey 1958
Algerine class			
FORT FRANCIS	Survey ship 1949	ROCKCLIFFE	BU 1960
KAPUSKASING	Survey ship 1959	SAULTE STE MARIE	BU 1959
NEW LISKEARD	Survey ship 1959	WALLACEBURG	To Belgium 1959
OSHAWA	Survey ship *Portage* 1958	WINNIPEG	To Belgium 1959
PORTAGE	BU 1960	DRUMMOND-VILLE	Sold 1958

Bangor class
All these were officially stricken, but were re-acquired from the Crown Assets Disposal Corporation in 1952. *Brockville*, *Digby*, *Granby*, *Lachine*, *Melville*, *Noranda*, *Transcona* and *Truro* were transferred to the Marine Section of the Royal Canadian Mounted Police in 1950 and renamed *Macleod*, *Perry*, *Colonel White*, *Starnes*, *Cygnus*, *Irvine*, *French*, *MacBrien* and *Herchmer*. In 1951 *Macleod*, *Perry* and *Colonel White* were returned to the RCN and assumed their former names. All were re-designated coastal escorts (FSE) in 1953. At the end of 1957 ten were sold to Turkey: *Blairmore*, *Fort William*, *Kenora*, *Kentville*, *Mahone*, *Medicine Hat*, *Nipigon*, *Sarnia*, *Swift Current* and *Westmount*, renamed *Beycoz*, *Bodrum*, *Bandirma*, *Bartin*, *Beylerbeyi*, *Biga*, *Bafra*, *Buyukdere*, *Bozcaada* and *Bornova*. A Hedgehog spigot mortar replaced the 122pdr gun in most.
Algerine class
Re-rated as coastal escorts (FSE) in 1953. *New Liskeard* and *Kapuskasing* were lent to the Department of Mines and Technical Surveys. *Kapuskasing* was sunk as a target on 3.10.78. In 1959 *Wallaceburg* and *Winnipeg* were sold to Belgium and renamed *Georges Lecointe* and *A F Dufour*, replacing ex-RN sisters of the same names.

There were also the 'Flower' class corvette *Sackville* (minelayer 1956 – see main entry below) and the 'Isles' class minelaying trawler *Whitethroat* (survey ship 1960).

PENNANT NUMBERS
Pennant numbers carried by the principal ships in this section were as follows:

Cruisers			
CCL 31	UGANDA	CCL 32	ONTARIO
Destroyers			
DDE 215	HAIDA	DDE 213	NOOTKA
DDE 216	HURON	DDE 219	ATHABASKAN
DDE 217	IROQUOIS	DDE 225	SIOUX
DDE 218	CAYUGA	DDE 228	CRUSADER
DDE 214	MICMAC		
Frigates			
FFE 315	NEW GLASGOW	FFE 318	JONQUIERE
FFE 304	NEW WATERFORD	FFE 305	LA HULLOISE
FFE 310	OUTREMONT	FFE 321	LANARK
FFE 324	ST CATHERINE'S	FFE 322	LAUZON
FFE 306	SWANSEA	FFE 307	PENETANG
FFE 301	ANTIGONISH	FFE 316	PRESTONIAN
FFE 303	BEACON HILL	FFE 309	STE TERESE
FFE 314	BUCKINGHAM	FFE 311	STETTLER
FFE 317	CAP DE LA MADELEINE	FFE 302	STONETOWN
		FFE 313	SUSSEXVALE
FFE 312	FORT ERIE	FFE 319	TORONTO
FFE 308	INCHARRON	FFE 320	VICTORIAVILLE
Minesweepers			
FSE 180–199	BANGOR class	FSE 168–177	ALGERINE class

MAJOR SURFACE SHIPS

Laid down on 29 July 1943 and launched on 16 November 1944, *Magnificent* was returned to the RN in June 1957 and was put into reserve for eight years before being sold for scrapping. While in the RCN she was designated CVL 21 and had the number '21' painted at each end of the flight deck. In 1954 she was re-classified as RML 21 and served mainly as a ferry carrier and transport. Various USN radars replaced the British sets, but she was otherwise similar.

Ex-British MAJESTIC class *aircraft carrier*

Particulars: As Australian *Sydney*

No	Name	Builder	Acquired	Fate
CVL 21	MAGNIFICENT	Harland & Wolff	21.5.48	Returned 14.6.57

Bonaventure 12.4.57 RCN

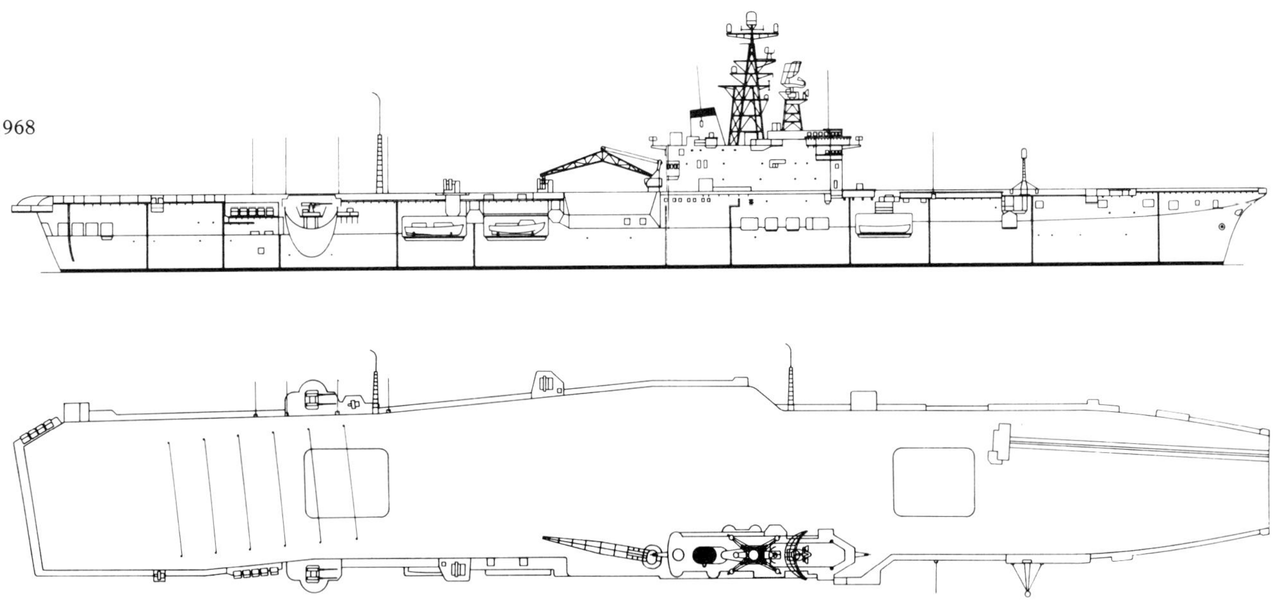

Bonaventure 1968

Ex-British MAJESTIC class *aircraft carrier*

Displacement: 16,000t standard; 20,000t full load
Dimensions: 704ft oa × 128ft max × 25ft
214.6m × 39.0m × 7.6m
Machinery: 2-shaft geared steam turbines, 4 Admiralty boilers, 40,000shp = 24.5kts. Oil 3200t
Armour: Mantlets over magazines
Armament: 8–3in/50 Mk 33 (4×2), 4–3pdr saluting, 21–24 aircraft
Complement: 1370

No	Name	Builder	Acquired	Fate
CVL 22	BONAVENTURE (ex-*Powerful*)	Harland & Wolff	17.1.57	BU 1971

Laid down on 27 November 1943 and launched on 27 February 1945, the former *Powerful* was modernised between 1952 and 1957 and sold to the Royal Canadian Navy as HMCS *Bonaventure* (RML 22). She was completed with an angled deck, but aircraft, radar and armament were supplied from the United States. Although similar to her sisters she had a distinctive appearance, with a tall lattice mast, a raked funnel and a big sponson for her twin 3in AA guns.

The original air group consisted of sixteen F-2H Banshee fighters and eight S-2F Tracker ASW aircraft, but it was subsequently changed to an entirely ASW mission with eight Trackers and thirteen HO4S-3 Whirlwind helicopters, and in time the Whirlwinds gave way to CHSS-2 Sea King helicopters. During her 1966–67 modernisation the ship was given new radars and improved accommodation and aircraft handling, and the two forward 3in gun mountings were removed to save weight. The new radars were the Dutch Signaal type.

The ship's designation changed twice, to RRSM 22 in 1957, then to CVL 22, and she carried the number '22' for deck recognition. She was paid off in April 1970 and was laid up for disposal.

Both these ships were given a Canadian variant of the British Type 15 frigate conversion. It followed the same lines but with a much bigger square bridge and American electronics. *Algonquin* had an RN-pattern twin 4in gun mounting aft and a USN-pattern twin 3in/50 forward, whereas *Crescent* had the twin 4in forward and the 3in aft. Both ships also had two single 40mm Boffin abreast of the funnel, and *Crescent* had a prominent cap to her funnel. Conversion was undertaken at Esquimalt DYd in 1952–56.

ALGONQUIN class *destroyer escorts*

Particulars: As British *Rapid* class

No	Name	Yard	Reconstructed	Fate
DDE 224	ALGONQUIN (ex-*Valentine*, ex-*Kempenfelt*)	Esquimalt DYd	1952–56	BU 1971
DDE 226	CRESCENT	Esquimalt DYd	1952–56	BU 1971

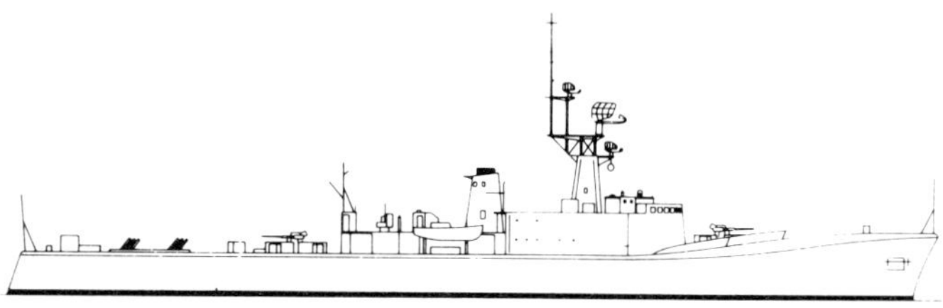

Ottawa 1957

In the late 1940s there was a pressing NATO requirement for anti-submarine vessels, and funds were voted for seven destroyer escorts or frigates for the RCN. The British *Whitby* design was still on the drawing-board, and as the USN had nothing which met the exacting RCN Staff Requirement it was decided to design a new class of ship in Canada. Using the services of the Montreal naval architects, German & Milne, and under the direction of a senior constructor, Rowland Baker, seconded from the Director of Naval Construction, work started in 1949. Baker produced a design basically similar to the *Whitby*, but incorporating several ideas of his own. The machinery was identical with the British ships and *St Laurent*'s was in fact imported from the UK, but electronics and electrical gear were of American and Canadian design. To flatter Canadian susceptibilities, Baker was careful to make the appearance as different as possible from the *Whitby*, but the ship which resulted was virtually a Type 12 specification (albeit with a different hull form) translated by a different design team.

During 1961–66 all seven were modernised, with variable-depth sonar (VDS) on the stern, one Limbo Mk 10 mortar removed and a helicopter platform added. The funnel was replaced by twin uptakes to allow a hangar to be fitted between, and the two 40mm guns were removed. Thereafter they were rated as helicopter-carrying destroyers (DDH) and shipped one CHSS-2 Sea King. *St Laurent* was laid up in 1974 but not sold until 1979; on 12 January 1980 she foundered in tow to the breakers. *Skeena*, *Ottawa* and *Fraser* were given major overhauls in 1977–79; the 'Destroyer Life Extension' (DELEX) refit was applied to *Saguenay*, *Margaree* and *Assiniboine* in 1980, and to the other three in 1981.

Assiniboine paid off in 1989 to become harbour TS at Halifax, and *Saguenay* was paid off for disposal in 1990. The others will go in the mid-1990s as the 'City' class join the Fleet.

ST LAURENT class *destroyer escorts*

Displacement: 2000t normal; 2600t deep load
Dimensions: 371ft wl, 366ft oa × 42ft × 13ft 2in
113.1m, 111.6m × 12.8m × 4.2m
Machinery: 2 shaft English-Electric geared steam turbines, 2 Babcock & Wilcox boilers, 30,000shp = 28kts
Armament: 4–3in/50 DP (2×2), 2–40mm/60 Boffin (2×1), 2 Limbo Mk 10 ASW mortars
Sensors: Radar SPS-12, SPS-10, SPG-48; sonar SQS-501, SQS-502, SQS-503, SQS-504
Complement: 290

No	Name	Builder	Laid down	Launched	Comp	Fate
DDE 205	ST LAURENT	Canadian Vickers	22.11.50	30.11.51	29.10.55	Sold 1979
DDE 206	SAGUENAY	Halifax SYd	4.4.51	30.7.53	15.12.56	Sold 1990
DDE 207	SKEENA	Burrard, Vancouver	1.6.51	19.8.52	30.3.57	For disposal 1994
DDE 229	OTTAWA	Canadian Vickers	8.6.51	29.4.53	10.11.56	Sold 1992
DDE 230	MARGAREE	Halifax SYd	12.9.51	29.3.56	5.10.57	BU 1994
DDE 233	FRASER	Burrard, Vancouver	11.12.51	19.2.53	28.6.57	For disposal 1994
DDE 234	ASSINIBOINE	Marine Ind, Sorel	19.5.52	12.2.54	16.8.56	HS 1989

Fraser 1.10.85 L & L van Ginderen

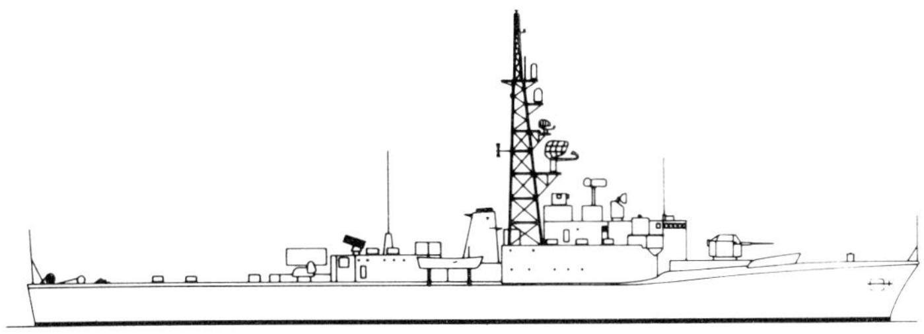

Gatineau 1972

RESTIGOUCHE class *destroyers*

Developed from the original *St Laurent* class and ordered in 1952. On the same hull it was possible to mount the British twin 3in/70 Mk 6 as designed for the previous class, but a USN-pattern twin 3in/Mk 22 was retained aft.

Four ships – *Restigouche, Gatineau, Kootenay* and *Terra Nova* – were modernised in 1966–73, with VDS on the stern and an ASROC anti-submarine missile launcher aft. This necessitated the removal of one Limbo Mk 10 mortar and the 3in Mk 22 twin mounting, and a new lattice foremast was stepped. *Chaudière, Columbia* and *St Croix* paid off into reserve in 1974, but in 1979 *St Croix* was reduced to harbour service at Halifax, and she and *Chaudière* were deleted in 1988. *Columbia* reduced to harbour training at Esquimalt in 1974. In September 1990 *Terra Nova* was hurriedly refitted for the Gulf crisis, with Mk 15 Phalanx CIWS on the quarterdeck in place of the Mk 10 mortar, two quadruple Harpoon missile launchers in place of the Asroc missile system, two 40mm/60 Bofors Boffin guns in place of the boats, and 12.7mm MGs, plus upgraded EW and communications. Early in 1991 she was relieved by *Restigouche*, which had a similar refit. The four survivors will be replaced by the 'City' class.

Displacement: As *St Laurent* class
Dimensions: As *St Laurent* class
Machinery: As *St Laurent* class
Armament: 2–3in/70 Mk 6 DP (2×2), 2–3in/50 Mk 22 DP (1×2), 2 Limbo Mk 10 ASW mortars, 2–40mm Boffin (2×1)
Sensors: Radar SPS-12, SPS-10, Sperry Mk 2, SPG-48; sonar SQS-501, SQS-503
Complement: 290

No	Name	Builder	Laid down	Launched	Comp	Fate
DDE 257	RESTIGOUCHE	Canadian Vickers	15.7.53	22.11.54	7.6.58	For disposal 1994
DDE 235	CHAUDIERE	Halifax SYd	30.7.53	13.11.57	14.11.59	Scuttled 1992
DDE 236	GATINEAU	Davie SB, Lauzon	30.4.53	3.6.57	17.2.59	Extant 1995
DDE 256	ST CROIX	Marine Ind, Sorel	15.10.54	17.11.57	4.10.58	HS 1974, BU 1991
DDE 258	KOOTENAY	Burrard, Vancouver	21.8.52	15.6.54	7.3.59	Extant 1995
DDE 259	TERRA NOVA	Victoria Machinery	11.6.53	21.6.55	6.6.59	Extant 1995
DDE 260	COLUMBIA	Burrard, Vancouver	11.6.53	1.11.56	7.11.59	HS 1974

Restigouche 19.8.93 L & L van Ginderen

Saskatchewan 20.10.82 L & L van Ginderen

MACKENZIE class *destroyers*

Four more 'repeat *Restigouches*' were ordered in 1957, but the opportunity was taken to improve habitability and provide better pre-wetting and better bridge and weatherdeck fittings to cope with extreme cold. Originally they were rated as DDs, like the *Restigouche* class, but they carry DDE pennant numbers. They all mounted two triple 12.75in Mk 32 launchers for ASW torpedoes, and in *Yukon* and *Qu'Appelle* the British 3in/70 mount forward was replaced with a US 3in/50. DELEX refits (SQS-505 in lieu of SQS-503 and new navigational radar) were carried out at Esquimalt in 1982–85. From 1992 they were replaced in service by the 'City' class frigates.

Particulars: As *Restigouche* class

No	Name	Builder	Laid down	Launched	Comp	Fate
DDE 261	MACKENZIE	Canadian Vickers	15.12.58	25.5.61	6.10.62	Sold 1995
DDE 262	SASKATCHEWAN	Victoria Machinery	16.7.59	1.2.61	16.2.63	For disposal 1995
DDE 263	YUKON	Burrard, Vancouver	25.10.59	27.7.61	25.5.63	For disposal 1993
DDE 264	QU'APPELLE	Davie SB, Lauzon	14.1.60	2.5.62	14.9.63	Sold 1992

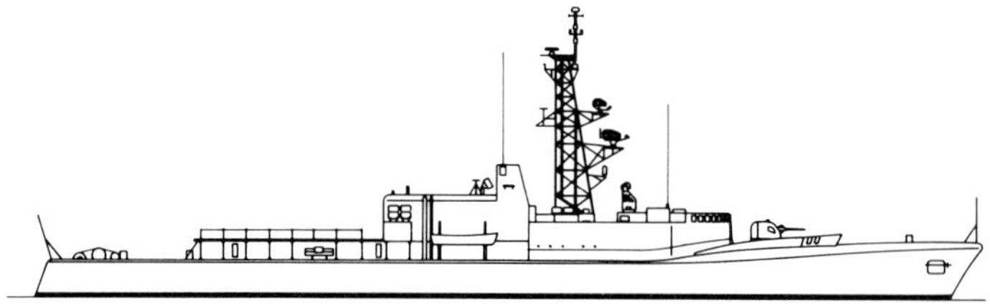

Annapolis 1989

The final development of the original *St Laurent* ordered in 1959 incorporated all the improvements approved for the earlier ships, particularly the Sea King helicopter deck and hangar. Because of the extra topweight the lighter USN-pattern 3in/50 Mk 33 was reintroduced in place of the British-pattern Mk 6. Both ships were modernised in 1977–79 and are now rated as DDHs. Further modernisations under the DELEX programme took place in 1982–85, and in 1987–88 the SQS-501 variable-depth sonar was replaced by the SQR-501 CANTASS towed sonar array for trials and evaluation.

ANNAPOLIS class *helicopter destroyers*

Displacement: 2400t normal; 3000t deep load
Dimensions: As *St Laurent* class
Machinery: As *St Laurent* class
Armament: 2–3in/50 Mk 33 DP (1×2), 6–12.75in Mk 32 TT (2×3), 1 Limbo Mk 10 ASW mortar, 1 helicopter
Sensors: Radar SPS-12, SPS-10, Sperry Mk 2, Mk 48; sonar SQS-503, SQS-504, SQS-501
Complement: 228

No	Name	Builder	Laid down	Launched	Comp	Fate
DDE 265	ANNAPOLIS	Halifax SYd	July 1960	27.4.63	19.12.64	Extant 1995
DDE 266	NIPIGON	Marine Ind, Sorel	Apr 1960	10.12.61	30.5.64	Extant 1995

Annapolis Oct 1977 C & S Taylor

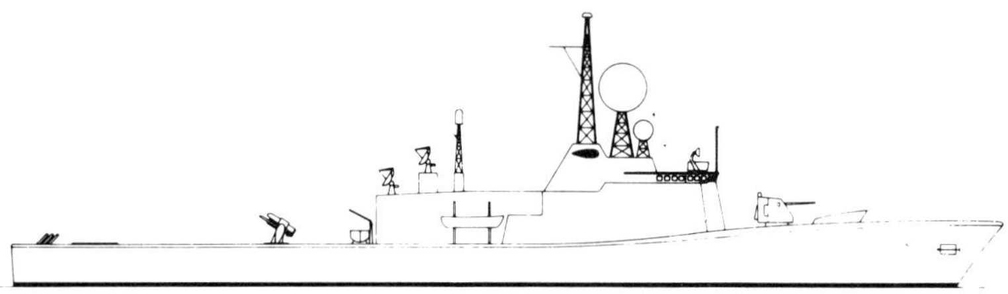

'Tribal' class design 1960

Eight large general-purpose frigates were announced in 1960; they were to be given 'Tribal' names, but in 1963 the project was cancelled before orders had been placed. The design was developed into the *Iroquois* class DDHs (see below).

'TRIBAL' class *missile frigates*

Displacement: As *Iroquois* class
Dimensions: As *Iroquois* class
Machinery: 2-shaft geared steam turbines, 36,000shp = 30kts
Armament: 1–5in/54 Mk 42, 1 Tartar RIM-24 SAM system, ASW TT
Complement: As *Iroquois* class

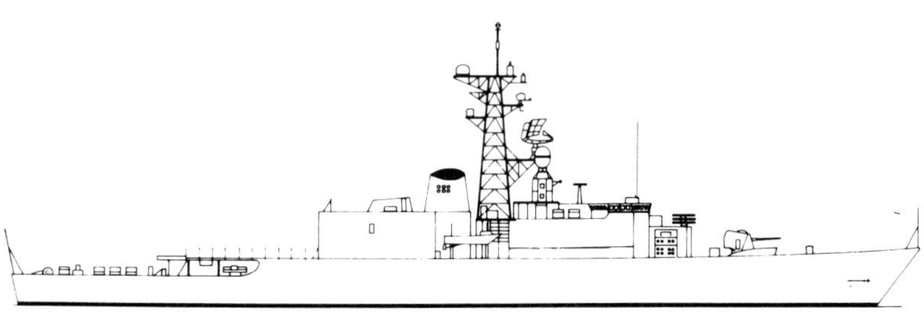

Huron as completed

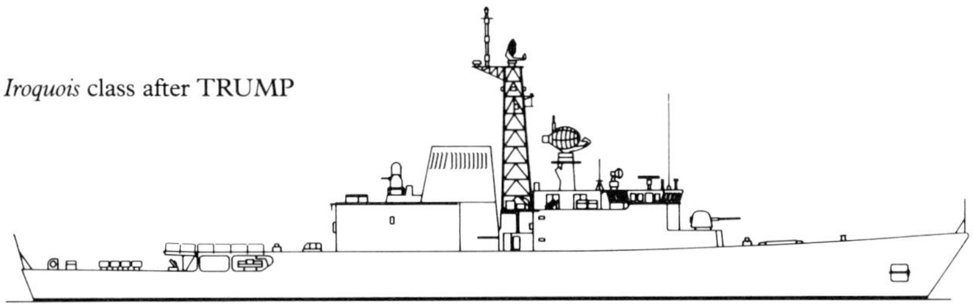

Iroquois class after TRUMP

Iroquois 20.9.85 L & L van Ginderen

Ordered in 1968 as a revised design on the same hull as the cancelled general-purpose frigates of 1963 (see above). The design allowed for two big helicopters (CHSS-2) to enhance their ASW capabilities, and they are still among the best equipped escorts in the world. The 5in gun is Italian, the radars Dutch.

In 1986 the long-deferred plans to build AAW ships came to fruition. A contract was awarded to Litton Systems Canada Ltd for the 'Tribal' class Update and Modernisation Programme (TRUMP). The plan involved total rebuilding of the superstructure, new propulsion and new weaponry and electronics.

The 5in gun is replaced by a Mk 41 vertical launch system for Standard SM-2 medium-range SAMs, a 76mm OTO Melara Super Rapid gun is added in 'B' position, and a Mk 15 Phalanx CIWS is positioned abaft the remodelled funnel. Internally, the electronics suite is completely overhauled and new GM Allison 570 KF cruising turbines replace the Pratt & Whitney units. The electronic warfare equipment is also upgraded: Shield decoy-launchers, SLQ-504 Canews ESM, and ULQ-6 ECM.

Algonquin was modernised by MIL-Davie at Lauzon, Quebec, in 1987–91; *Iroquois* taken in hand November 1988, and two remaining ships were finished in 1993. The unmodernised *Athabaskan* was sent to the Persian Gulf in September 1990, with Mk 15 Phalanx CIWS on the quarterdeck. From 1991 rated as DDG 280–283.

IROQUOIS class *helicopter destroyers*

Displacement: 3551t normal; 4700t deep load; 5100t deep load (modernised)
Dimensions: 398ft wl, 423ft oa × 50ft × 14.6in
121.3m, 128.9m × 15.2m × 4.4m
Machinery: 2-shaft COGOG, 2 Pratt & Whitney FT4A2 gas turbines, 50,000shp; 2 P&W FT12AH3 cruising GTs, 7400shp = 29kts (2 GM Allison 570KF cruising GTs (modernised)), endurance 4500nm at 20kts
Armament: 1–76mm/62 Super Rapid, 1 Mk 41 VL Standard SM-2 (MR) SAM system (29 rounds), 1 Mk 15 Phalanx 20mm CIWS, 6–324mm Mk 32 ASW TT (2×3), 2 CH-12A Sea King helicopters
Sensors: Radar LW-08, STIR 1.8; sonar SQS-501, SQS-505 (modernised)
Complement: 23 officers, 232 ratings + 30 aircrew

No	Name	Builder	Laid down	Launched	Comp	Fate
DDH 280	IROQUOIS	Marine Ind, Sorel	15.1.69	28.11.70	29.7.72	Extant 1995
DDH 281	HURON	Marine Ind, Sorel	15.1.69	3.4.71	16.12.72	Extant 1995
DDH 282	ATHABASKAN	Davie SB, Lauzon	1.6.69	27.11.70	30.11.70	Extant 1995
DDH 283	ALGONQUIN	Davie SB, Lauzon	1.9.69	23.4.71	30.9.73	Extant 1995

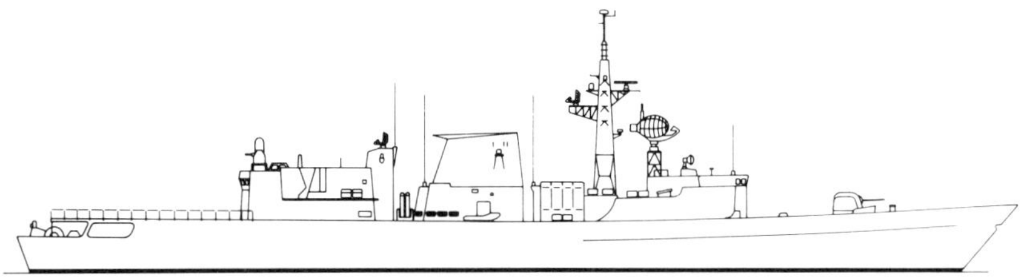

'City' class design

The decision was taken on 22 December 1977 to order six ships of a projected twenty vessel programme, but lengthy delays have afflicted the programme, and a contract to build the first six was not awarded to St John Shipbuilding Ltd, New Brunswick until June 1983. Design was shared between St John SB and Paramax Electronics, and three hulls were subcontracted to Marine Industries Ltd of Sorel, Quebec (now known as MIL-Davie). In December 1987 six more were ordered.

The lead-ship *Halifax* started trials in August 1990 and was accepted in June 1992. The last of class is planned to be in service by 1996, but there have been several delays.

'CITY' class *helicopter frigates*

Displacement: 4750t deep load
Dimensions: 408ft 6in pp, 440ft oa × 53ft 9in × 16ft 2in
124.5m, 134.1m × 16.4m × 4.9m
Machinery: 2-shaft COGOG, GE LM 2500 gas turbines, 46,000shp; 1 SEMT-Pielstick 20PA6-V280 diesel, 8800bhp = 28kts. Range 7100nm at 15kts
Armament: 1–57mm/70 SAK Mk 2, 8 Harpoon RGM-84C SSMs, 2 Mk 48 VL Sea Sparrow SAM systems (16 rounds), 1 Mk 15 Phalanx 20mm CIWS, 4–324mm Mk 32 ASW TT (2×2), 8–12.7mm MGs
Helicopter: CH-124A Sea King or EH.101 Merlin
Sensors: Radar SPS-49(V)5, Sea Giraffe HC 150; sonar SQS-505(V)6, SQR-501 CANTASS
Complement: 225

No	Name	Builder	Laid down	Launched	Comp	Fate
FFH 330	HALIFAX	St John SB	19.3.87	30.5.88	29.6.92	Extant 1995
FFH 331	VANCOUVER	St John SB	19.5.88	8.7.89	23.8.93	Extant 1995
FFH 332	VILLE DE QUEBEC	MIL-Davie	16.12.88	16.5.91	14.7.94	Extant 1995
FFH 333	TORONTO	St John SB	22.4.89	15.6.91	29.7.93	Extant 1995
FFH 334	REGINA	MIL-Davie	6.10.89	25.11.91	30.7.94	Extant 1995
FFH 335	CALGARY	MIL-Davie	15.6.91	28.8.92	25.9.94	Extant 1995
FFH 336	MONTREAL	St John SB	9.2.91	28.2.92	21.7.94	Extant 1995
FFH 337	FREDERICTON	St John SB	25.4.92	26.6.93	10.9.94	Extant 1995
FFH 338	WINNIPEG	St John SB	19.3.93	25.6.94	30.9.94	Extant 1995
FFH 339	CHARLOTTETOWN	St John SB	17.12.93	1.10.94	14.5.95	Extant 1995
FFH 340	ST JOHN'S	St John SB	24.8.94	30.4.95	Dec 1995	Extant 1995
FFH 341	OTTAWA	St John SB	27.4.95	Dec 1995	Jun 1996	Building

Vancouver 19.8.93

L & L van Ginderen

Ordered October 1991, but given town names early 1995. Intended originally as mine countermeasures vessels with a limited peacetime patrol and training function, their MCM capability has been drastically reduced to cut costs.

KINGSTON class *maritime coastal defence vessels*

Displacement: 962t full load
Dimensions: 160ft 9in pp, 181ft 4in oa × 37ft × 11ft
49m, 55.3m × 11.3m × 3.4m
Machinery: 2-shaft Wärtsilä diesel-electric, 3000bhp = 15kts. Range 4000nm at 12kts
Armament: 1–40mm/60 Mk 5C Boffin, 2–12.7mm MG (2×1)
Sensors: Radar KH 1006; sonar towed sidescan
Complement: 17

No	Name	Builder	Laid down	Launched	Comp	Fate
700	KINGSTON (ex-*Frontenac*)	Halifax SY	12.8.95	Aug 1995	Oct 1995	Extant 1995
701	NANAIMO	Halifax SY	–	–	–	Building
702	EDMONTON	Halifax SY	–	–	–	Building
703	WHITE HORSE	Halifax SY	–	–	–	Building
704	YELLOWKNIFE	Halifax SY	–	–	–	Building
705	GOOSE BAY	Halifax SY	–	–	–	Building
706	SUMMERSIDE	Halifax SY	–	–	–	Building
707	MONCTON	Halifax SY	–	–	–	Building
708	BRANDON	Halifax SY	–	–	–	Ordered
709	SASKATOON	Halifax SY	–	–	–	Ordered
710	GLACE BAY	Halifax SY	–	–	–	Ordered

SUBMARINES

An ex-USN *Balao* class (ex-radar picket) fleet submarine, launched 18 June 1943, on loan to the RCN for two (extended to five) years for ASW training at Esquimalt, BC, and commissioned there on 11 May 1961. In May 1966 the loan was extended indefinitely, but she was returned on 2 October 1969.

Ex-US BALAO class *submarine*

Particulars: As US *Balao* class

No	Name	Builder	Acquired	Fate
SS 71	GRILSE (ex-*Burrfish*, SSR 312)	Electric Boat	May 1961	Returned to USN 1969

Ex-USN *Tench* class (Fleet Snorkel conversion) launched 1 October 1944 and bought to replace *Grilse*. Commissioned at Esquimalt 2 December 1968 and returned to USN for scrapping 31 December 1974.

Ex-US TENCH class *submarine*

Particulars: As US *Tench* class

No	Name	Builder	Acquired	Fate
SS 75	RAINBOW (ex-*Argonaut*, SS 475)	Portsmouth N Yd	Dec 1968	Returned to USN 1974

British OBERON type *submarines*

Particulars: As British *Oberon* class

No	Name	Builder	Laid down	Launched	Comp	Fate
SS 72	OJIBWA (ex-*Onyx*)	Chatham DYd	Sep 1962	29.2.64	23.9.65	Extant 1995
SS 73	ONANDAGA	Chatham DYd	June 1964	25.9.65	22.6.67	Extant 1995
SS 74	OKANAGAN	Chatham DYd	Mar 1965	17.9.66	22.6.68	Extant 1995

Identical with British *Oberon* class, but armed with Mk 37-C ASW torpedos, not anti-ship weapons (twenty-two carried). *Ojibwa* began 'Submarine Operational Update Programme' (SOUP) in 1980, *Onandaga* in 1983 and *Okanagan* in 1982–86. This involved a new fire control system (Singer-Librascope Mk 1 mod O), completely revised sonar arrangements Marconi Type 2051/Triton with enlarged bow dome for the passive array and the active transducer in the fin; Type 2007 to be retained), and weapon launching and fire control systems were modified from 1987 to handle Mk 48 Mod 4 torpedoes and Sub-Harpoon missiles. *Onandaga* transferred to the Pacific in 1994.

MINE WARFARE VESSELS

SACKVILLE *minelayer*

Displacement: 1085t normal; 1350t deep load
Dimensions: 205ft oa × 33ft 2in × 14ft 6in
62.5m × 10.1m × 4m
Machinery: 1-shaft VTE, 2 boilers, 2750ihp = 16.5kts. Oil 230t
Armament: None
Complement: 38

The only 'Flower' class corvette retained in Canadian service, *Sackville* was laid up until 1953–56 when she was converted to a minelayer (AN 113) for laying 'loops' or controlled minefields. She was lent to the Naval Research Laboratory for oceangraphic work and finally in 1964 was designated AGOR 113 when converted for survey work.

(For illustration see under France)

GASPE ('BAY') class *coastal minesweepers*

Displacement: 390t normal; 412t deep load
Dimensions: 140ft pp, 152ft oa × 28ft × 8ft 6in max
42.7m, 46.3m × 8.5m × 2.6m
Machinery: 2 shafts, 2 GM 12cyl diesels, 2400bhp = 15kts. Oil 52t
Armament: 1–40mm/60 Mk 7
Complement: 40

No	Name	Builder	Launched	Fate
MCB 143	GASPE	Davie SB	12.11.51	To Turkey 1958
MCB 144	CHALEUR (i)	Port Arthur	21.6.52	To France 1954
MCB 145	FUNDY (ii)	St John DD	Apr 1953	To France 1954
MCB 146	COMOX	Victoria Machinery	24.4.52	To Turkey 1958
MCB 147	COWICHAN (i)	Davie SB	12.11.51	To France 1954
MCB 148	UNGAVA	Davie SB	20.5.53	To Turkey 1958
MCB 149	QUINTE	Port Arthur	8.8.53	Sold 1966
MCB 150	MIRAMICHI (i)	St John DD	Oct 1952	To France 1954
MCB 151	FORTUNE	Victoria Machinery	14.4.53	Sold 1966
MCB 152	JAMES BAY	Canadian Yarrow	12.3.53	Sold 1966
MCB 153	THUNDER (i)	Canadian Vickers	17.7.52	To France 1954
MCB 154	RESOLUTE	Kingston SYd	20.6.53	Sold 1966
MCB 156	CHIGNECTO (i)	Marine Industries	21.6.52	To France 1954
MCB 157	TRINITY	Davie SB	31.7.53	To Turkey 1958
MCB 159	FUNDY (ii)	Davie SB	14.6.56	Extant 1995
MCB 160	CHIGNECTO (ii)	Davie SB	17.11.56	Extant 1995
MCB 161	THUNDER (ii)	Port Arthur	27.10.56	Extant 1995
MCB 162	COWICHAN (ii)	Canadian Yarrow	26.2.57	Extant 1995
MCB 163	MIRAMICHI (ii)	Victoria Machinery	22.2.57	Extant 1995
MCB 164	CHALEUR (ii)	Marine Industries	11.5.57	Extant 1995

Four coastal minesweepers (AMC, later MCB) were ordered in 1950, followed by ten more in 1951; they were based closely on the British 'Ton' class, with wood planking and aluminium framing. In 1954 *Chignecto*, *Chaleur*, *Fundy*, *Miramichi*, *Cowichan* and *Thunder* were sold to France and renamed *La Bayonnaise*, *Le Dieppoise*, *La Dunkérquoise*, *La Lorientaise*, *La Malouine* and *La Paimpolaise*; six replacements were built for them. In 1958 four more, *Ungava*, *Trinity*, *Comox* and *Gaspe*, were sold to Turkey and renamed *Tekirdag*, *Terme*, *Tirebolu* and *Trabzon*. Designation was changed from MCA to MCB in 1954; in 1972 the six survivors were redesignated small patrol escorts (PFL) and since 1979 have been used for training with PB pennant numbers.

Bras D'Or 1984 L & L van Ginderen

MISCELLANEOUS

BRAS D'OR *experimental hydrofoil*

Displacement: 180t normal; 237.5t deep load
Dimensions: 150ft 10in oa × 21ft 6in × 15ft (hull depth)
46.0m × 6.5m × 4.6m
Machinery: 2 shafts, 2 Pratt & Whitney FT4A gas turbines, 2000shp = 15kts
Armament: (Designed) 12–12.75in Mk 32 TT (4×3)
Complement: 17

Name	Builder	Launched	Fate
BRAS D'OR	Marine Industries, Sorel	Jul 1968	Sold 1982

This vessel was jointly funded by the Canadian and British governments and designed by De Havilland (Canada). She ran a series of trials in 1969–71, but although moderately successful she was expensive to operate and was laid up in 1971. She remained in a state of preservation in the dockyard at Halifax until sold in 1982.

'BIRD' class *submarine chasers*

Displacement: 66t full load
Dimensions: 92ft × 17ft × 5ft 4in
30.2m × 5.6m × 1.7m
Machinery: 2-shaft diesels, 1200bhp = 14kts
Armament: 1–20mm, Hedgehog, DCs
Complement: 21

Class: *Loon* (PCS 780), *Cormorant* (PCS 781), *Bleu Heron* (PCS 782), *Mallard* (PCS 783)

Built in 1955, of wood and aluminium, they were designed for harbour patrol and ASW training. *Blue Heron* was transferred to the Marine Section of the Royal Canadian Mounted Police in 1956, and the remaining three were paid off on 23 May 1963.

BLUETHROAT *loop-layer*

Displacement: 785t normal; 870t deep load
Dimensions: 150ft 8in pp, 157ft oa × 33ft × 10ft
45.9m, 47.8m × 10m × 3m
Machinery: 2-shaft diesels, 1200bhp = 13kts
Armament: None
Complement: 25

Ordered in 1951 as a loop-layer and completed in 1955. Redesignated AGOR 114 in 1964 and converted for survey duties. *Bluethroat* was sold to United States interests and delivered to them at Halifax in December 1991.

Cape Verde

This republic of nine major islands, some 250 miles off the coast of Senegal, by referendum became independent from former Portuguese Guinea-Bissau in 1975. From January 1979 the Soviet Union supplied its left-wing government with two 'Shershen' class fast attack craft (configured as gunboats, the usual torpedo tubes not being fitted) and three 'Zhuk' class coastal patrol craft. The 'Shershen' class boats are no longer in service as of 1995, and a total of eight 51ft patrol craft built by Peterson, USA, have been ordered in their place. Three in service 1995.

Chile

The last ten years have been very stressful for the Chilean Navy. Chile has been beset by international and domestic problems. Relations with Argentina and Peru have become increasingly strained. Argentina and Chile have had a long-standing border dispute at the tip of the continent. Rumours that the disputed area might contain oil deposits have increased the tensions caused by this dispute. Oil is a critical problem for Chile. The nation can meet only a fraction of the national demand through indigenous production. Relations with Peru have been strained since the War of the Pacific (1879–83), when Chile crushed Peru. The significant increase in strength of the Peruvian Navy *vis-à-vis* that of Chile in recent decades has added to the pressure upon the Chilean Navy. Thus, today the Chilean Navy faces reputedly strong fleets to both the north and south.

Chile's international problems, which directly affect the navy, do not end here. The fall of the Allende government in 1973 and the subsequent conflict between his exiled supporters and the new government have strained United States–Chilean relations to breaking point. Thus, the major supplier of warships and spare parts to Latin America has cut off Chile. This impacts very heavily on the maintenance of the *Fletcher* class destroyers and other US-built ships in the Chilean fleet. The reactivation of the cruiser *O'Higgins* in 1978 and the submarine *Simpson* in 1977 evidenced the need for fleet units. Also, Chile had been unable to purchase new warships from abroad or to secure a bi-national construction agreement as Argentina and Peru have done.

During the Falklands War, the Chilean fleet was moved south, adding pressure to the Argentine Navy. Chile's acquisition problems improved during the early 1980s. Chile was able to order two Type 209 submarines and built three small landing ships under a bi-national construction agreement with France. During the 1980s the Chilean Navy managed to acquire four ex-British *County* class guided missile destroyers. These very seaworthy ships are well suited for the harsh conditions around the Chilean coasts, acting as flagships for smaller destroyers and frigates. Their acquisition made it possible to retire the old cruiser *O'Higgins* (ex-*Brooklyn*) which has become a museum ship in the USA. The main naval base is at Talcahuano where there is also a major repair yard. Other bases are Puerto Montt, Punta Arenas, Iquique and Puerto Williams. In 1995 personnel numbered 22,800 including 2000 officers and 5000 Marines. Conscription is for a 12-month period. During 1982 there were persistent rumours that the British carrier *Hermes* was on offer to Chile for transfer after her final commission in the Royal Navy but these came to nothing.

For many decades the Chilean Navy, unit for unit, has been reported to be the best in Latin America. Evidence suggests that every effort is being made to retain this quality. Soon, however, the adverse material position of the navy must have an impact.

FLEET STRENGTH 1947

BATTLESHIPS

Name	Launched	Disp	Fate
HUASCAR	1865	1130t	Extant 1995 as museum
ALMIRANTE LATORRE	17.11.13	28,000t	Discarded 1959

Huascar was captured from the Peruvians on 8 October 1879 during the War of the Pacific.

Almirante Latorre had her main engines and auxiliary machinery overhauled in 1950 and she was equipped with radar. In 1951 she suffered an engine room disaster; she remained inactive throughout the remainder of her career.

CRUISERS

Name	Launched	Disp	Fate
CHACABUCO	1898	4300t	Discarded 1959
O'HIGGINS	1898	8500t	Discarded 1959

Chacabuco was extensively rebuilt in 1942 and was actively employed in 1950.

DESTROYERS

Name	Launched	Disp	Fate
Serrano class			
ALDEA	29.11.28	1090t	Discarded 1957
HYATT	21.7.28	1090t	Discarded 1963
ORELLA	8.3.28	1090t	Discarded 1967
RIQUELME	28.5.28	1090t	Discarded 1963
SERRANO	25.1.28	1090t	Discarded 1967
VIDELA	16.10.28	1090t	Discarded 1960

SUBMARINES

Name	Launched	Disp	Fate
FRESIA	1915	364t/435t	Discarded 1953
GUACOLDA	1915	364t/435t	Discarded 1949
GUALA	1915	364t/435t	Discarded 1953
ALMIRANTE SIMPSON	15.1.29	1540t/2020t	Discarded 1957
CAPITAN O'BRIEN	2.10.28	1540t/2020t	Discarded 1957
CAPITAN THOMPSON	15.1.29	1540t/2020t	Discarded 1958

MINOR WARSHIPS

Name	Acquired	Disp	Fate
British 'River' class frigates			
COVADONGA (ex-*Seacliff*, ex-*Megantio*)	3.3.46	1445t	Sold 1968
ESMERALDA (ex-*Glace Bay*, ex-*Lauzon*)	3.1.46	1445t	Sold 29.11.68
IQUIQUE (ex-*Joliette*)	3.1.46	1445t	Sold 29.11.68
British 'Flower' class corvettes			
CASMA (ex-*Stellarton*)	1946	980t	Sold 2.7.69
CHIPANA (ex-*Strathray*)	1946	980t	Sold 2.7.69
PAPUDO (ex-*Thorlock*)	1946	980t	Sold 17.5.67

Both classes were Canadian wartime construction. *Esmeralda* was renamed *Baquedano* in 1952. *Chipana* was employed as a survey ship.

Other vessels on strength in 1947 included three ex-US tug type patrol craft *Lautaro, Lientur* (*Lientur* scrapped 1986, *Lautaro* c1990) and *Leucoton* (700t, 1944) and the depot ship *Araucano* (6438t, 1929). In 1947 Chile had just acquired 4 ex-US LSM type landing ships, named *Aspirante Goicolea, Aspirante Izaza, Aspirante Morel* and *Guardiamarina Contreras*; and six ex-US LCIL type landing craft: *Cabo Bustos, Grumete Diaz, Eduardo Llanos, Grumete Tellez, Grumete Bolados* and *Soldado Canaves*; there were also eight LCPs, twenty-two LCVPs and two LCPLs transferred from the USN at the same time. The larger landing craft were withdrawn from service from 1958 onwards, the last vessel being stricken in 1967.

MAJOR SURFACE SHIPS

Ex-US BROOKLYN class *cruisers*

No	Name	Builder	Acquired	Fate
02	O'HIGGINS (ex-*Brooklyn*)	New York N Yd	1951	Decomm 1992, foundered on way to breakers
03	PRAT (ex-*Nashville*)	New York SB	1951	Decomm 1984, BU 1985

O'Higgins and *Prat* cost Chile $37m apiece – 10 per cent of the original cost. Both refitted in the US 1957–58. *O'Higgins* was damaged in a collision in 1974 and was laid up as an accommodation hulk; later repaired and recommissioned in 1978. Light AA comprised 28–40mm (6×4, 2×2) and 12–20mm (6×2), and the principal radars were SPS-6 and SPS-4. A Jet-Ranger helicopter was embarked.

O'Higgins 1981 Chilean Navy

Göta Lejon before transfer as the *Latorre* — Swedish Navy

Almirante Williams 1971 — C & S Taylor

Ex-Swedish TRE KRONOR class *cruiser*

No	Name	Builder	Acquired	Fate
04	LATORRE (ex-*Göta Lejon*)	Eriksbergs	15.7.71	Deleted 1984. Sold for scrap Aug 1986

Reconstructed 1951–52 and modernised 1958. The 6in guns were high angle, automatic anti-aircraft weapons with 70° elevation. AA armament comprised 4–57mm and 11–40mm (all single) and the main radars were Dutch LW-03 and British Types 277 and 293.

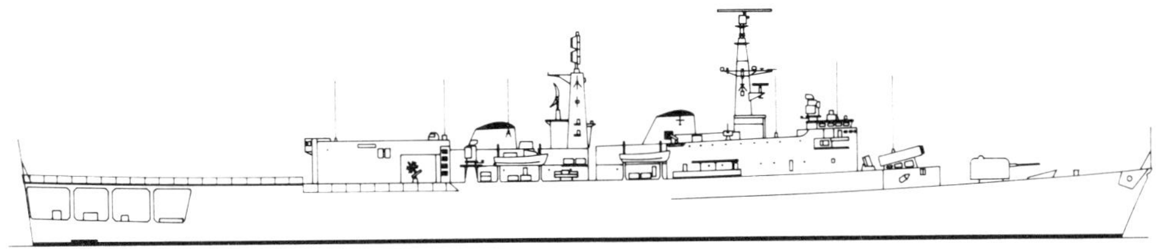

Blanco Encalada 1988

Ex-British COUNTY class *destroyers*

No	Name	Builder	Launched	Acquired	Fate
11	PRAT (ex-*Norfolk*)	Swan Hunter	16.11.67	6.4.82	Extant 1995
12	COCHRANE (ex-*Antrim*)	Fairfield	19.10.67	22.6.84	Extant 1995
14	LATORRE (ex-*Glamorgan*)	Vickers	9.7.64	3.10.86	Extant 1995
15	BLANCO ENCALADA (ex-*Fife*)	Fairfield	9.7.64	12.8.87	Extant 1995

All four destroyers purchased and transferred from United Kingdom. Completed extensive refits after transfer. All named after senior officers, however the official titles are not used. *Blanco Encalada* and *Cochrane* had their Seaslug missiles and associated fire control gear removed and were converted into helicopter carriers with a large landing deck and hangar capable of accommodating two Super Puma helicopters. Other two unconverted ships will serve as flagships. Seaslug Mk 2 SAM missiles are in use only in Chile which bought the remaining missiles from the UK in 1986. All of the class are to get a short range missile system, possibly the Israeli Barak I and new command, control and communications facilities as well as new ECM/ESM equipment and optronic directors. For more details see under United Kingdom.

Ex-US FLETCHER class *destroyers*

No	Name	Builder	Acquired	Fate
14	BLANCO ENCALADA (ex-*Wadleigh*)	Bath Iron Wks	1963	Deleted 1983
15	COCHRANE (ex-*Rooks*)	Seattle-Tacoma	1963	Deleted 1983

Armed with 4–5in (4×1), 6–3in/50 (3×2), 5–21in TT (1×5), two Hedgehogs, two torpedo racks, six K-guns. Initially three more *Fletchers* were to have been transferred in 1966. However, four *Charles Lawrence* class units were substituted.

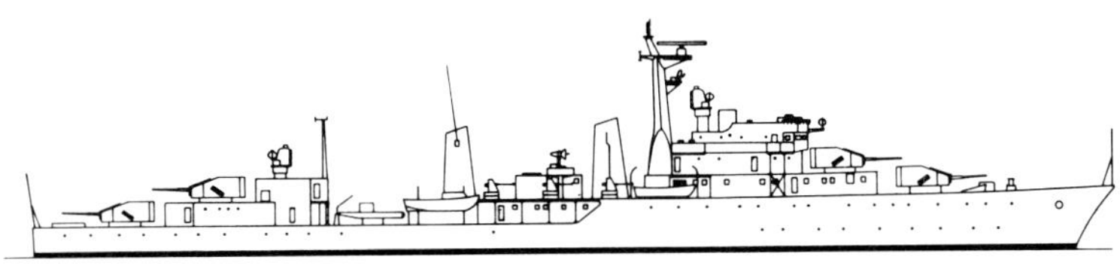

Almirante Williams as completed

ALMIRANTE class *destroyers*

Displacement: 2730t standard; 3300t full load
Dimensions: 402ft × 43ft × 13ft 4in
122.5m × 13.1m × 4.0m
Machinery: 2-shaft Parsons Pametrada geared turbines, 2 Babcock & Wilcox boilers, 54,000bhp = 34.5kts. Range 6000nm at 16kts
Armament: 4–4in/60 (4×1), 6–40mm/70 (6×1), 5–21in TT (5×1), 2 Squid ASW mortars
Sensors: Radar Plessey AWS-1 (see notes)
Complement: 266 (17 officers)

No	Name	Builder	Laid down	Launched	Comp	Fate
18	ALMIRANTE RIVEROS	Vickers, Barrow	12.4.57	12.12.58	31.12.60	Extant 1995
19	ALMIRANTE WILLIAMS	Vickers, Barrow	20.6.56	5.5.58	26.3.60	Extant 1995

Air-conditioned quarters. Two Seacat SAM (2×4) in lieu of 2–40mm fitted in 1964 at Talcahuano. Modernised in Britain between 1971–75 when four Exocet MM38 SSMs were fitted in place of the TT; 6–324mm Mk 32 ASW TT were fitted at the same time. Two of the Exocet SSMs were transferred to the *Allen M Sumner* class destroyers in 1980. Sensors now comprise Plessey AWS-1, Decca 629, Marconi SNW-10, SWW-20, SGR-102, SNG-20 with Dutch M-4 directors for Seacat. Sonar is Type 164B. The Vickers-designed main armament is unique; max range is 12,500 yards (11,400m) and max elevation is 75°. Like many Chilean warships, they are camouflaged.

CHILE

Ex-US ALLEN M SUMNER class *destroyers*

Both units received a FRAM II modernisation while in US service. VDS removed in 1980. Carried two Exocet SSMs between the funnels.

No	Name	Builder	Acquired	Fate
16	MINISTRO ZENTENO (ex-*Charles S Sperry*)	Federal, Kearny	8.1.74	Deleted 1990
17	MINISTRO PORTALES (ex-*Douglas H Fox*)	Todd-Pacific, Seattle	8.1.74	Deleted 1990

Ex-US CHARLES LAWRENCE class *destroyers escorts*

Riquelme was used for spare parts and was not commissioned. Constructed as destroyer escorts, these units were converted to high speed transports while in US service. Chile employed these ships as escorts. These four were transferred instead of three *Fletcher* class units. Only *Uribe* is active as a training ship.

No	Name	Builder	Acquired	Fate
27	ORELLA (ex-*Jack C Robinson*)	Dravo, Pittsburg	25.11.66	Deleted 1984
28	RIQUELME (ex-*Joseph E Campbell*)	Bethlehem, Hingham	1.12.66	See notes
26	SERRANO (ex-*Odum*)	Consolidated, Orange	25.11.66	Deleted 1984
29	VIRGILIO URIBE (ex-*Daniel Griffin*)	Bethlehem, Hingham	1.12.66	TS 1995

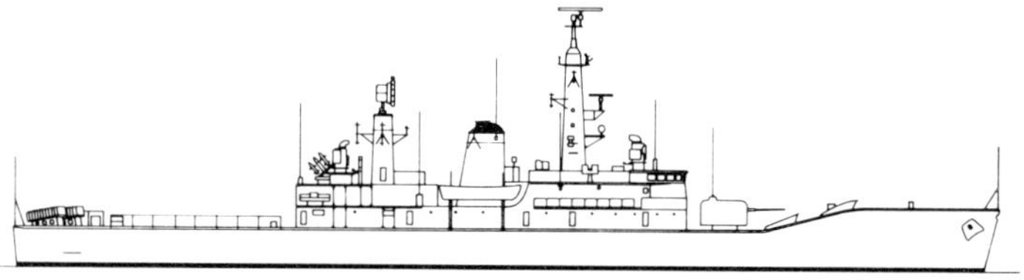

Condell 1977

British LEANDER type *frigates*

Initially *Condell* was to have been named *Latorre*. These Chilean *Leanders* have taller foremasts than their British near-sisters. Exocet MM38 (4×1) fitted in the late 1970s, is mounted on the fantail as opposed to replacing the 4.5in gun as in the British Exocet-equipped *Leanders*. These Chilean units do not have VDS or Limbo ASW mortar. *Ministro Zenteno* was acquired in 1990 and *General Baquedano* in 1992. See under United Kingdom for full details.

No	Name	Builder	Laid down	Launched	Comp	Fate
07	ALMIRANTE LYNCH	Yarrow, Scotstoun	6.12.71	6.12.72	25.5.74	Extant 1995
06	CONDELL	Yarrow, Scotstoun	5.6.71	12.6.72	21.12.73	Extant 1995
08	MINISTRO ZENTENO (ex-*Achilles*)	Yarrow, Scotstoun	1.12.67	21.11.68	9.7.70	Extant 1995
09	GENERAL BAQUEDANO (ex-*Ariadne*)	Yarrow, Scotstoun	1.11.69	10.9.71	10.2.73	Extant 1995

Lynch 1974 C & S Taylor

SUBMARINES

Ex-US BALAO class *submarines*

Simpson served until 1975 when she was decommissioned; she was reactivated in 1977. *Thomson* had an overhaul in the US in 1966 when a new conning tower was installed; reduced to a training hulk in late 1970s.

No	Name	Builder	Acquired	Fate
21	SIMPSON (ex-*Spot*)	Mare Island N Yd	1962	Deleted 1982
22	THOMSON (ex-*Springer*)	Mare Island N Yd	23.1.61	See notes

British OBERON class *submarines*

Both units were ordered in 1969. They suffered delays in fitting out due to recabling; see comments concerning *Oberons* in Brazilian service. *Hyatt* suffered a minor explosion in January 1966. Chile has developed the capacity to overhaul these units in her own yards. For further details see under United Kingdom.

No	Name	Builder	Laid down	Launched	Comp	Fate
23	HYATT (ex-*Condell*)	Scott-Lithgow	10.1.72	26.9.73	27.9.76	Extant 1995
22	O'BRIEN	Scott-Lithgow	17.1.71	21.12.72	15.4.76	Extant 1995

THOMSON class (209 Type 1300) *submarines*

Two submarines ordered in 1980 from the HDW yard at Kiel to the enlarged and improved Type 209 design. Two more were projected in 1988, but the plans were dropped. Maximum snorkelling speed is 12kts, diving depth 250m and have sail and masting larger than the other ships of this class to cope with wave size off Chilean coasts. Utilised components from cancelled Iranian order. Refitted in Chile in 1990–91. Currently a requirement for up to four new submarines of 1400t exists. First may be ordered from abroad and remainder from local ASMAR yard at Talcahuano.

Displacement: 1285t surfaced; 1390t submerged
Dimensions: 59.5m × 6.2m × 5.5m
Machinery: 1 shaft diesel-electric, 4 MTU gensets, 1760kW, 1 Siemens main electric motor, 4600shp, 11kts surfaced/21.5kts submerged. Fuel 110t. Range 16/400nm at 21.5/4kts submerged and 8200nm at 8kts
Armament: 8–533 TT (bow, 14 AEG SUT torpedoes)
Sensors: Radar Thomson-CSF Calypso II; sonar Krupp Atlas CSU 3
Complement: 31 (5 officers)

No	Name	Builder	Laid down	Launched	Comp	Fate
20	THOMSON	Howaldtswerke	1.11.80	28.10.82	31.7.84	Extant 1995
21	SIMPSON	Howaldtswerke	15.2.82	29.7.83	18.9.84	Extant 1995

AMPHIBIOUS WARFARE VESSELS

Ex-US LST type *tank landing ships*

Class (no, former name/no):
Comandante Hemmendinger (88, ex-*New London County*), *Comandante Araya* (89, ex-*Nye County*), *Aguila* (91, ex-*Aventinus*, ex-*LST 1092*), *Comandante Toro* (97, ex-*LST 277*).

Transferred in 1973. *Comandante Toro* stricken 1977, *Aguila* 1980 (after grounding).

Ex-US LSM type *landing ship*

Class (no, acquired):
Aspirante Morel (92, 2.9.60).

Ex-US *Aloto*, ex-*LSM 444* transferred to replace earlier LSM of the same name. Stricken 1977.

Ex-US LCU type *landing craft*

Class (no, acquired):
Grumete Diaz (96, Jun 1970).

Ex-US *LCU 1396*, previously designated an LCT (6). Stricken 1977.

French BATRAL type *landing ships*

No	Name	Builder	Comp	Fate
91	MAIPO	ASMAR, Talcahuano	1.1.82	Extant 1995
92	RANCAGUA	ASMAR, Talcahuano	8.8.88	Extant 1995
93	CHACABUCO	ASMAR, Talcahuano	15.4.86	Extant 1995

French design and French equipment, armed with 2–40mm and 1–81mm mortar. Laid down 1980. See under France for further details.

ELICURA class *landing craft*

Displacement: 290t light; 750t full load
Dimensions: 145ft × 34ft × 12ft 9in
44.2m × 10.4m × 3.9m
Machinery: 2 shafts, diesels, 900bhp = 10.5kts. Oil 77t. Range 2900nm at 9kts
Armament: 3–20mm
Complement: 20

No	Name	Builder	Comp	Fate
90	ELICURA	Talcahuano	10.12.68	Extant 1995
94	OROMPELLO	Dade Dry Dock Co, Miami	5.9.64	Extant 1995

Two further LCTs built in Chile are of similar design.

In 1995 the Chilean Navy also operated eleven ex-US LCVPs.

SMALL SURFACE COMBATANTS

Ex-US CHEROKEE class *patrol vessel*

No	Name	Builder	Acquired	Fate
63	SERGENTO ALDEA (ex-*Arikara*)	Charleston SB	1.7.71	Extant 1995

Ex-US fleet tug, built 1943.

Ex-Israeli SAAR 4 (RESHEV) class *fast attack craft (missile)*

Class (former name, transfer date):
Casma (ex-*Romach*, 1979), *Chipana* (ex-*Keshet*, 1979)

Both refitted before transfer. Pennant 30 and 31. They are stationed in the Beagle Channel. Harpoon missiles were removed prior to transfer. Both units extant 1995.

Ex-Israeli SAAR 3 class *fast attack craft (missile)*

Class (no, former name, transfer date):
Iquique (32, ex-*Hamit*, 1988), *Covadonga* (33, ex-*Hetz*, 1988).

Purchased in 1988 and transferred in 1989. Harpoon missiles removed prior the transfer. Acquired in lieu of local construction of missile boats and in lieu of additional units of larger *Saar 4* class. Both extant 1995.

GUACOLDA class *fast attack craft (torpedo)*

Displacement: 134t full load
Dimensions: 111ft 6in wl, 118ft 7in oa × 18ft 4in × 5ft 5in max
34.0m, 36.2in × 5.6m × 1.68m
Machinery: 2 shafts, 2 Mercedes-Benz MB 839Bd diesels. 4800bhp = 32kts. Range 1500nm at 15kts
Armament: 2–40mm (2×1), 4–21in TT (4×1) British Mk IV
Sensors: Radar Decca 505
Complement: 20

No	Name	Builder	Launched	Fate
81	FRESIA	Bazan, Cadiz	9.12.65	Extant 1995
82	GUACOLDA	Bazan, Cadiz	30.7.65	Extant 1995
83	QUIBORA	Bazan, Cadiz	1966	Extant 1995
84	TEGUALDA	Bazan, Cadiz	1966	Extant 1995

Lürssen designed craft, these units serve in the Beagle Channel. The TT are British Mk 4.

PAPUDO class *large patrol craft*

Displacement: 313t normal; 417t full load
Dimensions: 173ft 6in × 23ft × 10ft 2in
52.9m × 7.0m × 3.1m
Machinery: 2 shafts, 2 GM diesels, 2800bhp = 19kts. Range 5000nm at 18kts
Armament: 1–40mm, 4–20mm (2×2)
Complement: 69

Class (no): *Papudo* (P 37, ex-*PC 1646*), *Abatao* (P 36), *Pisagua* (P 38).

Basically similar to the US PC design (and the postwar Turkish 'Hisar' class), the name ship was built with US 'offshore' funds by ASMAR, Talcahuano, and was commissioned 27 November 1971. The two other craft were cancelled. *Papudo* extant 1995.

MARINERO FUENTEALBA class *large patrol craft*

Displacement: 215t
Dimensions: 80ft × 21ft × 9ft
24.4m × 6.4m × 2.7m
Machinery: 1 Cummins diesel, 340bhp = 9kts. Range 2600nm at 9kts
Armament: 1–20mm, 3–12.7mm MG
Complement: 19

No	Name	Builder	Comp	Fate
75	MARINERO FUEN-TEALBA	ASMAR, Talcahuano	22.7.66	Extant 1995
76	CABO ODGER	ASMAR, Talcahuano	21.4.67	Extant 1995

Fishing boat type hull. Operated by Coast Guard.

MICALVI class *large patrol craft*

Displacement: 518t full load
Dimensions: 149ft 3in × 27ft 10in × 9ft 6in
45.5m × 8.5m × 2.9m
Machinery: 2 shafts, 2 Caterpillar 3512TA diesels, 2560bhp = 15kts. Range 4200nm at 12kts
Armament: 1–40mm/60 Bofors, 2–20mm Oerlikon (2×1)
Complement: 23

No	Name	Builder	Comp	Fate
PSG 71	CONTRAMAESTRE MICALVI	ASMAR	30.3.93	Extant 1995
PSG 72	CONTRAMAESTRE ORTIZ	ASMAR	15.12.93	Extant 1995
PSG 73	ASPIRANTE ISAZA	ASMAR	30.1.94	Extant 1995
PSG 74	ASPIRANTE MOREL	ASMAR	Jun 1994	Extant 1995

A class of large patrol craft built under the design name *Taitao.* Multipurpose vessels with good seakeeping qualities for work in the Chilean EEZ. Lightly armed and fitted for service as a small transport. Can also service navigational aids. Fitted with 2.5t capacity crane on after deck. Can carry 35t of cargo in hold and 18t in containers on deck.

PILLAN class *coastal patrol craft*

Displacement: 43t full load
Dimensions: 61ft × 17ft 4in × 5ft 7in
18.6m × 5.3m × 1.7m
Machinery: 2 MTU diesels, 816bhp = 30kts. Range 700 at 15kts
Armament: 2–20mm, 2 DCs
Complement: ?

Class (no):
Pillan (GC 1801), *Tronador* (GC 1802), *Rano Kau* (GC 1803), *Villarrica* (GC 1804), *Corcovado* (GC 1805), *Llaima* (GC 1806), *Antuco* (GC 1807), *Osorno* (GC 1808), *Choshuenco* (GC 1809), *Copahue* (GC 1810)

Ten coastal patrol craft built by Maclaren, Niteroi in Brazil on GRP hulls. Ordered in 1977 and delivered in 1979–82. Belong to Coast Guard.

Ex-Israeli DABUR class *coastal patrol craft*

Class (no):
Grumete Diaz (1814), *Grumete Bolados* (1815), *Grumete Salinas* (1816), *Grumete Tellez* (1817), *Grumete Bravo* (1818), *Grumete Campos* (1819).

Class of six transferred from Israel and commissioned on 3 January 1991. Used for coastal patrol and anti-smuggling activities. Extant 1995.

There are also other patrol craft used by Coast Guard:
Two *Protector* class *Alacalufe* (LEP 1603) and *Hallef* (LEP 1604) built in 1989 by ASMAR, Talcahuano, with option for six more (107t full load, 32.2m, 5200bhp = 20kts).

Ona (LEP 1601) and *Yagan* (LEP 1602) built by Asenav and commissioned in 1980 (79t full load, 24.6m, 1300bhp = 22kts, 2–12.7mm MG).

Fifteen inshore patrol craft completed by Asenav in 1982–83 named *Maule, Lauca, Aconcagua, Rapel, Isluga, Loa, Maulin, Copiapo, Cau-Cau, Pudeto, Rio Rinihue, Chadmo, Caspana, Petrohue* and *Rio Bueno* (LPM 1901–1915, 14t full load, 1300bhp = 18 kts, 1–12.7mm MG).

China 1947–1949

At the end of the Second World War no one group ruled all of China. The official government, the Kuomintang (Nationalists) controlled by Generalissimo Chiang Kai-shek were based at Chungking on the Yangtze. They ruled over about half the land area, but this was mainly in the south-west of the country and did not include any of the major ports, which along with most of the rest of the coastline and the entire northeast were still in Japanese hands. The rest was dominated by the Communists under Mao Tse Tung. They were based in Yunnan but had troops and sympathisers scattered throughout the country.

The Japanese pulled out of Inner Mongolia, Manchuria, Taiwan (Formosa) and Hainan in September 1945, and with the help of the Americans the Kuomintang forces regained most of the major towns. However, the countryside was mainly Communist, and although the Kuomintang had several successes (the fragile truce between the Kuomintang and the Communists, dating back to 1936, had broken down well before the end of the Second World War), culminating in the taking of the Communist capital in Yunnan in March 1947, the Communists gradually gained the upper hand. China and the United States had signed a treaty of friendship on 4 November 1946, but American aid and military forces were gradually disengaged.

The Communists gained so much ground that on 1 September 1948 they proclaimed the North China People's Republic, and after a series of major battles they crossed the Yangtze into the Kuomintang's heartland in April 1949. On 16 July 1949 the Kuomintang commanders decided to withdraw to Taiwan and the other islands under their control. This massive evacuation was completed on 8 December. Meanwhile the People's Republic of China had been proclaimed at Peking on 1 October 1949.

All of the Chinese Navy's major warships had either been sunk or captured by the Japanese in 1937; the only ships to remain under the control of the Government were six gunboats and an MTB. Most of the latter remained at the Kuomintang capital of Chungking throughout the Second World War, and their numbers were swollen by one American and three British river gunboats that were also trapped at Chungking and which were transferred to the Chinese Navy in February 1942. The major prewar ships that had been taken over by the Japanese were all sunk by 1945, and the only vessels that were returned were two more gunboats.

Despite the political confusion in China, she was still thought of as a major power, especially by the Americans. It was therefore decided to build up her navy with British and American help. Britain supplied two frigates and several smaller craft, and a cruiser was purchased to act as the flagship. The United States provided rather more, including six destroyer-escorts, and large numbers of patrol boats, minesweepers and landing vessels. American aid to the Kuomintang forces temporarily ceased on 5 August 1949 though US vessels assisted with the evacuation.

China also claimed a number of Japanese warships as reparations. Because of the US desire to rebuild her navy and the justifiable claim that Japan had destroyed it, China obtained the lion's share of the surviving Japanese warships. Three destroyers, together with smaller numbers of destroyer-escorts and smaller craft, were transferred in 1946–47 after they had completed the repatriation of POWs and Japanese troops. In addition, a number of gunboats and other small warships were captured by Chinese forces in 1945, and these were also incorporated.

When the Chinese Navy was expanded after the Second World War it suffered badly from a lack of trained sailors among the 40,000 oficers and men. In addition, although the Allied ships were operational when they were handed over, most of the Japanese vessels, including all the larger ones, were disarmed at the end of the war (almost all in Fleet Strength table). The country was in such disorder that none of them was rearmed prior to the collapse of the Kuomintang Government. The surviving dockyards had neither the trained men nor the equipment to put things right. On top of all of this, some of the officers and many of the men were sympathetic to the Communists, and a number of the Chinese Navy's ships, including the only cruiser, defected to the Communists during the last stages of the Civil War. Many more were captured either because they could not be moved or because they were cut off by the crossing of the Yangtze in April 1949. Finally, several ships were lost by accident, and several more were sunk in action with Communist land forces.

FLEET STRENGTH 1947

Ex-Japanese DESTROYERS

Name	Disp (std)	Acquired	Fate
Minekaze class: launched 1922			
SHEN YANG (ex-*Namikaze*)	1345t	3.10.47	To Taiwan 1949, BU 1960s
Kagero class: launched 1939			
TAN YANG (ex-*Yukikaze*)	2050t	1947	To Taiwan 1949, BU 1971
Akitsuki class: launched 1944			
FEN YANG (ex-*Yoitsuki*)	3485t	29.8.47	To Taiwan 1949, BU 1963
Ex-Japanese escort destroyers			
Matsu class: launched 1944			
HENG YANG (ex-*Kaede*)	1262t	6.7.47	To Taiwan 1949, BU *c*1960
HUI YANG (ex-*Sugi*)	1262t	31.7.47	To Taiwan 1949, BU *c*1960
Tachibana class: launched 1944–45			
HSIN YANG (ex-*Hatsuyume*)	1289t	6.7.47	To Taiwan 1949, BU 1964
HUA YANG (ex-*Tsuta*)	1289t	31.7.47	To Taiwan 1949, BU *c*1950

Ex-US DESTROYER ESCORTS

Name	Disp (std)	Acquired	Fate
'GMT' class: launched 1942			
TAI KANG (ex-*Wyffels*)	1150t	Aug 1945	To Taiwan 1949, deleted 1972
TAI PING (ex-*Decker*)	1150t	Aug 1945	To Taiwan 1949, sunk 14.11.54

Ex-British Frigate

Name	Disp (std)	Acquired	Fate
'Flower' class			
FU PO (ex-*Petunia*)	1020t	Jan 1946	Sunk 19.3.47

Fu Po sank in collision with the merchant ship *Hai Ming* in the Formosa Strait. *Matsu* class rearmed after 1950 with ex-Japanese weapons by Taiwan. *Tai Ping* torpedoed and sunk by Communist torpedo boats off the Tachen Islands. *Hsin Yang* rearmed with 2–120mm (2×1), 8–25mm (4×2), 4–13mm, 36 DC after 1949. *Hua Yang* sailed to the Pescadores, was never re-armed and acted as a hulk. *Tai Kang* took Chiang Kai-shek on his second exploratory visit to Taiwan in July 1949. See under Taiwan for *Tan Yang*.

Ex-Japanese ESCORTS

Name	Disp (std)	Acquired	Fate
Etorofu (Type A) class: launched 1942–43			
KU AN (ex-*Oki*)	870t	29.8.47	To PRC 1949, stricken *c*1986
LIN AN (ex-*Tsushima*)	870t	31.7.47	To Taiwan 1949, deleted 1963
Ukuru (Type B) class: launched 1944			
HUI AN (ex-*Shisaka*)	940t	6.7.47	To PRC 1949, stricken 1986
Mikura (Type B) class: launched 1944			
WEI TAI (ex-*Yashiro*)	940t	29.8.47	To Taiwan 1949
Kaibokan (Type C) class: launched 1944–45			
MUKDEN (ex-*No 81*)	745t	1947	To PRC 1949, renamed *Shen Yang* and again, deleted *c*1980
CHI AN (ex-*No 85*)	745t	1947	To PRC 1949, renamed 1960s, deleted *c*1980
HSIN AN (ex-*No 205*)	745t	31.7.47	To Taiwan 1949, hulked *c*1960
CHAO AN (ex-*No 107*)	745t	21.8.47	To Taiwan 1949, deleted 1963
JUI AN (ex-*No 67*)	745t	June 1947	To Taiwan 1949, deleted 1963
LIAO HAI (ex-*No 215*)	745t	1945	To Taiwan 1949, hulked *c*1960
Kaibokan (Type D) class: launched 1944–45			
CHENG AN (ex-*No 40*)	740t	29.8.47	To Taiwan 1949, deleted 1963
CHIEH 5 (ex-*No 14*)	740t	6.6.47	To PRC 1949, renamed *Wu Jang*, stricken *c*1986
CHIEH 6 (ex-*No 194*)	740t	6.7.47	Renamed *Wei Hai* 1948, to PRC 1949, stricken *c*1986
CHIEH 12 (ex-*No 118*)	740t	31.7.47	To PRC 1949, renamed *Jangsha*, stricken *c*1986
CHIEH 14 (ex-*No 198*)	740t	1947	To PRC 1949, renamed *Shian*, stricken *c*1986
TAI AN (ex-*No 104*)	740t	29.8.47	To Taiwan 1949, deleted 1960
TUNG AN (ex-*No 192*)	740t	31.7.47	To Taiwan 1949, deleted 1960

Etorofu Type A class
Lin An rearmed after 1953 with 1–120mm, 1–76mm/50 and 8–25mm and fitted for minelaying. Reclassified as a frigate, deleted 1963. *Ku An* renamed by Communists – *Chang Pai* (230) and rearmed with 3–100mm (3×1) and 3–37mm (3×1) Soviet guns in 1950s, also fitted as minelayer. Extant with S Sea Fleet until stricken *c*1986.
Ukuru Type B class
Hui An not renamed by Communists. Rearmed in late 1960s or early 1970s with 2–120mm (2×1) and 6–12.7mm MG. In early 1980s was armed with 3–100mm (3×1) and 4–37mm (4×1) in E Sea Fleet. Stricken 1986.
Mikura (Type B) class
Wei Tai went to Taiwan in 1949, and was renamed *Hsueh Feng*. Rearmed in 1954 with 2–120mm (2×1) and 2–40mm (2×1) ex-Japanese guns – having been disarmed since August 1945. Renamed *Cheng An* in the late 1950s and deleted in 1963.
Kaibokan (Type C) class
Hsin An was rearmed with 2–76mm (2×1) and 6–25mm (6×1) but was wrecked on 26 September 1954 in Pescadores Islands. Laid up non-operational and renamed *Chang An*, and hulked *c*1960. *Chao An* was rearmed with 2–76mm (2×1) and 2–40mm (2×1). Deleted in 1963 as was *Jui An* which had been rearmed with 2–120mm (2×1), 1–76mm and 8 MG. *Liao Hai* was rearmed as in data and assigned to the army. She was hulked and renamed *Tsing Hai c*1960 or possibly earlier.
Kaibokan (Type D) class
Cheng An and *Tai An* rearmed with 2–120mm (2×1), 11–20mm and 2 MG. Served as gunboats until deletion in 1963. *Tung An* never rearmed. The four Communist units armed with 2–76mm US Mk 26 (2×1), 8–37mm (8×1), 4/8–25mm (4×1 or 4×2) in 1955, served until stricken *c*1986, two each with N and E Sea Fleets.

GUNBOATS

Name	Launched	Disp (std)	Fate
YI HSIEN (ex-*Yat Sen*, ex-*Atada*)	12.11.30	1520t	To Taiwan 1949, deleted 1960s
Chiang Chen class			
CHIANG YUAN	1905	550t	To PRC 1949, deleted 1963
Chu class			
CHU KUAN	1907	740t	To Taiwan 1949, deleted 1960s
CHU TUNG	1906	740t	To PRC 1949, deleted 1960s
Ming Cheun class			
MING CHEUN	1929	420t	To PRC 1949, memorial *c*1984
Yung Chien class			
YUNG CHI (ex-*Hsai Hsing*, ex-*Asuga*, ex-*Yung Chi*)	1915	860t	To PRC 1949, deleted *c*1970
Yung Feng class			
YUNG HSIANG	1912	830t	To Taiwan 1949, deleted 1950s
Yung Sui class			
YUNG SUI	1929	650t	To PRC 1949, deleted 1970s

These were the only large ships left of the pre-1937 Navy. The sloop size *Yi Hsien* was the prewar *Yat Sen* retaken from the Japanese at Shanghai 25 August 1946. In the early 1950s her Japanese poop deck was removed and she was refitted as a TS which she remained until deleted. Her post-1946 armament was 1–120mm, 1–76mm, 4–40mm (2×2), 5–25mm and 6–7.7mm MG. Radar and sonar were fitted. *Ming Cheun* was first renamed *Hsien Ning* then *Chang Chiang* by the Communist Navy. *Yung Chi* and *Yung Sui* were sunk by Communist forces in the Yangtze in April 1949 and salved by them.

RIVER GUNBOATS

Name	Launched	Disp (std)	Fate
Ex-British Peterel class			
YING SHAN (ex-*Gannet*)	10.11.27	310t	To PRC 1949, deleted 1960s
Falcon class			
YING TEH (ex-*Lung Huan*, ex-*Falcon*)	18.5.31	354t	To PRC 1949, deleted 1960s
Sandpiper class			
YING HAO (ex-*Sandpiper*)	9.6.33	185t	To PRC 1949, deleted 1960s
Ex-US Wake class			
TAI YUAN (ex-*Tatara*, ex-*Wake*, ex-*Guam*)	28.5.27	370t	To PRC 1949, deleted 1960s
MEI YUAN (ex-*Tutuila*)	14.6.27	370t	To PRC 1949, deleted 1960s

All except *Tai Yuan* were transferred to the Chinese Navy by the RN and USN at Chungking in February 1942. *Tai Yuan* had been captured by the Japanese at Shanghai on 8 December 1941 and renamed *Tatara*; she was handed back in 1946. *Lung Huan* was renamed *Ying Teh* in 1948.

There were also *Wei Ning*, a 400t customs vessel launched in 1934, captured by the Japanese in 1937 and renamed *Bunsei*; she was recaptured by the Chinese in 1945, given back her original name, and used as a patrol boat. She was sunk in 1949 and salved by the Communists, being deleted in the 1950s. *Ku Ai 102* was a German-built 60t MTB, launched in 1937, taken over by the Communists in 1949 and deleted in the 1960s

Ex-Japanese SUBMARINE CHASERS

Name	Acquired	Disp (std)	Fate
Ch 28 class: launched 1943			
YAI LUNG (ex-*Ch 49*)	3.10.47	420t	To Taiwan 1949, renamed *Hai Hung* then *Chu Chiang*, deleted 1964
Cha 1 class			
KAO MING (ex-*Cha 220*)	Aug 1945	130t	To PRC 1949
KUANG KUO (ex-*Cha 223*)	1947	130t	To PRC 1949
Ch 7 class: launched 1938			
FU LING (ex-*Ch 9*)	3.10.47	291t	*Hai Ta* 1948, to Taiwan renamed *Min Chiang*, deleted 1960s

Ex-Japanese COASTAL MINESWEEPERS

Name	Acquired	Disp (std)	Fate
Natsushima class: launched 1914			
CHIEH 29 (ex-*Kuroshima*)	14.11.47	450t	To Taiwan 1949, BU *c*1960
Wa 1 class: launched 1942–43			
No 200 (ex-*Wa 4*)	3.10.47	215t	To PRC 1949, deleted 1976
No 201 (ex-*Wa 14*)	3.10.47	215t	To PRC 1949, deleted 1976
No 202 (ex-*Wa 19*)	3.10.47	215t	To Taiwan 1949, renamed *Chiang Yung*, deleted *c*1970
No 203 (ex-*Wa 22*)	3.10.47	215t	To Taiwan 1949, renamed *Chiang*, deleted *c*1970

Ex-Italian MINELAYER

Name	Acquired	Disp (std)	Fate
Ostia class: launched 1927			
HSIEN NING (ex-*Okitsu*, ex-*Lepanto*)	1945	615t	To Taiwan 1949, deleted 1956

Ex-Axis RIVER GUNBOATS

Name	Acquired	Disp (std)	Fate
Italian: launched 1918			
KIANG KUN (ex-*Narumi*, ex-*Ermanno Carlotto*)	1946	180t	To PRC 1949, deleted 1960s
Japanese: launched 1911			
YANG CH'I (ex-*Toba*)	1945	250t	To PRC 1949, deleted 1960s
Japanese: launched 1922			
AN TUNG (ex-*Ataka*)	1945	725t	To PRC 1949, deleted 1970s
CHANG TEH (ex-*Seta*)	1945	305t	To PRC 1949, deleted 1960s
Atami class: launched 1929			
YUNG PING (ex-*Atami*)	1945	206t	To PRC 1949, deleted 1960s
YUNG AN (ex-*Futami*)	1945	206t	To PRC 1949, deleted 1960s
Fushima class: launched 1939			
KIANG SHIH (ex-*Fushima*)	1946	320t	To PRC 1949, renamed *Chiang Feng*, deleted 1960s then *Fu Chiang*
NAN CHANG (ex-*Sumida*)	1946	320t	To PRC 1949, deleted 1960s
Ex-Portuguese Macau class: launched 1909			
WU FENG (ex-*Maiko*, ex-*Macau*)	1946	95t	To PRC 1949, deleted 1960s
Hashidate class: launched 1939			
CHANG CHI (ex-*Uji*)	1945	999t	To PRC 1949, deleted 1986

Hashidate class
Chang Chi renamed *Nan Chang* (53/224) and rearmed in 1950s with 2–130mm (2×1) *Gordyy* class destroyer guns and 6–37mm (6×1). Served with E Sea Fleet until stricken *c*1986.

Ex-Japanese LANDING SHIP

Name	Acquired	Disp (std)	Fate
T 103 class: launched 1944			
LU SHAN (ex-*T172*)	3.10.47	887t	To Taiwan 1949, wrecked and BU *c*1954–55

Ex-Japanese MINELAYER

Name	Acquired	Disp (std)	Fate
Hirashima class: launched 1941			
YUNG CHING (ex-*Saishu*)	3.10.47	720t	To Taiwan 1949, BU 1964

Hirashima class
Yung Ching was soon rearmed with 1–105mm, 1–40mm, 7–25mm, 1–20mm and 120 mines. At first classified as light minelayer, as before, but by 1953 reclassified as a frigate. This indicated the shortage of suitable frigates in the early 1950s.

MAJOR SURFACE SHIPS

Ex-British ARETHUSA class *cruiser*

Name	Builder	Launched	Acquired	Fate
CHUNG KING (ex-*Aurora*)	Portsmouth DY	20.8.36	19.5.48	To PRC 1949

The ex-RN cruiser *Aurora* was sold to China, fetched by a crew of 682, and defected to the Communists in Shanghai on 2 March 1949 when she was renamed *Huang Ho*. The Nationalists claimed to have sunk her at Taku with bombers but she was salved and recommissioned by late 1951 undergoing another name change (allegedly) *Pei Ching*. China's largest warship was laid up *c*1954. BU in 1960s.

Ex-British 'HUNT' class (Type 1) *frigate*

Name	Builder	Launched	Acquired	Fate
LIN FU (ex-*Mendip*)	Swan Hunter, Wallsend	9.4.40	19.5.48	Returned to RN 27.5.49

Former RN wartime escort destroyer, with Type 285 and 290 radar, rerated AA frigate (F 160) in 1947, transferred to China for eight years but returned to RN at Hong Kong due to the Nationalist collapse on the mainland. Then transferred to Egypt as *Mohamed Ali El Kebir*. See under Egypt.

Ex-US 'DET' class *destroyer escorts*

Class (no, former name/no):
Tai Chao (ex-*Carter*, DE 112), *Tai Ho* (ex-*Thomas*, DE 102), *Tai Hu* (ex-*Breeman*, DE 104), *Tai Tsang* (25, ex-*Bostwick*, DE 103)

Four ex-USN 'long hull' destroyer escorts, all handed over on 31 December 1948. All sailed to Taiwan in 1949. *Tai Chao* and *Tai Tsang* deleted in 1973, *Tai Ho* in 1972 and *Tai Hu* in 1975.

AMPHIBIOUS WARFARE VESSELS

Ex-US LST class *tank landing ships*

At least thirty were transferred from the USN to China during 1946–48; eleven went to Taiwan including *Chung Shu* (ex-LST 520), and sixteen to the Communists. Thirteen extant 1995. Some are used as stationary accommodation or submarine depot ships. Deployments in 1995: *355*, *361* South Sea Fleet, *901–907* North Sea Fleet and *921–926* East Sea Fleet. In addition, the repair ship *Hsing An* (ex-USS *Achilles*, APL 41, ex-*LST 455*) was transferred in September 1947. She became the Communist *Taku Shan*, after salvage following grounding and fire in 1949. Taiwan units rebuilt late 1960s and thirteen were extant 1995. See under Taiwan. Two mainland LSTs transferred to N Vietnam as tankers, others used as merchantmen or fitted with RL.

Ex-US LSM class *medium landing ships*

About thirty were transferred from the USN to China during 1946–48; fifteen went to the Communists and eleven to Taiwan. *Mei Lo* (ex-*LSM 157*) damaged by Communist gunfire and beached on Quemoy Island, 8 September 1958. All fifteen mainland (some merchantmen and most minelayers) had been deleted by 1993 and two of the Taiwan LSMs were extant in 1982. Mainland deployments in 1982: *394–396* (N Sea), *511* and *931–936* (E Sea), *352–354* and *393* (S Sea).

Ex-US LSMs at Shanghai 1984 Courtesy Ross Gillett

Ex-US LCT (5) and LCT (6) *tank landing craft*

More than fifteen of these two classes were transferred from the USN during 1946–48. Built during 1942–43. The LCT(6) were modified versions of the LCT(5) with the bridge offset to starboard and bow and stern loading. Ten or eleven went to the Communists and six to Taiwan. The six Taiwanese units were extant 1995, but the Chinese vessels were all deleted *c*1986.

Ex-US LCI class *landing craft*

About twenty-five USN LCI of various types, including at least seven LCI(G) and an LCI(M) were transferred to China during 1946–48. All built 1943–45. Fourteen, including at least three LCI(G), went to the People's Republic (some fitted with RL or minesweeping gear) and five to Taiwan, including an LCI(M) and four LCI(G). Four mainland ships were extant in 1982 but have since been deleted.

SMALL SURFACE COMBATANTS

Ex-US PC (173ft) class *submarine chasers*

Class (former no):
Chien Fang (ex-*PC 1549*), *Huang Pu* (ex-*PC 492*), *Hung Tse* (ex-*PGM 26*, ex-*PC 1557*), *Kung Kiang* (ex-*PC 1233*), *Kan Tang* (ex-*PGM 15*, ex-*PC 1091*), *Chialing* (ex-*PC 1247*), *Pao Ying* (ex-*PGM 20*, ex-*PC 1551*), *Tung Ting* (ex-*PGM 13*, ex-*PC 1089*), *Wu Sung* (ex-*PC 490*), *?* (ex-*PC 593*), *?* (ex-*PC 595*), *?* (ex-*PC 598*), *?* (ex-*PGM 12*, ex-*PC 1088*), *?* (ex-*PGM 14*, ex-*PC 1090*)

Fourteen transferred to the Chinese Navy between 1946 and 1949, six of which had been rebuilt 1944–45 as PGM type gunboats. The PGMs had 1–60mm mortar and RLs in addition to guns, and the PCs had differing numbers of light AA guns. The first eleven except *Kan Tang* went to Taiwan in 1949, and the other four were taken over by the Communists. Three were deleted in the 1970s but one was still in service in 1982 as a S Sea Fleet training ship rearmed with 2–76mm (2 × 1) and 3–37mm (3 × 1) but has since been stricken. Taiwan fates and numbers were as follows: *Chien Fang* and *Wu Sung* deleted 1951–52, ex-*PC 598* deleted 1954, *Tung Ting* sunk by Communist torpedo boats 10 January 1955, *Pao Ying* renamed *Ying Chiang* so badly torpedoed 20 January 1955 that BU, ex-*PC 593* deleted 1955, *Hung Tse* and *Chialing* (renamed *To Kiang*, 104) deleted 1964, *Kung Kiang* (113) and *Huang Pu* (105, renamed *Fu Kiang*) deleted *c*1970.

Ex-US SC (110ft) class *submarine chasers*

Ten transferred 1946–49, ex-*SC 518*, ex-*SC 637*, ex-*SC 648*, ex-*SC 698* (renamed *Chu Chien 103*), ex-*SC 703*, ex-*SC 708* (*Chu Chien*, *102*), ex-*SC 722*, ex-*SC 722*, ex-*SC 723*, ex-*SC 735*. All sailed to Taiwan and all deleted *c*1970 except *101 Chu Chien* *c*1955.

Ex-British *harbour defence motor launches*

Eight ex-RN HDMLs were transferred to China after the Second World War. They were named *Fang 1–8* (ex-*ML 1033*, *1047*, *1058*, *1059*, *1068*, *1390*, *1405* and *1406* respectively). They went to Taiwan in 1949. *Fang 3* and *Fang 8* were sunk in 1950 and the remainder were deleted *c*1970.

MINE WARFARE VESSELS

Ex-US ADMIRABLE class *minesweepers*

Class (former name/no, transferred):
Yung Ning (ex-*Magnet*/AM 260, Aug 1945), *Yung Sheng* (ex-*Lance*/AM 257, Aug 1945), *Yung Shun* (ex-*Logic*/AM 258, Aug 1945), *Yung Ting* (ex-*Lucid*/AM 259, Aug 1945), *Yung Chun* (ex-*Gavia*/AM 363, May 1946), *Yung Feng* (ex-*Prime*/AM 279, May 1946), *Yung Ho* (ex-*Delegate*/AM 217, May 1946), *Yung Hsing* (ex-*Embattle*/AM 226, May 1946), *Yung Kang* (ex-*Elusive*/AM 225), *Yung Chang* (ex-*Refresh*/AM 287, Jun 1948), *Yung Chia* (ex-*Implicit*/AM 246, Jun 1948), *Yung Hsiu* (ex-*Pinnacle*/AM 274, Jun 1948), *?* (ex-*Nimble*/AM 266, Jun 1948), *?* (ex-*Reform*/AM 286, Jun 1948), *?* (ex-*Phantom*/AM 273, Jun 1948), *Yung Shou* (ex-*Pivot*/AM 276, Aug 1948)

Last two PCE type gunboats. All sixteen sailed to Taiwan 1949. Three, ? ex-USS *Nimble* (AM 266), ? ex-USS *Reform* (AM 286) and ? ex-USS *Phantom* (AM 273) were deleted almost immediately. Eight remained as minesweepers. Of these, one, *Yung Ting*, ex-USS *Lucid* (AM 259) was renamed *Yung Ming* in 1964 and later converted into a survey ship. She was deleted in 1976. Two more, *Yung Hsing* (ex-USS *Embattle*/AM 226) and *Yung Chia* (ex-USS *Implicit*/AM 246) were converted into coastguard vessels (early 1970s) and were stricken in *c*1986. The former was renamed *Hung Hsing*. Of the five other minesweepers, *Yung Ning* (ex-USS *Magnet*/AM 260) was deleted in 1963 and *Yung Hsiu* (ex-USS *Pinnacle*/AM 274), *Yung Sheng* (ex-USS *Lance*/AM 257), *Yung Shou* (ex-USS *Pivot*/AM 276) and *Yung Shun* (ex-USS *Logic*/AM 258) were deleted in the early 1970s. One, *Yung Feng* (ex-USS *Prime*/AM 279) was converted into a coastal minelayer. She was deleted *c*1970. The other four were converted into gunboats with 6–40mm (3 × 2) and 1–76mm. *Yung Chang* (ex-USS *Refresh*/AM 287) was sunk by a Communist escort off the South China coast on 14 November 1965.

Yung Ho (ex-USS *Delegate*/AM 217) and *Yung Kang* (ex-USS *Elusive*/AM 225) were scrapped in 1964 and *Yung Chun* (ex-USS *Gavia*/AM 363) was deleted *c*1970. The two ex-USN PCE class gunboats (which differ from the *Admirable* class mainly by not having minesweeping gear aft) were fitted with a second 76mm gun aft in 1955. *Yung Tai* (ex-USS *PCE 867*) was damaged in action with Communist escorts off the South China coast on 14 November 1965 and was deleted soon after. *Yung Hsiang* (ex-USS *PCE 869*) was renamed *Wei Yuan*, and was deleted *c*1970.

Ex-US YMS type *motor minesweepers*

Five ex-USN units transferred to the Chinese in 1948. Ex-*YMS 339*, *346*, *367* and *393* were built in 1942–43 for the USN, but ex-*YMS 2017* (ex-*YMS 17*) had served with the RN 1943–45. All were taken over by the People's Republic in 1949; ex-*YMS 339* was deleted in 1963 and the remainder in 1976.

China

PEOPLE'S REPUBLIC

Two connected forces have driven China during this century: the desperate need to create a modern society, and the need to reverse the long history of foreign domination. The first need caused a series of revolutions culminating in the current People's Republic. The second has often determined the legitimacy of the Chinese government. The People's Republic in particular felt the tensions between the two imperatives. To Mao and his colleagues, the triumph of communism was the triumph of a modern social order over the corrupt past. Mao's view was that once the social order had been reformed, the other elements of modernisation, such as the creation of heavy industry, would inevitably follow. The rub was that, virtually everywhere else, communism had triumphed in the wake of the Soviet Red Army, as an extension of the power of the Soviet Union. Mao was as much a product of China and its continuing revolution as of Communist ideology. He was definitely not Stalin's man. Indeed, he was well aware that, in China as in Europe, Stalin had been quite happy to sacrifice local Communists to further Soviet foreign policy.

Mao's ideology told him that the only true enemy was the capitalist West, which after all had long exploited his country. Although the Western powers had given up their special rights (*eg*, in ports such as Shanghai) during the Second World War, Mao could hope to eject the West from other Asian colonies (which had been in the historic Chinese sphere of influence). On the other hand, Mao's ideology did not explain the fact that Stalin was running a Russian, rather than an internationalist, policy, and that despite its Communist trappings Stalin's Russia continued to harbour ambitions in China. For example, as recently as 1944 Stalin had tried to annex the large Chinese province of Sinkiang. As for the Chinese revolution, Stalin had, as usual, been more interested in controlling it than in supporting it. He had expected the Kuomintang to win after the war, and indeed had supported them until Mao's victory seemed nearly inevitable. Only then had Mao enjoyed much Soviet support.

Mao went to Moscow in 1950 to negotiate a treaty with Stalin, in the knowledge that only the Russians would provide him with the support he so desperately needed, but also in the knowledge that they were quite as rapacious as the Western enemy he more openly attacked. However, Mao was also painfully aware of the weakness of his own position. His forces had probably nearly exhausted the weapons they had captured during the Civil War. In February 1950 he signed a 30-year mutual assistance treaty with Stalin. Within the year his depleted forces would intervene in Korea, suffering terrible losses in support of Stalin's Korean puppet.

In January 1950 Mao's navy consisted of Kuomintang ships which had either defected or which had been captured in port or on the Yangtze: a cruiser, seven frigates, twenty-one smaller escorts, five minesweepers, thirteen river gunboats, six patrol boats, and about sixty landing craft. Few were in a usable condition. Many ex-Japanese ships had not yet been rearmed. Furthermore, the retreating Kuomintang did their best to sink defecting or captured ships; their B-24 bombers managed to sink the one cruiser soon after her defection. The Communists also managed to acquire some British and Canadian escorts which had been demilitarised as small freighters after the Second World War. By 1954, Mao had enough small attack craft to contest the waters of the Formosa Strait, and by 1958 he had a big enough navy to attempt, unsuccessfully, to seize Quemoy.

For Mao, victory was incomplete in 1950. The Kuomintang enemy retained Taiwan and several offshore islands, such as Quemoy. Mao did seize Hainan in 1950. Through the next decades, the two Chinese navies, Communist and Nationalist, would clash periodically. With the outbreak of the Korean War, the US Navy began to operate in the Formosa Strait to preclude any seaborne invasion of Taiwan. However, US policy was also to convince those on Taiwan not to seek an early return match on the mainland.

With the outbreak of war in Korea, and particularly with Chinese intervention, it became impossible for the US government to reconcile with the People's Republic. Trade and recognition were both blocked, though by the late 1950s the European powers had all recognised the People's Republic.

When Stalin died in 1953, Mao pointed out that he, rather than any Russian of the new generation, was the natural leader of the world revolution. The Russians' shocked reaction must not really have surprised Mao: the Russians saw world revolution mainly as a way of winning a Russian war against the West. They were not interested in an ideological crusade for its own sake. However, their power rested at least partly on leadership in a world revolutionary movement, and they could not afford to lose that leadership.

Furthermore, it cannot have escaped Mao that, despite much talk about committing forces in Korea in 1950, the Soviets let the Chinese fight for them, at a very high human and material cost. Later, one reason the Soviets had to provide so much support to the Vietnamese was fear that the Chinese would exploit any failure to do so as proof that the Soviets had abandoned the world revolution (which the Chinese could then taken over).

The beginning of the Sino-Soviet split was probably the Soviet proposal, in 1954, that China should contribute her fledgling navy to a joint Sino-Soviet Pacific Fleet run by the Soviets. The problem was papered over at the time, and Soviet material aid continued. Its initial fruits were Soviet warships, mainly built prewar: four *Gordyy* class destroyers, thirteen submarines, eight large patrol craft, two minesweepers, and fifty-five 'P 4' and twelve 'P 6' class MTBs. Then the Soviets provided parts for more modern surface ships and submarines, which were assembled in China. Finally the Chinese began to build these classes for themselves, eventually adapting them to new weapons, some of them of Western origin. These Soviet designs still form the great bulk of the Chinese fleet. Probably the last examples of Soviet naval aid were small numbers of Styx missiles, which the Chinese eventually adapted as Silkworm.

By about 1957 the Chinese wanted something much more important: assistance in developing nuclear and thermonuclear weapons, the keys to true great-power status. There was already considerable political friction. In 1956 Khrushchev denounced Stalin and his 'cult of personality'. To Mao, who had his own 'cult of personality', and who had also ordered mass murders like Stalin's, this was a personal, if slightly indirect, attack. Mao also feared that Khrushchev's policy of peaceful coexistence with the West was the beginning of some sort of united front against China. By 1960, Khrushchev had withdrawn Soviet technicians and was publicly attacking Mao. His successors found that the split was not, as they had imagined, merely a matter of Khrushchev's personality and blunders. The logic of great power politics was at work: Mao really was more enemy than friend. Khrushchev provided parts for a single 'Golf' class submarine and perhaps for its missiles, but no warheads. After 1960 all military assistance was cut off, and Soviet specialists left China.

Ultimately, the logic of the Chinese Revolution required any Chinese government to try to take back what had been seized over the long centuries of Chinese weakness. Russia had been by far the most rapacious of the foreign powers, and the Chinese recalled a long history of 'unequal treaties'. By the mid-1960s, they demanded that the Russians admit their historical crimes and 'rectify' the border. By 1969 the two powers were clashing on the Amur River (problems there justified a wave of new construction of Soviet river gunboats). The Russians began hinting to the United States that they were interested in a pre-emptive nuclear strike against the nascent Chinese nuclear industry and forces. Apparently the United States refused to acquiesce; among other things it pointed out that one consequence of any such attack would probably be devastating fallout on Japan.

By the mid 1950s, then, Mao was keenly aware that he could not depend on the Russians to help modernise his country. He had won his revolution very largely with home-grown resources, and he imagined, it seems, that sufficient ideological fervour would make industrialisation possible. In 1958, then, he announced the 'great leap forward': China would pull herself up by her bootstraps. As in Russia a few decades before, those responsible for the 'leap' told Mao what he wanted to hear: inspired by him, the peole had accomplished what experts said could not be done. As in Russia, ideology could not change the facts of life, and the 'great leap' largely failed. It also caused a devastating famine, since effort that should have gone into agriculture went instead into primitive industrial work (for example, into backyard blast furnaces to make pig iron).

Even had it worked as advertised, the leap would not have given China the sort of industrial base needed to develop modern armed services. Mao was forced to chose priorities. He chose to concentrate on the one decisive weapon, the only one the Russians would have to respect: the atomic, and then the hydrogen, bomb.

By the early 1960s Mao was in an increasingly uncomfortable position. He had no reliable foreign partner, and the industrial base assembled with Russian help was relatively weak. The failure of the 'leap' had greatly strengthened the hands of experts and bureaucrats. Mao himself was losing control. On the other hand, in 1965 his nuclear team succeeded in exploding a nuclear bomb. He now had a deterrent, however feeble. Whatever weakness he had to accept in non-nuclear forces, no foreign power would lightly intervene in Chinese affairs. Indeed, Mao was happy to say that, leading the most populous country on earth, he alone could afford to fight a nuclear war.

Now Mao could take on the experts and bureaucrats he blamed for the failure of the 'great leap'. He initiated a cultural revolution, in which education itself became a sin. Again, the result was disastrous. China lost a generation of trained people, because the schools closed and the teachers were sent to work in the countryside (or killed off). Progress in developing weapons, such as naval missiles and torpedoes, effectively stopped, because as the revolution grew the military design teams could not be altogether insulated from it. In many cases industry could not produce what was needed. For example, the quality of metallurgy in diesel engines declined drastically (it was reported that the engines of the submarines sold to Egypt wore out during the delivery voyage).

From the point of view of the navy, the combination of the Sino-Soviet split and the cultural revolution virtually froze technology at the level of the mid-1950s. In some key categories, such as torpedoes, production of even obsolescent equipment could not begin until the 1970s.

Just as China was a key player in, and much affected by, the Second World War, the cold war greatly shaped her postwar history. In 1949–50 the US position towards Mao was determined in large part by fear of Soviet expansion; it was assumed that Mao was part of that assault on the West. The 'loss of China' became an important issue in the United States, and internal politics (largely General Douglas McArthur's personal ambitions, presumably for the 1952 Presidential Nomination) probably led McArthur to proclaim in 1950 that the Korean War was not merely a fight against the North Korean attack on the South but also a fight against communism in Asia. Given that proclamation, the Communist Chinese entered the war as United Nations forces moved deep into North Korea. They might well have done so in any case, since the defeat of North Korea might have undermined the new Chinese regime, but the events in Korea certainly cemented Mao's hostility towards the West, and thus his alliance with the Soviets.

By the mid-1950s Mao was competing with the Soviets in revolutionary fervour. He, rather than the Russians, could claim credit for helping Ho Chih Minh to victory over the French in 1954. Similarly, the Malayan revolutionaries (1948–60) were Chinese-inspired. Mao's most spectacular coup was gaining leverage over the Sukarno regime in Indonesia. At the same time, the Chinese could present themselves to emerging Third World countries, for example in Africa, as a more realistic model for development than the Russians. They had come from a position much closer to those their clients occupied.

By this time it was clear to the Soviet leadership that, with the advent of nuclear weapons on both sides, direct military action in Europe was unlikely to be profitable. That did not end their interest in assaulting the West. Khrushchev pointed out that the West remained vulnerable in the Third World, from which it obtained vital raw materials. Leadership in the world revolutionary movement therefore offered the Russians valuable leverage against the West. Conversely, losing that leadership to the Chinese was doubly bad. First, it would mean the loss of the Soviets' best weapon and secondly, history suggested that the Chinese would eventually turn on the Russians, remembering their own historical crimes. From about 1960 on, Russians began to talk about the gloomy prospect of war against the human mass Mao controlled.

A decade after Korea, the United States intervened in Vietnam partly for domestic reasons: the Kennedy Administration feared that it would suffer the political effects of 'losing Vietnam', particularly since it expected to face a hawkish Republican challenger in 1964. There was also a real fear that Vietnam would be the first of a string of successes for a Sino-Soviet policy of backing Third World revolutionaries in crucial places such as Indonesia. From a Chinese perspective, the Vietnamese Communists obviously had to be supported, and in this case that meant a sort of truce with the Russians (the Russians later complained that the Chinese had used the occasion to seize arms shipments passing through China by rail).

For the United States, China was never the main cold war issue. The question was always how to contain the Soviets. By the early 1950s, some US experts knew that whatever alliance the Russians had built with the Chinese could not last; Mao had made his own revolution, and he would never function for long as Stalin's puppet. Therefore there was always a possibility that conflict between China and the Soviet Union could be exploited. Some argued that continued pressure on both countries would tend to exacerbate the tension between them, rather than welding them together. The memory of Korea, and the Democrats' fear of Republican charges, made any approach to China impossible until the 1970s.

When Richard Nixon, a Republican, was inaugurated in 1969, the United States was still deeply engaged in Vietnam against forces supported by the Chinese. However, Nixon realized that the United States could benefit from the intensifying hostility between China and the Soviet Union. As the Vietnam War wound down in 1971, he opened relations with Mao. The deepening relationship between the United States and the People's Republic was cemented by the Soviet invasion of Afghanistan, which the Chinese viewed as evidence of Soviet aggressiveness ('hegemonism'). They feared that a Soviet victory in Afghanistan would destabilise other southwest Asian states, such as Pakistan.

By that time Mao, and his cultural revolution, had died. Mao's successor, Deng Xiaoping, took a far more pragmatic approach to the two linked questions of the Chinese revolution. He proclaimed the 'Four Modernisations,' which he hoped would finally realize China's enormous potential. He also made a radical announcement: the object was to modernise (which included 'getting rich'), and the way to that object was less important. Capitalism was no longer an enemy; it was merely an alternative way to the same goal. China was strong enough that she could afford to open herself to Westerners and their industry.

In the list of the four vital modernisations, building up the Chinese military machine came last, not first. Deng had drawn a valuable lesson from the Soviet experience: the Soviets had put military modernisation first, and had been unable to keep up with Western powers for which military production was not the main priority. The West had been able to draw on a very creative civilian sector to produce military innovation. The Soviets had enjoyed no such resource, and had suffered accordingly.

It was not that Deng doubted the importance of military power. Quite the opposite: he faced a real threat in the form of the Soviets. However, the opening to the United States made it far less likely that the Soviets would strike. Moreover, the United States would want to help build up Chinese military power, and that might make modernisation much easier. During the 1980s China was able to import considerable Western hardware, mainly French and Italian, generally in small quantities. Production licenses were discussed but they generally were not bought. However, it is clear that considerable reverse-engineering took place. The Chinese were fortunate to be entering the naval market at this time, just before systems came to be dominated by software control. It is far easier to reverse-engineer hardware than to tease proprietary software out of the chips into which it has been embedded.

United States interest in naval cooperation rose during the 1980s, and for a time it appeared that China would licence-produce the Mk 46 torpedo, the LM-2500 gas turbine, and possibly even Phalanx. Assistance was suspended after the Tiananmen Square massacre in the summer of 1989, and by the time military relations were resumed the Chinese were no longer very interested in these programmes. The Soviet Union had collapsed, and modern Soviet technology could be purchased at a relatively low price. The most outstanding example is the new programme to buy and licence-produce the 'Kilo' class submarine. There is also a reported plan to licence-produce *Sovremennyy* class missile destroyers. For a time it appeared that China would buy and complete the unfinished Russian carrier *Varyag*, but that never occurred.

Western friendliness also helped build up a substantial domestic computer and software industry, which would later contribute to Chinese naval power.

None of this is to say that China became an American or Western ally (although the Russians claimed that was the case often enough). To the Americans, China was a valuable card which tied down strong Soviet forces in Asia, where they could not contribute to a war in Europe. To the Chinese, the West was a valuable card which limited any Russian concentration on China, and which thus helped protect the country. For example, when the Chinese invaded Vietnam in 1979, they feared a possible Soviet attack in central Asia; they sent women and children in that area into shelters. No such attack materialised, possibly because the Soviets were already busy elsewhere (this was before the invasion of Afghanistan).

From a Chinese point of view, it was best to let the two superpowers concentrate on each other, rather than on China. It was vital, then, to support individual countries against 'hegemonism'. The Chinese saw nuclear proliferation as a guarantee against both superpowers, and at least through the 1980s they argued that the more nuclear states, the less the chance of war (they were fond of pointing out that the Soviets would never lightly have invaded a nuclear-armed Afghanistan). This idea was reflected in Chinese arms exports throughout the 1990s. Reportedly, for example, the Pakistani bomb developers used a Chinese plan. From a naval point of view, this type of weapons trade bought badly-needed time.

Another important development in the 1980s was vastly increased naval exports, particularly of missiles. Iran and Iraq both bought Chinese missiles during their long and inconclusive war; presumably their money helped fund new weapons development. Because China was willing to charge its friends very low prices, Chinese warships (frigates and submarines) were successfully exported to Bangladesh, Egypt, Pakistan and Thailand. There was some question as to their quality (for instance, the Thai frigates apparently lack truly watertight doors and damage control equipment), and their sensors and weapons are largely obsolete (except for missiles). Thailand has tried to solve the problem by buying hulls built in China but equipped with Western systems.

In 1995, the Chinese Navy is at a point of transition. By far the bulk of the fleet consists of ships of obsolete design, equipped with the unintegrated systems of the 1950s. That does not preclude the development of considerable firepower, using fire-and-forget missiles (such as Silkworm), but it does mean that the present Chinese fleet has little ability to defend itself against air attack. It is, therefore, essentially a coastal fleet which must operate under the umbrella of land-based aircraft. Because it seems unlikely that this fleet has much in the way of data links to the shore, it will be difficult for the fleet to coordinate with land-based strike aircraft (it is too easy for the aircraft to hit friendly ships by mistake).

The Chinese are well aware of the problem; their goal is to extend the reach of their fleet to well offshore. One aspect of this programme is the determination to acquire an aircraft carrier. Another is the decision to modernise the fleet. A few ships already incorporate the new technology, such as a combat data system derived from the French Tavitac 2000.

The Chinese economy is booming, driven by the considerable degree of capitalism which Deng welcomed. Some analysts believe that, if the present rate of growth continues, China will have the world's largest economy some time in the coming century. By then the 'four modernisations' may have provided enough industrial muscle to convert some units of the fleet into a modern deep-water force.

That rosy forecast avoids some major questions. The most important is whether China will be stable enough to weather the transition from the current mixed economy to something more clearly capitalist. When Deng accepted that capitalism was a legitimate road to the future, he opened up to question the legitimacy of the Chinese Communist Party. If the party were not needed as the vanguard of the people, why should the economy pay for its perquisites? Why should it exist at all? When the question was raised, in effect, at Tiananmen

Square, Chinese troops proved that the party still had the ability to maintain its position, whatever the rationale.

Six years later, the free-market part of the Chinese economy is doing quite well, although a high rate of inflation probably exaggerates its success. The state-controlled economy, which includes heavy industry, is still grossly inefficient. Party hacks still have their non-performing jobs. It is still virtually impossible to fire workers. The free-market sector is being milked to support the older Stalinist industrial sector. As it happens, most of the free market is in the southeast; most of the heavy industry is in the northeast, in places such as Beijing and Harbin. To make matters more interesting, local military units have invested heavily in local industry, and have profited heavily. Army involvement has already had some odd consequences. For example, some years ago the Party Central Committee wanted to outlaw satellite television dishes, which were being used by many citizens to receive subversive programming from outside China. The General Staff vetoed the idea. It was making far too much money out of the manufacture and sale of the dishes.

It must, then, occur to many in the southeast that the northeast (and other, less industrialised, areas) are acting as a drag to their own growth and prosperity. They must wonder whether a less centralised China might be a happier one, at least for them. The fading of the Soviet threat may make them more adventurous. That was certainly the effect of the fading of the Soviet threat on those who wanted to break out of federal Yugoslavia a few years ago.

National military strategy can have a unifying effect. The People's Republic of China still evinces interest in taking back Taiwan and the other Nationalist islands, but for the moment it seems most unlikely that force will be used. Indeed, there is some evidence that the two governments will make common cause on the main international issue, the Spratly Islands.

Further south there are more serious issues. China once regarded southeast Asia as her sphere of influence, before the West (initially the British) broke down her power from the 1840s on. A regime bent on reviving Chinese power (*ie*, any legitimate Chinese regime) would also probably want to revive the informal power relationship to the south. The locals disagree, and there has long been a dispute with Vietnam (the ethnic side of this problem led to massacres and ejection of ethnic Chinese in the 1970s). The Chinese unsuccessfully invaded Vietnam in 1979, and during the 1980s they supported the Pol Pot regime in Cambodia largely as a way of sapping Vietnamese strength.

The Spratly Islands are a more obvious issue. It is generally assumed that a sea of oil lies under these South China Sea islands. They are claimed by China, Indonesia, Malaysia, the Philippines and Vietnam. As long ago as January 1974 Chinese warships engaged South Vietnamese warships asserting a claim in the South China Sea. In 1995 Filipino warships destroyed some Chinese claim markers. Clearly China would like to enforce its claim. The opposition seems feeble. However, in the absence of an aircraft carrier and a fleet capable of long-range operation, any Chinese naval assault in the area would be quite dangerous. For some time China is said to have maintained a war chest specifically for the Spratlys (these funds were supposed to have paid for the *Varyag*).

It seems likely, then, that the Spratlys problem will spur the Chinese to invest more heavily in their fleet, especially since the major land problem, the former Soviet Union, has declined so dramatically.

NAVAL ORGANISATION

At present, the Chinese fleet is still organised as a coastal force, in three separate groups: the North Sea Fleet (HQ at Tsingtao), the East Sea Fleet (HQ at Shanghai), and the South Sea Fleet (HQ at Chan Chiang). The East Sea Fleet faces Taiwan. There are also coastal districts. During the 1980s the Northern Fleet, facing the Soviet Pacific Fleet, was the strongest, including two submarine divisions (about fifty boats). The South Sea Fleet included only one submarine division, and many small craft. Presumably it will be built up as tension increases in the Spratly Islands.

In common with the other Chinese forces, the navy has been drawn down in recent years in order to improve its quality. Thus in 1982 personnel strength was 284,000 (excluding about 38,000 for the naval air arm and a similar strength for coast defense). A decade later the official figure was 176,000, including 6,000 naval infantry, 28,000 in coast defence, and 25,000 in naval aviation.

Current naval air arm strength (1993) is: about thirty H-6 (Badger) bombers, some of which can launch two HY-4 (Silkworm) missiles; about eighty H-5 (Beagle) bombers, some of which are armed with rocket-propelled stand-off torpedoes (the Soviet RAT-52). Some may be capable of launching single HY-4 missiles; about seventy Nanchang Q-5 Fantan strike aircraft; about seventy J-8 Finback (modified MiG-21) interceptors; abou 280 J-6 interceptors (modified MiG-19); about 110 J-5 interceptors (modified MiG-17); about ten Be-6 ASW amphibians; three Y-8 maritime patrol aircraft; seven SH-5 ASW amphibians; about twenty Z-5 ASW helicopters (modified Mi-4); twelve Super Frelon helicopters; about ten Zhi-9 (copy of French Dauphin) light helicopters.

Note the weakness of the long-range strike force (only a very few H-6s definitely carry stand-off missiles) and the obsolescence of many of the interceptors on which the fleet must depend. Also, none of them has much range. In 1994 the Chinese began buying Russian SU-27s, which would greatly extend the reach of the shore-based air component. At least some of them may also carry modern Russian air-to-surface missiles, such as the supersonic AS-17 (Kh-31).

DESIGNS

Like the Soviets, the Chinese number their designs. The currently known list, with Western nicknames, is:

010	'T-43' class Minesweeper
021	'Huangfeng' missile boat (Osa class)
024	'Houku' missile boat (export EM1A; Homa is EM1B)
025, 026	'Huchuan' missile hydrofoil
033	'Ming' submarine (export designation ES3E)
	'Romeo' submarine (export designation ES3B)
037	'Hainan' sub-chaser
051	'Luda' destroyer
051HT	'Luhu' destroyer
053HT	'Jianghu II' frigate
053HT(H)	'Jianghu IV' frigate
053K	'Jiangdong' frigate
055	'Jiangwei' frigate
062	'Shanghai II' sub-chaser
065	'Jiangnan' frigate
067	'Yunnan' landing craft
073	'Yudao' landing ship
079	'Yuling' utility landing craft
312	Drone minesweeper
343M	'Houxin' missile boat
520T	'Houjian' missile boat

MAJOR SURFACE SHIPS

Ex-Soviet Type 7 *destroyers*

Other particulars: See 1922–46 volume
Armament: 4 HY-2 SSM (2×2), 4–130mm (4×1), 8–37mm (4×2), 60 mines. See notes
Sensors: Radar Cross Bird, Square Tie, Mina, Neptune; sonar hull mounted
Complement: 197

No	Name	Builder	Acquired	Fate
101	ANSHAN (ex-*Razyaschy*)	Dalzavod, Nikolayev	1954–55	Deleted 4.5.92
102	CHANG CHUN (ex-*Reshitelny*)	Dalzavod, Nikolayev	1954–55	Deleted *c*1991
103	CHI LIN (ex-*Retivy*)	Komsomolsk, Nikolayev	1954–55	Deleted *c*1991, museum ship
104	FU CHUN (ex-*Rezky*)	Komsomolsk Nikolayev	1954–55	TS 19.9.91

Two transferred in December 1954 and two in July 1955. Until 1971 armed with 4–130mm (4×1), 2–76mm (1×2), 4–37mm (4×1), 1–20mm and 6–533mm TT (2×3). Then during 1971–74 all had their TT replaced by twin SSM launchers and the single 37mm by twin mountings. The HY-2 missile is a Chinese derivative of the Soviet SS-N-Z. Despite this modernisation, they were already obsolete and, given that they were lightly built, it is a tribute to the Chinese Navy that they were kept in service for so long. By the end of their careers, their best speed was 30kts, down from an original 37kts. All had been stricken by 1992.

'Luda' class

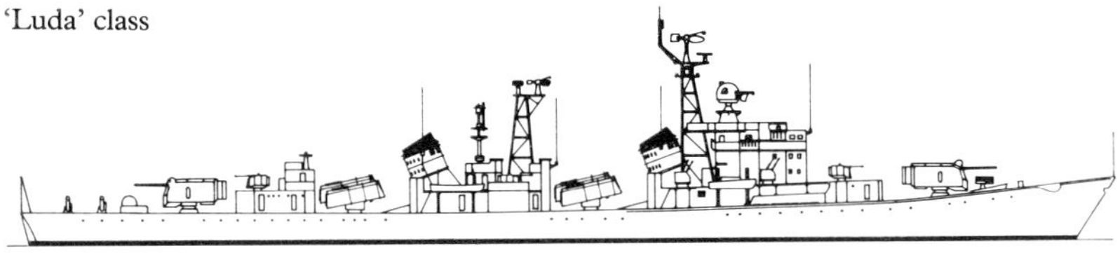

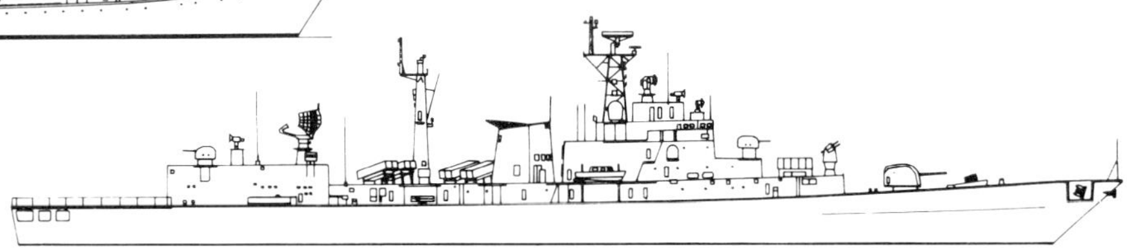

Zhanjiang as completed

'LUDA' class *destroyers*

Displacement:	3760t standard; 3960t full load
Dimensions:	433ft 1in × 42ft × 14ft 5in *132.0m × 12.8m × 4.39m*
Machinery:	2-shaft geared steam turbines, 4 boilers, 72,000shp = *c*32kts. Range 3200nm/4000nm at 18kts/15kts
Armament:	6 HY-2 SSM (2×3), 4–130mm (2×2), 8–57mm or 37mm (4×2), 4–25mm (2×2), 2 FQF 2500 ASW RL, 2/4 DCT, 2 DC racks, mines
Sensors:	Radar Cross Slot, Wasp Head, Post Lamp, Square Tie, Neptune; sonar hull mounted
Complement:	*c*300

No	Name	Laid down	In service	Fate
103	JINAN	1966	1972	Extant 1995
106	XIAN	1968	1972	Extant 1995
107	YINCHUAN	1968	1972	Extant 1995
108	XINING	1969	1973	Extant 1995
109	KAIFENG	–	1974	Extant 1995
110	DALIAN	–	–	Extant 1995
131	NANJING	1970	1974	Extant 1995
132	HEFEI	1970	1975	Extant 1995
133	CHONGQING	–	–	Extant 1995
134	ZUNYI	–	–	Extant 1995
160	–	–	–	Lost Aug 1978
161	CHANGSHA	1971	1975	Extant 1995
162	NANNING	1977	1980	Extant 1995
163	NANCHANG	–	1980	Extant 1995
164	GUILAN	–	1980	Extant 1995
165	ZHANJIANG	–	1981	Extant 1995

The first ocean-going warships built by the PRC, these vessels are derived from the Soviet *Neustrashimmy* class (NATO 'Tallinn') of which only one was built, despite their outward resemblance to the later 'Kotlin' class. They substitute triple SSM launchers for the TT of the Soviet ships. Hull design, weaponry and sensors compare very badly with other contemporary destroyers, or even with Taiwan's refitted ex-US ships. One unit of this class (identity unknown) was lost due to an explosion off Zanjiang in August 1978. FQF 2500 ASW RL is derived from Soviet RBU-1200. They have had underway replenishment capability since the early 1980s. In 1987, 105 was refitted with a helicopter hangar and flight deck aft, replacing the after gun mountings and the DCT.

There have been several attempts by the Chinese Navy to obtain Western armament and electronics for these ships. Plans for British Aerospace and Vosper Thornycroft to equip the entire class with new operations rooms and Sea Dart SAM fell through in the early 1980s, reportedly due to Chinese dissatisfaction with British missile performance in the Falklands War, and the arms embargo in 1989 following the Tiananmen Square massacre put an end to the proposed acquisition of US technology, reported to include Phalanx CIWS and helicopters. Since then, 109 has been fitted with a French Crotale octuple SAM launcher plus its controlling radar, and other vessels of the class have reportedly received other French electronics. The class is split almost equally between the three fleets.

'LUDA II' class *destroyers*

Armament:	8 YJ-1 SSM (2×4), 4–130mm (2×2), 8–37mm (4×2), 6–324mm ASW TT (2×3), 2 ASW RL, 2 DC racks, 38 mines
Sensors:	See notes
Other particulars:	See 'Luda' class

No	Name	Builder	In service	Fate
166	ZHUHAI	Dalian SY	1992	Extant 1995
–	–	Dalian SY	1992	Extant 1995

Two units of this improved version of the 'Luda' class entered service in 1992. The YJ-1 SSM is a wholly Chinese design, and the ships can also carry the CY-1 ASW missile in place of some of the SSMs. They also carry Western-built sonars (Thomson-Sintra bow-mounted and VDS (DSBV 23/43)) and ECM (Thomson-CSF DR-2000S intercept, Alligator jammer and Dagaie decoy launchers). These two ships are reportedly fitted with the French-supplied Tavitac 2000 combat direction system. China is now producing this system under licence (designated CCS-3 for export).

Zhanjiang 16.3.95 — L & L van Ginderen

'LUHU' class *destroyer*

Displacement:	5700t full load
Dimensions:	492ft 2in × 52ft 6in × 16ft 5in *150.0m × 16.0m × 5.0m*
Machinery:	2 shafts (CP propellers) CODOG, 2 GE LM-2500 gas turbines, 53,600shp = 30kts, 2 MTU 12V1163 TB83 diesels, 8840bhp = 20kts. Range 5000nm at 16kts
Armament:	8 YJ-1 SSM (2×4), 1 Crotale SAM (1×8), 2–100mm (1×2), 8–37mm (4×2), 1 CY-1 ASW missile system, 6–324mm ASW TT (2×3), 2 helicopters
Sensors:	Radar Type 756, MX-902, Thomson-CSF Sea Tiger, Hai Ying, Thomson-CSF DRBC-32F, Type 347G; sonar bow mounted, VDS
Complement:	300

No	Name	Builder	Laid down	Launched	Comm	Fate
112	ZHANJIANG	Qiuxin SY	25.4.90	Jun 1991	2.2.93	Extant 1995

Enlarged version of 'Luda' class. Originally planned to be fitted with US weapons and equipment, and GM gas turbines were exported to China for this class. Helicopter deck and hangar aft, and gun mounts are of a new enclosed type. Following the 1989 embargo, future units of this class will have to be equipped with different gas turbines, possibly of Russian design. This vessel is probably equipped with the Chinese-built version of the French Tavitac 2000 CDS. Given her Western electronics and enhanced capabilities, this is a powerful ship by PRC standards.

Xian in 1980 — Courtesy Ross Gillett

Ch'eng Tu after 1970s reconstruction

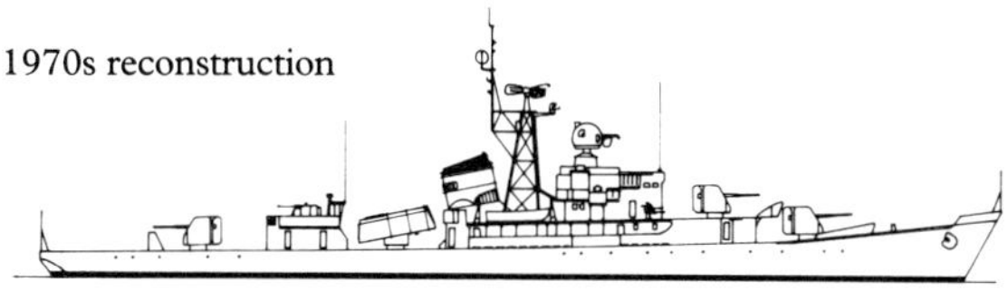

Soviet 'RIGA' class *frigates*

No	Name	Builder	Launched	Fate
507	CHENGDU	Hutung, Shanghai	28.4.56	See notes
505	GUIYANG	Hutung, Shanghai	26.9.56	See notes
508	GULIN	Kuang Chu, Canton	1957	See notes
506	KUNMING	Kuang Chu, Canton	1957	See notes

These ships were assembled from Soviet components in Chinese shipyards and were of the Soviet 'Riga' class design. First three laid down 1955, completed in pairs 1958 and 1959. Two were completed with modified superstructures. All later rebuilt with a larger bridge and lattice mast. As originally built, they had 3–37mm (3×1) with 3–533mm TT (1×3) abaft the funnel. From 1971 onward the TT were replaced by a twin SSM launcher and the AA armament was modified to 4–37mm (2×2). Although this eliminated some of the worst weaknesses from the armament, it was still very poor for a modern escort. The ASW armament of four DCT was particularly weak. All served with the S Sea Fleet. All deleted 1990–93.

'Jiangnan' class

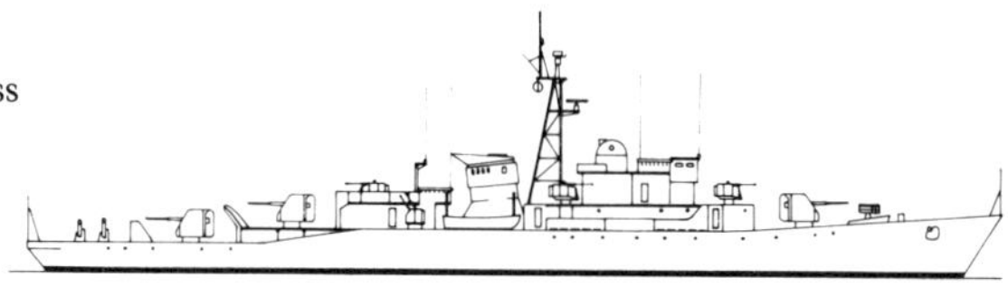

'JIANGNAN' class *frigates*

Displacement: 1350t standard; 1600t full load
Armament: 3–100mm (3 × 1), 8–37mm (4 × 2), 4–14.5mm (2 × 2), 2 MBU 1800 ASW RL, 4 DCT, 2 DC rails, fitted for minelaying
Sensors: Radar Ball Gun, Wok Won, Neptune; Sonar?
Other particulars: As Soviet 'Riga' class

No	Builder	Laid down	Comp	Fate
209	Kiangnan, Shanghai	1965	1967	Extant 1995
214	Kiangnan, Shanghai	1965	1967	Extant 1995
231	Kiangnan, Shanghai	1966	1968	Extant 1995
232	Kuang Chu, Canton	1966	1968	Extant 1995
233	Kuang Chu, Canton	1967	1969	Extant 1995

The first major warships, also spelt as 'Kiangnan' and named after their Shanghai builders, to be designed and built in China since 1945, even though they were an adaptation of the Soviet 'Riga' class design. Only the first ship's launch date (January 1966) is known. They use the same machinery and basic hull form, but the forecastle has been raised and extends three-quarters the length of the ship. This gives more room for living space and equipment. The armament has been rearranged, with one 100mm gun forward and two aft. They are efficient ships, but their sensors are obsolescent and their AA surface and ASW armament are all weak. Unlike the 'Rigas', they were not built with TT so it would be more difficult to find room for SSMs to improve the surface armament. However, their armament is so weak compared with probable opponents that it is surprising that something has not been tried, particularly as one of the class engaged S Vietnamese warships in 1974. It may be that there is inadequate stability due to the enlarged forecastle to make such an alteration to the armament. Four are in S Sea Fleet, one with E Sea Fleet. These vessels are thought to be approaching the end of their active lives.

Yintang as completed

'JIANGDONG' class *frigates*

Displacement: 1800t standard; 2200t full load
Dimensions: 337ft 9in oa × 39ft 4in × 13ft 2in
103m × 12.8m × 4.3m
Machinery: 2 shafts, 2 Pielstick diesels; 24,000hp = 28kts. Range 4000nm at 15kts
Armament: 4 HQ-61 SAM (2 × 2), 4–100mm (2 × 2), 8–37mm (4 ×2), 2 MBU 1800 ASW RL, 2 DCT, 2 DC rails
Sensors: Radar ?
Complement: 195

No	Name	Builder	Launched	Fate
531	YINTANG	Hutong, Shanghai	1973	Extant 1995
532	ZHONGDONG	Hutong, Shanghai	1974	Discarded incomplete 1982

This class, also spelt 'Kiang Tung', utilises an enlarged 'Riga'-type hull, with two large diesels (one per shaft) instead of four small ones. Laid down in 1971 and 1972, they were an attempt to provide an AA escort, something the Chinese Fleet badly needed. However, the SAM system caused many problems and did not become operational until the mid-1980s, after some 20 years of development. Subsequently vessels using the same hull were completed as *Kiang Hu* class SSM armed frigates. The second unit, *Zhongdong*, was discarded incomplete in 1982. The later 'Jianghu' class frigates were built on the same hulls, with SSM armament replacing the SAM (see below).

'Jianghu I' class

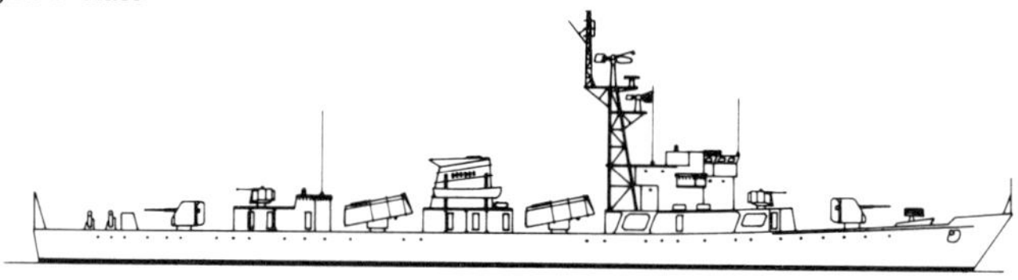

'JIANGHU I and II' class *frigates*

Displacement: 1425t standard; 1702t full load
Dimensions: 337ft 9in × 39ft 4in × 13ft 2in
103.0m × 12.8m × 4.3m
Machinery: 2 shafts, 2 SEMT-Pielstick 12 PA6 280BTC diesels, 16,000bhp = 25.5kts. Range 3000nm at 18kts
Armament: 'Jianghu I': 4 HY-2 SSM (2×2), 2–100mm (2×1), 12–37mm (6×2), 2 RBLL-1200 ASW RL, 4 DCT, 2 DC racks, mines
'Jianghu II': 4 HY-2 SSM (2×2), 4–100mm (2×2), 8–37mm (4×2), 2 RBU-1200 ASW RL, 4 DCT, 2 DC racks, mines
Sensors: Radar Type 765, Eye Shield, Square Tie; sonar EH-5 hull mounted; ECM Jug Pair, Square Head, High Pole
Complement: 195

No	Name	Builder	Fate
Jianghu I			
509	CHANG DE	Jiangnan SY	Extant 1995
510	SHAOXING	Jiangnan SY	Extant 1995
511	NANTONG	Jiangnan SY	Extant 1995
512	WUXI	Jiangnan SY	Extant 1995
513	HUAYIN	Jiangnan SY	Extant 1995
514	ZHENJIANG	Jiangnan SY	Extant 1995
515	XIAMEN	Jiangnan SY	Extant 1995
516	JIUJIANG	Jiangnan SY	Extant 1995
517	NANPING	Jiangnan SY	Extant 1995
518	JIAN	Jiangnan SY	Extant 1995
519	CHANGZHI	Jiangnan SY	Extant 1995
520	KAIFENG	Jiangnan SY	Extant 1995
555	ZHAOTONG	Jiangnan SY	Extant 1995
556	XIANGTAN	Jiangnan SY	Extant 1995
Jianghu II			
533	NINGBO	Hudong SY	Extant 1995
534	JINHUA	Hudong SY	Extant 1995
543	DANDONG	Hudong SY	Extant 1995
545	TIANSHAN	Hudong SY	Extant 1995
552	YIBIN	Hudong SY	Extant 1995
553	SHAOGUAN	Hudong SY	Extant 1995
554	ANSHUN	Hudong SY	To Bangladesh 1990
557	JISHOU	Hudong SY	Extant 1995

First unit launched in 1975, and construction continued until 1993, despite this being an obsolete design. 'Jianghu II' type replace the single 100mm mounts with auto-loading twin mounts, two 37mm mounts being removed in compensation. Electronics have been updated since 1982. Two units, with twin 57mm replacing 100mm, were sold to Egypt in 1984–85, and *Anshun* was transferred to Bangladesh in 1990.

Zhenjiang 1984 Courtesy Ross Gillett

Huangshi as completed

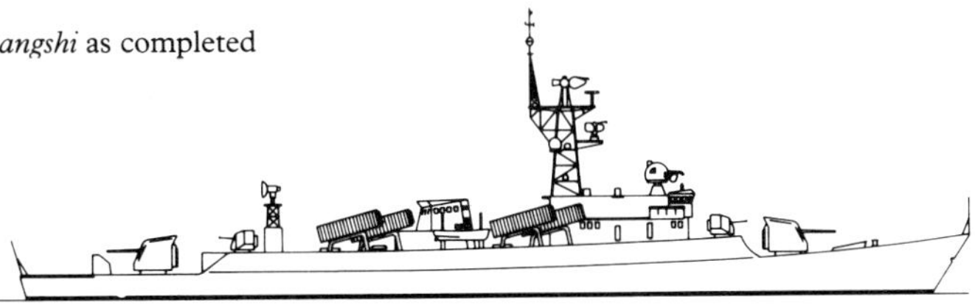

'JIANGHU III' class *frigates*

Displacement: 1865t full load
Dimensions: 338ft 7in × 35ft 6in × 10ft 2in
103.2m × 10.83m × 3.10m
Machinery: As 'Jianghu I/II' class
Armament: 8 YJ-1 SSM (4×2), 4–100mm (2×2), 8–37mm (4×2), 2 RBU-1200 ASW RL, 2 DCT
Sensors: Radar Type 765, Eye Shield, Rice Lamp, Square Tie, Sun Visor; sonar Type EH-5; ECM Elettronica Newton-Beta suite
Complement: 180

No	Name	Builder	Comm	Fate
535	HUANGSHI	Hudong SY	1986	Extant 1995
536	WUHU	Hudong SY	1987	Extant 1995
537	ZHOUSHAN	Hudong SY	1992	Extant 1995
538	–	Hudong SY	–	?

Improved version of 'Jianghu II' class, with a full shelter deck amidships supporting four twin YJ-1 SSM launchers. They carry the same twin autoloading 100mm mounts as their predecessors. Four vessels of this class were built for Thailand 1988–91, two of which were completed with a helicopter pad and hangar in place of the after gun mount.

Siping as completed

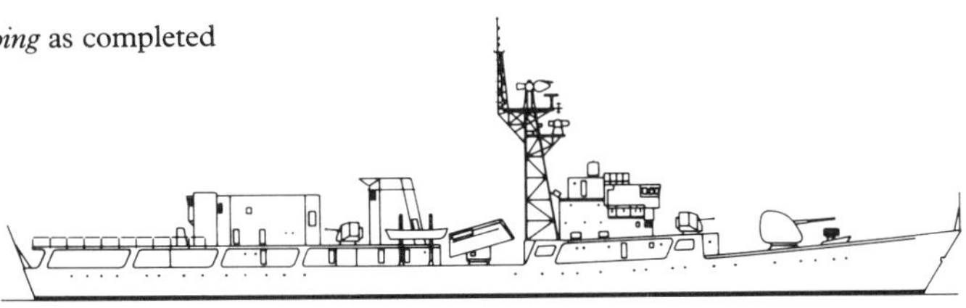

'JIANGHU IV' class *frigate*

Displacement: 1820t full load
Dimensions: 338ft 7in × 33ft 6in × 10ft
103.2m × 10.2m × 3.05m
Machinery: As 'Jianghu I/II' class
Armament: 2 HY-2 SSM (1×2), 1–100mm/55 Creusot-Loire Compact, 8–37mm (4×2), 6–324mm ILAS-3 ASW TT, 2 RBU-1200 ASW RL, 1 helicopter
Sensors: Radar Type 956, Eye Shield, Square Tie; sonar Type EH-5 hull mounted; ECM Jug Pair, Square Head, High Pole, 2 Mk33 decoy launchers
Complement: 185

No	Name	Builder	Launched	Comp	Fate
544	SIPING	Hudong SY	Sep 1985	Nov 1986	Extant 1995

This was the first Chinese warship to be built with helicopter capability, with a hangar and pad aft for one Zhi-9 helicopter (a Chinese-built version of the French SA-365N Dauphin), replacing the after gun mount and two SSM launchers of the other 'Jianghu' classes. Also equipped with a French main gun and ASW torpedoes and US decoy launchers. It is believed that no further units of this class are planned.

Anqing as completed

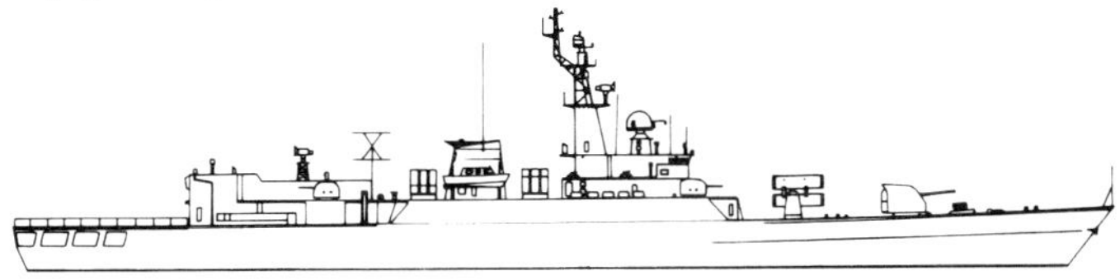

'JIANGWEI' class *frigates*

Displacement: 2250t full load
Dimensions: 377ft 4in × 45ft 11in × 13ft 1in
115.0m × 14.0 × 4.0m
Machinery: 2 shafts (CP propellers), 4 MTU 20V1163 TB92 diesels, 21,460bhp = 28kts. Range 2500nm at 18kts
Armament: 6 C-802 SSM (2×3), 6 HQ-61 SAM (1×6), 2–100mm (1×2), 8–37mm (4×2), 2 RBU-1200 ASW RL, 1 helicopter
Sensors: Radar Type 765, Eye Shield, air search, 2 Rice Lamp, Sun Visor; sonar S-07 bow-mounted; ECM RWD-8, NJ8-3, 2 decoy launchers
Complement: 180

No	Name	Builder	Comp	Fate
539	ANQING	Hudong SY	Dec 1991	Extant 1995
540	HUINAN	Hudong SY	Dec 1992	Extant 1995
541	–	Hudong SY	?	?
542	–	Hudong SY	?	?
543	–	Hudong SY	?	?
544	–	Hudong SY	?	?

First frigates to mount the Chinese-built HQ-61 SAM since the troubled *Yintang* (see above), in a sextuple mount forward of the bridge, without reloads. The 37mm guns are in the later enclosed mounts. ASW armament is limited to two RBU-1200 RLs. Two units are under construction for Thailand, which will however carry different equipment and power plants, for delivery in 1995–96. A version of this class was offered for export in the early 1990s, with French-built SAMs and medium guns, as well as some Western electronics, but has not been taken up by any navy.

Ex-British 'CASTLE' class *frigates*

Several that had served in the Royal Canadian Navy were sold as merchant ships to China in 1947. After the Communist takeover of the mainland two were rearmed, at much the same time as the Nationalists also rearmed two. The ex-mercantile SS *Shih Lin*, ex-mercantile SS *Hsi Ling*, ex-mercantile *Ta Tung* (all 1947), ex-HMCS *Coppercliff* (1944), ex-British *Hever Castle*; and the ex-mercantile SS *Yuan Pei*, ex-mercantile SS *Ta Shun* (both 1947) ex-HMCS *Bowmanville* (1944) ex-HMS *Nunnery Castle* were both rearmed with 2–130mm (2 × 1), 1–45mm and 5–37mm (5 × 1). One was renamed *Kuang Chou*, but the name of the other is not known. There is some doubt about which one was renamed *Kuang Chou*. The other ship was discarded in the 1960s but *Kuang Chou* (602) remained in service until 1986, when she was stricken.

Ex-British 'FLOWER' class *frigates*

Name	Builder	Acquired	Fate
KAI FENG (ex-*Cloverlock*, ex-*Clover*)	Fleming & Ferguson, Paisley	1949	Extant 1982
LIN I (ex-*Ziang Teh*, ex-*Heliolock*, ex-*Surprise*, ex-*Heliotrope*)	Crown	1949	Deleted 1979
? (ex-*Maw Hwa*, ex-*Tenacity*, ex-*Candytuft*)	Grangemouth		

Several were sold as merchant ships to China by the RN after the Second World War. During the last stages of the 1949 struggle for mainland China two, or possibly three, of these were rearmed by the People's Republic. By the early 1950s *Kai Feng* (formerly the mercantile SS *Cloverlock* ex-HMS *Clover*) and *Lin I* (ex-mercantile SS *Ziang Teh*, ex-mercantile *Heliolock*, ex-USS *Surprise*, ex-British HMS *Heliotrope*) were armed with Soviet weapons. In 1974–75 *Kai Feng*, 2–100mm (2 × 1), 1–45mm, 4–37mm (2 × 2), was disarmed and refitted as a survey ship in which form she remained in service until *c*1989. *Lin I*, 2–100mm (2 × 1), 2–37mm (2 × 1), was deleted in 1979. The mercantile SS *Maw Hwa*, ex-USS *Tenacity*, ex-British *Candytuft* may also have been rearmed, but if so she was no longer in service by the mid-1950s.

SUBMARINES

Ex-Soviet Series VI 'M IV' class *coastal submarine*

One (built 1933–34) was transferred from the Soviet Pacific Fleet *c*1955. She was used for training (only 2–533mm TT and 18 crew) and deleted in 1963.

Ex-Soviet Series VI bis 'M II' class *coastal submarine*

Ex-*M 39* (ex-*M 53*), launched 1935, was transferred from the Soviet Pacific Fleet in 1955. Used for training and deleted 1963.

Ex-Soviet Series XV 'MV' class *coastal submarines*

Four of this late 1930s design (built 1946–47) transferred *c*1955. They were numbered *M 200*, *M 201*, *M 202* and *M 203*. They were used mainly for training. *M 200* was deleted in 1963 but the other three survived until the late 1970s. None remain in service.

Ex-Soviet SHCH class *submarines*

Four of these late 1930s units transferred in 1954–55. Their design was out of date when they were first built, and by the mid-1950s they were totally obsolete. All deleted in 1963.

Ex-Soviet Series IX bis STALINETZ class *submarines*

Class (former name):
S 400 (ex-*S 51*), *S 401* (ex-*S 52*), *S 402* (ex-*S 53*), and *S 403* (ex-*S 57*). Four transferred 1955. Launched 1939–40 by Dalzavod Yd, Vladivostok. Obsolete when transferred, but all remained in service until the 1970s.

Soviet 'WHISKEY' class *submarines*

Class:
119, 120, 122, 123, 127, 129, 131, 201–207, 221, 241, 243, 244, 265–267

Several transferred in the mid-1950s. The USSR supplied components for five more to be assembled in Chinese shipyards until eventually the Chinese built the entire submarine. A total of twenty-one entered service during 1956–64, most being built at Kiang Chou, Canton and Kiangnan, Shanghai. Obsolete by the 1980s, six had been stricken by 1989, with the remainder either laid up in reserve or serving as stationary harbour training boats. All had been disposed of by 1995.

Soviet 'ROMEO' class *submarines*

Class:
126, 140, 142, 143, 153, 172, 176, 208–212, 227–229, 245, 248, 249, 254, 267–270, 281–283 (plus 52 others)

Four supplied by the USSR prior to the Sino-Soviet split and between the early 1960s and late 1970s a total of eighty-four were built at the rate of nine a year in the Kiang Chou, Canton and Kiangnan, Shanghai shipyards. Six of these Chinese-built 'Romeos' were transferred to North Korea, (1973–75) which built more of them. Another four went to Egypt (first pair in 1983, second pair in 1984). Approximately seventy of these boats were reported extant with the Chinese Navy in 1995, but only twenty of these were actually operational, suggesting that the modernisation programme proposed *c*1983 may not have taken place. Now an obsolete class.

'Romeo' class — Courtesy Ross Gillett

Soviet 'GOLF' class *strategic submarine*

The design for this Soviet submarine were supplied to China prior to the expulsion of the Soviet advisers in 1960, but although *200* was completed at Luda in 1964 with the three vertical tubes for SLBMs at the aft end of the conning tower the SLBMs themselves were not provided. One of the missile tubes was removed *c*1974 when she was modified for trials of the new Chinese-built ballistic missile. In August 1981 an unnamed 'Golf' class submarine was reported lost off the Chinese coast during an 'underwater test firing', presumably an early launch of the CSS-NX-3 SLBM, as the SS-N-4 carried by Soviet 'Golf' class boats was launched on the surface. However, later reports stated that the first successful launch of the Chinese SLBM was made from *200* on 12 October 1982. She is extant 1995, as a trials ship.

'Han' class

'HAN' class *nuclear attack submarines*

Displacement: 5000t submerged
Dimensions: 328ft (*403–405* 354ft) × 36ft × 28ft
100.0m (108m) × 11.0m × 8.5m
Machinery: 1 shaft nuclear: 1 pressurised-water reactor, turbo-electric drive, 15,000shp = 25kts/30kts
Armament: 6–533mm TT (bow)
Sensors: Radar navigational; sonar DUUX-5
Complement: 75

Class (launched):
401 (1970), *402* (1977), *403* (1983), *404*, *405* (1990)

The first nuclear-powered vessels and the first submarines designed and built in China, by the Huludao Shipyard. Considerable help with the powerplant was received from West Germany. They have an *Albacore*-type teardrop hull. *401* underwent prolonged trials, and construction of further units was delayed by powerplant problems. The reliability of the earlier units has been called into question. The last three units of the class are approximately 26ft (*8m*) longer than the original boats, which was thought to be due to the fitting of cruise-missile tubes abaft of the sail, but this is now considered unlikely. All five were extant 1995, serving with the N Fleet.

'Ming' class

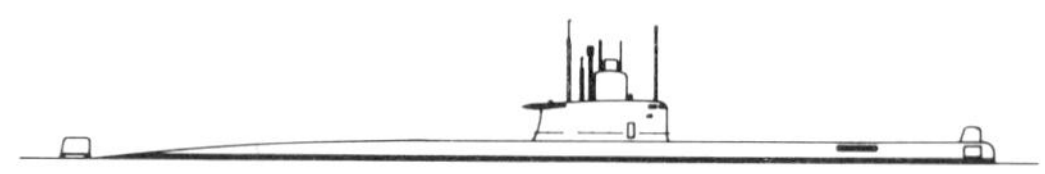

'MING' class *attack submarines*

Displacement: 1584t surfaced; 2113t submerged
Dimensions: 249ft 4in × 24ft 11in × 16ft 9in
76.0m × 7.6m × 5.1m
Machinery: 2 shafts diesel-electric, 2 diesels, 2 electric motors, 2500hp = 15kts/18kts. Range 8000nm at 9kts surfaced
Armament: 8–533mm TT (6 bow, 2 stern, 16 torpedoes or 32 mines)
Sensors: Radar Snoop Plate; sonar Hercules, Feniks
Complement: 58

Class (launched):
232 (1975), *233* (1975), *341* (1982), *342, 343, 353, ?*

This is a Chinese-built derivative of the Russian 'Romeo' class with slightly more engine and motor power and a redesigned outer hull. The Chinese wanted greater power, so in the 1970s they built two units with triple screws (three diesels, three motors accommodated in a more voluminous after hull). This version was a failure and the boats were broken up. The standard production 'Ming' gains about ¼kt on a 'Romeo', and can be recognised by an outer hull enlarged forward of the sail. A modified version (Project 035) has French sonar, ESM and torpedo fire control. A single unit (Project 039), launched in 1994, reverted to the original triple-screw configuration, and has a redesigned sail housing a sonar. The Chinese designated the original Russian 'Romeo' Project 031; the licence-built version is Project 033. Project 034 was the unsuccessful triple-screw variant, and Project 036 was the proposed cruise missile boat (see below), of which none have been reported built.

Xia 1987

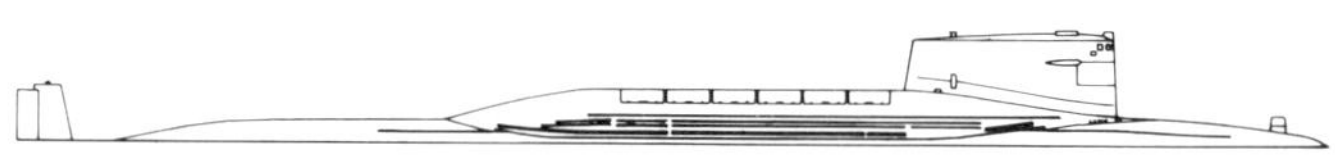

'XIA' class *ballistic missile submarine*

Displacement: 8000t submerged
Dimensions: 394ft × 33ft × 26ft
120m × 10m × 8m
Machinery: 1 shaft nuclear: 1 pressurised-water reactor, turbo-electric drive = *c*20kts submerged
Armament: 12 SLBM, 6–533mm TT (bow)
Complement: 100

Name	Builder	Laid down	Launched	Comp	Fate
406	Huludao SY	1978	Apr 1981	1987	Extant 1995

The first (and thus far only) Chinese SSBN, built by the same yard as the 'Han' class SSN. Equipped with twelve tubes for the CSS-N-3 ballistic missile, first launched from the 'Golf' class trials boat in 1981. The first launch from *406* took place in July 1988, but no further launches have been announced. Reports of further construction of units of this class have been vague, with possibly three more boats planned, with up to four more missile tubes and equipped with an improved SLBM, but as of 1995 no more boats have been reported extant.

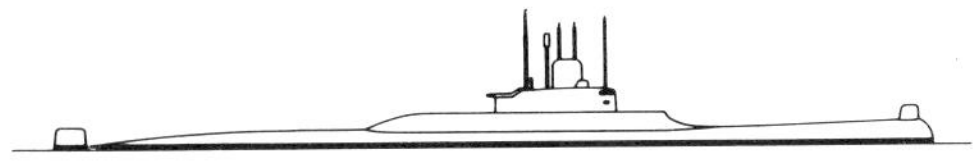

'Wuhan' class

'WUHAN' class *cruise missile submarine*

Class (in service):
351 (1987)

A converted 'Romeo' class boat, fitted with six tubes for trials of the surface-launched C-801 cruise missile, which was to arm the proposed 'E5SG' class (see under 'Ming' class above). Extant 1995.

AMPHIBIOUS WARFARE VESSELS

YU LING class *medium landing ships*

About ten of this 1500t, 250ft design have been built since 1971. Enlarged versions of US Second World War LSMs.

YUKAN class *tank landing ships*

Displacement: 3110t full load
Dimensions: 392ft 1in × 51ft 2in × 9ft 3in
119.5m × 15.6m × 2.82m
Machinery: 2 shafts, 2 SEMT-Pielstick 12 PA6 V280 diesels, 10,000bhp = 18kts. Range 3000nm at 14kts
Armament: 8–57mm (4×2), 8–25mm (4×2)
Sensors: Radar 2 Type 756
Complement: 109

Class:
927, 928, 929, 934, 990, 991

Built as replacements for the ex-US wartime LSTs since 1978 by the Zhonghua Shipyard, being both larger and faster. Can carry a load of 500t. *934, 990* and *991* are a later version with a helicopter deck aft. All extant 1995.

'YUDAO' class *medium landing ships*

Displacement: 1460t full load
Dimensions: 269ft 3in × 41ft 4in × 10ft 2in
82.07m × 12.6m × 3.10m
Armament: 8–25mm (4×2)
Sensors: Radar Type 756
Complement: 60

Class:
972, 975, 984, 985

Replacements for the ex-US LSMs (all deleted by 1993), entering service from 1980. Smaller version of LST design. *975* now a museum ship at Qingdao, remainder extant 1995.

YUNNAN class *utility landing craft*

Displacement: 133.2t full load
Dimensions: 90ft 2in oa, 78ft 9in wl × 17ft 8in × 4ft 6in
27.5m, 24m × 5.4m × 1.4m
Machinery: 2 shafts, 2 diesels = 10kts
Armament: 2/4–14.5mm MG (1 or 2×2)
Complement: 6

Class:
3313, 3321, 3344, 5526, 7566, 7588 and many more

Hangzhou Yd built 300 during 1968–72 and all are extant 1995. The 49ft × 13ft (*15m × 4m*) cargo deck can hold 46t or a tank. A stubbier 100t LCU *7568* was seen at Canton in April 1980.

OTHER MINOR LANDING CRAFT

'Yuqin' class landing craft (110t full load, 79ft 1in × 17ft 1in × 3ft 7in, *24.1m × 5.2m × 1.1m*, 2 diesels = 11.5kts, 2–14.5mm MG (2×1); built 1968–1972. Can carry 110 troops. Two transferred to Bangladesh 1984. Fifty extant 1995.

'Yuchai' class landing craft (70t full load, 65ft 7in × 14ft 1in × 3ft 3in, *20.0m × 4.3m × 1.0m*, 2 diesels = 10kts, 4–14.5mm MG (2×2)): built during the 1960s. Some employed as transports rather than as landing craft proper. Thirty extant 1995.

'Dagu-A' air cushion vehicle (61t, 55kts); one unit apparently built as an experimental prototype *c*1980. Can carry 16t cargo. Bow ramp. Extant 1995.

SMALL SURFACE COMBATANTS

Ex-Australian BATHURST class *ocean minesweeper (escort)*

No	Name	Builder	Acquired	Fate
53/220	LOYANG (ex-*Cheung Hing*, ex-*Bendigo*)	Cockatoo D Yd	1949	Stricken 1986

This ex-RAN minesweeper (launched 1 March 1941) was sold on 5 May 1947 and became the Chinese mercantile *Cheung Hing*. The Communist forces took her over and reconverted her from a merchant ship. She was originally rearmed with 2–130mm (2×1) and 2–37mm (2×1) but was later fitted with 2-100mm (2×1) and 4–37mm (2×2). She was classed as an escort and served with the E Sea Fleet.

Ex-British 'ISLES' class *patrol trawlers*

Two which had been disarmed and sold to Chinese owners in 1946 for mercantile work were taken over by Communist forces and rearmed with 1–76mm and 3–20mm in 1949–51. Ex-HMS *Hoxa* (launched 15.1.41) became the mercantile SS *Sung Hwei* in 1946. The other may have been HMS *Balta* which became the mercantile SS *Ching Hai* on 12.7.46. One ex-Canadian *Basset* class patrol trawler (dimensionally identical to the 'Isles' class) was also rearmed. This was either SS *Sung Ming* (ex-HMCS *Comox*), SS *Sung Li* (ex-HMCS *Gaspe*) or SS *Sung Ling* (ex-HMCS *Nanoose*, ex-*Nootka*). All three units were deleted in 1967.

Ex-Soviet 'KOMAR' class *fast attack craft (missile)*

Seven or eight transferred in the early 1960s. Two or four reported as in service in the early 1980s but since discarded.

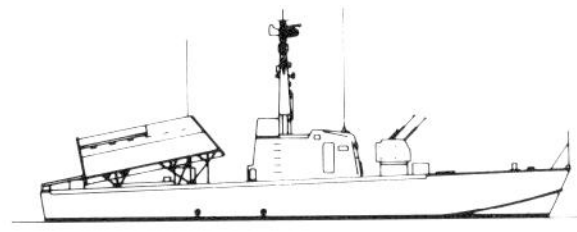

'Hoku' class

'HOKU' class *fast attack craft (missile)*

Displacement: 68t standard; 80t full load
Dimensions: 91ft 10in × 21ft × 6ft
28m × 6.3m × 1.8m
Machinery: 4 shafts, 4 M50 diesels, 4800 bhp = 39kts. Range 400nm/500nm at 30kts/25kts
Armament: 2 SS-N-2 SSM (2×1), 2–25mm (2×1)
Sensors: Radar Square Tie
Complement: 18

Class:
1109 series, *3125* series

Chinese-built development of the Soviet 'Komar' with a steel rather than a wooden hull. The SSM launchers are farther inboard and they have a pole mast. Otherwise they are very similar apart from a slightly modified hull form. Anything from 80 to 110 'Hokus' were built since the late 1960s. Their value has gradually been eroded by the widespread adoption of ECM by potential enemies. Transfers: four to Albania 1976–77, four to Bangladesh 1983, four to Egypt 1984, four to Pakistan 1981–82. Approximately seventy remained in service in 1995. A single hydrofoil variant of this class, *Homa* (*c*100t, 95ft/*29m*, 4–25mm), was built in the 1970s but not repeated.

'HUANGFENG' class *fast attack craft (missile)*

Class:
3100 series

Chinese version of the Soviet 'Osa I' class. Anything from 80 to 112 'Huangfeng' class were built in Chinese shipyards (Kiangnan, Shanghai) since *c*1966–67. The first four (Soviet transfers 1965–67) differed in having 4–30mm (2×2) in place of 25mm. In the early 1980s, some units were rearmed with 30mm guns and new SSM. Transfers: four to Pakistan 1984, four to Bangladesh 1988. In 1994 five modified versions of this class were sold to Iran, with five more to follow (see under Iran). Approximately eighty units were extant 1995.

'Huangfeng' class — Courtesy Ross Gillett

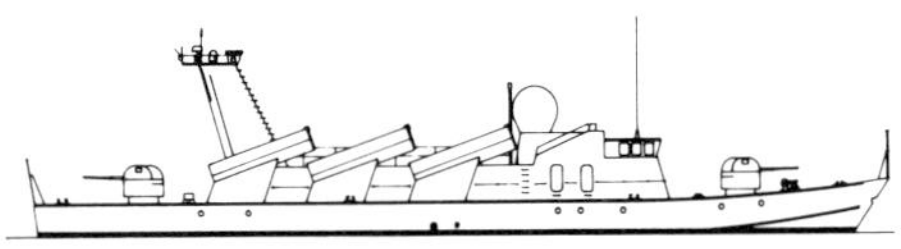

'Hola' class
NB 1/750 scale

'HOLA' class *fast attack craft (missile)*

Displacement: 185t standard; *c*200t standard
Dimensions: 137ft 10in oa × 26ft 6in × 6ft 6in
42m × 8.1m × 2m
Machinery: 3 shafts, 3 diesels, 12,000bhp = 36kts
Armament: 4 or 6 SS-N-2 SSM (6×1 or 4×1), 4–25mm (2×2)
Sensors: Radar Square Tie

Enlarged and lengthened version of 'Osa I' class with greater striking power. Built *c*1970 and numbered *5100*. She can be distinguished by her aft mounted radome and folding mast. Only one has been built. Armament reduced to 2 SS-N-2 launchers by the 1980s.

'HOUXIN' class *fast attack craft (missile)*

Displacement: 430t full load
Dimensions: 203ft 5in × 23ft 7in × 7ft 4in
62.0m × 7.2m × 2.24m
Machinery: 4 shafts, 4 diesels, 13,200bhp = 32kts. Range 2000nm at 14kts
Armament: 4 C-801 SSM (2×2), 4–37mm (2×2), 4–14.5mm MG (2×2)
Sensors: Radar Type 756, Square Tie, Rice Lamp

Class: *751, 752, 753, 754, 755* + 1–3 others

Built by the Qixin Shipyard, Shanghai, from 1989, with the first unit entering service in June 1991. FAC(M) variant of 'Haiju' class patrol boat. Six to eight units extant 1995.

'HOUJIAN' class *fast attack craft (missile)*

Displacement: 520t ful load
Dimensions: 214ft 7in × 27ft 7in × 7ft 10in
65.4m × 8.4m × 2.38m
Machinery: 3 shafts, 3 SEMT-Pielstick 12PA6V 280 MPC diesels, 17,280bhp = 33.5kts. Range 1800nm at 18kts
Armament: 6 C-801 SSM (2×3), 2–37mm (1×2), 4–30mm (2×2)
Sensors: Radar Type 756, Square Tie
Complement: 75

Class (in service):
770 (Jul 1991)

Single unit built by the Huangpo Shipyard, Shanghai, as an example of a design offered for export with Western weapons. Extant 1995.

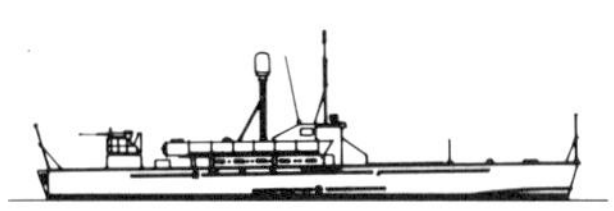

Chinese 'P4' class
NB 1/750 scale

Ex-Soviet 'P 4' class *fast attack craft (torpedo)*

Class:
5220 series

Over seventy Soviet 'P 4s' were supplied to China from 1952. Fifty-five were supplied plus twelve P-6 (eighty built in China). Six were transferred to Albania in 1956 followed by six Chinese-built units in 1965 and four to Bangladesh in 1983. The class was gradually deleted from 1976, with ten boats being stricken, then twenty in 1977 and five each in 1978 and 1981. By the mid-1980s it was estimated that perhaps fifty vessels remained in reserve, with only a few still operational. None are reported extant in 1995.

Chinese 'P6' class
NB 1/750 scale

Soviet 'P 6' class *fast attack craft (torpedo)*

Class:
5200 series

Twelve Soviet-built were transferred to PRC. About eighty built in China (mainly Shanghai) 1956–66. Six transferred to North Vietnam 1967. Approximately ten extant 1995, probably non-operational.

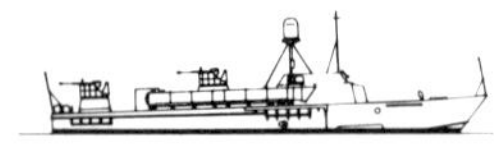

'Huchuan' class
NB 1/750 scale

'HUCHUAN' class *torpedo hydrofoils*

Displacement: 39t standard; 45t full load
Dimensions: 71ft 6in oa × 16ft 6in × 3ft 3in
21.8m × 4.9m × 1m
Machinery: 3 shafts, 3 M50 12-cyl diesels; 3600hp = 55kts. Range 500nm at 20kts
Armament: 2–533mm TT (2×1), 4–14.5mm MG (2×2)
Sensors: Radar Skin Head
Complement: 20

Class:
205, 207, 209, 248, 3206, 3214, 6218, 7230 and many more

Nearly 200 of these successful torpedo armed hydrofoils, the world's first operational foil-borne warships, have been built since 1966 by Hutung of Shanghai. By 1968 at least twenty-five were in service with S Sea Fleet. There are two versions, one with the bridge well forward and both twin MG mountings aft, and the other version with the bridge farther aft and the MGs fore and aft. Transfers: thirty-two to Albania (1968–74), four to Pakistan, four to Tanzania (1975), three to Romania which has also built her own version, four to Bangladesh in 1988, and four to Zaire. Approximately 120 were built for the Chinese Navy, of which eighty were extant 1995.

Ex-Soviet BO 2 type *submarine chasers*

Six of these Second World War 'Artillerist' class units transferred in the mid-1950s. Replaced by Chinese 'Hainan' class in mid-1960s and deleted 1967.

Soviet 'KRONSTADT' class *large patrol craft*

Class (deployment):
251–258 (N Sea Fleet at Tsingtao and Lushin), *262, 263, 633–635* (E Sea Fleet in Chou Shan Islands), *651–656* (S Sea Fleet in Hainan)

Six transferred from the USSR in 1956–57. Fourteen more were built at Shanghai and Canton from Soviet components in 1956–57. Armed in Chinese service with 6–14.5mm MG (3×2) instead of 2 MG. Units of this class, despite their age, remained in service until 1990, but have since all been stricken.

Ex-Soviet 'SO 1' class *submarine chasers*

Two transferred 1960. Neither remains in service.

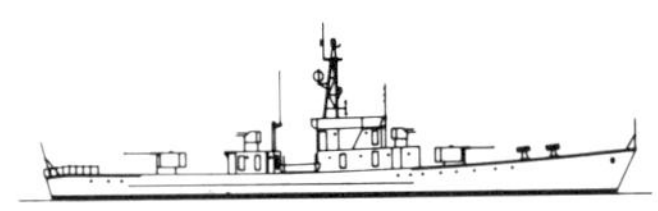

'Hainan' class

'HAINAN' class *large patrol craft*

Displacement: 360t standard; 400t full load
Dimensions: 193ft 8in oa × 24ft × 7ft 10in
59m × 7.3m × 2.4m
Machinery: 4 shafts, 4 diesels, 8000bhp = 30.5kts. Range 1300nm at 15kts
Armament: 4–57mm (2×2) (some 2–76mm (2×1)), 4–25mm (2×2), 4-MBU 1800 ASW RL, 2 DCT, 2 DC rails, mine rails
Sensors: Radar Pot Head (most), Skin Head (remainder)
Complement: *c*60

Class:
200 series, *300* series, *600* series, *800* series (total ninety-five units)

Developed from the Soviet 'SO 1' class but have a considerably larger hull. Early ships were armed with single US Mk 26 76mm guns fore and aft, but on later vessels twin 57mm are standard. Built at an average rate of four per year since 1964, they are still the standard Chinese large patrol craft. Transfers: seven to Algeria (1990–91), two to Bangladesh (one in 1982 and one in 1985), eight to Egypt (two in 1983 and six in 1984), ten to Myanmar (six in 1991 and four in 1993), six to North Korea (three pairs in 1975, 1976 and 1978 respectively), and four to Pakistan in 1981. In 1995, there were approximately ninety-five boats of this class in Chinese service.

'Hainan' class boats *c*1980 Courtesy Ross Gillett

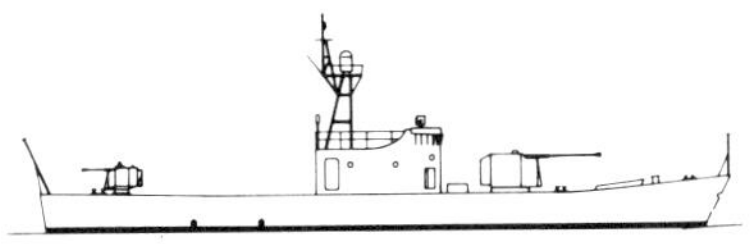

'Shanghai I' class
NB 1/750 scale

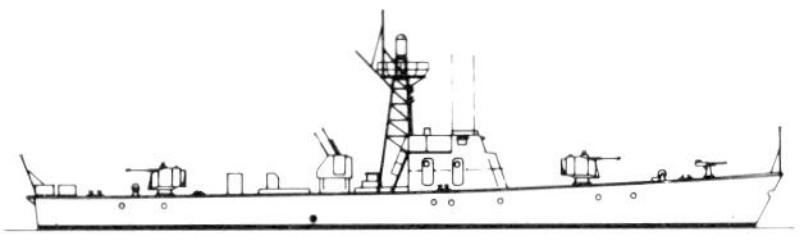

'Shanghai II' class
NB 1/750 scale

'SHANGHAI' class *large patrol craft*

Displacement: 120t standard; 155t full load ('Shanghai I': 100t standard)
Dimensions: 128ft oa × 18ft × 5ft 8in ('Shanghai I' 120ft oa/*35.1m*)
39m × 5.5m × 1.7m
Machinery: 4 shafts, 4 M50 F-4 diesels, 4800hp = 28kts. Range 800nm at 17kts
Armament: (Type II – see notes) 4–37mm (2×2), 4–25mm (2×2), 1 ASW RL
Sensors: Radar Pot Head or Skin Head; sonar hull mounted (some have VDS)
Complement: 25

Class:
1300 series, *4300* series, *8300* series

Very simple and economical coast patrol vessels. Twenty-five 'Shanghai I' were built from 1959, of which eight were transferred to North Korea and ten deleted in 1978. These had a shorter hull than later versions and were armed with 2–57mm (1×2) forward and 2–37mm (1×2) aft. Some had 2–457mm TT (1×2) but these were later removed. The longer-hulled Types II, III, IV and V were built from 1961. The Type III have 57mm in place of the 37mm and the other two types have more minor differences. Some have 2–75mm Army RRs (1×2) mounted forward. Over 350 were built. Transfers: Albania (six), Bangladesh (six), Cameroun (two), Congo (three), Guinea (six), Pakistan (twelve), Sierra Leone (three), Sri Lanka (ten), Tanzania (eight), Tunisia (two), North Vietnam (eight) and Zaire (four). In addition the Romanians have built their own versions. About 280 of the later type remain in service with the Chinese Fleet, but the first to be built have been discarded.

'HAIKOU' class *large patrol craft*

Displacement: 160t standard; 175t full load
Dimensions: 150ft oa × 21ft × *c*7ft
45.7m × 6.4m × c2.1m
Machinery: 4 shafts, 4 diesels = *c*30kts. Range 850nm at 20kts (est)
Armament: 4–37mm (2×2), 4–25mm (2×2)
Other particulars: Not known

Enlarged 'Shanghai' class. Six built in the late 1960s. Only three extant in 1982, since deleted.

HAIJUI class *large patrol craft*

Displacement: 450t full load
Dimensions: 205ft 5in
62m
Machinery: 4 shafts, 4 diesels, 8800bhp = 28kts. Range 750nm at 18kts
Armament: 4–57mm (2×2), 4–30mm (2×2), 2 RBU-1200 ASW RL, 2 DCT, 2 DC racks, mines
Sensors: Radar Pot Head, Round Ball; sonar hull-mounted (see notes)
Complement: 70

Class:
688, *693*, *694*, *697*

Class entering service since 1987. Lengthened version of the 'Hainan' class with more modern AA weapons. *688* and *697* have a Thomson-Sintra SS 12 VDS in place of the after 57mm mount. Four units extant 1995.

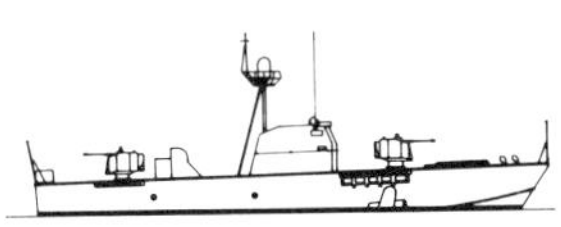

'Shantung' class hydrofoil
NB 1/750 scale

SHANTOU class *coastal patrol craft*

Displacement: *c*70t standard; 80t full load
Dimensions: 84ft oa × 20ft × 6ft 6in
25.7m × 6.1m × 2m
Machinery: 4 shafts, 4 diesels, 3000hp = 28kts. Range 500nm/750nm at 28kts/15kts
Armament: 4–37mm (2×2), 2–12.7mm (2×1), 8 DC
Sensors: Radar Pot Head or Skin Head
Complement: 17

Steel hull, gun-armed Chinese version of Soviet 'P 6' class FAC(T). Approximately eighty were built 1956–60. Transfers: North Korea (eight in 1968) and North Vietnam (twelve). Over half of these ageing boats were deleted between 1976 and 1982, the remainder having been reported as no longer in service *c*1988.

An unsuccessful hydrofoil variant of this class ('Shantung') was built in the 1960s (2 shafts, 2 diesels, 3800bhp = 40kts, 4–37mm (2 × 2)). Only three units of this class were in service *c*1980 and have since been stricken.

'HUANGPU' ('WHAMPOA') class *river patrol craft*

Displacement: 42t standard; 50t full load
Dimensions: 88ft 7in × 13ft × 5ft
27m × 4m × 1.5m
Machinery: 2 shafts, 2 diesels, 1000bhp = 14kts. Range 400nm at 9kts
Armament: 4–25mm (2×2) (see notes)
Sensors: Radar Skin Head
Complement: 25

Built in Shanghai and Canton 1950–55. Low freeboard and inadequate propulsion; ten deleted 1977, five in 1978 and seven in 1981. Some armed with 2–37mm (1×2) and 2–12.7mm MG (1×2). Remainder deployed in 1982: twenty-five in E Sea, eight in N Sea and seven S Sea.

'YULIN' class *river patrol craft*

Displacment: 10t full load
Dimensions: 40ft × 9ft 6in × 3ft 6in
12.8m × 2.9m × 1.1m
Machinery: 1 shaft, 1 diesel, 300bhp = 20–24kts
Armament: 2–12.7mm MG (2×1)
Complement: 10

Built in Shanghai 1964–68. Transfers: four to Congo (1966), three to Cambodia and four to Tanzania. In 1982 forty were extant but are being phased out.

OTHER COASTAL PATROL CRAFT

Thirty 'Behai' or 'Pen Hai' class (80t full load, 90ft, 3 diesels, 900bhp = 18kts, 4–25mm in twin mounts), built in 1950s, most manned by Naval Militia and all now deleted.

Thirty 'Fukien' or 'Fuijan' class (65ft, 2–12.7mm MG, 1–7.62mm MG): river patrols in E Sea Fleet area. All now deleted.

'Ying Kou' class (30t, 70ft, 1 diesel, 300bhp = 16kts, 4–12.7mm MG in twin mounts); built early 1960s. All now deleted.

Four 'Taishan' class (60t, 95ft, diesels, 2–37mm, 2–12.7mm); built 1950s and 1960s. All now deleted.

MINE WARFARE VESSELS

Soviet 'T 43' class *ocean minesweepers*

Class (deployment):
341 (E Sea), *342* (E Sea), *364–366* (N Sea), *375–380* (E Sea), *386–389* (S Sea), *396–399* (S Sea), *801–803* (N Sea), *807* (N Sea)

Four Soviet units were transferred in 1954–55, two of which were returned in 1960. Another twenty-one were built in Chinese shipyards at Shanghai and Canton from 1956. Last pair completed in 1966. Some early ships had a slightly smaller hull. In the late 1970s three ships were converted to survey vessels and three more to civilian research ships. Production resumed in the mid-1980s. Approximately forty-three extant 1995.

WOOSUNG class *ocean minesweeper*

One *c*250t–300t ship (2–37mm, 4–14.5mm MG) was built in 1970–72, but she (pennant number 535) was unsuccessful. No more were built and she was transferred to North Vietnam in 1974.

WOSAO class *inshore minesweeper*

Displacement: 310t full load
Dimensions: 146ft 11in × 20ft 4in × 7ft 5in
44.8m × 6.2m × 2.3m
Machinery: 2 shafts, 2 diesels, 2000bhp = 15kts
Armament: 4–25mm (2 × 2), mines
Sensors: Radar Type 756
Complement: 17

Class (completed):
4422 (1988)

One unit built by the Wusung Shipyard in Shanghai in 1988. Possibly four others have since been built. Steel-hulled. Design has been offered for export. *4422* extant 1995.

FUSHUN class *coastal minesweepers*

Particulars: See 'Shanghai II' class above.

Twenty units built since 1976. Version of 'Shanghai II' class large patrol boat with sweeping gear replacing after gun mount. Number extant 1995 unclear, but several may have been discarded.

FUTI class *drone minesweepers*

Displacement: 47t full load
Dimensions: 68ft 8in × 13ft 9in × 4ft 3in
20.9m × 4.2m × 1.3m
Machinery: 1 shaft, 1 diesel = 11kts
Armament: None
Complement: 3 (for ferrying)

Approximately sixty built since 1984. Remote-controlled, but can be manned. Few in service 1995, as reportedly poor seaboats. Can sweep for both magnetic and acoustic mines.

NOTE
Up to eighty auxiliary minesweepers (modified fishing boats) in service 1995.

Siette de Agosto as completed — Navarret Collection

Colombia

Colombia has traditionally possessed one of the smallest navies in South America. The nation's river flotilla is as important as the deep water fleet. The short service time of many units suggests the Navy is experiencing maintenance problems. There are 7200 officers and men serving with the Colombian Navy and Colombia also has 6800 marines. National service is for a period of two years. The country's main naval base is at Cartagena which also has a naval training school. Balanguero, Barranquilla, Santa Marta and Buenaventura, also naval bases, are less important.

With all the older destroyers retired, the biggest and most powerful Colombian surface warships are the four frigates of the FS 1500 type built in Germany. These handy, helicopter-capable ships are to be better armed (SAM, CIWS, possible even SSM) if the funds become available. Colombia also expressed interest in two 600t missile corvettes built for Iraq in Italy and it was planned to acquire these ships in 1994–95, budget permitting.

FLEET STRENGTH 1947

DESTROYERS

Name	Launched	Disp	Fate
Vouga class			
ANTIQUOIS (ex-*Douro*)	1933	1383t	Discarded 1961
CALDAS (ex-*Tejo*)	1933	1383t	Discarded 1961

Both units were completely refitted 1952–53 and 1954–55 respectively.

RIVER GUNBOATS

Name	Launched	Disp	Fate
Cartagena class			
BARRANQUILLA	10.5.30	142t	Discarded 1970
CARTAGENA	26.3.30	142t	Discarded 1987
SANTA MARIA	16.4.30	142t	Discarded Dec 1962

Also in service in 1947 were the gunboat *Mariscal Sucre* (125t, 1909), and the French-built coast guard patrol craft *Jumin, Carabobo* and *Pinchincha* (120t, 1925). The river flotilla operated three small gunboats, *Fernandez Madrid* (150t, 1926), *Demetrio Salamanca* (120t) and *Presidente Mosquera* (230t, 1912).

MAJOR SURFACE SHIPS

Siette de Agosto 1961

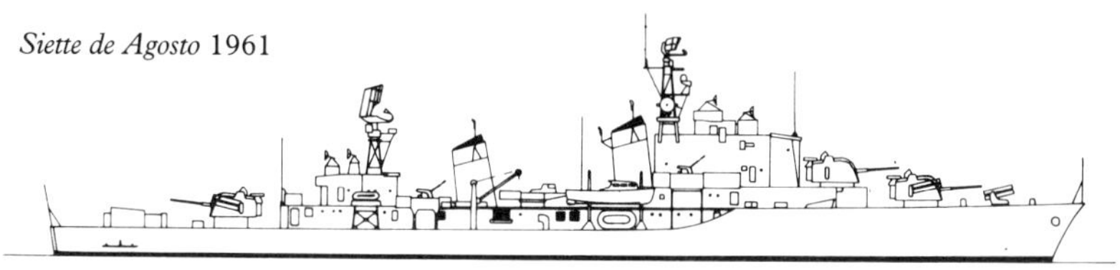

Ordered in 1954, these units were modified *Halland* class destroyers. They mounted a different armament, more electronics, and different interior arrangements than their near sisters in Sweden. In July 1957 *Trece de Junio*'s name was changed to *Siette de Agosto*. *Siette de Agosto* was extensively refitted in the US in 1975–76, including a machinery overhaul – nevertheless 25kts reported as best recent speed, although they made 35kts on trials. Dutch electronics, with M4 type radar directors for gun armament.

Swedish modified HALLAND class *destroyers*

Displacement: 2650t standard; 3300t full load
Dimensions: 397ft 2in pp × 40ft 7in × 15ft 4in max
121.1m × 12.4m × 4.7m
Machinery: 2-shaft De Laval double reduction geared turbines, 2 Penhoët boilers, 55,000shp = 32kts. Range 450nm/3000nm at 32kts/18kts
Armament: 6–120mm (3×2), 4–40mm (4×1), 4–533mm TT (1×4), 1–375mm Bofors ASW RL (1×4)
Sensors: Radar LW-03, DA-02
Complement: 248

No	Name	Builder	Laid down	Launched	Comp	Fate
D 06	SIETTE DE AGOSTO (ex-*Trece de Junio*)	Götaverken, Gothenborg	Nov 1955	19.6.56	31.10.58	Discarded 1986
D 05	VEINTE DE JULIO	Kockums, Malmö	Oct 1955	26.6.56	15.6.58	Discarded 1984

Ex-US FLETCHER class *destroyer*

No	Name	Builder	Acquired	Fate
D 01	ANTIOQUIA (ex-*Hale*)	Bath Iron Wks	Dec 1960	Discarded 20.12.73

Santander received a FRAM II modernisation while in US service.

Ex-US ALLEN M SUMNER class *destroyers*

No	Name	Builder	Acquired	Fate
D 02	CALDAS (ex-*Willard Keith*)	Bethlehem, San Pedro	1.7.72	Discarded 1977
D 03	SANTANDER (ex-*Waldron*)	Federal, Kearny	30.10.73	Discarded 1986

Tayrona Mar 1982 L & L van Ginderen

Almirante Padilla 15.5.83 L & L van Ginderen

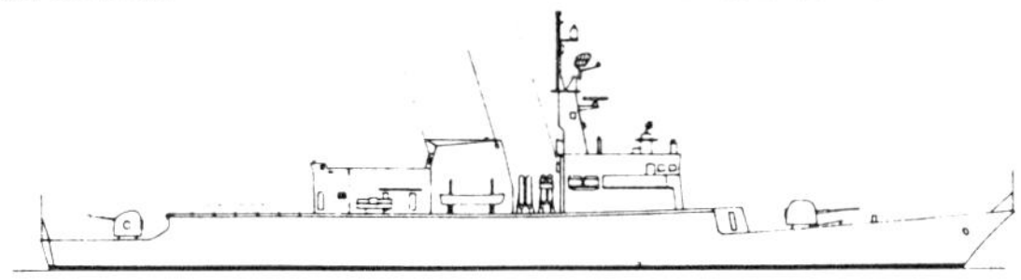

FS 1500 type

Four frigates ordered in 1980 from German yard to the FS 1500 design. Fitted with fin stabilisers. Provided with hangar and landing pad and equipped with MBB BO 105 CB light helicopter. Combat data system Thomson-CSF TAVITAC, fire control Thomson-CSF Vega II and two Canopus optronic directors. Fitted with CSEE Dagaie chaff launcher. Empty space before the bridge is to be filled by SAM launcher in the future. Two ships of modified type FS 1500 serve with Malaysian navy.

West German FS 1500 type *frigates*

Displacement: 1500t standard; 2100t full load
Dimensions: 295ft 3in × 37ft × 11ft 1in
93.5m × 11.3m × 3.7m
Machinery: 2 shafts, 4 MTU 20V 1163 TB92 diesels, 23,400bhp = 27kts. Range 7000nm at 14kts
Armament: 8 MM40 Exocet (2 × 4), 1–76mm OTO Melara Compact, 2–40mm Breda (1 × 2), 6–324mm Mk 32 ASW TT (2 × 3), 1 helicopter
Sensors: Radar Thomson-CSF Sea Tiger, Castor IIB; sonar Krupp Atlas ASO 4–2
Complement: 90

No	Name	Builder	Laid down	Launched	Comp	Fate
51	ALMIRANTE PADILLA	Howaldtswerke, Kiel	17.3.81	6.1.82	31.10.83	Extant 1995
52	CALDAS	Howaldtswerke, Kiel	14.6.81	23.4.82	14.2.84	Extant 1995
53	ANTIOQUIA	Howaldtswerke, Kiel	22.6.81	28.8.82	30.4.84	Extant 1995
54	INDEPENDIENTE	Howaldtswerke, Kiel	22.6.81	21.1.83	24.7.84	Extant 1995

Cordoba was decommissioned in 1980 and is used as a museum. She was the only one of the above to receive a FRAM II modernisation.

Ex-US CROSLEY/BUCKLEY class *destroyer escorts/high speed transports*

No	Name	Builder	Acquired	Fate
DT 07	ALMIRANTE BRION (ex-*Barber*)	Norfolk N Yd	8.12.68	Discarded 1974
DT 03	ALMIRANTE PADILLA (ex-*Tollberg*)	Bethlehem, Hingham	14.8.65	Discarded 1973
DT 04	ALMIRANTE TONO (ex-*Bassett*)	Consolidated, Orange	6.9.68	Discarded 1977
DY 15	CORDOBA (ex-*Ruchamkin*)	Philadelphia N Yd	24.11.69	Museum 1980

An ex-US destroyer escort of the *Dealey* type, originally built 1955–57.

Ex-US DEALEY class *destroyer escort*

No	Name	Builder	Acquired	Fate
D 07	BOYACA (ex-*Hartley*)	New York SB	8.7.72	Extant 1995

All served in the Korean War.

Ex-US TACOMA class *patrol frigate*

No	Name	Builder	Acquired	Fate
FG 14	ALMIRANTE BRION (ex-*Burlington*)	Consolidated Steel, San Pedro	1953	Discarded 1968
FG 11	ALMIRANTE PADILLA (ex-*Groton*)	Walter Butler, Superior	Mar 1947	Discarded 1965
FG 12	CAPTAIN TONO (ex-*Bisbee*)	Consolidated Steel, San Pedro	Feb 1952	Discarded 1962

SUBMARINES

Both units ordered 1971; see Greek *Glavkos* class for details.

West German Type 209/1200 *submarines*

No	Name	Builder	Laid down	Launched	Comp	Fate
SS 28	PIJAO	Howaldtswerke, Kiel	1.4.72	10.4.74	17.4.75	Extant 1995
SS 29	TAYRONA	Howaldtswerke, Kiel	1.5.72	16.7.74	18.7.75	Extant 1995

These units (similar to those ordered by Pakistan and Taiwan) can carry two submarine vehicles and eight frogmen, with 2050g of explosives.

Italian Type SX 506 *midget submarines*

Displacement: 58t/70t
Dimensions: 75ft 5in × 6ft 7in × 13ft 3in
23.0m × 2.0m × 4.0m
Machinery: Diesel electric, 300bhp = 8kts/6kts. Range 1200nm at 7kts
Complement: 5

No	Name	Builder	Comp	Fate
SS 20	INTREPIDO	Cosmos, Livorno	1972	Extant 1995
SS 21	INDOMABLE	Cosmos, Livorno	1972	Extant 1995
SS 23	RONCADOR	Cosmos, Livorno	1974	Stricken *c*1981
SS 24	QUITO SUENO	Cosmos, Livorno	1974	Stricken *c*1981

MORROSQUILLO class *landing craft*

Class (no):
Morrosquillo (TM 246), *Uraba* (TM 247), *Bahia Honda* (TM 248), *Bahia Portete* (TM 249), *Bahia Solano* (TM 250), *Bahia Cupica* (TM 251), *Bahia Utria* (TM 252), *Bahia Malaga* (TM 253)

Eight LCU 1466A landing craft built for the US Army in 1954 and transferred 1991–92 (347t full load, 3 diesels = 7kts, 2–12.7mm MG). Used as transports. All extant 1995.

SMALL SURFACE COMBATANTS

Ex-US CHEROKEE class *patrol vessels*

Class (no, former name, acquired):
Pedro de Heredia (RM 72, ex-*Choctaw*, 1961), *Sebastian de Belalcazar* (RM 73, ex-*Carib*, 15.3.79), *Rodrigo de Bastedas* (RM 74, ex-*Hidatsa*, 15.3.79), *Bahia Solano* (RM 75, ex-*Jicarilla*, 15.3.79).

Pedro de Heredia was purchased outright in 1978. Ex-US tugs used for patrol and rescue duties, armed only with 1–3in gun. *Bahia Solano* deleted 1987, remainder extant 1995.

Ex-US ASHEVILLE class *patrol gunboats*

The *Quita Suevo* (111, ex-*Tacoma*) and *Albuquerque* (112, ex-*Welch*) were reported leased to Colombia in 1981. Transferred to Customs Service 1992. Both extant 1995.

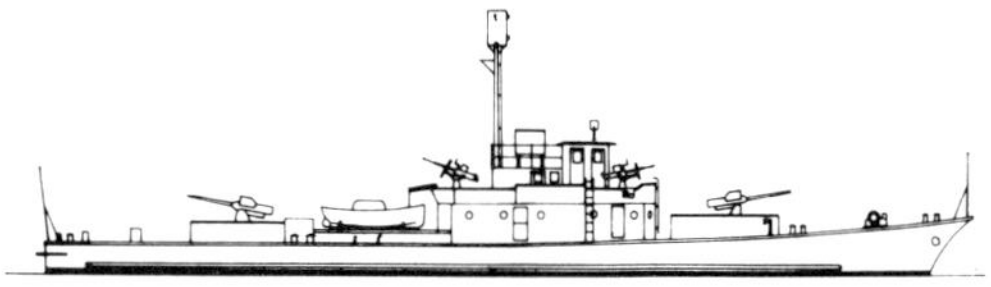

Arauca 1957
NB 1/750 scale

ARAUCA class *river gunboats*

Displacement:	184t
Dimensions:	163ft 3in oa × 23ft 6in × 2ft 9in max *49.8m × 7.2m × 0.8m*
Machinery:	2 shafts, 2 Caterpillar diesels, 916bhp = 13 kts
Armament:	2–76mm (1×2), 4–20mm (4×1)
Complement:	43

Class:
Arauca, Leticia and *Rio Hacha* built by Union Industrial de Barranquilla; launched 1955; completed 1956; extant 1995. *Leticia* converted to a hospital boat in the early 1970s and armament removed. Extant 1995.

OTHER PATROL CRAFT

Colombia also operated two coastal patrol craft (*Espartana*, 50t, commissioned 1950, stricken 1993) and *Capitan R D Binney* (23t, commissioned 1947, extant 1995) both built by Ast Naval, Cartagena and six river patrol craft. Colombia also has a coast guard and a separate customs service.

Six river patrol craft *Juan Lucio, Alfonso Vargas, Fritz Hagale, Humberto Cortes, Calibio, Carlos Galindo* (LR 122–124, 126–128) delivered in 1952–54 (33t full load, 23.2m, 280bhp = 13kts, 2–20mm). Extant 1995.

Three patrol craft *Pedro Gual, Esteban Jaramillo, Carlos E. Restrepo*, delivered 1964 by Schurenstedt KG (85t, 32.9m, 2450bhp = 26kts, 1–20mm). First two deleted in 1984 and third in 1986.

Three Finnish *Ruissalo* class, *Carlos Alban, Jorge Soto del Corval, Nito Restrepo* delivered in 1971 (130t full load, 33m, 2500bhp = 17kts, 2–20mm). Second deleted in 1984 and third in 1990.

Two *Swift* 105ft class *Rafael del Castillo y Rada* (GC 102), *Olaya Herrera* (AN 201) delivered 1983 (103t full load, 31.5m, 5320bhp = 25kts, 1–40mm Bofors, 2–12.7mm), *Olaya Herrera* was badly damaged in 1986. Extant 1995.

Two *Swift* 110ft class *Jose Maria Palas* (GC 103), *Medardo Monzon* (GC 104) commissioned 1989–90 (95t full load, 33.5m, 2 diesels, 25kts, 1–40mm, 1–12.7mm, 2 MG). Extant 1995.

Two US Mk III PBs *Jaime Gomez* (GC 105) and *Nepomuceno Pena* (GC 106) delivered in 1990 (34t full load, 690bhp = 28kts, 1–40mm, 1–20mm), original guns replaced by 2–12.7mm, 2 MG and 1 Mk 19 grenade launcher. Extant 1995.

Two large patrol craft *Juan Nepomuceno Eslava* (113) and *Jose Maria Garcia de Toledo* (114) built by Bender Marine in 1994. Extant 1995.

Nine *Tenerife* class river patrol craft built by Bender Marine (12.4m, 3 diesels, 850bhp = 29kts, 3–12.7mm MG, 2–7.62mm MG, 1 grenade launcher). Acquired 1993, all extant 1995.

Two *Octubre* class river patrol craft built by Swiftships (18ft full load, 13.7m, 2 diesels = 32kts, 2–12.7mm MG, 2–7.62mm MG). Acquired 1992, both extant 1995.

Five *Rio* class river patrol craft (ex-US PBR Mk II) built 1970 and acquired 1989 (9.8m, 2 diesels = 24kts, 2–12.7mm MG, 1–7.62mm MG, 1–60mm mortar). All extant 1995.

There are also two 'Rotork' type and twelve *Delfin* class patrol craft of 12.7m and 7.9m respectively.

Comoros

This four–island group NW of Madagascar declared unilateral independence from France in 1975; France later recognised the transfer of power, but allowed one of the islands, Mayotte, to remain French. The French LCT 9061 (ex-HMS *Buttress*, LCT(8) 4099) was transferred in 1976, as in inter-island transporter. Two Japanese Maritime Safety Agency *Yamazuri* class patrol boats, *Kasthala* and *Ntringhui*, were delivered in October 1981.

Congo

The history of Congo (formerly Middle Congo, of French Equatorial Africa) since independence in 1960 has been turbulent and confused. Another coup in 1979 returned the People's Republic to its version of Marxism-Leninism (which it renounced in July 1990). There were in the 1980s about 100 Soviet advisers and a similar number of East Germans in the country; several hundred Cubans, many of them military advisers, occupy strategic sites in the country.

Congo (population 2,346,000 million), like Guinea and Angola, is not, however, a Soviet satellite in the eastern European sense. She retains freedom of action in foreign and economic affairs and has improved relations with China. The main trade and development partner is France (Congo remains a member of the franc zone) with assistance also from the USA and European countries for specific projects.

The Congolese armed forces have expanded rapidly in the past few years, and the army has been largely involved in policing civil disturbances brought about by allegations of vote-rigging in national elections. The French and Chinese military aid formerly received was mostly replaced by the technical and other assistance from Cuba already mentioned. The navy (formed 1963) operates four fast attack craft from its base at Pointe-Noire, and has also some smaller Brazzaville-based patrol craft for the very long Congo and Ubargi river frontier with Zaire.

Ex-Soviet 'SHERSHEN' class *fast attack craft (gun)*

Three transferred in December 1979 without TT. All discarded by 1989.

Ex-Chinese 'SHANGHAI' class *fast attack craft (gun)*

Three transferred in March 1975 and numbered P 401 to P 403. Renumbered P 201 to P 203. Out of service 1995.

Spanish PIRANA class *large patrol craft*

Displacement:	125t standard; 134t full load
Dimensions:	118.7ft × 19ft × 6.2ft *36.2m × 5.8m × 1.9m*
Machinery:	2 shafts, 2 Bazán/MAN V8V 16/8 TLS diesels, 3200bhp = 25kts. Range 1100nm at 17kts
Armament:	1–40mm Breda-Bofors (144rds), 2–20mm Oerlikon (1000rds), 2–12.7mm Browning M2 MG
Sensors:	Radar Raytheon RN-1220
Complement:	19

Class (no):
Marien Ngouabi (ex-*L'Intrepide*, P 601) launched 7.7.82 and completed Nov 1982, *Les Trois Glorieuses* (ex-*L Vaillant*, P 602), *Les Maloango* (ex-*Terrible*, P 603).

Very similar to Spanish *Barcelo* class FAC. Ordered from Bazán, Cadiz, in August 1980. All arrived Pointe-Noire 3 April 1983. Only P 602 was operational as of 1995.

MISCELLANEOUS

The French seaward patrol craft *VC 4* (P 754) purchased in 1962 and renamed *Reine N'Galifourou*. Returned to France in 1965, and subsequently transferred to Senegal.

There are three ex-Soviet 'Zhuk' class coastal patrol craft (*V 301* to *V 303*) extant 1995, one ex-Chinese *Yulin* class patrol boats (transferred 1966) out of service 1995 and about twelve other small craft for river patrol duties.

Costa Rica

Although Costa Rica possesses both Pacific and Caribbean coastlines and covers an area of some 19,653 square miles, she has no military forces of any significance. The Civil Guard, numbering about 1200 men, is responsible along with a police force of some 3000 for internal security, the Army having been abolished in 1948. In an area of political upheaval and uncertainty (Costa Rica is bordered by Nicaragua to the north and Panama to the south) the country remains stable and operates only ten patrol craft. These include a 105ft 'Swift' class fast patrol boat, five smaller 65ft 'Swift' class patrol boats, all six built in 1979 and three coastal patrol craft built in the mid-1950s (10t), one ex-USCG 'Cape' class craft and one ex-USCG 'Point' class craft. There are also two 36ft and one 42ft 'Swift' inshore patrol craft.

Cuba

The pre-Castro Cuban Navy was totally dependent upon the United States. Since independence at the turn of the century, Cuba, politically and militarily, had been dominated by American policies. The fleet reflected this. In 1959 the Cuban Navy was composed of ex-US warships, most of which had been built during the Second World War. Many of these were comparatively larger types such as *Tacoma* class frigates; their equipment was old and the ships were not well maintained. Many of these ships were built as ASW platforms and much of the US training aid was in this speciality. This force was ill-suited for a war against Castro's guerrilla forces and reflected the ultra-conservative doctrines of both the Juan Batista military and the American advisers, although the Cuban Navy had a history of political involvement. The assumption and resumption of power by Juan Batista brought with it purges of the officer corps of all services and a resulting loyalty to the regime – as well as a lack of military professionalism and poor leadership. During the years immediately preceding Fidel Castro's emergence the navy was increasingly restless, and accordingly harassed by the Batista forces. Like all Latin American navies, the Cuban Navy was conservative. Officers were generally from the middle or lower upper class. Naturally, a group with this background must come to odds with the Castro leadership. In July 1959, Raul Castro, Commander-in-Chief of the Armed Forces, assigned thirty-six politically indoctrinated army officers to the navy. In September the Marine Corps was ordered to be dissolved. The cohesiveness of the navy was diluted by assigning personnel, including ships' crews, to various construction projects, mainly the improvement of transit facilities.

The new Cuban Navy was as dependent upon the Soviet Union as the old one had been upon the United States. Since 1962 the backbone of the fleet has been guided missile patrol boats. In fact, these were the first such craft in the Western Hemisphere. These missile boats have been supplemented by other patrol craft. Soviet naval aid took on far greater dimensions with the transfer of submarines, beginning in 1979 and continued with Soviet financial aid. A 'Koni' class frigate was transferred in 1981, with two more following in 1984 and 1988. Cuba's greatest maritime asset is her strategic location. The long-rumoured Soviet submarine base at Cienfuegos brought this painfully into focus. This key location married to Fidel Castro's professed desire to export revolution and his willingness to act as a Soviet surrogate, made the Cuban Navy one of the most important in the Western Hemisphere, despite its relatively small size. Main naval bases are at Canasi, Cienfuegos, Havana, Mariel and Punta Ballenatos; 13,500 officers and men, including 1000 marines, serve with the Cuban Navy and there is a three-year period of national service.

Currently probably all acquisitions of new material for the navy will be stopped owing to the country's grave economic crisis, and difficulties will be encountered with maintenance of the existing ex-Soviet equipment, as no aid from Russia is expected. All this will undoubtedly lower the level of efficiency of the naval forces.

FLEET STRENGTH 1947

Name	Launched	Disp	Fate
Cuba (sloop)			
CUBA	10.8.11	2055t	Discarded 1971
Patria (sloop)			
PATRIA	10.8.11	1200t	Discarded 1955
Baire (gunboat)			
BAIRE	1906	500t	Discarded 1948

Patria operated as a training ship for the latter years of her service. There were also a number of coast guard patrol craft: the Cuban-built *Leoncio Prado* (80t, 1946); the ex-US type *Camaguey Habana*, *Las Villas*, *Oriente* and *Pinar del Rio*; and the 45t ex-US Coast Guard *GC 11–14*, *22*, *32–34*. The older *Donativo* (130t, 1932) and *Matanzas* (80t, 1912) were also on strength. Three US-built Vosper type MTBs (*R 41–R 43*) were employed as air-sea rescue vessels.

MAJOR SURFACE SHIPS

Ex-Soviet 'KONI' (Type 1159) class *frigates*

Class: 350, 353, 356

Three 'Koni' (Type 1159) frigates were transferred on 24 September 1981, 8 February 1984 and 10 April 1988 respectively. These ships are said to have no names, however the name *Mariel* was mentioned in connection with the first unit. They are similar to the Algerian version of the type with continuous superstructure. Two of them are usually based at Mariel. These are the most important units of the small Cuban Navy. All three extant 1995. See under Algeria for details.

Ex-Soviet 'FOXTROT' (Type 641) class *submarines*

Class: 725, 727, 729

Three large patrol submarines of Soviet Type 641 transferred in 1979, 1980 and 1984 respectively. Submarines No 725 and 727 were under refit at Havana as of 1991. The Cuban navy also received one 'Whiskey' (Type 613) submarine in 1979 which is non-operational and used as battery charging station. All extant 1995 but likely to be disposed of in the near future.

AMPHIBIOUS WARFARE VESSELS

Ex-Soviet 'POLNOCNY B' (Type 771) class *landing ships*

Class: 690, 691

Two 'Polnocny B' (Type 771) landing ships built by Stocznia Polnocna in Gdansk, Poland were transferred by the Soviet Navy in September and December 1982. Extant 1995.

Ex-Polish 'EICHSTADEN' (Type 709) class *landing craft*

Two LCPs of Polish Type 709 transferred from Poland in 1990, probably to replace the last craft of the Soviet 'T 4' class.

Cuba operates seven 'T 4' class LCMs which are used for harbour duties.

SMALL SURFACE COMBATANTS

Ex-US TACOMA class *patrol escorts*

Class (transfer date):

Antonio Maceo (ex-*Peoria*, 1947), *Jose Marti* (ex-*Eugene*, 1947), *Maximo Gomez* (ex-*Grand Island*, 1947)

All units modernised 1956 at Key West, Florida; all discarded 1975. *Jose Marti* fitted as a flagship and sunk as a target 1975.

Ex-US PCE class *patrol escort*

Class (transfer date):

Caribe (ex-*PCE 872*, 1947), *Siboney* (ex-*PCE 893*, 1947)

Both units modernised 1956 at Key West, Florida. *Caribe* had box deckhouse removed 1953. Discarded 1976 and 1973 respectively.

Ex-Soviet 'PAUK II' (Type 1241.2) class *corvette*

One 'Pauk II' class corvette built by Yaroslav SY was transferred in May 1990. Transfer of following units slowed or halted probably due to the financial difficulties of both parties. Similar to the ships serving in Indian Navy. Extant 1995.

Ex-Soviet 'KOMAR' (Type 183-R) class *fast attack craft (missile)*

Eighteen units transferred – twelve in 1962, four during 1963–66, two in 1967; four discarded in 1979. Nine deleted 1982–84 and the rest by 1990.

Ex-Soviet 'OSA I' (Type 205) class *fast attack craft (missile)*

Six units transferred – two in 1972, two in 1973 and two in 1974. One unit deleted in 1981. Five extant 1995.

Ex-Soviet 'OSA II' (Type 205-M) *fast attack craft (missile)*

Thirteen units transferred – two in 1976, one in 1978, four in 1979, two in 1981 and four in 1982. All units extant 1995.

Ex-US PC type *submarine chaser*

Class:
Baire (ex-*Tribesman*, ex-*PC 790*). Acquired from Honduras in 1956 and re-armed; sank at dock, Isle de Pines, 17 April 1961 during an abortive coup.

Ex-Soviet 'KRONSTADT' (Type 122-bis) class *large patrol craft*

Six units towed by merchant ships to Cuba in February 1962; two units discarded in 1979 and the other four in 1984.

Ex-Soviet 'SO 1' (Type 210–M) class *large patrol craft*

Twelve units; six transferred in 1964 and six in 1967; two discarded 1979, one in 1981, three in 1984, two in 1985, two in 1990 and two by 1992.

Soviet 'P 6' (Type 183) class *fast attack craft*

Twelve units towed by merchant ships to Cuba in 1962; six units discarded 1979; six discarded 1985. Depth charge and mine carrying capabilities.

Ex-Soviet 'P 4' class *fast attack craft (torpedo)*

Twelve units transferred 1962–64. Seven deleted in 1984, five in 1985.

Ex-Soviet 'TURYA' (Type 206–M) *fast attack craft (hydrofoil)*

Class: 101, 102, 108, 112, 130, 165, 178, 180, 193.

Two units transferred in 1979, two in 1980, two in 1981, two in 1983 and the last one in November 1983. Fitted with four torpedo tubes. All extant 1995.

Ex-Soviet ZHUK (Type 199) class *coastal patrol craft*

A total of forty of these small patrol craft were transferred from Soviet Union since 1971, the last in 1989. Few have been deleted since. About twenty-seven extant 1995. Used by Coast Guard, pennant numbers painted in red.

Ex-Soviet STENKA (Type 205–E) class *fast attack craft (patrol)*

Two of these craft were transferred from the USSR in February 1985 and one in August 1985. Similar to the units used by KGB, but with TTs and sonar removed. Extant 1995.

MINE WARFARE VESSELS

Ex-Soviet 'SONYA' (Type 1265) class *minesweepers/hunters*

Two units transferred in August and December 1980, two in January and December 1985. Pennant numbers 560, 561, 571, 578. Extant 1995.

Ex-Soviet 'YEVGENYA' (Type 1258) class *inshore minesweeper*

Class: 501–504, 507, 509, 510–514, 538.

Two transferred in 1977, one in 1978, two in 1979, two in 1980, two in 1981, one in 1982 and four in 1984. These boats are organised into two squadrons. All extant 1995.

Ex-Polish 'K 8' (Type 361-T) class *minesweeper*

One unit transferred in 1978. Deleted 1982.

Cyprus

The Republic of Cyprus, third largest island in the Mediterranean, became independent of British rule in 1960 and has been strife-torn ever since. Civil war between the 80 per cent Greek population and the Turkish minority together with a wider war between Greece and Turkey were only just averted in the period 1963–67. A UN peacekeeping force (UNFICYP) was introduced in March 1964 while Britain continues to hold the two Sovereign Base Areas of Akrotiri and Dhekelia, totalling 99 square miles, in the south of the island.

Both these outside forces were powerless in the face of the Turkish invasion in July 1974, precipitated by the Cypriot National Guard's overthrow of Archbishop Makarios. Makarios returned as president only to find that 38 per cent of the island had become the 'Turkish Federated State of Cyprus' with settlers from the mainland (only 50 miles away) as well as 40,000 troops reinforcing the Turkish minority.

This enforced partition remains the position today and prevents the creation of a significant Cypriot Navy despite the island's large, although mainly 'flag of convenience', merchant marine.

Currently two small Cypriot naval forces exist side by side, one of the Republic of Cyprus and the other of the Turkish Republic of Cyprus, where independence was declared in 1983, but is recognised only by Turkey. The patrol force of about 320 officers and ratings from the Republic of Cyprus has acquired a few patrol craft from Germany and Yugoslavia.

Ex-Soviet 'P 4' class *fast attack craft (torpedo)*

For a considerable period six ex-Soviet 'P 4' class MTBs served in Cypriot waters, four transferred October 1964 (?1962) and two February 1965 (probably two extra engines have been supplied since that time). However the majority were sunk during the Turkish invasion of July 1974; one undamaged craft was beached and captured (now preserved in a marine museum at Istanbul).

French *fast attack craft (missile)*

Two (80t, four SS12 SSM, 2–20mm AA) ordered from CN de l'Esterel, Cannes, were not taken up and were transferred to Greece as *Kelefstis Stamou* and *Diopos Antoniou*. Both commissioned 1975. Cypriot names unknown. Wooden hull. For further details see under Greece.

French '32L' type *patrol craft*

Displacement:	96t full load
Dimensions:	32.1m oa × 6.5m × 0.9m
Machinery:	2 shafts, 2 SACM UD33 V12M5 diesels, 4000bhp = 30kts. Range 1500nm at 15kts
Armament:	1–40mm/70 Breda, 1–12.7mm MG
Complement:	22

Class:
Salamis (P 01, in service 1982)

This fast patrol craft was built in 1981–82 by CN de l'Esterel, Cannes, for Naval Command of National Guard. Belongs to the Republic of Cyprus. Extant 1995.

Ex-German R41 group *motor minesweepers (patrol craft)*

Three units (launched 1943), Cypriot and ex-German names unknown. Were taken up 1962–63 from mercantile use and rearmed as patrol boats with 1–40mm and/or 2–20mm. One destroyed by Turkish air attack on 8 August 1964 at Xeros (beached near Xeros habour). One (with radar) had pennant no 15 early 1970s. Further fates of two ships unknown – probably lost during the Turkish operations of July 1974 (one stricken 1973?).

COASTAL PATROL CRAFT

Two FAC-23 Jet patrol craft, *Evagoras* (PV 20), *Posidon* (PV 21), built in Yugoslavia by Brodotehnika SY, in service 1995 (60t full load, 80ft 9in, 3560bhp, KaMeWa waterjets, 45kts, 1–20mm Oerlikon).

Two patrol craft acquired in 1992 from Germany, possible names *Agathos* and *Panagos*.

Two 58ft patrol craft *Aphrodite* (PL 1), *Kimon* (PL 2) built by CN de l'Esterel, in service 1995. Controlled by Cyprus Maritime Police.

Two Yugoslav-built patrol craft *Adonis*, *Kiniris*, delivered in 1991.

PATROL CRAFT USED BY TURKISH REPUBLIC OF CYPRUS

Raif Denktas (74), transferred from Turkish Coast Guard in 1988.

Two patrol craft *102*, *103* delivered by Taskizak Yard in 1990 and 1991 (25t full load, 47ft 11in, 18kts).

MISCELLANEOUS

Several of ten former fishing boats (*c*50t, 1–2 MG or 20mm) were probably lost July 1974. Some unconfirmed reports give purchase of three Type '101' MTBs in early 1970s from Yugoslavia.

Five SAB 12 type harbour patrol craft (ex-GDR MAB 12 class) bought from Germany in 1992. All extant 1995.

Czechoslovakia

Czech Republic and Slovakia 1993 on

After the Second World War Czechoslovakia, freed from Axis occupation mainly by the Red Army, was rebuilt within the pre-1938 boundaries (with the exception of Transcaucasian Ukraine which was incorporated into the Soviet Union) while her internal affairs were remodelled on the Soviet pattern.

As a result of attempts to introduce political reforms Czechoslovakia was invaded in 1968 by most of the Warsaw Pact countries to restore the Communist regime. From this time Soviet troops were permanently stationed in the country.

Being landlocked, Czechoslovakia does not maintain a navy, but there is a river patrol force operated by 1200 Border Guard personnel wearing naval type uniforms. In 1990 this force consisted of about 18 armed launches.

Following the overthrow of the Communist regime at the end of 1989 and the break-up of the Warsaw Pact, Czechoslovakia successfully negotiated the withdrawal of Soviet troops by mid-1991. In December 1991 the country became an associate member of the European Community and applied to become a member of NATO. On 1 January 1993, following a referendum, the country was divided into the Czech Republic and Slovakia. It is believed that the Czech Republic has retained control of the river force.

Denmark

In the 1930s the majority of the Danish people and at least two political parties were in agreement that military defence of the country was unfeasible. The role of the armed forces was relegated to that of border watch and the troops were ordered to offer no resistance when the Germans invaded on 9 April 1940. Officially, Denmark was now under German 'protection', but the Danish Government was forced to 'lend' six of the most modern torpedo-boats to Germany, and more such loans were expected. Resistance and sabotage increased and by 1943 relations between the Danes and the occupiers had deteriorated. On 29 August the Germans launched a *'coup'* and the occupation became official. As had been planned, the Danish Navy scuttled most of its ships, but a few small boats escaped to neutral Sweden. When the war ended, the Danish Navy had ceased to exist. Several German warships were allocated to Denmark as war reparations, and she received other vessels, chiefly from Britain, on loan.

With Norway, Denmark became a member of NATO and gradually enlarged her navy with ships built at Copenhagen Navy Yard and various civilian yards, and with vessels lend-leased from the USA. Together with the new German Federal Republic, Denmark found herself responsible for the protection of the narrow waters of the Baltic – a role which laid heavy emphasis on motor torpedo-boats, submarines, minelayers and minesweepers. Some frigates and corvettes were built, and various patrol boats for use off Greenland. Despite some financial difficulties, the Danish Navy has kept up a programme of progressive modernisation, either refitting existing vessels (as in the case of the submarine force and the *Falster* class minelayers) or commissioning new classes, such as the multi-role *Flyvefisken* class and the *Thetis* class patrol frigates. In 1995 the navy had 4769 regular personnel, plus 5500 reservists and 4030 naval Home Guard. Bases are at Fredrikshaven (submarines) and Korsor (surface vessels), with a small facility at Gronnedal in Greenland.

FLEET STRENGTH 1947

FRIGATES

Name	Launched	Disp	Fate
British 'River' class			
HOLGER DANSKE (ex-*Monnow*)	4.12.43	1445t	Stricken 1959
NIELS EBBESEN (ex-*Annan*)	29.12.43	1445t	Stricken 1963

Both purchased 1945 and became TS.

CORVETTE

Name	Launched	Disp	Fate
THETIS (ex-*Geranium*)	23.4.40	898t	Stricken 1963

Purchased from Britain 1945

TORPEDO-BOATS

Name	Launched	Disp	Fate
HAVKATTEN	1919	108t	Stricken Nov 48
Ex-German T 4	1938	962t	BU 1950
Ex-German T 19	18.12.41	853t	BU 1951
Najaden class			
HUITFELDT (ex-*Nymfen*)	22.6.43	782t	Stricken 1966
WILLEMOES (ex-*Najaden*)	17.3.43	782t	Stricken 1966

*Huitfeldt c*1963 Royal Danish Navy

SUBMARINES

Name	Launched	Disp	Fate
Havmanden class			
HAVMANDEN	19.6.37	340t	Discarded 1949–50
HAVRUEN	6.11.37	340t	Discarded 1949–50
HAVKALEN	3.3.38	340t	Discarded 1949–50
HAVHESTEN	11.7.40	340t	Discarded 1949–50
Ex-British 'U' class			
SPRINGEREN (ex-Polish *Dzik*, ex-*P52*)	11.10.42	540t	Returned Oct 1957, BU
STOREN (ex-*Vulpine*)	28.12.43	545t	Returned Jan 1959, BU
SAELEN (ex-French *Morse*, ex-*Vortex*)	19.8.44	545t	Returned Jan 1958, BU

All the *Havmanden* class were scuttled 29 August 1943, but were salvaged. The 'U' class boats were chartered from the Royal Navy at the end of the war; they were originally known by their pennant numbers only, as *U 1*, *U 2* and *U 3* respectively.

DENMARK

MINELAYERS

Name	Launched	Disp	Fate
LINDORMEN	30.3.40	614t	Stricken 1969
LAALAND	1941	350t	Stricken 1974
LOUGEN	1941	350t	Stricken 1974

They later wore the NATO pennant numbers (in the above order) N 39, N 40 and N 41.

MINESWEEPERS

Name	Launched	Disp	Fate
NARHVALEN (ex-*TB*)	1917	110t	Stricken 1949
Söbjörnen class (classified as patrol boats 1957)			
SOHESTEN	30.4.42	274t	Stricken 1962
SOHUNDEN	16.5.42	274t	Stricken 1959
SOLOVEN	3.12.42	274t	Stricken 1959
SORIDDEREN	11.4.42	274t	Stricken 1962
MS I class			
MS I SORTE SARA	1941	70t	Stricken 1956
MS 2 ASKO	1941	70t	Stricken 1971
MS 3 BAAGO	1941	70t	Stricken 1968
MS 5 ENO	1941	70t	Stricken 1971
MS 6 FANO	1941	70t	Stricken 1960
MS 7 HJORTO	1941	70t	Stricken 1968
MS 8 LYO	1941	70t	Stricken 1971
MS 9 MANO	1941	70t	Stricken 1971
MS 10 STRYNO	1941	70t	Stricken 1956

In 1957 the six remaining boats were reclassified as patrol boats. Other small minesweepers were: ex-German *MR 26* (1938, 100t, stricken 1949); ex-German *MR 152*, MR 150 *Asnæs*, MR 155 *Bognæs*, MR 156 *Drynæs*, MR 157 *Egenæs*, *MR 160*, MR 167 *Helgenæs*, MR 168 *Lynæs*, *MR 170*, *MR 173*, MR 174 *Rinkenæs*, *MR 175*, *MR 176*, *MR 214* (1940–43, 130t, stricken 1950–57); ex-German *MR 226*, *MR 229*, MR 230 *Stigsnæs*, *MR 233*, MR 236 *Trellenæs*, MR 242 *Vornæs* (1943, 150t, stricken 1950–57); *ME 26*, *ME 32*, *ME 36*, *ME 83*, *ME 84*, *ME 86*, *ME 152–157*, *ME 160*, *ME 173*, *ME 263*, *ME 307* (*c* 1940–42, 240t, on loan from RN, ten returned 1948, remaining six in 1950); *ME 1016*, *ME 1038*, *ME 1042*, *ME 1044* (1943–44, 360t, on loan from RN, returned 1950); MSK 1 *Aalholm*, MSK 2 *Birkholm*, MSK 3 *Ertholm* (1944–45, 70t, later patrol boats, stricken 1973, 1970 and 1976 respectively); MSK 4 *Fryholm*, MSK 5 *Græsholm*, MSK 6 *Lindholm* (1945, 68t, later patrol boats, stricken 1970, 1968 and 1974 respectively); ML 1 *Graadyb*, ML 2 *Klördyb*, ML 3 *Vejdyb* 1945, 21t, stricken 1956, 1969 and 1969 respectively).

*Lindormen c*1965 Royal Danish Navy

*Stigsnaes c*1955 Royal Danish Navy

PATROL BOATS

Name	Launched	Disp	Fate
HEIMDAL	1.2.35	705t	Stricken 1961
TERNEN	1937	82t	Stricken 1957
FREJA	23.12.38	332t	Stricken 1968

Ternen was designed for use off Greenland. *Heimdal* and *Freja* carried NATO pennant numbers A 542 and A 541 respectively.

ICEBREAKERS

Name	Launched	Disp	Fate
MJOLNER	1890	515t	Stricken 1958
ISBJORN	1923	1675t	Stricken 1969
LILLEBJORN	1926	1000t	Stricken 1975
STOREBJORN	1931	2540t	Stricken 1976

*Esbern Snare c*1962 Royal Danish Navy

MAJOR SURFACE SHIPS

British 'Hunt' Type 2 escort destroyers, originally launched in 1941. Initially lent for 4 years, extended for further four years. Laid down 10 February 1941, 12 June 1940 and 7 June 1940 respectively; completed 14 April 1942, 11 December 1941 and 18 October 1941 respectively. Reconstructed 1953–54, and then carried 6–4in (3×2), 3–40mm (3×1), 2 DCT.

Ex-British 'HUNT' class *frigates*

Particulars: As British 'Hunt' Type 2

No	Name	Builder	Acquired	Fate
F 341	ESBERN SNARE (ex-*Blackmore*)	Stephen	1953	BU 1966
F 342	ROLF KRAKE (ex-*Calpe*)	Swan Hunter	1953	BU 1966
F 343	VALDEMAR SEJR (ex-*Exmoor*)	Swan Hunter	1953	BU 1966

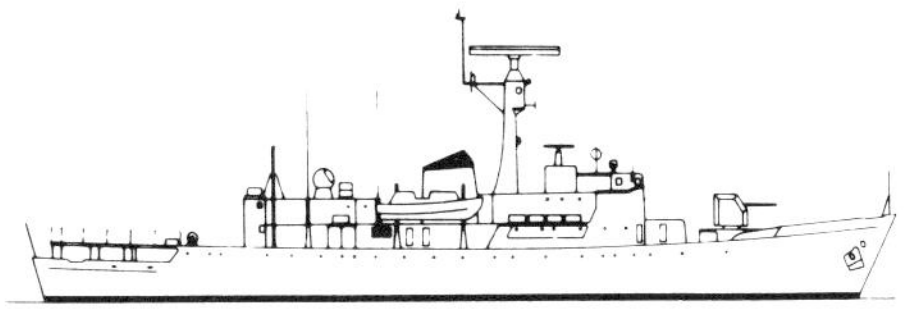

Hvidbjörnen 1974

Officially rated as 'inspection vessels'. Ordered 1960–61. Specially designed and equipped for fishery protection in Greenland, Faroe and North Sea waters. Originally carried an Alouette III helicopter; later replaced with Lynx. CWS-1 radar replaced by AWS-6 in large radome atop mast: *Hvidbjörnen* and *Vaedderen* so equipped in 1987 and the remaining pair in 1988.

HVIDBJORNEN class *frigates*

Displacement: 1345t standard; 1650t full load
Dimensions: 219ft 8in pp, 238t 2in oa × 38ft × 16ft
67.0m, 72.0m × 11.5m × 5.0m
Machinery: 1 shaft, 4 General Motors diesels, 6400bhp = 18kts. Range 6000nm at 13kts
Armament: 1–3in DP, DCT, 1 helicopter
Sensors: Radar CWS-1, NWS-1; sonar MS-26
Complement: 85

No	Name	Builder	Laid down	Launched	Comp	Fate
F 348	HVIDBJORNEN	Aarhus Flydedok	4.6.61	23.11.61	15.12.62	Stricken 1992
F 349	VAEDDEREN	Aalborg	30.10.61	6.4.62	19.3.63	Stricken 1992
F 350	INGOLF	Svendborg	5.12.61	27.7.61	27.7.63	Survey duties 1982; stricken 1992
F 351	FYLLA	Aalborg	27.6.62	18.12.62	10.7.63	Stricken 1991

Vaedderen as completed

Navarret Collection

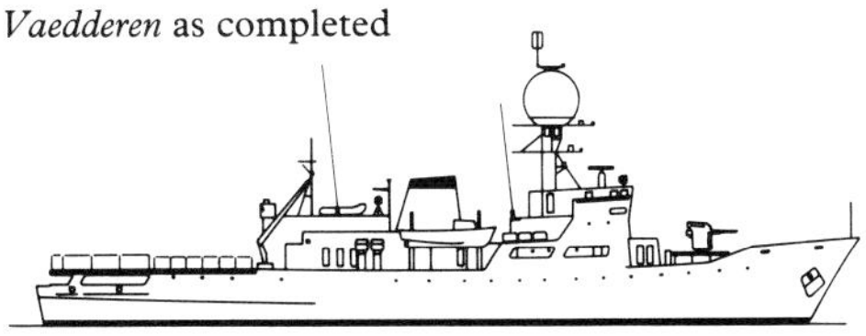

Beskytteren 1986

Officially rated as 'inspection vessel' (fishery protection). Modified *Hvidbjörnen* class, specially designed and equipped for Greenland service. Alouette III helicopter was later replaced by Lynx. Received same AWS-6 radar as *Hvidbjörnen* class. Planned for refitting for continued service.

BESKYTTEREN *frigate*

Displacement: 1540t standard; 1970t full load
Dimensions: 239ft 6in wl, 244ft oa × 41ft × 14ft 9in
73.0m, 74.4m × 12.5m × 4.5m
Machinery: 1 shaft, 3 Burmeister & Wain Alpha diesels, 7440bhp = 18kts. Range 4500nm at 16kts, 6000nm at 13kts
Armament: 1–3in DP, 1 helicopter
Sensors: Radar CWS-2, NWS-1, NWS-2; sonar MS-26
Complement: 60

No	Name	Builder	Laid down	Launched	Comp	Fate
F 340	BESKYTTEREN	Aalborg	15.12.74	27.5.75	27.2.76	Extant 1995

Beskytteren as completed

Royal Danish Navy

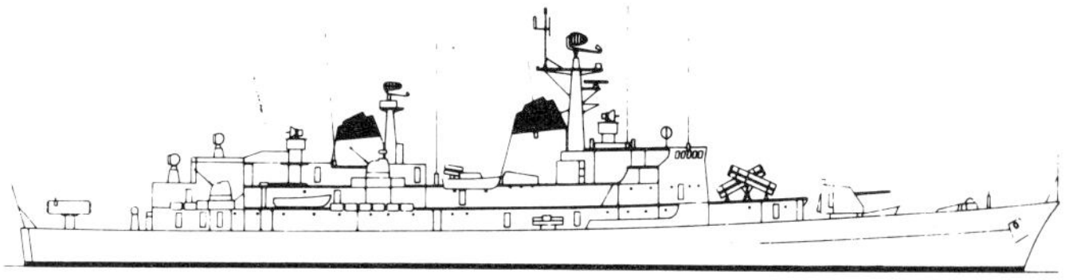

Herluf Trolle 1977

Indigenous Danish design, but built with US 'Offshore' funds. The original triple TT mounting was replaced by two twin sided amidships, and the Terne ASW missile provided for in the design was never fitted. Modernised 1976–78, when the 'B' 5in twin mount and TTs were replaced by Harpoon SSM launchers (2×4), a Sea Sparrow SAM launcher (1×8) and 6–324mm TT (2×3) for Swedish wire-guided ASW torpedoes installed. *Herluf Trolle* suffered a serious engine-room fire in 1982, returning to service after repairs in 1983. In 1987 both ships were placed in reserve with only maintenance crews, and were later stricken.

PEDER SKRAM class *frigates*

Displacement: 2200t standard; 2720t full load
Dimensions: 354t 3in pp, 396ft 5in oa × 39ft 5in × 14ft 1in deep
108.0m, 112.6m × 12.0m × 4.3m
Machinery: 2-shaft CODOG: 1 MTU 20 V 956 diesel, 4800bhp = 18kts; General Electric LM 2500 gas turbine, 18,400shp = 28kts. Range 800nm at 28kts or 2500nm at 18kts
GG4A-3 gas turbines, 37,000shp = 28kts (44,000shp = 32kts max)
Armament: 4–5in/38 (2×2), 4–40mm Bofors (4×1), 3–21in (533mm) TT (1×3), DC
Sensors: Radar CWS-2, CWS-3, NWS-1, NWS-2, M-66, Mk 91; sonar MS-26
Complement: 200

No	Name	Builder	Laid down	Launched	Comp	Fate
F 352	PEDER SKRAM	Helsingör	25.9.64	20.5.65	30.6.66	In reserve 1987; stricken
F 353	HERLUF TROLLE	Helsingör	18.12.64	4.9.65	16.4.67	In reserve 1987; BU 1995

Herluf Trolle c1970

Ambrose Greenway

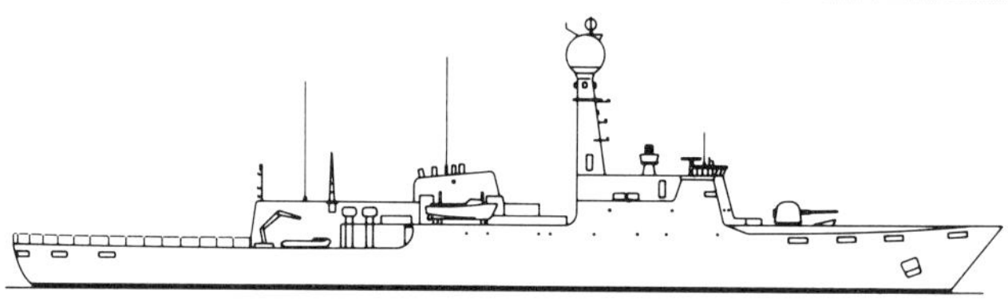

Thetis as completed

'Stanflex 2000' design ordered in October 1987 to replace the *Hvindbjörnen* class. Employed principally for fishery protection and survey duties. In 1991 *Thetis* was equipped to serve as an oil survey ship off Greenland, under lease to the Nunaoil consortium. The class has a helicopter deck and hangar for one Sea Lynx helicopter. Provisional plans were laid to arm these ships to frigate standard, with Harpoon SSM (2×4), two octuple Sea Sparrow VLS and enhanced sensors and ECM. However financial difficulties have meant that these plans are increasingly unlikely to bear fruit.

THETIS class *frigates*

Displacement: 2600t standard; 3500t full load
Dimensions: 327t 3in pp, 369ft 1in oa × 47ft 3in × 19ft 8in
99.75m, 112.5m × 14.4m × 6.0m
Machinery: 1 shaft, 3 MAN-Burmeister & Wain 12V 28/32 diesels, 12,000bhp = 21.5kts. Range 8500nm at 15.5kts
Armament: 1–76mm/62 OTO-Melara, 1–20mm, 2 DC racks, 1 helicopter
Sensors: Radar Mil 009, FR 1505, AWS-6; sonar CTS-36 hull-mounted, Salmon HF VDS (see notes); ECM Sabre intercept, Scorpion jammer
Complement: 61 (11 officers)

No	Name	Builder	Laid down	Launched	Comp	Fate
F 357	THETIS	Svendborg	10.10.88	14.7.89	1.7.91	Extant 1995
F 358	TRITON	Svendborg	27.6.89	16.3.90	2.12.91	Extant 1995
F 359	VAEDDEREN	Svendborg	19.3.90	21.12.90	9.6.92	Extant 1995
F 360	HVIDBJORNEN	Svendborg	2.1.91	11.10.91	30.11.92	Extant 1995

Hvidbjörnen 13.9.93
L & L van Ginderen

Niels Juel 14.6.88
L & L van Ginderen

All vessels built at part of MDAP; sisters of the Italian *Albatros* class. Originally classified as submarine chasers, from 1954 they were referred to as corvettes, although they carry a frigate's pennant number.

BELLONA class *corvettes*

Displacement: 800t standard; 900t full load
Dimensions: 250ft 4in wl, 259ft 2in oa × 31ft 9in × 9ft 10in deep
76.3m, 79.0m × 9.6m × 3.0m
Machinery: 2 shafts, 2 Ansaldo Fiat 409T diesels, 4400bhp = 20kts. Range 3000nm at 18kts
Armament: 2–3in (2×1), 1–40mm, 2 Hedgehog, 4 DCT
Sensors Radar NWS-1, CWS-1; sonar QCU-2
Complement: 109

No	Name	Builder	Laid down	Launched	Comp	Fate
F 344	BELLONA	Navalmeccanica, Castellammare	1954	9.1.55	31.1.57	Stricken 1981
F 345	DIANA	Cantiere del Tirreno, Riva Trigoso	1953	19.12.54	30.7.55	Stricken 1974
F 346	FLORA	Cantiere del Tirreno, Riva Trigoso	1953	25.6.55	28.8.56	Stricken 1978
F 347	TRITON	Cantiere Navali di Taranto	1953	12.9.54	10.8.55	Stricken 1981

Olfert Fischer as completed

NIELS JUEL class *corvettes*

Displacement: 1190t standard; 1320t full load
Dimensions: 262ft 6in pp, 275ft 6in oa × 33ft 9in × 13ft deep
80.0m, 84.0 × 10.3m × 4.0m
Machinery: 2-shaft CODOG: 1 MTU 20 V 956 diesel, 4800bhp = 18kts; 1 General Electric LM 2500 gas turbine, 18,400shp = 28kts. Range 800nm at 28kts or 2500nm at 18kts
Armament: 1–76mm, 8 Harpoon SSM (2×4), 1 Sea Sparrow SAM system (1×8), DCs
Sensors: Radar AWS-5, Skanter 009, Philips 3cm, 9LV-200, 1 Mk 91 Mod 1; sonar MS-26
Complement 90

No	Name	Builder	Laid down	Launched	Comp	Fate
F 354	NIELS JUEL	Aalborg	20.10.76	17.2.78	3.1.79	Extant 1995
F 355	OLFERT FISCHER	Aalborg	29.11.78	10.5.79	29.2.80	Extant 1995
F 356	PETER TORDENSKJOLD	Aalborg	5.12.79	30.4.80	30.3.81	Extant 1995

Contracted 5 December 1975. An interesting design incorporating equipment from many countries – the 76mm gun is Italian (OTO Melara compact), missile systems and gas turbines are American, the electronics are British and Dutch, etc. The planned ASW torpedo system was never fitted, leaving the class with extremely limited ASW capability. It was originally planned to fit them with the US RAM (Rolling Air Frame) SAM system (two 10-round launchers per ship), but in 1991 the programme was opened up to other competitors, including the French-built SADRAL SAM and the British VISRAD system (using Starstreak missile). On 6 April 1988 *Olfert Fischer* was damaged in a collision with a ferry, and in 1991 she served with UN forces in the Gulf. *Niels Juel* also suffered collision damage in March 1992. In 1995 the class was to receive a refit, with the Sea Sparrow system to be re-equipped with the later RIM-7P missile and improved electronics.

SUBMARINES

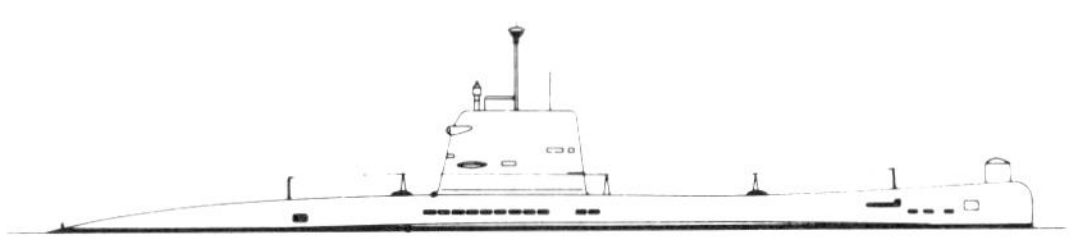

Delfinen 1960 **NB 1/750 scale**

First Danish-designed submarines built since the war. *Springeren* was built with US 'Offshore' funds. Replaced in service by the ex-Norwegian *Kobben* class boats.

DELFINEN class *submarines*

Displacement: 595t standard; 643t submerged
Dimensions: 178ft 10in × 15ft 5in × 13ft
54.5m × 4.7m × 4.0m
Machinery: 2-shaft Burmeister & Wain 12cyl V-diesels plus Brown-Boveri electric motors, 1200bhp/1200shp = 15kts/15kts. Range 4000nm at 8kts
Armament: 4–533mm TT (bow)
Sensors: Sonar, active and passive types
Complement: 33

No	Name	Builder	Laid down	Launched	Comp	Fate
S 326	DELFINEN	Copenhagen N Yd	1.7.54	4.5.56	16.9.58	Decomm 1983, stricken 11.9.84
S 327	SPAEKHUGGEREN	Copenhagen N Yd	1.12.54	20.2.57	27.6.59	Discarded 1989
S 328	TUMLEREN	Copenhage N Yd	22.5.56	22.5.58	15.1.60	Stricken Aug 1981
S 329	SPRINGEREN	Copenhagen N Yd	3.1.61	26.4.63	22.10.64	Discarded 1989

DENMARK

Licence-built boats of German Type 205 (West German *U 4*), modified for Danish needs. They are to be modernised to bring them up to the same standard as the ex-Norwegian *Kobben* class boats (see below), with the Atlas PSU-83 sonar suite, new EW equipment and new batteries. *Narhvalen* modernised 1993–94, *Nordkaperen* to begin in 1995.

NARHVALEN class *submarines*

Displacement: 370t standard; 500t submerged
Dimensions: 146ft × 15ft 4in × 12ft 6in
44.5m × 4.6m × 3.8m
Machinery: 1 shaft, 2 Mercedes-Benz diesels, 2 electric motors, 1500bhp/1500shp = 12/17kts
Armament: 8–533mm TT (bow)
Sensors: Sonar; active and passive types
Complement: 22

No	Name	Builder	Laid down	Launched	Comp	Fate
S 320	NARHVALEN	Copenhagen N Yd	16.2.65	10.9.68	27.2.70	Extant 1995
S 321	NORDKAPEREN	Copenhagen N Yd	20.1.66	18.12.69	22.12.70	Extant 1995

Three units purchased 1986 as replacements for *Delfinen* class. A further two units were wanted by the Danish Navy, but were not acquired due to lack of funds. The class was refitted at Bergen during 1987–91, receiving the Krupp-Atlas PSU-83 sonar suite, new fire-control system and a propulsion overhaul. The refit resulted in the hull being lengthened by 5ft 3in (*1.6m*). The ex-Norwegian *Kya* was purchased in 1989 to replace the original third unit *Stadt*, which was damaged beyond repair in 1987. *Saelen* sank whilst under tow to Denmark, but was raised and returned to service in 1993.

Ex-Norwegian KOBBEN class *submarines*

Particulars: See Norwegian *Kobben* class (and notes below)

No	Name	Builder	Transferred	Fate
S 322	TUMLEREN (ex-*Utvaer*)	Rheinstahl Nordseewerke	1989	Extant 1995
S 323	SAELEN (ex-*Uthaug*)	Rheinstahl Nordseewerke	1990	Extant 1995
S 324	SPRINGEREN (ex-*Kya*)	Rheinstahl Nordseewerke	1991	Extant 1995

Springeren 4.11.92 L & L van Ginderen

SMALL SURFACE COMBATANTS

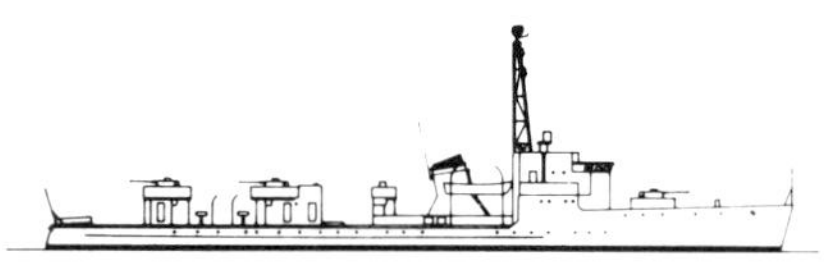

Buhl 1954

BILLE class *torpedo-boats*

Displacement: 392t standard; 400t full load
Dimensions: 210ft × 21ft 1in × 7ft 6in
64.0m × 6.4m × 2.3,
Machinery: 2-shaft Danish Atlas geared steam turbines, 6900shp = 29kts
Armament: 3–40mm (3×1), 2–20mm (2×1), 2 DCT, 6–456mm TT (2×3), mines
Complement: 65

No	Name	Builder	Launched	Fate
T 11	BILLE	Copenhagen N Yd	21.9.46	Stricken 1960
T 12	BUHL	Copenhagen N Yd	10.9.47	Stricken 1960
T 13	HAMMER	Copenhagen N Yd	23.6.48	Stricken 1960
T 14	HOLM	Copenhagen N Yd	5.2.48	Stricken 1960
T 15	KRABBE	Copenhagen N Yd	27.6.46	Stricken 1960
T 16	KRIEGER	Copenhagen N Yd	4.5.46	Stricken 1960

Planned 1942, drawings and material ready at armistice 1945. Reclassified as patrol boats 1952 with pennant numbers P 570-P 575. Rebuilt 1953–55, TTs exchanged for ASW weapons.
(For illustrations see under Germany in 1922–46 volume)

Ex-German S-BOAT type *torpedo-boats*

Displacement: 112–121t standard
Dimensions: 107ft 4in pp, 114ft 5in oa × 17ft 5in × 5ft 1in
32.7m, 34.9m × 5.3m × 1.7m
Machinery: 3 shafts, 3 Daimler-Benz 20cyl 4-stroke diesels, 7500bhp = 39/42kts. Range 700/750nm at 30/35kts. Oil 17t
Armament: 1–37mm, 3–20mm (3×1), 2 MG, 2–533mm TT (2×1)
Complement: 24/40

Class:
T 51 (ex-*S 316*), *T 52* (ex-*S 107*), *T 53* (ex-*S 216*), *T 54* (ex-*S 133*), *T 55* (ex-*S 206*), *T 56* (ex-*S 127*), *T 57* (ex-*S 305*), *T 58* (ex-*S 79*), *T 59* (ex-*S 197*), *T 60* (ex-*S 97*), *T 61* (ex-*S 207*), *T 62* (ex-*S 68*)

These twelve ex-German *Schnellboote* were purchased 1947 and 1948, also ex-*S 15* and ex-*S 122* in 1947 for 'cannibalising' with spare parts for the other twelve. In 1951 six boats were transferred from Norway: *Lyn (ex-S 64) Brand* ex-*S 10*), *Storm* (ex-*S 85*), *Blink* (ex-*S 302*), *Tross* (ex-*S 117*), *Kjekk* (ex-*S 195*). In 1951 all eighteen boats received new pennant numbers and names in the following order: P 551 *Glenten*, P 552 *Gribben*, P 553 *Havörnen*, P 554 *Haerfuglen*, P 555 *Högen*, P 556 *Isfuglen*, P 557 *Jagtfalken*, P 558 *Musvaagen*, P 559 *Raagen*, P 560 *Ravnen*, P 561, *Skaden*, P 562 *Stormfuglen*, P 563 *Taarnfalken*, P 564 *Tranen*, P 565 *Falken*, P 566 *Hejren*, P 567 *Lommen*, P 568 *Viben*. The gun armament was later reduced to 1–40mm, 1–20mm. *Haefuglen* stricken 1955, *Högen*, *Isfuglen* and *Musvaagen* 1956, *Havörnen*, *Raagen* and *Taarnfalken* 1957, *Glenten*, *Gribben*, *Skaden*, *Falken* and *Lommen* 1961, *Jagtfalken*, *Ravnen*, *Stormfuglen* and *Tranen* 1963, *Hejren* and *Viben* 1965.

FLYVEFISKEN class *torpedo-boats*

Displacement: 110t standard
Dimensions: 118ft 2in pp, 120ft 2in oa × 18ft × 5ft 11in
36.0m, 36.6m × 5.5m × 1.8m
Machinery: 3-shaft Mercedes-Benz diesels, 7500bhp = 40kts
Armament: 1–40mm, 1–20mm, 2–533mm TT
Complement: 23

Class:
Flyvefisken (P 500), *Hajen* (P 501), *Havkatten* (P 502), *Laxen* (P 503), *Makrelen* (P 504), *Sværdfisken* (P 505)

All built at Copenhagen N Yd. Voted 1952, laid down 1953, launched 1954–55. Development of German S-boat type. All stricken 1974 and scrapped 1976.

Flyvefisken as completed Royal Danish Navy

Glenten c1970 — Royal Danish Navy

FALKEN class *torpedo-boats*

Displacement: 119t standard
Dimensions: 118ft × 17ft 10in × 6ft
36.0m × 5.4m × 1.8m
Machinery: 3 shafts, MTU diesels, 9000bhp = 40kts
Armament: 1–40mm, 1–20mm, 4–533mm TT (4×1)
Sensors: Radar NWS-1
Complement: 22

Class:
Falken (P 506), *Glenten* (P 507), *Gribben* (P 508), *Högen* (P 509)

Laid down, launched, commissioned respectively 1.11.60, 19.12.61, 4.10.62 – 3.1.61, 15.3.62, 15.12.62 – 15.5.61, 18.7.62, 26.4.63 – 1.9.61, 4.10.62, 6.6.63. All built at Copenhagen N Yd as part of US Military Aid Programme. All stricken 1978–79.

Söridderen on trials — Vosper Thornycroft

SOLOVEN class *torpedo-boats*

Displacement: 100t standard; 120t full load
Dimensions: 96ft 2in pp, 99ft 2in oa × 26ft 3in × 8ft 3in
29.3m, 30.2m × 8.0m × 2.5m
Machinery: 3-shaft CODOG: 3 Bristol Proteus gas turbines, 12,750shp = 50kts; 2 GM 6V 71 diesels (on outer shafts), ?bhp = 10kts
Armament: 2–40mm (2×1), 4–533mm TT (4×1), mines
Sensors: Radar NWS-1
Complement: 21

Class:
Söloven (P 501), *Söridderen* (P 511), *Söbjornen* (P 512), *Söhesten* (P 513), *Söhunden* (P 514), *Söulven* (P 515)

Laid down, launched, commissioned respectively 27.8.62, 19.4.63, 12.2.65 – 4.10.62, 22.8.63, 10.2.65 – 9.7.63, 19.8.64, Sep 65 – 5.9.63, 31.3.65, June 1966 – 18.8.64, 12.1.66, Dec 1966 – 30.3.65, 27.4.66, Mar 1967. The design is a combination of British *Brave* type hull form and *Ferocity* style construction. *Söloven* and *Söridderen* built at Vosper, Portsmouth, remaining four at Copenhagen N Yd. A pair of TT was removed when the after gun was given an enclosed shield. All in reserve by 1988, and all stricken by 1992 as *Flyvefisken* class came into service.

Laxen Aug 1993 — L & L van Ginderen

Norby 30.5.89 — L & L van Ginderen

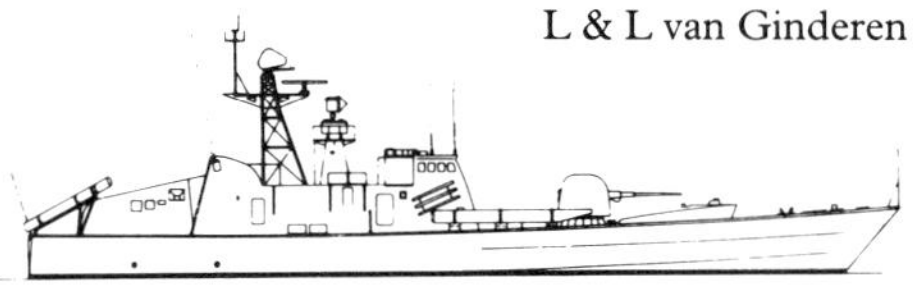

Sehested as completed
NB 1/750 scale

WILLEMOES class *fast attack craft*

Displacement: 240t standard; 260t full load
Dimensions: 144ft 5in, 151ft × 24ft 4in × 7ft 3in
44.0m, 46.0m × 7.4m × 2.2m
Machinery: 3-shaft CODOG: 3 Rolls-Royce Proteus 52M-544 gas turbines, 12,750shp = 40kts; 3 GM V 71 diesels for cruising, 800bhp = 12kts
Armament: 4 Harpoon SSM (2×2), 1–76mm/62, 2–533mm TT (2×1), 6 RFL, mines.
Sensors: Radar NWS-3, 9LV-200
Complement: 24

No	Name	Builder	Comp	Fate
P 540	BILLE	Fredrikshavn Vaerft	Jun 1976	Extant 1995
P 541	BREDAL	Fredrikshavn Vaerft	21.7.77	Extant 1995
P 542	HAMMER	Fredrikshavn Vaerft	1.4.77	Extant 1995
P 543	HUITFELDT	Fredrikshavn Vaerft	15.6.77	Extant 1995
P 544	KRIEGER	Fredrikshavn Vaerft	22.9.77	Extant 1995
P 545	NORBY	Fredrikshavn Vaerft	22.11.77	Extant 1995
P 546	RODSTEEN	Fredrikshavn Vaerft	16.2.78	Extant 1995
P 547	SEHESTED	Fredrikshavn Vaerft	Mar 1978	Extant 1995
P 548	SUENSON	Fredrikshavn Vaerft	Jun 1978	Extant 1995
P 549	WILLEMOES	Fredrikshavn Vaerft	7.10.76	Extant 1995

Based on Swedish *Spica II* class. Armament variations include: 8 Harpoon, no TT; or 4 TT, no Harpoon. First order for four boats, then eight and finally ten. *Willemoes*, the prototype boat, was completed in the latter arrangement. *Bille* was the first series production boat and was modified following the *Willemoes* trials. Three CP propellers are fitted, and normal endurance is 36 hours. The gun is the OTO Melara Compact, and 20 mines can be carried in lieu of TT and missiles. All are to be fitted with a point-defence SAM system, with US Stinger, Swedish RBS 70, French SADRAL or British Starstreak under consideration. In 1989, *Norby* carried out tests with Mistral missiles mounted on the after superstructure.

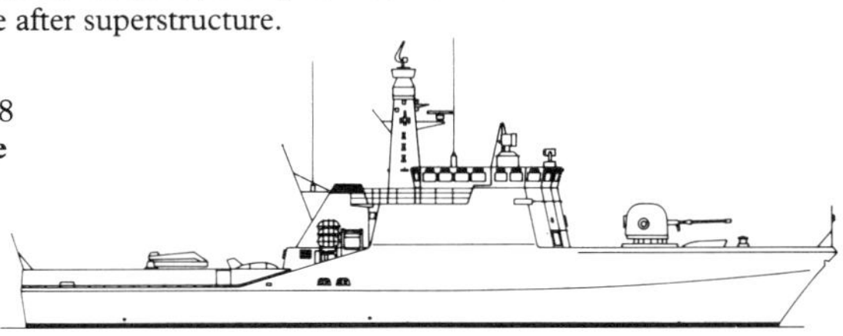

Flyvefisken 1988
NB 1/750 scale

FLYVEFISKEN class *fast attack craft/patrol boats*

Displacement: 320t full load
Dimensions: 164ft 1in pp, 177ft 1in oa × 29ft 6in × 9ft 10in
50.0m, 54.0m × 9.0m × 3.0m
Machinery: 3 shaft CODAG: 1 GE-Fiat LM-500 gas turbine, 5680shp = 35kts; 2 MTU 16V396 TB94 diesels, 6960bhp = 19kts; hydraulic drive slow-speed system, 1 GM 12V71 diesel, 500bhp = 9kts; bow thruster
Armament: Patrol boat and MCM configuration: 1–76mm/62 OTO Melara, 2–12.7mm MG (2×1)
FAC: 4 Harpoon SSM (2×2), 1–76mm/62 OTO Melara, 2–533mm TT (2×1). See notes
Sensors: Radar Terma Pilot, AWS-6; sona TSM 2640 MF hull-mounted; ECM Sabre intercept, Cygnus jammer, 1 decoy RL (1×4)
Complement: Patrol boat: 17 (4 officers)
FAC: 19 (4 officers)

No	Name	Builder	Comp	Fate
P 550	FLYVEFISKEN	Aalborg Vaerft	19.12.89	Extant 1995
P 551	HAJEN	Aalborg Vaerft	19.7.90	Extant 1995
P 552	HAVKATTEN	Aalborg Vaerft	1.11.90	Extant 1995
P 553	LAXEN	Aalborg Vaerft	22.3.91	Extant 1995
P 554	MAKRELEN	Aalborg Vaerft	1.10.91	Extant 1995
P 555	STOREN	Aalborg Vaerft	Apr 1992	Extant 1995
P 556	SVAERDFISKEN	Aalborg Vaerft	1.2.93	Extant 1995
P 557	GLENTEN	Aalborg Vaerft	1993	Extant 1995
P 558	GRIBBEN	Aalborg Vaerft	1994	Extant 1995
P 559	LOMMEN	Aalborg Vaerft	–	Building
P 560	RAVNEN	Aalborg Vaerft	–	Building
P 561	SKADEN	Aalborg Vaerft	–	Building
P 562	VIBEN	Aalborg Vaerft	–	Projected
P 563	–	Aalborg Vaerft	–	Projected

Multipurpose craft of the 'Stanflex-300' design, with interchangeable equipment and armament modules for patrol boat, fast attack craft, MCM and proposed ASW roles. The prototype and first group of six production units was ordered in July 1985, and a further six in June 1990. A total of sixteen vessels was originally planned, but budget cuts have reduced this to fourteen. This class was designed to replace the *Daphne* class patrol boats, the *Söloven* class FAC(T) and the existing minesweepers. It is also planned for them to fulfil auxiliary roles such as fishery protection and oceanographic research. In the patrol boat configuration four Harpoon SSM and two 533mm TT are fitted. In the MCM role, six pairs of Bofors Double Eagle minehunting ROVs have been ordered, and *Havkatten* operates two MRF-1 class minesweeping drones (see below). A further six minesweeping modules are on order.

The original AWS-6 air search radar has been replaced with the Telefunken TRS-3D set, and in November 1993 four Sea Sparrow SAM systems were ordered for this class, with an option for ten more, fitted in standard modular weapon units.

Nymphen c1970 — Royal Danish Navy

DAPHNE class *patrol boats*

Displacement: 150t standard; 170t full load
Dimensions: 121ft 5in × 22ft 4in × 8ft 6in
37.0m × 6.8m × 2.6m
Machinery: 3 shafts, 2 12cyl Maybach diesels, plus 1 Foden FD-6 diesel for cruising, 2600/100bhp = 20kts
Armament: 1–40mm, 2 ASW rocket launchers, 2 DCT, mines
Sensors: Radar NWS-3; sonar MS-26
Complement: 23

Class:
Daphne (P 530), *Dryaden* (P 531), *Havmanden* (P 532), *Havfruen* (P 533), *Najaden* (P 534), *Nymfen* (P 535), *Neptun* (P 536), *Ran* (P 537), *Rota* (P 538)

Built at Copenhagen N Yd, laid down 1959–62, launched 1960–64, commissioned 1961–65. *Havmanden* was stricken in 1979. *Daphne, Najaden* and *Neptun* were disarmed in 1982, and all had been stricken by 1992 as the *Flyvefisken* class came into service.

MINE WARFARE VESSELS

BESKYTTEREN class *minelayers*

Displacement: 753t standard; 1095t full load
Dimensions: 203ft 6in oa × 34ft 2in × 8ft 6in
62.0m × 10.4m × 2.6m
Machinery: 2 shafts, 2 diesels, 280bhp = 13kts. Range 2500nm at 12 kts
Armament: 8–40mm (4×2), mines
Complement: 70

No	Name	Builder	Acquired	Fate
N 50	BESKYTTEREN	Charleston N Yd	1954	Stricken 1966
N 51	VINDHUNDEN	Charleston N Yd	1954	Stricken 1966

Ex-US *LSM 390* and *LSM 392* originally built in 1944–45, reconstructed as minelayers in USA. *Beskytteren* transferred 8 June 1954.

Langeland 1971 **NB 1/750 scale**

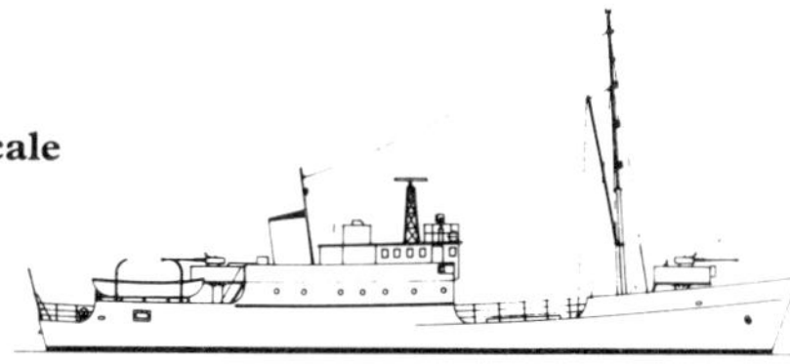

Langeland 1971 (1/750sc)

LANGELAND *minelayer*

Displacement: 310t standard; 330t full load
Dimensions: 134ft 6in pp, 144ft 4in oa × 23ft 8½in × 7ft 3in
41.0m, 44.0m × 7.2m × 2.2m
Machinery: 2 shafts, 2 Burmeister & Wain diesels, 770bhp = 12kts
Armament: 2–40mm (2×1), 2–20mm, mines
Complement: 37

No	Name	Builder	Launched	Fate
N 42	LANGELAND	Copenhagen N Yd	17.5.50	In reserve 1982; stricken

Commissioned 1951. Layer of controlled minefields.

FALSTER class *minelayers*

Displacement: 1800t standard; 1900t full load
Dimensions: 238ft pp, 252ft 6in oa × 41ft × 13ft
73.0m, 77.0m × 12.5m × 4.0m
Machinery: 2 shafts, 2 GM 16-56703 diesels, 4800bhp = 17kts
Armament: 4–3in (2×2), 4–20mm Oerlikon (4×1), 400 mines
Sensors: Radar CWS-2, NWS-2, NWS-3, M-46
Complement: 120

No	Name	Builder	Launched	Fate
N 80	FALSTER	Nakskov	19.9.62	Extant 1995
N 81	FYEN	Fredrikshavn	3.10.62	Extant 1995
N 82	MOEN	Fredrikshavn	6.3.63	Extant 1995
N 83	SJAELLAND	Nakskov	14.6.63	Extant 1995

First two ordered 1960, remainder 1962, commissioned 1963 and 1964 respectively. Of a special NATO design, 'Scandinavian' type of minelayer. *Fyen* also acts as a training ship for midshipmen, and *Sjælland* as a depot ship for submarines and torpedo-boats. The 3in gun is the US Mk 33. All are to be refitted to remain in service until 2010. They are to receive point-defence SAM systems (Stinger or Mistral) and updated combat data systems.

LINDORMEN class *minelayers*

Displacement: 420t standard; 570t full load
Dimensions: 145ft 4in pip, 147ft 8in oa × 26ft 3in × 8ft 3in
44.3m, 45.0m × 8.0m × 2.5m
Machinery: 2 shafts, 2 Wichmann diesels, 4200bhp = 14kts
Armament: 2-20mm (2×1), 60 mines
Sensors: Radar NWS-3
Complement: 27

No	Name	Builder	Launched	Fate
N 43	LINDORMEN	Svendborg	9.9.77	Extant 1995
N 44	LOSSEN	Svendborg	11.11.77	Extant 1995

Laid down 20 January 1977 and 22 June 1977 respectively. Commissioned 26 October 1977 and 30 January 1978 respectively. Primarily layers of controlled minefields but *Lossen* also submarine depot ship.

Falster c1970

Lossen Jan 1993 L & L van Ginderen

Gronsund 25.5.92 L & L van Ginderen

Ex-US AMS/MSC type *coastal minesweepers*

Particulars: As US *Bluebird* class

Class (no, former name, transferred):
Aarösund (M 571, ex-*AMS 127*, 24.1.55), *Alssund* (M 572, ex-*AMS 128*, 5.4.55), *Egernsund* (M 573, ex-*AMS 129*, 3.8.55), *Grönsund* (M 574, ex-*AMS 256*, 21.9.56), *Guldborgsund* (M 575, ex-*AMS 257*, 11.11.56), *Omösund* (M 576, ex-*AMS 221*, 20.6.56), *Ulvsund* (M 577, ex-*AMS 263*, 20.9.56), *Vilsund* (M 578, ex-*AMS 264*, 15.11.56)

NATO coastal minesweepers of the MSC 60 type, built in the USA, launched 1954–56. *Aarösund* stricken and *Omösund* deleted for cannibalisation in 1981. *Alssund, Egernsund* and *Ulvsund* stricken 1988. *Guldborgsund* was fitted with a deckhouse between the funnel and the bridge for survey duties and was stricken in 1992. *Grönsund* and *Vilsund* received the same deckhouse and were extant 1995.

ASVIG class *inshore minesweepers*

Displacement: 180t standard; 200t full load
Dimensions: 106ft 8in pp, 116ft 2in oa × 23ft 8in × 5ft 5in
32.5m, 35.4m × 7.2m × 1.70m
Machinery: 2 shafts, 2 diesels, 1100bhp = 14kts
Armament: 1–20mm
Complement: 14

Class (no, commissioned):
Asvig (M 579, 6.9.61), *Mosvig* (M 580, 25.10.61), *Sandvig* (M 581, 1.2.62), *Saelvig* (M 582, 30.4.62)

Built at Copenhagen Navy Yard 1959–60. All stricken 1977.

MRF 1 class *drone minesweepers*

Displacement: 38t full load
Dimensions: 59ft 9in × 15ft 7in × 3ft 11in
18.2m × 4.75m × 1.2m
Machinery: 1 shaft, 1 diesel, 350bhp = 12kts
Armament: None
Sensors: Radar navigational; sonar TSM 2054 sidescan
Complement: 4 (ferrying)

Class (in service):
MRF 1 (Mar 1991), *MRF 2* (Dec 1991)

Built by Danyard, Alborg for operations in pairs with the *Flyvefisken* class vessels in minesweeper configuration (see above). GRP construction. A further four units have been ordered, out of a proposed ten. Both extant 1995.

MRD 2 Sep 1993 L & L van Ginderen

MISCELLANEOUS

FISHERY PROTECTION VESSELS

Since 1945 Denmark has built a number of trawler-type vessels for fishery patrols in Greenland and Faroes waters. They are not true warships, but since they are operated by the Danish Navy, their principal characteristics are listed below. All were extant in 1995 unless otherwise noted.

Skarven (Y 382), *Teisten* (Y 383) – 125t; 82ft × 20ft 9in × 10ft, *25.0m × 6.3m × 3.0m*; 1 Alfa diesel, 180bhp = 9kts; 1–20mm; launched 1951; *Skarven* grounded 7.5.66 and was stricken; *Teisten* stricken 1978.

Maagen (Y 384), *Mallemukken* (Y 385) – 190t full load; 88ft × 23ft × 10ft, *27.0m × 7.0m × 3.0m*; 1 diesel, 385bhp = 11kts; 2–20mm; launched 1960. Based in Greenland.

Barsö (Y 300), *Drejö* (Y 301), *Romsö* (Y 302), *Samsö* (Y 303), *Thurö* (Y 304), *Vejrö* (Y 305), *Farö* (Y 306), *Læsö* (Y 307), *Römö* (Y 308) - 155t standard; 84ft × 20ft × 10ft, *25.6m × 6.0m × 3.0m*; 1 diesel, 385bhp = 11kts; 2–20mm; launched 1969 (first six), 1973 (last three). All extant 1995.

Agdlek (Y 386), *Agpa* (Y 387), *Tulugaq* (Y 388) 390t full load; 103ft × 26ft × 19ft, *31.4m × 8.0m × 5.8m*; 1 B × W diesel, 800bhp = 12kts, 2–20mm; launched 1974 (except *Tulugaq* 1978). All extant 1995.

There are also large numbers of small patrol boats; *Y 375, Y 376* – 12t, 26kts, 1 MG, built 1974; *Y 377–Y 379*, 7.5t, 26kts, 1 MG, built 1974–75; and about forty inshore patrol craft manned by the Home Guard and numbered between *MHV 1* and *MHV 95*. They were built from the mid-1950s onwards.

The Home Guard also operates six patrol boats of the MHV 800 class: *Aldebaran* (MHV 801), *Carina* (MHV 802), *Aries* (MHV 803), *Andromeda* (MHV 804), *Gemini* (MHV 805), *Dubhe* (MHV 806) – 83t full load, 77ft 9in × 18ft 4in × 6ft 7in, *23.7m × 5.6m × 2.0m*, 13kts, 2–7.62mm MG (2 × 1), crew 8. Replacements for previous Home Guard craft, total of twenty-eight planned. Six in service 1995.

The Ministry of Fisheries operated the patrol vessel *Havörnen*, built to the British 'Osprey' design by Fredrikshavn Vaerft in 1979 – 3680t; 164 × 34ft × 9ft, *50.0m × 10.5m × 2.8m*; 2 diesels; 4640bhp; 1 helicopter. She was built to mercantile specifications, and was unarmed, but is included because the 'Osprey' design has been promoted as a possible warship. Transferred to Namibia November 1992.

The Danish Navy also has the use of four icebreakers operated by the Ministry of Trade and Shipping. They are *Elbjörn* (893t standard, 12kts, launched in 1932); *Danbjörn* and *Isbjörn* (3425t standard, 17kts, launched in 1966 and 1969); and *Thorbjörn* (1500t standard, 16kts, launched 1980). They are employed by the navy for survey work during the summer months.

DANNEBROG *royal yacht*

Displacement:	1130t standard
Dimensions:	246ft × 35ft 9in × 12ft 2in *75.0 × 10.9 × 3.7m*
Machinery:	2 shafts, 2 diesels, 1800bhp = 14kts
Armament:	2–37mm
Complement:	57

Built at Copenhagen N Yd, launched and commissioned 1931. 1979–80 refitted, new machinery. Pennant number A 540. Extant 1995.

Dannebrog Jul 1994 L & L van Ginderen

Djibouti

This tiny state (formerly French Somaliland) independent since 1977 and outlet for half Ethiopia's trade, relies on French land, sea and air forces for its protection. A French Navy minesweeping flotilla has been based there since November 1980. Ninety personnel operate two 35t 'Mousa Ali' class patrol craft donated by France in 1985–86, as well as five Iraqi-built speedboats and other small craft.

Dominican Republic

During this era the numerical strength of the Dominican Navy peaked under the regime of Rafael Trujillo. Trujillo was assassinated in 1961, and many political changes took place. The navy declined in size, and most of the units were renamed in 1962. There are just over 3900 officers and men serving with the Dominican Navy the main bases being at Santo Domingo, Las Calderas and Haina.

FLEET STRENGTH 1947

FRIGATES

Name	Launched	Disp	Fate
British 'River' class			
MELLA (ex-*Presidente Trujillo*, ex-*Carlplace*)	1944	1400t	Extant 1995
British 'Flower' class			
CRISTOBAL COLON (ex-*Lachute*)	1944	980t	Discarded 1978

Ex-Canadian units. *Mella* modified for use as a presidential yacht, renamed *Mella* in 1962; used for training and as a flagship. Transferred in 1946.

Cristobal Colón also cited as *Colón*. Transferred in 1946.

The only other vessels on strength in 1947 were: three 100t launches (*LR 101–103*); *LR 102* lost 1956, survivors renamed *Capitan Alsina*, extant 1995, and *Capitan Maduro*, discarded 1968), launched in 1944; three 47t patrol craft (*GC 9–11*) built in the USA in 1943, and eight smaller launches.

MAJOR SURFACE SHIPS

Ex-British 'H' class *destroyer*

Class:

Duarte (ex-*Trujillo*, ex-*Hotspur*, 1936–48). Purchased November 1948; pennant no D 101. Renamed *Duarte* in 1962; discarded 1972.

Ex-British 'F' class *destroyer*

Class:

Sanchez (ex-*Generalissimo*, ex-*Fame*, 1935–49). Purchased February 1949, pennant no D 102, renamed *Sanchez* in 1962; discarded 1968.

Ex-US ASHEVILLE class *frigate*

Class:

Juan Pablo Duarte (ex-*Natchez*, ex-*Annan*). Purchased July 1947; pennant no F 102; lost 1958.

Generalissimo 1968 L & L van Ginderen

Ex-US TACOMA class *frigates*

Class (former names):
Capitan General Pedro Santana (ex-*Presidente Peynado*, ex-*Pueblo*), transferred September 1947, discarded 1979; *Gregorio Luperon*, ex-*Presidente Troncoso*, ex-*Knoxville*, transferred September 1947, discarded 1979. Pennant nos F 104 and F 103. Both renamed in 1962. Deleted 1982.

Ex-Canadian 'FLOWER' class *corvettes*

Class (former name, transfer date, fate):
Gerardo Jansen (ex-*Peterborough*, 1947, discarded 1972), *Juan Alejandro Acosta* (ex-*Louisbourg*, 1947, discarded 1978), *Juan Bautista Cambiaso* (ex-*Belleville*, 1947, discarded 1972), *Juan Bautista Maggiolo* (ex-*Riviere du Loup*, 1947, discarded 1972), ? (ex-*Asbestos*, 1947, lost en route from Canada; Dominican Navy name not available). One additional unit, *Cristobal Colon*, had been transferred in 1946 and is listed among those ships on hand in 1947. She was discarded in 1978. They carried the pennant nos C 104, C 102, C 103, C 105 and C 101 respectively.

AMPHIBIOUS WARFARE VESSELS

Ex-US LSM type *landing ship*

Class (former name, transfer date, fate):
Sirio (ex-*LSM 483*, March 1958, deleted 1989)

Two further vessels of this type were transferred after the war and served in an auxiliary capacity as *San Rafael* and *Antares*; the former was discarded in 1960 and the latter was sold in 1959.

MINOR LANDING CRAFT

Three ex-US LCIL type served in the Dominican Navy: *BDI 101* and *102* were transferred *c*1948; stricken 1957. A further unit was acquired from the merchant service in 1952, renamed *Paraiso* and served until 1960.

Two LCT(6) type were transferred from the USA *c*1948 and served as *17 de Julio* and *Maria Josefina* until stricken in 1956.

Two LCU type landing craft were built locally by the Astilleros Navales Dominicanos in 1957: *Enriquillo* (ex-*17 de Julio*), and *Samana*. The latter was lost in heavy weather, but replaced by a similar ship of the same name, extant 1995.

SMALL SURFACE COMBATANTS

Ex-US PC type *submarine chasers*

Class (former names, transfer date, fate):
Patria (ex-*Capitan Wenceslao Arvelo*, ex-*PC 1207*, 1947, discarded 1962), *Constitucion* (ex-*Cibas*, ex-*PC 1497*, 1947, discarded 1968), *27 de Febrero* (ex-*PC 613*, 1947, discarded 1968)

Two units were renamed in 1962. Pennant nos P 102, P 103, P 101.

Ex-US COHOES class *corvettes*

Class (pennant number, former name, transfer date):
Cambiaso (P 207, ex-*Etlah*, 29.9.76), *Separacion* (P 208, ex-*Passaconaway*, 29.9.76), *Calderas* (P 209, ex-*Passaic*, 29.9.76).

Former net tenders, but now armed with 2–3in and 3–20mm guns and used as patrol craft. Pennant nos P 207–209. All extant 1995.

Ex-US ADMIRABLE class *corvettes*

Class (pennant number, former name, transfer date):
Prestol Botello (BM 454, ex-*Separacion*, 13.1.65), *Tortuguero* (BM 455, ex-*Signet*, 13.1.65).

Originally minesweepers. Both extant 1995.

Ex-US Coast Guard 165B class *large patrol craft*

Class (former names, transfer date, fate):
Independencia (ex-*Icarus*, 1948, deleted 1993), *Libertad* (ex-*Rafael Atoa*, ex-*Thetis*, 1948, deleted 1993), *Restauracion* (ex-*Galathea* 1948, deleted 1993)

US Coast Guard *Argo* class 165ft cutters (335ft full load), armed with 1–3in, 1–40mm and 1–20mm; *Libertad* was renamed in 1962.

US PGM 71 class *large patrol craft*

Class:
Betelgeuse (ex-US *PGM 77*), pennant number is GC 102, built by Petersen, USA and transferred under the Military Aid Programme (14 January 1966). Extant 1992.

The Dominican Navy also operates the following patrol craft: one large patrol craft *Capitan Alsina* (ex-RL 101) GC 105, built in 1944 and acquired in 1957 (100t, 1000bhp = 17kts, 2–20mm), extant 1995.

Four *Bellatrix* class coastal patrol craft *Procion* (GC 103), *Aldebaran* (GC 104), *Bellatrix* (GC 106) and *Capella* (GC 108) delivered by Sewart in 1967–68 and 1972 (60t, 896bhp = 18.7kts, 3–12.7mm), extant 1995.

Two Swiftships 110ft class large patrol craft *Canopus* (GC 107) and *Orion* (GC 109) delivered by Swiftships in 1984 (93.5t, 1530bhp = 23kts, 1–40mm, 2–12.7mm), extant 1995. *Canopus* renamed *Cristobal Colón*.

Ecuador

Until the acquisition of *Guayas* in 1947, Ecuador possessed a few ex-yachts and small coast guard units. Although Ecuador acquired a few frigate-sized units, most of the equipment remained obsolescent until the mid-1970s. Beginning in 1975, Ecuador began to acquire first class fighting units for the first time. Naval patrol boats are employed primarily for Ecuador's 200-mile coastal zone and it is a common occurrence for American fishing boats to be held up and fined. This action affected the transfer of ships from the US, in particular, the *Gearing* class destroyer *Southerland* in 1981. During the past decade the Ecuadorian Navy was much reinforced by the acquisition of two ex-British *Leander* class frigates, which however are in need of armament update. These ships are well complemented by six Italian-built missile corvettes of the *Esmeraldas* class. Naval bases are at Guayaquil, Manta, San Lorenzo and the Galapagos Islands, Guayaquil being the main base. The Galapagos Islands are also the base for Ecuador's naval infantry (about 1900 men); 600 officers and men are employed by the navy and national service is selective.

MAJOR SURFACE SHIPS

Ex-US GEARING class *destroyer*

No	Name	Builder	Acquired	Fate
DO 01	PRESIDENTE ELOY ALFARO (ex-*Holder*)	Consolidated, Orange	1.9.78	Deleted 1991

Although a FRAM I reconstruction, ASROC was deleted before transfer. Overhauled at Norfolk Shipbuilding, Virginia, 1978–79.

Ex-US TACOMA class *frigate*

No	Name	Builder	Acquired	Fate
E 21	GUAYAS (ex-*Covington*)	Globe, Superior	Aug 1947	Discarded 1972

Pennant no originally E 01.

Ex-British 'HUNT' class (Type 1) *frigates*

No	Name	Builder	Acquired	Fate
D 01	PRESIDENTE ALFARO (ex-*Quantock*)	Scotts	18.10.54	Discarded 1978
D 02	PRESIDENTE VELASCO IBARRA (ex-*Meynell*)	Swan Hunter	18.10.54	Discarded 1978

Both units refitted by J Samuel White, Cowes, Isle of Wight. Commissioned 16 August 1955 and August 1955 respectively

Ex-British LEANDER class *frigates*

No	Name	Builder	Launched	Acquired	Fate
FM-01	PRESIDENTE ELOY ALFAR (ex-*Penelope*)	Vickers	17.8.62	25.4.91	Extant 1995
FM-02	MORAN VALVERDE (ex-*Danae*)	HM Dockyard, Devonport	31.10.65	25.4.91	Extant 1995

Both frigates acquired from UK in 1991 as replacements for discarded ex-US destroyer and destroyer escort. Both are Batch 2 Exocet conversions completed in 1980 and 1982. Torpedo tubes removed before transfer. Transferred without Seacat and Exocet missiles. Biggest units of Ecuadorian Navy, but already over 30 years old.

Ex-US CHARLES LAWRENCE class *destroyer escort*

Class:
Moran Valverde (ex-*Veinticinco de Julio*, ex-*Enright*)

Transferred 14 July 1967; purchased 30 August 1978. Constructed as a destroyer escort, this unit was converted to a high speed transport while in US service; she could carry 162 troops. Ecuador employed her primarily as an escort and she carried the pennant no D 01. She had a small helicopter deck added in the late 1970s. Deleted 1989.

SUBMARINES

West German Type 209 (Type 1300) class *submarines*

No	Name	Builder	Launched	Fate
SS 12	HUANCAVILVA	Howaldtswerke, Kiel	18.3.77	Extant 1995
SS 11	SHYRI	Howaldtswerke, Kiel	8.10.76	Extant 1995

Similar to units sold to Greece, Turkey, Argentina and Peru, both were ordered in 1974 and completed 1 June 1978 and 16 March 1978 respectively. See Greek *Glavkos* class for details. New pennant nos S 101 and S 102. Refitted in Germany in 1983 and 1984.

AMPHIBIOUS WARFARE VESSELS

Ex-US LST type *tank landing ship*

One unit, *Hualcopo* (ex-*Summit County*, LST 1148) transferred in February 1977 (1650t, 11.6kts, built by Chicago Bridge). Pennant no T 61. Extant 1995.

Ex-US LSM type *landing ship*

Two units, *Jambelli* and *Tarqui*, transferred in 1958 (513t, 12.5kts, built 1945). Pennant nos T 51 and T 52. Deleted 1983, 1987.

Ecuador also operates six British Rotork 'Sea trucks', LF 91–96.

SMALL SURFACE COMBATANTS

Esmeraldas class design

Italian modified ESMERALDAS class *corvettes*

Displacement: 605t standard; 685t full load
Dimensions: 189ft 7in pp, 204ft 5in oa × 30ft 6in × 9ft 3in max
57.8m, 62.3m × 9.3m × 2.8m
Machinery: 4 shafts, 4 MTU 20V956 TB92 diesels, 24,400bhp = 37kts. Range 1200nm/4000nm at 31kts/18kts
Armament: 6 MM40 Exocet SSM (2×3), 1 Albatros SAM (1×4), 1–76mm OTO Melara Compact, 2–40mm (1×2) 6–324mm Mk 32 ASW TT (2×3), 1 light helicopter
Sensors: Radar SMA 3RM-20, RAN-105, Orion 10X; sonar Diodon; ECM Gamma system
Complement: 51

No	Name	Builder	Launched	Fate
CM 14	EL ORO	CNR Tirreno	9.2.81	Extant 1995
CM 11	ESMERALDAS	CNR Tirreno	1.10.80	Extant 1995
CM 15	GALAPAGOS	CNR Tirreno	4.7.81	Extant 1995
CM 16	LOJA	CNR Tirreno	27.2.82	Extant 1995
CM 13	LOS RIOS	CNR Tirreno	27.2.81	Extant 1995
CM 12	MANABI	CNR Tirreno	9.2.81	Extant 1995

Esmeraldas class corvettes were ordered in 1978. Generally similar to the *Wadi M'ragh* class built for Libya. Laid down and completed: CM 11 (27.9.79, 7.8.82), CM 12 (19.2.80, 21.6.83), CM 13 (5.12.79, 9.10.83) CM 14 (20.3.80, 11.12.83), CM 15 (4.12.80, 26.5.84), CM 16 (25.3.81, 26.5.84). *El Oro* was out of commission for two years from 1985 after a bad fire. Fitted with combat data system Selenia IPN 10. There is a helicopter landing deck but no hangar. Very powerful and useful units well armed against surface and aerial threats.

Ex-US PCE type *patrol vessel*

Class:
Esmeraldas (ex-*Eunice*), *Manabi* (ex-*Pascagoula*).

Transferred 29 November 1960 and 5 December 1960 respectively. Discarded 1970s. Pennant nos changed from E 03 and E 02 to E 22 and E 23 to P 22 and P 23.

MANTA class *fast attack craft*

Class (no, comp):
Manta (LM 24, 11.6.71), *Tulcan* (LM 25, 25.4.71), *Nueva Rocafuerte* (ex-*Tena*, LM 26, 23.6.71).

Similar to Chilean *Guacolda* class but one extra diesel makes them 3kts faster. Originally armed with 1–40mm gun, an Oerlikon twin RL, 2–21in TT, but rearmed in 1979 with 2–30mm Emerlec (1×2) and in 1981 four Gabrield SSMs replaced the TT. All extant 1995.

QUITO class *fast attack craft (missile)*

Displacement: 250t standard; 265t full load
Dimensions: 154ft 2in × 23ft × 7ft 10in
47.0m × 7.0m × 2.4m
Machinery: 4 shafts, 4 MTU 16V538 diesels, 14,000bhp = 35kts. Range 600nm at 30kts
Armament: 4 MM38 Exocet SSMs, 1–76mm/62 OTO Melara Compact, 2–35mm (1×2)
Sensors: Radar Triton, Vega fire control, Pollux, Racal Decca
Complement: 34

Class (no, comp):
Quito (LM 31, 13.7.76), *Guayaquil* (LM 32, 22 December 1977), *Cuenca* (LM 33, 17.7.77).

All built by Lürssen, Vegesack. Named after cities. Extant 1995.

Quito 1979 USN

Ex-US PGM type *large patrol craft*

Class (no, acquired):
Veintecinco de Julio, ex-*Quito* (LC 71, 30.11.65), *24 de Mayo*, ex-*Diez de Agosto*, ex-*Guayaquil* (LC 72, 30.11.65).

US *PGM 71* class patrol craft built by Peterson, Sturgeon Bay. Transferred to coast guard 1980. New numbers LGC 31, 32. Extant 1995.

OTHER SMALL CRAFT

Six 35t launches, *LP 7–12*, built in 1954 by Schürenstedt, Bardenfleth, West Germany, deleted 1960.

Six 64t full load launches, *LSP 1–6* built 1954–55 by Schürenstedt. They were given the pennant nos *LP 81–86*, and *LSP 4–6* were stricken 1976. The survivors were renamed *10 de Agosto*, *9 de Octubre* and *3 de Novembre* respectively, and later *Baha Hoyo*, *Pichincha* and *Portoviejo*; extant 1995.

Three 34t patrol craft built by Halter, New Orleans in 1976: *Comandancia de Balao*, *Comandancia de Guyaquil*, *Comandancia de Salinas*; all were extant 1995.

Six Halter Marine Rio *Puyango* class patrol craft *Rio Puyango*, *Rio Matage*, *Rio Zarumilla*, *Rio Chone*, *Rio Daule*, *Rio Babahoyo* (LGC 40–45) delivered in 1986 and last four by Astinave 1988 (17t, 26kts, 1–12.7mm, 2 MG), extant 1995.

Two *Espada* large patrol craft *5 de Agosto* and *27 de Febrero* built by Moss Point Marine, delivered in 1991 (190t full load, 112ft, 34.1m, 2790bhp = 27kts, 1–40mm Bofors, 2–12.7mm), extant 1995.

Two Swiftships type river patrol craft completed 1992 (17t, 22kts, 2–12.7mm MG, 2–7.62mm MG). Both extant 1995.

Egypt

The Royal Egyptian Navy was negligible in 1947, its most important unit being the weakly-armed sloop *El Amir Farouk*. Plans to acquire three sloops, two 'Castle' class frigates, four *Algerine* class minesweepers and smaller craft were interrupted by the 1948–49 Arab-Israeli War, in which the Egyptian Navy succeeded only in losing *El Amir Farouk*, and when implemented in 1949–50 were modified, the actual ships acquired being detailed below.

Further warship purchases were made by Nasser's new republic in 1955–56. Two old British destroyers were delivered in July 1956 as tension over the Suez Canal was increasing, while from the Soviet Union came two new destroyers, four minesweepers and twelve MTBs. Despite of these reinforcements, the navy was badly mauled in the Suez Operations (29 October–7 November), losing three frigates, one LST and about six MTBs in the face of overwhelming Anglo-French strength. Egyptian claims that two officers sank the French battleship *Jean Bart* by unspecified means were unfounded, and duly ignored by the non-Arab world. On the other hand the navy's blocking of the Suez Canal was efficiently done.

The naval losses were soon made good by the Soviet Union, which also transferred submarines for the first time. During the 1967 Six Day War, despite it being precipitated by Egypt's blockade of the Gulf of Aqaba at Sharm el Sheikh, the larger Egyptian units were kept in port, and the fighting was left to motor torpedo-boats. No successes were reported, and four Egyptian boats are believed to have been sunk, two by Israeli boats and two by the destroyer *Eilat*.

Between 1962 and 1967 the Soviet Union transferred twelve 'Osa I' and eight 'Komar' class fast attack craft, armed with SS-N-2 Styx surface-to-surface missiles (SSM). In a celebrated encounter on the night of 21 October 1967, one or more of these boats, moored at Port Said, launched Styx missiles at *Eilat*, patrolling some 13 miles offshore. *Eilat* took evasive action and opened fire on the incoming missiles, but was totally disabled by two hits in her machinery spaces, and sank an hour later after a third hit. This incident attracted widespread attention in Western navies, which then had no operational SSM, and no modern shipborne defence against them.

Eilat was avenged physically when Israeli jets sank the Egyptian destroyer *El Qaher* in May 1970; the Israelis also rapidly developed counter-measures to deal with the missile threat. The results were seen in the October 1973 (War of Ramadan), in which for the first time opposing sides fought with SSMs, and engagements were almost entirely between fast attack craft; of these, there were nineteen Egyptian and six Syrian, armed with Styx, to fourteen Israeli, armed with the new Gabriel I, which had half the range of the Styx and a much smaller warhead. The Egyptians operated their boats piecemeal, and did not make the best of their missile superiority, often electing to fire their Styx at distances beyond the Gabriel's range and then withdrawing to harbour. Out of an estimated fifty Styx fired, none hit, because of Israeli ECM, high-speed manoeuvring and gun attacks on incoming missiles, while Gabriels are believed to have sunk between three and five Egyptian boats.

The USSR did not replace these losses (coming after the 1972 expulsion of 20,000 Soviet advisers and followed by the ending of Red Fleet Egyptian port facilities in April 1976), so Egypt was compelled to rely more on Western shipbuilders and repairers. The reopening of the Suez Canal in 1975 was a moment of glory for the navy and a commercial blessing, but the canal had been cleared by British and American mine warfare ships. The proposed construction of two *Agosta* class submarines in France and two *Lupo* class frigates in Italy was abandoned. In 1978 the frigates *Lincoln* and *Salisbury* were purchased from Britain, with the guided missile destroyer *Devonshire* and the submarine *Cachalot* to follow, but Cairo had second thoughts and cancelled the transaction (although not before *Salisbury* reached Gibraltar), probably because the ships were too old. Two elderly US *Gearing* FRAM I destroyers were subsequently also offered and rejected. There was no way that Egypt could afford to replace most of the obsolete force of ex-British and Soviet destroyers, or maintain the large numbers of submarines, and for over a decade it had to rely almost entirely on its coastal forces.

With nearly 20,000 men, the Egyptian Navy is both large (in regional terms) and well-trained. It has held frequent exercises with the USN, and has recently benefited from the stance Egypt took in the Gulf conflict. However, it is still difficult for the navy to obtain up-to-date equipment of the quality it would like, so that it can deal with such incidents as the mining of the Red Sea in 1984 without needing outside help.

The Egyptian Navy's Mediterranean bases are Alexandria, Port Said, Sollum and Mersa Matruh while Port Tewfik, Hurghada, Safaqa and Ras Banas ($106m modernisation being funded by the US) cover the Red Sea. Since 1980 a unique Otomat missile coastal defence system has been installed; 30 truck-mounted batteries with mobile radar to replace the 1960s Soviet 'Samlet' system.

FLEET STRENGTH 1947

The sloop *El Amir Farouk* built by Hawthorn Leslie (Newcastle-upon-Tyne) was sunk by Israeli action off Gaza 22 November 1948, possibly by limpet mines or an explosive motor boat.

There were also the transport *El Amira Fawzia* (1929, 2640t, 2-3pdr, renamed *El Quseir c*1953, discarded 1967); the veteran royal yacht *Mahroussa* (built 1865 by Samuda, London, 3417t, renamed *El Horria c*1953), still active in 1995 as a training ship and occasional presidential yacht after by far the longest career of any sea-going ship in a modern navy); and the small motor boats *Darfeel*, *Noor el Bahr*, *El Sarea* and *Raqib* (used by the RN as harbour defence boats during the Second World War and discarded 1948–*c*1953).

MAJOR SURFACE SHIPS

Ex-British 'Z' class *destroyers*

No	Name	Builder	Acquired	Fate
833	EL FATEH (ex-*Zenith*, ex-*Wessex*)	Denny	July 1956	Extant 1995
	EL QAHER (ex-*Myngs*)	Vickers-Armstrong, Tyne	July 1956	Sunk 16.5.70

Purchased from the RN in 1955 and refitted before delivery. Modernised in the UK during 1963–64 (TT removed and short mainmast replaced by a tall thin lattice mast with Type 960 search radar). *El Qaher* was sunk in the Red Sea by Israeli aircraft at Ras Banas (Berenice). *El Fateh* is still used for training.

Ex-Soviet 'SKORIY' class *destroyers*

No	Name	Builder	Acquired	Fate
	EL NASSER (i)	USSR	11.6.56	Returned USSR 1967
822	EL ZAFFER	USSR	11.6.56	Discarded 1985
	DAMIET (i)	USSR	Jan 1962	Returned USSR 1967
888	SUEZ	USSR	Jan 1962	Discarded 1985
666	6 OCTOBER (ex-*Al Nasser*)	USSR	Apr 1967	Discarded 1985
844	DAMIET (ii)	USSR	Apr 1967	Discarded 1985

El Nasser (i) and *El Zaffer* (one possibly ex-*Stroikii*) transferred at Alexandria in June 1956; one reported damaged by British aircraft 1 November 1956 and the other set on fire and damaged by French aircraft on 2 November 1956. *Damiet* and *Suez* (one possibly ex-*Bessmennyi*) transferred in 1962. In 1967 *El Nasser* and *Damiet* returned to the Soviet Union and were replaced by modified *Skory* class ships, given the same names. *El Nasser* (ii) renamed after the day Egypt crossed the Suez Canal in 1973 Arab-Israeli War. *El Zaffer* modernised as a guided missile destroyer; 2–85mm (3.4in) and 2–37mm AA guns replaced by two SS-N-2 SSMs, probably taken from a 'Komar' class FAC (1978). *Damiet* (also given as *Domiat*) commemorated the frigate sunk in the 1956 Suez operations.

Ex-British BLACK SWAN class *frigate*

No	Name	Builder	Acquired	Fate
555 ex-42	EL TARIK (ex-*El Malek Farouk*, ex-*Whimbrel*)	Yarrow	Nov 1949	Extant 1995

Modified *Black Swan* class sloop, purchased from the RN in late 1949. Now, she is used as a training ship.

El Tarik 10.4.94 L & L van Ginderen

Ex-British 'FLOWER' class *frigate*

No	Name	Builder	Acquired	Fate
	El SUDAN (ex-*Mallow*, ex-*Partizanka*, ex-*Nada*, ex-*Mallow*)	Harland and Wolff	28.10.49	Discarded *c*1975

This ex-'Flower' class corvette, launched in 1940 and loaned to Yugoslavia 1943–49, was transferred to Egypt. Used for training from *c*1970.

The frigate *Misr* (ex-mercantile *Malrouk*), sunk by collision S of Suez during the night of 16–17 May 1973, is listed as a sister ship of *El Sudan*. However, her original 'Flower' name has not been located, and she may have been of another class (numerous 'Castle' class corvettes and *Bangor* class minesweepers were also sold to commercial interests after the war).

Ex-British 'HUNT' class (Type 1) *frigates*

No	Name	Builder	Acquired	Fate
525 (ex-11)	PORT SAID (ex-*Mohammed Ali*, ex-*Ibrahim el Awal*, ex-*Cottesmore*)	Yarrow	July 1950	Hulked 1986
	IBRAHIM EL AWAL (ex-*Mohamed Ali*, (ex-*Mendip*, ex-*Lin-Fu*, ex-*Mendip*)	Swan Hunter	Nov 1949	To Israel 1956

Former escort destroyers, launched in 1940, transferred from the RN. They exchanged names about 1951–52. *Ibrahim el Awal* surrendered off Haifa on 31 October 1956 after coming under fire from the French destroyer *Kersaint*, and subsequently became the Israeli *Haifa* (see under Israel). *Port Said* lost her 5–2pdr guns in the 1960s, and her remaining light AA guns were replaced by 2–37mm and 2–25mm Russian-made weapons *c*1972. She had only one navigation radar and no ASW equipment.

Ex-British 'RIVER' class Frigates

No	Name	Builder	Acquired	Fate
	ABIKIR (ex-*Usk*)	Smith's Dock	Apr 1950	Scuttled 1.11.56
	DOMIAT (ex-*Nith*)	Robb	May 1950	Sunk 1.11.56
511(ex-43)	RASHID (ex-*Spey*)	Smith's Dock	Dec 1949	Extant 1995

These ships, launched 1941–43, were purchased from the RN in 1948 and refitted before delivery. *Abikir*, scuttled as a blockship in the Suez Canal, was refloated on 8 April 1957, beached and abandoned. *Domiat* was sunk by gunfire from the cruiser HMS *Newfoundland* 8 miles S of Suez; there were sixty-nine survivors. *Rashid* has been used as a submarine support ship since *c*1975, and has a large deckhouse aft.

Ex-Spanish DESCUBIERTA class *frigates*

No	Name	Builder	Launched	Comp	Fate
961	EL SUEZ (ex-*Centinela*)	Bazan	6.10.79	21.8.84	Extant 1995
946	EL ABOUKIR (ex-*Serviola*)	Bazan	20.12.79	27.10.84	Extant 1995

Both laid down for the Spanish Navy, and purchased by Egypt in 1982. Fitted with eight Harpoon SSM (4×2).

Chinese JIANGHU class *frigates*

No	Name	Builder	Acquired	Fate
951	NAJIM AL ZAFIR	Jiangnan, Shanghai	27.10.84	Extant 1995
956	EL NASSER	Jiangnan, Shanghai	16.4.85	Extant 1995

Both ordered from China in 1982. Have 4-57mm (2×2) in place of 100mm and fitted with Elettronica EW suite.

Ex-US KNOX class *frigates*

No	Name	Builder	Acquired	Fate
961	DAMIETTA (ex-*Jesse L Brown*)	Avondale	1994	Extant 1995
966	RASHID (ex-*Moinster*)	Avondale	1994	Extant 1995

Two units transferred from USN 1994 on five-year lease.

EX-US TACOMA CLASS
In addition to the above, the ex-US *Tacoma* class (British 'Colony' class) frigates *Papua* (ex-*Howett*) and *Tobago* (ex-*Hong Kong*, ex-*Holmes*) were acquired by Egypt in 1950 as mercantile passenger vessels. They were taken over by the Egyptian Navy on 17 May 1953 for refitting as warships, but so far as is known no work had been done when they were scuttled as blockships in the Suez Canal on 1 November 1956. They were later refloated, beached and abandoned.

SUBMARINES

Ex-Soviet 'WHISKEY' class *submarines*

Class:
415, 418, 421, 432, 455, 477

Eight transferred (four in June 1957, three on 24 January 1958 and one in January 1962), but two returned to the Soviet Union in 1966 and were replaced by 'Romeo' class boats. Two more exchanged in 1971–72 at Leningrad. *455* is in reserve, and may be used for spares; the rest remain in service.

Ex-Soviet 'MV' class *submarine*

One (number known) transferred in 1957 and deleted 1971.

Ex-Soviet 'ROMEO' class *submarines*

Class:
711, 722, 733, 744, 755, 766

Five transferred in 1966 (two replacing two 'Whiskey' class boats returned to the USSR) and one in 1969. Two (*733* and *744*) may have been cannibalised for spares. Two remain nominally operational in 1995.

Ex-Chinese 'ROMEO' class *submarines*

Class:
831, 842, 852, 858

Two units were delivered in 1983 and a further two in 1984. The Egyptian Navy planned to refit them with US fire control equipment and torpedoes in 1989–91 with the assistance of Tacoma SB Co. Delays in obtaining US Congressional approval for the deal set this programme back by three years.

AMPHIBIOUS WARFARE VESSELS

LST *Aka* (ex-US *LST 178*, acquired 1950) was sunk as a blockship near Lake Timsah in the Suez Canal on 1 November 1956. Refloated in February 1957, beached and abandoned. Three ex-Soviet 'Polnocny I' class LCTs (acquired 1974) were in service 1985. Numbered *910, 912* and *915* originally, renumbered *301, 303* and *306*.

Nine ex-Soviet 'Vydra' class LCUs (acquired 1968–9) were in service in 1995; numbered *640, 643, 645, 648, 649, 682, 684, 685* and *688*. Some temporarily refitted as gunboats after the 1973 war, with 122mm 8-barrelled rocket launchers and 37mm or 40mm guns. All extant *c*1994. Four Soviet

'SMB-1' class LCUs (acquired 1965) were in service in 1982. All since stricken.

Twenty very small LCTs of the LCM type were acquired in 1948 and numbered 1 to 20. Nos 15 and 20 were sunk in the 1956 Suez operations, eight have been discarded and ten remained in service as harbour craft in 1982. All since stricken.

SMALL SURFACE COMBATANTS

Ex-Soviet 'KOMAR' class *fast attack craft (missile)*

Eight transferred to Egypt in 1962–67. Four believed to have been sunk (one by Israeli aircraft on 16 May 1970 and two in the October 1973 War). Refitted with Italian CRM-12 D/S-2 diesels. Remainder all discarded by 1985.

Ex-Soviet 'OSA I' class *fast attack craft (missile)*

Twelve transferred to Egypt in 1966. Four believed to have been sunk (some in the October 1973 War) and eight, numbered *301*, *310*, *323*, *341*, *356*, *378*, *389* and *390*, were in service in 1982. All carry SA-N-7 'Grail' shoulder-launched missiles, new Kelvin Hughes 1006 surveillance and Decca 916 navigation radars. The renumbered *631*, *635*, *637* and *639* were no longer in service by 1990. *633*, *641* and *643* are extant 1995.

Ex-Soviet 'SHERSHEN' class *fast attack craft*

Class:
310, *321*, *332*, *343*, *354* and *365*

Transferred 1967–68. Two were configured as torpedo boats, with 4–21in TT, and four as gunboats, with tubes replaced by 122mm BM21 multiple rocket launchers and one SA-N-5 'Grail' SAM. They were renumbered *751*, *753*, *755*, *757*, *759* and *761*. One discarded, remainder extant 1995.

'OCTOBER' class *fast attack craft (missile)*

Displacement:	71t standard, 82t full load
Dimensions:	84ft oa × 20ft × 5ft *25.5m × 6.1m × 1.3m*
Machinery:	4 shafts, 4CRM-18 DS/2 diesels, 5400bhp = 40kts. Range 400nm at 30kts
Armament:	2 Otomat Mk 1 SSM, 4–30mm A32 (2×2)
Sensors:	Radar Decca AC 1226, Marconi S810; ECM Decca Cutlass RDL-1, 2 Protean decoy launchers
Complement:	20

Class:
207—212

Launched 1975–76 at Alexandria, with Soviet 'Komar' class wooden hulls and Italian machinery. Weapons and electronics installed in 1979–81 by Vosper Thornycroft, Portchester. Last pair (*207*, *212*) reached Alexandria 25 April 1981. The Marconi-Sperry Sapphire radar/tv fire control system is fitted. They are very cramped, access to the engine room hatches being by a removable GRP operations room. They were renumbered *781*, *783*, *785*, *787*, *789* and *791*. 791 was lost overboard on her delivery voyage, salvaged and repaired. First two discarded 1989, remainder extant 1995.

RAMADAN class *fast attack craft (missile)*

Displacement:	258t light; 307t normal; 350t full load
Dimensions:	170ft 7in oa × 25ft × 6ft 7in mean *52.0m × 7.6m × 2.0m*
Machinery:	4 shafts, 4 MTU 20cyl 20V-538 TB 91 diesels, 18,000bhp = 37kts. Range 2000nm at 15kts. Oil 43.3t
Armament:	4 Otomat Mk 1 SSM (2×2), 1–76mm 62 OTO Melara Compact, 2–40mm Breda Bofors (1×2)
Sensors:	Radar Marconi S820, S810, 2 ST802, Decca AC 1226; ECM Decca-Racal Cutlass, 2 Protean MEL decoy launchers
Complement:	40

No	Name	Builder	Launched	Fate
561	RAMADAN	Vosper	6.9.79	Extant 1995
562	KHYBER	Vosper	12.6.80	Extant 1995
563	EL KADESSEYA	Vosper	19.2.80	Extant 1995
564	EL YARMOUK	Vosper	12.6.80	Extant 1995
565	HETTEIN	Vosper	17.6.81	Extant 1995
566	BADR	Vosper	25.11.80	Extant 1995

Laid down and completion dates: *Ramadan* (23.11.78, 20.7.81), *Khyber* (26.2.79, 15.9.81), *El Kadesseya* (23.4.79, January 1982), *El Yarmouk* (16.5.79, March 1982), *Hettein* (29.9.79, June 1982), *Badr* (March 1980, June 1982). Ordered from Vosper Thornycroft, Portchester 4 September 1977 for £153 million to a design derived from the RN *Tenacity* class FPBs. Renumbered 670, 672, 674, 676, 678 and 680 until 1982. *Ramadan* and *Khyber* arrived at Alexandria 9 November 1981. They have the very advanced Marconi-Sperry Sapphire fire control system, 2 + 1 OFD trackers and Ferranti CAAIS action information system.

'October' class 210 on trials — Vosper Thornycroft

Ex-Chinese 'HOKU' class *fast attack craft (missile)*

Class:
401–406

Six acquired in September 1984. Since renumbered *611–616*. Two no longer operational, remainder in service 1995.

Ex-Soviet 'SO 1' class *large patrol craft*

Class:
211, *217*, *222*, *228*, *230*, *233*, *239*, *244*, *251*, *255*, *262*, *266*

Eight transferred in 1962–67, and four after 1970. Armament varied: six had BM21 rocket launchers, others 2–21in TT or SA-N-5 'Grail' SAMs. Decca 916 radar was fitted. All discarded by 1986.

Ex-Chinese 'HAINAN' class *patrol craft*

Two transferred in 1983 and six more in 1984. Four to be fitted with 6–324mm ASW TT (3×2). All extant 1995.

Ex-Chinese 'SHANGHAI II' class *patrol craft*

Four transferred 1984. Reportedly rearmed with Egyptian-built 23mm guns on original mountings. All extant 1995.

MINE WARFARE VESSELS

Ex-British BANGOR class *minesweepers (corvettes)*

No	Name	Builder	Launched	Fate
	MATROUH (ex-*Stornoway*)	Robb	1949	?Sunk 1968–69
	NASR (ex-*Bude*)	Lobnitz	1949	Expended *c*1972
	SOLLUM (ex-*Wedgeport*)	Dufferin	1949	Lost 7.4.53

Former minesweepers, launched 1940–41. Details of *Matrouh*'s loss are not available. Transferred from the RN and rated as corvettes. *Nasr* sunk as a target for Styx SSMs, and *Sollum* foundered off Alexandria in heavy weather.

Ramadan on trials — Vosper Thornycroft

Ex-US YMS type *coastal minesweepers*

Class (former BYMS no):
Arish (2028), *Darfur* (2041), *Gaza* (2013), *Kaisaria* (2075), *Kordofan* (2212), *Malek Fuad* (2035), *Naharia* (2069), *Rafah* (2149), *Tor* (2175)

Wooden-hulled boats launched 1942–43 and acquired 1949–50; all single-funnelled except *Arish*, which had two stacks. *Gaza* blew up off Mersa Matrouh in a fuel tank explosion on 26 July 1950, *Darfur* and *Tor* transferred to Algeria in 1962, rest hulked or discarded *c*1967–70.

Ex-Soviet 'T 43' class *ocean minesweepers*

Class (no, former no):
Assiut (675, ex-303), *Bahaira* (659, ex-301), *Charkieh* (650, ex-304), *Dakhla* (672, ex-305), *Gharbia* (656, ex-306), *Miniya* (302), *Sinai* (679).

Four transferred from the USSR in 1956–59; *Assiut* and *Sinai* after 1970. *Miniya* sunk in the Gulf of Suez by Israeli aircraft on 6 February 1970. Only *Sinai* operational as of 1993.

Ex-Soviet 'T 301' class *inshore minesweepers*

Class (no):
El Fayoum (708), *El Manufieh* (619)

Transferred 1962; later in harbour service. Discarded 1991.

Ex-Soviet 'YURKA' class *ocean minesweepers*

Class (no):
Aswan (695), *Giza* (690), *Qena* (696), *Sohag* (699)

Transferred from the USSR in 1970–71 and slightly modified for Egyptian service (with hull scuttles and no 'Drum Tilt' radar), and all remain in service in 1995. Plans for modernisation, including new engines and side-scan sonar, have been set aside.

US 110ft *coastal minesweepers*

Displacement:	175t standard; 188t full load
Dimensions:	110ft oa, 102ft wl × 27ft × 5ft *33.53m, 31.09m × 8.23m × 1.53m*
Machinery:	2 MTU 12V183 TC82 diesels, 2 Schottel Z-drives, 1,034bhp = 12.4kts. 1 White Gill thruster, 300shp = 4kts
Armament:	2–12.7mm MG (1×2)
Sensors:	Radar Sperry Rastar; sonar Thomson-Sintra/Thoray TSM 2022 HF
Complement:	25

Order placed for six units December 1990 with Swiftships Inc, Louisiana. GRP hull. Can also be used as route survey craft in peacetime. Three in service 1995.

MISCELLANEOUS

The following light craft and amphibious force units have seen service with the Egyptian Navy:

Five Fairmile 'B' type motor launches acquired from Britain 1948–50. Two (names not known) sunk *c*1949–50; *Hamza* (ex-*ML 134*), *Sab el Bahr* and *Saker el Bahar*, hulked in the late 1960s.

Two Fairmile 'D' type motor launches acquired from Britain in 1950 and named *El Naser* and *El Zafer*. Both discarded *c*1965.

Six Yugoslav '108' type MTBs (similar to US Higgins boats) bought in 1956. Four discarded in 1975 and two relegated to targets.

Thirty-six Soviet 'P 6' type MTBs were acquired 1956–70 (a few may have been built at Alexandria). At least ten have been sunk, two in the 1956 Suez operations, four in the Six Day War and four by Israeli aircraft in 1969–70, and four discarded (in 1975); twenty, numbered from *201*, were believed to be in service in 1982. Armament varied; two boats had 4–21in TT, some had the aft 25mm mounting and all TT replaced by a 122mm 8-barrelled 40 tube RL and 2–12.7mm MG. All were discarded by 1986.

Four ex-Soviet 'P 4' type MTBs were transferred from Syria in 1970. All stricken by 1983.

Two ex-Soviet 'K 8' class minesweeper boats, named *Safaga* and *Abou El Ghoussoum*, were delivered 1968 and deleted 1980.

Three British SRN-6 hovercraft, delivered 1976–79, may be minelayers.

The large patrol craft *Nisr, Nimr* and *Thar* were built at Port Said in 1963 for the Coast Guard (110t, 1–20mm). The Coast Guard also operates six GRP MV 70 class coastal patrol craft (70ft, 33t, 2–30mm, 1–20mm, 2 MG) and has six more on order from Italy. It received six *Timsah* class Egyptian-built patrol craft in 1981–84 (99t, 25kts, 1–30mm, 13 men) and six improved *Timsah II* (99t, 27kts, 2–20mm (1×2), 13 men) in 1988–89.

Eire

Ireland officially left the Commonwealth in 1948 and the Republic of Eire was proclaimed in April 1949. She joined the EEC in January 1973 but has maintained the policy of neutrality adopted before the Second World War. The Irish Navy or Naval Service was created in 1947 from the disbanded Wartime Marine Service and Port Control Service. Until the 1970s the Irish Navy was a coastal fishery protection service, but the increased fishing of ever smaller fish stocks, the adoption of a 200-mile EEZ and the exploitation of offshore minerals and oil and gas have meant that the area the Irish Navy has to patrol has increased many-fold. Modern Irish warships have to be capable of carrying out long patrols in the extremely inclement North Atlantic in mid-winter. This has meant a considerable expansion in numbers and an emphasis on much larger, more capable vessels. The main base is Haulbowline Island, Cork, which may be enlarged. The navy strength is made up of about 1000 volunteers including reserves. Shortage of money has meant that the navy has been forced once again to buy second-hand vessels — the *Peacock*s – rather than purpose-designed ships suited to the navy's needs.

FLEET STRENGTH 1947

FRIGATES

Name	Launched	Disp	Fate
CLIONA (ex-*Bellwort*)	11.8.41	1020t	Stricken Jan 1971
MACHA (ex-*Borage*)	22.11.41	1020t	Stricken Jan 1971
MAEV (ex-*Oxlip*)	28.8.41	1020t	Stricken Mar 1972

Three ex-RN 'Flower' class frigates (formerly corvettes) were acquired by the Irish Navy in 1946. *Cliona* and *Maev* were in the original short forecastle form. *Macha* had been fitted with an extended forecastle during the Second World War. One of her messdecks was converted into cabins for the flotilla commander, and she was used as leader. When purchased all were armed with 1–4in and 2–20mm (2×1), plus a Hedgehog ASW mortar in *Cliona* and *Maev* and a 2pdr (40mm) in *Cliona* and *Macha*.

Their pole masts were replaced by lattice masts in 1953, and in 1966–67 *Cliona* and *Macha* were refitted for fishery protection duties. Armament was reduced to 1–4in. *Maev* was placed in preservation in reserve and retained her original armament.

OTHER VESSELS

Name	Launched	Disp	Fate
FORT RANNOCH	1936	258t gross	Stricken 1947
MTBs 1—5	1939	32t	Stricken *c*1948

SMALL SURFACE COMBATANTS

Ex-British 'TON' class *coastal minesweepers*

Class (no, former name):
Banba (CM 11, ex-*Alverton*), *Fola* (CM 12, ex-*Blaxton*) and *Gráinne* (CM 10, ex-*Oulston*)

Ex-RN 'Ton' class coastal minesweepers, built during 1952–56. Purchased by the Irish Navy in 1971 to replace the worn out 'Flower' class frigates. Used for fishery protection duties. *Gráinne* had to use Bofors and small arms fire to arrest a 3277grt Soviet trawler in September 1976. Movable minesweeping gear had been landed, but the fixed gear remained on board. *Banba* stricken 1983, *Fola* in 1986 and *Gráinne* in 1985.

Deirdre 1979 L & L van Ginderen

DEIDRE *patrol vessel*

Displacement:	972t full load
Dimensions:	184ft 4in pp, 205ft 2in oa × 34ft 2in × 14ft 5in *56.2m, 62.5m × 10.4m × 4.4m*
Machinery:	1 shaft (CP propeller), 2 Polar SF 112 VS-F diesels, 4200bhp = 18kts. Oil 150t. Range 3000nm/5000nm at 15.5kts/12kts
Armament:	1–40mm, 2–2in RFL
Sensors:	Radar Decca type; sonar Simrad sidescan type
Complement:	46

No	Name	Builder	Launched	Fate
P20	DEIDRE	Verolme, Cork	29.12 71	Extant 1995

The first warship designed and built in Eire. Intended to be the prototype for a number of fishery protection vessels (entered service 19 June 1972), but her seakeeping, though adequate for most navies, was not as good as had been hoped. Subsequent vessels have been built to a slightly elongated and modified design. *Deidre* is fitted with Vosper stabilisers.

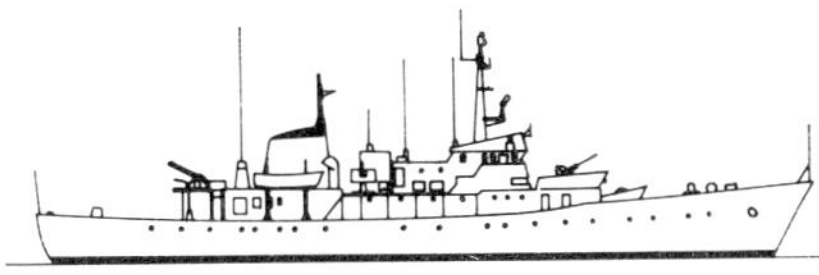

Emer 1979

EMER class *patrol vessels*

Displacement:	1020t full load
Dimensions:	192ft 1in pp, 213ft 11in oa × 33ft 6in × 15ft *58.5m, 65.2m × 10.2m × 4.6m*
Machinery:	1 shaft (CP propeller), 2 SEMT-Pielstick 6 PA6L-280 diesels, 4800bhp = 18kts. Oil 172t. Range 4500nm/6750nm at 18/12kts
Armament:	1–40mm, 2–20mm (2×1)
Sensors:	Radar Decca type; sonar Simrad sidescan type
Complement:	55

No	Name	Builder	Launched	Fate
P 21	EMER	Verolme, Cork	1977	Extant 1995
P22	AOIFE	Verolme, Cork	12.4.78	Extant 1995
P 23	AISLING	Verolme, Cork	3.10.79	Extant 1995

This class was developed from *Deidre*; they completed on 18 January 1978, 21 November 1979, 21 May 1980 respectively. They have a 7ft 8in (*2.4m*) longer hull with a raised forecastle to improve seakeeping, more powerful machinery to maintain the speed, better watertight subdivision and improved accommodation. *Aoife* and *Aisling* differ from *Emer* by having a satellite navigation system, and improved noise levels in the control rooms and accommodation to make it easier for them to patrol the new 200 mile EEZ limit. They also have a bow thruster and a new KaMeWa propeller.

Eithne Jul 1990 L & L van Ginderen

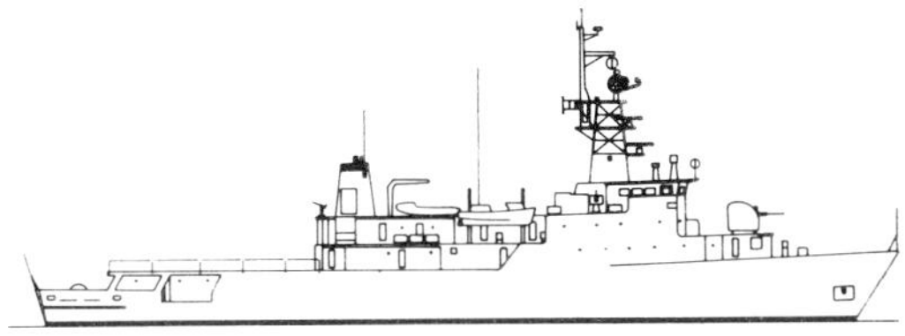

Eithne 1989

EITHNE class *patrol vessel*

Displacement:	1915t full load
Dimensions:	265ft × 39ft 4in × 13ft 8in *80.8m × 12.0m × 4.3m*
Machinery:	2 shafts, 2 Ruston Paxman diesels, 7200hp = 19kts. Range 7000nm at 15kts
Armament:	1–57mm Bofors, 2–20mm Rheinmetall (2×1), 2–7.62mm MG (2×1), 1 helicopter
Sensors:	Radar Decca TM 1229, Decca AC1629C, HSA DA-05; sonar Plessey
Complement:	82

No	Name	Builder	Launched	Fate
P31	EITHNE	Verolme, Cork	19.12.83	Extant 1995

Enlarged version of *Emer* class with a longer range, more powerful armament and a helicopter platform and hangar over the wide transom stern. Designed for long patrols at the limit of the 200-mile EEZ, leaving inshore work to the *Emers* and *Deidre*. A class of two was planned but the second unit was deferred due to problems with the shipyard and lack of money. Built as a flagship, *Eithne* completed on 7 December 1984.

Ex-British PEACOCK class *patrol vessels*

No	Name	Builder	Acquired	Fate
P41	ORLA (ex-*Swift*)	Hall Russell	8.10.88	Extant 1995
P42	CIARA (ex-*Swallow*)	Hall Russell	8.10.88	Extant 1995

Two units from Hong Kong, now surplus, entered Irish service on 21 November 1988.

MISCELLANEOUS

The Danish trawler *Ferdia* was temporarily hired by the Irish Navy to act as a fishery protection vessel until *Emer* entered service. In the mid-1970s the only other vessel the Irish Navy possessed that could police the extended EEZ was *Deidre*. *Ferdia* was returned to her owners in 1978.

El Salvador

A mountainous and densely populated state in which economic and social conditions are one of the main causes of the dissent and tension within the country. Successive repressive governments have encouraged the growth of guerrilla and terrorist groups and a string of *coups* has added to the republic's instability. In 1979 General Humberto Romero was overthrown by rebel officers and thus began a state of civil war which came to an end in 1992. Military rule came to an end in March 1982 when a constituent assembly was elected. During the civil war over 40,000 lives were lost. El Salvador has received much international aid to rebuild its war-damaged infrastructure and services. Much military aid comes to El Salvador from the United States although this is utilised mainly for the training and re-equipping of army and air force units.

army and air force units.

El Salvador has a 160-mile Pacific coastline and operates a small navy (2200 officers and men including naval infantry) employing patrol boats. The main operational ports are Acajutla, La Libertad and La Union.

One ex-British HDML patrol boat (46t, commissioned 1942, purchased 1959, deleted 1988).

One American-built patrol craft (33t, commissioned 1967, 3 GM diesels, built by Sewart, Louisiana). Carries pennant no GC 5. Sunk 1990.

Three Camcraft type patrol boats (100t, 3 GM diesels), carry pennant nos GC 6, GC 7 and GC 8, commissioned October 1975, December 1975 and November 1975 respectively. Extant 1995.

One Swiftships 77ft patrol craft (GC 11), delivered in 1985 (48t full load, 23.5m, 2325bhp = 26kts, 2–12.7mm, 1–81mm mortar). Extant 1995.

One Swiftships 65ft patrol craft (GC 10), delivered 1984 (36t full load, 1550bhp = 23kts, 1–20mm, 2–12.7mm). Extant 1995.

Ten *Protector* class patrol boats (LP 01–10), delivered in 1988–89 (9t, 680bhp = 28kts, 2–12.7mm, 2 MG). Extant 1995.

Six *Piranha* class patrol boats (LOF 1–6), delivered 1987 (8.2t, 680bhp = 26kts, 2–12.7mm, 2 MG). Extant 1995.

One LCM 6 (LD 1) and LCM 8 (LD 2, LD 3) transferred from USA in 1986 and 1987. Both extant 1995.

The navy also operates two ex-US CG type patrol boats (10.5t, commissioned 1950), a tug *Libertad* and twenty-five small outboard-propelled craft.

Equatorial Guinea

Equatorial Guinea, sandwiched between Gabon and Cameroun, became independent from Spain in 1968. West Africa's smallest state (1991 estimated population 356,000) received a Soviet 'P 6' class torpedo boat and a 'Poluchat' class patrol boat in 1974. All ex-Soviet equipment was discarded by 1986 and replaced with the ex-Nigerian P20 class *Rio Wele*, and the ex-US 68ft patrol craft *Isla de Bioko*, built in 1987–88. The 120 officers and men are based at the capital Malabo (Fernando Po), and at Bata (Rio Muni). They received some training from China and Spain.

Ethiopia (Eritrea)

The history of Ethiopia from 1947 shows a sharp discontinuity in September 1974, when the so-called 'creeping revolution' finally deposed the venerable Emperor, Haile Selassie, the effective ruler of the country since 1916 (except for the years of Italian occupation), and replaced his paternal feudalism by the Marxist military rule of the Dergue. The immediate cause of the revolution was the disastrous drought and famine of 1973, which cost up to 400,000 lives, but the Imperial Government had long been unable and perhaps unwilling to face the task of unifying and modernising their gigantic, primitive and ramshackle country. Eritrea, an ex-Italian territory forcibly amalgamated with Ethiopia in 1962 and providing its outlet to the Red Sea, was a battleground for vigorous independence movements, while to the south and east Somali irredentism in the Ogaden Desert also threatened national unity.

In international affairs the changes between 1974 and 1977 ended the pro-Western stance of Imperial days and left the Ethiopian government as the most pro-Soviet in Africa (and the most dependent on Soviet and Cuban military support). Arms shipments from the USA ceased in 1977, shortly before deliveries from the USSR began. The swift Ethiopian defeat of Somalia in the Ogaden during early 1978 could not have happened without assistance from Soviet and Cuban advisers and Cuban troops; these 'volunteers' helped shortly afterwards to roll back to the Sudan frontier the Eritrean liberation movements which, by 1977, had come to control nearly all the province, almost isolating Addis Ababa from the sea. In November 1978 Ethiopia and the USSR formalised their new alliance into a 20-year treaty of friendship and co-operation.

The Ethiopian armed forces were greatly expanded, the army totalling 244,500 in 1982. Military efficiency and political reliability were at a low, the regime looking ultimately to the 13,000 Cubans and 1000–2000 Soviet and East German advisers for protection against internal and external opponents. The Imperial Ethiopian Navy was founded in 1955, and a naval base and college were established at Massawa in 1956. The first ship was received there in 1957. The modest naval force supplied by the USA, the Netherlands, France and Yugoslavia was supplemented since 1977 by ex-Soviet fast attack craft.

The naval base at Assab, scene of the 1977–78 Soviet arms sealift and close to the strategic Bab el Mandeb Strait, was expanded to include repair facilities, while a small Soviet-constructed forward base in the Dahlak Islands was used as an anchorage for Russian warships. The effective revolution in Tigray province, as well as in most of the rest of the country, resulted by 1991 in the separation of Eritrea (which became fully independent in May 1993) and the loss to Ethiopia of its entire coastline. Some vessels were destroyed – especially at Massawa – but many escaped to internment in Djibouti. The fate of most of them is still unclear.

Ex-Soviet 'PETYA II' class *frigates*

Class (transferred):
F 1616, F 1617

Two Project 159A 'Petya II' arrived at Massawa July 1983 and March 1984 respectively. *F 1617* interned at Djibouti.

Ex-Soviet 'OSA II' class *fast attack craft (missile)*

Four transferred 1978, September 1980, 13.1.81 and 1982. Based in the Dahlak Is with Soviet advisers presumed to be aboard. One survived to be interned at Djibouti.

Ex-Soviet 'MOL' class *torpedo-boats*

Three transferred 1978, may not have TT.

Ex-Soviet 'ZHUK' class *coastal patrol craft*

Two transferred 1982 and a further two in 1990. Two lost during civil war.

Ex-US PC type *large patrol craft*

Class (former no):
Belay Deress (ex-*PC 1616*)

Built for W Germany at Brest, France, during 1953–55 under MDAP, but not delivered, being transferred to Ethiopia in 1957 under the MAP as her first warship. Found to be too advanced for the then state of Ethiopian naval training, and returned to the US in 1958, subsequently becoming the Italian *Vedetta* (discarded 1977).

Ex-US 'CAPE' and 'PGM 53' class *large patrol craft*

Class (former no):
PC 11 (ex-USCG WVP 95304), *PC 12* (ex-*WVP 95310*), *PC 13* (ex-USN *PGM 53*), *PC 14* (ex-*PGM 54*), PC 15 (ex-*PGM 18*)

The two ex-US Coast Guard cutters transferred in 1958, the three ex-USN patrol gunboats in 1961–62. *PC 12* sunk by Ethiopian aircraft in April 1977 while her crew were defecting; the other four were discarded 1984.

Ex-Netherlands WILDERWANK class *large patrol craft*

Class (former name and no):
MS 41 (ex-*Eist*, M 829)

Acquired from the Netherlands in 1971. Used for patrol duties, but minesweeping fittings remain. May have been fitted with SS 12 missiles. Discarded 1990.

Ex-Yugoslav 'KRALJEVICA' class *large patrol craft*

PBR 509 was transferred in 1975; her Ethiopian number is 51. Out of service 1984.

US 'SWIFT' type *large patrol craft*

Displacement:	118t full load
Dimensions:	105ft oa × 23ft 7in × 6ft 6in *32.0m × 7.2m × 2.0m*
Machinery:	2 shafts, 2 MTU MD 16V538 TB90 diesels, 7000bhp = 30kts. Range 1200nm at 18kts
Armament:	4–30mm Emerlec (2×2)
Sensors:	Radar Decca RM 916
Complement:	21

Class:
P 201, P 202, P 203, P 204

Six ordered from Swiftships, Louisiana in 1976; four delivered in April 1977 before US arms deliveries to Ethiopia ceased. *P 203* and *P 204* have 4–23mm (2×2) and 2–12.7mm MG (2×1) Soviet armament. *P 202* defected to Somalia in 1984.

AMPHIBIOUS WARFARE SHIPS

Four ex-US landing craft (two LCM type, purchased 1962, and two LCVP type, purchased 1971).

Two French 'EDIC' LCU class (L 1035 and L 1036, launched May 1977). L 1035 lost 1990. Six ex-Soviet 'T 4' class LCVP (transferred May 1979). Four were discarded by 1990. Two ex-Soviet 'Polnocny B' class landing ships transferred, LTC 1037 in November 1981. LTC 1038 in January 1983. One ex-Soviet 'Natya' class MCM ship and one ex-Soviet 'Sonya' class minesweeper were transferred in early 1991. Both were interned in Yemen later in 1991.

Fiji

Fiji was annexed by Britain in 1874 and this South Pacific archipelago became an independent state and member of the Commonwealth in 1970. The Fijian Naval Force was created on 12 June 1974 for coastal patrol, fishery protection, coastguard and survey duties. Following a coup in 1987, an order for four Australian ASI-315 'Pacific Forum' class patrol boats was cancelled, three of the four being completed for Tonga. A second coup took place on 25 September 1987, and Fiji was declared a republic on 7 October. Fiji's membership of the Commonwealth subsequently lapsed. Following a few years of political unrest, a coalition government was formed in 1994.

Ex-US 'BLUEBIRD' class *coastal minesweepers*

Class (no, former name/no):
Kikau (204, ex-*Woodpecker*/MSC 209), *Kula* (206, ex-*Vireo*/MSC 206), *Kiro* (205, ex-*Warbler*/MSC 205)

First two transferred on 14 October 1975 having served with USN since 3 February 1956 and 7 June 1955 respectively. Third unit transferred June 1976 having first been commissioned on 23 July 1955. Magnetic sweeper gear removed for deckhouse installation with helicopter platform. *Kiru* was discarded in 1985, and the other two in July 1990.

Ex-Israeli DABUR class *patrol boats*

Class:
301–304

Four purchased in 1991 and all extant 1995.

FIJI-built *landing craft*

Class:
Vasua, *Yaubula*

Both built at Suva and in service since 1978. Both powered by two GM diesels at a maximum of 8kts but displacement and dimensions differ. *Yaubula* is 500t full load. Both able to carry 200–220t of cargo

Finland

The second armistice in 1944 and the subsequent Treaty of Paris brought problems for the Finnish Navy; a maximum of 10,000t and 4500 men were allowed, and submarines and torpedoes were forbidden. The remaining coast defence ship *Väinämöinen* was taken as 'reparations' by the Soviet Union, the submarines were scrapped and the MTBs deprived of their tubes and reclassified as patrol boats.

Throughout the cold war, Finland was in a very difficult position, both geographically and politically, but managed to survive as a free nation, albeit one whose policies were strongly influenced by the need not to offend the Soviet Union. The navy received only small grants from the restricted defence budget and found it difficult to replace the losses to Soviet reparations. The fleet is principally composed of mine warfare vessels, missile boats and patrol vessels, mostly of Finnish construction. The world-famous Wärtsilä yard was the world's most successful builder of icebreakers, exporting to the former Soviet Union and Sweden.

In 1990 the abrogation of various clauses in the Treaty of Paris left Finland free to acquire submarines and torpedo-armed vessels, but financial limitations mean that there is likely to be little construction in the short term.

In 1995 there were 1800 personnel, plus 600 Border Guards. The fleet was organised into four flotillas (gunboats (Turku), missile craft, mine warfare vessels and patrol units (all at Helsinki)), until 1993 when it was reformed into the Archipelago Fleet, based at the main naval base at Turku, and the Gulf of Finland Fleet at Helsinki.

FLEET STRENGTH 1947

The coast defence ship *Väinämöinen* was ceded to the Soviet Union as part of reparations and renamed *Vyborg*, capital of the conquered southeast province of Finland. The submarines *Vetehinen, Vesihiisi, Iku-Turso, Saukko* and *Vesikko* were all stricken, the last preserved as a museum boat at Sveaborg, outside Helsinki.

SLOOPS

Name	Launched	Disp	Fate
Hämeenmaa class			
HAMEENMAA	1917	400t	BU 1953
UUSIMAA	1917	400t	BU 1953
Karjala class			
KARJALA	1918	342t	BU 1953
TURUNMAA	1918	342t	BU 1953

MINELAYERS

Name	Launched	Disp	Fate
LOIMU, LIESKA	1915–16	60t	Stricken 1953
POMMI, MIINA	1917	80t	Stricken 1953
RUOTSINSALMI	1940	310t	Stricken 1975

PATROL VESSELS

Name	Launched	Disp	Fate
AURA	1907	350t	Stricken 1971
VMV 1, 2, 5, 6	1930–31	30t	Stricken 1950–60
VMV 9, 11, 13, 15, 16	1935	30t	Stricken 1960–70, VMV 11 museum boat
TURSAS	1938	360t	Stricken 1975
J 1–4 (ex-Italian MAS 526–529), JYLHA, JYRY, JYSKE, JYMY	1939	22t	Stricken 1961
H 1–5, HYOKY, HIRMU, HURJA, HYRSKY, HAIJY	1943	20t	Stricken 1963
T 2–8, TAISTO, TYRSKY, TUIMA, TUISKO, TUULI T 7, T 8	1942–46	22t	Stricken 1962–77
VMV 19, 20	1943	21t	Stricken 1970

MINESWEEPERS

Name	Launched	Disp	Fate
RAUTU	1917	165t	Stricken 1950
Pukkio class			
PUKKIO, PORKKALA, PANSIO	1939–47	162t	Stricken 1974–79
Ahven class			
1 AHVEN, 2 KIISKI, 3 MUIKKO, 4 SARKI, 5 KUORE, 6 LAHNA	1936–37	17t	Stricken 1960–62
SM 1, 2, 4	1939–40	20t	Stricken 1951–54
AJONPAA, KALLANPAA	1941	52t	Stricken 1960 and 1962
Ex-US (purchased 1946)			
DR 2 PYHTAA, DR 7 PIRTTISAARI, DR 10 PURHA	1943–44	150t	DR 2 stricken 1978, others 1980
KUHA 1, 2, 4, 5, 7–18	1941–46	17t	Stricken 1957–62

Porkkala sunk 28 November 1941, was raised and rebuilt.

ICEBREAKERS

Name	Launched	Disp	Fate
MURTAJA	1898	1000t	Stricken 1959
SAMPO	1898	1850t	Stricken 1960
APU	1899	900t	Stricken 1959
TARMO	1907	2300t	Stricken 1969
OTSO	1936	800t	Stricken 1967
LOUHI (ex-*Sisu*)	1938	2012t	From 1975 HQ ship

FINLAND

MAJOR SURFACE SHIPS

Ex-British 'Bay' class *frigate*

No	Name	Builder	Acquired	Fate
–	MATTI KURKI (ex-*Porlock Bay*, ex-*Loch Seaforth*, ex-*Loch Muick*)	Charles Hill & Sons, Bristol	19.3.62	Stricken 1975

Ex-Soviet 'RIGA' class *frigates*

No	Name	Builder	Acquired	Fate
01	UUSIMA	Soviet N Yd	14.5.64	Stricken 1979
02	HAMEENMAA	Soviet N Yd	14.5.64	Stricken 1985

Hämeenmaa was converted to a minelayer 1978–79; no TT, 50 mines.

AMPHIBIOUS WARFARE VESSELS

Five *Kave* class LCUs (27t standard, 160t full load; 59ft × 16ft 5in × 4ft 4in, (*18m* × *5m* × *1.4m*); 2 Valmet diesels, 360bhp = 9kts; 1–20mm; 3 crew). Built by Hollming, Rauma, completed 19 December 1960. *Kave 5* lost in tow 15 December 1960. *Kave 1* completed by Haminan Konepaja 16 November 1956. All stricken *c*1990.

Six *Kala* class LCUs (60t standard, 200t full load; 88ft 7in × 26ft 2in × 6ft (*27m* × *8m* × *1.8m*); machinery and gun as *Kave* class; 10 crew). Built by Rauma Repola, *Kala 1* completed 16 November 56, *Kala 6* 4 December 1959. Can lay up to 34 mines. *Kala 2* and *Kala 5* stricken 1992, remainder extant 1995.

Kampela class LCUs (90t standard, 260t full load; 106ft 7in × 26ft 2in × 4ft 10in, (*32.5m* × *8m* × *1.5m*); 2 Scania diesels, 460bhp = 9kts; 2–20mm, mines; 10 crew. *Kampela 1* (in service 29 July 1976), *Kampala 2* (in service 21 October 1976) built by Enso Gutzeit, Savonlinna, *Kampala 3* (in service 23 October 1979) built by Finnmekano, Teija. All extant 1995. There are also five *Valas* class, six *Hauki* class and two *Pukkio* class, officially rated as artillery and personnel ferries.

SMALL SURFACE COMBATANTS

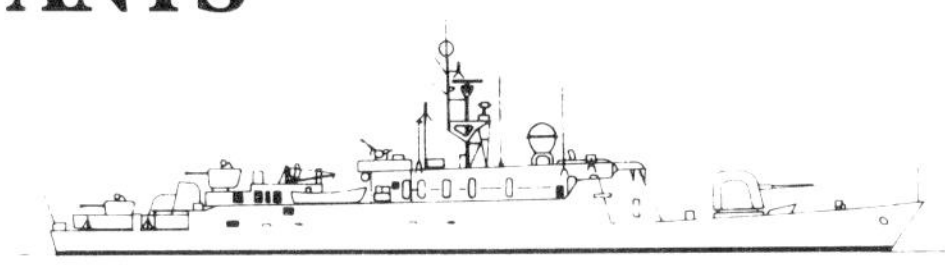

Turunmaa 1978

TURUNMAA class *corvettes*

Displacement: 650t standard; 770t full load
Dimensions: 229ft 7in oa, 243ft 1in wl × 25ft 7in × 7ft 11in
70.0m, 74.0m × 7.8m × 2.4m
Machinery: 3-shaft CODOG (3 CP propellers): 1 Olympus TM3B gas turbine, 16,180shp = 35kts, plus 3 MTU diesels, 4050bhp = 17kts. Oil 120t. Range 2500nm at 14 kts
Armament: 1–120mm Bofors, 2–40mm Bofors, 2–23mm (1×2), 2 RBU 1200 ASW RL (2×5), 2 DCT, 6–103mm RFL on main gun
Sensors: Radar HSA, M22, navigational
Complement: 70

No	Name	Builder	Launched	Fate
03	TURUNMAA	Wärtsilä, Helsinki	11.7.67	Extant 1995
04	KARJALA	Wärtsilä, Helsinki	16.8.67	Extant 1995

Ordered 18 February 1965. Laid down March 1967, completed 29 August 1968 and 21 October 1968 respectively, commissioned 1969. Engine exhausts taken astern either side of the quarterdeck to discharge as one 50ft astern, thus providing an IR decoy and allowing the ships to steam at 12 kts without using the propellers. Can cruise 17kts on diesels alone. Refitted 1984–86 by Wärtsilä, Turku, and fitted with new electronics.

Turunmaa 10.6.83 L & L van Ginderen

Isku 1973 Karl-Erik Westerlund

ISKU *experimental missile patrol boat*

Displacement: 115t standard; 140t full load
Dimensions: 86ft 6in wl, 91ft 10in oa × 28ft 7in × 6ft 5in
26.4m, 28.0m × 8.7m × 1.8m
Machinery: 4 shafts, 4 Soviet M50 diesels, 3600bhp = 15kts
Armament: 2–30mm (1×2), 4 MTO/68 (SS-N-2A Styx) SSM
Sensors: Radar 1 navigational, 1 Square Tie
Complement: 15

No	Name	Builder	Launched	Fate
16	ISKU	Reposaaren Konepaja, Björneborg	4.12.69	Extant 1995

Laid down November 1968 and commissioned 1970 for training and trials. Landing craft hull designed for missile armament and more powerful engines. In 1989–90 she was lengthened by 23ft (*7m*) and a lengthened deckhouse, mine rails and a crane fitted.

Ex-Soviet 'OSA II' class *fast attack craft (missile)*

Class (no):
Tuima (11), *Tuisko* (12), *Tuuli* (14), *Tyrsky* (15).

Purchased 1974–75 from the USSR. Finnish electronics. In 1993 *Tuima* was converted to a fast minelayer. SSM were removed and mine rails added. It is believed that the rest of the class will be similarly converted as their SSM are past their useful life. All extant 1995.

Tuima Apr 1994 L & L van Ginderen

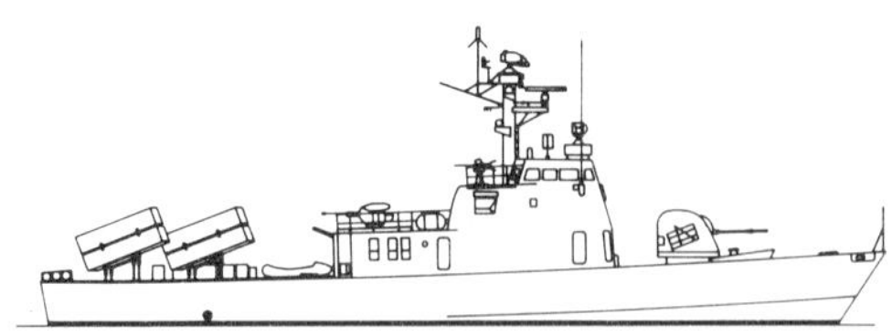

Helsinki 1990
NB 1/750 scale

HELSINKI class *fast attack craft (missile)*

Displacement: 280t standard; 350t full load
Dimensions: 147ft 7in × 29ft 3in × 9ft 9in
45.0m × 8.5m × 3.0m
Machinery: 3 shafts, 3 MTU 16V 538 TB92 diesels, 11,000bhp = 32kts
Armament: 8 RBS 15 SSM (4×2), 1–57mm Bofors Mk 1, 2–23mm (1×2), 2 Barricade chaff RL, mines
Sensors: Radar navigational, 9GA208, 9L V225; sonar SS 304 hull-mounted
Complement: 30

No	Name	Builder	Launched	Fate
60	HELSINKI	Wärtsilä, Helsinki	5.11.80	Extant 1995
61	TURKU	Wärtsilä, Helsinki	1985	Extant 1995
62	OULU	Wärtsilä, Helsinki	1985	Extant 1995
63	KOTKA	Wärtsilä, Helsinki	1986	Extant 1995

Prototype ordered 5 October 1978, with a further three ordered on 13 January 1981. A class of twelve was planned, but further construction was halted in favour of the *Rauma* class (see below). Aluminium hull. The three production models had a different-shaped pilot house to the prototype *Helsinki* as completed, but she was later modified to the same configuration. Armed with up to eight Swedish RBS 15 SSM (ordered 7 March 1983), but only four usually carried. Minelaying role requires removal of SSM tubes and DC cracks.

Turku 5.6.94 L & L van Ginderen

Rauma as completed
NB 1/750 scale

RAUMA class *fast attack craft (missile)*

Displacement: 215t standard; 248t full load
Dimensions: 134ft 6in pp, 157ft 6in oa × 26ft 3in × 2ft 10in (hull)
41.0m, 48.0m × 8.0m × 0.87m
Machinery: 2 shafts, 2 MTU 16V538 TB 93 diesels, 2 Riva Calzone IRC 155 waterjets, 8000bhp = 30kts
Armament: 6 RBS 15 SSM (2×2, 2×1), 1 Sako SAM launcher (1×6, Mistral missiles), 1–40mm/70 Bofors, 2–12.7mm MG (2×1), 4 Elma ASW RL (4×1), mines
Sensors: Radar navigational, 9GA208, 9LV225; sonar SONAC TPA towed array, ECM 9EW300 intercept, 2 decoy RL
Complement: 19 (5 officers)

Class (no, in service):
Rauma (70, 18.10.90), *Raahe* (71, 20.8.91), *Porvoo* (72, 27.4.92), *Naantali* (73, 23.6.92)

Shorter, shallower-draft successor to *Helsinki* class, ordered February 1987. Originally first unit to have been named *Luokka*. Two further groups of four are planned. SAM launcher is a Finnish designed conversion of a standard 23mm AA gun mount. Guns and missiles are interchangeable. All four extant 1995.

British 'DARK' class *fast attack craft (gun)*

Class:
Vasama 1 and *Vasama 2*.

Built 1955–57 by Saunders Roe, Anglesey, UK. Different from RN armament, 2–40mm, 4 DCs and mines. *Vasama 1* stricken 1977, *Vasama 2* stricken 1979.

NUOLI class *fast attack craft (gun)*

Displacement: 47t standard; 53t full load
Dimensions: 72ft 2in × 21ft 8in × 5ft
22.0m × 6.6m × 1.5m
Machinery: 3 Soviet M50 diesels, 2700bhp = 40kts
Armament: 1–40mm, 1–20mm
Sensors: Radar Decca 707
Complement: 15

Class (no, launched):
Nuoli 1–4 (31–34, 1961), *Nuoli 5–8* (35–38, 1962), *Nuoli 9* (39, 1963), *Nuoli 10–12* (40–42, 1964), *Nuoli 13* (43, 1966).

All built at Laivateollisuus, Turku (Åbo). *Nuoli 9* deleted 1979 and *Nuoli 4* in 1980. *Nuoli 1–3, 6* and *9* discarded 1984. Remainder modernised 1982–83 with new armament: 2–23mm/60 (1 × 2), 1–12.7mm MG, 4DC. *Nuoli 5* stricken 1992, remaining five extant 1995.

Nuoli 2 c1975 Karl-Erik Westerlund

Rymattyla 1975 Karl-Erik Westerlund

RIHTNIEMI class *large patrol craft*

Displacement: 90t standard; 110t full load
Dimensions: 101ft 8in × 18ft 8in × 5ft 11in
31.0m × 5.6m × 1.8m
Machinery: 2 shafts, 2 MTU diesels, 2500bhp = 15kts
Armament: 1–40mm, 1–20mm (from 1976 4–23mm (2×2)), 2 RBU 1200 ASW RL (2×5)
Complement: 20

No	Name	Builder	Launched	Fate
51	RIHTNIEMI	Repola, Rauma	1956	Extant 1995
52	RYMATTYLA	Repola, Rauma	1956	Extant 1995

Ordered June 1955. Commissioned 21 February and 20 May 1957 respectively. Modernised 1980–81. Also minesweepers.

RUISSALO class *large patrol craft*

Displacement: 110t standard; 130t full load
Dimensions: 108ft 11in × 18ft 6in × 5ft 11in
32.0m × 6.0m × 1.8m
Sensors: Radar Decca, hull sonar
Other particulars: As *Rihtniemi* class

No	Name	Builder	Launched	Fate
53	RUISSALO	Laivateollisuus, Turku	16.6.59	Extant 1995
54	RAISIO	Laivateollisuus, Turku	2.7.59	Extant 1995
55	ROYTTA	Laivateollisuus, Turku	2.6.59	Extant 1995

Ex-*R 3–5*, completed 11 August 1959, 12 September 1959, 14 October 1959 respectively; 53 modernised 1976, other two 1980. Also minesweepers. In 1991 *Ruissalo* conducted trials with a lightweight VDS system in place of the after gun mounting.

Ruissalo c1970 Karl-Erik Westerlund

*Keihassalmi c*1975 Karl-Erik Westerlund

MINE WARFARE VESSELS

KEIHASSALMI class *minelayer*

Displacement:	290t standard; 360t full load
Dimensions:	184ft 9in × 25ft 4in × 6ft 6in *56.0m × 7.7m × 2.0m*
Machinery:	2 shafts, 2 Wärtsilä diesels, 1600bhp = 15kts
Armament:	2–40mm, 2–20mm (2×1), up to 100 mines
Complement:	60

No	Name	Builder	Launched	Fate
05	KEIHASSALMI	Valmet, Helsinki	16.3.57	Stricken 1992

Commissioned 1958. The 2–40mm changed to Soviet 4–30mm (2 × 2) in 1975.

Pohjanmaa as completed

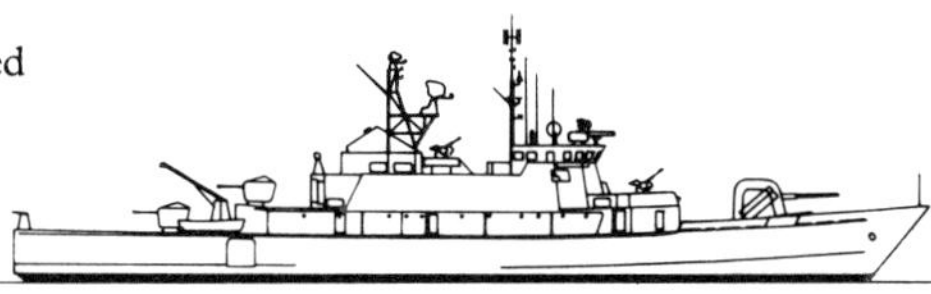

POHJANMAA class *minelayer*

Displacement:	1000t standard; 1100t full load
Dimensions:	255ft 9in × 37ft 8in × 9ft 10in *78.0m × 11.5m × 3.0m*
Machinery:	2 shafts, 2 Wärtsilä-Vasa 16V22 diesels, 5800hp = 20kts. Range 3500nm at 17kts
Armament:	1–120mm, 2–40mm (2×1), 8–23mm (4×2), 2 RBU 1200 ASW RL (2×2), up to 120 mines
Sensors:	Radar navigational, air search, Philips 9LV 100 fire control; sonar 2 sets; ECM 2 decoy RL
Complement:	77–90 (as training ship)

No	Name	Builder	Launched	Fate
01	POHJANMAA	Wärtsilä, Helsinki	22.8.78	Extant 1995

Designed in 1976 and laid down May 1978. Also TS, as such maximum fifty-six men in Portacabins on mine deck. Commissioned 8 June 1979. Helicopter pad on quarter-deck. Bow thruster. In 1992 two 12.7mm MG replaced two twin 23mm mounts forward.

HAMEENMAA class *minelayers*

Displacement:	1000t standard; 1300t full load
Dimensions:	228ft 4in pp, 252ft 7in oa × 38ft 1in × 9ft 10in *69.6m, 77.0m × 11.6m × 3.0m*
Machinery:	2 shafts (CP props), 2 Wärtsilä-Vasa 12V 22MD diesels, bow thruster, 6400bhp = 20kts
Armament:	1 Sako SAM launcher (1×6, Mistral missiles), 2–40mm/70 Bofors (2×1), 4–23mm/60 (2×2), 2 RBU-1200 ASW RL (2×1), 2 DC racks, 4 mine rails (150 mines)
Sensors:	Radar navigational; sonar hull-mounted
Complement:	45

No	Name	Builder	Launched	Fate
02	HAMEENMAA	Finnyards, Rauma	11.11.91	Extant 1995
05	UUSIMAA	Finnyards, Rauma	Jun 1992	Extant 1995

Originally ordered from Wärtsilä, Helsinki, in July 1989 as replacements for *Keihässalmi*, but had to be re-ordered in December of that year following the bankruptcy of the original yard, entering service in 1992 and 1993 respectively. They were built with bow and stern ramps so as to be able to serve as logistic transports. Strengthened hulls and icebreaker bows for ice navigation.

Pohjanmaa 30.5.92 L & L van Ginderen

Hameenma Jul 1994 L & L van Ginderen

PANSIO class *coastal minelayers*

Displacement:	450t full load
Dimensions:	144 ft 4in oa, 128ft 7in wl × 32ft 10in × 6ft 7in *44.0m, 39.2m × 10.0m × 2.0m*
Machinery:	2 shafts, 2 MTU 12V 183 diesels, 1500bhp = 10kts
Armament:	2–23mm/60 (1×2), 1–12.7mm MG, 50 mines
Sensors:	Radar navigational
Complement:	12

Class (no, in service):
Pansio (876, 25.9.91), *Pyhäranta* (475, 26.4.92), *Porkkala* (777, 29.10.92)

Ordered in May 1990 from Olkiluoto Telakka. Multi-role vessels built as minelayers, pollution control ships and landing craft/cargo carriers to support coastal artillery facilities. Bow and stern ramps, and a vehicle deck accommodating up to two trucks, with mine rails fitted. All three extant 1995.

Ex-US BYMS type *inshore minesweepers*

Class (no):
Katanpää (1), *Purunpää* (2), *Vahterpää* (3), *Tammenpää* (4)

Ex-US BYMS 2049, 2032, 2044, 2047. Launched 1941–42, purchased 1948, scrapped 1958–60.

KUHA class *inshore minesweepers*

Displacement:	90t standard
Dimensions:	87ft 2in × 22ft 8in × 6ft 6in *26.6m × 6.9m × 2.0m*
Machinery:	1 shaft, 2 Cummins NT-380M diesels, 600bhp = 11kts
Armament:	2–20mm (2×1) or 2–23mm (1×2) and 1–20mm
Sensors:	Radar Decca
Complement:	15

Class:
Kuha 21–26

Ordered 1972 and launched at Laivateollisuus, Turku, 1974–75. Hulls constructed of GRP. Active rudders.

Kiiski 1 as completed

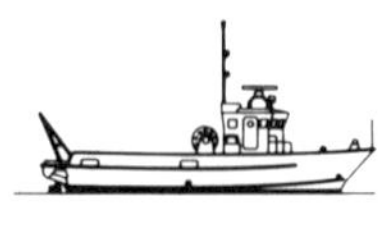

KIISKI class *coastal minesweepers*

Displacement:	20t full load
Dimensions:	49ft 10in × 13ft 5in × 3ft 11in *15.18m × 4.1m × 1.2m*
Machinery:	2 waterjets, 2 Valmet 611 CSMP diesel, 340bhp = 10kts
Armament:	None
Sensors:	Radar navigational
Complement:	4

Class (nos, in service):
Kiiski 1–7 (521–527, 1984–85)

GRP hulled, built by Fiskar's Turun, Turku. Originally designed to be operated in both manned and drone modes (controlled from the *Kuha* class minesweepers), but in the event operate in manned mode only. Equipped to counter both magnetic and acoustic mines. All seven extant 1995.

ICEBREAKERS

All built by Wärtsilä, Helsinki, with pairs of propellers fore and aft. Extant 1995 unless otherwise stated.

Voima (4415t standard; 274ft × 63ft 88in × 23ft, *83.5m × 19.4m × 6.8m;* 6 diesels, 10,500bhp = 16.5kts, 58 crew). Launched 1952, in service 1953. Rebuilt and re-engined 1978–79 to serve until 1994.

Four *Karhu* class (3540t standard; 243ft × 57ft × 21ft, *74.2m × 17.4m × 6.4m;* 4 diesels, 7500bhp = 16kts, 53 crew), *Karhu* (launched 22.10.57, in service December 1958), *Murtaja* (22.9.58, in service 1959), *Sampo* (launched and in service 1960), *Hansa* (W German-owned, in service 25.11.66) has a Finnish crew and summers off Finland. *Karhu, Murtaja* and *Sampo* stricken *c*1985, replaced by the *Otso* class (see below).

Three *Tarmo* class (4890t standard; 284ft × 71ft × 23ft, *86.5m × 21.7m × 6.8m;* 4 diesels, 12,000bhp = 16.5kts, 58 crew). *Tarmo* (in service 1963), *Varma* (1968), *Apu* (25.11.70). *Tarmo* and *Varma* deleted 1993.

Two *Urho* class (as Swedish *Atle* class). *Urho* (launched 1974, in service 5.3.75), *Sisu* (1975, in service 28.1.76). A helicopter but no other armament, 45 crew.

Two *Otso* class (8500t, 324ft 10in × 80ft 3in × 26ft 3in (*99.01m × 24.46m × 8.0m*), 20,400shp = 18.5kts, 28 crew): *Otso* (launched 12.7.85, in service 30.1.86, *Kontio* (launched 30.7.86, in service 29.1.87). Replacements for *Karhu* class.

Two *Fennica* class (12,500t, 380ft 7in × 85ft 4in × 27ft 7in (*116.0m × 26.0m × 8.4m*), 16kts, 16 crew): *Fennica* (launched 10.9.92, in service March 1993), *Nordica* (launched 18.6.93, in service January 1994). Helicopter deck plus hangar forward.

Otso Apr 1994 — L & L van Ginderen

France

At the end of the Second World War little remained of the powerful fleet which France had built during the 1930s. Two battleships – one incomplete – and a handful of cruisers, *contre-torpilleurs* and submarines had escaped the German occupation in 1940. Most had fled to North Africa and had later rejoined the Allied forces, seeing heavy service in all theatres especially during the latter part of the war. The remainder of the French fleet had been scuttled at Toulon in 1942.

France therefore found herself bereft of a fleet that could satisfy her traditional maritime status, her extensive naval infrastructure, and the demands of a far-flung empire which she was in the process of retrieving from the Allies and from the Japanese. Industry had been crippled by the war and the economy was in ruins. Consequently France was unable either to construct or to finance a new fleet in the immediate postwar period.

Inevitably, recourse was made to her British and American allies, who not only provided war-built ships surplus to their own requirements, but were particularly generous as regards the allocation of German and Italian prizes. Many of the latter were modern, powerful vessels which had seen little war service and were in excellent condition. By 1947 the Marine Nationale looked quite respectable – at least on paper.

Logistics and maintenance, however, soon became a nightmare. Gun calibres included 155mm (French), 152mm (French), 150mm (German), 138.6mm (French), 135mm (Italian), 127mm (German), 120mm (Italian), 105mm (German), 102mm (British), 100mm (French), 90mm (French), 88mm (German), 76mm (US), 75mm (French), 40mm (US), 37mm (French/German/Italian), 20mm (various), 13.2mm (French) and the British 2pdr. Propulsion machinery was equally difficult to maintain. Attempts were made to rationalise weaponry by the adoption of a German-pattern 105mm mounting to replace the British 4in and the Italian 135mm, and the standardisation of light AA guns around the 40mm and 20mm favoured by the British and Americans. But this hastily assembled 'fleet' could clearly do little more than tide the French over until such time as a new navy could be constructed, and the 1950s saw a steady procession of ex-German and ex-Italian units to the breakers' yards. They served their purpose, nevertheless, and the ex-German submarines proved of particular value as experimental vessels, exercising considerable influence on postwar submarine construction.

THE NEEDS OF THE EMPIRE

After the war France was anxious to re-establish herself as an imperial power. She clung tenaciously to her possessions in the Far East in the face of a growing wave of nationalism in that region. The result was the protracted conflict in Indo-China, in which the navy played a considerable part. US-built patrol craft were purchased for use around the coast and in the rivers and estuaries, and there were successive deployments by the carriers *Dixmude* and *Arromanches,* and later by the *Bois Belleau* and *Lafayette.*

The acquisition of these two ex-American carriers on a long-term loan was largely due to the problems experienced in sustaining a French carrier presence off Indo-China. They brought with them newer and more effective aircraft – Hellcats, Helldivers, Corsairs and Avengers – to replace the ageing Seafires and Dauntlesses, and enabled the former escort carrier *Dixmude* to be relegated to support.

Once the usefulness of carriers as an instrument for the projection of French maritime power was established they quickly became the central element of the fleet, and the two big battleships – one of which was completed at great expense in the early postwar years – went into an early retirement. *Jean Bart* was present at Suez but only as a fire support ship, while the main Anglo-French naval presence was a five-carrier task force which included *Lafayette* and *Arromanches.*

NATO

In 1949 France became a founder-member of NATO. Apart from national security needs there were many advantages accruing from membership of the alliance. During the immediate postwar period much of the French fleet was of British or American construction, and not only were there clear benefits in continuing this association from a logistics point of view, but France was now given access to the latest tactical thinking and military technology. There were also to be extensive financial benefits from the American-funded Mutual Defense Assistance Program (MDAP).

Commitments to French national status as an imperial power on the one hand, and to the security of the North Atlantic on the other, made demands of a different nature on the Marine Nationale. In the first postwar review a naval statute established a requirement for (a) 400,000t of escorts, patrol vessels, minesweepers and small craft for NATO; and (b) 360,000t of carriers, submarines and amphibious vessels for national requirements. Financial and technological difficulties resulted in the failure of the French Parliament to commit itself to the programme, but the project was saved by allied aid amounting to 40 per cent of the NATO commitment. Between 1951 and 1957 30,000t was laid down per year.

In order to conform to the new thinking, the prewar system of classification – *croiseur léger, contre-torpilleur, torpilleur, aviso de 1ère/ 2e classe* – was abandoned. The allocation of all destroyer/frigate sized ships to the protection of the carriers (in the national defence role) or of the convoy (in the NATO role) led to the adoption of the term *escorteur*, to which the word *rapide* would be added for a vessel capable of fleet speed, together with a sub-classification to indicate ASW or AA protection as the primary role of the ship. In 1951 this system of classification was in its turn abandoned in favour of '1st class escort' (30kts plus) and '2nd class escort'. This appears to have provided inadequate differentiation of the ship's mission, and the classification system was changed again in 1953 and 1955. The *Surcouf* and *Le Corse* classes, both authorised in 1949, were the prototypes of the new escorts and underwent all these changes in classification while building:

Surcouf (T 47)
(AAW – carrier escort)
1949: *Escorteur Rapide Anti-Aérien*
1951: *Escorteur de 1ère classe*
1953: *Escorteur Rapide*
1955: *Escorteur d'Escadre* (fleet escort)

Le Corse (E 50)
(ASW – convoy escort)
1949: *Escorteur Rapide Anti-Sous-marin*
1951: *Escorteur de 2e classe*
1953: *Escorteur*
1955: *Escorteur Rapide*

Other vessels were modified to fit in with these new concepts. The former light cruiser *De Grasse*, which had been laid down prewar and launched in 1946, was redesigned as an anti-aircraft cruiser, and fitted with the same combination of 127mm and 57mm guns as the *Surcouf* class. She and her near-sister, *Colbert*, laid down in 1953, were to provide command facilities and AA protection at the centre of a carrier task force. The two former Italian light cruisers of the *Attilio Regolo* class were rebuilt 1951–54 and rearmed on the lines of the new AAW escorts; they would later serve as flotilla leaders for the latter.

Curiously, the prewar organisation of destroyer-sized ships into three-ship *Divisions* persisted and is still in existence today. There was also a continuing need for the colonial sloop, designed to patrol alone on colonial stations – hence the *escorteur d'union française* designation conferred on the frigates of the *Commandant Rivière* class commenced in the late 1950s.

Submarines continued to be divided into 1st class (1200t standard) and 2nd class (600t) boats, although a new element was a small 'hunter-killer' submarine (*Aréthuse* class) designed to prevent enemy submarines from disrupting French mercantile traffic in the Mediterranean.

After a number of false starts the new carriers and amphibious units on which France's ability to project her maritime power was to be based were authorised. The loan or purchase of British and American vessels in the immediate postwar period had given the Marine Nationale a breathing-space in which to acquire the necessary operating experience and technical expertise to ensure that the designs were in line with the latest developments elsewhere.

Finally, with the help of American financial aid, large numbers of mine countermeasures and patrol craft were acquired to provide France's contribution to the security of the seaports of western Europe.

A CHANGE OF POLICY

When General de Gaulle came to power in 1958 he expressed immediate dissatisfaction with France's position in the NATO command structure, and in particular with what he regarded as inadequate representation in the AFMED command, to which the bulk of the fleet was allocated. His claims received an unsympathetic hearing from the Americans, who were anxious not to upset their other allies, and in March 1959 France duly informed the North Atlantic Council that the French Mediterranean Fleet would henceforth remain under national control. From this moment on the withdrawal of French forces from the NATO command structure was a continuous process: the remaining naval units were withdrawn from the Channel and Atlantic commands, the land and air forces followed in 1966, and finally in 1969 notice was served on all NATO HQs and other installations on French soil.

The political aspirations of De Gaulle produced far-reaching changes in French military strategy and consequently in the structure of the fleet itself. The desire to create a foreign policy independent of Washington and Moscow brought with it a complete reorientation of defence priorities. At the beginning of the 1960s De Gaulle defined these as: 'To deter – to intervene – to defend'. France could no longer live under the protection of the American nuclear umbrella. Immediately plans were put in hand to develop a submarine-based *Force de Dissuasion* similar in concept to Polaris.

France already possessed a considerable intervention capability based on three carriers and a new generation of amphibious units and colonial sloops, and this capability would be maintained and improved through modernisation.

'Defence' now came third in the list of priorities, and in view of the limited finance available now that the Marshall Plan was at an end would continue to rely heavily on the ships procured during the 1950s.

In size and capability the French SSBN force was to be similar to that of the Royal Navy. The British, however, acquired Polaris 'on the cheap' from their American allies; they paid no development costs, had unrestricted access to American technological expertise in the design of the submarines, and were permitted to use the firing ranges off the coast of Florida for tests and trials. France, on the other hand, had to begin from scratch: missiles and warheads had to be developed and tested, firing ranges constructed. The United States had refused to supply France with enriched uranium or with a reactor. The design of a French pressurised water-cooled reactor was therefore undertaken in collaboration with the CEA – the civilian atomic energy authority; because the Americans would supply enriched uranium only for land experimental use, a nuclear propulsion plant was built ashore at Cadarache and tested to simulate typical SSBN patrols.

Caissons were built for test-firings of the missiles. In 1967 this role was taken on by a specially constructed submarine, *Gymnote*, fitted with the same guidance and inertial navigation system as the new SSBNs (SNLE in French parlance). The first missile was launched at sea at the Landes Trials Centre at the end of 1968. Observations of the missile's flight were made by the *Henri Poincaré*, a former Italian tanker rebuilt at Brest 1964–67 and equipped with long-range tracking antennae for monitoring the re-entry phase, accompanied by three *escorteurs rapides* specially converted for the purpose.

Simultaneously nuclear warheads were being tested at the Experimental Centre at Mururoa Island in the Pacific, the tests being supervised by a whole fleet of ships including the cruiser *De Grasse* – partially disarmed and fitted as a command ship – and the two new landing ships *Orage* and *Ouragan*.

The cost of all this was enormous. In 1965, when the SNLE programme was in full swing, the only other ships under construction or on order for the Marine Nationale were the two *frégates* of the *Suffren* class, two submarines of the *Daphné* class, and the landing ship *Orage*. These would be the only major vessels outside the SNLE programme completed during the second half of the decade. Not a single unit in the former NATO category – convoy escorts, MCM vessels, patrol craft – was built.

THE 'PLAN BLEU'

In 1969 De Gaulle retired from French political life; there was, however, to be no significant change in foreign policy. Nobody in the ruling Gaullist Party was prepared to engineer a *rapprochement* with the United States and thereby throw into question the independence of France's position in world affairs. In June 1970 the new Premier, Michel Debré, reaffirmed: 'We shall either have a national defence or no defence at all'. The development of the French sea-based nuclear deterrent force continued to take priority over all naval commitments.

In some quarters of the Marine Nationale, however, concern was beginning to be voiced regarding the state of France's conventional naval forces. The large surface fleet built up during the 1950s was now ageing fast, despite extensive mid-life conversions to many of the large fleet escorts, and few replacements were being built. Successive five-year plans, vulnerable as they were to piecemeal deletions and postponements when money was short, had failed to deliver the goods. Inevitably recourse was made to a massive long-term plan, in which the future size and structure of the fleet was strictly determined. The *Plan Bleu*, as it was to be known, was formulated in 1972, and provided for the following fleet structure by 1985: five SNLE, two aircraft carriers, two helicopter carriers, thirty *frégates/corvettes*, two replenishment tankers, twenty SSK/SSN, fifty LRMP aircraft, two assault ships, LCU, LST, etc, thirty-five *avisos*, thirty FPB, thirty-six MHC/MSC and 85,000t of logistics support and maintenance ships.

The objectives of the naval policy which this fleet was intended to sustain were: (a) the maintenance of an effective sea-based nuclear deterrent force; (b) the ability of the fleet to operate on the high seas; (c) a defence capability for the French maritime approaches; and (d) a defence capability for French overseas territories.

The first three categories paralleled exactly De Gaulle's 'to deter –

Blue Water navy: *Isère* refuels the carrier *Clemenceau* and the escort *La Bourdonnais* Navarret Collection

to intervene – to defend'. The fourth, however, marked an interesting modification of previous policy in that it established the need for a French naval *presence* (as opposed to a distant intervention capability) in the overseas territories. French interest in these territories appeared to have ended during the 1960s following the withdrawal from southeast Asia and from the colonies in North Africa. The 1970s, however, saw a steady build-up of French naval forces in the Indian Ocean – now for the first time a focal point in world affairs – and the basing of ships at Djibouti, Mayotte and La Réunion. It was to provide such a presence that the new *avisos* of the A 69 class – designed for coastal ASW and patrol – and the FPBs were included in the programme. By 1980 the French Indian Ocean squadron regularly numbered twenty ships.

The *Plan Bleu* decreed an increase in the carrier force from three to four ships, and plans were already in hand for the additional vessel, an ASW/assault helicopter carrier. The old fleet escorts of the T 47/53/56 classes were to be replaced by a scaled-down version of the *Tourville* (F 67) design, the C 70 *corvette*. The latter class would be built in two versions, one specialising in ASW and the other armed with surface-to-air missiles.

French naval plans have an unhappy record, and the *Plan Bleu* was no exception. In 1977 one of the four carriers, eight of the *avisos* and eight of the submarines were deleted from the plan. The surface ship programme was subject to constant delays, with orders being placed piecemeal as and when funds were available. Whenever the opportunity for export sales arose, the French naval construction programme was slowed or interrupted to ensure early delivery to the customer.

The underlying reason for these budgetary problems was the disproportionate increase in spending on the strategic missile submarine force. In 1982 FOST accounted for 26 per cent of the naval budget. The corresponding figure for 1990 was 37.4 per cent – an increase of more than 11 per cent over eight years. The French strategic force not only enjoyed absolute priority over other naval programmes, but had become relatively more expensive because of the increasingly sophisticated technology employed both in the missiles and the submarine platforms. This steadily ate into the overall budget, and was responsible for the delays and cancellations in surface ship programmes which were characteristic of French naval development throughout the 1970s and 1980s.

By the late 1980s concern was being expressed about the Marine Nationale's ability to fulfil its worldwide policing commitments, and new frigate programmes were put in train as replacements for the ageing *avisos-escorteurs* of the *Commandant Rivière* class. In order to finance the maximum number of hulls on a still-tight budget, the Marine Nationale resorted to the novel expedient of a utility 'patrol frigate' built to commercial standards and ordered from Chantiers de l'Atlantique, St Nazaire (the *Floréal* class). These ships will now complement the sophisticated first-line frigates of the *La Fayette* class, the construction of which has been subject to serious delays.

Reductions in the defence budget following the end of the Cold War have resulted in further cuts and delays in the construction programme; completion of the carrier *Charles de Gaulle* has been delayed by three years, the last two SSNs of the *Améthyste* class have been cancelled, two further LPDs of the *Foudre* class have been deferred to a later programme, and the BAMO Ocean MCM programme has been abandoned. There are many over-aged ships still active in the current fleet, and some of these will have to remain in service until urgently-needed replacements are completed. Problems extend to the Aéronavale, which will be obliged to soldier on with refurbished F-8 Crusader interceptors until the Rafale ACM fighter enters service at the end of the 1990s. However, the proportion of the naval budget taken by FOST has now declined to 25 per cent; construction of the new SSBNs of the *Le Triomphant* class has been slowed, and the last two ships projected have been cancelled, leaving a future deterrent force of only four boats.

COMMAND STRUCTURE

The command structure of the Marine Nationale is divided into four maritime zones. The Atlantic zone is under the command of CEC.LANT, with its HQ at Brest, and the Mediterranean zone is commanded by CEC.MED, based at Toulon. These two metropolitan commands are complemented by two overseas commands: the C-in-C in the Indian Ocean is designated AL.INDIEN, and exercises command from a flagship; the C-in-C Pacific is designated AL.PACI, and is based at Papeete.

Since the Second World War Metropolitan France has been divided into three maritime regions: the *1ère Préfecture Maritime* covers the Channel area, with its headquarters in Cherbourg; the *2e Préfecture Maritime*, covering the area from the Cotentin Peninsula down to the border with Spain, with its HQ in Brest; and the *3e Préfecture Maritime*, covering the Mediterranean region, with its HQ in Toulon.

In the aftermath of the Cold War it was decided that the French Armed Forces should be subjected to a radical review, entitled '*Armées* 2000'. As a consequence of this review it was decided to base all naval forces in the Atlanic on Brest, and to create a new command structure which would better reflect current French naval missions. The Atlantic and Mediterranean squadrons were disbanded, the new command structure reflecting the various components of the Marine Nationale rather than geographical operational boundaries. The principal force components comprise: the *Force d'Action Navale* (FAN), constituted by the carriers, the major amphibious units and the nuclear attack submarines, supported by nine modern frigates (four AAW, five ASW), based at Toulon; *Aviation Embarqué* (AE); the *Groupe d'Action Sous-Marine* (GASM), comprising six first-line ASW frigates plus ten of the smaller *avisos*, based at Brest; the *Force Océanique Stratégique* (FOST), currently with five SSBNs and based on the Île Longue in the Brest roadstead; and the *Force de Guerre des Mines*, comprising all MCM vessels, including the three units of the *Eridan* type based at Toulon, but with its HQ in Brest.

MAJOR RADARS OF FRENCH MANUFACTURE

Designation	Description	In service
DRBV 11	Air/surface surveillance	1954
DRBV 20A	Medium-range air surveillance	1954
DRBV 22A/C/D/E	Medium-range air surveillance	1956
DRBV 20C	Long-range air surveillance	1961
DRBV 23B/C	Long-range air surveillance	1961

DRBV 50	Low-altitude target designation	1961
DRBV 13	Surveillance and tracking	1973
DRBV 26A/C/D	Long-range air surveillance	1974
DRBV 51A/B/C	Low-altitude target designation	1976
DRBV 15A/C	Target designation	1985
DRBV 21A	Medium range air surveillance	1992
DRBI 10	Height-finder	1957
DRBI 23	3-D air surveillance/tracking	1967
DRBJ 11B	3-D air surveillance/tracking	1991
DRBR 51 A/B	Tracker/illuminator for Masurca	1967
DRBC 11	127mm fire control	1954
DRBC 30	57mm fire control	1954
DRBC 31B	127mm fire control	1959
DRBC 31A	57mm fire control	1959
DRBC 32A/B	100mm fire control	1961
DRBC 32C/D/E	100mm fire control (digital)	1972
DRBC 33A	100mm fire control	1986

FLEET STRENGTH 1947

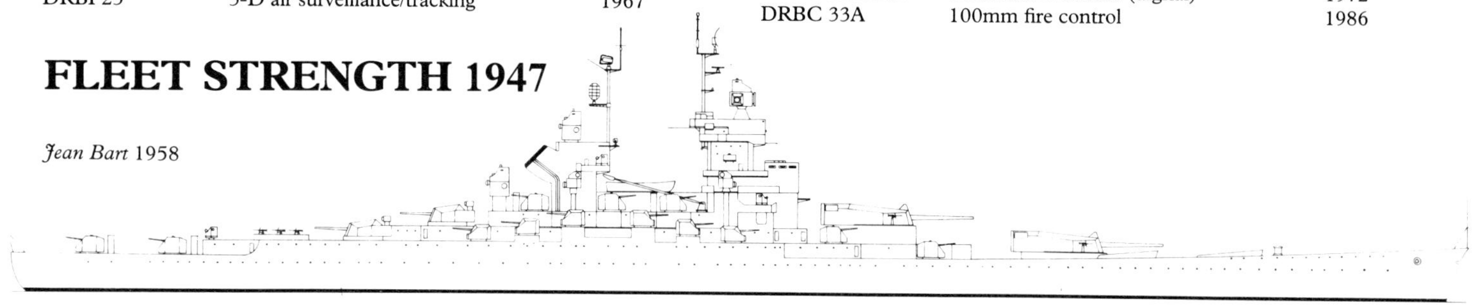

Jean Bart 1958

BATTLESHIPS

Name	Launched	Disp (std)	Fate
RICHELIEU	17.1.39	38,500t	Hulked 1959, BU 1968
JEAN BART	6.3.40	42,800t	Stricken 1960, BU 1970

Richelieu remained largely unaltered, except for the removal of some of the smaller AA weapons, until placed in reserve in 1956. *Jean Bart,* which was incomplete and had been heavily damaged at Casablanca, was transferred to Brest for completion. Work was delayed because of economic difficulties and, but for the strength of public opinion, construction would probably have been abandoned. The first sea trials took place in January 1949, but the new AA armament was not installed until the winter of 1951–52. As finally completed *Jean Bart* carried 24–100mm in twin turrets, disposed symmetrically in four groups on either side of the superstructure. There were six of the new 57mm twin mountings amidships, with four abeam 'B' turret and a further four on the quarterdeck. Modifications were made to the turret mast, and the latest surveillance and fire control radars fitted. To compensate for the increase in topweight *Jean Bart* was fitted with a tear-shaped torpedo bulge which increased beam to 33.5m. She was present as a fire support ship at Suez in 1956, but was subsequently used as an accommodation ship at Toulon.

AIRCRAFT CARRIER

Name	Launched	Disp (std)	Fate
DIXMUDE (ex-*Biter*)	18.12.40	12,850t	To US 1966

Transferred from Britain in 1945, *Dixmude* saw active service off Indo-China in 1947, flying SBD Dauntless dive-bombers. In 1949 she was relegated to the rôle of aviation transport and served as such until 1960, when she replaced the *Jules Verne* as base ship for the amphibious forces at St Mandrier.

CRUISERS

Name	Launched	Disp (std)	Fate
SUFFREN	3.5.27	9940t	BU 1974
JEANNE D'ARC	14.2.30	6500t	BU 1966
EMILE BERTIN	9.5.33	5890t	BU 1959
La Galissonnière class			
GLOIRE	28.9.35	7600t	BU 1958
MONTCALM	26.10.35	7600t	BU 1970
GEORGES LEYGUES	24.3.36	7600t	BU 1959
Le Fantasque class			
X 01 LE TERRIBLE	29.11.33	2570t	BU 1962
X 02 LE MALIN	17.8.33	2570t	BU 1964
X 03 LE FANTASQUE	15.3.34	2570t	Stricken 1957
X 04 LE TRIOMPHANT	16.4.34	2570t	Stricken 1954

Suffren was the only surviving French heavy cruiser to see active service in the immediate postwar period. She served in the Far East before paying off into reserve in 1949. The older cruisers *Duquesne* and *Tourville* were used only for training and were hulked in 1950.

Of the prewar light cruisers, *Gloire, Montcalm* and *Georges Leygues* all had active postwar careers until their disposal in the late 1950s. *Emile Bertin* became a sea-going gunnery ship in 1947 and was placed in reserve in 1952. The older light cruiser *Duguay Trouin* was used as a depot ship until her disposal in 1952. *Jeanne d'Arc* continued to serve as a school ship until 1964, when she was replaced by the new helicopter carrier of the same name.

The four survivors of the *Le Fantasque* class of *contre-torpilleurs* were redesignated light cruisers and saw extensive service, especially in the Far East during the early 1950s. Even by this time they could sustain a maximum speed of 35kts. *Le Triomphant* was broken up earlier than the others and cannibalised for spares. *Le Malin* was hulked in 1956 and the following year *Le Terrible* became a school ship.

DESTROYERS AND TORPEDO-BOATS

No	Name	Launched	Disp	Fate
T 05	TIGRE	2.8.24	2130t	BU 1954
T 06	ALBATROS	28.6.30	2440t	BU 1959
Ex-German vessels				
T 01	MARCEAU (ex-*Z 31*)	15.5.41	2660t	Stricken 1958
T 02	HOCHE (ex-*Z 25*)	16.3.40	2660t	BU 1962
T 03	KLEBER (ex-*Riedel*)	22.4.36	2310t	BU 1958
T 04	DESAIX (ex-*Jacobi*)	24.3.36	2310t	BU 1951
T 07	ALSACIEN (ex-*T 23*)	Nov 1941	1100t	BU 1955
T 08	LORRAIN (ex-*T 28*)	Nov 1942	1100t	BU 1960
T 09	DOMPAIRE (ex-*T 14*)	1939	850t	BU 1952
T 10	BIR HAKEIM (ex-*T 11*)	1.3.39	850t	BU 1952
T 11	BACCARAT (ex-*T 20*)	Oct 1941	850t	BU 1952

Five ex-German destroyers and six torpedo-boats were transferred to the Marine Nationale in 1946. Of the large 'Narvik'-type vessels, *Z 39* was in such bad condition that she was cannibalised to provide spares for the other two units, renamed *Hoche* and *Marceau.* The former was in reserve 1949–51 and was then refitted and served as an ASW trials ship from 1953 until 1956. Her armament at this time consisted of three single 150mm, 8–40mm (8×1) and six 550mm torpedo tubes (2×3) of the same model as that installed in the *T 47* class. *Marceau* underwent a refit 1948–50 during which US radar was installed and the AA armament was altered to 8–40mm (8×1). She saw active service until 1953, when she was placed in reserve.

Desaix was refitted 1946–47 and had US radar installed. She then served as a weapons trials ship until 1949. Her sister *Kléber* was refitted 1948–51, when US radar and 6–40mm were installed. She served until 1953 when she was placed in reserve. Of the large torpedo-boats, *T 35* was dismantled for spares. *Lorrain* was refitting until 1949, and was then used for ASW training at Toulon; *Alsacien* was similarly employed until 1952, when she was placed in reserve.

All the larger units except *Desaix* were given NATO 'D' numbers in 1951, but the three smaller torpedo-boats were disposed of before the new system of classification came into existence.

The former *contre-torpilleurs Tigre* and *Albatros* were employed as sea-going tenders, the latter being assigned to the gunnery school. *Albatros* had her forward boilers and funnels removed, speed being reduced to 24kts.

Georges Leygues 5.9.56 L & L van Ginderen

SUBMARINES

No	Name	Launched	Disp	Fate
S 01	ARCHIMEDE	6.9.30	1380t	BU 1952
S 02	LE GLORIEUX	29.11.32	1380t	BU 1952
S 03	CASABIANCA	2.2.35	1380t	BU 1952
S 04	LE CENTAURE	14.10.32	1380t	BU 1952
S 05	JUNON	15.9.35	600t	BU 1955
S 06	IRIS	23.9.34	600t	BU 1950
S 07	RUBIS	30.9.31	760t	Stricken 1949
Ex-German submarines				
S 08	MILLE (ex-*U 471*)	6.3.43	770t	Stricken 1963
S 09	LAUBIE (ex-*U 766*)	Apr 1943	770t	Stricken 1961
S 10	BLAISON (ex-*U 123*)	2.3.44	1050t	Stricken 1957
S 11	BOUAN (ex-*U 510*)	4.9.41	1050t	BU 1958
S 12	ROLAND MORILLOT (ex-*U 2518*)	4.10.44	1620t	BU 1968

By 1947 only four submarines of the prewar *1500-tonne* ocean-going type, two of the '2nd class' *600-tonne* type, and the minelaying submarine *Rubis* remained. With the transfer of the ex-German boats even these few were quickly paid off, with the exception of *Le Glorieux* and *Junon*, the latter being employed for training.

In 1946 France received six U-boats: two from the UK and four that had been seized in French ports. *Blaison* was a Type IXB, *Bouan* a Type IXC. A second Type IXB, *U 129*, was broken up and cannibalised for spares. *Millé* and *Laubie* were both Type VIIC boats. The fifth boat, a small Type XXIII (*U 2326*), was lost accidentally in 1946. The sixth, *Roland Morillot*, was a Type XXI, and was to have a considerable influence on subsequent French construction. At first received on loan from the Royal Navy, then transferred, this boat was used extensively as a trials vessel until her disposal in the late 1960s. All the ex-German boats except *Bouan* needed repairs before entering service. Guns were removed and all received a snorkel. In the early 1950s *Bouan* was given a new streamlined fin.

Besides the larger U-boats, four Type XXVII (*Seehund*) midget submarines fell into French hands at the end of the war. Nos *74*, *90*, *107* and *365* became *S 621–24* respectively. They underwent extensive testing at Toulon, and provided the inspiration for a French 30t design. Two units (*Q 224–25*) were to have been built under the 1956–57 programmes but the orders did not materialise. Two of the German midgets were stricken in 1954 and the other pair in 1959.

AVISOS

No	Name	Launched	Disp	Fate
1st class				
A 01	LA GRANDIERE	22.6.39	1970t	BU 1959
A 02	DUMONT D'URVILLE	21.3.31	1970t	BU 1958
A 03	SAVORGNAN DE BRAZZA	18.6.31	1770t	BU 1957
2nd class				
A 05	BISSON	6.3.46	660t	Stricken 1964
A 06	COMMANDANT DE PIMODAN	1941	660t	Stricken 1966
A 07	COMMANDANT AMYOT D'INVILLE	1941	660t	Stricken 1965
A 08	GAZELLE	17.6.39	650t	BU 1961
A 09	ANNAMITE	17.6.39	650t	To Morocco 1961
A 10	CHEVREUIL	17.6.39	650t	To Tunisia 1959
A 11	COMMANDANT BORY	26.1.39	630t	BU 1953
A 12	COMMANDANT DELAGE	25.2.39	630t	BU 1960
A 13	LA MOQUEUSE	21.1.40	630t	BU 1965
A 14	LA GRACIEUSE	30.11.39	630t	BU 1958
A 15	COMMANDANT DOMINE	2.5.39	630t	BU 1960
A 16	LA CAPRICIEUSE	19.4.39	630t	BU 1964
A 17	COMMANDANT DUBOC	16.1.38	630t	BU 1963
A 18	LA BOUDEUSE	Sep 1939	630t	BU 1958
A 19	ELAN	27.7.38	630t	BU 1958
Avisos Hydrographiques				
A 32	LA PEROUSE (ex-*Sans Peur*)	1940	1370t	Stricken 1969
A 33	BEAUTEMPS-BEAUPRE (ex-*Sans Souci*)	1940	1370t	Stricken 1969

The three large 1st class *avisos* served largely unaltered until the late 1950s. The more numerous 2nd class *avisos* of the *Elan* and *Annamite* classes ended the war with two main guns of 90mm, 100mm or British 4in (102mm) calibre, a single 40mm and four or six 20mm. The main calibre was standardised around the single 105mm mounting in the early postwar years. Many ships were fitted with two initially, but by the mid-1950s only a single mounting was carried. The construction of the seventh vessel of the *Annamite* class, the *Commandant Ducuing*, was abandoned in 1948. Several of the small *avisos* saw service in Indo-China. In 1957 *Gazelle* became a training ship.

FRIGATES

No	Name	Launched	Disp	Fate
Ex-US destroyer escorts				
F 01	ALGERIEN (ex-*DE 107*)	27.11.43	1300t	Stricken 1965
F 02	SENEGALAIS (ex-*DE 106*)	11.11.43	1300t	Stricken 1965
F 03	SOMALI (ex-*DE 111*)	12.2.44	1300t	Disarmed 1956
F 04	HOVA (ex-*DE 110*)	22.1.44	1300t	Stricken 1960
F 05	MAROCAIN (ex-*DE 109*)	1.1.44	1300t	Stricken 1960
F 06	TUNISIEN (ex-*DE 108*)	17.12.43	1300t	Stricken 1960
Ex-British frigates				
F 07	L'AVENTURE (ex-*Braid*)	30.11.43	1400t	BU 1961–62
F 08	LA SURPRISE (ex-*Torridge*)	16.8.43	1400t	To Morocco 1964
F 09	L'ESCARMOUCHE (ex-*Frome*)	1.6.43	1400t	BU 1961–62
F 10	CROIX DE LORRAINE (ex-*Strule*)	8.3.43	1400t	BU 1961–62
F 11	TONKINOIS (ex-*Moyola*)	27.8.42	1400t	BU 1961–62
F 12	LA DECOUVERTE (ex-*Windrush*)	18.6.43	1400t	BU 1959

The six ex-British frigates of the 'River' class were purchased 1943–44. The original 4in mountings were replaced postwar by two single 105mm mountings. In 1953 *Tonkinois* was renamed *La Confiance*; *L'Escarmouche* was renamed *L'Ailette*.

The six ex-American *Cannon* class DEs were purchased in 1944. Eight further units were transferred 1950–52 (see under Major Surface Ships). *Algérien* was later renamed *Oise*, and *Sénégalais* was renamed *Yser*. In 1956 *Somali* was disarmed and became an experimental vessel. She was renumbered A 607, and in 1968 was renamed *Arago*.

MINESWEEPERS

No	Name	Fate	No	Name	Fate
Ex-German minesweepers (launched 1937–41, 685t)					
A 20	AILETTE (ex-*M 24*)	To W Germany 1963	A 26	BELFORT (ex-*M 205*)	To W Germany 1963
A 21	ANCRE (ex-*M 252*)	BU 1955	A 27	PERONNE (ex-*M 251*)	BU 1950
A 22	SOMME (ex-*M 9*)	BU 1955	A 28	VIMY (ex-*M 253*)	To W Germany 1963
A 23	MEUSE (ex-*M 28*)	BU 1948	A 29	BAPAUME (ex-*M 35*)	BU 1950
A 24	OISE (ex-*M 38*)	BU 1955	A 30	LAFFAUX (ex-*M 81*)	To W Germany 1963
A 25	YSER (ex-*M 85*)	To W Germany 1963	A 31	CRAONNE (ex-*M 202*)	BU 1950
Ex-German minesweepers (launched 1942, 543t)					
	SUIPPE (ex-*M 432*)	BU 1953		AISNE (ex-*M 452*)	BU 1949
	MARNE (ex-*M 442*)	BU 1953			

In addition to the above, the German minesweepers *M 4*, *M 12*, *M 21*, *M 275*, *M 277*, *M 404*, *M 408*, *M 434*, *M 454*, *M 475*, *M 476* and *M 495* were transferred to France, but none saw active service and they were broken up by 1949. The remaining units were reclassified *avisos de 2ème classe* and were employed as training and patrol vessels.

There were also thirty small wooden ex-American minesweepers of the YMS type. Names were:
Amarante, *Anémone*, *Armoise*, *Asphodèle*, *Aubépine*, *Balsamine*, *Basilic*, *Belladone*, *Campanule*, *Capucine*, *Clématite*, *Dahlia*, *Digitale*, *Genêt*, *Gentiane*, *Géranium*, *Glycine*, *Héliotrope*, *Hortensia*, *Jasmin*, *Jonquille*, *Lotus*, *Marjolaine*, *Myosotis*, *Perce-neige*, *Pétunia*, *Pimprenelle*, *Primevère*, *Tiare* and *Zinnia*.

In the immediate postwar period they carried the pennant numbers D 01–30. *Zinnia* was renamed *Lotus* and *Pétunia* became *Tiare*. Ten continued in service into the 1960s.

SUBMARINE-CHASERS

Thirty 280t sub-chasers of the American PC type were acquired in 1944 and designated *escorteurs*:
L'Attentif, *Carabinier*, *Cavalier*, *Cimeterre*, *Coutelas*, *Dague*, *Dragon*, *L' Effronté*, *L'Emporté*, *L'Eveillé*, *Fantassin*, *Franc-tireur*, *Goumier*, *Grenadier*, *Hussard*, *L'Indiscret*, *Javelot*, *Lancier*, *Lansquenet*, *Légionnaire*, *Mameluk*, *Le Résolu*, *Pique*, *Le Rusé*, *Sabre*, *Spahi*, *Tirailleur*, *Le Vigilant*, *Le Volontaire* and *Voltigeur*. Three more, *Hué*, *Luang-Prabang* and *Pnom-Penh* were purchased postwar for service in Indo-China. There were also forty-nine SC type wooden sub-chasers of 95t.

NATO PENNANT NUMBERS
Carried from 1950 onwards by ships in this section:

Destroyers

D 601	MARCEAU	D 610	LE FANTASQUE
D 602	HOCHE	D 611	LE TERRIBLE
D 603	KLEBER	D 612	LE MALIN
D 604	ALSACIEN	D 613	LE TRIOMPHANT
D 605	LORRAIN		

Submarines

S 605	JUNON	S 611	BLAISON
S 609	MILLE	S 612	BOUAN
S 610	LAUBIE	S 613	ROLAND MORILLOT

Frigates

F 701	ALGERIEN	F 708	LA SURPRISE
F 702	SENEGALAIS	F 709	L'ESCARMOUCHE
F 703	SOMALI	F 710	CROIX DE LORRAINE
F 704	HOVA	F 711	TONKINOIS
F 705	MAROCAIN	F 712	LA DECOUVERTE
F 706	TUNISIEN	F 731	LA GRANDIERE
F 707	L'AVENTURE	F 732	DUMONT D'URVILLE
F 733	SAVORGNAN DE BRAZZA	F 742	COMMANDANT DOMINE
F 734	ANNAMITE	F 743	COMMANDANT DUBOC
F 735	CHEVREUIL	F 744	LA BOUDEUSE
F 736	GAZELLE	F 745	LA CAPRICIEUSE
F 737	BISSON	F 746	LA GRACIEUSE
F 738	COMMANDANT AMYOT D'INVILLE	F 747	LA MOQUEUSE
F 739	COMMANDANT DE PIMODAN	F 748	ELAN
F 740	COMMANDANT BORY	F 750	LA PEROUSE
F 741	COMMANDANT DELAGE	F 751	BEAUTEMPS BEAUPRE
		F 752	AMIRAL MOUCHEZ

Minesweepers

M 601	MEUSE	M 649	PIMPRENELLE
M 602	OISE	M 650	PRIMEVERE
M 603	SOMME	M 651	GENET
M 604	YSER	M 652	ZINNIA
M 605	AILETTE	M 653	BALSAMINE
M 606	BELFORT	M 654	CAPUCINE
M 607	LAFFAUX	M 655	DIGITALE
M 608	VIMY	M 656	MARJOLAINE
M 609	MARNE	M 657	AUBEPINE
M 610	SUIPPE	M 658	BELLADONE
M 641	AMARANTE	M 659	HORTENSIA
M 642	ARMOISE	M 660	JONQUILLE
M 643	ASPHODELE	M 661	CAMPANULE
M 644	BASILIC	M 662	HELIOTROPE
M 645	DAHLIA	M 663	JASMIN
M 646	PETUNIA	M 664	PERCE-NEIGE
M 647	CLEMATITE	M 665	ANEMONE
M 648	GENTIANE	M 666	GERANIUM

MAJOR SURFACE SHIPS

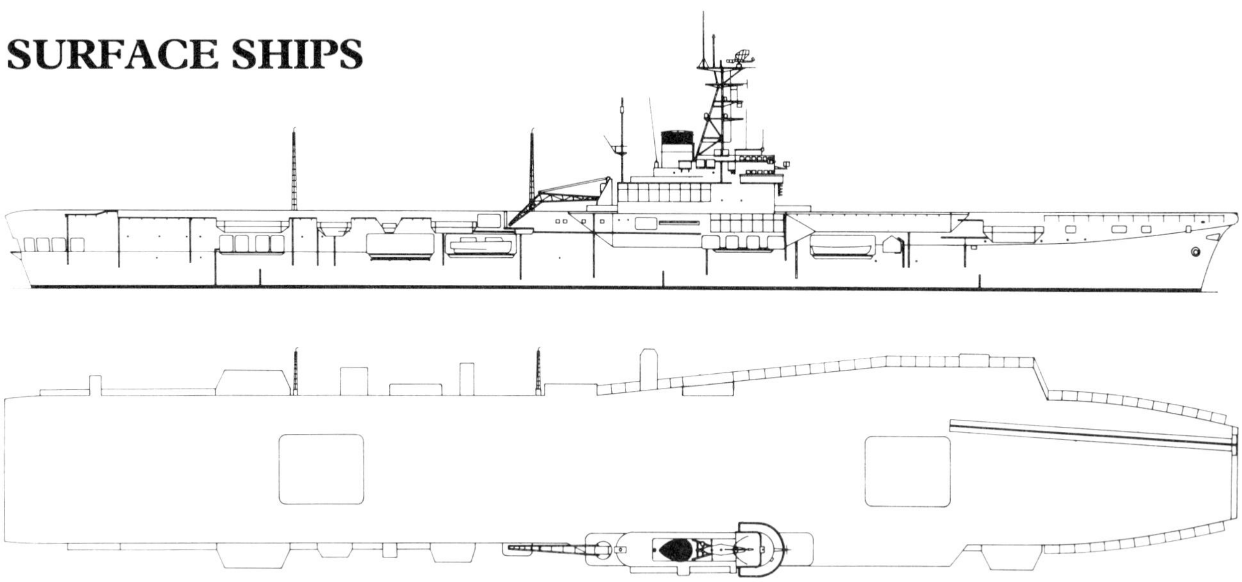

Arromanches 1971

Ex-British COLOSSUS class *aircraft carrier*

Displacement: 14,000t standard; 17,900t full load
Dimensions: 630ft pp, 693ft oa × 80ft wl, 112ft fd × 23ft
192.0m, 211.2m × 24.5m, 34.2m × 7.2m
Machinery: 2-shaft Parsons geared steam turbines, 4 boilers, 40,000shp = 25kts. Range 12,000nm at 14.6kts
Armament: 24–2pdr (6×4), 19–40mm (19×1), 24 aircraft. See notes.
Complement: 824 (1400 with aircrew)

No	Name	Builder	Acquired	Fate
R 95	ARROMANCHES (ex-*Colossus*)	Vickers-Armstrong, Tyne	6.8.46	Stricken 1974

Transferred from Britain in 1946 for a five-year loan period, *Arromanches* was purchased outright when the loan expired in 1951. By this time she had already made two deployments to Indo-China, flying SBD Dauntless dive-bombers and Seafire Mk XV fighters on the first occasion, and Helldivers and Hellcats on the second. Two further deployments followed before the collapse of the French military position in 1954. She then operated in the Mediterranean where she participated in the Suez operation in 1956, flying Corsairs and Avengers.

In 1957–58 *Arromanches* underwent a major modernisation in which she received a 4° angled flight deck and a mirror landing sight. The single catapult was removed, as was the entire AA armament, and she was fitted with a DRBV 22A air surveillance radar. The angled deck added 6ft (1.8m) to the overall beam.

After this first modernisation *Arromanches* became a training carrier, operating the new Alizé ASW aircraft in addition to the Fouga CM175M Zéphyr, a naval version of the Magister trainer. She thus helped to build up an adequate supply of trained pilots for the jet squadrons which would operate from *Clemenceau* and *Foch*.

In 1962 *Arromanches* embarked a squadron of HSS-1 helicopters and took on the assault role in addition to the training role. After a further major refit in 1968 she was redesignated a helicopter carrier, with multiple missions including the intervention and ASW roles, the fast operational transport role, and training. She retained these missions until withdrawn from service in 1974, at which time it was envisaged that her place would be taken by a new purpose-built helicopter carrier (see PH 75).

Arromanches 12.6.48 CMP

PA 28 design

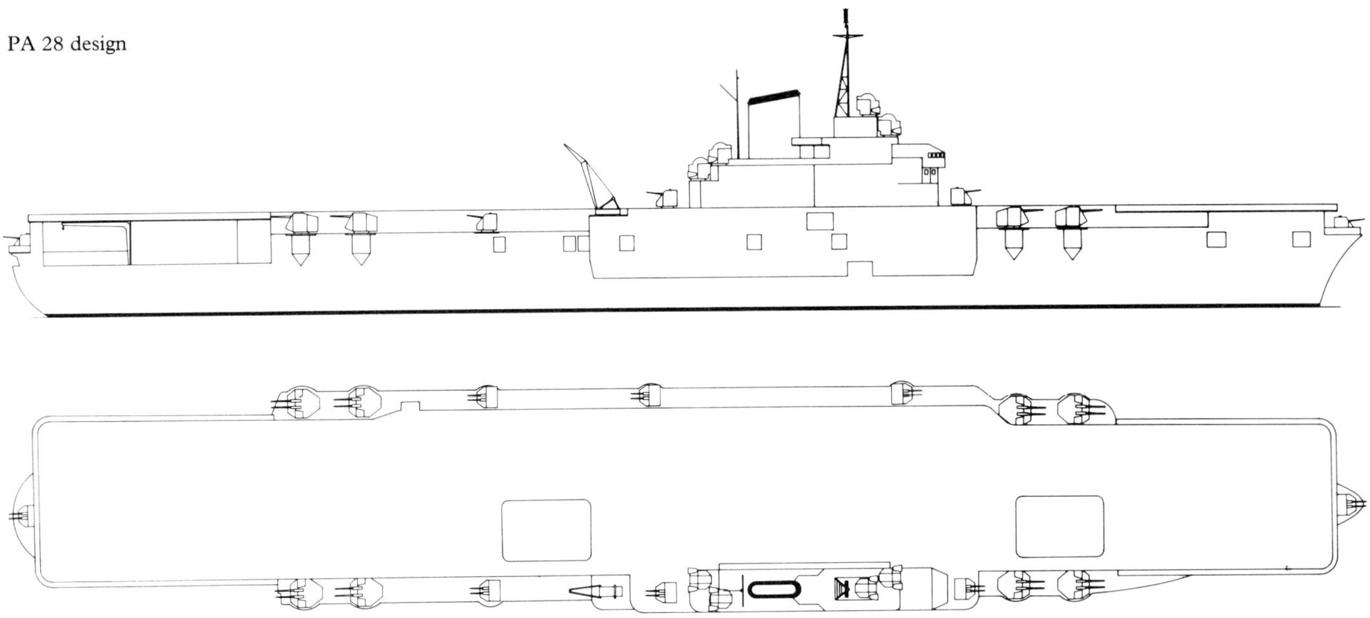

PA 28 *aircraft carrier*

Displacement: 15,700t standard; 20,000t full load
Dimensions: 705ft pp, 755ft oa × 83ft wl, 118ft fd × 21ft
215.0m, 230.0m × 25.4m, 36.0m × 6.5m
Machinery: 2-shaft Parsons geared steam turbines, 4 boilers, 105,000shp = 32kts
Armament: 16–100mm (8×2), 16–57mm (8×2), 45 aircraft
Complement: 1800

No	Name	Builder	Laid down	Launched	Comp	Fate
—	CLEMENCEAU	Arsenal de Brest	—	—	—	Project abandoned 1950

Designs for future carriers had continued throughout the Occupation and in August 1947 the construction of a new carrier, designated PA 28, was approved. She was to be named *Clemenceau* and would be laid down at Brest.

Of similar size and configuration to the light fleet carriers built by the Royal Navy during the 1940s, PA 28 was unusual in that while the flight deck was the strength deck – as in contemporary British construction – the ends of the ship were open like those of US carriers.

As in the prewar *Joffre* design, the large hangar, 164m × 24m, was offset to port. Hangar and flight deck were served by two lifts, each 15m × 10m with a capacity of 12t, and these were placed forward and aft of the island and offset in such a way as to place them on the starboard edge of the hangar itself. This permitted aircraft movement to port of the after lift, enabling the hangar to be extended farther aft without interruption. Seven arrester wires were to be fitted aft, and the PA 28 is credited with two catapults (although it is difficult to see how a second would have been accommodated on the axial flight deck, given the position of the forward lift). The propulsion system was similar to that installed in the *Volta* class *contre-torpilleurs*, resulting in a 7kt advantage over the British light fleet carriers.

The projected AA armament was exceptionally powerful: a pair of 100mm twin mountings disposed in each quadrant, three 57mm twin mountings on either side amidships, and a further two 57mm mountings at the open ends of the ship. The 57mm mounting was of a new design which was to see service in all major French surface ships of the 1950s. Four large HA fire control directors were mounted on the island.

Clemenceau was to have carried a new generation of aircraft. It was planned to accommodate twenty-two large twin-engined torpedo-bombers of the NC1070 type then under development and five SE582 fighters (a naval version of the land-based SE580) in the hangar. A further twenty-two SE582 fighters were to be suspended from the hangar roof.

Following intense debate concerning the military value of aircraft carriers, the PA 28 project was abandoned in 1950 and none of the carrier aircraft under development went beyond the prototype stage.

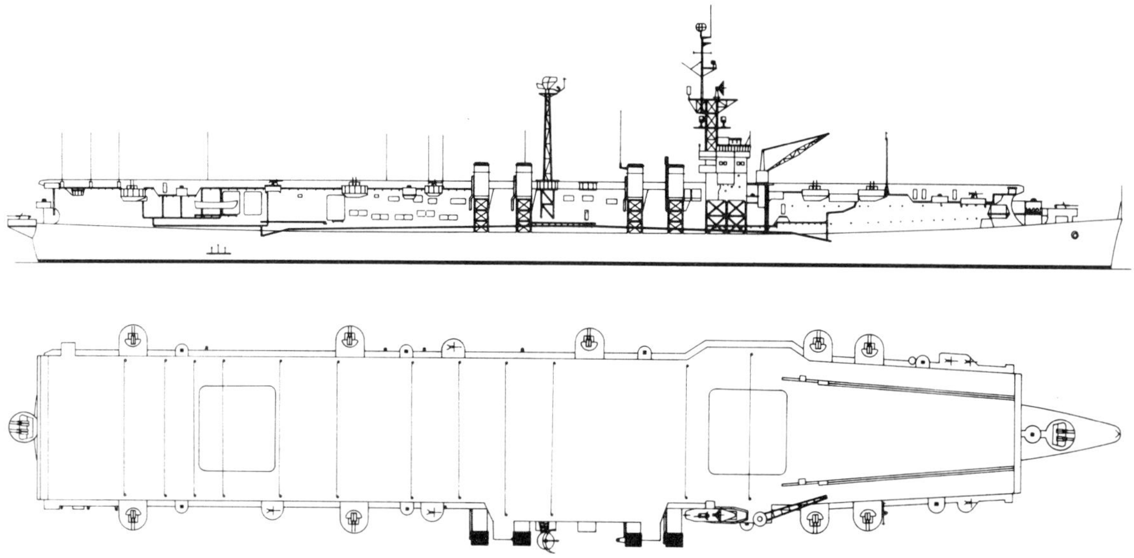

Lafayette 1959

In 1950 the USS *Langley* was obtained from the United States as part of MDAP, and she was delivered the following year as *Lafayette*. In 1953 she was joined by *Belleau Wood*, which was secured on a five-year loan and renamed *Bois Belleau*. Both carriers initially operated Hellcats, Helldivers and Avengers purchased from the United States in 1950–53.

Lafayette served off Indo-China in 1953, to be replaced in 1954 by her sister-ship. Both underwent refits on their return to France, at which time the single catapult was replaced and a tall lattice mast added forward of the second pair of funnels topped by a DRBV 22A air surveillance radar. In a later refit (1958–59) *Lafayette* received a DRBI 10 height-finder in place of the US antenna which previously occupied the foremast.

From the mid-1950s both ships operated F4 Corsairs and were based at Toulon. *Lafayette* took part in the Suez operation in 1956. In 1960 *Bois Belleau* was employed as an aircraft transport between France and the United States before being returned to the US Navy.

Ex-US INDEPENDENCE class *aircraft carriers*

Displacement:	11,000t standard; 15,800t full load
Dimensions:	600ft wl, 623ft oa × 72ft wl, 104ft fd × 24ft *182.9m, 189.7m × 21.9m, 31.7m × 7.2m*
Machinery:	4-shaft geared steam turbines, 4 boilers, 100,000shp = 32kts. Range 11,000nm at 15kts
Armament:	26–40mm (2×4, 9×2), 6–20mm (3×2), 26 aircraft
Sensors:	Radar SK-2, SP
Complement:	1400

No	Name	Builder	Acquired	Fate
R 96	LAFAYETTE (ex-*Langley*)	New York SB	1950	Returned USA 1963
R 97	BOIS BELLEAU (ex-*Belleau Wood*)	New York SB	Sep 1953	Returned USA 1960

Bois Belleau 12.6.57 USN

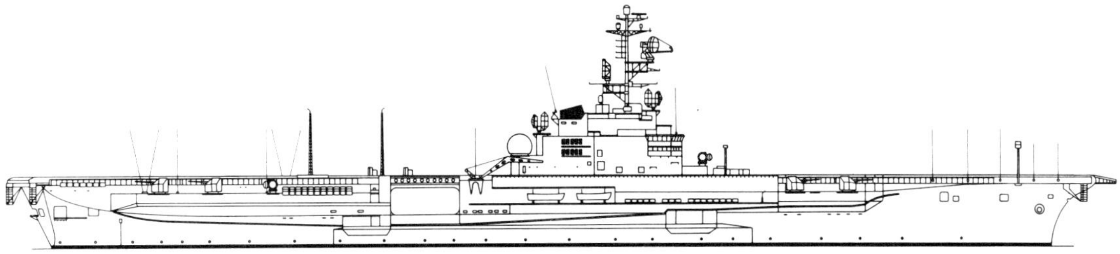

Clemenceau 1976
NB 1/1750 scale

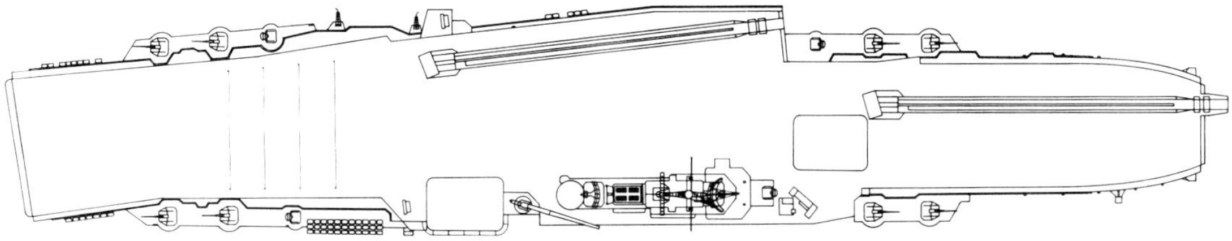

CLEMENCEAU class *aircraft carriers*

Displacement:	22,000t standard; 32,780t full load
Dimensions:	781ft pp, 870ft oa × 104ft wl, 168ft fd × 28ft *238.0m, 265.0m × 31.7m, 51.2m × 8.6m*
Machinery:	2-shaft Parsons geared steam turbines, 6 boilers, 126,000shp = 32kts. Range 7500nm at 18kts
Armament:	8–100mm Mod 1953 (8×1), 40 aircraft
Sensors:	Radar DRBV 20C, DRBV 23B, DRBV 50, DRBI 10, 4 DRBC 32A, NRBA 50; sonar SQS-503
Complement:	1338 (as aircraft carrier)/984 (as helicopter carrier)

No	Name	Builder	Laid down	Launched	Comp	Fate
R 98	CLEMENCEAU	Arsenal de Brest	Nov 1955	21.12.57	12.11.61	Extant 1995
R 99	FOCH	Penhoët Loire/ Arsenal de Brest	Feb 1957	28.7.60	15.7.63	Extant 1995

Clemenceau, allocated the project number PA 54, was the first carrier designed as such to be completed in France. Considerably larger than PA 28, she was designed to incorporate all the advances made in carrier aviation during the early 1950s. A fully-angled flight deck was provided, together with a mirror landing sight and an altogether more comprehensive outfit of air search and tracking radars.

The length/displacement ratio is exceptional. The flight deck measures 165.5m × 29.5m and is angled at 8° to the ship's axis. The forward lift, 17m × 13m, is forward of the island and offset to starboard. The after lift, 16m × 11m, was positioned on the deck-edge in order to clear the angled flight deck and to increase hangar space. Both lifts had a designed capacity of 15t, which has since been increased to 20t. The hangar, as in previous French designs, is offset to port, and has an overall length of 180m (of which 152m are usable) and a width of 22–24m, with 7m clearance overhead. Two 52m Mitchell-Brown steam catapults are fitted. In the original design both would have been located forward, but one was moved to the angled deck in order to create a deck parking area to starboard. The catapults can accept aircraft weights of up to 20t and can produce a speed of 140kts with a 30kt wind-over-deck.

Armour was provided for the flight deck (45mm) with an armoured box 30mm–50mm over the machinery and other vital parts.

The two-shaft propulsion system is remarkable for a ship of this size; the figure of 63,000shp per shaft has been surpassed only by the 'super carriers' of the US Navy.

Originally *Clemenceau* was to have been armed with twelve 57mm twin mountings, but these were subsequently replaced by the new single 100mm, and the ship was finally completed with only eight 100mm single mountings.

A capacity of up to sixty aircraft was contemplated, but by 1960 this figure had fallen to forty – a reflection of the increased size and weight of the new jets. A new generation of aircraft was specially designed to operate from the new carriers. Two flights of Étendard IVM ground support fighters (into which was incorporated a handful of IVP tactical reconnaissance/tanker aircraft) were embarked, together with a flight of Alizé turboprop ASW aircraft. As a temporary measure the ships operated the Aquilon (Sea Venom) interceptor, but in 1963 forty-two F-8E Crusaders were purchased, and a flight of these was attached to each carrier after minor modifications in 1966. Helicopters, first American and then of French origin, were operated for plane-guard duties and later for ASW. The Étendard IV was replaced by the Super Étendard in the late 1970s, but the small size and light construction of the ships has made it difficult to find a suitable replacement for the Crusader.

Foch had the distinction of being launched twice: once in July 1959 by Chantiers de l'Atlantique, who built the hull up to hangar deck level together with the propulsion machinery, and again in July 1960 by Arsenal de Brest, where the ship was fitted out. She was completed with bulges, which proved so successful that they were retro-fitted to *Clemenceau* during her 1966 refit.

Fire control for the 100mm was performed by DRBC 32A and DRBC 31D radars in *Clemenceau*, and by DRBC 32C and DRBC 31C radars in *Foch*.

In 1977–78 *Clemenceau* underwent a major overhaul in which the catapults were strengthened, new auxiliary boilers were installed, and a closed-circuit TV system was fitted. She also received a SENIT 2 tactical data system, removed from the deactivated T 53 destroyer *Jauréguiberry*. At the time the magazines were modified to enable AN-52 tactical nuclear weapons to be carried, and a central inertial guidance system was fitted which transfers position data to the inertial guidance system of each of her Super Étendard strike aircraft. *Foch* underwent a similar refit in 1980–81, being fitted with a SENIT 2 system from the destroyer *Tartu*.

Both ships received Syracuse Satcom terminals in 1984, and in 1985–86 *Clemenceau* underwent a second major refit, during which four 100mm guns and their respective fire control radars were removed and replaced by two Crotale EDIR point-defence missile systems. The Crotale installations are *en echelon*, and have required extensive remodelling of the sponsons on which they are located. The remaining four 100mm guns have had their original DRBC 32A fire control radars replaced by the lightweight digital DRBC 32C model. New catapult mechanisms have been fitted. Similar modifications were extended to *Foch* in 1987–88, which in addition received the prototype CSEE Dallas (Deck Approach and Landing Laser System), together with modifications to ordnance magazines and handling gear to enable her to accommodate the ASMP nuclear missile.

Major problems have been experienced in maintaining a viable air group for these ships. Since 1975 only one ship has been retained in full commission, the other serving as a helicopter carrier with a reduced complement. *Clemenceau* is now to remain in service until 1999, when she will be replaced by *Charles de Gaulle*. *Foch* is scheduled to decommission in 2003; she is currently being refitted to enable her to operate the Rafale M interceptor and will receive two Sadral SAM systems in 1996.

Foch 21.7.88 L & L van Ginderen

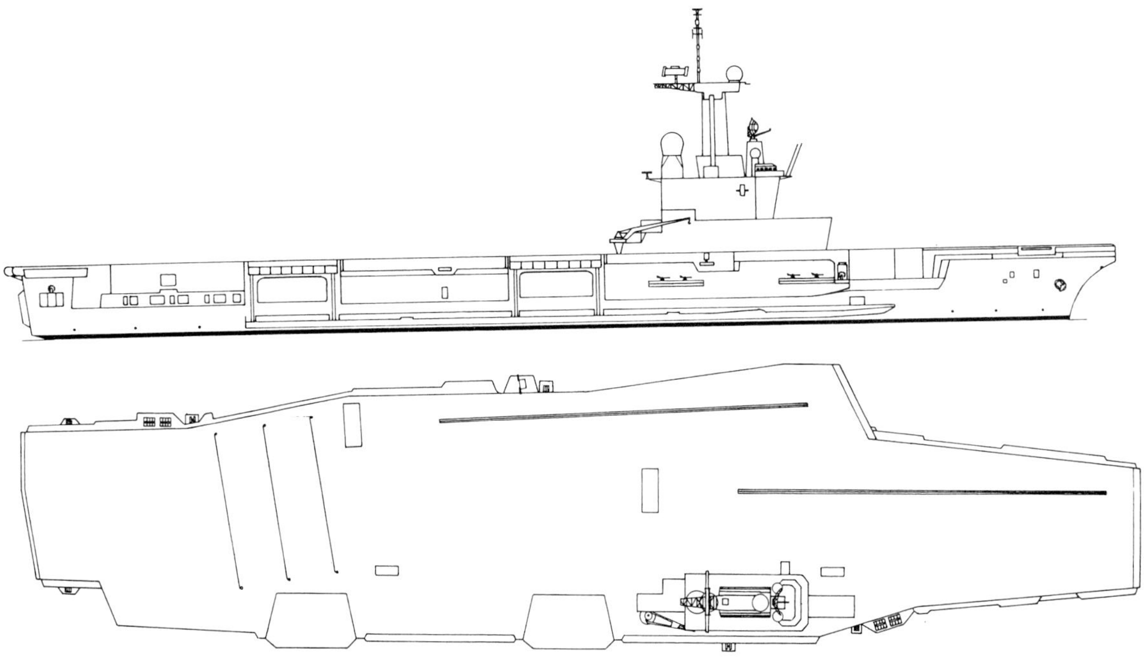

Charles de Gaulle design

CHARLES DE GAULLE class *aircraft carriers*

Displacement:	34,500t standard; 36,600t full load
Dimensions:	781ft wl, 858ft oa × 103ft wl, 211ft fd × 28ft *238.0m, 261.5m × 31.5m, 64.4m × 8.5m*
Machinery:	2-shaft nuclear: 2 150MW Type K15 PWR, double-reduction geared steam turbines, 82,000shp = 27kts
Armament:	4 v/l SAAM systems (4×8; 32 missiles), 2 SADRAL SAM (2×6), 8–20mm (8×1), 4 Sagaie chaff r/l, about 40 aircraft
Sensors:	Radar DRBV 26D, DRBV 15C, DRBJ 11D/E, Arabel FC
Complement:	1750

No	Name	Builder	Laid down	Launched	Comp	Fate
R 91	CHARLES DE GAULLE	Arsenal de Brest	14.4.89	14.5.94	—	Building
R	Unnamed	Arsenal de Brest	—	—	—	Projected

The PAN programme dates from 1980, when the Defence Council established future requirements as two nuclear-powered carriers of 35,000t with catapults and arrestor gear capable of handling aircraft with a maximum weight of 22t, the first ship to enter service in 1992. Construction was delayed for financial reasons, and the order for *Charles de Gaulle* was not placed until 1986.

Hull dimensions are virtually identical to those of *Clemenceau*, this being the maximum that can be accommodated in the building dock at Brest Naval Dockyard. However, the requirement to operate heavier aircraft led to the adoption of a broad United States Navy-style flight deck with substantial overhang, served by two deck-edge lifts to starboard. Each of the lifts has a capacity of 36t, enabling two aircraft to be transferred simultaneously between the flight deck and the hangar. The latter has a length of 138m and is divided into two bays. It can accommodate about half the air group, the remaining aircraft being kept permanently on deck.

The angled deck is 193m in length and is set at 8° to the ship's axis. The two US C13-3 75m catapults, capable of accelerating a 22t aircraft to a take-off speed of 140kts have been largely built by Ruelle. The three 4.25in arrestor wires and their Mk 7 Mod 3 hydraulic brakes have been built in The Netherlands. One catapult serves the angled deck, while the other serves the forward end of the flight deck. The forward catapult is offset to port, and intrudes on to the angled deck. It is therefore not possible to conduct take-off and landing operations simultaneously; the advantage of this arrangement is that it frees a large parking area to starboard which can accommodate six aircraft. The adoption of nuclear propulsion has enabled the island to be located forward of the two deck-edge lifts, thereby preventing mutual interference between the latter and take-off operations using the forward catapult.

The propulsion system is based on two PWR reactors derived from the model developed for the SSBNs of the *Le Triomphant* class. Concern has been expressed regarding the relatively low maximum speed of 27kts, but the only alternative was to instal a third reactor in what was already a tight design. The secondary steam generators supply steam to the catapults and to four turbo-alternators each of 4000kW, which provide power to two independent switchboards. There are also four diesel alternators each of 1100kW to provide power when alongside or in emergencies. A SATRAP active fin stabilisation system is fitted to ensure a steady platform for flight operations. The propulsion machinery and ordnance magazines are housed within a protective armoured box, designed to survive a 90° collision with an escorting frigate at full speed.

The Aster SAAM is a radar-guided anti-missile system being developed jointly by France and Italy. The missile is launched vertically from modules located forward of the island to starboard, and outboard of the flight deck to port. There are also two Sadral launchers, firing the Mistral IR point-defence missile, and four Sagaie chaff launchers.

The major problem experienced with the PAN project has been the provision of suitable aircraft. Development of the Rafale ACM interceptor has been delayed, and the completion of *Charles de Gaulle* has been put back to 1998.

Charles de Gaulle 14.5.94 L & L van Ginderen

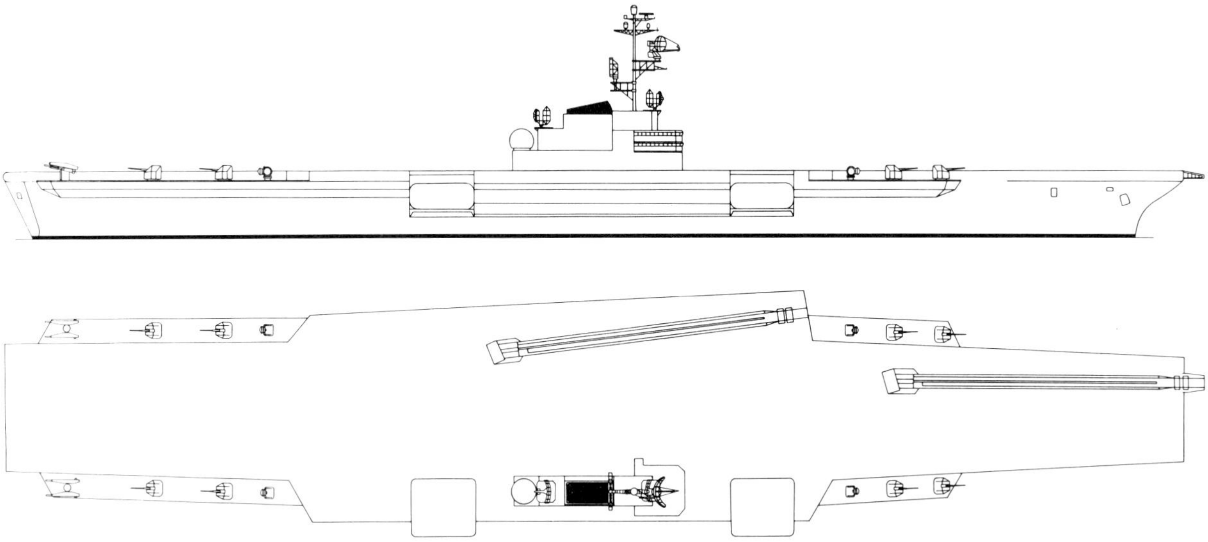

PA 58 design
NB 1/1750 scale

In 1958, while *Clemenceau* and *Foch* were still under construction, a third carrier was approved. The PA 58 design was larger and heavier, and although it bore a superficial resemblance to PA 54, was clearly influenced by the new 'super carriers' built by the US Navy. The shape of the flight deck, which was given considerable overhang on either side amidships, enabled the island to be positioned farther outboard. Parking space amidships was thereby increased, and the 200m hangar was positioned centrally. Two deck-edge lifts, each 17m × 14m, were sited forward and aft of the island to starboard. The length of the catapults was increased to about 75m so that larger aircraft could be handled. They were, however, positioned in the same way as those on *Clemenceau* so that there was a clear separation between flying operations, which were confined to the port side of the ship, and parking and handling operations. The angled deck measured 192m and was angled at 8° as on *Clemenceau.*

PA 58 *aircraft carrier*

Displacement:	35,000t standard; 45,000t full load
Dimensions:	860ft pp, 939ft oa × 112ft wl, 190ft fd *262.0m, 286.0m × 34.0m, 58.0m*
Machinery:	4-shaft geared steam turbines, 200,000shp = 33kts
Armament:	2 Masurca SAM (2×2), 8–100mm Mod 1953 (8×1)
Complement:	?

The significant increase in power needed to sustain fleet speed led to a corresponding increase in the number of shafts from two to four. Protection of flight deck and machinery was on a similar pattern to *Clemenceau,* but the thickness of the armour was slightly increased.

The eight single 100mm of *Clemenceau* were to be retained but, in addition, it was planned to fit a twin launcher for the new Masurca SAMs on either side of the flight deck aft (*cf* the US Navy's *Kitty Hawk*).

In addition to Alizé ASW aircraft and Étendard fighter-bombers, PA 58 was to operate Mirage IVM heavy strike aircraft. The latter, under development for the Navy since 1956, had a length of 19m, a wing-span of 12m, and a take-off weight of 20t.

PA 58, possibly to have been named *Verdun,* was delayed by financial problems, and the Defence Staff considered a smaller design, derived from *Clemenceau,* in which the after guns would have been replaced by Masurca before the project was finally abandoned in 1961.

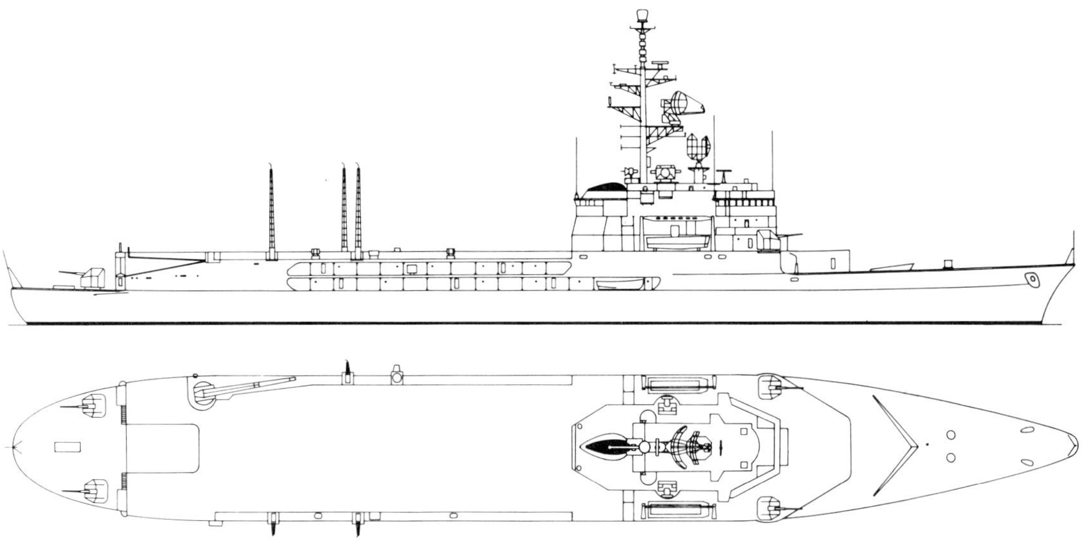

La Resolue as completed

In the mid-1950s the need arose to replace the old training cruiser *Jeanne d'Arc*. Various proposals were considered, including the use of a flotilla of *avisos escorteurs*, but in 1956 it was decided to construct a specialised helicopter cruiser which in wartime could be employed for ASW operations, for amphibious assault, or as a troop transport capable of lifting a battalion of 700 men.

The resulting PH 57 design adopted a hull form based on that of

JEANNE D'ARC *helicopter carrier*

Displacement;	10,000t standard; 12,365t full load
Dimensions:	564ft pp, 597ft oa × 73ft wl, 79ft fd × 24ft *172.0m, 182.0m × 22.0m, 24.0m × 7.3m*
Machinery:	2-shaft Rateau-Bretagne geared steam turbines, 4 boilers, 40,000shp = 26.5kts. Range 6000nm at 15kts
Armament:	4–100mm Mod 1953 (4×1), 4–8 helicopters (see notes)
Sensors:	Radar DRBV 22D, DRBV 50, DRBI 10, 3 DRBC 32A; sonar SQS-503
Complement:	627 (inc 183 officer cadets)

No	Name	Builder	Laid down	Launched	Comp	Fate
R 97	JEANNE D'ARC (ex-*La Résolue*)	Arsenal de Brest	7.7.60	30.9.61	30.6.64	Extant 1995

Jeanne d'Arc 1985 French Navy

the anti-aircraft cruiser *Colbert*. A conventional cruiser superstructure forward accommodated all command and control facilities, together with the boiler uptakes, while the after part of the ship was dominated by a helicopter deck (62m × 21m) beneath which were located the hangar and aviation facilities. The ship was named *La Résolue* as a temporary measure, adopting the name *Jeanne d'Arc* when her predecessor was withdrawn from service in 1964.

High speed was not a tactical requirement, so the size of the propulsion machinery was considerably reduced compared with *Colbert*. There are two main machinery spaces, each with two boilers and a turbine, separated by a single bulkhead. After the ship's sea trials, the funnel had to be heightened in order to clear the bridge structure.

Initially it was planned to fit four 100mm single mountings around the superstructure with a further two on the stern, controlled by three FC directors, but the two mountings abeam the superstructure were deleted.

Electronic equipment is similar to that aboard *Clemenceau*. Sonar was provided, and the ship was to have carried a quadruple 305mm ASW mortar on the forecastle, but this was never fitted. It was also planned to replace the mortar by a Masurca SAM launcher at a later date, but this plan, too, was abandoned. In 1974, however, six MM38 Exocet launchers were installed forward of the bridge. A Crotale SAM system and a SENIT 2 tactical data system were to have been installed during the 1980s, but these and other projected modifications were abandoned because of budget pressures. The DRBI 10 height-finding radar was removed in 1983–84.

The bridge structure, in addition to the navigation bridge, contains an action information centre, a control centre for amphibious landing operations and, at the after end, a helicopter control bridge.

Hangar and flight deck are served by a single centreline lift with a capacity of 12t. Originally this was to have projected from the after end of the flight deck, but during construction it was decided to enclose it by extending the hangar structure aft in order to provide repair and inspection workshops and handling rooms for aircraft ordance.

The hangar can accommodate four Super Frelon helicopters, which were introduced into service at about the same time as *Jeanne d'Arc* and can serve in both the ASW and assault roles. In wartime, hangar space could be doubled by removing some of the officer cadet accommodation, allowing a further four helicopters to be embarked. Five helicopter spots are marked on the flight deck, and there is sufficient space for two simultaneous take-offs. In peacetime, *Jeanne d'Arc* has tended to operate smaller helicopters such as the HSS-1, the Alouette III and the Lynx.

Following her commissioning *Jeanne d'Arc* carried the President of France across the Atlantic on a visit to the West Indies. She has served as flagship of the training squadron, being accompanied by a single frigate/destroyer escort, the latest of which is the newly-built *Germinal*. The squadron leaves Brest every autumn for a six month cruise taking in the Atlantic, the Mediterranean, and sometimes the Indian Ocean and the Pacific.

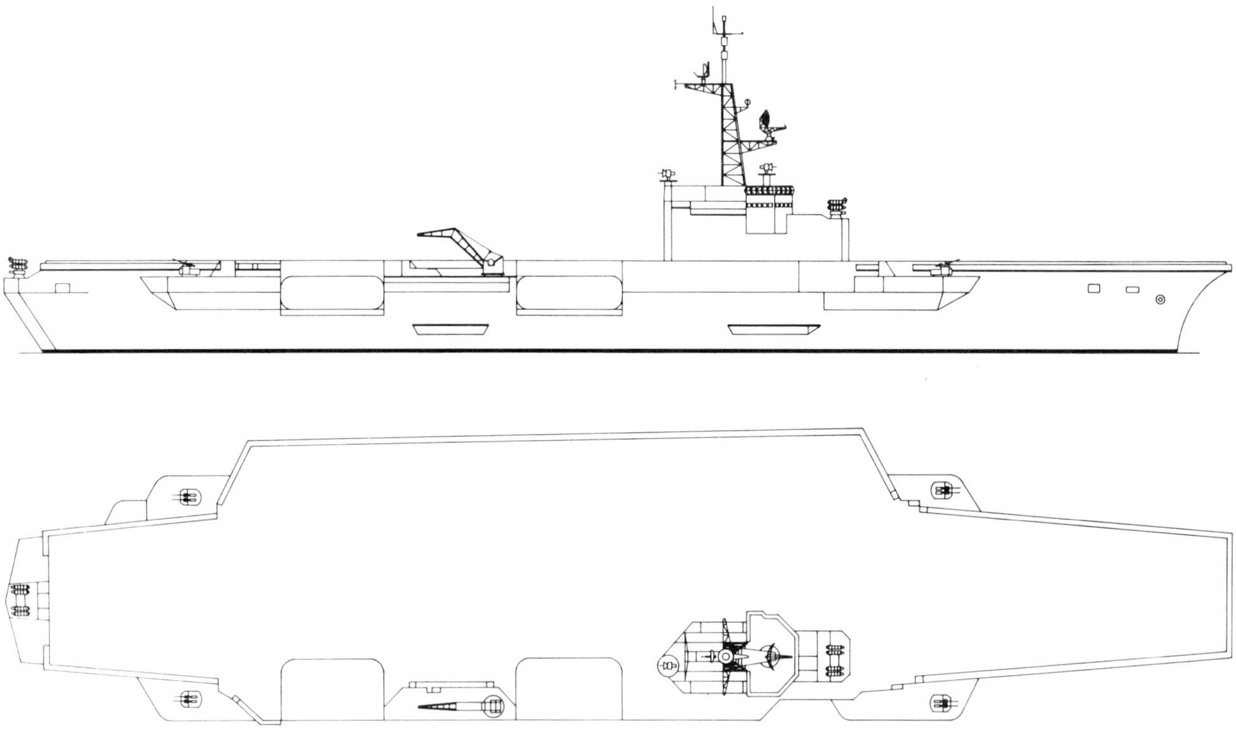

PH 75 design

PH 75 *helicopter carrier*

Displacement:	16,400t standard; 18,400t full load
Dimensions:	682ft oa × 87ft wl, 157ft fd × 21ft *208.0m × 26.4m, 46.0m × 6.5m*
Machinery:	2-shaft nuclear: 1 CAS-230 reactor, 2 turbo reduction-condenser groups, 65,000shp = 28kts; 2 AGO standby diesels
Armament:	2 Crotale SAM (2×8), 2–100mm Mod 1968 (2×1), 2 Sagaie chaff r/l, 10–25 helicopters
Sensors:	Radar DRBV 26, DRBV 51, DRBC 32; sonar DUBA 25
Complement:	890 (+1000 troops)

In 1970, with *Arromanches* fast approaching the end of her active life, studies were undertaken to provide for a replacement. A number of possible designs were produced, all of which shared the 55,000shp steam plant of the F 67 frigates, a through deck for helicopters, and a uniform armament of four 100mm mountings.

The first three versions displaced 20,000–22,000t, had a waterline length between 187m and 200m and a beam of 28m, with a maximum speed of 27kts. The first two versions incorporated a docking well aft, 40–50m long and 14m wide. The docking well of the first variant could accommodate four LCMs (26t), while that of the second variant, which was longer and higher, could accommodate two LCMs and two CTMs (56t). (The LCM has a cargo capacity of

30t, and the CTM of 90t, and both can carry a light or medium tank.) The flight deck configuration of the first variant was clearly influenced by the US Navy's *Iwo Jima* class; the deck itself was rounded off at its forward edge, and deck-edge lifts were staggered to port and starboard. In the second variant the flight deck had a broad centre section not unlike that of the US Navy's CVAs, with lifts on either side of the island to starboard incorporated into the overhang. A similar arrangement was adopted in the third variant, which dispensed altogether with the docking well.

These first three variants all had a hangar of about 90m length with a capacity of eight Super Frelons and eighteen Lynx. There was accommodation for 600 men and their equipment, together with light vehicles.

The fourth variant was a smaller vessel, 170m × 25m, with a displacement of 15,000t, and a consequent increase in speed to 28.5 kts. Flight deck configuration was similar to that of *Clemenceau*, with a centreline lift forward and a deck-edge lift to starboard, aft of the island. Helicopter capacity was reduced to seven Super Frelons and sixteen Lynx, and troop capacity to 450 men.

In the event none of these designs was proceeded with, but they served as the basis of an even more ambitious project, PH 75. Similar in conception, size and configuration to the third variant of the 1970 designs, PH 75 was to have nuclear propulsion to give her unlimited range for distant intervention. Like *Jeanne d'Arc*, she would be able to assume the ASW role as an alternative to that of amphibious assault, and was also designed for disaster relief. She was, therefore, given extensive hospital facilities consisting of three main wards, an X-ray ward, an intensive care ward, an infectious diseases ward, two dental surgeries and a laboratory.

In the intervention role she would carry a special landing force of units from the *Forces Terrestres d'Intervention* (FTI) and their supporting air units (CAFI), plus a Helicopter Movement Command Centre. One thousand troops could be accommodated in designated quarters, with a further 500 in supplementary spaces in the hangar.

The ship would embark Super Frelon and Lynx helicopters in the ASW role, or Puma helicopters for the assault mission. Hangar dimensions were 84m × 21m × 6.5m and there were two deck-edge lifts, each with a capacity of 15t. There was one fixed and one mobile crane, and extensive helicopter support facilities including workshops, munitions-handling rooms, magazines, and fuel tanks for 1000m^3 of aviation fuel.

PH 75 was to be fitted as a command ship, with an action information centre, a communications centre, an ASW centre and an amphibious operations centre. She would carry 1250t of FFO for refuelling escorting warships. As with other French nuclear-powered vessels there were emergency diesel propulsion units capable of powering the ship for 3000nm at 18kts.

PH 75 was originally intended to complete in 1981, but financial problems delayed construction, and the design was successively re-designated PA 75, PA 78, PA 82, and PA 88. The change in classification from 'PH' (*Porte-Hélicoptères*) to 'PA' (*Porte-Aéronefs*) indicated an intention to provide for the embarkation of VTOL aircraft. The project was finally abandoned in favour of a new, revised design: the PAN *Charles de Gaulle* (see below).

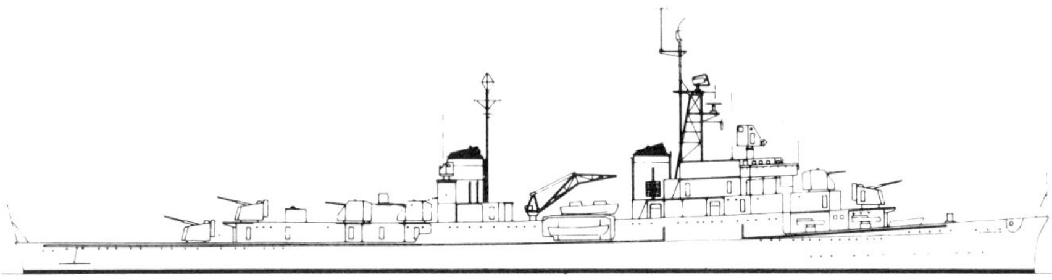

Guichen 1955

Ex-Italian ATTILIO REGOLO class *cruisers*

Displacement: 3680t standard; 5500t full load
Dimensions: 455ft pp, 466ft oa × 47ft × 13ft
138.7m, 142.2m × 14.4m × 4.1m
Machinery: 2-shaft Belluzzo geared steam turbines, 4 boilers, 110,000shp = 39kts. Range 3600nm at 18kts
Armament: 6–105mm (3×2), 10–57mm Mod 1951 (5×2), 12–550mm TT (4×3)
Sensors: Radar DRBV 20A, DRBV 11, DRBC 11, 2 DRBC 30
Complement: 353

No	Name	Builder	Acquired	Fate
D 606	CHATEAURENAULT (ex-*Attilio Regolo*)	OTO, Leghorn	1948	Stricken 1962
D 607	GUICHEN (ex-*Scipione Africano*)	OTO, Leghorn	1948	Stricken 1961

Acquired under the Peace Treaty of 1948, these two cruisers of the *Attilio Regolo* class had seen so little service that it was decided that they should undergo an extensive conversion to fit them as fast ASW/AA cruisers. They were converted 1951–54 at La Seyne Dockyard and re-rated *escorteurs d'escadre* – the same classification as the T 47 class destroyers; details in the table are as reconstructed. The original armament was completely removed. German-pattern 105mm twin mountings were fitted forward and aft, with one of the new 57mm twin mountings in 'B' position, and a further four 57mm mountings grouped around the after deckhouse. The main fire control director was located above the bridge and the secondary director aft. Four triple banks of ASW torpedo tubes were fitted at upper deck level forward of the bridge, and a short lattice foremast carried DRBV 11 and DRBV 20A surface and air search antennae.

With the completion of the new *escorteurs d'escadre* of the T 47 and T 53 classes, *Châteaurenault* and *Guichen* were fitted out in the late 1950s as command ships and flotilla leaders, the after 105mm mounting and two banks of torpedo tubes being removed to provide extra accommodation.

Chateaurenault 20.10.61
L & L van Ginderen

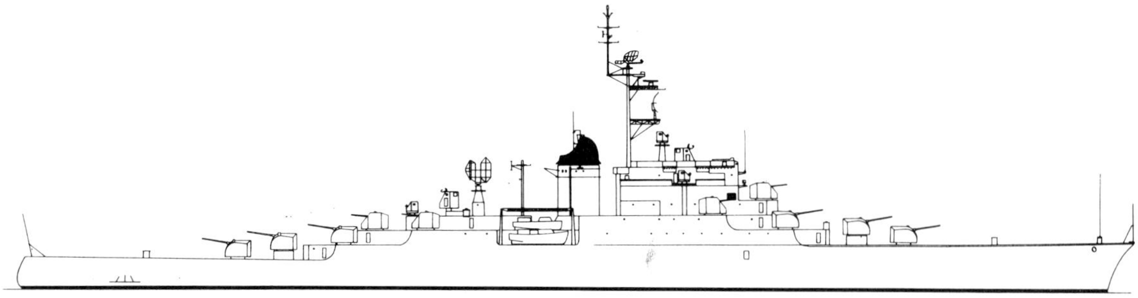

De Grasse 1961

DE GRASSE *cruiser*

Displacement:	9380t standard; 11,545t full load
Dimensions:	592ft pp, 518ft oa × 61ft × 18ft *180.5m, 188.3m × 18.6m × 5.5m*
Machinery:	2-shaft Rateau-Bretagne geared steam turbines, 4 boilers, 110,000shp = 33kts
Armament:	16–127mm Mod 1948 (8×2), 20–57mm Mod 1951 (10×2)
Sensors:	Radar DRBV 20A, DRBV 11, DRBI 10, 4 DRBC 11, 4 DRBC 30
Complement:	950 (983 as flagship)

No	Name	Builder	Laid down	Launched	Comp	Fate
C 610	DE GRASSE	Arsenal de Lorient	Nov 1938	11.8.46	3.9.56	BU 1976

De Grasse had been laid down at Lorient in 1938 as a conventional light cruiser armed with nine 6in (152mm) guns. Construction had been interrupted by the war and was not resumed until it ended. The hull was duly launched in 1946, but construction was again suspended to enable studies to be made of a new AA armament. It was finally decided to complete the ship as an anti-aircraft cruiser. Construction was resumed in January 1951. Completion was delayed when the ship sank in June 1954 after she had been floated out with sea-cocks open.

The armament selected was based on the 127mm and twin 57mm weapons which were to constitute the main armament of the new fleet escorts of the T 47 class. The mountings were disposed in a symmetrical pattern forward and aft, with the 127mm at upper deck and 01 deck levels and the 57mm grouped around the superstructure. Fire control arrangements were exceptionally complete: no fewer than four directors for the 127mm mountings were provided, disposed in a lozenge arrangement, and each was paired with a smaller 57mm director. All guns were radar-controlled with stabilised gunlayers for automatic tracking. Six 550mm torpedo tubes in two triple banks were also to have been fitted, but were dropped from the final plans.

De Grasse was equipped as a command ship with the ability to exercise radar control over air strikes. Besides the air and surface surveillance radars carried on her foremast she was fitted with the new DRBI 10 height-finding radar.

De Grasse after 1966 Navarret Collection

The machinery grouping and single funnel of the original design was retained. The main structural alterations involved lengthening and reshaping the original transom stern and the redesign of the bridge structure. When completed *De Grasse* had a standard displacement 1400t greater than envisaged in the 1938 plans. Possibly too much was attempted, as the innermost four 57mm mountings, together with the two main fire control directors abeam the bridge structure, were removed in a refit in 1962. At the same time the original DRBV 11 radar was replaced by a DRBV 20A.

In 1966 *De Grasse* underwent major reconstruction to equip her as flagship of the Pacific Experimental Centre. The signals department was considerably enlarged and a tall lattice mast carrying communications aerials was fitted aft. The after pair of 127mm mountings and all 57mm mountings, together with all directors except the main director above the bridge, were removed. The two air search radars were replaced by a single more powerful model, DRBV 23, and the ship was given protection against fall-out. These modifications resulted in a lower standard displacement (9000t), and the complement was reduced to 560 with accommodation for 120 engineers and technicians. *De Grasse* served in her new role until stricken in 1973.

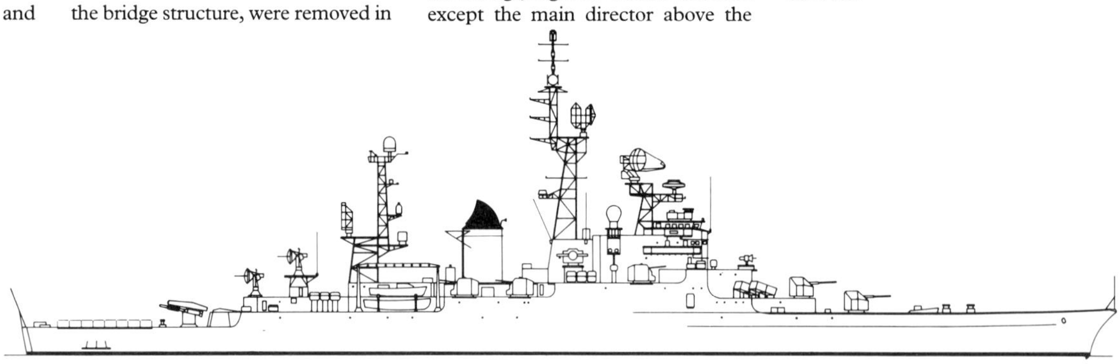

Colbert 1989

COLBERT *cruiser*

Displacement:	9085t standard; 11,100t full load
Dimensions:	593ft oa × 66ft × 21ft *180.5m × 20.3m × 6.5m*
Machinery:	2-shaft CEM Parsons geared steam turbines, 4 boilers, 87,000shp = 33kts. Range 4000nm at 25kts
Armament:	16–127mm Mod 1948 (8×2), 20–57mm Mod 1951 (10×2)
Sensors:	Radar DRBV 20A, DRBV 22A, DRBI 10, 4 DRBC 31B, 4 DRBC 31A; sonar DSBV 1
Complement:	977

No	Name	Builder	Laid down	Launched	Comp	Fate
C 611	COLBERT	Arsenal de Brest	Dec 1953	24.3.56	5.5.59	Stricken 1991

Developed from the *De Grasse*, *Colbert* incorporated a number of modifications, including a shortened transom stern, an increase in beam, a knuckle in the hull forward, and a new system of protection (50mm deck; 50mm–80mm belt). These changes appear to have improved stability, as *Colbert* retained her full designed armament until she was rebuilt in 1970. A reduction in installed horsepower was accompanied by the adoption of higher-pressure boilers, which contributed to a lower displacement. The funnel was moved farther aft to clear the bridge structure, and a lattice mast was stepped immediately forward of it.

Like *De Grasse*, *Colbert* was equipped as a fleet command ship, but in addition there was provision for her use as a fast transport, with accommodation for 2400 men and their equipment. The adoption of a transom stern allowed for a helicopter landing pad aft.

Armament and fire control as completed were identical with *De Grasse*, although early drawings show an additional pair of 57mm mountings on the stern. Provision was made in the original design for eventual missile conversion. The large planar DRBV 20C air search radar was not ready in time and, as a temporary measure, *Colbert* was fitted with the older 'A' model, which she carried until her 1962–63 refit.

From April 1970 until October 1972 *Colbert* underwent a major refit to equip her with guided missiles for fleet air defence. The original refit plan involved the suppression of all 127mm and 57mm mountings in favour of Masurca surface-to-air missiles and six single 100mm guns, but financial difficulties led to the retention of three 57mm mountings on either beam in place of four of the 100mm guns, resulting in a saving of 80m francs from the original estimate of 350m. The new main armament, therefore, consisted of two 100mm single mountings forward, together with a DRBC 32C fire control director, and a twin Masurca launcher aft with two DRBR 51 tracker/illuminators. Two of the original DRBC 31 directors were retained abeam the after end of the bridge structure to serve the remaining 57mm mountings. Forward of the bridge, bedplates were installed for four MM38 Exocet SSMs. The bridge structure was rebuilt and major modifications made to the electronics suite, which then comprised DRBV 20, DRBV 23C, DRBV 50, DRBI 10, DRBI 51 (2), DRBC 32C and DRBC 31 (2). Electrical power was increased to 5000kW and air-conditioning provided throughout the ship. Complement was reduced to 560. Following conversion, the original Masurca Mk 2 Mod 2 beam-riding missiles were replaced by the Mod 3 semi-active homer, but plans to fit Crotale were never implemented. A SENIT 1 action information system enabled *Colbert* to exercise control of the air and surface situation at the centre of a widely dispersed formation.

Following her missile conversion in 1970–72 *Colbert* served as flagship of the Mediterranean Squadron until her demise. In 1980 four MM38 Exocet SSMs were installed forward of the bridge, and in a 1981–82 refit a Syracuse Satcom system was fitted and the Masurca missile system was updated. She was intended to serve until 1997, but decommissioned in 1991.

Colbert 1972 Navarret Collection

Ex-Italian ORIANI/'SOLDATI' class *destroyers*

These four ex-Italian destroyers, launched 1936–41, were acquired under the Peace Treaty of 1948. They remained largely unaltered until their disposal, and saw little active service.

Displacement: 1690t (*D'Estaing* 1675t) standard; 2530t (*D'Estaing* 2254t) full load
Dimensions: 333ft pp, 350ft oa × 34ft × 12ft
101.6m, 106.7m × 10.2m × 3.4m
Machinery: 2-shaft Parsons (*Gravière*–Belluzzo) geared steam turbines, 3 boilers, 48,000shp = 36kts. Range 2350nm at 14kts
Armament *Gravière*: 5–120mm (2×2, 1×1), 12–20mm (6×2), 6–533mm TT (2×3); *Duperré, Duchaffault*: 4–120mm (2×2), 2/3–37mm (2/3×1), 8–20mm (4×2), 3–533mm TT (1×3); *D'Estaing*: 4–120mm (2×2), 4–37mm (4×1), 6–20mm (3×2), 3–533mm TT (1×3)
Complement: 157

No	Name	Builder	Acquired	Fate
T 12	JURIEN DE LA GRAVIERE (ex-*Mitragliere*)	CNR, Ancona	15.7.48	Stricken 1956
T 13	DUPERRE (ex-*Velite*)	OTO, Leghorn	24.7.48	Stricken 1961
T 14	DUCHAFFAULT (ex-*Legionario*)	OTO, Leghorn	15.8.48	Stricken 1956
T 15	D'ESTAING (ex-*Alfredo Oriani*)	OTO, Leghorn	8.8.48	Stricken 1954

Casabianca 18.1.84 MG Photographic

Larger than other contemporary European destroyers, the T 47 class, authorised 1949–52, were clearly in direct line of succession to French prewar construction. In mission, layout, and in the calibre adopted for their main armament, they bear a particularly close resemblance to the *Le Hardi* class.

Designated *escorteur d'escadre*, the T 47 was designed to give AAW protection to the new carriers and other fleet units. A 127mm (5in) calibre was adopted for the main armament to enable the ships to use standard US ammunition, and this was backed up by a heavy secondary battery of 57mm twin AA mountings of a new pattern. The inclusion of a heavy AA battery in addition to the dual-purpose main armament was a response to the weakness of French prewar-built ships in this respect, and was also responsible for the increase in length compared with *Le Hardi*.

In the original plans little provision was made for ASW operations beyond the traditional depth-charge racks. A single quadruple bank of 550mm anti-ship torpedo tubes was to be mounted on the centreline between the after groups of 127mm and 57mm mountings. After considering the American Hedgehog and the British Squid it was finally decided to fit the ship with four triple banks of tubes, mounted along the deck edge on either side, of which the forward pair would fire L3 anti-submarine homing torpedoes and the after pair either L3 ASW or K2 anti-ship torpedoes. Immediately abaft each bank of ASW tubes was a ready-use locker containing three reloads. Hull sonars of French design were fitted.

In recognition of the increased threat from the air the T 47 was to carry a new generation of French radars. Initially it was intended to fit the ships with a tall British-pattern lattice mast carrying the DRBV 20A air search antenna, but in the event twin tripods with lattice supports were fitted, the second of which carried a DRBV 11 surface/air search radar. A single FC director was fitted for the main armament, with a second director aft for the 57mm.

As the new fleet escorts were designed to operate in company with the carriers and were not expected to engage in independent operations against hostile surface units, high speed was a less important consideration than it was for the prewar *torpilleurs* and *contre-torpilleurs*. The hull was entirely welded and light alloy was used extensively in the upperworks to reduce top-weight. The ships were assembled using eighty-four prefabricated sections.

In the early 1960s *Surcouf*, *Cassard* and *Chevalier Paul* had their forward 57mm mounting removed to allow the bridge to be extended, and became command ships, replacing the two former Italian light cruisers in that role. In 1962 *Cassard* was fitted experimentally with a helicopter deck.

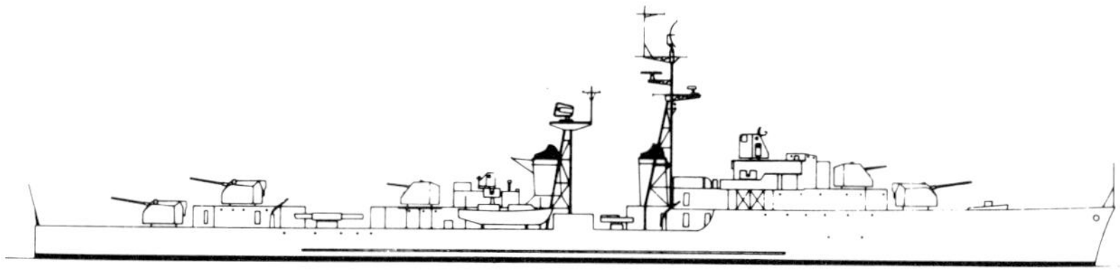

Vauquelin as completed

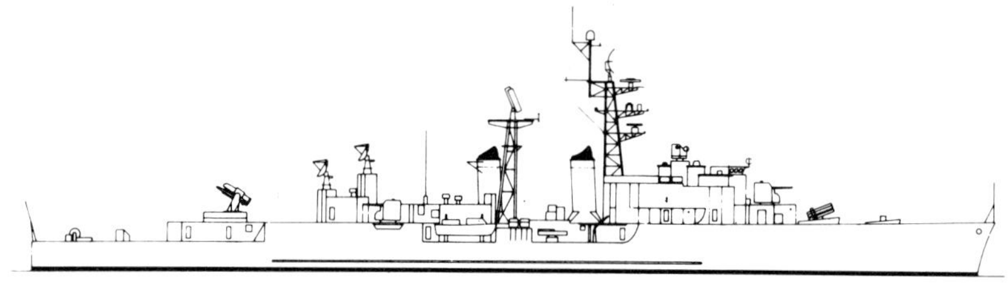

Dupetit Thouars 1971 (Tartar conversion)

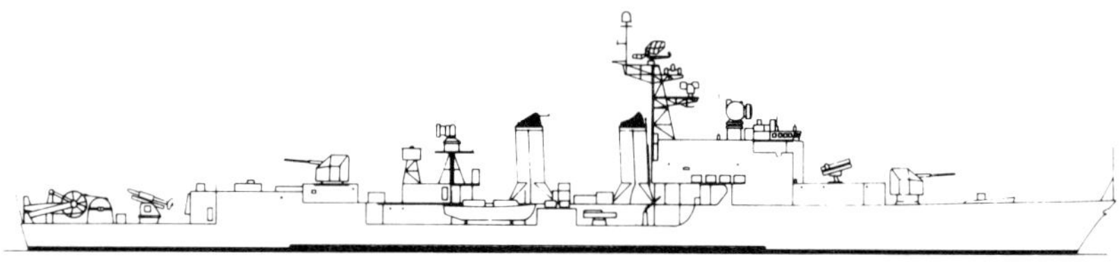

Maille Brézé 1971 (ASW conversion)

SURCOUF (T 47) class *destroyers*

Displacement;	2750t standard; 3740t full load
Dimensions:	422ft oa × 42ft × 18ft *128.6m × 12.7m × 5.4m*
Machinery:	2-shaft Rateau geared steam turbines, 4 boilers, 63,000shp = 34kts. Range 5000nm at 18kts
Armament:	6–127mm Mod 1948 (3×2), 6–57mm Mod 1951 (3×2), 4–20mm (4×1), 12–550mm TT (4×3; 18 K2/L3 torpedoes)
Sensors:	Radar DRBV 20A, DRBV 11, DRBC 11, DRBC 30; sonar DUBV 1, DUBA 1
Complement:	347

No	Name	Builder	Laid down	Launched	Comp	Fate
D 621	SURCOUF	Arsenal de Lorient	Jul 1951	3.10.53	1.11.55	BU 1971
D 622	KERSAINT	Arsenal de Lorient	Nov 1951	3.10.53	20.3.56	Stricken 1983
D 623	CASSARD	A C Bretagne	Nov 1951	12.5.53	14.4.56	Stricken 1974
D 624	BOUVET	Arsenal de Lorient	Jun 1952	25.9.54	13.5.56	Stricken 1982
D 625	DUPETIT-THOUARS	Arsenal de Brest	Mar 1952	4.2.54	15.9.56	Stricken 1988
D 626	CHEVALIER PAUL	F C de la Gironde	Feb 1952	28.7.53	22.12.56	Stricken 1971
D 627	MAILLE-BREZE	Arsenal de Lorient	Oct 1953	2.7.55	4.5.57	Museum Ship 1988
D 628	VAUQUELIN	Arsenal de Lorient	Mar 1953	2.7.55	3.11.56	Stricken 1986
D 629	D'ESTREES	Arsenal de Brest	May 1953	27.11.54	19.3.57	Stricken 1985
D 630	DU CHAYLA	Arsenal de Brest	Jul 1953	27.11.54	4.6.57	Stricken 1991
D 631	CASABIANCA	A C Bretagne	Oct 1953	13.11.54	4.5.57	Stricken 1984
D 632	GUEPRATTE	F C de la Gironde	Aug 1953	8.11.54	6.6.57	Stricken 1985

Tartar conversions

Between 1962 and 1965 *Dupetit-Thouars, Kersaint, Bouvet* and *Du Chayla* were rearmed with the American Tartar missile system. A single Mk 13 launcher replaced the after pair of 127mm mountings, the twin SPG-51 tracker-illuminators being mounted on a raised deckhouse between the after pair of 57mm mountings. The forward 127mm mounting was replaced by a sextuple 375mm Bofors ASW r/l, allowing the DRBC 31 director to be moved to the bridge top in place of the main fire control director. Only the forward (ASW) torpedo tubes were retained. The masts and funnels were heightened and an SPS-39A 3-D antenna fitted in place of the DRBV 11 atop the mainmast.

From 1968 onwards the Tartar conversions were updated: an SPS-39B radar (with the planar SPA-72 antenna) replaced the 'A' model, and a SENIT 2 action information centre was installed in the after end of the bridge structure. In 1979 *Dupetit-Thouars* and *Du Chayla* had their original air search radar replaced by a DRBV 22A. Complement was reduced to 277. *Bouvet* was stricken in 1982 and her missile system removed for refurbishment and installation on the first of the C 70 (now F 70) AA destroyers, *Cassard*. *Kersaint* decommissioned the following year to provide a second missile system for *Cassard*'s sister-ship, *Jean Bart*. Her hull was subsequently used to test the effect of fires on various types of cabling and other standard shipboard equipment in the wake of the Falklands conflict. *Dupetit-Thouars* remained in service until the completion of *Cassard*. *Du Chayla* decommissioned in 1991, when *Jean Bart* entered service. All four ships of the class served in the Atlantic Squadron until their demise.

ASW conversions

Following the experimental installation of a bow sonar and VDS on *D'Estrées* in the early 1960s the remaining five ships underwent an extensive ASW conversion 1968–70 involving replacement of their entire armament and sensor outfit. Many of the systems had already seen service on *La Galissonière*. A single Malafon ASW missile launcher was installed aft with its magazine immediately forward of it. Single 100mm mountings (Mod 1953) controlled by a DRBC 32A director were fitted forward and aft, and a sextuple 375mm Bofors ASW r/l mounted in 'B' position. As with the Tartar conversions, only the forward set of tubes was retained. A DRBV 22A air search antenna replaced the original model atop the single tripod foremast, with a DRBV 50 air/surface radar beneath it; DUBV 23 and DUBV 43 sonars were fitted. The installation of a new bow sonar involved a reshaped clipper bow with a stem anchor, increasing overall length to 132.5m. Damage control was updated and air conditioning was extended to all living spaces for a reduced ship's complement of 260.

The five ASW conversions of the T 47 class continued in service into the mid-1980s, being steadily replaced by new purpose-built ASW destroyers of the *Georges Leygues* (C 70) class. *D'Estrées* and *Guépratte* served with the Mediterranean Squadron, the others with the Atlantic Squadron. Shortly after decommissioning in 1988 *Maillé-Brézé* became a museum ship at Nantes.

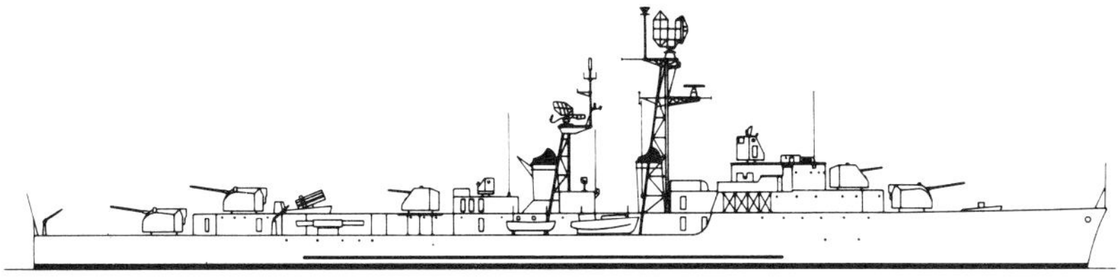

Forbin 1959

DUPERRE (T 53) class *destroyers*

Displacement: As T 47
Dimensions: As T 47
Machinery: As T 47
Armament: 6–127mm Mod 1948 (3×2), 6–57mm Mod 1951 (3×2), 2/4–20mm (2/4×1), 1–375mm Bofors ASW RL (1×6), 6–550mm TT (2×3; K2 or L3 torpedoes)
Sensors: Radar DRBV 22A, DRBI 10, DRBC 11, DRBC 30; sonar DUBV 1, DUBA 1
Complement: 346

No	Name	Builder	Laid down	Launched	Comp	Fate
D 633	DUPERRE	Arsenal de Lorient	Nov 1954	23.6.56	8.10.57	Stricken 1992
D 634	LA BOURDONNAIS	Arsenal de Brest	Aug 1954	15.10.55	3.3.58	Stricken 1976
D 635	FORBIN	Arsenal de Brest	Aug 1954	15.10.55	1.2.58	Stricken 1981
D 636	TARTU	A C Bretagne	Nov 1954	2.12.55	5.2.58	Stricken 1979
D 637	JAUREGUIBERRY	F C de la Gironde	Sep 1954	5.11.55	15.7.58	Stricken 1977

Forbin after 1973 ECPA

The T 53 design was a follow-on from the T 47. Five ships were authorised in 1953; the sixth was completed to a revised design.

The main difference as compared with the T 47 was an increased capability for tracking and controlling aircraft. The air search radar – an improved model – was moved to the mainmast so that a DRBI 10 heightfinder could be installed on the foremast. ASW capability was increased by the addition of a 375mm Bofors ASW r/l aft; the forward banks of ASW tubes were deleted in compensation.

In 1967 *Duperré* was disarmed and became a trials ship (pennant no A 633) for a large variable-depth sonar. In 1972–74 she underwent an extensive ASW conversion. Instead of Malafon, a hangar and flight deck for a Lynx were installed aft, the helicopter recovery system being a simpler version of that in the F 67 and C 70 classes. A single 100mm (Mod 68) was fitted forward; four single MM38 Exocet launchers were installed abreast the fore funnel, and Syllex chaff dispensers abeam the hangar. Radars and sonars were DRBV 22A, DRBV 51A, DRBC 51A, DRBC 32E, DUBV 23, DUBV 43 VDS. All the original torpedo tubes were removed, but two catapults for eight L5 torpedoes were installed immediately abaft the bridge structure. The ship's complement was reduced to 272. The DRBC 32E fire control was not installed until the late 1970s.

The rebuilt bridge structure houses command spaces and a SENIT 2 action information system. *Duperré* served as Flagship of the Atlantic Fleet until she ran aground in 1978. Because of the extent of her modernisation it was decided to repair her using parts from her sister-ship *Jauréguiberry*. She served in the Atlantic Squadron except for a brief period in the Mediterranean Squadron in 1986–89.

Between 1969 and 1971 *Tartu, La Bourdonnais* and *Jauréguiberry* received a SENIT 2 action information centre, and TACAN aerials were added to the foremast. In 1973 *Forbin* had her forward 57mm and after 127mm mountings removed and a flight deck fitted aft. She then served alongside *Jeanne d'Arc* as a school ship until her demise in 1981.

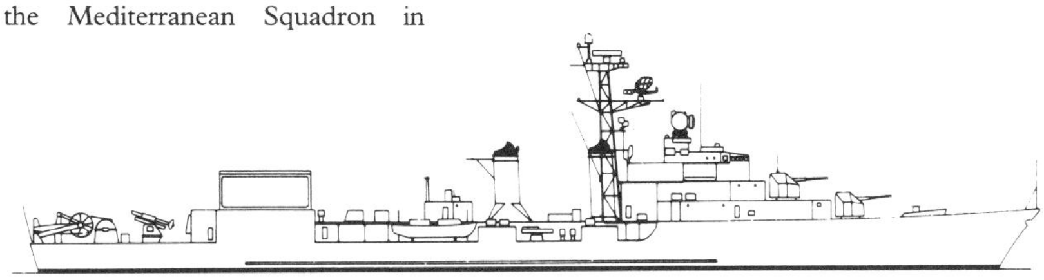

La Galissonnière 1976

LA GALISSONNIERE (T 56) *destroyer*

Displacement: As T 47
Dimensions: As T 47 except length 436ft
132.8m
Machinery: As T 47
Armament: 2–100mm (2×1), 1 Malafon ASW (13 missiles), 6–550mm TT (2×3; 12 L3 torpedoes), 1–305mm ASW mortar (1×4); 1 helicopter
Sensors: Radar DRBV 22A, DRBV 50, DRBC 32A; sonar DUBV 23, DUBV 43 VDS
Complement: 333

No	Name	Builder	Laid down	Launched	Comp	Fate
D 638	LA GALISSONNIERE	Arsenal de Lorient	Nov 1958	12.3.60	11.7.62	Stricken 1990

Of the same basic design as the T 47/53 series, *La Galissonnière* was completed as an experimental vessel fitted with an entire new generation of weapons and sensors. Originally it was intended that she should be armed with three of the new automatic 100mm DP mountings and 2–30mm AA, but the design was recast to incorporate the Malafon ASW missile launcher, and her armament as completed leaned heavily towards anti-submarine warfare. She provided the necessary experience for the design of a new generation of ASW fleet escorts.

La Galissonnière was the first French escort to operate a helicopter. In order to make the most economical use of deck space, the top of the Malafon magazine formed the centre of the flight deck and a collapsible hangar was fitted above it, the sides folding down to provide additional width. This arrangement proved less

La Galissonnière as completed Navarret Collection

than satisfactory and was not repeated in subsequent ships equipped with helicopters. The first helicopter operated was an Alouette II; this was later superseded by an Alouette III.

Immediately forward of the helicopter deck a quadruple 305mm ASW mortar was fitted. Only the forward banks of tubes were retained from the original design. A new generation of sonars was fitted and both the bow and the stern of the ship had to be reshaped to accommodate them, resulting in an increase in length. The low-frequency DUBV 23 bow sonar and DUBV 43 VDS are a matching pair with the same basic characteristics, and after trials aboard *La Galissonnière* were adopted as the standard sonars of subsequent ASW construction.

In the late 1960s, following the completion of weapon and sonar trials, *La Galissonnière* was incorporated into the active fleet and became a squadron leader. The 305mm mortar was removed. She served in the Mediterranean until 1981 when she was transferred to the Atlantic.

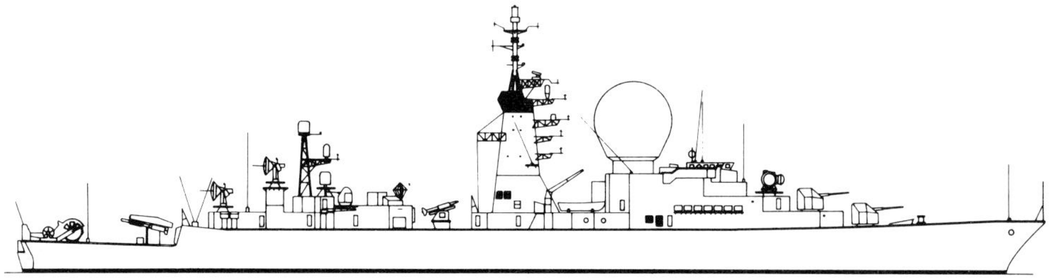

Suffren 1970

SUFFREN (FLE 60) class *guided missile destroyers*

Displacement: 5090t standard; 6090t full load
Dimensions: 486ft pp, 517ft oa × 51ft × 24ft
148.0m, 157.6m × 15.5m × 7.25m
Machinery: 2-shaft Rateau geared steam turbines, 4 boilers, 72,500shp = 34kts. Range 5100nm at 18kts
Armament: 1 Masurca SAM (1×2; 48 missiles), 2–100mm Mod 1964 (2×1), 4–20mm, 1 Malafon ASW (13 missiles), 4 catapults for 10 L5 torpedoes, 2 Syllex chaff r/l
Sensors: Radar DRBI 23, DRBV 50, 2 DRBR 51, DRBC 32A; sonar DUBV 23, DUBV 43 VDS
Complement: 355

No	Name	Builder	Laid down	Launched	Comp	Fate
D 602	SUFFREN	Arsenal de Brest	Dec 1962	15.5.65	20.7.67	Extant 1995
D 603	DUQUESNE	Arsenal de Lorient	Nov 1964	12.2.66	1.4.70	Extant 1995

Built under the 1960–65 Plan, these ships marked a radical departure from previous practice both in terms of armament and architecture. Classified as *frégates lance-engins* (now *lance-missiles*, FLM), they were designed to provide area defence and anti-submarine protection to the carriers *Foch* and *Clemenceau*, and are therefore similar in conception to the US Navy 'frigates' of the late 1950s and early 1960s, and the Royal Navy's Type 82 destroyer. They accompanied the French carriers to the Mediterranean Fleet when the latter were transferred from the Atlantic in 1975. Three ships were initially projected, with more to follow, but the programme was cut short by financial pressures related to the SSBN programme.

The Masurca area defence system, which was of French design and manufacture, was fitted only in these ships and the cruiser *Colbert*. Masurca was initially a beam-rider, but the Mod 3 variant currently in use employs semi-active guidance. The system was upgraded in 1982–85 to prolong its service life into the late 1990s. The DRBI 23 3-D air search and tracking radar is housed within a distinctive radome which dominates the ship's silhouette. Data from the ship's sensors are co-ordinated by a SENIT 1 tactical data system, and the ships are fitted with three pairs of non-retractable fin stabilisers to ensure a steady weapons platform.

The ASW systems are essentially those trialled on *La Galissonnière* during the early 1960s. The magazine for the single Malafon launcher is housed within the after deckhouse. *Suffren* was the first ship in the Marine Nationale to employ fixed catapults to launch ASW torpedoes. The four catapults are housed in the deckhouse between the mast and the bridge structure.

Exocet SSMs were fitted in 1977–79. Both ships received the Syracuse Satcom system in the mid-1980s, and in 1985 *Duquesne* had the DRBV 50 target designation radar atop the central mast replaced by the more advanced DRBV 15A. *Suffren* underwent a major refit in 1989-90 which focused on upgrading her electronics: the DRBV 50 was replaced by DRBV 15A, and the DRBC 32A gunfire control radar by the digital DRBC 33A, which has the DIBC 1A Piranha III IR/optronic attachment. A new integrated ECM outfit on a par with that of the new F 70 AA destroyers was installed, the Syllex chaff r/l being replaced by Sagaie, and two single 20mm guns were added abeam the DRBC 33 FC radar. *Duquesne* was similarly modified in 1990–91.

Duquesne 1977 C & S Taylor

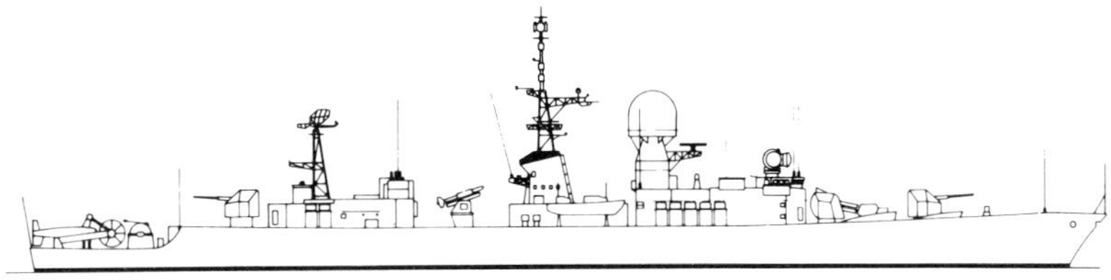

Aconit 1977

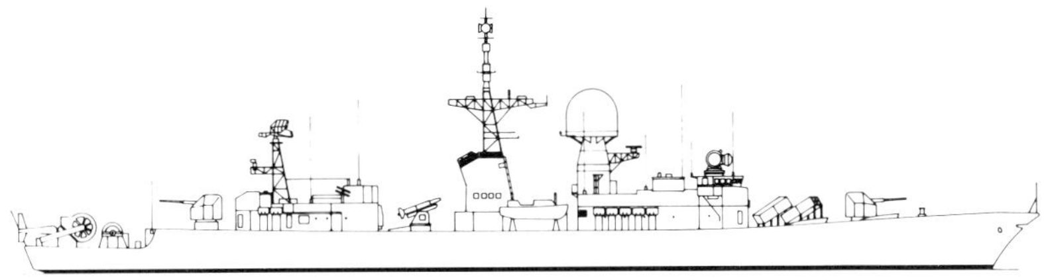

Aconit 1989

ACONIT (C 65) *destroyer*

Displacement: 3500t standard; 3900t full load
Dimensions: 417ft oa × 44ft × 19ft
127.0m × 13.4m × 5.8m
Machinery: 1-shaft Rateau geared steam turbines, 2 boilers, 28,650shp = 27kts. Range 5000nm at 15kts
Armament: 2–100mm Mod 1968 (2×1), 1 Malafon ASW (13 missiles), 1–305mm ASW mortar (1×4), 2 catapults for 10 L5 torpedoes, 2 Syllex chaff R/L
Sensors: Radar DRBV 13, DRBV 22A, DRBC 32B; sonar DUBV 23, DUBV 43 VDS
Complement: 228

No	Name	Builder	Laid down	Launched	Comp	Fate
F 703	ACONIT	Arsenal de Lorient	Jan 1967	7.3.70	30.3.73	Extant 1995

Aconit was the prototype for a series of five ASW *corvettes* designed for convoy escort work in the North Atlantic. Similar in size and conception to the ocean escorts of the US *Knox* class, she shares with them a single-shaft steam propulsion system which makes her unique in the Marine Nationale. Although well-equipped for ASW by the standards of the day, the absence of a helicopter capability limited the potential of the design, and the four projected sister-ships were abandoned in favour of the larger, more capable C 67 (later F 67) class.

In terms of architecture she is a scaled-down version of the *Suffren*, with a block of superstructure forward surmounted by a prominent radome, a combined mast and funnel amidships, and a lower superstructure block aft. The ASW systems are essentially those trialled by the *La Galissonnière*. The Malafon reload magazine is housed within the after superstructure block. The SENIT 3 tactical data system was designed specifically for this class, and is also a feature of the F 67.

As completed, *Aconit* was fitted with an automatically-reloading quadruple 305mm ASW mortar immediately forward of the bridge, but in 1984–85 this was replaced by two quadruple launchers for MM40 Exocet SSMs. At the same time the DRBV 13 multi-mode pulse-doppler radar housed within the radome was replaced by a DRBV 15A.

Aconit has served in the Atlantic Squadron since completion. She received the DSBV 62C towed array in 1991, and the original fire control radar was replaced by the all-digital DRBC 32D. Syracuse satcom is now fitted.

Aconit 13.5.92 L & L van Ginderen

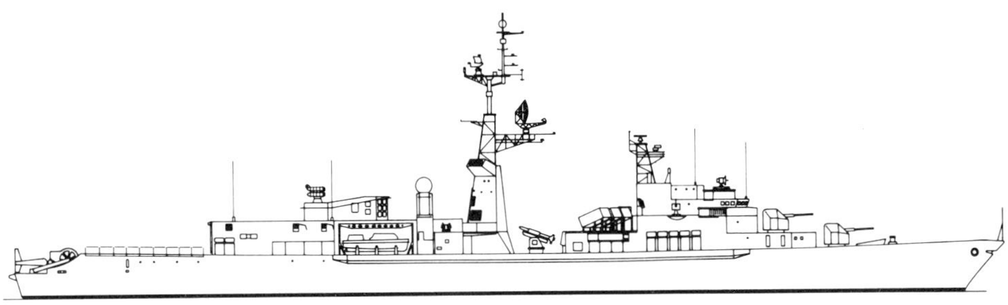

Tourville 1989

TOURVILLE (F 67) class *guided missile destroyers*

Displacement: 4850t standard; 5800t full load
Dimensions: 466ft pp, 501ft oa × 50ft × 21ft
142.0m, 152.5m × 15.3m × 6.5m
Machinery: 2-shaft Rateau geared steam turbines, 4 boilers, 54,400shp = 31kts. Range 5000nm at 18kts
Armament: 6 MM38 Exocet SSM, 1 Crotale SAM (1×8; 26 missiles), 1 Malafon ASW (13 missiles), 2–100mm Mod 1968 (2×1), 2–20mm (2×1), 2 catapults for 10 L5 torpedoes, 2 Lynx helicopters, 2 Syllex chaff R/L
Sensors: Radar DRBV 26, DRBV 51B, DRBC 32D; sonar DUBV 23, DUBV 43 VDS
Complement: 282

No	Name	Builder	Laid down	Launched	Comp	Fate
D 610	TOURVILLE	Arsenal de Lorient	16.3.70	13.5.72	21.6.74	Extant 1995
D 611	DUGUAY-TROUIN	Arsenal de Lorient	26.2.71	1.6.73	17.9.75	Extant 1995
D 612	DE GRASSE	Arsenal de Lorient	14.6.72	30.11.74	1.10.77	Extant 1995

Essentially an enlargement of the C 65 design, the F 67 incorporated from the outset a double hangar capable of accommodating two Lynx ASW helicopters, and MM38 Exocet SSMs. There was also a doubling-up of installed horsepower and the ships reverted to a conventional two-shaft machinery installation.

As originally conceived *Tourville* was to have received the same sensor outfit as *Aconit*. However, while the new ships were under construction it was decided to fit the new DRBV 26 air surveillance radar, and replace the complex DRBV 13 with a DRBV 51B (initially DRBV 50) target designation radar. A new lightweight variant of the DRBC 32 fire control radar was fitted. The ships retained the SENIT 3 tactical data system trialled in *Aconit*.

Tourville and *Duguay-Trouin* were completed with a third 100mm mounting atop the hangar, but it was decided at an early stage to replace this by a Crotale point-defence SAM system. *De Grasse* was completed with only the two forward guns in place. In early 1979 *Duguay-Trouin* received the first production model of Crotale, and the associated DRBV 51 target designation radar was moved to the mainmast to give improved performance against low-flying targets. *Tourville* was similarly modified in 1980, with *De Grasse* following in 1981.

These were the first French ships of destroyer size designed from the outset to carry helicopters. The large double hangar is served by a SPHEX traverse system, and in the centre of the helicopter deck there is a grille for use with the Harpoon securing device. Two pairs of non-retractable stabilisers are fitted.

De Grasse incorporated a number

of improvements over the earlier ships, including superior habitability. Her boilers were modified to enable her to burn distillate fuel, and this modification has since been extended to her two sisters.

The Syracuse satcom system was fitted during the mid-1980s, and all three ships received the DSBV 61C towed array in 1990. From 1993 they began to receive the new SPLASM ASW suite. Modifications include an improved DSBV 62C towed array, silencing of the auxiliary machinery, an upgrade of the SENIT 3 tactical data system and the removal of Malafon (to be replaced by the Murène ASW torpedo, and possibly by the OTO/Matra Milas ASW missile at a later date). They will also receive Dagaie in place of their Syllex chaff r/l, together with the Syracuse II satcom system. Since completion all three ships have served in the Atlantic, where they currently form the backbone of GASM.

De Grasse 19.7.80 L & L van Ginderen

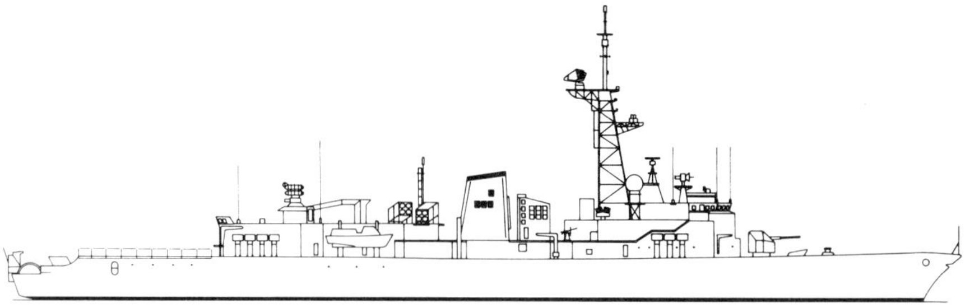

La Motte-Piquet 1989

GEORGES LEYGUES (C 70 ASW) class *guided missile destroyers*

Displacement: 3550t standard; 4350t full load
Dimensions: 423ft pp, 456ft oa × 46ft × 19ft
129.0m, 139.1m × 14.0m × 5.7m
Machinery: 2-shaft CODOG: 2 Olympus TM-3B gas turbines, 52,000hp = 30kts; 2 SEMT-Pielstick 16 PA 6 CV280 diesels, 10,400bhp = 21kts. Range 9500nm at 17kts
Armament: 4 MM38 Exocet (8 MM40 from D 642), 1 Crotale SAM (1×8; 26 missiles), 1–100mm Mod 1968, 2–20mm (2×1), 2 catapults for 10 L5 torpedoes, 2 Lynx helicopters, 2 Syllex/Dagaie chaff r/l
Sensors: Radar DRBV 26, DRBV 51C, DRBC 32E; sonar DUBV 23B, DUBV 43B VDS (see notes)
Complement: 216

No	Name	Builder	Laid down	Launched	Comp	Fate
D 640	GEORGES LEYGUES	Arsenal de Brest	16.9.74	17.12.76	10.12.79	Extant 1995
D 641	DUPLEIX	Arsenal de Brest	17.10.75	2.12.78	16.6.81	Extant 1995
D 642	MONTCALM	Arsenal de Brest	5.12.75	31.5.80	28.5.82	Extant 1995
D 643	JEAN DE VIENNE	Arsenal de Brest	26.10.79	17.11.81	25.5.84	Extant 1995
D 644	PRIMAUGUET	Arsenal de Brest	19.11.81	17.3.84	5.11.86	Extant 1995
D 645	LA MOTTE-PIQUET	Arsenal de Brest	9.2.82	6.2.85	18.2.88	Extant 1995
D 646	LATOUCHE-TREVILLE	Arsenal de Brest	31.5.85	19.3.88	1990	Extant 1995

Prompted by the requirement to replace the destroyers of the T 47 and T 53 classes, the C 70 project aimed to provide a common hull and propulsion system for two distinct classes; one to provide ASW protection for the fleet, and the other to provide area air defence. In the event it proved difficult to achieve the degree of commonality desired, and the final designs share only a common hull.

Six ASW vessels were laid down before construction was begun on the first of the AAW ships. The destroyers of the *Georges Leygues* class incorporate many of the features of the F 67 type, but on a reduced displacement. They have adopted the full-width hangar and have identical helicopter handling facilities. They also carry the same outfit of sensors, although a SENIT 4 tactical data system, using more advanced computers and all-French components, superseded the SENIT 3 system of their predecessors. The reduction in length meant only a single 100mm gun mounting forward, and Malafon had to be abandoned because of the adoption of gas turbine main propulsion machinery, which required large air intakes and exhaust uptakes amidships.

The adoption of gas turbines was influenced by developments in the Royal Navy and the Royal Netherlands Navy. However, only the Olympus half of the Olympus/Tyne COGOG package adopted by the latter navies was acceptable, as it was felt that the Tyne lacked sufficient power for running astern, particularly in the tropics, and was also expensive to run at cruising speed. SEMT-Pielstick diesels were adopted instead, and coupled with the Olympus main turbines in a CODOG arrangement.

On completion *Georges Leygues* was

Georges Leygues as completed Navarret Collection

fitted with the second prototype Crotale launcher (the first was fitted in the trials ship *Île d'Oléron*). The missiles can be fired either from the ship's command centre or from an FC compartment beneath the launcher. The last three units of the class have the upgraded Crotale EDIR variant, in which an infrared tracker has been added to the launcher/director to counter SSMs. A DRBV 15A target designation radar, with improved performance against low-flying targets, has replaced DRBV 51C, and DRBV 26 is not fitted. These modifications will be extended to the early ships of the class during the 1990s.

The last three ships also have a modified sonar outfit. The DUBV 43C VDS is an improved model with a maximum operating depth of 700m and a theoretical range of 20,000m; it is paired with the more advanced DUBV 24C hull sonar. There is also an LF towed linear hydrophone array, the DSBV 61. Earlier ships will receive a full SLASM upgrade from 1997.

Originally an eighth ship was projected, but this was cancelled as an economy measure. D 645 and D 646 were launched at Brest, but were towed to Lorient for fitting-out. Five ships currently serve in the Mediterrean and two in the Atlantic.

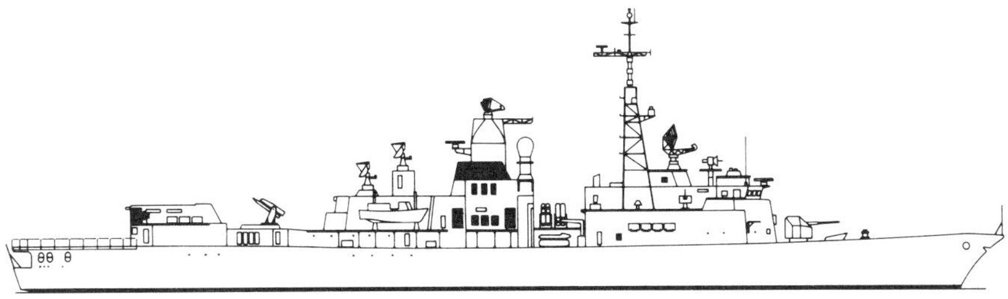

Cassard 1989

CASSARD (F 70 AA) class *guided missile destroyers*

Displacement: 3900t standard; 4500t full load
Dimensions: 423ft pp, 456ft oa × 46ft × 18ft
129.0m, 139.1m × 14.0m × 5.6m
Machinery: 2 shafts, 4 SEMT-Pielstick 18 PA 6 V280 BTC diesels, 42,300bhp = 29.6kts. Range 8000nm at 17kts
Armament: 1 Standard SAM (40 SM-1 missiles), 2 SADRAL SAM (2×6), 1–100mm Mod 1968, 2–20mm (2×1), 2 catapults for 10 L5 torpedoes, 1 Lynx helicopter, 2 Dagaie, 2 Sagaie chaff r/l
Sensors: Radar DRBV 26C, DRBJ 11B, 2 SPG-51C, DRBC 33A; sonar DUBA 25A (D614), DUBV 24C (D615)
Complement: 225

No	Name	Builder	Laid down	Launched	Comp	Fate
D 614	CASSARD	Arsenal de Lorient	3.9.82	6.2.85	29.7.88	Extant 1995
D 615	JEAN BART	Arsenal de Lorient	12.3.86	19.3.88	1.9.91	Extant 1995
D 616	COURBET	Arsenal de Lorient	—	—	—	Cancelled
D 617	Unnamed	Arsenal de Lorient	—	—	—	Cancelled

Cassard as completed — French Navy

The destroyers of the C 70 AA type (re-designated FAA – *frégates anti-aériennes* – in June 1988) were intended to replace the four T 47 AAW conversions. Four ships were to have been built, and the third and fourth of the class were indeed authorised in 1983. However, the decision by the United Stated Navy to terminate production of the SM 1 Standard missile in favour of the more advanced SM 2 resulted in their cancellation in the late 1980s.

Cassard shares a common hull with the *Georges Leygues*, but the original intention to adopt the same CODOG propulsion system was abandoned at an early stage in the design. The hot exhaust gases from the Olympus turbines were thought to be incompatible with the major electronic arrays required for the air defence rôle. Propulsive power is provided instead by four SEMT-Pielstick 18 PA fast diesels, which feature double supercharging. Flexible engine mountings have reduced the level of noise emissions as compared with traditional diesel installations.

Above decks there are few similarities between the two types – a reflection of their very different missions. The superstructures are of lightweight aluminium alloy, close attention being paid to resistance to fire and corrosion. The after part of the *Cassard* is dominated by the Standard missile system. The Mk 13 single-arm launchers and SPG 51 tracker/illuminators were removed from the *escorteurs d'escadre Bouvet* and *Kersaint*, and refurbished. A DRBV 26 provides long-range air surveillance, as in the *Georges Leygues*, and the ships are fitted with the advanced DRBJ 11B pulse-coded target-designation radar, housed within a distinctive GRP radome amidships. Development of this radar was delayed by technical problems, and a DRBV 15 was installed as a temporary measure on the first ship, *Cassard*. The SENIT 6 tactical data system has been developed specifically for this class. It is based on six 15M 125X computers of French design and manufacture, serving twelve display consules and two horizontal plots. It is currently envisaged that Standard will be replaced by the Aster SAAM system at a future refit.

The prolonged gestation period for these ships resulted in the design being recast several times. Originally, they were to have had a second 100mm mounting on the quarterdeck. However, this was subsequently replaced by a hangar for a small helicopter, with positions for SADRAL point-defence missile launchers on either side. The Lynx helicopter can be used both for ASW and for anti-ship strike, using AS.15 missiles. The SAMAHE 210 helicopter handling system can be adopted to various types of undercarriage, including skids, and features munitions trucks which position missiles, depth charges and torpedoes beneath the helicopter to assist rearming.

ASW provision is austere. However, it is envisaged that the ships will eventually be fitted with the lightweight DSBV 62C towed array.

Anti-missile protection has received particular attention in these ships, which have the first fully-integrated ECM suite in the Marine Nationale. Passive countermeasures include the ARBR 17 radar warning antenna and the DIBV 1A Vampir infrared surveillance antenna, while for active countermeasures there are two ARBB 33 jammers, together with Dagaie and Sagaie chaff r/1. The SADRAL point defence missile system is specifically designed for use against sea-skimming missiles. Each launcher houses six Mistral missiles which combine optronic guidance systems with infrared homing.

Both ships serve with the *Force d'Action Navale* (FAN).

FRANCIS GARNIER *aviso*

Displacement: 2165t standard; 3068t full load
Dimensions: 285ft pp, 318ft oa × 44ft × 15ft
87.0m, 96.9m × 13.3m × 4.7m
Machinery: 2 shafts, 2 Fiat diesels plus diesel-electric drive, 7800bhp + 1300hp = 20kts. Range 5000nm at 15kts
Armament: 4–120mm (2×2), 4–37mm (2×2), 4–20mm (2×2)
Complement: 234

No	Name	Builder	Acquired	Fate
A 04	FRANCIS GARNIER (ex-*Eritrea*)	CCM	12.2.48	Stricken 1966

This ex-Italian colonial sloop was reclassified *aviso de 1ère classe* on her acquisition under the Peace Treaty of 1948. In the 1950s she became F 730.

Acquired under MDAP, six of these ships were transferred in 1950, with *Malgache* and *Arabe* following in 1952. At first designated *torpilleurs d'escorte*, they became *escorteurs de 2e classe* in 1953. In the mid-1960s *Malgache* had her ASW weapons removed and became command ship of the amphibious training centre in place of *Dixmude*.

Ex-US CANNON class *destroyer escorts*

Armament: 3–76mm (3×1), 2–40mm (1×2), 8 to 12–20mm, 1 Hedgehog, 8 DCTs
Complement: 150
Other particulars: As US *Cannon* class (DET type)

No	Name	Builder	Acquired	Fate
F 717	ARABE (ex-*Samuel S Miles*)	Federal, Newark	Aug 1950	Stricken 1958
F 718	KABYLE (ex-*Riddle*)	Federal, Newark	Aug 1950	Stricken 1959
F 719	BAMBARA (ex-*Swearer*)	Federal, Newark	Sep 1950	Stricken 1959
F 720	SAKALAVE (ex-*Wingfield*)	Federal, Newark	Sep 1950	Stricken 1959
F 721	TOUAREG (ex-*Bright*)	Western Pipe & Steel, San Pedro	Nov 1950	Stricken 1958
F 722	SOUDANAIS (ex-*Cates*)	Tampa SB	Nov 1950	Stricken 1959
F 723	BERBERE (ex-*Clarence L Evans*)	Dravo, Wilmington	Mar 1952	Stricken 1969
F 724	MALGACHE (ex-*Baker*)	Federal, Newark	Mar 1952	Stricken 1969

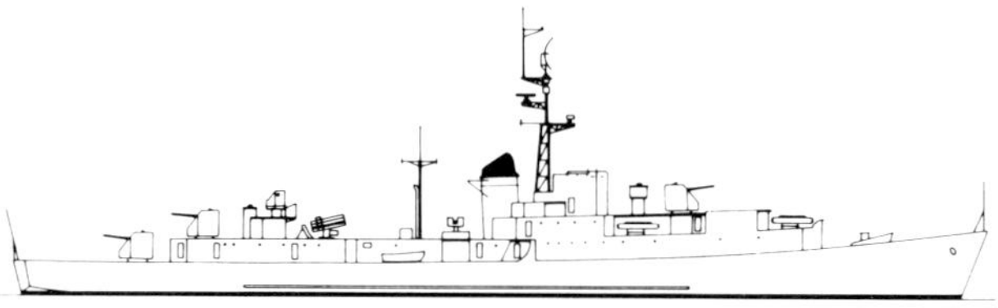

Le Corse as completed

LE CORSE (E 50) class *frigates*

Displacement: 1250t standard; 1702t full load
Dimensions: 312ft pp, 327ft oa × 34ft × 14ft
95.1m, 99.7m × 10.3m × 4.3m
Machinery: 2-shaft Rateau geared steam turbines, 2 boilers, 20,000shp = 28kts. Range 4000nm at 15kts
Armament: 6–57mm Mod 1951 (3×2), 2–20mm (2×1), 1–375mm ASW r/l (1×6), 2 DC mortars, 1 DC rack, 12 TT (4×3; 24 K2/L3 torpedoes)
Sensors: Radar DRBV 20A, DRBC 30; sonar DUBV 1, DUBA 1
Complement: 174

No	Name	Builder	Laid down	Launched	Comp	Fate
F 761	LE CORSE	Arsenal de Lorient	Oct 1951	5.8.52	15.4.55	Stricken 1975
F 762	LE BRESTOIS	Arsenal de Lorient	Nov 1951	16.8.52	19.1.56	Stricken 1975
F 763	LE BOULONNAIS	A C de la Loire	Mar 1952	12.5.53	5.8.55	Stricken 1977
F 764	LE BORDELAIS	F C de la Méditerranée	May 1952	11.7.53	7.4.55	Stricken 1975

The E 50 class was designed to produce sea-going convoy escort vessels with a large radius of action. They were the first French ships to be built specifically for convoy escort. Although clearly influenced by the latest American thinking (they were of similar size to the contemporary *Dealey* class DEs, with which they shared an American-style flush-decked hull form with a transom stern), they were entirely of French design. Two were authorised under the 1949 Programme, and the second pair in 1950. The first three ships were financed under MDAP.

The original ASW armament comprised four triple banks of ASW torpedo tubes, all mounted forward at 01 deck level. Immediately abaft each bank was a ready-use locker carrying three more reloads, the torpedoes being transferred tail-first into the muzzle of the tubes. Initial drawings show a German-pattern single 105mm gun aft, with two of the new 57mm AA mountings in 'B' and 'X' positions. In the event, however, the 105mm was replaced by a third 57mm mounting for a homogeneous AA battery. There was a fire control position forward of the bridge with a director aft above the second mounting. As with the T 47 destroyers, initial drawings show a single tall lattice foremast on the British pattern, topped by a DRBV 20A air search radar, but a tall tripod with lattice support was eventually adopted. *Le Bordelais* was completed with a modified Strombos-Velensi funnel cap.

It was decided at a late stage to fit one of the new sextuple Bofors ASW r/l. but the only position available – immediately forward of the DRBC 31 director – was not ideal because of its poor arcs fore and aft. This defect was corrected in the E 52 class.

The designed maximum speed was 26kts, but this was exceeded by an average of 3kts on trials. In convoy escort operations the E 50s were expected to maintain 14kts for maximum fuel economy.

In the early 1960s *Le Brestois* had a single 100mm fitted in place of her after 57mm mounting for experimental purposes and a DRBC 32A fire control director replaced the Bofors ASW r/l. At the same time the foremast was modified on the pattern of the latter E 52, and a DRBV 22A antenna replaced the DRBV 20A. The other three vessels were placed in reserve in 1964–66 but were later reactivated.

Le Vendéen (date unknown)
EPCA

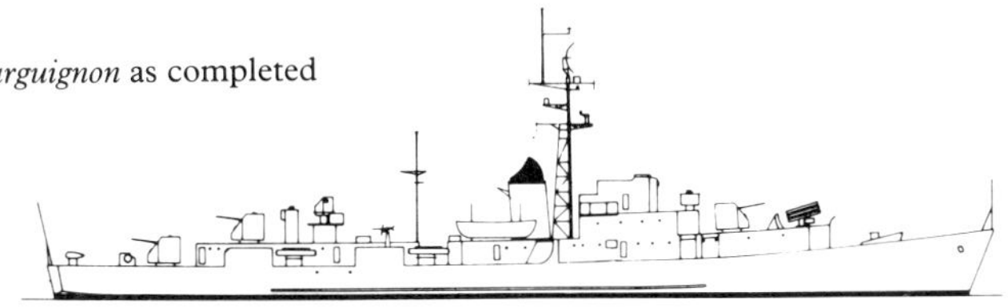

Le Bourguignon as completed

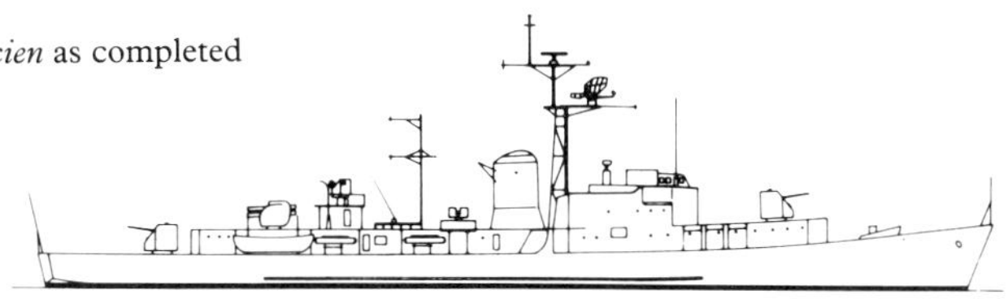

L'Alsacien as completed

LE NORMAND (E 52) class *frigates*

Displacement: As E 50
Dimensions: As E 50
Machinery: As E 50 but Parsons turbines in some. Range 4500nm at 15kts
Armament: As E 50 except E 52B type (*L'Asacien, Le Provençal, Le Vendéen*) 4–57mm (2×2), 2–20mm (2×1), 1–305mm ASW mortar (1×4), 2 DC mortars, 1 DC rack, 12–550mm TT (4×3; 12 K2/L3 torpedoes)
Sensors: Radar DRBV 22A, DRBC 31; sonar DUBV 1, DUBA 1
Complement: 175

No	Name	Builder	Laid down	Launched	Comp	Fate
F 765	LE NORMAND	F C de la Méditerranèe	Jul 1953	13.2.54	3.11.56	Stricken 1980
F 766	LE PICARD	A C de la Loire	Nov 1953	31.5.54	20.9.56	Stricken 1979
F 767	LE GASCON	A C de la Loire	Feb 1954	23.10.54	29.3.57	Stricken 1977
F 768	LE LORRAIN	F C de la Méditerranée	Feb 1954	19.6.54	1.1.57	BU 1981
F 769	LE BOURGUIGNON	Penhoët	Jan 1954	28.1.56	11.7.57	Stricken 1976
F 770	LE CHAMPENOIS	A C de la Loire	May 1954	12.3.55	1.6.57	Stricken 1975
F 771	LE SAVOYARD	F C de la Méditerranée	Nov 1953	7.5.55	14.6.56	Stricken 1980
F 772	LE BRETON	Arsenal de Lorient	Jun 1954	2.4.55	20.8.57	Stricken 1977
F 773	LE BASQUE	Arsenal de Lorient	Dec 1954	25.2.56	18.10.57	Stricken 1980
F 774	L'AGENAIS	Arsenal de Lorient	Aug 1955	23.6.56	14.5.58	Stricken 1985
F 775	LE BEARNAIS	Arsenal de Lorient	Dec 1955	23.6.56	18.10.58	Stricken 1980
F 776	L'ALSACIEN	Arsenal de Lorient	Jul 1956	26.1.57	27.8.60	Stricken 1981
F 777	LE PROVENÇAL	Arsenal de Lorient	Feb 1957	5.10.57	6.11.60	Stricken 1980
F 778	LE VENDEEN	F C de la Méditerranée	Mar 1957	27.7.57	1.10.60	Stricken 1981

E 52 was a follow-on design from the E 50 incorporating a number of improvements. The first seven units, ordered in 1952, were financed under MDAP. The remaining ships were paid for by France; two were ordered in 1953, two in 1954 and three in 1955.

The major difference in layout compared to E 50 was the relocation of the ASW tubes, together with their ready-use lockers, from 01 deck forward to the upper deck amidships. Besides reducing topweight this had the effect of freeing the forward position for the Bofors ASW r/l with a consequent improvement in arcs.

The first vessels completed had a similar radar fit to E 50, but later units were fitted from the start with the new DRBV 22A air search radar, which was carried at a lower level than the DRBV 20A. All ships were later brought up to the same standard.

In place of the tiered bridge structure of the earlier ships, *L'Agenais* and *Le Béarnais* was given a large single block extending well forward, the fire control position being relocated on top.

The last three ships of the class incorporated a number of important modifications, and were known as the Type E 52B. The 57mm mounting in 'X' position was replaced by a large quadruple 305mm ASW mortar of a new design. The Bofors ASW r/l was deleted and the forward 57mm mounting moved into its place to improve arcs. Reload magazines for the ASW torpedoes were removed, suggesting that the early vessels of the E 50 and E 52 classes may have suffered from excessive topweight. The E 52B class units had the larger bridge structure introduced by the previous two ships and in addition had the Strombos-Velensi funnel cap. Two further vessels of this type were included in the 1957 naval estimates but were cancelled due to financial problems.

In the mid-1960s *Le Savoyard, Le Breton* and *Le Basque* were assigned to the Naval Group of Trials and Measurements and were involved in initial testing of the MSBS missile. The latter two ships had their bridge structure modified on the pattern of the later ships of the class, and all three had tracking equipment installed in place of the third 57mm mounting. In the late 1970s *Le Basque* was fitted with FC equipment aft for trials of the OTOMAT missile. *L'Agenais* had a VDS fitted in place of her after 57mm mounting and the forward sets of ASW torpedo tubes were removed. She became trials ship (A 784) fitted with a towed linear passive hydrophone array (Flute ETBF). She was replaced in the sonar trials rôle by *Commandant Rivière*. In 1981 *Le Vendéen* was also reclassified as a trials ship (A 778) and briefly ran trials for a stern-mounted 550mm tube for wire-guided torpedoes during the same year before paying off in December.

E54 Type

In 1954 plans were drawn up for a new class of austere convoy escorts. Dimensions were 89.0m × 9.5m on a displacement of 1000t standard. ASW armament was to be on a par with the E 50, but AA armament was restricted to four 40mm. The first three ships were to have been laid down at Lorient but the design was not proceeded with.

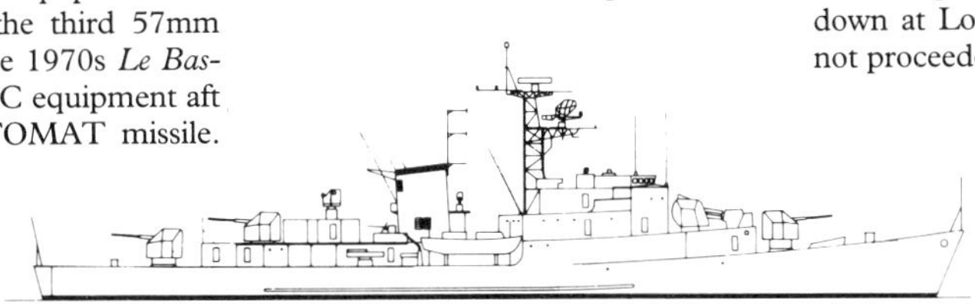

Commandant Rivière 1961

COMMANDANT RIVIERE class *frigates*

Displacement: 1750t (*Balny* 1650t) standard; 2230t (*Balny* 1950t) full load
Dimensions: 322ft pp, 338ft oa × 38ft × 14ft
98.0m, 103.0m × 11.5m × 4.3m
Machinery: 2 shafts, 4 SEMT-Pielstick 12cyl diesels (except *Balny, Bory*; see notes), 16,000bhp = 25kts. Range 7500nm at 16kts
Armament: 3–100mm Mod 1953 (3×1), 2–30mm/40mm (2×1), 1–305mm ASW mortar (1×4), 6–550mm TT (2×3; 6 L3 torpedoes)
Sensors: Radar DRBV 22A, DRBC 32A; sonar DUBA 3, SQS 17
Complement: 166

No	Name	Builder	Laid down	Launched	Comp	Fate
F 725	VICTOR SCHOELCHER	Arsenal de Lorient	Oct 1957	11.10.58	15.10.62	To Uruguay 1988
F 726	COMMANDANT BORY	Arsenal de Lorient	May 1958	11.10.58	5.3.64	Extant 1995
F 727	AMIRAL CHARNER	Arsenal de Lorient	Nov 1958	12.3.60	14.12.62	To Uruguay 1991
F 728	DOUDART DE LAGREE	Arsenal de Lorient	Mar 1960	15.4.61	1.5.63	Stricken 1991
F 729	BALNY	Arsenal de Lorient	Mar 1960	17.3.62	1.2.70	Stricken 1994
F 733	COMMANDANT RIVIERE	Arsenal de Lorient	Apr 1957	11.10.58	4.12.62	Trials ship 1986
F 740	COMMANDANT BOURDAIS	Arsenal de Lorient	Apr 1959	15.4.61	10.3.63	To Uruguay 1990
F 748	PROTET	Arsenal de Lorient	Sep 1961	7.12.62	1.5.64	Stricken 1992
F 749	ENSEIGNE DE VAISSEAU HENRY	Arsenal de Lorient	Sep 1962	14.12.63	1.1.65	Stricken 1994

The *Commandant Rivière* was a dual-purpose design intended to perform the traditional rôle of a colonial *aviso* in peacetime and the NATO rôle of convoy escort in wartime. At first christened *escorteurs d'union française* these ships were redesignated *avisos escorteurs* in 1959. The first of the class was ordered in 1955, followed by six units in 1956 and the final pair in 1957.

Three 100mm guns were fitted, with the provision that the third could be replaced by a helicopter deck. The DRBC 32A FC director was located above the after two mountings. Two single 30mm AA of a new pattern were fitted amidships. The 305mm ASW mortar in 'B' position reloaded from a small glacis forward of the mountings, and could fire a special shore bombardment round. Two triple banks of ASW torpedo tubes were fitted, but in contrast to the E 50/52 series there was no provision for reloading. There was accommodation either for a flag officer and staff, or for an eighty-man commando team, and two 9m LCPs each with a capacity of twenty-five men were carried on davits.

Because of the requirement for long range and fuel economy, a two-shaft diesel installation was adopted. *Commandant Bory* was initially fitted with Sigma free-piston generators driving gas turbines, but these were removed in 1974–75 and replaced by the standard diesel installation. *Balny* trialled an even more radical CODAG propulsion plant in which a Turbomeca M38 gas turbine (a modified Atar-8 aero-engine downrated from 15,000hp to 11,500hp) was combined with two 3600bhp AGO V16 diesels for cruising. The gas turbine and the diesels could be clutched together to drive the single shaft, which was fitted with a 3.6m diameter controllable-pitch propeller. The compactness of the installation enabled her to carry 100t more fuel than her sisters.

From 1975 until 1979 all except *Balny* had 'X' 100mm gun mounting replaced by four MM38 Exocet launchers. All had a DRBN 32 navigation radar fitted in place of the original DRBV 50 target acquisition radar, and all except F 725, F 729, F 748 and F 749 had their single 30mm AA guns replaced by standard 40mm mounts.

In 1984–85 *Commandant Rivière* underwent an extensive conversion to enable her to serve as a trials ship for the prototype DSBV 61 towed array (ETBF). The entire stern was gutted and two caissons 20m × 1m were added to the sides at upper deck level. The main armament was removed, 'B' gun mounting being replaced by a single 40mm and the DRBC 32 FC radar by a crane. She received the DUBV 43 VDS mechanism removed from the T 47 destroyer *Casabianca*, and her original SQS-17 hull sonar was replaced, probably by the prototype DUBV 24 intended for the later C 70 ASW destroyers.

Commandant Bory 19.2.83 L & L van Ginderen

Ensigne de Vaisseau Jacoubet 24.9.85 L & L van Ginderen

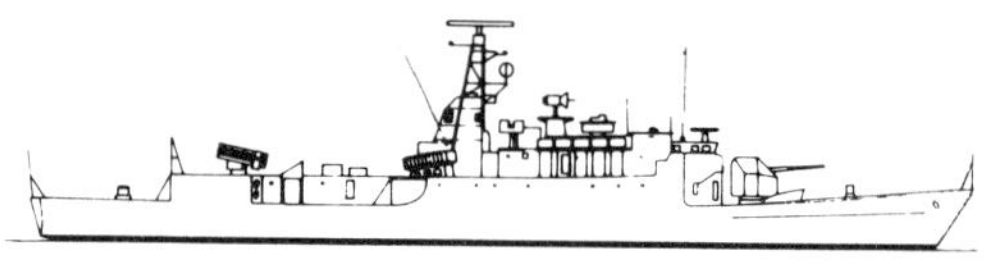

D'Estienne D'Orves 1975

D'ESTIENNE D'ORVES (A 69) class *corvettes*

Displacement: 1100t standard; 1250t full load
Dimensions: 249ft pp, 262ft oa × 34ft × 17ft
76.0m, 80.0m × 10.3m × 5.3m
Machinery: 2 shafts, 2 SEMT-Pielstick 12 PC 2 V400 diesels, 11,000bhp = 23.5kts. Range 4500nm at 15kts
Armament: 2 MM38 Exocet SSM, 1–100mm Mod 1968, 2–20mm (2×1), 1–375mm ASW r/l, 4 catapults for L3/L5 torpedoes
Sensors: Radar DRBV 51A, DRBC 32E; sonar DUBA 25
Complement: 105

No	Name	Builder	Laid down	Launched	Comp	Fate
F 781	D'ESTIENNE D'ORVES	Arsenal de Lorient	1.9.72	1.6.73	10.9.76	Extant 1995
F 782	AMYOT D'INVILLE	Arsenal de Lorient	1.9.73	30.11.74	13.10.76	Extant 1995
F 783	DROGOU	Arsenal de Lorient	1.10.73	30.11.74	30.9.76	Extant 1995
F 784	DETROYAT	Arsenal de Lorient	15.12.74	31.1.76	4.5.77	Extant 1995
F 785	JEAN MOULIN	Arsenal de Lorient	15.1.75	31.1.76	11.5.77	Extant 1995
F 786	QUARTIER-MAITRE ANQUETIL	Arsenal de Lorient	1.8.75	7.8.76	4.2.78	Extant 1995
F 787	COMMANDANT DE PIMODAN	Arsenal de Lorient	1.9.75	7.8.76	20.5.78	Extant 1995
F 788	SECOND-MAITRE LE BIHAN	Arsenal de Lorient	1.11.76	13.8.77	7.7.79	Extant 1995
F 789	LIEUTENANT DE VAISSEAU LE HENAFF	Arsenal de Lorient	Mar 1977	16.9.78	13.2.80	Extant 1995
F 790	LIEUTENANT DE VAISSEAU LAVALLEE	Arsenal de Lorient	11.11.77	29.5.79	16.8.80	Extant 1995
F 791	COMMANDANT L'HERMINIER	Arsenal de Lorient	7.5.79	7.3.81	19.1.86	Extant 1995
F 792	PREMIER-MAITRE L'HER	Arsenal de Lorient	15.12.78	28.6.80	15.12.81	Extant 1995
F 793	COMMANDANT BLAISON	Arsenal de Lorient	15.11.79	7.3.81	28.4.82	Extant 1995
F 794	ENSEIGNE DE VAISSEAU JACOUBET	Arsenal de Lorient	Apr 1979	26.9.81	23.10.82	Extant 1995
F 795	COMMANDANT DUCUING	Arsenal de Lorient	1.10.80	26.9.81	17.3.83	Extant 1995
F 796	COMMANDANT BIROT	Arsenal de Lorient	23.3.81	22.5.82	14.3.84	Extant 1995
F 797	COMMANDANT BOUAN (ex-*Commandant Levasseur*)	Arsenal de Lorient	12.10.81	23.4.83	11.5.84	Extant 1995

Ordered as replacements for the 1950s-built *escorteurs* of the E 50 and E 52 series, the *avisos* of the A 69 class illustrate the changes in French maritime philosophy which had taken place by the late 1960s. They are austere vessels designed for coastal (not ocean) ASW and for patrol on colonial stations. There is additional berthing for a troop detachment of one officer and seventeen men.

The *aviso* rôle demanded a dual-purpose gun, good endurance and manoeuvrability. Since endurance was a more important consideration than speed, a 2-shaft diesel installation was adopted, producing a maximum speed of about 23.5kts. The two diesels are mounted side by side in a single engine room, and are controlled from an acoustically insulated and air-conditioned machinery control room immediately abaft the engine room. The *Commandant L'Herminier* has SEMT-Pielstick 12 PA 6 BTC diesels with infrared suppression features; protracted trials delayed her entry into operational service. All ships of the class are fitted with fin stabilisers except for *Ducuing* and *Birot*, which have a 'dynamic' stabilisation system. Early ships of the class experienced problems with funnel gases, and the funnel was heightened from the *Jean Moulin* onwards; this modification was subsequently extended to all ships of the class.

ASW capabilities are strictly limited. The austere high-frequency DUBA 25, the transducer for which is retractable but housed in a fixed dome, is designed for acquisition and attack in coastal waters down to 220m. The magazine for the 375mm r/l is beneath the after deckhouse and contains thirty reloads, the rounds being hoisted by a hydraulic lift. The r/l is remote-controlled from the operations centre. Inside the deckhouse itself are four catapults for ASW torpedoes angled at 30 degrees to the centreline on a forward bearing; no reloads are carried. It was intended to provide a hangar and flight deck for a Dauphin helicopter in *Blaison* and *Jacoubet*, but the scheme was abandoned.

Originally the class was to have been divided into two sub-groups, designated A 69 and A 70 respectively. Ships belonging to the latter sub-group would have been fitted with Exocet SSMs. In the event bedplates for Exocet were fitted in all ships, and the missiles are always carried by ships deploying overseas. Fire control for the single 100mm is provided by a DRBC 32E monopulse radar served by a semi-analogue, semi-digital computer.

The A 69 class underwent a mid-life update programme from the mid-1980s. The first unit to be refitted, *Amyot d'Inville*, emerged in late 1986 with a new 100mm gun, a US SLQ-25 Nixie torpedo decoy system, an upgraded sonar, and L5 catapult-launched torpedoes in place of the ageing L3 model. Most have been retro-fitted with the *Installation de Tir Léger* (ITL) for the MM 38/40 SSM, and those without ITL are being fitted with the *Installation de Tir Standard* (ITS) to enable them to serve on overseas stations. In 1993 F 793 and F 794 had the Bofors r/l replaced by a Syracuse II satcom terminal.

The ruggedness, simplicity and cheapness of the design has appealed to a number of foreign buyers. In 1976 the original *Le Hénaff* and *L'Herminier* were purchased by South Africa while under construction, but following the UN arms embargo on that country they were transferred to Argentina in 1978. They were replaced by two vessels carrying the same name and pennant number.

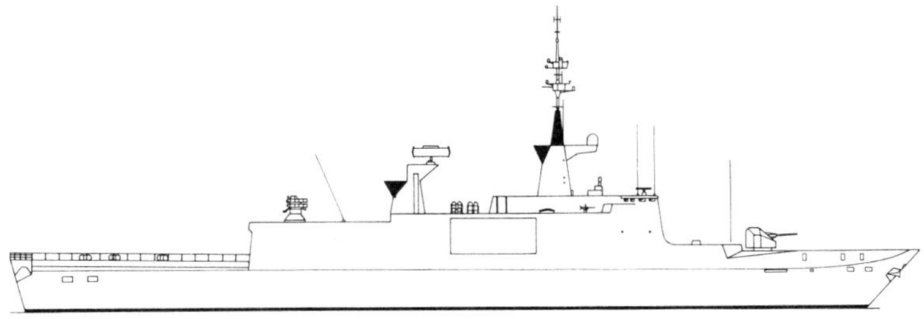

Lafayette as completed

LAFAYETTE class *frigates*

Displacement:	3280t full load
Dimensions:	377ft pp, 410ft oa × 51ft × 13ft (hull) *115.0m, 125.0m × 15.4m × 4.0m*
Machinery:	2-shaft CODAD; 4 SEMT-Pielstick 12 PA 6V BTC diesels, 17,600bhp = 25kts. Range 7000nm at 15kts
Armament:	8 MM40 Exocet SSM (2×4), 1 Crotale EDIR SAM (1×8), 1–100mm Mod 1968, 2–20mm (2×1), 2 Dagaie chaff r/l, 1 AS 565 Panther helicopter
Sensors:	Radar DRBV 15C, DRBC 33A
Complement:	156

No	Name	Builder	Laid down	Launched	Comp	Fate
F 710	LAFAYETTE	Arsenal de Lorient	15.12.89	13.6.92	Dec 1994	Extant 1995
F 711	SURCOUF	Arsenal de Lorient	6.7.92	3.7.93	—	Building
F 712	COURBET	Arsenal de Lorient	15.9.93	Aug 1994	—	Building
F713	JAUREGUIBERRY	Arsenal de Lorient	—	—	—	Ordered
F714	GUEPRATTE	Arsenal de Lorient	—	—	—	Ordered
F715	RONARC'H	Arsenal de Lorient	—	—	—	Ordered

The *frégates légères* of the *Lafayette* class are multi-purpose frigates designed to perform a variety of missions, which include crisis intervention and defence of the sea approaches of both Metropolitan France and her overseas territories. They could also be integrated into an intervention task force based on an aircraft carrier and/or amphibious ships.

A modular approach has been adopted to their construction in order to give the necessary flexibility of weapon fit for possible future missions and to appeal to the export market. Each ship comprises about seventy modules, two of which are large propulsion modules housing the diesel engines. Many of the modules are prefabricated off-site and assembled on the building way at Lorient Naval Dockyard.

The other novel feature in the design is the emphasis on acoustic and electro-magnetic discretion. The diesel engines will be mounted on suspended rafts in order to reduce noise emissions. The sides of the hull and the superstructures are inclined at about 10 degrees, and an inclined bulwark forward shields the single gun mounting. The superstructures and masts are coated with radar-absorbent paint, and there is a composite radar-absorbent cladding on the forecastle and quarterdeck. The superstructures are 'clean', the ship's boats being concealed behind doors amidships.

The armament initially fitted will be biased towards the overseas patrol mission. There is a single 100mm CADAM gun mounting, MM40 Exocet missiles, together with a lightweight Crotale EDIR modular mounting for short-range air defence. A TAVITAC 2000 tactical data system is installed. The capacious hangar and flight deck can accommodate a helicopter in the 9–10t range, and there is magazine stowage beneath the flight deck for both AS15 and AM39 missiles. Initially an AS 565 Panther helicopter will be carried, but this will eventually be superseded by the NH-90. The ECM outfit will be on a par with that of the *Cassard* class air defence ships, and they will be fitted with the Syracuse 2 satcom system.

Forward of the bridge, space has been reserved for the installation of two 8-missile SAAM modules as and when the system becomes available. SAAM will be fitted from the outset in the second batch of three ships. Although anti-submarine warfare is a low priority for the new frigates, they have the potential to accommodate the new DSBV 62C towed array, Murène homing torpedoes, and the SLAT anti-torpedo system.

The first three ships were ordered in July 1988, but construction was delayed by the 1989 budget cuts. The second batch of three was ordered in 1992.

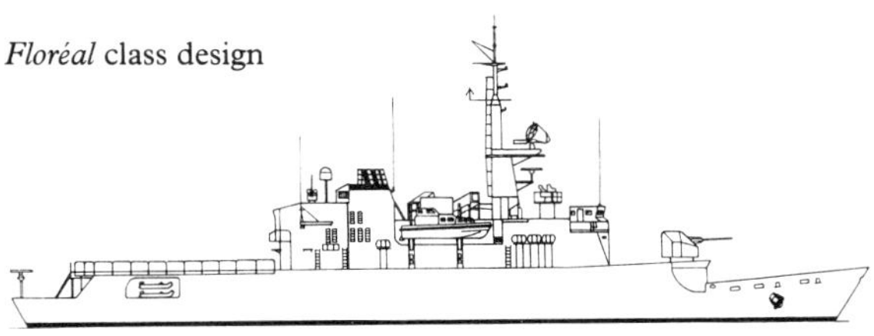

Floréal class design

Nivose 23.8.92 L & L van Ginderen

FLOREAL class *corvettes*

Displacement:	2600t trials; 2950t full load
Dimensions:	280ft pp, 307ft oa × 46ft × 14ft *85.2m, 93.5m × 14.0m × 4.3m*
Machinery:	2 shafts, 4 SEMT-Pielstick 6PA 6L280 diesels, 8800bhp = 20kts. Range 9000nm at 15kts
Armament:	2 MM 38 Exocet SSM, 1–100mm Mod 1968, 2–20mm (2×1), 1 AS 565 Panther helicopter
Sensors:	DRBV 21A, Najir FC
Complement:	80 + 24 commandos

No	Name	Builder	Laid down	Launched	Comp	Fate
F 730	FLOREAL	Chantiers de l'Atlantique	2.4.90	6.10.90	27.5.92	Extant 1995
F 731	PRAIRAL	Chantiers de l'Atlantique	11.9.90	23.3.91	20.5.92	Extant 1995
F 732	NIVOSE	Chantiers de l'Atlantique	16.1.91	10.8.91	16.10.92	Extant 1995
F 733	VENTOSE	Chantiers de l'Atlantique	28.6.91	14.3.92	5.5.93	Extant 1995
F 734	VENDEMIARE	Chantiers de l'Atlantique	17.1.92	29.8.92	20.10.93	Extant 1995
F 735	GERMINAL	Chantiers de l'Atlantique	17.8.92	13.3.93	Apr 1994	Extant 1995

A response to the requirement for a 'cheap frigate' capable of performing general patrol duties in low-threat areas, the *Floréal* class are the first French naval vessels of their size to be ordered from a private shipyard since the late 1950s.

They have been built to mercantile standards except in respect of the aviation facilities, damage control, and munitions stowage. The reduction in cost was such that three hulls could be built for the price of a single frigate of the *Lafayette* class; the use of modular construction techniques reduced construction time to 24 months per vessel. The hull and superstructures are of welded steel, with transverse bulkheads. The design has an exceptionally broad beam for its length to enable a hangar and flight deck for a 9–10 ton helicopter to be worked in aft. Fin stabilisers provide an exceptionally stable platform.

The twin shafts are each powered by two diesels, and have controllable pitch propellers. Twin rudders combine with the latter to provide exceptional manoeuvrability. Maximum speed is adequate only for patrol duties, but at a 7kt cruising speed endurance is virtually unlimited. Contrary to standard naval practice, all four diesels are located in a single spacious machinery room for ease of maintenance and surveillance. Control and surveillance is exercised from a fully-automated machinery control centre. The three electrical generators are disposed fore and aft of the engine room, the two main switchboards being located in separate damage control areas. Diesel fuel and TR5 aviation fuel can be received by the abeam or astern replenishment method, although there is only a single transfer station as compared with two for first-line naval vessels.

The ships can operate a helicopter up to the size of the AS 332F Super Puma. It was initially envisaged that they would receive the minimum armament compatible with the surveillance mission, but the final installation is on a par with that of the *avisos* of the A 69 class.

Communications systems are also on a par with the A 69 class, and there is provision for the installation of a Syracuse 2 satcom system. A detachment of twenty-four commandos can be accommodated in addition to the crew of 100 and medical facilities include a consultation room together with hospital beds. These features enable the ships to undertake both 'policing' duties and humanitarian missions.

Following sea trials, the ships were delivered to Lorient Naval Dockyard for weapon installation and trials of the combat system. They have now deployed in support of the French overseas territories, replacing the *Commandant Rivière* class in this rôle.

SUBMARINES

In 1951 four demilitarised 'S' class submarines were lent by the Royal Navy for ASW training. All had their torpedo tubes closed off and two, *Saphir* and *Sultane*, had a modified hull form and streamlined fin, giving them an increase in speed of 3kts when dived. *Sibylle* was accidentally lost near Toulon on 23 September 1952. The initial loan period of four years was extended for the other three.

SAPHIR class *training submarines*

Displacement: 715t standard; 990t submerged
Dimensions: 217ft oa × 24ft × 14ft
66.1m × 7.2m × 4.2m
Machinery: 2 shafts, 2 8cyl Admiralty diesels plus 2 electric motors, 1900bhp/1300shp = 14.5kts/9kts. Range 6000nm at 10kts
Armament: See notes
Complement: 44

No	Name	Builder	Acquired	Fate
S 616	SAPHIR (ex-*Satyr*)	Scotts	1951	Returned UK 1961
S 614	SULTANE (ex-*Statesman*)	Cammell Laird	1951	Returned UK 1959
S 615	SIRENE (ex-*Spiteful*)	Scotts	1951	Returned UK 1958
S 614	SIBYLLE (ex-*Sportsman*)	Chatham DYd	1951	Lost 23.9.52

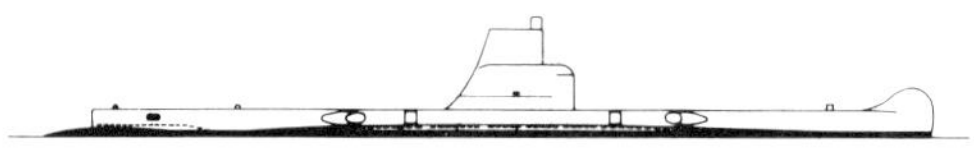

L'Artemis 1955

LA CREOLE class *patrol submarines*

Displacement: 970t surfaced; 1250t submerged
Dimensions: 241ft oa × 21ft × 14ft
73.5m × 6.5m × 4.2m
Machinery: 2 shafts, 2 Sulzer (*L'Africaine* Schneider) diesels plus 2 electric motors, 3000bhp/1400shp = 15.5kts/9.3kts. Range 8800nm at 10kts. See notes
Armament: 10–550mm TT (4 bow, 4 amidships, 2 stern); *La Créole, L'Africaine, L'Astrée* also 1–88mm, 2–20mm (2×1)
Complement: 62

No	Name	Builder	Laid down	Launched	Comp	Fate
S 606	LA CREOLE	A C Augustin-Normand	Dec 1937	8.4.40	1949	BU 1961
S 607	L'AFRICAINE	A C de la Seine-Maritime	Dec 1937	7.12.46	Oct 1949	BU 1963
S 608	L'ASTREE	A C Dubigeon	Nov 1938	3.5.46	Apr 1949	BU 1966
S 601	L'ANDROMEDE	A C Dubigeon	Nov 1938	17.11.49	Apr 1953	BU 1966
S 603	L'ARTEMIS	A C Augustin-Normand/ A C Dubigeon	May 1939	28.6.42	Feb 1954	BU 1967

Five submarines of the *Aurore* type remained uncompleted – and relatively undamaged – at the end of the Second World War. *La Créole*, which had been towed to the UK at the time of the occupation, was put back on the slipway at Le Havre and launched a second time (on 8 May 1946). She and two of her sisters, *L'Astrée* and *L'Africaine*, were completed to a modified design, with a German-model 88mm gun mounted in the forward end of an enlarged fin, and two 20mm AA at its after end. There were ten torpedo tubes in place of the original nine. As well as the four internal bow tubes and two stern tubes there were pairs of external tubes forward and aft of the fin.

L'Artémis and *L'Andromède*, whose construction was less advanced, were redesigned to give faster diving and a higher submerged speed (10.3kts) and were given a fin incorporating a snorkel similar to that of the US 'Guppies'. The completion of *L'Artémis* was delayed in order to incorporate new fire control equipment destined for the *Narval* class. *La Créole* was fitted with a snorkel at the same time as the later two boats.

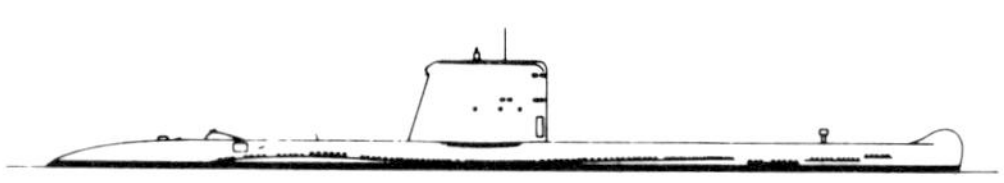

Requin 1976

NARVAL class *patrol submarines*

Displacement: 1635t standard; 1910t submerged
Dimensions: 257ft oa × 26ft × 17ft
78.4m × 7.8m × 5.2m
Machinery: 2 shafts, 2 Schneider 7cyl diesels plus 2 electric motors, 4400bhp/5000shp = 16kts/18kts. Range 15,000nm at 8kts
Armament: 8–550 TT (6 bow, 2 stern)
Sensors: Sonar DUUA 1
Complement: 63

No	Name	Builder	Laid down	Launched	Comp	Fate
S 631	NARVAL	Arsenal de Cherbourg	Oct 1951	11.12.54	1.12.57	Stricken 1986
S 632	MARSOUIN	Arsenal de Cherbourg	Nov 1951	21.5.55	1.10.57	Stricken 1986
S 633	DAUPHIN	Arsenal de Cherbourg	Jan 1952	12.9.55	1.8.58	Stricken 1992
S 634	REQUIN	Arsenal de Cherbourg	Feb 1952	3.12.55	1.8.58	Stricken 1985
S 637	ESPADON	A C Augustin-Normand	Dec 1955	15.9.58	2.4.60	Stricken 1985
S 638	MORSE	A C de la Seine-Maritime	Feb 1956	10.12.58	2.5.60	Stricken 1986

In 1947 the STCAN was asked to produce plans for a new class of submarine of 1200 tonnes standard. Requirements included a speed of over 16kts submerged and a range of 15,000nm with snorkel. Long range was the overriding consideration because the continued existence of a French empire demanded the ability to make a lengthy transit and still be able to undertake a patrol of 7–14 days.

In drawing up the plans the constructors leaned heavily on experience with the ex-German Type XXI boat *Roland Morillot* and the two reconstructed 1500t boats of French origin (*L'Andromède* and *L'Artémis*).

In addition Dinechin, who was responsible for the basic design, visited the USA in order to make a detailed study of current US Navy thinking.

The result was an improved Type XXI, almost identical in dimensions and displacement but with superior performance. Two boats were ordered in 1949, two in 1950, and the final pair in 1954.

Problems were experienced in developing a high-tensile steel capable of resisting pressures at the 200m maximum diving depth, and the increased pressures also necessitated many changes in other areas, notably more powerful emergency blowing systems. An exceptionally strong all-welded double hull was adopted. Hull-form, propellers, rudder and driving planes were all the subject of intensive studies. The Cherbourg-built boats were assembled in prefabricated sections each 7m × 10m – the first French submarines constructed by this method.

The action information and fire control centre was based on that developed in *L'Artémis*, as was the fin with its raised section aft. The latest detection equipment was fitted. In addition to the six bow tubes, two tubes were fitted aft outside the pressure hull; fourteen torpedoes were carried.

Problems were experienced in finding a suitable diesel, as few French engineering firms were willing to undertake a series of only twelve engines. Schneider 2-stroke diesels were chosen in preference to 2/4-stroke Sulzer and 4-stroke MAN diesels. When fast diesels later became available it was decided to replace the machinery by a diesel-electric installation. Three SEMT-Pielstick PA 4 12cyl diesels, each rated at 750bhp, replaced the Schneider units. New electric motors each of 2500hp were installed, and to these were added a pair of electric cruise motors each of 40hp. All six boats were taken in hand from 1965 onwards, and besides the replacement of the propulsion system major improvements were made to the weapons and detection equipment. The two stern tubes were removed

Marsouin as completed Navarret Collection

and the number of torpedoes carried was increased to twenty. The fin was rebuilt.

When first completed the *Narval* class were subjected to extensive trials. In 1958 *Dauphin* established a record of forty-two days snorkelling. Only the propulsion system proved less than satisfactory. But the reconstruction during 1965–70 was a complete success. *Marsouin* had a serious fire on 4 August 1978 but was subsequently repaired.

The submarines of the *Narval* class continued in service with the Atlantic Squadron into the mid-1980s. *Narval* was fitted with a swimmer delivery vehicle in an open-topped container fitted to the after deck casing. *Requin* was modified for trials of the SM-39 missile in 1980. *Marsouin* was decommissioned in November 1982 to provide spares for the submarines remaining in service, and *Espadon* became a museum exhibit after paying off in 1985. In 1986 *Dauphin* was fitted with the prototype bow section for the *Améthyste* class SNA, which extended her length overall by 2.97m.

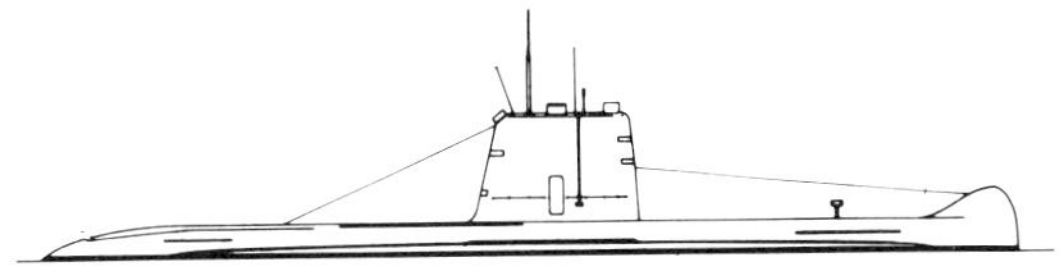

Arethuse 1971 **NB 1/750 scale**

ARETHUSE class *attack submarines*

Displacement: 543t surfaced; 669t submerged
Dimensions: 163ft oa × 19ft × 13ft
49.6m × 5.8m × 4.0m
Machinery: 1 shaft, 2 SEMT-Pielstick 12cyl diesels plus 1 electric motor, 1060bhp/1300shp = 12.5kts/16kts
Armament: 4–550mm TT (bow)
Sensors: Sonar DUUA 1
Complement: 39

No	Name	Builder	Laid down	Launched	Comp	Fate
S 635	ARETHUSE	Arsenal de Cherbourg	Mar 1955	9.11.57	23.10.58	Stricken Apr 1979
S 636	ARGONAUTE	Arsenal de Cherbourg	Mar 1955	29.6.57	11.2.59	Stricken 1982
S 639	AMAZONE	Arsenal de Cherbourg	Dec 1955	3.4.58	1.7.59	Stricken Jul 1980
S 640	ARIANE	Arsenal de Cherbourg	Dec 1955	12.9.58	16.3.60	Stricken Mar 1981

From their earliest conception the *Aréthuse* class were designated *sous-marins de chasse*. They were, therefore, the world's first hunter-killer submarines designed as such, and marked a significant departure from the ocean-going *Narval* class. Designed specifically for the Mediteranean, they were to be based at Mers-el-Kebir to prevent enemy submarines infiltrating the extensive mercantile traffic between metropolitan France and the colonies in North Africa.

The staff requirements demanded a 100 per cent submarine with a small silhouette and excellent manoeuvrability. They also specified the ability to dive deep and fast, to carry out operations at snorkel depth, and to maintain high speed for an hour when dived. Advanced sensors were to provide data for a well-equipped action information centre.

Although in conception the *Aréthuse* class was clearly influenced by the German Type XXIII design, they were bigger boats with higher submerged speed and a heavier torpedo armament – four tubes plus four reloads.

They were the first French submarines to adopt diesel-electric propulsion. Two groups of electro-generators producing 337kW drove a single large electric motor of 1300hp. The generators and auxiliaries had spring suspension, while the electric motor was coupled directly to the shaft. This resulted in near-silent operation.

All four boats served in the Mediterranean from completion.

The hull of *Argonaute* was transferred to Paris in 1989 to be refurbished as a museum exhibit.

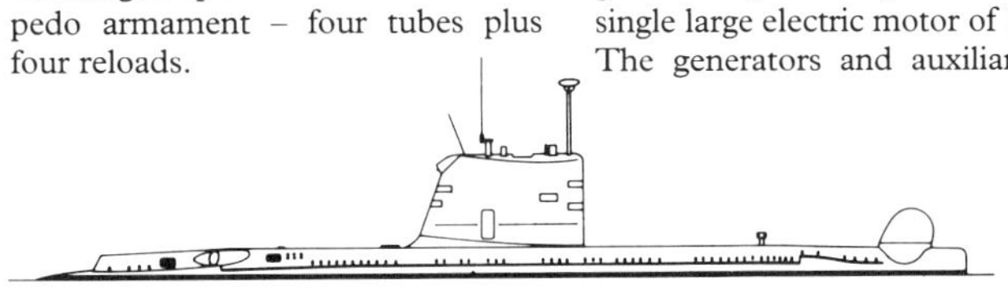

Doris 1975 **NB 1/750 scale**

DAPHNE class *patrol submarines*

Displacement: 869t surfaced; 1043t submerged
Dimensions: 190ft oa × 22ft × 17ft
57.8m × 6.8m × 5.25m
Machinery: 2 shafts, 2 SEMT-Pielstick diesels plus 1 electric motor, 1300bhp/1600shp = 13.5kts/16kts. Range 4500nm at 5kts
Armament: 12–550mm TT (8 bow, 4 stern)
Sensors: Sonar DUUA 1
Complement: 45

No	Name	Builder	Laid down	Launched	Comp	Fate
S 641	DAPHNE	A C Dubigeon	Mar 1958	20.6.59	1.6.64	Stricken 1989
S 642	DIANE	A C Dubigeon	Jul 1958	4.10.60	20.6.64	Stricken 1988
S 643	DORIS	Arsenal de Cherbourg	Sep 1958	14.5.60	26.8.64	Stricken 1994
S 644	EURYDICE	Arsenal de Cherbourg	Jul 1959	19.6.60	Sep 1964	Lost 4.3.70
S 645	FLORE	Arsenal de Cherbourg	Sep 1958	21.12.60	21.5.64	Stricken 1989
S 646	GALATEE	Arsenal de Cherbourg	Sep 1958	22.9.61	25.7.64	Stricken 1991
S 647	MINERVE	A C Dubigeon	May 1958	31.5.61	10.6.64	Lost 27.1.68
S 648	JUNON	Arsenal de Cherbourg	Jul 1961	11.5.64	25.2.66	Extant 1995
S 649	VENUS	Arsenal de Cherbourg	Aug 1961	24.9.64	1.1.66	Stricken 1990
S 650	PSYCHE	Arsenal de Brest	May 1965	28.6.69	7.6.70	Extant 1995
S 651	SIRENE	Arsenal de Brest	May 1965	28.6.69	3.9.70	Extant 1995

Developed from the small hunter-killer submarins of the *Aréthuse* class, the *Daphné* was designed as a second-class counterpart to the ocean-going submarines of the postwar *Narval* class. Double-hull construction was used, and staff requirements emphasised manoeuvrability, low noise, a small crew and ease of maintenance. The latter requirement influenced the adoption of no fewer than twelve torpedo tubes (eight bow, four stern) with no reloads. Further reductions in complement were effected by the adoption of a maintenance system based on modular replacement. The last two boats incorporated a number of improvements, and have the Type 12 PA4-135 diesel in place of the original PA1 model.

Beginning in 1971 all except *Daphné* underwent an extensive modernisation, being fitted with a DUVA 2 search and attack sonar in a prominent bow dome and the DLT D3 torpedo FC system.

Two sisters were lost in the western Mediterranean: *Minerve* (S 647) in 1968 and *Eurydice* (S 644) in 1970. The remaining units have served since completion in the Mediterranan Squadron, with the exception of *Psyché* and *Sirène*, which are based at Lorient. The class has proved to be a popular export design. Ten have been built in France for Portugal (four), Pakistan (three) and South Africa (three), and a further four have been built under licence in Spain.

Psyché 10.7.92
L & L van Ginderen

Gymnote as completed
Navarret Collection

In 1954 studies were made of a torpedo-armed nuclear-powered submarine; the boat was even laid down at Cherbourg in 1956 as hull No Q 244. A surface displacement of 4000t on a length of 120m was projected. The inability of the French to produce enriched uranium at this time meant, however, that a heavy-water reactor using unrefined uranium was the only possible propulsion plant. This proved too heavy and construction was halted in 1958. Negotations with the Amercians for the purchase of a reactor failed, and even the purchase of enriched uranium was refused except for land experimental use. The project was abandoned in 1959.

GYMNOTE class *ballistic missile trials submarine*

Displacement: 3000t surfaced; 3250t submerged
Dimensions: 276ft oa × 35ft × 25ft
84.0m × 10.6m × 7.6m
Machinery: 2 shafts, 4 SEMT-Pielstick diesels plus 2 electric motors, ?/2600shp = 11kts/10kts
Armament: 4 SLBM tubes
Complement: 78

No	Name	Builder	Laid down	Launched	Comp	Fate
S 655	GYMNOTE	Arsenal de Cherbourg	17.3.63	17.3.64	17.10.66	Stricken 1986

With the decision to develop an independent sea-based deterrent force, it was decided to complete the hull (now designated Q 251) as an experimental submarine for the testing of the new ballistic missiles. A prominent casing abaft the fin housed four vertical launch tubes. *Gymnote* was also fitted with the prototype guidance and inertial navigation system intended for *Le Redoutable.* She launched her first M-1 missile at the end of 1968 at the Landes Trials Centre. She later conducted trials for the other missiles in the series, and in 1977–79 underwent a refit in which the missile tubes were enlarged to accommodate the new M-4 missile due to enter service in 1985. *Gymnote* was involved in test-firings of the M-4 until she was stricken in 1986.

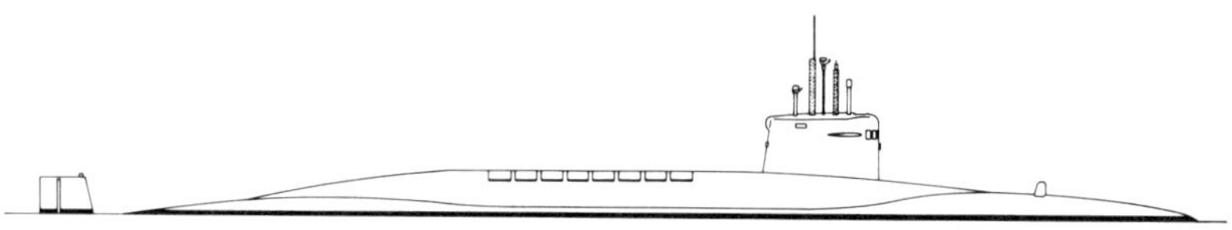

L'Inflexible 1985

Owing to the political circumstances in which the *Force de Dissuasion* was conceived (see Introduction), no US assistance in the design and construction of the SNLEs was forthcoming. The first reactor constructed, the PAT 1, was produced in collaboration with the CEA (the French Atomic Energy Authority) and tested extensively ashore at Cadarache. Simultaneously development of the missiles took place, the first MSBS being launched from the experimental submarine *Gymnote* at the Landes Trials Centre at the end of 1968.

Construction of the submarines posed a number of technical prob-

LE REDOUTABLE class *ballistic missile submarines*

Displacement: 8045t surfaced; 8940t submerged
Dimensions: 422ft × 35ft × 33ft
128.0m × 10.6m × 10.0m
Machinery: 1-shaft nuclear: 1 pressurised water-cooled reactor, 2 turbines plus 2 turbo-alternators, 1 electric motor, 15,000shp = 20kts/25kts. Auxiliary diesel 2670bhp
Armament: 16 SLBM tubes, 4–550mm TT (bow; 18 L5 torpedoes). See notes
Sensors: Radar DRUA 33; sonar DUUV 23, DUUX 2
Complement: 135

No	Name	Builder	Laid down	Launched	Comp	Fate
S 611	LE REDOUTABLE	Arsenal de Cherbourg	30.3.64	29.3.67	1.12.71	Stricken 1991
S 612	LE TERRIBLE	Arsenal de Cherbourg	24.6.67	12.12.69	1.1.73	Extant 1995
S 610	LE FOUDROYANT	Arsenal de Cherbourg	12.12.69	4.12.71	6.7.74	Extant 1995
S 613	L'INDOMPTABLE	Arsenal de Cherbourg	4.12.71	17.8.74	23.12.76	Extant 1995
S 614	LE TONNANT	Arsenal de Cherbourg	Oct 1974	17.9.77	3.4.80	Extant 1995
S 615	L'INFLEXIBLE	Arsenal de Cherbourg	27.3.80	23.6.82	1.4.85	Extant 1995

lems such as the development of high-tensile steel to cope with the increase in diving depth (up to 300m), the adaptation of the hull to nuclear propulsion, and the development of suitable steering mechanisms. The cylindrical hull was constructed using prefabricated all-welded sections each of 200t.

In addition to the main propulsion machinery there is an emergency diesel capable of bringing the submarine home in the event of a failure in the nuclear plant; it has a range of about 5000nm. The diving planes are on the fin, as in US boats. Four torpedo tubes are fitted forward, a total of eighteen torpedoes being carried. The submarine's position can be accurately determined by reference to three inertial navigation centres (CIN) ashore, and there is a periscope for celestial navigation.

The first MSBS was the M-1, an 18t two-stage missile of similar length and diameter to the US Poseidon, but carrying only a single 500-kiloton warhead and with a range of only 2500km. The M-1 missile was installed in the first two boats, but was then superseded by the 3000km M-2, which went into service aboard the third boat, *Le Foudroyant.* From 1976 onwards, both the M-1 and the M-2 were replaced by the M-20, which had a one-megaton thermonuclear warhead and an upgraded re-entry vehicle system.

L'Indomptable and *Le Tonnant* have a metallic reactor core in place of the oxide cores of earlier boats. *L'Inflexible*, the sixth boat, is an intermediate design between the *Le Redoutable* class and the new-generation SNLE of the *Le Triomphant* class, and incorporates a number of improvements. She was fitted from the outset to fire the M-4 missile, which carries six TN-70 150kt MIRVs out to a range of 2150nm (4000km). She also has the new DSUX 21 multifunction sonar array, DUUX 5 passive ranging, a DSUV 61 towed array, and improved inertial navigation systems. The diving planes are mounted higher on a more streamlined fin, and diving depth has been increased from 200m to 300m. Considerable attention was given to reducing noice emissions.

Le Redoutable was the first to decommission in 1991. The other units, however, have recently completed major refits to bring them up to the same standard as *L'Inflexible.* The M-20 single warhead missile has been replaced by the M-4, and the DSUX 21 multifunction sonar has been fitted. *Le Tonnant* was refitted at Cherbourg in 1985–87, followed by *L'Indomptable* and *Le Terrible*. The refit of *Le Foudroyant* was completed in 1993. *L'Inflexible* will convert to the 3250nm (6000km) M-5 missile around the year 2000.

Le Redoutable as completed — Navarret Collection

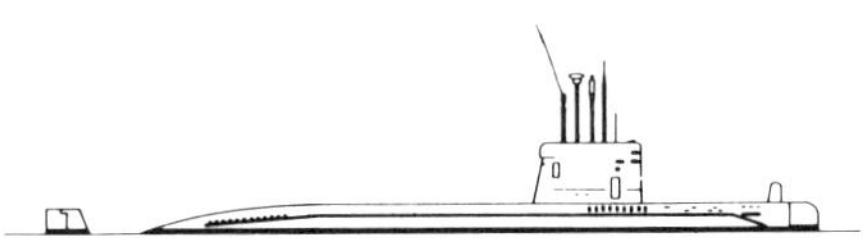

Agosta 1977

AGOSTA class *attack submarines*

Displacement: 1490t surfaced; 1740t submerged
Dimensions: 222ft oa × 22ft × 18ft
67.6m × 6.8m × 5.5m
Machinery: 1 shaft, 2 SEMT-Pielstick 320-16 PA 4 185 diesels, 1 electric motor, 3600bhp/4600hp = 12kts/17.5kts. Range 8500nm at 9kts
Armament: 4–550mm TT (bow; 20 L5/F17 torpedoes)
Sensors: Radar DRUA 33; sonar DUUA 2A/D, DSUV 22, DUUX 2A
Complement: 54

No	Name	Builder	Laid down	Launched	Comp	Fate
S 620	AGOSTA	Arsenal de Cherbourg	1.11.72	19.10.74	28.7.77	Extant 1995
S 621	BEVEZIERS	Arsenal de Cherbourg	17.5.73	14.6.75	29.7.77	Extant 1995
S 622	LA PRAYA	Arsenal de Cherbourg	1974	15.5.76	9.3.78	Extant 1995
S 623	OUESSANT	Arsenal de Cherbourg	1974	23.10.76	27.7.78	Extant 1995

These ocean-going submarines are of double-hull construction, the space between the two hulls being used for ballast, fuel tanks, auxiliary tanks and for the sensing heads of most of the acoustic equipment. The outer shell is cylindrical, graduating to an oval form forward to accommodate the sensors. The profile is more streamlined than that of earlier French submarines, and influenced the design of the SNA of the *Rubis* class. All deck fittings retract into recesses in the outer casing to improve waterflow. The forward diving planes are close to the bow, and are complemented by cruciform control surfaces aft. Maximum diving depth is 300m.

The propulsion system features a 320-cell battery with twice the capacity of the *Daphné* class. The maximum underwater speed of 17.5kts can be sustained for one hour, and can be boosted to 20.5kts for 5 mins to enable the submarine to take evasive action. In addition to the main electric motor there is a small 32hp 'creep' motor for silent operations, capable of driving the submarine at 3.5kts.

The sensor outfit comprises: the DUUA 2 search and attack sonar; a DSUV 22 passive array, comprising a crown of thirty-six hydrophones, to provide a bearing on enemy noise emissions; and a DUUX 2 passive acoustic telemeter to provide range. Since completion the *Agosta*s have been fitted with the DSUV 62A towed array.

The reduction to only four torpedo tubes was made possible by the high hit probability of modern wire-guided torpedoes. The tubes employ a pneumatic ram to discharge the torpedo, enabling firing to take place at any speed or depth. They can accept both 550mm and 533mm torpedoes. The DLA D3 fire control is centralized in one computer bank. *Bévéziers* was the trials boat for the SM39 submarine-launched variant of the Exocet SSM, and the remaining three units of the class have since been retro-fitted with this capability.

All four boats served in the Mediterranean Squadron until the mid-1980s, when they transferred to the Atlantic to replace the elderly *Narval* class. The *Agosta* has proved a popular export design; two units ordered for South Africa in the mid-1970s were subsequently transferred to Pakistan, and four have been built under licence in Spain.

Rubis design

The first French *sous-marin nucléaire de chasse* was authorised in 1964. Reportedly named *Rubis*, it was to have displaced 4000t and was to be armed with four torpedo tubes. The order was cancelled in 1968, the year in which construction was due to begin, and it was stated that SSNs would be included in the new programme.

When the new design materialised it was for a submarine smaller than any SSN in service with the major world navies. In effect it was a nuclear-powered adaptation of the *Agosta* design, the rationale being that a large programme would be economically feasible only if the SSNs were equipped with weapons and sensors already in production.

The comparatively small size of the SNA 72 was made possible by the development of a compact, integrated reactor-exchanger with turbo-electric drive. Natural circulation is sufficient at lower speeds, thereby reducing noise emissions. There are two turbo-alternators, the larger unit generating power for the main electric motor while the smaller of the two provides power for the boat's auxiliary systems. A small emergency motor powered by a diesel generator is capable of driving the submarine for 50nm in the event of reactor failure.

Unlike the *Agosta*, the *Rubis* is of single-hull construction, with high-tensile Marel steel ensuring a diving depth of 300m. The corresponding increase in the diameter of the pressure hull has enabled three decks to be worked in beneath and immediately abaft the fin.

The forward diving planes are carried high on the conning tower. Operation of all control surfaces is by hydraulic-electric remote control; the one-man console has an autopilot facility with manual override. A DSUV 62 towed array has been added since completion. *Saphir* was the first boat to deploy the SM39 Exocet SSM; *Rubis* was subsequently retro-fitted to fire the missile. All four boats are currently being upgraded to *Améthyste* standard to improve their ASW capabilities.

Endurance is rated at forty-five days (sixty days maximum). Unusually for SSNs, these boats have a twin-crew system similar to that of their ballistic missile counterparts. The annual operating cycle allows for 240 days active service. All four units of the class serve in the Mediterranean. In July 1993 *Rubis* was involved in a collision with a tanker and has undergone major repairs.

RUBIS (SNA 72) class *nuclear attack submarines*

Displacement: 2385t surfaced; 2670t submerged
Dimensions: 237ft oa × 25ft × 23ft
72.1m × 7.6m × 6.9m
Machinery: 1-shaft nuclear: 1 CAS 48 48MW integrated reactor-exchanger with turbo-electric drive, 1 electric motor, 9500hp = 25kts (sub). Jeumont-Schneider auxiliary diesel
Armament: 4–550mm TT (bow; 14 L5/F17 torpedoes or SM39 Exocet SSM)
Sensors: Radar DRUA 33; sonar DUUA 2B, DSUV 22, DUUX 5, DSUV 62A towed array
Complement: 66

No	Name	Builder	Laid down	Launched	Comp	Fate
S 601	RUBIS (ex-*Provence*)	Arsenal de Cherbourg	11.12.76	7.7.79	23.2.83	Extant 1995
S 602	SAPHIR (ex-*Bretagne*)	Arsenal de Cherbourg	1.9.79	1.9.81	6.7.84	Extant 1995
S 603	CASABIANCA (ex-*Bourgogne*)	Arsenal de Cherbourg	21.9.81	22.12.84	21.4.87	Extant 1995
S 604	EMERAUDE	Arsenal de Cherbourg	1.3.83	12.4.86	15.9.88	Extant 1995

The *Rubis* class proved to be generally successful, but were significantly noisier than their British and US counterparts, which limited their effectiveness against other submarines. The reduction of noise emissions was therefore the priority for the second group of SNA. *Améthyste* is not only the name of the lead-submarine of the class, but is an acronym for *AME*lioration *T*actique, *HY*drodynamique, *S*ilence, *T*ransmission, *É*coute (= tactical, hydrodynamic, silencing and sonar improvement).

The hull has an improved body-of-revolution configuration with a lengthened bow section constructed of composites. This bow is being retro-fitted to the *Rubis* class, which will be brought up to the same acoustic standard as the *Améthyste*; *Saphir* began her refit in 1989.

The pressure hull is of high-tensile HLES 80 steel. *Améthyste* has the new DMUX 20 multi-function search and attack sonar, improved communications and a new tactical data system. The propulsion system is essentially that of *Rubis*, but with greater attention given to the flexible mounting of machinery.

Construction was delayed by budgetary problems, and the third and fourth boats of the class have been cancelled.

AMETHYSTE class *nuclear attack submarines*

Displacement: 2400t surfaced; 2670t submerged
Dimensions: 241ft oa × 25ft × 21ft
73.6m × 7.6m × 6.4m
Machinery: As *Rubis*
Armament: As *Rubis*
Sensors: Radar DRUA 33; sonar DMUX 20, DUUX 5, DSUV 62C towed array
Complement: 66

No	Name	Builder	Laid down	Launched	Comp	Fate
S 605	AMETHYSTE	Arsenal de Cherbourg	11.10.84	14.5.88	20.3.92	Extant 1995
S 606	PERLE	Arsenal de Cherbourg	27.3.87	22.9.90	7.7.93	Extant 1995
S 607	TURQUOISE	Arsenal de Cherbourg	—	—	—	Cancelled
S 608	DIAMANT	Arsenal de Cherbourg	—	—	—	Cancelled

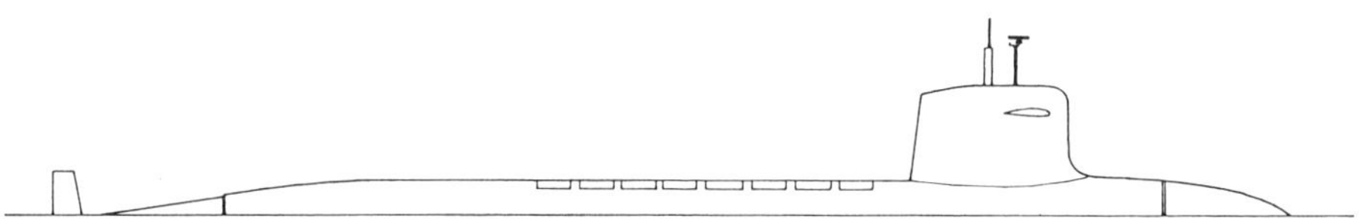

Le Triomphant design

This class will replace the first-generation SNLE of the *Le Redoutable* class. They have a superior hull-form and a reconfigured fin. The adoption of NLES 100 high-tensile steel gives them a 500m diving depth. Considerable attention has been paid to reducing noise emissions. The main propulsion machinery is raft-mounted, and the auxiliary machinery will be suspended and isolated from the hull. There is a single large-diameter, low-revolution propeller. The 150MW K15 reactor is essentially the same model which will power the carrier *Charles de Gaulle*. It uses slightly enriched uranium, and must therefore be refuelled more frequently than US- and British-type PWRs which use highly enriched uranium. This helps to preserve French independence of foreign uranium enrichment facilities.

The sensor outfit includes the new DMUX 80 multi-function sonar, comprising passive bow and flank arrays. The first two units will enter service with the M-4 missile. Later boats will be armed with the advanced M-5 missile currently under development, which will carry 8–12 TN-75 MIRVs. *Le Triomphant* began trials in 1994. Construction of the other boats has been delayed, and there will now be only four instead of the six initially projected. A special construction facility had to be built at Cherbourg to accommodate the large hull.

LE TRIOMPHANT class *ballistic missile submarines*

Displacement: 12,700t surfaced; 14,200t submerged
Dimensions: 453ft oa × 41ft × 41ft
138.0m × 12.5m × 12.5m
Machinery: 1-shaft neclear; 1 Type K15 150MW PWR, turbo-electric drive, 1 electric motor, 41,500shp = 20kts/25kts. Auxiliary diesel
Armament: 16 M4/M5 SLBM, 4–533mm TT (12 torpedoes or SM39 Exocet SSM)
Sensors: Radar DRUA 33; sonar DMUX 80, DUUX 5, DSUV 62 towed array
Complement: 111

No	Name	Builder	Laid down	Launched	Comp	Fate
S 616	LE TRIOMPHANT	Arsenal de Cherbourg	9.6.89	13.7.93	—	Building
S 617	LE TEMERAIRE	Arsenal de Cherbourg	—	—	—	Building
S 618	LE VIGILANT	Arsenal de Cherbourg	—	—	—	Ordered
—	Unnamed	Arsenal de Cherbourg	—	—	—	Projected
—	Unnamed	Arsenal de Cherbourg	—	—	—	Cancelled
—	Unnamed	Arsenal de Cherbourg	—	—	—	Cancelled

AMPHIBIOUS WARFARE VESSELS

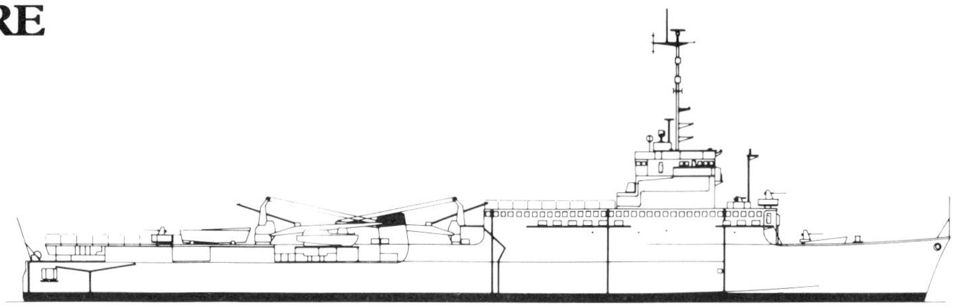

Ouragan 1975

Ex-US LST type *tank landing ships*

Displacement: 1625t standard; 4080t full load
Dimensions: 316ft wl, 328ft oa × 50ft × 10ft
96.3m, 100.0m × 15.2m × 3.0m
Machinery: 2 shafts, 2 diesels, 1800bhp = 11kts. Range 24,000nm at 9kts
Armament: 8–40mm (2×2, 4×1), 12–20mm
Complement: 111 (+163 troops)

CLASS:
Laïta (L 9001, ex-*LST 177*), *Orne* (L 9002, ex-*LST 508*), *Vire* (L 9003, ex-*LST 347*), *Rance* (L 9004, ex-*LST 223*), *Odet* (L 9005, ex-*LST 815*), *Chéliff* (L 9006, ex-*LST 874*), *Adour* (L 9007, ex-*LST 860*), *Golo* (L 9008, ex-*LST 973*).

A number of ex-American LSTs were transferred to the Marine Nationale 1949–51. *Adour* was badly damaged through sinking in the early 1950s, but was raised and subsequently became an accommodation ship. Five units were disposed of 1957–62; *Laïta* became a depot ship in 1962, and *Odet* and *Chéliff* continued in service until 1969–70.

FOUDRE type (ex-US) *dock landing ship*

Displacement: 4500t standard; 7930t full load
Dimensions: 454ft pp, 458ft oa × 72ft × 16ft
138.4m, 139.5m × 22.0m × 4.8m
Machinery: 2-shaft geared steam turbines, 2 boilers, 7400shp = 17kts. Range 7400nm at 15kts
Armament: 1–105mm, 2–120mm mortars, 4–40mm (2×2), 4–20mm (4×1)
Complement: 212

Built by Newport News as *LSD 12* (launched 29 December 1943) and serving in the Royal Navy as *Oceanway*, this vessel was returned to the USA in 1947, and was subsequently transferred to France under MDAP in 1952. As first completed she carried a single 3in (76mm) DP gun, 4–2pdr (1×4) and 16–20mm (16×1), but she was completely re-armed by the French. The docking well could accommodate two LCT(3) landing craft. She gave the Marine Nationale much needed experience in amphibious operations and influenced the design of the *Ouragan* class TCDs. She was stricken in 1969.

ISSOLE *landing craft*

Displacement: 600t full load
Dimensions: 161ft oa × 23ft × 7ft
48.7m × 7.0m × 2.1m
Machinery: 2 shafts, 2 diesels, 100bhp = 12kts

Issole (L 9097) was built at Toulon 1957–58. The design was an adaptation of a mercantile coaster, with bow doors and a ramp. In the early 1970s she became a coastal transport and was given the pennant number A 734. She was stricken in 1979.

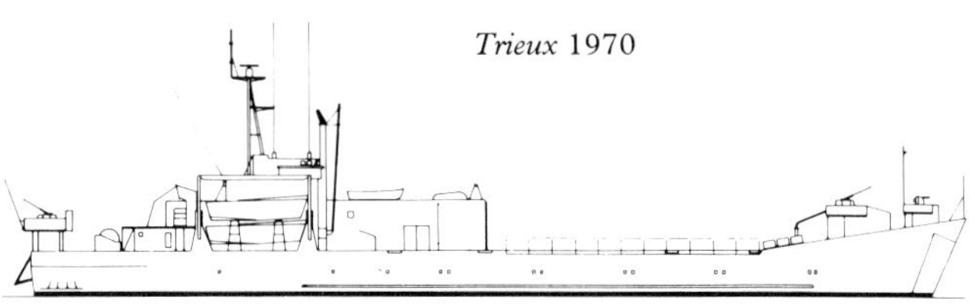

Trieux 1970

TRIEUX class *tank landing ships*

Displacement: 1750t standard; 4225t full load
Dimensions: 327ft pp, 335ft oa × 51ft × 11ft
96.6m, 102.1m × 15.5m × 3.2m
Machinery: 2 shafts, 2 SEMT-Pielstick diesels, 2000bhp = 11kts. Range 18,500nm at 10kts
Armament: 1–120mm mortar, 3–40mm (3×1); *Trieux*, *Blavet* 2–40mm (2×1), 2–20mm (2×1)
Complement: 85 (+170 troops)

No	Name	Builder	Launched	Fate
L 9007	TRIEUX	A C de Bretagne	6.12.58	Stricken 1989, BU 1992
L 9003	ARGENS	A C de Bretagne	7.4.59	Stricken 1985
L 9009	BLAVET	A C de Bretagne	15.1.59	Stricken 1985
L 9008	DIVES	A C de la Seine-Maritime	29.6.60	Stricken 1986
L 9004	BIDASSOA	A C de la Seine-Maritime	30.12.60	Stricken 1986

Derived from the US Navy's LST, of which several were operated by the Marine Nationale in the postwar period, the *Trieux* class were designated BDC (*Bâtiments de Débarquement de Chars*). They carried four LCVPs beneath davits on each side of the superstructure. There was fixed accommodation for a maximum of 329 troops, and a further 320 could be carried over short distances. In the logistics role they could carry 1000t of freight, 400t of fresh water and 270t of diesel fuel. MacGregor-type loading hatches were fitted.

In the late 1960s *Blavet* and *Trieux* were fitted with a hangar for two Alouette helicopters immediately forward of the bridge. They served on distant colonial stations as logistics ships, *Blavet* being based in French Polynesia in the Pacific, being superseded in that role by the BATRALs of the *Champlain* class. *Bidassoa* was allocated to the amphibious training centre at Lorient.

OURAGAN class *dock landing ships*

Displacement: 5800t standard; 8500t full load
Dimensions: 474ft pp, 489ft oa × 71ft × 18ft min
144.5m, 149.0m × 21.5m × 5.4m
Machinery 2 shafts, 2 SEMT-Pielstick 12PC 2V diesels, 8640bhp = 17kts. Range 8000nm at 15kts
Armament: 2–120mm mortars, 4–40mm (4×1) *Ouragan* only
Sensors: Sonar SQS-17 (*Ouragan*)
Complement: 211 (+470 troops)

No	Name	Builder	Launched	Fate
L 9021	OURAGAN	Arsenal de Brest	9.11.63	Extant 1995
L 9022	ORAGE	Arsenal de Brest	22.4.67	Extant 1995

Designated TCD (*Transports de Ćhalands de Débarquement*) these two vessels were designed with a dual function in mind. They can lift a full commando and its support troops and put them ashore using helicopters; and they can put ashore personnel, tanks, vehicles and supplies using landing craft. While the design was clearly influenced by the American LSD concept, an increase in flexibility was obtained, particularly with regard to the operation of helicopters and the accommodation of vehicles, by the use of removable sections of deck both externally and internally.

The docking well is 120m long, and floods to a depth of 3m. The stern gate measures 14m × 5.5m. Two EDICs, each with a capacity of eleven light tanks, or eighteen LCM(6)s can be accommodated. Movement of the water control sluices and valves is automatic, using pumps controlled from a central command post. A temporary deck 90m long in fifteen sections can be laid out to increase the stowage space for vehicles or cargo, but the well is then reduced to half its original size. In the logistics role, 1500t of equipment can be carried, the cargo being handled by two 35t cranes mounted above the docking well. Alternative loads include eighteen Super Frelons, eighty Alouettes, 120 AMX-13 tanks, eighty-four DUKWs, 340 jeeps or twelve 50t barges.

Instead of the conventional LSD superstructure there is a small island to starboard which accommodates the navigation bridge and a combined centre for directing amphibious and helicopter operations. The main helicopter deck was designed to operate either three Super Frelon troop-carrying helicopters or ten Alouettes. A further Super Frelon or three Alouettes can be operated from a removable deck in six sections which covers the after part of the well for 36m. A total of 349 troops can be transported (470 over short distances), and three LCVPs are carried topsides.

The TCDs were intended to serve as repair and maintenance ships on distant deployments. They can dock a 400t ship and are equipped with hull and machinery repair shops, and electrical and ordnance workshops.

Ouragan was laid down in June 1962 and completed in January 1965; her sister was laid down in June 1966 and completed in March 1968. On completion *Orage* was assigned to the Pacific Experimental Centre. In this rôle she was unarmed, and the surgical compartments and inboard removable decks were replaced by a modular structure within the well deck which enabled her to serve as a floating HQ. Great attention was paid to the provision of decontamination areas. She was due to decommission in 1993, but her service life will now be extended due to delays with the *Foudre* class. Under the 1992 reorganisation both ships have been allocated to the FAN, with Toulon as their base port.

EDIC/EDA type *tank landing craft*

Displacement: 635/736t full load
Dimensions: 194ft × 39ft × 4ft 6in
59.0m × 12.0m × 1.3m
Machinery: 2 shafts, 2 MGO diesels, 1000bhp = 8kts. Range 1800nm at 8kts
Armament: 2–20mm (2×1), 2–12.7mm MG in some
Complement: 17

No	Fate	No	Fate
L 9091	Stricken 1988	L 9070	Extant 1995
L 9093	Stricken 1988	L 9072	Stricken 1993
L 9094	Stricken 1991	L 9073	Stricken 1986
L 9096	Stricken 1988	L 9074	Extant 1995
L 9082	To Madagascar 1985	L 9051	Extant 1995
L 9083	Stricken 1986	L 9052	Extant 1995
L 9084	Extant 1995		

The EDIC (*Engins de Débarquement Infanterie Chars*) were built in three groups. *L 9091–96* were launched in 1958, *L 9070–74* in 1967–69, and *L 9051–52* in 1987. They are conventional landing craft with a capacity of eleven trucks or five LVTs. The last two units (EDIC III) are of a modified design, and have SACM M60 Type V12 ASM diesels, which produce 1400bhp for a maximum speed of 12kts.

Between the two groups four EDA (*Engins de Débarquement Ateliers*) were constructed. Launched in 1964 and numbered *L 9081–84*, these were specially fitted as repair and logistics craft to serve on overseas stations. *L 9084* is classified as an 'auxiliary electrical stores ship' (BAME).

L 9051 and *L 9052* are based in the Pacific.

CDIC type *tank landing craft*

Displacement: 380t light; 710t full load
Dimensions: 195ft × 39ft × 6ft
59.4m × 11.9m × 1.8m
Machinery: 2 SACM UD30 V12 diesels, 1080bhp = 10.5kts. Range 1000nm at 10kts
Armament: 2–20mm (2×1), 2–12.7mm MG
Complement: 12

Designed to operate from the TCDs of the *Foudre* class, the CDIC (*Chaland de Débarquement Infanterie Chars*) is essentially an improved EDIC. The wheelhouse can be lowered to facilitate docking. Like the EDIC III, the CDICs were built at SFCN, Villeneuve-la-Garenne; *L 9061* was completed in October 1988 and *L 9062* in March 1989. Further projected orders have been postponed because of delays in the TCD programme.

Champlain 6.11.92 L & L van Ginderen

CHAMPLAIN class *landing ships*

Displacement: 750t/770t standard; 1330t/1386t full load
Dimensions: 223ft pp, 262ft oa × 43ft × 10ft
68.0m, 80.0m × 13m × 3m
Machinery: 2 shafts, 2 SACM V12 diesels, 3600bhp = 16kts. Range 3500nm at 13kts
Armament: 2–40mm/20mm (2×1), 2–81mm mortars (2×1)
Sensors: Radar DRBN 32
Complement: 44 (+138/180 troops)

No	Name	Builder	Launched	Fate
L 9030	CHAMPLAIN	Arsenal de Brest	17.11.73	Extant 1995
L 9031	FRANCIS GARNIER	Arsenal de Brest	17.11.73	Extant 1995
L 9032	DUMONT D'URVILLE	At Français de l'Ouest	27.11.81	Extant 1995
L 9033	JACQUES CARTIER	At Français de l'Ouest	28.4.82	Extant 1995
L 9034	LA GRANDIERE	At Français de l'Ouest	11.12.85	Extant 1995

Designated BATRAL (*BÂtiment de TRansport Léger*), these vessels are designed for overseas service. They carry a landing company (*guépard*), comprising five officers, 133 men, six *Léopard* armoured personnel carriers and six lorries. One LCVP and one LCP are carried on the upper deck. A maximum load of 330t can be carried, and cargo can be transferred ashore either by the 10t hydraulic crane or via the bow ramp, which has a maximum capacity of 40t. The three ships of the second series have improved accommodation for a total of 180 troops, and a higher bridge structure. The last unit, *La Grandière*, was built for export and subsequently acquired by the Marine Nationale; she has a longer flight deck than her sisters. A similar unit was built for Gabon, while three earlier vessels serve with the Moroccan Navy.

La Grandière currently serves with the FAN at Toulon; *Champlain* serves in the Indian Ocean, *Francis Garnier* in the French West Indies, and the remaining two ships are based in the Pacific.

Bouganville 1988

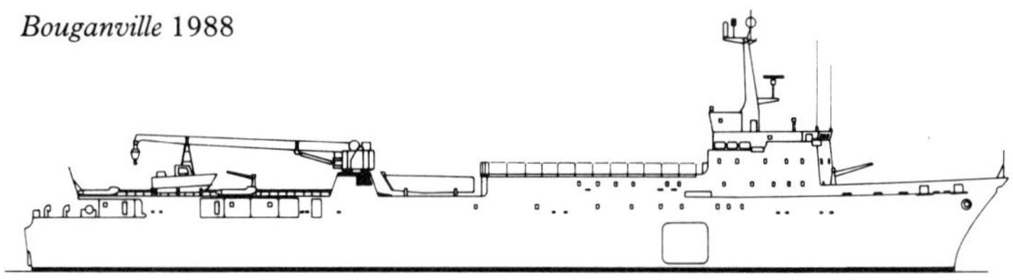

BOUGAINVILLE *dock landing ship*

Displacement: 4200t standard; 4870t full load
Dimensions: 344ft pp, 372ft oa × 56ft × 14ft min
105.0m, 113.5m × 17.0m × 4.2m
Machinery: 2 shafts, 2 SACM MGO 195 V12 RVR diesels, 4800bhp = 14.6kts. Range 6000nm at 12kts
Armament: 2 Super Puma helicopters
Sensors: Radar DRBN 32
Complement: 53 (+ 50 passengers/500 troops)

No	Name	Builder	Launched	Fate
L 9077	BOUGAINVILLE	Dubigeon, Nantes	3.10.86	Extant 1995

Designated BTMS (*Bâtiment de Transport Moyen et de Soutien*), *Bougainville* was ordered in 1984 for the Directorate of Nuclear Experimentation, to serve as a logistics vessel between Papeete and the Mururoa Test Centre.

She is essentially a scaled-down version of the TCD 90 (see below), with a docking well 78m × 10.2m able to accommodate three CTM-type landing craft or forty 20ft containers. The stern door measures 10m × 6m. A removable deck in three 6m sections can be fitted above the docking well to increase vehicle/cargo capacity. The dock area is served by a fixed 37t crane to starboard and by a travelling crane with a capacity of 25t. Vehicles can also be embarked via a 6m side ramp to starboard. Vessels up to the size of the 400t patrol vessels of the *L'Audacieuse* class can be accommodated in the docking well for maintenance and repair.

The flight deck is amidships and measures 26m × 17m. Particular attention was given to providing a stable platform in adverse sea conditions. Two Super Puma helicopters can be operated, but there are no hangar facilities.

The priority given to logistics and support operations is reflected in high endurance (forty-five days maximum) and a relatively low speed. Twin bow thrusters provide manoeuvrability in narrow anchorages, and steadiness when embarking and disembarking landing craft. Fifty civilians can be accommodated in double berths in addition to the crew, and 500 troops can be transported in an emergency.

Shortly after the ship was launched by Dubigeon the shipyard was forced to close down, and *Bougainville* was completed by Chantiers de l'Atlantique.

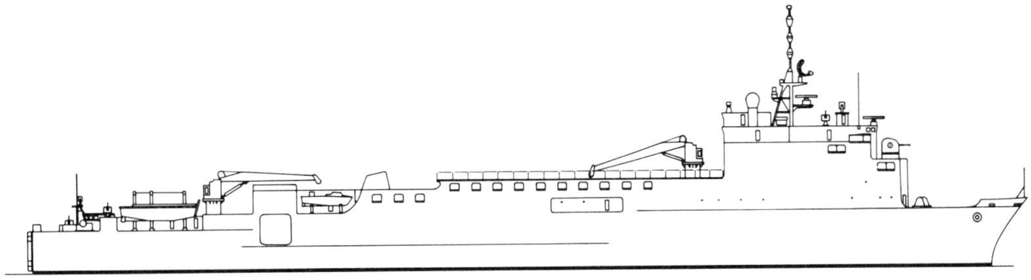

Foudre design

FOUDRE (TCD 90) class *dock landing ships*

Displacement: 9300t standard; 11,880t full load
Dimensions: 525ft pp, 551ft oa × 77ft × 17ft min
160.0m, 168.0m × 23.5m × 5.2m
Machinery: 2 shafts, 2 SEMT-Pielstick 16 PC 2.5-V400 diesels, 21,600bhp = 21kts. Range 11,000nm at 15kts
Armament: 2 Simbad SAM (2×2), 1–40mm, 2–20mm (2×1), 4 Super Puma Helicopters
Sensors: DRBV 21A (not in L 9011)
Complement: 210 (+ 350 passengers/1200 troops)

No	Name	Builder	Launched	Fate
L 9011	FOUDRE	Arsenal de Brest	19.11.88	Extant 1995
L 9012	SIROCO	Arsenal de Brest		Ordered
L 9013	unnamed			Projected
L 9014	unnamed			Projected

The TCDs of the *Foudre* class were designed as replacements for the *Ouragan* class. Three were to have been ordered under the 1984–88 naval programme, but budgetary problems resulted in construction of the second and third ships being deferred. The second ship was authorised in 1994 for completion in 1998.

These ships are larger than *Ouragan* and *Orage*, and are designed to embark a 350-man mechanised regiment of the *Force d'Action Rapide*, together with 1080t of combat vehicles and cargo. Like their predecessors, they will have a secondary rôle as mobile logistics and support ships. Improvements over the *Ouragan* class include: a higher transit speed; a hangar for four Super Puma helicopters; an elevator with a capacity of 52t connecting the docking well with the vehicle/cargo decks; a side ramp for rapid embarkation of vehicles from the dockside; permanent accommodation for 470 troops (*Ouragan* could embark this number only in an emergency); and advanced hospital facilities.

The docking well measures 122m × 13.5m and can accommodate a maximum of ten 24m CTM landing craft, or two of the new CDIC tank landing craft (see below). Two LCVP are suspended from davits abeam the after part of the docking well. The docking well can accept any of the standard French overseas patrol and support craft, including the 400t Super Patra, the BSR (*Bâtiment de Soutien Régional*) of the *Chamois* and RR 2000 types, and coastal tugs. It is served by a fixed 37t crane to starboard and transfer mechanisms to lift machinery and heavy equipment. The workshops can handle a wide variety of repair and maintenance tasks.

Two Super Puma helicopters can operate simultaneously from the 50m flight deck amidships, and the fitting of a removable deck in five 6m sections above the docking well provides either an additional helicopter spot or a 40 per cent increase in vehicle parking space. Abeam the after end of the hangar, to port, a second 37t crane serves the flight deck area. The hospital facilities are located directly beneath the flight deck to facilitate the transfer of casualties from the helicopters.

The main propulsion machinery is identical to that of the fleet oilers of the *Durance* class. This facilitates maintenance and gives the ships a speed advantage of 4kts over the *Ouragan* class. Twin bow thrusters are fitted to keep the ships steady when embarking and disembarking landing craft. The computer-based amphibious command centre, which is housed in the forward superstructure, is served by an advanced communications suite which includes the Syracuse satcom system.

The second and third ships will have a larger flight deck, a more extensive AA armament comprising single Breda-Mauser 30mm guns and Simbad SAM launchers, and the DRBV 21 radar for controlling aircraft over land.

Foudre 21.7.93 — L & L van Ginderen

SMALL SURFACE COMBATANTS

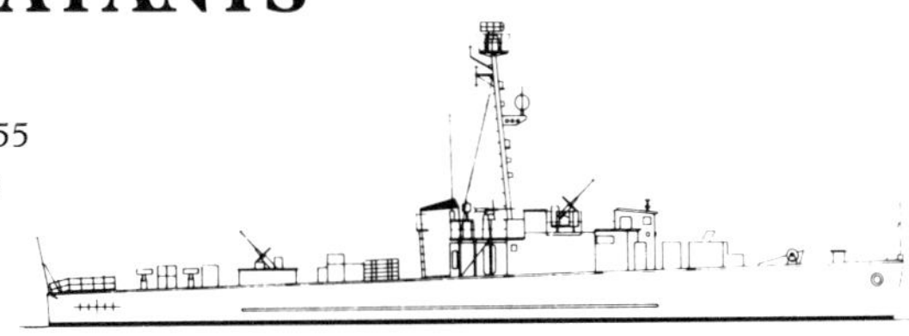

Le Fougueux 1955
NB **1/750 scale**

LE FOUGUEUX class *submarine-chasers*

Displacement: 325t standard; 400t full load
Dimensions: 174ft oa × 24ft × 10ft
53.1m × 7.3m × 3.1m
Machinery: 2 shafts, 4 SEMT-Pielstick fast diesels, 3240bhp = 19kts. Range 3000nm at 12kts
Armament: 2–40mm (2×1), 2–20mm (2×1), 1–120mm ASW mortar, 2 DCT, 1 DC rack (*Le Fougueux, L'Opiniâtre, L'Agile* 1 Hedgehog, 4 DCT, 2 DC racks)
Sensors: Sonar QCU-2
Complement: 62

No	Name	Builder	Launched	Fate
P 630	L'INTREPIDE	CMN, Cherbourg	12.12.58	Stricken 1977
P 635	L'ARDENT	A C Augustin-Normand	17.7.58	Stricken 1980
P 637	L'ETOURDI	F C de la Méditerranée	5.2.58	Stricken 1978
P 638	L'EFFRONTE	A C Augustin-Normand	27.1.59	Stricken 1978
P 639	LE FRONDEUR	CMN, Cherbourg	26.2.59	Stricken 1977
P 640	LE FRINGANT	F C de la Méditerranée	6.2.59	Stricken 1982
P 641	LE FOUGUEUX	A C Dubigeon	31.5.54	Stricken 1975
P 642	L'OPINIATRE	F C de la Méditerranée	4.5.54	Stricken 1975
P 643	L'AGILE	A C Provence	26.6.54	BU 1978
P 644	L'ADROIT	Arsenal de Lorient	6.9.58	Stricken 1979
P 645	L'ALERTE	Arsenal de Lorient	5.10.57	Stricken 1979
P 646	L'ATTENTIF	Arsenal de Lorient	10.7.58	Stricken 1978
P 647	L'ENJOUE	Arsenal de Lorient	5.10.57	Stricken 1978
P 648	LE HARDI	CMN, Cherbourg	17.9.58	Stricken 1977

The first three ships were funded under MDAP. The other eleven vessels were purchased by France: five were authorised in 1955 and six in 1956. Other similar units were built for Yugoslavia, Tunisia, Portugal and Ethiopia.

After the first three ships changes were made in the shape of the bridge structure and in the armament. *L'Intrépide* had a single torpedo tube fitted above the stern.

LA COMBATTANTE *patrol boat*

Displacement: 180t standard; 202t full load
Dimensions: 148ft × 24ft × 7ft
45.0m × 7.4m × 2.5m
Machinery: 2 shafts, 2 SEMT-Pielstick 8 PA 4 200 VG DS diesels, 3600bhp = 23kts. Range 2000nm at 12kts
Armament: 4 SS.11 SSM (1×4) 1–40mm, 2–12.7mm MG
Complement: 25

Authorised in 1960, *La Combattante* (P 730) was to have been the prototype of a new class of *patrouilleurs garde-côtes*. She was built by CMN, Cherbourg and launched 20 June 1963. Construction was of laminated wood and plastics, giving a low magnetic signature. She conducted trials of the Exocet missile and served as the basis for a whole series of FPBs built for export.

A second 40mm mounting was added aft in the 1970s, but was replaced by a launcher for 14 flares before *La Combattante* left for the Indian Ocean in 1975. She could carry a group of 80 men and their equipment for a short passage.

In 1978–79 the original diesels were replaced by the new PA 4 VG DS engine. The latter had a very high power/mass ratio, and employed two-stage supercharging and a variable-geometry precombustion chamber. Power was increased to 4000bhp. *La Combattante* was initially deployed to stations in the Indian Ocean and the Pacific, but was subsequently based at Toulon and manned by the Gendarmerie Nationale. She was stricken in 1994.

La Combattante c 1974 Navarret Collection

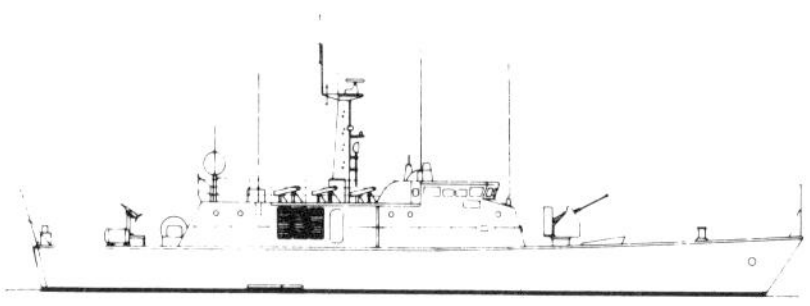

Pertuisane 1980

TRIDENT (PATRA) class *patrol boats*

Displacement: 115t standard; 148t full load
Dimensions: 126ft pp, 134ft oa × 19ft × 5ft
38.5m, 40.7m × 5.9m × 1.6m
Machinery: 2 shafts, 2 AGO 195 V12 CZSHR diesels, 5000bhp = 28kts. Range 1500nm at 15kts
Armament: 6 SS.12 SSM (6×1), 1–40mm, 1–12.7mm MG
Complement: 19

No	Name	Builder	Launched	Fate
P 670	TRIDENT	Auroux, Arcachon	31.5.76	Extant 1995
P 671	GLAIVE	Auroux, Arcachon	27.8.76	Extant 1995
P 672	EPEE	CMN, Cherbourg	31.3.76	Extant 1995
P 673	PERTUISANE	CMN, Cherbourg	2.6.76	Extant 1995

Intended for the surveillance and control of territorial waters, these four vessels were to have been the first of a series of thirty *patrouilleurs* projected under the *plan bleu*, of which sixteen were for overseas service. In the event they proved too small for the latter task, and after four boats had been completed the programme was abandoned in favour of the larger Super PATRA.

As completed there were six single launchers for SS.12 SSMs atop the superstructure. These were later removed and replaced by a second 12.7mm MG. All four boats are based in Metropolitan France, and are manned by the Gendarmerie Nationale; *Trident* and *Epée* are on the Atlantic Coast, and *Glaive* and *Pertuisane* at Toulon.

L'Audacieuse 8.1.93 L & L van Ginderen

STERNE *patrol boat*

Displacement: 270t standard; 380t full load
Dimensions: 161ft oa × 25ft × 9ft
49.0m × 7.5m × 2.8m
Machinery: 2 shafts, 2 SACM V12 CZSHR diesels, 4200bhp = 20kts. Range 4900nm at 12kts
Armament: 2–12.7mm MG
Sensors: Radar DRBN 32
Complement: 18

Sterne (P 680) was built to mercantile standards for fishery protection duties within the 200-mile economic zone. Construction funds were provided by the ship owners, but the ship is manned by naval personnel. She was launched by A & C de la Perrière, Lorient, on 31 October 1979. She is equipped with a large infirmary, and features an active tank stabilsation system. An electro-hydraulic drive system coupled to the starboard shaft provides an economical cruise speed of 6.5kts. A second unit was to have been ordered in 1987, but was cancelled.

La Fougueuse 1988

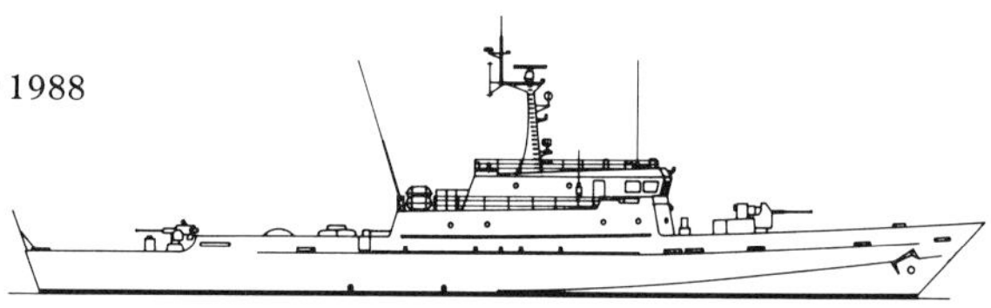

L'AUDACIEUSE (Super PATRA) class *patrol boats*

Displacement: 406t standard; 454t full load
Dimensions: 179ft oa × 26ft × 8ft
54.5m × 8.0m × 2.5m
Machinery: 2 shafts, 2 Alsthom/SEMT-Pielstick 16 PA 4 V200 VGDS diesels, 8000bhp = 24.5kts. Range 4200nm at 15kts
Armament: 1–40mm, 1–20mm, 2–12.7mm MG
Sensors: Radar DRBN 32
Complement: 24 + 20 passengers

No	Name	Builder	Launched	Fate
P 682	L'AUDACIEUSE	CMN, Cherbourg	21.3.84	Extant 1995
P 683	LA BOUDEUSE	CMN, Cherbourg	21.5.84	Extant 1995
P 684	LA CAPRICIEUSE	CMN, Cherbourg	31.10.84	Extant 1995
P 685	LA FOUGUEUSE	CMN, Cherbourg	17.12.84	Extant 1995
P 686	LA GLORIEUSE	CMN, Cherbourg	25.1.85	Extant 1995
P 687	LA GRACIEUSE	CMN, Cherbourg	26.3.85	Extant 1995
P 688	LA MOQUEUSE	CMN, Cherbourg	8.4.86	Extant 1995
P 689	LA RAILLEUSE	CMN, Cherbourg	2.9.86	Extant 1995
P 690	LA RIEUSE	CMN, Cherbourg	17.10.86	Extant 1995
P 691	LA TAPAGEUSE	CMN, Cherbourg	16.2.87	Extant 1995

An improved class of *patrouilleurs* designed for overseas service. They are altogether larger and more seaworthy than their predecessors of the *Trident* class. Four units were originally to have been part of the *Force de Service Public* under the designation SP 400, and were to have been fitted with fire-fighting, search-and-rescue, and anti-pollution equipment. This project was abandoned, and all units are now identical. The programme was delayed by propulsion problems experienced with the first ship, *L'Audacieuse*. In order to minimise infrared emissions the exhaust gases from the diesels were discharged underwater, and the escape valves were found to admit water at high temperatures. In 1990 *L'Audacieuse* was given conventional twin funnels to resolve the problem, and this modification has now been extended to all other units of the class.

Two large holds, served by a hydraulic crane, are provided to enable these vessels to transport equipment and stores between the overseas islands, and there is accommodation for twenty passengers. Provision was made for the later addition of two Exocet missiles and a fire control radar, and *La Glorieuse* carried out trials of a Thomson-CSF VDS-12 sonar in 1985. P 683 and P 690 currently serve in the Indian Ocean, P 684 and P 685 are based in French Guiana, and five units (P 686–689 and P 691) are in the Pacific; P 682 is at Cherbourg.

GREBE class *patrol boats*

Displacement: 300t standard; 410t full load
Dimensions: 499ft oa × 32ft × 7ft
52m × 9.8m × 2.3m
Machinery: 2 shafts, 2 SACM-RVR UD33 V12 diesels, 4800bhp = 24kts, 2 90kW auxiliary electric motors. Range 4500nm at 12kts
Armament: 2–12.7mm MG (2×1)
Sensors: Radar Decca 2690
Complement: 19

No	Name	Builder	Launched	Fate
P 679	GREBE	SFCN, Villeneuve La Garenne	16.11.89	Extant 1995
P –	–	CMN, Cherbourg	–	Building
P –	–	CMN, Cherbourg	–	Building
P –	–	CMN, Cherbourg	–	To Mauretania

Ordered to replace the former minesweeper *Mercure* in the fishery protection rôle, *Grèbe* was based on the commercial *Espadon 50* design, employing a deep-vee hull form capable of operating at speed in Sea State 4. She has a stern embarkation ramp for an EDL 700 7-metre rigid inflatable, and accommodation for up to 22 passengers. A 400m³/hr fire-fighting monitor is installed, and there are two $7m^3$ capacity spill recovery holding tanks.

Three additional units were ordered in 1992, but construction was delayed by the closure of the original contract yard. The fourth unit is being completed by LNI, Lorient, for Mauritania.

MINE WARFARE VESSELS

(For illustrations see under United States)

US AGILE class *ocean minesweepers*

Particulars: As US *Agile* class

CLASS:
Bellingham Shipyards, Washington – *Alençon* (M612), *Autun* (M 622), *Berlaimont* (M 620), *Berneval* (M 613), *Bir Hacheim* (M 614), *Cantho* (M 615), *Dompaire* (M 616), *Garigliano* (M 617), *Mytho* (M 618), *Origny* (M 621), *Vinh Long* (M 619).
Tacoma Boatbuilding, Washington – *Baccarat* (M 623).
Peterson Builders, Wisconsin – *Narvik* (M 611), *Colmar* (M 624), *Ouistreham* (M 610).

In 1952 the USA agreed to transfer fifteen new ocean minesweepers of the *Agile* class to the *Marine Nationale* under MDAP. Launched 1953–54 and transferred 1954–57, they had wooden hulls and diesels of non-magnetic stainless steel alloy. *Autun, Baccarat, Berlaimont, Colmar, Narvik, Origny* and *Ouistreham* had taller funnels than the other ships.

Origny was modified 1961–62 as an ocean research vessel and was renumbered A 640. *Mytho* became a tender in 1965. *Narvik* (A 769) was converted in 1976 for trials of the AP4 sweep and a lenticular sonar. *Autun* and *Colmar* were stricken 1978 and broken up in 1985–86, and *Bir Hacheim* was returned to the USA in 1970 and later transferred to Uruguay.

Cantho, Dompaire, Garigliano, Mytho and *Vinh Long* were converted to minehunters 1975–79. They received a new block bridge structure, the PAP-104 minehunting vehicle and a bow thruster. The remaining five units underwent a similar conversion in 1978–81, but were fitted only with the DUBM 41 side-scan sonar.

Dompaire and *Mytho* were stricken in 1987, *Vinh Long* and *Garigliano* in 1988 and *Cantho* in 1989; they were replaced by the purpose-built minehunters of the *Eridan* class.

Berneval was stricken 1987, *Berlaimont* in 1989, *Baccarat* and *Alençon* in 1993 and *Ouistreham* in 1994. The minesweeping rôle of these vessels was to have been taken over by the new BAMO catamaran.

Eglantine c 1960 Navarret Collection

US ADJUTANT class *coastal minesweepers*

Particulars: As US *Adjutant* class

Class:
Harbor Boat Building, California – *Aconit* (M 640), *Azalée* (M 665), *Camélia* (M 671).
Frank L Sample, Maine – *Acacia* (M 638), *Acanthe* (M 639), *Ajonc* (M 667).
Stephen Bros, California – *Bégonia* (M 669), *Coquelicot* (M 673), *Giroflée* (M 677), *Laurier* (M 681), *Magnolia* (M 685), *Glaïeul* (M 678), *Pavot* (M 631), *Pivoine* (M 633), *Réséda* (M 635).
Tacoma Boatbuilding, Washington – *Lobélia* (M 684), *Muguet* (M 688), *Eglantine* (M 675), *Glycine* (M 679), *Liseron* (M 683), *Mimosa* (M 687), *Chrysanthème* (M 672), *Gardénia* (M 676), *Jacinthe* (M 680).
National Steel & SB, San Diego – *Lilas* (M 682), *Marguerite* (M 686), *Cyclamen* (M 674).
Pacific Boatbuilding, Tacoma, Washington – *Bleuet* (M 670).
South Coast Co, Newport Beach, California – *Pervenche* (M 632), *Renoncule* (M 634).

In 1952 the USA agreed to allocate thirty-six coastal minesweepers of the *Adjutant* class to the *Marine Nationale* under MDAP. Thirty were delivered 1953–54, but of the remaining six, four were returned to the USA after their delivery to Indo-China (two were subsequently transferred to Japan) and two were re-allocated to Spain.

They were constructed of wood and non-magnetic alloys. *Eglantine, Lobélia* and *Pivoine* became tenders in 1965. In 1968 *Jacinthe* was modified as a minelayer and renumbered A 680. *Marguerite, Pavot* and *Reconcule* were returned to the USA 1969–70 and subsequently transferred to Uruguay and Turkey; *Bégonia* and *Glaïeul* were returned in 1974. *Coquelicot* and *Aconit* (renamed *Marjolaine* 1967) were transferred to Tunisia in 1973 and 1977 respectively. In 1974 *Ajonc* became a training vessel for divers at St Mandrier and was renumbered A 701. Three further vessels, *Gardénia* (A 711), *Liseron* (A 723), and *Magnolia* (A 770) became base ships for mine demolition teams and were stricken in 1986–88. *Giroflée, Bleuet* and *Chrysanthème* were stricken in 1977, the latter two being cannibalised for spares. *Pervenche, Camélia* and *Laurier* followed in 1980, *Acanthe, Lilas, Muguet, Mimosa* and *Lobélia* in 1981, and *Jacinthe* in 1982. *Ajonc* was stricken in 1988; *Réséda, Acacia* and *Mimosa* in 1982; *Pivoine, Cyclamen, Glycine* and *Lobélia* in 1983. *Eglantine* and *Azalée* were broken up in 1985.

Ex-Canadian 'BAY' class *coastal minesweepers*

Particulars: As Canadian 'Bay' class

Class:
St John DD – *La Dunkerquoise* (M 726, ex-*Fundy*), *La Lorientaise* (M 731, ex-*Miramichi*).
Port Arthur SB – *La Dieppoise* (M 730, ex-*Chaleur*).
Canadian Vickers – *La Paimpolaise* (M 729, ex-*Thunder*).
Marine Industries – *La Bayonnaise* (M 728, ex-*Chignecto*).
Davie SB – *La Malouine* (M 727, ex-*Cowichan*).

These six coastal minesweepers of the Canadian 'Bay' class were launched 1951–53 and transferred to France in 1954 under MDAP. They had wooden hulls and aluminium frames and decks. In the early 1960s they were modified for service as colonial patrol boats. Sweep gear was removed, air conditioning installed, and they were renumbered P 653, P 652, P 655, P 657, P 654 and P 651 respectively. *La Bayonnaise* and *La Malouine* were stricken in 1976 and 1977 respectively. *La Lorientaise, La Dunkerquoise* (both stricken in 1986), *La Dieppoise* and *La Paimpolaise* (both stricken 1987) were stationed in the Pacific until the mid-1980s, when they were replaced by purpose-built patrol craft of the 'Super Patra' class.

Sirius (left) and *La Bayonnais*
Navarret Collection

British 'HAM' class *inshore minesweepers*

Particulars: As British 'Ham' class

Class:
White, IOW – *Tulipe* (M 771), *Oeillet* (M 774), *Pâquerette* (M 775)
Blackmore, Bideford – *Hortensia* (M 783), *Jasmin* (M 776)
Harris, Appledore – *Violette* (M 773)
McLean, Renfrew – *Capucine* (M 782), *Pétunia* (M 789)
Taylor, Shoreham – *Armoise* (M 772)
Brooke Marine – *Aubépine* (M 781), *Myosotis* (M 788)
Fairlie Yacht Co – *Jonquille* (M 787)
McGiver – *Géranium* (M 784)
Vosper, Gosport – *Hibiscus* (M 785)
Taylor, Chertsey – *Dahlia* (M 786)

Built by shipyards in the UK, these inshore minesweepers of the 'Ham' class were transferred 1954–55 under MDAP. In the mid-1970s all ships of the class were reclassified as auxiliary or patrol craft. *Myosotis* became a diving tender (A 710), while *Hibiscus, Dahlia, Tulipe, Capucine, Oeillet, Hortensia, Armoise* and *Pâquerette* were allocated the numbers A 735–42. *Jasmin* (P 661), *Violette* (P 788), *Géranium* (P 784) and *Jonquille* (P 787) became patrol boats of the *Gendarmerie Maritime* at Lorient, La Pallice, Cherbourg and Toulon respectively. *Pâquerette* (P 789) was similarly fitted in 1976 and allocated to Toulon. *Pétunia* became a buoy tender (*patrouilleur baliseur*) and was given the number P 662. *Armoise* and *Aubépine* were stricken during the 1970s. All but two were stricken 1984–86. *Hibiscus* and *Jasmin* were BU 1989.

Vega 1954
NB 1/750 scale

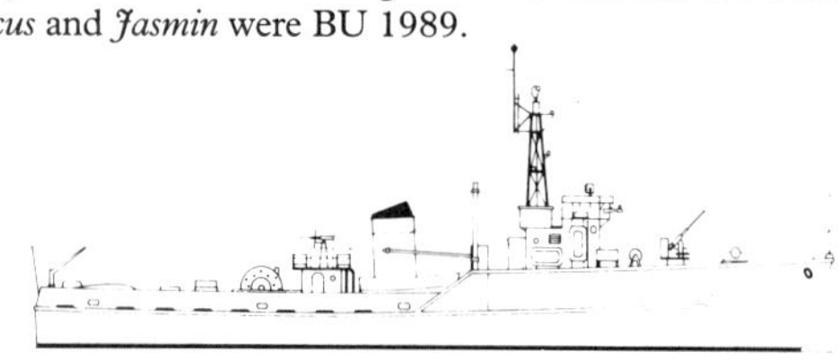

SIRIUS class *coastal minesweepers*

Displacement: 365t standard; 424t full load
Dimensions: 140ft pp, 152ft oa × 28ft × 8ft
42.7m, 46.4m × 8.6m × 2.5m
Machinery: 2 shafts, SIGMA free piston generators and Alsthom or Rateau-Bretagne gas turbines, 2000hp = 15kts. 1 diesel-electric aux engine, 500hp. Range 3000nm at 10kts
Armament: 1–40mm, 1 or 2–20mm (1 or 2×1)
Complement: 38

No	Name	Builder	Launched	Fate
M 701	SIRIUS	Arsenal de Cherbourg	6.10.52	Stricken 1972–74?
M 702	RIGEL	Arsenal de Cherbourg	13.5.53	Stricken 1972–74?
M 703	ANTARES	Arsenal de Cherbourg	21.1.54	Stricken 1978
M 704	ALGOL	A C Augustin-Normand	15.4.53	Stricken 1977
M 705	ALDEBARAN	A C Augustin-Normand	27.6.53	BU 1973
M 706	REGULUS	Penhoët	18.11.52	Stricken 1972–74?
M 707	VEGA	Penhoët	14.1.53	BU 1982
M 708	CASTOR	A C Augustin-Normand	19.11.53	Stricken 1972–74?
M 709	POLLUX	A C Augustin-Normand	16.7.54	Returned USA 1970
M 710	PEGASE	A C Augustin-Normand	21.6.56	Stricken 1975
M 734	CROIX DU SUD	A C de la Seine-Maritime	13.6.56	To Seychelles 1979
M 735	ETOILE POLAIRE	A C de la Seine-Maritime	5.3.57	Stricken 1981
M 736	ALTAIR	Arsenal de Cherbourg	27.3.54	Stricken 1981
M 737	CAPRICORNE	Arsenal de Cherbourg	8.8.56	Stricken 1988
M 740	CASSIOPEE	Penhoët	16.11.53	Stricken 1977
M 741	ERIDAN	Penhoët	18.5.54	BU 1982
M 742	ORION	A C de la Seine-Maritime	20.11.53	Returned USA 1970
M 743	SAGITTAIRE	A C de la Seine-Maritime	12.1.55	BU 1982
M 744	ACHERNAR	CMN, Cherbourg	12.8.54	Returned USA 1971
M 745	PROCYON	CMN, Cherbourg	12.12.54	Returned USA 1970
M 746	ARCTURUS	CNF	12.3.54	Stricken 1981
M 747	BETELGEUSE	CNF	12.7.54	Stricken 1987
M 748	PERSEE	CMN, Cherbourg	23.5.55	BU 1973
M 749	PHENIX	CMN, Cherbourg	23.5.55	Stricken 1992
M 750	BELLATRIX	A C Augustin-Normand	21.7.55	Stricken 1975
M 751	DENEBOLA	CMN, Cherbourg	12.7.56	Stricken 1975
M 752	CENTAURE	Penhoët	8.3.55	Returned USA 1971
M 753	FORMALHAUT	Penhoët	24.4.55	Returned USA 1970
M 754	CANOPUS	A C Augustin-Normand	31.12.53	Stricken 1986
M 755	CAPELLA	CMN, Cherbourg	6.9.55	Stricken 1987
M 756	CEPHEE	CMN, Cherbourg	3.1.56	BU 1991
M 757	VERSEAU	CMN, Cherbourg	26.4.56	BU 1991
M 758	ARIES	Penhoët	13.3.56	To Morocco 1974
M 759	LYRE	Penhoët	3.5.56	Stricken 1981

Of the thirty-four French minesweepers completed belonging to this group, twenty-eight were funded under MDAP. Three further vessels were built for Yugoslavia. They had the same general characteristics as the British 'Ton' class and the design was developed in close collaboration with Thornycroft and the Royal Navy, but there were differences in the construction of the hull. The latter was of laminated wood and aluminium, producing a strong, rigid, light body; the keel and stern were of heavy wood. They were fitted with mechanical, magnetic and acoustic sweep gear. *Capricorne* was given greater degaussing treatment than the other ships. *Croix du Sud, Etoile Polaire, Arcturus, Bételgeuse, Phénix, Canopus, Capella, Céphée, Verseau, Aries* and *Lyre* were fitted with SEMT-Pielstick diesels instead of gas turbines.

In 1960, *Altair, Arcturus* and *Croix du Sud* became station ships in the West Indies and were renumbered P 656, P 650 and P 658 respectively; all sweep gear was removed. Six further units were reclassified as patrol boats in 1973 and modified for fishery protection and surveillance duties: *Canopus*, *Etoile Polaire*, *Véga*, *Antarès*, *Eridan* and *Sagittaire* became P 659, P 660, P 707, P 703, P 741 and P 743 respectively. Six ships were returned to the USA 1970–71. In 1977 *Lyre* and *Véga* were placed in reserve; at the same time *Bételgeuse* became a trials ship for the DUBM 41 side-scan sonar and was renumbered A 474, and *Sagittaire* and *Aldébaran* (ex-*Eridan*) were allocated to the Missile Test Centre. *Canopus* performed general surveillance and fishery protection duties off the Atlantic coast until being stricken in 1986.

Capella was stricken prematurely following a collision in April 1987. *Bételgeuse*, was also retained into the late 1980s. *Phénix* continued to operate out of Cherbourg on fishery protection duties until her demise in 1992.

MERCURE *coastal minesweeper*

Displacement: 365t standard; 400t full load
Dimensions: 146ft oa × 27ft × 8ft 6in
44.4m × 8.3m × 4.0m
Machinery: 2 shafts, 2 MTU diesels, 4000bhp = 15kts. Range 3000nm at 15kts
Armament: 2–20mm (1×2)
Complement: 48

Known as Type DBI, *Mercure* was to have been the first of a new class of French coastal minesweeper based on the ex-American MSCs of the *Acacia* class. She was ordered from CMN, Cherbourg, under MDAP and launched on 21 February 1957. Six further units were later built for the Federal German Navy.

The hull was of laminated wood assembled or glued under heavy pressure, with anti-magnetic longitudinal metal bonding. The navigation bridge was of light alloy. All propulsion and generating machinery was flexibly mounted for shock resistance, and during sweep operations the ship's engines could be operated from a small cabin just beneath the upper deck.

The original diesels did not meet expectations and *Mercure* spent a number of years in reserve. Finally funds were allocated for modification as a fishery protection vessel at Cherbourg Dockyard. The original shafts and the KaMeWa CP propellers were retained, but MGO 2500 diesels replaced the original German units. Fuel tanks were enlarged to give a radius of 6200nm at 10kts. The sweep gear was removed and a new deckhouse containing medical facilities was installed aft. She was renumbered P 765, and replaced *Le Fringant* in this role in 1981. *Mercure* performed general surveillance and fishery protection duties off the Atlantic Coast until stricken in 1990.

Mercure as completed — Navarret Collection

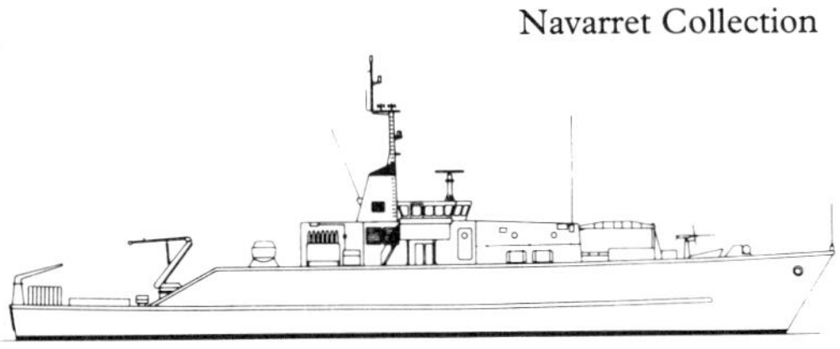

Clio as completed

CIRCE class *minehunters*

Displacement: 460t standard; 510t full load
Dimensions: 153ft pp, 167ft oa × 29ft × 12ft
46.5m, 50.9m × 8.9m × 3.6m
Machinery: 1 shaft, 1 MTU diesel, 1800bhp = 15kts; twin active rudders (7kts). Range 3000nm at 12kts
Armament: 1–20mm, 2 PAP 104
Sensors: Radar Decca 1229; sonar DUBM 20
Complement: 45

No	Name	Builder	Launched	Fate
M 712	CYBELE	CMN, Cherbourg	2.3.72	Extant 1995
M 713	CALLIOPE	CMN, Cherbourg	20.10.71	Extant 1995
M 714	CLIO	CMN, Cherbourg	10.6.71	Extant 1995
M 715	CIRCE	CMN, Cherbourg	15.12.70	Extant 1995
M 716	CERES	CMN, Cherbourg	10.8.72	Extant 1995

The world's first purpose-built minehunters, the *Circé* class were designed round the DUBM 20 minehunting sonar and the PAP 104 minehunting vehicle, which is handled by the large articulated davit mounted on the stern. The system can detect mines laid to a depth of 60m at a maximum range of 500m from the ship. No sweep gear is fitted. The sonar and operations control room are housed in the superstructure block forward of the bridge. *Cérès* was fitted with the prototype EVEC automatic plotting table, which was subsequently retro-fitted to the other units. All ships received the upgraded DUBM 20A with coherent processing during refits in the mid-late 1980s, and this has now been further upgraded to DUBM 20B.

A low magnetic signature and silent operation were design priorities. As insufficient expertise with GRP had been acquired it was decided that the hull should be of sandwich construction, with blocks of wood and foam inside an outer skin of wood. The outer skin was coated with a film of GRP to ease maintenance. The deckhead and superstructure are a composite of wood and glass resin.

There are two independent propulsion systems, one specifically for minehunting operations. The latter comprises two 'active rudders' each with a small crew propeller mounted at its base, energised by a 260hp electric motor. All propulsion machinery can be operated either from the bridge or from a soundproof control room located above the main deck. The five units of the *Circé* class moved from Cherbourg to Brest in 1993.

ERIDAN class *minehunters*

Displacement: 562t standard; 595t full load
Dimensions: 169ft oa × 29ft × 11ft 6in
51.6m × 9m × 3.5m
Machinery: 1 shaft, 1 Brons-Werkspoor A-RUB 215 V12 diesel, 1900bhp = 15kts; twin active rudders (7kts) and bow thruster. Range 3000nm at 12kts
Armament: 1–20mm, 2–12.7mm MG, 2 PAP 104
Sensors: Radar Decca 1229; sonar DUBM 21B
Complement: 55

No	Name	Builder	Launched	Fate
M 641	ERIDAN	Arsenal de Lorient	2.2.79	Extant 1995
M 642	CASSIOPEE	Arsenal de Lorient	28.9.81	Extant 1995
M 643	ANDROMEDE	Arsenal de Lorient	22.5.82	Extant 1995
M 644	PEGASE	Arsenal de Lorient	23.4.83	Extant 1995
M 645	ORION	Arsenal de Lorient	6.2.85	Extant 1995
M 646	CROIX DU SUD	Arsenal de Lorient	6.2.85	Extant 1995
M 647	AIGLE	Arsenal de Lorient	8.3.86	Extant 1995
M 648	LYRE	Arsenal de Lorient	15.11.86	Extant 1995
M 649	PERSEE	Arsenal de Lorient	9.3.88	Extant 1995
M 650	SAGITTAIRE(i)	Arsenal de Lorient	9.11.88	To Pakistan 1992
M650	SAGITTAIRE(ii)	Arsenal de Lorient	—	Building

Developed from the *Circé* class minehunter, the *Eridan* class was part of a collaborative project (the 'Tripartite' MH) between France, the Netherlands and Belgium. Under the terms of the agreement each country built its own GRP hulls to a standard design; the French were responsible for the minehunting systems and electronics, the Netherlands for the main propulsion system, and Belgium for the minehunting propulsion system and the electrical generating plant.

The hull, decks and partitions are formed by a single GRP skin 20–40mm thick, stiffened by trapezoid-section formers, the hull/former connections being reinforced by fibreglass pins. The hull is moulded in a steel shell.

The propulsion system is similar to that of the *Circé* class with the addition of a pair of bow thrusters for additional steadiness during minehunting operations. The electric motors for the 'active rudders' and bow thrusters are driven by a group of three Astazou IVB gas turbine alternators, each of 150kW. A fourth diesel-drive alternator of 160kW supplies the ship with power at cruise speed and while alongside. The main shaft drives a five bladed CP propeller, enabling the blades to be feathered when the auxiliary propulsion system is in use. The main propulsion machinery is controlled from the bridge,

Cassiopeé 1.9.94 — L & L van Ginderen

but the auxiliary propulsion system can also be controlled from the operations room.

The DUBM 21B minehunting sonar is a smaller, more advanced model which can detect and classify ground and moored mines to a depth of 80m. It is retracted into a well in the hull for cruising. Unlike the *Circé* class these ships have a limited minesweeping capability; light mechanical sweep gear was fitted from the outset, and the AP4 acoustic sweep from 1985. The PAP 104 magazine has a capacity of twenty-seven explosive charges. The minehunting control centre, which is housed within the superstructure, features the EVEC 20 plotting table and advanced inertial navigation equipment. Decca Hi-Fix and Syledis radio precision navigation equipment is fitted. A 5t container can be embarked stored for various tasks, including research, patrol, extended diving and drone control.

Fifteen units were originally projected for the Marine Nationale, but this was reduced by budgetary pressures to ten. The first six units, together with the diving tender *Styx*, are based at Brest, their primary mission being to keep clear the channels used by the French SSBN force. The remainder are based at Toulon, where they have replaced the converted US *Agile* class minehunters of the *Cantho* sub-group. *Sagittaire* was sold to Pakistan in 1992, and is to be replaced by a new hull.

VULCAIN class *diving tenders*

Displacement: 409t standard; 490t full load
Dimensions: 136ft oa × 25ft × 10ft 6in
41.6m × 7.5m × 3.2m
Machinery: 2 shafts, 2 SACM MGO V16 AFHR diesels, 2200bhp = 13.7kts; 750hp bow thruster. Range 7400nm at 9kts
Armament: 1–12.7 MG
Sensors: Radar Decca 1226
Complement: 15 + 12 divers

No	Name	Builder	Launched	Fate
M 611	VULCAIN	La Perrière, Lorient	17.1.86	Extant 1995
M 612	PLUTON	La Perrière, Lorient	13.5.86	Extant 1995
M 613	ACHERON	CMN, Cherbourg	19.11.86	Extant 1995
M 614	STYX	CMN, Cherbourg	3.3.87	Extant 1995

Derived from the local support tenders of the *Chamois* class, these ships have replaced the former US *Adjutant* class converted minesweepers in the diving support rôle. A team of up to fifteen divers can be accommodated. Two semi-rigid inflatables are carried abaft the funnels. The single hydraulic crane can lift 5t at 6m radius, 3.5t at 10m. The Syledis precision navigation system is fitted.

Vulcain is based at Cherbourg, *Styx* at Brest, and *Pluton* at Toulon. *Achéron* is training ship for the diving school at Toulon.

Narvik class design

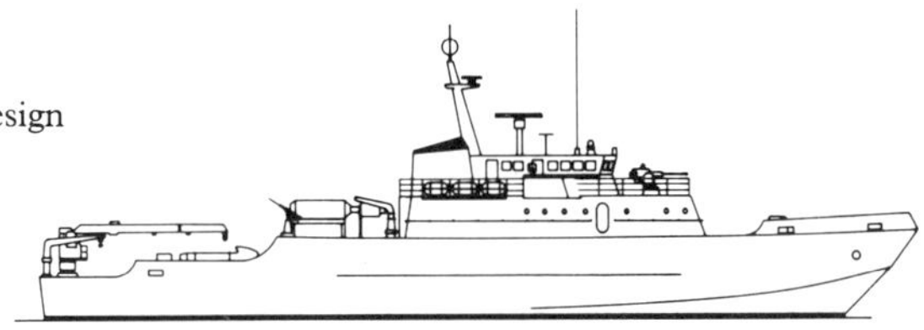

NARVIK (BAMO) class *ocean mine countermeasures ships*

Displacement: 830t trials; 905t full load
Dimensions: 171ft × 49ft × 12ft
52m × 14.8m × 3.6m
Machinery: 2 shafts, 2 ? diesels, 2720bhp = 15kts; bow and stern thrusters. Range 5000nm at 10kts
Armament: 1–20mm, 2–12.7mm MG, 1/2 PAP 104
Sensors: Radar ?; sonar DUBM 42 (with DUBM 60B)
Complement: 46

No	Name	Builder	Launched	Fate
M 660	NARVIK	Arsenal de Lorient	22.3.91	Extant 1995
—	AUTUN	Arsenal de Lorient	—	Cancelled
—	BIR HAKEIM	Arsenal de Lorient	—	Cancelled
—	COLMAR	Arsenal de Lorient	—	Cancelled
—	GARIGLIANO	Arsenal de Lorient	—	Cancelled
—	BERLAIMONT	Arsenal de Lorient	—	Cancelled
—	Unnamed	—	—	Cancelled
—	Unnamed	—	—	Cancelled
—	Unnamed	—	—	Cancelled

Designed by DCN, the BAMO (*Bâtiment Anti-Mines Océanique*) was intended to provide a conventional minesweeping capability in deeper waters, a mission previously performed by the former ocean minesweepers of the US *Agile* class. A catamaran hull constructed of GRP was selected in order to create the deck space needed for mechanical, magnetic and acoustic (DCN AP4) sweeps. It was claimed that a catamaran would provide improved seakeeping, stability and manoeuvrability in deepwater minesweeping operations.

Narvik was funded under the 1990–93 programme, but all other projected units were cancelled in 1992 due to budgetary problems. *Narvik* will now serve as a trials ship. She will have the Lagadmor mine warfare system based on the DUBM 42 sonar array. The latter comprises two side-looking sonars together with an ahead-looking sonar (DUBM 60B), and can be towed at 6.5kts to search a swathe 200m wide at depths of up to 300m. Two PAP 104 Mod 5 remotely-controlled submersibles will be carried, and the vessels will be fitted with Syledis and Navstar precision navigation systems.

ANTARES class *mine countermeasures vessels*

Displacement: 340t full load
Dimensions: 93ft oa × 25ft × 12ft 6in
28.3m × 7.7m × 3.8m
Machinery: 1 shaft, 1 Baudouin diesel, 800bhp = 10kts. Range 3600nm at 10kts
Armament: None
Sensors: Sonar DUBM 41B
Complement: 25

No	Name	Builder	Launched	Fate
M 770	ANTARES	Socarenam, Boulogne	30.5.93	Extant 1995
M 771	ALTAIR	Socarenam, Boulogne	1994	Extant 1995
M 772	ALDEBARAN	Socarenam, Boulogne	1994	Extant 1995

These three ships were built to replace the elderly ocean minesweepers of the *Agile* class for navigation training, and for route surveillance in the approaches to Brest. The design is based on a trawler hull. The DUBM 41B towed sonars were removed from the MSOs. A mechanical sweep is also fitted. The ships are classified BINRS (*Bâtiments d'Instruction à la Navigation Remorqueurs de Sonars*).

Gabon

The Republic of Gabon, formerly part of French Equatorial Africa, became independent in 1960, and for many years was politically stable, with a right-of-centre autocracy firmly in control and black Africa's highest per capita income; a one-party state was formally instituted in 1968. The country's first multi-party elections were held in 1990. This has been followed by much unrest amid allegations of vote rigging and fraud. French military intervention quickly ended the one coup, in February 1964.

Gabon, like Ivory Coast, follows a 'moderate' foreign policy, usually taking the French line; a flirtation with Libya in 1973–74 has now cooled off. She consistently broke trade sanctions against Rhodesia, and was accused of complicity in the guerrilla attack on Cotonou in Benin (Dahomey) in January 1977. The continued French interest is mainly the consequence of Gabon's mineral wealth (iron, manganese and uranium); under 1974 bilateral defence arrangements, France keeps four Jaguar aircraft and 900 marines at Libreville, and has another 1500 men near at hand in the landlocked Central African Republic.

These forces protect Gabon from external threats, and give France effective control over her 1,300,000 inhabitants, as the Gabonese Army is only of reinforced battalion strength. The 480-volunteer navy, based at Port Gentil, south of Libreville and just below the Equator, has a French commanding officer and operates a few fast attack craft and patrol craft along the 500-mile coastline. Port Gentil (opened March 1968) was enlarged in 1981.

GENERAL NAZAIRE BOULINGUI *fast attack craft (missile)*

Displacement: 150t full load
Dimensions: 138ft oa × 25ft 4in × 6ft 6in
42.0m × 7.7m × 1.9m
Machinery: 3 shafts, 3 MTU 20V 672 TY90 diesels, 9450bhp = 38.5kts. Oil 28.4t, range 1500nm at 15kts
Armament: 4 SS12M SSM (2×2), 1–40mm Bofors, 1–20mm
Sensors: Radar Decca RM 1226
Complement: 20

Class (no, former name):
General Nazaire Boulingui (P 10, ex-*President El Hadj Omar Bongo*)

Triple-skinned mahogany-hulled boat launched by CNE, Cannes, on 21 November 1977 and commissioned 7 August 1978, and in service 1995. Reached 40kts (10,500bhp) on trials. Wire-guided optically fired missiles. Has been re-engined with three SACM diesels giving 5,400bhp for a speed of 32kts.

General Nazaire Boulingui (ex-*President El Hadj Omar Bongo*) as completed
Navarret Collection

N'GOLO class *fast attack craft (gun)*

Displacement: 65t standard; 88t full load
Dimensions: 88.5ft oa × 22.3ft × 6.9ft
27.3m × 6.8m × 2.1m
Machinery: 2 shafts, 2 MTU 16V538 diesels, 7000bhp = 40kts. Range 600nm/1200nm at 40kts/35kts
Armament: 1–40mm Bofors, 2–20mm Rheinmetall RM 202
Sensors: Radar Decca RM 916
Complement: 16

N'golo (GC 04) ordered from Intermarine, Sarzana, Italy, laid down in June and launched December 1976. She was completed in March 1977 but burnt out on 13 March 1980. Replaced in 1981 by craft of same name and number. Largest high-speed GRP hull in service. *N'golo* reached 43.8kts (7960bhp) on measured distance trials off Italy. Galileo fire control systems. Discarded in 1989.

French SUPER PATRA class *fast attack craft*

Class (no, launched):
Generale d'Armée Ba Oumar (P 07, 18.12.87), *Colonel Djoue d'Bany* (P 08, 23.3.90)

Two built by CMN, Cherbourg, both extant 1995. P 07 armed with 1–57mm Bofors and 1–20mm Oerlikon. P 08 only fitted with 2–20mm to save money.

N'GUENE *large patrol craft*

Displacement: 90t standard; 118t full load
Dimensions: 95ft 9in pp, 105ft 7in oa × 22ft 4in × 7ft 7in
29.2m, 32.2m × 6.8m × 2.3m
Machinery: 3 shafts, GM 16V 149TE diesels, 4800bhp = 27kts. Range 825nm/1400nm at 25kts/15kts
Armament: 1–40mm Bofors, 1–20mm, 2–12.7mm MG (1×2)
Sensors: Radar Deccca RM 916
Complement: 21

N'guéné (GC 03) is an aluminium-hulled boat built by Swiftships, Morgan City, USA (1974–75) and in service since February 1976. Armament 2–40mm when delivered. Discarded in 1989.

MISCELLANEOUS

The French seaward patrol craft *VP 775* (ex-*VP* 25, ex-British *HDML 1021*) was acquired as Gabon's first warship in 1961 and named *Bouet-Willaumez*. She was hulked in February 19[illegible]8 and deleted in 1975. One French *Champlain* class landing ship *President El Hadj Omar Bongo* (L 05), launched 1984, extant 1995. In service are the patrol craft *Colonel Djoue Dabany* (ex-*President Albert Bongo*, (GC 02, 1972, CNE, Cannes, commissioned March 1972, 80t, 2–20mm, 30kts, 17 men) and the Gabon-built *President Léon M'ba* (GC 01, launched 16 January 1968, 85t, 2–12.7mm MG, 12.5kts, 16 men), a Senegalese-built LCM *Manga* (two more were ordered from Bazan, Cadiz, in August 1981) and a coast guard of about ten small coast patrol craft and launches.

German Democratic Republic 1949–1990

The German Democratic Republic was proclaimed on 7 October 1949 and the Soviet Military Administration formally transferred power to it, being replaced by the Soviet Control Commission. In February 1950 the military *Seepolizei* (Sea Police) was formed under the Ministry of the Interior and in May that year the first flotilla, consisting of the former Danish fishery protection vessel *Hvidbjörnen*, six *R*-boats and probably some *Flugsicherungsboote* returned by the Soviets, was commissioned. However, the German Communists had already been planning to re-establish the navy, the 'KS' (or 'Sperber') class patrol boats having been ordered on 1 March 1949, seven months before the formal establishment of the GDR.

A reduction of reparations by 50 per cent and admission to the COMECON in 1950 benefited the country economically. An ambitious five-year plan was drawn up in 1951, designed to double industrial production by 1955. In 1952 all military police units were united into the *Kasernierte Volkspolizei* (Quartered People's Police) and the *Seepolizei* became a part of that organisation. The *Seepolizei*'s first large ships, namely six 'Habicht' class minesweepers, were ordered and series construction of over fifty *Delphin*/*Tümmler* class patrol boats was begun. In 1952, in great secrecy, plans were formulated for a submarine force, to be equipped with former *Kriegsmarine* Type VIIC U-boats, then in Soviet hands, or Soviet 'M' class boats. These plans were abandoned after the July 1953 riots, the official (though still secret) reason being economic difficulties. Meanwhile *U 1308* was raised in 1953, and *U 2344* two years later, but both boats were eventually broken up.

After Stalin's death in 1953 the Control Commission was abolished which encouraged the Germans, already under pressure to fulfil the plan, to rise against the authorities in June. Only intervention by Soviet troops restored order; nevertheless the government decided to revise economic policy. The Soviets announced an end to the collection of reparations as well as cancellation of debts and reduction of the cost of the Soviet occupying forces. In March 1954 it was announced by the Soviet Union that East Germany was a sovereign state and the following year she was included in the Warsaw Pact. At that time the *Seepolizei* comprised six 'Habicht' class minesweepers and about 100 patrol boats built locally (1950–53) plus ships returned by the Soviets. After completion of this initial programme a further six improved 'Habicht' class minesweepers and fifty 'Schwalbe' class minesweeping boats were ordered. After the creation of the *National Volksarmee* (NVA) in January 1956 the *Seepolizei* was transformed into the *Seestreikrafte* (Naval Force). By 1960 this force was increased considerably by the inclusion of four 'Riga' class frigates, twenty-seven 'P 6' type MTBs and sixteen 'SO 1' type subchasers delivered by the Soviet Union as well as ten 'Krake' class minesweepers. From 1957 onwards ships of the *Seestreikrafte* participated in combined naval exercises with units of the Polish Navy and the Soviet Baltic Fleet.

In November 1960 the *Seestreikrafte* was transformed into the *Volksmarine* (People's Navy) and ships were given names as well as the numbers originally carried. These numbers were occasionally changed, but unfortunately the same numbers were often used on different ships at various times. For example, the number '822' was first used on a *Delphin* class patrol boat, then on the 'P 6' class FAC(T)s *Hans Beimler*, then *Anton Saefkow*, *Julius Adler* and *Bruno Kühn*, and finally by the 'Shershen' class *Fritz Behn*. The *Volksmarine* was now experiencing shortages of trained sailors. Its role widened in 1961 when, after the erection of the Berlin Wall, the navy was given full responsibility for direct protection of the sea border and patrol craft of the Frontier Guard were transferred to it. In 1962 an 18-month conscription period had to be introduced. The construction programme could be realised without interruption and by the mid-1960s Peenewerft at Wolgast had delivered about six 'Robbe' class landing ships and twelve 'Labo' landing craft. The enlargement of the *Volksmarine* was completed with transfer of twelve 'Osa' class missile craft from the Soviet Union.

The *Volksmarine* began its first modernisation programme in 1969. With the exception of missile craft (another three added in 1976) almost all types were replaced by new ones. At first new construction included thirty 'Libelle' class MTBs, fifty-three small 'Kondor' class minesweepers, eighteen 'Bremse' class patrol craft and fourteen 'Frosch' class landing ships (three further units were under construction at that time).

By 1982 the minesweepers, MTBs, patrol craft and landing ships were replaced by new vessels. The ASW programme was started in 1978 to replace the ex-Soviet 'Riga' class frigates by 'Koni' class units and the *Hai* class submarine-chasers by domestically-built *Parchim* class corvettes. The corvette construction programme was concluded in 1985 and three frigates were purchased by 1986. The last stage of the *Volksmarine* modernisation started in 1984 and comprised updating of the strike force, which was composed of several, mostly outdated, missile and torpedo craft. The objective of this programme was to produce all missile-armed squadrons as the FAC(T)s were regarded as unsuitable for modern warfare. The programme consisted of the purchase of ex-Soviet 'Tarantul' class missile craft and the construction of much smaller, locally designed 'Balcom 10' missile craft fitted with a new generation of missiles and sensors.

Thanks to this programme the East Germans were to have a more balanced and homogeneous coastal force than any other Soviet satellite state. The navy was to have a considerable strike force, coastal ASW capability, large amphibious lift capacity and numerous mine warfare vessels.

From the early 1980s, together with the Soviet and Polish navies, the *Volksmarine* took part in the annual joint cruises and exercises of the Warsaw Pact Baltic Joint Squadron, usually deploying one or two frigates accompanied by auxiliaries. The three-week exercises included North Sea deployment. The supply ship force was strengthened in order to provide adequate logistic support for sustained operations in the North Sea and Norwegian Sea.

In 1980 the *Volksmarine* comprised the 1st Flotilla (landing ships, ASW ships, minesweeping and support ship divisions), 4th Flotilla (frigates, ASW ships, minesweeping and support ship divisions), 6th Flotilla (one missile and two torpedo FACs brigades). The *Grenzbrigade Küste* (Coastal Border Brigade) of 4100 men, in three divisions was subordinated to the Ministry of the Interior in 1989. In 1990 the navy had a total of 12,300 men, of whom 4700 were afloat, 4100 in coastal defence (three battalions each of six SS-C-3 'Styx' missile launchers and five coastal batteries armed with thirty-five 6in guns) and 1000 in naval aviation. Ratings were on 1-year conscription. The main bases were Peenemünde (1st Flotilla), Warnemünde (4th Flotilla) and Dranske-Bug (6th Flotilla), with smaller bases at Sassnitz, Wolgast and Tarnewitz (HQ coastal defence). The naval aviation division, with bases in Parow, Peenemünde (both ASW squadrons) and Laage (fixed wing), consisted of twenty Mi-8 'Hip' and eight Mi-14 'Haze' helicopters, as well as twenty Su-22 'Fitter' fighter-bombers. Naval infantry was established in 1985 at Prora on Rügen Island and comprised 1250 men of the 29th Motorised Rifle Regiment, which was replaced in 1987 by the 28th Regiment (part of the 8th Motorised Rifle Division).

In the 40 years from its foundation the German Democratic Republic appeared to be the most strictly pro-Moscow of the Soviet satellite states. The support of Soviet troops stationed there, the Communist Party's hold on a well-disciplined society and a relatively prosperous economy gave the impression that East Germany was the 'most reliable link in the chain of the socialist commonwealth'. However, the relaxation of Soviet control and the increasing strength of independence movements in neighbouring eastern European states prompted a similar development in the DDR in the summer of 1989. Within a few months the Communist regime was overthrown, and in their first free elections since 1933 East Germans voted for reunification with West Germany. The reunification came about on 3 October 1990. As with the rest of the apparatus of the East German state, the former *Volksmarine*, with three frigates, twelve landing ships, sixteen corvettes, nineteen FAC(M)s, twenty FAC(T)s and twenty-seven minesweepers, which until 1990 had been under strict Soviet operational control, became the property of the united Germany.

It was announced originally that no *Volksmarine* vessels would be taken over, but eventually fifteen combat ships and fifteen auxiliaries were incorporated into the *Bundesmarine* for a short time. However most of the former *Volksmarine* vessels have since been disposed of, being either given or sold to civilian companies or various foreign navies. The largest such transfer was the sale of thirty-nine ships to Indonesia at a price of US$314m. Thus, for the fourth time in modern history the German navy, or a part of it, was disposed of *en bloc*.

The numbers given in the tables are those with which the ships first commissioned. Exceptions are given in the notes.

MAJOR SURFACE SHIPS

Ex-Danish HVIDBJORNEN *fishery protection vessel*

Armament:	1–85mm, 2–37mm (1×2), 2–25mm (2×1) (after refit)
Complement:	40 + 80 cadets
Other particulars:	As Danish *Hvidbjörnen*

No	Name	Builder	Acquired	Fate
	ERNST THAELMANN (ex-*Hvidbjörnen*)	Copenhagen N Yd	12.6.50	BU 1968

Built for the Royal Danish Navy as the fishery protection vessel *Hvidbjörnen* (launched 1928). Scuttled 29 August 1945 in the Great Belt, she was raised and repaired by the *Kriegsmarine*, and was found damaged at Warnemünde after the war. Take over by the GDR on 12 June 1950, she was refitted by Peenewerft, Wolgast (new bridge, armament 4–37mm, 2–25mm) and entered *Volksmarine* service on 10 December 1952 as a training ship. In 1956, after re-boilering and rearming (see table), she was transferred to the Frontier Guard. She was renamed *Albin Köbis* in 1961 when her original name was transferred to a 'Riga' class frigate. She was hulked in 1961 and scrapped in 1968.

Ex-Danish KVINTUS class *minelayers*

No	Name	Acquired	Fate
844	FÜRSTENBURG (ex-*Kvintus*)	12.6.50	BU 1959
845	PRENZLAU (ex-*Sixtus*)	12.6.50	BU 1973

Two former Danish minelayers found damaged in Rostock in 1945. They were taken over by the GDR in June 1950, and commissioned as training ships for the *Seepolizei* in April 1952 (199t, 1–37mm). Both were deleted in 1959, *Fürstenburg* being broken up while *Prenzlau* became the training ship *Freundschaft* for the *Gesellschaft für Sport und Technik* (a paramilitary youth organisation) until 1973. Another former Danish minelayer, *Lossen*, was rebuilt as the salvage tug *Wismar* and was deleted in 1962.

Ex-Soviet 'RIGA' class *frigates* (Project 50)

No	Name	Acquired	Fate
I-61	ERNST THÄLMANN (ex-*Olen*)	15.12.56	Stricken 29.8.77
I-62	KARL MARX (ex-*Tur*)	15.12.56	Stricken 31.8.76
103	KARL LIEBKNECHT	10.10.59	Stricken 10.10.68
104	FRIEDRICH ENGELS	10.10.59	Stricken 10.10.68

A pair of brand-new 'Riga' class frigates were bought from the Soviet Union for 24m marks in 1956 and commissioned as the coast defence ships I-61 and I-62. A second pair of older, but recently modernised (new masts, electronics and ASW), ships were acquired three years later, when the earlier two were renumbered 101 and 102 respectively, and were themselves modernised in 1963. *Thälmann* was seriously damaged by a fire in the engine room and in 1967 *Karl Marx*'s engines were damaged in a storm. They were later renumbered 121–124, while in the 1970s the two surviving ships bore the numbers 141 and 142. All had been broken up by 1977.

Ex-Soviet 'KONI I' class *frigates* (Project 1159)

No	Name	Builder	Commissioned	Fate
141	ROSTOCK	Zelenodolsk	25.7.78	Deleted Aug 1991
142	BERLIN	Zelenodolsk	10.5.79	For disposal 1991
143	HALLE	Zelenodolsk	28.1.86	Deleted Aug 1991

Officially classed as *Küstenschutzschiffe 2 Ranges* (2nd Rate Coastal Defence Ships), the first pair were the first of the class built in the Soviet Union for export. They were armed with 2 SA-N-4 'Gecko' SAM (1×2), 4–76mm (2×2), 4–30mm (2×2), 2 RBU-6000 ASW RL (2×12), 22 mines, 2 chaff RL (2×16). Dummy 16-tubes launchers were mounted originally on *Halle*, the others had them fitted in 1987. Full official name of 142 was *Berlin Haupstadt der DDR*. The acquisition of a fourth ship, to restore the Coastal Defence Division to its previous strength, was cancelled. Their SA-N-4 systems were removed by the Soviets before German reunification in 1990. Upon unification *Rostock* and *Halle* were taken over by the *Bundesmarine* as *F 224* and *F 225* respectively, while *Berlin* was immediately put up for disposal. Both ships were decommissioned by the *Bundesmarine* after six months' service. Algeria proposed to purchase all three in 1991, but the offer was declined.

AMPHIBIOUS WARFARE VESSELS

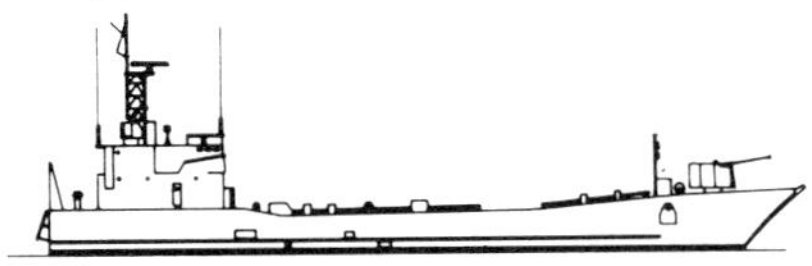

'Robbe' class early 1970s

'ROBBE' class *landing ships* (Project 47)

Displacement: 600t standard; 709t normal; 1263t full load
Dimensions: 210ft × 38ft × 8ft 2in
64.0m × 11.6m × 2.5m
Machinery: 2 shafts, 2 40DM diesel engines, 4800bhp = 14kts. Range 1060nm at 12kts
Armament: 2–57mm (1×2), 4–25mm (2×2), 14 tanks
Sensors: Radar KSA3 (later TSR333)
Complement: 32

Class (nos, commissioned:
Schwedt (501/650, 10.4.64), *Eisenhüttenstadt* (502/618, 20.5.64), *Grimmen* (503/619, 20.5.64), *Lübben* (504/616), *Hoyerswerda* (505/619, 20.5.64), *Eberswalde* (506/614, 20.10.64)

Six built by Peenewerft, Wolgast. Originally numbered 501–506, later renumbered in 610 series, not in sequence. Cargo capacity 500t. Deleted between 23 August 1975 (*Hoyerswerda*) and 6 October 1978 (*Eisenhüttenstadt* and *Grimmen*), and broken up, except for *Hoyerswerda* which was used as a barracks ship at Peenemünde for a time.

'Frosch' class as completed

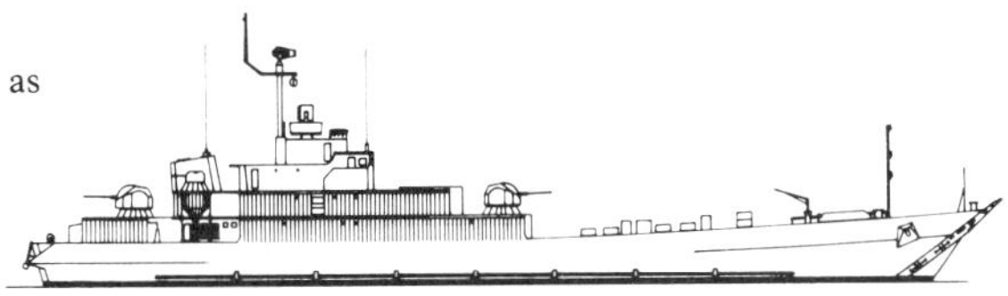

'FROSCH' class *landing ships* (Project 108/109)

Displacement: ('Frosch I') 1774t normal; 1950t full load
('Frosch II') 1530t normal
Dimensions: 297ft 7in × 36ft 5in × 11ft 2in
90.7m × 11.1m × 3.4m
Machinery: 2 shafts, 2 61B diesel engines, 9500bhp = 18kts. Range 2000nm
Armament: 'Frosch I': 4–57mm/70 (2×2), 4–30mm/65 (2×2), 40 mines, 7 tanks
'Frosch II': 4–57mm/70 (2×2), 4–25mm (2×2), 36 mines, 650t cargo
Sensors: 'Frosch I': Radar Muff Cob, Strut Curve, Don, Square Head, High Pole
'Frosch II': Radar Don
Complement: 'Frosch I': 42
'Frosch II': 35

Class (nos, commissioned):
Hoyerswerda (611, 12.11.76), *Hagenow* (612/632, 1.12.76), *Frankfurt/Oder* (613, 2.2.77), *Eberswalde-Finow* (614/634, 28.5.77), *Eisenhüttenstadt* (615, 4.1.79), *Grimmen* (616, 15.6.79), *Lübben* (631/632, 15.3.78), *Schwerin* (632/612, 19.10.77), *Neubrandenburg* (633, 28.12.77), *Cottbus* (634/614, 26.5.78), *Anklam* (635, 14.7.78), *Schwedt* (636, 7.9.79), *Nordperd* (E-35/E-171, 30.10.79), *Südperd* (E-36/E-172, 26.2.80)

Built 1975–78 by Peenewerft, Wolgast, under designation 'Project 108' as replacements for the earlier classes, these were the largest ships in the *Volksmarine*, and could carry up to twelve armoured vehicles (total military lift of 600 tonnes). They are fitted with a bow ramp, but have no bow doors and no stern doors or ramp. Superimposed above the forward twin 57mm mounting a reserve site is provided for a close-in fire support system of two 40-barrel MBU-2500 122mm RLs, which were actually fitted on *Schwerin*. The last two were built 1978–80 as armed support ships under the designation 'Project 109' ('Frosch II') with the combat outfit reduced but with an 8-tonne capacity crane fitted amidships. All deleted 1 October 1990. All transferred to Indonesia August 1993.

'Frosch' class — Twardowski Collection

'LABO' class *landing craft* (Project 46)

Displacement: 150t standard; 232t normal; 285t full load
Dimensions: 134ft 9in × 22ft 7in × 3ft 7in
41.1m × 6.9m × 1.1m
Machinery: 2 shafts, 4 6NVD26A diesels, 1080bhp = 10kts. Range 540nm at 10kts
Armament: 4–25mm (2×2)
Complement: 15

Class (no):
601–605, *Gerhard Prenzler* (606), *Rolf Peters* (607), *Heinz Wilkowski* (608), *609–612*

Built by Peenewerft, Wolgast and entered service between October 1960 and December 1962. Could carry two tanks. Three vessels received names on 1 March 1967. All were renumbered on various occasions in 630, 650 and 670 series. 601–604 and 606 deleted 22 November 1976 and the remainder on 24 October 1977. All broken up except for *Rolf Peters* which became a supply tender with the number E-30, and in 1981 she was transferred to civilian use, being refitted in Poland and renamed *Neuendorf*.

SMALL SURFACE COMBATANTS

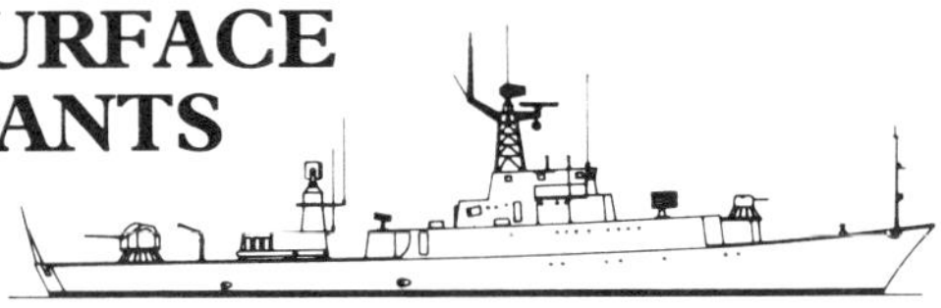

Parchim as completed

PARCHIM class *corvettes* (Project 133)

Displacement: 793t normal; 854t full load
Dimensions: 226ft 5in wl, 246ft 9in (oa) × 32ft 2in × 9ft
69.0m × 75.2m × 9.8m × 2.73m
Machinery: 3 shafts, 3 M504 A-3 diesels, 14,250bhp = 24.7kts. Oil fuel 55t. Radius 2100nm/14kts
Armament: 2 SA-N-5 SAM (2×4), 2–57mm (1×2), 2–30mm (1×2), 2 RBU-6000 ASW RL (2×12), 4–400mm ASW TT (4×1), 1 DCT, 2 chaff RL (2×16), 60 mines
Sensors: Radar Strut Curve, TSR-333, High Pole, Muff Cob; sonar M/F hull mounted MG-322, MG-329, dipping sonar; ECM Watch Dog
Complement: 57

Class (no, commissioned):
Gadebusch (211, 31.8.84), *Grevesmühlen* (212, 21.9.84), *Bergen* (213, 1.2.85), *Angermünde* (214, 26.7.85), *Lübz* (221, 12.2.82), *Bad Doberan* (222, 30.6.82), *Güstrow* (223, 10.11.82), *Waren* (224, 23.11.82), *Prenzlau* (231, 11.5.83), *Ludwigslust* (232, 4.7.83), *Ribnitz-Damgarten* (233, 29.10.83), *Teterow* (234, 27.1.84), *Wismar* (241, 9.7.81), *Parchim* (242, 9.4.81), *Perleberg* (243, 19.9.81), *Bützow* (244, 30.12.81)

Ribnitz-Damgarten 9.10.93 — L & L van Ginderen

These vessels were built 1977–85 by Peenewerft, Wolgast under the designation 'Project 133' in two series as replacements for the *Hai* class submarine-chasers, but with diesel engines only instead of the CODAG power plant of their predecessors. Officially classed as *Küstenschutzschiffe 3 Ranges* (3rd Rate Coastal Defence Ships), they had the NATO code 'Balcom 4'. Reportedly poor seaboats: not fitted with anti-rolling systems. VDS dipping sonar is fitted on the starboard side, though the ship must stop in order to lower the sonar's transducer. Based on this design, a series of twelve 'Parchim II' class corvettes were built 1985–90 by Peenewerft for the Soviet Navy but with armament and sensors upgraded. *Gadebusch*, *Grevesmühlen*, *Lübz*, *Teterow* and *Wismar* were integrated in the *Bundesmarine* with numbers P 6167, P 6164, P 6169, P 6168 and P 6170 respectively. Remainder paid off 3 October 1990 and offered for sale. In August 1993 all sixteen were transferred to Indonesia.

Ex-Soviet 'TARANTUL I' *missile corvettes* (Project 1241)

Class (no, commissioned):
Albin Köbis (771, 31.10.84), *Rudolf Egelhoffer* (772, 16.10.85), *Fritz Globig* (773, 5.10.85), *Paul Eisenschneider* (774, 12.2.86), *Hans Beimler* (775, 28.10.86)

These vessels were built by the 341 Yard, Rybinsk (USSR), 1983–85 as a partial replacement of the ageing 'Osa I' class FAC(M)s. They were probably intended as leaders for the smaller 'Balcom-10' FAC(M)s of local design. *Köbis* and *Egelhoffer* transferred on 24 September 1984, *Globig* in September 1985 and the last pair in October 1985. In April 1987 they were renumbered 571–575 in the same order. *Rudolf Egelhoffer* was incorporated into the *Bundesmarine* and renamed *Hiddensee* (P 6166). She was acquired by the USN in 1991, renamed *Beagle* and employed as a survey ship. The remainder were deleted on 3 October 1990. Offers in 1991 from Malta for the purchase of *Köbis* and *Globig*, and from Poland for *Eisenschneider* and *Bemler* were reportedly turned down.

FORELLE class *experimental MTBs*

Displacement:	65t full load
Dimensions:	90ft 10in × 23ft 7in × 3ft 3in *27.7m × 7.2m × 1.0m*
Machinery:	2 shafts, 2 MB 511 diesels, 5000bhp = 38kts. Range 1000nm
Armament:	2–12.7mm MGs, 2–53cm TT (planned)
Complement:	12

Class:
I, G-01, G-02

Experimental, light alloy hulled boats built by Rosslauer Schiffswerft, Rosslau. The first, unnamed, boat was ready for trials in October 1955. Following an accident off Peenemünde, she was rebuilt in 1956 and used for further trials until she was deleted in late 1958. A further two boats were built, commissioning in 1961. They were fitted with four Soviet M50F diesels with a total output of 4800hp. Following trials, both were commissioned with the Frontier Guard on 13 August 1963. Both deleted on 10 July 1965, and broken up.

Ex-Soviet 'P 6' type *fast attack craft (torpedo)* (Project 183)

Class (nos, stricken):
Artur Becker (S-68, later 801/826, 1.11.67), *Edgar Andre* (S-62, later 802/821, 31.5.68), *Julius Adler* (S-61, later 803/822, 1.11.67), *Fritz Behn* (S-69, later 804/841, 14.10.68), *Rudolf Breitscheid* (S-63, later 805/842, 14.10.68), *Rudi Arndt* (S-64, later 806/824, 1.11.67), *Bernhard Bästlein* (S-67, later 807/827, 1.11.67), *Herbert Balzer* (S-65, later 808/825, 1.11.67), *Willi Bänsch* (S-66, later 809/844, 14.10.68), *Fritz Reidel* (811/825, 21.10.70), *Anton Saefkow* (812/826, 5.9.70), *Ernst Schneller* (813/864, 20.1.69), *Arvid Harnack* (814/821, 4.8.70), *Bruno Kühn* (815/822, 4.8.70), *Erich Kuttner* (816/824, 16.10.69), *Heinz Kapelle* (817/823, 12.10.70), *Fritz Lesch* (818/825, 10.10.69), *Wilhelm Leuschner* (819/824, 21.10.70), *Josef Römer* (821/829, 10.10.69), *Hans Beimler* (102, later 822/862, 1.11.67), *Hans Coppi* (105, later 823/863, 1.11.67), *Max Roscher* (104, later 824/843, 14.10.68), *Hanno Günther* (825/846, 14.10.68), *Fritz Heckert* (826/866, 1.11.67), *Willi Sachse* (827/863, later *Fiete Schulze*, 31.7.69), *Adam Kuckhoff* (109, later 828/845, 14.10.68), *Wolfgang Thies* (829/866, 31.7.69).

S-68–S-66 transferred on 8 October 1957, S-67–S-69 on 19 October 1957 and 102, 105, 825 and 826 on 30 October 1958. The remainder (with Pot Head instead of Skin Head radars) were transferred on 7 September 1960. Received names in 1960. Afterwards frequently renumbered (some boats as many as six times), although always in the 800 series. On 31 August 1968, *Bänsch* collided with the Danish vessel *Drottningen* in the Baltic and sank.

Most of this class were broken up immediately following deletion, but *Andre*, *Arndt* and *Behn* were used as target boats until 1971, *Breitscheid* and *Kuttner* were fitted with enclosed bridges and renumbered as rescue boats *R-71* and *R-72* (broken up 1970 and 1975 respectively) and *Beimler*, *Günther* and *Roscher* were refitted and transferred to the Frontier Guard where they served as *G-81* to *G-83* until 1971. *Lesch* became the trials vessel *V41* and was broken up in 1973, and *Becker*, *Coppi* and *Heckert* received enclosed bridges and were sold to Tanzania 1974–75. In the 1970s the former *Fiete Schulze* and *Wolfgang Thiess* impersonated wartime German 'S39' *schnellboote* in the GDR film 'Rotterknechte'.

As this class was often stationed in isolated havens, the former salvage ship *Grobian* was rebuilt as an accommodation ship, remaining in service until 1981, and four torpedo recovery craft were built on 'Schwalbe' class hulls and named *Sperber 1–4*.

Ex-Soviet 'OSA I' class *fast attack craft (missile)* (Project 205)

Class (nos, commissioned):
Max Reichspietsch (701/751, later 711, 20.11.62), *Albin Köbis* (702, later 712, 26.11.62), *Josef Schares* (712/753, 6.10.71), *Rudolf Egelhoffer* (713/751, 14.1.62), *Richard Sorge* (714/713, 14.1.64), *Otto Tost* (714/731, 28,9.71), *August Lüttggens* (713, later 731/732, 24.9.64) *Paul Eisenschneider* (714, later 732/713, 16.10.64), *Heinrich Dorrenbach* (732/711, 3.9.71), *Karl Meseberg* (733/733, 21.10.64), *Walter Kramer* (734/712, 5.12.64), *Paul Schulze* (751/753, 24.12.64), *Paul Wieczorek* (752/754, 24.9.65), *Fritz Gast* (753/714, 29.11.65), *Albert Gast* (754/734, 23.12.65)

The three boats transferred in 1971 replaced the oldest of the class, *Reichspeitsch*, *Köbis* and *Sorge*, which were relegated to training duties. *Köbis* was deleted on 22 September 1981, *Eisenschneider* on 6 November 1981 and *Egelhoffer* on 26 November 1981 – the last two with cracks in their engine seats – and two of the training vessels were restored to active duty to replace them. From 1984 the class was gradually being replaced by ex-Soviet 'Tarantul' class FAC(M)s. The remaining boats were fitted with chaff launchers in 1987. They were all deleted on 1 October 1990 and scheduled for disposal. In August 1993 the disarmed *Dorrenbach*, *Tost* and *Schares* were transferred to the Luthuanian Navy, whilst in early 1994 *Wieczorek*, *Fritz Gast* and *Albert Gast* went to Latvia and *Reichspietsch*, *Sorge*, *Lüttgens*, *Meseberg*, *Kramer* and *Schulze* went to Estonia.

'BALCOM-10' type *fast attack craft (missile)* (Project 151)

Displacement:	347t full load
Dimensions:	160ft 5in × 28ft 3in × 7ft 3in *48.9m × 8.6m × 2.2m*
Machinery:	3 shafts, 3 M520 diesels, 16,200hp = 35kts. Range 1600nm
Armament:	8 SS-N-25 SSM (2×4) 4 SA-N-5 SAM (1×4) 1–76mm, 1–30mm Gatling, 2 chaff launchers (2×16)
Sensors:	Radar Plank Shave, Bass Tilt, SRN, Square Head, Salt Pot
Complement:	33

Class (no, commissioned):
Sassnitz (591, 31.7.90), *Ostseebad Sellin* (592, 2.10.90), *Ostseebad Binz* (593, 23.12.90), *594* (10.5.90), *595* (launched 7.7.90), *596–598* (cancelled)

A local design intended as replacements for the ageing 'Osa I' class, and built by Peenewerft, Wolgast. Up to fifty units were planned for the USSR, Poland and the GDR (which was to receive twelve). The political changes in eastern Europe in 1990 prevented the USSR from delivering the SS-N-25 systems and although *Sassnitz* had her missile canisters fitted for trials, she commissioned without them. On reunification *Sassnitz* was taken over by the *Bundesmarine* as *P 6165*. In 1992–93 she was re-engined, refitted with larger superstructure and Western electronics, and entered service with the *Bundesgrenschutze* (Federal Border Patrol) as *BG 22*. *Sellin* and *Binz* were also taken into *Bundesgrenschutze* service as *BG 23* and *BG 24*, but were not rebuilt. The incomplete hulls of *594–596* were sold to Poland in October 1990.

'ILTIS' class *fast attack craft (torpedo)* (Project 63)

Displacement:	17.4t full load
Dimensions:	48ft 10in × 11ft 2in × 3ft 5in *14.9m × 3.4m × 1.1m*
Machinery:	2 shafts, 2 'M50F4' diesels, 2400bhp = 52kts. Range 230nm
Armament:	2–21in TTs
Sensors:	Radar KSA 5 (later TSR 222)
Complement:	3

Class:
I, II, III, V-01, V-05, V-06, V-08, V-09, V-13, 911–925, 941-945, 951-955, 957-959, 960, 965

The design for this class was based on the German war-built *LS* class light MTBs intended primarily to operate from commerce raiders, and successfully employed by the *Kriegsmarine* in the Aegean in 1944. The geographical and hydrometeorological conditions in the western part of the Baltic lend themselves to surprise operations carried out by small, ultra-fast FACs both in defensive and offensive roles. The prototypes were ordered simultaneously in June 1958 in four yards, using a variety of hull materials and construction techniques. Peenewerft (design codenamed 'Iltis' or 'Project 63') and Rosslauer Schiffswerft (design of hydrofoil boat 'Project 81') prepared hulls with aluminium alloy AlMg 4.5, and Yachtwerft, Berlin, built wooden hulls with aluminium alloy frames (design codenamed 'Hydra'). The fourth hull, designed by Boots und Reparaturwerft, Griefswald, was to be made of GRP (codename 'Möwe', Project 58) but did not get further than the planning stage. The first three experimental boats were built by Peenewerft in 1959 with smaller dimensions (16t, 45ft 3in × 11ft 10in × 1ft 8in, *13.8m × 3.6m × 0.52m*) and were deleted *c*1964. They were followed by six pre-production boats, initially numbered in the *V* series, and commissioned between August 1963 and May 1964. *V-01* was deleted 31 March 1966 and put on exhibition at Wolgast, whilst the remaining five were deleted between February 1968 and May 1972. The thirty boats of the production series were commissioned between July 1964 and January 1966. They were frequently renumbered (*916* and *954*, for example, nine times), in *900* or *9000* series. They were all stricken April 1973–March 1977 and broken up, except for *923* which became a memorial. Replaced by the 'Libelle' class.

'HYDRA' class *fast attack craft (torpedo)* (Project 68)

Displacement: 19.2t full load
Dimensions: 56ft 1in × 11ft 10in × 4ft 11in
17.1m × 3.6m × 1.5m
Machinery: 2 shafts, 2 'M50F4' diesels, 2400bhp = 50kts
Armament: 3–21in TTs; could carry mines
Sensors: Radar KSA 5 (later TSR 222)
Complement: 3

Class:
I, II, III, W-13, W-17, W-18, 930–931, 933–936, 939, 941–946, 948, 952–955, 957–958 and two uncompleted boats

This design was prepared by Yachtwerft, Berlin, under the designation 'Project 68', competing with the 'Iltis' class; they were larger and slower, but more powerfully armed and more suited to general purpose use. The wooden hull backed by aluminium alloy frames was stepped aft to accommodate three single stern-fired tail-first TTs which could be removed to carry mines or accommodate landing parties. Two unnamed prototypes (one shorter, 42ft 8in, *13.0m*) were built 1958–60 and deleted in 1963. There then followed a series of twenty-four boats, commissioned between November 1964 and October 1968. Frequently renumbered in the 900 series, they were deleted January 1968 to July 1975 and replaced by the 'Libelle' class. One unnamed boat was broken up shortly after trials in 1964. *941* was put on display in a museum in Dresden, *930* was put ashore and became a training boat and *943* was hulked at Wiek and renamed *Diana*.

Ex-Soviet 'SHERSHEN' class *fast attack craft (torpedo)* (Project 206)

Class (nos, commissioned):
Rudolf Breitscheid (821/852, 1.8.70), *Fritz Behn* (822/851, 5.9.70), *Ernst Schneller* (823/851,12.10.70), *Anton Saefkow* (824/853, 28.11.70), *Fiete Schulze* (825/854, 23.12.70), *Artur Becker* (841/816, 14.10.68), *Heinz Kapelle* (841/811, 6.10.71), *Hans Coppi* (842/812, 9.11.68), *Bernhard Bästlein* (843/836, 14.12.68), *Arvid Harnack* (843/813, 16.7.71), *Edgar Andre* (844/811, 18.1.69), *Fritz Heckert* (845/834, 24.7.69), *Willi Bäntsch* (861/831, 16.8.69), *Walter Husemann* (862/856, 19.9.69), *Bruno Kühn* (862/832, 26.8.71), *Adam Kuckhoff* (863/814, 15.10.69), *Wilhelm Florin* (864/815, 13.12.69), *Ernst Grube* (865/833, 16.7.70)

These ships took over the names of the 'P 6' class vessels they replaced. In 1977, *Becker, Bästlein* and *Husemann* became training boats, but were returned to active service when the four oldest boats (*Coppi, Bänsch, Andre* and *Heckert*) were deleted on 1 September 1984. However, *Becker* was deleted a month later. *Harnack* was stricken 15 August 1985, *Bänsch, Breitscheid* and *Behn* on 1 October 1986, and *Husemann* and *Schulze* on 20 October 1986. The last six were stricken on 28 February 1990. All were broken up on deletion.

'LIBELLE' class *fast attack craft (torpedo)* (Project 131)

Displacement: 28t full load
Dimensions: 62ft 4in × 14ft 9in × 5ft 7in
19.0m × 4.5m × 1.7m
Machinery: 3 shafts, 3 'M50F3' and 'M50F7' diesels, 3600bhp = 48kts. Range 350nm
Armament: 2–21in TTs, 2–23mm AA (1×2), 2 mine launchers
Sensors: Radar TSR 333
Complement: 5

Class:
I–III, V-11, (later *956*), *934–935, 943–945, 951–955, 961–965, 980–984, 990–999*

Replacements for the 'Iltis' and 'Hydra' classes. Of aluminium alloy construction, the main armament consisted of two single stern-launch tail-first 21in TTs in hull recesses on both sides. The deck space on both sides of the CT were occupied by removable launchers for KMD type ground mines which could be replaced by portable windlasses and seats for the transport of commandos. The hulls were assembled on a production line in the Rechlin shipyard, then fitted by Peenewerft, Wolgast. Three unnamed experimental boats were built 1971–1973, followed in 1974 by *V-11*, which was displayed at the celebrations of the 25th anniversary of the GDR. Series production of thirty boats began in 1972, with the first unit commissioning on 20 December 1974 and the last three on 31 March 1977. Vessels of this class were stricken between 1 July 1984 (fifteen boats) and 31 May 1990 (last two boats). *951* sank following a collision in 1986; although salvaged, she was stricken on 30 June 1986. *997, 999, 952,* and *963* were acquired by museums in Bremerhaven, Hamburg, Rostock and Dresden respectively.

'Hydra' class — Twardowski Collection

Ex-Soviet 'SO 1' type *submarine chasers* (Project 201M)

Class (nos, stricken):
Kormoran (411/424, G-44/G-46, 3.10.76), *Albatros* (412, G-64/G-46, 29.4.72), *Sperber* (413/421, G-41/G-43, 20.5.75), *Habicht* (414/422, G-42/G-44, 3.10.75), *Elster* (431, G-61/G-43, 29.4.72), *Kranich* (432, G-32/G-44, 29.4.72), *Möwe* (433, G-63/G-45, 14.10.72), *Falke* (434, G-16/G-41, 29.4.72), *Reiher* (451, G-17/G-42, 14.10.72), *Bussard* (452, G-43, 8.6.71), *Weihe* (453/423, G-18/G-41, 20.5.75), *Adler* (454, G-15/G-21, 29.4.72)

Six commissioned 7 December 1959, a further three on 29 December 1959 and the final pair on 15 August 1960. Received names in 1961. In 1965 transferred to Frontier Guard and renumbered in G series. In 1972, *Sperber, Habicht, Weihe* and *Kormoran* returned to *Volksmarine* service for a time. All broken up following deletion.

'Hai I' as completed
NB 1/750 scale

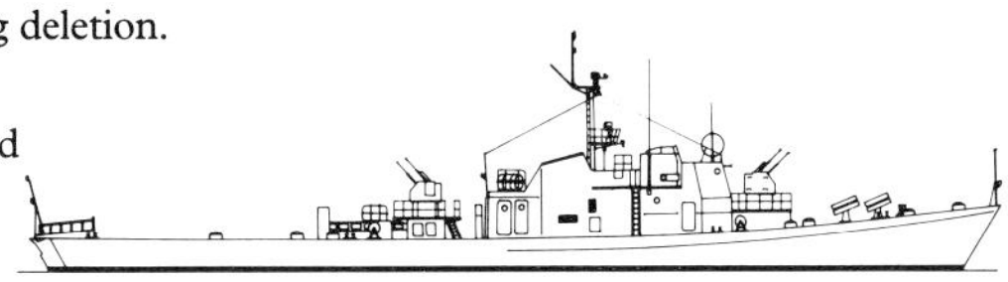

'HAI I' *experimental submarine chaser* (Project 12.1)

Displacement: 278t
Dimensions: 159ft 9in × 19ft 8in × 6ft 7in
48.7m × 6.0m × 2.0m
Machinery: 3 shafts, 2 6KVD43 and 1 Dora 40 diesels, 4600hp = 19kts. Range 1500nm
Armament: 4–25mm (2×2), 4 RBU-1200 ASW RL, 4 DCTs, 2 DC racks (see notes)
Complement: 11

Designed since 1958, she was built by Peenewerft, Wolgast, during 1961–63, as experimental craft to develop the 'Hai' class for which she was the prototype and commissioned without name on 26 November 1963 for trials. She was completed with two large flat funnels abreast. Initially she carried no armament, later being fitted with guns only and finally ASW weapons were introduced. She could also carry a 45mm gun aft. Relegated to training duties after conclusion of trials and experiments, named *S 01*. Stricken 2 November 1970.

'HAI II' *experimental submarine chaser* (Project 12.3)

Displacement: 316t
Dimensions: 169ft 7in × 21ft 8in × 8ft 6in
51.7m × 6.6m × 2.6m
Machinery: 3 shafts, CODOG, 1 D40 diesel and 2 'Pirna 014' gas turbines, 12,200shp = 32kts
Armament: 4–25mm AA (2×2), 4 RBU-1200 ASW RL, 2 DC racks
Sensors: Radar Pot Head, Square Head; sonar Tamir
Complement: 29

Laid down on 1 October 1961 by Peenewerft, Wolgast, under the designation 'Project 12.3', this vessel was fitted with the CODOG experimental power plant, as the second 'Hai' class prototype. The planned twin 57mm AA mounting was not fitted on the afterdeck because of stability problems. She was commissioned on 11 April 1963 as the experimental vessel *V 81*, and renamed *Werdau* in 1973. Stricken 23 April 1983.

'Libelle' class — Twardowski Collection

Bad Doberan as completed Twardowski Collection

'HAI III' class *submarine chasers* (Project 12.4)

Displacement: 320t full load
Dimensions: 169ft 7in × 21ft 8in × 8ft 6in
51.7m × 6.6m × 2.6m
Machinery: See 'Hai II' class. Range 1000nm at 18kts
Armament: 4–30mm (2×2), 4 RBU-1800 ASW RL (4×5), 2 DC rails (24 DC), mines
Sensors: Radar Pot Head, Drum Tilt, Square Head, High Pole A; sonar Tamir-II
Complement: 29

Class (nos, stricken):
Perleberg (414/241, 3.4.81), *Gadebusch* (421/211, 2.8.83), *Wismar* (422/214, 2.5.80), *Sternberg* (423/215, 28.2.84), *Parchim* (424/213, 31.10.80), *Ludwigslust* (431/215, 1.11.82), *Lübz* (433/244, 8.10.81), *Bad Doberan* (451/242, 29.12.81), *Bützow* (452/243, 10.7.81), *Ribnitz-Damgarten* (453/214, 15.6.83), *Teterow* (454/213, 10.1.83), *Grevesmühlen* (461/212. 21.10.83)

First unit – *Grevesmühlen* – laid down 8 August 1964 at Peenewerft, Wolgast, and commissioned on 5 July 1965 together with *Gadebusch, Wismar, Sternberg* and *Parchim. Ludwigslust* commissioned 1 December 1965, *Perleberg* and *Bad Doberan* on 8 January 1966, and the remainder on 4 November 1966. Later renumbered, first in a 400 and then in a 200 series. Gas turbines used only for short bursts of hunting speed. All broken up after deletion except for *Grevesmühlen* which was used for explosion tests in 1984.

'SPERBER I' class *coastal patrol craft*

Displacement: 72t
Dimensions: 87ft 11in × 15ft 1in × 4ft 7in
26.8m × 4.6m × 1.4m
Machinery: 3 shafts, 3 Jumo 205D diesels, 1600hp = 22kts. Range 400nm
Armament: 3–20mm (1×1, 1×2)
Complement: 20

Class (nos, stricken):
Freundschaft (111, later G 111/G-61, 1.10.66), *Pionier* (112, later G 112/G-61, 20.5.63), *Solidarität* (113, later G 121/G-41, 5.9.64), *114* (later G 122/ G-65, 10.7.65), *115* (later G 131/G-61, 5.9.64), *116* (later G 132/G-25, 20.4.67)

Twenty boats of this class were ordered on 1 March 1949, before the formal foundation of the GDR. However, only six were built by Yachtwerft, Berlin, armed at Wolgast and commissioned in December 1950 (named boats) and 1951 (other three). In 1962 they were transferred to the Frontier Guard and renumbered in a G series. In the early 1960s they were rearmed with either 4–12.7mm MG or 4–14.5mm, and re-engined with two 6NVD26A diesels of 540hp each, coupled to the outer shafts only. Speed fell to 12kts. All broken up once stricken.

Another boat of similar dimensions was built by Peenewerft. Fitted with two 200hp diesels and 1–20mm, she operated as training boat *852* 1951–56. She was later transferred to the *Gesellschaft für Sport und Technik* and renamed *Patriot.*

'SPERBER II' class *coastal patrol craft*

Displacement: 78t
Dimensions: 94ft 6in × 15ft 9in × 4ft 11in
28.8m × 4.8m × 1.5m
Machinery: 3 shafts, 3 Jumo 205D diesels, 1600hp = 22kts. Range 400nm
Armament: 2–20mm Oerlikon (1×2)
Complement: 20

Class:
111–118, 121–126, 131–136

Improved version of 'Sperber I' design, with lengthened hull. Six built July 1951–April 1952 by Peenewerft, Wolgast, and the remainder by Yachtwerft, Berlin. The first boats were armed with two or three 20mm Flak 38, but the majority received wartime 20mm Oerlikons supplied to the Soviet Union under Lend-Lease. In June some of these had their Oerlikons replaced with 4–12.7mm MG. The remainder were converted to submarine chasers in 1955, receiving a new enclosed bridge, a Tamir 10 sonar and two DC racks (6 DCs each). Receiving same re-engining as 'Sperber I' class (see above) in the early 1960s. All deleted 1963–1969. Three were disarmed and transferred to the *Gesellschaft für Sport und Technik*. Renamed *Partisan, Pionier* and *Ernst Schneller*, they were used for youth training.

'DELPHIN' class *coastal patrol craft*

Displacement: 44t
Dimensions: 73ft 6in × 14ft 5in × 4ft 7in
22.4m × 4.4m × 1.4m
Machinery: 2 shafts, 2 Jumo 205C diesels, 600bhp = 13kts. Range 130nm
Armament: 1 or 2–12.7mm MG
Complement: 10

Class:
821–826, 831–836

Twelve boats built by Yachtwerft, Berlin and commissioned between 16 November 1953 and 24 July 1954. Officially classed as 'roads protection boats'. In 1957 all transferred to the Frontier Guard and given 'G' numbers. In the early 1960s they were fitted with 6NDV26A diesels and light tripod masts. Deleted between October 1964 and May 1969 and broken up, save for one boat transferred to the *Gesellschaft für Sport und Technik.*

'TUMMLER I and II' class *coastal patrol craft*

Displacement: 38t
Dimensions: 73ft 6in × 14ft 5in × 4ft 2in
22.4m × 4.4m × 1.3m
Machinery: 'Tummler I': 2 shafts, 2 Jumo 205C diesels, 600bhp = 13kts. Range 130nm
'Tummler II': 2 shafts, 2 IFA EM-6-20 diesels, 240hp = 11kts
Armament: 4–12.7mm MG (2×2)
Complement: 9

Class:
'Tummler I': *181–186*
'Tummler II': *G-111, G-112, S-21, S-22, S-24, S-25*

Twelve boats built by Yachtwerft, Berlin, as 'roads protection boats' with higher, enclosed bridge. Commissioned between July and August 1954 ('Tummler I') and August and November 1956 ('Tummler II'). All deleted and broken up 1963–66.

'SAS' class *coastal patrol craft*

Displacement: 90t full load
Dimensions: 68ft 11in × 18ft 8in × 9ft 2in
21.0m × 5.5m × 2.79m
Machinery: 1 shaft, 1 '4NVD 36' diesel, 150bhp = 11kts
Complement: 7

Class (former name, final number):
G-91 (ex-*Braunschweig, G-761*) *G-92* (ex-*Thuringen, G-61*) *G-93* (ex-*Lübeck, G-63*), *G-94* (ex-*Gross Berlin, G-762*), *G-95* (ex-*Hannover, G-64*), *G-96* (ex-*Holstein, G-65*), *G-97* (ex-*Bremen*)

Built in the late 1940s by Kremer & Sohn for the Sassnitz fishing company. Transferred to the Frontier Guard in 1965, having been refitted with new bridges and masts. *G-97* deleted April 1976 and the remainder between April 1985 and May 1988. All were subsequently broken up except for *G-94* which was sold for civilian use.

'BREMSE' class *coastal patrol craft* (Project GB-23)

Displacement: 38t
Dimensions: 75ft 9in × 15ft 5in × 4ft 7in
23.1m × 4.7m × 1.4m
Machinery: 2 shafts, 2 6VD18/15HA diesels, 500hp = 17kts. Range 300nm
Armament: none
Sensors: Radar TSR 222
Compliment: 6

Class (commissioned):
G-30–G-33 (27.11.71), *G-34–G-39* (17.3.72)

Ten built by Yachtwerft, Berlin, as replacements for the 'Delphin' and 'Tummler' classes. Later renumbered *G-730–G-739*. After reunification four were retained by the Federal Border Guard and renamed *Prignitz* (BG 61), *Uckermark* (BG 62), *Altmark* (BG 63) and *Börde* (BG 64). In 1992 five were sold to Tunisia, two to Malta and two to Jordan.

In 1973, four boats of 'Project TB-23 (1582)' (armed variant with enclosed bridge) were sold to Tanzania.

MINE WARFARE VESSELS

'Habicht I' class
NB 1/750 scale

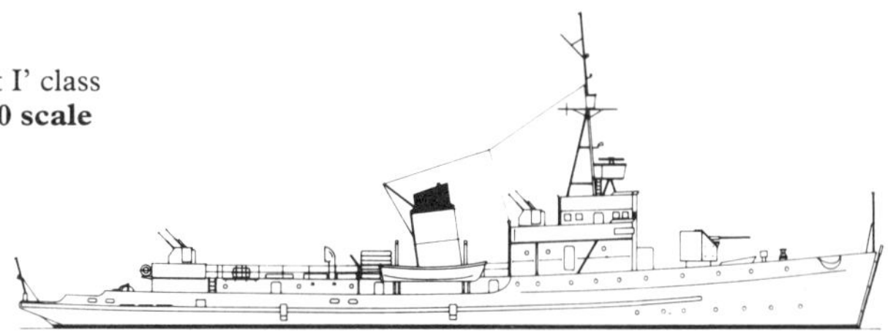

'HABICHT' class *minesweepers/minelayers*

Displacement: 510t ('Habicht II' 670t)
Dimensions: 194ft 11in ('Habicht II' 213ft 7in) × 26ft 3in × 7ft 10in ('Habicht II' 8ft 2in)
59.1m (65.1m) × 8.0m (8.1m) × 2.4m (2.5m)
Machinery: 2 shafts, 2 diesels, 2500bhp = 18kts ('Habicht II' 17kts). Range 3700nm
Armament: 1–85mm, 8–25mm (4×2), 2 DCT (22 DC), 18 mines ('Habicht II' 20 mines)
Sensors: Radar navigational; sonar Tamir II
Complement: 38

Class (nos, commissioned, fate):
'Habicht I': *Suhl* (6-31/213, 27.4.53, deleted 19.11.67), *Schwerin* (6-32/214, 7.8.53, deleted 10.11.67), *R-21* (6-33, 1953, deleted 1.10.68), *Neubrandenburg* (6-34/251, 8.12.53, deleted 10.12.65), *R-11* (6-25, 1954, deleted 28.10.70), *Cottbus* (6-36/212, 2.2.54, deleted 27.2.65)
'Habicht II': *Bergen* (6-11/213. 2.8.55, deleted 15.10.68), *Frankfurt/Oder* (6-12/212, 6.8.55, deleted 28.10.70), *Stralsund* (6-13/211, 12.9.55, deleted 28.10.70), *Sassnitz* (6-14/252, 15.11.55, deleted 10.12.65), *Wolgast* (6-15/273, 15.11.55, deleted 27.2.65), *Greifswald* (6-16/214, 20.12.55, deleted 15.10.68)

Ordered on 1 January 1951 from Volkswerft, Stralsund. Design apparently heavily dependant on wartime 'M40' minesweepers, though diesel propulsion employed instead of steam. Last six boats completed with additional 20ft section amidships as a result of operational experience. Original armament on 'Habicht I' comprised 1–37mm and 4–25mm (4×1); 'Habicht II' class completed with armament listed above and earlier ships were later rearmed to same standard.

In 1957 *6-33* ran aground off Sassnitz, and was refloated and rebuilt, with one other of the class, as salvage ships *R-11* and *R-21*. Received names in 1960. Some later fitted with tripod masts and improved electronics, while *Frankfurt/Oder* and *Stralsund* had their 85mm replaced with 2–57mm in the mid-1960s. Most broken up on deletion, although *Stralsund* was hulked as *S-91* and not broken up until 1974, and *R-11* was transferred to the Maritime School as the training vessel *Ernst Thälmann*. A few boats were scuttled off Peenemünde Haken as targets for naval aircraft.

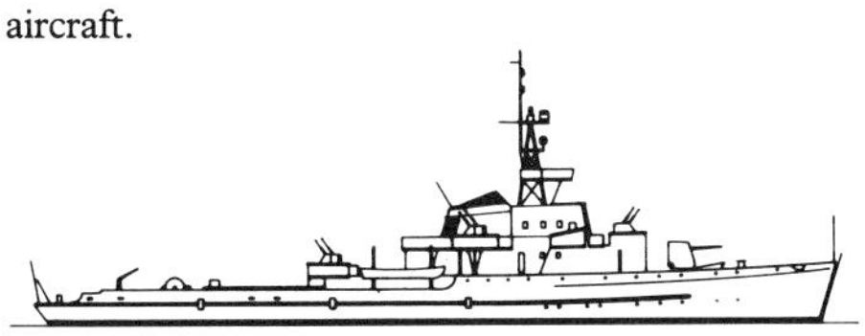
'Krake' class 1971

'KRAKE' class *minesweepers/minelayers*

Displacement: 553t normal; 741t full load
Dimensions: 216ft 10in × 27ft 7in × 8ft 3in
66.1m × 8.4m × 2.5m
Machinery: 2 shafts, 2 6KVD43A diesels, 2400bhp = 16.5kts. Range 2460nm
Armament: 1–85mm, 10–25mm (5×2), 2 DCT, 30 mines
Sensors: Radar navigational; sonar Tamir II
Complement: 38

Class (nos, commissioned, fate):
Halle (220/221, 21.9.57, deleted 14.10.72), *Dresden* (221/225, 9.11.57, deleted 26.10.73), *Rostock* (223, 28.11.57, deleted 16.10.73), *Gera* (224/242, 20.12.57, deleted 14.10.72), *Karl-Marx-Stadt* (225/224, 14.5.58, deleted 25.10.76), *Leipzig* (226/222, 31.10.58, deleted 1.10.76), *Berlin* (227/245, 16.8.57, deleted 13.10.67), *Erfurt* (228/243, 28.3.58, deleted 14.10.72), *Magdeburg* (229/223, 6.3.58, deleted 14.10.72)

Development of 'Habicht' class, work beginning in 1955 at Peenewerft, Wolgast. Reported that first four units built for Poland. *Dresden*, *Berlin*, *Leipzig* and *Karl-Marx-Stadt* served as training ship before deletion.

'KONDOR I' class *coastal minesweepers* (Project 89.100)

Displacement: 339t standard; 377t full load
Dimensions: 178ft 7in × 23ft 4in × 7ft 10in
52.0m × 7.1m × 2.4m
Machinery: 2 shafts, 2 40D diesels, 4000bhp = 20kts. Range 1900nm
Armament: 2–25mm (1×2)
Sensors: Radar TSR-333; sonar KLA58m
Complement: 24

Class (nos, commissioned, fate):
V-31 (28.11.70, deleted 16.12.81), *Greifswald* (301/316, 29.5.69, deleted 2.4.90, to Guinea-Bissau, 1990), *Bergen* (302, 29.5.69, deleted 20.5.81, extant 1994 as fishery protection vessel), *Anklam* (303, 30.5.69, deleted 1.12.76, to GST as *Ernst Thälmann*, sold 1990), *Ueckermünde* (304, later G-11/G441, 1.7.69, 1990 to BGS as *GS01*, to Malta 1992), *Bansin* (315, later G-46/G446, 24.12.70, deleted 28.2.90, BU), *Templin* (335, later G-16/G442, 20.12.69, 1990 to BGS as *GS06*, to Tunisia 1992), *Neustrelitz* (336/316, later G-41/G412, deleted 28.2.90, BU), *Vitte* (341, later G-21/G421, 18.2.70, deleted 6.2.90, BU), *Zingst* (342, later G-22/G444, deleted 2.3.89, BU), *Prerow* (343, later G-23/G413, 15.5.70, deleted 2.3.89, BU), *Graal-Müritz* (344, later G-24/G413, 15.5.70, deleted 2.3.89, BU), *Kühlungsborn* (345, later G-25/G445, 3.6.70, 1990 to BGS as *GS07*, extant 1995), *Ahrenshoop* (346, later G-26/G415, 8.8.70), 1990 to BGS as *GS08*), *Demmin* (351/331, later G-12/G442, 1990 to BGS as *GS02*, to Tunisia 1992), *Kirchdorf* (351/311, later G-42/G425, 19.9.70, deleted 15.8.90, BU), *Malchin* (352/332, later G-13/G441, 18.10.69, 1990 to BGS as *GS03*, to Tunisia 1992), *Boltenhagen* (352/312, later G-43/G443, 1990 to BGS as *GS09*, extant 1995), *Klütz* (353/313, later G-43/G443, 3.10.70, deleted 28.2.90, BU), *Altentrepow* (353/333, later G-14/G414, 5.9.69, 1990 to BGS as *GS04*, to Tunisia 1992), *Pasewalk* (354/314, later G-45/G426, 1990 to BGS as *GS05*, to Malta 1992), *Rerik* (354/314, later G-45/G426, 12.11.70, deleted 28.2.90, BU)

Replacements for 'Habicht' and 'Krake' classes. These smaller, more numerous boats reflected changes in *Volksmarine* mine warfare operational concepts. First boat, after prolonged trials, entered service as experimental ship *V-31*. First production boat laid down October 1967 and twenty-one were built by Peenewerft, Wolgast. In the mid-1970s all were transferred to the Frontier Guard (retaining their sweeping gear), and renumbered in 'G' series, apart from *Anklam*, *Bergen* and *Greifswald* which became training ships. Upon deletion *Bergen* became the fishery protection vessel *Warnemünde* (with enlarged superstructure) and was retained in this role after reunification by the German Ministry of Agriculture and Fisheries. Three remain in service with the Federal Border Guard, the remainder having been disposed of (disarmed) to Tunisia, Malta and Guinea-Bissau (see above).

In addition, two torpedo recovery vessels (*Libben* and *Strelasund*, 380t) and two intelligence-gathering craft (*Komet* and *Meteor*, 359t) were built in 1971–72 on 'Kondor I' hulls. *Strelasund* was broken up in 1990, whilst *Libben* was transferred to Lithuania. *Komet* and *Meteor* were both transferred to Estonia in 1994.

'Kondor II' class
NB 1/750 scale
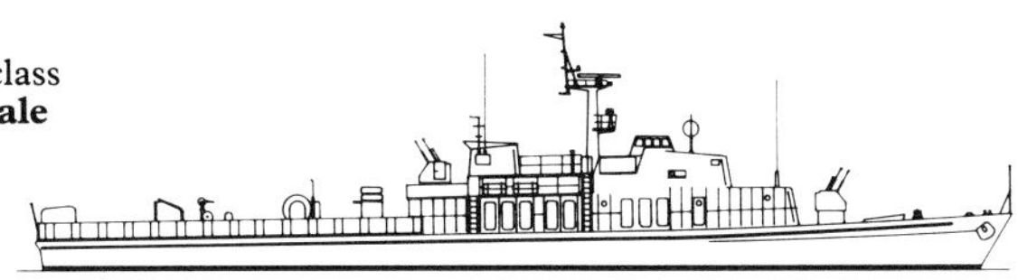

'KONDOR II' class *ocean minesweepers* (Project 89.200)

Displacement: 436t normal
Dimensions: 179ft 7in × 23ft 3in × 7ft 10in
54.7m × 7.7m × 2.4m
Machinery: 2 shafts, 2 40D diesels, 4000bhp = 18kts. Range 2000nm
Armament: 6–25mm (3×2)
Sensors: Radar TSR-333; sonar KLA58m
Complement: 30

Class (nos, commissioned, fate):
Wolgast (V-32/V811, 1.6.71, deleted 3.10.90, to Indonesia 1993), *Rosslau* (V383/V813, 12.9.72, deleted 3.10.90), *Kamenz* (S-24, 24.7.71, deleted 1.12.81, to Latvia 1993), *Sömmerda* (311, 7.4.73, to *Bundesmarine* as *M2670* 1990, to Indonesia 1993), *Eisleben* (312, 24.5.73, to *Bundesmarine* as *M2671* 1990, to Uruguay 1993), *Gransee* (313/333, 30.6.73, deleted 3.10.90), *Stralsund* (S-25, later 314, 14.8.71, deleted 3.1.91, for sale), *Zeitz* (314, 11.7.73, deleted 1.12.81, preserved), *Wittstock* (S-26, later 315, 11.9.71, deleted 20.4.90, BU), *Hettstedt* (315, 9.8.73, deleted 1.12.81 for preservation, to Indonesia 1993), *Altenburg* (315/335, 11.9.73, deleted 2.4.90, BU), *Kyritz* (321, 2.10.71, deleted 14.3.90, BU), *Neuruppin* (322/342, 30.10.71, deleted 10.4.90, BU), *Strasburg* (323/346, 12.11.71, deleted 3.10.90, BU), *Röbel* (324, 1.12.71, deleted 3.10.90, to Latvia 1993), *Pritzwalk* (325, 22.12.1971, deleted 3.10.90, to Indonesia 1993), *Rathenhow* (326, 24.3.72, deleted 12.2.90, BU), *Schönebeck* (372/334, 29.2.73, deleted 20.3.90, BU), *Dessau* (331, 3.5.72, deleted 3.10.90), *Bitterfeld* (332, 26.6.72, to *Bundesmarine* as *M2672* 1990, to Indonesia 1993), *Tangerhütte* (333/313, to *Bundesmarine* as *M2669* 1990), *Genthin* (334, later V382/V812, 7.8.72, deleted 24.10.88, BU), *Zerbst* (335/316, 7.8.72, deleted 3.10.90, to Indonesia 1993), *Grimma* (336, 10.11.73, deleted 3.10.90, to Indonesia 1993), *Oranienburg* (341, 30.9.72, deleted 3.10.90, to Indonesia 1993), *Jüterbog* (342, 1.11.72, deleted 3.10.90, to Indonesia 1993), *Bernau* (343, 1.12.72, to *Bundesmarine* as *M2673* 1990, to Uruguay 1991), *Eilenburg* (344, 22.12.72, to *Bundesmarine* as *M2674* 1990, to Uruguay 1991), *Riesa* (345/322, 3.2.73, deleted 3.10.90, to Uruguay 1991), *Wilhelm-Pieck-Stadt-Guben* (346/323, 5.7.73, deleted 29.1.90, BU)

Development of 'Kondor I' class with hull lengthened by 4.7m to accommodate additional armament and modern minesweeping gear. Class of twenty-eight (first unit laid down November 1969), all built by Peenewerft, Wolgast. Modernised since 1985, receiving SA-N-5 SAM launcher forward of the bridge and an additional deckhouse aft for long-range FRG-1 magnetic minesweeping gear. Following reunification, six boats were taken into *Bundesmarine* service while the rest were laid up. Disposals: four to Uruguay 1991 (guns removed but sweeping gear intact); nine to Indonesia in July 1993; and two to Latvia August 1993 (guns removed but sweeping gear intact). There have been rumours of further planned sales to Surinam and Portugal.

R511–516 *minesweeping boats*

Six war-built *R-boats* of the *R 401* group were returned by the USSR in 1950 and commissioned as *R1–R6* on 29 May 1950 with the *Seepolizei*. In 1955 the numbers were changed to *811–816* and in 1956 to *7-61* to *7-66*. All deleted 1956 and broken up, apart from *816* which was transferred to the *GST* as the training ship *Freundschaft*, serving until 1959.

'SCHWALBE' class *minesweeping boats*

Displacement: 84t
Dimensions: 93ft 10in × 14ft 9in × 4ft 11in
28.6m × 4.5m × 1.5m
Machinery: 2 shafts, 2 6VD-224 (Series III: 6NVD26A) diesels, 300hp (Series III: 540hp) = 11.3kts. Range 500nm
Armament: 2–25mm (1×2) or 2–12mm MG (1×2)
Complement: 14

Class (nos):
Series I: *411–416*
Series II: *421–426*, *Aue* (431/311), *Zwickau* (432/312), *Freiburg* (433/321), *Senftenberg* (434/511), *Forst* (435/341), *Waren* (436/322, later G-17), *Meissen* (441/311), *Görlitz* (442/312), *Kamenz* (443), *Prenzlau* (444/321), *Eisenach* (445), *Gotha* (446/322), *Jena* (451/377), *Greiz* (452/331), *Possneck* (453/332), *Eisleben* (454/353), *Kothen* (455/334), *Zeitz* (456/331)
Series III: *Torgau* (321, later G-33/G-13), *Calbe* (322, later G-34/G-14), *Stendal* (323, later G-35/G-15), *Burg* (324, later G-36/G-16), *Borna* (325, later G-31/G-11), *Wurzen* (326/345, later G-32/G-12), *Weisswasser* (342/354), *Weimar* (341/353), *Luckenwalde* (343/323), *Brandenburg* (344/374), *Nauen* (345/373), *Güstrow* (361/364), *Ilmenau* (363/362), *Meiningen* (364/363), *Hagenow* (365/321, later G-19), *Sonneberg* (365/234), *Sternberg* (422), *Ankalm* (622).

Design work commenced in January 1952. A total of forty-eight boats were built in three series. Series I was built by 'E. Thälmann' Yard in Brandenburg/Havel: first three ships of series I entered service on 16 November 1953 and last three on 19 May 1954. Further, slightly improved ships were built by Yachtwerft, Berlin. Unnamed boats of Series II were commissioned in June 1956. *Aue*, *Zwickau*, *Freiberg*, *Senftenberg*, *Forst* and *Waren* were armed with 2–12.7mm DShK MGs, but stability problems occurred. Further boats had improved stability either through changed hull shape or additional ballast and fitting of 2–25mm was possible. Series III were commissioned between April 1956 and January 1957. Upon arrival of ex-Soviet 'P 6' class in 1957 the *412*, *413*, *415* and *416* (Series I) were converted to torpedo recovery vessels and renumbered *B-62* to *B-64*. Other two became survey ships. Series I boats were deleted between April 1967 and September 1971. All were broken up, apart from *416* transferred to *GST* and used as a training boat. *421–425* of Series II were deleted between June 1964 and December 1968, transferred to militarised *Seehydrographischen Dienst* together with two of Series I, renumbered with *D* superior and broken up later. Rest of Series II were deleted between September 1965 and October 1986. *Prenzlau* was transferred to Tanzania in 1966 together with *Ilmenau* of Series III. *426* and *Eisleben* became trial ships *V-72* and *V-89* respectively. Later *V-72* was cut into sections, transported inland and reerected into training ship. The rest were broken up. Of Series III seven were taken over by *Grenzbrigade-Küste* and renumbered with *G* superior; they were broken up 1971–73. *Sternberg* sank after being rammed by British freighter *Abbotsford* off Warnemünde on 30 September 1963 and was subsequently salvaged and scrapped. Others were broken up between October 1964 and August 1974.

In addition to four converted 'Schwalbes' four further torpedo recovery vessels were built on 'Schwalbe' hulls and named *Sperber 1* (commissioned 25 August 1958) to *Sperber 4*. Deleted 1963–69.

German Federal Republic 1947–1990: Germany 1990–

THE FIRST DECADE: 1945 to 1955

The unconditional surrender of Germany in 1945 brought to an end the history of the *Kriegsmarine*. Disposal of the surviving naval units posed a delicate problem. Stalin wanted one-third because the Soviet Union's naval forces were ill-equipped; the US was not interested in acquiring German fighting units, while Great Britain's chief interest lay with the U-boats which she wished to study and then destroy. Under the final terms of the contentious Potsdam Conference signed on 1 August 1945, the Soviet Union received one-third of all surviving warships; the remainder were distributed among the other Allied nations. The US and Great Britain proposed that all German merchant shipping be placed under the control of the Combined Shipping and Adjustment Board and The United Maritime Authority until the end of the war with Japan. A further US proposal was that all riverine and coastal vessels remain German property to assist in industrial regeneration. The US feared that in the event of a complete breakdown of German postwar industry she would end up effectively paying war reparations to the Soviet Union indirectly via her Marshall Aid Plan.

Meanwhile there remained the problem of clearing the extensive minefields. The Berlin Declaration of June 1945 charged Germany with this task. In July the German Mine Sweeping Administration (GMSA) was formed; a force of 755 vessels, including 440 ex-*Kriegsmarine* minesweepers, manned by 16,000 German seamen under British command, which swept a total of 581,000 mines, 126,000 of which were German. By December 1947 the GMSA had been disbanded and replaced by the Minesweeping Unit of the Customs Inspection, Cuxhaven. This subordinate arm of the British Frontier Inspection Service (FIS) was supervised by the RN in the last stage of the mine clearance and was disbanded in June 1951.

In March 1951 the German Frontier Inspection Authority was created, and this included a naval frontier inspection unit which at first deployed vessels of the former minesweeping force, and included four patrol boats, three fast boats, ten war-built fishing vessels, one sailing boat, one tanker, one tug, two tenders and other small craft. The formation of this unit coincided with the establishment of the new Federal Republic of Germany.

In July 1951 the US Navy helped to establish the Labour Service Unit (LSU) which had two sub-units: LSU(C) consisting of small boats of the Rhine and Weser River Patrols, and LSU(B) consisting of ex-GMSA and ex-Customs, Cuxhaven minesweepers plus other vessels. The total was six 1940 and 1943 type German minesweepers, twenty-six *R-Boote* (fast motor minesweepers), two tenders, three *S-Boote*, one submarine-chaser, three air-sea rescue boats, one salvage barge, one tanker, two accommodation ships and six patrol craft. LSU(C) consisted of twenty-two LCMs and four LCTs.

THE BIRTH OF THE GERMAN FEDERAL REPUBLIC

As early as 1946 the Germans had been allowed their own government; regional parliaments and local governments had been elected, and in January 1947 the US and British occupation zones had become linked economically. The Marshall Aid Plan provided the first funds for the re-establishment of the new Republic. In September 1948 a parliamentary council had been appointed to work out the constitution of a future Germany. In April 1949 the occupation zones of the three Western Allies, the United States, Great Britain and France, were linked economically and all three accepted the new constitution which came into force on 23 May 1949.

On 14 August 1949 the first German *Bundestag* (Parliament) was elected and in September Konrad Adenauer became the first Federal Chancellor of the German Federal Republic.

FIRST STEPS TO GERMAN REARMAMENT

As early as 1949 the first talks had been held between the High Commissioners and the Federal German Government concerning a German role in Western defence policy, in the light of the worsening international situation between the Soviet Union and her former allies. Steps towards this development had been the so-called 'Dunkirk Pact' of 1945 between Great Britain and France, which was followed in 1948 by the 'Brussels Contract', when The Netherlands, Belgium and Luxembourg joined together to form the 'Western European Union' under the stimulus of the Communist Party's seizing control of Czechoslovakia. The Communist blockade of Berlin in 1948–49 and the Communist rise to power in China led to the foundation of NATO.

With the advent of the Korean War, Sir Winston Churchill proposed the building of a 'European Defence Community' with German participation. The possibilities and consequences were examined by the so-called '*Amt Blank*' (literally, Blank Office, after

PROPOSED GERMAN NAVAL BUILD-UP PRIOR TO GERMANY'S JOINING NATO

Type	Autumn 1951	EDC plan*	Oct 1954	Sep 1955	End 1955
Mining vessels	3	2	2	2	2
Destroyers	12	–	12	12	18
Submarine chasers	12	18	18	6	–
Escorts	12				10
MTBs	36	60	36	40	40
Submarines (coastal)	24	–	12	12	12
Landing craft	36	36	30	12	–
Minesweepers (coastal)	24	24	16	24	
Minesweepers (fast)	36	36	36	30	54
Patrol vessels	36	10	10	10	–
Naval aircraft	204	54	54	58	58
Personnel	?	?	?	35,000	?

* plan of the European Defence Community
? no figure available

the name of its chief) and it was decided that the basic prerequisites for such participation were changes to the Occupation Statute, a legal declaration that the Second World War was officially over, and an end to the dismantling of German industry. On 26 May 1952 the German Contract was signed and on May 27, the Contract of the Western European Defence Union. The latter was rejected by the French National Assembly in 1952, but British initiatives brought about the signing of the Paris Contract in October 1954, so that on 5 May 1955 – ten years after the capitulation – Germany became a member of NATO and was now independent in all questions of internal and foreign politics.

THE NEW FEDERAL GERMAN NAVY

On 7 June 1955 the former '*Amt Blank*' became the new Federal Defence Ministry. Its first order, dated 1 January 1956, initiated the so-called 'Programme 6000', consisting of a naval instruction division, with four officers, twenty-four NCOs and 140 ratings. In July 1956 the Frontier Inspection Authority was disbanded and all its vessels and some personnel were transferred to the new navy which now numbered four patrol boats, two fast motor boats, ten coast guard boats, two tenders, one tanker, one tug, six MMS type boats and one training boat. LSU(B) and LSU(C) were also disbanded and their vessels, with many crews, were handed over: six minesweepers, twenty-six *R-Boote*, one submarine chaser, nine patrol boats, three tugs, one accommodation ship, one tanker, one salvage barge, two tenders and other small craft.

To meet the first urgent requirements of the new navy, many ships were purchased or taken on lease from abroad, and some war reparations ships were returned to Germany. These included six ex-British frigates, six Lease-Lend US destroyers, four purchased LSMs, two LSMRs, one LSD, seven LSTs and a number of minor vessels. In addition, five former 1935 type minesweepers were purchased from France.

In 1950 the navy envisaged its requirements as: twelve torpedo-boats, thirty-six MTBs, twenty-four submarines, twelve escorts, thirty-six LSIs, thirty-six minehunters, twelve submarine chasers, thirty-six patrol boats, twenty-four minesweepers and 114 naval aircraft.

The first intake of ships was a hotch-potch of types and all the vessels were so outdated that they could never have been made fit for active service, but they did fulfil the vital training role. In 1955 the first naval building programme was defined and was sanctioned by the *Bundestag* in May 1956. The second programme was sanctioned in 1958 and the third programme, of 1960, was sanctioned in 1960–61. It envisaged the strength of the Federal German Navy as: twelve destroyers, six escorts (frigates), forty MTBs, twelve submarines, fifty-two minesweepers, two training ships, twelve landing craft, ten patrol boats, 120 auxiliaries and fifty-eight naval aircraft.

In 1965 the *Bundestag* fixed the final strength of the navy as: six destroyers (Type 119), four destroyers (Type 101), six frigates (Type 120), fifty MTBs (Types 140, 141 and 142; forty built), thirty submarines (Types 201 and 205; twelve built), twenty patrol boats (Types 360, 361 and 362), twenty-four coastal minesweepers (Types 320 and 321), thirty fast minesweepers (Types 340 and 341), thirteen tenders (Types 401, 402 and 403), twenty inshore/riverine minesweepers (Type 390; only one built), six landing craft (Type 502; none built), twelve tankers, twelve supply ships, eight transports, twenty tugs and numerous small craft for trials and training, plus two large training ships. This was later increased by: three destroyers (Type 103), twenty MTBs (Type 148, replacing Type 140), ten MTBs (Type 143, replacing Type 141).

In 1995 the Federal German Navy numbered 108 fighting ships, sixty-six auxiliaries and 119 naval aircraft.

INTEGRATION WITH NATO

The Federal German Navy today is fully integrated with NATO's naval forces, her ships being able to fulfil all their duties within the NATO mandate. The *Lütjens* class destroyers and *Bremen* class frigates, together with the new generation of *Brandenburg* class frigates, serve in turn with the Standing Naval Forces Atlantic (STANAVFORLANT) and in the Adriatic with the Standing Naval Forces Mediterranean (SNFM or STANAVFORMED). The *Lindau* class minesweepers/minehunters and the *Hameln* and *Frankenthal* class ships serve with the Standing Naval Forces Channel (STANAVFORCHAN). In addition, the navy's ships take part in all bi- and multi-national NATO manoeuvres in order to maintain a standardised training level. The function of the Federal German Navy is to protect the frontiers of the Republic, and to defend the Baltic and North Sea entrances, in cooperation with NATO's naval, ground and air forces. Based on the doctrine of 'forward defence', the navy's main operations theatre is the Baltic and the North Sea.

PROPULSION AND WEAPONS

With the exception of three older types of destroyer with steam propulsion, the navy has followed the example of the *Kriegsmarine* and used diesel propulsion with CP propellers for all major units. The hulls of submarines and minesweepers are of non-magnetic materials.

Weapons are NATO standard, except in the older units which retain their original armament. The 100mm/55 and the 40mm/70 guns are to be found in almost all the NATO navies, as is the new 76mm/62 OTO Melara gun. All missiles except the French MM38 Exocet are of US origin. All ships down to minesweepers can carry mines: in the major units minerails are fixed on deck; in the smaller ships part of the mine deck equipment is removable.

GERMAN SHIPYARDS

At the end of the war Germany's shipbuilding industry had been virtually destroyed and any remaining machinery was dismantled. The first objective was the rebuilding of the mercantile fleet. All warship construction was strictly prohibited; naval yards did not exist and it was, and is, not intended to re-start them. Today the German shipbuilding industry has a high international reputation, and domestic and foreign warship construction is a valuable part of it. Lürssen Werft has more than 100 years experience in the building of small fighting boats, and concentrates mainly on MTBs. As early as 1953–54 the yard received special permission to build *Plejad* class fast MTBs for the Swedish Navy. Apart from the German Navy's *Jaguar* type MTBs, the yard has produced more than fifty fast fighting craft for foreign navies, including improved *Jaguar* types and entirely new designs. Lürssen designs are built under licence in many countries.

During the 1950s Abeking & Rasmussen built a number of R-boat types for Indonesia and are building a similar type for Brazil, in addition to the *Schütze* type minesweepers. Their most recent design is the SAR 33 type, a hard-chine planing boat, to be built in large numbers under licence in Turkey.

Blohm & Voss entered the field of warship export with the MEKO 360 type frigates for Nigeria and the Argentine Republic. Argentina is now building the Blohm MEKO 140 A 16 type frigate under licence.

Howaldtswerke-Deutsche Werft, Kiel, built the FS 1500 corvette, but are primarily engaged in building submarines. Under the management of Professor Gabler, Ingenieurkontor Lübeck (IKL) designed the successful types built by Howaldtswerke and Nordseewerke, Emden (now Thyssen-Nordseewerke). Norway has ordered fifteen Type 207, and Denmark has built two Type 205 under licence. More than thirty-five Type 209s have been built to foreign orders. After the lifting of the displacement limits on 21 July 1980, larger submarines have been built, such as the SSK Type 1500 for India and the Types TR 1700 and 1400 for Argentina.

A large number of small yards are building small patrol vessels to foreign orders.

REUNIFICATION

In the mid-1980s it became apparent that the understanding reached between the Soviet Union and the USA had lead to sweeping changes in the political climate. Both parties agreed to end the arms race by prioritising a policy of comprehensive disarmament. Following this, cracks began to appear in the Eastern Bloc as a result of the catastrophic economic situation its member countries found themselves in, a situation which the German Democratic Republic, so

Ship	Fate
Koni class frigates	
ROSTOCK (F 224)	Laid up 1991
HALLE (F 225)	Laid up 1991
Sassnitz class	
SASSNITZ (P 6165)	Laid up 1991. Renamed *Neustrelitz* (BG 23) and in service with the Border Police
Tarantul 1 class missile corvette	
HIDDENSEE (P 6166) (ex-*Rudolf Egelhofer*)	To USA 1991
Parchim class corvettes	
GADEBUSCH (P 6167)	Laid up 1991. To Indonesia
TETEROW (P 6168)	Laid up 1991. To Indonesia
LÜBZ (P 6169)	Laid up 1991. To Indonesia
WISMAR (P 6170)	Laid up 1991. To Indonesia
Kondor class ocean minesweepers	
TANGERHÜTTE (M 2669)	Laid up 1991
SÖMMERDA (M 2670)	Laid up 1991. To Indonesia
EISLEBEN (M 2671)	Laid up 1991. To Uruguay 13.11.91
BITTERFELD (M 2672)	Laid up 1991. To Indonesia
BERNAU (M 2673)	Laid up 1991. To Uruguay 13.11.91
EILENBURG (M 2674)	Laid up 1991. To Uruguay 13.11.91

long seen as the strongest of the Warsaw Pact countries and the Soviet Union's most reliable ally, was far from immune.

The Berlin Wall and the 1000km of mined border fence came down in November 1989 and Germany was formally reunited on 3 October 1990. On that date the *Nationale Volksarmee* (the GDR's 'National People's Army') and with it the *Volksmarine* (the 'People's Navy') were taken over by the *Bundeswehr* (the armed forces of the Federal Republic).

This take-over was purely symbolic in nature, as one of the conditions under which reunification was sanctioned by the former wartime allies was a reduction in German armed forces – within a context of general international force reductions – to a total *Bundeswehr* strength of 370,000 personnel by 1994. As a result of this, all units of the *Volksmarine* were decommissioned the day before reunification and only a small number of these vessels hoisted the Federal flag the next day. All of these vessels were either sold or otherwise disposed of by the end of 1991. The table below shows the *Volksmarine* units briefly taken over by the *Bundeswehr* and their eventual fates. In 1995, the strength of the *Bundeswehr* (including *Bundesmarine*) is to be further reduced to *c*320,000 personnel.

MAJOR SURFACE SHIPS

Ex-US FLETCHER class *destroyers*

Armament: 4–5in/38 Mk XII (4×1), 6–3in/50 Mk XXII (3×2), 5–533mm (21in) TT (1×5), 2–533mm ASW TT(2×1), 2 Hedgehog
Sensors: Radar SPS-6, SPS-10, Mk 25, Mk 36, Mk 35; sonar SQS-29
Other particulars As US *Fletcher* class

No	Name	Builder	Acquired	Fate
D 170	ZERSTORER 1 (ex-*Anthony*)	Bath Iron Wks	17.1.58	Sunk as target 16.5.79
D 171	ZERSTORER 2 (ex-*Ringold*)	Federal, Kearny	14.7.59	To Greece Sep 1981
D 172	ZERSTORER 3 (ex-*Wadsworth*)	Bath Iron Wks	6.10.59	To Greece Oct 1980
D 178	ZERSTORER 4 (ex-*Claxton*)	Federal, Kearny	15.12.59	To Greece Feb 1981
D 179	ZERSTORER 5 (ex-*Dyson*)	Consolidated, Orange	23.2.60	To Greece Feb 1982
D 180	ZERSTORER 6 (ex-*Charles Ausburne*)	Consolidated, Orange	12.4.60	Sold for BU Oct 1968

During 1958–60 these six US *Fletcher* class destroyers were transferred under MDAP for a lease period of five years; they did not receive German names. Before being handed over they were modernised with new weapons, new electronics and additional German command systems. During later refits the bridge structure was altered.

In 1974 *Zerstörer 4* carried a containerised 76mm/62 OTO Melara gun for trials purposes. From 1973 the quintuple TT were removed from all ships.

Zerstörer 6 was sold for scrap and *Zerstörer 1* was sunk in the Mediterranean as a torpedo target. In 1976 the remaining ships were purchased by the Federal German Navy and handed over to the Greek Navy in 1981–82 as part of the military aid programme (see Greece).

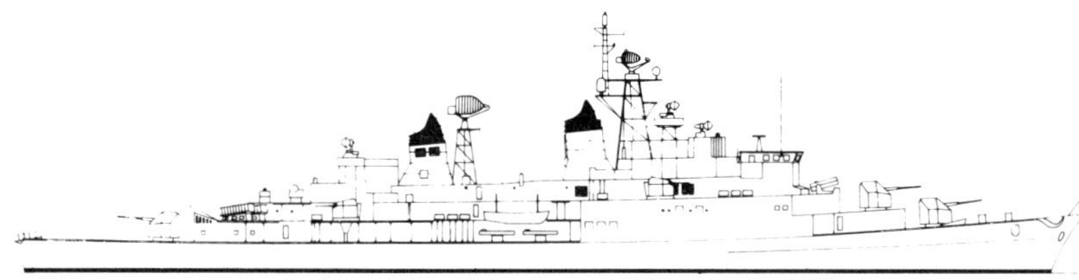

Hamburg 1979

HAMBURG class *destroyers*

Displacement: 3340t standard; 4330t full load
Dimensions: 420ft wl, 438ft 9in oa × 44ft × 17ft
128m, 133.7m × 13.4m × 5.2m
Machinery: 2-shaft Wahodag turbines, 4 Wahodag boilers, 72,000shp (52,992kW) = 36kts. Oil 800t. Range 6000nm at 13kts
Armament: 3–100mm/55 (3×1), 4 MM38 Exocet SSM (2×2), 8–40mm/70 (4×2), 4–533mm ASW TT (4×1), 2–375mm ASW mortars (2×4)
Sensors: Radar KH 14/9, DA-08, LW-04, SGR-103, M-45; sonar ELAC 18V
Complement: 284

No	Name	Builder	Laid down	Launched	Comp	Fate
D 181	HAMBURG	Stülcken, Hamburg	29.1.59	26.3.60	23.3.64	Decomm 24.2.94
D 182	SCHLESWIG-HOLSTEIN	Stülcken, Hamburg	20.8.59	20.8.60	12.10.64	Decomm 15.12.94
D 183	BAYERN	Stülcken, Hamburg	14.9.60	14.8.62	6.7.65	Decomm 16.12.93, sold for BU 25.4.95
D 184	HESSEN	Stülcken, Hamburg	15.2.61	4.5.63	8.10.68	Decomm 23.9.90, sold for BU 8.9.91

Originally the German Navy wanted twelve Type 101 destroyers and, in fact, ordered them in August 1957, but it transpired that they could not be built on the planned displacement of 2500t. So the Western European Union by a *post facto* decision raised the displacement to 6000t.

All four ships were modernised 1974–77: the boilers were modified to burn light oil, and the lattice mainmast was removed. The 'X' position 100mm was replaced by two MM38 Exocet twin launchers. In 1978 all ships received an enclosed bridge. During subsequent refits the superstructure and funnel caps were modified and the electronics were updated. The five TTs were replaced by four short ASW TTs. The data table refers to the ships' 1982 state.

Hessen was sold for breaking up 8 September 1991, the remaining three following in 1994–95. This class will be replaced by the Type 123 frigates.

Hessen 29.11.80
L & L van Ginderen

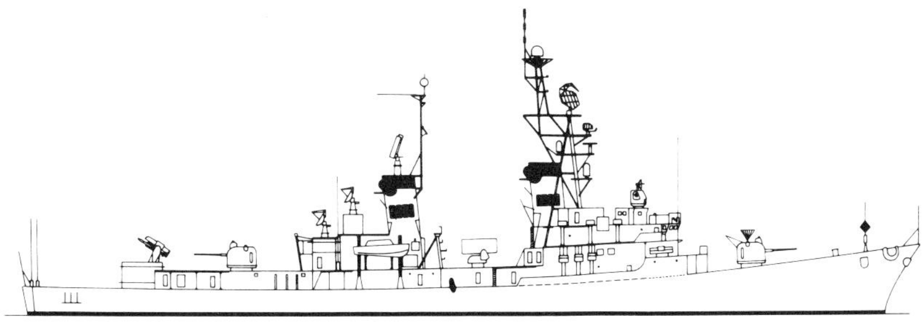

Lütjens 1975

These Type 103 destroyers are based on the USN family of 'Tartar guided-missile destroyers SCB 155' (*Charles F Adams* class Mod 14 – *Goldborough*, DDG 20 and following vessels). They stemmed from a US/German agreement of 11 May 1964; the USN ordered them 1 April 1965 as DDG 28–DDG 30.

They differ from the original *Charles F Adams* class in having modified funnels with side exhausts, and different masts. Originally it was intended to build six ships in German shipyards, but three were ordered from Bath Iron Works, USA. At the end of the 1970s their boilers were modified to burn light oil; at the beginning of the 1980s their weapons and electronics were updated. *Mölders* completed her refit on 29 March 1984, *Rommel* on 26 July 1985 and *Lütjens* on 16 December 1986. Following this updating the ships were reclassified Type 103B. See US section, *Charles F Adams* class.

LUTJENS class (US CHARLES F ADAMS type) *guided missile destroyers*

Displacement: 3370t standard; 4717t full load
Dimensions: 420ft 4in wl, 441ft 3in oa × 47ft 3in × 14ft 9in
128.1m, 134.4m × 14.4m × 4.5m
Machinery: 2-shaft General Electric turbines, 4 boilers type D-V2M, 70,000shp (51,520kW) = 36kts. Oil 930t. Range at 4500nm at 20kts
Armament: 1 Tartar Mk 13 (40 SM-1MR SAM and Harpoon SSM missiles), 2–5in/54 Mk XVI (US Mk 42; 2×1), 1 ASROC ASW (1×8), 6–324mm Mk 32 ASW TT
Sensors: Radar SPS-40, SPS-10, SPS-52, KH 14/9, SPG-51C, SPG-53; sonar SQS-23
Complement: 333

No	Name	Builder	Laid down	Launched	Comp	Fate
D 185	LUTJENS	Bath Iron Wks	1.3.66	11.8.67	22.3.69	Extant 1995
D 186	MOLDERS	Bath Iron Wks	12.4.66	13.4.68	20.9.69	Extant 1995
D 187	ROMMEL	Bath Iron Wks	22.8.67	1.2.69	2.5.70	Extant 1995

Mölders 8.8.94
L & L van Ginderen

Gneisenau 1961
L & L van Ginderen

These seven frigates belonged to the Royal Navy's *Black Swan*, modified *Black Swan* and 'Hunt' Types 2 and 3 classes, and all had seen service during the Second World War. They were purchased in 1957, renamed and designated Type 138. They were modernised in British yards and all except *Scheer* had their original armament when transferred. *Scharnhorst* and *Gneisenau* were later rebuilt: their bridges, masts and funnels were changed and they received standard NATO weapons. During a refit *Scheer* received additional radar, and the armament of *Raule* and *Brommy* was changed; only *Raule* received an ASW component.

During their entire service with the Federal German Navy they served as training ships: *Gneisenau* and *Scharnhorst* for gunnery, *Raule* and *Brommy* for underwater weapons, *Scheer* for radar, and *Hipper* and *Graf Spee* for cadet training. Towards the end of their active lives it was envisaged that *Hipper* and *Graf Spee* be converted to air-sea rescue/pilot vessels, but this was dropped because of the cost. All were sold for scrap 1967–71 except *Scharnhorst* which served as a training hulk for damage control until being sold for BU in 1990. The acquisition dates relate to their commissioning in the Federal German Navy.

Ex-British HUNT class *frigates*

Displacement: 1050t standard; 1610t full load (*Gneisenau*)
1087t standard; 1490t full load (*Raule, Brommy*)
Dimensions: 272ft pp, 264ft 3in wl, 282ft 9in oa × 31ft 6in × 12ft 6in (*Gneisenau*)
83.0m, 80.6m, 86.1m × 9.6m × 3.8m
272ft wl, 281ft 6in oa × 31ft 6in × 10ft 3in (*Raule, Brommy*)
83.0m, 85.9m × 9.6m × 3.1m
Machinery: 2-shaft Parsons turbines, 2 Admiralty boilers, 19,000shp (13,984kW) = 27kts. Oil 280t. Range 2000nm at 12kts (*Gneisenau*)
Armament: 6–4in/45 (3×2), 6–2pdr (1×4, 2×2); after conversion 1–100mm/55, 3–40mm/70 (1×2, 2×1) (*Gneisenau*)
4–4in/45 (2×2), 5–2pdr (1×4, 1×1), 2–533mm TT, 2 Hedgehog; after conversion 1–40mm/70, 14–375mm ASW mortar (1×4), 2–533mm ASW TT (2×1) (*Raule, Brommy*)
Complement: 83 (*Gneisenau*), 170 (*Raule, Brommy*)

No	Name	Builder	Acquired	Fate
F 212	GNEISENAU (ex-*Oakley*)	Yarrow	18.10.58	BU 1977
F 217	RAULE (ex-*Albrighton*)	John Brown	14.5.59	BU 1971
F 218	BROMMY (ex-*Eggesford*)	White	14.5.59	BU 1979

Ex-British BLACK SWAN class *frigates*

Displacement: 1470t (1580t *Scheer*) standard; 1925t full load
Dimensions: 283ft 3in wl, 299ft 9in oa × 40ft × 11ft 6in
86.4m, 91.3m × 12.2m × 3.5m (*Scharnhorst, Hipper*)
283ft 3in wl, 299ft 9in oa × 37ft 6in × 11ft
86.4m, 91.3m × 11.4m × 3.4m (*Graf Spee*)
322ft 4in oa × 41ft × 11ft 2in
98.2m × 12.4m × 3.4m (*Scheer*)
Machinery: 2-shaft General Motors/Parsons turbines, 2 Admiralty (Foster Wheeler *Scharnhorst*) boilers, 4200shp(3134kW) = 18.2kts. Oil 440t, later 341t (*Scharnhorst*), 460t (*Graf Spee*), 387t (*Scheer*). Range 3200nm (2783nm *Scheer*) at 12kts
Armament: 6–4in/45 (3×2), 12–20mm (6×2); after conversion 2–100mm/55 (2×1), 4–40mm/70 (2×1, 1×2) (*Scharnhorst*)
6–40mm/70 (2×2, 2×1), 4–40mm/70 (2×2 *Scheer*), 2 Hedgehog (*Hipper, Graf Spee*)
Complement: 147 (*Scharnhorst*), 147 + 40 (*Hipper, Graf Spee*), 60 (*Scheer*)

No	Name	Builder	Acquired	Fate
F 213	SCHARNHORST (ex-*Mermaid*)	Denny	28.5.59	BU 1990
F 214	HIPPER (ex-*Actaeon*)	Thornycroft	10.1.59	BU 1967
F 215	GRAF SPEE (ex-*Flamingo*)	Yarrow	21.2.59	BU 1967
F 216	SCHEER (ex-*Hart*)	Stephen	21.5.59	BU 1971

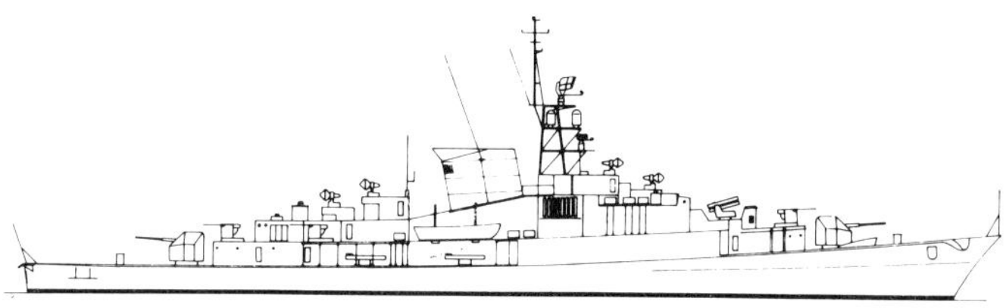

Emden as completed

These Type 120 frigates were ordered in March 1957 having been designed in 1955–57 as 'Geleitboot 55' (Escort 55). They were the first German warships to receive CODAG propulsion. During numerous refits they received increased hull protection and stability, and updated electronics. The original two 533mm TT were subsequently replaced by four tubes for short 533mm ASW torpedoes. *Köln*, *Emden* and *Karlsruhe* were replaced by Type 122 frigates of the same name. *Köln* was converted into a stationary training hulk for damage control and fire-fighting exercises in 1989, the remaining ships being sold to Turkey 1983–89.

KÖLN class *frigates*

Displacement: 2090t standard; 2750t full load
Dimensions: 344ft 6in wl, 360ft 4in oa × 36ft 1in × 15ft 1in
105.0m, 109.8m × 11.0m × 4.6m
Machinery: 2-shaft (CP propellers) CODAG: 4 MAN diesels, plus 2 Brown Boveri gas turbines, 12,000bhp + 24,000shp = 32kts. Oil 360t. Range 3450nm at 12kts
Armament: 2–100mm/55 (2×1), 6–40mm/70 (2×2, 2×1), 4–533mm ASW TT (4×1), 2–375mm ASW mortars (2×4)
Sensors: Radar DA-02, SGR-103, KH 14/9, M44, M45; sonar PAE/CWE hull-mounted M/F set
Complement: 238

No	Name	Builder	Laid down	Launched	Comp	Fate
F 220	KÖLN	Stülcken, Hamburg	21.12.57	6.12.58	15.4.61	TS 1989
F 221	EMDEN	Stülcken, Hamburg	15.4.58	21.3.59	24.10.61	To Turkey 23.9.83
F 222	AUGSBURG	Stülcken, Hamburg	29.10.58	15.8.59	7.4.62	Decomm 30.3.88, BU 1989
F 223	KARLSRUHE	Stülcken, Hamburg	15.12.58	24.10.59	15.12.62	To Turkey 28.3.83
F 224	LÜBECK	Stülcken, Hamburg	28.10.59	23.7.60	6.7.63	To Turkey 1.12.88
F 225	BRAUNSCHWEIG	Stülcken, Hamburg	28.7.60	3.2.62	16.6.64	To Turkey 4.7.89

Braunschweig 1971
C & S Taylor

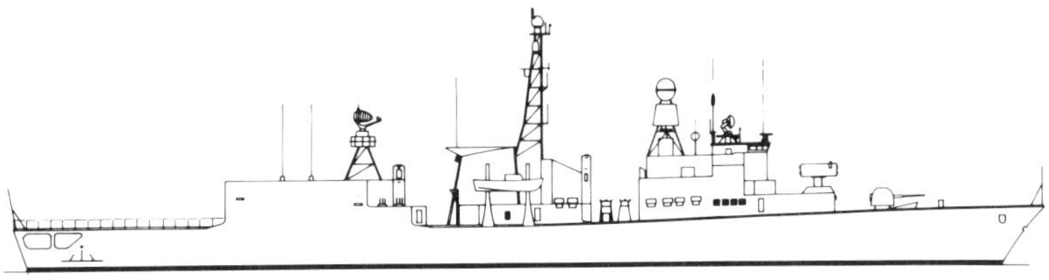

Bremen as completed

BREMEN class *frigates*

Displacement: 3700t standard; 3800t full load
Dimensions: 399ft 7in wl, 440ft oa × 47ft 10in × 19ft 9in
121.8m, 128.0m × 14.6m × 6.0m
Machinery: 2-shaft (CP propellers) CODOG: 2 GE LM 2500 gas turbines, 50,000shp (36,800kW) = 30kts, plus 2 MTU diesels, 10,400bhp (7654kW) = 18kts. Range 4000nm at 18kts
Armament: 8 Harpoon SSM (2×4), 1–76mm/62, 1 Sea Sparrow SAM (1×8), 4 Mk 32 ASW TT (2×2), 2 helicopters
Sensors: Radar 3RM-20, LW-08, WM-25, HSA STIR; sonar SQS-21B(Z)
Complement: 189

No	Name	Builder	Laid down	Launched	Comp	Fate
F 207	BREMEN	Bremer Vulkan	9.7.79	27.9.79	7.5.82	Extant 1995
F 208	NIEDERSACHSEN	AG Weser	9.11.79	9.6.80	15.10.82	Extant 1995
F 209	RHEINLAND-PFALZ	Blohm & Voss	25.9.79	3.9.80	9.5.83	Extant 1995
F 210	EMDEN	Thyssen-Nordseewerke	23.6.79	17.12.80	7.10.83	Extant 1995
F 211	KÖLN	Blohm & Voss	16.6.80	29.5.81	19.10.84	Extant 1995
F 212	KARLSRUHE	Howaldtswerke, Kiel	10.3.81	8.1.82	19.4.84	Extant 1995
F 213	AUGSBURG	Bremer Vulkan	4.4.87	17.9.87	3.10.89	Extant 1995
F 214	LÜBECK	Thyssen-Nordseewerke	1.6.87	15.10.87	19.3.90	Extant 1995

Koln 24.9.86 L & L van Ginderen

Preparatory work on the design of new fighting ships began with the institution of the new navy in 1955–56, but until the 1960s all design was based on the knowledge and experience of wartime officers and civil servants who had re-enlisted. The German shipbuilding industry was faced with the problem of closing a ten-year gap in design experience. The design of the *Hamburg* and *Köln* classes was commendable given the short space of time in which it was produced, but like other contemporary designs they suffered from the WEU's displacement limit, which resulted in too many components for the displacement. This is why all the units of the first shipbuilding programme look so voluminous and have such high superstructures.

The role of Types 101 and 120 were very clearly defined: escorting convoys against submarines and aircraft, repelling of surface attack plus minelaying ability. At this time the submarine, the chief opponent, was diesel-powered, and the aircraft menace was represented by fighter-bombers. The weaponry of this first generation of German surface combatants clearly matched these requirements: panoramic aircraft-warning radars plus gun-directors for quick reaction. The 100mm turrets plus the 40mm produced the necessary gun component. The destroyers' propulsion consisted of conventional steam turbines, the frigates had diesels plus gas turbines for dash speed. But the tremendous development in engineering technology, nuclear propulsion, new weapons and electronic guidance systems, made conventional warships of this first generation quickly obsolete.

During the 1960s preliminary studies for a follow-up type were produced. The first design was for an AA corvette of 1200t displacement, ten of which were envisaged. But the increasing submarine menace necessitated an ASW component which forced the displacement up to 2500t. Because of the restricted defence budget of the German Federal Republic at this time, the project was dropped and in place of the new AA/ASW corvettes both the *Lütjens* class destroyers and the ten boats of the *Zobel* class were modernised.

The next design stage was the 'Frigate 70' which was to have incorporated all the required components. Although the budgetary board had accepted on 23 January 1969 and the defence board of the Bundestag had accepted this Type on 8 February 1969, the orders were cancelled because the design stage had not been completed and the displacement had risen to 3600t while the building costs had soared.

So design studies started again from scratch by comparing various foreign frigate designs of which the Netherlands' 'Standard' frigate of the *Kortenaer* class turned out to be optimal. On 17 July 1975, a 'Memorandum of Understanding' with the Netherlands was signed, followed by acceptance by the German Defence Board on 28 January 1976.

On 26 November 1977 six units were ordered, designated the Type 122, and a further two on 6 December 1985. The general project responsibility lay with the Bremer Vulkan shipyard, the other units being built by Blohm & Voss, Thyssen-Nordseewerke, AG Weser and HDW, Kiel. The final fitting out was the job of Bremer Vulkan. The NATO Sea Sparrow system employs a Mk 21 launcher; twenty-four missiles are carried. The helicopters are WG-13 Lynx with DAQS-13D dipping sonar and the Canadian Bear Trap system is fitted to the helicopter deck. The US RAM (Rolling Airframe Missile) point defence SAM is to be fitted atop the hangar in two 24-round launchers, when the system becomes operational. The new frigates replaced the Type 119 destroyers (*Fletchers*) and three units of the *Köln* class. Although the fighting potential of the navy's surface combatants has risen, the number of destroyers and frigates decreases.

Brandenburg 20.6.95
L & L van Ginderen

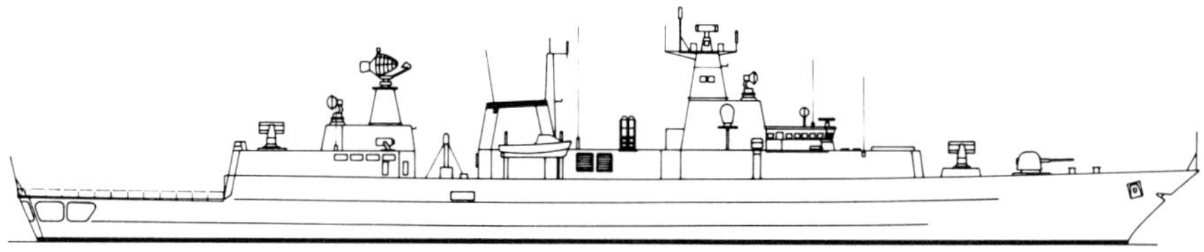

Brandenburg design

These frigates replace the destroyers of the *Hamburg* class. First units ordered 28 June 1989 from a consortium headed by Blohm & Voss, with both Thyssen Noordseewerke and HDW, Kiel, also participating. The VLS for the RAM missiles was designed so that it could be expanded to take double the number of missiles. The MM38 Exocet SSM were taken from the *Hamburg* class frigates.

BRANDENBURG class *frigates*

Displacement: 3660t standard; 4275t full load
Dimensions: 413ft 9in wl, 455ft 8in oa × 57ft 1in × 20ft 8in
126.1m, 138.9m × 16.7m × 6.3m
Machinery: 2-shafts (CP propellers): 2 gas turbines, 25,480shp (38,000kW) = 29kts, plus 2 MTU diesels, 5630bhp (8400kW) = 18kts
Armament: 4 MM38 Exocet SSM, 1–76mm OTO Melara, 2 RAM (21 missiles each), 1 Sea Sparrow SAM VLS (16 missiles), 4 Mk 32 ASW TT (2×2), 2 Düppel/IR launchers Breda, 2 Sea Lynx helicopters
Sensors: Radar; sonar; ECM
Complement: 219

No	Name	Builder	Laid down	Launched	Comp	Fate
F 215	BRANDENBURG	Blohm & Voss	11.2.92	28.8.92	14.10.94	Extant 1995
F 216	SCHLESWIG-HOLSTEIN	Bremer Vulkan	1.7.93	8.6.94		Completing
F 217	BAYERN	Thyssen-Nordseewerke	16.12.93	30.6.94		Completing
F 218	MECKLENBURG-VORPOMMERN	HDW, Kiel	23.11.93	23.2.95		Completing

SUBMARINES

HAI class (ex-Type XXIII) *coastal submarines*

Displacement: 233.9t surface; 275t submerged
Dimensions: 118ft 6in oa × 9ft 9in × 12ft 3in
36.1m × 3.0m × 3.7m
Machinery: 1-shaft diesel-electric: 1 Mercedes-Benz diesel, plus 1 electric motor, 635shp (467kW) = 9.7kts/12.5kts. Oil 18t. Range 1350nm at 9kts surfaced
Complement: 17

No	Name	Builder	Acquired	Fate
S 170	HAI (ex-*U 2365*)	Deutsche Werft, Hamburg	15.8.57	Foundered 14.9.66
S 171	HECHT (ex-*U 2367*)	Deutsche Werft, Hamburg	1.10.57	BU 1969

Belonging to the Type 240, these boats were former *Kriegsmarine* submarines of the coastal Type XXIII. *U 2365* was commissioned on 2 March 1945 and lost 8 May 1945 after a bomb hit, while *U 2367* was commissioned 17 March 1945 and sank after a collision on 5 May 1945. In 1956 they were raised and recommissioned as *UW 20 Hai* and *UW 21 Hecht*. They were used as training boats for the weapons school. Both received numerous modifications and subsequently received equipment envisaged for the new Type 201. They were enlarged, with a new conning tower and the old MWM diesels were replaced by Mercedes-Benz. From now on they had – like the follow-up types – pure electric propulsion; the diesels were used solely for charging the batteries, the propellers were driven by electric motors only. *Hai* was lost during a storm in 1966.

WILHELM BAUER (ex-Type XXI) *experimental submarine*

Displacement: 1620t surfaced; 1820t submerged
Dimensions: 247ft 5in wl, 251ft 6in oa × 21ft 7in × 22ft 6in
75.4m, 76.7m × 6.6m × 6.8m
Machinery: As Type 206. Range 11,000nm/490nm at 12kts/3kts submerged
Armament: 4–533mm TT
Complement: 57

This was the former *U 2540* of the *Kriegsmarine* built by Blohm & Voss, launched 13 January 1945, commissioned 24 February 1945 and lost near the Flensburg lightship on 3 May 1945 after bomb hits. Raised 3 November 1958, she was commissioned 1 September 1960 after a total refit as *Wilhelm Bauer* (Y 888). Incorporated as the only Type 241 boat, she served for experimental purposes only, first with a naval crew, after her rebuilding at Blohm & Voss, and then with a civilian crew acting for the Trials Command 71 of the *Bundeswehr*. The data above relates to the submarine in her final guise, with a trials version of the Type 206 propulsion unit. She was never considered an operational unit and was decommissioned on 15 March 1982.

Type 201 *coastal submarines*

Displacement: 395t surfaced; 433t submerged
Dimensions: 139ft wl, 142ft 9in oa × 15ft × 13ft
42.4m, 43.5m × 4.6m × 4.0m
Machinery: 1-shaft electric drive: 2 Mercedes-Benz diesels, plus 1 electric motor, 1200bhp/1500shp = 10kts/17.5kts. Range 3800nm at 10kts
Armament: 8–533mm TT (bow)
Complement: 21

No	Name	Builder	Launched	Fate
S 180	U 1 (i)	Howaldtswerke, Kiel	21.10.61	BU 1971
S 181	U 2 (i)	Howaldtswerke, Kiel	25.1.62	BU 1971 (?)
S 182	U 3	Howaldtswerke, Kiel	7.5.62	BU 1971

Originally consisting of twelve boats, this class of coastal submarines belonged to the Type 201 and was first designed as '*U-Boot 55*' by Ingenieurkontor Lübeck (IKL). The boats were to be section-built and were ordered in March 1959. In November 1958 it was decided that the hulls should be non-magnetic, but the material used quickly showed corrosive defects. In June 1962 *U 1* showed the first signs of structural weakness which quickly increased so that *U 1* and *U 2* had to be decommissioned. Later they were entirely rebuilt (see Type 205).

U 3 was first commissioned for the Royal Norwegian Navy as a training boat, as Norway wanted to have fifteen boats of this type. Norway's coastal waters had no corrosive effect on the hull. In June 1964 the boat was handed back to Germany and was commissioned as a training boat. At the end of her service career she was used for shock-testing trials and scrapped in 1971. For test purpose *U 1* for a short time shipped an above-surface stern torpedo tube.

Type 202 *coastal submarines*

Displacement: 100t surfaced; 137t submerged
Dimensions: 72ft 3in wl, 75ft 9in oa × 11ft 3in × 9ft
22.0m, 23.1m × 3.4m × 2.7m
Machinery: 1-shaft electric drive: 2 Mercedes-Benz diesel. 330bhp (243kW) = 6kts/13kts. Range 400nm at 4kts dived
Armament: 2–533mm TT (bow)
Complement: 6

No	Name	Builder	Launched	Fate
S 172	HANS TECHEL	Atlaswerke, Bremen	15.3.65	BU after 1970
S 173	FRIEDRICH SCHURER	Atlaswerke, Bremen	10.11.65	BU after 1970

Type 202 were originally built in an endeavour to develop a Type on this displacement. They made extensive trials, first with naval crews, and later civilian. They were ordered in February 1959. *Friedrich Schürer* had a Kort jet on her vertical rudder. Hopes for them were not realised and they were soon decommissioned and scrapped.

Hai 1957
CMP

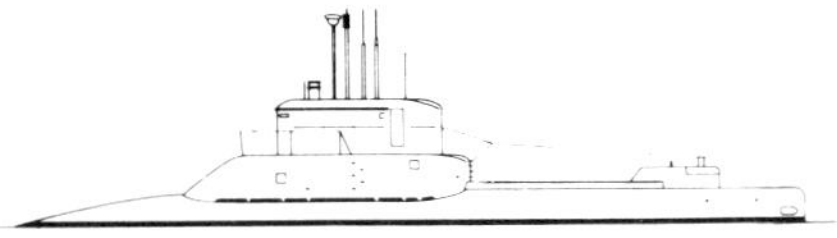

U 12 1971
NB 1/750 scale

Type 205 *coastal submarines*

Displacement: 419t surfaced; 455t submerged
Dimensions: 141ft 3in pp, 144ft 5in (*U 1, 2, 9*) 143ft 9in (*U 10*) 150ft 4in (*U 11, 12*) oa × 15ft × 13ft 9in
43.3, 44.0m (43.8m) (45.8m) × 4.5m × 4.2m
Sensors: Radar Calypso; sonar SRS-M1H, GHG AN5039A1
Other particulars: As Type 201

No	Name	Builder	Launched	Fate
S 183	U 4	Howaldtswerke, Kiel	25.8.62	BU 1977
S 184	U 5	Howaldtswerke, Kiel	20.11.62	BU 1975
S 185	U 6	Howaldtswerke, Kiel	30.1.63	BU 1977
S 186	U 7	Howaldtswerke, Kiel	10.4.63	BU 1976
S 187	U 8	Howaldtswerke, Kiel	19.6.63	BU 1977
S 188	U 9	Howaldtswerke, Kiel	20.10.66	Museum ship 19.8.93
S 189	U 10	Howaldtswerke, Kiel	5.6.67	Decomm 16.2.93, museum ship Wilhelshaven
S 190	U 11	Howaldtswerke, Kiel	9.2.68	Extant 1995
S 191	U 12	Howaldtswerke, Kiel	10.9.68	Extant 1995
S 180	U 1 (ii)	Howaldtswerke, Kiel	17.2.67	Decomm 29.11.91 and sold to Thyssen-Nordseewerke, Emden, BU 1993
S 181	U 2 (ii)	Howaldtswerke, Kiel	15.7.66	BU 1993

With the exception of *U 1* (ii) and *U 2* (ii), all boats were originally to have been of the Type 201. The order was placed in March 1959 to the particulars of this class. Because of the corrosive defects in *U 1* and *U 2*, the hull of the following boats, *U 4–U 8*, were covered with tin, and suffered severe operational restrictions. Construction of *U 9–U 12* was stopped until a newly developed non-magnetic steel could be produced. *U 1* and *U 2* were entirely re-hulled with this steel, their inner construction remaining intact. They could carry sixteen mines in lieu of their eight torpedoes. They differed from the Type 201 in having a totally enclosed conning tower whose shape differed to form four sub-types: *U 7* differing from *U 4–U 6*, *U 8* and from *U 9–U 12* and *U 11* and *U 12* also had a different rudder arrangement. *U 4–U 8* belonged to the training group and were scrapped after decommissioning.

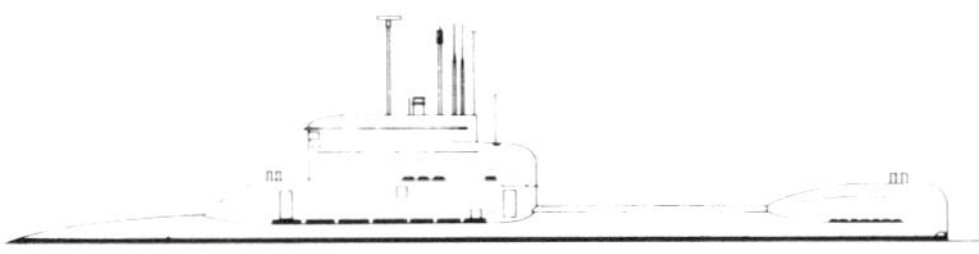

U 28 1976
NB 1/750 scale

Type 206 *coastal submarines*

Displacement: 456t surfaced; 500t submerged
Dimensions: 159ft 6in oa × 15ft × 14ft
48.6m × 4.6m × 4.3m
Machinery: As Type 201. Oil 23.5t
Armament: 8–533mm TT (16 torpedoes), 24 mines
Sensors: Radar Calypso; sonar AN410A4, AN5039A1, DBQS-21D
Complement: 21

U 25 3.8.92 L & L van Ginderen

No	Name	Builder	Launched	Fate
S 192	U 13	Howaldtswerke, Kiel	28.9.71	Extant 1995
S 193	U 14	Rheinstahl-Nordseewerke, Emden (RNSW)	1.2.72	Extant 1995
S 194	U 15	Howaldtswerke, Kiel	15.6.72	Extant 1995
S 195	U 16	RNSW, Emden	29.8.72	Extant 1995
S 196	U 17	Howaldtswerke, Kiel	10.10.72	Extant 1995
S 197	U 18	RNSW, Emden	31.10.72	Extant 1995
S 198	U 19	Howaldtswerke, Kiel	15.12.72	Extant 1995
S 199	U 20	RNSW, Emden	16.1.73	Extant 1995
S 170	U 21	Howaldtswerke, Kiel	9.3.73	Extant 1995
S 171	U 22	RNSW, Emden	27.3.73	Extant 1995
S 172	U 23	RNSW, Emden	22.5.73	Extant 1995
S 173	U 24	RNSW, Emden	26.6.73	Extant 1995
S 174	U 25	Howaldtswerke, Kiel	23.5.73	Extant 1995
S 175	U 26	RNSW, Emden	20.11.73	Extant 1995
S 176	U 27	Howaldtswerke, Kiel	21.8.73	Extant 1995
S 177	U 28	RNSW, Emden	22.1.74	Extant 1995
S 178	U 29	Howaldtswerke, Kiel	5.11.73	Extant 1995
S 179	U 30	RNSW, Emden	26.3.74	Extant 1995

The Type 206 boats were designed by the IKL in 1964–65, *U 25–U 30* being the replacement units for the Type 205 boats *U 4–U 8*. In April 1969 the orders for *U 13–U 24* and in autumn 1970 those for *U 25–U 30* were placed. The general interior arrangement resembles that of the Types 201 and 205, the only visual difference being a rounded bow with sonar dome. Improvements were the use of wire-guided torpedoes, decreased underwater noise, better steering and higher underwater manoeuvrability. Submerged range is 200nm at 5kts.

In addition to the sixteen mines that can be carried in lieu of the torpedoes, this class can carry twenty-four extra mines in a pair of external GRP containers. This ingenious system was developed by IKL to allow the submarines to embark both torpedoes and mines simultaneously if required.

In 1988–92 the following were converted to Type 206A submarines: *U 15–U 18, U 22, U 23, U 24, U 25, U 26, U 28, U 29*. Conversion entailed fitting of updated electronics, weapons, periscopes, navigational equipment and improved crew accommodation.

AMPHIBIOUS WARFARE VESSELS

Ex-US LSM type *medium landing ships*

Displacement: 830t standard; 1080t full load
Dimensions: 196ft 6in wl, 203ft 7in oa (*Krokodil* after conversion 191ft 7in, 201ft 10in) × 34ft 6in × 8ft 3in
59.8m, 62.1m, (58.4m, 61.5m) × 10.5m × 2.5m
Machinery: 2 shafts, 2 GMC diesels, 2800bhp (2061kW) = 12.5kts. Oil 195.8t. Range 2500nm at 12kts
Armament: 2–40mm/56 (1×2)
Complement: 50

No	Name	Builder	Acquired	Fate
L 750	KROKODIL (ex-*LSM 537*)	Brown, SB Houston	15.8.58	Sold 10.5.76
L 751	EIDECHSE (ex-*LSM 491*)	Brown, SB Houston	15.8.58	Sold 10.5.76
L 752	SALAMANDER (ex-*LSM 553*)	Charleston N Yd	15.8.58	Sold 22.5.70
L 753	VIPER (ex-*LSM 558*)	Charleston N Yd	15.8.58	Sold 22.5.70

Krokodil 23.8.58 L & L van Ginderen

Ex-US LSMR type *support landing ships*

Displacement: 994t standard; 1084t full load
Dimensions: 204ft 3in wl, 210ft 3in oa × 34ft 6in × 9ft 4in
62.2m, 64.1m × 10.5m × 2.8m
Machinery: 2 shafts, 2 GMC diesels, 2800bhp (2061kW) = 12.5kts. Oil 164t. Range 2500nm at 12kts
Armament: 1–5in/38, 4–40mm/56 (2×2), 20–5in (127mm) rocket projectors (10×2), after conversion: 1–5in/38, 4–40mm/70 (2×2), 16–5in (127mm) rocket projectors (8×2)
Complement: 100

No	Name	Builder	Acquired	Fate
L 754	OTTER (ex-*Smyrna River*, LSMR 532)	Brown SB, Houston/ Charleston N Yd	15.8.58	Sold 8.9.71
L 755	NATTER (ex-*Thames River*, LSMR 534)	Brown SB, Houston/ Charleston N Yd	15.8.58	Sold 8.9.71

In 1958 the Federal German Navy bought four former LSMs and two former LSMRs from the USN, all built in 1944–45. The vessels were taken over 15 August 1958 at Charleston, Va and commissioned 5 September 1958. They were incorporated as the Types 550 (*Eidechse* and others) and 551 (*Otter*, *Natter*). During their service they remained unmodified, with the exception of the armament as shown in the particulars. *Krokodil* was designed as a command and hospital craft and therefore received a helicopter platform over the stern and a new hinged bow-ramp while the folding doors were removed. No other unit of this class was converted to this scheme. *Krokodil* was sold 10 May 1976 and served as a pipeline carrier for Pauses Shipping, as did *Eidechse*. *Salamander* was scrapped and *Viper* was sold to the Norfolk Line, Scheveningen, and renamed *Duchess of Holland*.

BUTT class *utility landing craft*

Displacement: 166t standard; 397t full load
Dimensions: 120ft 5in wl, 131ft 5in oa × 29ft × 7ft 3in
36.7m, 40.0m × 8.8m × 2.2m
Machinery: 2 shafts, 2 MWM diesels, 1320shp (971kW) = 12kts. Oil 40t
Armament: 2–20mm Mk 20 (2×1)
Complement: 15

Class (no, launched, fate):
Flunder (L 760, 6.1.66, extant 1995), *Karpfen* (L 761, 5.1.66, decomm 30.1.92), *Lachs* (L 762, 17.2.66, extant 1995), *Plötze* (L 763, 18.3.66, extant 1995), *Rochen* (L 764, 18.3.66, decomm 7.2.92, to Greece 20.10.92), *Schlei* (L 765, 17.5.66, extant 1995), *Stör* (L 766, 18.5.66, decomm 16.9.92), *Tümmler* (L 767, 14.6.66, decomm 16.9.92), *Wels* (L 768, 15.6.66, decomm 11.12.92), *Zander* (L 769, 13.7.66, extant 1995), *Butt* (L 788, 28.3.65, decomm 4.12.92), *Brasse* (L 789, 28.3.65, decomm 16.4.92), *Barbe* (L 790, 26.11.65, decomm 26.9.91, to Greece 13.3.92), *Delphin* (L 791, 25.11.65, decomm 26.9.91, to Greece 20.2.92), *Dorsch* (L 792, 17.3.66, decomm 26.9.91, to Greece 13.3.92), *Felchen* (L 793, 19.4.66, decomm 25.10.92, to Greece 31.1.93), *Forelle* (L 794, 20.4.66, decomm 1.11.91, to Greece 31.1.92), *Inger* (L 795, 14.7.66, decomm 9.9.92), *Makrele* (L 796, 22.8.66, decomm 8.11.91, to Greece 20.2.92), *Muräne* (L 797, 23.8.66, decomm 14.2.92, to Greece 20.10.92), *Renke* (L 798, 22.9.66, to Greece 16.11.89), *Salm* (L 799, 23.9.66, to Greece 16.11.89).

All built by Howaldtswerke, Hamburg. These multi-purpose landing craft of the type MZL 63, being an improved version of the US *LCU 1640* class, were ordered in 1964 and belong to the Type 520. Shortage of crews and changing concepts meant that most of them were transferred to the reserve immediately after delivery, laid up and cocooned. With the exception of *Renke* and *Salm* all were finally commissioned in 1973. From 5 August 1974 to 9 September 1992, *Inger* served as training boat for the Seamen Training Group.

All were refitted in the 1970s, receiving an enclosed bridge and a modern AA armament consisting of 2–20mm Rheinmetall guns. All were designed to carry minerails for the transport and laying of mines. Stern and stem were designed in such a manner that the boats could form a chain, so as to produce a massive floating bridge.

OTHER LANDING CRAFT

Apart from these twenty-two boats, there are twenty-eight LCM type boats (*LCM 1–28*). Carrying a five-man crew they serve for transport and supply purposes, but are unarmed. *LCM 1–11* transferred to Greece April 1991; *LCM 12, 13, 15–21, 23, 27, 28* decommissioned 1993; *LCM 14, 22, 24–26* extant 1995.

From 1959 to 1968 a former US LCT was in commission. This was *LCU 1* ex-*L 7981*, ex-*LCU 779*, ex-*LSU 779*, ex-*LCT 779*. With only one MG she had no fighting value and served as headquarters ship of ten LCAs of the beachmaster company. She was sold 19 September 1968.

SMALL SURFACE COMBATANTS

French LE FOUGUEUX type *submarine chaser*

Particulars: As French *Le Fougueux* class

Originally ordered by the USA as a part of a NATO order in France, where a whole series of this type was built, *UW 12* was built by Dubigeon, Nantes. Belonging to the *La Fougueux* class she was commissioned in 1955 by the USN together with a sister ship and was given the pennant number P 1618. Taken over by Germany in 1957, she was commissioned on 12 March 1957 by the navy, pennant number *UW 12* as Type 179 and was used as training boat for the underwater weapons school. She was decommissioned 15 September 1967 and sold to the Tunisian Navy which commissioned her on 16 July 1970 as *Sakiet Sidi Youssef* (P 313).

The second boat, *P 1616*, was not taken over by the navy, but went to Ethiopia under MDAP, where she was renamed *Belay Deress*. She was soon handed back to the USN who lent her to Italy 3 February 1959. She was commissioned 3 December 1959 under her new name *Vedetta* (F 597).

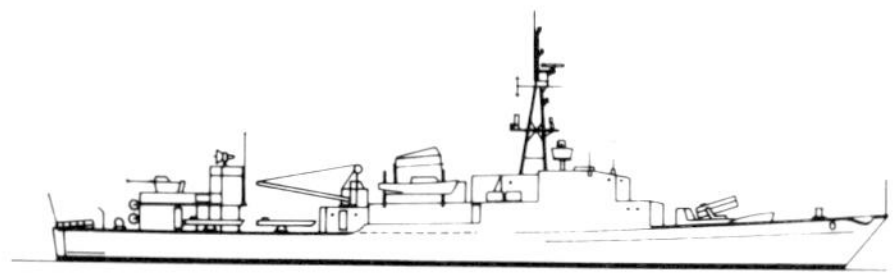

Najade 1964

THETIS class *corvettes*

Displacement: 599t standard; 732t full load
Dimensions: 215ft wl, 229ft oa × 27ft × 13ft 10in
65.5m, 69.8m × 8.2m × 4.2m
Machinery: 2 shafts (CP propellers), 2 MAN diesels, 6000bhp (4416kW) = 24kts. Oil 78t. Range 2760nm at 15kts
Armament: 2–40mm/70 (1×2), 1–375mm ASW mortar (1×4), 4–533mm ASW TT
Sensors: Radar KH 14/9, TRS-N; sonar ELAC 1BV
Complement: 68

No	Name	Builder	Launched	Fate
P 6052	THETIS	Rolandwerft, Bremen	22.3.60	Decomm 6.9.91, to Greece
P 6053	HERMES	Rolandwerft, Bremen	9.8.60	Decomm 7.9.92, to Greece
P 6054	NAJADE	Rolandwerft, Bremen	6.12.60	Decomm 6.9.91, to Greece
P 6055	TRITON	Rolandwerft, Bremen	5.8.61	Decomm 7.9.92, to Greece
P 6056	THESEUS	Rolandwerft, Bremen	20.3.62	Decomm 30.4.92, to Greece

Originally designed in 1958–59 as 'Large B Type' torpedo-recovery vessels, their official designation until 1974 was fleet service craft, but now they are referred to as the Type 420. It soon turned out that they were unable to fulfil their planned role in torpedo-recovery and they were used for various purposes, mainly as training units of the weapons school. During the 1970s they were modified and received a powerful ASW component. The crane amidships and the two boats were removed, the original 2–533mm TT for ASW torpedoes were replaced by 4–533mm TT. In addition they received the necessary sonar systems.

Najade differed from her sisters in having an enlarged bridge. Because they were three times reclassified their pennants changed: P 6111–6115, A 1430–1434, and finally P 6052–6056. All transferred to Greece 1991–92.

Triton 16.8.82 — L & L van Ginderen

HANS BURKNER *corvette*

Displacement: 983t standard; 1347t full load
Dimensions: 265ft 3in oa × 30ft 10in × 9ft 3in
81.0m × 9.4m × 2.8m
Machinery: 2 shafts, 4 MAN diesels, 13,600bhp (10,010kW) = 24kts. Range 2180nm at 15kts
Armament: 2–40mm/70 (1×2), 1–375mm ASW mortar (1×4), 2–533mm ASW TT
Complement: 60

No	Name	Builder	Launched	Fate
Y 879	HANS BURKNER	Atlas-Werke, Bremen	16.7.61	Decomm 26.10.90

Originally rated as a 'Large B Type' torpedo-recovery vessel, *Hans Burkner* was manned by the navy who first commissioned her in May 1963. However she was turned over to Trials Command 71 and operated under the pennant number A 1449 with a reduced civilian crew. All armament except the Bofors ASW mortar was removed, and the ship served as a trials ship for ASW equipment and underwater experiments.

SILBERMÖWE class *fast attack craft*

Displacement: 109t standard; 155t full load (*Seeschwalbe* as *UW 9* 135.6t)
Dimensions: 111ft 9in wl, 116ft 9in oa × 17ft × 5ft 3in
34.0m, 35.6m × 5.2m × 1.6m
Machinery: 3 shafts (*Seeschwalbe* CP propellers), 3 Mercedes-Benz (*Seeschwalbe* Maybach) diesels, 9000bhp (6624kW) = 42kts. Oil 13t
Armament: 2–20mm (2×1), later 1–40mm/70 (*Eismöwe, Raubmöwe, Seeschwalbe* 2–40mm (2×2), 2–533mm TT (2×1))

No	Name	Builder	Launched	Fate
P 6052	SILBERMÖWE (ex-*Silver Gull*)	Lürssen	21.1.55	To Greece Nov 1968
P 6053	STURMMÖWE (ex-*Storm Gull*)	Lürssen	7.12.54	To Greece Nov 1968
P 6054	WILDSCHWAN (ex-*Wildswan*)	Lürssen	23.3.55	To Greece Nov 1968
P 6055	EISMÖWE	Lürssen	15.9.55	To Greece Nov 1968
P 6056	RAUBMÖWE	Lürssen	27.10.55	To Greece Nov 1968
P 6057	SEESCHWALBE	Lürssen	13.9.56	Sold 29.1.74

These boats were a modified repeat of the *Kriegsmarine*'s wartime S-boats. They were designated as Type 149. The first three were ordered in February 1952 by the then Federal Home Office for the Federal Frontier Guard. But the projected speed of 43kts was prohibited under the terms of the Potsdam Conference and construction was stopped by the British Military Security Department. From the spring of 1953 construction of these boats continued, but officially on behalf of the Royal Navy. On 3 May 1955, after renewed negotiations, the Military Security Department gave permission for the construction of three more boats. The first three served with German crews under British command until their incorporation into the new navy on 29 May 1956. The fourth and fifth boats were commissioned by the Federal Frontier Guard (Naval) as *S 1* and *S 2* on 15 December 1955 and 2 February 1956 respectively. *S 1* and *S 2* were taken over by the navy on 1 July 1956 together with all the vessels of the Federal Frontier Guard (Naval). The last boat *Seeschwalbe* was commissioned on 23 April 1957. The first five boats were transferred to the Greek Navy in November 1968 as military aid, but *Seeschwalbe* was deleted on 1 October 1960 and became a training vessel *UW 9* (W 63). In 1964 she was rebuilt and served under her new name *Wilhelm Laudahn* (Y 839) for the Trials Command 73 especially carrying out trials of new engines with a civilian crew. She was frequently modified and rebuilt.

WARTIME S-BOATS

Also belonging to the Type 149 were three further units: *UW 10* (W 49) ex-*S 130*, *UW 11* (W 50) ex-*S 208* and the former *S 116*.

UW 10 was commissioned for the *Kriegsmarine* on 21 October 1943 as *S 130*, taken by the British in 1945 and served in the Royal Navy as experimental craft *FPB 5130*. From 1948 she served, together with *Silbermöwe* and *Sturmmöwe*, as a fishery protection vessel, with a German crew. Commissioned 7 May 1957 as *UW 10*, she served as a training boat for the underwater weapons school. On 15 August 1963 she was reclassified as *EF 3* (Y 840) and served with a civilian crew for Trials Command 71, until being decommissioned and sold 13 January 1991.

UW 11 was commissioned by the *Kriegsmarine* on 28 September 1944 as *S 208*, became a British prize in 1945 and served in the Royal Navy as experimental craft *MTB 5208*. From 1951 she served as a fishery protection vessel. Commissioned on 12 March 1957 as *UW 11*, she too served as a training boat for the underwater weapons school. After its winding-up she was used as a target boat.

S 116 was commissioned by the *Kriegsmarine* on 9 July 1942. Little is known about her post war fate. In 1952 she was purchased by the Federal Frontier Guard (Naval) to be used as a training hulk. In 1957 the navy took over the hulk and used her for the damage control training group. She was destroyed by fire on 25 May 1965.

While *S 116* remained unmodified, *UW 10* and *UW 11* were disarmed. During their service with the Royal Navy they had foreign engines. On their return they were re-engined with Mercedes-Benz diesels. *EF 3* (ex-*UW 10*) has only her middle shaft remaining.

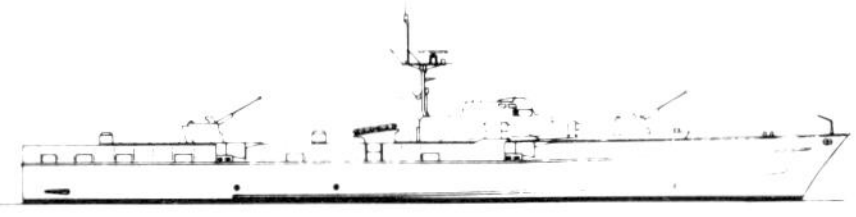

Elster 1982
NB 1/750 scale

JAGUAR class *fast attack craft*

Displacement: 183.6t (P 6068–P 6077 194.6t) standard; 210t (P 6068–P 6077 221t) full load
Dimensions: 130ft 7in wl, 139ft 10in oa × 23ft 4in × 7ft 6in (P 6068–P 6077 8ft)
39.8m, 42.6m × 7.1m × 2.3m (2.4m)
Machinery: 4 shafts, 4 Mercedes-Benz (P 6068–P 6077 4 Maybach) diesels, 12,000bhp (8832kW) = 43kts. Range 500nm/1000nm at 39kts/32kts
Armament: 2–40mm/70 (2×1), 4–533mm TT (4×1)
Complement: 39

No	Name	Builder	Launched	Fate
P 6058	ILTIS	Lürssen	15.8.57	Sold 3.6.78
P 6059	JAGUAR	Lürssen	12.6.57	To Turkey Feb 1976
P 6060	LEOPARD	Lürssen	4.1.58	Sold 15.9.75
P 6061	LUCHS	Lürssen	2.11.57	Sold 27.10.75
P 6062	WOLF	Lürssen	21.9.57	To Turkey Nov 1975
P 6063	TIGER	Kröger	21.4.58	To Turkey Jan 1976
P 6064	PANTHER	Kröger	18.10.58	Sold 14.12.78
P 6065	LOWE	Lürssen	8.11.58	To Turkey Oct 1975
P 6066	FUCHS	Lürssen	20.12.58	Sold 3.6.78
P 6067	MARDER	Lürssen	21.3.59	Sold 27.10.75
P 6068	SEEADLER	Lürssen	1.2.58	To Greece Mar 1977
P 6069	ALBATROS	Lürssen	20.3.58	To Greece 1977
P 6070	KONDOR	Lürssen	17.5.58	To Greece Mar 1977
P 6071	GREIF	Lürssen	28.6.58	To Greece Oct 1976
P 6072	FALKE	Lürssen	30.8.58	To Greece Oct 1976
P 6073	GEIER	Lürssen	1.10.58	To Greece Oct 1976
P 6074	BUSSARD	Kröger	29.11.58	To Greece 1976
P 6075	HABICHT	Kröger	21.2.59	To Greece 1977
P 6076	SPERBER	Kröger	4.4.59	To Greece Jun 1976
P 6077	KORMORAN	Kröger	16.7.59	To Greece May 1977
P 6082	WEIHE	Lürssen	20.6.59	To France as target 1973
P 6083	KRANICH	Lürssen	15.8.59	Museum ship Bremerhaven, 1974
P 6084	ALK	Kröger	19.9.59	To Turkey 1976
P 6085	STORCH	Lürssen	16.11.59	To Turkey Jun 1975
P 6086	PELIKAN	Kröger	12.12.59	To Turkey Jun 1975
P 6087	HAHER	Lürssen	9.1.60	To Turkey Jan 1976
P 6088	ELSTER	Lürssen	26.3.60	Sold 7.1.78
P 6089	REIHER	Lürssen	7.5.60	To Turkey Feb 1976
P 6090	PINGUIN	Lürssen	4.7.60	To Turkey Aug 1975
P 6091	DOMMEL	Lürssen	20.8.60	Sold 7.1.78

Designated 'Schnellboot 55' by Lürssen Werft, Vegesack, these composite-built units belonged to the Type 140/141. The only difference between the classes was in the choice of engine. This type was a further development of the well-known wartime 'S-boats' (E-boats) of the *Kriegsmarine*, but incorporating the latest advances.

The boats could carry mines, when the aft TTs were removed, the minerails being fixed on deck.

Originally *Iltis* carried 2–20mm/85, but later she received the standard 2–40mm/70. Experiments with *Pelikan* and *Geier*, to cover the hulls with a radar-absorbing coat of rubber and plastic, were unsuccessful.

For experimental purposes *Pelikan* received a conical stub mast. *Geier* was rebuilt several times, receiving a conical stub mast and later a telescopic mast. Finally she received additional superstructure aft of the bridge for ABC warfare experiments. The Type 149 boat *Wilhelm Laudahn* (Y 839) ex-*Seeschwalbe* was modified to the same standards. *Kormoran* was used for trials of stern TTs which were later part of the armament of the 142 and 149 Types. After they had been stricken they were transferred as military aid in groups of ten each to the Greek and Turkish navies. The others were sold for scrap. *Weihe* went to France as a 'fast target' and *Kranich* is exhibited at the German Maritime Museum in Bremerhaven.

HUGIN class *fast attack craft*

Displacement: 64.8t standard; 75.5t full load
Dimensions: 75ft 6in wl, 80ft 5in oa × 24ft 8in × 3ft 8in
23.0m, 24.5m × 7.5m × 1.1m
Machinery: 2 shafts, 2 Napier-Deltic diesels, 6280bhp (4622kW) = 45kts. Oil 10t
Armament: 2–40mm/70 (2×1), 4–533mm TT (4×1)
Complement: 34

No	Name	Builder	Launched	Fate
P 6191	HUGIN	Batservice Verft, Mandal	26.3.60	To Turkey Aug 1964
P 6192	MUNIN	Batservice Verft, Mandal	26.6.60	To Turkey Aug 1964

Based on a private design study by the Norwegian shipbuilder who designed the prototype vessel *Nasty* for the Norwegian Navy. Later the *Tjeld* class was developed from this design. The navy ordered two units (designated Type 152) in May 1959 to find out whether a cheap hard-chine planing boat would be of use in narrow coastal waters. Trials showed that the design did not match requirements and no more were ordered. The two existing boats were handed over to Turkey as military aid.

(For illustration see under Greek *Astrapi*)

PFEIL and STRAHL *fast attack craft*

Displacement: 75t standard; 85t full load (*Strahl* 96.5t; 101.6t)
Dimensions: 90ft 8in wl, 95ft oa × 24ft × 7ft
(*Strahl* 97ft 2in, 99ft 2in × 23ft 10in × 6ft 10in)
27.6m, 29.0m × 7.3m × 2.1m (29.6m, 30.2m × 7.3m × 2.1m)
Machinery: 2-shaft CODOG: 2 Proteus gas turbines, 8500shp = 54kts; 2 diesels, 300bhp = ?kts. Oil 21t (*Strahl* 3-shaft Proteus gas turbines, 10,500shp = 55kts)
Armament: 2–40mm/70 (2×1), 2–533mm TT (2×1)
Complement: 22

No	Name	Builder	Launched	Fate
P 6193	PFEIL	Vosper	26.10.61	To Greece Jan 1967
P 6194	STRAHL	Vosper	10.1.61	To Greece Jan 1967

Both designated as the Type 153, these two hard-chine planing boats were ordered 22 August 1960 as experimental units for comparative tests. *Strahl* was of the British *Brave* type, while *Pfeil* was of the improved *Ferocity* type. Both boats had their gas turbine exhausts situated in the stern. After trials they were handed over to the Greek Navy as military aid.

ZOBEL class *fast attack craft*

Displacement: 172.5t standard; 190t full load (after conversion: 212.5t standard; 219t full load)
Dimensions: 130ft 8in wl, 139ft 10in oa × 23ft 4in × 7ft 6in
39.8m, 42.6m × 7.1m × 2.3m
Machinery: As Mercedes-engined *Jaguar* class (after conversion 38.5kts)
Armament: As *Jaguar* class (after conversion 2–533mm TT)
Sensors: Radar KH 14/9, M-20
Complement: 39

No	Name	Builder	Launched	Fate
P 6092	ZOBEL	Lürssen	28.1.61	Decomm 7.9.82, sold to private owner 28.12.83
P 6093	WIESEL	Lürssen	14.3.61	Decomm 6.3.84, to Turkey 8.10.84
P 6094	DACHS	Lürssen	10.6.61	Decomm 6.12.83, to Turkey 14.5.84
P 6095	HERMELIN	Kröger	5.8.61	Decomm 12.1.83, trials vessel *Kor 2* 1985–87, sold to private owner 25.5.88
P 6096	NERZ	Lürssen	5.9.61	Decomm 8.7.82, sold to private owner 14.4.87
P 6097	PUMA	Kröger	26.10.61	Decomm 17.12.81, TS Brake Naval Technical College 1983, sold 1993
P 6098	GEPARD	Lürssen	14.4.62	To Turkey 14.5.84
P 6099	HYANE	Kröger	31.3.62	Decomm 5.6.84, to Turkey 8.10.84
P 6100	FRETTCHEN	Lürssen	20.11.62	Decomm 9.8.83, to Turkey 14.5.84
P 6101	OZELOT	Lürssen	4.2.63	Decomm 10.1.84 to Turkey 8.10.84

The Lürsen-designed 'Schnellboot 60', now Type 142, was a modified version of the Type 140/141. They differed from the foregoing classes in having additional superstructure aft of the bridge, which was tested earlier in *Geier*, a more streamlined bridge and masts of a different shape. A further ten projected boats were not built.

Nerz was used for weapons trials in 1963–64 including the 'Tartar-Bullpup' guided missile system, which necessitated a spherical radome situated on a four-pod mast. The system was too heavy for these boats, the project was dropped, and continued with the British Seacat SAM instead. The Seacat missile was tried on a launch frame on the forward 40mm/70, but these experiments were also dropped.

At the end of the 1960s all the boats were substantially modernised, receiving a new mast with a spherical radome including a target indicating radar. The 4–533mm TT were removed and 2–533mm TT firing Seal wire-guided torpedoes were fitted in the stern. (See also the experiments with *Kormoran*, under *Jaguar* class.) The 40mm/70 was now fully radar-directed. Since 1981 the boats were successively decommissioned and are being replaced by the new Type 143A boats. Their M-20 fire control radar systems were installed in the new *Hameln* class fast minesweepers.

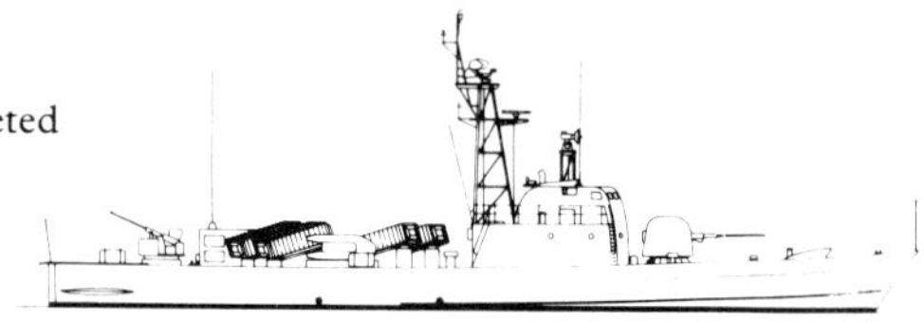

Tiger (S 41) as completed
NB 1/750 scale

S 41 class *fast attack craft*

Displacement: 234t standard; 263.7t full load
Dimensions: 144ft 5in pp, 150ft 8in wl, 154ft 3in oa × 23ft × 7ft 3in
44.0m, 46.0m, 47.0m × 7.0m × 2.2m
Machinery: 4 shafts, 4 MTU diesels, 12,000bhp (8832kW) = 38.5kts. Oil 33t. Range 570nm/1600nm at 30kts/15kts
Armament: 4 MM38 Exocet SSM (2×2), 1–76mm/62, 1–40mm/70
Sensors: Radar 3RM20, Triton, Pollux
Complement: 30

No	Name	Builder	Launched	Fate
P 6141	TIGER (*S 41*)	CMN, Cherbourg	27.9.72	Extant 1995
P 6142	ILTIS (*S 42*)	CMN, Cherbourg	12.12.72	Decomm 12.10.92, to Greece
P 6143	LUCHS (*S 43*)	CMN, Cherbourg	7.9.73	Extant 1995
P 6144	MARDER (*S 44*)	CMN, Cherbourg	5.5.73	Decomm 25.5.94, to Greece 16.3.95
P 6145	LEOPARD (*S 45*)	CMN, Cherbourg	3.7.73	Extant 1995
P 6146	FUCHS (*S 46*)	CMN, Cherbourg	21.5.73	Extant 1995
P 6147	JAGUAR (*S 47*)	CMN, Cherbourg	20.9.73	Extant 1995
P 6148	LOWE (*S 48*)	CMN, Cherbourg	10.9.73	Extant 1995
P 6149	WOLF (*S 49*)	CMN, Cherbourg	11.1.74	Extant 1995
P 6150	PANTHER (*S 50*)	CMN, Cherbourg	10.12.73	Extant 1995
P 6151	HAHER (*S 51*)	CMN, Cherbourg	26.4.74	Decomm 24.6.94 to Greece 16.3.95
P 6152	STORCH (*S 52*)	CMN, Cherbourg	25.3.74	Decomm 12.11.92, to Greece
P 6153	PELIKAN (*S 53*)	CMN, Cherbourg	4.7.74	Extant 1995
P 6154	ELSTER (*S 54*)	CMN, Cherbourg	8.7.74	Extant 1995
P 6155	ALK (*S 55*)	CMN, Cherbourg	15.11.74	Extant 1995
P 6156	DOMMEL (*S 56*)	CMN, Cherbourg	30.10.74	Extant 1995
P 6157	WEIHE (*S 57*)	CMN, Cherbourg	13.2.75	Extant 1995
P 6158	PINGUIN (*S 58*)	CMN, Cherbourg	26.2.75	Extant 1995
P 6159	REIHER (*S 59*)	CMN, Cherbourg	15.5.75	Extant 1995
P 6160	KRANICH (*S 60*)	CMN, Cherbourg	26.5.75	Extant 1995

The steel-hulled Type 148 missile-firing attack craft are a modified French 'La Combattante II' type, which in turn is based on a Lürssen design. Orders were placed on 18 December 1970, based on a German-French contract of 23 October 1970. These boats were to replace the *Jaguar* class (Type 140). Under the project managment of the Direction Technique des Constructions Navales (DTCN) of the French Ministry of Defence, the main shipyard employed was Constructions Mécaniques de Normandie (CMN).

S 46, S 48, S 50, S 52, S 54, S 56, S 58 and *S 60* were fitted out by Lürssen, the hulls having been brought from Cherbourg to Germany. In 1981 in addition to their numbers, the boats were given the traditional names of their forerunners.

Alk May 1993 L & L van Ginderen

S 61 class *fast attack craft*

Displacement: 380t standard; 393.7t full load
Dimensions: 178ft 6in pp, 188ft 9in oa × 26ft 3in × 8ft 3in
54.4m, 57.5m × 8.0m × 2.5m
Machinery: 4 shafts, 4 MTU diesels, 16,000–18,000bhp (11,776–13,248kW) = 33–38kts. Oil 43t. Range 600nm/1600nm at 30kts/16kts
Armament: 4 MM38 Exocet SSM (2×2), 2–76mm/62 (2×1), 2–533mm TT
Sensors: Radar 3RM20, WM-27
Complement: 40

No	Name	Builder	Launched	Fate
P 6111	ALBATROS (*S 61*)	Lürssen	22.10.73	Extant 1995
P 6112	FALKE (*S 62*)	Lürssen	21.3.74	Extant 1995
P 6113	GEIER (*S 63*)	Lürssen	18.9.74	Extant 1995
P 6114	BUSSARD (*S 64*)	Lürssen	14.4.75	Extant 1995
P 6115	SPERBER (*S 65*)	Kröger	15.1.74	Extant 1995
P 6116	GREIF (*S 66*)	Lürssen	4.9.75	Extant 1995
P 6117	KONDOR (*S 67*)	Kröger	6.3.75	Extant 1995
P 6118	SEEADLER (*S 68*)	Lürssen	17.11.75	Extant 1995
P 6119	HABICHT (*S 69*)	Kröger	5.6.75	Extant 1995
P 6120	KORMORAN (*S 70*)	Lürssen	14.4.76	Extant 1995

The Type 143 were designed to replace the Type 141 (P 6068–6077). The design of the composite hulls is by Lürssen, the military requirements were defined in October 1966, the contract was signed 7 July 1972 and the orders placed 13 July 1972. The TT are aft-launching, and fire Seal wire-guided torpedoes. Since the end of 1982 the boats bear not only their original pennant numbers, but also the traditional names of their forerunners.

Type 143A design
NB 1/750 scale

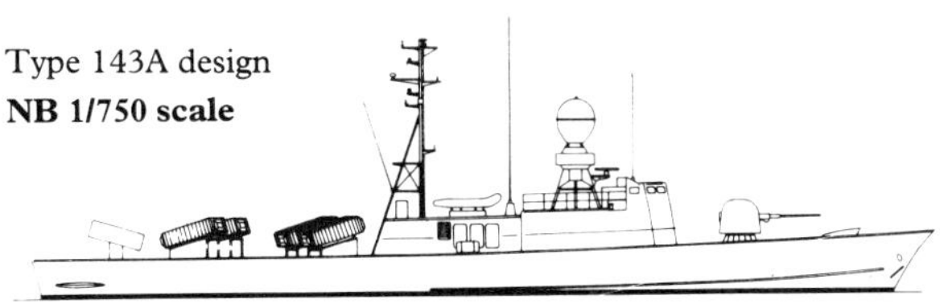

S 71 class *fast attack craft*

Displacement: As *S 61* class, but 397.5t full load
Dimensions: 189ft oa × 25ft 6in × 7ft 3in
57.6m × 7.8m × 2.2m
Machinery: As *S 61* class
Armament: 4 MM38 Exocet SSM, (2×2), 1–76mm/62, RAM ASMD System (1×24)
Sensors: As *S 61* class
Complement: 35

No	Name	Builder	Launched	Fate
P 6121	GEPARD (*S 71*)	Lürssen	25.9.81	Extant 1995
P 6122	PUMA (*S 72*)	Lürssen	8.2.82	Extant 1995
P 6123	HERMELIN (*S 73*)	Kröger	8.12.81	Extant 1995
P 6124	NERZ (*S 74*)	Lürssen	18.8.82	Extant 1995
P 6125	ZOBEL (*S 75*)	Kröger	30.6.82	Extant 1995
P 6126	FRETTCHEN (*S 76*)	Lürssen	26.1.83	Extant 1995
P 6127	DACHS (*S 77*)	Kröger	14.12.82	Extant 1995
P 6128	OZELOT (*S 78*)	Lürssen	7.6.83	Extant 1995
P 6129	WIESEL (*S 79*)	Lürssen	8.8.83	Extant 1995
P 6130	HYANE (*S 80*)	Lürssen	5.10.83	Extant 1995

Officially designated Type 143A these boats replaced the Type 142. The general layout resembles Type 143, but has no stern TT. The US RAM point-defence system is intended to be fitted in place of the Type 143s' after 76mm gun. The orders were placed 15 December 1978.

Ex-KFK type *patrol craft*

Displacement: 110t standard; 112t full load
Dimensions: 67ft 6in–70ft 6in wl, 73ft 3in oa × 21ft × 9ft
20.5m–21.5m, 22.3m × 6.4m × 2.7m
Machinery: 1 shaft, 1 Demag diesel, 150bhp (11kW) = 11kts. Oil 3.4t. Range 1200nm at 7kts
Armament: 1–20mm
Complement: 16

Class (no, builder, fate):
KW 1 (W 1, ex-*Vs 1441*, Burmester, Swinemünde, sold 12.12.68), *KW 2* (Y 828, ex-*M 3253*, Burmester, Swinemünde, 15.8.75, to Greece), *KW 3* (Y 829, ex-*K 561*, Burmester, Swinemünde, decomm 22.5.90), *KW 4* (W 4, ?, to Tanzania 30.8.64), *KW 5* (W 5, ?, to Tanzania 30.8.64), *KW 6* (Y 836, Eschler & Gabler, Lübeck, sold 1976), *KW 7* (W 7, ?, to Greece 22.12.67), *KW 8* (Y 831, ?, to Greece 15.8.75), *KW 9* (W 9, ?, to Tanzania 30.9.64), *KW 10* (W 10, ?, to Tanzania 30.9.64). All were commissioned 1.7.56.

These were former *Kriegsmarine* series-built naval fishing vessels (*Kriegsfischkutter*-KFKs) which were built in great numbers during the Second World War to serve as auxiliary patrol craft and reconnaissance units. The original design came from Maierform GmbH and was developed with a view to their future peacetime usage as fishing vessels. With the exception of the three boats *KW 1–KW 3*, which had been part of the GMSA postwar, they were repurchased by the Federal Frontier Guard (Naval) and modified to their needs. The modification was accepted by the Federal Home Office 21 September 1951, the boats being commissioned as *W 13–W 19*.

When taken over by the new navy the ten boats formed a harbour guard squadron. In 1963 they were decommissioned with the exception of *KW 2* and *KW 8*, which served as training boats for the Seamen Training School.

KW 3 was converted to a radio-control boat of the Wilhelmshaven naval arsenal, still serving with civilian crew. Four boats were sent as a gift to Tanzania and three to the Greek Navy. All the others were sold or scrapped.

Kw 15 class *patrol craft*

Displacement: 51.4–51.6t standard; 71.5–77.8t full load
Dimensions: 84ft 10in–88ft 8in wl, 89ft–95ft 3in oa × 15ft–15ft 6in × 4ft 6in
25.8m–27.0m, 27.1m–29.0m × 4.6m–4.7m × 1.4m
Machinery: 2 shafts, 2 Mercedes-Benz diesels, 2000bhp (1472kW) = 25kts. Oil 7t
Armament: 1/2–20mm (1/2×1) or 4–20mm (2×2)
Complement: 17

Class (no, builder):
KW 15 (Y 827, Schweers, Bardenfleth), *KW 16* (Y 830, Lürssen Lemwerder), *KW 17* (Y 845, Schürenstedt, Bardenfleth), *KW 18* (Y 832, Abeking & Rasmussen, Lemwerder). *KW 19* (Y 833, Schweers, Bardenfleth), *KW 20* (Y 846, Lürssen, Lemwerder). All commissioned 30 November 1956.

These six units of the Type 369 were former USN patrol boats, which served under the designation *USN 54–59* with the LSU's Weser River Patrol. They resembled the four patrol boats *P 1–P 4* of the Frontier Guard, which were commissioned 1 July 1956 for the navy as *KW 11–KW 14* and finally became the Type 909 air-sea rescue boats *FL 5–FL 8*.

They were launched in 1951–52 and commissioned with the USN in 1952–53. They were unarmed except during their navy career as training boats.

KW 15, *KW 16*, *KW 17* and *KW 20* were decommissioned in 1964 and transferred to the new inspection unit who commissioned them as *BG 1–BG 4*. After their return in 1971 they were disarmed like their sisters and commissioned as security vessels with civilian crews for the Territorial Command. All decommissioned 1993–94. Since the 1960s *KW 19* has served as a dispatch boat in the North Sea area with a civilian crew. She was decommissioned 22 December 1981.

MINE WARFARE VESSELS

BAMBERG class (ex-US LST type) *minelayers*

Displacement: 3640t standard; 4140t full load
Dimensions: 316ft wl, 332ft 8in oa × 51ft 10in × 9ft
96.3m, 101.4m × 15.8m × 2.7m
Machinery: 2 shafts, 2 GMC diesels, 2650bhp (1950kW) = 11kts. Oil 600t. Range 15,000nm at 9kts
Armament: 6–40mm/56 (after conversion 6–40mm/70) (2×2, 2×1)
Complement: 60

No	Name	Builder	Acquired	Fate
N 120	BOCHUM (ex-*Rice County*)	American Bridge	1960	To Turkey 13.12.76
N 121	BOTTROP (ex-*Saline County*)	Missouri Valley	1960	To Turkey 12.12.76
A 1403	BAMBERG (ex-*Greer County*)	Jefferson-ville	1960	Sold 15.7.70

In 1960 a total of seven US wartime LST type tank landing ships was purchased and in the summer of 1961 they arrived with German crews at Wilhelmshaven: *Rice County*, LST 1089; *Saline County*, LST 1101; *Greer County*, LST 799; *Montgomery County*, LST 1401; *Millard County*, LST 987; *Ulysses*, ARB 9, ex-*LST 967* and *Diomedes*, ARB 11, ex-*LST 1119*.

The last two were converted into the Type 726 maintenance ships *Odin* (A 512) and *Wotan* (A 513). The planned conversion of LST 987 and LST 1041 to maintenance ships was dropped and both were sold for scrapping.

The first three units arrived at Wilhelmshaven with the pennants A 1403–A 1405 and were converted to Type 370 minelayers in 1962. During this conversion the bow ramp was removed and the bow doors welded up. Superstructure and interior equipment was adapted for the future use, six rails with launching ramps were fixed on the deck, the platform and lower deck also had minerails and two launching ramps with hatches. *Bochum* and *Bottrop* came into service on 9 April 1964 and 6 February 1964; they decommissioned on 14 April 1971 and 28 September 1971, were refitted and handed over to the Turkish Navy. The conversion of *Bamberg* was 60 per cent finished when this was stopped and the ship was sold for scrap.

Sachsenwald 1970

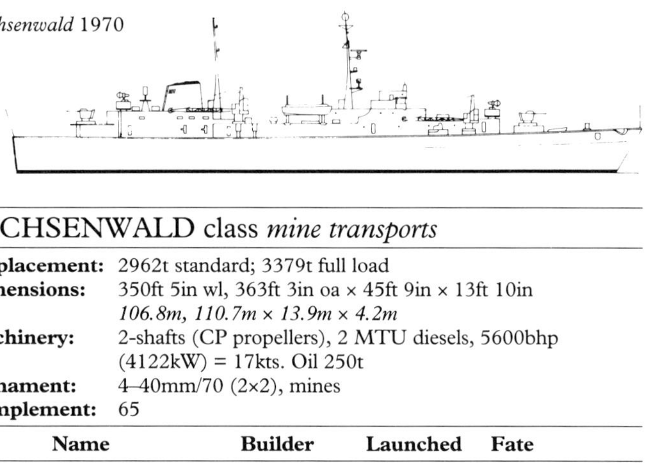

SACHSENWALD class *mine transports*

Displacement: 2962t standard; 3379t full load
Dimensions: 350ft 5in wl, 363ft 3in oa × 45ft 9in × 13ft 10in
106.8m, 110.7m × 13.9m × 4.2m
Machinery: 2-shafts (CP propellers), 2 MTU diesels, 5600bhp (4122kW) = 17kts. Oil 250t
Armament: 4–40mm/70 (2×2), mines
Complement: 65

No	Name	Builder	Launched	Fate
A 1437	SACHSENWALD	Blohm & Voss	10.12.66	Decomm 26.9.91, sold for BU 9.2.95
A 1438	STEIGERWALD	Blohm & Voss	10.3.67	Decomm .11.93, sold for BU 3.5.95

Projected under the auxiliary shipbuilding programme of 1959, both units were designed in 1963 and ordered in 1965. Officially, they were Type 762 mine transports, but they had full minelaying capabilities. All decks had minerails, the stern had four hatch-covered launching ramps and two rails on deck lead to the sideward launching positions. Their crane equipment enabled them to supply other ships with mines.

Ex-1935 type *ocean minesweepers*

Displacement: 685t (*Wespe* 682t) standard; 878t (*Wespe* 874t) full load
Dimensions: 216ft 6in wl, 224ft 5in (*Wespe* 223ft 5in) oa × 28ft 6in × 8ft 6in
66.0m, 68.3m (68.1m) × 8.7m × 2.6m
Machinery: 2-shaft reciprocating steam engines, 2 Wagner boilers, 3500shp (2576kW) = 18.3kts. Oil 155t. Range 1440nm at 9kts
Armament: 1956: 2–105mm/45 (2×1)
1958: 1–105mm/45, 1 Hedgehog
1960: 2–40mm/70 (2×1), 4–20mm (2×2), 1 Hedgehog
Complement: 100

No	Name	Builder	Acquired	Fate
F 207	BIENE (ex-*M 205*)	A G Neptun, Rostock	1956	BU 1974
F 208	BREMSE (ex-*M 253*)	Deutsche Werft, Hamburg	1956	BU 1976
F 209	BRUMMER (ex-*M85*)	Lübecker Maschinenbau-gesellschaft, Lübeck	1956	BU 1974
F 210	HUMMEL (ex-*M81*)	Lübecker Maschinenbau-gesellschaft, Lübeck	1956	BU 1976
F 211	WESPE (ex-*M 24*)	Flender-Werke, Lübeck	1956	Sunk as target 25.10.73

Designated Type 319 these minesweepers were former *Kriegsmarine* minesweepers of the 1935 type launched in 1940–41. Being a new design they had – in contrast to earlier minesweeper types – oil-burning high-pressure boilers and double-compound Lenz-type steam machinery.

After the war they became Allied prizes and were first used for GMSA duties. Later they were handed over to the French Navy to serve as minesweepers and fishery protection vessels under the following names: *Biene* as *Belfort* (M606), *Bremse* as *Vimy* (M 608), *Brummer* as *Yser* (M 604), *Hummel* as *Laffaux* (M 607) and *Wespe* as *Ailette* (M 605).

Repurchased by Germany, the five were transferred to Wilhelmshaven in December 1956 and January 1957 and handed over to the navy, and wore the following pennant numbers: *Biene* (Q 74), *Hummel* (Q 75) and *Wespe* (Q 76).

In February 1957 they recommissioned after being refitted, the armament was partially changed to new weapons and the interiors were refurbished to German standards. After that they formed the First Escort Squadron with an F pennant number, but only served as officer training ships.

In 1963 they were decommissioned and transferred to the Wilhelmshaven naval arsenal and modified to target ships, in which role they served for a short time. In 1966 *Biene* and *Brummer* went to the damage control training group as training hulks. In 1967 *Hummel* was handed over to the Trials Command 71 as a target ship and scrapped in 1976. *Wespe* was sunk in 1973 by *Zerstörer 4*, while *Bremse* was scrapped in August 1976. During their service career they received standardised weapons and their funnel caps were changed repeatedly.

Ex-1940 and 1943 type *ocean minesweepers*

Displacement: 543t (*Seeschlange* 582t) standard; 775t (*Seeschlange* 821t) full load
Dimensions: 189ft (*Seepferd* 179ft 2in; *Seeschlange* 204ft 5in) wl, 204ft 5in
(*Seepferd* 204ft; *Seeschlange* 207ft) oa × 28ft–29ft 3in × 8ft 6in–8ft 10in
57.6m (54.6m; 62.3m), 62.3m (62.2m; 63.1m) × 8.5m–9.0m × 2.6m–2.7m
Machinery: 2-shaft triple expansion, 2 Navy (*Seeschlange* Navy-Wagner boilers, 2700shp, (1987kW) = 17.2kts (*Seeschlange* 16.7kts).
Coal 162t (*Seeschlange* 180t). Range 4000nm at 10kts
Armament: 1–3in (76.2mm), 4–40mm (2×2), 3–20mm (3×1)
Complement: 70

No	Name	Builder	Acquired	Fate
M 187	SEEHUND (ex-*M 388*)	Elsflether Werft	17.7.56	Sold for BU 7.9.73
M 188	SEEIGEL (ex-*M 460*)	Nederlandsche Dok Mij, Amsterdam	30.8.56	Sold 4.5.73
M 189	SEELOWE (ex-*M 441*)	P Smit, Rotterdam	17.7.56	BU 1969
M 190	SEEPFERD (ex-M 294)	Lindenau, Memel	30.8.56	BU 1966
M 191	SEESCHLANGE (ex-*M 611*)	A G Neptun, Rostock	15.8.56	BU 1967
M 192	SEESTERN (ex-*M 278*)	Rickmers Werft, Wesermünde	15.8.56	BU 1966

Also designated Type 319, these were *Kriegsmarine* minesweepers of the 1940 and 1943 types, launched 1942–45. Designed in 1939–40 they showed a resemblance to the 1914–15 minesweeper of the Imperial German Navy. The only *Bundesmarine* coal-burning warship with VTE engines. In 1945 they became Allied prizes and served with the GMSA. From 1951 on they belonged, disarmed, to the LSU(B) carrying the pennant numbers M 201–M 206. Before being handed over to the Federal German Navy they refitted and commissioned with letter codes, but in 1957 they received standard NATO 'M' pennant numbers. After decommissioning in 1960 they were partially modified to accommodation ships and received a block-shaped superstructure on the stern. As *WBM I–IV* (*W*ohn*b*oot *M*inensucher = accommodation boat minesweeper) they served as floating quarters for the crews of ships fitting out. *Seeigel* was modified once more and served *Torpedoklarmachstelle II* (torpedo clearance establishment). All scrapped 1966-73; *Seeigel* sold in 1973.

LINDAU class *coastal minesweepers*

Displacement: 362t standard; 380t full load (after conversion 387t standard; 402t, as minehunter 465t, full load)
Dimensions: 137ft 9in (after conversion 142ft 9in) wl, 149ft 6in (first six 147ft 8in; after conversion all 154ft 6in) oa × 26ft 5in × 8ft 3in
42.0m (43.5m), 45.6m (45.0m; 47.1m) × 8.0m × 2.5m
Machinery: 2 shafts (CP propellers), 2 Maybach diesels, 4000bhp (2944kW) = 17kts. Oil 42t. Range 1950m at 12kts
Armament: 1–40mm/70
Sensors: Radar TRS-N or KM 1419; sonar DSQS-11 (DSQS-11A in Type 351; Type 193M in Type 331)
Complement: 45

Class (no, launched, fate):
Göttingen (M 1070, 1.4.57, 19.1.79 minehunter, extant 1995), *Koblenz* (M 1071, 6.5.57, 21.6.78 minehunter, extant 1995), *Lindau* (M 1072, 16.2.57, 10.2.78 minehunter, extant 1995), *Schleswig* (M 1073, 2.10.57, 19.3.81 Troika, extant 1995), *Tübingen* (M 1074, 12.8.57, 20.3.78 minehunter, extant 1995), *Wetzlar* (M 1075, 24.6.57, 6.10.78 minehunter, decomm 30.6.95), *(Paderborn* (M 1077, 4.2.57, 17.9.81 Troika, extant 1995), *Weilheim* (M 1076, 5.12.58, 17.11.78 minehunter, decomm 15.6.95), *Cuxhaven* (M 1078, 11.3.58, 6.6.79 minehunter, extant 1995), *Düren* (M 1079, 12.6.58, will become Troika), *Marburg* (M 1080, 4.8.58, 28.6.79 minehunter, extant 1995), *Konstanz* (M 1081, 30.8.58, 24.5.82 Troika, extant 1995), *Wolfsburg* (M 1082, 10.12.58, 4.3.82 Troika, extant 1995), *Ulm* (M 1083, 10.2.59, 11.11.81 Troika, extant 1995), *Flensburg* (M 1084, 7.4.59, 12.8.72 minehunter, decomm 26.6.91, museum ship), *Minden* (M 1085, 9.6.59, 31.5.78 minehunter, extant 1995), *Fulda* (M 1086, 19.8.59, 1969 minehunter, decomm 26.3.92), *Völkingen* (M 1087, 20.10.59, 15.7.79 minehunter, extant 1995)

The Type 320 coastal minesweepers were built under the first naval programme and were a German version of the American-designed NATO standard AMS/MSC type. They were built of wood and non-magnetic materials. The first unit, *Lindau*, was the first ship to be built for the Federal German Navy. All were built by Burmester, Bremen. After completion of the first six boats it turned out that the extremely high bridge structure affected stability, so the next units were built with one deck less in the bridge; the first boats were modified to the same bridge. Another problem was the stern section, the boats having an upright transom stern, which resulted in insufficient working space on the poop, so the hulls were enlarged by 6m by the addition of a stern compartment. During the same refit all boats received a fixed bulwark from bow to stern, the hull plating in the midship section was reinforced, an 18t lead keel and bilge keels were added. Aft of the funnel an ABC proofed space was erected and the original loading derricks abreast the funnel were replaced by cranes.

In 1968–69 *Fulda* was converted to a minehunter, as was *Flensburg* in 1970–72. Instead of the original sonar equipment they received special minehunting sonar (Plessey Type 193M), the minesweeping equipment was removed as mines were now detected and deactivated by scuba-divers. The forecastle superstructure was extended aft. To improve manoeuvrability they were equipped with retractible Schottel-propellers in the transom (in fact, powerful electric outboard motors). This propulsion permits running at extremely slow speeds or manoeuvring on the spot. Between 1976 and 1979 a further ten boats were modified to similar standards with the exception of the Schottel-propulsion, in place of which they received a special gearing on the shaft to produce the same manoeuvrability. On the quarterdeck they received a hydraulic telescoping crane and two PAP 104 (*Poisson Auto Propulsé*) working in connection with the minehunting sonar Type 193M. Later, *Fulda* and *Flensburg* were also modified to the same standards, but retained their Schottel-propulsion. Now they are distinguished as the Types 331A (*Fulda, Flensburg*) and 331B (*Lindau, Tübingen, Minden, Koblenz, Göttingen, Cuxhaven, Weilheim, Marburg* and *Völklingen*).

From 1979 to 1983 the remaining six units (*Schleswig, Paderborn, Düren, Ulm, Konstanz* and *Wolfsburg*) were converted to control ships for the 'Troika'

Marburg 4.6.82 L & L van Ginderen

Seehund 3, a 'Troika' drone 18.6.93 L & L van Ginderen

system, being designated Type 351. The Troika system consists of a radio control minesweeper of the Type 351 and three self-propelled drones. The drones are of the HFG-F 1 type and are named *Seehund 1–18*. All eighteen *Seehunde* form a pool from which craft are drawn as necessary.

The system was introduced after a year of experiments and trials with *Niobe* (see below) and the prototype drones *Walross* and *Seekuh 1–3*.

Seehund 1–18 were designed by Blohm & Voss and built by MaK, Kiel. The drones have a boat-shaped hull, are fully non-magnetic and can be controlled both manually (during the passage) and by remote control (during the minehunting process). They also operate conventional sweep gear and can be used against acoustic mines. Their principal particulars are: 78ft 4in wl, 81ft 9in oa × 15ft 2in × 6ft (*23.9m, 25.0m × 4.6m × 1.8m*); 1-shaft Schottel propulsion, diesel-electric drive, 1 MWM diesel 320bhp (235.5kW) = 9.4kts. Range 520nm at 8.8kts. Complement 3.

VEGESACK class *coastal minesweepers*

Machinery: 2 shafts (CP propellers), 2 Mercedes-Benz diesels, 1320bhp (972kW) = 14.5kts. Oil 34.8t. Range 5000nm at 12 kts
Other particulars: As French *Mercure*

Class (no, launched):
Vegesack (M 1250, 21.5.59), *Hamelin* (M 1251, 20.8.59), *Detmold* (M 1252, 17.11.59), *Worms* (M 1253, 30.1.60), *Siegen* (M 1254, 29.3.60), *Passau* (M 1255, 25.6.60).

These Type 321 coastal minesweepers were built at CMN, Cherbourg, as members of the *Mercure* class. This was a French version of the NATO standard AMS/MSC type built to German order. The wooden-hulled boats were built during the first naval building programme and replaced the obsolete Type 319 coal-burners.

The boats had a remarkably short service career: after their decommissioning they were in the reserve fleet from 1963 to 1974 and were only put into service for mobilisation exercises. After their deletion in December 1973 five units were handed over to the Turkish Navy as military aid during June–October 1975. It was planned to convert *Passau* to an oceanography vessel; when this plan was dropped she also went to Turkey in 1978.

Herkules 1974
NB 1/750 scale

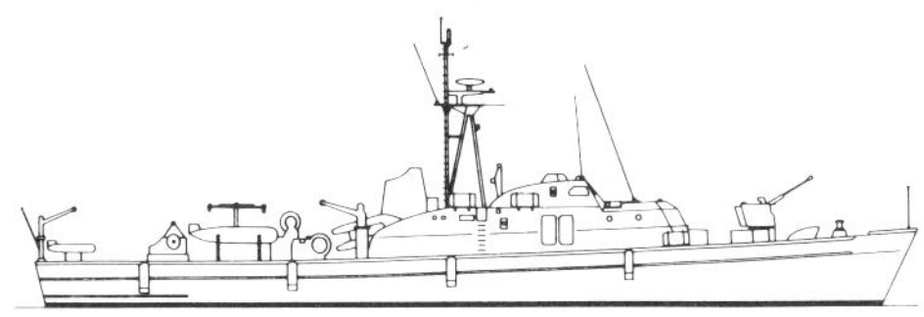

SCHUTZE class *inshore minesweepers*

Displacement: 241t standard; 266.5t full load
Dimensions: 144ft 9in wl, 155ft 9in oa × 22ft 10in × 7ft 6in
44.1m, 47.4m × 6.9m × 2.3m
Machinery: 2 shafts (CP propellers), 2 Mercedes-Benz diesels, 4500bhp (3312kW) = 25kts. Oil 26.4t. Range 1000nm at 18kts. See notes
Armament: 1–40mm/70 (as patrol boats 2–40mm/70)
Complement: 31

Class (no, launched):
Abeking & Rasmussen – *Mira* (M 1050, 16.12.59), *Castor* (M 1052, 12.7.62), *Krebs* (M 1052, 26.1.59), *Orion* (M 1053, 10.8.61), *Pollux* (M 1054, 15.9.60), *Sirius* (M 1055, 15.3.61), *Rigel* (M 1056, 2.4.62), *Regulus* (M 1057, 18.12.61), *Mars* (M 1058, 1.12.60), *Spica* (M 1059, 25.5.60), *Scorpion* (M 1060, 21.5.63). *Stier* (M 1061, 30.10.58), *Schütze* (M 1062, 20.5.58), *Waage* (M

Regulus 9.6.85 L & L van Ginderen

1063, 9.4.59), *Algol* (M 1068, 23.1.63), *Wega* (M 1069, 10.10.62), *Steinbock* (M 1091, 25.8.58), *Fische* (M 1096, 14.7.59), *Gemma* (M 1097, 6.10.59), *Capella* (M 1098, 26.2.60)

Schurenstedt, Bardenfleth *Deneb* (M 1064, 11.9.61), *Jupiter* (M 1065, 15.12.61), *Pluto* (M 1092, 9.8.60), *Widder* (M 1094, 12.3.59), *Uranus* (M 1009, 15.3.60)

Schurenstedt, Bardenfleth-*Deneb* (M 1064, 11.9.61), *Jupiter* (M 1065, 15.12.61), *Pluto* (M 1092, 9.8.60), *Widder* (M 1094, 12.3.59), *Uranus* (M 1009, 15.3.60)

These fast minesweepers of the Types 340/341, designed by Abeking & Rasmussen 1957–59, were built of wood and non-magnetic materials. They resemble the last R-boat type of the *Kriegsmarine* and at first were designated '*SM-boot 55*'. They had subsidiary roles as coastal patrol craft and minelayers. As planned the boats were to have had Voith-Schneider propellers, but results with the first prototypes *Schütze* and *Krebs* were unsatisfactory and all boats were given two shafts and CP propellers. All boats from M 1060 onwards have two Maybach diesels of 4760bhp (3530kW) and are referred to as the Type 341.

In the mid-1960s the superstructure of some of the boats had to be renewed because of corrosion. *Schütze*'s hull was given an experimental radar-absorbing plastic coating, but it was unsuccessful as it tended to come loose at high speeds. The original derricks abreast the funnel were replaced by swivelling cranes. As commissioned, *Gemma*, *Pegasus*, *Waage*, *Deneb*, *Regulus* and *Altair* had no minesweeping gear, but 2–40mm/70 in singles were fitted in order to test the patrol boat role; later they were equipped to the general standard.

During their construction period, the pennants changed a number of times until they received those noted above. *Castor* carried out experiments with the Seacat SAM in 1968. For this purpose she got launching frames at the 40mm/70 position and a director on the bridge. In 1964 *Stier* was relieved from the active squadron and was converted in 1966 to a mine clearance diver boat; she now belongs to the Type 732 under her new pennant Y 849. During the conversion the minesweeping gear was removed and replaced with a decompression chamber. Decommissioned 31 March 1995.

Mira was deleted 2 February 1978 and attached to the Naval Technical College as a stationary training boat, sold 1983. On 18 January 1973 *Algol* became an exercise hulk for the damage control training group, sold for BU 1993. *Pegasus* was sold on 7 May 1974 to HDW, Kiel, and served as a security vessel for the Type 209 submarines building for foreign contractors. She was sold to a private owner on 26 March 1974 and became an ocean-going yacht. On 5 November 1989 she was seized and confiscated by the German Border Police for involvement in drug smuggling. After deletion *Krebs* (22 June 1976), *Orion* (3 June 1979), *Steinbock* (5 November 1976), *Uranus* (1973), *Capella* (1976), *Pluto* (1 July 1987), *Neptun* (28 February 1990) and *Fische* (20 April 1989) were handed over to local sections of the German Naval League, to serve as floating clubrooms. *Widder* decommissioned 14 July 1989 and became a stationary training hulk at the Seaman Training Group at Borkum. The remaining ships were decommissioned and sold to private owners 1989–92.

Ex-R-BOAT type *inshore minesweepers*

Displacement:	*Capella* group 150t standard; 155t full load *Aldebaran* group 125t standard; 130t full load *UW 6* group 140t standard; 148t full load
Dimensions:	*Capella* group 126ft 8in wl, 134ft 9in oa × 18ft 10in × 5ft 3in *38.6m, 41.1m × 5.7m × 1.6m* *Aldebaran* group 120ft 9in wl, 127ft oa × 18ft 5in × 5ft *36.8m, 38.7m × 5.6m × 1.5m* *UW 6* group 120ft 9in wl, 130ft oa × 18ft 10in × 5ft 9in *36.8m, 39.7m × 5.7m × 1.7m*
Machinery:	2 Voith-Schneider propellers, 2 MAN diesels, 1800bhp (1325kW) = 18kts. Oil 11–18t. Range 900nm at 15kts (*UW 6* group: 2 shafts, 2 MWM diesels, 2800bhp (2061kW) = 24kts. Oil 14.7t. Range 1000nm at 15kts)
Armament:	1 or 2–20mm
Complement:	27–34

Class (no, commissioned, fate):

Capella group – *Capella*, ex-*R 133* (M 1050, 19.6.56, sold 1972), *Spica*, ex-*R 147* (M 1059, 31.7.56, sold 1970), *Castor*, ex-*R 138* (M 1051, 19.6.56, BU 15.6.78), *Mars*, ex-*R 136* (M 1052, 19.6.56, BU after 1966), *Orion*, ex-*R 132* (M 1053, 5.6.56, hulk DMB 20.6.68), *Pollux*, ex-*R 140* (M 1054, 19.6.56, sold 1970), *Regulus*, ex-*R 142* (M 1055, 31.7.56, BU 14.3.68), *Rigel*, ex-*R 135* (M 1056, 5.6.56, BU after 1966), *Saturn*, ex-*R 146* (M 1057, 31.7.56, sold 2.11.72), *Sirius*, ex-*R 144* (M 1058, 5.6.56, BU after 1970), *Spica*, ex-*R 147* (M 1059, 31.3.56, sold 1970), *Jupiter*, ex-*R 137* (M 1065, 31.7.56, hulk DMB 1969), *Merkur*, ex-*R 134* (M 1066, 5.6.56, sold 1.8.70), *UW 5*, ex-*R 150* (W 47, 11.12.56, sold 6.11.79).

Aldebaran group – *Aldebaran*, ex-*R 91* (M 1060, 16.11.56, sold 1.7.72), *Algol*, ex-*R 99* (M 1061, 30.10.56, hulk DMB 4.3.70), *Arkturus*, ex-*R 128* (M 1062, 16.11.56, BU 14.3.68), *Altair*, ex-*R 76* (M 1063, 30.11.56, sold 15.12.70), *Deneb*, ex-*R 127* (M 1064, 30.11.56, hulk DMB 20.6.68), *Wega*, ex-*R 67* (M 1069, 30.10.56, hulk DMN 1966), *Pegasus*, ex-*R 68* (M 1067, 16.11.56, sold 9.3.70), *Skorpion*, ex-*R 120* (M 1068, 16.11.56, hulk DMB 1.7.74), *UW 4*, ex-*R 101* (Y 870, 11.12.56, BU March 1981)

UW 6 group – *UW 6*, ex-*R 408* (W 48, 11.12.56, BU 1977), *OT 1*, ex-*R 406* (W 52, 21.1.57, sold after 1959), *AT 1*, ex-*R 266* (W 61, 21.1.57, BU 1968), *AT 2*, ex-*R 407* (W 62, 21.1.57, target 1963)

These former *Kriegsmarine* R-boats were all built by Abeking & Rasmussen. They became Allied prizes after the war and saw service with the GMSA and the German Minesweeping Unit. After those forces were disbanded they were commissioned by the USN for the LSU(B), carrying the pennant numbers USN 130–USN 155.

Before being handed over to the new Federal German Navy they were substantially refitted; they commissioned from June 1956. While the first boats retained their old USN pennant numbers without the prefix USN, subsequent ships commissioned had a pennant number formed of a letter-group. From 1957 they received the NATO prefix 'M' and were designated Type 359. During their *Bundesmarine* career they were unaltered except for a lattice mast and some minor modifications to the training boats.

After decommissioning most served as accommodation boats carrying the designation WBR (abbr. for *Wohnboot Räumboot* = accommodation boat minesweeper) and formed the Type 730. After their final deletion *Orion*, *Deneb*, *Skorpion* and *Jupiter* became floating quarters for German Navy League local sections. *UW 4*, *UW 5*, *Algol*, *Altair*, *Merkur* and *Wega* were handed over to the Navy League, but have since been scrapped. Until 1972 *Saturn* was used as an exercise hulk for the damage control training group and *Aldebaran*, after minor modifications, served as a training boat for mine divers (pennant Y 850, Type 732). *UW 6* was used as a floating electric generator, *Stromer*, at the Naval Arsenal, Kiel. *Jupiter* became the training boat *OT 1* (ii) and *Regulus* the training boat *AT 1* (ii).

During the period 1946 to 1951 some boats were leased from the USN by private owners: *Arkturus* served as *Annegret*, *Wega* as *Hilligenlei* and *Rüstringen*, *UW 6* as *Hansa IV* and *OT 1 (i)* as *Arngast*.

HANSA *inshore minesweeper*

Displacement:	155t standard; 175t full load
Dimensions:	111ft 8in, 115ft 5in oa × 21ft 10in × 6ft 5in *34.0m, 35.2m × 6.6m × 1.9m*
Machinery:	1 shaft, 1 Mercedes-Benz diesel, 1000bhp (736kW) = 14kts. Oil 14t. Range 1500nm at 12kts
Armament:	1–40mm/70
Complement:	20

NIOBE *inshore minesweeper*

Displacement:	169.7t standard; 171t full load
Dimensions:	110ft 9in wl, 115ft 5in oa × 22ft 5in × 5ft 8in *33.8m, 35.2m × 6.8m × 1.7m*
Machinery:	2 shafts, 2 Mercedes-Benz diesels, 2000bhp (1472kW) = 16kts. Oil 11t. Range 800nm at 12kts
Armament:	1–40mm/70
Complement:	20

Designed in 1955–56 as coast guard boats of the Types 361 and 360, they differed in the propulsion arrangement: *Hansa* had one shaft, *Niobe* two. Both were built by Krögerwerft and were launched on 18 November 1957 and 8 August 1957 respectively. These wooden boats were the first to use predominantly non-magnetic materials. In 1968 they were reclassified as riverine minesweepers Types 392 and 391, their pennant number changing from a 'W' to 'M' prefix.

Since 1969 *Hansa* has served as a mine diver boat, pennant number Y 806, of the Type 732. During refit the minesweeping gear was removed and a cabin was erected on the quarterdeck. Pennant number changed to M 1052 in 1987. Decommissioned 16 January 1992 and sold 1994.

Niobe was transferred as early as 1964 as a trials boat for the mine warfare experimental department. In 1968 she was converted to a mine warfare RC-control boat, the pennant prefix changing to 'Y'. In December she was modified yet again, changed her Type to 740, and served with a civilian crew at Trials Command 71. During modifications her minesweeping gear was removed and a hut was erected on the poop, and extensive electronic equipment was installed. The completely disarmed boat served for prototype trials and tests of the 'Troika' system using the towed-skid prototypes *Walross* and *Seekuh 1–3*. Particulars of the Troika prototypes are:

Walross: 84.7t full load; 88ft wl × 11ft 6in × 7ft (*26.8m × 3.5m × 2.1m*); 1 shaft Schottel propeller, diesel-electric drive, 290bhp (213.4kW) = 9kts. Oil 5.8t.
Seekuh: 90.1t full load (*Seekuh 1*); 78ft 2in wl, 88ft 8in oa × 11ft 8in × 7ft 4in (*23.8m, 27.0m × 3.5m × 2.2m*); *Seekuh 2, 3* 25ft (*7.6m*) wl, 1 shaft Schottel propeller, diesel-electric drive, 1 diesel 320bhp (235.5kW) = 9kts. Oil 6.2t.

ARIADNE class *inshore minesweepers*

Displacement:	199t–205t standard; 252t full load
Dimensions:	116ft 2in wl, 124ft 9in oa × 25ft 2in × 26ft 4in × 7in *35.3m, 38.0m × 7.6m–8.0m × 2.1m*
Machinery:	2 shafts, 2 Mercedes-Benz diesels, 2000bhp (1472kW) = 14.5kts. Oil 16t. Range 830nm at 12kts
Armament:	1–40mm/70
Complement:	24

Class (no, launched, fate):

Ariadne (M 2650, 23.4.60, decomm 13.12.91), *Freya* (M 2651, 25.6.60, decomm 5.5.92), *Vineta* (M 2652, 17.9.60, decomm 20.12.91), *Hertha* (M

2653, 27.2.61, decomm 7.5.92), *Nymphe* (M 2654, 20.9.62, decomm 18.6.92), *Nixe* (M 2655, 3.12.62, decomm 2.4.92 and sold 1994), *Amazone* (M 2656, 27.2.63, decomm 9.4.92), *Gazelle* (M 2657, 14.8.63, decomm 2.7.92)

These wooden-hulled coast guard boats were designed in 1958–59 as 'KW 55-1', using mainly non-magnetic material in view of the successful outcome with *Hansa* and *Niobe*. All units were built by Krögerwerft, and were commissioned with the pennant prefix 'W' and belonged to the Type 362.

From 1959 they were reclassified as riverine minesweepers of the Type 393 and pennant prefix was changed to 'M'. The eight units were built in two groups which differed from each other in the shape of the main frame. From 1968 to 1974 they were in reserve with pennant prefix 'Y' although, with the exception of *Gazelle*, they never carried it. In 1974 they were recommissioned as an active minesweeping squadron. All decommissioned 1991–92.

FRAUENLOB class *inshore minesweepers*

Displacement: 237.8t standard; 246.4t full load
Dimensions: 116ft 3in wl, 124ft 9in oa × 26ft 4in × 7ft
35.4m, 38.0m × 8.0m × 2.1m
Machinery: 2 shafts, 2 Mercedes-Benz diesels, 2000bhp (1472kW) = 14.5kts. Oil 13.3t. Range 993nm at 12 kts
Armament: 1–40mm/L70
Complement: 24

Class (no, launched):
Frauenlob (M 2658, 26.2.65), *Nautilus* (M 2659, 19.5.65), *Gefion* (M 2660, 19.6.65), *Medusa* (M 2661, 25.1.66), *Undine* (M 2662, 16.5.66), *Minerva* (M 2663, 25.8.66), *Diana* (M 2664, 13.12.66), *Loreley* (M 2665, 14.3.67), *Atlantis* (M 2666, 20.6.67), *Acheron* (M 2667, 11.10.7).

These coast guard boats ('KW 60') of the Type 362 were reclassified in 1967 to riverine minesweepers of the Type 394. They resemble the previous *Ariadne* class. The first seven boats commissioned with the pennant prefix 'W'; from 1968 they carried the prefix 'Y' and finally since 1974, 'M'. *Nautilus* decommissioned 28 April 1994, *Minerva* and *Diana* 16 February 1995 and *Atlantis* and *Acheon* 20 March 1995; remainder extant 1995.

HOLNIS class *inshore minesweepers*

Displacement: 150t standard; 180.9t full load
Dimensions: 113ft 3in wl, 119ft 10in oa × 24ft 4in × 5ft 8in
34.5m, 41.7m × 7.4m × 1.7m
Machinery: 2 shafts, 2 Mercedes-Benz diesels, 2000bhp (1472kW) = 16.5kts. Oil 14.7t
Armament: 1–20mm (later removed)
Complement: 25

No	Name	Builder	Launched	Fate
Y 1400	HOLNIS	Abeking & Rasmussen	20.5.65	Extant 1995

Another type of inshore riverine minesweeper was planned (Type 390), but only the prototype unit, *Holnis*, was commissioned on 31 March 1966. She was launched on 20 May 1965 at Abeking & Rasmussen who were to have built the whole class, but the project was cancelled.

After commissioning, *Holnis* carried the pennant prefix 'M'. In 1968 she was converted to a communications trials boat. Guns and minesweeping gear were removed and a hut was erected on the poop. She now belongs to the Type 740, and has a civilian crew. Original pennant number Y 836 changed to A 836 in 1985 and again in 1987 to Y 1400.

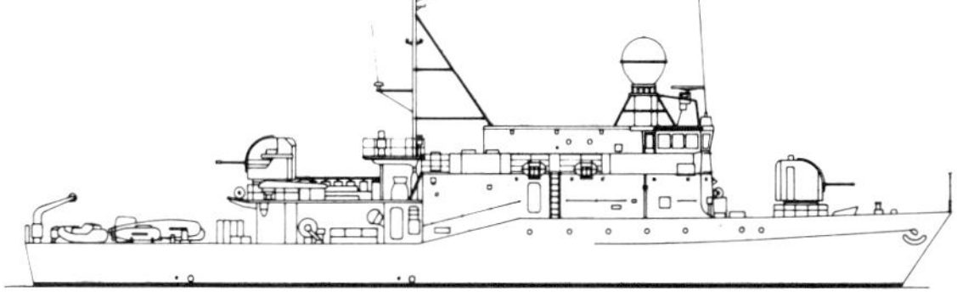

Hameln 1989

HAMELN class *fast minesweepers*

Displacement: 450t standard; 590t full load
Dimensions: 167ft 4in wl, 178ft 6in oa × 30ft 2in × 8ft 2in
51m, 54.4m × 9.2m × 2.5m
Machinery: 2 shafts (CP propellers), 2 MTU diesels, 5540–6100hp 4080–4480kW) = 18kts
Armament: 2–40mm/L 70 (2×1), SAM, 2 Hot Dog/Silver Dog chaff launchers
Sensors: Radar M 20/2 fire control, MIRGA 6, PALIS; sonar ?; ECM DR 2000 radar warning
Complement: 37

Kulmbach 12.9.94 L & L van Ginderen

No	Name	Builder	Launched	Comp	Fate
M 1090	PEGNITZ	Lürssenwerft	13.3.89	9.3.90	Extant 1995
M 1091	KULMBACH	Abeking & Rasmussen	15.6.89	24.4.90	Extant 1995
M 1092	HAMELN	Lürssenwerft	15.3.88	29.6.89	Extant 1995
M 1093	AUERBACH	Lürssenwerft	18.6.90	7.5.91	Extant 1995
M 1094	ENSDORF	Lürssenwerft	8.12.89	25.9.90	Extant 1995
M 1095	ÜBERHERRN	Abeking & Rasmussen	30.8.88	19.9.89	Extant 1995
M 1096	PASSAU	Abeking & Rasmussen	1.3.90	18.12.90	Extant 1995
M 1097	LABOE	Krögerwerft	13.9.88	7.12.89	Extant 1995
M 1098	SIEGBURG	Krögerwerft	14.4.89	17.7.90	Extant 1995
M 1099	HERTEN	Krögerwerft	22.12.89	26.2.91	Extant 1995

The Type 343 fast minesweepers, called 'mine combat boats' replaced the *Schütze* class fast minesweepers. Ten units were ordered on 3 July 1985. They are built of non-magnetic steel and have semi-buried forecastles. The two 40mm guns are mounted on the forecastle and a platform at the end of the superstructure respectively. Their M 20/2 fire control systems were taken from the decommissioned *Zobel* class FACs. Five boats are to be converted to minehunters, and the remaining five to control ships for *Seehund*.

Frankenthal design

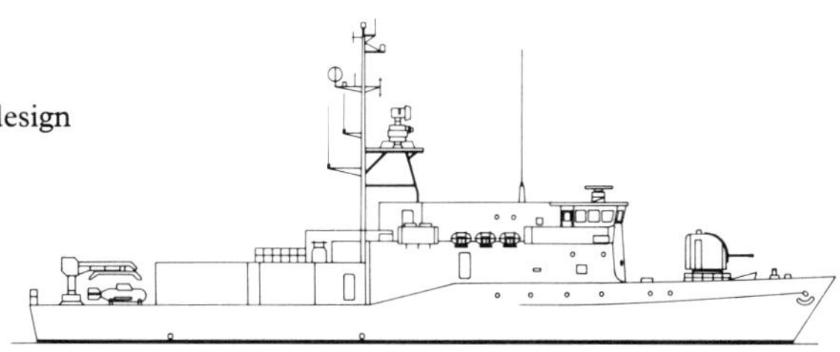

FRANKENTHAL class *minehunters*

Displacement: 650t standard
Armament: 1–40mm/L 70, 2 MBB-Pinguin-B3 mine disposal vehicles
Other particulars: As *Hameln* class

No	Name	Builder	Launched	Comp	Fate
M 1060	WEIDEN	Abeking & Rasmussen	14.5.92	30.3.93	Extant 1995
M 1061	ROTTWEIL	Krögerwerft	12.3.92	7.7.93	Extant 1995
M 1062	SULZBACH-ROSENBERG	Lürssenwerft	27.4.95	—	Completing
M 1063	BAD BEVENSEN	Lürssenwerft	21.1.93	9.12.93	Extant 1995
M 1064	GRÖMITZ	Krögerwerft	22.4.93	23.8.94	Extant 1995
M 1065	DILLINGEN	Abeking & Rasmussen	26.5.95	25.4.95	Extant 1995
M 1066	FRANKEN-THAL	Lürssenwerft	6.2.92	16.12.92	Extant 1995
M 1067	BAD RAPPENAU	Abeking & Rasmussen	3.6.93	19.4.94	Extant 1995
M 1068	DATTELN	Lürssenwerft	27.1.94	1.12.94	Extant 1995
M 1069	HOMBURG	Krögerwerft	21.4.94	_	Completing

Not fitted with M 20 fire control system.

Neckar 3.6.83 L & L van Ginderen

MISCELLANEOUS

RHEIN (Type 401) class *support ships*

Displacement: 2370t standard; 2749–3016t full load
Dimensions: 304ft 6in wl, 322ft 2in oa × 38ft 10in × 17ft 2in
92.9m, 98.2m × 11.8m × 5.2m
Machinery: 2 shafts (CP propellers), 6 MTU diesels, 11,400bhp (8390kW) = 22kts. Oil 330t
Armament: 2–100mm/55 (2×1), 4–40mm/L70 (2×2)
Sensors: Radar KM 14/9, SGR-105, M45
Complement: 122

No	Name	Builder	Launched	Fate
A 58	RHEIN	Schlieker, Hamburg	10.12.59	Decomm 26.6.92, sold for BU 25.11.94
A 61	ELBE	Schlieker, Hamburg	5.5.60	To Turkey Mar 1993
A 62	WESER	Elsflether Werft	11.6.60	To Greece 6.7.76
A 63	MAIN	Lindenau, Kiel	23.7.60	Decomm 25.11.93
A 64	RUHR	Schlieker/Blohm & Voss	8.8.60	To Turkey 9.11.76
A 66	NECKAR	Lürssen	26.6.61	Decomm 30.11.89, sold for BU
A 68	WERRA	Lindenau, Kiel	26.3.63	Decomm 21.3.91, sold for BU 1993
A 69	DONAU	Schlichting, Lübeck-Travemünde	26.11.60	Decomm 1.12.94, to Greece 13.3.95

Saar 1971

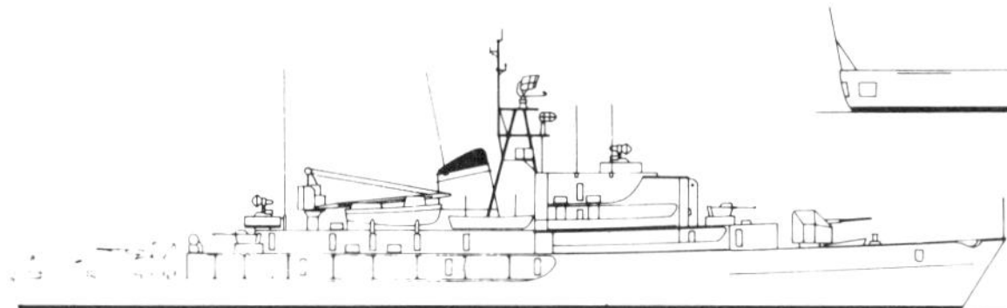

MOSEL (Type 402) class *support ships*

Displacement: 2330t standard; 3010t full load
Dimensions: 304ft 6in wl, 323ft 4in oa × 38ft 10in × 17ft 2in
92.9m, 98.5m × 11.8m × 5.2m
Machinery: 2-shafts diesel-electric: 6 MTU diesels, 12,000 bhp (8832kW) = 22.5kts. Oil 441t. Range 2500nm at 16kts
Complement: 99
Other particulars: As *Rhein* class

No	Name	Builder	Launched	Fate
A 54	ISAR	Blohm & Voss,	14.7.62	To Turkey 30.9.82
A 65	SAAR	Norderwerft, Hamburg	11.3.61	Decomm 15.2.91, sold for BU 31.8.90
A 67	MOSEL	Schlieker/Blohm & Voss	15.12.60	Decomm 30.6.89, sold for BU 30.12.91

LAHN (Type 403) class *support ships*

Displacement: 2633t standard; 2886t full load
Dimensions: 292ft pp, 304ft 6in wl, 327ft oa × 38ft 10in × 17ft 2in
89.0m, 92.8m, 99.7m × 11.8m × 5.2m
Machinery: As *Mosel* class, but oil 452t
Armament: 4–40mm/70 (2×2)
Sensors: Radar SGR-103
Complement: 114

No	Name	Builder	Launched	Fate
A 55	LAHN	Flender-Werke, Lübeck	21.11.61	Decomm 26.4.91, sold for BU 30.12.91
A 56	LECH	Flender-Werke, Lübeck	4.5.62	Decomm 30.6.89, sold for BU 31.8.90

Under the first building programme a total of eleven tenders (depot ships) was ordered, to a design produced in 1957–59. They were designed primarily to act as support vessels for particular types of warship. Officially, therefore, they are not fighting units, but considering their displacement, speed and armament they are real multi-role ships which can serve as command ships for their squadron, as accommodation ships (capacity for an additional 200 men), as escorts, minelayers and so on.

The tenders are sub-divided into three classes: Type 401 fast attack craft tenders, Type 402 fast minesweeper tenders, Type 403 submarine tenders. The Type 401 has pure diesel propulsion while the Types 402 and 403 have diesel-electric propulsion. The tenders of the Types 401 and 402 have identical armament, while the Type 403 has only 40mm guns.

Not all the tenders were actually allocated to flotillas; for example, *Weser*, *Ruhr* and *Donau* served as cadet training ships for a short time during the 1960s.

Because of the serious shortage of manpower, *Isar*, *Weser*, *Ruhr*, *Lech* and *Donau* were placed in reserve from 1968 to 1971. *Donau* was recommissioned in 1970 and *Lech* in 1975, while *Weser* was handed over to the Greek Navy, and *Ruhr*, *Elbe* and *Isar* to the Turkish Navy as military aid.

Deutschland 1963

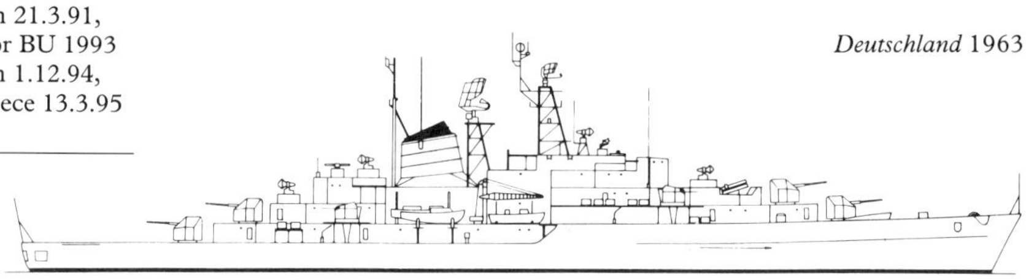

DEUTSCHLAND *training ship*

Displacement: 4880t standard; 5684t full load
Dimensions: 424ft 6in wl, 453ft 6in oa × 52ft 9in × 16ft 8in
130.0m, 138.2m × 16.1m × 5.1m
Machinery: 3-shaft (CP propellers), Wahodag steam turbines, 2 Wahodag boilers, plus 4 MTU diesels, 16,000shp (11,776kW) = 22kts. Fuel and diesel oil 643t. Range 3800nm at 12kts
Armament: 4–100mm/55 (4×1), 6–40mm/70 (2×2, 2×1), 2–533mm TT (stern), 4–533mm ASW TT (4×1), 2–375mm ASW mortars (2×4)
Sensors: Radar LW-08, SGR-114, SGR-105, SGR-103, M-45; sonar ELAC 1BV
Complement: 172

No	Name	Builder	Launched	Fate
A 59	DEUTSCHLAND	Nobiskrug, Rendsburg	5.11.60	Decomm 28.6.90, sold for BU Oct 1993

The training ship *Deutschland* was of the Type 440 and was not strictly a fighting unit although she was designed in 1957–58 as a multi-purpose ship in case of war for troop transport, hospital, escort and minelaying duties. The extraordinary displacement required special permission from the WEU, and limits to speed and armament were imposed.

Deutschland served as a cadet training ship only and for this reason she had a number of unique features, in combination not found in any other ship. The propulsion was mixed: the central shaft steam-turbine driven, the outer shafts by two different types of diesel: one Maybach and one Mercedes-Benz. In 1981 this was rationalised to Mercedes diesels because of the maintenance costs. The armament included all Federal German Navy types except missiles.

In addition to a nucleus crew of 172 men, a further 250 cadets could be accommodated. The stern TTs were removed in the mid-1970s.

Deutschland as completed Navarret Collection

NEUSTADT class *patrol craft*

Displacement: 191.4t standard; 218t full load
Dimensions: 118ft wl, 129ft oa × 23ft × 7ft 2in
36.0m, 39.3m × 7.0m × 2.1m
Machinery: 3 shafts, (1 shaft for cruising, 1 MWM diesel 685bhp (504kW), 2 shafts, 2 MTU diesels 6000bhp (4416kW) for high speed) = 9kts and 30kts. Oil 15.5t. Range 450nm at 27kts
Armament: 2–40mm/70 (2×1)
Complement: 23

Class (no, launched):
Neustadt (BG 11, 27.2.69), *Bad Bramstedt* (BG 12, 2.4.69), *Uelzen* (BG 13, 25.7.69), *Duderstadt* (BG 14, 3.6.69), *Eschwege* (BG 15, 16.9.69), *Alsfeld* (BG 16, 11.11.69), *Bayreuth* (BG 17, 9.1.70), *Rosenheim* (BG 18, 12.3.70).

The so-called Federal Frontier Guard (*Bundesgrenzschutz*) was instituted on 15 February 1951 and was taken over by the newly-created navy (*Bundeswehr*) on 1 July 1956. Its personnel could choose to enter the service or remain as civil servants. The entire function of the Federal Frontier Guard (Naval) ('Bundesgrenzschutz/See') was assumed by the navy when the former unit was disbanded.

The duties of the Federal Frontier Guard are: prevention of illegal entrants, passport control and security of the border. The strength of the Federal Frontier Guard was originally fixed by the Allied High Commissioners at 10,000 men, including 500 naval personnel. The unit is subordinated to the Federal Ministry of Home Affairs.

In 1963, because of continual border incidents, but also in order to release the navy from frontier control duties – especially on the German Democratic Republic's border – the Ministry of Home Affairs and the Federal Ministry of Defence agreed to re-institute a small naval frontier inspection unit. As a first step four former Type 369 coast guard boats were handed over to the Guard in 1964. Meanwhile the first order for new boats had been given and on 27 February 1969 the first new vessel was launched. The Type consisted of eight boats designed and built by Lürssen. In contrast to the usual grey paint, they have a blue hull and white superstructure. In 1990 *Uelzen* was sold to Mauretania.

Ghana

Ghana under Kwame Nkrumah became in March 1957 the first British black African territory to achieve independence, largely because of her relatively advanced electorate and strong cocoa-based economy. The initial promise was not, however, maintained; within a few years, strains appeared in the carefully imported pseudo-Westminster constitutional arrangements, and, as a result of extravagance, mismanagement and a catastrophic fall in the world cocoa price, the economy almost collapsed.

The first democratic experiment ended with Nkrumah's overthrow by the Army in February 1966, since when Ghana has had military rule rather than not and an economy that has not recovered. Internationally, the fledgeling state became the centre of pan-African nationalism and a vigorous denouncer of 'neo-colonialism', but by 1966, as Nkrumah's outlook moved left, the official policy had become an ill-defined amalgam of Pan-Africanism and 'African Socialism', entitled 'Nkrumahism'. Since 1966 domestic matters have swept these 'isms' out of sight.

The volunteer armed forces, initially derived from the modest colonial army, gradually expanded in the years of independence and as much as 12 per cent of the gross national product was spent on the armed forces in the 1970s, a very high figure for a country under no threat from her neighbours, and emphasising the privileged role of the military in tropical Africa. Despite this huge sum, the training of the armed forces deteriorated and their development stagnated, affecting the soldiers' professionalism and dedication to duty. Less than 1 per cent of GNP is now spent on the armed forces.

The Ghanian Navy, very much a junior partner, has been built up into a modest coast defence force split into Western and Eastern Commands, mainly useful for coast guard purposes and fishery protection. By the end of 1981 fish piracy was so rife that Eastern Command issued a shoot to sink warning. The volunteer personnel, planned in 1960 for 500, reached 800 in 1970 and 1360 in 1982, but by 1993 had fallen to 750. The first ships, two ex-RN inshore minesweepers, reached Sekondi-Takoradi (Tema, east of Accra, is the other main base) in November 1959, and subsequent construction and purchases have been very moderate, except during the last years of Nkrumah, when there were grandiose plans for a sizeable fleet. Two corvettes of a novel type completed in 1964 and 1965 were to have been followed by two more, and by two frigates and a support ship but of the latter five, only one frigate materialised, and then not in Ghanaian service. The chronic shortage of funds has badly affected the navy.

Kromantse on trials — Vosper Thornycroft

BLACK STAR *frigate*

Particulars: See British *Mermaid*

Modified RN Type 41 frigate, to serve also as presidential yacht, ordered in 1965 as *Black Star*, but cancelled after President Nkrumah's deposition in 1966. The British Government accepting responsibility for the hull, the ship was completed by the builders. Acquired by the RN in 1972 after six years lying idle in the Clyde and commissioned in 1973 as *Mermaid*. See under United Kingdom (Part I) and Malaysia (Part II). Another such frigate was projected, but never begun.

Kromantse as completed
NB 1/750 scale

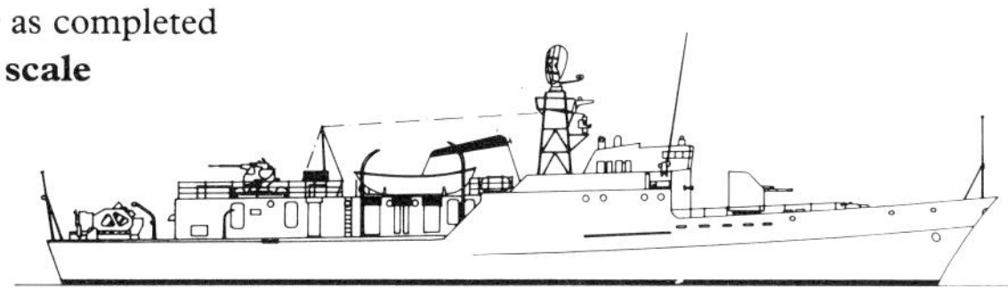

KROMANTSE class *corvettes*

Displacement: 380t light; 440t standard; 500t full load
Dimensions: 177ft oa × 28ft 6in × 13ft
54.0m × 8.7m × 4.0m
Machinery: 2 shafts, 2 Bristol Siddeley Maybach 16cyl diesels, 7100 bhp = 20kt. Fuel 60t. Range 1100nm/2900nm at 14 kts/18kts
Armament: 1–4in/40 Mk 23, 1–40mm/60 Mk 9, 1 Squid Mk 4, 2–2in RFLs
Sensors: Radar Plessey AWS-1, Type 978 navigation set; Sonar Type 164
Complement: 54

No	Name	Builder	Launched	Fate
F 17	KROMANTSE	Vosper, Portsmouth	5.9.63	Discarded 1989
F 18	KETA	Vickers, Tyne	18.1.65	Discarded 1989

Vosper Mk 1 corvettes, designed in conjunction with Vickers, and the first in the series of Vosper (now Vosper Thornycroft) corvette and frigate designs. Laid down 10 December 1962 and 25 October 1963. Completed 27 July 1964 and 18 May 1965. Novel, compact and relatively cheap ASW patrol vessels. Both refitted 1974–75. Two more units were projected, but never begun.

Yogaga as completed
NB 1/750 scale

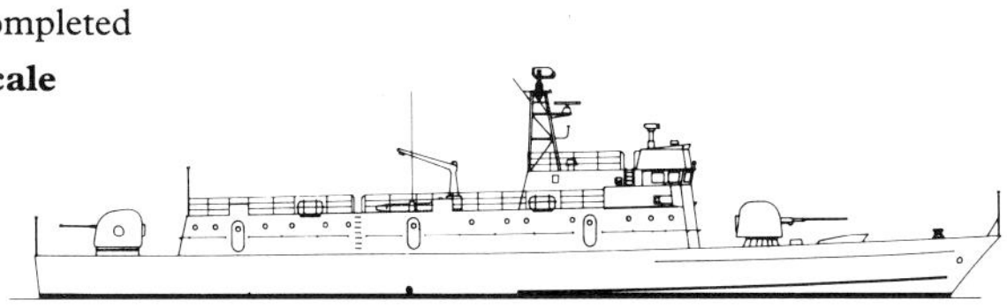

ACHIMOTA class *fast attack craft (gun)*

Displacement: 380t standard; 410t full load
Dimensions: 190ft 7in oa × 25ft × 9ft 2in
58.1m × 7.6m × 2.8m
Machinery: 3 shaft, 3 MTU 16V 538 TB91 diesels, 11,400bhp = 33kts. Range 3300nm/5200nm at 16kts/12kts
Armament: 1–76mm/62 OTO Melara Compact (250rnds), 1–40mm/70 Bofors (750rnds)
Sensors: Radar navigational, Thomson-CSF Canopus A
Complement: 40

Class (no):
Achimota (P 28), *Yogaga* (P 29)

PB 57 type, launched 14 March 1979 and delivered 27 March 1981 by Lürssen, Vegesack. Larger and faster than *Dzata* class. In service 1995.

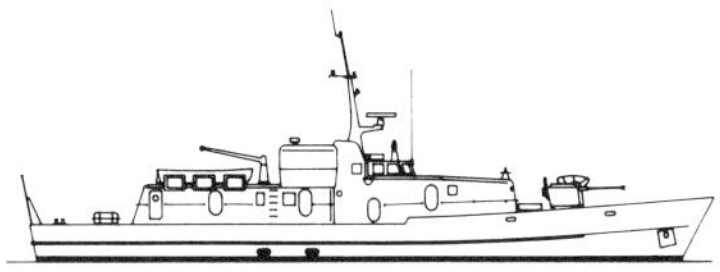

Sahene 1976
NB 1/750 scale

DELA class *large patrol craft*

Displacement: 160t full load
Dimensions: 115ft 6in × 21ft 4in × 5ft 11in
35.2m × 6.5m × 1.8m
Machinery: 2 shafts, 2 MTU MD 16V 538 TB90 diesels, 3000bhp = 30kts. Range 1000nm at 30kts
Armament: 1–40mm Bofors (RFLs attached)
Complement: 32

Class (no):
Dela (P 24), *Sahene* (P 25)

Built 1973–76 for fishery protection by Ruthoff Werft, Mainz, West Germany. The builders' 1975 bankruptcy prevented construction of four further units. Refitted by Lürssen 1981. Stricken *c* 1992.

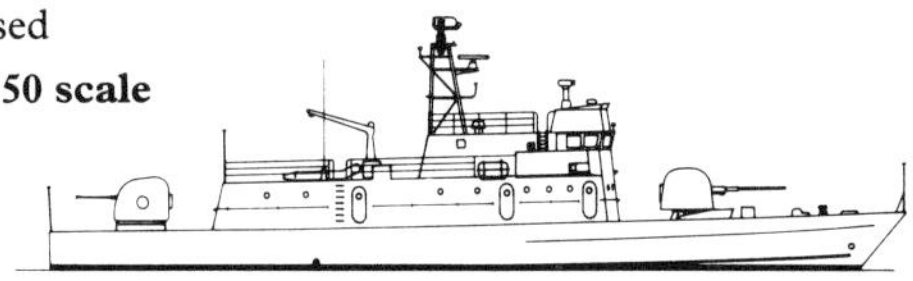

Dzata as originally proposed with 76mm gun **NB 1/750 scale**

DZATA class *large patrol craft*

Displacement: 212t standard, 255t full load
Dimensions: 147ft 4in × 23ft × 7ft 6in
44.9m × 7.0m × 2.3m
Machinery: 2 shafts, 2 MTU diesels, 7200bhp = 27kts. Range 1100nm/2000nm at 25kts/15kts
Armament: 2–40mm/70 Bofors, 2–7.62mm MG
Sensors: Radar navigational, Thomson-CSF Canopus A
Complement: 356

Class (no, in service):
Dzata (P 26, 4.12.1979), *Sebo* (P 27, 2.5.80)

TNC 45 type, laid down January 1978 and launched 19 September 1979 by Lürssen, Vegesack, and in service 1995.

British 'FORD' class *seaward defence boats*

Class (no, comp):
Elmina (P 13, 18.10.62), *Komenda* (P 14, 17.5.62)

As RN ships, but built for Ghana and commissioned in November 1962. Both discarded by 1985.

Ex-British 'HAM' class *inshore minesweepers*

Class (no):
Afadzato (M 12), Yogaga (M 11)

Formerly *Ottringham* and *Malham* respectively, transferred from the RN in 1959. Fitted with funnels. The Foden diesel and centre shaft were removed in 1967. Discarded 1977.

Ex-British 'TON' class *coastal minesweeper*

Ejura (M16) formerly HMS *Aldington* (M 1171) loaned by the RN in 1964 and transferred permanently in 1974. Discarded 1979.

MISCELLANEOUS

The British maintenance repair craft *MRC 1122*, a converted LCT (4) type, was purchased in 1965 and named *Asuantsi*. Used as a base workshop and for sea training, she was discarded in 1970.

Four 'Poluchat I' class coastal patrol boats, numbered *P 20*, *P 21*, *P 22* and *P 23*, were transferred from the USSR in 1967. *P 22* discarded 1970, the others 1973. There are also four coastal patrol craft of the 'Spear 2' class, built by Fairey Marine in 1978, and two Rotork service craft.

Greece

In 1947 the civil war which had split the country since 1944 was nearing its end, though it did not actually cease until 1949. It left a country already ruined by the German occupation even worse off. The navy had succeeded in retaining its identity throughout the Second World War, but it had been almost completely re-equipped with ships on loan from Britain. In 1947, however, America took over from Britain as the supporter of the Greek governments, and the Truman Doctrine, which announced that America would support the non-communist countries in southwest Europe, meant that large quantities of US arms, including warships, were supplied to Greece from the late 1940s. The first new ships to be ordered for the Greek Navy since the 1930s were the *Tjeld* type MTBs, ordered from Norway in 1965.

PROBLEMS WITH TURKEY

Since the wars of independence the traditional Greek enemy has been Turkey, and despite both countries joining NATO in 1952, little has changed. The Greek islands off the Turkish mainland in the Aegean have been a traditional source of quarrels, intensified by the search for oil and minerals on the continental shelf in the 1970s. Greece claims a continuous EEZ from the Greek mainland to the islands off Turkey, while Turkey claims that large areas of the Aegean belong to her Exclusive Economic Zone (EEZ).

An even greater source of discord than the Aegean has been Cyprus. From the beginnings of the EOKA struggle for independence against the British, Greece favoured *Enosis*, or union with Greece, while Turkey held out for the independence of Cyprus. The formation of the Republic of Cyprus in 1960 did nothing to solve the problem, and open civil war in Cyprus between the Greek and Turkish populations led to intervention by the armed forces of both countries.

The intercession of the UN separated Greece and Turkey before open warfare could break out, but when Turkey invaded Cyprus in 1974 and Greece was prevented from intervening, she left NATO. The decision to withdraw was influenced by the association of NATO, and in particular the United States, with the military 'Colonels' regime which ruled Greece from 1967 until it was toppled in November 1973. Greece retained some ties with NATO, and eventually rejoined in 1981.

THE GREEK NAVY

The main determinant of the strength of the Greek Navy since 1947 has been the strength of the Turkish Navy. Although both countries received considerable numbers of wartime ex-USN destroyers, the main strength of the Greek Navy in the 1970s and 1980s lay in its force of modern German-built submarines and its fast attack craft, well-suited to operations in the restricted waters of the Aegean. The multitude of narrow channels and islands also explains the large mine warfare and amphibious forces maintained throughout the postwar period. However, by the 1980s it was becoming imperative that replacements be found for the ageing major surface units, which were becoming increasingly difficult to maintain. Attempts to establish a local shipbuilding industry have been hampered by financial problems and the general weakness of the Greek economy, which lead to the abandonment in the mid-1980s of plans to build Netherlands *Kortenaer* class frigates in Greek yards and major delays to later frigate, LST and FAC construction programmes. Indeed, it was not until Netherlands and USN destroyers and frigates became available for transfer following the end of the cold war that the older units could finally be retired. However, the amphibious forces were still largely dependent upon Second World War-vintage ships in 1995. The navy also still suffers its perennial problem of shortage of skilled manpower, which has improved little since the early 1960s, when the cruiser *Helle* was discarded mainly because she absorbed almost all the trained ratings needed to man the destroyers and frigates.

Despite the acquisition of the *Kortenaer* frigates and a considerable quantity of German-built equipment (beginning in the 1960s), the Greek Navy remains largely dependent on the USA both for ships and weaponry, and in 1991 an agreement was signed continuing the US presence at Suda Bay in Crete. The Greek Navy has benefited greatly from the 'cascading' of ships by the USN since the end of the cold war, as Turkey has also.

SPECIAL NOTE
In the following tables, in the ship names
A = *Antipliarchos* I = *Ipopliarchos* P = *Plotarchos*

FLEET STRENGTH 1947

CRUISERS

Name	Launched (full load)	Disp	Fate
GIORGIOS AVEROFF	Mar 1910	9956t	Discarded 1951

Averoff was not finally taken out of service until the ex-Italian cruiser *Helle* was commissioned. *Averoff* is now a national monument and is moored at Poros.

DESTROYERS

Name	Launched	Disp	Fate
'B' class			
SALAMIS (ex-*Boreas*)	18.7.30	1360t	Discarded 1951
'E' class			
NAVARINON (ex-*Echo*)	16.2.34	1375t	Discarded 1956

Salamis was the ex-British 'B' class destroyer *Boreas*. She was on loan to the Greek Navy during 1944–51, and was scrapped in April 1952. *Navarinon* was the ex-British 'E' class destroyer *Echo*, on loan during 1944–56. She was scrapped in April 1956.

FRIGATES

Name	Launched	Disp	Fate
'Hunt' Type 2			
AEGEON (ex-*Lauderdale*)	5.8.41	1050t	Discarded 12.12.59, BU 1960
KRITI (ex-*Hursley*)	25.7.41	1050t	Discarded 12.12.59, BU 1960
THEMISTOCLES (ex-*Bramham*)	29.1.42	1050t	Discarded 12.12.59, BU 1960
'Hunt' Type 3			
ADRIAS (ex-*Tanatside*)	30.4.42	1087t	Discarded 1963, BU 1964
HASTINGS (ex-*Catterick*)	22.11.41	1087t	Discarded 1963, BU 1963
KANARIS (ex-*Hatherleigh*)	18.12.41	1087t	Discarded 12.12.59, BU 1960
MIAOULIS (ex-*Modbury*)	13.4.42	1087t	Discarded 1960, BU 1960
PINDOS (ex-*Bolebroke*)	5.11.41	1087t	Discarded 1960, BU 1961
'Flower' class			
APOSTOLIS (ex-*Hyacinth*)	19.8.40	1020t	Discarded 1961, BU 1962
KRIEZIS (ex-*Coreopsis*)	23.5.40	1020t	Discarded 1952, BU 1952
SAKHTOURIS (ex-*Peony*)	4.6.40	1020t	Discarded 1951, BU 1952
TOMPAZIS (ex-*Tamarisk*)	28.7.41	1020t	Discarded 1952, BU 1963

The ex-British 'Hunt' Types 2 and 3 were originally rated as escort destroyers; they were later reclassified as frigates. Of the 'Hunt' Type 2, *Kriti* and *Themistocles* were lent by the British in 1943, and *Aegeon* was lent in 1946. The 'Hunt' Type 3 *Kanaris*, *Miaoulis* and *Pindos* were lent in 1942 and *Adrias* and *Hastings* in 1946. *Hastings* was named after a British officer who had fought in the Greek War of Independence. The ex-British 'Flower' class were originally rated as corvettes, but were later reclassified as frigates. *Tompazis* had been modified with a long forecastle, but the rest retained their original short forecastles. All four were lent by Britain in 1943. A further 'Flower' class, *St Lykoudis*, ex-mercantile *Chania*, ex-*Nasturtium*, ex-French *La Paimpolaise*, was sold to Greece as the merchantman *Chania* in 1948. She was later purchased by the Greek Navy and renamed *St Lykoudis*, and was employed as an unarmed lighthouse tender until she was discarded in 1976.

SUBMARINES

Name	Launched	Disp	Fate
'U' class (2nd Series)			
AMFITRITI (ex-*Untiring*)	20.1.43	545/740t	Discarded 1952
XIFIAS (ex-*Upstart*)	24.11.42	545/740t	Discarded 1952
'V' class			
ARGONAFTIS (ex-*Virulent*)	23.5.44	545/740t	Discarded 3.10.58
DELFIN (ex-*Vengeful*)	20.7.44	545/740t	Discarded 1957, BU 1958
PIPINOS (ex-*Veldt*)	19.7.43	545/740t	Discarded 10.12.57, BU 1958
TRIAINA (ex-*Volatile*)	20.6.44	545/740t	Discarded Oct 1958, BU 1959

The ex-Britsh 'U' class submarines *Amfitriti* and *Xifias* were both lent to Greece by Britain in 1945. Their hulks were expended as targets in July 1957 off the English S coast. The ex-British 'V' class submarine *Pipinos* was lent in 1943. *Delfin* was lent in 1944 and *Argonaftis* and *Triaina* in 1946. *Argonaftis* left Malta November 1958 in tow for breakers on R Tyne, broke adrift and stranded on Spanish coast and was sold in 1961.

PATROL VESSEL

Name	Launched	Disp	Fate
PC type			
VASSILEFS GEORGIOS II (ex-*PC 622*)	1942	280t	Discarded *c*1960

The ex-USN PC boat *Vassilefs Georgios II* was transferred to Greece under lease-lend in June 1944.

MINESWEEPERS

Name	Fate	Name	Fate
BYMS type: launched 1942–45, 215t			
AFROESSA (ex-*BYMS 2185*)	Discarded 1973	SIMI (ex-*BYMS 2190*)	Discarded 1966
ITHAKI (ex-*BYMS 2240*)	Discarded 1966	ZAKYNTHOS (ex-*BYMS 2209*)	Discarded 1973
KALYMNOS (ex-*BYMS 2033*)	Discarded 1973	AURA (ex-*BYMS 2054*)	Discarded 1967
KARTERIA (ex-*BYMS 2065*)	Discarded 1973	ANDROMEDA (ex-*BYMS 2261*)	Discarded 1967
KEFALLINIA (ex-*BYMS 2171*)	Discarded 1966	KLEIO (ex-*BYMS 2152*)	Discarded 1967
KERKYRA (ex-*BYMS 2172*)	Discarded *c*1960	LAMBADIAS (ex-*BYMS 2182*)	Discarded 1968
LEFKAS (ex-*BYMS 2086*)	Discarded 1966	PIGASSOS (ex-*BYMS 2221*)	Discarded 1968
LEROS (ex-*BYMS* 2186)	Discarded 1966	PROKYON (ex-*BYMS 2076*)	Discarded 1968
PAPALOS (ex-*BYMS 2066*)	Discarded 1973	THALIA (ex-*BYMS 2252*)	Discarded 1967
PATMOS (ex-*BYMS 2229*)	Discarded 1966	ARIADNE (ex-*BYMS 2058*)	Discarded *c*1975
PAXI (ex-*BYMS 2056*)	Discarded 1972	VEGAS (ex-*BYMS 2058*)	Discarded *c*1975
SALAMINIA (ex-*BYMS 2067*)	Discarded *c*1960		
MMS type: launched 1940–44, 165t. All discarded *c*1960			
ANDROS (ex-*MMS 310*)		MIKONOS (ex-*MMS 5*)	
ARGYROKASTRON (ex-*MMS 58*)		SYROS (ex-*MMS 313*)	
CHIMARA (ex-*MMS 1*)		TEPELENI (ex-*MMS 46*)	
KORYTSA (ex-*MMS 53*)		TINOS (ex-*MMS 144*)	

BYMS type
These ex-British BYMS were transferred to Greece prior to 1947. *Afroessa* to *Zakynthos* were used as coastal minesweepers, *Aura* to *Thalia* as coastal patrol boats and *Ariadne* and *Vegas* as coastal survey craft.

MMS type
Ex-Britsh inshore minesweepers transferred to the Greek Navy prior to 1947.

PATROL CRAFT

Name	Name
Fairmile B type: launched 1942–44, 73t. All discarded *c*1960	
DOLIANA (ex-*ML 840*)	KARPENISSI (ex-*ML 867*)
DRAMA (ex-*ML 341*)	KASSOS (ex-*ML 554*)
ELEFTERON (ex-*ML 478*)	KASTELORIZON (ex-*ML 840*)
HALKI (ex-*ML 578*)	KOS (ex-*ML 565*)
KALAMBAKA (ex-*ML 483*)	NISSIROS (ex-*ML 864*)
KARPATHOS (ex-*ML 561*)	TILOS (ex-*ML 569*)
HDML type: launched 1940–44, 54t. All discarded *c*1960	
BIZANI (ex-*HDML 1221*)	KARIA (ex-*HDML 1307*)
DAVLIA (ex-*HDML 1032*)	KASRTAKI (ex-*HDML 1375*)
DISTRATON (ex-*HDML 1292*)	KLISSOURA (ex-*HDML 1149*)
FARSALA (ex-*HDML 1252*)	PORTARIA (ex-*HDML 1051*)

The ex-British MLs and HDMLS were all transferred prior to 1947.

LANDING SHIPS

Name	Fate	Name	Fate
LST 1-510 class			
CHIOS (ex-*LST 31*)	Discarded 1977	SAMOS (ex-*LST 35*)	Discarded 1997
LEMNOS (ex-*LST 36*)	Discarded 1977		

These three ex-USN *LST 1–510* class tank landing ships were transferred to the Greek Navy under lease-lend in 1944.

PENNANT NUMBERS

Although allocated NATO numbers Greek warships usually wear their own totally separate pennants. Since these national numbers are the ones that are usually painted up, they are quoted in the tables following. However, even these were occasionally altered.

MAJOR SURFACE SHIPS

Ex-Italian DUCA D'AOSTA class *cruiser*

Particulars: As Italian *Duca D'Aosta* class

No	Name	Builder	Acquired	Fate
C 94	HELLE (ex-*Eugenio di Savoia*)	Ansaldo	1951	Discarded 1964

Originally launched in 1935, *Eugenio di Savoia* was transferred to Greece under the terms of the Italian Peace Treaty to compensate Greece for the sinking of the previous *Helle* in 1940 by an Italian submarine. *Helle* was finally delivered to Greece in 1951, but she absorbed most of the skilled technicians that were needed to run the smaller Greek warships so she saw little service. She was the Greek flagship during 1951–64.
(For illustration see under Italy in 1922–46 volume)

Ex-US GLEAVES class *destroyers*

Particulars: As US *Gleaves* class

No	Name	Builder	Acquired	Fate
20	DOXA (ex-*Ludlow*)	Bath Iron Wks	21.5.51	Discarded 1972
65	NIKI (ex-*Eberle*)	Bath Iron Wks	18.4.51	Discarded 1972

Doxa and *Niki* were built as USN *Gleaves* class destroyers. *Doxa* was transferred on 21 May 1951 and *Niki* in April 1951. They were given a refit in 1962 when the 6–20mm were removed, as were the 5–21in (533mm) TT (1 × 5). A tripod foremast replaced the original polemast.
(For illustration see under United States in 1922–46 volume)

Ex-US FLETCHER class *destroyers*

Particulars: As US *Fletcher* class

No	Name	Builder	Acquired	Fate
06	ASPIS (ex-*Conner*)	Boston N Yd	15.9.59	Stricken 1991
56	LONCHI (ex-*Hall*)	Boston N Yd	15.9.59	Stricken 10.10.90
63	NAVARINON (ex-*Brown*)	Bethlehem, San Pedro	27.9.62	Stricken 1981
85	SFENDONI (ex-*Aulick*)	Consolidated, Orange	21 8.59	TS, stricken
28	THYELLA (ex-*Bradford*)	Bethlehem, San Pedro	27.9.62	Stricken 1981
16	VELOS (ex-*Charette*)	Boston N Yd	15.9.59	Stricken 1991
65	NEARCHOS (ex-*Z 3*, ex-*Wadsworth*)	Bath Iron Wks	30.10.80	To Logistics Command 18.9.81
42	KIMON (ex-*Z 2*, ex-*Ringold*)	Federal SB	18.9.81	BU 1993

These were all built 1942–43 as USN *Fletcher* class destroyers. *Sfendoni* was transferred on loan to Greece under MDAP on 21 August 1959, *Aspis*, *Lonchi* and *Velos* on 15 September 1959, and *Navarinon* and *Thyella* on 27 September 1962. All the loans were renewed in March 1970 and all six were purchased on 25 April 1977. *Navarinon* and *Thyella* had 5–5in (127mm) (5×1), 10–40mm (2×4, 1×2) and no TT; the rest had 4–5in (127mm) (4×1), 6–3in (76mm) (2×3) and 5–21in (533mm) TT (1×5) and 6–324mm) (12.75in) Mk 32 ASW TT (2×3). Ex-West German *Fletcher* class *Z 1*, ex-USS *Anthony*, was acquired for spares in 1979 after being used as a target. Further units were later acquired for spares from the same source: ex-*Z 4* in 1981, and ex-*Z 5* in February 1982. Ex-*Z 2* and ex-*Z 3* had 2–533mm ASW TT in lieu of the Mk 32 and were equipped for minelaying.

A plan to modernise three units of this class, reportedly D 42, D 65 and one of the ex-West German units, announced in May 1985, came to nothing and the remaining five ships of this class were all in reserve by 1987, the ex-West German units being cannibalised for spares. *Sfendoni* served as a stationary training ship until she was stricken. All had been stricken by 1991.

Kimon (as German *Z 2* in 1977) USN

Ex-US ALLEN M SUMNER class *destroyers*

Miaoulis was built 1943–44 as a USN *Allen M Sumner* class destroyer, and received a FRAM II modernisation in the 1960s. She was transferred to Greece on 16 July 1971, and operated an Alouette III helicopter from the DASH flight deck. In 1986 she underwent modernisation at the Eleusis Shipyard, including the fitting of a new telescoping hangar for a larger AB-212 ASW helicopter. She was deleted in 1992 on the arrival of the four *Charles F Adams* class destroyers.

(For illustration see under United States in 1922–46 volume)

Particulars: As US *Allen M Sumner* class (FRAM II type)

No	Name	Builder	Acquired	Fate
D 211	MIAOULIS (ex-*Ingraham*)	Federal, Kearney	16.7.71	Deleted 1992

ex-US GEARING class *destroyers*

Particulars: As the US *Gearing* class (FRAM I and FRAM II types)

No	Name	Builder	Acquired	Fate
212	KANARIS (ex-*Stickell*)	Consolidated, Orange	1.7.72	Deleted 1994
213	KOUNTOURIOTIS (ex-*Rupertus*)	Bethlehem, Quincy	10.7.73	Deleted 1994
214	SACHTOURIS (ex-*Arnold J Isbell*)	Bethlehem, Staten I	4.12.73	Stricken 1993
210	THEMISTOCLES (ex-*Frank Knox*)	Bath Iron Wks	30.1.71	Deleted 1992
215	TOUMBAZIS (ex-*Gurke*)	Todd-Pacific, Seattle	17.3.77	Deleted 1994
216	APOSTOLIS (ex-*Charles P Cecil*)	Bath Iron Wks	2.8.80	Deleted 1993
217	KREIZIS (ex-*Myles C Fox*)	Consolidated, Orange	8.7.81	Deleted 1994

All were built 1944–46 as USN *Gearing* class destroyers. *Themistocles* was given a FRAM II radar picket modernisation in the 1960s. She was transferred on 30 January 1971 and given a minor refit in 1976–77 when a hangar and flight deck for an Alouette III helicopter replaced her radar picket facilities. In 1980 she was fitted with 4–30mm Emerlec AA (2×2) in place of her single 20mm, and in 1987 she received the same upgraded helicopter facilities as *Miaoulis* (see above). She was deleted in 1992 when the four *Charles F Adams* class destroyers were acquired.

The remaining six were FRAM I modernisations. *Kanaris* was transferred 1 July 1972, *Kountouriotis* 10 July 1973 (purchased 11 July 1978), *Sachtouris* 4 December 1973 (purchased 11 July 1978), *Toumbazis* was purchased on 17 March 1977, *Apostolis* on 2 August 1980 (initially intended for cannibalisation) and *Kreizis* on 8 July 1981. All were refitted 1987–88, receiving 1–76mm OTO Melara on the helicopter deck aft. 212–215 also received 4 Harpoon SSM (2×2) abaft the hangar. *Apostolis* and *Kreizis* did not receive Harpoon, but were fitted with 1–40mm Bofors just forward of the bridge.

Two further *Gearing* class destroyers (ex-*Corry* DD 817 and ex-*Dyess* DD 880) were acquired in 1980 for cannibalisation for spares. All had been deleted by 1995.

Kreizis (as US *Corry*) USN

Ex-US CHARLES F ADAMS class *destroyers*

Particulars: As US *Charles F Adams* class

No	Name	Acquired	Fate
218	KIMON (ex-*Semmes*, DDG 18)	13.9.91	Extant 1995
219	NEARCHOS (ex-*Waddell*, DDG 24)	1.10.92	Exant 1995
220	FORMION (ex-*Joseph Strauss*, DDG 16)	1.10.92	Extant 1995
221	THEMISTOCLES (ex-*Berkeley*, DDG 15)	1.10.92	Extant 1995

Ex-DDG 18 was leased to Greece on 13 September 1991, recommissioning a year later. The other three were transferred at San Diego on 1 October 1992, recommissioning in Greece 30 April 1994. A fourth unit (ex-*Richard E Byrd*, DDG 23) was acquired on 1 October 1992 for use as a stationary training ship and for spares. They were probably not transferred with Harpoon missiles, but these may be acquired later. It is reported that the Greek Navy intends to upgrade their electronics, fitting a new combat data system to replace the obsolete NTDS, new sonar and an improved SAM control system.

Ex-US CANNON class *destroyer escorts*

Particulars: As US *Cannon* class (DET type)

No	Name	Builder	Acquired	Fate
01	AETOS (ex-*Slater*)	Tampa SB	15.3.51	Stricken 1991
31	HIERAX (ex-*Ebert*)	Tampa SB	15.3.51	Stricken 1991
54	LEON (ex-*Eldridge*)	Federal, Newark	15.1.51	Stricken 1991
67	PANTHIR (ex-*Garfield Thomas*)	Federal, Newark	15.1.51	Stricken 1992

All four were built 1943–44 as US DET type destroyer escorts, but were later reclassified as frigates. *Leon* and *Panthir* were transferred to Greece under MDAP on 15 January 1951 and *Aetos* and *Hierax* on 15 March 1951. Their 3–21in (533mm) TT (1×3) were removed, and they were fitted with 6–324mm (12.75in) Mk 32 ASW TT (2×3). This class was well overdue for replacement by the 1980s but was not finally disposed of until 1991–92. *Hierax* and *Leon* were retained as training hulks.

GREECE

Ex-West German RHEIN class *frigate/support ships*

Particulars: See Federal German Navy *Rhein* class

No	Name	Builder	Acquired	Fate
03	AEGEON (ex-*Weser*)	Elsflether Werft	6.7.76	Stricken 1991

The West German depot ship *Weser* was transferred to Greece on 6 July 1976 and renamed *Aegeon*. These *Rhein* class depot ships were designed to be used as frigates if necessary, and were armed accordingly. Because of the Greek Navy's pressing shortage of major modern warships, *Aegeon* was therefore used as a frigate, as well as acting as a support ship for the missile patrtol boats and MTBs. On the acquisition of the *Kortenaer* and MEKO 200 frigates, she was stricken.

Ex-Netherlands KORTENAER ('STANDARD') class *frigates*

Armament: 8 Harpoon SSM (2×4), 1 Aspide SAM (1×8; 24 missiles), 2–76mm OTO Melara, 2–20mm Phalanx CIWS (2×1), 4–324mm Mk 32 ASW TT (2×2), 2 helicopters
Other particulars: See Netherlands *Kortenaer* class

No	Name	Acquired	Fate
F 450	ELLI (ex-*Pieter Florisz*)	26.6.81	Extant 1995
F 451	LIMNOS (ex-*Witte De With*)	26.6.81	Extant 1995
F 460	AEGEON (ex-*Banckert*)	14.5.93	Extant 1995
F 459	ADRIAS (ex-*Callenburgh*)	30.3.94	Extant 1995
F 460	NAVARINO (ex-*Van Kinsbergen*)	1.2.95	Extant 1995

In an attempt to reduce its dependency on the USN, the Greek Navy decided to acquire frigates of the Netherlands *Kortenaer* class, the first two units being ordered in 1980 and 1981 respectively. In order to ensure rapid delivery, the sixth and seventh units being built for the Netherlands Navy were purchased and replacements ordered for the Dutch. It was originally planned for a third unit to be built by the Hellenic Shipyard at Skaramanga, with Dutch assistance, but this was later abandoned due to financial problems. Instead a further three ships were acquired from the Netherlands 1993–95. It is also planned for more ships of this class to be transferred in 1996 as they are retired by the Netherlands Navy.

These ships have been fitted with lengthened hangars to accommodate the larger AB-212 ASW helicopters operated by the Greeks. Normally only one helicopter is carried. *Elli* and *Limnos* were rearmed with Aspide SAMs replacing the NATO Sea Sparrow carried aboard the Dutch ships, two Phalanx CIWS port and starboard and double the original number of Harpoon SSM. This extra equipment has raised questions as to the stability of these ships. The later three acquisitions were transferred without Harpoon or the after 76mm gun, but are to be brought up to the same standard as the first two ships, using guns taken from the deleted *Gearing* class destroyers.

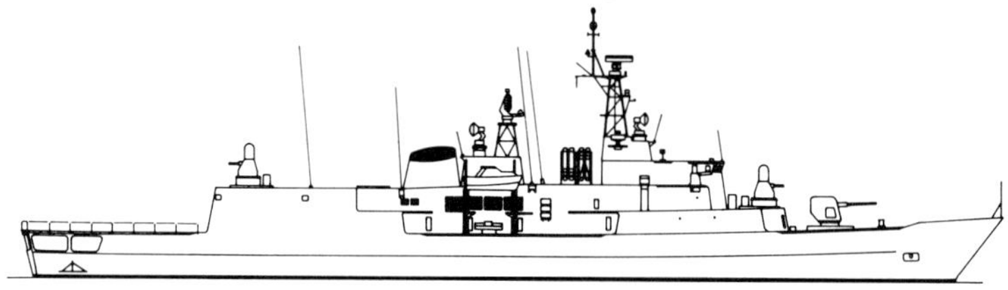

MEKO 200HN design

MEKO 200 Mk 3 class *frigates*

Displacement: 2710t standard; 3200t full load
Dimensions: 385ft 6in oa, 359ft 3in pp × 48ft 7in × 19ft 8in
117.5m, 109.5m × 14.8m × 6.0m
Machinery: 2 shafts (CP propellers) CODOG; 2 GE LM-2500-30 gas turbines, 60, 656shp = 31kts, 2 MTU 20V956 TB82 diesels, 10,400bhp = 21kts. Range 4100nm at 18kts
Armament: 8 Harpoon SSM (2×4), 1 Mk 48 VLS (16 Sea Sparrow SAM), 1–5in/54 Mk 45, 2–20mm Phalanx CIWS (2×1), 6–324mm Mk 32 ASW TT (2×3), 1 helicopter
Sensors: Radar Decca 2690BT, HSA MW-08, HSA DA-08, 2 HSA STIR-18; sonar SQS-56 hull mounted; ECM Argo AR 700 intercept, Argo APECS-II jammer, 2 decoy launchers
Complement: 173 (22 officers)

No	Name	Builder	Laid down	Launched	Comp	Fate
F 452	HYDRA	Blohm & Voss, Hamburg	17.12.90	25.6.91	12.11.92	Extant 1995
F 453	SPETSAI	Hellenic SY, Skaramanga	11.8.92	9.12.93	–	Completing
F 454	PSARA	Hellenic SY, Skaramanga	9.12.93	–	–	Building
F 455	SALAMIS	Hellenic SY, Skaramanga	–	–	–	Building?

In the mid-1980s plans were discussed for the purchase of a frigate design, to be built in Greece with foreign assistance. Among those considered were the Italian *Lupo* class and a modernised version of the British *Leander* class, but on 18 April 1988 it was announced that the German MEKO 200 class had been chosen, the first unit to be built by Blohm & Voss, Hamburg and the remaining three in Greece. Similar ships have been purchased by Portugal and Turkey. The ships are fitted with US-built weapons and some electronics, and are to receive Sikorsky S-70B Seahawk helicopters.

The first, German-built, ship *Hydra*, was completed in 1992 but did not formally commission until 30 April 1993, following a serious fire aboard her in January 1993 at Portland while she was on trials. The construction of the three Greek-built units has been greatly delayed by much the same sort of financial difficulties as affected the proposed *Kortenaer* construction programme in the 1980s. In an attempt to speed up progress, sections of F 453 were prefabricated in Germany and she was reportedly launched towards the end of 1993. Once again, Greek attempts to build warships in her own yards have been plagued with difficulties.

Hydra 2.7.92
L & L van Ginderen

Three *Knox* class frigates were transferred on lease in 1992. *Ipiros* remained in an American shipyard until early 1993, undergoing repairs to her sonar systems, and all three were formally commissioned into the Greek Navy on 30 April 1993. While in USN service, all three had been fitted with spray strakes and bow bulwarks to improve sea-keeping. They are also equipped to fire Harpoon SSM from their ASROC launchers (four missiles carried in the ASROC magazine), and carry one AB-212 ASW helicopter. A further unit was to be transferred. Turkey has also acquired ships of this class.

Ex-US KNOX class *frigates*

Particulars: See US *Knox* class

No	Name	Transferred	Fate
F 456	IPIROS (ex-*Connole*)	30.8.92	Extant 1995
F 457	THRAKI (ex-*Trippe*)	30.7.92	Extant 1995
F 458	MAKEDONIA (ex-*Vreeland*)	25.7.92	Extant 1995

SUBMARINES

Both these boats were built 1942–43 as USN *Gato* class submarines, and fitted with a streamlined fin postwar. *Poseidon* was transferred on loan under MDAP to Greece on 8 August 1957 and *Amfitriti* on 21 April 1958. *Poseidon* was purchased in April 1976 for spares to keep the remaining ex-US Second World War submarines operational.

(For illustration see under United States)

Ex-US GATO class *submarines*

Particulars: As US *Gato* class

No	Name	Builder	Acquired	Fate
S 78	AMFITRITI (ex-*Jack*)	Electric Boat	21.4.58	Discarded *c*1972
S 09	POSEIDON (ex-*Lapon*)	Electric Boat	8.8.57	Stricken 1976

Triaina was built 1943–44 as a US *Balao* class submarine, and was given a streamlined fin postwar. She was transferred to Greece on loan under MDAP on 26 February 1965, and was purchased in April 1976. She is no longer operational, being used for alongside training.

Ex-US BALAO class *submarine*

Particulars: As US *Balao* class (Fleet Snorkel type)

No	Name	Builder	Acquired	Fate
S 86	TRIAINA (ex-*Scabbardfish*)	Portsmouth N Yd	26.2.65	TS 1982

(For illustration see under United States)

This submarine was built 1943–44 as a US *Balao* class boat and modernised in the early 1950s to the GUPPY IIA configuration. *Papanikolis* was purchased on 26 July 1972 and was mainly used for training, but could have been used operationally if necessary.

Ex-US BALAO class *submarine*

Particulars: As US *Balao* class (GUPPY IIA type)

No	Name	Builder	Acquired	Fate
S 114	PAPANIKOLIS (ex-*Hardhead*)	Manitowoc	26.7.72	Stricken 1993

(For illustration see under United States)

This boat was built 1945–46 as a US *Tench* class submarine. She was first converted to GUPPY II configuration in the early 1950s, then to GUPPY III in 1961–62. Greece purchased her on 29 October 1973. Mainly used for training, she was stricken in 1993.

Ex-US TENCH class *submarine*

Particulars: As US *Tench* class (GUPPY III type)

No	Name	Builder	Acquired	Fate
S 115	KATSONIS (ex-*Remora*)	Portsmouth N Yd	29.10.73	Stricken 1993

Greece was the first country to adopt the German-designed and -built Type 209 single-hulled submarine, which has since been purchased by Turkey and many South American navies. She can dive to more than 650ft (200m), and has a powerful battery of torpedo tubes (plus reload torpedoes) and a comparatively small crew. The second group, ordered in 1975–76, have a longer hull to give more internal volume. The first group had proved to be rather too cramped to take all the equipment, electronics and crew. The Type 209s have replaced almost all the old Second World War American submarines.

On 5 May 1989 a contract was placed in Germany for the modernisation of the first group of this class, to bring them up to German Type 206A standards. *Triton* was refitted at Howaldtstwerke, Kiel, completing in July 1993 whilst it is hoped to complete the other three, refitting at Salamis, by 1996. The programme includes a new torpedo-launch system (with Sub-Harpoon SSM capability), new sonar and electronics. As a result of this upgrading, complement will increase by five. The second group are scheduled for a similar modernisation once this programme is completed.

GLAVKOS class *submarines*

Displacement: 1105t/1230t (2nd group 1180t/1300t) normal
Dimensions: 178ft 6in (2nd group 183ft 5in) oa × 20ft 4in × 17ft 11in
54.4m (55.9m) × 6.2m × 5.5m
Machinery: 1 shaft, 4 MTU diesels plus 1 Siemens electric motor, 2250bhp/3600shp = 11kts/21kts. Range 25nm/230nm/400nm at 20kts/8kts/4kts submerged
Armament: 8–533mm TT (bow; 14 torpedoes)
Sensors: Radar?; sonar?
Complement: 31

No	Name	Builder	Laid down	Launched	Comp	Fate
First group						
S 110	GLAVKOS	Howaldtswerke, Kiel	1.9.68	15.9.70	5.11.71	Extant 1995
S 111	NEREUS	Howaldtswerke, Kiel	15.1.69	7.6.71	10.2.72	Extant 1995
S 112	TRITON	Howaldtswerke, Kiel	1.6.69	19.10.71	23.11.72	Extant 1995
S 113	PROTEUS	Howaldtswerke, Kiel	1.10.69	1.2.72	8.8.72	Extant 1995
Second group						
S 116	POSEIDON	Howaldtswerke, Kiel	15.4.76	21.3.78	22.3.79	Extant 1995
	AMFITRITI	Howaldtswerke, Kiel	14.9.76	14.6.78	14.8.79	Extant 1995
S 118	OKEANOS	Howaldtswerke, Kiel	1.10.76	16.11.78	15.11.79	Extant 1995
S 119	PONTOS	Howaldtswerke, Kiel	15.1.77	22.3.79	29.4.80	Extant 1995

Glavkos as completed
Official

AMPHIBIOUS WARFARE VESSELS

(For illustration see under countries of origin in this and 1922–46 volume)

Ex-US LSD type *dock landing ships*

Particulars: As US LSD type

No	Name	Builder	Acquired	Fate
L 153	NAFKRATOUSSA	Newport News	1953	Discarded 1971
L 153	NAFKRATOUSSA (ex-*Fort Mandan*)	Boston N Yd	1971	Extant 1995

The first ex-USN LSD to be acquired, *Hyperion* (LSD 9, built 1943), was taken over by the Greek Navy in 1953 and acted as the HQ Ship for the Greek amphibious forces. She was replaced in 1971 by ex-US *Fort Mandan* (LSD 21, built 1945), also renamed *Nafkratoussa*, which had received a FRAM I modernisation. She had an SPS-6 radar added *c*1982 but this has since been removed. She remains the flagship of the amphibious forces.

Ex-British LST (3) type *tank landing ships*

Particulars: As British LST (3) type

Class:
Axios, *Strymon*, *Alfios*, *Acheloos*, *Pinios*, *Aliakmon*

Six ex-British LST (3)s were transferred to the Greek Navy in 1947. *Axios* (ex-*LST 3007*) and *Strymon* (ex-*LST 3502*) were returned in June 1962. *Alfios* (ex-*LST 3020*) was returned in August 1963, *Acheloos* (ex-*LST 3503*) was discarded in 1964 and *Pinios* (L 104) ex-*LST 3506* and *Aliakmon* (L 171) ex-*LST 3002* in 1968–69.

Ex-US LSM 1 class *medium landing ships*

Particulars: As US *LSM 1* class

Class:
I Grigoropoulos (L 161), *I Tournas* (L 162), *I Daniolos* (L 163), *I Roussen* (L 164), *I Krystalidis* (L 165), *I Merlin* (I 166)

Six ex-USN *LSM 1* class LSMs were transferred by the USN to Greece in 1958. *I Merlin* (ex-*LSM 557*) and *I Krystalidis* (ex-*LSM 541*) on 30 October 1958 and *I Daniolos* (ex-*LSM 227*), *I Grigoropoulos* (ex-*LSM 45*), *I Roussen* (ex-*LSM 399*) and *I Tournas* (ex-*LSM 102*) on 3 November 1958. *I Merlin* was sunk in collision with a supertanker on 15 November 1972. *I Tournas* stricken 1990. *I Grigoropoulis* and *I Daniolos* stricken 1993. Remainder extant 1995.

Ex-US LST I class *tank landing ships*

Particulars: As US LST type (*LST 1–510 group*)

Class:
Lesbos (L 172), *Rodos* (L 157), *Syros* (L 144)

Three ex-USN *I–510* type LSTs were transferred to Greece in the 1960s to add to the three transferred under lease-lend in 1944. They were *Lesbos* (ex-*Boone County*, ex-*LST 389*), and *Rodos* (ex-*Bowman County*, ex-*LST 391*) transferred on 9 August 1960, and *Syros* (ex-*LST 325*) transferred on 29 May 1964. They were added to the three LSTs of this type transferred under lease-lend in 1944. *Lesbos* was stricken *c*1994, and the remainder are to follow as the new *Chios* class LSTs enter service.

Ex-US LST 511 class *tank landing ships*

Particulars: As US LST type (*LST 511–1152* group)

Class:
Ikaria (L 154), *Kriti* (L 171)

Ikaria (ex-*Potter County*, ex-*LST 1086*) was transferred from the USN to Greece on 9 August 1960. *Kriti* (ex-*Page County*, ex-*LST 1076*) was transferred in March 71 and purchased by Greece on 11 July 1978. Both extant 1995, but will be replaced by the *Chios* class LSTs.

Ex-US TERREBONNE PARISH class *tank landing ships*

Particulars: As US *Terrebonne Parish* class

Class:
Inouse (L 104), *Kos* (L 116)

Two ex-USN *Terrebonne Parish* class LSTs *Inouse* (ex-*Terrell County*, ex-*LST 1157*) and *Kos* (ex-*Whitefield County*, ex-*LST 1169*), were purchased by Greece on 17 March 1977. Both were built in 1953, and they were modernised in Greece after purchase. The transfer of two more was projected, but came to nothing. Both extant 1995.

CHIOS class *tank landing ships*

Displacement: 4400t full load
Dimensions: 380ft 4in × 50ft 2in × 11ft 4in
115.9m × 15.30m × 3.44m
Machinery: 2 shafts, 2 diesels, 10,600bhp = 17kts.
Armament: 1–76mm/62 OTO Melara, 2–40mm Bofors (2 × 1), 4–20mm (2 × 2)
Sensors: Radar Kelvin-Hughes Type 1007, Thomson-Sintra Triton V, Thomson-CSF Castor
Complement: 108 (+ 303 troops)

No	Name	Builder	Launched	Fate
L 173	CHIOS	Eleusis SY	16.12.88	Extant 1995
L 174	SAMOS	Eleusis SY	6.4.89	Completing
L 175	IKARIA	Eleusis SY	5.7.90	Completing
L 176	LESBOS	Eleusis SY	–	Building
L 177	RODOS	Eleusis SY	–	Building

Ordered from the Eleusis Shipyard on 15 May 1986 as replacements for the ex-US LSTs (see above), which they generally resemble. The programme was to have completed by September 1990, but financial problems at the builders have delayed it considerably. Carry two LCVPs and have a large helicopter deck aft, capable of handling Sea King helicopters. *Chios* completed for trials in May 1993.

Ex-British LCT (4) type *tank landing craft*

Particulars: As British LCT (4) type

Twelve ex-British LCTs of the LCT (4) type were acquired by Greece from Britain in 1946. *Amorgas* (ex-*LCT 1301*), *Anafi* (ex-*LCT 1293*), *Kandados* (ex-*LCT 587*), *Kithera* (ex-*LCT 1198*), *Malakassi* (ex-*LCT 619*), *Milos* (ex-*LCT 1300*), *Palaiochori* (ex-*LCT 620*), *Serifos* (ex-*LCT 1227*), *Sofades* (ec-*LCT 594*), *Thira* (ex-*LCT 1297*), *Vraghni* (ex-*LCT 607*) and one other were all discarded (most to Greek commercial interests) prior to 1968, *Kithera* and *Milos* were re-acquired by the Greek Navy and were extant 1995, but soon to be deleted.

Ex-US LCU 501 class *utility landing craft*

Particulars: As US *LCU 501* class

Class:
Kassos (L 145), *Karpathos* (L 146), *Kimolos* (L 147), *Kea* (L 148), *Kithnos* (L 149), *Sifnos* (L 150), *Skopelos* (L 151), *Skiathos* (L 152).

Eight ex-US *LCU 501* class (ex-LCT (6) originally built in 1944) were acquired between 1959 and 1962. *Skopelos* (ex-*LCU 852*) and *Skiathos* (ex-*LCU 827*) were transferred in 1959; *Kea* (ex-*LCU 1229*), *Kithnos* (ex-*LCU 763*), and *Sifnos* (ex-LCU 677) were transferred in 1961; *Kassos* (ex-LCU 1382), *Karpathos* (ex-*LCU 1379*) and *Kimolos* (ex-*LCU 971*) were transferred in 1962. All except *Kithnos* deleted 1991–92. *Kithnos* extant 1995 but approaching the end of her service life.

Ex-German BUTT class *utility landing craft*

Class (no, former name, transferred):
Ios (L 167, ex-*Barbe* 31.1.92), *Sikinos* (L 168, ex-*Dorsch*, 31.1.92), *Irakleia* (L169, ex-*Forelle*, 31.1.92), *Folengandros* (L 170, ex-*Delphin*, 31.1.92), *Naxos* (L 178, ex-*Renke*, 16.11.89), *Paros* (L 179, ex-*Salm*, 16.11.89), L 180 (ex-*Rochen*, 20.10.92), L 181 (ex-*Mürane*, 20.10.92)

Built by Howaldtswerke, Hamburg in the 1960s, to a design based on the US LCU 1626 class. Enclosed bridge and armament of 2–20mm (2 × 1), ramps bow and stern. Normal load 141t. All extant 1995.

Ex-German KLASSE 521 *landing craft*

Class (former name):

ABM 20 (ex-*Seenelke*), *ABM 21* (ex-Hering), *ABM 22* (ex-*Orfe*), *ABM 23* (ex-*Maräne), ABM 24* (ex-*Saibling*), *ABM 25* (ex-*Stint*), *ABM 26* (ex-*Äsche*), *ABM 27* (ex-*Hummer*), *ABM 28* (ex-*Krille*), *ABM 29* (ex-*Krabbe*), *ABM 30* (ex-*Auster*).

Built by Rheinwerft, Walsum in the 1960s. *ABM 20* transferred 5 March 1991, the remainder on 25 April 1991. Based on the US LCM (8) design. Cargo 60t. All extant 1995.

OTHER LANDING CRAFT

LCM
Thirteen LCM were transferred from the USN prior to 1965.

LCA
Thirteen ex-British LCA were on loan from 1947 to the early 1960s.

LCVP
Thirty-four ex-USN LCVP were transferred for use with the LSTs.

LCP
Fourteen LCP have been built in Greece from 1977.

SMALL SURFACE COMBATANTS

(For illustration see under countries of origin in this and 1922–46 volume)

Ex-British ALGERINE class *corvettes*

Particulars: As British *Algerine* class

No	Name	Builder	Acquired	Fate
M 12	ARMATOLOS (ex-*Aries*)	Toronto Shipyard	1947	Discarded 1977
M 58	MACHITIS (ex-*Postillion*)	Redfern	1947	Discarded 1976
M64	NAVMACHOS (ex-*Lightfood*)	Redfern	1947	Discarded 1976
M 74	POLEMISTIS (ex-*Gozo*)	Redfern	1947	Discarded 1977
M 76	PYROPOLITIS (ex-*Arcturus*)	Redfern	1947	Discarded 1977

These five corvettes were built 1942–43 as British *Algerine* class flcct minesweepers in Canada for the USN, and supplied to the RN under lease-lend. All were transferred to Greece in 1947. By the late 1960s they were being used mainly as transports; *Armatolos* was also employed as a lighthouse tender. *Machitis* was disarmed as a training ship, *Polemistis* was also used for training and *Pyrpolitis* was a command ship for amphibious operations.

Ex-US PGM 9 class *patrol vessels*

Particulars: As US *PGM 9* class (ex-PC type)

Class:

A Laskos (P 53), *A Pezopoulos* (P 75), *P Meletopoulos* (P 57), *P Arsanoglou* (P 14), *P Chadzikonstandis* (P 96), *P Blessas* (P 61)

Six ex-USN *PGM 9* class patrol vessels, originally built in 1943–44, were acquired from the USA in August 1947. All six had been launched in 1943–44. In 1963 all except *P Blessas* were fitted with a Hedgehog ASW mortar, and 2–40mm (2 × 1) were removed. *P Blessas* (ex-*PGM 28*) was discarded in 1963, *A Laskos* (ex-*PGM 16*) and *P Meletopoulos* (ex-*PGM 22*) in 1971, *A Pezopoulos* (ex-*PGM 21*) in 1977 and *P Arsanoglou* (ex-*PGM 25*) and *P Chadzikonstandis* (ex-*PGM 29*) in 1979.

Ex-US LSSL type *patrol vessels*

Particulars: As US LSSL type

Class:

P Maridakis (P 94), *P Vlachavas* (P 95)

Two ex-USN LSSL type patrol vessels, both built in the USA in 1945, were transferred to Greece: *P Maridakis* (ex-*LSSL 65*) in June 1958 and *P Vlachavas* (ex-LSSL 45) on 12 August 1957. Both were discarded in 1976.

Norwegian TJELD type *motor torpedo-boats*

Particulars: As Norwegian *Tjeld* class

Class:

Andromeda (P 21), *Iniohos* (P 22), *Kastor* (P 23), *Kykonos* P 24), *Pigasos* (P 25), *Toxotis* P 26)

An order for six MTBs was placed with Mandal, Norway in 1965. These were identical with the Norwegian *Tjeld* (improved *Nasty*) class except that

Astrapi (as the German *Strahl* in 1962) CMP

they have 2–40mm (2 × 1) in place of the Norwegian boats' 1–40mm and 1–20mm. All were delivered in 1967. These were the first new warships to be built for Greece since the 1930s. *Iniohos* was discarded in 1972, and the remainder, renumbered P 196–P 200 respectively, were deleted in 1983. However, *Andromeda*, *Kykonos*, *Pigasos* and *Toxotis* were re-engined and refurbished to re-enter service in 1989. All four extant 1995.

British BRAVE type *motor torpedo-boat*

Particulars: See German Federal Republic *Strahl*

Class:

Astrapi (P 20)

The West German MTB *Strahl*, built to the British *Brave* class design and launched by Vosper on 10 January 1962, was transferred to Greece and renamed *Astrapi* in April 1967. She was refitted by Vosper in 1968 and discarded in 1979.

British FEROCITY type *motor torpedo-boat*

Particulars: See German Federal Republic *Pfeil*

Class:

Aiolos (P 19)

The West German MTB *Pfeil*, launched on 26 October 1961 and built to the British *Ferocity* design by Vosper, was purchased by Greece in April 1967 and renamed *Aiolos*. She was refitted by Vosper in 1968 and discarded in 1976.

Ex-West German SILBERMÖWE class *motor torpedo-boats*

Particulars: See German Federal Republic *Silbermöwe* class

Class:

Delphin (P 15), *Drakon* (P 16), *Polikos* (P 17), *Polidefkis* (P 18), *Foinix* (P 27)

Five West German *Silbermöwe* class MTBs, built 1951–56, were transferred to Greece on 17 December 1968. *Delfin* (ex-*Sturmöwe*), *Drakon* (ex-*Silbermöwe*), *Foinix* (cx-*Eismöwe*), *Polikos* (cx-*Raubmöwe*) and *Polidefkis* (ex-*Wildschwan*) were all discarded in 1974.

Ex-West German JAGUAR class *motor torpedo-boats*

Particulars: See German Federal Republic *Jaguar* class

Class:

Hesperos (P 196), *Kataigis* (P 197), *Kentauros* (P 198), *Kyklon* (P 199), *Laiaps* (P 228), *Scorpios* (P 229), *Tyfon* (P 230)

Seven ex-West German *Jaguar* class MTBs have been commissioned in the Greek Navy. *Hesperos* (ex-*Seeadler*) was transferred on 24 March 1977, *Kataigis* (ex-*Falke*) on 12 December 1976, *Kentauros* (ex-*Habicht*) on 22 May 1977, *Kyklon* (ex-*Greif*) on 12 December 1976, *Laiaps* (ex-*Kondor*) on 24 March 1977, *Scorpios* (ex-*Kormoran*) on 22 May 1977 and *Tyfon* (ex-*Geier*) on 12 December 1976. In addition, ex-*Albatros*, ex-*Bussard* and ex-*Sperber* were transferred in 1976–77 to provide spares for the operational boats. All the active boats were refitted before being transferred. During 1980 all were renumbered P 50–P 56 in the above order. *Kataigis* was discarded late in 1981 and *Kentauros* in 1985 to provide spares for the active boats. Remainder extant 1995.

KELEFSTIS STAMOU class *fast attack craft*

Displacement:	80t standard; 115t full load
Dimensions:	105t oa × 19ft × 5ft 4in *32.0m × 5.8m × 1.6m*
Machinery:	2 shafts, 2 MTU diesels, 2700shp = 30kts. Range 1500nm at 15kts
Armament:	4 SS 12 SSM (4 × 1), 1–40mm, 1–20mm
Sensors:	Radar ?
Complement:	17

Class:

Diopos Antoniou (P 29), *Kelefstis Stamou* (P 28)

Built by CN de l'Esterel 1974–75 for Cyprus, but the Cypriot Government was not allowed to take delivery of them (this was at the time of the Turkish invasion of Cyprus), so they were taken over by the Greek Navy instead. They have wooden hulls, and can be armed with 2–20mm (2×1) in place of 1–40mm and 1–20mm. Renumbered P 287 and P 288 respectively in 1980. Both extant 1995.

GREECE

Ipopliarchos Konidis as completed
NB 1/750 scale

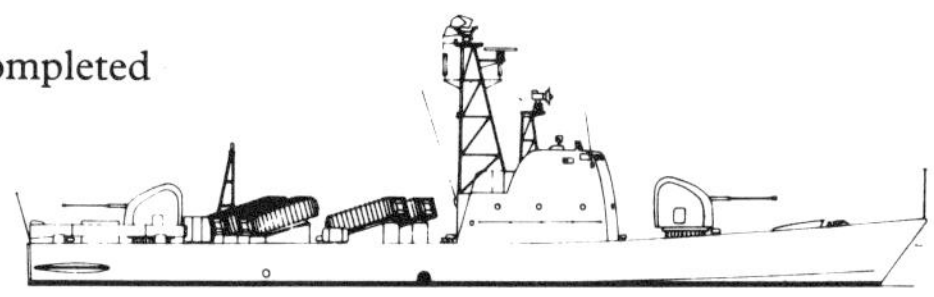

French 'LA COMBATTANTE II' type *fast attack craft*

Displacement:	234t standard; 255t full load
Dimensions:	154t 3in oa × 23ft 4in × 8ft 3in *47.0m × 7.1m × 2.5m*
Machinery:	4 shafts, 4 MTU diesels, 12,000bhp = 36.5kts. Range 8500nm at 25kts
Armament:	4 MM38 Exocet SSM (4×1), 4–35mm (2×2), 2–21in (533mm) TT (2×1)
Sensors:	Radar Thomson types Pollux, Triton
Complement:	40

Class:
A Anninos (P 14), *I Arliotis* (P 15), *I Konidis* (P 16), *I Batsis* (P 17)

Four steel-hulled 'La Combattante II' type missile patrol boats similar to the Israeli *Saar* class were ordered in 1969. They were built by CM de Normandie at Cherbourg, and it was originally intended to build six more in Greece, but the more capable 'La Combattante III' type were built instead. *A Anninos* (ex-*Navisithoi*) was launched on 8 September 1971, *I Arliotis* (ex-*Evniki*) on 26 April 1971, *I Batsis* (ex-*Kalypso*) on 26 January 1971 and *I Konidis* (ex-*Kymothoi*) on 20 December 1971. They were completed in 1971–72 and renamed in 1973. They were originally numbered (in the above order) P 55, P 56, P 53 and P 54. All four were extant 1995.

French 'LA COMBATTANTE III' type *fast attack craft*

Displacement:	359t standard; 425t (2nd group 429t) full load
Dimensions:	184ft × 26ft 3in × 8ft 3in *56.2m × 7.9m × 2.5m*
Machinery:	4 shafts, 4 MTU diesels, 18,000bhp (2nd group 15,000bhp) = 36.5kts (32.5kts). Range 700nm/2000nm at 32kts/15kts
Armament:	4 MM38 Exocet SSM (4×1) (2nd group 6 Penguin SSM, (6×1), 2–76mm/62 (2×1), 4–30mm (2×2), 2–533mm TT (2×1)
Sensors:	Radar Thomson types Castor, Pollux, Triton (2nd group TM-1226, D-1280)
Complement:	42

Class:
1st group – *A Laskos* (P 20), *A Blessas* (P 21), *A Troupakis* (P 22), *A Mykonios* (P 23); 2nd group – *Simeoforos Kavaloudis* (P 24), *A Kostakos* (P 25), *I Deyiannis* (P 26), *Simeoforos Xenos* (P 27), *Simeoforos Simitzopoulos* (P 28), *Simeoforos Starakis* (P 29)

Save for the armament these closely resemble the 'La Combattante III' type large missile patrol boats in other navies including Israel's. The first group of four boats, *P Blessas* (launched 10 November 1976), *A Laskos* (launched 6 July 1976), *I Mykonios* (launched 5 May 1977) and *I Troupakis* (launched 6 January 1977) were ordered in September 1974 and all built by CM de Normandie at Cherbourg. They have four MM38 Exocet SSMs. The SST-4 torpedoes are wire-guided, and the 3in (76mm) is an OTO Melara compact. The second group of six boats were ordered from the Hellenic Shipyard at Skaramanga in Greece, and were built under French supervision. *Simeoforos Kavaloudis* was launched on 10 November 1979 and completed on 14 July 1980. The remainder were completed (in the above order) on 9 September 1980, December 1980, 24 March 1981, June 1981 and 12 October 1981. They were ordered in 1978 and have Penguin 2 SSM in place of Exocet, and less expensive machinery and electronics. All were extant 1995.

Antipliarchos Laskos as completed — Official

Navmachos Aug 1994 — L & L van Ginderen

'OSPREY 55' class *fast attack craft (missile)*

Displacement:	515t full load
Dimensions:	179ft 8in × 34ft 5in × 8ft 4in *54.75m × 10.5m × 2.55m*
Machinery:	2 shafts, 2 MTU 1163 TB 93 diesels, 9870bhp = 24.7kts. Range 500nm at 22kts
Armament:	1–40mm Bofors, 2–20mm Rheinmetall (2×2), mines
Sensors:	Radar navigational
Complement:	36

Class (no, launched):
Armatolos (P 18, 19.12.89), *Navhamos* (P 19, 16.5.90)

Lengthened version of design licensed from Frederikshavn Vaerft, Denmark. Original order for two units (with an option for two more) placed in March 1988. It was intended to arm them with four Harpoon SSM, a twin Bofors mount plus associated radars. Only two boats were built, the further ten projected vessels being cancelled in favour of the locally-designed 'P 100' class (see below). They currently serve as patrol vessels, and in this role can carry up to twenty-five troops. It is reported that they are to receive SSMs and 76mm guns from decommissioned *Gearing* class destroyers. Both boats were extant 1995.

Ex-US ASHEVILLE class *large patrol vessels*

Class (no, former name):
Ormi (P 229, ex-*Beacon*), *Tolm* (P 230, ex-*Green Bay*)

Two boats transferred 22 November 1989, having been in USN reserve since 1977. Refitted in Greece prior to commissioning in June 1990. Gas turbines removed prior to transfer. Both extant 1995.

Ex-German THETIS class *corvettes*

Class (no, former name):
Niki (P 62, ex-*Thetis*), *Doxa* (P 63, ex-*Najade*), *Elevtheria* (P 64, ex-*Triton*), *Karteria* (P 65, ex-*Hermes*)

Built 1959–60 by Roland Werft, Bremen. Transferred in pairs on 6 September 1991 and 7 September 1992, being formally commissioned on 30 April 1993. A fifth unit (ex-*Theseus*) was transferred in December 1992 for cannibalisation for spares. Torpedo tubes now modified to fire US 324mm torpedoes, and current best speed only 19kts. All four extant 1995.

Ex-German S 41 class *fast attack craft (missile)*

Two boats (ex-*Iltis*, P 6142 and ex-*Storch*, P 6152), transferred late 1992. Both extant 1995.

P 100 class *large patrol craft*

Displacement:	555t full load
Dimensions:	185ft 4in × 32ft 10in × 8ft 2in *56.5m × 10.0m × 2.5m*
Machinery:	2 shafts, 2 MTU 1163 TB 93 diesels, 9870bhp = 24kts. Range 2200nm at 22kts
Armament:	1–76mm/61 OTO Melara, 2–40mm Breda (1×2), 2–20mm Rheinmetall (2×1), mines
Sensors:	Radar navigational, Triton, Selenia RTN-10X
Complement:	36 (+ 25 troops)

Class (no, launched):
Pyropolitis (P 20, 16.9.92), *Polemistis* (P 21)

Modified version of 'Osprey 55' design (see above), built by the Hellenic Shipyard, Skaramanga. Order for two units (with option for a third) made on 19 February 1989, and the first was laid down in January 1991, completing for trials in 1993. The class was originally planned to be armed with four Harpoon missiles. Financial problems have greatly delayed this programme. One (possibly two) extant 1995.

COASTAL PATROL BOATS

The following small patrol craft were also in service:

KFK class – Arhikelefstis Maliopoulis (ex-*W 8*), *Arhikelefstis Stassis* ex-*W 2*), *Anemos* (ex-*W ?*)

German MFV type patrol boats of the Second World War, transferred from the Federal German Navy in 1969 (*Anemos*) and 30 August 1975 (the other two). *Anemos* (discarded *c*1975) and *Maliopoulis* were used as survey craft. Now stricken.

Dilos class – Dilos (P 267), *Lindas* (P 269), *Knossos* (P 268)

An 86t design by Abeking & Rasmussen, used by the Greek Navy for air sea rescue, built 1977–81. More have been built for the Customs and Coast Guard. They were built by the Hellenic Shipyard, Skaramanga, and were launched 1977–81. All extant 1995.

N I Goulandris class – N I Goulandris I (P 289), *N I Goulandris II* (P 290)

Coastal patrol boats of 39t, armed with 2–20mm, designed by the German firm of Abeking & Rasmussen, built in Greece 1975–77 and paid for by the Greek shipowner whose name they carried. P 289 blew up and sank 24 June 1983. P 290 deleted *c*1992.

E Panagopoulos class – E Panagopoulos I (P 61), *E Panagopoulos II* (P 70), *E Panagopoulos III* (P 96)

As with the previous class, named after a prominent shipowner. P 61 stricken 1991, remainder stricken 1993.

MINE WARFARE VESSELS

Ex-US MMC type *coastal minelayers*

Particulars: As US MMC type (ex-*LSM I* class)

Class:
Aktion (N 04), *Amvrakia* (N 05)

Aktion (ex-*MMC 6*, ex-*LSM 301*) and *Amvrakia* (ex-*MMC 7*, ex-LSM 303) were built in the USA in 1943–45 as *LSM 1* class medium landing ships. They were completely rebuilt as MMC prior to being transferred to Greece on 1 December 1953. The armament consists of 8–40mm (4 × 2), 6–20mm (6 × 1) and 100–130 mines, handled by a pair of derricks fore and aft. Both were in service in 1995.

Ex-US ADJUTANT class *coastal minesweepers*

Particulars: As US *Adjutant class*

Class:
Atalanti (M 202), *Antiopi* (M 205), *Faedra* (M 206), *Thalia* (M 210), *Niovi* (M 254)

Five USN *Adjutant* class coastal minesweepers, originally built in the USA for Belgium under MDAP, were subsequently returned by Belgium to the USA and re-transferred to Greece. *Atalanti* (ex-Belgian *St Truiden*) and *Antiopi* (ex-Belgian *Herve*) were transferred to Greece on 29 July 1969 and *Faedra* (ex-Belgian *Malmedy*), *Niovi* (ex-Belgian *Laroche*) and *Thalia* (ex-Belgian *Blankenberge*) were transferred to Greece on 26 September 1969. These all have a small funnel and open bridge. All were in service in 1995.

US FALCON class *coastal minesweepers*

Particulars: As US *Falcon* class (MSC 294 type)

Class:
Aedon (M 248), *Klio* (M 213, ex-*Argo*), *Aigli* (M 246), *Alkyon* (M 211), *Avra* (M 214), *Dafni* (M 267), *Kichli* (M 241), *Kissa* (M 242), *Pleias* (M 240)

All were built by Peterson, and have an enclosed bridge and a large, prominent funnel. Both this and the previous class have wooden hulls. Extant 1995, although overdue for replacement.

Grenada

Independence for this tiny island state (133 square miles, population 84,000) came in 1974 and was followed in 1979 by a *coup* in which Maurice Bishop, heading the New Jewel Movement seized power from Sir Eric Gairy. Political unrest and the death of the prime minister in 1983 prompted intervention by Caribbean and US forces. A general election was held in December 1984 and phased withdrawal of US military personnel was completed in June 1985, since when the state has enjoyed political stability.

The country operates a patrol force consisting of:

One Guardian class coastal patrol craft *Tyrrel Bay* (PB 01) built at Lantana, delivered 1984 (90t full load, 32.0m, 2000bhp = 24kts, 3–12.7mm, 2 MG). Extant 1995.

Two Brooke Marine patrol craft delivered 1972 (10t, 22kts, 3 MG), deleted in 1990.

Three Fairey Spear class delivered 1978, deleted 1988–90.

Two Boston Whalers acquired 1988–89. Both extant 1995.

Guatemala

The republic of Guatemala, bordered by Mexico, Belize, Honduras and El Salvador, covers an area of 42,042 square miles and is traversed by mountain ranges. The country's political turbulence is mirrored by its geographical location – afflicted, as it often is, by earthquakes; a major eruption in 1976 demolished several towns and caused about 25,000 deaths. In March 1982 Guatemala experienced a military *coup* and the country's constitution has been temporarily suspended. This is the most recent in a wave of overthrows which began in 1954; terrorist activity is rife throughout Guatemala. There have been several confrontations with the neighbouring state of Belize, over which Guatemala claims sovereignty. In 1972 and again in 1977 Guatemala sent troops to the border with Belize and Britain responded by reinforcing the existing military forces in Belize (then still a British colony). Civilian rule was restored in 1984 and a new constitution promulgated in 1985. However, further political unrest has seen further changes in leadership, and it was not until January 1994 that a measure of stability returned following a referendum.

The Guatemalan Navy was established in 1959 primarily for carrying out coast guard work. There are bases at Santo Tomas de Castillas on the Atlantic coast and Sipacate on the Pacific coast. There are 1245 men serving with the Navy (including 700 Marines) and Guatemala operates a two-year period of nationel service. Most of the country's vessels come from the United States which paid out a total of $23.6m aid 1955–77.

Swedish JAGAREN class *patrol vessel*

The Swedish patrol vessel (ex-minesweeper) *Snapphanen* was transferred in 1959 to become the first Guatemalan warship. Renamed *José Francisco Barrundia*, she was rearmed with 2–3in (2×1) and 2–25mm guns, she was laid up in an inoperative state by 1964.

28ft *coastal patrol craft*

Class (no):
Xucuxuy (P 281), *Camalote* (P 282)

Transferred in 1961 and deleted in 1979.

'BROADSWORD' class *coastal patrol craft*

Displacement: 90.5t standard; 110t full load
Dimensions: 105ft × 20ft 5in × 6ft 4in
32m × 6.2m × 1.9m
Machinery: 2 shafts, 2 GM 16V 149TI diesels, 3200hp = 32kts. Range 1150nm at 20kts
Armament: 1–12.7mm, 1–75mm recoilless, 1–81mm mortar, 4 MGs
Complement: 20

No	Name	Builder	Comp	Fate
P 1051	KUKOLKAN	Halter Marine	4.9.76	Extant 1995

Rearmed with 2–20mm Oerlikon in 1989.

Sewart 85ft type *coastal patrol craft*

Displacement: 60t
Dimensions: 85ft × 18ft 8in × 3ft
25.9m × 5.7m × 0.9m
Machinery: 2 GM diesels, 2200bhp = 23kts. Range 400nm at 12kts
Armament: 2 MGs
Complement: 12

No	Name	Builder	Comp	Fate
P 851	UTATLAN	Sewart Louisiana	May 1967	Extant 1995
P 852	SUBTENIENTE USORIO SARAVIA	Sewart, Louisiana	1972	Extant 1995

Halter 'CUTLASS' class *coastal patrol craft*

Class (no, commissioned):
Tecunuman (P 651, 1972), *Kaibilbalam* (P 652, 1972), *Azumanche* (P 653, 1972), *Tzacol* (P 654, 10.3.76), *Bitol* (P 655, 4.4.76), *Gucumaz* (P 656, 1981)

Built by Halter Marine, New Orleans. Extant 1995.

Ex-US AVR type *coastal patrol craft*

Class (no, commissioned):
Cabrakan (P 631, 1945), *Hunahpu* (P 632, 1945)
Transferred 1965 and 1964 respectively. US Coast Guard 63ft, 32t type. Deleted 1982.

In 1979 Guatemala ordered three coastal patrol craft from Abeking and Rasmussen, Lemwerder.
Thirty river patrol craft are being or have been built in Guatemala by Trabejos Baros SY in 1979. Wooden construction. Extant 1995.

Ex-US LCM (8) type *landing craft*

Class:
Chinaltenango (E 81)
Transferred December 1965; 115t full load. Deleted 1987.

Ex-US LCU 1466 type *landing craft*

One craft of this type in service; 360t full load. Deleted 1982.

MACHETE class *transport craft*

Class (no, commissioned):
Picuda (P 361, 4.4.76), *Barracuda* (P 362, 4.4.76).
Built by Halter Marine, 8.3t, 36ft × 12ft 6in × 2ft (*11m × 3.8m × 0.6m*), can carry 20 troops. Extant 1995.

Guinea

Guinea became independent in 1958, when her people voted against continued association with France in the referendum held in French colonies and territories. For thus spurning de Gaulle's offer of participation in the great Francophone community, the new republic was cut off without a sou; the estrangement lasted 20 years until reconciliation in December 1978. Traditional agriculture accounts for about half the gross domestic product, but the most important and dynamic area is mining, Guinea having about 30 per cent of the world's crude bauxite, and iron ore deposits of an estimated eight billion tons.

Internationally, Guinea, although a one-party state and leftward-leaning under Sékou Touré, was not a Soviet satellite despite arms deliveries and advisers from Cuba going back to 1960. Since 1975 there has been a generally steady improvement and normalisation of relations with the West and Guinea's neighbours; in 1980, refuelling facilities formerly granted for Soviet naval long-range reconnaissance aircraft at Conakry airport were withdrawn, another straw in the wind. Guinea strengthened her 1971 defence agreement with neighbouring Sierre Leone into a Mutual Defence Pact in August 1981. A few days after the death of Touré in March 1984, a military *coup* took place, but following a referendum a military-civilian government was established in January 1991. Since then numerous political parties have emerged, although only a presidential election has so far been held.

On past performance, both the efficiency and loyalty of the Guinean military must be considered questionable. On 11 November 1970 a four-ship attack on Conakry by 350–400 Portuguese-led armed dissidents from the then Portuguese Guinea (now Guinea-Bissau) achieved complete surprise and reached its objective before being defeated in three days fighting. Soon after the first Soviet warships arrived at Conakry which became the Red Navy's most visited Atlantic port, surpassing all of Cuba's combined. Its peak importance came during the 1975–76 Angolan Civil War. Until the early 1980s, most of the navy's vessels had been supplied by the USSR, but since then some US-built equipment has been introduced, as well as more ex-Soviet craft.

Ex-Soviet 'T58' class ocean minesweeper transferred May 1979 with all sweeping gear removed and renamed *Lamine Sadji Kaba*. Discarded 1987.
Three ex-Soviet 'Shershen' class FAC (T); two transferred 1978, one in December 1979 minus TT. Discarded 1977.
Six ex-Chinese 'Shanghai II' class FAC (G) transferred in pairs, 1973, 1974 and 1976. Pennant numbers 733–738. All discarded by 1987.
Four ex-Soviet 'P 6' class FAC(T) transferred 1971–72 and deleted 1981.
US-built 65ft patrol boat *Vigilante* (P 300) in service June 1986, extant 1995.
US-built 77ft patrol boat *Intrepid* (P 328) in service December 1987, extant 1995.
Two ex-Soviet 'Bogomol' class patrol boats. P 320, P 321, transferred 1989 and 1991 respectively. Both extant 1995.
French-built coastal patrol craft delivered by CNE in 1979.
Three ex-Soviet 'Poluchat I' class coastal patrol craft.
Two ex-Soviet 'MO VI' class coastal patrol craft transferred 1972–73.
Two ex-Soviet LCUs.

Guinea-Bissau

The independence of this republic (formerly Portuguese Guinea) was recognised in 1974, ending a guerrilla war (since 1959) against the Portuguese Army. Her navy, based at Bissau, has 250 volunteers who operate five French and Spanish-built coastal patrol boats, two ex-Soviet 'T 4' landing craft and smaller units. A Netherlands-built patrol craft was delivered in May 1981 and named *Naga*. She was discarded in 1991. In 1978 the Soviets transferred one 'Shershen' class FAC(T) minus TT, and one 'P 6' torpedo-boat. Both were discarded by 1988. Two ex-Chinese 'Shantou' class patrol boats were transferred in 1983, and discarded and replaced by two more in 1986. Two ex-Soviet 'Bogomol' class patrol boats were transferred in 1988–89.

Guyana

A poor country (although there are large bauxite and manganese deposits) bordered by Brazil, Surinam and Venezuela, Guyana has a 270-mile coastline and covers an area of 83,000 square miles, much of it rain forest and some areas have yet to be explored fully. The country was granted independence from Britain in 1966 (previously known as British Guiana). Guyana has had border clashes with all of her neighbours; Surinam has laid claim to 6000 square miles of Guyanan territory in which there is much bauxite to be mined. In 1969 there was a brief armed encounter with Surinam over this claim which was settled temporarily when both sides agreed to withdraw. Guerrilla activity took place along the Brazilian border in 1972 but again there was no serious confrontation.

Guyana operates a Defence Force, which is responsible for the operation of the country's patrol boats. The Defence Force itself numbers about 2000 volunteer personnel (including 60 women) and of these 100 are employed on naval duties. Bases are at Georgetown and New Amsterdam.

Two ex-North Korean patrol boats (names and numbers unknown) 35t, 40kts, armed with 2–14.5mm MGs, 2 shafts, 2 petrol engines, 9 crew. Both units transferred 1980. Deleted 1988.
One 103ft Vosper Thornycroft type large patrol craft, *Peccari* (DF 1010), 96t standard, 27kts, armed with 2–20mm, 2 Paxman-Ventura diesels, 22 crew. Launched 26 March 1976, commissioned 26 January 1977. Discarded 1992. Guyana ordered a second identical patrol craft in 1977; the order was subsequently cancelled. See Trinidad and Tobago for details.
Three 40ft Vosper Thornycroft coastal patrol craft, *Jaguar, Margay, Ocelot*. Numbers unknown. Commissioned 28 April 1971, 21 May 1971 and 22 June 1971 respectively, 10t, 19kts, 1–7.62mm MG, 2 Cummins diesels, 6 crew. Constructed with glass fibre hulls.
Four ex-US 45ft class coastal patrol craft, *Camoudie, Labana, Rattler* and one name unknown. Deleted 1992.
One LCU; *Kimbla*, 600t, 14kts, 2 Caterpillar diesels, built by Damen, Netherlands. Launched 21 August 1981. Deleted 1991.
Guyana also operates two ex-shrimp boats, and a lighter (YFN 960).

Peccari on trials — Vosper Thornycroft

Haiti

Haiti occupies the western section of the island of Hispaniola, the eastern section being occupied by the more wealthy Dominican Republic, and together with several smaller offshore islands covers an area of 10,700 square miles, and has a population of 6,625,000. A former French colony, Haiti became an independent state in 1804. Deployment of US marines on the island came to an end in 1947, having begun in 1915. There has been much political instability in recent decades. In 1986 the Duvalier regime fell, and in 1991 the democratically-elected President Aristide was himself ousted by a military coup. After increasing international pressure and growing violence against Aristide's supporters by the military regime, US troops landed in 1994 to restore the legitimate president. The president commands all three armed forces in Haiti; there is no navy as such but a Coast Guard (165 men) which operates a number of small patrol craft. Main base is a Port au Prince.

Two ex-US *Sotoyomo* class tugs, *Henri Christophe* (MH 20, ex-*Samoset*), ? (MH 21, ex-*Keywadin*). Transferred 18 September 1978 and 28 March 1981 respectively. Latter unit to be employed in oceanographic research. Launched 14 July 1944 and 9 April 1945 respectively. Out of service 1993.

Three 65ft commercial cruiser type coastal patrol craft, *Jean Claude Duvalier* (MH 21), *MH 22* and *MH 23*, 33t, 25kts, 1–20mm, 2–12.7mm MG, 3 shafts, 3 GM 8V71 diesels. Built by Sewart, Louisiana and in service 1976. *MH 22, MH 23* deleted 1990. *Jean Claude Duvalier* now stricken.

Nine US 3812 VCF class patrol boats, *Le Maroon* (MH 11), *Oge* (MH 12), *Chavannes* (MH 13), *Capois* (MH 14), *Bauckman* (MH 15), *Makandal* (MH 16), *Charlemagne Perrault* (MH 17), *Sonothax* MH 18 and *Bois Road Tonnerre* MH 19. 8.5t, 1–12.7mm MG, 1–7.62 MG 2 shafts, 2 GM 6V 71N diesels, 4 crew. Built by Monark, Monticello, Arkansas. These units came into service 1980–81. MH 17–MH 19 extant 1995, remainder stricken for cannibalisation.

Haiti also operates one, possible two, US 'Enforcer' class patrol boats, *MH (?)* and *MH 6*, 9.5m, built by Bertram, Miami.

Honduras

Lying between Nicaragua and Guatemala with El Salvador to the west, Honduras is one of the larger of the Central American republics (43,277 square miles) and also the poorest. Her Caribbean coastline extends for roughly 375 miles while she has a 63-mile strip of Pacific coastline bordering the Gulf of Fonseca. Honduras has recently (1982) emerged from a long spell of military government during which time repressive action was taken by military leaders against agricultural unions; discontent was due to the slow pace with which agrarian reform was taking place. In 1982 a civilian government led by Roberto Suazo Cordora came into power. Honduras, like many Latin American countries, has experienced border disputes with all the countries adjacent to her. In 1969 the so-called 'Soccer War' took place with El Salvador, occasioned by disagreements over a football match. The dispute continued intermittently until 1976 when both countries agreed to talks. More recently Honduras found herself the thoroughfare for arms supplies to El Salvadorian guerrillas coming from Nicaragua and consequently has been exposed to pressure from the US government in an effort to reduce these supplies of arms. The terrorist violence experienced by Honduras' neighbours now appears to be establishing itself in Honduras; in 1982 measures were taken by the military against terrorist groups.

The republic's Coast Guard force employs 900 men, a percentage of whom are carrying out the 24 months' national service. The main base is at Puerto Cortes on the Caribbean coast.

Three 'Swift' type 105ft class patrol boats, *Guaymuras* (FN 1051), *Honduras* (FN 1052), *Hibures* (FN 1053). Built by Swiftships, Morgan City, Louisiana, in service April 1977 (FN 1051) and March 1980 the latter units. 103t, 32kts, 2 shafts, 2 MTU diesels, 17 crew. Armament 1–20mm Gatling, 2–12.7mm, or 6–20mm (2×3). Extant 1995.

One US 85ft (commercial cruiser type) patrol boat; *Chamelecon* (FN 8502), built by Swiftships, Louisiana, in service date unknown. 50t, 23kts, 2 shafts, 2 GM diesels, 10 crew. Armament 1–20mm, 2–12.7mm. Extant 1995.

Five US 65ft commercial cruiser type patrol boats; *Aguan* (ex-*Gral*, FN 6501), *Goascoran*, ex-*J T Cabanas* (FN 6502), *Petula* FN 6503), *Ulua* (FN 6504), *Chuluteca* (FN 6505). Built by Swiftships, Morgan City, Louisiana, in service December 1973 (FN 6501), January 1974 (FN 6502), 1980 remainder. 33t, 25kts, 3 shafts, 3 GM diesels, armed with 2–12.7mm MG, 5 crew. Haiti ordered FN 6501 and FN 6502 originally, transferred to Honduras 1977; three later units fitted with 3 MTU diesels (36kts), armed with 1–20mm, 2–12.7mm. Extant 1995.

Two inshore patrol craft; FN 2501 and FN 2502, names unknown. Built by Ampela Marine, Honduras, entered service 1981–82. 3t, 24kts, 25ft × 8ft 10in × 1ft 4in (*7.6m × 2.7m × 0.4m*), 1 Chrysler diesel, waterjet drive, armed with 1–12.7mm MG, 1–7.62mm MG, 4 crew. Now stricken.

Two Guardian coastal patrol craft *Copan* (FNH 106), *Tegucigalpa* (FNH 107) delivered by Lantana Boatyard in 1983 and 1986 (94t, 2070bhp = 30kts, 1–20mm Gatling, 3–20mm Hispano Suiza (1×3), 2–12.7mm). Extant 1995.

Twelve *Piranha* class river patrol craft delivered 1986–90, one landing craft (LCU) *Punta Caxinas* (FNH 1491) delivered 1988 by Lantana Boatyard (625t full load, 14kts), extant 1995.

Honduras also operates six ex-fishing boats; FN 7501, FN 7502, FN 7503, FN 7504, FN 7505, FN 7506. Extant 1995.

Hong Kong

Hong Kong was seized by the British from China in 1839 during the first Opium War, and British ownership was confirmed in 1842 by the Treaty of Nanking. Kowloon was acquired in 1860, and in 1898 a 99-year lease was taken on the New Territories. Hong Kong remains a British Crown Colony, and its force of small patrol boats are operated by volunteers of the Marine District of the Royal Hong Kong Police Force. The Royal Navy also maintained a presence at Hong Kong. This diminished from a cruiser squadron in the early 1950s to the frigate HMS *Mermaid* in the mid-1970s. In 1982 it consisted of five 'Ton' class large patrol craft and two SRN-6 hovercraft and from 1984 the intention was that these would be replaced by the five *Peacock* class patrol vessels being 75 per cent funded by the Hong Kong Government. The RN presence is being withdrawn, and Hong Kong will be returned to China in 1997.

Two Command vessels (222t full load, 111ft 4in × 24ft × 10ft 6in, *33.9m × 7.3m × 3.2m*, 2 Cummins diesels, 6746bhp = 11.8kts, 5200nm range, crew 25 or 29 plus two platoons of troops, 1–12.7mm MG): *Sea Lion* (PL 1), *Sea Tiger* (PL 2), built HK United DYd 1965. Both discarded 1993.

Two Command patrol boats (450t full load, 131ft 3in × 27ft 11in × 10ft 5in, *40m × 8.5m × 3.2m*, 2 Caterpillar 3512 diesels, 2,350bhp = 14kts, range 1,300nm at 14kts, 2–12.7mm (MG (2 × 1), crew 29 or 33 plus two platoons of police): *Sea Panther* (PL 3), *Sea Horse* (PL 4), built 1986–88. Both extant 1995.

Ten Damen type coastal patrol craft (86t standard, 85ft 10in × 19ft 5in × 5ft 10in, *26.2m × 5.9m × 1.8m*, 2 MTU 12V 396 TC 82 diesels, 2600bhp = 23kts, 1 MAN D2566, 1956hp = 2–8kts cruising, 1400nm at 8kts, 1–12.7mm MG, Decca 150 radar, crew 14): PL 60–69 built by Chung Wah of Kowloon. First three in service February 1980, next four by 8 September 1980, last three by early 1981. PL 65 named *Cetus*, PL 66 *Dorado*.

Fifteen Damen Mk 3 coastal patrol craft (85t, 87ft × 19ft × 6ft, *26.5m × 5.8m × 1.8m*, 2 diesels = 25kts, 1–7.62mm MG, 17 crew): PL *70–84*, built 1984–85. All extant 1995.

Eight 78ft Vosper Thornycroft type (82t full load, 78ft 6in × 17ft 2in × 5ft 6in, *23.9m × 5.2m × 1.7m*, 2 Cummins diesels, 1500bhp = 20.7kts, 4000nm at 20kts, 1–12.7mm MG, crew 16): *Sea Cat* (PL 50), *Sea Puma* (PL 51), *Sea Leopard* (P 52), *Sea Eagle* (PL 53), *Sea Hawk* (PL 54), *Sea Lynx* (PL 55), *Sea Falcon* (PL 56); all built at Singapore and in service since May 1972–May 1973. All discarded 1993. *Sea Horse* (PL 4) is a 1958, 15.5kt, 21-man predecessor still in service.

Six Australian 'Pacific Forum' class patrol boats (170t full load, 107ft × 30ft × 5ft 3in, *32.6m × 8.2m × 1.6m*, 2 diesels, 1 waterjet = 26kts, 1–12.7mm MG, 2–7.62mm MG (2 × 1), crew 19): *Protector* (PL 51), *Guardian* (PL 52), *Defender* (PL 53), *Preserver* (PL 54), *Rescuer* (PL 55), *Detector* (PL 56). Built 1992–93. Replaced Vosper Thornycroft *Sea Cat* type (above). All extant 1995.

Nine 70ft HK-built patrol craft (52t full load, 70ft × 17ft × 5ft 2in, *21.3 × 5.2m × 1.6m*, 2 diesels, 215bhp = 10kts, 1600nm range, 1–12.7mm MG, crew 12): *Sea Rover* (PL 26), *Sea Farer* (PL 27), *Sea Roamer* (PL 28), *Sea Rider* (PL 29), *Sea Nomad* (PL 30), *Sea Wanderer* (PL 31), *Sea Raker* (PL 32), *Sea Rescuer* (PL 33), *Sea Ranger* (PL 34). PL 26–28 built 1954, rest 1955. All now discarded.

One 65ft patrol craft (48t full load, 1 diesel, 152hp = 10.5kts, 1400nm at 9kts, crew 11): *Islander* (PL 6), built *c*1960.

Ten 45ft converted wooden tugs built in Australia 1944–45, two discarded 1975 rest discarded by 1995.

Three 40ft *Jetstream* class (17t, 2 diesels, 740bhp = 24kts, 380nm range, 5 men), built by Cheoy Lee SY 1971.

Nine 30ft Fairey Marine 'Spear' class in service since 1981 with 1–7.62mm MG.

Eleven 22ft personnel launches built by Cheoy Lee SY 1970.

There are other logistic craft.

Hungary

The armistice signed in January 1945 in Moscow ordered the return of territories seized during 1938-40, 300 million dollars reparations and access to the country's territory by units of the Soviet Army. Hungary was proclaimed a republic in 1946 and in 1947 the communists seized power in the country following the disbanding of the Allied armistice commissions after signing the peace treaty in the same year. Internal affairs were modelled on the Soviet pattern and in 1955 the country became a member of the Warsaw Pact. In 1956 the Hungarian Revolution forced the government to establish a multi-party system and to announce on 1 November withdrawal from the Warsaw Pact. This resulted in Soviet invasion and a permanent Soviet presence which went unopposed by Western powers at the time. The political reforms were abolished but thorough economic reform now became possible and the country became the most prosperous of the COMECON states.

Being a landlocked country after the First World War, Hungary maintained a flotilla on the Danube which had been totally annihilated by the closing stages of the Second World War. Reconstruction of the Danube Flotilla was started in 1948 with the transfer of ex-Soviet river minesweepers. Then eight old craft (the four *Kecskemet* class 133t ships were stricken in the 1950s, *Baja* in 1969, and the 140t *Sopron* of 1918 also in the 1950s, as were the three *Honved* class 17t minesweeping boats of 1916) from the Austro-Hungarian Danube Flotilla were restored. Since the mid-1950s they have been replaced either by reconstructed ships of the inter-war period or by new construction.

In 1969 it was officially stated that the flotilla had been disbanded; however, the country continues to operate an independent maritime brigade (400 officers and men; ratings on one and half years service) of the army, and older patrol craft have been gradually disposed of and replaced by Yugoslav-built river minesweepers. About 13 craft were in active service in 1990, and others were in reserve. Several auxiliaries are operated, including troop transports up to 1000t, two transport barges which can be used as LCTs, and five LCUs.

Hungary, which together with Poland led the wave of political reform in eastern Europe, successfully pressed for the disbanding of the Warsaw Pact and was able to negotiate withdrawal of Soviet troops from its territory. In December 1991 Hungary became an associate member of the European Community and has applied for membership of NATO.

PN11 *river patrol boat*

Displacement: 87t
Dimensions: 108ft 11in × 12ft 9in × 3ft 7in
33.2m × 3.9m × 1.1m
Machinery: 2 shafts, Lang MP-35 diesels, 420bhp = 13.5kts
Armament: 2–37mm AA (2×1), 1–85mm howitzer
Armour: 30mm
Complement: –

Built in Budapest in 1944 as *PN 3*, and scuttled in the winter of 1945, this vessel was raised in 1948 and reconstructed with a 30mm casemate added, which affected her stability. Became *PN 31* and *PN 34*. Stricken 1973.

PN 21 type *river patrol boat*

Two 52t boats of the former Hungarian Danube Flotilla were raised in the late 1940s and entered service as *PN 21* (1950) and *PN 22*, armed with 1–37mm AA gun. They had 30mm casemate armour added and both were excessively top heavy. This resulted in the sinking of *PN 21* during manoeuvres in 1951, while *PN 22* was lengthened and her tonnage increased to 74t. *PN 22* was stricken in 1973.

PN 31 type *river patrol boat*

Displacement: 92t
Dimensions: 108ft 3in × l5ft 9in × 39ft 4in
33.0m × 4.8m × 1.0m
Machinery: 2 shafts, Endracek diesels, 900bhp = 13.4kts
Armament: 2–37mm AA (2xl), 1–12.7mm MG
Complement: –

Class:
PN31–33
Built 1952–56; scrapped 1973.

Yugoslavian 'NESTIN' class *river minesweepers*

Class:
Baja (AM 12), *Dunafoldvar* (AM 32), *Dunaujvaros* (AM 31), *Obuda* (AM 22), *Szass halombatta* (AM 21), *Ujpest* (AM 11), *AM 14* (reserve), *AM 24* (reserve)

These vessels were built by Brodotehnika, Belgrade, commissioned 1979–80. *AM 14* and *AM 24* deleted 1993, remainder extant 1995.

Ex-Soviet *MSBs*

Twelve wooden minesweepers of 10t were transferred in July 1948 to form the nucleus of the Danube Flotilla. They were all stricken in the 1960s.

AN-2 type MSB *patrol craft*

Displacement: 10.5t
Dimensions: 44ft × 12ft 6in × 2ft
13.4m × 3.8m × 0.6m
Machinery: 2 shafts, Csepel D-613 diesels, 170bhp = 13kts
Armament: 2–12.7mm DShK MG (1×2), 2–4 mines
Complement: –

These aluminium-hulled boats were built in 1940 for the Lake Balaton Flotilla. Two war losses were salved in 1948 and recommissioned with totally reconstructed superstructures.

A series based on this type were built commencing in 1953 for service on the Danube, and forty-five boats with numbers *542–001* to *542–053* were extant in 1995.

Iceland

Although a member of NATO, Iceland has no navy. However, there is a coast guard service which is mainly employed in fishery protection and offshore patrol duties. Although its patrol craft were regularly referred to as 'gunboats' during the 1975–76 'Cod War' with the UK, they have only nominal armament (originally 1–57mm gun: vessels in service in 1990 were rearmed with 1–40mm Bofors).

Aegir, 1929, 507t, 14kts, BU 1968
Saebjorg, 1937, 98t, 10kts, stricken Aug 1965
Gautur (ex-*Odinn*), 1938, 72t, 11kts, stricken 1.1.63
Hermadur, 1947, 208t, 13kts, foundered 17.7.59
Maria Julia, 1950, 138t, 11.5kts, sold 1969
*Thor**, 1951, 920t, 17kts, stricken
Albert, 1956, 200t, 12kts, stricken
*Odinn**, 1959, 1000t, 18kts, extant 1995
Arvakur, 1962, 716t, 12kts, stricken
*Aegir**, 1968, 1150t, 19kts, extant 1995
Baldur, 1974, 740t, 15kts, stricken
*Tyr**, 1974, 1150t, 19kts, extant 1995

* have sonar, helicopter deck and hangar.

India

'The future of India will undoubtedly be decided on the sea' wrote K M Panikkar in 1945. This lone oceanic visionary was striving even before the subcontinent's independence to counteract the overwhelmingly land-based attitudes and preoccupations of its rulers. India lacks a continuous or general maritime tradition yet it was seaborne Europeans who colonised her in the eighteenth and nineteenth centuries. The evolution of the Indian Navy to its present status of seventh largest in the world and third largest in Asia (with the region's only aircraft carrier) is inseparable from its pre-independence background.

In 1939 the Royal Indian Navy, having almost disappeared in the economies of the 1920s, stood at eight ships and 1600 men, all based at Bombay. By the end of 1944 the Fleet consisted of 117 warships including seven sloops, nine corvettes, fourteen ocean minesweepers and forty-nine coastal craft as well as the Landing Craft Wing. Shore establishments included six ports, gunnery, radar, torpedo, signals, engineering and tactical training schools while personnel including nearly 1000 Indian Wrens (WRINS) totalled almost 35,000. The RIN contribution to Allied victory was at least comparable with the Royal Australian Navy's. Four of the major ships had Indian captains in 1945 but expansion had been so headlong that British officers were needed long after independence. The unexpected early defeat of Japan led to rapid demobilisation down to 11,000 men by the end of 1946. Poor shore conditions and Nationalist agitation sparked a five-day mutiny at Bombay in February 1946. It spread to the Andamans, Hong Kong and Aden but was over almost as soon as it started after ten sailors were killed. Only the sloop *Hindustan* exchanged fire with the Indian army at Karachi.

Partition split the navy as it did all other institutions. Before 1939 most sailors had been Muslim and only wartime expansion brought in large numbers of Hindus. Over half the fleet's ships sailed to the two halves of Muslim Pakistan. Neither force took part in the 1948–49 war between the new states, nor indeed in the 1965 fighting which was again a localised frontier dispute. Ambitious expansion plans went ahead. Postwar acquisition of cruisers and destroyers had been proposed as early as mid-1943 and these were acquired in 1948–49 despite the manning problem. On 26 January 1950 the Royal Indian Navy was officially retitled the Indian Navy and in 1952 the first Indian admiral took command. Three classes of the latest British frigates were ordered in 1954–56 and new minesweepers acquired. Most ambitiously of all, India bought the light fleet carrier HMS *Hercules* in 1957 and had her rebuilt as *Vikrant*. The Fleet Air Arm trained a new naval aviation service to operate from her with Sea Hawk jets and helicopters. The 'Indian Flotilla' became the Indian Fleet on 30 December 1957.

In 1961 the navy played its part in the almost bloodless annexation of Portuguese Goa, from which European seapower had first dominated the Indian Ocean. By 1964 manpower had reached 19,500 when it had been only 8800 in 1958. The navy now shared in the general defence expansion after the shock of China's limited invasion in 1962. Between 1966 and 1970 manpower doubled and the fleet expanded from thirty-two warships to fifty-five. Traditional reliance on Britain was supplemented by the purchase of cheap Soviet ships on easy terms from 1967. The Soviet Union also helped to expand the naval base at Visakhapatnam. India had her first submarines (four) by the time she went to war for a third time with Pakistan and now the navy could show its teeth.

The carrier *Vikrant* headed a task force of seven destroyers and frigates, two submarines, a minesweeper, five fast attack craft and three landing ships in the Bay of Bengal during December 1971. Their blockade of doomed East Pakistan was totally effective. More than that, the navy trained *Mukhti Bahini* Bengali guerrillas to plant limpet mines on ships and bridges in the waterways of the Pakistan state before Indian's conventional invasion. The Indian Western Fleet also demonstrated its superiority over Pakistan's main force. The submarine *Ghazi* was sunk by ASW surface forces off Visakhapatnam while trying to ambush *Vikrant*.

Bangladesh's creation meant that the navy need no longer worry about a two-sea war and could concentrate on long term expansion and indigenous warship building. The August 1971 mutual assistance treaty with the USSR did not lead to a Soviet predominance in naval matters. Bombay's Mazagon Dockyard built six British *Leander* type frigates between 1966 and 1981, fitting Western electronics and expanding their helicopter facilities. A number of indigenous designs have been produced, expanding both Western and Soviet designs, and building under licence. Both German and Russian types have been acquired, the former being built at Bombay.

Some attempts at expansion have proved disappointing. The plans to build an aircraft carrier in time for the fiftieth anniversary of Independence is running very late, and plans have been recast more than once. The collapse of the Soviet Union casts doubt on India's ability to continue supporting the largely Soviet-built Eastern Fleet, and recent events suggest that maintenance of hulls and equipment has been neglected, with disastrous consequences.

An Indian Coast Guard to patrol the country's 200-mile EEZ was formed on 1 February 1977 as part of the navy. It became independent in August 1978 with a rear-admiral as director-general. The force started with two former navy frigates but two home-built classes of armed patrol vessels are in service.

India's prompting of the UN declaration of the Indian Ocean as 'a zone of peace' in 1971 has not reduced 'outside' naval activity since Britain's withdrawal. The US Navy, based mainly on Diego Garcia, wnet into South Asian waters for the first time in a permanent way to counter the Soviet Indian Ocean Squadron, but the eclipse of Iran leaves India unchallenged as the main regional sea power. *Vikrant*'s life is being extended by the acquisition of Sea Harriers.

Since 1977 the navy has up to 55,000 volunteer sailors including a naval air arm of 5000 men with thirty-five combat aircraft and thirty-six helicopters. In 1995 they manned an active fleet of eighty-six warships and thirty auxiliaries.

FLEET STRENGTH 1947

FRIGATES (ex-sloops)

Name	Fate	Name	Fate
Modified Hastings class			
HINDUSTAN	To Pakistan 1948		
Modified Bittern class			
JUMNA	Stricken 1983	GODAVARI	To Pakistan 1948
SUTLEJ	Stricken 1983	NARBADA	To Pakistan 1948
Modified Black Swan class			
CAUVERY	Paid off 1979	KISTNA	Stricken 1982
'River' class			
TIR (ex-*Bann*)	Paid off 1979	SHAMSHER (ex-*Nadder*)	To Pakistan 1948
DHANUSH (ex-*Deveron*)	To Pakistan 1948	KHUKRI (ex-*Trent*)	Survey ship 1951

Hindustan was transferred from the former Royal Indian Navy but was worn out and of little military value. *Godavari* and *Narbada* became the Pakistani *Sind* and *Jhelum*, having previously been handed over from the ex-RIN; *Jumna* and *Sutlej* formed the 12th Frigate Squadron but from 1955 *Sutlej* served as a surveying vessel, followed by *Jumna* in 1957. *Jumna* was renamed *Jumuna* c 1976.

Kistna had her after 4in guns replaced by a deckhouse c1962. *Cauvery* acted as temporary Fleet Flagship in the mid-1950s.

Neza (ex-Trent) was returned to the RN in April 1947 but four sisters were retained. *Dhanush* and *Shamsher* went to Pakistan in 1948, the former retaining her name but the latter becoming *Zulfiquar*. *Tir* was converted to a midshipmen's training ship in 1948; *Khukri* was refitted as a survey ship in 1951 and renamed *Investigator*.

MINESWEEPERS

Name	Fate	Name	Fate
Ex-Bangor class			
ROHILKAND	Sold 1960	KATHIAWAR	To Pakistan 1948
KONKAN	For disposal 1973	OUDH	To Pakistan 1948
RAJPUTANA	BU 1961	MALWA	To Pakistan 1948
BALUCHISTAN	To Pakistan 1948		
Ex-Bathurst class			
BENGAL	BU 1967	MADRAS	For disposal 1962
BOMBAY	For disposal 1962		

Kathiawar, Oudh and *Malwa* became the Pakistani *Chittagong, Dacca* and *Peshawar* in 1948

Vikrant 1961
Navarret Collection

MAJOR SURFACE SHIPS

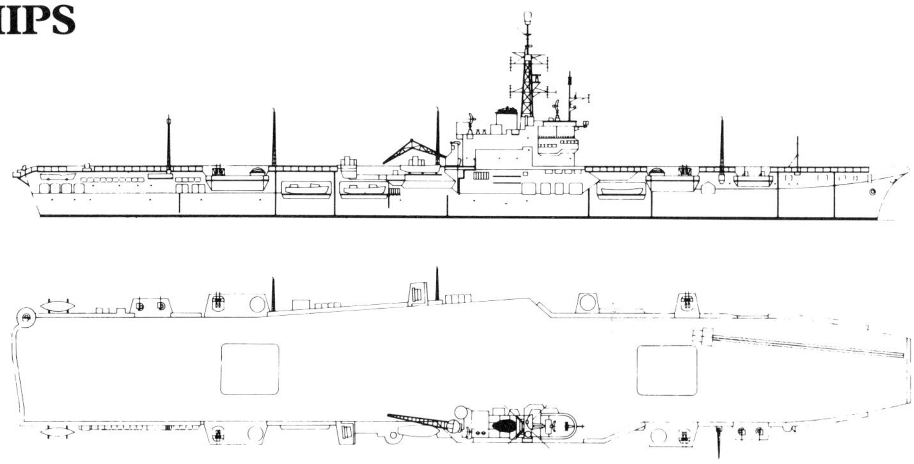

Vikrant as completed
NB 1/750 scale

VIKRANT *aircraft carrier*

The last of the light fleet carriers to be completed, HMS *Hercules* was bought in January 1957 by the Indian Navy, building to a recast design resumed at Harland & Wolff's yard in Belfast three months later. When she was ready in 1961 she had received a steam catapult, an angled deck, air conditioning and new radar, similar to HMAS *Melbourne*.

The air group comprised ten Sea Hawk jet fighters, four Alizé ASW aircraft and two Alouette helicopters, but in 1979 it was announced that re-equipment with Sea Harrier V/STOL aircraft would take place. Delivery of eight began in January 1983. During the 1971 Indo-Pakistan War *Vikrant* operated in the Bay of Bengal and flew off several strikes against naval and military targets including Chittagong. In 1973–74 she was extensively overhauled, again in 1979–82, and in 1987–89 she was rebuilt with a ski-jump. The delay in starting work on a replacement means that she will have to continue beyond the turn of the century despite being in poor condition.

Displacement: 16,000t standard; 19,500t full load
Dimensions: 700ft oa × 128ft max × 24ft
213.4m × 39.0m × 7.3m
Machinery: 2 shaft geared steam turbines: 40,000shp = 23kts. Oil 3200t. Range 12,000nm/6200nm at 14kts/23kts
Armament: 15–40mm/60 Bofors (4×2, 7×1), 16 aircraft
Sensors: Radar 960, 278, 293, 963
Complement: 1075 (peace), 1345 (war)

No	Name	Builder	Acquired	Fate
R 11	VIKRANT (ex-*Hercules*)	Vickers-Armstrong, Tyne	4.3.61	Extant 1995

(For illustration see UK *Hermes*)

VIRAAT *aircraft carrier*

Handed over to Indian control 14 November 1986 and underwent a limited overhaul before recommissioning at Devonport 12 May 1987. Formally commissioned into on arrival in India 15 February 1989. Operates twelve Sea Harrier Mk 51 and nine Sea King Mk 42B/C. Seacat SAM system may be replaced with Russian CADS-1 CIWS. Engine room flooded in September 1993.

Particulars: See *Hermes* in UK section

No	Name	Builder	Acquired	Fate
R 22	VIRAAT (ex-*Hermes*)	Vickers-Armstrong, Barrow	19.4.86	Extant 1995

New aircraft carrier class

In 1989 it was announced that a new class of three carriers was planned, two to replace *Vikrant* and *Viraat*, and a third to allow a carrier task force to operate permanently with both Eastern and Western Fleets. Initially a French design from DCN was selected, a 28,000t ship based on *Charles de Gaulle*, but conventionally powered. It was hoped to operate new Soviet fixed-wing aircraft as well as STOVL, so the ships would need catapults and arrester gear.

Severe industrial troubles were encountered at Cochin Shipyard, the only yard with capacity to build such a large ship, and it has been admitted officially that the first CV cannot be ready in time for the Independence jubilee in 1997. In 1991 there was another change, this time selecting the Italian 12,000t *Giuseppe Garibaldi* design to save money (perhaps also reflecting a disagreement with the French over financial arrangements). Work has now switched to an 'indigenous' design (possibly based on a merchant hull). At the time of writing work has not started, and the new carrier is unlikely to appear before the early years of the next century.

Purchase of the damaged Russian aircraft carrier *Admiral Gorshkov* was reportedly under consideration in 1994 but the deal seems unlikely to succeed.

Ex-British LEANDER class *cruiser*

Delhi arrived at Bombay in September 1948, having been refitted. Apart from playing herself (*Achilles*) in a 1953 film of the Battle of the River Plate, the only further action she saw was in 1961, when Goa was seized by a small naval force. Although on the active list at the time of the Indo-Pakistan War in 1971 she did not get to sea, having been replaced by *Mysore* as Fleet Flagship in 1957. TT were removed in 1958. In 1969 she visited Australia, New Zealand and Fiji; thereafter she was based at Cochin as a harbour TS, with the 40mm and 4in guns removed to provide additional space for cadets. In July 1977 she returned to Bombay under her own power to pay off. Although preservation was out of the question steps were taken to display a twin 6in turret at the Defence Academy in Poona.

Displacement: 7114t standard; 9740t full load
Dimensions: 522ft pp, 554ft 6in oa × 55ft 8in × 20ft
159.1m, 169.0m × 17.0m × 6.1m
Machinery: 4 shaft Parsons geared turbines: 6 Admiralty 3-drum boilers, 72,000shp = 32.5kts. Oil 1800t. Range 12,000nm at 14kts.
Armament: 6–6in/50 (3×2), 8–4in/45 Mk 16 (4×2), 8–40mm/60 Mk 5 (4×2), 6–40mm/60 Mk 7 (6×1), 4–3pdr saluting, 6–21in TT (2×3)
Sensors: Radar 274, 285, 277Q, 960, 293
Complement: 680 (800 as TS)

No	Name	Builder	Acquired	Fate
C 74	DELHI (ex-*Achilles*)	Cammell Laird	5.7.48	For disposal May 1978

Ex-British FIJI class *cruiser*

Bought from Britain on 8 May 1954 for £300,000. Refitted and rebuilt by Cammell Laird. 'X' 6in turret and all six TT were removed, the bridge was modified, two lattice masts stepped, electrics were renewed and machinery refitted. Handed over at Birkenhead on 29 August 1957 and relieved *Delhi* as flagship. She took part in the 1971 Indo-Pakistan War and in May 1978 relieved *Delhi* once more, this time as flagship of the Training Squadron. She was in two major collisions, the second being in 1972 with the frigate *Beas*. *Mysore* was decommissioned at the end of 1981 and sold for scrapping in 1986.

Displacement: 8700t standard; 11,040t full load
Dimensions: As British *Fiji* class
Machinery: As British *Fiji* class
Armament: 9–6in/50 Mk 23 (3×3), 8–4in/45 Mk 16 (4×2), 10–40mm/60 Mk 5 (5×2), 2–40mm/60 Mk 7 (2×1)
Sensors: Radar 274, 275, 960, 277Q, 293
Complement: 800

No	Name	Builder	Acquired	Fate
C 60	MYSORE (ex-*Nigeria*)	Vickers-Armstrong, Tyne	29.8.57	BU 1986

Ex-British 'R' class *destroyers*

All given lattice foremast in place of tripod, new funnel ventilating trunks for tropical service, retubed boilers, TT removed and new radar prior to transfer. The new top hamper was counterweighted with 30t of permanent pig iron ballast. Arrived in Indian waters January 1950 as 11th Destroyer Flotilla.

No	Name	Builder	Acquired	Fate
D 115	RANA (ex-*Raider*)	Cammell Laird	9.9.49	For disposal 1976
D 41	RAJPUT (ex-*Rotherham*)	John Brown	29.7.49	For disposal 1976
D 209	RANJIT (ex-*Redoubt*)	John Brown	4.7.49	For disposal 1979

Rajput 1981

Soviet 'KASHIN II' class *destroyers*

New construction and unlike in modified Soviet units the SS-N-2C SSM are mounted forward. The stern 76mm turret has been deleted to provide a helicopter hangar reached by a ramp. They have twin 30mm instead of singles.

No	Name	Builder	Acquired	Fate
D 51	RAJPUT	Nikolayev	30.9.80	Extant 1995
D 52	RANA	Nikolayev	28.6.82	Extant 1995
D 53	RANJIT	Nikolayev	24.11.83	Extant 1995
D 54	RANVIR	Nikolayev	28.8.86	Extant 1995
D 55	RANVIJAY	Nikolayev	15.1.88	Extant 1995

DELHI class *destroyers*

The first of four Project 15 destroyers was ordered in March 1986 followed by a second, and a third in 1992. An eventual class of six is planned. The design is officially described as a stretched 'Kashin', built with Russian technical assistance. The first of class has Russian gas turbines, but later ships will have LM-2500 GTs made under licence by HAL.

Displacement: 6500t standard
Dimensions: 524ft 10in × 55ft 9in × 21ft 3in
160.0m × 17.0m × 6.5m
Machinery: 2-shaft CODAG, 2 AM-50 gas turbines, 54,000shp, 2 MTU diesels = 30kts(?)
Armament: 2–76mm/60 (1×2), 4 SS-N-22 'Sunburn' SSMs, 2 SA-N-7 'Gadfly' SAMs or Trishul, possibly CAD-N-1 CIWS, 6–324mm ASW TT (2×3)
Sensors: Not known
Radar RALW-002, RAWS 003; sonar Spherion
Complement: Not known

No	Name	Builder	Laid down	Launched	Comp	Fate
–	DELHI	Mazagon Dock	14.11.87	1.2.91	4.6.93	Fitting out
–	MYSORE	Mazagon Dock	2.2.91	4.6.93	–	Fitting out
–	BANGALORE	Mazagon Dock	14.12.92	–	–	Building
–	–	Mazagon Dock		–	–	Projected

Ex-British 'HUNT' class (Type 2) *frigates*

Lent by RN initially for three years and rerated as escort destroyers. They were sold to the Indian Navy in April 1959.

No	Name	Builder	Acquired	Fate
D 94	GANGA (ex-*Chiddingfold*)	Scotts	Apr/May 1953	For disposal 1975
D 92	GODAVARI (ex-*Bedale*)	Hawthorn Leslie	27.4.53	For disposal 1979
D 93	GOMATI (ex-*Lamerton*)	Swan Hunter	27.4.53	For disposal 1975

British BLACKWOOD class *frigates*

Slightly modified Type 14, ordered in 1954. The 4–21in TT were not fitted. *Khukri* was torpedoed and sunk by the submarine *Hangor* during the 1971 Indo-Pakistan War, the first warship so sunk since 1945; 191 of the 288 crew were lost. *Kirpan* (means 'Sword') and *Kuthar* were transferred to the new Coast Guard Service in 1978.

No	Name	Builder	Laid down	Launched	Comp	Fate
F 49	KHUKRI	White	29.12.55	20.11.56	15.7.58	Sunk 9.12.71
F 44	KIRPAN	Stephen	5.11.56	19.8.58	Jul 1959	CG 1978
F 46	KUTHAR	White	19.9.57	14.10.58	Nov 1959	CG 1978

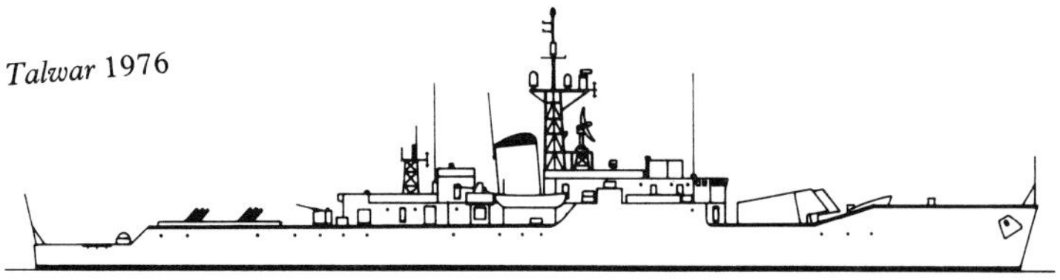

Talwar 1976

Two slightly modified Type 12 frigates were ordered in 1956. They differed mainly in having a twin 40mm Bofors Mk 5 with a CRBF director in place of the STAAG Mk 2. In 1974–75 *Talwar* had her 4.5in Mk 6 replaced by three SS-N-2 'Styx' SSM with 'Square Tie' radar above the bridge. In 1977–78 *Trishul,* flagship of 15th Frigate Squadron, was similarly refitted.

British WHITBY class *frigates*

No	Name	Builder	Laid Down	Launched	Comp	Fate
F 40	TALWAR	Cammell Laird	7.6.57	18.7.58	26.4.59	BU 1992
F 43	TRISHUL	Harland & Wolff	19.2.57	18.6.58	13.1.60	Extant 1995

Three Type 41 frigates were ordered in 1954 and the RN made the *Panther* available from its own programme. *Brahmaputra* is now a TS, with the after 4.5in guns and CRBF director replaced (1978) by classrooms. *Beas* modernised in 1980, her 40mm GFCS radar being removed.

British LEOPARD class *frigates*

No	Name	Builder	Laid Down	Launched	Comp	Fate
F 31	BRAHMAPUTRA (ex-*Panther*)	John Brown	20.10.55	15.3.57	31.3.58	BU 1986
F 37	BEAS	Vickers, Tyne	29.11.56	9.10.58	24.5.60	BU 1992
F 38	BETWA	Vickers, Tyne	29.5.57	15.9.59	8.12.60	BU 1988

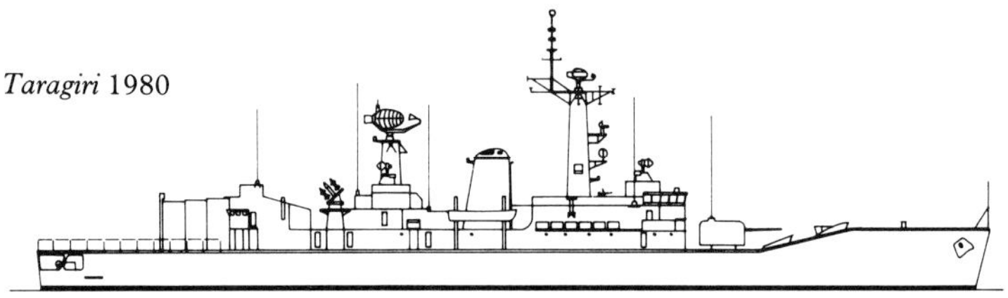

Taragiri 1980

Although originally intended to be repeats of the *Leander* design, progressive improvements have been made, and the later ships are a hybrid Anglo-Dutch design. Only *Nilgiri* and *Himgiri* have British electronics; the other two have Dutch Signaal radars built under licence by Bharath Electronics in India, and all but *Nilgiri* have M-4 directors for the Seacat SAMs. The Alouette III helicopter is too long for the original *Leander* hangar and so a telescoping type is fitted. The last two have been completed to a modified design, with a longer flight deck to permit a Sea King helicopter to be carried. This involved removing the Mk 10 mortar (replaced by 1–375mm Bofors ASW RL on fo'c's'le) and installing a large telescopic hangar much nearer the stern. The hull side beneath the flight deck is open to facilitate the handling of mooring wires etc. The six ships form the 14th Frigate Squadron.

British 'broad beam LEANDER' type *frigates*

No	Name	Builder	Laid down	Launched	Comp	Fate
F 33	NILGIRI	Mazagon Dock, Bombay	Oct 1966	12.10.68	3.6.72	Extant 1995
F 34	HIMGIRI	Mazagon Dock, Bombay	1967	6.5.70	23.11.74	Extant 1995
F 36	DUNAGIRI	Mazagon Dock, Bombay	14.9.70	24.10.72	18.2.76	Extant 1995
F 35	UDAYGIRI	Mazagon Dock, Bombay	Jan 1973	9.3.74	5.5.77	Extant 1995
F 37	TARAGIRI	Mazagon Dock, Bombay	1974	25.10.76	16.5.80	Extant 1995
F 38	VINDHYAGIRI	Mazagon Dock, Bombay	1975	12.11.77	8.7.81	Extant 1995

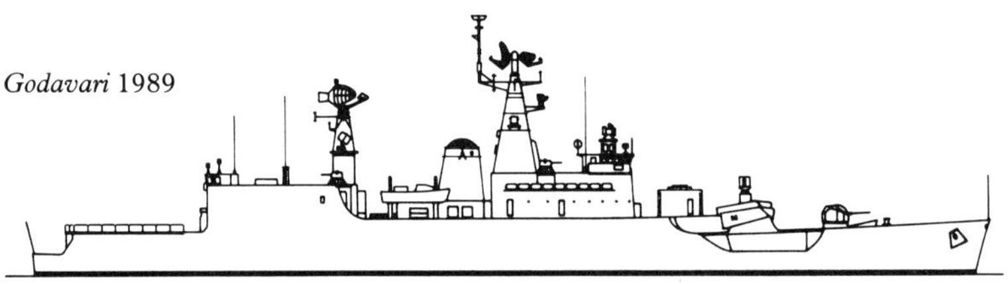

Godavari 1989

By broadening and lengthening the basic *Leander* hull the designers at Mazagon Dock were able to incorporate an extra Sea King helicopter. With four 'Styx' missiles and SAN-4 missiles forward they are the first examples of a Western hull design incorporating Russian weapons. Two 30mm 'Gatlings' are mounted on the hangar roof. A second group of three will have 30kts machinery.

GODAVARI class *frigates*

Displacement: 3500t normal; 4100t deep load
Dimensions: 397ft wl, 414ft 9in oa × 47ft 11in × 29ft 6in (over sonar)
121.0m, 126.4m × 14.6m × 9.0m
Machinery: 2-shaft geared steam turbines, 2 Babcock & Wilcox boilers, 30,000shp = 27kts. Range 4500nm at 12kts
Armament: 2–57mm/62 DP (1×2), 8–30mm/65 (4×2), 4 SS-N-2C 'Styx' SSMs, SA-N-4 'Gecko' SAM, 6–324mm ILAS-3 ASW TT (2×3), 2 Sea King Mk 42B or 1 Sea King and 1 Chetak helicopters
Sensors: Radar LW-08, ZW-06, Head Net-C, Drum Tilt, Pop Group, Muff Cob; sonar 184M or APSOH, 162M, VDS
Complement: 51 officers, 262 ratings

No	Name	Builder	Laid down	Launched	Comp	Fate
F 20	GODAVARI	Mazagon Dock, Bombay	2.6.78	15.5.80	10.12.82	Extant 1995
F 21	GOMATI	Mazagon Dock, Bombay	1981	20.3.84	10.12.85	Extant 1995
F 22	GANGA	Mazagon Dock, Bombay	1980	21.10.81	16.4.88	Extant 1995

Udaygiri Jun 1977 G Arra

Godavari 1983 RAN

Three Type 16A 'Improved *Godavari*' class frigates are planned. Although originally planned to to be only slightly modified versions of the preceding ships, several major improvements have been made. Machinery has been reported to be Russian AM-50 gas turbines or a CODOG arrangement, but other reports suggest the same geared steam turbines as the original *Godavaris*. Shikari combat data system is a locally-built version of the Italian SADOC-1. Malachit-E is the export version of the Russian SS-N-9 Siren.

IMPROVED GODAVARI class *frigates*

Displacement: 4400t normal; 5400t deep load
Dimensions: 426ft 6in × 52ft 6in × 16ft 5in
130.0m × 16.0m x 5.0m
Machinery: See notes
Armament: 4 P-120 Malachit-E SSM, 1 Trishul SAM system, 1–76mm/62 OTO Melara, 4–30mm AK630 CIWS, 6–324mm NST-58 ASW TT, 1 Sea King or Chetak helicopter
Sensors: Radar Rashmi, RAWS-03, RAWL-02, Aparna; sonar HUS-001; ECM/ESM Indra; combat data system Shikari
Complement: Not known

No	Name	Builder	Laid down	Launched	Comp	Fate
F 23	BRAMAPUTRA	Mazagon Dock, Bombay	1989	29.1.94	–	Fitting out 1995
F 24	–	Mazagon Dock, Bombay	–	–	–	Projected
F 25	—	Mazagon Dock, Bombay	–	–	–	Projected

SUBMARINES

Soviet 'FOXTROT' class *submarines* (Project 641)

Class (no, transferred)

Kursura (S 20, Dec 1970), *Karanj* (S 21, Oct 1970), *Kandhera* (S 22, Jan 1969), *Kalvari* (S 23, 16.7.68), *Vela* (S 40, 31.8.73), *Vagir* (S 41, 3.11.73), *Vagli* (S 42, 10.8.74), *Vagsheer* (S 43, 26.12.74)

India's first submarines, built at Leningrad and Severodvinsk. Their base is at Vishakhapatnam. *Kandhera* out of service 1990, followed by *Kalvari* in 1992. Rest extant 1995. One may be sold or leased to Singapore in 1995.

Shishumar class

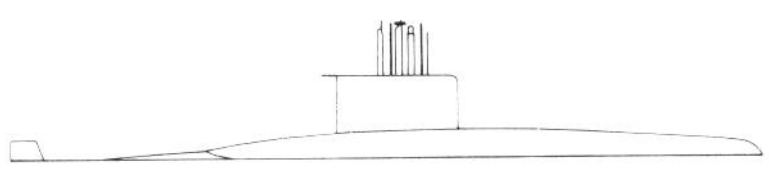

SHISHUMAR class *submarines*

Displacement: 1690t/1850t
Dimensions: 211ft 2in (oa) × 21ft 4in × 19ft 8in
64.4m × 6.5m × 6.0m
Machinery: 1 shaft diesel-electric, 4 MTU 12V493 diesels, 2400bhp, 1 electric motor, 5000hp = 11kts/22kts. Range 8000nm at 8kts (snorkel)
Armament: 8–533mm TT (bow, 14 SUT torpedoes)
Sensors: Radar Calypso; sonar CSU-83, DUUX-5
Complement: 40 (8 officers)

No	Name	Builder	Laid down	Launched	Comp	Fate
S 44	SHISHUMAR	HDW, Kiel	1.5.82	13.12.84	28.9.86	Extant 1995
S 45	SHANKUSH	HDW, Kiel	1.9.82	11.5.84	20.11.86	Extant 1995
S 46	SHALKI	Mazagon	5.6.84	30.9.89	7.2.92	Extant 1995
S 47	SHANKUL	Mazagon	3.9.89	21.3.92	28.5.94	Extant 1995

A Howaldtswerke-Deutsche Werft of Kiel design. The Indian Navy finally signed a contract for four boats on 11 December 1981, with the first pair built in Germany and the second pair to be assembled at Mazagon, Bombay. The project suffered lengthy delays because all materials had to be imported and workers had to be trained. These delays lead to plans to build a furthet two boats being dropped and Soviet-built 'Kilo' class being acquired instead.

These vessels are unique in being fitted with the IKL rescue sphere, capable of evacuating the entire crew.

Shalki Aug 1994 L & L van Ginderen

Russian 'KILO' class *submarines* (Project 877EM)

Class (no, commissioned):

Sindhugosh (S 55, 30.4.86), *Sindhudhvaj* (S 56, 12.6.87), *Sindhuraj* (S 57, 20.10.87), *Sindhuvir* (S 58, 26.8.88), *Sindhuratna* (S 59, 22.12.88), *Sindhukesari* (S 60, 16.2.89), *Sindhukirti* (S 61, 4.1.90), *Sindhuvijay* (S 62, 6.3.91)

All built by Sudomekh, Leningrad/St Petersburg, and have reportedly been very sucessful. Six ordered originally, then increased to ten following delays with *Shishumar* class, but acquisition of final pair (possibly 'leaders') since dropped. All extant 1995. Reports of plans to acquire ten 'Improved Kilo' class in 1994 appear to have been premature.

Chakra 1988

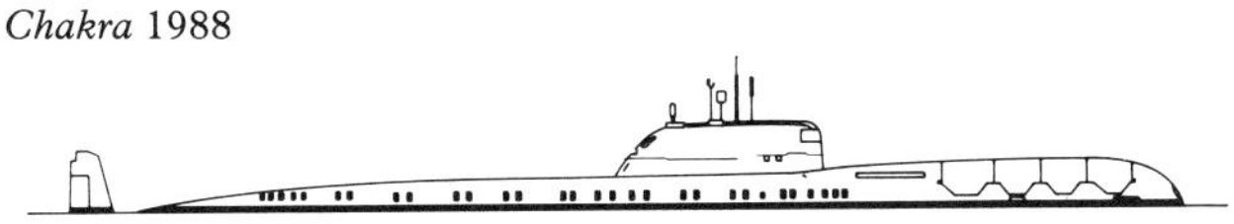

Soviet 'CHARLIE I' class *cruise missile submarine*

Class:

Chakra

Nuclear-powered SSGN leased without missiles 5 January 1991 and returned 1991, without ever becoming operational. Reportedly plagued with technical problems and plans to lease a second boat (possibly to be called *Chitra*) never materialised.

Acquired as part of India's own long-term SSN plans for trials and evaluation. Development of indigenous nuclear power plant has been under way since 1974, but is still at least ten years from completion.

AMPHIBIOUS WARFARE VESSELS

Ex-British LST 3 class *tank landing ship*

HMS *Avenger (LST 3001)* became INS *Magar* (L 11) in 1951 and was broken up in 1988.

Ex-Soviet 'POLNOCNY' class *landing ships*

Class (no, transferred):
Gharial (L 12, 1966), *Guldar* (L 13, 1966), *Ghorpad* (L 14, 1975), *Kesari* (L 15, Sep 1975), *Shardul* (L 16, Dec 1975), *Sharabh* (L 17, Mar 1976), *Cheetah* (L 18), *Mahish* (L 19), *Guldar* (L 21), *Khumbir* (L 22)

Two 'A' class (L 12–13) since scrapped, followed by four 'C' and four 'D's.

VASCO DA GAMA class *utility landing craft*

Displacement: 500t full load
Dimensions: 175ft 10in × 26ft 2in × 4ft
53.6m × 8.0m × 1.2m
Machinery: 3 shafts, 3 Kirloskar-MAN W8V 17.5/22 AMAL diesels, 1245bhp = 9kts. Range 1000nm at 8kts
Armament: 2–40mm (2×1)
Complement: ? + 120 troops or 250t cargo

Class (no):
Vasco da Gama (L 34), L 35–L 37, *Midhur* (L 38), *Mangala* (L 39), L 40.

Mk 3 LCU, two built by Hooghly Dock & Engineering, remainder at Goa SY, during 1980–87.

MAGAR class *tank landing ships*

Class (no):
Magar (L 20), *Gharial* (L 21)

Based on British *Sir Lancelot* design, delivered 1987–92. Originally planned to be a class of eight.

SMALL SURFACE COMBATANTS

Soviet 'PETYA III' class *frigates (corvettes)*

Class (no):
Arnala (P 68), *Androth* (P 69), *Anjadip* (P 73), *Andaman* (P 74), *Amini* (P 75), *Kamorta* (P 77), *Kadmath* (P 78), *Kiltan* (P 79), *Kavaratti* (P 80), *Katchal* (P 81), *Kanjar* (P 82), *Amindivi* (P 83)

New export-version ships, the five 'K' names being delivered in 1969 followed by five more with 'A' names in 1974 and, slightly later, by *Amindivi* and *Kanjar*.

Unlike Soviet ships have four RBU-2500 ASW RL and 3–21in TT as well as simpler radio equipment. It is reported that some of them needed attention to their machinery en route from Vladivostok to India. Despite having 'P' (patrol) pennant numbers the ships form the 31st and 32nd Frigate Squadrons. *Andaman* sank in heavy weather in the Bay of Bengal on 22 August 1990. *Kavarati, Kiltan, Katchal, Kanjar* and *Amandivi* paid off 1986–88 and were scrapped. Remainder scheduled for disposal 1994 but at least two were still running at the end of that year.

'DURG' class (Soviet 'Nanuchka II') *missile corvettes*

Class (no):
Vijaydurg (K 71), *Sindhudurg* (K 72), *Hosdurg* (K 73)

Soviet Project 1234E, export variant of 'Nanuchka' type built at Leningrad and delivered 1976–78. Identical to Soviet type, except armed with SS-N-2C Styx rather than SS-N-9 Siren missiles and carries Square Tie radar. All extant 1995, based at Bombay, but one laid up to provide spares for other two.

VEER class (Soviet 'Tarantul I') *missile corvettes*

Displacement: 480t normal; 580t deep load
Dimensions: 186ft 9in × 34ft 4in × 8ft 2in
56.5m × 10.5m × 2.5m
Machinery: 2-shaft COGOG, 2 NK-12MV gas turbines, 24,200shp plus 2 cruising turbines, 6000shp = 36kts. Range 2000nm at 20kts
Armament: 4 SS-N-2C Styx SSM, 1 SA-N-5 SAM, 1–76mm/60, 2–30mm CIWS
Sensors: Radar Plank Shave, Pechora, Bass Tilt
Complement: 40

Class (no, commissioned):
Veer (K 40, 12.5.87), *Nirbhik* (K 41, Jan 1988), *Nipat* (K 42, Jan 1989), *Nishank* (K 43, 12.9.89), *Nirghat* (K 44, Jun 1990), *Vibhuti* (K 45, 3.6.91), *Vipul* (K 46, 16.3.92), *Vinash* (K 47, 20.11.93), *Vidyut* (K 48, 16.1.95), *Nashak* (K 49, Mar 1995), *K 50, K 51.*

Soviet Project 1241E design missile corvettes armed with P 20/P 21 SS-N-2C Styx missiles. K 40–K 44 built by Volodarskiy Shipyard, Russia, K 45–K 46, 49, 51 by Mazagon Dockyard, Bombay and K 47–48, 50 by Goa Shipyard. All extant 1995. Further nine vessels projected.

Andaman 1975 L & L van Ginderen

Khukri as completed

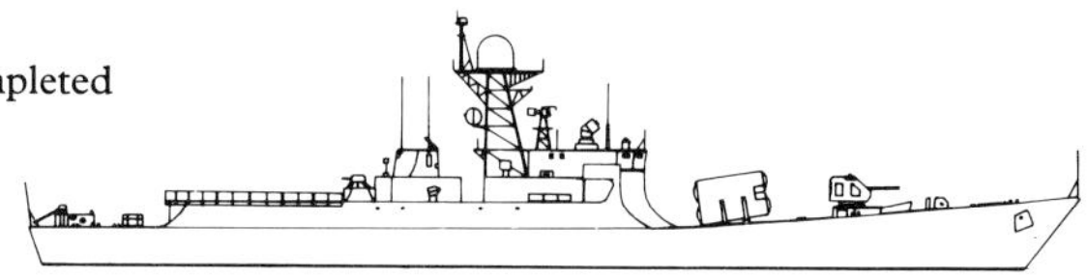

KHUKRI class *missile corvettes*

Displacement: 1350t deep load
Dimensions: 298ft 8in × 34ft Sin × 8ft 3in
91.0m × 10.5m × 2.5m
Machinery: 2-shaft SEMT-Pielstick 18PA6 diesels, 14,400bhp = 25kts. Range 4000nm at 16kts
Armament: 4 SS-N-2C Styx SSM, SA-N-5 Grail SAM, 1–76mm/60 DP, 4–30mm (2×2)
Sensors: Radar Positive E, Plank Shave, Bass Tilt, 1245
Complement: 79 (10 officers)

Class (no, commissioned):
Khukri (P 49, 23.8.89), *Khutar* (P 46, 7.6.90), *Kirpan* (P 44, 12.1.91), *Khanjar* (P 47, 22.10.91), *Kora* (P?, 1996), *Kirch* (P?, 1997)

Indigenous design possibly derived from Russian 'Koni' class hull, with forecastle deck extended aft to form helicopter deck. Class of twelve planned to replace 'Durg' class. P 46 and P 49 built by Mazagon Dockyard, Bombay, the rest by Garden Reach Shipyard, Calcutta. Five extant 1995, with three units building.

ABHAY class *ASW corvettes*

Displacement: 480t normal; 530t deep load
Dimensions: 189ft 7in × 30ft 9in × 7ft 6in
57.8m × 9.4m × 2.3m
Machinery: 2-shaft M517 radial diesels, 16,000bhp = 32kts. Range 2000nm at 20kts
Armament: 1–76mm/60, 1–30mm AK630 CIWS, 4–533mm ASW ITT
Sensors: Radar Strut Curve, Bass Tilt, Pechora; sonar Rat Tail VDS
Complement: 32

Class (no, commissioned):
Abhay (P 33, Mar 1989), *Ajay* (P 34, 24.1.90), *Aksay* (P 35, Jan 1991), *Agray* (P 36, Feb 1991)

Russian project 1241PE, a variant of the 'Pauk' class design built by Volodarskiy Shipyard at Rybinsk, ordered 1987. Differ from original in having VDS and ASW torpedo tubes. All extant 1995.

Ex-Soviet 'OSA I and II' class *fast attack craft (missile)*

Class (no):
'Osa I': *Veer* (K 82), *Vidyut* (K 83), *Vijeta* (K 84), *Vinash* (K 85), *Nipat* (K 86), *Nashat* (K 87), *Nirbhik* (K 88), *Nirghat* (K 89).

'Osa II: *Prachand* (K 90), *Pralaya* (K 91), *Prabal* (K 92), *Pratap* (K 93), *Chamak* (K 94), *Chapal* (K 95), *Chapak* (K 96), *Charag* (K 97)

First eight boats transferred 1971, K 84 and K 87 lost their missile tubes (three each) to the frigates *Talwar* (late 1975) and *Trishul* (1978), and used as patrol craft. K 87 converted to carry frogmen in 1980. 'Osa II' boats in service 17 February 1976 and 5 November 1976 (four each). K 82–89 scrapped in the 1980s, remainder replaced as *Veer* and *Vibhuti* classes delivered.

SDB Mk 2 class *large patrol craft*

Displacement: 203t full load
Dimensions: 123ft × 24ft 7in × 5ft 10in
37.5m × 7.5m × 1.8m
Machinery: 2 shafts, 2 Deltic 18–42K diesels, 7,000bhp = 32kts, 1 Cummins NH-220 cruise diesel = 14kts. Range 1000nm at 14kts
Armament: 1–40mm/60 Bofors, 2 DCT, 2 DC rails (18 Mk 7 and 10 Mk 12 DCs)
Complement: 30

Class (in service)
T 51 (17.11.78), *T 52* (3.9.77), *T 53* (12.4.78), *Rajhans* (T 54, Jan 1981), *T 55* (Feb 1981)
Indian design derived from the RN 'Ford' class. Built by Garden Reach SYd of Calcutta for the new Coast Guard. *T 51* launched 31 December 1975 and began trials September 1977. Construction way behind schedule. Named ships operated by Coast Guard.

British 'FORD' class *seaward defence boats*

Class (no, in service):
Abhay (P 3135, 13.11.61), *Ajay* (?, 21.9.60), *Akshay* (P 3136, 8.1.62), *Amar* (P 3137)
Built by Hooghly Docking & Eng Co, Calcutta. *Amar* transferred to Mauritius in 1974. *Akshay* and *Ajay* transferred to Bangladesh 1973 and 1974. Larger crew (35) than RN vessels. All scrapped by mid-1980s.

Ex-Soviet 'POLUCHAT I' class *coastal patrol craft*

Class (No):
Panaji (501/P 249), *Panvel* (502/P 246), *Pamban* (503/P 247), *Puli* (504/P 248), *Pulicat* (505/P 250)
Transferred 1967–69. Originally numbered in P sequence. Originally Home Dept units in service with the Coast Guard since 1977. All disposed of in 1980s.

SUKANYA class *offshore patrol vessels*

Displacement:	1650t normal; 1890t deep load
Dimensions:	334ft 7in × 37ft 9in × 11ft 3in *102.96m × 11.5m × 3.4m*
Machinery:	2-shaft Kirloskar-Pielstick 16PA6 V289 diesels, 12,800bhp = 21kts. Range 7000nm at 17kts
Armament:	1–40mm/60 Bofors Mk 7, 1 helicopter
Sensors:	Radar Selenia air search, Racal-Decca navigation
Complement:	145 (12 officers)

Class (no, commissioned):
Sukanya (P 50, 31.8.89), *Subhadra* (P 51, 25.1.90), *Suyarna* (P 52, 4.4.91), *Savitri* (P 53, 27.11.90), *Sarayu* (P 54, 8.10.91), *Shargda* (P 55, 27.10.91), *Sujatta* (P 56, 3.11.93)
First three (P 50–P 52) built by Korea Tacoma, Masan to their 'Neptune' design, remainder by Hindustan Shipyard at Visakhapatnam. Three more built for Coast Guard. Fitted with hangar and flight deck for Chetak light helicopter.

Vikram class

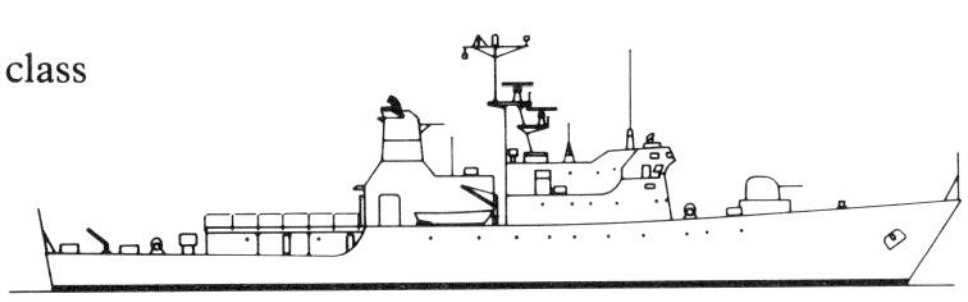

VIKRAM class *offshore patrol vessels*

Displacement:	940t normal; 1223t full load
Dimensions:	243ft 1in × 37ft 5in × 10ft 6in *74.1m × 11.4m × 3.2m*
Machinery:	2 shafts, Pielstick 16PA6 V280 diesels, 12,800bhp = 22kts. Range 4000nm at 16kts
Armament:	1–40mm/60 Bofors Mk 7, 1 helicopter
Sensors:	Radar 2 Racal-Decca 1226
Complement	96 (11 officers)

Class (no):
Vikram (T 33), *Vijaya* (T 34), *Veera* (T 35), *Varuna* (T 36), *Vajra* (T 37), *Vivek* (T 38), *Vigraha* (T 39), *Varad* (T 40), *Varaha* (T 41)
Designed with technical assistance from NEVESBU in the Netherlands. Order placed with Mazagon Dockyard, Bombay, in 1979 and ships delivered December 1983–March 1991. Operated by the Coast Guard.

Other patrol craft operated include:
Seven *Jija Bai* class 44m inshore patrol craft, three built by Sumidagawa, remainder by Garden Reach and delivered 1983–85: *Jija Bai* (T 64), *Chand Bibi* (T 65), *Kittur Chinnama* (T 66), *Rani Jindan* (T 67), *Habbakhatum* (T 68), *Ramadevi* (T 69), *Avvayar* (T 70). Operated by Coast Guard.
Seven SDB Mk 3 large patrol craft: built by Garden Reach and Goa yards and delivered 1984–86. Slightly larger version of SDB Mk 2. Some reported to be operated by Coast Guard.
Six *Tara Bai* class 46m inshore patrol craft, designed by Lürssen: two ordered from Singapore Shipbuilding and Engineering, four from Garden Reach, delivered 1987–90: *Tara Bai* (T 71), *Ahalya Bai* (T 72), *Lakshmi Bai* (T 73), *Akka Devi* (T 74), *Naiki Devi* (T 75), *Ganga Devi* (T 76).
Eight C 01 class 20m inshore patrol craft built in Korea 1990–92 and taken over from India Oil Corporation.
Ten P 2000 type inshore patrol craft ordered to Angram design from Anderson Marine, Chincalim, for delivery from 1992.

MINE WARFARE VESSELS

Ex-British 'TON' class *coastal minesweepers*

Class (no, former name, fate):
Cuddalore (M 90, ex-*Wennington*), *Cannanore* (M 91, ex-*Whitton*), *Karwar* (M 92, ex-*Overton*), *Kakinada* (M 93, ex-*Durweston*)
Transferred from RN in 1956 and named after small Indian ports. In reserve from early 1978 when Soviet 'Natya' class arrived. All scrapped in 1980s.

Ex-Soviet 'NATYA' class *ocean minesweepers*

Class (no, transfer):
Pondicherry (M 61, Apr 1978), *Porbandar* (M 62, Apr 1978), *Bedi* (M 63, Jul 1979), *Bhavnagar* (M 64, Jul 1979), *Alleppey* (M 65, Aug 1980), *Ratnagiri* (M 66, Aug 1980), *Karwar* (M 67, Sep 1986), *Cannanore* (M 68, Nov 1987), *Cuddalore* (M 69, Nov 1987), *Kakinada* (M 70, May 1987), *Kozhikoda* (M 71, Dec 1986), *Konkan* (M 72, Dec 1988).
Last of class transferred early in 1989. Lack stern ramp seen in their Russian sister-ships. Project 266 ME design.

British 'HAM' class *inshore minesweepers*

Class (no, former name):
Bimlipathan (M 79, ex-*Hildersham*), *Bassein* (M 80, ex-*Littleham*), *Bhatkal* (M 81), *Bulsar* (M 82)
First two transferred from RN in 1955. Last two built by Mazagon of Bombay being launched May 1967 and 17 May 1969 respectively. All extant 1995.

Ex-Soviet 'YEVGENYA' class *inshore minesweepers*

Class (no):
Mahe (M 83), *Malvan* (M 84), *Mangalore* (M 85), *Malpe* (M 86), *Mulki* (M 87), *Magdala* (M 88)
Three delivered in 1983, followed by three more a year later.
It is hoped to build a class of ten minehunters at Goa, but no design has yet been selected.

Indonesia

After Indonesia finally gained independence from the Netherlands on 27 December 1949 the young republic started with a moderate build-up of her own naval forces. The first vessels of the Indonesian Navy were two former Australian *Bathurst* class minesweepers which were commissioned as corvettes in December 1949. In 1950 two sister ships and two former US subchasers as well as several Netherlands customs and governmental vessels which had survived the war in the Netherlands East Indies were added to the fleet. During 1949–52 seventeen patrol boats of the *Alkai* class were built by Netherlands shipyards for Indonesia and from 1951 the former destroyer *Tjerk Hiddes* served for many years as the flagship. She was named *Gadjah Mada*.
In 1954 a programme for new construction was set up which included two destroyers and two corvettes ordered in Italy. Ten minesweepers and sail training vessel *Dewarutji* were built by German shipyards and in 1958 six patrol boats were received from Yugoslavia. Indonesia's approach to the Eastern Bloc led to a massive naval build-up. From 1959 to 1964 about one hundred naval vessels, including a cruiser, seven destroyers, seven frigates and fourteen submarines were received from the Soviet Union. None of these ships was involved in the 1963–66 Borneo Confrontation with Britain and Malaysia; only Indonesian marines and river craft took part.
In October 1965 when a military coup in Indonesia led by General Raden Suharto ended President Sukarno's pro-Soviet policy, the situation changed completely. As the new regime refused to pay the $800m debt to Russia and many thousands of communists were massacred, supply of spares for the above mentioned ships was ended. For this reason and the lack of funds for maintenance, the Indonesian government in July 1970 announced its decision to sell all ships of Russian origin for scrapping. In 1973 a decision was taken to reduce Indonesia's armed forces: the army from 300,000 to 200,000, air force and navy from 100,000 to 30,000 each. The marine corps which had been reduced from 16,000 to 5000 was however increased in subsequent years to 12,000.

During this period of rundown the navy found itself involved in the expansion of Indonesian territory. In August 1975 civil war broke out in the Portuguese colony of East Timor. By mid-September thirty Indonesian warships, including three to five frigates, were offshore and on 7 December they bombarded in support of marine and paratroop (1000 men each) landings to seize the island capital Dili. With the aid of local anti-communist guerrillas the coastal towns of Maubara and Baucau, the Ocussi Ambeno enclave and the offshore island of Atauro were all occupied before the year was out. East Timor became the sprawling archipelago's twenty-seventh province.

After all ex-Soviet destroyers and most frigates had been sold for scrap new ships were ordered. During 1977–79 three new corvettes (light frigates) of the *Fatahillah* class were built in Netherlands yards. Four frigates, twelve minesweepers and several auxiliary ships were bought from the United States. The South Korean shipyard Korea Tacoma, Masan won a contract for four missile-armed fast attack craft.

During the 1980s the Indonesian Navy was reinforced with no fewer than nine frigates, three former British 'Tribals' and six modernised Netherlands *Van Speijk*s. Future acquisition of Netherlands *De Ruyter* class missile destroyers is also planned when these are phased out of service with the Netherlands Navy. There are also new fast attack craft of the Lürssen 'FPB 57' type built for the navy and the customs service. Two modern 'Alkmaar' type minesweepers were commissioned as well as six Korean-built landing ships of the 'Teluk' class.

In 1993 Indonesia purchased sixteen corvettes, twelve landing ships and nine minesweepers of the former East German *Volksmarine* from Germany, one of the largest bloc resales of used warships in recent years. The transaction has been plagued by cost overruns. The deal includes extensive training of Indonesian personnel in Germany. The names of all Indonesian ships are preceded by the designation KRI (*Kapalperang Republik Indonesia*) which means 'Warship of the Republic of Indonesia'.

In 1995 some 1000 men served in the naval air arm with twenty-three aircraft and twenty-eight helicopters. Currently there are over forty-six helicopters (Alouette III, Bo-105, AS-332 Puma, Westland Wasp), many based on ships, and as many as fifty-four aircraft. According to the recent reports the current personnel strength of the Indonesian Navy is 41,000 officers and men, including 13,000 marines and 1000 naval aviation personnel.

Its principal bases are Gorontalo, Kemayoran (Jakarta) and Surabaya on Java. A new naval base was built at Teluk Ratai on South Sumatra, costing a reported £3bn to replace that at Surabaya and completing in 1991. The nine local shipyards are able to build small surface units.

MAJOR SURFACE SHIPS

Ex-Soviet SVERDLOV class *cruiser* (Project 68-bis)

No	Name	Builder	Acquired	Fate
201	IRIAN (ex-*Ordzhonikidze*)	Admiralty	Sep 1962	BU

Sold to Indonesia in 1962 where she arrived in October 1962 with pennant no 201. A second Soviet cruiser was to have been acquired by the end of 1963 according to Indonesian sources but was never delivered. There were also talks about an aircraft carrier built on the *Sverdlov* class hull but eventually nothing came out of it. *Irian* was modified to serve in the tropical climate but still was ill suited to the tasks of the Indonesian Navy. She was put on a disposal list in 1972 because of lack of spare parts needed to keep this impressive ship running.

Ex-British 'N' class *destroyer*

No	Name	Builder	Acquired	Fate
	GADJAH MADA (ex-*Tjerk Hiddes*, ex-*Nonpareil*)	Denny	1.3.51	BU 1961

RN 1941-built destroyer transferred to the Royal Netherlands Navy in 1942. Served as Indonesian flagship. Stricken and scrapped in 1961.

Ex-Soviet SKORYY class *destroyers* (Project 30 bis)

No	Name	Builder	Acquired	Fate
201	SILIWANGI	USSR	1959	Stricken 1971
202	SINGAMANGA-RADJA	USSR	1959	Stricken 1971
203	SANDJAJA	USSR	1959	Stricken 1971
204	SAWUNGGALING (ex-*Srawadjala*)	USSR	1959	Stricken 1971
304	SULTAN ISKANDARMUDA	USSR	1962	Stricken 1971
306	DIPONEGORO	USSR	1964	Stricken 1973
307	BRAWIDJAJA	USSR	1964	Stricken 1973

All 1950–53 built warships. *Sandjaja*, *Siliwangi*, *Singamangaradja* (means 'gannet') and *Sultan Iskandarmuda* were stricken in 1971 and used for spare parts for remaining vessels; *Brawidjaja* and *Diponegoro* (named after the prince who fought the 1825–30 Great Java War against the Dutch) deleted in 1973. *Diponegoro* was probably renamed *Sultan Badarudin*.

ALMIRANTE CLEMENTE type *frigates*

No	Name	Builder	Acquired	Fate
250	IMAN BONDJOL	Ansaldo, Leghorn	5.5.56	Stricken 1978
251	SURAPATI	Ansaldo, Leghorn	5.5.56	Stricken 1978

Two fast frigates of the type designed and built (both laid down 8 January 1956, completed 19 May and 28 May 1958) earlier for Venezuela as *Almirante Clemente* class. The light AA armament differed in being 6–30mm (3×2) and 6–20mm (3×2). They were never really modernised and were deleted in 1978. Pennant numbers 355 and 356 replaced 250 and 251.

Ex-Soviet 'RIGA' class *frigates* (Project 50)

No	Name	Builder	Acquired	Fate
351	JOS SUDARSO	USSR	1964	Stricken 1986
352	SLAMET RIJADI	USSR		Stricken 1973
353	NGURAH RAI	USSR		Stricken 1974
355	MONGINSIDI	USSR		Stricken 1971
357	LAMBUNG MANG-KURAT	USSR	1964	Stricken 1986
539	KALIALI	USSR		Stricken *c*1985
360	NUKU	USSR		Stricken 1981
358	HANG TUAH	USSR		Stricken 1971

These seven ships of the 'Riga' class (built 1954–57) were transferred from the USSR in 1962–64: two (original pennant nos 405 and 406) in 1962 with cruiser *Irian*, two in 1963 and the last three in 1964. *Monginsidi* was placed in reserve in 1969.

Ex-US CLAUD JONES class *frigates*

No	Name	Builder	Acquired	Fate
341	SAMADIKUN (ex-*John R Perry*)	Avondale	20.2.73	Extant 1995
342	MARTADINATA (ex-*Charles Berry*)	Avondale	31.1.74	Extant 1995
343	MONGINSIDI (ex-*Claud Jones*)	Avondale	16.12.74	Extant 1995
344	NGURAH RAI (ex-*McMorris*)	Avondale	16.12.74	Extant 1995

These four ships modelled on *Dealey* class destroyer (ocean) escorts were fitted with diesel propulsion instead of steam plant and were considered not too successful by US Navy and as surplus to their needs were sold to Indonesian Navy. *Monginsidi* took part in the 1975 E Timor operations. In 341 and 342 two Soviet 37mm (1 × 2) replaced a US 76mm and 2–25mm (1 × 2) were added to the fo'c'sle break. No 341 lost her ESM domes. SPS-10 and SPG-52 radar have been added. Sonar is now SQS-42. *Samadikun* has been Fleet flagship. All refitted at Subic Bay, Philippines, 1979–82.

Nala 27.3.80 — L & L van Ginderen

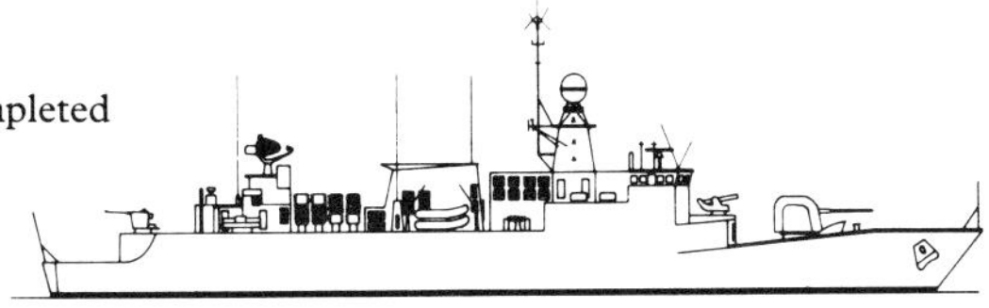

Malahayati as completed

FATAHILLAH class *frigates*

Displacement: 1200t standard; 1450t full load
Dimensions: 276ft oa × 36ft 5in × 10ft 9in
83.9m × 11.1m × 3.4m
Machinery: 2-shaft CODOG (2 CP propellers): 1 gas turbine Rolls-Royce Olympus TM 3B 28,000shp = 30kts, plus 2 diesels MTU 16V956 TB91, 6000bhp = 21kts. Range 4250nm at 16kts
Armament: 4 MM38 Exocet SSM (2×2), 1–120mm/46 Bofors (400rds), 1–40mm/70 Bofors (3000rds), 2–20mm (2×1), 2–375mm
Bofors SR-375A ASW RL (1 × 2); 54 Neilli and Erica rockets, Mk 32 324mm ASW TT (2 × 3, 6 spare torpedoes)
Sensors: Radar Decca AC 1229, HSA DA-05/2, WM-28; sonar PHS-32; ECM SUSIE-1 system, 2 Vickers Mk 4 chaff RL (50rds)
Complement: 89

No	Name	Builder	Launched	Fate
361	FATAHILLAH	Wilton-Fijenoord	22.12.77	Extant 1995
362	MALAHAYATI	Wilton-Fijenoord	19.6.78	Extant 1995
363	NALA	Wilton-Fijenoord	11.1.79	Extant 1995

These compact ships of small frigate type were ordered from the Netherlands in August 1975 and designed by NEVESBU which belonged to Rhine-Schelde-Verolme. *Fatahillah* laid down 31 January 1977 and completed 16 July 1979. *Malahayati* laid down 28 July 1977, completed 21 March 1980. Air-conditioned living spaces and NBC warfare citadel. DAISY computer system. Officially they are classed as corvettes. *Nala* (laid down 21 January 1978, completed August 1980) differs from sisters having a helicopter (BO 105) facility and collapsible hangar for one small helicopter and her light AA consists of two Bofors 40mm/70 abreast the hangar.

Ex-Netherlands VAN SPEIJK class *frigates*

No	Name	Builder	Acquired	Fate
351	AHMAD YANI (ex-*Tjerk Hiddes*)	Nederlandsche DSM	10.10.86	Extant 1995
352	SLAMET RIYADI (ex-*Van Speijk*)	Nederlandsche DSM	1.11.74	Extant 1995
353	YOS SUDARSO (ex-*Van Galen*)	De Schelde	2.11.87	Extant 1995
354	OSWALD SIHAAN (ex-*Van Nes*)	De Schelde	1.11.88	Extant 1995
355	ABDUL HALIM PERDANAKASUMA (ex-*Evertsen*)	De Schelde	1.11.89	Extant 1995
356	KAREL SATSUITUBUN (ex-*Isaac Sweers*)	Nederlandsche DSM	1.11.90	Extant 1995

All six ships belong to the Netherlands *Van Speijk* class based on the British *Leander* class, but were originally fitted with broader enclosed bridge and two Sea Cat systems. New infrared suppression stack caps added early 1980s. First two transferred under the agreement signed on 11 February 1986, and under which the ships were provided with spare parts but not helicopters. Provision was made for eight Harpoon SSM missiles per ship, but normally only two are carried. US-made SQR-18A towed passive sonar array was removed from the last two ships before transfer. Fitted with SEWACO II data system.

Ex-British 'TRIBAL' class *frigates*

No	Name	Builder	Acquired	Fate
331	MARTHA KRISTINA TIYAHAHU (ex-*Zulu*)	Stephen, Govan	2.5.85	Extant 1995
332	WILHELMUS ZAKARIAS YOHANNES (ex-*Gurkha*)	Thornycroft	16.10.85	Extant 1995
333	HASANUDDIN (ex-*Tartar*)	HM Dockyard Devonport	3.4.86	Extant 1995

These ships are the only survivors of the class of seven general purpose frigates built for the RN and fitted with COSAG type machinery. Purchased by Indonesia on 16 April 1984 and refitted by Vosper Thornycroft before transfer. Variable depth sonar of Type 199 removed from *Gurkha* before transfer.

Wilhelmus Zakarias Johannes Mar 1989 RAN

HSAR DEWANTARA *training frigate*

Particulars As Iraqi *Ibn Khaldoum*

No	Name	Builder	Launched	Fate
364	HSAR DEWANTARA	Titograd, Yugoslavia	11.10.80	Extant 1995

Ordered 14 March 1978 from Yugoslavia, where the hull (laid down 11 May 1979) was built and engines fitted. Armament and electronics were fitted in the Netherlands and Indonesia. She was completed 20 August 1981. In peacetime used for training and troop transport.

SUBMARINES

Ex-Soviet 'WHISKEY' class *submarines* (Project 613)

Class (no):
Tjakra, *Trisula* (402), *Nagabanda* (403), *Nagarangsang* (404), *Hendradjala* (405), *Alugoro, Nanggala, Tjandrasa* (408), *Widjajadanu* (409), *Pasopati* (410), *Tjundamani, Bramastra* (412); two without Indonesian names.

Nanggala and *Tjakra* (bought from Poland) were transferred in 1959, the former being overhauled at Surabaya, and others followed. The four additional submarines which arrived in Indonesia on 28 June 1962 brought the total numbers of this class transferred to fourteen units. Some were to be used only for spare parts and others kept in reserve. *Nanggala* and *Tjakra* were deleted from the list in 1972; *Alugoro, Hendradjala, Nagarangsang, Tjandrasa, Tjjundamani, Trisula, Widjajadanu* in 1974, *Naga Banda* in 1976. *Bramastra* stricken 1981. *Pasopati*, the sole named survivor, had new British batteries but was stricken in November 1990.

German Type 209 *submarines*

No	Name	Builder	Launched	Fate
401	CAKRA	Howaldtswerke, Kiel	10.9.80	Extant 1995
402	NANGGALA (ex-*Candrasa*)	Howaldtswerke, Kiel	10.9.80	Extant 1995

German export boats ordered on 2 April 1977, and laid down 25 November 1977 and March 1978 respectively. They were completed 18 March 1981 and 6 July 1981. There were plans to build two more boats of the 209/1300 type.

AMPHIBIOUS WARFARE VESSELS

Ex-US LST 511 class *landing ships*

Class (no, former names):
Teluk Bajur (502, ex-*LST 616*), *Teluk Bone* (511, ex-*Iredell County, LST 839*), *Teluk Kau* (504, ex-*LST 562*), *Teluk Langsa* (501, ex-*LST 1128*), *Teluk Manado* (505, ex-*LST 657*) *Teluk Ratai* (509, ex-*Inagua Shipper*, ex-*LST*...), *Teluk Saleh* (510, ex-*Clarke County, LST 601*), *Teluk Tomini* (508, ex-*Bledsoe County, LST 356*), *Tandjung Nusanie* (ex-*Lawrence County, LST 887*), *Tandjung Radja* (ex-*LST 1090*), *Teluk Mentawi* (595, ex-*LST*...)

Former US LST type landing ships transferred in March 1960 (505), June 1961 (502, 510, 511) and July 1970 (504 and 501). Built in 1943. 'Teluk' means 'bay'. *Tandjung Nusanie* stricken in 1974. Some armed with ex-Soviet 37mm guns. *Tandjung Radja* stricken 1966. Nos 502 and 508 have been in Military Sealift Command since 1978. *Teluk Manado* stricken 1983. Remainder due for scrapping in 1995.

Japanese type LST *landing ship*

Class:
Teluk Amboina, no 869 (changed to 503)

Built in Sasebo, Japan, launched 17 March 1961 and transferred to Indonesian Navy in June 1961. A copy of US LST 511 class. Transferred to Military Sealift Command. Extant 1995.

TELUK SEMANGKA class *landing ships*

Displacement: 1800t; 3770t full load
Dimensions: 328ft 1in × 50ft 6in × 13ft 9in, 9ft 10in mean
100.0m × 15.4m × 4.2m, 3.0m
Machinery: 2 shafts, 2 diesels, 6860bhp = 15kts. Range 7500nm at 13kts
Armament: 2–40mm, 2–20mm
Complement: 13 officers, 104 men + 202 troops

Class (no, completed):
Teluk Semangka (512, 20.1.81), *Teluk Penyu* (513, 20.1.81), *Teluk Mandar* (514, July 1981), *Teluk Sampit* (515, 1981), *Teluk Banten* (516, May 1982), *Teluk Ende* (517, 2.9.82)

All built by Korea Tacoma, Masan. Four ships ordered in 1979 and two more helicopter-capable in 1981. Project based on the US 'LST 542' class design. Beaching load is 690t (up to 17 tanks). Maximum cargo capacity is 1800t (incl 17 main battle tanks). First four ships have helicopter landing pads amidships and aft. The last two ships in the series include a larger hangar built into the superstructure and helicopter deck is one level above the main deck. All ships are fitted with davits for two LCVP landing craft. 516 and 517 have increased command facilities and can act as flagships and can carry three big AS 332 Super Puma helicopters. All extant 1995.

Ex-GDR 'FROSCH I' class *medium landing ships* (Project 108)

No	Name	Acquired	Fate
531	TELUK GILIMANUK (ex-*Hoyerswerda*)	1993	Extant 1995
532	TELUK CELUKAN BAWANG (ex-*Hagenow*)	1993	Extant 1995
533	TELUK CENDRAWASIH (ex-*Frankfurt/Oder*)	1993	Extant 1995
534	TELUK BERAU (ex-*Eberswalde-Finow*)	1994	Extant 1995
535	TELUK PELENG (ex-*Lübben*)	1994	Extant 1995
536	TELUK SIBOLGA (ex-*Schwerin*)	1994	Extant 1995
537	TELUK MANDAO (ex-*Neubrandenburg*)	1994	Extant 1995
538	TELUK HADING (ex-*Cottbus*)	1994	Extant 1995
539	TELUK PARIGI (ex-*Anklam*)	1994	Extant 1995
540	TELUK LAMPUNG (ex-*Schwedt*)	1994	Extant 1995
541	TELUK JAKARTA (ex-*Eisenhüttenstadt*)	1994	Extant 1995
542	TELUK SANGKULIRANG (ex-*Grimmen*)	1994	Extant 1995

Twelve former GDR fast medium landing ships of the 'Frosch I' class built by VEB Peenewerft, Wolgast, to an indigenous East German design 1976–79. Officially transferred to Indonesia 25 August 1993. Refitted in German yards prior to transfer, where all armament was removed, to be replaced later after delivery. A very valuable addition to the Indonesian Navy's landing forces in view of the block obsolescence of the ex-US LST 511 type landing ships.

Ex-GDR 'FROSCH II' class *logistic support ships* (Project 109)

No	Name	Transferred	Fate
543	TELUK CIREBON (ex-*Nordperd*)	25.8.93	Extant 1995
544	TELUK SABANG (ex-*Südperd*)	25.8.93	Extant 1995

Two logistic support ships of the 'Frosch II' class transferred to Indonesia 25 August 1993. Built by VEB Peenewerft, Wolgast, as a development of the 'Frosch I' landing ships. Originally commissioned in the GDR Volksmarine on 3 October 1979 and 26 February 1980. Fitted with an 8t crane amidships and a bow ramp as on the 'Frosch I' class. All armament removed prior to transfer; to be rearmed in Indonesia.

Ex-Yugoslav LCT type

Class:
Teluk Katurai, Teluk Wadjo, Teluk Weda, Teluk Wori

Transferred from Yugoslavia on 1 November 1958. Nos 862, 860, 861 and 863 respectively. *Teluk Wadjo* and *Teluk Weda* in reserve in 1969. All stricken by 1974.

Ex-US LCI type *infantry landing craft*

Class (former names):
Amahai (864, ex-*Tropenvogel*, ex-*LCI 467*), *Marich* (866, ex-*Zeemeeuw*, ex-*LCI . . .*), *Piru* (868, ex-*Zeearend*, ex-*LCI 420*), *Baruna* (ex-*Vogel*, ex-*LCI 948*), *Namlea* (ex-*Stormvogel*, ex-*LCI 588*)

Former US infantry landing craft, built in 1943. Handed over by the Netherlands East Indies Government on formation of Indonesian Navy in 1950. *Baruna* and *Namlea* were rerated as pilot ship and lightship in 1961. Remaining three deleted in 1974.

Ex-US LCU type *infantry landing craft*

Class:
Dore, Amurang, Banten

Transferred from USN. Built by Korneuburg SY, Austria, in 1968. *Dore* extant 1995, other two now civilian-operated.

KUPANG class *utility landing craft*

Displacement: 200t full load
Dimensions: 99ft 5in × 32ft 1in × ?
30.2m × 9.8m × ?m
Machinery: 2 diesels, ?bhp = 11kts. Range 700nm at 11kts
Armament: Nil
Complement: 17

Class: (in service):
Kupang (3.11.78), *Dili* (27.2.79), *Nusantara* (1980)

Indonesian-built and designed small landing craft type based on US LCU 1610 class. Built in Surabaya by Naval Training Centre. All in Military Sealift Command. Extant 1995.

SMALL SURFACE COMBATANTS

Ex-Australian BATHURST class *minesweepers (corvettes)*

Class (former names):
Bateng (ex-*Ambon*, ex-*Cairns*), *Hang Tuah* (ex-*Morotai*, ex-*Ipswich*), *Pati Unus* (ex-*Tidore*, ex-*Tamworth*), *Radjawali* (ex-*Banda*, ex-*Wollongong*)

Built in Australia as ocean minesweepers and launched in 1941–42. *Hang Tuah* and *Pati Unus* transferred from Royal Netherlands Navy, 28 December 1949, *Banteng* and *Radjawali* transferred 6 April 1950. *Hang Tuah* was reported sunk by rebel aircraft off Balikpapan, East Borneo on 28 April 1958; *Banteng* and *Randjawali* arrived at Hong Kong for scrapping in April 1968; and *Pati Unus*, latterly as TS, was disposed of in 1969.

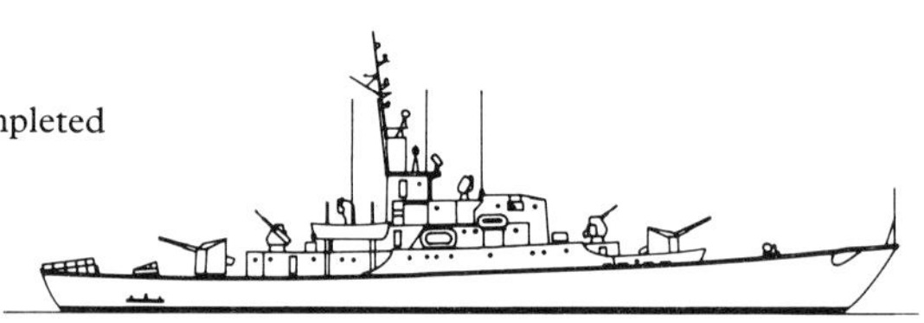

Pattimura as completed

PATTIMURA class *corvettes*

Displacement: 950t standard; 1200t full load
Dimensions: 270ft 2in oa, 246ft pp × 34ft × 9ft
82.4m × 75.0m × 10.4m × 2.7m
Machinery: 3 shafts, 3 Ansaldo-Fiat diesels, 6900bhp = 22kts. Oil 100t. Range 2400nm/18kts
Armament: 2–76mm/40 (2 × 1), 2–30mm/70 (1 × 2), 2 Hedgehogs, 4 DCT, DC rail
Complement: 110

No	Name	Builder	Launched	Fate
252	PATTIMURA	Ansaldo, Leghorn	1.7.56	Stricken 1980s
253	SULTAN HASANUDIN	Ansaldo, Leghorn	24.3.57	Stricken 1979

Italian *Albatros* class corvettes with diesel engines. Such ships were built for Denmark, one for Netherlands and three for Italy. All sub-types vary in details. Indonesian ships were ordered together with fast frigates of *Surapati* class. Rearmed with Soviet 2–85mm and 4–25mm (2 × 2), 4–14.5mm MG (2 × 2) from deleted FAC 1976–77. New Pt. Nos 371 and 372 from *c*1975 instead of 252, 253. Both laid down 8 January 1956 and completed 28 January and 8 March 1958 respectively.

Ex-GDR PARCHIM class *corvettes* (Project 133.1)

No	Name	Acquired	Fate
371	KAPITAN PATIMURA (ex-*Prenzlau*)	23.9.93	Extant 1995
372	UNTUNG SUROPATI (ex-*Ribnitz*)	23.9.93	Extant 1995
373	NUKU (ex-*Waren*)	15.12.93	Extant 1995
374	LAMBUNG MANGKURAT (ex-*Angermünde*)	1994	Extant 1995
375	CUT NYAK DIEN (ex-*Lübz*)	25.2.94	Extant 1995
376	SULTAN THAHA SYAIFUDDIN (ex-*Bad Doberan*)	25.2.94	Extant 1995
377	SUTANTO (ex-*Wismar*)	1994	Extant 1995
378	SUTEDI SENOPUTRA (ex-*Parchim*)	1994	Extant 1995
379	WIRATNO (ex-*Parleberg*)	1994	Extant 1995
380	MEMET SASTRAIRIA (ex-*Bützow*)	1994	Extant 1995
381	TJIPTADI (ex-*Bergen*)	1994	Extant 1995
382	HASAN BASRI (ex-*Güstrow*)	1994	Extant 1995
383	IMAN BONJOL (ex-*Teterow*)	1994	Extant 1995
384	PATI UNUS (ex-*Ludwigslust*)	1994	Extant 1995
385	TEUKU UMAR (ex-*Gravesmühlen*)	1994	Extant 1995
386	CUT MEUTIA (ex-*Gadebusch*)	1994	Extant 1995

Former *Volksmarine* corvettes built by VEB Peenewerft, Wolgast and originally commissioned 1981–84. Following the reunification of Germany a few remained in service with the *Bundesmarine* for a short period. Formally transferred to Indonesia on 25 August 1993. All refitted prior to transfer, with increased range and air-conditioning fitted in accommodation spaces, retaining original armament. Good sea boats with a good turn of speed, but their multi-cylinder Soviet engines have proved a constant source of trouble. Similar vessels were built in the GDR for the Soviets and still serve in the Russian Navy.

German JAGUAR type *fast attack craft (torpedo)*

Class:
Adjak, Anoa, Biruang, Harimau, Madjan Kumbang, Madjan Tutul, Serigala, Singa

'TNC 45' boats built by Lürssen, Vegesack, in 1959–60. The first four boats had wooden hulls, but the second four were built of steel. Pennant nos 601–608. *Madjan Tutul* sunk on 15 January 1962 by Netherlands warships off Borneo while trying to land infiltrators in New Guinea, her two sisters fled. *Adjak* and *Singa* deleted in 1974, *Serigala* in 1977, *Madjan Kumbang* in 1979. *Biruang* (652) and *Harimau* (654) deleted since 1982.

Ex-Soviet 'P 6' class *fast attack craft (torpedo)* (Project 183)

Class:
Angin Bohorok, Angin Badai, Angin Taufan, Angin Gending, Angin Prahara, Angin Pujuh, Angin Pasat, Angin Wambrau, Angin Brubu, Angin Tonggi, Angin Grenggong, Angin Wamandais, Angin Kumbang, Angin Ribut and ten boats without names.

Former Soviet interchangeable gun/torpedo boats. A total of 24 reported delivered since 1961, including eight in 1961 and six in 1962. Pennant numbers in the 1600 range. Probably the last ten were intended for cannibalisation to provide spare parts for operational boats. All stricken by 1975.

Ex-Soviet KOMAR class *fast attack craft (missile)* (Project 183-R)

Class (no):
Guawidjaja (612), *Hardadali, Kalamisani* (602), *Kalanada, Katjabola, Kelaplintah, Pulanggeni* (605), *Nygapasa* (611), *Surotama* (608), *Sarpamina, Sarpawasesa* (603), *Tritusta*

Transferred in three batches: six in 1961–63, four in September 1964 and two in 1965. Ten ships were in commission and two in reserve. Missiles now of probably doubtful capability. *Hardadali* sank early 1976 after hitting an underwater obstacle. *Surotama, Kalamisani, Nagapasa* stricken in 1979, *Pulanggeni* in 1974, *Katjabola, Tritusta* in 1976. Last five stricken 1979–81.

Badak 22.8.93 L & L van Ginderen

PSK Mk 5 class *fast attack craft (missile)*

Displacement:	290t full load; 250t standard
Dimensions:	175ft 10in × 26ft 2in × 5ft 2in *53.6m × 8.0m × 1.6m*
Machinery:	2-shaft CODOG (2 CP propellers): 1 gas turbine GE LM 2500, 25,000shp = 41kts, 2 diesels MTU 12V331 TC81 2240bhp = 17kts. Range 2000nm/17kts
Armament:	4 MM38 Exocet SSM (4 × 1), 1–57mm/70 Bofors Mk 1, 1–40mm/70 Bofors, 2–20mm (1 × 2)
Sensors:	Radar Decca AC 1229, HSA WM-28
Complement:	32

Class (no, completed):
Mandau (621, 20.7.79), *Rencong* (622, 20.7.79), *Badak* (623, Feb 1980), *Keris* (624, Feb 1980)

All built by Korea-Tacoma. First ship laid down May 1977. Modified US *Asheville* class FAC design. All extant 1995.

Ex-Soviet KRONSTADT class *large patrol craft* (Project 122-bis)

Class (no):
Barakuda (817), *Kakap* (816), *Katula* (811), *Landjuru, Lapai, Lumba-Lumba, Madidihang, Momare, Palu* (818), *Pandrong* (814), *Sura* (815), *Tohok* (829), *Tongkol, Tjutjut.*

Fourteen former Soviet submarine chasers. Built 1951–54 and transferred to the Indonesian Navy on 30 December 1958 (eight) and later *Landjuru, Lapai, Lumba-Lumba, Madidihang, Momare* and *Tongkol* were withdrawn from active service in 1970 and used for spares for the rest of the type. Stricken 1975 – *Landjuru, Lapai, Lumba-Lumba, Madidihang, Tongkol, Tjutjut*; 1976 – *Katula, Momare, Tohok*; 1981 – *Kakap*; 1986 – *Barakuda, Sura, Pandrong, Palu.*

Ex-US PC (173ft) type *submarine chasers*

Class (no, former name):
Hui (805, ex-*Malvern*, PC 580), *Tenggiri* (ex-*PC 1183*), *Tjakalang* (807, ex-*Pierre*, PC 1141), *Torani* (ex-*Manville*, PC 581), *Alu Alu* (ex-*PC 787*)

Former US submarine chasers of the steel-hulled PC type built in 1942–43. *Pierre* transferred in October 1958, *Malvern* and *Manville* in March 1960. *Alu Alu* was removed from the effective list in 1961. *Tenggiri* later had letters TGR painted on her bows instead of the no 309. Probably other boats had their markings also changed. *Torani* stricken in 1979, *Tenggiri* in 1974.

Ex-Yugoslav 'KRALJEVICA' Type 509 *large patrol craft*

Class (no, former name):
Lajang (819, ex-*PBR 515*), *Lemadang* (820, ex-*PBR 517*), *Krapu* (821, ex-*PBR 518*), *Dorang* (822, ex-*PBR 513*), *Todak* (823, ex-*PBR 514*), *Bubara* (ex-*PBR 516*)

Six bought and transferred on 27 December 1958. Original pennant nos 310–312, 314–316 until *c*1976. *Bubara* and *Lajang* stricken 1970, *Krapu* 1976, and *Lemadang* 1984. *Dorang* and *Todak* in reserve from 1989.

KELABANG class *large patrol craft*

Class:
Kalahitam, Kelabang, Kompas

So called 'Mawar' class submarine-chasers built in 1966–70 in Indonesia at Surabaya; 150t, 127ft 11in × 18ft 8in × 5ft 10in, *39m × 5.7m × 1.8m*, 1–40mm, 1–20mm, 21kts. Early armament: 4–12.7mm MG (2 × 2). All stricken 1981.

Lürssen 'PB 57' (Nav I and II) type *large patrol boats*

Displacement: 447t full load (428t, 801, 802)
Dimensions: 178ft 6in wl, 190ft 7in oa × 25ft × 9ft
54.4m, 58.1m × 7.62m × 2.73m
Machinery: 2 shafts, 2 MTU 16V956 TB92 diesels, 8260bhp = 27kts. Range 2200nm at 27kts, 6100nm at 15kts.
Armament: 1–57mm/70 Bofors Mk 2, 1–40mm Bofors, 2–20mm Rheinmetall (2 × 1), 2–533mm TT (2 × 1), SUT wire-guided torpedoes with 2 reloads, (801 and 802: 1–57mm Bofors Mk 2, 1–40mm Bofors, 2–20mm Rheinmetall (2 × 1), 2–7.62mm MG.
Sensors: Radar Racal Decca 2459, Signaal WM 22 (Nav I only); sonar Signaal PMS 32 (Nav I only); ESM DR 2000S3
Complement: 3 officers + 39 men (801, 802: 9 + 31)

Class (no, completed):
Singa (651, Apr 1988), *Ajak* (653, Apr 1989), *Pandbong* (801, 1990), *Sura* (802, 1991)

All four built by PT, PAL SY, Surabaya. Mid-body for the first unit was sent in 1984 from Lürssen Werft, which also supervised the whole programme. First two are in ASW version. The second two are practically identical to the units originally destined for the Maritime Security Agency and fitted for general patrol and SAR duties. All extant 1995.

Lürssen 'PB 57' (Nav III and IV) type *large patrol boats*

Displacement: 423t full load
Dimensions: 190ft 8in × 25ft × 9ft
58.1m × 7.62m × 2.73m
Machinery: 2 shafts, 2 MTU 16 V956 TB92 diesels, 8260bhp = 30.5kts (sust 28.1kts). Range 6100nm/15kts
Armament: 1–40mm Bofors, 2–7.62mm, 1 Boch 105 helicopter
Complement: 9 officers, 31 men (17 spare berths)

Class (no, in service):
Kakap (811, Oct 1988), *Kerapu* (812, 5.4.89), *Tongkol* (813, 1991), *Brevang* (814, 1992)

These ships ordered from West Germany in 1982 are SAR, transport and patrol version of the combat vessels of the 'FPB 57 Nav I' and 'II' type. All built by PT PAL SY, Surabaya with the first two shipped from Lürssen and the second pair assembled by PT PAL. Have raised helicopter deck at the stern, carry 1000 rounds of 40mm ammunition. All fitted with 70m range water cannon of 294m^3/h capacity for fire fighting. Two rescue launches are stowed aft. Originally destined for Maritime Security Agency. As transports can carry two rifle platoons. All extant 1995.

860 type *patrol boats*

Displacement: about 150t
Dimensions: 105ft
32.0m
Machinery: 2 shafts, 2 diesels, 20kts
Armament: 1–40mm Mk 3, 2–12.7mm
Complement: 22

Class:
860, 861

Indonesian design apparently based on the Australian *Attack* class patrol boat, but with lower freeboard, bigger superstructure and without bow bulwarks. Two ships were completed in 1983 by PT, PAL SY, Surabaya. Fitted with Decca navigation radar. Both extant 1995.

Ex-US PGM 39 class *patrol craft*

Class (no, former names):
Bentang Silungkang (572, ex-*PGM 55*), *Bentang Waitatiri* (571/841, ex-*PGM 56*), *Bentang Kalakuan* (570/843, ex-*PGM 57*)

Originally intended as amphibious control craft but now used for normal patrol duties. All transferred in January 1962. Built in 1958–60 by Peterson Boatbuilders. *Silungkang* stricken 1981, others 1982.

Ex-Australian ATTACK class *patrol craft*

Class (no, former name, transfer):
Sibaru (847, ex-*Bandolier*, 16.11.73), *Suliman* (848, ex-*Archer*, 1974), *Sigalu* (857, ex-*Barricade*, 22.4.82), *Silea* (858, ex-*Acute*, 6.5.83), *Siribua* (859, ex-*Bombard*, 1983), *Siada* (862, ex-*Barbette*, Feb 1985), *Sikuda* (863, ex--*Attack*, 24.5.85), *Sigurot* (864, ex-*Assail*, 30.1.86)

Transferred from RAN after refit. Light-alloy superstructure. All extant 1995.

Australian CARPENTARIA type *coastal patrol craft*

Class (no, in service):
Sabola (853, Oct 1976), *Sadarin* (855, Dec 1976), *Salmaneti* (856, Jul 1977), *Samadar* (851, Aug 1976), *Sasila* (852, Sep 1976), *Sawangi* (854, Nov 1976)

Built by de Havilland Marine, Australia in 1975–77. Grant-aided purchase. See under Burma for data. All six transferred to sea police in 1980s. All extant 1995.

MINE WARFARE VESSELS

Ex-Netherlands DJEMBER class *coastal minesweepers*

Class:
Djombangh, Djampea, Enggano, Flores

Four transferred to Indonesia in 1951. They were scuttled in 1942 at Tandjong Priok (where built) and repaired by Japanese. After the war returned and transferred to Indonesia. All stricken in 1974.

German R-BOAT type *inshore minesweepers*

Class:
Pulau Rangsang, Pulau Rau, Pulau Rindja, Pulau Raas, Pulau Rempang, Pulau Rengat, Pulau Rupat, Pulau Rusa, Pulau Roma, Pulau Roti

Built by Abeking & Rasmussen, Lemwerder in 1954–57. These boats had a framework of metal covered with wood. *Pulau Ras* and *Pulau Roti* stricken 1969; *Pulau Rau, Pulau Rempang, Pulau Rusa, Pulau Roma* stricken 1974 and *Pulau Rindja* 1975, *Pulau Rengat* 1979, *Pulau Rupat* 1981.

Ex-Soviet 'T 43' class *ocean minesweepers* (Project 254)

Class (no):
Pulau Rani (701), *Pulau Roon* (703), *Pulau Radja* (705), *Pulau Ratewo* (702), *Pulau Rorbas* (704), *Pulau Rondo* (706)

Four transferred in 1962 and two in 1964. Some had Decca 110 radar. *Pulau Roon* deleted 1979, *Pulau Rondo* 1976, *Pulau Radja* and *Pulau Rorbas* 1983, *Pulau Rani* and *Pulau Rondo* 1988. 701 and 702 reportedly still in service as patrol craft.

Ex-US BLUEBIRD class *coastal minesweepers*

Class (no, former name):
Pulau Alor (717, ex-*Meadowlark*), *Pulau Anjer* (719, ex-*Limpkin*), *Pulau Atang* (721, ex-*Frigate Bird*), *Pulau Aru* (722, ex-*Falcon*), *Pulau Aruan* (718, ex-*Jacana*), *Pulau Impalasa* (720, ex-*Humming Bird*)

Transferred from USN in 1971; all have wooden hulls with low magnetic signature. Stricken 1976.

PULAU RENGAT class *minehunters*

Displacement: 510t; 586t full load
Dimensions: 169ft × 29ft 2in × 8ft 7in
51.5m × 8.9m × 2.62m
Machinery: 2 shafts, 2 CP propellers, 2 MTU 12V 396 TCD51 diesels, 1900 bhp = 15.5kts, 2 × 75hp bow thrusters, 2 × 120hp Schottel rudder propellers = 7kts. Range 3500nm at 10kts
Armament: 2 × 20mm Rheinmetall
Complement: 45

No	Name	Builder	Launched	Fate
711	PULAU RENGAT (ex-*Willemstad*)	Van der Giessen de Noord	23.7.87	Extant 1995
712	PULAU RUPAT (ex-*Vlaardingen*)	Van der Giessen de Noord	27.8.87	Extant 1995

New ships ordered on 29 March 1985 (laid down 22 July 1985) and 30 August 1985 from the shipyard and taken from Royal Netherlands Navy production. Carry two PAP 104 Mk 5 remote-controlled submersibles. GRP construction. Sweep equipment includes magnetic sweep, acoustic gear, mechanical sweep. Minehunting system is the French Thompson-CSF IBIS V type. Construction of up to ten units in Indonesia is planned, but funds are lacking.

Ex-GDR KONDOR II class *minesweepers* (Project 89.2)

No	Name	Acquired	Fate
721	PULAU ROTE (ex-*Wolgast*)	Jul 1993	Extant 1995
722	PULAU RAAS (ex-*Hettstedt*)	Jul 1993	Extant 1995
723	PULAU ROMANG (ex-*Pritzwalk*)	Jul 1993	Extant 1995
724	PULAU RIMAU (ex-*Bitterfeld*)	Jul 1993	Extant 1995
725	PULAU RONDO (ex-*Zerbst*)	Jul 1993	Extant 1995
726	PULAU RUSO (ex-*Oranienburg*)	Jul 1993	Extant 1995
727	PULAU RANGSANG (ex-*Jüterbog*)	Jul 1993	Extant 1995
728	PULAU RAIBU (ex-*Sömmerda*)	Jul 1993	Extant 1995
729	PULAU REMPANG (ex-*Grimma*)	Jul 1993	Extant 1995

Nine former *Volksmarine* minesweepers built by VEB Peenewerft, Wolgast and originally commissioned 1971–73. Laid up after German reunification, they were acquired by Indonesia in 1993 and shipped there in the Russian dockship *Trans-shelf*. Armament was removed before transfer. They were rearmed in Indonesia and are to receive new sonars and additional minesweeping equipment, although their MCM role is to take second place to EEZ patrol duties.

MISCELLANEOUS

Two Fairey Marine 'Spear' class patrolcraft bought in 1973–74.

One Boeing Jetfoil hydrofoil, *Bima Samudera I*, launched 22 October 1981. Arrived Indonesia 1982 for naval/civil evaluation.

The Customs Service is now undergoing an expansion and many patrol boats are being constructed at home and western European shipyards. Many of these ships are of typically naval construction, and include thirty-five Lürssen 'FPB 28' class patrol boats (68.5t full load, armament 1–12.7mm), built in the 1980s to replace older units of similar characteristics. The Customs Service also has seven 'BC 2001' class (built CMN, Cherbourg, 1980–81, 70.3t full load, 1–12.7mm); seven 'BC 3001' class (built Chantiers Navals de l'Esterel, Cannes, 1979–81, 71t full load, 1–12.7mm) and three 'BTC 1001' class (built l'Esterel 1975, 56t full load, 1–12.7mm). There are still about twenty 'BC 401' class patrol boats built by Lürssen in 1960–62 of very similar characteristics to the new FPB 28 series.

The Maritime Police has under its command numerous seagoing patrol boats with potential for naval use:

Ten 'DKN 504' class, built Japan 1963–64 (444t, 15.3kts, 1–20mm, 2–12.7mm)

Nine 'DKN 908' class, built Italy 1961–64 (159t, 21 kts, 3–20mm)

Iran

Iran's navy is the largest in the Persian Gulf area and possesses considerable missile capability. From the early 1960s the Imperial Iranian Navy was trained in the West, mainly by Great Britain and the USA. Iran also had the advantage of being a member of CENTO, and took part in the MIDLINK series of exercises with the US, Pakistan and Royal navies. Senior officers therefore had a thorough grounding in sound naval practice and, although a number were lost as a result of the 1979 revolution, a reasonable reservoir of expertise still remains.

Even before the revolution, however, it was reported that the general standard of ship and equipment maintenance was well below that of Western navies, and this situation worsened as a result of the departure of Western advisers, disruption of naval administration and difficulty in obtaining spares. The most effective units are the 'Combattante II' design (*Kaman* class) fast attack craft with the US-supplied Harpoon missile followed by the *Saam* class frigates fitted with the Italian Sea Killer missile.

There is no reliable information about losses in the eight-year war between Iran and Iraq apart from Iraqi reports that two frigates had been sunk by them on 22 September 1980. At the same time Iraq claimed that fifteen vessels, described as patrol boats and gunboats, were reported sunk by Iraq on 20–22 September. Main bases in the Persian Gulf are at Bandar Abbas on the Strait of Hormuz, Bushire, Kharg Island and Khorramshahr; in the Indian Ocean at Chah Bahar and in the Caspian Sea at Bandar. There were 18,000 officers and men (two-year conscripts) serving in the navy in the early 1990s.

Currently the naval personnel is estimated at 16,000 officers and ratings, a substantial decrease from the earlier levels. There is a tendency to underestimate Iranian strength after the protracted eight-year war with Iraq, but the country boasts a 3 per cent population growth rate and the total population is likely to top 100 million early in the next century (in 1992 the population stood at 61.2 million); Iran clearly has the human and natural resources to play a significant role as a regional power. Some elements in Pakistan, for example, are looking to Iran as a counterweight to India. Iranian naval forces have diminished since the Shah's era but remain important. During the past few years they have held exercises in the open sea, and there is also a considerable repair capacity in Iran, as was clearly shown by the repairs effected in 1989–90 to the heavily damaged light frigate *Sabalan*.

Generally old warships inherited from the Shah's navy are still in service. Few fighting ships have been acquired during the past decade, except midget submarines, Chinese fast attack craft, two landing ships, and a replenishment oiler from the United Kingdom.

Iran remains hostile to the West, and now is acquiring modern diesel-electric submarines. Soviet Project 877 'Kilo' class submarines were selected and crews are being trained in the Baltic. This represents a dangerous boost in weapons proliferation. The submarines would be based outside the Gulf at Chah Bahar. However, submarine expertise is not acquired quickly, and it is possible that foreign experts (Russian or Chinese) will be employed. Iran is suffering from a Western embargo on arms sales, and has been forced to arm some of its ships with Chinese SSM, as old stocks of the Harpoon SSM were exhausted. The same applies to other types of armament and equipment (AA guns from Russia, anchor mines from North Korea, etc).

Despite being a big oil producer, Iran has little cash to rebuild its rundown navy. According to the published 1990 Iranian estimates, rebuilding of the army after the lengthy conflict with Iraq will cost $15–25bn; the navy has much lower priority than the army and in time may receive one third or less of the funds assigned to the army.

FLEET STRENGTH 1947

PATROL CRAFT

Name	Launched	Disp	Fate
CHAROGH (ex-*Nilam*)	26.7.31	331t	BU *c*1953
CHANBAAZ (ex-*Hira*)	12.9.31	331t	BU *c*1953
SIMORGH (ex-*Lal*)	3.8.31	331t	BU *c*1953

Charogh and *Simorgh* built by CNR, Palermo. *Chanbaaz* built by OCP, Naples.

MOTOR LAUNCHES

Name	Launched	Disp	Fate
BABOLSAR	1935	*c*30t	Stricken *c*1972
GORGAN	1935	*c*30t	Stricken *c*1972
SEF INDREUDE	1935	*c*30t	Stricken *c*1972

These were stationed in the Caspian Sea armed with 1–47mm Skoda and 1 MG. Built by CNR, Palermo.

MAJOR SURFACE SHIPS

Artemiz 1975

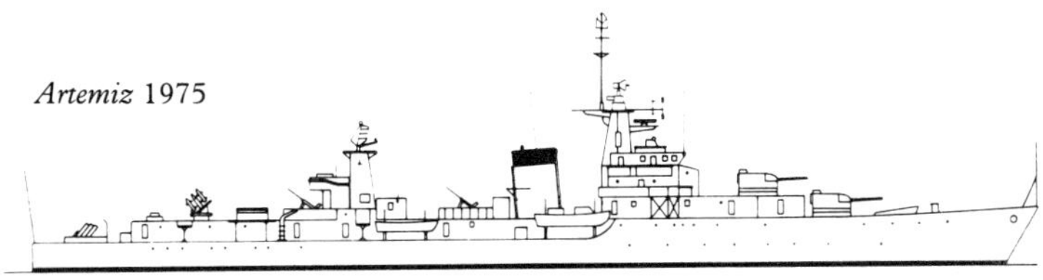

Ex-British 'BATTLE' class *destroyer*

Armament: 4–114mm/45 QF Mk 4 (2×2), 4–40mm/60 Bofors (4×1), 1 Seacat SAM launcher (1×4; est 16 missiles), 1 Squid ASW mortar (1×3). See notes

Complement: 270

Other particulars: As British 'Battle' class (1st group)

No	Name	Builder	Acquired	Fate
D 5	ARTEMIZ (ex-*Sluys*)	Cammell Laird	26.1.65	Extant 1995

The former RN destroyer HMS *Sluys* (D 60) was transferred to Iran at Southampton and handed over to the Iranian Navy following a three-year modernisation by Vosper Thornycroft. Later given pennant no 51. Often referred to as having 8–40mm in early 1970s. Four Standard SSM launchers (eight missiles) added after refit at Cape Town, South Africa in 1975–76. Now has only two–40mm (2×1) aft. Plessey AWS-1 air surveillance radar with on-mounted IFF, two Contraves Sea Hunter fire control radars, Plessey MS-26 sonar and Decca RDL-1 ESM.

The ship was renamed *Damavand* in 1985 and had her Sea Cat missile launcher removed, at the same time being rearmed with two twin Soviet 23mm/80 AA guns. *Damavand* was still regularly seen at sea in 1989 and is credited with only 23kts maximum speed.

Artemiz May 1969 Vosper Thornycroft

Ex-US ALLEN M SUMNER class *destroyers*

No	Name	Builder	Acquired	Fate
61	BABR (ex-*Zellars*)	Todd Pacific Shipyards	19.3.71	Extant 1995
62	PALANG (ex-*Stormes*)	Todd Pacific Shipyards	16.2.72	Extant 1995

FRAM II conversions at Norfolk, Virginia in FY 61. *Stormes* replaced the outdated *Gainard* (stricken from USN 1970). However, they are far from being sister ships. For instance, the main deckhouse on *Babr* ('Tiger') was widened during FRAM refit as far forward as amidships, but only half as far forward on *Palang* ('Leopard'). Both ships underwent a full refit as well as conversion at Philadelphia N Yd before sailing in mid-1974 for Iran. The air-conditioning was improved, replacing two 100kW diesel generators by two 500kW units, B gun-mount was removed with its hoists and magazines (which became store rooms since these ships will operate away from suppliers of spare parts), accommodation was altered, a Canadian telescopic hangar similar to the US *Knox* class frigates (AB-204 ASW helicopter), was fitted, the four Standard SSM launchers (eight missiles, which can be used as SAM) were sited on a platform between stacks, VDS was rigged on the 01 level forward of the bridge, and two Hedgehogs were fitted in B position. *Babr* carried an SQS-43 sonar system that included an SQS-31B hull-mounted sonar and SQA-10 VDS. *Palang*'s VDS was removed to provide spares for *Babr* which has now lost hers. New torpedo stowage was mounted between the funnels (to make room in the hangar) and the landing platform was widened out to the ship's sides while the old hangar was removed and scrapped.

As *Babr* had the small ECM house and mast atop the hangar and *Palang* had the large one, *Palang*'s house was transferred to *Babr* as she was further advanced and due for completion first; a new house and mast was then built for *Palang*. These old ships with their additional SSM/SAM and ASW armament were more effectively armed than many warships currently in the USN.

On both ships VDS and Hedgehogs were removed and one twin Soviet 23mm/80 AA gun mounting installed aft in place of the VDS. Both ships were active in 1992 and while at sea had the second digit of the pennant numbers painted out to make identification more difficult.

Ex-US DD 713 *Kenneth D Bailey* and DD 881 *Bordelon* (both *Gearing* class) were purchased 13 January 1975 and 1 February 1977 for spares.

US Modified SPRUANCE class (DD 993 class) *destroyers*

Class:
Kouroush, *Daryush*, *Ardeshir*, *Nader*, *Shapour*, Anoushirvan

The original order for these superb AA ships placed with the US Navy by the Imperial Iranian Government in 1974 was for six. They were assigned the US Navy hull numbers DD 993–998 for accounting purposes; Iranian pennant nos 11–16 respectively and named after emperors. Two (DD 995 and DD 997) were cancelled in June 1976 before the order from Ingalls Shipbuilding Corp for the remaining four was issued on 23 April 1978. On the same day DD 996 and DD 998 were reclassified to DD 995 and DD 996. The Iranian Navy planned to classify them as cruisers. During the Iranian Revolution DD 995 and DD 996 were cancelled on 3 February 1979 (French sources give DD 993 and DD 994). On 31 March 1979 the two remaining ships DD 993 and DD 994 were cancelled (French sources give DD 995 and DD 996). The four were officially acquired by the US Navy on 25 July 1979 with the signing of the FY79 supplementary budget request. On 8 August 1979 DD 993–996 were reclassified DDG 993–996. These ships are now the USN *Kidd* class. US names: *Kidd* (ex-*Kouroush*), *Callaghan* (ex-*Daryush*), *Scott* (ex-*Nader*) and *Chandler* (ex-*Anoushirvan*). For full details see under United States.

Ex-British LOCH class *frigate*

Babr (ex-*Derby Haven*, ex-*Loch Assynt*) completed as depot ship, acquired in July 1949. Served as depot ship for landing craft. No pennant no. Armed with 2–102mm (2×1). Paid off 30 October 1969 and stricken in 1972.

Nagdi Jul 1994 L & L van Ginderen

Nagdi as completed

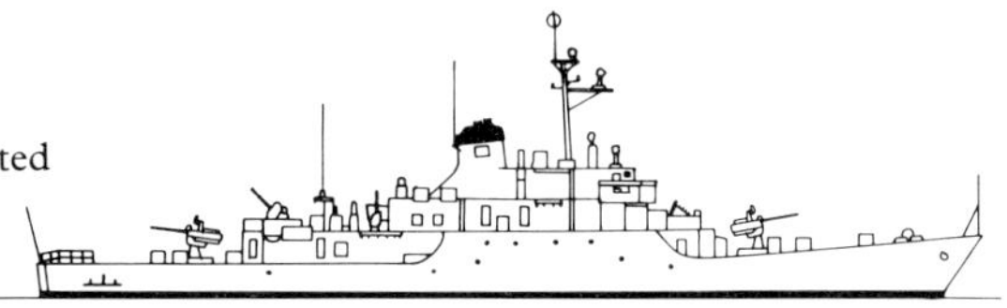

BAYANDOR class (US PF 103 class) *frigates*

Displacement: 900t standard; 1135t full load
Dimensions: 299ft 2in × 32ft 11in × 10ft
83.8m × 10.0m × 3.0m
Machinery: 2 shafts, 4 Fairbanks-Morse 38D81/8 diesels, 6000bhp = 20kts. Oil 110t. Range 3000nm at 15kts
Armament: 2–76mm/50 DP Mk34 (2×1), 2–40mm/60 (1×2), 1 Hedgehog, 4 DCT, 2 DC racks, 60 DCs. See notes
Sensors: Radar SPS-6, Raytheon, SPG-34; sonar SQS-17
Complement: 133–140

No	Name	Builder	Launched	Fate
F 25	BAYANDOR (ex-*PF 103*)	Levingston SB	7.7.63	Extant 1995
F 26	NAGHDI (ex-*PF 104*)	Levingston SB	10.10.63	Extant 1995
F 27	MILANIAN (ex-*PF 105*)	Levingston SB	4.1.68	Sunk 1982
F 28	KAHNAMUIE (ex-*PF 106*)	Levingston SB	4.4.68	Sunk 1982

Laid down and completion dates: *Bayandor* (20.8.62, 15.5.64); *Naghdi* (12.9.62, 22.7.64); *Milanian* (1.5.67, 13.2.69); *Kahnamuie* (12.6.67, 13.2.69).

Transferred from the USA to Iran under Mutual Assistance Programme (MAP). Built in two phases, five years apart. Design similar to the Italian-built *Pattimura* class built for Indonesia. Sometimes classified as corvettes. Named after naval officers killed in action with the British in 1941. Pennant nos changed to 81–84 in 1976.

Twin Soviet 23mm AA were bought and have been added forward of the bridge in place of the single Hedgehog ASW mortar. Now reported to have 4–23mm AA (2×2) instead of 2–23mm AA (1×2). Mid-life conversion planned (but now probably cancelled) to include 76mm DP OTO Melara guns. Two lost in the war with Iraq in 1982. Two sister ships; see under Thailand.

In 1988 *Naghdi* completed the reconstruction of accommodation and change of main engines. The Soviet twin 23mm/80 AA guns and depth charge racks were replaced by two 20mm Oerlikon GAM BO1 guns in 1990.

Saam 1972

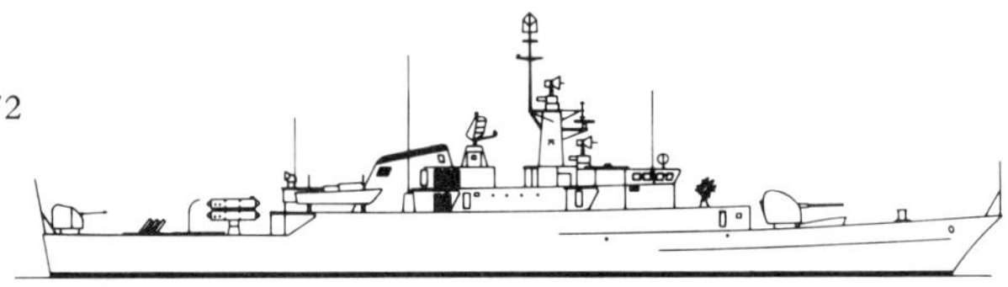

SAAM class (Vosper Thornycroft Mk 5 class) *frigates*

Displacement: 1250t standard; 1540t full load
Dimensions: 290ft pp, 310ft oa × 34ft 5in × 10ft 6in–11ft 6in
88.4m, 94.5m × 10.5m × 3.2m–3.5m
Machinery: 2 shaft CODOG: 2 Rolls-Royce Olympus TM 3A gas turbines, 46,000shp = 40kts, plus 2 Paxman 16cyl Ventura diesels for cruising, 3800bhp = 17.5kts. Oil 150t (overload 250t). Range 5000nm at 15kts
Armament: 1 Sea Killer Mk 2 SSM launcher (1×5), 1 Seacat SAM launcher (1×3; est 9 missiles), 1–114mm/45 DP Mk5 (1–114mm/55 DP Mk 8 in *Rostam* and *Faramarz*), 2–35mm/90 Oerlikon-Bührle GDM-A (1×2), 1 Limbo Mk10 ASW mortar (1×3). See notes
Sensors: Radar Plessey AWS-1, 2 Contraves Sea Hunter 4; sonars Types 170 and 174
Complement: 125–135 (room for 146)

No	Name	Builder	Launched	Fate
DE 12	SAAM	Vosper Thornycroft	25.7.68	Extant 1995
DE 14	ZAAL	Vickers, Barrow	4.3.69	Extant 1995
DE 16	ROSTAM	Vickers, Newcastle	4.3.69	Extant 1995
DE 18	FARAMARZ	Vosper Thornycroft	30.7.69	Sunk 19.4.88

Laid down and completion dates: *Saam* (22.5.67, 20.5.71); *Zaal* (3.3.68, 1.3.71); *Rostam* (10.12.67, 26.5.72); *Faramarz* (25.7.68, 28.2.72).

Since 1977 new pennant nos 71–74 respectively. It was announced on 25 August 1966 that Vosper Ltd, Portsmouth, received an Iranian order for four 'destroyers'. Three factors made the design possible. The first was the development of the Olympus marine gas turbine with a rating in excess of 22,500hp – a great advance on the AEI G6 turbine (7500shp) fitted in the British 'County' and 'Tribal' classes. This made it possible for the gas turbine to become the main source of power instead of merely a boost turbine used in conjunction with conventional steam turbines. Diesel engines for cruising could therefore be added in a CODOG arrangement which considerably simplified gearing. The second factor was the advent of a new generation of smaller and lighter but more powerful weapons systems that did not require such a large ship to carry them. The third factor was the reduced crew needed not only thanks to gas turbines and the higher degree of weapon automation but also to the

Zaal May 1977
C & S Taylor

increased reliability accruing from each of these developments and the consequent reduction in maintenance. Unit replacement methods associated with both gas turbines and modern electronics would ensure faster repairs than would have previously been possible.

As the Mk 5 'destroyer' project which gave birth to the *Saam* class was a private design, the Vosper designers were able to reduce weight and cost – which at £6.5m was about two-thirds the price of similarly equipped vessels built to the standards demanded by Western navies – by a careful consideration of how far these standards were applicable to the role envisaged for these particular ships. They considered that the standards of shock-resistance laid down by the Admiralty were unnecessarily high, being based on the high probability of near-misses for an older, and less accurate, generation of weapons. Savings were also made by lowering the standards of engine mountings considered essential to reduce underwater noise in a vessel whose primary function was ASW. As a defence against submarine attack the class was to rely on high speed and the manoeuvrability bestowed on them by their KaMeWa CP propellers. *Rostam* completed at Barrow. Ships named after heroes of the Shah Nameh, the national epic. Air-conditioned throughout. Retractable Vosper fin stabilisers. *Saam* and *Zaal* were at HM Dockyard Devonport July–August 1975; major refit including replacement of older Mk 5 114mm gun by Mk 8 took place. Completed May 1977. Hull-mounted sonar at high speeds retracts into the hull and is covered by doors to protect it against excessive pressures. All weapons and sensors are integrated in a computerised operations room, and all machinery is operated from a similar centralised control room with full instrumentation. Sometimes credited with different range figures 550nm at 36kts (on gas turbines) or 3200nm at 18kts (on diesels); with additional fuel 800 and 4100nm respectively.

All four ships were renamed in 1985 as follows: *Alvand* (ex-*Saam*), *Alborz* (ex-*Zaal*), *Sabalan* (ex-*Rostam*) and *Sahand* (ex-*Faramarz*). *Sahand* was sunk on 19 April 1988 by three Harpoons from USS *Joseph Strauss* and cluster bombs from aircraft based on USS *Enterprise*. At the same time *Sabalan* was badly damaged, with her back broken by a laser guided bomb, but was repaired in 1989–90 and became operational in 1991. In 1988 the Sea Cat missiles were replaced by twin Soviet 23mm/80 guns, and these in turn together with both boats had been replaced by three 20mm Oerlikon GAM BO1 guns in 1990.

NATO Standard type *guided missile frigates*

Displacement: 2900–3000 (?3500)t standard; 3750–3800t full load
Dimensions: 419ft 11in–427 2in × 47ft 3in–47ft 11in × 14ft 5in–15ft 1in (19ft–19ft 8in props)
128.0–130.2m × 14.4–14.6m × 4.4m–4.6m (5.8–6.0m)
Machinery: 2-shaft CODOG (CP): 2 General Electric LM2500 gas turbines, 50,000shp = 30kts, plus 2 MTU 20V956 TB92 diesels = 10,400bhp = *c*18–20kts. Range 4000nm at 18–20kts
Armament: 8 Harpoon SSM (2×4), 1 Mk 13 Mod 4 missile launcher (total *c*42 Harpoon SSM and Standard SM1–MR) or 1 Sea Sparrow SAM Mk 29 launcher (1×8) instead, 1–127mm/54 DP Mk 45, 2–20mm Vulcan-Phalanx Mk 15 CIWS (2×6), 4–324mm Mk 32 ASW TT (2×2) fixed for Mk 46 torpedoes,
2 helicopters (considered SH-60B LAMPS III or WG13 Sea Lynx or Puma
Complement: 189

Four units (to be built at West German yards with Bremer-Vulkan as prime contractor), eight units (to be built at Holland, Rhine-Schelde-Verolme).

These ships were ordered in 1978; after the successful trials of the Netherlands *Kortenaer* and the contract for the German Type 122. Although Iran specified that the basic design for the new class should be the same as the NATO Standard Frigate, compatibility was needed with other US systems within the Fleet. This would reduce costs, simplify training and speed up commissioning. Apart from one or two alterations, hull layout was to remain as *Kortenaer* class. The propulsion system was to be the same as in the German Type 122 (*Bremen* class).

The main operational requirement for Iran was to possess vessels with an exceptionally high standard of accommodation for service in the Persian Gulf. With the need to operate over an extended radius of action, far from supporting bases and out of the range of aerial cover, it was essential that the Iranian frigates carried a strong defensive armament backed up by a comprehensive EW system.

The main fire control for the 127mm gun, Harpoon and Standard SM1–MR were to be two sets of Lockheed Mk 86 systems (a total of two SPG-60 and once SPQ-9 radars). Long range early warning would be provided by an SPS-48C three-dimensional radar. The other EW system was to consist of an SLQ-32 (V2 or possibly V3). Sonars were to consist of a bow-mounted 80 set with the German VDS set in a well aft. The ships were also to be equipped with NTDS, Tacan, navigational radar, chaff system and Link 11, 14 and 4A.

The contract was cancelled as a result of the Iranian Revolution of 1979.

SUBMARINES

West German Type 209 *submarines*

Six West German (export) Type 209 submarines (1400t) had been ordered on March 1978 from Howaldtswerke-Deutsche Werft (HDW), Kiel. These were cancelled early in 1979, by which time considerable progress had probably been made in the West German yard. First planned to be launched in April 1981, and all planned for delivery at the end of 1983.

Ex-US TANG class *submarines*

Agreement on the transfer of three ex-*Tang* class submarines (launched 1951) from USN reached in 1975 to provide training for the establishment of a larger submarine force. At the time of the revolution in Iran these three ships were earmarked for transfer. *Kousseh* (Iranian pennant no 101 or S 101; ex-US SS 566 *Trout*) had been transferred after extensive renovation on 19 December 1978 at New London. Her acquisition was officially cancelled on 3 February 1979. She lay at New London until late 1979, with a token crew, when she was towed to Philadelphia and laid up in reserve (handed back to the USN May 1979). The vessel is reported to be for scrap in 1992 following an Iranian announcement that it has dropped all claims to military hardware which remains undelivered.

The transfers of the other two, US SS 563 *Tang* as Iranian *Dolfin* (some sources incorrectly refer to her as SS 567 *Gudgeon*) and SS 565 *Wahoo* as Iranian *Nahang*, were cancelled on 3 February 1979 (to Turkey as *Piri Reis* 1980) and 1979 respectively. *Wahoo* was half-way through a two year overhaul when Iran cancelled the project on 31 March 1979. Still in a dismantled condition, she was finally decommissioned on 31 March 1980 and laid up in reserve.

Ex-Russian 'KILO' class *submarines* (Project 877EKM)

Class (no, in service):
Tareq (901, 21.11.92), *Noor* (902, 6.6.93), *I* (904, 1994)

It was officially announced in 1990 that the Iranian Navy was seeking to acquire fully-fledged diesel-electric submarines and that Iranian sailors were being trained at Riga. Three submarines of the 'Kilo' class (Project 877 EKM) were then acquired from the Admiralty Yard at St Petersburg, being launched in 1991, 1992 and 1993 respectively. The first boat was transferred to Iran under Russian colours with a largely Russian crew. They are based at Chah Bahar on the northern shore of the Gulf of Oman. For details see under Soviet Union (and successors). These are very capable vessels which will have a considerable effect on the balance of military power in the region, and they pose a new potential threat to commercial shipping in the Gulf.

Midget submarines

The Iranian Navy has also shown interest in midget submarines, with one submarine assembled in Bandar Abbas and completed in 1987. It proved unsuccessful and a second of different design was purchased in North Korea and delivered in 1988. In 1992 the possibility of buying submarines from Pakistan was considered.

It has been reported that a total of nine midget submarines had been imported from North Korea by 1993, and that twenty-four such vessels were planned. The North Korean boats have a reported displacement of 76t surfaced/90t dived with a speed of 12kts/8kts. There is a 'wet and dry' compartment for divers.

AMPHIBIOUS WARFARE VESSELS

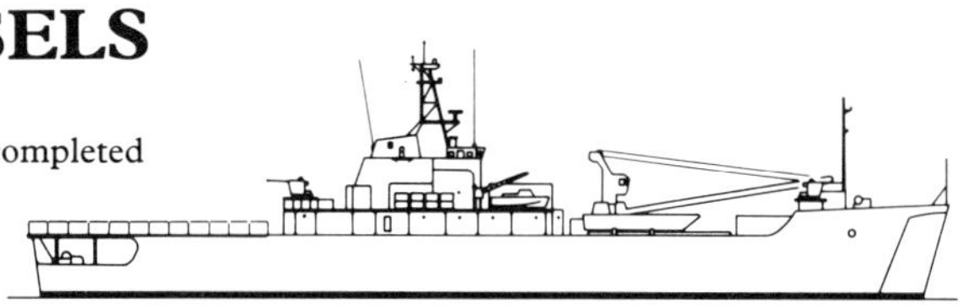

Hengam as completed

HENGAM class *tank landing ships*

Displacement: 2540t full load
Dimensions: 284ft 9in wl, 305ft 1in oa × 49ft × 7ft 3in
86.8m, 93.0m × 14.9m × 2.2m
Machinery: 2 shafts, 2 CP propellers, 4 Paxman Ventura 12CM diesels, 5600bhp = 14.5kts or 4 MTU 16V 652 TB81 diesels, 5800bhp = 14.5kts (second pair). Range 4000nm at 12kts
Armament: 4–40mm/60 Bofors (4×1) or 8–23mm/80 Soviet mountings, (4×2) for *Tonb* and *Lavan*, 1 BM-21 multiple RL, 2 British Mk 5 rocket flare launchers, 1 helicopter
Sensors: Radar Decca 1229
Complement: 80 + 227 troops

No	Name	Builder	Launched	Fate
511	HENGAM	Yarrow, Clyde	27.9.73	Extant 1995
512	LARAK	Yarrow, Clyde	7.5.74	Extant 1995
513	LAVAN	Yarrow, Clyde	27.2.79	Extant 1995
514	TONB	Yarrow, Clyde	6.12.79	Extant 1995

Lack the through tank deck of the larger British *Sir Lancelot* class. Flight deck for one helicopter aft. Carry up to nine tanks depending on size (one Chieftain abreast or two T54/55). Cargo capacity of 700t includes 300t of vehicle fuel. Vehicle deck is 140ft × 30ft 9in (42.7m × 9.4m) and can accommodate according to some sources twelve Soviet T55 or six British Chieftain tanks. Named after Persian Gulf islands. An upper deck 10t crane can handle two Uniflote cargo lighters and twelve Z-boat LCPs. In the pure logistics role 10–20t or 30–10t containers would be loaded.

First two ordered 25 July 1972 at Yarrow (Shipbuilders) Ltd, Clyde; four more ordered 20 July 1977. *Hengam* (original no 51) laid down late 1972 and completed 12 August 1974. *Larak* laid down 1973 and completed 12 November 1974. *Lavan* laid down 6 February 1978. Some payments made over, construction continuing slowly. Yarrow had already ordered the material for the last two ships by the time of their cancellation in March 1979. *Tonb* was completed for trials in October 1984 followed shortly by *Lavan*. Both were released in 1985 as 'hospital ships' without any armament. *Tonb* and *Lavan* commissioned on 21 February 1985 and 16 January 1985.

Two 36001 LST are reportedly under construction at Bushire for completion during 1995–96.

IRAN AJR class *tank landing ships*

Displacement: 2274t full load
Dimensions: 176ft 2in oa × 35ft 5in × 9ft 10in
53.7m × 10.8m × 3.0m
Machinery: 2 shafts, 2 diesels, 2200bhp = 12.5kts
Armament: 1–12.7mm MG
Complement: 30

Class:
Iran Ajr (ex-*Arya Rakhsh*), *Iran Asir* (ex-*Arya Akian*), *Iran Ghaydr* (ex-*Arya Dokht*), *Iran Bahr* (ex-*Arya Sahand*), *Iran Badr* (ex-*Arya Boum*)

Five roll-on/roll-off landing craft built by Teraoka, Japan, in 1978–79 and acquired by the Iranian Navy in 1980 for use primarily as minelayers. *Iran Bahr* and *Iran Badr* sunk by Iraq in 1980 while *Iran Ajr*, loaded with moored mines, was captured in September 1987 by US Navy and after examination was scuttled on 26 September 1987. *Iran Asir* and *Iran Ghaydr* extant 1995.

Ex-US LSIL type *landing ships*

Class (former name):
Hengam (ex-French *LSIL 9037*, ex-US *LSIL 768*), *Larak* (ex-US *LSIL 710*)

First transferred from USN to Iran in 1957 under MDAP (during 1953–57 in French navy). Second was loaned to Iran December 1958. Pennant nos 41 and 42 respectively. Fates uncertain – *Hengam* probably stricken early 1970s, *Larak* mid-1970s.

Ex-US LCU type *landing craft*

Queshm (or *Quesham*, some sources give *Gheshme* or *Gheshne*; ex-US *LCU 1431*) was transferred to Iran by USA September 1964 under MAP. Pennant no probably LCU 47, now 501. French references report that she was sunk 1978 as a Harpoon missile target, but British and German sources reported her still extant in the early 1980s. Now stricken. For further details see under United States.

CIVILIAN LANDING SHIPS USED BY THE MILITARY

Three *Iran Hormuz 24* class, named *Iran Hormuz 24*, *25* and *26*, built South Korea 1985–86; fitted with large bow doors (2014t full load, 240ft oa, 2400bhp = 12kts), extant 1995.

Three *Iran Hormuz 21* class, named *Iran Hormuz 21*, *22* and *23*, built Netherlands 1984–85, fitted with bow door and ramp (1400t, 213ft oa, 730bhp = 9kts), extant 1995.

One 'Foque' class LSL, named *Foque 1*, launched Bandar Abbas 1988 (250t full load, 125ft oa, military load 120t), extant 1995.

Fourteen 'Rotork' class LCU (type 412) built in UK (9t, 41ft 8in oa, 240bhp = 28kts, military load 30 troops), extant 1995.

British BH 7 'WELLINGTON' class *hovercraft*

Displacement: 33t empty; 50t full load
Dimensions: 75ft 6in–78ft 5in × 44ft 11in × 34ft 1in/42ft height
23.9m × 13.7m × 10.4/12.8m
Machinery: 1–6.4m diameter prop, 1 Rolls-Royce Proteus Gnome 15M549 (15M541 in Mk 5) gas turbine, 4250shp = 60kts (speed is reduced to 35kts in a 1.4–metre sea). Oil 9–10t. See notes
Armament: 2 Browning MG
Sensors: Radar Decca 914
Complement: ?

Class (in service):
101 (November 1970), *102* (March 1971), *103* (mid-1974), *104* (mid-1974), *105* (late 1974), *106* (early 1975)

Built by British Hovercraft Corporation. First pair are logistics-support Mk 4 version, with a 14t payload, able to carry sixty troops. French reports give speed 65kts and range 400nm at 56kts. The four Mk 5 units have an SSM recess for two or four Harpoon but missiles are not mounted. Extant 1995.

British SR N6 'WINCHESTER' class *hovercraft*

Displacement: 10.9t full load
Dimensions: 48ft 7in × 25ft 3in × 12ft 6in/15ft 9in height
14.8m × 7.7m × 3.8/4.8m
Machinery: 1 Gnome Model 1050 gas turbine, ?–1050shp = 52–58kts. One Peters diesel as auxiliary unit. Range 170nm/54kts
Armament: 1 or 2 MG. See notes
Complement: 3

Class:
01–08

Built by British Hovercraft Corporation. Credited with one or 2–12.7mm MG or only 1–7.62mm MG. The Iranian Navy had the world's largest fully operational hovercraft squadron for coastal defence and logistic duties. Building dates uncertain – ordered 1970–72 and commissioned 1973 (three), 1974 (two) and 1975 (three). Lack of spares must be starting to tell by now and there were no reports of hovercraft action in the war with Iraq. Some refitted in the UK in 1984. Can carry four 500kg mines. Probably less than half operational in 1995.

SMALL SURFACE COMBATANTS

KAMAN class ('Combattante II' design) *fast attack craft (missile)*

Displacement: 249t standard; 275t full load
Dimensions: 154ft 2in × 23ft 4in × 6ft 3in
47.0m × 7.1m × 1.9m
Machinery: 4 shafts, 4 MTU 16V538 TB91 diesels, 14,400bhp = 37.5kts. Oil 41t. Range 2000nm/700nm, 15kts/33–33.7kts
Armament: 4 Harpoon SSM launchers (2×2), 1–76mm/62 DP OTO Melara, 1–40mm/70 Bofors
Sensors: Radar Signaal WM 28, Racal Decca 1226; ECM Thomson-CSF Alligator 5–A; ESM Thomson-CSF DR 2000 Dalia
Complement: 31

No	Name	Builder	Launched	Fate
P 221	KAMAN	CMN, Cherbourg	8.1.76	Extant 1995
P 222	ZOUBIN	CMN, Cherbourg	14.4.76	Extant 1995
P 223	KHADANG	CMN, Cherbourg	15.7.76	Extant 1995
P 224	PEYKAN	CMN, Cherbourg	12.10.76	Sunk 1980
P 225	JOSHAN	CMN, Cherbourg	21.2.77	Sunk 1988
P 226	FALAKHON	CMN, Cherbourg	2.6.77	Extant 1995
P 227	SHAMSHIR	CMN, Cherbourg	12.9.77	Extant 1995
P 228	GORZ	CMN, Cherbourg	28.12.77	Extant 1995
P 229	GARDOUNEH	CMN, Cherbourg	23.2.78	Extant 1995
P 230	KHANJAR	CMN, Cherbourg	27.4.78	Extant 1995
P 231	NEYZEH	CMN, Cherbourg	5.7.78	Extant 1995
P 232	TABARZIN	CMN, Cherbourg	15.9.78	Extant 1995

Gorz 1978 C & S Taylor

Contracted 19 February 1974 (first six) and 14 October 1974 at CMN, Cherbourg. Meaning of names: bow, javelin, arrowhead, arrow, boiling oil, sling, scimitar, mace, roulette, dagger, spear and battleaxe. The later boats have not yet received their missiles – only 12 Harpoon missiles were delivered but these may have sunk Iraqi 'P 6' FAC and a 'Polnocny' landing ship. Laying down/in service dates: *Kaman* (5.2.75/12.8.77), *Zoubin* (4.4.75/12.9.77), *Khadang* (20.6.75/15.3.78), *Peykan* (15.10.75/31.3.78), *Joshan* (5.1.76/23.3.78), *Falakhon* (15.3.76/31.3.78), *Shamshir* (15.5.76/31.3.78), *Gorz* (5.8.76/22.8.78), *Gardouneh* (18.10.76/11.9.78), *Khanjar* (17.1.77), *Neyzeh* (12.9.77), *Tabarzin* (24.6.77). Last three units, all commissioned 1 August 1981, left Cherbourg with approval of French Government (previously embargoed at Cherbourg April 1979 due to non-payment of cash). On way to Iran P 232 seized by a pro-Royalist group on 13 August 1981 near Cadiz and briefly returned to Toulon.

To avoid any further problems these boats were delivered to Iran by merchant ship as deck cargo. The last three boats were not fitted with Harpoon tubes on delivery. Presently all Harpoon missiles are believed to have been expended and replaced by Chinese C 801 SSMs. One some there may be portable SA-7 launchers. *Peykan* was sunk by an Iraqi 'Osa' class FAC using SS-N-2 missiles in November 1980 while *Joshan* was sunk by the US Navy on 19 April 1988.

North Korean CHAHAO class *fast attack craft (gun)*

Built in North Korea and transferred to Iran in 1987. Called *Zafar* class by the Iranians. Re-engined and speed may be reduced. Hull shape based on old Soviet 'P 6' class MTBs (Project 183). In addition at least one Iraqi 'Osa II' FAC and one *Bogomol* patrol craft, from Gulf conflict of 1991, are in Iranian hands.

Chinese HUDONG class *fast attack craft (missile)*

Displacement: 171t standard; 205t full load
Dimensions: 110ft 9in × 24ft 11in × 8ft 10in
33.6m × 7.6m × 2.7m
Machinery: 3 shafts, 3 diesels, 8000bhp=35kts. Range 800nm at 30kts
Armament: 4 YJ-1 Eagle Strike/C-801 SSM, 2–30m/65 AK-230 (2×1)
Complement: 28 (3 officers)

Class (in service):
301–305 (Sep 1994)

FAC(M) of the new Chinese 'Hudong' class designed on the standard 'Huangfen' class hull (itself based on the Russian 'Osa I'). Negotiations with the Chinese for the purchase of such craft began in 1991, but were held up for some time by arguments over the choice of SSM. Five craft were finally delivered on board a Chinese docking vessel in September 1994, with five more to follow. It is reported that they may be fitted with the improved C-802 tupe missile. This acquisition enhances the Iranian Navy's coastal attack capability.

US Improved PGM 71 *large patrol craft*

Displacement: 105t standard; 146t full load
Dimensions: 100ft 1in × 22ft × 10ft 2in
30.5m × 6.7m × 3.1m
Machinery: 2 shafts, 8 General Motors 8–71 diesels, 2000bhp = 15kts. Range 1500nm/10kts
Armament: 1–40mm, 2–20mm, 2–12.7mm MG. See notes
Sensors: Radar Decca 303; sonar SQS 17B
Complement: 20

Class (no, former name, launched, in service):
Parwin (PGM 67, ex-*PGM 103*, 1967), *Bahram* (PGM 66, ex-*PGM 112*, *c*1969), *Nahid* (PGM 67, ex-*PGM 122*, 1970)

All built by Peterson, Sturgeon Bay, Wisc. French sources give 1–40mm, 4–20mm (2×2), 2 Mk22 Mousetrap, 2 DC racks. Names mean Mercury, Mars and Venus. Pennant numbers in 1992, 211–213. Extant 1995.

US 'CAPE' class (USCG 95ft class) *large patrol craft*

Displacement: 98t standard; 148t full load
Dimensions: 94ft 10in × 20ft 4in × 5ft 11in–6ft 7in
28.9m × 6.2m × 1.8–2.0m
Machinery: 2 shafts, 4 Cummins diesels, 2200bhp = 20kts. Range 1500nm/15kts
Armament: 1–40mm, 2 Mk 22 Mousetrap ASW RL, 2 DC racks (8–300lb DC)
Complement: 15

Class (no, former no, in service):
Keyvan (201, PC 61, 14.1.56), *Tiran* (202, PC 63, 1957), *Mehran* (203, PC 62, 1959), *Mahan* (204, PC 64, 1959)

All built by US Coast Guard Yard, Curtis Bay, Maryland. Transferred under MAP-US Contract Nos 95CGMDAI, 95CGMDA3, 95CGMDA8 and 95CGMDA5 respectively. Names derive from islands in the Gulf. *Keyvan* sometimes referred as *Keywan*, *Kayvan* or *Kaivan*. Fitted with navigation radar. *Mehran* lost 1980–81 in Iraq war. Mousetrap replaced by twin Soviet 23mm/80 guns, or 20mm Oerlikon. Extant 1995.

COASTAL PATROL BOATS

Twenty-six US 50ft Mk II class (22t full load, 2 diesels, 26kts, 4–12.7mm MG, 6 crew) built by Peterson for Iranian Coast Guard 1976–77. Numbered *1201–1220*. Six extant 1995.

Two ex-British Admiralty HDMLs: *SDML 1389* transferred in early 1950s as *FDB 58*. Then renamed *Tahmadou* (pennant no FDB65). British *SML 323* (employed on survey duties; ex-*HDML 1081*) transferred 21 June 1956 as *Asalon*. Both employed as despatch boats. It was officially stated in 1971 that they were taken out of service.

Nine *Azar* class (90ft, 65t std, 2 diesels, 22kts, 2 MG) built by CN Lamar, La Spezia 1954–55: *Azav*, *Chahab*, *Darakhsh*, *Navak*, *Peykan*, *Tondar*, *Tondbad*, *Toufan*, *Tousan*. Probably all stricken 1970s.

Two ex-ASR craft (ex-British MTBs) were handed over to the Customs Guard in 1953. Probably named *Favour* and *Larak* (1942, 25t standard, 40kts).

Three *Gohar* class (70t std, 75ft, 2 diesels, 27–28kts, 1–12.7mm MG, 19 crew): *Gohar*, *Shahpar*, *Shahram*. Built by Abeking & Rasmussen, Lemwerder, West Germany in 1969–71 as fishery protection units for the Iranian Navy, transferred to the Iranian Coast Guard 1975 and to Sudan the same year as *Skeikan* (? or *Shekan*), Kadir (? or *Kader*) and *Karari* respectively. German sources give 95ft 2in × 16ft 5in × 4ft 11in (*29.0m × 5.0m × 1.5m*). Decca 202 navigation radar.

Twelve US 40ft (*Sewart* type) class (10t std, 2 diesels, 30kts, 2–7.62mm MG, about 7 crew): *Mahmavi-Hamraz*, *Mahmavi-Taheri*, *Mahmavi-Vanedi*, *Mardjan*, *Mordarid*, *Sadaf* and six others (names unknown – they are possibly *Lovon*, *Nanang*, *Schanak*, *Genbod*, *Kosay*, *Dolphin* (Pennant nos 51–56; ex-US Contract Nos *4OCGA01*, *4OCGB01*, *4OCGC01*, *4OGGD01*, *4OCGE01*, *4OCGF01* – built by USCG Yard, Curtis Bay, Maryland). Three given to Sudan 1975. Six stricken since.

All served in Iranian Coast Guard. Launches used for port duties, built by Sewart, Morgan City in 1963 and 1970s as Sewart standard 40ft type (according to British reports six were transferred June 1963). Now six in Iranian Coast Guard (pennant nos 5001 and above, some serve in the Caspian Sea), four (built 1970) given to Sudan in 1975 (?1978) and two stricken.

Twenty US 'Mk III' class (41.6t, 3 diesels, 690bhp = 30kts, 3–12.7mm, crew 5) delivered in 1975–76; about 50 were ordered in 1976 to be shipped out and completed in Iran. It is not known how many were finally assembled, six lost in Gulf wars and others scrapped. Ten extant 1995.

Fifty-one 'Boghammar' type patrol craft (6.4t, 42ft 8in, 714bhp = 46kts, 1–12.7mm MG, 1 RPG 7 RL or 106mm recoilless rifle, 1–12 barrelled 107mm RL, crew 5–6), delivered 1984–85 from Sweden for Islamic Revolutionary Guard; thirty extant 1995.

MINE WARFARE VESSELS

US MCS class *coastal minesweepers*

No	Name	Builder	Launched	Fate
32	SHAHBAZ (ex-*MSC 275*)	Bellingham Shipyards, Wash	22.11.58	Lost through fire 1975
31	SHAHROKH (ex-*MSC 276*)	Bellingham Shipyards, Wash	*c*1959	Extant 1995
33	SIMORGH (ex-*MSC 291*)	Tacoma Boat-building, Wash	3.3.61	Sunk 1981
34	KARKAS (ex-*MSC 292*)	Peterson Builders, Wisc	1962	Extant 1995

Of wooden construction. Transferred from the USA under MAP in 1959–62. According to British reports *Shahbaz* stricken in 1974 after collision damage. *Shahrokh* is the Iranian Navy's largest unit in the Caspian Sea. Decca 707 navigation radar and UQS 1 sonar. Three later units have pennant nos 301–303 respectively, and probably 2–20mm AA (1×2). *Simorgh* sunk by Iraq 1981 but reported to have been returned to service in 1992.

US COVE class *inshore minesweepers*

Class (no, former no, transferred):
Kahnamuie (301, ex-*MSI 13*, 3.9.64), *Riazi* (302, ex-*MSI 14*, 15.10.64)

Built by Tacoma Boatbuilding, delivered under MAP – transferred at Seattle, Washington. Sometimes referred to as *Cape* class. In August 1967 *Kahnamuie* was renamed *Harischi* as the name was required for one of the new Iranian frigates built in USA (ex-*PF 106*). Navigation radar. Given new pennant nos 311 and 312 respectively. *Harischi* sunk by Iraqi forces 1980–81, but reportedly returned to service in 1992. *Riazi* extant 1995.

Ex-British ALGERINE class *minesweeper*

Palang (ex-British *Fly*) acquired in 1949. Served as frigate with 2–102mm (2×1). Without pennant no. Paid off in December 1966 and stricken 1972.

Iraq

Although the Iraqi coastline is very short – just 58km – she maintained a relatively powerful navy until the Gulf conflict of 1991. Her geographical location at the head of a narrow and restricted body of water (the Persian Gulf) with an exit outside her control made the February 1981 order placed with Italian shipyards for four *Lupo* class fast frigates, six corvettes and a *Stromboli* class replenishment ship seem nonsensical – a deep sea navy without a sea to sail on. With hindsight it can be seen that this enlarged and modernised navy was intended to pose a military and political threat throughout the Gulf region; occupation of the disputed islands of Bubiyan and Warbah held by Kuwait would in any case have improved access to Iraq's main naval base, Umm Qasr.

Naval forces played a relatively minor part in the Iran-Iraq war, which lasted for most of the 1980s. Information on naval operations is still sparse, though both Iran and Iraq were reported to have lost eleven minor warships (including four Iraqi patrol boats claimed sunk by Iran on 21 September 1980). The Iraqi navy, with some 3000 men, remained the smallest of her armed services, and the order for the training frigate *Ibn Khaldoum* from a Yugoslav yard reflected the difficulty of training adequate numbers of men to crew the Italian-built ships, whose delivery was held up pending the end of the war. Furthermore, Iraq's main port, Basra, was effectively out of action throughout the conflict.

Following the ending of the Iran-Iraq war, delivery of the Italian-built warships was again held up by Iraq's occupation of Kuwait in late 1990. Iraq suffered a severe defeat at the hands of Allied armies and navies in Operation Desert Storm in 1991, and the navy was one of the major casualties. All Iraqi naval vessels which ventured into the open sea during Desert Storm were either sunk or severely damaged, and others were destroyed or damaged in port. It is likely that all the missile boats, landing ships and minesweepers were sunk. Personnel strength of the navy had increased to 4600 men by 1989, but is impossible to assess accurately in the wake of the Desert Storm defeat. The Iraqi Navy has effectively ceased to exist as an operational force.

FLEET STRENGTH 1947

Four 'Thornycroft' type patrol vessels, disp 67t, were launched in 1937 and, when no longer serviceable, were used as barracks. Numbered *1–4*, renamed *Abd al Rahman*, *Al Ghazi*, *Dat Al Diyan* and *Janada*.

Ibn Khaldoum design

IBN KHALDOUM *training frigate*

Displacement:	1850t full load
Dimensions:	317ft 4in oa, 295ft 3in pp × 36ft 8in × 14ft 10in *96.7m, 90.0m × 11.2m × 3.6/4.5m*
Machinery:	2 shaft CODOG (CP propellers): 1 Rolls-Royce Olympus TM-3B gas turbine, 22,000shp = 26kts, plus 2 MTU 16V 956 TB 91 diesels, 7500bhp = 20kts. Range 4000nm at 20kts
Armament:	1–57mm/70 Bofors, 1–40mm/70,8–20mm AA (4×2), 2 ASW TT and ASW mortar, DC rack
Sensors:	Radar Decca 1226, 1229, 1 search Philips 9LV 200 Mk II; sonar Plessey MS 26
Complement:	93 + 100 trainees

No	Name	Builder	Launched	Fate
507	IBN KHALDOUM	Titograd, Yugoslavia	1978	Extant 1995

Training ship with frigate characteristics of Yugoslav design. Laid down 1977 and in service 21 March 1980. The first of three such ships ordered – one for the Yugoslav Navy which was never completed and the other named *Hasr Dewantara* by Indonesia. Ships differ in machinery and armament (no helicopter pad). These are handsome ships of similar appearance to the Vosper Thornycroft frigates of the Iranian, Brazilian and Nigerian navies. *Ibn Khaldoum* could carry four SSM. Used as a transport in the war with Iran. Renamed *Ibn Marjid*. Severely damaged February 1991, but remains afloat.

Italian LUPO class *frigates*

Displacement:	2213t standard; 2525t full load
Dimensions:	370ft 1in oa, 347ft 9in pp × 39ft 4in × 12ft 7in *112.8m, 106.0m × 11.98m × 3.84m*
Machinery:	2-shaft (CP propellers) CODOG, 2 Fiat/GE LM 2500 gas turbines, 50,000shp=35kts, 2 GMT A230–20M diesels, 7800bhp = 20.5kts. Range 900nm at 35kts, 3450nm at 20.5kts
Armament:	8 Otomat Mk II SSM, 8 Albatros SAM (1×8) with no reloads, 1–127mm OTO Melara, 4–40mm Breda Dardo (2×2), 6–324mm ASW TT, 2 AB 212 helicopters
Sensors:	Radar 3RM20, RAN–11X, RAN–10S, 2–Orion RTN–10XRCT, 2–Orion RTN–20X; hull sonar Edo 610E or Raytheon 1160B, 2–SCLAR chaff launchers
Complement:	185

No	Name	Builder	Launched	Fate
F 14	HITTIN	Fincantieri, Ancona	27.7.83	Not delivered
F 15	THI QUAR	Finantieri, Ancona	19.12.84	Not delivered
F 16	AL QADISSIYA	Fincantieri, Ancona	31.3.84	Not delivered
F 17	AL YARMOUK	Fincantieri, Riva Trigoso	20.6.85	Not delivered

These ships were ordered in 1981 at Cantieri Navali Riuniti (CNR) and are very similar to the frigates built in Italy for the Peruvian and Venezuelan navies, fitted with a fixed helicopter hangar. Building was delay by US objections to the sale of major components of LM 2500 gas turbines. Finally the ships were completed in 1985-87 (two in 1985, one each in 1986 and 1987) but not delivered to Iraq because of a trade embargo imposed during the Iraq-Iran war. They were laying at Muggiano and were finally released for delivery in 1990, but delivery did not take place because of Iraqi aggression against Kuwait. These frigates were taken over by Italian Navy using funds provided outside the existing navy budget.

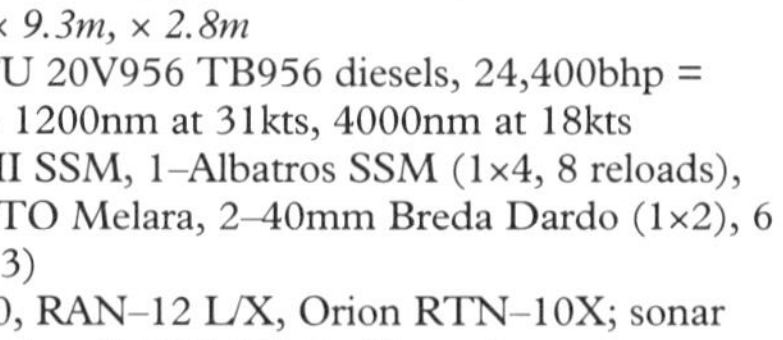

Mussa ben Nussair 1987

ABI SERH class *missile corvettes*

Displacement:	600t standard; 675t full load
Dimensions:	204ft 5in oa, 189ft 8in pp × 30ft 6in × 9ft 2in *62.3m, 57.8m × 9.3m, × 2.8m*
Machinery:	4 shafts, 4 MTU 20V956 TB956 diesels, 24,400bhp = 37.5kts. Range 1200nm at 31kts, 4000nm at 18kts
Armament:	6 Otomat Mk II SSM, 1–Albatros SSM (1×4, 8 reloads), 1–76mm/62 OTO Melara, 2–40mm Breda Dardo (1×2), 6 ASW TT (2 × 3)
Sensors:	Radar 3RM–20, RAN–12 L/X, Orion RTN–10X; sonar Thompson Diodon, 2–SCLAR chaff launchers
Complement:	51

No	Name	Builder	Launched	Fate
F 210	MUSSA BEN NUSSAIR	Fincantieri, Muggiano	16.12.82	Not delivered
F 212	TARIQ IBN ZIYAD	Fincantieri, Muggiano	8.7.83	Not delivered
F 214	ABDULLAH IBN ABI SERH	Fincantieri Marghera	5.7.83	Not delivered
F 216	KALID IBN AL WALID	Fincantieri, Marghera	5.7.83	Not delivered
F 218	SAAD IBN ABI WAKKAD	Fincantieri, Marghera	30.12.83	Not delivered
F 220	SALAH ALDIN AYOOBI	Fincantieri, Marghera	30.3.84	Not delivered

Six corvettes based on the 'Wadi M'ragh' design for Libya were ordered in 1981 by Iraq in two versions. The basic version (four ships F 214–220) is described above. Two other corvettes similar to the Ecuadorian 'Esmeraldas' class are fitted with telescopic hangar and landing pad for helicopter at a cost of four of the six Otomat SSMs. All these ships were completed in 1986–87 but never delivered as was the case with the *Lupo* type frigates. They are still in Italian hands, and although it had been proposed that the Italian Navy take them over, two were planned to be sold to Morocco and another two to Colombia, although by 1995 the sales have not yet taken place.

Polish 'POLNOCNY D' class *medium landing ships* (Project 773)

Class (delivery):
Atika (1977), *Janada* (1977), *Nouh* (1978), *Ganda* (Sep 1979)
Four ships of 'Polnocny D' type built in Poland in 1976–80 by Naval Shipyard, Gdynia. Deck structure amidships acts as a helicopter platform and a stores lift, closed during helicopter manoeuvres. One lost to Iranian Harpoon missiles 1980–81, two sunk 1991, one escaped to Iran.

Ex-Soviet 'OSA I' class *fast attack craft (missile)* (Project 205)

Class:
Hazirani, Kanun Ath-Thani, Nisan, Tamuz, I, II
Six delivered in 1972–74, two probably since stricken. Some names are not reliable. One craft defected to Iran 1991, one extant 1995.

Ex-Soviet 'OSA II' class *fast attack craft (missile)* (Project 205M)

Class:
Sa'd, Khalid Ibn, Al Walid, I–IV
Eight boats of this Soviet type were delivered in pairs in 1974, 1975, 1976 and the last pair since then. Some names may be suspect. Two sunk by Iranian warships and two by Iranian aircraft since September 1980. Two replacements delivered 1984. One defected to Iran 1991, remainder lost.

Ex-Soviet 'P 6' class *fast attack craft (torpedo)* (Project 183)

Class:
Al Adrisi, Al Bahi, Al Shaab, Al Tami, Alef, Ibn Said, Lamaki, Ramadan, Shulab, Tamur, Tareq Ben Zaid, I
Twelve units transferred from the USSR; two in 1959, four in November 1960 and six in January 1961. Stricken as worn out. Pennant numbers included 217–222. Six sunk by Iranian forces since 1980, some by Harpoon missiles.

Ex-Soviet 'SO 1' class *large patrol craft* (Project 201M)

Three ships transferred in 1962. Employed as large patrol craft. Carry pennant numbers 310–312. All sunk 1991.

Ex-Soviet 'POLUCHAT I' class *large patrol craft*

Two ships of *Poluchat* type were transferred from USSR in late 1960s. Used as patrol craft or for torpedo recovery. One sunk 1991, two extant 1995.

Ex-Soviet 'ZHUK' class *coastal patrol craft* (Project 199)

Five transferred to 1975. Three sunk 1991, two extant 1995.

Ex-Soviet 'NYRYAT' II class *patrol craft*

Two multi-purpose craft possibly employed as diving craft. There are also two ex-Soviet *PO 2* (*Pozarnyy 2*) type fire boats of very similar design.

Ex-Soviet 'T 43' class *ocean minesweepers*

Class (no):
Al Yarmouk (465), *Al Kadisia* (467)
Built in mid-1950s and transferred to Iraq in early 1970s. Pennant numbers changed to 412 and 417. Both sunk 1991.

Ex-Soviet 'YEVGENYA' class *inshore minesweepers*

Three of these GRP-hulled inshore minesweepers were delivered in 1975 ostensibly as 'oceanographic craft'. Armed with 2–25mm. One lost 1991, two extant 1995.

Yugoslav NESTIN class *river minesweepers*

Four transferred in 1980. Three extant 1994–5.

Israel

The Israeli Navy is a small, but powerful, force centred around three widely separated operational bases at Haifa (the main base where the majority of units and certainly the most modern missile boats are sited), Ashdod (established May 1967) and the Red Sea port of Eilat. Sharm el Sheikh was returned to Egypt in April 1982.

Israel has a vital strategic location on the boundary of two important seas: the Mediterranean and the Red Sea. She is surrounded by largely hostile neighbours and faces continuing internal disruption stemming from the unresolved Palestinian problem. The peace treaty signed with Egypt at Camp David, USA, is of great importance to Israeli industrial and military needs. Among other things it has made it possible for Israeli ships to pass safely through the Suez Canal and so Israeli naval units can be sent from the Mediterranean to the Red Sea in two to three days instead of sailing round Africa.

THE ARAB-ISRAELI WARS 1948–82

The navy has always been the Cinderella of the Israeli Defence Forces. In the 1948–49 War of Independence it consisted of about 350 sailors with RN and/or 'illegal' immigrant boat experience. Their flagship was the 2150t ex-US Coast Guard cutter *Northland*, renamed *Elath* (better known as *Eilat*). They also had a few motor boats and naval frogmen, who sank the Egyptian sloop *El Amir Farouk* off Gaza. In 1956 the spectacular capture of the Egyptian frigate *Ibrahim el Awal* off Haifa by Israel's two newly acquired destroyers was owed largely to the air force, and the same Israeli destroyers' shore bombardment of Rafah the next night was no help to the Army.

Before the 1967 Six Day War the navy had added submarines and eight home-built landing craft to the inventory of three destroyers, eight MTBs and an ASW ship, but Israel's lightning victory left the fleet little to do. Three torpedo fast attack craft reached Sharm el Sheikh from Eilat to find the Egyptian naval blockade that had precipitated war no longer existed. Raids by naval frogmen on Alexandria and Port Said achieved psychological but no physical damage.

The 1967–70 'War of Attrition' gave more scope to Israel's warships. Their vulnerability to modern missiles was dramatically illustrated by the sinking of her flagship *Eilat*, a lesson that was to be learnt with a vengeance. Meanwhile in the Gulf of Suez tank landing craft carried an armoured battalion into Egypt for a raid in depth (September 1969) that cost the Egyptian naval C-in-C his job.

Israeli naval intelligence was less surprised than most by the 1973 Yom Kippur War and events were dramatically to vindicate the fleet's new emphasis on missile and electronic warfare. Thirteen Israeli missile fast attack craft with sixty-three Gabriel launchers defeated twenty-seven Arab 'Osa' and 'Komar' boats mounting eighty-four longer-ranged Styx rockets. Off Latakia (Syria), on the first night of the war and 200 miles from base, five Israeli units fought the first missile versus missile surface action in naval history. They evaded two Styx salvoes of six missiles each, shooting down one with 76mm gun fire, and eighteen minutes later, having closed the range, fired a Gabriel salvo that hit all three Syrian ships. Two blew up and the third was finished off with 40mm fire. Two nights later six Israeli craft dodged four Styx volleys (twelve missiles) to destroy three out of four Egyptian 'Osa' boats off Damietta, the third vessel again being sunk by gunfire. Another twenty-eight Arab missiles, fired in later actions, all missed. The Gabriel Mk I missile was 85 per cent effective.

The relatively protracted Yom Kippur operations allowed the navy to come into its own. Constant sorties tied down an armoured brigade on the Syrian coast. The Egyptians, whose Suez Canal crossing stunned the world, failed entirely off Sinai and in the Gulf of Suez; in one typical action five Israeli patrol boats sank nineteen armed fishing boats in the Ras Ghareb anchorage. For a loss of three killed and twenty-four wounded the navy sank nineteen Arab regular warships.

In the course of the war the Israeli Navy used new coastal warfare tactics; fast attack craft instead of escort ships were employed. The enemy was generally attacked in his own bases, rarely on the open sea. The Israelis deployed their fast attack craft together with helicopters providing mid-flight guidance for missiles and decoys for enemy missiles. Such deployment was a novelty at the time, and gave excellent results.

Israeli naval operations in the 1980s showed a daring only made possible by total air supremacy. In June 1981 fast attack craft hit the PFLP HQ at Tripoli in Lebanon. A year later the navy made a maximum effort in amphibious and gunfire support of the army coastal thrust that reached Beirut within four days. The Palestinian military evacuation to Cyprus was effectively regulated by Israel's warships patrolling off the Lebanese capital.

Already in possession of the best trained and equipped fast attack craft group in the Mediterranean, Israel formed a second such unit for deployment on the Red Sea during the 1980s. The rapid build-up of the Libyan Navy prompted the Israeli Navy to order a new type of fast attack corvette with good seakeeping qualities and designed purely for attack. In her present economic situation Israel cannot afford to build a defensive navy and so her requirements are for attack craft, such as missile armed hydrofoils and small guided-missile patrol boats.

To counter infiltration by PLO terrorist groups along the 150-mile coastline there are many coastal radar surveillance stations, and patrols by 'Dabur' and 'Yatush' class craft. These radars and patrols also provide early warning of low-level aerial attacks coming in over the sea.

Israeli Aircraft Industries (IAI) is also involved in the development and production of military equipment for the navy. There are even some small warships built by IAI such as fast attack craft of the 'Dabur' or 'Dvora' (missile) classes.

During the 1980s the Israeli Navy commissioned only a small number of new warships such as 'Saar 4.5' class boats and three missile hydrofoils of the US 'Flagstaff 2' class. However, during this time ambitious submarine and corvette plans both received the go-ahead. Two new ocean-going diesel-electric submarines of the 'Dolphin' class were ordered while three 'Eilat' class corvettes of very advanced design are being built in the USA. These ships, called by their builders 'pocket cruisers', will extend Israel's maritime reach significantly. There are also plans to acquire six fast attack craft (gun) and two LCTs.

MISSILES

Surface-to-surface Gabriel type guided missiles designed and produced in Israel are well-known internationally. There are two versions, Mk 1 and Mk 2, and the much improved Gabriel Mk 3 is now entering service. Gabriel Mk 1 carried out trials in the late 1950s and during the early 1960s it was decided to adopt the concept of the missile armed fast attack craft as the main unit for future deployment with the Israeli Navy. The Israeli Navy's concept of the missile/gun armed FPB to replace the destroyer and frigate was vindicated by the success of the Gabriel Mk 1 in the 1973 Yom Kippur War, despite its disadvantages such as limited range (20km) due to the solid fuel rocket motor and target handling (resulting from the semi-active homing system which permits only one target to be selected for engagement). During the early 1970s the Gabriel Mk 2 was developed to carry increased fuel of a more modern type, which extended its range to 36km. The range was in fact greater than 36km but the effective range was limited by the extent of coverage of the search and fire control radars then operational. The Gabriel Mk 2 still has the semi-active homing missile. IAI has therefore developed the Gabriel Mk 3 missile, which uses a fully active homing head, enabling the operator to fire the missile and forget it. In fact the Mk 3 missile can operate in three modes: fire and forget (possible to engage four targets simultaneously); fire and update – the missile's computer is updated while in flight to assimilate a target's changing parameters, which enables the homing seeker in the head to be activated at the latest possible moment to overcome ECM; and fire and control – this incorporates the capability of the Mk 1 and Mk 2 missiles to control the missile throughout the whole of its flight and thus to concentrate on one target alone.

The latest concept in missile tactics to gain favour is that of the 'Cocktail', designed to overcome and confuse ECM systems: different types of missiles are fired at the target simultaneously and it is possible that a Mk 2 and Mk 3 Gabriel might be fired simultaneously using different modes of attack. Furthermore it is to be noted that the *Reshef* class are now equipped with the US Harpoon missile so providing even greater possibilities for 'Cocktail' mixes.

Work on the proposed Gabriel Mk 4 anti-ship swept-wing missile has been abandoned. It was to have weighed is 960kg (150–200kg warhead) with a speed of Mach 0.85.

Reportedly up to the present 2500–3000 Gabriels in all versions have been manufactured. It is interesting to note the current cost of one Gabriel missile: Mk 1/2 – $400,000, Mk 3 – $450,000, Mk 3 A/S – $550,000. These costs have risen dramatically since 1971, when a Gabriel Mk 1 SSM cost only $85,000–$95,000. The Gabriel Mk 2 is being manufactured in South Africa under licence, while the Taiwanese 'Hsiung Feng I' is an unlicensed copy.

In recent years Israeli industry has also produced a surface-to-air missile called Barak, designed for small fast attack craft. It achieves a very short reaction time (time to maximum range less than 30 seconds). It was originally conceived to be used with an elevatable/trainable eight-cell box launcher, but now will use a thirty-two-cell VLS. The missile weighs 98kg, is fitted with a warhead of 22kg and its range is 500–10,000m. One missile costs about $200,000. It entered Israeli service in 1991–92. It was exported to Chile in 1988 and also armed the new Singaporean 'Victory' class.

GUNS

In addition to developing their missile technology, IAI has developed a new twin 30mm gun mount for point defence, the TCM 30, using the Hispano-Suiza 831 30mm. The mount is a low, relatively heavy rigid structure supporting the muzzles and which is designed to help overcome muzzle droop and vibration etc, leading to small dispersion, an essential feature in a gun anti-missile system. The fully automatic system is a completely above deck system with a magazine of 150 rounds per gun fired in bursts of 5, 10, 15 or 20 rounds. Target speeds up to 250–300m/sec can be handled and the barrels have an elevation of -50° to +85°. The mount has undergone satisfactory trials in a *Reshef* class ship and was expected to equip the *Reshef* class and vessels then under construction, with subsequently a possible retrofit for the older 'Saar' class boats. In the end it was fitted only to the *Shimrit* class hydrofoils.

In practice the Israeli Navy adopted mostly foreign systems. All bigger fighting ships are equipped with US Vulcan Phalanx Mk 15 CIWS and OTO Melara automatic 76mm/62 guns. Phalanx CIWS is planned to be replaced with Barak SAM. The new corvettes will receive the new American Sea Vulcan 25mm close-in weapon system.

TRAINING AND MANPOWER

The navy's high-quality personnel are its greatest asset and this quality is assured by its own electronics school, training boys for two years from the age of sixteen before national service. Most officers and senior petty officers are regulars, while the 3500 conscripts in the total of 6600 sailors in 1995 also join before their three year period. Naval commando and frogman training is done by 750 men, women take on many shore jobs and there are about 4000 reserves who do up to sixty days annual training. Command is vested in a rear-admiral or commodore.

In 1995 the Israeli Navy had 7000 men in active service (900 officers). The compulsory national service is three years for Jews and Druses.

MAJOR SURFACE SHIPS

PROJECTED *destroyers*

In the mid-1950s construction of two destroyers (1100/1200t, 32kts) was considered.

Ex-British 'Z' class *destroyers*

No	Name	Builder	Acquired	Fate
40	ELATH (ex-*Zealous*)	Cammell Laird	15.7.55	Sunk 21.10.67
42	YAFFA (ex-*Zodiac*)	Thornycroft	15.7.55	Stricken 1972

Transferred at Cardiff Docks. Renamed *Elath* and *Yaffa* (often referred to as *Yaffo*). Refitted before going to Israel in 1956: *Elath* by Harland & Wolff, Langton Dock, Liverpool; *Yaffo* by Crichton, Trafalgar Dock, Liverpool. *Elath*, better know as *Eilat*, sank on 21 October 1967 after being hit by three Styx SSMs 14nm off Port Said. The missiles were fired from 'Komar' class Egyptian FAC *inside* the harbour. Casualties aboard the Israeli Navy Flagship were forty-seven killed or missing and ninety wounded out of 199. This was the first time in history that a warship was sunk by a ship-launched missile. In July 1967 *Elath* and two torpedo FAC had sunk two Egyptian torpedo FAC off Rumani, Sinai.

Ex-British 'RIVER' class *frigates*

No	Name	Builder	Acquired	Fate
?28	MIVTAKH (ex-*Violetta*, ex-*Orkney*)	Canadian Yarrow	1950	To Ceylon 1959
32	MISNAK (ex-*Sharon*, ex-*Hallowell*)	Canadian Vickers	1950	To Ceylon 1959
?30	MISGAV (ex-?, ex-*Strathadam*)	Canadian Yarrow	1951	BU 1959–60

Three Canadian-built RN frigates of 1943–44 sold to merchant service 1947, 1945 and 1947 respectively from which the Israeli Navy acquired them. Additional 4in gun mounted in *Misgav*, 1956. First two sold in August 1959 to Royal Ceylon Navy; third scrapped 1959–60 or stricken only in 1961.

Ex-British 'Modified FLOWER' class *frigates*

Name	Builder	Acquired	Fate
WEDGEWOOD (ex-*Colon*, ex-*Beauharnois*)	Morton	1950	BU 1956
HAGANAH (ex-*Balboa*, ex-*Norsyd*)	Morton	1950	BU 1956

Ex-RCN corvettes sold into merchant service and acquired by the Israeli Navy.

Ex-British HUNT class (Type 1) *frigate*

No	Name	Builder	Acquired	Fate
38	HAIFA (ex-*Ibrahim el Awal*, ex-*Mohamed Ali el Kebir*, ex-*Lin Fu*, ex-*Mendip*)	Swan Hunter	31.10.56	Stricken 1972

Captured off Haifa in the only major naval action of the 1956 Arab-Israeli War. The Egyptian frigate sailed from Port Said and, undetected in the darkness off the Israeli coast, fired 200rds of 4in from 0330 at Haifa naval installations causing some damage but no casualties. The French destroyer *Kersaint* fired 64 rounds back but did not pursue as the Israeli destroyers *Yaffa* and *Eilat* were across the Egyptian's line of escape. They fired 436 rounds from 0527 at ranges under 9000yds but it was two rocket-firing Ouragan jets which at dawn disabled *Ibrahim* and caused her surrender at 0710 with two killed and eight wounded. She failed to scuttle due to rusty valves and *Eilat* towed her into Haifa. Commissioned into Israeli Navy January 1957. Served as hulk from about 1970 until stricken.

Eilat as completed

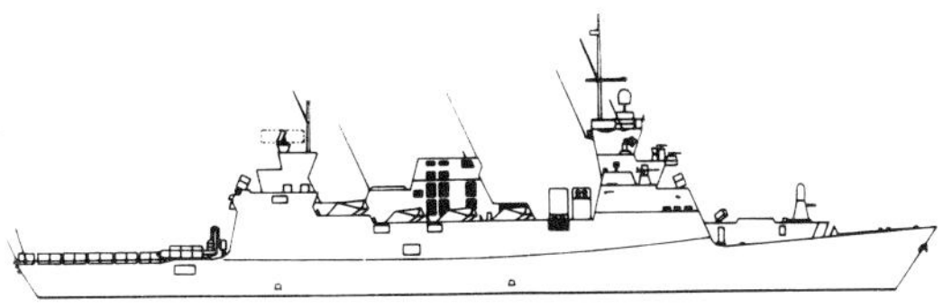

'EILAT' ('SAAR 5') class *missile corvettes*

Displacement: 1075t standard; 1200t full load
Dimensions: 283ft 6in oa, 251ft 4in wl × 39ft × 10ft 6in
86.4m, 76.6m × 11.9m × 3.2m
Machinery: 2 shafts, 2 KaMeWa CP propellers, CODOG, 1 LM 2500 gas turbine, 23,000shp=33kts, 2 MTU 12V 1163 TB82 diesels, 6600bhp=20kts. Range 3500nm at 17kts
Armament: 8 Harpoon SSM launchers (2×4), 8 Gabriel II SSM launchers (8×1), 2 Barak I SAM VLS (2×32), 1–20mm Vulcan-Phalanx CIWS, 2 Sea Vulcan 25mm CIWS (2×5), 6–324mm TT ASW Mk 32 (Mk 46 torpedoes), 1 helicopter Dauphin SA 366G or Kaman SH-2F or Hellstar RPV.
Sensors: Radar Elta EL/M 2218S, Cardion SPS 55, 1 navigational set; fire control 3 Elta EL/M 2221 GM STGR, 3 Elop MSIS optronic directors; sonar EDO Type 796 Mod 1, VDS; ESM/ECM Elisra NS 9003/5, 4 chaff launchers, Nixie SLQ 25 towed torpedo decoy; combat data system Elbit NTCCS
Complement: 64 (16 officers) + 10 (4 officers) aircrew

No	Name	Builder	Laid down	Launched	Comp	Fate
501	EILAT	Ingalls	22.2.92	6.2.93	Feb 1994	Extant 1995
502	LAHAV	Ingalls	25.9.92	20.8.93	May 1994	Extant 1995
503	HANIT	Ingalls	5.4.93	May 1994	Nov 1994	Extant 1995

Three of these compact, fast and powerful missile corvettes were ordered on 8 February 1989 to be built by Ingalls, Pascagoula, USA, to a design jointly prepared by Israeli Shipyards at Haifa, the Israeli Navy and Ingalls. There were originally to have been eight, then four, and now the option for the fourth unit will not be taken up. They were financed using the FMS fund. Especially designed with the suppression of radar, noise and heat signatures in mind, these ships have steel hulls with machinery resiliently mounted and are fitted with the Prairie Bubbler Masker System. These ships appear top-heavy, as a result of the traditional Israeli lack of concern about the top weight of their ships. This may lead to stability problems in service and force a reduction in the ships' formidable armament, which earned them the nickname 'pocket cruisers' from their builders.

Combat installation is being done in Israel. The planned 76mm OTO Melara gun was replaced by a 20mm Vulcan-Phalanx as the navy considers it much more useful for air defence than the heavier weapon. The numerous SSM can be used against surface targets and these ships are not expected to operate alone in high-threat areas. They were conceived as flotilla leaders and command ships for groups of FAC(M)s. The first unit, *Eilat*, finished her initial sea trials in October 1994.

Eilat on trials (Ingalls)

SUBMARINES

Ex-British 'S' class *submarines*

No	Name	Builder	Acquired	Fate
71	TANIN (ex-*Springer*)	Cammell Laird	9.10.58	Cannibalised 1968
72	RAHAV (ex-*Sanguine*)	Cammell Laird	Oct 1958	Stricken 1972

RN boats launched 1945, both refitted in Great Britain before delivery to Israel in December 1959 (*Tanin*) and May 1960 (*Rahav*). Originally very handy craft, capable of making a 'crash dive' in 30 seconds, and both fitted with 'Snort' mast and sonar domes. *Tanin* put frogmen into Alexandria on 5 June 1967. They did not return. *Rahav*, unable to submerge, was stricken after serving as an ASW ship in the Six Day War as worn out in 1968 and was cancelled to keep *Tanin* in commission, and *Tanin* herself was awaiting disposal in 1972. Used for training.

Ex-British 'T' class (1941 and 1942 Programmes) *submarines*

No	Name	Builder	Acquired	Fate
75	LEVIATHAN (ex-*Turpin*)	Chatham DYd	19.5.67	Deleted 1975
77	DAKAR (ex-*Totem*)	Devonport DYd	10.11.67	Lost 25.1.68
77	DOLPHIN (ex-*Truncheon*)	Devonport DYd	9.1.68	Deleted 1977

Acquisition of these 1943–44 built RN subs by Israel was announced in November 1964. Refitted before handover. *Dakar* lost with all hands in the E Mediterranean on passage from Britain. Modernised as British boats: *Turpin* lengthened by 12ft, *Totem* and *Truncheon* by 20ft. All guns and external TT were removed.

German Type 206 *coastal submarines*

Displacement: 420t normal/600t submerged
Dimensions: 147ft 8in × 15ft 5in × 12ft 2in
45.0m × 4.7m × 3.7m
Machinery: 1 shaft, 2 MTU 12V493 TY60 diesels and AEG generators plus 1 electric motor, 1200bhp/1800shp = 11kts/17kts. See notes
Armament: 8–533mm TT (bow sub; 2 spare torpedoes). See notes
Sensors: Radar Plessey, fire control Tios System; sonar Plessey; ESM Elisra NS 9034
Complement: 22

Name	Builder	Launched	Fate
GAL	Vickers Ltd, Barrow	2.12.75	Extant 1995
TANIN (ex-*Gur*)	Vickers Ltd, Barrow	25.10.76	Extant 1995
RAHAV	Vickers Ltd, Barrow	8.5.77	Extant 1995

A contract was signed for the building of these boats by Vickers in April 1972. Designed by the well-known German submarine designer Prof Gabler from Lübeck; also known as IKL/Vickers Type 206 or T540. *Gal* (laid down 1973 and commissioned December 1976) ran aground on delivery voyage, but was repaired. *Tanin* (laid down 1974) and *Rahav* (laid down 1976) were commissioned June and 18 December 1977 respectively. Differ from the German Navy's version in having a different conning tower and other changed characteristics. Designed to be equipped with 1 SLAM SAM launcher (1×6, retractable) intended to destroy enemy ASW helicopters. The first type of submarine since 1945 to be designed with armament that cannot be used in standard TT (apart from SLBMs). Tropicalised.

These small submarines, which are extremely cramped by modern standards and really only suitable for short endurance patrols, are excellent ASW units. They are small and quiet and in the variable Mediterranean environment are extremely difficult to detect. They operate as defensive units providing early warning, surveillance and the landing of sabotage teams. It is admitted that three submarines are not really sufficient to meet Israeli Navy requirements and they are to some extent limited by short endurance and relatively slow torpedoes. The vessels are armed with a British weapon system incorporating eight bow TT of the swim out type, the torpedoes being wire guided and controlled by a FCS linked to a one-man operated passive sonar system, with an active system available if required. The tear-drop hull is of single compartment design with a single escape tower in the fin. The forward hatch in the hull is used for battery removal and reloading torpedoes.

SLAM SAM system not fitted. Sub Harpoon and associated fire control equipment installed in 1983. Obsolete Mk 37 torpedoes replaced with NT 37E type in 1987–88. Fire control system to be updated locally starting from 1994.

'DOLPHIN' class *submarines*

Displacement: 1550t surfaced; 1720t submerged
Dimensions: 187ft × 22ft 4in × 20ft 4in
57.0m × 6.8m × 6.2m
Machinery: 1 shaft, 3 MTU 16V396SE diesels, 1 Siemens electric motor, 4243bhp/3875shp = 11kts/20kts. Range 8000nm at 8kts surfaced, 420nm at 8kts dived
Armament: 6–533mm TT (bow, 8 reloads)
Complement: 35

Two submarines of similar design to the German '212' type were projected in 1988 when Ingalls Shipbuilding Division of Litton Industries Corp was chosen as prime contractor. Boats were to be built in Germany with FMS funds, approved in July 1989 with effective contract date of January 1990. Cancelled in November 1990 due to the lack of adequate defence funding. In April 1991 after the Gulf conflict, German funding was secured and two ships will be built by HDW, Kiel, and Thyssen Nordseewerke, Emden. First steel cut in April 1992, projected delivery in 1997. There are plans to order a third boat. The 'Dolphin' class is to be fitted with Krupp Atlas or Litton TCS fire control system and Krupp Atlas sonar. Diving depth was set at 250m and endurance at 30 days. Equipped with 'wet and dry' compartment for frogmen. Fitted for Sub Harpoon anti-ship missiles.

AMPHIBIOUS WARFARE VESSELS

Ex-US LSM class *landing ships*

Three purchased in 1972 from civil sources. Armed with 4–20mm. Israeli and all ex-names unknown. Stricken since 1982.

ETZION GUEBER class *tank landing craft*

Displacement: 182t standard; 230t full load
Dimensions: See notes
Machinery: 2 shafts, 2 diesels, 1280bhp = 10kts
Armament: 2–20mm (2×1)
Complement: 12

Class (no):
Etzion Gueber (or *Ezion Geber*, 51), *Shikmona* (or *Shiqmona*, 53), *Kessaraya* (55)

All built by Israel Shipyards, Haifa. Commissioned 1965. Often credited with other dimensions: 100ft × 19ft 4in × 4ft 3in (*30.5m × 5.9m × 1.3m*) or 120ft × 23ft 4in × 4ft 7in (*36.6m × 7.1m × 1.4m*). All three and *Ashdod* (see *Ash* class) transported overland to Eilat in the Six Day War by day and back into the desert by night to tie down Egyptian warships in the Red Sea by simulating a build up for the capture of Sharm el Sheikh. All stricken 1991.

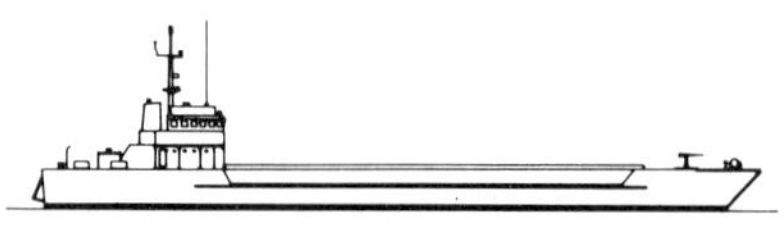

Ashdod 1971

ASH class (LCT type) *tank landing craft*

Displacement: 400t standard; 730t full load
Dimensions: 206ft 9in × 32ft 10in × 5ft 11in
62.7m × 10.0m × 1.8m
Machinery: 3 shafts, 3 MWM diesels, 1900bhp = 10.5kts. Oil 37t
Armament: 2–20mm (2×1). See notes
Complement: 20

Class (no, in service):
Ashdod (61, 1966), *Ashkelon* (63, 1967), *Achziv* (65, 1967)

All built by Israel Shipyards, Haifa, 1966–67 (one of the latter two units due for completion in August was finished in five days for the Six Day War) and extant 1995. Helicopter pads fitted aft. *Ashdod* was an early trials ships for Barak VLS.

BAT SHEVA

Displacement: 900t standard; 1150t full load
Dimensions: 312ft × 36ft 9in × ?26ft 11in
95.1m × 11.2m × ?8.2m
Machinery: 2 shafts, 2 diesels, ?bhp = 10kts
Armament: 4–20mm, 4–12.7mm MG
Complement: 26

Bat Sheva built in Netherlands 1967 and bought from S Africa in 1968. Sometimes classified as a transport. Has bow ramp. Extant 1995.

In addition to the above, the Israeli Navy also operated a number of other landing craft. Reported to be a mixed flotilla of ex-US LCTs (including ex-*LCT 640* and ex-*LCT 673*), LCI and LCM types, ex-British vessels of the LCT(6) type and ex-German craft (referred to as *Siebel Ferries* or MFP type). Landing craft of the LCT, LCI and other types, were taken out of commission for disposal before the early 1970s (with the exception of one LCT, which was given to the Israeli National Museum in Haifa). According to the latest German reports three LCMs are in service. Probably all since stricken.

SMALL SURFACE COMBATANTS

'Saar 5' design
NB 1/750 scale

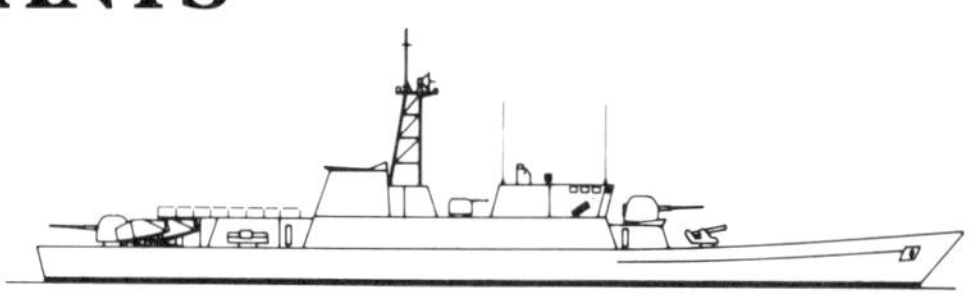

'SAAR 5' class *missile corvettes*

Displacement: 850t full load
Dimensions: 253ft 3in × 30ft 2in (?27ft 8in) × 10ft–10ft 10in
77.2m × 9.2m (?8.4) × 3.0–3.3m
Machinery: 2-shaft (CP propellers) CODAG: 1 General Electric LM2500 gas turbine plus 2 MTU turbocharged diesels, 20,000–25,000shp plus 8000bhp = 40kts (42kts max; 18–20kts on diesels). Range 4500nm/20kts on diesels (5500nm with additional fuel), endurance 20 days
Armament: 4 Gabriel SSM (4×1), 2–76mm/62 DP OTO Melara (2×1), 6–30mm/75 AA Hispano-Suiza A32 (3×2), 1–375mm Bofors ASW RL (1×2), 6–324mm AS TT Mk 32 (2×3), 1 helicopter (type unknown). See notes
Complement: 45

A British source reported two building early 1981 and probably further six planned, but French reported incorrectly: 'this design is a project offered by the shipyard but not yet ordered by the Israeli Navy'. 'QU-09-35' (or 'Q9-O35') known also as 'Corvette 850' was an interim design between 500t missile boats (high speed, manoeuvrability) and corvettes over 1000t (range, firepower, electronics, CIC). Planned to have a mix of Harpoon and Gabriel SSM launchers. These ships may well have been intended for over-the-horizon target data acquisition for Harpoon-fitted *Reshefs* ('Saar 4' and 'Saar 4.5' classes) with command and control facilities, particularly since the trials of a low-mounted helicopter platform in *Tarshish* proved unsatisfactory. Details on planned armament uncertain – sometimes credited with eight Harpoon SSM (2×4), 4–30mm/85 Emerlec (2×2), 1 short-range SAM system, 1 single (or twin) 40mm/70 Breda or one single 57mm/70 Bofors, without any TT.

Electronic equipment was to have comprised: long range search/surveillance, air search, surface search and navigation radars; gun fire control radar; active/passive sonar; ECM. These corvettes were not built. The *Eilat* was ordered in 1989 instead (see above).

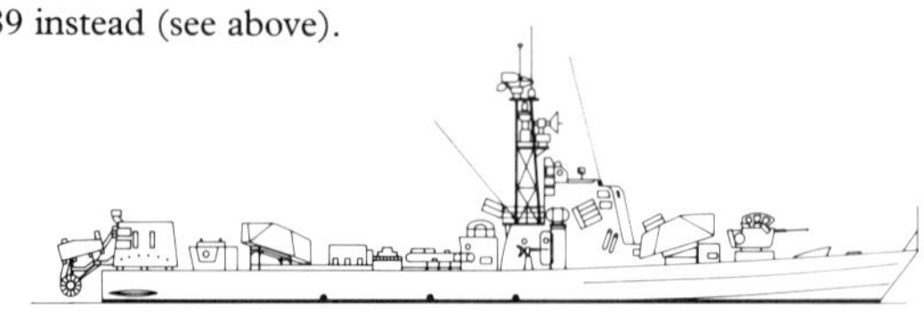

Mifgav 1983
NB 1/750 scale

'SAAR 1' class *fast attack craft (missile)*

Displacement: 220t standard; 250t full load
Dimensions: 146ft 7in × 23ft × 5ft 9in mean, 8ft 2in deep
44.9m × 7.0m × 1.7m, 2.5m
Machinery: 4 shafts, 4 Maybach-Mercedes MD872 (later known as MTU 16V 538 TB90) diesels, 13,500bhp = 42kts. Oil 30t. Range 2500nm/1600nm/1400nm/1000nm at 15kts/20kts/25kts/30kts

Armament: 'Saar 1' class: 3–40mm/70 Breda Model 58/11 (3×1), 4–32mm Mk 32 ASW TT (4×1; Mk 46 torpedoes), 2–12.7mm MG. As 'Saar 2' see notes
'Saar 3' class: 6 Gabriel SSM (2×3 trainable mounts amidships), 1–76mm/62 DP OTO Melara, 2–12.7mm MG, 4, 8 or 12 large chaff RL
Sensors: Radar Thomson-CSF Neptune, Selenia Orion; ECM VHF DIF and intercept gear
Complement: 35–40 (5 officers)

Class (no, launched):
Mirtach (311, 11.4.67), *Miznag* (312, 1967), *Mifgav* (313, 1967), *Eilath* (321, 14.6.68), *Haifa* (322, 14.6.68), *Akko* (323, 1968)
Saar (331, 25.11.68), *Soufa* (332, 4.2.69), *Gaash* (333, 24.6.69), *Herev* (341, 20.6.69), *Hanit* (342, 1969), *Hetz* (343, 14.12.69)

All built by CMN, Cherbourg, although designed by Lürssen Werft of Bremen. First six ordered in 1965 had 3–40mm guns (configured like the Italian *Freccia* class patrol boats). Classified as 'Saar 1' class. Next six ordered in 1966 with 1–76mm OTO Melara. First three units and *Haifa* delivered to Israel unarmed, the former therefore not operational for the June 1967 Six Day War. Last five escaped to Israel from Cherbourg on 24 December 1969 nominally as Norwegian oilfield supply craft named *Starboat I–V* under Panamanian flag despite the French arms embargo. Arrived in Israel January 1970 after world publicity for their endurance and seaworthiness. *Akko* and *Saar* delivered after the embargo was lifted. Only 'Saar 1' class fitted with searchlight sonar ELAC and it became 'Saar 3' after Gabriel SSM fitted. 'Saar 2' 76mm gun and magazine (interim fit before missiles available) left no space for ASW equipment. Armament varies from 1–40mm (fwd) and eight Gabriel SSM (2×3 trainable launchers amidships and aft, 2×1 fixed mounts forward or five Gabriel SSM (1×3 amidships, 2×1 fwd) and 2–40mm (2×1 fwd and aft) as 'Saar 2' to 3–40mm and 4 ASW TT as all-gun configuration 'Saar 1'. Armaments are now fairly standardised, but triple Gabriel launchers can be interchanged with the after 40mm mounts. 'Saar 3' group can now probably mount two Harpoon SSM (2×1) instead of one triple Gabriel. A radar controlled missile system, aided by optical sights, launching a 150lb HE warhead at a low altitude to a range of 12.5nm (22nm in the later versions). Eventually 2–12.7mm MG can be removed. Cost $2m (late 1960s prices) per unit. *Haifa* was reportedly previously named *Misnak*.

On 'Saar 3' class their superstructure is slightly different forward in order to accommodate the larger gun. This group also has small bridge wings added and relocated inflatable life rafts. 'Saar 2' and 'Saar 3' (and their armament) were not only the most sophisticated vessels in Israel in the early 1970s, but were among the most powerful warships in the Mediterranean. They sank seven Arab missile FAC without loss in October 1973. Extant 1995. *Saar*, *Gaash* and *Herev* stricken 1990. *Hanit* and *Hetz* sold to Chile 1988. *Soufa* stricken 1993. Remainder stricken 1994.

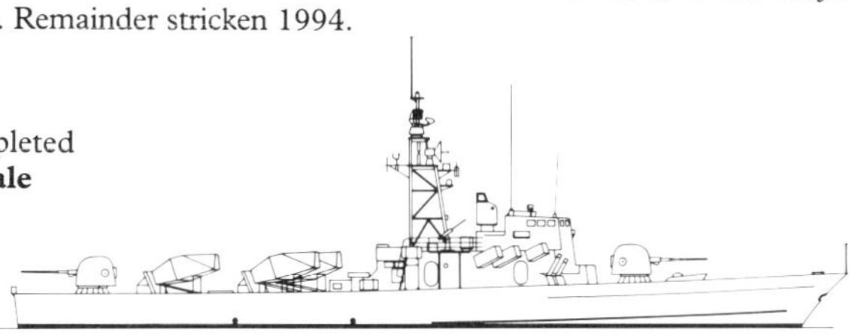
Yaffo as completed
NB 1/750 scale

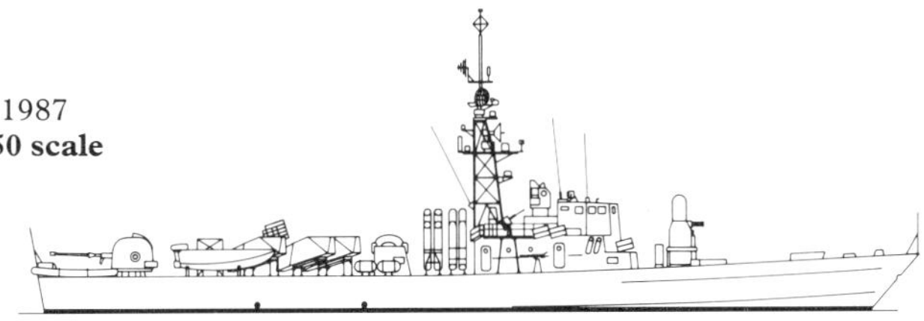
Modelet 1987
NB 1/750 scale

RESHEF ('SAAR 4') class *fast attack craft (missile)*

Displacement: 415t standard; 450t full load
Dimensions: 190ft 7in × 24ft 11in × 7ft 10in
58.1m × 7.6m × 2.4m
Machinery: 4 shafts, 4 MTU 16V 956 TB91 diesels, 14,000bhp = 32kts, sustained power 10,680bhp. Range 4000nm at 17.5kts, 1600nm at 30kts
Armament: 'Saar 4' class: 6 Gabriel SSM (6×1 all fixed; designed with 7: 1×3, 4×1), 2–76mm/62 DP OTO Melara (2×1), 2–20mm Oerlikon (2×1) or 4–12.7mm MG (2×2)
Sensors: Radar Thomson-CSF Neptune TH-D1040, Selenia Orion RTN-10X; ECM MN-53 intercept; sonar EDO 780
Complement: 45

Class (launched, in service):
Reshef (19.2.73, Apr 1973), *Keshet* (2.8.73, Oct 1973), *Romah* (Jan 1974, Mar 1974), *Kidon* (4.7.74, Sep 1974), *Tarshish* (17.1.75, Mar 1975), *Yaffo* (3.2.75, Apr 1975), *Nitzahon* (10.7.78, Sep 1978), *Atsmout* (3.12.78, Feb 1979), *Moledet* (22.3.79, May 1979), *Komemiut* (19.7.78, Aug 1980)

The development of *Reshefs* began towards the end of 1968. Chaff launchers (four large single and six small 12-barrelled, for long-distance cluttering of radar screens); Israeli-made missiles and electronics and US missiles. An expansion of the building slips at Haifa enabled rapid construction of the next six units ordered in January 1975. Four Harpoon SSM fitted to 'Saar 4' in early 1978 in addition to the reduced number of Gabriel launchers. Israeli homing systems attached to Harpoons. Gabriel II system, fitted with a TV camera capable of transmitting a picture to the firing ship, was installed after the ships were built.

Tarshish has carried a temporary helicopter deck in place of the after 76mm gun for experimenting with over-the-horizon targeting for Harpoon. 76mm guns have been specially adapted for shore bombardment.

Yaffo deployed in Mediterranean, other five of 'Saar 4' class in Red Sea. Quarters are air-conditioned. ELAC Sonar fitted in ships deployed in the Red Sea. VHFD/F and radar warning device.

Excellent sea-keeping qualities and endurance and at cruising speed has a very long range. Four ships made the passage from Israel to the Red Sea round Africa via Gibraltar relying entirely on refuelling at sea. Two sailed to New York in July 1976. The first pair was successfully engaged in the October 1973 Arab-Israeli War. *Reshef* sank a Syrian 'T 43' minesweeper with Gabriel off Latakia.

Three have been built in Haifa for S Africa and nine more in Durban. Two others bought late 1979 and 1981 by Chile as *Casma* (ex-Israeli *Romach*) and *Chipana* (ex-Israeli *Keshet*).

The 1995 armament fit is: two to four Harpoon SSM, four to six Gabriel II or III SSM, one or two 76mm/62 OTO Melara, 2–20mm Oerlikon, one Vulcan Phalanx Mk 15 20mm CIWS (1×6), 2–12.7mm MG. It is now planned to replace the Vulcan Phalanx with Barak SAM VLS. VDS sonar can be fitted. Total designed armament weight for *Reshef* class is 40t.

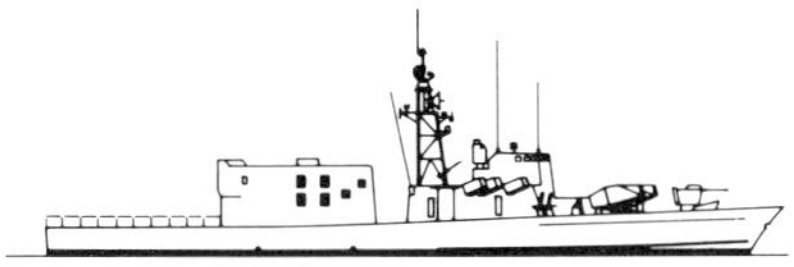
Aliya class design

ALIYA/ROMAT/HETZ ('SAAR 4.5') class *fast attack craft (missile)*

Displacement: 488t full load (500t *Aliya*)
Dimensions: 202ft 5in oa × 25ft × 8ft 2in
61.7m × 7.6m × 2.5m
Machinery: As *Reshef* class except *Hetz*: 4 MTU 16V 538 TB93 diesels 16,600bhp = 31kts. Range 3000nm at 17kts, 1500nm at 30kts
Armament: 8 Harpoon SSM (2×4), 6–8 Gabriel III SSM (6–8×1), (4 Harpoon and 4 Gabriel in *Aliya*, *Geoula*), 1 Barak I SAM VLS (1×32) (*Hetz*), 1–76mm/62 OTO Melara (except *Aliya*, *Geoula*), 2–20mm Oerlikon (2×1), 1 Vulcan Phalanx Mk 15 20mm CIWS (1×6), 2 or 4–12.7mm MG, 1 SA 366G Dauphin helicopter for over the horizon targeting or Hellstar RPV
Sensors: Radar Thomson-CSF TH-D 1040 Neptune, fire control Selenia Orion RTN-10X (*Hetz*: Elta EL/M-2221 GM STGR), ESM/ECM Elisra suite, Chaff RL
Complement: 45 (53 in *Aliya*, *Geoula*, *Hetz*)

Class (launched, in service):
Aliya (11.7.80, Aug 1980), *Geoula* (Oct 1980, 31.12.80), *Romat* (30.10.81, 1981), *Keshet* (Oct 1982, Nov 1982), *Hetz* (ex-*Nirit*, Oct 1990, Feb 1991)

These boats, built by Israel Shipyards in Haifa, employ lengthened *Reshef* class hull for additional armament or helicopter facilities. *Aliya* and *Geoula*, built with landing pad and a big boxy hangar for one helicopter, used for OTH targeting. Can serve as command boats for groups of other FAC. In the case of *Romat* and *Keshet* additional armament was substituted for helicopter and its hangar. The Vulcan Phalanx Mk 15 CIWS is mounted in the bow position in all 'Saar 4' and 'Saar 4.5' boats so fitted. The last boat in the series *Nirit* (later renamed *Hetz*), started construction in 1984 as private speculation by Israel Shipyards and after languishing on the ways was picked by the government. Building was resumed in 1989 and she was fitted out as combat-worthy trials ship for systems to be used in the *Eilat* (so called 'Saar 5') class corvettes. She has Barak SAM VLS installed aft in place of some Gabriel launchers and new big tower mast similar to that planned for the *Eilat* class.

'DVORA' class *fast attack craft (missile)*

Displacement: *c*38t standard; 47t full load
Dimensions: 70ft 11in × 18ft 4in × 2ft 11in/6ft (props)
21.6m × 5.5m × 0.9/1.8m
Machinery: 2 shafts, 2 MTU 12V331 TC81 diesels, 2720hp = 36kts. Range 700nm at 27 or 32kts
Armament: 2 Gabriel SSM (2×1), 2–20mm (2×1), 2–12.7mm MG
Sensors: Radar Decca 926
Complement: 8–10

One unit (name not known)

IAI design, similar to the 'Dabur' class but smallest ever missile craft. Constructed of aluminium alloy. The *Dvora* (meaning 'Seagull') because of her small size, can be transported by land to the required place. Can also be stored on land to ease maintenance. Trials from December 1977. Two 'Dvora' class were also ordered by Nicaragua from Israel. Price $3–4m per unit (15 per cent cost of large missile FAC). Stricken 1991.

'Dvora' class IAI

'SUPER DVORA' class *fast attack craft (gun)*

Displacement: 38t light; 54t full load
Dimensions: 73ft 5in oa × 18ft × 3ft 3in (5ft 11in over props)
22.4m × 5.5m × 1.0m (1.8m)
Machinery: 2 shafts, 2 MTU 12V 396 TB 93 diesels 3260bhp in Mk I version or 2 Detroit 16V-92TA diesels 2800bhp - 36kts in Mk II or 3 Detroit 16V-92TA diesels 4200bhp - 46kts in Mk III. Range 1200nm/17kts
Armament: 2–20mm/80 Oerlikon (2×1), 2–12.7mm or 7.62mm MG, 1–84mm RL, 2 DC racks
Complement: 9 (1 officer)

Enlarged and improved version of earlier 'Dvora' class. Built by Israeli Aircraft Industries (RAMTA). Ordered in March 1987, first started trials in November 1988 and the first two delivered in June 1989. Deliveries at two or four per year. Currently twelve (pennant No 810–821) are in service. SSM, torpedoes or 130mm MRL can be fitted if required. Fitted with Raytheon radar and Elop optronic director. First ten units are the Mk I and Mk II versions.

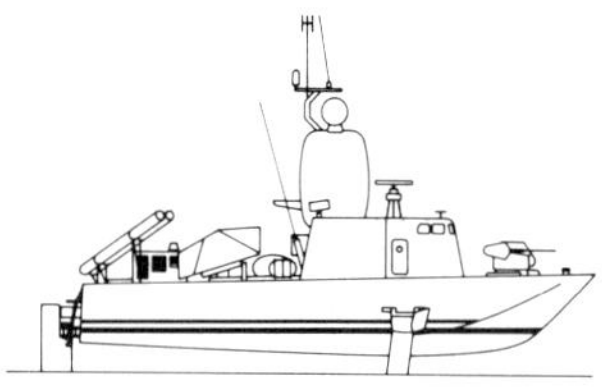

Shimrit class as built
NB 1/750 scale

SHIMRIT (US FLAGSTAFF 2) class *missile hydrofoils*

Displacement: 71t light; 103–105t full load
Dimensions: 104ft 3in foils retracted, 84ft hull oa × 42ft 6in foils, 24ft hull × 5ft 3in hull, 15ft 7in foils at rest, 6ft 4in foiling
31.79m, 25.6m × 12.95m, 7.3m × 1.6m, 4.75m, 1.93m
Machinery: 1 shaft, 1 CP main propeller, 1 Allison 501-KF gas turbine, 5400shp - 45–48kts, 2 GM 6V53 auxiliary diesels driving retractable 80hp hydraulic motors for hull-borne propulsion. Range 750nm/42kts, 2600nm/8kts
Armament: 4 Harpoon SSM (2×2), 2 Gabriel III SSM (2×1), 2–30mm Oerlikon TCM30 (1×2), 2–12.7mm MG (2×1)
Complement: 15

Class (no, in service):
Shimrit (M 161, Jul 1982), *Livnit* (M 162, 1983), *Sanpirit* (M 163, Jun 1985).
The *Shimrit* class (name means 'guardian') ordered from Grumman Lantana Yard, USA. It was intended to build up to twelve such craft in Israel (some sources report even fifteen units). Eventually only *Livnit* and *Sanpirit* were built by Israel SYs at Haifa. Building terminated after three units. Programme encountered many delays. Considerable teething troubles in *Shimrit* were finally overcome, but on trials she did not reach the designed speed of 52kts, attaining only 47–48kts. Turning radius 200m at 45kts. Endurance described as three to five days. Aluminium construction. Pineapple-shaped radome conceals radar, fire control and countermeasures aerials. Three laid up 1993.

Ex-British *motor torpedo-boats*

Class (no):
Lilitt (209), *Shaldagg* (210), *Tinshemett* (212)
At least three (possibly nine) British 70ft Vosper MTBs were purchased late 1940s. Ex-names not known. Reported about 1967 to be 'no longer in service'.

AYAH class *fast attack craft*

Displacement: 62t standard
Dimensions: 85ft 4in × 20ft 8in × 4ft 11in
26.0m × 6.3m × 1.5m
Machinery: 2 shafts, 4 12cyl Arsenal-Marine-Otto petrol engines, 4600bhp - 42kts max. Range 600/300nm at 29/42kts
Armament: See notes
Complement: 14 (After modernisation 15)

Class:
Ayah, Baz, Daya, Peress, Tahmass, Yasoor
Built by Chantiers de Meulan, France. Ordered 1949 and delivered 1951–52. Pennant nos T200 series. Wooden hull. Older sources give *Netz* instead of *Yasoor*. Armed as designed: 1–40mm, 4–20mm and 2–457mm TT. Served rather as MGBs without TT. After about ten years service were modernised. Two new Napier Deltic diesels with total 5000bhp replaced the old petrol engines. This also increased significantly safety on board. Superstructure altered and mast with radar added. Armed with 1–40mm, one or two 20mm and 2–457mm TT. Later 1–20mm and TT probably removed. It appears that in early 1970s at least two boats mounted 20mm guns forward as well as aft. Stricken 1972–74.

OPHIR class *fast attack craft*

Displacement: 40t standard
Dimensions: 70ft 1in × 17ft 1in × 5ft 1in
21.3m × 5.2m × 1.5m
Machinery: 2 shafts, 2 Packard-Otto high octane petrol engines, 4000bhp - 40kts
Armament: 1–40mm, 2–20mm, 2–457mm TT
Complement: ?

Class (no):
Ophir (T 150), *Shva* (T 151), *Tarshich* (T 152)
Built by Cantieri Baglietto, Varazze, Italy in 1956–57. Wooden hull. Armament variable, torpedoes later probably removed and 1–20mm added. Probably stricken mid-1970s.

YAR class *large patrol craft*

Displacement: 96t standard; 109t full load
Dimensions: 100ft × 20ft × 5ft 11in–6ft 3in
30.5m × 6.1m × 1.8–1.9m
Machinery: 2 shafts, 2 MTU diesels, ?bhp - 22kts
Armament: 2–20mm (2×1)
Complement: 16

Class (no):
Yarden (42), *Yarkon* (44)
Often incorrectly referred to as *Jarkon Echad* or *Yarkon Echad*.
Coast guard cutters, built by Yacht & Bootswerft Burmester, Bremen-Burg 1956–57. *Yarden* became a non-naval training craft. *Yarkon* stricken 1976.

KEDMA class *coastal patrol craft*

Displacement: 32t standard
Dimensions: 66ft 11in × 15ft 1in × 4ft 11in
20.4m × 4.6m × 1.5m
Machinery: 2 shafts, 2 diesels, 1540bhp - 25kts
Armament: 2–20mm. See notes
Complement: 10

Class (no):
Kedma (46), *Yama* (48), *Negba* (52), *Tzafona* (50)
Built in Japan 1968. Handy boats of the small seaward defence type. Steel hull. Used for coast guard and police work in peace time. Later probably all armed only with two MG – *eg Negba* has 2–7.62mm MG. Stricken 1992.

'DABUR' class *coastal patrol craft*

Displacement: 25–26t standard; 35t–39t full load
Dimensions: 65ft × 19ft × 2ft 8in
19.8m × 5.8m × 0.8m
Machinery: 2 shafts, 2 GM 12V-71TI diesels, 960bhp - 21kts (more powerful GE engines in some 1900bhp - 30kts). Range 450nm/13kts (older sources give 1200nm/17kts)
Armament: 2–20mm (2×1), 2 or 4–12.7mm MG (2×1 or 2×2). See notes
Sensors: Radar Decca 101 or 926; small active sonar
Complement: 6–9 depending on armament

Class:
34 units
Twelve built by Sewart Seacraft (USA), others by IAI (RAMTA), 1973–77. Aluminium hull, quarters air-conditioned and spacious. Good rough weather performance. Differences in armament between units. In some boats two 324mm ASW TT are fitted forward. Deployed in the Mediterranean and Red Seas; possible as these craft were designed for overland transport. Extant 1995.
Currently only eight remain in Israeli service, while others were transferred to Argentina (four units 1978), Nicaragua (four 1978), Sri Lanka (two

'Dabur' class IAI

1984), Fiji (four 1991) and Chile (six 1991). Five were given to Lebanon Christian Militia but later returned. Some boats fitted with more powerful GE main engines for more speed as the original version was considered not fast enough to cope with modern terrorist boats. Fitted with Elop optronic director. Being sold out of service as 'Super Dvora' boats enter service.

'YATUSH' class (ex-US PBR type) *patrol boats*

Ex-USN 6.5t GRP river craft bought since 1974 and up to twenty-eight were operational in 1982. Two given to Lebanese Christians 1975–76. Armament variously given as 2–3 MG (7.6mm or 12.7mm and/or 1–60mm mortar). Last nine units stricken 1990–91.

Ex-British FAIRMILE 'B' *motor launch*

One vessel (ex-*M 17*, ex-*ML...*) transferred in early 1950s and renamed *Haportzim*. Armed with 3–20mm and 12 DC. Stricken 1961.

Ex-British *harbour defence motor launches*

Class (no):
Dror (21), *Saar* (35), *Tirtsa* (25)

In early 1950s three vessels transferred. Original British nos not known. Armed with 1–40(?47)mm, 1–20mm and 8 DC. *Saar* stricken between 1965–68, others armed with 1–47mm, 1–40mm AA and 8 DC or only 2–20mm AA and 8 DC; used for coast guard and police work. *Dror* and *Tirtsa* stricken 1974–76.

MINE WARFARE VESSELS

WHALECATCHER type *minesweepers*

Displacement: 300t
Dimensions: 130ft × 25ft × 13ft
40m × 7.6m × 3.9m
Machinery: 1 shaft, 1 reciprocating engine, 500ihp - 10kts
Armament: 1–76mm, 1–20mm
Complement: ?

Class:
Drom A I, Drom A II

Built in South Africa, probably in 1940s. Broken up in the early 1950s.

MISCELLANEOUS

Ex-US PC class *submarine chaser*

PC 1188 (often incorrectly referred to as *PC 16*) transferred in early 1950s and renamed *Nogah* (since 1960s pennant no 22). Classified as patrol vessel; in mid-1970s served as target for Gabriel missiles. Further fate unknown.

Ex-US NORTHLAND *Coast Guard cutter*

USCG cutter WPG 49 *Northland* (launched 1927) acquired in 1948 (between 1947 and 1948 mercantile *Medina Ivrit*) and renamed *Eilath*. Sloop type, of exceptionally strong construction with icebreaker bows specially built for Bering Sea. She rammed an RN destroyer during her immigrant running days and had a 60mm mountain gun lashed to the deck during the 1948–49 war. Served in Israeli Navy as TS (mainmast and forward 76mm gun removed) and later as depot ship for MTBs. In the late 1950s renamed *Matzpen*. Sold for scrap in 1962.

Ex-US MAYFLOWER *Presidential Yacht/Coast Guard training ship*

Mayflower was one of three very large and luxurious steam yachts very similar in size and appearance, all designed by G L Watson. Built by J & G Thompson in 1897, she was bought by the USN in 1899 and armed. In 1902 she was recommissioned as the US Presidential Yacht (in 1905 Roosevelt arranged for the treaty ending the Russo-Japanese War to take place on board). Later became pennant no PY 1. President Hoover withdrew her from service as the Presidential Yacht in 1929. She was laid up in Philadelphia N Yd when she was badly burned in June 1931. The hull was sold to, and refitted by, H L Gielow. The new owner had to sell her, so she went into South American trade as SS *Butte*. The Maritime Commission acquired her in 1942, and after being refitted, armed with 1–127mm and 2–76mm and renamed WPG 183 *Mayflower*, she entered the US Coast Guard where she served as a TS until 1946. Decommissioned and sold, she caught fire and was towed to Baltimore. Purchased and put under the Panamanian flag, she was named *Mala* in 1947 and carried refugees from France to Isreal. Purchased by the Israeli Navy *c*1950; served as a patrol craft and TS *Maoz*. Broken up 1955. *Maoz* often incorrectly referred to as ex-USCGC *Tampa* (launched 1921).

Italy

The Italian Navy operates in two very different regimes: the blue water of the two Mediterranean basins, and the narrow seas of the Adriatic and the choke point of the Sicilian straits. Coastal traffic is also important, so the navy has to be able to convoy it. Until 1960, the Strait of Otranto were also an important choke point, since the Soviets had a submarine base in Albania. Certainly from 1969, the Italian government also had to take into account Libyan hostility and the possibility that, in wartime, Libya would be a major Soviet base. With the end of the Cold War, instability in North Africa has become a major worry for all of the Mediterranean European powers, and as this is written the San Marco battalion carried by Italian amphibious warships is being enlarged to regimental strength (by 1997). Conversely, with the end of the Cold War the light ships bought to control the Sicilian straits against Soviet incursions are no longer important, and they will not be replaced as they wear out.

Particularly during the Cold War, the Mediterranean was an extremely complex theatre of operations. The massed airfields of the southern Soviet Union were not far away, and from 1969 the Soviets also enjoyed base rights directly across the Mediterranean in Libya (their forces in Egypt, in place until 1973, were only slightly further away). For air cover, the Italian fleet had to depend on land-based aircraft. As computer control of such aircraft developed, it became more and more important for major fleet units to be able to share the air tactical picture used by the air defence authorities ashore. That was one factor encouraging the adoption of elaborate computer combat direction systems (SADOC) at sea.

Yet another new factor in Italian naval construction is the new treaty on the Law of the Sea which defines a 200-mile Exclusive Economic Zone (EEZ) around each country. Smuggling has always been popular in the Mediterranean; the advent of the EEZ makes it easier for para-naval forces (Coast Guard, in US parlance) to try to deal with smugglers and also with polluters. Although not an EEZ function, control of illegal immigrants crossing the Mediterranean is important to Italy, as it is to the other European Mediterranean nations. As in the United States, corvettes built for EEZ enforcement are potential wartime reinforcements for the fleet. Since they are often paid for out of non-naval funds, their purchase has little impact on more specialised naval programmes.

Italy ended the Second World War with a substantial fleet and a very ambiguous position as a co-belligerent of the Allies. The peace treaty was negotiated as relations between the wartime Allies collapsed, and a final settlement was not reached until 1949. Under the treaty, Italy was not permitted to possess aircraft carriers, modern battleships (two old units, *Andrea Doria* and *Caio Duilio*, were retained as training ships), submarines, or torpedo boats (the famous 'MAS'). Many of the best surviving ships were transferred to Allied countries, including the Soviet Union. No new warships could be laid down until 1 January 1950.

As it happened, Italy joined the new NATO alliance soon after the treaty had been completed. The United States and Britain were already hotly engaged with the civil war in Greece, so by the late 1940s there was every interest in maintaining an effective Italian Navy. Thus surviving submarines were retained as battery-charging hulks, and wartime torpedo boats had their tubes removed to become motor gunboats. The prohibitions against submarines and torpedo boats were dropped in 1952.

The Italian economy could not support naval reconstruction, but the United States was willing to pay for considerable work in European shipyards. A five-year building programme laid out in 1950 called for a 6000t cruiser, two AA cruisers (about 4000t), six destroyers (AAW/ASW) of about 2500t (the first two provisionally named *San Giorgio* and *San Marco*), an ASW escort of about 1500t, an experimental vedette, several fast MGBs, and coastal patrol craft.

The 4000-tonner was expected to be an improved version of the wartime 'Capitani Romani' class, armed with US-type 5in/38 guns. The destroyers would have been armed either with 5in/38s or with the 135mm guns of the 'Capitani Romani' class in two twin mounts. The escort would have been armed with four 90mm guns and four 40mm, plus ASW weapons.

In fact, instead of building a new cruiser, the Italians rebuilt four existing ones: the two largest of the surviving wartime ships, the *Giuseppe Garibaldi*s, and two of the old 'Capitani Romani', which became super-destroyers or destroyer leaders rather than AA cruisers. Two new destroyers were built, their hulls based on wartime practice. Instead of the other four, four big frigates (*Canopo* class) were built. That left enough money to build three escorts (*Albatros* class); five sisters were built for other NATO accounts.

The ships of the five-year programme were armed mainly with US-supplied weapons, and equipped with US sensors, but they introduced a new Italian-developed ASW mortar. Many ships were to have been equipped with 21in ASW torpedo tubes firing wire-guided torpedoes, but these weapons were not ready in time, and the tubes apparently were not installed.

This programme also included the first new light attack craft, which were intended to form the basis for a force capable of blocking the Sicilian straits and the Strait of Otranto. They were considerably larger than the wartime 'MAS' and Allied MTBs. Ultimately they led to Italian interest in missile-armed hydrofoils.

US transfers filled out the programme: two submarines converted to GUPPYs were transferred in 1954–55. They were the first truly modern Italian submarines, capable of high underwater speed. Two *Benson* class destroyers and three destroyer escorts were transferred in 1951.

Finally, many existing ships were modernised. Three surviving war-built submarines were refitted and somewhat streamlined, although apparently they were not fitted with snorkels. Two prewar destroyers were modernised, and many of the prewar and war-built corvettes remained in service as interim escorts.

Meanwhile, an Italian shipyard, Ansaldo, developed light corvette and light destroyer/frigate designs. NATO bought both the corvette (for Denmark and the Netherlands) and the frigate (for Portugal). Venezuela bought the frigate (as the *Almirante Clemente* class). This success was repeated with the later *Lupo* class frigate and its smaller companion, the *Wadi M'Ragh* class corvette in the 1970s and early 1980s. The last major customer was Iraq, which bought a large fleet in 1981. When that fleet could not be delivered (or paid for), the effect on Italian shipyards was disastrous. Since 1981, the only sales of complete ships have been *Lerici* class minehunters, although the Italian naval industry has been quite successful in selling weapons and sensors.

The combination of the big five-year programme and transfers of US warships provided Italy with a modern fleet. A second five-year programme was apparently laid down in 1956. It brought the Italian Navy into the missile age. The first year of the programme (1956–57) included the first Italian missile destroyer, a Tartar-armed version of the *Impetuoso* class analogous to the contemporary US *Charles F Adams* class, as well as the first postwar submarine (*Enrico Toti*), and four fast corvettes (CV 2 or *Bergamini* class). The distinction between corvettes and the earlier frigates was that the former were diesel-powered escorts; the latter were somewhat faster steam (later, gas turbine) ships. Corvettes were intended primarily for coastal convoy work. For 1957 the Italian Navy planned a helicopter carrier. The second half of the programme was expected to include two big missile-armed cruisers, a second missile destroyer (*Intrepido*), and a second submarine. This programme also included the conversion of the big cruiser *Giuseppe Garibaldi* to a missile ship.

The last two years of the programme (1959–60 and 1960–61) included four more steam escorts (*Circe* class, described as improved *Centauros*). Only the first pair were built, as the *Alpino* class.

As in 1951–55, the programme was considerably modified. The helicopter carrier was never built, but the two big missile cruisers (the *Andrea Doria* class) had large helicopter hangars and flight decks; in effect they shared the capability of a small helicopter carrier. A third missile cruiser, tentatively named *Enrico Dandolo*, planned for 1959–60, was deferred (the larger cruiser *Vittorio Veneto* was bought under the 1961–62 programme). The two submarines had to be deferred; they were not ordered until 1962. However, one existing submarine, *Pietro Calvi*, was radically modernised for high underwater speed, in analogy with the US GUPPY programme and the British T class conversions.

A more spectacular submarine project, a nuclear submarine to be named *Guglielmo Marconi*, was included in the 1959–60 programme. Throughout the early 1960s the Italian Navy annually reported ongoing studies (the submarine was to have been based on the US *Skipjack* class), but the project was ultimately dropped. It is not clear when (or whether) the US Navy was approached for the necessary technology.

Again, US transfers were significant. A US 'fleet snorkel' submarine was provided in 1960, presumably to help make up for the delay in building the *Toti* class. The United States also supplied four ocean minesweepers and seventeen coastal sweepers, and financed the construction of many coastal and inshore sweepers in Italian yards. By 1957, Italy had a total of thirty-six modern coastal minesweepers and twenty inshore units.

In 1959–60 the national military electronics combine, Selenia, was formed. It produced a series of major radars and missiles (such as Aspide), and it was also responsible for the hardware of the Italian Navy's SADOC computerised combat direction system, which was based on the US NTDS. Italy adopted Link 11, which in the early 1960s was almost entirely restricted to high-value US and French units (the British and Netherlands navies adopted the more limited Link 10). This choice, which at the time was costly in hardware terms, was presumably necessary given the close relationship between the Italian Navy and the Sixth Fleet. Italy established a naval programming centre to produce the required software. Later, an export version of SADOC was incorporated in many Italian export frigates.

It seems unlikely that a third five-year programme was formally set. The early 1960s were occupied in filling in gaps left in the early plans. Some programmes stopped for lack of funds, such as the *Bergamini* and *Toti* classes, were built at this time. Four corvettes (the *De Cristofaro* class) were ordered in 1962.

The modern Italian Navy which emerged from the two five-year plans included three helicopter-carrying cruisers (one replacing the cruiser flagship *Giuseppe Garibaldi*), two destroyer leaders (ex-cruisers), four destroyers (two armed with area defence missiles), six fast ASW frigates (of a planned eight), twelve corvettes (four of them so small as to be obsolescent, fit mainly for fishery protection and other gunboat duties), and eight snorkel submarines (one modernised wartime unit, three US-supplied Fleet Snorkels, and four *Toti*s). Two US-supplied *Benson*s might be considered substitute fast frigates, to make up the total of eight, and three US-supplied destroyer escorts were, in effect, corvettes. As late as 1967, there were also seventeen war-built *Ape* class corvettes.

In 1967 the new construction programme (presumably for 1966–70) included two new large submarines (1370t, in effect replacing the abortive *Marconi*), a fourth missile cruiser (tentatively named *Trieste*), two new missile destroyers (the *Audace* class, announced April 1966), two general purpose frigates (the two *Alpino*s which had not yet been ordered), one corvette (a fifth, improved, *De Cristofaro*, powered by 20,000shp gas turbines, ordered in 1967), three landing ships (*Quarto* class), and a nuclear-powered fleet replenishment ship of new design. Still under construction were the missile cruiser *Vittorio Veneto* (which would replace the old modernised cruiser *Giuseppe Garibaldi*) and four small ASW submarines (*Toti* class). Two more US fleet snorkel submarines were transferred in 1966 to replace the two old and relatively unmodified war-built submarines. Including the *Pietro Calvi* and the *Toti*s, by 1970 Italy had ten modern submarines. That strength became a force goal.

Much of this programme had to be cancelled in 1968: the two submarines (restored in 1972 to become the first pair of *Nazario Sauro*s), the missile cruiser (never restored), the frigates (restored in 1972 as the first pair of *Lupo*s, officially described as improved *Alpino*s), the corvette, two of the three landing ships (with another, projected, pair cancelled), and the fast replenishment ship. About 1970, however, the navy was able to order a prototype missile hydrofoil, *Sparviero*. Also, about 1973 a second pair of *Lupo*s was ordered, in effect replacing the fifth *De Cristofaro*.

The *Lupo*s were significant improvements on the ships they, in theory, replaced. As in other navies, with these ships it became obvious that the distinction between destroyers and frigates had blurred; henceforth Italian destroyers would be area-defence missile ships, while frigates would be fast ASW ships capable only of defending themselves against air attack. The first pair of *Lupo*s finally filled out the 1959 plans for the *Circe*/*Alpino* class. The other two in effect replaced the first pair of postwar destroyers.

The 1972 restorations did little to replace the substantial postwar fleet, which was already ageing. A 1974 White Paper argued that without a formal replacement programme, the fleet would be more than halved by 1985. It recommended: two more *Audace* class missile destroyers to replace *Impavido* and *Intrepido*; eight *Lupo*s, in addition to the four already on order (they became the *Maestrales*; four *Sparviero* class (60t) missile hydrofoils; two 250t hydrofoils; two *Sauro* class submarines in addition to the two already under construction; ten minehunters; one underway replenishment ship of about 8000t; one hydrographic ship.

It would be necessary to lay down an amphibious dock ship and a small carrier to replace, towards the end of the 1980s, the existing amphibious ship (*Quarto*) and the two *Andrea Doria*s. That is, just as the two cruisers had in effect replaced one helicopter carrier, one helicopter carrier and a pair of area air defence ships could replace the cruisers.

The number of projected minehunters did not approach the number of ageing minecraft, but no navy could have afforded replacement on a one-for-one basis, as unit cost rose so quickly with the new minehunting technology. In addition, several existing ships would have to be modernised.

A 1975 Navy Law enshrined the 1974 recommendations and laid out a ten-year replacement programme. Most of the programme was executed, including the carrier and the amphibious dock ship. The planned pair of big hydrofoils would have been the Italian version of the NATO hydrofoil then in the design stage (the only ships of this type actually built were the US PHMs). They were never built; instead the Italian Navy received another pair of *Sparvieros*.

As in the past, parts of the programme had to be deferred due to shortages of money: the two missile destroyers, two of the fast frigates, six of the minehunters, and the planned LPD. In addition, the projected purchase of thirty-six AB-212 shipboard helicopters was cut to twenty-seven.

About this time, also, the Italian Navy received one last series of ex-US warships: two *Tang*s, two more GUPPY-type submarines, and a pair of *Fletcher* class destroyers (one of which had been converted in the 1950s for ASW). The submarines made it possible for the Italian Navy to realise a force goal of ten modern snorkel submarines, a standard it maintained until 1993, when it was cut to eight. The destroyers made up for a temporary shortfall in modern frigates due to the 1968 cancellation.

In 1981, official Italian naval policy called for a fleet comprising two light carriers, six missile destroyers, twenty frigates/corvettes, one large training ship, ten–twelve fast attack craft, twenty-four mine countermeasures craft, three underway replenishment ships, three maritime patrol squadrons, and eight submarines. Presumably this meant two battle groups, one for each Mediterranean basin, each built around a carrier and a fast underway replenishment ship, supported by three missile destroyers and several ASW frigates. This basic fleet configuration survived into the 1990s, but the strength per battle group declined to two missile destroyers and, probably, four frigates (*ie*, ASW ships of battle group speed). Other frigates were expected to operate independently, *eg* around choke points, to stiffen lighter forces. Plans ultimately called for each of the two main battle units to be a carrier like the new *Giuseppe Garibaldi* (as of 1992, the projected second carrier was to be named *Giuseppe Mazzini*).

The balance of battle group escorts between AAW and ASW reflected the high cost of AAW units. The hope for at least three AAW escorts per carrier remains: for the future, the Italian Navy plans six high-capacity AAW ships, 'Horizon' class frigates, to replace two AAW destroyers and four ASW frigates.

By this time the Italian Navy was consciously moving from small fast narrow-seas combatants to larger ships, such as the *Minerva* class corvettes, which could be used either near choke points or in the wider Mediterranean. In 1992 long-range planning envisaged using non-naval funds to buy four more Coast Guard-type corvettes (the *Cassiopeia* class), but that project has never materialised. If four more corvettes are in fact built, they may be of much more capable *Minerva* type.

As for the submarine force, throughout the 1980s it seemed likely that a wholly new design, 'S 90', would enter production at some point. Meanwhile four evolved versions of the original *Sauro* class followed the first four, the expectation being that the first pair of 'S 90's would be ordered in 1993 for completion in 1997. At that point the four very small coastal ASW submarines of the *Toti* class would have been replaced, and Italy would have a fully ocean-going force of modern diesel-electric units. In 1993 defence funding was cut drastically, and the submarine force goal fell to eight units; the construction of the 'S 90's was put off until about the turn of the century, when the first two *Sauro*s will be due for replacement.

The main programmes of the period after the 10-year programme were those deferred in the early 1980s (listed above) and two new series of corvettes, the *Minerva*s to replace the more capable coastal escorts, and four *Cassiopeia*s to replace the *Albatros* class. The planned run of minehunters was increased to twelve units, to keep the specialist yard alive until the first of a new class of six deep hunters could be ordered. The latter were conceived to meet a Soviet threat, and thus may no longer be so attractive (the French cancelled an analogous programme). However, they are still carried in the Italian long-range programme. For the present, the main programmes in view are the second carrier and the big 'Horizon' class frigate/destroyer.

Italy was one of the first countries to establish an independent air force, and as a consequence until very recently all fixed-wing aviation was in air force hands. Many Italian naval officers considered this arrangement responsible for Second World War disasters, and during the 1950s a very sympathetic US Navy tried to help the Italian Navy establish its own air arm. Success was limited to shipboard helicopters, which were bought in substantial numbers and which carried not only ASW weapons but also anti-ship missiles.

The maritime air component ashore consists of strike fighters and maritime patrol aircraft, the latter with naval crews under naval command (but air force ownership). The 1949 peace treaty forbade Italy from possessing bombers; the air arm was limited to 200 fighters. When restrictions were relaxed, the United States supplied maritime aircraft, initially Helldivers (SB2C) and Harpoons (PV-2 bombers) which were organised into ASW squadrons. About 1955 they were followed by S2F-1 (S-2A) Tracker ASW aircraft. By the mid-1970s, eighteen Trackers formed one squadron alongside two squadrons (eighteen aircraft) of Breguet Atlantics ordered in October 1968 (delivery was completed in 1973). Another fourteen Atlantique Mk 2 were planned under the 1975 programme to replace the ageing Trackers, but they were never ordered, and the Trackers were retired without replacement. Then the question of replacing the Atlantiques arose. It appeared that Italy would buy the P-7, the US replacement for the P-3 Orion, but that programme was cancelled. Instead, existing Atlantiques are being modernised to Mk 2 standard.

Italy currently operates four squadrons of Tornados, one of G-91Ys, and three of AMX strike aircraft. Some Tornados carry Kormoran stand-off missiles, and at least some of the other strike aircraft are assigned anti-ship missiles. It is not clear to what extent crews specialise in this role.

WEAPONS AND SENSORS

Italy built up a major naval weapons industry in the 1950s, and many systems were exported. At the end of the Second World War, the major firms were Odero-Terni-Orlando (OTO) and Breda for guns and mountings, Whitehead for torpedoes and mines, and Officine Gallileo for optics, including periscope optics. The Italian government itself created a national electronics industry, including a microwave radar company, MicroLambda. After producing several radars in the 1950s, MicroLambda became a component of Selenia, the major Italian military electronics company. Another company, Elettronica, produced most of the Italian naval electronic warfare systems (Selenia systems were used mainly for export). Breda produced some important chaff rocket launchers. Italy also produced its own submarine sonars, although surface ship sonars were generally licence-produced versions of US types.

During the 1960s, OTO merged with Melara to become OTO-Melara; this firm specialised in calibres from 76mm up. Breda limited itself to 40mm and smaller calibres, generally using Bofors or Oerlikon guns in its own mountings (in the 1980s it added a Mauser 30mm gun to its product line). By the early 1990s the two gun-making firms had merged to form the present OTOBreda. In addition to guns, OTO-Melara developed a major anti-ship missile, Otomat, in conjunction with the French company, Engins Matra. Most Otomats in current service worldwide are the Italian version, 'Teseo' (the letters stand for one of two alternative Otomat fire control systems). In addition to Otomat itself, a torpedo-carrying derivative, MILAS, will arm French and Italian warships; it can be fired from an Otomat launcher.

Italian naval helicopters are armed with the Marte Mk 2 missile, a derivative of the earlier Sistel Sea Killer with an enlarged nose housing an Otomat active radar seeker. Sea Killer itself was a beam-riding anti-ship missile rejected by the Italian Navy but supplied to Iran and Libya. It appears that the Israeli Gabriel used the same fire control system for mid-course guidance (unlike Sea Killer, it has a semi-active terminal seeker).

ITALY

For the systems listed below, the export designation is given first, with the official naval designation in parentheses.

Weapons

Aspide	Surface-to-air and air-to-air missile loosely based on the US Sparrow; launchers can fire either weapon. The associated ship fire control system is Albatros. Unlike all versions of Sea Sparrow, Aspide allows for reloads, using an elaborate frame which lifts out behind the launcher. It carried two sets of reloads (sixteen missiles in addition to the eight on the launcher).
Sea Killer	Beam-riding anti-ship missile by Contraves, tested by the Italian Navy as the Nettuno system, using the airframe of an anti-air missile, similar in size to Sea Sparrow or Aspide (20cm × 3.9m, 168kg, range 10km). The Mk 2 version was exported to Iran and Libya, and used in the Iran–Iraq War.
Marte	Sea Killer adapted for helicopters, using an active seeker in its current Mk 2 version.
Otomat	Surface-to-surface missile with an active seeker. Mk 1 range was 60–80km. Mk 2 can accept mid-course guidance from a helicopter; maximum range is about 150km. Dimensions: 46cm × 4.82m, 750kg with 200kg warhead. Turbojet power. The Italian system is also called Teseo.
Kangaroo (Canguro)	Interim heavy torpedo for use until A-184 was available; it comprised a German G-7e with wire guidance carrying a Mk 44, which it could fire at a target.
A-184	Heavy wire-guided torpedo developed in the 1970s for both surface ships (fired aft from tubes above the transom) and from submarines.
A-244	Alternative to the US Mk 44, widely exported. A-244S is an improved version with better shallow-water capability. A follow-on A-290 was abandoned in favour of a Franco-Italian project, Impact.
Menon (K-113)	The standard Italian postwar ASW mortar, equivalent to Limbo and to Weapon Alfa, was installed in both single- and triple-barrelled versions. The 320mm (4.6m long) barrel or barrels were fired at 45deg elevation, range (400m to 900m) being set by varying the amount of firing gas let into the barrel. Reloading (at 90deg elevation) was automatic. The projectile weighed 160kg. Typically Menon was ripple-fired to produce a 21-round pattern covering 80yds × 180yds; the projectile had a contact fuse with hydrostatic back-up. Larger ships had the triple-barrel version. According to the Italian official history, Menon was derived from the British Limbo. There was also a short triple-barrel Menon mortar, used extensively as a depth-charge mortar postwar (Menon 42). For example, the *Centauro* class had four (two either side) aft. Maximum range was 140m.

The most prominent postwar guns have been: the OTO-Melara 5in/54 lightweight and various 76mm/62s; the current Super Rapido is credited with 120 rounds/min and is advertised as a close-in weapon to compete with Phalanx and Goalkeeper. The argument is that the Gatling guns control fire in what amounts to 10-round bursts, so rates like 6000 rounds/min are really equivalent to 600 bursts/min. Since the proximity-fuzed 76mm shell need not actually hit the target to neutralise it, it can be argued further that a lower rate of fire will suffice, say the 240 rounds/min put out by a pair of 76mm Super Rapidos. The alternative is Breda's Fast Forty, a twin 40mm mount firing 300 rounds/min per barrel, or 600 per minute together (and these rounds are proximity-fused). Breda was a major 40mm supplier to the postwar German Navy; it provided a powered feed (from overhead) for single 40mm guns used by many navies.

During the 1950s OTO developed an unusual 3in mount in which the two barrels were mounted one atop the other, the elevating gear lying between them. This mounting appeared only in the *Centauro* class escorts, and was apparently unsuccessful, being replaced by the predecessor to the current single-barrel 76mm/62.

Radars

MLA-1	The first major Italian radar, an L-band air search set made by MicroLambda for export; it was widely used in the Spanish Navy, and on board the Italian-built *Almirante Clemente* class export frigates (for Indonesia, Portugal, and Venezuela). Presumably it was the basis for RAN-3L, which was made by Selenia, the successor company to MicroLambda.
MLN-1	MicroLambda surface search radar.
MLT-4	MicroLambda improved version of the US Mk 39. MLT-3 was a version of Mk 26.
SPQ-2 (SPQ-701, SPS-702)	Dual-purpose surface- and short-range air-search radar for small craft; used as a surface search set in frigates and destroyers. SPQ-701 is a developed version used on board *Sparviero* class hydrofoils; SPS-702 is for *Maestrale* class frigates. These radars were made by SMA.
RAN-3L (SPS-768)	Selenia's standard L-band air search radar for large ships, development of which began in May 1966. In Italian service it replaced the US SPS-12 and SPS-40.
RAN-10S (SPS-774)	This S-bank air search radar is used on board *Lupo* and *Maestrale* class frigates.
RAN-11L/X and -12L/X	Multi-band radars for small ships such as hydrofoils.
EMPAR (SPY-790)	Search radar for Anglo-French-Italian 'Horizon' class using the Aster missile; will probably be used in French and Italian ships.
RTN-7X	Fire control radar for OG 3 director
RTN-10X (SPG-70)	Orion fire control radar for Argo NA-10 system; also used by Royal Navy as Type 912 in *Amazon* (Type 21) class
RTN-20X (SPG-74)	Fire control radar for Dardo system
RTN-30X (SPG-75)	Fire control radar for Dardo-E system

(Note that the Orion radars can control short-range missiles (Aspide) as well as 40mm and 76mm guns.)

Electronic Warfare

SLQ-A,B,C, D (Newton)	Predecessor to Nettuno, using directional jammers in spherical radomes. It also uses non-directional jammers in stabilised cylindrical radomes. Widely exported.
SLQ-732 (Nettuno)	Current phased array system, also used in Spain as Nettunel
SLQ-747 (INS-3)	Selenia system, mainly for export, but used on *Minerva* class corvettes

Combat Direction

SADOC (IPN)	Computer system derived from the US NTDS, but using Italian hardware and eventually Italian software. The current version is SADOC 2 (IPN-20).

Sonars

IP-64	Analog sonar suite of the *Toti* class, comprising a conformal passive chin array, and an active circular array on deck.
IPD 70	Sonar suite of the *Sauro* class, using much IP-64 technology and thus largely analog. IPD 70/S is a digital version. It employs a low frequency chin array, a high-frequency cylindrical array in the upper part of the bow, and a passive ranger along the sides of the ship. The passive ranger is MD 100.

FLEET STRENGTH 1947

(including ships transferred under the terms of the Peace Treaty)

BATTLESHIPS

Name	Launched	Disp (full load)	Fate
Cavour class			
GIULIO CESARE	15.10.11	29,032t	To USSR 15.12.48
Dorio class			
ANDREA DORIA	30.3.13	28,882t	Discarded 1.11.56
CAIO DUILIO	24.4.13	29,391t	Discarded 15.6.56
Littorio class			
ITALIA (ex-*Littorio*)	22.8.37	45,236t	Discarded 1.6.48
VITTORIO VENETO	22.7.37	45,029t	Discarded 1.2.48

Cavour class

Cesare (reconstructed during October 1933–October 1937) was assigned to the USSR as war reparations as *Z 11*. The Russians renamed her *Novorossiisk*, and she was probably mined and sunk at Sevastopol on 29 October 1955 (see Soviet Union section).

Doria class
Doria (rebuilt during 8.4.37–26.10.40) and *Duilio* (rebuilt during 1.4.37–15.7.40) were both used as training ships postwar except for a period in 1947–49 when *Duilio* was the flagship of the C in C of the Italian Navy. *Doria* was laid up in reserve from June 1953, followed by *Duilio*. Both were scrapped during 1957–61.

Littorio class
Italia and *Veneto* were returned to Italy in 1946. *Italia* was assigned to the USA and *Veneto* to the UK as War Reparations. They were ordered to be scrapped and were broken up during 1948–55 (*Italia*) and 1948–51 (*Veneto*).

CRUISERS

Name	Launched	Disp (full load)	Fate
Cadorna class			
LUIGI CADORNA	30.9.31	700lt	Discarded 1.5.51
Montecuccoli class			
RAIMONDO MONTECUCCOLI (C 552)	2.8.34	8853t	Discarded 1.6.64
Duca D'Aosta class			
EMANUELE FILBERTO DUCA D'AOSTA	22.4.34	10,374t	To USSR 2.3.49
EUGENIO DI SAVOIA	16.3.35	10,672t	To Greece 1.7.51
Abruzzi class			
LUIGI DI SAVOIA DUCA DEGLI ABRUZZI (C 550)	21.4.36	11,575t	Discarded 30 4.61
GIUSEPPE GARIBALDI (C 551)	21.4.36	11,117t	Discarded Jan 1972
'Capitani Romani'			
GIULIO GERMANICO	26.7.41	5334t	Renamed *San Marco* (D 563), discarded 1971
POMPEO MAGNO	24.8.41	5334t	Renamed *San Giorgio* (D 562), discarded 1980)
SCIPIONE AFRICANO	12.1.41	5334t	To France 15.8.48
ATTILIO REGOLO	28.8.40	5334t	To France 15.8.48

Cadorna class
Cadorna was employed for a short period as a training ship, but was laid up in reserve before 1950.

Montecuccoli class
When she was converted to a cadet training ship during 1947–mid-1949, the 152mm/53 twin turret in 'B' position was removed as was the 100mm/47 twin mounting in 'Q' position. Her light AA armament became 8–40mm (4 × 2) and 4–20mm (4 × 1). She was also fitted with radar. Two boilers were removed during the refit, so that she developed 75,000shp giving a maximum speed of 29kts. She was replaced as cadet training ship by *San Giorgio*.

Duca D'Aosta class
Both these ships were transferred to foreign navies as war reparations. *D'Aosta* went to the USSR and was renamed *Stalingrad*. She was later renamed *Kerch* and was discarded in the late 1950s. *Savoia* was transferred to Greece and was renamed *Helle*. She was discarded in 1964.

Abruzzi class
These two ships formed the backbone of the Italian Navy during the 1950s. Both were rebuilt during 1950–53. The secondary armament was reduced to 4–100mm/47 (2 × 2) and the light AA guns altered to 24–40mm (4 × 4, 4 × 2). Two of the eight boilers were removed, reducing the power to 85,000shp and the maximum sea speed to 29kts. The bridge was rebuilt.

Garibaldi was modernised at La Spezia in 1949–51 and *Abruzzi* was somewhat more completely modified in 1951–53. *Garibaldi* emerged with US radars: an SO-8 navigational set in a small radome atop her foremast, and a big SK-2 atop her mainmast. *Abruzzi* had her bridge structure considerably cut down and a tripod foremast stepped; it carried surface search radar (on a topmast) and the SK-2 dish (which was replaced by SPS-6C about 1955). About 1955 a height-finder was installed atop the heavy mainmast (which had carried a British Type 281 radar before the ship was modernised). A fire control radar was installed atop the main director forward. *Abruzzi* was decommissioned 30 April 1961 and discarded without further modification.

Garibaldi was rebuilt as a Terrier missile cruiser in 1957–61, design work having begun in 1954. Her forward turrets were replaced by twin rapid-fire (80 rounds/min) 135mm/45 (replaced by 135mm/53 in 1968) mounts and eight single 76mm/62 were fitted, four on each side. The Terrier launcher replaced the after turrets. A lattice foremast carried an SPS-39 3-D radar at its head, and SPS-6 and a surface search radar on its forward face. A new lattice mainmast carried a huge Italian-built ARGO-5000 long-range air-search radar. The usual pair of SPG-55 missile guidance radars were mounted aft, near the missile launcher. An Italian-built radar-equipped director was mounted above the bridge, with two more on each side of the superstructure to control the 76mm guns. Terrier missiles were stowed below the twin launcher aft, as in the first pair of US cruiser conversions, *Boston* and *Canberra* (the later Terrier CLGs and other Terrier ships all had their missiles stowed horizontally, rammed horizontally rather than vertically on to the launcher). If the systems were identical, magazine capacity was seventy-two weapons. Four Polaris ballistic missile tubes were sunk into the deck abaft the Terrier magazine. Presumably they were part of a larger US programme to distribute Polaris among NATO cruisers. At this time the United States had placed Jupiter ballistic missiles in Italy, and Polaris may have been considered a sea-going equivalent. After the 1962 Cuban Missile Crisis, the Jupiters were withdrawn, and the Polaris idea probably also lapsed. The ship tested dummy missiles, but never carried real ones. The two funnels were cut to one, the number of boilers was cut to four large and two small, so power fell to 85,000shp (30kts rather than 33kts). Electrical generating power had to be increased to 4500kW.

Garibaldi was decommissioned 20 February 1971 and stricken 16 November 1976.

'Capitani Romani' class
Germanico had been scuttled on 29 August 1943 before she was completed, but she was salvaged in 1947. Both she and *Magno* were rebuilt during 1951–55 with American weapons and radar. Their existing armament was replaced by 6–127mm/38 (3 × 2) in 'A', 'X' and 'Y' positions and 20–40mm AA. In 'B' position they had an Italian Menon triple ASW mortar. They were renamed *San Marco* and *San Giorgio* respectively and were recommissioned in 1956. In 1957 they were re-rated from cruisers to scouts and in 1958 from scouts to destroyer leaders. Both received US-supplied sensors: SPS-6 and SG-6B radars and SQS-11 sonar, the same sets as in the first postwar destroyers and escorts. The 5in fire control system was the US Mk 37, with Mk 25 radar. They then carried the number D 562 (*San Giorgio*) and D 563. *San Marco* was scrapped in this condition, but *San Giorgio* was completely rebuilt as a cadet training ship during 1963–65. She was fitted with CODAG machinery (2 Tosi Metrovick G 6 gas turbines, each of 7500shp, and 4 Fiat-Tosi diesels each of 4000bhp) on two shafts, giving a maximum speed on diesels alone of 20kts and combined of 28kts. She had new funnels, the second being set much farther aft, and was given accommodation for 130 cadets. 'X' position 127mm/38 mount was replaced by a single 76mm/62 and two more single 76mm/62 were fitted amidships, all the 40mm being removed. Two triple 324mm TT were

Giuseppe Garibaldi c 1965 Navarret Collection

San Giorgio c 1960 Navarret Collection

also added. As a training ship, *San Marco* had the new Italian SPQ-2 radar and three of the new Orion RTN-7X fire control radars for her 76.2mm guns.

Africano and *Regolo* were transferred to France as war reparations and were renamed *Guichen* and *Châteaurenault* respectively. They were discarded in 1961 and 1962 respectively.

DESTROYERS

Name	Launched	Disp (full load)	Fate
'Navigatori' class			
NICOLOSO DA RECCO	5.1.30	2580t	Discarded 15.7.54
Maestrale class			
GRECALE (D 552)	17.6.34	2207t	Discarded 31.5.64
Oriani class			
ALFREDO ORIANI	30.7.36	2254t	To France 8.8.48
'Soldati' class (1st group)			
ARTIGLIERE (ex-*Camicia Nera*)	8.8.37	2250t	To USSR 21.2.49
CARABINIERE (D 551)	23.7.38	2250t	Discarded 18.1.65
FUCILIERE	31.7.38	2250t	To USSR 17.1.50
GRANATIERE (D 550)	24.4.38	2250t	Discarded 1.7.58
(2nd group)			
LEGIONARIO	16.4.41	2500t	To France 15.8.48
MITRAGLIERE	28.9.41	2500t	To France 15.7.48
VELITE	31.8.41	2500t	To France 24.7.48

'Navigatori' class
By 1950 *Recco*'s best sea speed was 28kts. She was scrapped without being rebuilt.

Maestrale class
In 1949 *Grecale* was fitted with a British destroyer-type bridge and a lattice mast, and was given new radars. She had a light AA armament of 3–37mm guns. She was converted to a fast ASW escort in 1952–53. Her TT were removed and her light AA guns were replaced by 6–40mm. She was re-rated as a frigate on 10 April 1957. In 1959–60 she was rebuilt as a command ship. All her armament except for 2–40mm was landed and she was given a CIC and extensive communications equipment and extra accommodation to enable her to act as the fleet flagship until *Garibaldi* re-entered service.

Oriani class
Oriani was transferred to France as war reparations and was renamed *D'Esta-ing*. She was discarded in 1954.

'Soldati' class (1st group)
Artigliere and *Fuciliere* was transferred to the USSR as war reparations as *Z 12* and *Z 20* respectively. They were discarded in the late 1950s.

Carabiniere and *Granatiere* (D 551 and D 552 respectively) were rebuilt as fast ASW escorts in 1953–54. Their TT were removed and their light AA armament was replaced by 6–40mm. They were subsequently re-rated as frigates. *Carabiniere* was used as a trials ship for the Menon ASW mortar and was later used to test new Italian naval weapons. The was again re-rated as an auxiliary experimental ship (A 5314) after 1960.

(2nd group)
Legionario, Mitragliere and *Velite* were transferred to France as war reparations and renamed *Duchaffault, Jurien de la Gravière* and *Duperré* respectively. They were discarded in 1954, 1956 and 1961 respectively.

TORPEDO BOATS

Name	Launched	Disp (full load)	Fate
Pilo class			
GIUSEPPE CESARE ABBA	25.5.15	885t	Discarded 1.9.58
ANTONIO MOSTO	20.5.15	885t	Discarded 15.12.58
ROSOLINO PILO	24.3.15	885t	Discarded 1.10.54
La Masa class			
GIACINTO CARINI	7.11.17	837t	Discarded 31.12.58
NICOLA FABRIZI	8.7.18	837t	Discarded 1.2.57
Curtatone class			
MONZAMBANO	6.8.23	925t	Discarded 1.2.57
Spica class			
CASSIOPEA (F 553)	22.11.36	995t	Discarded 31.10.59
SAGITTARIO (F 557)	21.6.36	985t	Discarded 1.7.64
SIRIO (S 554)	14.11.35	985t	Discarded 31.10.59
ARETUSA (F 556)	6.2.38	1030t	Discarded 1.8.58
CALLIOPE (F 551)	16.4.38	1030t	Discarded 1.8.58
CLIO (F 555)	3.4.38	1030t	Discarded 31.10.59
LIBRA (F 552)	3.10.38	1030t	Discarded 1.4.64
Pegaso class			
ORIONE (F 554)	21.4.37	1575t	Discarded 1.1.65
ORSA (F 558)	21.3.37	1575t	Discarded 1.7.64
Ciclone class			
ALISEO	20.9.42	1625t	To Yugoslavia 3.5.49
ANIMOSO	15.4.42	1625t	To USSR 16.3.49
ARDIMENTOSO	27.6.42	1625t	To USSR 28.2.49
FORTUNALE	18.4.42	1625t	To Yugoslavia 28.4.49
INDOMITO	6.7.43	1625t	To Yugoslavia 28.4.49
Ariete class			
ARIETE	6.3.43	1110t	To Yugoslavia 30.4.49
BALESTRA	4.10.47	1110t	To Yugoslavia 1948

Pilo class
These ships were built as destroyers but reclassified as torpedo-boats on 1 October 1929. *Pilo* was scrapped without having been modified, but *Abba* and *Mosto* were converted to fast minesweepers in 1953–54. Their armament was altered to 1–102mm and 2/4–20mm AA guns, and they were given radar.

La Masa class
They were reclassified from destroyers to torpedo-boats on 1 October 1929. Both were converted to fast minesweepers in 1952–54. *Carini* was armed with 2–102mm (2 × 1) and 2–20mm, *Fabrizi* had only 1–102mm. Both were fitted with radar. *Fabrizi* was scrapped after being discarded but *Carini* was used as a training hulk, and renumbered GM 517.

Curtatone class
Monzambano was scrapped without having been rebuilt.

Spica class
(*Climene* group) *Cassiopea* was modernised as a fast corvette in 1950–52. Her TT were removed, and a Hedgehog added. (*Perseo* group) *Sirio* was rebuilt as a fast corvette in 1951–52. Her TT were removed, a Hedgehog added and only one 100mm/47. By 1958 the 100mm/47 had been replaced by 2–40mm (2 × 1). *Sagittario* was converted to an armament trials ship (F 557) in 1952–53, and was equipped with a variety of weapons. (*Alcione* group) *Aretusa* and *Calliope* were converted to fast corvettes in 1952–53, *Clio* in 1951–52 and *Libra* in 1950–52. Their TT and one 100mm/47 gun were removed and a Hedgehog and new radar were fitted. By 1958 the remaining 100mm/47 had been replaced by 2–40mm (2 × 1).

Pegaso class
Orione was rebuilt as a fast corvette in 1953–54 and *Orsa* in 1954–55. As with the other conversions, a Hedgehog and new radar was fitted and the aft 100mm/47 gun and the TT were removed. In about 1958 the other 100mm/47 gun was also removed.

Ciclone class
Aliseo and *Indomito* were transferred to Yugoslavia as war reparations and renamed *Biokovo* and *Triglav* respectively. *Animoso, Ardimentoso* and *Fortunale* were transferred as war reparations to the USSR as *Z 16, Z 19* and *Z 17* respectively.

Ariete class
Ariete was transferred to Yugoslavia as war reparations and renamed *Durmitor*. *Balestra* was captured by the Germans on her slip and renamed *TA 47*. She was taken over and completed by the Yugoslavs and renamed *Ucka*. Their sister *Fionda* was also captured by the Germans on her slip (and renamed *TA 46*). She was taken over by the Yugoslavs, but although they renamed her *Valebit* and started to complete her, she had been very badly damaged by the Germans at the retreat and proved to be not worth completing.

Pietro Calvi 1962

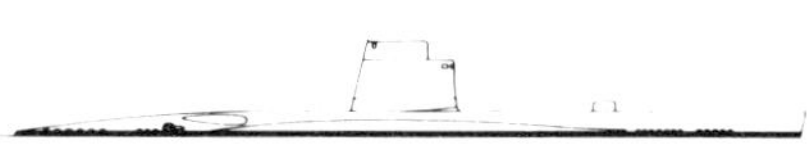

SUBMARINES

Name	Launched	Disp (surface)	Fate
'H' class			
H 1	16.10.16	355t	Discarded 23.3.47
H 2	19.10.16	355t	Discarded 23.3.47
H 4	17.4.17	355t	Discarded 23.3.47
Glauco class			
OTARIA (ex-*Espadarte*)	20.3.35	1054t	Discarded 1.2.48
Foca class			
ATROPO	20.11.38	1305t	Discarded 23.3.47
ZOEA	5.12.37	1305t	Discarded 23.3.47
Marcello class			
DANDOLO	20.11.37	1043t	Discarded 23.3.47
Brin class			
BRIN	3.4.38	1000t	Discarded 1.2.48

Name	Launched	Disp (surface)	Fate
Cagni class			
AMMIRAGLIO CAGNI	20.7.40	1653t	Discarded 1.2.48
Romolo class			
R 3	7.9.46	2155t	BU *c*1947
R 4	30.9.46	2155t	BU *c*1947
R 7	21.10.43	2155t	BU *c*1947
R 8	28.12.43	2155t	BU *c*1947
R 9	27.2.44	2155t	BU *c*1947
R 10	12.7.44	2155t	BU *c*1947
R 11	6.8.44	2155t	Discarded 1947
R 12	29.9.44	2155t	Discarded 1947
Mameli class			
GIOVANNI PROCIDA	1.4.28	810t	Discarded 1.2.48
GOFFREDO MAMELI (ex-*Masaniello*)	9.12.26	810t	Discarded 1.2.48
TITO SPERI	25.5.28	810t	Discarded 1.2.48
Pisani class			
VETTOR PISANI	24.11.27	866t	Discarded 23.3.47
Bandiera class			
FRATELLI BANDIERA	7.2.29	925t	Discarded 1.2.48
LUCIANO MANARA	5.10.29	925t	Discarded 1.2.48
CIRO MENOTTI	29.12.29	925t	Discarded 1.2.48
Squalo class			
SQUALO	15.1.30	920t	Discarded 1.2.48
Bragadin class			
MARCANTONIO BRAGADIN	21.7.29	965t	Discarded 1.2.48
FILIPPO CORRIDONI	30.3.30	965t	Discarded 1.2.49
Settembrini class			
RUGGIERO SETTIMO	29.3.31	938t	Discarded 23.3.47
Argonauta class			
JALEA	15.6.32	650t	Discarded 1.2.48
Sirena class			
GALATEA	5.10.33	680t	Discarded 1.2.48
Perla class			
DIASPRO	5.7.36	680t	Discarded 1.2.48
ONICE	15.6.36	680t	Discarded 23.3.47
PERLA	3.5.36	680t	Captured 9.7.42, became Greek *Matrozos*
TURCHESE	19.7.36	680t	Discarded 1.2.48
Adua class			
ALAGI	15.11.36	680t	Discarded 23.5.47
Acciaio class			
BRONZO	28.9.41	697t	Captured 12.7.43, became French *Narval*
GIADA (ex-*PV 2*, ex-*Giada*) (S 501)	10.7.41	697t	Discarded 1.1.66
NICHELIO	12.4.42	697t	To USSR 7.2.49
PLATINO	1.6.41	697t	Discarded 1.2.48
Flutto class (Type I)			
MAREA	10.12.42	930t	To USSR Feb 49
NAUTILO	12.4.42	930t	Became Yugoslav *Sava*
VORTICE (ex-*PV 1*, ex-*Vortice*) (S 502)	23.2.43	930t	Discarded 1.8.67
(Type II)			
PIETRO CALVI (ex-*UIT 7*, ex-*Bario*) (S 503)	23.1.44	913t	Discarded Jan 1972
CM class			
CM 1	5.9.43	90t	Discarded 1.2.48
CM 2	Apr 1945	90t	Museum exhibit 1951
CB class			
CB 8	1.8.43	42t	BU 1948
CB 9	1.8.43	42t	BU 1948
CB 10	1.8.43	42t	BU 1948
CB 11	24.8.43	42t	BU 1948
CB 12	28.8.43	42t	BU 1948
CB 19	1943	42t	BU 1947
CB 20	1943	42t	To Yugoslavia?
CB 22	1943	42t	Museum exhibit 1951

By 1947 most Italian submarines were in poor condition or totally obsolete or both, and these boats were discarded in March 1947. The Peace Treaty prohibited Italy from building or commissioning any submarines, so almost all the remainder were disposed of in February 1948. Only the USSR took submarines as war reparations, the *Acciaio* class boat *Nichelio* being transferred as the *Z 14*, and the *Flutto* class *Marea* as the *Z 13*. *Nautilo* of the same class was salvaged by the Yugoslavs and renamed *Sava*.

No *Romolo* class submarine saw service in the postwar Italian Navy, but the two incomplete boats *R 11* and *R 12*, which had been scuttled in 1945, were salvaged in 1946 and used as oil hulks for more than 20 years. Two small submarines, *CM 2* and *CB 22*, became permanent museum exhibits in 1951.

Only two Italian submarines remained in servcie, the *Acciaio* class *Giada* and *Flutto* class *Vortice* (S 502). These were used as training boats, but evaded the terms of the Peace Treaty by being classed as battery-chargers with their names changed to *PV 2* and *PV 1* respectively. The 'no submarines' clause expired in 1952, and they were returned to the navy list on 1 March 1951 (*Giada*) and 1 January 1952 (*Vortice*), resuming their original names. Between 1951–53 both were rebuilt as training boats with no gun and a streamlined fin.

The incomplete *Bario* of the *Flutto* (*Type II*) class, which had been scuttled in 1945, was salvaged in the same year. She was rebuilt during 1957–61, being relaunched on 21 June 1959, and she entered service on 16 December 1961. She differed very considerably from her half-sister *Vortice*, having a new tear-drop bow and stern, with only one shaft. As rebuilt her data was:

Displacement: 905t/1107t
Dimensions: 216ft 6in oa × 23ft × 13ft 3in (*66.0m × 7.0m × 4.0m*)
Machinery: 1 shaft, 2 MAN diesels plus 3 electric motors, 2700bhp/?shp = 14kts/14kts
Armament: 4–533mm TT (4 × 1)
Complement: 60

She was renamed *Pietro Calvi* (S 503) in March 1961, and was effectively a new submarine with a greatly improved underwater performance.

COLONIAL SLOOP

Name	Launched	Disp (full load)	Fate
ERITREA	28.9.36	3068t	To France 12.2.48

She was transferred to France as war reparations and renamed *Francis Garnier*. She was discarded in 1960.

CORVETTES

Name	Launched	Disp (full load)	Fate
Gabbiano class			
APE (F 567)	22.11.42	660t	Discarded 1965
BAIONETTA (ex-*Partigiana*) (F 578)	5.10.42	660t	Discarded Jan 1972
BOMBARDA (ex-*UJ 206*, ex-*Bombarda*) (F 549)	31.8.42	660t	Discarded 1976–77
CHIMERA (F 569)	30.1.43	660t	Discarded 1976–77
CORMORANO (F 575)	20.9.42	660t	Discarded Jan 1972
CRISALIDE (F 547)	8.12.47	660t	Discarded 1972
DANAIDE (F 563)	21.10.42	660t	Discarded 1968
DRIADE (F 568)	7.10.42	660t	Discarded 1.8.66
FARFALLA (F 548)	4.1.48	660t	Discarded Jan 1972
FENICE (F 577)	1.3.43	660t	Discarded 1.10.65
FLORA (F 572)	1.12.42	660t	Discarded *c*1970
FOLAGA (F 576)	14.11.42	660t	Discarded 1965
GABBIANO (F 571)	23.6.42	660t	Discarded Jan 1972
GRU (F 566)	23.12.42	660t	Discarded *c*1970
IBIS (F 561)	12.12.42	660t	Discarded Jan 1972
MINERVA (F 562)	5.11.42	660t	Discarded 1969
PELLICANO (F 574)	12.2.43	660t	Discarded *c*1970
POMONA (F 573)	18.11.42	660t	Discarded 1965
SCIMITARRA (F 564)	16.9.42	660t	Discarded Jan 1972
SFINGE (F 579)	9.1.43	660t	Discarded 1976–77
SIBILLA (F 565)	10.3.43	660t	Discarded Jan 1972
URANIA (F 570)	21.4.43	660t	Discarded Jan 1972

The *Gabbiano* class were numerically the largest Italian ship class to survive the war. Three, *Bombarda, Crisalide* and *Farfalla*, were scuttled incomplete by the Germans in April 1945 and completed postwar.

These ships were completed with 1–100mm/47, 3–20mm/70 (3 × 1), 4–20mm/65 (2 × 2), 2–450mm (17.7in) TT, 8 DCT, 2 depth charge tracks (6 charges each), and 2 towed ASW charges. Torpedo tubes were landed postwar and then re-installed in 1952–53; they fired the Whitehead W 200 torpedo (45cm × 575cm, 930kg). A major reconstruction programme was conducted in 1952–56. A new British-style bridge structure with flat angled sides provided a CIC and an open bridge. In 1956 the corvettes divided into two main groups:

Group I: convoy AA escorts: *Ape, Chimera, Cormorano**, *Danaide**, *Fenice***, *Flora***, *Folaga***, *Pellicano, Pomona***, *Sibilla***, *Sfinge**. Starred units had SPS-6 air search radars; the others had the US-supplied SO-13 surface search set or the Italian (SMA)-made NSM-8 surface search set (double stars). Main armament was 4–40mm/56 (2 × 2) controlled by one Mk 51 forward and another aft. The ASW armament was one Mk 10 Hedgehog abaft the forward

40mm mount, four short Menon mortars, a single depth charge track aft, and a pair of 450mm torpedo tubes. *Cormorano* and *Danaide* (with SPS-6 radars) had a prominent deckhouse forward of the bridge, and lacked the Hedgehog otherwise mounted there. *Sfinge*, the other SPS-6 unit, mounted a 100mm gun forward instead of 40mm guns; she also lacked the Hedgehog (later she received the Hedgehog and 40mm guns forward). Similarly, when newly completed *Driade* and *Flora* had a 100mm gun forward, but by 1956 they had Hedgehogs. In 1956 *Ape* was largely disarmed as a commando/frogman transport; she was left with only 2–20mm guns forward of her superstructure. The standard ASW armament included 144 Hedgehog rounds and sixty-eight Menon bombs and depth charges. The sonar was a US-supplied QCU-2. All but *Pomona* and *Sibilla* later were fitted with heavy double-layer depth charge tracks aft which could hold 12 Mk 6 depth charges.

Group II: convoy ASW escorts: *Baionetta***, *Driade, Farfalla, Gru, Ibis, Minerva, Scimitarra, Urania, Crisalide.* Double-starred units had the Italian-made NSM-8 surface search radar; the others had SO-13. Main gun armament was 3–40mm (1 × 1, 1 × 2 aft). They carried a Mk 15 trainable stabilised Hedgehog forward of the single 40mm gun forward, plus the torpedo tubes and Menon projectors of the AAW ships. However, most of these ships eventually were fitted with massive multi-layer depth charge tracks aft, which resembled wine racks. At least as newly converted in 1953, *Farfalla* mounted a 100mm gun forward and lacked the Hedgehog (it had been installed by 1956).

Gabbiano and *Bombarda* were also ASW escorts; after 1956 they had no 40mm forward at all, just a Mk 15 Hedgehog, with the same twin 40mm mount aft, and 2–20mm/70 (2 × 1) alongside the bridge structure. Like the other ships, they had four Menon depth charge throwers aft, but they had two short depth charge tracks rather than the 'wine racks'. When newly completed, both ships temporarily mounted a 100mm gun forward (and no Hedgehog).

GUNBOAT

Name	Launched	Disp (full load)	Fate
Bafile class			
ERNESTO GIOVANNINI	11.3.22	206t	Discarded 1.12.50

Designed as a small minelayer but used as patrol boat.

MINESWEEPERS

Name	Fate	Name	Fate
'RD 6' type: launched 1916, 193t			
RD 6	To Yugoslavia 19.8.47		
'RD 16' type: launched 1917–18, 197–198t			
RD 16	To Yugoslavia 25.8.48	RD 27	To Yugoslavia 27.8.48
RD 21	To Yugoslavia 25.8.48	RD 28	To Yugoslavia 14.9.48
RD 25	To Yugoslavia 1948	RD 29	To Yugoslavia 14.9.48
'RD 32' type: launched 1919, 179t			
RD 32	Discarded 30.6.56	RD 34	Discarded 30.6.56
'RD 40' type: launched 1919, 200t			
RD 40	Discarded 15.4.56	RD 41	Discarded 1.1.53
Azalea class: launched 1942–43, 215t			
AZALEA (ex-*BYMS 2142*, ex-*YMS 142*)	Discarded 1966	MAGNOLIA (ex-*BYMS 2206*, ex-*USS 206*)	Discarded 1966
BEGONIA (ex-*BYMS 2173*, ex-*YMS 173*)	To Customs Apr 1966	ORCHIDA (ex-*BYMS 2137*, ex-*YMS 137*)	Discarded 1966
DALIA (ex-*BYMS 2141*, ex-*YMS 141*)	To Customs Apr 1966	PRIMULA (ex-*BYMS 2278*, ex-*YMS 278*)	Discarded 1966
FIORDALISO (ex-*BYMS 2277*, ex-*YMS 277*)	Discarded 1966	TULIPANO (ex-*BYMS 2194*, ex-*YMS 194*)	Discarded 1966
GARDANIA (ex-*BYMS 2150*, ex-*YMS 150*)	Discarded 1966	VERBENA (ex-*BYMS 2280*, ex-*YMS 280*)	Discarded 1966
GLADIOLO (ex-*BYMS ?* ex-*YMS*)	Discarded 1966		
Anemone class: launched 1942, 207t			
ANEMONE (ex-*BYMS 2009*, ex-*YMS 9*)	Discarded 1966	MUGHETTO (ex-*BYMS 2023*, ex-*YMS 23*)	Discarded 1966
BIANCOSPINO (ex-*BYMS 2012*, ex-*YMS 12*)	Discarded 1966	NARCISCO (ex-*BYMS 2024*, ex-*YMS 24*)	Discarded 1966
GERANIO (ex-*BYMS 2014*, ex-*YMS 14*	Discarded 1966	OLEANDRO (ex-*BYMS 2027*, ex-*YMS 27*)	Discarded 1966
Dance class: launched 1940–41, 530t			
DR 307 (ex-*Minuet*)	Discarded 1962	DR 312 (ex-*Gavotte*)	Discarded 1962
DR 308 (ex-*Twostep*, (ex-*Tarantella*)	Discarded 1962	DR 316 (ex-*Hornpipe*)	Discarded 1962
Shakespearian class: launched 1941, 545t			
DR 310 (ex-*Othello*)	Discarded 1962		
Isles class: launched 1940–43, 545t			
DR 301 (ex-*Burra*)	Discarded 1962	DR 309 (ex-*Grain*)	Discarded 1962
DR 302 (ex-*Cumbrae*)	Discarded 1962	DR 311 (ex-*Mousa*)	Discarded 1962
DR 303 (ex-*Unst*)	Discarded 1962	DR 313 (ex-*Foula*)	Discarded 1962
DR 304 (ex-*Staffa*)	Discarded 1962	DR 314 (ex-*Ensay*)	Discarded 1962
DR 305 (ex-*Filla*)	Discarded 1962	DR 315 (ex-*Stroma*)	Discarded 1962
DR 306 (ex-*Egilsay*)	Discarded 1962		
'DV' class: launched 1944–45, 110t			
DV 401 (ex-*DV 111*)	Discarded 1957	DV 409 (ex-*DV 116*)	Survey Ship c1960
DV 402 (ex-*DV 112*)	Discarded 1959	DV 411 (ex-*DV 121*)	Discarded 1959
DV 403 (ex-*DV 113*)	Discarded 1959	DV 412 (ex-*DV 122*)	Discarded 1959
DV 404 (ex-*DV 115*)	Discarded 1959	DV 413 (ex-*DV 123*)	Discarded 1959
DV 405 (ex-*DV 131*)	Discarded 1957	DV 414 (ex-*DV 124*)	Discarded 1959
DV 406 (ex-*DV 132*)	Discarded 1957	DV 415 (ex-*DV 125*)	Discarded 1957
DV 407 (ex-*DV 134*)	Discarded 1957	DV 133	Discarded 1953
DV 408 (ex-*DIDV 114*)	Survey ship c1960	DV 135	Discarded 1953

The first seven went to Yugoslavia as war reparations and were renamed *ML 301–307.* The remainder served in the Italian Navy until they were scrapped.

The *Azalea* and *Anemone* class coastal minesweepers were USN YMS that had been transferred to the Royal Navy during the Second World War. They were acquired by the Italian Navy in April 1947. The main difference between the two classes was that the *Azalea*s had one funnel and the *Anemone*s had two. They were allocated the pennant numbers M 5400–M 5416. Some ships of both classes had their 76mm gun replaced by a 20mm gun in 1956. The almost identical *Dance, Shakespearian* and *Isles* class trawlers were transferred from the British to the Italian Navy between January and March 1946. The Italians classified them as minesweepers, and gave them the pennant numbers M 5301–M 5316. In 1956 they all had their 76mm gun replaced by a 20mm, giving them an armament of 3–20mm (3 × 1). In 1962 they were removed from the active list and reclassified as general purpose auxiliaries. The 'DV' class coastal minesweepers were reclassified as motor patrol boats in *c*1955 and used for coastal patrols. Some were refitted with lattice masts, *DV 408* and *DV 409* were disarmed in *c*1960 and used as survey vessels. They were discarded in *c*1967.

LIGHT CRAFT

Name	Fate	Name	Fate
MAS 430 class: launched 1930, 14t			
MAS 433	Discarded 1.8.49	MAS 434	Discarded 1.8.49
MAS 438 class: launched 1934–35, 40t			
ME 38 (ex-*AS 25* ex-*MAS 438*)	Discarded 1.7.50	ME 41 (ex-*AS 28*, ex-*MAS 441*)	Discarded 1.6.50
ME 40 (ex-*AS 27*, ex-*MAS 440*)	To USSR 6.7.49		
MAS 501 class: launched 1936–37, 21.5t			
MAS 1 (ex-*MEB 1*, ex-*MAS 510*)	Discarded 28.6.50	MAS 3 (ex-*MEB 3*, ex-*MAS 520*)	Discarded 1.7.50

Name	Fate	Name	Fate
MAS 2 (ex-*MEB 2*, ex-*MAS 514*)	Discarded 1.7.50	NE 109 (ex-*MAS 4*, ex-*MEB 4*, ex-*MAS 521*)	Discarded 1.7.50
MAS 516	To USSR 6.7.49	MAS MT 523 (ex-*MAS 523*)	Discarded 1949
MAS 519	To USSR 6.7.49	MAS 9 (ex-*MEB 9*, ex-*MAS 525*)	Discarded 1950
MAS 526 class: launched 1939, 25.2t			
MAS 5 (ex-*MEB 5*, ex-*MAS 538*)	Discarded 1.7.50	MAS 6 (ex-*MEB 6*, ex-*MAS MT 545*, ex-*MAS 545*)	Discarded 1.7.50
MAS 543	To France 26.9.49	MAS 7 (ex-*MEB 7*, ex-*MAS 547*)	To Customs 1.7.50
MAS 555 class: launched 1941, 28t			
MAS 8 (ex-*MEB 8*, ex-*MAS 562*)	Discarded 1.7.50		
MS First series: launched 1942, 62.4t			
MS 471 (ex-*MS 611*, ex-*MV 611*, ex-*MS11*)	Discarded 1.5.65	MS 473 (ex-*MS 613*, ex-*MV 613*, ex-*MS 31*)	Discarded 1975
MS 472 (ex-*MS 612*, ex-*MV 612*, ex-*MS 24*)	Discarded 1974	MS 35	To France 15.12.48
MS Second Series: launched 1942–43, 67t			
MS 52	To USSR 18.5.49	MS 65	To USSR 6.7.49
MS 53	To USSR 18.5.49	MS 483 (ex-*MS 617*, ex-*MV 617*, ex-*MS 72*)	Discarded 1.3.63
MS 474 (ex-*MS 614*, ex-*MV 614*, ex-*MS 54*)	Discarded 1978	MS 484 (ex-*MS 618*, ex-*MV 618*, ex-*MS 73*)	Discarded 1.3.63
MS 481 (ex-*MS 615*, ex-*MV 615*, ex-*MS 55*)	Discarded 1978	MS 475 (ex-*MS 619*, ex-*MV 619*, ex-*MS 74*)	Discarded 1965
MS 482 (ex-*MS 616*, ex-*MV 616*, ex-*MS 56*)	Discarded 1.3.63	MS 75	To USSR 6.7.49

Name	Fate	Name	Fate
MS 61	To USSR 6.7.49		
VAS First series: launched 1942, 68t			
VAS 711 (ex-*VAS 201*)	Discarded 1953	VAS 493 (ex-*VAS 714*, ex-*VAS 218*)	Discarded 1957
VAS 491 (ex-*VAS 712*, ex-*VAS 204*)	Discarded 1957	FAS 715 (ex-*VAS 222*)	Discarded 1953
VAS 716 (ex-*VAS 205*)	Discarded 1953	VAS 494 (ex-*VAS 721*, ex-*VAS 224*)	Discarded 1953
VAS 492 (ex-*VAS 713*, ex-*VAS 211*)	Discarded ? 1953	? (ex-*VAS 228*)	Discarded 1953
VAS Second series: launched 1942–43, 68t			
VAS 495 (ex-*VAS 722*, ex-*VAS 233*)	Discarded 1953	VAS 497 (ex-*VAS 726*, ex-*RA 266*, ex-*VAS 241*)	Discarded 1957
VAS 723 (ex-*VAS 235*)	Discarded 1953	VAS 245	To USSR 30.6.49
VAS 496 (ex-*VAS 724*, ex-*VAS 237*)	Discarded 1957	VAS 245	Burnt out 21.8.47
VAS 725 (ex-*RA 265*, ex-*VAS 240*)	Discarded 1953	VAS 248	To USSR 1949

Under the terms of the Peace Treaty the torpedo launching gear had to be removed from all Italian torpedo-boats. They were reclassified as *Motovedette* (MV) and rearmed with 2–6 x 20mm guns. All the small torpedo-boats (MAS) were either transferred to foreign navies as war reparations or discarded by the end of 1950, though many remained in use for a number of years as harbour craft. Those large torpedo boats (MS) that were not transferred as war reparations, however, were refitted with TT in 1952–53. *MS 472* and *473* were rebuilt as convertible MTB/MGB in 1960 and *MS 474* and *481* were similarly rebuilt in 1961. They were given new machinery developing a total of 4500bhp = 33kts. As MGBs they were armed with 2–40mm (2 × 1) and as MTBs 1–40mm and 2–450mm TT. They could also carry mines. The VAS (anti-submarine) boats that were not transferred as war reparations were used mainly as coastal minesweepers.

MISCELLANEOUS

In 1947 the Italian Navy possessed a number of ex-Allied and ex-Axis landing craft. Of the ten ex-German (Italian-built) and four ex-Italian MFP-D class landing craft in service in 1947, nine remained in 1981. All had been deleted by 1995. Five of the ten ex-British LCT (3) are still in service, though by 1982 three were lighthouse tenders and seven were repair craft. Five were deleted 1990–93, and five extant 1995.

Giuseppe Garibaldi 1985 — Fincantieri

MAJOR SURFACE SHIPS

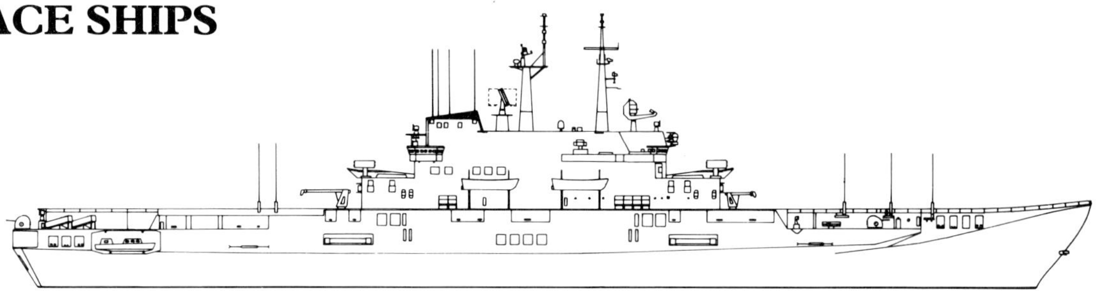

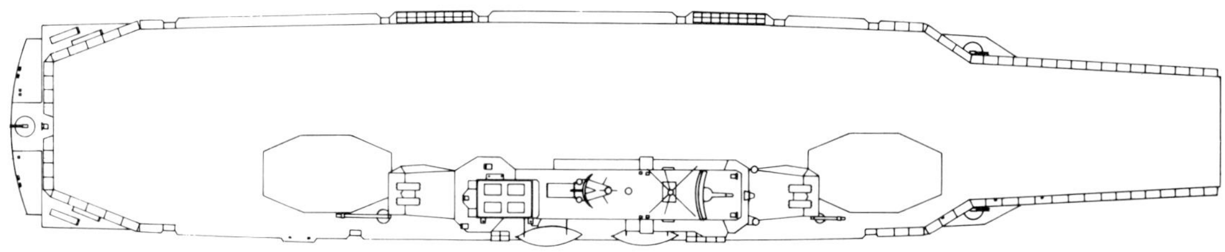

Giuseppe Garibaldi as completed

The first Italian aircraft carrier, this ship was originally limited in its capabilities by a law dating from 1923 preventing the Italian Navy from operating combat aircraft from ships. When this law was changed in 1992, sixteen AV-8B+ were purchased. The design envisaged VSTOL aircraft operations, since it ultimately included a ski-jump ramp forward. The hangar can accommodate either twelve SH-3D Sea King helicopters or ten AV-8B and one Sea King. *Garibaldi* was ordered on 20 February 1978, and began sea trials on 3 December 1984.

In about *c*1990 the original fit of four Otomat SSMs was replaced by eight Otomat 2 SSMs (the folding-wing type). Only four missiles are usually carried, however. She is equipped as a fleet flagship and also serves as an 'interpreter' between the on-shore Italian air defence system and deployed ships (their electronic data links are not compatible with the land-based system). Like the British *Invincible* class she can carry troops (up to 600).

A second, larger, ship is planned as of 1995. Possible names are *Giuseppe Mazzini* or *Conti di Cavour*. This project may be merged with the British *Invincible* replacement and the Spanish project for a second carrier, some time early in the next century.

GIUSEPPE GARIBALDI *aircraft carrier*

Displacement: 10,000t standard; 13,850t full load
Dimensions: 591ft 2in oa × 99ft 9in (78ft 1in wl) × 22ft (26ft 9in over sonar)
180.2m × 30.4m (23.8m) × 6.7m (8.2m)
Machinery: 2 shafts, 4 GE/Fiat LM-2500 gas turbines, 80,000shp = 29.5kts. Range 7000nm at 20kts
Armament: 4 Otomat SSM (4×1), 2 Albatros SAM launchers (2×8, 48 Aspide missiles), 6–40mm Breda Dardo (3×2), 6–324mm ASW TT (2×3), aircraft see notes
Sensors: Radar SPS-768, SPS-774, SPS-52C, SPG-74, SPG-75; sonar DE 1160 LF bow-mounted; ECM SLQ-732, SADOC 2 combat direction system
Complement: 550 plus 230 air group plus 45 staff

No	Name	Builder	Laid down	Launched	Comp	Fate
C551	GIUSEPPE GARIBALDI	Italcantieri, Monfalcone	26.3.81	4.6.83	30.9.85	Extant 1995

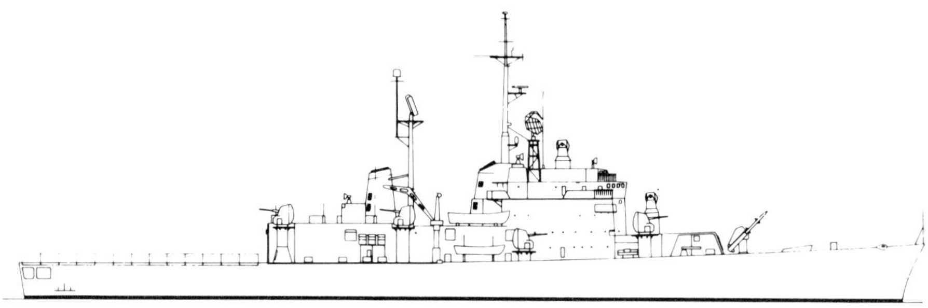

Andrea Doria 1978

Both ships were ordered under the 1957–58 new construction programme. They were designed to carry Terrier SAMs and three Sea King ASW helicopters to provide long range AA and ASW protection to task groups and convoys. An enlarged version of the *Impavido* class destroyer design was used, with a considerably increased beam to allow a hangar and a 98ft 6in × 52ft 6in (*30m × 16m*) flight deck to be worked in aft. In practice they proved to be too small to operate Sea Kings, so they carried four smaller ASW helicopters instead. The flight deck was cantilevered out at the extreme stern to provide more operating space.

This class was to have been armed with the SMP 3 76mm/62 gun used in the *Albatros* class. Due to its failure, the new OTO-Melara weapon was substituted. These guns were adopted because the Italian Navy decided in 1958 that nothing short of a 76mm weapon would be adequate for point defence against aircraft. These were the first Italian warships with automated combat data systems (SADOC, equivalent to the contemporary US NTDS).

The official sketch published in 1959 showed a lattice foremast like that of the newly-modernised *Garibaldi*, with SPS-39 at its top and SPS-6 or SPS-12 on a platform further down; there was no mainmast. In fact they were completed with two pole masts, the foremast carrying SPS-12 and the mainmast SPS-39A. The gun fire control system was the new Italian-made NA-9 Orion with RTN-10X (SPG-70) radar.

Doria was modernised in 1976–78. *Duilio* was converted into a training ship (to replace *San Marco*) in lieu of full modernisation in 1979–80. The most important change was replace-

DORIA class *helicopter cruisers*

Displacement: 5000t standard; 6500t full load
Dimensions: 472ft 7in pp, 489ft 9in oa × 56ft 5in × 16ft 5in
144.0m, 149.3m × 17.3m × 5.0m
Machinery: 2-shaft geared steam turbines, 4 Foster Wheeler boilers, 60,000shp = 30kts. Range 5000nm at 7kts
Armament: 1 Terrier SAM (1×2), 8–76mm/62 (8×1), 6–324mm TT (2×3), 4 helicopters. See notes
Sensors: Radar SPS-12, SPS-39A, SPQ-2; sonar SQS-39
Complement: 485

No	Name	Builder	Laid down	Launched	Comp	Fate
C 553	ANDREA DORIA	CNR, Riva Trigoso	11.5.58	27.2.63	23.2.64	Stricken 19.7.91
C 554	CAIO DUILIO	Castellammare	16.5.58	22.12.62	30.11.64	In reserve for stricking 30.9.92

ment of the Terrier missile by SM-1(ER). Presumably the existing SADOC system was modernised. *Doria* received updated sensors: SPS-40 2-D air search radar in place of SPS-12, SPS-52 in place of SPS-39A (*Duilio* received the new flat array antenna, but not fully updated electronics), SPG-55C in place of the earlier SPG-55A missile fire control radar and SQS-23 sonar in place of SQS-39 (a variant of SQS-4). In *Duilio*, classrooms replaced the former helicopter hangar and the after pair of 76mm guns and their NA-9 director were removed. By 1980, both ships had Italian RAN-3L (SPS-768) long-range air search radars in place of the earlier US sets. Both also received new EW suites (SPR-4 intercept, SLQ-D jammers).

A third unit, to have been named *Enrico Dandolo*, was deferred; the larger *Vittorio Veneto* was built instead.

Caio Duilio as completed — Navarret Collection

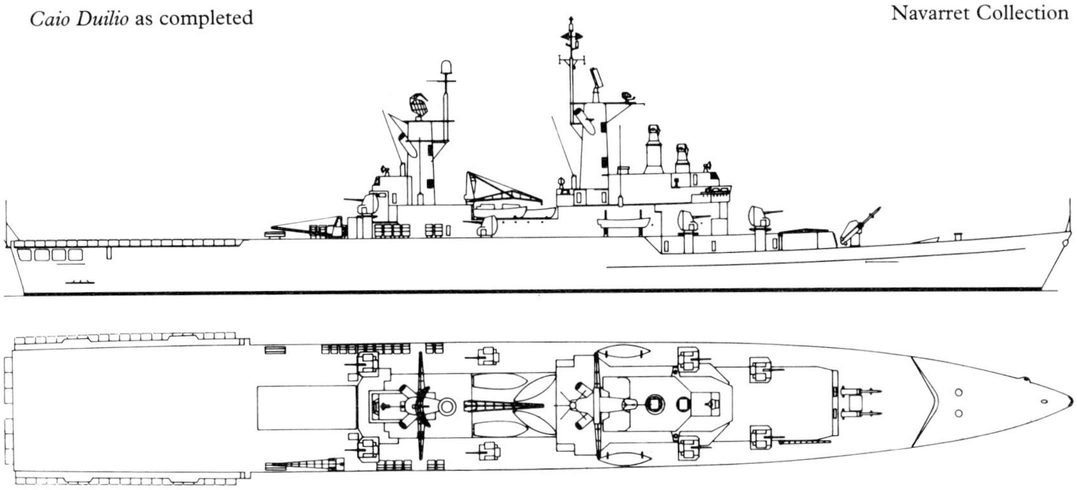

Vittorio Veneto 1974

VITTORIO VENETO *helicopter cruiser*

Displacement:	7500t standard; 8850t full load
Dimensions:	560ft pp, 589ft 3in oa × 63ft 7in × 19ft 9in *170.6m, 179.6m × 19.4m × 6.0m*
Machinery:	2-shaft geared steam turbines, 4 Foster Wheeler boilers, 73,000shp = 30.5kts. Range 5000nm at 17kts
Armament:	1 Terrier/ASROC SAM/ASW (1 × 2), 8–76mm/62 (8 × 1), 6–324mm TT (2 × 3), 9 helicopters. See notes
Sensors:	Radar SPS-40, SPS-52, SPQ-2B, SPG-55B, RTN-10X; sonar SQS-23
Complement:	550

No	Name	Builder	Laid down	Launched	Comp	Fate
C 550	VITTORIO VENETO	Castellammare	10.6.65	5.2.67	12.7.69	Extant 1995

This larger cruiser was built instead of a third *Andrea Doria*. On the enlarged hull, the hangar could be worked in below the flight deck aft instead of forward of it. The enlarged flight deck (60ft 8in × 13ft 3in, *18.49m × 4.03m*) can accommodate six SH-3D or nine AB-212 helicopters, compared to four AB-212 on an *Andrea Doria*. The deeper hull forward accommodates a third missile drum below the usual pair; the Mod 7 version of the Mk 10 launcher in this ship (Aster) fires both the Terrier anti-aircraft weapon and the ASROC ASW missile (*Vittorio Veneto* was the only ASROC-firing ship in the Italian Navy). Missile capacity rose from forty to sixty.

Vittorio Veneto was modernised in 1981–84, 4 Otomat Mk 2 SSM and 2 Dardo point defence systems (with 3 twin 40mm guns) being added. An Italian RAN-3L (SPS-768) radar replaced the US-supplied SPS-40, and RTN-20X (SPG-74) fire control radars were added for the Dardo systems. The earlier NA-10 systems (with RTN-10X radars) controlling the 76.2mm guns remained. Terrier missiles were replaced by the successor SM-1 (ER).

Vittorio Veneto served as fleet flagship until replaced by *Giuseppe Garibaldi*; she now (1995) serves as a training ship, and is likely to remain in service until about the turn of the century. With the retirement of the US Terrier/SM(ER) ships, she is the last ship in the world armed with the two-stage Standard Missile.

A projected second ship would have been named *Italia*. When she dropped out of the long-range programme, a larger missile cruiser/carrier, *Trieste*, was projected, but she, too, was cancelled. Ultimately the Italian Navy chose to separate the carrier and area air defence roles, building a light carrier.

Vittorio Veneto 1985
Italian official

Ex-US BENSON/GLEAVES class *destroyers*

Particulars: As US *Benson/Gleaves* class

No	Name	Builder	Acquired	Fate
D 553	ARTIGLIERE (ex-*Woodworth*)	Bethlehem	11.6.51	Discarded 1971
D 554	AVIERE (ex-*Nicholson*)	Boston NY	11.6.51	Sunk as target 1975

Artigliere was the ex-USN *Benson* class destroyer USS *Woodworth* and *Aviere* was the ex-USN *Gleaves* class destroyer USS *Nicholson*. *Aviere* could be distinguished from her half-sister by having round funnels, a higher bridge and no shield on 'X' 5in (127mm) gun mounting. Both were transferred to the Italian Navy under MDAP on 11 June 1951. They had their 5–21in (533mm) TT (1 × 5) removed. When they were replaced by the ex-*Fletcher* class *Fante* and *Geniere*, *Artigliere* was discarded straight away, but *Aviere* was converted to an experimental ship in 1970. She was first used to develop the OTO Melara lightweight 127mm/54 (5in; mounted in 'B' position) and 76mm/62 (3in; in 'X' position), and then the Albatros SAM.

Ex-US FLETCHER class *destroyers*

Particulars: As US *Fletcher* class

No	Name	Builder	Acquired	Fate
D 561	FANTE (ex-*Walker*)	Bath Iron Wks	2.7.64	Discarded 1977
D 555	GENIERE (ex-*Pritchett*)	Tacoma	7.1.70	Discarded 1975

The two ex-USN *Fletcher* class destroyers were transferred to Italy on 2 July 1969 (*Fante*) and on 7 January 1970 (*Geniere*). Both were built in 1943 and modernised postwar. *Fante* became an ASW escort (DDE), with 2–5in guns and Weapon Alfa in 'B' position. *Geniere* was a 4-gun *Fletcher*, with a twin 3in/50 replacing her 'Q' 5in gun.

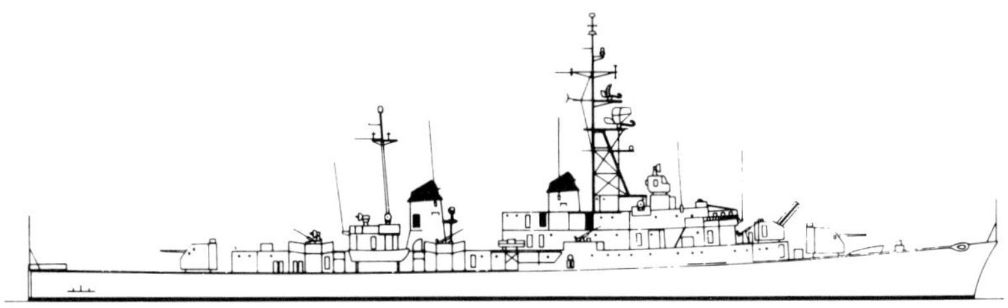

Impetuoso c 1978

IMPETUOSO class *destroyers*

Displacement: 2775t standard; 3810t full load
Dimensions: 405ft pp, 418ft 9in oa × 43ft 2in × 14ft 9in
123.4m, 127.6m × 13.2m × 4.5m
Machinery: 2-shaft geared steam turbines, 4 Foster Wheeler boilers, 65,000shp = 34kts. Range 3000nm at 16kts
Armament: 4–5in (127mm)/38 (2 × 2), 16–40mm/56 (2 × 4, 4 × 2) 1 triple ASW mortar, 2–533mm TT
Sensors: Radar SPS-6, SG-6B, SPG-34, Mk 25; sonar SQS-11
Complement: 315

No	Name	Builder	Laid down	Launched	Comp	Fate
D 558	IMPETUOSO	CNR, Riva Trigoso	7.5.52	16.9.56	25.1.58	Stricken 1983
D 559	INDOMITO	Ansaldo	24.4.52	9.8.55	23.2.58	Stricken 3.11.80

These two ships were ordered in February 1950, and were the first postwar Italian destroyers. The hull design was based on that of the wartime 'Commandante' class. The powerplant was probably that of the US *Gearing* class. Armament and sensors broadly matched those of contemporary US ASW destroyers, except that the ASW weapon was the Italian Menon mortar. A major reconstruction proposed in the late 1960s would have rearmed them with a single lightweight 5in/54 forward and Tartar aft. As in the US Navy, the short-range SQS-11 sonar initially installed was eventually replaced by the lower-frequency SQS-4, and the fixed 533mm TT were replaced by triple 12.75in tubes firing, in this case, Mk 44 torpedoes. By 1980, *Impetuoso* had apparently been fitted with a US WLR-1 ESM system, and with an Elettronica noise jammer (SLQ-A). Her sister did not show the characteristic radomes of these systems.

Impetuoso c 1975 Navarret Collection

Impavido 1970 Navarret Collection

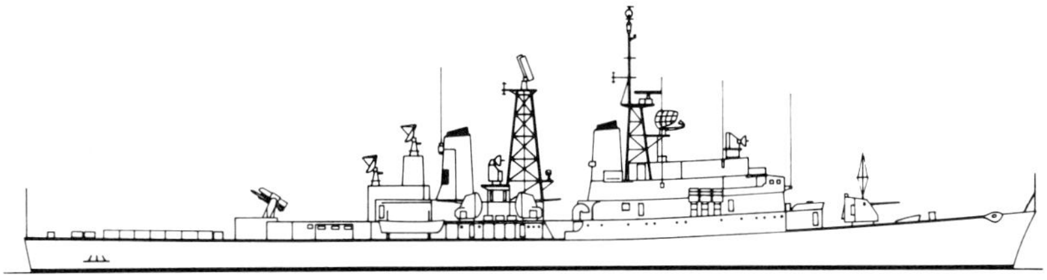

Impavido 1976

These are essentially *Impetuoso*s with the after twin 5in/38 replaced by a Tartar missile launcher, in rough analogy with the way the US *Charles F Adams* class was derived from the *Forrest Sherman* class. They also have the much longer-range sonar of the US ships (the *Impetuoso*s had modernised Second World high-frequency sets), and they lack the Menon mortar of the earlier design. Single 76mm/62s (two on each side) with Italian directors replaced the 40mm guns of the earlier ships, the Italian Navy having decided in about 1958 that the 40mm could not deal adequately with modern aircraft. Consideration was given to installing 135mm guns (as on the modernised *Giuseppe Garibaldi*) instead of US-type 5in/38s. These ships were ordered in, respectively, 1957 and 1959 and modernised in 1976–77 and 1974–75, respectively. The Tartar missile was replaced by SM-1(MR), as in other ships equipped with this system. The SPS-39 3-D radar was modernised (it received the later flat planar array antenna) and an Elettronica EW system installed. The original gun fire control system was replaced by an Argo NA 10 with RTN-10X radars and a computer combat direction system was fitted.

IMPAVIDO class *destroyers*

Displacement: 3201t standard; 3990t full load
Dimensions: 430ft 11in oa × 44ft 9in × 14ft 6in
131.3m × 13.7m × 4.4m
Machinery: 2-shaft geared steam turbines, 4 Foster Wheeler boilers, 70,000shp = 33kts. Range 3300/1500nm at 20kts/30kts
Armament: 1 Tartar SAM, 2–5in (127mm)38 (1 × 2), 4–76mm/62 (4 × 1), 6–324mm TT (2 × 3)
Sensors: Radar SPS-12, SPS-39, SPQ-2, SPG-51, RTN-10X; sonar SQS-23
Complement: 340

No	Name	Builder	Laid down	Launched	Comp	Fate
D 570	IMPAVIDO	CNR, Riva Trigosa	10.6.57	25.5.62	16.11.63	Stricken 30.6.92
D 571	INTREPIDO	Ansaldo	16.5.59	21.10.62	28.7.64	Stricken 31.8.91

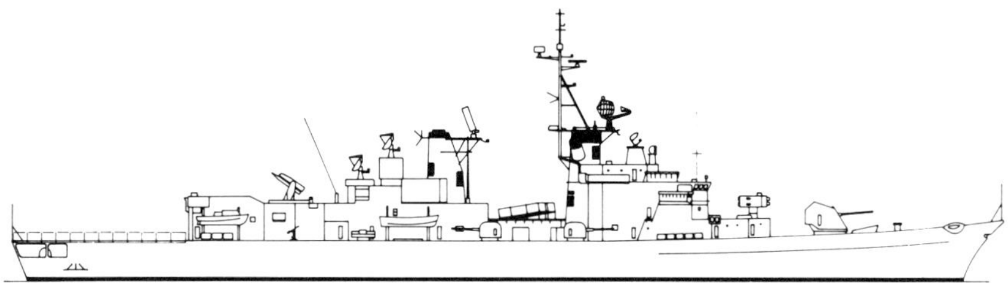

Audace after modernisation

Although initially described as 'Improved *Impavidos*', these ships showed little resemblance to their predecessors. They incorporated the new SADOC computer combat direction system, which enormously increased their capability, and a large helicopter hangar aft (for two AB-212 or one SH-3D). The gun battery consisted of the new Italian-made lightweight 5in/54 and the new Compact 76mm/62. The four tubes for 21in wire-guided torpedoes fired aft over the stern. *Audace* was apparently to have been fitted with a US-supplied SPS-40, but received SPS-12 instead; her sister probably was completed with the Italian RAN-3L, and both ships soon had this radar.

Both ships were modernised: *Ardito* was completed in March 1988, and her sister was completed early in 1991. They now have a Dardo-E close-in defensive system employing a short-range missile (Aspide, in an 8-round box launcher in 'B' position replacing a 5in/54 gun) and 76mm Super Rapido guns. The SM-1 area defence missile system has been upgraded to Mk 13 Mod 5 Weapons Direction System and Mk 74 Mod 13 missile fire control system, with SPG-51C fire control radars. In the course of the refit, the aft-firing 21in torpedo tubes were removed and the combat direction system upgraded. The main search radars are now SPS-768 and SPS-774 (in effect, target indication radar for Dardo-E), with the US-supplied 3-dimensional SPS-52C.

AUDACE class *destroyers*

Displacement: 3600t standard; 4400t full load
Dimensions: 448ft 3in oa × 47ft 3in × 15ft 1in
136.6m × 14.4m × 4.6m
Machinery: 2-shaft geared steam turbines, 4 Foster Wheeler boilers, 73,000shp = 33kts. Range 3000nm at 20kts
Armament: 1 Tartar SAM, 2–127mm/54 (2 × 1), 4–76mm/62 (4 × 1), 4–533mm TT (4 × 1), 6–324mm TT (2 × 3), 2 helicopters
Sensors: Radar SPS-768 (RAN-3-L), SPS-52, SPQ-2, SPG-51B, RTN-10X; sonar CWE-610
Complement: 380

No	Name	Builder	Laid down	Launched	Comp	Fate
D 551	AUDACE	CNR, Riva Trigoso	27.4.68	2.10.71	16.11.72	Extant 1995
D 550	ARDITO	Castellammare	19.7.68	27.11.71	5.12.72	Extant 1995

Ardito 1977 C & S Taylor

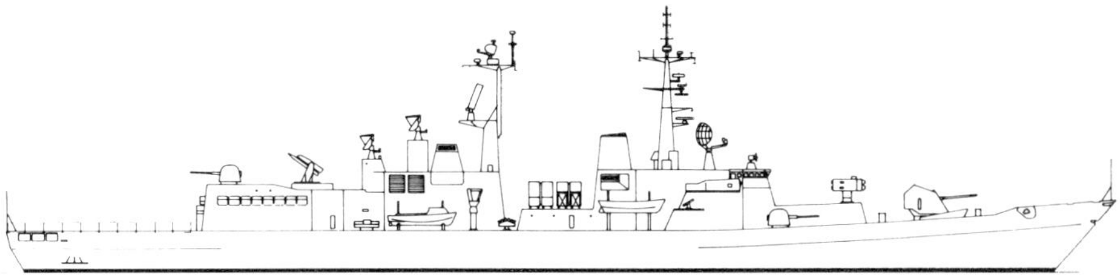

Animoso as completed

LUIGI DURAND DE LA PENNE class *destroyers*

These destroyers were ordered in March 1986 to replace the old *Impavido* class. Originally named *Animoso* and *Ardimentoso*, they were renamed in 1992 to honour Second World War heroes. Their 127mm guns came from the 'B' positions of the modernised *Audace* class, which now carry a similar armament. The 76mm guns, Aspide missiles and four SPG-76 directors form an integrated close-in defensive system. The helicopter deck is large enough to accommodate the planned EH-101 as well as the large SH-3D currently in Italian service.

A second pair was cancelled when Italy joined the Anglo-French 'Horizon' frigate programme. Four such ships were to have replaced the *Audace* class and the stricken *Andrea Doria* class cruisers. However, in December 1993 the second pair were cancelled to provide money to buy the four existing *Lupo* class frigates already built for Iraq. As of 1995, Italy was considering buying as many as six 'Horizons', including four to replace the four elderly *Lupo*s already in service. But naval funding was shaky (see also the problems of the S90 submarine project below) and the future of the Italian part of the programme was in some doubt.

Displacement: 4500t standard; 5400t full load
Dimensions: 484ft 7in × 52ft 10in × 16ft 5in (22ft over sonar)
147.7m × 16.1m × 5.0m (6.7m)
Machinery: 2 shafts (CP propellers) CODOG: 2 GM/Fiat LM-2500 gas turbines, 55,000shp = 31.5kts, 2 GMT BL230-20DVM diesels, 12,600bhp=21kts. Range 7000nm at 18kts (diesels)
Armament: 8 Otomat SSM (4 × 2), 1 Albatros SAM (1 × 8, Aspide missiles), 1–127mm/54, 3–76mm/62 OTO Melara (3 × 1), 6–324mm ASW TT (2 × 3), 2 helicopters
Sensors: Radar SPS-774, SPS-768, SPS-52C, 2 SPG-51D, 4 SPG-76; sonar DE 1164, DE 1167; ECM SLQ-732, SADOC 2 combat direction system
Complement: 400 (35 officers)

No	Name	Builder	Laid down	Launched	Comp	Fate
D 560	LUIGI DURAND DE LA PENNE (ex-*Animoso*)	Fincantieri	26.7.86	29.10.89	Dec 1993	Extant 1995
D 561	FRANCESCO MIMBELLI (ex-*Ardimentoso*)	Fincantieri	3.12.89	13.4.91	Dec 1993	Extant 1993

Francesco Mimbelli
18.7.94
L & L van Ginderen

Ex-US CANNON class *frigates*

Particulars: As US *Cannon* class (DET type)

No	Name	Builder	Acquired	Fate
F 590	ALDEBARAN (ex-*Thornhill*)	Federal, Newark	10.1.51	Discarded 1976
F 591	ALTAIR (ex-*Gandy*)	Tampa	10.1.51	Discarded 1971
F 592	ANDROMEDA (ex-*Wesson*)	Federal, Newark	10.1.51	Discarded 1970

Originally built in 1943, all three were transferred from the USN to the Italian Navy in 1951. Modernisation was limited to installation of a new SPS-6 air search radar, on a tripod mast that replaced the earlier pole, in refits beginning in 1954. The original US surface search radar was replaced by an Italian SMA type. The existing IFF system was replaced. Unlike ex-US destroyer escorts in other navies, they were never fitted with modern scanning sonars (they retained their old QGB searchlights to the end of their careers). However, the new A/S Attack Plotter (ASAP) was installed. *Altair* was blown up as a target in 1971.

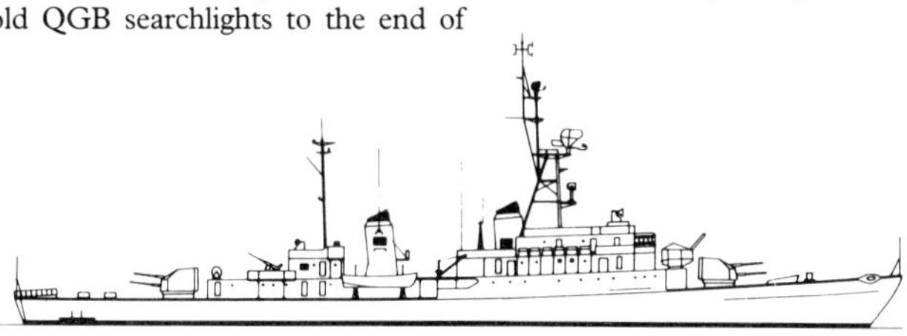

Castore 1961

CANOPO class *frigates*

Castore and *Cigno* were financed by US 'off-shore procurement' MDAP funds. Initially they had Micro-Lambda search radars (MLA-1), but by 1960 all four had SPS-6. The guns and fire control were Italian; sensors were American (the 76mm fire control radar was MLT-4, an Italian version of the US Mk 39, made by MicroLambda, and the main director was the Officine Galileo OG 2, comparable to the US Mk 57; the forward director was an enlarged version carrying an operator's cab. The 40mm guns were controlled by OG 1, comparable to the US Mk 51. The main ASW weapon was the new Italian Menon

Displacement: 1807t standard; 2250t full load
Dimensions: 306ft 3in pp, 338ft 4in oa × 39ft 4in × 12ft 6in
93.3m, 103.1m × 12m × 3.8m
Machinery: 2-shaft geared steam turbines, 2 Foster Wheeler boilers, 22,000shp = 26kts. Range 3000nm at 20kts
Armament: 4–76mm/62 (2 × 2), 4–40mm/70 (2 × 2), 1 Menon mortar, 2–533mm TT (2 × 1)
Sensors: Radar SPS-6, SG-6B; sonar SQS-11
Complement: 207

No	Name	Builder	Laid down	Launched	Comp	Fate
F 531	CANOPO	CNT	15.5.52	20.2.55	1.4.59	Stricken 1981
F 533	CASTORE	CNT	14.3.55	8.7.56	14.7.57	Decomissioned 30.11.80; stricken
F 554	CENTAURO	Ansaldo	31.5.52	4.4.54	5.5.57	Stricken 1984
F 555	CIGNO	CNT	10.2.54	20.3.55	7.3.57	Stricken 1983

mortar, which was also mounted in the contemporary *Impetuoso* class destroyer. The two torpedo tubes were mounted angled forward on either side between the funnels. In 1963–64 two of the four Menon short range mortars (depth charge mortars) were removed. At least some units probably had their original SQS-11A sonars (versions of the late-war US QHB) replaced by the lower-frequency SQS-4. Reportedly all were ultimately fitted with SQS-36.

The unusual over-and-under 76mm mounting (with the elevating gear between the barrels) was unsuccessful. It was complex to load, difficult to maintain, and the mount carried few ready-service rounds. It was superseded by a single mount, capable of firing 60 rounds/min, the lineal ancestor of the current 76mm/62 Compact and Super Rapido. *Castore* was rebuilt in 1966–67 with 3–76mm/62 (3 × 1) replacing these mountings and the pair of 40mm guns, and with Mk 32 triple ASW torpedo tubes firing Mk 44 torpedoes. The OG 2 directors were replaced by OG 3 (comparable to the US Mk 56), carrying a new Orion RTN-7X (predecessor to Orion 10, RTN-10X) radar. The new SPQ-2 was installed. An EW suite was later installed (probably the US WLR-1 intercept set and the Italian Elettronica SLQ-A noise jammer on the foremast). A new SQS-36 sonar was installed. The others were also rebuilt, the last (after about 1972) being *Cigno*. *Castore* was apparently unique in having a second tracking radar (with a larger dish) on a short lattice mast atop her bridge; it had been removed by 1972. *Castore* and *Centauro* had their gun fire control directors forward; *Canopo* had hers aft, overlooking the two 76mm/62 mounts there.

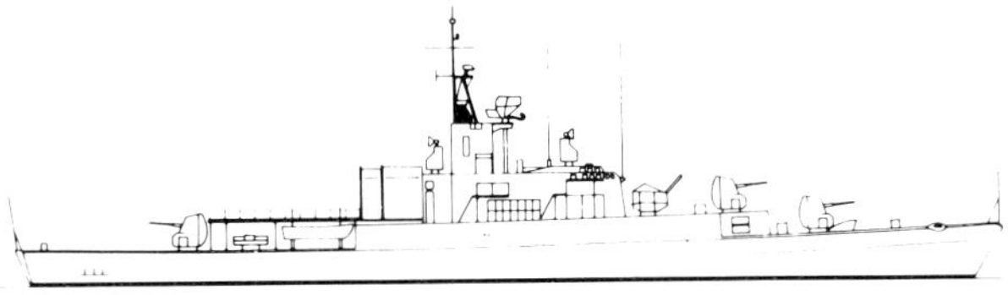

Luigi Rizzo 1962

BERGAMINI class *frigates*

Displacement:	1410t standard; 1650t full load
Dimensions:	283ft 11in pp, 308ft 4in oa × 37ft 3in × 10ft 2in *86.5m, 94.0m × 11.4m × 3.1m*
Machinery:	2 shafts, 4 diesels, 16,000bhp = 25kts. Range 3000nm at 18kts
Armament:	2–76mm/62 (2 × 1), 1 Menon ASW mortar, 6–324mm TT (2 × 3), 1 helicopter. See notes
Sensors:	Radar SPS-12, SPQ-2, RTN-10; sonar SQS-40
Complement:	163

No	Name	Builder	Laid down	Launched	Comp	Fate
F 593	CARLO BERGAMINI	CRDA	19.7.59	16.6.60	23.6.62	Stricken Nov 1981
F 594	VIRGILIO FASAN	Castellammare	6.3.60	9.10.60	10.10.62	Stricken 7.12.88
F 595	CARLO MARGOTTINI	Castellammare	26.5.57	12.6.60	5.5.62	Stricken 31.5.88
F 596	LUIGI RIZZO	Castellammare	26.5.57	6.3.60	15.12.61	Stricken 30.11.80

Although originally described as corvettes ('Corvette Veloci Tipo 2', or fast corvette Type 2), these ships were much closer to the faster frigates in ASW capability, and they were all re-rated as frigates. Even so, they were in effect replaced by the *Minerva* class corvettes. The last two ships listed were announced in June 1956 as part of the second five-year naval programme. Another pair was added that November. The official laying-down dates presumably have little real significance. Work on *Bergamini* was suspended for a time in 1959 due to financial problems.

The main distinction between these 'fast corvettes' and the earlier frigates was their diesel powerplant, which limited their speed. *Bergamini* and *Rizzo* had Tosi high-speed (900rpm) diesels; the other pair have Fiats. Engines are in two adjacent compartments.

The design initially called for three 76mm/62s ('A', 'B', 'X' positions), for a twin 40mm gun in 'Y' position, for two Menon single-barrel mortars side by side abaft 'B' position, and for three aft-firing 21in (533mm) torpedo tubes on the fantail. The fire control system was the new OG 3 with its Orion RTN-7X radar. In 1958, however, the Italian Navy decided that 76mm was the minimum calibre for a useful anti-aircraft weapon, and the twin 40mm gun was eliminated; one of the two Menon mortars replaced it. Then in 1960 the navy decided that an ASW helicopter, armed with a homing torpedo, was necessary to deal with modern fast submarines. The Menon aft was eliminated, the uptakes merged with the mast to form a 'mack', and a helicopter pad and telescopic hangar placed abaft the bridge structure. The fixed torpedo tubes were replaced by a pair of lightweight Mk 32 triple tubes (the first in the Italian Navy) for Mk 44 torpedoes. By this time the planned SQS-11 sonar had been superseded by the longer-range SQS-40 (a variant of SQS-4). The two single-barrel mortars abaft 'B' mount were replaced by a single mortar, which could lay down only a 7-round pattern. To make helicopter operation easier, the ships were fitted with fin stabilisers.

The US WLR-1 electronic intercept system (or its SLR-2 predecessor) seems to have been planned, but apparently the more compact Elettronica SPR-A was substituted.

Then the Italian Navy adopted a larger shipboard helicopter, the AB-212, to replace the earlier AB-204. It required a larger pad and hangar, so the after 76mm gun was eliminated. *Margottini* was modified in 1968, *Fasan* in 1969, *Bergamini* in 1970, and *Rizzo* in 1971.

Carlo Bergamini as completed — Navarret Collection

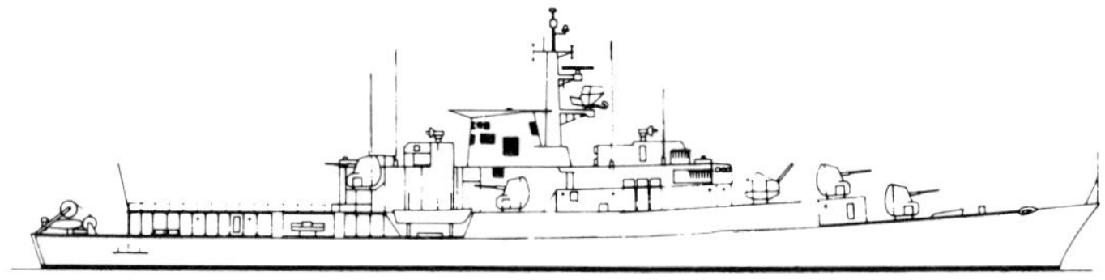

Alpino as completed

ALPINO class *frigates*

Displacement:	2000t standard; 2700t full load
Dimensions:	349ft 2in pp, 371ft 10in oa × 43ft × 12ft 4in *106.4m, 113.3 × 13.1m × 3.8m*
Machinery:	2-shaft CODAG; 2 gas turbines plus 4 diesels, 15,000shp + 16,800bhp = 29kts. Range 3500nm at 18kts
Armament:	6–7mm/62 (6 × 1), 1 Menon ASW mortar, 6–324mm TT (2 × 3), 2 helicopters
Sensors:	Radar SPS-12, SPQ-2, RTN-30X, RTN-10X; sonar SQS-43, SQA-10 VDS
Complement:	263

No	Name	Builder	Laid down	Launched	Comp	Fate
F 580	ALPINO (ex-*Circe*)	CNR, Riva Trigosa	27.2.63	10.6.67	14.1.68	Extant 1995
F 581	CARABINIERE (ex-*Climene*)	CNR, Riva Trigoso	9.1.65	30.9.67	28.4.68	Extant 1993

These ships were conceived as improved *Centauros*; two each were included in the 1959–60 (*Circe* and *Climene*) and 1960–61 (*Perseo, Polluce*) programmes. In fact only the first two were ever built and their design was radically recast to parallel that of the *Bergamini* class. In 1963–65 the decision was taken to replace the planned steam plant (27,000shp) with a CODAG plant, based on experience with the modified destroyer leader *San Giorgio* and with a fast motor gunboat. Each shaft was powered by a pair of Tosi diesels and a Tosi-Metrovick G6 gas turbine; cruise speed on diesels alone is 24kts.

An official drawing of the original design for these ships shows single 76mm/62 mounts fore and aft, a single-barrel Menon mortar in 'B' position, and a pair of 40mm guns alongside a hangar for an AB-204 helicopter. The hull form is recognisably that of the earlier *Centauro*, with its dory bow. Unlike a *Centauro*, it has only a single funnel, which suggests that much higher steam conditions (or pressure firing) were planned. By this time (see above) the 40mm was being rejected, and by 1962 announced particulars of the design showed three 76mm guns, presumably two of them alongside the helicopter hangar and one forward. The gun abaft the hangar would have been eliminated to allow for enlargement, as in the *Bergaminis*. The ships emerged in 1968 with sharply raked bows (the first in Italian frigate practice), and with no fewer than six 76mm guns: two superfiring forward (with the mortar abaft B position, as in the *Bergamini* class), and two on each side amidships. The director was the new NA-9, with three Orion radars (RTN-10X: one forward, one each side aft atop the hangar). The EW outfit was SPR-A. No jammer seems to have been installed initially, but the ships were completed with platforms for jammers on either side of their massive uptakes, and the usual cylindrical noise jammers (SLQ-A type) were installed some time in the 1970s.

These ships introduced variable-depth sonar to Italian frigate practice. Given the shallowness of the surface layer in summer, VDS is essential. The sonar selected, the US SQS-43, uses the same electronic 'stack' to drive both hull and towed units.

Both were extensively refitted in the 1980s (*Alpino* was completed July 1985) with new sonars (two independent DE 1164s, for hull and VDS), an improved combat direction system (SADOC, as in the *Lupos* and *Maestrales*), and a modern EW system (SLQ-747, with its pair of spherical jammer radomes (for directional jammers) replacing the earlier cylinders alongside the big uptakes.

Both ships were renamed (with destroyer names); the second pair were never built. In effect, their place in the rolling construction programme was taken by the first two *Lupos*.

Carabiniere was reduced to auxiliary status (reclassified January 1993) to serve as weapons trials ship in place of *Quarto*. Her Menon mortar was replaced by a single Milas (ASW missile) box launcher. She also conducted trials with the new EMPAR radar. As of 1994, *Alpino* was scheduled for retirement in 1997, *Carabiniere* in 1996.

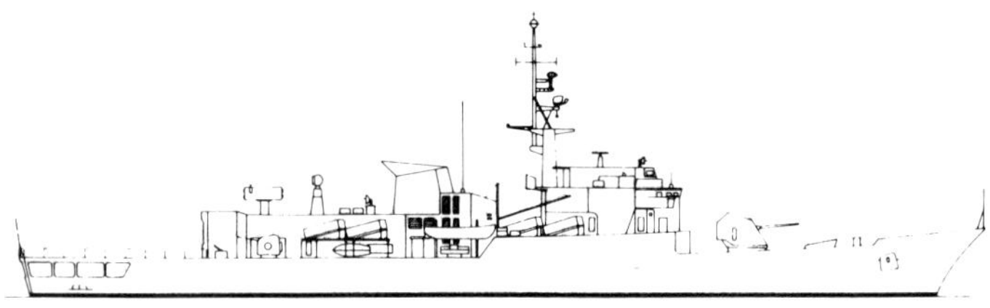

Lupo as completed

LUPO class *frigates*

Displacement:	2208t standard; 2500 full load
Dimensions:	347ft 10in pp, 370ft 2in oa × 39ft 4in × 12ft *106.0m, 112.8m × 12.0m × 3.7m*
Machinery:	2-shaft CODOG: 2 gas turbines, 50,000shp = 35kts, plus 2 diesels, 8490bhp = 20.5kts. Range 4350nm at 16kts on diesels
Armament:	8 Otomat SSM (8 × 1), 1–127mm/54, 1 Sea Sparrow SAM (1 × 8), 4–40mm/70 (2 × 2), 6–324mm TT (2 × 3), 1 helicopter
Sensors:	Radar RAN-10s, RAN-11, SPQ-2, RTN-10X, RTN-20; sonar DE-1160B
Complement:	185

No	Name	Builder	Laid down	Launched	Comp	Fate
F 564	LUPO	CNR	11.10.74	29.7.76	12.9.77	Extant 1982
F 565	SAGITTARIO	CNR	4.2.76	22.6.77	18.11.78	Extant 1982
F 566	PERSEO	CNR	28.2.77	12.7.78	3.8.79	Extant 1982
F 567	ORSA	CNR	1.8.77	1.3.79	13.2.80	Extant 1982
F 582	ARTIGLIERE (ex-*Hitteen*)	Fincantieri, Ancona	31.3.82	27.7.83	Mar 1985	Extant 1995
F 583	AVIERE (ex-*Thi Qar*)	Fincantieri, Ancona	Sep 1982	19.12.84	1985	Extant 1995
F 584	BERSAGLIERE (ex-*Al Qadissiya*)	Fincantieri, Ancona	15.4.83	31.3.84	1986	Extant 1995
F 585	GRANATIERE (ex-*Al Yarmouk*)	Fincantieri, Riva Trigoso	12.3.84	20.6.85	Apr 1987	Extant 1995

These ships were originally described as modified *Alpinos*, taking the place of the two units planned but never built. A single lightweight 127mm/54 gun replaces the two forward 76mm/62s of the earlier ship, and close-in air defence is provided by a combination of missiles (the US Sea Sparrow system, rather than the later Albatros with the Aspide missile) and twin 40mm guns. Presumably the Italian Navy thought, at this time, that the missiles could deal with the aircraft threat, but that a very high rate of fire, not attainable by a 76mm gun, was needed to shoot down missiles. Later this position would reverse, with the appearance of the 76mm Super Rapid gun, firing twice as fast as earlier weapons. With the appearance of the helicopter, the Menon mortar was no longer needed, and it was eliminated. As in *Alpino*, there is a helicopter hangar aft. This class and the contemporary *Sparviero* attack boat introduced the Otomat SSM to Italian service. Originally there were eight separate Otomat box launchers, but with the advent of Otomat Mk 2 that is reduced to four pairs, alongside the forward and after superstructures.

Also unlike *Alpino*, this class has an automated combat direction system (SADOC) and can thus be integrated with a battle group using Link 11. In this sense it is a considerable departure from previous frigates, and might be considered an ASW fleet escort. The EW system comprises SLR-4 (intercept) and SLQ-D (jammer: one set on each side). After refits beginning in 1991, ships have SMA CORA radars in big radomes atop their bridges. Presumably CORA is a track-while-scan surface search/anti sea skimmer set, replacing SLQ-2.

The Italian Navy normally operates these ships with one rather than two helicopters. Unlike the *Alpinos*, this ship has a CODOG powerplant in which LM-2500 gas turbines suffice for high speed (speed on diesels is about 20kts).

Although clearly designed for the Italian Navy (as evidenced by the history running from the CIRCE class through the 1972 re-ordering of the ships originally planned for 1968), the *Lupo* design enjoyed considerable export success, being sold to Venezuela (six ships), Peru (four), and Iraq (four). Of the the Italian ships, all but *Orsa* (Muggiano) were built at Riva Trigoso.

Four ships were ordered for Iraq in February 1981; delivery was held up by the Iran–Iraq War, then by lack of cash, and then by the embargo on Iraq due to the invasion of Kuwait and the subsequent conflict. In January 1992 the Italian Council of Ministers decided to buy them for the Italian Navy, but the navy protested that they would not fit its requirements, and the sale was held up until July 1993 (by which time refit work had already begun). At that time it was hoped that their appearance would make it possible for the *Maestrales* to be withdrawn for badly-needed refits. A plan to build four new frigates was dropped. Unfortunately, the ships had deteriorated while they were idle, and it turned out that virtually none of their equipment was comparable to that in Italian service (for example, the EW suit was SLQ-747, and the sonar was a Krupp-Atlas type). Initially it seemed likely that they would be completely refitted with standard Italian equipment, but that is not being done. Instead, the ships are being reconditioned and, reportedly, their sonars and EW equipment are being removed. They will serve as gunboats, equivalent to US Coast Guard cutters.

Carabiniere c 1970 USN

Perseo 28.6.83 L & L van Ginderen

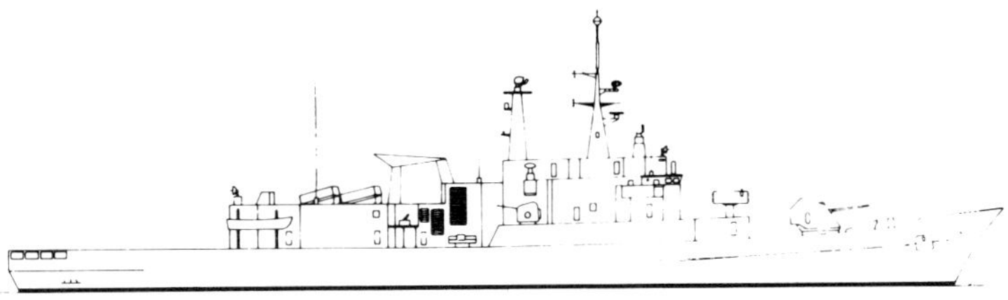

Maestrale as completed

This is a modified *Lupo* design, about 10 per cent larger, for better sea-keeping at the cost of about 2.5kts in calm water. Hangar space for two helicopters was bought by halving the number of Otomat missiles. The anti-aircraft missile system is Albatros, with a mechanized reloader (total twenty-four missiles, eight of them in the box launcher); the Sea Sparrow of the earlier *Lupo* has no means of power reloading, and whatever extra missiles are carried must be brought up from below by hand. The fire control system for the gun and Aspide missile is NA-30A (*Lupo* had NA-10 for the gun plus Mk 91 for the missile), using the RTN-30X (SPG-75) radar (rather than the earlier RTN-10X/SPG-70). Also unlike the *Lupos*, they have a VDS to supplement their hull sonar (DE 1164 rather than the DE 1160B of the *Lupo*). The VDS tow cable is to be lengthened from 600m to 900m, and a passive array streamed from the VDS 'fish', as in the U.S. SQR-18 system (*Maestrale* ran trials in 1989). Note that, unlike a *Lupo*, *Maestrale* carries a pair of stern tubes for large-calibre torpedoes (A-184) in addition to her lightweight weapons. To some extent these torpedoes must make up for the loss of the four Otomats.

Of the eight ships, the last two were cancelled in 1977 and then re-ordered in 1980.

MAESTRALE class *frigates*

Displacement: 2990t standard; 3250t full load
Dimensions: 374ft 2in pp, 402ft 9in oa × 42ft 9in × 27ft 5in
114.0m, 122.7m × 12.9m × 8.4m
Machinery: 2-shaft CODOG: 2 Fiat LM 2500 gas turbines, 50,000shp = 33kts, plus 2 diesels, 11,000bhp = 21kts. Range 6000nm at 16kts
Armament: 4 Otomat SSM (4 × 1), 1–27mm/54, 1 Albatros SAM (1 × 8), 4–40mm/70 (2 × 2), 2–533mm TT (2 × 1), 6–324mm TT (2 × 3), 2 helicopters
Sensors: Radar RAN-10S, SPQ-2D, RTN-30X, RTN-20X; sonar DE-1164, DE-1164 VDS
Complement: 232

No	Name	Builder	Laid down	Launched	Comp	Fate
F 570	MAESTRALE	CNR, Riva Tregoso	8.3.78	2.2.81	7.3.82	Extant 1995
F 571	GRECALE	CNR, Maggiore	21.3.79	12.9.81	5.2.83	Extant 1995
F 572	LIBECCIO	CNR, Riva Tregoso	1.8.79	7.9.81	5.2.83	Extant 1995
F 573	SCIROCCO	CNR, Riva Tregoso	26.2.80	17.4.82	20.9.83	Extant 1995
F 574	ALISEO	CNR, Riva Tregoso	26.2.80	29.10.82	20.9.83	Extant 1995
F 575	EURO	CNR, Riva Tregoso	15.4.81	25.3.83	7.4.84	Extant 1995
F 576	ESPERO	CNR, Riva Tregoso	1.8.82	19.11.83	4.5.85	Extant 1995
F 577	ZEFFIRO	CNR, Riva Tregoso	15.3.83	19.5.84	4.5.85	Extant 1995

Scirocco 1991
L & L van Ginderen

SUBMARINES

(For illustration see under United States)

Ex-US GATO class *submarines*

Particulars: As US *Gato* class (Fleet Snorkel type)

No	Name	Builder	Acquired	Fate
S 510	LEONARDO DA VINCI (ex-*Dace*)	Electric Boat	15.12.54	Discarded 1973
S 511	ENRICO TAZZOLI (ex-*Barb*)	Electric Boat	31.1.55	Discarded 1973

These two *Gato* class submarines originally built in 1942–43, both underwent GUPPY 1B modernisation in 1953–54. Ex-USS *Dace* was transferred on 15 December 1954 and renamed *Leonardo da Vinci* and ex-USS *Barb* was transferred on 31 January 1955 and renamed *Enrico Tazzoli*. They were originally lent for five years, but in 1959 this was extended by another five years and it was later extended again. They formed the backbone of the Italian submarine force in the late 1950s and early 1960s. A 1972 photograph of *Tazzoli* shows a prominent passive sonar array wrapped around the fairing of her periscope shears, presumably for test purposes.

(For illustration see under United States in 1922–46 volume)

Ex-US BALAO class *submarines*

Particulars: As US *Balao* class

No	Name	Builder	Acquired	Fate
S 507	ALFREDO CAPPELLINI (ex-*Capitaine*)	Electric Boat	5.3.66	Discarded 1977
S 512	EVANGELISTA TORICELLI (ex-*Lizardfish*)	Manitowoc SB	9.1.60	Discarded 1976
S 514	FRANCESCO MOROSINI (ex-*Besugo*)	Electric Boat	31.3.66	Discarded 15.11.75

Ex-USS *Lizardfish* was transferred on 9 January 1960 and renamed *Evangelista Torricelli*, ex-USS *Capitaine* was transferred on 5 March 1966 and renamed *Alfredo Cappellini* and ex-USS *Besugo* was transferred on 31 March 1966 and renamed *Francesco Morosini*. These boats were all Fleet Snorkels, substantially less capable than the GUPPYs transferred earlier. *Cappellini* was fully streamlined but retained her 5in/25 gun aft; the other pair were semi-streamlined but apparently did not retain their guns. By the early 1970s *Torricelli* was no longer operational and was used for training and experiments.

(For illustration see under United States)

(For illustration see under United States in 1922–46 volume)

Ex-US BALAO class *submarines*

Particulars: As US *Balao* class (GUPPY III type)

No	Name	Builder	Acquired	Fate
S 502	GIANFRANCO GAZZANA PRIAROGGIA (ex-*Pickerel*)	Boston N Yd	18.8.72	Stricken 1981
S 501	PRIMO LONGOBARDO (ex-*Volador*)	Portsmouth N Yd	18.8.72	Stricken 31.1.80

These ex-USN *Balao* class submarines, originally built in 1944–46, had been given the GUPPY II modernisation in 1952–54 and the very extensive GUPPY III modernisation in 1961–63. These were the first foreign transfers of the US PUFFS (BQG-4) passive underwater fire control system.

(For illustration see under United States in 1922–46 volume)

Ex-US TANG class *submarines*

Particulars: As US *Tang* class

No	Name	Builder	Acquired	Fate
S 515	LIVIO PIOMARTA (ex-*Trigger*)	Electric Boat	10.7.73	Stricken 28.2.86
S 516	ROMEO ROMEI (ex-*Harder*)	Electric Boat	20.2.74	Stricken 31.5.88

The *Tang*s were the immediate successors to the GUPPYs and Fleet Snorkels which the Italian Navy had received. They were equipped with the PUFFS passive ranging/fire control array. Two sisters survive in the Turkish Navy, and one unit (*Trout*) is to be converted (1995) into an unmanned mobile underwater sonar target for the US Navy.

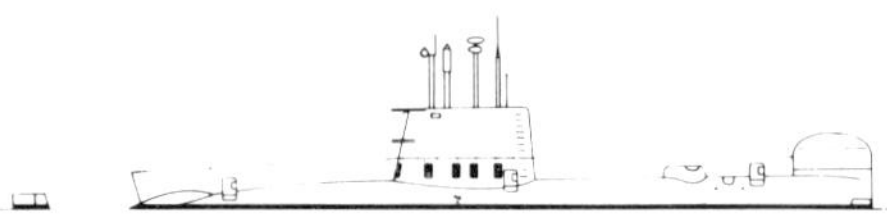

Enrico Toti as completed **NB 1/750 scale**

TOTI class *attack submarines*

Displacement: 535t/591t
Dimensions: 151ft 8in oa × 15ft 5in × 13ft 1in
46.2m × 4.7m × 4.0m
Machinery: 1 shaft, 2 diesels plus 1 electric motor, 2200hp = 9.7/14kts. Range 3000nm at 5kts (surfaced)
Armament: 4–533mm TT bow, 8 torpedoes)
Sensors: Radar RM-20/SMG; sonar IP-64
Complement: 26

No	Name	Builder	Laid down	Launched	Comp	Fate
S 505	ATTILIO BAGNOLINI	CRDA	15.4.65	26.8.67	16.6.68	Stricken 4.7.91
S 506	ENRICO TOTI	CRDA	15.4.65	12.3.67	22.1.68	Stricken 1.4.92
S 513	ENRICO DANDOLO	CRDA	10.3.67	16.12.67	25.9.68	Stricken 1993
S 514	LAZZARO MOCENIGO	CRDA	12.6.67	20.4.68	11.1.69	Decomm 15.10.93

The small *Toti*s were the first Italian-designed postwar submarines. Apparently the design began with a US-sponsored NATO project for a very small submarine, possibly inspired by US work on small ASW submarines about 1955 (the nuclear *Tullibee* was bought instead). The German Type 205/206 may also have come out of this programme. The first pair were included in the 1956 programme, and apparently the design changed considerably between then and 1964, when they were formally re-ordered. As described in 1958, they were rather larger: 800t/1000t, with 2-shaft diesels-electric powerplants, 1400hp = 13.5/8kts, and four torpedo tubes. By 1960, the Italians had decided on a single-shaft powerplant for higher underwater speed: 850t std, 1020t surfaced and 1230t submerged; 214ft 7in × 19ft × 17ft 1in, *65.4m × 5.8m × 5.2m*, 2 diesels, 1 electric motor, 2200hp = 13.515kts, with four torpedo tubes (eight torpedoes). At this time the wartime submarine *Pietro Calvi* was being rebuilt with a single screw for high underwater performance. Then the design was cut drastically (1964), to 380/480t, 160ft × 15ft 3in × 15ft 3in, *48.77m × 4.65m × 4.65m*, with the same 2200hp powerplant (for 9/13kts), and with only three torpedo tubes. Another account of this small design gives diesel power as 1360bhp and motor power (single motor, single screw) as 850hp, for a speed of 10kts surfaced and 14kts submerged. The final design was slightly smaller but presumably with a fuller hull form, since it accommodated a 2200hp powerplant. Speed for one hour was 15kts.

While these boats were being designed, the Italian Navy planned to buy one or more nuclear submarines for deep-ocean work. After that project was dropped, a longer-range diesel submarine (*Sauro*) was substituted. It would seem to follow that the cuts in size accompanied a decision to build a two-tier (inshore/deep water) force.

When commissioned, boats carried four wire-guided Canguro (Kangaroo) ASW torpedoes and four antiship torpedoes; later they were armed with six A-184s. About 1972 they were fitted with three stub arrays projecting up from the hull on either side, for passive ranging. Diving depth was 180m (590ft).

Enrico Dandolo Aug 1990 — L & L van Ginderen

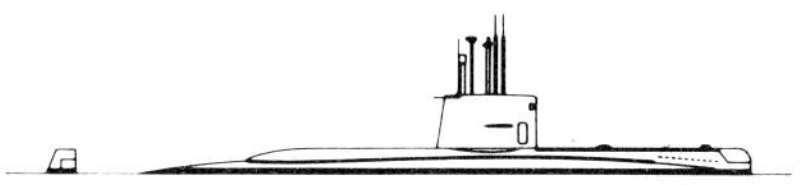

Nazario Sauro as completed

SAURO class *attack submarines*

Displacement: 1456t/1641t
Dimensions: 209ft 7in oa × 22ft 5in × 18ft 9in
63.9m × 6.8m × 5.7m
Machinery: 1 shaft, 3 diesels plus 1 electric motor, 3210bhp/3650shp = 12/20kts. Range 7000nm (surfaced), 100nm at 4kts (submerged)
Armament: 6–533mm TT (bow, 12 torpedoes)
Sensors: Radar RM-20/SMG; sonar IPD-70, Velox M-5
Complement: 45

No	Name	Builder	Laid down	Launched	Comp	Fate
S 518	NAZARIO SAURO	CRDA, Monfalcone	26.6.74	9.10.76	1.3.80	Extant 1995
S 519	CARLO FECIA DI COSSATO	CDRA Monfalcone	15.7.76	16.11.77	1.3.80	Extant 1995
S 520	LEONARDO DA VINCI	Italcantieri, Monfalcone	1.7.76	20.10.79	6.11.82	Extant 1995
S 521	GUGLIELMO MARCONI	Italcaniteri, Monfalcone	23.10.79	20.9.80	16.10.82	Extant 1995

This class was delayed by both financial and technical problems. *Sauro* and *Cossato* were originally ordered in 1967, but budget difficulties caused them to be cancelled and they were not reordered until 1972. Their completion was then delayed because their batteries did not work properly, and different batteries had to be purchased from Sweden. The first two did not commission until March 1980. These were the first Italian submarines built of HY-80 steel; operating depth is 250m (820ft). They were armed with the new A-184 wire-guided torpedo. The MD 100 S passive ranger consists of three arrays spaced along the casing above the pressure hull. The Velox sonar intercept receiver is a very widely-used French type, with its array in a small dome on deck forward; the other sonars are Italian.

Mid-life updates, to improve habitability and replace batteries, were conducted from 1990 onwards. In addition, an integrated combat system (BSN-716), which replaced the earlier SISU fire control system, was installed. It can control Sub-Harpoon missiles (which had not, however, been ordered by Italy as of 1994).

The class was to have been followed by a new S90 design (*ie*, the submarine for the 1990s), but delays led to construction of the two series of improved *Sauro*s described below.

Salvatore Pelosi 1988

SALVATORE PELOSI class *submarines*

Displacement: 1476t/1662t
Dimensions: 211ft 2in × 22ft 5in × 18ft 7in
64.36m × 6.83m × 5.66m
Machinery: Speed 11kts/19kts
Complement: 50 (7 officers)
Other particulars: As *Sauro* class

No	Name	Builder	Laid down	Launched	Comp	Fate
S 522	SALVATORE PELOSI	Fincantieri, Monfalcone	23.7.86	29.11.86	14.7.88	Extant 1995
S 523	GIULIANO PRINI	Fincantieri, Monfalcone	30.7.87	12.12.87	11.11.89	Extant 1995

This is a slightly lengthened, improved version of the original *Sauro* class, bought because the new *S90* was not yet ready. The two units were ordered 7 March 1983. They are capable of diving deeper (normal depth 300m (984ft), crush depth 600m (1968ft)) and are capable of firing US Sub-Harpoon missiles (but none had been purchased as of 1995). A single pressure-proof bulkhead was added to provide a pressure-proof internal space to shelter the crew in the event the submarine sank.

PRIMO LONGOBARDO class *submarines*

Displacement: 1653t/1826t
Dimensions: 217ft 8in × 22ft 5in × 19ft 8in
66.35m × 6.83m × 6.0m
Other particulars: As *Salvatore Pelosi* class

No	Name	Builder	Laid down	Launched	Comp	Fate
S 524	PRIMO LONGOBARDO	Fincantieri, Monfalcone	19.12.91	20.6.92	Dec 1993	Extant 1995
S 525	GIANFRANCO GAZZANA PRIAROGGIA	Fincantieri, Monfalcone	12.11.91	26.6.93	May 1995	Extant 1995

This is a further evolution of the *Sauro* design, slightly lengthened and with an improved hull form, carrying more fuel. In 1992 they were described as lengthened to fire Sub-Harpoon, although this missile can also be fired from earlier submarines. These two boats were ordered on 28 July 1988, again because the follow-on *S90* class had been delayed.

This is the projected follow-on to the *Sauro* class. The first two of this class were authorised but orders for them were not placed as planned in 1993 (for delivery in 1997). The next pair are to replace the first two boats of the *Sauro* class. Unlike previous Italian submarines, this design shows a long parallel-body midships section to support passive ranging and flank arrays. In addition there is a bow array and a chin array below the TT. During the 1980s, the Italian Navy hoped that the sonars and combat system for these submarines would be bought on a multi-national basis, presumably with the French Navy. Reported planned diving depth was 400m (*c*1300ft).

S90 class *submarines*

Displacement:	2500t/2780t
Dimensions:	228ft 8in × 26ft 11in × 20ft 8in *69.7m × 8.2m × 6.3m*
Machinery:	1 shaft diesel-electric: 3 GMT A210 18NM diesels, 3600bhp, driving 3895kW generators; 1 electric motor, 6000hp peak, 3200hp sustained = 11kts/22kts
Armament:	6–533m TT (24 torpedoes or Sub-Harpoon SSM)
Sensors:	See notes
Complement:	50 (8 officers)

AMPHIBIOUS WARFARE VESSELS

QUARTO class *landing ships*

Displacement:	764t standard; 980t full load
Dimensions:	226ft 6in oa × 31ft 4in × 5ft 11in *69.0m × 9.6m × 1.8m*
Machinery:	3 shafts, 3 diesels, 2300bhp = 13kts
Armament:	4–40mm (2 × 2). See notes
Sensors:	Radar SPQ-2, RAN-3L
Complement:	?

Class:
Quarto, Marsala, Capera

Built at Taranto and launched on 18 March 1967, *Quarto* was intended to be the first of a large classe of landing ship, intermediate in size between LSTs and LSMs. The two planned sister ships (*Marsala* had been laid down 18 March 1967) fell victim to the 1968 cuts. Two projected ships, *Lombardo* and *Piemonte*, were cancelled altogether. Reportedly *Quarto* was less than successful; she became a missile test ship, her 40mm guns removed. She was finally stricken 31 February 1992.

Ex-US DE SOTO COUNTY class *landing ships*

Particulars: As US *De Soto County* class

Following the failure of the *Quarto* class landing ships, two *De Soto County* class LSTs were acquired from the USN. Both were transferred in July 1972; ex-USS *York County* (launched at Newport News on 5 September 1957) was renamed *Caorle* (L 9891) and ex-USS *De Soto County* (launched at Boston N Yd on 28 February 1957) was renamed *Grado* (L 9890). Both were purchased outright in 1981. *Grado* and *Caorle* were stricken on 15 January 1988 and 15 January 1989 respectively.

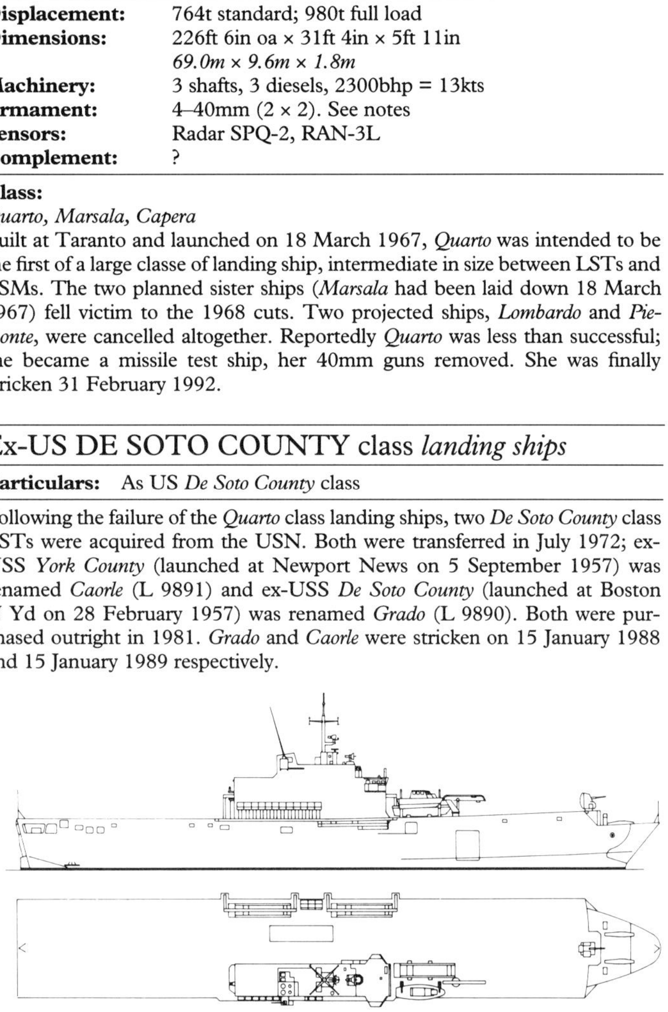

San Marco 1988 **NB 1/750 scale**

SAN GIORGIO class *dock landing ships*

Displacement:	5000t standard; 7665t full load
Dimensions:	437ft 4in × 67ft 3in × 17ft 3in *133.3m × 20.5m × 5.25m*
Machinery:	2 shafts, 2 GMT A420.12 diesels, 16,800bhp = 21kts
Armament:	1–76mm/62, 5 helicopters
Sensors:	Radar SPS-702, SPG-70
Complement:	163 (17 officers) plus 345 troops

No	Name	Builder	Launched	Fate
L 9892	SAN GIORGIO	Fincantieri, Riva Trigoso	25.2.87	Extant 1995
L 9893	SAN MARCO	Fincantieri, Riva Trigoso	10.10.87	Extant 1995
L 9894	SAN GIUSTO	Fincantieri, Riva Trigoso	3.12.93	Extant 1995

San Giorgio was ordered in March 1984. *San Marco* was financed by the Ministry of Civil Protection for disaster relief, and has more extensive medical facilities; she served as a hospital ship during the 1991 Gulf conflict. *San Giusto* is a modified version intended as a peacetime training ship; she had a larger island superstructure and lacks the bow vehicle ramp of the other two ships. Each ship can operate five SH-3D Sea King or CH-47 Chinook helicopters and carries three LCM, launched via the well deck. The first two units also carry four LCVP in davits on the port side. *San Giorgio* and *San Marco* can beach to unload vehicles, up to thirty of which are carried in a below-decks hangar (*San Giusto* can carry thirty-six vehicles). The first two ships are armed with old-model 76mm guns salvaged, with their fire control systems, from scrapped frigates; *San Giusto* has a 76mm OTO Melara Compact.

Ex-US LCM type *landing craft*

Particulars: As US LCM(3) and LCM(6) type

Twenty ex-USN LCM (3) and LCM (6) numbered between MTM 9901 and MTM 9929, were transferred to the Italian Navy in 1952–53. One was discarded in 1967 and on in 1980. Fifteen ex-US LCM were left in 1994. In addition, Fincantieri built eleven new LCM (MTM 217 class) out of GRP for the *San Giorgio* class (three per ship plus two spares): 62t, 60ft 8in × 16ft 9in × 2ft 11in, *18.50m × 5.10m × 0.9m*, 2 FIAT diesels, 560bhp = 9kts, with a capacity of 20t.

Ex-US LCVP type *landing craft*

Particulars: As US LCVP Type

Thirty LCVP were transferred from the USN to the Italian Navy during 1952-72. Twenty-two have since been discarded. A further twenty-eight have been built in Italy. The whole class was numbered in the MTP 9703–MTP 9754 range. Six were left by 1994. Eleven new MTP 96 class personnel landing craft, with bow ramps, were built for the *San Giorgio* class (three per ship plus two spares). Each can carry forty-five troops.

Ex-US LSSL type *gunboats*

Particulars: As US LSSL type

Class:
Alano (L 9851), *Bracco* (L 9852), *Mastino* (L 9853), *Molosso* (L 9854), *Segugio* (L 9855), *Spinone* (L 9856)

Six ex-USN LSSL were transferred to Italy under MDAP on 25 July 1951. They were renamed *Alano* (ex-*LSSL 34*), *Bracco* (ex-*LSSL 38*), *Mastino* (ex-*LSSL 62*), *Molosso* (ex-LSSL 63), *Segugio* (ex-*LSSL 64*) and *Spinone* (ex-*LSSL 118*). They were rated as support gunboats in the Italian Navy. All were discarded during 1969–71.

San Marco Jan 1995 L & L van Ginderen

SMALL SURFACE COMBATANTS

Ex-British 'FLOWER' class *corvette*

Particulars: As British 'Modified Flower' class

No	Name	Builder	Acquired	Fate
A 5307	STAFFETTA (ex-*Elbano*, ex-*Prudent*, ex-*Privet*)	Morton	1949	Discarded 1970

She was built in 1942 as the British 'Modified Flower' class corvette (later frigate) *Privet*, and was transferred to the USN as *Prudent* (PG 96) in 1943. After the Second World War she was returned to the RN and was transferred to Italy as the *Elbano* in 1949. She was renamed *Staffetta* in 1951. In 1953 she was converted to a hydrographic survey ship, armed with 2–20mm (2 × 1).

Ex-British ALGERINE class *minesweeper (corvette)*

Particulars: As British *Algerine* class

No	Name	Builder	Acquired	Fate
F 560	ALABARDA (ex-*Eritrea*, ex-*Ammiraglio Magnaghi*, ex-*Larne*)	Lobnitz	1946	Stricken 1.7.68

This ship was initially used as a colonial corvette, and in April 1950 was reclassified as an ASW corvette. She was armed with 1–4in/47 forward, 4–37mm (4 × 1), 4 Menon ASW mortars, and 2 DC racks (4 DC each). In 1954 the 37mm guns were replaced with 4–40mm/56 (2 × 1, 1 × 2). The 4in gun was replaced by a deckhouse in 1956, when she was converted into a divisional flagship for corvettes.

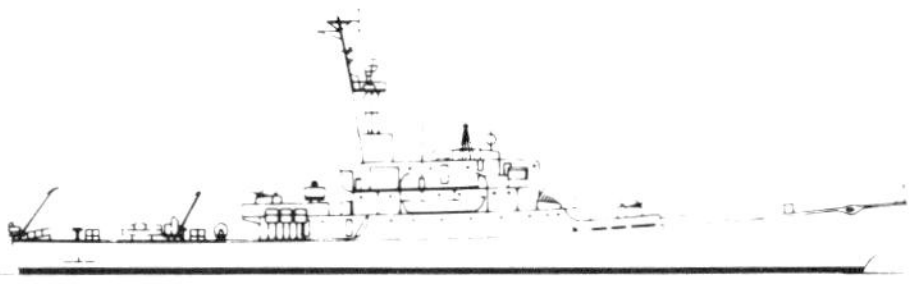

Albatros 1974

ALBATROS class *corvettes*

Displacement: 800t standard; 950t full load
Dimensions: 228ft pp, 250ft 5in oa × 31ft 8in × 9ft 2in
69.5m, 76.3m × 9.7m × 2.8m
Machinery: 2 shafts, 2 Fiat diesels, 5200bhp = 19kts. Range 5000nm at 18kts
Armament: 2–40mm/70 (2 × 1), 2 Hedgehog ASW mortars (2 × 1), 6–324mm TT (2 × 3). See notes
Sensors: Radar MLN-1A; sonar QCU-2
Compliment 110

No	Name	Builder	Launched	Fate
F 542	AQUILA (ex-*Lynx*)	Breda, Marghera	31.7.54	Stricken 1991
F 543	ALBATROS	Castellammare	18.7.54	Hulk 30.4.86, stricken 30.4.89
F 544	ALCIONE	Castellammare	19.9.54	Stricken 1991
F 545	AIRONE	Castellammare	21.11.54	Stricken 1991

Albatros as completed Navarret Collection

Eight ships of this type were built in Italy in the early 1950s under MDAP funding. The Italian Navy were given three (*Albatros* class), four went to Denmark (*Triton* class) and one (*Lynx*) went to the Netherlands Navy. *Lynx* was ceded to the Italian Navy on 18 October 1961 and was renamed *Aquila*. The Italian ships were originally to have been armed with 90mm guns. As completed, they had 2–76mm/62 (2 × 1), 2–40mm/70 (1 × 2), 2 Mk 11 Hedgehogs, 4 Menon ASW mortars and 1 depth charge track. The SMP 3 guns (made by Stabilimenti Meccanici di Pozzuoli) could fire single rounds or bursts (at 50 rounds/min), using a 14-round drum magazine. The gun re-loaded at 90 deg elevation. The same gun was planned for the *Andrea Doria* class cruisers. A radar fire control director, probably OG 2, was mounted atop the bridge.

SMP 3 was the first Italian automatic 76mm gun. In service it was a nightmare; the revolving drum feed was too large, and the gun was described as 'fatally complex'. It gave way to the OTO Melara series which culminated in the current 76mm/62 Super Rapido. The *Albatros* class was rearmed in 1963.

As rearmed, these ships had 4–40mm (1 single forward, 1 single and 1 twin aft). Eventually triple lightweight torpedo tubes replaced the Menon bomb-throwers aft. The original MLN-1 radar was replaced by SPQ-2.

Pietro de Cristofaro 1966

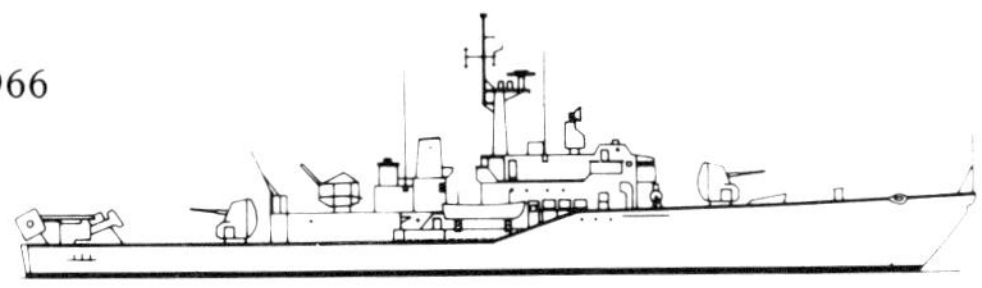

DE CRISTOFARO class *corvettes*

Displacement: 850t standard; 1020t full load
Dimensions: 246ft 2in pp, 263ft 3in oa × 32ft 10in × 8ft 3in
75.0m, 80.2m × 10.0m × 2.5m
Machinery: 2 shafts, 2 diesels, 8400bhp = 23kts. Range 4000nm at 16kts
Armament: 2–76mm/62 (2 × 1), 1 Menon ASW mortar, 6–12.75in (324mm) TT (2 × 3)
Sensors: Radar SPQ-2, RTN-10; sonar SQS-36, SQS-36 VDS
Complement: 131

No	Name	Builder	Launched	Fate
F 540	PIETRO DE CRISTOFARO	CNR, Riva Trigoso	29.5.65	Terminal reserve 31.10.92
F 541	UMBERTO GROSSO	Ansaldo	12.12.64	Terminal reserve 1993
F 546	LICIO VISINTINI	CRDA	30.5.65	Terminal reserve 1.5.92
F 550	SALVATORE TODARO	Ansaldo	24.10.64	Stricken 1994

For the first time in postwar Italian practice, these ships were given a forecastle. They also had a transom stern, a departure from previous practice. As in the somewhat later *Alpinos*, they had a VDS, in their case a separate SQS-36 handled by a crane mounted on deck. There was also a medium-frequency hull sonar, reportedly a separate SQS-36. The 76mm gun was the successful OTO-Melara type, controlled by an OG 3 director (Orion RTN-7X radar), with an optical director (probably OG 1) aft. The single-barrel Menon mortar was mounted aft, in 'X' position. Sonars and ASW weapons were removed when these ships were reclassified as patrol ships for fishery protection in 1989.

Pietro de Cristofaro 20.11.85 L & L van Ginderen

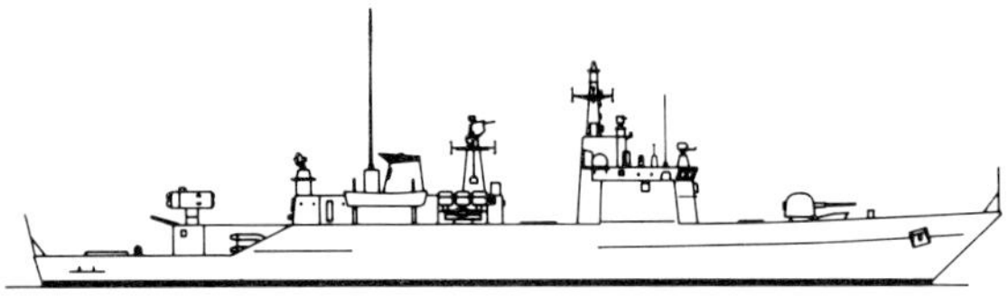

Minerva 1989

MINERVA class *corvettes*

Displacement:	1029t standard; 1285t full load
Dimensions:	284ft 1in × 34ft 5in × 10ft 4in (15ft 9in max) *86.6m × 10.5m × 3.16m (4.8m)*
Machinery:	2 shafts, 2 GMT BM 230.20 DVM diesels, 11,000bhp = 25kts. Range 3500nm at 18kts
Armament:	1 Albatros SAM (1 × 8, Aspide missiles), 1–76mm/62 OTO Melara, 6–324mm ASW TT (2 × 3)
Sensors:	Radar SPS-774, SPG-76; sonar DE 1167; ECM SLQ-747
Complement:	113 (7 officers)

No	Name	Builder	Comp	Fate
F 551	MINERVA	Fincantieri, Riva Trigoso	10.6.87	Extant 1995
F 552	URANIA	Fincantieri, Muggiano	10.6.87	Extant 1995
F 553	DANIADE	Fincantieri, Muggiano	13.2.88	Extant 1995
F 554	SFINGE	Fincantieri, Riva Trigoso	13.2.88	Extant 1995
F 555	DRIADE	Fincantieri, Riva Trigoso	7.9.91	Extant 1995
F 556	CHIMERA	Fincantieri, Riva Trigoso	15.1.91	Extant 1995
F 557	FENICE	Fincantieri, Riva Trigoso	11.9.90	Extant 1995
F 558	SIBILLA	Fincantieri, Riva Trigoso	16.5.91	Extant 1995

These corvettes are designed for later upgrades. The first four were authorised in 1982 and ordered in 1983; the next four were authorised in 1985 and ordered in 1987. As of 1995 a third group of four seems possible, depending on finances. Their peacetime roles are surveillance, fishery protection, training and search-and-rescue; in effect they are more sophisticated equivalents to the contemporary *Cassiopeia* class (see below). Space and weight have been reserved for Aspide reloads, for four Otomat SSM and for a VDS (the Otomat launchers can fire MILAS ASW missiles). The combat direction system (SADOC 2) is compatible with that of larger Italian frigates, destroyers and the carrier. Unusually for ships this small, they have Link 11 capability.

The first sketches released (in 1983) showed a 'mini-Lupo' design (1504t full load) with a helicopter hangar aft and frigate-type electronics, with a speed of 30kts. Note that this class have 'F' pennant numbers in the frigate series.

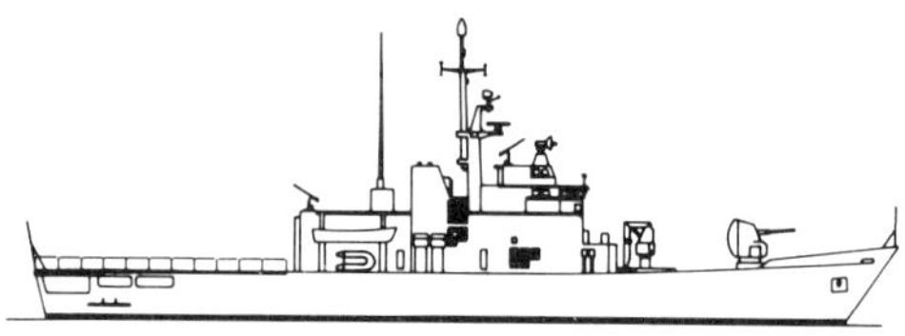

Cassiopeia 1989

CASSIOPEIA class *corvettes*

Displacement:	1110t standard; 1475t full load
Dimensions:	261ft 10in × 38ft 9in × 11ft 10in *79.8m × 11.8m × 3.6m*
Machinery:	2 shafts, 2 GMT BL 230.16 diesels, 8800bhp = 21kts. Range 3300nm at 17kts
Armament:	1–76mm/62, 2–20mm (2 × 1), 1 helicopter
Sensors:	Radar SPS-702, SPG-70
Complement:	60 (6 officers)

No Comp Fate	Name	Builder	Laid down	Launched
P 401	CASSIOPEIA	Fincantieri, Muggiano	21.10.89	Extant 1995
P 402	LIBRA	Fincantieri, Muggiano	23.3.91	Extant 1995
P 403	SPICA	Fincantieri, Muggiano	23.3.91	Extant 1995
P 404	VEGA	Fincantieri, Muggiano	23.3.91	Extant 1995

These corvettes are equivalent to US Coast Guard cutters. Their main armament and its fire control system were taken from the scrapped *Bergamini* class frigates. The class was authorised on 31 December 1982, and the first four units were paid for by the Ministry of the Merchant Marine. Plans for a further four did not materialise; the fifth unit (to have been named *Orione*) and the sixth were cancelled in 1991.

Sibilla Apr 1993 — L & L van Ginderen

VEDETTA *patrol boat*

Particulars: As French MDAP PC type

She was built at Brest as *PC 1616* under MDAP for Germany, and was launched on 30 September 1954. However, she was delivered instead to Ethiopia in 1957 as *Belay Deress*. She was too sophisticated for the fledgling Ethiopian Navy, so she was again transferred under MDAP to Italy. She was delivered on 3 February 1959 and was renamed *Vedetta* (F 597). After refitting she was classed as a corvette, and was mainly used as a fishery protection vessel until being discarded in 1977. Note that this was *not* the standard US PC design, but a 174ft equivalent (with very different appearance) designed in France for construction to US account.

FULMINE *patrol boat*

Displacement:	300t standard; 340t full load
Dimensions:	154ft pp, 163ft oa × 21ft 8in × 7ft *46.9m, 49.7m × 6.6m × 2.1m*
Machinery:	2 shafts, 4 diesels, 9000shp = 30kts.
Armament:	1–76mm/62, 2–40mm/56 (1 × 2), 2–450mm TT (2×1). See notes
Complement:	60

Fulmine was ordered in 1952 from CRDA as *VAS 470*. She was launched on 14 November 1955 and completed in 1956. She was designed as a submarine chaser and was classed as a corvette. The initial armament consisted of 2–40mm/56 (1 × 2), 2–450mm (17.75in) torpedoes with launchers (2 × 1), a Hedgehog Mk 10 ASW mortar and 2 DCT. She was used as a leader for the torpedo-boat flotilla, and was renamed *Sentinella* (F 598). In 1965 she was reclassified as a gunboat and renamed *Fulmine*. In 1967 a 76mm/62 (3in) was mounted in place of the Hedgehog forward. She was discarded in *c*1970.

Ex-US ELCO VOSPER type *motor torpedo-boats*

Particulars: As US Elco Vosper type

Class: *SIS 811–SIS 814, SIS 821–SIS 824, SIS 831–SIS 834*

Twelve ex-USN Vosper type MTBs were acquired by the Italian Navy in 1948 and reconstructed during 1949–52. They were later renumbered *MS 461–64, 421–24* and *431–34* respectively. All were armed with 5–20mm/65 and 2–450mm (17.75in) torpedoes (2 × 1) in launchers. *MS 421, 423–4* and *434* were discarded in 1958, *MS 433, 461* and *464* in 1959, *MS 422, 462* and *463* in 1960 and *MS 431, 432* in 1961.

Ex-US HIGGINS type *torpedo-boats*

Particulars: As US Higgins type

Class:
GIS 841–GIS 844, GIS 851–GIS 854

Eight ex-USN Higgins type PT Boats were acquired in March 1948 and refitted during 1949–53. In the mid-1950s they were renumbered *MS 441–44* and *451–54* respectively. *MS 854* was discarded in 1959. They were armed with 1–40mm/56, 2- or 3–20mm (2 or 3 × 1) and 2–450mm (17.75in) torpedoes on launchers (2 × 1). In the early 1960s they were reconstructed with SPS-21 radar and ASM/185/CRM petrol engines in place of the original Packards. Their maximum speed after reconstruction was 34kts. *MS 444* was discarded in 1966 and in the same year *MS 442, 451* and *452* were transferred to the Customs service. *MS 441* and *453* had their armament reduced to 2–20mm (2 × 1) forward and were used for frogmen support. *MS 441, 443* and *453* were stricken *c*1985.

Ex-German S-BOAT type *motor gunboat*

Particulars: As German *S 701* group

Class:
MS 621

An ex-German *S 701* series S-Boat, which had been converted to the merchant ship *Toros*, was purchased by the Italian Navy and refitted in 1953–54. She was rearmed with 2–40mm/56 (2 × 1) and 2–21in (533mm) TT (2 × 1), and was numbered *MS 621*. She was later renumbered *MC 485* and was discarded in 1965.

Folgore c 1970 — Navarret Collection

FOLGORE *patrol boat*

Displacement: 160t standard; 190t full load
Dimensions: 129ft 4in × 19ft × 5ft 7in
39.4m × 5.8m × 1.7m
Machinery: 4 shafts, 4 diesels, 10,000bhp = 36kts
Armament: 2–40mm (2 × 1), 2–533mm TT
Complement: 39

Class:
Folgare (P 490)

She was authorised in 1950 and built by CRDA. Her original number was *MC 490*, but she was later renamed *Folgore*. The design was a convertible MGB/MTB and she could carry up to 4–533mm (21in) TT in launchers (4 × 1). *Folgare* had a very rapid acceleration from cruising speed to full speed. She was discarded in 1977.

LAMPO class *patrol boats*

Displacement: 170t standard; 196t full load
Dimensions: 141ft 2in oa × 20ft 8in × 4ft 11in
43.0m × 6.3m × 1.5m
Machinery: 3-shaft CODAG: 1 Metrovick gas turbine plus 2 MTU 518D diesels, 4500shp + 7200bhp = 39kts
Armament: 3–40mm/70 (3 × 1). See notes
Sensors: Radar 3ST7-250
Complement: 33

Class:
Lampo (P 491), *Baleno* (P 492)

The *Lampo* class are improved versions of *Folgore* (1957 programme). They were designed as convertible boats, with 2– or 3–40mm/70 (2 × or 3 × 1) as MGBs, and 1–40mm/70 and 4–17.7in (450mm) torpedoes in launchers (4 × 1) or 2–40mm/70 (2 × 1) and 2–21in (533mm) torpedoes (2 × 1) as MTBs. They were the first Italian warships to have gas turbines. *Lampo* (P 491, ex-*MC 491*) was built at Taranto and launched on 22 November 1960. *Baleno* (P 492, ex-*MC 492*) was also built at Taranto and was launched on 10 May 1964. She was modified in 1967–69. Both had their original Fiat diesels replaced by MTU diesels in 1975. Two further vessels of this type, *MC 493* and *MC 494* were projected under the 1960 programme but were not built. Both stricken 31 March 1983.

Freccia 1967 **NB 1/750 scale**

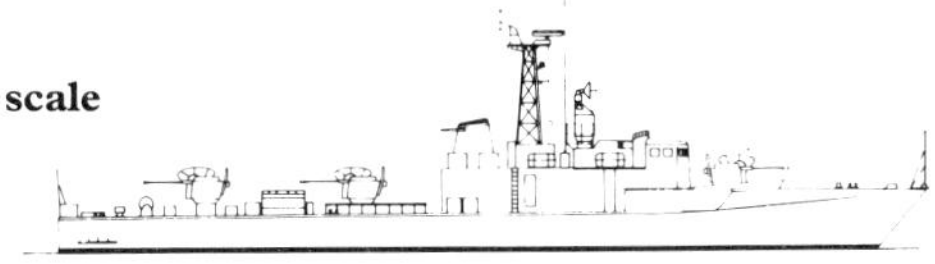

FRECCIA class *patrol boats*

Displacement: 175t standard; 205t full load
Dimensions: 151ft 4in oa × 23ft 8in × 5ft 1in
46.1m × 23.6m × 5.1m
Machinery: 3-shaft CODAG: 1 Proteus gas turbine plus 2 Fiat diesels, 4250shp + 7600bhp = 40kts. Range 800nm at 27kts
Armament: 3–40mm/70 (3 × 1). See notes
Sensors: Radar RTN-150 (*Saetta* only)
Complement: 37

Class (no):
Freccia (P 493), *Saetta* (P 494), *Dardo* (MC 592)

The *Freccias* are enlarged and improved *Lampos*, with a Proteus gas turbine in place of the *Lampos*' Metrovick. Unlike the *Lampo*s, they retain their Fiat diesels. These ships were built under the 1960 programme. The 1962 programme envisaged five Type 590 (*Freccia* class) MGB and three type 491 (*Lampo* class) MGB, of which only two *Freccia* were ordered; they were dropped officially from the programme in January 1972 when the *Sparviero*s were clearly in prospect. *Freccia* (P 493, ex-*MC 590*), built at CNR, Riva Trigoso, was launched on 9 January 1965. *Saetta* (P 494, ex-*MC 591*) built by CRDA, was launched on 11 April 1965. *Dardo*, ex-*MC 592* was laid down on 10 May 1964 at Taranto, but she was not completed, and *Strale*, ex-*MC 593* was never started.

They were designed to be convertible, in one day, to an MGB (2– or 3–40mm/70 (2 × or 3 × 1), an MTB (1–40mm/70, 4–21in (533mm) torpedoes on launchers, 4 × 1) or a fast minelayer (1–40mm/70), 8 mines). They can also carry missiles, and *Saetta* has been used for trials with the Sea Killer SSM. They could be distinguished from the earlier boats by their raised forecastles. Deleted 15 March 1984 and 1985 respectively.

SPARVIERO class *guided missile hydrofoils*

Displacement: 62.5t full load
Dimensions: Foilborne: 75ft 4in oa × 39ft 7in (foils) × 4ft 9in
23.0m × 12.1m × 1.5m
Hullborne: 80ft 7in oa × 23ft (hull) × 6ft 2in
24.6m × 7.0m × 1.9m
Machinery: Foilborne: 1 Proteus gas turbine via waterjet, 4500shp = 50 kts
Hullborne: 1 shaft, 1 diesel, 180bhp = 8kts. Range 400nm/1200nm at 45 kts/8kts
Armament: 2 Otomat SSM (2 × 1), 1–76mm/62
Sensors: Radar 3RM-7, RTN-10X
Complement: 10

Class (no, launched):
Sparviero (P 420, 9.5.73), *Nibbio* (P 421, 29.2.80), *Falcone* (P 422, 27.10.80), *Astore* (P 423, 20.7.81), *Grifone* (P 424, 1.12.81), *Gheppio* (P 425, 24.6.82), *Condore* (P 426, 25.1.83)

Development of conventional Italian patrol boats ceased in 1964 when a consortium of the Italian Government's research branch, Carlo Rodriguez – which built commercial hydrofoils – and the American firm of Boeing, which was building its first jetfoil, came into existence. This consortium, Alinave, designed a small hydrofoil for the Italian Navy based on Boeing's *Tucumcari*. The prototype, *Sparviero* (P 420), was built during 1971–74, but no more were built until 1977. Six more were completed during 1980–83. *Sparviero* has a single-cell launcher whereas the later boats use the Teseo system.

They utilise the Boeing jetfoil system, with one foil forward and two aft. Foilborne power comes from a Proteus gas turbine driving a waterjet, while hullborne power consists of a GM diesel. The boats are built entirely of aluminum. A much larger NATO hydrofoil based on this foil system proved too expensive; it was adopted only by the US Navy (as the PHM). Indonesia bought the Jetfoil system but failed to buy the large number of fast gunboat/transports envisaged. Japan adopted this design for a series of missile patrol boats laid down from 1991 on. *Sparviero* was stricken 30 September 1991; the others are to be retired in 1999–2002.

Sparviero as completed — S Cioglia

MINE WARFARE VESSELS

DAINO class *minesweepers*

Particulars: As German 1940 type and 1943 type minesweepers

Class:
Antilope, Gazzella, Daino (M 5537–M 5539)

Three ex-German coal-burning minesweepers were purchased by Italy on 20 July 1949. *B 1* ex-*M 328*, launched on 12 June 1943 was of the 1940 type and was slightly smaller than the other two. She was assigned to the USN in 1947 under the terms of the Peace Treaty. *B 2* ex-*M 803*, launched on 19 October 1944, and *B 3* ex-*M 801*, launched on 9 September 1944, were both of the 1943 type and were assigned to the USN in 1945. They were originally classed as auxiliary ships, but were converted to oil burning, reclassified as patrol ships and armed with 1–3.9in/47 (100mm) aft, and 3–40mm/56 (3 x 1), plus a DCT. *B 1* was renamed *Antilope*, *B 2* was renamed *Daino* and *B 3* was renamed *Gazzella*. *Daino* was later fitted with a Hedgehog ASW mortar. They were classified as minesweepers in 1954 and as coastal escorts in June 1956 with various changes of pennant numbers. *Antilope* was discarded in 1958, and *Daino* and *Gazzella* were reclassified as corvettes in 1959. In 1960 *Daino* was converted to a survey ship and *Gazzella* to a cadet training ship. Both were discarded in 1966.

Ex-US AGGRESSIVE class *ocean minesweepers*

Particulars: As US *Agile/Agressive* class

Class:
Salmone (M 5430), *Sgombro* (M 5432), *Squalo* (M 5433), *Storione* (M 5434)

Four ex-USN *Aggressive* class ocean minesweepers were transferred under MDAP to the Italian Navy while under construction in 1956–57. *Salmone* (ex-*MSO 507*) was transferred in June 1956, *Sgombro* (ex-*MSO 517*) was transferred in June 1957, as was *Squalo* (ex-*MSO 518*). *Storione* (ex-MSO 506) was transferred in February 1956. *Sgombro* was discarded in 1990, and *Salmone* (used for fishery patrol since 1989) in 1991. The other two were to have been scrapped, but instead they were stripped of minesweeping gear to serve as fisheries patrol craft.

Ex-US ADJUTANT/AGAVE classes *coastal minesweepers*

Particulars: As US *Adjutant* class

Seventeen ex-USN *Adjutant* class minesweepers were transferred to Italy while the class was under construction in 1953–54. *Abete* (ex-*AMS 72*), *Acacia* ex-*AMS 73*), *Betulla* (ex-*AMS 74*), *Castragno* (ex-*AMS 75*), *Cedro* (ex-*AMS 76*), *Ciliego* (ex-*AMS 79*), *Faggio* (ex-*AMS 80*), *Frassino* (ex-*AMS 81*), *Gelso* (ex-*AMS 82*), *Larice* (ex-*AMS 88*), *Noce* (ex-*AMS 89*), *Olmo* (ex-*AMS 90*), *Ontano* (ex-*AMS 133*), *Pino* (ex-*AMS 134*), *Pioppo* (ex-*AMS 135*), *Platano* (ex-*AMS 136*) and *Quercia* (ex-*AMS 137*) all have wooden hulls, as does *Mandorlo* (ex-*Salice*, ex-*MSC 280*) transferred from the USN on 16 December 1960. *Mandorlo* (*M 5519*) differs from the earlier seventeen (numbered M 5501–M 5517 in the above order) by having a modified bridge and funnel.

The nineteen otherwise identical *Agave* class, *Agave, Allora, Edera, Gaggia, Gelsomino, Giaggiolo, Glicine, Loto, Mirto, Timo, Trifoglio, Vischio, Bambu, Ebano, Mango, Mogano, Palma, Rovere* and *Sandalo* (M 5531–M 5542, M 5521–M 5527) were all built in Italian yards, with composite wood and alloy hulls, from 1955.

Pioppo of the *Adjutant* class and *Mirto* of the *Agave* class were converted to survey ships. Following the successful use of *Mandorlo* as a minehunter, six more ships of both classes were converted to minehunters from 1978. *Acacia, Betulla* and *Ciliego* of the *Adjutant* class and *Rovere* of the *Agave* class were scrapped in 1974, *Abete* of the *Adjutant* class was scrapped in 1977, *Faggio, Ontano, Pino, Gaggia* and *Glicine* were scrapped in 1980; *Queria* and *Trifoglio* were stricken in 1981 *Ebano, Gelsomino, Olmo* and *Noce* were all discarded in 1982–83. *Edera* was stricken 1987, *Larice* 30 June 1988, *Sandalo* 30 October 1988, *Agave* 30 September 1990, *Giaggiolo* 30 November 1990, *Timo* and *Vischio* 1991, *Frassino* 30 November 1992, *Loto* 31 October 1992, *Cedro* 1993, and *Mandorlo* 30 November 1993 (had been used for hydrographic survey). In 1994 only *Castagno, Gelso*, and *Platano*, all minehunters, were still in use as minecraft. Four units (*Bambu, Mango, Mogano* and *Palma*) are patrol boats assigned to the United Nations force in the Red Sea. *Mirto* and *Pioppo* serve as hydrographic survey craft. *Alloro*, which had been converted to a training ship in 1986, was stricken 31 August 1992.

(For illustration see under United Kingdom)

ARAGOSTA class *inshore minesweepers*

Particulars: As British 'Ham' class

Twenty NATO 'Ham' class inshore minesweepers were constructed in Italian yards during 1955–57. They are similar to the British 'Ham' class. Although they were fitted to mount 1–20mm, none has carried it in service. They were built under MDAP, *Arsella* (M 5451), *Attina* (M 5453), *Calamaro* (M 5454), *Conchiglia* (M 5455), *Dromia* (M 5456), *Ostrica* (M 5460), *Paguro* (M 5461), *Seppia* (M 5467), *Tellina* (M 5468) and *Totano* (M 5469) were discarded in 1974. *Gambero* (M 5457), *Pinna* (M 5462), *Scampo* (M 5466), *Grancio* (M 5458) and *Riccio* (M 5465) were stricken in 1979–80. *Aragosta* (M 5450), *Astice* (M 5452), *Mitilo* (M 5459), *Polipo* (M 5463, ex-*Polpo*) and *Porpora* (M 5464) were reclassified as tenders and as navigational training craft in 1984–85. Two others became ferries, and another, the stricken *Scampo* (renamed *Murena*), became a torpedo trials and retriever craft. All still existed in 1994. *Aragosta* is a combat swimmer training boat, with a decompression chamber in a deckhouse aft.

Milazzo Sep 1994 — L & L van Ginderen

Sapri 1989
NB 1/750 scale

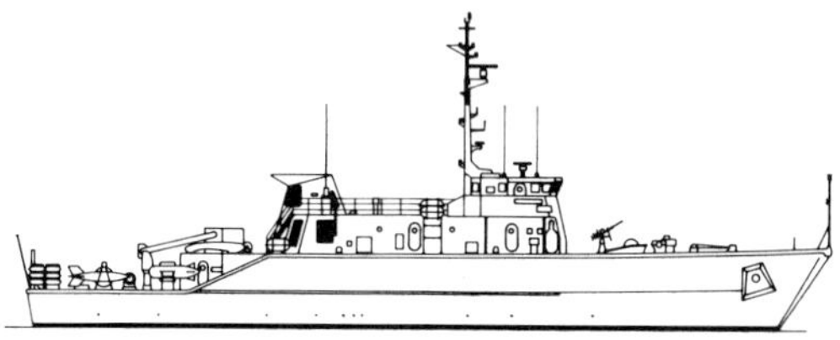

LERICI class *minehunters*

Displacement:	488t standard; 520t full load
Dimensions:	164ft x 31ft 4in x 8ft 10in *49.98m x 9.56m x 2.7m*
Machinery:	1 shaft, 1 GMT B230-SM diesel, 1840bhp = 15kts Range 2500nm at 12 kts
Armament:	1–20mm/80 Oerlikon
Sensors:	Sonar SQQ-14; command system SSN-714
Complement:	47 (4 officers)

Class (no, launched):
Lerici (M 5550, 3.9.82), *Sapri* (M 5551, 5.4.84), *Milazzo* (M 5552, 4.1.85), *Vieste* (M 5553, 18.4.85)

These ships were ordered in April 1978; sister ships are in the Malaysian and Nigerian fleets, and the US *Cardinal* class is a modified version. The US-supplied SQQ-14 sonars were replaced by a substantially modified FIAT-built version, SQQ-14 (IT), which incorporates the British Type 2048 side-scanning sonar (Speedscan) for route survey work. Each minehunter supports up to seven divers, as well as MIN-77 and a Pluto remote controlled submersible. All four extant 1995. Another six ships were deleted from the 1975 ten-year plan; they were revived as the *Gaeta* class (see below).

GAETA class *minehunters*

Displacement:	665t standard; 697t full load
Dimensions:	172ft 1in x 32ft 5in x 9ft 8in *52.45m x 9.87m x 2.95m*
Machinery:	1 shaft, 1 GMT BL230-BN diesel, 1985bhp = 14.3kts Range 2500nm at 12kts
Armament:	1–20mm/80 Oerlikon
Sensors:	Sonar SQQ-14 (IT); command system SSN-714
Complement:	47 (4 officers)

Class (no, launched):
Gaeta (M 5554, 28.7.90), *Termoli* (M 5555, 15.12.90), *Alghero* (M 5556, 4.5.91), *Numana* (M 5557, 26.10.91), *Crotone* (M 5558, 11.4.92), *Viareggia* (M 5559, 3.10.92), *Chioggia* (M 5560, 9.4.94), *Rimini* (M 5561, 1997)

Improved and lengthened *Lerici* class, with a new thruster unit for precise positioning and the Italian-modified version of the US SQQ-14 sonar. Although names were announced as early as 1980, the first six ships were not ordered until April 1988. The last two units were cancelled in August 1992, but reinstated in 1993 to keep the Intermarine Yard going until the next series of minehunters could be ordered. Seven extant 1995.

In 1995 the long-term shipbuilding plan still calls for building six deep ocean minehunters, but the project may be dropped in the absence of the Soviet deep-mining threat (the French have already virtually abandoned their own analogous deep minehunter, the BAMO or *Narvik* class). These deep hunters would have the British Type 2095 (improved 2093) deep minehunting sonar and would be capable of sweeping to depths of 400m (1312ft).

MISCELLANEOUS

Ex-US KENNETH WHITING class *support ship*

Particulars: As US *Kenneth Whiting* class

The ex-US seaplane tender *St George* (AV 16) was transferred on 11 December 1968 and was converted to a support ship for special forces named *Andrea Bafile* (L 9871). Long in unmaintained reserve, she was stricken for scrap on 31 May 1995.

Ex-US BARNEGAT class *support ship*

Particulars: As US *Barnegat* class

The USN MTB tender *Oyster Bay* (laid down as a *Barnegat* class seaplane tender), was purchased by the Italian Navy under MDAP on 23 October 1957. She was renamed *Pietro Cavezzale* (A 5301) and was used as a frogmen support ship and for miscellaneous duties. Stricken 31 March 1994.

Ivory Coast

The Ivory Coast under Houphouët-Boigny, a political veteran, obtained independence from France in 1960, and had been a model of stability and prosperity for tropical Africa. Her gross domestic product in 1988 far exceeded that of neighbouring Ghana and this from only two-thirds the population.

The conservative and anti-Communist policies followed externally have made her sometimes into France's mouthpiece in Africa, and have kept the country out of the most important black African councils and groupings. Similarly, Ivory Coast was one of the few countries to recognise breakaway Biafra, and granted asylum to her defeated leader in 1970.

President Houphouët-Boigny died in December 1993 and has been replaced by the parliamentary speaker Henri Konan-Bédié, who will serve until new elections in October 1995. There has been some political unrest following allegations of fraud in the 1990 elections.

The selective service defence forces are small, with an Army of about 4500 (including a marine battalion) and a Ministry of Marine-controlled navy of 700 men (150 in 1974) manning a coastal and river patrol force. The ports of Abidjan (shipyard), Sassandra, Tabou and San Pedro are all used for the navy's work which includes fire fighting and anti-pollution patrols. The Ivorian forces are, and are likely to remain, almost entirely trained, supplied and maintained by France; a 450-strong French marine infantry battalion is stationed near Abidjan.

Ex-US SC type *large patrol craft*

Wooden-hulled submarine chaser, formerly US *SC 1337*, transferred to France in 1943 (as *CH 71*, re-rated as patrol vessel *P 699* in 1951), and from France to Ivory Coast as *Patience* in 1961. Discarded *c*1972.

Le Vigilant as completed
NB 1/750 scale

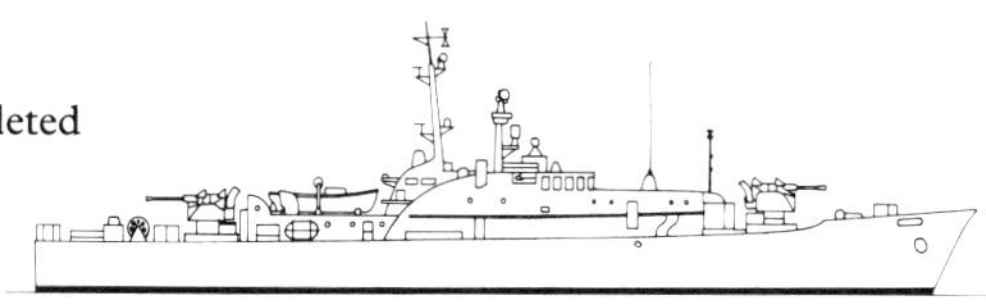

LE VIGILANT class *large patrol craft (missile)*

Displacement: 235t full load (*Le Valeureux* 250t)
Dimensions: 155ft 10in oa (*Le Valeureux* 157ft 6in) × 23ft 7in × 7ft 6in
47.5m (48.0) × 7.0m × 2.3m
Machinery: 2 shafts, 2 MGO diesels, 2400bhp = 18.5kts (*Le Valeureux* 4200bhp = 23.0kts). Range 2000nm at 16kts
Armament: 4 MM40 Exocet SSM, 2–40mm/60 (1 × 2)
Sensors: Radar DRBN-32, Decca (*Le Valeureux*)
Complement: 34

Name	Builder	Launched	Fate
LE VIGILANT	SFCN	23.5.67	Extant 1995
LE VALEUREUX	SFCN	8.3.76	Extant 1995

Franco-Belge type P 48 boats ordered in 1966 and 1974. Laid down 20 January 1967 and 28 October 1975, completed 29 February 1968 and 25 September 1979 respectively. *Vigilant* refitted at Brest, France 1981 when both vessels received Exocet. *Le Valeureux* re-engined in 1987. Speed now 18.5kts.

L'ARDENT class *large patrol craft (missile)*

Class:
L'Ardent, L'Intrépide

Ordered from Auroux, Arcachon in 1977, launched 21 July 1978, and in service since 6 October 1978. Differ from the French *Trident* class in having a 20mm gun instead of a 12.7mm, and an increased crew of 19. Plans to refit them with four Exocet SSM were not carried out.

MISCELLANEOUS

The French 70t seaward patrol craft *La Persevèrance* VC 9 (P 759) was acquired 26 April 1963 and named *Sine Saloum*. The 450t training ship *Locodjo*, built in W Germany in 1953 and seized in 1970 when gun-running to Biafra. Both were stricken in 1979.

There are also four coastal and river patrol craft, the 750t 'Batral' type *Elephant* landing ship launched 1975 and commissioned 21 February 1977 and refitted at Brest 1981, similar to the French *Champlain*), one LCVP (built in Abidjan, 1970) and one LCPS. All extant 1995.

Jamaica

The island state of Jamaica gained independence within the Commonwealth from Britain in 1962. Lying to the south of Cuba it covers an area of some 4411 square miles and has a population of 2,460,700. A democratic country, following a foreign policy of non-alignment, Jamaica has remained a stable and untroubled nation since independence. The Jamaican Defence Force is headed nominally by the prime minister but in practice by a major general. The Coast Guard, Maritime Arm of the Defence Force, is based at HMJS *Cagway*, Port Royal; 18 officers command 180 men while the Coast Guard Reserve employs a total of 46 officers and men.

FORT CHARLES *patrol boat*

Displacement: 103t standard; 130t full load
Dimensions: 103ft 4in × 18ft 9in × 6ft 10in
31.5m × 5.7m × 2.1m
Machinery: 2 shafts, 2 MTU MB 16V538 TB90 diesels, 6000hp = 32kts. Range 1200nm at 18kts
Armament: 1–20mm, 2–12.7mm MG
Complement: 20

Class:
Fort Charles (P 7) was built by Teledyne Sewart in 1974.

Refitted 1979–80 at Jacksonville, Florida. Capable of carrying 24 troops; converts to floating hospital. Extant 1995.

There are also three 85ft commercial cruiser class patrol craft: *Discovery Bay* (P 4), *Holland Bay* (P 5), *Manatee Bay* (P 6). Built by Sewart Seacraft, Berwick, Louisiana 1966–67, 60t, 30kts, *25.9m × 5.6m × 1.8m*, 3 shafts, 3 MTU diesels, 3–12.7mm MG, 11 crew. Engines refitted on two occasions, last time in 1975–77. Extant 1995.

One Hero class patrol craft *Paul Bogle* (P 8) delivered by Lantana Boatyard 1985 (93t, 30kts, 1–20mm, 2–12.7mm MG). Extant 1995.

Japan

The Japanese Navy did not completely die after its defeat in 1945. It lived on in a twilight existence that it suited both the Japanese and the occupying Americans to play down, both at the time and later. There is, however, continuity between the old Imperial Navy and the modern Maritime Self Defence Force (MSDF) both in ships and personnel. Variously designated as the General Affairs Bureau of the Second Demobilisation Ministry, the Maritime Safety Agency, the Maritime Safety Force and finally the Maritime Self Defence Force (MSDF), the Japanese Navy always remained operational and eventually grew once more into one of the world's major naval forces.

On 1 December 1945 the old Navy Ministry became the Second Demobilisation Ministry. The Military Affairs Bureau became a General Affairs Bureau, headed by Rear Admiral Yashio Yamamoto, former head of the Bureau's Military Affairs Section. Yamamoto controlled two sections, one of which organised the huge repatriation task and looked after ship maintenance. The other section was Captain Kuzo Tamura's minesweeping organisation tasked with clearing the more than 100,000 mines strewn around Japan's coast. Disarmed warships, including the aircraft carriers *Hosho* and *Katsuragi* and the 'Akizuki' class destroyers *Hanatzuki*, *Yoizuki* and *Hatzuki* were used for repatriation, but by the end of 1947 these had all been turned over to Allied navies or scrapped. The minesweeping force's task was less easy to complete and although it was steadily reduced in size it continued to operate a fleet, the backbone of which was provided by wooden vessels of the 'Cha-1' and 'Pa No. 1' type originally built as auxiliary submarine chasers or patrol vessels. A few larger 'Ukuru' class escorts were also retained for use as both minesweepers and mother ships, along with former mercantile auxiliaries that were useful for mine destructor duties against pressure mines.

Article 9 of the 1946 Constitution renounced 'war as a sovereign right of the nation and the threat or use of force as a means of settling international disputes'. In order to accomplish this aim 'land, sea and air forces' were never to be maintained. A maritime nation such as Japan could not, however, do without a constabulary force to maintain good order in the seas around its long coastline. Advised by a US Coast Guard captain, the Transportation-Communications Ministry set up a new organisation 'for the purpose of protecting life and property and preventing, detecting and suppressing violation of law at sea'. On 28 August 1947 the Demobilisation Ministry (now called the 'Second Bureau Demobilisation Board') allocated the new body twenty-eight small vessels for use in patrol work, and on 1 May 1948 the Maritime Safety Board, usually translated Maritime Safety Agency (MSA) officially came into being. It took over the former General Affairs Bureau, Tamura's minesweeping section becoming the Fairway Safety Office, and the former repatriation section the Patrol and Rescue Division. The MSA was a controversial organisation and, perhaps to diminish its 'naval' aspect, the 'Ukurus' became weather ships with the Home Affairs Department (although they returned to the MSA later).

The MSA was initially limited to 10,000 personnel and 125 vessels of a combined tonnage no greater than 50,000t. Its ships were not to exceed 15 knots and the only weapons allowed were personal small arms. It was only to operate in 'the high seas adjacent to Japan', but this allowed use in intelligence-gathering in the vicinity of Sakhalin and the other northern islands recently taken over by the USSR. The MSA's minesweepers were the best and most competent source of MCM assets when the Communist use of mines in Korean waters disrupted operations off the coast of Korea. Virtually the entire minesweeping capability of the MSA, forty-six minesweepers and a mine destructor ship were sent, under 'Mr' Tamura, to operate off Wonsan, Kunsan, Inchon, Haiju and Chinampo. They swept about 200 miles of channel and over 600 square miles of anchorages. A 'Cha-1' minesweeper, *MS 14*, was sunk by a mine off Wonsan and a larger 'Pa No. 1', *MS 28* went aground and was also lost off Kunsan. The operation began on 2 October 1950 and the Korean detachment was officially dissolved to return to normal duties on 15 December. A 'Well Done' was transmitted to the MSA by Vice Admiral Turner Joy, US Naval Commander Far East, nine years to the day after Pearl Harbor.

The Korean War helped change American attitudes that Japan required something more military than the MSA (which was enlarged to 18,000 personnel) to defend Japanese security interests at sea. Encouraged by Rear Admiral Arleigh Burke, Joy's Chief of Staff, former Imperial Navy flag officers were called back to what was now the 'Second Demobilisation Liquidation Bureau' to plan for a new naval force to be based around former Soviet American lend-lease frigates in storage in Japan. For political and financial reasons it was decided to form the new force as an autonomous 'district' of the MSA. Training of officer instructors began in January 1952 and the 'Coastal Security Force' came into existence on 26 April the same year with an original establishment of 6038. Already, however, the political background had changed. Japan had promised to 'increasingly assume responsibility for its own defence against direct and indirect aggression'. In August 1952, therefore, the 'Coastal Security Force' left the MSA to become the 'Maritime Safety Force', a part of the new National Safety Agency, the original name of what was redesignated after two years the 'Self Defence Force'. The newly independent navy took over the Fairway Safety minesweeping organisation from the MSA, which continued in a purely coastguard role responsible for fisheries patrol, search and rescue, aids to navigation, pollution protection, traffic supervision and the interdiction of illegal goods.

The new title 'Maritime Self Defence Force' reflected the passage of the Defence Agency Establishment and Self Defence Forces Laws on 1 July 1954. The Self Defence Forces Law stated their primary mission to be 'to defend the nation against direct and indirect aggression for the purpose of preserving peace and the independence of the nation, and maintaining national security'. The Maritime Self Defence Force was tasked with the conduct of operations 'chiefly at sea'. On foundation it consisted of 115 units totalling 58,293t. Its main combatant assets were the patrol frigates supplemented by amphibious fire support ships supplied in 1953 to be employed as gunboats, all on loan from the USA. Two American 'Gleaves' class destroyers appeared, however, before the year was out, also on long-term loan. Backed by the US Navy, the MSDF fought and won a successful campaign to possess its own air arm, although until 1969 it had to accept joint training with the Air SDF. The USN provided the MSDF with 217 aircraft as military aid and partially funded the construction of maritime patrol aircraft in Japan. As late as 1968 some 40 per cent of the MSDF's aircraft were still American owned.

The MSDF structure has remained basically unchanged since its inception. The Chief of the Maritime Staff reports to the Director General Defence Agency and is supported by a Maritime Staff Office. Subordinate to the Maritime Staff Office are three major components, (a) the Self Defence Fleet (the major operational units), (b) the five Regional Districts (Yokosuka, Kure, Sasebo, Maizuru and Ominato) and (c) the support, training and medical units. The Self Defence Fleet was originally composed of two escort flotillas and a patrol flotilla for allocation to the Regional Districts as required; the latter controlled minesweeping forces and air stations. As the MSDF grew in size, however, the Regional Districts acquired their own flotillas of older and smaller craft and the first-line aircraft and minesweepers passed to the main fleet command. By 1972 the Self Defence Fleet looked much as it does today with a Fleet Escort Force of four flotillas, the Fleet Air Force (in 1995 three fixed wing maritime patrol wings and two ASW helicopter wings), two Mine Countermeasures Flotillas, a Submarine Flotilla (later joined by a second) and a Fleet Training Command.

Japan had begun construction of new 'high speed escort vessels' by the end of 1954. These were destroyer type ships optimised for ASW and for a time the designator DDK was used. They were supplemented by three gun-armed 'DDAs' in the later 1950s. In all twelve destroyer type ships were included in the original MSDF programmes. Then in 1957 in the context of a new 'Basic Policy for National Defence' a First Defence Build-up Plan was drawn up. Financial constraints meant this was a disappointment for the MSDF with only three new destroyers planned; in the event only two were constructed and two more obtained on loan from the USA. Money became a little more available in the decade 1962–72 with the Second and Third Buildup Programmes. These consisted of six smaller DDKs supplemented by a larger destroyer, one air defence DDG and four larger ASW ships. The Japanese relied on the Americans for weapons, including the less than successful Weapon Alfa and DASH systems. Both were operated with a little greater success in the MSDF but neither became fully satisfactory. Instead the Japanese turned to Canadian techniques of operating the large Sea King helicopter and under the Third Build-up Programme laid down a pair of innovative large helicopter destroyers (DDHs) capable of carrying three of these aircraft each.

In 1972 the Cabinet adopted a Fourth Defence Build-up plan that placed special emphasis on strengthening defence capabilities in the sea areas around Japan. The plan called for fifty-four vessels including two more helicopter destroyers and two DDGs, one air defence and one surface-to-surface. The programme was much delayed and amended in the contemporary atmosphere of *détente* and eventually became a ten-year programme including three air defence guided missile ships, the last of which completed in 1982. A feature of this period, however, was the adoption of indigenous Japanese electronics.

The 1983 Defence White Paper announced that Japan had been building 'its maritime defence capability year-by-year, by setting its goal at having the ability to provide protection of sea lines of communication in sea waters extending to about 1000 nautical miles or so from its shores in the event of establishing sea routes covering several hundred nautical miles in the surrounding waters of the country when an armed attack occurs against Japan'. The sea routes in mind were to the southwest and southeast. Such was the rationale of the '8–8' Fleet plan that has provided the escort flotillas each with eight modern surface escorts operating eight ASW helicopters. As from the start in 1954 the emphasis in the Escort Fleet is still on anti-submarine warfare but it is the aim that each flotilla should deploy a pair of DDGs to provide organic air defence. This has still not been achieved but it will be when the four Aegis destroyers are available, together with less capable AAW ships.

The end of the Cold War provides a challenge to traditional Japanese attitudes as to the size and role of the MSDF. Although Japan only spends just over 1 per cent of GNP on defence, that GNP is now so large that she has built up what must be regarded as one of the world's most powerful navies. Only a lack of carriers and nuclear-powered submarines prevents the MSDF being number three in the world naval league. The Japanese constitutional and cultural postwar aversion to things military still reflects itself in an unwillingness to build carrier type ships (despite MSDF desire to do so under the guise of landing vessels) and a reticence to send ships to take part in distant operations even under UN auspices. However the deployment in 1991 of the minesweepers *Yurishima, Hikoshima, Awashima* and *Sakushima* to the clearing-up operations in the Gulf probably marked a watershed in the emergence of the MSDF as a significant means by which re-emerged Japan plays the role that her economic position as the world's number two – and most sea-dependant – power implies.

JAPANESE SENSORS

The Japanese have employed foreign (usually US) designed equipment domestically produced with their own designations as well as sensors of their own design.

Radars: OPS-1 (a version of the US SPS-12 air search set with an SPS-6 type antenna); OPS-2 (US SPS-6B air search); OPS-4 (probably US Raytheon 1404, surface search); OPS-9 (based on the British 978, surface search); OPS-10 (small ship surface search); OPS-11 (the first Japanese designed air search radar, although probably based on the US SPS-40); OPS-12 (planar array air search); OPS-13 (small ship surface search); OPS-14 (air search); OPS-15 (surface search); OPS-16 (surface search); OPS-17 (surface search); OPS-18 (surface search); OPS-19 (navigational); OPS-22 (small ship surface search); OPS-24 (planar array air search); OPS-28 (surface search); OPS-29 (small ship surface search); OPS-35 (surface search); OPS-36 (surface search); OPS-37 (US SPS-5B, surface search); ZPS-4 (submarine search); ZPS-6 (submarine search).

Sonars: surface ship: OQS 1/2 (licence built US SQS-4/29); OQS-3 (licence built US SQS-23); OQS-4; OQS-12 (modified US SQS-31 with OQA-1 VDS and licence built SQS-35J); OQS-101; OQS-102 (US SQS-53); OQR-1 (US SQR-19 towed array); OQR-2 (US SQR-19A towed array).

Submarine: ZQQ-1, ZQQ-2, ZQQ-3 (with licence built SQS-36J active sets); ZQQ-4 (with licence built SQS-36J active sets and US supplied BQR-15 towed arrays); ZQQ-5/5B (including towed and fink arrays).

Minehunting: ZQS-1 (modified US UQS-1); ZQS-2 (licence built British 193); ZQS-3.

WARSHIPS OF THE FORMER IMPERIAL JAPANESE NAVY – REMAINING IN SERVICE AFTER 1 MAY 1948

ESCORT DESTROYER

TACHIBANA class *escort destroyer*

No	Name	Launched	Disp	Fate
DE 261	WAKABA (ex-*Nashi*)	17.1.45	1530t	Discarded 1972

One of the numerous *Tachibana* class, an emergency ship design constructed at the end of the Second World War. She was sunk 28 July 1945 by aircraft off Hitakiri Point, Inland Sea, but in 1955 was raised and purchased by the MSDF.

Reconstructed at Kure Zosen, completing 12 May 1956; renamed and re-commissioned 30 May 1956. Used for training 1956–57 she was converted 10 September 1957–26 March 1958 to a radar picket, receiving a lattice foremast, a tripod mainmast and a large air search radar mount atop her bridge. She was reclassified as an experimental ship in 1968. On recommissioning her dimensions were 321ft 6in pp × 30ft 7in × 11ft max (*98.0m × 9.3m × 3.4m*). Her machinery was rated at 14,000shp and maximum speed was 27kts. Range 4680nm at 16 kts.

Armament: 2–3in/50 (1×2), 1 Hedgehog, 6 K-guns, DC racks.
Complement: 206.

ESCORTS/LARGE PATROL VESSELS

UKURU *class*

No	Name	Launched	Disp (std)	Fate
PL 102	OKIJA (ex-*Ikuna*)	4.9.44	940t	Discarded 1963
PL 103	ATSUMI (ex-*Chikubu*)	24.11.44	940t	Discarded 1962
PL 104	SATSUMA (ex-*Ukuru*)	15.5.44	940t	Discarded 1966
PL 105	TSUGARU (ex-*Shinnan*)	19.5.44	940t	Discarded 1966
PL 106	KOJIMA (ex-*Shiga*)	9.2.45	940t	Discarded 1964

All but *Shiga* minesweeper tenders 1945–47; *Chikubu* also used as minesweeper. *Shiga* used as both minesweeper and as a ferry between Pusan and Hakata. Transferred in May 1948 for use as weather ships; *Shiga* used as depot ship. Commissioned with new names and numbers as Maritime Safety Agency large patrol vessels in 1955; armament fitted 1–3in/50, 2–20mm.

MINESWEEPERS/SMALL PATROL VESSELS

'CHA 1' *type*

Original No.	1948 No.	Name (1955–57 No.)	Fate
Cha 1	PS 18	CHIDORI (PS 134)	Discarded 1960
Cha 4	PS 20 (1953: NS 81)	KIJI	Discarded 1959
Cha 27	PS 23		Discarded 1952
Cha 57	PS 21	KAWASEMI (PS 136)	Discarded 1957
Cha 58	PS 32	UZURA (PS 140	Discarded 1960
Cha 68	PS 27	KAMOME (PS 139)	Discarded 1961
Cha 71	PS 39	HO'OJIRO (PS 143)	Discarded 1964
Cha 72	MS 07	UMITSUBAME (MSI 697)	Harbour service 1961
Cha 78	PS 28 (1953: MS 82)	OTORI	Discarded 1959

Original No.	1948 No.	Name (1955–57 No.)	Fate
Cha 79	MS 08	YUHIBARI	Discarded 1959
Cha 80	PS 08	SEKIRE (PS 128)	Discarded 1964
Cha 86	MS 09	HAYATAKA	Discarded 1959
Cha 88	MS 10	YUKARI (MSI 692)	Harbour service 1961
Cha 90	MS 11	HAHUO	Discarded 1959
Cha 93	MS 12	MIYAKODORI	Discarded 1959
Cha 99	MS 13	IWATSUBAME (MSI 698)	Harbour service 1961
Cha 154	PS 06	UGUSUI (PS 126)	Discarded 1960
Cha 155	PS 26	SHIGI (PS 155)	Discarded 1957
Cha 157	PS 37	AOSAGI (PS 141)	Discarded 1961
Cha 159	PS 38	HINOZURU (PS 142)	Discarded 1963
Cha 161	PS 24	KOTAKA (PS 137)	Discarded 1956
Cha 166	PS 17	HATO (PS 133)	Discarded 1963
Cha 168	MS 57	NISHIKIDORI	Discarded 1959
Cha 169	PS 11	YAMASEMI (PS 130	Discarded 1956
Cha 171	MS 01	CHIYOZURU (MSI 695)	Harbour service 1961
Cha 173	PS 40	SHIRATAKA (PS 144)	Discarded 1959
Cha 175	PS 01	KASASAGI (PS 121)	Discarded 1959
Cha 179	PS 41	YAMAGARA (PS 145)	Discarded 1960
Cha 181	PS 16	HIBARI (PS 132)	Discarded 1961
Cha 183	PS 1	BENIZURU	Discarded 1955
Cha 185	PS 19	HAYABUSA (PS 135)	Discarded 1958
Cha 186	PS 13	HATSUKARI (PS 131)	Discarded 1960
Cha 187	PS 04	YAMADORI (PS 124)	Discarded 1958
Cha 194	PS 42	MIZUTORI (PS 146)	Fate unknown
Cha 196	PS 25 (1953: MS 83)	WAKATAKA	Discarded 1959
Cha 198	PS 02	WAKASAGI (PS 122)	Discarded 1964
Cha 202	MS 14		Sunk 1950
Cha 203	PS 22 (1953: MS 84)	HAYODORI (MSI 700)	Discarded 1964
Cha 214	MS 15	HAYATORI (MSI 699)	Discarded 1964
Cha 215	MS 16	TOMOZURU (MSI 693)	Discarded 1963
Cha 217	PS 07	TSUGUMI (PS 127)	Discarded 1960
Cha 219	PS 14 (1953: MS 85)	SHIRASAGI	Discarded 1959
Cha 221	MS 02	YOSHIKIRI (MSI 690)	Harbour service 1961
Cha 222	MS 03	HATSUTAKA (MSI 696)	Discarded 1964
Cha 227	PS 05	OSHIDORI (PS 125)	Discarded 1957
Cha 231	MS 17	SHIRATORI (MSI 694)	Discarded 1963
Cha 232	PS 09 (1953: MS 85)	OTAKA	Discarded 1963
Cha 236	PS 15	KOMADORI	Discarded 1954
Cha 239	PS 10	KARI (PS 129)	Discarded 1955
Cha 241	MS 04	FURUTAKA	Discarded 1959
Cha 245	PS 03	MANAZURU (PS 123)	Discarded 1961
Cha 246	MS 05	YAMABATO	Discarded 1959
Cha 249	MS 06	KAMAZURU (MSI 691)	Discarded 1963

'Cha 1' type launched 1941–44, 130t. Auxiliary submarine chasers built of wood; based on harbour tug design. Converted to minesweepers during war. Transferred to Maritime Safety Agency 1948 as minesweepers (MS) and Small Patrol Vessels (PS). Minesweepers to Maritime Safety Force (later MSDF) 1 August 1952. MSA patrol vessels named and renumbered 1953. MSDF ships named 1956 and eleven re-rated Inshore Minesweepers (MSI) in 1957.

Pa No. 1 class

Original No.	1948 No.	1956 Name	Fate
Pa No. 31	MS 18	UKISHIMA	Harbour service 1962
Pa No. 84	MS 19	TSURUSHIMA	Harbour service 1963
Pa No. 134	MS 20	OTOSHIMA	Discarded 1962
Pa No. 135	MS 21	MATSUSHIMA	Harbour service 1963
Pa No. 136	MS 22	HIMESHIMA	Discarded 1962
Pa No. 138	MS 23	AWASHIMA	Harbour service 1962
Pa No. 139	MS 24	KARUSHIMA	Harbour service 1963
Pa No. 152	MS 25	KAMOSHIMA	Discarded 1962
Pa No. 153	MS 26	TAKASHIMA	Harbour service 1963
Pa No. 154	MS 27		Sunk 1949
Pa No. 175	MS 28		Wrecked 1950
Pa No. 179	MS 29	OSHIMA	Discarded 1963

Launched 1945. 238t. Auxiliary patrol vessels built under 1943–44 programme with wooden hulls intended for eventual conversion to fishing vessels. Used as minesweepers post war and transferred to Maritime Safety Board 1948. To MSDF on its creation; named 1956.

300t aircraft rescue boats. Became MSA Small Patrol Vessels *Kawachidori* (PS 101), *Hamachidori* (PS 102), *Asachidori* (PS 103), *Miachidori* (PS 104), *Tomachidori* (PS 105), *Okichidori* (PS 106), *Sawachidori* (PS 107), *Wakachidori* (PS 108), *Murachidori* (PS 109), *Shimachidori* (PS 112), *Haruchidori* (PS 115); and *Yuchidori* (MS 62) used as mine destructor vessel. (Reclassified 'minesweeping boat' ASM 71, 1961)

189t aircraft rescue boat minesweeper *Okichidori* (MS 68) (Reclassified ASMN 72, 1961)

All above discarded 1963–69.

The 2,850t 2TM tanker *Soei* (former merchantman *Soei Maru*) was used as a mine destructor vessel and numbered MS 32. She was also used as a minesweeper tender. Discarded 1963. Another large merchantman, MS 31 of 6 888t was discarded in 1955. The 3 800t icebreaking transport *Soya*, requisitioned by the IJN in 1939, was used as repatriation ship and then by the MSA from 1948 as a lighthouse supply ship and navigational aid vessel. Converted to Antarctic Research Ship (PL 107) in 1956 and more extensively in 1957 and again in 1958. Assigned to guard rescue duties as patrol vessel (PL 107) 1963. Discarded 1978.

MAJOR SURFACE SHIPS

Both these ships (launched 1941) were transferred under MDAP and were the first destroyers to serve in the MSDF. During their lengthy stay with the MSDF, new electronics were added and the forward superfiring 5in gun was removed to compensate the added torpedo weight. Both ships were returned to the United States in 1969 but were later transferred to Taiwan, *Hatakaze* as *Hsien Yang* and *Asakaze* for spares. See under United States and Taiwan for full details.

Ex-US GLEAVES class *destroyers*

Particulars: As US *Gleaves* class

No	Name	Builder	Acquired	Fate
DD 181	ASAKAZE (ex-*Ellyson*)	Federal, Kearny	19.10.54	Returned USA 1969
DD 182	HATAKAZE (ex-*Macomb*)	Bath Iron Wks	19.10.54	Returned USA 1969

Ex-US FLETCHER class *destroyers*

Particulars: As US *Fletcher* class

No	Name	Builder	Acquired	Fate
DD 183	ARIAKE (ex-*Heywood L Edwards*)	Boston N Yd	10.3.59	Deleted 1974, BU 1976
DD 184	YUGURE (ex-*Richard P Leary*)	Boston N Yd	10.3.59	Deleted 1974

Both launched 6 October 1943; towed to Japan 1959. Both ships had been reconstructed by March 1963. Changes included improved bridges, larger CICs, new radars and a tripod mast. No 3 5in mounting was removed from both ships to compensate added topside weight, while No 2 5in mounting was removed from *Ariake* and replaced by Weapon Alfa. See under United States for full details.

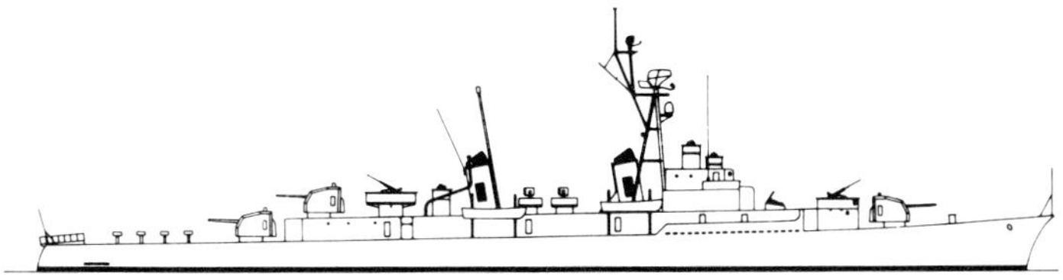

Harukaze as completed

HARUKAZE class ('A type') *destroyers*

Displacement: 1700t standard; 2340t full load
Dimensions: 347ft 9in wl, 348ft 9in oa × 34ft 5in × 12ft 2in max
106.0m, 106.3m × 10.5m × 3.7m
Machinery: *Harukaze*: 2-shaft Mitsubishi/Escher Weiss geared turbines, 2 Hitachi/Babcock boilers, 30,000shp = 30kts. Oil 557t
Yukikaze: 2-shaft Westinghouse geared turbines, 2 Combustion Engineering boilers, 30,000shp = 30kts. Oil 557t. Range 6000nm at 18kts
Armament: 3–5in/38 (3 × 1), 8–40mm (2 × 4), 2 Hedgehog, 8 K-guns, 1 DC rack
Sensors: Radar SPS-5, SPS-6; sonar SQS-29
Complement: 240

No	Name	Builder	Laid down	Launched	Comp	Fate
DD 101	HARUKAZE	Mitsubishi Zosen, Nagasaki	15.12.54	20.9.55	26.4.56	TS 27.3.81, deleted 1985
DD 102	YUKIKAZE	Shin Mitsubishi, HI, Kobe	17.10.54	20.8.55	31.7.56	TS 27.3.81, deleted 1985

The *Harukaze* class was authorised under the 1953 fiscal year programme as 'A type' high-speed escort vessels. The class were conservative in size and armament, and electric welding was extensively used in hull construction. As with the destroyer escorts built under the 1953 programme, different propulsion plants were used for study purpose. All weapons and electronics were US produced. In March 1959 both units were fitted with ASW torpedo launchers. In the mid-1970s, *Yukikaze*'s after mast, 5in mounting, K-guns, and DC rack were removed and a towed passive sonar array was installed. They were transferred to Fleet Training and Development Command on 27 March 1981 as Special Use Auxiliaries ASU 7002 and ASU 7003. In 1980 *Yukikaze* was further modified to mount 1–5in/38 and 4–40mm only, in order to carry out trials with a later model towed passive sonar array. *Harukaze* is moored as a training hulk at the Etajima Naval Academy.

Harukaze as completed
Navarret Collection

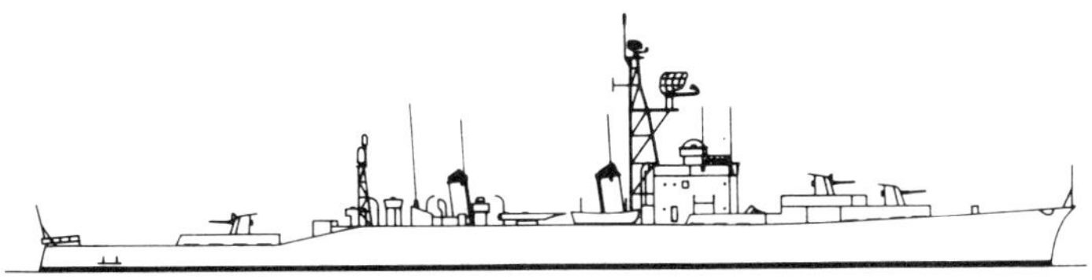

Ayanami as completed

AYANAMI class *destroyers*

Displacement: 1720t standard; 2500t full load
Dimensions: 357ft 7in wl × 35ft 2in × 12ft 2in light
109.0m × 10.7m × 3.7m
Machinery: 2-shaft Mitsubishi/Escher Weiss turbines, 2 Mitsubishi-Nagasaki CE type boilers, 35,000shp = 32kts. Range 6000nm at 18kts
Armament: 6–3in (3 × 2), 4–21in TT (1 × 4), 2 ASW torpedo racks, 2 Hedgehog, 2 Y-guns
Sensors: Radar OPS-1, OPS-15 (DD 103–106, OPS-2); sonar OQS-12 or 14 (DD 103, 104, 110, also OQA-1 VDS)
Complement: 220

No	Name	Builder	Laid down	Launched	Comp	Fate
DD 103	AYANAMI	Mitsubishi Zosen, Nagasaki	20.11.56	1.6.57	12.2.58	TS 1982, deleted 25.12.86
DD 104	ISONAMI	Shin-Mitsubishi, Kobe	14.12.56	30.9.57	14.3.58	TS 1975, deleted 1.7.87
DD 105	URANAMI	Kawasaki, Tokyo	1.2.57	29.8.57	27.2.58	TS 1982, deleted 25.12.86
DD 106	SHIKINAMI	Mitsui Zosen, Tamano	14.12.56	25.9.57	15.3.58	TS 1976, deleted 1.7.87
DD 110	TAKANAMI	Mitsui Zosen, Tamano	8.11.58	8.8.59	30.1.60	TS 1985, deleted 1.3.89
DD 111	ONAMI	Ishikawajima HI, Kobe	20.3.59	13.2.60	29.8.60	TS 1987, deleted 1.3.90
DD 112	MAKINAMI	Iino HI, Maizuru	20.3.59	25.4.60	28.10.60	TS 1987, deleted 1.3.90

The *Ayanami* class were designed principally for ASW warfare and originally designated DDKs. Four ships were built under the 1955 programme, one under the 1957 and two under the 1958 programmes. The 3in guns were US Mk 33s but the shields were uniquely Japanese. Later units of the class had a bulwark aft of the hull break. Reload torpedoes for the 21in tubes were carried in racks on deck, which was standard practice in Japanese destroyers during the war. *Ayanami* was equipped with eight wash-down systems for removing radio-active fallout. The units built under the 1957 and 1958 programmes were equipped with ECM, an improved sonar, and improved air conditioning. VDS was mounted in *Ayanami* (1965), *Isonami* (1966), and *Takanami* (1967). These three units had their original ASW torpedo system, the Mk 32 torpedo thrown overboard by the Mk 8 'Poor Boy' launcher, replaced by Mk 32 TT at the same time. *Isonami and Shikinami* had their 21in TT mount replaced by a deckhouse when converted to TS. The electronics were largely Japanese versions of US sets (OPS-1 is US SPS-6, and OPS-2 is SPS-12, for example). Transferred to Fleet Training and Development Command as Special Use Auxiliaries ASU 7004–7007, ASU 7009, ASU 7013 and ASU 7014 respectively between 1975 and 1987.

Uranami May 1968 L & L van Ginderen

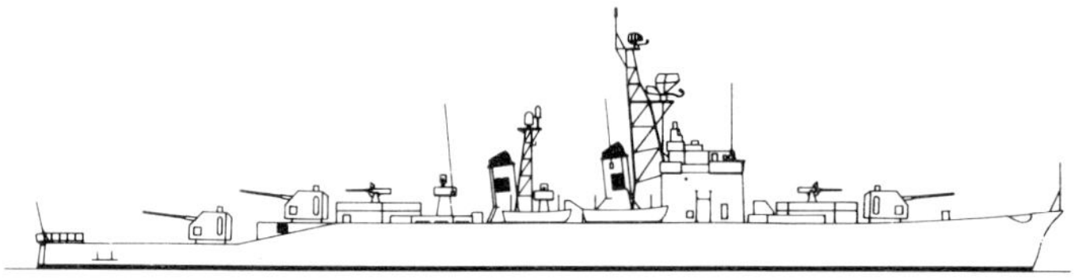

Murasame 1959

The *Murasame* class were designed principally for AAW and originally were designated DDAs. Two ships were built under the 1956 programme and the other under the 1957 programme. The 5in guns mounted in this class were Mk 39s removed from the US *Midway* class aircraft carriers, the only ships then fitted with this gun; this 5in/54 was to have been the successor to the ubiquitous 5in/38. Actually, their rate of fire was disappointing, making them less than the ideal AA weapon. The handling rooms and magazines for these guns required additional space, thus making the *Murasame*s much more crowded ships than the *Ayanami*s. Weapons constitute 18 per cent of the weight of the *Murasame* class units. *Murasame* and *Harusame* fitted with 6–12.75in Mk 32 ASW TT, replacing torpedo racks, Y-guns and DC, in 1975 and 1979 respectively. *Murasame* became ASU 7006 in 1984, used as experimental vessel for deep-sea minesweeping equipment; *Harusame* became submarine depot ship ASU 7008 in 1985.

MURASAME class *destroyers*

Displacement: 1838t (*Harusame* 1800t, *Yudachi* 1840t) standard
Dimensions: 360ft 9in wl × 36ft × 12ft 2in light
110.0m × 11.0m × 3.7m
Machinery: *Harusame*: 2-shaft Ishikawajima geared turbines, 2 Ishikawajima Foster Wheeler boilers, 35,000shp = 32kts
Murasame: 2-shaft Mitsubishi Escher Weiss geared turbines, 2 Mitsubishi CE type boiler, 35,000shp = 32kts
Yudachi: 2-shaft Kanpon-Ishikawajima geared tubines, 2 Ishikawajima FW-D type boilers, 35,000shp = 32kts. Range 6000nm at 18kts
Armament: 3–5in/54 (3 × 1), 4–3in/50 (2 × 2), 2 ASW torpedo racks, 1 Hedgehog, 2 Y-guns, 1 DC rack
Sensors: Radar OPS-1, OPS-15; sonar SQS-29 (DD 109, also OQA-1 VDS)
Complement: 220

No	Name	Builder	Laid down	Launched	Comp	Fate
DD 107	MURASAME	Mitzubishi Zosen, Nagasaki	16.12.57	31.7.58	31.3.59	Auxiliary 1984, deleted 23.3.88
DD 108	YUDACHI	Ishikawajima HI, Tokyo	16.12.57	31.7.58	16.3.59	Auxiliary 1984, deleted 1987
DD 109	HARUSAME	Uraga Dock, Yokosuka	17.6.58	18.6.59	15.12.59	Auxiliary 1985, deleted May 1989

Murasame on trials
Navarret Collection

Akizuki 25.9.63
CMP

The *Akizuki* class were built under the 1957 US Military Aid Programme and carried the 5in/54 as in the *Murasames*. They featured a long forecastle hull and were designed as flotilla leaders. Weapon Alfa, an ASW mortar developed in the United States, proved to be unsatisfactory. Weapons and ASW sensors were modernised on *Teruzuki* in 1976–77 and on *Akizuki* in 1977–78, with a 375mm Bofors ASW mortar replacing Weapon Alfa and 6–12.75in Mk 32 ASW TT replacing the original launchers; VDS was added and SQS-29 sonar was replaced with SQS-23. Torpedo tubes and VDS removed on conversion to auxiliaries. *Akizuki* (ASU 7010) had after gun turret replaced by trial towed sonar array in 1989. *Teruzuki* (ASU 7012) re-rated TU 3504 as submarine depot/target ship 20 June 1991.

AKIZUKI class *destroyers*

Displacement: 2388t standard; 2890t normal
Dimensions: 387ft 2in oa, 377ft 2in pp × 39ft 4in × 13ft 1in light
118.0m, 115.0m × 12.0m × 4.0m
Machinery: *Akizuki*: 2-shaft Mitsubishi Escher Weiss geared turbines, 4 Mitsubishi CE type type boilers, 45,000shp = 32kts
Teruzuki: 2-shaft Westinghouse geared turbines, 4 Mitsubishi/CE type boilers, 45,000shp = 32kts
Armament: 3–5in/50 (3 × 1), 4–3in/50 (2 × 2), 1 Weapon Alfa, 2 Hedgehog, 2 Y-guns, 2 ASW torpedo, 4–21in TT (1 × 4, 8 torpedoes)
Sensors: Radar OPS-1, OPS-15; sonar SQS-23, OQA-1 VDS
Complement: 330

No	Name	Builder	Laid down	Launched	Comp	Fate
DD 161	AKIZUKI	Mitsubishi Zosen, Nagasaki	31.7.58	26.6.59	13.2.60	Auxiliary 1985, deleted Dec 1993
DD 162	TERUZUKI	Shin Mitsubishi, Jyuko, Kobe	15.8.58	24.6.59	29.2.60	Auxiliary 1986, TS 1.7.87, deleted Sep 1993

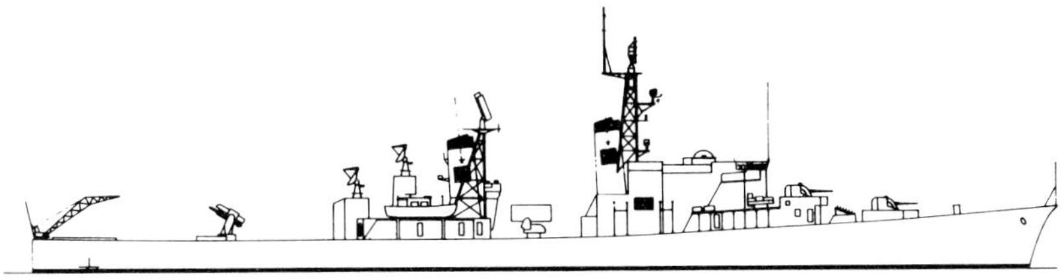
Amatsukaze as completed

Amatsukaze was the first guided missile ship built in Japan. She required a little over two years to build, a remarkably short time. The Tartar (Standard) system was applied from the United States. Her flush deck design gives a clean appearance, and because of her clean quarterdeck many publications mistakenly credit her with the capacity to handle a helicopter. She was refitted in 1967 when the Mk 32 TT and the SPS-52 radar were installed. A further refit with OTO Melara 76mm compact mountings and improved fire control was planned but was cancelled.

AMATSUKAZE *destroyer*

Displacement: 3050t standard; 4000t full load
Dimensions: 429ft 9in pp × 43ft 11in × 13ft 9in max
131.0m × 13.4m × 4.2m
Machinery: 2-shaft Ishikawajima/GE geared turbines, 2 Ishikawajima/Foster Wheeler boilers, 60,000shp = 33kts. Oil 900t. Range 7000nm at 18kts
Armament: 1 Mk 13 launcher (40 Standard MR-1(MR) missiles), 4–3in/50 (2 × 2), 1 ASROC ASW system (1 × 8), 2 Hedgehog, 6–12.75in Mk 32 ASW TT (2 × 3)
Sensors: Radar OPS-17, SPS-29, SPS-52, SPG-51, SPG-34; sonar SQS-23
Complement: 290

No	Name	Builder	Laid down	Launched	Comp	Fate
DDG 163	AMATSUKAZE	Mitsubishi, Nagasaki	29.11.62	5.10.63	15.2.65	To decommission 1995

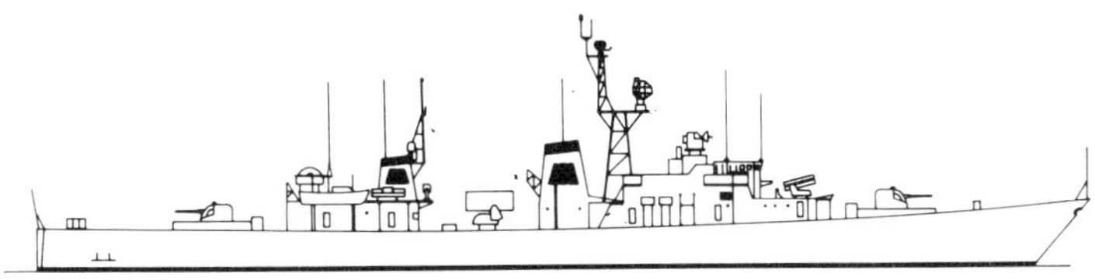
Asagumo 1974

The *Yamagumo* class carry an impressive ASW armament, and being diesel-powered, the ships have a small complement for their size and potential. They were classified as 'escort destroyers' (DDK). Originally, *Yuugumo* was to have been the lead ship of a new class. The fact that this last unit was laid down 12 years after the first reflects well upon the success of the design. They were authorised as follows: *Yamagumo* – 1962 Programme; *Makigumo* – 1963; *Asagumo* – 1964; *Aokumo* – 1965; *Akigumo* – 1967; *Yuugumo* – 1967. *Yamagumo* and *Makigumo* have a raised stern, which houses a VDS; the other units received the VDS during construction and, therefore, have a flat sterm. On conversion to TS first pair had a lecture room built under the ASROC launcher and a chart room added on the signal deck. *Asagumo* was converted for use as a training support ship for the submarine force.

YAMAGUMO class *escort destroyers*

Displacement: 2100t standard; 2700t full load
Dimensions: 376ft 11in pp × 38ft 9in × 13ft 2in max
114.9m × 11.8m × 4.0m
Machinery: 2 shafts, 6 Mitsubishi (Mitsui in DDK 113 and 114) diesels, 26,500bhp = 27kts. Range 7000nm at 20kts
Armament: 4–3in/50 (2 × 2), 1 ASROC ASW system (1 × 8), 1–375mm ASW rocket launcher (1 × 4), 6–12.75in Mk 32 ASW TT (2 × 3)
Sensors: Radar OPS-11, OPS-17; sonar SQS-23 (OQS-3 in DDK 119–121), SQS-35(J) VDS (not in DDK 115)
Complement: 210

No	Name	Builder	Laid down	Launched	Comp	Fate
DDK 113	YAMAGUMO	Mitsui, Tamano	23.3.64	27.2.65	29.1.66	TS 20.6.91
DDK 114	MAKIGUMO	Uraga, Yokosuka	10.6.64	26.7.65	19.3.66	TS 20.6.91
DDK 115	ASAGUMO	Maizuru Jukogyo, Maizuru	24.6.65	25.11.66	29.8.67	ASU 7018 18.10.93
DDK 119	AOKUMO	Sumitomo, Uraga	2.10.70	30.3.72	25.11.72	Extant 1995
DDK 120	AKIGUMO	Sumitomo, Uraga	7.7.72	23.10.73	24.7.74	Extant 1995
DDK 121	YUUGUMO	Sumitomo, Uraga	4.2.76	21.5.77	24.3.78	Extant 1995

Amatsukaze on trials Navarret Collection

Makigumo Aug 1985 L & L van Ginderen

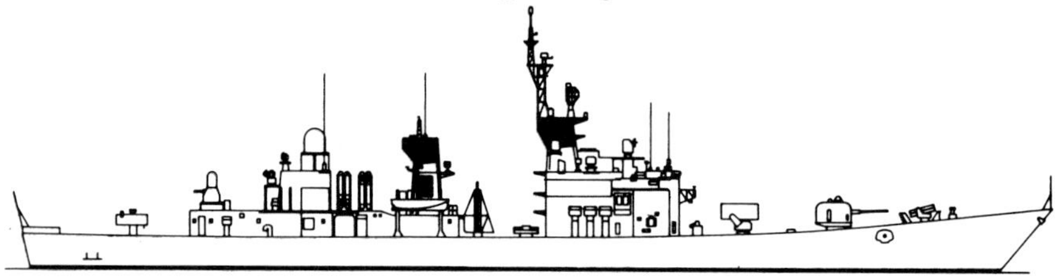

Kikuzuki 1987

The *Takatsuki* class have a typical 'American' appearance because of their 5in guns and ASROC. The ships were designed for the US ASW DASH system (three drones), which proved to be unsuccessful and was removed in 1977. *Mochizuki* and *Nagatsuki* have a knuckle in the hull sides forward. *Kikuzuki* is the only unit fitted with fin stabilisers. VDS has been mounted in *Takatsuki* (1970), *Kikuzuki* (1972), *Mochizuki* (1972) and *Nagatsuki* (1972). All units are fitted with gravity davits and telescopic high-line equipment. A substantial modernisation programme was carried out on *Takatsuki* (April 1984 to October 1985) and *Kikuzuki* (May 1985 to December 1986). The after 5in/54 and disused DASH hangar were replaced by Sea Sparrow SAM (1 × 8, Mk 29 launcher), eight Harpoon SSM (2 × 4), provision for Phalanx CIWS, improved electronics and the addition of the American SQS-18A TACTASS towed array sonar. Modernisation for the rest of the class was cancelled and the ships were allocated to district flotillas. *Nagatsuki* served for a time as a cadet training ship.

TAKATSUKI class *destroyers*

Displacement: 3100t standard; 4500t full load
Dimensions: 446ft 2in oa × 43ft 11in × 14ft 5in max
136.0m × 13.4m × 4.4m
Machinery: 2-shaft Mitsubishi geared turbines, 2 Mitsubishi/CE boilers, 60,000shp = 32kts. Range 7000nm at 20kts
Armament: 2–5in (2 × 1), 1 ASROC ASW system (1 × 8), 1–375mm ASW rocket launcher (1 × 4), 6–12.75in Mk 32 ASW TT (2 × 3), 3 DASH (now removed)
Sensors: Radar OPS-11B, OPS-17; sonar SQS-23 (OQS-3in DD 166 and 167), SQS-35(J) VDS (not in DD 166)
Complement: 270

No	Name	Builder	Laid down	Launched	Comp	Fate
DD 164	TAKATSUKI	Ishikawajima-Harima	8.10.64	7.1.66	15.3.67	Extant 1995
DD 165	KIKUZUKI	Mitsubishi	15.3.66	25.3.67	27.3.68	Extant 1995
DD 166	MOCHIZUKI	Ishikawajima-Harima Tokyo	25.11.66	15.3.68	25.3.69	Extant 1995
DD 167	NAGATSUKI	Mitsubishi HI, Nagasaki	2.3.68	19.3.69	12.2.70	Extant 1995

Nagatsuki 1976
USN

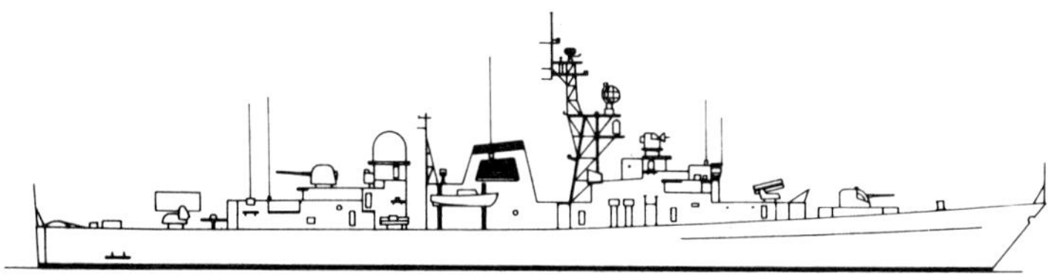

Murakumo 1979

This class paralleled the *Yamagumo* DDKs, substituting DASH for ASROC; the DASH system was a failure and is no longer carried. In 1978 *Murakumo* was fitted with 1–76mm OTO Melara compact, and an ASROC ASW system; the ASROC was placed on what had been the flight deck for DASH and a 76mm OTO Melara gun replaced the aft 3in mount. The other units were refitted with ASROC but retained their after 3in Mk 32s.

MINEGUMO class *escort destroyers*

Displacement: 2100t standard; 2750t full load
Dimensions: 376ft 11in pp × 38ft 9in × 12ft 6in max
114.9m × 11.8m × 3.8m
Machinery: 2 shafts, 6 Mitsubishi IZUEV 30/40 diesels, 26,500bhp = 27kts. Range 7000nm at 20kts
Armament: 4–3in/50 (2 × 2), 1–375mm ASW rocket launcher (1 × 4), 6–12.75in Mk 32 ASW TT (2 × 3). See notes
Sensors: Radar OPS-11, OPS-17, SPG-34; sonar OQS-3 (plus SQS-35(J) VDS in DDK 118)
Complement: 205

No	Name	Builder	Laid down	Launched	Comp	Fate
DD 116	MINEGUMO	Mitsui, Tamano	14.3.67	16.12.67	21.8.68	Extant 1995
DD 117	NATSUGUMO	Uraga, Yokosuka	26.6.67	25.7.68	25.4.69	Extant 1995
DD 118	MURAKUMO	Maizuru HI	19.10.68	15.11.69	21.8.70	Extant 1995

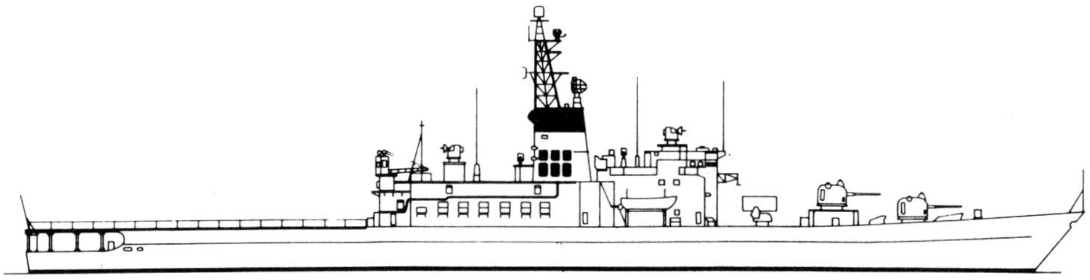

Haruna as completed

The *Haruna* class and the follow-on *Shirane* are the only destroyer-size ships in the world which carry three large ASW (HSS-2) helicopters. They have a strong ASW armament but were originally weak against air and surface targets. A Sea Sparrow SAM launcher was added on the hangar roof during FRAM refits (*Haruna* 31 March 1986–31 October 1987, *Hiei* 31 August 1987–30 March 1989). Two Phalanx CIWS and chaff launchers were also fitted. These units have very large bridge areas (100m²). The funnel is located somewhat off-centre towards the port side of the ship, thus permitting space for the third helicopter in the hangar. Fin stabilisers are installed at the bilge keel, and the 'bear trap' landing system is installed. The ships are used as flagships for escort flotillas.

HARUNA class *helicopter destroyers*

Displacement: 4700t standard; 6300t full load
Dimensions: 501ft 11in pp × 57ft 5in × 16ft 8in max
153.0m × 17.5m × 5.1m
Machinery: 2-shaft geared turbines, 2 boilers, 70,000shp = 32kts
Armament: 2–5in/54 (2 × 1), 1 ASROC ASW system (1 × 8), 6–12.75in Mk 32 ASW TT (2 × 3), 3 helicopters
Sensors: Radar OPS-11, OPS-17 (replaced by OPS-28); sonar OQS-3
Complement: 340

No	Name	Builder	Laid down	Launched	Comp	Fate
DD 141	HARUNA	Mitsubishi, Nagasaki	19.3.70	Dec 1971	22.3.73	Extant 1995
DD 142	HIEI	Ishikawajima-Harima, Tokyo	8.3.72	13.8.73	27.12.74	Extant 1995

Hiei 29.7.85 L & L van Ginderen

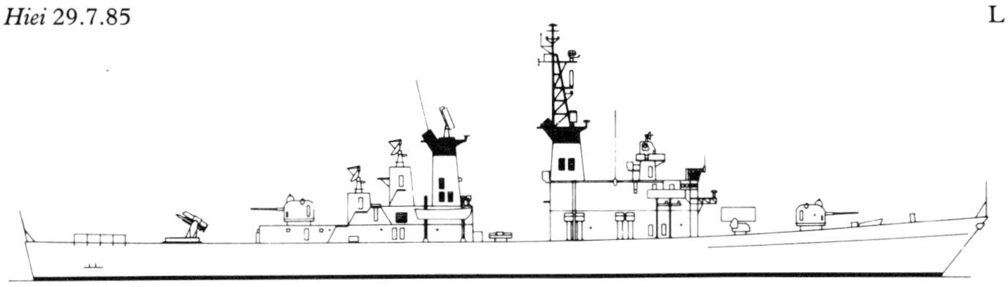

Tachikaze as completed

The *Tachikaze* class greatly improved the MSDF's anti-aircraft capabilities. The ships were fitted with progressively more advanced combat data systems: OYQ-1B in *Tachikaze*, OYQ-2B in *Asakaze* and OYQ-4 in *Sawakaze*. The propulsion plant is identical with that fitted in the *Haruna* class. Eight Harpoon SSM (2×4) and two Phalanx CIWS were added to *Tachikaze* in 1983 and to the others in 1987.

TACHIKAZE class *destroyers*

Displacement: 3850t standard; 4800t full load
Dimensions: 469ft 2in pp × 46ft 10in × 15ft 1in max
143.0m × 14.3m × 4.6m
Machinery: 2-shaft geared turbines, 2 boilers, 70,000shp = 32kts
Armament: 1 Standard SAM (1 × 1 Mk 13 launcher, SM-1(MR)missiles), 2–5in/54 (2 × 1), 1 ASROC ASW system (1 × 8), 6–12.75in Mk 32 ASW TT (3 × 3)
Sensors: Radar OPS-11, SPS-52B, SPG-51C; sonar OQS-3
Complement: 277

No	Name	Builder	Laid down	Launched	Comp	Fate
DD 168	TACHIKAZE	Mitsubishi, Nagasaki	19.6.73	12.12.74	26.3.76	Extant 1995
DD 169	ASAKAZE	Mitsubishi, Nagasaki	27.5.76	15.10.77	27.3.79	Extant 1995
DD 170	SAWAKAZE	Mitsubishi, Nagasaki	14.9.79	4.6.81	Feb 1982	Extant 1995

Shirane c 1985
JMSDF

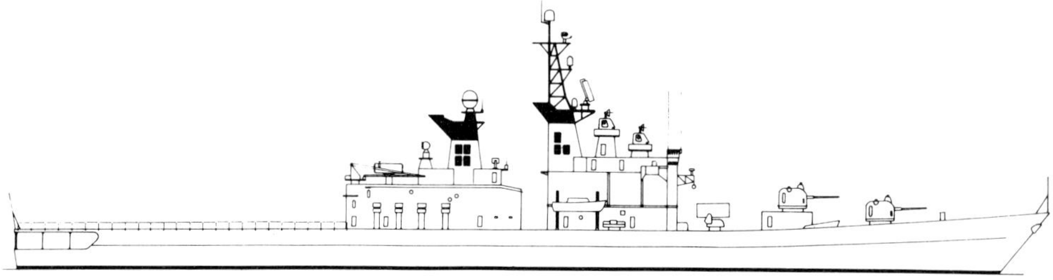

Shirane as completed

SHIRANE class *helicopter destroyers*

These two units are improved *Haruna* type with better AAW capabilities. Both HSS-2B helicopters originally carried have now been replaced by Sea Hawks. Both received the American SQR-18A TACTASS towed passive sonar array, and 2–20mm Phalanx CIWS (2 × 1) by mid-1990. The after funnel is set to starboard and the forward one to port. The ships are fitted with Vosper Thornycroft stabilisers. They are used as flagships for escort flotillas.

Displacement: 5200t standard; 6800t full load
Dimensions: 521ft pp × 57ft 5in × 17ft 5in max
158.8m × 17.5m × 5.3m
Machinery: 2-shaft geared turbines, 2 boilers, 70,000shp = 32kts
Armament: 1 Sea Sparrow SAM (1 × 8), 2–5in/54 (2 × 1), 1 ASROC ASW system (1 × 8), 6–12.75in Mk 32 ASW TT (2 × 3), 3 helicopters
Sensors: Radar OPS-12, OPS-28; sonar OQS-101, SQS-35(J) VDS, SQR-18A towed array
Complement: 370

No	Name	Builder	Laid down	Launched	Comp	Fate
DDH 143	SHIRANE	Ishikawajima-Harima, Tokyo	25.2.77	18.9.78	17.3.80	Extant 1995
DDH 144	KURAMA	Ishikawajima-Harima, Tokyo	1977	20.9.79	27.3.81	Extant 1995

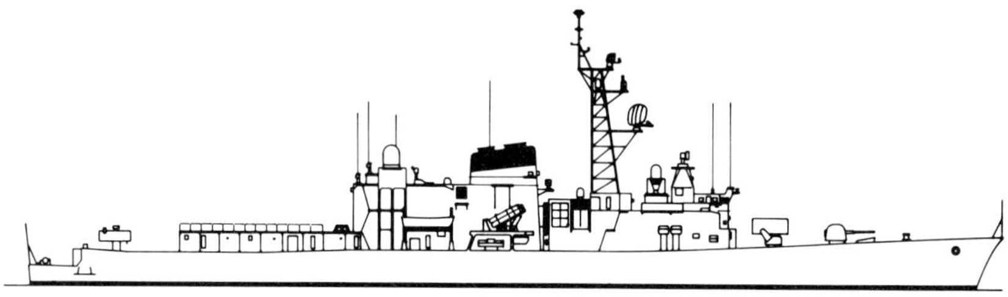

Yamayuki 1987

HATSUYUKI class *destroyers*

These units are multi-purpose destroyers with a balanced anti-air, surface and sub-surface capability. The Sea King helicopters are operated on the 'Bear Trap' system. Two Phalanx CIWS were fitted to all ships of the class by mid-1994. The higher displacement of the later ships is caused by the substitution of steel superstructure for the original aluminium alloy. *Matsuyuki* was the first of the class to be fitted with OQR-1 towed array sonar, in 1990; others are being fitted.

Displacement: 2850t standard; 3700t full load (from DD 129 on 3050t/3800t)
Dimensions: 432ft 4in pp × 44ft 11in × 14ft 3in max
131.7m × 13.7m × 4.3m
Machinery: 2-shaft COGOG (CP propellers): 2 Kawasaki Rolls-Royce Olympus TM-3B gas turbines, 56,780shp = 30kts; 2 Tyne RM-1C gas turbines, 10,860shp = ?kts
Armament: 8 Harpoon SSM (2 × 4), 1 Sea Sparrow SAM (1 × 8), 1–76mm OTO Melara, 1 ASROC ASW system (1 × 8), 6–12.75in Mk 32 ASW TT (2 × 3), 1 helicopter
Sensors: Radar OPS-18, OPS-14B; sonar OQS-4 (see notes)
Complement: 190

No	Name	Builder	Laid down	Launched	Comp	Fate
DD 122	HATSUYUKI	Sumitomo, Uraga	14.3.79	7.11.80	23.3.82	Extant 1995
DD 123	SHIRAYUKI	Hitachi, Maizuru	3.12.79	4.8.81	8.2.83	Extant 1995
DD 124	MINEYUKI	Mitsubishi, Nagasaki	7.5.81	19.10.82	26.1.84	Extant 1995
DD 125	SAWAYUKI	Ishikawajima-Harima, Tokyo	22.4.81	21.6.82	15.2.84	Extant 1995
DD 126	HAMAYUKI	Mitsui, Tamano	4.2.81	27.5.82	18.11.83	Extant 1995
DD 127	ISOYUKI	Ishikawajima-Harima, Tokyo	20.4.81	19.9.83	23.1.85	Extant 1995
DD 128	HARUYUKI	Sumitomo, Uraga	11.3.82	6.9.83	14.3.85	Extant 1995
DD 129	YAMAYUKI	Hitachi, Maizuru	25.2.83	10.7.84	3.12.85	Extant 1995
DD 130	MATSUYUKI	Ishikawajima-Harima, Tokyo	7.4.83	25.10.84	19.3.86	Extant 1995
DD 131	SETOYUKI	Mitsui, Tamano	26.1.84	3.7.85	11.12.86	Extant 1995
DD 132	ASAYUKI	Sumitomo, Uraga	22.12.83	16.10.85	20.2.87	Extant 1995
DD 133	SHIMAYUKI	Mitsubishi, Nagasaki	8.5.84	29.1.86	17.2.87	Extant 1995

Shimayuki 30.9.88
L & L van Ginderen

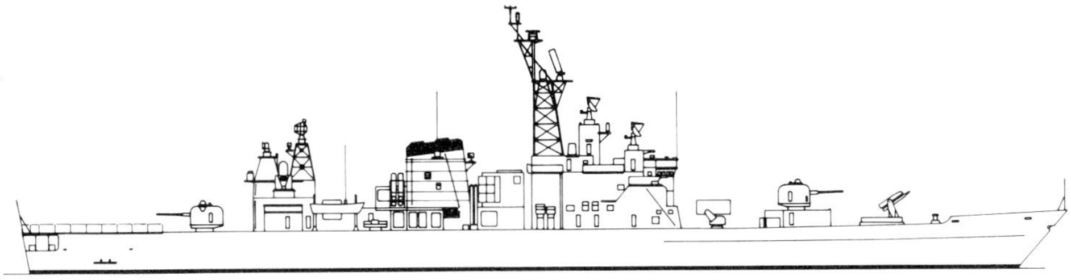

Shimakaze 1989

HATAKAZE class *destroyers*

First unit ordered under 1981 budget, second under 1983 programme. A third unit was requested in 1985 but refused. Although both fitted with helicopter platform (HHS-2B or Sea Hawk operated), they have no hangar. They provide AAW defence to the Escort Flotillas to which they are allocated.

Displacement: 4600t standard: 5500t full load
Dimensions: 492ft × 53ft 9in × 15ft 8in
150.0m × 16.4m × 4.8m
Machinery: 2-shaft COGAG: 2 Olympus TM 38 plus 2 Spey SMIA gas turbines, 76,050shp = 30kts
Armament: 2–5in/54 (2 × 1), 8 Harpoon SSM (2 × 4), 1 Standard SAM (1 × 1; 40 missiles), 2–20mm Phalanx CIWS (2 × 1), 1 ASROC ASW system (1 × 8), 6–12.75in Mk 32 ASW TT (2 × 3), helicopter platform
Sensors: Radar SPS-52C, OPS-28B, OPS-11C, SPG-51C; sonar OQS-4
Complement: 260

No	Name	Builder	Laid down	Launched	Comp	Fate
DD 171	HATAKAZE	Mitsubishi, Nagasaki	20.5.83	9.11.84	27.3.86	Extant 1995
DD 172	SHIMAKAZE	Mitsubishi, Nagasaki	30.1.85	30.1.87	23.3.88	Extant 1995

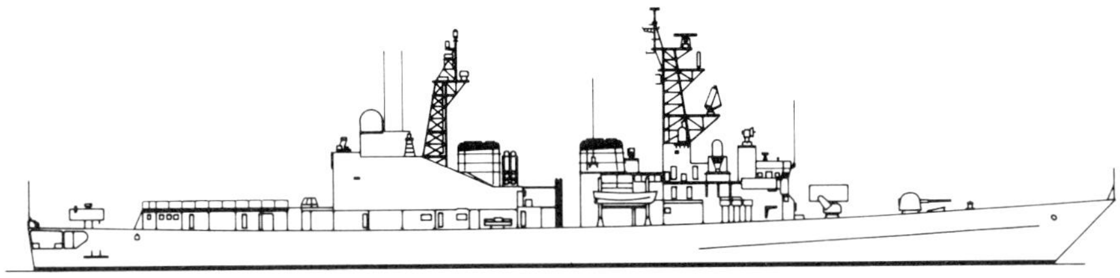

Hamagiri 1989

ASAGIRI class *destroyers*

An improved *Hatsuyuki* with upgraded electronics and a new propulsion plant. The last four were completed with further improved radar and helicopter data link; originally the helicopters carried were Sea Kings, now replaced by Sea Hawks. *Umigiri* was the first to be fitted with towed-array sonar, which is being retrofitted to the rest of the class. The after funnel and mast arrangements had to be altered because of gas turbine exhaust problems.

Displacement: 3500t standard; 4200t full load
Dimensions: 449ft 5in × 48ft × 14ft 7in
137.0m × 14.6m × 4.5m
Machinery: 2-shaft COGAG (CP propellers); 4 Rolls-Royce Spey SM-1A gas turbines, 53,300shp = 30kts
Armament: 8 Harpoon SSM (2 × 4), 1 Sea Sparrow SAM (1 × 8), 1–76mm OTO Melara, 2–20mm Phalanx CIWS (2 × 1), ASROC ASW system (1 × 8), 6–12.75in ASW TT (2 × 3), 1 helicopter
Sensors: Radar OPS-14C (DD 151–154), OPS-24 (DD 155–158), OPS-28C; sonar OQS-4A, OQR-1 TACTASS
Complement: 220

No	Name	Builder	Laid down	Launched	Comp	Fate
DD 151	ASAGIRI	Ishikawajima-Harima, Tokyo	13.2.85	19.9.86	17.3.88	Extant 1995
DD 152	YAMAGIRI	Mitsui, Tamano	5.2.86	8.10.87	25.1.89	Extant 1995
DD 153	YUUGIRI	Sumitomo, Uraga	25.2.86	21.9.87	28.2.89	Extant 1995
DD 154	AMAGIRI	Ishikawajima-Harima, Tokyo	3.3.86	9.9.87	17.3.89	Extant 1995
DD 155	HAMAGIRI	Hitachi, Maizuru	20.1.87	4.6.88	31.1.90	Extant 1995
DD 156	SETOGIRI	Sumitomo, Uraga	9.3.87	12.9.88	14.2.90	Extant 1995
DD 157	SAWAGIRI	Mitsubishi, Nagasaki	14.1.87	25.12.88	6.3.90	Extant 1995
DD158	UMIGIRI	Ishikawajima-Harima, Tokyo	31.10.87	9.11.89	12.3.90	Extant 1995

Yamagiri 11.10.92
L & L van Ginderen

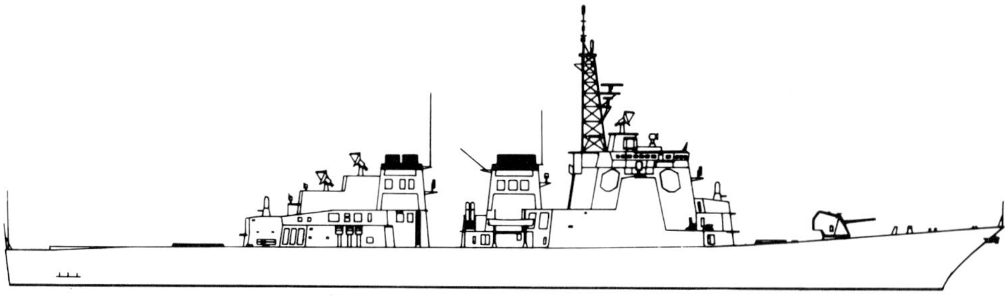

Kongo class

Based on the US *Arleigh Burke* design, these large 'Aegis destroyers' are the most sophisticated warships in the world outside the USN. They are intended to provide both air cover for escort groups and air defence (including anti-tactical ballistic missile defence) for the Japanese homeland. The programme was slowed by US reluctance to release Aegis technology and the planned class of eight ships has been reduced to four for reasons of cost. *Myoko* is due to complete in March 1996, and the fourth unit two years later. Equipped with a platform aft for a Sea Hawk helicopter, but have no hangar.

KONGO class *destroyers*

Displacement: 7250t standard; 9485t full load
Dimensions: 528ft 2in × 68ft 7in × 20ft 4in
161.0m × 21.0m × 6.2m
Machinery: 2-shaft COGAG (CP propellers): 4 GE LM-2500 gas turbines, 102,160hp = 30kts
Armament: 8 Harpoon SSM (4 × 2), 2 Mk 41 VLS groups (29 cells forward, 61 aft for Standard SM2-MR block 2 SAM and vertically-launched ASROC), 1–5in/54 OTO Melara, 2–20mm Phalanx CIWS (2 × 1), 6–12.75in ASW TT (2 × 3)
Sensors: Radar SPY1D/SPG-62, OPS-28C, OPS-20; sonar OQS-102, OQR TACTASS
Complement: 300

No	Name	Builder	Laid down	Launched	Comp	Fate
DD 173	KONGO	Mitsubishi, Nagasaki	8.5.90	26.9.91	25.3.93	Extant 1995
DD 174	KIRISHIMA	Mitsubishi, Nagasaki	7.4.92	19.8.93	Mar 1995	Extant 1995
DD 175	MYOKO	Mitsubishi, Nagasaki	8.4.93	6.10.94	Mar 1996	Completing 1995
DD 176	–	Ishikawajima-Harima, Tokyo	Apr 1995	–	–	Building 1995

Kongo 1993
JMSDF

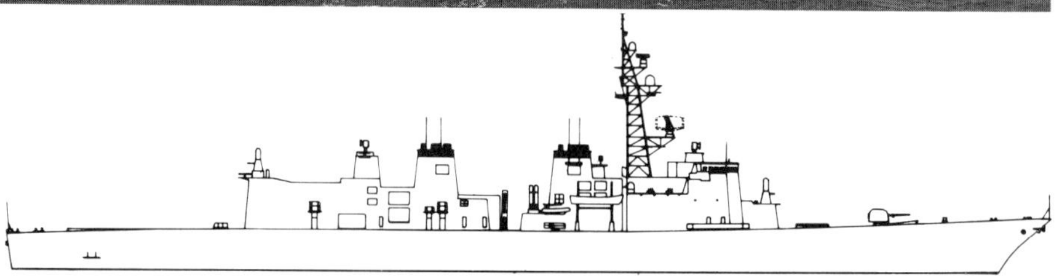

Murasame design

Eight of these ships are planned to complement the *Kongo* class destroyers. The VLS cells are positioned in two groups, one before the bridge and the other aft of the second funnel. The first unit was approved in 1991. They are much larger than the original *Asagiri*s and are a cross between that class and the *Kongo*s in both machinery and armament. The VLS may be adapted to launch Standard SAMs, which would make the ships extra missile magazines for the *Kongo*s. Sixteen Sea Sparrow SAMs can be controlled simultaneously by the ships' own fire control systems. Sea Hawk helicopters will be carried to provide enhanced ASW capability. Four more ships have been approved in Fiscal Years 1994–95, DD 103 to be built by Sumitomo, Uraga and DD 104 by Mitsubishi, Nagasaki. *Murasame* is due to commission in March 1996 and the second a year later.

MURASAME class *destroyers*

Displacement: 4400t standard; 5100t full load
Dimensions: 495ft 5in × 55ft 7in × 17ft 1in
151.0m × 17.0m × 5.2m
Machinery: 2-shaft COGAG (CP propellers): 2 Rolls-Royce Spey SMIC gas turbines, 2 General Electric LM 2500 gas turbines, 84,630hp = 30kts
Armament: 8 Harpoon SSM (2 × 4), 2 Mk 49 VLS groups (29 cells forward, 61 cells aft; 90 Sea Sparrow SAM or Vertical-Launch ASROC), 1–76mm OTO Melara, 2–20mm Phalanx CIWS (2 × 1), 6–12.75in ASW TT (2 × 3), 1 helicopter
Sensors: Radar OPS-24, OPS-28C; sonar OQS-4A, OQR-1 TACTASS
Complement: 160

No	Name	Builder	Laid down	Launched	Comp	Fate
DD 101	MURASAME	Ishikawajima-Harima, Yokohama	18.8.93	Sep 1994	–	Completing 1995
DD 102	–	Mitsui, Tamano	24.8.94	Aug 1995	–	Building 1995

Launched 1943 and transferred on loan in 1953. They were named after trees, as were IJN destroyers. *Keyaki* had a deckhouse added abaft the mainmast and served as a flotilla flagship; *Kaede* received a similar deckhouse. All ships were technically returned on 28 August 1962, but were immediately transferred outright to the Japanese government the same day. *Kusu* was converted to a drone target carrier in 1964. The following units were reclassified from escort vessels to training ships (moored): *Buna* (1.2.65), *Kashi* (1.4.65), *Momi* (1.4.65), *Tochi* (1.4.65), *Ume* (1.4.65), *Kaede* (31.3.66), *Maki* (31.3.66), *Matsu* (31.3.66), *Nara* (31.3.66), *Sakura* (31.3.66). *Tsuge* was discarded in 1968; *Nire* and *Shii* were returned to the USA on 31 March 1970; and *Kiri*, *Keyaki* and *Sugi* were decommissioned 31 March 1970.

Ex-US TACOMA class *frigates*

Particulars: As US *Tacoma* class

No	Name	Builder	Acquired	Fate
PF 281	KUSU (ex-*Ogden*)	Consolidated, San Pedro	1953	Deleted 1972
PF 282	NARA (ex-*Machias*)	Froemming	1953	Deleted 1972
PF 283	KASHI (ex-*Pasco*)	Kaiser, Richmond	1953	Deleted 1972
PF 284	MOMI (ex-*Poughkeepsie*)	Butler	1953	Deleted 1972
PF 285	SUGI (ex-*Coronado*)	Consolidated, San Pedro	Jan 1953	Decommissioned 31.3.70
PF 286	MATSU (ex-*Bath*)	Froemming	Dec 1953	Deleted 1972
PF 287	NIRE (ex-*Sandusky*)	Froemming	1953	Returned 31.3.70
PF 288	KAYA (ex-*San Pedro*)	Consolidated, San Pedro	1953	Deleted 1972
PF 289	UME (ex-*Allentown*)	Froemming	Apr 1953	Deleted 1972
PF 290	SAKURA (ex-*Carson City*)	Consolidated, San Pedro	Apr 1953	Deleted 1972
PF 291	KIRI (ex-*Everett*)	Kaiser, Richmond	Mar 1953	Decommissioned 31.3.70
PF 292	TSUGE (ex-*Gloucester*)	Butler	1953	Deleted 1958
PF 293	KAEDE (ex-*Newport*)	Butler	1953	HS 31.3.68
PF 294	BUNA (ex-*Bayonne*)	American SB	1953	Deleted 1972
PF 295	KEYAKI (ex-*Evansville*)	Letham D Smith	Oct 1953	Decommissioned 31.3.70
PF 296	TOCHI (ex-*Albuquerque*)	Kaiser, Richmond	1953	Deleted 1972
PF 297	SHII (ex-*Long Beach*)	Consolidated, San Pedro	1953	Returned 31.3.70
PF 298	MAKI (ex-*Charlottesville*)	Butler	Jan 1953	Deleted 1972

Originally built in 1943, both ships were transferred on 14 June 1955 under MDAP and returned to the United States in 1975.

Ex-US CANNON class *frigates*

Particulars: As US *Cannon* class (DET type)

No	Name	Builder	Acquired	Fate
DE 262	ASAHI (ex-*Amick*)	Federal, Newark	14.6.55	Returned 1975
DE 263	HATSHUI (ex-*Atherton*)	Federal, Newark	14.6.55	Returned 1975

Akebono was authorised under the 1953 Programme as one of three 'B type' escort vessels. She was built with geared turbines and her near-sisters, *Ikazuchi* and *Inazuma* with diesels. There were many differences between the two designs, the most noticeable being *Akebono*'s twin funnels in contrast to the one funnel of the near-sisters. All weapons and electronics were US produced. In March 1959 her original gun armament was removed and she was fitted with 2–3in quick-firing guns. Four K-guns and a DC rack were also removed.

AKEBONO ('B type') *escort*

Displacement: 1075ft standard; 1350t full load
Dimensions: 301ft pp × 27ft 9in × 10ft 11in max
91.8m × 8.5m × 3.4m
Machinery: 2-shaft Ishikawajima geared turbines, 2 Ishikawajima/Foster Wheeler boilers, 18,000shp = 28kts
Armament: 2–3in/50 (2 × 1), 4–40mm (2 × 2), 1 Hedgehog, 8 K-guns, 1 DC rack. See notes
Complement: 193

No	Name	Builder	Laid down	Launched	Comp	Fate
DE 201	AKEBONO	Ishikawajima Jukogyo, Kobe	10.12.54	30.10.55	20.3.56	Discarded 1976

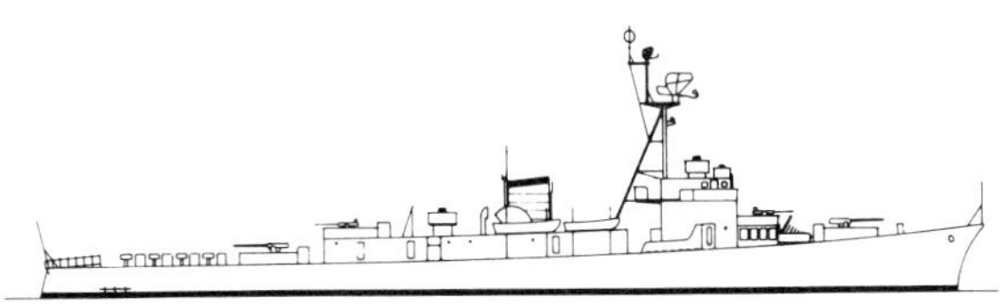

Inazuma as completed

Both units were authorised under the 1953 Programme as two further 'B type' escort vessels, powered by diesels. The *Ikazuchi* class had one funnel whereas *Akebono* had two – a prominent recognition difference. All weapons and electronics were US produced. The original 2–3in guns and 1–40mm gun were removed in 1959 and replaced by 2–3in quick-firing guns. Four K-guns and one DC rack were also removed. Due to projected vibrations from the diesels, mild steel plates were used and an aluminium alloy was employed in the superstructure.

IKAZUCHI class ('B type') *escorts*

Displacement: 1080t standard
Dimensions: 287ft 1in pp, 288ft 9in oa × 28ft 6in × 10ft 3in max
87.5m, 88.0m × 8.7m × 3.1m
Machinery: 2 shafts, 2 Mitsubishi diesels, 12,000bhp = 25kts
Armament: 2–3in/50 (2 × 1), 4–40mm (2 × 2), 1 Hedgehog, 8 K-guns, 1 DC rack
Complement: 145

No	Name	Builder	Laid down	Launched	Comp	Fate
DE 202	IKAZUCHI	Kawasaki, Kobe	18.12.54	6.9.55	29.5.56	Discarded 1976
DE 203	INAZUMA	Mitsui, Tamano	25.12.54	4.8.55	1.4.56	Discarded 1976

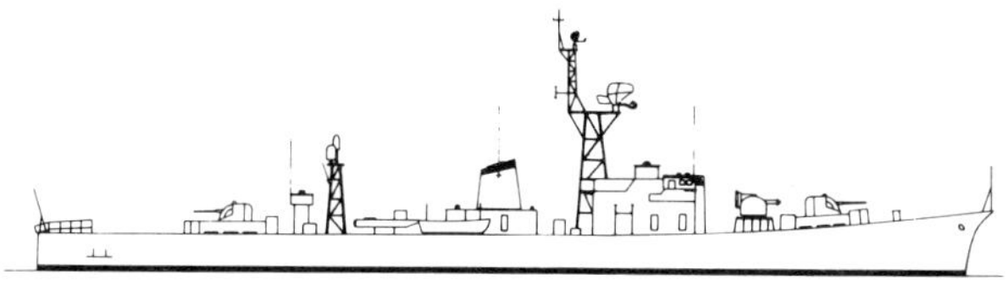

Mogami as completed

Two ships were constructed under the 1959 and two under the 1961 programme. Weapons and machinery varied considerably throughout these units (*Isuzu* and *Mogami* had Weapon Alfa instead of a rocket launcher, and no DCT) and they served as test platforms. VDS was mounted on *Mogami* in 1966 and *Kitakami* in 1968. All had different machinery, but *Isuzu* had only two engines, and consequently a smaller funnel. On conversion to auxiliary/training status TT (and VDS in DE 212 and 213) removed.

ISUZU class *escorts*

Displacement: 1490t standard; 1700t full load
Dimensions: 308ft 5in wl, 318ft 3in oa × 34ft 1in × 11ft 6in max
94.0m, 97.0m × 10.4m × 3.5m
Machinery: 2 shafts, 4 diesels, 16,000bhp = 26kts
Armament: 4–3in/50 (2 × 2), 4–21in TT (1 × 4), 1–375mm ASW rocket launcher (1 × 4), 6–12.75in Mk 32 ASW TT (3 × 2), 1 DCT, 1 DC rack. See notes
Sensors: Radar OPS-1, OPS-16; sonar OQS-12 or OQS-14 (OQA-1 VDS in DE 212 and 213)
Complement: 183

No	Name	Builder	Laid down	Launched	Comp	Fate
DE 211	ISUZU	Mitsui, Tamano	16.4.60	17.1.61	29.7.61	Auxiliary (ASU 7105) 8.4.88 deleted Mar 1992
DE 212	MOGAMI	Misubishi, Nagasaki	4.8.60	7.3.61	28.9.61	TS (TV3505) 1.7.87, deleted Jun 1991
DE 213	KITAKAMI	Ishikawajima-Harima, Tokyo	7.6.62	21.6.63	1.2.64	Auxiliary (ASU 7016) 31.3.90 deleted Oct 1993
DE 214	OHI	Maizuru Jukogyo, Maizuru	10.6.62	15.6.63	1.2.64	Auxiliary (ASU 7017) 31.1.90, deleted Feb 1993

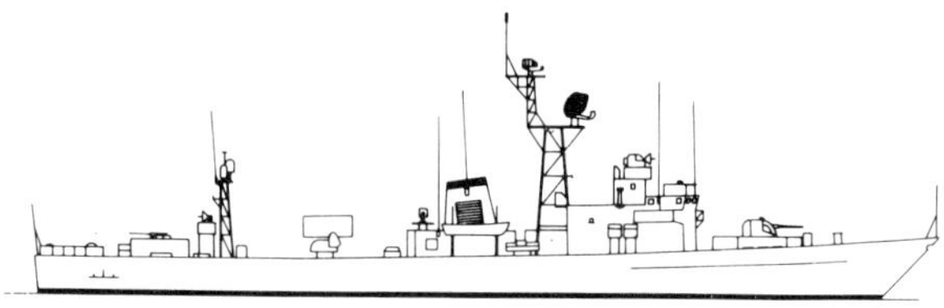

Chikugo as completed

Chikugo units are the smallest ships in the world to be armed with ASROC, and most units are equipped with VDS. Particular attention was given to reduce vibration and noise. Because of the large superstructure area for ships of this size, they were given greater beam to improve stability. Early units are 1470t standard; from *Teshio* onwards they are 1530t. The Mitsui diesels are fitted to DE 216, DE 220, DE 222 and DE 224.

CHIKUGO class *escorts*

Displacement: 1470–1530t standard; 1700–1800t full load
Dimensions: 305ft 2in pp × 35ft 6in × 11ft 6in max
93.0m × 10.8m × 3.5m
Machinery: 2 shafts, 4 Mitsubishi-Burmeister & Wain UEV 30/40 or Mitsui 28 VBC-30 diesels, 16,000bhp = 28kts. Range 10,700nm/12,000nm at 12kts/9kts
Armament: 2–3in/50 (1 × 2), 2–40mm (1 × 2), 1 ASROC ASW system (1 × 8), 6–12.75in Mk 32 ASW TT (2 × 3)
Sensors: Radar OPS-16, OPS-14; sonar OQS-3 (plus SQS-35(J) VDS in most ships)
Complement: 165

No	Name	Builder	Laid down	Launched	Comp	Fate
DE 215	CHIKUGO	Mitsui, Tamano	9.12.68	13.1.70	31.7.70	Extant 1995
DE 216	AYASE	Ishikawajima-Harima, Tokyo	5.12.69	16.9.70	20.7.71	Extant 1995
DE 217	MIKUMO	Mitsui, Tamano	17.3.70	16.2.71	26.8.71	Extant 1995
DE 218	TOKACHI	Mitsui, Tamano	11.12.70	25.11.71	17.5.72	Extant 1995
DE 219	IWASE	Mitsui, Tamano	6.8.71	29.6.72	12.12.72	Extant 1995
DE 220	CHITOSE	Hitachi, Maizuru	7.10.71	25.1.73	21.8.73	Extant 1995
DE 221	NIYODO	Mitsui, Tamano	20.9.72	28.8.73	8.2.74	Extant 1995
DE 222	TESHIO	Hitachi, Maizuru	11.7.73	29.5.74	10.1.75	Extant 1995
DE 223	YOSHINO	Mitsui, Tamano	28.9.73	22.8.74	6.2.75	Extant 1995
DE 224	KUMANO	Hitachi, Maizuru	29.5.74	24.2.75	19.11.75	Extant 1995
DE 225	NOSHIRO	Mitsui, Tamano	27.1.76	23.12.76	31.8.77	Extant 1995

Kitakami as completed Navarret Collection

Yoshino as completed Robert L Scheina Collection

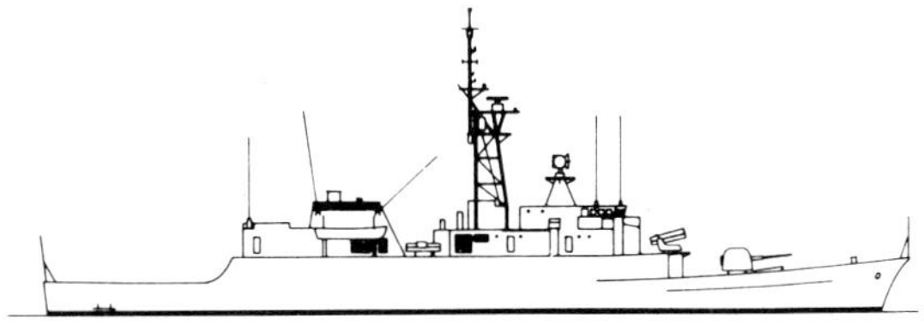

Ishikari during trials

Two units of the *Ishikari* class were originally authorised, one in 1977–78 and the second in 1979–80, but only one was built. As many as twenty were planned. This design is smaller and less sophisticated than the preceding *Chikugo* class, but too much seems to have been attempted on the displacement – hence the curtailment of the programme, and the evolution of the *Yubari* design (see below).

ISHIKARI class *escort*

Displacement: 1200t standard; 1450t full load
Dimensions: 277ft 3in × 32ft 9in × 11ft 6in
84.5m × 10.0m × 3.5m
Machinery: 2-shaft CODOG (CP propellers): 1 Kawasaki/Rolls-Royce Olympus TM-3B gas turbine, 28,390shp = 25kts; 1 Mitsubishi diesel, 5000bhp = 19kts
Armament: 8 Harpoon SSM (2 × 4), 1–76mm OTO Melara compact, 1–375mm ASW rocket launcher (1 × 4), 6–12.75in Mk 32 ASW TT (2 × 3)
Sensors: Radar OPS-28; sonar SQS-36
Complement: 90

No	Name	Builder	Laid down	Launched	Comp	Fate
DE 226	ISHIKARI	Mitsui, Tamano	17.5.79	18.3.80	30.3.81	Extant 1995

Ishikari 1981
JMSDF

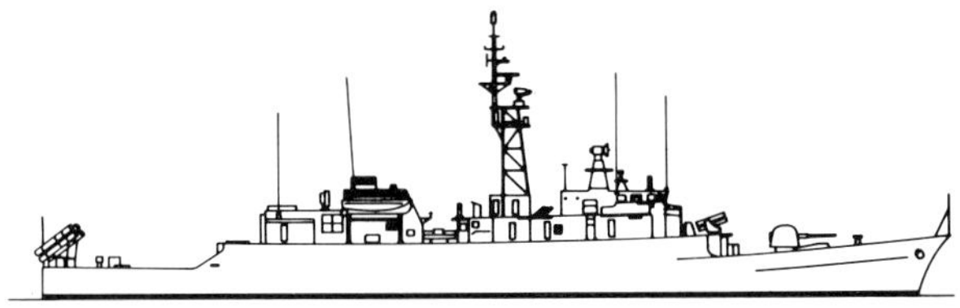

Yubari 1983

An improved and slightly enlarged *Ishikari*, with improved accommodation and increased fuel capacity. The second ship was ordered under the FY80–81 programme, but no further ships appeared. Phalanx CIWS was part of the original design, but has not as yet been fitted.

YUBARI class *escorts*

Displacement: 1470t standard; 1690t full load
Dimensions: 298ft 6in × 35ft 5in × 11ft 6in
91.0m × 10.8m × 3.5m
Machinery: 2-shaft CODOG: 1 Kawasaki/Rolls-Royce Olympus TM-3B gas turbine, 28,400shp = 25kts; 1 Mitsubishi diesel, 4650bhp = ?kts
Armament: 8 Harpoon SSM (2 × 4), 1–76mm/62 OTO Melara compact, 1–375mm ASW rocket launcher (1 × 4), 6–12.75in Mk 32 ASW TT
Sensors: Radar OPS-28; sonar SQS-36J
Complement: 98

No	Name	Builder	Laid down	Launched	Comp	Fate
DE 227	YUBARI	Sumitomo, Uraga	9.2.81	22.2.82	18.3.83	Extant 1995
DE 228	YUBETSU	Hitachi, Maizuru	14.1.82	25.1.83	14.2.84	Extant 1995

Abukuma 1993
L & L van Ginderen

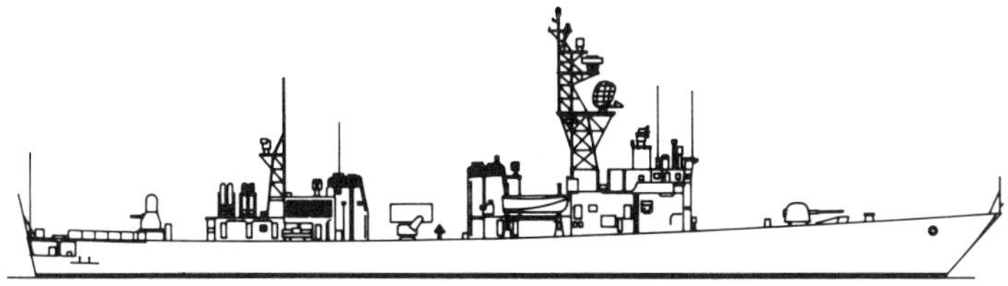

Jintsu 1989

A much larger and more capable class of escorts, comparable to the earlier smaller destroyers. Two ordered in successive estimates 1987–89. Fitted for towed-array sonar.

ABUKUMA class *escorts*

Displacement: 2050t standard; 2550t full load
Dimensions: 357ft 7in × 44ft × 12ft 6in
109.0m × 13.4m × 3.8m
Machinery: 2-shaft CODOG: 2 Rolls-Royce Spey SM1A gas turbines, 26,650shp = 27kts; 2 Mitsubishi diesels, 6000bhp = 27kts
Armament: 8 Harpoon SSM (2 × 4), 1–76mm/62 OTO Melara, 1–20mm Phalanx CIWS, 1 ASROC ASW system, 6–12.75in Mk 32 ASW TT (2 × 3)
Sensors: Radar OPS-14C, OPS-28; sonar OQS-8
Complement: 115

No	Name	Builder	Laid down	Launched	Comp	Fate
DE 229	ABUKUMA	Mitsui, Tamano	17.3.88	21.12.88	12.12.89	Extant 1995
DE 230	JINTSU	Hitachi, Maizuru	14.4.88	31.1.89	28.2.90	Extant 1995
DE 231	OHYODO	Mitsui, Tamano	8.3.89	19.12.89	23.1.91	Extant 1995
DE 232	SENDAI	Sumitomo, Uraga	14.4.89	26.1.90	15.3.91	Extant 1995
DE 233	CHUKUMA	Hitachi, Maizuru	14.2.91	22.1.92	24.2.93	Extant 1995
DE 234	TONE	Sumitomo, Uraga	8.2.91	6.12.91	8.2.93	Extant 1995

SUBMARINES

The US *Gato* class submarine *Mingo* (launched 30 November 1942) was transferred in August 1955 under MDAP. Her crew was trained at the US Submarine Base, New London, Connecticut. *Kuroshio* was used primarily as a training boat until decommissioned on 31 March 1966.

Ex-US GATO class *submarine*

Particulars: As US *Gato* class

No	Name	Builder	Acquired	Fate
SS 501	KUROSHIO (ex-*Mingo*)	Electric Boat	15.8.55	Decommissioned 31.3.66

Oyashio at launch 25.5.59
CMP

Oyashio was built under the 1956 programme and was the first submarine built in Japan since the Second World War. Although equipped with a snorkel, she was a very conservative design. She had a whale-shaped hull with an internal frame and her 480 main storage batteries were arranged in four groups within two rooms. The diving planes were housed inside the hull, between the internal and external frames. She served as a TS from 1 April 1975.

OYASHIO *submarine*

Displacement: 1139t surfaced; 1420t submerged
Dimensions: 258ft 6in pp × 22ft 11in × 15ft max
78.8m × 7.0m × 4.6m
Machinery: 2-shaft, 2 diesels plus 2 electric motors, 2700bhp/5960shp = 13kts/19kts. Range 10,000nm at 10kts
Armament: 4–21in TT (bow)
Complement: 65

No	Name	Builder	Laid down	Launched	Comp	Fate
SS 511	OYASHIO	Kawasaki, Kobe	25.12.57	25.5.59	30.6.60	Discarded 30.9.76

Constructed under the 1959 programme, these boats and their near-sisters, the *Natsushio* class, were small and limited in capability. They were air-conditioned and had good habitability and a large safety factor. They were modelled after the US *Barracuda* and were somewhat shorter and thicker than *Oyashio*. In order to reduce hull resistance, an effort was made to produce a perfect double hull; external framing was used in order to improve internal space. Attention was paid to improving underwater manoeuvrability, and a 'joystick' was installed for steering. The class used a water pressure system for launching torpedoes, thus eliminating the emission of air bubbles. The main storage batteries were water-cooled.

HAYASHIO class *submarines*

Displacement: 650t surfaced; 800t submerged
Dimensions: 193ft 6in oa × 21ft 4in × 13ft 6in max
59.0m × 6.5m × 4.1m
Machinery: 2-shafts, 2 diesels plus 2 electric motors, 1350bhp/1700shp = 12kts/15kts
Armament: 3–21in TT (bow)
Complement: 43

No	Name	Builder	Laid down	Launched	Comp	Fate
SS 521	HAYASHIO	Shin Mitsubishi, Kobe	6.6.60	31.7.61	30.6.62	Discarded 1979
SS 522	WAKASHIO	Kawasaki, Kobe	7.6.60	28.6.61	10.8.62	Discarded 23.3.79

Hayashio as completed Navarret Collection

The *Natsushio* class was authorised in 1961 and, like their near-sisters, the *Hayashio*s, were small and limited in capability. They were air-conditioned and had good habitability and a large safety factor.

NATSUSHIO class *submarines*

Displacement: 690t surfaced; 850t submerged
Dimensions: 200ft 1in pp × 21ft 2in × 13ft 6in max
61.0m × 6.5m × 4.1m
Machinery: 2-shafts, 2 diesels plus 2 electric motors, 1350bhp/1700shp = 12kts/15kts
Armament: 3–21in TT (bow)
Complement: 43

No	Name	Builder	Laid down	Launched	Comp	Fate
SS 523	NATSUSHIO	Shin-Mitsubishi Jyuko, Kobe	5.12.61	18.9.62	30.6.63	Deleted 20.3.78
SS 524	FUYUSHIO	Kawasaki Jyuko, Kobe	6.12.61	14.12.62	31.1.64	Deleted 20.6.78

Asashio as completed

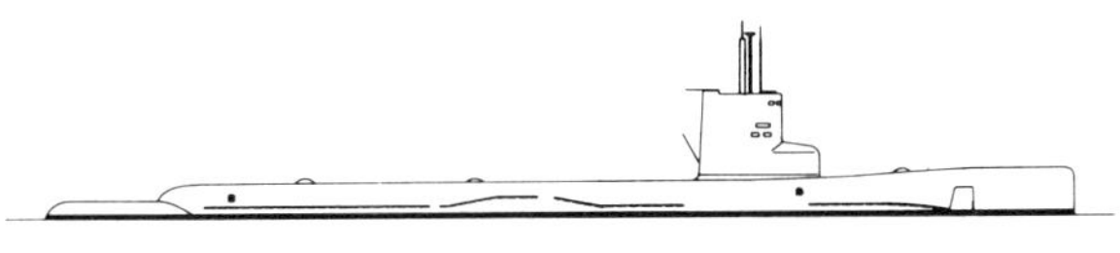

These boats are often cited as two distinct classes with the lead unit being segregated, since she had a different shaped bow and less sophisticated sonar. These were the first fleet submarines constructed by the Japanese since the Second World War, being double the displacement of the *Hayashio* and *Natsushio* classes, and were specially equipped to serve as 'targets' in ASW exercises. The bow TT were for US Mk 54 and the aft tubes were for 'swim out' torpedoes.

OSHIO class *submarines*

Displacement: 1650t (*Oshio* 1600t) standard; 2150t (*Oshio* 2100t) submerged
Dimensions: 288ft 8in pp × 26ft 11in × 16ft (*Oshio* 15ft 5in)
88.0m × 8.2m × 4.9m (4.7m)
Machinery: 2 shafts, 2 Kawasaki-MAN diesels plus 2 electric motors, 2300bhp/6300shp = 14kts/18kts
Armament: 8–21in TT (6 bow, 2 stern)
Complement: 80

No	Name	Builder	Laid down	Launched	Comp	Fate
SS 561	OSHIO	Mitsubishi, Kobe	29.6.63	30.4.64	31.3.65	Deleted 20.8.81
SS 562	ASASHIO	Kawasaki, Kobe	10.10.64	27.11.65	13.10.66	Deleted 30.3.83
SS 563	HARUSHIO	Mitsubishi, Kobe	12.10.65	25.2.67	21.12.67	Deleted 30.3.84
SS 564	MICHISHIO	Kawasaki, Kobe	26.7.66	5.12.67	29.8.68	Deleted 28.3.85
SS 565	ARASHIO	Mitsubishi, Kobe	5.7.67	24.10.68	25.7.69	Deleted Mar 1986

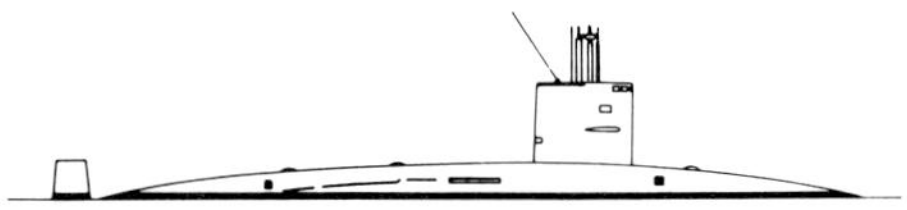

Uzushio as completed

The first teardrop-hull submarines produced by Japan. These boats had double hulls, a bow sonar array, and torpedo tubes amidships; NS-63 high-tensile steel was used in the hull shell in order to permit greater diving depth (200m). The class was equipped with a separate emergency blowing system which enables the submarine to suface rapidly. A three-dimensional automatic steering system, consisting of a combination of automatic depth-maintenance system and automatic direction-maintenance system is installed. The ATSS (Auxiliary Training Submarines) are used for experimental trials as well as training.

UZUSHIO class *submarines*

Displacement: 1850t surfaced; 3600t submerged
Dimensions: 236ft 3in pp × 32ft 6in × 24ft 7in max
72.0m × 9.9m × 7.5m
Machinery: 1 shaft, 2 Kawasaki-MAN V8/V24–30 diesels plus 2 electric motors, 3600bhp/7200shp = 12kts/20kts
Armament: 6–21in TT (amidships)
Complement: 80

No	Name	Builder	Laid down	Launched	Comp	Fate
SS 566	UZUSHIO	Kawasaki, Kobe	25.9.68	11.3.70	21.1.71	Deleted 24.3.87
SS 567	MAKISHIO	Mitsubishi, Kobe	21.6.69	27.1.71	2.2.72	Deleted Mar 1988
SS 568	ISOSHIO	Kawasaki, Kobe	9.7.70	18.3.72	25.11.72	ATSS 8001 Oct 1989, deleted Mar 1992
SS 569	NARUSHIO	Mitsubishi, Kobe	8.5.71	22.11.72	28.8.73	ATSS 8002 Jun 1990, deleted Mar 1993
SS 570	KUROSHIO	Kawasaki, Kobe	5.7.72	22.2.74	27.11.74	ATSS 8003 May 1991, deleted Dec 1993
SS 571	TAKASHIO	Kawasaki, Kobe	6.7.73	30.6.75	30.1.76	ATSS 8004 Jul 1992, extant 1995
SS 572	YAESHIO	Kawasaki, Kobe	14.4.75	19.5.77	7.3.78	ATSS 8005 Dec 1993, extant 1995

Nadashio 1985

A development of the *Uzushio* class, with a deeper diving capability and improved electronics. Regular replacement programme of one boat per year. Carry ZQQ-4 or ZQQ-5 bow sonar, a modified version of BQS-4. SS 574 and SS 577–582 carry Sub-Harpoon, fired from TT. *Nadashio* was the first unit to be fitted with ZQR-1 towed array sonar in 1987, since fitted to rest of class.

YUSHIO class *submarines*

Displacement: 2200t surfaced (SS 574, 577–582 2250t, SS 576 2300t)
Dimensions: 249ft 4in pp × 32ft 6in × 24ft 7in max
76.0m × 9.9m × 7.5m
Machinery: 1 shaft, 2 Kawasaki-MAN V8/V24 diesels, plus electric motors, 3400bhp/7200shp = 13kts/20kts
Armament: 6–21in TT (amidships)
Sensors: See notes
Complement: 80

No	Name	Builder	Laid down	Launched	Comp	Fate
SS 573	YUSHIO	Mitsubishi, Kobe	3.12.76	29.3.79	26.2.80	Extant 1995
SS 574	MOCHISHIO	Kawasaki, Kobe	28.4.78	12.3.80	5.3.81	Extant 1995
SS 575	SETOSHIO	Kawasaki, Kobe	17.4.79	12.2.81	17.3.82	Extant 1995
SS 576	OKISHIO	Kawasaki, Kobe	17.4.80	5.3.82	1.3.83	Extant 1995
SS 577	NADASHIO	Mitsubishi, Kobe	10.4.81	27.1.83	6.3.84	Extant 1995
SS 578	HAMASHIO	Kawasaki, Kobe	8.8.82	1.1.84	5.3.85	Extant 1995
SS 579	AKISHIO	Mitsubishi, Kobe	15.8.83	22.1.85	5.3.86	Extant 1995
SS 580	TAKESHIO	Kawasaki, Kobe	3.4.84	9.2.86	3.3.87	Extant 1995
SS 581	YUKISHIO	Mitsubishi, Kobe	11.5.85	23.1.87	11.3.88	Extant 1995
SS 582	SACHISHIO	Kawasaki, Kobe	11.5.86	17.2.88	24.3.89	Extant 1995

A natural progression from the previous *Yushio* class with improved noise reduction and ZQR-1 towed array sonar. Hull sonar is the Hughes/Oki ZQQ-5B. Can fire Sub-Harpoon from the torpedo tubes. Built one per year to replace the *Uzushio*s.

HARUSHIO class *submarines*

Displacement: 2450t surfaced
Dimensions: 252ft 7in × 32ft 10in × 25ft 4in
77.0m × 10.0m × 7.7m
Machinery: 1 shaft, 2 Kawasaki 12V25/25S diesels, plus electric motors, 5520bhp/7200shp = 12kts/20kts
Armament: 6–21in TT (amidships, 20 Type 89 torpedoes or Sub-Harpoon missiles)
Sensors: See notes
Complement: 75

No	Name	Builder	Laid down	Launched	Comp	Fate
SS 583	HARUSHIO	Mitsubishi, Kobe	21.4.87	26.7.89	30.12.90	Extant 1995
SS 584	NATSUSHIO	Kawasaki, Kobe	8.4.88	20.3.90	20.3.91	Extant 1995
SS 585	HAYASHIO	Mitsubishi, Kobe	9.12.88	17.1.91	25.3.92	Extant 1995
SS 586	ARASHIO	Kawasaki, Kobe	8.1.90	17.3.92	17.3.93	Extant 1995
SS 587	WAKASHIO	Mitsubishi, Kobe	12.12.90	22.1.93	1.3.94	Extant 1995
SS 588	FUYUSHIO	Kawasaki, Kobe	12.12.91	16.2.94	Mar 1995	Extant 1995
SS 589	–	Mitsubishi, Kobe	24.12.92	Jun 1995	Mar 1997	Completing

Harushio Jul 1993
L & L van Ginderen

The next incremental step in Japanese submarine design, the increased displacement of which (approx. 2700t surfaced) will be due to the adoption of large flank sonar arrays. There are reports that this class may employ an auxiliary air-independent propulsion system, although it is unlikely to be ready for SS 590. First unit approved in the 1993 Budget; second in 1994, third in 1995. To replace the *Yushio* class at the standard rate of one per year.

'Improved HARUSHIO' class *submarines*

No	Name	Builder	Laid down	Launched	Comp	Fate
SS 590	–	Kawasaki, Kobe	26.12.93	Nov 1996	Mar 1998	Building
SS 591	–	Mitsubishi, Kobe	1995	1997	1999	Authorised

Nemuro as completed JMSDF

AMPHIBIOUS WARFARE VESSELS

Ex-US LSSL type *large support ships/gunboats*

Particulars: As US LSSL class

Class:
Ajisai (ex-*LSSL 88*), *Akane* (ex-*LSSL 100*), *Aio* (ex-*LSSL 98*), *Ayame* (ex-*LSSL 115*), *Azami* (ex-*LSSL 27*), *Bara* (ex-*LSSL 78*), *Botan* (ex-*LSSL 129*), *Ezogiku* (ex-*LSSL 24*), *Fuji* (ex-*LSSL 75*), *Fuyo* (ex-*LSSL 110*), *Hagi* (ex-*LSSL 130*), *Hamagiku* (ex-*LSSL 87*), *Hamayu* (ex-*LSSL 68*), *Hasu* (ex-*LSSL 89*), *Hiiragi* (ex-*LSSL 114*), *Himawari* (ex-*LSSL 102*), *Himeyuri* (ex-*LSSL 20*), *Hinagiku* (ex-*LSSL 83*), *Isogiku* (ex-*LSSL 106*), *Iwagiku* (ex-*LSSL 120*), *Kaido* (ex-*LSSL 76*), *Kanna* (ex-*LSSL 109*), *Karukaya* (ex-*LSSL 67*), *Keito* (ex-*LSSL 60*), *Keshi* (ex-*LSSL 114*), *Kiku* (ex-*LSSL 57*), *Kikyo* (ex-*LSSL 119*), *Nogiku* (ex-*LSSL 22*), *Oniyuri* (ex-*LSSL 13*), *Ran* (ex-*LSSL 104*), *Renge* (ex-*LSSL 126*), *Rindo* (ex-*LSSL 79*), *Sasayuri* (ex-*LSSL 52*), *Sawagiku* (ex-*LSSL 84*), *Sekichiku* (ex-*LSSL 12*), *Shida* (ex-*LSSL 90*), *Shiragiku* (ex-*LSSL 72*), *Shobu* (ex-*LSSL 96*), *Suiren* (ex-*LSSL 94*), *Sumire* (ex-*LSSL 14*), *Susuki* (ex-*LSSL 58*), *Suzuran* (ex-*LSSL 25*), *Tsuta* (ex-*LSSL 85*), *Tsutsuji* (ex-*LSSL 101*), *Yagaruma* (ex-*LSSL 103*), *Yamabuki* (ex-*LSSL 116*), *Yamagiku* (ex-*LSSL 82*), *Yamayuki* (ex-*LSSL 18*), *Yuri* (ex-*LSSL 104*)

Wartime built fire support ships, originally rated as LCS(3)Ls; rocket launchers were removed before all were transferred in 1953. They were named after flowers, are armed with either 1–3in/50 and 4–40mm, or three twin 40mm, and all were disposed of between 1957 and 1961. See under United States in 1922–46 volume for full details.

Ex-US LSM type *medium landing ship*

Particulars: As US LSM type

Class:
Yorikutei (LSM 3001)

The US *LSM 125* was transferred to the French Navy in 1954 as *LSM 9013* for use in Vietnam, and later passed on to Japan (in 1958). She was deleted in 1974.

Ex-US LST type *tank landing ship*

Particulars: As US LST type

Class (no, former name and no):
Oosumi (LST 4001, ex-*Daggett County*, LST 689), *Shimokita* (LST 4002, ex-*Hillsdale County*, LST 835), *Shiretoko* (LST 4003, ex-*Nansemond County*, LST 1064)

US wartime built LSTs transferred under MDAP and commissioned 1 April 1961. The first two were deleted in 1974 and 1975 respectively, and the last was returned to the USA in 1976 for transfer to the Philippines.

ATSUMI class *tank landing ships*

Displacement: 1550 standard; 2400t full load
Dimensions: 292ft × 42ft 8in × 8ft 10in
89.0m × 13.0m × 2.7m
Machinery: 2 shafts, 2 Kawasaki/MAN diesels, 440bhp = 13kts. Range 4300nm at 12kts
Armament: 4–40mm (2 × 2)
Sensors: Radar OPS-9
Complement: 100

Class (no, launched):
Atsumi (LST 4101, 13.6.72), *Motobu* (LST 4102, 3.8.73), *Nemuro* (LST 4103, 16.6.77)

Atsumi is only 1480t standard and is a knot faster. All were built by Sasebo. Troop capacity is 120 plus twenty vehicles; they carry two LCVPs in davits and can carry an LCM on deck. All three were in service in 1995.

MIURA class *tank landing ships*

Displacement: 2000t standard; 3200t full load
Dimensions: 308ft 4in pp, 321ft 6in oa × 45ft 11in × 9ft 10in
94.0m, 98.0m × 14.0m × 3.0m
Machinery: 2 shafts, 2 Kawasaki/MAN diesels, 4400bhp = 14kts. Range 4300nm at 12kts
Armament: 2–3in/50 (1 × 2), 2–40mm (1 × 2)
Sensors: Radar OPS-14, OPS-16
Complement: 18

Class (no, launched):
Miura (LST 4151, 13.8.74), *Ojika* (LST 4152, 2.9.75), *Satsuma* (LST 4153, 12.5.76)

All were built by Ishikawajima-Harima, Tokyo. They can carry 180 troops and 1800t cargo; two LCVPs are slung under davits, and a travelling gantry handles two LCMs carried on the foredeck. The US Mk 33 3in mounting is forward and the 40mm mounting is aft. *Satsuma* has run trials with a 76mm OTO Melara compact in place of the 3in mount. All were in service in 1995.

Projected tank landing ship/landing ship dock

Displacement: 8900t standard
Dimensions: 557ft 8in × 75ft 6in × 24ft 7in
170.0m × 23.0m × 7.5m
Machinery: 2 shafts, 2 diesels = 22kts
Complement: 130

This ship has its origins in a 5500t LST requested in 1989–90. The plan was not approved, partly on account of the proposed ship's resemblance to an aircraft carrier, with its large helicopter deck. A new request was presented in Fiscal Year 1993 for a larger ship and this was approved. The vessel is to carry US-built LCACs (Landing Craft Air Cushion) and will also operate large helicopters (Chinooks). She will be the nearest thing to a power-projection asset the MSDF has deployed and as such is likely to be the subject of political controversy. The above details are therefore provisional at best. Due to be laid down November 1995 and launched December 1996.

YURA class *utility landing craft*

Displacement: 500t standard; 590t full load
Dimensions: 190ft 4in × 31ft 3in × 5ft 8in
58.0m × 9.5m × 1.7m
Machinery: 2 shafts, 2 Fuji diesels, 3000bhp = 12kts
Armament: 1–20mm Gatling
Complement: 32

Class (no, launched):
Yura (LSU 4171, 10.8.80), *Noto* (LSU 4172, 1.11.80)

Both built by Sasebo Heavy Industries and completed 27 March 1981. They are small LSTs with a bow door and ramp. Both were in service in 1995.

YUSOTEI class *utility landing craft*

Displacement: 420t standard
Dimensions: 170ft 7in × 28ft 6in × 5ft 2in
52.0m × 8.7m × 1.6m
Machinery: 2 shafts, 2 Mitsubishi 516MTK diesels, 3040bhp = 12kts
Armament: 1–20mm Gatling
Complement: 28

Class (no, launched):
Yusotei-Ichi-Go (LSU 2001, 1.10.87), *Yusotei-Ni-Go* (LSU 2002, 15.5.91)

Also numbered LC 01–02. Built at Sasebo and in service March 1988 and March 1992 respectively. Third unit cancelled.

MINOR LANDING CRAFT

Japan also operated six ex-US LCUs (*LCU 2001–2006*), forty-two ex-US LCMs (*LCM 1001–1042*), and eight LCVPs. The last thirteen LCMs were transferred under MDAP in 1961, the remainder having been taken over on June 1995. Most were deleted in the 1970s although some LCUs survived into the 1990s as exercise minelayers.

To replace them the Japanese built fifteen 24t craft of the US LCM (6) design and twenty-two 13ft LCVPs (mostly with GRP hulls). These are carried by LSTs. Other craft of these designs have been built for use as miscellaneous service launches (YFs). In 1995 there were about ten LCM YFs and seventeen LCVP YFs in service. Two US-built LCACs (landing-craft air cushion) have been authorised for delivery in 1995.

SMALL SURFACE COMBATANTS

KAMOME class *submarine chasers*

Displacement: 330t (*Kari, Kiji, Taka, Washi* 310t) standard
Dimensions: 177ft 2in pp × 21ft 4in × 6ft 7in max
54.0m × 6.5m × 2.0m
Machinery: 2 shafts, Mitsui/Burmeister & Wain (*Kari, Kiji, Taka,* Kawasaki/MAN) diesels, 4000bhp = 21kts. Oil 21.5t. Range 2000nm at 12kts
Armament: 2–40mm (2 × 2), 1 Hedgehog, 2 K-guns (not in PC 305–307), 2 DC racks
Complement: 60

Class (no, launched):
Uraga, Yokosuka – *Kamome* (PC 305, 3.9.56), *Misago* (PC 307, 1.11.56)
Fujinagata, Osaka – *Kari* (PC 301, 26.9.56), *Taka* (PC 303, 17.11.56)
Iino, Maizuru – *Kiji* (PC 302, 11.9.56), *Washi* (PC 304, 12.11.56)
Kure – *Tsubame* (PC 306, 10.10.56)

All authorised under the 1954 programme, laid down between December 1955 and March 1956, and completed January–March 1957. In effect they were up-dated versions of the wartime US PC concept. All were scrapped 1977–78.

HAYABUSA *submarine chaser*

Displacement: 370t standard
Dimensions: 190ft 3in × 25ft 7in × 6ft 6in max
58.0m × 7.8m × 2.0m
Machinery: 3-shaft, CODAG: 2 Burmeister & Wain diesels, 4000bhp, plus 1 gas turbine, 6000shp = 26kts
Armament: 2–40mm (1 × 2), 1 Hedgehog, 2 DCT, 2 DC racks
Complement: 80

Built by Mitsubishi, Nagasaki, laid down 23 May 1956, launched 20 November 1956 and completed 10 June 1957. Authorised under the 1954 programme, *Hayabusa* (PC 308) had a unique machinery arrangement with a gas turbine on the centre shaft. She was converted to a yacht (ASY 91) at Yokohama Yacht Co in 1977–78, when she was disarmed and new enlarged superstructure added. Painted white overall, she was deleted in 1987.

UMITAKA class *submarine chasers*

Displacement: 440t standard; 480t full load
Dimensions: 196ft 9in pp × 23ft 3in × 7ft 9in max
60.0m × 7.1m × 2.4m
Machinery: 2 shafts, Mitsui/Burmeister & Wain diesels, 4000bhp = 21kts. Range 3000nm at 12kts
Armament: 2–40mm (1 × 2), 1 Hedgehog, 2 DCT, 2 DC racks
Sensors: Radar OPS-35; sonar SQS-11A
Complement: 80

Class (no, launched):
Ōotaka (PC 309, 3.9.59), *Umitaka* (PC 310, 25.7.59)

Both were authorised under the 1957 programme, *Ōotaka* being built by Kure and *Umitaka* by Kawasaki, Kobe. They were later fitted with 2 Mk 3 launchers for Mk 32 torpedoes in place of the DCT. They were transferred to training duties in 1980 and 1981 respectively, and renumbered ASU 86 and 88. Both were deleted in 1984.

MIZUTORI class *submarine chasers*

Displacement: 420t–450t standard; 450–480t full load
Dimensions: 196ft 9in pp × 23ft 3in × 7ft 8in max
60.0m × 7.1m × 2.4m
Machinery: 2 shafts, 2 MAN diesels, 3800bhp = 20kts. Oil 24.5t. Range 3000nm at 12kts
Armament: 2–40mm (2 × 2), 1 Hedgehog, 1 DC rack, 2 ASW torpedo racks
Sensors: Radar OPS-35 or -36; SQS-11A
Complement: 70

Class (no, launched):
Kawasaki, Kobe – *Mizutori* (PC 311, 22.9.59)
Fujinagata, Osaka – *Yamadori* (PC 312, 22.10.59), *Kasasagi* (PC 304, 31.5.60), *Kumataka* (PC 318, 21.10.63)
Kure – *Ōotori* (PC 313, 27.5.60), *Wakataka* (PC 317, 13.11.62)
Sasebo – *Hatsukai* (PC 315, 24.6.60), *Umidori* (PC 316, 15.10.62), *Shiratori* (PC 319, 8.10.64), *Hiyodori* (PC 320, 29.9.65)

The first two were authorised in the 1958 programme, with three in 1959, two in 1961 and one each in 1962, 1963 and 1964. All but one were reclassified as auxiliaries (ASU) and given a training role: PC 311, 312 and 314 on 27 March 1981; PC 313, 315, 317 and 318 by March 1982, PC 316 in 1984 and PC 319 in 1986. *Hiyodori* was reconstructed in 1986–87 as Auxiliary Special Service Yacht (ASY 92), and is extant 1995. Remainder now deleted.

TYPE 1 *fast patrol craft*

Displacement: 70t standard
Dimensions: 85ft 4in pp × 22ft 4in × 3ft 7in max
26.0m × 6.8m × 1.1m
Machinery: 2 shafts, 2 diesels, 14,000bhp = 31kts
Armament: 1–40mm, 2–21in TT (2 × 1)
Complement: 18

Class:
Hitachi Zosen – *No 1, No 2*
Mitsubishi Zosen – *No 3, No 4*
Azuma Zosen – *No 5, No 6*

All units were authorised in the 1953 programme. Nos *801–806* were assigned on 1 September 1957. These boats had an ASW capability, sonar domes being installed on the hull bottoms, but at high speeds the sonars offered considerable resistance. All units were deleted 1971–72.

TYPE 7 *fast patrol craft*

Displacement: 104t standard
Dimensions: 109ft 5in pp × 24ft 7in × 3ft 11in max
34.1m × 7.5m × 1.2m
Machinery: 3 shafts, 3 Mitsubishi diesels, 6000bhp = 33kts
Armament: 2–40mm (2 × 1), 4–21in TT (4 × 1)
Complement: 24

Class (launched):
Mitsubishi Zosen – *No 7* (2.2.57), *No 8* (20.7.57)

Authorised under the 1954 programme, laid down August 1956 and launched 31 October 1957 and 15 November 1957 respectively. Numbers *807* and *808* were assigned 1 September 1957. This class was fitted with an improved sonar dome which offered less resistance at high speed than the type used in the Type 1. From January 1968 until December 1972, *No 7* was fitted with an Ishikawajima-Harima T64 marine gas turbine for test purposes. Both units were deleted in 1973.

No 9 *fast patrol craft*

Displacement: 64t standard
Dimensions: 71ft 4in oa × 19ft 9in × 6ft max
21.7m × 6.2m × 1.8m
Machinery: 2 shafts, 2 Napier Deltic diesels, 5000bhp = 40kts
Armament: 1–40mm, 2–21in TT (2 × 1)
Complement: 15

Type 9 craft, built by Saunders-Roe, Beaumaris, GB, delivered 29 July 1957 and accepted 2 September 1957. Similar to the British 'Dark' class. Number *809* was assigned 1 September 1957. Deleted 1972.

No 10 *fast patrol craft*

Displacement: 120t standard
Dimensions: 105ft pp × 27ft 8in × 3ft 10in max
32.0m × 8.5m × 1.2m
Machinery: 3 shafts, 3 diesels, 9000bhp = 40kts
Armament: 2–40mm (2 × 1), 4–21in TT (4 × 1)
Complement: 20

Type 10 craft, built by Mitsubishi, Shimonoseki, laid down 30 September 1962, launched 28 July 1961 and completed 25 May 1962. Authorised under the 1960 programme; Number *810* was assigned. This boat was built as a test platform for the follow-on, mass produced Type II and was fitted with dipping sonar and sonobuoys. Deleted 1975.

PT 13 as completed JMSDF

TYPE 11 *fast patrol craft*

Displacement: 110t standard; 125t full load
Dimensions: 116ft 6in pp × 30ft 3in × 3ft 11in max
25.5m × 9.2m × 1.2m
Machinery: 3-shaft CODAG: 2 Mitsubishi diesels plus 2 IHI gas turbines, 11,000shp = 40kts
Armament: 2–40mm (2 × 1), 4–21in TT (4 × 1)
Sensors: Radar OPS-13
Complement: 28

Class (no, laid down, commissioned):
PT 11 (811, 17.3.70, 27.3.71), *PT 12* (812, 22.4.71, 28.3.72), *PT 13* (813, 28.3.72, 16.12.72), *PT 14* (814, 23.3.73, 15.2.74), *PT 15* (815, 23.4.74, 10.7.75)

All were built by Mitsubishi, Shimonoseki and were deleted 1990–94.

'PG 1' class *fast attack hydrofoil*

Displacement: 50t standard
Dimensions: 71ft 6in × 22ft 11in × 5ft 7in
21.8m × 7.0m × 1.7m
Machinery: 1 pump jet: 1 GE/IHI LM500 gas turbine, 5522hp = 46kts. Also diesel-driven retractible propeller = 8kts
Armament: 4 SSM-1B SSM (2 × 2), 1–20mm Sea Vulcan Gatling
Sensors: Radar surface search
Complement: 11

Class (no, commissioned):
PG 01 (821, 22.3.93), *PG 02* (822, 22.3.93)

Italian-type hydrofoils built by Sumitomo, Uraga with assistance from Fincantieri. Built for interceptor duties. Four more projected but third not authorised for 1995 Fiscal Year.

OTHER SMALL CRAFT

Fifteen 18t patrol boats were transferred to Japan under MDAP in 1958, and numbered *Shokai* (patrol boat) *1–7* and *11–18*. They were deleted in the early 1970s.

Nine new 18t boats, *PB 19–PB 27*, were built by Ishikawajima-Harima, Yokohama, between 1971 and 1973. They are GRP-hulled craft, capable of 20kts, and armed with 1–20mm. Four deleted 1992, one in 1994 and one in 1995, remainder extant 1995.

MINE WARFARE VESSELS

ERIMO *minelayer/minesweeper*

Displacement: 630t standard; 670t full load
Dimensions: 210ft × 26ft × 8ft
64.0m × 7.9m × 2.4m
Machinery: 2 shafts, diesels, 2500bhp = 18kts
Armament: 1 40mm, 2 20mm (2 × 1), 1 Hedgehog, 2 K-guns, 2 DC racks
Complement: 80

Erimo (AMC 491) was built by Uraga Dock and launched 12 July 1955. She was designed as a multipurpose minesweeper, minelayer and submarine-chaser. She was converted to an Explosive Ordnance Disposal tender (YAS 69) (for mine clearance divers) in 1975–76, with an armament of 2–40mm (1 × 2), 2–20mm (2 × 1) and a single Mk 10 Hedgehog. Deleted 1982.

TSUGARU *minelayer/cable layer*

Displacement: 950t standard
Dimensions: 216ft 4in × 34ft 2in × 11ft
66.0m × 10.4m × 3.4m
Machinery: 2 shafts, diesels, 3200bhp = 16kts
Armament: 1–3in/50, 2–20mm (2/1), 4 K-guns, 40 mines
Sensors: Radar OPS-16
Complement: 100

Tsugaru (ARC 481) was built by Yokohama Shipyard and launched 19 July 1955; she was designed as a dual purpose coastal minelayer and cable layer. She was lengthened and widened to 338ft × 47ft 11in (*103.0m × 14.6m*) by Nippon Kokan, Tsurumi in 1969–70 and converted to a specialised cable ship (ASU 7001), retaining only 2–20mm guns. She was deleted in 1990.

HAYASE *mine warfare support ship/minelayer*

Displacement: 2000t standard; 3050t full load
Dimensions: 324ft 9in × 42ft 8in × 12ft 6in
99.0m × 13.0m × 3.8m
Machinery: 2 shafts, 4 Kawasaki/MAN diesels, 6400bhp = 18kts
Armament: 2–3in/50 (1 × 2), 2–20mm (2 × 1), 6–12.75in Mk 32 ASW TT (2 × 3), 200 mines
Sensors: Radar OPS-17; sonar SQS-11A
Complement: 85

No	Name	Builder	Launched	Fate
MST	HAYASE	Ishikawajima-Harima	21.6.71	Extant 1995

A multi-role warfare ship, with five rails for minelaying, a platform aft for landing a minesweeping helicopter, and the facilities to support mine countermeasures operations. Originally fitted with OPS-14 air search radar. The design was developed into the *Souya* type (see below). *Hayase* was the flagship for the historic out-of-area development to the Gulf in 1991. Flagship First MCM Flotilla.

Souya 1971

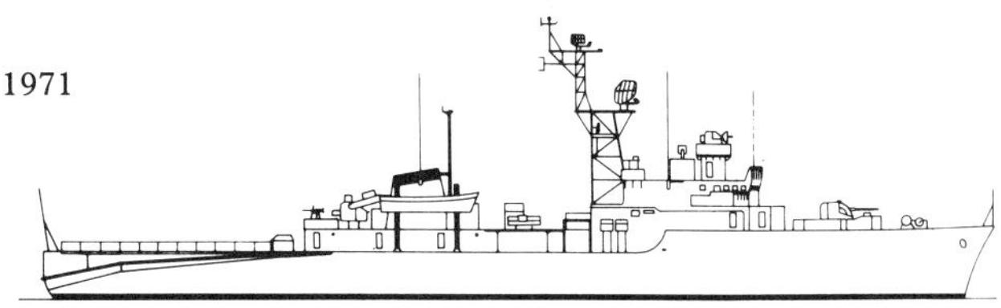

SOUYA *minelayer*

Displacement: 2150t standard; 3250t full load
Dimensions: 324ft 9in × 49ft 3in × 13ft 9in
99.0m × 15.0m × 4.2m
Sensors: Radar OPS-14, OPS-16; Sonar SQS-11A
Other particulars: As *Hayase*

No	Name	Builder	Launched	Fate
MMC 951	SOUYA	Hitachi, Maizuru	31.3.71	Extant 1995

Laid down on 9 July 1970 and commissioned on 30 September 1971, *Souya* has six mine rails, four in the stern, and two externally. She is also equipped to operate a minesweeping helicopter, although there is no hangar. Flagship Second MCM Flotilla. Both *Souya* and *Hayase* (above) are to be replaced by new 5600t minesweeper support ships/minelayers authorised 1994 and due in service 1996–98. *Souya* will be the first to be replaced.

Ex-US ALBATROSS class *coastal minesweepers*

Particulars: As US YMS type

Class (no, former no):
Etajima (MSC 656, ex-*AMS 10*), *Moroshima* (MSC 663, ex-*YMS 372*), *Ninoshima* (MSC 662, ex-*YMS 376*), *Nuwajima* (MSC 657, ex-*AMS 18*), *Ogishima* (MSC 659, ex-*AMS 32*), *Ujishima* (MSC 655, ex-*AMS 5*), *Yakushima* (MSC 658, ex-*AMS 28*), *Yugeshima* (MSC 660, ex-*AMS 36*), *Yurishima* (MSC 661, ex-*AMS 40*)

Belonging to the wartime TMS type (renamed the *Albatross* class postwar), they were transferred under MDAP in 1955 (except *Moroshima* and *Ninoshima* in 1959). They had short careers in the JMSDF, most being deleted in 1957, although *Etajima* and *Ujishima* served until 1966, and *Yakushima* was returned to the USA in 1970.

Ex-US AMS/MSC type *coastal minesweepers*

Particulars: As US AMS/MSC type

Class (no, former no, acquired):
Yashima (MSC 651, ex-*AMS 144*, 16.12.54), *Hashima* (MSC 652, ex-*AMS 95*, 3.6.55), *Tsushima* (MSC 653, ex-*AMS 255*, 18.7.56), *Toshima* (MSC 654, ex-*MSC 258*, 18.7.56)

Belonging to the numerous US-built AMS type (reclassified MSC in February 1955), they were transferred on completion to Japan. MSC 651 and 652 were reclassified as auxiliaries in 1971, and the other two shortly afterwards. MSC 682 was deleted in 1976 and the remainder in about 1978.

YASHIRO *coastal minesweeper*

Displacement: 230t standard; 255t full load
Dimensions: 118ft pp × 22ft 8in × 6ft 2in
36.0m × 6.9m × 1.9m
Machinery: 2 shafts, diesels, 1200bhp = 13kts
Armament: 1–20mm

Yashiro (MSC 601) was built by Nippon Kokan, Tsurumi and launched 26.3.56. She was transferred to an auxiliary role as a diving tender (YAS 58) and was deleted in 1981.

ATADA class *class minesweepers*

Displacement: 240t standard; 260t full load
Dimensions: 118ft pp, 123ft 4in oa × 21ft × 6ft 9in
36.0m, 37.6m × 6.4m × 2.1m
Machinery: 2 shafts, 2 diesels, 1200bhp = 13kts
Armament: 1–20mm

Class (no, launched):
Atada (MSC 601, 12.3.56), *Itsuki* (MSC 602, 11.3.56)

Both built of wood and lightweight metals by Hitachi, Kanagawa. Both were transferred to auxiliary roles as gunnery tenders (YAS 56 and 57) and were deleted in 1981.

Katsura 1979 JMSDF

KASADO class *minehunters/minesweepers*

Displacement: 330t standard; 360t full load
Dimensions: 140ft × 27ft 6in × 7ft 6in
45.7m × 8.4m × 2.3m
Machinery: 2 shafts, 2 Mitsubishi diesels, 1200bhp = 14kts
Armament: 1–20mm
Sensors: Radar OPS-9; sonar ZQS-9
Complement: 43

Class (no, launched):
Amani (MSC 623, 31.10.66), *Chiburi* (MSC 620, 29.11.63), *Habushi* (MSC 609, 19.6.59), *Hario* (MSC 618, 10.12.62), *Hirado* (MSC 613, 3.10.60), *Hotaka* (MSC 616, 23.10.61), *Ibuki* (MSC 628, 2.12.67), *Kanawa* (MSC 606, 22.4.59), *Karato* (MSC 617, 11.12.62), *Kasado* (MSC 604, 19.3.58), *Katsura* (MSC 629, 18.9.67), *Koozu* (MSC 609, 12.11.59), *Koshiki* (MSC 615, 9.11.61), *Kudako* (MSC 622, 8.12.64), *Mikura* (MSC 612, 14.3.60), *Minase* (MSC 627, 10.1.67), *Mutsure* (MSC 619, 16.12.63), *Ootsu* (MSC 621, 5.11.64), *Rebun* (MSC 624, 7.12.65), *Rishiri* (MSC 625, 22.11.65), *Sakito* (MSC 607,22.4.59), *Shikine* (MSC 613, 22.7.60), *Shisaka* (MSC 605, 20.3.58), *Tatara* (MSC 610, 14.1.60), *Tsukumi* (MSC 611, 12.1.60), *Urume* (MSC 626, 12.11.66)

Wooden coastal minesweepers, similar to many other classes in Western navies. The even-numbered vessels were built by Hitachi, Kanagawa, and the remainder by Nippon Kokan, Tsurumi. By 1982 all had been transferred to auxiliary roles. MSC 612 and 613 became tenders to the gunnery school; MSC 621 and 609 became MCM support ships (MST 474 and 473); MSC 604, 608, 610, 614 and 618 became hydrographic ships; and the rest were reclassified YAS, most serving as mine clearance diver support ships. All were deleted between 1981 and 1990.

TAKAMI class *minehunters/minesweepers*

Displacement: 380t standard
Dimensions: 170ft 7in × 28ft 10in × 7ft 10in
52.0m × 8.8m × 2.3m
Machinery: 2 shafts, 2 Mitsubishi diesels, 1440bhp = 14kts
Armament: 1–20mm
Sensors: Radar OPS-9; sonar ZQS-2
Complement: 45–47

Class (no, launched):
Takami (MSC 630, 15.7.69), *Iou* (MSC 631, 12.8.69), *Mikaye* (MSC 632, 3.6.70), *Utone* (MSC 633, 6.4.70), *Awaji* (MSC 634, 11.12.70), *Toushi* (MSC 635, 13.12.70), *Tewi* (MSC 636, Oct 1971), *Murotsu* (MSC 637, Oct 71), *Tashiro* (MSC 638, 2.4.73), *Miyato* (MSC 639, 3.4.73), *Takane* (MSC 640, 8.3.74), *Muzuki* (MSC 641, 5.4.74), *Yokose* (MSC 642, 21.7.75), *Sakate* (MSC 643, 5.8.75), *Oumi* (MSC 644, 28.5.76), *Fukue* (MSC 645, 12.7.76), *Okitsu* (MSC 646, 4.3.77), *Hashira* (MSC 647, 8.11.77), *Iwai* (MSC 648, 8.11.77)

Wooden minesweepers (developed from the *Kasado* class) which also carried divers for mine clearance duties. The electronics were largely Japanese-produced versions of British equipment. The even-numbered vessels were built by Nippon Kokan, Tsurumi, and the remainder by Hitachi, Kanagawa. *Miyato* was the first to be deleted on 31 March 1982, almost all being converted to YAS over the next twelve years. In 1995 *Fukue* was extant as MST 426, a mothership for the MSB inshore minesweepers. All but six of the YASs had been deleted 1995.

HATSUSHIMA class *minehunters/minesweepers*

Displacement: 440t standard
Dimensions: 180ft 5in × 30ft 10in × 7ft 10in
55.0m × 9.4m × 2.4m
Machinery: 2 shafts, 2 Mitsubishi diesels, 1440bhp = 14kts
Armament: 1–20mm
Sensors: Radar OPS-9; sonar ZQS-2B
Complement: 45

Class (no, launched):
Nippon Kokan, Tsurumi – *Hatsushima* (MSC 649, 30.10.78), *Miyajima* (MSC 651, 18.9.79), *Enoshima* (MSC 652, 25.7.80), *Miijima* (MSC 655, 2.6.81), *Yakushima* (MSC 656, 22.6.82), *Toroshima* (MSC 659, 23.6.83), *Hahajima* (MSC 660, 27.6.84), *Etajima* (MSC 663, 17.6.85), *Kamishima* (MSC 664, 20.6.86), *Moroshima* (MSC 667, 11.6.87), *Yurishima* (MSC 668, 13.5.88)
Hitachi, Kanagawa – *Ninoshima* (MSC 650, 9.8.79), *Ukishima* (MSC 653, 11.7.80), *Oshima* (MSC 654, 17.6.81), *Narushima* (MSC 657, 7.6.82), *Chichijima* (MSC 658, 18.7.83), *Takashima* (MSC 661, 18.6.84), *Newajima* (MSC 662, 5.6.85), *Himeshima* (MSC 665, 10.6.86), *Ogishima* (MSC 666, 10.6.87), *Hikoshima* (MSC 669, 2.6.88)

Enlarged versions of the *Takami* type, and also wooden hulled. Besides conventional sweep gear, they carry Type 5.4 minehunting vehicles. *Ukishima* carried out trials with a 20mm JM61-MB Gatling-type gun, which was adopted for the rest of the class. *Yurishima* and *Hikoshima* went to the Gulf in 1991. *Hatsushima* converted to YAS 1995. Remainder extant 1995.

Hamajima Jul 1993 L & L van Ginderen

AWASHIMA class *minehunters/minesweepers*

Displacement: 490t standard
Dimensions: 189ft 4in × 30ft 10in × 7ft 11in
57.7m × 9.4m × 2.4m
Machinery: 2 shafts, 2 Mitsubishi diesels, 1400bhp = 14kts
Armament: 1–20mm Gatling
Sensors: Radar OPS-9 (from MSC 674, OPS-39); sonar ZQS-2B (from MSC 674, ZQS-3)
Complement: 45

Class (no, completed):
Hitachi, Kanagawa – *Awashima* (MSC 670, 13.12.89), *Ieshima* (MSC 673, 19.12.90), *Tsukishima* (MSC 674, 18.3.93), *Maejima* (MSC 675, 15.12.93), *Makishima* (MSC 677, 12.12.94), *MSC 679* (1996)
Nippon Kokan, Tsurumi – *Sakushima* (MSC 671, 13.12.89), *Uwajima* (MSC 672, 19.12.90), *Kumejima* (MSC 676, Dec 1994), *Tobishima* (MSC 678, 10.3.95), *MSC 689* (1996)

A lengthened version of the *Hatsushima* class, to provide improved accommodation. Built of wood, like their predecessors. *Awashima* and *Sakushima* participated in the 1991 deployment to the Gulf.

YAEYAMA class *ocean minehunters/minesweepers*

Displacement: 1000t standard
Dimensions: 219ft 10in × 38ft 8in × 10ft 2in
67.0m × 11.8m × 3.1m
Machinery: 2 shafts, 2 Mitsubishi diesels, 2400bhp = 14kts
Armament: 1–20mm Gatling
Sensors: Radar OPS-39; sonar SQQ-32 VDS
Complement: 60

Class (no, launched):
Hitachi, Kanagawa – *Yaeyama* (MSO 301, 28.8.91)
Nippon Kokan, Tsurumi – *Tsushima* (MSO 302, 11.9.91), *Hachijyo* (MSO 303, 15.12.92)

Three wooden-hulled ocean minehunters intended to be a class of six. First two projected in the 1989 Budget. Fitted with the S-7 deep-sea autonomous minehunting vessel and the S-8 deep-sea mine disposal vehicle.

MSB type *inshore minesweepers*

Class (no):
Nana-Go (MSB 707), *Hachi-Go* (MSB 708), *Kyuu-Go* (MSB 709), *Jyuu-Go* (MSB 710), *Jyuu-Ichi-Go* (MSB 711), *Jyuu-Ni-Go* (MSB 712)

Six Wooden-hulled boats were built by Hitachi, Kanagawa (odd numbers) and Nippon Kokan, Tsurumi in 1973–75. They were 74ft craft of 58t full load and during minesweeping operations they were supported by the *Ootsu* and then by *Fukue*. MSB 707–708 deleted 1992; 709–710 in 1995.

MISCELLANEOUS

AZUMA *training support ship*

Displacement: 1950t standard; 2400t full load
Dimensions: 308ft 5in pp, 321ft 6in oa × 42ft 7in × 12ft 6in max
94.0m, 98.0 × 13.0m × 3.8m
Machinery: 2-shaft Kawasaki/MAN V8V 23/30 ATL diesels, 4000bhp = 18kts
Armament: 1–30/50in, 2 Mk 4 ASW torpedo racks
Sensors: Radar OPS-15, SPS-40; sonar SQS-11A
Complement: 185

Laid down 13 July 1968, launched 14 April 1969 and completed 21 October 1969 by Hitachi, Maizuru. A multi purpose support ship, *Azuma* carries drones for anti-aircraft practice. In service (ATS 4204) in 1995.

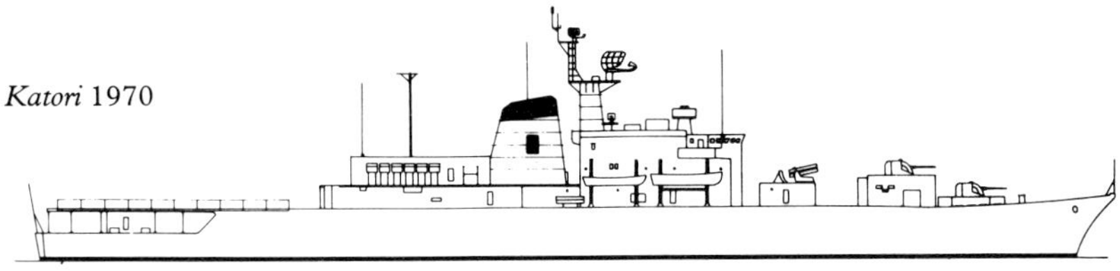

Katori 1970

KATORI *training ship*

Displacement: 3372t standard; 4100t full load
Dimensions: 400ft 3in pp, 418ft 2in oa × 49ft 1in × 14ft 4in max
122.0m, 127.5m × 15.0m × 4.3m
Machinery: 2-shaft sets of Ishikawajima geared turbines, 20,000shp = 25kts
Armament: 4–3in/50in (2 × 2), 1 ASW rocket launcher (1 × 4), 6–12.75in Mk 32 ASW TT (3 × 2), 1 helicopter
Sensors: Radar SPS-12, OPS-15; sonar OQS-3
Complement: 460 (inc trainees)

Katori was built by Ishikawajima-Harima Tokyo, laid down 8 December 1967, launched 19 November 1968 and completed 10 September 1968. She is capable of serving as a training ship in peacetime and as an escort command ship in wartime. She was ordered under the 1966 programme. In service 1995 and about to be replaced by *Kashima*.

KUROBE *training support ship*

Displacement: 2270t standard; 3200f full load
Dimensions: 331ft 5in × 54ft 1in × 13ft 1in
10.0m × 16.5m × 4m
Machinery: 2 shafts, 2 Fuji 8L27 5XF diesels, 8700bhp = 20kts
Armament: 1–76mm OTO Melara
Sensors: Radar OPS-14, OPS-18
Complement: 156

Laid down by Nippon Kokan, Tsurumi 31 July 1987, launched 23 May 1988 and commissioned 23 March 1989. Carries drones for AAW operations training.

KASHIMA *training ship*

Displacement: 4060t standard
Dimensions: 469ft 2in × 59ft 1in × 15ft 1in
143.0m × 18.0m × 4.6m
Machinery: 2 shaft CODOG: 2 Rolls-Royce Spey gas turbines, 26,650hp = 26kts; 2 Mitsubishi 616U diesels, 8000bhp = ?kts
Armament: 1–76mm OTO Melara, 2–40mm saluting guns (2 × 1), 6–12.75in Mk 32 ASW TT (2 × 3), helicopter platform
Sensors: Radar air and surface search
Complement: 389 (including trainees)

Built by Hitachi, Maizuru to replace *Katori*. Laid down 20 April 1993 and launched 23 February 1994. Due into service 1995. The project was delayed for a year. Wartime role as command ship and escort.

MARITIME SAFETY AGENCY

After the setting-up of the MSDF, the MSA continued to carry out coast guard duties with armed patrol vessels. These are operated by the Guard and Rescue Department. In addition, there are Hydrographic and Aids to Navigation Departments. Like the MSDF, the Agency is organised into districts with eleven regional Maritime Safety offices. The MSA's duties include those which in other countries would fall to the navy, a recent example being the escort of a shipload of nuclear material from Europe. A large fleet of offshore patrol vessels of various sizes has been built up since the Agency's foundation in 1948 with vessels left over from the Second World War (see above).

MUROTO class *large patrol vessels*

Displacement: 750t standard
Dimensions: 200ft × 30ft 6in × 10ft 3in
60.9m × 9.3m × 3.1m
Machinery: 2 shafts, 2 diesels, 1,500bhp = 12kts
Armament: 1–3in, 2–20mm (2 × 1)

No	Name	Builder	Launched	Fate
P 201	MUROTO	Uraga, Tokyo	5.12.49	Deleted 1975
P 202	DAIO	Uraga, Yokyo		Deleted 1974

Japan's first postwar-built large patrol vessel, *Muroto* was delivered to the Maritime Safety Board on 20 March 1950. She made 15.37kts on trials.

NOJIMA class *large patrol vessels*

Displacement: 950t standard
Dimensions: 226ft 6in × 30ft 3in × 10ft 6in
69.1m × 9.2m × 3.2m
Machinery: 2 shafts, 2 Uraga/Sulzer 6MD 42 diesels, 3000bhp = 17.5kts
Armament: None

No	Name	Builder	Launched	Fate
PL 11	NOJIMA	Uraga, Tokyo	12.2.62	Deleted 1989
PL 12	OJIKA	Uraga, Tokyo	10.6.63	Deleted 1991

Built for training and ocean rescue work and used as patrol and weather ships.

KOJIMA class *large patrol vessel*

Displacement: 1100t
Dimensions: 228ft 4in × 33ft 10in × 10ft 6in
69.6m × 10.3m × 3.2m
Machinery: 1 shaft, 1 Uraga/Sulzer 7MD 51 diesel, 2600bhp = 17kts
Armament: 1–3in, 1–40mm, 1–20mm
Complement: 107 (including 47 cadets)

No	Name	Builder	Launched	Fate
PL 21	KOJIMA	Kure	21.5.64	Deleted 1993

Maritime Safety Academy training ship. Armament changed to 1–40mm, 1–20mm and 1–12.7mm MG in early 1980s.

ERIMO class *large patrol vessels*

Displacement: 1009t
Dimensions: 251ft 4in × 30ft 2in × 9ft 11in
76.6m × 9.2m × 3.0m
Machinery: 2 shafts, 2 Burmaster and Wain 635V 2BV 45 diesels, 4800bhp = 19.8kts
Armament: 1–3in, 1–20mm

No	Name	Builder	Launched	Fate
PL 13	ERIMO	Hitachi, Mukaishima	14.8.65	Deleted 1994
PL 14	SATSUMA	Hitachi, Mukaishima	Apr 1966	Extant 1995

Erimo specially strengthened for service in waters north of Japan.

DAIO class *large patrol vessels*

Displacement: 1194t
Dimensions: 251ft 4in × 31ft 6in × 10ft 8in
76.6m × 9.6m × 3.3m
Machinery: 2 shafts, 2 diesels, 7000bhp = 20kts
Armament: 1–40mm, 1–20mm

No	Name	Builder	Launched	Fate
PL 15	DAIO	Hitachi, Maizuru	19.6.73	Extant 1995
PL 16	MUROTO	Naiki, Tagama	5.8.74	Extant 1995

Modified version of *Erimo* class, slightly enlarged with more powerful engines.

IZU class *large patrol vessels*

Displacement: 2080t normal
Dimensions: 313ft 4in × 38ft × 12ft 10in
95.6m × 11.6m × 3.9m
Machinery: 2 shafts, 2 SEMT-Pielstick 2 PC2 SV400 diesels, 12,000bhp = 20kts
Armament: 1–40mm
Complement: 72

No	Name	Builder	Launched	Fate
PL 31	IZU	Hitachi, Maizuru	Jan 1967	Extant 1995
PL 32	MIURA	Hitachi, Maizuru	Oct 1968	Extant 1995

Originally long-range weather ships equipped with weather observation radar. Strengthened against ice. Armament added later for patrol work.

SOYA (ii) class *large patrol vessel*

Displacement: 3700t
Dimensions: 323ft 5in × 51ft 2in × 17ft 1in
98.6m × 15.6m × 5.2m
Machinery: 2 shafts, 2 SEMT-Pielstick 12PC2 SU 400 diesels, 12,000bhp = 21kts
Armament: 1–40mm, 1–20mm. 1 helicopter
Complement: 71

No	Name	Builder	Launched	Fate
PLH 01	SOYA (ii)	Nippon Kokan	3.7.78	Extant 1995

Built to replace the icebreaking PL of the same name. First designated PL 01; current designation adopted 13 December 1985. Bell 212 helicopter carried.

TSUGARU class *large patrol vessels*

Displacement: 3730t
Dimensions: 344ft 5in × 49ft 2in × 16ft 1in
105.0m × 15.0m × 4.9m
Machinery: 2 shafts, 2 SEMT-Pielstick 12PC2 SV 400 diesels, 12,000bhp = 22kts
Armament: 1–40mm or 35mm, 1–20mm, 1 helicopter
Complement: 69

No	Name	Builder	Launched	Fate
PLH 02	TSUGARU	IHI, Tokyo	6.12.78	Extant 1995
PLH 03	OOSUMI	Mitsui, Tamano	1.6.79	Extant 1995
PLH 04	URAGA	Hitachi, Maizuru	12.12.79	Extant 1995
PLH 05	ZAO	Mitsubishi, Nagasaki	29.10.81	Extant 1995
PLH 06	CHIKUZEN	Kawasaki, Kobe	18.3.83	Extant 1995
PLH 07	SETTSU	Sumitomo, Oppama	21.4.84	Extant 1995
PLH 08	ECHIGO	Mitsui, Tamano	4.7.89	Extant 1995

Development of the *Soya* class for ocean patrol and rescue work. Designated PLH from 13 February 1985. Bell 212 helicopters carried.

Uraga 11.5.84 L & L van Ginderen

SHIRETOKO class *large patrol vessels*

Displacement: 1290t
Dimensions: 255ft 10in × 31ft 6in × 10ft 6in
78.0m × 9.6m × 3.2m
Machinery: 2 shafts, 2 Fuji 8 540B or Niigata 8MA 40 diesels, 7000bhp = 20kts
Armament: 1–40mm or 35mm (PL 101–105, 127–128 1–20mm additionally)
Complement: 41

No	Name	Builder	Launched	Fate
PL 101	SHIRETOKO	Mitsui, Tamano	13.7.78	Extant 1995
PL 102	ESAN	Sumitomo	Aug 1978	Extant 1995
PL 103	WAKASA	Kawaju, Kobe	Aug 1978	Extant 1995
PL 104	YAHIKO	Mitsubishi, Shimonoseki	Aug 1978	Extant 1995
PL 105	MOTOBU	Sasebo	Aug 1978	Extant 1995
PL 106	RESHIRI	Shikoku	27.3.79	Extant 1995
PL 107	MATSUSHIMA	Tohohu	11.4.79	Extant 1995
PL 108	IWAKI	Naikai	28.3.79	Extant 1995
PL 109	SHIKINE	Usuki	27.4.79	Extant 1995
PL 110	SURUGA	Kurushima	20.4.79	Extant 1995
PL 111	REBUN	Narasaki	Jun 1979	Extant 1995
PL 112	CHOKAI	Nihonkai	Jun 1979	Extant 1995
PL 113	ASHIZURI	Sanoyasu	Jun 1979	Extant 1995
PL 114	OKI	Tsuneishi	Jun 1979	Extant 1995
PL 115	NOTO	Miho	Jul 1979	Extant 1995
PL 116	YUNAKUNI	Hayashikane, Nagasaki	Jun 1979	Extant 1995
PL 117	KURIKOMA (ex-*Kudaka* (ex-*Daisetsu*)	Hakodate	22.8.79	Extant 1995
PL 118	SHIMOKITA	Ishikawajima, -Harima Kakoki	Sep 1979	Extant 1995
PL 119	SUZUKA	Kanasashi	4.10.79	Extant 1995
PL 120	KUNISAKI	Kouyo	8.10.79	Extant 1995
PL 121	GENKAI	Oshima	Sep 1979	Extant 1995
PL 122	GOTO	Onomichi	Oct 1979	Extant 1995
PL 123	KOSHIKI	Kasado	Sep 1979	Extant 1995
PL 124	HATERUMA	Osaka	Nov 1979	Extant 1995
PL 125	KATORI	Tohoku	May 1980	Extant 1995
PL 126	KUNIGAMI	Kanda	28.3.80	Extant 1995
PL 127	ETOMO	Naikai	30.9.81	Extant 1995
PL 128	MASHU	Shikoku	14.10.81	Extant 1995

Mass-produced development of the *Erimo* class for EEZ patrol duties.

MIZUHO class *large patrol vessel*

Displacement: 4900t
Dimensions: 426ft 6in × 50ft 10in × 17ft 8in
130.0m × 15.5m × 5.4m
Machinery: 2 shafts, 2 SEMT-Pielstick 14PC3-SU400 diesels, 18,200hp = 23kts
Armament: 1–35mm, 1–20mm Gatling, 2 helicopters
Complement: 100 (+ 30 aircrew)

No	Name	Builder	Launched	Fate
PLH 21	MIZUHO	Mitsubishi, Nagasaki	5.6.85	Extant 1995
PLH 22	YASHIMA	Nippon Kokan, Tsurumi	20.1.88	Extant 1995

Large PLH class built primarily for search and rescue work more than 200 miles from the Japanese coast. S-62 or Fuji-Bell 212 helicopters carried.

Suruga Nov 1981 L & L van Ginderen

NOJIMA class *large patrol vessel*

Displacement: 950t
Dimensions: 285ft 5in × 34ft 5in × 11ft 6in
87.0m × 10.5m × 3.5m
Machinery: 2 shafts, 2 Fuji 8540B diesels, 7000bhp = 19kts
Armament: 1–20mm Gatling

No	Name	Builder	Launched	Fate
PL 01	NOJIMA	Ishikawajima-Harima, Tokyo	30.5.89	Extant 1995

Surveillance and rescue command ship. Helicopter platform but no hangar.

SHIKISHIMA class *large patrol vessel*

Displacement: 6500t
Dimensions: 492ft 1in × 55ft 10in × 19ft 8in
150.0m × 17.0m × 6.0m
Machinery: 2 shafts, 2 SEMT-Pielstick 16PC2.5 diesels, 20,800bhp = 25kts
Armament: 4–35mm (2 × 2), 2–20mm Gatling (2 × 1), 2–12.7mm MG (2 × 1), 2 water cannon, 2 helicopters

No	Name	Builder	Launched	Fate
PLH 31	SHIKISHIMA	Ishikawajima-Harima, Tokyo	27.6.91	Extant 1995

Built to escort ships carrying plutonium from Europe to Japan. This job was given to the MSA because of the reluctance to deploy MSDF units out of area. Equipped with two water cannon for fire-fighting and for possible use against protestors. Helicopters are either Bell 212 or Super Puma. Built instead of the third *Mizuho* class unit.

OJIKA class *large patrol vessels*

Displacement: 1883t
Dimensions: 299ft 8in × 36ft 1in 11ft 6in
91.4m × 11.0m × 3.5m
Machinery: 2 shafts, 2 Fuji 8540B diesels, 7000bhp = 18kts
Armament: 1–35mm, 1–20mm Gatling
Complement: 38

No	Name	Builder	Launched	Fate
PL 12	OJIKA	Mitsui, Tamano	28.4.91	Extant 1995
PL 13	SATSUMA	Hakodate DY	Jun 1994	Extant 1995
PL 14	–	–	–	Building 1995

Development of the *Nojima* class of 1989. Search and rescue command ships.

KOJIMA (ii) class *large patrol vessel*

Displacement: 2650t
Dimensions: 377ft 4in × 45ft 11in × 16ft 5in
115.0m × 14.0 × 5.0m
Machinery: 2 shafts, 2 diesels, 8000bhp = 18kts
Armament: 1–35mm, 1–20mm Gatling, 1–12.7mm MG
Complement: 118

No	Name	Builder	Launched	Fate
PL 21	KOJIMA (ii)	Hitachi, Maizuru	10.9.92	Extant 1995

New training vessel built as direct replacement for ship of same name and number deleted in 1993. Commissioned 11 March 1993.

MEDIUM PATROL VESSELS

NB Classes built in parallel with large patrol vessels. Twin-screw, diesel-powered ships unless otherwise stated.

Three *Awaji* class (510t, 1–3in, 1–20mm): *PM 01–PM 03*, completed 1950. All deleted 1974–75.

Fourteen *Rebun* class (450t, 1–3in, 1–20mm): *PM 04–PM 17*, completed 1951. All deleted 1979.

Five *Chifuri* class (465t, 1–3in, 1–20mm): *PM 18–PM 22*, completed 1952. All deleted 1979.

Four *Kunashiri* class (350–M 3 type) (498t, 1–20mm): *PM 65–PM 68*, completed 1969–73. All extant 1995.

Three *Miyake* class (350t, 1–20mm): *PM 70–PM 72*, completed 1972–73. All extant 1995.

Twenty *Bihuro* class (350–M4 type) (636t, 1–20mm): *PM 73–PM 88, PM 90–PM 93*, completed 1972–73. All extant 1995.

Two *Takatori* class (634t, unarmed): *PM 89, PM 94*, completed 1978–79. Salvage/fire fighting boats. Both extant 1995.

Fourteen *Teshio* clas (630t, 1–20mm Gatling); *PM 01–PM 13*, completed 1980–88. All extant 1995.

Three *Anami* class (230t, 1–20mm Gatling): *PM 95*, completed 1992. Extant 1995, PM 96–97 building 1995.

One 550t class *PM 97* (1–20mm Gatling), with icebreaker bow launched May 1994. To replace penultimate ship of the *Matsuura* class in 1995.

SMALL PATROL VESSELS

Seventeen *Kuma* class (258t, 1–40mm): *PS 01–PS 17*, completed 1971. Deleted 1974–77.

Three *Nagara* class (260t, 1–40mm): *PS 18–PS 20*, completed 1952. Deleted 1976–77.

Two *Tokachi* class* (336t/324t, 1–40mm): *PS 51–PS 52*, completed 1954. Deleted 1980.

One *Teshio* class* (422t, 1–40mm): *PS 53*, completed 1955. Deleted 1980.

Seven *Yahagi* class* (350 type) (333t, 1–40mm): *PS 54–PS 59, PS 69*, completed 1956–61 (*PS 69* completed 1970 for Okinawa local government, transferred to MSA 1972, extant 1995). Remainder deleted 1981–86.

Five *Matsuura* class* (350–M2 type) (420t, 1–20mm gun): *PS 60–PS 64*, completed 1961–67. Two extant 1995, remainder deleted since 1986.

Fourteen *Hidaka* class (165t, single screw, unarmed): *PS 32–PS 39, PS 41*, PS 43–PS 46, completed 1961–68. Three extant 1995.

Seven *Akagi* class (126t, 1–12.7mm MG): *PS 101–PS 107*, completed 1980–84. All extant 1995.

Six *Mihashi* class (180 type) (180t, 1–12.7mm MG): *PS 01–PS 03*, completed 1988–94. Extant 1995.

Four *Takatsuki* class (115t, 1–12.7mm MG): *PS 108–PS 111*, completed 1992–94. Extant 1995.

*Re-rated medium patrol vessels (PM) with same numbers *c* 1970. Type 350 developed as standard PM of the 1970s.

Jordan

It was officially stated in 1969 that the Hashemite Kingdom of Jordan had no naval force, but the Jordanian Coast Guard, sometimes called the Jordan Sea Force, takes orders directly from the Army Director of Operations at GHQ. A Dead Sea flotilla no longer exists. Jordan's Sea Force base is at Aqaba.

Currently the Coast Guard personnel strength is 750 officers and ratings. In the late 1980s the king decided that three small fast attack craft should be ordered for patrolling the 12-mile maritime border on the Gulf of Aqaba. Three boats were ordered from Vosper in the United Kingdom and delivered in 1991, a considerable increase in Jordan's naval forces.

HAWK class *fast attack craft (gun)*

Displacement: 124t full load
Dimensions: 100ft oa × 22ft 8in × 4ft 11in
30.5m × 6.9m × 1.5m
Machinery: 2 shafts, 2 MTU 16V 396 TB94 diesels, 5800bhp = 32kts. Range 750nm at 15kts, 1500nm at 11kts
Armament: 2–30mm Oerlikon GCM-A03 (1×2), 1–20mm Oerlikon GAM-B01, 2–12.5mm MG, 2 Wallop Stockade chaff RL
Sensors: Radar Kelvin Hughes 1007, fire control Radamec Series 2000 optronic; combat data system Racal Cane 100
Complement: 16 (3 officers)

Class (no, in service):
Al Hussein (101, Dec 1989), *Al Hussan* (102, Mar 1990), *Abdullah* (103, 1991)

Ordered in December 1987 from Vosper Thornycroft. GRP construction. Up to seven were sought. Transported to Aqaba in September 1991 on a merchant vessel. All three extant 1995.

PATROL CRAFT

Three Rotork class patrol craft (9t, 41ft 8in, 240bhp = 28kts, 1–7.62mm MG): *Al Hashim, Al Faisal, Al Hamza*; delivered late 1990, all extant 1995.

Four US-built 38ft Bertram 'Enforcer' type coastal patrol craft (8t, 25kts, 3 MG) acquired in August 1974: *Faysal, Han, Hasayu* and *Muhammed.* All extant 1995.

Two US-built 30ft Bertram type coastal patrol craft (6.5t, 25kts, 3 MG, 8 crew): *Ali* and *Abdullah*, GRP hulls. Acquired 1974, both stricken 1992.

Two ex-GDR 'Bremse' class inshore patrol craft (42t, 74ft, 14kts, 1–12.7mm MG): *Shorouq I* (ex-*G 30/GS 5*), *Shorouq II* (ex-*G 31/GS 42*), built 1971–72, transferred from Germany 1991, both extant 1995.

Kenya

This East African colony became independent in December 1963 and a republic within the Commonwealth one year later. The new state was led from 1963 to 1978 by Jomo Kenyatta, who, while setting up a one-party state, preached moderation and conciliation; the legacy of Mau Mau was forgotten, white farmers and some Asian traders allowed to remain, and in general the regime came to be accepted as one of the most stable and pro-Western in tropical Africa. Although tribal politics have occasionally led to uncertainty and unrest, the transition to Daniel Moi's leadership after Kenyatta's death was peaceful.

The volunteer armed forces have traditionally been small in numbers and lightly equipped. Kenya receives military assistance from Britain, and is re-equipping her air force with US credits; she could almost certainly rely on overt military support from these powers if threatened by a neighbour. US naval forces made routine visits to Mombasa, making use of the usual supply facilities that were available to other countries, including the Soviet Union from 1968. More recently, Kenya has responded to the crises in the Gulf and the Horn of Africa with a 1980 mutual defence and assistance agreement allowing US forces access to Kenyan airfields, and making Mombasa, where an 18,000-ton dry dock was opened in 1978, a possible storage and revictualling depot for the US Navy.

Plans for the Kenyan Navy were announced in March 1964, and it was inaugurated on 12 December 1964, the first anniversary of independence; training, supply and maintenance have been almost entirely British (initial equipment worth 3,640,000 dollars). The present strength of about 4000 men and six large patrol craft, four of which were retrofitted with Israeli Gabriel missiles, is adequate for patrolling Kenya's 350-mile coastline and for limited offshore duties.

SIMBA class *large patrol craft (gun)*

Displacement:	96t standard; 109t full load
Dimensions:	95ft wl, 103ft oa, ×19ft 10in × 5ft 10in *28.9m, 31.3m × 6m × 1.8m*
Machinery:	2 shafts, 2 Paxman Ventura 12YJCM diesels, 2800bhp = 24kts. Oil 14t. Range 1500nm at 16kts
Armament:	2–40mm/60 Mk 7 (2×1), 2–2in RFL Mk 5
Complement:	23

Class (no, launched, in service):
Simba (P 3110, 9.9.65, 23.5.66), *Chui* (P 3112, 25.11.65, 7.7.66), *Ndovu* (P 3117, 22.12.65, 27.7.66)

Vosper type ordered 28 October 1964 from Vosper, Portsmouth. All sailed 22 August 1966 and arrived Mombasa 4 October 1966. Similar to Malaysian *Sri* class. All stricken by 1988.

MAMBA *large patrol craft (missile)*

Displacement:	125t standard; 160t full load
Dimensions:	123ft oa × 22ft 6in × 5ft 2in *37.5m × 6.9m × 1.6m*
Machinery:	2 shafts, 2 Paxman Ventura 16YJCM diesels, 4000bhp = 25kts. Range 3000nm at 12kts
Armament:	4 Gabriel Mk 2 SSM (4×1), 4–30mm (2×2)
Sensors:	Radar Decca RM-916
Complement:	25

Class (no, builder, launched):
Mamba (P 3100, Brooke Marine, Lowestoft, 6.11.73).

In service since 7 February 1974. Original 2–40mm/60 Mk 7 replaced in 1981 by Israeli guns and missiles plus EW gear. Refitted by Vosper Thornycroft in 1989–90. Extant 1995.

Simba as completed Vosper Thornycroft

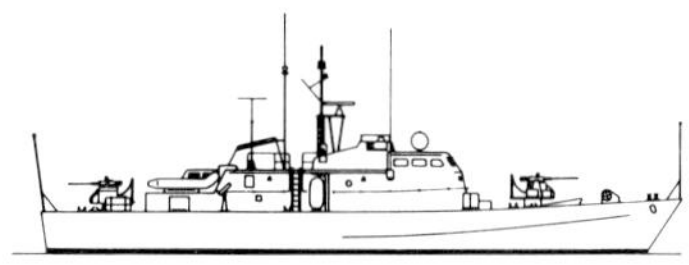

Harambee 1976 **NB 1/750 scale**

MADARAKA class *large patrol craft (missile)*

Displacement:	120t standard; 145t full load
Dimensions:	107ft oa × 20ft × 5ft 7in *32.6m × 6.1m × 1.7m*
Machinery:	2 shafts, 2 Paxman Valenta 16RP 200M diesels, 5400bhp = 25kts. Range 2300nm at 12kts
Armament:	4 Gabriel Mk 2 SSM (1×4), 2–40mm/60 Mk 7
Sensors:	Radar Decca RM-916
Complement:	21

Class (no, launched):
Madaraka (P 3121, 28.1.75), *Jamhuri* (P 3122, 14.3.75), *Harambee* (P 3123, 2.5.75)

Brooke Marine type ordered 10 May 1973 from Brooke Marine, Lowestoft. First two in service since 16 June 1975, *Harambee* since 28 June 1975. Israeli missiles fitted 1981–82. P 3121 refitted by Vosper Thornycroft 1989–90. All extant 1995.

Nyayo as completed

NB 1/750 scale

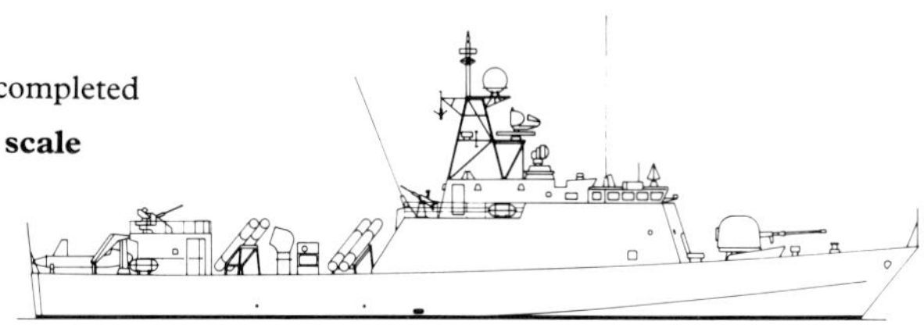

PROVINCE class *fast attack craft (missile)*

Armament:	4 Otomat Mk II SSM (2×2), 1–76mm OTO Melara, 2–30mm BMARC (1×2), 2–20mm (2×1)
Sensors:	Radar Plessey AWS-4, Decca AC 1226
Other particulars:	As Oman *Province* class

Class (no, launched):
Nyayo (P 3126, 20.8.88), *Umoja* (P 3127, 5.3.87)

Two ordered from Vosper Thornycroft in September 1984. Generally similar to craft built for Oman and Egypt. Both extant 1995.

Ex-British 'FORD' class *seaward defence boat*

Nyati (P 3102) Formerly HMS *Aberford*, RN loaned in 1964 and transferred permanently in 1967. Sold 1971.

ROYAL EAST AFRICAN NAVY 1950–64

Mention must be made of the Royal East African Navy, the colonial forerunner of the navies of independent Kenya and Tanzania.

At the end of the Second World War, the Kenyan RNVR operated the *Rosalind*, a 'Shakespearian' class minesweeping trawler loaned by the RN as a training ship at Mombasa, while the Zanzibari RNVR, which combined for certain purposes with the Kenyan, used the *Ndovu* and *Alhathari*, two tugs fitted for minesweeping (380t, 1–3in, 2 Lewis guns, 1 DCT, 10kts, built Paisley 1928). *Ndovu* was discarded in 1948 and *Alhathari* in 1951.

The East African Naval Force was formed in 1950 as successor to the Kenyan RNVR, with additional financial contributions from Tanganyika and (in later years) Uganda and Zanzibar. In 1953 the new force was designated the Royal East African Navy. Ships in service were *Rosalind* at Mombasa and the *MFV 206* at Kilindini (the *Mvita*, the base ship at Kilindini, seems to have been the MFV 204). In addition to training, the REAN performed usefully in famine relief operations. The inshore minesweeper *Bassingham* was acquired on loan in June 1958 to replace *Rosalind*, by then nearing the end of her useful life, but proved unsuitable and was returned to the RN in October 1961. No replacement had been transferred when the decision to disband the REAN was announced in December 1961. The government of newly independent Tanganyika considered the REAN unsuitable in its existing form and the other three governments were unable to meet the extra costs of modernising the force. *Rosalind* was discarded in 1963 and *Bassingham* was on the sales list in 1964. The fate of the MFV 204 *Mvita* has not been traced.

Kiribati (Gilberts)

This Central Pacific island group, formerly the British Gilberts, achieved independence as a Commonwealth republic in 1979. The 33 islands, half uninhabited, dot 2 million square miles of the SW

Central ocean. Tarawa, the capital, has 40 per cent of the population (56,452 in 1978) and its Marine Training School teaches seamen for employment with overseas shipping companies. Kiribati ordered a 17-metre GRP patrol craft of 22t (23.6kts, 1–7.62mm MG, 7 men) from Cheverton of Cowes (UK) in 1980 but it was not delivered, being bought by the Maldives instead.

In 1992 two Australian ASI 315 patrol boats (165t full load, 2lkts, 17 crew) were ordered. One, *Teanoai*, entered service on 22 January 1994 and is extant 1995. Two more may be ordered.

Korea

Allied troops entered Korea – Soviets from the North and Americans from the South – to take the Japanese surrender in 1945. Korea was divided into Soviet and US areas at the 38th Parallel. This resulted finally in the establishment of two Korean states – the Communist North and the anti-Communist South. Soviet troops were reportedly withdrawn in 1948 and US in 1949. South Korea was invaded from the North on 25 June 1950 and the resulting Korean War lasted more than three years. Most of the North's sixteen patrol craft were sunk or run aground within a week by US and British warships. More than a million people were killed, 2.5 million were left homeless and property damage was estimated at over a billion dollars. Hostilities did not end until 27 July 1953. South Korea was helped by troops and ships from fifteen UN countries, while North Korea got help from China and the Soviet Union. South Korea, originally the more agricultural and less developed part of the country, is officially called the Republic of Korea, and the 20 per cent larger but half as populous North became the Democratic People's Republic of Korea.

The mainstay of the North Korean Navy has long been a force of 'Romeo' and 'Whiskey' class patrol submarines, of which the latter at least are now nearing the end of their useful service lives. The surface forces comprise a nucleus of four locally built frigates of the 'Najin' class and fast missile craft supported by several hundred gun/ torpedo units of various types. Four of these made headlines on 23 January 1968 firing at and arresting the intelligence ship USS *Pueblo* in international waters. The ship was taken into Wonsan and eighty-two crew endured eleven months captivity before being released on North Korea's terms. These units are probably intended for coastal operations against more numerous South Korean large surface ships. Principal bases are Wonsan (east coast) and Nam (west coast) with fourteen minor harbours. Two 'Samlet' coast defence missile sites exist.

North Korea has acquired only one locally built experimental catamaran corvette/frigate, some 'Romeo' type submarines and numerous small craft in the course of the 1980s. Generally the North Korean Navy is of limited combat value, but the 'Romeo' type submarines remain a force to be reckoned with. Personnel strength is reported to be 41,500 officers and men plus about 40,000 reserves and the national service in the navy is five years.

The South Korean Navy had 7500 men by September 1949 and about thirty-four small ex-US and Japanese ships had been acquired before the Korean War began though many had not been delivered. Five frigates and landing ships were added during hostilities and in July 1954 took the disputed Takeshima islands. They were active afterwards seizing Japanese fishing boats infringing the 'Rhee Line' in a two-year dispute. The first two destroyers were received in 1963.

South Korea is, in terms of indigenous naval construction, one of the most advanced maritime nations in Asia. The merchant fleet was doubled in numbers and quadrupled in tonnage since 1974. Many warship designs have been produced, with an eye not only on home requirements, but also for increasing export in the region to Thailand, Indonesia, Taiwan, and the Philippines. Recent construction programmes range from *Ulsan* class frigates and large sophisticated missile-armed fast attack craft to small patrol boats. For its part the South Korean Navy seems to be emphasising large missile craft designed for defence against larger ships or small fast patrol boats. Eight *Paek Ku* type missile boats (called PSMM-5) have been constructed. The *Ulsan* class are very important as they were built to replace old, worn out ex-American destroyers as the chief surface units.

During the 1980s many more ships were built including twenty-six modern small frigate-sized *Po Hang* and *Dong Hae* class ships, numerous patrol boats and craft and a new class of minesweepers/ minehunters. Many more large patrol ships were built for the Coast Guard. South Korea is actively pursuing its policy of building its own warships and buying the equipment and armament for them abroad. There are ambitious plans for building in Korean shipyards 3500–4000t frigates. Submarines were ordered from Howaldtswerke Deutsche Werft (a variant of the popular Type 209).

Fleet HQ is at Chinhae. The six naval commands (under a rear-admiral or commodore) are based at Inchon, Mokpo (both west coast), Cheju island south of the peninsula, Mukho, Pohang and Pusan (both east coast). The navy flies over twenty US Tracker ASW aircraft from Kimhae near Pusan. Since October 1973 the Marine Corps has been under the direct command of the Vice Chief of Naval Operations.

South Korea

MAJOR SURFACE SHIPS

Ex-US FLETCHER class *destroyers*

No	Name	Builder	Acquired	Fate
911	CHUNG MU (ex-*Erben*)	Bath Iron Wks	1.5.63	Stationary TS 1989
912	SEOUL (ex-*Halsey Powell*)	Bethlehem, Staten I	27.4.63	Stricken 1982
913	PUSAN (ex-*Hickox*)	Federal, Kearny	15.11.68	Stricken Apr 1989

Former USN destroyers built in 1942–43 and bought on 31 January 1977. Original Korean pennant nos 91–93.

Ex-US GEARING (FRAM II) class *destroyers*

No	Name	Builder	Acquired	Fate
915	CHUNG BUK (ex-*Chevalier*)	Bath Iron Wks	5.7.72	Extant 1995
916	JEONG BUK (ex-*Everett F* Larson)	Bath Iron Wks	30.10.72	Extant 1995

Built in 1944–45 and converted in American service to radar picket destroyers (DDR) in 1949 and subsequently modernised under the FRAM II programme in the 1960s. Fitted with a small hangar and flight deck. Armament includes an Alouette III helicopter. *Chung Buk* received reportedly in 1976 a 20mm Vulcan-Phalanx CIWS which was fitted on the hangar roof. *Jeong Buk* was armed with a twin 30mm Emerlec gun mounting. Both fitted with eight Harpoon SSM (2×4). Both loaned in 1972 and bought on 31 January 1977. Originally had Korean pennant nos 95 and 96.

Ex-US ALLEN M SUMNER (FRAM II) class *destroyers*

No	Name	Builder	Acquired	Fate
917	DAE GU (ex-*WallaceLLind*)	Federal, Kearny	4.12.73	Extant 1995
918	IN CHEON (ex-*De Haven*)	Bath Iron Wks	5.12.73	Stricken 1993

Fitted with small flight deck and hangar intended for DASH drone but now are to be fitted with LAMPS (with light helicopter). Transferred in 1973 and sold in January 1977. Fitted with eight Harpoon SSM (2 × 4).

Ex-US GEARING (FRAM I) class *destroyers*

No	Name	Builder	Acquired	Fate
922	KANG WON (ex-*William R Rush*)	Federal, Kearny	1974	Extant 1995
921	KWANG JU (ex-*Richard E Kraus*)	Bath Iron Wks	23.2.77	Extant 1995
919	TAEJON (ex-*New*)	Consolidated, Orange	23.2.77	Extant 1995
925	JEONG JU (ex-*Rogers*)	Consolidated, Orange	11.8.81	Extant 1995
923	KYONG KAI (ex-*Newman M Perry*)	Consolidated, Orange	27.7.81	Extant 1995

Former US *Gearing* FRAM I class destroyers built in 1944–46. All modernised in 1960s under FRAM I programme. Fitted with enlarged helicopter hangar and landing deck for a small Alouette III helicopter. 919, 921, 922 fitted with eight Harpoon SSM (2×4). *Kang Won* was loaned initially and bought on 1 July 1978.

Ex-US TACOMA class *patrol frigates*

No	Name	Builder	Acquired	Fate
61	DU MAN (ex-*Muskogee*)	Consolidated Steel, San Pedro	5.11.50	Stricken *c*1974
62	APNOK (ex-*Rockford*)	Consolidated Steel, San Pedro	5.11.50	Sunk as target 1963
63	TAE TONG (ex-*Tacoma*)	Kaiser Richmond	8.10.51	Stricken *c*1974
65	NAK TONG (ex-*Hoquiam*)	Kaiser, Richmond	8.10.51	Stricken *c*1974
66	IM CHIN (ex-*Sausalito*)	Kaiser, Richmond	Sept 1952	Stricken *c*1974

Former USN frigates modelled on the British 'River' class. Built in 1943–44. These ships were amongst the twenty-eight ships of this class transferred to USSR during summer 1945. Survivors were returned to US Navy in Japan in 1949. First large ships loaned to Korean Navy. Two more, ex-*Pasco* (PF 6) and *Gloucester* (PF 22) were towed from storage in Japan to Korea for spare part cannibalisation in January 1969. *Apnok* was damaged in collision on 21 May 1952; returned to USA and sunk as target in 1963. Replaced by *Im Chin* transferred in September 1952. Note that pennant no 64 was not assigned as Koreans consider 'four' as an unlucky number. Second pair transferred at Yokosuka.

Ex-US CANNON ('DET') class *frigates*

No	Name	Builder	Acquired	Fate
71	KYONG KI (ex-*Muir*)	Tampa SB	Feb 1956	Stricken 28.12.77
72	KANG WON (ex-*Sutton*)	Tampa SB	Feb 1956	Stricken 28.10.77

Former US destroyer escorts of Cannon ('DET') class. Built in 1944 and transferred at Boston to Korean Navy. Named after Korean states. Taken over by the Philippines.

Ex-US APD type *frigates/high-speed transports*

No	Name	Builder	Acquired	Fate
822	KYONG NAM (ex-*Cavallaro*)	Defoe, Bay City	Oct 1959	Stricken 1984
823	AH SAN (ex-*Harry L Corl*)	Bethlehem, Hingham	Jun 1966	Stricken 1984
825	UNG PO (ex-*Julius A Raven*)	Bethlehem, Hingham	Jun 1966	Stricken 1984
826	KYONG PUK (ex-*Kephart*)	Charleston N Yd	Aug 1967	Stricken 30.4.85
827	JONNAM (ex-*Hayter*)	Charleston N Yd	Aug 1967	Stricken 1986
828	CHI JU (ex-*William M* Hobby)	Charleston N Yd	Aug 1967	Stricken 1989

Former US high-speed transports (APD) begun as destroyer escorts (DE) and converted during construction. Built in 1943–45. In Korean service four latter ships were originally rated as gunboats (PG 83, 85–87); changed in 1972 to APD. First pair originally APD 81 and 82. Fitted for about 160 troops and can carry four LCVPs. These ships are in two different configurations with high and low bridges. Fitted with tripod masts. All finally bought outright 15 November 1974. All stricken in 1980s.

Ex-US RUDDENROW ('TEV') class *frigate*

No	Name	Builder	Acquired	Fate
827	CHUNG NAM (ex-*Holt*)	Defoe, Bay City	19.6.63	Stricken 1984

Built in 1943–44 and bought outright on 15 November 1974. Stricken 31 January 1984.

Keong Buk Dec 1993 Courtesy Ross Gillett

Ulsan 1983

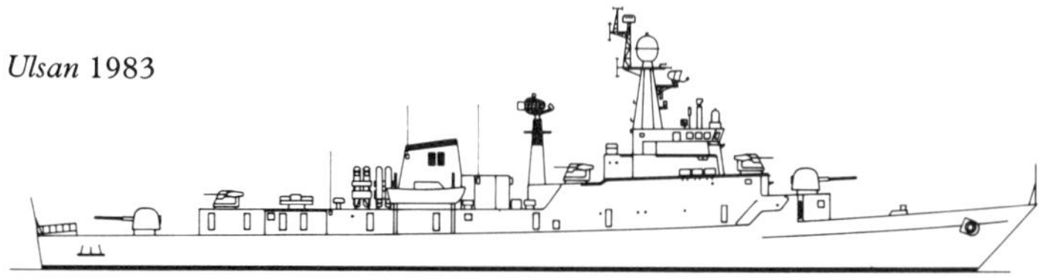

ULSAN class *frigates*

Displacement: 1495t light; 1600t standard; 2180t full load (2300t for 957, 958, 959, 961)
Dimensions: 344ft 6in oa, 321ft 6in pp × 39ft 4in × 11ft 6in
105.0m, 98.0m × 12.0m × 3.5m
Machinery: 2-shaft CODOG (CP propellers): 2 GE LM 2500 gas turbines, 54,400shp=35kts, 2 MTU 12V956 TB82 diesels, 7200bhp=20kts. Range 900nm/34kts, 400nm/18kts
Armament: 8 Harpoon SSM (2×4), 2–76mm/62 OTO Melara (2×1), 8–30mm/85 Emerlec (4×2) or 6 Breda 40mm/70 (3×2 in 956–961), 6–324mm Mk 32 ASW TT (2×3), 2 DC racks
Sensors: Radar HSA DA-05, HSA ZW-06 (Marconi S 1810 in 957–961), HSA WM-28 (Marconi ST 1802 in 957–961), Raytheon SPS 10C; sonar HSA PHS-32
Complement: 25 officers, 120 ratings

No	Name	Builder	Launched	Fate
951	ULSAN	Hyundai, Ulsan	8.4.80	Extant 1995
952	SEOUL	Korea SB, Pusan	24.4.84	Extant 1995
953	CHUNG NAM	Korea SB, Pusan	1985	Extant 1995
955	MASAN	Korea Tacoma, Masan	26.10.84	Extant 1995
956	KEONG BUK	Daewoo, Okpo	1985	Extant 1995
957	CHON NAM	Hyundai, Ulsan	19.4.88	Extant 1995
958	CHE JU	Daewoo, Okpo	1988	Extant 1995
959	BUSAN (or *Pusan*)	Hyundai, Ulsan	20.2 92	Extant 1995
961	CHUNG JU	Daewoo, Okpo	20.3.92	Extant 1995

Korean-designed, well-armed, fast general purpose frigates built locally at various shipyards. After the first three the next three units ordered in 1981 and the last in 1984. Steel hull with alloy superstructure. These ships lack any helicopter facilities. Netherlands and British electronic equipment including the two HSA LIOR optronic directors. Combat data system in 957 and 958: Samsung/Ferranti WSA 423. All equipment and armament bought off-the-shelf from European and American manufacturers. Beginning from 956 Emerlec gun mountings replaced by three 40mm Breda twins. There are some variations in the basic equipment and it differs slightly on some ships. All have SLQ-25 Nixie torpedo decoys. *Ulsan* previously listed as *Ulsan Ham*, while the class is sometimes called type HDF-2000. Plans exist for building of an additional series of up to eight units of the class to replace old ex-American destroyers of Second World War vintage. Big offshore patrol ships designed on the hull of *Ulsan* class but with different engines and armament were built for India.

New *frigates*

Displacement: 3900t full load
Dimensions: 444ft 3in × 46ft 7in × 13ft 9in
135.4m × 14.2m × 4.2m
Machinery: 2 shafts CODOG, 2 GE LM 2500 gas turbines, 58,200shp = 30kts, 2 MTU or SEMT Pielstick diesels, *c* 8000bhp. Range 4000nm at 18kts
Armament: 8 Harpoon SSM (2×4), 1 Sea Sparrow SAM VLS Mk 48 (16 missiles) (later ships 1 Standard SAM VLS Mk 41), 1–127mm/54 OTO Melara, 2–30mm Goalkeeper CIWS, 6–324mm ASW TT (2×3), 1 helicopter
Sensors: Radar Plessey AWS-6A, SPS 49(V)5; fire control Contraves TMX-KA; ECM/ESM Argo AR 700/APECS, 4 chaff launchers, 1 SLQ-25 Nixie torpedo decoy; combat data system Atlas/Contraves COSYS 200
Complement: 170 (15 officers)

This is the much delayed programme planned for many years. The first keel was to be laid down by Daewoo, Okpu, in mid-1992 for completion in 1996. It had been planned to lay down a new ship each year from 1993. These will be fully-fledged frigates as used by the Western navies. Well armed and equipped with a helicopter (Westland Super Lynx) these ships will be the most powerful surface units in the Korean Navy. The final electronic suit will be chosen during the construction of the prototype. This new class is sometimes called HDF 3500 or KDX 2000 or DW 4000.

SUBMARINES

CHANG BOGO class (Type 209/1200) *submarines*

Displacement: 1100t surfaced; 1290t submerged
Dimensions: 184ft 1in × 20ft 4in × 18ft 1in
56.1m × 6.2m × 5.5m
Machinery: 1 shaft diesel-electric, 4 MTU 12V 396 SE diesels plus 1 electric motor, 3800bhp/4600shp = 11kts/21.5kts. Range 7880nm at 8kts surfaced
Armament: 8–533mm TT (bow, 14 torpedoes or 28 mines)
Sensors: Radar Atlas Elektronik ISUS 83TFCS; sonar Atlas Elektronik CSU 83 (active/passive)
Complement: 69 (11 officers)

No	Name	Builder	Launched	In service	Fate
061	CHANG BOGO	HDW, Kiel	Jun 1992	2.6.93	Extant 1995
062	YI CHON	Daewoo	14.10.92	1994	Extant 1995
063	CHOI MUSON	Daewoo	Sep 1993	1994	Extant 1995
065	–	Daewoo	1994	1995	Completing
066	–	Daewoo	1995	1996	Building
067	–	Daewoo	1996	1997	Building

First three boats ordered in February 1988 and the next batch of three in October 1989. A further three boats, planned to be ordered in 1993, have been delayed. The programme called for a total of eighteen submarines, to be split between the three fleets of the South Korean navy, but it now seems unlikely that funding will be available for such a large class of vessels. Prototype built in Germany but later units being built at Daewoo's Opko yard with far greater Korean input. Maximum diving depth is 1050ft (320m). Boats fitted with two 240-cell, 11,500Amp/hr batteries weighing a total of 257t. Armed with German System Technik Nord (STN) SUT Mod 2 wire-guided torpedoes, or can alternatively carry up to 28 mines. These are the first fully-fledged submarines in the South Korean Navy, in contrast to the North which has maintained a submarine force since the 1960s (see under North Korea).

KSS–1 'TOLGORAE' class *small submarines*

Displacement: 150t/175t
Dimensions: ?
Machinery: 1 or 2 diesel generator sets
Armament: 2–406mm TT (or 533mm TT?)
Complement: ?

Three small submarines of the 'Tolgorae' class were built by Hyundai, Ulsan, launched 1983–87. The last boat was delivered in 1988. They are fitted with Krupp-Atlas passive sonar. There are probably also in service up to eight Italian-built 'Cosmos' midget submarines of 70t displacement.

AMPHIBIOUS WARFARE VESSELS

Ex-US LST type *tank landing ships*

Class (no, former names, transfer):
Ex-LST 120, ex-LST 213, *Dan Yang* (ex-LST 343), ex-LST 378, ex-LST 380, *Ryong Pi* (ex-LST 388), *An Tong* (ex LST 491), ex-LST 536, ex-LST 594, *Chon Po* (ex-LST 595), *Ulsan* (ex-LST 608), ex-LST 624, *Ryong Hwa* (ex-LST 659), *Lyung Wha* (ex-LST 805), *Un Bong* (LST 671, ex-LST 1010, Feb 1955), *Duk Bong* (LST 672, ex-LST 227, Mar 1955), *Bi Bong* (LST 673, ex-LST 218, May 1955), *Kae Bong* (LST 675, ex-*Berkshire County* LST 288, March 1956), *Wee Bong* (LST 676, ex-*Johnson County* LST 849, Jan 1959), *Su Yong* (LST 677, ex-*Kane County* LST 853, Dec 1958), *Buk Han* (LST 678, ex-*Lynn County* LST 900, Dec 1958), *Hwa San* (LST 679, ex-*Pendar County*, LST 1080, Oct 1958).

All twenty-two transferred in 1950s. *An Tong* was wrecked and scrapped on 24 January 1952. All ships except the last eight stricken by the end of 1959. Some never got Korean names. Some were probably renamed while in Korean service: *Ryong Hwa* (ex-*Chong Ho*), *Chon Po* (ex-*Chochiwan*), *Ryong Pi* (ex-*Samlangjin*). There was also *Jung Su* (ex-*APL 59*, ex-*LST 53*), a barrack ship transferred in 1955 and since stricken. *Duk Bong* stricken 1989 after grounding.

ALLIGATOR class *tank landing ships*

Displacement: 4200t full load
Dimensions: 350ft 9in × 50ft 2in × 9ft 10in
106.9m × 15.3m × 3.0m
Machinery: 2 shafts, 2 SEMT Pielstick 16PA6 V280 diesels, 12,800bhp = 16kts. Range 10,000nm at 12kts
Armament: 2–40mm/70 Breda (2×1), 2–20mm Oerlikon (2×1)
Complement: 120

Class (in service):
I (1994), II (building 1995)

Tank landing ships of the standard HDL-400 type ordered in June 1990 from Korea Tacoma, Masan. First of the class began trials in 1993. Very similar ships were built for Venezuela and Indonesia. Design developed from US LST 1 and LST 511 classes of the Second World War. New ships of improved design are characterised by substantially higher speed.

Ex-US LSM type *medium landing ships*

Class (no, former no):
Tae Cho (LSM 601, ex-LSM 546), *Tyo To* (LSM 602, ex-LSM 268), *Tok To* (LSM 603, ex-LSM 419), *Ka Tok* (LSM 605, ex-LSM 462), *Ko Mun* (LSM 606, ex-LSM 30), *Pian* (LSM 607, ex-LSM 96), *Pung To* (LSM 608, ex-LSM 54), *Wol Mi* (LSM 609, ex-LSM 57), *Ki Rin* (LSM 611, ex-LSM 19), *Nung Ra* (LSM 612, ex-LSM 84), *Sin Mi* (ex-LSM 316), *Ul Rung* (ex-LSM 17).

Twelve former USN ships, built in 1944–45. First four transferred in 1955 the rest in 1956. *Sin Mi* and *Ul Rung* had served as French L 9014 and L 9017 in Indo-China 1954–55. All finally bought 15 November 1974. *Tok To* was scrapped in 1963, *Pung To* serves as mine force flagship, fitted with minelaying rails and designated LSML, others LSM. The number LSM 604 was not assigned. *Toe Cho, Tyo To, Ka Tok* stricken 1982, *Pung To* 1984. Seven extant 1995. Pennant nos. changed to LSM 652–661 (with nos. 654 and 660 unassigned).

Ex-US LSSL type *large support landing ships*

Class (no, former no):
Chung Jin (LSSL 105, ex-LSSL 1056), *Yung Huang Man* (LSSL 107, ex-LSSL 77), *Kang Hwa Man* (LSSL 108, ex-LSSL 91), *Po Song Man* (LSSL 109, ex-LSSL 54), *Yong 11 Man* (LSSL 110, ex-LSSL 84).

Chung Jin was a former US infantry landing ship (LSIL, ex-LCI/L). The rest were employed as gunboats. Transferred in 1950s; *Chung Jin* and *Yong Il Man* were stricken in 1962; *Yung Huang Man* and *Kang Hwa Man* were decommissioned in 1960 and *Po Song Man* in 1962.

Ex-US LSMR type *fire support landing ship*

Class:
Si Hung (LSMR 311, ex-*St Joseph River* LSMR 527)

Transferred on 15 September 1960. Renumbered LSMR 650. Stricken 1984.

US LCU 501 type *utility landing craft*

Class:
Mulkae 71 (LCU 1, ex-LCU 531)

First unit built in 1943 as LCT(6) 631 and transferred to Korea in 1960. Renamed LCU 1 in Korean service. Stricken 1987.

US LCU 1610 type *utility landing craft*

Class:
Mulkae 72, Mulkae 73, Mulkae 75, Mulkae 76, Mulkae 77, Mulkae 78.

Six copies of this design were built in S Korea and brought into service 1979–81. Extant 1995.

Ex-US LCM (8) class *landing craft*

Ten ex-US Army units transferred September 1978. Extant 1995.

SMALL SURFACE COMBATANTS

Ex-US AUK class *corvettes (former ocean minesweepers)*

Class (no, former name):
Shin Song (PCE 711, ex-*Ptarmigan* MSF 376), *Sunchon* (PCE 712, ex-*Speed* MSF 116), *Koje* (PCE 713, ex-*Dextrous* MSF 341)

Former USN fleet minesweepers built in 1941–44. Originally designated AM in USN. Original Korean numbers PCE 1001–1003. The minesweeping gear was removed before the transfer and a second 76mm gun was fitted aft, ASW was augmented. *Shin Song* was transferred in July 1973, *Suchon* in November 1967 and *Koje* in December 1967. All stricken 1984.

PO HANG class *fast corvettes*

Displacement: 950t; 1180–1300t full load
Dimensions: 288ft 9in oa, 273ft 11in wl × 32ft 10in, 32ft 2in wl × 9ft 6in
88.0m, 83.5m × 10.0m, 9.8m × 2.9m
Machinery: 2-shaft CODOG, 2 CP propellers, 1 GE LM 2500 gas turbine, 27,200shp=31kts, 2 MTU 12V956 TB82 diesels, 6200bhp (according to some sources these are SEMT Pielstick 12PA6V280 diesels, total 4800bhp). Range 4000nm/15kts, 800nm/31kts
Armament: 2 MM 38 Exocet SSM (756–765), 1–76mm/62 OTO Melara (2–76mm/62 OTO Melara from 766 onwards), 4–30mm/85 Emerlec (2×2, in 756–765), 4–40mm/70 Breda (2×2, from 766 onwards), 6–324mm ASW Mk 32 TT (from 766 onwards), DC racks
Sensors: Radar SPS-64(V) or Marconi 1810, HSA WM-28 or Marconi S-1802; sonar PHS-32 (from 766 onwards)
Complement: 95 (10 officers)

No	Name	Builder	Commissioned	Fate
756	PO HANG	Korea SB, Pusan	18.12.84	Extant 1995
757	KUN SAN	Korea Tacoma, Masan	18.12.84	Extant 1995
758	KYONG JU	Hyundai, Ulsan	1986	Extant 1995
759	MOK PO	Daewoo, Okpo	1986	Extant 1995
761	KIM CHON	Korea SB, Pusan	1987	Extant 1995
762	CHUNG JU	Korea Tacoma, Masan	1987	Extant 1995
763	JIN JU	Hyundai, Ulsan	1988	Extant 1995
765	YO SU	Daewoo, Okpo	1988	Extant 1995
766	AN DONG	Korea SB, Pusan	1989	Extant 1995
767	SUN CHON	Korea Tacoma, Masan	Jun 1989	Extant 1995
768	IRI	Hyundai, Ulsan	Jun 1989	Extant 1995
769	WON JU	Daewoo, Okpo	1989	Extant 1995
771	JE CHON	Korea Tacoma, Masan	1989	Extant 1995
772	CHEON AN	Korea SB, Pusan	1989	Extant 1995
773	SONG NAM	Daewoo, Okpo	1990	Extant 1995
775	BU CHEON	Hyundai, Ulsan	1990	Extant 1995
776	DAE CHON	Korea Tacoma, Masan	1990	Extant 1995
777	SIN HAE	Korea SB, Pusan	1990	Extant 1995
778	SOK CHO	Daewoo, Okpo	1991	Extant 1995
779	YONG YU	Hyundai, Ulsan	1991	Extant 1995
781	NAM WON	Korea Tacoma, Masan	1991	Extant 1995
782	KWAN MYONG	Korea SB, Pusan	1991	Extant 1995
783	–	Korea Tacoma, Masan	1992	Extant 1995

Small fast general purpose corvettes designed in Korea and fitted with off-the-shelf Western equipment and armament. First laid down in 1983 at Korea SB in Pusan. Have broad superstructure amidships. Besides the twenty-two named units listed above, two more were to be completed in 1992. Class was sometimes called KCX. First eight units have Exocet missiles. Beginning from *An Dong* (766) the ships are fitted with two twin 40mm Breda gun mountings instead of Emerlec, have no SSM, mount the Marconi S1810 surveillance radar at the masthead and have two Rademac optronic director for 40mm guns. Some of the later units in the series (completed in 1988 or later) are said to be fitted with all-diesel propulsion plant with four MTU engines (16,000bhp) for much lower speed. First eight ships (756–765) are equipped with HSA Sewaco ZK combat data system while the rest received the Ferranti WSA 423 system. All fitted with HSA Lirod or Radamec 2400 optronic director.

DONG HAE (HDC 800) type *fast corvettes*

Displacement: 800t normal; 1076t full load
Dimensions: 256ft 3in oa × 31ft 6in × 8ft 6in
78.1m × 9.6m × 2.6m
Machinery: 2-shaft CODOG, 2 CP propellers, 1 GE LM 2500 gas turbine, 27,800shp=31kts, 2 MTU 12V956 TB82 diesels, 6200bhp. Range 4000nm at 15 kts
Armament: 1–76mm/62 OTO Melara, 1–40mm Bofors, 4–30mm/85 Emerlec (2×2), 6–34 Mk 32 ASW TT, DC racks
Sensors: Radar SPS-64(V), HSA WM-28; sonar HSA PHS-32 or Edo 768
Complement: 95 (10 officers)

No	Name	Builder	In service	Fate
751	DONG HAE	Korea SB, Pusan	1982	Extant 1995
752	SU WON	Korea Tacoma, Masan	1983	Extant 1995
753	KANG REUNG	Hyundai, Ulsan	1983	Extant 1995
755	AN YANG	Daewoo, Okpo	1983	Extant 1995

This is a reduced version of the *Dong Hae* class. Four had been ordered in 1980, each from a different major shipyard. Have lattice mast. Ships proved too small for their designed duties and only four were built commissioning in 1982–83. At the aft end of the superstructure there was place for two SSM launchers which is now occupied by twin 40mm Bofors gun mounting of Second World War vintage.

HAN KANG class *patrol corvette* (Type HDC 1150)

Displacement: 1000t normal; 1200t full load
Dimensions: 288ft oa × 32ft 10in × 9ft 6in
87.8m × 10.0m × 2.88m
Machinery: 2-shaft CODOG, 1 GE LM 2500 gas turbine, 27,800shp=31kts, 2 MTU 12V956 TB82 diesels, 6200bhp. Range 4000nm at 15kts
Armament: 1–76mm/62 OTO Melara, 1–40mm Mk 3, 2–20mm Vulcan (2×1)
Sensors: Radar Raytheon SPS-64(V), HSA WM-28
Complement: 72

No	Name	Builder	In service	Fate
1005	HAN KANG	Korea SB, Pusan	Dec 1985	Extant 1995

This class uses the same hull as the *Po Hang* class fast corvettes. These ships are used by the Coast Guard but in the event of war will be taken over by the navy. Their armament is similar, but the ASW TT were not fitted. Have optronic HSA LIOD director as a backup for WM-28.

'MAZINGER' class *patrol ships* (Type HDP 1000)

Displacement: 1200t standard; 1450t full load
Dimensions: 264ft 1in × 32ft 2in × 13ft 9in
80.5m × 9.8m × 3.2m
Machinery: 2 shafts, 2 SEMT-Pielstick 12PA6 V280 diesels 9600bhp = 21.5kts. Range 700nm at 18kts
Armament: 1–40mm Bofors, 2–20mm Oerlikon (2×1)
Complement: 69 (11 officers)

Class (in service):
1001 (29.11.81), 1002 (1982), 1003 (1982)

Three 'Mazinger' class patrol ships ordered on 7 November 1980 from Korea Tacoma and Hyundai. Steel construction. Ship numbered 1001 used as command vessel. Used by Coast Guard for offshore surveillance and general patrol duties. Only three built.

Coast Guard also has at its disposal six 'Sea Whale' class patrol ships (*501–503, 505–507*, in service 1979–82, 580t full load, 24kts), twenty-two 'Sea Shark' class patrol boats (280t full load, 28kts) and eighteen 'Sea Gull' class (80t, 30kts), and two new 'Bukhansan' class (371t full load, 28kts, modified 'Sea Wolf' class) *Bukhansan* (278), *Chulmasan* (279).

Two large patrol craft No 402, 403 (430t full load, 53.7m, 2 diesels = 19kts, 1–20mm Vulcan Gatling, 4–12.7mm MG) built by Hyundai. Delivered 1990, 1991 respectively. Both extant 1995.

Ex-US PCE (185ft) type *patrol vessels*

Class:
Ko Jin (ex-*Report* MSF 289), *Ro Ryang* (ex-PCEC 882), *Myong Ryang* (ex-PCEC 896), *Han San* (ex-PCEC 873), *Ok Po* (ex-PCEC 898), *Tang Po* (ex-*Maria* PCE 842), *Pyok Pa* (ex-*Dania* PCE 870), *Ryul Po* (ex-*Somerset* PCE 892), *Sa Chon* (ex-*Batesburg* PCE 903); No PCEC 50–53, 55, PCE 56–59.

Nine USN patrol craft based on the *Admirable* class minesweeper hull built in 1943–44. Four were modified in US service as control ships (PCEC) for operation with landing craft, being fitted with additional radio equipment in an enlarged bridge area. Transfers: *Ro Ryang* and *Myong Ryang* in February 1955, *Han San* and *Ok Po* in September 1955 and the rest in 1961. *Tang Po* was sunk by N Korean coastal guns on 19 January 1967. *Han San, Ok Po, Ro Ryang, Myong Ryang* stricken 1977 and the rest in 1979.

Ex-US PC (173ft) type *submarine chasers*

Class:
Pak Tu San (ex-*Ensign Whitehead*, ex-PC 823), *Kum Kang San* (ex-PC 810), *Sam Kak San* (ex-PC 802), *Chiri San* (ex-PC?), *Han La San* (ex-PC 485), Myo Hyang San (ex-PC 600), *O Tae San* (ex-*Winnemussa* PC 1145), *Kum Chong San* (ex-*Grosse Point* PC 1546), *Sol Ak* (ex-*Chadron* PC 546)

Nine USN units built in the Second World War, *Pak Tu San*, the Navy's first ship, was purchased in September 1949 for $18,000 and delivered in early 1950; next four (three at Honolulu when war began) transferred during Korean War; *Chiri San* was mined and sunk off Wonsan, Korea on 26 December 1951; *Kum Chong San* and *O Tae San* transferred in 1960; *Sol Ak* transferred in 1964. Korean pennant numbers PC 701–709. *Han La San* was sunk in a typhoon at Guam in November 1962 and although raised was scrapped in 1964. *Kum Kang San* and *Sam Kak San* were decommissioned in 1960 and scrapped; *Myo Hyang San* scrapped in 1968; *O Tae San, Sol Ak* and *Kum Chong San* stricken in 1970s.

Ex-US PT type *fast attack craft (torpedo)*

Class (no, former no):
Kalmaeki (PT 23, ex-PT 616), *Kiroki* (PT 25, ex-PT 619), *Olpamei* (PT 26, ex-PT 613), *Jebi* (PT 27, ex-PT 620)

Former US MTB of plywood hull construction built by Electric Boat Co in 1945 and transferred in January 1952. *Olpamei* destroyed in September 1952. The rest deleted in 1969. Former PT 616 was returned to USA for use as a memorial.

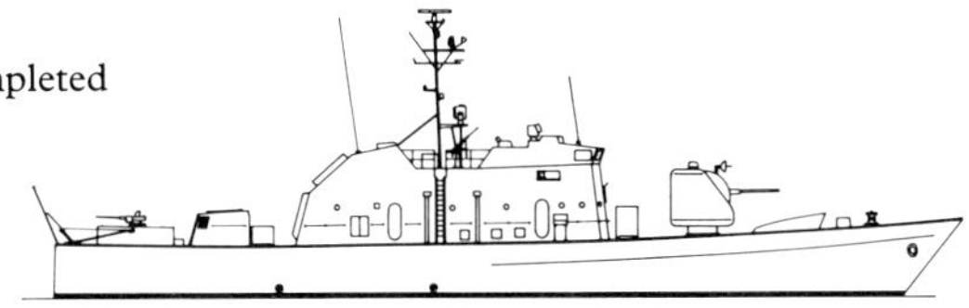

Paek Ku 12 as completed
NB 1/750 scale

PAEK KU (PSMM type) class *fast attack craft (missile)*

Displacement:	240t standard; 268t full load
Dimensions:	165ft pp, 176ft 2in oa × 24ft × 9ft 6in *50.3m, 53.7m × 7.3m × 2.9m*
Machinery:	COGAG, 2 shafts, (CP propellers): 6 Avco Lycoming TF 35 gas turbines, 16,800shp = 40 + kts. Range 2400nm at 18kts
Armament:	2 Standard SSM (1 reload each), 1–76mm/50 Mk 34, 1–40mm, 2–12.7mm MG. See notes
Sensors:	LN-66HP, SPG-50. See notes
Complement:	32

Class (no, former name, completed):
Paek Ku 52 (PGM 352, ex-*Paek Ku 12,* 14.3.75), *Paek Ku 53* (PGM 353, ex-*Paek Ku 13,* 14.3.75), *Paek Ku 55* (PGM 355, ex-*Paek Ku 15,* 1.2.76), *Paek Ku 56* (PGM 356, ex-*Paek Ku 16,* 1.2.76), *Paek Ku 57* (PGM 357, ex-*Paek Ku 17,* 1.2.76), *Paek Ku 58* PGM 358, ex-*Paek Ku 18,* 1977), *Paek Ku 59* (PGM 359, ex-*Paek Ku 19,* 1977), *Paek Ku 61* (PGM 361, 1978)

Based on the US *Asheville* design and built for Korean Navy. Considered as multi-mission patrol ships (PSMM), by their builders. First three Tacoma, USA, the other five at Tacoma, S Korea. Aluminium hulls. Korean name *Paek Ku* means 'seagull'. The six TF 35 gas turbines turn two propeller shafts and one or two or three turbines can be selected to provide each shaft with variety of power settings. Korean-built units (*Paek Ku 56 – Paek Ku 61*) are armed with one automatic 76mm/62 OTO Melara gun forward, one twin 30mm/85 Emerlec gun mounting and two 12.7mm machine guns as well as with two twin launchers for Harpoon SSM. Fitted with Westinghouse M-1200 fire control system, using inputs from LN-66HP radar and optical director, while American-built units use the US Mk 63 fire control system with SPG-50 radar. Prototype launched on 17 February 1975 and others in 1975–77. All eight extant 1995.

Ex-US ASHEVILLE class *fast attack craft (missile)*

US FAC (gun) ex-*Benicia* (PG 96), fitted in Korean service with Standard SSM. Built in 1969–70. Transferred to Korea on loan 15 October 1971, became *Paek Ku 2.* Now *Paek Ku 5* (PGM 351, formerly 11 and 101). Arrived S Korea January 1972. No ASW weapons fitted. Missiles fitted in 1975–76. Extant 1995.

KILURKI class *fast attack craft (gun)*

Particulars: As US CPIC type

Class:
Kilurki 11 (ex-*Gireogi*)

Built in USA and designated CPIC (Coastal Patrol and Interdiction Craft). *Gireogi* conducted extensive trials in US Navy. She was launched in 1973 by Tacoma Boatbuilding Co, Tacoma, Wash.; other five to be constructed in Korea by Korea Tacoma International. Of the planned next five boats none were built. *Kilurki II* returned to USA.

Ex-US PCS (136ft) type *large patrol craft*

Class:
Su Song (ex-*PCS 1426), Kum Seong* (ex-*PCS 1445), Mok Seong* (ex-*PCS 1446*), *Hwa Seong* (ex-*PCS 1448*), *Kae Seong* (ex-Korean AMS 504), *Kilchu* (ex-Korean AMS 514), *Yong Kung* (ex-Korean AMS 518)

Seven former USN submarine chasers of wooden construction built on YMS type hull in 1941–44. Acquired from USN in 1952. *Su Song* returned to USA in 1963, *Kae Seong, Kil Chu* decommissioned in 1955–56; the rest were returned to US custody and scrapped in 1970s.

Ex-US 95ft type *coast guard patrol craft*

Class (former name, no):
PB 3 (ex-*Cape Rosier,* WPB 95333), *PB 5* (*Cape Sable,* WPB 95334), *PB 6* (*Cape Providence,* WPB 95335), *PB 7* (*Cape Florida,*WPB 95325), *PB 8* (*Cape Porpoise,* WPB 95327), *PB 9* (*Cape Falcon,* WPB 95330), *PB 10* (*Cape Trinity,* WPB 95331), *PB 11* (*Cape Darby,* WPB 95323), *PB 12* (*Cape Kiwanda,* WPB 95329)

Nine former USCG steel-hulled patrol craft transferred to Korean Navy in September 1968. *PB 7* was stricken after grounding in May 1971. All others stricken 1984.

Kilurki class
NB 1/750 scale

KILURKI type *large patrol craft*

Displacement:	120t; 140t full load
Dimensions:	108ft × 26ft 3in × 3ft 7in *32.9m × 8.0m × 1.1m*
Machinery:	3 shafts, 3 MTU diesels, 10,200hp = 35 kts. Range 1000nm at 20kts
Armament:	1–40mm, 2–12.7mm, 2 Exocet SSM
Complement:	26

Class:
Kilurki 71 (PKM 271), *Kilurki 72* (PKH 272)

Built by Korea Tacoma, Masan in 1971–72. Extant 1995

'SEA DOLPHIN' class *patrol boats*

Displacement:	113t normal; 144t full load
Dimensions:	108ft 7in × 22ft 8in × 5ft 9in *33.1m × 6.9m × 1.75m*
Machinery:	2 shafts, 2 MTU 16V538 TB90 diesels, 6,000bhp=34kts. Range 1000nm at 20kts
Armament:	1–40mm, 2–30mm/85 Emerlec (1×2), 2–20mm (2×1), 2–12.7mm or 2–30mm/85 Emerlec (1×2), 2–20mm Gatling Vulcan
Complement:	29

Class:
Kilurki (or PKM) *211–213, 215–219, 221–223, 225–229, 231–233, 235–239, 251–253, 255–259*

Thirty-two 'Sea Dolphin' class patrol boats designed for 38kts; can make 32kts sustained. Built in the late 1970s at Korea Tacoma, Chinhae. According to some reports it is the PKM pennant numbers which run in series while individual names are *Kilurki 11, Kilurki 12,* etc. All extant 1995.

CHEBI class *coastal patrol boats*

Displacement:	70t normal; 78t full load
Dimensions:	84ft 4in × 17ft 9in × 3ft 11in *25.7m × 5.4m × 1.2m*
Machinery:	2 shafts, 2 MTU 16V538TD90 diesels, 5200bhp=40kts. Range 500nm at 20kts
Armament:	1–40mm, 1–20mm, 4–12.7mm (2×2)
Complement:	25

Class (no):
Chebi 51 to *Chebi 89* (PK 151 to PK 189)

Thirty-nine small fast patrol boats of the *Chebi* (Schoolboy) class built by Korea SB, Pusan, in the late 1970s and 1980s. Armament varies. Sometimes classed as 'Sea Hawk' or 'Sea Fox' according to the commercial type names given by the building yards. Six transferred to the Phillipines in 1993, and a further six in 1994. Others extant 1995.

Ex-US 40ft Sewart type *coastal patrol craft*

Class:
SB 1–3, 5

Aluminium-hulled boats built in USA by Stewart and transferred to Korean Navy in 1964. No 4 was not assigned. Stricken 1982.

Ex-US 65ft Sewart type *coastal patrol craft*

Class:
FB 1–3, 5–10

Built in USA by Stewart and transferred to Korean Navy in 1967. Referred to as *Toksuuri No 1* etc by Koreans. No 4 being considered unlucky was not assigned. All stricken 1982.

MINE WARFARE VESSELS

Ex-US YMS type *coastal minesweepers*

Class:
Kang Jim (ex-*YMS 354*), *Kyoung Chu* (ex-*YMS 358*), *Kwang Chu* (ex-*YMS 413*), *Kae Song* (ex-*YMS ?*), *Kimhae* (ex-*YMS ?*), *Kangkae* (ex-*YMS ?*), *Kang Nung* (ex-*YMS 463*), *Kang Wha* (ex-*YMS 245*), *Ka Pyong* (ex-*YMS 210*),

Kang Kyong (ex-*YMS 330*), *Kaya San* (ex-*YMS 423*), *Kupo* (ex-*YMS 323*), *Kim Chon* (ex-*YMS 258*), *Kil Chu* (ex-*BYMS 5*), *Ko Yung* (ex-*Ue Nam*, ex-*BYMS 55*), *Kong City* (ex-*BYMS 6*), *Ko Won* (ex-*YMS ?*), *Yong Kung* (ex-*BYMS 8*), *Kum Hwa* (ex-*Curlew* MSCO 8, ex-*AMS 8*, ex-*YMS 218*), *Kim Po* (ex-*Kite* MSCO 22, ex-*AMS 22*, ex-*YMS 375*), *Ko Chang* (ex-*Mockingbird* MSCO 27, ex-*AMS 27*, ex-*YMS 419*).

Twenty-one wooden Second World War-built units transferred 1948–56. Korean pennant numbers MSC 501–521. Fate: MSC 501 BU 1959, MSC 502 BU 1962, MSC 503 stricken 1972, MSC 504–506 (not known). MSC 507 BU 1959, MSC 508 lost 1959, MSC 509 sunk 1950, MSC 510 BU 1963, MSC 511 lost 1949, MSC 512 BU 1956, MSC 513 BU 1968, MSC 514 BU 1959, MSC 515 BU 1959, MSC 516 sunk 1950, MSC 517 lost 1948, MSC 518 BU 1955, MSC 519, 520 and 521 stricken in 1977.

US BLUEBIRD class *coastal minesweepers*

Class (no, former no, transfer):
Ha Dong (MSC 556, ex-MSC 296, Aug 1963), *Ko Hung* (MSC 552, ex-MSC 285, Sep 1959), *Kum Kok* (MSC 553, ex-MSC 286, Oct 1959), *Kum San* (MSC 551, ex-MSC 284, Jun 1959), *Nam Yang* (MSC 555, ex-MSC 295, Aug 1963), *Sam Kok* (MSC 557, ex-MSC 316, Jul 1968), *Yong Dong* (MSC 558, ex-MSC 320, 2.10.75), *Ok Cheon* (MSC 559, ex-MSC 321, 2.10.75).

Built by Peterson, Wisconsin, USA specially for transfer to Korea under MAP. Wooden hulled with non-magnetic metal fittings. Used to be numbered MSC 522–530. All eight extant 1995.

Ex-US MSB type *minesweeping boat*

Former USN MSB 2 transferred 1 Dec 1961, renamed *Pi Bong* (MSB 1) and bought 2 July 1975. Stricken 1982.

KANG KEONG 'Swallow' class *minehunters*

Displacement: 470t normal; 520t full load
Dimensions: 164ft × 27ft 3in × 8ft 6in
50.0m × 8.3m × 2.6m
Machinery: 2 Voith-Schneider vertical cycloidal propellers, 2 MTU diesels, 2040bhp = 15kts. Range 2000nm at 10kts.
Armament: 1–20mm Oerlikon, 2–12.7mm MG
Sensors: Radar Raytheon; sonar GEC/Marconi 193M (Mod 1 or Mod 3) minehunting; combat data system Racal MAINS 500
Complement: 44 (5 officers) + 4 divers

No	Name	Builder	In service	Fate
561	KANG KEONG	Kangam Shb.	Dec 1986	Extant 1995
562	KANG JIN	Kangam Shb.	May 1991	Extant 1995
563	KO REONG	Kangam Shb.	Nov 1991	Extant 1995
565	KIM PO	Kangam Shb.	Apr 1993	Extant 1995
566	KO CHANG	Kangam Shb.	Oct 1993	Extant 1995
567	KUM WHA	Kangam Shb.	Apr 1994	Extant 1995

Design of these ships is based on the Italian *Lerici* class. Prototype was to be delivered in 1986, but it was two years late. Six completed and two more are planned. Plans envisage building of sixteen units. Fitted with two Gaymarine Pluto submersibles and GEC-Marconi 193M Mod 1 sonar as well as Racal-Decca MAINS plotting gear. GRP hull. Fitted with bow thruster. Some reports suggest they are illegal copies of the original Italian design.

1993 *minelayer design*

Displacement 3300t full load
Dimensions: 340ft 6in × 49ft 3in × 11ft 2in
103.8m × 15.0m × 3.4m
Machinery: 2 shafts CODAD, 4 diesels, 17,200bhp = 22kts. Range 4500nm at 15kts
Armament: 1–76mm/62 OTO Melara, 2–40mm/70 Breda (2 x 1), 6–324mm Mk 32 ASW TT, mines, 1 helicopter
Sensors: Radar air and surface search, fire control, navigation; sonar bow-mounted high frequency (active search and attack type)
Complement: 160

In 1991 Hyundai was commissioned to draw up a design for a new class of large, relatively fast minelayers for the South Korean Navy. This was completed in 1993 and it was expected that the first ship would be ordered in 1994. Designed with large stern helicopter platform but no hangar. These ships will have twin rudders and fin stabilisers.

North Korea

MAJOR SURFACE SHIPS

'Najin' class

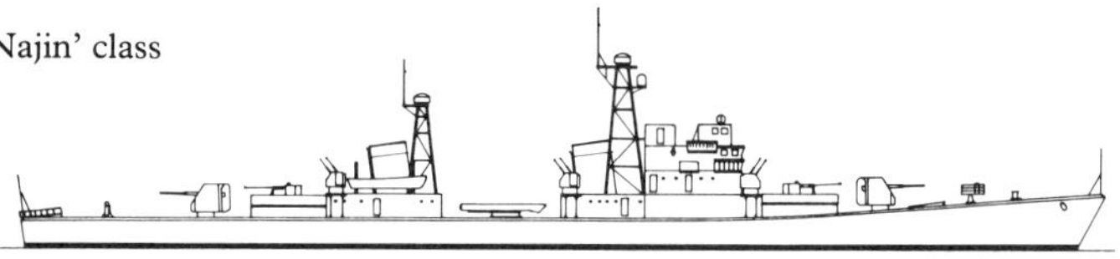

'NAJIN' class *frigates*

Displacement: 1200t standard; 1500t full load
Dimensions: 328ft × 32ft 6in × 8ft 10in
100.0m × 10.0m × 2.7m
Machinery: 2 shafts, 2 diesels, 15,000bhp = 26kts. Range 4000nm at 14kts
Armament: 2–100mm/50 (2×1), 4–57mm (2×2), 4–25mm (2×2), 8–14.5mm (4×2), 3–533mm TT (1×3), 2–RBU-1200, 2 DC racks, 2 ASW mortars, mines
Sensors: Radar Skin Head, Pot Head, Slim Net, Drum Tilt; IFF: Ski Pole and hull sonar
Complement: 155 or 180

No	Builder	Launched	Comp	Fate
3025	N Korea	1972	1973	Extant 1995
3026	N Korea	1973	1975	Extant 1995

Two 'Najin' class frigates of indigenous construction are of flush deck construction reminiscent of old Soviet 'Kola' class destroyers; however Korean ships are diesel powered. Built by Nanjin SY. In the space between funnels occupied by torpedo tubes the two launchers for SS-N-2A were fitted in 1980s making these ships missile frigates. This is a crude installation using the same launchers as fitted on the 'Osa I' (Projekt 205) type FACs. All other armament is aged; all guns are of Soviet Second World War design, while most of the electronics were introduced into service during the 1950s (the only exception is the Drum Tilt of 1961 normally associated with AK-230 gun mountings). These ships are of little combat value. One of these ships now carries pennant number 531 instead of 3000 series (probably the second is 639). Names unknown. Number of 25mm guns increased to twelve, whilst 14.5mm guns removed.

'SOHO' class *experimental frigate*

Displacement: 1600t standard; 1850t full load
Dimensions: 246ft 1in oa × 49ft 3in × 12ft 6in
75.0m × 15.0m × 3.8m
Machinery: 2 shafts, 2 diesels, 15000bhp = 27kts
Armament: 4 SS-N-2A Styx SSM (1×4?), 1–100mm/56, 2–37mm (1×2), 4–25mm (2×2), 2–RBU-1200, DC racks
Sensors: Radar navigation, Square Tie; hull sonar
Complement: 190–200

Probably catamaran design layout as suggested by the hull dimensions with large helicopter deck aft. Built by Najin SY, reportedly launched in 1980 or 1981 and completed in 1983. Occasionally the name *Mayang* has been quoted for this ship. Design evidently is not a success as no more were built.

SUBMARINES

Ex-Chinese and North Korean 'ROMEO' class *submarines* (Project 633)

Seven Chinese-built 'Romeo' class submarines were transferred in 1973 (two), 1974 (two) and 1975 (three). Since then approximately fifteen such submarines were constructed in North Korea at Mayang DY and Sinpo. One of these submarines was reported lost during exercises in February 1985. Programme probably terminated in 1992.

Ex-Soviet 'WHISKEY' class *submarines* (Project 613)

Four submarines transferred from the USSR in 1960s. These were stricken and in 1974 another four 'Whiskey' type submarines were transferred. Now all probably near the end of their useful lives and without any real combat value. Used for training and diving restricted to periscope depth due to age.

Midget submarines and submersibles

There are probably nineteen to twenty (some sources quote higher numbers in the region of sixty) locally built midget submarines mainly intended for commando operations off the South Korean coast. One is reported to be of 41m length (built in 1970s) but other much smaller (17–25t, 65ft 7in, *20m*, 10/4kts, crew 2 + 6–7 divers). These midget submarines are carried on merchant mother ships. There are also some eight so-called semi-submersibles for saboteur delivery, built at Wonsan in the late 1980s: 5t, 50kts, 28ft 3in × 8ft 3in × ?ft (*8.6m × 2.5m × ?m*), carrying six persons. They travel fast on the surface and then ballast down to the awash condition just before delivery. Sighted off Pusan on 20 December 1985.

AMPHIBIOUS WARFARE VESSELS

'NAMPO' class *fast assault landing craft*

Displacement: 82t full load
Dimensions: 84ft 2in × 20ft × 6ft
27.7m × 6.1m × 1.8m
Machinery: 4 shafts, 4 M50 diesels, 4800bhp = 40 kts. Range 375nm at 40kts
Armament: 6–14.5mm MG (3×2)
Complement: 20

A large class of 100 small fast assault ships based on the P 6 type MTB hull. Built in Korea since *c*1975. All have retractable bow ramp for landing operations. Nine 'Hanchon' class LCU 18 home-built LCM, and four 'Hantae' 50m class (built 1980s) are also in service.

'HANCHON' class *landing craft (LCM)*

Displacement: 145t full load
Dimensions: 117ft 2in × 25ft 11in × 3ft 11in
35.7m × 7.9m × 1.2m
Machinery: 2 shafts, 2 Type 3D-12 diesels, 630bhp = 10kts. Range 600nm/6kts
Armament: 2–14.5mm/93 (1×2)
Complement: 15

A new class of locally designed landing craft (LCM) currently building in North Korean shipyards. Currently seven are in service while the programme continues. Fitted with Shin Head radar and can carry military load of two tanks or 200 troops. Fitted with bow ramp for loading and unloading operations. There are also total of sixteen landing craft of 'Hungnam' and eight of 'Hante' class capable of lifting 4–5 medium tanks. Up to fifty-two 'Songjong' small hovercraft type landing craft were built in the 1980s. In various versions they can carry 35–55 troops.

SMALL SURFACE COMBATANTS

Ex-Soviet TRAL class *coastal minesweepers (corvettes)* (Project 53)

Class:
Ex-*Strela*, *Tros Provodnik*, *Podsekatel*, *Paravan*, *Kapsul*, *Viecha*, *Ĉeka* (T1–8)

Eight old Soviet ships built in 1938–42, transferred in 1950s and used as patrol vessels. There were only two left in service (ex-*Strela*, *Paravan*) by 1993.

'Sariwan' class

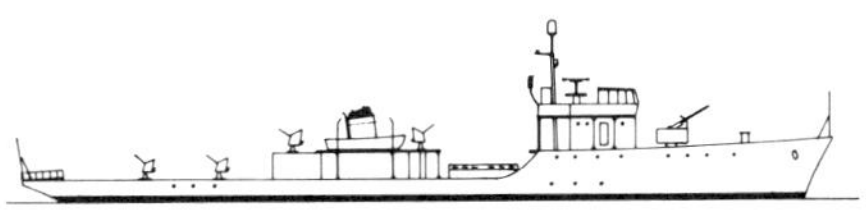

'SARIWAN' class *corvettes*

Displacement: 475t standard; 600t full load
Dimensions: 203ft 8in × 23ft 11in × 7ft 10in
62.1m × 7.3m × 2.4m
Machinery: 2 shafts, 2 diesels, 3000bhp=21kts
Armament: 1–100mm, 2–37mm (1×2), 16–14.5mm MG (4×4), DC rails, 30 mines
Sensors: Radar Skin Head, Dori 2; IFF Ski Pole or Yard Rake
Complement: 65

Class:
725, 726, 727

Large patrol craft/escort ships built in Korea in the mid-1960s and closely modelled on the ex-Soviet *Tral* class minesweepers. They had the same silhouette. Armament also given as 1–76mm, 2–57mm, 4–2mm (2×2). Four units sometimes reported. Extant 1995.

Ex-Soviet 'OSA I' class *fast attack (missile)* (Project 205)

Twelve transferred from the USSR in 1968 and four more in 1972–83. In late 1980s four were deleted. Four Chinese 'Huang Fen' (version of 'Osa I) transferred in 1982. Sixteen of the twelve North Korean version of 'Osa I' named 'Soju' reported completed in 1981–95.

Soviet 'KOMAR' class *fast attack craft (missile)* (Project 183R)

Six delivered by Soviets and four built locally. There are reports of the existence of 'Sohung' class, a Korean version of 'Komar'. Ten were completed in 1980–95.

'SIN HUNG' and KU SONG' class *fast attack craft (torpedo)*

Displacement: 25t full load
Dimensions: 60ft × 11ft × 5ft 6in
18.3m × 3.4m × 1.7m
Machinery: 2 shafts, 2 diesels, 2400hp=40kts
Armament: 2–14.5mm MG, 2–533mm or 457mm TT
Sensors: Radar navigation
Complement: ?

MTBs built in N Korea from mid-1950 to 1970. They resemble Soviet D-3 class MTBs of 25 years ago. 120 active in 1995.

Ex-Soviet 'P 6' class *fast attack craft (torpedo)* (Project 183)

These boats built in the late 1950s in USSR (45) and China were transferred upon completion to Korea. Remainder built of steel in N Korea and called 'Sinpo' class, six of which had no TT. Some are armed with 2–37mm (2×1) guns instead original 4–25mm (2×2). Forty-six extant in 1995. Pennant numbers in 400 sequence.

Ex-Soviet 'P 4' class *fast attack craft (torpedo)*

Small MTBs with aluminium hulls built in Korea and China in 1951–57, all now stricken.

'IWON' class *fast attack craft (torpedo)*

MTBs built in Korea in late 1950s. Modelled on the Soviet 'P 2' type, all since stricken.

Ex-Soviet 'SHERSHEN' class *fast attack craft (torpedo)* (Project 206)

Four transferred in 1968.

'AN JU' class *fast attack craft (torpedo)*

Displacement: 35t full load
Dimensions: 65ft × 12ft × 6ft
19.8m × 3.7m × 1.8m
Machinery: 4 shafts, 4 M50 diesels, 4800bhp = 50kts. Range 1200nm/20kts
Armament: 2–25mm (1×2), 2–533mm TT
Sensors: Radar navigation set
Complement: 20

TBs designed and built in Korea in 1960s, six extant in 1995.

Ex-Soviet 'MO IV' class *fast attack craft (gun)*

Twenty of these 1945–47 built units transferred to Korea in 1957–60 and extant 1982. Four MO-I type submarine chasers transferred in 1954 were deleted 1971–72.

'Chaho' class
NB 1/750 scale

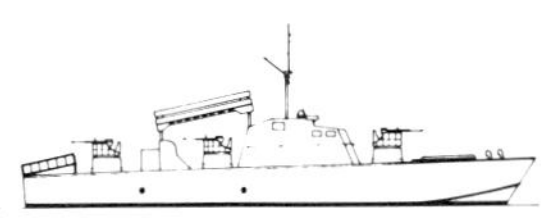

'CHAHO' class *fast attack craft (gun)*

Displacement: 80t full load
Dimensions: 90ft 11in × 20ft × 6ft
27.7m × 6.1m × 1.8m
Machinery: 4 shafts, 4 M50 diesels, 4800bhp = 38kts
Armament: 8–23mm (4×2), 40-tube RL, 4–14.5mm MG (2×2)
Sensors: Radar navigation set
Complement: *c*25

Korean designed FAC built since 1974 on 'P 6' type wooden hulls, sixty extant 1995. There are also thirty 'Hong Jin' class FAC similar to 'Chaho' built since 1975; armed with 1–85mm tank gun and 4–14.5mm MG (2×2).

'K 48' class *large patrol craft*

Displacement: 110t full load (est)
Dimensions: 125ft × 18ft 1in × 4ft 11in
38.1m × 5.5m × 1.5m
Machinery: 2 shafts, 2 diesels, 4000bhp=24kts
Armament: 1–76mm/50, 3–37mm (3×1), 6–14.5mm MG (3×2)
Sensors: Radar Skin Head
Complement:

Korean designed ships built in Korea in late 1950s, now deleted.

Ex-Soviet 'ARTILLERIST' class *large patrol craft* (Project 122)

Two transferred in 1950s and stricken in 1971–72. Built in 1943–44.

Soviet 'SO 1' class *submarine chasers*

Eight transferred from USSR in 1957–61 and a further eleven were built in N Korea. All extant 1995.

'CHODO' class *large patrol craft*

Displacement: 130t (est)
Dimensions: 140ft 19ft × 8ft 6in
42.7m × 5.8m × 2.6m
Machinery: 2 shafts, 2 diesels, 6000bhp = 25kts. Range 2000nm/10kts
Armament: 4–37mm (4 × 1), 4–25mm (2 × 2)
Sensors: Radar Skin Head, Don
Complement: 40

Korean built in mid-1960s, four extant 1982 one numbered 971, three extant 1995.

Ex-Chinese 'SHANGHAI II' class *large patrol craft*

Fifteen transferred from China since 1967. One deleted 1988, one in 1990, remainder extant 1995.

Ex-Chinese 'HAINAN' class *large patrol craft*

Six transferred in pairs in 1975, 1976 and 1978. Extant 1995.

'TAE CHONG I' and 'II' class *large patrol craft*

Displacement: 385t standard; 410t full load (I), 425t full load (II)
Dimensions: 196ft 3in, 199ft 6in (II) × 23ft 8in × 7ft 3in
59.8m, 60.8m × 7.2m × 2.2m
Machinery: 4 shafts, 4 40-D diesels, 10,000bhp = 30kts. Range 2000nm at 12kts
Armament: 1–100mm/56 (II), 2–57mm/70 (1×2), 1–37mm/63 (I), 2–25mm (1×2), 4–14.5mm/93 (2×2), 2–RBU 1220, 2 DC racks
Sensors: 1 Pot Head radar; sonar Stag Ear
Complement: 80

North Korean design of mid 1970s developed from the Chinese 'Hainan' class patrol craft. There are eight craft of the 'Tae Chong I' class and four of the improved 'Tae Chong II'. 'Tae Chong II' are still in production, joining the fleet at about one per year. The 'Tae Chong II' class is also referred to as 'Mayang' class.

'SOMAN' class *large patrol craft*

Displacement: 190t full load
Dimensions: 91ft 6in × 19ft × 6ft 3in
27.9m × 5.8m × 1.9m
Machinery: 1 shaft, 1 Type 3D–12 diesel, 315bhp=10kts. Range 1200nm/9kts
Armament: 4–25mm/60 (2×2), 4–14.5mm/93 (2×2), 16 mines
Complement: 50 (7 officers)

Single trawler type patrol craft used as command vessel, auxiliary gunboat and small minelayer.

Ex-Chinese 'SWATOW' class *coastal patrol craft*

Eight transferred from China in 1968. All now stricken.

Soviet 'KM 4' class *coastal patrol craft*

Small coastal patrol boats of 10t built in Korea to Soviet design, ten extant in 1982, all since stricken.

MINE WARFARE VESSELS

Ex-Soviet 'T 43' class *ocean minesweepers* (Project 255)

Built in 1954. Transferred from USSR in 1960s. Now deleted.

'YUKTO I' and 'II' class *coastal minesweepers*

Displacement: 60t full load
Dimensions: 78ft 9in, 68ft 11in (II) × 13ft 1in × 5ft 7in
24m, 21m × 4m × 1.7m
Machinery: 2 shafts, 2 diesels, 18kts
Armament: 1–37mm/63 or 2–25mm/80 (1×2), 2–14.5mm/93 (1×2), 4 mines
Complement: 22 (4 officers)

A new class of locally designed wooden coastal minesweepers built in the 1980s to replace the obsolete ex-Soviet 'KN 14' class. Currently there are 19 'Yukto I' and four 'Yukto II' boats in service. There is a new class of inshore minesweepers in existance which are called 'Pipa-Got'. Details unavailable.

Kuwait

The Kuwaiti navy was primarily a coast guard force, controlled by the Ministry of the Interior and consisting of lightly armed patrol craft (delivered from 1966 onwards) and a number of small landing craft. Naval personnel totalled about 2100 in 1989, almost double the total in 1982.

Both the navy and the naval air force were taken by surprise by the Iraqi invasion in August 1990, and all but two fast attack craft were captured intact by the invaders. The captured vessels were used by the Iraqis during Operation Desert Storm, and all were badly damaged or sunk by the Allied forces, many by Sea Skua missiles; the ex-Kuwaiti vessels were virtually defenceless as Iraqi personnel were not familiar with their systems.

In general, the Gulf War has shown that small fast attack craft are less effective than was hitherto almost universally assumed. They are very vulnerable to helicopter missile attack, and at high speed their masts vibrate so much that their pencil-beam radars have difficulty locking on to targets. The only obvious solution to the problems of such craft is to fit them with fire-and-forget SAM missiles for close-in air defence.

Kuwaiti losses in August 1990 included five Lürssen 'TNC 45' and one 'FPB 57' fast attack craft; five Seagull, fifteen Thornycroft, one Halter Marine and seven Magnum Sedan patrol craft; four Loadmaster and six Vosper Singapore type landing craft; ten tugs and launches; twelve Super Puma helicopters; and three Azimut launches of the Customs service.

Some of the vessels captured by Iraq which survived the Gulf War afloat were returned to their rightful owners, but in their damaged and unserviceable condition they are of little value and it is probable that most or all of them will be scrapped after the removal of any useful equipment remaining. The Kuwaiti Navy is actively pursuing its new construction plans, which include the acquisition of two corvettes and two 34m landing craft in 1995, and for the future three minehunters, two diving support vessels and a training ship. Eight large patrol vessels are to be ordered in France.

Sabhan after being hit by Sea Skua missiles Feb 1994 MoD

Al-Saadi 1984
NB 1/750 scale

Lürssen 'TNC 45' type *fast attack craft (missile)*

Displacement: 228t standard; 268t full load
Dimensions: 139ft 9in pp, 147ft 4in oa × 24ft 3in × 7ft 7in
42.3m, 44.9m × 7.4m × 2.3m
Machinery: 4 shafts, 4 MTU 16V538 TB92 diesels, 14,400bhp = 40.5kts. Range 500nm/1600nm at 38kts/16kts
Armament: 4 MM40 Exocet SSM (2×2), 1–76mm/62 DP OTO Melara (250rds), 2–40mm/70 Breda-Bofors Compatto (1×2; 1800rds)
Sensors: Radar Decca 1226, HSA WM-28; ECM Decca Cutlass RDL-2 Dagaie chaff RL
Complement: 40

Class (no, in service):
Werija (K 451, 26.4.84), *Mashuwah* (K 452, 26.4.84), *Jalboot* (K 453, 26.4.84), *Istiqlal* (K 455, 9.8.84), *Al Ahmadi* (P 4509, 9.8.84), *Al Mubareki* (P 4511, 9.8.84)

Ordered at Lürssen Werft 1980 to the standard TNC 45 design. All launched 1982–83. First four and the sixth renamed and renumbered soon after completion as *Al Boom* (P 4501), *Al Betteel* (P 4503), *Al Sanbouk*(P 4505), *Al Saadi* (P 4507) and *Al Abdali* (P 4511). *Al Sanbouk* escaped to Bahrain 1990 upon Iraqi invasion of Kuwait, but the others were captured and taken over by the Iraqi Navy, only to be sunk or heavily damaged in February 1991 during Operation Desert Storm. Only *Al Sanbouk* extant 1995.

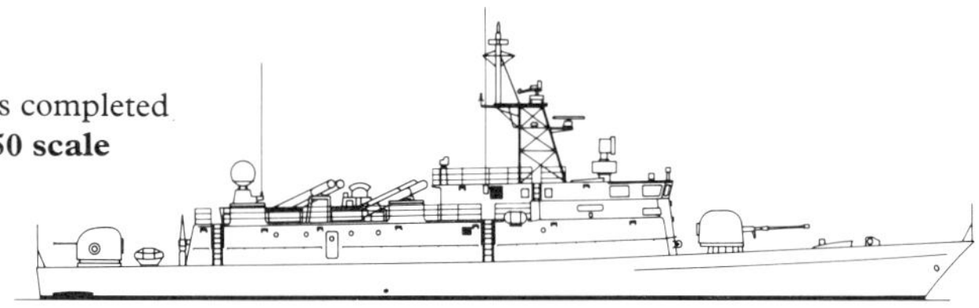

Istiqlal as completed
NB 1/750 scale

Lürssen 'FPB 57' type *fast attack craft (missile)*

Displacement; 350t standard, 425t full load
Dimensions: 190ft 7in oa × 25ft × 8ft 10in
58.1m × 7.6m × 2.7m
Machinery: 4 shafts, 4 MTU 16V 956 TB91 diesels, 15,000bhp = 36kts. Range 1300nm at 30kts, 700nm at 35kts
Armament: 4 MM40 Exocet SSM (2×2), 1–76mm/62 OTO Melara, 2–40mm/70 Breda (1×2), fitted for minelaying, Dagaie chaff RL
Sensors: Radar Marconi S 810, Decca TM 1226C, fire control PEAB 9LV228 (with 9LV200 search and tracking radars); ESM: Racal Cutlass
Complement: 40 (4–5 officers)

Class (no, in service):
Istiqlal (P5702, Aug 1983), *Sabhan* (P 5704, 9.8.84)

Two modified FPB 57 fast attack craft ordered 1980 at Lürssen Werft, Vegesack, Germany. Launched 1982–83. Originally scheduled to enter service 1982 and 1983 respectively. Had much larger superstructure amidships than typical FPB 57 type boats. Functioned as leaders for six TNC 45 type boats. Fitted with CSEE Lynx optical sight. Early in their careers armed with MM38 missiles. *Istiqlal* escaped to Bahrain August 1990 when Iraq invaded Kuwait. *Sabhan* was captured and sunk February 1991 during Operation Desert Storm.

ASI 315 type *large patrol craft*

Displacement: 150t full load
Dimensions: 88ft 11in pp, 123ft oa × 21ft 4in × 6ft 7in
27.1m, 37.5m × 6.5m × 2.0m
Machinery: 2 shafts, 2 MTU 16V 396 TB94 diesels, 5800bhp = 28kts; 1 Hamilton 422 waterjet for cruising, 1 MTU 8V 183 TB 94 diesel, 750bhp. Range 300nm at 28kts
Armament: 1–20mm Oerlikon, 1–7.62mm MG
Sensors: Radar navigational, surface search
Complement: 11 (3 officers)

Class (no, in service):
Inttisar (P 301, 20.1.93), *Aman* (P 302, 20.1.93), *Maimon* (P 303, Jun 1993), *Mobark* (P 304, Jun 1993)

First two were ordered from Australian Shipbuilding Industries in 1991, second pair July 1992. Built to proven ASI 315 design widely exported to small Pacific nations. Durable steel hull and aluminium superstructure. Small-engine and waterjet propulsion used only for cruising during extended patrol work.

Thornycroft 78ft type *coastal patrol boats*

Displacement: 40t standard
Dimensions: 78ft × 15ft 6in × 4ft 6in
25.5m × 4.7m × 1.4m
Machinery: 2 shafts, 2 8cyl Rolls-Royce V8 marine diesels, 1340bhp max, 1116bhp = 20kts. Range 700nm at 15kts
Armament: Reported 1 MG or unarmed
Complement: 12

Class (launched):
Al Salemi (30.6.66), *Al Shurti* (1972), *Al Mubaraki* (16.7.66), *Aman* (Mar 1968), *Intisar* (1972), *Marzook* (1969), *Mashoor* (1969), *Maymoon* (Apr 1968), *Murshed* (1970), *Wathah* (1970)

Two built by Thornycroft before their merger with Vosper and eight after the merger. *Al Salemi* and *Al Mubaraki* shipped to Kuwait on 8 September 1966 and the last pair *Al Shurti* and *Intisar* in 1972.

Welded steel hulls, aluminium alloy superstructures. Twin hydraulically operated rudders, Decca type D202 radar. Modified superstructure and no funnel on later. Designed to have 1–40mm.

Vosper 56ft *coastal patrol boats*

Displacement: 25t
Dimensions: 56ft 1in × 16ft 1in × 4ft 3in/7ft 7in
17.1m × 4.9m × 1.3m/2.3m
Machinery: 2 shafts, 2 MTU MB 6V331 diesels, 1350bhp = 26kts for first two and 1800bhp = 29–30kts for last 3 or 4 boats. Range 320nm at 20kts
Armament: 1–20mm, 2 MG
Complement: 8

Class:
Dastoor, Kasar, Qahir, Sagar, Salam, and probably one with unknown name.

First two ordered September 1973 and built by Vosper Thornycroft Ltd, Singapore, laid down 31 October 1973 and commissioned June 1974. Steel hulls and aluminium superstructure. Three more ordered in 1978. Built by Vosper Private Ltd and commissioned 1979. French source gives two types – first two units built 1974 armed as above and later four with 2–20mm AA (2×1). Second pair ordered 10 March 1977, in service December 1978 and third pair ordered July 1978, in service 1979. Lost to Iraq 1990.

Vosper 46ft *coastal patrol craft*

Displacement: 21.5t
Dimensions: 46ft 3in × 14ft 5in × 3ft 11in, 7ft 2in max
14.1m × 4.4m × 1.2m, 2.2m
Machinery: Probably 2 shafts, 2 Rolls-Royce C8M-410 diesels, 780bhp = 21–21.7kts
Armament: Can mount 2–20mm AA (2×1), 2–7.62mm MG
Complement: 5

Class: *Mahroos*

Built by Vosper Thornycroft Ltd, Singapore. Ordered October 1974, delivered 1975 and commissioned January 1976. Hull is of welded steel construction with aluminium superstructure. Originally not armed. Extant 1990.

Vosper 36ft *coastal patrol craft*

Displacement: 6.8t full load
Dimensions: 36ft 5in oa × 10ft 10in × 5ft 3in
11.1m × 3.3m × 1.6m
Machinery: 2 shafts, 2 Sabre 210 turbocharged diesels, 420bhp = 27kts
Armament: 4–7.62mm MG (2×2)
Complement: 4

Class:
Antar, Al Salmi II, Al Sebbah, Istiqlal II, Qarah, Warbah and eight unnamed

All built by Vosper Thornycroft, Singapore. Wooden construction with nylon sheathed hulls. First four unnamed delivered in 1972, second four unnamed in service 1973. Six stricken 1987. Remaining two extant 1990. First three named boats in service 1976 and last three on 2 December 1977. All named boats lost to Iraq 1990.

South Korean 'SEAGULL' type *patrol craft*

First five of these boats ordered 1985 at Korea SB & Eng, Pusan, and delivered 1986. All captured by Iraq 1990 and destroyed February 1991. Second series of five boats ordered 1988 but not delivered before Iraqi invasion. Locally known as 'Sawahil 35' class.

'LOADMASTER Mk II' type *logistics landing craft*

Displacement: 175t light; 350t normal; 420t full load
Dimensions: 108ft 3in oa × 33ft 6in × 5ft 9in
33.0m × 10.2m × 1.75m
Machinery: 2 shafts (Kort nozzles), 2 Caterpillar 3412 DITA diesels, 1214bhp = 10.5kts. Range 1000nm at 10kts
Armament: none
Complement: 7 (1 officer)

Class (no):
Al Saffar (L 101), *Al Seef* (L 102), *L 103*, *Al Baldani* (L 104)
Ordered from Cheverton 1982. Original design taken over by Fairey Marine, Cowes. First pair completed October-November 1984, second pair September 1985. Can carry 60t liquid cargo and 90t dry cargo or accommodate two 60t main battle tanks. There are 50 places for seated passengers. Bow ramp four metres wide. Lost to Iraq August 1990. There were also six SRN6 Mk8 type hovercraft operated by the Coast Guard, built by British Hovercraft, Cowes, which were operational in 1990.

Vosper type *logistic landing craft*

Displacement:	170t/320t
Dimensions:	105ft 11in × 24ft 7in × 8ft 2in *32.3m × 7.5m × 2.5m*
Machinery:	2 shafts, 2 Rolls-Royce C8M-41D diesels, 750bhp = 9.5kts. Range 1500nm at 9kts
Armament:	None
Complement:	?

Class:
Hadiya, Ceriff, Al Jahra
All built by Vosper Private Ltd, Singapore. Ordered July 1978 and commissioned 1979. Unusual design with stern landing ramp. Lost to Iraq 1990.

Vosper type *landing craft*

Displacement:	88t standard; 170t full load
Dimensions:	88ft 7in × 22ft 8in × 4ft 3in *27.0m × 6.9m × 1.3m*
Machinery:	Probably 2 shafts, 2 Rolls-Royce C8M-41D diesels, 752bhp= 10kts. Range 1500nm at 9kts
Armament:	None
Complement:	9 (can carry 8 passengers)

Class:
Waheed, Fareed, Regga
Built by Vosper Thornycroft Ltd, Singapore. First two ordered 1970 and third October 1974 by Kuwait Ministry of Interior. Able to carry 6400 gallons oil, 9400 gallons water and 40t deck cargo; a 2.5t derrick employed. Supports landing-parties working on islands offshore. Lost to Iraq 1990.

AL TAHADDY class *logistic landing craft*

Displacement:	130t standard; 250t full load
Dimensions:	141ft 1in × 32ft 10in × 5ft 11in *43.0m × 10m × 1.8m*
Machinery:	2 shafts, 2 MTU diesels = 13kts
Armament:	None
Complement:	?

Class (no, in service):
Al Tahaddy (L 401, 1994), *Al Soumood* (L 402, 1994)
Two logistic landing craft ordered after the Gulf War to replace lost Loadmaster Mk II type from Singapore Shipbuilding and Engineering Ltd, delivered 1994. Capacity 80t cargo plus 5 containers (5 TEU). Standard design with bow ramp, open cargo deck and machinery and superstructure aft. Fitted with small articulated cargo crane.

OTHER SMALL CRAFT
Twelve Simmoneau Standard 12 type small patrol boats (10t, 45ft 11in × 12ft 6in × 2ft 4in, *14.0m × 3.8m × 0.7m*, 2 inboard Caterpillar engines, 850bhp = 40kts. 1–12.7mm, 1–7.62mm MG, complement 4): built by Simmoneau Marine, delivered 1993.
Seventeen Cougar type fast inshore patrol boats: three Cat 900 (32ft), three Cat 1000 (33ft), three Predator 1100 (35ft), four Type 1200 (38ft), four Type 1300 (41ft): all racing hulls acquired 1991–92 as a stopgap measure. All armed with 1–7.62mm MG.
Thirty-three Al-Shaali type inshore patrol boats: thirteen 33ft, twenty 28ft: built by Al-Shaali Marine, Dubai 1992.

Laos

Laos has no coastline and her navy consists of small gunboats and landing craft operating mainly on the Mekong River. Laos was formed from several different parts of French Cochin China as an independent members of the French Union in 1949 and achieved true independence in 1954. She had already been invaded by the Vietminh in 1953 and was soon split between the Communist Pathet Lao in the East and the Royal Laotian Government in the West. In 1962 a 14-nation conference at Geneva decided to recognise this split and neutralise Laos, but the fighting continued with the Pathet Lao gradually winning. During the 1950s and 1960s the United States and France supplied enough gunboats and landing craft to make up four squadrons, and in the early 1970s there were seven ex-USN LCM 6 and twenty-eight gunboats of 6–21t, plus seven small transports.

After the fall of South Vietnam the Pathet Lao took over Laos and on 2 December 1975, the King having abdicated, declared a People's Democratic Republic, which was then completely overrun by Vietnamese troops (about 45,000 deployed in 1982). There are about 600 personnel and forty river craft in service in 1995.

Lebanon

Lebanon maintains a naval force which consists mainly of small patrol boats and craft supported by former US utility landing craft. The strength of this navy has remained constant over the years and there are no serious plans for a development programme. Israel gave seven patrol boats to her Lebanese Christian allies in 1975–76. A strong customs service rather than a navy is required. Some of a ten-year $955m programme to re-equip the republic's armed forces may have found its way to the navy. The main base is at Beirut/Jounieh; there are a total of 200–400 officers and men serving in the navy commanded by a colonel.

One large French-built patrol craft (105t standard, 2 diesels, 27kts, 2–40mm, 2–12.7mm MG): *Tarablous* (31, laid down by CNE June 1958, launched June 1959). Deleted 1991.
Two *Jihad* class large patrol craft ordered in 1975 but sold to Libya January 1978 when Lebanon unable to pay.
Two ex-RN Fairmile B motor launches: *Djounieh* (ex-*ML ?*) and *No 41*. Stricken 1975.
Three *Byblos* class motor launches (28t standard, 66ft, 2 diesels, 18kts, 1–20mm, 2 MG): *Byblos* (11), *Sidon* (12), *Beyrouth* (13) built 1954–55 by CNE of France. Deleted 1990.
Two Fairey Marine 'Tracker Mk2' class patrol boats (31.5t full load, 2 diesels, 29kts, 11 crew, 1-twin 23mm of Soviet origin). In service 21 January 1980 and 8 February 1980 (names unknown).
One ex-French 'EDIC' type landing craft: *Sour* (ex-L 9096) loaned on 7 November 1983 and since stricken.
Two French 'EDIC III' type landing craft: *Sour* (21) and *Damour* (22) in service 1985–86, built by SFCN, Villeneuve-la-Garonne: ordered in 1983, can carry 11 trucks or five armoured personnel carriers. These craft were damaged in 1990 but were repaired in 1991. Extant 1995.
Six 'Aztec' class 9-metre coastal patrol craft: *CP 1001–1006*, GRP customs boats supplied by Crestitalia, La Spezia, 1980. *CP 1006* deleted 1991.
Five ex-RN 'Attacker' class patrol craft (38t, 65ft, 21kts, 1–20mm Oerlikon, 2 MG): *L 01–L 05* (ex-*Attacker, Hunter, Chaser, Fencer*) three acquired 1992, two in 1993. All extant 1995.
One ex-US LCU 1466 utility landing craft: *Sour* (ex-*LCU 1474*) launched 1955 and transferred November 1958. Stricken 1980.

Liberia

Although Liberia has been independent since her foundation in 1847, there is no long-standing naval tradition, her naval forces having generally been negligible or non-existent. In 1982 the 445-man volunteer Coast Guard (set up *c*1956) operated from Monrovia and three other bases, three CG 27 class Swedish-built 87ft patrol boats (50t, in service 27.9.80, 3 MGs, 25kts), two US-built Swiftship 65ft boats (38t, in service 22.7.76, 81mm mortar and 3 MGs, 24kts) and a US-built 42ft craft. The large patrol boat *Alert* (ex-US *PGM 102*, transferred 1967) was deleted in 1981. Two ex-USCG motor launches (*ML 4001* and *ML 4002*, sold in 1957) were discarded in 1976. The 1980 coup by Army NCOs ended 133 years of civilian rule and completely upset the power structure in the country. In 1987 the Coast Guard was renamed the Liberian Navy. Civil war has destroyed much of the country, including two of the three CG 27 class patrol craft, *Master Sergeant Samuel K Doe* (ex-*Nuah River*) and *General Thomas Quiwoulepe* on 20 July 1990. The remaining one is named *Albert Porte*. Despite a pan-African intervention force, there is no sign of the war ending.

Libya

After the Second World War, the provinces of Libya were placed provisionally under British and French administration until the country became independent in December 1951, the first independent state created by the UN. The conservative, pro-Western rule of King Idris was bolstered by an American and British military presence, but this did not prevent its overthrow by a military coup in September 1969. The new Arab Republic reversed previous policies – Western bases and troops were evacuated by 1970 – and has subsequently steered a somewhat maverick course internationally, a course apparently often dependent on the whims of her erratic yet mercurial leader – Colonel Muammar el Gaddafi.

In the world arena, oil-rich Libya has developed love-hate relationships with both Western and Eastern blocs. Nearer home, the main enemies are Israel and Zionism, but realistic appreciations of Libya's current military capabilities and her oil installations' vulnerability have prevented direct armed conflict. As a fervent exponent of pan-Arab Socialism, she has proposed (and occasionally affected) unions with other Arab states, but these have always proved transitory. No money or effort have been spared in attempts to destabilise less revolutionary Arab leaders, or even revolutionary leaders following different roads to socialism. Libya replaced Algeria as the main arms supplier to the Polisario Front in the Western Sahara, and also began economic and military penetration southwards, alarming the fragile and vulnerable states of the Sahel and tropical Africa; attempts in 1979 to shore up Idi Amin in Uganda and Bokassa in the Central African Empire brought only ignominy and ridicule, but the Libyan intervention in Chad in 1980 successfully ended traditional French influence.

Libya's armed forces are numerically weak, as expected for a population of only 3,250,000, but make up for this by being lavishly equipped with the most modern weapons, which look good on paper but cause formidable problems of training and maintenance; in major July 1977 border clashes with Egypt, Libyan forces performed badly. Some 1800 Soviet advisers were reported in the country in 1982, but the USSR never secured bases in avowedly anti-Communist Libya.

The navy was set up in November 1962 by a British Naval Mission, with technical assistance and training that began in Malta. Since 1974 it has been expanded rapidly with defence spending rising 250 per cent; the volunteer personnel increased from 2000 to 3000 between 1974 and 1980. Libya's merchant marine quadrupled over the same period. The Fleet, which has been obtaining up-to-date ships from both West and East, now includes submarines, missile-armed corvettes and fast attack craft, an amphibious force in theory able to land 2000 troops plus units with a minelaying capability. The bases are Tripoli (5000-ton floating dock), Benghazi, Derna, Tobruk and Bandiyah.

In the 1980s there were a number of armed clashes between the Libyan Navy and Air Force and both Italian and US forces. These culminated in Operation El Dorado Canyon, when carrier-borne USN A-6 Intruders destroyed two Libyan missile boats.

MAJOR SURFACE SHIPS

Dat Assawari 1973

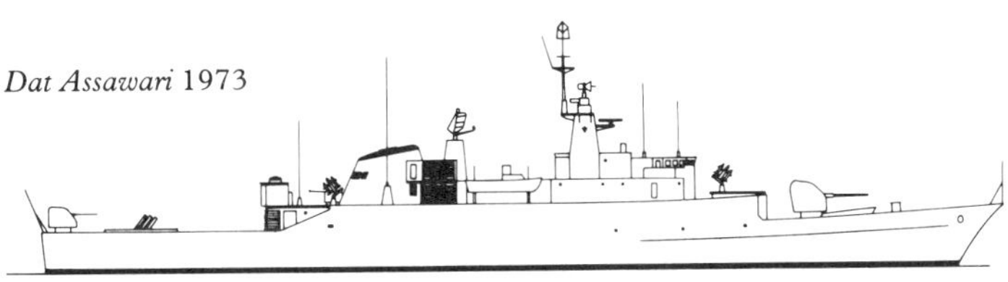

Al Hani 29.7.91 L & L van Ginderen

Vosper Thornycroft Mk 7 frigate, ordered in February 1968, laid down 27 September 1968 and completed 1 February 1973, slightly larger than the two Iranian Mk 5s and has a different armament. Vibrates strongly at speeds over 27kts. *Dat Assawari* was refitted (1979–81) by CNR, Riva Trigosa (Genoa) despite bomb damage on 29 October 1980 with the following changes: 4 Otomat Mk 1 SSM (4 × 1) added aft, 4 Albatros SAM (1 × 4) in place of Seacat forward, 6–12.75in ASW TT (2 × 3), Decca TM 1229 radar, Diodon sonar and Selenia RAN-12/X combat data system added. Italian RAN-10S search radar gives a reported 50 per cent range improvement on the old British set. Now numbered 211 (ex-101). She underwent a major refit in 1989–9.

DAT ASSAWARI *frigate (missile)*

Displacement: 1325t standard; 1625t full load
Dimensions: 330ft oa, 310ft pp × 36ft × 11ft 2in
94.5m, 100.6m × 11.0m × 3.4m
Machinery: 2-shaft CODOG: 2 Rolls-Royce Olympus gas turbines, 46,400shp = 37.5kts, plus 2 Paxman Ventura 16Y JCM diesels, 3500bhp = 17kts. Range 5700nm at 17kts
Armament: 1–4.5in/55 Mk 8, 2–40mm/70 (2×1), 2–35mm/90 Oerlikon (1×2), 2 Seacat SAM (2×3), 1 Limbo Mk 10 ASW mortar. See notes
Sensors: Radar Plessey ASW-1, RDL-1; sonar Types 185, 170, 174
Complement: 132

No	Name	Builder	Launched	Fate
F 01	DAT ASSAWARI	Vosper Thornycroft	13.10.69	Extant 1995

Ex-Soviet 'KONI III' class *frigates*

No	Name	Builder	In service	Fate
212	AL HANI	Zelenodolsk	28.6.86	Extant 1995
213	AL GHARDANIA	Zelenodolsk	23.10.87	Extant 1995

Eleventh and twelfth units of 'Koni III' class built for export. Extended deckhouse to accommodate additional air-conditioning equipment.

Dat Assawari 10.9.83 L & L van Ginderen

SUBMARINES

Soviet 'FOXTROT' class *submarines*

Class (no):
Al Badr (311), *Al Fatah* (312), *Al Ahad* (313), *Al Matrega* (314), *Al Khyber* (315), *Al Hunyan* (316)

Constructed on a re-activated Sudomekh SY building line in Leningrad; *Al Badr* reached Tripoli 27 December 1976. Rest delivered late 1977, early 1978, 17 March 1981 and February 1982. Libyan crews were trained in the USSR and up to twelve Soviet advisers served aboard each boat until 1992. Only one submarine fully operational as of 1995.

Yugoslav MALA class *midget submarines*

Two ordered 1979. First delivered January 1982.

AMPHIBIOUS WARFARE VESSELS

IBN OUF class *landing ships*

Displacement: 2200t standard; 2800t full load
Dimensions: 326ft 5in × 51ft 2in × 7ft 10in
99.5m × 15.6m × 2.4m
Machinery: 2 shafts, 2 SEMT-Pielstick diesels, 5340bhp = 15.4kts. Range 4000nm at 14kts, 30-day endurance
Armament: 6–40mm/70 Breda AA (3 × 2), 1–81mm mortar
Sensors: Radar Decca 1226, Thomson-CSF Triton
Complement: 35 (+ 240 troops)

No	Name	Builder	Launched	Fate
130	IBN OUF	CNIM, La Seyne	22.10.76	Extant 1995
131	IBN HARITHA	CNIM, La Seyne	18.10.77	Extant 1995

Can carry 570t cargo inc eleven tanks. Stern helicopter platform. 130 laid down 1 April 1976 and completed 11 March 1977, 131 laid down 18 April 1977 and completed 10 March 1978. Renumbered 132 and 134. Possibly non-operational 1995.

Ibn Haritha 1979 L & L van Ginderen

Soviet 'POLNOCNY-C' class *landing ships*

Class (no):
Ibn al Hadrani (112), *Ibn al Qis* (113), *Ibn Omaya* (116) and *Ibn el Farat* (118)
Number 112 transferred December 1977, 116 and 118 in June 1979. 113 lost by fire at sea 14/15 September 1978 during landing exercise. Like Iraq's export versions with helicopter deck fwd of superstructure.

Ex-Turkish 'C 107' class *large landing craft*

Class (former name):
Ras el Hillel (ex-*C 132*), *El Kobayat* (ex-*C 133*), *Ibn al Idrissi* (ex-*C 130*), *Ibn Marwhan* (ex-*C 131*)
Transferred 7 December 1979. Up to fifty originally planned but only four transferred. Built by Taskizak and Gölcük N Yds. Cargo 300 troops, five tanks. All extant 1995.

SMALL SURFACE COMBATANTS

TOBRUK *corvette*

Displacement: 440t standard; 500t full load
Dimensions: 177ft oa × 28ft 6in × 13ft
54.0m × 8.7m × 4.0m
Machinery: 2 shafts, 2 Paxman Ventura 16 YJCM diesels, 3800bhp = 18kts. Oil 60t. Range 2900nm at 14kts
Armament: 1–4in/40 Mk 23 LA, 4–40mm/60 Mk 9 (4 × 1), 2–2in RFL, 2–12.7mm MG
Sensors: Radar surface warning
Complement: 63

No	Name	Builder	Launched	Fate
C 411	TOBRUK	Vosper, Portsmouth	29.7.65	Hulked 1989

Type 1 corvette, laid down 22 February 1965 and completed 30 March 1966, arrived Tripoli 15 June 1966. Generally similar to *Keta* and *Kromantse* built for Ghana but with less powerful machinery, and AA armament. Two 40mm guns removed *c*1970. Apartments for use as state yacht. Vosper anti-roll fins and air conditioning.

Wadi M'Ragh as completed

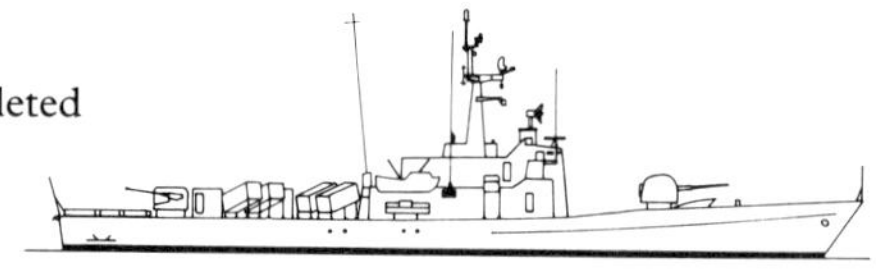

WADI M'RAGH class *missile corvettes*

Displacement: 545t light; 670t full load
Dimensions: 189ft 7in pp, 202ft 5in oa × 30ft 6in × 8ft 10in
57.8m, 61.7m × 9.3m × 2.7m
Machinery: 4 shafts, 4 MTU type MA 16V 956 TB91 diesels, 14,000bhp = 31.5kts. Oil 126t. Range 1400nm/4150nm at 33kts/18kts, 15 day endurance
Armament: 4 Otomat Mk 1 SSM (4 × 1), 1–76mm/62 OTO Melara Compact, 2–35mm/90 Oerlikon (1 × 2), 6–12.75in ASW TT (2 × 3), 16 mines on 2 rails (as minelayer)
Sensors: Radar Selenia RAN-11L/X, Decca TM-1226, RTN-10X Orion; Sonar Thomson-CSF Diodon; ECM Selenia ISN-1 intercept
Complement: 58

No	Name	Builder	Launched	Fate
C 412	ASSAD AL TADJIER (ex-*Wadi M'Ragh*)	Muggiano, La Spezia	29.9.77	Deleted 1993
C 413	ASSAD AL TOUGOUR (ex-*Wadi Majer*)	Muggiano, La Spezia	20.4.78	Deleted 1993
C 414	ASSAD AL KALIJ (ex-*Wadi Marseat*)	Muggiano, La Spezia	14.12.79	Deleted 1993
C 415	ASSAD AL HUDUD (ex-*Wadi Magrawa*)	Muggiano, La Spezia	21.6.79	Deleted 1993

Missile corvettes ordered in 1974: 412 and 413 laid down 25 May 1976; 414 on 26 October 1977; 415 on 25 May 1978. 412 entered service on 14 September 1979, 413 on 12 February 1980, 414 and 415 on 28 March 1981. All renamed 1981. Four more ordered May 1980. Slower than comparably armed FAC, but have much greater range and sea-keeping qualities. NBC protection. Six similar ships were built for Ecuador and four for Iraq. Only one unit operational 1990, and all discarded by 1995.

Soviet 'NANUCHKA II' class *missile corvettes*

Class (no, transferred):
Tariq Ibn Ziyad (416, Oct 1981), *Ain al Galaza* (417, 1983), *Ain Zaara* (418, 1983), *Ain Zaquit* (419, 1983)
419 sunk by US aircraft 24 March 1986, and 416 damaged next day, and repaired in the USSR 1990. Survivors extant 1995.

SUSA class *fast attack craft (missile)*

Displacement: 95t standard; 114t full load
Dimensions: 100ft oa × 25ft 6in × 7ft
30.5m × 7.8m × 2.1m
Machinery: 3-shaft CODOG: Bristol Siddeley Proteus gas turbines; 12,750shp = 54kts; 2 GM 6–71 cruising diesels, 190hp
Armament: 8 SS12 SSM (2 × 4), 2–40mm/60 Mk 9 (2 × 1), 2 RFL
Complement: 20

Class (no, launched):
Susa (P 512, 31.8.67), *Sirte* (P 513, 10.1.68), *Sebha* (ex-*Sokna*) (P 514, 29.2.68)
Wooden, nylon sheathed boat ordered from Vospers 12 October 1966 and modelled on Vosper's Danish *Sölöven* class. All commissioned January 1969 as the West's first missile-armed FAC after Vosper and Nord Aviation had designed a sighting turret for the wire-guided SS12. All completely refitted 1977 in Italy and fitted with new engines 1985–86. Extant 1995, but may be out of action.

Susa on trials Vosper Thornycroft

Ex-Soviet 'OSA II' class *fast attack craft (missile)*

Class (no):
Al Katum (205), *Al O'Work* (206), *Al Rwae* (207), *Al Baida* (208), *Al Nabhaa* (209), *Al Fikar* (210), *Al Safhaa* (952), *Al Zakab* (954), *Al Zuara* (956), 953 and 513 plus one other.

Libya has twelve, the first transferred in October 1976, four in 1977, one in 1978, three in October 1979 to Tobruk and three in 1980. Fitted with new engines in Italy 1984–85. All extant 1995.

BEIR GRASSA class *fast attack craft (missile)*

Displacement:	258t standard; 311t full load
Dimensions:	160ft 8in oa × 24ft 11in × 7ft 11in *49.0m × 7.6m × 2.4m*
Machinery:	4 shafts, 4 MTU 20V538 TB91 diesels, 18,000bhp = 40kts. Range 1600nm at 15kts
Armament:	4 Otomat Mk 1 SSM (4 × 1), 1–76mm/62 OTO Melara Compact, 2–40mm/70 Breda Bofors (1 × 2)
Sensors:	Radar Thomson-CSF: Triton search, Castor track, Vega II fire-control; ECM Thomson-CSF
Complement:	27

No	Name	New name	Launched	Fate
518	BEIR GRASSA	*Sharara*	28.6.79	Extant 1995
519	BEIR GZIR	*Shehab*	22.1.80	Extant 1995
520	BEIR GTIFA	*Wahg*	20.5.80	Extant 1995
521	BEIR GLULUD	*Waheeg*	30.9.80	Sunk 24.3.86
522	BEIR ALGANDULA	*Shouaae*	14.1.81	Extant 1995
523	BEIR KTITAT	*Shouala*	22.4.81	Extant 1995
524	BEIR ALKARIM	*Shafak*	23.6.81	Extant 1995
525	BEIR ALKARDMEN	*Bark*	23.6.81	Extant 1995
526	BEIR ALKUR	*Raad*	30.11.81	Extant 1995
527	BEIR ALKUESAT	*Laahibe*	28.1.82	Extant 1995

'Combattante IIG' type boats ordered from CMN, Cherbourg in May 1977 at an estimated cost of about £300 million. Laying down dates: 518 (13.3.78), 519 (10.6.78), 520 (30.1.79), 521 (20.10.79), 522 (12.9.79), 523 (17.12.79), 524 (11.3.80), 525 (9.6.80), 526 (20.10.80) and 527 (20.1.81). *Beir Grassa* was to have been delivered in February 1981 but the French embargo on arms sales to Libya, imposed because of Libyan intervention in Chad, delayed her handover till 5 February 1982, that of *Beir Gzir* until 3 April 1982. 520 and 521 were delivered in July 1982 and 522 in September, 525 on 11 March 1983. Renamed, and renumbered, 1983. Survivors reported non-operational as of 1995.

GARIAN class *large patrol craft*

Displacement:	120t standard; 159t full load
Dimensions:	106ft oa × 21ft 2in × 5ft 6in *32.3m × 6.5m × 1.7m*
Machinery:	2 shafts, 2 Paxman 12cyl Ventura 10YJCM diesels, 3600bhp = 24kts. Range 1800nm/2300nm at 13kts/11kts
Armament:	1–40mm/60 Mk 9, 1–20mm (690rds for both), 2–2in RFL
Complement:	15–22

Class (no, launched):
Garian (PC 1, 21.4.69), *Khawlan* (PC 2, 29.5.69), *Merawa* (PC 3, 30.9.69), *Sabratha* (PC 4, 25.10.69)

Ordered from Brooke Marine, Lowestoft, England, in December 1967. First two commissioned 20 August 1969, second two early 1970. At least one has Soviet BM21 122mm 20-barrel MRL instead of the 20mm gun. All extant 1995.

FARWA class *large patrol craft*

Displacement:	100t full load
Dimensions:	100ft oa × 21ft × 5ft 6in *30.5m × 6.4m × 1.7m*
Machinery:	3 shafts, 3 Rolls-Royce DV8TLM diesels, 1740bhp = 18kts. Range 1800nm at 14kts
Armament:	1–20mm Oerlikon Mk 7
Complement:	18

Class (commissioned):
Ar Rakib (4.5.67), *Farwa* (4.5.67), *Benina* (29.8.69), *Misurata* (29.8.69), *Akrama* early 1969), *Homs* (early 1969)

Vosper Thornycroft type, for customs and fishery protection, launched 1967–68, *Ar Rakib*, *Akrama* and *Farwa* transferred to Malta in 1978; rest extant 1995.

MINE WARFARE VESSELS

Ex-British 'HAM' class *inshore minesweepers*

Class (former name):
Brak (ex-*Harpham*), *Zuara* (ex-*Greetham*)

Loaned by the RN in 1963 as the new Libyan Navy's first ships. Transferred permanently and renamed in September 1966. Both discarded 1973.

Soviet 'NATYA' class *ocean minesweepers*

Class (no):
Al I'sar (ex-*Ras el Gehis*, 111), *Al Tayyar* (ex-*Ras Hadad*, 113), *Ras al Hamman* (115), *Ras al Falluga* (117), *Ras al Oula* (119), *Ras al Dawar* (121), *Ras Massad* (123), *Ras al Hani* (125)

Eight delivered 1981–86.

As with India's six they lack the stern ramp of Soviet Navy units. Two minefields laid 8 miles from Tripoli in June 1973, several since using Italian-built ro-ro ship *El Timsah* (acquired 1979).

MISCELLANEOUS

One 78ft Thornycroft coastal patrol craft (built Singapore 1962), the maintenance repair craft *Zleiten* (ex-British *MRC 1013*), bought 5 September 1966 and now a hulk, the 2200t logistic support ship *Zeltin* (Vosper, launched 29 February 1968, 15kts) can dock FAC of up to 120ft, fourteen SAR 33 class customs patrol boats (150t, 40kts, 1–40mm and 2–12.7mm MG) of W German design ordered early 1980 from Taskizak N Yd, Istanbul, two customs patrol boats *Jihad* and *Salam* (120t, 27kts, 4–20mm Hispano-Suiza A32) bought from Müller of W Germany in January 1978 when Lebanon was unable to pay. Six Yugoslav-built PB 90 class, in service 1985–86. All extant 1995.

Madagascar

Madagascar regained independence as the Malagasy Republic in 1960, after 64 years of French rule. The new state, under a right-wing autocratic regime, maintained strong economic and cultural links with France and joined the usual Francophone African organisation. Military rulers (since 1972) renegotiated the cooperation agreement with France, French air and naval bases were evacuated by 4000 troops in 1973. A policy of 'active neutralism', *ie* non-alignment, denies the use of the important naval base of Diego Suarez to foreign powers. The country is too remote and insignificant to influence African affairs appreciably; traditionally, the Malagasys (8,900,000 in 1982) have kept aloof from the mainland, perhaps because of indifference or a feeling of superiority.

The Malagasy armed forces, including the Gendarmerie, total nearly 30,000 18-month conscripts, and have been incorporated into the People's Armed Forces. Formed on 15 February 1961, the 530-man navy (including a company of 120 marines) has doubled since 1974 and operates a coast patrol force of light craft from Diego Suarez (shipyard transferred from French ownership 1975), Tamatave and Majunga. Lesser bases in the south of the island are Tulear, Nossi-Be, Fort Dauphin and Manakara. Madagascar's warships have been supplied by France, W Germany and N Korea. In 1991 the navy and air force combined as the Malagasy Aeronaval Force.

MALAIKA *large patrol craft*

Particulars: As Ivory Coast *Le Vigilant*

P 48 type boat ordered from SFCN in 1966. Laid down 5 December 1966 and launched 22 March 1967. In service since 11 April 1968 still with original armament of 2–40mm. Extant 1995, non-operational.

Ex-US YMS type *coastal minesweepers*

Class (former name, no):
Tanamasoandro (ex-*Marjolaine*, ex-D 337, ex-YMS 69), *Jasmine* (ex-*Jasmin*, ex-D 385, ex-YMS 29)

Wooden-hulled 1942 US-built ships and transferred to France in 1954. *Tanamasoandro*, bought 18 February 1961 by Madagascar as her first warship and used as a patrol vessel, was discarded in 1968; *Jasmine*, acquired in 1965 and rated as a tender, was deleted in 1976.

MISCELLANEOUS

The French seaward patrol craft VC 8 (P 758) was acquired in 1963 and named *Malaika*. Returned to France in 1967.

Five Maritime Police small coastal W German-built patrol craft (commissioned 1962, 46t, 22kts, 1–40mm). Since stricken.

'Batram' class LCT *Toky* (in service October 1974, intermediate in size between the French 'Batral' and EDIC types and paid for by France although built at Diego Suarez: she is reportedly armed with 8 SS12 SSM as well as guns.

Four ex-N Korean 'Nampo' class LCAs (acquired February and June 1979, rampless and used only as patrol craft), three sank in a storm April 1984, fourth non-operational 1995, and two ex-US LCMs.

One French EDIC class LCT transferred September 1985. Only operational vessel 1995.

The Coast Guard and training ship *Fanantenana* (1040t, 12kts, 1040t) is the ex-trawler *Richelieu*, Bremen-built in 1959 and refitted Diego Suarez December 1966–July 1967.

Malawi

Malawi (independent 1964) has no sea coast, but some small patrol craft (100 men) are in service at Chipoka on Lake Malawi (formerly Lake Nyasa) which is a quarter of the country. A Fairey Marine 'Spear' class boat joined the other three British-built vessels (acquired since 1968) in 1976 and a French-built 36t patrol boat *Kasanga* (ex-*Chikala*) entered service on 17 December 1984.

Malaysia

In 1950 the Malayan Naval Force consisted of the minelayer *Laburnum* (ex-Japanese *Wakataka* transferred to the Royal Navy at Singapore on 17 October 1947), the tank landing craft *Pelandok,* three seaward defence motor launches with four more awaiting transfer, the motor fishing vessel *Panglima* and the torpedo recovery vessel *Simbang.* No additions were made while the Royal Navy shielded the new Malayan Federation during the long years of the 'Emergency' up to 1960, even though independence came in 1957. On 16 September 1963 the Federation was enlarged by the addition of Singapore, Sabah and Sarawak (both states of North Borneo) and became Malaysia.

The Royal Malaysian Navy now had the role of patrolling a 3000 mile coastline and the vital 500-mile South China Sea link between the two halves of the Federation; a task complicated by Singapore's secession in 1965. Fortunately for the fledgling service the Royal Navy maintained a massive deployment for the Borneo 'Confrontation' with Indonesia as new warships were delivered and worked up. In 1964 the first frigate joined the minesweepers, as did the 103ft Vosper patrol craft which formed the simple and reliable basis of the Federation's new fleet. Fast attack craft were added in 1967 and by the early 1970s missile boats and landing craft were joining in numbers. In 1970 volunteer personnel numbered 4000. Defence spending increased sixfold between 1975 and 1981 in the uncertain era after Britain's final withdrawal and the fall of South Vietnam although Malaysia remains covered by the Five-Power Defence Arrangement of 1 November 1971, under which Britain and Australasia would offer aid if her security or Singapore's were threatened. RAAF fighter squadrons are still based in the Federation.

In 1995 the service numbered 12,700 with 1000 regular reserves and an RMNVR of 900 men and women, with ambitious expansion plans after a decade that saw a tripling of the Malaysian merchant fleet and a revival of piracy in the South China Sea. Shore bases are principally KD (*Kapal Diraja* = 'royal ship') *Malaya* at Woodlands on the Johore Strait and KD *Labuan* in Borneo. A new port was completed at Kuantan on the east coast in 1981 and Lumut (Telok Muroh) became KD *Pelandok* in 1984, main Fleet base, dockyard and training centre on the Strait of Malacca. Commodores head Malaysia's two naval areas, covering the mainland and Borneo.

Acquisition plans include OPVs, new FACs and, eventually, six submarines. It is also hoped to equip several ships with helicopter hangars.

MAJOR SURFACE SHIPS

Ex-British 'LOCH' class *frigate*

No	Name	Builder	Acquired	Fate
F 433	HANG TUAH (i) (ex-*Loch Insh*)	Robb	2.10.64	Stricken 1977

In 1964 HMS *Loch Insh* was transferred from the RN, refitted at Portsmouth DY and renamed *Hang Tuah* after a fifteenth century Malay admiral. She was stricken in 1977 when replaced by HMS *Mermaid*, and scrapped.

Ex-British MERMAID *frigate*

No	Name	Builder	Acquired	Fate
F 76	HANG TUAH (ii) (ex-*Mermaid*)	Yarrow	May 1977	Extant 1995

In May 1977 the RN frigate *Mermaid* was bought to replace the elderly *Hang Tuah*, and took over the same name when commissioned on 22 July the same year. It was hoped to replace the twin 4in with a GIAT 100mm Compact but this was dropped, and she remains virtually unaltered. Refitted 1991-92 for training duties. Diesels to be replaced 1996.

Rahmat 1972

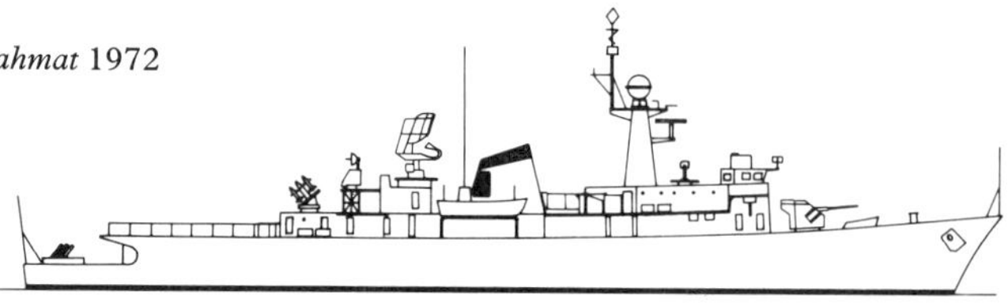

RAHMAT *frigate*

Displacement: 1250t standard; 1600t full load
Dimensions: 300ft pp, 308ft oa × 34ft 1in × 14ft 10in
91.4m, 93.8m × 10.3m × 4.5m
Machinery: 2-shaft CODOG (CP propellers): 1 Rolls-Royce Olympus TMIB gas turbine, 19,500shp = 26kts, plus Crossley-Pielstick diesels, 38.50bhp = 16kts. Range 6000nm/1000nm at 16kts/26kts
Armament: 1–4.5in/45 Mk 5, 2–40mm/70 Bofors (2×1), 4 Seacat SAM (1×4), 1 Limbo Mk 10
Sensors: Radar Decca 626, Signaal M-20, M-44, LW-02; sonar 170B, 174
Complement: 140

No	Name	Builder	Launched	Fate
F 24	RAHMAT (ex-*Hang Jebat*)	Yarrow, Scotstoun	18.12.67	Extant 1995

On 11 February 1966 Malaysia ordered a new frigate to be called *Hang Jebat*. She was laid down on 1 June 1967 and completed on 13 September 1971. Emphasis was laid on simplicity, with a hand-loaded gun forward, single Bofors guns amidships and a depth-charge mortar and a Seacat close-range missile system aft. There is also provision for a helicopter, with hatch covers over the Limbo well forming the flight deck. She has an unusual machinery layout, with three engines driving into one gearbox which, in turn, drives two propeller shafts.

The ship was renamed *Rahmat* before completion and is similar to the Thai Navy's *Makut Rajakumarn*. Modernised 1982–83 and Seacat SAM replaced by additional 40mm/70.

Rahmat on trials — Vosper Thornycroft

Lekir 1988

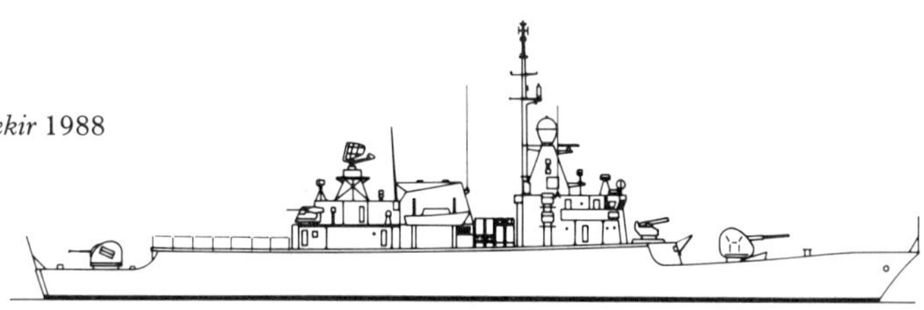

KASTURI class *frigates*

Displacement: 1500t standard; 1850t deep load
Dimensions: 319ft 1in oa × 37ft 1in × 11ft 6in
97.3m × 11.3m × 3.5m
Machinery: 2-shaft MTU 20V1103 TB92 diesels, 28,000bhp = 28kts. Range 3000nm at 18kts
Armament: 1–100mm Compact, 1–57mm/70 Mk 1, 4–30mm Emerlec (2×2), 1–375mm ASW rocket-launcher
Sensors: Radar DA O8, WM 22, TM 1226C; sonar DSQS-21C
Complement: 13 officers, 111 ratings

No	Name	Builder	Comp	Fate
25	KASTURI	HDW, Kiel	15.8.84	Extant 1995
26	LEKIR	HDW, Kiel	15.8.84	Extant 1995

Two FS 1500 design were ordered in February 1981 but plans to build another pair were shelved. Both laid down 3 January 1983 and launched 14 May 1983. May in due course be fitted with telescopic hangars for the newly acquired Wasp HAS 1.

Kasturi May 1993 L & L van Ginderen

Lekui class

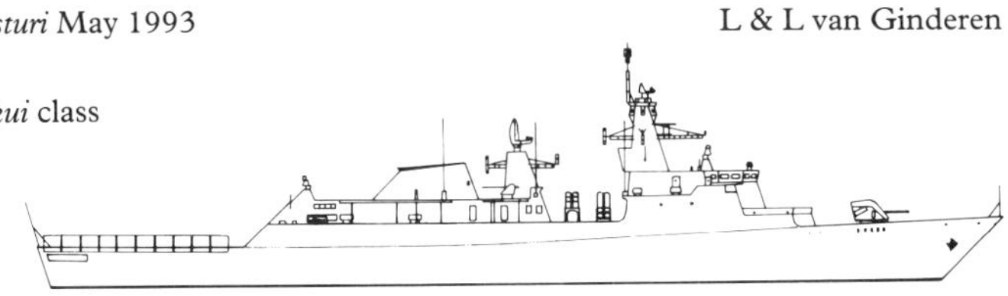

LEKIU class *frigates*

Displacement: 1845t normal; 2270t full load
Dimensions: 319ft 10in wl, 346 oa × 42ft × 11ft 9in
105.0m, 97.5m × 12.8m × 3.6m
Machinery: 2-shaft MTU 20V1183 TB 93 diesels, 40,000bhp = 28.5kts
Armament: 8 MM40 Exocet SSM, 1VLS Seawolf SAM (16 missiles), 1–57mm Bofors Mk 2, 2–30mm Mk 1 (2×1), 1 helicopter
Sensors: Radar DA-08, Sea Giraffe 150HC, 1802; sonar Spherion
Complement: 146

No	Name	Builder	Laid down	Launched	Fate
30	LEKIU	Yarrow SB	Dec 1992	3.12.92	Fitting out 1995
29	JEBAT	Yarrow SB	Nov 1994	May 1995	Building

Negotiations started with Yarrow in the late 1980s for the construction of two 80m corvettes, but the design was expanded into a frigate. The contract was worth £400m when the ships were ordered on 31 March 1992. Nautis-F command system, Mentor-2(V)1 electronic warfare system fitted.

Marikh 1988

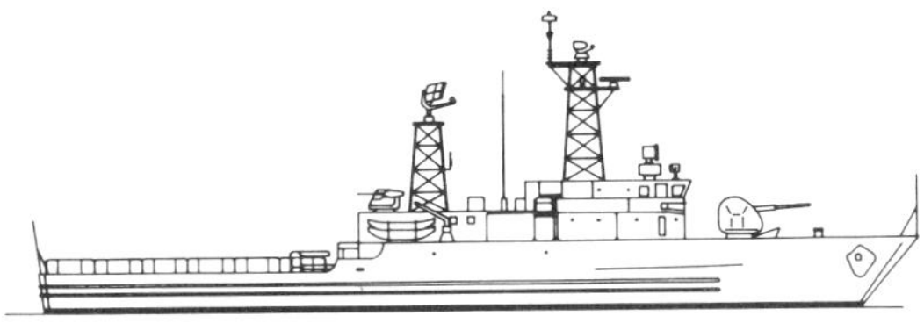

MUSYTARI class *offshore patrol vessels*

Displacement: 1300t deep load
Dimensions: 246ft oa × 35ft 5in × 12ft 1in
75m × 10.8m × 3.7m
Machinery: 2-shaft Pielstick diesels, 12,720bhp = 20kts. Range 6000nm at 20kts
Armament: 1–100nm/55 Compact, 2–30mm Emerlec (1×2)
Sensors: Radar DA-05, 9GA-600, TM 1226
Complement: 76 officers and ratings

No	Name	Builder	Launched	Fate
160	MUSYTARI	Korea SB, Pusan	20.7.84	Extant 1995
161	MARIKH	Malaysia SB & Eng, Johore	21.1.85	Extant 1995

Ordered June 1983. For patrol duties. A proposed third unit was not built.

New Generation Patrol Vessel Project

In 1993 the RMN announced that it was inviting bids for a large new class of offshore patrol vessels. This has since been designated the New Generation Patrol Vessels (NGPV) Project, but at the time of writing the final selection had not yet been made. The Australian shipbuilder Transfield is proposing a joint RMN/RAN design (see Australia section), but competition from other shipbuilders is intense. Broad requirements are for a 1000t hull armed with a medium-calibre gun, carrying a light helicopter and with provision for fitting SSMs and SAMs in wartime. Up to twenty-four may be built but the first batch will be twelve hulls.

SUBMARINES

In the late 1980s the Naval Staff formulated a requirement for submarines. Setting up a submarine force would take place in two stages, leasing or buying two refurbished second-hand boats as a preliminary to ordering six modern boats. Various proposals have been examined, notably British, French and Swedish, but the priorities were changed in 1990–1991. OPVs and corvettes will be acquired before submarines, putting the programme back to the end of the 1990s at least.

AMPHIBIOUS WARFARE VESSELS

Ex-US *tank landing ships*

Class (no, former name/nos):
Sri Langkawi (A 1500, ex-*Hunterdon County*, AGP 838, ex-LST 838), *Sri Banggi* (A 1501, ex-*Henry County*, LST 824), *Rajah Jerom* (A 1502, ex-*Sedgwick County*, LST 1123).

A 1500 loaned by USN on 1 January 1971, sold 1 August 1974, other two sold 7 October 1976. A 1500 was converted to small craft tender in US service. All have midships helicopter pad. A 1500 scrapped in the 1980s, and other two largely reduced to harbour service by 1992, supporting light forces.

SRI INDERA SAKTI class *logistic support ships*

Displacement: 1800dwt; 4300t full load
Dimensions: 328ft × 49ft 2in × 15ft 8in
100m × 15m × 4.8m
Machinery: 2 shafts, 2 Deutz S/BMV6 540 diesels, 5986bhp = 16.5kts. Endurance 60 days
Armament: 1–40mm
Sensors: Radar navigation set, Naja optronic FCS
Complement: 215

No	Name	Builder	Launched	Fate
A 1503	SRI INDERA SAKTI	Bremer Vulkan	1.7.80	Extant 1995
A 1504	MAHAWANGSA	Korea Tacoma	1.7.80	Extant 1995

A 1503 ordered 24 October 1979, laid down 15 February 1980 and completed 24 October 1980 as multi-purpose ship. A 1504 ordered in February 1981 and delivered May 1983. Usable as TS, FAC/MCM tender, C & C ship and troop transport. Bow thruster, stern helicopter deck and midships 15t crane all fitted. Large ops room, vehicle holds and divers' compression chamber are included. Up to 1300t of fuel and 200t water can be loaded with 3230sq ft (300sq m) of provisions. Ten 20ft containers can be carried on deck.

Sri Indera Sakti 1988 L & L van Ginderen

Ex-US NEWPORT class *tank landing ship*

Particulars: As US *Newport* class

Class (no, former name):
Sri Indera Putera (1505, ex-*Spartanburg County*, LST 1192)
Transferred from the USN in late 1994.

MINOR LANDING CRAFT

Five US LCM (6) class: *LCM 1–5* built in Australia and transferred *c*1970.

Nine RCP class LCVP (15t std, 30t full load, 17kts, 1–20mm, 35 troops or vehicle): *RCP 1–9* built Hong Leong–Lürssen SY, Butterworth, Malaysia. All in service 1974.

Fifteen LCP class (19t full load, 48ft 4in × 14ft × 3ft 3in, *14.6m × 4.3m × 1.0m;* 2 Cummins diesels, 400bhp = 16kts): built in Australia and transferred 1965–66. Light armour on midships pilot house.

Two *Jernih* class LCU (290t full load, 124ft 8in/*38m*, 2 Caterpillar D343 T diesels, 730bhp = 8kts: *Jernih* and *Terijah* built at Brooke DY, Malaysia 1977–78 to supply Sarawak. Another longer LCU, *Meleban*, was launched at the same yard on 15 October 1977.

SMALL SURFACE COMBATANTS

PERKASA class *fast attack craft*

Displacement: 95t normal; 114t deep load
Dimensions: 90ft wl, 99ft 9in oa × 24ft × 7ft max
27.4m, 30.4m × 7.3m × 2.1m
Machinery: 3 shafts: Protetus gas turbine plus 6cyl GM diesels, 12,750shp + 380bhp = 54kts. Fuel 24.25kt
Armament: 1–40mm/60 Bofors Mk 9, 1–20mm Oerlikon, 4–21in side-launched torpedoes or 10 mines
Sensors: Radar navigation set only
Complement: 24

Class (no, launched):
Perkasa (P 150, 26.10.65), *Handalan* (P 151, 18.1.66), *Gempita* (P 152, 6.4.66), *Pendekar* (P 153, 24.6.66)

Four modified versions of the RN 'Brave' class were ordered in October 1964 and shipped out from England in 1967. P 150 achieved 57kts on trials. Generally similar to the RN craft but the hull was made of bonded wood, as in the private-venture boat *Ferocity*. Air conditioned quarters. Armed with 8 SS 12 SSM (2 × 4) in 1972. Another improvement was the provision of cruising diesels on the wing shafts to improve endurance. All were built by Vosper at Southampton and all were stricken in 1977.

PERDANA class *fast attack craft*

Armament: 4 MM38 Exocet SSM (4×1), 1–57mm Bofors, 1–40mm/70 Bofors, 6–103mm RFL on 57mm mount
Complement: 35
Other particulars: As under Greece

Class (no, in service):
Perdana (P 3501, 21.12.72), *Serang* (P 3502, 31.1.73), *Ganas* (P 3503, 23.2.73), *Ganyang* (P 3504, 20.3.73)

Four 'Combattante II' type missile strike craft were ordered from CMN, Cherbourg in 1970. Although equipped with Swedish guns the fire control is the French Vega type normally associated with the 'Combattante II' FAC series. All four were launched in 1971–72 and delivered in 1973 after sailing for Malaysia on 2 May. All extant 1995.

HANDALAN class *fast attack craft*

Displacement: 240t normal; 268t full load
Dimensions: 143ft 1in oa × 23ft × 12ft 2in aft
42.6m × 7.0m × 2.3m
Machinery: 3 shafts, 3 MTU diesels, 10,800bhp = 34.5kts. Range 1850nm at 14kts
Armament: 4 MM38 Exocet SSM (4×1), 1–57mm/70, 1–40mm/70 Bofors
Sensors: Radar PEAB 9LV200
Complement: 39

Class (no, in service):
Handalan (P 3511, 26.10.79), *Perkasa* (P 3512, 26.10.79), *Pendikar* (P 3513, 26.10.79), *Gempita* (P 3514, 26.10.79).

Four modified versions of the Swedish *Spica* class were ordered on 13 August 1976 to replace the *Perkasa* class torpedo-boats, whose names were transferred. The bridgework is farther forward than in the original *Spica* design, no torpedoes are mounted, and diesels replace the gas turbines. All four built by Karlskrona Varvet. Extant 1995.

Pendikar 1990 — L & L van Ginderen

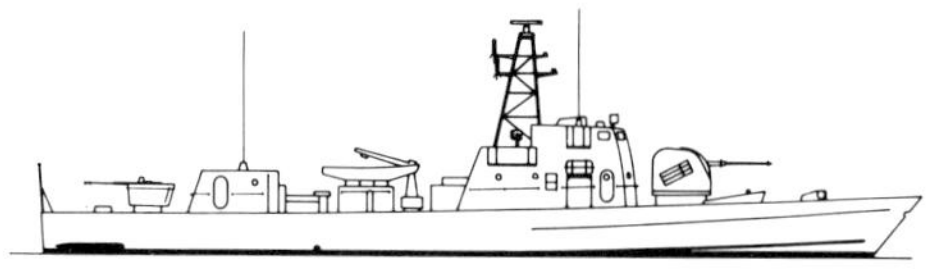

Jerong as completed
NB 1/750 scale

JERONG class *fast attack craft*

Displacement: 210t normal; 255t full load
Dimensions: 147ft 8in oa × 23ft × 8ft 2in aft
44.9m × 7.0m × 2.5m
Machinery: 3 shafts, 3 MTU diesels, 9900bhp = 32kts. Range 2000nm at 15kts
Armament: 1–57mm Bofors, 1–40mm/70 Bofors
Sensors: Radar Decca 1226, Naja optronic FCS
Complement: 35

Class (no, in service):
Jerong (P 3505, 21.3.76), *Tudak* (P 3506, 16.6.76), *Paus* (P 3507, 18.8.76), *Yu* (P 3508, 15.11.76), *Baung* (P 3509, 11.1.77), *Pari* (P 3510, 23.3.77)

Six B 45 type gunboats were ordered in 1973 from Lürssenwerft in Bremen-Vegesack but to be built at a new yard (Hong Leong) at Butterworth near Singapore. They are a Lürssen design somewhat smaller than the 'Combattante II' type, with Swedish armament.

SRI KEDAH class *large patrol craft*

Displacement: 96t normal; 109t deep load
Dimensions: 95ft wl, 103ft oa × 19ft 10in × 5ft 6in
31.4m × 6m × 1.7m
Machinery: 2 shafts, 2 Bristol Siddeley MTU MD665/18 diesels, 3500bhp = 27kts. Range 1400nm at 14kts
Armament: 2–40mm/70 Bofors (2×1)
Sensors: Radar
Complement: 22

Class (no, launched, in service):
Sri Kedah (P 3138, 4.6.62), *Sri Selangor* (P 3139, 17.7.62–25.3.63), *Sri Perak* (P 3140, 30.8.62–Jun 1963), *Sri Pahang* (P 3141, 15.10.62), *Sri Kelantan* (P 3142, 8.1.–12.11.63), *Sri Trengganu* (P 3143, 12.12.62–16.12.63), *Sri Sabah* (P 3144, 30.12.63–2.9.64), *Sri Sarawak* (P 3145, 20.1.-30.9.64), *Sri Negri Sembilan* (P 3146, 17.9.–28.9.64), *Sri Melaka* (P 3147, 25.2.–2.11.64). *Kris* (P 3148, 1.1.–11.3.66), *Sundang* (P 3149, 22.5.–29.11.66), *Badek* (P 3150, 8.5–15.12.66), *Renchong* (P 3151, 22.6.–17.1.67), *Tombak* (P 3152, 20.6.66–2.3.67), *Lembing* (P 3153, 22.8.66–12.4.67), *Serampang* (P 3154, 15.9.66–19.5.67), *Panah* (P 3155, 10.10.66–27.7.67), *Kerambit* (P 3156, 20.11.66–28.7.67), *Beludu* (P 3157, 11.1.–12.9.67), *Kelawang* (P 3158, 31.1–4.10.67), *Rentaka* (P 3159, 15.3–22.9.67), *Sri Perlis* (P 3160, 26.5.67–24.1.68), *Sri Johore* (P 3161, 21.8.67–14.2.68)

Six patrol craft were ordered from Vosper, Portsmouth, in September 1961 for delivery in 1963; the next four were ordered in March 1963 for delivery in 1965, followed by fourteen more in 1965 for delivery in 1966–68. Various minor improvements have been made in each group (Second group has greater range of 1600nm at 14kts) but the basic design is the same – an all-welded steel hull with aluminium superstructure. The last group were subsequently renumbered P 34, 36–47 and 49 respectively. *Sri Kedah* and *Sri Pahang* were stricken in 1976. Nineteen extant 1995.

'PZ' class *large patrol craft*

Displacement: 221t full load
Dimensions: 126ft 4in × 22ft 10in × 5ft 10in
38.5m × 7m × 1.8m
Machinery: 2 shafts, 2 MTU 20V 538 TB92 diesels, 10,000bhp = 31kts. Range 550nm/1200nm at 31kts/24kts
Armament: 1–40mm (1000rds), 1–20mm (2000rds), 2–7.62mm MG, 2 RFL
Sensors: Radar Kelvin Hughes
Complement: 38

Class (no):
Lang Hitam (PZ 1), PZ 2–12

Ordered from Hong Leong-Lürssen, Butterworth, Malaysia in 1979. First delivery August 1981, five in late 1981 and remainder by mid-1982. Royal Malaysian Marine Police ships.

Ex-British *seaward defence motor launches*

Class (former no, fate):
Sri Kedah (SDML 3501, BU 1959), *Sri Trengganu* (SDML 3502), *Sri Pahang* (SDML 3505, BU 1965), *Sri Negri Sembilan* (SDML 3506), *Sri Perak* (SDML 3507), *Sri Kelantan* (SDML 3508, BU 1965), *Sri Selangor* (SDML 3509, BU 1961)

Originally 7 SDMLs made up the 200th Seaward Patrol Craft Squadron based at Singapore. Only SDML 3502 remained in service by 1970. SDML 3506 and SDML 3507 were offered for sale in 1966.

'PX' class *coastal patrol craft*

Displacement: 85t full load
Dimensions: 87ft 6in × 19ft × 4ft 10in
26.7m × 5.8m × 1.5m
Machinery: 2 shafts, 2 MTU MB 820 Db diesels, 2460bhp = 25kts. Range 700nm at 15kts
Armament: 2–20mm (2×1)
Complement: 15

Class (no):
Mahkota (PX 1), *Temenggong* (PX 2), *Hulubalang* (PX 3), *Maharajasetia* (PX 4), *Maharajalela* (PX 5), *Pahlawan* (PX 6), *Bentora* (PX 7), *Perwira* (PX 8), *Pertanda* (PX 9), *Shahbandar* (PX 10), *Sangsetia* PX 11), *Lakasamana* (PX 12), *Pekan* (PX 13), *Kelang* (PX 14), *Kuala Kangsar* (PX 15), *Arau* (PX 16), *Sri Gumantong* (PX 17), *Sri Labuan* (PX 18)

Eighteen built by Vosper of Singapore 1963–70. Named after islands. All extant 1995 and operated by Royal Malaysian Marine Police except last two which are in Sabah government use.

Improved 'PX' class *coastal patrol craft*

Class (no):
Alor Star (PX 19), *Kota Bahru* (PX 20), *Kuala Tregganu* (PX 21), *Johore Bahru* (PX 22), *Sri Menanti* (PX 23), *Kuching* (PX 24), *Sri Gaya* (PX 25), *Sri Kudat* (PX 26), *Sri Tawau* (PX 27)

Nine built by Vosper of Singapore 1972–73 as 92t units, 91ft long (27.8m) with longer range (750nm) and larger crew (18). All extant 1995 in police service.

PX 28 class *coastal patrol craft*

Displacement: 114t
Dimensions: 95ft 1in × 19ft 9in × 5ft 7in
29.0m × 6.0m × 1.7m
Machinery: 2 shafts, 2 Paxman Valenta CM diesels, 2250bhp = 36kts. Range 1200nm at 24kts
Armament: 2–20mm (2×1)
Complement: 20

Class (no):
Sangitan (PX 28), *Sabahan* (PX 29), *Dungun* (PX 30), *Tioman* (PX 31), *Tumpat* (PX 32), *Segama* (PX 33)

Six built by Penang SB Co between 1979–83. All extant 1995 in police service.

MINE WARFARE VESSELS

Ex-British 'TON' class *coastal minesweepers*

Class (no, former name, transferred):
Mahamiru (M 1127, ex-*Darlaston*, 1960), *Kinabulu* (M 1134, ex-*Essington*, 1964), *Ledang* (M 1143, ex-*Hexton*, Oct 1963), *Tahan* (M 1163, ex-*Lullington*, Apr 1966), *Jerai* (M 1168, ex-*Dilston*), *Brinchang* (M 1172, ex-*Thankerton*, May 1966.

Six coastal minesweepers were transferred from the RN in 1960–66 starting with *Mahamiru*. All underwent modernisation at Singapore in 1972–73, but in 1977 *Jerai* was stricken, and the other three were deleted in the 1980s.

Ex-British 'HAM' class *inshore minesweepers*

Class (no, former name, in service):
Langka Suka (M 2606, ex-*Bedham*, 1958–67), *Sri Johore* (M 2602, ex-*Altham*, 1959–67), *Sri Perlis* (M 2604, ex-*Asheldham*, 1959–67), *Temasek* (M 2612, ex-*Brantingham*, 1958–66), *Jerong* (M 2627, ex-*Boreham*, Jan 1966), *Todak* (M 2610, ex-*Felmersham*, Mar 1966).

Six RN units transferred in two groups, four in 1958–59 and two in 1966. Some had 2–20mm aft instead of sweeping gear.

MAHAMIRU class *minehunters*

Class (no):
Mahamiru (11), *Jerai* (12), *Ledang* (13), *Kinabalu* (14)

Four modified *Lerici* type GRP minehunters were ordered from Intermarine in Italy on 20 February 1981. They arrived at Lumut on 26 March 1986, and all wereextant in 1995. They differ from Italian sisters in propulsion and electronics.

Maldives

This republic of about 2000 coral islands and islets (only 210 inhabited) lies 450 miles SW of Sri Lanka and extends more than that distance in twelve major atolls from north to south. No island is larger than 5 square miles or higher than 8ft above sea level. Britain granted independence in July 1965. The hereditary sultanate became a republic by referendum in November 1968. Britain retained an RAF staging post on Gan in the southernmost ADDU Atoll until March 1976. The almost entirely Muslim population (143,469 in 1978) are expert fishermen and traders with a merchant marine totalling forty-three ships of 180,117 grt in 1982.

The republic's navy was created in 1976 by the departing RAF handing over a 68ft target towing launch, a 63ft pinnace and four utility landing craft. A 45ft Fairey Marine customs launch had been bought the year before. Three ex-Soviet trawlers (2–25mm guns) were handed over by the USSR after being confiscated for illegal fishing, but an attempt to rent Gan in 1977 as a fishing base was rejected. There are about 250 personnel, five patrol boats and an LCU in service in 1995. A 17m Cheriton design patrol boat originally built for Kiribati was purchased in 1984. A 21m British *Tracker* type patrol boat was ordered in 1985 and commissioned in 1987, and three similar ex-British Customs and Excise boats were transferred in July 1989. The LCU was built in Singapore in 1991.

Mali

This landlocked desert republic (independent 1960) has five small river patrol craft (50 men) on her long stretch of the upper Niger, operating from Timbuktu, Barnako (capital), Segou and Mopti.

Malta

A Maltese coastal patrol force of small craft was formed in 1973. Malta achieved independence in 1964 but a Royal Navy presence was prolonged until 1979 under a NATO arrangement concluded in 1972. In September 1980 Malta declared her neutrality, banning any foreign troops or bases, including Soviet warship docking facilities (the first Soviet Navy courtesy visit was in 1978). Italy, France and Algeria have all since guaranteed the island republic's status.

Personnel strength of the Maltese coastal patrol force has slowly increased. In 1989 there were 150 officers and ratings and in 1992 there were 190. During the past few years numerous older patrol craft were decommissioned, but new acquisitions were few and insignificant. Some existing craft badly need replacement. In 1988 the government gave authorisation to seek bids for three or four new patrol boats and, according to some reports, Malta showed an interest in two ex-East German 'Tarantul I' missile corvettes built in the USSR. Such new ships would have given some bluewater capability to the Maltese naval force. In the event only two 'Kondor I' class patrol craft were acquired. In 1992 all pennant nos. were changed from 'C' to 'P'.

Malta-built customs launch (25t, 19kts, 2–12.7mm MG, 8 crew). Launched 1969, bought 1973, stricken 1979.

Three ex-W German customs launches (90–100t, 9–16kts, 1–12.7mm MG, 7–9 crew) bought 1974. *C 29* (ex-*Kondor*) built by Lürssen, stricken 1979. *C 27* (ex-*Brunsbüttel*) built Buschmann, Hamburg, 1953, stricken 1989. *C 28* (ex-*Geier*) built Bremen Burg, stricken 1979.

Two ex-Libyan (Yugoslav-built) customs launches (100t, 21kts, 1–12.7mm MG, 12 crew). *C 25* (ex-*Arraid*) and *C 26* (ex-*Tariq*) transferred 16 January 1974. *C 25* stricken 1990, *C 26* in 1991.

Two ex-US 'Swift' class PCF type patrol boats (22.5t full load, 25kts, 1–81mm mortar, 4–12.7mm MG, 6–8 crew). *C 23* (ex-US *C 6823*), *C 24* (ex-US *C 6824*) built by Sewart Seacraft 1967, bought or donated Jan–Feb 1971. Extant 1995.

Three ex-Libyan *Farwa* class large patrol craft transferred 1978. *C 30* (ex-*Farwa*) sank 1981; *C 29* (ex-*Akrama*) stricken 1989; *C 28* (ex-*Ar Rakib*) extant 1992.

Two 10t customs launches launched 1979. *Aphrodite, Kiklan* built by Guy Couacher. Fitted with two 7.62mm MG and Decca 110 radar.

Two ex-Yugoslav Type 131 coastal patrol craft transferred 31 March 1982. *President Tito* (C 38, ex-*Cer*) and *Ganni Bonnici* (C 39, ex-*Dom Mintoff, ex-Durmitor*). Stricken 1993.

Two British RAF RTTL Mk 2 class rescue launches (34.6t, 68ft, 30kts) transferred early 1970s. *C 68* (ex-*2768*), *C 71* (ex-*2771*); both deleted 1988.

Two US NOAA patrol craft acquired 1991 (35t, 59ft, 20kts, 2–7.62mm MG). *C 25* (ex-*1255*), *C 26* (ex-*1257*). Extant 1995.

Two ex-GDR 'Kondor I' class patrol craft *P 30* (ex-*Ückermünde*) and *P 31* (ex-*Pasewalk*) transferred from Germany 1992. Sonar removed, armament 1–14.5mm, 2–12.7mm MG.

Marshall Islands

Formerly a dependency of the United States, the Marshall Islands created their navy on independence in 1987. It has about sixty personnel and operates two patrol boats as of 1995. *Ionmeto 1* is an ex-US oilfield supply boat purchased in 1987, the *Ionmeto 3* is an Australian-built ASI 315 'Pacific Forum' class patrol boat delivered in 1991. *Ionmeto 2* was an ex-USCG 'Cape' class patrol boat transferred in 1987 and sold in 1992.

Mauritania

Mauritania, an independent Islamic republic since 1960, is larger in area than Nigeria but mainly desert, with a population of only 1.5 million, uneasily divided between the 80 per cent Arabs and 20 per cent black Africans. There are few natural resources, and the country has been exhausted by her 1975–79 support for Morocco against the Polisario Front guerrillas of the Western Sahara. Several palace revolutions since 1978 have caused a change of sides, and the French and Moroccan troops that formerly assisted her weak army have left. The Islamic Navy was formed on 25 January 1966. The country's fishing fleet of French and locally built trawlers is aiming to catch 250,000 tons a year. South Korea is to help build a naval shipyard.

The 500 volunteer naval personnel man a growing force of patrol craft based at Port Etienne, Nouadhibou, on Cape Blanc adjoining Western Sahara. In the south are the Senegal river ports of Rosso, Boghé and Kaedi.

Spanish BARCELO class *fast attack craft (gun)*

Class (no, in service):
El Vaiz (P 362, Dec 1979), *El Beg* (P 363, May 1979), *El Kenz* (P 364, Jan 1981).

Ordered 21 July 1976 from Bazán, Cadiz. First two delivered late due to trials collision in December 1978. All extant 1995.

French PATRA class *patrol boat*

Class (no, former name, in service):
El Nasr (P 411, ex-*Le 10 Juillet*, ex-French *Rapière*, 1.11.81)

Build by CNA as a private venture, purchased initially by the French Navy and then transferred to Mauritania. Not fitted with rumoured SSM armament. Renamed in 1988. Extant 1995.

Ex-British JURA class *offshore patrol vessel*

Class (former name, acquired):
N'Madi (ex-*Criscella*, ex-*Jura*, Jul 1989)

Former Scottish fisheries patrol vessel built Hall Russell 1975. Very similar to British *Isles* class except fitted with two Polar SP112 diesels. Sold commercially in 1988 and leased to Mauritania. Extant 1995.

TICHITT class *coastal patrol craft*

Class (in service):
Tichitt (Apr 1969), *Dar el Barba* (Sept 1969).

Built by CNE. First refitted at Dakar 1980, second in the Canaries 1982. Same as Tunisian *Istiklal* and Moroccan *El Sabiq* patrol craft. Both extant 1995.

French 18-metre *coastal patrol craft*

Class (in service):
Imrag'ni (Nov 1965), *Sloughi* (May 1968)

CNE built 20t units (21kts, 2 diesels, 1–12.7mm MG, 8 men) given to Mauritania. *Sloughi* deleted 1986, *Imrag'ni* out of service *c*1990.

MISCELLANEOUS

Ex-Spanish patrol vessels *Centinela* and *Serviola* (1953, 270t, 2–37mm, 12kts) served as *Tekane* and *Keur Macenf* from 5 March 1977, two ex-Soviet whalers, the *Boulanouar* and *Idini* (1956, 847t, 2–30mm, 1 MG), converted as patrol vessels and support ships. They were all sold to a Las Palmas (Canaries) shipbroker 12 January 1981.

One ex-East German *Nieastadt* class border patrol craft *Z'Bar* (ex-*Uelzen*) given by the German government in March 1990.

Mauritius

Mauritius became independent within the Commonwealth in March 1968. The densely populated main island has 90 per cent of the million inhabitants with 30,000 more on Rodrigues 350 miles to the east and a few hundred on the satellite islands. A defence agreement with Britain was not renewed when it expired in 1976, and the British naval communication base was closed down. The Mauritius Naval Volunteer Force, 111 strong, was disbanded in June 1968. Instead of a regular army, there is a security force of about 3500 men, including paramilitary elements trained by French advisers. Defence spending was only 4 million dollars in 1980.

In April 1974 India transferred as a gift a large patrol craft of the *Abhay* class (Indian-built version of the British 'Ford' class seaward defence boats); this ship kept her name of *Amar* and is extant 1995. Two ex-Soviet 'Zhuk' class coastal patrol craft were acquired in 1989, and re-named *Rescuer* and *Retriever*. There are also nine 15t patrol boats built in India 1989–90. The Soviet Union laid naval mooring buoys near Mauritius and used Port Louis as a fishing base. This fine harbour was also the seventh most visited by the Soviet Indian Ocean Squadron in the period 1962–80. Warships of all the major powers and India helped with cyclone relief operations in February 1975.

Mexico

Although the fifth largest navy in Latin America behind those of Argentina, Brazil, Chile and Peru, all of the large ships and most of the small craft are old, former US, equipment. Mexican naval forces are concerned mainly with patrolling fishing grounds as poaching is frequent; there have been brushes with Guatemala and the US. Mexico's offshore oilfields, recently developed, have necessitated greater vigilance by the navy and the raising of the defence budget from $567m in 1979 to $1150m in 1981 is proof of this. The navy operates a Gulf Command and a Pacific Command both of which are divided into zones. Main bases are at Veracruz, Tampico, Acapulco, Puerto Cortes, Manzanillo, Guaymas, Cuidad del Carmen and Islas Mujeres; 37,650 officers and men are employed – this figure includes 8650 marines and members of the naval air force.

FLEET STRENGTH 1947

CRUISER-TRANSPORT

Name	Launched	Disp	Fate
PROGRESO	1907	1590t	Discarded 1947

PATROL SLOOPS

Name	Launched	Disp	Fate
GUANAJUATO	29.5.34	1300t	Extant 1995
QUERETARO	29.6.34	1300t	Discarded 1975
POTOSI	24.8.34	1300t	Discarded 1975

GUNBOAT-TRANSPORT

Name	Launched	Disp	Fate
DURANGO	28.6.35	1600t	Extant 1995

GUNBOATS

Name	Launched	Disp	Fate
G 20	1922	130t	Discarded 1954
G 21	1922	130t	Discarded 1954
G 22	1922	130t	Discarded 1956
G 23	1922	130t	Discarded 1954
G 25	1922	130t	Discarded 1956
G 26	1922	130t	Discarded 1954
G 27	1922	130t	Discarded 1954
G 28	1922	130t	Discarded 1963
G 29	1922	130t	Discarded 1955

The only other significant vessels in service in 1947 were three trawler-type coast guard auxiliaries, *Mazatland*, *Acapulco* and *Veracruz* (486t, 1918). They were withdrawn from service after 1949.

MAJOR SURFACE SHIPS

Ex-US FLETCHER class *destroyers*

No	Name	Builder	Acquired	Fate
E 01	CUATHEMOC (ex-*Harrison*)	Consolidated, Orange	Aug 1970	Deleted 1982
E 02	CUITLAHUAC (ex-*John Rodgers*)	Consolidated, Orange	Aug 1970	Extant 1995

Ex-US GEARING class *destroyers*

No	Name	Builder	Acquired	Fate
IE 03	QUIETZALCOATL (ex-*Vogelgesang*)	Bethlehem, Staten I	Feb 1982	Extant 1995
IE 04	NETZAHUALCOYOL (ex-*Steinaker*)	Bethlehem, Staten I	Feb 1982	Extant 1995

FRAM I modernisation in US service. Renumbered E 10, E 11 respectively. *Quietzalcoatl* renamed *Ilhuicamina*.

Vincente Guerrero 1993 L & L van Ginderen

Ex-US CHARLES LAWRENCE/CROSLEY class *destroyer escorts*

Class (former name, transfer date):
California (ex-*Belet*, 12.12.63), *Chihuahua* (ex-*Barber*, 1971), *Coahuila* (ex-*Renour*, Jun 1967), *Papaloapan (ex-Earhart*, 12.12.63), *Tehuantepec* (ex-*Joseph M Auman*, 12.12.63), *Usmacinta* (ex-*Don O Woods*, 12.12.63)

Constructed as destroyer escorts, these units were converted to high speed transports while in US service. They could carry 162 troops. Used as patrol vessels while in Mexican service. *California* stranded and lost on Bahia, California, 16 January 1972; *Papaloapan* discarded 1976; *Tehuantepec* discarded 1989; all others extant 1995. *Usmacinta* renamed *Miguel Hidalgo*, *Coahuila* renamed *Vincente Guerrero*, *Chihuahua* renamed *Jose Maria Morelos Y Pavon* and *Tehuatepec* renamed *Zacatacas*.

Ex-US TACOMA class *frigates*

No	Name	Builder	Acquired	Fate
	CALIFORNIA (ex-*Hutchinson*)	Consolidated Steel, San Pedro	Nov 1947	Discarded Jun 1994
	PAPALOAPAN (ex-*Gladwyne*)	Globe, Duluth	Nov 1947	Discarded Jun 1974
	TEHUANTEPEC (ex-*Bangor*)	American SB, Lorain	Nov 1947	Discarded Aug 1964
	USUMACINTA (ex-*Annapolis*)	American SB, Lorain	Nov 1947	Discarded Aug 1964

Ex-US EDSALL class *frigate*

No	Name	Builder	Acquired	Fate
A 06	MANUEL AZUETA (ex-*Hurst*)	Brown SB, Houston	1.10.73	Extant 1995

Gulf Fleet command training ship.

Ex-US BRONSTEIN class *frigates*

No	Name	Builder	Acquired	Fate
E 40	NICOLAS BRAVO *(ex-McCloy)*	Avondale	16.11.93	Extant 1995
E 42	HERMENEGILDO GALENA (ex-*Bronstein*)	Avondale	16.11.93	Extant 1995

Two relatively modern frigates (by Mexican Navy standards) transferred from the US Navy in 1993. These were the first 'second generation' US postwar frigates. For further details see US section.

AMPHIBIOUS WARFARE VESSELS

Ex-US LST type *tank landing ships*

Class (former name):
Panuco (ex-*Park County*), *Manzanillo* (ex-*Clearwater County*)

Transferred 20 September 1971 and 25 May 1972, and employed as rescue ships, numbered A 01 and A 02. Extant 1995.

SMALL SURFACE COMBATANTS

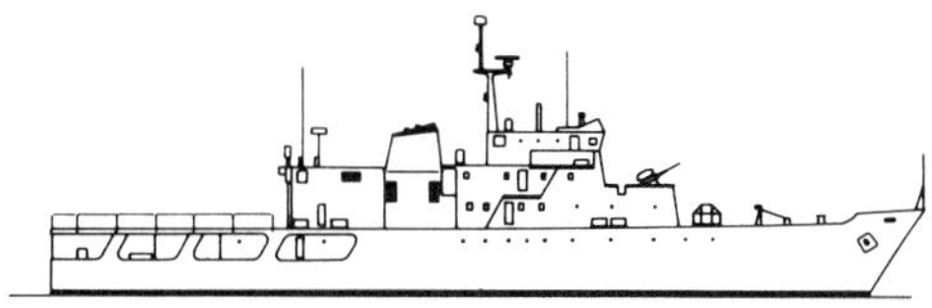

Halcon class as built

Spanish HALCON type *corvettes*

Displacement: 767t normal; 910t full load
Dimensions: 220ft oa, 206ft 9in × 34ft 6in × 10ft
67.0m, 63.0m × 10.5m × 3.08m
Machinery: 2 shafts, 2 MTU-Bazan 16V 956 TB91 diesels, 7500bhp = 21kts. Range 5000nm at 18kts
Armament: 1–40mm Breda, 1 MBB BO 105C helicopter
Sensors: Radar Decca AC 1226, Tacan, optronic director CSEE Naja
Complement: 52 (10 officers)

No	Name	Builder	Comp	Fate
GH 01	CADETE VIRGILIO URIBE	Bazan, Cadiz	2.6.82	Extant 1995
GH 02	TENIENTE JOSÉ AZUETA	Bazan, Cadiz	30.8.82	Extant 1995
GH 03	CAPITAN DE FREGATA PEDRO SÁINZ DE BARANDA	Bazan, Cadiz	30.10.82	Extant 1995
GH 04	COMODORO CARLOS CASTILLO BRETÓN	Bazan, Cadiz	4.11.82	Extant 1995
GH 05	VICEALMIRANTE OTHÓN BLANCO	Bazan, Cadiz	16.11.82	Extant 1995
GH 06	CONTRAALMIRANTE ANGEL MORTIZ MONASTERIO	Bazan, Cadiz	17.12.82	Extant 1995

Ordered in 1980. Generally similar to the ships built for Argentina. Contracts for further eight have been shelved. Delivery of GH 04 delayed by accident. GH 05 and GH 06 reportedly delivered only on 24 February 1983 and 24 March 1983. According to some reports main engines of 20V 956 TB 91 type capable of 13,300bhp for speed of 22–32kts. Large landing deck and hangar for helicopter. Used for EEZ patrol. Pennant nos. changed to C 11–C 16.

Vicealmirante Othóne P Blanco 1994 L & L van Ginderen

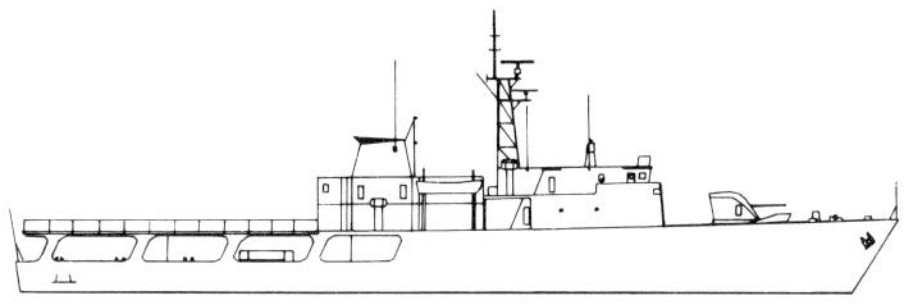

Uxmal 1991

AGUILA class *corvettes*

Displacement: 910t normal; 1175t full load
Dimensions: 244ft × 34ft 4in × 8ft 4in
74.4m × 10.5m × 2.5m
Machinery: 2 shafts, 2 MTU 20V 956 TB92 diesels, 11,700bhp = 22kts. Range 3820nm at 18kts
Armament: 1–57mm/70 Bofors Mk 2, 1 helicopter MBB BO 105C
Sensors: Radar Raytheon SPS 64 (V), fire control Elsag NA 18 optronic director
Complement: 75 (11 officers)

No	Name	Builder	Comp	Fate
GA 01	UXMAL	Tampico NSY	Nov 1991	Extant 1995
GA 02	MITLA	Veracruz NSY	1992	Extant 1995
GA 03	PETEN	Tampico NSY	1993	Extant 1995
GA 04	ANAHUAC	Veracruz NSY	1993	Extant 1995

Originally one ship ordered each from Tampico NSY and Veracruz. Improved version of *Halcon* class. Fitted with hangar and landing deck for helicopter. Programme slowed down considerably due to financial problems. The first unit renamed on completion *Sebastian José Holzinger* (C 01) and the others, respectively, *Capitan de Navio Blas Godinez Brito* (C 02), *Brigadier Jose Maria de la Vega Gonzalez* (C 03), *General Filipe B Berriozabal* (C 04). On completion, C 01 and C 02 were temporarily armed with a 40mm gun. To be used for EEZ patrol duties.

Ex-US AUK class *minesweepers (corvettes)*

Class (no, former name):
Leandro Valle (IG 01, ex-*Pioneer*), *Guillermo Prieto* (IG 02, ex-*Symbol*), *Mariano Escobedo* (IG 04, ex-*Champion*), *Ponciano Arriaga* (IG 03, ex-*Competent*), *Manuel Doblado* (IG 05, ex-*Defense*), *Sebastian Leido de Tejada* (IG 06, ex-*Devastator*), *Santos Degollado* (IG 07, ex-*Gladiator*), *Ignacio de la Llave* (IG 08, ex-*Spear*), *Juan N Alvarez* (IG 09, ex-*Ardent*), *Melchior Ocampo* (IG 10, ex-*Roselle*), *Valentin G Farias* (IG 11, ex-*Starling*), *Ignacio Altamirano* (IG 12, ex-*Sway*), *Francisco Zarco* (IG 13, ex-*Threat*), *Ignacio L Vallarta* (IG 14, ex-*Velocity*), *Jésus G Ortega* (IG 15, ex-*Chief*), *Gutierriez Zamora* (IG 16, ex-*Seater*), *Juan Aldarma* (IG 18, ex-*Pilot*), *Hermenegildo Galeana* (IG 19, ex-*Sage*)

All units transferred 1973. Reclassified as corvettes after minesweeping and ASW gear removed. All extant 1995.

Ex-US ADMIRABLE class *minesweepers (corvettes)*

Class (no, former name):
DM 01 (ID 01, ex-*Jubilant*), *DM 02* (ID 02, ex-*Hilarity*), *DM 03* (ID 03, ex-*Execute*), *DM 04* (ID 04, ex-*Specter*), *DM 05* (ID 05, ex-*Scuffle*), *DM 06* (ID 06, ex-*Eager*), *DM 10* (ID 10, ex-*Instill*), *DM 11* (ID 11, ex-*Device*), *DM 12* (ID 12, ex-*Ransom*), *DM 13* (ID 13, ex-*Knave*), *DM 14* (ID 14, ex-*Rebel*), *DM 15* (ID 15, ex-*Crag*), *DM 16* (ID 16, ex-*Dour*), *DM 17* (ID 17, ex-*Diploma*), *DM 18* (ID 18, ex-*Invade*), *DM 19* (ID 19, ex-*Intrigue*)

All units transferred 1 October 1962, except *DM 04*, February 1973. Reclassified as corvettes after minesweeping and ASW gear removed. *DM 02, 06, 10* and *16* deleted 1986. All others extant 1995.

Ex-US PCE type *submarine chasers*

Class (former no, fate):
Blas Godinez (ex-PCE 844, discarded 1965), *David Porter* (ex-PCE 847, discarded 1965), *Pedro Sainz de Baranda* (ex-PCE 868, discarded 1965), *Tomas Marin* (ex-PCE 871, discarded 1972), *Virgilio Uribe* (ex-PCE 875, discarded 1965)

All units purchased 1947. Armed with 1–3in, 6–40mm, 4–20mm, 4 DCT (*Uribe* no DCT; *Godinez* 3–40mm, 5–20mm).

Ex-US PC type *submarine chasers*

Class (former no, fate):
CG 30 (ex-*PC 820*, discarded 1966), *CG 31* (ex-*PC 608*, discarded 1964), *CG 32* (ex-*PC 614*, discarded 1964), *CG 33* (ex-*PC 813*, discarded 1966), *CG 34* (ex-*PC 794*, discarded 1964), *CG 35* (ex-*PC 824*, discarded 1966), *CG 36* (ex-*PC 1224*, discarded 1964), *CG 37* (ex-*PC 819*, discarded 1966), *CG 38* (ex-*PC 1210*, discarded 1971)

All units purchased 1947. Many were not armed while in Mexican service.

'Azteca' class FPB — ABMTM

'AZTECA' class *patrol boats*

Displacement: 148t full load
Dimensions: 111ft 10in oa, 101ft 4in pp × 28ft 3in × 6ft 6in
34.1m, 30.9m × 8.6m × 2.0m
Machinery: 2 shafts, 2 Ruston Paxman Ventura diesels, 3600bhp = 24kts. Range 2400nm at 12kts
Armament: 1–40mm, 1–20mm, 2 MG
Complement: 24

Class (no):
Andres Quintana Roo (P 01), *Matias de Cordova* (P 02), *Miguel Ramos Arizpe* (P 03), *José Maria Izazago* (P 04), *Juan Bautista Morales* (P 05), *Ignacio Lopez Rayon* (P 06), *Manuel Crescencio Rejon* (P 07), *Antonio de la Fuente* (P 08), *Leon Guzman* (P 09), *Ignacio Ramirez* (P 10), *Ignacio Mariscal* (P 11), *Heriberto Jara Corona* (P 12), *José Maria Mata* (P 13), *Feliz Romero* (P 14), *Fernando Lizardi* (P 15), *Francisco J Mujica* (P 16), *Pastor Rouaix José Maria* (P 17), *José Maria del Castillo Velasco* (P 18), *Luis Manuel Rojas* (P 19), *José Natividad Macias* (P 20), *Esteban Baca Calderon* (P 21), *Ignacio Zaragoza* (P 22), *Tamaulipas* (P 23), *Yucatan* (P 24), *Tabasco* (P 25), *Veracruz* (P 26), *Campeche* (P 27), *Puebla* (P 28), *Margarita Maza de Juarez* (P 29), *Leona Vicario* (P 30), *Josefa Ortiz Dominiquez* (P 31)

Designed by a British agency and ordered from Associated British Machine Tool Makers which sub-contracted the construction of the first twenty-one (from 1974 onwards) and assisted with the building in Mexico of a further ten (at Vera Cruz and Salina Crus). They are not entirely satisfactory in service, being subject to maintenance problems.

Mexico also operates the following:
Two ex-US 'Point' class patrol craft *Punta Morro* (P 60, ex-US *Point Verde*) and *Punta Mastun* (P 61, ex-P 46, ex-US *Point Herron*) built in the early 1960s and acquired 1991. Extant 1995.

Three ex-US 'Cape' class patrol craft *Cabo Corrientes* (P 42, ex-*Jalisco*, ex-US *Cape Carter*), *Cabo Corzo* (P 43, ex-*Nayarit*, ex-US *Cape Hedge*) and *Cabo Catoche* (P 44, ex-US *Cape Hatteras*) built between 1953–59 and acquired 1990–91. All extant 1995.

Four *Isla Coronado* class fast attack craft built by Equitable Shipyards 1994 (52t full load, 3 diesels, 16,200bhp = 50kts, 1–12.7mm MG, 2–7.62mm MG, 9 crew): *Isla Coronado* (P 51), *Isla Lobos* (P 52), *Isla Guadalupe* (P 53), *Isla Cozumel* (P 54). All extant 1995.

Micronesia

The Federated States of Micronesia became independent from the US on 10 May 1989. It comprises the Caroline Islands archipelago islands of Truk, Yap, Pohnepei and Kosrai. A patrol force with 100 personnel was set up in 1990, although the USA still retains responsibility for the defence of the islands from outside threats. Vessels extant in 1995 were:

Two ASI-35 'Pacific Forum' class patrol boats: *Palakir* (FSM 01), *Micronesia* (FSM 02). Built in Australia and transferred 7 June 1990 and 25 January 1991 respectively. Differ from other boats of the same class in service with other Pacific states in having spray strakes forward.

Three ex-USCG 'Cape' class patrol boats: *FSM 03* (ex-*Cape Cross*), *Constitution* (FSM 04, ex-*Cape Corwin*), *FSM 05* (ex-*Cape George*). All transferred 30 March 1990.

Morocco

The Sherifian Empire (now Kingdom) of Morocco obtained independence in 1956 from France and Spain, which had ruled it as a protectorate since 1912. Internally, initial and unsuccessful attempts at parliamentary government gave way to direct rule by King Hassan II. To obtain a national consensus for intervention in the Western Sahara since 1975, a four-party democratic system and parliament has been reintroduced, but ultimate power remains firmly in royal hands.

Moroccan foreign policy in the Middle East and Africa has been pro-Western since independence. French interventions in the Shaba province of Zaire in 1977 and 1978 were supported, Moroccan soldiers being airlifted to put down the rebel forces. Morocco also publicly approved the Camp David agreements between the United States, Egypt and Israel, but modified her attitude to retain essential subsidies from the Saudis and the Gulf States. Her main external problem in recent years, however, has been the West Sahara War. Morocco and Mauritania jointly occupied the former Spanish Sahara when Spain withdrew in 1975 to secure the world's richest nitrate deposits, but faced growing resistance from the Polisario Front, sponsored by Algeria and more recently Libya. In spite of French and Moroccan military support, Mauritania gave up the struggle in July 1979 and transferred her portion of the territory to Morocco; a military stalemate resulted, with Morocco controlling the main towns and the phosphate mines, and the Polisario Front making sporadic guerrilla raids from its bases near the Algerian border. The Western Sahara extends Morocco's 2100-mile coastline by half as much again.

Military, Morocco has concentrated her efforts on her Army and Air Force, with France as the main arms supplier. The United States has also supplied arms and in May 1982 concluded an agreement allowing the Rapid Deployment Force to use Moroccan bases. The self-esteem and prestige of the armed forces is high, as often happens when a right wing regime lacks a strong political base, but their loyalty to the throne has sometimes been justly suspect.

The Royal Moroccan Navy only dates from 1960 and is relatively small (personnel 7000 in 1995 including 1500 marines), but expanding, with most warships and naval equipment from France, although important recent orders have gone to Spain. An analysis of the 18-month conscript sailors suggests that they are reliable politically and adequately trained, but operationally inexperienced.

The main Atlantic bases are Casablanca, Safi, Agadir, Kenitra and Tangier; none near the Western Sahara. The age-old Spanish colonial ports of Ceuta and Melilla continue to dominate Morocco's Mediterranean coastline. A fishing dispute in 1973 caused a brief Moroccan-Spanish naval action. Morocco has been trying to acquire new frigates since the mid-1980s but lacks adequate funds.

Ex-British 'RIVER' class *frigate*

No	Name	Builder	Acquired	Fate
31	EL MAOUNA (ex-*La Surprise*, ex-*Torridge*)	Blyth	1964	Discarded 1975

Ex-RN frigate bought by France in 1944 and by Morocco in June 1964. Transferred in 1965 after conversion as flagship and royal yacht, extra accommodation and a helicopter landing deck having been added aft.

Modified Spanish DESCUBIERTA class *frigate*

No	Name	Builder	Launched	Fate
51	COLONEL ERRHAMAN (ex-*El Maouna*)	Bazán, Ferrol	26.2.82	Extant 1995

Ordered 14 June 1977 and laid down 20 March 1979. Strikes delayed delivery till 10 March 1983. Full details of modifications are not available, but Exocet MM38 or MM40 SSM will probably be carried instead of Harpoon in the Spanish ships. Particulars are as for the Spanish *Descubierta* class.

French CHAMPLAIN class *landing ships*

No	Name	Builder	Comp	Fate
42	DAOUD BEN AICHA	Dubigeon, Normandy	28.5.77	Extant 1995
43	AHMED ES SAKALI	Dubigeon, Normandy	June 1977	Extant 1995
44	ABOU ABDALLAH EL AYACHI	Dubigeon, Normandy	Dec 1978	Extant 1995

Ordered 12 March 1965 (first two) and 19 August 1975.

Abou Abdallah El Ayachi Jan 1983 L & L van Ginderen

French EDIC type *landing craft*

Lieutenant Malghagh (41, ex-21) ordered 1963 from CN Franco-Belges. Laid down 17 September 1963, launched 10 January 1964 and commissioned 1965. Extant 1995.

Ex-French CHAMOIS class *corvette (minesweeping sloop)*

No	Name	Builder	Acquired	Fate
33	EL LAHIQ (ex-*Annamite*)	Arsenal de Lorient	1961	Returned France 1967

CORMORAN class *fast attack craft (missile)*

Displacement: 363t standard; 355t full load
Dimensions: 194ft 6in oa × 24ft 7in × 6ft 3in
59.3m × 7.5m × 1.9m
Machinery: 2 shafts, 3 MTU-Bazán 16V956 TB 91 diesels, 13,500bhp = 36kts. Range 700nm/2500nm at 27kts/15kts
Armament: 4 Exocet MM38 SSM, 1–76mm/62 OTO Melara Compact (300rds), 1–40mm/70 Breda Bofors (1472rds), 2–20mm/90 Oerlikon GK204 (2 × 1, 3000rds)
Sensors: Radar Raytheon 1620, HSA WM-25 fire control, Decca TM-1229C
Complement: 41

Class (no, completion):
El Khattabi (P 35, 30.6.81), *Boutouba* (P 36, 11.12.81), *El Harty* (P 37, 25.2.82), *Azouggarh* (P 38, 2.8.82)

Ordered 14 June 1977 from Bazán, Cadiz, as modified Spanish *Lezaga* class (derived from Lürssen Type 143 design). *El Khattabi* launched 1 July 1980. All extant 1995.

Okba as completed **NB 1/750 scale**

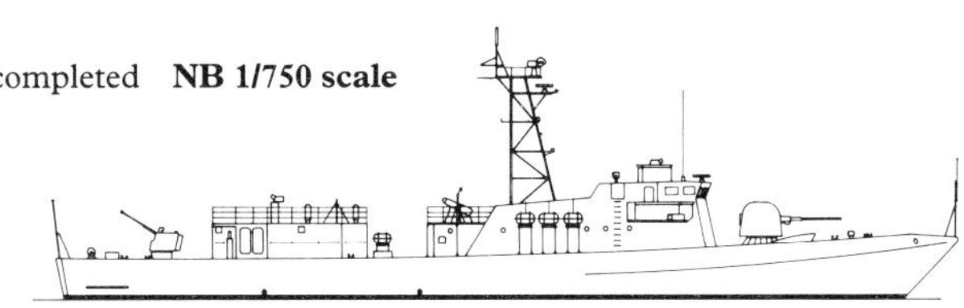

OKBA class *fast attack craft*

Displacement: 375t standard; 445t full load
Dimensions: 177ft 1in pp, 188ft 9in oa × 25ft × 7ft 1in
57.5m, 54.0m × 7.6m × 2.1m
Machinery: 4 shafts, 4 AGO V16 diesels, 11,040bhp = 28kts. Range 2500nm at 16kts, 20 days endurance
Armament: 1–76mm/62 OTO Melara Compact, 1–40mm/70 Breda
Sensors: Radar Vega system
Complement: 53

Class (no, launched):
Okba (33, 10.10.75), *Triki* (34, 1.12.76)

PR 72 type boats ordered from SFCN June 1973. Laid down 28 April 1975 and 22 November 1974, completed 16 December 1976 and 12 July 1977. Renumbered 302, 303. 302 refitted 1985. Both extant 1995.

OSPREY 55 class *patrol craft*

Displacement: 420t normal; 500t full load
Dimensions: 180ft 5in × 26ft 7in × 9ft
55.0m × 8.1m × 2.75m
Machinery: 2 shafts, 2 MAN/B & W diesels, 4960bhp = 18kts. Range 4500nm at 16kts
Armament: 1–40mm, 2–20mm (2 × 1)
Sensors: Radar Decca 1226
Complement: 15

Class (no, in service):
El Lahiq (308, Nov 1987), *El; Talifiq* (309, Feb 1988), *El Hamiss* (316, Feb 1990), *El Karib* (317, Feb 1990)

Four units built by Danyard, Denmark 1987–90. Intended for fishery protection. Similar vessels built for Greece (with more powerful armament) and Senegal. All extant 1995.

El Talifiq 7.5.92 L & L van Ginderen

VIGILANCE class *patrol boats*

Machinery:	2 shafts, 2 MTU TB 82 diesels, 7600bhp = 22kts
Armament:	1–40mm, 2–20mm (2 × 1)
Other particulars:	As Spanish *Lazaga* class

Class (no, in service):
Lieutenant de Vaisseau Rabhi (310, Sep 1988), *Errachiq* (311, Dec 1988), *El Akid* (312, Apr 1989), *El Maher* (313, Jun 1989), *El Majid* (314, Sep 1989), *El Bachir* (315, Dec 1989)

Six units built by Bazan, Spain. Reduced power and armament than Spanish *Lazaga* class for fishery protection duties.

Ex-US PC type *patrol vessel*

Name	Builder	Acquired	Fate
AGADIR (ex-*Goumier*, (ex-USS *PC 545*)	Defoe	1960	Returned France 1964

Transferred from the USN to France in 1944 and from France to Morocco in 1960, as the first ship in the new Moroccan Navy. Returned to France in 1964 and sold in 1965.

Modified French LE FOUGUEUX class *large patrol craft*

Lieutenant Riffi (32)

Built by CMN, Cherbourg, laid down May 1963, launched 1 March 1964 and in service May 1964. Armament reduced to 2–40 by 1982. Discarded 1990.

AL BACHIR *large patrol craft*

Displacement:	125t light; 154ft full load
Dimensions:	124ft 8in pp, 133ft 2in oa × 20ft 10in × 4ft 8in *40.6m, 38.0m × 6.4m × 1.4m*
Machinery:	2 shafts, 2 SEMT-Pielstick diesels, 3600bhp = 25kts. Oil 21t. Range 3000nm at 12kts
Armament:	2–40mm Bofors, 2–12.7mm MG
Complement:	23

Numbered 22 (ex-12), ordered in 1964 from CMN, Cherbourg, laid down June 1965 and launched 25 February 1967. In service since April 1967.

Ex-French SIRIUS class *coastal minesweeper*

Tawfic (ex-*Aries*). Transferred on 4-year loan 28 November 1974 (renewed 1978). In 1982 patrol duties as P 659. Loan made permanent 1980. Discarded 1986.

MISCELLANEOUS

There are ten coastal patrol craft: one ex-French seaward defence motor launch (ex-P 762, VC 12), transferred free 15 November 1960 and named *El Sabiq*, discarded 1980; six Albatros CMN P 32 type (built 1975–76) numbered 23–28 – with six more planned – and 3 Arcor 31 type).

Mozambique

This Portuguese territory became independent in June 1975, when Samora Machel's Frelimo guerrilla movement, which had risen against the colonial government in 1964, took power. Compared with less populous but richer Angola, the period of transition was peaceful; nevertheless, the departure of some 80,000 Portuguese settlers, about 80 per cent of the total, coupled with the immediate and almost complete socialisation of the economy, left the country in ruins.

A Treaty of Friendship and Co-operation was signed with the Soviet Union in March 1977. A succession of incursions, first by Rhodesia and then by South Africa, made it impossible for the Mozambique government to impose order on the country, despite aid from the USSR and other countries.

The *c*400-man Navy (volunteer whereas the Army and Air Force are 2-year conscripts) operated a force of small coast and lake patrol craft; the personnel were trained by Tanzania and the USSR. Their four main bases, Maputo (HQ, formerly Lourenço Marques, a 4500-ton capacity Soviet 'Klaipeda' floating dock was installed there in October 1981), Beira, Nacala and Pemba (Porto Amelia), cover the long coastline while Metangula serves Lake Nyasa. The country is bisected by the mighty Zambezi. The end of white minority rule in South Africa may help Mozambique rebuild its shattered economy and its navy.

In 1982 the Mozambican Navy consisted of the following coastal patrol and other units:

Six ex-Soviet 'Zhuk' class transferred 1979 (2), Oct 1980 (2) and Oct 1981 (2). Three extant 1995.

One ex-Soviet 'Poluchat' class transferred 1977, discarded since 1982.

Ex-Portuguese *Sirius* and *Vega* of the *Bellatrix* class. Since discarded.

Three ex-Portuguese *Jupiter* class (which operate on lake Nyasa) and the smaller ex-Portuguese *Antares*. Since discarded.

Ex-Portuguese survey ship *Almirante Lacerda* (ex-RCN *Bangor* class minesweeper *Caraquet*), transferred in 1975 and reported as a training and support hulk in Maputo.

Two ex-Soviet 'SOi' class patrol boats, transferred 1985. Both extant 1995.

Two ex-Soviet 'Yevgenya' class MCM craft, transferred 1985. Both extant 1995.

Ex-Portuguese *Bombarda* class LDG (LCT) *Cimitarra* and 2 ex-Portuguese LDM 100 type landing craft. A report that the last-named has been wrecked is not confirmed. All since discarded.

Myanmar (Burma)

The Republic of Myanmar as an exceptionally insular, almost 'forgotten' country, one quarter bigger than France, follows a policy of non-alignment and neutrality. Bordered to the north by China, and to the east by Thailand and west by Bangladesh, the 35 million Burmese are in a truly isolated position. Maintaining an isolated policy in the face of turbulence to east, north and south is a particularly difficult task. Up to the present this small navy established on 4 January 1948 has relied upon ex-British and American vessels of principally Second World War vintage supported by some Yugoslav-built craft.

With the policy of neutrality the Burmese Navy does not require a vast force, just sufficient craft to police the 1900 mile coastline together with a number of small craft for patrolling the numerous rivers and waterways. These forces are maintained mainly for a counter-insurgency role, a necessary task as there have been a number of insurrections in the country and, as much of the countryside is a dense jungle, the main means of communication is by water. In the 1970s the navy embarked on a modest modernisation programme to police the 200-mile EEZ, receiving from Australia six *Carpentaria* patrol craft, and from Denmark three offshore patrol vessels of the so called British-designed *Osprey* class/Danish *Havörnen*. It is unlikely that Myanmar will consider any extensive expansion programme and any new construction will be centred on light craft and the above mentioned OPV type vessels to replace overaged ex-British 'River', *Algerine* and US PCE ships.

During the 1980s Burma acquired a few small patrol boats and a group of six chinese 'Hainan' class vessels, Myanmar is being increasingly isolated from its traditional arms suppliers by the repressive actions of its government, and will probably go to China for replacements of its ageing fleet. Naval aviation consists of three maritime patrol aircraft of Fokker F-27M Friendship type, ten Aerospatiale SA 316B Alouette III and ten Kawasaki-Bell 47G-3 helicopters.

There are three naval regions with seven major bases: Yangon (Rangoon), Bassein (Irrawaddy Delta), Sittwe (Akyab), Moulmein in the Salween estuary, Mergui in the far south Tenasserim State, Sinmalaik and Seikyi. A 1958 strength of 3000 sailors with forty-two ships had become 10,000 (including 800 marines) with 106 ships by 1982.

Currently the personnel strength is reported as 7000 officers and men including 800 marines; some sources mention however 12000 officers and men. Service is by voluntary enlistment for two years and a rear-admiral (Vice Chief of Staff, Defence Forces, Navy) commands.

Ex-British 'RIVER' class *frigate*

Name	Builder	Acquired	Fate
MAYU (ex-*Fal*)	Smith's Dock	1947	Stricken 1979

Ex-RN frigate (F 266) bought in 1947 and re-named March 1948. Had RN Type 974 radar.

Ex-British LCG (M) type *landing craft*

Class:
Indaw, Inlay, Inma, Inya
Former British LCG(M) type landing craft, gun medium. Employed as gunboats. All stricken: *Indaw* in 1976, *Inma* and *Inya* in 1980–82, *Inlay* in 1983, and gave their names to OPVs built in Denmark.

Ex-US LCU 1610 type *utility landing craft*

Ex-US *LCU 1626*, cargo capacity 168t, transferred on completion at Rangoon under MDAP in June 1967 (launched 1963). Used as transport. Renamed *Aiyar Lulin* (603), early 1980s. Extant 1995. Second vessel reported built by Burma Naval DY.

AIYAR MAUNG class *landing craft*

Displacement: 250t full load
Dimensions: 125ft 6in × 30ft × 4ft 6in
38.2m × 9.1m × 1.4m
Machinery: 2 shafts, 2 Kubota diesels, 560bhp = 10kts
Complement: 10

Class:
Aiyar Maung (605), *Aiyar Min Thar* (607), *Aiyar Mai* (604), *Aiyar Min Tha Mee* (606)
Four landing craft launched March 1969 by Yokohama Yacht. Can carry 100t of cargo. Used as transports. All extant 1995.

SINDE class *landing craft*

Displacement: 220t full load.
Dimensions: 96ft 8in × 22ft × 4ft 6in
29.5m × 6.7m × 1.4m
Machinery: 2 shafts, 2 Kubota diesels, 300bhp = 10kts
Complement: ? + 30 passengers

Class:
Sinde, Htonbo
Two small landing craft built by Yokohama Yacht in 1978. Can carry 50t of cargo. Used as transports.

Ex-US LCM 3 type *landing craft*

Class:
LCM 701–708
Eight US-built 52t units (2 Gray Marine diesels, 4500bhp = 9kts, 2–20mm) used as transports able to carry 30t of stores. Extant 1995. *LCM 709–710* were added recently.

'OSPREY' class *offshore patrol vessels (corvettes)*

Displacement: 385t light; 505t full load
Dimensions: 164ft × 34ft 6in × 9ft
50m × 10.5m × 2.8m
Machinery: 2 shafts (2 CP propellers), 2 Burmeister & Wain Alpha 16V23L-VO diesels, 4640bhp = 20kts. Range 4500nm at 16kts
Armament: 2–40mm (1×2), 1–20mm
Complement: 20 (accommodation for 35)

Class (no, completed):
Indaw (FV 55, 30.5.80), *Inma* (FV 56, Dec 1981), *Inya* (FV 57, 1982)
Ordered in Denmark at Frederikshavn Vaerft in 1978. Identical with Danish *Havörnen.* Built in 1979–82. Mercantile design of helicopter-carrying small fishery protection vessel. Sometimes known as FV 710 type. Carry a David Still 25kt boarding boat in an inclined ramp at the stern. Sailed by the navy for the People's Pearl and Fishery Dept. *Indaw* operated effectively in her first six months against pirates, smugglers and illegal fishermen. All extant 1995.

Chinese 'HAINAN' class *large patrol boats*

Class:
Yan Sit Aung (43), *Yan Myang Aun* (44), *Yan Nyein Aung* (45), *Yan Khwin Aung* (46), *Yan Min Aung* (47), *Yan Ye Aung* (48), *Yan Paing Aung* (447), *Yan Win Aung* (448), *Yan Aye Aung* (449), *Yan Zwe Aung* (450)
A class of later variant of the Chinese 'Hainan' class large patrol boats fitted with tripod masts. First batch of six (43–48) delivered January 1991. Four more (447–450) acquired 1993.

Indigenous design *fast attack craft (gun)*

Displacement: 213t
Dimensions: 147ft 8in × 23ft × 8ft 2in
45.0m × 7.0m × 2.5m
Machinery: 2 shafts, 2 Mercedes Benz diesels = 30kts
Armament: 2–40mm Bofors (2×1)
Complement: 34 (7 officers)

Two FACs (gun) laid down 1991 at the Naval Engineering Depot at Rangoon. Completed 1994. Typical patrol craft layout with central superstructure and guns fore and aft.

Yugoslav PB 90 class *patrol boats*

Displacement: 80t standard
Dimensions: 89ft 11in × 21ft 8in × 7ft 3in
27.4m × 6.6m × 2.2m
Machinery: 3 shafts, 3 diesels, 4920bhp=32kts. Range 400nm at 25kts
Armament: 8–20mm (2×4), 1–128mm launcher for illuminants
Complement: 17

Class:
336, 337, 338
These fast ships were built by the Yugoslav yard Brodotechnika for an African country and completed in 1986–87, but not delivered. Sold to Burmese Navy and delivered in October 1990. Renumbered *424, 425, 426.* All extant 1995.

PGM 412 class *patrol boats*

Displacement: 128t full load
Dimensions: 109ft 11in × 22ft × 6ft 7in
33.5m × 6.7m × 2.0m
Machinery: 2 shafts, 2 Deutz SBA 16MB216 LLKR diesels, 2720bhp = 17kts. Range 1400nm at 14kts.
Armament: 2–40mm (2 × 1)
Complement: 17

Class:
PGM 412–415
Built in Burma by Burma Naval Dockyard, Rangoon, first two completed in 1983–84. Last two boats delivered to Coast Guard in 1993 but may be taken over by the navy.

RPC 11 class *river patrol boats*

Displacement: 30t
Dimensions: 49ft 10in × 14ft 1in × 3ft 7in
15.2m × 4.3m × 1.1m
Machinery: 2 shafts, 2 Thornycroft RZ6 diesels, 250bhp=10kts. Range 400nm cruising
Armament: 1–20mm
Complement: 8

Class:
RPC 11–15, I–IV
A new class built locally in the 1980s. An additional four boats of similar characteristics (37t) were built by the Naval Engineering Depot in 1990–1.

Ex-US PCE 827 class *patrol vessel*

Yan Taing Aung (41 ex-*Farmington* PCE 894) transferred 18 June 1965. Built 1942–44 by Willamette Iron & Steel. Fitted with SPS-5 radar and QCU-2 sonar. Extant 1995.

Ex-US ADMIRABLE class *ocean minesweeper (corvette)*

Yan Gyi Aung (42, ex-*Creddock, MSF 356*) transferred at San Diego 31 March 1967 with minesweeping gear removed. Built by Willamette Iron & Steel 1943–44. Fitted with SPS–5 radar and QCU–2 sonar. Extant 1995.

US 'SWIFT' type *large patrol craft*

Class:
PGM 421–423
See under Ethiopia for data. Armament differs being 2–40mm (2 × 1), 2–20mm. MTU diesels are 12V331 TC81, crew 25. Built by Swiftships, Morgan City, USA. Launched 1976, completed 1977–78 acquired 1980, extant 1992.

US 83ft type *motor gunboats*

Class:
MGB 101–110
Former USCG 83ft type patrol boat engines with new hulls built at Rangoon DYd, Burma. Completed in 1960. Three (*MGB 103, 107, 109*) are reported to have been sunk. Originally launched in 1938–42 and transferred by USCG under MDAP in 1951. *101, 104–106, 108* stricken 1983. Remainder reported extant 1995.

Ex-British *motor launches*

Class:
ML 391, 414, 415, 418, 437 (Fairmile type, launched in 1946), *ML 1102, 1103, 1104* (smaller Thornycroft type loaned in 1946)
All eight stricken *c* 1957–58.

Ex-British HDML type *motor launches*

Class:
MGB 1272, 1299, 1304, 1306, 1369, 1456, 1462, 1467, 1477, 1478, 1486
Eleven former RN HDMLs of same numbers transferred 1948–50. *MGB 1299, 1369, 1467, 1477* and *1486* stricken in 1957; others followed.

US PGM type *patrol gunboats*

Class:
PGM 401–406 (ex-US *PGM 43–46, 51, 52*)
Built by Marinette Marine, USA, 1959–61, austere export version, USCG 95ft type. Raytheon 1500 radar in *405* and *406*, EDO–320 in rest. Extant 1995.

Australian CARPENTARIA class *coastal patrol craft*

Displacement: 27t full load
Dimensions: 51ft 6in × 15ft 8in × 4ft 4in
15.7m × 4.8m × 1.3m
Machinery: 2 shafts, 2 GM 12V 71TI diesels 1120bhp = 27kts. Range 700nm/950nm at 22kts/18kts
Armament: 2–20mm
Sensors: Radar Decca 110
Complement: 10

Class:
01–06
Six built by de Havilland, Sydney. Aluminium construction. Ordered December 1978 for fishery protection, launched 1979–80, completed 1980–81. First two delivered 1979. Extant 1992.

British DARK class *fast attack craft*

Class:
PTS 101–105 (renamed during construction *T 201–205*)
Five interchangeable MTBs/MGBs built by Saunders-Roe and completed in 1956–57. Of aluminium construction, with riveted skin and aluminium alloy framework. The Saunders Roe slow-speed electric drive was fitted to facilitate manoeuvring in the confined inland waters where the craft were required to operate. Total cost of this order was £1.8m. All stricken 1975.

Y 301 class *river gunboats*

Displacement 120t
Dimensions: 104ft 9in × 24ft × 3ft
31.7m × 7.3m × 0.8m
Machinery: 2 shafts, 2 MTU diesels, 1000bhp = 13kts
Armament: 2–40mm (2×1), 1–2pdr
Complement: 29

Class:
Y 301–310
All ten launched in 1957 and completed in 1958 by Yugoslav yard Uljanik in Pula. Built to specific Burmese order. Extant 1995.

Nawarat as completed
NB 1/750 scale

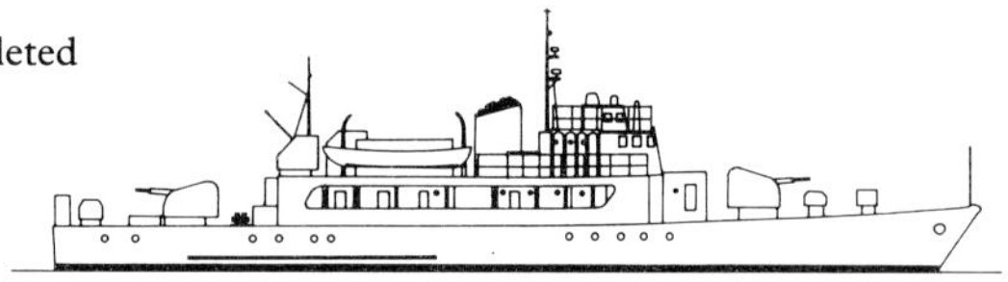

NAWARAT class *river gunboats*

Displacement: 400t standard, 450t full load
Dimensions: 163ft × 26ft × 4ft 7in
50.0m × 8.1m × 1.7m
Machinery: 2 shafts, 2 Paxman-Ricardo turbo-charged diesels, 1160bhp = 12kts
Armament: 2–25pdr (Army), 2–40mm
Complement: 43

Class (in service)
Nawarat (25.4.60), *Nagakyay* (3.12.60)
Built at the Government Dockyard Dawbon, Rangoon. Launched 1959 and completed 1960. Extant 1995.

Y 311 class *river gunboats*

Displacement: 250t
Dimensions: 121ft 4in × 24ft × 3ft 6in
37.0m × 7.3m × 1.1m
Machinery: 2 shafts, 2 MTU diesels, 1000bhp = 14kts
Armament: 2–40mm, 2–20mm (both 1×2)
Complement: ?

Class:
Y 311, Y 312
These improved Y 301 class vessels were built in Burma by Simalak Dockyard in 1967–69. Extant 1995.
There are also six PBR 211–216 river patrol craft of American PBR type acquired in 1978; 8.9t, 25kts, 3 MG (1 × 2, 1 × 1).

Ex-British ALGERINE class *ocean minesweeper*

Name	Builder	Acquired	Fate
YAN MYO AUNG (ex-*Mariner*, ex-*Kincardine*)	Port Arthur	18.4.58	Stricken 1982

Used also as an escort vessel (corvette). Canadian-built in 1943–45. Handed over to Burmese Navy in London. Fitted for minelaying and carried sixteen mines, eight on each side. Stricken 1982.

Ex-British MMS type *motor minesweepers*

Class (stricken):
MMS 197 (1957), *MMS 201* (1956)
Wooden units transferred from RN at Rangoon in Oct 1946. Both scrapped.

MISCELLANEOUS
Eight armed river transports (98t, 160bhp = 12kts, 1–40mm, 3–20mm. Earlier just 6–20mm): *Hintha, Saban, Sagu, Seinda, Setkaya, Setyahat, Shwepazun, Shwethida* and *Sinmin*. Six stricken 1988.
Boom defence vessel *Barstoke* served on loan from RN 1946–59.
There are three army landing craft (*LCM 001–003*) extant 1995.

Namibia

Namibia achieved independence from South Africa on 20 March 1990. In 1993 the Danish government transferred one 'Osprey' class fishery protection vessel (ex-*Havornen* –see Denmark section) to Namibia, with Danish officers aboard to train local crews. Several other small patrol craft and fisheries research ships are also operated.

Netherlands

The havoc wrought on the Netherlands Navy and its supporting defence industries by the Second World War was considerable. Most of the naval units still under construction at the time of the German occupation in May 1940 had been scuttled or completed for the German Navy; a few had escaped to Britain, where many items of equipment – especially armament and fire control – had to be installed, modified or replaced, either because the ships had left without them or because those already fitted were incompatible with available munitions. A number of older vessels also fled to Britain, where they saw heavy war service, while the bulk of the powerful East Indies Squadron had been destroyed by the Japanese in the early months of 1942. Some of the older ships which served in the European theatre and proved difficult to maintain had been replaced by British construction.

As if this were not enough, the Germans, in the face of the irresistible advance of the Allied armies through northern France and Belgium in September 1944, had embarked on the systematic destruction of the installations in Netherlands seaports and shipbuilding yards in the western part of the country. Practically all the tools of these yards were dismantled and taken to Germany, together with all the items of equipment on which the Germans could lay hands. Ships were scuttled in the narrow waterways and harbour entrances: the incomplete hull of the cuiser *De Zeven Provincien*, for example, was launched at Schiedam with the intention of using it for this purpose.

RECONSTRUCTION

The necessity of reconstructing the Royal Netherlands Navy in the face of a European political situation which was still far from stable was immediately recognised, and the Naval Staff accordingly drew up a new Fleet Plan which aimed at the construction of a new fleet combining the latest theories of naval warfare with the practical lessons learned from the participation of Netherlands naval units in the Second World War and the losses experienced.

By 1950 authorisation had been given for the purchase of a light fleet carrier from Britain, for the completion of the two large cruisers of the *Provincien* class, and for the construction of twelve large ASW destroyers and four submarines. Until these ships could be completed the Netherlands Navy would use its motley collection of prewar Netherlands-built and wartime British-built vessels to build up the expertise of its seamen, particularly in the new speciality constituted by ASW operations. Further submarines and escorts were acquired on loan from Britain and the USA for this purpose.

The difficulties inherent in the implementation of the Fleet Plan were two-fold: the lack of experience in designing and building ships as a result of the German occupation which had taken its toll of the design teams of the Netherlands United Shipbuilding Bureaux, and the near-total destruction of the Netherlands defence industries caused inevitable delays in the construction and fitting out of the vessels. In particular, the decision to develop a home-based electronics industry capable of supplying all the necessary radars and fire control systems resulted in a number of ships being completed with empty mast platforms. The problems involved in designing a ship on the basis of predicted antenna dimensions and weights supplied by HSA while the radars concerned were still at the development stage will be self-evident to the reader. Nevertheless, the decision to develop a Netherlands defence electronics industry rather than rely on the purchase of US or RN equipment has been a great success for the Netherlands, not only in terms of the quality of the equipment produced, but also because of foreign sales.

The design problems involved in the reconstruction of the Netherlands Navy during the immediate postwar period were eased by close collaboration with British constructors, who provided particular assistance in the revision of the plans of the *Provincien* class cruisers, and whose hand can also be seen in the hull form and layout of the *Holland* class destroyers.

NATO

The abandonment of colonial commitments in the Far East and membership of NATO in 1949 confirmed the reorientation of Netherlands naval strategy towards the European theatre in general, and anti-submarine warfare in particular. The mission of the Royal Netherlands Navy could now be divided into two clear areas of responsibility. The navy would provide blue-water units to NATO's East Atlantic command for open-ocean ASW, and would, in addition, have a major responsibility for the defence of shipping and harbours in the North Sea against submarines, aircraft and mines. The new fleet units under construction were therefore organised into squadrons of ASW destroyers with a cruiser or carrier flagship to serve as hunting groups. The single carrier, *Karel Doorman*, was modernised and had her air complement changed from that of a conventional attack carrier to that of an ASW carrier on the American model.

For the North Sea role, small diesel powered frigates were acquired from the United States and large numbers of minesweepers were built both in US and Netherlands yards. Many of these ships were funded under the Mutual Defense Assistance Program (MDAP).

By the end of the 1950s this naval programme was complete, and most of the prewar and war-built ships had been stricken or returned.

The formulation of the new Fleet Plan was accompanied by a decision to develop Den Helder as the principal base of the Royal Netherlands Navy. This was a gigantic task because all existing naval installations there had been damaged or destroyed by Allied or German air attacks, and the prewar harbour, even if restored, was totally inadequate for the envisaged fleet. A new harbour was therefore constructed in typical Dutch fashion, using the mudflats east of the existing port to form a large central basin with finger piers of varying sizes projecting from its sides. The old harbour, which now lay to the west, was closed off at its southern end and land was provided for various new installations.

All the large ocean-going units and submarines were to be based on Den Helder under the Commander Netherlands Task Group, while the smaller patrol craft and minesweepers would be divided among Amsterdam, Rotterdam and Flushing.

During the 1960s the structure of the Navy remained basically unchanged. The British *Leander* design was adopted to provide replacements for the American DEs purchased in 1950; an ocean-going combat support ship, *Poolster*, was built for operations in the eastern Atlantic; and the carrier *Karel Doorman*, badly damaged by a boiler fire in 1968. was sold to Argentina and replaced by eight French Atlantique long-range maritime patrol aircraft.

THE 1974 FLEET PLAN

The Fleet Plan formulated in 1974 provided for twenty-three major surface ships organised into three task groups, a single squadron of six submarines, and thirty MCM craft operating as three groups. The composition of the anti-submarine task groups was to be as follows:–

–Two task groups of one guided missile flagship *(Tromp* class), six ASW frigates ('Standard' class) and a combat support ship to operate in the eastern Atlantic.

–A task group comprising a smaller guided-missile flagship (an AAW variant of the 'Standard' frigate) and the six *Van Speijk* class ASW frigates allocated to Channel Command.

Replacements for the ageing submarines of the *Dolfijn* class would be ordered in two batches of two, while a third batch would eventually be ordered to replace the comparatively modern boats of the *Zvaardvis* class.

Two mixed MCM groups would operate off the Netherlands coast while a third smaller group would be provided for operations in the North Sea under CINCHAN.

In 1981 this plan was modified when two frigates of the 'Standard' class were sold to Greece while on the stocks. In view of the increased aerial threat in the eastern Atlantic, English Channel and North Sea areas it was decided that the two replacements ordered would be air defence variants of the 'Standard', and that the original 'Standard AD' would be replaced by the first of a new series of 'M' class frigates designed to operate in the North Sea.

THE 1984 FLEET PLAN

In 1984 a new ten-year Fleet Plan was prepared. Replacements were urgently required for the elderly ASW corvettes of the *Roofdier* class. Moreover, the navy's policy of ordering modern surface combatants in large batches in order to reduce building costs was creating major problems for the country's shipbuilders, as by the early 1980s the last of the 'Standard' frigates were fitting out, with no prospect for further orders until 1987, when the 'M' class was due to come on line. The collapse of the Rhein-Schelde-Verolme Group in 1983 led the Netherlands Government to reconsider its shipbuilding programme.

It was decided to bring forward construction of the 'M' class; the first batch of four was ordered in February 1984 to replace the *Roofdier* class, which were stricken immediately to help finance the programme, and options were taken on a further four ships as replacements for the *Van Speijk* class. Orders for the second batch of two *Walrus* class submarines were placed at the same time.

It was planned that the two guided-missile frigates of the *Tromp* class would receive an extensive mid-life update from 1988 onwards, and that they would be followed by the 'Standard' frigates of the *Kortenaer* class. The pace of construction for the fifteen 'Tripartite' minehunters of the *Alkmaar* class was increased, and a new class of fifteen minesweepers was to be ordered from 1988 onwards.

This plan had to be modified in 1986. Budget cuts reduced the surface fleet to twenty ships, of which only sixteen would be operational. Four of the *Van Speijk* class were reduced to reserve status and subsequently sold to Indonesia, and five of the fifteen minehunters were placed in reserve. It was decided that of the two guided-missile flagships only *Tromp* would be modernised, and that two of the *Kortenaers* would not undergo the planned 'capability upkeep programme' and a further two would not receive the Goalkeeper CIWS. In compensation, the 'M' class frigate had by this time grown in both size and capability, and would now be capable of operations beyond the North Sea.

Further economies were to follow. Under the 1989 Modified Ten-Year Plan neither of the two flagships was to be modernised, and the 'capability upkeep programme' planned for the *Kortenaers* was limited to six units. The third batch of *Walrus* class submarines was cancelled. The Defence White Paper of 1991 proposed radical changes in the organisation of the fleet. By 1996 there were to be only two large task groups, each comprising a flagship of the *Tromp* class, an air defence frigate of the *Heemskerck* class, four 'M'-class and three *Kortenaer* class anti-submarine frigates, three submarines, and a fast combat support ship. A third reserve task group would comprise four mothballed *Kortenaers*, to be decommissioned by 2000. Only six of the fifteen *Alkmaar* class minehunters would remain operational, the others being placed in reserve.

Even this reduced fleet proved to be unsustainable in a much-changed political and economic climate. In order to preserve planned construction programmes such as the new air defence frigates (*Tromp* replacements) and the amphibious transport dock for the Marines the navy opted for premature disposal of the *Kortenaers*. Three ships were sold to Greece in 1993 and a further three are scheduled for sale to the same country in 1996.

The new White Paper published in January 1994 established requirements in terms of sustainable peacetime naval deployments rather than force structures. Thus the navy will deploy ships on a continuous basis as follows: two frigates to NATO standing forces, one to the Netherlands West Indies, and a fourth as guard-ship in home waters. It has been calculated that these deployments can only be sustained with a force of twelve operational frigates (in addition to those in major refit).

COMMAND STRUCTURE

The Netherlands currently holds the NATO post of COMBENECHAN, and one of the three senior positions in Channel Command. The national command structure in the postwar era has been based on four maritime regions – Ijmond (HQ: Amsterdam), Rijmond (HQ: Rotterdam), Texel (HQ: Den Helder), and Schelde (HQ: Vlissingen).

The only 'national' naval responsibility is the Netherlands West Indies (HQ: Curaçao).

MAJOR RADARS OF NETHERLANDS MANUFACTURE

Designation	Description	In service
LW-01	Long-range air surveillance	1956
LW-02/03	Long-range air surveillance	1956
LW-08	Long-range air surveillance	1978
SMART-L	Long-range air surveillance	projected
DA-01/02	Medium-range air/surface surveillance	1956
DA-05	Medium-range air/surface surveillance	1979
DA-08	Medium-range air/surface surveillance	1986
SMART	Medium-range target designation	1992
SPS-01	3-D air surveillance/tracking	1975
M25	152mm fire control	1953
M45	120mm (and smaller) fire control	1953
M44	Seacat fire control	1967
WM25	Gun/missile fire control	1975
STIR-180	Gun/missile fire control	1978
STIR-240	Tracker/illuminator for Standard	1986

FLEET STRENGTH 1947

AIRCRAFT CARRIER

Name	Launched	Disp	Fate
KAREL DOORMAN (ex-*Nairana*)	20.5.43	13,825t	Returned RN 1948

Acquired on loan in 1946, this vessel was returned to the Royal Navy in 1948, her place being taken by the former HMS *Venerable* (see under Major Surface Ships).

CRUISERS

Name	Launched	Disp	Fate
JACOB VAN HEEMSKERCK	16.9.39	4150t	Stricken 1958
TROMP	24.5.37	4200t	Stricken 1958

After several years of active service these two cruisers became accommodation ships and were renumbered A 879 and A 878 respectively.

DESTROYERS

No	Name	Launched	Disp	Fate
J 1	BANCKERT (ex-*Quilliam*)	29.11.41	1750t	Stricken 1957
J 2	EVERTSEN (ex-*Scourge*)	8.12.42	1796t	Stricken 1962
J 3	VAN GALEN (ex-*Noble*)	17.4.41	1760t	Stricken 1957
J 4	PIET HEIN (ex-*Serapis*)	25.3.43	1796t	BU 1961
J 5	TJERK HIDDES (ex-*Nonpareil*)	25.6.41	1760t	To Indonesia 1951
J 6	KORTENAER (ex-*Scorpion*, ex-*Sentinel*)	26.8.42	1796t	Stricken 1962
HX 4	MARNIX (ex-*Garland*)	24.10.35	1335t	Stricken 1964

Tjerk Hiddes and *Van Galen* were transferred in 1942, and the other war-built destroyers in 1945–47. *Evertsen*, *Kortenaer* and *Piet Hein* were reconstructed as fast frigates at the Rijkswerf Willemsoord 1957–58, when they were fitted with a helicopter platform aft. *Marnix* was transferred from the Royal Navy in 1947. Following a refit 1948–49 she became an ASW training ship.

Van Galen 27.3.52 L & L van Ginderen

SUBMARINES

No	Name	Launched	Disp	Fate
–	O 21	21.10.39	962t	BU 1958
–	O 23	5.12.39	962t	BU 1949
–	O 24	18.3.40	962t	Stricken 1956
–	O 27 (ex-*UD 5*)	26.9.41	962t	Stricken 1959
O 28	TIJGERHAAI (ex-*Tarn*)	29.11.44	1320t	Stricken 1966
O 29	ZWAARDVIS (ex-*Talent*)	17.7.43	1320t	Stricken 1963

From the mid-1950s the surviving three Netherlands-built boats were all used for training. *Zwaardvis*, transferred in 1943, and *Tijgerhaai*, transferred 1945, were fitted with a snorkel and had a sonar dome between the bow tubes. Two further 'T' class boats were acquired on loan in 1948 (see below).

FRIGATE

No	Name	Launched	Disp	Fate
JO 1	JOHAN MAURITS VAN NASSAU (ex-*Ribble*)	23.4.43	1463t	BU 1960

Johan Maurits Van Nassau was transferred in 1943. She served postwar as an ASW and gunnery training ship. She had single 105mm mountings in place of the British 4in guns.

GUNBOATS

No	Name	Launched	Disp	Fate
N 1	FLORES	24.10.35	1734t	Depot ship 1955
N 3	VAN KINSBERGEN	5.1.39	2095t	Stricken 1959
N 5	VAN SPEIJK (ex-*K 3*)	22.3.41	1365t	BU 1960

Van Speijk had been completed by the Germans during the war. In 1953 she was re-engined with diesels from the submarine *O 23*, resulting in a 3kts increase in speed. She became an accommodation ship and was renumbered A 877. *Van Kinsbergen* was laid up in 1945, and in 1951–52 was renamed, the original 4–120mm being replaced by two single German-pattern 105mm. In 1955 she became a depot ship; *Flores* was broken up in 1968.

MINELAYER

Name	Launched	Disp	Fate
WILLEM VAN DER ZAAN	15.12.38	1267t	Stricken 1963

Willem Van Der Zaan was reclassified as a frigate in 1953. In 1961 she became an accommodation and repair ship for minesweepers at Flushing and was renumbered A 880. The former minelayer *Jan Van Brakel* was converted to a survey ship at Soerabaya in 1950 and renumbered A 906.

CORVETTES

Name	Launched	Disp	Fate
TERNATE (ex-*Kalgoorlie*)	7.8.41	790t	Stricken 1958
BATJAN (ex-*Lismore*)	10.8.41	790t	Stricken 1958
BOEROE (ex-*Toowoomba*)	26.3.41	790t	Stricken 1958
CERAM (ex-*Burnie*)	25.10.40	790t	Stricken 1958
AMBAN (ex-*Cairns*)	5.7.41	790t	To Indonesia 1950
BANDA (ex-*Wollongong*)	5.7.41	790t	To Indonesia 1950
MOROTAI (ex-*Ipswich*)	11.8.41	790t	To Indonesia 1949
TIDORE (ex-*Tamworth*)	14.3.42	790t	To Indonesia 1949

These eight Australia *Bathurst* class minesweepers were acquired in 1946 and served as patrol ships in the Far East, four being transferred to Indonesia. In the early 1950s the single 4in gun was replaced by a 105mm mounting.

MINESWEEPERS

Seven ex-British wooden MMS of 255t (launched 1943–44) were operated until disposed of 1949–52; *Duiveland* (MV 13), *Overflakkee* (MV 14), *Schokland* (MV 15), *Tholen* (MV 16), *Voorne* (MV 17), *Wieringen* (MV 19) and *Ijsselmonde* (MV 20). *Duiveland* and *Ijsselmonde* were transferred to Indonesia. The remainder were broken up.

There were also ten BYMS type vessels of 223t; *Borndiep*, *Deurloo*, *Hollandsch Diep*, *Marsdiep*, *Oosterschelde*, *Texelstroom*, *Vliestroom*, *Volkerak*, *Westerschelde* and *Zuiderdiep*. They were numbered from MV 34 to MV 43 respectively (changed in 1950 to NATO numbers M 831–840). In the late 1950s *Zuiderdiep*, *Borndiep*, *Marsdiep*, *Deurloo* and *Vliestroom* became diving vessels and were renumbered A 904–906, A 921 and A 922 respectively. The unconverted ships were disposed of in 1958 and the diving vessels followed in 1962.

Eight MMS type minesweepers (219t) were operated until 1958 (except *Vlieland*, sunk New Guinea 1951): *Ameland*, *Beveland*, *Marken II*, *Putten*, *Rosenburg*, *Terschelling*, *Texel* and *Vlieland*. They were initially numbered MV 5–12, and were later given the NATO numbers M 861–867. Four prewar Netherlands-built minesweepers (460t, launched 1936–37) were employed as patrol and gate vessels: *Abraham Crijnsson*, *Abraham Van Der Hulst*, *Jan Van Gelder* and *Pieter Florisz*. Initially numbered MV 1–4, they were later given the numbers A 925–928. They were stricken in 1962.

NATO PENNANT NUMBERS

Carried from 1950 onwards by ships in this section:

Cruiser

C 803	JACOB VAN HEEMSKERCK	C 804	TROMP

Destroyers

D 801	BANCKERT	D 805	PIET HEIN
D 802	EVERTSEN	D 806	TJERK HIDDES
D 803	VAN GALEN	D 807	MARNIX
D 804	KORTENAER		

Submarines

S 801	O 21	S 812	TIJGERHAAI
S 804	O 24	S 814	ZWAARDVIS
S 807	O 27		

Frigates

F 802	JOHAN MAURITS VAN NASSAU	F 814	BOEROE
F 803	FLORES	F 815	CERAM
F 804	VAN KINSBERGEN	F 816	TERNATE
F 805	VAN SPEIJK	F 824	WILLEM VAN DER ZAAN
F 813	BATJAN		

Karel Doorman Oct 1963 L & L van Ginderen

MAJOR SURFACE SHIPS

KAREL DOORMAN *aircraft carrier*

Displacement: (all details 1958) 15,892t standard; 19,896t full load
Dimensions: 630ft pp, 693ft oa × 80ft wl, 121ft fd × 25ft
192.0m, 211.3m × 24.4m, 37.0m × 7.6m
Machinery: 2-shaft Parsons geared steam turbines, 4 boilers, 40,000shp = 23.5kts. Range 12,000nm at 14kts
Armament: 12–40mm (12 × 1), 21 aircraft. See notes.
Sensors: Radar LW-01, LW-02, DA-01, 2 VI-01, ZW-01
Complement: 1462

No	Name	Builder	Acquired	Fate
R 81	KAREL DOORMAN (ex-*Venerable*)	Cammell Laird	1.4.48	To Argentina 1969

Laid down on 3 December 1942, launched on 30 December 1943 and completed on 17 January 1945, the British light fleet carrier *Venerable* was purchased in 1948 and given the same name as the former escort carrier she replaced. In her early years she operated Sea Fury FB 11 fighter-bombers and Firefly Mk 4/5 ASW aircraft purchased from the UK; twenty-three Sea Fury FB 51s were later built in the Netherlands under licence.

In the early 1950s the Dutch decided that the excellent condition of the hull and machinery justified a full modernisation to incorporate the latest developments in carrier aviation. This was duly undertaken by Wilton-Fijenoord, the refit lasting from 1955 until 1958. A heavier, modified, angled flight deck, 165.8m long was fitted, together with a mirror landing sight and a new steam catapult to port forward; a new type of arrester gear was installed. The island superstructure was completely redesigned to incorporate the new generation of Dutch radars. A tall funnel, of similar configuration to that of the cruisers and the new destroyers, carried an LW-02 air search antenna, while an even taller quadruped lattice mast carrying a long-range LW-01 air search and a DA-01 combined air/surface search radar replaced the former tripod. Atop either end of the island was a VI-01 'nodding' height-finder, and the sensor fit was completed by a ZW-01 navigation radar above the bridge. The original AA armament of six quadruple pompoms and single 40mm was removed and replaced by twelve single 40mm mountings, two of which occupied each quadrant with the remaining two pairs at either end of the island. Soon after the ship recommissioned, however, the port forward pair was removed to clear the end of the angled deck. A number of major internal compartments were redesigned, and the main and auxiliary machinery received a complete overhaul.

After modernisation *Karel Doorman* operated a squadron of Sea Hawk FGA 6 jet fighters and a squadron of TBM-3 Avengers. These were replaced in the early 1960s by eight S2F-1 Tracker ASW aircraft and six HSS-IN Seabat helicopters, the ship becoming the centre of a submarine hunter-killer group.

In 1965–66 *Karel Doorman* was reboilered with units removed from the uncompleted British carrier *Leviathan* of the same class. She was badly damaged by a boiler-room fire on 29 April 1968, and was subsequently sold to Argentina, being first refitted by Wilton-Fijenoord. In this refit she also received the turbines from *Leviathan.*

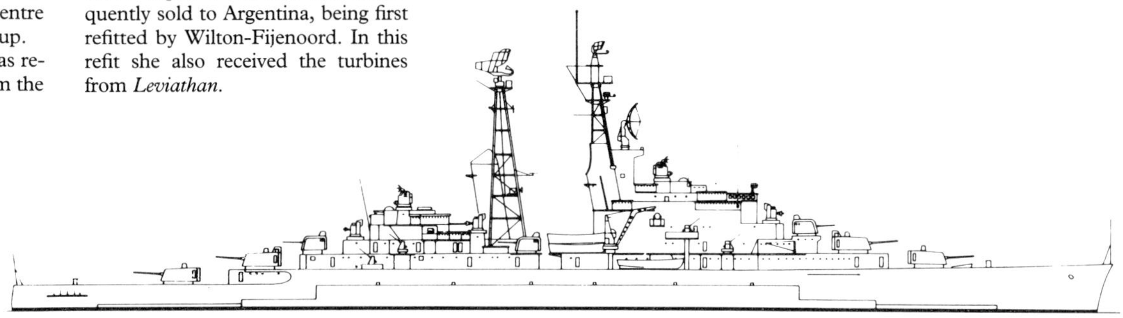

De Ruyter 1960

DE RUYTER class *cruisers*

Displacement: 9529t standard; 11,850t full load
Dimensions: 591ft pp, 615ft (*De Zeven Provincien* 609ft) oa × 57ft × 22ft
182.4m, 187.3m (*185.7m*) × *17.3m* × *6.7m*
Machinery: 2-shaft De Schelde-Parsons geared steam turbines, 4 boilers, 85,000shp = 32kts
Armament: 8–152mm (4 × 2), 8–57mm (4 × 2), 8–40mm (8 × 1). See notes
Sensors: Radar LW-01, DA-01, VI-01, ZW-01, 2 M25, 2 M45. See notes
Complement: 926

No	Name	Builder	Laid down	Launched	Comp	Fate
C 801	DE RUYTER (ex-*De Zeven Provincien*)	Wilton-Fijenoord	5.9.39	24.12.44	18.11.53	To Peru 1973
C 802	DE ZEVEN PROVINCIEN (ex-*Eendraght*, ex-*Kijkduin*)	Rotterdam DD	19.5.39	22.8.50	17.12.53	To Peru 1976

De Zeven Provincien after Terrier SAM conversion CMP

The hulls of the two *Provincien* class cruisers – an enlargement of the prewar *De Ruyter* design – were still intact at the end of the Second World War, although little work had been done on them during the German occupation. *De Zeven Provincien* (which had been given an 'Atlantic' bow with more pronounced rake and flare, on the instructions of the German navy), had been launched with a view to using her to block the port of Rotterdam, while *Eendragt* was still on the stocks. All of the equipment intended for the ships had been removed by the Germans; much of this was in any case obsolete, so construction was postponed until 1947 when completely new plans were drawn up with the aid of the British.

The armament was revised to take account of the increased threat from the air. Eight 152mm DP guns with an elevation of 60°, partly manufactured by Wilton-Fijenoord under licence from Bofors, replaced the former battery of ten 150mm. The new weapons had a rate of fire of 15 rounds per minute and were fully automatic and radar-controlled, as was the new secondary battery of twin-mounted 57mm guns in quick revolving triaxially stabilised turrets. Eight Bofors 40mm/70 in single mounts and a 103mm flare launcher were later added.

War experience favoured the abandonment of the original layout of the propulsion machinery – three boiler rooms followed by three engine rooms – in favour of the unit system. The whole plant was therefore divided into two independent units each consisting of a boiler room and an engine room.

These two major modifications resulted in considerable changes in the internal layout of the ships because of the different arrangement of magazines, hoists, and boiler trunking. The adoption of the unit system for the propulsion machinery also necessitated a second funnel, while the need to accommodate an action information centre, new fire control stations, navigational aids and air and surface search antennae resulted in major changes in the bridge structure – which doubled in size – and the provision of two large tripod masts. In order to preserve deck space the forefunnel trunking was incorporated into the bridge structure and the foremast built around it, and the mainmast was rebuilt around the second funnel following sea trials. Welding was used wherever practicable and aluminium alloy employed extensively in order to keep topweight within limits. Despite this the ships turned out over 1000t heavier than the original design, while retaining the same dimensions.

Other modifications made as a result of war experience included the dispersal of the electric generators (AC replacing DC), the dispersal of living quarters, and the incorporation of the latest damage-control techniques. The elimination of scuttles and the accompanying need for more ventilation and air-conditioning placed additional demands on electrical generating capacity, which had to be greatly increased. On completion the two ships, which by now had been renamed, conducted their trials in Arctic and Equatorial waters respectively.

It was planned to fit both ships with Terrier for operations in the northeast Atlantic, but the cost of reconstruction was considerable, and it was felt that the remaining useful life of *De Ruyter* was too short to justify the expense involved. The conversion of *De Zeven Provincien* involved plating up the main deck to the forecastle deck aft because of the extra depth needed to accommodate the missile magazine rings; small openings in the stern were needed for line handling. The after 152mm and 57mm mountings and the forward 40mm mountings were removed; armament as reconstructed was thus one Terrier Mk 10 SAM launcher (40 missiles), 4–152mm (2 × 2), 6–57mm (3 × 2) and 4–40mm (4 × 1). A new mainmast carrying the 3-D SPS-39 antenna was sited abaft the second funnel, the LW-01 air search antenna being relocated on the forward edge of the funnel itself. ZW-01 was retained, as were one each of the two M25 and M45 sets, and two SPG-55s were fitted for missile guidance. In a later modification (1971–72), the obsolescent VI-01 heightfinder was removed and the DA-01 scanner relocated in its place in order to reduce topweight. The original SPS-39 scanner was replaced by an AN/SPA-72 planar antenna, and Corvus chaff dispensers were added at 01 level abreast the bridge.

Both cruisers served as squadron flagships throughout their careers, being succeeded in that role by the *Tromp* class missile destroyers. The Terrier installation on *De Zeven Provincien* had to be removed and returned to the USA when she was sold to Peru.

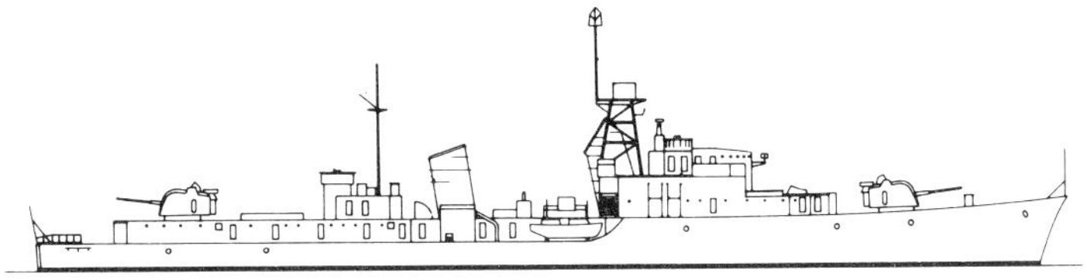

Holland as completed

HOLLAND class *destroyers*

Displacement: 2215t standard; 2765t full load
Dimensions: 361ft pp, 371ft oa × 37ft × 17ft
109.9m, 113.1m × 11.4m × 5.1m
Machinery: 2-shaft Werkspoor-Parsons geared steam turbines, 4 boilers, 45,000shp = 32kts. Range 4000nm at 18kts
Armament: 4–120mm (2 × 2), 1–40mm, 2–375mm ASW RL (2 × 4), 2 DC racks
Sensors: Radar LW-02, DA-01, ZW-01, M45; sonar Types 170B, 162
Complement: 247

No	Name	Builder	Laid down	Launched	Comp	Fate
D 808	HOLLAND	Rotterdam DD	21.4.50	11.4.53	31.12.54	To Peru 1978
D 809	ZEELAND	Royal Schelde	12.1.51	27.6.53	1.3.55	BU 1979
D 810	NOORD BRABANT	Royal Schelde	1.3.51	28.11.53	1.6.55	Scrapped following collision 1974
D 811	GELDERLAND	Wilton-Fijenoord	10.3.51	19.9.53	17.8.55	Hulked 1973

Specifically designed for ocean ASW operations, the *Holland* class were the first European fleet destroyers to be completed without anti-ship torpedoes. They were required to protect task forces or convoys against submarine attack, to hunt submarines either singly or in ASW hunter-killer groups, to assist in the air defence of task forces or convoys, and to defend the latter against attack by light surface forces. To equip them for the latter two roles they were to carry a heavy dual-purpose armament; the 120mm Bofors twin mounting chosen was fully automatic and radar-controlled, capable of 45 rounds per minute and with an elevation from – 10° to + 85°. Additional back-up in the anti-air role was to be provided by five 57mm, while a number of ASW mortars (including the British Squid) were considered before the adoption of the four barrelled Bofors 375mm R/L.

Six ships were ordered in 1948 for completion in 1952, and six more for completion in 1953–5, but it proved impossible to meet these dates largely because of the damage caused to Netherlands defence industries by the war. In order to speed construction the first four units were built using equipment salvaged from destroyers of the *Isaac Sweers* class and stored away in sheds during the German occupation. The other eight units then became a separate subgroup (see below).

The *Holland* class were fitted with propulsion machinery which had been designed for much smaller destroyers, and this imposed some restrictions on size and maximum speed compared with their successors. Two of the sets of turbines had been built prewar for the uncompleted units of the *Isaac Sweers* class, while a further two similar sets had been ordered by the Germans. In spite of this, 40.3kts were achieved on trials, albeit without a number of important items of equipment on board.

The hull form is reminiscent of contemporary British destroyers. Unusually, for ships of their size, side armour as well as deck protection was incorporated in order to reduce vulnerability to near-misses from aerial bombing, and great attention was paid to watertight subvision. The high-tensile A52 steel which had proved so successful in prewar vessels was used for plating and all structural work. In order to reduce topweight, aluminium alloy was used wherever possible, the entire after deckhouse being so constructed. Electric welding was used throughout – the most extensive use to date of this technique by Netherlands shipyards.

Wind tunnel tests on a forefunnel enclosed in a quadruped lattice mast led to the adoption of a common funnel design for the destroyers, the two cruisers and *Karel Doorman* (on refit). The air search radars under development were not ready at the time of completion of the *Holland* class, and all four ships were initially fitted with a pole mainmast. This was changed to a short lattice structure in 1955, but the DA-01 and LW-02 antennae were not fitted until 1957–58. The additional topweight involved may have been responsible for the reduction of the light AA armament to a single 40mm fitted forward of the second funnel.

In the late 1950s it was proposed that the after 120mm be placed by guided missiles, but the conversion never materialised and the ships continued in service without any major modification until disposed of in the mid-1970s. The 120mm mountings from *Gelderland* were installed in the new *Tromp* class.

Holland 4.7.76 L & L van Ginderen

Groningen 1980 C & S Taylor

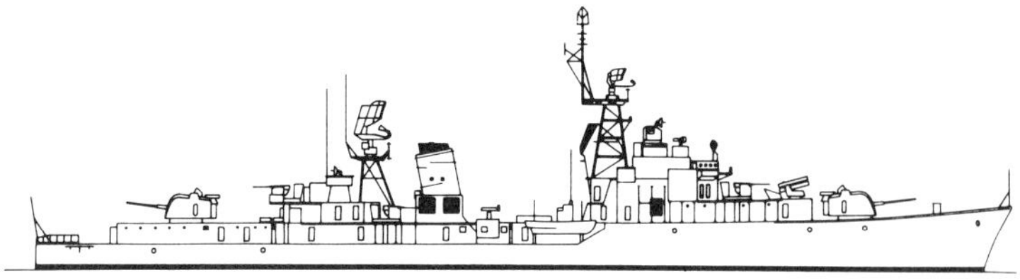

Friesland as completed

The major difference between the *Friesland* class and their predecessors lay in the increased size and power of their propulsion machinery, which was identical to that of the US *Gearing* class; 42.8kts was attained on trials. The increase in length and displacement enabled them to carry a heavier battery of light AA guns: four single 40mm were grouped around a built-up after structure carrying the lattice mainmast, and a further two were fitted forward of the bridge.

In 1960 *Utrecht* was fitted with eight 533mm anti-submarine torpedo tubes, four on each side of the upper deck. *Overijssel* followed in 1961, but the project was dropped – as was a similar one in the Royal Navy – because of new developments in ASW, and the tubes were removed.

The two forward 40mm were removed in the mid-1960s, and the fire control radar for the remaining 40mm in 1977–78. In the early 1970s the British sonars originally fitted were replaced by a CWE-610 – the same model fitted in the *Tromp* class frigates.

From their completion the *Friesland* and *Holland* classes constituted the backbone of the ASW hunting groups until replaced in that role by the 'Standard' frigates.

FRIESLAND class *destroyers*

Displacement: 2497t standard; 3070t full load
Dimensions: 370ft pp, 381ft oa × 38ft × 17ft
112.8m, 116.0m × 11.7m × 5.2m
Machinery: 2-shaft Werkspoor geared steam turbines, 4 boilers, 60,000shp = 36kts. Range 4000nm at 18kts
Armament: 4–120mm (2 × 2), 6–40mm (6 × 1), 2–375mm ASW RL (2 × 4), 2 DC racks
Sensors: Radar LW-02, DA-01, ZW-01, M45; sonar Types 170B, 162
Complement: 284

No	Name	Builder	Laid down	Launched	Comp	
D 812	FRIESLAND	Amsterdam	17.12.51	21.2.53	22.3.56	BU 1979
D 813	GRONINGEN	Amsterdam	21.2.52	9.1.54	12.9.56	To Peru 1980
D 814	LIMBURG	Royal Schelde	28.11.53	5.9.55	31.10.56	To Peru 1980
D 815	OVERIJSSEL	Wilton-Fijenoord	15.10.53	8.8.55	4.10.57	To Peru 1982
D 816	DRENTHE	Amsterdam	9.1.54	26.3.55	1.8.57	To Peru 1981
D 817	UTRECHT	Royal Schelde	15.2.54	2.6.56	1.10.57	To Peru 1980
D 818	ROTTERDAM	Rotterdam DD	7.1.54	26.1.56	28.2.57	To Peru 1981
D 819	AMSTERDAM	Amsterdam	26.3.55	25.8.56	10.4.58	To Peru 1980

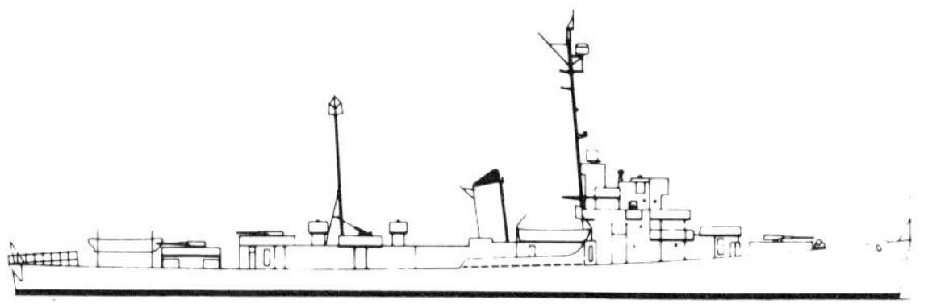

De Zeeuw 1966

These ex-American DEs of the *Cannon* class were purchased with MDAP funds in 1950; the first four were delivered in 1950 and the last pair in 1951. *Van Ewijck, Dubois, De Zeeuw* and *Van Zijll* retained their triple bank of torpedo tubes when transferred, but these, together with most of the 20mm AA, were later removed. All six ships were returned to the US Navy in December 1967 for scrapping.

Ex-US CANNON class *destroyer escorts*

Armament: 3–3in (3 × 1), 6–40mm (3 × 2), 8 to 10–20mm (8 to 10 × 1), 1 Hedgehog, 4 DCT, 2 DC racks, 3–21in TT (1 × 3) except *Van Amstel, De Bitter*
Other particulars: As US *Cannon* class (DET type)

No	Name	Builder	Acquired	Fate
F 806	VAN AMSTEL (ex-*Burrows*)	Dravo, Wilmington	Jun 1950	Returned USA Dec 1967, BU
F 807	DE BITTER (ex-*Rhinehart*)	Federal, Newark	Jun 1950	Returned USA Dec 1967, BU
F 808	VAN EWIJCK (ex-*Gustafson*)	Federal, Newark	Oct 1950	Returned USA Dec 1967, BU 1968
F 809	DUBOIS (ex-*O'Neill*)	Federal, Newark	Oct 1950	Returned USA Dec 1967, BU
F 810	DE ZEEUW (ex-*Eisner*)	Federal, Newark	May 1951	Returned USA Dec 1967, BU
F 811	VAN ZIJLL (ex-*Stern*)	Federal, Newark	May 1951	Returned USA Dec 1967, BU

Vos 1970 **NB 1/750 scale**

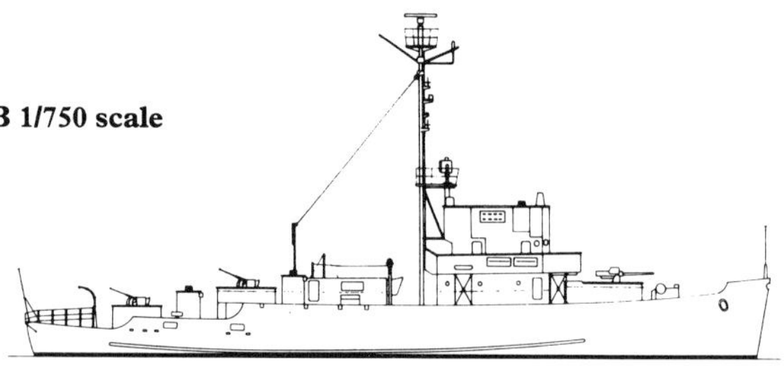

Fret May 1981
L & L van Ginderen

The six 'Roofdier' class frigates were all built in the USA with MDAP funds. They were designed to escort slow coastal convoys in Channel and North Sea areas and were operated as a single squadron by the Royal Netherlands Navy.

They were slow, lively ships fitted with Second World War-pattern anti-aircraft weapons and detection equipment. They proved useful for a number of peacetime tasks, especially fishery protection, and some were retained in this role in the North Sea until the mid-1980s, when they were decommissioned to release funds for new construction. Many of the 20mm AA guns were removed during the 1970s.

'ROOFDIER' class *frigates*

Displacement: 808t standard; 975t full load
Dimensions: 180ft pp, 185ft oa × 33ft × 10ft
54.9m, 56.2m × 10.0m × 2.9m
Machinery: 2-shafts, 2 General Motors 12–567 ATL diesels, 1600bhp = 15kts. Range 4300nm at 10kts
Armament: 1–3in, 6 or 4–40mm (6 × 1; *Panter, Jaguar* 4 × 1), 8–20mm, 1 Hedgehog, 2 (*Panter, Jaguar* 4) DCT
Sensors: Sonar QCU-2
Complement: 96

No	Name	Builder	Laid down	Launched	Comp	
F 817	WOLF	Avondale	15.11.52	2.1.54	26.3.54	Sold 1985
F 818	FRET	General SB	18.12.52	30.7.53	4.5.54	BU 1988
F 819	HERMELIJN	General SB	2.3.53	6.3.54	5.8.54	BU 1987
F 820	VOS	General SB	3.8.52	1.5.54	2.12.54	BU 1988
F 821	PANTER	Avondale	1.12.52	30.1.54	11.6.54	BU 1987
F 822	JAGUAR	Avondale	10.12.52	20.3.54	11.6.54	BU 1988

Built under MDAP, the frigate *Lynx* was similar to the Italian *Albatros* and Danish *Triton* classes. She served for only five years in the Royal Netherlands Navy before being transferred to Italy in 1961. For further details see under *Albatros* class corvettes (Italy).

LYNX *frigate*

Displacement: 800t standard; 950t full load
Dimensions: 228ft pp, 250ft 5in oa × 31ft 8in × 9ft
69.5m, 76.3m × 9.7m × 2.8m
Machinery: 2 shafts, 2 Fiat diesels, 5200bhp = 19kts. Range 2400nm at 18kts
Armament: 2–3in (2 × 1), 2–40mm (1 × 2), 2 Hedgehogs, 2 DCT
Sensors: Sonar QCU-2
Complement: 109

No	Name	Builder	Laid down	Launched	Comp	
F 823	LYNX	Breda Marghera	Jul 1953	31.7.54	2.10.56	To Italy Oct 1961

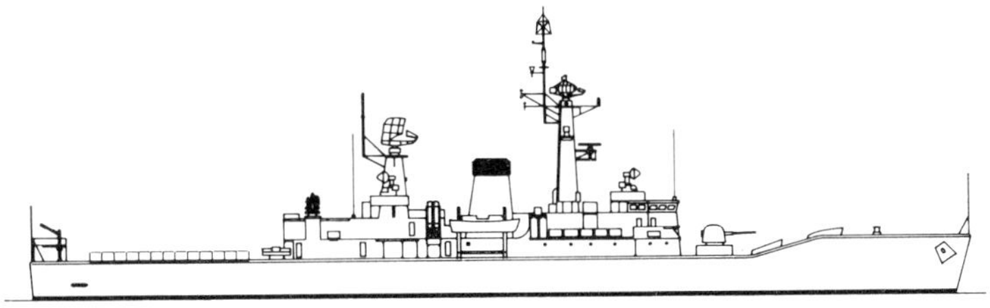

Isaac Sweers 1988

VAN SPEIJK class *frigates*

Displacement: 2200t standard; 2850t full load
Dimensions: 360ft wl, 372ft oa × 41ft × 18ft
109.7m, 113.4m × 12.5m × 5.8m
Machinery: 2 shaft Werkspoor-English Electric geared steam turbines, 2 boilers, 30,000 shp = 28.5kts. Range 4500nm at 12kts
Armament: 2–4.5in (1 × 2), 2 Seacat SAM (2 × 4), 1 helicopter, 1 ASW mortar Mk 10 (1 × 3). See notes
Sensors: Radar LW-03, DA-02, M45, M44; sonar Types 170B, 162. See notes
Complement: 251

No	Name	Builder	Laid down	Launched	Comp	Fate
F 802	VAN SPEIJK	Amsterdam	1.10.63	5.3.65	14.2.67	To Indonesia 1986
F 803	VAN GALEN	Royal Schelde	25.7.63	19.6.65	1.3.67	To Indonesia 1987
F 804	TJERK HIDDES	Amsterdam	1.6.64	17.12.65	16.8.67	To Indonesia 1986
F 805	VAN NES	Royal Schelde	25.7.63	26.3.66	9.8.67	To Indonesia 1988
F 814	ISAAC SWEERS	Amsterdam	5.5.65	10.3.67	15.5.68	To Indonesia 1989
F 815	EVERTSEN	Royal Schelde	6.7.65	18.6.66	21.12.67	To Indonesia 1989

Van Speijk 1970 L & L van Ginderen

Built to replace the elderly DEs of the *Van Amstel* class, the *Van Speijks* were an adaptation of the British *Leander* class. In order to facilitate rapid construction few changes were made in the basic design, and even the armament of the British ships was adopted in its entirety, thus adding the 114mm gun to the 120mm standard in other Netherlands vessels. The stipulation that as much equipment of Netherlands manufacture as possible be incorporated led to a number of changes in other areas, in particular the surveillance radars and fire control systems, which were provided by HSA. The size advantage of the small M44 director provided for Seacat over the British model enabled a second launcher to be worked in on the hangar roof.

Four ships were ordered in 1962, with the other two following in 1964. They were the first Netherlands escorts to operate helicopters, and on completion were quickly integrated into the ASW squadrons alongside the *Holland* and *Friesland* classes.

Plans were drawn up in 1974 for an extensive mid-life modernisation which aimed at improving the ships' operational efficiency and making them compatible with the new generation of frigates coming into service. Major changes in armament were the first priority, and as far as possible systems installed in the 'Standard' frigates were adopted. The 114mm twin mounting was replaced by an OTO-Melara 76mm compact gun, and Harpoon canisters were added between the funnel and the mainmast. Triple Mk 32 ASW tubes (2 × 3) were placed abreast the mainmast (in later conversions they were moved to the upper deck), and the hangar was extended aft to accommodate the new Lynx helicopter, the well for the Limbo mortar well being plated over to form part of the flight deck. Seacat was retained, and two Corvus chaff launchers were installed.

The operations room was completely rebuilt and a SEWACO-V data and weapons control system installed. The DA-02 search radar atop the foremast was replaced by a DA-05, and communications facilities were improved. The British sonars were replaced by the same US-manufactured model fitted in the *Tromp* class frigates. The full sensor outfit now consisted of LW-03, DA-05, M45(1), M44(2), CWE-610 and PDE-700.

Significant reductions in the complement of the ships (to two hundred and thirty-one) were achieved by increased automation of the propulsion machinery. The reduction in complement enabled improvements to be made in accommodation and messing facilities.

The modernisation took place between 1976 and 1982. As refitted the *Van Speijks* were intended to make up the third ASW group envisaged in the 1974 Defence Plan, with operations centred on the North Sea and English Channel areas. However, the class became a victim of the 1986 defence cuts, with four ships being paid off in that year. All four were purchased by Indonesia and transferred between 1986 and 1988. The other two units, which were fitted with the US SQR-18 towed array, were retained until 1989–90 when they too were sold to Indonesia. The towed arrays were removed prior to transfer.

Tromp as completed

C & S Taylor

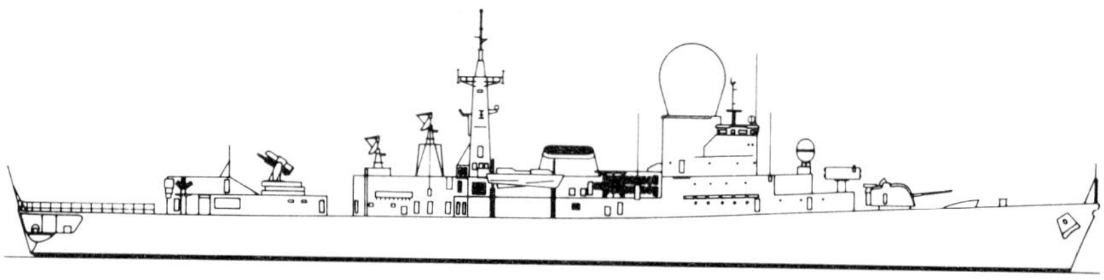

Tromp 1975

TROMP class *guided missile frigates*

Displacement: 3665t standard; 4308t full load
Dimensions: 430ft pp, 453ft oa × 49ft × 22ft
131.0m, 138.2m × 14.8m × 6.6m
Machinery: 2 shaft COGOG: 2 TM 3B Olympus gas turbines, 44,000shp = 28kts, plus 2 RM 1A Tyne gas turbines, 8200shp = 18kts. Range 5000nm at 18kts
Armament: 1 Standard SAM (40 SM-1 missiles), 8 Harpoon SSM (2 × 4), 1 NATO Sea Sparrow SAM (1 × 8, 16 missiles), 2–120mm (1 × 2), 2 Corvus chaff RL, 6–324mm Mk 32 TT (2 × 3), 1 WG–13 Lynx helicopter
Sensors: Radar SPS-01, WM-25, SPG-51C; sonar CWE-610, Type 162
Complement: 306

No	Name	Builder	Laid down	Launched	Comp	Fate
F 801	TROMP	Royal Schelde	4.8.71	4.6.73	3.10.75	Extant 1995
F 802	DE RUYTER	Royal Schelde	22.12.71	9.3.74	3.6.76	Extant 1995

These two large missile frigates were designed as replacements for the postwar 6in gun cruisers of the *De Ruyter* class. Staff requirements stipulated that they be able to protect a task force or convoy against aircraft and guided weapons, and incorporate sophisticated command facilities; they would have a secondary ASW and anti-ship capability. Since completion *Tromp* and *De Ruyter* have served as flagships of two of the three anti-submarine task groups established in accordance with the 1974 Fleet Plan.

The broad hull-form, high freeboard and non-retractable fin stabilisers of these ships combine to give them excellent sea-keeping qualities, together with the necessary internal volume for an extensive electronics outfit and a high level of habitability. The construction of the hull and the superstructure below the radome is of steel, the remaining upperworks being of sea-water resistant aluminium alloy.

Initially they were to have had steam propulsion, but the desire to reduce crew numbers led to the adoption of a COGOG gas turbine system identical to that installed in contemporary British vessels, but with gearing designed and supplied by the shipbuilders. The Olympus turbines were down-rated to improve gas-generator life and reduce maintenance. Twin spade rudders are fitted, each with its own electro-hydraulic steering gear. The rudders are electrically synchronised but can be operated independently. Stringent noise and shock requirements were laid down; all machinery is resiliently mounted, and there are silencers in the uptakes and intakes. The entire machinery installation is monitored from a highly-automated machinery control room, which is situated near the operations room to ensure good communications. The ventilation and air-conditioning system allows individual control over all living and electronic spaces, and the ships are designed for complete NBC close-down.

Originally the ships were to have been fitted with the Sea Dart SAM in return for British adoption of the large HSA 3-D radar for the CVA-01 and the Type 82 escorts. However, Sea Dart proved to be volume-intensive, and the more compact US Tartar/Standard system was adopted instead. The 120mm gun mountings were removed from the old destroyer *Gelderland* and refurbished. The SEWACO–I tactical data system is based on DAISY–1 computers and display systems using equipment of predominantly Netherlands manufacture.

In 1980 both ships received new plastic radomes for their SPS-01 radar, and in 1984-85 a number of further modifications were made, including the installation of RAMSES (Reprogrammable Advanced Multi-mode Shipboard ECM System) and Link 11, and the substitution of four US Super RBOC chaff launchers for the original British Corvus. In a major mid-life modernisation planned to begin in 1988 the new SMART 3–D radar would have replaced the SPS–01. These plans were first modified and then cancelled altogether in 1989. However, as a result of the Gulf crisis the modernisation was partially reinstated, and both ships have now received a Goalkeeper CIWS and Super RBOC chaff launchers. It is now envisaged that these ships will eventually be replaced by two air defence vessels developed in collaboration with Germany and Spain.

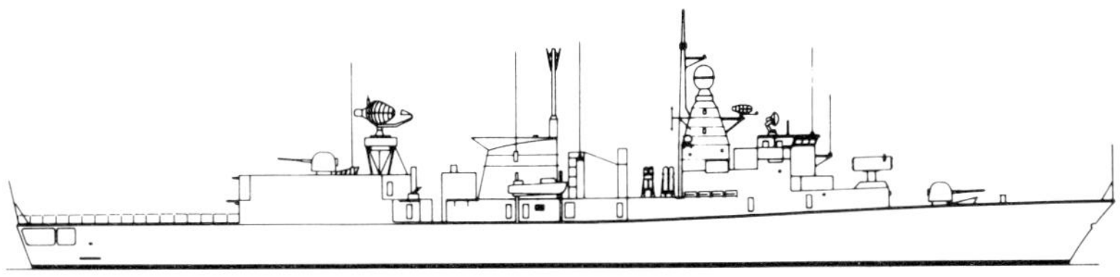

Kortenaer 1978

KORTENAER ('STANDARD') class *guided missile frigates*

Displacement: 3000t standard; 3785t full load
Dimensions: 400ft pp, 427ft oa × 47ft × 20ft
121.8m, 130.2m × 14.4m × 6m
Machinery: 2 shaft COGOG: 2 Olympus TM 3B gas turbines, 51,600shp = 30kts, plus 2 Tyne TM 1C gas turbines, 9800shp = 20kts. Range 4000nm at 20kts, 4700nm at 16kts
Armament: 8 Harpoon SSM (2 × 4), 1 NATO Sea Sparrow SAM (1 × 8, 24 missiles), 1–76mm, 1–40mm or 1–30mm Goalkeeper CIWS (see notes), 4–324mm TT (4 × 1), 2 Corvus/Super RBOC chaff RL, 2 WG–13 helicopters
Sensors: Radar LW-08, WM-25, STIR-180, ZW-06; sonar SQS-505
Complement: 200

No	Name	Builder	Laid down	Launched	Comp	Fate
F 807	KORTENAER	Royal Schelde	8.4.75	18.12.76	26.10.78	Extant 1995
F 808	CALLENBURGH	Royal Schelde	30.6.75	26.3.77	26.7.79	To Greece 1994
F 809	VAN KINSBERGEN	Royal Schelde	2.9.75	16.4.77	24.4.80	To Greece 1995
F 810	BANCKERT	Royal Schelde	25.2.76	30.9.78	29.10.80	To Greece 1993
F 811	PIET HEIN	Royal Schelde	28.4.77	3.6.78	14.4.81	Extant 1995
F 812	PIETER FLORISZ(i)	Royal Schelde	1.7.77	15.12.79		To Greece
F 813	WITTE DE WITH(i)	Royal Schelde	13.6.78	27.10.79		To Greece
F 816	ABRAHAM CRIJNSSON	Royal Schelde	25.10.78	16.5.81	26.1.83	Extant 1995
F 823	PHILIPS VAN ALMONDE	Wilton-Fijenoord	1.10.77	11.8.79	2.12.81	Extant 1995
F 824	BLOYS VAN TRESLONG	Wilton-Fijenoord	1.5.78	15.11.80	25.11.82	Extant 1995
F 825	JAN VAN BRAKEL	Royal Schelde	16.11.79	16.5.81	14.4.83	Extant 1995
F 826	PIETER FLORISZ(ii), (ex-*Willem van der Zaan*)	Royal Schelde	21.1.81	8.5.82	1.10.83	Extant 1995

Designed as replacements for the twelve postwar ASW destroyers of the *Holland* and *Friesland* classes, the frigates of the *Kortenaer* class were intended to have the maximum degree of commonality of weapons, electronics and propulsion systems with other contemporary NATO frigates. Attempts were initially made to develop a common hull-form with the Royal Navy, but these and approaches to other NATO navies foundered. The Netherlands Navy therefore elected to develop its own 'Standard' design, so-called because it was planned to build ASW and AAW variants on a standard, common hull, and because the design conformed to NATO requirements for ocean escorts.

Eight were ordered in 1974, with a further four in 1976. However, the original *Pieter Florisz* (F 812) and *Witte de With* (F 813) were sold to Greece while fitting out, and it was decided that these should be replaced by two AAW variants, which became the *Jacob van Heemskerck* class (see below).

The hull-form is similar to that of contemporary French construction, with continuous upper and main decks, a single rudder of the semi-balanced type, and a clipper bow with 'negative' sheer. The desire for good performance in a seaway resulted in a length/beam ratio less favourable than is generally accepted, and careful design was necessary to ensure that the ships met speed requirements.

The propulsion system and machinery layout are identical to those of the *Tromp* class. All machinery is resiliently mounted and there is a single pair of non-retractable fin stabilisers. A high degree of automation means that the total engine room complement is only twenty-nine. The ship can be fought and navigated from the operations room, which is fitted with a SEWACO-II tactical data system. On the main deck there is a central passageway running the entire length of the ship, and personnel spaces are grouped in such a way as to minimise traffic.

Two Lynx helicopters can be accommodated, but the ships carry only one during peacetime operations. Similarly, only two or four Harpoon SSM are fitted in peacetime, although this figure would be increased to eight in wartime. The weapons systems are all in widespread service with other NATO navies, but the electronics systems, with the sole exception of the hull sonar, are of Netherlands design and manufacture. The first two units were initially fitted with a second 76mm mounting atop the hangar, but this was replaced by a single 40mm Bofors in later ships. *Kortenaer* had her second 76mm mounting replaced by a single 40mm in 1982, and *Callenburgh* was fitted with the prototype Goalkeeper CIWS in 1984. This has now been fitted in all ships remaining in service with the RNIN, but a projected modernisation which included installation of the SMART 3-D radar and the SQR-18A towed array was subsequently abandoned. Three ships were sold to Greece in 1992, and a further three (F 807, F 811 and F 816) will be transferred in 1996.

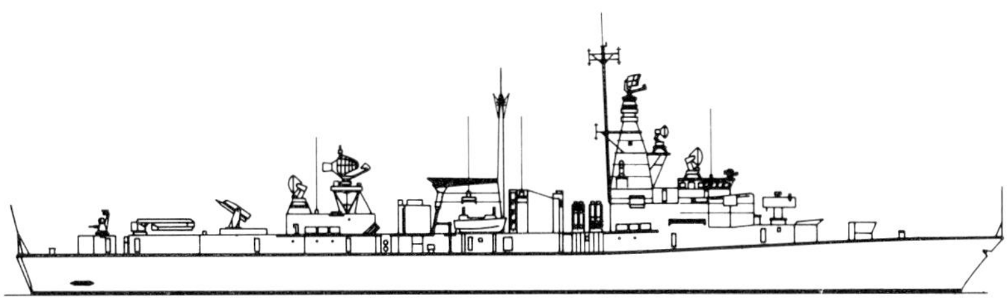

Jacob van Heemskerck 1989

JACOB VAN HEEMSKERCK ('L') class *missile frigates*

Displacement: 3000t standard; 3750t full load
Dimensions: As *Kortenaer* class
Machinery: As *Kortenaer* class
Armament: 1 Standard SAM (40 SM-1 missiles), 8 Harpoon SSM (2 × 4), 1 NATO Sea Sparrow SAM (1 × 8, 24 missiles), 1–30mm Goalkeeper CIWS, 4–324mm TT (4 × 1), 2 Super RBOC chaff RL
Sensors: Radar LW-08, DA-08, 2 STIR-240, STIR-180, ZW-06; sonar PHS-36
Complement: 197

No	Name	Builder	Laid down	Launched	Comp	Fate
F 812	JACOB VAN HEEMSKERCK	Royal Schelde	21.1.81	5.11.83	15.1.86	Extant 1995
F 813	WITTE DE WITH	Royal Schelde	15.12.81	25.8.84	17.9.86	Extant 1995

The air defence variant of the 'Standard' frigate has much in common with its ASW counterparts. Not only does it share a common hull-form and propulsion system with the latter, but it features a number of common weapons and electronics systems. The major external differences are the after deckhouse for the US Standard missile system which replaces the helicopter hangar, the suppression of the 76mm gun mounting forward, and the heightened foremast with its DA-08 target tracking radar.

Pieter Florisz 11.7.83 L & L van Ginderen

Witte de With 24.2.86 L & L van Ginderen

It is envisaged that the latter will eventually be replaced by the SMART (Signal Multi-beam Acquisition Radar for Targeting) 3-D tracking radar. The fire control radar for the US Standard SAM is not the standard US SPG-51, but a modified STIR radar of Netherlands design and manufacture.

The arrangements for reloading the eight-cell box launcher for the NATO Sea Sparrow SAM have been considerably modified. Reloads are carried in a tall magazine structure projecting from the forward end of the bridge and are run up manually on an angled ramp connecting the magazine doors with the after end of the launcher. ECM capabilities are on a par with the later *Kortenaers*, with the RAMSES active/passive ECM system complemented by a pair of US Super RBOC chaff launchers. A SEWACO-II tactical data system is fitted, together with NATO Link 11 communications. Besides SMART, these ships may receive a towed array at a future refit. Both ships have accommodation for a task group commander and staff.

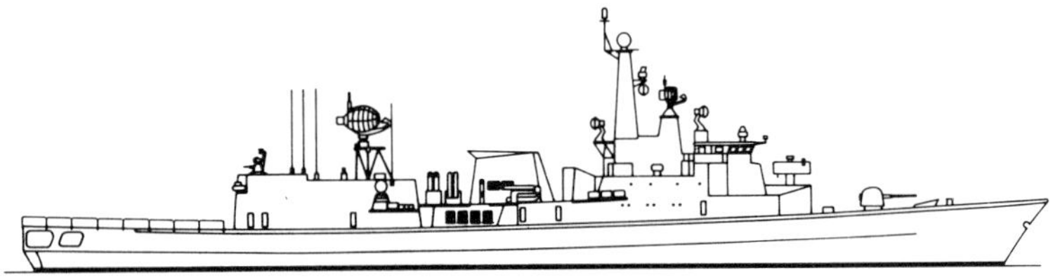

Karel Doorman design

KARREL DOORMAN ('M') class *guided missile frigates*

Displacement: 2800t standard; 3320t full load
Dimensions: 375ft pp, 401ft oa × 47ft × 20ft
114.4m, 122.3m × 14.4m × 6.1m
Machinery: 2 shaft CODOG: 2 Rolls-Royce SM-1A/C Spey gas turbines, 37,550hp/48,970hp = 29/30kts, 2 Stork-Werkspoor 12 SWD 280 V-12 diesels, 8,450bhp = 21kts. Range 5000nm at 18kts
Armament: 8 Harpoon SSM (2 × 4), 1 VLS Sea Sparrow SAM (16 missiles), 1–76mm, 1–30mm Goalkeeper CIWS, 2–20mm (2 × 1), 4–324mm TT (4 × 1), 2 Super RBOC chaff RL, 1 WG-13 Lynx helicopter
Sensors: Radar LW-08, SMART, 2 STIR-180, Decca 1690; sonar PHS-36, DSBV 61A towed array
Complement: 154

No	Name	Builder	Laid down	Launched	Comp	Fate
F 827	KAREL DOORMAN	Royal Schelde	26.2.85	20.4.88	31.5.91	Extant 1995
F 829	WILLEM VAN DER ZAAN	Royal Schelde	6.11.85	21.1.89	28.11.91	Extant 1995
F 830	TJERK HIDDES	Royal Schelde	28.10.86	9.12.89	3.12.92	Extant 1995
F 831	VAN AMSTEL	Royal Schelde	3.5.88	19.5.90	27.5.93	Extant 1995
F 832	ABRAHAM VAN DER HULST	Royal Schelde	8.2.89	7.9.91	15.12.93	Extant 1995
F 833	VAN NES	Royal Schelde	10.1.90	16.5.92	24.6.94	Extant 1995
F 834	VAN GALEN	Royal Schelde	7.6.90	21.11.92	Dec 1994	Extant 1995
F 828	VAN SPEIJK	Royal Schelde		1.10.91	23.3.89	Completing

This class was originally intended to replace the six elderly US-built frigates of the *Roofdier* class. Four ships were to have been built, and the design was essentially a scaled-down version of the 'Standard' frigate, with a standard displacement of 1900t and a length overall of 111.8m (367ft). Intended to operate primarily in the North Sea, the ships were to be fitted for peacetime surveillance and fishery protection duty, but were to be capable of operating with the 'Standard' in the North Atlantic in the anti-submarine role.

With the completion of the 'Standard' construction programme there was a crisis at the De Schelde Shipyard, and in order to provide continuity of work the first batch was ordered three years earlier than planned. The contract was signed in June 1985, after the first hull had been laid down. Contracts for the weapons, electronics and command systems were placed later in the same year. A further four ships were ordered in April 1986 to replace the frigates of the *Van Speijk* class, which were retired early to finance the programme.

The final design is significantly larger, the lengthened stern section reflecting the decision to provide hangar/flight deck facilities for a helicopter larger than the current Westland Lynx (probably the NATO NH-90). To port of the hangar are vertical launch tubes for NATO Sea Sparrow, which have replaced the conventional eight-cell box launcher. Other changes reflect the lessons of the Falklands crisis and advances in electronics. In particular, the original DAISY computer bank has been superseded by a combat system based around a triple Ethernet-based databus, which allows the physical distribution of processors and databases over the whole area of the ship. The sides of the hull and superstructure have been angled to maximise radar discretion, and a passive 'rudder-roll' stabilisation system is fitted in place of the traditional fins.

Otherwise the ships reflect many of the original requirements. Three semi-rigid inflatables are carried for boarding ships and oil rigs, and there is accommodation for a detachment of thirty marines. Extensive automation has enabled the crew size to be reduced to 154 officers and men.

The decision to order the ships earlier than planned has resulted in the hulls being ready in advance of much of their equipment. *Karel Doorman* was completed with the less powerful Spey SM-1A gas turbine, and has run trials lacking the SMART 3-D radar, the US Argo APECS-II EW suite, and the Thomson-CSF DSBV 61 Anaconda towed array. She has an interim SEWACO VII A combat system and the integrated monitoring and control system (IMCS), which looks after propulsion, power generation, accommodation, and NBC facilities, has yet to be installed. The SMART radar ran trials on the *Jan van Brakel* of the *Kortenaer* class, and was installed only in 1992. The first ship to have the APECS-II EW suite, the DSBV 61 towed array and the upgraded SEWACO VII B combat system was the fifth unit *Abraham van der Hulst*. These systems will eventually be retro-fitted to all ships.

In 1987 the names were reshuffled, *Van Speijk* becoming the last ship; she retains her original pennant number.

Tjerk Hiddes 1995 L & L van Ginderen

It is planned to order two air defence frigates in 1998 to replace the *Tromp* class. The new frigates will be a collaborative project between the Netherlands, Germany and Spain. A project definition contract was awarded to Royal Schelde in December 1993. It is hoped to have the first ship in service by 2001.

The new ships will have the advanced SMART-L air surveillance radar, and the APAR fixed planar array. An area defence missile is under consideration to complement Evolved Sea Sparrow. The sonar system is being developed by Thomson-Sintra of France, and may feature the TNO-FEL ALF combination of a hull-mounted projector and two 130m towed receiver arrays. The propulsion system will be based on an advanced intercooled, recuperative-cycle gas turbine.

Projected guided missile frigates

Displacement: 4400t full load
Dimensions: 414ft wl, 442ft oa × 48ft × 15ft
126.4m, 134.8m × 14.6m × 4.5m
Machinery: 2-shaft CODOG; 2 WR-21 gas turbines = 28-30kts; 2 diesels. Range 5000nm at 18kts
Armament: VLS Mk 41 SAM (32 Evolved Sea Sparrow missiles), 8 Harpoon SSM (2×4), 1–127mm/54 Mk 45, 1–30mm Goalkeeper CIWS, 2–20mm (2×1); 2–4 chaff RL; 4–324mm Mk 32 TT (4×1), 1 NH-90 helicopter
Sensors: Radar SMART-L, APAR; sonar hull-mounted LF
Complement: 200

No	Name	Builder	Laid down	Launched	Comp	Fate
2	projected					

SUBMARINES

These two ex-British 'T' class boats, launched on 27 June 1942 and 21 August 1944 respectively, were lent to the Netherlands from 1948 until 1953. They were unmodified except for the fitting of a snorkel.

Ex-British 'T' class *submarines*

Particulars: As British 'T' class (1940–42 programmes)

No	Name	Builder	Acquired	Fate
S 811	DOLFIJN (ex-*Taurus*)	Vickers-Armstrong, Barrow	1948	Returned RN Dec 1953
S 813	ZEEHOND (ex-*Tapir*)	Vickers-Armstrong, Barrow	1948	Returned RN 1953

These two submarines were former *Balao* class boats acquired on loan from the US Navy, initially for a period of five years. The loan period was later extended to ten years and then to fifteen. They were specially modified prior to their transfer to a GUPPY IA configuration. This involved the streamlining of the hull form and fin, in which a snorkel was incorporated, and the installation of improved batteries for a higher submerged speed.

Walrus (launched 20 February 1944) and *Zeeleeuw* (9 January 1944) were extensively employed as ASW targets from the time of their transfer in 1953 until their disposals in 1970-71.

Ex-US BALAO class *submarines*

Displacement: 1830t surfaced; 2440t submerged
Dimension: 307ft oa × 27ft × 17ft
93.6m × 8.2m × 5.2m
Machinery: 2 shafts, 4 2-stroke diesels plus electric motors, 6500bhp/4600 shp = 17kts/5kts. Range 12,000nm at 10kts
Armament: 10–21in TT (6 bow, 4 stern; 24 torpedoes)
Complement: 79

No	Name	Builder	Acquired	Fate
S 802	WALRUS (ex-*Icefish*)	Manitowoc SB	1953	Returned USN 1971, BU
S 803	ZEELEEUW (ex-*Hawksbill*)	Manitowoc SB	1953	Returned USN 1970, BU

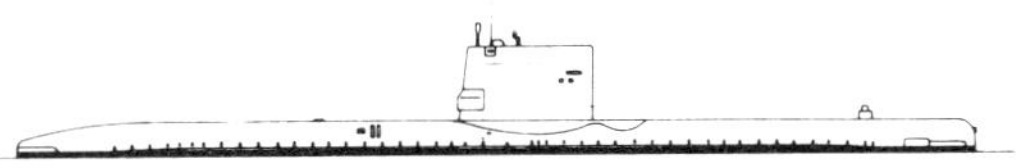

Potvis 1976

These were the first submarines to be built postwar for the Royal Netherlands Navy, and were of a novel triple-hulled design. Three cylinders were arranged in a triangular shape within an outer casing. The upper cylinder accommodates the crew, control spaces and armament, while the shorter lower cylinders each housed an independent propulsion unit comprising a twelve cylinder diesel, an electric motor, and a 168-cell battery.

The order for the second pair was postponed while the possibility of nuclear propulsion was considered. It was reinstated in 1962, and *Potvis* and *Tonijn* were completed with a number of modifications which were subsequently extended to the earlier boats. The early boats had a length of 257ft (78.3m) and retained their original MAN 12-V6V 22/30 diesels (each 1400bhp) until their demise; *Potvis* and *Tonijn* had theirs replaced by SEMT-Pielstick units in 1978-79.

In the early 1970s *Zeehond* had a new sonar fitted in a dome atop the casing, but this was subsequently removed. She was to have paid off in 1987, but the fire which delayed completion of *Walrus* led to a decision to extend her service life until 1990. Since decommissioning she has been used by RDM as a trials ship for an air-independent propulsion system.

DOLFIJN (ii) class *submarines*

Displacement: 1494t surfaced; 1826t submerged
Dimensions: 261ft oa × 26ft × 16ft
79.5m × 7.9m × 5.0m
Machinery: 2 shafts, 2 MAN 12-V6V 22/30 diesels, 2 electric motors, 2800bhp/4000hp = 14.5kts/17kts
Armament: 8–533mm TT (4 bow, 4 stern)
Sensors: Radar Type 1001; sonar?
Complement: 64

No	Name	Builder	Laid down	Launched	Comp	Fate
S 808	DOLFIJN	Rotterdam DD	30.12.54	20.5.59	16.12.60	BU 1985
S 809	ZEEHOND	Rotterdam DD	30.12.54	20.2.60	16.3.61	Test hulk 1990
S 804	POTVIS	Wilton-Fijenoord	17.9.62	12.1.65	2.11.65	Stricken 1992
S 805	TONIJN	Wilton-Fijenoord	27.11.62	14.6.65	24.2.66	Stricken 1991

Dolfijn as completed

Navarret Collection

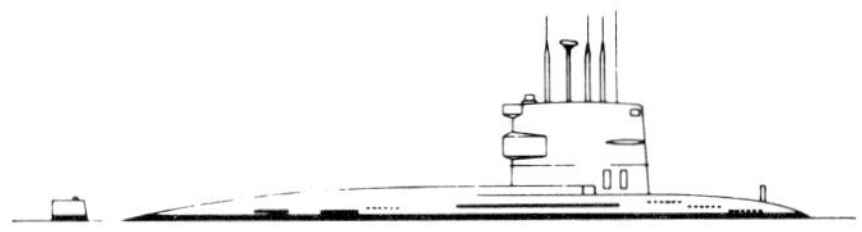

Zwaardvis as completed

ZWAARDVIS class *attack submarines*

Displacement: 2408t surfaced; 2640t submerged
Dimensions: 219ft oa × 28ft × 23ft
66.9m × 8.4m × 7.1m
Machinery: 1 shaft, 3 Werkspoor RVB 215X 12 diesels, 1 Holec electric motor, 4200bhp/5100hp = 13kts/20kts
Armament: 6–533mm TT (bow, 20 Mk 48/37 torpedoes)
Sensors: Radar Type 1001; sonar (as refitted) Eledone Octopus, Type 2026 towed array, DUUX 5
Complement: 67

No	Name	Builder	Laid down	Launched	Comp	Fate
S 806	ZWAARDVIS	Rotterdam DD	14.7.66	2.7.70	18.8.72	Stricken 1994
S 807	TIJGERHAAI	Rotterdam DD	14.7.66	25.5.71	20.10.72	Stricken 1995

These boats were closely based on the US *Barbel* class. They are single-hulled, with three decks amidships housing the control spaces, accommodation for the crew, and the batteries respectively. The forward section accommodates the sonars, torpedo tubes and the torpedo handling room. The after section houses the machinery: three diesel generators with a single large electric motor abaft them, coupled to a single shaft. All machinery is mounted on a false deck with spring suspension. In accordance with contemporary US practice, these boats have cruciform after control surfaces and fin-mounted diving planes.

As much Netherlands-manufactured equipment as possible was incorporated. The HSA M8 Mod 7 fire control system, which uses a digital computer, permits the simultaneous launching of two torpedoes, one of which may be wire-guided. Torpedoes currently in use include the US Mk 48 'heavyweight', and the re-worked NT-37D ASW torpedo.

In the late 1980s *Zwaardvis* and *Tijgerhaai* received a limited midlife modernisation to extend their service lives into the late 1990s, when they were due to be replaced by the third batch of *Walrus*-class boats (now cancelled). A Thomson-CSF Eledome sonar was fitted, together with a British Type 2026 towed array. Both paid off in 1994–95.

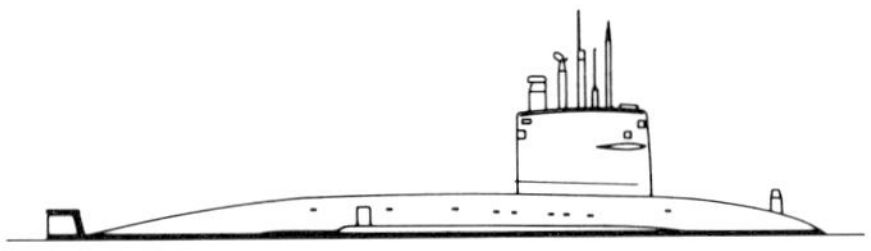

Zeeleeuw 1988

WALRUS class *submarines*

Displacement: 2450t surfaced; 2800t submerged
Dimensions: 222ft oa × 28ft × 23ft
67.7m × 8.4m × 7m
Machinery: 1 shaft, 3 SEMT-Pielstick 12 PA 4V 200 diesels, 1 Holec electric motor, 6300bhp/6910hp = 13kts/20kts
Armament: 4–533mm TT (bow, 20 Sub-Harpoon SSM or Mk 48/37 torpedoes)
Sensors: Radar ZW-07; sonar Eledone Octopus, Type 2026 towed array, DUUX 5
Complement: 52

No	Name	Builder	Laid down	Launched	Comp	Fate
S 801	WALRUS	Rotterdam DD	11.10.79	28.10.85	25.3.92	Extant 1995
S 802	ZEELEEUW	Rotterdam DD	24.9.81	20.6.87	25.4.90	Extant 1995
S 808	DOLFIJN	Rotterdam DD	12.6.86	25.4.90	29.1.93	Extant 1995
S 810	BRUINVIS	Rotterdam DD	18.4.88	26.4.92	July 1994	Extant 1995
	unnamed	Rotterdam DD				Cancelled
	unnamed	Rotterdam DD				Cancelled

These submarines were designed to replace the ageing *Dolfijn* class. Nuclear propulsion was initially considered but was rejected on financial and technological grounds. Approaches were also made to the British Royal Navy with a view to developing a joint design but these failed, despite almost identical staff requirements, and the Netherlands finally opted for an improved *Zwaardvis*.

The two classes have similar dimensions and a similar internal layout. The most important difference in external configuration is the adoption of 'X'-shaped after control surfaces similar to those fitted experimentally in USS *Albacore* in 1961 and subsequently adopted by the Swedish navy for its own submarines. These are particularly attractive for shallow-water operations and for manoeuvring alongside, and microprocessor control permits very fine control of the boat's movement at all speeds and depths.

Werkspoor, who had supplied the diesels for *Zwaardvis*, had taken that engine out of production, but suitable diesels similar to those adopted for generating electrical power aboard the 'Standard' frigates were purchased from SEMT-Pielstick. The third and fourth boats of the class reverted to the Brons-Werkspoor diesel. All electronic components, including the main electric motor, are of Netherlands manufacture. The control gear is designed to give uniform speed continuity between 1.5kts and full speed. Manpower has been reduced by fifteen by the adoption of a single control room housing the machinery control and surveillance consoles and an attack centre with a SEWACO-VIII tactical data system based on two main computers and seven GYPSY multi-purpose display consoles.

The waterslug tubes can fire both torpedoes and the Sub-Harpoon SSM. The Thomson-CSF medium-frequency attack sonar has a passive capability as well as a wide-band active mode, and there is a British Type 2026 towed array. The submarines are also fitted with Sperry Mk 29 and SATNAV position-fixing equipment.

In order to ensure continuity of work at RDM, construction of the first pair was brought forward by three years, the first steel being cut prior to the signing of the official contract in June 1979. This created major problems when the Netherlands Navy subsequently demanded a 50 per cent increase in diving depth (300m). This has been achieved by the use of French MAREL high-tensile steel, at the expense of considerable redesign, a two-year delay in completion and a 40 per cent increase in cost. The change to high-tensile steel required a reduction in the number of welded joints and hull apertures, while the increase in diving depth necessitated machine intercooling (in place of high-pressure cooling systems), duplex valves, a pressurised fuel system, a 'wet' exhaust system, sophisticated hydraulics and additional emergency blowing systems.

A further major delay to the programme occurred in August 1986, when the hull of the first boat, *Walrus*, was severely damaged by fire. Following reconstruction she was relaunched 13 September 1989; fitting out was expedited by diverting components intended for the third and fourth boats. The latter were ordered in 1984, but a third pair, intended to replace the *Zwaardvis* class, was cancelled in 1989.

Zeeleeuw 1993
L & L van Ginderen

AMPHIBIOUS WARFARE VESSELS

Official artist's impression of *Rotterdam*

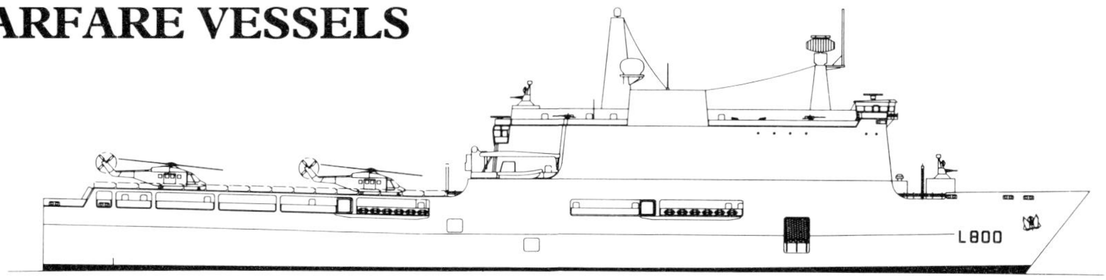

ROTTERDAM *amphibious transport dock*

Displacement: 1200t full load
Dimensions: 476ft wl, 523ft oa × 82ft × 19ft 4in
145m, 160m × 25m × 5.9m
Machinery: 2-shafts, 4 diesels, ?bhp = 20kts. 2 600hp electric motors for slow speeds. Range 6000nm at 12kts
Armament: 2–30mm Goalkeeper CIWS, 4–20mm (4×1)
Sensors: Radar DA-08
Complement: 113–127 + 600 troops

No	Name	Builder	Launched	Fate
L 800	ROTTERDAM	Royal Schelde	–	Building

The RNI Navy's first amphibious warfare vessel is being built in a collaborative project with Spain. The first steel was cut in February 1994, and the ship is due to enter service in 1997.

Rotterdam can transport a fully-equipped battalion of 600 Marines, together with 170 APCs or 30 MBTs. The 750m² docking well can accommodate four LCU/LCM or six LCVP. The capacious hangar can support up to four EH-101 or six NH-90 transport helicopters, and there is a two-spot flight deck aft. The ship will also have a 100-bed hospital, and will be capable of undertaking environmental and disaster relief tasks.

MINE WARFARE VESSELS

Ex-US ADJUTANT class *coastal minesweepers*

Particulars: As US *Adjutant* class

Class:
Tampa Marine, Florida – *Beemster* (M 845), *Bedum* (M 847), *Borculo* (M 849), *Borne* (M 850), *Breukelen* (M 852)
Broward Marine, Florida – *Bolsward* (M 846), *Beilen* (M 848), *Brummen* (M 851), *Blaricum* (M 853)
G W Kneass Co, San Francisco – *Breskens* (M 855), *Boxtel* (M 857), *Brouwershaven* (M 858)
Puget Sound Navy Yard – *Brielele* (M 854), *Bruinisse* (M 856)

These fourteen coastal minesweepers of the *Adjutant* class were transferred 1953–54 under MDAP. They were constructed of wood and non-metallic alloys.

Bolsward, *Breukelen* and *Bruinisse* were returned to the USA in 1972. *Brummen* and *Brouwershaven* were stricken in 1974, and the remainder in 1975.

Beemster c 1960 Navarret Collection

Onversaagd 1968
NB 1/750 scale

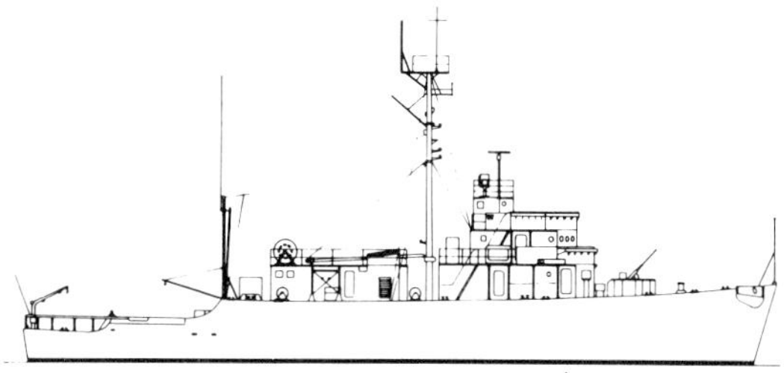

Ex-US AGILE class *ocean minesweepers*

Particulars: As US *Agile* class

Class:
Astoria Marine Construction Co – *Onversaagd* (M 884), *Onbevreesde* (M 885), *Onvervaard* (M 888)
Peterson, Wisconsin – *Onverschrokken* (M 886), *Onvermoeid* (M 887), *Onverdrotten* (M 889)

These six *Agile* class ocean minesweepers were transferred 1954–55 under MDAP. They had wooden hulls, and diesels of non-magnetic stainless steel alloy.

In 1966 they were transferred to escort duties and renumbered A 854–59. Two years later *Onbevreesde*, *Onvervaard* and *Onverdrotten* became MCM Group Command and Support Ships. In 1972 *Onverschrokken* became a torpedo trials ship and was renamed *Mercuur*. *Onvermoeid* was returned to the USA in 1972, and after being refitted provisionally as a hydrographic ship *Onversaagd* was sold in 1979. *Onbevreesde* was stricken in 1987, and broken up in 1989 and *Onvervaard* was stricken in 1982 and broken up in 1985. *Onverdrotten* was stricken in 1982.

Dokkum 1971
NB 1/750 scale

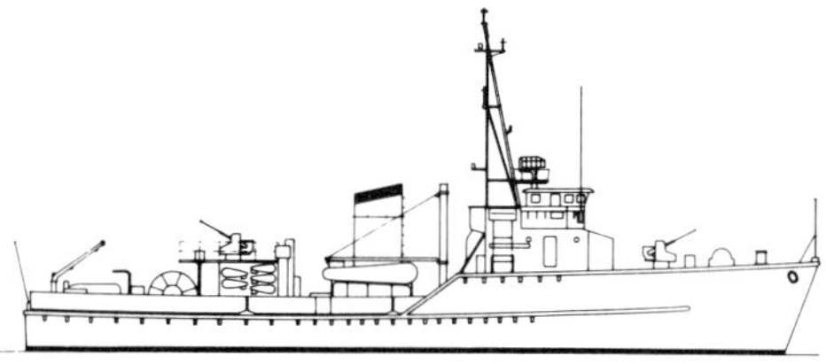

DOKKUM class *coastal minesweepers*

Displacement: 373t standard; 453t full load
Dimensions: 153ft oa × 29ft × 8ft
46.6m × 8.8m × 2.3m
Machinery: 2 shafts, 2 Fijenoord MAN or Werkspoor diesels, 2500bhp = 14kts. Range 2500nm at 10kts
Armament: 2–40mm (2 × 1); later 1–20mm
Sensors: Radar Decca TM 1229C
Complement: 38

No	Name	Builder	Launched	Fate
M 801	DOKKUM	Wilton-Fijenoord	12.10.54	Trials ship 1984
M 802	HOOGEZAND	Werf Gusto, Schiedam	22.3.55	Stricken 1992
M 806	ROERMOND	Wilton-Fijenoord	13.8.55	Stricken 1987
M 809	NAALDWIJK	Werf de Noord	1.2.55	Stricken 1993
M 810	ABCOUDE	Werf Gusto, Schiedam	2.9.55	Stricken 1993
M 812	DRACHTEN	Niestern SU	24.3.55	Stricken 1993
M 813	OMMEN	J & K Smit, Kinderdijk	5.4.55	Stricken 1993
M 815	GIETHORN	L K Smit & Zn, Kinderdijk	30.3.55	Stricken 1992
M 817	VENLO	Arnhemse SM	21.5.55	Stricken 1992
M 818	DRUNEN	Werf Gusto, Schiedam	24.3.56	Stricken 1984
M 820	WOERDEN	Haarlemse SM	28.11.56	Stricken 1991
M 823	NAARDEN	Wilton-Fijenoord	27.1.56	Extant 1995
M 827	HOOGEVEEN	Werf de Noord	8.5.56	Stricken 1992
M 828	STAPHORST	Werf Gusto, Schiedam	21.7.56	Stricken 1984
M 830	SITTARD	Niestern SU	26.4.56	Stricken 1987
M 841	GEMERT	J & K Smit, Kinderdijk	13.3.56	Stricken 1992
M 842	VEERE	L K Smit & Zn, Kinderdijk	9.2.56	Stricken 1984
M 844	RHENEN	Arnhemse SM	31.5.56	BU 1986
Wildervank sub-group				
M 803	WILDERVANK	De Vries Lentsch	24.2.55	BU 1977
M 804	STEENWIJK	Haarlemse SM	1955	Stricken 1970
M 805	GIETEN	De Haan & Oerlemans	1955	Stricken 1970
M 807	WAALWIJK	Verschure, Amsterdam	6.4.55	BU 1977
M 808	AXEL	De Groot & Van Vliet	5.3.55	To Oman 1974
M 811	AALSMER	Gebr Pot, Bolnes	23.4.55	To Oman 1974
M 814	MEPPEL	Haarlemse SM	19.7.56	Stricken 1974
M 816	LOCHEM	E J Smit & Zn, Westerbroek	1955	Stricken 1970
M 819	GOES	De Vries Lentsch	23.3.56	Stricken 1974
M 822	LEERSUM	De Haan & Oerlemans	11.8.56	BU 1977
M 824	SNEEK	Verschure, Amsterdam	1956	Stricken 1970
M 826	GRIJPSKERK	De Groot & Van Vliet	10.3.56	Harbour training 1973
M 829	ELST	Gebr Pot, Bolnes	21.3.56	To Ethiopia 1971
M 843	LISSE	E J Smit & Zn, Westerbroek	1956	Stricken 1970

The *Dokkum* class was based on the British 'Ton' class, eighteen vessels being financed under MDAP, and a further fourteen – known as the *Wildervank* sub-group – being purchased by the Netherlands. Construction was similar to that of the British vessels. The fourteen units of the *Wildervank* sub-group had Werkspoor diesels in place of the Fijenoord units on the *Dokkums*.

Waalwijk, *Woerden*, *Leesum*, *Lisse* and *Rhenen* were fitted as diving vessels 1962–65, and *Roermond* followed in 1968. Complement was reduced to 32. *Dokkum*, *Drunen*, *Veere* and *Staphorst* were converted to minehunters in 1968–73, when a British Type 193 minehunting sonar and an ARNAS marker buoy automated display system were installed. These vessels were the first of the class to de-commission, being replaced by the purpose-built minehunters of the *Alkmaar* class. *Woerden* was fitted out as a general-purpose tender (A 882) to serve in the Antilles.

All 14 vessels of the *Wildervank* class were disposed of between 1970 and 1977. Six of the *Dokkum* class (M 809–810, M 812–813, M 823 and M 830) underwent modernisation during the late 1980s. Four paid off in 1993 (M 809–10, M 812–13), but will be retained in reserve until 1997.

Houtepen as completed — Navarret Collection

VAN STRAELEN class *inshore minesweepers*

Displacement: 151t standard; 171 full load
Dimensions: 109ft oa × 22ft × 6ft
33.1m × 6.7m × 1.8m
Machinery: 2 shafts, 2 Werkspoor diesels, 1100bhp = 13kts
Armament: 1–20mm
Complement: 12

No	Name	Builder	Launched	Fate
M 868	ABLAS	Werf de Noord	26.9.59	BU 1986
M 869	BUSSEMAKER	De Vries Lentsch	27.2.60	Sold 1986
M 870	LACOMBLE	Arnhemse SM	6.2.60	Stricken 1984
M 871	VAN HAMEL	De Vries Lentsch	28.5.60	To Peru 1984
M 872	VAN STRAELEN	Arnhemse SM	17.5.60	Sold 1986
M 873	VAN MOPPES	Werf de Noord	10.5.60	Stricken 1982–83
M 874	CHOMPFF	Werf de Noord	10.5.60	BU 1988
M 875	VAN WELL GROENEVELD	Arnhemse SM	1.10.60	Sold 1986
M 876	SCHUILING	De Vries Lentsch	30.6.60	To Sea Cadets 1986
M 877	VAN VERSENDAAL	Werf de Noord	Aug 1961	To Sea Cadets 1986
M 878	VAN DER WEL	De Vries Lentsch	May 1961	To Peru 1984
M 879	VAN'T HOFF	Werf de Noord	15.3.61	Sold 1985
M 880	MAHU	Werf de Noord	15.3.61	Stricken 1982–83
M 881	STAVERMAN	De Vries Lentsch	Jul 1961	BU 1986
M 882	HOUTEPEN	Arnhemse SM	Jun 1961	Sold 1986
M 883	ZOMER	Arnhemse SM	4.3.61	For disposal 1986

Eight of these inshore minesweepers were funded under MDAP and the remaining eight were purchased by the Netherlands. Although similar in size, conception and construction to the British 'Ham' class, they were strictly a national design, and could be distinguished from other European inshore minesweepers by the configuration of their funnel and mast.

The sixteen vessels of this class were disposed of in 1982–84. *Lacomble* was transferred to the Sea Cadets at Den Helder in July 1984, and *Van Hamel* and *Van der Wel* were sold to Peru in February 1984 for use as survey craft.

Makkum 15.1.88 — L & L van Ginderen

ALKMAAR ('Tripartite') class *minehunters*

Displacement: 510t standard; 544t full load
Dimensions: 169ft oa × 29ft × 9ft max
51.6m × 9m × 2.6m
Machinery: 1 shaft, 1 Brons-Werkspoor A-RUB 215X V12 diesel, 1900bhp = 15kts; twin active rudders (7kts) and bow thruster. Range 3000nm at 12kts
Armament: 1–20mm, 2 PAP 104
Sensors: Radar Decca TM 1229C; sonar DUBM 21B
Complement: 29–42

No	Name	Builder	Launched	Fate
M 850	ALKMAAR	Van der Giessen de Noord, Alblasserdam	18.5.82	Extant 1992
M 851	DELFZIJL	Van der Giessen de Noord, Alblasserdam	29.10.82	Extant 1995
M 852	DORDRECHT	Van der Giessen de Noord, Alblasserdam	26.2.83	Extant 1995
M 853	HAARLEM	Van der Giessen de Noord, Alblasserdam	9.7.83	Extant 1995
M 854	HARLINGEN	Van der Giessen de Noord, Alblasserdam	9.7.83	Extant 1995
M 855	SCHEVENINGEN (ex-*Hellevoetsluis*)	Van der Giessen de Noord, Alblasserdam	2.12.83	Extant 1995
M 856	MAASLUIS	Van der Giessen de Noord, Alblasserdam	5.5.84	Extant 1995
M 857	MAKKUM	Van der Giessen de Noord, Alblasserdam	27.9.84	Extant 1995
M 858	MIDDELBURG	Van der Giessen de Noord, Alblasserdam	23.2.85	Extant 1995
M 859	HELLEVOETSLUIS (ex-*Scheveningen*)	Van der Giessen de Noord, Alblasserdam	18.7.85	Extant 1995
M 860	SCHIEDAM	Van der Giessen de Noord, Alblasserdam	20.12.85	Extant 1995
M 861	URK	Van der Giessen de Noord, Alblasserdam	2.5.86	Extant 1995
M 862	ZIERIKZEE (ex-*Veere*)	Van der Giessen de Noord, Alblasserdam	4.10.86	Extant 1995
M 863	VLARDINGEN	Van der Giessen de Noord, Alblasserdam	6.8.88	Extant 1995
M 864	WILLEMSTAD	Van der Giessen de Noord, Alblasserdam	27.1.89	Extant 1995

The 'Tripartite' minehunter was developed as a collaborative project with France and Belgium. Each navy built its own hulls, while other equipment was subcontracted to each of the three countries involved. The Netherlands was responsible for the main propulsion system, which underwent extensive testing following the determination of acoustic performance specifications at the Netherlands Ship Model Basin. The Werkspoor supercharged diesel, the reduction gearing, the Lips CP propeller, the machinery control and the air-conditioning system are all of Netherlands manufacture. (For further technical details see under France.)

A 5t modular van abaft the superstructure can contain a decompression station, communications equipment, or drone control gear. All units are fitted with OD–3 mechanical sweep gear. Two single 12.7mm, satcom, and portable US Stinger missiles were added to M 856 and M 859 when they deployed to the Gulf in 1987. The first three units were placed in reserve in 1987. *Dordrecht* was subsequently replaced by *Scheveningen*, which was damaged in a grounding incident. The original *Vlaardingen* and *Willemstad* were sold to Indonesia while under construction in 1985 and replaced by two new hulls. *Middleburg* and *Hellevoetsluis* were to have been transferred to Egypt as *Mecca* and *Medina*, but the transaction fell through.

Following the cancellation of the joint Belgian-Dutch programme for a new coastal minesweeper in January 1993, it was decided to modify seven of the *Alkmaar* class as follows: four will receive a propelled variable depth sonar (PVDS) to enable them to detect mines buried beneath the sea bed; three others will serve as control ships for MCM drones (cf. German 'Troika' system) due to enter service in 1998. All seven will receive new minehunting sonars and command/control systems.

New class *minesweepers*

For details see Belgium

No	Name	Builder	Launched	Fate
8 projected		Van der Giessen de Noord		cancelled

These ships were to have been built in a collaborative project with Belgium. The memorandum was signed in April 1989, and in November 1990 a design contract was awarded to Van der Giessen de Noord Marinebouw in conjunction with Béliard Polyship NV; initial plans were completed in 1992. Originally sixteen hulls were to be built for the Netherlands Navy and ten for Belgium, but these figures were subsequently reduced to eight and six respectively. In January 1993 the Netherlands pulled out of the project due to budgetary constraints.

MISCELLANEOUS

Balder 1961
NB 1/750 scale

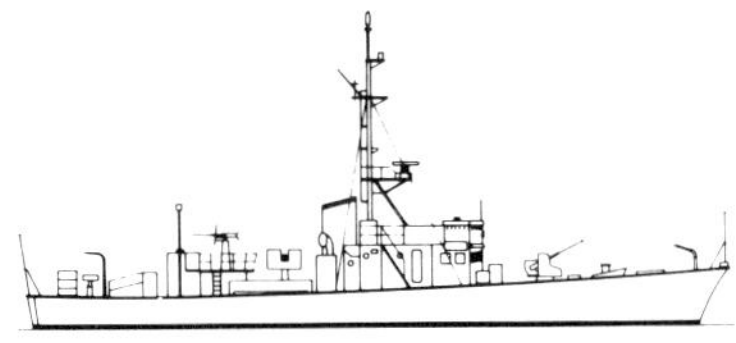

BALDER class *submarine chasers*

Displacement: 149t standard; 225t full load
Dimensions: 115ft pp, 119ft oa × 20ft × 6ft
35.0m, 36.3m × 6.2m × 1.9m
Machinery: 2 shafts, 2 diesels, 1050bhp = 15.5kts. Range 1000nm at 13kts
Armament: 1–40mm, 3–20mm (3 × 1), 2 DCT, Mousetrap
Complement: 27

Class:
Balder (P 802), *Bulgia* (P 803), *Freyr* (P 804), *Hadda* (P 805), *Hefring* (P 806).

The five small vessels of the *Balder* class were built in the Netherlands under MDAP. They were laid down at the Rijkswerf Willemsoord (Den Helder) 1953–54, and all launched in 1954; the first three were placed in reserve in 1975.

In 1986 *Bulgia* was renumbered A 880, and employed as a training ship for midshipmen. *Hadda* (stricken 1986) was retained to provide spares support. *Balder* and *Hefring* were stricken in 1985 and *Freyr* in 1986.

New Zealand

Unlike Australia, New Zealand does not have the industrial base to consider building her own major warships, and until recently remained dependent upon the United Kingdom for naval equipment. Few ships have been built for the Royal New Zealand Navy since 1947 and these have been little-modified versions of standard RN designs – initially the *Rothesay* type, and then the *Leanders*. Local capability has now advanced from basic to advanced refits of the *Leanders* and industrial participation in the *Anzac* programme.

When the two *Rothesays* were due for replacement in the early 1980s, the purchase of two Type 24 frigates from Yarrow was considered, but two second-hand *Leander* class were bought from the Royal Navy in October 1981 instead. To replace the four frigates it was hoped to build four *Anzac* class in collaboration with the Australians, but only two were ordered.

Four small patrol craft were ordered in 1982 to replace the ageing HDMLs in the training role. A political row with the United States over visits of nuclear-capable warships led to the abrogation of the ANZUS Treaty. The New Zealand Government sees itself as the protector of small island states in the SW Pacific, a role which implies rather more power-projection capability than before, but funds are scarce. Plans to build a logistic support ship to carry equipment for the Army's Rapid Reaction Force have finally been realised, with the conversion of a ro-ro ferry purchased at the end of 1994.

FLEET STRENGTH 1947

CRUISERS

Name	Launched	Disp (deep load)	Fate
BLACK PRINCE (C 81)	27.8.42	7410t	BU 1962
BELLONA (C 63)	29.9.42	7410t	BU 1959

In 1946 these two ships replaced *Gambia* and *Achilles*, on loan to the RNZN. *Black Prince* became the flagship while *Bellona* was used for harbour training at Auckland before being returned to the RN in April 1956. *Black Prince* left RNZN service in November 1961.

Other vessels in service with the RNZN in 1947 included the modified 'Flower' class corvettes *Arabis* and *Arbutus*; these were returned to the RN in 1950 and scrapped the following year. The 'Bird' class ASW and MS trawlers *Kiwi* and *Tui* were both laid up in 1946, but *Tui* later recommissioned for oceangraphic service, not paying off until 1967. *Kiwi* was sold in 1963. The 'Castle' class trawlers *Hinau* and *Rimu* were sold in 1955 and the 'Isles' class *Inchkeith*, *Killegray*, *Sanda* and *Scarba* were all for disposal in 1958.

There were also the British HDML type motor launches *SDML 3551–3556* and *3562–3567*. Five were given names in the mid-1950s: *Mako* (ex-*SDML*

3551), *Paea* (ex-*SDML 3552*), *Viti* (ex-*SDML 3555*), *Takapu* (ex-*SDML 3556*) and *Tarapunga* (ex-*SDML 3566*). The remainder were named c1960. In the above order the names were *Tamaki*, *Irirangi*, *Olphert*, *Pegasus*, *Toroa*, *Wakefield*, *Managa* and *Viti* was renamed *Ngapona*. They were given pennant numbers prefixed with 'P', but retained the old numbers. They were used for survey duties, fishery protection and training. Some were again renamed in about 1970, and were stricken as follows: *Maroro* 1972, *Kahawai*, *Makro*, *Parore* and *Tamure* 1975, and *Manga* 1980. In 1995 the remaining vessels were *Paea*, *Kupara* (ex-*Pegasus*), *Kowa* (ex-*Toroa*), *Kahawai* (ex-*Takapu*) and *Mako* (ex-*Tarapunga*).

MAJOR SURFACE SHIPS

Ex-British Modified DIDO class *cruiser*

Displacement: 5950t standard; 7410 deep load
Dimensions: 485ft pp, 512ft oa × 50ft 6in × 17ft 11in mean deep load
147.8m, 156.1m × 15.4m × 5.5m
Machinery: 4-shaft Parsons geared turbines, 4 Admiralty 3-drum boilers, 62,000shp = 32kts. Oil 1100t
Armour: Side 3in, bulkheads 1in
Armament: 8–5.25in/50 QF Mk 1 (4×2), 8–40mm (3×2, 2×1), 6–2in TT (2×3)
Complement: 530

No	Name	Builder	Acquired	Fate
C 89	ROYALIST	Scotts	1954	BU 1968

Royalist was modernised at Devonport DYd in 1954–56, with her former close-range armament replaced by three twin 40mm STAAG Mk 2 automatic AA mountings (one in 'Q' position and two abreast the second funnel) and TT removed. The bridge was rebuilt, lattice masts replaced tripods and Mk 6M directors were installed forward and aft. With Flyplane predictors she now had one of the most effective AA defence systems of any British-built cruiser, and she replaced *Bellona* when that ship returned to Britain to pay off at the end of 1955. She damaged her boiler tubes when her feedwater became contaminated with seawater, and paid off prematurely in July 1966. She was returned to the RN and scrapped in Japan.

Ex-British LOCH class *frigates*

Displacement: 1435t standard; 2260t deep load
Dimensions: 307ft × 38ft 7in × 12ft 4in mean deep load
93.6m × 11.8m × 3.8m
Machinery: 2-shaft VTE, 2 Admiralty 3-drum boilers, 5500ihp = 19.5kts. Oil 724t
Armament: 1–4in/45 QF Mk V HA, 4–2pdr pompom (1×4), 2 Squid ASW mortars
Complement: 114

No	Name	Builder	Acquired	Fate
F 422	HAWEA (ex-*Loch Eck*)	Smiths Dock	1.10.48	BU 1965
F 426	KANIERE (ex-*Loch Achray*)	Smiths Dock	7.7.48	BU 1967
F 424	PUKAKI (ex-*Loch Achanalt*)	Robb	3.9.48	BU 1966
F 625	ROTOITI (ex-*Loch Katrine*)	Robb	7.7.49	BU 1967
F 421	TAUPO (ex-*Loch Shin*)	Swan Hunter	3.9.48	BU 1962
F 517	TUTIRA (ex-*Loch Morlich*)	Swan Hunter	1.4.49	BU 1962

Acquired from the Royal Navy. Virtually no change was made to them during their RNZN service.

British ROTHESAY type *frigates*

Particulars: As British *Rothesay* class

No	Name	Builder	Laid down	Launched	Comp	Fate
F 111	OTAGO (ex-*Hastings*)	Thornycroft	5.9.57	11.12.58	22.6.60	Stricken 1982
F 148	TARANAKI	J S White	27.6.58	19.8.59	28.3.61	Stricken 1982, sold

Otago was ordered in February 1956 for the RN, but in February 1957 the contract was taken over by the RNZN, and a second ship was ordered at the same time. Apart from internal modifications they were identical with RN ships, and received quadruple Seacat GWS22 SAM systems. They also carried 6–12.75in (324mm) Mk 32 ASW TT (2 × 3). In 1978–79 *Taranaki* was converted to a training ship, minus Seacat and Limbo mortars, with a single 40mm Mk 7 on the after superstructure. The Type 277 radar was removed from *Otago*. In 1980 it was announced that *Taranaki* would be converted to a training ship/offshore resources protection vessel. For this purpose she would retain the 4.5in guns forward but would be re-engined, with twin-shaft Rolls-Royce Tyne gas turbines in place of the steam turbines. This interesting project was dropped after the British Defence Review in June 1981, when *Leander* class frigates became available at a cheap price. Both ships were withdrawn when the new *Leanders* were acquired.

NOTE
The *Whitby* class frigate *Blackpool* was lent by the RN for five years from June 1966 until HMNZS *Canterbury* was complete. She returned to Great Britain in 1971 and her ship's company turned over to the new frigate. For full details see under Major Surface Ships, United Kingdom.

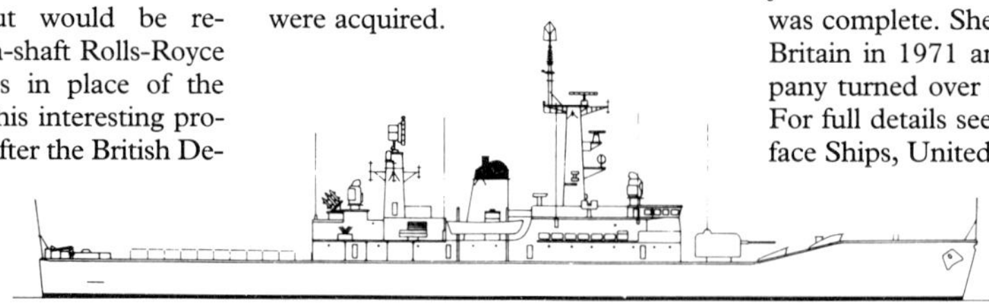

Canterbury 1976

British LEANDER type *frigates*

Particulars: As British *Leander* class

No	Name	Builder	Laid down	Launched	Comp	Fate
F 55	WAIKATO	Harland & Wolff	10.1.64	18.2.65	16.9.66	Extant 1995
F 148	CANTERBURY	Yarrow	12.4.69	6.5.70	22.10.71	Extant 1995
F 69	WELLINGTON (ex-*Bacchante*)	Vickers-Armstrong, Tyne	Acquired Oct 1991			Extant 1995
F 104	SOUTHLAND (ex-*Dido*)	Yarrow	Acquired Oct 1991			Stricken 1995

In June 1963 one of the new *Leander* class was ordered from the RN, but when a second frigate was approved five years later the modified or 'broad-beam' version was ordered (August 1968). They are similar to the RN ships apart from a modified funnel cap and USN-pattern Mk 32 TT abaft the boats port and starboard. *Waikato*'s hangar was enlarged to accommodate a Lynx, but the others still operate Wasps. *Wellington* and *Canterbury* had their MRS 3 fire control and ADAWS-5 combat system replaced by RCA's R-76 and MUSL's Nautis-F between 1984 and 1992, and *Canterbury* received a Signaal LW–08 surveillance radar in place of Type 965 *Wellington* is similarly fitted.

In October 1981 it was announced that the Ikara-armed *Leander*, HMS *Dido*, and the broad-beam *Bacchante* would be purchased from the Royal Navy in 1981–82. HMNZS *Wellington* (ex-*Baccante*) sailed to New Zealand on 11 October 1982, but *Southland* did not sail until December 1983, when her refit was completed. She paid off in March 1995.

When the RAN ordered the *Anzac* class on 14 August 1989, the RNZN was given an option to buy up to four, but in September 1989 a decision was made to build only two. Offsets included fabrication of superstructure modules in New Zealand. *Tekaha* scheduled to be delivered in February 1997, the second unit in November 1998.

Australian ANZAC class *frigates*

Particulars: As Australian *Anzac* class

No	Name	Builder	Laid down	Launched	Comp	Fate
–	TEKAHA	Transfield	19.9.94	–	–	Building
–	–	Transfield	Feb 1995	–	–	Building

SMALL SURFACE COMBATANTS

Ex-Australian BATHURST class *minesweepers*

Displacement: 650–790t standard
Dimensions: 186ft 2in oa × 31ft 2in × 9ft 10in mean deep load
56.7in × 9.5m × 3.0m
Machinery: 2-shaft VTE, 2 3-drum boilers, 1550–2000ihp = 15–16kts. Oil 124–153t
Armament: 1–4in/45, 1–40mm
Complement: 80–90

Class (fate):
Echuca (BU 1967), *Inverell* (BU 1979), *Kiamia* (BU 1979), *Stawell* (BU 1968)

Acquired from the RAN in 1952. Despite their age they replaced the 'Loch' class on training duties in 1965–66, with sweep gear removed and the 4in gun replaced by a 40mm. For pennant numbers see under Australia.

PUKAKI class *patrol craft*

Displacement: 105t standard; 135t full load
Dimensions: 107ft × 20ft × 11ft 10in
32.8m × 6.1m × 3.6m
Machinery: 2 shafts, 2 Paxman YCJM diesels, 3000bhp = 25kts. Range 2500nm at 12kts
Armament: 2–12.7mm MG (1×2), 1–81mm mortar/12.7mm MG
Sensors: Radar Decca 916
Complement: 21

Class (no, launched):
Pukaki (P 3568, 1.3.74), *Rotoiti* (P 3569, 8.3.74), *Taupo* (P 3570, 25.7.74), *Hawea* (P 3571, 9.9.74)

All built by Brooke Marine, Lowestoft, UK; all put up for sail 1991.

Taupo as completed — Courtesy Ross Gillett

MOA class *patrol craft*

Displacement: 91.5t standard; 105t full load
Dimensions: 88ft × 20ft × 7ft 3in
26.8m × 6.7m × 2.2m
Machinery: 2 shafts, 2 Cummins KT-1105M diesels, 710bhp = 12kts. Range 1000nm at 11kts
Armament: 1–12.7mm MG
Sensors: Radar Racal-Decca 916; sonar Klein 595 Tracpoint towed sidescan
Complement: 19

Class (no):
Moa (P 3553), *Kiwi* (P 3554), *Wakakura* (P 3555), *Hinau* (P 3556)

Construction of four units approved 1982, and contract awarded to Whangarei Engineering and Construction Co. All commissioned November 1983 to October 1985. Fitted with sidescan sonars 1993–94 to give them MCM capability. All extant 1995.

MISCELLANEOUS
Charles Upham (ex-*Merchandian Queen II*; 7200t) acquired 1994. Former ro-ro ferry purchased for conversion to a military sealift ship. Extant 1995.

Nicaragua

The largest of the Central American republics, Nicaragua has had an unstable and strife-torn history. Until 1979, government was in the hands of General Somoza, commander of the National Guard, who had become president in 1936. The declaration of martial law in 1973 had lead to the emergence of a strong, communist-backed guerrilla organisation, the Sandanista Liberation Movement, which forced Somoza to resign in 1979, following the withdrawal of US aid for the regime by the Carter Administration. Fears of the Sandanistas' links to the Soviet Union (a $166.8m aid deal was signed in 1982) and of Nicaragua serving as a base for further Communist subversion in Central America caused the Reagan government to back the Contra rebels against the Sandanistas, and a ten-year civil war ensued, until the Sandanistas lost their parliamentary majority in the February 1992 elections. However, political unrest since then has seen the re-emergence of the Sandanistas in a coalition government and in 1992–94 government forces fought disgruntled Sandanista and former Contra guerrillas dissatisfied with the failure of economic reforms.

Nicaragua has no navy as such; patrol boats are operated by the Marine Section of the National Guard (1200 officers and men). Main bases are at Corinto (the main port), Puerto Cabezas (Caribbean), Puerto Somoza and San Juan del Sur.

Fleet strength:

Four Israeli 'Dabur' type patrol craft *CG 10*, *CG 11*, *CG 12*, *CG 13* delivered in 1978; one lost in 1985, second severely damaged in 1987. Non-operational.

Ten North Korean 'Sin Hung' type patrol craft *400*, *402–410*, delivered in 1983 (two), 1984 (two) and 1989 (six); two deleted.

Four North Korean 'Kim Jin' type patrol craft *306*, *308* and two more, delivered in 1980s, two deleted in 1990, remainder non-operational.

Ten ex-Soviet 'Zhuk' type patrol craft *304*, *305*, *307*, *309*, *311*, *313*, *315*, *317* and two more, delivered in 1982–89; two lost in the hurricane of October 1989, one deleted 1991.

Two French 'Vedette' type patrol craft *300*, *302* delivered in 1983 (57t full load, 28.2m, 1520bhp = 24kts, 2–14.5mm).

Two 'Tuyacan' class patrol craft *301*, *303*, (32.0m, 2–37mm), non-operational.

Rio Kuringwas (GC 7) patrol craft of Sewart type (60t, 26kts, 3 MG) delivered in 1972, reported captured by Honduras.

Six American 'Hatteras' type patrol craft transferred in 1980; two 24.5m patrol craft *GC 6*, *GC 7* acquired in 1960, four 27.5m patrol craft GC 2, GC 3, GC 4, GC 5 acquired in 1947; all deleted.

Four ex-Polish 'K 8' (Type 361T) type minesweeping boats transferred in 1984, extant 1995.

Four ex-Soviet 'Yevgenya' class inshore minesweepers *501*, *503*, *508*, *510*, transferred in pairs 1984 and 1986; *503* sunk in hurricane in 1989.

There are also two tank landing lighters (45.8m, 9kts) built in 1985 by Damen, Netherlands

Nigeria

The Federal Republic of Nigeria, with her population of around 112 million and vast oil revenues, is by far the most populous African country and much the wealthiest and most powerful country in black Africa; her gross domestic product exceeds South Africa's and is four times that of Egypt. The merchant marine expanded by fifty ships in the 1970s and 1980s.

Nigeria's political history since independence in 1960 has often been troubled and turbulent. The original parliamentary government evolved into quasi-dictatorial rule by a corrupt elite, which manipulated elections and censuses. Increasing polarisation along tribal lines, followed by mounting chaos, led to the January 1966 Army coup. This temporarily eased tensions, but seemed after a short time to be leading to Ibo (Easterner) domination; it was followed after six months by a second coup, led by Northern elements in the Army, which put 31-year-old Lt-Colonel Yakubu Gowon into power. Massacres of Ibos in the north ensued. Attempts to keep the East in the Federation failed, and the proclamation of the 'Republic of Biafra' in May 1967 was followed by civil war.

The navy (a frigate and five small combatants) remained loyal and, even though its former minister (later Biafra's foreign minister) joked 'What navy?', a naval blockade of the 200-mile Biafran coast at the outset effectively isolated the rebel republic from outside recognition or help (except by air). Amphibious landings at Bonny (to capture its ocean oil terminal) and Calabar in the end proved major springboards to Federal victory.

Biafra collapsed in January 1970, and General Gowon started a successful programme of national reconciliation. He was, however, unable to cope with the rise in world oil prices in 1973 and 1974. There was general relief when Gowon was ousted in July 1975; a new dynamic military ruler corrected the worst abuses of his regime, and set Nigeria firmly back on the road to civilian rule. A new constitution, loosely based on the American was carefully drawn up and took effect on 1 October 1979.

The navy has grown slowly from its 200-man pre-independence beginnings in 1956, volunteer personnel having increased from 555 in 1960 to 1720 in 1970 and about 5000 in 1990. Until 1964 British officers were in command. The build-up of ships from 1970 was steady, the ex-British warships originally transferred being discarded as they wore out, and replaced by new construction from the UK, Netherlands, West Germany and France. The navy is far ahead materially of other tropical African forces; it operates two frigates and four corvettes, and has the only ship-borne helicopters, as well as the only admiral, between the Mediterranean and South Africa. However, this expansion came to a halt in the 1980s, and since then the navy has stagnated. Nigeria has not been able to afford a navy of this size, and the ships are in poor condition and some have been held abroad awaiting payment for work done on them. The only recent additions to the fleet have been two Italian MCM vessels.

The training of its personnel is considered less satisfactory than its material state although India, USA, Australia, France and West Germany all help. A shortage of skilled and experienced technicians has caused mishandling of delicate equipment and a corresponding reduction in operational efficiency. The leadership, although young, is of uncertain quality, lacking initiative and a sense of devotion to duty; promotion in the past has often depended on tribal factors and nepotism.

The Fleet is split between Flotilla Command at Lekki (for the Niger and its delta), Western Naval Command (based on Apapa-Lagos where a second dockyard is building) and Eastern Naval Command at Calabar and Port Harcourt (home of Nigerian Shipbuilders Ltd). These and the main naval staff sections are normally headed by a commodore. A Fleet Air Arm was set up in 1981 for which Westland delivered the first of three Lynx helicopters. From 1976 Decca installed a chain of four coastal radars that covers the country's 400 nautical mile coastline.

To conclude, Nigeria's navy is impressive on paper, but it has so far failed to tackle the revival of piracy off West Africa and Lagos in particular. It is not, at present, equal to a properly led, trained and maintained force of similar paper strength.

MAJOR SURFACE SHIPS

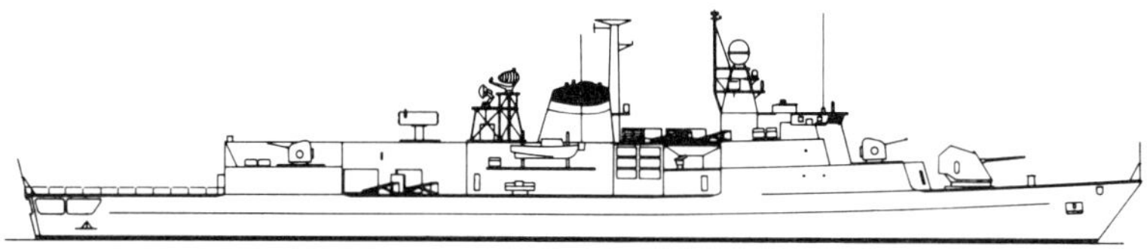

Aradu during trials period, May 1981

Ordered 3 November 1977 from Blohm & Voss, Hamburg, to the MEKO 360H design (see Argentina), the first frigate in the world using extensive modular prefabrication and 'containerised' weapons (46 modules) and sensors (76) to speed construction. One Lynx helicopters is carried. Renamed (1 November 1980) after launch. *Aradu* means 'Thunder' and she arrived at Lagos 21 December 1981 to become the Nigerian Navy flagship at a cost of $84.5 million or 10 per cent of the 1981 defence budget. She was damaged in two collisions in 1987 and from 1991 to 1994 was refitted at Lagos.

ARADU *frigate*

Displacement: 3600t standard; 4200t full load
Dimensions: 390ft 5in wl, 413ft oa × 49ft 3in × 14ft 2in
119.0m, 125.9m × 15.0m × 4.3m
Machinery: 2-shaft CODOG (2 CP propellers): 2 Rolls-Royce Olympus TM3B gas turbines, 50,000shp = 30.5kts plus 2 MTU Type 20 V 956 TB92 diesels 10,000bhp = 18kts. Fuel 440t. Range 6500nm at 18kts. Endurance 90 days
Armament: 8 Otomat Mk 2 SSM (8 × 1), 1 Apside Mk 2 Mod 9 SAM (1 × 8, 24 missiles), 1–127mm/54 OTO Melara (460rds), 8–40mm/70 Breda Bofors (4 × 2, 10,752rds), 6–12.75in STWS-13 TT (2 × 3; 18 torpedoes), 2 helicopters
Sensors: Radar AWS-5A, WM-25, Decca 1226 navigation; sonar PHS 32; ECM Elsag/Breda 105mm chaff RL (2 × 20; 120rds) Decca RCM0-2
Complement: 200–237

No	Name	Builder	Laid down	Launched	Comp	Fate
F 89	ARADU (ex-*Republic*)	Blohm & Voss	2.5.79	25.1.80	4.9.81	Extant 1995

Aradu 26.8.81

L & L van Ginderen

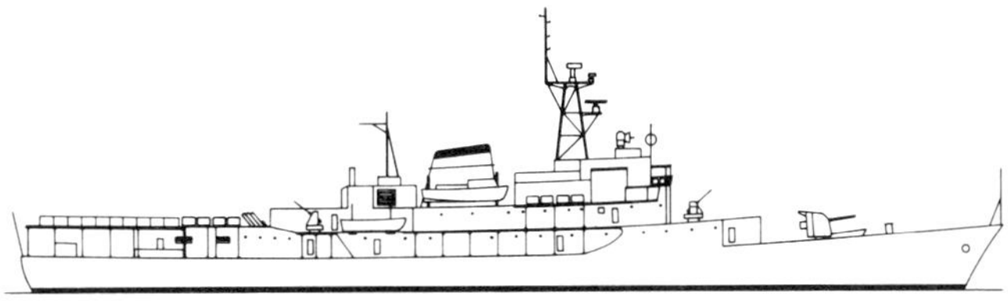

Nigeria 1975

Ordered at original price of £3.5 million from the Netherlands Shipbuilding Bureau (NEVESBU) as combined escort, training ship and flagship; armed with RN-pattern armament. Fitted with a helicopter pad aft. Refitted Birkenhead (UK) 1973 and Schiedam (Holland) 1977. Renamed *Obuma* in 1981. An intended refit was abandoned in 1983 because of lack of money and she is not currently able to go to sea.

NIGERIA *frigate*

Displacement: 1724t standard; 2000t full load
Dimensions: 341ft 3in pp, 360ft 3in oa × 37ft × 11ft 6in
104.0m, 109.8m × 11.3m × 3.5m
Machinery: 2 shafts, 4 MAN 16 V8B cyl diesels, 16,000bhp = 26kts. Range 3500nm at 15kts
Armament: 2–4in/45 Mk 19 (1 × 2), 4–40mm/60 Mk 7 (4 × 1), 1 Squid ASW mortar
Sensors: Radar Types 293, 275, Plessey AWS-4; sonar Types 170, 174 (removed 1977)
Complement: 216

No	Name	Builder	Laid down	Launched	Comp	Fate
F 87	OBUMA (ex-*Nigeria*)	Wilton-Fijenoord	9.4.64	12.4.65	16.9.65	Extant 1995

AMPHIBIOUS WARFARE VESSELS

West German type RO-RO 1300 *landing ships*

Displacement: 1470t standard; 1750t full load
Dimensions: 242ft 1½in pp, 285ft 5in oa × 45ft 11in × 7ft 6in
74.5m, 97.0m × 14.0m × 2.3m
Machinery: 4 shafts, 2 MTU 16V 956 TB92 diesels, 6694bhp (4920kW) = 17kts. Range 5560nm at 14kts
Armament: 1–40mm Bofors (300rds), 2–20mm Oerlikon (2 × 1; 3000rds), 81mm mortar (optional)
Sensors: Radar Decca 1226
Complement: 56 (plus 220–1000 troops)

No	Name	Builder	Launched	Fate
L 1312	AMBE	Howaldtswerke, Hamburg	7.7.78	Extant 1995
L 1313	OFIOM	Howaldtswerke, Hamburg	7.12.78	Extant 1995

Built as Type 502 landing ships, a design not bought by the *Bundesmarine*. Laid down 3 March 1978 and 15 September 1978, completed 11 May 1979 and July 1979. Bow and stern ramps, former can unload 52t tanks. Typical loading could be 400t of vehicles plus troops. Names mean 'crocodile' in different Nigerian languages. Cost $30.8 million each.

Ambe 1979 L & L van Ginderen

SMALL SURFACE COMBATANTS

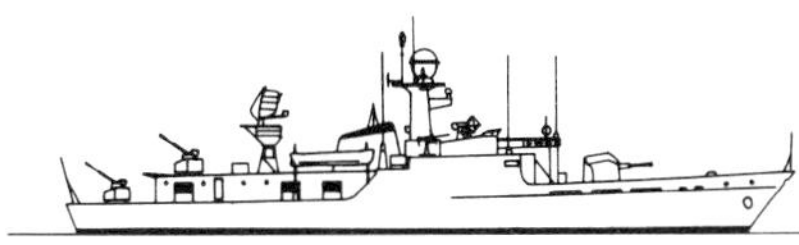

Dorina 1972

DORINA class *corvettes*

Displacement: 500t standard; 650t full load
Dimensions: 185ft pp, 202ft oa × 31ft × 11ft 4in
56.4m, 61.6m × 9.5m × 3.3m
Machinery: 2 shafts, 2 MAN type V8V 24 30B diesels, 8000bhp = 22kts. Oil 68t. Range 3500nm at 14kts
Armament: 2–4in/45 Mk 19 (1 × 2), 2–40mm/60 (2 × 1), 2–20mm Mk 7a (2 × 1)
Sensors: Radar Plessey AWS-1, M-22, Decca TM-626; sonar Plessey MS-22; ECM Decca
Complement: 67

No	Name	Builder	Launched	Fate
F 81	DORINA	Vosper Thornycroft	16.9.70	See notes
F 82	OTOBO	Vosper Thornycroft	25.5.71	Stricken 1987

Vosper Thornycroft Mk 3 corvettes, ordered 28 March 1968, larger and faster than the Mk 1 versions built for Ghana and Libya. Laid down 26 January 1970 and 28 September 1970, completed 21 July 1972 and 10 November 1972. Both were refitted by Vosper 1975–76. The 4in guns are hand-loaded and there is no ASW armament to exploit the sonar carried. *Dorina* sank at her moorings on 16 April 1987 and raised a month later. In April 1988 she was shipped to Italy as deck cargo for refitting as a patrol boat with new machinery and armament of 1–40mm. No work has been done as of 1995 because of lack of funds.

Otobo Apr 1976 C & S Taylor

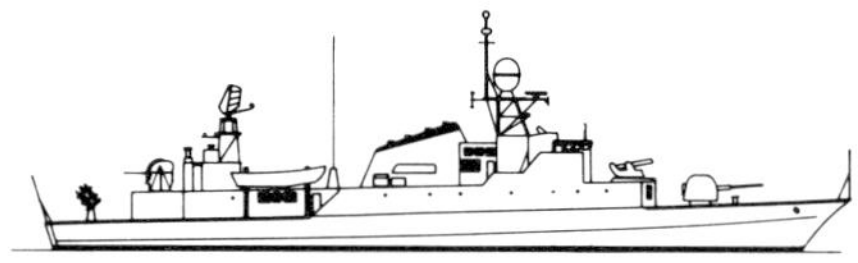

Erin'mi 1980

ERIN'MI class *corvettes*

Displacement: 700t standard; 850t full load
Dimensions: 210ft pp, 226ft oa × 31ft 6in × 9ft 10in
64.0m, 69.0m × 9.6m × 3.0m
Machinery: 2 shafts, 4 MTU type 20V956 TB92 diesels (2 CP propellers), 17,600bhp = 28kts. Range 2200nm at 14kts
Armament: 1–76mm/62 OTO Melara Compact (750rds), 1–40mm/70 Bofors, 2–20mm Oerlikon GAM 204 GK (2 × 1), Seacat SAM (1 × 3; 15 missiles), 2–375mm Bofors ASW RL (1 × 2; 24 rockets)
Sensors: Radar Plessey AWS-2, WM-24, TM-1226; sonar PMS-26; ECM Decca Cutlass system
Complement: 90

No	Name	Builder	Launched	Fate
F 83	ERIN'MI	Vosper Thornycroft	20.1.77	Extant 1995
F 84	ENYIMIRI	Vosper Thornycroft	9.2.78	Extant 1995

Vosper Thornycroft Mk 9 corvettes ordered 22 April 1975 for $40 million. Laid down 14 October 1975 and 11 February 1977, completed 25 February 1980 and 14 July 1980. Funnels were heightened during trials so that smoke would not blind the Bofors fire-control system at the stern. Names, like those of the *Dorina* class, mean 'hippopotamus' in different Nigerian languages. F 83 sailed for Nigeria May 1980. Can maintain 20kts on 2 diesels. Both were refitted by Vosper Thornycroft in the early 1980s. Neither is operational as of 1995.

Erin'mi 1979 Vosper Thornycroft

Damisia as completed
NB 1/750 scale

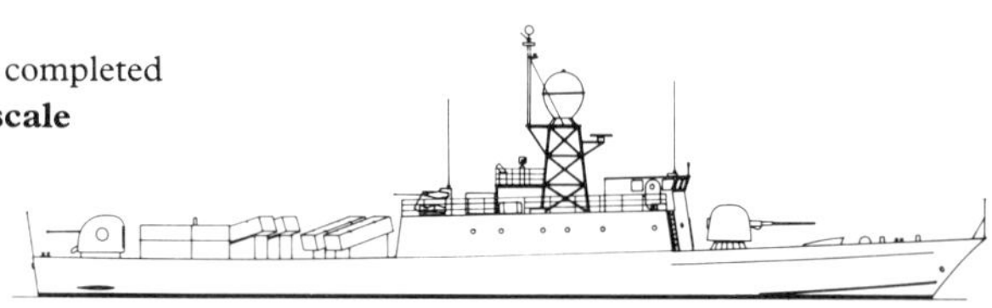

EKPE class *fast attack craft (missile)*

Displacement: 295t standard; 378t full load
Dimensions: 187ft wl, 200ft oa × 25ft 7in × 7ft 11in
57.0m, 60.9m × 7.79m × 2.41m
Machinery: 4 shafts, 4 MTU 16cyl 956 TB92 diesels, 20,840bhp (15,298kW) = 38kts. Range 1300/3300nm at 30/16kts
Armament: 4 Otomat Mk 1 SSM (4 × 1), 1–76mm/62 OTO Melara Compact, 2–40mm/70 Breda Compact (1 × 2), 4–30mm Emerlec (2 × 2), 2–533mm TT
Sensors: Radar Decca TM-1226C, HSA WM-28; ECM Decca RDL-1 intercept
Complement: 40

Class (no, launched):
Ekpe (P 178, 17.12.79), *Damisia* (P) 179, 25.5.80), *Agu* (P 180, 7.11.80)

S 143 type W German boats ordered with different armament from Lürssen, Vegesack, 14 November 1977. Laid down 17 February 1979 and arrived Lagos together 29 September 1981. They are fitted with Dutch WM-28 fire control. TT are aft, firing Seal wire-guided torpedoes. All extant 1995, held at a shipyard in France, where they had been sent for refitting 1986–88, because of non-payment for work done.

SIRI class *fast attack craft (missile)*

Displacement: 385t standard; 430t full load
Dimensions: 184ft 5in oa × 26ft 3in × 6ft 11in
56.2m × 8.0m × 2.1m
Machinery: 4 shafts, 4 MTU 16V956 TB92 diesels, 20,840bhp (15,298kW) = 37kts. Range 2000/800nm at 15/30kts
Armament: 4 Exocet MM38 SSM (2 × 2), 1–76mm/62 OTO Melara Compact (250rds), 2–40mm/70 Breda Compact (1 × 2, 1600rds), 4–30mm Emerlec (2 × 2; 1970rds)
Sensors: Radar Decca 1226, Thomson-CSF Triton and Castor ESM Decca RDL-1
Complement: 42

Class (no, launched, completed):
Siri (P 181, 3.6.80, 19.2.81), *Ayam* (P 182, 10.11.80, 11.6.81), *Ekun* (P 183, 11.2.81, 18.9.81)

'Combattante IIIB' type boats ordered from CMN, Cherbourg, 14 November 1977. Laid down 15 May 1979, 7 September 1979 and 14 November 1979. Fitted with 2 CSEE Panda fire control directors. Names mean 'tiger' in the three main Nigerian languages. Did not leave France until 9 May 1982 when Nigeria completed payment. Commissioned 6 February 1982. Nigeria ordered eighteen Exocet MM38 missiles from Aérospatiale in 1980. All three extant 1995.

Ex-US PC type *patrol vessel*

Name	Builder	Acquired	Fate
OGOJA (ex-*Queen Wilhelmina*, ex-*PC 468*)	Lawley & Sons	1965	Lost Oct 1969

Old patrol vessel, presented to the Netherlands in 1942 and laid up from 1953, was lent to Nigeria while the frigate *Nigeria* was building. Transferred permanently *c*1966 and wrecked off Brass (Niger Delta) in October 1969 during the civil war.

ARGUNGU class *large patrol craft*

Displacement: 90t full load
Dimensions: 95ft 1in oa × 18ft × 5ft 2in
32.0m, 29.0m × 5.5m × 1.6m
Machinery: 2 shafts, 2 Paxman 8cyl Ventura YJCM diesels, 2200bhp = 20kts
Armament: 1–40mm/60 Mk 9, 1–20mm, 2–2in R(F)L (see notes)
Complement: 25

Class (no, launched):
Argungu (P 165, 4.7.73), *Yola* (P 166, 12.6.73), *Bras* (P 169, 12.1.76), *Epe* (P 170, 9.2.76)

Ordered from Abeking and Rasmussen, Lemwerder (FRG) in pairs (1971 and 1975). *Argungu* laid down 4 July 1973 and units completed in pairs (October 1974 and March 1976). Rearmed with 4–30mm Emerlec (2 × 2) in 1978. All extant 1995.

MAKURDI class *large patrol craft*

Displacement: 115t standard; 143t full load
Dimensions: 107ft oa × 20ft × 11ft 6in
32.6m × 6.1m × 3.5m
Machinery: 2 shafts, 2 Paxman Ventura 12cyl diesels, 3000bhp = 20.5kts. Oil 18t. Range 2300nm at 12kts
Armament: 2–40mm/60 Mk 9 (960rds). See notes
Complement: 21

Class (no, launched):
Makurdi (P 167, 21.3.74), *Hadejia* (P 168, 25.5.74), *Jebba* (P 171, 1.12.76), *Oguta* (P 172, 17.1.77)

First pair ordered June 1972 and second October 1974 from Brooke Marine, Lowestoft (GB). P 167 and P 168 commissioned 14 August 1974, P 171 and P 172 29 April 1977. First pair refitted with 4–30mm Emerlec (2 × 2) by Brooke Marine 1981, second pair likewise in Nigeria. All extant 1995.

Hadejia after 1982 refit Brooke Marine

Ex-British 'FORD' class *seaward defence boats*

Class (former name):
Benin (ex-*Hinksford*), *Bonny* (ex-*Gifford*), *Enugu* (–), *Ibadan* (i) (ex-*Montford*), *Ibadan* (ii) (ex-*Bryansford*), *Kaduna* (ex-*Axford*), *Sapele* (ex-*Dubford*)

Enugu was the first warship built for Nigeria 1960–61. *Benin, Ibadan* (i) and *Kaduna* bought from the RN in 1966, *Bonny* and *Sapele* (1967–68) and *Ibadan* (ii) in 1968. *Ibadan* (i) served as the Biafran *Vigilance* May–September 1967; sunk by the Nigerian Navy at Port Harcourt on 10 September 1967, she was salved and later broken up. *Kaduna* and *Ibadan* (ii) were deleted in 1965; *Benin* (P 03), *Bonny* (P 04), *Enugu* (P 05) and *Sapele* (P 09) were deleted in the 1980s.

Ex-British ALGERINE class *minesweeper*

Name	Builder	Acquired	Fate
NIGERIA (ex-*Hare*)	Harland & Wolff	July 1959	BU 1962

A war-built minesweeper, transferred from the RN; returned to UK in 1962 with boiler and other defects, and scrapped as beyond economic repair.

Italian LERICI class *MCM vessels*

Class (no, in service):
Ohue (M 371, Apr 1988), *Maramba* (M 372, Apr 1988)

First unit ordered 9 April 1984, second in May 1984 from Intermarine, Sarzana. Option for two more not taken up. Delay in delivery caused by difficulty in obtaining export licence. Both extant 1995.

MISCELLANEOUS

Besides the above, the following minor fighting vessels were acquired from the RN in 1959: the seaward defence motor launch *Kaduna* (ex-*SDML 3515*) deleted 1965, the Fairmile 'B' type minesweeping motor launches *Calabar* (ex-*MSML 2223*) deleted 1969, and *Sapele* (ex-*MSML 2217*), discarded 1967, and the landing craft *Lokoja* (ex-*LCT(4) 1213*), refitted in 1966–67 and deleted in 1978.

Three ex-Soviet 'P 6' class MTBs, acquired in 1967 and assigned the names *Ekpen, Ekun* and *Elole*, were discarded in 1975.

The Nigerian Coast Guard operates fifteen GRP coastal patrol boats (21t, 33kts, 1–20mm, 2 MG) ordered from Intermarine, Italy, in October 1978, delivered July 1981–May 1982 for the anti-smuggling role. It also has a Fairey 'Tracker' and two Fairey 'Spear' boats for fishery protection (since 1978). The Marine Police operate a large force of fast launches (including six small hovercraft, delivered 1982) in the Niger Delta and on Lake Chad.

Norway

When the Second World War ended Norway had a larger and more modern navy than she had had in April 1940, when the Germans invaded. Only a few units escaped to Britain, but the Norwegian government in exile built up a new navy, mostly of ships borrowed or purchased from Great Britain and Canada. In 1947 the Norwegian Navy consisted of six destroyers (one incomplete), four torpedo boats, five submarines, two frigates, three corvettes and two fishery protection vessels, one coastal minelayer, ten minesweepers, one submarine-chaser, ten motor torpedo-boats and three motor launches.

During the immediate postwar years, Norway, like all the European countries, had difficulty finding money for defence measures. The navy had received a number of German warships in 1945, but all these vessels were in bad condition and were soon scrapped. In 1951, the USA delivered ten motor torpedo-boats.

The first step to building a new navy was taken in 1950 with the ordering of motor torpedo-boats from a domestic yard. Thereafter Norway constructed, launched and commissioned series of ships and boats for the navy, built at naval and private yards, in cooperation with other NATO navies and with help from the United States under the Mutual Defense Assistance Program.

The proliferation of offshore establishments in the North Sea oilfields brought an increased need for patrol vessels, and in the 1980s the *Nordkapp* class of multipurpose 'offshore protection vessels' was commissioned. The Norwegian Navy has continued to update its fleet, acquiring new submarines and developing a class of surface-effect MCM vessels. However, there have been problems, with the new *Ula* class submarines reportedly causing dissatisfaction due to excessive machinery noise, and the first surface-effect vessel not entering service until a year after its due date. Further programmes are planned, including the acquisition of new frigates to replace the *Oslo* class (an indigenous design or Dutch *Kortenaer* class ships both being considered), new fast attack craft and further modernisations for other classes.

FLEET STRENGTH 1947

DESTROYERS

Name	Launched	Disp (standard)	Fate
Bergen class			
BERGEN (ex-*Cromwell*) (D 304)	6.8.45	1710t	Stricken 1.1.67
OSLO (ex-*Crown*) (D 303)	19.12.45	1710t	Stricken 1966, BU
STAVANGER (ex-*Crystal*) (D 306)	12.2.45	1786t	Stricken 1.1.67,
TRONDHEIM (ex-*Croziers*) (D 305)	19.9.44	1786t	Stricken 1.5.67
Stord class			
STORD (ex-*Success*) (D 300)	3.4.43	1796t	Stricken 1959, BU
Aalesund class			
AALESUND	1941	1220t	Scrapped 1950
Odin class			
ODIN (F 302)	17.1.39	632t	Stricken 1959
TOR (F 303)	9.9.39	632t	Stricken 1959
BALDER (F 304)	11.10.39	632t	Stricken 1959
Sleipner class			
SLEIPNER (F 300)	7.6.36	597t	Stricken 1956
GYLLER (F 301)	7.7.38	597t	Stricken 1959

Bergen class
In Norwegian service from respectively, 6.8.45, 19.12.45, 6.8.45 and 19.9.45.

Stord class
In Norwegian service from 3.4.43. *Stord*'s sister-ship *Svenner* (ex-*Shark*) was sunk by a German S-boat off Normandy 6.6.44.

Aalesund class
Two sister-ships, laid down in Apr 1939, were both captured by the Germans on 9.4.40, and were renumbered *ZN 4* and *ZN 5*. The hull of the second was broken up. *Aalesund* was never completed; her construction was abandoned in 1950.

Odin class
Improved *Sleipner* class. *Odin* surrendered to the Germans at Marvika naval depot, Kristiansand on 11.4.40, and was renamed *Panther*. *Tor* and *Balder* were completed by the Germans and renamed *Tiger* and *Leopard*. All three were retro-ceded in May 1945 and converted to frigates 1954, 1954 and 1952 respectively.

Sleipner class
Sleipner was the largest Norwegian warship to get to Britain in April 1940. *Gyller* surrendered at Kristiansand on 11.4.40 and was renamed *Löwe*. Both were retro-ceded in May 1945. *Gyller* was converted to a frigate in 1957. Her sistership *Aeger* was bombed on 9.4.40 at Hundvaagöy, Stavanger, and wrecked.

FRIGATES

Name	Launched	Disp (standard)	Fate
ARENDAL (ex-*Badsworth*) (F 310)	17.3.41	1317t	Stricken 1.5.61
NARVIK (ex-*Glaisdale*) (F 309)	5.1.42	1230t	Stricken 1.5.61, BU

These two *Hunt* class escort destroyers, *Badsworth* (Type 2) and *Glaisdale* (Type 3) were purchased by Norway in 1946. Wore 'D' pennant numbers until 1956.

CORVETTES

Name	Launched	Disp (standard)	Fate
Andenes class			
ANDENES (ex-*Acanthus*)	26.5.41	925t	Stricken 1956
NORDKYN (ex-*Buttercup*)	10.4.41	925t	Stricken 1956
SOROY (ex-*Eglantine*)	11.6.41	925t	Stricken 1956
Nordkapp class			
NORDKAPP	18.8.37	243t	Stricken 1956, later fishing vessel
SENJA	25.8.37	243t	Stricken 1956, later fishing vessel

NORWAY

Andenes class
These British 'Flower' class corvettes were transferred to the Norwegian flag during the war. Other British corvettes transferred were: *Montbretta*, sunk 1942; *Rose*, sunk 1944; *Potentilla*, returned to the RN in 1944; *Tunsberg* (ex-*Shrewsbury Castle*), mined 1944.

Nordkapp class
Rated as fishery protection vessels. *Nordkapp* escaped to Britain in Apr 1940. *Senja* was captured by the Germans on 9.4.40, and was used as a patrol vessel, renamed *Löwe*, later *H 2*, *DW*, *V 6735* and *V 6315*, retro-ceded 1945.

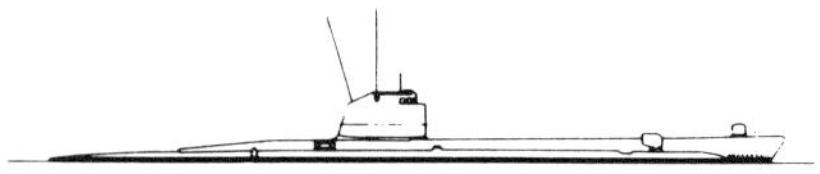

Uthaug 1957

SUBMARINES

Name	Launched	Disp (standard)	Fate
Ula class			
ULA (ex-*Varne*) (S 300)	22.6.43	630t	Stricken Jul 1964
UTSTEIN (ex-*Venturer*) (S 302)	4.5.43	545t	Stricken Jan 1964
UTVÆR (ex-*Viking*) (S 303)	5.5.43	545t	Stricken Dec 1964
UTSIRA (ex-*Variance*) (S 301)	22.5.44	545t	Stricken Dec 1962
UTHAUG (ex-*Votary*) (S 304)	21.8.44	545t	Stricken Oct 1965

British *'U'* class, purchased from Great Britain 1943–46, modernised in 1955–56, with external streamlining, snorkel and sonar dome.

MINELAYER

Name	Launched	Disp (standard)	Fate
LAUGEN	Oct 1917	335t	Scrapped 1948

Captured by the Germans Apr 1940, designated *NN 05*, later *M 22*, retro-ceded May 1945.

MINESWEEPERS

Name	Fate	Name	Fate
Glomma class: launched 1940, 590t			
GLOMMA (ex-*Bangor*)	Stricken 1.12.61	TANA (ex-*Blackpool*)	Stricken 1.5.61
Orkla class: launched 1943–45, 350t			
ORKLA (ex-*MMS 1085*)	Stricken 1964	VEFSNA (ex-*MMS 1086*)	Stricken 1962
Alta class: launched 1944–45, 232t			
VINSTRA (ex-*NYMS 247*)	Stricken 1959	BEGNA (ex-*NYMS 381)*	Stricken 1961
GAULA (ex-*NYMS 305*)	Stricken 26.6.63	RANA (ex-*NYMS 406*)	Stricken Mar 1962
DRIVA (ex-*NYMS 377*)	Stricken 1961	VORMA (ex-*NYMS 480*)	Stricken 1959
ALTA (ex-*NYMS 379*)	Stricken 1959		
Otra class: launched 1939, 335t			
OTRA	Stricken Apr 1963	RAUMA	Stricken Apr 1963

Glomma class
British *Bangor* class diesel-engined minesweepers, purchased in 1946.

Orkla class
British MMS type, purchased in June 1946. From 1951 they were used as X-ray ships for torpedo-boat testing.

Alta class
US YMS type, transferred under Lease-Lend in 1945.

Otra class
Captured by the Germans on 9.9.40 and renamed *Togo* and *Kamerun*. Retro-ceded in May 1945. They were also used as coastal minelayers. Four ex-whalecatchers, *Penang*, *Polarnacht*, *Rotger* and *Sölöven*, were also used as minesweepers in 1947.

SUBMARINE CHASER

Name	Launched	Disp	Fate
KING HAAKON VII (ex-*PC 457*)	29.4.42	280t	Stricken 1974

Used as a fishery protection vessel from *c*1950; TS 1953.

MOTOR TORPEDO-BOATS

Faulk (ex-*MTB 704*), *Hauk* (ex-*MTB 711*), *Jo* (ex-*MTB 713*), *Lom* (ex-*MTB 719*), *Ravn* (ex-*MTB 720*), *Skarv* (ex-*MTB 721*), *Stegg* (ex-*MTB 722*), *Teist* (ex-*MTB 723*), *Tjeld* (ex-*MTB 716*). All ex-British Fairmile 'D' type MTBs, launched 1944, 128t. *Hauk* was lost by fire on 23.11.48, *Jo* was stricken in 1958, the remaining seven were stricken in 1959.

MOTOR LAUNCHES

ML 125, *ML 213*, *ML 573*. All ex-British MLs, stricken in 1947.

NATO PENNANT NUMBERS
Carried from 1950 by ships in commission. Numbers for major ships are quoted after the name in the above tables, but the following were also allocated.

Minesweepers			
M 309	GLOMMA		
M 319	DRIVA	M 318	GAULA
M 310	TANA	M 320	ALTA
M 313	ORKLA	M 322	BEGNA
M 314	VEFSNA	M 330	RANA
M 317	VINSTRA	M 321	VORMA
Minelayers			
N 33	RAUMA	N 34	OTRA
Motor torpedo-boats			
P 343	FAULK	P 347	LOM
P 344	RAVN	P 348	TEIST
P 345	STEGG	P 349	JO
P 346	SKARV	P 350	TJELD

MAJOR SURFACE SHIPS

Taken over at the end of 1952 on a four-year loan, but bought outright in July 1956, when the pennant numbers were changed from 'D' to 'F'. They had been refitted in 1954 with a lattice mast and two Squid ASW mortars in place of 'X' twin 4in, the armament then consisting of 4–4in (2×2), 4–40mm.

Ex-British 'HUNT' (type 2) class *escort destroyers*

Particulars: As British 'Hunt' Type 2

No	Name	Builder	Acquired	Fate
F 311	TROMSO (ex-*Beaufort*)	Yarrow	end 1952	Stricken 1965
F 312	HAUGESUND (ex-*Zetland*)	Cammell Laird	end 1952	Stricken 1965

Originally Canadian 'River' type, modernised for ASW in 1953-54. Lent to Norway in 1956 and purchased outright in 1959. *Garm* was converted to a depot ship for MTBs in 1964 and renamed *Valkyrien*, and *Troll* to a depot ship for submarines in 1965 and renamed *Horten*. They were given the pennant numbers A 535 and A 530.

Ex-Canadian 'RIVER' class *frigates*

Particulars: As Canadian 'River' class

No	Name	Builder	Acquired	Fate
F 313	DRAUG (ex-*Penetang*)	Davie SB	25.1.56	BU 1966
F 314	TROLL (ex-*Prestonian*)	Davie SB	Mar 1956	Stricken 1972
F 315	GARM (ex-*Toronto*)	Davie SB	Mar 1956	Stricken 1977

Trondheim 1988

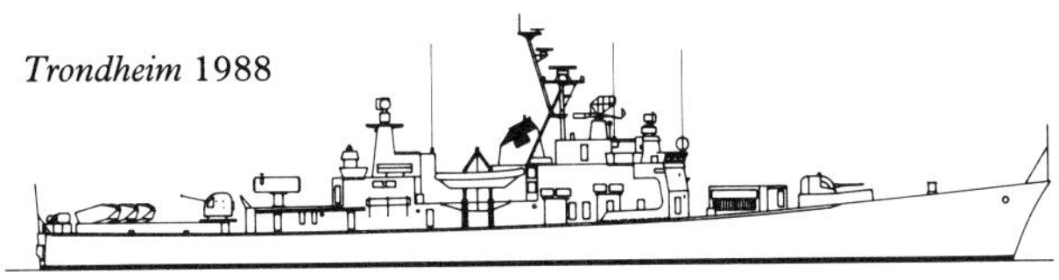

Built under a five-year naval programme voted in 1960, half the cost being carried by the United States. Based on the design of US *Dealey* class destroyer-escorts, but with many modifications, including increased freeboard forward to suit sea conditions off the Norwegian coast. They were modernised in the late 1970s, when the SSM, SAM and TT systems were installed. The 3in guns are US Mk 33s, the Sea Sparrow launcher a US Mk 29, with a Mk 91 mod O radar director. The electronics fitted in the 1970s were of mixed origin, the search radar being French, the Decca TM–1226 navigation set British, and the HSA M 22 tactical and fire control radar being Dutch. From 1988 to 1991 the class received a further modernisation. The aft twin 3in mount and two Penguin launchers were removed to make room for Thomson-Sintra TSM 2633 hull and VDS sonars replacing the US SQS-36, and decoy rocket launchers were fitted. *Bergen* was placed in reserve in 1993, but returned to active service to replace *Oslo*, which sank on 25 January 1994 after running aground.

OSLO class *frigates*

Displacement: 1450t standard; 1760t full load
Dimensions: 308ft pp, 318ft oa × 36ft 7in × 17ft 4in
93.9m, 96.6m × 11.2m × 5.3m
Machinery: 1-shaft STAL-de Laval double reduction geared turbines, 2 Babcock & Wilcox boilers, 20,000shp = 25kts. Range 4500nm at 15kts
Armament: 6 Penguin SSM (2×3), 1 NATO Sea Sparrow SAM (1×8; 24 missiles), 4–3in/50 (2×2), 1 Terne III ASW RL (1×6), 6–324mm Mk 32 TT (2×3)
Sensors: Radar DRBU–22, TM–1226, M22; sonar Terne Mk III attack, SQS–36
Complement: 150

No	Name	Builder	Laid down	Launched	Comp	Fate
F 300	OSLO	Horten N Yd	1963	17.1.64	29.1.66	Sunk 25.1.94
F 301	BERGEN	Horten N Yd	1964	23.8.65	15.6.67	Extant 1995
F302	TRONDHEIM	Horten N Yd	1963	4.9.64	2.6.66	Extant 1995
F 303	STAVANGER	Horten N Yd	1965	4.2.66	1.12.67	Extant 1995
F 304	NARVIK	Horten N Yd	1964	8.1.65	30.11.66	Extant 1995

Stavanger Nov 1979
C & S Taylor

SUBMARINES

German VIIC type U-boats, originally launched in 1943. They were transferred in 1950 from Great Britain.

Ex-German TYPE VIIC *submarines*

Particulars: As German Type VIIC

No	Name	Builder	Acquired	Fate
S 309	KAURA (ex-*U 995*)	Blohm & Voss, Hamburg	1950	Stricken Jan 1963
S 307	KYA (ex-*U 926*)	Neptun-Werft, Rostock	1950	Stricken Mar 1964
S 308	KINN (ex-*U 1202*)	Schichau, Danzig	1950	Stricken 1.6.61

German XXIII type U-boat, originally launched in January 1945. Turned over to Norway in 1950 by Great Britain.

Ex-German TYPE XXIII *submarine*

Particulars: As German Type XXIII

No	Name	Builder	Acquired	Fate
–	KNERTER (ex-*U 4706*)	Germaniawerft, Kiel	1950	Stricken 1953

Voted July 1959 under a five-year programme, with half the cost being paid by United States. Based on the German Type 207, but with a stronger hull for deeper diving, and revised equipment. They can fire the US Mk 37 wire-guided torpedo. For training purposes the German *U 4* of the same design was lent 1962–64 for the building period and under the Norwegian flag she carried the name *Kobben* and pennant number S 310. *Svenner*, equipped for officer training, has a second periscope and is 1m longer.

Six boats of this class were modernised with new fire control equipment and other electronics 1989–1991 and remain in Norwegian service as of 1995. *Sklinna* was renumbered S 314 in 1988. Three units were transferred to Denmark 1989–91. *Stadt* was to have been one of these, but was damaged beyond repair in 1987 after grounding, and was replaced by *Kya*. In October 1991 *Kaura* was transferred to Denmark to provide spares for the sunken *Saelen* (ex-*Uthaug*) (see Denmark).

KOBBEN class *submarines*

Displacement: 370t standard; 530t submerged
Dimensions: 149ft oa × 15ft × 14ft
45.4m × 4.6m × 4.3m
Machinery: 1 shaft, 2 Maybach diesels plus 1 electric motor, 2400bhp/1200shp = 10kts/17kts
Armament: 8–533mm (21in) TT (bow; 8 torpedoes)
Complement: 18

No	Name	Builder	Laid down	Launched	Comp	Fate
S 315	KAURA	Nordseewerke, Emden	1961	16.10.64	5.2.65	To Denmark Oct 1991
S 316	KINN	Nordseewerke, Emden	1960	30.11.63	8.4.64	Stricken 1982
S 317	KYA	Nordseewerke, Emden	1961	20.2.64	15.6.64	To Denmark 1991
S 318	KOBBEN	Nordseewerke, Emden	1961	25.4.64	17.8.64	Extant 1995
S 319	KUNNA	Nordseewerke, Emden	1961	16.7.64	1.10.64	Extant 1995
S 300	ULA	Nordseewerke, Emden	1962	19.12.64	7.5.65	To USA May 1991
S 301	UTSIRA	Nordseewerke, Emden	1963	11.3.65	1.7.65	Stricken 1991
S 302	UTSTEIN	Nordseewerke, Emden	1962	19.5.65	9.9.65	Stricken 1991
S 303	UTVAER	Nordseewerke, Emden	1962	30.6.65	1.12.65	To Denmark 1989
S 304	UTHAUG	Nordseewerke, Emden	1962	8.10.65	16.2.66	To Denmark 1990
S 305	SKLINNA	Nordseewerke, Emden	1963	21.1.66	17.8.66	Extant 1995
S 306	SKOLPEN	Nordseewerke, Emden	1963	24.3.66	17.8.66	Extant 1995
S 307	STADT	Nordseewerke, Emden	1963	10.6.66	15.11.66	Stricken 1987
S 308	STORD	Nordseewerke, Emden	1964	2.9.66	9.2.67	Extant 1995
S 309	SVENNER	Nordseewerke, Emden	1965	27.1.67	1.7.67	Extant 1995

Kuna Jan 1995 L & L van Ginderen

Ula 1989

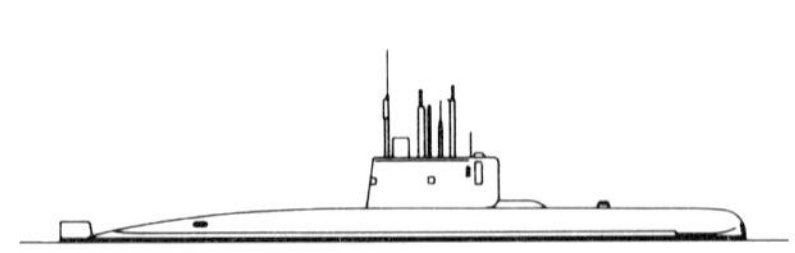

Six units ordered from Thyssen Nordseewerke, Emden on 30 September 1982. An option for a further two units was not taken up. Diving depth 250m. Reported to suffer from noise problems with their machinery. *Ula* was damaged by a practice torpedo during trials in 1989, *Uredd* was involved in a docking accident in March 1991 and suffered a control room fire in February 1992.

ULA class *submarines*

Displacement: 940t standard; 1150t submerged
Dimensions: 193ft 7in × 17ft 9in × 15ft 1in
59.0m × 5.40m × 4.6m
Machinery: 1 shaft: 2 MTU 12V93 AZ diesels plus 1 electric motor 1700shp = 12kts/17kts. Range 5000nm at 8kts snorkelling
Armament: 8–533mm TT (bow; 14 torpedoes)
Sensors: Radar Type 1007; sonar DBQS-21F
Complement: 19 (3 officers)

No	Name	Builder	Laid down	Launched	Comp	Fate
S 300	ULA	Thyssen Nordseewerke	29.1.87	28.7.88	27.4.89	Extant 1995
S 301	UTSIRA	Thyssen Nordseewerke	15.6.90	21.11.91	30.4.92	Extant 1995
S 302	UTSTEIN	Thyssen Nordseewerke	6.12.89	25.4.91	14.11.91	Extant 1995
S 303	UTVAER	Thyssen Nordseewerke	8.12.88	19.4.90	8.11.90	Extant 1995
S 304	UTHAUG	Thyssen Nordseewerke	15.6.89	18.10.90	7.5.91	Extant 1995
S 305	UREDD	Thyssen Nordseewerke	23.6.88	22.9.89	3.5.90	Extant 1995

Utstein 3.8.92 L & L van Ginderen

AMPHIBIOUS WARFARE VESSELS

Ex-US LCU type *utility landing craft*

Particulars: As US *LCU 1466* class

In 1958 US *LCU 1478* was transferred to Norway as *Tjeldsund* (A 30). The vessel was completed in 1953, originally intended to be rebuilt as a minelayer but this was not carried out. Stricken 1975.

KVALSUND class *utility landing craft*

Displacement: 569t standard; 590t full load
Dimensions: 164ft × 33ft 6in × 5ft 10in
50.0m × 10.2m × 1.8m
Machinery: 2 shafts, 2 Maybach diesels, 1350bhp = 11.5kts
Armament: 2–20mm (2×1)
Complement: 10

Class:
Kvalsund (L 4500, ex-A 31), *Raftsund* (L 4501, ex-A 32)

Built by Hjöivold, Kristiansand in 1970. Could carry five tanks and 80–180 troops: rails for 120 mines. Both stricken 1992.

REINOYSUND class *utility landing craft*

Displacement: 560t standard; 596t full load
Dimensions: 167ft 4in wl, 171ft oa × 34ft × 6ft
51.0m, 52.1m × 10.3m × 1.8m
Machinery: 2 shafts, 2 Maybach diesels, 1350bhp = 11.5kts
Armament: 3–20mm (3×1), 2–12.7mm MG
Complement: 10

Class:
Reinöysund (L 4502, ex-A 33), *Söröysund* (L 4503, ex-A 34), *Maursund* (L 4504, ex-A 35), *Rotsund* (L 4505, ex-A 36), *Borgsund* (L 4506, ex-A 37).

Built at Mjellum & Karlsen, Bergen, launched 1970–72, commissioned 1972–73. Modified version of *Kvalsund* class, with similar capabilities. All extant 1995. It is planned to replace this class *c*1999.

SMALL SURFACE COMBATANTS

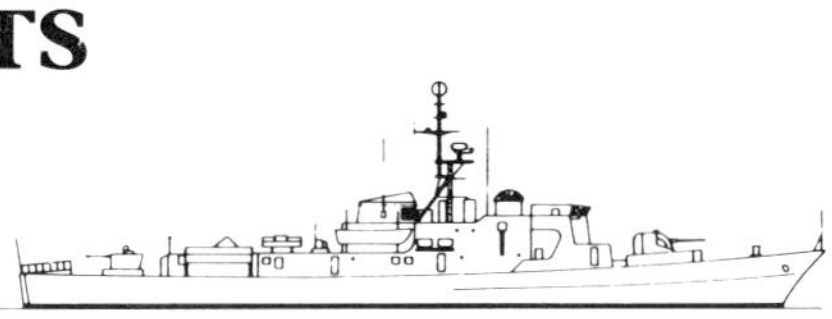

Sleipner 1966

SLEIPNER class *corvettes*

Displacement: 600t standard; 780t full load
Dimensions: 227ft 4in × 26ft × 7ft 10in
69.3m × 7.9m × 2.4m
Machinery: 2 shafts, 4 Maybach diesels 8800bhp = 22kts
Armament: 1–3in, 1–40mm AA, 1 Terne III ASW system, 6–324 Mk 32 TT (3×2), 1 DC rack
Sensors: Radar Decca 202, TM–1226; sonar Terne Mk III attack, SQS-36
Complement: 63

No	Name	Builder	Launched	Fate
P 950	SLEIPNER	Nylands	9.11.63	Stricken 1993, BU
P 951	AEGER	Akers, Oslo	24.9.65	Stricken 1993, BU

In 1959 the decision was taken to build five of this class, but three were cancelled. *Sleipner* was laid down 1963 and completed 29 April 1965. *Aeger* (originally named *Balder*) was laid down 1964 and completed 31 March 1967. Both were specially equipped as submarine chasers in 1972 when the ASW TT were fitted; later renumbered F 310 and F 311. The 3in gun was a US Mk 34, and the original Mk 63 fire control system was replaced with two Swedish TVT–300 optronic systems. In 1988 the SQS-36 sonar was replaced by a Thomson-CSF TSM2633 set. In the 1980s they served principally as training ships. Both scrapped 1993.

VADSO *corvette*

Displacement: 631t normal; 900t full load
Dimensions: 169ft 4in × 29ft 6in × 18ft
51.6m × 9.0m × 5.5m
Machinery: 1 shaft, 1 MAK 8BM451 diesel, 1400bhp = 12kts. Range 15,000nm at 12kts
Armament: 1–40mm, 1 DC rack
Sensors: Radar navigational
Complement: 20

Vadsö (P 340), built at Stord Verft as a whaler, launched 1951, purchased and rebuilt 1976. Sold commercially 1985.

Ex-US SC type *patrol boats*

Armament: 1–40mm (originally 1–6pdr), 4–20mm, 2 MG
Other particulars: As US SC type

Class:
Hessa (ex-*SC 683*), *Hitra* (ex-*SC 718*), *Vigra* (ex-*SC 1061*)

Transferred in 1947 from USA. Pennant numbers were P 391, P 392, P 393 respectively. Stricken 1959.

Ex-German S-BOAT type *fast attack craft*

Armament: 1–40mm, 3–20mm, 2–533mm (21in) TT
Other particulars: As German S-boats (*S 38* and *S 100* groups)

Class:
Kvikk (ex-*S 98*), *Lyn* (ex-*S 64*), *Rapp* (ex-*S 174*), *Snar* (ex-*S 210*), *Storm* (ex-*S 85*), *Tross* (ex-*S 117*)

Launched 1943, 1941, 1944, 1944, 1942 and 1942 respectively. All taken over in 1947; *Storm*, *Tross* and *Lyn*, were handed over to Denmark in 1951. The other three were stricken 1952.

Ex-German S-BOAT type *fast attack craft*

Armament: 1–40mm, 1–20mm, 2 MG, 2–533mm (21in) TT
Other particulars: As German S-boats (*S 6* and *S 10* groups)

Class:
Blink (ex-*S 9*), *Brand* (ex-*S 10*), *Kjekk* (ex-*S 12*)

Launched 1934–35. All three were ceded to the Allies in 1947; *Brand* was transferred to Denmark in 1951, and the other two were discarded shortly afterwards. Norway also received *S 21*, *S 48*, *S 76*, *S 195* and *S 302*, and although the last pair carried the names *Kjekk* and *Blink* for a short time before transfer to Denmark in 1951, none was operational.

Ex-US 'ELCO' type *fast attack craft*

Armament: 1–40mm, 1–20mm, 4–12.7mm MG, 2–533mm (21in) TT
Other particulars: As US 'Elco' type

Class:
Hai (P 956), *Hauk* (P 957), *Hval* (P 958), *Hvass* (P 959), *Sel* (P 950), *Sild* (P 951), *Skrei* (P 952), *Snar* (P 953), *Snögg* (P 954), *Springer* (P 955)

Ex-US 'Elco' type PT-boats *PT 602–PT 606*, *PT 608–PT 612*, launched in 1945. Four were transferred under MDAP on 16 March 1951 and the remaining six later in the same year. In 1955 *Hauk*, *Hvass*, *Snar* and *Snögg* were renamed becoming *Laks*, *Delfin*, *Lyr* and *Knurr* respectively. *Skrei* was BU in 1960, *Delfin*, *Knurr* and *Sel* were stricken 1 December 1961, *Lyr* on 1 March 1962, *Sild* in December 1962, and the remaining four (whose names had been transferred to the *Tjeld* class) in 1966.

RAPP class *fast attack craft*

Displacement: 72t standard
Dimensions: 87ft × 23ft × 5ft
26.5m × 7.0m × 1.5m
Machinery: 2 shafts, 4 Packard petrol engines, 4800bhp = 41kts (32kts sea speed)
Armament: 1–40mm, 1–20mm, 4–533mm TT
Complement: 18

Class:
Rapp (P 351), *Rask* (P 352), *Kvikk* (P 353), *Kjapp* (P 354), *Snar* (P 355), *Snögg* (P 356)

Norwegian construction, built at Westermoens Baatbyggeri, Mandal. *Rapp* launched 7 May 1952, completed 18 November 1952; remaining five launched 1953–55 and completed 1953–56. Wooden hull construction. All stricken 1970.

NORWAY

NASTY *fast attack craft*

Displacement:	69t standard; 76t full load
Dimensions:	75ft pp, 80ft 4in oa × 24ft 6in × 6ft 9in *22.9m, 24.5m × 7.5m × 2.1m*
Machinery:	2 shafts, 2 Napier Deltic diesels, 5000bhp = 43kts
Armament:	2–40mm, 4–533mm (21in) TT
Complement:	22

Experimental boat, built at Westermoens Baatbyggeri, Mandal, as private venture, launched and completed 1958, used as prototype boat for *Tjeld* class. Wooden hull construction. Stricken 1967.

TJELD class *fast attack craft*

Displacement:	70t standard; 76t full load
Dimensions:	75ft 6in pp, 80ft 4in oa × 24ft 6in × 6ft 9in *23.0m × 24.5m × 7.5m × 2.1m*
Machinery:	2 shafts, 2 Napier Deltic 18cyl turbo-charged diesels, 6200bhp = 45kts. Range 450nm/1600nm at 40kts/25kts
Armament:	1–40mm, 1–20mm, 4–533mm TT (or as gunboat 2–40mm, 2–533mm TT)
Complement:	22

Class:
First group – *Tjeld* (P 343), *Skarv* (P 344), *Teist* (P 345), *Jo* (P 346), *Lom* (P 347), *Stegg* (P 348), *Hauk* (P 349), *Falk* (P 350), *Ravn* (P 357), *Gribb* (P 388), *Geir* (P 389), *Erle* (P 390)
Second group – *Skrei* (P 380), *Hai* (P 381), *Sel* (P 382), *Hval* (P 383), *Laks* (P 384), *Knurr* (P 385), *Delfin* (P 386), *Lyr* (P 387)

All were built by Westermoens Baatbyggeri, Mandal. The first series was ordered in 1957, launched 1959–60, and commissioned during 1960–62. The second series was ordered under a new five-year programme in 1962, launched 1962–63 and commissioned 1963–66. They were developments of the *Nasty* design intended for series production. Two similar vessels went to West Germany (and eventually Turkey), two to the USA and six to Greece.

Skarv was stricken in 1978 and *Teist*, *Lom*, *Jo*, *Hauk*, *Falk* and *Sel* in 1979. The remainder were laid up in reserve, and by 1982 only eight survived; all stricken by 1995.

STORM class *fast attack craft*

Displacement:	100t standard; 125t full load
Dimensions:	120ft oa × 20ft 6in × 5ft *36.6m × 6.2m × 1.5m*
Machinery:	2 shafts, 2 Maybach diesels, 7200bhp = 32kts. Range 550nm at 32kts
Armament:	6 Penguin SSM (6×1), 1–76mm, 1–40mm, DCT
Sensors:	Radar TM-909, WM-26
Complement:	19

Class:
Storm (P 960), *Blink* (P 961), *Glimt* (P 962), *Skjold** (P 963), *Trygg* (P 964), *Kjekk* (P 965), *Djerv** (P 966), *Skudd* (P 967), *Arg* (P 968), *Steil** (P 969), *Brann* (P 970), *Tross* (P 971), *Hvass** (P 972), *Traust* (P 973), *Brott* (P 974), *Odd** (P 975), *Pil* (P 976), *Brask* (P 977), *Rokk** (P 978), *Gnist* (P 979)

In 1959 twenty-three MGBs were voted under the next five-year programme, but the number was reduced to twenty. Built at Bergens Mekaniske Verksteder (except those marked * by Westermoens, Mandal). A first *Storm* was built as a prototype, launched 8 February 1963 and completed 31 May 1963. This boat underwent many modifications during 1963–65 and was eventually scrapped. First series boat *Blink* launched 28 June 1965 and completed 18 December 1965. Last boat commissioned 1968. Later fitted with TVT-300 optronic tracker and laser rangefinder abaft the mast. *Pil* was stricken in 1982 and used for fire-fighting training. *Glimt*, *Arg*, *Brann*, *Tross*, *Traust*, *Brott*, *Odd* and *Rokk* were deleted 1991–92, three being sold disarmed to Latvia, Lithuania and Estonia respectively in December 1994. Remainder extant 1995.

SNOGG class *fast attack craft*

Displacement:	125t standard; 138t full load
Dimensions:	120ft × 20ft 6in × 5ft *36.5m × 6.2m × 1.3m*
Machinery:	2 shafts, 2 Maybach diesels, 7200bhp = 32kts. Range 550nm at 32kts
Armament:	4 Penguin SSM (4×1), 1–40mm, 4–533mm (4×1) TT, 2 DC racks
Sensors:	Radar TM-626
Complement:	19

Class:
Snögg (P 380, ex-*Lyr*), *Rapp* (P 981), *Snar* (P 982), *Rask* (P 983), *Kvikk* (P 984), *Hjapp* (P 985)

All built at Westermoens Baatbyggeri, Mandal, launched 1969–70, commissioned 1970–71. Generally similar to *Storm* class. Have PEAB TORI fire control system; TT fire T–61 wire-guided torpedoes. It is planned for this class to receive new electronics and a Mistral SAM launcher. All six extant 1995.

Gribb Jul 1970 L & L van Ginderen

Hvas Jun 1985 L & L van Ginderen

Skarv Jul 1994 L & L van Ginderen

HAUK class *fast attack craft*

Displacement:	120t standard; 155t full load
Dimensions:	120ft × 20ft × 5ft *36.5m × 6.2m × 1.5m*
Machinery:	2 shafts, 2 Maybach diesels, 7300bhp = 35kts. Range 440nm at 34kts
Armament:	6 Penguin 2 SSm (6×1), 1–40mm, 1–20mm, 2–533mm TT
Sensors:	Radar TM-1226; sonar SQ3D/SF
Complement:	22

Class:
Hauk (P 986), *Orn* (P 987), *Terne* (P 988), *Tjeld* (P989), *Skarv* (P 990), *Teist* (P 991), *Jo* (P 992), *Lom* (P 993), *Stegg* (P 994), *Falk* (P 995), *Ravn* (P 996), *Gribb* (P 997), *Geir* (P 998), *Erle* (P 999)

Ten boats ordered 12 June 1975 from Bergens Mekaniske Verksteder, four (P 996–P 999) from Westamarine, Mandal (the new name for Westermoens Baatbyggeri). *Hauk* laid down May 1976, launched 21 February 1977 and commissioned August 1977. *Erle*, the last boat, commissioned on 10 December 1980. The 40mm is a Bofors, the 20mm is by Rheinmetall, and they also carry 2–50mm rocket flare launchers. Fitted with MSI–80S fire control system, using the two Decca TM-1226 radars, a TVT–300 optronic tracker and an Ericsen laser rangefinder. All extant 1995.

OTHER SMALL CRAFT

There are also the patrol and inspection boats *Tarva* (TSD 1) and *Welding* (OSD 1), launched 1974 at Fjellstrand Aluminium Yachts, Omastrand; commissioned 1 December 1974 and 1 November 1974 respectively. Principal particulars were: 27.5t; 53ft 6in × 17ft 4in × 3ft 10in, (*16.3m × 5.3m × 1.2m*); 1–12.7mm MG; 2 shafts, 2 General Motors diesels = 15kts; complement 4. Both now stricken.

MINE WARFARE VESSELS

Ex-US LSM type *minelayers*

Particulars: See Danish *Beskytteren* class

No	Name	Builder	Acquired	Fate
N 45	VALE (ex-*LSM 492*)	Brown SB, Houston	Oct 1952	To Turkey 1.10.60
N 46	VIDAR (ex-*LSM 493*)	Brown SB, Houston	Oct 1952	To Turkey 1.10.60

Ex-US *LSM 1* class landing ships rebuilt as coastal minelayers in the USA during 1952, and transferred to Norway. On 1 October 1960 they were returned to the US Navy and later turned over to Turkey.

VARGSUND class *minelayers*

Displacement: 600t standard
Dimensions: 185ft 6in oa × 38ft 9in × 4ft 3in
56.5m × 11.8m × 1.3m
Machinery: 2 shafts, 2 Paxman Ricardo diesels, 500bhp = 8kts. Oil 43t
Armament: 2–20mm AA
Complement: 20

Class:
Vargsund (N 35), *Reinöysünd* (N 36)
Ex-British LCTs, built *c*1945, converted to minelayers 1952. Stricken 1960.

Ex-US AUK class *coastal minelayers*

Displacement: 180t standard; 1250t full load
Dimensions: 215ft wl, 221ft 2in oa × 32ft 2in × 11ft
65.5m, 67.4m × 9.8m × 3.4m
Machinery: 2 shafts, 2 GM diesels, 2070bhp = 16kts
Armament: 1–3in/50, 4–20mm, 2 Hedgehog ASW mortars, 3 DCT
Complement: 83

No	Name	Builder	Acquired	Fate
N 47	TYR (ex-*Sustain*, MMC 2)	American SB Cleveland	1959	Stricken 1976
N 48	GOR (ex-*Strive*, MMC 1)	American SB, Cleveland	1959	Stricken 1976
N 49	BRAGE (ex-*Triumph*, MMC 3)	Associated SB, Seattle	1960	Stricken 1978
N 50	ULLER (ex-*Seer*, MMC 4)	American SB, Cleveland	1959	Stricken 1978

Originally US *Auk* class steel-hulled ocean minesweepers launched in 1941–43. *Gor* and *Tyr* were converted to coastal minelayers at Charleston N Yd in 1959 and *Brage* in 1960; *Uller* was similarly converted at Horten N Yd in Norway. All were transferred under MDAP, and were numbered AM (later MSF) 119, 117, 323 and 112 in the order of the above table. *Uller*'s armament was different: 1–3in/50, 1–40mm, 1 Terne ASW system, 1 DCT.

BORGEN *controlled minelayer*

Displacement: 282t standard
Dimensions: 102ft 6in oa × 26ft 3in × 11ft
31.2m × 8.0m × 3.4m
Machinery: 2 shafts, 2 GM diesels, 660bhp = 9kts
Armament: 1–20mm AA, mines
Complement: 18

No	Name	Builder	Launched	Fate
N 51	BORGEN	Horten N Yd	29.4.60	Extant 1995

Coastal minelayer of Swedish 'Mul 12' type commissioned in 1961.

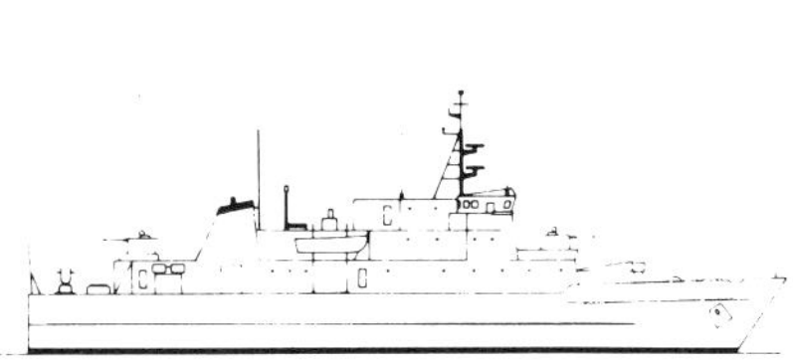

Vidar as completed

VIDAR class *coastal minelayers*

Displacement: 1150t standard; 1673t full load
Dimensions: 212ft 6in × 39ft 4in × 14ft 1in
64.8m × 12.0m × 4.3m
Machinery: 2 shafts, 2 Wichmann 7AX diesels, 4200bhp = 15kts. Oil 247t
Armament: 2–40mm (2×1), 320–400 mines
Sensors: Radar TM-1226; sonars SQ3D
Complement: 50

No	Name	Builder	Launched	Fate
N 52	VIDAR	Skaaluren SB, Rosendal	18.3.77	Extant 1995
N 53	VALE	Mjellum & Karlsen, Bergen	5.8.77	Extant 1995

Laid down March 1976 and February 1976 respectively; commissioned 21 October 1977 and 10 February 1978 respectively. Both were completed at Bergen. They are multi-role ships, capable of serving as minelayers (for which they have three decks with automated hoists between), torpedo-recovery vessels, transports, fishery protection vessels, or even limited ASW escorts (for which they have been fitted with 6–324mm Mk 32 ASW TT and 2 DC racks).

Ex-German 1940 type *coastal minesweepers*

Particulars: As German 1940 type

Class:
M 261, M 272, M 302, M 306, M 321, M 322, M 323, M 326, M 361, M 362, M 364, M 365, M 436
Britain transferred thirteen ex-German minesweepers to Norway in 1948. *M 321* and *M 326* were BU in 1951 and the remaining eleven during the following years.

US ADJUTANT type *coastal minesweepers*

Armament: 2–20mm
Sensors: Radar RM-707; sonar UQS-1
Other particulars: As US *Adjutant* class

Class:
Sauda (M 311, ex-*AMS 102*), *Sira* (M 312, ex-*MSC 132*), *Tana* (M 313, ex-*Roesleare*, ex-*MSC 103*), *Alta* (M 314, ex-*Arlon*, ex-*MSC 104*), *Ogna* (M 315), *Vosso* (M 316), *Glomma* (M 317, ex-*Bastogne*, ex-*MSC 151*), *Tista* (M 331), *Kvina* (M 332), *Utla* (M 334)
Sauda and *Sira* launched 1953 at East Boothbay, Maine, USA. *Alta*, *Glomma* and *Tana* were transferred in March 1966 from Belgian Navy in exchange for *Lagen* and *Namsen*. Five built in Norway; at Mandal, Risör (*Kvina* and *Utla*) and Rodendal (*Tista*), launched: *Ogna* 18 June 1954, *Vosso* 16 May 1954, *Tista* 1 June 1954, *Kvina* 21 July 1954 and *Utla* 2 March 1955. In 1977 *Tana* was converted to a minehunter, with British Type 193M sonar, TM-1226 radar, and PAP 104 remote-controlled minehunters. All have Rheinmetall 20mm guns. *Sauda* and *Ogna* stricken 1986, *Sira*, *Vosso*, and *Glomma* in 1992. Remainder used as trials vessels for equipment for new *Oskoy*/*Alta* class (see below), and are due to be disposed of as the new class comes into service. Only *Alta* and *Kvina* extant 1995.

Sira c 1979 Royal Norwegian Navy

Ex-US AGGRESSIVE class *ocean minesweepers*

Particulars: As US *Aggressive* class

Class:
Lagen (M 950, ex-*MSO 498*), *Namsen* (M 951, ex-*MSO 498*)
Launched 13 August 1954 and 15 October 1954 at Bellington, USA, taken over 27 September 1955 and 1 November 1955 respectively. Transferred to Belgian Navy in 1966 in exchange for *Alta*, *Glomma* and *Tana* of *Sauda* (US *Adjutant*) class.

OKSOY/ALTA class *minehunters/minesweepers*

Displacement: 335t standard; 367t full load
Dimensions: 170ft 7in pp, 177ft 10in oa × 44ft 5in × 7ft 1in (2ft 10in on cushion)
52.0m, 54.2m × 13.55m × 2.15m (0.87m)
Machinery: 2 waterjets; 2 MTU 12V396 TE94 propulsion diesels, 3840bhp, 2 MTU 8V396 TE 54 lift-fan diesels, 1880bhp = 30kts. Range 1200nm at 20kts
Armament: 1 Sadral SAM (1×6, Mistral missiles), 2–20mm (2×1),2–12.7mm MG
Sensors: Radar RN 80 navigational; sonar (minesweepers) SA-950 mine avoidance, (minehunters) TSM-2023N variable-depth hf.
Complement: 37 (14 officers)

Class (no, in service):
Minehunters: *Oksoy* (M 340, 15.8.94), *Karmoy* (M 341, 12.12.94), *Maaloy* (M 342, 24.3.95), *Hinnoy* (M 343, 30.6.95)
Minesweepers: *Alta* (M 350, Dec 1995), *Otra* (M 351, Apr 1996), *Rauma* (M 352, Aug 1996), *Orkla* (M 353, Jan 1997), *Glomma* (M 354, Jul 1997)

Surface-effect minehunters/minesweepers ordered from Kvaerner Mandal in November 1989. Rigid-sidewall air-cushion design. First four built as minehunters and carry two remotely-operated Pluto Plus submersibles, whilst the next five are to be minesweepers. All machinery mounted at main-deck level to reduce acoustic and magnetic signatures. Four minehunters extant 1995, with minesweepers to enter service in 1996–97.

Oskoy 1.4.95 L & L van Ginderen

MISCELLANEOUS

HAAKON VII *training vessel*

Ex-US seaplane tender *Gardiner Bay*, launched 2 December 1944. Taken over 17 May 1958 and converted to training vessel for midshipmen (A 537). Displacement 1766t standard, 2800t full load. Rearmed with 1–5in (127mm), 8–40mm and 4–20mm. 2 diesels, 6080bhp = 18kts. Complement 215 plus 86 cadets and petty officer apprentices. Stricken 1974.

Horten as completed

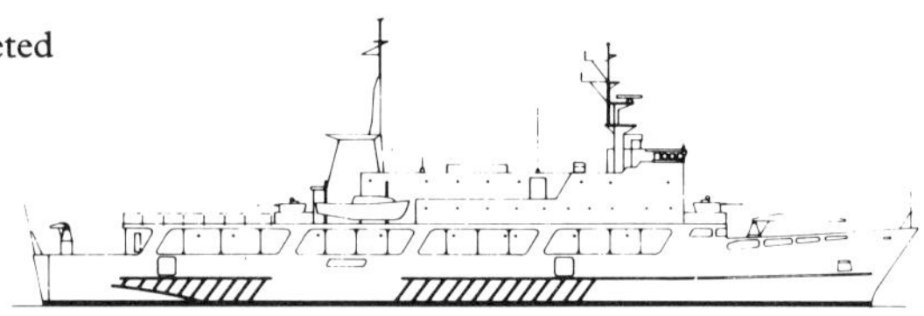

HORTON *support ship*

Displacement: 2500t full load
Dimensions: 269ft pp, 285ft 5in oa × 42ft 7in × 23ft
82.0m, 87.0m × 13.0m × 7.0m
Machinery: 2 shafts, 2 Wichmann diesels, 4200bhp = 16.5kts
Armament: 2–40mm (2×1), mines
Complement: 85

Horten (A 530) was built at Horten Werft, launched 12 August 1977, and commissioned in April 1978. A general purpose logistical support and depot ship for submarines and fast attack craft, but has some limited offensive capability. Served as Royal Yacht 1985–86 whilst *Norge* undergoing repairs (see below). Extant 1995.

Nordkapp as completed

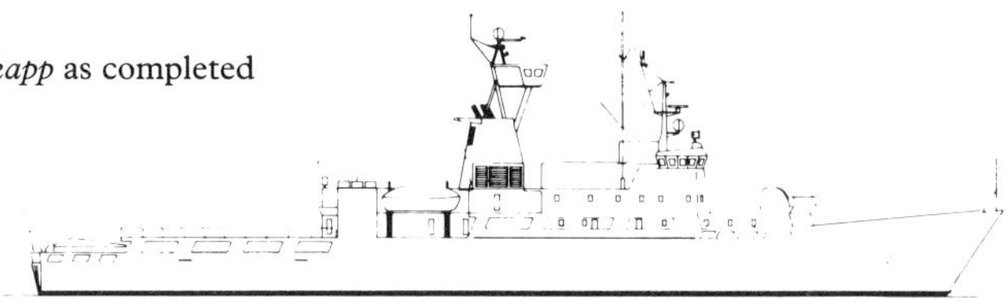

NORDKAPP class *patrol vessels*

Displacement: 2700t standard; 3200t full load
Dimensions: 319ft 10in wl, 344ft 6in oa × 47ft 11in × 16ft 1in max
97.5m, 105.0m × 14.6m × 4.9m
Machinery: 2 shafts, 4 Wichmann diesels, 14,400bhp = 23kts. Range 7500nm at 15kts (on 2 diesels)
Armament: 1–57mm, 4–20mm (4×1), 6–324mm Mk 32 TT (2×3), 1 DC rack (6 DCs), 1 helicopter
Sensors: Radar TM-1226, RM-914, ASW-4, GLF-218; sonar SS-105
Complement: 55 peace; 76 war

No	Name	Builder	Launched	Fate
W 320	NORDKAPP	Bergens Mek Verk	14.5.80	Extant 1995
W 321	SENJA	Horten Verft	16.3.81	Extant 1995
W 322	ANDENES	Haugesund Mek Verk	21.3.81	Extant 1995

In 1977 the Naval Material Command requested two new types of vessel to patrol the EEZ, three of an ice-strengthened type and four of normal construction. However, the Coast Guard budget was cut and only the above three Type 320 ships were built. Only *Nordkapp* is strengthened for ice navigation. The first two were ordered in July 1978 and the third in November 1979. They are versatile ships capable of patrol, fire-fighting, anti-pollution work, meteorological reporting, and escort duties in wartime. For this last function they can be fitted with six Penguin 2 SSMs (6×1) and chaff launchers. The guns are Bofors (57mm) and Rheinmetall: they have a NAVKIS AIO system, and carry a Sea Lynx helicopter. *Andenes* served in the Gulf conflict 1991 as support ship for the Danish corvette *Nils Juel*.

NORGE *royal yacht*

Ex-*Philante*, launched at Camper & Nicholson, Gosport, England, 17 February 1937. A gift from the Norwegian people to the late King Haakon VII after the Second World War. Naval pennant number A 533. Severely damaged in a fire on 8 March 1983 and did not return to service until mid-1986. Displacement 686t standard. Dimensions 250ft wl, 263ft oa × 28ft × 15ft 3in (*76.3m, 80.2m × 8.5m × 4.6m*). 8cyl diesels, 3000bhp = 17kts.

COAST GUARD

The Norwegian Coast Guard (Kystvakt) was set up in 1976 to protect Norway's burgeoning offshore interests, and to patrol the new 200nm Exclusive Economic Zone (EEZ). At first the service used ex-naval fishery protection vessels and chartered trawler-type merchant ships, but the *Nordkapp* class patrol vessels would make useful warships, and therefore full details are included. The following Coast Guard vessels might also be considered to have some naval value:

Andenes (W 303), *Senja* (W 304), *Nordkapp* (W 305): 400t standard; 183ft 9in × 31ft 3in × 16ft (*56.0m × 9.5m × 4.9m*), 16kts, 1–76mm, launched 1957 in the Netherlands, purchased 1965 and converted to patrol vessels. *Nordkapp* later A 531, *Senja* A 536. Stricken 1981.

Farm (W 301), *Heimdal* (W 302), 600t gross, 177ft 3in × 27ft × 16ft (*54.0m × 8.2m × 4.9m*), launched 1962, two diesels, 2700bhp = 16kts, 1–40mm, complement 29. Modernised in 1979–80. Ex-navy. Both extant 1995.

Nornen (W 300), 1060t standard, 203ft 5in × 32ft × 15ft 9in (*62.0m × 10.0m × 4.8m*), 17kts, 1–76mm, launched and completed 1963 at Mjellum & Karlsen, Bergen. Ex-navy. Modernised 1977. Extant 1995.

Oman

The Sultanate of Oman, a little smaller than Britain in area and with a population (1991 estimate) of 1,559,000, is situated at the southern end of the Arabian Peninsula and holds a strategically very important position, controlling the approaches to the Persian Gulf through the Gulf of Oman. With the additional small area under her control on the Mussandam Peninsula she also virtually controls the narrow Strait of Hormuz which is the only means of entry to the Gulf. It was possible to create a modern navy in the 1970s only with the help of Great Britain and the navy employs a number of expatriates mainly from the UK. The commander of the navy and a few senior officers are on secondment from the Royal Navy. More Omanis are gradually being recruited into the navy and all officers in the Patrol Boat Squadron apart from the commanding officers are Omanis. In 1982 the first Omani officer to take command of a patrol boat took up his post, but the emerging navy is in grave need of trained officers and men.

Most Omani ships are modern and potent but there is an evident lack of support vessels and auxiliaries. A British firm, Brooke Marine Ltd, has been a main supplier of ships as has Vosper Thornycroft to a lesser extent.

The Omani Navy has proved its worth in its involvement in the control of shipping in the Strait of Hormuz Shipping Separation Zones which pass through Omani waters. Requiring two patrol vessels on duty supported by a supply vessel, this is a relatively heavy commitment for such a small force and has been carried out with success since November 1979 and reflects well on the navy's ability. Patrol boats regularly fired at land targets in an effort to disrupt camel trains during the Dhofar War of 1965–75, although the success of this method was never confirmed. The navy's role in defeating the communist led insurgency in the sultanate's southern province was mainly an unglamorous but essential one of logistic support along the barren coastline.

Currently the personnel strength of the Omani Navy is about 3500 officers and men (including civilian employees). In 1987 it was 2400 officers and men. Naval service is voluntary. The navy received some valuable ships in the course of the 1980s, such as four large missile boats of the *Dhofar* class, the landing ship *Nasr Al Bahr* and some smaller ships and craft. In March 1992 an order was placed for two large fast corvettes of the *Muheet* class, which when delivered will be the largest fighting ships in the Omani Navy apart from the landing ship *Nasr Al Bahr.* Plans to acquire a US *Knox* class frigate were abandoned. In 1992 naval aviation included four aircraft of Short SMS–3M Seavan type; two of Dornier Do 228–100 type and two Pilatus Porter. The main naval bases are at Muscat (main base dockyard and slipway), Wudam, Mina Raysut, Jazirat Ghanam and Musandam.

Quahir al Amwaj class design

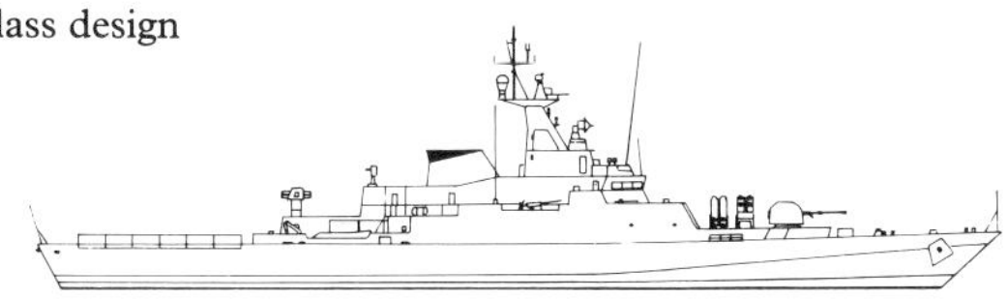

QUAHIR AL AMWAJ class *fast corvettes*

Displacement: 1420t full load
Dimensions: 249ft 4in wl, 274ft 7in oa × 37ft 9in × 11ft 6in
76.0m, 83.7m × 11.5m × 4.7m
Machinery: 2 shafts (CP propellers) CODAD, 4 Crossley-SEMT-Pielstick 16PA6V 280STC diesels, 28,160bhp = 25kts. Range 4000nm at 10kts
Armament: 8–MM40 Exocet SSM (2×4), 8–Crotale NG SAM (1×8), 1–76mm/62 OTO Melara, 2–20mm Oerlikon, BMARC GAM-B01, 1 medium helicopter
Sensors: Radar Signaal MW 08, Kelvin Hughes 1007; sonar see notes; fire control radar Signaal STING, 1 Signaal optronic director, 2 optical directors; command and control Signaal SEWACO-FD with Thomson Sintra TACTICOS, Link Y and SATCOM; ESM/ESM Thomson CSF DR 3000, 2 Barricade chaff launchers
Complement: 80

No	Name	Laid down	Launched	Fate
Q 31	QUAHIR AL AMWAJ	21.5.93	21.9.94	Completing
Q 32	AL MUA'ZZAR	Apr 1994	–	Building

A new class of two fast corvettes ordered in March 1992 from Vosper Thornycroft. First delivery expected in February 1996. Based on 'Vigilance' class design. Class is also known as the *Muheet* class and Project Type 83. Equipped with a landing pad for medium helicopter but have no hangar. These ships can also be fitted with Thomson Sintra hull sonar and BAe Sema ATAS which add 8t to displacement but will not affect longitudinal stability and helicopter operations. Capable of cruising at 20kts on one shaft powered by two engines. These powerful combatants will greatly enhance the capabilities of the Oman Navy. The real worth of these high-technology acquisitions will depend in the short term on the maintenance of close links with the West and in the long term on the training of Omani personnel.

AL SAID *armed yacht (escort vessel)*

Displacement: 785t standard; 930t full load
Dimensions: 188ft 4in pp, 203ft 5in oa × 35ft 1in × 9ft 10in
57.4m, 62.0m × 10.7m × 3.0m
Machinery: 2 shafts, 2 Paxman Ventura 12-YJCM diesels, 2470bhp = 17kts
Armament: 1–40mm
Sensors: Radar Decca TM-626
Complement: 32 + 7 staff (plus 32 troops)

Al Said was built by Brooke Marine Ltd, ordered in 1969, launched 7 April 1970 and commissioned as the navy's first ship in 1971. Designed as a yacht for the Sultan of Muscat and Oman, she was converted to a dual purpose ship with a gun on her forecastle and serves as flagship of the sultanate navy. A Fairey Marine Spear patrol boat is carried and she is also fitted with a helicopter deck. Could be used as an escort vessel. Converted to training ship 1983–84 and renamed *Al Mabrukah* (A1). Two 20mm guns and Decca 1226 radar added. Extant 1995.

Al Munassir as completed

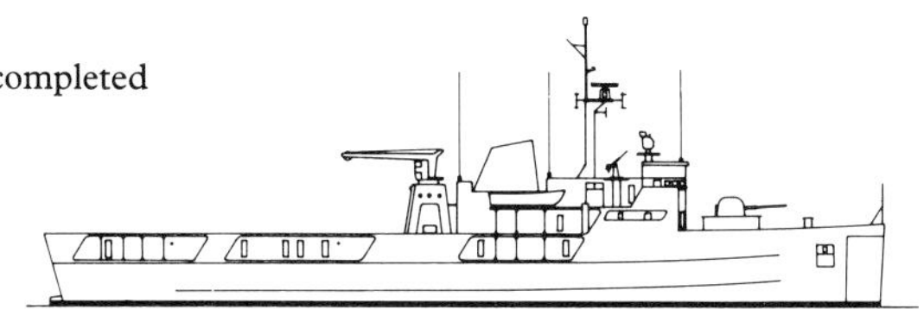

AL MUNASSIR *logistic support ship*

Displacement: 200t full load
Dimensions: 266ft 6in pp, 276ft oa × 49ft × 7ft 4in
81.25m, 84.0m × 15.0m × 2.23m
Machinery: 2 shafts (CP propellers), 2 Mirrlees-Blackstone ESL8MGR diesels, 2400bhp = 12kts. Range 2000nm at 12kts
Armament: 1–76mm/62 OTO Melara, 2–20mm (2×1)
Sensors: Radar Decca TM-1229
Complement: 45 (plus 188 troops)

This unusual ship was built by Brooke Marine Ltd, Lowestoft. Laid down on 4 July 1977, launched on 25 July 1975, completed on 3 April 1979. Equipped with bow doors and ramp for beaching. Bluff-bowed hull shape and slow speed of only 12kts. Can carry 550t of cargo or eight heavy tanks. Carries a travelling crane of 16-ton capacity which spans helicopter deck; one Westland Sea King or smaller helicopter can be carried. Fitted with Sperry Sea Archer optical fire control. The ship can also be employed as a command vessel. She carries hull number *L 1.* Now in reserve.

Nasr al Bahr 1984

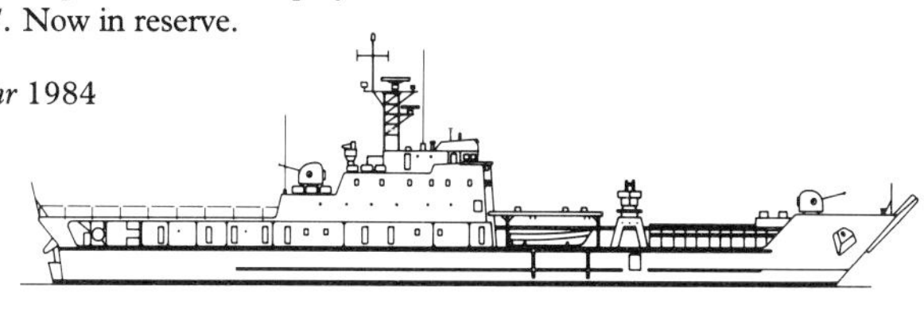

'KALAAT' type *landing ship*

Displacement: 2500t full load
Dimensions: 262ft 6in pp, 305ft 1in oa × 50ft 10in × 7ft 6in
80.0m, 93.0m × 15.5m × 2.3m
Machinery: 2 shafts, 2 CP propellers, 2 Paxman Valenta 18RP200CM diesels, 7800bhp = 16kts. Range 4000nm at 13kts
Armament: 4–40mm/70 Breda (2×2), 2–20mm (2×1)
Sensors: Radar Decca 1226, Decca 1290, PEAB 9LV200 fire control system and CSEE Lynx electro-optical gunsight
Complement: 13 officers and 68 men (plus 240 troops)

The logistic landing ship *Nasr Al Bahr* of the Algerian 'Kalaat' type ordered on 18 March 1982, launched on 16 May 1984 and completed on 13 February 1985 by Brooke Marine Lowestoft. Vehicle deck 246ft × 24ft (75m × 7.4m), bow ramp (18m) long, beaching cargo load 450t or seven main battle tanks. Equipped with two 'Sea Truck' LCVP craft. At the stern a flight deck for one Sea King helicopter is arranged but no hangar. There is a 16t travelling cargo crane on the deck forward of the superstructure. Maximum cargo load is 650t. Two thousand rounds of ammunition for 40mm guns are provided. The ship carries hull number L 2. Extant 1995.

Nasr al Bahr 15.4.85 L & L van Ginderen

OMAN

SABA AL BAHR class *utility landing craft*

Displacement: 230t full load
Dimensions: 98ft 5in oa, 84ft pp × 26ft 3in × 4ft
30.0m, 25.6m × 8.0m × 1.2m
Machinery: 2 shafts, 2 Caterpillar 3408TA diesels, 1840bhp=8kts. Range 1800nm at 8kts
Armament: –
Complement: 11

Class (no, laid down, launched, completed)
Saba Al Bahr (C 8, launched 30.6.81), *Al Doghas* (C 9, 8.9.82, 12.11.82, 10.1.83) *Temsah* (C 10, 8.9.82, 15.12.82. 12.2.83).

Built by Vosper (Singapore) Pte Ltd. First ordered in April 1981 and the other two on 8 May 1982. Can carry 100t vehicles or stores. C 9 and C 10 are longer than C 8 with length increased to 108ft 3in oa (33.0m). Extant 1995.

The following twin diesel LCUs were extant in 1995 except *Al Doghas* (stricken 1981), and *Al Dhaibah* (C 6) (stricken 1982).

Al Sansoor (C 4), (*Al Doghas* C 5, ex-*Kinzeer al Bahr*), 'Loadmaster' class launched by Cheverton, Cowes in 1974, in service January 1975; 60t (130t full load), 6kts, 60ft × 20ft × 3ft 7in, (*18.0m × 6.0m × 1.2m*), crew 6.

Sulhafa al Bahr (C 3), 'Loadmaster' class launched in 1975 by Cheverton, Cowes; 30dwt, 45t, 8.5kts, 45ft × 15ft × 3ft (*13.7m × 4.6m × 1.7m*).

Al Dhaibah (C 6), launched in 1975 by Impala Marine, Twickenham; 75dwt, 122grt, 9kts, 75ft × 23ft × 5ft (*22.9m × 7.0m × 1.5m*), crew 8.

Kasab Kumzar, I, II, III, built in 1977 by Impala Marine, Twickenham; 122grt, 9kts, 84ft × 25ft × ?) (*25.6m × 7.6m × ?*)

Al Neemran (C 7), I, launched in 1979–81 by Lewis Offshore, Stornoway; 85t, 8kts (*25.5m × 7.4m × 1.8m*), crew 8.

Zara 16, *Zara 20*, *Zara 22* built in 1981, 11t standard, 23t full load, 20kts (*18m × 3.8m × 0.5m*), 2–12.7mm MG (2x1), crew 4. *Zara 22* is 52ft 6in oa (16m).

SMALL SURFACE COMBATANTS

Dhofar as completed
NB 1/750 scale

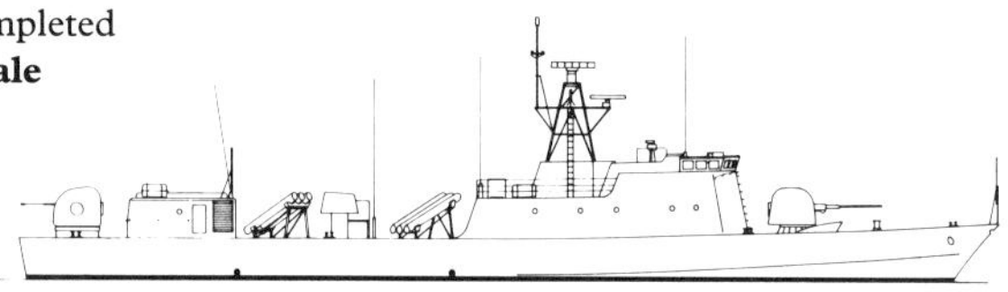

DHOFAR class *fast attack craft (missile)*

Displacement: 311t light; 394t full load
Dimensions: 170ft 7in pp, 186ft oa × 26ft 10in × 7ft 11in (6ft 11in hull)
52.0m, 56.7m × 8.2m × 2.4m (2.1m)
Machinery: 4 shafts, 4 Paxman Valenta 18RP200 diesels, 18,200bhp = 40kts. Range 2000nm/15kts
Armament: 6 (8 in B 11, 12, 14) MM40 Exocet SSM (2×3), 1–76mm/62 OTO Melara, 2–40mm/70 Breda (1×2), 2–12.7mm (2×1)
Sensors: Radar in *Dhofar* (Plessey AWS-4, Decca 1226), other ships Decca TM 1226C
Complement: 45 (5 officers) plus 14 cadets

Class (no, laid down, launched, in service):
Dhofar (B 10, 30.9.80, 14.10.81, 7.8.82), *Al Sharqiyah* (B 11, Oct 1981, 2.12.82, 5.12.83), *Al Bat'nah* (B 12, 9.12.81, Nov 1982, 18.1.84), *Mussandam* (B 14, 8.10.87, 19.3.88, 31.3.89)

Also referred to as the 'Province' class. Designed by Vosper Thornycroft (UK) Ltd as a modified version of the Egyptian 'Ramadan' class FAC. B 10 ordered in 1980, B 11 and B 12 in 1981 and B 14 in 1986. B 10 is fitted with Sperry Archer Mk 2 fire control system while others have PEAB 9LV300 system. These ships are fitted with two 100hp electric motors for manoeuvring. Extant 1995.

Al Bat'nah 12.9.83 L & L van Ginderen

Project MAWJ *fast attack craft (gun)*

Displacement: 475t full load
Dimensions: 178ft 10in × 26ft 3in × 8ft 10in
54.5m × 8.0m × 2.7m
Machinery: 2 shafts (CP propellers), 2 SEMT-Pielstick 16 PA4 V200 diesels, 8000bhp = 24.5kts. Range 4200nm at 15kts
Armament: 1–76mm/62 OTO Melara, 2–30mm (2×1)
Sensors: Radar surface search; fire control optronic director
Complement: 26 (3 officers) plus 20 spare berths

In September 1993 three craft, with the option for five more, were ordered from the CMN shipyard, Cherbourg. Design is a modified French 'P 400' type large patrol craft with twin funnels instead of the underwater exhaust for the main engines. First unit is due to be delivered in late 1995 without armament, which will be fitted at the Muscat dockyard. The 76mm guns will probably come from the second-group *Al Bushra* class large patrol craft (see below), which these class will replace in service.

Al Mansur 1978
NB 1/750 scale

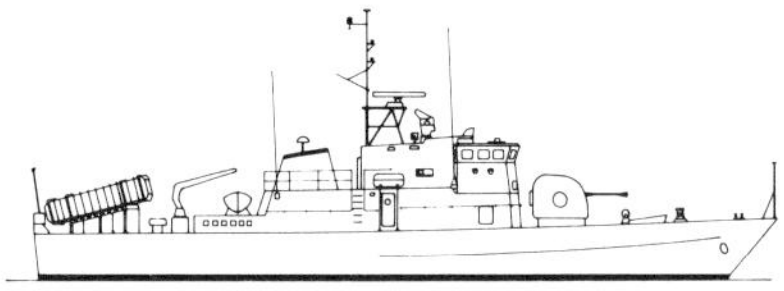

AL BUSHRA class *large patrol craft (missile and gun)*

Displacement: 140t standard; 185t full load (B 1–3), 125t standard; 160t full load (B 4–7)
Dimensions: 123ft oa × 22ft 8in × 5ft 7in
37.5m × 6.9m × 1.7m
Machinery: 2 shafts, 2 Paxman Ventura 16 RP-200 diesels, 4800bhp = 28kts. Range 3230nm at 13kts (B 1–3), 3300nm at 12kts (B 4–7)
Armament: 2 MM38 Exocet SSM (2×1), 2–40mm/70 Breda-Bofors (1×2), 2–7.62mm MG (in B 1–3); 1–76mm/62 AA OTO Melara (130rds), 1–20mm, 2–7.62mm MG (in B 4–7)
Sensors: Radar Decca TM-916
Complement: 27

Class (no, in service):
Al Bushra (B 1), *Al Mansur* (B2, 26.3.73), *Al Nejah* (B 3, 13.5.73), *Al Wafi* (B 4, 24.3.77), *Al Fulk* (B 5, 24.3.77), *Al Mujahid* (B 6, 6.10.77), *Al Jabbar* (B 7, 6.10.77)

All 37.5m boats built by Brooke Marine Ltd. First three ordered on 5 January 1971 and commissioned 1973, B 4–7 ordered April 1974 and delivered 1977. B 1–B 3 rearmed with Exocet missiles and Sperry Sea Archer fire control systems by builders, November 1977–November 1978. *Al Bushra* lost on 28 November 1978 when she was washed off the delivery ship while in the Bay of Biscay. A replacement boat was ordered in 1980 and is a *Dhofar*. B 4–7 fitted with Lawrence-Scott optical director. B 1–3 originally armed with two single 40mm gun fore and aft. *Al Mansur* stricken 1986. Remainder extant 1995.

Ex-Netherlands DOKKUM class *coastal minesweepers*

Class:
Al Nasiri (ex-*Axel*), *Al Salihi* (ex-*Aalsmeer*)

Pennant nos 1 and 2. These ex-Netherlands minesweepers (launched 1955) were bought by Oman in 1974. Air-conditioned and fitted as patrol boats. Similar to British 'Ton' class. Their minesweeping gear was removed and the third 40mm gun placed on the main deck aft. Stricken in 1981.

HARAS 1–5 *patrol boats*

Displacement: 45t full load
Dimensions: 75ft × 20ft × 5ft 10in
22.9m × 6.1m × 1.8m
Machinery: 2 shafts, 2 Caterpillar D438 diesels, 1840bhp = 24.5kts. Range 600nm at 20kts, 1000nm at 11kts
Armament: 2–20mm
Sensors: Radar Decca 101
Complement: 11

These boats named *Haras 1–5* were built in 1975 by Vosper Thornycroft, Singapore. First four operational from 22 December 1975. GRP hulls. *Haras 5* delivered in November 1978. Extant 1995.

HARAS 7 *patrol boat*

Displacement: 80t full load
Dimensions: 94ft × 17ft × 3ft 3in
28.7m × 5.2, × 1.0m
Machinery: 2 shafts, 2 MTU 8V331 IC82 diesels, 1866bhp = 25kts
Armament: 2–20mm (2×1)
Sensors: Radar Decca 1226C
Complement: 13

Small patrol boat built in 1981 by Karlskronavarvet AB. Swedish type CG 29 (improved CG 27). GRP hull. Sister ship *Haras 9* and *10* also built by Karlskrona, Sweden. Extant 1995.

HARAS 8 *patrol boat*

Displacement: 30t standard; 33t full load
Dimensions: 63ft × 15ft 8in × 4ft
19.2m × 4.8m × 1.2m
Machinery: 2 MTU 8V331 diesels, 1200/1600bhp = 30kts. Range 1650nm/2300nm at 17kts/12kts
Armament: 2–12.7mm MG
Complement: 10

Built in 1981 by Le Comte, Netherlands. Type PT 1903 Mk III. In service August 1981. Extant 1995.

HARAS 9 *patrol boat*

Displacement: 53t
Dimensions: 78ft 9in × 18ft 1in × 6ft 2in
24.0m × 5.5m × 1.9m
Machinery: 2 shafts, 2 MTU 12V331 diesels, 2660bhp = 25kts
Armament: 1–20mm Oerlikon
Complement: 11

Patrol boat of CG27 type built in 1980 by Karlskronavarvet on GRP hull. Operated by the police.
Royal Oman Police also operates:
Dheeb Al Bahar I in service 1984, built by Watercraft Ltd, 80t full load, 38kts, 68ft 3in oa × 19ft × 4ft 11in, (*20.8m × 5.8m × 1.5m*), armament 1–20mm, 6–7.62mm (6 × 1).
Dheeb Al Bahar II. Dheeb Al Bahar III in service 1988, built at Yokohama, 65t full load, 36kts, 75ft 6in × 17ft 1in × 3ft 11in (*23.0m × 5.2m × 1.2m*), armament 1–7.62mm.

SEEB class *patrol craft*

Displacement: 75t full load
Dimensions: 82ft 10in oa, 75ft 6in pp × 19ft × 5ft 2in
25m, 23m × 5.8m × 1.6m
Machinery: 3 shafts, 2 MTU 12V331TC–92BW255 diesels, 3072bhp = 26kts. Range 200nm at 8kts. 1 Cummins 200bhp cruise diesel for 8 kts
Armament: 1–20mm
Sensors: Radar Decca
Complement: 13

Class (no:)
Al Seeb (B 20), *Al Shinas* (B 21), *Al Sadh* (B 22), *Al Khasab* (B 23)
Built by Vosper, Singapore; all commissioned August 1981. GRP hulls. Extant 1995.

Other patrol craft.
Five Watercraft and Emsworth type inshore patrol craft (16/18t, 45ft/52.5ft, 32kts) built 1981 and 1987: *Zahra 14, 15, 17, 18, 21.*

Pakistan

For the first twenty-four years of its existence the Pakistan Navy faced the impossible task of maintaining maritime links between the two halves of history's most geographically absurd state, round 3500 miles of hostile coastline. It inherited only two of the old RIN training establishments in August 1947. If that was not bad enough, the service came definitely last in the new Muslim state's defence priorities. In the air and on the ground Pakistan might hope to offset India's permanent numerical advantage; never at sea. The Royal Pakistan Navy's warships, like India's, were Royal Navy transfers until the early 1960s. A smaller cruiser acquired in 1956 was joined by a pair of 'Battle' class destroyers, more powerful and newer than India's units, but on the other hand Pakistan could not match her neighbour's seaborne air power. She did acquire the subcontinent's first submarine in 1964, an ex-US *Tench* class.

There was no maritime dimension to the 1965 Indo-Pakistan War though, according to the then Pakistan Air Force C-in-C, Indonesia offered to launch diversionary naval attacks. The service grew steadily in manpower from 7700 in 1963 to 9870 in 1970. But this build-up, including four submarines, could not match India's and in the hectic days of December 1971 the navy's strategic and numerical inferiority was glaringly revealed. In East Pakistan Rear-Admiral Shariff's four large patrol craft and seventeen MG-armed civilian boats were neutralised by Indian carrier and land-based aircraft on the first day. Thereafter the sailors became soldiers in a lost cause. Off Karachi on the same day India's Western Fleet sank the destroyer *Khaibar* and a minesweeper in the largest surface action since 1945 before shelling and rocketing the naval base. This treatment was repeated on 9 December with three foreign merchant ships being hit as well. Pakistan's only naval consolation was the submarine *Hangor*'s sinking of the Indian frigate *Khukri* in the Arabian Sea.

In defeat the navy turned to China for fast attack and patrol craft, receiving sixteen in 1972–73. By the end of 1980 the submarine force had been doubled and modernised US destroyers were being acquired. Volunteer manpower reached 22,000 in 1995 year with 5000 reserves. A naval air arm of three Breguet Atlantic patrol aircraft and ten helicopters (both armed with AM-39 Exocet) was set up in 1975.

MAJOR SURFACE SHIPS

Ex-British BELLONA class *cruiser*

No	Name	Builder	Acquired	Fate
C 84	BABUR (i) (ex-*Diadem*)	Hawthorn Leslie	29.2.56	BU 1985

Refitted at HM Dockyard, Portsmouth, with new bridge, radar and lattice masts. Renamed *Babur,* after the Mogul emperor on 5 July 1957. Adapted to TS during 1961 refit and reduced to harbour service until replaced in 1982 by the new *Babur,* and renamed *Jahanqir.* C prefix dropped from pennant number in 1963.

Babur 1957 L & L van Ginderen

Ex-British 'O' class *destroyers*

No	Name	Builder	Acquired	Fate
D 21	TARIQ (i) (ex-*Offa*)	Fairfield	30.11.49	BU 1959
260	TIPPU SULTAN (i) (ex-*Onslow*)	John Brown	30.9.49	Stricken 1980
261	TUGHRIL (i) (ex-*Onslaught*)	Fairfield	3.3.51	Stricken 1977

Refit and conversion of *Tippu Sultan* and *Tughril* to Type 16 frigates with US MDAP funds agreed in London on 29 April 1957. Both ships were converted at Liverpool but *Tariq* was handed back to RN in 1959. *Tughril* was disarmed in 1977 and used as midshipmen's naval academy training ship. *Tippu Sultan* was stripped 1980 with her hulk being used as a work platform.

Ex-British 'Ch' and 'Cr' class *destroyers*

No	Name	Builder	Acquired	Fate
D 129	TAIMUR (i) (ex-*Chivalrous*)	Denny	29.6.54	BU 1961
D 160	ALAMGIR (ex-*Creole*)	White	29.2.56	BU 1982
D 162	JAHANGIR (ex-*Crispin*, ex-*Craccher*)	White	29.2.56	BU 1982
D 164	SHAH JAHAN (i) (ex-*Charity*)	Thornycroft	16.12.58	BU 1982

Taimur was lent by RN after 22 October 1953 agreement, handed over at Liverpool, and returned to UK in 1960. *Alamgir* and *Jahangir* bought in 1956 and modernised with US MDAP funds by Thornycroft being handed over to the Pakistan Navy on 18 March and 20 June 1958 respectively at Southampton. *Shah Jahan* bought by US under MDAP and handed over at Cowes after refit by White.

Ex-British 'BATTLE' class *destroyers*

No	Name	Builder	Acquired	Fate
D 161	BADR (i) (ex-*Gabbard*)	Swan Hunter	29.2.56	BU 1989
D 163	KHAIBAR (i) (ex-*Cadiz*)	Fairfield	29.2.56	Sunk 5.12.71

Modernised and refitted with US MDAP funds, *Badr* at Yarrow being handed over on 24 January 1957 and sailing for Karachi from Portsmouth on 17 February 1957 *Khaibar* refitted by Gowan, Glasgow, and handed over 1 February 1957. Both named after early Islamic victories. *Khaibar* sunk off Karachi by Indian 'Osa' FAC 5 December 1971. *Badr* became HQ ship of the new Maritime Safety Agency in 1985.

Ex-US GEARING (FRAM I) class *destroyers*

No	Name	Builder	Acquired	Fate
D 160	ALAMGIR (ex-*Cone*)	Bethlehem, Staten I	1.10.82	Extant 1995
D 164	SHAH JAHAN (ii) (ex-*Harold J. Ellison*)	Bethlehem, Staten I	1.10.83	For disposal 1994
D 165	TARIQ (ii) (ex-*Wiltsie*)	Federal, Kearny	29.4.77	MSA 1990
D 166	TAIMUR (ii) (ex-*Epperson*)	Federal, Kearny	29.4.77	Extant 1995
D 167	TUGHRIL (ii) (ex-*Henderson*)	Todd-Pacific, Seattle	30.9.80	Extant 1995
D 168	TIPPU SULTAN (ii) (ex-*Damato*)	Bethlehem, Staten I	30.9.80	For disposal 1994

First pair refitted by Campbell Industries, San Diego for $7.43m in 1977–78. *Tariq* finished 2 June 1978, *Taimur* 16 February 1978. *Tariq* replaced *Badr* as flagship of Maritime Safety Agency 25 January 1990 and renamed *Nazim* (D 156).

Ex-British 'COUNTY' class *guided missile destroyer*

No	Name	Builder	Acquired	Fate
C 84	BABUR (ii) (ex-*London*)	Swan Hunter	Feb 1982	Extant 1995

Bought February 1982 and refitted, Seaslug SAM system removed from stern. Handed over 24 March 1982. Flight deck extended to permit Sea King Mk.45 to be operated. Mk 15 Phalanx CIWS added on former hangar, and three twin 37mm/45 gun mountings added.

Ex-British modified BITTERN class *frigates*

No	Name	Builder	Acquired	Fate
F 40	JHELUM (ex-*Narbada*)	Thornycroft	1948	BU 1959
F 32	SIND (ex-*Godavari*)	Thornycroft	1948	For disposal c1960

Transferred by India 1948. *Sind* refitted by original builders at Southampton 1949–50 touring Australasian ports 1951. *Jhelum* refitted 1950–51, retained her tripod mast until also replaced by lattice in Karachi 1955. Also fitted as flagship.

Ex-British 'RIVER' class *frigates*

No	Name	Builder	Acquired	Fate
F 265	ZULFIQUAR (ex-*Dhanush*, ex-*Deveron*)	Smiths Dock	1948	BU 1983
F 392	SHAMSHER (ex-*Nadder*)	Smiths Dock	1948	Stricken 1960

Transferred by India in 1948, *Shamsher* became a training ship. *Zulfiquar* became a survey ship in 1951 with her aft 4in removed to fit a charthouse. She surveyed for the new port of Chalna, East Pakistan. Renumbered 262 in 1963.

Ex-British 'LEANDER' class *frigates*

No	Name	Acquired	Fate
F 262	ZULFIQUAR (ex-*Apollo*)	14.10.88	Extant 1995
F 263	SHAMSHER (ex-*Diomede*)	15.7.88	Extant 1995

Two broad-beamed *Leanders* were bought from the Royal Navy in 1988. The only change made was to add an Alouette III helicopter.

Ex-US 'GARCIA' class *frigates*

No	Name	Acquired	Fate
F 264	SAIF (ex-*Garcia*)	31.1.89	Returned Jan 1994
F 265	ASLAT (ex-*O'Callahan*)	8.2.89	Returned 14.11.93
F 266	KHAIBAR (ii) (ex-*Brumby*)	31.3.89	Returned 14.11.93
F 267	SIQQAT (ex-*Koelsch*)	31.5.89	Returned 1993

Four ASW frigates were acquired from the US Navy and arrived at Karachi in July 1989. Mk 15 Phalanx CIWS to be fitted 1991–92, but towed array removed before *Saif* was handed over.

Ex-US BROOKE class *missile frigates*

No	Name	Acquired	Fate
F 161	BADR (ex-*Julius A Furer*)	31.1.89	Returned 11.12.93
F 162	HARBAH (ex-*Brooke*)	8.2.89	Returned 1993
F 163	TABUK (ex-*Richard Page*)	31.3.89	Returned Jan 1994
F 164	HUNAIN (ex-*Talbot*)	31.5.89	Returned 11.12.93

Four AAW frigates were acquired from the US Navy in 1989, all arriving at Karachi July 1989. *Badr* received Mk 15 Phalanx CIWS on flight deck in 1990, others may follow.

Ex-British AMAZON class *frigates*

Armament: 4 RGM-84 Harpoon SSM, 1–4.5in/55 Mk 8, 2–30mm (1 × 2), 1–20mm Phalanx CIWS, ? Tp 45 ASW TT, 1 Lynx helicopter
Sensors: Radar DA-08, RTN-10X; sonar ATAS(V)2
Other particulars: As British *Amazon* class

No	Name	Acquired	Fate
F 181	TARIQ (iii) (ex-*Ambuscade*)	28.7.93	Extant 1995
F 182	BABUR (iii) (ex-*Amazon*)	30.9.93	Extant 1995
F 183	KHAIBAR (iii) (ex-*Arrow*)	1.3.94	Extant 1995
F184	BADR (iii) (ex-*Alacrity*)	1.3.94	Extant 1995
F 185	SHAH JEHAN (iii) (ex-*Active*)	23.9.94	Extant 1995
F 186	TIPPU SULTAN (iii) (ex-*Ambuscade*)	23.9.94	Extant 1995

Purchase of all six Type 21 frigates from the RN was announced on 27 April 1993, but the deliveries were phased over 14 months. To be fitted with new Dutch radar and active towed array sonar (ATAS), French DR-3000S electronic warfare suite and Swedish 9LV Mk 3 combat data system.

SUBMARINES

Ex-US TENCH class *submarine*

No	Name	Builder	Acquired	Fate
S 130	GHAZI (i) (ex-*Diabolo*)	Portsmouth N Yd	1.6.64	Sunk 4.2.71

Loaned by USN after Fleet Snorkel conversion at Philadelphia N Yd. Commissioned *Ghazi* ('Defender of the Faith') at USN Submarine Base, New London, Connecticut. Sunk by Indian ASW surface ships off Visakhapatnam while trying to torpedo the carrier *Vikrant* at the start of the third Indo-Pakistan War.

HANGOR class *submarines*

No	Name	Builder	Commissioned	Fate
S 131	HANGOR	Arsenal de Brest	12.1.70	Extant 1995
S 132	SHUSHUK	CN La Ciotat	12.1.70	Extant 1995
S 133	MANGRO	CN La Ciotat	8.8.70	Extant 1995
S 134	GHAZI (ii) (ex-*Cachalote*)	A C Dubigeon	25.1.69	Extant 1995

French *Daphne* class with some internal modification for Pakistani service. First pair laid down 1 December 1967 and in service 12 January 1970. *Mangro* laid down 8 July 1968 and in service 8 August 1970. *Ghazi* bought from Portugal in December 1975. *Hangor* torpedoed and sank the Indian frigate *Khukri* on 9 December 1971, the first such attack since 1945.

HASHMAT class *submarines*

No	Name	Builder	Commissioned	Fate
S 135	HASHMAT (ex-*Astrant*)	AC Dubigeon	7.2.79	Extant 1995
S 136	HURMAT (ex-*Adventurous*)	A C Dubigeon	6.2.80	Extant 1995

When France cancelled an order for ??*Agosta* class submarines destined for South Africa they were quickly bought by Pakistan in November 1978. They had been built in 1976–79, and *Hashmat* (completed 17 February 1979) arrived at Karachi 31 October 1979. *Hurmat* was completed on 17 February 80, and arrived Karachi 11 August 1980. SM-39 Exocet not fitted.

French AGOSTA-90B type *submarines*

Displacement: 1510t/1760t
Dimensions: 221ft × 22ft 4in × 17ft 9in
67.57m × 6.8m × 5.4m
Machinery: 1-shaft diesel-electric=12.5kts/20.5kts
Armament: 4–533mm TT (20 F17P torpedoes plus SM39 Exocet)
Sensors: Sonar DUUA-1D, DUUA-2A/B, DUUX-2A
Complement: 36

No	Name	Builder	Commissioned	Fate
–	–	DCN Cherbourg	Projected 1999	Building
–	–	Karachi Naval Dyd	Projected 2000	Ordered
–	–	Karachi Naval Dyd	–	Projected

An agreement worth US$ 950m was signed 20 September 1994 for four boats. First of class laid down early 1995, to complete 1998–99. Second unit will be built in sections at Cherbourg and assembled at Karachi. The third unit will be fitted with an 8m 'plug' containing a 200kW MESMA air-independent propulsion system, which will be retro-fitted to the two earlier boats. To be equipped with the SUBTICS command system developed from that fitted in the new French SSBNs. They are the first export submarines to be equipped with SM39 sub-launched Exocet.

Italian SX 404 type *midget submarines*

Displacement: 40t
Dimensions: 52ft 5in oa × 6ft 7in
16.97m × 2.0m
Machinery: 1-shaft electric = 11kt/6½kts. Range 1200nm surface, 60nm underwater
Complement: 4

A commercial rather than naval design of midget submersible capable of carrying 12 passengers. Built 1972–73 by Cosmos, Livorno, and five out of six still in service. Diving depth 330ft (100m). One lost at sea 27 December 1976. See under Taiwan. Replaced by three new midgets in 1990s, probably of SX 756 design.

SMALL SURFACE COMBATANTS

Ex-British BASSET class *trawlers*

Class (no, former name, builder, fate):
Bahawalpur (P 149, ex-*Baroda*, ex-*Lucknow*, *Shalimar*, for disposal *c*1960), *Lahore* (P 12, ex-*Rampur*, ex-*Barisal*, Burn, for disposal *c*1960).
Two transferred from India 1948.

Ex-Chinese 'HUCHWAN' class *torpedo hydrofoils*

Class:
HDF 01–04
Transferred from China 1973 and all four sold 1992.

Ex-Chinese 'HUKU' class *fast attack craft (missile)*

Four transferred from China 1981–82. All extant 1995.

'TOWN' class *large patrol craft*

Displacement: 115t standard; 143t full load
Dimensions: 107ft × 20ft × 11ft
36.6m × 6.1m × 3.4m
Machinery: 2 shafts, 2 MTU 12V 538 diesels, 3400bhp = 24kts
Armament: 2–40mm/70 (600rds)
Complement: 19–24

Class (no):
Rajshahi (P 140), *Jessore* (P 141), *Comilla* (P 142), *Sylhet* (P 143).
Ordered from Brooke Marine on 5 October 1963 and named after East Pakistan towns. *Jessore* and *Comilla* commissioned 20 May 1965, *Rajshahi* and *Sylbet* 2 August 1965. Specially strengthened hulls with sea resistant aluminium alloy superstructures. All were attacked by Indian aircraft on 4 December 1971. Off Chittagong *Rajshahi* was hit six times by four Hunter jets, set on fire, losing one killed and six wounded but regained port after the crew checked flooding and fire in the engine room. Extant 1995. Nine aircraft sank *Comilla* in the same action. *Jessore* and *Sylhet* destroyed in air strikes from the carrier *Vikrant* the same day at Khulna to the West. *Jessore* salvaged by Bangledeshi Navy.

Lahore 6.11.85 L & L van Ginderen

Ex-Chinese 'SHANGHAI II' class *large patrol craft*

Class (no):
Lahore (P 140), *Quetta* (P 141), *Mardan* (P 143), *Gilgit* (P 144), *Pishin* (P 145), *Sukkur* (P 147), *Sehwan* (P 148), *Bahawalpur* (P 149), *Bannu* (P 154), *Kalat* (P 156), *Larkana* (P 159), *Sahival* (P 160).
Eight transferred 1972, four in 1973. All named after Pakistani cities. Four transferred to Maritime Security Agency: only P 144, P 145 and P 149 extant 1995.

Ex-Chinese 'HAINAN' class *large patrol craft*

Class (No):
Baluchistan (P 155), *Sind* (P 159), *Sarhad* (P 161), *Punjab* (P 197)
First pair transferred mid-1976, second April 1980. *Baluchistan* for disposal 1992, remainder stricken 1993–94.

Ex-Chinese 'HEGU' class *fast attack craft (missile)*

Class (no)
Haibat (P 1021), *Jalalat* (P 1052), *Jurat* (P 1053), *Shujaat* (P 1054)
First pair transferred 5 May 1981, second pair October 1981. Extant 1995

Ex-Chinese 'HUANGFEN class *fast attack craft (missile)*

Class (no)
Azmat (P 1025), *Dehshat* (P 1026), *Himmat* (P 1027), *Guwwat* (P 1028)
Four transferred from China April 1984. All extant 1995.

COASTAL PATROL CRAFT
Eighteen MC 55 class (22.8t full load, 54ft 1in × 17ft × 2ft 10in, *16.5m × 5.2m × 0.8m*, 2 V6 diesels, 1600hp = 30kts, 1–14.5mm MG, 5 crew): *P 551–568* built by Crestitalia and delivered 1979–80. GRP units, *P 553* named *Vaqar.*
One Fairey Marine 'Spear' class.
Four ex-British SDMLs: *SDML 3518* and *2519* BU 1965. *SDML 3517* and *3520* stricken 1975–77.

MINE WARFARE VESSELS

Ex-British BANGOR class *minesweepers*

Class (no, former name):
Dacca (M 245, ex-*Oudh*), *Peshawar* (M 55, *ex-Malwa*), *Baluchistan* (M 182, ex-*Greenock*), Chittagong (ex-*Kathiawar*, ex-*Hartlepool*)
Transferred from India 1948, second pair were turbine-powered. *Dacca* and *Baluchistan* for disposal *c*1960; *Chittagong* sold 1956; *Peshawar* sold 1959.

Ex-US BLUEBIRD class *coastal minesweepers*

Class (no, former no, transfer):
Mahmood (M 160, MSC 267, Apr 1957), *Momin* (M 161, MSC 293, Jul 1962), *Mubarak* (M 162, MSC 262, Jan 1957), *Muhafiz* (M 163, AMS 138, 25.2.55), *Mujahid* (M 164, MSC 261, Oct 1956), *Mukhtar* (M 165, MSC 274, 25.6.59) *Munsif* (M 166, MSC 273, 25.6.59), *Moshal* (M167, MSC 294, 13.7.63)
Momin and *Moshal* have lower bridges than the other six. All built under MAP. *Mujahid and Mubarak* arrived Karachi on 5 December 1957. *Munsif* stricken 1979. *Muhafiz* sunk off Karachi by Indian Navy on 5 June 1957. *Momin, Moshal* and *Mubarak* stricken in 1980s, *Mujahid* in 1990; only *Mahmood* and *Mukhtar* extant 1995, and to be replaced by the new minehunters.

French 'ERIDAN' type *minehunters*

Class (no, former name):
Munsif (166, ex-*Sagittaire*), *Muhafiz* (167), *Mahmood* (168)
Three Tripartite GRP minehunters ordered from DCN January 1992, one transferred in November 1992 from the French Navy, one launched by DCN Lorient on 8 July 1995 and one to be built at Karachi Naval Dockyard with DCN assistance.

Panama

Panama, the most southerly of the Central American republics, lies between Colombia (from which country independence was gained in 1903) and Costa Rica and is bisected by the Panama Canal which came under the joint control of Panama and the United States in 1977–78. By 1999 Panama will assume sovereignty over the entire canal. A military coup in 1968 led by Omar Torrijos took the task of government out of civilian hands until 1978 when Dr Aristides Royo became president.

An attempt in February 1988 by President Delvalle to remove General Noriega as commander of the armed forces failed. Noriega ousted Delvalle, who was replaced by Manuel Palma. Presidential elections were held in May 1989, the result of which, appearing to give victory to the opposition, was annulled by Noriega, who assumed power as head of state on 15 December 1989. On 20 December US troops invaded Panama to oust Noriega, and Guillermo Endara, probable winner of the elections in May, took over as president.

There is a large US military presence (roughly 10,000 troops) in the Canal Zone on the Caribbean side of the canal, as the US Southern Command is based here. The Panamanian Coast Guard, divided between the Pacific and Caribbean coastlines, is responsible for maritime defence and employs approximately 280 officers and men.

Vosper 103ft type *large patrol craft*

Particulars: See under Trinidad and Tobago

No	Name	Builder	Comp	Fate
GC 10	PANQUIACO	Vosper, Portsmouth	Mar 1971	Extant 1992
GC 11	LIGIA ELENA	Vosper, Portsmouth	Mar 1971	Deleted 1989

Launched 22 July 1970 and 25 August 1970 respectively. Armament varies from Trinidad and Tobago's units, being 2–20mm. Fitted with fin stabiliser equipment. New Pennant nos P 301 and P 302.

Ex-US AVR class *coastal patrol craft*

Class (no, commissioned, transfer date):
Ayanasi (GC 12, 1943, 1965), *Zarti* (GC 13, 1943, 1966).

Fitted with Raytheon 1500B radar. Have wooden hulls coated with fibreglass. *Ayanasi* deleted, *Zarti* sunk 1989.

Ex-US Coast Guard 40ft type *coastal patrol craft*

Class (no, commissioned, transfer date):
Marti (GC 14, 1950, 1962), *Jupiter* (GC 15, 1950, 1962). No radar fitted. *Marti* sunk 1989. *Jupiter* deleted 1988.

Ex-US Coast Guard 40ft type *coastal patrol craft*

Class (former name, no):
Tiburon (ex-*Smokey Hill River*, GC 10).

Launched 7 July 1945 and purchased by Panama 14 March 1975 from a shipping company. Can carry 400t cargo. Employed on logistic support duties. Sold 1988.

Other patrol craft include:
One ex-USCG 'Point' class large patrol craft *Tres de Noviembre* (P 204, ex-*Point Barrow*), built in early 1960s and acquired 1991. Extant 1995.

One US MSB 29 type patrol craft/minesweeper built in 1954 and acquired 1993. Extant 1995.

Two ex-US MSB 5 type coastal patrol craft/minesweepers (ex-MSB 25, MSB 28 and MSB 41) built in 1952–56 and acquired 1993. Both extant 1995.

Panama also operates:
Five ex-US LCM (8) class landing ships; *GNT 1*, *GNT 2*, *GNT 3*, *Colba* and *Cebaco*. Transferred 1972 (three) and 1989 (two). Employed on logistic support duties. Two deckhouses fitted in place of the wheelhouse. One deleted 1986, two in 1989.

One YF 852 class lighter (ex-*YF 886*), launched 25 May 1945, by Defoe SB, Bay City, Michigan, transferred May 1975. Employed in logistic support duties.

One 65ft patrol craft, *GNT 8*, Panamanian-built, 60t, 8kts, 1 GM diesel, 6 crew. Deleted.

One former shrimp boat, 65ft, 11kts, purchased 1976 also employed on logistics support duties.

Two Swiftships 65ft patrol craft *Commandante Torrijos* (P 201), *Presidente Porras* (P 202) built by Swiftships, delivered 1982 (35t full load, 1020bhp = 23kts, 1–12.7mm, 1 MG). P 202 sunk in 1989, P 201 extant 1995.

Papua New Guinea

Papua New Guinea (PNG) received her independence from Australia in September 1975, but strong links remain and all the navy has been transferred from and trained by the RAN. HQ is at Port Moresby with the patrol boat base at Lombrun on Manas in the Admiralties. It is a very small force to patrol the 3200 mile coastline and rivers as well as six island groups. The 500 volunteer sailors are part of the Defence Force.

Four ASI 315 'Pacific Forum' patrol boats: *Tarangua* (P 01), *Dreger* (P 02), *Seeadler* (P 03), *Basilisk* (P 04). Built 1986–89 and all extant 1995.

Five ex-Australian *Attack* class large patrol craft: *Aitape* (P 84), *Samarai* (P 85), *Ladava* (P 92), *Lae* (P 93), *Madang* (P 94). *Madang* extant 1995. Remainder stricken for spares from 1981.

Two ex-Australian *Balikpapan* utility landing craft: *Salamaua* (31) and *Buna* (32). Both extant 1995.

Seven Australian personnel landing craft: *Kokuba*, *Kaiapit*, *Kandep*, *Kiunga*, *Kuniawa*, *Kukipi*, *Kiunga*. Built and delivered 1975. Two extant 1995.

Four landing craft: *Burfoam* (launched May 1981), *Burcrest* (launched 20.7.81), *Bursea* (launched 6.4.82) and *Burwave* (launched 18.5.82) built by Sing Koon Seng of Singapore. All extant 1995.

Paraguay

Paraguay is one of two South American nations without a coastline. However, navigable rivers form part of her international boundaries and penetrate deep into her interior. Paraguay has traditionally maintained a small river force. Most of the navy's ships are extremely old and in need of replacement. Bases are at Puerto Sajonia (Asuncion) where there are naval repair facilities, Chaco Island (naval air base) and Bahia Negra. Navy personnel includes 3664 officers and men including Coast Guard and 500 marines.

FLEET STRENGTH 1947

Paraguay postwar

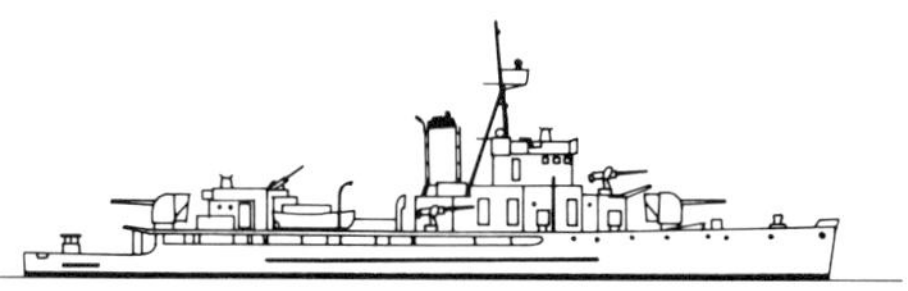

ARMOURED RIVER GUNBOATS

Name	Launched	Disp	Fate
HUMAITA	1930	745t	Extant 1995
PARAGUAY	1930	745t	Extant 1995

There were also six ex-US Coast Guard patrol craft, P 1–6 (45t, acquired 1944), the old *Capitan Cabral* (180t, 1907) and three smaller river craft.

SMALL SURFACE COMBATANTS

Ex-Argentine BOUCHARD class *minesweepers (patrol craft)*

No	Name	Builder	Acquired	Fate
M 1	NANAWA (ex-*Seaver*)	Rio Santiago NY	13.3.64	Extant 1995
M 2	CAPITAN MEZA (ex-*Bouchard*)	Sanchez, San Fernando	5.3.68	Extant 1995
M 3	TENIENTE FARINA (ex-*Py*)	Rio Santiago NY	5.3.68	Extant 1995

Constructed as minesweepers, are being used as river patrol craft. Armament changed to 4–40mm/60 (1×4) 2–12.7mm, MG. Pennant nos changed to P 01, P 02, P 03 and then P 02, P 03, P 04.

Brazilian RORAIMA class *river gunboat*

Class (Pt No):
Itaipu (P 2)

One river gunboat ordered in 1982 from Brazilian yard Arsenal de Marinha, Rio de Janeiro. Laid down on 3 March 1983, launched on 16 March 1984 and completed on 2 April 1985. Equipped with small helicopter landing deck. Pt No changed to P 05 in late 1980s. Extant 1995.

Paraguay also operates a large patrol craft, *Capitan Cabral* (A 1), (180t, launched 1907, wooden hull); two US-built CG type coastal patrol craft (*P 1, P 2*) and six ex-US 701 class coastal patrol craft, *P 101–2P 106.*

Five river patrol craft *P 07, P 08, P 09, P 10, P 11*, built in Paraguay 1989–91 (18t full load, 200bhp = 12kts, 2–12.7mm MG). Extant 1995.

Peru

During the early 1960s, the Peruvian Navy began to mature into a balanced force, once again comparable to those of Argentina, Brazil, and Chile. The decades of deprivation had forged a formidable submarine tradition. This was now complemented by the acquisition of powerful surface units.

The acquisition of elderly cruisers by the Peruvian Navy might mislead some into underestimating the capabilities of the fleet. Peru's chief rival is Chile. Should there be a conflict, the navy which can maintain station the longest might well be the victor. Large cruisers, armed with Exocet missiles, are formidable station platforms. Although many of the navy's surface units are old, most are armed with Exocet missiles and they are kept in fighting trim.

Tension between Peru and the US has been brought about by the intrusion of American tuna boats into the 200-mile zone of waters extending from Peru's coastline. As a result of the impounding of tuna boats, arms and other military equipment being sold to Peru by America was curtailed for a time. A separate coast guard was formed in 1975 and a naval air arm earlier.

Personnel in 1995 totalled 2500 officers and 26,500 men, plus 2500 marines (two years service). Main naval bases are at Callao and San Lorenzo (for submarines); Iquitos is the HQ of the Amazon flotilla and other inland bases are Puno (on Lake Titicaca) and Madre de Dios.

FLEET STRENGTH 1947

CRUISERS

Name	Launched	Disp	Fate
ALMIRANTE GRAU	1906	3200t	Discarded 1958
CORONEL BOLOGNESI	1906	3200t	Discarded 1958

Both units were converted from coal to oil, 1923–25, at Balboa, Canal Zone. *Almirante Grau* served as the fleet flagship. Italian Giradelli AA fire directors were added in the 1930s. By the late 1940s, these ships saw very limited service.

Palacios (left) and *Villar* Sep 1973 USN

DESTROYERS

Name	Launched	Disp	Fate
GUISE	1917	1350t	Discarded *c*1947
VILLAR	1917	1200t	Discarded *c*1954

Guise and *Villar* were rushed to the Peruvian Amazon in 1933, when acquired by Peru. They were part of a hastily gathered fleet sent to confront Colombia.

SUBMARINES

Name	Launched	Disp	Fate
ARICA (ex-*R 4*)	10.5.23	576t/755t	Discarded 1960
CASMA (ex-*R 2*)	29.3.26	576t/755t	Discarded 1960
ISLAY (ex-*R 1*)	12.7.26	576t/755t	Discarded 1960
PACOCHA (ex-*R 3*)	21.4.28	576t/755t	Discarded 1960

All 'R' class submarines received major refits 1935–36 and 1955–56. They acquired their new names in April 1957, when those of all submarines were changed to honour famous Peruvian naval battles.

RIVER GUNBOATS

Name	Launched	Disp	Fate
AMERICA	1904	240t	Inactive
AMAZONAS	1934	250t	Extant 1995
LORETA	1934	250t	Extant 1995

America lies at Iquitos, Peru, in total disrepair. There were also the smaller *Napo* (98t, 1920), *Portillo* (49t, 1902), and the ancient *Iquitos* (50t, 1875, stricken 1967).

In 1947 Peru acquired two ex-US YMS type minesweepers *Bondy* (ex-*YMS 25* and *San Matin* (ex-*YMS 35*), stricken 1974; and four ex-US LCT/6 type landing craft, numbered *BT 1–4*; stricken mid-1950s. There were also six ex-US Coast Guard cutters (45t type) built in 1943 and transferred 1943–44; stricken 1958–61.

MAJOR SURFACE SHIPS

Ex-British FIJI class *cruisers*

No	Name	Builder	Acquired	Fate
81	CAPITAN QUINONES (ex-*Almirante Grau*, ex-*Newfoundland*)	Swan Hunter	30.12.59	Discarded 1979
82	CORONEL BOLOGNESI (ex-*Ceylon*)	Stephen	9.2.60	Discarded May 1982

Originally named *Almirante Grau* by the Peruvians, her name was changed to *Capitan Quinones* in 1973, when Peru acquired the former Dutch cruiser *De Ruyter*. Since the War of the Pacific (1879–83), the name *Almirante Grau* has been reserved for the fleet's flagship. She was transferred on 30 December 1959 at Portsmouth, England. She had been reconstructed in 1951–53 at Devonport. Two lattice masts, a new bridge, and improved AA were fitted. Her torpedo tubes were removed at that time. *Capitan Quinones* was fitted with two lattice while *Coronel Bolognesi* was fitted with one lattice and one tripod mast. *Coronel Bolognesi* was transferred on 9 February 1960 at Portsmouth, England. In 1955–56 she had been refitted with a lattice foremast and a covered bridge; her torpedo tubes were removed at that time. As transferred *Coronel Bolognesi* mounted 9–6in (3×3), 8–4in (4×2), and 18–40mm (5×2, 8×1) – this being six more 40mm than her sister, *Capitan Quinones*. The principal electronics were British radar Types 960, 277; 293 and 274, with US Mk 34 fire control for two of the 4in mounts.

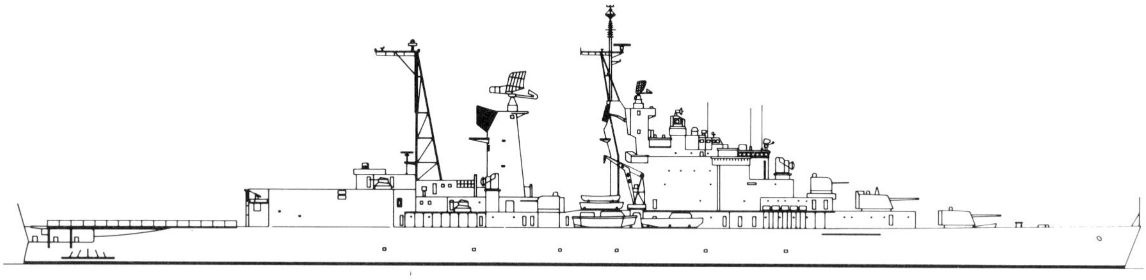
Aguirre 1988

For early history see under Netherlands. In 1962–64 *De Zeven Provincien* was converted to a Terrier missile ship by Rotterdamsche Droogdok Mij, Rotterdam; *De Ruyter* was to have been converted as well, but there was a shortage of funds. After sale to Peru *Aguirre* had her missile system removed and she was converted into a helicopter cruiser with flight deck (35m × 17m) and large hangar (20.5m × 16.5m) capable of operating up to three Agusta ASH-3D Sea King helicopters armed with AM 39 Exocet missiles. Conversion was completed on 31 October 1977. Electronics now comprise radars LW 02, ZW 03, SGR 103, fire control Mk 45, M 25, sonar CWC 10N. In 1986 *Aguirre* was renamed *Almirante Grau* and former *De Ruyter* become *Proyecto 01*, while refitting in Amsterdam. Original Peruvian names were assigned once more as soon as *Grau* resumed her post-refit trials in 1987. There is always an active cruiser named after *Almirante Grau* in the Peruvian Navy. *Aguirre* was refitted and her boilers retubed in 1986. *Grau* was brought back to Peru with incomplete armament fit due to lack of funds. She is fitted to receive 8 SSMs of Otomat or Exocet type and had all her 57mm guns removed, instead 40mm guns are to be installed. Electronic fit of *Grau* is as follows: radars DA 08, Racal Decca 1226, WM 25, Signaal STIR, 2 Lirod 8 optronic directors, combat data system SEWACO PE. She is also fitted with 2 Dagaie and 1 Sagaie chaff launchers. To be armed with SAM launcher. Can carry 3250 rounds for 152mm guns and 16,000 rounds 40mm. Both had Pt Nos changed to CH 81 and CH 84.

Ex-Netherlands DE RUYTER class *cruisers*

Displacement: 9850t standard; 12,250t full load (*Almirante Grau* 9529t standard; 12,165t full load)
Dimensions: 598ft 5in pp, 614ft 7in oa × 56ft 8in × 22ft max
182.4m, 187.3m × 17.2m × 6.7m
Machinery: 2-shaft De Schelde-Parsons geared turbines, 4 Werkspoor-Yarrow boilers, 85,000shp = 32kts. Range 7000nm at 12kts
Armour: Belt 76–102mm, decks 20–25mm
Armament: 4–6in (2×2), 6–57mm (3×2), 4–40mm (2×2), 3 helicopters (*Almirante Grau* 8–6in (4×2), 8–57mm (4×2), 8–40mm (4×2))
Complement: 953

No	Name	Builder	Acquired	Fate
84	AGUIRRE (ex-*De Zeven Provincien*)	Rotterdam Drydock Co	Aug 1976	Extant 1995
81	ALMIRANTE GRAU (ex-*De Ruyter*)	Wilton-Fijenoord, Schiedam	7.3.73	Extant 1995

Initially two more *Fletcher* class units were to have been transferred to Peru, but this was never carried out. In 1975 both units had helicopter landing pads constructed over their fantails.

Ex-US FLETCHER class *destroyers*

No	Name	Builder	Acquired	Fate
72	GUISE (ex-*Isherwood*)	Bethlehem, Staten I	15.12.60	Discarded 1981
71	VILLAR (ex-*Benham*)	Bethlehem, Staten I	8.10.61	Discarded 1980

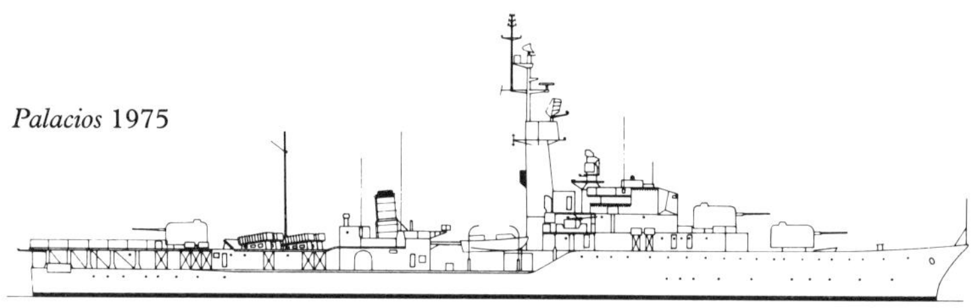
Palacios 1975

Both units were refitted in 1970–73 by Cammell Laird, Birkenhead. Foremast was enclosed and a Plessey AWS-1 radar installed and eight Exocet SSMs (4×2) fitted. *Palacios* commissioned February 1973 and *Ferre* April of that year. A helicopter deck was added in 1975 but in 1977–78 a fuller substantial refit replaced the after 4.5in mount with a helicopter hangar (later the hangar was removed and the 4.5in mount replaced), the second funnel was raised and streamlined and two twin Breda Bardo 40mm mounts replaced the singles in the bridge wings (with Selenia NA-10 director aft); the Squid ASW mortar and sonar were removed.

Ex-British DARING class *destroyers*

No	Name	Builder	Acquired	Fate
DM 74	FERRE (ex-*Decoy*)	Yarrow	Dec 1969	Extant 1995
DM 73	PALACIOS (ex-*Diana*)	Yarrow	1.12.69	Discarded 1993

No modifications carried out before transfer.

Ex-Netherlands HOLLAND class *destroyer*

No	Name	Builder	Acquired	Fate
75	GARCIA Y GARCIA (ex-*Holland*)	Rotterdam DD	2.1.78	Deleted 1986

No modifications carried out before transfer. *Bolognesi* and *Castilla* paid off but still used for spare parts. Pennant nos changed to DM 70, 71, 72, 76–79.

Ex-Netherlands FRIESLAND class *destroyers*

No	Name	Builder	Acquired	Fate
71	CASTILLA (ex-*Utrecht*)	Royal Schelde	6.10.80	Deleted 1990
76	CAPITAN QUINONES (ex-*Limburg*)	Royal Schelde	27.6.80	Deleted 1991
77	VILLAR (ex-*Amsterdam*)	Amsterdam	23.5.80	Deleted 1991
78	GALVEZ (ex-*Groningen*)	Amsterdam	2.3.81	Deleted 1991
79	DIEZ CANSECO (ex-*Rotterdam*)	Rotterdam DD	29.6.81	Deleted 1991
72	GUISE (ex-*Drenthe*)	Amsterdam	3.6.81	Deleted 1985
70	COLONEL BOLOGNESI (ex-*Overijssel*)	Wilton-Fijenoord	14.6.82	Deleted 1990

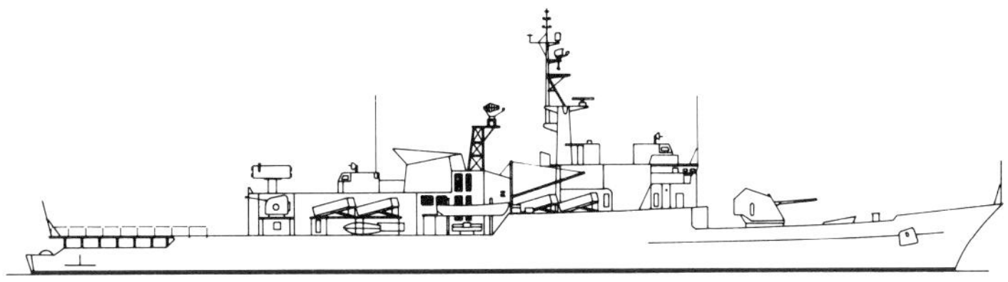

Meliton Carvajal as completed

Italian modified LUPO class *frigates*

Peruvian-built units differ considerably from the standard *Lupo* design and two sisters built in Italy. They have two masts, a standard hangar, not a telescopic one, are fitted, and their Albatros reloading system is hand powered. Armament: 8 Otomat SSMs (8×1), 1 Albatros SAM (1×8, with reloads), 1–127mm/54, 4–40mm/70, 2–20 barrelled 105mm rocket launcher, 6 Mk 32 ASW TT (2×3); have fixed hangars. 53 and 54

No	Name	Builder	Laid down	Launched	Comp	Fate
51	MELITON CARVAJAL	CNR, Riva Trigoso	8.8.74	17.11.76	5.2.79	Extant 1995
52	MANUEL VILLAVICENCIO	CNR, Riva Trigoso	6.10.76	7.2.78	25.8.79	Extant 1995
53	MONTERO	SIMAC, Callao	Oct 1978	8.10.82	25.7.84	Extant 1995
54	MARIATEGUI	SIMAC, Callao	1979	8.10.84	10.10.87	Extant 1995

are first major warships built on the Pacific coast of South America. Pennant nos changed to FM 51–FM 54. See under Italy for further details.

Ex-US BOSTWICK class *destroyer escorts*

Class (former name):
Aguirre (ex-*Waterman*), *Castilla* (ex-*Bangust*), *Rodriquez* (ex-*Weaver*)
All three units transferred 26 October 1951 under MDAP, arriving in Peru 24 May 1952. Prior to transfer, TTs were removed and the ships were reconditioned. Pennant nos D 2, D 1 and D 3, changed in 1959 to DE numbers and in 1960 to 62, 61 and 63 respectively. *Aguirre* sunk in 1974 in Exocet test. *Castilla* and *Rodriguez* discarded in 1979.

Ex-US TACOMA class *frigates*

Class (former name):
Calvez, (ex-*Woonsocket* 1944–48)
Purchased 1948, she was overhauled at Norfolk Navy Yard in 1952. Name often incorrectly cited as *Teniente Calvez*. Pennant no F 1. Discarded 1961 and broken up.

Ex-Canadian 'RIVER' class *frigates*

Class (former name):
Ferre (ex-*Poundmaker*), *Palacios* (ex-*St Pierre*)
Both units were purchased in 1947. Modernised with new guns and fire control at the New York Naval Shipyard in 1952. Their names are often incorrectly cited as *Teniente Ferre* and *Teniente Palacios*. Pennant nos F 3 and F 4, became FE numbers in 1959 and in 1960 they were changed to 66 and 65. Both discarded in 1966.

SUBMARINES

Dos del Mayo 1970

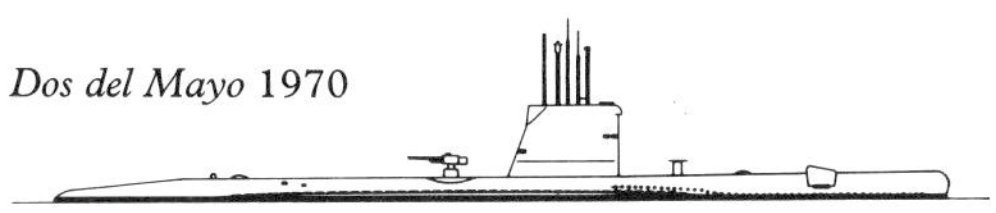

ABTAO class *submarines*

The *Abtao* class is a modification of the wartime US *Mackerel* design. Pennant nos were originally 5–8 and then SS 1–SS 4. *Abtao* and *Dos de Mayo* carry a 5in/25 deck gun abaft the fin. In April 1957 names were changed to those of famous Peruvian naval battles. *Dos de Mayo* and *Abtao* were refitted at Groton, Connecticut, in 1965 and the other pair were refitted in 1968. New batteries shipped in 1981. Eldour sonar fitted.

Displacement: 825t surfaced; 1400t submerged
Dimensions: 243ft oa × 22ft × 14ft max
74.1m × 6.7m × 4.3m
Machinery: 2-shaft diesel-electric: 2 General Motors single acting, Type 278A diesels, plus 2 electric motors, 2400hp = 16kts/10kts. Range 5000nm at 10kts
Armament: 6–21in TT (4 bow, 2 aft)
Sensors: Radar SS-2A; sonar BQR-3, BQA-1A
Complement: 40

No	Name	Builder	Laid down	Launched	Comp	Fate
S 42	ABTAO (ex-*Tiburon*)	Electric Boat	12.5.52	27.10.53	20.2.54	Extant 1995
S 43	ANGAMOS (ex-*Atun*)	Electric Boat	27.10.55	5.2.57	1.7.57	Deleted 1990
S 41	DOS DE MAYO (ex-*Lobo*)	Electric Boat	12.5.52	6.2.54	14.6.54	Extant 1995
S 44	IQUIQUE (ex-*Merlin*)	Electric Boat	27.10.55	5.2.57	10.10.57	Deleted 1993

Dos del Mayo Jun 1966
L & L van Ginderen

Name of S 49 changed after a few weeks, both commissioned 1975 after refit. Ex-*Tench* also purchased 16 September 1976 for spares. SS 48 sunk in collision 1988, salvaged 1989 and used for spares.

Ex-US GUPPY 1A type *submarines*

No	Name	Builder	Acquired	Fate
S 48	PACOCHA (ex-*Atule*)	Portsmouth N Yd	31.7.74	Sunk 1988
S 49	LA PEDRERA (ex-*Pabellon de Pica*, ex-*Sea Poacher*)	Portsmouth N Yd	1.7.74	Deleted 1995

First two were ordered in 1969. Units 3 and 4 ordered 12 August 1976 and 5 and 6 ordered 23 March 1977. Last four are slightly larger at 56.0m and 1000t surfaced (so-called Type 209/1200).

West German Type 209 *submarines*

Particulars: As Greek *Glavkos* class

No	Name	Builder	Laid down	Launched	Comp	Fate
S 32	ANTOFAGASTA	Howaldtswerke, Kiel	3.10.77	19.12.79	20.2.81	Extant 1995
S 36	ARICA	Howaldtswerke, Kiel	1.11.71	5.4.74	21.1.75	Extant 1995
S 33	PISAGUA (ex-*Blume*)	Howaldtswerke, Kiel	15.8.78	19.10.80	12.7.83	Extant 1995
S 31	CASMA	Howaldtswerke, Kiel	15.7.77	31.8.79	19.12.80	Extant 1995
S 35	ISLAY	Howaldtswerke, Kiel	15.3.71	11.10.73	28.8.74	Extant 1995
S 34	CHIPANA (ex-*Pisagua*)	Howaldtswerke, Kiel	1.11.78	19.5.81	20.9.82	Extant 1995

AMPHIBIOUS WARFARE VESSELS

Ex-US LST type *landing ships*

Class (no, former name):
Chimbote (34, ex-m/s *Rawhiti*, ex-*LST 283*), *Paita* (35, ex-*Burnett County LST 512*), *Salaverry* (140, ex-m/s *Iquitos*, ex-m/s *Carpine*, ex-*LST?*).

Acquired in 1951, 1957 and 1977, first and second served commercially before bought by Peruvian Navy. *Chimbote* and *Paita* renumbered 142 and 141. *Paita* deleted 1983, *Chimbote* c1984 and *Salaverry* 1986.

Ex-US TERREBONNE PARISH class *landing ships*

Class (no, former name):
Paita (DT 141, ex-*Walworth County* LST 1164), *Pisco* (DT 142, ex-*Waldo County* LST 1163), *Callao* (DT 143, ex-*Washoe County* LST 1165), *Eten* (DT 144, ex-*Traverse County* LST 1160)

Leased from US Navy for five years on 7 August 1984, lease extended to 1994. All recommissioned on 4 March 1985. Unarmed at the time of the transfer. Extant 1995.

Ex-US LSM type *landing ships*

Class (no, former no):
Lomas (36, ex-*LSM 396*), *Atico* (37, ex-*LSM 554*)

Both units transferred July 1959, deleted 1986.

Ex-US AUK class *corvettes*

Class (no, former name):
Galvez (68, ex-*Ruddy*), *Diez Camesco* (69, ex-*Shoveler*)

War-built US fleet minesweepers transferred under MDAP on 1 November 1960. The minesweeping gear was replaced with sonar and a Hedgehog ASW mortar and the ships are employed as patrol craft. Transferred to the Coast Guard in 1975, and stricken 1981.

SMALL SURFACE COMBATANTS

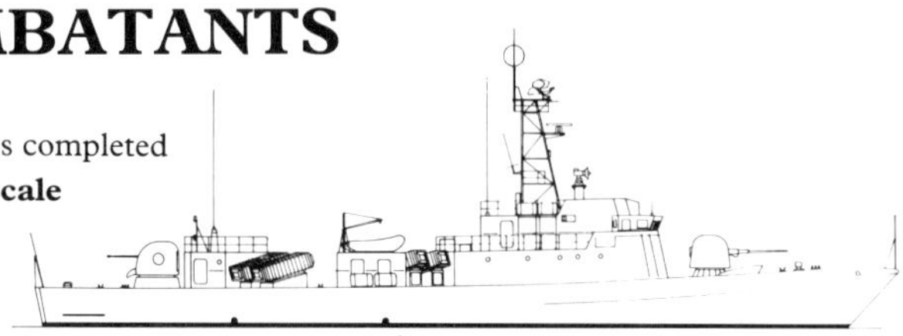

Santillana as completed
NB 1/750 scale

French PR-72P type *missile corvettes*

Displacement:	560t normal; 610t full load
Dimensions:	210ft oa, 193ft 6in pp × 27ft 3in × 8ft 6in *64.0m, 59.0m × 8.3m × 2.6m*
Machinery:	4 shafts, 4 SAGM AGO 240 V16 diesels, 20,000hp = 37kts. Range 700nm/2000nm at 30kts/16kts
Armament:	4 MM38 Exocet SSM (2×2), 1–76mm OTO Melara DP, 2–40mm Breda Bofors (1×2), 2–20mm (2×1)
Sensors:	Thomson-CSF Triton, Vega fire control, Decca 1226
Complement:	45

No	Name	Builder	Launched	Fate
21	VELARDE	Lorient DY	16.9.78	Extant 1995
22	SANTILLANA	SFCN	11.9.78	Extant 1995
23	DE LOS HEROS	Lorient DY	16.9.78	Extant 1995
24	HERRERA	SFCN	16.2.79	Extant 1995
25	LARREA	Lorient DY	20.5.79	Extant 1995
26	SANCHEZ CARRION	SFCN	28.6.79	Extant 1995

All units ordered 1976 and completed in 1980 (first three) and 1981. Designed by SFCN at Villeneuve-la-Garonne. Pennant nos changed to CM 21–CM 26.

Ex-US PGM type *gunboat*

Rio Sama (PC 11, ex-*PGM 78*) was transferred under MDAP in September 1966. A second vessel, *Rio Chira* (ex-*PGM 11*) was built at SIMA, Callao with US assistance, and completed in June 1972. Both were transferred to the Coast Guard in 1975 and renumbered 222 and 223. *Rio Sama* deleted 1990, *Rio Chira* extant 1995.

VELARDE class *fast patrol craft*

Displacement:	100t normal; 130t full load
Dimensions:	109ft 7in oa, 103ft 4in wl × 21ft × 5ft 7in *33.4m, 31.5m × 6.4m × 1.7m*
Machinery:	2 shafts, 2 Napier-Deltic diesels, 6200bhp = 30kts
Armament:	2–20mm (2×1)
Complement:	25

Class (no, launched):
Velarde (21, 10.7.64), *Santillana* (22, 24.4.64), *De Los Heros* (23, 18.11.64), *Herrera* (24, 26.10.64), *Larrea* (25, 18.2.65), *Sanchez Carrion* (26, 18.2.65).

Designed and built by Vosper, Portsmouth, as multi-purpose patrol, rescue, fishery protection vessels, able to carry guns, torpedoes (side-launching racks), twin RL, mines or DCs (sonar was fitted). Transferred to Coast Guard in 1975 and renamed *Rio Vilor* (229), *Rio Ica* (228), *Rio Chicawa* (224), *Rio Pativilca* (225), *Rio Huaroa* (226), and *Rio Locumba* (227) respectively. 225 and 227 extant 1992. 224 deleted 1986, 229 in 1990, 228 in 1982, 226 in 1983.

Santillana 1964 Vosper Thornycroft

RIO CANETE class *patrol craft*

Displacement: 296t full load
Dimensions: 167ft 3in oa × 24ft 3in × 5ft 6in
50.98m oa × 7.4m × 1.7m
Machinery: 2 shafts, 4 Bazan MTU diesels, 5640bhp = 22kts. Range 3000nm at 17kts
Armament: 1–40mm Bofors, 1–20mm Oerlikon
Complement: 30 (4 officers)

Class (no, in serv.):
Rio Nepena (PC 243, 1981), *Rio Tambo* (PC 244, 1982), *Rio Ocona* (PC 245, 1983), *Rio Huarmey* (PC 246, 1984), *Rio Zana* (PC 247, 1985), *Rio Canete* (PC 248, ex-234, 1976).

Built by SIMA at Chimbote (PC 248 at Callao). Designed speed of 25kts was never achieved. Steel hulls and aluminium superstructure. Prototype scrapped in 1990. All others extant 1995. Belong to Coast Guard.

Maranon as completed
NB 1/750 scale

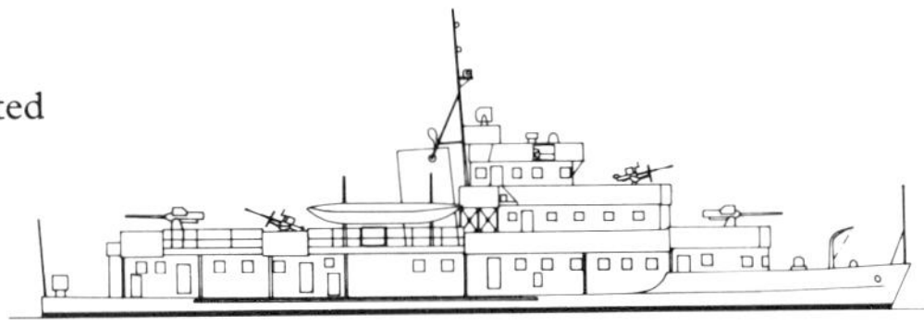

UCAYALI class *river gunboats*

Displacement: 365t full load
Dimensions: 154ft 9in wl × 32ft × 4ft max
47.1m × 9.7m × 1.2m
Machinery: 2 shafts, 2 British Polar diesels Type M441, 800bhp = 12kts. Range 5000nm at 10kts
Armament: 2–3in (2×1), 7–20mm (2×2, 1×3)
Complement: 40

No	Name	Builder	Launched	Fate
13	MARANON	Thornycroft	Apr 1951	Extant 1995
14	UCAYALI	Thornycroft	7.3.51	Extant 1995

Stationed at Iquitos on the Amazon River. They have a shallow draft which is necessitated by the river environment. Specially designed for operation in the tropics.

There are also a number of smaller vessels employed on inland waters:

La Pedrera and *Rocafuerte* (Yarrow, 1954) of 150t full load, used as troop transports on the Amazon.

Four Italian-built 37t, 18kt patrol craft for the River Salto built in 1960: *Rio Zarumila*, *Rio Tambes*, *Rio Piwa* and *Rio Salta*; the last was stricken 1966, *Rio Zarumila* deleted 1990, the rest extant 1995.

Six patrol craft built in 1980–82 by McLaren, Niteroi (43t full load, 18.6m, 2 diesels = 25kts, 2–20mm Oerlikon): *La Punta*, *Rio Chillon*, *Rio Santa*, *Rio Viru*, *Rio Lurin*, *Rio Surco* (PP 230–232, 235–236). All extant 1995.

Four 14t launches patrol Lake Titicaca, and there are also four ex-US Coast Guard 40ft utility boats.

Philippines

The Philippines are an extensive archipelago of eleven large and 7079 other islands spreading over an area of some 115,800 square miles with a total coastline of about 14,000 miles. The archipelago (population 62,868,000 in 1991) lies to the north of Indonesia and bisects the sea routes between China, Japan, Korea, Asian Russia and Indonesia and routes to the Indian Ocean. Strategically, therefore, the Philippines hold an important position as regards maritime strategy in South East Asia. The country has ninety-four ports and several thousand vessels involved in inter-island traffic and fisheries.

The Filipino Government has maintained close ties with the USA since the republic's full independence on 4 July 1946. The US Navy had a twenty-five-year lease on its major base at Subic Bay and withdrew in 1992. This resulted in extensive military aid, the majority of warships being fairly sizeable, but now greatly overage, ex-American units (over forty-nine acquired by 1950). Certainly the age of the present seventeen-ship escort force indicates a definite need for reliefs.

A planned modernisation programme, which was due to be completed in 1983, called for the construction of twelve missile-armed patrol combatants, three to be of the South Korean *Paek Ku 12* class, as replacements for the old ex-US frigates. It was also reported that interest was shown in the acquisition of up to two submarines, probably German Type 209s, but nothing came of these plans.

There are many coastal patrol craft and larger patrol vessels (the Coast Guard was formed as a branch of the navy in October 1967) designed and used to patrol the large territorial seas as the Filipino Government claimed a 300nm limit. The Philippines Navy also has some old landing ships used mainly as transports for landing stores on harbourless coasts or in small fishing ports. Both these categories have been used against Muslim secessionists on Mindanao and in the Sulu Archipelago since 1965 in the bloody Moro Rebellion. There were only two small coastal motor minesweepers (US *Bluebird* class) probably only for training purposes.

Selective service manpower, for an active fleet of ninety-one surface combatants and one hundred and two amphibious vessels, totalled 13,560 (12,000 reserves) in 1995 (4000 in 1958, 8000 in 1970) with a marine corps of 7500 organised in two seven-battalion brigades. The main base is Sangley Point. The Coast Guard had 3400 service personnel in 1995. There are six main commands held by captains or commodores; Naval Defence Forces; Coast Guard Sealift; Amphibious Command; Training; Ready Force and Fleet Support.

Due to the poor state of the Philippines economy all 1980s–'90s naval building programmes were effectively stopped or cancelled, and only some old ships were refitted for further service. Some patrol boats were built, and an ambitious plan to build twelve fast attack craft was implemented. The first of these were to be three boats of the Spanish Cormorán type, but lack of funds has stalled this programme. Planned earlier boats of the Korean PSMM-5 type were never built due to lack of funds. The naval aviation has at its disposal three Fokker F-27M maritime patrol aircraft (one reported lost), ten Britten-Norman BN-2 Defender light maritime patrol aircraft and four to ten MBB Bo-105 light helicopters.

MAJOR SURFACE SHIPS

Ex-US BUCKLEY ('TE') class *frigate*

No	Name	Builder	Acquired	Fate
PF	RAJAH SOLIMAN (ex-*Bowers*)	Bethlehem, San Francisco	21.4.61	BU 1966

Former US high speed transport converted from destroyer escort built in 1943. Transferred under MAP. Fitted as command ship for the Philippine Navy. Sunk in typhoon in June 1962, raised but stricken on 3 December 1964.

Ex-US EDSALL (FMR') class *frigate*

No	Name	Builder	Acquired	Fate
PF4	RAJAH LAKANDULA (ex-*Tran Hung Dao*, ex-*Camp*)	Brown SB, Houston	5.4.76	Stricken 1988

USN destroyer escort converted to radar picket after the Second World War, redesignated DER instead. Transferred to South Vietnamese Navy on 6 February 1971. Acquired by Philippines in 1975 and formally transferred the next year. Extant 1995 as stationary HQ ship.

Ex-US CASCO class *cutters*

Class (no, former names):
Andres Bonifacio (PF 7, ex-*Ly Thoung Kiet*, ex-*Chincoteague* WHEC 375, ex-*AVP 24*), *Georgio de Pilar* (PF 8, ex-*Ngo Kuyen*, ex-*McCulloch* WHEC 386, ex-*Wachapreague* AGP 8, ex-*AVP 56*), *Diego Silang* (PF 9, ex-*Tran Quang Khai*, ex-*Bering Strait* WHEC 382, ex-*AVP 34*), *Francisco Dagohoy* (PF 10, ex-*Tran Binh Trong*, ex-*Castle Rock* WHEC 383, ex-*AVP 35*)

Built by L Washington S Yd as *Barnegat* class seaplane tenders for USN. All four transferred to Coast Guard in 1946–48, initially on loan designated WAVP and then on permanent transfer except ex-*McCulloch* transferred outright from USN to CG, subsequently redesignated as high endurance cutters (WHEC). Transferred from CG to S Vietnam in 1971–72 and officially transferred to Philippines in 1976. *Francisco Dagohoy* commissioned in Philippines Navy in 1979. Two other cutters *Yakutat* WHEC 380 (ex-*AVP 32*) and *Cook Inlet* WHEC 384 (ex-*AVP 36*) transferred in 1976 for spares. All stricken 1986.

Ex-US CANNON ('DET') class *frigates*

Class:
Datu Kalantiaw (PF 76, ex-*Booth*), *Datu Siratuna* (PF 77, ex-*Asahim*, ex-*Amick*), *Rajah Humabon* (PF 78, ex-*Hatsuhi*, ex-*Atherton*), I (ex-*Kyong Ki*, ex-*Muir*), II (ex-*Kang Won*, ex-*Sutton*)

USN destroyer escorts type built in 1943–44. *Datu Kalantiaw* transferred 15 December 1976, and sold on 31 August 1978; *Datu Siratuna* and *Rajah Humabon* served in Japanese Navy in 1955–75 and were transferred to Philippines in 1976 and purchased at scrap value, towed to S Korea in October 1978 for overhaul and modernisation. The last two ships without new names were taken over from Korean deletion list and were probably cannibalised for spares for active units. *Datu Kalantiaw* lost on 20 September 1981 in Typhoon 'Clara'. *Datu Siratuna* stricken in 1984. *Rajah Humabon* due to strike soon.

AMPHIBIOUS WARFARE VESSELS

Ex-US LST type *landing ships*

Class (no, former names, transferred):
Cutabato (ex-*LST 75*), *Pampanga* (ex-*LST 842*), *Bulacan* (ex-*LST 843*), *Albay* (ex-*LST 865*), *Misamis Oriental* (ex-*LST 875*). *Bataan* (ex-*Caddo Parish*, LST 515), *Cagayan* (LT 86, ex-*Hickman County*, LST 825), *Ilcos Norte* (LT 97, ex-*Madera County*, LST 905), *Zamboanga del Sur* (LT 86, ex-*Cam Ranh*, ex-*LST 975*), *Mindoro Occidental* (LT 93, ex-*LST 222*, 15.7.72), *Surigao del Norte* (LT 94, ex-*LST 488*), *Surigao del Sur* (LT 95, ex-*LST 546*, 15.7.72), *Tarlac* (LT 500, ex-*LST 47*, 13.9.76), *Laguna* (LT 501, ex-*LST 230*), *Samar Oriental* (LT 502, ex-*LST 287*, 13.9.76), *Lanao del Sur* (LT 503, ex-*LST 491*, 13.9.76), *Lanoa del Norte* LT 504, ex-*LST 566*, 13.9.76), *Leyte del Sur* (LT 505, ex-*LST 607*, 13.9.76), *Davao Oriental* (LT 506, ex-*Daggett County*, LST 689), *Benguet* (LT 507, ex-*Davies County*, LST 692, Sep 1976), *Sierra Madre* (AL 57, ex-*Dumagat*, ex-*My Tho*, ex-*Harnett County* APG 821, ex-*LST 821*), *Aurora* (LT 508, ex-*Harris County*, LST 822, Sep 1976), *Cavite* (LT 509, ex-*Hillsdale County*, LST 835), *Samar del Norte* (LT 510, ex-*Nansemond County*, LST 1064), *Cotabato del Norte* (LT 87, ex-*Orleans Parish*, LST 1069, ex-*MCS 6*, Oct 1976), *Tawi-Tawi* (LT 512, ex-*LST 1072*, Sep 1976), *Agusan del Sur* (ex-*Nha Trang*, ex-*Jerome City*, *LST 848*).

Twenty-three US tank landing ships built in 1942–45. Names after Filipino island and straits. Several have undergone major refits including replacement of frames and plating as well as engines and electrics. First five ships transferred in 1948 and the next three in November 1969. Other ships acquired from S Korea and S Vietnam in the 1970s (LT 93–95 unarmed July 1972) especially after the fall of S Vietnam (LT 54 and 87). LT 510 bought 24 September 1977. Vietnamese ships were bought at scrap value on 5 April 1976. LT 500 series recommissioned 8 August 1978 and 18 October 1978 after Japan refits. *Cutabato* and *Pampanga* were scrapped. *Bataan* renamed *Maquindanao*. All reported as being in poor condition. In 1995, only LT 86, 501, 504, 507, 516 remained in service.

Ex-US LSM type *medium landing ships*

Class (no, former names, transferred):
Isabela (LP 41, ex-*LSM 463*, Mar 1961), *Batanes* (i) (ex-*LSM 236*), *Oriental Mindoro* (LP 68, ex-*LSM 320*, Apr 1962), *Batanes* (ii) (ex-*Huong Giang*, ex-*Oceanside*, ex-*LSM 175*), *Western Samar* (LP 66, ex-*Hat Giang*, ex-*LSM 9011*, ex-*LSM 335*)

Two transferred to the Philippines in 1961–62. *Batanes* (i) ran aground 1971, stricken 1972. LSM 175 acquired from S Vietnam in 1975. Also LSM 110 (17 November 1975) for spares cannibalisation. All stricken in early 1990s.

BACOLOD CITY class *landing ships (logistics)*

Displacement:	4265t full load
Dimensions:	272ft 8in × 60ft × 12ft 2in *83.1m × 18.3m × 3.7m*
Machinery:	2 shafts, 2 GM EMD 16-645E2 diesels, 3900bhp = 11.6kts. Range 6000nm/11kts
Armament:	None
Complement:	30 (6 officers)

No	Name	Commissioned	Fate
LC 550	BACOLOD CITY	1.12.93	Extant 1995
LC 551	CAYAGAN DE ORO CITY	Mar 1994	Extant 1995

Two ships ordered from the Trinity Marine 3 April 1993 to US *Frank S Besson* class design currently produced for the US Army. Typical ro-ro design with 10,500sq ft cargo deck. Military load of 2280t (only 990t for amphibious operations). Can also accommodate up to 150 fully equipped troops. Ships are fitted with 250bhp bow thruster for ease of manoeuvring at low speeds. Helicopter platform astern.

Ex-US LSSL 1 class *fire support landing ships*

Class (no, former names):
Camarines Sur (LF 48, ex-*Nguyen Duc Bong*, ex-*LSSL 129*), *Sulu* (LF 49, ex-*LSSL 96*), *La Union* (LF 50, ex-*Doan Ngoc Tang*, ex-*LSSL 9*)

La Union was transferred to France in 1951 (*Hallebarde* L 9023), transferred to Japan 1956–64; returned and transferred to S Vietnam in 1965 for spares. Acquired by Philippines in 1975 as was *Lulu Phu Tho* (ex-LSSL 101) for spares. *Camarines Sur* and *Sulu* transferred on 17 November 1975. Ex-US LSSL 68 and 87 transferred from Japan September and November 1975 but not commissioned. Armament: 1–76mm, 4–40mm, 4–20mm, 4 MG. Stricken 1991.

Ex-LSIL type *infantry landing ships*

Class:
Marinduque (LF 36, ex-*LSIL 875*), *Seorsogon* (LF 37, ex-*Thien Kich*, ex-*L 9038*, ex-*LSIL 872*), *Camarines Norte* (LF 52, ex-*Loi Cong*, ex-L 9034, ex-*LSIL 699*), *Misamis Occidental* (LF 53, ex-*Tam Set*, ex-*L 9033*, ex-*LSIL 871*)

Transferred to France and later to S Vietnam and in 1975 acquired by Philippines. There is also LSIL 476 transferred at the same time for spares. First three stricken 1980, last unit 1979.

MINOR LANDING CRAFT
Eleven ex-US LCM (8) type transferred 1973–75.

Fifty ex-US LCM (6) type transferred 1955 (1), 1971–73 (13), 1973–75 (36).

Seven ex-US LCVP type transferred in 1955–56 (2), 1965 (1), 1971 (2), 1973 (2).

Three ex-US LCU 1466 type transferred from Japan 17 November 1975.

SMALL SURFACE COMBATANTS

Ex-US ADMIRABLE class *ocean minesweepers (corvettes)*

Class (no, former names):
Pagasa (ex-APO 21, ex-*Quest* AM 281), *Samar* (ex-*Project* AM 278), *Magat Salamat* (PS 20, ex-*Chi Lang II*, ex-*Gayety* MSF 239)

Minesweeping gear removed upon transfer in 1948. *Pagasa* renamed several times: *Santa Maria*, *Corregidor* and *Mount Samat*; served as presidential yacht and had her stern rebuilt, later used as command vessel. *Samar* used as patrol vessel, armament: 1–76mm, 4–40mm (2 × 2), 4–20mm; APO 21: 1–76mm, 4–20mm. *Samar* later served as survey vessel. *Magat Salamat* was built in 1943–45 and transferred to S Vietnam in April 1962 and to Philippines in November 1975. Minesweeping gear removed and ship used as large patrol ship (1–76mm, 2–40mm, 6–20mm). *Magat Salamat* extant 1995.

Ex-US PCE type (185ft) *patrol craft (corvettes)*

Class:
Cebu (PS 28 ex-*PCE 881*), *Negros Occidental* (PS 29, ex-*PCE 884*), *Leyte* (PS 30, ex-*PCE 885*), *Pangasinan* (PS 31, ex-*PCE 891*), *Iloilo* (PS 32, ex-*PCE 897*), *Miguel Malvar* (PS 19, ex-*Ngoc Hoi*, ex-*Brattleboro* PCER 852), *Sultan Kundarat* (PS 22, ex-*Dong Da II*, ex-*Crestview*, *PCER 895*), *Datu Marikudo* (PS 23, ex-*Van Kiep II*, ex-*Amherst PCER 853*)

Built in 1943–45 and transferred in July 1948. Originally numbered E 28–32. Armament: 1–76mm/50, 3 or 6–40mm, 4–20mm (4x1), 2–234mm Mk 32 ASW TT (3x1), DC, Hedgehog. *Leyte* ran aground in 1979. Last three transferred from Vietnamese Navy in 1975–76. PCER 852, 895, 853 originally fitted as rescue ships. Extant 1995.

Ex-US AUK class (MSF type) *ocean minesweepers (corvettes)*

Class:
Rizal (PS 69, ex-*Murrelet*, MSF 372), *Quezon* (PS 70, ex-*Vigilance*, MSF 324)

Former US Navy minesweepers (originally designated AM) built in 1943–45 and transferred to Philippines on 18 June 1965 and 19 August 1967 respectively. Upon transfer the minesweeping gear was removed and a second 76mm gun fitted aft until replaced by raised helicopter pad. SPS-5 radar and SQS-17B sonar; armament: 2–76mm/50 (2x1), 4–40mm (2x2), 4–20mm (2x2), 3–324mm Mk 32 ASW TT, 1 Hedgehog, 2 DCT, 2 DC rails. Used as patrol vessels. *Quezon* stricken 1994, *Rizal* extant 1995.

Ex-US PC (173ft) type *submarine chasers (large patrol craft)*

Class:
Camarines Sur (PS 21, ex-*PC 1121*), *Bohol* (PS 22, ex-*PC 1131*), *Zamboango del Sur* (PS 23, ex-*PC 1133*), *Batangas* (PS 24, ex-*PC 1134*), *Nueva Ecija* (PS 25, ex-*PC 1241*), *Negros Oriental* (PS 26, ex-*PC 1563*), *Capiz* (PS 27, ex-*PC 1564*), *Nueva Viscaya* (PS 80, ex-*Altus*, ex-*PCE 568*), *Negros Oriental* (PS 29, ex-*E 312*, ex-*L'Inconstant*, ex-*PC 1171*)

Eight transferred in 1947–48, 1958 and March 1968. *Camarines Sur* stricken in 1953, *Negros Oriental* sunk in typhoon at Guam in November 1962 (raised and stricken in 1963), *Bohol* scrapped in 1969, *Capiz* and *Batangas* stricken in 1979. Others: *Nueva Ecija* (1977) and *Zamboango del Sur* also stricken (16 October 1956). PC 1171 escaped from Cambodia to the Philippines and was bought by the latter in December 1976. PS 29 stricken 1989, PS 80 lost in typhoon 13 November 1989.

Ex-US PCS type *submarine chasers (large patrol craft)*

Class:
Tarlac (PG 11, ex-*PCS 1399*, ex-*YMS 450*), *Laguna* (LG 12, ex-*PCS 1403*)

Built in 1942 and transferred in January 1948. Both scrapped in 1969. Armament: 1–76mm, 1–40mm, 4–20mm.

Ex-US SC (110ft) type *submarine chasers (large patrol craft)*

Class:
Cagayan (ex-*SC 731*), *Mountain Province* (ex-*SC 736*), *Ilocus Sur* (ex-*SC 739*), *Alert* (ex-*SC 1267*), *Surigao* (ex-*SC 747*), *Isabella* (ex-*SC 750*), *Cavite* (ex-*SC 981*), *Ilocus Norte* (ex-*SC 1274*)

Built in 1942–43 and all eight transferred in 1946–48. *Ilocus Norte* renamed *Malampay Sound*. *Ilocus Sur*, *Cagayan* and *Isabella* stricken in 1956. *Alert* was sunk in 1956. Others since stricken.

US SC type (110ft) *(large patrol craft)*

Class:
Antique (ex-*PGM 36*), *Camarines Sur* (ex-*PGM 33*), *La Union* (*PGM 35*), *Masbate* (ex-*PGM 37*), *Misamis Occidental* (ex-*PGM 38*), *Sulu* (ex-*PGM 34*)

Six built under MAP for Philippines; four transferred in 1955 and two (*Masbate* and *Misamis Occidental*) in 1956. Design based on the 110ft wooden submarine chasers of the Second World War. *Masbate* sunk 1965. *Camarines Sur, Sulu, La Union, Antique, Misdamis Occidental* decommissioned in 1973–74.

US PGM type *(large patrol craft)*

Class (no):
Agusan (PG 61, ex-*PGM 39*), *Catanduanes* (PG 62, ex-*PGM 40*), *Palawan* (PG 64, ex-*PGM 42*), *Romblon* (PG 63, ex-*PGM 41*), *Yachi* (ex-*PGM 60*), *Yanga* (ex-*PGM 59*), *Yundi* (ex-*PGM 57*), *Basilan* (ex-*Hon Troc*, ex-*PGM 83*)

Steel-hulled patrol gunboats, longer versions of the US Coast Guard 95ft 'Cape' class built under MAP in US by Tacoma Boatbuilding for Philippines and transferred upon completion in 1960 (first pair in service March; second in June). *Yachi*, *Yanga* and *Yundi* intended for Indonesia transferred to Philippines in 1965. *Basilan* acquired from S Vietnam in December 1975. PG 61–64 extant 1995.

KATAPANGAN class *large patrol craft*

Displacement: 135t standard; 150t full load
Dimensions: 100ft 4in × 18ft 7in × 5ft
30.6m × 5.7m × 1.5m
Machinery: 2 shafts, 2 MTU MB 820 D61 diesels, 2050bhp = 16kts
Armament: 2–30mm/85 Emerlec, 4–12.7mm, 2–7.62mm MG
Complement: 30

Class (no):
Katapangan (P 101), *Bangong Lakas* (P 102), *Kagitingan* (P 103), *Bagong Silang* (P 104)

Designed and built by W Müller, Hamelin, W. Germany. Prototype delivered for trials 11 October 1978. Both in service 9 February 1979 but plans to build more at Cavite Navy Yd and Boseco, Bekan, in the Philippines abandoned after building two more (P 103, P 104), due to their inadequacy. Last two completed 1982. *Katapangan* stricken 1989 for spares, remainder extant 1995.

Ex-British *seaward defence motor launches*

Class:
SDB 1323, 1326, 1328, 1329

Former British HDML type motor launches acquired from Australia in 1959 and commissioned without name change. Stricken *c*1963.

Ex-US 'SWIFT' Mk 1 and 2 type *coastal patrol craft)*

Class:
PCF 300-317 (ex-US *PCF 6633, 6634, 35–38, 681–688, 6911–6913*)

Eighteen inshore patrol craft built in the USA. *PCF 302–305* served in USN prior the transfer, rest built under MAP. *PCF 300–305* transferred in March and August 1966, *PCF 306–313* in February 1968 and *PCF 314–316* in July 1970. *PCF 317* was Filipino built (ferro concrete) and used as a yacht for the president's wife until discarded 1976. *PCF 304* and *306* stricken then. *PCF 308* stricken 1979, *PCF 303* in 1988; fifteen extant in 1992.

ABRA class *coastal patrol craft*

Displacement: 40t standard; 72t full load
Dimensions: 87ft × 9ft × 5ft
26.5m × 2.7m × 1.5m
Machinery: 2 shafts, 2 MTU diesels, 2400bhp = 25kts
Armament: 2–20mm (2×1), 1–12.7mm MG
Complement: 15 or 16

Class (no, in service):
Abra (FB 83, 8.1.70), *Bukindon* (FB 84, 1971), *Tablas* (FB 85, 1975)

First ship built by Vosper at Singapore, other two at Cavite N Yd in the Philippines. Similar to Malaysian 'PX' class. All deleted in early 1990s

US 'SWIFT' Mk 3 type *coastal patrol craft*

Displacement: 33t full load
Dimensions: 65ft × 16ft 1in × 3ft 3in
19.8m × 4.9m × 1.0m
Machinery: 3 shafts, 3 GM 8V 71 TI diesels, 1590bhp = 30kts. Range 500nm at 30kts
Armament: 2–12.7mm (2×1), 2–7.62mm MG (2×1)
Sensors: Radar LN-66
Complement: 8

Class:
PCF 318–323, 333, 334, 336–340, 341–352

Built in USA by Sewart for Philippines Navy. First six delivered January-June 1972, *PCF 333* and *334* in April 1975, *PCF 338* in July, *336* in November, *339* and *340* in December. Surviving boats renamed *DF 325–32* belong to Coast Guard whilst *DF 353, 354* operated by the navy.

De Havilland Marine series 9209 *coastal patrol craft*

Displacement: 16.5t full load
Dimensions: 45ft 10in × 14ft 6in × 3ft 4in
14m × 4.4m × 1m
Machinery: 2 shafts, 2 Caterpillar D348 diesels, 740bhp = 25kts. Range 500nm at 12kts
Armament: 2–12.7mm (2×1)
Complement: 8

Class:
PC 326–331

Built by De Havilland Marine of Sydney, Australia with GRP hulls. In service 20 November 1974–8 February 1975. Four extant 1995 as *DF 326, DF 328, DF 330, DF 331*.

MARCELO type *coastal patrol craft*

Displacement: 21.75t full load
Dimensions: 46ft × 14ft 1in × 3ft 4in
14m × 4.3m × 1m
Machinery: 2 shafts, 2 MTU 8V-331 TC80 diesels, 1800bhp = 46kts. Range 200nm at 36 kts
Armament: 4–12.7mm (1 × 2, 2 × 1)
Sensors: Radar LN-66

Class:
PSB 411, PSB 414, PSB 417–434

Eighty ordered August 1975 from Marcelo of Manila but fifteen of twenty-five completed that year destroyed by fire, halting production. Turning MG set into fo'c'sle. Ten extant 1995.

Spanish 'CORMORÁN' type *patrol boats*

Three modified 'Cormorán' type patrol boats were ordered in 1991 in Spain. Empresa Nacional Bazán, San Fernando, built the first boat in 28 months, delivered in 1994, and the next two will be built in the Philippines with technical assistance from Bazán. These new boats displace 380t and reach speed of 35kts on a three-shaft 11,250bhp diesel installation. According to French reports they will be armed with MM 40 Exocet SSMs apart from the guns (1–76mm/62 OTO Melara, 2–40mm/70 Breda twin). The planned programme announced in 1990 include six missile-armed boats followed by six gun-armed versions. Construction has not yet begun.

'LAUNCESTON' type *large patrol craft*

Displacement: 396t full load
Dimensions: 187ft × 26ft 11in × 7ft 11in
57.0m × 8.2m × 2.4m
Machinery: 3 shafts, CP propellers, 3 MTU 16V 956 TB91 diesels, 11,250bhp = 30kts. Range 3500nm/18kts
Armament: 1–76mm/62 OTO Melara, 2–40mm/70 Breda (1×2), 2–25mm Oerlikon (2×1)
Sensors: Radar HSA WM 22
Complement: –

Six ships were to be built by Launceston Marine, Tasmania, Australia. Negotiations continued by Australian Submarine Corporation and contract signed on 21 October 1991 for three units. Originally planned CODAG machinery arrangement and water-jet propulsion. First unit to be built by the former Carrington Slipways for delivery in 1996.

GENERAL AMILIO AGUINALDO class *large patrol craft*

Displacement: 279t full load
Dimensions: 144ft 4in × 24ft 3in × 5ft 3in
44.0m × 7.4m × 1.6m
Machinery: 2 shafts, 4 Detroit 12V92TA diesels, 2040bhp = 20kts. Range 1100nm/18kts
Armament: 2–40mm Bofors, 2–20mm Oerlikon, 4–12.7mm
Complement: 58 (6 officers)

Class (no, commissioned):
General Amilio Aguinaldo (PG 140, 21.11.90), *General Antonio Luna* (PG 141, Mar 1994)

Aguinaldo was launched 23 June 1984 at Cavite NSY but no work was carried out on it for some time. Was to be armed with four SSM. Steel hull. Designed speed of about 20kts would be rather low for a missile boat. Armament is to be upgraded with missiles and 1–76mm/62 OTO Melara gun. Current plans foresee four more craft of this class being built.

PSB 436 class *patrol craft*

Displacement:	24.6t full load
Dimensions:	54ft 10in oa × 16ft 5in × 4ft 3in *16.7m × 5.0m × 1.3m*
Machinery:	2 shafts, 2 MTU 8V396 TB93 diesels, 2400bhp = 36kts
Armament:	1–12.7mm

Class:
PSB 436–439

Considered as an improved version of the earlier Philippines patrol craft, GRP hull, ordered in 1982 at Cavite Navy SY and completed for service in 1985. Originally fifty-five units planned, but construction terminated after only four were completed.

US 'Halter' PFC 70 type *patrol boats*

Displacement:	56t full load
Dimensions:	77ft 1in × 20ft × 6ft 7in *23.5m × 6.1m × 2.0m*
Machinery:	2 shafts, 2 GM 16V92 TA diesels, 1380bhp = 28kts. Range 1200nm at 12kts
Armament:	4–12.7mm M3 (4×1), 2–7.62mm (2×1)
Complement:	8

Four boats ordered in September 1989 for $9.4m at Trinity–Equitable SY, New Orleans. Were to be armed with one 40mm gun. Second batch of three ordered September 1990. Completed for service in 1990–91. Plans called for another thirty-one boats to be built in the Philippines under FMS programme. It is not known if they actually will be built.

Ex-South Korean SEA HAWK class *coastal patrol craft*

Class (no):
Conrado Yap (PG 840), *Jose Artiga Jnr* (PG 844), *Leon Tadina* (PG 845), *Leopoldo Regis* (PG 847), *Loreto Danipog* (PG 849), *Sulpicio Fernandez* (PG 853), *PG 842, PG 848, PG 851, PG 852*

Ten fast coastal patrol craft of South Korean Sea Hawk class (Type PK 181) built by Korea Tacoma in Hyundai 1975–78 and transferred to Philippines 19 June 1993. Some differences in armament. It is planned to name all these boats in due course. With this acquisition, it is likely that further construction of Halter PCF 70 type craft will be much reduced.

MINE WARFARE VESSELS

Ex-US BLUEBIRD class *coastal minesweepers*

Class (no, former no):
Zambales (PM 55, ex-*MSC 218*), *Zamboanga del Norte* (PM 56, ex-*MSC 219*)

Built by US yards specifically for transfer under MAP. Transferred 7 March 1956 and 23 April 1956. Wooden hull with non-magnetic metal fittings. Stricken in 1979.

Ex-US AGILE class *ocean minesweepers*

Class (no, former name):
Davao del Norte (PM 91, ex-*Energy* MSO 436), *Davao del Sur* (PM 92, ex-*Firm* MSO 444)

Built in 1953–54 and transferred on 5 July 1972. Designated PM (patrol minesweepers). Wooden hull. Stricken in 1976. To US for disposal 1977.

MISCELLANEOUS

Two Italian 68ft Rodriguez type patrol hydrofoils (28t, 38kts 400nm range, 1–20mm, 1–TT, 15 men): *Camiguin* (H 72), and *Siquisor* (H 73) completed at Messina April 1965. Deleted 1979.

Two Japanese 68ft Hitachi *Zozen* type patrol hydrofoils (32t, 37.8kts, 400nm range, MG fore and aft, 15 men): *Bontoc* (M 74) and *Balek* (M 75) built at Osaka in 1966. Stricken 1979.

Poland

The Polish Navy was re-established between 1945 and 1948 with the destroyer *Blyskawica*, three submarines, four minesweepers, two patrol craft and a sail training ship (all of Polish origin), three BYMS type minesweepers acquired in Britain and nine minesweepers, twelve submarine chasers and two MTBs from the Soviet Union in lieu of war reparations from the *Kriegsmarine*. The units thus assembled were insufficient for the defence of over 500km of coastline, and in 1947 a Naval Estimate was prepared for the construction of a 42,000t navy (eighteen submarines, twenty-four MTBs, nine escorts, forty-four patrol craft, sixty-six minesweepers), over 20 years. This programme reflected adoption of the Soviet naval doctrine of that time, which confined the navy to covering the flanks of large land armies with extensive use of minefields guarded by coastal artillery while attacks on enemy communications would be executed by small combatants and naval aviation. The 1947 estimate evidently exceeded what was possible from a war-shattered economy and was substituted by a 1950–55 plan for construction of five MTBs, five patrol craft, five minesweepers and five minesweeping boats. This plan was abandoned in 1950 by the Soviet officers in command, who saw the Polish Navy as a local detachment of the Soviet Baltic Fleet. In 1951 the 'Plan for Reinforcing Defences' was authorised, proposing intensive modernisation of the army, air force and AA defence. The navy's role was confined to the defence of the main base at Gdynia, and further development was restricted to the placing in service of suitable ships. As a temporary measure, a landing craft flotilla was formed, comprising former German and US boats, and the ageing destroyer *Burza* was modernised. The necessity of replacing ageing naval craft was only acknowledged four years later in the 'Military Development Plan for 1955–60', which proposed increasing the navy to five destroyers, nine frigates, twelve submarines and ninety-six MTBs (all to be imported from the Soviet Union), in addition to sixty minesweepers and eight submarine chasers which were to be built locally.

This enforced rearmament programme brought about a sharp decline in the national economy, the first sign of which was an acute food shortage in 1952. However, only the death of Stalin and the easing of tension, both internationally (at the Geneva conference of July 1955) and internally after the June 1956 riots in Poznań made it possible to halt both the Sovietisation of Poland and the rearmament programme. This resulted in the withdrawal of Soviet officers from certain command posts and the return of Polish military uniforms. The military expansion programme was halted, though by then the navy had already purchased from the Soviet Union two *Skory* class destroyers, six 'MV' class submarines and twenty-seven fast craft, while twelve 'T43' type and seven 'TR40' type minesweepers were built locally under Soviet licence.

The Khrushchev regime forced Poland to undertake a new armaments programme; approved in March 1961 this provided for, within a period of seven years, a navy comprising a destroyer flotilla (three ships), a missile boat flotilla (six craft), a submarine brigade (seven boats), an MTB brigade (nineteen boats), a subchaser flotilla (eight boats), a minesweeper brigade (twenty-four craft), a river minesweeper flotilla (seven boats) and a landing ship brigade (twenty-eight ships). As well as the minesweepers, the Polish shipbuilding industry was to construct the subchasers, MTBs, landing ships and auxiliary ships. The desired force (at least on paper) was reached by the late 1960s, but numerous craft of this force needed replacement.

The decline of the Polish planned economy, burdened with military expenditure and deprived of modern technology, created a crisis in the second half of the 1960s and an explosion of social unrest in December 1970. The ailing economy caused delays in the replacement of ageing ships and these were made worse by the attitude of the Polish Ministry of Defence, which emphasised the development of the army and air forces designated for participation in Warsaw Pact strategic planning. Therefore the basic combat potential planned for the Polish Navy in 1961 was never reached and at the end of the decade the Polish Navy comprised one 'Kotlin SAM' class guided missile destroyer, four 'Whiskey V' class submarines, fifty small surface combatants, twenty-three landing ships and twenty-four minesweepers though some of the ships were by that time obsolete and the naval aviation division had not been modernised. Thus the Polish Navy, which had been the largest of the non-Soviet naval forces of the Warsaw Pact, lost this position to East Germany in the early 1980s. In 1980 a modernisation programme aimed at replacement of the majority of combat ships was commenced: this included the acquisition of submarines, surface escorts, FACs, MCVs and assault force craft. The surface ships were to be built locally and to be equipped with Soviet sensors and weaponry. However, a wave of strikes which swept the country in the summer of 1980 led to the creation of political organisations independent of the Communist Party, and the latter was able to regain control only after the imposition of martial law 18 months later. These events caused a further decline in the frail Polish economy and enforced a reduction in military expenditure. The naval programme was therefore pursued at a much slower pace. The first new naval vessels to emerge from local shipbuilders were the 'Noteć' class GRP minehunters, but a major element of the programme – a class of seven surface escorts – proved to be a failure as only one, *Kaszub*, was built, and this was not operational until 1990. Therefore after the withdrawal of the 30 year-old guided missile destroyer *Warszawa* in 1986, the Poles were unable to contribute a major surface ship to the Warsaw Pact Baltic Joint Squadron, until a 'Kashin mod' class ship was lent by the Soviet Union in January 1988. Construction of FAC(M)s was not started until 1988 in co-operation with East Germany, and only the replacement landing craft was pursued more vigorously; even so, only five of the *Lublin* class have been completed. Additional purchases from the Soviet Union proved a considerable extra strain on the economy. They included four 'Tarantul I' class FAC(M)s replacing some of the dozen 'Osa I' class boats. Noteworthy was the transfer of the 'Kilo' class submarine *Orzel* in 1986, and the smaller and cheaper 'Foxtrot' class followed.

Following economic restraints brought about as a result of martial law, the Polish economy once again showed signs of crisis. Hardships suffered by the people caused much social unrest in the summer of 1988, and the Communist Party, discouraged by Moscow's unwillingness to compromise its new image by supporting an unpopular regime, decided to negotiate with leaders of the hitherto illegal opposition. This resulted in the peaceful removal of communist rule after elections in June 1989 and triggered a similar process in other eastern European countries.

Poland was the first country to join the 'Partnership for Peace' programme and the most eager to join NATO, making the best possible use of the time given by the temporary weakness of the Russian Empire. An estimate to create a balanced and independent naval force was drawn up and provided for five submarines, twelve surface escorts, eighteen FACs, twenty MCVs and a naval aviation force of eighty aircraft. This programme was however beyond the capacity of the Polish economy, driving forcefully from the centralised system to the open market. The Polish Navy in 1995 comprised, on paper, three submarines, a missile destroyer, a surface escort, thirteen FAC(M)s, nineteen subchasers, twenty-four minesweepers and five landing ships. However most of these ships are without practical combat value either because of obsolescence (destroyer, submarines, subchasers) or lack of weapon systems (escort, FACs). The only units of practical value are one 'Kilo' class submarine, four 'Tarantul' class FACs, five landing ships and sixteen MCVs. The Coast Guard was separated from the navy in 1991. The fleet headquarters are at Gdynia, and bases were at Gdynia Oksywie (destroyer *Warszawa*, submarines, FACs), Hel (*Kaszub*, 'Obluze' class subchasers, ocean minesweepers and 'Mamry' class MCVs), Swinoujscie (landing force and 'Gardno' class MCVs) and Kolobrzeg ('Pilica' class subchasers). The naval aviation comprises 34th Fighter Wing (Gdynia Babie Doly) of thirty-eight MIG-21 fighters, four TS-11 Iskra training aircraft, two An-2 aircraft and two Mi-2 'Hoplite' helicopters, 7th Training Wing (Siemirowice) of fifteen TS-11, eight An-2 and two An-28 aircraft, 40th Helicopter Squadron (Darlowo) of fourteen Mi-14 'Haze' and five Mi-2 'Hoplite', 18th Despatch and Rescue Squadron of four RW Anakonda, three W-3 Sokol, ten Mi-2 helicopters and two An-28 aircraft.

FLEET STRENGTH 1947

DESTROYERS

Name	Launched	Disp	Fate
Wicker class			
BURZA	16.4.29	1540t	BU 1977
Grom class			
BLYSKAWICA	1.10.36	2144t	Preserved 1976

Wicker class
Burza (N 52, 52, 272) served with the Polish Naval Detachment in Britain from 1939 to 1946. She was handed over to the RN in 1946 and not returned to Poland until 1951 because of her poor condition. Rebuilt 1952–55 in the Stocznia im Komuny Paryskiej yard, Gdynia, to serve as an AA defence ship, she was rearmed with 4–100mm (4×1), 8 (later reduced to 6–37mm) (4×2, later 2×2 + 2×1), 4 DCT and 1 double DC rack. The bridge was widened, the two after funnels were trunked into a single uptake, the TTs were removed and a large fire control platform was built up amidships. *Burza* was stricken from the active list on 24 February 1960 and converted into a museum ship. In 1976 she was replaced in this role by *Blyskawica* and subsequently scrapped.

Grom class
Blyskawica (N 51, 51, 271) served during the war with the Polish Naval Detachment in Britain and was handed over to the RN in 1946. She was returned to Poland (Gdynia) in 1947, overhauled 1949–50 and had her 4in barrels replaced by 100mm guns to allow Soviet ammunition to be used. In addition she received 10–37mm (4×2, 2×1), 3–21in TT (1×3), 2 DCT (replaced by four new-pattern units during a 1957–61 refit) and 2 DC racks. Served as navy flagship until the early 1960s.

SUBMARINES

Name	Launched	Disp	Fate
Wilk class			
RYS	22.4.29	980t	BU 1954
WILK	12.4.29	980t	BU 1952
ZBIK	14.6.29	980t	BU 1954
Orzel class			
SEP	17.10.38	1100t	Stricken 1969

Wilk class
Minelaying submarines. *Rys* (B 12) and *Zbik* (B 13) were interned by Sweden in September 1939 and returned to Gdynia on 25 October 1945. *Wilk* ('Wolf') succeeded in reaching British waters in 1939 and remained there until 1951 when she was towed to Poland and scrapped owing to her poor condition.

Orzel class
Sep (B 11, P 11, 291) returned to Gdynia from Swedish internment on 25 October 1945. During a 1946–47 refit the unfinished work was completed and her 105mm gun was replaced with a 100mm weapon to take Russian ammunition. The gun was removed in the late 1950s and *Sep* was subsequently relegated to training duties.

MINESWEEPERS

Name	Launched	Disp	Fate
Jaskolka class			
CZAJKA	10.4.35	183t	Stricken 1970
MEWA	10.1.35	183t	Stricken 1970
RYBITWA	26.4.35	183t	Stricken 1970
ZURAW	22.8.38	185t	Hulked 1971
Ex-Soviet T 371 class	1944–45	146t	Stricken *c*1959

Jaskolka class
Returned from German service on 12 March 1946 with rebuilt sterns and bridges and an armament comprising 4 to 8–20mm guns. The boats returned to their previous names while *Zuraw* was reclassified as a surveying ship in 1948 (renamed *Kompas* in 1951) and rebuilt 1959–63. Her armament was removed and a raised forecastle and tripod mast were fitted. The remaining boats served as patrol craft from July 1949 (with numbers *D 45–D 47*, since 1960 *325–327*) and were later rearmed with 2–37mm (1×2), 2–0.5in MG (1×2) and 2 DC racks.

Blyskawica c 1955 — Marek Twardowski Collection

Ex-Soviet T 371 class
The Leningrad-built craft *T 225, T 228, T 231, T 241, T 243, T 244, T 246, T 465* and *T467* transferred to Poland on 5 April 1946 to represent her quota of German ships seized by the Allies. They were renamed traditionally after birds and given two-letter bow identification codes, later changed to numbers: *Albatros* (AL, later T 21), *Czapla* (CP, later CL, T 11), *Jaskolka* (JK, T 14), *Jastrzab* (JS, JT, T 23), *Kania* (KN, T 25), *Kondor* (KD, T 12), *Kormoran* (KR, KM, T 24), *Krogulec* (KG, T 13) and *Orlik* (OL, OK, T 22). They were in service until 1958–59, when *Albatros, Kzapla* and *Kondor* were sold to an inland waterways enterprise while the others were relegated to auxiliary duties.

MOTOR BOATS

Ex-Soviet D 3 type MTBs
TP 1 and *TP 2* formerly *TK 76* and *TK 116* (32t) transferred together with the minesweepers. Launched *c*1944 and rebuilt in 1954 and used as training boats later on, renumbered *ST* pennant letters were changed to *KT 81* and *82* respectively. Stricken 1958–59.

Ex-Soviet 'OD-200' type submarine chasers
Bezwzgledny (BW), *Bystry* (BS), *Dziarski* (DR), *Dzielny* (DL), *Karny* (KR), *Niedoscigly* (ND), *Nieuchwytry* (NW), *Odwazny* (OW), *Smialy* (SM, SL), *Sprawny* (SP), *Szybki* (SZ), all with two-letter bow markings, later replaced by numbers S 51–56, S 61–65.

Formerly *MO 325* to *MO 329, MO 368* to *MO 372, MO 552* transferred on 5 April 1946 with the other boats. Launched 1944, 32t. All stricken 1959.

Ex-Soviet BMO type submarine-chasers
Blyskawiczny (BL, later S 71) transferred from the Soviet Union on 5 April 1946, formerly *BMO 552.* Launched 1944 (74t). Stricken early 1960s.

Nieuchwytny class river patrol boat
The river AMGB *Nieuchwytny,* launched 1933 (38.5t), scuttled on 10 September 1939, raised by the Germans and commissioned as *Pionier.* Scuttled again in 1944, raised 1947 and recommissioned with the Polish Navy as *Okon.* During 1955–57 served with the Frontier Guard. Stricken 1957.

MISCELLANEOUS
Other small units included the frontier patrol boat *Hel* (former customs cutter *Batory,* renamed *7 Listopada* then later *KP 1*), four ex-German *Pionierlandungsboote* and ten (later increased to nineteen) ex-US LCP(L)s used for landing or patrol purposes as well as minor craft including nine Frontier Guard patrol boats adapted from fishing boats and commissioned during 1947–50. All these smaller craft stricken during 1957–62.

MAJOR SURFACE SHIPS

Ex-Soviet SKORYY class *destroyers*

No	Name	Builder	Acquired	Fate
53	GROM (ex-*Smetlivy*)	Zhdanov Yd	15.12.57	Stricken 1975
54	WICHER (ex-*Skoryy*)	Zhdanov Yd	29.6.58	Stricken 1975

Two units, which included the class prototype, were transferred from the USSR. Numbers changed to 273 and 274 respectively in 1960. Designated Project 30-bis.

Ex-Soviet 'KOTLIN SAM II' class *missile destroyer*

No	Name	Builder	Acquired	Fate
275	WARSZAWA (ex-*Spravedlivyy*)	Zhdanov Yd	25.6.70	Stricken 31.1.86

Built in Leningrad in 1956, and converted to missile destroyer in 1968. Flagship of the Polish Navy, decommissioned allegedly because of boiler breakdown. Replaced by the 'Kashin Mod' class missile destroyer of the same name. Designated Project 56E.

Grom c 1958 Marek Twardowski Collection

Ex-Soviet 'KASHIN MOD' class *missile destroyer*

No	Name	Builder	Acquired	Fate
271	WARSZAWA (ex-*Smelyj*)	61 Kommunar Yd	3.1.88	Extant 1995

Built at Nikolayev in 1969 and converted in the mid 1970s. Lent by the Soviet Navy after a lengthy refit at Leningrad, to replace the 'Kotlin SAM' missile destroyer of the same name. First Polish warship equipped with helicopter platform and towed VDS. Fleet flagship. Designated Project 61MP. Polish navigational radars SRN-7534 and SRN-207 installed in 1988. Purchased outright in 1993.

Kaszub as completed

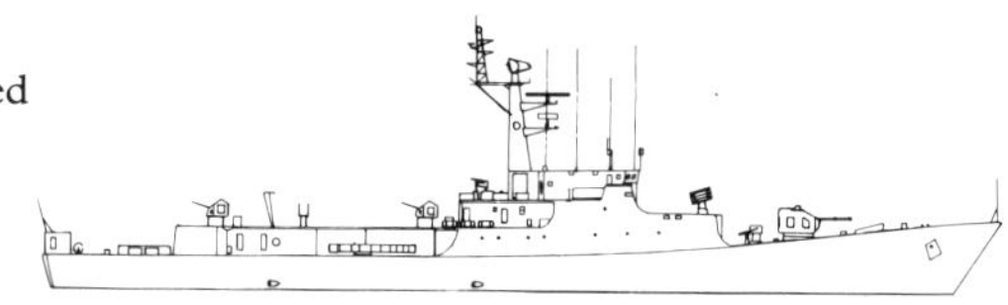

KASZUB *corvette*

Displacement: 1051t standard; 1183t full load
Dimensions: 270ft 2in × 32ft 9in × 9ft 2in
82.3m × 10.0m × 2.8m
Machinery: 4 shafts CODAD, 4 Sulzer-Cegielski 16 AS-V 25/30 diesels, 17,500bhp = 28kts. Range 2800nm at 14kts
Armament: 2 SA-N-5 Strela SAM (2 × 4), 1–76mm AK 176, 6–23mm ZU-23-2-M (3 × 2), 2–53cm TTs (1 × 2), 2 RBU-6000 ASM (2 x 12), 2 DC rails, chaff launchers
Sensors: Radar Strut Curve (MR-302), Bass Tilt (MR 123), SNR 7453 TSR 333 (navigational), Square Head, Salt Pot, sonar MR 322T, MG-16, MG-329M
Complement: 67

No	Name	Builder	Laid down	Launched	Comp	Fate
240	KASZUB	Polnocna Yd, Gdansk	9.6.84	4.10.86	15.3.87	Extant 1995

NATO codename 'Balcom 6', Polish designation Project 620. Hull design based on the Soviet 'Grisha' class but with diesel engines instead of turbines. First ocean-going combat warship ever built in Poland. In all seven ships were planned but series construction was postponed because of technical problems and budgetary cuts. New modified design is scheduled to start construction. The planned twin 57mm (ZIF-72) gun mounting was not fitted aft because of stability problems, while installation of the SA-N-4 missile system with Peel Group director was abandoned because of the system's propensity to failure in ships of this size due to vibration at high speeds. The present armament is entirely provisional, with the 76mm gun lacking its fire control radar Bass Tilt (MR-123). Because of technical problems she was rarely seen at sea before 1990. In 1990 *Kaszub* was operated by the Coast Guard, being transferred to the navy in 1991.

Kaszub 7.6.94 L & L van Ginderen

SUBMARINES

Ex-Soviet 'MV' class *submarines*

Class (no):
Kaszub (M 100, P 100, 301), *Kujawiak* (M 104, P 104, 305), *Kurp* (M 105, P 105, 306), *Krakowiak* (M 102, P 102, 303), *Mazur* (M 101, P 101, 302), *Slazak* (M 103, P 103, 304)

Four, built 1948–52 in Soviet yards, were delivered to Poland in 1954 and two in 1955. The last one reportedly ran aground in the late 1950s and was badly damaged but repaired later and supposedly recommissioned as *Mazowsze.* The 45mm gun was removed during 1958–59 from all units. All stricken by February 1970.

Ex-Soviet 'WHISKEY V' class *submarines*

Class (no. fate)

Orzel (292, stricken 30.12.83), *Sokol* (293, stricken 12.12.87), *Kondor* (294, stricken 1986), *Bielik* (295, stricken 1989)

Built at Leningrad in the late 1950s and transferred to the Polish Navy on 30 December 1962, 24 October 1964 and 1969 (last two), as replacements for the ex-Soviet *M* class coastal submarines. Designated Project 613. They were used for northern and Norwegian sea deployments. Replaced by the 'Kilo' and Foxtrot' classes.

Ex-Soviet 'KILO' class *attack submarine*

No	Name	Builder	Acquired	Fate
291	ORZEL	Sudomekh Yd, Leningrad	21.6.86	Extant 1995

Built for the Polish Navy as the lead ship of a group of four to replace obsolete 'Whiskey' class. Further transfers were postponed after delivery of *Orzel* because of financial problems Designated Project 877E. Former Soviet codename 'Varshavianka'.

Orzel Jun 1994 L & L van Ginderen

Ex-Soviet 'FOXTROT' class *attack submarines*

No	Name	Builder	Acquired	Fate
292	WILK	Sudomekh Yd, Leningrad	3.11.87	Extant 1995
293	DZIK	Sudomekh Yd, Leningrad	10.12.88	Extant 1995

Loaned from the Soviet Navy to maintain strength of the submarine force after postponement of purchase of the 'Kilo' class boats. Designated Project 647. *Wilk* refitted 1993.

AMPHIBIOUS WARFARE VESSELS

Ex-German MFP type *landing ships*

Class:

BDD 1–3, *BDD 5–7* (later *863–865*)

Six were pressed into the navy in 1950 (two hitherto in civilian service, four incomplete, one damaged). The pennant letters were later changed to *ODD* (*ODD* stands for *Okret Desantowy Duzy* = landing ship, large). Stricken 1957–63.

Ex-Italian MZ type *landing ship*

One recommissioned by the Polish Navy in 1950 as *BDD 4*, later changed to *ODD 4*. Stricken *c*1960.

Ex-US LCT 5 type *landing ships*

Class:

BDS 50–60

Eleven acquired from the civilian service in 1950. The pennant letters were changed later to *ODS* (*ODS* stands for *Okret Desantowy Sredni* = landing ship, medium). *BDD 56* was converted to salvage tender in the early 1950s, the remainder were scrapped during 1957–60.

GDYNIA class *landing ships*

Displacement: 775t standard
Dimensions: 239ft 6in × 28ft 3in × 6ft 1in
73.0m × 8.6m × 1.9m
Machinery: 2 shafts, 2 diesels, 4400bhp = 19kts. Range 3300nm
Armament: 2–30mm AK-230 (1 × 2), 2 MW-18M RL (2 × 18)
Sensors: Radar Don 2, Square Head, High Pole
Complement: 35

Class (no):

Gdynia (888), *Glogow* (889), *891–893*, *San* (894), *895–897*, *Warta* (898), *899*

Built by the Polnocna Yard, Gdansk during 1965–68 under the designation 'Project 770D'. NATO designation 'Polnocny A'. Modified versions of this class were built in a large series for the Soviet Union as 'Project 771'. Can carry up to five tanks and 120 troops. SA-N-5 Strela SAM quadruple launchers were added on some in the 1980s. All deleted 1991–92.

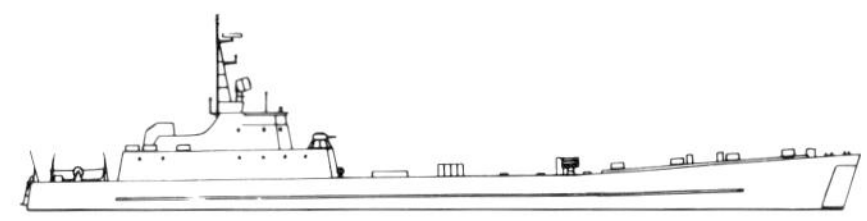

Lenino as completed

LENINO class *landing ships*

Displacement: 800t standard
Dimensions: 242ft 9in × 28ft 3in × 6ft 1in
74.0m × 8.6m × 1.9m
Machinery: 2-shafts, 2 diesels, 4400bhp = 19kts. Range 3300nm
Armament: 4–30mm AK-230 (2 × 2), 2 MW-18M RL (2 × 18)
Sensors: Radar Drum Tilt (MR-104), Don 2, Square Head, High Pole
Complement 40

Class (no):

Lenino (801), *Studzianki* (802), *Siekierki* (803), *Brda* (804), *Polichno* (805), *Rablow* (806), *Janow* (807), *Narwik* (808), *Odra* (809), *Falaise* (810), *Budziszyn* (890)

Built at the Polnocna Yard, Gdansk, 1968–71 as Project 771. NATO designation 'Polnocny B'. Modified version of *Gdynia* class. Can carry up to five tanks and 120 troops. SA-N-5 Strela SAM quadruple launchers were added on some in the 1980s. All deleted 1991–92.

GRUNWALD *command ship*

Displacement: 1253t full load
Dimensions: 266ft 8in × 30ft 6in × 7ft 9in
81.3m × 9.3m × 2.4m
Machinery: 2 shafts, 2 Type 40DM diesels, 5000bhp = 17kts. Range 2600nm at 12kts
Armament: 2–30mm AK-230 (1 × 2), 2 MW-18M RL (2 × 18)
Sensors: Radar Drum Tilt (MR-104), Don 2, Square Head, High Pole
Complement: 45

Class (no):

Grunwald (811)

Project 776, NATO designation 'Polnocny C', built by the Polnocna Yard in Gdansk 1971–73 as the prototype of a class of twenty-seven ships built for export to the Soviet Union (nine), Libya (four), India (eight) and Iraq (five). Fitted as a command ship for a flotilla of landing ships, she can accommodate fifty-four staff and in addition to extra communications and data transmission facilities was fitted with Decca navigation system. Fitted with four quadruple SA-N-5 Strela SAM launchers in 1980s. Extant 1995.

Grunwald 7.9.92 L & L van Ginderen

LUBLIN class *landing ships*

Displacement: 1400t standard; 1745t full load
Dimensions: 314ft 4in oa, 265ft 9in pp × 35ft 5in × 7ft 6in
95.8m, 81.0m × 10.8m × 2.3m
Machinery: 3 shafts, Sulzer-Cegielski 6ATL 25 diesels, 2900bhp = 16.5kts. Range 1400nm at 16kts
Armament: 8–23mm (4×2), up to 134 mines
Sensors: Radar SRN 7453, SRN 443, Square Head
Complement: 37

Class (no):
Lublin (821), *Gniezno* (822), *Krakow* (823), *Poznan* (824), *Torun* (825)

NATO code 'Balcom 11', designed and built during 1988–91 by the Polnocna Yard, Gdansk (Project 767). Design work started in 1979 to achieve replacement of the 'Polnocny' class combat landing ships, while the prototype was completed ten years later without features necessary for typical offensive combat landings. The 12,000-strong 7th Sea Landing Division specially trained for assault landings has been reduced to a brigade size of 5000 men and should no longer be equipped with 'PT 76' amphibious tanks. The prototype was completed in August 1989 and commissioned on 12 October 1989. Officially classed as transport/minelaying ships, these were built as replacements for the 'Polnocny' class landing ships. Roll-on/roll-off type ferry design with a continuous, open hold deck, spanned by the superstructure; fitted with a bow ramp with folding end and an aft ramp. Can carry ten tanks or 369t of military lift. Lack of support rocket launchers is significant. A landing force detachment of 135 men may be accommodated. Sixth unit not built due to financial problems. All extant 1995.

Ex-US LCM 3 type *landing craft*

Class:
BDM 100–102 (later *880–882*)

Three craft of this type were acquired from civilian service in 1950. The pennant letters were later changed to *ODM* (*ODM* stands for *Okret Desantowy Maly* = landing ship, small). Possibly stricken in the early 1960s.

'EICHSTADEN' class *landing craft*

Displacement:	25t standard; 30t full load
Dimensions:	52ft 6in × 13ft 1in × 2ft 8in *16.6m × 4.0m × 0.8m*
Machinery:	2 shafts, 2 diesels, 900bhp = 18kts
Armament:	None
Complement:	3

Class:
857–871

Fifteen craft of the LCP(L) type were built by the Naval Yard in Gdynia 1962–64. Designated Project 719 and modernised in 1969 by resiting of the bridge and plating up of the hatch. Firing slots are provided forward to enable troops to use light weapons when closing a shore. Deleted 1990–93.

'MARABUT' class *landing craft*

Displacement:	60t full load
Length:	70ft × 13ft 9in × 3ft 3in *21.3m × 4.2m × 1.0m*
Speed:	2 shafts, 2 diesels = 10kts
Armament:	2–14.5mm MG (1×2)

Class:
516–519 (later *872–875*)

Built by the Naval Yard, Gdynia, in 1973. Designated Project 719. Can carry a tank or thirty infantry. Plastic hull, not successful design, stricken 1987.

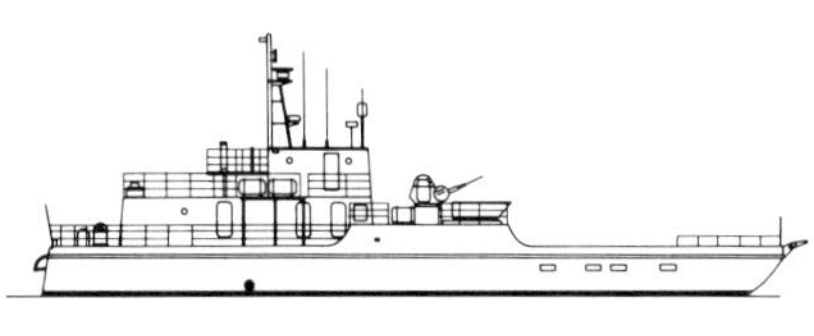

Deba class

DEBA class *landing craft*

Displacement:	164t standard; 170t full load
Dimensions:	122ft × 23ft 10in × 5ft 6in *37.2m × 7.3m × 1.7m*
Machinery:	3 shafts, 3 Type M 401A diesels, 2900bhp = 20kts
Armament:	2–23mm ZU-23-2 (1 × 2), 2 PW-LWD line-charge projectors
Sensors:	Radar SRN 207A (navigational), IFF Chrom-K, navigational system AD-2, Garus. Echo-sounder 4-501.
Complement:	12

Class:
851–853

Built during 1987–91 by the Naval Yard in Gdynia under designation Project 716 as replacements for Project 709 ('Eichstaden' class). Previously planned class of twelve was reduced to four of which only three were completed because of budgetary cuts. Can carry up to two amphibious tanks or fifty fully-equipped troops. All extant 1995.

SMALL SURFACE COMBATANTS

Ex-Soviet 'P 6' type *fast attack craft (torpedo)*

Class:
KT 71–80, KT 83–92

Twenty boats transferred from the Soviet Union November 1956 to 1958. (*KT* stands for *Kuter Torpedowy* = MTB). All boats were renumbered *401* to *420* in the early 1960s and twelve were deleted during 1973–75. At least three of the remaining boats were converted into target boats. Only these three were extant in 1995.

Ex-Soviet 'OSA I' class *fast attack craft (missile)*

Class (no, fate):
Hel (421, transferred to Coast Guard), *Gdansk* (422, transferred to Coast Guard), *Gdynia* (423, converted to Coast Guard patrol boat *SG 301* in 1990), *Kolobrzeg* (424, converted to Coast Guard patrol boat *SG 304* in 1990), *Szczecin* (425, converted to Coast Guard patrol boat *SG 302* in 1990), *Elblag* (426, converted to Coast Guard patrol boat *SG 303* in 1990), *Puck* (427, extant 1995), *Ustka* (428, extant 1995), *Oksywie* (429, extant 1995), *Darlowo* (430, extant 1995), *Swinoujscie* (431, extant 1995), *Dziwnow* (432, extant 1995), *Wladyslawowo* (433, extant 1995)

Soviet Project 205, transferred from the Soviet Union during 1963–68 (first batch of eight) and 1971–75 (remaining five). Boats converted to the Coast Guard patrol craft had their missile launchers as well as Soviet-made radars removed while the twin 30mm AK-230 gun mounting was replaced with a twin 25mm mounting.

BLYSKAWICZNY *fast attack craft (torpedo)*

Displacement:	94t standard
Dimensions:	79ft 8in × 25ft 7in × 4ft 11in *24.3m × 7.8m × 1.5m*
Machinery:	4 shafts CODAG, 4 Type M-50 diesels, 4800bhp, 1 Type TM-1 gas turbine, 7200shp = 55.8kts
Armament:	2–25mm 2-M-23 (1 × 2), 4–53cm TT (4 × 1)
Sensors:	Radar Pot Drum (Reja)

Class (no):
Blyskawiczny (451)

Design work originated in late 1950s. Designated 'Project 663D', built by the Polnocna Yard, Gdansk as experimental boat for development of a new class of fast attack craft. Completed on 11 June 1965, deleted in the 1970s.

BITNY class *fast attack craft (torpedo)*

Displacement:	94t standard; 128t full load
Dimensions:	81ft 4in × 25ft 7in × 4ft 11in *24.8m × 7.8m × 1.5m*
Machinery:	4 shafts CODAG, 4 Type M-50 diesels, 4800bhp, 1 Type TM-1 gas turbine, 7200shp = 59.2kts
Armament:	2–30mm AK-230 (1 × 2), 4–53cm TT (4 × 1), mines
Sensors:	Radar Pot Drum (Reja)
Complement:	11

Class (no):
Bitny (452), *Dzielny* (453), *Dziarski* (454), *Bystry* (455), *Szybki* (456), *Sprawny* (457), *Odwazny* (458), *Odporny* (459)

NATO designation 'Wisla'. Series version of the previous design, codename 'Project 664'. Very compact and highly-automated by Eastern bloc standards. Aluminium alloy hull and deckhouses, welding used throughout. On diesels they can attain 14kts. Gas turbine (developed from aircraft engines) was used for high speed, but range limited because of high fuel consumption. Hampered by vibration and excessive material fatigue caused by dynamic loads accumulated in forward part of the hull All deleted by 1984.

'PROJECT 665' *fast attack craft (missile)*

Displacement:	460t
Dimensions:	168ft 4in *51. 3m*
Machinery:	4 shafts, 4 Type M-504B diesel engines, 20,000bhp = 36kts
Armament:	4 SSM, 1 57mm AA (1 × 2), 2–23mm AK-230 (2×2)

Design of new class of FAC was prepared by Polnocna Yard in Gdansk (57mm gun was to be replaced by 76mm when available) and construction started in the late 1970s, but abandoned in 1978 . The reason was refusal by the Soviet Union to deliver the missile systems, because of the Warsaw Pact decision to emphasise naval shipbuilding in East Germany.

Ex-Soviet 'TARANTUL' class *missile corvettes*

Class (no):
Gornik (434), *Hutnik* (435), *Metalowiec* (436), *Rolnik* (437)

Built during 1982–87 by the 341 Yard, Rybinsk (USSR), as replacements for the 'Osa' class FAC(M). Designated Project 1241RE. First pair delivered in 1983–84, second pair in 1988–89. Plans to acquire a further pair from the former East German Navy have been abandoned. All extant 1995.

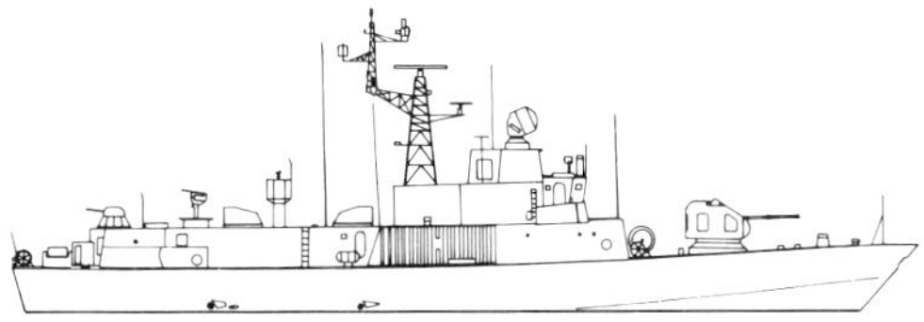

Piorun 1995

ORKAN class *fast attack (missile)*

Displacement: 311t standard; 369t full load
Dimensions: 163ft 5in oa, 147ft 8in wl × 28ft 7in × 7ft 3in
49.8m, 45.0m × 8.7m × 2.2m
Machinery: 3 shafts, 3 Type M520 diesels, 16,200bhp = 39kts. Range 2400nm at 13kts
Armament: 1 SA-N-5 Strela SAM (1 × 4), 1–76mm AK 176, 1–30mm AK 630 Gatling, 4 chaff launchers (4 × 8)
Sensors: Radar Bass Tilt (MR 123), 2 SNR 7453 (navigational), Square Head, Salt Pot
Complement: 33

Class (no):
Orkan (421), *Piorun* (422), *Grom* (ex-*Huragan*) (423)

'Balcom 10' design, Polish designation Project 660, East German designation Project 151, designed for the East German Navy as the 'Sassnitz' class. Six ships of this type have been planned for the Polish Navy, instead of the abandoned 'Project 665' class. Following preparation works in Polnocna Yard in Gdansk, in order to speed delivery, the hulls under construction in the former East German Peenewerft Yard, Wolgast (launched 1990) were purchased and completed during 1992–95 by the Polnocna Yard, Gdansk. Delivery of further hulls have been frustrated. The two quadruple SS-N-25 (Soviet codename Uran) SSM launchers planned were not delivered. At present purchase of Western systems is being considered (Swedish Robot RBS 15), but plans have not been finalised because of financial problems.

BITNY *submarine chaser*

Displacement: about 50t
Dimensions: 92ft 2in pp × 15ft 5in × 4ft 3in
28.7m × 4.7m × 1.3m
Machinery: 2 Hispano-Suiza diesels, 1800bhp
Armament: 2–37(?)mm, 2 DC racks

Built 1951–53 on the hull of the ex-German *Polizeikampfboote* which was found in Torun after the war. Completed in Gdynia and commissioned with number S 72, used for experimental purposes, fate unknown.

Ex-Soviet 'KRONSTADT' class *submarine chasers*

Class (no):
Czujny (361), *Nieugiety* (362), *Zawziety* (363), *Zwrotny* (364), *Zwinny* (365), *Zreczny* (366), *Wytrwaly* (367), *Grozny* (368)

First four units acquired in 1955; last four on 15 December 1957. Designated Project 122 bis. Deleted 1969–72. Originally numbered DS 41–DS 48 respectively.

OP 201 *large patrol craft*

Displacement: 140t
Dimensions: 128ft 7in × 18ft 8in × 4ft 11in
39.2m × 5.7m × 1.5m
Machinery: 2 shafts, 2 diesels = 20kts
Armament: 2–37mm, 2 DC racks
Sensors: Radar Skin Head

Built on the hull of the ex-German *R-boat* found after the war in Swinoujscie and launched on 6 October 1956. Probably belonged to the *R 272* series; she retained her armoured bridge. Used by Frontier Guard, deleted *c*1980.

'OKSYWIE' class *large patrol craft*

Displacement: 130t standard; 150t full load
Dimensions: 114ft 10in × 19ft × 5ft
35.0m × 5.8m × 1.5m
Machinery: 2 shafts, 2 diesels, 2400bhp = 19.5kts
Armament: 2–37mm (2×1), 4–12.7mm MG (2×2), 2 DC racks
Sensors: Radar Skin Head; sonar Tamir

Class:
OP 202–OP 204

Improved version of the *OP 201* type (*OP* stands for *Okret Patrolowy* = patrol craft). Launched 1957 and later reconstructed and fitted with the new mast and RN 231 type radar. Renumbered *302* to *304*; used by Frontier Guard. Deleted 1980–81.

'GDANSK' class *large patrol craft*

Displacement: 193t full load
Dimensions: 129ft 3in × 19ft 4in × 5ft 7in
39.4m × 5.9m × 1.7m
Machinery: 2 shafts, 2 Type 6KVD-43 diesels, 1200bhp = 19kts
Armament: 2–37mm (2 × 1), 2–12.7mm MG (1 × 2), 2 DC racks
Sensors: Radar Skin Head (Zarnica); sonar Tamir
Complement: 33

Class (no):
Neptun (OP 205), *Saturn* (OP 206), *Pegaz* (OP 207), *Orion* (OP 208), *Kastor* (OP 209), *Cirus* (OP 210), *Jupiter* (OP 211), *Wega* (OP 212), *Minor* (OP 212)

Built by the Naval Yard, Gdynia during 1959–60 under designation 'Project 902' as a development of the 'Oksywie' class. Reconstructed by the Naval Yard, Gdynia during 1971–75, and fitted with new mast and SNR 231 radar. Renumbered 311–319. Used by the Frontier Guard. Deleted 1983–86.

'OBLUZE' class *large patrol craft/submarine chasers*

Displacement: 210t standard; 238t full load
Dimensions: 135ft 4in × 20ft 8in × 6ft 3in
41.3m × 6.3m × 1.9m
Machinery: 2 shafts, 2 Type 40 DM diesel engines, 5000bhp = 24kts. Range 1200nm at 16kts. Fuel 24t
Armament: 2–30mm 30 (1 × 2), 2 DC racks (12 DC), mines
Sensors: Radar Pot Head (patrol craft only), SRN 231, Drum Tilt (MR 104) (subchasers only), Square Head (subchasers only), High Pole; sonar hull mounted active attack high frequency MG-1 1–265
Complement: 12

Class (no):
Patrol craft: *Fala* (OP 321), *Szkwal* (OP 322), *Zefir* (OP 323), *Zorza* (OP 324), *Tecza* (OP 325)
Subchasers: *Wytrwaly* (351), *Grozny* (352), *Zawziety* (353), *Zwinny* (354), *Zwrotny* (355), *Zreczny* (356), *Nieugiety* (357), *Czujny* (358)

Polish designation Project 912. Developed from the 'Gdansk' class with steel hulls and aluminium alloy deckhouses. Built by the Naval Yard in Gdynia in two series as patrol craft of the Frontier Guard (1965–68) and, following delivery of the MR-104 radar systems from the Soviet Union, as subchasers (1969–72) – designated Project 912M. Some of the patrol boats have had their aft 30mm twin mounting removed and have been renumbered SG 321–SG 325. The subchasers are without practical combat value now and are unable to detect modern submarines. All extant 1995. A replacement class, designated Project 924, was developed by the Polnocna Yard, Gdansk in 1989, but further works have been postponed until an alternative armament supplier to the former Soviet Union has been found.

'KAPER' class *large patrol craft*

Displacement: 376t full load
Dimensions: 139ft 5in × 27ft 7in × 9ft 10in
42.5m × 8.4m × 3.0m
Machinery: 2 shafts, 2 Cegielski-Sulzer 8ATL25/30 diesels, 4720bhp = 17.6kts. Range 2800nm at 14kts
Armament: None
Sensors: Radar navigational
Complement: 11

Class:
SG 311, SG 312

Type SKS 40. Fishery protection vessels, built in 1991 by the Wisla Yard in Gdansk. Operated by Coast Guard. Both extant 1995.

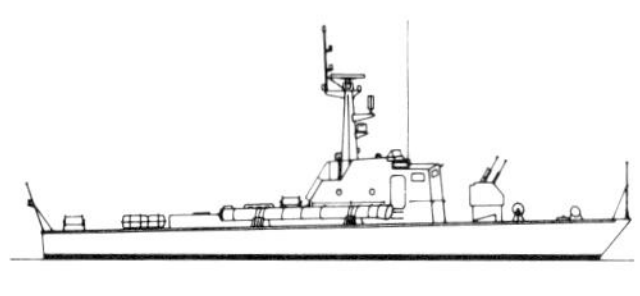

'Pilica' class
NB 1/750 scale

'PILICA' class *patrol craft/submarine chasers*

Displacement: 89t standard; 100t full load
Dimensions: 93ft 10in × 18ft 11in × 4ft 4in
28.6m × 5.8m × 1.3m
Machinery: 3 shafts, 3 type M50-F5 diesels, 3600bhp = 27.3kts. Range 430nm. Fuel 7t
Armament: 2–25mm 2-M-3 (1 × 2), (*166* to *176* 2–53cm TT)
Sensors: Radar RN 231, IFF; sonar dipping VDS (subchasers)
Complement: 15

Class:
Patrol craft: *161–165*
Subchasers: *166–176*

Polish designation Project 918. Built during 1973–74 by the Naval Yard in Gdynia as patrol craft (87t) and Project 918M built during 1973–82 as subchasers, the latter fitted with TTs and dipping sonar aft. All later rearmed with 2–23mm ZU-23-2. Patrol craft renamed *SG 161–SG 166* in 1991. All extant 1995.

KP 100 class *inland patrol craft*

Displacement: 27t full load
Dimensions: 69ft 3in × 8ft 10in × 4ft 3in
21.1m × 2.7m × 1.3m
Machinery: 2 shafts, 2 Skoda diesels, 360bhp = 14kts. Range 120nm
Armament: 1–12.7mm MG
Complement: 8

Class:
KP101–KP107

Built in Hungary at the Balaton-Fured Yard during 1953–54 for service with the Frontier Guard. Designated Project 101. Stricken 1974; five scrapped, two laid up.

'WISLOKA' class *coastal patrol craft*

Displacement: 40t full load
Dimensions: 76ft 5in × 16ft × 5ft 3in
23.3m × 4.9m × 1.6m
Machinery: 2 shafts, 2 diesels, 600bhp = 14kts. Fuel 2.5t
Armament: 2–12.7mm MGs (1×2)
Sensors: Radar SRN 231
Complement: –

Class:
KP141–KP152

Frontier Guard patrol craft built during the late 1970s as Project 90. Extant 1995.

MINE WARFARE VESSELS

Ex-US YMS type *minesweepers*

Class:
Delfin (DF, later T 31; ex-*BYMS 2211*), *Foka* (FK, later T 32; ex-*BYMS 2257*), *Mors* (MS, later T 33; ex-*BYMS 2282*)

Four *BYMS* motor minesweepers were purchased in Britain and three of them were commissioned on 19 April 1948 with the Polish Navy. Stricken in 1955–57.

Soviet 'T43' class *ocean minesweepers*

Class (no):
Zubr (601), *Tur* (602), *Los* (603), *Dzik* (604), *Bizon* (605), *Bobr* (606), *Rosomak* (607), *Delfin* (608), *Foka* (609), *Mors* (610), *Rys* (611), *Zbik* (612)

First three (601–603, 570t, 193ft 3in, 58m, 14kts) built under Soviet licence during 1955–58 by Stocznia im Komuny Paryskiej and designated Project 254K. The others were built during 1957–62 as modified design Project 254M (604–612, 600t, 193ft 7in, 59m, 15kts) to fit Polish made equipment. *Tur* refitted as a radar picket in the late 1970s, losing her aft 37mm mounting and sweeping gear for quadruped mast carrying the Big Net air warning antenna. All deleted 1987–93.

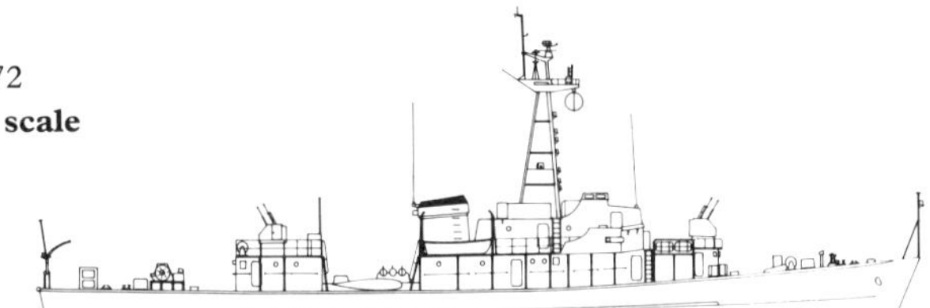

Tukan 1972
NB 1/750 scale

KROGULEC class *ocean minesweepers*

Displacement: 424t standard; 503t full load
Dimensions: 190ft 11in × 26ft 2in × 6ft 11in
58.2m × 7.9m × 2.12m
Machinery: 2 shafts, 2 FIAT 2312 SS diesel engines, 3600bhp = 18.4kts (some fitted with 2 Cegielski-Sulzer 6AL 25/30 diesel engines 2200bhp). Range 2000nm/17kts
Armament: 6–25mm 2-M-3 (3 × 2, see notes), 16 mines, 2 DCTs (12 DC). Sweeps: MT-2 contact, TEM-52M electromagnetic, BGAT acoustic
Sensors: Radar RN-231, IFF; sonar hull mounted high frequency MG-11-265A
Complement: 49

Class (no, fate):
Orlik (613, deleted 1989), (614, deleted 1989), *Jastrzab* (615, deleted 1990), *Kormoran* (616, deleted 1993), *Czapla* (617, deleted 1990), *Albatros* (618, extant 1995), *Pelikan* (619, deleted 1993), *Tukan* (620, extant 1995), *Flaming* (621, extant 1995), *Rybitwa* (622, extant 1995), *Mewa* (623, extant 1995), *Czajka* (624, extant 1995)

Designation Project 206F, planned as series of fifteen ships with construction scheduled for 1957. Design modified as compared with the T43 type with more compact layout, flush deck and close range weapons added. However because of financial and technical problems (short delivery of Fiat engines) only twelve ships were built during 1962–67 by the Gdynia Yard. Since 1980s some have two twin 23mm AA ZU-23-2 fitted insted of old fashined 25mm guns and two quadruple SA-N-5 SAM launchers added on bridge wings. Completely outdated in 1990s, to be replaced by Project 257 (*Lodowka* class) minehunters when budget resources would allow.

GOPLO class *coastal minesweepers*

Displacement: 203t standard; 225t full load
Dimensions: 125ft 4in × 22ft 11in × 5ft 10in
38.2m × 7.0m × 1.8m
Machinery: 2 shafts in nozzles, 2 Type M 401A diesels, 2000bhp = 14kts. Range 1100nm at 9kts
Armament: 2–23mm ZU-23-2 (1 × 2), 6–24 mines
Sensors: Radar RN 231 (navigational), IFF, navigational system Szop; sonar SHL 100, SHL 200
Complement 5

Class (no):
Goplo (630), *Gardno* (631), *Bukowo* (632), *Dabie* (633), *Jamno* (634), *Mielno* (635), *Wicko* (636), *Resko* (637), *Sarbsko* (638), *Necko* (639), *Naklo* (640), *Hancza* (642), *Druzno* (641)

Goplo designated Project 207D, others Project 207P. Low-magnetic field minesweepers with GRP hull. Built by the Naval Yard in Gdynia as replacement for the ageing minesweeping force. The prototype *Goplo* was built during 1981–82 (she is used as experimental vessel) and carried at first an old twin 25mm 2-M-3 mounting which was later substituted by a 23mm one (Polish designation Wrobel). Series construction took place during 1983–91. Named after lakes. All extant 1995.

MAMRY class *coastal minesweepers*

Displacement: 203t standard; 225t full load
Dimensions: 125ft 4in × 22ft 11in × 5ft 10in
38.2m × 7.0m × 1.8m
Machinery: 2 shafts in nozzles, 2 Type M 401A diesels, 2000bhp = 14kts. Range 1100nm at 9kts
Armament: 2–23mm ZU-23-2MR (1 × 2), 6–24 mines
Sensors: Radar RN 231 (navigational), IFF, navigational system Szop; sonars SHL 100, SHL 200
Complement: 5

Class (no):
Mamry (643), *Wigry (644), Sniardwy* (645), *Wdzydze* (646, building 1995)

Modified versions of the *Goplo* class with changes to the armament, deckhouses and fittings. Planned as a class of eight ships, designated Project 207M, built by the Naval Yard in Gdynia since 1991 with three completed by 1995.

'LENIWKA' class *coastal minesweepers*

Displacement: 300t full load
Dimensions: 84ft 7in × 23ft 7in × 8ft 10in
25.8m × 7.2m × 2.7m
Machinery: 1 shaft Cegielski-Sulzer 6AL20/24 diesel, 570bhp = 11kts
Armament: None
Sensors: Radar navigational; sonar hull mounted
Complement: –

Class:
625, 626

Ex-fishing stern trawlers of 'B410' type built at the Ustka Yard, Poland, and commissioned as coastal minesweepers in 1984 for experimental purposes. Extant 1991.

Soviet 'TR 40' type *river minesweepers*

Class:
TR 41–47

Seven built during 1954–56 at the Wisla yard, Gdansk, under Soviet licence as 'Project 151' (*TR* stands for *Tralowiec Rzeczny* = river minesweeper), later renumbered *811–817*. Stricken 1969–70.

Soviet 'K 8' type *minesweeping boats*

Class:
KTR 931–964

A total of thirty-six boats were built during 1956–60 by the Polnocna Yard Gdansk, under Soviet licence as Project 361T while an additional ten were built as patrol boats under designation Project 361S. (KTR stands for *Kuter tralowy* = minesweeping boat). All stricken by 1986.

Portugal

Portugal's strategic position, and the fact that she eventually sided with the Allies during the Second World War, brought considerable benefits to her armed forces from the mid-1940s, in which the navy shared. Her fragile economy would not have supported a naval re-equipment programme, but American aid started soon after the end of the war. In 1949, Portugal was a founder member of NATO and in 1951 she signed a treaty with the United States for establishing American bases in Portuguese territory, including the Azores. War-built ships were transferred, but from the 1950s a modest programme of new building commenced.

THE COLONIES

In the 1960s, Portugal still had an extensive overseas empire, and most of her warships, including a large number of small frigates and patrol vessels, were designed to patrol their waters. India invaded and annexed the colonies of Goa, Daman and Diu on the Indian subcontinent in December 1961, and the Portuguese Navy lost a frigate and two patrol boats in a futile effort to prevent this. In the face of widespread unrest in the African colonies where independence movements sprang up in Angola, Mozambique and Portuguese Guinea (now Guinea-Bissau), the navy was hurriedly expanded to assist the other services. Large numbers of simply equipped small frigates, patrol boats and landing craft were built in a few years.

THE POST-COLONIAL ERA

The independence movements gradually won more and more territory. The end of Dr Salazar's dictatorship appeared to change none of Portugal's attitudes, but when his successor, Dr Caetano, was toppled by a military *coup* in April 1974, Portugal at last had leaders who were prepared to abandon the colonies and the drain they represented on the Portuguese economy and armed forces. The African colonies achieved their independence in 1974, 1975 and 1976, and most of the small craft built to serve in their waters were either transferred to Angola or Mozambique or scrapped.

PORTUGAL'S NAVY

A few medium-sized frigates were built in the 1960s to American and French designs, but the problems of the Portuguese economy have prevented any large-scale modernisation of these ageing ships and most of them are now suitable only for fishery protection duties. Many of the navy's smaller vessels have been built in Portugal, but plans in the early 1980s to build new frigates locally were abandoned. Indeed, the acquisition of the three German-built MEKO 200 frigates in 1990 was possible only because a consortium of NATO countries provided 60 per cent of the funding.

In 1995, the Portuguese Navy had 15,000 personnel plus 2600 marines. In 1990 five Sea Lynx helicopters were ordered to form Portugal's first seaborne aviation force. These are carried by the MEKO 200 frigates.

FLEET STRENGTH 1947

DESTROYERS

Name	Launched	Disp (full load)	Fate
Douro class			
DAO	27.7.34	1563t	Discarded 29.11.60
DOURO	16.8.35	1563t	Discarded Dec 1959
LIMA	29.5.33	1563t	Discarded 16.10.65
TEJO	4.5.35	1563t	Discarded 9.2.65
VOUGA	25.1.33	1563t	Discarded 3.6.67

They were refitted during 1946–49 in Britain. The machinery was overhauled, the aft funnel shortened, the original funnel casings replaced by aluminium ones to reduce topweight, and a tripod foremast installed. The AA armament was increased to 3–40mm (3 × 1) and 3–20mm (3 × 1) and radar and sonar was installed. They were again refitted in 1957 to enhance their ASW capabilities. Armament then consisted of 2–4.7in (120mm), 5–40mm (1 × 2, 3 × 1), 3–20mm, 1 Squid ASW mortar, 2 DC racks, 4–21in (533mm) TT.

SUBMARINES

Name	Launched	Disp	Fate
Delfim class			
DELFIM	1.5.34	1092t	Discarded *c*1950
ESPARDARTE	30.5.34	1092t	Discarded *c*1950
GOLGINHO	30.5.34	1092t	Discarded *c*1950

These Vickers-built submarines were replaced by ex-British 'S' class boats in 1948.

SLOOPS

Name	Launched	Disp	Fate
Albuquerque class			
ALFONSO DE ALBUQUERQUE	28.5.34	2440t	Sunk 18.12.61
BARTOLOMEU DIAS	10.10.34	2440t	Depot Ship Feb 1967, discarded *c*1969
Velho class			
GONCALVES ZARCO	28.11.32	1414t	Discarded *c*1969
GONCALO VELHO	3.8.32	1414t	BU June 1961
Nunes class			
PEDRO NUNES	17.3.34	1220t	Discarded 1976
JOAO DE LISBOA	21.5.36	1220t	Discarded 17.8.66

Albuquerque class

These first-class sloops were re-rated as frigates. *Albuquerque* was sunk by the Indian cruiser *Mysore* and Indian destroyers during the Indian invasion and annexation of Goa in December 1961. *Dias* was converted to a depot ship in February 1967 and renamed *Sao Cristovão*. All her armament except for 'A' and 'Y' position 4.7in (120mm) guns were removed.

Velho class

Second-class sloops. These were reclassed as frigates, as were the second-class sloops of the *Nunes* class.

Nunes class

Nunes was converted to a survey ship in 1956 and *Lisboa* in 1961. Her 20mm were replaced by 4–40mm guns (4 × 1).

GUNBOATS

Name	Launched	Disp	Fate
Zaire class			
DIO	Oct 1929	397t	Discarded *c*1970
ZAIRE (ex-*Goa*)	26.2.25	397t	Discarded *c*1960
Faro class			
FARO	1927	295t	Discarded *c*1960
LAGOS	1930	295t	Discarded *c*1960
TETE	1918	100t	Discarded *c*1975
Azevia class			
AZEVIA	1941–42	270t	Discarded 1975
BICUDA	1941–42	270t	Discarded 1976
CORVINA	1941–42	270t	Discarded 1975
DOURADA	1941–42	270t	Discarded 1975
ESPILHA	1941–42	270t	Discarded 1971
FATACA	1941–42	270t	Wrecked 22.1.49

Zaire class

Diu was used as a training ship from *c*1965.

Faro class

Faro and *Lagos* were employed as fishery protection vessels. *Tete* was a river gunboat on the River Zambesi. She was re-rated as a patrol boat in 1960.

Azevia class

The *Azevia* class were all used as fishery protection vessels. *Fataga* became a total loss after grounding on 22 January 1949.

Alfonso de Albuquerque Apr 1960 L & L van Ginderen

PORTUGAL

MAJOR SURFACE SHIPS

Ex-British 'RIVER' class *frigates*

Particulars: As British 'River' class

No	Name	Builder	Acquired	Fate
F 331	DIOGO GOMES (ex-*Awe*)	Fleming & Ferguson	May 1949	Discarded 20.4.69
F 332	NUNO TRISTAO (ex-*Avon*)	Charles Hill	May 1949	Discarded *c*1972

These two ex-British 'River' class frigates, launched in 1943, were purchased by Portugal in 1948 and transferred in May 1949. Both were refitted in 1959 when their ASW armament was modernised. The Hedgehog ASW mortar on the forecastle was removed and the 4in (102mm) gun moved farther forward. Two Squid ASW mortars (2 × 3) were then positioned side by side between the gun and the bridge. The DCT were also removed. The armament then consisted of 2–4in (2 × 1), 6–40mm, 2 Squid ASW, 2 DC racks.

Ex-US JOHN C BUTLER class *destroyer escorts*

Particulars: As US *John C Butler* class (WGT type)

No	Name	Builder	Acquired	Fate
F 333	DIOGO CAO (ex-*Formoe*)	Federal	7.2.57	Discarded 19.11.68
F 334	CORTE REAL (ex-*McCoy Reynolds*)	Federal	7.2.57	Discarded 21.10.68

Formoe was originally to have been renamed *Zaire* and *Reynolds*, *Zambeze*, but when the two ex-USN 'WGT' type DE (launched 1944) were transferred under MDAP on 7 February 1957 they were actually renamed *Cao* and *Real* respectively. Both had their 3–21in (533mm) TT (3 × 1) removed, leaving an armament of 2–5in/38 (2 × 1), 10–40mm (1 × 4, 3 × 2), 1 Hedgehog, 8 DCT, 2 DC racks. They were rated as frigates by the Portuguese.

Ex-British 'BAY' class *frigates*

Particulars: As British 'Bay' class

No	Name	Builder	Acquired	Fate
F 336	ALVARES CABRAL (ex-*Burghead Bay*)	Hill	Apr 1959	Discarded 1971
F 479	D FRANCISCO DE ALMEIDA (ex-*Morecambe Bay*)	Pickersgill	May 1961	Discarded 1970
F 478	VASCO DA GAMA (ex-*Mounts Bay*)	Pickersgill	May 1961	Discarded 1971
F 337	PACHECO PEREIRA (ex-*Bigbury Bay*)	Hall Russell	Apr 1959	Discarded 1970

Ex-British 'Bay' class AA frigates, launched in 1944–45. *Cabral* and *Pereira* were purchased in April 1959 and transferred to Portugal on 11 May 1959, and *De Almeida* and *Da Gama* were purchased in May 1961 and transferred after a refit on 3 August 1961. A half-sister, *Alfonso de Albuquerque*, ex-*Dalrymple*, ex-*Luce Bay*, built by Wm Pickersgill, launched 12 April 1945 and completed in 1949 as an unarmed survey ship, was purchased by Portugal in April 1966. She was employed as a survey ship and was stricken in 1983.

PEDRO ESCOBAR *frigate*

Displacement: 1250t standard; 1600ft full load
Dimensions: 321ft 6in oa, 306ft 9in wl × 35ft 6in × 10ft
98.0, 93.5m × 10.9m × 3.1m
Machinery: 2-shaft geared steam turbines, 2 boilers, 24,500shp = 32kts. Range 2800nm at 13.5kts
Armament: 2–76mm/50 DP (2 × 1), 2–40mm (1 × 2), 3–533mm (21in) TT (1 × 3). See notes
Complement: 165

No	Name	Builder	Laid down	Launched	Comp	Fate
F 335	PEDRO ESCOBAR	Castellammare	7.1.55	25.9.55	July 1957	Discarded 1975

Pedro Escobar had the same basic hull and machinery as the Italian-designed Venezuelan *Almirante Clemente* class, but had a considerably lighter armament to reduce topweight. She was modernised in 1968–69 to improve her AA and ASW armament, and was rearmed with 4–3in (76mm) (2 × 2), 2 Squid ASW mortars (2 × 3) and 6–12.7in (324mm) TT (2 × 3). She had MLT-3A and MLA-1B radar and SQS-29 sonar fitted, and this armament made her equivalent to the *Da Silva* class frigates. *Pedro Escobar* was built to a NATO order.

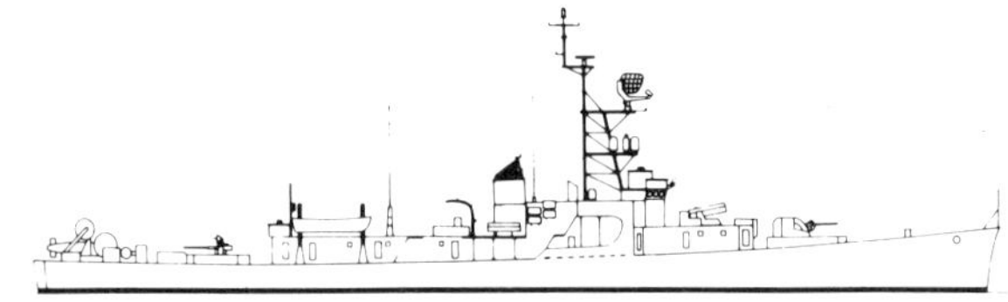

Almirante Pereira da Silva as completed

US DEALEY type *frigates*

Particulars: As US *Dealey* class

No	Name	Builder	Laid down	Launched	Comp	Fate
F 472	ALMIRANTE PEREIRA DA SILVA	Lisnave	14.6.62	2.12.63	20.12.66	Stricken *c*1989
F 473	ALMIRANTE GAGO COUTINHO	Lisnave	2.12.63	13.8.65	29.11.67	Stricken *c*1989
F 474	ALMIRANTE MAGALHAES CORREA	Viana do Castelo	30.8.63	26.4.65	4.11.68	Stricken *c*1989

These three ships were built to a slightly modified version of the US *Dealey* class DE design. The ships were fitted for prolonged service in tropical waters and they had 2–14.75in (375mm) Bofors ASW rocket launchers (2 × 4) in 'B' position in place of Weapon Alfa, plus 6–12.75in (324mm) Mk 32 ASW TT (2 × 3). The complement was 166 and they were fitted with MLA-1B search, Type 978 tactical, SPG-34 fire control and RM-316P navigational radars and SQS-30 (*Da Silva*), SQS-31 (*Coutinho*), SQS-32 (*Correa*) search, SQA-10A VDS and DUBA-3A attack sonar. They were built in Portugal under MDAP. A proposed modernisation programme was abandoned and they were all deleted *c*1989.

Almirante Gago Coutinho 1977 L & L van Ginderen

Commandante Joao Belo 9.7.84 L & L van Ginderen

The *Belo* class are very similar to the French *Rivières*, but are fitted for tropical service and have 2–40mm/70 in place of the *Rivières*' 2–30mm (2 × 1). The *Belos* have a complement of 200 and have DRBV-22A search, DRBV-50 tactical, DRBC-31D fire control and RM-316 navigational radars, with SQS-17A search and DUBA-3A attack sonars. In the mid-1980s it was planned for this class to undergo a major refit, with one or both of the after 100mm turrets to be replaced with a helicopter deck and hangar on two ships, and with SSM on the other two, but this was abandoned due to cost. Instead, their ASW capability is to be modernised, in a programme due to complete at the end of 1995. A new SQS-510(V) sonar is to be fitted, along with US Mk 32 ASW TT and Mk 46 torpedoes and the SLQ-25 Nixie torpedo decoy system.

French COMMANDANT RIVIÈRE type *frigates*

Particulars: As French *Commandant Rivière* class

No	Name	Builder	Laid down	Launched	Comp	Fate
F 480	COMANDANTE JOAO BELO	AC de Nantes	6.9.65	22.3.66	1.7.67	Extant 1995
F 481	COMANDANTE HERMENGILDO CAPELO	AC de Nantes	13.5.66	29.11.66	26.4.68	Extant 1995
F 482	COMANDANTE ROBERTO IVENS	AC de Nantes	13.12.66	8.8.67	23.11.68	Extant 1995
F 483	COMANDANTE SACADURA CABRAL	AC de Nantes	10.8.67	15.3.68	25.7.69	Extant 1995

Joao Coutinho as completed

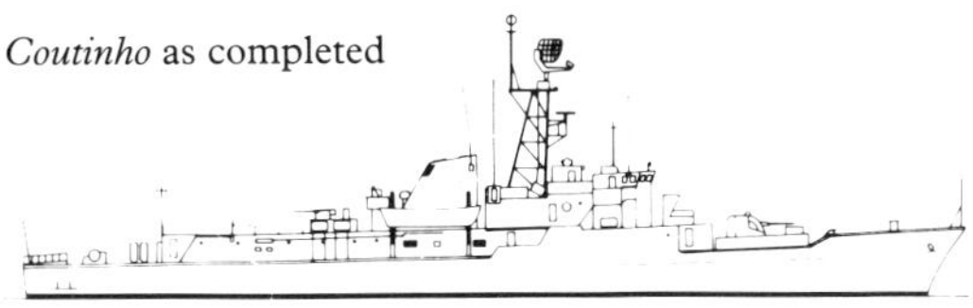

Baptiste de Andrada 1979

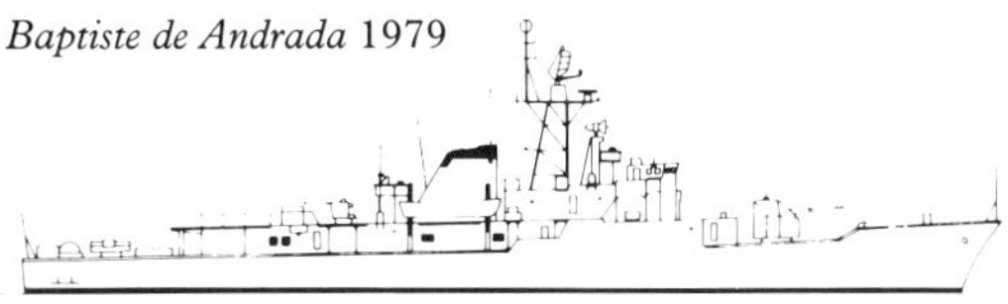

The *Joao Coutinho* class were designed by Blohm and Voss, Hamburg and built by them and by Bazan, Spain. They were relatively cheap and sparsely equipped light frigates, which could be acquired by Portugal in sufficient numbers to replace the old second-class sloops, corvettes, and gunboats. They can carry thirty-four marines, and their chief function until 1975 was patrolling off Portugal's colonies. There is provision for a light helicopter to land aft. As with the *Belo* class, a proposed modernisation including the fitting of SSM and SAM was cancelled for financial reasons. All ASW armament and sonar had been removed from the ships of this class by 1987, and the obsolete MLA-1B air-search radars had been replaced by a Kelvin-Hughes surface search set, more suited to their current role as fishery protection vessels.

The *De Andrada* class utilised the same hull and machinery as the *J Coutinhos*, but had a more sophisticated armament and sensors. They have a helicopter platform but no hangar. After the withdrawal from the colonies, all four were to be sold to Colombia in 1977, but the deal was not completed. Lack of funds has prevented any modernisation plans for this class, and *Baptiste de Andrada* was placed in reserve in 1991, and *Alfonso Cerqueira* was stricken in 1994. Neither of these two classes is particularly well armed or equipped and are only really usable for coastal patrol or fishery protection duties.

JOAO COUTINHO and DE ANDRADA classes *frigates*

Displacement: 1252t standard; 1401t full load
Dimensions: 265ft 10in pp, 277ft 8in oa × 33ft 10in × 10ft 10in
81m, 84.6m × 10.3m × 3.3m
Machinery: 2 shafts, 2 OEW-Pielstick diesels, 10,560bhp = 24.4kts. Range 5000nm at 18kts
Armament: *J Coutinho* class 2–3in/50(76mm) (1 × 2), 2–40mm/70 (1 × 2), 1 Hedgehog ASW mortar, 2 DCT, 20 DC in racks
De Andrada class 1–3.9in (100mm), 2–40mm/70 (2 × 1), 6–324mm (12.75in) TT (2 × 3), 1 helicopter
Sensors: Radar *J Coutinho* class MLA-1B, TM-626, SPG-50
De Andrada class AWS-2, TM-626, Pollux
Sonar *J Coutinho* class QCU-2, *De Andrada* class Diodon
Complement: *J Coutinho* class 93, *De Andrada* class 113

No	Name	Builder	Laid down	Launched	Comp	Fate
J Coutinho class						
F 475	JOAO COUTINHO	Blohm & Voss, Hamburg	Sep 1968	2.5.69	7.3.70	Extant 1995
F 476	JACINTO CANDIDO	Blohm & Voss, Hamburg	Apr 1968	16.6.69	10.6.70	Extant 1995
F 477	GENERAL PEREIRA D'ECA	Blohm & Voss, Hamburg	Oct 1968	26.7.69	10.10.70	Extant 1995
F 478	AUGUSTO DE CASTILHO	Bazan	Aug 1968	5.7.69	14.11.70	Extant 1995
F 485	HONORIO BARRETO	Bazan	July 1968	11.4.70	15.4.71	Extant 1995
F 471	ANTONIO ENES	Bazan	Apr 1968	16.8.69	18.6.71	Extant 1995
De Andrada class						
F 486	BAPTISTE DE ANDRADA	Bazan	1972	Mar 1973	19.11.74	In reserve 1991
F 487	JOAO ROBY	Bazan	1972	3.6.73	18.3.75	Extant 1995
F 488	ALFONSO CERQUEIRA	Bazan	1973	6.10.73	26.6.75	Stricken 3.1.94
F 489	OLIVEIRA E CARMO	Bazan	1972	Feb 1974	Feb 1976	Extant 1995

Antonio Enes Jul 1994 L & L van Ginderen

Alvares Cabral Jul 1994 L & L van Ginderen

The history of Portuguese plans for new frigates has been somewhat complicated. In 1982 it was expected that three frigates based on the Netherlands *Kortenaer* class and modified to suit Portuguese requirements would be built in a programme funded by a consortium of NATO countries, with the first unit to be built in the Netherlands and the remainder in Portugal. However, these plans were cancelled in August 1984. After further consideration of several plans, it was finally decided in 1986 to buy German-built MEKO 200 type frigates. See below.

Netherlands KORTENAER class *frigates*

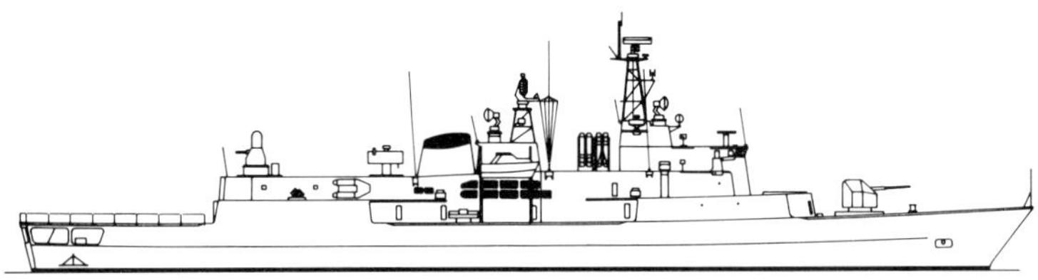

Vasco da Gama as completed

On 25 July 1986 three MEKO 200 type frigates were ordered from Blohm & Voss, Hamburg (one) and Howaldtswerke, Kiel (two). These were the ships finally chosen by the Portuguese Navy, after considering the Netherlands *Kortenaer* type (see above) and a Spanish proposal. The programme was 60 per cent funded by a NATO consortium comprising the USA, Canada, West Germany, Norway and the Netherlands. Similar ships have been built for Greece and Turkey. These ships give the Portuguese Navy its first seaborne aviation capability, each carrying one Sea Lynx helicopter.

MEKO 200 type *frigates*

Displacement: 2920t standard; 3200t full load
Dimensions: 380ft 3in × 48ft 7in × 19ft 7in
115.9m × 14.8m × 5.97m
Machinery: 2 shafts (CP props) CODOG, 2 GE LM-2500-30 gas turbines, 60,000shp = 32kts, 2 MTU 12V1163 TB83 diesels, 8840bhp = 18kts. Range 4100nm at 18kts (diesels)
Armament: 8 Harpoon SSM (2 × 4), 8 Sea Sparrow SAM (1 × 8), 1–100mm/55 CADAM, 1–20mm Phalanx CIWS, 6–324mm Mk 32 Mod 5 ASW TT (2 × 3), 1 helicopter
Sensors: Radar Kelvin-Hughes 1007, HSA MW-08 Mod 3, HSA DA-08, 2 HSA STIR-18; sonar SQS-510 (V); ECM APEX II/AR-700, SLQ-25 Nixie towed torpedo decoy
Complement: 182 (23 officers)

No	Name	Builder	Laid down	Launched	Comp	Fate
F 330	VASCO DA GAMA	Blohm & Voss, Hamburg	2.2.89	26.6.89	20.11.90	Extant 1995
F 331	ALVARES CABRAL	Howaldtswerke, Kiel	2.6.89	2.5.90	18.1.91	Extant 1995
F 332	CORTE REAL	Howaldtswerke, Kiel	20.10.89	2.5.90	22.11.90	Extant 1995

SUBMARINES

These ex-British 'S' class submarines, built in 1943–45, were all purchased in 1948. The light AA guns (1–20mm, 3–0.303in (7.7mm)) were landed in 1961. The 'S' class boats were replaced by the *Daphnes*.

Ex-British 'S' class *patrol submarines*

Particulars: As British 'S' class (1942/43 programmes)

No	Name	Builder	Acquired	Fate
S 160	NARVAL (ex-*Spur*)	Cammell Laird	Nov 1948	Discarded 1.10.69
S 161	NAUTILO (ex-*Saga*)	Cammell Laird	11.10.48	Discarded 25.1.69
S 162	NEPTUNO (ex-*Spearhead*)	Cammell Laird	Aug 1948	Discarded 1.9.67

These French-designed *Daphne* type submarines were ordered in 1964 and are slightly modified to suit Portuguese requirements. They are fitted for operating in tropical waters, have a complement of fifty and DRUA-31 radar and DUUA-1 active and DSUV passive sonar. They are scheduled *Cachalote* was puchased by Pakistan in December 1975 and renamed *Ghazi*.

French DAPHNE type *patrol submarines*

Particulars: As French *Daphne* class

No	Name	Builder	Laid down	Launched	Comp	Fate
S 163	ALBACORA	Dubigeon	6.9.65	15.10.66	1.10.67	Extant 1995
S 164	BARRACUDA	Dubigeon	19.10.65	24.4.67	4.5.68	Extant 1995
S 165	CACHALOTE	Dubigeon	27.10.66	16.2.68	25.1.69	To Pakistan Dec 1975
S 166	DELFIM	Dubigeon	12.5.67	23.9.68	1.10.69	Extant 1995

Albacora 1983
L & L van Ginderen

AMPHIBIOUS WARFARE VESSELS

LANDING CRAFT

Name	Launched	Disp (full load)	Fate
LCT type			
ALFANGE (LDG 101)	1964	635t	To Angola 1975
ARIETE (LDG 102)	1964	635t	To Mozambique? 1975
BACAMARTE (i) (LDG 103)	1964	635t	Discarded 1976
CIMITARRA (LDG 104)	1965	635t	To Mozambique 1975
MONTANTE (LDG 105)	1965	635t	Discarded 1976
BOMBARDA (LDG 201)	1969	635t	Extant 1995
ALABARDA (LDG 202)	1970	635t	Extant 1995
BACAMARTE (ii) (LDG 203)	1985	635t	Extant 1995
LCM type			
LDM 101–121	1963–72	50t	3 extant 1995
LDM 201–205	1963–65	50t	None extant 1995
LDM 301–313	1964–67	50t	None extant 1995
LDM 401–424	1964–67	56t	5 extant 1995

LCT type
The LCT were based on the British LCT (4) Type. The last two differ slightly from the first five. They have Decca RM-316 radar. Built by Mondego S Yd. Three extant 1995.

LCM type
The LCM are based on the USN LCM type. The four series differ slightly from one another; all built by Mondego S Yd. Most were discarded in 1975–76; *LDM 424* was deleted in 1982 and *LDM 421* in 1993, and both transferred to Guinea-Bissau in October 1993. Only *LDM 119–121, 406, 418, 420, 422* and *423* extant 1995.

Other minor landing craft
A number of 12t assault landing craft designated LD (later LDP) were built at Mondego from 1961 onwards in the three series *LDP 101–LDP 109, LDP 201–LDP 217, LDP 301–LDP 304.* Most were discarded in 1975, leaving only *LDP 216* in service in 1982, which has since been deleted.

SMALL SURFACE COMBATANTS

Ex-British BANGOR class *corvettes*

Particulars: As British *Bangor* class

No	Name	Builder	Acquired	Fate
F 470	CACHEU (ex-*Comandante Almeida Carvalho*, ex-*Fort York*, ex-*Mingan*)	Dufferin, Toronto	1950	Discarded 1971
A 525	ALMIRANTE LACERDA (ex-*Caraquet*)	N Vancouver	1946	Discarded 1975

The *Bangor* class fleet minesweepers *Caraquet* and *Fort York* (ex-*Mingan*) were built in Canada in 1941 and served in the Canadian Navy before being sold to the Portuguese Navy on 29 June 1946 and 26 September 1950 respectively and renamed *Almirante Lacerda* (A 525) and *Comandante Almeida Carvalho* (A 527). Both were originally employed as survey ships, but in 1965 *Carvalho* was renamed *Cacheu* and reclassified as a corvette. Both ships were armed with 1–3in (76mm) and 2–20mm.

Ex-British 'ISLES' class *patrol vessels*

Particulars: As British 'Isles' class

Four British 'Isles' and 'Tree' class minesweeping trawlers were lent to Portugal from October 1943 to July 1945, *Bruray* as *P 1*, *Mangrove* as *P 2*, *Hayling* as *P 3* and *Whalsay* as *P 4*. They were later purchased by Portugal: *Bruray* on 11 February 1946 as *Sao Miguel*, *Mangrove* on 11 February 1946 as *Faial*, *Hayling* on 11 June 1946 as *Terceira* and *Whalsay* on 11 June 1946 as *Santa Maria*. *Saltarelo* ('Dance' class) was purchased by Portugal on 4 August 1946 and renamed *Salvador Correia*, and *Ruskholm* was purchased on 2 September 1949 and renamed *Baldaque da Silva*. The latter was used first as a patrol vessel, then as a minesweeper and then as a survey ship. The others were used mainly as patrol vessels, *Correia* being stationed in Angola. *Sao Miguel* was discarded in 1956, *Terceira* in 1957, *Correia* in 1961, *Faial* in 1967 and *Santa Maria* in 1968. When the first *Correia* was discarded in 1961, *Da Silva* was renamed *Salvador Correia* and took the preceding ship's place in Angola. She was discarded in the early 1970s. They were all armed with 1–3in (76mm) and 2–20mm. *Correia* (ex-*Da Silva*)'s 3in (76mm) was removed in 1964.

Ex-US PC type *patrol vessels*

Particulars: As US PC type

Class:
Principe (P 581, ex-*Flores*, ex-*PC 812*), *Madeira* (P 582, ex-*PC 811*), *Santiago* (P 583, ex-*PC 1257*), *Sal* (P 584, ex-*PC 809*), *Sao Tome* (P 585, ex-*PC 1256*), *Sao Vicente* (P 586, ex-*PC 1259*)

Six ex-USN PCs were purchased by Portugal in 1948 under MDAP. They were originally employed mainly for air-sea rescue patrols off the Portuguese, Azores and Madeira coasts. In 1957 they were rearmed to bring them up to the standards of the new *Le Fougueux* type. Their 3in (76mm) and 2–20mm guns were removed and the new armament was 1–40mm, 3–20mm (3 × 1), 1 Hedgehog ASW mortar and 4 DCT. *Santiago* was discarded in July 1967 and the remainder were discarded during 1968–69.

French LE FOUGUEUX type *patrol vessels*

Particulars: As French *Le Fougueux* class

Class (no, launched):
Dubigeon – *Maio* (P 587, ex-*Funchal*, 27.9.54)
Normand – *Porto Santo* (P 588, 9.2.55), *Sao Nicolau* (P 589, 7.6.55)
Viana do Castelo – *Brava* (P 590, 2.5.56), *Fago* (P 591, 2.5.56)
Mondego – *Boavista* (P 592, 10.7.56)
Alfeite D Yd – *Santo Antão* (P 593, 8.6.56), *Santa Luzia* (P 594, 17.1.57)

ARGOS class *patrol vessels*

Displacement: 180t standard; 210t full load
Dimensions: 131ft 3in pp, 136ft 8in oa × 20ft 6in × 7ft
40.0m, 41.7m × 6.2m × 2.1m
Machinery: 2 shafts, 2 Maybach diesels, 1200bhp = 17kts. Oil 16t
Armament: 2–40mm (2 × 1)
Sensors: Radar?
Complement: 24

Class:
Argos (P 372), *Cassiopeia* (P 373), *Dragão* (P 374), *Escorpião* (P 375), *Hidra* (P 376), *Lira* (P 361), *Orion* (P 362), *Pegaso* (P 379), *Centauro* (P 1130), *Sagitario* (P 1134)

Six *Argos* class were built at Alfeite and four at Viana do Castelo between June 1963 and September 1965. They were intended for use in the Portuguese colonies. *Centauro*, *Escorpião*, *Lira*, *Orion* and *Pegaso* were transferred to Angola in 1975, with *Argos* and *Dragão* for spares, while *Cassiopeia* was discarded in 1975 and *Hidra* and *Sagitario* in 1976.

CACINE class *patrol vessel*

Displacement: 292t standard; 310t full load
Dimensions: 144ft 5in × 25ft 2in × 7ft 3in
44.0m × 7.7m × 2.2m
Machinery: 2 shafts, 2 MTU diesels, 4400bhp = 20kts. Range 4400nm at 12kts
Armament: 2–40mm (2 × 1), 1–37mm RL (1 × 32)
Sensors: Radar KH-975
Complement: 33

Class:
Cacine (P 1140), *Cunene* (P 1141), *Mandovi* (P 1142), *Rovuma* (P 1143), *Quanza* (P 1144), *Geba* (P 1145), *Zaire* (P 1146), *Zambesi* (P 1147), *Limpopo* (P 1160), *Save* (P 1161)

The first four ships were built to this design at Alfeite D Yd during 1967–69. The next six were built at Mondego, during 1968–71 and *Limpopo* and *Save* in 1973. They replaced the earlier PC and *Le Fougueux* type patrol vessels. All were extant in 1995.

Zaire 1980 L & L van Ginderen

COASTAL PATROL BOATS

Name	Fate	Name	Fate
Fairmile 'B' type: launched 1942–44, 85.6t full load			
ALTAIR (L 1)	BU 1955	DENEB (L 4)	Discarded 1959
BELATRIX (L 2)	Discarded *c*1959	ESPIGA (L 5)	Discarded Nov 1959
CANOPUS (L 3)	Discarded Sep 1959	FOMALHAUT (L 6)	Discarded 1959
Albatroz class: launched 1974–75, 45t full load			
ALBATROZ (P 1162)	Extant 1995	AGUIA (P 1165)	Extant 1995
ACOR (P 1163)	Extant 1995	CONDOR (P 1166)	Extant 1995
ANDORHINA (P 1164)	Extant 1995	CISNE (P 1167)	Extant 1995
Dom Aleixo class: launched 1967, 67.7t full load			
DOM ALEIXO (P 1148)	Extant 1995	DOM JEREMIAS (P 1149)	Extant 1995
Alvor class: launched 1966–67, 35.7t full load			
ALVOR (P 1156)	Discarded 1975	ALJELUR (P 1158)	Discarded 1975
ALBUFEIRA (P 1157)	Discarded 1975		
Jupiter class: launched 1944–65, 49t full load			
JUPITER (P 1132)	See notes	SATURNO (P 1136)	See notes
MARTE (P 1134)	To Angola 1975	URANO (P 1137)	See notes
MERCURIO (P 1135)	See notes	VENUS (P 1133)	To Angola 1975
Belatrix class: launched 1961–62, 29t full load			
ALDEBARAN	Discarded 1975	FOMALHAUT (P 367)	Survey ship 1975
ALTAIR (P 377)	To Angola 1975	POLLUX (P 368)	To Angola 1975
ARCTURUS	Discarded 1975	PROCION (P 1153)	Discarded 1975
BELATRIX (P 363)	Discarded 1975	RIGEL (P 378)	To Angola 1975
CANOPUS (P 364)	Discarded 1975	SIRIUS	To Mozambique 1975
DENEB (P 365)	Discarded 1975	VEGA	To Mozambique 1975
ESPIGA (P 366)	To Angola 1975		
Algol class: launched 1964, 24t full load			
ALGOL (P 1138)	Discarded *c*1975		
Castor class: launched 1964, 22t full load			
CASTOR (P 580)	Discarded 1976		
Antares class: launched 1959, 18t full load			
ANTARES (P 360)	To Mozambique 1975	SIRIUS	Sunk Dec 1961
REGULUS (P 369)	Discarded 1975	VEGA	Sunk Dec 1961
Rio Minho class: launched 1957, 14t full load			
RIO MINHO (P 370)	Extant 1995		

Fairmile 'B' type
The six Fairmile 'B' type motor launches were purchased from Britain in 1946. They were disarmed and used as air-sea rescue launches. A large deckhouse was built aft.

Albatroz class
Six were built at Alfeite in 1974–75 and eight more have been built since. They have RM-316P radar, and are armed with 1–20mm and 2–0.5in (12.7mm) (1 × 2).

Dom Aleixo class
These are armed with 1–20mm. *Dom Jeremias* is in use as a survey craft.

Alvor class
These were all built at Alfeite, and were armed with either 1–20mm or 2 MG.

Jupiter class
These were armed with 1–20mm and were built by Mondego. Two were transferred to Angola; of the remainder, three went to Mozambique and one was discarded, but it is not clear which.

Belatrix class
All were built by Beyerische Schiffbau, Germany. They are armed with 1–20mm.

Algol was built by Argibay, Lisbon, and armed with 2 MG.

Castor was built by Mondego and armed with 1–20mm.

Antares class
Designed by James Taylor, Shoreham, who built *Antares* and the glass-fibre hulls of the other three. *Sirius* and *Vega* were sunk in action with Indian forces during the invasion and annexation of Goa in 1961.

Rio Minho patrols the River Minho on the Spanish-Portuguese border.

MINE WARFARE VESSELS

US AGILE type *ocean minesweepers*

Particulars: As US *Agile* class

No	Name	Builder	Launched	Fate
M 418	CORVO	Burger Boats	28.7.54	Discarded 1973
M 417	GRACIOSA	Burger Boats	19.11.53	Discarded 1973
M 416	PICO	Bellingham	18.6.54	Discarded 1973
M 415	SAO JORGE	Bellingham	30.4.54	Discarded 1973

USN *Agile* type MSOs built in the USA for Portugal under MDAP. While building they were assigned the USN designations MSO 487, 486, 479 and 478 respectively. They were all laid down in 1953 and commissioned between 24 April 1955 and 23 November 1955.

British 'TON' type *coastal minesweepers*

Particulars: As British 'Ton' class

Class:
Sao Roque (M 401), *Ribeira Grande* (M 402), *Lagoa* (M 403), *Rosario* (M 404)

Four 'Ton' class coastal minesweepers were built in the CUF Shipyard, 1954–57. *Lagoa* and *Sao Roque* were funded by the USA under MDAP, and *Ribeira Grande* and *Rosairo* were paid for by Portugal. They had been ordered in 1954. Their 1–40mm gun was removed in 1972, leaving them with 2–20mm. All were extant in 1995, but are now used as patrol vessels.

Ex-US ADJUTANT class *coastal minesweepers*

Particulars: As US *Adjutant* class

Class:
Ponta Delgada (M 405), *Horta* (M 406), *Angra do Heroismo* (M 407), *Vila do Porto* (M 408), *Santa Cruz* (M 409), *Velas* (M 410), *Lajes* (M 411), *Sao Pedro* (M 412)

Eight USN *Redwing* type MSC were built in the USA 1951–53 and transferred under MDAP to Portugal 1953–55. *Ponta Delgarda* (ex-*Adjutant*, *AMS 60*), *Santa Cruz* (ex-*AMS 92*) and *Angra do Heroismo* (ex-*AMS 62*) were discarded in 1973, *Santa Cruz* (ex-*AMS 92*) and *Lajes* (ex-*AMS 146*) were discarded in 1975 and *Horta* (ex-*AMS 61*), *Vila do Porto* (ex-*AMS 91*) and *Velas* (ex-*AMS 145*) were discarded in 1976.

Qatar

Qatar is a small but prosperous Gulf State (4247 square miles, coastline 350 miles, population (1992 UN estimate 485,000). More than half the population reside in the capital, Doha. The state's naval forces evolved from a coast guard force in the 1970s.

Until 1971 Qatar was one of nine independent emirates in the Arabian Gulf which had special treaty relations with the UK. In that year, with the withdrawal of British forces from the area, the treaties were terminated. The state does rely to some extent on Western assistance to operate its armed forces, which have total personnel of about 7000.

The 1994 strength of the Naval Force was 850 officers and men, including the Marine Police. The main base is at Doha and a second base is located at Halul Island. The navy has received no new craft since three 'Combattante III M' fast attack craft were delivered in 1982–83, but in 1992 funds were allocated for four fast attack craft to be built by Vosper Thornycroft (UK) Ltd. Truck-mounted Exocet missiles are deployed for coast defence.

'COMBATTANTE III M' type *fast attack craft (missile)*

Displacement: 345t standard; 395t full load
Dimensions: 173ft 10in pp, 183ft 9in oa × 26ft 9in × 7ft 1in
53m, 56m × 8.1m × 2.1m
Machinery: 4 shafts, 4 MTU 20V538 TB93 diesels, 15,000bhp = 38.5kts. Range 2000nm at 15kts
Armament: 8 MM40 Exocet SSM (2×4), 1–76mm/62 DP OTO Melara, 2–40mm/70 Breda-Bofors Compatto (1×2), 2–30mm/85 Emerlec (2×2), Dagaie chaff RL
Sensors: Radar Decca 1226, Thomson-CSF Pollux, Castor; ESM/ECM Racal Cutlass/Cygnus
Complement: 41 (6 officers)

Class (laid down, launched, in service):
Damsah (6.6.81, 17.6.82, 10.11.82), *Al Ghariyah* (26.8.81, 23.4.82, 10.2.83), *R'Biva* (27.10.81, 22.12.82, 11.5.83)

Ordered from CMN, Cherbourg October 1980. In 1980 Qatar ordered about 100 MM40 Exocet missiles for these boats and land batteries. Two CSEE Panda directors for fire control. Very similar to Nigerian *Siri* class. Pennant numbers Q 01, Q 02, Q 03 respectively. Correct name for Q 03 may be *Ryigah*. Extant 1995.

Vosper Thornycroft 'VITA' type *fast attack craft*

Displacement: 376t full load
Dimensions: 183ft 9in × 29ft 6in × 7ft 7in
53.6m × 9.0m × 2.3m
Machinery: 4 shafts, 4 MTU 20V 538 TB93 diesels, 18,740bhp = 35kts. Range 1800nm at 12kts
Armament: 4 or 8 MM40 Exocet SSM (2 × 2 or 2 × 4), 1–76mm/62 OTO Melara, 1–30mm Goalkeeper CIWS or Thomson-CSF Crotale SAM, 2–12.7mm
Sensors: Radar air/surface search, navigational; fire control Signaal STING, IRESCAN electro-optical tracker; combat data system SEWACO FD with Thomson-CSF TACTICOS, Link Y; ECM/ESM Thomson CSF DR3000S, Dassault ARBB33, CSM Dagaie chaff launcher
Complement: 47 (7 officers)

A contract for four big FACs of a modified 'Vita' type to be built by Vosper Thornycroft (UK) Ltd was announced on 4 June 1992. First steel was cut on 20 July 1993 and the prototype boat, named *Bazan*, was launched on 1 April 1995. Other three craft to be completed at six-monthly intervals thereafter. Not all data has been released officially and therefore the information given above may change. Vessels are characterised by a long raised forecastle for better seakeeping. The side-plating forward will feature a pronounced anti-spray knuckle. Proposed replacement of Goalkeeper CIWS with Crotale SAM is a result of Gulf conflict experience (see comments in introduction to Kuwait section). These four craft, when completed, will greatly enhance the capabilities of the Qatar Navy.

Vosper Thornycroft 110ft type *large patrol craft*

Displacement: 120t
Dimensions: 110ft × 21ft × 5ft 6in
33.5m × 6.4m × 1.7m
Machinery: 2 shafts, 2 Paxman Valenta 16cyl diesels, 6000–6250bhp = 27kts
Armament: 2–20mm (1×2)
Complement: 25

Class (no, in service):
Barzan (Q 11, 13.1.75), *Hwar* (Q 12, 30.4.75), *That Assuari* (Q 13, 3.10.75), *Al Wusaail* (Q 14, 28.10.75), *Fateh al Khair* (Q 15, 22.1.76), *Tariq* (Q 16, 1.3.76)

Ordered in 1972–73. All laid down by Vosper Thornycroft, Portsmouth, between September 1973 and November 1974. Some sources give displacement 100t standard, 140t full load and power 4000bhp. Others refer to these ships as 103ft type with 31.1m pp, 32.4m oa × 6.3m × 1.6m and armed with 2–30mm (1×2) and 2–20mm. Extant 1995.

Barzan 1975 Vosper Thornycroft

British 75ft *coastal patrol craft*

Two 60t units (names unknown) built by Whittingham & Mitchell, Chertsey, England. In service 1969 with 2–20mm. Stricken by 1990.

Keith Nelson 45ft type *coastal patrol craft*

Displacement: 13t
Dimensions: 44ft 3in × 12ft 6in × 3ft 7in
13.5m × 3.8m × 1.1m
Machinery: 2 Caterpillar diesels, 800bhp = 26kts
Armament: 1–12.7mm MG, 2–7.62mm MG (2×1)
Complement: 6

Two units (names unknown) built early 1970s by Keith Nelson, Bembridge, a division of Vosper. A third vessel of this group has been converted into a pilot cutter. Extant 1995.

Fairey Marine 'SPEAR' class *coastal patrol craft*

Displacement: 4.3t/10t
Dimensions: 29ft 10in × 9ft 2in × 2ft 8in
9.1m × 2.8m × 0.8m
Machinery: 2 shafts, 2 Perkins T6–354 diesels, 580bhp = 26kts
Armament: 3–7.62mm MG
Complement: 4

Class:
Q 70–95

All built by Fairey Marine 1974–77. Seven ordered early 1974. Delivered June 1974–February 1975. Contract for further five assigned December 1975. Three more delivered; two on 30 June 1975 and one on 14 July 1975. Order for ten further craft (four Mk 1, six Mk 2) received October 1976 and delivery effected April 1977. Extant 1995.

MISCELLANEOUS

There are seven British-built Watercraft P 1200 type coastal patrol craft. (12.7t, 2-MGs, 660hp, 2 shafts = 29kts) with a complement of four men each. Built 1980.

Also, two Fairey Marine 'Interceptor' class fast assault/rescue craft (1.25t, 270bhp = 35kts with a range of 150nm at 30kts). Delivered 28 November 1975.

Six Damen 'Polycat 1450' type coastal patrol craft, *Q 31–36* (18ft full load, 47ft 7in oa, 840bhp = 26kts, 1–20mm Oerlikon), delivered in 1980. GRP construction.

Four Crestitalia MV-45 type (17t full load, 47ft 7in oa, 1270bhp = 32kts, 1–20mm Oerlikon, 2–7.62mm MG), delivered in 1989. GRP construction.

Romania

Although Romania changed to the Allied side in the closing stages of the Second World War, this did not excuse the country from postwar restrictions placed upon her by the USSR.

The peace treaty of 10 February 1947 required a reduction of the Romanian Army to 120,000 men, the Air Force to 150 aircraft while the Navy was allowed to maintain 15,000t overall tonnage. At the time that tonnage was barely reached as the Romanians operated two old destroyers, one submarine, two escorts converted from the old torpedo-boats, two old gunboats and five MTBs returned by the Soviets. This force was supplemented by more modern units returned by the Soviet Union by the early 1950s namely two destroyers, two submarines and a minelayer. At that time the Romanians began the river craft programme which resulted in the completion of nineteen such craft for service on the Danube.

During the late 1940s and early '50s the country was remodelled on the Soviet pattern, a process which had started in September 1944 due to the presence of Soviet troops. The increasing dominance of the local communists, backed by Soviet military power, forced King Michail I to abdicate in 1947, and this ended any likelihood of organised opposition. In 1948 the Communist Party took absolute power and enforced radical collectivisation and industrialisation programmes (1949–62).

Romania's membership of the Warsaw Pact improved her navy's situation but not to the same extent as with neighbouring Bulgaria. Four 'MV' type coastal submarines, four ex-German *M-boats* (1940 type), three 'Kronstadt' class patrol craft and twenty-one T 301 class inshore minesweepers were acquired from the Soviet Union during the late 1950s. In addition eight incomplete hulls of the 'TR 40' type river minesweepers were transferred from Poland for completion in Romania and twelve landing craft were laid down in a local yard. In June 1958 Soviet forces left the country.

The early 1960s saw slow but steady development as twelve 'P 4' type MTBs were transferred from the Soviet Union and transfer of the 'Osa' class missile boats was started. Further transfer of Soviet warships was interrupted suddenly in 1964 when Romania refused to follow the Soviet Union's example in relations with Red China or accept COMECON plans for economic development. This rift was widened when Romania took an independent line in her trade relations and refused to participate in the invasion of Czechoslovakia in 1968. Such disobedience resulted in a considerable reduction of Soviet supplies and left the Romanian Navy, except for three 'Poti' class corvettes, acquired in 1970, with only five modern combat naval craft ('Osa' class) in service at the beginning of the 1970s.

Therefore Chinese assistance was willingly invited and licence was bought for construction of torpedo hydrofoils of the 'Huchwan' class and patrol craft of the 'Shanghai' class. Over twenty-five craft of the former and thirty of the latter type were commissioned during the later 1970s. Apart from these boats the Romanians themselves began construction of three types of river craft.

Romania's somewhat divergent foreign policy was sufficient to satisfy Bucharest's ambitions for independence without provoking open conflict with the Kremlin. This in turn allowed President Ceaucescu the political freedom for an unparalleled concentration of power into his own hands and the pursuit of a series of major projects, increasingly erratic in conception and increasingly at odds with the capacity of the Romanian economy to support them. Among these was a most spectacular naval expansion programme undertaken in the early 1980s. A huge fleet (compared with the country's resources) of missile destroyers (officially classed as battlecruisers), four frigates, twelve torpedo craft and two large auxiliaries were to be built domestically. The first submarine, after a 25-year absence of this type in the Romanian Navy, was purchased in 1986 from the Soviet Union.

The results, however, were far from satisfactory. The major surface ships were too large for coastal defence and it was hard to conceive of their use in other waters. Moreover, they were of rather rudimentary design, fitted with outdated weapons and sensors purchased from the Soviet Union. The ASW forces were inadequate as were the MCM capabilities, amphibious capacity was negligible and the auxiliaries were too small.

Construction of such a large fleet, together with other grandiose projects, was attained at a cost unequalled elsewhere in Europe, and this was at a time of declining living standards, decapitalisation of machinery and devastation of the environment. Despite strenuous sacrifices imposed on the Romanian people, most of the extravagant programmes had to be postponed due to the collapse of the economy. Naval shipbuilding was stopped in 1987, and since then no new ships or craft have been laid down. At the end of the 1980s Romania possessed quite a large fleet by eastern European standards, but this force had 'operetta and parade ground' value rather than true operational worth.

The increasing hardship and repression suffered by the Romanian people led in 1989 to an unprecedented level of unrest, and news of events in other eastern European countries sparked a popular revolt, which was at least partly orchestrated by the army and the Communist Party and resulted in the execution of President Ceaucescu. Partial disintegration of the apparatus of the state and of the Communist Party itself, combined with the continuing struggle of non-communist political groups for power, accelerated the collapse of the economy. Following the break-up of the Warsaw Pact, Romania has pursued a policy of seeking membership of NATO and associate membership of the European Community.

The Romanian navy has a total of 10,000 men, of whom 3400 are afloat, 2000 are in coastal defence (one SSC2b 'Samlet' missile battalion and gun batteries totalling one hundred 130mm and 6in pieces, as well as radar stations along the coast) and 100 personnel in naval aviation. Ratings are on 15-months' conscription. The fleet headquarters are at Mangalia, and the main bases are at Mangalia and Constanţa. The Danube River Flotilla has headquarters at Giurhiu and has bases at Giurhiu, Sulina, Galati and Dulcea. The naval aviation division, which was established in 1983, has its headquarters and base at Constanţa and consists of six Mi14 'Haze' land-based and six license-built French designed IAR-316 'Alouette III' shipborne helicopters. Because of lack of oil fuel some of the larger warships have been non-operational since 1990.

FLEET STRENGTH 1947

DESTROYERS

Name	Launched	Disp	Fate
Marasti class			
D 11	26.3.17	1391t	Stricken mid-1950s
D 12	30.1.18	1391T	Stricken mid-1950s

Former destroyers *Marasti* and *Marasesti* which were seized by the USSR in August 1944 and commissioned by them as *Lovky* and *Logkij* respectively. Returned on 12 October 1945.

SUBMARINE

Name	Launched	Disp	Fate
Delfinul class			
ex-DELFINUL	22.6.30	650t	Stricken 1957

Seized by the USSR on 27 August 1944. Commissioned into the Soviet Black Sea Fleet on 14 September 1944 as TS-3. Returned on 12 October 1945 to Romania and renamed *Delfinul*. Stricken 1957 and BU.

ESCORT VESSELS

Name	Launched	Disp	Fate
Naluca class			
E 1	11.8.14	262t	BU 1960
Sborul class			
E 2	6.8.14	257t	BU *c*1960

Former torpedo-boats *Smeul* and *Sborul* converted for escort duties during the war. Both seized by Soviet troops in Constanţa on 29 August 1944 and 27 August 1944 respectively. Commissioned into Soviet Navy on 14 September 1944 as *Toros* and *Musson* respectively. Returned 12 October 1945 and reverted to previous names. Both deleted mid-1950s and BU.

GUNBOATS

Name	Launched	Disp	Fate
Capitan Dumitrescu class			
D 61	1916	355t	Extant 1995
D 62	1917	355t	Extant 1995

Dumitrescu, *Ghigulescu* and *Stihi Eugen* seized by the USSR 29 August 1944 and commissioned into the Black Sea Fleet as *Araks*, *Angara* and *Akhtuba* respectively. Araks was sunk by a mine in Odessa harbour on 10 January 1945, raised nine days later and BU. Remaining two returned to Romania on 14 October 1945 and served until *c*1960, then converted for surveying service and became *Stihi* (113) and *Constanta* (112).

MOTOR TORPEDO BOATS

The 72ft Vosper type *Viscolul* returned from the Soviet Black Sea Fleet (*TKA 955*) on 22 September 1945 and was broken up after 1947.

The 470ft *Vantul* class 32t boats (ex-*TKA 951–954*) likewise.

RIVER PATROL BOATS

Three 51t *Captain Nicolae Lascar Bogdan* class units of *c*1906 returned from Soviet service (*SKA 754–756*) in 1945 and were broken up after 1947.

MAJOR SURFACE SHIPS

REGELE FERDINAND class *destroyers*

Name	Builder	Launched	Fate
D 21 (ex-*Letuchy*, ex-*Regina Maria*)	Pattison, Naples	2.3.29	Stricken late 1950s
D 22 (ex-*Likhoi*, ex-*Regele Ferdinand*)	Pattison, Naples	2.12.28	Stricken late 1950s

Distrugatoáre ('destroyers') seized by the USSR in August 1944. Both returned to Romania 3 July 1951. *Regina Maria* renamed *D 21* (later *D 10*) and *Regele Ferdinand* renamed *D 22* (later *D 11*). Both deleted late 1950s and subsequently BU.

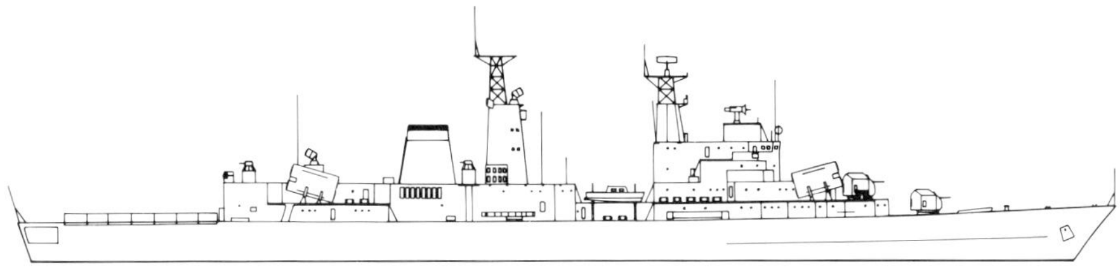

Muntenia as completed

MUNTENIA *missile destroyer*

Displacement: 4500t standard; 5790t full load
Dimensions: 474ft 4in × 48ft 6in × 23ft
144.6m × 14.8m × 7.0m
Machinery: 2 shafts, 4 diesels, 31,000bhp = 32kts
Armament: 8 SS-N-2C SSM(4×2)
4–76mm (2×2), 8–30mm (4×2), 6–533mm TT (2×3), 2 RBU-1200 ASW RL, 2 IAR-316 Alouette III and 1 IAR-330 helicopters
Sensors: Radar Strut Curve, Drum Tilt, Hawk Screech, High Pole; hull sonar ECM Watch Dog
Complement: 270

No	Name	Builder	Laid down	Launched	Comm	Fate
111	MUNTENIA	Mangalia Yard	1979	1982	3.6.85	Extant 1995

Designed in Romania and based on available Soviet machinery, weapons and sensors; the latter, however, are inadequate for a vessel of this size, which carries armament and radar fits equivalent to those of two 'Tarantul' class missile corvettes. Classified as a 'battle cruiser'. Despite helicopter capability – typical of modern ocean escorts – she is not fitted with modern ASW systems. Non-operational since June 1988 to conserve funds. Modernised 1990–92 to improve stability: masts reduced in size, SSM launchers lowered by one deck and RBU-1200 replaced by RBU-6000. Renamed *Marasesti.* Extant 1995, reportedly for sale.

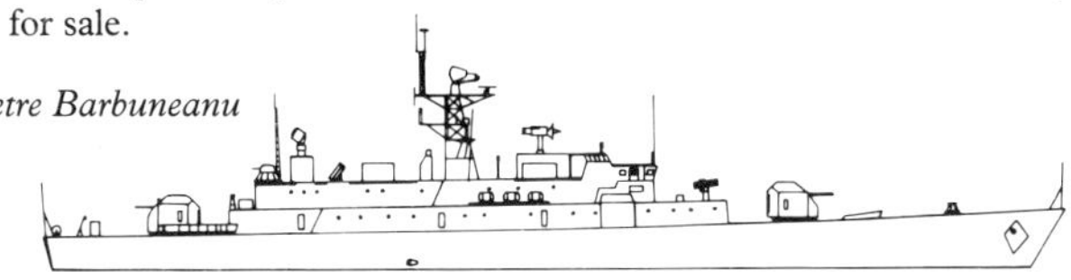

Admiral Petre Barbuneanu 1983

TETAL class *frigates*

Displacement: 1440t full load
Dimensions: 303ft 1in × 38ft 4in × 9ft 8in
92.4m × 11.7m × 3.0m
Machinery: 2 shafts, diesels, 20,000bhp = 24kts
Armament: 4–76mm (2×2), 4–30mm (2×2), 2–14.5mm MG (1×2), 4–533mm TT (2×2), 2 RBU 2500 ASW RL
Sensors: Radar Strut Curve, Drum Tilt; hull sonar; ECM Watch Dog
Complement: 98

No	Name	Builder	Launched	Comm	Fate
260	ADMIRAL PETRE BARBUNEANU	Mangalia Yard	1981	4.2.83	Extant 1995
261	VICE-ADMIRAL VASILE SCODREA	Mangalia Yard	1982	3.1.84	Extant 1995
262	VICE-ADMIRAL VASILE URSEANU	Mangalia Yard	1983	3.1.85	Extant 1995
263	VICE-ADMIRAL EUGENIU ROSCA	Mangalia Yard	1985	23.4.87	Extant 1995

Small AS escorts designed in Romania and based on Soviet weapons and sensors.

Modified TETAL class *frigates*

Displacement: 1500t full load
Dimensions: 303ft 1in × 38ft 4in × 10ft
92.4m × 11.7m × 3.1m
Machinery: As *Tetal* class
Armament: 1–76mm, 4–30mm (4×1), 4–533mm ASW TT (2×2), 2 RBU-6000 RL, 1 IAR-316 Alouette III helicopter
Sensors: Radar Strut Curve, Drum Tilt, Spin Trough, High Pole; sonar hull-mounted
Complement: 95

No	Name	Builder	Launched	Comm	Fate
265	CONTRE ADMIRAL EUSTATIU SEBASTIAN	Mangalia Yard	1988	30.12.89	Extant 1995
266	–	Magnalia Yard	1989	1992	Extant 1995

Improved version of *Tetal* class, with rearranged superstructure and a helicopter deck aft.

SUBMARINES

REQUINUL class *submarine*

Seized by the USSR on 29 August 1994 at Constanta and commissioned into Black Sea Fleet as *TS-1*. Renamed *N-39* in August 1947 and later *S-39* in June 1949. Returned to Romania July 1951, reverting to previous name. Deleted in the late 1950s and BU.

A near sister *Marsuinul* was also seized and incorporated into the Soviet Navy and renamed *TS-2* (later *N-40* and *S-40*). Sank at Poti on 20 February 1945 following a torpedo explosion, but was raised and repaired. Deleted 28 November 1950 and BU December 1950.

Ex-Soviet 'M V' type *submarines*

Four boats were transferred from the USSR in 1957 and served with the Romanian Navy until stricken in 1967. Names or numbers not known.

Ex-Soviet 'KILO' class *submarine*

No	Name	Acquired	Fate
521	DELFINUL	Dec 1986	Extant 1995

Transferred in December 1986, and was the first Romanian submarine commissioned since 1967. Too large for operational use in the Black Sea. Transfer of a second boat cancelled for financial reasons.

AMPHIBIOUS WARFARE VESSELS

BRAILA class *utilty landing craft*

LCU type landing craft, a dozen built in Romania probably in the late 1950s. Numbers in *400* series; nine were stricken 1983–90. Extant 1990: *420–421*.

SMALL SURFACE COMBATANTS

Ex-Soviet 'POTI' class *corvettes*

Class (no, commissioned):
Contre Admiral Nicolae Cristescu (31, 5.12.68), *Contre Admiral Nicolae Negru* (32, 1.4.69), *Contre Admiral Irimescu* (33, 20.10.69)

Built at Zelenodolsk, USSR, and transferred on completion. Extant 1995. Have 2–53cm TT instead of 400cm and RBU-2500 ASW RL instead of RBU-6000.

Ex-Soviet 'P 4' type *fast attack craft (torpedo)*

Class:
87–92

A dozen such craft were transferred from the USSR probably at the beginning of the 1960s. Only half of them remained in service in the mid-1970s carrying numbers *87* to *92*. All stricken by 1987.

Ex-Soviet 'OSA 1' class *fast attack craft (missile)*

Class (no, commissioned):
Soimul (194, 3.11.64), *Vulturul* (195, 3.11.64), *Uliul* (196, 31.12.64), *Acvila* (197, 27.12.64), *Eretele* (198, 1.1.65), *Albatrosul* (199, 25.2.81)

Six units purchased in the USSR during 1961–64. One boat stricken 1981 and replaced by the Romanian-built *Albatrosul*. Remainder extant 1995.

Chinese 'HUCHWAN' class *torpedo hydrofoils*

Class:
VT 51–77, 320–325

Three boats were acquired from China in the early 1970s, and others built locally under Chinese licence at Mangalia Shipyard, at a rate of about two per year during 1974–83 (fifteen of *51–77* series) and 1988–90 (*320–325*). Two of this class, *Mars* and *Jupiter*, were built without TTs and foils, and used as diving tenders. Ex-Chinese boats deleted 1993, remainder extant 1995.

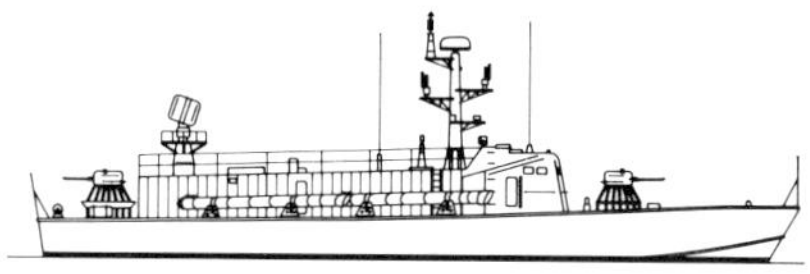

Epitrop class

'EPITROP' class *fast attack craft (torpedo)*

Displacement: 215t full load
Dimensions: 120ft 9in × 24ft 11in × 6ft
36.8m × 7.6m × 1.8m
Machinery: 3-shafts, M-503A diesels, 12,000bhp = 36kts. Range 500nm at 35kts
Armament: 4–30mm (2×2), 4–21in TTs
Sensors: Radar Pot Drum, Drum Tilt, High Pole
Complement: –

Class (no, commissioned):
Naluca (201, 19.5.79), *Smeul* (202, 25.10.79), *Viforul* (203, 14.1.80), *Vijelia* (204, 7.2.80), *Viscolul* (205, 30.4.80), *Virtejul* (206, 1.9.80), *Fulgerul* (207, 30.12.80), *Vintul* (208, 25.5.81), *Vulcanul* (209, 26.10.81), *Furtuna* (210, 13.1.82), *Trasnetul* (211, 15.6.82), *Tornada* (212, 5.10.82)

Composite design of the 'Osa' class FAC(M) hull and machinery but with torpedo armament. Built in Mangalia, the prototype appeared 1981. Extant 1995.

Ex-Soviet 'TARANTUL I' class *fast attack craft (missile)*

Class (no, commissioned):
Sborul (188, Dec 1990), *Lastunul* (189, Feb 1992), *Pescarusul* (190, Feb 1992)

Built at Petrovsky Shipyard, USSR, and later transferred to Romania. All extant 1995.

Ex-Soviet 'KRONSTADT' class *large patrol craft*

Class:
V1–V3

Purchased from the USSR 1956. Of 310t standard, rearmed 1980 to 2–37mm, 2–14.5mm MGs, 2 RBU-1200 ASW RLs (2×5), 2 DCTs, 2 DC chutes. Fitted with Ball End surface radar. Two deleted 1990, one in 1991.

Chinese 'SHANGHAI II' class *large patrol craft*

Class:
VP 20–40 (patrol version), *VS 41–44* (ASW version)

Three boats (*VP 22, 24, 25*) of this type were purchased in China in the early 1970s, and the others were built in Mangalia at a rate of about two per year since 1973. The patrol version is armed with 1–57mm, 2–37mm (1×2) while the ASW version is fitted with 1–37mm, 4–14.5mm MGs, 2–RBU 1200 ASW RLs (2×5), 2 DC racks. Two additional craft of this class, *Venus* and *Saturn*, were fitted with different superstructure and used as harbour security craft, and later as diving tenders. All extant 1995, though four are in reserve and gradually being cannibalised.

BRUTAR class *river monitors*

Displacement: 320t full load
Dimensions: 150ft × 26ft 4in × 4ft 9in
45.7m × 8.0m × 1.5m
Machinery: 2 shafts, 2 diesels, 2700bhp = 16kts
Armament: 1–100mm, 2–30mm (1×2), 10–14.5mm MG (2×4, 2×1), 1–122mm RL (1×40)
Sensors: Radar navigational
Complement: –

Class (no, commissioned):
Grivita (94, 21.11.86), *Rahova* (95, 14.4.88), *Opanez* (177, 24.7.90), *Smirdan* (178, 24.7.90), *179* (14.5.92), *180* (30.6.93)

Built at Mangalia. Operated by the Danube Flotilla. The main gun armament consists of a T-54/55 tank turret, and MGs are also turreted. Extant 1995.

VG class *river patrol craft*

Displacement: 40t full load
Dimensions: 52ft 5in × 14ft 4in × 4ft
16.0m × 4.4m × 1.2m
Machinery: 2 shafts, 2 3D12 diesels, 600bhp = 18kts
Armament: 1–20mm, 1–7.9mm MG
Complement: 10

Class:
VG 10–13 (stricken 1988), *VG 14–17* (stricken *c*1992)

Ten craft built in Galati in 1954 with steel hulls for service with the Danube Flotilla.

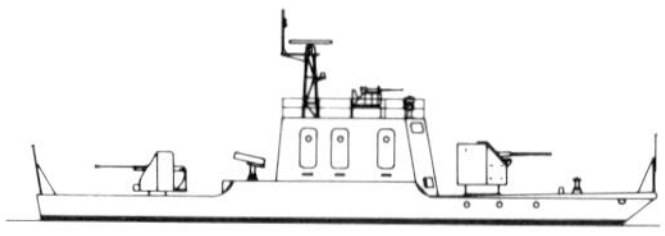

VB 93 1978 **NB 1/750 scale**

VB class *river patrol craft*

Displacement: 127t full load
Dimensions: 106ft 3in × 16ft × 3ft
32.4m × 4.8m × 0.9m
Machinery: 2 shafts, 2 diesels, 2400bhp = 17kts
Armament: 1–76mm, 4–14.5mm MG (2×2), 1–81mm mortar
Complement: 25

Class:
VB 76–93

Built at Dulcea Shipyard since 1973 at a rate of about two per year for the Danube Flotilla. Extant 1995.

SM class *river patrol boats*

Displacement: 22t
Dimensions: 40ft × 9ft 10in × 3ft
12.2m × 3.0m × 0.9m
Machinery: Diesels, 12kts
Armament: 2 MGs
Complement: –

Class:
SM 161–164 (stricken 1989), *SM 165–173* (stricken 1990)

Built in Romania 1954–56.

SD class *river patrol boats*

Class:
SD 270, 274, 275, 277, 278

Began building late 1970s in Romania. Fitted with minesweeping gear. All deleted 1990.

O.23E type *hovercraft*

One experimental 14t, 20ft 1in × 12ft × 6ft 2in (*6.1m × 3.6m × 1.9m*), air cushion vehicle powered by one M-337 engine of 210bhp, completed 1987. Has endurance of 4.5hrs and a crew of three.

NOTE:
Five Romanian river monitors (*Ardeal, Besarabia, Bratianu, Bucovina, Lahovari*) seized by the USSR in 1944 were returned on 3 July 1951. Renamed *M 207, M 206, M 202, M 205* and *M 201* respectively, they were all deleted 1959–60.

MINE WARFARE VESSELS

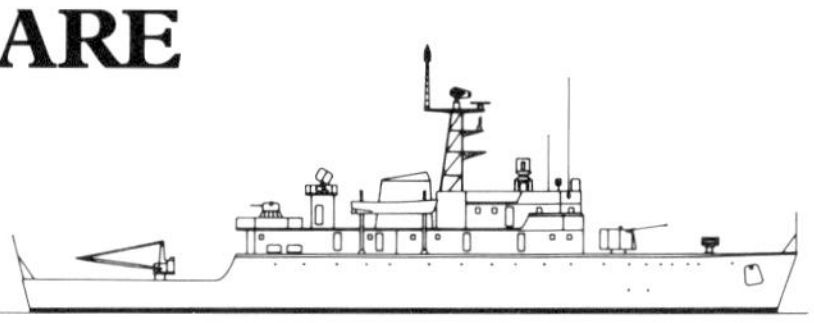

V-A Constantin Balescu as completed

COSAR class *minelayers*

Displacement: 1200t standard; 1450t full load
Dimensions: 259ft 2in × 34ft 9in × 11ft 6in
79.0m × 10.6m × 3.5m
Machinery: 2 shafts, diesels, 6400hp = 19kts
Armament: 1–57mm, 4–30mm (2×2), 8–14.5mm MG (2×4), 2 RBU-1200 ASW RL (2×5), 200 mines
Sensors: Radar Strut Curve, Don-2, Muff Cob, Drum Tilt; hull mounted sonar; ECM Watch Dog
Complement: 75

Class (no, commissioned):
Vice Admiral Ioan Murgescu (271, 30.12.80), *Vice Admiral Constantin Balescu* (274, 16.11.81)

Minelaying/MCM support ships designed and built at Mangalia Shipyard. *Murgescu* fitted with helicopter platform aft for IAR-316 Alouette III. Both extant 1995.

Ex-German 'M 40' type *coastal minesweepers*

Class:
Democratia (13), *Descatusaria* (14), *Desrobirea* (15), *Dreptatea* (16)

German Second World War minesweepers of *M Boot* 1943 design, 543t standard displacement. Built at Galati as coal-burners; converted to oil 1951, rearmed with 6–37mm (3×2), 4 MGs, DCTs. Silhouettes have changed considerably due to conversions 1976–83, which included fitting of navigation radar, two RBU-1200 RLs and a helicopter platform aft. The bridge was reshaped and a lattice mast fitted. Rerated as corvettes and renamed in 1993 as *Vice Admiral Mihai Gavrilescu, Vice Admiral Ioan Balanescu, Vice Admiral Emil Grecescu* and *Vice Admiral Ioan Georgescu* respectively. All extant 1995.

MUSCA class *minesweepers*

Displacement: 790t full load
Dimensions: 199ft 4in × 31ft 1in × 9ft 2in
60.8m × 9.5m × 2.8m
Machinery: 2 shafts, 2 diesels, 4800bhp = 17kts
Armament: 4–30mm (2×2), 16–14.5mm MG (4×4), 2 RBU-1200 ASW RL
Sensors: Radar Krivach, Drum Tilt; sonar hull-mounted
Complement: –

Class (no, commissioned):
Lieutenant Remus Lepri (24, 23.4.87), *Lieutenant Lupu Dinescu* (25, 6.1.89), *Lieutenant Dimitrie Nicolescu* (29, 7.12.89), *Sub-Lieutenant Alexandru Axente* (30, 7.12.89)

Built at Mangalia Shipyard. Probably have minelaying capability. All extant 1995.

Ex-Soviet 'T 301' class *inshore minesweepers*

Class (no, commissioned):
Captain Constantin Dumitrescu (4, 1956), *Lieutenant Constantin Virtosu* (5, 1957), *Lieutenant Gheorghe Niculescu* (6, 1957), *Lieutenant Moise Pauta* (7, 1957), *Lieutenant Nicolae Stoicescu* (8, 1957), *Lieutenant Alexandru Pompilian* (9, 1958), *Commodore Dumitru Lupascu* (17, 1958), *Commodore Vasile Toescu* (18, 1959), *Commodore Constantin Micoescu* (19, 1959), *Captain Adrian Vasiliu* (26, 1960), *Captain Constantin Istrate* (27, 1960), *Captain Alexandru Constantinescu* (28, 1960), *DR 1–3*, *DR 11–16*, *DR 21–25*, *DR 29*

Twenty-seven of these 146t standard displacement steel-hulled vessels were transferred from the USSR 1955–60. Individual craft have been stricken since the mid-1970s, and the remainder were laid up in the mid-1980s. Six deleted in 1990, and three more in 1991. Named ships above modernised since 1986 at Magnalia Shipyard and are extant 1995.

Soviet 'TR 40' type *river minesweepers*

Class:
VD 421–428

Eight incomplete hulls built 1955–60 were purchased in Poland and transferred to Romania. All completed there by 1960. Stricken 1983–85.

VD 141 class *minesweeping boats*

Displacement: 97t full load
Dimensions: 109ft × 15ft 7in × 2ft 8in
33.2m × 4.8m × 0.8m
Machinery: 2 shafts, 2 'M50' diesels, 1200bhp = 13kts
Armament: 4–14.5mm MG; can carry mines
Complement: –

Class:
VD 141–165

Built at Dobrata Severin Shipyard during 1976–84. All extant 1995.

NOTE:
The minelayer *Amiral Murgescu* was seized by the USSR in Constanţa on 29 August 1944 and commissioned in the Soviet Navy as *Don* on 14 September 1944, becoming a training ship in April 1945. Until May 1956 she served as a headquarters ship, and was then hulked and renamed *PKZ-107*. Converted to a floating workshop in January 1958 and renamed *PM-76* (*PMR-76* since June 1966). Deleted 27 May 1988 and BU July 1988.

St Kitts

The two islands which make up the state of St Kitts-Nevis form part of the Leeward group of islands in the eastern Caribbean. St Christopher is the larger of the two (68 square miles) and also contains the capital, Basseterre. The official status was a State in Association with Britain from February 1967 and the state became an independent nation in September 1983. The islands' governor-general is appointed by the British sovereign and Britain remains responsible for defence. The police force operates two patrol boats and one patrol craft.

One 'Spear' class patrol boat, *Ranger I* entered service 10 September 1974, built by Fairey Marine. 4.3t, 29ft 10in in length, 30kts, 2 diesels, 360hp, 2–7.62mm MGs, 2 crew. Extant 1995.

One Swiftships 110ft patrol craft *Stalwart* (C 253) delivered in 1985 (99t full load, 35.5m, 2400bhp = 21kts, 2–12.7mm, 2 MG). Extant 1995.

St Lucia

A mountainous and heavily forested island state, St Lucia in the Windward group of islands was a British colony until February 1979; the British sovereign is still the country's head of state. Twenty-seven miles long, the island covers 238 square miles and has a population of about 153,000. The Customs Service operates a patrol boat from Castries, a port that is highly regarded in the area because of the shelter it affords in heavy weather.

One 40ft Brooke Marine type patrol craft, *Helen*, completed 20 July 1970 and built by Brooke Marine, Lowestoft. 14t, 22kts, 2 Cummins diesels, 370hp, 2–7.62mm MGs. Sold 1981.

One Swift 65ft patrol boat *Defender* (P 02) delivered 1983 (42t, 22kts). Extant 1995.

St Vincent

One of the Windward Islands, lying in the eastern Caribbean, the state of St Vincent incorporates several of the Grenadines and covers an area of 150 square miles (388 sq km). In 1979 independence was granted and the head of state is the British sovereign, represented by a governor-general. The Police Force operates a Marine Wing based at Kingstown employing one patrol craft.

One 45ft Brooke Marine type patrol craft; *Chatoyer*, 15t, 21kts, 2 Cummins diesels, 370hp, 3 MG. Lost in a hurricane 1979.

One 75ft Vosper type patrol craft; *George McIntosh* (SVG 05), entered service March 1981, 70t, 24.5kts, 2 Caterpillar diesels, 1–20mm, 11 crew, built by Vosper, Portchester. Replacement for *Chatoyer*. Extant 1995.

One Swiftships 120ft patrol craft *Captain Mulzac* (SVG 01) delivered by Swiftships in 1987 (101t, 36.6m, 2700bhp = 21kts, 2–12.7mm, 2 MG). Extant 1995.

Saudi Arabia

Of all the Gulf States the Kingdom of Saudi Arabia is certainly the largest and most powerful with a coastline bordering both the Persian Gulf and Red Sea. With the 1355-mile coastline, much of which is unpopulated, divided between two seas, Saudi Arabia has a considerable problem in maintaining adequate surveillance over such an area, as well as presenting a cohesive naval presence to meet a threat in any one particular area. And in addition she formed the Gulf Co-operation Council with the region's smaller states in 1981. The Kingdom has a substantial merchant fleet including several modern roll-on/roll-off vessels, containerships, and passenger vessels.

Until a few years ago the Royal Saudi Navy (founded with one ship in 1960) was little more than a token force of three MTBs of German origin and small patrol boats. In 1991 Saudi Arabia announced a record budget of 143 billion rials ($38.1bn), 36 per cent going to defence.

A US expansion programme undertaken by the Saudis was planned as a ten year project (1972–82) and probably undertaken to match the Iranian Navy. This programme included: four MSC 322 type coastal minesweepers, four *Badr* class missile corvettes and nine *Al Siddiq* class missile fast attack craft. In 1980 a new programme was launched, this time involving the French naval industry. With these two programmes completed the Saudis have a well-balanced, medium-sized fleet. Ships ordered in France under the 'Sawari' programme included four helicopter-carrying frigates and two large underway fleet replenishment ships similar to the French *Durance* class. In addition twenty-four helicopters were ordered. Twenty of these were armed with anti-ship missiles and four equipped for the SAR role.

These expansion programmes, initiated even before a May 1981 announcement that defence spending would be at a record level, came about because of Saudi unease at the emergence of Iraq as a major military force in the region. This unease was increased by the spread of Soviet influence, the religious revolution in Iran and the bitter Iran–Iraq war.

The Saudi Government envisaged problems with the upkeep of equipment from difference sources (US and France) and so US ships were provided on the basis of a ten year naval co-operation agreement which included an extensive training programme; a similar 'total package' agreement had been concluded with France. The main naval bases are at Jiddah and Jubail with minor ones at Yanbu and on the Gulf at Ras Tannurah, Dammam and Ras al-Mishab.

In 1979–80 the Saudi Navy still had fewer than 2000 men and little more than 1000t of fighting vessels. By 1982 this had gone to 3500 volunteer officers and ratings and by 1985 personnel strength had grown to 4500, in 1995 to 11,400 officers and men including 1200 marines, and in 1992 10,800 officers and men (including 1200 marines).

Expansion plans still very much underway with six British *Sandown* class minehunters being built as the *Al Jawf* class, and three 3800t frigates planned (probably modified French *La Fayette* class) together with up to eight submarines. Six submarines may be ordered in the near future in a French yard. If such a deal went through Saudi personnel would be trained at Lorient using the *Daphne* class submarine *Flore*. Orders are however unlikely until the mid-1990s. It was also reported that up to two midget submarines of 225t and six smaller KSS-1 Tolgorae boats of 150/175t may be acquired from South Korea.

Saudi personnel are still familiarising themselves with some of the new equipment acquired by the navy in the 1980s. Interestingly the Saudi Navy had little involvement in the international naval activities in the Gulf in 1990–91. Saudi naval aviation currently has at its disposal twenty-three Aerospatiale SA 365F Dauphin 2 type helicopters, six Aérospatiale SA 322/BI transport helicopters and six US Lockheed P-3C Orion aircraft. There are plans to acquire six more P-3Cs.

MAJOR SURFACE SHIPS

For some years Saudi Arabia had planned to acquire fully fledged AAW frigates from France. An outline agreement was signed in 1989 for three to be built in Lorient, according to the design based on the *La Fayette* class. Two frigates of this class will be built by DCN under the new 'Sawari II' contract signed on 19 November 1994.

F 30005 class (Improved French LA FAYETTE class) *frigates*

Displacement: 3700 full load
Dimensions: 423ft 3in × 50ft 6in × 13ft 5in
129.0m × 15.4m × 4.1m
Machinery: 2 shafts, 4 SEMT-Pielstick 16 PA6 280BTC diesels, 38,400bhp = 28kts
Armament: 8 MM 40 Exocet SSM (2×4), 1–VLS for Aster SAM (1×16) or Crotale SAM, 12 Matra Sadral (Mistral) SAM (2×6), 1–100mm/55 Creusot Loire, 2–30mm/85 Emerson Emerlec (1×2), 4–324mm ASW TT (2×2)
Complement: 160

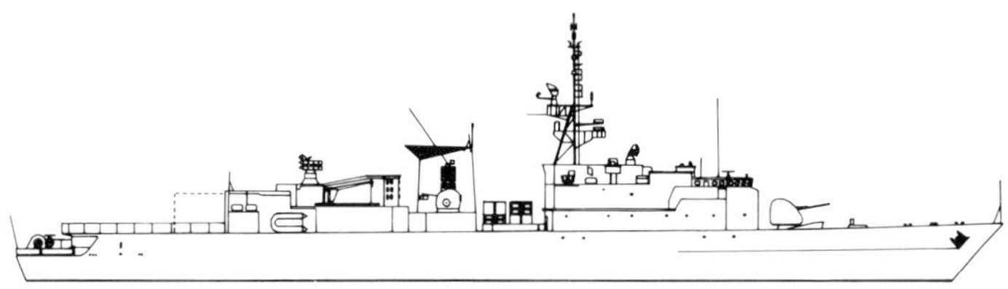

Al Madinha as completed

These four helicopter-carrying frigates were orded under the 'Sawari' programme in October 1980 from France: the first contract for frigate-type warships won by the French naval industry after the Second World War. The ships are fitted with modern armament and electronics and are capable of operating independently, greatly enhancing the Saudi Navy's capability.

Although the frigates are well armed, emphasis so far has been on anti-surface vessel capability which in view of the worthwhile targets which the Saudi Navy presents to those with hostile intentions, should be made good at the earliest opportunity.

AL MADINAH class *frigates*

Displacement: 1990t standard; 2610t full load
Dimensions: 349ft 5in wl, 377ft 4in oa × 41ft × 15ft 4in
106.5m, 115.0m × 12.5m × 3.4/4.7m
Machinery: 2 shafts, 4 SEMT-Pielstick diesels, 16 PA 6 BTC diesels, 32,500bhp = 30kts. Range 8000nm at 15kts or 6500nm at 18kts
Armament: 8 Otomat Mk 2 SSM (2×4), 1–100mm/55 Compact, 4-40mm/70 Breda-Bofors (2×2), 1 8-cell Crotale SAM launcher (24 missiles), 4–550mm ASW TT (F17P torpedoes), 1 SA 365F Dauphin helicopter with 2–4 AS-15 missiles
Sensors: Radar 2 Decca 1226, Thompson-CSF Sea Tiger DRBV-15, DRBC-32E, Castor II FCS, 3 CSEE Naja optronics FCS; Sonar Thompson-CSF Diodon hull mounted sonar, Diodon/Sorel VDS; ECM: Thompson-CSF DR 4000ESM, 2 CSEE Dagaie chaff launchers, SENIT VI CCS
Complement: 179

No	Name	Builder	Laid down	Launched	Comp	Fate
702	AL MADINAH	Lorient Naval Dockyard	15.10.81	23.4.83	4.1.85	Extant 1995
704	HOFOUF	CNIM, La Seyne (Toulon)	14.6.82	24.6.83	31.10.85	Extant 1995
706	ABHA	CNIM, La Seyne (Toulon)	17.12.82	23.12.83	4.4.86	Extant 1995
708	TAIF	CNIM, La Seyne (Toulon)	1.3.83	25.5.84	29.8.86	Extant 1995

Al Madinha Mar 1985
L & L van Ginderen

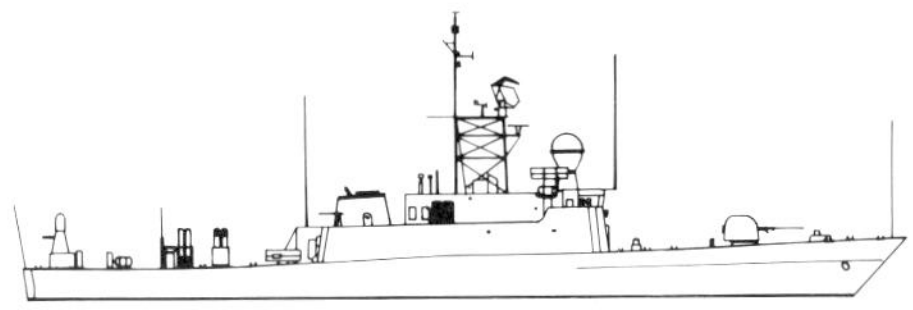

Badr 1990

These ships ordered on 30 August 1977 from Tacoma Boatbuilding, Tacoma, USA belong to the light frigate/corvette type. Named after early Islamic victories. They are powerfully armed for their size, and equipped with modern electronic gear. Fitted with fin stabilisers, the ships were completed well behind schedule and are considered to be overweight by over 25 per cent.

BADR class (US PCG type) *missile corvettes*

Displacement: 903t standard; 1038t full load
Dimensions: 245ft oa × 31ft 6in × 14ft 7in
74.7m × 9.6m × 4.5m
Machinery: 2-shaft CODOG: 1 General Electric LM 2500 gas turbine, 23,000shp = 30kts, plus 2 MTU diesels, 3000hp = 20kts
Armament: 8 Harpoon SSM (2×4), 1–76mm/62 OTO Melara, 2–20mm (to get Vulcan-Phalanx CIWS), 1–81mm mortar, 2–40mm army grenade launchers, 6–324mm Mk 32 ASW TT (2×3)
Sensors: Radar SPS-40B, SPS-55, Mk 92 FCS; sonar SQS-56, EW SLQ–32
Complement: 58 (7 officers, 51 ratings)

No	Name	Builder	Laid down	Launched	Comp	Fate
612	BADR (ex-*PCG 1*)	Tacoma Boatbuilding	9.4.79	6.10.79	30.11.80	Extant 1995
614	AL YARMOOK (ex-*PCG 2*)	Tacoma Boatbuilding	20.8.79	13.5.80	1.9.81	Extant 1995
616	HITTEN(ex-*PCG 3*)	Tacoma Boatbuilding	7.1.80	5.9.80	2.9.82	Extant 1995
618	TABUK (ex-*PCG 4*)	Tacoma Boatbuilding	22.9.80	18.6.81	10.1.83	Extant 1995

AMPHIBIOUS WARFARE VESSELS

AL QIAQ class (US LCU 1646) *utility landing craft*

Class:
Al Qiaq (ex-*SA 310*), *As Sulayel* (ex-*SA 311*), *Al Ula* (ex-*SA 312*), *Afif* (ex-*SA 313*), Nos 212, 214, 216, 218

Belong to the US LCU 1646 type. Built by Newport SY, Newport, USA, in 1975/76. Temporary US designations SA 310–313 (SA = Saudi Arabia). Note only even pennant numbers. Extant 1995.

DHEBA class (US LCM(6) type) *landing craft*

Class:
Al Leeth (224), *Al Quonfetha* (226), *Haqil, Dheba* (220), *Umlus* (222), *212, 214, 216, 218*

These eight US LCM(6) type were transferred in two batches in 1977. Numbered boats have been stricken.

SMALL SURFACE COMBATANTS

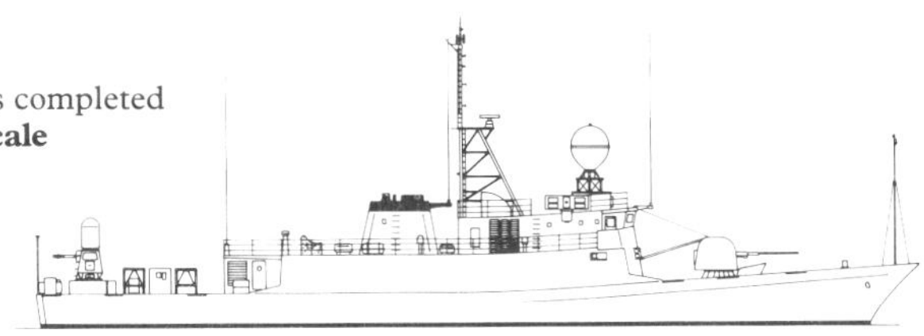

Al Farouq as completed
NB 1/750 scale

AL SIDDIQ class (US PGG type) *fast attack craft (missile)*

Displacement: 425t standard; 495t full load
Dimensions: 190ft 4in oa × 26ft 6in × 6ft 5in
58.02m × 8.08m × 1.95m
Machinery: 2-shaft CODOG: 1 GEM LM 2500 gas turbine, 6,500shp = 38kts, plus 2 MTU diesels, 3000bhp = 18kts
Armament: 4 Harpoon SSM (2×2), 1–76mm/62 OTO Melara, 2–20mm (2×1), 1 Vulcan-Phalanx CIWS (20mm), 1–81mm mortar, 2–40mm Army grenade launchers
Electronics Radar SPS–55, Mk 92 FCS
Complement: 35

Class (no, former name, in service):
Al Siddiq (511, ex-*PGG1*, 15.12.80), *Al Farouq* (513, ex-*PGG2*, 22.6.81), *Abdul Aziz* (515, ex-*PGG3*, 23.11.81), *Faisal* (517, ex-*PGG4*, 23.11.81), *Khalid* (519, ex-*PGG5*, 11.1.82), *Amir* (521, ex-*PGG6*, 21.6.82), *Tariq* (523, ex-*PGG7*, 16.8.82), *Oqbah* (525, ex-*PGG8*, 18.10.82), *Abu Obaidah* (527, ex-*PGG9*, 6.12.82)

Ordered on 16 February 1972 from Peterson Builders, Sturgeon Bay, Wisconsin, USA. They have the same machinery as the larger *Badr* class corvettes and the same armament (minus four Harpoon SSMs and ASW TTs). Heavily overweight, displacing 495t full load. Very powerful vessels for their size. Named after members of the royal family. Note lower speed on the diesels compared with the larger *Badr* class. Completed in 1980-82; the first boat was delivered on 8 December 1980 last in November 1982. All extant 1995.

Tariq 10.6.93 L & L van Ginderen

DAMMAM class (German JAGUAR type) *fast attack craft (torpedo)*

Class:
Dammam, Khabar, Maccah

German *Jaguar* class (Type 141) ordered in 1968 and built by Lürssen, Vegesack in 1968–69. All three were refitted during 1976–77. Currently used for training. Extant 1995.

AL JOUF class *large patrol craft*

Displacement: 210t full load
Dimensions: 126ft 8in × 26ft 3in × 6ft 3in
38.6m × 8.0m × 1.9m
Machinery: 3 shafts, 3 MTU 16V538 TB93 diesels, 11,265bhp = 38kts. Range 1700nm at 15kts
Armament: 2–20mm Oerlikon (2 × 1), 2–12.7mm (2 × 1)
Complement: 20

Class (no):
Al Jouf (351), *Turaif* (352), *Hail* (353), *Najran* (354)

Four fast patrol craft ordered on 18 October 1987 from Blohm & Voss and completed in pairs on 15 June and 20 August 1989 respectively. Steel hull and light alloy superstructure. Operated by Coast Guard. Extant 1995.

SEA GUARD type *fast attack craft (missile)*

Displacement: 53t standard
Dimensions: 73ft 10in × 18ft 4in × 5ft 7in
22.5m × 5.6m × 1.7m
Machinery: 2 shafts, 2 MTU diesels, 2920bhp = 35kts
Armament: 2 Aérospatiale AS-15TT SSM (2×1), 2–12.7mm (2×1)
Complement: ?

Class:
Al Riyadh, Zulrab

Two small FACs built by Simoneau Marine in 1992. Operated by Coast Guard. Extant 1995.

MINE WARFARE VESSELS

ADDRIYAH class (US Bluebird type) *coastal minesweepers*

Class (no, former name):
Addriyah (412, ex-*MSC 322*), *Al Quysymah* (414, ex-*MSC 323*), *Al Wadeeah* (416, ex-*MSC 324*), *Safwa* (418, ex-*MSC 325*)

Ordered 30 September 1975 from Peterson Builders, Wisconsin. Laid down 1976–77, launched 1976–77 and delivered in 1978–79. Built in USA under International Logistics Programme. Note only even pennant numbers. Extant 1995.

AL JAWF class (UK Sandown type) *single role minehunters*

Displacement:	450t standard; 480t full load
Dimensions:	172ft 9in × 34ft 5in × 6ft 11in *52.7m × 10.5m × 2.1m*
Machinery:	2 Voith-Schneider cycloidal propellers, 2 Paxman 6RP200E diesels, 1500bhp = 13kts. Range 3000nm at 12kts
Armament:	2–35mm/85 Emerson Emerlec (1 × 2), 2 Loral Hycor SRBOC Mk36 Mod 1 decoys
Sensors:	Radar Kelvin Hughes 1007; fire control Contraves TMEO optronic director; sonar Plessey Nautis M
Complement:	34 (7 officers)

Class (no, former name, in service):
Al Jawf (420, ex-*Inverness*, 12.12.91), *Shaqra* (422, 7.2.93), *Al Kharj* (424, 1994), *Onaizah* (426), *Al Rass* (428), *Al Bahan* (430)

Six ships of the British *Sandown* class were ordered on 3 December 1988 to be built by Vosper Thornycroft at Woolston and Portsmouth. First vessel, taken from the Royal Navy building line, delivered 1991, the second being launched on 15 May 1991 and delivered 1993. Option for the last three has yet to be taken up. GRP construction. Differ from RN version in having the twin 35mm mount raised one-half deck. Fitted with two Schottel bow thrusters for improved manoeuvrability. Silent electric drive gives speed of 6kts. Carry two PAP 104 Mk 5 mine disposal ROVs.

OTHER PATROL CRAFT

One ex-USCG 95ft large patrol craft: *Riyadh* (former US contract designation 95CGMDA9) launched in 1959 as one of the US CG 'Cape' type was built and transferred under MDAP from the USA in 1960. Steel hulled. Belongs to the Coast Guard. Built by US Coast Guard Yard, Curtis, Maryland, USA. Stricken 1990.

Sixteen SRN-6 type hovercraft. Eight acquired from the British Hovercraft Corporation Ltd February–December 1970, eight more delivered 1981–83; 10t, 58kts, 1050bhp, 1–7.62mm MG. Seven extant 1995.

There are also sixty-eight small patrol boats of under 10m length.

Two large patrol craft *Badr* and *Yarmouk* built by Bayerische Schiffsbau in the early 1970s: 65t full load, 2 GM diesels = 25kts, 2–12.7mm. Extant 1995.

Twelve *Rapier* class (Nos 127–138), launched 1976–77 by Halter Marine, New Orleans; 24t, 28kts, 1300bhp, 2–7.62mm MG. Extant 1995.

Twenty-five *Scorpion* class (Nos 140–164), launched 1979–81 by Bayerische Schiffsbaugesellschaft (twenty) and Arminiuswerft, Bodenwerder (five); 33t, 25kts, 1300bhp. Extant 1995.

Two *CGV-26 Explorer* type large patrol craft *Al Jubatel* and *Salwa* built by Abeking and Rasmussen, completed April 1987: 95t full load, 85ft 11in × 19ft × 6ft 9in, *26.2m × 5.8m × 2.1m*, 2 MTU diesels, 5800bhp = 35kts, 1–20mm Oerlikon GAM-B01, 2–12.7mm. Operated by Coast Guard. Extant 1995.

Seventeen Halter type coastal patrol craft delivered by Halter Marine in 1991–92; 56t full load, 2 Detroit diesels, 1380bhp = 28kts, 2–12.7mm MG, 2–7.62mm MG. Extant 1995.

Forty Simoneau 51 type inshore patrol craft built by Simoneau Marine, delivered in 1989–92; 22t full load, 2 diesels, 2400bhp = 33kts, 1–20mm GIAT, 2–7.62mm MG. Mostly used by naval commandos. Extant 1995.

Eight SRN6 Mod 8 hovercraft acquired in mid-1980s from British Hovercraft Corp; 17t normal, 55kts, military lift 30 troops, 1–7.62mm MG. Extant 1995.

Sixteen Slingsby SAH 2200 hovercraft supplied from UK in 1990. There are plans for thirty of such craft; 10.6m, 40kts, military lift 24 troops, 1–7.62mm MG. Extant 1995.

One Pegasus type hydrofoil delivered in 1985 by Boeing, Seattle; 115t full load, 27.4m, 2 gas turbines, 8660shp = 46kts, 2–20mm GE Sea Vulcan (Gatling). Mostly used as escort and tender to the Royal Yacht. Extant 1995.

Senegal

This West African republic, 80 per Muslim, gained independence from France in 1960, and has since maintained a fair degree of political freedom and stability under Léopold Sédar Senghor, one of Africa's best known and most distinguished leaders. Senghor retired at the end of 1980, the first recorded voluntary withdrawal from office of a civilian black African president. Senegal has tried various parliamentary arrangements, and is at present successfully operating a multi-party system. On 1 February 1982 she united with her tiny riverine neighbour The Gambia to form the Confederation of Senegambia, but this was later dissolved on 30 September 1989.

In foreign affairs, demonstrations of solidarity with liberation movements in Africa and the Middle East have increased; thus Senegal, like most African countries, broke diplomatic ties with Israel in October 1973, during the Yom Kippur War. Relations with the former colonial power have generally been good, with numerous cooperation agreements in operation. The March 1974 defence agreement allows France to keep a reinforced marine infantry battalion near Dakar and a French Navy Breguet Atlantic maritime patrol aircraft there, emphasising its strategic location on the South Atlantic and South American sea routes.

A border dispute with Mauritania was defused early in 1992, diplomatic relations were restored and the border reopened. There is an insurgent separatist movement in the Casamance region in the south, and although a ceasefire was agreed between the government and that movement in July 1993, there is still no long-term political agreement.

Senegal's French trained and equipped armed forces total about 9700. The small Navy (formed in 1961) of 350 two-year conscripts operates a coast patrol force craft from the capital, Dakar. The other Atlantic base is St Louis at the mouth of the Senegal river which is patrolled from Rosso, Podor and Kaedi. Kaolack is another river base.

NJAMBUUR *fast attack craft*

Particulars: As Moroccan *Okba* class

PR 72M type ordered from SFCN in 1978, launched 23 December 1980 and delivered September 1981. She entered service in July 1982 as P 773. Fire control by two CSEE Naja optical directors. Armament also differs from Moroccan boats being 2–76mm/62 OTO Melara Compact and 2–20mm F2. Could be fitted with Exocet. Extant 1995.

OSPREY 55 type *large patrol boat*

Particulars: As Moroccan *Osprey 55* type

Class (in service):
Fouta (1.6.87) Built by Danyard, Denmark for fishery protection duties. Extant 1995.

ST LOUIS class *large patrol craft*

Particulars: As Ivory Coast *Le Vigilant*

Class (launched):
St Louis (5. 8.70), *Popenguine* (22.3.74), *Podor* (20.7.76)

P 48 type. *St Louis* was ordered from Ch Franco-Belges, other two from SFCN. Laid down 20 April 1970, 20 November 1973, 5 February 1976, completed 5 August 1970, 22 March 1974, 20 July 1976. Decca navigation radar. All extant 1995.

Ex-Chinese 'SHANGHAI II' *large patrol craft*

Class:
Banjul, *Cuajur*

Both transferred by China on 2 February 1989 after refit with 6–25mm (2 × 3) and new radar. Both extant 1995.

MISCELLANEOUS

Wooden-hulled submarine chaser, formerly the US SC 1344, transferred to France in 1943 (as *CH 24*, re-rated as patrol vessel P 700 in 1951) and from France and Senegal in 1961 as the first Senegalese warship – *Sénégal*. Discarded 1974.

The French seaward patrol craft *VC 5* (P 755) and *VC 4* (P 754) (ex-Congolese *Reine N'Galifourou*) were acquired in 1963 and 1965, and renamed *Casamance* and *Siné-Saloum* respectively. Both discarded 1979.

There are also three coastal patrol craft of the 62ft Canadian-built 'Interceptor' class named *Senegal II, Casamance II* and *Siné-Saloum II* (in service since 1979) used for fishery protection, an ex-French LCT of the EDIC type *Falémè* (transferred 7 January 1974) returned to France and replaced by the newly-built *Karakene* 23 June 1986 and two ex-US LCM (6) boats, transferred July 1968 and named *Djomboss* and *Douloulou*. Both discarded 1990.

The Gambia added three British-built coastal patrol boats (20mm guns, delivered 1974–78) and 50 men to Senegal's Navy, which returned to the Gambia in 1989. Senegal also received two 16-metre 20kt Spanish-built patrol craft on 10 February 1982.

Seychelles

The Seychelles Indian Ocean archipelago of ninety-two granite and coral islands, independent in 1976 and with a left-wing non-aligned government from June 1977, has a 1000-strong defence force trained by Tanzanian military advisers, and an embryo naval force (organic to the Army) consisting of 150 sailors, the ex-French *Sirius* class coastal minesweepers *Croix du Sud*, transferred as a gift on 20 January 1979 as a patrol boat (without sweeping gear) and renamed *Topaz* (for particulars, see France section), and the 350t LCT *Cinq Juin* (ordered 12 December 1977) used as an inter-island transport. *Topaz* was discarded in 1987 but refitted in 1990. *Cinq Juin* is also extant 1995. The small patrol boat *Junon* was discarded in 1991. Two ex-Soviet 'Zhuk' class were transferred, *Constance* in 1981, and *Fortune* in 1982, and the ex-Soviet 'Turya' class *Zoroaster* in 1986. The 42m FPB class patrol boat *Andromache* was built in Italy in 1983 and refitted in 1985–86. These four also remain in service 1995.

Sierra Leone

Sierra Leone, which became independent in 1961, acquired her first naval units in 1973 when three 'Shanghai II' class patrol boats were transferred from China. These were discarded in the early 1980s, and two more – *Moa* and *Maimbana* – were transferred in 1987. These are currently in service alongside three *Pompoli* class 634t landing ships (*Pompoli, Gulama* and *Kallondo*) built in Japan in 1980 and a Swift 110ft patrol boat *Farandugu*, purchased from the USA in 1987. The 160 man force covers the 250 mile coastline, including the excellent natural anchorage at Freetown.

Singapore

Singapore became internally self-governing in 1959. She had been a British possession almost uninterruptedly since 1819; first ruled by the East India Company, then controlled from India during 1826–67, and finally as the separate colony of the Straits Settlements. Britain retained control of the naval base until a final withdrawal in March 1976. Singapore joined Malaysia (formerly Federation of Malaya) when it was set up in September 1963, but pulled out following disputes and Chinese–Malay clashes in August 1965. She then set up her own armed forces, replacing expatriates with islanders, but retained links with Malaysia. Singapore is now a member of ASEAN (the Association of South East Asian Nations), which is gradually taking on aspects of a military alliance. In 1980 the Soviet Pacific Fleet lost the Singapore repair facilities it had used heavily in the 1970s, indeed more than any other foreign Asian port.

Almost all trade to and from the Far East passes the port of Singapore, giving it great strategic significance and making it one of the world's major trading centres. Her spectacular economic growth is today attested by a merchant fleet of 988 ships totalling 7,664,229grt. This has assisted Singapore's drive to self-sufficiency – most of her warships have been built in the two specialist warship yards and she now exports warships to other navies. The country only consists of Singapore Island itself, the size of the Isle of Wight, plus a small archipelago of fifty-four other islands, but despite her small size she has a powerful and efficient force of gun- and missile-armed patrol boats. These are well-suited to policing the restricted waters around Singapore, and coping with pirates and infiltrators, but other navies would have to be called on to cope with large-scale mining or submarines. In 1994 plans were announced to create a submarine force and a programme was begun to replace the *Seawolf* class FACs.

In 1994 the service's 4000 men (2–3 year conscripts and regular volunteers) were commanded by a colonel.

SUBMARINES

Ex-German Type 206 *submarines*

Particulars: See German Type 206

Late in 1994 the Ministry of Defence announced that it hopes to acquire up to six redundant Type 206 diesels-electric submarines. These will undoubtedly be upgraded to Type 206A or similar standard before being recommissioned.

Ex-Indian 'FOXTROT' class *submarine* (Project 641)

Particulars: See Russian 'Foxtrot' class

As part of the preparations for establishing a submarine force the Singaporean Navy will lease or buy a Russian-built 'Foxtrot' class boat from the Indian Navy in 1995.

AMPHIBIOUS WARFARE VESSELS

Ex-US LST 542 type *tank landing ships*

Class (no, former name):
Endurance (L 201, ex-*Holmes County*, LST 836), *Excellence* (L 202, ex-*LST 629*), *Intrepid* (L 203, ex-*LST 759*), *Perseverance* (i) (L 206, ex-*LST 623*), *Persistence* (L 205, ex-*LST 613*), *Resolution* (L 204, ex-*LST 649*)

Nine ex-US LSTs, all 1944-built, were acquired by Singapore, but three were sold without entering naval service. *Endurance* was loaned from the USN on 1 July 1971 and purchased on 5 December 1975. Ex-USS *LST 276* and *LST 532* were transferred in June 1974 but were sold commercially. The other six were originally numbered A 81 to A 86. *Endurance* has been leased commercially since 1976. *Excellence, Intrepid* (only a stern 40mm gun each, former has helicopter pad) and *Resolution* were rebuilt in 1977 with a derrick forward of the bridge and a pole mast in place of the lattice mast. *Perseverance* (i), *Persistence* and *Resolution* discarded 1992. Remainder extant 1995.

Ex-British SIR LANCELOT class *tank landing ship*

Class (no, former name, commissioned):
Perseverance (ii) (L ?, ex-*Lowland Lancer*, ex-*Sir Lancelot* 21.11.92)

Former RN logistic ship, sold commercially 1989 and converted to car ferry. Bought October 1992 and converted back to military use. Extant 1995.

AYR CHAWAN class *landing craft*

Displacement:	150t full load
Dimensions:	88ft 6in × 22ft × 4ft *27.0m × 6.9m × 1.3m*
Machinery:	2 shafts, 2 diesels, 650bhp = 10kts
Armament:	None
Complement:	?

Class:
Ayr Chawan, Ayer Merban and two others

Built by Vosper Thornycroft, Singapore 1968–69 and extant 1995.

MINOR LANDING CRAFT

Two *Brani* class LCU (56t full load, 55ft 8in × 14ft × 4ft 7in, *17m × 4.3m × 1.4m*, 2 diesels, 460bhp = 9kts): *Brani* and *Berlayer* built in Australia 1955–56 and stricken *c*1991.

SMALL SURFACE COMBATANTS

VICTORY class *missile corvettes*

Displacement:	550t normal; 600t full load
Dimensions:	196ft 6in pp, 206ft 6in oa × 30ft 6in × 8ft 6in *59.9m, 62.95m × 2.6m*
Machinery:	4-shaft, 4 MTU 20V538 98 diesels = 35kts
Armament:	8 AGM-84 Harpoon SSM, 1 Barak 2 VLS (32 missiles), 1–76mm L/62 OTO Melara, 4–12.7mm (4 × 1), 6–324mm ILAS-3 ASW TT
Sensors:	Radar Sea Giraffe 150HC, Type 1007; sonar Salinon VDS
Complement:	49

Class (no, in service):
Victory (P 88, 18.8.90), *Valour* (P 89, 18.8.90), *Vigilance* (P 90, 18.8.90), *Valiant* (P 91, 25.5.91), *Vigour* (P 92, 25.5.91), *Vengeance* (P 93, 25.5.91)

Variant of Lürssen's MGB 62 design (see Bahrain) but with different electronics and weapons. Design has not proved successful because of excessive topweight. The big mast, similar to that on the Israeli *Hertz* class, holds Israeli EW equipment. It may be replaced by two lighter masts to solve topweight and windage problems. Barak 2 VLS installed in 1994.

Victory May 1990 L & L van Ginderen

Sea Dragon as completed
NB 1/750 scale

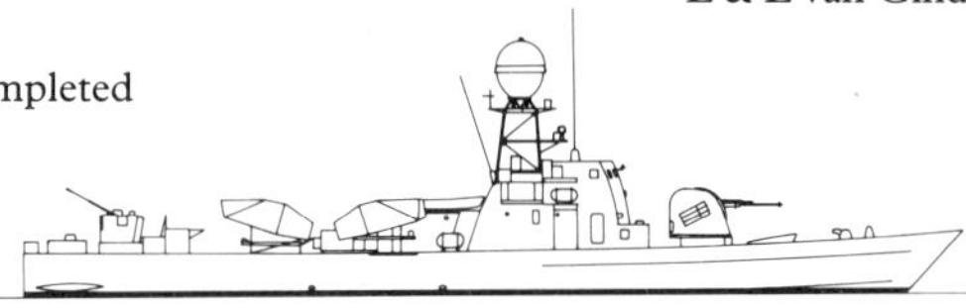

SEA WOLF class *fast attack craft*

Armament:	5 Gabriel SSM (1 × 3, 2 × 1), 1–57mm (504rds), 1–40mm/70 (1008rds), 2 RFL on 57mm
Other particulars:	See Bahrain and Kuwait

Class (no):
Sea Wolf (P 76), *Sea Lion* (P 77), *Sea Dragon* (P 78), *Sea Tiger* (P 79), *Sea Hawk* (P 80), *Sea Scorpion* (P 81)

Six Lürssen TNC 45 FACs ordered in 1970. Apart from machinery and armament they are similar to those supplied since to the Gulf navies. Armed with the first Gabriel SSMs to be exported. *Sea Dragon*, *Sea Hawk*, *Sea Scorpion* and *Sea Tiger* all built at Singapore with assistance from Lürssen. First pair commissioned 1974, next pair 1975 and the final two in 1977. Re-equipped with Harpoon SSMs and new electronics 1985–92. All extant 1995.

Vosper 110ft Type A and B *fast attack craft*

Displacement:	100t standard; 130t full load
Dimensions:	103ft 8in wl, 109ft 8in oa × 21ft × 5ft 8in *31.6m, 33.5m × 6.4m × 1.8m*
Machinery:	2 shafts, 2 MTU 16V 38 diesels, 7200hp = 32kts. Range 1100nm at 15kts
Armament:	Type A: 1–40mm/70, 1–20mm Type B: 1–76mm70, 1–20mm
Sensors:	Type A: Radar Decca TM 626 Type B: Radar HSA M 26, Decca TM 626
Complement:	Type A: 22 Type B: 19

Class (no, launched):
Independence (P 69, 15.7.69), *Freedom* (P 70, 18.11.69), *Sovereignty* (P 71, 25.11.69), *Justice* (P 72, 20.6.70), *Daring* (P 73, 1970), *Dauntless* (P 74, 6.5.71)

Six Vosper Thornycroft 110ft patrol boats were ordered on 21 May 1968. Three have a 40mm gun forward (Type A) and three a 76mm set further aft and a shorter superstructure (Type B). Type B also have Signaal M 26 Radome atop the short lattice mast. Steel hull and aluminium alloy superstructure, developed from the Malaysian 103ft type. First unit of each type built at Vosper Thornycroft's Portsmouth Yard; *Independence*, Type A, delivered 8 July 1970, and *Sovereignty*, Type B, delivered February 1970. Remainder built at Vosper Thornycroft's Singapore yard; *Freedom* and *Justice* (Type A) delivered 11 January 1971 and 23 April 1971 respectively, and *Daring* and *Dauntless* (Type B) delivered September and December 1971 respectively. All extant 1995, but are to be replaced by *Fearless* class.

Sovereignty (Type B) as completed Vosper Thornycroft

Independence (Type A) Vosper Thornycroft

Lürssen 'FPB 57' *fast attack craft*

Particulars: See Kuwait

Three ordered in 1980 from Singapore SB, Jurong. First laid down June 1980. Armament of 1–76mm OTO Melara (300rds) and 2–35mm Oerlikon GDM-A (1 × 2, 2750rds), possibly SSM as well. Susie-1 passive intercept ESM. Three more may be built for export. Similar to German Type 143 (S 61) class.

FEARLESS class *fast attack craft*

Displacement:	?
Dimensions:	147ft 8in wl, 180ft 5in oa *45.0m, 55.0m*
Machinery:	?
Armament:	4/6 Gabriel SSM, 1–76mm L/62, 6–324mm ILAS-3 ASW TT
Sensors:	?
Complement:	?

Class:
Fearless, 2–6

Six new missile-armed FACs ordered in 1994, and first of class building at Singapore SB, Jurong.

British 'FORD' type *large patrol boat*

Panglima (P 68) was built at United Engineers, Singapore and launched on 14 January 1956 to the design of the British 'Ford' class seaward defence boats but with 1–20mm and fifteen crew. Commissioned May 1956, transferred to Royal Malaysian Navy in 1963 when Federation formed. Returned to Singapore 1967 and now used as TS.

ENDEAVOUR *large patrol craft*

Displacement:	250t full load
Dimensions:	135ft × 25ft × 8ft *40.9m × 7.6m × 2.4m*
Machinery:	2 shafts, 2 Maybach diesels, 2600bhp = 20kts. Range 800nm at 8kts
Armament:	2–20mm (2 × 1)
Complement:	24

Endeavour (P 75) was built by Schiffswerft Oberwinter, W Germany in 1955 and purchased from the Netherlands on 30 September 1970. Used as TS and diving tender.

MINE WARFARE VESSELS

Ex-US MSC type *minesweepers*

Class (no, former name):
Jupiter (M 101, ex-*Whippoorwill*), *Mercury* (M 102, ex-*Thrasher*)

Former US coastal minesweepers built in 1955 and acquired by Singapore in 1975. Rearmed with new 20mm in 1980. Relegated to auxiliary duties. *Jupiter* discarded 1990, *Mercury* extant 1995.

BEDOK class *minehunters*

Displacement:	310t normal; 360t full load
Dimensions:	147ft 8in wl, 155ft 10in oa × 31ft 6in × 7ft 6in *45.0m, 47.5m × 9.6m × 2.30m*
Machinery:	2 shafts, 2 Saab-Scania diesels, 1440bhp = 15kts
Armament:	1–40mm/70 Bofors, 2–7.62mm MG
Sensors:	Radar WM 20; sonar TSM 2002
Complement:	39

Class (no, completed):
Bedok (M 105, 1994), *Kalau* (M 106, 1995), *Katang* (M 107, 1995), *Punggol* (M 108, 1996)

Original order for two modified *Landsort* type from Karlskronavarvet, Sweden made April 1991, followed by an order for a further two. M 108 left Sweden 17 August 1993 as deck cargo. Remainder fitting out at Singapore SB.

MISCELLANEOUS

Twelve 'Swift' class fast patrol craft (45t, 34kts, 1–20mm, 1–7.62mm MG, crew 13); P 10–P 21 built 1980–81 by Singapore SB, Jurong, based on Australian De Havilland 'Capricornia' design. All formally commissioned 20 October 1981. Operated by Singapore Marine Police.

Four Vosper 'PX' type coastal patrol craft: in service since 1969. Twenty PC 32 class speedboats in service since 1978–79 and nineteen larger craft (PX 14–33) building at Sembawang SY.

Solomon Islands

The Solomon Islands, best known for the 1942–43 US-Japanese naval campaign, became an independent state within the Commonwealth in July 1978. The island group covers 249,000 sq nm in the southwest Pacific with six major islands in the 900-mile chain. An Australian-built *Carpentaria* class coastal patrol craft, *Tulagi*, policed the archipelago since 30 March 1979. The 26m Australian P-150 class *Savo* (ex-*Pioneer*) was added in 1984, and two ASI 315 'Pacific Forum' patrol boats. *Lata* and *Auki*, were ordered in October 1985 and built in 1988–89. All are extant 1995.

Somalia

The Somali Republic was formed in 1960 by the union of the former UN Trust Territory (and ex-Italian colony) of Somalia and the former British Somaliland. The country is wretchedly poor and under-developed; since time immemorial a sparse, predominantly nomadic population has scratched a subsistence living from arid savannah and barren uplands.

Politically, the corrupt and divided civilian government since independence was overthrown in 1969 by an authoritarian military regime. China provided aid for major civilian projects, while the Army came to rely on Soviet equipment and training; in return, the Soviet Navy developed a small base at Berbera, in the north adding a floating pier and drydock, barracks, and missile store as well as expanding oil storage and the airfield. The Russian Indian Ocean Squadron used this port more than any other in the period 1962–80. In 1974 the creeping transformation of the Ethiopian Government to military socialism was accompanied by spreading internal chaos and led to increasing demands by the Ogaden Somalis for self-determination. Soviet and Cuban support for Ethiopia caused Somalia to break her Communist ties and expel all Russians in November 1977, realigning herself with the West.

Open fighting flared in the Ogaden in 1977 and 1978; initial Somali advances were halted around Harar and Diredawa, and an Ethiopian counter-attack, spearheaded by Cubans, drove the invaders from the country. Guerrilla warfare in the Ogaden continued for many years, and an estimated 1.3 million Somali refugees have fled from the fighting to a number of camps.

The Somali Army doubled in size after 1977. The 1974 treaty with the USSR was abrogated in 1977, and Soviet military aid ceased. In August 1980 the USA signed an agreement to provide Somalia with $17.4 million worth of defensive weapons and equipment in return for port facilities in Berbera and Mogadishu (Kismayu is the other main naval base), but has held back on more military aid because of continued Somali involvement in the Ogaden. The revolution in January 1991 overthrew Major-General Siad Bare. This was followed in 1992 by the three-year American and UN intervention, which failed to resolve the situation. Most of the navy has been destroyed, and the surviving vessels are not operational.

The following were in service between 1960 and 1991:

Two ex-Soviet 'Osa II' class fast attack craft (missile) transferred in December 1975.

Four Soviet 'MOL' class fast attack craft, new units transferred in 1976; two without TT.

Four ex-Soviet 'P 6' class fast attack craft (torpedo) transferred 1968.

Five ex-Soviet 'Poluchat I' class large patrol craft transferred 1965 (two) and 1966 (three).

Ex-Soviet 'Polnocny A' class landing ship transferred December 1976.

Four ex-Soviet 'T 4' class landing craft transferred 1968–69.

One Swift 105ft patrol boat P202 defected from Ethiopia in 1984.

South Africa

The South African Naval Service was founded in 1922, and between the wars operated two minesweeping trawlers and a survey ship; most of the seamen were South Africans while the Royal Navy provided the officers. From this very modest beginning, in a period when the RN's South Atlantic Squadron, based at Simonstown, still shielded South Africa, the SANS expanded enormously during the Second World War, providing personnel for RN ships and manning armed trawlers and whalers for coastal patrol work. In 1944–45 the RN transferred three 'Loch' class frigates, in addition to the two 'River' class frigates and some eighty smaller craft in service. Personnel totalled over 8000 by July 1945, and 3000 more were serving in the RN.

The South African Navy was officially founded in 1946, but with rapid demobilisation the new navy kept only the three 'Loch' class frigates, a minelayer, eleven harbour defence motor launches and two boom defence vessels. The RN protection remained, but the Nationalist Party government elected in 1948 found itself increasingly at odds with Britain, and took steps to increase its independence in defence matters; to its navy, already strengthened by the acquisition of two fleet minesweepers in 1947, were added two destroyers in 1950–52 and two coastal minesweepers in 1954.

Under the 1955 Simonstown Agreement, Britain handed over her base to South Africa (1 April 1957) and the SAN was expanded to take responsibility for the Cape sea route. Six fast anti-submarine frigates (three conversions, later reduced to one, and three new construction), eight coastal minesweepers and four seaward defence boats would be purchased from Britain, as well as Buccaneer strike aircraft and Shackleton maritime patrol aircraft. Britain and her allies would in return be able to use the extensive fuelling, storage, repair and communications facilities at Simonstown, even in a war which did not involve South Africa. Joint training and exercises with the RN and attendance of naval personnel at courses in the UK kept the SAN up to date.

On becoming a republic in 1961, South Africa was compelled to leave the Commonwealth, the first stage towards her status of pariah among nations. Militarily, she increasingly began 'looking to her moat', but arms procurement became steadily harder; Britain first refused to provide submarines as offensive weapons, and then started making difficulties about defensive weapons, although supplying a survey ship in 1969. France, always alert to a profitable arms deal, was more accommodating, and three *Daphne* class boats were ordered from Dubigeon-Normandie. A new submarine base was built at Simonstown, with an enclosed repair yard and a Synchrolift marine elevator, capable of lifting any South African warship except the replenishment ship *Tafelberg*.

In 1974 the British Labour Government abrogated the Simonstown Agreement, giving up base facilities and ending all naval co-operation. Increasingly isolated, South Africa tried to replace obsolete tonnage and augment her naval strike capability by ordering two *Agosta* class submarines and two type A 69 corvettes in France in 1975–76, but the 1977 mandatory UN arms embargo prevented their delivery, and they were sold to Pakistan and Argentina respectively. An important order for six missile-armed fast attack craft placed with Israel in 1974 was not affected by the embargo, which is not observed by Israel. Naval personnel reached 6758 (including 1574 two-year conscripts) by 1981 (plus 10,000 reservists), a notable increase on the figures of 2370 in 1960 and 3440 in 1970. A Marine Corps was formed for harbour defence in July 1979 but was disbanded in 1990. The Advokaat radar system scans the republic's 2100 nautical mile coasts and monitors 600,000 tons of shipping passing the Cape every day. The principal naval base is Simonstown and forces have been withdrawn from Durban as an economy measure.

The initial enthusiasm for fast attack craft soon waned when their operational limitations were exposed by the demanding South Atlantic sea conditions. As early as 1981 it was predicted that locally-built corvettes would replace them, and this plan began to take formal shape in 1991.

The submarine force was modernised, using local resources, but there was an acute need for two more hulls to ease the pressure on personnel. The French embargo on the *Agosta* class was a bigger blow than losing the two A 69s in 1978.

The end of the war in Angola and Namibian independence led to severe cutting of defence budgets. Inevitably the SAN was severely affected, being forced to defer modernisation plans and to cut personnel. In 1995 total personnel stood at 5034. The new ANC government is willing to support the armed forces' requests for modernisation, but the weakened state of the economy threatens navy plans.

FLEET STRENGTH 1947

FRIGATES

No	Name	Launched	Disp (std)	Fate
F 432	GOOD HOPE (ex-*Loch Boisdale*)	5.7.44	1435t	Scuttled 1976
F 10	NATAL (ex-*Loch Cree*)	19.6.44	1435t	Survey ship 1957
F 602	TRANSVAAL (ex-*Loch Ard*)	2.8.44	1435t	Scuttled 1976

Presented by RN in 1944–45. In 1955 *Good Hope* was modified as a despatch vessel/training ship, with 2–4in/45 Mk 16 (1×2) forward, a raised funnel and a reception platform aft. *Transvaal* modernised along similar lines in 1962, but with the forecastle extended aft for extra accommodation. Both ships sunk as breakwaters in 1976; *Natal* sunk as a target in September 1972.

MISCELLANEOUS
There were also the following minor fighting vessels; the harbour (later seaward) defence motor launches *1197–1207*, built in South Africa during the Second World War and discarded during 1953–75; the controlled minelayer *Spindrift* (later *Skilpad*); an ex-German trawler, sold 1958; and the 'Bar' class boom defence vessels *Barcross* (later *Somerset*) and *Barbrake* (later *Fleur*), respectively deleted 1986 and sunk as a target in October 1965.

MAJOR SURFACE SHIPS

Ex-British 'W' class *destroyers*

No	Name	Builder	Acquired	Fate
D 278	JAN VAN RIEBEECK (ex-*Wessex*, ex-*Zenith*)	Fairfield	29.3.50	Sunk as target May 1980
D 237	SIMON VAN DER STEL (ex-*Whelp*)	Hawthorn Leslie	1952	BU 1976

These destroyers, transferred from the RN, underwent modified Type 16 frigate conversions in 1962–66, their particulars being altered as follows: displacement 2105t standard, 2750t full load; draught (max) 17ft (5.2m); armament 4–4in/45 Mk 16 (2×2, in 'B' and 'Y' positions), 4–40mm/60 (4×1), 4–3pdr sauluting, 4–21in TT (1×4), two DCT, two DC racks, two Wasp ASW helicopters; complement 192. The 21in TT were removed *c*1972, and 6–12.75in Mk 32 ASW TT (2×3) added. *Jan van Riebeeck*, sunk as a target in May 1980, was hit by a Skorpioen (Gabriel II) SSM from FAC *Jim Fouche* and subsequently torpedoed by submarine *Maria van Riebeeck*.

Ex-British RAPID class *frigates*

No	Name	Builder	Acquired	Fate
F 157	VRYSTAAT (ex-*Wrangler*)	Vickers-Armstrong, Barrow	29.11.56	Sunk as target Apr 1976

Type 15 conversion by Harland & Wolff of former 'W' class destroyer, purchased from the RN. Arrived Cape Town 19 February 1957. Two further Type 15 purchases (one was to have been HMS *Roebuck*) were cancelled in 1956. *Vrystaat* was torpedoed in a submarine training exercise.

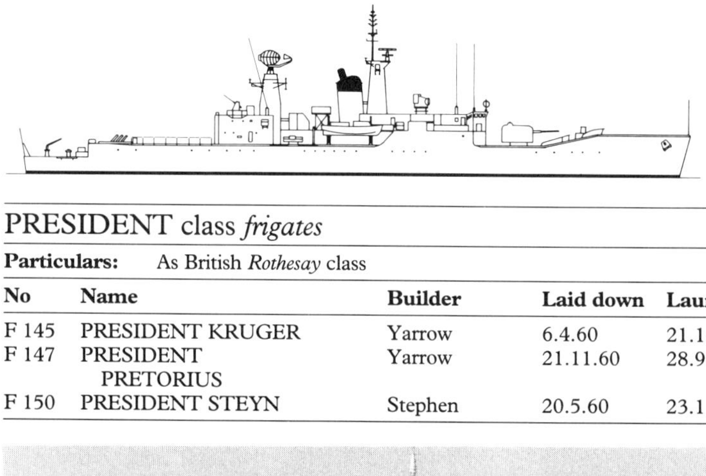

President Kruger 1979

PRESIDENT class *frigates*

Particulars: As British *Rothesay* class

No	Name	Builder	Laid down	Launched	Comp	Fate
F 145	PRESIDENT KRUGER	Yarrow	6.4.60	21.10.60	1.10.62	Sunk 18.2.82
F 147	PRESIDENT PRETORIUS	Yarrow	21.11.60	28.9.62	4.3.64	Sold 1990
F 150	PRESIDENT STEYN	Stephen	20.5.60	23.11.61	26.4.63	Sold 1990

President Steyn 4.8.72 L & L van Ginderen

As built, almost identical with the RN *Rothesay* class, although having a twin 40mm Mk 5 at main deck level in place of the STAAG Mk 2. All three refitted at Simonstown (*President Pretorius*, 1968–70, *President Steyn*, 1969–71, *President Kruger*, 1971–77) with a plated foremast, French Jupiter surveillance radar on a new plated mainmast, and Italian Elsag NA9C fire control in place of the Mk 6M director (the latter not in *President Kruger* until 1979). A small hangar for a Wasp helicopter added abaft the mainmast, with 2–40mm/60 singles on top, and the forward Limbo well was plated over to form a flight deck. Six 12.75in Mk 32 ASW TT (2×3) were added *c*1972. *President Kruger* was given a modified funnel cap to carry smoke clear of the foremost radar arrays *c*1975. *President Steyn* was decommissioned in August 1980 to be cannibalised for her sisters. The two survivors were laid up at Simonstown for some years stripped of armament and sensors. *President Kruger* was lost 80 miles southwest of the Cape on 18 February 1982 with thirteen of her crew in a night collision with the replenishment ship *Tafelberg* (A 243).

Replacements for the old 'Loch' class frigates with the same names ordered from France in 1976, two A69 units building for the French Navy being made available. Differed slightly from standard French design, having USN-pattern Mk 32 triple ASW TT on the quarterdeck. In September 1978, when both ships were almost ready, France cancelled the contract, because of the UN arms embargo on South Africa. Sold to Argentina on 25 September 1978 and renamed *Drummond* and *Guerrico* respectively.

GOOD HOPE class *corvettes*

Particulars: As French *d'Estienne d'Orves* class

No	Name	Builder	Laid down	Launched	Comp	Fate
F 432	GOOD HOPE (ex-*Lieutenant de Vaisseau Le Hénaff*)	Arsenal de Lorient	12.3.76	16.3.78	Mar 1978	See notes
F 102	TRANSVAAL (ex-*Commandant l'Herminier*)	Arsenal de Lorient	1.10.76	Sep 1977	Nov 1978	See notes

During the 1980s strenuous efforts were made to find replacements for the *President* class, including clandestine construction or purchase, in defiance of the UN embargo. Currently plans are for four light frigates displacing about 1500t, and armed with a 76mm gun, electronics and

Light frigates/Offshore patrol vessels

Skerpioen missiles to be removed from stricken *Minister* class vessels. Yarrow Shipbuilders in the UK and EN Bazán in Spain were shortlisted in November 1994. The project was deferred in 1995 because of funding problems and political opposition within the new ANC government. Plans for smaller patrol craft have been dropped.

SUBMARINES

Similar to those built in France for the French, Pakistani and Portuguese Navies, and in Spain for the Spanish Navy.

All three underwent a mid-life overhaul, with upgraded sonar and displays, but original French torpedoes retained. *Emily Hobhouse* recommissioned July 1988, *Johanna van der Merwe* in late 1990, *Maria van Riebeeck* in 1992.

Efforts were made to buy two *Daphne*s from Portugal in the early 1980s.

MARIA VAN RIEBEECK class *submarines*

Particulars: See French *Daphne* class

No	Name	Builder	Laid down	Launched	Comp	Fate
S 97	MARIA VAN RIEBEECK	A C Dubigeon	14.3.68	18.3.69	22.6.70	Extant 1995
S 98	EMILY HOBHOUSE	A C Dubigeon	18.11.68	24.10.69	25.1.71	Extant 1995
S 99	JOHANNA VAN DER MERWE	A C Dubigeon	24.4.69	21.7.70	21.7.71	Extant 1995

Maria van Riebeeck 1.4.92

L & L van Ginderen

Ordered in June 1975, but in 1978 France cancelled the contract because of UN arms embargo on South Africa. Became Pakistani *Hashmat* and *Hurmat* respectively. In the mid-1980s rumours circulated of a feasibility study into building two submarines, probably to a German Type 209 design, but almost certainly using Israeli technical assistance.

French AGOSTA class *submarines*

Name	Builder	Laid down	Launched	Comp	Fate
ASTRANT	A C Dubigeon	15.9.76	14.12.77	17.2.79	See notes
ADVENTUROUS	A C Dubigeon	1976	1.12.78	17.2.80	See notes

SMALL SURFACE COMBATANTS

P W Botha 1994 L & L van Ginderen

MINISTER class *fast attack craft (missile)*

Particulars: As Israeli *Reshef* class

Class (no, completed):
Jan Smuts (P 1561, 18.7.77), *P W Botha* (P 1562, 2.12.77), *Frederick Creswell* (P 1563, 6.4.78), *Jim Fouché* (P 1564, 22.12.78), *Frans Erasmus* (P 1565, 27.7.79), *Oswald Pirow* (P 1566, 4.3.80), *Hendrik Mentz* (P 1567, 11.2.83), *Kobie Coetzee* (P 1568, 11.2.83), *Magnus Malan* (P 1569, 4.7.86)

First three boats ordered from Haifa Shipyard, Israel, in late 1974. Material was supplied for three more to be built by Sandock Austral Navy Yard, Durban, and a further six at 50m rand ($56m) were ordered from Sandock Austral (three later cancelled) each under a contract signed on 15 November 1977. All named after previous Ministers of Defence. Identical with the Israeli *Reshefs*, apart from using a South Africa produced version of the Gabriel II SSM called Skerpioen (Scorpion). *P W Botha, Kobie Coetzee* and *Oswald Pirow* recommissioned in 1986–89 after overhauls. The remaining six were scheduled for replacement by corvettes, but the continued deferment of those plans led to a major service life extension programme. Major elements include a new combat system, upgrading of the Skerpioen missile system, new engine-room monitoring system, new target-designation system and improved 76mm ammunition. *Frederick Cresswell* was the first to undergo this modernisation, recommissioning in 1995. Some of the class are likely to be laid up and cannibalised for spares.

British 'FORD' class *seaward defence boats*

Class (no, former name, transfer/launched/in service):
Gelderland (P 3105, ex-*Brayford*, 30.8.54), *Nautilus* (P 3120, ex-*Glassford*, 23.8.55), *Rijger* (P 3125, 5.2.58, Dec 1958), *Haarlem* (P 3126, Aug 1959), *Oosterland* (P 3127, 27.1.59, 6.12.59)

Two transferred from the RN in 1954–55, three built in Britain for the SAN in 1957–59. All for disposal 1987. (*Haarlem* served as an inshore survey craft from 1963.)

MINE WARFARE VESSELS

Ex-British ALGERINE class *minesweepers*

No	Name	Builder	Acquired	Fate
M 439	BLOEMFONTEIN (ex-*Rosamund*)	Port Arthur	1947	Sunk as target June 1967
M 291	PIETERMARITZBURG (ex-*Pelorus*)	Lobnitz	1947	Hulked 1976

Purchased from the RN in 1947, *Pietermaritzburg* converted to a midshipmen's TS in 1961–62, with new armament of 2–4in/45 Mk 16 (1×2), 2–40mm/60 Bofors, and accommodation added aft.

Ex-British 'TON' class *coastal minesweepers*

Class (no, former name, transferred):
Johannesburg (M 1207, ex-*Castleton*, 1958), *Kimberley* (M 1210, ex-*Stratton*, 1958), *Port Elizabeth* (M 1212, ex-*Dumbleton*, 27.10.58), *Mosselbaai* (M 1213, ex-*Oakington*, 1958), *Walvisbaai* (M 1214, ex-*Packington*, 1958), *East London* (M 1215, ex-*Chilton*, 27.10.58), *Windhoek* (M 1498, launched 28.6.57), *Durban* (M 1499, launched 12.6.57), *Pretoria* (P 1556, ex-*Dunkerton*, Aug 1955), *Kaapstad* (P 1557, ex-*Hazleton*, Sep 1955)

Eight transferred from the RN in 1955–59, two built in Britain for the SAN in 1957. *Kimberley* (1977–78) and *Port Elizabeth* (1979–80) converted to minehunters – with Type 193M sonar, Type 1006 radar and 2 PAP-104 remote mine disposal vehicles. M 1207, M 1212–13, M 1499, M 1556–57 decommissioned in 1987 for lack of trained personnel, and scrapped. Four survivors extant 1995 are undergoing major hull renewal, with upgraded MCM capability.

'RIVER' class *inshore minehunters*

Class (no):
Umkomaas, Umhloti, Umzimkulu, Umgeni
Four 48m, 380t craft built by Dorbyl (formerly Sandock Austral) to a German design, and commissioned 1986–87. Wooden-hulled and equipped with Klein sidescan sonars. Another four may be ordered, but a larger design is needed to replace the coastal MCM craft.

Soviet Union 1947–1991: Russian Federation and Successor States 1991–

When the first edition of this book was published in 1982, the Soviet Union was a nuclear superpower. She had built up the world's second most powerful blue-water fleet, and in some measures had even surpassed the United States. Now much of that fleet is in ruins, and with it much of the power formerly attributed to the Soviet Union. During 1991 the nationalist stresses, which had always plagued the country, became so strong that the USSR itself was officially dissolved at the end of the year. It was replaced, at least in theory, by a Commonwealth of Independent States (CIS; Russian equivalent SNG). When this was written, in early 1995; it was impossible to say whether the former Soviet unitary state was really dead. The newly independent republics, first Ukraine, then Russia and the others, have all claimed their own military forces, but the Russian President, Boris Yeltsin, argued that the central forces, including the navy, should survive in unitary form, controlled by some sort of CIS government. He finally consented to removing the navy from the 'strategic forces' in 1992, so that the fleet could eventually be divided. Only Russia and Ukraine have very significant coasts, so they will share the bulk of the former Soviet naval forces. Ukraine includes the major centre for building large surface warships, Nikolayev. Late in 1991 warships there included an incomplete but launched carrier, *Varyag*, and portions of the nuclear-propelled carrier *Ulyanovsk*, which were later broken up. Whether *Varyag* will ever be completed is unclear. Ukraine has formed its own navy, although it is unclear what portion of the former Soviet Black Sea Fleet it will get.

Nor is it clear that all the old republics will continue to be fully independent. Their economies were tightly integrated, to the extent that, for example, Russia depends heavily on food from Ukraine. It is by no means certain that a Ukraine bent on independence, and more interested in exporting its products for hard currency than on feeding Russia, will ultimately be tolerable to a much larger and better-armed Russia.

The collapse of Soviet military (including naval) production was quite sudden. Naval production rates, particularly those for submarines, remained high throughout the 1980s, including the early years of Mikhail Gorbachev's tenure. By 1988, Gorbachev was announcing major cuts, but it was by no means certain that he could (or would) enforce them on a military production machine bound by the inertia of a planned economy which had been built up largely for its benefit. It appears that the 1986–90 Five-Year Plan showed very little in the way of production cuts or of conversion of military equipment plants to civilian purposes. The most impressive change was the well-publicised scrapping of obsolete warships. Westerners were treated to the sight of barge-loads of 'Whiskey' class submarines going to foreign shipbreakers (largely because so few Soviet breaking facilities existed: old Soviet warships often had simply been beached and abandoned rather than scrapped; in some cases this practice led to a Western belief that no ships were discarded, but the beached units could not have been recommissioned). Numerous surface warships, including some surprisingly recent units, were discarded. However, the rate of new ship production, particularly submarine production, seemed unchanged.

The economic reforms and political uncertainty of 1991, however, deeply affected military production. From early 1991 there was no new Five-Year Plan (1991–95), largely because no such plan could be written for the emerging mixed economy. Gorbachev had largely emasculated the one means of economic and political control he had, the Communist Party, but he had not produced any substitute. As suppliers faltered, the shipyards slowed their own production. Western projections of Soviet warship launchings during 1991 proved wildly optimistic. Individual yards began to scramble for commercial business in hopes of obtaining hard currency from abroad, and thus of feeding and clothing their many workers. For example, the vast submarine yard at Severodvinsk found itself building commercial barges for Germany.

The failure of an attempted coup in August 1991 showed that the Party had lost its cohesion (the coup plotters called for a government of national unity, not a return to Communism) and also that Gorbachev himself had largely lost the initiative. It was seized by Boris Yeltsin, who was more interested (at least for the time) in the power of his own republic than in the cohesion of the Soviet Union as a whole. Hence the formal dissolution of the Union at the end of 1991.

By that time, the command economy was clearly in ruins; the former Soviet Union was suffering hyperinflation and, among other ills, hoarding was causing severe shortages of essentials. In the scramble for hard currency, much of the fleet was being sold off, either for scrap or for use abroad. This sale of Soviet military equipment will probably strongly effect the balance of power in much of the Third World for the next five or ten years, before the ships wear out and the lack of follow-on servicing and spares becomes evident.

For some time, nationalism within the old republics had been growing openly. Many of the republics are themselves multi-national, so that it is by no means clear that they are natural coherent units. It seems very likely that dissolution, or at least attempts at dissolution, will continue for some time, and that the historical process begun in the late 1980s is far from over. The most appropriate analogy might be the Russian Civil War, during which Lenin was initially quite willing to let the peripheral regions (such as Ukraine) go, gaining them back once he had consolidated his power in the primary area, Russia.

The drive for naval power was probably among the causes of the collapse of the old system. Throughout the 1970s, the Soviet military economy seemed to outperform that of the West. Some sceptics did guess that the quality of Soviet military equipment, particularly electronics, did not approach that common in the West, but this view was not widely held. Similarly, maintenance of the mass conscription system (and little or no success in building a corps of long-term enlisted men) made it increasingly difficult to maintain more and more sophisticated weapons and related systems. This problem, not at all apparent to outside observers, was revealed only as Soviet military writers became more open. It seems to have been particularly felt in the submarine service, but it is now apparent that the surface fleet also suffered its share of embarrassing accidents.

From 1979 on, the USSR faced two great strains. One was the difficulty of warfare in Afghanistan. Military failure was a political problem because it undermined the sense of omnipotence which kept the Party in power. Throughout Russian history, failure in peripheral wars (which failed to stimulate Russian patriotism) had led to political convulsions. For example, disaster in the Crimea led to the call for modernisation which caused the liberation of the serfs in 1861. Failure in the Far East in 1904–5 led to a failed revolution.

The war in Afghanistan became even more difficult when the incoming US Administration of Ronald Reagan decided to support the rebels with arms, particularly with modern anti-aircraft missiles.

The other great strain on the USSR was imposed by a US administration determined to make up for the relative decline of the 1970s, a decline most publicly evident in the naval balance (though perhaps even more significant in the 'hollow army'). The new US administration began a programme, the Strategic Defense Initiative (SDI, 'star wars'), which aimed ultimately to overcome the apparent superiority of Soviet strategic forces. The Soviet Union was already credited with an aggressive strategic defence programme of its own, but SDI promised much greater effectiveness, building partly on Western superiority in cheap electronics to multiply missile-killing weapons.

In the end, the Soviet system began to collapse when the central, military power of the state was revealed by the failure of the war in Afghanistan to be much more limited than most Soviet citizens had imagined. The failure of Soviet industry to match the US SDI programme added to this perception of unsuspected weakness, and the heavy investment in production (including naval production) which was necessary to enable the backward and inflexible industrial complex to compete with the new US initiative effectively squeezed out all capital which might otherwise have gone into reform and modernisation of the Soviet system itself.

Gorbachev had first attempted to reform the system by making use of the perception that under Brezhnev it had become far too corrupt. The Brezhnev years, which incidentally had been the years of rapid naval growth, were labelled the 'era of stagnation'. Modernisation, first under Yuri Andropov and then under Gorbachev, was associated in the naval sphere with the ending of heavy investment in warships; the fall of Admiral Gorshkov, the creator of the modern Soviet Navy, was an important symbol in this process.

Reform of the Soviet system through a much publicised restoration of legal norms failed. Gorbachev perceived that he could proceed only by breaking the Party bureaucracy. His encouragement of individuals outside the Party meant sanctioning criticism of the Party system itself, and led directly to a measure of electoral democracy which allowed no role for the Party at all.

Modernisation of the state, in Gorbachev's view, could be paid for only by tapping the one large pool of capital in his system, the capital tied up in military production. He had to cut these costs. This meant publicly admitting failure in Afghanistan. It also required withdrawal from the expensive Eastern European empire (in effect, another public admission of failure) and from commitments in the Third World. The withdrawal from Eastern Europe, culminating in the collapse of the Warsaw Pact, required the USSR to withdraw very large numbers of military personnel, and housing and feeding those personnel became further major drains on the economy.

These withdrawals removed much of the logic of naval expansion. By the late 1980s, Soviet officials were renouncing traditional offensive military concepts in favour of an ill-defined defensive doctrine; forces would be reduced to 'defensive sufficiency'. Gorbachev also pursued *détente* with the West, to a much greater degree than had his predecessors. In the past, *détente* abroad had always been accompanied by internal Soviet propaganda intended to maintain the tension necessary for internal stability. This time Gorbachev's other reforms made such a contradiction impossible. Soviet citizens began to appreciate how much they were paying for defence, and how little foreign threat there really was.

The effect of collapse was that the bankrupt state could no longer support its vast military (including naval) production machine. During 1991 Soviet officials declared that existing facilities would be converted to civilian production. It remains unclear whether this can be done, or whether, like the pre-1917 Russian military industrial base, the existing base will collapse and will have to be rebuilt in the distant future. For example, there is no guarantee that the key personnel will choose to stay within the CIS; they will probably find much more lucrative employment overseas. Many of the barriers to the emigration of such personnel will disappear with the collapse of the central authority.

All of this makes the present chapter more of an obituary than a description of a developing naval power. By early 1992 the former Soviet Navy had been reduced to minor operations. It had virtually left the Mediterranean. Many of its auxiliaries were being used to carry fuel and food back to the naval bases; the main concern had become how to feed the machines and house the personnel, not how to use it for foreign policy purposes. Yet the ships and the men remain. Under different circumstances, both might easily be revived and turned to more traditional ends. Some part of the industrial base itself may be saved if the present owners manage to turn it to the export market.

TACTICS

The old Soviet Navy was characterised by a consistent tactical style, originally developed before the Second World War for its coast defence forces. It was necessary to develop a new tactical style after the Revolution because so much of the traditional pre-Revolutionary fleet (and its industrial base) had disappeared. The remnants of the fleet had to face substantial foreign fleets. Furthermore, operations at all levels had to be centrally controlled, initially for fear that too many of the commanders, who were ex-Czarist officers, would prove disloyal. Finally, the new Communist regime claimed that its more 'scientific' form of warfare would prevail. The goal would generally be sea denial (initially, coast defence) rather than sea control.

A central commander (ideally, outside the battle) would direct dispersed assets against a common target; the attackers would fire from maximum range so as to avoid damage to themselves. Soviet tactics showed themselves in operating procedures and in the designs of many classes of warships. For example, the concept of co-ordinated group operations externally controlled applied to submarines as well as to surface ships. Since the submarines receive their target data from external sources, they must remain within radio range, *ie*, near the surface, despite their high underwater speed and deep diving depth. The desire to execute group attacks from long range probably explains Soviet interest in stand-off weapons delivery *eg*, by submarine-launched missiles.

Centrally-controlled tactics place a heavy burden on systems to detect and track the target while the attack is set up, and then to inform the attackers of the precise location of the target. Initially the Soviet relied on land HF/DF stations to detect and locate enemy warships approaching within about a thousand miles of their territory. Later the HF/DF net was supplemented by passive satellites, which could detect ships' line-of-sight radar (and presumably, satellite communication) emissions. All of these systems contributed to a central plot of possible target movements.

A second (mainly active radar) surveillance layer was devoted to targeting. For example, reconnaissance versions of the 'Badger' (Tu-16) bomber were intended to precede strikes by land-based 'Badger' missile-bomber formations. 'Bear-D' (Tu-95RTs) reconaissance bombers supported submarines and surface ships armed with SS-N-3 missiles. Destroyers were converted to 'tattletales' intended to shadow enemy surface units, particularly in the Mediterranean. This category also included the now-defunct active radar satellite (RORSAT) system which supported submarines ('Oscars') and surface ships firing SS-N-19 missiles. Unlike the passive layer, the active surveillance layer reported directly to missile shooters.

Eventually, the active radar satellite system was supplemented by a passive system (EORSAT, in Western terminology) which also reported directly to the shooters. Unlike a land-based HF/DF net, a satellite-based passive system could locate a target on the surface of the sea, since it points in three rather than in two dimensions. At the time of writing, the RORSAT system has died, but the EORSATs have survived, though not in sufficient numbers to support the sort of continuous world-wide surveillance the Soviets had in mind.

From the early 1960s on, the Soviets also deployed specialised intelligence-gathering ships (AGIs), intially converted trawlers but later specially-built. The AGIs were intended mainly to collect electronic intelligence rather than as part of the real-time operational intelligence system. However, their information could be used to support the passive detection systems, since they made Western radio and radar emissions recognisable. AGIs off important naval bases could also report sorties, and thus could feed into the central plot of Western naval movements. In some cases AGIs were probably expected to detect sorties by Western ballistic missile submarines. The Russians hoped to dead-reckon these ships across to their waters, estimating when they would approach waters near the Soviet Union.

Clearly the Soviet plot of Western naval movements, which was the basis for many of their naval weapons systems, also benefitted enormously from espionage coups such as the betrayal of US naval codes by the Walkers. It is not clear to what extent the Soviets assumed that they could rely on such information. Certainly they and their Tsarist predecessors were extremely competent at collecting and using human intelligence sources, and such sources could make up for enormous technical difficulties. For example, it seems likely that throughout the early 1970s John Walker personally provided the Soviets with something close to a plot of US submarine positions. To

A 'Zulu IV' (left) and three 'Foxtrot' class diesel attack submarines on 23.2.68 following the 50th anniversary of the Soviet Armed Forces CMP

gain similar insight into Soviet submarine operations, the West had to build up a massive net of underwater detectors (SOSUS) and signals intelligence sites.

Initially central direction meant direction from land headquarters. However, the big 'Okean' exercise of 1970 apparently demonstrated that no such central direction could be altogether satisfactory. The Soviets therefore decided to decentralise: in most cases seagoing flagships would control groups of ships and, in some cases, aircraft. The flagships themselves would enjoy data-link communication with headquarters ashore.

From the late 1960s on, the Soviets developed a series of computer combat direction systems broadly comparable to Western ones such as the US NTDS and the British ADAWS, but reflecting their own concerns. These 'Second Captains' form a link between the commanding officer and the ship's weapons systems. The computer forms a compiled tactical picture and applies Soviet-style doctrinal rules to it to determine an optimum course of action. In the earliest version (in the carrier *Moskva* and the submarine tender/flagship *Volga*) output was in textual form, and the command saw only a Second World War-style grease pencil compiled plot. Modern 'Second Captains' provide synthetic video on a bridge or command centre (BIP, battle information post) work station. The 'Second Captain' differed dramatically from its Western counterparts in that it embodied a series of tactical rules (a very Soviet idea) and in that it proferred advice (and recorded deviations therefrom).

Only very recently have some Western combat direction systems incorporated tactical advice, and even then it is clearly advice rather than instruction. More generally, Western users would have considered the capacity of Soviet command system computers insufficient to carry both a complex tactical picture and data from which advice could be developed. Presumably the Soviets did not expect any individual ship to have to cope with so much information. Very unlike a Western combat direction system, the 'Second Captain' seems to devote little effort to sorting out data from alternative sources. Target identification is probably mainly by ESM, the data from which pass through the 'Second Captain'.

The main effect of emphasising tactical doctrine (which was the case throughout the Soviet military) was presumably a rigidity foreign to Western navies. When 'Second Captains' first appeared, a senior Soviet officer was said to bemoan the likely loss of nearly all initiative; he was probably right. In a very pessimistic society, doctrine is always welcome, because the individual following it faithfully cannot be blamed for tactical failure.

Roughly contemporary with the first 'Second Captains' were the first group ASW systems, in which a group of ships were linked to fire their long-range mortars (RBU-6000s) together into an area in which a computer (fed by their sonars) determined a submarine was to be found. The 'Fig Jar' data link initially commanded sub-chasers from shore; in the mid-1960s the computer moved out to sea.

These two ideas came together. In the 1970 'Okean' exercise the Soviets tested their new naval technologies, including the early 'Second Captains' and long-range anti-ship missiles. They found that direct command from ashore was impractical: turn-around time for information was far too long. They needed some intermediate command level, a sort of flagship. Like the distributed RBU-6000 system, this one would use a data link: it would collect information from the ships of the group, then send them commands. The commands would go to their 'Second Captains', which would pass them on to the ships' weapon systems. The new system might be called a 'Second Admiral'.

Much the same idea, of remotely controlling a ship's weapons, was applied to coastal units carrying anti-ship missiles. A 'Tarantul' does not have a computer, but its missile fire control system can accept commands via a one-way data link. The link allows a shore command post to concentrate the fire of several boats against a single target.

In 1970 the Soviet Navy was concerned with two distinct surface-ship missions, attacking ships (which might be called the rocket cruiser, or RKR, role) and strategic ASW (mainly pro-SSBN, or PSW). It quickly developed 'Second Admirals' for each role. They were tested on board two *Sverdlov* class cruisers: *Zhdanov* for RKR, *Admiral Senyavin* for PSW. These ships were tested in the 'Okean' 1975 exercise. Specially-built flagships followed: the *Kirov* class for RKR, the *Kiev* class for PSW (the latter were modified from an existing design, hangar space being surrendered for 'Second Admiral' facilities).

Eventually the 'Second Captain' and higher-level concepts were considerably extended. For example, the 'Second Captain' on board a 'Bear F' Mod 4 maritime patrol aircraft can control surface units. As designed, the *Kuznetzov* command system was expected to accept radar data from the ship's fighters – and to order them to fire their air-to-air missiles. Submarines now incorporate 'Second Captains' (and, reportedly, combat direction systems copied from some Western ones) and some of them act as brigade flagships.

The 'Second Captain' had very visible consequences for Soviet surface warship design. Typically a ship has one long-range radar feeding it directly. Air targets chosen on the basis of its tactical picture are passed to the ship's weapon system, which has its own shorter-range search radar. It is apparently *not* the case, as many have imagined, that the pairing of long- and short-range radars reflects fear that the primary unit may fail; they have different purposes. The recent *Sovremennyy* class is unique in having only a single air search set, whose picture is presumably shared by the ship's 'Second Captain' and her weapons systems.

Tactical style was a great constant in Soviet naval development, cutting across a very considerable variation in naval missions and even in favoured war scenarios. The other great constant was the Soviet economic system, geared to a series of five-year plans. Soviet industrial managers were rewarded for their success in fulfilling their 'norms' under the plan, and such fulfilment was easiest when production rates remained constant throughout the plan. Thus there was strong resistance to any change in the classes of ships to be produced, and in some cases quite obsolete programmes and concepts survived well past what should have been a natural demise. Planning also affected the design of new weapons and new platforms, since efforts must always be made to assure the compatibility of all components: changes at any stage are extremely difficult. (This one consideration may well explain much of the Soviet preference for deck-mounted box missile launchers which had minimal impact on the overall ship design, but which were often dismissed in the West as operationally inefficient.) Other elements of the Soviet economic system included a relatively backward electronic industry, so that some ships were commissioned and deployed abroad with important equipment not yet installed, and there have been reports of ships on foreign stations with empty missile tubes.

Ships and their systems were all designed by bureaus (TsKBs [central design bureaus] for ships, OKBs for aircraft and missiles), the best-known being such aircraft organisations as MiG and Tupolev. The main ship design units, numbered during all or most of the Soviet period, emerged in the 1990s with names and with attempts to sell their products abroad. Important postwar ship design KBs were: TsKB-16 (hived off from TsKB-17 to develop battleships and battlecruisers in the 1950s, then assigned to develop ballistic-missile launching units), TsKB-17 (major surface combatants), TsKB-18 (original submarine KB; assigned in the 1950s to develop cruise-missile launching submarines; now the Rubin bureau), TsKB-19 (small combatants; now the Almaz bureau), TsKB-53 (destroyers and probably frigates; now the Northern Planning and Design Bureau), SKB-112 (submarine KB created at the Sormovo yard, Gorkiy; now Lazurit), SKB-143 (Walter and then nuclear submarines; created May 1948; may have merged with TsKB-16; now Malakhit).

The bureaus had less independence than is generally imagined. As in the West, the main naval staff developed requirements (TTZs, tactical-technical requirements). In the cases of ships and weapons, a naval studies organisation (a higher naval institute, an NII) developed feasability studies (sketch designs) to check whether the TTZs could be met. It obtained technological data from a central studies agency, such as the Krylov Institute for hydrodynamics, or TSAGI for aerodynamics. The design bureau began work on the basis of the feasibility study; policy issues generally entered at the staff level when the TTZs were set. To a limited extent, however, a design bureau might sell its ideas up the line (just as, in the West, an aircraft company will sometimes develop an airplane as a private venture, and hope that the prospective buyer bites). Competition between bureaus was relatively rare, although in some cases (as in the Berezhniak and Chelomey bureaus) areas of responsibility might overlap, causing competition.

POSTWAR STRATEGY TO 1953

Within the limits imposed by the Soviet system, the navy was the subject of violently shifting political direction. Its modern history can be traced to Stalin's determination to build up a large ocean-going fleet, a decision he announced as part of the 1937–42 Five-Year Plan. Previously a poverty-stricken Soviet Navy had evolved a 'combined-arms' strategy in which shore-based aircraft, coast defence guns, submarines, and torpedo-boats would guarantee the security of the Soviet shoreline against foreign attack. These relatively inexpensive coastal forces responded to the principal requirement of Soviet (and Russian) governments: that the coast be closed to foreigners trying to infiltrate in peacetime or to invade in war, and that it also be sealed against Soviet citizens trying to leave. Thus many naval-type coastal units had always been under the control of the KGB, the border police. The coastal navy has always enjoyed political support, and its existence was a constant Soviet naval development. Latterly, it was perhaps best characterised by a variety of missile craft ranging from 'Osas' up through 'Nanuchkas', by very large numbers of submarine-chasers and minesweepers, and by the land-based bombers. All of these forces were controlled from shore, and the efficacy of shore-based target acquisition systems limited their effective offshore range.

In 1936–37 Stalin resolved to build a second, blue-water fleet. The overall shape of this 'Large Shipbuilding Programme' determined the sort of naval industrial base he needed, and it influenced the postwar programme. Only recently have details become available. In August 1937 Stalin formally requested a report on fleet size; it called for an increase of 630,000t over the fleet laid out in a 26 June 1936 government resolution. Stalin particularly favoured battleships and heavy cruisers (battlecruisers). A 1936 draft called for twenty-four battleships, twenty cruisers, and about 350 submarines and other ships, but it was not approved. The next version, which also was not approved, cut the number of capital ships in favour of cruisers and destroyers. It added two carriers (for the Northern and Pacific Fleets). As announced in January 1938, the ten-year programme (1938–47) included fifteen battleships, thirty-five cruisers (probably fifteen heavy and twenty light), 607 medium ships (destroyer leaders, destroyers, minesweepers, SKRs), 336 submarines, and 348 torpedo boats. Then it was cut to eight years (1940–47). The plan was still quite ambitious; one version called for twelve battleships, sixteen heavy cruisers, thirty-two light cruisers, two carriers, 198 destroyers and leaders, and 433 submarines, all to be built by 1948. The final version of the eight-year plan, submitted by Kuznetzov on 10 August 1939, omitted the carriers. By 1 January 1947 the navy was to have had twenty-four heavy ships, plus ten cruisers, seventeen destroyer leaders, 182 destroyers, and 344 submarines. These numbers presumably included ships already on hand. None was even remotely practicable.

Stalin never officially approved the programme, but he did order its implementation. One problem was that he kept it top secret, not even allowing fleet commanders to see its full extent. He personally examined designs, at least for the heavy ships. Recent accounts show that there were considerable delays as attempts were made to speed construction before sufficient planning had been done.

Finally the navy managed to obtain a 1938–42 plan (third Five Year Plan) calling for laying down six battleships, two aircraft carriers, four heavy and seventeen light cruisers, twelve destroyer leaders, 197 submarines, and 271 combat boats. It is not clear whether destroyers, SKRs, or minesweepers were planned. The light cruisers were the *Chapaevs*. Planned deliveries during this period were eleven cruisers (including the first of those laid down under the 1938–42 plan), seven destroyer leaders, forty-six destroyers, ten SKRs, eighty-nine minesweepers, 359 combat boats, and 201 submarines. Presumably the two carriers, if they survived at all, were to have been laid down during the next Five Year Plan (1943–47).

In fact the shipbuilding industry worked to annual approved plans based on an unapproved draft of the Five Year Plan; ships were laid down as slips filled during the previous Five Year Plan were vacated. Annual plans for 1938–41 called for laying down four *Sovetskiy Soyuz* class battleships (Project 23), two *Kronstadt* class heavy cruisers (Project 69), seven *Chapaev* class (Project 68) and two *Maxim Gorkiy* class (Project 26bis) light cruisers, three destroyer leaders (Project 48 class), thirty destroyers (the last two Project 7U, and twenty-eight Project 30), 100 submarines (five *K*-class, thirty-one *S*-class, six *L*-class, eleven *Shch*-class, forty-five *M*-class, two experimental), fifteen *Yastreb* class SKRs (Project 29), ten large submarine-chasers, thirty-six minesweepers, ten net- and minelayers, and three monitors, plus fast attack boats. However, this plan could not be met, as there was no approved programme and hence little industrial coordination. For example, between 1938 and the first half of 1941 only nineteen submarines (two *K*, two *Shch*, and fifteen *M* class) were placed in service.

As war approached, it became painfully obvious that the sheer industrial cost of the programme could not be borne. A 19 October 1940 decree made drastic cuts. Kuznetzov wanted to stop the two heavy cruisers under construction at Leningrad in order to speed constructiion of light units, but Stalin would not allow his two favourite ships to die. Of the ships under construction, the battleship *Sovetskaya Rossiya* and the two heavy cruisers would be pressed forward, in hopes that the battleship could be launched in June 1943 and the heavy cruisers in 1942. Another battleship (*Sovetskaya Belorussiya*) was to be broken up on the slip. Work would proceed on smaller ships, from light cruisers downward. The outbreak of war on 22 June 1941 ended these plans.

In retrospect, Stalin seems to have decided to build a big blue-water fleet (actually, fleets, given the geographical separation of Soviet fleets) almost because he considered it an essential trapping of a major power; a recently published account of the travails of the programme gives no rationale for the numbers of ships involved. One of the first indications of Stalin's big-fleet ambitions was his elevation of the Soviet Pacific Squadron to Fleet status.

Stalin's big-fleet programme built the industrial base on which the postwar Soviet Navy rose. Each of the four fleets (Baltic, Northern, Black Sea and Pacific) was to have its own primary yard for new construction. Major shipbuilding centres already existed at Leningrad and at Nikolayev on the Black Sea. Now Molotovsk (renamed Severodvinsk postwar) was built to serve the Northern Fleet; Komsomolsk-on-Amur served the Pacific Fleet. Neither had any considerable industrial hinterland, and each was intended as a satellite of one of the established centres. Thus Molotovsk was to assemble material fabricated at Leningrad, while Komsomolsk worked with Nikolayev. In addition Stalin ordered the construction of a submarine factory at Gorkiy on the Volga, east of Moscow. Submarines built there could run their trials on the Caspian Sea, and could then travel up the Volga and through the Soviet canal system to the White Sea and the Northern Fleet. However, that canal system in itself imposed limits on the size of submarines built at Gorkiy and so reduced the value of the yard postwar as the size of the standard submarine grew.

These original yards continued in the 1980s to dominate Soviet warship production. Severodvinsk produced nuclear submarines for the Northern Fleet, as Komsomolsk did for the Pacific Fleet. Designed for year-round operation, it has covered and heated sheds. The Leningrad complex includes the Admiralty yard (nuclear ships and submarines; it built *Kirov*), the Sudomekh submarine works, the Zhdanov yard (which served as lead yard in postwar destroyer and light cruiser programmes), and the Baltic yard, responsible for *Sverdlov* class cruisers but not for more recent warships. There is also a light warship yard on Petrovski Island. On the Black Sea, Nikolayev is a similar concentration. The South Yard built heavy warships such as the *Kiev* and *Moskva* classes, the North Yard, cruisers such as the 'Karas'. Inland, on the Volga, there is a light craft yard at Zelenodolsk to supplement the old mass production facility at Gorkiy. There are in addition major fleet bases: Severomorsk for the Northern Fleet, the Leningrad-Kronstadt complex for the Baltic Fleet, Sevastopol for the Black Sea Fleet, and Vladivostok for the Pacific Fleet, with a large base also at Petropavlovsk on the Kamchatka Peninsula.

The Soviet industrial system, including the base for the navy, was badly damaged by the Second World War. Leningrad was besieged and its factories and shipyards partly destroyed. Nikolayev was occupied by the Germans. The satellite yards, cut off from their sources of material, were largely idle in wartime, although they were not themselves damaged. In 1945 some of their slips were still occupied by the remains of Stalin's prewar programme, but there was little hope of immediately resuming work on it. All that could be done was the completion of prewar hulls from light cruisers (*Chapayev* class) down.

However, although the Soviet Navy had suffered terribly at the hands of the German invaders, it did derive several important benefits from the Red Army's victory. In East Prussia it seized the Schichau yard at Elbing, a specialist builder of light surface craft; the town of Königsberg, where the yard was located, became the Soviet city of Kaliningrad. The former Walter works fell into Soviet hands, as did several examples of the new and revolutionary Type XXI submarine, and several sets of snorkel gear. In East Germany the Soviets found scientists and engineers who had worked on the German Navy's advanced torpedoes and sonars, as well as on air-launched anti-ship missiles. All of these formed major elements of the postwar Soviet fleet programme. At the same time the Soviet fleet benefited from Allied Lend-Lease, which provided it with modern radars, sonars and ASW weapons.

In 1945, with the prewar Soviet Fleet and its industrial basis in ruins, Stalin faced enormous costs if he wished to revive his prewar plans. Existing ships were obsolete at best, although the technology for a more modern fleet had fallen into Soviet hands. It could, moreover, be argued that with all the claims on Soviet resources for industrial reconstruction, and indeed with the Army predominating, funds would best not be spent on naval expansion.

Apparently during 1944–45 all three Soviet services developed elaborate studies of the military lessons of the Second World War. Each then worked out an ideal postwar programme, on the assumption that the wartime allies would soon become enemies.

None of these programmes was particularly practical, and a special Politburo commission was formed to develop a compromise programme achievable, in theory, within Soviet industrial resources. Stalin was directly involved in decisions as to both total numbers and characteristics of each type.

Stalin was certainly interested in reviving both the sea-going fleet and the older coastal fleet. He apparently envisaged two stages: first, construction of modified versions of prewar designs (such as the *Sverdlov* class cruisers and the *Skoriy* class destroyers), then construction of ships of the new types reflecting wartime experience, such as new heavy cruisers (which became the *Stalingrad* class, Project 82) and the new destroyers ('Tallinn' and then 'Kotlin' classes). Presumably the modified prewar types were associated mainly with the 1946–50 Five Year Plan, and the new designs with the 1956–60 Plan.

In March 1945 the main naval staff proposed a ten-year programme: as of 1 January 1956 the navy would have four battleships, ten heavy cruisers, thirty cruisers, fifty-four light cruisers, six large aircraft carriers, six small carriers, 132 large destroyers, 226 destroyers, 168 large submarines, 204 medium submarines, 123 small submarines, and lesser craft. These figures presumably included ships already in existence or under construction. Apparently the light cruisers were a new design (Project 65). Stalin reviewed the plan in September 1945. Those present included the then navy C-in-C Admiral Kuznetzov (soon to be exiled to the Far East as commander of the Pacific Fleet) and the head of the shipbuilding industry, I I Nosenko.

Kuznetzov cut the carrier proposal to four large and four small ones. Stalin wanted to avoid building them for the present, on the grounds that the fleet could get by with land-based aircraft when operating in the Baltic and Black Seas, or near the Kuriles in the Pacific. However, he did agree that the Northern and Pacific Fleets needed carriers, and for a time he approved building two small ones. They seem to have been rejected, in the end, on the grounds of insufficient yard capacity.

Kuznetzov's views of cruiser roles seem reminiscent of those current before the Second World War showed how dominant aircraft could be. He thought that heavy cruisers would deal with the US *Alaska* class large cruisers, and heavy cruisers, and would help pin down enemy capital ships (as the British heavy cruisers had shadowed the *Bismarck*). Light cruisers would escort major combatants (carriers and battleships), dealing with destroyer attacks and leading Soviet destroyers. A special anti-aircraft cruiser seems also to have been considered.

As he would do again in 1951, Kuznetzov tried to cut the heavy cruiser (Project 82) programme; Stalin replied that he would rather cut battleships. However, he would not abandon them altogether (not least because of what he saw as their prestige value). He therefore asked for three: the ship already on the slip at Molotovsk (and therefore not damaged by war), with two more to be laid down in 3–4 years. Stalin was inclined to demand that the heavy cruisers outgun any foreign equivalents, and hence that they be armed with 12in guns, but Kuznetzov convinced him that a new 9in gun, with a higher rate of fire, would be better. Within a few years Stalin had returned to his earlier view, and the heavy cruiser programme bifurcated into 9in (Project 66) and 12in (Project 82) classes. Stalin was also dissatisfied that the heavy cruisers would not be ready before the end of the programme; but Nosenko promised to have four complete and three in the water by 1956, presumably with three more on the slip.

The figure of thirty cruisers was set by building capacity. Initially it seems that they would have been a new type, intermediate between very small cruisers suited to leading destroyer formations, and the heavy cruisers. The proposed intermediate design was probably Project 65, which by 1947 was being considered in both 6in and 7.1in versions. The alternative was a simpler design, Project 68bis, based on the prewar Project 68 (*Chapaev*). Stalin opted for quick production over greater sophistication, and killed Project 65 in favour of 68bis (*Sverdlov*). It seems likely that the five incomplete *Chapaev*s were included in the thirty cruiser figure. According to a recent account of the *Sverdlov* class, the plan for thirty was cut to twenty-five to allow for heavy cruiser construction, but that is not consistent with the yard capacity argument cited here. The fifty-four small cruisers seem to have been dropped due to limited yard capacity.

Kuznetzov wanted at least a few (he said at least ten large destroyers [about 3500t; Project 40]); Stalin objected that it would be better to build light cruisers and standardise on a 2500t destroyer. Ultimately he decided that 250 destroyers would be built; the special large type was apparently dropped. Stalin was well aware that changing designs would slow the programme. He wanted to continue building the existing type (Project 30bis, *Skoriy*) until about 200 had been completed, partly because the shift to dual-purpose guns in the new Project 41 would delay production. Again, he considered quantity far more important than quality. In the end, the planned number of destroyers was cut to 188; the run of *Skoriy* production ended at sixty-eight hulls.

Nosenko convinced the navy to cut the number of planned SKRs to 200 (the original figure is unknown). The fleet also agreed to a programme of thirty fleet and 400 coastal minesweepers.

Submarines were more controversial. Stalin objected to the number of different design projects, wondered whether so many submarines were needed at all, and asked why large submarines were needed. Kuznetzov managed to keep forty large submarines in the programme, and convinced Stalin that the rest of the programme would not slow down heavy-ship construction because it would be executed mainly at Sormovo, at Gorkiy on the Volga. The number of separate design projects was also cut.

Overall, compared to the prewar plan, the number of heavy ships was cut to about a quarter, cruisers and submarines were little changed, and destroyers, SKRs, and minesweepers were roughly doubled.

It seems that the long-range plan shifted drastically about 1950. The battleships were cancelled, but the cruiser programme survived. Stalin seems to have seen the battlecruisers as raiders, useful to hold Western naval attention (as he was well aware the German capital ships in Norway had done during the Second World War) and so to reduce the Western naval threat, which at that time he apparently considered mainly an amphibious threat. It appears that the submarine programme was very considerably expanded at this time.

Submarines were a difficult proposition. During 1942–44 work proceeded on successors to the prewar designs; Russian writers have identified Project 608 as a successor to the earlier *S* and *Shch* medium classes. The *K*-class was apparently modified; the improved *M*-class was actually built as Series XV. By 1944, it was painfully obvious that these designs were obsolete. For example, examination of a raised Type VIIC U-boat, *U 250*, showed that the existing structural design could no longer be considered modern. For the moment, boats of existing design were completed: seven *S*-boats at Gorkiy (Sormovo) in 1946–48 and fifty-three *M-V* class small submarines at Sudomekh in 1947–53. Nearly all of the *M*-boats were built in sections in the Leningrad area, assembled at Sudomekh, and shipped by rail to the Far East and the Baltic. Some boats may eventually have been fitted with snorkels, but there was no programme analogous to the US or British postwar modernisation.

In 1945, German snorkels and Type XXI U-boats fell into Soviet hands. A series of new TTZs was issued in 1946, for all three submarine categories: Projects 611 ('Zulu', long-range), 612 (small), and 613 ('Whiskey', medium-range). Project 614 was an abortive programme to complete Type XXIs.

Projects 611 and 613 entered production. They were not quite equivalent to the wartime German fast submarines, since (at least as initially built) they had extensive gun batteries and lacked snorkels. Project 612 was abandoned in favour of a closed-cycle diesel submarine, Project 615 ('Quebec'); it was the culmination of prewar experiments with such powerplants.

TTZs for second-generation medium- and long-range submarine designs ('Romeo', Project 633, and 'Foxtrot', Project 641) were issued about 1952–53, presumably for construction under the 1956–60 Five Year Plan. They were deeper divers and incorporated snorkels, a new low-frequency bow array, Feniks-M (roughly equivalent to the US BQR-4), and a new torpedo fire control system which did not require periscope input (Leningrad, using the new Arktika tracking sonar). Unlike the earlier designs, they never had provision for deck guns.

The table below gives reported numbers actually planned (probably *c*1953) plus the numbers actually completed. Plans were reportedly based on tactical organisation: six-boat brigades for long-range 'Zulus', eighteen-boat divisions for other types expected to operate inshore. The small number of 'Zulus' reflects both Soviet industrial capacity, which in 1946–50 was far better able to build small than large submarines, and the higher priority presumably associated with coast defence than with oceanic operations.

Type	Plan	Completed
'Zulu' (611)	96	26
'Whiskey' (613)	288	236 (including some in China)
'Quebec' (615)	36	31

Most of the units actually built (except for 'Quebecs') were built during the 1951–55 plan period. The cuts apparent for 'Zulu' and

'Whiskey' may actually indicate that these designs were stopped at about the end of the plan period in favour of their design successors, 'Foxtrot' and 'Romeo'. Published claims of major cuts ordered in 1953, on Stalin's death, are not plausible, given the actual pace of 'Whiskey' production. However, published figures do make it possible that cuts were ordered (by Khrushchev) in 1955–56.

A report of planned numbers of 'Foxtrot' and 'Romeo'-class submarines probably dates from about 1955 and was probably planned for completion during the 1956–60 plan period:

Type	Plan	Completed
'Romeo' (633)	336	21
'Foxtrot' (641)	224	62 (75 including transfers)

The increased number of 'Foxtrots' may reflect Admiral Kuznetzov's increased interest in oceanic warfare. 'Foxtrot' numbers were reportedly based on seven-boat brigades, possibly meaning a leader boat plus the usual brigade.

It seems likely that as of 1955 plans called for switching to production of closed-cycle submarines (along with the new nuclear submarines) under the 1961–65 and later plans. Two parallel series of closed-cycle plants were under development at this time: the Russian Kreislauf diesel and the German-derived Walter.

Also about 1955, the Russians apparently planned two parallel nuclear programmes: pressurised-water and liquid-metal. The sheer research effort required (particularly after they began to experience problems with liquid-metal reactors about 1957) presumably made it difficult to continue work on closed-cycle plants. They were abandoned in 1960, leaving the diesel submarine programme, which had been considered an interim effort, alive.

Stalin was also interested in direct nuclear attacks on Western rear areas. At first Soviet nuclear bombs were far too large for delivery by anything but long-range bombers. However, in 1952 work began on a naval weapon, a very long-range torpedo 150cm in diameter, which could be delivered by submarine. The 'November' (Project 627) class was conceived for this purpose, as was an (as yet undescribed) diesel submarine. In 1949 some experimental work began on naval ballistic missiles. The Soviets reportedly tried, without any success, to build the submarine-towed missile container the Germans had conceived during the Second World War. Compared to a shipboard installation, it had the important advantage of accommodating the sort of immense weapon needed to lift a large warhead. Reportedly several hundred such containers were built in series of ten or twenty.

In 1951 Admiral NG Kuznetzov, the prewar architect of the major Soviet fleet programmes, had returned to Moscow from exile as commander of the Pacific Fleet. It is tempting to associate his return with the cancellation of the big battleship (Project 24). It is known that Kuznetzov revived the carrier issue. Stalin still had not ruled one way or the other when he died, and then Kuznetzov apparently formally revived the carrier programme (Project 72). A recent Russian account recalls with some awe that Kuznetzov had the courage to tell Stalin that the Project 82 raider was worthless in the face of naval aircraft. It was cancelled at Stalin's death. Kuznetzov was apparently determined to build up an integrated fleet. For example, he enthusiastically supported the large Project 66 cruiser, a carrier escort. Given the cost of the major units, Kuznetzov may also have decided that some of the early postwar designs called for overly expensive ships: the Project 41 ('Tallinn') destroyer and the Project 42 ('Kola') 'guardship' (SKR). He, or some other senior naval official (current Russian authors cannot say for sure), ordered them replaced by smaller and less expensive designs, which became Projects 56 ('Kotlin') and 50 ('Riga'), respectively.

The prewar interest in coastal defence continued. Well offshore were medium submarines (the prewar *Shch* and *S* classes, the postwar *W* class). Closer in were 'guardships' (SKR, *Storozhevoi Korabl*), coastal submarines (*M* class), and shore-based aircraft. The inshore layer comprised small gunboats, MTBs (TK, *Torpedniye Kater*), coastal ASW craft, and coastal batteries (some of them afloat in the form of monitors and coastal defence ships).

SKRs were often mistakenly described in the West as frigates, but they were coastal pickets rather than convoy escorts. In peacetime they were seen largely as a means of keeping foreigners out of the Soviet Union, and Soviet citizens in. This gunboat function explains the heavy gun and torpedo armament of the two postwar classes ('Kola and 'Riga'). The SKR category lapsed with the 'Riga' class, but was revived in 1978 for 'Krivaks' (which had been designated large ASW ships, BPKs).

Except for its innermost layer, this coastal force was primarily intended to deal with enemy surface ships. Reportedly Stalin was particularly concerned with the threat of seaborne invasion (SKRs did have some limited ASW armament). Efforts to adapt the wartime German technology of anti-ship bomber-launched missiles to Soviet naval aircraft were pressed forward. The first such weapon was probably Shchuka, tested in 1951–52. The Soviets also adopted the wartime German radio-guided 'Fritz-X' bomb. They may have operated a version of the German Hs-293, which was produced in Czechoslovakia postwar.

The inner layer included gunboats (AK, *Artilleristiy Kater* or BO, *Bolshoi Okhotnik*, large hunters) like the 'Artillerist' and 'Kronstadt' classes (Projects 122 and 122bis), which might be considered a sort of inshore SKR. Smaller sub-chasers (MO, *Morskoi Okhotnik*, or 'sea hunters') were intended for harbour defence (they corresponded to wartime British HDMLs, for example). There were also numerous minesweepers (both ocean [MT, *Morskoi Tralchik*] and base [BT, equivalent to the US YMS]). There were also river forces comprising monitors and armoured gunboats (BK, *Broniye Kater*).

Finally, Stalin retained the amphibious craft (German-designed MFPs) left over from the Second World War. He showed no interest, however, in anything more, and the MFPs (similar to British LCMs) were unsuited to rough weather.

When Stalin died in March 1953, Kuznetzov survived in office. Late in 1953, probably at the suggestion of the missile designer V N Chelomey, his staff began to consider placing cruise or ballistic missiles on board warships, probably in the context of the 1956–60 Five Year Plan. On 26 January 1954 the Council of Ministers formally approved a plan (Project Volna) to place long-range cruise and ballistic missiles on board submarines. Presumably there was a parallel surface ship project. Apparently the last of the Project 82 battlecruisers, building at Molotovsk (*Severodvinsk*) was re-ordered as a missile ship (Project 82R); three more were planned. The existing *Shchuka* was ordered adapted to a destroyer-size platform (Project 57bis, 'Krupny'), and work began on a patrol boat weapon (which became Styx). A converted air-launched weapon, AS-1, was adapted for 1956 tests on board the cruiser *Admiral Nakhimov*. The 'Krupnys' were designated RK, *Raketnyy Korabl* (rocket ship). The rocket-armed patrol boat became a 'rocket cutter' (*Raketnyy Kater*).

An existing army ballistic missile (Scud, R-11) was adapted to submarine use (as soon as it was clear that this adaption was practical, the big strategic torpedo was cancelled and Project 627 reoriented to a conventional submarine role). A surfaced 'Zulu' class submarine fired a Scud (R-11FM) ballistic missile on 16 September 1955. It was the first such firing in the world. It was apparently accepted that any submarine-borne strategic missile would be fired after the submarine had surfaced (the US Navy fired its 'Regulus' cruise missile from surfaced submarines).

Khrushchev fired Admiral Kuznetzov in October 1955, nominally because he had failed to avoid the disastrous accidental sinking of the ex-Italian battleship *Novorossiysk*, but much more likely because he rejected Khrushchev's decision to cut back drastically on major surface ships. Later Russian supporters of a large-ship carrier-oriented navy recalled Admiral Kuznetzov warmly and rejected Khrushchev's dictate as 'voluntarism'. Kuznetzov was replaced by Admiral Sergei G Gorshkov, who would remain commander-in-chief until 1985. Gorshkov's initial task was to implement Khrushchev's cuts; Khrushchev later described him (incorrectly) as a submariner who would be happy to create the desired submarine-oriented fleet. However, Gorshkov probably benefitted far more from his Second World War connection with the Black Sea and, therefore, with the Ukrainian Khrushchev.

Gorshkov's Black Sea connection probably revived the Soviet amphibious force. In the Second World War, Gorshkov had mounted numerous small operations around the flanks of German forces. He soon began a programme of freighter conversions, and he was probably responsible for the 'MP 2' to 'MP 8' programmes. These largely unarmed ships were suitable for administrative, as opposed to contested, landings. It is clear in retrospect that Khrushchev had very different ideas. He abolished the separate naval infantry units, and declared that the army could do well enough without special amphibious units.

KHRUSHCHEV'S 'REVOLUTION IN MILITARY AFFAIRS'

Stalin avoided any admission that the existence of nuclear weapons would lead to radical changes, and indeed he forbade the discussion of the implications of such weapons.

In effect, then, nuclear weapons were forbidden fruit, suddenly accessible within the Soviet military establishment upon Stalin's death. Within two years, the Soviet leadership was to proclaim the 'Revolution In Military Affairs' under which new doctrine they were to be considered decisive in future war. Khrushchev would then announce that he was uninterested in any weapon save 'nuclear-missile' weapons, and a round of inter-service rivalry would begin. The stage for this round was set by general dissatisfaction with the cost of Stalin's fleet. Khrushchev would later characterise the naval shipyards as 'metal-eaters', and the big surface ships as 'coffins'. Khrushchev decided to cancel the projected 1956–60 Five Year Plan in favour of a completely revised programme. The new plan was worked out during 1955–57 and emerged as a new Seven Year Plan (1959–65). To enforce the accompanying shift, Khrushchev humiliated the senior level of the Soviet military. He also emphasised his break with Stalin by detailing many of the dictator's crimes in the famous 'Secret Speech' of 1956. The combination of industrial, military and political disruption (the loss of Communist Party prestige) led to Khrushchev's demise in 1964.

To begin with, Khrushchev apparently demanded plans showing how each service would accommodate the new military outlook. In October 1955 he convened a high-level meeting in Sevastopol to discuss the future of the Soviet fleet in light of the revolution in military affairs. This meeting shaped the 1956–65 Soviet shipbuilding programme.

Khrushchev proposed that submarines and land-based naval aircraft form the basis of the fleet; he wondered whether surface ships were still viable. Given the obvious power of missiles and nuclear weapons, he stopped work on heavy naval artillery (including coast defence weapons) and on armour, including underwater protection. He did press for development of stronger structural materials, presumably the titanium which would soon be incorporated into the design of the 'Alfa' class submarines.

Khrushchev called for the construction of seagoing ASW ships with sonar and rocket ASW weapons. Presumably they were conceived as counters to the new generation of strategic attack submarines (such as the new Soviet boats and US Regulus boats). This statement seems to explain the origin of the 'Petya' class. It would also explain the demise of the surface-oriented SKRs. Khrushchev also argued that destroyers needed both improved ASW weapons and effective AAW weapons (it is not clear why such ships were needed at all, if the major surface units were no longer wanted). The new destroyer became the 'Kashin' class.

Given the threat of nuclear attack, the fleet had to be dispersed: Khrushchev wanted a new auxiliary fleet to support dispersed basing. He also called for new facilities to support rapid ship construction. Khrushchev may have had in mind the quick reconstruction of the fleet after initial nuclear damage.

As for carriers, Khrushchev agreed that they would be essential to protect sea communications, but he considered them a project only for the distant future. There would be time enough to build a prototype when it was needed.

Marshal Zhukov, the Minister of Defence, supported Khrushchev. There was no point in competing quantitatively with the West, which was dependent on sea communications. It would therefore be better to emphasise submarines, providing them with nuclear powerplants and more powerful weapons. Construction of new surface ships should be aimed at supporting the submarine fleet and cooperating with ground forces; preference should go to light high-speed missile-armed cruisers, destroyers with short-range missiles, patrol combatants, submarine hunters and minesweepers. He agreed that AAW ships were needed for open theatres, but rejected carriers in the near term. Nor did he see any need for amphibious ships.

Navy representatives tried to temper Khrushchev's radical views. According to a recent Russian account, they argued that 'it is impossible to imagine a navy of the future without aircraft carriers and carriers of fighter and strike aviation'. Together with light carriers, they wanted cruisers and destroyers, with ships specialised for ASW and AAW. They also wanted to retain existing cruisers, refitting them with missile systems. The navy pressed for fast nuclear submarines armed with missiles and long-range torpedoes (apparently the naval staff agreed that all future submarines should be nuclear).

The naval staff listed important new technologies: sonar, navigation systems (presumably delivering strategic missiles), ASW torpedoes and missiles, ASW helicopters, and gas turbines for surface ship propulsion. It seems significant that anti-ship missiles were not mentioned: work on them was already well underway. The staff also pointed out that the base infrastructure was not being built up fast enough to support the new fleet of submarines and surface combatants.

Kuznetzov had probably just been fired (the Russian article describing the meeting does *not* name him as the navy representative). With him went the big surface ships which would have comprised a Soviet battle group: the Project 82R missile-battlecruisers, the Project 72 aircraft carrier, Project 75 missile conversion of the *Sverdlov* class cruisers, and remaining 'Kotlin' class destroyers (Project 56). Cancellation of Project 75 led to immediate suspension of work on the incomplete *Sverdlovs*. In addition, remaining 'Riga' class 'guardships' (SKRs: Project 50) were cancelled, probably to free yards to build the new offshore ASW craft.

Khrushchev reportedly wanted to cancel all non-nuclear submarines, presumably including the peroxide (Walter) and Kreislauf designs. His naval staff pleaded that an all-nuclear submarine force was unaffordable. Moreover, some conventional boats would have to be built to operate in shallow water and for export. That kept the closed-cycle development and design programme alive until 1960. It also kept the 'Romeo' and 'Foxtrot' building programmes alive, albeit curtailed to release resources for the new missile submarines. That proved fortunate, since production of more advanced submarines did not meet expectations: any 'Foxtrots' then scheduled for transfer went into Soviet service (only later boats were transferred). About 1959–60 it was decided that short-range submarines were no longer worthwhile, and 'Romeo' production was ordered stopped; the production tooling was sold to China.

Presumably as a consequence of the October 1955 meeting, basic directives for new types of combat ships were approved at 17 and 25 August 1956 meetings of the Council of Ministers. The new types certainly included the 'Kashin' class. The 'Petya' class may already have been approved as a direct outgrowth of earlier sub-chasers and as a test-bed for gas turbines (its history has not yet been described in detail). These directives probably also approved both the liquid-metal reactor programme and the 'Alfa' class submarine. Plans were made to develop pressurised-water and liquid-metal reactor submarines in parallel.

Almost certainly the naval representatives at the October meeting had managed to convince Khrushchev to approve four small 'Kynda' class cruisers as one-for-one replacements for the four 82Rs, although their arguments favouring carriers failed altogether. Note that the 'Kyndas' were always designated cruisers (RKR, *Raketnyy Kreyser*) in distinction to the 'Krupnys' (RK).

Khrushchev also approved twenty 'Kashins', which were explicitly conceived as escorts. At this time other missile-armed destroyers ('Krupnys') were being built; the 'Kashins' may also have been intended to escort them, *ie* to protect them against possible attack by NATO carrier aircraft. This mission may also have applied to the new offshore ASW units. The 'Kashins' were to be the last Soviet warships designated destroyers (EM, *Eskadry Mimenosets* or *Esminets*, meaning fleet mine [torpedo] craft) until the advent of the *Sovremennyy* class. It may have been significant that the first few were initially designated SKRs; about 1956–60 there may have been a feeling that a *combination* of anti-ship and ASW units covered by AAW missile ships would be the natural replacement for the old SKR offshore screen.

Also about August 1956, Khrushchev approved a proposal to expand the strategic submarine programme (in response to a February 1956 proposal by the Naval Scientific-Technical Committee). Until that time, it had apparently been limited to conversions of existing 'Whiskey' and 'Zulu' class diesel submarines. Two types of missile, cruise (P-5, NATO's Shaddock, SS-N-3) and ballistic (R-13, NATO's Sark, SS-N-4) were developed. Work was probably already proceeding on a much longer-range cruise missile developed by Ilyushin (probably P-20). Like P-5, it had to be launched from the surface. There may also have been a much longer-range surface-launched ballistic missile.

It seems likely that the 1956 directive called for (i) new classes of strategic submarines, both nuclear- and diesel-powered, to be built from the keel up; and (ii) cruise and ballistic missiles that could be fired from submerged submarines, hence which would be more viable in the face of improving US defences, such as SOSUS (as an alternative, a much longer-range weapon could be fired at US targets from beyond SOSUS coverage). The new ballistic missile was presumably SS-N-5 (R-21). The new cruise missile, Bazalt (SS-N-12), reportedly proved difficult to develop, and it never entered service as planned.

There was probably also a directive to accelerate the anti-ship missile programme. The new P-15 Styx was initially associated with a new missile boat, which became Project 205 'Osa'. It was, however, tested on board a specially modified torpedo boat in 1957. Similarly, the big Shchuka was tested on board a modified 'Kotlin', but it seems to have been intended for a new destroyer-size platform, the Project 57bis 'Krupny'. The likely proposal was to adapt the test platforms as *operational* platforms, in the form of modified torpedo boats (Project 183R 'Komar') and destroyers (Project 56M 'Kildin' class). The 'Kildins' were probably hulls stopped by the October 1955 directive.

These directives did not constitute a fleet plan; they were only a demand for new kinds of ships. Nor did they supersede ongoing projects such as the new strategic submarines (see below). The naval staff began work on a fleet plan reflecting Khrushchev's ideas concerning the new kind of war. Apparently it was completed and approved in 1958. No details have been published , but a revised 1964 version has been reported. The 1958 plan probably projected a fleet organisation parallelling that of the Army, with four fronts: two major (Northern and Pacific) and two lesser (Baltic, Black Sea/ Mediterranean). Each front (Fleet) would be organised into specialised armies (*eg* strike, anti-ship, ASW, logistics) and each army into divisions and then into brigades. For example, the Northern Fleet anti-ship army would consist of aircraft, surface ship, and submarine divisions. At least later on, each division consisted of one heavy and two light brigades, typically seven ships or submarines to a brigade (three pairs and one leader). At least in the case of anti-ship weapons, heavy meant long-range, light, short-range (in theory the heavy units would smash enough of the enemy's defences to allow the more numerous shorter-range weapons to attack). The seven-ship brigade structure presumably originated with the 1958 plan. Coastal forces were probably organised in six-unit brigades and in three-brigade divisions, all the units in a division often being of the same type.

The reader is cautioned that, despite this neat organisation, in practice brigades often operated with odd numbers of ships or submarines; building programmes rarely if ever delivered units quickly enough to form the desired homogenous formations. Apparently major units often operated outside any brigade structure.

At this time the Soviets were concerned with Western strike carriers and with Western bombers. The navy could usefully attack both the carriers and many of the bomber bases. According to the published account of the 1955 meeting, the main naval weapons to be used against the carriers would be submarines and naval bombers, the latter armed with the earliest Soviet air-launched missiles.

It seems likely that about this time a radical underwater-launched anti-ship missile was proposed. Work began on a fast submarine to carry it, a companion to the torpedo-firing 'Alfa' (it became Project 661, 'Papa'). It seems likely that parallel torpedo- and missile-firing closed-cycle designs were begun at about the same time. For example, Project 660 was an abortive closed-cycle diesel submarine; the proximity of its design number to that of 'Papa' suggests that it may have been an analogous closed-cycle type.

The fleet plan probably dramatised the need for large numbers of ships and submarines, just as the unit cost of revolutionary new types was becoming painfully evident. Reportedly it revived plans to produce 'Foxtrot' class diesel submarines for the Soviet Navy. The parallel 'Romeo' programme was presumably also revived, but it soon died, probably because it was not so clearly useful. Khrushchev may well have seen the new fast jet bombers and small missile combatants as a better offshore screen, particularly since the threat was far more likely to be an air or missile strike from well offshore than an amphibious operation closer in.

Apparently quite separate from the fleet plan was the new strategic attack naval mission. The navy initially avoided interservice problems because of the failure of contemporary bomber and long-range missile programmes; in 1955 the navy was the only Soviet service capable of delivering Khrushchev's decisive blow against North America. Then the situation changed. In 1957 the first Soviet ICBM, the massive SS-6, was successfully launched. Soviet ground-based missiles were operated by the army artillery arm, which had great political influence. With effect from 1959, Khrushchev reallocated roles and missions among the Soviet services.

A new arm, the Strategic Rocket Forces (SRF), was created out of the former missile arm of the Soviet artillery. The navy was ejected from the strategic role. At the same time, all anti-ship missile bombers were assigned to the navy. All of the navy's defensive fighters, which had been used to protect bases, were reassigned to the air defence force (*PVO-Strany*).

The reassignment fitted Khrushchev's new emphasis on strategic nuclear warfare. In a quick nuclear war, there was little point in concentrating defensive resources on naval bases. Better to allow the PVO to concentrate as required against attacking US bombers. However, the navy did have a well-defined defensive mission: protecting the Soviet Union by striking at NATO nuclear-attack platforms: carriers and missile submarines, both those firing the existing Regulus cruise missile and the new ones firing the Polaris ballistic missile.

Polaris could be fired from about twice the range of Regulus, 1200nm, and from the beginning the US Navy planned to use it from even greater distances. The planned Soviet counter, a small helicopter carrier capable of maintaining two ASW helicopters aloft throughout her patrol, became the *Moskva*. According to the official requirement for this ship, she was expected to operate as part of a surface hunting force. Her flagship role would have justified installation, on board *Moskva*, of the prototype 'Second Captain' automated command system. Two ships were built; they were probably intended for a single fleet, most likely the Northern. It seems likely that the longer-range programme called for a second pair for the Pacific Fleet. By the time *Moskva* was in service, the Mediterranean had become an important theatre of operations, and the two ships often operated there. That would have opened a requirement for a third pair. The two later pairs became the *Kiev* class. The *Moskva* project may well have been associated with the decision to provide twelve 'Kotlins' (six per *Moskva*) with improved ASW systems. By the time the ships actually entered service, operating concepts had changed drastically (see below).

The Russians probably thought that the *Moskva* hunting group would need anti-ship back-up, not least to deal with US carriers which might cover Polaris operating areas (the Soviets later became intensely interested in using surface and naval air forces to assist their own SSBNs). It is not clear to what extent the design of the anti-ship cruisers was associated with the *Moskva* project (no design histories of the Russian missile cruisers had been published at the time of writing), but it is known that an anti-ship version of *Moskva* was cancelled and broken up on the slip. It is tempting to think of the successor design, *Kiev*, as an enlarged *Moskva* incorporating the anti-ship weapons of the abortive companion ship.

There was also another reason for the Soviets to have become interested in dealing with US carriers at much greater distances. Polaris roughly coincided with the successful deployment of the long-range A-3 Skywarrior, which could reach targets about 1500nm from its carrier (earlier nuclear strike bombers could not reach much beyond 500nm). The announcement of the much more capable A-5 Vigilante may also have had some effect. In either case, it could be argued that shore-based aircraft would no longer suffice. Without much reconnaissance support even fast submarines might well fail to intercept the ships. Even so, in 1959 a directive was issued to design a fast submarine specifically to intercept fast carriers, 'Papa' (Project 661). She would be armed with missiles launched from underwater.

The need to deal with carriers seems to have made missile-firing cruisers (in Russian parlance, 'rocket cruisers', RKR) more attractive. Planning for RKR divisions probably began about 1959; construction of a new 'Kresta' class began about 1962.

From a bureaucratic point of view, the anti-carrier requirement would have been most welcome. With the SRF granted an effective monopoly on long-range strikes, Admiral Gorshkov and his colleagues presented the existing strategic submarines in a new light: they could fire their missiles at carriers. Probably the navy conceived a two-step programme for both ballistic and cruise missiles. Initially the weapons would be fired at preselected points through which carriers could be expected to pass. Later they would be provided with anti-ship guidance mechanisms. The second stage was already underway in the form of SS-N-3. In the case of the ballistic weapons, it was not until 1973–74 that a guided version, SS-N-13, was tested, and by that time it was no longer wanted. SS-N-6, the first really long-range Soviet submarine-launched ballistic missile, apparently began life in 1960 as an anti-carrier weapon.

SS-N-6 may have been conceived as the long-range ballistic companion to the big Ilyushin missile. Although its full range would not have been appropriate to anti-carrier attack, the high speed corresponding to that range may well have been attactive because it so limited time from launch to target. Similar reasoning would not applied to the big Ilyushin cruise missile, which was presumably doomed by the shift from strategic attack to anti-carrier attack.

As the navy tried to preserve its ballistic force by shifting it to an anti-ship role, the Strategic Rocket Force seems to have argued that its own land-based missiles could take over the *naval* role. It seems to have advertised missiles based in the southern Soviet Union as a means of dealing with NATO naval forces in the Mediterranean.

Work on the next naval building programme (presumably 1966–75, split into Five Year Plans to allow mid-term changes) probably began about 1962. That year Khrushchev was shown a menu of future naval technology, including a submarine-launched missile targeted by satellite (SS-N-19) and a new small-boat missile (probably the extended-range Styx). In approving a range of projects, Khrushchev demanded that ship endurance (in days and in ammunition) and operating range be increased to match; presumably he was pushing out the defensive perimeter the Soviet Navy was intended to protect.

A new RKR, Project 934 (ultimately 'Kresta I', Project 1134) had been designed. Plans called for one RKR division in the Northern Fleet, to comprise one heavy brigade (seven Project 934, armed with new SS-N-12 missiles) and two light brigades (each of seven Project 934A, armed with shorter-range missiles which became SS-N-9). The corresponding submarine division would have consisted of a heavy brigade of 'Papa' class (Project 661) missile-firing submarines and two light brigades of very fast torpedo submarines ('Alfa' class). The Pacific Fleet would have received the next RKR division, somewhat advanced over those in Northern Fleet: a Project 934B heavy brigade and two light brigades analogous to Project 934A. For the generation beyond, new heavy and light RKR were conceived. Probably work began only on Project 964, which ultimately became *Slava* (Project 1164). This entire programme died when the emphasis shifted to ASW.

The heavy brigades of the three Baltic Fleet RKR divisions would have consisted of anti-ship rocket (SS-N-9) versions of an existing SKR ('Riga' successor) design, which later became the 'Krivak' class frigate; the light divisions would have been 'Nanuchkas'. Given the heavy/light distinction, the 'Nanuchkas' may have been conceived to carry SS-N-2 missiles. The Black Sea Fleet would have had two 'Krivak'/'Nanuchka' RKR divisions. According to the fleet plan, the 'Krivaks' would have attacked NATO surface forces blocking the Danish Straits, and 'Nanuchkas' would have broken through into the North Sea and the English Channel. At least some of them would have attacked land targets.

The numbers ordered did not quite meet these targets. The plan for the Western fleets required a total of five inshore divisions, ten brigades of 'Nanuchkas' (seventy boats). In fact six brigades (one ultimately for export) were ordered built in Leningrad, and another brigade built at Vladivostok (eight boats were exported, plus three built specially for export, and one boat was scrapped incomplete).

In both the Baltic and the Black Sea, the ASUW submarines would have been diesel boats. The closed-cycle projects were all terminated in 1960, and no nuclear submarines were planned for these enclosed seas. That raises the question of whether there was a projected diesel missile boat equivalent to 'Papa'.

Khrushchev approved the plan with one important exception: he rejected seaborne flagships. He thought the brigades could be controlled adequately from shore headquarters. He would be proven wrong only in the 'Okean' 1970 exercise mentioned above. Even so, the numbers apparently were not affected: seven Project 934 and fourteen Project 934A were ordered.

About 1961 it became obvious that neither 'Papa' nor 'Alfa' could be mass-produced, at least not in the near-term. Fortunately a torpedo submarine with a conventional reactor, 'Victor' (Project 671), was ready for production. The missile system planned for 'Papa' was also in trouble, so an adapted Styx was installed in a hurriedly-redesigned submarine, 'Charlie' (Project 670) which used one (rather than two) 'Victor' reactors. It later turned out that although 'Victor' had been conceived as a torpedo submarine to attack surface ships, it was easily adapted to fire the SS-N-15 ASW missile derived from the US SUBROC.

The position of the longer-range missile submarines ('Echo II' and 'Juliett' armed with SS-N-3 and the ballistic missile units) was somewhat anomalous; these refugees from the aborted strategic programme did not fit the tactical anti-ship divisional structure. However, it is possible to imagine them as the long-range element of a submarine equivalent to the long/short-range surface ship ASUW force (the short-range portion might be missile- or torpedo-armed submarines).

As for the amphibious forces, about 1963, with his power already waning, Khrushchev allowed the revival of the special naval infantry. At about the same time the Soviets began to design landing ships, such as the 'Alligator', which could provide a measure of fire support while the troops landed; note that the earlier 'Polnocny' was initially little more than an updated 'MP 10' lighter. The *Ivan Rogov* was the ultimate step in this series, capable of launching air-cushion vehicles. The latter may also have been associated with the possible needs of warfare in the Far East.

THE POST-KHRUSHCHEV ERA

With the departure of Khrushchev, the SRF must have lost its greatest champion, and the older Soviet forces saw their way back to a more balanced strategy. Perhaps most significantly, Admiral Gorshkov had a personal connection with Khrushchev's successor, Leonid Brezhnev, probably dating back to his Black Sea Fleet days during the Second World War. Later some Soviet army officers would complain that Gorshkov had been able to get virtually anything he wanted, often at the expense of the army (some officers who made such complaints apparently forfeited advancement in their careers). The charge echoes that often levelled at the Kaiser's navy: that the cost of building the German battleship fleet had just equalled the cost of the extra army corps which might have assured victory in 1914.

With Khrushchev's departure, the navy saw its way back into strategic warfare, still the most prestigious of all Soviet war-fighting functions, given the 'revolution in military affairs'. The ongoing 'Yankee' (Project 667) programme for a new ballistic missile submarine became a programme for a strategic SSBN (reportedly details of the design show how hastily it was completed). Soviet naval doctrine did show a certain hesitancy in returning to strategic affairs. Well into the 1970s Admiral Gorshkov would argue that his submarines were a reserve force, a modern version of the old fleet-in-being, which would be available for use, or for threatened use, after the primary strategic force, the SRF, had been expended.

Given the relatively short range of its missile, a 'Yankee' had to cross the Atlantic or Pacific to come within range of US targets. To be sure, some 'Yankees' would generally be stationed off the US coast in peacetime, if only as insurance against surprise attack, but they would represent a far smaller fraction of the 'Yankee' force than would have been the case in the US Navy (the Soviets typically had only about 10 per cent of their SSBNs at sea at any one time).

The transit concept was part of a new phased war strategy. Khrushchev had believed that, because they were so effective, nuclear weapons would be used from the outset. The doctrine of the post-Khrushchev era apparently envisaged a non-strategic or even non-nuclear opening phase, during which each side would try to destroy the nuclear weapons of the other, possibly by non-nuclear attacks. In this context one might imagine a Soviet naval war plan in which the initial blow would fall on US sea control forces, leaving the way open for hunter-killers to attack US SSBNs, well before any escalation to the strategic level.

Another important influence on the Soviet Navy of this period was the move into the Mediterranean, beginning in 1967. The Black Sea Fleet had gone from great importance under Stalin to a relatively minor position under Khrushchev, as the Northern Fleet became dominant. With the rise of the Soviet naval presence in the Mediterranean (largely to protect the Soviet position in the Middle East), the Black Sea Fleet became extremely important, as it provided the Mediterranean Squadron surface units (the submarines came from the Northern Fleet).

The phased strategy was demonstrated in October 1973, when the Soviet Mediterranean Squadron was hurriedly reshuffled in a crisis. The most modern ships, that is the ASW fleet, were moved back into the Black Sea where they came under the protection of land-based aircraft; the older anti-carrier ships were moved into the Mediterranean to form classical Soviet anti-carrier groups. Presumably the ASW fleet would have moved back out to sweep the Mediterranean of Polaris submarines, had the carriers been destroyed.

The Soviet Navy became very interested in supporting SSBNs attempting to reach the open sea. A new pro-SSBN role (PSW) was incorporated in a new fleet plan drawn up in 1964. A separate army of each fleet corresponded to each major warfare area: anti-ship (ASUW), ASW, PSW. For example, an ASUW army would include one or more RKR divisions, one or more attack bomber divisions, and one or more submarine divisions, plus support facilities. Similar elements would be included in an ASW army. This was *not* a division by ship type: ASUW submarines are not under the same command as ASW or PSW submarines, though presumably there is some coordination at the fleet level.

The new PSW role was clearly more urgent than the RKR role. To fill it, most RKRs were redesigned. The big new RKR, Project 934, was redesignated Project 1134 ('Kresta I', which retained its RKR main armament but was redesignated a BPK, a large ASW ship). The next three and the 934As were all completed as ASW ships, their SS-N-9 missiles replaced by SS-N-14s (they became Project 1134As, 'Kresta IIs'). The second brigade of 1134As was reordered as gas turbine powered 1134Bs ('Karas'). Similarly, the 'Krivaks' were redesigned with SS-N-14s instead of SS-N-9s. Production continued slightly beyond the thirty-five presumably envisaged in 1964.

It would appear that the 'Nanuchkas' were redesigned to carry the SS-N-9s which could not go aboard the 'Krivaks'. One indication of this switch is their type designation, MRK, now typically translated small (*maly*) rocket 'cutter', but clearly inappropriate, since the much smaller 'Osa' and its ilk are rocket cutters (RKAs). MRK *would* be logical if it really meant 'small rocket ship'.

The first post-Khrushchev Five Year Plan (1966–70) included

The Cold War: *Kiev* being shadowed by HMS *Danae* as she makes for the Shetland/Faeroes gap, Aug 1976 CMP

development of a much longer-ranged submarine missile, SS-N-8, which could hit American targets from waters the Soviet Navy could hope to control. Once this weapon had entered service (1973), PSW changed meaning. It now became essential to protect waters ('bastions') within which SSBNs would operate. To the extent that the SSBNs were a vital reserve force, that protection might have to last throughout a war. During the 1980s, the US Navy would realise that pressure on the 'bastions' could tie down Soviet forces which might otherwise have attacked NATO naval and merchant shipping.

Two new surface ship types were designed: the *Kiev* class helicopter/VTOL carrier (Project 1143) and the *Udaloy* class ASW cruiser/destroyer (Project 1155). *Kiev* began as a follow-on to *Moskva* for other fleets. *Udaloy* was essentially an improved 'Kara' without an area defence missile system. It seems likely that the *Kiev*/*Udaloy* combination was conceived in rough analogy to the earlier combination of a *Moskva* and ASW destroyers. The heavy ASW unit of the main fleet, then, would be a *Kiev* operating with six *Udaloys*; presumably a total of twenty-four were planned (but only fourteen were ever laid down). In fact such ideas would soon be thrown out, due largely to the experience of the 1970 'Okean' exercise.

The shift to ASW also affected the submarine force. A long-range anti-submarine missile was developed by copying the US SUBROC (as SS-N-15). Production of 'Juliett' ASUW submarines was cut, and some of the components released went into a new class of ASW submarines ('Tango', Project 641BUKI). The new nuclear attack submarine ('Victor', Project 671) was adapted to fire SS-N-15.

The RKR role remained. In 1964–66 the decision was made ultimately to build next-generation ships for both fleets: *Slava*, *Kirov*, *Sovremennyy*. TTZs for the latter pair were probably written in 1969 in tandem with TTZs for new-generation missiles (SS-N-19 and SS-N-22); a preliminary design for *Slava* (which fired the previous-generation SS-N-12) seems already to have existed. It appears *Slava* (which has the suggestive code-name *Atlant*), was the Northern Fleet heavy RKR. Probably the Pacific Fleet heavy RKR was Project 954, which ultimately emerged as the Project 1144 *Kirov* nuclear command cruiser. The light RKR for both fleets became Project 956 *Sovremennyy*. The tactical concepts involved soon collapsed. For example, *Kirov*s were redesigned to serve as flagships. However, note that out of seven *Slava*s presumably originally planned, four were built; out of seven *Kirov*s, five were laid down (three completed, one still under construction in 1995). *Sovremennyy* numbers will probably match the twenty-eight (four brigades) probably originally conceived.

Presumably the Project 949 heavy missile submarine ('Oscar') was part of this programme. It appears that a folding-fin version of SS-N-22 was planned for new submarines. The combination of 'Oscars' and SS-N-22-armed submarines would form an underwater equivalent of the surface heavy/light brigade structure.

The 1970 exercise proved that both PSW and RKR forces required at-sea flagships. The ongoing *Kiev* design was revised to become a PSW flagship (some hangar space was sacrificed). *Udaloy*s became divisional flagships, presumably intended to lead groups of 'Krivaks' or 'Krivak' successors (*Neustrashimmyy* class). They would also work with the new long-range ASW aircraft and with ASW submarines. The new heavy cruiser became an RKR flagship.

Plans seem to have called for ASW and RKR flagships and back-up flagships in each of the two main fleets (the two *Moskva*s would have functioned as ASW flagships in the Black Sea/Mediterranean). Back-ups were essential because the two primary fleets were so far from the building/repair yards (St Petersburg on the Baltic for the *Kirov*s, Nikolaev on the Black Sea for the *Kiev*s). Indeed, both *Kiev*s in the Pacific ultimately had to be scrapped because they could not be brought to Nikolaev for major repairs. Presumably another pair of carriers (which became the *Kuznetzov*s) would eventually have been built for the Northern Fleet, and a pair of *Kiev*s moved to the Black Sea to replace the obsolete *Moskva*s. Reports of an abortive fifth *Kirov* suggest that ultimately one or two would have operated with the Black Sea/Mediterranean fleet. In fact none of these plans every completely matured. The Baltic Fleet would use shore-based headquarters.

At the same time, Admiral Gorshkov began to write about another, very different Soviet fleet mission. The basis of Soviet ideology was the constant struggle between forces of 'progress' (led by the Soviet Union) against the capitalist world. In the post-Khrushchev period it was accepted that one vital role of the Soviet armed forces was to deter 'reactionary' assault on the centre of 'progressive' movements, *ie* upon the Soviet Union. This concept was soon extended to shielding pro-Soviet insurrections in the Third World. Admiral Gorshkov argued that the first example of such an operation was the use of Soviet warships off Angola in 1975–76. He called it securing Soviet state interests in peacetime. This new concept was extremely attractive for a navy which had always been subject to intense political attack, often on the basis of the apparent obsolescence of its function. It was formulated as Soviet leaders came to believe that the balance of world power (the 'correlation of forces') was tipping in their direction, hence that they could safely risk more direct intervention in the Third World.

Admiral Gorshkov and his associates also argued, about 1979–80, for a more substantial wartime role, involving a distant 'oceanic' theatre of operations which would be wholly naval. The carrier decision, the centrepiece of a distant-ocean strategy, had already been taken, about 1975, for construction during 1978–85 (*ie* during the 1981–85 Five Year Plan). The other services recognised the oceanic theatre idea as a distinct threat, and the idea was roundly denounced in the main Soviet military journals. However, construction of a new generation of carriers proceeded. They were justified internally as a means of supplementing land-based air defences. They may also have been seen as a means of launching fighter escorts to protect bombers attacking US carriers in areas such as the Norwegian Sea. The first justification was clearly defective, since the carriers could not quickly move to counter incoming strategic bombers. However, it may have been designed to overcome resistance by the Soviet air forces. The air defence justification was also used to help explain how the carriers did not violate Gorbachev's new doctrine of 'defensive sufficiency'.

The 1980s were not kind to the Soviet Union or to its fleet. It was soon apparent that the West, particularly the United States, did not share the Soviet view of the state of the 'correlation of forces'. Where the Soviets had hoped to engage Western naval forces far out to sea, the new American (later NATO) Maritime Strategy assumed that the West would take the naval initiative. Soviet naval exercises came to emphasise defense against approaching Western battle groups rather than the offensives of the past.

Three other developments of this period deserve mention. KGB ship requirements began to expand, apparently to the point of taking over naval hulls conceived for other purposes, particularly 'Krivak IIIs' and the final *Udaloys*. In both cases, the vital requirement may have been the ability to operate helicopters (earlier KGB ships, modified corvettes, lacked any such capacity).

A second new development was the emergence of a Soviet equivalent to the US Tomahawk torpedo-tube launched strategic/anti-ship missile, SS-N-21. A new submarine, 'Akula' (Project 971), was designed specifically to carry it. Although 'Akula' looked like an attack submarine, in fact it was conceived as a special-purpose type, and designated accordingly.

Finally, there was a sudden improvement in Soviet submarine silencing, initially in the 'Victor III' (Project 671asRTMxs) class and then in the follow-on 'Sierra' (Project 945) and 'Akula' (Project 971). It was about the right time for a generational change in Soviet submarine design ('Victor' dated from about 1960). Furthermore, during the mid-1970s the Soviet had become aware of NATO success in tracking their existing submarines, through the efforts of the Walker spy ring. By this time the Soviets were probably quite aware of the details of NATO submarine-detection technology, such as Lofar (passive narrow-band processing). They knew what had to be done; the only question was whether to accept the associated industrial cost. Presumably the illegal acquisition of Western propeller-cutting technology was part of the silencing programme.

Ironically, the new submarines surprised the West mainly because the previous generation ('Victor' and 'Charlie') had been *louder* than expected. It had been assumed that the Soviets lagged behind the West by five to eight years in silencing, that the goals generally matched those of the West. Thus there was considerable concern that the next generation, due in the mid-1970s, would negate such vital passive detection systems as SOSUS and the principal sonobuoys. When the 'third generation' acoustic threat did not materialise on time, Western analysts assumed that the Soviets had decided to favour quantity over quality, that they would always be loud. Then came 'Victor III'. Existing strategic submarines began to emerge from refits significantly quieter.

In retrospect, it seems likely that the Soviet leadership decided to improve silencing drastically after the Walkers' revelations forced it to confront the fact that NATO could track virtually all of their existing submarines. John Walker himself clearly had access to the plot of submarine positions at Norfolk, and intercepted and decoded US radio traffic would have provided Soviet intelligence with the sort of operational information US and NATO commanders had concerning the current locations of Soviet submarines. Moreover, the Walker ring was run by the KGB, rather than by Soviet military intelligence (GRU). Its information was probably used by Yuri Andropov, the KGB chief, to discredit Admiral Gorshkov, a prop of the existing Brezhnev regime, who presumably preferred to claim that existing silencing was satisfactory, largely to avoid the industrial turmoil associated with a new and aggressive silencing campaign.

Industrial productivity had never been altogether satisfactory, particularly in such new fields as electronics. By the early 1980s, many new warships were going to sea without key electronic systems, and sometimes even without portions of their armament. The Soviets did continue to develop innovative prototypes, and sometimes they were able to acquire electronic components from the West. Overall, however, it seemed unlikely that the system could compete with the Western powers in the ongoing electronic revolution. Unfortunately, the Soviets had no Khrushchev who could envisage some radical military reaction comparable to the 'revolution in military affairs'. The post-Khrushchev settlement, in which all segments of the Soviet system were allowed to develop much as they liked, precluded that. In effect, the costs of maintaining standing forces and building the sort of forces already in production, could not be sustained.

Mikhail Gorbachev's attempt to solve the problem by reforming the system doomed it.

The collapse, and particularly the break-up of the old Soviet Union, had important implications. Suddenly the very large military sector of the old Soviet economy no longer enjoyed the sort of resources formerly available. It badly needed cash, and what amounted to a 'fire sale' began. Reportedly Iran benefitted when it bought 'Kilo' class submarines at drastically reduced prices. In the late spring of 1995 a major scandal at the Russian arms export agency included the charge that plans for a carrier had been sold to China at the absurdly low price of a million dollars. At the time of writing (June 1995) despite aggressive marketing, the Russians have had little success selling naval systems apart from submarine sales to Iran and to China (in the latter case, probably including production tooling for 'Kilo' class submarines).

Quite aside from the drastic loss of resources previously available for defence, no successor state enjoyed the combination of industrial and base resources which had been available to the old Soviet Navy. For example, the main gas turbine plant is in Ukraine. That may explain why Russia is building steam-powered *Sovremenny* class destroyers, rather than gas turbine *Udaloys*. Other important sub-systems were built in plants scattered among the successor republics. This distribution reflects a long-standing policy to maintain coherence by forcing plants to deal with each other over the borders of the republics. Now, with the republics trading with one another as separate countries, it is no longer possible for the Ministry of Defence to order components across the same borders.

Similarly, the carrier-building yard, at Nikolaev, is no longer in Russian hands (it is Ukrainian). One consequence was that major accidents to the machinery of the *Kiev*-class carriers could no longer easily be repaired; these ships are being broken up prematurely. Similarly, the future of the second *Kuznetzov* class carrier, *Varyag*, now depends on economic relations between Russia and Ukraine, rather than merely on the state of the Russian defence budget.

The Black Sea Fleet faces the loss of its major base, Sevastopol in the Crimea. Novorossisk may be built up as an alternative. Ironically, the Crimea was Russian until Khrushchev, a Ukrainian, transferred it to the Ukraine in 1954. At the time of writing, it is still possible that the Crimea, whose population is overwhelmingly ethnic Russian, will eventually become a Russian–Ukrainian flashpoint. It is also possible that the industrialised Eastern Ukraine, which has a largely Russian population, will eventually move back into Russia, largely restoring direct access to the scattered parts of the Soviet military industrial base.

For the present, conflict between Russia and Ukraine over the Black Sea Fleet is having some odd effects. For example, Ukraine has protested bitterly as obsolete ships are sold for scrap, arguing that it is not enjoying any of the income being generated. As a consequence, some ships of obsolete types survive in the Black Sea. Some ships also survive because they are, in effect, the only housing available for their crews.

SOME GENERAL OBSERVATIONS

This sketch of the evolution of the primary role of the Soviet Fleet emphasises discontinuities in naval doctrine. The effect of industrial inertia continued past programmes reflecting earlier emphases, whether or not those emphases continued to be important to the Soviet naval leadership. Existing ships were, moreover, pressed into service to meet the new requirements. For example, the 'Kashin' class missile destroyers, probably the last 'traditional' destroyers ever built by any navy, were initially designated 'fleet torpedo craft' in the Tsarist tradition. With the turn towards ASW/SSBN operations, they were redesignated large ASW ships (BPK) and employed as command ships for group ASW operations by smaller submarine-chasers. It is possible that this involved the installation of specialised computers for sonar data reduction, but the physical changes were small. Similarly, the 'Foxtrot' and 'Tango' class diesel submarines were direct descendants of the 'Zulus' designed for classical submarine warfare in the North Atlantic. However, their relative silence made them effective ASW platforms, and that appeared to be their mission. Thus, in effect, the Soviet Navy of any one point was an overlay of several quite different navies, and so was the Soviet naval production programme.

It is worth noting that although the Soviet system of industrial planning emphasised long-range programmes and thus imposed considerable rigidity, the Soviets were also very willing to copy their adversaries' equipment, allowing them in selected cases to shift warship missions and characteristics much more quickly than simple, industrial patterns might suggest. Examples of such copies would include the 'Atoll' air-to-air missile, a copy of the US Sidewinder, and quite possibly the SS-N-15 SUBROC equivalent and the SS-N-3 surface-to-surface guidance system. One might contrast the typical Western attitude that if it is 'not invented here' it is not worth having.

It is important, too, to recognise that the warships were only the most visible elements of a much more comprehensive establishment. Reference is frequently made to the large force of land-based strike bombers, at first 'Badgers', and later including many 'Backfire' supersonic bombers, as well as some 'Blinders'.

DESIGNATIONS

The Soviets assigned a Project number to each new design (or, at least, to each design selected for production); in some cases (such as ballistic missile submarines) a project, embracing both a hull and associated weapons, was assigned a code name. The system began before the Second World War. Postwar, it appears that all warship designs, and possibly all merchant ship designs, were given project numbers within a unified Shipbuilding Ministry series. There is ample evidence (*eg* in the text below) that the numbers are not sequential. To some extent they may have been assigned in type-specific blocks, *eg* a 600–900 series for submarines, or a 100–200 series for small combatants. It is also possible that some ships were reassigned numbers in new series. In some cases the number may be in two parts, *eg* Project 1135 is sometimes written 11–35.

The project-number system was also adopted by Poland and by East Germany. Their series were distinct from those adopted by the Soviets, but one navy would probably refer to a ship bought from another country by that country's project number (as in the cases of 'Polnocnys' and 'Parchim IIs').

Until quite recently, both project numbers and most individual ship names were highly classified; now, however, the Russians have released many of them. They are provided in the text, for both prewar and postwar ships. In some cases the project numbers for postwar ships show an obvious relationship to prewar designs. In a few cases, it appears that project numbers were erroneously used twice. Furthermore, some classes had their project numbers changed in the concept stage. Changes have been indicated wherever they are known.

Suffixes were applied, unfortunately not always consistently, to indicate variations. 'Bis', which is often rendered 'B', indicated a second version; it could also indicate an armament change. The letter 'A' generally meant the first major modification ('B' indicated the second); but 'A' was also often used merely to mean a modification, and thus often to stand for some other Russian letter or letters. 'K' indicated either a 'corrected' design or a command system change. 'M' generally indicated a modernisation (and could also apply to missile and gun system designators). 'U' might mean improved or extended (range) or a flagship (as in Project 68U) or even a training version (as in aircraft versions). 'E' generally meant export, but 'Eh', which is often rendered 'E', would mean experimental, as in *Admiral Nakhimov*, Project 67Eh. 'P' generally meant a version (*pogranichniy*) for Maritime Border Guards. 'T' could mean torpedo, but it could also mean minesweeper (*tralchik*) version, and it clearly had other meanings as well. Prefixes generally mean experimental prototypes (see Project 611, for example); a 'P' or 'V' prefix generally meant missile-related.

In the absence of quotable project numbers or actual ships' names, NATO assigned its own code names for reporting purposes. Initially (when the actual names of surface ships were known) that applied mainly to submarines, which were given the phonetic equivalents of letters, *eg* 'Whiskey' for W or 'Alfa' for A. Letters appear to have been assigned randomly, the first three being W, Q, and Z ('Whiskey', 'Quebec', and 'Zulu'). Surface ships were generally given Slavic-sounding names, beginning with K for major combatants. Minecraft were given Russian diminutive proper names, such as *Natya* or *Alesha*. Numerical designations were applied to smaller units. For example, the P-series designations of early postwar Soviet torpedo craft, such as P-6, are all of NATO origin; they have nothing to do with Soviet designators. The same is true of the designations of postwar landing craft.

Ships themselves have permanent hull numbers (in a single S- series generally related to building yard construction sequence), but the numbers are not clearly sequential, and most of them are not as yet available. They are not given in the lists which follow.

SOVIET NAVAL WEAPONS AND SENSORS

NUCLEAR WEAPONS

Many Soviet naval weapons had alternative nuclear and conventional versions. Since the late 1950s Soviet military doctrine consistently emphasised nuclear weapons as the single decisive element in modern warfare, and it is sometimes assumed that such weapons would be used at the outset of a naval war. The detection of nuclear torpedoes aboard the elderly 'Whiskey' class submarine which grounded in Swedish home waters in October 1981 underlined this belief. However, more recently it has been reported that the 'Kynda' class cruiser which sortied into the Mediterranean during the October 1973 crisis was not carrying the standard outfit of nuclear-armed SS-N-3 missiles: one might suspect that, in a period of intense crisis, the Soviet leadership would have been chary of unintended nuclear use. One might even go further, and observe that Soviet ideology asserted that the primary war scenario was a *Western* surprise attack on the Soviet Union. Soviet naval forces deployed only in relatively small numbers in peacetime, and one might go so far as to think of the nuclear weapons on board as an emergency 'equaliser' – to be removed in the event of lengthy limited-war warning. The reader is cautioned that such analysis is speculative, but it still seems worth keeping in mind.

GUN SYSTEMS

406.4mm/50 MK-1, the gun designed for the prewar and postwar battleships. It fired a 1108kg (2442lb) shell at 850m/sec to a maximum range of 46.2km (50,000yds). Turret weight was 2364t. Elevation/depression 45/4 degrees. Rate of fire was 2.5rpm. The crew per turret was one hundred.

304.8mm/54 MK-15 of 1938 planned for the prewar *Kronstadt* (Project 69) class. The 471.0kg shell had an initial velocity of 900m/sec; elevation/depression was 45/3 degrees. Roller path diameter was 9.7m. Protection: 305mm face, 125mm side and top. Maximum time between rounds: 19–24sec. The SM-31 gun of the postwar *Stalingrad* (1950) was a 304.8mm/61 firing a 467.0kg shell at 950m/sec; elevation/depression was 50/5 degrees. Maximum range was 49.3km (54,000yds). Roller path diameter was 9.9m. Protection: 240mm face, 225mm side, 125mm top. Maximum time between rounds: 18–22sec. Like the contemporary *Sverdlov* turret, SM-31 had a radar ranger on its right-hand side. By way of comparison, the old 305mm/52 on board *Gangut* class dreadnoughts fired a 470.9kg shell at 762m/sec. Elevation/depression was 25/5 degrees, and roller path diameter was 9.1m. The Mk-3-12 mount (presumably a modified *Gangut* type) fired a 410.9kg shell at 777.9m/sec to a maximum range of 29km (31,700yds); elevation/depression was 40/5 degrees, and rate of fire was 2rpm. Crew per turret (955t) was 62. There was also a coast defence version and a railroad version (TM-3-12).

220mm/65 This weapon was planned for the abortive Project 66 heavy cruiser. It fired a 176kg shell at 985m/sec at a rate of 3–5rpm (compared to 2–3rpm for Project 82); maximum range was 47.5km (52,000yds).

180mm/57 MK-3-180, the main armament of the *Kirov* class, in 274.4t triple turrets, with a rate of fire of 5.5rpm and elevation/depression of 50/4 degrees. Maximum range was 37.5km (41,000yds) (97.5kg shell fired at 920m/sec). The earlier cruiser *Krasni Krim* carried single 180mm/60 guns (B-1-K) with similar characteristics.

152.4mm/57 MK-5, the main weapon of the *Chapaev* and *Sverdlov* classes (the latter used MK-5bis). There was also an MK-4. The triple turret was designed in 1938, and the gun tube may have been designated B-38. The gun fired 55kg SAP and HE shells at 950m/sec to a maximum range of 30.7km (33,600yds). Elevation/depression was 45/5.5 degrees. Effective range was reportedly 18km (19,685yds). Reportedly a nuclear shell existed for these guns. Rate of fire was 7.5rpm. Armour, in a *Sverdlov*: turret face 175mm, back 60mm, side 65mm, top 75mm, barbette 130mm. This turret carried a Shtag-B rangefinding radar.

130mm Standard prewar and postwar destroyer calibre. The old 130mm/50 was based on a Vickers semi-automatic 5.1in/55 VSM Mk 'B' delivered before the First World War for Russian destroyers and cruisers; it appeared in both single (B-2-LM, 12.5t) and twin (B-13, 49t) mounts, with a maximum rate of fire per gun of about 11rpm (total 23rpm in the twin). The gun fired a 33.4kg shell at 870m/sec to a maximum range of 26km (28,400yds) (reported effective range 14–15km); elevation/depression were 45/5 degrees. The version in the *Skoryy* class, B-13-2C, had 10mm (0.4in) protection. A new

130mm/58 gun, 56-SM, appeared in a new triaxially stabilised mount in 1953, aboard *Neustrashimyy* and the 'Kotlin' class. The gun and mount were also manufactured in China. The mount suggests German wartime and prewar practice, in that the guns elevate through its roof (to 80 degrees). They are credited with 10rpm per barrel and with a maximum range of 28km (30,600yds). A new twin 130mm/54 (AK-130, Lev) was introduced in 1981 on *Sovremennyy* and *Slava* class cruisers and *Frunze*. It is liquid-cooled. Maximum rate of fire is 86rpm per barrel, the shell weighs 32kg, and maximum range is 29.5km (32,300yds). Muzzle velocity is 850m/sec (2790ft/sec). Elevation/depression are 80/10 degrees. The mount weighs 94t. This gun fires laser-guided shells. An automated version, comparable to the US 5in Mk 45 or to the British 4.5in Mk 8, was apparently developed for the abortive nuclear missile destroyer (Project *Anchar*).

100mm Standard frigate and cruiser secondary battery calibre. 'Kola' and 'Riga' classes had the 1947-model single 100mm/56 (B-34), elevating/depressing to 85/5 degrees; it fired a 15.6kg shell at 900m/sec to 22km (24,000yds), and had a maximum rate of fire of 12rpm. The mount weighed 7.7t and had a crew of seven.

A shorter 51-calibre was designed for *Kirov* class secondary batteries; as B-24 it originally had a maximum elevation of 40 degrees, but was modernised for dual-purpose use in 1947. A twin version was designed in 1940 for the abortive Soviet battleship programme. B-24DL had a muzzle velocity reduced to 872m/sec (range 21km [23,000yds]). The *Sverdlovs* and *Chapayevs* had a twin 100mm/70 gun in a triaxially-stabilised German-style mount (SM-5-1S), the guns elevating through the roof to 85 degrees and depressing to 8 degrees. In these ships, maximum surface range was 25.6km (28,000yds), maximum altitude 20km (about 33,000ft). Muzzle velocity was 1000m/sec (3280ft/sec). Rate of fire: 18rpm. This calibre was reintroduced in 1976 in 'Krivak II' class frigates and in the cruiser *Kirov* in a single water-cooled 100mm/59 mount (AK-100) firing a 15.6kg shell at 880m/sec at 60, 45, or 15rpm, with a maximum range of 21.3km (effective range 8km/8750yds). Standard ammunition stowage is 322 rounds. Elevation/depression 85/10 degrees. River monitors are armed with 100mm tank guns (not the 115mm weapons previously reported) designated D-10T2S, capable of firing 7 rounds/min out to a range of 14,600m (16,000 yds). Some earlier monitors had 76.2mm weapons (see below).

85mm The 1943-pattern twin 85/52 in *Skoryys* (92-K, mount HAM 39-3) fired a 9.2kg shell at 800m/sec to 15.5km (17,000yds); mount weight was 17.8t, comparable to that of a single 130mm gun. Rate of fire was 15–18rpm per gun. Elevation/depression was 85/5 degrees. Crew was fifteen. An earlier 90-K was a single mount (5.3t). Note that contemporary Western sources rated this weapon at 950m/sec muzzle velocity.

76.2mm Note that the calibre was 3in rather than some round metric number. The standard prewar and early postwar Soviet version was a 76.2mm/55, in single open (34-K), single shielded (81-K), and twin (39-K) versions. In each case the gun fired a 6.61kg shell at 813m/sec to a maximum range of 13km (14,200yds), at a rate of 20rpm per gun. Mount weights were, respectively, 4.9t, 12t, and 12.5t. Elevation/depression 85/5 degrees. In 1961 this calibre was revived in the form of the twin 76.2mm/59 AK-726 (mount ZIF-67), a successor to the twin 57mm then in service: It fires a 5.9kg shell at 980m/sec to a maximum range of 15.7km (17,000yds); the crew is nine. Elevation/depression limits are 84/2 degrees. The firing rate is 90rpm per gun (maximum continuous burst 40–45 rounds). Ammunition supply is 138 rounds per mount. Total weight is 26t. Work on this mount began in 1954, and the TTZ was issued in 1956. The single AK-176 (76mm/59) entered service in 1979. It is found on board 'Grisha-V', 'Parchim II', 'Nanuchka III', 'Pauk', and 'Tarantul' class craft, controlled by a Bass Tilt radar. Although it is designated a 76mm rather than a 76.2mm gun, it fires the same shell as the AK-726 at the same muzzle velocity; typical ammunition stowage is 152 rounds (120rpm). Elevation/depression limits are 80/10 degrees. A successor AK-176M uses a slightly shorter (54-calibre) gun (with the same muzzle velocity); elevation/depression limits are 85/15 degrees. The mount weighs 11.2t. There is also a 76.2mm/48 gun in the turret of a PT-76 tank, used on board river monitors (D-56TS).

57mm The old open 57mm/70 existed in single, twin (ZIF-31B), and quadruple mounts (the latter with the pairs of guns superimposed): 150rpm per barrel. Effective vertical range: 4.5km (4920yds; 2.8kg shell, muzzle velocity 900–1000m/sec). The twin water-cooled 57mm/75 of 1961 (AK-725, ZIF-72) was the first gun of this type without an on-mount crew: 2.8kg shell, 1020m/sec, maximum range 14.5km (15,740yds), 120rpm, standard stowage 1100 rounds, elevation 85 degrees.

45mm A quadruple 45mm/85 mount (SM-20-EIF) was widely used in the early 1950s: 75 rounds/bbl/min, effective vertical range 4km (4370yds); 2.2kg shell fired at 900m/sec. Earlier single versions were 21-K and 21-KM (respectively, 46 and 68.6 calibres long, firing 1.41 and 1.43kg shells at 760 and 870m/sec, for ranges of 9.9km (9920yds) and 10.7km (11,100yds); maximum rates of fire were 30 and 40rpm, and the mounts weighed 3t and 5t; elevation/depression angles were 82 degrees and 10 degrees for both).

37mm Twin water-cooled 37mm/67.5 (V-47M) mount resembling the Western Bofors, developed from a Bofors 25mm built under license. It fired a 0.782kg shell at 880m/sec to a maximum range of 8.1km (effective range 4km [4370yds]) at 180rpm per gun, and had a crew of seven; mount weight was 3.2t and elevation/depression were 90/15 degrees. The corresponding air-cooled single and twin mounts were 70-K and V-11M. V-11M was the mount in 'Rigas' and *Sverdlovs*, for example.

30mm A twin remote-controlled automatic 30mm/65 mount for light craft (AK-230) was introduced in 1959: it fired a 0.36kg shell at 1050m/sec to a maximum range of 6.7km (7327yds), and elevated to 85 degrees. A typical mount stowed 1000 rounds of ammunition and fired at 2100rpm. A six-barrel Gatling (AK-630) appeared in 1975. It fires a slightly heavier shell (0.39kg) at a lower velocity (900m/sec), and is credited with a maximum range of 8.1km (8850yds). The peak rate of fire is 5000rpm, and the mount stows 2000 rounds. AK-630M (1980) stows 3000 rounds per mount. A new single-barrel AK-306 also appeared in 1980; it fires 900rpm and stows 500 rounds per mount. There is also a BP-30 of 1976, a one-man single mount firing a lighter-weight shell (0.28kg) at low velocity (185m/sec). It fires 400rpm and the standard mount carries 400 rounds. Maximum range is 1.75km (1914yds).

25mm The single 25mm/83 mount, 8A-KM, was a Bofors gun built under license; it fired a 0.288kg shell at 900m/sec to a maximum range of 7km (7655yds); maximum firing rate was 240rpm, and it had a crew of three (the mount weighed 1.1t). Elevation/depression were 85/10 degrees. The postwar twin over-and-under 25mm/60 (2M-3) was fitted initially to large numbers of light craft, then many other types of ship. Maximum rate of fire was 150-200rpm, and the weapon was always optically controlled. Effective range against air targets is reportedly 3km (9800ft) (0.34kg shell, 900 m/sec). This weapon appeared in 1952, and was superseded by the AK-230.

23mm A twin 23mm/87 gun, in a mount similar to that of AK-230, appeared on 'Pchela'-class border patrol hydrofoils in 1965. Rate of fire was 1600rpm; muzzle velocity was 970m/sec, and maximum elevation was 90 degrees. Effective range was 3km (3280yds). This mount may later have been used on board amphibious craft.

14.5mm 2M-7 heavy machine gun (14.5mm/93), firing at 600rpm per barrel (150 rounds in the ready-use belt) at 1000m/sec, with a maximum range of 7km (7655yds). It replaced the earlier DHSK 38 12.7mm machine gun on a one-for-one basis some time after 1945. There is also a modern twin 14.5mm twin enclosed mount in 'Zhuk' class patrol boats.

12.7mm Beginning in 1976, some ships carried twin over-and-under machine guns to defend against small boats. UTES-M fires a 0.05kg bullet at 840m/sec; the rate of fire is 1600rpm and the mount stows 200 rounds. Maximum range is 6.1km (6671yds).

ANTI-AIRCRAFT MISSILES

NOTE: Anti-aircraft missile systems are listed by their NATO designations in bold.

M-1 Volna System **SA-N-1A:** missile V-600 (4K90), launchers ZIF-101 and ZIF-102; command system Yatagan and radar director Parus; in Projects 58 ('Kynda'), 1134 ('Kresta I'), 61, and 61M ('Kashin' and 'Kashin mod'). M-1M (missile V-601 or RZ-61; production designation 4K91) is in these types plus 56AE and 61ME ('Kashin' for India); the NATO designation is **SA-N-1B**. There may

SA-N-1 aboard the 'Kashin' class destroyer *Obratsovy* 1976. Note the Yatagan (Peel Group) FC radar atop the bridge C & S Taylor

also have been a M-1P (Volna-P) version limited to the ZIF-101 launcher, using a Yatagan-P control system. This was a naval version of the land-based SA-3 Goa. Work began in 1955, and the prototype appeared in 1959 aboard the modified 'Kotlin' class destroyer *Bravyy*. Fire control: Yatagan (Peel Group). Range was originally 15km (16,400yds) but the system was later developed to a reported range of 31,500m (17nm); it is capable of interception between 300ft and 50,000ft. A recent Russian article on the 'Kashin' class notes that this system was considerably developed, making it capable of engaging low-flying aircraft and surface ships; it became the first general-purpose SAM. The Soviet Navy apparently considered but rejected an alternative, the long-range semi-active air-to-air R-4RR (NATO AA-5) missile developed for the Tu-128 interceptor. Dimensions: 18in/45cm (body), 27in/70cm (booster) × 19ft (5.9m); span 4ft/1.2m. Launch weight is about 1300lb/600kg (warhead 132lb/60kg). The ZIF-101/102 launcher is stabilised. Magazine capacity (ZIF-101 in 'Kashin' or SAM-Kotlin' or 'Kynda') is 16 in two vertical cylinders or 32 in the ZIF-102 (with four cylinders) on a mod-'Krupny'.

M-2 Volhov-M system was the **SA-N-2**, a navalised SA-2, firing the V-753 missile, in the converted cruiser *Dzerzhinski*. It required a unique height-finding radar, High Lune, in addition to its Fan Song command fire control radar. Range is reportedly 44,000m (23.7nm); it has a 290lb/130kg warhead. Dimensions: 19in/0.5m (missile), 27in/0.7m (booster) × 35ft/10.7m, span 5ft 6in/1.7m (missile), 7ft 2in/2.2m (booster), launch weight 4750lb/2150kg. Reportedly magazine capacity was only 10 in *Dzerzhinskiy*.

M-11 Shtorm was the **SA-N-3** Goblet system, firing the V-611/4K60 missile from B-187 and B-192A launchers, using the 4R60 Grom command system with Head Lights radar. There is also a Shtorm-M (4K65 missile, 4R60M Grom-M control system; NATO **SA-N-3A**). Introduced in 1967, this is the only naval missile without a land-based equivalent. Like SA-N-1, -2, and -4, it is command-guided (there were incorrect reports that it was semi-active). Reported range is 30,000m (16nm); the improved version reaches out to 55,000m (29nm). Magazine capacity is 22 (in *Moskva*) or 24 (B-187A launcher) or 36 (B-192 launcher).

Osa-M (RZ-13/4K33 missile, 4S33 (ZIF-122) launcher, 4R33 'Baza' control system is NATO's **SA-N-4** Gecko. Osa-2M (4K33M, 9M33M3, 9M33M5 missiles) is a SA-N-4B. There is also Osa-MA (9M33M missiles) on Project 1159TP only. This point-defence missile corresponds roughly to Sea Sparrow. Range is about 9000m (5nm); probably it increased to 14,800m (8nm) in the later Osa-2M version. Dimensions: 8in/21cm × 10ft/3.2m, span 25in/64cm, weight 420lb/190kg, warhead 110lb/50kg. Reportedly the 20-round magazine of the ZIF-122 launcher (in which missiles are stowed nose down for extraction by the twin-arm launcher (in which missiles are stowed nose down for extraction by the twin-arm launcher proper) leaks. The successor system is Kinshal (NATO **SA-N-9** Gauntlet).

Strela-2 (9M32 missile, 9P43 launcher), for submarines and surface ships, is **SA-N-5**, the naval version of the SA-7 Grail hand-launched IR-guided missile. Strela-2M uses the 9M32M missile and the 9P58 launcher. Strela-3 (**SA-N-8**) uses the 9M36 missile and the MTU-4S launcher. It is the naval version of SA-14 (Gremlin), successor to SA-7. Strela-3M is a modified version using the 9M36M missile. 9P58 is the single-tube launcher. The others are presumably Russian designations of the East German-designed 4-round FASTA-4M launcher. Another reported designation, FAM-14, may be a corruption of SAM-14. Some reports equate SA-N-8 with the next generation 9M39 (Igla-M) missile (SA-X-18 Grouse, system designation 9K38).

S-300F Fort (Ryf for export) system (*Kirov* and *Slava* classes) is a naval version of the land-based SA-10 (S-300PMU); it uses the 3M41 missile, the 3S41 launcher, and probably the 3R41 Volna command system; there is also an S-300M Fort-M, probably using the 48N6 missile. The NATO designation is **SA-N-6** Grumble. Guidance is by track-via-missile. Missiles are carried in revolver-type 8-round vertical launchers, one every 4–5 sec. They are loaded in their canisters, and are considered good for 10 years before overhaul. The missiles ignite only after they clear the ship's deck. Each Top Dome can handle six targets simultaneously, controlling twelve missiles. Maximum slant range against targets at 6600ft (2000m) and higher altitudes in 49nm (90km); against low-altitude targets (altitudes 80ft [25m] and less), 13.7nm (25km). Maximum speed at interception is 2296kts/4200km/h.

M-22 Uragan (Shtil for export) uses the 3M90 missile, the 3S90 launcher, and the 3R90 Orekh control system; the NATO designation is **SA-N-7** Gadfly; it is the naval version of the land-based SA-11. Each single-arm launcher has a 24-round magazine; the *Sovremennyy* version has six channels for forty-eight missiles (two launchers). Like the US SM-2, this missile has a commandable autopilot. Commands are sent up after burnout, to avoid interference from the motor plume. The launcher fires a missile every 12 sec. Missile dimensions: 16in/40cm × 18ft 9in/555cm (fin span 2ft 9in/86cm); weight 1521lbs/690kg (warhead 154lbs/70kg). Missiles maneuvre up to 20 G; maximum speed is Mach 3.

Igla uses the 9M39 or 9M313 missile (9P516 or 9P519 launcher) on board submarines and surface ships. Igla-1 uses the 9M313 missile (9P519 launcher); this is the **SA-N-10** system (naval version of SA-16). 9P519 is the single tube launcher. 9P516 is presumably the power-operated twin-arm launcher for IR-guided missiles (4/launcher), carried by the nuclear-powered auxiliary *Urals* (SSV-33) and reportedly by some Russian 'Kilo' class submarines. These launchers are reloaded from below; a 'Kilo' carries 8 reloads. The two launcher arms elevate separately, so that each can line up with the reloader.

Kinshal (Klinok) uses the 3M95 missile (3S95 launcher) and the 3R95 Mars-Passat control system; this is the **SA-N-9** Gauntlet system. This is the vertically-launched successor to SA-N-4, controlled by the Cross Swords director. Missiles are carried in 8-round groups feeding vertical launchers; they may be stacked vertically, for a total of 16 per group. Each director can guide up to 8 missiles against four targets in a 60×60 degree sector. Response time is 8 to 24 sec, and missiles are fired at the rate of one every 3 sec (20/min). Maximum range is 12km (6.6nm); minimum is 1.5nm. Altitude limits are 10m and 6000m. Target maximum speed is 700m/sec (acceleration 300m/sec). Missile maximum speed is 850m/sec. The single-stage (dual-thrust) missile is 9M330 (length about 9ft/2.8m, 364lbs/165kg with 33lbs/15kg warhead).

Kortika (Kashtan export version) uses the 3M87 and 3M88 missiles, the 3S87 launcher, and the 3R87 control system. The missile is presumably a naval version of the 9M311 missile associated with the corresponding land-based SA-19 Grison system (the designation 9M311 has been associated with the naval system). The NATO designation is **SA-N-11** Grison; it is part of the CADS-N-1 gun/missile system introduced on the cruiser *Kalinin*, which carries eight tube-launchers (auto-reloaded from a 48-round magazine below the mount) and a pair of 30mm Gatling guns (probably longer single-barrel ones in a later version) and its own Hot Flash (3P87) radar. 9M311 is a two-stage rocket with a larger diameter booster. The associated gun is usually a 30mm Gatling, but it can be single-barrelled. Effective system range is 500–8000m (guns down to 500m, missiles from 1500 to 8000m [8750yds]); altitude limits are 15m (0m with guns) and 4000m (about 13,000ft), and maximum target speed is 500m/sec. Targets are tracked both by radar and electro-optically.

Yozh is the **SA-N-12** system (naval version of the land-based SA-17, successor to SA-11). The land version fires the 9M38M2 missile (SA-11 is the 9M38 missile). Photographs of SA-N-7 on its launcher show the missile in a somewhat awkward position; the launcher may have been designed to accommodate the larger SA-17 equivalent.

ANTI-SHIP AND STRATEGIC CRUISE MISSILES

Soviet ship-launched missile development followed three very distinct paths: long-range air-breathing cruise missiles, short-range rocket cruise missiles, and ballistic missiles. Each category was associated mainly with one design bureau (OKB). The air-breathers are all due to V N Chelomey, who formed a new design bureau in 1944 to develop a Soviet equivalent to the German V-1.

Chelmomey's bureau was shut down in February 1953 and its efforts transferred to the MiG bureau. It was revived in August 1954, within the navy research directorate (Chelomey had been appointed assistant to Admiral P G Kotov, Assistant Chief Commander of the Soviet Navy for Armaments, and Kotov sold the idea of a naval missile design bureau). The first product of the new bureau was SS-N-1 (KSShch Shchuka, or 'Pike'). Despite problems with SS-N-1, Chelomey was given his own design bureau about 1958–59 (it is now NPO Mashinostroyeniye). Chelomey has been credited publicly with SS-N-3, -12 and -19. The naval guided rockets (Styx and its successors) were developed by a second bureau, Berezhniak's (now Raduga). Berezhniak had designed the wartime Soviet rocket fighters.

Most of the ballistic missiles are credited to a Makayev OKB; two experimental weapons (D-3 and D-7 systems) were developed by the Yangel' bureau, which was the second Soviet ballistic missile design organisation (after Korolev, who designed the early land-based missiles, including the huge SS-6). Some writers have credited SS-N-4 and -5 to Yangel', on the theory that SS-N-6 was so different that it must have had different designers.

Work on anti-ship missiles began in 1948, with the Shchuka missile derived by V N Chelomey from the German V-1. In 1944 Chelomey had been assigned to adapt V-1 technology for all the Soviet services, within a main design bureau (OKB); other KBs were assigned to nuclear weapons and to defensive missiles. Chelomey's 1944 version of the V-1 was designated 10-Kh; it was dropped by a Tu-2 or Pe-8 bomber. Between 1947 and 1953 Chelomey experimented with a pair of straight-winged twin pulse-jets, 14-Kh and 16-Kh (14Kh may have flown late in 1946). There was also a ground-launched version of 10-Kh, 10-KhN.

None of the pulse-jets was particularly satisfactory, so in 1948 Chelomey became interested in the liquid rocket-powered Shchuka, which was designed to be dropped by a long-range bomber, ultimately an Il-28 ('Beagle'); in 1949 work on a coast defence version, Shtorm, as well as A and B variants, began.

The V-1 had flown a straight course from its launcher to a selected point (the only adjustment was in the duration of the flight, as measured by revolutions of a small propellor driven by the airstream). For Shchuka, Chelomey developed a new autopilot which flew the missile to preset geographical coordinates (hard-wired at the factory), so that a single launcher could attack multiple targets (albeit not too far from its heading). When the missile reached the set coordinates, its engine cut out (as in a V-1). Presumably the coordinates were coded. One security advantage of such a system (compared, say, to one in which the range and bearing were set) was that no battery commander could choose to fire at anything but the assigned target.

Planned Shchuka launch altitude was about 2km, range was 15–20km, and speed was about 300m/sec (in 1951–52 trials, missiles were launched at up to 10km, reaching a range of up to 60km). The missile approached its target at an altitude of 10–12m, dropping its warhead near the target so that it would pass under the target's keel. Shchuka A used a radio command guidance, probably for manoeuvres after it reached the selected aim point. Shchuka B had the first Soviet active radar seeker, presumably turned on when the missile reached its selected aim point. The land- and air-launched Shchuka apparently was unknown to the West; it never received any code-name. However, Shtorm ramps were seen both in Europe and in the Soviet Far East from about 1950. It is not clear whether the air-launched version ever became operational.

These missiles were quite slow. Chelomey's next step, 15-Kh, was to substitute a turbojet with a ventral intake, flanked by a pair of rocket boosters. Missiles still needed a short guide rail, which projected from the missile container. 15-Kh entered service late in 1948 or early 1949, and remained in ground service through the late 1960s; it was probably deployed on the Chinese border. By this time there was interest in attacking targets further afield. This missile was a point-to-point attack weapon, not a radar-guided anti-ship missile.

About 1949, Chelomey began work on a redesigned 15-Kh with the swept wings that the central aerodynamic institute, TsAGI, then espoused. The new missile, 20-Kh, had a longer fuselage, a tapered nose, and sharply swept (rather than straight) wings. The vertical fin was moved below the fuselage and the horizontal tail was eliminated. Because it had the same powerplant, it was not much faster than 15-Kh. However, it eventually became the basis for the long-range P-5 (Shaddock, SS-N-3) missile.

In a parallel programme, the MiG design bureau developed the Komet (NATO code-name AS-1 Kennel) air-launched missile, based on a Mig-15 fuselage, with beam-riding guidance. The engineer responsible, Berezhniak, later left to form his own design bureau, now known as Raduga. Raduga later developed Soviet short-range anti-ship missiles, beginning with P-15 (NATO designation Styx, SS-N-2).

Chelomey's design bureau was shut down in 1953, but later that year he managed to convince the navy to support a ship-launched anti-ship version of 15-Kh. Apparently naval studies of ship-launched missiles officially began at the end of 1953. Chelomey's KSShch (15-Kh) missile (NATO code-name **SS-N-1** Scrubber) was officially developed by an OKB headed by M Orlov. KSShch was carried in armoured 8-round hangars, and assembled and warmed up before launching. The warm-up container may originally have been designed for 'Whiskey' class submarines (a total of twenty shipboard launchers entered service).

Conversion of the prototype launch platform, the 'Kotlin' class destroyer *Bedoviy*, began in 1955; she was accepted on 1 January 1957. Meanwhile an army truck was modified as an experimental launcher system (SM-59). The first, unsuccessful, shipboard test was on 2 February 1957; the missile (without a warhead) functioned properly two weeks later (against the former destroyer leader *Yerevan*). The first successful warhead test (against a radio-controlled torpedo-boat making 30kts at 22km) was on 6 September 1957. The missile was also tested against the incomplete hull of the cruiser *Stalingrad*, penetrating 70mm of deck armour on its final test (27 December 1957). It was then accepted for service, even though of twenty-seven tests eight had been complete failures, nineteen dropped their warheads but only two were entirely successful. As a consequence, two alternatives were considered: making the warhead integral with the missile or arming it with a homing torpedo. In the end, the original warhead-dropping missile was deployed on board a total of twelve ships: four modified 'Kotlins' (Project 56M: one converted after completion, three converted while building) and eight newly-built 'Krupnys' (Project 57bis).

Ships carrying KSShch had conventional gun fire control systems. It seems likely that the missile was fired to preset coordinates; only when it reached them did it switch on its radar seeker (it has also been credited with radio mid-course guidance, but a recent Russian account calls it a fire-and-forget weapon). Each missile was tracked out so that the presets of later shots could be corrected (in effect, spotted, in gun fire control terms). The system was quite cumbersome: missiles were stowed without their wings, and they took about 40 minutes to prepare for firing, in a small container from which the launch rail projected. The launcher was designated SM-59-1A. KSShch was 25ft 3in/7.69m long, with a body diameter of 6ft 3in/1.9m and an estimated wing span of 15ft/4.6m; at launch it weighed 6305lb/2860kg (with a 1367lb/620kg warhead). It flew at an altitude of 60m, with a maximum range of 55nm/100km; speed was 260–280m/sec.

Meanwhile, late in 1953 Berezhniak began work on a ship-fired version of his Komet, Strela (arrow: this designation may also have applied to KSShch). In 1954 he began development of a coast defence version, Sopka S-2 (4K87; NATO SSC-2 Salish; the Russians also designated the original version FKR-1). A Strela launcher replaced the two forward 6in turrets aboard the *Sverdlov*-class cruiser *Admiral Nakhimov* for 1956–57 tests, including a demonstration for Marshal Tito of Yugoslavia. However, the system was rejected, possibly because beam-riding guidance limited its range to the horizon. The coast defence and air-launched versions did see service.

An August 1994 Russian article reports that in 1955 Berezhniak began work on a small rocket-powered missile intended specifically for coastal combatants, P-15 (Termit system; 4K40 and 4K40T missiles; prototype designation 50-Kh; 4S30 launcher; NATO **SS-N-2A**, Styx). The powerplant was a rocket, to permit very rapid launch, on the theory that an attacking boat would not be able to remain within attack range for very long. There was no nuclear warhead, because the range was so short (initially 13nm/25km) that it was expected that the blast would damage the launch platform. This is a fire-and-forget missile, but a former Soviet naval missile officer claims that, at least at first, there was a mid-course guidance phase while the missile flew at 400m, before it turned on its active seeker. Production missiles were designed to switch on their active (or, in later versions, IR or radar) seekers at an estimated range of 12km from the target.

The effect of the booster on the launching boat could be enormous. When missiles were launched, the crew of an 'Osa' could not occupy the machinery spaces, the afterpeak, or the upper deck, and special shields were raised over the windows of the pilot house. Missiles could be launched at any speed over 15kts in Sea State 4; in Sea State 5 they could be launched at 15kts to 30kts depending on the angle of the sea to the boat's course.

This was the first modern anti-ship missile used in combat, sinking the Israeli destroyer *Eilat* in October 1967. Termit was also used successfully by the Indian Navy against Pakistan in 1970–71, by the Chinese Navy in the Paracels in 1974, and by the Iraqi Navy against Iran from 1980 on. However, boats armed with this weapon were badly beaten in the October 1973 Middle East War, reportedly partly because of the considerable time required to line up gyros before firing.

A missile boat, Project 205 (NATO 'Osa') was apparently designed in parallel with P-15, and the missile was also adapted to be fired from a converted torpedo-boat (Project 183R, NATO designation 'Komar'). P-15 was first tested in October 1957, the first operational 'Komars' appearing in 1959, and the first 'Osas' in 1960. P-15 was formally accepted for service by the Soviet Navy in 1960.

The missile had a maximum range of 22nm/40km and a minimum range of 8km. Cruise altitude was 100–200m, and speed was 320m/sec. Dimensions: diameter 30in/75cm (31in/77.54cm in P-20 versions), length 21ft 4in/6.5m, wingspan 9ft/2.75m. Launch weight was 4685lb/2125kg (warhead 1058lb/480kg).

The Square Tie targeting radar can detect a destroyer under favourable conditions at up to 22nm/40km (theoretical range of the radar is 48nm/88km), and fast patrol boats can be detected at about 10nm/18.25km. The operator manually sets a range gate around the target, and missile firing data are computed for target ranges of 3 to 15nm (5.5–27km). At greater ranges, the impact point has to be calculated manually on a manoeuvring board, which significantly reduces hit probability. The missile platform has to be turned into the direction of the impact point, and in theory the number of missiles launched varies with the target: seven to eight for a cruiser, four for a destroyer. Hit probability against fast missile boats is lower due to their limited radar cross-section, and the doctrine is to fire two to four. The missile is launched at 15 degrees to the vertical, then climbs at 45 degrees to an altitude of about 137m. The seeker, operating at one of six preset frequencies, is switched on 6nm/11km from the estimated impact point. It seeks the strongest echo, and (at least in early versions) cannot distinguish fixed from moving targets. The terminal manoeuvre is a dive.

Early Soviet estimates were that it would take two hits to sink a destroyer (roughly consistent with the later *Eilat* attack) and that twelve shots (three boats) were needed to assure this level of damage. The six boats of a brigade were needed to insure that three would reach the target. Missile installation and arming initially took 15hr, but later was reduced to 12; by 1962 an 'Osa' could be outfitted with four missiles in 4 to 6 hrs.

Maximum range was 25nm/46km, but effective range was 16nm/29.25km. The missile's altitude can be preset at 100, 150, 200, 250 or 300m (the missile uses a barometric altimeter).

From 1961 on Berezhniak developed an improved P-15U (Termit-U system, 40K40U missile; KT67B, KT-67ER launchers; NATO **SS-N-2B**), with folding wings (so that it could be fired from a canister, rather than from a bulkier hangar) and with improved guidance. Carried by modified 'Osas' (Projects 205U, 205ER), it entered service in 1965.

P-20 (P-20 Missile,KT-97BE launcher) was the corresponding export weapon, for 'Nanuchkas' and 'Kashins'.

P-15U was adopted in preference to Chelomey's P-25, designed in 1961 and tested on board a simulated 'Osa'. Unlike a Styx, P-25 had a pair of boosters strapped alongside its body.

Termit-M (P-15M) was the next modified version, using the P-15M missile, with an auxiliary IR seeker to supplement its radar seeker. The missile may also have been called Rubezh. It probably originated in a 1961 project for a 50–70km patrol boat missile. P-15M is 4K40M (later redesignated 3M40M). The corresponding launchers are KT-138 and KT-97B. P-15M has the NATO designation **SS-N-2C**. With a radio (rather than barometric) altimeter it could cruise at 25 or 50m rather than at the 100–400m of the earlier weapon.

Range increased to 44nm/80km. Termit-M therefore had a nuclear version with a 15kT warhead.

This missile is considerably longer than P-15/P-15U (21ft 6in/6.55m rather than 19ft/5.8m). It was carried on board modified destroyers ('Kildin' and mod 'Kashin' classes) and aboard small combatants (Projects 205M [modified 'Osa'], 206MP ['Matka'], and 1241.1 ['Tarantul 2']). This missile was different enough from P-15 that NATO originally assigned it a new designation, SS-N-11.

P-20L was a version modified for submarine launch, as a stop-gap for the failure of the specially-designed Ametyst missile (see below). Carried by 'Charlie' (Project 670) class submarines, its NATO code-name was **SS-N-7**. Dimensions: 31in/0.78m × 23ft/7m; span 8ft/2.4m. Launch weight: 5291lb/2400kg. In a submarine, the Klon fire control system relied mainly on radar target detection. Compared to the surface-launched version' P-20L has a plastic nose cap and different tail fins, to improve its hydrodynamics.

The P-15M system was modified to use a pair of new missiles, P-21 and P-22 (production designation 4K51, later 3M51); their NATO designation is **SS-N-2D**. It retained its Klon FCS, using the Square Tie acquistion radar. Termit-R is probably the designation of a replacement system using the Korrell FCS (with Plank Shave target acquisition radar; Korrell may also have incorporated a data link command mode). P-21 has a radar seeker; P-22 has an IR seeker. This version has a Doppler radar altimeter to reduce missile drift in altitude; it also uses a different fuel to increase range (to about 55nm/100.5km). The designation P-20K may also be associated with these systems.

A 'Tarantul' typically carries three radar and one IR missile. The Russians estimated that three hits would destroy a light cruiser, one to two hits would destroy a destroyer or frigate. Assuming that the missile itself is 95 per cent mechanically reliable, the hit probability depending on the radar target, the Russians estimated that one missile boat carrying four weapons (typically three radar and one IR) could destroy one missile destroyer or missile frigate; one or two missile patrol boats, landing ships or unprotected minelayers; or two to three unescorted transports. The IR version, which homes on the target's IR plume, has to be launched more than 30 degrees from the direction of the sun and is ineffective in visibility less than 100m (*ie*, thick fog and drizzle or snow).

The export equivalent of Termit-M (P-15M) is P-20M. It appeared on board export 'Tarantuls' (Project 1241E), 'Kashins' (Project 61ME), and 'Koni' (Project 1159TR). Versions of P-15/P-20 were widely exported, and also made under license in India and Iraq. The Chinese derivative, Silkworm, is also made in North Korea.

P-27 is a new (*c*1990) version with a new higher-frequency (V-band) seeker and possibly with a data link back to the launching platform. Its seeker and data link may be those associated with the P-270 (3M30) missile. Some 'Tarantuls' carry this weapon.

Meanwhile work proceeded on long-range weapons, both cruise and ballistic, for strategic attack from submarines. Chelomey's big turbojet powered missile was designated P-5 Pityorka (Student); a competing missile, developed by Beriev, was P-10. The Chelomey missile had folding wings (probably developed for an abortive application to a big surface ship, the Project 82R battlecruiser) and could, therefore, be carried by a relatively small submarine, a 'Whiskey'.

Beriev's larger P-10 was also stowed with its wings folded, but they had to be unfolded on its zero-length launcher after it left its canister (forward of the submarine's sail). The missile was successfully tested on land in 1956 and at sea the following year. P-10 had to be tested on board a large 'Zulu', and despite its superior performance was judged too massive to be worthwhile. Each P-10 container was about twice as long as a P-5 container, and P-5 required far less time to prepare and fire. The few P-10s made became target drones.

P-5 (4K34 and 4K95 production designations) entered service in 1959 aboard 'Whiskey Twin Cylinder' class submarines (NATO called it **SS-N-3B**, Shaddock). P-5 carried a 200kT warhead. It took 35 to 40min (far more than Western experts have estimated) to set up the missile, *eg* to run up gyros, during which the launching platform had to run a steady course. Over its production run the missile was refined aerodynamically, and the thrust of its turbojet roughly doubled, so that maximum speed increased from about 470kts (about 540mph) to better than 600kts (about 690mph, close to Mach 1; some reports have the missile barely exceeding Mach 1). A modified version, P-5D (4K95D), had a radar (Doppler) rather than a barometric altimeter, which made it possible to fly higher. This version has also been described as supplied with means of registering the path covered and drift due to winds, for better accuracy. There was also a P-5K version. A modified version with the later anti-ship seeker described below (SSC-1b Sepal) was S-35 (4K95). This weapon remained in service as late as the mid-1970s.

At about this time (1956–60) Ilyushin developed a longer-range cruise missile, P-20, for a modified version of the 'November' class (Project P627A) nuclear attack submarine. It never entered service (it may never have flown). The P-20 designation was later re-used for an SS-N-2 version.

Chelomey apparently convinced the navy to combine the P-5 airframe with the guidance system perfected in the KSShch programme. A submarine-launched version, P-6 Progress (20-Kh, 4K48; NATO **SS-N-3A**), was carried aboard 'Juliett' (Project 651) and 'Echo II' (Project 675) class submarines. It is not clear when development began. P-7 (4K77) was a land-attack version, presumably with the flexibility to hit assigned, rather than pre-planned, targets. It apparently had an autopilot into which ship and target position coordinates could be set (P-5 had to be launched from a preset position to hit the target hard-wired into the missile). Although P-7 proved too inaccurate for the planned mission (and therefore did not enter production), subsequent dedicated anti-ship cruise missiles, particularly the long-range types, often had a reserve shore-attack mode.

In 1956 Chelomey began work on a next-generation weapon, code-named Bazalt, to be launchable from underwater. It would be powered by a new rocket-ramjet engine, and hence would be about twice as fast as P-5. NATO eventually designated this weapon **SS-N-12** Sandbox. Its production designation was 4K80. Ironically, it was never launched from underwater.

SS-N-12 aboard *Kiev* Jul 1976 RAF

This project included a new anti-ship seeker which could be remote-controlled by a launching ship. To do that effectively, the launch ship had to track the outgoing missiles, using a new track-while-scan radar, Scoop Pair (two missiles per set, controllable within an 80 degree sector). Missiles could also send their radar seeker video back to the launching ship. Radar data sent back by the first weapon a ship fired would be used to aim succeeding weapons. Missiles could be fired either in remote-controlled or in self-guided mode. The radar video receiving antenna was code-named Plinth Net. A system which tracked outbound missiles (and which could correct their paths) did not necessarily require the radar video receiving element.

There is evidence that a simple version of the system. which did not require Plinth Net, was conceived for the Project 82R battlecruiser (a sketch of the ship reportedly shows Scoop Pair) and then for the Project 58 ('Kynda') cruiser which succeeded it. It may have been the P-7 (4K77) mentioned above. The missile system incorporating both guidance and radar video talkback was apparently conceived in 1956, and was designated P-35. Apparently the associated rocket-ramjet missile was not particularly successful, so ultimately the guidance system was probably installed in a P-5 type airframe, designated P-35 (4K44). There were also a P-35B missile (the differences from the original P-35 are unknown), a P-35M version (3M44 missile), and an S-35 coastal version (see above).

P-35 could cruise at any of three altitudes: 400m (range 44nm/80km), 4000m (range 82–109nm/150–200km), and 7000m (range 191nm/350km); in the terminal phase it flew at 100m. A single Scoop Pair radar could guide two missiles against targets within an 80 degree arc. Although a 'Kynda' had two such radars, one report suggets that she had only a single guidance package, so that only one could be used at a time. In self-guided mode, she could salvo all eight of her ready-use weapons. There were two seeker modes, one for anti-ship and the other for land attack. P-35 was 32ft 4in/9.88m long. Launch weight was 12,346lb/5600kg (warhead approx 1100lb/500kg); range was 44–191nm/80–250km (cruise speed 450–480m/sec).

The P-35 project included radar video down-link from a targeting aeroplane, for over-the-horizon attack: initially a Badger (the previously unknown Tu-16RTs version), soon after from a Bear-D (Tu-95RTs), and later from a helicopter (Ka-25RTs). This system was accepted into service in 1956. About 1962 Khrushchev was shown a proposal for a satellite targeting system, probably in connection with the later Granit (SS-N-19) missile system. It became operational with the P-35 successor, Bazalt, in 1978.

P-35 trials began in the Caspian Sea in 1961; targets included a 2500t destroyer leader (Project 48 type), which sank in three minutes after a hit by an inert round.

For any of these long-range anti-ship systems, the great problem was that the missile might well arrive near the target so long after information had been received that the target had moved so far that the missile would be unable to acquire it. For example, in the case of the US Tomahawk, elaborate measures had to be taken to estimate where the target would be when the missile arrived. The Russians apparently hoped that doubling the speed of the missile (for Bazalt) would quadruple its chance of hitting the target. Therefore the next-generation rocket cruiser, Project 934 (ultimately 1134, 'Kresta I') was designed to carry only four weapons, rather than the sixteen of its 'Kynda' predecessor. As it happened, when Bazalt development stalled, 'Kresta I' was built with four subsonic P-35s.

P-35 development included an improved guidance package in which the missile could be instructed to fly a radar search pattern around the pre-selected radar turn-on point. There is also an ARM version. When Dazalt did enter service, it had a more sophisticated autopilot which could be programmed for waypoints and other mid-course manoeuvres.

The concept of command control with radar video data link was used in many later missile systems, such as those NATO designated SS-N-9, and SS-N-19, and SS-N-22. The missile-tracking radar for SS-N-9 and -22 was generally concealed by a radome code-named Band Stand (several different radars were mounted within such radomes, on board different classes of ship). Data links were Fish Bowl (on 'Nanuchkas') and Light Bulb (on, for example, *Sovremennyys* and some 'Tarantuls'). One important advantage of the radar down-link was that a particular target could be selected from within a formation. Eventually the Chelomey OKB developed a technique for transmitting information that could be built up on board the launch ship into an ISAR image of a potential target, and compared to a shipboard target library.

It appears that Chelomey developed a short-range jet-powered companion to P-35, P-50. In the late 1950s P-50 was scheduled for deployment on both a cruiser (Project 934A, which eventually became 1134A, 'Kresta II') and on board a frigate, which ultimately became Project 1135, 'Krivak'. This weapon probably armed at least some 'Nanuchkas'. The larger surface ships for which it was planned were reordered as ASW ships (reportedly a few early 'Krivaks' were nearly completed carrying P-50 launchers). In theory, a mixed group of 934s and 934As would attack a naval formation, the long-range Bazalts being used to break up defences so that the numerous P-50s could finish off the group. P-50 dimensions: 7ft × 29ft/2.1m × 8.8m (forebody diameter 31in/0.78m, aft body depth abaft jet intake, 34in/0.87m), span 7ft/2.1m. Launch weight: 6509lb/2952.5kg. The NATO designation for the 'Nanuchka' weapon was SS-N-9 Siren.

In 1959 staff requirements for a new submarine (Project 661, 'Papa') to be armed with submerged-launch anti-ship missiles were approved. The planned missile was Ametyst (P-70, 4K66). This weapon was probably developed by Chelomey. Apparently it encountered serious development problems, and a version of Styx (P-20L) was adopted instead for the 'Charlie' (Project 670) class built when the 'Papa' programme stalled. NATO called both P-70 *and* P-20L **SS-N-7** (P-70 may never have been seen).

In 1962, Berezhniak (Raduga) began work on the P-120 (Malakhit, 4K85) solid-fuel rocket missile (NATO **SS-N-9** Siren), which could be fired either by a surface ship or a submarine. It had two seekers (radar and IR), and a maximum range of 66nm/120km (it could be fired at higher speed to reach a shorter range). In 1972 P-120 was accepted for service on board 'Nanuchkas' (Project 1234) and 'Charlie II' class submarines (Project 670M). The standard triple surface-ship tube is designated KT-84. An export Malakhit-E version was later sold to India for the new *Dehli* (Project 15) class. P-120 formed the basis for the ASW missile which superseded P-50 on board larger ships, 85R (NATO SS-N-14). Note that P-50 and P-120 apparently shared the same NATO designation, since they could both be fired from the same tube.

Apparently a TTZ for a supersonic P-120 successor and a new cruiser to carry it (Project 956, *Sovremennyy*) was approved in 1969. Both Chelomey and Berezhniak seem to have developed weapons. Since both fit the same container (KT-190 launcher), and since neither was seen until the fall of the Soviet Union, NATO assumed they were a single weapon, **SS-N-22** Sunbeam. It now appears that some of the first fourteen *Sovremennyys* and the two prototype 'Tarantul' class missile boats were armed with Chelomey's rocket-powered P-80, incorporating the ISAR talk-back link. Reportedly a folding-fin submarine-launched version of P-80 has entered service as P-100 (it is fired from 65cm torpedo tubes); it may also be fired out of the catapults of the frigates of the *Neustrashimmyy* class. P-100 may be called Moskit-M, and it is associated with a KT-152M launcher. A folding-fin version, P-100, can be fired from 65cm torpedo tubes and from the torpedo catapults of the *Neustrashimmy* class. There is probably also a shore version. P-80 dimensions: 20in × 20ft 3in/0.52 × 9.4m (fin span 6ft/1.8m); the *Sovremmennyy* canister is 6ft × 31ft 6in/1.85 × 9.6m. A Chelomey OKB poster (at Moscow in 1992) gave range as 130km (70nm).

Later destroyers, later 'Tarantuls', the 'Sivuch' type surface-effect boat, and the WIGs (Ekranoplans) all had Berzhniak's rocket-ramjet P-270 Moskit (3M80), which has a simpler data-link. P-270 (3M80) dimensions: 4ft 3in × 30ft 9in/0.13m × 9.38m (with wings folded), 8708lb/3950kg with a 661lb/300kg warhead, speed 2800km/hr (1510kts), range 49nm/90km (as in SS-N2C), flight altitude 7–20m, capable of 60 degree turn after launch. The characteristics suggest that this was the intended successor to Raduga's Styx. Raduga claims that 1.2 missiles suffice for a mission kill of a destroyer, or 1.5 for a transport. The seeker can be switched to an ARM mode. A cruise speed of 3600km/hr has been claimed for this missile.

Chelomey developed two further supersonic missiles of this type:

the ramjet Yakhont (Oniks system, 3M55 missile; the system may be designated 3K55 and is associated with a SM-312 launcher) and Alfa (Biryuza P-10 system, 3M54 missile). Yakhont seems to be intended to arm modified 'Nanuchka' (Project 1234.2), *Neustrashimmyy* (Project 1154), and modified *Sovremennyy* with conventional or vertical launch tubes (Project 956M and 956U). Alfa is apparently planned only for 956U. Neither had entered service when the Soviet Union collapsed, but both have been offered at arms shows held since then.

Current policy seems to envisage replacing both medium-range and long-range weapons with a single common weapon, perhaps Alfa or a derivative. This may be the P-1000 Vulcan missile installed experimentally in a pair of 'Echo II' class submarines (K-1 and K-22, Project 675MKV).

Bazalt itself was not accepted into service until 1975 (dimensions: 5ft 3in/1.6m (forebody 2ft 6in/0.78m) × 38ft 4in/11.7m; span 10ft 6in/3.2m; launch weight just over 11,023lb/5000kg). In 1961, Admiral Gorshkov proposed to Khrushchev a new cruise missile which a submerged submarine could launch against surface ship targets at a range of up to 600km (370nm); surface combatant targeting range was to have been extended to 800km (430nm). Gorshkov proposed that the missile be targeted from space. This may have been either a last gasp of the Bazalt project or a glimmer of the later Granit (SS-N-19).

Given serious problems with the Bazalt airframe, Chelomey began work on a new missile similar in configuration to P-5/P-35. A new higher-powered turbojet became available in 1964. Reportedly formal work began in 1969, perhaps in parallel to that on P-80. The new missile, Granit (P-500, 3M45, NATO **SS-N-19** Shipwreck) was the first to be vertically-launched, by carriers, *Kirov* class cruisers and 'Oscar' (Project 949) class submarines. Granit was accepted into service in 1983. There may also be a P-700 version. Granit dimensions: 3ft 1in/0.96m × 33ft 6in/10.2m; span 10ft 6in/3.2m.

As carried by *Kirov* class missile cruisers and by *Kuznetzov* class carriers, P-500 can be tracked and controlled via the ship's main air search radar. The video down link provides ISAR data to the firing ship. There are also conventional radar and ARM versions (the latter said to be intended specifically to attack Aegis ships), and the standard radar seeker can switch to ARM mode in flight (though in that case it is less sensitive than the dedicated ARM version). Warheads: nuclear (200 or 350kT), 1000kg (2205lb) unitary shaped charge high explosive, and bomblet (primarily for anti-ship attack, but also usable against land targets: 750 × 1kg, a mix of incendary, AP, HE, which can be varied to meet requirements).

SS-N-19 is so fast (and probably has so good a search pattern) that it can function without mid-course guidance. An 'Oscar' class missile submarine can, therefore, fire the missile submerged, on the basis of information provided by a satellite via the Punch Bowl down-link antenna.

The next missile of this size and supersonic performance is Meteorit (3M25; NATO **SS-N-24** Scorpion); the only known installation was on board a modified 'Yankee' class submarine (Project 667M). There was also a Meteorit-M, 3M25M. This weapon was originally described only as a strategic cruise missile with intercontinental range, but P-750 Grom may have been an anti-ship version. It is tempting to think of this weapon as part of a programme, the other element of which is Chelomey's Yakhont or Alfa. Meteorit may have been designed by the Chelomey OKB. Raduga developed an air-launched equivalent, NATO designated AS-19 Kola (no Soviet designation is known). Both were cancelled about 1990. There was also a ground version, now prohibited by treaty.

During the late 1970s or early 1980s the Russians decided to develop equivalents of the US Harpoon anti-ship missile (Kh-35 Uran; 3K24 system, 3M24 missile, NATO **SS-N-25**) and of the US Tomahawk strategic cruise missile (S-10/3M10 missile, 3K10 Granat system, NATO **SS-N-21** Sampson).

Initially the East German Zeiss organisation was expected to develop an electro-optical seeker for Kh-35. Well before the collapse of the Soviet Union, financial problems had aborted this plan, and the missile ended up with either a radar or an imaging IR seeker. Unlike the missiles described above, this one was developed by the OKB responsible for tactical air-to-surface weapons, Zvezda. The plan for a optical seeker suggests that in at least some versions there is a down-link, allowing the operator to select an aim point. Kh-35 has been shown in both ship- and air-launched versions and for coast defence. There is also a submarine version, the Russian equivalent of Sub-Harpoon. The planned ship platform was the KKR 'Uran'; drawings of a 'Tarantul' carrying four quadruple launchers have been displayed. Kh-35 is now being mounted on modernised 'Krivak' class frigates. A triple Kh-35 launcher can replace a single SS-N-2 on board an 'Osa', and a sextuple launcher can replace the usual triple SS-N-9 launcher on board a 'Nanuchka'. Some boats may have been refitted. The canister was first seen on board the East German attack boat *Sassnitz*. Dimensions: 17in/0.42m × 14ft 6in/4.4m (1058lb/600kg, including 320lb/145kg warhead). Fin span (diagonal) is 3ft/93cm. Air-launched IIR and active radar versions are 12ft 4in/3.75m long (1058lb/480kg), with a range of 3–71nm/5–130km, launch altitude 200–5000m, and flight altitude 5–10m (speed 300m/sec).

Lyulev's 3M10 (S-10 System, Granat), code-named **SS-N-21** Sampson by NATO, was broadley equivalent to the US Tomahawk. Apparently the 'Akula' class submarine was designed specifically to carry this 51cm diameter weapon; 'Yankee Notch' was modified for it. Range is about 1600nm/2960km. A shorter-range anti-ship version was almost certainly either under development or in service.

A ship-launched version of Zvezda's Ovod has also been announced. Ovod (Kh-59, NATO AS-18 Koala) is an electro-optical missile with a link back to the launching aircraft. The naval version, Ovod-M, may have been the missile planned for the single 'Nanuchka' modified with open launch rails.

A Russian list of anti-ship missiles includes a previously unknown Bolid system (3M15 missile; the system may be designated 3K15) associated with the projected (abortive) submarine Project 881.

ASW MISSILE SYSTEMS

At the key October 1955 meeting, ASW missiles were emphasised as a major new area of development. The requirements for the 'Kashin' (Project 61) class initially included such weapons. However, development was clearly slow, since they were soon replaced by an ASW helicopter. Presumably the Soviet helicopter was analogous to the British MATCH system, in which the ship vectored it out to a target using its own sonar (the helicopter carried a torpedo or depth bomb but no sonar of its own).

The Soviet missiles were all included in an RPK series.

RPK-1 Vikhr' system (missile 81R, launcher MS-32, command system Sprut; NATO **FRAS-1**) in *Moskva* and *Kiev* classes; probably by Novator OKB. The nuclear warhead rocket (an adapted army FROG) was unguided, with a range of 5–14nm/10–25km; it could be used against submarines at depths to 3200ft (1000m) using a ship's sonar or helicopter sonar data. This weapon was accepted for service in 1968.

RPK-2 Viyoga system (missile 82R, fired from 53cm torpedo tubes; NATO **SS-N-15** Starfish) in submarines, by Novator OKB. This weapon was a Soviet copy of the US SUBROC, originally armed only with a nuclear depth bomb. Now there is an alternative torpedo warhead (originally the UMGT, now the APR-2E). Maximum range is 25nm (45.7km). A surface version fired from 53cm torpedo tubes arms the missile cruisers *Frunze* and *Kalinin*. In these ships it is fired like a torpedo, dives into the water, stabilises itself, and then ignites. In effect it replaces SS-N-14. It was apparently chosen to arm 'Krivak III' and modified *Udaloy*, although the appropriate torpedo tubes may not have been fitted. This system was accepted for service in 1969. It could be launched from depths down to 160ft (50m), and range was 5–22nm (10–40km).

RPK-3 Metel system (missile 85R, fired from KT-100 launcher with command system Musson in Project 1135 ['Krivak'] and from KT-106 launcher with command system Grom-M [which also controls SA-N-3] in Project 1134A/B ['Kresta 2' and 'Kara']; NATO **SS-N-14** missile), based on the P-120 (SS-N-9) anti-ship missile.

SS-N-14 ASW missiles aboard the 'Udaloy' class cruiser *Admiral Kharmolov* 8.7.94 L & L van Ginderen

This command-guided rocket-powered missile is broadly equivalent to Malafon or Ikara. It carries a homing torpedo (originally AT-2U) or a nuclear depth bomb. Reportedly the original nuclear depth bomb version was 60R; the original torpedo version (carrying a 40cm homing weapon) was 70R; a new torpedo version (carrying a short 53cm weapon) was 80R. The 'universal' (*ie*, dual-purpose) URPK-3 version of the system (85R missile) also carries a warhead in its body, so that it can be used as an anti-ship weapon. There is also a coast defence version (which may not have been deployed by the Russians). In 'Krivak III' and modified *Udaloy* (as planned) and in *Frunze* and *Kalinin* as built, this weapon was superseded by a torpedo-tube launched SS-N-15 (90R). Against targets in direct-path range, detected and tracked by hull sonars, the missile is commanded to drop its torpedo immediately above the submarine position. 'Krivaks' are limited to this mode. Against a more distant target (*eg*, detected by an aeroplane cooperating with the ship, or in the convergence zone), the attacking ship sends a helicopter to the estimated submarine position. The helicopter uses sonobouys and a dipping sonar to locate the submarine, hovering almost directly over it. The missile is fired directly at the hovering helicopter, which orders the torpedo dropped at the appropriate moment (presumably via a link back to the launching ship). Without the helicopter, attack range must be limited roughly to the horizon, about 10nm (18.3km). The missile typically flies at 400m; maximum range is 27nm (50km) (speed Mach 0.95). Dimensions (85R): 1ft 10in × 23ft 6in/0.57m × 7.2m (full depth, including underslung weapon, is 4ft 5in/1.35m); weight is 8818lb (4000kg). Note that when 'Krivaks' first appeared carrying the trainable launcher for this missile, it was assumed that they were intended for anti-ship work; the unseen missile in the container was tentatively designated SS-N-10. Presumably this designation was not shifted to the ASW missile that actually occupied the container because of a belief that an alternate (non-existent) anti-ship weapon could occupy the launch container.

Modified versions; URPK-5 Rastrub-A system (85RU with KT-100U launcher in Project 1135M ships, 1980) and probably URPK-4 Rastrub-B system (85RU with KT-106U launcher and probably Drakon command system in Project 1155 (*Udaloy* class ships).

RPK-5 Leevyen ('heavy rain') fires the 89R missile from the KT-129 6-barrel launcher (NATO RBU-10000) in one Project 1124 ship ('Grisha' class) and from a 10-barrel launcher (NATO RBU-12000) in larger ships. The 10-barrel system may be called Udav. Range: 216m to 10,000m. There are both ASW and anti-torpedo (hard- and soft-kill) rounds. The Zapad system (90R missile fired from an RBU-6000 rocket launcher) in Project 1154 (*Neustrashimmyy*) is probably related to RPK-5. In both cases, the round is guided acoustically, as in the new air-launched S3V guided depth bomb.

RPK-6 is a 65cm successor to RPK-2: the Vodopod system (83R missile) uses a torpedo warhead and the Vodopei system (84R missile) presumably uses a nuclear depth bomb warhead. RPK-6 entered service in 1979. This is probably the system NATO code-named **SS-N-16** Stallion. An improved version is URPK-6 (Vodopod/Vodopei-MK, using 83RN and 84RN missiles). The associated surface ship systems are Veder (86R missile, torpedo) and Vspletsk (86R missile, presumably with a nuclear depth bomb). RPK-6 may have been the first Soviet 65cm weapons, responsible for the shift from 53cm to 65cm torpedo tubes. Larger size bought about twice the range of the earlier weapon. SS-N-16 seems to have been conceived in alternative surface ship (Veder) and submarine (Vodopod) missiles, both carrying the same torpedo (Orlan) to 54–65nm (100–120km). The current payload is the short 53cm torpedo. Thus far only definite surface ship application is the *Neustrashimy*, which has six large torpedo catapults (similar to those in French ships) rather than the usual 53cm tubes (she carries sixteen reloads). According to a recent Russian book, this weapon is fired from the SUW-N-1 launcher of the *Moskva* class.

STRATEGIC BALLISTIC MISSILES

The Soviets developed strategic (*ie*, nuclear) naval cruise and ballistic missiles in parallel. Modified versions of the German V-1 and V-2 were tested both in towed submarine barge form and on board submarines. V-2 barges (with which the US Navy experimented unsuccessfully) were reportedly built in experimental series (perhaps 300 were made), but no more than about twenty were on hand at any one time, and none was successful. As in small-scale US experiments, missile exhaust blew apart the barge as the missile fired, throwing it off course.

Apparently the first attempt to build a naval strategic nuclear weapon was a massive torpedo, T-15 (1550mm [diameter] × 23m, 40 tons [with a 3.54 ton warhead], to be fired at 29kts at 30 or 40km stand-off range). A submarine could carry only one, in a special large-diameter tube. The first Russian nuclear submarine, 'November' (Project 627), was conceived to carry T-15; there was also a diesel submarine project (perhaps the 641A cancelled in October 1955). The torpedo idea seems to have arisen after attempts to build heavy nuclear artillery shells failed, probably about 1952. T-15 died when 1955 experiments showed that a submarine could fire a ballistic missile from a considerably greater distance. A conventional-size torpedo, T-5, was then developed (53cm diameter, 40kts/10km). It entered service in 1959.

The experimental single-missile submarine installations were 'Whiskey Single Cylinder' (firing P-5/SS-N-3) and a modified 'Zulu' (B-67, Project 611A) carrying a Scud (R-11FM: first fired 16 September 1955). Work on this weapon, and on a submarine to fire it, formally began with a January 1954 directive. S P Korolev decided to have the submarine surface to fire, reportedly partly because an attempt to start a missile in a tube had failed (this may refer to an unsuccessful missile barge experiment). Naval officers attached to the project disagreed (they managed to fire model missiles from both dry and water-filled tubes). In the end, Korolev compromised: the missile would be carried fuelled, and prepared underwater for launch, to minimise the submarine's exposure on the surface. Compared to R-11, the naval version had kerosene fuel and nitric acid oxidiser, which could be stored in the missile. The usual combination of alcohol and liquid oxygen evaporated constantly and therefore had to be drained periodically, the tanks then being refilled. A special test stand, to simulate ship motion, was built at Kapustin Yar.

The next step was two missiles per submarine ('Whiskey Twin Cylinder') and a two-tube 'Zulu' (Project AV611, carrying R-13/SS-N-4). Ballistic missiles were raised to the top of the sail, fuelled, and then fired. Apparently they had to be kept well clear of the surface of the sea, *eg*, to avoid icing. Khrushchev witnessed a Pacific Fleet firing by B-58 in October 1958; he then emphasised the need for the missile revolution then underway.

The step beyond was series production of diesel and nuclear platforms for the new cruise and ballistic missiles: 'Whiskey Long Bin' (four SS-N-3) and 'Echo I' (Project 659: six SS-N-3); and 'Golf' (Project 629) and 'Hotel' (Project 658) class submarines, each firing three SS-N-4 ballistic missiles, in sail tubes above a necked-down pressure hull section, so that missiles could be hoisted for surface firing. Parallel diesel and nuclear designs were necessary mainly because nuclear reactor production was limited.

The next missiles (R-21/SS-N-5 and Bazalt/SS-N-12) were designed to be fired submerged. SS-N-5 was popped out of its tube by gas or air before igniting. Because it did not have to be raised well above the water for firing, it no longer had to be carried in the submarine's sail.

Again parallel nuclear (Projects 667/668) and diesel (Projects 651/652) submarines were designed. Most of them did not materialise, at least in the planned form. Project 668 carried six Bazalts between its outer and pressure hulls, angled up at 45 degrees; Project 667 carried its six SS-N-5 missiles on either side, alongside its necked-down pressure-hull section. Project 652 reportedly carried four shorter-range ballistic missiles, of a type which never entered production (it was presumably one of the experimental types listed below). The diesel submarines in this programme may have been intended specifically to operate in narrow seas, such as the Baltic, to hit fixed targets in NATO rear areas such as the United Kingdom; nuclear strategic submarines would have been used against the United States.

As it happened, Bazalt failed and Khrushchev's hostility to the naval strategic mission put off work on Project 667; ultimately it emerged as 'Yankee' (Project 667A), the basis for most later Soviet strategic submarines. Meanwhile SS-N-5 went to sea on board the existing 'Golf' and 'Hotel' class submarines, and Project 651 was redesigned to launch SS-N-3s on the surface; it became the 'Juliett' class. The nuclear-powered 'Echo' was redesigned with two more missile launchers. All new submarines firing SS-N-3 ('Juliett' and 'Echo II') were provided with missile-guidance equipment appropriate to the new anti-ship role. Project 652 died, possibly because its short-range missile failed (or because the growing force of Soviet land-based ballistic missiles could carry out its mission of attacking fixed targets less expensively).

The Soviet Navy gained a new strategic attack role after Khrushchev's fall in 1964. Now the submarines were advertised largely as a reserve force. Once the primary (land-based) strategic force had been exhausted, the existence of an intact reserve might well determine the outcome of a war. In fact, however, submarines were also maintained as a deterrent, on patrol off the US coast.

The Soviet strategic submarine force gradually became a more and more important element of the Soviet strategic force. Soviet naval strategy came to emphasise its protection. The maritime strategy espoused by the United States during the 1980s used their significance to gain leverage: by threatening the 'bastions' within which submarines would operate, the US Navy would force the Soviets to concentrate their naval energies on protecting the submarines, rather than attacking NATO sea communications.

By that time the Soviets had developed a new strategic attack concept: a surprise attack against Western command and control centres, many of which were near seacoasts. It appears that the nuclear version of SS-N-16, perhaps with a more powerful warhead or warhead setting, could be used for quick attacks of this type. The Soviets probably considered such attacks equivalent to the threatened attacks against their own command centres by European-based Pershing missiles.

They also returned to the earlier idea of the strategic reserve force. 'Typhoon' (Project 941) class submarines were conceived as long-endurance underwater missile bases, capable of remaining at sea for as long as a year.

RSM-designations previously published were apparently invented specifically to be given in START treaty documents. Missile *systems* all had D-numbers.

D-1 was the modified Scud (R-11FM) fired by 'Zulu' and early 'Golf' class submarines; it was developed by S P Korolyev (OKB-1) under an August 1953 directive. Reportedly the earliest version of this army-derived weapon was non-nuclear, with an 1764lb/800kg warhead. Reportedly a 'Zulu' could not fire in sea conditions beyond wind force 5. The submarine surfaced, opened both tubes, lifted the missiles on their platforms in the tubes, and turned them to align the missile so that the planes of two of the three gyros pointed in the direction of fire. This preparation took several minutes, and after that the missile could be fired only when the submarine was on an even keel. The missile was held by four brackets on the platform (the hoof-and-claw) while its motor built up sufficient thrust. D-1 was accepted for service in 1959. Initial work for the naval adaptation was done by a special KB under Makayev, who designed the later submarine ballistic missiles. Range was 82nm/150km (CEP 4nm/8km) and dimensions were 2ft 11in × 33ft 11in/0.88m × 10.34m (fin span 5ft 9in/1.75m; weight empty 1.68t/full 5.47t, warhead 2150lb/975kg). Engine thrust was 8.26t. D-1 remained in service through 1967. Of seventy-seven launches in 1958–67, fifty-nine were successful (three missile failures, seven due to ship or equipment, eight to other causes).

D-2 (R-13 missile, production designation 4K50, NATO **SS-N-4** Sark) was developed by the Makayev OKB to arm the 'Golf' and 'Hotel' classes. The draft project was developed between late 1955 and early 1956; development of the D-2 system was authorised in August 1956. Flight trials at Kapustin Yar (June 1959 – March 1960) were followed by test firings from a 'Golf' class submarine; D-2 was accepted for service in October 1961. This liquid-fuelled missile had to be fired by a surfaced submarine. Each weapon was raised to the top of the submarine's sail by its own elevator, and held there by a 'hoof-and-claw'. Missiles were initially carried empty, with fuel in special tanks. Reportedly the Soviets switched to carrying them fuelled after several poisoning incidents due to seals broken by vibration. Prelaunch checkout was done submerged, after which the battery in the missile lasted about 1hr. Preparation time was 10–15min, and the missile had a 10-man crew. The liquid fuel rocket had a single main combustion chamber with four others clustered around it. Maximum speed was 2050m/sec, maximum altitude 145km (compared to 25km for the naval Scud). Flight time was 7min 5sec. After reaching apogee, the missile pitched over into a pre-programmed dive, and the warhead separated and flew on, reaching a terminal speed of 700m/sec. Reportedly submarines carrying three R-13 in 1960–61 had one or two practice weapons with dummy warheads. Dimensions: 4ft 3in × 38ft 10in/1.3m × 11.84m, weight 3.73t empty/13.75t at launch, throw weight 3522lb/1597.5kg. Engine thrust 27.72t. Yield was reportedly 1MT, design range 350nm/650km, and CEP about 4km. This system was in service 1961–73: 311 were launched, 225 successfully (thirty-eight missile failures, thirty-eight failures due to various known causes, ten due to unknown causes). The missile could be launched at speeds of up to 15kts, in sea states up to 5.

A recent Russian book lists a two-stage solid-fuel Maket (dummy) missile with a range of 355nm (650km); dimensions were 5ft 9in × 47ft 6in/1.75m x 14.5m (launch weight 19t). This weapon, presumably unsuccessful, was paraded in Moscow and advertised as R-13 (the contemporary intelligence phrase was, 'those which don't fly, march').

D-3 (R-15 missile) was a 1960 experimental missile by the Yangel OKB. No details are available. It may have been the failed weapon planned for the abortive Project 652 submarine. Work on this weapon was authorised by a government resolution of 17 August 1956; it was stopped by a 3 December 1958 resolution which ordered the OKB to switch to the R-21. The OKB was withdrawn from naval work in May 1959, so that it could work on land-based R-14 and R-16 missiles.

D-4 (R-21, 4K55 missile) was developed by Makayev to arm modified submarines: Projects 629A, 629M, 658M; the NATO designation was **SS-N-5** Mod 1 Serb. Mod 2 was probably designated D-4A (R-21A). Development was authorised on 17 March 1959. This successor to SS-N-4 was the first Soviet submerged-launch ballistic missile to enter service. A 'Whiskey' (Project 613) submarine was specially modified as a 'wet launch' test bed. In addition, a Northern Fleet 'Zulu', *B-67* (which had launched the first R-11FM), was modified to wet-launch a modified R-11FM (izd. S4.7: the first launch attempt, which failed, was on 14 August 1959; a launch in October 1960 succeeded). Tests of the R-21 prototype, K1.1, began in the Black Sea in May 1960. A 'Golf' (K-102) was modified (as Project 629B) to test-fire R-21; it first fired on 24 February 1962 (there were twenty-seven test shots in all). The missile could be launched at submarine speeds of 2–4kts, at depths of 40–60m, and in sea states up to 5. After the missile left the submarine, compressed air propelled the missile to 30m above the surface before it fired. Dimensions: 4ft 3in × 46ft 7in/1.3m × 14.2m, launch weight 19.65t. Throw weight was 2599lb/1179kg. Warhead yield was 1MT in Mod 1, 800kT in Mod 2. Designed ranges were 755nm/1400km (CEP 2.7–3.6km) for Mod 1, 862nm/1600km (CEP 1.3–3.0km) for Mod 2. Of 228 service test launches in 1963–82, 193 succeeded (nineteen were missile failures).

D-5 (R-27, 4K10 missile, Zyb) was developed by Makheev to arm 'Yankee' class submarines (Project 667A); it also armed Project 605. The treaty designation was RSM–25; the NATO designation is **SS-N-6** Sawfly. This system was authorised 24 April 1962, to arm the Project 667A ('Yankee' class) submarine. The SS-N-6/D-5 combination is Navaga. According to a Soviet émigré, the missile concept dates from the beginning of the Seven Year Plan in 1958. That year the Soviet MoD authorised Leningrad Research Institute 4 (NII-4) of Northern Fleet's 4th Missile/Gunnery Directorate to develop a TTZ for a third generation SLBM as an anti-carrier weapon. It wanted a single-stage, liquid-fuel underwater launch missile 1.8m × 10m (4m shorter than SS-N-4); weight to be less than 19,000kg. Inertial guidance for 1.0km CEP, low-overpressure thin wall fuel and oxidizer tanks which increased range to 2400km (vs 1300km for SS-N-5). The single RV carried a 0.7MT yield warhead. Although liquid-fuelled like its predecessors, D-5 used the 'Ampule' system which allowed it to remain fuelled for up to a year. The design incorporated numerous safety features to prevent premature fuel ignition. As built, R-27 is a 2-stage missile with a small powered upper stage. Using mainly 'off the shelf' technology, the missile completed the four preliminary test stages by 1967. The first sea launches were from a 'Whiskey' class submarine. The system was accepted on 13 March 1968. Missiles could be fired in salvoes of eight each. There are three versions: Mod 1 (1968), 1300nm/2400km; Mod 2 (1973), 1600nm/3000km (probably D-5U/R-27U); Mod 3 (1974), 2 MRVs, also 1600nm range (probably D-5A/R-27A). The Mod 1 warhead yield was 1MT. Accuracy: CEP about 1.85km. Reportedly about 1000 were built. Dimensions: 4ft 11in × 29ft 2in/1.5m × 8.89m; launch weight 14.2t (1st stage 13.5t); thrusts 23t and 3t; throw weight 1433lb/650kg.

The post-Khrushchev 8th and 9th Five Year Plans both incorporated a new series of SSBN and 4th-generation SLBM for introduction in the mid and late 1970s. As an interim, SS-N-6 was modified for coutervalue strikes against North American coastal targets. The missile achieved 550km more range at the expense of reducing warhead yield to 0.65MT (a Russian source claims that yield was not reduced); it was tested at Severomorsk in 1972. Then the introduction of SS-N-8 in 1972–73 led to the final version, with two RVs (Mod 3), flight tested in 1973, and retrofitted to all 'Yankee' class submarines. RV yields are variously reported as 200 and 350kT. Reported CEPs (Russian source) are 1.3–3.0km (Mod 1), and 1.3km in the other versions.

No D-6 has been reported. Logic suggests that it may have been reserved for the abortive SS-NX-13.

D-7 (RT-15M missile) was a 1963 experimental system by Yangel. This was a 2-stage solid-fuel missile. Reported range was 1292nm/2400km (as in R-27/SS-N-6); the 4ft 11in × 34ft 5in/1.5m ×

10.5m weapon carried a 1MT warhead. RT-15M seems to have been an unsuccessful R-21/SS-N-5 successor.

D-8 was a 1967 experimental missile. Reported characteristics: 2-stage liquid fuel, single warhead, range 4200nm/7800km, dimensions 5ft 4in × 37ft 9in/1.65m × 11.5m, launch weight 20t. The range matches that of the contemporary R-29/SS-N-8; this missile seems to have been an unsuccessful R-27/SS-N-6 successor.

D-9 (R-29, 4K75 missile, Vysota; NATO **SS-N-8** Mod 1 Sawfly; treaty designation RSM-40) was developed by Makeyev. SS-N-8 Mod 2 is D-9D (R-29D, 4K75D), D-9DU (R-29DU, 4K75DU, Volna) is **SS-N-18** Mod 1 Stingray (RSM-50). D-9R (R-29R) is **SS-N-18** Mod 2 (RSM-50). D-9RM (R-29RM) is **SS-N-23** Skiff (RSM-54).

R-29 was designed for the 'Delta' (Project 667B/667BD) class SSBN, using stellar-inertial guidance to obtain useful accuracy even when firing from 'bastions' which Soviet naval forces could protect. It entered service in 1973. Two versions: Mod 1, 4200nm/7800km range; Mod 2, 4900nm/9100km; one RV (about 800kT) in each case. Accuracy CEP 1.6km in Mod 2. Reportedly the stellar-inertial system was not always reliable, and initial reports of spectacular accuracy were exaggerated. Such systems, for example, sometimes fix on the wrong star or set of stars. Dimensions (2-stage missile): 5ft 11in × 42ft 8in/1.8m × 13.0m (without warhead); launch weight 33.3t; throw weight 2425lb/1100kg. The Delta/SS-N-8 (Project 667B) system is Murena; Delta II (667BD) is Murena-M.

D-9U/R-29U (4K75DU Volna, NATO **SS-N-18** Stingray) was the first Soviet MIRV submarine missile, using stored liquid propellant as in R-27 and R-29 (SS-N-6/8). Versions: Mod 1, 3 MIRV, 3500nm/6480km; Mod 2, one RV, 4300nm/7960km; Mod 3, 7 RV, 3500nm/6480km. CEP is reportedly 0.8km (a Russian source claims 0.9-1.4km). Two stages. Dimensions: 4ft 11in × 46ft 3in/1.8m × 14.1m (without warhead), 35.3t, throw weight 1650kg. Mod 1 was introduced in 1977, Mods 2 and 3 in 1979. Yields: Mod 1, 200kT, Mod 2, 450kT. CEPs 1.4km in Mods 1 and 2. The Delta III/SS-N-18 system is Kalmar.

R-29RM (Shtil-1 -1N, -2, -3A, and -3N; NATO **SS-N-23** Skiff) is the 3-stage liquid fuel missile of the Delta IV class. Dimensions: 6ft 3in × 48ft 7in/1.9m × 14.8m (without warhead), 40.3t, throw weight 6173lbs/2800kg. It carried 4 × 100kT warheads. The Delta IV/SS-N-23 system is Delfin. This missile became operational in 1985. Range is 4469nm/8300km, as in SS-N-20 (CEP 0.5 – 0.9km).

No D- or R-designation is known for **SS-NX-13** (4K18), a guided (anti-radar) ballistic missile tested on board a Project 629 or 658 submarine in 1973. Reported range was 700–1100 km; presumably much less would have been used in the anti-ship role, to minimise time delay between launch and arrival. Dimensions were about 1.5m × 10.0m.

No D- or R-designation is known for **SS-N-17** Snipe (treaty designation RSM-45). It was a limited-production 2-stage solid-fuel missile to replace SS-N-6; in one 'Yankee' conversion (Navaga-M system). Range 2100nm/3890km (CEP 1.4km). First tested 1975, it went to sea in the sole 'Yankee II' in 1977. It has a MIRV-type post-boost vehicle but only one RV. Dimensions: 5ft 1in × 34ft 9in/1.54m × 10.6m (without warhead); launch weight 26.9t, throw weight 992lb/450kg. This missile has been attributed to a previously unknown design bureau. KB ZiF. This missile shares some technology with the land-based SS-16 and SS-20; all three reportedly have similar post-boost vehicles. SS-17 may be related to the 3780nm/7000km solid-fuel missile proposed to Khrushchev in 1961.

There is no indication of D-numbers for strategic rockets between 9 and 19. However, it is possible that these numbers were reserved for non-strategic systems. In connection with modifications to 'Kotlin' class destroyers, a Russian article gave a D-13 designation for the ship's RBU-2500 Smerch system.

D-19 Taifun (R-39, 3M65 missile, Rif-M, treaty designation RSM-52) is NATO's **SS-N-20** Sturgeon, carried by 'Typhoon' class submarines (Project 941). This 3-stage solid-fuel rocket carries six to ten × 100kT RVs. Dimensions: 7ft 10in × 52ft 5in/2.4m × 16.0m (without warhead), 90t, throw weight 5622lb/2550kg. Note that launch weight includes components jettisoned after launch. Without them, the missile weighs 84t at launch (first stage 52.8t). R-39 was first fired in 1982. Range is 4469nm/8300km (CEP 0.5 – 1.9km). At least one of the solid-fuel rocket engines was developed by the 'Yuzhnoe' (ex-Korolyev) OKB following a September 1973 directive (which sets the timing for SS-N-20 development). The 3D65 rocket engine first flew in December 1981.

As the Soviet Union collapsed, two more missile systems were in development, one (RSM-54M) for 'Delta IV' class submarines, and one (RSM-52M) for the 'Typhoon' class (Project 941).

ASW ROCKET LAUNCHERS

A full ASW rocket system (RPS) consists of the RBU (launcher) plus the RGB (rocket), the UDB (depth fuse), and the PUSB (firing control device). The figure in the designation is usually the nominal range in metres.

MBU-600: this copy of the wartime Allied Hedgehog was widely used in the early 1950s; it was roughly contemporary with MBU-900. In each, the implied range was far greater than that of the equivalent wartime Allied weapon.

RBU-600: six barrels, 11.8in/30cm calibre, in two vertical rows, only in modified 'Kotlin' destroyers; NATO formerly called it MBU-4500. All six rockets fired simultaneously (198lb/90kg rockets with 121lb/55kg warheads). Introduced in 1962–63.

RBU-900: the Soviet equivalent to the wartime US Mousetrap rocket projector, consisting of two pairs of rocket tracks, one above the other. It does not train. This weapon appeared in some early 'Kola' and 'Riga' class SKRs.

RBU-1000: same barrels and, originally, projectiles (RGB-10 rockets) as RBU-600 in horseshoe formation, introduced 1962–63. It loads automatically, barrel by barrel, vertically, and the barrels fire in sequence. It is trainable. This is now a dedicated anti-torpedo weapon. Loadouts: 60 RGB-14 rounds (5 salvoes/launcher) in a 'Kara' or 'Kresta II', 48 rounds (4 salvoes/launcher) in a 'Kashin'.

RBU-1200 (Uragan system): two horizontal rows of 9.8in/25cm barrels, three above two. It elevates for range but does not train. It fires RGB-21 (50 rounds: 252mm × 1228mm; weight 158lb/71.5kg, including 71lb/32kg of explosives). Lethal radius is 6m. Sinking speed is 6.85m/sec (depth settings are 10–330m). Flight time is 3–16.3sec. At the maximum range, 1586yds/1450m, bombs are spread over a 70m × 150m area. Mounted in minesweepers and SO-1 class sub-chasers.

RBU-1800: five barrel 25cm rocket launcher (three above, two below) introduced 1955–56 as the first entirely Soviet-designed ASW projector, based on an army weapon. Manually loaded, on board early 'Kronstadt' and SO-1 class sub-chasers and on board minesweepers.

RBU-2500 (Katyusha; Smerch system) has two horizontal rows of tubes (total 16) to fire the RGB-25 round (312mm × 1340mm, weight 187lb/85kg, including 57lb/26kg of explosives). Range is 550–2700m. Lethal radius is 5m. Sinking speed is 11m/sec (depth settings 10–320m). Flight time is 3–25sec. The load-out in a 'Riga' class frigate was 128 rounds (4 salvoes/launcher). The launcher is manually reloaded. Introduced 1957–58. Smerch may laso be coded as D-13.

A Russian list includes an RBU-4500 30cm system (six barrels).

RBU-6000 (formerly designated MBU-2500A by NATO): twelve barrels in horseshoe arrangement, fired in pairs, reloaded automatically in vertical position, barrel by barrel. Introduced 1960–61. Range is 5700–6000m. 'Kresta II' and 'Kara' have 144 RGB-60 rounds (6 salvoes/launcher), 'Kashin' has 192 rounds (8 salvoes/launcher). The bomb can be fitted with the guidance system of the S3V depth bomb described below. This weapon can be used for shore bombardment.

Two RBU-6000 ASW RL aboard *Neukrotimy* 8.6.90 L & L van Ginderen

RPK-5 (see below) is designated RBU-12000 by NATO. There are both 10- and 6-barrel versions, firing both ASW and anti-torpedo rounds.

BOMBARDMENT ROCKET LAUNCHERS

The shore bombardment rocket launchers aboard amphibious ships and the 'Shmel' class river gunboats are adapted army weapons. The 17-barrel 140mm (5.5in) weapon coded BM-14/17 by NATO entered land service in 1954. BM-14/18 was designed as a naval weapon from the outset; it was mounted aboard 'Polnocny' class landing ships and some 'P 4' type torpedo boats. Both fire 70lb/32kg rockets to 4.8nm/9km. Some 'Shershens' had a 40-barrel BM-21 (122mm, 101lb/46kg rockets, range 8nm/15km) in place of their torpedo tubes.

TORPEDOES

Torpedoes are generally designated by calibre (cm) and by year of introduction. Calibres are 40cm (16in), 45cm (18in), 53cm (21in), and 65cm (25.6in). NATO designations follow Soviet practice but often do not match Soviet designations; wherever possible both designations are given below. In at least some cases the Soviets used drastically abbreviated designations which did not include calibre. Air-launched weapons used a different format. E indicates electric propulsion; U indicated a 'universal' torpedo suitable to attack both surface and submerged targets. GTs were 'evasion devices', torpedoes to be fired back at her puruers by a submarine under attack (M, *maliy*, for small, B, *bolshoi*, for large).

65cm

The first 65cm torpedoes were **65-73** (1102lb/500kg warhead) and **65-76** (nuclear); both run at 14m depth (46ft) and achieve a range of 54,680yds/50km at 50kts. There is probably also a newer 65cm torpedo carrying a 1984lb/900kg warhead (NATO **65-80**, 29ft/9.1m long, with an alternative 50kT nuclear warhead). A 65cm ASW torpedo has also been reported. A DT long-range thermal torpedo described in 1993 is 65cm × 1100cm (433in), 9920lb/4500kg with a 992lb/450kg warhead with proximity and contact fuses. It can be fired either as a wake-follower or as a straight-runner. The turbine powerplant uses a steam-gas (presumably Walter) mixture. The NATO designation for the 65cm dual-purpose (ASW and anti-ship) submarine-fired torpedo is E65-83.

53cm

53-27L carried postwar by North Korean motor torpedo boats: 21in × 276in (3800lb, 550–600lb warhead), 45kts/3980yds.

53-38: standard torpedo in service 1941 (21in × 284in; 3560lb with 660lb/300kg warhead; -38U version was 293in long with 880lb/400kg warhead; both 44.5kts/4000m, 34.5kts/8000m, 30.5kts/10,000m). A magnetic proximity fuse entered service about 1943. This torpedo was presumably developed from the 53-36 of 1936, the first two-speed Soviet torpedo (43.5kts/4000m and 32kts/8000m, with a 300kg warhead (265kg in 53-27L) in the same dimensions as 53-27L but slightly lighter (1700kg vs. 1710kg).

ET-80: The first Soviet electric torpedo, 53cm × 7.5m (294.6in), 1800kg with a 400kg (880lb) warhead: 29kts/4000m. It entered service in 1943. It was apparently adapted from the German G-7e, details of which were obtained under the Hitler-Stalin Pact of 1939–41. An improved postwar version, ET-46, had greater range and speed: 31kts/6000m, and carried a heavier warhead (450kg/990lbs). It may have been based on the improved German torpedoes captured when *U 250* was raised in September 1944.

53-39: improved torpedo which entered service July 1941; more powerful engine and increased air and fuel capacity. Suitable for angled fire (21in × 295in, 3923lb, 317kg warhead, 51kts/400m, 39kts/8000m, 34kts/10,000m). 53-39PM was a pattern-running version.

53-51: first postwar submarine torpedo, straight- and pattern-running (21in × 307in, 880lb/400kg warhead, 51kts/4000yds, 39kts/10,000yds).

53-56V (submarines) **-56VA** (torpedo-boats) are standard air-steam powered export torpedoes, replacing 53-39. Both have pattern-running modes based on the wartime German FAT (53-56VA is preset by the torpedo FCS of 'Shershen' class or later torpedo-boats). The exploder is an inertia-pendulum with 4 degrees of freedom, so that it can work at any impact angle. 53-56VA: 533mm × 7490mm, 1750kg (21in × 295in, 3859lb), 51kts/4000m, 39kts/8000m, 310kg warhead. 53-56V: 1900kg (4188lb), 400kg warhead, 41kts/8000m. The mechanical FCS on 'Romeos' and 'Foxtrots' is designed to work with 53-56, and probably cannot easily be adapted to alternatives. These torpedoes are spindle-set.

53-57 (NATO 53-56): first torpedo with HTP engine, a modified 53-51 (51kts/9000m, 45kts/18,000m, run depth 2–14m, 533mm × 7688mm, weight 2000kg, warhead 305kg). This weapon could be fired by both surface ships and submarines. It was the standard straight-running anti-ship torpedo of the 1960s. The first operational Soviet nuclear torpedo was a version of 53-57, carrying a 15kT warhead. Its development designation was T-5; it entered service in 1959 or 1960. Reported performance was 50kts/11,000m or 32kts/21,000m. The Soviet designation of this nuclear torpedo is unknown.

53-61: submarine anti-ship torpedo; improved 53-57 with two settings (55kts/19,000m, 35kts/22,000m) for straight-, or pattern-, or circular-running. The improved -61M of 1963 added a passive homer.

53-65: this torpedo replaced 53-61. It introduced a new semi-closed cycle (oxygen) powerplant and wake homing for attacks against ships; it was the first Soviet thermal homing torpedo; wake-following is largely immune to self-noise (21in × 307in, 65kts/12,000m, 40kts/24,000m, run depth 14m, warhead 300kg). Operational 1968. -65M of 1969 had an improved homer. 53-65K (1969) is a 53-61 upgraded to -65 standards and usable by surface ships as well as submarines (45kts/19,000m, run depth 12m, warhead 304kg). The export version is 53-65KE: 533mm × 7945mm, 2070kg, 307.6kg HE, up to 45kts at 4–12m run depth, to 19,000m. NATO designated a nuclear version 53-68 or 53-68N; it was a modernised version of the nuclear 53-57, operational in 1970 (21in × 284in, 45kts/15,000yds, 20kt warhead).

53-70 (NATO designation): submarine and surface ship passive thermal pattern-runner. 53-70VA is a NATO designation for a torpedo with a 560kg warhead.

53-83 (NATO designation): successor to 53-65 wake-follower, using Mk 48 propulsion, for launch by ships and submarines.

BGT is the straight-running 53cm nuclear evasion weapon (55kts/15,000m).

SAET-50: the first Soviet homing torpedo, a modification of the earlier electric torpedo (ET-80) of 1942. This passive anti-ship weapon was developed 1946–50 as SAET-2, and accepted into service in 1950 as SAET-50. Range at 23kts was either 4.6km with a V-6-III battery or 6km with a V-6-IV battery. A modified SAET-50M appeared in 1955. SAET-50 was probably developed partly from the German T-5 captured on board *U 250* in 1944.

SET-53 (NATO ET-80A): the first Soviet ASW homing torpedo. Work began in 1950; it was intended to engage targets moving at not less than 9kts and at 20m to 200m depths. This weapon also had an electronic proximity fuse (range 5m). It was tested in Lake Ladoga in 1954–55, with formal state trials between August and October 1957; it was accepted into service in 1958 as SET-53. SET means *Somonavodyashchayasya Electricheskaya Torpeda*, or 'self-guiding electric torpedo'. Dimensions: 21in × 307in. Range at 23kts was 6000m with a V-6-IV battery or 7500m with a T-7 battery. This torpedo could acquire a cavitating submarine (21dB at 24.8kHz) at about 700m. Warhead weight was only 100kg, presumably because electronics were so bulky. The formal designation was presumbably SET-53-58.

SET-53M was an improved SET-53 adopted in 1964. Warhead and operating depth did not change, but performance was improved to 29kts/14,000m. This version may have incorporated pirated British Mk 20 active/passive technology. Dimensions: 533.4mm × 7800mm, total weight 1480kg.

SAET-60 (probably actually SAET-53-60): the second Soviet 53cm electronic homing anti-ship torpedo (1961, 42kts/13,000m, 400kg warhead). SAET-M (1969, 42kts/12,800m, 35kts/14,600, 300kg), probably actually -60M, is an improved version. There was also a nuclear version, which NATO designated ET-80A(66) or ET-80-66N; it replaced the nuclear version of 53-57 described above. Maximum running depth was 1000ft; a deep run would tend to enlarge the area affected by the nuclear burst (20kT in this case) and thus would reduce the effect of any aiming error. Dimensions: 21in × 303in; performance 35kts/10,000m, 20kts/40,000m.

SET-65 (NATO ET-80A): passive acoustic submarine torpedo, successor to SET-53/SET-53M with 1.7 times the speed, twice the range of SET-53: 40kts/15,000m and 200m maximum depth; 1850kg with 250kg warhead (533mm × 7800mm). With active/passive guidance, this was the first serious Soviet ASW torpedo. The true designation was probably SET-53-65. SET-65Ch of 1972 (reportedly 21in × 307in, 45kts/15,000m, 460m depth, 250kg warhead) was a converted 53-65. Performance was also reported as 35kts/10,000m and 24kts/20,000m. Details of an export SET-65E were published in 1993: 53.3cm × 780cm (307.1in), 1700kg with 205kg warhead. There was an improved version, SET-65M (NATO ET-80A(76)).

E53-66: electric gyro torpedo (straight- and pattern-runner).

TEST-68: the first Soviet submarine-launched wire-guided torpedo (1968); TEST means *Teleupravlyayannaya Torpeda*, 'teleguided torpedo'. Performance: 29kts/24,000m, maximum depth 250m, 200kg warhead. No NATO designation is known; this torpedo may not have been known to NATO when it entered service.

TEST-71 (NATO E53-71): 1971 submarine-launched wire-guided anti-submarine torpedo, successor to TEST-68 (35kts/25km, maximum depth 400m, 200kg warhead). TEST-71ME was described publicly in 1993 (53.3cm × 793.5cm/312.4in, 1820kg with 205kg warhead).

SET-72: dual-purpose torpedo launched from both surface ships and submarines (1972): performance 41kts/8000m, maximum depth 750m, warhead 80kg (probably nuclear).

TEST-E: 1977 surface ship ASW torpedo (40kts/20,000m, 400m depth, 200kg warhead); the proper designation is probably TEST-53-77E. This is a surface ship version of TEST-71. It may have a nuclear version. Reported characteristics: 53cm × 7.8m, 35kts/15,000m, 270kg warhead. For surface ships (probably including 'Grisha V') and submarines. This was presumably NATO's ET-80A(76).

USET-80 (NATO E53-83): submarine-launched dual-purpose wire-guided torpedo, 1980: 48kts/18,000m (also given as 40kts/15,000), maximum depth 10m/330ft (also given as 1000m depth), 80kg warhead. This weapon may be designated **SAET-80** or **UEST-80**.

TEST-96: new electric surface ship/submarine wire-guided torpedo for use against submarines or surface ships: it follows surface ships wakes in its terminal phase, or homes actively or passively against submarines (53.3cm × 800cm/315in, 1800kg with 250kg warhead).

AT-2 (NATO 53-72): missile warhead, 1973; performance 40kts/3800m, 400m maximum depth, 100kg warhead. This short 21in torpedo arms SS-N-14 and probably -16. NATO listed two follow-on weapons, E53-72A of 1975 and E53-79A.

Shkval is a recently-developed rocket torpedo which a submarine is expected to fire back when it detects an oncoming weapon. It is in effect the ultimate BGT. The cone-shaped Shkval travels within a supercavitating bubble which it produces with its own exhaust. It is fired as a conventional torpedo, and ignites its rocket after travelling 50m from the submarine. Range is given as 6nm at 200kts. Running depth is 400m. Dimensions: 53cm × 8.2m. The warhead is nuclear, with a preset burst range. The published designations are VA-111 and M-5.

RAT-52 was a special air-launched rocket torpedo, dropped by Tu-14s and Il-28s and exported to China, which produced its own version. Once in the water it made 58–68kts over a range of 520m.

45-12 may have been supplied to allied navies postwar. Dimensions: 45cm × 5.7m (224in), weight 860kg with 140kg warhead, 38kts/3000m, 29kts/6000m. There was also a 45-15.

45-36NU/45-36ANU: standard Second World War torpedo-boat and air-dropped torpedo. The reinforced ANU could be dropped at speeds of up to 200kts (360km/h). Dimensions: 18in × 224in. Performance: 45kts/4400yds (air version 41kts/3280yds). Weight 2000lb (warhead 440lb).

45-54VT (1075kg) and **45-56NT** (1035kg) were the standard anti-ship torpedoes carried in pairs by Il-28 naval bombers in the 1950s, and probably still used by the PRC on its H-5 (Il-28 copy) bombers.

AT-1 (NATO E45-70A): the original payload of the SS-N-14. This torpedo executes an up-and-down helical search, down to about 1000ft (300m). Dimensions: 18in × 154in; performance: 28–30kts/10,000m. Warhead: 198lb. This torpedo may have superseded SET-40 on board some small ASW ships; at least one 'Mirka' was armed with 45cm torpedoes in the late 1960s or early 1970s.

UMGT-1 (NATO E45-75A) is the later payload for SS-N-14 and -16 missiles, presumably replacing AT-2. It became operational in 1981. This updated AT-1 can operate down to 500m. Within the same dimensions it achieves 41kts/8000m; warhead is 60kg (132lbs). There was a later lengthened and modernised version.

MGT (probably NATO SET-40-62): 1961 submarine-launched evasion torpedo, a passive-homing anti-escort weapon to be fired while evading (28kts/6000m [also given as 25kts/6000m]; run depth 10m/33ft, warhead 80kg [also given as 50kg]). MGT was apparently designed for missile-firing submarines. It was *not*, as often stated, an ASW torpedo; it loitered to find an approaching escort, and its run depth limit protected the firing submarine.

SET-40 (NATO E-40-65A): first Soviet surface ship lightweight torpedo, 1962; possibly also for firing from submarines. Work began in the second half of the 1950s, probably originally on a 345mm weapon. Like the US Mk 44, it uses an active pinger (acquisition range 640yds); 29kts/7500m, maximum operating depth 200m (660ft), 80kg (176lb) warhead. Dimensions: 16in × 177in.

SET-40M: modernised deeper-diving SET-40, 1968: 400m depth, 80kg warhead.

SET-72 (NATO E40-75): first Soviet multi-role torpedo (anti-ship/anti-submarine), using an active/passive seeker; for use by both submarines and surface ships. Dimensions match those of SET-40; reported performance is 30kts/14,000yds. NATO lists E40-75 (0.4m × 4.5m, 30kts/12,000m) as the current standard shipboard lightweight ASW weapon. The designation SET-73 may apply to this weapon.

USET-95 (0.4m × 4.7m, 650kg, 80kg HE), announced in 1993, can be launched by both surface ships and submarines; it follows surface ships wakes and homes actively and passively on submarines.

APSET-96, announced in 1993, is suitable for aircraft or helicopter launch, with a drag parachute (0.4m × 3.84m, 720kg with 60kg warhead, active-passive homing in both planes, with proximity and contact fuses).

APR-2E, a new solid-fuel ASW torpedo, was shown (in poster form) at the 1992 Moscow Air Show. APR means *Aviatsiya Protivo Raketny*, 'airborne ASW rocket', but in this case rocket probably only means solid-fuel. It is dropped by aircraft and ASW helicopters (CFE documents released in 1990 show that it is the standard weapon of the Mi-14 [Hazel] ASW helicopters). The 'E' probably means export version. APR-2E is smaller-diameter than previous homing torpedoes, and probably arms the conventional version of SS-N-15. Dimensions: 0.35m × 3.7m (14in ×146in), weight 575kg (1273.3lb), maximum speed 115km/h (62.9kts). This weapon is in effect a Soviet Mk 46, but it is very different in concept: it loiters, listens, and then takes off after its target. The Soviets apparently also developed a piston-engined torpedo, which NATO designated 40-79, which would have been a Mk 46 equivalent. No Soviet designation is known.

MINES

The Soviet Navy was well-known as an advocate of mine warfare. Many surface combatants were equipped to lay mines, and minelaying in enemy harbours was one of the major roles of Soviet naval aircraft as early as the 1930s. The later missile attack regiments were direct successors of early mine-torpedo regiments, presumably retaining the mining responsibility. In the 1980s there was speculation that Soviet aerial mining would be the best tactic to use against NATO European ports in wartime. The reader may recall that the Soviets supplied the North Koreans with contact and ground mines that closed the port of Wonsan for two weeks in 1950.

Western interest in deep minehunting was inspired by initial reports, about 1975, of two new mine types: the rocket-propelled RVM, and the anti-submarine UEP mine. The Soviets later introduced a torpedo mine similar in concept to the US Captor and also a submarine-fired mobile mine.

A small stockpile of 5-20kT nuclear mines has been reported.

Reports of total reserves vary. In 1987 the US DoD claimed a total of 300,000, including about 100,000 moored contact mines. Many were exported to Soviet client states, such as Egypt, Libya, North Korea, and Vietnam. Inventory figures below are estimates published unofficially in 1988, and probably derived from an official report then about a decade old. Designations given below are mainly those applied by NATO; the letters follow generic Soviet mine descriptions for the various types.

Major current types are:

AMD/KMD/KMK Series/MKD Air/surface/submarine-laid ground mines, the number (-500 or -1000, but -1000 only for submarine mines) indicating the weight in kilograms. Known influence mechanisms are (i) magnetic, (ii) acoustic, (iii) pressure, and (iv) combination. There may be others; a Libyan-laid KMD was recovered in the Red Sea in 1984, carrying a new influence mechanism. It may have carried a modular electronic firing mechanism replacing the earlier plug-in type. The original AMD appeared during the Second World War. It was the air-dropped version of the earlier MKD magnetic ground mine, the first modern Soviet ground mine. AMD-II was reported in 1950 (KMD is a surface-laid equivalent, an AMD strapped to a dolly). There was no AMD-III, but AMD-IV appeared in 1954. The AMD designations are NATO, not Soviet. Claimed inventories: 18,000 each of AMD-500 and -1000, 35,000 AMD-IV. In 1988 a projected successor, AMD-P-V, was listed, with an estimated inventory of 12,000. No separate inventories of the other series were given.

KRAB Moored acoustic influence mine, with a charge of 250kg. It entered service in 1949, and an improved version appeared in 1960. The mine can be moored in 600m of water, its case at up to 150m (500ft). Estimated inventory: 10,000.

MAG Moored ASW antenna mine, a modified MKB. It can also carry horns. Maximum laying depth is 1490ft, and the antennas are 24m long (upper) and 35m (lower), so a single mine can sweep out a depth interval of about 60m (about 200ft). Case depth is 8ft to 146ft. Estimated inventory: 10,000.
MKB-3 Standard moored contact mine, the successor to the M-08, a North Korean-made version of which the Iranians used in the Persian Gulf. It has five chemical horns. Charge weight is 200kg. Estimated inventory: 30,000.
RVM (Cluster Bay) Rocket-propelled rising mine. The mine listens for an approaching ship or submarine, then turns on its active sonar to determine depth and speed. The rocket goes straight up, with a preset fuze to burst it at estimated target depth. This weapon, more than any other, was responsible for Western interest in deep minesweeping. It appeared in 1969, but Western navies seem to have become aware of it only about 1973. An air-dropped version (A-RVM) was also reported. The mine is moored, with a maximum case depth of 300m (about 1000ft) and a maximum mooring depth of 750m. Estimated inventory: 15,000. A later torpedo-bearing mine was designated G-RVM.
UDM Universal influence mine against surface ships and submarines, laid by surface ships and aircraft (and possibly also by submarines). Total weight is about 1000kg (charge weight 650kg). In its magnetic version, the mine releases a sensor which floats closer to the surface. This mine was used by the Iraqis during the Gulf conflict. This may be the mine otherwise described as sectionalised. Estimated inventory: 10,000.
UEP (Cluster Gulf) This is a stand-off mechanism. Underwater Electric Potential (UEP) is due to the corrosion current flowing between dissimilar metals in a ship or submarine hull. UEP was known during the Second World War, but apparently the Soviets were the first to apply it to an ASW mine. The current version has a case weight of about 480kg (charge 30kg). The Soviets may have used UEP mines of their own as early as 1944. Estimated inventory: 10,000.

AIR-SEARCH RADARS

NOTE; Most names are NATO ones usually based on antenna appearance. In some cases the Soviet designations (MR- or a code name) are known.
Big Net (MR-500 Fut-N or Kliver) L-band air search radar (850MHz, 250/550pps, 3.5/2.0 microsec pulses) introduced in the 'Kresta I' and 'Kashin' classes. It was developed for cruisers and destroyers in 1948-55 and tested in 1955; it entered service in 1957 and is contemporary with Top Trough. Big Net was the last major Soviet two-dimensional long-range air search radar; Top Sail is, in effect, its successor.
Boat Sail Air search radar on Whiskey-class submarines converted to radar pickets.
Cheese Cake Small-ship search radar.
Cross Bird (Guis-2M) P-band radar (210MHz, 400-445pps, 2-5 microsec pulses, beam 25-30 degrees) derived from the British Type 291, and carried by cruisers and destroyers. The specification was issued in August 1946 and the radar was first tested on board the cruiser *Molotov* in 1948. The original Guis-1M had two Yagis and a reflector; -2M, the most used version, had four paired dipoles, as in Type 291. In the initial tests, -1M detected aircraft at 14-29nm and a cruiser at 11.5nm; resolution in bearing was 4 degrees.
Cross Round (Positive-E) Surface/air search radar, enclosed in a large radome. It appears on board the new *Neustrashimyy*, the German-built 'Parchim IIs' and the Indian *Khukri*. This is the acquisition element of the Cross Sword fire control system.
Don Series Standard X-band surface-search sets (80kW, 800/1600pps, 1/0.5 microsec pulses, 1.1 x 20 degrees beam). Being superseded by Palm Frond.
Flat Spin Similar to, but flatter than Hair Net.
Hair Net S-band air search set (1950: 420-425pps, 1 – 1.5 microsec pulses, beam width 30 degrees). Slim Net was a modified version with a much larger antenna, for a much narrower beam.
Head Net (MR-310 Angara) S-band small-ship radar. Head Net A (MR-300 Angara), now extinct, used a single antenna. Head Net B had two back-to-back, for higher net data rate. Head Net C (MR-310U Angara M) has two back-to-back antennas with one canted 30 degrees to the horizon. Comparison betweeen the apparent positions of the same target as seen by the two radars gives target altitude (V-beam height-finding). Head Net C entered service in 1963 on board the 'Kashin' and 'Kynda' classes. There may also have been a very similar radar called Sea Net. This radar is being superseded by Top Plate.
High Sieve/Low Sieve (Ryf-1M and -2M) surface target indicator and fire control radar, developed in parallel with Cross Bird and Knife Edge under the 1946-48 three-year radar plan. Ryf was tested with Knife Edge in 1948 on board the cruiser *Molotov*. S-band (427 pulses/sec, 1 microsec pulses, scan rate 6rpm; 150kW; 7 degrees beam). High Sieve was first seen in 1954 on the *Sverdlov* and *Skoryy* classes. Low Sieve was first seen in 1957 on board missile cruisers, cruisers, and *Neustrashimyy*. See also Ball End and Ball Gun.
Kivach 3 Current small-combatant navigational radar.
Knife Rest A (P-8)/Spoon Rest P-band (75MHz, 100kW, 50pps, 5-12 microsec pulses, 20 × 25 degrees beam) long-range air search radar, originally for ground use. Knife Rest, developed under the 1946-48 radar plan, appeared in 1952 aboard *Sverdlovs* (replacing Cross Bird) and on some T-43 radar pickets. Spoon Rest used a much larger antenna and operated at higher frequency (147-151MHz); it appeared in 1965, and was fitted on the after superstructures of some *Sverdlov* class cruisers.
Mius X-band surface search radar.
Neptune X-band navigational radar of the 1950s (42kW, 2700/675pps, 0.25/1.0 microsec, 1.6 degrees beam).
Nyada (MR-212) X-band surface-search radar.
Palm Frond (MR-320) Standard surface-search radar, replacing Don 2 and Don Kay from the 'Krivak' class onwards. The version in an *Udaloy* is MR-320M Podkat.
Peel Cone (Positiv) Air/surface search radar on 'Pauk' and 'Stenka' classes, superseding Pot Drum.
Peel Pair X-band surface search radar of the late 1960s; it looks like two Don-2 antennas mounted back to back. This radar may be called Dubrava.
Plank Shave (Garpun) Surface search (missile fire control) radar superseding Square Tie in the 'Tarantul' class. The Russian name means Harpoon.
Plate Steer Modified Top Steer, in which the cylindrical 3D antenna is replaced by one of the flat plates of Top Plate.
Pot Drum (MR-102)/Pot Head (Reya) Standard small combatant X-band radars, superseding Skin Head. Pot Drum came first (1958: 150kW, pulse widths 0.5-1.0 microsec); Pot Head (1960) had better resolution, *eg*, for periscope detection (100kW, 1600-1650 and 3200-3300pps, pulse widths 1.0 and 0.25 microsec). These radars were superseded by Square Tie.
Skin Head (Zarnitsa) Derivative of US SO series S-band radars for patrol boats; developed 1946-48 and tested 1948 (80kW, 400pps, 1 microsec pulse). Superseded by Pot Drum/Pot Head.
Sky Watch (Mars-Passat) Phased-array radar for carriers, first seen on board the last of the *Kiev* class, *Baku* (now renamed *Gorshkov*). It replaced the Top Pair of earlier ships of this class. Reportedly it has been less than altogether successful.
Slim Net (FUT-N) S-band air/surface search radar (420-425 pps, 1 – 1.5 microsec pulses, 1 degree beam) for cruisers, destroyers, and frigates (1957); it was a slightly modified Hair Net. It was on all 'Kolas' and eventually on board all but one 'Riga'. A modified version, with a solid rather than lattice reflector, is made in China as MX-902 (Eye Shield).
Snoop Head (Tobol) X-band submarine search radar with an ESM radome (Bald Head) on the same mast, above it. The reflector is slotted. This antenna appeared in the late 1970s on board 'Alfa' and Oscar class submarines.
Snoop Pair (Tobol) X-band submarine search radar using a pair of back-to-back antennae. The two feeds appear somewhat different, so it may be designed to produce two overlapping beams. The reflector is solid and resembles that of Snoop Slab, albeit with feed from above rather than from below. An earlier radar using much the same antenna (*eg*, on 'Yankees') was Albatros.
Snoop Plate (Nakat) First postwar submarine search radar, superseding British-supplied Type 267. Many components duplicate those of Neptune, Don, and Post Lamp (X-band, 80kW, 1618pps, 1-2 microsec).
Snoop Slab (Flag) Submarine search radar which replaced Snoop Plate from 1957 on; it has a massive parabolic reflector and very noticeable motor housing.
Snoop Tray (Albatros) Submarine search radar introduced in 1962, with a nearly flat and almost rectangular reflector, tipped over at about a 25 degrees angle. It succeeded Snoop Slab in production (initially on board 'Tango', 'Golf', 'November', 'Hotel', and 'Echo' classes).
Spin Trough (Volga) Surface search radar in 'Krivaks' and in auxiliaries, small craft, etc, as an alternative to Don 2.
Square Tie (Rangout) Surface search radar for missile fire control in Komar and Osa classes. Superseded by Plank Shave (although Square

Tie continued in production for export). This radar is manufactured in China as Type 331, so its Soviet designation may be MR-331.
Strut Curve (MR-302 Rubba)/Strut Pair (Rubba-M) S-band small ship air search radar, contemporary with the Head Net series, on board 'Petya' and 'Mirka' class frigates and 'Poti' and 'Grisha' class corvettes. It operates at the upper end of S-band and is unusual in lacking a balancing vane behind its reflector; that may mean that it is not stabilised. Strut Pair is two Strut Curves back to back for high data rate; it uses pulse compression.
Top Plate/Half Plate (MR-700 Fregat) Top Plate was the first Soviet planar-array frequency-scanning 3D radar; it supersedes Head Net C. Both back-to-back arrays are canted to the horizontal at opposite (*ie*, analogous) angles. In theory canting would correct for a small bearing error as the radar rotates while the beam scans vertically. Half Plate is a single Top Plate array; there are two versions (A and B).
Top Sail (MR-600 Voskhod)/Top Pair (Flag) Top Sail, introduced in the *Moskva* class, was the first Soviet frequency-scanning radar. Top Pair is a Top Sail back-to-back with Big Net; presumably both share the same waveguide and thus operate at the same frequency.
Top Steer (Fregat) This consists of an S-band frequency-scanned radar (similar in form to the US SPS-39, but with a lattice reflector) back-to-back with Strut Curve; the two share a common waveguide. This radar preceded Top Pair into service, the *Kiev* carrying Top Sail (rather than Top Pair) and Top Steer. There is also a Top Steer/Top Plate in which the 3D antenna is replaced by the Top Plate planar array.
Top Trough L-band target indicator radar analogous to the British Type 983, probably using the same transmitter as Big Net. It was installed on board some *Sverdlovs* and some destroyers.

GUN AND TORPEDO FIRE CONTROL RADARS

These radars went through several generations. The early postwar surface search/target indicator radars used feeds which scanned rapidly across their reflectors to produce a detailed picture of an area around the target (the US Navy used a similar technique for splash-spotting in its wartime Mk 13 main battery radar). In the next generation, Hawk Screech/Owl Screech, a rapidly-spinning feed scanned across the reflector, and the operator had a special console on which he could designate a target. He could indicate when the target reappeared, to measure its movement in range and bearing. The console transmitted these rates to a fire control computer. This was track-while-scan (TWS) operation. It provided no means of automatic tracking; the operator periodically checked target position against projected (rate-aided) position. The external mark of TWS operation is the cylindrical radar feed, incorporating a spinner switching among beams.

The next step was to add auto-tracking, in the form of a tilting plate which could be used to track the designated target (Drum Tilt, Bass Tilt). The next generation, Kite Screech, presumably has the scanning and tracking mechanisms moved into the radar enclosure.

Missile guidance radars used much the same TWS technology introduced in Hawk Screech, although some form of monopulse has been reported for Pop Group. It may be a modified version of scanning called Scan With Compensation (SWC), which is mathematically almost equivalent to monopulse.
Ball End/Ball Gun Same radar transmitter as High and Low Sieve.
Bass Tilt (MR-123 Vympel) X-band fire control radar, for 76.2mm guns on recent small combatants; it appeared in 1972. It is similar to, but simpler than, Drum Tilt. MR-123/76 controls only a 76mm gun (AK-176). MR-123-02/76 controls both the 76mm gun and 30mm gun(s) (AK-630s).
Cross Sword Control radar/director for the SA-N-9 missile system, combining back-to-back search/acquisition radars on top with a phased array target/missile tracker.
Drum Tilt (MR-104, Rys) X-band fire control radar controlling 30mm guns. It appeared about 1961. The housing is tipped up at a fixed angle of 25 degrees.
Egg Cup (SHTAT-5) Range-only S-band radar for 6in and 3.9in guns, developed 1946-48. Detection range 12nm (destroyer).
Eye Bowl S-band tracker (secondary radar) for the SS-N-14 ASW missile, which is command-guided. Ships equipped with Head Lights use that system instead.
Fan Song E (Volkhov) SA-2 fire control radar on *Dzerzhinski.*
Fire Iron 550-610 MHz (700-900 pulses/sec, pulse width 1 – 2.5 microsec). Effective range 22-23nm. Presumably this was a developed Swordfish.
Front Dome (Orekh) Tracker-illuminator for the SA-N-7 missile system. It resembles Bass Tilt but is larger.
Front Door/Front Piece Secondary radar to track SS-N-3 missiles, receive their radar video, and transmit commands. On board 'Echo II' and 'Juliett' class submarines and *Slava* class cruisers. The *Kiev* and *Kirov* classes have an alternative antenna, Front Door C, originally called Trap Door. The submarine antenna is now called Front Door A. The SS-N-12 version of the tracker is Argon. It is also used for the Vulcan (P-1000) missile on the two converted 'Echo IIs'.
Half Bow (Zalp) X-band radar (65-70kW, beam width 0.8 degrees) developed 1948-50 specifically to give accurate bearings for blind firing against surface targets; it had a stabilised antenna. It could coordinate with Post Lamp (Zarya). Tested September–November 1950 on the destroyer *Bestrashnyy*, and then in 1951 on the cruiser *Sverdlov*.
Hawk Screech (Yakor or FUT-B)) Radar director for close-range AA fire; used with 57mm quad mount. Maximum range is probably about 6500yds (18,000ft altitude). This was the first Soviet auto-tracking system; it was stabilised and could be used for blind-fire. X-band (1245-1270 pulses/sec, 0.6 degree pulses). Hawk Screech was probably conceived in 1949; it appeared in 1954. This radar introduced the new TWS (rate-aided tracking) technique.
Head Lights (Grom) Control radars for the SA-N-3 system. The two large (4m) dishes track the target, the two small dishes (1.8m) the target. A fifth dish transmits guidance commands. Two dishes are needed for tracking because one is scanned horizontally and one vertically. There are three versions: Head Lights A, the original; Head Lights B, which adds SS-N-14 control capability; and Head Lights C, for the *Kiev* class, with greater power for longer effective range.
Kite Screech (MR-114 Lev, MR-145 Drakon, and MR-184) Kite Screech A (mid 1970s: MR-114 and -145) is a C/X/Ka-band radar controlling 100mm guns. Kite Screech B (MR-184, mid-1980s) is an X/Ka-band radar controlling 130mm guns. In each case the Ka-band radar is a precision tracker; the target is acquired by the C/X or X-band TWS radar. MR-114 is on board 'Krivak II' class frigates and *Kirov* class cruisers. MR-145 is the successor system on board, *eg*, *Udaloys*. MR-114 may also be called Yakhond.
Long Bow X-band (1100, 1650, 3000-3300 pulses/sec; pulse widths 0.25 microsec at high PRF, 0.75 microsec at low PRFs). Replacement for Post Lamp; first observed 1957.
Muff Cob (MR-103, Bars) X-band fire control for twin 57mm guns. Unlike the other two nacelle fire control radars (Bass and Drum Tilt), it can elevate. It first appeared about 1962.
Owl Screech (MR-105 Turel) An improved Hawk Screech (1961) for 76.2mm dual-purpose guns. Peak power is 250kW.
Peel Group (Yatagan) SA-N-1 missile fire control radar director, using orange-peel shaped sections over which horizontal and vertical beams scan (a large pair for target tracking, a smaller pair for missile tracking, and a command dish). The target is tracked in S-band, the missile in X-band.
Pop Group (Osa-2M, MPZ-301) SA-N-4 missile fire control radar director, topped by an acquisition radar (2-D). The face of the director carries a large antenna for missile tracking and a smaller one for target tracking; both are probably inverse-Cassegrain antennas covered by flat plates, with monopulse or SWC clusters behind them. The search antenna operates at 6-8 GHz, the target tracker at 14.2-14.8 GHz. Note that the same designation applies to both missile system and fire control.
Post Lamp (Zarya) X-band (488 pulses/sec), horizontal beam 0.8 degree. Destroyer and cruiser torpedo fire control radar developed 1949-50, later modified as a gun-control set but then relegated again to torpedo control. The feed rotates to scan the antenna for what amounts to TWS operation. Range: 20nm on a large ship. This radar was modified to control 3.9in – 6in coastal artillery.
Scoop Pair (Binom) SS-N-3 fire control system, presumably a secondary radar. It tracks the outgoing missile and probably receives its radar video. Presumably it is equivalent to Front Door/Front Piece. Each consists of stabilised upper and lower antennae, each of which has dual horns and a pair of superimposed antennae (presumably with different polarisations).
Sun Visor (Yakor-2M) X-band fire control radar (two versions, A and B); pulse rates 2500, 3500, 1600-4500pulses/sec; pulse width 0.2 microsec; scans at 40 cycles/sec. Horizontal beam 0.8 degree wide. Effective range 22-23nm on surface targets, 40nm on air target. Sun Visor appeared about 1953.
Sword Fish 500-520 MHz (700 pulses/sec, pulse width 1 microsec). Considered almost obsolete in 1959. Effective range 22-23nm. This radar was derived from the wartime British Type 285.
Top Bow X-band target indication and surface fire control radar for *Sverdlov* and *Skoryy* classes, first seen 1952; probably a modified Half Bow.

Top Dome SA-N-6 phased-array missile guidance radar. The same antenna, then, presumably time-shares target tracking, missile tracking, and missile commands.

ECM/ESM

Bell Clout/Bell Slam/Bell Tap (Zaliv) is the ESM system associated with the Side Globes jammer (different radomes house DF antennas working at different frequencies, probably, respectively, S/C,C/X and K). Bell Tap was a targeting system in 'Osa' class missile boats. Zaliv first appeared in the contemporary 'Kynda' class (without the Side Globes jammer). Later missile cruisers have the full Zaliv/ Side Globes combination. Bell Nest is a special higher-frequency antenna carried by 'Nanuchkas', which also have Bell Tap. The separate elements of Zaliv were designated MRP-11 through -14. For example, mod 'Kyndas' had MRP-11 and -14 (they retained their Top Hat jammers).

Bell Thump/Bell Bash is the ESM system associated with the Wine Flask directional jammer in large cruisers. This combination replaces Zaliv/Side Globes. In *Kuznetzov, Admiral Nakhimov, Petr Velikiy, Novorossiysk, Minsk.*

Bell Nip (Kursk) is a new generation Ka/Ku-band short-range self-protection jammer using a pair of cones atop paired arrays of cavity-backed spirals. Bell Push (*Kuznetzov*) is a successor.

Rum Tub (MR-404) ESM system for targeting long-range anti-ship missiles such as SS-N-19, using intermittent radar ducting. The successor is Wine Glass.

Bell Shroud/Bell Squat (Start) was a destroyer/frigate combination designed to counter the new Ku-band aircraft radars of the 1960s, initially on board mod 'Kashins' and then on board 'Krivaks' 'Kashins' and *Udaloy*s had Start-1; *Sovermennys* have Start-2.

Side Globe (MR-401 Gorzuf) is the jammer on major surface combatants of the 1970s and early 1980s. Wine Flask is its successor. The MR-401 designation applies to the *Slava* class. Earlier versions probably had different numbers, *eg* MR-150 for the *Moskva* class.

Top Hat (Gafel)was the first Soviet jammer; it is triggered by Watch Dog. In a 'Krupny', it was designated Krab-11 and -12 (there are both deception and noise jammers). The jammer on a *Sverdlov* class cruiser was *Machta* class.

Watch Dog (Bizan) was the first Soviet ESM set, associated with the Top Hat jammer. Stop Light is the submarine version. Reportedly the Soviets considered Watch Dog/Top Hat equivalent to the US WLR-1/3/ ULQ-6 combination. In a 'Krupny', Watch Dog was designated Bizan'-4A.

Wine Flask was the jammer to replace Side Globes. Half Hat is the small-combatant equivalent. NATO earlier designated this system Modified Football. This long-range electronically-steered system first appeared on board the cruiser *Frunze.*

Wine Glass (formerly Football) is the replacement for Rum Tub, on board *Kirovs, Sovremennyys,* and *Neustrashimyys,* to target long-range missiles (SS-N-19 and -22).

DATA LINKS

Band Stand: missile tracker supporting mid-course guidance for SS-N-9, SS-N-22, and possibly non-export SS-N-25. Probably a beacon tracker. The Band Stand/Light Bulb combination to target Moskit (SS-N-22) is Monolit. The radar in Band Stand is Mustang.

Bell Crown: data link between flagship and slave command systems. The slaves transmit sensor data; the flagship sends back commands to be executed through the slaves' 'Second Captain' command systems.

Fish Bowl: missile guidance data link on board some 'Nanuchka' and the sole 'Sarancha'. Two mounted, one each side. The Band Stand/ Fish Bowl combination to target SS-N-9 in 'Nanuchkas' is Titanit.

Light Bulb missile guidance data link for SS-N-9, -22, and possibly non-export SS-N-25. It may also function as a general-purpose data link.

Plinth Net: missile video receiver for SS-N-3.

Punch Bowl (Kasatka): satellite-to-ship data link to support SS-N-12 and SS-N-19. The space reconnaissance system was originally designated US. Procurement was formally authorised on 21 July 1967; satellites were launched using R-36 (SS-9) missiles.

SONARS

Modern sonars are designated by code-name and by an MG-number; earlier ones were numbered in a GAS (Russian initials for sonar) series. NATO also assigns code-names. Sonar 'komplexes' have MGK-numbers. Since NATO identifications must be mainly by reception of sonar signals, the same NATO name can cover several different sets operating at similar frequencies and producing similar signals.

The first generation of postwar submarines was equipped with an HF active searchlight set, Tamir-SLS (NATO code-name Pike Jaw), probably copied from British sets supplied in wartime. First generation postwar submarines ('Whiskey', 'Zulu', and 'Quebec') carried Tamir-5LS in the forefoot. They also had Feniks, probably a Russian adaptation of the wartime German 'Balkon' (Balcony), under the hull, near the breeches of the torpedo tubes. Feniks was broadly equivalent to the contemporary US BQR-2, presumably with less effective amplifiers. Some boats had Mars, a passive LF set similar to that used by the Germans during the Second World War, using flat semi-circular arrays on each side of the bow, applied to the side of the hull. LF array was used to bring the boat into contact with the target. The combination of Tamir and Mars equipped the prototype 'November' class (Project 627) nuclear submarine, *K-3.*

The next generation submarine system substituted a large bow array, Feniks-M (MG-10, NATO Trout Cheek) based on the German late-war GHG for the earlier Feniks or Mars. Feniks-M was tested on board two 'Whiskeys' in 1958, and was deployed on board 'Foxtrots', 'Romeos', and production 'Novembers' (Project 627A; the 627 prototype had the earlier combination). A recently-published inboard profile of a 'Foxtrot' class submarine shows a cylindrical transducer above the MG-10 array, inside the same bow dome. It is probably a 'Hercules' scanning sonar (MG-15). 'Romeos' had no such set.

The accompanying active pinger in the forefoot, Arktika (MG-100, NATO Pike Jaw; Arktika-M was MG-200), was associated with a mechanical fire control system, Leningrad. Unlike contemporary Western systems, it turns torpedoes through two gyro angles in sequence. The torpedoes of a spread therefore emerge in parallel, and thus can hit despite errors in target range (the single gyro angle applied by Western systems causes the torpedoes of a spread to diverge, so much beyond the estimated range the target is unlikely to be hit). Russian writers indicate that Arktika and Leningrad finally provided their submarines with the option of firing torpedoes without exposing periscope; presumably Arktika was the first Russian sonar capable of tracking a target.

'Romeos' were also fitted with a sonar intercept set (probably MG-23) in a small dome above their bows. 'Foxtrots' had additional sonar 'windows' on their sails, presumably for medium- and lower-frequency intercept arrays used in conjunction with the bow array. These boats also all had an underwater communications set (MG-29, NATO Fez) in a narrow dome above the bow dome.

The sonar suite of a 'Foxtrot', 'Juliett', or 'Echo II' comprised Feniks-M (MG-10), Arktika-M (MG-200), Hercules (MG-15), the sonar intercept set (MG-23; also MG-13 in a 'Foxtrot') and the MG-29 underwater communications set. 'Golf' had a simpler suit: Feniks-M (MG-10), Arktika (MG-100) and Hercules (MG-15).

Large cylindrical active sonars were designed for the next generation of nuclear submarines, the 'Alfa', 'Papa', 'Victor', and 'Charlie' classes. Earlier submarines had separate passive and active search and tracking sonars; these units had single sonars capable of both search and track. Hence the use of MGK designations.

'Alfa' (Project 705) introduced a new 4m-diameter Kerch sonar, the MGK-100 'complex'. Its beam could be steered both vertically and horizontally. A recently-published inboard profile of a 'Delta' shows a transducer in the shape of a truncated cone, with the narrow part at the bottom; much the same shape appears in drawings of the new 'Kilo' (Project 877) class. MGK-100 became the standard second generation nuclear submarine sonar, appearing in such classes as 'Victor I', 'Charlie', 'Tango', 'Yankee' and 'Delta I/II/III'. Some modernised 'Echo II' class submarines may also have been fitted with this suite. The version in 'Delta III' is called Omega. These sonar suites probably also included a medium-frequency active sonar (MG-14 in the case of 'Charlie I' and probably others). The Soviets considered sonars like Kerch broadly equivalent to the US BQQ-2.

NATO code-named the full MGK-100 suite Shark Teeth. Late production 'Foxtrots' reportedly had only the passive element, code-named Shark Teeth E, retaining their earlier active pingers. However NATO applied a different code-name, Squid Arm, to the 'Alfa' sonar suite; it included a Mouse Roar HF obstacle-avoidance sonar, which was common to many later classes. Presumably 'Alfa' lacked the medium-frequency active sonar of the full Shark Teeth system.

'Papa' had a new 6m-diameter cylindrical sonar, Rubin. It appears not to have been used on board any contemporary submarine, presumably because it was far too massive. No MGK designator has appeared.

'Delta IV' (Project 667BRDM) submarines have an MGK-200 active/passive sonar. The full suite is designated Skat-2; NATO designates the full system Shark Gill. 'Delta IV' adds a separate sonar set, Skat-2. Skat itself was introduced in the 'Victor II' (Project 671RT) class. It may be related to the Barracuda spherical array reported on board 'Victor III' (Project 671RTM) class submarines. Skat is also reported on board 'Mike' (Project 685) and on board 'Oscar' and 'Typhoon' class submarines.

Barracuda was probably derived from the US spherical bow sonars, BQS-6 and BQR-7 (wrapped around the BQS-6 sphere), which were presumably compromised when the Soviets gained access to the associated SUBROC missile (which led to development of their SS-N-15). Its array is a sphere (of about 1000 elements, about two-thirds the size of the US sphere) operating at 0.5–5khz (3.5kHz active) wrapped in a 27-stave (20 double, 7 single) linear array operating passively at 3.5kHz, the standard US and British active frequency. These submarines also have a flank array (Akula), about a third as long as the submarine, probably inspired by the British Type 186/2007. It was reportedly the first such array on board a Soviet submarine. There is also an unnamed passive ranger, presumably equivalent to the French DUUX-2.

Finally, this class introduced a towed array, Pithon (NATO Shark Tail), 0.08m x 80m long, with 50 hydrophones, operating at 20–200Hz; it is streamed on a 756m cable (37.5mm diameter). In a 'Victor III', it has full DEMON processing. Probably the tail and the flank array share a processor, as in some US submarines.

The MGK-300 series may have been reserved for surface ships.

'Kilos' have MGK-400 (which NATO also designates Shark Gill). Export versions apparently lack the active component. The main transducer, like that of MGK-200, is a truncated cone, with the smaller end pointed down. Apparently there is no passive flank array nor, generally, is there a towed array. The associated mine detection sonar (possibly Mouse Roar) is MG-519; the sound velocity meter is MG-553, and the cavitation detector is MG-512. These auxiliary sets are probably also included in the Skat system described above.

'Akulas' (Project 971) and 'Sierras' (Project 945) have a new sonar suite, Skat (MGK-503). Reportedly 'Sierra' has her bow tubes splayed out to clear the bow sphere.

Reportedly the new Project 885 has a double bow array something like that of the new US *Seawolf* (separate transmitting and receiving elements), with her tubes moved back to clear it. The designation is probably MGK-500.

MG-29 (Khost) was an underwater communications and IFF set probably introduced in the 1960s, for surface ships and submarines. MG-26 was another such set.

The first postwar surface ship sonar was the Tamir-5N searchlight, on board large surface ships from the 'Kolas' to the *Sverdlovs*. Tamir-5L, for submarines, was introduced in 1947–50; Tamir-10, for small surface ships, appeared in 1950, Tamir-11 and .11M following in 1952. The next stage was to tilt the searchlight transducer (as in contemporary Western attack sonars associated with ahead-thrown weapons). Tamir-11 seems to have been MG-11, a searchlight hull sonar used on board Projeet 201M ('SO 1') class sub-chasers and minesweepers *(eg,* MG-11–265 in Project 265 sweepers).

The successor high-frequency searchlight was Pegas-2 (GAS-572, introduced 1955, and installed in the 'Tallinn' (Project 41) and 'Kotlin' (Project 56) classes and the reconditioned 'Skoriy' class, Project 31). Pegas-2M was installed on board 'Kotlins' modified for ASW (Project 56PLO).

The Soviets trailed the Western shift to scanning sonars by about a decade: their first such set, Herkules, appeared in 1957. It was installed in the new surface combatants then being designed; 'Kynda' (Project 58) and 'Kresta I' (Project 1134) as well as in some modernised 'Kotlins' (Project 56K AAW missile destroyers and Project 56M SSM missile destroyers equipped with the SS-N-1 weapon. In the later ships it replaced the Pegas searchlight.

'Kashin' (Project 61) and 'Petya' (Project 159) introduced a new combination of scanning search/target indication (Titan, MG-312: NATO Bull Nose) and attack sonars (Vychegda, MG-311, NATO Wolf Paw) to support the new RBU-6000; active range was 5km (about 5500 yds). Bull Nose also appeared on board the new BPKs ('Kresta' and 'Kara' classes). It was first tested in 1960, and probably operates at 7.5kHz with a 7500yds/6858m keying interval. This sonar may have been derived from the compromised British Type 184. Some ships, such as 'Krupnys' converted for ASW in the late 1960s, received Titan-2 (MG-332) sonars.

When 'Kashins' were rebuilt (as Project 61M), their sonars were replaced by a new Platina system (Platinum; MG-335; NATO Bull Horn). In these ships the sonar could search out to 130 degrees to either side of the bow. Claimed ranges were 4–6km in direct path and 25–30km in convergence zone. Platina was also installed on board modified 'Kotlins' (Project 56U, converted to carry P-15M Styx missiles).

'Krivaks' have a related sonar (also code-named Bull Nose by NATO), Piranha (MGK-335MS). The K in the designator suggests that Piranha includes both the bow sonar and the Mare Tail VDS. The bow array operates at 4.5, 5.0, 5.5kHz bandwidth (CW and FM pulses); pulse widths 30, 60, 120msec. Typically the same consoles control RBU-6000 and process data from its smaller sonar (which is in a separate dome about 60ft abaft the main dome). 'Krivak III' has a solid-state version. Estimated ranges: 3nm passive (6000yds), 5nm active (10,000yds). These figures correspond to SQS-23 performance. 'Grishas' are credited with an MGK-335MS sonar system, working with an Elk Tail VDS. The *Sovremennyy is* credited with an MG-335 sonar, almost certainly actually MGK-335.

MG-322, the main hull sonar in 'Konis' (MG-322T version) and in German (and probably Russian) 'Parchims', is probably a related type. 'Parchims' also have MG-329 (presumably a searchlight fire control set) and possibly an MG-16 mine-avoidance set. Presumably MG-322T makes up for the absence of the separate searchlight set by providing an elevating beam for attack, as earlier US SQS-4 series.

The first Soviet VDS appeared on board a 'Petya I' class trials ship in 1966; it became operational two years later on board *Moskva.*

The *Moskva* class introduced a much larger low-frequency sonar, Orion (NATO Moose Jaw) in a big retractable sonar dome. It was probably associated with the SUW-N-1 weapon, and was always credited with convergence-zone capability (instrumented range is 21.6nm/40km). It may have been based on the compromised British Type 177. The associated VDS, Vega (NATO Mare Tail), is probably also based on Type 177 (at a different frequency range). Pike (also code-named Mare Tail, hence presumably using a similar pinger) is the version on board 'Krivaks' (Pike-MK is on a 'Krivak III').

Horse Jaw (NATO name), the successor to the big Orion/Moose Jaw, is in the last two *Kirovs* and the last two *Kievs*. It is apparently an all-round scanner rather than a sector scanner, and is credited with instrumented ranges of 30/60/120km, *ie,* with multiple convergence zone range. The associated VDS is Horse Tail. Both (as well as the contemporary destroyer sonar, Horse Tail) are reportedly based on compromised French technology, which in turn is based on US commutator practice, very different from the British preformed-beam technique in Types 177 and 184.

Neustrashimy has a new sonar, Zvezda M-1, broadly comparable to SQS-53C, with digital reception *and* transmission. A drawing of the export version of an *Udaloy* class destroyer shows a sonar window in the forefoot and a side panel further aft, presumably analogous to the US SQQ-23 PADLOC (passive location by triangulation).

Like larger combatants, 'Mirkas' and 'Petyas' had the Titan/Vichegda sonar combination. The next size down, represented by a 'Poti', had Hercules, a higher-frequency scanning sonar. The successor combination replacing Titan/Vichegda in an early 'Grisha', is Shelon (hull) and a helicopter dipper (Argon). The smaller ASW units (*eg,* 'Pauks') have a next-generation hull sonar, Bronza (the sole 'Babochka' had Sheksna).

Small-combatant dipping sonars are Elk Tail for the 'Grisha' class and Ka-25 helicopters, and Rat Tail (14-l5kHz), used by Ka-27 helicopters. Unlike a ship with a conventional VDS, a ship using a dipper must stop, dip, and only then run. Such craft are best suited to 'sprint and drift' operation, and they are designed to retrieve their dippers very rapidly.

MG-69 and -79 are minehunting sonars in the 'Natya' and 'Sonya' classes. The current type is the 49kHz MG-89; range 1000 1200m on moored mine, 350–400m for bottom mine, localisation error up to 1.5 degrees and 1.5 per cent of range. MG-7 is a 300kHz anti-swimmer sonar lowered from a forward davit (*eg,* on board a 'Krivak III' or a 'Kashin'). It is used as a minehunting sonar on board inshore craft.

Most surface combatants apparently have underwater sound communication systems. An example is MG-35E, on board the export version of the Project 1265 minesweeper ('Sonya'). It is a non directional HF/LF set for voice and morse, which also measures ranges by a DME (transponder) technique. MG 35E operates alternately with the ship's sonar and echo sounder. It is roll- and pitch-stabilised. Presumably there is a submarine equivalent.

Along with sonars, the Soviets experimented extensively with non-acoustic submarine detectors, apparently with no great success. For example, the standard surface ship suite of the late 1960s included an MI-llOR infra-red wake detector (there was no NATO code name). There may also be a turbulence (wake) detector; many submarines have protrusions from their sails which appear to be non-acoustic wake detectors.

SOVIET UNION

PROJECT NUMBERS

1 *Leningrad* class large destroyers

2 *Uragan* class TB-SKR Series 1

3 *Fugas* class minesweeper Series 1

4 *Uragan* class TB-SKR Series 2 (for Pacific Fleet: *Metel* class)

6 *Dekabriest* class submarine Series 1, 2, 11, 13

7 *Gordiy* class destroyer; 7-U is *Storozhevoy* class; 7-UE is *Strashniy* class

9 *Stalinets* (S-class) submarine Series 9, 9bis

19 *Kirov* (Italian-built SKR)

20 *Tashkent* class destroyer leader

23 Prewar battleship *Sovetskii Soyuz* class/23NU revision (first postwar)

24 Postwar battleship cancelled 1950

25 Tactical A/C-carrying cruiser (TAKR) (flight-deck cruiser); but also used for 26,000t battleship

26 *Kirov* class cruiser; 26bis is *Maxim Gorkiy*

29 *Yastreb* class SKR

30 *Ognevoi* class destroyer; 30K was *Vnushitelniy*; 30bis was *Skoriy*; 30B, 30BA, and 30K were export *Skoriy* types; 30T was a targetdrone version of *Skoriy* (also described as tropicalised, however). There was also 30AV modernisation of the early 1970s. 30A was a version of 30 planned prewar

31 *Skoriy* modernised

32A Missile test bed converted from 7-U

33 Cruiser *Voroshilov* modified with OS-24 on board (test bed)

35 Destroyer by TsKB-32, sketch design April 1941. Later 'Mirka I' class sub-chaser; 35M is 'Mirka II'

37 Small destroyer by TsKB-17, sketch design May 1942, technical project November 1942. Later became SKR by TsKB-32, February 1944.

38 *Minsk* class destroyer leader (mod *Leningrad)*

39 *Uragan* class SKR Series 3 (*Molniya* class, for Pacific Fleet)

40 Fleet destroyer: sketch design June 1944, technical project November 1944; number also reported for conversion of Tsarist yacht *Marti* to minelayer

41 *Neustrashimyy* ('Tallinn' class, for Pacific Fleet)

42 *Kola* (postwar TB-SKR)

43 *Zhemchug* (mod 3 as SKR)

45 *Opitniy* (high speed experimental destroyer)

47 Destroyer leader (sketch design) by TsKB-32, October 1941

48 *Kiev* (improved *Tashkent* type destroyer leader, abortive)

50 'Riga' (postwar TB-SKR, successor to 'Kola')

52 'Purga' (armed icebreaker)

53 *Fugas* class coastal minesweeper Series 2; 53-U was Series 3; 53-ETU was Series 4

56 'Kotlin' class postwar destroyer; 56-PLO was ASW version; 56K was prototype of a SAM 'Kotlin'; 56A was series production of the SAM 'Kotlin'; 56AE was a SAM 'Kotlin' for Poland; 56E was the prototype 'Kildin' (armed with the SS-N-l missile); 56M was a 'Kildin' converted from a ship under construction; 56U was a 'Kildin' converted after completion; 56EM was an export version of 'Kotlin' for China (abortive)

57bis *Boikiy* ('Krupnyy') class missile destroyer (SS-N-l missiles); 57AM Kanin, also 57PLO; see also 1157

58 'Kynda'/58.1 Modified 'Kynda'

59 *Polukhin* class minesweeper

60 Possible abortive *Skoriy* modernisation project c 1960

61 'Kashin' class AA missile destroyer; also 61E; 61M and 61MP (mod 'Kashin'); 61ME for India

63 First Soviet project for a nuclear-powered cruiser

64 Battleship 'B' (unbuilt)

65 Abortive postwar light cruiser, designed in both 6in and 7.1in versions, dropped in favour of 68bis

66 Prewar heavy cruiser, revived postwar

67EP *Admiral Nakhimov* (cruiser converted as missile test ship)

68K *Chapaev* class cruiser; 68bis *Sverdlov* class cruiser; 68A modernised *Sverdlov*; 68U1 *Zhdanov*; 68U2 *Admiral Senyavin*

69 *Kronstadt* class prewar battle cruiser with 12in guns; 69.1 redesigned with German 15in guns

70 *Dzerzhinski*, to 70E (SAM conversion of cruiser)

71 Prewar carrier, number later used for 68bis mod

72 Postwar carrier cancelled 1955

73 *Vladimir Polukhin* class fleet minesweepers; modified version was 73KP

75 This number was used for a *Sverdlov* missile conversion, presumably not *Dzerzhinski.* It may have been an SSM ship

80 'Blokha' TKA M-400, 1941 experimental MTB; 35.3m, 2 × 45cm TT

82 Postwar heavy (12in) cruiser; 82R was a cruise missile version armed with SS-N-3

83 *Petropavlovsk* (ex-German *Lutzow,* handed over 1940) and cruiser by TsKB-17

85 Postwar light carrier

92 Icebreaker *Lenin*

95 Malodki (small submarine) Series III and VI; but it may apply only to the closed-cycle M-401 of Series XII

96 *M*-class submarine (Series XV)

96 bis Modified *M*-class for Caspian Sea by TsKB-18, July 1942; number later used for an *M*-class submarine with twin TT, December 1943; technical project June 1944, detail design 1944–45

97 Modified ('corrected') *S*-class submarine by TsKB-18, November 1942

97AP *Dobryna Nikitich* (armed icebreaker) /97A, E, D, AD, M all AGBs; *Ivan Susanin* is 97P. AP and P are armed, others are not; they are probably civilian units (the P designator indicates a KGB patrol ship)

98 Small submarine for closed sea operations, by KB-112 (Sormovo yard at Gorkiy); sketch design March 1941, technical project design October 1941

99 Minelaying submarine by TsKB-18, sketch design August 1941, technical project design April 1942

101 Cruiser submarine ('KU' class) by TsKB-18; sketch design August 1941, technical project design April 1942

106 MP-10/SMB-1 landing craft; 106K is 'Vydra' class landing craft

116 G-5 MTB; there is also a 116bis

122 'Artillerist' class sub-chaser; 122bis was 'Kronstadt' class

123 Postwar small MTB ('P-4'); also 123B and 123K with radar

125 'P-8'/'P-10' torpedo boats; 125P Ktylatyy is 'Pchela' class hydro foil coastal patrol craft

130 *Berezha* class degauss/deperm

133 Antares: Muravey (this number [133.1M] also applies to the German-built 'Parchim' class – in that case it is an East German number

141 'Kashtan' class

151 TR-40 river sweeper

152 Sub-chaser by TsKB-51 1942

157 MTB with closed superstructure by TsKB-50; technical project November 1941

158 Type L MTB by TsKB-50, sketch design April 1941

159 Sub-chaser by TsKB-50, sketch design December 1941. Later 'Petya'; 159M mod 'Petya I'; 159A 'Petya II'; 159AM mod 'Petya II'; 159AE 'Petya III'

161 MBKa; Second World War armoured boat (BK-501 class)

163 Large MTB/sub-chaser, technical design February 1942

165 DB motor launch (PAB-1220 coastal battery class)

183 'P-6' and 'P-8/10' class torpedo boats; 183R was 'Komar' class missile boat; 183TK had boost gas turbine; 183A had Arctilite covering; 183TS was target boat

186 MKP class Second World War armoured boat (BK-521 class): sea-going gunboat by KB-194; sketch design March 1944, technical project June 1944

188 MP-8 class landing craft

191 Postwar river gunboat

192 Small postwar armoured boat; also used for Stividor; MB-70 class small tug

194 BMO sub-chaser (MO-318 class)

199 MO-VI sub-chaser; erroneously applied to 'Zhuk'

200 TM-200 MTB (wartimc), also OD-200 sub-chascr; 'P-2' class and MO-211 class SC

201 '501-1' class sub-chaser (and 201M, 201P)

202 (A-202) 'Roslavl' class tugs

204 'Poti' class sub-chaser

205 'Osa' class missile boat; 205M is 'Osa II'; 205P 'Tarantul' is Stenka; 205MA and 205PE are Slepen; other 'Osa II' variants are 205MR and 205Ch. 205A is probably 205 rebuilt to 205M standard, 205K is the boat with P-15M missiles and new diesels. 205M has P-20 missiles. 205MR has P-20K missiles, 205Ch has P-20M missiles (and 400Hz AC power supply). 205.4 is 'Osa II' rearmed with 12 SS-N-25

206 'Shershen'; 206M Shtorm is 'Turya'; 206MR Vikhr' is 'Matka'; 206E is 'Mol'' (but see above claim). 'Shershen' is also listed as 206T. 206.5 is Almaz, a light FAC with 4 × SS-

N-25. 206E is Mol', the export torpedo boat; 206EP is the 'Bogomol'' export patrol boat. 206E was probably redesignated 206ET to distinguish it from 206EP. 206E and its relatives are sometimes called 205E *etc*, but the 206 designation seems far more likely

220 'Sekstan' class

229 'Khabarov' class

233 'Kolomna' class

253 T-222 class minesweeper (253-L is NATO 'T-301' class); 225 built 1946–56

254 'T-43' class ocean minesweepers; 254M is 'T-43' radar picket; there is also 254K

225 Coastal minesweeper ('T-301' successor)

257 'Vanya' class coastal minesweepers; 257D is 'Vanya I', 257DM is 'Vanya II'; 257DT is a minehunter variant. Total 72 built

264 'T-58' class ocean minesweepers; 264M is 'T-58' radar picket (SKR)

265 'Sasha' class coastal minesweepers

266 'Yurka' class ocean minesweepers (and 266A); 266M Akvamarin is 'Natya' class (266DM Akvamarin-2 is 'Natya II'; also given as 266ME)

268 Steel MTB TsKB-32, sketch design April 1942, technical project August 1942

286 *Nakhimovya* class

301M *Oskol* class repair ships

303 Small river gunboat by TsKB-51, pre-design study October 1954

304 *Amur I* class repair ships, 304M is *Amur II*

306 'T-4' class landing craft

310 Batur: 'Don' class tender/command

311 Monitor by TsKB-51

317 'Alesha' class minelayer

320 Ob' class hospital ship (may be B-320)

323 'Lama' class cruise missile transports (including 323B) (2001-M?)

361T K-8 class

364 'Pozharnyy-I' class fireboat

368 'Poluchat-I' class torpedo retriever and coastal patrol craft

376U *Yaroslav* class (post security boat); I-376 is *Yaroslavets* class, the *Ilyusha* class special minesweeper; also listed as *Nytyat-3* class diving tender, also 376U and 376G; same hull as PO-2 utility launch

393 'Bakan' class

394 'Krim' class stern-haul trawler; 394B 'Primo'rye' class AGI

419 'Sura' class

437 'Seyma' class, 437M is 'Khobi' class coastal tanker

450 'MP-4' class landing craft

498 Prometey: *Anton Mazhin* class (also 498T) harbour tugs

502M 'Mayak' class armed trawler; 502R is AGI version; 502U is training ship version. 502 is 'Kurs' class

503 Stern-haul trawler; 503M 'Al'pinist' class AGI

512 'MP-6' class; may be just the ammo transport version

513 'T-43' radar picket; 513M is environmental monitor version

514 Nevelskoi

522 Nyryat-1 diving tender (see also 378)

527 'Prut' class rescue ship; also 527M

530 'Nepa' class rescue ship; also given as 'Karpaty'

532A 'Valday' class ('T-58' as rescue ship)

535M 'Yelva' class ('Krab') diving tender

536 *Pionier Moskvyy* class salvage submersible tenders (listed as 05360, 05361). Modified standard timber/carrier container ships

537 'Elbrus' class rescue ship

550 'Amguzma' class

561 'Voda' class water tanker

563 'Kazbek' class (unrep)

567 *Leninski Komsomol* class (unrep)

572 SChS-150 minesweeping boat

577 'Uda' class (unrep)

596P 'Vytegrales' class timber carriers, mod as 1918

601 Modified S-class submarine by TsKB-18, October 1992; later 'Golf III' class modified missile submarine test bed; number earlier used for S-class submarines completed after Second World War (Series IXbis): S-21 through S-26, S-35

602 Modified M-class submarine with twin 40cm TT by TsKB-18; sketch design May 1942, technical design July 1942

605 'Golf IV'

607 Transport submarine by TsKB-18, 1944; detail design October 1945

608 Closed-sea submarine by TsKB-18; sketch project August 1942, technical project 1943. Number later used for medium-tonnage submarine by TsKB-18, December 1943. Sketch design March 1944, technical project October 1944. Cancelled in favour of new designs

609 *L*-class submarine by TsKB-18, technical project February 1943

610 'Keyla' class cargo ship; one converted to AGI

611 'Zulu' class submarine. Versions: 611M (streamlined, gun removed), 611A (Scud-'Zulu'), 611B (611RU (special training), A611 (weapons tests, probably torpedoes), AB611 (weapons test, probably torpedoes), AB611U (trial version with extended range), AB611K (experimental flotilla leader), AB611E (possible new engine test bed), AB611D (penetrating minelayer, to get through defensive fields to plant mines inside; failed), AB611RA (test for new anti-ship missile), AB611RE (test bed, no ID), AB611S (Spetznaz experimental landing sub), B611 (experimental oiler sub). 611E may be civilian oceanographic boat.

612 Postwar coastal submarine, not built (M successor)

613 'Whiskey' class submarine. Versions: 613E export, also 613 with a different E – engineering?; also 613A armament, 613AD, 613M modified version, 613S, 613RB, 613Kh (test bed with Kh-15, precursor to 'Single Cylinder'), EP613, B613 (probably experimental armament version). Prefix means test bed. 613S: salvage version. 613AD numerous minelayer conversions. 613V is torpedo test version. 613Ts is torpedo target version. 613B is extended range

614 Type XXI submarine (German design, not built)

615 'Quebec' class submarine with 2 TT/615A modified, 4 TT/A615 is Kreislauf test bed. There is also a 615A. A Russian article indicates that A615 was the prototype, and that M615 was the production version

616 Type XXVI (German Walter boat, not built)

617 Walter submarine ('Whale'); 617M improved version

618 Abortive closed-cycle (kreislauf) submarine project: medium submarine

619 'Golf V' class modified missile submarine

626 Possible diesel submarine to fire big nuclear torpedo; abortive (number speculative)

627 'November' class nuclear submarine; 627A was production variant; P627A was abortive missile variant

628 Salvage version of 'Golf' (abortive)

629 'Golf' class missile submarine. Variants: 629B was test ship for SS-N-5 missiles; 629M is 'Golf II'; 629KS (possibly 629KO) is the SSQ variant (three units). As of 1988, six E 629M in Baltic and seven in Pacific. There is also a 629R. 629M may also have been denoted 629A

630 Abortive diesel submarine *c*1960

631 Abortive diesel submarine *c*1960

633 'Romeo' class diesel submarine; 633A was a missile version, possibly predecessor to 'Whiskey Long Bin'. 633RV was the trials boat for the 65cm torpedo, with a big casing built on the bows

635 Walter submarine, dropped 1960

636 Modified 'Kilo' class diesel submarine (probably error for 1636). This number seems to have been used for an abortive diesel submarine about 1960

637 'Quebec' with safer powerplant, designed 1954–55

639 Abortive SSBN carrying a Yangel missile, either R-15 or R-21

640 'Whiskey Canvas Bag' (SSV; also 640V and 640U)

641 'Foxtrot' class diesel submarine; also 641K (different diesel); 641BUKI is 'Tango' class; 641A was a missile version. Was it parallel to P627A, to fire the big P-20 missile?

643 Abortive closed-cycle submarine project; planned production version of 617

644 'Whiskey Twin Cylinder'

645 'November' class submarine with liquid-metal powerplant

647 Walter submarine, probably production version of 617, dropped 1960

650 'Telbnovsk' class (unrep)

651 'Juliett' (SSG)

652 Abortive ballistic missile equivalent to 651

653 Abortive P-20 strategic submarine

656 Number sometimes incorrectly associated with 'November' class liquid metal variant (see 645)

658 'Hotel' class strategic submarine (SSBN); 658U is 'Hotel II'. Number also given as 658M
659 'Echo I' class missile submarine (SSGN); 659T is conversion to torpedo attack submarine; may be code-named Chaika
660 Abortive closed-cycle (kreislauf) submarine project
661 'Papa' class missile submarine
665 'Whiskey Long Bin' SSG conversion
666 Abortive cruise missile (SS-N-12) submarine parallel to 'Yankee'; number also used for modified 'Whiskey' salvage/rescue submarine
667 Strategic submarine (SSBN): 667A Navaga is 'Yankee', 667AU is modified 'Yankee' with SINS to fire SS-N-6 Mod 2 (R-27U), 667B Murena is 'Delta I', 667BD Murena-M is 'Delta II', 667BDR Kal'mar is 'Delta III', 667BDRM Del'fin is 'Delta IV'. 667M Navaga-M is 'Yankee' modified to fire SS-N-17 Missile; 667AR (also listed as 667AT and 667E) is Andromeda, 'Yankee' modified to fire SS-N-24 cruise missile ('Yankee Sidecar')
668 Abortive cruise missile (SS-N-12) submarine parallel to 'Yankee'.
669 Abortive large torpedo submarine to follow 'November'
675 'Echo II' class missile submarine (SSGN); version with SS-N-12 Bazalt missile is 675M
685 'Mike' class experimental attack submarine
688 Bolzva class (688 is Bolzva-3)
690 Kefal': 'Bravo' class target submarine
701 'Hotel III'
705 Lira: 'Alfa' class attack submarine; also 705K, 795ZhMT (possibly refitted with pressurised-water reactor)
706 'Neptun' class underwater reactor
712 'Sliva' class salvage tug
714 'Goryin' class tugs
728 'Kayra' class survey launches
730 'Tigyr' class
733 Goliat: 'Okhtenskiy' class (also 733S) tugs
734 'Dnepr' class submarine repair ship
740 *Yonyi Partisan* class
741 Number possibly associated with 'Papa' (661)
742 *Yunty Partisan* class naval cargo ship; 742B is ammunition ship
745 'Sorum' class tugs, also 'Sidehole' class small tugs
754 Nepa
770 'Polnocny-A' (Polish project number)
771 'Polnocny-B (Polish project number)
772 *Bryza* class training cutter (may be Polish project number)
774 'Polnocny-C' (may be Polish project number)
775 'Ropucha' (may be Polish project number); 775M may be 'Ropucha II'
802 'Tomba' class generator ships
821 Number possibly associated with Beluga (1710)
850 *Nikolai Zubov* class oceanographic ship; 850M is AGI version (may be Polish project number)
852 *Akademik Krylov* class
856 *Akademik Krylov* class
860 'Samar' class
861 'Moma' class AGI (also 861 M)
862.1 Yug class AGI (also 862, 862 11); 862 is Yug class oceanographic 862.2 ship
864 'Vishnaya' class AGI
865 Pirhanya: Losos-type midget submarine *BUT* this number was also applied to the *Sibiryakov* class oceanographic ship
870 'Kamenka' class
871 'Biya' class
872 'Finik' class (PNR B-91)(may be Polish number); 873 is 'Finik-2' and 874 is modified 'Finik'
873 'Siberyakov' class
877 Varshavsyanka: 'Kilo' class diesel submarine; 877M fitted for wire-guided torpedoes; 877KM has new command system; 877E is export version; 877EM is export with 2 wire-guided torpedoes; 877EKM has new command system and possibly longer hull
881 Abortive missile submarine, designed late 1980s
885 New attack submarine, 1994
887 *Smolnyi* class training ship
888 'Mod. Wodnik' class for training (may be Polish number)
902 Lun': 'Utka' type WIG
904 'Orlan' type troop-carrying WIG (Orlenok)
934 Original project for 'Kresta' as anti-ship cruiser armed with SS-N-12; 934K was an abortive command version, and 934A had shorter-range missiles (P-50); renumbered 1135 when switched to ASW role
935 Reported designation of the SS-N-P version of 'Krivak', renumbered 1135 in analogy to 934/1134
940 Lenok: 'India' class rescue submarine; see also 1847
941 Akula: 'Typhoon' class strategic submarine (SSBN)
943 Possible original project number for 'Mike' class; see 685 above
945 'Sierra' class attack submarine; 945B is 'Sierra II'
949 Antey: 'Oscar' class missile submarine (SSGN); 949A Antey II is 'Oscar II'. Granit (which is the project name for SS-N-19) was also associated with 949 954 Original project number for *Kirov* class before redesignation as 1144
956 *Sovremmenny* class; 956A *(Bespokoynyy* and later) have longer cruise missile tubes and probably also SA-17 anti-aircraft missile. Was this originally 965, with typo?
964 Reported designation of *Slava* class before redesignation as 1164
970 Sibiryakov
971 Barts: 'Akula' cruise missile submarine (for SS-N-21)
976 *Akhaziya* class oceanographic ship
1041 'Svetlyak'; export version 1041.2 without ASW equipment, and as 1041.1 with 8 SS-N-25 missiles
1123 *Moskva* class
1124 Albatros: 'Grisha I' (also 1124.1), 1124A is 'Grisha II'; 1124M/1124P are 'Grisha III'; 1124.4 and 1124K are trials ships (both designated 'Grisha IV'); 1124MU and 1124MEh are 'Grisha V'. Another source: 'Grisha III' is 1124M, and 'Grisha V' is 1124EM. Number originally assigned to Second World War river gunboat
1125 Second World War armoured river gunboat (with tank turret)
1127 Abortive missile destroyer armed with P-15 (Styx)
1128 *Sibir* class space events ship
1129 *Sibir* class space events ship
1130 *Desna* class space events ship
1134 'Kresta I' class; 1134A Berkut-A is 'Kretsa II', 1134B Berkut-B is 'Kara', 1134BF is a modified 'Kara' *(Azov)* for SA-N-6 trials
1135 Burevestnik: 'Krivak' class SKR; 1135.2 is 'Krivak IV'; 1135M Burevesknik-M is 'Krivak II'; 1135.5 (also reported as 1135P Nerey) is 'Krivak III' 1140 Svetlyak
1141 Sokol: 'Baboochka' class hydrofoil ASW boat; 1141.1 Sokol is 'Baboochka II' (Mukha)
1143 Carriers: 1143/1143.2 were *Kiev* class, 1143.4 was *Gorshkov,* 1143.5 was *Kuznezov* class (Orel), 1143.7 was *Ulyanovsk*
1144 Orlan: *Kirov* class
1145 MPK *(eg,* MPK 215, 220)(small ASW ships, Sokol-2)(?) 1149 Alternative designation for 'Matka' (206MP)
1154 Yastreb: *Neustrashimyy* class; 1154.1 is modified version (fourth and later ships)
1155 *Udaloy* class; 1155.1 has cruise missile vice SS-N-14; ll55R *Adm Zakharov)* was the first unit completed with an operational SA-N-9missilesystem. Maybe ll55A, ll55B, 1155.1 for KGB
1157 Possible redesignation of mod 'Kanin' class
1159 'Koni' class export corvette
1163 Export corvette; 'Grisha' successor
1164 Atlant: *Slava* class
1166.1 'Gepard' export frigate
1171 Tapir (also Nosirog): 'Alligator' (is this also 1170?)
1172 'Emba' class cable layers
1174 Yedinorog: *Ivan Rogov* class
1175 Biriusa (Emba-2) class cable layers
1176 'Akula': Ondatra landing craft
1190 Second World War armoured river gunboat (monitor) *Khasan* class
1204 'Shmel' class river gunboat
1205 Skat: surface effect craft 'Gus' class
1206 Surface effect craft: 1206T Kasafka is Pelikan MCM craft, 1206.1 Murena is Tsaplya, 1206 Kal'mar is Lebed'
1208 Videlsya: 'Yaz' class river gunboat
1209 'Utenok' class surface effect craft
1212 'Piyavka' class river craft
1232 Surface effect craft: 1232.1 Dzheryan is 'Aist', 1232.2 Zubr is 'Pomornik'
1234 Burya: 'Nanuchka'; 1234E is 'Nanuchka II'; 1234.1 is 'Nanuchka III'; 1234.2 is 'Nanuchka IV'
1239 Sivuch: 'Dergach' class surface effect boat

1240 Uragan: 'Sarancha' class missile hydrofoil
1241 Molniya: 'Tarantul' class; 1241RZ is 'Tarantul III'; there are also 1241.1 ('Tarantul I') and 1241.7 ('Tarantul II'); 1241RE is export version. 1241.2 Molniya-2 is 'Pauk'; 1241 PE is the export version. Export versions with new missiles: 12418 (16-SS-N-25) and 12421 (4-SS-N-22). 1241.M is 'Tarantul II'; 1241. IMP is 'Tarantul III'; 1241. IRE is 'Tarantul IV', new export version with Kortika
1244 Svetlyak (1244.1)(see also 1140)
1248 Moskit: 'Vosh' class river gunboat
1252 Almaz: 'Zhenya' (also given as 1263
1253 1253B is 'Vanya' modernised (see also 257)
1256 Topaz: 'Andryusha' class magnetic sweeper
1258 Korund: 'Yevgenya' class coastal minesweeper
1259 Malakhit: 'Olya' class MSB; 1259.2 Sapfir is Lida class mine sweeper
1260 'Gorya' class minehunter
1265 Yakhont: 'Sonya' class coastal minesweeper
1274 'Klaz'ma' class cable layer
1288 'Bambuk' class AGI (1288.4 is Kamchatka class)
1300 Nel'ma: Ilyusha class minesweeping drone
1380 'Baltika' class converted fishing boat/MCM
1388 'Shelon' class torpedo retriever; 1388M environmental monitor
1392 'Osa-I' class converted to target boat; 1392V is target control boat
1400 'Zhuk' class; 1400ME is a lengthened version; 1400.6 offered for export (199 erroneously applied to this boat)
1408.1 'Saygak' class river Patrol launch (also 14081)
1415 'Tanya': 'Flamingo' class diving tender
1431.0 'Mirage' export patrol boat, 1990
1452 'Ingul' salvage tug; also given as 'Pamir'
1453 'Ingul' class
1454 'Sorum' class armed tugs and AGI
1537.1 Polyusclass oceanographic ship
1539 'Alpinist' class
1541 'Luza' class special-liquids tanker
1545 'Baskunchak' class (unrep for Maritime Border Guards)
1549 'Manych' class water tankers
1552 'Sofia' class (unrep)
1559B *Boris Chilikin* class (unrep); 1559 is Desna class
1593 *Boris Butoma* (unrep)
1636 Possible original designation of 636 improved 'Kilo'
1710 Makrel': 'Beluga' class research submarine
1783 'Vala' class special liquids-tanker
1784 Late production T4 (see also 306)
1791 'Amga' class ballistic missile transports
1799 'Pelym' class degauss/deperm
1806 Midget subnarine; BUT also listed as Onega environmental monitor
1823 'Muna' class small munitions transport
1824 'Muna' class (also 1824B) coastal survey ships
1826 Aslya: 'Bal'zam' class AGI
1828 Lira: *Aleksandr Brykin* ballistic missile transport
1832 Midget submarine
1837 Midget submarine
1840 'Lima' class research submarine
1844 Kair: 'Toplivo' class fuel lighters (and 1844D
1847 Rescue submersible of 'India' class (940)
1850 'Andizhan' class cargo ship to ammunition ship
1851 'X-Ray' class research submarine (as-l l)
1859 *Berezhina* class underway replenishment
1861 Large air-cushion landing craft
1884 'Ugra' class training ship
1886.1 'Ugra' class command tenders
1889U Alternative reported number for Ugra training ship
1893 'Katun' class seagoing fre boat
1896 'Nyiryat' class inshore survey craft and diving tenders
1908 *Akademik Sergei Korolev* space events ship
1909 *Kosmonaut Yuri Gagann*
1910 'Uniform' class research submarine (AS-15)
1914 *Marshal Nedelin* class space events ship
1918 'Vitegrales' class command/logistics tenders
1929 *Kosmonaut Pabel Beliyaev* class
1930 *Akademik Nikolai Plyugin* space events ship
1941 Titan: *Ural* (Kapusta) nuclear-powered AGI
1968 'Belyanka' special-liquids tankers (also given as 19680)
1993 'Katun' class seagoing fre boat (see also 1893)
2020 'Malina' class submarine tenders

NOTES TO TABLES

Designations have been *transliterated* from the Russian. Thus, the Cyrillic C is rendered S and the Cyrillic P is rendered R (the Cyrillic symbol for P is rendered P).

Launch dates are *christening* dates, which sometimes did not coincide with actual launchings. Completion dates are delivery dates, after builders' trials. When only one date is given, unless otherwise indicated, it is a delivery date. I think the strike dates listed are dates of withdrawal from service; for example, *Kiev* was officially stricken on 31.8.94.

FLEET STRENGTH 1947

The data which follows should be taken with considerable caution. It was common in Western accounts of the Soviet Fleet to assume that all Soviet warships were maintained constantly in active status, so that the fleet list equated to the active fleet. That was unrealistic; the Soviets maintained a considerable reserve fleet of uncertain availability, apparently with skeleton crews. Thus it is difficult to determine which of the ships which survived the Second World War were active postwar or, for that matter, when they were discarded. Soviet ships rarely operated outside home waters through the late 1950s, limiting Western observation of fleet status. The list which follows is not exhaustive, but it includes all ships which reportedly survived the war, plus ships transferred under Lend-Lease and under the Peace Treaties.

Major Lend-Lease units were returned in 1949 in exchange for Italian ships, under the Italian peace treaty. Lend-Lease small craft were scrapped or destroyed in 1955–56 as part of agreements winding up the Soviet Lend-Lease debt; note purchases made at this time.

Note the large-scale scrappings of elderly warships in 1958–60 as part of Krushchev's programme of reducing the size of the Soviet conventional forces. Presumably many of these ships had not been active for some years.

Note that only a few scrapping dates are available; the data on Soviet ship disposals, at least in the unclassified literature, is extremely sparse. However, the dates which are given below should be considered indicative of the periods during which classes were discarded.

BATTLESHIPS

Name	Launched	Disp (std)	Fate
Gangut class			
OKTYABRSKAYA REVOLUTSIYA	7.10.11	23,000t	BU 1956–59
SEVASTOPOL	29.6.11	23,000t	BU 1957
Ex-British Royal Sovereign class			
ARKHANGELSK (ex-*Royal Sovereign*)	29.4.15	29,150t	To RN 4.2.49, BU
Ex-Italian Cavour class			
NOVOROSSIYSK (ex-*Giulio Cesare*)	15.10.11	24,300t	Sunk 29.10.55

Oktyabrskaya Revolutsiya operational 1950, at Kronstadt from at least Nov 1954; superstructure and turrets being removed Jul 1956. *Sevastopol* BU at her name port. Also reported in existence but inactive 1958. Reportedly the loss of the last ship, a war prize replacing *Royal Sovereign* afte being handed over on 15 December 1948, on a German ground mine while at anchor in the Black Sea, was the occasion for the removal of Admiral Kuznetzov as Soviet C-in-C; he was replaced by Admiral Gorshkov.

COAST DEFENCE SHIP

Name	Launched	Disp (std)	Fate
Ex-Finnish Ilmarinen class			
VYBORG (ex-*Väinämöinen*)	28.12.30	3900t	Stricken *c*1958, BU by 1965

Flagship of the Second World War Finnish Navy ceded as part of the 1947 war reparations. At Kronstadt summer 1956, miscellaneous auxiliary (pennant no 98) 1960.

CRUISERS

Name	Launched	Disp (std)	Fate
Admiralty type			
KRASNY KRIM	21.11.15	6833t	BU 1960
Modified Admiralty type			
KRASNY KAVKAZ	21.6.16	7650t	Target 1956
Omaha class			
MURMANSK (ex-*Milwaukee*)	24.3.21	7050t	To USN 8.3.49
Kirov class (Project 26)			
KIROV	30.11.36	7880t	Stricken 1975
VOROSHILOV	28.6.35	7970t	BU 1960s

Name	Launched	Disp (std)	Fate
Maxim Gorky class (Project 26B)			
MAXIM GORKY	30.4.38	8177t	Hulk 1960
MOLOTOV	23.2.39	8177t	Stricken 1970s
KALININ	Apr 1943	8177t	BU 1960s
KAGANOVITCH	Oct 1943	8177t	BU 1960s
Chapaev (Project 68)			
KUIBISHEV	31.1.41	11,300t	BU 1965
FRUNZE	31.12.40	11,300t	BU 1960–61
CHAPAEV	1940	11,300t	BU 1964
ZHELEZNIAKOV	1940	11,300t	Stricken 1976
CHKALOV	1948	11,300t	BU 1981
Ex-German Nürnberg class			
ADMIRAL MAKAROV	6.12.34	6520t	BU 1959
Italian Duca d'Aosta class			
STALINGRAD	22.4.34	8317t	Auxiliary 1960s

Admiralty type
TS from 7 March 1945, decommissioned 1958.

Modified Admiralty type
TS from 12 May 1947, decommissioned. Expended as SS-N-1 missile trials target.

Kirov class
Name ship overhauled 1950, 1953. Reported at Leningrad 'under overhaul' (ONI) Nov 1954–Feb 1957, then non-operational. Laid up at Krylov Ship-Leningrad 1959–60; hulk by early 1960. *Voroshilov* was converted to serve as trials ship for the *Moskva* class weapon system, with SA-N-3 forward and a helicopter hangar aft.

Maxim Gorky class
Name ship BU at Leningrad. *Molotov* renamed *Slava* (1958) and TS from 1960s. *Kagenovitch* renamed *Petropavlovsk* (1957).

Chapaev class
Seventeen ships were planned for the 1938–42 Five Year Plan (five to be laid down 1939, one in 1940, six in 1941, and five in 1942). The original TTZ called for a modified *Maxim Gorkiy* with the new MK-5 triple 152mm/57 turret in place of the earlier triple 180mm, plus a new enclosed 100mm/56 secondary mount (B-54: total eight guns) in place of the earlier open B-34 mount. Twelve twin 37mm mounts (66-K type) would replace the six twin open mounts of the earlier type. Side armour was doubled to 100mm; deck armour remained at 50mm. The design was to incorporate a new gun fire control system. Standard displacement would be limited to 8300t. Trial speed would be at least 35kts. After the sketch design was reviewed, a fourth triple 152mm turret was added, at a cost in endurance and protection. The technical design approved on 13 July 1939, called for a standard displacement of 10,620t and a trial speed of 33.5kts (using the same machinery as the *Kirovs*). Project 68I (imported) was an abortive 1939 version with German 150mm guns (abandoned when the Russians decided their own gun was better); 86S was a version with Skoda guns (stopped 1940 when the Germans stopped work).

Ships were completed postwar to a 'corrected' (68-K) design with radar and better AA guns (37mm water-cooled rather than air-cooled) as part of the 1946–55 Ten Year Plan. Aircraft and torpedo tubes were eliminated. The standard displacement rose to 11,450t (normal 12,187t, full load 14,138t); speed fell to 32.6kts.

Only the 1939 ships were completed: *Chapaev* (16 May 1950), *Chkalov* (5 November 1950), *Zhelezniakov* (19 April 1950), *Frunze* (15 December 1950), *Kuibyshev* (22 December 1959). The first three were built at Leningrad, the others at Nikolayev. *Frunze* was the first launched (30 December 1940). *Chkalov* was the only one launched postwar (25 October 1947). Ships in the water at Nikolayev were towed away before the Germans seized the yard. Two more were laid down but not launched: *Ordzhonikidze* (laid down 31 December 1940) and *Sverdlov* (31 January 1941). They were broken up incomplete. Ordered but not laid down were: *Lenin*, *Avrora*, *Parhomenko*, *Kotovski*, *Shchors* and *Lazo* plus three more planned. The last two named ships were to have been built at Komsomolsk for the Pacific Fleet.

Name ship hulked 1960, BU 1964. *Chkalov* renamed *Komsomolets* (1960). Served as TS after transfer to Northern Fleet with *Zhelezniakov* in 1950. They were BU 1981 and 1976, respectively. *Frunze* was BU 1961, and *Kuibyshev* 1966.

Ex-German Nürnberg class
Handed over to USSR Jan 1946 as war prize. TS 1950, based at Kronstadt to 1956.

Ex-Italian Duca d'Aosta class
Handed over as war reparation 2.3.1949. Later renamed *Kerch*. Operational through late 1950s; auxiliary status early 1960s.

DESTROYERS

Name	Fate	Name	Fate
Leningrad class: (Project 1): launched 1933–38, 2150t std			
LENINGRAD	BU 1960s	TBILISI	BU 1960s
MINSK	BU 1960s	BAKU	BU 1967–68
Novik type: launched 1911–16, 1200–1440t			
KARL LIEBKNECHT	BU 1950s	URITZKY	BU 1953
KUIBISHEV	BU 1957	NEZAMOZHNIK	TS 1956
VOIKOV	BU 1950s	ZHELEZNIAKOV	BU 1956
STALIN	BU *c*1953		
Project 7 class: launched 1936–41, 1855t std			
BODRY	Stricken ? 1950s	RESHITELNY	To China 1955
BOIKY	BU 1958	RETIVY	To China 1955
GREMYASHCHIY	Stricken 1958	GROSYASHCHIY	Stricken early 1960s
GROMKY	Stricken ? 1960s	REZKY	To China 1955
GROZNY	Stricken early 1950s	REZVTY	Stricken early 1960s
STEREGUSHCHY	Stricken ? 1950s	REVNOSTNY	Stricken 1950s
RYANY	Auxiliary 1960	RAZARYONNY	A-bomb target 15.11.58
RAZTOROPNY	Target 1959	RAZUMNY	Foundered *c*1960
RAZYASHCHY	To China 1955	REKORDNY	Stricken ? 1950s
Project 7-U class: launched 1938–41, 2192t std			
STOROZHEVOI	BU 1959	SVIREPY	Stricken ? 1950s
SILNY	Stricken ? 1960s	STROGY	Stricken ? 1950s
VITSE ADMIRAL DROZD	BU 1959	STROINY	BU 1959
STRASHNY	Stricken ? 1950s	SOOBRAZITELNY	Museum ship ship 1966
SLAVNY	Stricken 1960s		
Experimental Type (Project 45): launched 1935, 1570t			
OPITNY	Decommissioned 1950, BU *c*1950		
Ognevoy (Project 30) class: launched 1940–47, 1800t			
OGNEVOY	Stricken? 1960s	ODARENYY	Stricken ? 1960s
OZARENYY	To Bulgaria 1956	STALIN	Stricken ? 1960s
		VNUSHITYELNYY	Target 1959
OSMOTRITELNYY	Stricken ? 1960s	VLASTNYY	Target 1959
OTLICHNYY	Stricken ? 1960s	VYNOSLIVYY	Stricken ? 1960s
OBRAZTSOVYY	Stricken ? 1960s	OTVAZHNYY	Stricken ? 1960s
Ex-British/US 'flush deckers': launched 1917–19, 1090t			
ZHARKY	To RN 28.2.49	DIERZKY	To RN 23.6.49
ZHOSTKY	To RN 8.9.52	DOSTOINY	To RN 28.2.49
ZHIGUCHI	To RN 30.1.50	DRUZHNY	To RN 23.8.52
ZHYIVUCHI	To RN 23.6.49	DOBLESTNY	To RN 4.2.49
Ex-Romanian Regele Ferdinand class: launched 1928–29, 1400t std			
LYETUCHI	To Romania 1953	LIKHOI	To Romania 1953
Ex-German destroyers: launched 1935–41			
PROVORNY (ex-*Z 33*)	BU 1960–62	PYLKIY (ex-*Z 15*)	BU 1960–61
PROCHNY (ex-*Z 20*)	BU 1950s	POSPESHNYI (ex-*Z 14*)	Deleted by 1953, BU
Ex-Japanese destroyers			
POSPESHNYY (ex-*Harutzuki*)	Target 1965	VOLNYY (ex-*Shin*)	?
VERNYY (ex-*Hibiki*)	?	VOZROZHDENYY (ex-*Kiri*)	?
VNIMATELNNY (ex-*Kin*)	?	VYRAZITELNYY (ex-*Hatsuzakara*)	?
VOLEVOI (ex-*Kaya*)	?	(ex-*Ushio*)	
Ex-German torpedo-boats			
PRIMERNYY (ex-*T 39*)	BU 1960	PORYVISTYY (ex-*T 17*)	BU 1957
PODVIZHNYY (ex-*T 12*)	BU 1957		
Ex-Italian 'Soldati' class: launched 1937–38, 1690–1820t std			
Z 12 (ex-*Artigliere*)	Stricken 1958	Z 20 (ex-*Fuciliere*)	Stricken 1958
Ex-Italian Ciclone class torpedo-boats; launched 1942, 910 std			
Ex-*Animosa*	Stricken ? late 1950s	Ex-*Ardimentoso*	Stricken late 1950s

Leningrad class
Minsk at Kronstadt 1955–57, auxiliary (pennant TSL-75) 1958–59, used as target-tower from 1959 and transferred from Baltic to Northern Fleet. *Tbilisi* (non-operational at Leningrad Nov 1954, auxiliary Apr 1956, TS Aug 1957) also used as target-tower from 1959 (TSL-50). *Baku* auxiliary early 1959, still on list 1965. Project 1 embraced only *Leningrad* and *Kharkov*. Project 38 was *Minsk* and *Moskva*; Project 38 bis was *Tbilisi* and *Baku*. Project 48 was the follow-on class, reportedly based on the *Tashkent* design (*Kiev, Yerevan, Stalinabad, Petrozavodsk, Ashkabad, AlmaAta, Ochakov, Perekop, Archangelsk, Murmansk*). These last ships were not completed, due to the outbreak of war (only the first three were laid down, the first two, at Nikolayev, being launched, towed away from Nikolayev, and broken up postwar). Although she was built abroad, the fast leader *Tashkent* was assigned project number 20. Two repeat *Tashkents* (project 'I') were ordered in Leningrad, but they were not built.

Project 7 class
Four of these Second World War ships went to China in pairs Dec 1954 and July 1955. *Razumny* became an Arctic electronic survey ship from 1958–59 and was beached in a storm off Murmansk. In all, thirty-three were ordered and twenty-nine laid down; four (*Legkii, Lovkii, Burnyy, Boyevoi*) were never laid down. One (*Reshatelnyy*) was lost through grounding before completion; seven were delivered after the outbreak of war. One account states that *Pronzitelnyi* and *Porazhaloshchii* were laid down in 1936 and scrapped incomplete in 1938, these names being applied to Project 7-U ships and then cancelled.

Project 7-U class
Slavny became a TS *c*1958, as probably did *Soobrazitelny* until she became a Black Sea museum. In all, twenty were ordered and eighteen laid down (not laid down: *Pronzitelnyy, Porazhaloshchii*). One unit (*Sovershenny*) was destroyed on the slip, eight being delivered after the outbreak of war. *Storozhevoy*, torpedoed in June 1941, had her fore end replaced by the fore end of the *Ognevoy* class destroyer *Ognennyy* in Leningrad.

Ognevoy class
The class was developed to a 15 November 1937 TTZ, to replace the Project 7 class, which was condemned for the vulnerability of its boiler/turbine arrangement and for its limited range. The shipbuilding industry offered Project 7-U as a simple alternative. During development of the new TTZ, the navy decided to mount the main guns and the machine guns in enclosed gunhouses. The new Project 30 design was approved 27 October 1939 for thirty ships. Its powerplant, developed from those of the *Leningrad* class and Project 7-U, was still somewhat unsatisfactory in that it employed forced-draft water-tube boilers (war experience would show that near-misses could easily open up the closed boiler rooms and so drastically reduce power). In July 1943 changes were ordered: installation of six 37mm AA (70-K) guns, radar, new sonar, stronger scantlings, and depth charge throwers; 60t of solid ballast had to be added. A recent Russian article mentions a prewar Project 30A variant intended to use an imported US powerplant.

Only one ship was completed to the original Project 30 design: *Ognevoy* (completed May 1945). Two more were blown up on the ways in Nikolayev in 1941: *Otmenyy, Obuchennyy*. Another sixteen were not completed (*Ognennyy, Ozhestochennyy, Ostryy, Oslepitelnyy, Ostrorozhnyy, Otchetlivyy, Otzyvchivyy, Otradnyy, Ozhivlennyy, Otlichitelnyy, Organizovannyy, Otbornyy, Otrazhayushchoiy, Veduschchiy, Vnezapnyy, Opasnyy*). Finally, ten were ordered but not laid down: *Otchayannyy, Obshchitelnyy, Zharkiy, Zhivoy, Zhutkiy, Zhyostkiy, Zhguchiy, Zhestokiy, Zorkiy, Zvonkiy* (another source says that the first two ships were laid down). The ten ships completed postwar were to a modernised Project 30-K design: *Otlichnyy, Obraztsovyy, Otvazhnyy, Odarenyy, Ozareny, Osmotritelnyy, Okhotnyy* (renamed *Stalin* 1949), *Vnushitelnyy, Vynoslivyy, Vlastnyy*.

Ex-German destroyers
Provorny served in the Baltic after being handed over by RN in summer 1945. The other three were all 1946 war prizes, all considered second line by 1955 and in auxiliary role by 1959. *Pylkiy* was active to 1956 becoming an accommodation ship in 1958.

Ex-Japanese destroyers
Harutzuki (2701t) was renamed *Pospeshnyy*, then *Oskol*, then *Vnezapnyy*, and became the target vessel *TSL-64* in 1965. In 1959 she had a Knife Rest radar amidships, and a helicopter landing area aft, but no weapons. *Hibiki* (1980t) was renamed *Vernyy* and later *Dekabrist*. *Kaya* (1262t), *Kin* (600t) and *Shin* were not reactivated. *Vyrazitelnyy* was laid up unarmed and then became a torpedo retriever and the target vessel *TSL-24* in 1959. She was in service up to about 1966.

Ex-Italian 'Soldati' class
Handed over as war reparations 21 February 1949 and 17 January 1950 respectively. In 1955 ONI reported that these units were being employed in type training. They had not yet been refitted with Soviet weapons. Both converted to target ships early 1960.

Ex-Italian Ciclone class
War reparations handed over 1949 and considered second line by 1955.

SUBMARINES

Lend-Lease British 'U' class
V 2, V 3, V 4 all returned 10 Feb 1949 and BU 1950.

OPERATIONAL SUBMARINES OF SOVIET PREWAR DESIGN, 1945–64 ACCORDING TO US NAVAL INTELLIGENCE
(Note that this includes postwar construction)

Year	M-V	Shch	L	S	K classes
1947	90	76	28	33	14
1948	97	77	28	34	14
1949	106	79	28	34	14
1950	113	78	22	34	14
1951	133	74	22	34	14
1952	143	70	22	34	14
1953	147	67	22	34	14
1954	134	67	20	34	14
1955	106	65	20	31	14
1956	89	56	19	29	14
1957	83	46	16	29	14
1958	79	32	12	24	14
1959	76	20	10	22	14
1960	75	13	9	20	14
1961	48	8	5	16	12
1962	46	0	2	6	9
1963	41	0	0	0	0
1964	34	0	0	0	0

Note that ONI was still carrying four old D (*Dekabrist*) class and two old P class (*B 1* and *B 31*) submarines as second-line units in 1955, even though later lists did not include them.

Many submarines were renamed in Oct 1949 in a single uniform B series; previously they had been named by class. Thus *P 3* became *B 1*; *D 2* became *B 2* (but *D 5* became *B 32*); *L 3* became *B 3* (but *L 2* became *B 31*, *L 4* became *B 34*, *L 5* became *B 35*, *L 7* became *B 23*, *L 8* became *B 24*, *L 9* became *B 19*, *L 10* to *18* became *B 10* through to *18*, *L 20* to *22* became *B 20* to *22*). The surviving K boats were also renamed: *K 21* became *B 4*, *K 51* to *56* became *B 6* to *9*. It is not clear whether *Shch*-class submarines were renamed. The U-boats retained their N numbers (*eg N 26* became *B 26*).

According to recently-published Russian sources, through the end of 1945 the Soviets completed: six D-class (Series I), six L-class (Series II), four *Shch*-class (Series III), three P-class cruiser (Series IV), twelve *Shch*-class (Series V), fourteen *Shch*-class (Series V-bis), thirteen *Shch*-class (Series V-bis 2nd series), thirty M-class (Series VI), twenty M-class (Series VI-bis), three S-class (Series IX), thirty-one S-class (Series IX-bis), thirty-two *Shch*-class (Series X), nine *Shch*-class (Series X-bis-, six L-class (Series XI), forty-five M-class (Series XII), seven L-class (Series XIII), five L-class (Series XIII-bis), eleven K-class (Series XIV), and four M-class (Series XV). At the outbreak of the Second World War they also had on hand five 'AG' class (US supplied H-boats of First World War vintage) and the ex-British L-55; the old Czarist *Pantera* (renamed *Komissar* in 1922 and *B-2* in 1934) had been discarded before the oubreak of war.

Thus total production runs through 1945 were: M-V-class, four (total of all M-class, seventy-nine) *Shch*-class, eighty-four; L-class, twenty-four; S-class, thirty-four; K-class, eleven. Many of these submarines did not survive the war, and in most cases the totals never matched those estimated in the West. Thus the US table above should be considered suggestive of the rate at which submarines were retired, rather than particularly precise.

Prewar submarine designs construction of which continued postwar: Project 96 (M-V-class, Series XV): *M-201* to *-219*, *M-234* to *-253*, *M-270* to *-294*, *M-299* (total sixty-five), of which *M-207* to *-213* were never launched. *M-209* was stopped 23 July 1946. Deliveries: *M-201* and *-202* in 1943, *M-203* and *-204* in 1944, *M-205/6* and *-214/18* in 1947, *M-234/8* in 1948, *M-239/249* in 1949, *M-250/276* in 1950, *M-277/284* in 1951, *M-285/291* in 1952, *M-292, -294, -299* (and possibly *-293*in 1953. All were designed by TsKB-18, and built by Sudomekh. Transfers: six to Poland 1954–55, four to China 1954–55, one to Egypt May 1957.

Project 601 (S-class, Series IXbis): *S-21* through and *S-38*, completed 1946–48. *S-21* was begun at Leningrad Sudomekh and worked on at a new yard at Astrakhan. *S-22* through *-26* were built at Gorkiy; *S-35* was built at Sudomekh. The S-class (Series IX comprised *S-4/26, S-31/38, S-45/48, S-50/62, S-101/108* (many were not completed). *S-1/3* were Series IX.

Contrary to published reports, no other submarines of prewar design were completed postwar. A single K-boat was scrapped on the slip in 1949; thus only eleven of a planned 100 (as of 1937) were completed. This particular K-boat was probably a modified version reported cancelled in 1949 after operating experience had revealed serious structural and other limitations.

WAR PRIZES
Ex-German Type IIB:
U 18, U 24
Reported in service (Black Sea) 1947, scrapped by 1960.
Ex-German Type VIIC:
S 81 (ex-*U 1057*), *S 82* (ex-*U 1058*), *S 83* (ex *U 1064*), *S 84* (ex-*U 1305*)
British designations *N 21, 22, 23, 24*; allocated to USSR 1947, probably broken up late 1950s. One other Type VIIC, *U 250*, was salvaged in Aug 1944

and taken to Kronstadt 15–16 Sept 1944. She was evaluated but probably was not placed in service.

Ex-German Type IXC:
B 26 (ex-*U 1231*)
British designation *N 26*; taken over 1947 and probably broken up late 1950s.

Ex-German Type XXI:
The Soviets captured twenty incomplete Type XXI U-boats when their troops overran Danzig on 30 March 1945: *TS-5* (ex-*U 3538*, in 1947 renamed *R-1*), *TS-6* (ex-*U 3539*, in 1947 renamed *R-2*), *TS-7* (ex-*U 3539*, in 1947 renamed *R-2*), *TS-8* (ex-*U 3541*, in 1947 renamed *R-4*), *TS-9* (ex-*U 3542*, in 1947 renamed *R-5*), *TS-10* (ex-*U 3543*, in 1947 renamed *R-6*), *TS-11* (ex-*U 3544*, in 1947 renamed *R-7*), *TS-12* (ex-*U 3545*, in 1947 renamed *R-8*), *TS-13* (ex-*U 3546*), *TS-15* (ex-*U 3547*), *TS-17* (ex-*U 3548*), *TS-18* (ex-*U 3549*), *TS-19* (ex-*U 3550*), *TS-32* (ex-*U 3551*), *TS-33* (ex-*U 3552*), *TS-34* (ex-*U 3553*), *TS-35* (ex-*U 3554*), *TS-36* (ex-*U 3555*), *TS-37* (ex-*U 3556*), *TS-38* (ex-*U 3557*). Contrary to much postwar speculation, none entered service; all were destroyed or scrapped in 1947–48. In addition four complete boats were turned over by the allies: *N-27* (ex-*U 3515*), *N-28* (ex-*U 2529*), *N-29* (ex-*U 3035*), *N-30* (ex-*U 3041*). *N-27* (later *B-27* and then *BSh-28*) served in the Baltic Fleet as an active submarine until Dec 1955. She then served as a pier trainer (renamed *UTS 3* Jan 1957) until 1972. The other three served in the Baltic Fleet through 1955 and were scrapped 1958–59.

The Soviets did capture four Italian 'CB' class midgets, but contrary to postwar reports, it appears they never saw active service.

Ex-German Type XXIII
M 51 (ex-*U 2353*)
U 2334 was reported salvaged in the Baltic in 1948, but she does not appear in recent Soviet lists. *U 2353* taken over by the British, designated *N 31*, was allocated to USSR 1947, and scrapped in 1963.

Ex-Italian Acciaio class:
Z 14 (ex-*Nichelio*)
British designation *N 42*; to USSR Feb 1949, stricken mid-1950s.

Ex-Italian Flutto class:
Z 13 (ex-*Marea*)
British designation *N 41*; to USSR Feb 1949, stricken mid-1950s.

Ex-Estonian Lembit class:
S 85 (ex-*Lembit*)
Experimental unit at Kronstadt 1956, memorial at Tallinn 1979.

Ex-Romanian Requinul class
TS 1
Returned 1951

Ex-Romanian Marsinul class
TS 2
Returned 1951

Ex-Romanian Delfinul class
TS 3
Returned 1951, stricken 1957; may not have been operational

Ex-British 'U' class
V 2 (ex-*Unbroken*), *V 3* (ex-*Unison*), *V 4* (ex-*Ursula*)
All returned 10 Feb 1949, BU 1950.

The Soviets also captured numerous midget submarines, including five of the Italian CB 1 class (captured from Romania 1944). All were BU about 1955.

FRIGATES

Name	Fate	Name	Fate
Uragan class (Project 39): launched 1929–35, 450t std			
URAGAN	BU 1958–59	MIETEL	BU 1958–59
TAIFUN	Stricken 1959	VYUGA	BU 1958–59
SMERCH	Stricken 1959	TUCHA	Stricken 1959
GROZA	BU 1958–59	GROM	Stricken 1959
VIKHR	Stricken 1959	ZARNITSA	Stricken 1959
SHTORM	Stricken 1959	MOLNYA	Stricken 1959
SKVAL	Stricken 1959		
Yastreb class (Project 29): launched 1940–41, 906t std			
YASTREB	BU 1958–59	SOKOL	Stricken early 1960s
ORYOL		KONDOR	
KORSHUN		ALBATROS	
VORON		KIROV	
BERKUT		DZERZHINSKI	

Yastreb class
A total of twenty were ordered. *Yastreb* was completed as planned, at Zhdanov. *Oryol, Korshun,* and *Zorkii* were completed postwar at Zhdanov to Project 29-K design. Five more laid down at Zhdanov were not completed: *Berkut, Sokol, Grif, Voron, Kondor.* Six were laid down at Nikolayev in 1939–40: *Tigr, Leopard, Rys, Yaguar, Kuguar, Pantera.* Project 29-P was intended for the Far East: *Albatros* (commissioned 1 February 1946) and *Burevestnik* (commissioned 12 February 1947) were completed postwar. *Fregat* and *Orlan* were not completed. A twentieth ship has not been named.

Ex-US Tacoma class
EK 1–28, transferred 1945, all returned 1949 and laid up in Japan; many were transferred to US allies: Japan, South Korea, Thailand, and Colombia. *EK 3* (ex-*Belfast*) foundered off Petropavlovsk 17 Nov 1948.

WAR PRIZES
Six ex-Romanian torpedo boats, used as minesweepers: six units, all returned in 1949; Soviet names are not known.

Two ex-Japanese *Ukuru* class: acquired 1947 ex-*Ikino* and ex-*Kozu*.

Eleven ex-Japanese 'Kaibokan' Type C escorts: acquired 1947 ex-Nos *34, 48, 52, 71, 76, 77, 78, 79, 105, 221, 227*

Most never reactivated, but one Kaibokan No 2 operated as frigate *EK 37* in 1957, and two Kaibokan No 1s were operating as surveying ships, one named *Sorakoram* (G 127); another, at Vladivostok, carried pennant G 111.

COASTAL CRAFT
The Soviets entered the Second World War with 269 MTBs, and reinstated twenty-four old units in wartime; another 166 were built. In addition, about twelve Romanian and three Bulgarian MTBs were obtained in 1944. Another 200 US PT-boats were received under Lend-Lease. However, 122 MTBs were lost in wartime, another fifty-two stricken. Of the Lend-Lease units, twelve were war losses, sixty-six were broken up, twenty-six were destroyed before US observers, and the rest were returned in 1954–58. The Soviets claimed that 393 MTBs were operational on 2 September 1945.

These boats were developed by the central aero- and hydrodynamic institute (TsAGI), initially under the direction of the famous aircraft designer A N Tupolev. The first thirty-six were ordered under the 1926–32 Five Year Plan. An unsuccessful GANT-3 prototype (named *Pervenets*: 8.9t, 56ft 9in, *17.3m*, two 600hp aircraft petrol engines = 55kts, 340nm cruising range, 1–450mm TT (stern trough), 3 men) was succeeded by GANT-4 (12t, 48–50kts, 2–45cm TT [troughs]); prototype built 1927–28, then fifty-nine built 1928–32 as Sh-4). A successor ANT-5 (built 1931–33) was the prototype of the main wartime class, G-5 (Project 116: 329 built 1935–43). It achieved 58kts on two 1000hp aircraft engines. G-5 characteristics: 17.8t, 62ft 8in, *19.1m*, two 1000hp GAM-34 aircraft engines = 48–54kts, 220nm range, 2–53cm TT [troughs]). The successor Project 123 design ('Komsomolets': thirty-one completed 1944–45) was developed further postwar: it had ahead-firing 45cm TT: 20t, 61ft 4in, *18.7m*, two 1200hp US supplied Packard engines = 48kts, range 345nm at 28.8kts. G-5 production comprised the prewar Series 7 – 9 (1934–36) and 10 (1937), and the wartime Series 11. Project 123 may have been Series 12 and 13.

There was also a parallel series of large wooden hard-chine boats. In effect they were prototypes of the postwar BTKA concept. D-3, which ran trials in 1939–40, carried two 53cm torpedoes in dropping cradles aft, firing forward: 36.1t, 73ft 6in, *22.4m*, 850–1000hp = 37–45kts. Another seventy-three were built in wartime, plus thirty-six ('PP-19-OK' type) built as MOs (sub-chasers) in 1942–44; many were designated SKA or SK: 38.9t, 72ft 6in, *22.1m*, 2200hp = 23.8kts (9.7kts cruising), 1–37mm, 2–12.7mm. Two were transferred to Poland in 1946.

The first five boats of a new TM-200 ('Yunga') class (see below) were completed before the end of 1945. So were eighty-eight of the corresponding ASW type, OD-200 (designated in an MO series: 47.2t, 79ft 9in, *23.4m*, 1800hp = 28kts, range 590nm at 9kts, 1–37mm or 25mm, 2–12.7mm, 2 DC, 19 men. NATO designated this type 'MO-V', and production probably continued postwar.

There were also several experimental MTB classes:

'Staliniy' (two built, as *S-1* and completed 1931): 28t, 77ft 1in, *23.5m*, 2250hp = 26kts, range 170nm, 3–45cm TT, 2–7.62mm MG, 6 men.

DTK: three built 1934–35, Nos 104, 114, 124, no details.

SM-3 (one built 1938): 34t, 68ft 3in, *20.8m*, 3600hp = 30kts, range 380nm at 24kts, 2–53cm torpedoes in cradles aft, 2–12.7mm MG, 9 men.

SM-4 (one boat, No 124, built 1939–41): 42t, 72ft 2in, *22m*, 4000hp = 30kts, 3–12.7mm MG, 9 men. This boat was later rerated as an MO.

G-6 (one boat, begun 1931 but not completed until 1939): 86t, 119ft 9in, *36.5m*, 7760hp = 49.8kts, range 435nm at 28.5kts; 3–53cm torpedoes, 1–45mm, 1–12.7mm, 4–7.62mm MG, crew 30.

G-8 (one boat, built 1937–40): 31.3t, 79ft 5in, *24.2m*, 4000hp = 32kts, range 350nm at 27kts, 3–53cm torpedoes in stern troughs, 3–12.7mm, 10 men.

D-2 (built 1938–40): 17t, two 1100hp engines, became an MO with 2 MG and 52 depth charges.

L-5 (four boats, TKL-1/4, begun 1936–39, in service 1939–40): 11.3t, 78ft 9in, *24m*, 2000hp = 70kts (probably using an air cushion), range 250nm, 2–45cm TT and 2–12.7mm, 5 men.

L-9 (three boats) *TKL-220/22*, in service 1941): 2.13t, no other details.

DD (one boat, *STK DD*, built at Baku and Poti 1942–43): 50t, 82ft, *25m*, 3000hp = 34kts, range 1000nm at 10kts, 4–45cm TT, all forward-firing, 3–12.7nm, 11 men.

US supplied Lend-Lease boats:
Type A-1 (Vosper): eighty-three units, *TKA-220* through *-271*, *TK-517* and *-518*, *-527* and *-528*, *-537* through *-570*, *-578* through *-590*, *-597*.

Type A-2 (Higgins): 52 units: *Nos. 21–22*, *Nos. 202–212*, *TK-213*, *TK-215* through *-219*, *-222*, *-851* through *-882*.

Type A-3 (Elco): 31 units: *TK-417* through *-447*.

There were also subchasers (MO, *Moskoi Okhotnik*, *ie*, sea hunters), built in peacetime mainly as guardcraft (PKA, PSKA) for the Maritime Border Guards. Large sub-chasers were numbered in a BO (*Bolshoi Okhotnik*, *ie*, large hunter) series. The MO-1 prototype was followed by the MO-2 class, construction of which stopped after one series, followed by the improved MO-4 class. Details: One MO-1 type (*PK-139*, later *SK-312*) built 1934–35: 51.1t, 85ft 4in, *26m*, 1130hp = 14kts, range 400nm at 8kts, 2–12.7mm, 21 men, 2 shaft powerplant.

Twenty-seven MO-2 type built 1935–36 as PSkas, later PKs; some later designated MOs. 50t, 85ft 11in, *26.2m*, 3 shafts, 2250hp = 26kts, range 285nm at 15kts, 2–45mm, 2–12.7mm, 2 DC racks, 4 mines, 21 men. Compared to MO-1, this version added a third engine.

Four MO-3 type (modified MO-2) were built 1935 as PKs. 55.4t, 85ft 11in, *26.2m*, 3 shafts, 2065hp = 24kts, 8kts cruise, 2–45mm, 2–12.7mm, 2 DC racks, 21 men. The single BK-3 was probably a related armoured type: *PK-224*, later *MO-411*, built 1934: 75t, 88ft 3in, *26.9m*, 2250hp = 18kts, range 285nm at 15kts, 2–45mm, 2–12.7mm, 2 DC racks, 4 mines, 20 men.

Two-hundred and nineteen MO-4 built 1937–44; some were PKs or SKAs; 56.5t, 88ft 3in, *26.9m*, 3 shafts, 2550hp = 25.5kts, range 367nm at 15kts, 2–45mm, 2–12.7mm, 2 DC, 4 mines, 21 men.

One MO-5, built 1939–41: 60t, 89ft 3in, *27.2m*, 3 shafts, 2550hp = 26kts, range 700nm at 15kts, 2–45mm, 2–12.7mm, 2 DC tracks, 21 men.

Four MO-6 (SKA series, built 1943–44): 66t, 87ft 11in, *26.8m*, 3600hp = 23kts, range 700nm at 8kts, 2–45mm, 2–12.7mm, 23 men.

The larger BO-2 (Project 194) class was built 1942–45; sixty-six were completed in wartime with three-shaft powerplants (two M-50 plus 1 auxiliary diesel). Most had BO numbers; some were SKs (guardships): 52.6t, 81ft 4in, *24.8m*, 2400hp = 21.2kts, range 495nm at 8.4kts, 1–37mm, 3–12.7mm.

Construction of Project 122 (*Artillerist* type) large sub-chasers (BMO) was begun under the 1938–42 (Third) Five Year Plan, but construction was delayed; eleven were to have been in service by 1 January 1939. Most completed in wartime were powered by GM diesels. Builders: Gorkiy (Sormovo), Zelenodolsk, Molotovsk (Severodvinsk), Vladivostok. In all, twenty-three were completed 1940–45. They had BO designations. 249t, 165ft 7in, *50.47m*, 3600hp = 22.3kts, range 1300nm at 9kts, 1–85mm, 2–37mm, 2–12.7mm, 2 DC racks, 49 men. The postwar 'Kronstadt' class was in successor type (Project 122bis).

Lend-Lease US PTC type: *MO 230–249*, *435–458*, 29t, 28kts, 1–20mm: 26 transferred 1943–44; five were war losses, nine were returned to the USN in 1956, seven were broken up postwar, and five were scuttled in the Barents Sea before US observers in 1958.

Lend-Lease RPC type: *SK 250–258*, SK 521–527, 36 transferred 1943–44, 15 were returned in 1955, one scuttled in the Barents Sea before US observers in 1956, and the remainder scrapped postwar.

Lend-Lease US SC type: *BO 202–246*, *BO 301–332*, transferred 1943–45; four lost, ten returned 1955, 20 scrapped, and 29 scuttled before US observers in 1955–56. The other five were purchased by the Soviet Navy.

WAR PRIZES

In 1945–49 the Soviets obtained twenty-three or more German R-301 class, twenty-six or more German R-218 class, fourteen German R-17 class motor minesweepers ('Raumboote'), and three Italian VAS. They also seized twenty-six or more German S-100 class MTBs, four early S-boats, four Italian MS-boats, and two Italian MAS 501 class MTBs. It is not clear when these boats were discarded.

MINE WARFARE VESSELS

Name	Fate	Name	Fate
Ex-Imperial Yacht, minelayer: 6200t			
MARTI (*Oka* 1957)	BU *c*1960		
Tral class coastal minesweepers (*Project 53-U*): launched 1935–40, 434t std			
STRELA	To N Korea 1955	GAK	
TROS	To N Korea 1955	RYM	
PROVODNIK	To N Korea 1955	T 215	
PODSEKATYEL	To N Korea 1955	KONTRADMIRAL KHOROSHIN	BU *c*1947–59
PARAVAN	To N Korea 1955	T 220	
KAPSUL	To N Korea 1955	TRAL	
VEKHA	To N Korea 1955	SHCHIT	
CHEKA	To N Korea 1955	ISKATEL	
FUGAS		MINA	
GAFEL		ARSENI RASSKIN	

Name	Fate	Name	Fate
Polukhin class ocean minesweepers (*Project 73-K*): launched 1940–46, 700t std			
VLADIMIR POLUKHIN	Stricken 1960s	TIMOFEI ULYANTSEV	Stricken 1960s
PAVEL KHAKHRELOV	Stricken 1960s	MIKHAIL MARTYNOV	Stricken 1960s
ALEKSANDR PETROV	Stricken 1960s	FEDOR MITROFANOV	Stricken 1960s
KARL ZEDIN	Stricken 1960s	LUKA PALKOV	Stricken 1960s
VASILI GROMOV	Stricken 1960s	PAVLIN VINOGRADOV	Stricken 1960s
ANDRIAN ZOSIMOV	Stricken 1960s	STEPAN GREDYSHKO	Stricken 1960s
VLADIMIR TREFOLEV	Stricken 1960s	SEMYON PALEIKOV	Stricken 1960s

Polukhin class

A total of twenty were ordered, of which seventeen were completed (fourteen of them at Leningrad). Ships completed in wartime at Leningrad were *Vladimir Polukhin* and *Vasili Gromov*. Ships completed postwar were: *Pavel Khakhrelov*, *Aleksandr Petrov*, *Karl Zedin*, *Andrian Zosimov*, *Vladimir Trefolev*, *Timofei Ulyantsev*, *Mikhail Martynov*, *Fedor Mitrofanov*, *Luka Palkov*, *Pavlin Vinogradov*, *Stepan Gredyshko*, and *Semyon Palikov*. Ships completed postwar (1948) at Nikolayev were: *Ivan Borisov*, *Sergei Shuvalov*, and *Ivan Sladkov*. Ships laid down at Nikolayev but not completed were *Pavel Golovkin*, *Nikolai Markin*, and *Semyon Roshal* (taken to Poti but not completed). Completion dates of the fifteen Project 73-K fast sweepers: *Pavel Khokhryakov* (T-251, 25.11.49), *Aleksandr Petrov* (T-252, 8.12.49), *Karl Zedin* (T-253, 18.11.49), *Andrian Zosimov* (T-255, 15.2.50), *Vladimir Trefolev* (T-256, 27.12.47), *Timofeiy Ulyantsev* (T-257, 4.1.49), *Mikhail Martinov* (T-258, 12.8.48), *Fedor Mitrofanov* (T-258, 7.11.47), *Luka Pan'kov* (T-260, 23.2.48), *Pavlin Vinogradov* (T-261, Sep 1949), *Stepan Gryadushko* (T-262, Sep 1949), *Semen Pelikhov* (T-263, Sep 1949), *Ivan Borisov* (T-451, Sep 1948), *Sergei Shuvalov* (T-452, Apr 1948), *Ivan Sladkov* (T-454, Apr 1948).

In addition, ninety-two small minesweepers (Project 253-L: *T-222* through *-249*, *T-351* through *-391*, *T-434* through *-441*, *T-459* through *-470*, *T-472* through *-479*) were built in wartime. Production continued postwar; NATO called the class 'T-301'. Postwar construction amounted to 103 'T-301' Class (Project 253-LP and 255): *T-229*, *T-330*, *T-301–316*, *T-328*, *T-340–350*, *T-364–369*, *T-392–400*, *T-436–458*, *T-480*, *T-481*, *T-494–500*, *T-528–533*, *T-651–655*, *T-726–731*, *T-741–746*. *T-313* was *Shakhin*; *T-500* was *Evgeniy Nikonov*.

Ex-US Admirable class ocean minesweepers: 625t

T 111–T 120, *T 272–T 282*, *T 521–T 526*, *T 593–T 598*. War losses: *T 114*, *T 115*, *T 117*, *T 118*, *T 120*, *T 278* (possibly postwar loss), *T 279*, *T 522*. Destroyed Barents Sea before US observers 1956: *T 111–T 113*, *T 119*. Certified unseaworthy and BU in Soviet Union 1954, *T 281*, *T 521*, *T 523–T 526*, *T 593–T 598*. Purchased by Soviets 1956: *T 272–T 277*.

Ex-British MMS(1) type: 165t, 11kts, 2–12.7mm MG: *T 108–T110*

Ex-British MMS(2) type: 255t, 10kts, 2–20mm: *T 121*, *T 122*, *T 193–T 202*

Ex-US YMS type: 203t, 14kts, 1–3in: *T 151–T 156*, *T 181–T 192*, *T 581–T 592*, *T 599–T 611*.

In 1954 the Soviets stated that *T 114*, *T 116*, *T 118*, *T 120*, *T 152*, *T 278*, *T 279*, and *T 610* had all been lost; that *T 181–T 192* in the Black Sea had been declared unseaworthy and BU; that *T 151*, *T 153*, *T 156*, *T 271–T 277*, *T 289*, *T 282*, *T 588*, *T 591*, *T 602* and *T 610* had all been scrapped in the Pacific. In addition, *T 522*, *T 581*, *T 586*, *T 587*, *T 604*, *T 605*, *T 606*, *T 608*, *T 609* and *T 611* in the Pacific were BU *c*1956. Finally, *T 111–T 113*, *T 115*, *T 119*, *T 120*, *T 154*, *T 155*, *T 218*, *T 521*, *T 523–T 526*, *T 583*, *T 585*, *T 589*, *T 590*, *T 592–T 601* and *T 603* were all destroyed before American observers in 1956.

WAR PRIZES

At the end of the Second World War, the Soviets obtained fourteen M-35, twenty-nine M-40, and seventeen or more M-43 class German ocean minesweepers. Some writers have suggested that the 'T 43'/'T 58' classes were derived from these ships.

AMPHIBIOUS WARFARE VESSELS

The Soviets appear not to have had any specialised landing ships of their own; they did use a wide variety of merchant ships as extemporised transports during the Second World War. Under Lend-Lease, the Soviet Pacific Fleet received thirty LCI, fifteen LCT, and fifty-four LCM(3) type landing craft; the Northern Fleet received two LCM(3), two LCS(S), two LCT, and two LCVP. At least five LCI and three LCM were lost in the Pacific.

WAR PRIZES

In 1955 ONI credited the Soviets with three Italian MZ class landing craft; one German MNL; eight German SF; sixteen German LT class MFP; twenty-nine German F class MFP, and fifteen MFP Type D class. The Soviets put these designs into production in the 1950s; see MP-2. The Soviets also obtained twenty-nine German AF type amphibious support ships, built on MFP hulls, and armed with 2–3.5in, 2–37mm and four quadruple 20mm, plus four RL; 330t (std), 355t (full), 154ft 2in × 21ft 4in × 5ft 3in (max), *47m* × *6.5m* × *1.6m*, and powered by three 375bhp diesels, with a speed of 10kts. They were coastal craft only.

RIVER CRAFT

Numerous riverine craft, mainly armoured cutters (BK), were built during the Second World War. As of 1945 the following had been completed: twenty 'MBK' (Project 161), eight 'MKL' (Project 186; another twenty were completed in 1946: *BK 529–559*). Ninety-seven Type 1124 (armed with two 76.2mm tank turrets), 151 Type 1125 (armed with one 76.2mm tank turret; another fifty-one were completed in 1946–47: *BK-153, BK-155–157, BK-187–199, BK-335–339, BK-425–430, BK-435–459*), and seven S-40 class (armed with one 76.2mm tank turret). In addition, two large (1900t) *Khasan* (Project 1190) class river monitors were completed in wartime, and a third was delivered in 1946.

War experience showed that these craft needed more protection and more defensive armament. However, a study of Project 1190 showed that any such improvement would increase its draft and overall size too much. As a consequence, plans for building large and small monitors under the 1946–55 programme were dropped. Instead, efforts were concentrated on replacing the 1124/1125 boats.

MAJOR SURFACE SHIPS

Aircraft carrier projects

The abortive 1938–42 Five Year Plan included two carriers (Project 71), for the Northern and Pacific Fleets, the first to be laid down in 1941, the second in 1942. A 13,000t hull would have accommodated thirty fighters and fifteen bombers. There was also a 10,600t light carrier design, based on the *Chapaev* class cruiser hull, for twenty aircraft (ten fighters). Both designs were soon considered obsolete. The main naval staff issued new carrier TTZs in January 1943, for carriers to accommod ate thirty, forty-five, or sixty aircraft. TsKB-4 was assigned to develop new plans for a battleship, a carrier and a cruiser. A November 1943 reorganisation shifted carrier work to a new TsKB-17, which was responsible for all later Soviet heavy combatants. The carrier project was redesignated Project 72, and at least some plans were ready by November 1944. At about the same time, the Naval Academy, which had been evacuated from Leningrad to Samarkand, produced a large study of wartime warship development trends.

Late in 1944 the Soviet government decided to redesign existing warships and to develop new ones based on war experience. A commission to submit proposals was formed in January 1945; it included a carrier branch. The latter proposed fleet carriers for the Northern and Pacific Fleets and small carriers for the Baltic and Black Sea Fleets. The commission suggested that specialists be sent abroad to gain experience on board Allied carriers, that a Liberty ship be converted into a training carrier, and that two American carriers be bought or borrowed for the Pacific and Northern Fleets. The naval staff doubted that the Allies would agree; work proceeded on home-grown designs. Ultimately the commission proposed light, escort, fleet, and heavy carriers; light and escort carriers would carry air defence fighters, multi-purpose ASW search aircraft, and AA guns; they would differ only in speed and armour. Fleet carriers would add medium multi-purpose aircraft (torpedo bombers) and light multi-purpose aircraft (dive bombers – attack fighters); they would have the same speed as the light carriers, but would be armoured like cruisers, with extra flight deck armour and extra armoured bulkheads near the elevators. Heavy carriers would have more torpedo bombers.

In all, thirty-three variants were developed, twenty-four for escorts, three for light, four for fleet and two for heavy carriers. For example, all the escort carriers accommodated twenty-four fighters and eighteen ASW aircraft; an 18,410t ship (armour 25mm flight deck, 50mm deck, 80mm sides and bulkheads, 25mm bulkheads) would have made 20kt on 33,500shp. There was also a much smaller unarmoured equivalent. The light carrier (armoured like the escort) would have displaced 30,560/37,580t (794ft × 97ft 1in × 26ft 10in/*242m × 29.6m × 8.18m*), accommodating thirty-six fighters and twenty-seven torpedo bombers; designed armament was 9 twin 100mm, 6 quad 45mm, and 14–23mm; she would have made 34kt on 242,000shp. The fleet carrier (35,270/39,670t, 898ft 11in × 105ft 10in/*274m × 32m × 10.3m*) would have accommodated twenty-four fighters plus twenty-four light and twenty-four medium attack aircraft; designed armament was 12 twin 100mm, 6 quad 45mm, and 24–23mm. Armour would have increased slightly; 60mm armoured deck, 100mm side. She would have made 34kt on 277,000shp. Finally, a heavy carrier (82,370/89,770t, 1148ft 4in × 134ft 6in × 35ft 1in/*350m × 41m × 10.7m*) would have accommodated sixteen more medium attack aircraft and would have had a thicker armour deck (150mm). Armament would have been lighter: 9 twin 100mm, 8 quad 45mm, 32–23mm. She would have required 384,000shp to make 34kt. Although Stalin apparently initially approved Admiral Kuznetzov's suggestion that two small carriers should be included in the postwar fleet, he soon concluded that they were not worthwhile, and the 1946–55 programme showed no carriers at all.

Kuznetzov (supported by Admiral A Golovko) revived the proposal soon after he was reinstated as head of the Soviet Navy in 1951. Stalin never ruled on this issue. Work on a carrier design did not resume until after Stalin's death in 1953. A forty-aircraft light carrier had been designed by 1955, but it was cancelled in October 1955. Two pre-project designs were developed in 1954, one by TsKB-17 and one by the navy's internal studies organisation, TsNII-45. Each would carry 40 aircraft (40 fighters) and would have had a range of 5000nm. The TsKB-17 version would have displaced 30,555t, with a speed of 34kts, 16–130mm guns, 100mm side armour, and 70mm deck armour. TsNII-45 suggested 21,000t, with a speed of 32.2kts, 16–76mm guns, and 50mm belt armour. The operational-technical requirements (preliminary TTZ) issued in 1955 called for a displacement of 23,000t to 24,000t, 40 aircraft, 32kts, a range of 5000 to 6000nm, armed with 16–100mm guns, with 50mm belt and 70mm deck armour, and 2 steam catapults. In 1989 Admiral Cherniavin claimed that 'it was decided to move from the simple to the complex', *ie* from helicopter carriers to full CTOL carriers. Work on the *Moskva* design began in 1959.

A tactical-technical requirement for a future CTOL carrier (code-named Orel, Eagle) was formulated in the early 1970s; it envisaged a 75,000t to 80,000t nuclear-powered ship to carry at least seventy aircraft, including fighters, attack aircraft, ASW and AEW aircraft. Like a US carrier, she would have had four steam catapults. Admiral Gorshkov decided to include anti-ship missiles as well. A design bureau was assigned to this project in 1973. Marshal Ustinov, the Minister of Defence, asked whether the third *Kiev* could not be modified with a catapult for MiG-23A fighters, a project called Naverku ('Upstairs'). It would have carried thirty-six airplanes and helicopters. Ustinov clearly was unwilling to accept the full Orel. A revised staff requirement for a smaller ship (60,000t, fifty aircraft, mainly fighters, with nuclear power) was issued (June 1974).

By 1975 the 1978–85 naval programme was being discussed (*ie* the latter portion of the 1976–80 Five-Year Plan and the 1981–85 Plan). Ustinov once more raised the VTOL *v* CTOL issue. Full-deck carrier plans existed, but none had been approved. The proposal was to include two Orels in the 1978–85 plan. Admiral Cherniavin stated in 1989 that the programme to provide fighters (rather than the Yak-38 attack aircraft flying from the *Kiev* class) at sea was adopted in 1976, *ie* for the 1976–80 Five-Year Plan. In 1979 the design was again revised down, to accommodate forty-two aircraft plus anti-ship missiles. Presumably this was the *Tbilisi* class. The *Ulyanovsk* was probably an attempt to return to the full Orel concept.

About 1978, when these carriers were being planned, Admiral N Amelko, formerly commander of the Soviet Pacific Fleet and then Deputy Chief of the Naval Staff, proposed a helicopter carrier as an alternative to the big *Tbilisi*. He suggested that the 'Atlantika' class ro-ro ship, then in series production, be modified as a low-cost helicopter carrier, as an alternative to the projected carriers. A design, code-named Khalzan, was developed in 1980. It was evaluated alongside a possible helicopter carrier based on the *Baku* and also alongside the new carrier. Khalzan had a very noisy twin-shaft gas turbine powerplant. It was also argued that she could not have operated the new Yak-141 fighters or the airborne early warning helicopters then in the design stage. Both the alternative designs could have operated both types of aircraft. Khalzan was also considered significantly less survivable against Harpoon and Mk 48 torpedo attack. For example, her power plant occupied only a single compartment, one hit on which would have disabled her. She lacked both anti-ship missiles of her own and torpedo tubes.

ULYANOVSK class *aircraft carrier*

Displacement: 65,000t standard; 85,000t full load
Dimensions: 1089ft oa × 124ft 6in wl, 246ft ext, × 38ft
332m × 38m, 75m, × 11.6m
Machinery: 4-shaft nuclear: 4 reactors = 200,000shp = 30kts
Armament: 12 SS-N-19 Granit, SA-N-12 Yezh, 8 CADS-N-I, 8 AK-630 'Gatlings', 80 aircraft

Name	Builder	Laid down	Launched	Comp	Fate
ULYANOVSK	Nikolayev	25.11.88	–	–	Cancelled 1.11.91, ordered BU 4.2.92

This was the projected nuclear-powered successor to the *Kuznetzov* class, presumably equipped with conventional steam catapults. She was laid down at Nikolayev in December 1988 and although her launch was expected during 1991, the economic collapse that year badly slowed work, and the split between Russia and Ukraine made essential parts such as the reactors undeliverable. By February 1992 this ship, which was 40 per cent complete, was being broken up. Reportedly Russia and Ukraine were unable to agree on her ownership, so they decided to share the proceeds of the scrap sale. She had been laid down on the slip vacated by *Varyag* (ex-*Riga*). A second unit was probably planned, but she was never laid down because *Ulyanovsk* was not launched. Prior to her cancellation, it was reported that she would be renamed *Petr Veliky* ('Peter the Great') if completed.

Few details are available, but reported length was about 320m (1050ft). It seems likely that the planned powerplant was a pair of *Kirov* plants (nuclear/steam turbine). Effective maximum power would depend on how much could usefully be transmitted through the propellers.

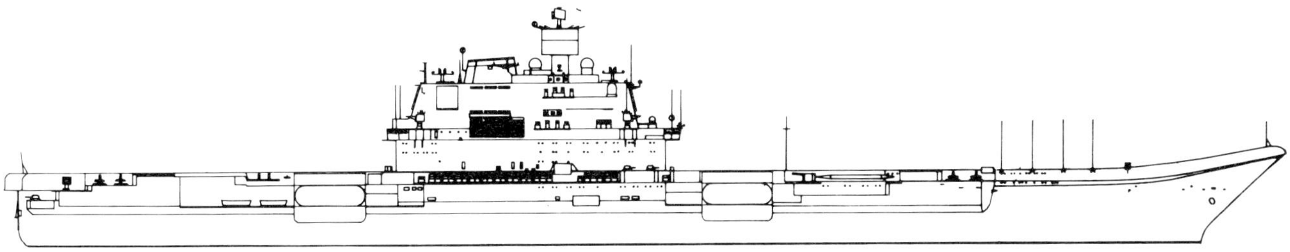

Admiral Kuznetzov as completed **NB 1/1750 scale**

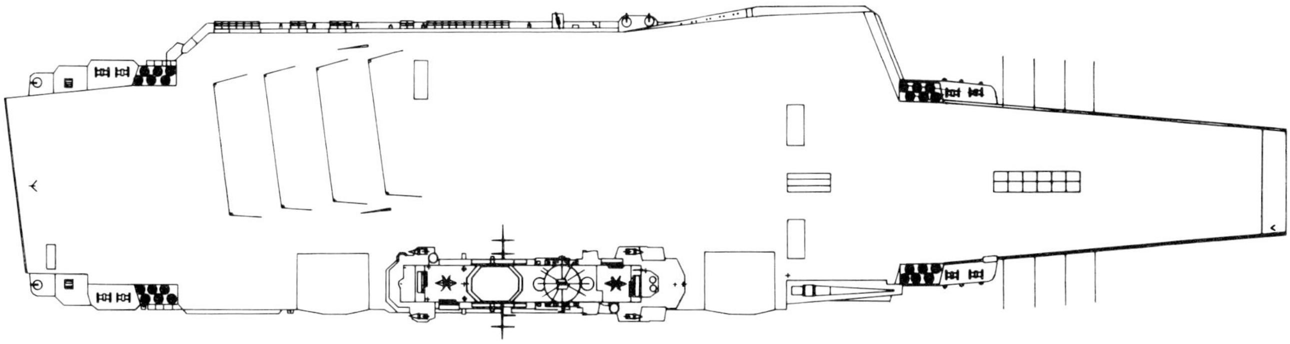

KUZNETZOV class *aircraft carriers* (Project 1143.5)

Displacement: 60,000t (67,000t full load)
Dimensions: 984ft × 124ft 8in × 36ft; 239ft 5in extreme width
300.0m × 37.8m × 11.0m; 73.0m
Machinery: 4 sets geared turbines, 200,000shp = 30kts; pressure-fired steam
Armament: 16 SS-N-19, 18 SA-N-9 launchers (8 missiles each), 8 CADS-N-I, 6 30mm Gatling, 2 RBU-12000, 18 fixed-wing aircraft (two 9-aircraft squadrons) and 12 helicopters (Helix-A and -C). Aircraft also reported as 24 Flanker (Su-27) and 4–6 Helix-A and -C helicopters
Sensors: Radar Sky Watch, 2 Strut Pair, 3 Palm Frond, 1 Plate Steer, 3 Cross Sword (FCS), sonar LF hull. EW of new types. 2 Punch Bowl satellite antennas.

Name	Builder	Laid down	Launched	Comp	Fate
ADMIRAL KUZNETZOV	Nikolayev	6.11.83	5.12.85	21.1.91	Extant 1995
VARYAG	Nikolayev	6.12.85	4.12 88		For sale 1992

Admiral Kuznetzov 1991 MoD

Admiral Kuznetzov was launched as *Tbilisi*; she ran some trials (including initial aircraft operations) under that name in October 1989. She was renamed on 4 October 1990 after the outbreak of anti-Soviet violence in Georgia. She was laid down as *Riga*, and renamed *Leonid Brezhnev* on 18 November 1982, but was renamed again in November 1988 after Mikhail Gorbachev had denounced Brezhnev and his 'era of stagnation'. *Varyag* was launched as *Riga*, but renamed after the Baltic States demanded their freedom. She was caught unfinished at Nikolayev when the Soviet Union broke up, and as of early 1992 her fate was uncertain. *Admiral Kuznetzov* slipped into the Mediterranean in December 1991 for the trip home to the Northern Fleet. She was still incomplete (in mid-1992), with many of her antennae not connected to any internal electronics.

Compared to the earlier *Kiev* class, these ships presumably represent a shift in Soviet naval policy. From the mid-1970s on, Admiral Gorshkov advertised the Soviet Navy as a 'sure shield' against Western intervention in 'wars of national liberation' in the Third World, *ie* against US carrier battle groups. Soviet policy seemed to be to create forces which could be interposed between local revolutionaries and counter-insurgency (Western) forces. True carriers would make Soviet distant-water operations in such contingencies far more effective. As planning for this class proceeded, Soviet comments about carrier operations became markedly more respectful. The earlier *Kiev* class, formerly identified as 'large ASW ships' or 'ASW cruisers' were reclassified as 'air-capable ships' in *Soviet* terminology.

While under construction, these ships were described as full-deck conventional carriers with steam catapults. It appears that Soviet attempts to develop such catapults (based in part on the steam catapults of the Indian carrier *Vikrant*) failed; the alternative chosen was a 12-degree ski-jump deck. That suffices to launch modern jet aircraft, which have thrust-to-weight ratios better than 1. Operations from such a deck broadly resemble those of pre-jet carriers, whose aircraft frequently made rolling take-offs. Unlike those carriers, these have jet-type angled decks (angle about 5.5 degrees). The disadvantage of the ski-jump are that it requires considerable wind over deck to function, that it may not support heavily-loaded aircraft, and that aircraft are not under positive control as they are launched (which may be significant in heavy seas). As of early 1992, *Kuznetzov* had experienced relatively little air activity, and it appeared that only a few pilots had been carrier-qualified. She has operated Su-27 Flanker, MiG-29 Fulcrum, and Su-25 Frogfoot aircraft. There is no indication that the Soviets had developed a suitable airborne early warning aircraft for this class (the one airplane sometimes so described, the An-72 MADCAP, was a private venture of the Antonov design bureau).

All aircraft are normally stowed below decks in a closed hangar. Hangar length is limited by the unusual (for a carrier) nest of SS-N-19 missiles under the flight deck, in launching tubes. This feature is reminiscent of the *Kirov* design. The Punch Bowl satellite link serves these SS-N-19s; these ships do not have the Front Door C (formerly Trap Door) of the previous class.

This project was first reported in 1979, *ie*, at the end of the 1976–80 Five-Year Plan. The name *Kuznetzov* recalls the Soviet admiral most closely identified with carriers and with an integrated blue-water fleet.

The second ship was renamed *Varyag* late in 1990 (by which time disturbances in Riga had made it clear that the original name was inappropriate). The new name was traditional in the Pacific Fleet, and on 15 December 1991 it was announced that the ship would go to the Pacific upon completion. By that time the ship, nearly complete (except for electronics and weapons), was ordered mothballed until Russian finances improved. However, in January 1992 it was reported that the incomplete *Varyag* was for sale for breaking up; the Russian Ministry of Defence claimed that it would cost $522m to complete her. At the same time it was reported that both the Indian and Chinese navies were interested in purchasing the ship for completion.

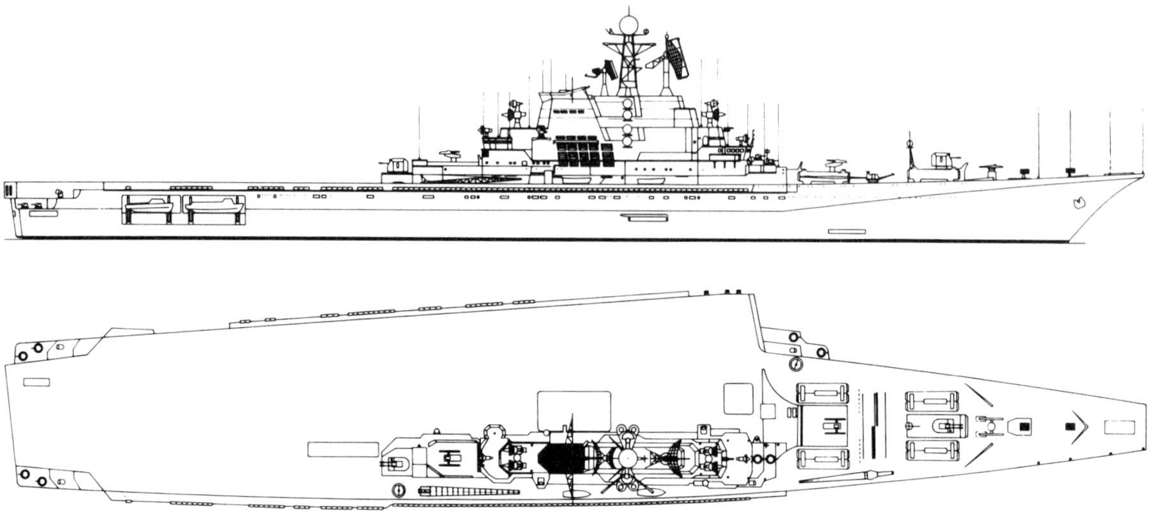

Kiev as completed
NB 1/1750 scale

The TTZ for this class was approved on 16 October 1968. They were designed by TsKB-17. Chronology suggests strongly that the *Kiev*s were associated with the 'Delta' class strategic submarines, which for the first time could fire from 'holding areas' which the Soviets could hope to control, such as the Norwegian Sea north of the GIUK Gap. It can be argued that the massive batteries of these ships were ideally suited to denying passage to NATO ASW forces, be they aircraft (which could be opposed by the Yak-38 fighter), surface ships (against which *Kiev* could release SS-N-12 cruise missiles), and submarines (against which she had a variety of weapons including her helicopters). As in the case of *Moskva*, this was hardly a traditional Western-type carrier function. Note, however, that the Yak-38 (twelve carried with eighteen to nineteen Hormone A and B helicopters) was observed primarily carrying ground-attack weapons, and appeared not to have any air intercept radar. If it was indeed intended to oppose NATO maritime patrol aircraft, especially the P-3, presumably it depended heavily on control by the ship; this would be in accord with standard Soviet fighter-control doctrine. In this context it was suggestive that the 'Forger' had an even design number, since the Soviets reportedly assigned odd numbers to fighters (*eg* MiG-21, -23, -25) and *even* ones to attack aircraft (*eg* Su-22).

Certainly among the most spectacular Soviet ships, these large cruisers or semi-cruisers combined the *Moskva* weapon system with the first reloadable SSM system since the *Kynda*s, and with the SA-N-4 point defence missile system. More importantly, to many analysts, they operated Yak-38 VTOL fighters. It was

KIEV class *air capable ships* (Project 1143 and versions)

Displacement: 36,000t standard; 42,000t full load

Dimensions: 902ft oa × 108ft wl, 154ft 10in ext × 26ft 11in
275.0m × 32.7m, 47.2m × 8.2m

Machinery: 4-shaft geared steam turbines, 140,000shp = 32kts. Fuel 7000t. Range 4000nm at 30kts, 13,500nm at 18kts

Armament: 4 S-N-12 (4×2, 24 missiles), 2 SA-N-3 (2×2, 72 missiles), 2 SA-N-4 (2×2, 40 missiles), 4–76mm (2×2), 8–30mm Gatling (8×1), 10–533mm TT (2×5), 1 SUW-N-1 ASW RL (1×2), RBU-6000 ASW RL (2×12), 31 aircraft. See notes

Sensors: Radar 2 Palm Frond, Top Sail, Top Steer, 2 Head Lights, 2 Pop Group, 2 Owl Screech, 4 Bass Tilt, Trap Door; sonar L/F hull, VDS; ECM 8 Side Globe, 12 Bell series, 4 Rum Tub, 2 chaff RL (2×2)

Complement: *c*1700

Name	Builder	Laid down	Launched	Comp	Fate
KIEV	Nikolayev	21.7.70	26.12.72	28.12.75	Stricken 30.6.93
MINSK	Nikolayev	28.12.72	30.9.75	27.9.78	Stricken 30.6.93
NOVOROSSIYSK	Nikolayev	30.9.75	26.12.78	14.8.82	Stricken 30.6.93
ADMIRAL GORSHKOV (ex-*Baku*)	Nikolayev	Dec 1978	1.4.82	11.12.87	Extant 1995

Minsk in the 1980s
RAF

difficult to divine the relative priorities of these systems: certainly the deep magazine for SS-N-12 missiles took up a lot of space and even length. Unlike other Soviet ASW ships of their period, they did not have the SS-N-14 torpedo-carrying stand-off missile, although they did have the Hormone/SUW-N-1 nuclear ASW system, and the helicopters could drop torpedoes, and some writers suggested that the *Kiev*s were intended to support Soviet amphibious operations. That seems unlikely; it is more probable that they were intended to support Soviet ballistic missile submarines by helping to bar NATO access to SSBN 'holding areas' in wartime. In such a role their large hulls improve sea-keeping, and their three-dimensional armament would have allowed them to oppose penetration of their patrol areas by NATO ASW aircraft, surface ship or submarines. However, it should be noted that the second ship to be completed, *Minsk*, went to the Pacific Fleet in mid-1979 instead of supplementing the *Kiev* in Northern Fleet. Moreover, only four ships were built, which hardly seemed sufficient to block the GIUK Gap. *Novorossiysk* first deployed to the N Atlantic mid-1983.

Kiev herself spent considerable time supporting Soviet amphibious exercises since her much publicised maiden voyage through the Mediterranean in July 1976 and many analyses of the Yak-38 suggested that close support or surface support was its design role. Certainly it was only marginally valuable as a fighter, given no airborne intercept radar – which in turn may be explainable in terms of ship-controlled intercept concepts or in terms of the weight an effective radar would consume. The problem is that the Soviets never made the anticipated heavy investment in amphibious assault ships, which might have explained the *Kiev* as a support ship for overseas operations. Nor can the Yak be explained as an anti-ship weapon, since it had no surface search radar (as in the Sea Harrier). To add to the confusion, the Soviets reclassified *Kiev* as a Tactical Aviation Cruiser (TAKR = *Taktycheskiy Avionosnyy Kreyser*) *c*1982.

A recent Soviet account suggests that *Kiev* was conceived as a lengthened *Moskva*, with additional helicopters. This enlargement could not overcome the basic defects of the design, and it had to be extensively rethought. The original ship had four helicopter take-off spots, and the new one was intended to have six. However, a seventh was needed for an alert search and rescue helicopter. There was no space for sufficient spots abaft the island, so they were lined up to form the characteristic angled through deck of the *Kiev* class. The larger ship could accommodate the additional armament which had been desired for the *Moskva* class. The powerplant was simply two *Moskva* plants. According to this account, the ship was not conceived as a carrier. Her anti-ship missiles were added in the draft design stage, the concept design stage during which detailed trade-offs are worked out. The fixed-wing air group was only decided upon about 1968, the Yak-36 VTOL prototype having interested naval personnel after its 1967 debut. A Yak-38 flew from *Moskva* for the first time on 18 November 1972.

The recent Soviet account states that the *Minsk* design (Project 1143M) was modified to provide for 50 per cent more aircraft, as well as a new sonar (Horse Jaw and Horse Tail in place of Moose Jaw, Bull Nose, and Moose Tail) and new electronic warfare equipment. Earlier Western reports suggested that the new sonars were introduced only in the third ship, *Novorossiysk*. According to the Soviet account, she followed the modified design. However, she lacked SA-N-4, and instead was fitted to take the successor SA-N-9 (six abaft the island and six forward to port). She lacked the Side Globes ECM system of her predecessors.

The fourth ship, *Baku* (later renamed *Admiral of the Fleet of the Soviet Union Gorshkov*, Project 1143.4) was again modified. She had a new phased-array radar (Sky Watch), a new fighter control system (Cake Stand, comparable to the Western Tacan), and more anti-ship missile launchers (twelve, to provide a larger salvo; but there were no reloads). She had SA-N-9 vertically-launched point defence missiles in place of the earlier SA-N-4s, but she lacked the medium-range SA-N-3s of the earlier ships. The 76mm dual-purpose guns were replaced by 100mm weapons. The earlier RBU-6000 system was replaced by a pair of new RBU-12000 anti-torpedo systems. The forward gun sponson before the angled deck was eliminated, and the superstructure frame made 'more dignified'. In effect the fourth ship introduced the electronic suite later used in the full carrier *Kuznetzov*. The mass of new systems took quite some time to install, and the ship's construction was quite protracted.

By late 1991 all surviving Yak-38 Forgers were out of service, considered either too worn out or too dangerous for use. The projected successor, the supersonic Yak-141 Freehand, had fallen victim to the Soviet economic collapse and seemed unlikely ever to enter production (the Yak design bureau had failed to find any foreign sponsors). One of the two flying prototypes had crashed.

By 1994 this class was virtually extinct. The two Pacific Fleet ships were stricken after accidents requiring major repairs at the building yard (Nikolaev), which was no longer part of Russia. *Kiev* was cannibalised to keep *Gorshkov* operating, then *Gorshkov* herself suffered a fire. She was reportedly offered to Indian in 1994, but no sale was completed. Reports that one or more ships would be sold to China also seem to have been incorrect, as both Pacific ships were sold for breaking up in Korea.

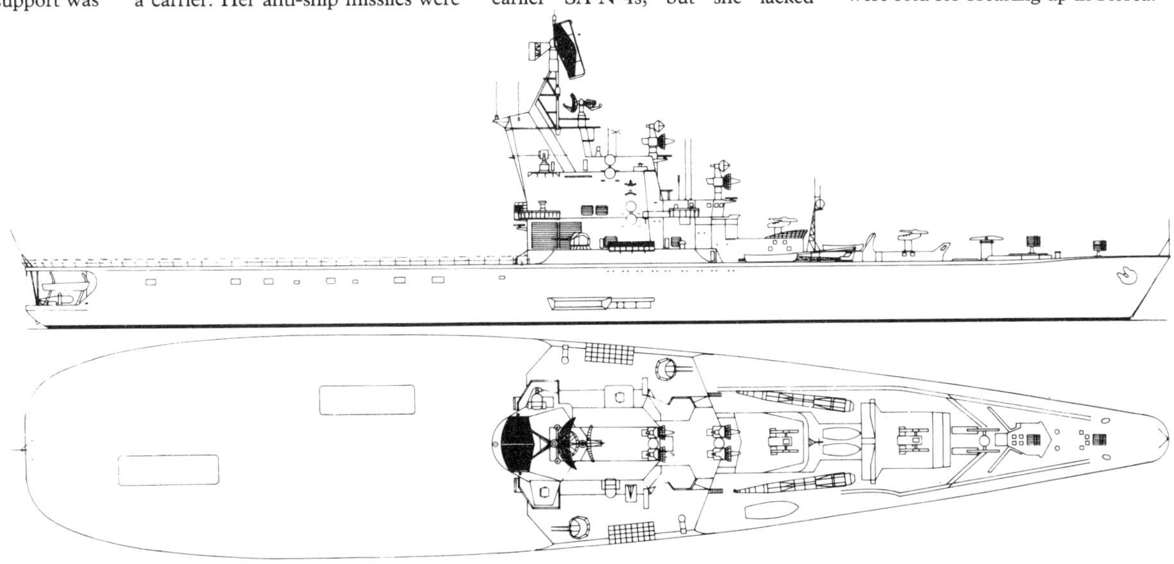

Moskva 1974

MOSKVA class *helicopter carriers*

Displacement: 11,200t light; 14,590t standard; 17,500t full load

Dimensions: 587ft 3in wl, 620ft 1in oa × 85ft 4in × 27ft 11in (full), 42ft 8in (over sonar dome)
179.0m, 189.0m × 23m × 8.5m, 13.0m

Machinery: 2-shaft geared turbines, 4 pressure-fired boilers, 100,000shp = 31kts; fuel 2600t. Range 14,000nm at 12kts or 4500nm at 29kts

Armament: 2 SA-N-3 (2×2, 44 missiles), 4–57mm (2×2, 4400 rounds), 1 SUW-N-1 launcher (1×2, 18 missiles), 2 RBU-6000, 10–533mm TT (removed mid-1970s), 14 helicopters (also given as 12–13; maximum 20)

Sensors: Radar Top Sail, Head Net-C, 2 Head Lights, 2 Muff Cob, 3 Don-2; sonar Moose Jaw bow, Mare Tail VDS; ECM 8 Side Globe, 8 Bell Series, 2 Chaff RL (2×2)

Complement: 850

Name	Builder	Laid down	Launched	Comp	Fate
MOSKVA	Nikolayev	15.12.62	14.1.65	25.12.67	Extant 1995
LENINGRAD	Nikolayev	15.1.65	31.7.68	2.6.69	Stricken 24.6.91

These two ships were designed specifically to counter Polaris submarines. According to a recent Russian account, naval staff studies showed that this mission required sea-based aircraft (helicopters) capable of searching large areas for a sustained period. Any submarines that were found would have to be dealt with by accompanying ships. This was not too different from the contemporary US CVS concept. Indeed, at just about the same time (1959) the US Navy was interested in building what it called a 'patrol escort' (PE), specifically to trail missile submarines off the US coast. The Soviet requirement, to maintain two helicopters continuously airborne on patrol, was matched by the US requirement. In both cases, the hope seems to have been to minimise ship size, providing just enough for helicopters and their reliefs (in the Soviet case, a total of eight).

Admiral Gorshkov approved the ship's operational requirement on 31 January 1959: it was to search for and destroy SSBNs and multi-purpose submarines in distant ASW zones as part of a ship search-strike force and in cooperation with other ships and ASW aircraft. The cruiser-design bureau (TsKB 17) offered two alternatives: a modified *Sverdlov* hull (several incomplete hulls still existed) and a small carrier. Although Defence Minister Malinovskiy approved both, only the small carrier was pursued.

As is typical of staff requirements, Gorshkov wanted the smallest possible hull to accommodate the eight helicopters. The design bureau (TsKB 17) argued that, since the helicopters would not be able to operate in bad weather, the ship ought also to have an all-weather ASW weapon system: a long-range sonar supporting an ASW missile system, as well as torpedoes and ASW rockets (RBUs). The bureau also wanted a self-defence missile system. The naval staff agreed, and the helicopter complement rose to ten–fourteen. Admiral Gorshkov approved the revised design specification on 25 January 1960. The chief of the Navy Main Shipbuilding Directorate emphasised the need to hold standard displacement down to 7000–8000t.

One way to do that might have been to work up from a destroyer rather than down from a cruiser. The destroyer design bureau (TsKB 53) was therefore asked for an alternate design. TsKB 17 itself tried eight alternatives, including a catamaran and an aluminium-magnesium hull; it soon concluded that only by using gas turbines (as yet untried) could the displacement limit be met in a steel hull. The navy turned down the exotic alternatives (and TsKB 53's design), opting for an 8500t vessel with pressure-fired boilers (the Project 58 'Kynda' class powerplant). On this basis the preliminary design was approved on 1 December 1960.

Additional studies were still required. The designers considered using new helicopters (which could operate in Sea State 6), reducing the crew to 300–350, cutting habitability to submarine or destroyer standards, and other economies. Thus in addition to the original eight, sixteen more variants were developed, the twenty-third (9300t) being approved in March 1961.

This was the first Soviet ship designed to operate a group of helicopters. In 1961 the air force announced that it would not provide the necessary technical advice, so the designers had to rely on the OKB designing the helicopters (Kamov's OKB-938). Worse, as the contract design approached completion, the air force announced that the chosen helicopter, the Ka-25, would not be adequate; a special expert commission had to be formed to decide the issue. By mid-1962 it seemed that the ship could be built on 9000t (29kts, 370 men). However, the navy's studies group wanted thirty-one more men. When the staff reviewed manning, it decided that the ship needed far more. Full three-shift manning during a prolonged patrol would require 497 men (which was cut to 415 by review). The contract design was approved on 25 January 1962.

Leningrad 1981 CMP

That did not end the process. The helicopter complement increased to fourteen (Ka-25 and Mi-8 ASW helicopters), radio equipment was changed, sea-keeping was improved (for operation in Sea States 6–7). It turned out that the hangar had to be expanded by 81 per cent (a second hangar for two helicopters had to be provided on the upper deck), and the flight deck area had to be increased by 77 per cent. Displacement grew to 10,000t and then to 10,750t; designs speed dropped to 28.5kts.

Meanwhile, Polaris range increased. During a 1962 summer visit to the Northern Fleet Khrushchev decided that the endurance (in time) and the range of surface combatants should be considerably increased. That meant more AAW weapons capacity and, as it turned out, a new radar (presumably Top Sail). Another twenty-six versions of the design were developed. Final figures were standard displacement 11,300t, design speed 29kts (using a more powerful version of the Project 58 plant), range 6000nm at 18kt, and crew 541. *Moskva* was laid down on 15 December 1962, for completion within two and a half years, *ie*, before the end of the 1959–65 Plan. However, changes in the design nearly doubled the building period. Thus she ran dock trials only between November 1966 and May 1967, began builders' trials only in May 1967, and official trials only in August 1967.

According to the Russians, the design was greatly complicated by the number of experimental and prototype weapons, many of them still in development, incorporated in it. These ships introduced a new low-frequency sonar in a massive (21m × 6.5m × 9m) retractable keel dome, the first Soviet VDS, a new SAM (SA-N-3), and a new ASW weapon (SUW-N-1, an adapted Army 'Frog' missile). As completed, the ships generally trimmed by the bow, suggesting that weights forward (*ie*, of the fixed weapons) were incorrectly estimated.

At least on board *Moskva*, there was considerable machinery trouble. Her powerplant had to be rebuilt twice, and she suffered a fire in 1973. Her crew was told (incorrectly) that she had been designed to make 33kts, and to search at 26kts. In practice she could barely make 30kts (and any sustained steaming at that speed caused severe damage), and she could sustain 24kts (maximum search speed) for no more than 2 or 3 hours. Her stabilisers were also unsuccessful. However, her ASW search system was considered quite successful. Her sonar ('Orion') filled a cylinder so large that the ship's hull bulged out around it (so that the cylinder could house when the ship entered a shallow harbour, such as Sevastopol); the cylinder interrupted the hull lines forward.

A recent Soviet account indicates that the *Moskva* design was considered unsuccessful. The air group was too small, and the ship pitched badly in heavy seas due to her full stern lines and her Y-shaped body, a shortcoming which had already been experienced in the prewar Project 7-U destroyers.

A July 1991 Russian account of Soviet carrier design states that a third, incomplete, anti-ship missile version of the *Moskva* design was broken up on the slip before the *Kiev* was laid down. Presumably it would have escorted the ASW ships; in effect the *Kiev* design combined the ASW role with the protective anti-ship missile battery. This ship was apparently 12m longer than the pair actually built. She may have carried a nest of missile tubes in the space devoted to a hangar on the two ships built. The original *Moskva* design may have incorporated SS-N-3 launchers alongside the superstructure (pointing forward on one side, aft on the other), but they could not be inlcuded in the ships as completed.

Moskva remained in commission in 1995; *Leningrad* was decommissioned 15 December 1991.

Work on this design continued until 1952. Reportedly Stalin insisted on building battleships rather than carriers, despite the advice of his main naval staff and his Navy Minister, Admiral N G Kuznetzov. The ten-year (1945–55) fleet programme drawn up in 1944 apparently envisaged construction of eight battleships, as well as battle cruisers and lesser ships. Project 24 was the suc-

Battleship (Project 24)

cessor to the prewar Project 23. Developed by TsKB-17. Pre-project estimates completed October 1944, sketch design 1945. Displacement 75,000t, maximum speed 30kts, cruising speed 24kts, economical speed 18kts (range 8000nm); guns 9–16in (3×3), 24–5.1in (12×2), 48–45mm (24×2), 60–25mm; protected against 16in shellfire and 1000kg bombs dropped from 5000m; 2 catapults and 6 aircraft. This project was initially assigned to TsKB-4, which completed a sketch design in March 1942. This TsKB continued work on the design through 1944. A photograph of a model supposedly showing the postwar battleship later proved to be a model of the *Stalingrad* class (Project 82).

Project 66 heavy cruiser design
NB 1/1750 scale

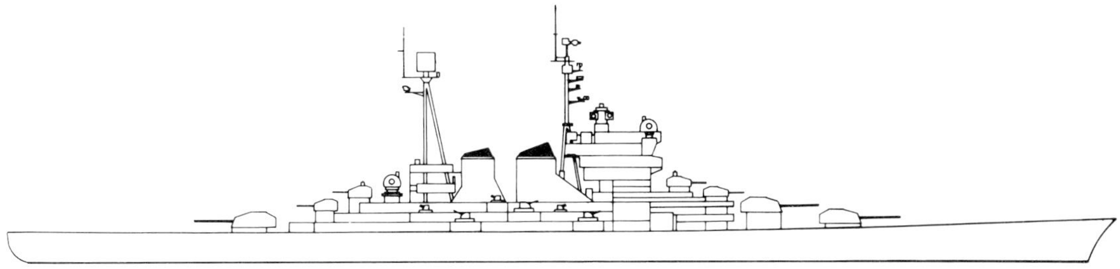

Project 82 was among the first true Soviet postwar warship designs. It began in 1941 as a heavy cruiser armed with 8in guns (the prewar programme also included a pair of 12in battlecruisers, the *Kronstadt* class [Project 69]). Reviewing the design in 1945, Stalin suggested that it be armed with 12in guns. Admiral Kuznetzov countered that he preferred a new 9in gun with a very high rate of fire. Studies were conducted in 1943–47, and at the end of 1947 TsKB-17 and the central naval design office began work on sketch designs with various main batteries, protection, displacement and speed. Apparently by this time Stalin himself had decided that the ships should be armed with 12in guns. Although the TTZ included a variety of fleet missions, Stalin always thought of the ships as raiders. When he reviewed the sketch designs in March 1948, he asked whether the ship could not be made fast enough to catch lesser enemy ships and to escape from more powerful ones, virtually the same terms applied before the Second World War to the German 'pocket battleships'. The other roles were stiffening of light forces, operating with a manoeuvering force, escorting convoys, destroying enemy cruisers, and attacks on coastal installations in support of the army's flanks. The sketch design was rejected, and the designer ordered to cut displacement (then 39,500t) in order to increase speed (from 32kts). A new sketch design was ready late in 1949. To obtain the desired 35kts, power had to be increased by nearly 30 per cent (to four 70,000shp turbines). Displacement was reduced partly by reducing side and transverse armour.

TsKB-17 was badly overloaded; it was also working on light cruisers (Projects 68K and 68bis) and battleships (Project 24). A new bureau, TsKB-16, was hived off to deal with the battleships and battlecruisers. The basic design formally approved on 25 March 1950 showed reduced dual-purpose and AA armament (compared to those in the approved TTZ), increased displacement (36,500t rather than 36,000t) and a speed of 35kts.

STALINGRAD class *battlecruisers* (Project 82)

Displacement: 36,500t standard; 42,300t full load
Dimensions: 852ft 8in wl, 897ft 4in oa × 105ft, 98ft 5in wl × 30ft 2in
260m, 332.6m × 32m, 30m × 9.2m
Machinery: 280,000shp = 35.3kts (24kts cruise). Range 5000nm at 18kts
Armour: Belt 7in (180mm), main turrets 91/2in (240mm)
Armament: 9–304mm (3×3), 12–130mm DP (6×2), 24–45mm (6×4), 40–25mm (10×4)
Complement: 1700 (provision for 30 staff)

Name	Builder	Laid down	Launched	Comp	Fate
STALINGRAD	Nikolayev	5.12.51	–	–	BU
MOSKVA	Baltic	Oct 1952	–	–	BU
KRONSTADT	Molotovsk	May 1955	–	–	BU

Stalin badly wanted these ships. To accelerate them, it was decided during 1950 that the detailed (technical) design would be completed in February 1951 without further sketch work. At that time the Ministry of Ship Production committed to begin two ships during the second quarter of 1951 (at Nikolaev and Leningrad) for delivery in 1954–55. In fact the technical design was approved only in June 1951, with parts of the basic TTZ not yet ready.

Armour was on a cruiser rather than a battleship scale: 180mm side sloped at 15 degrees (extending 1.7m below the design waterline) with 140mm (forward) and 125mm (aft) bulkheads; upper deck 50mm (8–20mm outside citadel); middle deck 70mm (50mm outside citadel); lower deck 15–20mm (10mm outside citadel); conning tower 260mm; main turrets 240mm face, 125mm side and roof (400–762mm rear, to provide enough weight to counterbalance the guns). The belt was expected to protect against 8in shells between 13,000 and 15,000yds (65 to 75 cables in Russian terms) and the deck against such shells beyond 34,000yds (170 cables). The rest of the hull was protected against 6in shells and against 500kg HE bombs dropped from 3000m (about 10,000ft). There was also a four-bulkhead torpedo protection system (air-liquid-air, thicknesses of internal bulkheads 8–15mm, 8–25mm, 50mm, 8–15mm from outer to inner), designed to resist a 400–500kg charge.

Stalin considered Project 82 a key part of the postwar building programme. At the end of 1951, a third ship was ordered from Molotovsk (later Severodvinsk). The programme was quite expensive; on 28 August 1951 Admiral Kuznetzov, who had only recently returned from exile in the Pacific Fleet, told Stalin that these ships were not worth building. Remarkably, he did not suffer for his audacity. No one else supported Kuznetzov, and every effort was made to speed up work. The builders promised to launch the lead ship on the eve of the anniversary of the Revolution, 6 November 1953; in fact by the end of 1952 it was clear that work was five to six months behind schedule. The ships were cancelled soon after Stalin died in March 1953.

In 1954 part of the hull of the lead ship, including the citadel, was launched for use as a target. Trials the following year apparently proved that the protective design would have worked.

It appears that Admiral Kuznetzov ordered the third ship, on which very little work had been done, redesigned as a missile platform (Project 82R). Another three ships were planned, but they were probably never laid down. Project 82R added hull plugs abaft B turret and forward of X turret, in each case accommodating a quadruple SS-N-3 launcher and two sets of reloads in the superstructure (total 24 missiles). The main and dual-purpose armament was unchanged, but light armament consisted of 24–3in (12×2) in the square mounts later adopted for the 'Kynda' class and, possibly, quadruple 57mm guns in place of the quadruple 45mm of Project 82. Standard displacement rose to 38,500t.

Project 82R was cancelled in October 1955.

Project 65 was a catch-all designation for a new light cruiser design to succeed the prewar Project 68 (*Chapaev*). In 1946 TsKB-17, which had designed all the prewar cruisers, and the Navy's Central Scientific Institute for Military Shipbuilding (TsNIIVK) began work on a new cruiser armed with 152mm guns (Project 82 covered 220mm and 305mm ships). The ten-year plan approved 27 November 1945 called for thirty such cruisers. They were intended for classic cruiser tasks: escorting a main fleet against enemy surface and air torpedo attacks; stiffening friendly torpedo forces; covering convoys; raiding enemy commerce; supporting landing operations; and planting minefields.

Project 65 *cruisers*

The approved TTZ called for a relatively small ship (8000–8500t), but initial sketch designs showed a displacement of about 18,000t. TsKB-17 then tried to pare down the ship to 14,500t, which it considered a lower limit for a truly modern cruiser. It also suggested substituting triple 180mm turrets (as in the prewar *Kirovs*) for 152mm, but the navy rejected this idea.

Presumably the great displacement was imposed by protection, particularly against underwater damage. Early in 1947 the Shipbuilding Ministry asked TsKB-17 to develop a series of variants: with 9–12 180mm or 152mm guns; with different levels of protection (100mm/50mm and 75mm/50mm side/deck armour); with different full speeds (32–35kts); with alternative powerplants (2 × 60,000shp or 3 × 50,000shp or 4 × 40,000shp; or, with high pressure and high temperature, 3 × 36,000shp or 3 × 50,000shp); and with alternative electrical systems (AC or DC) A further alternative would have been armed with nine dual-purpose 152mm guns. In all, forty versions (12,800t to 16,850t standard) were developed. They showed that adopting 180mm guns would not much affect displacement, so twelve 180mm were adopted. The nine-152mm (DP) version was kept alive because it provided superior firepower to contemporary foreign cruisers (the US *Worcester* and the British *Mauritius*).

The project was killed at a Spring 1947 Kremlin meeting; effort shifted to the Project 82 and Project 68-bis designs. Since Project 68-bis was a relatively simple redesign, the decision to kill off Project 65 was presumably an indication that Soviet resources could support only a single cruiser design, and that Stalin much preferred the Project 82 battlecruiser.

Project 66 *heavy cruiser*

Displacement:	24,800t standard; 30,750t full load
Dimensions:	828ft 2in × 100ft 4in × 29ft 6in *252.5m × 30.6m × 9m*
Machinery:	? = 32.5kts
Armour:	Belt 155mm, main turrets 135–170mm
Armament:	9–228mm/65 (3×3), 8–130mm/54 (4×2), 24–45mm (6×4), 24–45mm (6×4)
Sensors:	Not known
Complement:	Not known

Like Project 82, this design derived from the heavy cruiser component of the 1946–55 Ten Year Plan. Reviewing the fleet's cruiser strength in 1945, Admiral of the Fleet N G Kuznetzov, then the commander of the Soviet Navy, argued that existing and planned 6in and 7.1in cruisers could not face down the large US force of heavy cruisers, not to mention the *Alaska*s. In 1946 he ordered studies of heavy cruiser (Project 82) variants armed with 9in and 12in guns at the navy's scientific research institute and at the heavy combatant design bureau (TsKB-17). Work on a new rapid-fire 9in/65 gun had already begun at the central ordnance KB in Moscow (January 1945); the weapon was ready for trials in 1954. The 12in gun considered at this time was that developed prewar for Project 69. When Project 82 was shifted from 9in to 12in, the planned main battery was a new longer 12in gun.

Meanwhile the Naval Academy, which was responsible for operational analysis, proposed a new type of cruiser, heavily armed and very fast (36–38kts), lightly armoured to limit its displacement. TsKB-17 developed a pre-sketch design, LKT-22 (light cruiser with 22cm, *ie* 9in, guns). By the time it was ready in 1947, Kuznetzov had been exiled to the Far East; the design was approved by Admiral I S Yumashev, his successor.

Stalin showed no interest in LKR-22; he ordered efforts concentrated on the 12in gun Project 82. When Admiral Kuznetzov returned to head the Soviet Navy in the summer of 1951, he revived the 9in gun cruiser. TsKB-17 completed a preliminary sketch design early in 1952, and a formal sketch design, to a TTZ issued that April, was ready by the end of the year. Stalin now accepted the argument that medium cruisers armed with 9in guns were needed to face US 8in ships.

The TTZ included a demand that both main and secondary batteries be concentrated on the centreline for maximum arcs of fire, and that the ship be very heavily armoured. To the greatest possible extent, systems were to duplicate those on other ships, to make production easier. Protection and high speed were both quite expensive: the ship would have been quite large. The sketch design was reviewed in March 1953 and revisions were completed that August. On 23 December the naval staff asked the Naval Academy to review the TTZ, for completion by 1 March 1954.

The Academy doubted whether the new 9in/65 would really buy enough superiority over the US *Des Moines* type. Fire could not be controlled effectively at the guns' maximum range of 40,000 to 52,000yds (splashes could not be spotted, even by radar). Hit probability at 30,000 to 40,000yds would be so poor that the entire magazine (1080 rounds) could be exhausted without making enough hits to put a *Des Moines* out of action. At shorter range (10,000 to 30,000yds) the US ships would enjoy the advantages of a higher rate of fire and better accuracy. Project 66 was too large to be worthwhile. Work stopped before the Academy's suggestion, that speed be cut to 32kts so that armour could be thickened, could be adopted.

One peculiarity of the design seems to have been the use of a shorter 5.1in gun for the secondary battery (54 calibres long rather than the 58 calibres of Project 82 and of contemporary Soviet destroyers).

Aleksandr Suvorov Apr 1970 CMP

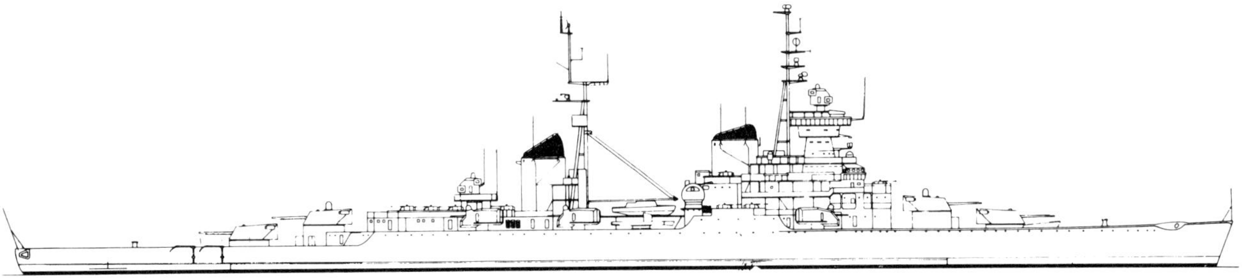
Sverdlov

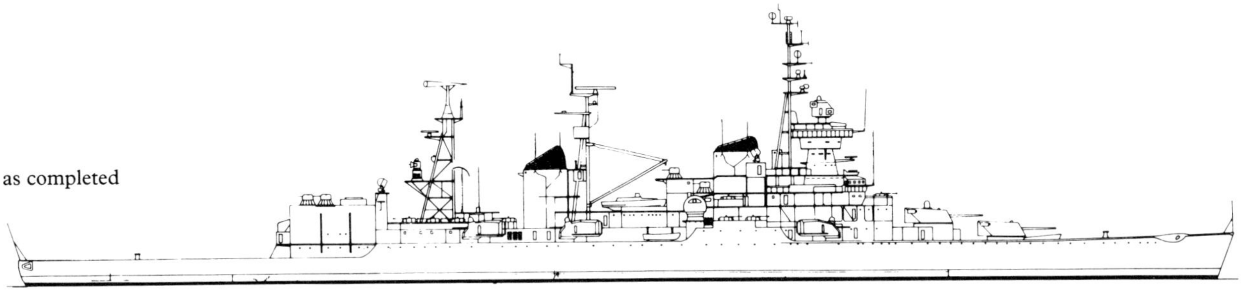
Admiral Senyavin as completed

SVERDLOV class *cruisers* (Project 68B)

Displacement: 13,600t standard; 16,640t full load
Dimensions: 672ft 4in wl, 689ft oa × 72ft × 22ft 7in
205m, 210m × 22.0m × 6.9m
Machinery: ? 110,000shp = 32.5kts. Range 9000nm at 18kts
Armour: Side 4in (100mm), bulkheads 4¾in (120mm), deck 2in (50mm), CT 6in (150mm)
Armament: 12–152mm/57 Mk 5 (4×3), 12–100mm/70 (6×2), 32–37mm (16×2), 10–533mm TT (2×5).
Sensors: Radar Sea Gull, Knife Rest A, Slim Net, Top Bow, Egg Cup (for both 152mm and 100mm mounts); sonar H/F hull
Complement: 1250

Name	Builder	Laid down	Launched	Comp	Fate
SVERDLOV	Baltic	15.10.49	5.7.50	15.5.52	Stricken 1989
ZHDANOV	Baltic	11.2.50	27.12.50	31.12.51	Stricken 1991
ADMIRAL USHAKOV	Baltic	31.8.50	29.9.51	8.9.53	Stricken 1987
ALEKSANDR SUVOROV	Baltic	26.2.51	15.5.52	31.12.53	Stricken 1990
ADMIRAL SENYAVIN	Baltic	31.10.51	22.12.52	30.11.54	Stricken 1991
DMITRY POZHARSKI	Baltic	31.3.52	25.6.53	31.12.54	Stricken 1987
KRONSTADT	Baltic	Oct 1953	11.9.54	–	BU 1961
TALLIN	Baltic	1953	11.9.54	–	BU 1961
VARYAG	Baltic	Dec 1952	5.6.56	–	BU 1961
ORDZHONIKIDZE	Admiralty	19.10.49	17.9.50	30.6.52	To Indonesia 1962
ALEKSANDRNEVSKI	Admiralty	30.5.50	7.6.51	31.12.52	Stricken 1989
ADMIRAL LAZAREV	Admiralty	6.2.51	29.6.52	30.12.52	Stricken 1986
SHCHERBAKOV	Admiralty	Jun 1951	17.3.54	–	BU 1961
DZERZHINSKI	Nikolayev	31.12.48	31.8.50	18.8.52	Stricken 1989
ADMIRAL NAKHIMOV	Nikolayev	27.6.50	29.6.51	27.3.53	Stricken 1961
MIKHAIL KUTUZOV	Nikolayev	23.2.51	29.11.52	30.2.54	?
ADMIRAL KORNILOV	Nikolayev	6.11.51	17.3.54	–	Hulk PKZ 130, 1957
OKTYABRSKAYA REVOLUTSIYA (ex-*Molotovsk*)	Severodvinsk	15.7.52	25.5.54	30.11.54	Stricken 1957
MURMANSK	Severodvinsk	28.1.53	24.4.55	22.9.55	Stricken 1992
ARKHANGELSK	Severodvinsk	1954	–	–	BU 1961
VLADIVOSTOK	Severodvinsk	1955	–	–	BU 1961

These large light cruisers were enlarged versions of the prewar *Chapaevs* (Project 68). They had the same machinery, main armament, and side protection, but had much increased fuel capacity (for an endurance of 9000nm; provisions sufficed for 30 days) and introduced an all-welded hull and a new type of underwater protection (incuding a double bottom over 75 per cent of their length, and twenty-three watertight bulkheads). These ships had new fire control radars (Ryf and Zalp for their main batteries, Yakor for their secondaries) as well as new types of directors (stabilised SPN-500s for the 100mm guns). Their 152mm turrets were a modified version of the MK-5 of the *Chapaev* class; the stabilised German-style 100mm/70 (SM-5-1) secondary mounts were new. The Zenit-68-bis AA fire control system could control both the secondary guns and 37mm light AA guns. The torpedo fire control system employed both optical rangefinders and a dedicated radar (Zarya). The Project 68bis design was formally approved on 27 May 1947, three months *before* the 68K version to which the *Chapaevs* were completed. The first three ships were named after *Chapaevs* abandoned incomplete in 1941: *Sverdlov*, *Ordzhonikidze*, *Dzerzhinski*.

On trials, *Sverdlov* made 114,000shp = 32.69kts and 121,700shp = 33.04kts. Reportedly the rated endurance of 9000nm was 1600nm above the design fiture. Reportedly, too, *Sverdlov* came out 370t light at 13,230t standard.

The postwar ten-year plan originally called for thirty cruisers. One account of the *Sverdlovs* claims that five were dropped in favour of the three Project 82s, but it seems more likely that the five *Chapaevs* counted as the first five postwar cruisers. At Stalin's death, the class was cut to twenty-one ships. After the first fifteen were laid down, the Soviet Navy ordered the remaining six (beginning with *Shcherbakov*; none was completed) protected against radioactive fallout as Project 68zif. It appears that plans were soon made to convert all these ships to missile units (Project 75). The early 1956 decision to stop construction presumably followed the October 1955 cancellation of Project 75. When work stopped, *Admiral Kornilov* was nearly complete, and she alone of the incomplete ships was hulked. In 1959 all the others were ordered broken up by the end of 1961.

In appearance these ships generally resembled the *Chapaevs*, but with their hull plating carried much further aft, with an additional pair of twin 100mm mounts near the bridge, and a taller cap to their second funnels. Torpedo tubes were all removed by the early 1960s. Active units were later fitted with either Top Trough or Big Net long-range air search radars. However, some retained their early Sea Gulls.

Standard Soviet practice was to pass these ships in and out of reserve, so that a ship could remain out of service for an extended period without going to the breakers. Even so, there were breakdowns: *Dzerzhinski* had to be towed home from the Mediterranean in 1979. Presumably the limited scale of Project 68A shows that these ships were not considered worth modernising.

Admiral Nakhimov was rebuilt in 1955–57 with an anti-ship missile (adapted from AS-1) in place of her forward turrets. She was stricken in February 1961, and used as a target. This conversion was probably Project 67EP, but may have been designated Project 71. *Dzerzhinski* was converted to an anti-aircraft missile cruiser in 1960–62, carrying the army's SA-2 missile (Project 70E). *Zhdanov* (Project 68-U-1) and *Admiral Senyavin* (Project 68-U-2) were both converted to command ships in 1970–72. Each had No. 3 triple 152mm replaced by command spaces, with a pop-up SA-N-4 missile launcher and four twin 30mm guns on top (*Senyavin* only), with another set of 30mm guns around the forefunnel; a new tripod mainmast carried the Vee Cone long-range communications antenna. *Senyavin* lost both after turrets and had a helicopter hangar aft. As rebuilt the ships were reclassified KU, *Korabl Upravleniy* (command ships), rather than KR, *Kreyser*.

Project 68A was a limited modernisation. *Oktybrskaya Revolutsiya* emerged from a refit in 1977 with an enlarged bridge. *Admiral Ushakov* and *Aleksandr Suvarov* were similarly altered in 1979. *Mikhail Kutuzov* was also similarly altered. All had their Egg Cup radars removed from 100mm mounts and exchanged two twin 37mm for four twin 30mm controlled by Drum Tilt radars. As modernised, *Admiral Ushakov* displaced 14,470t standard and 17,970t full load.

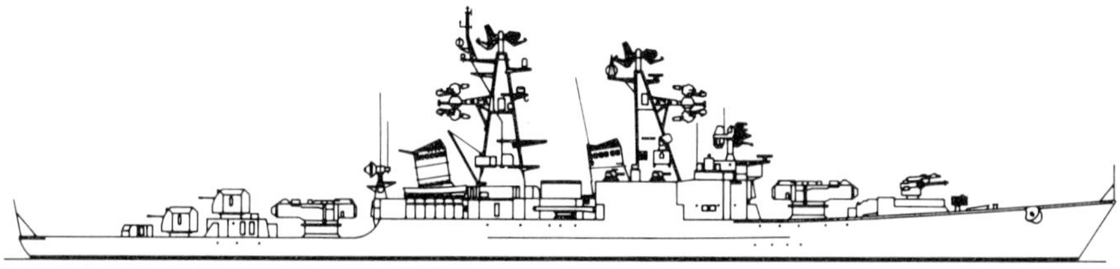

Varyag 1989

'KYNDA' class *cruisers* (Project 58)

Displacement: 4400t standard; 5600t full load
Dimensions: 464ft 9in oa × 51ft 10in × 17ft 5in (hull only)
141.7m × 15.8m × 5 3m
Machinery: 2-shaft geared steam turbines, 4 pressure-fired boilers, 100,000shp = 34kts. Range 2000nm at 34kts, 7000nm at 14.5kts
Armament: 2 SS-N-3 (2×4, 16 missiles), 1 SA-N-1 (1×12, 24 missiles), 4–76mm (2×2), 2 RBU-6000 ASW RL (2×12), 6–533mm TT (2×3)
Sensors: Radar 2 Don-2, 2 Head Net A or C, 2 Scoop Pair, Peel Group, Owl Screech, 2 Plinth Net (2 Bass Tilt in *Fokin*); sonar Hercules H/F hull; ECM 3 Bell series, 4 Top Hat
Complement: 375–390

Name	Builder	Laid down	Launched	Comp	Fate
GROZNY	Zhdanov	23.2.60	26.3.61	30.12.62	Stricken 24.6.91
ADMIRAL FOKIN (ex-*Vladivostok*, ex-*Admiral Fokin*, ex-*Steregushchiy*)	Zhdanov	5.10.90	19.11.61	28.12.64	Stricken 30.12.93
ADMIRAL GOLOVKO (ex-*Doblestnyi*)	Zhdanov	20.4.61	18.7.62	30.12.64	Stricken 1991
VARYAG	Zhdanov	13.10.61	7.4.63	20.8.65	Decomm 14.10.90

These four ships were Khrushchev's concession to Soviet naval advocates of surface ships. Like the 'Kashins', they were part of a new generation, development of which was ordered in 1956. The TTZ for this class was approved on 6 December 1956. In effect they replaced four Project 82R missile-firing battlecruisers, carrying the same quadruple missile mounts (but with one rather than two reloads per tube). Like other postwar surface combatants, they had the new pressure-fired steam plant. Like the contemporary 'Kashins', they had the new SA-N-1 defensive missile. According to a recent Russian publication, they were initially classed as guided missile cruisers (hence the destroyer names listed above), then reclassified as cruisers 31 October 1962. Of ten planned units, only four were built.

A recent Russian article associated the command-guided version of the SS-N-3 anti-ship missile (P-35) with this class. Ships were completed with the Scoop Pair missile-tracking radar (two missiles tracked per radar mounting) but without the 'Plinth Net' missile data link receiving antenna (but with the necessary platform). A helicopter pad aft could accommodate a 'Hormone' (Ka-25RTs) targeting helicopter. Plinth Net was added to *Groznyy* in 1973, and later to *Admiral Fokin*. The other two never received it.

The Soviets designated these ships 'rocket cruisers' (RKR) and they made a considerable impression in the West, with their main battery effective at up to 250nm, and their overall level of armament for their size, shorter (albeit beamier) than 'Kashins'. This classification later proved extremely embarrassing to the USN, which had numerous 'frigates' (DLG/DLGN) larger than Soviet cruisers of the

'Kynda' class 1964 CMP

'Kynda' and later classes, and suffered in comparative tables from its apparent lack of modern cruisers. The problem was solved only with the reclassification of most of the DLGs as cruisers in 1975.

Four 30mm 'Gatling' guns controlled by Bass Tilt were added to *Groznyy* 1980–82 and to *Varyag* in 1981. *Varyag* had a deckhouse fitted between her funnels. All were completed with Head Net A radars fore and aft; as of 1989 *Admiral Fokin* had one Head Net A and one Head Net C; *Varyag* had two Head Net C. *Groznyy* was assigned to the Baltic Fleet 1982. *Admiral Golovko* was in the Black Sea Fleet; the other two were in the Pacific. *Varyag* removed from operational service in 1990. She was to have become a museum, but as of 1995 this seems unlikely. Remainder since stricken.

'Kresta I' class 1975
CMP

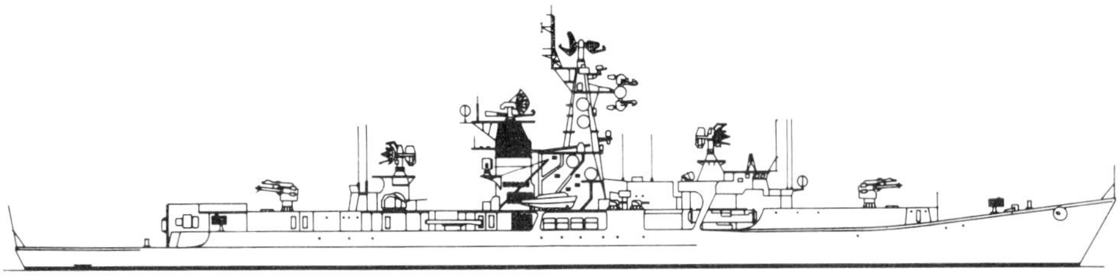

Vitse Admiral Drozd after reconstruction

'KRESTA I' class *cruisers* (Project 1134)

Displacement: 6000t standard; 7500t full load
Dimensions: 508ft 6in oa × 55ft 9in × 18ft 1in (hull only)
155.0m × 17.0m × 5.5m
Machinery: 2-shaft steam turbines, 4 pressure-fired boilers, 100,000shp = 34kts. Fuel 1150t. Range 2400nm at 32kts, 10,500nm at 14kts
Armament: 2 SS-N-3 (2×2, 4 missiles), 2 SA-N-1 (2×2, 44 missiles), 4–57mm (2×2), 2 RBU-6000 ASW RL (2×12), 2 RBU-1000 ASW RL (2×6), 10–533mm TT (2×5), 1 helicopter
Sensors: Radar Don Kay, Don-2, Big Net, Head Net C, 2 Plinth Net, Scoop Pair, 2 Peel Group, 2 Muff Cob (2 Bass Tilt in *Drozd*); sonar Hercules H/F hull set: ECM 8 Side Globe, 4 Bell series
Complement: 380

Name	Builder	Laid down	Launched	Comp	Fate
ADMIRAL ZOZULYA	Zhdanov	26.7.64	17.10.65	8.10.67	Extant 1995
VITSE-ADMIRAL DROZD	Zhdanov	26.10.65	18.11.66	27.12.68	Stricken 1.10.90
VLADIVOSTOK	Zhdanov	24.12.64	1.8.67	11.9.69	Stricken 1.1.91
SEVASTOPOL	Zhdanov	8.6.66	28.4.67	25.9.69	Stricken 19.4.90

These four ships are the remnant of a large anti-ship rocket-cruiser (RKR) programme, stopped when Soviet priorities shifted to ASW. Although the ships were still armed with SS-N-3 (P-35) anti-ship missiles, they were reclassified BPK (large ASW ships). The programme initially called for a brigade of seven heavy ASuW ships (armed with long-range missiles) plus two brigades (fourteen ships) of light ASuW ships (armed with shorter-range missiles). All were designated Project 934: the light-missile ships were 934A. Khrushchev rejected a proposed flagship version, presumably 934K (Kommandatur), but did not change the numbers (which would have amounted to a leader plus three pairs of ships).

The main battery was to have been a new ramjet Bazalt (SS-N-12) missile, which was considered four times as effective as the slower SS-N-3 (P-35). Unfortunately development was protracted, and P-35 had to be substituted. Compared to 'Kynda', this class showed increased emphasis on self-defence, with a second SA-N-1 launcher and twin 57mm guns. For the first time in Soviet practice, the targeting helicopter aft was provided with a hangar.

The abortive Project 934K would have been a stretched hull with an enlarged helicopter deck aft (about twice the length and width of the deck actually used), accommodating five helicopters (mainly to data-link beyond the horizon to other ships of the attack group). Armament would have matched that of Project 934.

The effect of the shift to ASW was to redesignate the first four ships and to stop construction of other ships of the series while they were heavily modified.

These ships were not extensively modified. *Vitse Admiral Drozd* was fitted with four 30mm 'Gatlings' abaft the SS-N-3 launchers and with a new deckhouse between them (1976); *Admiral Zozulya* was similarly refitted (completed 1991); *Sevastopol* received the deckhouse but not the 'Gatlings'. By 1992 only the recently-refitted *Admiral Zozulya* remained in service.

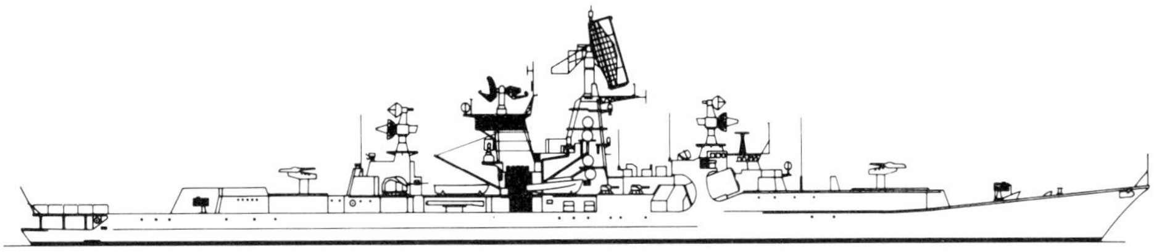

Admiral Oktyabr'skiy as completed

'KRESTA II' class *cruisers* (Project 1134A, Berkut A)

Displacement: 5600t standard; 6565t normal; 7535t full load
Dimensions: 485ft 7in wl, 521ft 6in oa × 55ft 1in, 54ft 10in wl × 17ft 5in
148.0m, 159.0m × 16.8m, 16.7m × 5.32m
Machinery: 2-shaft pressure-fired steam turbines (4 boilers), 91,000shp = 32kts; 1830t fuel oil. Range 5200nm at 18kts
Armament: 2–4 SS-N-14 (8 missiles), 2 SA-N-3 (2×2, 48 missiles), 2 RBU-6000 (2×12, 144 rockets), 2 RBU-1000 (2×6, 60 rockets), 4–57mm (2×2, total 2200 rounds), 4–AK-630 Gatling guns (total 8000 rounds), 2–5 21in TT, 1 Ka-25 helicopter
Sensors: Radar: 2 Don-Kay, Don-2, Top Sail, Head Net C, 2 Head Lights, 2 Muff Cob, 2 Bass Tilt; sonar Bull Nose; ECM 8 Side Globe, 7 Bell series
Complement: 343

Name	Builder	Laid down	Launched	Comp	Fate
KRONSTADT	Zhdanov	12.12.66	10.2.68	29.12.69	Stricken 24.6.91
ADMIRAL ISAKOV	Zhdanov	15.1.68	22.11.68	28.12.70	Stricken 30.6.93
ADMIRAL NAKHIMOV	Zhdanov	15.1.68	15.4.69	29.11.71	Stricken 31.1.91
ADMIRAL MAKAROV	Zhdanov	23.2.69	22.1.70	25.10.72	Stricken 3.7.92
MARSHAL VOROSHILOV	Zhdanov	20.3.70	8.10.70	15.9.73	Stricken 3.7.92
ADMIRAL OKTYABR'SKIY	Zhdanov	2.6.70	21.5.71	28.12.73	Stricken 30.6.93
ADMIRAL ISACHENKOV	Zhdanov	30.10.70	28.3.72	5.11.74	Stricken 3.7.92
MARSHAL TIMOSHENKO	Zhdanov	22.11.72	21.10.73	25.11.75	Stricken 3.7.92
VASILY CHAPAEV	Zhdanov	22.12.73	28.11.74	30.11.76	Stricken 30.6.93
ADMIRAL YUMASHEV	Zhdanov	17.4.75	30.9.76	30.12.77	Stricken 23.2.93

These were 'Krestas' redesigned for ASW service while under construction. The first three were to have been further 'Kresta Is'; the others were to have carried a shorter-range anti-ship missile battery (eight SS-N-9/P-50). The shorter-range missile was replaced by an ASW missile, probably adapted from the anti-ship weapon (SS-N-14), and a new bow sonar was fitted (the bow was more sharply raked to accommodate it). The surface-to-air missile was the more capable SA-N-3, requiring a better 3-D radar (Top Sail) than the Head Net C/Big Net combination of 'Kresta I'. SA-N-3, the only Soviet naval SAM with no land equivalent, reportedly had a substantial anti-ship capability, and would have been welcome in a ship otherwise restricted to ASW ordnance. These ships are designated BPK, large ASW ships.

Admiral Makarov allegedly bombarded the Angolan ports of Lobito and Benguela on 11–12 February 1976 in support of the MPLA during the Angolan Civil War. *Kronstadt* and *Admiral Isachenkov* were for disposal 1992.

The above metric data are from a Russian source and should be considered accurate.

Admiral Isakov Dec 1977

MoD

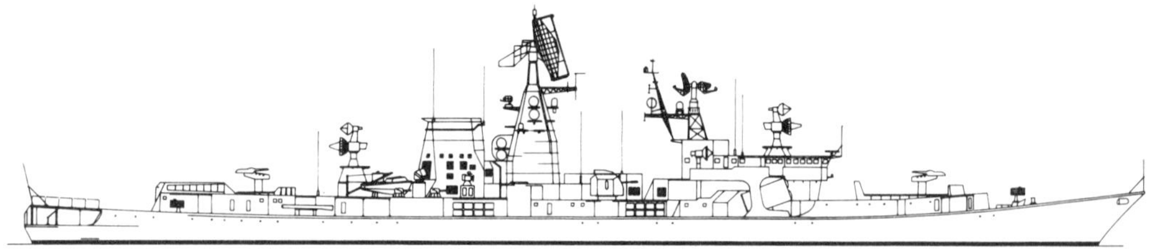

Ochakov as completed

'Karas' were redesigned versions of the version of Project 934 armed with short-range anti-ship missiles (Project 934A). Beyond the changes introduced in 'Kresta II', they were powered by gas turbines and were enlarged to provide substantially better command/control for associated ASW forces, such as aircraft. Reportedly the gas turbines were used partly because they were far quieter than the pressure-fired steam plant of the earlier ships. It is also possible that gas turbines were used partly because the main Soviet gas turbine plant was in southern Russia, near the building yard at Nikolaev on the Black Sea. These ships also added a short-range defensive missile (SA-N-4), a VDS, and 76mm rather than 57mm guns. The bow sonar is unofficially described as low- rather than medium-frequency, and one can speculate that the VDS was particularly attractive for use in the Mediterranean, where most of these ships served.

Azov (Project 1134BE) was test ship for the SA-N-6 missile system, with Top Dome fire control radar in place of Head Lights. She never left the Black Sea during her career. The name ship entered the Mediterranean on 2 March 1973. *Petropavlovsk* and *Tashkent* joined the Pacific Fleet in 1979.

'KARA' class *cruisers* (Project 1134B, Berkut B)

Displacement: 6700t standard; 7630t normal; 8565t full load
Dimensions: 531ft 6in wl, 568ft 9in oa × 60ft 8in, 55ft 1in wl × 17ft 5in
162.0m, 173.4m × 18.5m, 16.8m × 5.32m
Machinery: 2-shaft COGAG, 4 gas turbines (type M-5), 120,000shp = 32kts; 1830t fuel. Range 6500nm at 18kts
Armament: 2–4 SS-N-14 (8 missiles), 2 SA-N-3 (2×2, 72 missiles), 2 SA-N-4 (2×2, 40 missiles), 4–76mm (2×2, 4800 rounds total), 4 AK-630 Gatling (total 8000 rounds), 10 533mm TT (2×5, no reloads), 2 RBU-6000 (144 rounds), 2 RBU-1000 (60 rounds; not in *Petropavlovsk*), 1 Ka-25 helicopter. See notes.
Sensors: Radar: 2 Don-Kay, Don-2, Top Sail, Head Net C, 2 Head Lights, 2 Pop Group, 2 Owl Screech, 2 Bass Tilt; sonar: Bull Nose and Mare Tail, ECM 8 Side Globes, 7 Bell series (first 2 ships) or 2 and 4 Rum Tub
Complement: 380

Name	Builder	Laid down	Launched	Comp	Fate
NIKOLAYEV	Nikolayev	25.6.68	19.12.69	31.12.71	Stricken 29.10.92
OCHAKOV	Nikolayev	19.12.69	30.4.71	4.11.73	Extant 1995
KERCH	Nikolayev	30.4.71	21.7.72	26.12.74	Extant 1995
AZOV	Nikolayev	21.7.72	14.9.73	25.12.75	Extant 1995
PETROPAVLOVSK	Nikolayev	9.9.73	22.11.74	29.12.76	Extant 1995
TASHKENT	Nikolayev	22.11.74	5.11.75	21.12.77	Stricken 3.7.92
VLADIVOSTOK (ex-*Tallin*)	Nikolayev	5.11.75	5.11.76	31.12.79	Extant 1995

'Kara' class cruiser Feb 1976
MoD

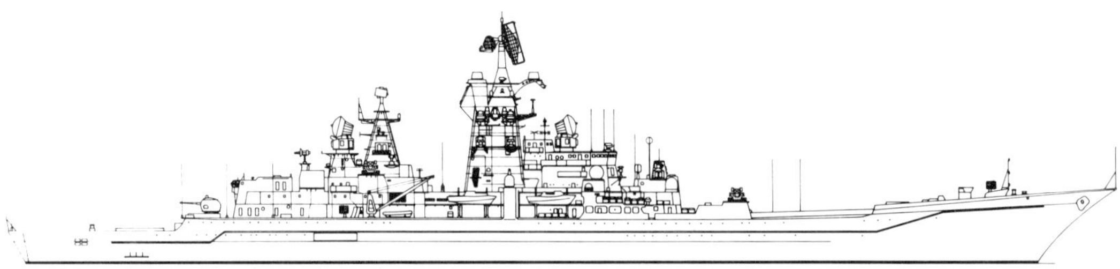

Kalinin 1989 **NB 1/1750 scale**

Kirov, the first Soviet nuclear-powered surface combatant, was sometimes called a 'battlecruiser'. She apparently employs a CONAS (combined nuclear and steam) powerplant, in which fossil-fuel boilers supply steam to turbines geared to the same shafts as the nuclear-powered turbines. The USN considered a somewhat similar scheme: it is difficult to achieve sufficient power for high speed in a compact enough nuclear plant. Combined nuclear and gas turbine and nuclear and steam turbine plants were considered but rejected on grounds of complexity. The Russians apparently saw things differently. Similar fossil-fuel boosters have been used in land-based Western power reactors.

KIROV class *cruisers* (Project 1144, Orlan)

Displacement: 24,000t standard; 28,000t full load
Dimensions: 754ft 7in wl, 813ft 8in oa × 78ft 9in wl, 91ft 10in ext × 24ft 7in hull
230.0m, 248.0m × 24.0m, 28.0m × 7.5m
Machinery: 2 nuclear reactors plus oil-fired superheaters, 150,000shp = 32–34kts. See notes
Armament: 20 SS-N-19 (20×1), 12 SA-N-6 (12×1, 96 missiles), 2 SA-N-4 (2×2, 40 missiles), 2-100mm (2×1), 8–30mm Gatling (8×1), 1 SS-N-14 (1×2, 14–16 missiles), 8–533mm TT (2×4), 1 RBU-6000 ASW RL (1×12), 2 RBU-1000 ASW RL (2×6), 3 helicopters
Sensors: Radar 2 Palm Frond, Top Pair, Top Steer, 2 Top Dome, 2 Pop Group, 2 Eye Bowl, Kite Screech, 4 Bass Tilt, Punch Bowl; sonar Horse Jaw and Horse Tail; ECM 8 Side Globe, 10 Bell series, 4 Rum Tub, 2 chaff RL (2×2)
Complement: 800

Name	Builder	Laid down	Launched	Comp	Fate
KIROV	Baltic, Leningrad	26.3.74	27.12.77	30.12.80	Extant 1995
FRUNZE	Baltic, Leningrad	26.7.78	26.5.81	31.10.84	Extant 1995
KALININ	Baltic, Leningrad	May 1983	26.4.86	30.12.88	Extant 1995
YURI ANDROPOV	Baltic, Leningrad	24.4.86	29.4.89	–	Completing
ADMIRAL KUZNETZOV	Baltic, Leningrad	–	–	–	Cancelled 4.10.90

Kirov 1980 MoD

Speed on reactors alone (90,000shp) would probably be about 24kts. Endurance at high speed, on both plants: 14,000nm at 30kts. On oil power alone: 1000nm at 17kts.

The TTZ for this class was approved on 25 May 1971. Design work on a nuclear-powered cruiser began in 1968. Initial studies were based on the contemporary 'singing frigate', which was armed with 6 to 8 medium-range (150km) anti-ship missiles (it was probably a pre-design for what became the *Sovremenny*). Admiral Gorshkov apparently personally rejected the design as too weak, in effect a nuclear-powered 'Kynda'. The design team thought that the ship should not displace more than 8,000t, and attempts to remain within that limit cost considerable time. They soon realised that the weapon systems they wanted (for anti-saturation AAW and for effective ASW, as well as good anti-ship firepower) would demand a displacement of at least 20,000t. Given existing submarine reactors, that in turn seemed to demand 4-6 reactors, which they rejected as cumbersome and dangerous. They were fortunate that a nuclear power design bureau came up with the elegant solution (nuclear base plant plus steam boost) which they adopted.

Kirov was conceived as a flagship to control anti-ship (RKR) formations, following the demonstrated failure of shore control in the 1970 Okean exercise. The design of later ships was further modified to provide more self-defence armament after the 1975 exercise showed how an enemy might well concentrate on destroying just such a flagship. The ship was probably based on an earlier nuclear RKR without flag facilities, possibly called Project 954, for the Pacific Fleet. The Russians have referred explicitly to an earlier nuclear cruiser design, Project 63, but no details have been published.

The main battery is SS-N-19 (Granit) in canted deck tubes. The Rum Tub ESM array is used to target these weapons passively. Defensive weapons are the new SA-N-6 vertically-launched missile and a reloadable SS-N-14 (using a large bow sonar and a new VDS). The second ship, *Frunze*, dispensed with SS-N-14 because she could fire SS-N-15 from her torpedo tubes. The space freed (and space aft, used for 'Gatlings' in *Kirov*) went into vertically-launched SA-N-9 missiles, but she retained two SA-N-4s (presumably as an interim measure, since SA-N-9 was not ready when the ship was completed). She substitued a twin 130mm mount for the two single 100mm of *Kirov*. The third ship, *Kalinin*, eliminated torpedo tubes in favour of the new RBU-12000 mount forward. She introduced a new close-in weapon, CADS-N-1. All three ships have different EW suites. *Kirov* has Side Globes. *Frunze* has Wine Flask. *Kalinin* has Foot Ball.

Protection: over reactor: 100mm side and deck over reactor, ends 35mm; over steering: side 70mm, deck 50mm, bulkheads 75mm. Magazines 80mm.

All four ships were renamed May 1992 as, respectively, *Admiral Ushakov*, *Admiral Lazarev*, *Admiral Nakhimov* and *Petr Velikiy*. A fifth, *Dzerzhinskiy*, was laid down 1989 but almost immediately stopped, and the components scrapped in November 1989. Another was probably planned but never ordered. In February 1993 it was officially stated that *Petr Velikiy* would be completed as part of a new ten-year naval programme; work began (albeit very slowly) by August 1993. Meanwhile two other units are reportedly effectively out of commission because they cannot be refuelled. As of early 1995, *Admiral Ushakov* had not deployed since experiencing a minor nuclear accident in the Mediterranean in January 1990; her propellers have been removed for repairs, and have not been replaced.

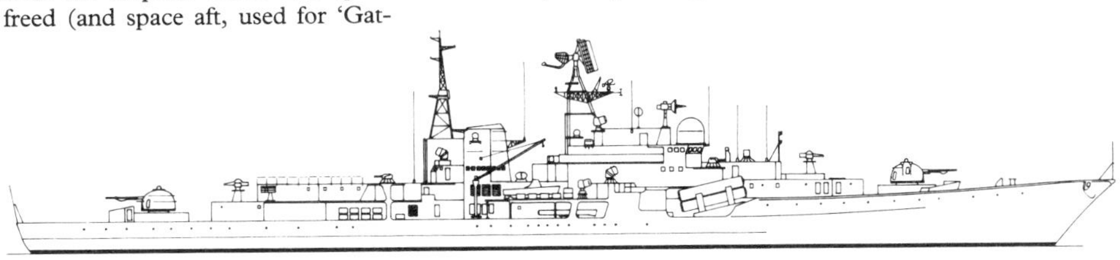

Sovremenny as completed

SOVREMENNYY class *cruisers* (Project 956 'Sarych')

Displacement: 6200t standard; 7800t full load
Dimensions: 511ft 10in oa × 56ft 9in × 21ft 4in
156.0m × 17.3m × 6.5m
Machinery: 2-shaft geared turbines, 4 pressure-fired boilers, 100,000shp = 35kts. Range 2400nm at 32kts, 10,500nm at 14kts
Armament: 2 SS-N-22 (2×4, 8 missiles), 2 SA-N-7 (2×1, ?48 missiles), 4–130mm DP (2×2), 4–30mm Gatling (2×2), 4–533mm TT (2×2), 2 RBU-1000 ASW RL (2×6), 1 helicopter
Sensors: Radar 3 Palm Frond, Top Steer, 6 Front Dome, 2 Bass Tilt, Kite Screech; sonar M/F bow; ECM 8 Bell series, 2 chaff RL
Complement: 350–380

Name	Builder	Laid down	Launched	Comp	Fate
SOVREMENNYY	Zhdanov	3.3.76	18.11.78	25.12.80	Extant 1995
OTCHAYANNYY	Zhdanov	4.3.77	29.3.80	30.9.82	Extant 1995
OTLICHNY	Zhdanov	22.4.78	21.3.81	30.9.83	Extant 1995
OSMOTRITELNYY	Zhdanov	27.10.78	24.4.82	30.9.84	Extant 1995
BEZUPRECHNYY	Zhdanov	1980	Aug 1983	6.11.85	Extant 1995
BOYEVOY	Zhdanov	26.3.82	4.8.84	28.9.86	Extant 1995
STOYKIY	Zhdanov	28.9.82	27.7.85	31.12.86	Extant 1995
OKRYLENNYY	Zhdanov	16.4.83	31.5.86	30.12.87	Extant 1995
VNUSHITELNY	Nikolayev	30.8.83	17.10.87	–	Hulked incomplete
BURNYY	Zhdanov	1984	Feb 1987	30.9.88	Extant 1995
GREMYASHCHIY (ex-*Vedushchyy*)	Zhdanov	23.11.84	30.5.87	30.12.88	Extant 1995
BYSTRYY	Zhdanov	29.10.84	28.11.87	30.9.89	Extant 1995
RASTOROPNYY	Zhdanov	15.8.86	4.6.88	30.12.89	Extant 1995
BEZBOYAZNENNYY	Zhdanov	1987	Mar 1989	28.11.90	Extant 1995
BEZUDERZHANNYY	Zhdanov	1987	Jun 1990	25.6.91	Extant 1995
BESPOKOYNYY	Zhdanov	1988	1990	28.12.91	Extant 1995
NASTOYCHIVY (ex-*Moskovskiy Komsomolets*)	Zhdanov	1989	1991	30.12.92	Extant 1995
BESSTRASHNYY	Zhdanov	–	1992	30.12.93	Extant 1995
VAZHNYY	Zhdanov	1989	–	1996	Building
VDUMCHIVYY	Zhdanov	1989	–	1997	Building

These SSM cruisers (designated EM, destroyers, by the Russians) are production successors to the 'Krestas' with very similar hulls and the same pressure-fired steam plants. They evolved out of a 1970–71 TTZ for a 5000t ship carrying three single-barrel guns, an air defence missile (presumably SA-N-4), and rocket launchers. It was intended primarily to support amphibious assaults, and was presumably associated with the large new amphibious ships, such as *Ivan Rogov*. The guns envisaged were presumably 130mm or larger calibre. Subsequent cost-benefit analysis led to drastic revision of the TTZ; the ship became a multi-purpose surface warship. This version of the design was completed in 1974, and shipyard work began in 1975.

As redesigned, the new destroyer fitted the earlier concept of the short-range element of an RKR division, armed with SS-N-22 missiles. Defensive armament comprises a new semi-active missile, SA-N-7, paired ASW torpedo tubes, and RBU-1000 anti-torpedo rockets. In effect the twin tubes are equivalent to the ubiquitous US Mk 32, and the Russians probably equate the medium-frequency sonar to the defensive SQS-56 of a US *Perry*. This class also introduced a new gun, a twin 130mm (5.1in).

Bezuprechnyy 9.7.90 L & L van Ginderen

Unlike earlier Russian designs, this one has only a single main radar mast, carrying a back-to-back unit. That suggests that both the ship's 'Second Captain' combat direction system *and* the missile system share a common radar picture, a considerable innovation.

Project 956A (beginning with *Bespokoynyy*) is a modified version, probably armed with the naval version of SA-17 (*ie*, with a new longer missile in place of SA-N-7). They may also have a different anti-ship missile. The first fourteen units were probably armed with P-80, the others with P-270 (3M80). Two versions have been offered for export: one with 16 SS-N-25 in place of the SS-N-22s, 2 CADS-N-1 flanking the end of the helicopter deck (and no AK-630s: 8700t) and one with a 24-missile VLS in place of the after 130mm gun, plus 4 CADS-N-1 (the forward pair replace the SS-N-22s: 9000t). The latter version is probably the improved version Admiral F Gromov, Russian Navy C-in-C, said would be phased into the programme. In December 1993 it was reported that the Russians planned a total of twenty-eight units (four brigades); as of 1994, sixteen had been completed and six more were building.

These ships are reportedly quite successful *Sovremennyy* began sea trials in the Baltic in August 1980; she entered the Black Sea in January 1982. About 1990–91 it was planned to transfer one ship to Cuba together with three 'Koni' class frigates, but that never occurred. In 1993 it was reported that China planned to build this class under licence. A variety of versions was offered that year at the major defence show at Dubai.

Two more units were ordered from the Nikolaev 61 Kommunar shipyard, but they were never completed. *Vnushitelnyy*, laid down in 1982 and launched 17 October 1987, was hulked. A second unnamed unit was never begun. *Nastoichy* suffered a serious fire while fitting out (3 October 1991). *Bespokoynyy*, the first 956A, suffered a serious fire at Baltysk, 13 August 1992.

Udaloy as completed

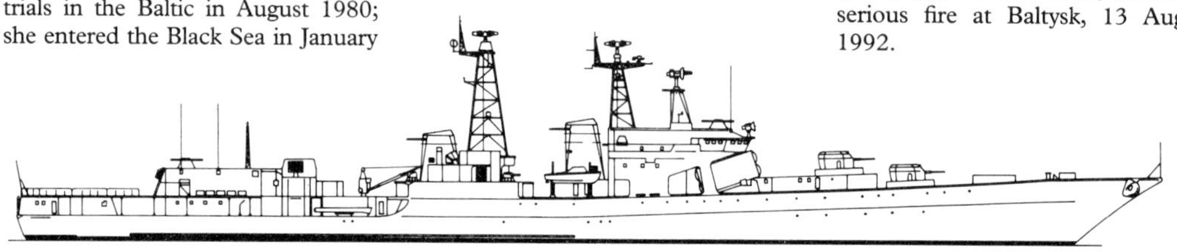

UDALOY class *cruisers* (Project 1155, 1155B, Fregat)

Displacement: 6200t standard; 7900t full load
Dimensions: 485ft 7in pp, 531ft 6in oa × 63ft 4in × 20ft 4in
148.0m, 162.0m × 19.3m × 6.2m
Machinery: 2-shaft COGAG (CP propellers): 4 gas turbines, *c*120,000shp = 35kts. Range 2400nm at 32kts, 10,500nm at 14kts
Armament: 2 SS-N-4 (2×4, 8 missiles), 8 vertical launch point defence (?48 missiles), 2–100mm (2×1), 4–30mm Gatling (4×1), 2 RBU-6000 ASW RL (2×12), 8–533mm TT (2×4), 2 helicopters
Sensors: Radar 2 Strut Pair, 2 Eye Bowl, 3 Palm Frond, Kite Screech, 2 Bass Tilt, 1 for helo control; sonar Horse Tail and Horse Jaw; ECM 8 Bell series, 2 chaff RL (2×2)
Complement: 300

Name	Builder	Laid down	Launched	Comp	Fate
UDALOY	Kaliningrad	23.7.77	5.2.80	31.12.80	Extant 1995
VITSE ADMIRAL KULAKOV	Zhdanov	4.11.77	16.5.80	29.12.81	Extant 1995
MARSHAL VASILYEVSKIY	Zhdanov	22.4.79	29.12.81	8.12.81	Extant 1995
ADMIRAL ZAKHAROV	Zhdanov	16.10.81	4.11.82	30.12.83	Extant 1995
ADMIRAL SPIRIDONOV	Kaliningrad	1981	1983	30.12.84	Extant 1995
ADMIRAL TRIBUTS	Zhdanov	19.4.80	26.3.83	30.12.85	Extant 1995
MARSHAL SHAPOSHNIKOV	Kaliningrad	1983	Jan 1985	30.12.85	Extant 1995
SIMFEROPOL* (ex-*Marshal Zhukov*)	Kaliningrad	12.6.84	24.12.85	30.12.87	Extant 1995
ADMIRAL LEVCHENKO	Zhdanov	27.1.82	21.2.85	30.9.88	Extant 1995
ADMIRAL VINOGRADOV*	Kaliningrad	5.2.86	4.6.87	30.12.88	Extant 1995
ADMIRAL KHARLAMOV	Kaliningrad	7.8.86	29.6.88	30.12.89	Extant 1995
ADMIRAL PANTALEYEV	Kaliningrad	1987	1988	19.12.91	Extant 1995
ADMIRAL CHABANENKO**	Kaliningrad	1988	1992	1994	Extant 1995
ADMIRAL BASISTIY	Kaliningrad	1990	–	–	Extant 1995

* Project 1155 bis. ** Project 1155.

These ships are in effect Soviet equivalents of the US *Spruance*. They are designed to act as ASW flotilla leaders, presumably working with frigates. In effect they are 'Kara' successors with paired helicopters aft and without area defence missiles (their SA-N-9s are, in effect, SA-N-4 successors). The TTZ for this class was approved on 17 October 1992.

The new SA-N-9 apparently encountered problems. As of 1992, the first two ships still lacked both missile launchers and directors. The third had launchers but no directors. The fourth, *Admiral Zakharov*, was the first to have the full system, including two directors (one of which was later removed for fitting to another ship). The next two ships had only single directors; the first to retain both directors was *Simferopol*. The first two ships were completed with Strut Pair 2-D radars; later ones all have 3-D Top Plates on their mainmasts. Ships with the full SAM system (fourth and later units) are Project 1155R. Sometimes the two versions are designated 1155A and 1155B.

The torpedo tubes normally carry four anti-ship and four ASW torpedoes; there are no reloads. These were the first Soviet ASW ships to carry two helicopters rather than one.

The final unit, *Admiral Panteleyev*, began trials in the Baltic in July 1991. *Admiral Tributs* suffered a serious shipyard fire at Vladivostok, 18 July 1991. *Admiral Zakharov* suffered a fire and explosion off Vladivostok, 17 February 1992, and was stricken; but she was reinstated and was under repair as of July 1993. *Admiral Kharlamov* had a major fire in 1993.

Two more ships (*Admiral Chebanenko*, and *Kucherov*) were ordered to a modified 1155.1 design, with a single 130mm mount forward, an RBU-12000, two CADS-N-1, four Gatling guns, and two quadruple SS-N-22 (rather than SS-N-14) mounts. They have no SS-N-14s. It seems likely that this was originally a modified version intended to fire SS-N-15 from torpedo tubes (hence not needing SS-N-14 tubes), comparable to the 'Krivak III' class. These ships probably have the same new sonar as the *Neustrashimyy* class. Like the 'Krivak III' class, these ships ultimately were to have been taken over by the KGB (now the Maritime Border Guards).

Admiral Tributs suffered a serious shipyard fire at Vladivostok on 18 July 1991. In 1991 *Admiral Ushakov* was so badly damaged by fire at Vladivostok (a gas turbine disintegrated) that she was being scrapped.

Admiral Kharmolov 8.7.94 L & L van Ginderen

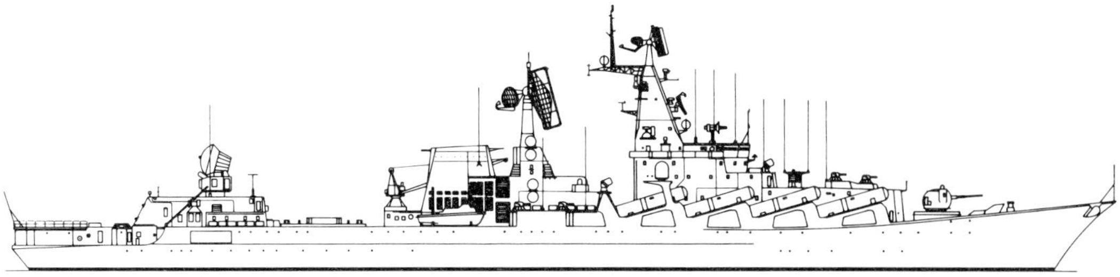

Marshal Ustinov 1989

This RKR is an evolved version of a design drawn in the early 1960s, incorporating the new supersonic cruise missile (SS-N-12 Bazalt). It would originally have resembled a stretched 'Kresta I' of about 12,000t, with SA-N-3 defensive missiles fore and aft and a hangar aft, and with a secondary battery of SA-N-4s. As in 'Kresta I', the missile tubes would have elevated. Later fixed tubes were substituted. The Bazalt programme stalled about 1962, and presumably this design was then put aside. The project name, Atlant, suggests that it was intended for the Northern Fleet, though in fact ships were distributed elsewhere when they were completed. When the RKRs were revived in the late 1960s, this design retained its original main battery but was provided with the new surface-to-air missile system, SA-N-6, aft. Presumably topweight precluded provision of an additional Top Dome forward. Since the single director covers only a 180 degree arc, these ships have a blind arc forward.

Defensive torpedo tubes are behind shutters near the stern.

Each uptake contains two boost turbine uptakes, one gas turbine uptake, and two gas turbine generator uptakes.

SLAVA class *cruisers* (Project 1164)

Displacement: 10,000t normal; 12,500t full load
Dimensions: 613ft 6in x 68ft 3in x 24ft 7in (hull)
187.0m x 20.8m x 7.5m
Machinery: COGOG (4 31,250shp boost turbines, two 12,000shp cruise turbines): 125,000shp = 34kts. Range 2000nm/30kts, 9000nm/15kts
Armament: 16 SS-N-12, 8 SA-N-6 (8 missiles per launcher), 2 SA-N-4 launchers (40 missiles), 2–130mm (1x2), 6–30mm Gatling, 10–533mm TT (2x5), 2 RBU-6000, 1 Hormone-B helicopter
Sensors: Top Pair, Top Steer, 3 Palm Frond, 1 Top Dome, 2 Pop Group, 1 Kite Screech, 3 Bass Tilt, 1 Front Door/Front Piece; sonar LF hull, MF VDS, EW Side Globe/Rum Tub; Punch Bowl satellite link
Complement: 600 (*Ustinov* 505)

Name	Builder	Laid down	Launched	Comp	Fate
SLAVA	Nikolayev	5.11.76	27.7.79	30.12.82	To Ukraine 1990
MARSHAL USTINOV (*ex-Admiral Lobov*)	Nikolayev	5.10.78	25.2.82	15.9.86	Extant 1995
CHERVONAYA UKRAINA	Nikolayev	31.7.79	28.8.83	25.12.89	Extant 1995
ADMIRAL LOBOV	Nikolayev	29.8.84	11.8.90	Mar 1993	Ukranian *Vilna Ukraina* 1995
ROSSIYA	Nikolayev	–	–	–	Cancelled 4.10.90
ADMIRAL GORSHKOV	Nikolayev	–	–	–	Cancelled 4.10.90

Although these ships probably lack flagship computer facilities, they are used in peacetime as flagships (*eg*, visiting foreign ports), in which capacity they embark up to 27 more officers and 24 more enlisted men. The TTZ for this class was approved in 1971, and the technical project (design) in August 1974.

Slava first deployed 15 September 1983. She was attached to the Black Sea Fleet, *Ustinov* to the Northern Fleet. *Lobov* was 74 per cent complete in February 1992; at 1 April of that year it was reported that she would be completed for the Ukraine. As of 1995, *Slava* was out of service; a refit begun December 1990 was stopped when 40 per cent complete. A fifth unit, to have been named *Oktyabrska Revolutsiya*, was cancelled.

Slava Nov 1983 RN

Stepenny 1954

(For photograph see under Poland)

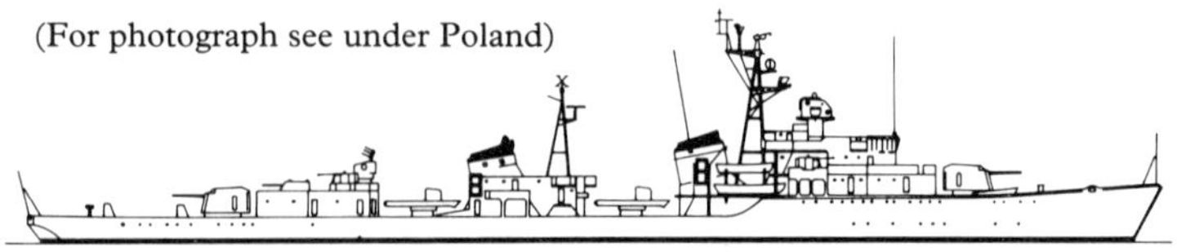

Ogenny after modernisation

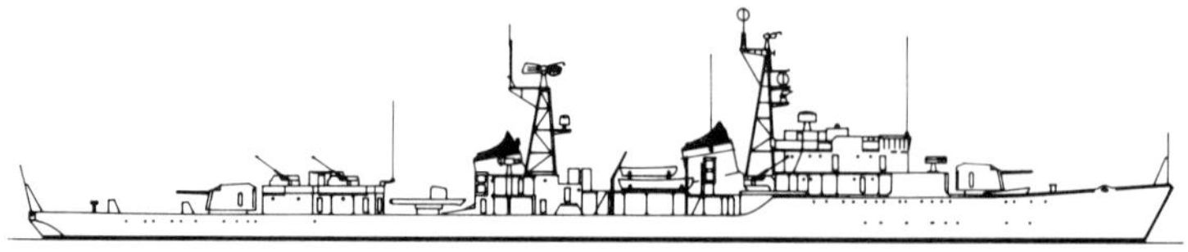

SKORIY class *destroyers* (Project 30B)

Displacement: 2316t standard; 3066t full load
Dimensions: 380ft 7in wl, 395ft 4in oa × 39ft 4in × 12ft 10in
120.5m, 116m × 12m × 3.9m
Machinery: 60,000shp = 36.5kts. Range 3500nm at 15.7kts
Armament: 4–130mm/50 (2×2: 150rpg), 2–85mm (1×2: 300rpg), 7–37mm (7×1), 10–533mm TT (2×5), 2 DC mortars, 2 DC tracks, (52 DC or 60 mines)
Sensors: Radar Gyus-1B, Ryf-1, Redan-2, Vympel-2; sonar Tamir-5H
Complement: 286

Name	Builder	Laid down	Launched	Comp	Fate
SMELIY	Zhdanov	16.5.48	29.9.48	21.12.49	Stricken 10.2.65, accommodation ship at Murmansk
STOIKII	Zhdanov	16.11.48	1.2.49	18.4.50	Stricken 22.2.80
SKORIY	Zhdanov	30.3.49	14.8.49	26.9.50	To Poland 4.8.58
SUROVYI	Zhdanov	17.8.49	1.10.49	31.10.50	Stricken 11.2.88
SERDITYI	Zhdanov	22.12.49	15.4.50	21.12.50	Stricken 14.3.75
SPOSOBNYI	Zhdanov	1.3.50	30.4.50	21.12.50	To Poland 26.12.57
STREMITEL'NYI*	Zhdanov	15.5.50	15.4.51	4.7.51	Decomm 25.6.84
SOKRUSHITELNYI	Zhdanov	15.9.50	30.6.51	28.11.51	Stricken 18.7.77
SVOBODNYI	Zhdanov	27.11.50	20.8.51	23.6.52	Stricken 19.1.83
STATNYI	Zhdanov	1.5.51	23.10.51	4.8.52	Stricken 8.7.82
SMETLIVYI	Zhdanov	24.5.51	17.11.51	5.8.52	To Egypt 1956
SMOTRYASHCHCHII	Zhdanov	21.6.51	19.2.52	4.11.52	Stricken 14.2.78
SVERSHCHENNYY	Zhdanov	16.7.51	24.4.52	20.12.52	Stricken 5.3.87
SER'EZHNIY	Zhdanov	25.10.51	13.7.52	18.12.52	Stricken 5.3.87
SOLIDNIY	Zhdanov	4.1.52	17.8.52	31.12.52	To Egypt 1956, returned 1968, stricken 15.4.87
STEPENNIY	Zhdanov	11.2.52	22.9.52	11.2.53	Stricken 14.3.86
BDITELNIY	Nikolayev	10.6.48	30.12.48	25.10.49	Planned for India: stricken 12.10.68
BEZUDERZHNIY	Nikolayev	20.7.48	31.3.49	30.12.49	Decomm 1.3.67
BUIYNIY	Nikolayev	15.4.49	23.9.49	29.8.50	Hulk at Sevastopol 1982–83, stricken 7.3.86
BEZUPRECHNIY	Nikolayev	15.7.49	31.12.49	9.9.50	Stricken 14.3.75
BESSTRASHCHNIY	Nikolayev	19.9.49	31.3.50	2.11.50	Decomm 2.6.76
BOEVOIY	Nikolayev	21.12.49	29.4.50	19.12.50	To Indonesia 1962
BYSTRIY	Nikolayev	20.2.50	28.6.50	19.12.50	Stricken 31.7.79
BURNIY	Nikolayev	18.5.50	29.8.50	4.6.51	To Egypt 1962
BESPOSHCHADNIY	Nikolayev	28.5.50	30.9.50	27.6.51	To Egypt 1962, returned 1968, stricken 9.2.88
BEZZHALOSTNIY	Nikolayev	12.7.50	30.12.50	6.7.51	To Indonesia 1964
BEZZAVETNIY	Nikolayev	28.9.50	30.3.51	11.11.51	To Indonesia 1959
BESSCHUMNIY*	Nikolayev	31.10.50	31.5.51	30.11.51	UTS-538 1.10.85; disarmed 15.6.79
BESPOKOINIY	Nikolayev	16.1.51	30.6.51	21.12.51	To Indonesia 1959
BEZBOYAZNENNIY*	Nikolayev	28.3.51	31.8.51	11.1.52	Decomm 12.8.77
BEZOTKAZNIY	Nikolayev	22.6.51	31.10.51	4.10.52	Stricken 25.4.85
BEZUKORIZNENNIY	Nikolayev	29.7.51	31.1.52	10.12.52	Stricken 3.1.67
BESSMENNIY	Nikolayev	12.9.51	31.3.52	10.12.52	To Egypt 1968
PYLKIY	Nikolayev	29.4.52	31.7.52	31.12.52	To Indonesia 1964
VSTRECHNIY*	Komsomolsk-on-Amur	29.4.48	20.5.49	7.12.49	Planned for India, to SM-302 1972, stricken
VEDUSHCHIY	Komsomolsk-on-Amur	31.7.48	21.8.49	26.12.49	Stricken 14.3.75
VAZHNIY	Komsomolsk-on-Amur	30.10.48	4.9.49	29.12.49	Stricken 14.3.75
VSPYL'CHIVIY	Komsomolsk-on-Amur	15.2.49	14.5.50	30.9.50	Stricken 12.3.74
VELICHAVIY	Komsomolsk-on-Amur	4.8.49	14.5.50	31.10.50	Decomm 12.8.77, stricken 10.2.78
VERTKIY	Komsomolsk-on-Amur	5.11.49	22.7.50	14.12.50	Decomm 1.9.78
VECHNIY	Komsomolsk-on-Amur	12.1.50	30.8.50	15.12.50	UTS-27 15.7.71, stricken 22.8.89
VIKHREVOIY*	Komsomolsk-on-Amur	28.2.50	19.5.50	27.12.50	Stricken 7.6.83
VIDNIY	Komsomolsk-on-Amur	27.5.50	17.5.51	21.12.51	UTS-263 12.2.71, decomm 30.5.83
VERNIY*	Komsomolsk-on-Amur	15.7.50	17.5.51	26.12.51	Stricken 21.3.81
VNEZAPNIY	Komsomolsk-on-Amur	23.9.50	14.6.51	28.12.51	To Indonesia 1959
VNIMATEL'NIY	Komsomolsk-on-Amur	31.10.50	2.8.51	26.12.51	Decomm 21.5.81, stricken 27.7.86
VYRAZIMITEL'NIY	Komsomolsk-on-Amur	14.12.50	26.8.51	29.12.51	To Indonesia 1962

Work on this design began on 8 October 1945. The armament, machinery and basic arrangement of the earlier Project 30 (*Ognevoy*) were retained to speed construction, and the basic design was approved 28 January 1947. The TTZ called for a maximum speed of 36kts and an endurance of 3500nm at 15.5kts (the figures roughly matched those for 30K; Project 30 had been designed for 30kts and 4080nm at 16kts). Armament was as installed, except that the TTZ required triple torpedo tubes and six rather than seven single 37mm guns. This was the first Soviet longitudinally-framed completely welded hull. Compared to Project 30, it had a heavily reinforced hull, the changes adding nearly 50 per cent to hull weight. To speed construction, ships were built in 101 prefabricated modular sections; one ship (*Serezhniy*) was completed in only one year.

This design had higher freeboard than the earlier Projects 7-U and 30, but not enough to avoid taking water. For example, on trials flooding through the boiler air intakes limited *Smeliy* to 28kts in rough weather (she exceeded 36kts in smooth water). Many guns could not be operated. The turning circle (8.5 lengths) was considered excessive, and bilge keels were later fitted.

From about 1952 on, the single-barrelled 70-K 37mm guns were replaced by four twin V-11 mounts and the heavy-calibre machine guns initially fitted were replaced by 2 or 6–25mm. A Pegas-2 sonar replaced the earlier Tamir-5N. The lead ship displaced 2316t, but 1952 ships added 35t (radar and hull reinforcement to prevent sagging due to the weight of water shipped over the bow).

In 1957 a TTZ for modernisation was issued. Project 31 was intended to carry out some new missions, which the Russians listed as ASW, AAW, anti-fast attack boat, patrol service and ELINT. The heavy 130mm director and one set of torpedo tubes were landed, the light armament was replaced by five single 57mm guns, and 2 RBU-2500 were mounted on a raised platform forward of the bridge.

Six were transferred to Egypt (Project 30-BA: two in 1956, two in 1962, two in 1968) and seven to Indonesia (Project 30-BK: four in 1959, one in 1962, two in 1964). Two more planned for India were not transferred. Habitability was improved, a ZIF-75 57mm mount replaced the forward group of AA guns (85mm and 37mm), and the Ryf radar was replaced by Fut-N. The Egyptian ships were fitted with Don radar and RBU-2500.

Name	Builder	Laid down	Launched	Comp	Fate
VOLEVOIY	Komsomolsk-on-Amur	1.3.51	11.9.51	29.12.51	To Indonesia 1959
VOLNYIY	Komsomolsk-on-Amur	12.6.51	4.6.52	31.12.52	Stricken 15.12.81
VKRADCHIVIY	Komsomolsk-on-Amur	14.7.51	4.6.52	31.12.52	Stricken 30.1.79
VDUMCHIVIY	Komsomolsk-on-Amur	5.11.51	31.7.52	31.12.52	Decomm 28.4.77
VRAZUMITELNIY	Komsomolsk-on-Amur	15.12.51	3.9.52	31.12.52	Stricken 7.2.77
OGNENNYIY*	Severodvinsk	14.8.48	17.8.49	28.12.49	Stricken 25.12.79, BU 1981
OTCHETLIVIY	Severodvinsk	29.10.48	14.9.49	28.12.49	Decomm 19.5.72
OSTRIY	Severodvinsk	21.12.48	16.4.50	25.8.50	Stricken 4.6.83
OTVETSTVENNIY	Severodvinsk	11.6.49	16.4.50	31.8.50	Decomm 12.4.63
OTMENNIY	Severodvinsk	8.10.49	18.6.50	6.11.50	Stricken 9.2.78
OTRYVISTIY	Severodvinsk	3.12.49	28.8.50	10.12.50	Stricken 7.2.77
OTRAZHAYUSHCHIY	Severodvinsk	3.3.50	3.10.50	17.12.50	Decomm 20.4.67
OTRADNIY	Severodvinsk	10.5.50	4.1.51	20.7.51	Stricken 12.6.84
OZHARENNIY	Severodvinsk	6.7.50	10.4.51	28.7.51	Decomm 31.7.80
OBEREGAYUSHCHIY*	Severodvinsk	23.9.50	14.5.51	28.10.51	Stricken 13.3.75
OKHRANYAYUSHCHIY	Severodvinsk	25.11.50	26.7.51	28.11.51	Stricken 27.2.87
OSTOROZHNIY	Severodvinsk	25.1.51	25.9.51	20.12.51	Stricken 15.12.81
OKRILENNIY	Severodvinsk	24.3.51	17.10.51	31.12.51	Stricken 25.12.78
OTZHIVCHIVIY	Severodvinsk	30.5.51	29.12.51	20.12.52	Stricken 18.7.77
OTCHAYANNIY	Severodvinsk	25.8.51	29.12.51	25.11.52	To Egypt 1968
OPASNIY*	Severodvinsk	20.10.51	1.6.52	9.12.52	Stricken 5.3.87
OZHIVLENNIY	Severodvinsk	12.1.52	4.8.52	24.1.53	Stricken 2.6.76
OZHESTOCHENNIY	Severodvinsk	3.4.52	26.9.52	4.3.53	Stricken 12.3.71

NB Units marked * modernised under Project 31

Neustrashimmy 1958 USN

NEUSTRASHIMYY (TALLINN) class *destroyer* (Project 41)

Displacement: 3100t standard; 3830t full load
Dimensions: 439ft 1in × 44ft 6in × 14ft 6in
133.83m × 13.57m × 4.42m
Machinery: 2-shaft geared turbines, 4 boilers, 66,000shp = 36kts
Armament: 4–130mm/58 (2×2; 425rpg), 8–45mm (4×2), 10–533mm TT (2×5), 2 4-barrel ASW RL, 6 DC throwers and tracks (105 DC), 48 mines
Sensors: Radar Fut-N air search, Ryf surface search; sonar Pegas
Complement: 305

Name	Builder	Laid down	Launched	Comp	Fate
NEUSTRASHIMYY	Zhdanov	5.7.50	29.1.51	31.1.55	Stricken 22.2.74

NB Ship began builder's trials 28.1.52 and official trials 1.8.54. Published sources do not explain this very protracted trials period.

This was the first true Soviet postwar destroyer, mounting dual-purpose guns intended to survive against radioactive fallout. The 188-destroyer 1946–55 Plan envisaged construction of 118 of these ships (the other seventy were *Skoriys*). Three postwar alternatives were considered. One was a modified Project 48 destroyer leader (2740t standard, 3460t full load, 418ft 8in × 40ft × 15ft, *127.6m × 12.2m × 4.58m*, 3 shafts, 81,000shp = 37.5kts). The large-ship design bureau, TsKB-17, designed Project 40 in both forecastle and flush-deck versions (3296t/4050t or 3250t/4000t, 442ft 11in × 42ft 8in × 14ft 7in or 14ft 6in, *135m × 13m × 4.45m* or *4.42m*, 2 shafts, 76,000shp = 35.8kts, range 4150nm at 18kts). TsKB-53, the destroyer bureau, designed Project 41 (3010t/3830t, 439ft 1in × 44ft 6in × 14ft 6in, *133.38m × 13.57m × 4.42m*, 2 shafts, 66,000shp = 36kts, range 5500nm at 14kts). Project 48 had three twin 130mm single-purpose mounts (plus one twin 85mm, three twin 37mm, and two twin 12.7mm machine guns, and two twin torpedo tubes). Project 40 had three twin dual-purpose mounts (130mm SM-2) plus six twin 45mm and two quintuple torpedo tubes. Project 41 had only two twin 130mm dual-purpose mounts, plus four twin 45mm and two quadruple 25mm; and two quadruple torpedo tubes. Project 48 was apparently rejected as obsolete, given its single-purpose guns (the two surviving ships of this type were broken up incomplete). Project 40 was rejected as too large for mass production; it was probably Admiral Kuznetzov's large destroyer of 1945.

Project 41, the smallest and simplest alternative, was selected. Ironically, it would soon be dropped as too large. When the TTZ was approved on 14 June 1947, production of 110 ships was planned. The draft design was approved 19 August 1948, and the technical design on 28 September 1949.

The two most important innovations in the design were the attempt to minimise structure above the weather deck, and the pressure-fired steam plant used in all later Soviet steam-powered warships. The ship also introduced AC electrical power into Soviet practice. Design work was complicated because the weapons were all new types, often without any TTZs. For example, Project 41 was the first Soviet destroyer with dual-purpose main guns (SM-2 stabilised twin 130mm/58) and with the new Pegas sonar. The main gun fire control system was similar to that for the *Sverdlov* class secondary battery. Like the *Sverdlov* guns, the twin 130mm carried their own range-only radars.

The flush-deck hull had two continuous internal decks, providing protected fore–aft access. It was nearly completely welded and was designed for prefabrication. The Project 41 destroyer was designed to remain afloat with three adjacent compartments flooded (two on a *Skoriy*). Splinter armour (8–10mm and 20mm thick) protected the fire controls, bridge, machinery-boiler casings, and weapons. Since all air ports were omitted (for nuclear protection), the design introduced radiator heating and air conditioning into Soviet destroyer practice.

The ship introduced pressure-fired KV-41 boilers producing high-pressure, high-temperature steam at 64kg/cmu22. Compared to the conventional steam plant on a *Skoriy*, this one was smaller and 100t lighter; at economical speed it consumed 20 per cent less fuel. Most importantly, it could be lit off without being warmed up. Trials showed, however, that the turbines were inefficient; at full power they required about 10 per cent more steam than expected. Essentially this steam plant was used in all the later major steam combatants: Projects 56, 57bis, 58, 1134/1134A, 1123, the carriers, and presumably also the *Slava* and *Sovremennyy* classes.

Trials were not altogether successful. The ship made only 33.5kts on 64,200shp, and endurance at 14kts was 5210nm rather than the 5500nm specified. The unfortunate bow form made for excessive spray forward, so the forward 130mm and 45mm mounts were difficult to operate. The ship vibrated badly aft. She did not answer her helm when backing down. It was soon clear that the arrangement of light AA guns on the sides of the ship made it difficult to engage targets forward or abaft the ship. The pre-production SM-16 45mm guns were unreliable, and later were replaced by quadruple SM-20-ZITs. Also, the obsolete depth charge launchers were eventually replaced by a pair of RBU-2500s.

However, the project died even before the ship departed for trials. Someone, it is not quite clear who, decided that the ship was too large; in June 1951 Project 41 was cancelled and incomplete hulls broken up. Later plans were sold to China to become the basis for the 'Luda' design. Recent accounts of Project 41 emphasise that the ship's size provided a valuable margin for modernisation. TsKB-53 was ordered to design a more compact version of Project 41, which became Project 56 ('Kotlin').

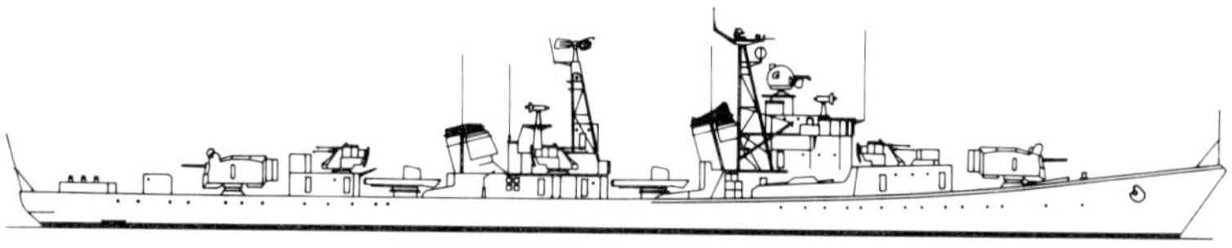

Plamenny as completed

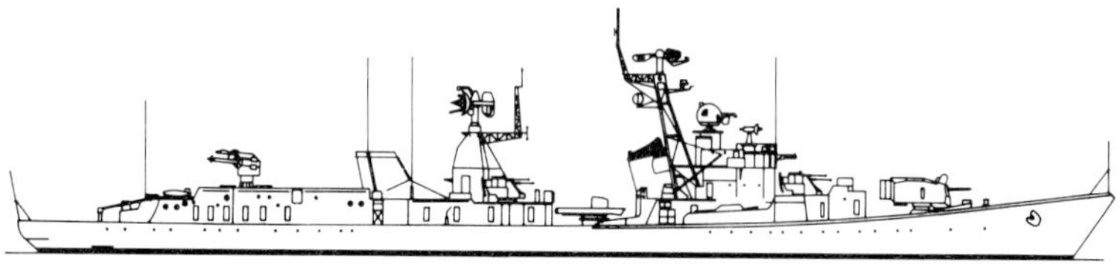

Bravy after missile conversion 1964

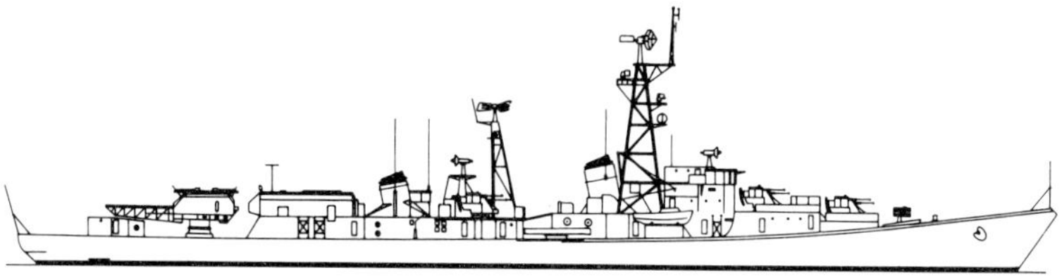

Prozorlivyy as completed

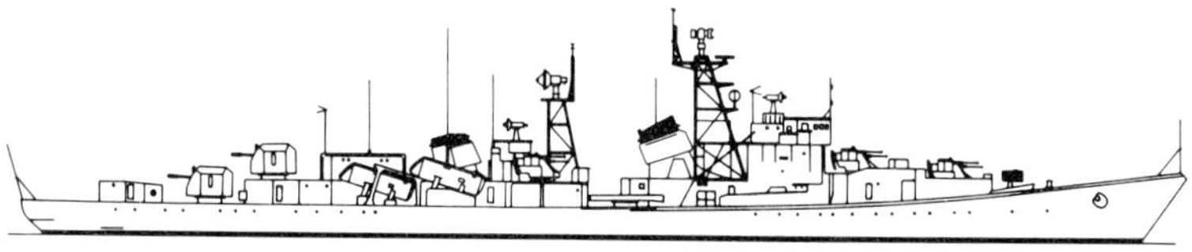

Bedovyy as completed

This 'Kotlin' (Russian *Spokoinyy*, Project 56) design was the shrunken version of Project 41. Work began under a 2 June 1951 decree on 'changes in the military characteristics of Project 41'. The main changes ordered were a reduction in full load displacement from 3770t to 3150t, full speed increased from 36kts to 39kts, cruising range (14kts) reduced from 5500nm to 4000nm, stores endurance halved to 10 days, and lighter torpedo tubes and unstabilised 45mm gun mounts (quadruple SM-20s instead of twin SM-16s, 750 rather than 1000 rounds/barrel). A production run of 100 ships (rather than the 110 planned Project 41s) was envisaged. In fact only twenty-seven were completed; production was stopped in October 1955. Four more hulls were then completed as missile ships (Project 56M, 'Kildin').

Presumably the 1951 figures represented a sketch design already available; TsKB-53 skipped the usual preliminary design stage to go directly to detailed design in hope of rescuing the production schedule. Work on the first two ships began in February 1952 (much prefabrication preceded the official keel laying). It was soon obvious that the preliminary figures had been too optimistic. However, the figures were officially confirmed in a TTZ issued 4 April 1954 (full load displacement was allowed to rise to 3220t, required speed to fall to 38.5kts, and the 25mm guns of Project 41 were eliminated). By this time the first unit was nearly ready to launch.

There were some differences from Project 41 armament: the director was a new type (Sfera-56; the analogous Sfera-50 was used on 'Riga'

'KOTLIN' class *destroyers* (Project 56) and 'KILDIN' class *destroyers* (Project 56M)

Displacement: 2662t standard; 3230t full load
Dimensions: 413ft 9in × 41ft 8in × 13ft 9in
126.1m × 12.7m × 4.19m
Machinery: 2-shaft geared turbines, 4 boilers, 72,000shp = 38kts.
Armament: 'Kotlin': 4–130mm/58 (2×2), 16–45mm (4×4), 10–533mm TT (2×5), 6 DC throwers plus 2 DC tracks (48 DC), 50 mines
'Kildin': see notes
Sensors: Radar Fut-N, Ryf; sonar Pegas-2
Complement: 284

Name	Builder	Laid down	Launched	Comp	Fate
Project 56 'Kotlin'					
SPOKOYNYY	Zhdanov	4.3.53	28.11.53	27.6.56	Stricken 19.4.90
SVETLYY	Zhdanov	4.3.53	27.10.53	17.9.55	Decomm 25.4.89
SPESHNYY	Zhdanov	30.5.53	7.8.54	30.9.55	Stricken 25.4.89
SKROMNYY**	Zhdanov	27.7.53	26.10.54	30.12.55	Stricken 25.4.89
SVEDUSHCHIY**	Zhdanov	7.12.53	17.2.55	31.1.56	Stricken 8.4.92
SMYSHLENYY*	Zhdanov	23.2.54	2.5.55	28.6.56	Renamed *Moskovskiy Komsomolets* 29.10.58, stricken 22.7.86
SKRYTNYY	Zhdanov	25.7.54	27.9.55	30.9.56	Stricken 25.4.89
SOZNATELNYY**	Zhdanov	25.9.54	15.1.56	31.10.56	Stricken 1.3.88
SPRAVEDLIVYY**	Zhdanov	25.12.54	12.4.56	20.12.56	To Poland 25.6.70
NESOKRUSHIMYY**	Zhdanov	15.6.55	20.7.56	30.6.57	Stricken 30.2.87
NAKHODCHIVYY**	Zhdanov	19.10.55	30.10.56	18.9.57	Stricken 25.4.89
NASTOYCHIVYY**	Zhdanov	3.3.56	23.4.57	30.11.57	Stricken 25.4.89
BLESTYASHIY*	Nikolayev	20.2.53	27.11.53	30.9.55	Stricken 30.7.87
BYVALVYY*	Nikolayev	6.5.53	31.3.54	21.12.55	Stricken 17.7.88
BRAVYY	Nikolayev	25.7.53	28.2.55	9.1.56	SAM 'Kotlin' prototype: stricken 30.7.87
BESSLEDNYY*	Nikolayev	1.4.54	5.11.55	31.10.56	Stricken 8.4.88
BURLIVYY*	Nikolayev	5.5.54	28.1.56	28.12.56	Stricken 25.5.89
BLAGORODNYY*	Nikolayev	5.3.55	30.8.56	18.7.56	Stricken 25.4.89
PLAMENNYY*	Nikolayev	3.9.55	26.10.56	31.8.57	Stricken 24.6.91
NAPORISTYY*	Nikolayev	17.8.55	30.12.56	31.10.57	Stricken 30.7.87
VYZYVAYUSHCHIY*	Komsomolsk-on-Amur	25.7.53	20.5.55	31.5.56	Stricken 25.4.89
VESKIY	Komsomolsk-on-Amur	30.1.54	31.7.55	30.3.56	Stricken 30.7.87
VDOKHNOVENNYY*	Komsomolsk-on-Amur	31.8.54	7.5.56	31.10.56	Stricken 5.3.87
VOZMUSHCHENNYY*	Komsomolsk-on-Amur	30.12.54	8.7.56	31.12.56	Stricken 5.3.87
VOZBUZHDENNYY**	Komsomolsk-on-Amur	29.7.55	10.5.57	31.10.57	Stricken 25.4.89

Name	Builder	Laid down	Launched	Comp	Fate
VLIYATELNYY	Komsomolsk-on-Amur	29.10.55	10.5.57	6.11.57	Stricken 17.7.88
VYDERZHANNYY	Komsomolsk-on-Amur	30.6.56	24.6.57	10.12.57	Renamed *Dalnevostochnyy Komsomolets* 20.2.67; stricken 25.4.92
Project 56M 'Kildin'					
BEDOVYY	Zhdanov	1.12.53	31.7.55	30.6.58	Project 56EM configuration; stricken 25.4.89
NEULOVIMYY	Zhdanov	23.2.57	27.2.58	30.12.58	Stricken 19.4.90
PROZORLIYY	Nikolayev	1.9.56	30.7.57	30.12.58	Stricken 24.6.91
NEUDERZHIMYY	Komsomolsk	23.2.57	24.5.58	30.12.58	Stricken 8.12.85
NEUKROTIMIY	–	–	–	–	Ordered 1955 but not built

NB * 56PLO ASW conversion; ** 56 AAW missile conversions

Svetlyy Jul 1976 RN

class SKRs) and the 45mm guns were arranged in a rhombus so that fire could be concentrated along the centreline. Unlike Project 41, Project 56 had no ASW rocket launchers forward.

To accommodate the armament and powerplant of Project 41 in a smaller hull, some volume was pushed up above the main deck. To limit topweight, aluminium-magnesium alloy was used in the superstructure. These ships may have had the first Soviet fin stabilisers.

Trials were not altogether satisfactory. Although the powerplant exceeded design power, *Spokoinyy* did not make her designed speed until major modifications had been made (a centreline rudder replaced the two abaft the propellors, and 4 bladed propellors replaced the earlier 3-bladed ones). Even brief periods of high-speed steaming greatly reduced the ship's endurance (one hour at full speed cost 184nm at economical speed, compared to 104nm in a *Skoriy*).

This class was very well liked in the Soviet Navy; their high bows made them dry, and they were very fast. *Blagarodnyy* reportedly made 42.9kts after a 1968 refit, and even the slowest of the class made 36kts about a decade after completion.

One incomplete ship, *Bedovyy*, was fitted out as a test platform for the new KSSHch (Shchuka, or SS-N-1) anti-ship missile. The Kiparis system, an adapted gunfire direction system, was used for missile control. Apparently it was then decided to accelerate the missile programme by making this test installation operational (the Project 57 'Krupny' class destroyer was already being developed specifically to fire the KSShch missile). Three more ships were completed to a new 56M missile configuration (NATO 'Kildin' class). One of them, *Neuderzhimyy*, had not been laid down to the Project 56 design, but presumably the material for her construction had already been assembled. These ships were armed with a new quadruple 57mm secondary gun (16 weapons, 4×4, two replacing the forward 130mm mount, two aft) and paired beam torpedo tubes (2×2) in place of the earlier quintuple mounts. Reportedly the armoured hangar aft contained six missiles. The existing Wasp Head (Sfera) director was retained. Three out of the four were modified in 1973–75 in rough analogy to 'mod Kashins', aft-firing SS-N-2s replacing SS-N-1s. They retained 57mm mounts, adding a pair of twin 76mm right aft in place of the earlier SS-N-1 hangar. *Neuderzhimyy* alone was not modified; reportedly she was used to launch her SS-N-1s as targets. All of these ships were disposed of in the late 1980s.

Twelve ships were modified in 1958 for better ASW firepower; they seemed to have parallelled the *Skoriy* ASW refit programme. The second torpedo tubes and depth charge throwers were removed, an ASW station (with improved sonar) was fitted, and two RBU-2500 were installed forward; RBU-600s replaced the depth charges aft. *Moskovskiy Kosomolets* had an RBU-6000 forward, no RBU-600s, and eventually a VDS. The remaining quintuple torpedo tube was adapted to fire either ASW or anti-ship torpedoes. Displacement increased to 3600t, and speed fell to 34kts.

Bravyy tested the new 'Volna' (SA-N-1) destroyer SAM in 1959–63 (Project 56-K). She retained only her forward 130mm and 45mm mounts; later she was fitted with RBU-2500s, twin torpedo tubes, and two more 45mm mounts. *Bravyy* was followed by eight 56A modernisations (begun 1966, completed 1969–71). All after weapons (including 3 quadruple 45mm), from the after tubes back were removed, together with the main mast; in their place was a SA-N-1 system with sixteen missiles in the magazine. Later two AK-230 guns were placed near the after funnel (*Nesokrushimyy*, *Soznatelnyy*, and *Vozbuzhdennyy* each eventually had four AK-230s). Two RBU-600 ASW rocket launchers were added forward (RBU-2500s in *Skromnyy*, which had three quadruple 45mm). The foremast carried an MR-310 3-D radar (in some cases, MR-300).

The remaining six ships were never modernised. *Svetlyy* had a stern helipad from which Ka-15 helicopter flights were practised for the first time. Two other ships also had helicopter decks. Most unmodified ships were eventually fitted with twin 25mm guns released by the disposal of lighter units.

Decommissionings began in 1986, the ships having operated very hard for about 13 years.

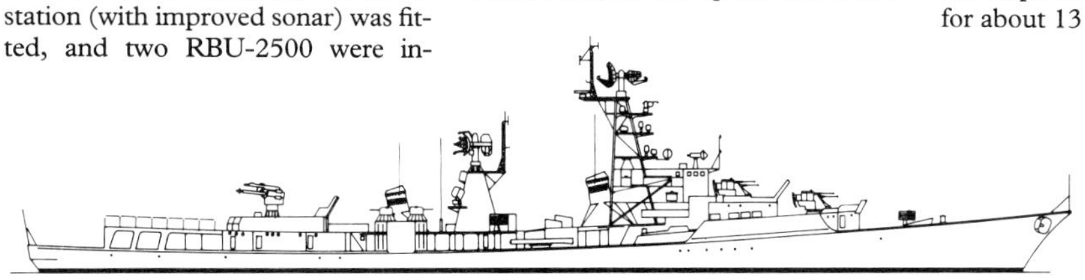

Zorkiy after 'Kanin' conversion

'KRUPNY' class *missile destroyers* (Project 57B)

Displacement: 3500t standard; 3850t normal; 4192t full load
Dimensions: 426ft 6in wl, 455ft 9in oa × 48ft 8in × 13ft 9in
130m, 138.9m × 14.84m × 4.2m
Machinery: 2 shaft geared turbines, 4 boilers, 72,000shp = 34.5kts. Range 3000nm at 18kts
Armament: 2 SS-N-1 (12 missiles; see notes), 16–57mm (4×4; 800rpb), 6–533mm TT (2×3), 2 RBU-2500 ASW RL (128 rounds)
Sensors: Radar Angara (Head Net), Zalp-Shch missile control, Neptun navigational; sonar Pegas-2
Complement: 31

Name	Builder	Laid down	Launched	Comp	Fate
GREMYASHCHIY	Zhdanov	25.2.58	30.4.59	30.6.60	Stricken 2.10.91
ZHGUCHIY	Zhdanov	23.6.58	14.10.59	23.12.60	Stricken 30.7.87
ZORKIY	Zhdanov	17.4.59	30.4.60	30.9.61	Stricken 30.6.93
DERZKIY	Zhdanov	10.10.59	4.2.60	30.12.61	Stricken 19.4.90
GNEVNYY	Nikolayev	17.12.57	30.11.58	10.1.60	Stricken 8.4.88
UPORNYY	Nikolayev	9.4.58	14.10.59	3.12.60	Stricken 29.6.93
BOYKIY	Nikolayev	2.4.59	15.12.60	26.6.61	Stricken 17.7.88
GORDYY	Komsomolsk	May 1959	24.5.60	6.2.61	Stricken 30.7.87
KHRABRYY	Komsomolsk	1959	1961	–	Cancelled 1963

These were the first Soviet missile warships. Their TTZ was signed by Admiral Kuznetzov on 25 July 1955. They were apparently designed from the beginning as missile ships; it is not clear why their designation was Project 57bis rather than Project 53. Project 57 may have been an alternative configuration, not yet described. Almost certainly there was never a corresponding classic destroyer design, which has sometimes been the subject of Western speculation ('Kruplin'). Basic design work was completed at the end of December 1956, but the new navy commander, Admiral Gorshkov, ordered some changes the following February. Note that this project predates the modified 'Kotlin' (56M). The design missions were to engage enemy surface ships and to attack shore targets, both using the new KSShch (SS-N-1) missile.

After a series of model tests of hulls large enough to accommodate the new missile system, a scaled-up 'Kotlin' was chosen, but modified to operate in a chemical or nuclear environment, with airtight remote-control cabs in the engine and boiler rooms. The turbines were modified for better efficiency.

Each of two SS-N-1 missile magazines held six weapons. In addition, one missile could be carried in the box launcher and another in the magazine, but in that case it was impossible to withdraw a faulty missile before launch. Compared to the Kiparis missile control system of Project 56M, the Topol' system in these ships had an improved computer and added a sighting/range-finding station; there was also an auxiliary control system, Kedr. Topol' used the same Zalp-Shch radar as the earlier Kiparis. It was to have incorporated the new Rangout (Square Tie) radar, but that was not ready in time (the radar ended up in Styx missile boats). Using an auxiliary system, the ship could engage two targets simultaneously, firing the missiles 5 to 15sec apart. Plans called for a television receiver (Grusha) by means of which the ship could receive target designation data from an aircraft beyond the horizon, but it was never reliable enough to be worth installing. Presumably it was the conceptual forebear of the system used with SS-N-3. The planned targetting aircraft was probably a Ka-15 helicopter, for which a pad was built aft, and for which the ship carried enough fuel for five flights. In fact the helicopter pad proved unsatisfactory (without the television, it could be used only for visual observation) and was little used; saluting guns were mounted on the pad.

As in the 56M missile destroyer, torpedo tubes were placed on the sides of the ship (but its greater beam made it possible to provide triple rather than twin tubes). The light anti-aircraft armament was arranged in a rhomboid similar to that of Project 56.

The ships were soon considered obsolescent; they lacked any effective means of self-defence and their KSShch missile was ineffective. On 30 April 1965 they were ordered converted to ASW ships, presumably as part of the shift towards ASW then in progress. The new version was Project 57PLO. The Bull Nose bow sonar (MG-332 Titan-2) was installed to control a pair of quintuple torpedo tubes (firing SET-65 ASW torpedoes) and three RBU-6000s (216 rounds); a helicopter pad aft supported a Ka-25PLO ASW helicopter (3 AT-1 torpedoes or 12 PLAB-250-120 or 24 PLAB-50 or 144 PLAB-MK depth bombs). There were also 72 sonobuoys, 12 marine markers, and 15 radar transponder buoys. The ship also had the MI-110R IR wake detector. For self-defence, the ship was fitted with SA-N-1 (32 missiles, Zif-102 launcher), two quadruple 57mm guns, and, in some cases, four AK-230s, as well as a pair of chaff launchers (ZiF-125). The diesel generator had to be upgraded from 300 to 500kW, and another 500kW generator added to supply the new sonar. Efforts were made to silence the ships, and their habitability was improved. Displacement increased to 3700t standard/4500t full load, and speed fell to about 32kts. To clear the new bow sonar, the bow was extended forward, so total length (overall) increased to 461ft 3in (140.6m). Complement increased to 318.

This modernisation proved so expensive as, apparently, to have deterred the Soviet Navy from any other attempt to rebuild a class of warship.

All eight were stricken in the late 1980s, the last four in 1987–88.

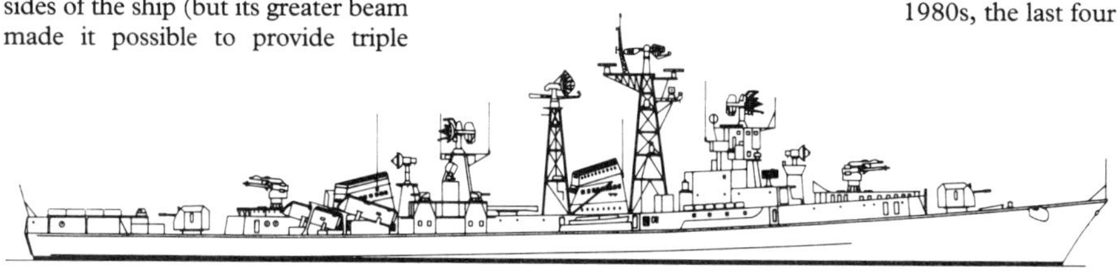

Slavny 1980

Krasnyy Krim 1974

Komsomolets Ukrainy was lead ship: laid down 15 September 1959, launched 31 December 1960, began trials 15 October 1962. She was laid down as SKR-26; *Soobrazitelnyy* was originally SKR-44.

This was one of the radical new classes designed in response to an August 1956 decree. It was described as a replacement not only for the 'Kotlin' class but also for the 'Riga' (Project 50) class, on the theory that neither could survive in the face of modern submarines and aircraft. The TTZ described this as an air defence/ASW patrol ship to defend warships and other vessels at sea against submarines, low-flying aircraft, and cruise missiles. As the design developed, the ASW role clearly became secondary. Admiral Gorshkov approved the operational requirement on 14 March 1956 and design-operational requirements in December 1956. The Naval Technical Institute considered seven alternative armament mixes and configurations, and included a gas turbine for the first time in a major Soviet warship. Detailed design was assigned to TsKB-53. The sketch characteristics were approved 30 April 1957; the preliminary design was submitted to Admiral Gorshkov late in June 1957 and approved 3 September.

The initial operational requirement was for four long-range SAM launchers (16–20 missiles each, with a single control system), three twin 76.2mm guns (1200 rounds), two (3) quadruple 57mm guns (1200rpg), gun fire control system, two ASW guided missile launchers (4–6 missiles), two RBU-6000 (80–96

'KASHIN' class *missile destroyers* (Project 61)

Displacement: 3400t standard; 4390t full load
Dimensions: 433ft 9in wl, 472ft 5in oa × 51ft 10in × 15ft 1in
132.2m, 144m × 15.8m × 4.6m
Machinery: 2-shaft CODAG; 4 gas turbines (M-3), 72,000shp = 34kts. Range 3500nm at 18kts
Armament: 2 SA-N-1 (32 missiles), 4–76mm (2×2; 600rpg), 5–533mm TT (1×5), 2 RBU-6000 ASW RL (96 rounds), 2 RBU-1000 ASW RL (24 rounds)
Sensors: Radar: Angara, Turel; sonar Titan, Vychegda
Complement: 266

Name	Builder	Laid down	Launched	Comp	Fate
OGNEVOY*	Zhdanov	9.5.62	31.5.63	31.12.64	SKR-31: conv Feb 1973; stricken 4.5.89, BU Oct 1990
OBRAZTSOVYY*	Zhdanov	29.7.63	23.2.64	29.9.65	SKR-2; stricken 30.6.93
ODARENNYY*	Zhdanov	22.1.63	11.9.64	30.12.65	Stricken 19.4.90
SLAVNYY*	Zhdanov	9.7.64	24.4.65	30.9.66	Conv. Sep 1975, stricken 24.6.91
STEREGUSHCHIY*	Zhdanov	9.7.64	20.2.66	21.12.66	Stricken 30.6.93
KOMSOMOLETS UKRAINY	Nikolayev	15.9.59	31.12.60	31.12.62	SKR-25; stricken 24.6.91
SOOBRAZITELNYY	Nikolayev	20.7.60	25.9.61	26.12.63	SKR-44; stricken 3.7.92
PROVORNYY	Nikolayev	10.2.61	23.3.62	25.12.64	SKR-37; stricken 21.8.90
OTVAZHNYY	Nikolayev	10.8.63	17.10.64	31.12.65	Renamed *Orel*; lost 30.8.74
STROYNYY	Nikolayev	20.3.64	28.7.65	15.12.66	Stricken 12.4.90, BU Jan 1991
KRASNYY KAVKAZ	Nikolayev	25.11.64	9.2.66	25.9.67	Extant 1995
RESHITELNYY	Nikolayev	25.6.65	30.6.66	30.12.67	Extant 1995
SMYSHLENNYY	Nikolayev	15.8.65	22.10.66	27.9.68	Stricken 22.2.93
STROIGIY	Nikolayev	22.2.66	29.4.67	24.12.68	Stricken 30.6.93
SMETLIVYY	Nikolayev	15.7.66	26.8.67	25.9.69	Extant 1995
SMELYY	Nikolayev	15.11.66	6.2.68	27.12.89	To Poland Dec 1987
KRASNYY KRIM	Nikolayev	23.2.68	28.2.69	15.10.70	Extant 1995
SPOSOBNYY	Nikolayev	10.3.69	11.4.70	25.9.71	Stricken 3.7.92
SKORYY	Nikolayev	20.4.70	26.2.71	23.9.72	Extant 1995
SDERZHANNYY*	Nikolayev	20.10.70	29.2.72	30.12.73	Extant 1995

NB * Project 61M modernisations. See notes

Smelyy 1981 L & L van Ginderen

rockets), one triple or two twin 345mm ASW TT (6 torpedoes); a speed of 36kts; range 300–3500nm at 18kts, and 15 days' stores endurance. The design-operational requirement changed the SAM battery to two twin launchers (12 missiles each) with two fire control systems, and cut the guns to 2–76.2mm (700 rounds). The planned ASW missiles were replaced by 1 Ka-25 ASW helicopter, but the two RBU remained. The 345mm TT gave way to one quintuple or two twin 400mm TT (a 40cm ASW torpedo was being developed). Displacement was expected to be 2800t standard.

Virtually all the planned equipment was being developed alongside the design. Thus by the preliminary design stage missile magazine capacity had been set at 16. Presumably as weight compensation, 76.2mm gun ammunition capacity was cut to 600 rounds. Problems with the new 40cm ASW torpedo led to a decision to replace it with 53cm weapons, initially in two triple mounts (as in Project 57bis 'Krupny') and then, in the contract design, in quintuple mount. The preliminary design added a pair of RBU-1000 anti-torpedo weapons (24 rounds). Displacement grew, to 3200t in the preliminary design and then to 3380t in the contract design. Expected trial speed had to be cut to 35kts in the contract design, and stores' endurance to 10 days.

The contract design was completed in March 1958. Weight growth included 11t added to the SAM magazine feed system between preliminary and contract design, 26t for the SAM control system and automatic units, and 25t in the main power plant.

Apparently initially a combined steam and gas turbine powerplant was planned, using a new 15,000shp M-2 gas turbine on each shaft. However, while design-operational requirements were being developed, the turbine developers offered a new 36,000shp M-3 unit (two 18,000shp turbines coupled together with reversing gears), with half the guaranteed service life of M-2 (initially 500hr rather than 1000hr, later 3000hr, comprising 100hr at 100 per cent power, 200hr at 80 per cent, and 2700hr at 50–60 per cent). The new destroyer became the first major gas turbine warship in the world; the M-3 powerplant weighed about half as much as the pressure-fired syeam plant in a 'Kotlin'. To produce 2.5 times as much electric power as a 'Kotlin' (2800kW) within a limited weight, the designers adopted gas turbine main generators. Using gas turbines, the ship could work up from rest to cruise speed in 10min, then to 30kts in 2–3min, and then full speed in another 3–5min; a steam plant would have taken about 12hr to do the same. As in later Western gas turbine warships, the gas turbines could be removed through the funnels. The ships infra-red signature was approximately halved by mixing air with the gas turbine exhaust in the funnels.

The M-3 turbine was built and tested in parallel with the lead ship *Komsomolets Ukrainy*. She was therefore initially limited to 65 per cent power (she turned out to need 12 per cent more than designed power at 30.3kts at 3790t). On full-power trials she later made 35.5kts. *Prozorlivvy* made 39.75kts on trials.

The new powerplant was so light that there was some fear that the usual topside weights would make the ship top-heavy; a new hull form was developed with knuckle from amidships to the transom, and with very sharp bow lines. Tumble-home in the hull was intended to reduce the ship's radar signature.

This was the first Soviet warship designed to withstand nuclear blast. To that end, the primary ship control station was separated from the enclosed bridge ('conning post') and buried in the hull, on the lower deck. Access fore and aft was by an enclosed passageway, with torpedo exits only through gastight airlocks.

The last ship in the class, *Sderzhannyy*, was completed to a modified design (Project 61M); another five ships (starred, above) were similarly modernised. Major improvements were better close-in air defence (the Start EW system with four 16-tube chaff launchers and four 30mm Gatlings with Vympel fire control system), and new radars (MR-300 and -500). The Gatlings replaced the two RBU-1000s originally fitted. Four P-15M (SS-N-2C Styx) cruise missiles, firing aft, were added. The Titan bow sonar was replaced by Platina, a related MF sonar including a towed VDS 'fish'. Stores endurance was increased to 25 days. Two superstructure tiers were added abaft the bridge to accommodate 7 extra officers (total complement increased to 320). Over 100t of solid ballast had to be added, and some fuel tanks had to kept permanently filled to maintain stability. Displacement rose to 4104t standard/4974t full load, speed fell to 32kts, and endurance increased to 4000nm at 18kts.

In 1974–76 *Provornyy* (Project 61E) was to test the new Uragan (SA-N-7) SAM system, which required new search radars. Further modernisations of this type were planned but not carried out. *Sposobnyy* had an unspecified major modification, presumably as a test ship.

Five more ships were built at Nikolayev for the Indian navy as Project 61ME: *Rajput*, the first, entered service in October 1980. The ships were built under Russian names: *Nadezhny, Gubitel'nyy, Lovkiy, Tverdyy, Tolkovoy*. This was a remarkable case of the revival of a relatively old design, the last 'Kashin' having being delivered in 1972.

Project ANCHAR (no NATO designation) *destroyer*

This nuclear destroyer project was cancelled in 1990; it was apparently intended to escort a carrier. It was described in 1992 as a multi-mission 12,000t ship powered by a nuclear gas turbine (nuclear for boost only). It would have had the new *Udaloy* class sonar system, a new 130mm gun system (under test in 1992), and a vertical launcher capable of firing ASuW, ASW, AAW, and land-attack missiles. The vertical launcher and the gun may appear on board late-production *Sovremennyy* class ships. The reference to ASuW and land-attack is suggested by literature advertising the new generation of dual-purpose cruise missiles.

Sovietsky Dagestan as completed

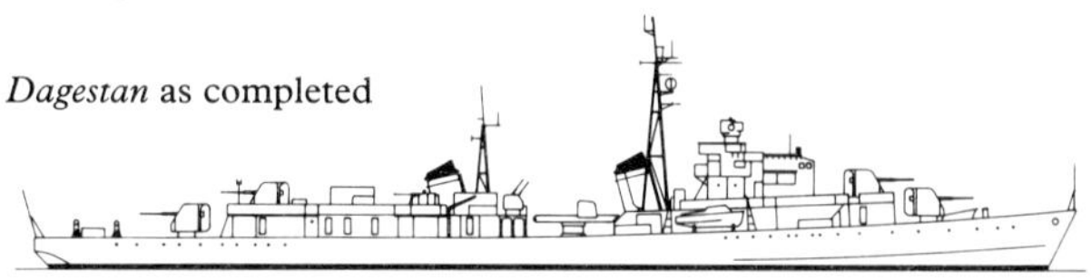

'Riga' class prototype rig

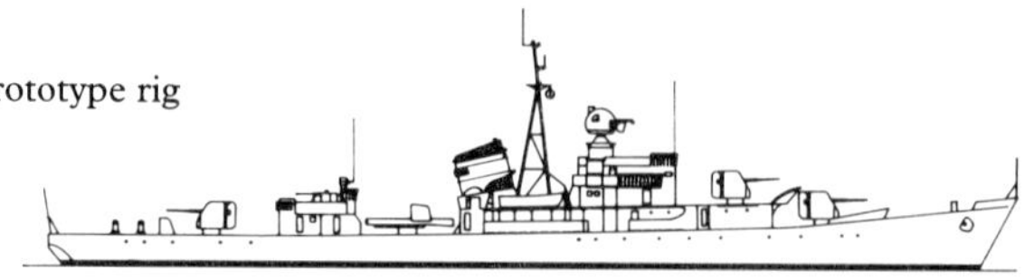

'KOLA' class *frigates* (Project 42)

Displacement: 1900t standard
Dimensions: 314ft 11in × 35ft 5in × 10ft 6in
96m × 10.8m × 3.2m
Machinery: 2-shaft, geared turbines, 2 boilers, 27,800shp = 30kts. Range 3300nm at 13.7kts
Armament: 4–100mm (4×1), 4–37mm (2×2), 3–533mm TT (1×3), 2 RBU-900 (later refitted with MBU-2500), 4 DCT (36 DC), 2 DC rails (12 DC)
Sensors: Radar Cross Bird and Ball Gun/Ball End; sonar H/F hull
Complement: 190

Class (completed):
Sokol (30.8.51; renamed *Sovetskiy Turkmenistan*), *Berkut* (10.5.52), *Kondor* (10.5.52), *Grif* (25.10.52; renamed *Sovetskiy Dagestan*), *Krechet* (18.12.52), *Orlan* (17.3.53; renamed *Sovetskiy Azerbaydzan*), *Lev* (17.3.53), *Tigr* (25.4.53).

Kondor (Northern Fleet) was lost by grounding 1962. *Krechet* may originally have been named *Leopard*. All were built at Kaliningrad.

This was the SKR counterpart to Project 41. Although a preliminary TTZ for a new SKR (Project 37) was set by the main naval staff on 21 August 1943, by 1945 effort was being concentrated on a redesign of the existing Project 29 (*Yastreb*). In 1945, Stalin argued that the *Yastreb* was the largest acceptable SKR, and that 900t would be a better limit on size. Although work on blueprints for the modified *Yastreb* was authorised on 30 January 1946, the ship was never built, and draft requirements for a new SKR were approved on 29 July 1946. The main roles were convoy escort, patrol work, the support of amphibious operations, and minelaying. Maximum displacement was set at 1800t, and required speed was 26–27kts; armament would comprise two twin 130mm mounts (SM-2, as in destroyers), two quad 45mm mounts, 12–25mm, a triple torpedo tube, four DC throwers, two DC tracks, and a Hedgehog (M-10).

For the first time since 1917, two design bureaus (TsKB-32 and TsKB-53) competed. Each produced versions with diesel and steam powerplants. The smallest, the TsKB-32 diesel design (two 6700bhp engines on each of two shafts) came out to 1860t, with an endurance of 5000nm at 15kts. The largest, the TsKB-53 steam design (13,000shp on each shaft), came out to 1925t.

As in other Soviet designs of this period, the draft requirements proved over-ambitious, and at the next stage (TTZ) a displacement limit of 1300t was imposed, to be achieved by cutting armament (to four single 100mm guns, two twin 37mm, the triple tubes, a pair of Mousetraps with 24 rockets each, and 2 depth charge tracks with 22 depth charges) endurance (from 5000nm to 2000nm), and stores capacity (from 20 to 10 days). As completed, the ships had four depth charge throwers in addition to the Mousetraps and depth charge tracks.

TsKB-32 (Leningrad) won the competition with a steam design, and the final TTZ was approved on 21 June 1947. The sketch design for this Project 42 escort was completed at the end of August 1947, but opinions on the ship's role were unsettled. Only late in 1947 did the naval staff agree to separate the SKR role from a planned minesweeping role. As built, Project 42 introduced prefabrication to Soviet practice (the hull was built in seven welded sections), and it had a dual-purpose main armament. This design also introduced DC current (which had not been used in the *Yastreb*s and *Skoriy*s). The lead ship, *Sokol*, was laid down 17 August 1949, launched 11 September 1950, and began builder's trials 27 March 1951; she was accepted on 30 June 1951. She was heavier than the allowed limit (1339t), but she exceeded her design speed on trials (she made 29.65kts). Seakeeping was superior to that of both *Yastreb* and, on paper, to that of the planned successor, Project 50 ('Riga'). Project 42 had twin rudders: tactical diameter at full speed was 4.7 lengths.

No formal record of the decision to discontinue production after eight ships has survived, but it seems likely that, as in the case of Project 41, the ships' size was considered unacceptable. The decision may have been made by Stalin himself.

'Kola' class — Marek Twardowski

'RIGA' class *frigates* (Project 50)

Displacement: 1160t standard; 1288t normal; 1416t full load; 1488t maximum (design displacement 1045t light, 1188t standard)
Dimensions: 295ft 3in × 33ft 6in × 10ft
91m × 10.2m × 3.16m
Machinery: 2-shaft, TV-9 geared turbines, 2 boilers, 21,000shp = 28kts. Range 1950nm at 14.5kts
Armament: 3–100mm (3×1, 200 rounds total), 4–37mm (2×2, total 4000 rounds), 4–25mm (added 1970s), MBU-600 (replaced by 2–RBU-2500, 128 rounds), 4 DCT (36 DC), 2 DC rails (12 DC), 2 or 3–533mm TT (1×2 or 3), 28 mines (max)
Sensors: Radar Neptune or Don 2, Slim Net, Sun Visor; sonar H/F hull (Herkules or Pegas)
Complement: 175

Class (by builder, completed):
Nikolayev (20 units): *Gornostay*, *Pantera* (21.5.54; renamed *Komsomolets Gruzhii*), *Rys'* (21.5.54), *Yaguar* (24.4.54), *Sarych* (31.8.54), *Puma* (31.8.54), *Volk* (31.10.54), *Kunitsa* (23.12.54), *Korsak* (30.12.54), *Norka* (30.4.55), *Grifon* (30.6.55), *SKR-51* (28.9.55), *SKR-53* (31.12.55), *SKR-57* (28.2.56), *SKR-58* (7.5.56), *SKR-63* (30.5.56), *SKR-66* (29.9.56), *SKR-67* (22.12.56), *Voron* (18.6.55), *SKR-52* (26.11.55)

Komsomolsk (7 units): *Zubr* (31.5.54), *Bizon* (30.6.54), *Aist* (27.8.54), *Laska* (25.10.54), *Pelikan* (30.11.54), *Pingvin* (31.12.54), *Gepard* (31.12.54)

Kaliningrad (41 units): *Leopard* (30.6.54), *Bars* (laid down 25.4.52, launched 25.6.53, completed 30.6.54), *Rosomakha* (30.4.54), *Sobol'* (13.10.54), *Barsuk* (15.9.54), *Kuguar* (31.4.54), *Enot* (30.10.54), *Filin* (9.12.54), *Lun'* (27.12.54), *Kobchik* (31.5.55), *Tur* (31.5.55), *Los'* (31.6.55), *Olen'* (27.8.55), *SKR-50* (26.10.55), *SKR-54* (1956), *SKR-56* (21.5.56), *SKR-59* (25.5.56), *SKR-60* (20.6.56), *SKR-61* (23.8.56), *SKR-62* (25.9.56; named *Irkutskii Komsomolets* 27.10.69), *SKR-64* (31.10.56; named *Komsomolets Litvii* 12.10.62), *SKR-65* (27.12.56), *SKR-68* (23.3.57), *SKR-69* (30.5.57), *SKR-70* (20.6.57), *SKR-71* (13.7.57), *SKR-72* (13.7.57), *SKR-73* (30.11.57), *SKR-74* (26.11.57), *SKR-55* (1957), *SKR-75* (30.12.57), *SKR-76* (15.6.58; named *Arkhangelskii Komsomolets* 25.10.68), *SKR-77* (29.6.58), *SKR-80* (31.7.58), *SKR-81* (31.8.58), *SKR-10* (21.10.58), *SKR-4* (13.12.58), *SKR-5* (31.12.58), *SKR-8* (31.12.58), *SKR-14*, *SKR-15*.

The lead ship, *Gornostay*, was laid down on 20 December 1951, launched 30 July 1952, and began builder's trials 13 January 1953.

Transfers: Bulgaria (*SKR-53* in 1957 [*Derzkiy*], *SKR-66* on 15 October 1958 [*Smeliy*], *Korchik* in 1986 [*Bodriy*]), East Germany (*Barsuk* [*Sassnitz*] 22

'Riga' class 1970 — CMP

November 1956; *Enot, Tur, Olen'* in 1957), Finland (*SKR-69, Filin* in 1957), Indonesia (*Aist, Bizon, Grifon, Zubr, Korsak, Puma, Pelikan, Sarich* 1964–66).

Just as the 'Kotlin' was the simpler follow-on to the big *Neustrashimyy*, this was the simpler follow-on to the sophisticated 'Kola', with its unit machinery. Apparently Stalin personally ordered the development of a smaller SKR, with a full-load displacement of 1200t. Development of a TTZ was ordered on 20 July 1950; the sketch design was to have been completed in September for presentation in October 1950. The technical design was completed and presented in March 1951, and the lead ship was laid down in the spring of 1951. The ship was designed by the new TsKB-820 (the design office associated with Yard 820, in Kaliningrad). To gain sufficient performance within the limited dimensions allowed, the ship used a pressure-fired steam powerplant, a type used in all subsequent Soviet steam combatants. For a time the designers also considered replacing single 100mm guns by twin enclosed mounts, and using 25mm guns instead of 37mm. In the end, Project 50 carried 15 per cent less ammunition than its predecessor, Project 42; and one fewer 100mm and one fewer torpedo tube (ultimately, however, a lightweight triple tube was developed, and installed in many ships).

This class introduced a new fire control system, SFERA-50 (Wasp Head: the analogous system on 'Kotlins' was SFERA-56) with remote power control of the guns (SSSP-50 system) and a stabilised optical sight (SVP-42-50). The sight worked with the Yakor' radar. The sight was effective against both air and surface targets, but apparently computing time was excessive (34–40sec). The prototype was tested in a 'Kola'. In 1983 an experimental unit carried a Hawk Screech director forward and Wasp Head aft. Some units had two twin small craft type 25mm mounts added in the 1970s.

The longitudinally-framed HTS hull was lightly armoured, with 7-8mm plating over the boiler and machinery spaces (7mm at the waterline, 8mm up to the main deck) and 7mm over gunshields and bridge. To hold down ship size, the unit machinery arrangement of the 'Kola' class was abandoned; both boilers were in one compartment; both turbines were in an adjacent compartment.

On trials the lead ship made 29.5kts at 1134t. In Sea State 6, the ship had to reduce speed to 23kts, and the guns could not be used at speeds of more than 16kts. Tactical diameter was 6.1 to 6.9 lengths at full speed. It is not clear why performance suffered so badly, compared to that of the 'Kola'.

Some ships were modernised for ASW, with two RBU-2500 instead of the original MBU-200, and a total of 128 RGB-25 rounds. Presumably they were also fitted with Hercules scanning sonars in place of their original Pegas sonars. Modernised ships also had solid ballast installed.

Production was cut short by Khrushchev's October 1955 cancellations. China built four units plus five improved versions ('Kiangnan' class). The Chinese ex-Russian units were unique in mounting two Styx missiles in place of their TT ('Kiangnans' had no missiles).

'Petya I' class

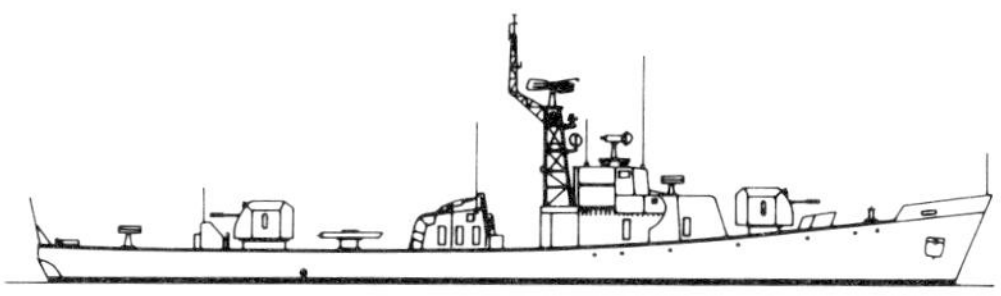

'Petya I' modified class

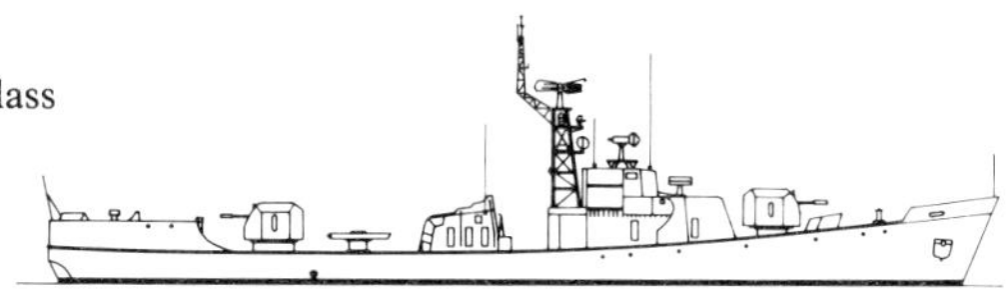

'Petya II' class

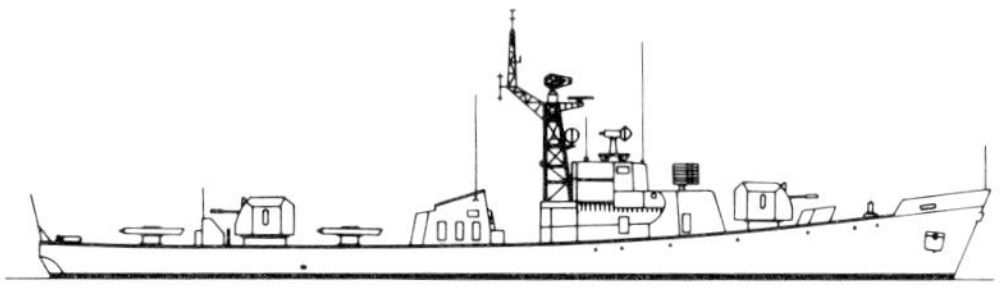

'Petya II' modified class

'PETYA' class *frigates* (Project 159)

Displacement: 950t standard; 1150t full load ('Petya II' 1160t)
Dimensions: 268ft 4in × 30ft 2in × 9ft 6in ('Petya II' 9ft 10in)
81.8m × 9.2m × 2.9m (3m)
Machinery: 2-shaft CODAG: 2 15,000shp gas turbines plus 1 6000bhp diesel, 36,000shp = 32kts. Range 4870nm at 10kts (diesel), 450nm at 29kts
Armament: 4–76mm, 2 RBU-2500 ASW RL, 2 DC racks, 5–406mm TT (1×5), 22 mines max ('Petya II' 2 RBU-6000, 10–406mm TT (2×5))
Sensors: Radar Don-2, Slim Net ('Petya II' Strut Curve), Hawk Screech; sonar H/F hull (Herkules), helicopter dipping sonar. M/F towed sonar in modified ships; ECM 2 Watch Dog
Complement: 90

Class (completed):
SKR-2, SKR-3, SKR-11, SKR-15, SKR-16, SKR-21 (named *Orlovskii Komsomolets* 12.3.79–18.3.92), *SKR-22* (named *Timofey Ulyantsev*), *SKR-23, SKR-27, SKR-33–36, SKR-38, SKR-40, SKR-41, SKR-43, SKR-46, SKR-47, SKR-78, SKR-82* (14.5.77), *SKR-90, SKR-92, SKR-96* (29.7.78), *SKR-98, SKR-103, SKR-110, SKR-112, SKR-115, SKR-120, SKR-126, SKR-128* (30.10.72), *SKR-130, SKR-133* (31.12.70), *SKR-135* (26.12.72), *SKR-138* (30.9.71), *SKR-141* (31.12.71) (37 units).

All by Kaliningrad (twenty-two including exports) and Khabarovsk (thirty-two including exports): lead ship by Kaliningrad 1960–61. Khabarovsk production amounted to: nine Project 159 'Petya I' completed 1961–66: *SKR-3, -11, -15, -21, -23, -41, -43, -46, -47*; eight Project 159A completed 1967–72; eleven Project 159AE completed 1968–77 for India; four Project 159M (completed 1973, 1974, 1976, 1982).

All remaining 'Petya I' believed discarded by 1989, all modified 'Petya I' 1990–91, all remaining 'Petya II' and 'III' by January 1994. 'Petya II' *SKR-16* to Azerbaijan July 1993, and *SKR-112* defected to Ukraine 21 July 1992.

Sixteen 'Petya I' completed 1961–64 (seven Project 159 with 2 RBU-2500, ten modified as Project 159M from 1973 on with 4 RBU-2500), twenty-seven 'Petya II' completed 1964–69 (Project 159A; one Project 159AM), 'Petya III' was the export version (4 RBU-25000, 1 triple 53cm TT): two to Ethiopia, ten to India, two to Syria, three to Vietnam; two were retained by the Soviet Navy for training. Ex-Soviet 'Petya II' were also exported: two to Ethiopia, three to Vietnam. They may have retained their 40cm TT. Designation changed from small ASW ship (PLK) to SKR 1964.

These were the first Soviet warships specially designated as ASW units. Like the 'Kashin', they were developed under the 1956 directive. The first large Soviet gas turbine combatant, 'Petya' was powered by a pair of M-2s plus a diesel. 'Mirka' (see below) is a similar hull in which a pair of jet engines in raised spaces exhausting directly aft rather than geared to propellor shafts.

'Petya II' has RBU-6000s and a second bank of 40cm torpedo tubes on the stern in place of the two after RBU-2500 of Project 159M. About 1966 one unit was fitted with an SUW-N-1 launcher in place of her forward 76.2mm mount. In 1967 one tested the VDS used on the *Moskva* class. In 1969 a unit tested a towed array. Twelve had various VDS (the modified 'Petya II' had one bank of TT removed, and the 'Petya I', lost two RBU-2500s). In each case the sonar was housed in a new raised stern deckhouse.

Recent disposals: three in 1987–88, eight in 1988–90, five in 1990–91.

'Mirka I' class

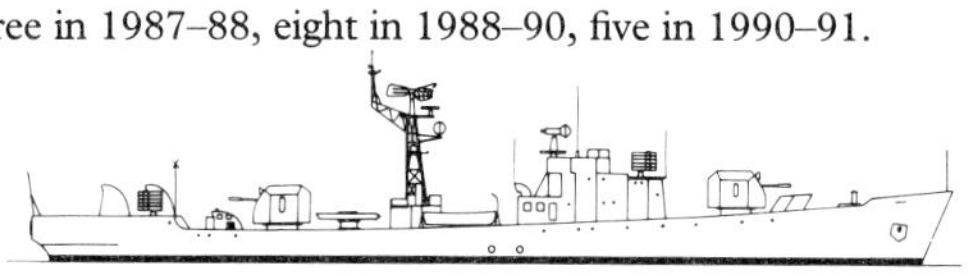

'Mirka II' class

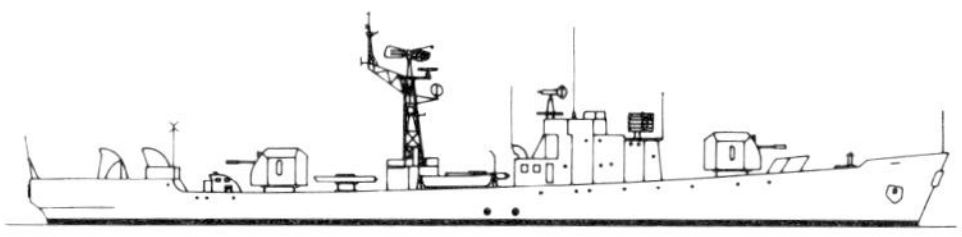

'MIRKA' class *frigates* (Project 35, 35M)

Displacement: 950t standard; 1150t full load
Dimensions: 270ft 4in × 30ft 2in × 9ft 6in
82.4m × 9.2m × 2.9m
Machinery: 2-shaft CODAG: 2 gas turbines plus 2 diesels, 30,000shp+ 12,000bhp = 32kts. Range 4800nm at 10kts, 500nm at 30kts
Armament: 4–76mm (2×2), 4 RBU-6000 ASW RL, 1 DC rack in some, 5–406mm TT (1×5) ('Mirka II' 2 RBU-6000, 2 sets of TT)
Sensors: Radar Don-2, Slim Net or Strut Curve, Hawk Screech; sonar H/F hull (Herkules or Pegas), dipping sonar in 'Mirka II'; ECM 2 Watch Dog
Complement: 98

Class:
Gangutets, Ivan Sladkov, SKR-6, SKR-7, SKR-9, SKR-12, SKR-13, SKR-19, SKR-20, SKR-22, SKR-26, SKR-32, SKR-48, SKR-83 (named *60 Let Komsomola Belorussii* 9.6.80), *SKR-84, SKR-86, SKR-87, SKR-117*

'Mirka I' had five banks of TT, and 4 RBU-6000 launchers (two on the forward superstructure and two aft), as in 'Petya I'. 'Mirka II' substituted a second bank of tubes for the after ASW launchers. Production extended between 1964 and 1966, and 'Mirka IIs' have been modernised with a helicopter-type dipping sonar in place of the former internal DC rack. The 'Mirka' machinery arrangement recalls that of the earlier 'Poti' class corvettes. The gas turbine arrangement aft precludes installation of a towed (as opposed to a dipping) sonar aft. Nine of each class, all built at Kaliningrad.

Recent disposals: five in 1989–90, three in 1990–91.

'Mirka II' class Marek Twardowski

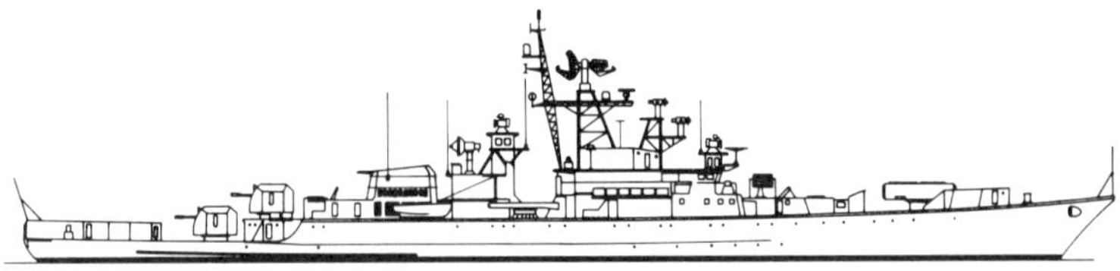

'Krivak I' *Dostoynyy* 1974

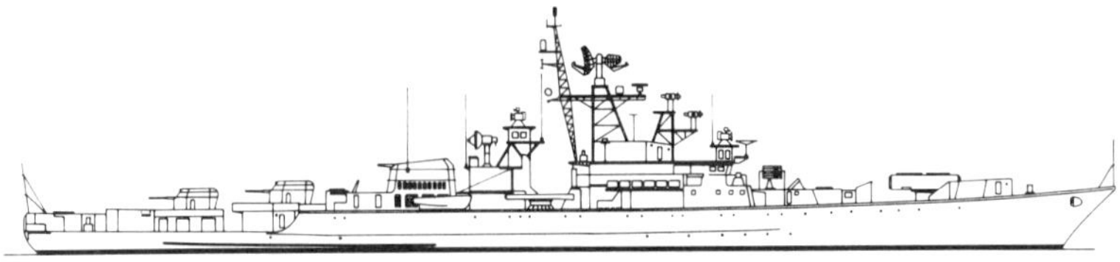

'Krivak II' *Rezvyy* as completed

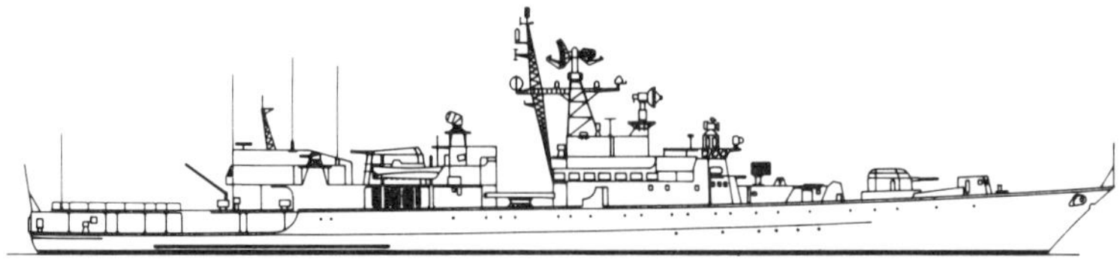

'Krivak III' class

'KRIVAK' class *missile frigates* (Project 1135)

Displacement: 3300t standard; 3575t full load
Dimensions: 405ft 3in oa × 46ft 3in × 15ft 1in
123.5m × 14.1m × 4.6m
Machinery: 2-shaft CODAG: total 72,000shp = 32kts (35kts on trial). Range 4600nm at 20kts, 1600nm at 31kts
Armament: 1 SS-N-14 (1×4, 4 missiles), 2 SA-N-4 (2×2, 40 missiles), 4–76mm (2×2), 2 RBU-6000 ASW RL, 8–533mm TT (2×4), 20 mines max ('Krivak II' single 100mm in place of twin 76mm
Sensors: Radar Don-2, Don-Kay or Palm Frond, Spin Trough Net C, 2 Eye Bowl, Kite Screech, 2 Pop Group; sonar Bull Nose and Mare Tail; ECM 2 Bell Shroud, 2 Bell Squat, 4 chaff RL (4×16)
Complement: 200

Class (by builder, commissioned):
'Krivak I' (Project 1135 Burevestnik):
Kaliningrad: *Bditelnyy* (1970), *Bodryy** (1971), *Svirepyy* (1971), *Storozhevoy* (1972), *Razyashchiy* (1973), *Rasumnyy* (1974), *Druznyy* (1975) (8 units)
Kamysch-Burun: *Dostoynyy* (1971), *Doblestnyy* (1973), *Deyatelnyy* (1973), *Bezzevetniy* (1.2.78), *Bezukoriznennyy* (Aug 1980), *Ladnyy* (Dec 1980), *Porvistiy* (Oct 1982) (7 units)
Zhdanov: *Zharkyy** (1975), *Retivyy* (1976), *Leningradskiy Komsomolets* (1976, renamed *Legkiy* Apr 1992)*, *Letuchiy* (1977)*, *Pylkiy** (1979), *Zadornyy* (Jul 1979) (6 units)
'Krivak II' (Project 1135M Burevestnik-M, all Kaliningrad): *Bessmennyy* (1978), *Gordelivyy* (Aug 1979), *Gromkiy* (Feb 1979), *Grozyashchiy* (Jul 1977), *Neukrotimyy* (Feb 1978), *Pitlivyy* (16.1.82), *Razitelnyy* (1977), *Revnostnyy* (Oct 1980), *Reskiy* (1976), *Rezvyy* (1975), *Ryanyy* (Apr 1980) (11 units)
'Krivak III' (Project 1135.5 Nerei or 1135P, all Kamysch-Burun): *Menzhinskiy* (1984), *Dzerzhinskiy* (1986), *Imeni XXVII Syezda KPSS* (1987; renamed *Orel*), *Imeni 70-Lettiya VUK KGB* (Jan 1988, ex-*Andropov*), *Imeni 70-Lettiya Progranvoysk* (Dec 1988), *Kedrov* (Dec 1989), *Vorovskiy* (Dec 1990), *Kirov*, *Krasny Vympel*. The last two units were taken over by Ukraine. *Kirov* was renamed *Latsis* (1991), then completed for Ukraine as *Hetman Petro Sagaidachnyy*. The last is to be completed for Ukraine as *Hetman Bayda Vyshnevetsky*.

Starred units were modified to 1135.2 ('Krivak 4') configuration.

This design began, probably about 1956, as a 'Riga' successor with a high forecastle hull (as in 'Kynda'), a main battery of three twin 76mm guns (one forward, two aft), two quadruple 57mm guns (fore and aft) and, presumably, a pressure-fired steam powerplant. Presumably the ship had the standard self-defence torpedo battery of this period, twin mounts on either side. There was also a design study of a radar picket version carrying a Top Sail (Voshkod) radar on a lattice mast forward of the funnel. By about 1963 a gas turbine version of the original SKR version was being offered for export, reportedly mainly to Indonesia, and by 1964 this version had been modified with SA-N-4 launchers in place of the earlier quadruple 57mm. About this time the Soviet fleet planners wanted a special narrow-sea rocket combatant for the Baltic and Black Seas, and the SKR was modified once more, with two triple SS-N-9 (P-50) launchers forward in place of the forward twin 76mm mount. At least a few ships were laid down to this Project 935 design. Then, with ASW once again paramount, the ships were redesigned (as Project 1135): SS-N-9 was replaced by a quadruple trainable SS-N-14 launcher. The ships were designated BPKs (redesignated SKRs in 1978).

Western navies were unaware of the ASW shift, and when the ships first appeared in 1970 they were assessed as short-range anti-ship shooters capable of 38kts (the reality was closer to 31kts).

'Krivak II' has a reshaped VDS housing aft to accommodate a different gun battery, single 100mm mounts in place of the earlier twin 76mm.

'Krivak III' was a major redesign to exploit the new TT-launched SS-N-15 ASW missile; the SS-N-14 mount was eliminated and the 100mm gun moved forward. That cleared space aft for a helicopter pad. Taken over by the KGB Maritime Border Guards, these ships were rated PSKR.

Modifications to 'Krivak I' class: *Zharkiy* and *Leningradskiy Komsomolets* had a new LF sonar suite (hull sonar and VDS). *Zarkiy* was the prototype. *Leningradskiy Komsomolets* completed modernisation refit in the Baltic May 1991: she received the new sonar suite, her Head Net C radar was replaced by Half Plate, and racks for eight SS-N-25 fitted in place of her RBU-6000s; she also received new ECM equipment. It seemed likely that other ships would be modified similarly, if resources permitted.

'Krivak I' class *Razyashchiy* (completed 1974) was stricken 1991.

On 9 November 1975 the poilitical commissar of *Storozhevoy* convinced part of the crew to mutiny at Kronstadt. Apparently he planned to go to Leningrad to protest what he saw as the corruption of the Soviet Communist movement. The ship was intercepted by naval bombers and warships, and the mutiny quashed. The ship was subsequently transferred to another fleet and (with a totally fresh crew) made many port visits, presumably to refute Western reports of the mutiny.

Razashchiy and *Doblestnyy* were stricken 1992 (*Razashchiy* towed for BU Aug 1993). *Svireppy*, *Silnyy*, and *Zharkiy* were all offered for BU 1994, although *Silnyy* had already apparently entered the yard at Kaliningrad for 1135.2 conversion, as indicated above. *Gordelivyy* began a refit 1992 but due to lack of funds had to be placed in reserve incomplete.

Bditelnyy Jul 1980 RAF

'IVAN SUSANIN' class *patrol icebreakers* (Project 97P and 97B)

Displacement: 3400t full load
Dimensions: 229ft 7in × 60ft × 21ft 4in
70.0m × 18.3m × 6.5m
Machinery: 2-shaft diesel-electric (3 13D100 diesel generators), 4800bhp = 14.5kts. Range 5500nm at 12kts, 13,000nm at 9.5kts
Armament: 2–76mm (twin mount), 2–30mm Gatling; may have SA-N-5
Sensors: 1 Strut Curve, 2 Don Kay, 1 Owl Screech
Complement: 140

Class:
Aisberg, *Dunaii*, *Imeni 26 Svezda KPSS*, *Ivan Susanin*, *Neva*, *Ruslan*, *Volga*, *Sadko*, *Vladimir Kavraiyskiiy* (Project 97B hydrographic ship)

Built at Admiralty, Leningrad 1975–81. Eight of a class of thirty-one modified as KGB border patrol ships, classified as PSKR. They differ from the naval icebreakers in being armed and in having enlarged superstructures and helicopter platforms (no hangars). The 30mm guns are optically controlled.

'Parchim II' class

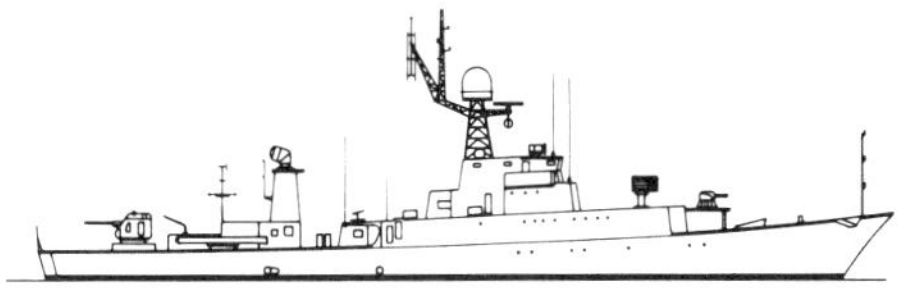

'PARCHIM II' class *corvettes*

Displacement: 800t normal; 950t full load
Dimensions: 237ft 10in × 30ft 10in × 9ft 10in, 15ft 1in (max)
72.5m × 9.4m × 3.0m, 4.6m
Machinery: 3 Type M504 diesels, 14,250bhp = 28kts
Armament: 2 SA-N-5/8 (2×4), 1–76.2mm, 1–30mm Gatling, 4–533mm TT (2×11), 2 RBU-6000, 2 DC racks (12 charges), mines
Sensors: Radar Positive-E, Spin Trough, Bass Tilt; sonar MF hull, HF dipping; ESM Watch Dog
Complement: 60

Class:
Yunga, MPK-67, MPK-99, MPK-105, MPK-192, MPK-205, MPK-213, MPK-216, MPK-219, MPK-224, MPK-228, MPK-229

Soviet version of a standard East German type, built at Peenewerft, equipped with a better air search radar and better weapons. The dipping sonar deploys through a door on the starboard side of the main deck superstructure. Inferior to the equivalent Soviet 'Grisha V', but presumably acceptable as payment in kind for Soviet aid. Twelve were delivered by Peenewerft in 1986–89 (launched 1985–89). The East German designation was Project 133.2.

NEUSTRASHIMY class *frigates* (Project 1154.0 Yastreb)

Displacement: 3500t full load
Dimensions: 426ft 6in × 50ft 11in × 18ft 5in
130m × 15.5m × 5.6m
Machinery: 2-shafts, 4 gas turbines, 110,000shp = 30kts
Armament: 1 SA-N-9 launcher, 2 CADS-N-1, 1 RBU-12000; 6 torpedo launchers (16 weapons); 1 helicopter; see notes
Sensors: LF bow sonar and VDS; 1 Top Plate; 2 Palm Frond; Cross Sword; 1 Kite Screech; Positive-E; EW is 2 Bell Shroud and 2 Bell Squat
Complement: –

Name	Builder	Laid	Launched	Comp	Fate
NEUSTRASHIMY	Kaliningrad	1986	May 1988	24.1.93	Extant 1995
NEPRISTUPNYI	Kaliningrad	1988	May 1991	1992	Extant 1995
TUMAN	Kaliningrad	1990			Building
NEPOKORNYI	Kaliningrad				

This is the 'Krivak' successor. Like 'Krivak III', it is intended to fire a surface-ship variant of the standard submarine ASW stand-off missile (in this case the 65cm SS-N-16). The elimination of special ASW missile tubes leaves space aft for helicopters. Missiles and torpedoes are fired from catapults like those on many French warships, which can be adapted to either 53cm or 65cm caliber. The torpedo-tube version of the SS-N-22 missile (P-100) can also be fired from these tubes, giving the ship considerable anti-ship firepower. The Positive-E radar aft controls the helicopter and also acquires targets for the two CADS-N-1 alongside the hangar.

SS-N-25 anti-ship missiles may eventually be mounted in the space between the superstructure blocks.

The sonar suite is the new one tested on the modernised 'Krivak I' frigates *Zharkiy* and *Leningradskiy Komsomolets*. Presumably it is also the new suite of the *Udaloy* follow-on. The very long hull dome accommodated both a bow array and a pair of flank arrays. There is also a VDS (but no towed array).

Reportedly this class was designed specifically to minimise radar and IR signatures. The hull has a flat flare over its whole length; superstructure levels are broken up to disperse reflections. The stacks are also shaped to reduce returns, and they have special exhaust eductors. However, the masts and radar antennas would probably provide sharp radar returns.

Neustrashimy began sea trials in December 1990; for a time NATO designated her Bal-Com 8. As of 1991, the second ship was scheduled for completion in 1992, and the third was to have been launched in 1993 for completion in 1994. *Nepokornyi* was intended as Project 1154.1. The current Ten Year Plan envisages continued production of this class. It was reported in December 1993 that seven units (1 brigade) are planned.

Neustrashimmy 1988 German Navy

'Koni' class 1980

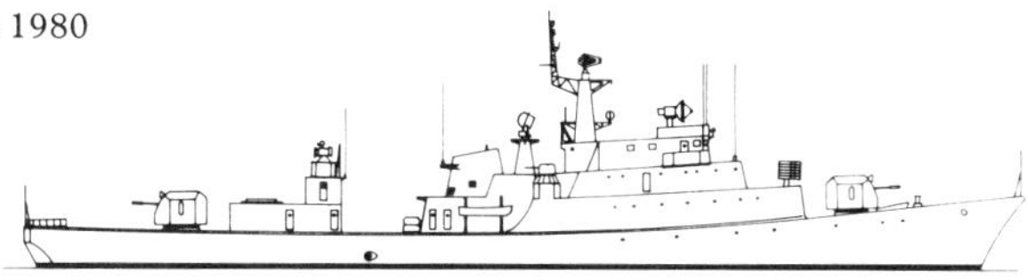

'KONI' class *frigate* (Project 1159)

Displacement: 1440t standard; 1660t full load
Dimensions: 311ft 8in oa × 42ft × 13ft 9in
95.0m × 12.8m × 4.2m
Machinery: 3-shaft CODAG: 1 gas turbine plus 2 diesels, 15,000shp = 15,000bhp = 27kts. Range 1800nm at 14kts
Armament: 1 SA-N-4 (1×2, 20 missiles), 4–76mm (2×2), 4–30mm (2×2), 2 RBU-6000 ASW RL, 2 DC racks, 20 mines max
Sensors: Radar Don-2, Strut Curve, Pop Group, Hawk Screech, Drum Tilt; sonar M/F hull; ECM 2 Watch Dog, 2 chaff RL (2×16)
Complement: 110

Delfin completed at Zelenodolsk 1978. The sole Soviet unit was transferred to Bulgaria in 1990. She had been retained for training and demonstration. Transfers: Algeria (three), Cuba (two), East Germany (three), Libya (two), Yugoslavia (two). These ships were presumably designed to carry SS-N-2 missiles. For example, the Algerian, Cuban, and East German ships have the Square Tie surface search radars used to control them, although they lack the missiles themselves. Yugoslav ships have four SS-N-2C missiles. 'Koni III' (Libya) is armed with 40cm TT (not otherwise exported) and SS-N-2C missiles, at the expense of one ASW rocket launcher. This is essentially an enlarged 'Grisha'.

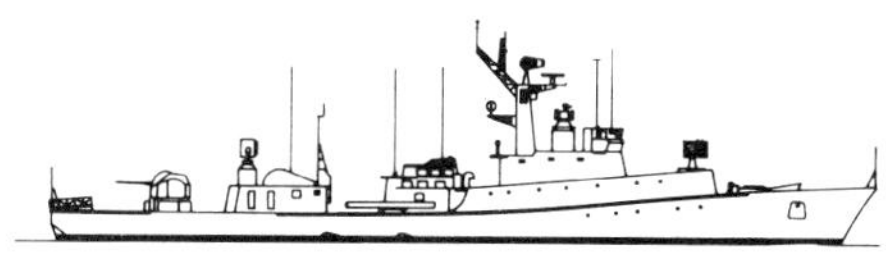

'Grisha I'

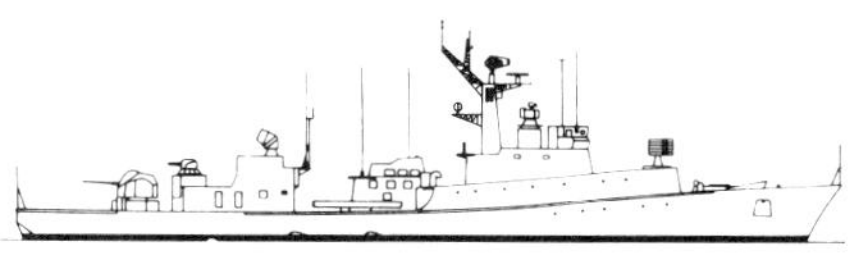

'Grisha II'

'Grisha V'

'GRISHA' class *frigates* (Project 1124, Albatros)

Displacement: 950t standard; 1200t full load
Dimensions: 234ft 11in × 32ft 2in × 12ft 2in
71.6m × 9.8m × 3.7m
Machinery: 3-shaft CODAG: gas turbines plus 4 diesels, 15,000shp or 19,000shp + 16,000bhp = 30kts. Range 4500nm at 10kts, 950nm at 27kts, 2000nm at 20kts
Armament: 1 SA-N-4 (1×2, 20 missiles), 2–57mm (1×2), 2 RBU-6000 ASW RL, 2 DC racks (12 DC), 4–533mm TT (2×2), 18 mines max ('Grisha II': second twin 57mm replaces SA-N-4, no Pop Group installed; 'Grisha III': 30mm Gatling aft in addition)
Sensors: Radar Strut Curve, Pop Group, Muff Cob, ('Grisha III' Bass Tilt); sonar Bull Nose and Elk Tail, M/F hull and dipping; ECM 2 Watch Dog
Complement: 60

Class (completed):
'Grisha I' (1124.1 Albatros, 16 units): *MPK-2–5, MPK-8, MPK-33, MPK-35, MPK-49, MPK-58, MPK-67* (11.9.68), *MPK-69, MPK-108, MPK-139, MPK-142, MPK-147, MPK-202. MPK-147* launched 13.10.68, accepted 10.10.70. *MPK-5* was the sole Project 1124.4 ('Grisha IV'). All completed

Neustrashimmy as completed

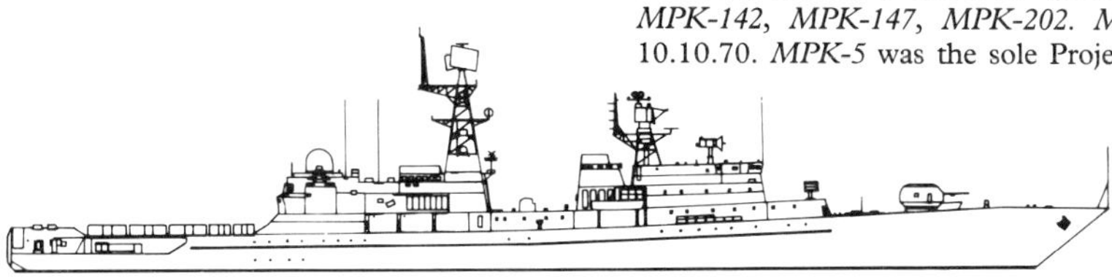

Grisha III (in Lithuanian service) 7.6.94 L & L van Ginderen

1968–75 at Kiev and Khabarovsk. Recognisable by a prominent plate, presumably for radiation hazard protection, forward of the Muff Cob radar controlling the 57mm gun aft.

'Grisha II' (1124A, also given as 1124P; 17 units): *Ametyst, Brilliant, Bezuprechniy, Bditelniy, Dozorniy, Zorkiy, Izumrud, Zhemchug, Izmail, Dnepr, Reshitelniy, Provorniy, Predanniy, Primernyy, Rubin, Sapfir, Smel'iy*. Version for KGB Maritime Border Guards, completed at Zelenodolsk 1973–76. A twin 57mm gun replaces the SA-N-4 launcher. The six KGB 'Grisha III' are sometimes lumped with this class. *Sapfir* lost in a storm 1967.

'Grisha III' (1124EhM, also given as 1124M, 34 units): *MPK-6, MPK-36* (31.12.71), *MPK-37, MPK-41* (31.12.72), *MPK-43* (*Odesskii Komsomolets* until 18.3.92), *MPK-44* (*Komsomolets Latvii* until 18.3.92), *MPK-59, MPK-81* (31.12.74), *MPK-101* (*Zaporozhskii Komsomolets* until 1992), *MPK-113, MPK-114, MPK-117* (31.12.73), *MPK-122* (31.12.75), *MPK-127* (*Komsomolets Gruzhyy* 26.8.80–18.3.92), *MPK-133, MPK-138, MPK-143* (31.12.76), *MPK-145* (completed 30.11.77), *MPK-152, MPK-155* (completed 30.9.80), *MPK-161, MPK-170* (14.10.78), *MPK-178* (21.12.84), *MPK-190, MPK-191* (21.11.85), *MPK-196, MPK-197, MPK-198, MPK-199* (*Komsomolets Armenii* until 18.3.92), *MPK-200* (*Primorskiy Komsomolets* until 1992, completed 25.12.87), *MPK-208, MPK-221, MPK-254* (*Komsomolets Beloryssii* until 1992), 1 other. Built at Kiev and Khabarovsk. AK-630 with Bass Tilt replaces Muff Cob. *MPK-44* and *MPK-108* were transferred to Latvia 6 November 1992. The six with adjectival names (as in destroyers) were for the KGB Maritime Border Guards.

MPK-104 (Project 1124K), which was probably laid down as a 'Grisha III', was test platform for the SA-N-9 missile (3 launchers, only 2 TT), and possibly also for an alternative powerplant. *MPK-5* ('Grisha IV': Project 1124.4) was used for trials of RBUs.

'Grisha V' (1124MU, 30 units): *MPK-7, MPK-10, MPK-52, MPK-56, MPK-69, MPK-82, MPK-85, MPK-104, MPK-107* (ex-*Irkutskiy Komsomolets*), *MPK-113, MPK-118* (ex-*Komsomolets Moldaviy*), *MPK-130* (ex-*Arkhangelskiy Komsomolets*), *MPK-134* (ex-*Kievskiy Komsomolets*), *MPK-203, MPK-207, MPK-214* (ex-*Lenninskaya Kuzhnitsa*), *MPK-217, MPK-291, Lutsk, Ternopol', L'vov, Lugansk, Zaporozhskaya Sech'*, 6 others. As in contemporary Soviet coastal combatants, this version has a 76mm gun in place of the earlier twin 57mm. It has a better air search radar (Strut Pair or Top Plate) and sonar (Platina rather than Shelon/Argon) but carries only one RBU-6000. Two launch positions for hand-held SAMs have been added at the break of the 01 level superstructure, just forward of the uptake. In at least one case, the 53cm TT were modified to launch wire-guided torpedoes. Some units, including the last two named, were built for the KGB Maritime Border Guards. Two units were taken over by the Ukrainian Navy, as U-200 *Lutsk* and *Ternopol'*. *MPK-82* is the 1124EMT export prototype, with Cross Dome radar rather than Half Plate; she also lacks the dipping sonar.

The following are divided between 'Grisha III' and 'Grisha V': *MPK-17, MPK-28, MPK-64, MPK-65, MPK-89, MPK-103, MPK-222.*

These small ASW ships (MPKs) are production successors to 'Poti', carrying much the same armament on a larger hull which adds an SA-N-4 missile launcher forward, both ASW RLs being moved into the superstructure. As in the earlier class, CODAG machinery is used because the ships are intended for 'sprint-and-dash' ASW tactics, dipping and then quickly retrieving their sonars. Early units had 15,000shp gas turbines, but 'Grisha III' has a 19,000shp unit. The diesels are M503s ('Grisha I' and some 'Grisha II') or M504 (later units).

Converted Trawlers

Two types are known, both presumably mobilisation prototypes for small ASW vessels used for peacetime training.

'Mayk' (Project 502GM): 920t standard, 1000t full load, 177ft 10in × 30ft 10in × 12ft 2in, *54.2m × 9.4 × 3.7m*, 1 shaft, 900hp = 14kts, range 9500nm at 12kts, 2–25mm (1×2), 2–400mm TT (2×1), 3 RBU-1200 ASW RL, 10 DC, complement 50. One unit (No. 225) converted 1989, served in Baltic Fleet 1992. At the same time, another seven were used as intelligence collectors (Project 502P), and three served as ASW training ships (Project 502U).

'Alpinist' (Project 1539): 1185t, 176ft 2in × 34ft 5in × 14ft 5in, *53.7m × 10.5m × 4.4m*, 1 shaft, 1320hp = ??, range 7200nm at 10kts, 2–25mm (2×1), 2–533mm TT (2×1), 10 DC, complement 35. One unit (No. 320) built at Yaroslav and converted 1989. In 1992 another five were in use as intelligence collectors with GS designations (Project 503M). Others are used for research.

SUBMARINES

Submarines were generally designated by letter and number: in October 1949 existing submarines, which had gone by individual class series, were designated as S (medium) or B (*bolshoi*, large). The M-series of small submarines was not affected. Thus existing L-, D-, and K-boats were redesignated *B-1* through *B-35* (later *K-51* through *-56* became *B-51* through *-56*, although *K-55* may have remained *B-8*). Large ex-U boats (*N-25* through *-30*) became *B-25* through *-30*. *S-* and *Shch*-boats were redesignated in an S-series, which also embraced Type VIIC U boats (*s-81* through *-84* and ex-*Lembit* (*s-85*). Later a K series began for nuclear submarines (though many were later designated or redesignated in the B-series, retaining their numbers). SS indicates a special service submarine. The big 'Typhoons' are numbered in a TK (presumably 'Tyhazely', heavy) series.

Note that numbers are *not* sequential. Yards were apparently assigned blocks of numbers in the S-, B-, and K-series. At times multiples of 100 seem to have been added to the numbers of early boats to make room for later classes.

'Whisky V' class

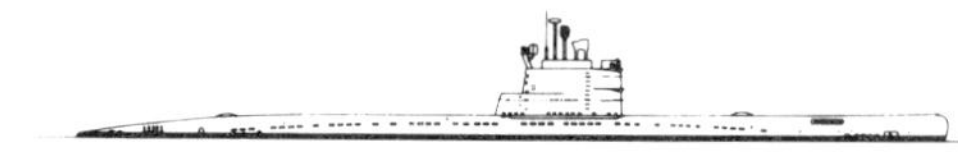

'Whiskey Canvas Bag' class

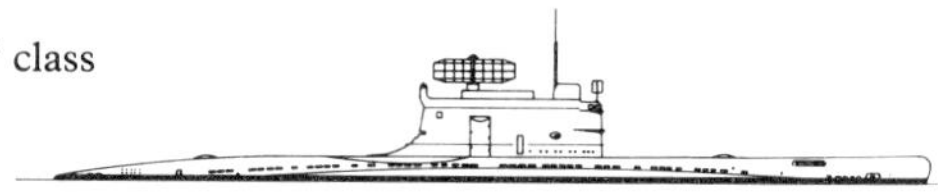

'Whiskey Long Bin' class

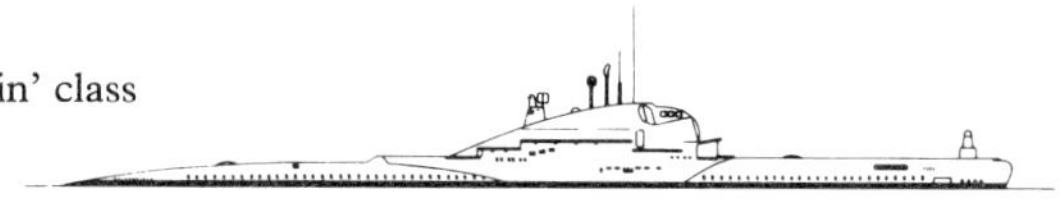

'WHISKEY' class *attack submarines* (Project 613)

Displacement: 1050t surfaced; 1340t submerged
Dimensions: 249ft 4in × 20ft 8in × 14ft 11in
76m × 6.3m × 4.55m
Machinery: 2-shafts diesel-electric; 2 Type 37D diesels, 4000bhp, 2 main motors, 2700hp, 2 creep motors, 100hp = 18.25kts/13.1kts. Range 8580nm at 10kts surfaced, 335nm at 2kts submerged (endurance 166hr submerged). Stores endurance 30 days
Armament: 6–533mm TT (4 bow, 2 stern, 12 torpedoes or 22 mines), 2–57mm (1×2, SM-24ZiF mounting), 2–25mm (1×2, 2M-8 mounting)
Sensors: Radar Feniks, Tamir; sonar Nakat, flag
Complement: 52

Class:

Builder:	Gorkiy	Nikolayev	Baltic	Komsomolsk
1951	1 (S-80)	–	–	–
1952	4 (S-43–46)	5 (S-61–65)	–	–
1953	16 (S-140–152, 155, 157,158)	11 (S-66–76	3 (S-153, 154, 156)	– –
1954	29 (S-159–186, 192)	14 (S-77–99, 86–91, 95–97, 100, 217)	–	1 (S-331)
1955	37 (S-193–200, 261–289)	18 (S-98, 218–234)	8 (S-187–191, 355–357)	4 (S-332–335)
1956	26 (S-290–297, 300, 325–329, 338–349)	15 (S-235–246, 250, 374, 376)	4 (S-358–361)	4 (S-336, 337, 390,391)
1957	–	9 (S-375, 377–384)	3 (S-362–364)	2 (S-392, 393)
1958	–	–	1 (S-365)	–
Totals:	113	72	19	11

Total built for Soviet navy, 215. Another twenty-one were assembled in China from Soviet-supplied components, 1956–64. The first boat, *S-80*, was laid down 13 March 1950, launched 21 October 1950 (70 per cent complete), and began dock trials 31 December 1950. Builder's trials followed between 27 April and 28 June 1951, and she was accepted on 2 December 1951. The last Sormovo submarine of this type, *S-349*, was accepted on 31 December 1956. Names with dates of naming: *Ulyanovskiy Komsomolets* (S-143, 17.9.76), *Pskovskiy Komsomolets* (S-191, 15.6.66), *Vladimirskiy Komsomolets* (S-263, 23.9.67), *Bryanskiy Komsomolets* (S-300). *S-74* sank Feb 1968.

S-178 of this class sank 24 October 1981 off Vladivostok after colliding with a refrigerator ship. *S-137* was the famous 'Whiskey on the rocks' which ran aground near the Swedish naval base of Karlskrona, 27 October 1981.

Transfers: Albania (two 1960, two seized from USSR 1961), Bulgaria (two 1958), China (five 1956–57), Cuba (one 1979), Egypt (six 1957–58, one 1962), Indonesia (fourteen 1959–62, of which two from Poland and two for spare parts), North Korea (four 1960s), Poland (two 1962, one 1964, two 1969), Syria (one 1986). Cuban and Syrian boats were battery-charging barges.

This was one of three submarine designs developed by TsKB-18 to TTZs formulated in 1946: Project 611 for a large submarine ('Zulu'), Project 612 for

S 137 aground off Karlskrona, Sweden 11.4.81 CMP

a small submarine (not built), and Project 613 for a medium submarine ('Whiskey'), the latter to succeed the earlier 'S' and 'Shch' classes. An earlier Project 608 for a follow-on 500–600t medium submarine was begun in 1939. Work was stopped by the outbreak of war, but by 1942 two versions (770t and 820t) were being considered. A new TTZ issued in September 1943 asked for a 640-tonner with a 120m test depth, and 660t and 687t versions were proposed. Project 608 was dropped in 1944 when the German *U 250* was raised and examined. The new TTZs (January 1946) required a diving depth of 200m. The Project 613 design was developed between October 1947 and August 1948. The Sormovo yard at Gorkiy was assigned to mass-produce these submarines, hopefully without interfering with major surface combatants. A special project office established there to work out detailed plans (SKB-112, for shipyard 112) became the Lazurit TsKB, responsible for submarines built at Gorkiy and at Kosomolsk-on-Amur.

Data above are for the original Project 613, with high-capacity batteries but also with guns (in and on deck abaft the sail) and without a snorkel. A 'Whiskey' tested the experimental snorkel for this class in the Caspian Sea in 1954. Apparently the first version seen in the West ('Whiskey I') had the 25mm gun but not the deck mount; 'Whiskey II' had both guns. 'Whiskey III' lacked gun armamnet but retained the deck platform. 'Whiskey IV' had the 25mm gun and a snorkel. 'Whiskey V' had all guns removed and was fully streamlined, with the step at the fore end of the sail faired over. This upgraded version was probably designated project 613M.

Although nominally reflecting German Type XXI practice, these boats had only slightly figure-eight form hulls at their battery sections fore and aft of the control room. The hull form aft resembled that of Type XXI. The Tamir active sonar was in the forefoot, with the Feniks passive sonar (similar to the earlier German Balkon) in a bulb at the fore end of the horizontal keel, under the breeches of the torpedo tubes.

As late as 1982, sixty boats were considered usable (forty-five active, fifteen in operational reserve); eighteen remained a decade later.

Variants:

613V: *S-66, 67, 70, 80, 88, 88?, 140, 141, 145, 150, 154, 160, 161, 168, 172, 176, 178, 181, 185, 194, 195, 197, 222, 224, 226, 294, 331* (extended range).

613Ts: *S-384* (torpedo target). *S-76* may also have been used as a target.

P-613: *S-146* (missile test, presumably 'Whiskey Single Cylinder'). Other reported missile test designation: 613Kh.

613D4: *S-229* (ballistic missile test for SS-N-5, with two tubes). A reported V613 designation may be related to this one. D-4 was SS-N-5.

613RV: *S-65* (torpedo test launch platform).

613S: *S-43* (salvage/rescue version; this may be the boat fitted with a diver lock-out chamber forward of her sail).

640: ('Whiskey Canvas Bag' radar picket): *S-62, 73, 144, 151.* There was also a modified Project 640V version. One boat reverted to attack submarine status and two or three were broken up.

644 ('Whiskey Twin Cylinder'): *S-44, 46, 69, 80, 158, 162*; two SS-N-3 launchers added abaft the sail. *S-80* of this class was lost with all hands (68 men) 27 January 1961, found in 1969 in 200m of water and raised 1970.

665 ('Whiskey Long Bin'): *S-61, 64, 142, 152, 155, 164.* Boats were converted at Leningrad 1961–63, to carry four SS-N-3 (P-5) missiles. Speed was reduced to 13.5/8kts.

666: *S-63* (experimental rescue submarine based on 613S).

Two (renamed *Sevryanka* [ex-*S-148*] and *Slavyanka*) were converted to fishery research craft in 1958.

Other reported variants: 613E (export), 613Eh (machinery), 613A (armament), 613AD (minelayer), EP613 (test bed).

'Whiskey Twin Cylinder' 1973 USN

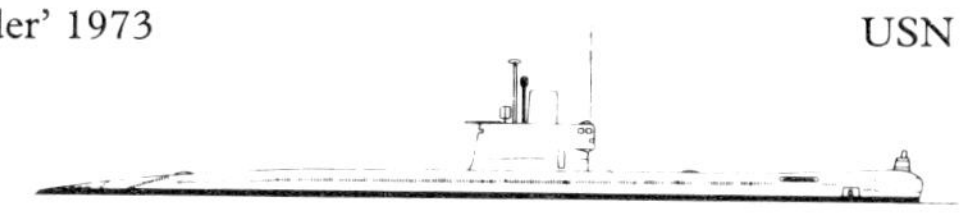

'Romeo' class

'ROMEO' class *attack submarines* (Project 633)

Displacement:	1330t surfaced; 1700t submerged
Dimensions:	252ft 8in × 22ft × 16ft 1in *77.0m × 6.7m × 4.9m*
Machinery:	2 shafts, 2 diesels, 4000bhp submerged = 15.5kts surfaced, 13kts submerged. Endurance 45 days. Range on snorkel 7000nm at 5kts and 16,000nm at 10kts (surfaced). Diving depth 270–300m
Armament:	8–533mm TT (6 bow, 2 aft, 14 torpedoes or 28 mines)
Sensors:	Radar Snoop Plate; sonar M/F (Herkules), passive array (Fenika); ECM Stop Light
Complement:	56

Class:

(All at Gorkiy, built 1957–62): *S-11* (later *SS-11*), *37, 128* (later *SS-128*), 139, 247 (later the training hulk *UTS-247*), *248, 249, 298, 299, 321, 330, 350, 351, 352, 353, 367, 368, 369, 370, 371, 372* (21 units).

This improved 'Whiskey (Project 613) could dive deeper and had greater range. The TTZ was reportedly issued in 1952–53. The Balcon-like passive array below the bow was replaced by a larger passive array on the upper part of the bow (Feniks-M). The Tamir attack sonar was replaced by an Arktika tracking sonar feeding a torpedo fire control system, Leningrad. Many units had small sonar domes on deck forward carrying a sonar intercept array, with underwater telephone (NATO 'Fez') above it. Unlike 'Whiskey', this submarine had a deeper bow carrying an extra pair of bow tubes. It appears that production of these boats, a large number of which were planned, was reinstated (with the 'Foxtrots') about 1956, then finally stopped a few years later and parts and tooling sold to China. Production in North Korea was apparently unauthorised by the Soviets. A Project 633A version (reportedly to carry a missile, probably similar to 'Whiskey Long Bin'), was cancelled in October 1955. The 'Whiskey Long Bin' may represent an application of this design to the earlier hull. Two built in 1958, three in 1959, five in 1960, nine in 1961, one in 1962. Transfers: Algeria (two 1982–83), Bulgaria (two 1971–72, two 1985–86), Egypt (five 1966, one 1969, plus four Chinese-built 1983–84), Syria (three 1985–86). Others were built in China (thirty-five, of which seven were transferred to North Korea in 1973–75: plus four to Egypt as noted) and North Korea (twelve, of which two were lost in 1985). The last units were stricken by 1987 and broken up by 1989. *SS-128* (Project 633RV) had a massive structure added atop her bow, carrying a pair of 65cm TT, to test the SS-N-16 missile and the new generation of 65cm torpedoes.

'Zulu III'Apr 1969 MoD

'Zulu I' class

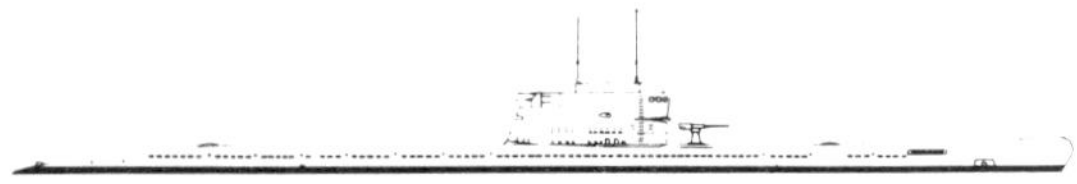

'Zulu II' class

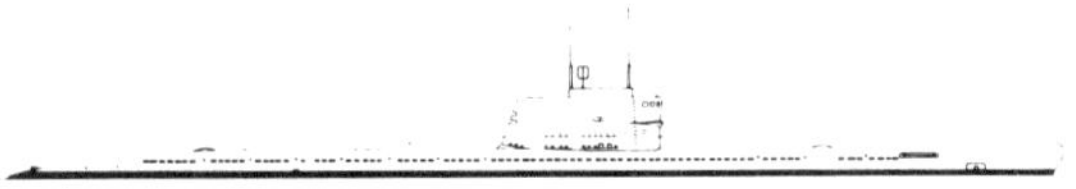

'Zulu III' class

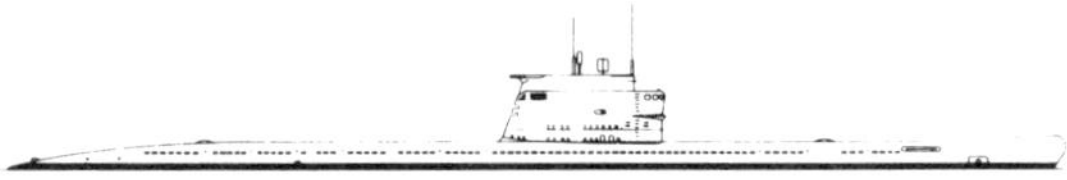

'Zulu IV' class

'Zulu V' class

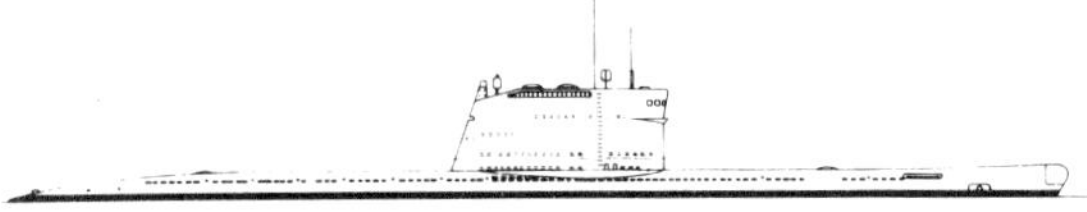

'ZULU' class *attack submarines* (Project 611)

Displacement: 1900t surfaced; 2350t submerged
Dimensions: 295ft 3in × 24ft 7in × 19ft 8in
90.0m × 7.5m × 6.0m
Machinery: 3 shafts, 2000bhp diesels, 5300shp submerged = 18kts surfaced, 16kts submerged. Endurance 70 days. Range 9500nm at 8kts on snorkel, 20,000nm surfaced
Sensors: Radar Snoop Plate; sonar M/F (Tamir), small passive array
Armament: 10–533mm TT (6 bow, 4 aft, 22 torpedoes or 44 mines), 2–57mm (1×2), 2–25m,m (1×2). See notes
Complement: 70

Class (by builder):
Sudomekh: reportedly one 1952, five 1953, six each 1954–55

Names: *B-61* through *-82*, *B-88* through *-91*. *B-61* was accepted 31 December 1953, *B-91* on 15 August 1958. *B-78* named *Murmanskiy Komsomolets* 1978. Missile conversions: P611 (one missile) was *B-67*; AV611 (two missiles each): *B-61*, *B-62*, *B-78*, *B-79*, *B-89*.

'Zulu' was the large-submarine equivalent to 'Whiskey', also designed to a 1946 TTZ. The project reached the design stage in 1948, and plans for the first boat were transferred to Sudomekh in Leningrad on 10 January 1951. Compared to the prewar K-boat, a 'Zulu' was about 400t larger. It could dive twice as deep (200m as opposed to 100m), had a much larger battery (4 × 112 cells as opposed to 2 × 120) and added a centre shaft carrying a diesel and a large electric motor (the two wing motors were only slightly more powerful than the 1200hp units on a K-boat) to increase submerged speed at the 1hr rate by about 50 per cent and low-speed endurance by a factor of nearly 3, from 160nm. Surface speed fell from 22.5kts, because even the three engines of this type were less powerful than the two 4200bhp units of the K-boat.

The first boats completed had deck guns and lacked snorkels. In 1956, with jet aircraft clearly immune to such weapons, deck guns were ordered removed. All boats were eventually converted to this Project 611M configuration, with snorkels.

Reported conversions, many for test: 611RU (special training), A611 and AB611 (weapons tests, probably torpedoes), AB611U (trial version with extended range), AB611K (experimental flotilla leader), AB611E (possible engine test bed), AB611D (penetrating minelayer, to get through defensive fields to plant mines inside; failed), AB611RE (test bed, otherwise unidentified; suffix suggests a missile), AB611S (salvage conversion), AB611D (experimental minelayer), P611 (experimental oiler sub). Two units were converted into the oceanographic research ships *Lara* and *Vega* (possibly Project 611RE). The special minelayer had a large sonar, intended to allow it to pass inside defensive fields to lay mines. Reportedly the failure of this version led to Soviet interest in submarine-launched mobile mines, which could be laid from outside the defensive fields.

In 1995 a 'Zulu' was on display at Den Helder, in the Netherlands.

This class was used to test the first Soviet submarine-launched ballistic missile. On 26 January 1954 a directive was issued to arm long-range submarines with ballistic missiles (the 'Volna' programme). TsKB-16 developed the necessary stabilised sail-top launcher from gun mounts originally designed for the Project 82 battlecruiser. The existing army R-11 missile (NATO 'Scud') was selected, and adapted as R-11FM. *B-67*, a Northern Fleet submarine launced in 1953, was converted to fire a single missile as Project V611. On 16 September 1955 she fired the world's first submarine ballistic missile. She was later used to test other new systems, such as new torpedo tubes, new periscopes, and the new sonar complex used on nuclear submarines.

Five boats (NATO 'Zulu V') were converted to take paired tubes (Project AV611). They formed the basis for the later Project 629. The first launch of a dummy missile from such a submarine was in 1958. The Scud-firing submarine has also been identified as Project 611A. These ships were eventually converted to fire SS-N-4 missiles (Project 611B). The Severodvinsk-built *B-78* was used for underwater-launch trials in 1964 (Project 611LB). Project AB611RA was reportedly converted to test a new anti-ship missile.

'Quebec' class

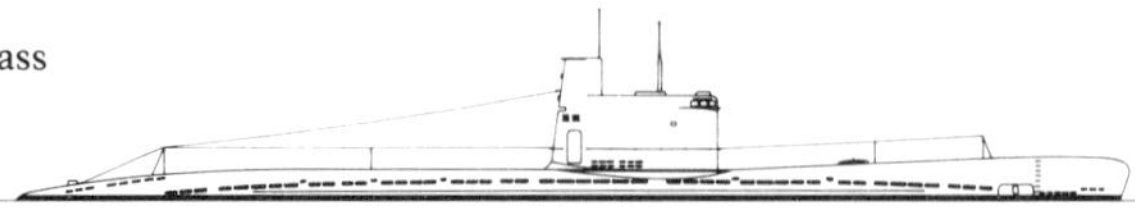

'QUEBEC' class *attack submarines* (Project 615)

Displacement: 460t surfaced; 540t submerged
Dimensions: 183ft 9in × 16ft 5in × 12ft 6in
56.0m × 5.1m × 3.8m
Machinery: 3 shafts, Kreislauf system; 2300bhp diesels, 2200shp submerged = 18kts surfaced, 16kts submerged. Range 7000nm cruising
Armament: 4–533mm TT (8 torpedoes), 2–25mm (1×2, later removed)
Sensors: Radar Snoop Plate; sonar Tamir, Feniks
Complement: 30

Class:
M-254 (prototype, Project 615; all others were A615; begun 17.3.50, launched 31.8.50, completed 30.5.53), *M-255–269*, *M-295–301*, *M-321*, *M-351–356*, *M-361* (completed 24.8.62).

M-255 through *M-259* delivered 1955, *M-260* through *M-268* and *M-351–354* in 1956, *M-269*, *M-295*, *M-297*, *M-298–300* and *M-355–6* in 1957, *M-301*, *M-321*, *M-354* in 1958. *M-361* tested the propulsion system planned for Project 637, the modified version of this design. *M-296* (redesignated *M-305*) became a memorial at Odessa in April 1984. One unit of this class was delivered on completion to the Naval Engineering Academy in Leningrad, and was never operational.

Note that the number *M-233* has also been given for the prototype. The prototype displaced 392t/493t; production boats were 405.8t/503.9t (also given as 411t/507.1t).

Completions all at Sudomekh: three in 1954, ten in 1955, twelve in 1956, five in 1957, for a total of thirty. Four were left in 1982. They were built in alternate groups of three, three, and then four. Launch dates for the first five groups: September 1954, then May, July, October 1955, and then April 1956. In 1956 US intelligence sources reported first completions in May 1955; the data above may refer to *launch* rather than to completion dates, in which case it would jibe with the more detailed 1956 report.

This closed cycle ('single propulsion system', in Soviet parlance) design was chosen in preference to the conventional Project 612 as successor to the earlier M-classes. The *M-615* was by TsKB-18; the production A-615 was by TsKB/16 (the chief designer, Kassatier, apparently transferred to TsKB-16). The prototype may have had two rather than four torpedo tubes. The Kreislauf powerplant, derived from prewar experiments, ran submerged on an internal supply of liquid oxygen; oxygen was added after exhaust was passed through a lime chemical absorbent (ED KhPI). The submarine could also snorkel. There were three engines: one 32D diesel (900bhp) on the centre shaft, plus a pair of 700bhp M-50P diesels (a type used on torpedo boats), with a 100hp creep motor on the centre shaft and a back-up diesel generator aft. Boats could run at low speed using only the centreline diesel.

Because liquid oxygen could be stored for only a limited period, the boat could not operate far from a base. To overcome this problem, in 1954–55 the Project 637 submarine was designed to use a new 'B-2 single product', consisting of solid granules containing bonded oxygen and a carbon dioxide absorbent. Exhaust gas passed through it would have its carbon dioxide leached out while its oxygen content was renewed. In 1958–59 *M-361*, then under construction, was retrofitted to test B-2, but in 1960 her trial programme was stopped and closed-cycle work ended because of the success of the nuclear programme. *M-361* may have been the boat seen at Sudomekh in July 1959 with a large 'shack' (about 26ft long × 14ft wide × 7ft 6in deep) built around her modified sail, with a 17ft vertical pipe at the after end of the shack. She (or a second such unit) was seen at Kronstadt in August 1960 with a second shack further aft.

Further otherwise unidentified closed-cycle designs, probably also using the Kreislauf system, were Projects 618 and 661. Given its place in the sequence, 618 was probably a medium submarine with this type of powerplant.

Prewar experiments apparently began with an M-boat, probably *M-92*, using liquid-oxygen as a stored oxidizer and exhausting overboard (the system was called YeD-VVD, a unified engine exhausting into water). Combustion could never be complete; the exhaust contained considerable oxygen, so the next stage (on *M-401*) was to extend underwater endurance by running the exhaust through a bed of lime to leach out the carbon dioxide before enriching it with more oxygen and feeding it back into the engine (this version was called YeD-KhPI, a unified engine with lime chemical absorbent). The project was also called REDO.

This Kreislauf system was dangerous: there were accidents prewar on M-class (Series XII) oxygen-fuelled (REDO) boats, during the Second World War on *M-401*, and postwar on *M-255*, *-256*, *-257*, *-259*, *-351* and *-352*; flooding after attempts to put out a fire sank *M-256* on 26 September 1957. Boats were modernised, probably during the late 1950s, to make the power plant more fire- and explosion-resistant, and these submarines continued to operate in the Baltic and Black Seas until they were stricken early in the 1960s. One (*M-305*) has survived as a memorial in Odessa.

'WHALE' *attack submarine* (Project 617)

Displacement: 950t surfaced
Dimensions: 204ft × 19ft 11in × 16ft 7in
62.18m × 6.1m × 5.06m
Machinery: 1 shaft; Walter powerplant, 7250shp = 11kts/20kts; 1 26-cell battery, electric motor, 540hp = 9.3kts; creep motor 140hp; aux powerplant 8-cylinder 4-stroke diesel, 450bhp (could be coupled either to main shaft or generator for propulsion or battery recharging). Range on Walter engine 120nm at 20kts, on electric motor 13.4nm at 9.3kts.
Armament: 6–533mm TT (bow; 12 torpedoes)
Sensors: Radar Nakat; sonar Tamir-5, Mars-24
Complement: 51

Name	Builder	Laid down	Launched	Commissioned
S-99	Sudomekh	5.2.51	5.2.52	6.6.52

Project 617 was the Soviet attempt to build an operational Walter submarine, based on technology captured in Germany in 1945. A special KB was set up in Germany in hopes of reviving the late-war Type XXVI design (22.5kts, 6 bow tubes firing forward and 4 more firing aft, without reloads). In 1946 TsKB-18 recreated the German Type XXVI design from German material; it was assigned Project number 616. This version soon proved unacceptable, so TsKB-18 began work on a new Project 617 to combine standard Soviet equipment with the 7500shp Walter turbine designed for Type XXVI. The preliminary design was completed by the end of 1947. A new SKB-143 was hived off from TsKB-18 specifically to develop Project 617. German Walter experts were used on the project until 1951. After the first (and, it turned out, only) boat was completed, SKB-143 was assigned to develop the first Soviet nuclear submarine. TsKB-18 took over the Walter project. The Sudomekh yard built a land-based prototype Walter powerplant, tests of which were completed in 1951.

Like the German Walter boats, this one carried her hydrogen peroxide in plastic bags in tanks slung below the main pressure hull. The hydrogen peroxide powerplant could start up in a little more than 2min (acceleration to full speed from a cold start took 9.5min). It lost propulsive power with depth, because exhaust had to be dumped overboard. During trials, the engine operated at 80m. At a depth of 95–130ft (30–40m), it produced 6050shp. Although the boat used Soviet equipment, in hull form it resembled its German relatives, its relatively short hull offering good maneuverability.

Between 1956 and 1959 *S-99* went to sea 98 times and covered over 6000nm on the surface and about 800nm submerged, including 315 using the Walter powerplant. On 19 May 1959 she suffered an explosion while starting her Walter plant at 80m depth; mud in the hull valve of her hydrogen peroxide supply pipe caused the peroxide to decompose and ultimately to blow an 8cm opening in the pressure hull, causing flooding. *S-99* was not repaired.

Project 643 was probably a long-range equivalent, work on which was stopped in 1960 when all the closed-cycle projects were closed down.

'Foxtrot' class

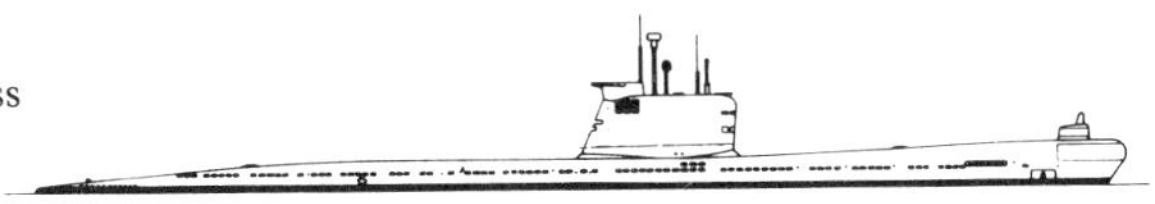

'FOXTROT' class *attack submarines* (Project 641)

Displacement: 1957t surfaced; 2484t submerged
Dimensions: 294ft 3in wl, 299ft 6in oa × 24ft 7in × 19ft 8in
89.7m, 91.3m × 7.5m × 6m
Machinery: 3 shafts; 3 diesels, 6000bhp = 16.6kts surfaced; 4 112-cell batteries, 3 motors (1–2700hp, 2–1350hp), 5400hp = 15.9kts submerged. Stores endurance 90 days
Armament: 10–533mm TT (6 bow, 4 stern)
Sensors: Radar Nakat-M; sonar Feniks-M, Arktika-M, Herkules, sonar intercept system
Complement: 75 (8 officers)

Class (completed):
B-2 (14.7.63), *B-4* (named *Chelyabinskiy Komsomolets* 18.1.63 – Apr 1992, completed 31.8.61), *B-6* (30.7.64), *B-7* (15.12.61), *B-8* (18.12.62), *B-9* (30.11.65), *B-15* (30.9.64), *B-21* (31.10.65), *B-25* (6.9.65), *B-26* (named *Yaroslavskiy Komsomolets* 15.2.65; completed 24.3.66), *B-28* (18.6.66), *B-29* (28.11.66), *B-31* (25.6.63), *B-33* (27.11.61), *B-34* (31.8.66), *B-37* (5.11.59), *B-38* (30.9.62), *B-39* (28.12.67), *B-40* (6.9.66), *B-41* (24.12.66), *B-46* (30.6.67), *B-49* (named *Vladimirskiy Komsomolets* 23.9.67; completed 30.6.67), *B-50* (8.12.62), *B-51* (26.10.67), *B-53* (5.11.62), *B-55* (20.9.63; renamed *B-855* in 1970), *B-57* (28.7.60), *B-59* (10.6.61), *B-85* (23.12.60), *B-94* (25.12.58), *B-95* (30.9.59), *B-98* (15.5.64), *B-101* (29.5.64), *B-103* (24.12.64), *B-105* (15.6.62), *B-107* (28.4.65), *B-109* (15.4.65), *B-112* (5.8.65), *B-116* (25.8.60), *B-130* (22.9.60), *B-133* (5.11.59 renamed *B-833* in 1970), *B-135* (5.11.59), *B-139* (15.3.60), *B-143* (5.11.60), *B-153* (30.9.61), *B-156* (18.4.61), *B-164* (16.10.61), *B-169* (28.7.62), *B-205* (28.12.69), *B-213* (11.8.70), *B-309* (10.12.78), *B-311* (11.2.78), *B-330* (27.10.77), *B-397* (31.12.67), *B-400* (25.9.68), *B-401* (19.9.68, yard N 402), *B-402* (5.9.69), *B-405* (2.7.69), *B-409* (10.9.71), *B-413* (25.12.68), *B-416* (4.12.69), *B-427* (4.12.71), *B-435* (6.11.70), *B-440* (25.12.70), *B-456* (10.10.72), *B-464* (16.12.73), *B-470* (5.7.73), *B-510* (20.10.83), *B-522* (20.9.74), *B-533* (8.10.76), *B-586* (30.11.79), *B-587* (25.10.81), *B-588* (28.9.82), *B-590* (21.11.80)

'Foxtrot' class Apr 1979 MoD

Note the re-use of low *B*-numbers from the 1949 series. Most were built at Sudomekh (three at Severodvinsk 1962). *B-94* was prototype: begun 3.10.57, launched 28.12.57.

Production for the Soviet Navy (62 units) ended in 1971.

Export production: total 12 (included in the list above). Transfers: Cuba (641K: *B-309* on 7.2.79, *B-586* in Mar 1980, *B-510* in Feb 1984), India (641I: *B-51* in Apr 1968, *B-401* in Mar 1969, *B-405* in Nov 1969, *B-402* in Feb 1970, *B-456* in Nov 1973, *B-470* in Dec 1973, *B-464* in Oct 1974, *B-522* in Feb 1975), Libya (641I: *B-311* in Dec 1976, *B-330* and *B-533* in Feb 1978, *B-587* in Feb 1981, *B-588* in Jan 1982, *B-590* in Feb 1983), Poland (ex-Soviet boats: 1987, 1988).

This was the large-submarine analog to 'Romeo', designed from the outset to snorkel. It had the same sonar suit as 'Romeo', with the new Leningrad torpedo FCS. A recently-published inboard profile of a 'Foxtrot' suggests that above the Feniks-M bow passive array (within the bow) there was a cylindrical active sonar, probably Hercules. There were also two rows of 'windows' for sonar transducers in the sail. Their dimensions suggest that they were medium- and low-frequency sonar-intercept arrays (presumably the portion in the dome above the bow was the high-frequency element). Compared to 'Zulu' (611), 'Foxtrot' could dive deeper (250m, maximum 280m) and had greater range. Project 641K had a new engine (2D42 rather than 37D). A planned Project 641A, probably a missile boat, was cancelled in 1955. Planned service life was 27 years; at least twenty had been discarded by 1989, five of them in 1987–88. *B-475* (completed 1967, decommissioned 1 April 1994) is now on display in London. In 1976 *B-2*, probably of this class, reportedly was so badly worn out that she could not leave her pier.

The original version had Type 37D diesels; Project 641K had faster-running Type 2D-42 diesels (1825bhp) geared to the shafts.

Abortive diesel submarine projects of this period included 630, 631, and possibly 636. No details are available.

'Tango' class

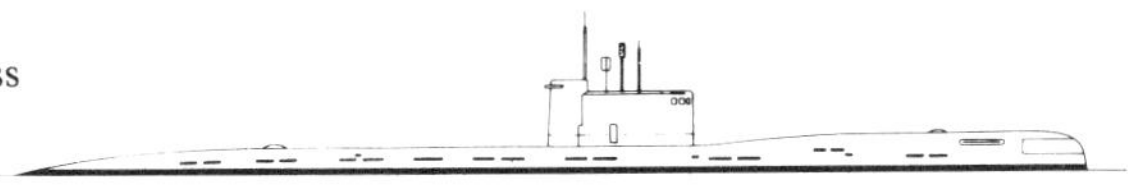

'TANGO' class *attack submarines* (Project 641 BUKI (Som))

Displacement: 3100t surfaced; 3900t submerged
Dimensions: 300ft 3in × 29ft 6in × 23ft
91.5m × 9.0m × 7.0m
Machinery: 3 shafts, diesels, 6000bhp = 20kts surfaced, 16kts submerged
Armament: 6–533mm TT (bow). Reportedly fitted for SS-N-15 torpedo tube-launched missiles
Sensors: Radar Snoop Tray; sonar Shark Teeth L/F
Complement: 72

Class (all built at Gorkiy):
B-30, B-97, B-215, B-225, B-290, B-303, B-307, B-312, B-319, B-380 (named *Gorkovskiy Komsomolets* until 18.3.92), *B-386, B-443, B-474, B-498, B-504, B-515, B-519, B-546*

Reported launch dates: two 1972, two 1974, two 1975, one 1976, two 1977, three 1978, two 1979, three 1980, two 1981

This ASW submarine was improvised from 'Foxtrot' components and surplus 'Juliett' hull sections (to gain about 50 per cent more internal volume; hull diameter grew from about 7.8m to 9m). The added volume was needed to accommodate the big new sonar and plotting facilities to support the associated SS-N-15 missile. There were three sonars in all: presumably the bow array, a sonar intercept set, and an underwater communications set. Although the class was first seen at the July 1973 Sevastopol Review, all units served with the Northern Fleet. The number built, eighteen, suggest that they were conceived as a bastion-defence division under shore control. The timing suggests an association with the development of the 'Delta' class, which was expected to operate in a bastion area. That the crew of 78 includes 17 officers suggests a more complex weapons system (a 'Foxtrot' has 75, including 8 officers).

The BUKI suffix to the project number means armament change (B)/enlarged (U)/new command system (K)/emergency design (I).

'Tango' class 1993 L & L van Ginderen

'November' class

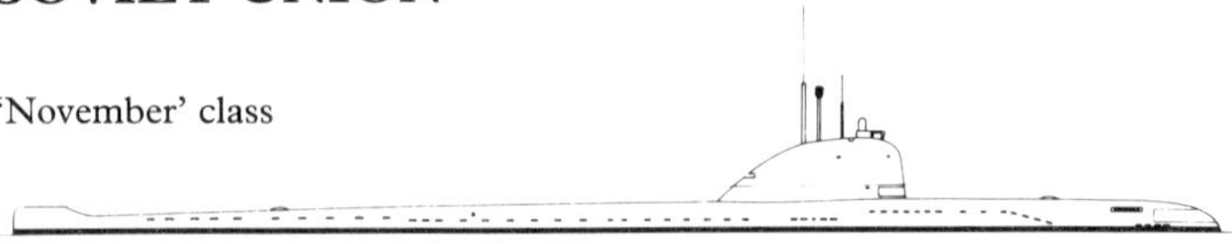

'NOVEMBER' class *attack submarines* (Project 627 Kit)

Displacement: 4500t surfaced; 5300t submerged
Dimensions: 363ft 11in × 29ft 6in × 25ft 3in
110.0m × 9.0m × 7.7m
Machinery: 2-shaft nuclear, 1 reactor, 30,000shp = 30kts
Armament: 8–533mm TT (bow, 24 torpedoes)
Complement: 80

Class (completed):
K-3 (named *Leninskiy Komsomol* 9.10.62; begun 15.5.54, launched 12.8.57, delivered 12.3.59), *K-14* (completed 31.12.59), *K-5* (prototype of 627A production series: completed 17.8.60), *K-8* (31.8.60), *K-52* (23.12.61), *K-21* (23.12.61), *K-11* (23.12.61), *K-133* (16.10.62), *K-181* (16.10.63), *K-115* (20.12.62), *K-152* (4.11.63), *K-27* (liquid-metal reactor version, Project 645: laid down 15.6.58, launched 1.8.62, completed 30.10.63), *K-42* (named *Rostovskiy Komsomolets* 8.5.81; completed 4.11.63), *K-50* (20.12.63)

All built at Severodvinsk. *K-3* began trials on 3 July 1958 and was first underway on nuclear power at 10:03 on 4 July 1958. Trials in the White Sea were completed on 1 December 1958. *K-8* was lost due to a fire off Cape Finisterre on 8 April 1970. She had suffered a steam-generator crack in the Arctic, 13 September 1961. *K-3* had steam generator trouble en route to the North Pole in June 1962; she suffered a fire on either 18 August or 8 September 1967, in which thirty-nine men died. Her reactors were later dumped in the Kara Sea. *K-11* had a severe reactor accident at Severodvinsk 12 February 1965. The reactor compartment had to be dumped in the Kara Sea. *K-5* also had a severe reactor accident.

This first Soviet nuclear submarine was conceived to deliver a strategic weapon, a 1550mm diameter T-15 torpedo (24m long) with a range of at least 40km. A parallel diesel submarine project (probably 630 or 631) was abandoned before the boat could be built. T-15 itself was abandoned when the alternative ballistic missile proved successful, and in 1954–55 the design was recast as a torpedo-attack submarine.

Under a 9 September 1952 directive, separate design groups were formed to develop the nuclear submarine and its powerplant. SKB-143 (Malakhit), which had been working on Walter submarines, was assigned to this project. It formally began the submarine design in March 1953; it would carry a single strategic-attack torpedo and a pair of 53cm torpedoes for self-defence. The submarine was provided with a targeting radar for use with T-15 (presumably mainly to confirm its position off an enemy coast). Each tube carried only a single torpedo.

Initial estimates (1952–53) called for a 2650–2700t (surfaced) submarine with an operating depth of 250–300m, capable of 22–24kts, with a stores endurance of 50–60 days and a crew of 70. The first sketch design (1953) showed 2896t/4500t, operating depth 275–300m, 24–25kts, 60 days' stores, and a crew of 75. The final version of the strategic attack submarine design (1954) showed 3050t/4746t, 300m depth, 24–25kts, 50–60 days' stores, and a crew of 76. Then the mission was changed; armament was switched to 8 bow tubes (in vertical rows) with 20 torpedoes. Displacement was almost unchanged (3050t/4750t)), as were speed, diving depth, and stores endurance; but crew increased to 85. Surfaced speed was 15.2kts.

Once the basic design had been recast, it was ordered developed into a production attack submarine, Project 627A (directive 22 October 1955). SKB-143 began work in August 1956. The prototype, *K-5*, was launched in September 1958 and was accepted on 27 December 1959. The most obvious difference between 627 and 627A was in their sonars. Probably mainly to achieve high speed, 627 lacked the Feniks chin array of contemporary diesel submarines, such as 'Zulu'. Instead, she had the simpler flush-mounted Mars-16KP, an earlier-generation set. Her Arktika searchlight sonar was behind a big window in the forward face of her sail (which was quite small by contemporary Russian standards). As yet there was no associated Leningrad system; the boat had a Prizma torpedo fire control system. The underwater communications set was Svet and the echo-sounder was Luch. Torpedoes could be fired at depths down to 100m. Unusually for Russian submarines, there was only a single periscope. Other masts carried two HF whips, the 'Park Lamp' HF/DF (actually for underwater radio reception), an ESM array, and radar.

On trials, *K-3* made 23.3kts at 65 per cent power, and 28–30kts at full power. Each of the two independent power trains comprised a 70MW VI-A reactor and a 17,500shp turbine set turning a motor-generator, as in a direct-drive diesel submarine. Also as in diesel boats, each shaft carried a 450hp creep motor. One unusual feature of these and later Soviet submarine reactors was a quench baffle within the reactor pressure vessel, to be used to quickly shut the reactor down (it was also used during refuelling). There was a substantial battery, needed partly to start up each reactor. The submarine could, therefore, either drive her propellers directly from the turbine or use one turbine, operating as a generator, to turn both shafts. Each reactor was connected to several heat exchangers. Reportedly heat exchangers were, initially at least, quite unreliable, so boats often ran on part of the output of only one reactor and rarely at full power. As these problems were solved, the operational speed of the class improved dramatically through the 1960s, one demonstrating her full power when chasing USS *Enterprise* in February 1968. Her unexpectedly good performance helped sell the *Los Angeles* class to the US Navy. Rated speed was 15kts/29kts. The same machinery arrangement was used on other first-generation submarines of the 'Echo' and 'Hotel' classes.

Because 627A was conceived from the outset as an attack submarine, it was provided with a full attack submarine sonar suit, Arktika and Feniks-M. Arktika was moved into a large chin fairing projecting below the keel, and Feniks-M was placed behind a sonar window on the upper part of the bow. The big sonar window in the sail was eliminated. A new narrow window covered elements of the high-frequency sonar intercept array (a recent drawing shows seven elements on the forward part of the sail, with one on either side of the after part). Medium-frequency intercept arrays were arranged to either side of the torpedo tubes, with a large window (presumably covering the low-frequency elements) on either side at the waterline. The submarine carried a Nakat radar and the RLK-1 ESM set. At 80 per cent power *K-5* made 28kts on trials.

The sixth unit of the class, *K-27*, was redesigned with a new liquid-metal (lead-bismuth) powerplant (as Project 645). Contract design began in 1955; because the hull very nearly duplicated that of Project 627, there was no need for a concept or preliminary design. The only important differences were that the bow of Project 645 was spherical (like that of 627) only below the waterline; it was conical above. Furthermore, unlike 627, 645 was designed to standard navy requirements. It added ship-service turbo-generators, like those of Western nuclear submarines. There was some internal rearrangement. The reactor room replaced the auxiliary machinery space of the 627; the turbo-generators and diesel generators replaced the reactor space of the 627.

Compared with Project 627A, this design needed 23 per cent less battery capacity because its higher-temperature reactor needed much less electricity to start up. The ship's combat system was also improved: a device for rapid torpedo loading was installed. It could simultaneously load any four tubes. Like a 627A, 645 had Arktika-M, Feniks-M, and the Leningrad system which allowed attacks without using a periscope. The usual second (air search) periscope was added. Non-magnetic steel was used in the outer hull, superstructure, and sail to reduce magnetic signature (degaussing weight and power were halved). However, as in contemporary German submarines, the new steel caused problems.

As built, the new submarine was slightly larger than a Project 627 (3414t/4370t, 360ft 3in × 27ft 3in × 19ft 2in, *109.8m × 8.3m × 5.85m*). Maximum speed was 30.2kt (range 35,400nm at this speed). Because the crew increased to 105, rated stores endurance fell to 50 days.

The contract design was completed in 1956, and *K-27* was laid down in 1958, launched 1 April 1962, and completed trials on 30 October 1963. Her lengthy building period was due to powerplant problems. However, she was able to begin a first long patrol sometime after 20 April 1964, and made a second between 15 July and 13 September 1965. During her next departure for sea, 24 August 1968, her port steam-generator broke down, as did some of the reactor thermal shielding. She returned and was decommissioned; sometime later her hulk was scuttled in the White Sea.

Project 627 was also adapted directly to fire a long-range missile. P627A was designed to carry a new P-20 long-range cruise missile developed by Ilyushin. It occupied a 82ft × 15ft (*25m × 4.6m*) cylindrical container faired into the after end of the sail, and took 6.5 minutes to erect. The hull was slightly enlarged, to 361ft 7 in × 30ft 2in × 20ft 10in (*110.2m × 9.2m × 6.34m*) (about 3950t), diving depth was limited to 285m, and speed would have fallen to 23–25.5kts. Like other contemporary Soviet strategic submarines, this one was provided with 40cm self-protection torpedoes, in this case in two tubes below the reduced nest of four full-size tubes; the load-out was one big torpedo per tube and 6–40cm torpedoes. A recently-published drawing suggests that the submarine would have had a new sonar suit. Arktika-M would have remained in the chin position, but there was a big sonar window in the fore part of the sail (presumably for Feniks-M) and a big window below the 53cm torpedo tubes (the 40cm tubes were outboard of it). The window may have been intended for the new large-diameter Kerch sonar introduced in the 'Alfa' and 'Victor' classes.

By early 1960 P-20 had been cancelled. A proposed torpedo conversion, PT-627A, was dropped in favour of completing the submarine as a standard 627A attack unit, *K-50*. Note that P-627 may have been considered a test bed for P-20 rather than its standard platform; reportedly another submarine, Project 653, was also designed to carry this weapon.

Leninsky Komsomol at the North Pole 1963 CMP

'Golf I' class

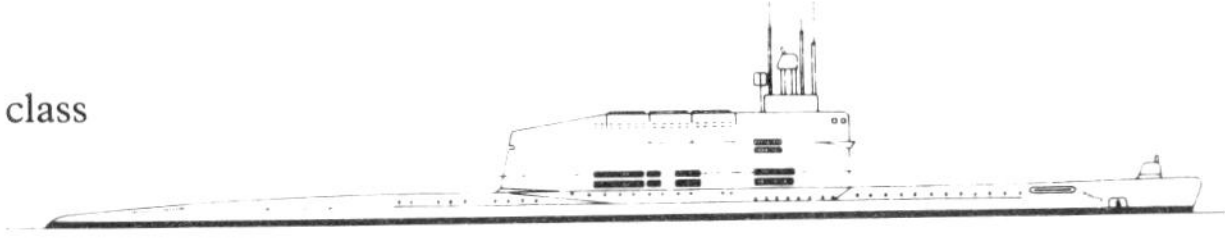

'Golf II' class

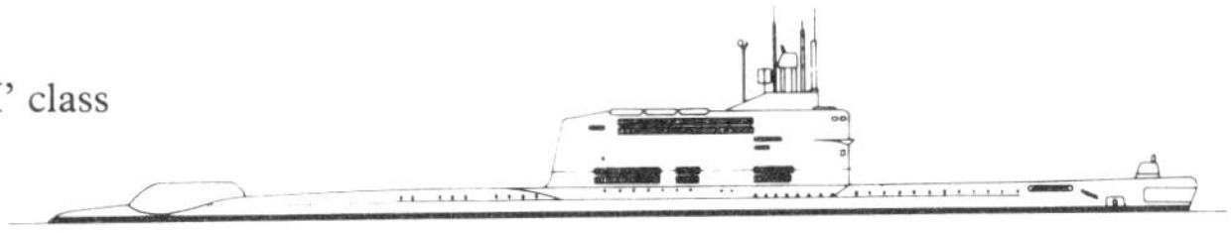

'Golf' class SSQ conversion

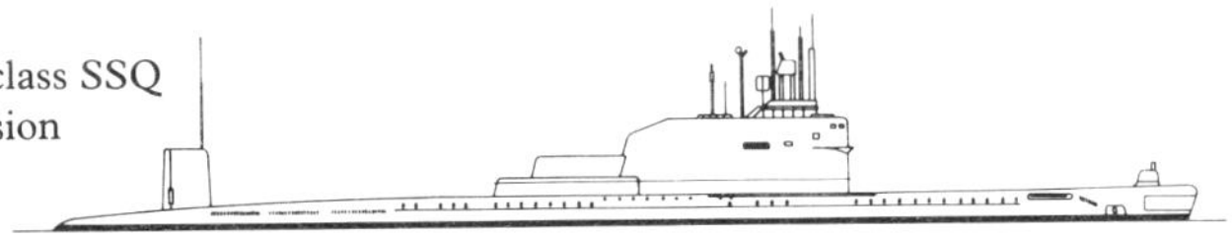

'Golf II' class Apr 1979 MoD

'Hotel II' class 1975 MoD

'GOLF' class *strategic submarines* (Project 628)

Displacement: 2500t surfaced; 2900t submerged
Dimensions: 328ft 1in × 27ft 11in × 21ft 8in
99m × 8.0m × 6.6m
Machinery: 3 shafts, 3 diesels of 2000bhp, 5300shp submerged = 12kts submerged. Endurance 70 days. Range 9000nm at 5kts
Armament: 3 SS-N-4, 10–533mm TT (6 fwd, 4 aft, 16 torpedoes)
Sensors: Radar Snoop Tray; sonar Artika, Fenik-M, Hercules
Complement: 59

Class (by builder):
Severodvinsk: *B-92* (renamed *K-96* 20.1.60), *B-40* (renamed *K-72*, then *K-372* in 1974), *B-41* (renamed *K-79*), *B-42* (renamed *K-83*, then *K-183* in 1974), *B-121* (renamed *K-102*), *B-125* (renamed *K-107*), *B-45* (renamed *K-88*), *B-61* (renamed *K-93*), *B-156* (renamed *K-113*), *K-118*, *K-36*, *K-91*, *K-110*, *K-153* (later *BS-153*), *K-142*. One completed 1958, two 1959, four 1960, five 1961, three 1962

Komsomolsk-on-Amur: *B-93* (renamed *K-126*), *B-103* (renamed *K-129*), *B-109* (renamed *K-136*), *B-113* (renamed *K-139*), *K-75*, *K-99*, *K-163*. One completed 1959, three 1960, two 1961, one 1962

Boats are listed in order of completion, beginning with *B-92* (laid down 1958, launched and completed 1959). *K-142* was completed in December 1962; *K-163* also in 1962.

No yard is listed for the 23rd unit, *K-129*.

One more unit was built in China from Russian-supplied components.

Boats originally with *B*-numbers were probably ordered as 'Foxtrots' and re-ordered as 'Golfs'.

K-102 (Project 629A) demonstrated an underwater launch of an R-21 (SS-N-5) missile before Khrushchev in February 1962.

K-129, lost in the Central Pacific in April 1968, became the subject of a massive CIA salvage attempt in August 1974, using the Hughes-built *Glomar Explorer*. It reportedly recovered a 38ft bow section, including eight dead crewmen (who were then buried at sea) and two nuclear torpedoes, from a depth of 17,000ft.

This ballistic-missile submarine with 'Foxtrot' components was built specifically to fire the new SS-N-4 (R-13) missile (some boats were equipped for a time with R-11FM ['Scud'] carried by converted 'Zulus'). Project 629V (*K-142*, 1967, 2 tubes) was the first submarine to test the submerged-launch D-4 missile. Project 629A (*K-129*) was the first submerged-launch boat with three tubes; *K-142* was later converted into this configuration. A total of thirteen units (Project 629M, 'Golf II') were converted. Project 601 ('Golf III': *K-118*) tested the D-9 (SS-N-8) missile system. She was lengthened to 361ft (*110m*), 2900t/3300t. Project 605 ('Golf IV': *K-102*, with 6 tubes) tested the D-5 (SS-N-6)/4K18 (SS-NX-13) system. The hull was lengthened by 60ft (*18.28m*). Project 619 ('Golf V': *K-153*, later renamed *BS-153*, 1978, with one tube) tested the D-19 missile system (SS-N-20). Project 629R was a communications relay version with missiles and torpedo tubes removed (NATO SSQ): *K-83* and *K-96* converted 1973–79 for Pacific and Black Sea, *K-107* for the Black Sea Fleet. Project 629I (*K-113*) was a 1974 minelayer conversion. At least one unit was converted into a torpedo submarine (Project 629KS).

Six 'Golf II' were deployed to the Baltic in the late 1970s, possibly as a response to the deployment of US Poseidon submarines in direct support of NATO. Another seven were in the Pacific. Disposals: one 'Golf II' 1987–88, one 'Golf SSQ' 1989, twelve 'Golf II' and one 'Golf V' 1989–90. The 'Golf III' had also been discarded by late 1989. The three SSQ remained in service in 1990.

'Hotel II' class

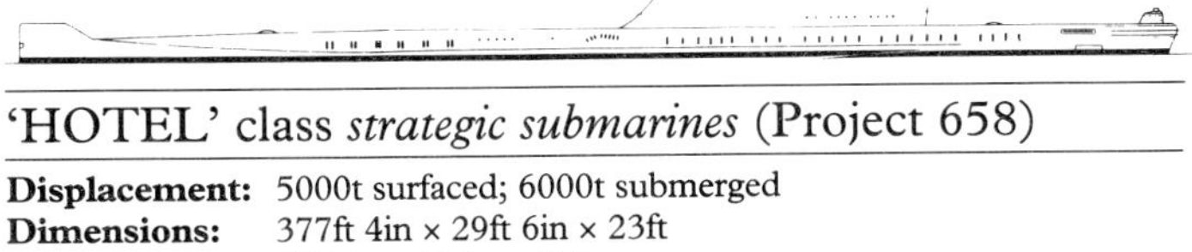

'HOTEL' class *strategic submarines* (Project 658)

Displacement: 5000t surfaced; 6000t submerged
Dimensions: 377ft 4in × 29ft 6in × 23ft
115.0m × 9.0m × 7.0m
Machinery: 2-shaft nuclear: 1 reactor, 30,000shp = 20kts/25kts (submerged)
Armament: 3 SS-N-4, 6–533mm TT (bow, 16 torpedoes), 2–406mm TT (stern)
Sensors: Radar Snoop Tray; sonar Arktika, Feniks-M, Herkules
Complement: 80

Class (completed):
K-19 (laid down 17.10.58, launched 8.8.59, completed 12.11.60), *K-33* (6.6.61), *K-55* (12.8.62), *K-40* (28.12.62), *K-16* (15.6.63), *K-145* (19.12.63), *K-149* (named *Ukrainskiy Komsomolets* in 1969, completed 12.2.64), *K-178* (30.6.64)

Built only at Severodvinsk: two completed 1959, one 1960, three 1961. One was stricken early in the 1980s. *K-19*, the lead ship (delivered November 1960), had a particularly unfortunate reputation in the Soviet Navy; she was nicknamed 'Hiroshima'. On 4 July 1961 she lost pressure in her primary coolant loop, causing ten radiation deaths; she had to be towed back to base. She collided with USS *Gato* 15 November 1969 at the entrance to the White Sea. In another accident, 24 February 1972, twenty-nine men died.

This design is a ballistic-missile equivalent of 'November' (Project 627); note that the hull-numbers of the two classes are interleaved. Like the contemporary 'Golf', this class was modified to fire SS-N-5 (Project 658M). Unlike 'Golf', this type had the new 40cm anti-escort defensive torpedo (MGT). One unit ('Hotel III', Project 701 *K-145*) was trials ship for SS-N-8: she was lengthened to 426ft (*129.8m*) 5500t/6500t during her 1965 conversion). Two were dismantled by March 1983 to conform to SALT provisions; one boat was disabled off Newfoundland and towed back to the Northern Fleet 1973. As of 1989 all surviving units had had their missile systems disabled under the SALT agreement. Two (Project 658U) were converted to communications relay boats (SSQ: *KS-1*, ex-*K-1*, and *KS-40*, ex-*K-40*).

The propulsion plant of this and the 'Echo' classes virtually duplicates that of 'November', and is therefore called a HEN plant.

'Echo I' class as converted early 1970s

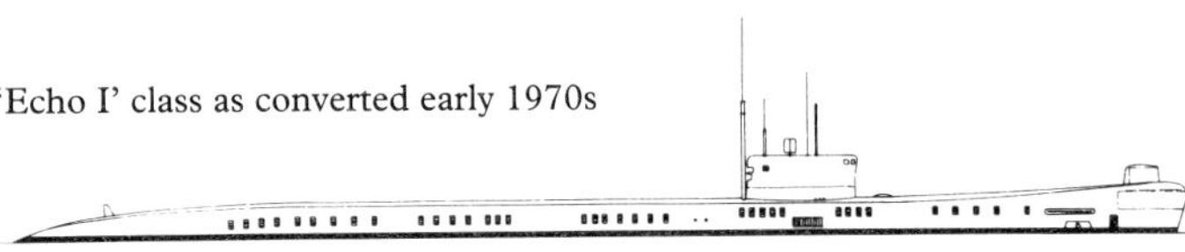

'ECHO I' class *cruise missile submarines* (Project 659)

Displacement: 4500t surfaced; 5500t submerged
Dimensions: 360ft 11in × 29ft 10in × 24ft 7in
110.0m × 9.1m × 7.5m
Machinery: 2-shaft nuclear: 1 reactor, 25,000shp = 20kts surfaced, 25kts submerged
Armament: 8 SS-N-3, 6–533mm TT (bow), 4–406mm TT (aft)
Sensors: Radar Snoop Tray; sonar M/F
Complement: 75

Class (completed):
K-45 (laid down 28.12.57, launched 12.5.59, in service 18.9.60, operational Jun 1961), *K-59* (10.12.61), *K-66* (10.12.61), *K-122* (13.4.62), *K-259* (Dec 1962).

All built at Komsomolsk and all converted to attack submarines under Project 659T. *K-45* suffered an underwater fire and was disabled off Okinawa in 1979. *K-122* suffered a similar accident (causing nine deaths) on 21 August 1983. *K-66* was damaged in 1980. Three remained in 1992; all are now gone. This type may have been code-named Chaika.

This design used the same 'HEN' powerplant as the 'Hotels'. Like them, it

'Echo I' class 1975 USN

carried the new 40cm anti-escort torpedo. These ships were designed for strategic attack; they could not accommodate the guidance radar needed to target SS-N-3 against ships. They were therefore converted into torpedo-firing submarines (Project 659T).

'Echo II' class

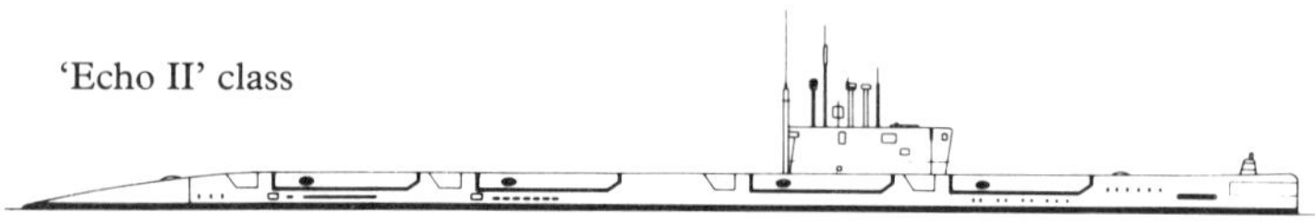

'ECHO II' class *cruise missile submarines* (Project 675)

Displacement: 5000t surfaced; 6000t submerged
Dimensions: 377ft 4in × 29ft 6in × 24ft 7in
115.0m × 9.0m × 7.5m
Machinery: 2-shaft nuclear: 1 reactor, 24,000shp = 20kts surfaced, 23kts submerged
Armament: 8 SS-N-3, 6–533mm TT (bow), 2–406mm TT (aft)
Sensors: Radar Front Door/Front Piece, Snoop Tray; sonar Arktika-M, Feniks-M, Herkules; ECM Stop Light
Complement: 90

Class:
K-1, *K-7* (renamed *K-127* in 1968), *K-10*, *K-22* (*Krasnogvardeets*), *K-23*, *K-28* (renamed *K-428*), *K-31* (renamed *K-431*), *K-34* (*Kefal'*, renamed *K-134*), *K-35*, *K-47*, *K-48*, *K-56*, *K-57* (renamed *K-557*), *K-74*, *K-86*, *K-90*, *K-94*, *K-104* (renamed *K-144*), *K-108*, *K-116*, *K-125*, *K-128*, *K-131*, *K-135*, *K-160*, *K-172*, *K-175*, *K-184*, *K-199*. (Total 29 units)

Builders: Komsomolsk: *K-10* completed 1962, one 1963, three 1964, two 1965, two 1966, two 1967

Severodvinsk: *K-166* and one other completed 1962, three 1963, four 1964, five 1965, four 1966

Accidents: One reportedly sank in the Kola Gulf 1968. *K-74* had a turbine destroyed due to the failure of its governor, 20 November 1965 in the Atlantic. *K-557* almost sank 20 June 1970 after a collision with USS *Tautog*. *K-56* collided with the research ship *Akademik Berg* near Nakhodka, 14 June 1973 (twenty-seven killed). *K-22* collided with USS *Voge* in the Mediterranean, 28 August 1976. *K-10* suffered an underwater collision in the Pacific in 1979 or 1980 and was nearly lost. *K-116* suffered a reactor meltdown at sea in July 1979. One of this class suffered an internal accident off Okinawa in August 1980, and had to be towed to Vladivostok. On 18 June 1984 *K-131* suffered a fire (thirteen men died). On 10 August 1985 one of *K-431*'s reactors exploded during refuelling near Vladivostok (ten killed). *K-175* had a serious reactor accident at Cam Ranh Bay, 1986. On 26 June 1989 one unit, reportedly *B-354* (probably actually *K-154*), suffered a nuclear accident (reportedly a leak in her primary reactor loop). The investigation which followed led to a decision to retire all Soviet first-generation (HEN powerplant) nuclear submarines.

This was an 'Echo' (Project 659) lengthened 16ft (*5m*) to accommodate mid-course guidance antennas (for anti-ship attack) and another pair of SS-N-3 launchers. Fourteen (Project 675M) were modified to fire the SS-N-12 Bazalt missile. Some units were fitted with the Punch Bowl (Kasatka-B) satellite targeting down-link. Apparently 675MK boats had both Kasatka-B and SS-N-12. To accommodate Punch Bowl, the sail was bulged on both sides. Equipment moved from the sail was accommodated in bulges on either side of the hull at the fore end of the missile tubes abeam the sail. Project 675MKV (*K-1*, *K-22*, and *K-35*) were further modified to fire the P-1000 Vulkan missile with its Argon-675KB control system. *K-86* was disarmed to work with deep-diving submersibles.

One unit was discarded in 1989, eleven more following in 1990–91. All unmodified 'Echo II' were gone by the end of 1993, and the remaining modified units were all discarded by the end of 1994.

'Echo II' class 1979 MoD

'Julliett' class

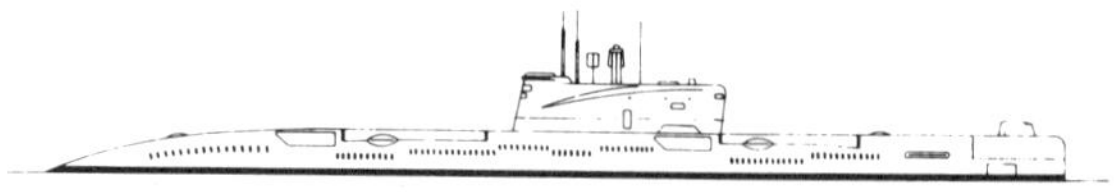

'JULIETT' class *missile submarines* (Project 651)

Displacement: 3000t surfaced; 3750t submerged
Dimensions: 295ft 3in × 32ft 10in × 23ft
90.0m × 10.0m × 7.0m
Machinery: 2 shafts, diesel-electric, 5000bhp = 16kts. Range 9000nm at 7kts, 15,000nm surface cruising
Armament: 4 SS-N-3, 6–533mm TT (bow, 16 torpedoes), 4–406mm TT (aft, 12 torpedoes)
Sensors: Radar Front Door/Front Piece, Snoop Slab; sonar Arktika, Feniks-M, Herkules
Complement: 78

Class (completed):
K-156 (renamed *B-156* in 1987; laid down 16.11.60, launched 31.7.62, completed 10.12.63), *K-85* (30.12.64), *K-24* (renamed *B-124*, completed 31.10.65), *K-68* (28.12.65), *K-63* (12.6.66), *K-70* (renamed *B-270*, completed 31.12.64), *K-77* (31.10.65), *K-58* (23.9.66), *K-81* (14.12.65), *K-73* (15.12.66), *K-67* (30.9.67), *K-78* (1.11.67), *K-203* (2.12.67), *K-304* (21.8.68), *K-318 (29.9.68)*, *K-120* (26.12.68). (Total 16 units)

All built at Gorki. Boats from *K-68* on had the Kasatka (Punch Bowl satellite targeting) system. This class was originally divided between the Northern and Black Sea Fleets, but all surviving Northern Fleet units were transferred to the Baltic during the 1980s, beginning with six in 1981, possibly for a theatre-strategic role using the nuclear version of SS-N-3. Two were discarded before 1989, and late that year another two from the Northern Fleet entered the Baltic to be scrapped. Another four were broken up during 1991; a fifth sank at her moorings, and early in 1992 a sixth awaited tow to the breakers. Late in January 1992 six remained in the Baltic and two in the Black Sea. All were gone by the end of 1994 (only two remained in mid-year).

In effect this was a diesel equivalent to 'Echo', carrying half the missiles. It was reportedly designed to carry SS-N-12 Bazalt, which would be fired when the submarine was submerged, in six inclined tubes fixed in the hull abaft the sail. A companion Project 652 would have carried a relatively short-range ballistic missile with a large warhead, also in six tubes buried in the hull. Work on Bazalt-armed submarines (both Project 651 and Project 666 [see 'Yankee', below]) stopped in 1962 due to problems with the missile. That left only SS-N-3, which had to be carried in elevating launchers which folded flat on deck, hence consuming considerable length. Project 651 could accommodate only four weapons. Eventually Bazalt matured into a useable weapon, and one boat was adapted to carry it (Project 651M). She could be distinguished by sail bulges to accommodate the associated Punch Bowl satellite down-link.

The big short-range ballistic missile was also unsuccessful. Project 652 was dropped altogether (possibly in favour of continued 'Golf' production), probably because it could not be stretched to provide the sort of salvo needed for the new anti-battlegroup mission.

The Project 651 sonar suit matches that of an 'Echo', with Arktika in the forefoot, Herkules surrounded by Feniks-M in the upper part of the bow, and a sonar intercept receiver in a sonar dome atop the bow. There is also a sonar window in the forward part of the sail, which swings aside to expose the submarine's Front Door/Front Piece missile guidance radar.

Unlike earlier three-engine Soviet diesel submarines, in this one the centreline diesel drove a generator. The generator engine could either drive the submarine at low speed through the motors on the two shafts, or it could charge the battery. During the 1980s the Soviets reportedly tried replacing this engine with much the same variety of air-independent powerplants as were then being tried in the West, including a small reactor.

'Juliett' class 1979 MoD

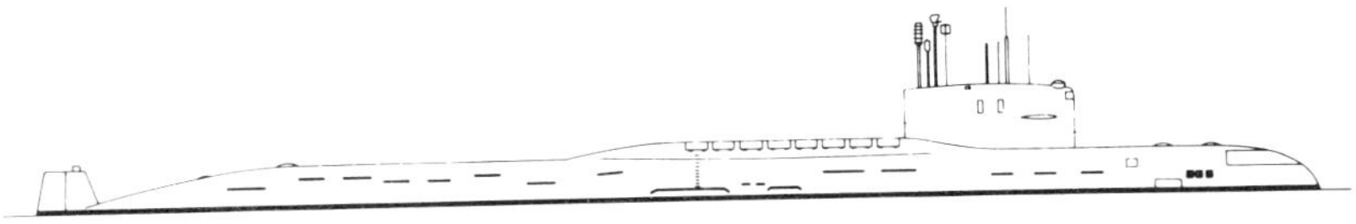

'Yankee' class as built

'YANKEE' class *strategic submarines* (Project 667A)

Displacement: 8000t surfaced; 9600t submerged
Dimensions: 426ft 6in × 39ft 4in × 29ft 10in
30.0m × 12.0m × 8.8m
Machinery: 2-shaft nuclear, 1 reactor, 50,000shp = 27kts
Armament: 16 SS-N-6, 4–533mm TT, 2–406mm TT (all bow)
Sensors: Radar Snoop Tray; sonar Shark Teeth
Complement: 120

Class (by builder, completed):
Severodvinsk: *K-137 Leninets* (laid down 9.11.64, launched 28.8.66, completed 5.11.67), *K-140* (30.12.67), *K-26* (3.9.68), *K-32* (26.10.68), *K-216* (27.12.68), *K-207* (30.12.68), *K-210* (6.8.69), *K-249* (27.9.69), *K-253* (28.10.69), *K-395* (5.12.69), *K-408* (25.12.69), *K-411* (31.8.70), *K-418* (22.9.70), *K-420* (29.10.70), *K-423* (13.11.71), *K-426* (22.12.70), *K-415* (30.12.71), *K-403* (20.8.71), *K-245* (16.12.71), *K-214* (31.12.71), *K-219* (31.12.71), *K-228* (31.12.72), *K-241* (23.12.71), *K-444* (9.12.72)

Komsomolsk: *K-399* (24.12.69), *K-434* (21.10.70), *K-236* (27.12.70), *K-389* (1970), *K-252* (1971), *K-258* (1971), *K-446* (1971), *K-451* (1971), *K-436* (1972), *K-430* (1972)

Total 34 units.

Severodvinsk units were Project Navaga; Komsomolsk units were Project Nalim. The technical project for this class was approved in 1964; the submarine was designed by Firma Rubin (TsKB-18).

K-219 suffered an accident in a missile tube 31 August 1973. Another such accident caused damage which sank her 6 October 1986. *K-140* had a reactor go out of control 23 August 1968 and one of her reactors was dumped in the Kara Sea; presumably she was selected for reconstruction (as the sole 'Yankee II') because she required a major refit in any case. *K-446* (Project 667AU) was the last Pacific Fleet unit in commission, in the autumn of 1994.

This was reportedly the first design intended to fire ballistic missiles submerged. Since missiles no longer had to be raised to the top of the sail before firing, their tubes could be moved into the hull abaft the sail, in two parallel rows. A parallel Project 668 was intended to fire the new Bazalt cruise missile submerged, from inclined hull tubes. The missiles could be packed relatively densely because the tubes were fixed in the hull. The pressure hull was necked-in abaft the control room to accommodate the missile tubes; the pressure hull necked-out further aft to take the machinery. About 1960 both designs were lengthened to provide a second set of tubes (total twelve ballistic or cruise missiles) abaft the machinery.

Work on Bazalt halted in 1962 due to aerodynamic problems, and Project 668 was abandoned. The only available cruise missile was SS-N-3, which could not be fired submerged, and which required an elevating launcher which stowed flat. That limited a submarine of this size to eight missiles, and there was no point in building a new type. Instead, a second batch of 'Echo IIs' was ordered. Eventually something like Project 666 materialised in the form of Project 949 ('Oscar'), which was armed with the more successful SS-N-19 Granit missile.

The ballistic missile version was redesigned in 1963–64 to carry sixteen SS-N-6 rather than twelve SS-N-5 missiles and, moreover, to carry them inside the pressure hull, as in contemporary Western SSBNs. It became the 'Yankee' class (Project 667A). The redesign seems to have been prompted by compromise of Western submarine construction or missile guidance technology.

Like earlier strategic submarines, these ships had 40cm self-defence (anti-escort) weapons, in their case in bow tubes flanking the four large caliber bow tubes.

A recent Russian account describes this design as highly automated for its time, with automatic prelaunch preparation of missiles and a computer to provide firing data. The modernised Project 667AU (which fired the R-27U missile [SS-N-6 Mod 2]) introduced an inertial navigation system. These ships could be distinguished from their sisters by bulges at the upper rear of their sails, possibly to accommodate receivers for the Soviet equivalent of the US Transit navigational satellite (to update the inertial system). Project 667AU units: *K-219*, *K-228*, *K-241*, *K-245*, *K-430*, *K-436*, *K-444*, *K-446* and *K-451*.

This class, 'Victor', and 'Charlie' all used second-generation Soviet machinery technology. The 'Yankee' reactor (and presumably that of the 'Victor' and 'Charlie' classes) was designated VM-4 ('Deltas' had VM-4SG). Apparently the strategic submarines used an OK-350 core (OK-350G is the current core type, 1995) and had five coolant pumps (presumably for five primary loops, and thus five heat exchangers). The attack submarines used four pumps (OK-300 cores) and thus had four heat exchangers; output is 72mW rather than 90mW. Rated core life was reportedly 5 years, but strategic submarines sometimes exhausted their cores in 3 years. In a 'Delta III' zirconium was added to the core, presumably to extend its lifetime.

All of these reactors apparently had higher thermal efficiencies than the first-generation plants: apparently in a 'Yankee', a 90mW reactor powered a 29,000shp turbine, whereas in a 'November', it took 70mW to power a 17,500shp turbine. On the other hand, the 177mW second-generation reactor of the 'Papa' powered a 40,000shp turbine, implying a thermal efficiency about midway between the two generations. That in turn implies higher steam conditions (temperature and pressure). At 45,000shp for 90mW, the third-generation reactors in 'Typhoons' apparently have about the same thermal efficiency as those of the 'Yankee'. In their case effort has gone into compactness (by using more highly-enriched uranium, some of it about twice as enriched as that in the second generation), reliability and silencing (of auxiliary equipment).

'Yankee I' class 1982 — Courtesy Ian Sturton

Boats were withdrawn from the Soviet strategic submarine force beginning about 1980, to remain within SALT treaty limits as 'Deltas' entered service. Missile sections were cut out and bows and sterns welded together. Later new sections were inserted into several units, to fit them for other roles. 'Yankee' patrols off the US coast ceased in 1987, though they may have resumed later. By mid-1991 only twelve ships (six Northern Fleet, six Pacific Fleet) were still operational as strategic submarines, and by 1992 only six were left. By 1993 all 'Yankee' strategic submarines were gone.

One unit (*K-140*: Project 667AM, Navaga-M) was converted to test the new SS-N-17 ballistic missile; it had twelve tubes. An ex-SSBN (*K-420*: Project 667M Andromeda) was converted to test the SS-NX-24 cruise missile; twelve missiles were carried in vertical tubes in a greatly-enlarged midbody (speed was reportedly reduced to 23kts and manoeuverability suffered). This submarine was relaunched in December 1982 and entered service in 1985; the programme was cancelled in 1988.

Four 'Yankee Notch': Project 667AT Grusha: *K-236*, *K-399*, *K-408*, *K-415* were converted in 1984–88 to fire SS-N-21 cruise missiles, launched from four horizontal tubes cut into each side of the former missile section (capacity is forty missiles). The sail was reconfigured and extended by 9ft 10in (*3m*), presumably to accommodate a communications buoy (as in the 'Akula' class). The two 40cm self-defence tubes may have been replaced by 53cm tubes. Three more projected conversions were cancelled.

One unit ('Yankee Pod', Project 09774, *K-403*) rebuilt in 1984 and lengthened by 13ft 1in (*4m*), was a sensor test platform. She had both a towed-array pod *and* a 'Delta' or 'Oscar' type towed array tube. What appeared to be torpedo front ends were mounted on either side of the leading edge of the sail, presumably for sensor tests. She was probably unarmed, a new sonar array filling her bow.

Another unit ('Yankee Stretch', Project 09780, *K-411*, known in Russia as SMPL/AS) completed conversion in 1990. This unit was lengthened to 524ft 11in (*160m*). The designator probably refers to support of small (*maliy*) special (S) submarines (PL) or to support of research submarines (AS series designators).

'DELTA I' class *strategic submarines* (Project 667B Murena)

Displacement: 9000t surfaced; 11,750t submerged
Dimensions: 459ft 4in × 39ft 4in × 28ft 7in
140.0m × 12.0m × 8.7m
Machinery: 2-shaft nuclear: 1 reactor, 50,000shp = 25kts
Armament: 12 SS-N-8, 4–533mm TT (bow, 12 torpedoes), 2–406mm TT (bow, 6 torpedoes)
Sensors: Radar Snoop Tray; sonar Shark Teeth
Complement: 120

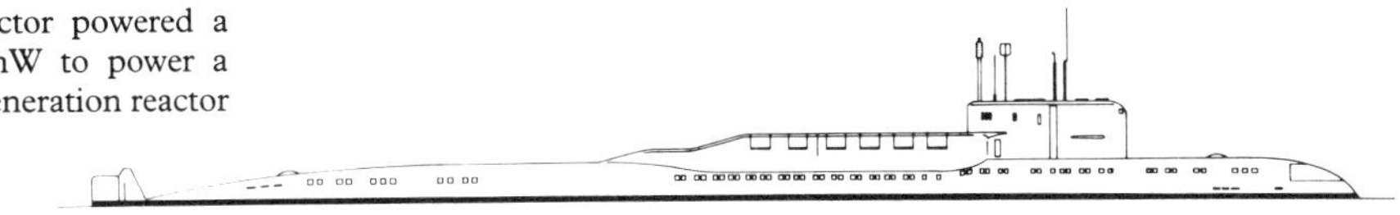

'Delta I' class

'Delta I' class 1973 MoD

Class (by builder, completed):
Severodvinsk: *K-279* (laid down 1971, launched Jan 1972, completed 22.12.72), *K-447* (1973), *K-450* (1973), *K-385* (1974), *K-457* (1974), *K-453* (1974), *K-460* (1975), *K-472* (1975), *K-475* (1975), *K-171* (1976)

Komsomolsk: *K-366* (1974), *K-417* (1974), *K-477* (1975), *K-497* (1975), *K-500* (1976), *K-512* (1976, 28.10.88–1991 named *70 Let VLKSM*), *K-523* (1977), *K-530* (1977)

A reference to a September 1977 accident mentions a Pacific Fleet unit, *K-171*; she may have been renamed.

This submarine was built on a 'Yankee' hull but carried twelve much longer-ranged missiles. The 'B' in the project number indicates a major armament change. The appearance of the new missile changed Soviet naval strategy because it made the defence of extensive wartime 'bastions', in which the submarines would operate, essential. The earlier 'Yankees' would have begun a war either on patrol off the US coast or in protected harbours; the navy's strategic support mission would have been to help them break out to sea.

This may have been the first Soviet ballistic-missile submarine with a 'Second Captain' combat data computer. In 1975 *Pravda* published an account of just such a system on board a ballistic missile submarine, in the context of the importance of the computer operator, who had to insert accurate data into the computer by means of which the CO of the submarine fought the ship.

By January 1994 two Northern Fleet and one Pacific Fleet unit had been retired. 'Delta' was designed by Firma Rubin (formerly TsKB-18). In 1994 Rubin was advocating a variety of civilian conversions of strategic submarines.

'DELTA II' class *strategic submarines* (Project 667 BD Murena-M)

Displacement:	10,000t surfaced; 12,750t submerged
Dimensions:	508ft 6in × 39ft 4in × 28ft 10in *155.0m × 12.0m × 8.8m*
Armament:	16 SS-N-8, 4–533mm TT (bow, 12 torpedoes), 2–406mm TT (bow, 6 torpedoes)
Complement:	126
Other particulars:	As 'Delta I' class

Class:
K-182 (laid down Apr 1973, launched Jan 1975, completed 20.9.75; named *Shestidesyatiletie Velikogo Oktyabrya* 4.11.77–1991), *K-92* (laid down 1973, launched 1975, completed 17.12.75), *K-193* (laid down 1974, launched 1975, completed 30.12.75), *K-421* (laid down 1974, launched 1975, completed 30.12.75).

All built at Severodvinsk 1974–75. This was a lengthened ('D' in the project number) 'Delta I' carrying four more missiles; underwater speed fell to 24kts. All will probably soon be retired.

This class introduced a new integrated sonar, Rubikon, in place of the Kerch of the 'Yankee' and 'Delta I' classes.

'Delta III' class MoD

'DELTA III' class *strategic submarines* (Project 667 BDR Kalmar)

Displacement:	10,500t surfaced; 13,250t submerged
Armament:	16 SS-N-18, 4–533mm TT (bow, 12 torpedoes), 2–406mm TT (bow, 6 torpedoes)
Other particulars:	As 'Delta I' class

Class (completed):
K-441 (laid down 1975, launched 1976, completed Dec 1976), *K-424* (1977), *K-449* (1977), *K-455* (1978), *K-490* (1978), *K-487* (1978), *K-44* (1979), *K-496* (1979), *K-506* (1979), *K-211* (1980), *K-223* (1980), *K-180* (1980), *K-433* (1981), *K-129* (1981).

All built at Severodvinsk: reported launchings: four in 1976, four in 1977, two in 1979, two in 1980, one on 30 April 1981, the last in December 1982. This is a 'Delta II' with a higher hump to accommodate the new SS-N-18 missile (presumably the 'R' in the project number). Reported normal operating depth is 350m (about 1200ft); test depth is 580m (1900ft). The missile compartment is 147.6ft (*45m*) long. The externally-framed pressure hull is 9.8m in diameter, *ie*, about the same diameter as the separate pressure hulls in a 'Typhoon'. The sonar suit, Rubikon, roughly matches that of a 'Victor I/II', but a recently-published inboard profile shows a small cylinder above the big bow array. Like the attack submarine, this one has a cylindrical array in the forward part of the sail. These ships are now receiving towed arrays, as in 'Delta IV'.

'DELTA IV' class *strategic submarines* (Project 667 BDRM Delfin)

Displacement:	10,800t surfaced; 13,500t submerged
Dimensions:	538ft × 39ft 4in × 28ft 6in *164.0m × 12.0m × 8.7m*
Armament:	16 SS-N-23, 6–533mm TT (bow, 18 torpedoes)
Complement:	130

Name	Builder	Laid down	Launched	Comp	Fate
K-51	Severodvinsk	23.2.84	1985	29.12.85	Extant 1995
K-84	Severodvinsk	Nov 1984	Dec 1985	Feb 1986	Extant 1995
K-64	Severodvinsk	Nov 1985	Dec 1986	Feb 1988	Extant 1995
K-114	Severodvinsk	Dec 1986	Sep 1987	Jan 1989	Extant 1995
K-117	Severodvinsk	Sep 1987	Sep 1988	Mar 1990	Extant 1995
K-18	Severodvinsk	Sep 1988	Nov 1989	Sep 1991	Extant 1995
K-407	Severodvinsk	Nov 1989	Jan 1991	20.2.92	Extant 1995

This is a redesigned 'Delta' (hence the 'M' suffix) with a new sonar system, Skat-2. The MGK-200 spherical array squeezes out the 40cm self-defence tubes, and there is a towed array and, presumably, a flank array. Presumably the entire suite is similar to that of a 'Victor III'. This class also has a new stern form approaching a 'double-Hoggner' (as in an 'Oscar'). There is also a new missile, SS-N-23. Hence the new code name for the submarine/missile combination. A follow-on missile is reportedly under development.

In January 1992 the Russian Ministry of Defence announced that another five units were being broken up incomplete. *K-51* was the 1000th submarine ordered for Soviet Union/Russia. She was named *Imeni 26 S'ezda KPSS* until 1992. *K-89* attempted to launch sixteen missiles in succession while in the White Sea, 7 December 1989; the third failed, and the falling missile apparently hit the submarine and injured thirteen men. One was damaged in collision with USS *Grayling*, 20 March 1993.

'Delta II' class

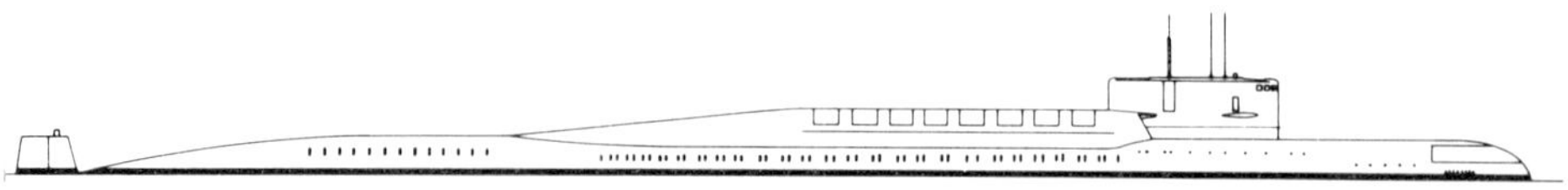

'Delta III' class

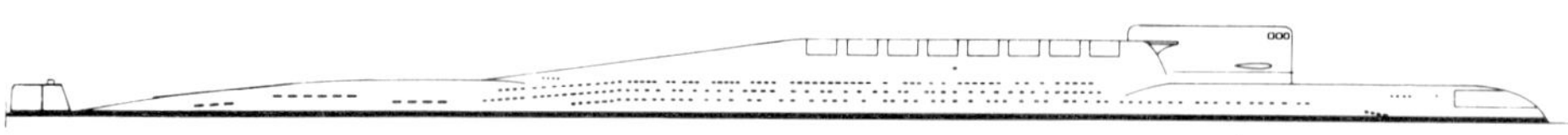

'Delta IV' class

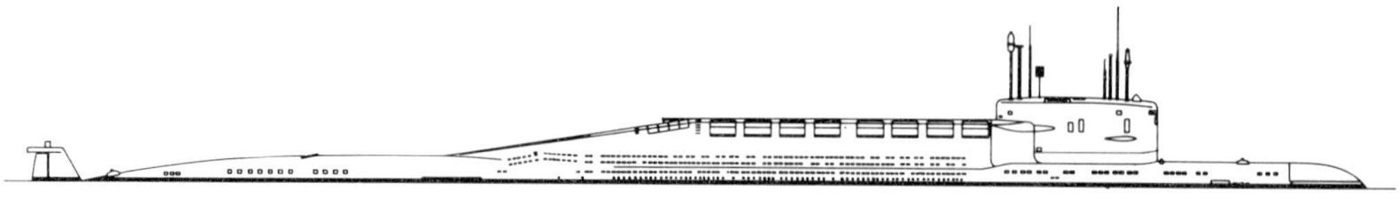

'Charlie I' class 1975 MoD

'Charlie I' class

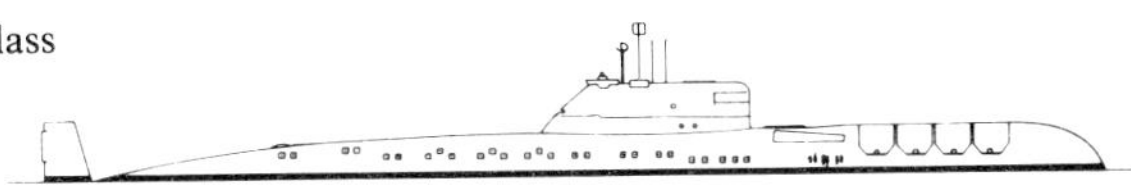

'Charlie II' class

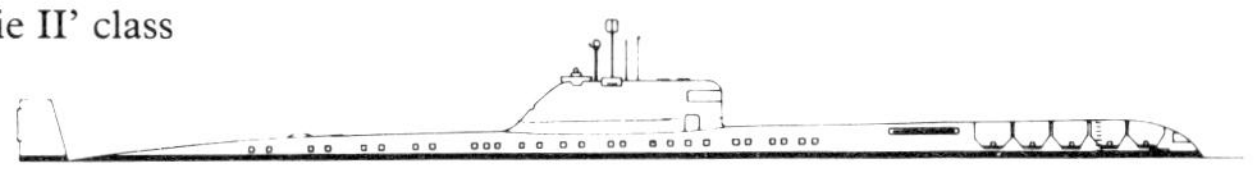

'CHARLIE' class *cruise missile submarines* (Project 670A Skat; Project 670M Skat M)

Displacement: 4000t surfaced; 4900t submerged ('Charlie II' 4300t/5100t)
Dimensions: 311ft 8in ('Charlie II', 340ft) × 32ft 10in × 26ft 3in
95.0m, 103.0m × 10.0m × 8.0m
Machinery: 1-shaft nuclear: 1 reactor, 15,000shp = 24kts
Weapons: 8 SS-N-7 (SS-N-9 in 'Charlie II'), 4–533mm TT, 2–406mm TT (bow)
Sensors: Radar Snoop Tray; sonar Shark Teeth L/F ('Charlie II' ECM Stop Light)
Complement: 100 (98 in 'Charlie II')

Class:
'Charlie I' (Project 670A Skat): *K-43*, *K-87* (renamed *K-212* in 1972), *K-25*, *K-121*, *K-313*, *K-308* (completed 20.9.70), *K-320*, *K-303*, *K-325*, *K-429*, *K-201*. *K-43* was the unit leased to India as *Chakra* (as Project 06709).

'Charlie II' (Project 670M Skat-M): *K-452* (named *Berkut* 1989), *K-458*, *K-479*, *K-503*, *K-508*, *K-209*

All built at Gorkiy. 'Charlie I' entered service at the rate of about two per year in 1968–73, followed by 'Charlie II' (launchings: one in each of 1973, 1975, 1977, 1978, 1979, 1980). First boat became operational 1968. *K-314* suffered a major reactor casualty (possibly a meltdown) December 1985. *K-429* sank east of Petropavlovsk 23 June 1983, reportedly because a ventilator was not closed before a post-refit dive. Raised and repaired, she sank at dockside. She was again raised but then was used only as a training hulk.

Project 670 was initially conceived as a mass-production nuclear submarine; hence the choice of a single-reactor powerplant at a time when the Soviet naval staff very strongly preferred two reactors and two propellers. The Lazurit KB (TsKB-112, associated with the Gorkiy yard) prepared the design, probably having won the competition described under 'Victor', below. Some time after the design had been begun, the submarine was redesigned to carry the anti-ship missile system which had been conceived for 'Papa'. Presumably that was because 'Papa' was so clearly going to be both late and overexpensive. There were also problems with the planned P-50 Ametyst missile. The boat was designed to accommodate the projected weapon, but for the first series a modified Styx, probably designated P-20L, was substituted. This missile is shorter than SS-N-9/P-50, so the tubes are plugged at their lower ends. Missiles are fired submerged on the basis of intermittent radar data; the fire control system is an adapted version of the surface ship Klon. A nuclear version can be fired (to first convergence zone) on the basis of sonar data. 'Charlie II' has the originally planned armament with a different fire control system.

The single-reactor powerplant does not offer the sort of speed required to keep up with fast carrier group targets; presumably it is left over from the earlier requirement for a submarine suited to mass production. In the late 1980s, 'collars' were added to the forward base of the sail, probably to reduce turbulent flow along the hull. Test depth is 400m (about 1300ft); crush depth is 600m.

K-452 was modified with a new sonar system, MGK-400 Rubikon, under Project 06704.

One unit was leased to India for 3 years as *Chakra* (5 January 1988). The Indians refused to renew the lease, reportedly because the Soviets never allowed any Indians in either the powerplant spaces or the missile control centre. This unit was in poor condition when returned, and has been discarded. However, it has also been reported that the Indians wish to obtain 'Charlie' plans and possibly components as the basis for their own future nuclear submarine. At least one Pacific Fleet 'Charlie I' was stricken 1992, and others during 1993; in 1994 only two remained (both in the Pacific Fleet). One 'Charlie II' was discarded in 1992; by 1994, four remained.

'Papa' class

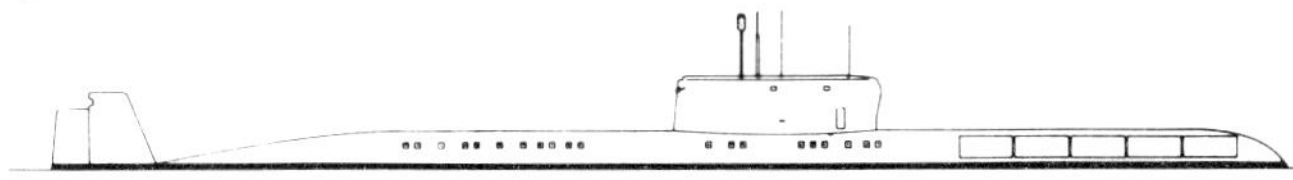

'PAPA' class *cruise missile submarine* (Project 661 Anchar)

Displacement: 5197t submerged
Dimensions: 350ft 9in × 37ft 9in × 26ft 3in
106.92m × 11.5m × 8m
Machinery: 2-shafts: 2 reactors, 2 152-cell batteries, 80,000shp = 16/42kts. Stores endurance 75 days
Armament: 10 SS-N-9, 4–533mm TT (bow, 12 torpedoes)
Sensors: Sonar Rubin
Complement: 82

Name	Builder	Laid down	Launched	Comp	Fate
K-222	Gorkiy	28.12.63	21.12.68	31.12.69	Discarded 1991; in storage at Severodvinsk 1995

'Papa' seems to have been conceived about 1958 as a missile-firing counterpart to 'Alfa', with a titanium hull (operating depth 400m) and very high speed (38kts was specified, but she had the higher rated speed quoted above). Design work was authorised in December 1959 as part of a programme to develop a new fast submarine, new types of powerplants, and other submarine R&D. Although a recent Russian article refers to the powerplant as part of the 'second generation', it is by no means the same as that in a 'Victor'. 'Papa' was designed on a priority basis by TsKB-16 (Malakhit). The new Ametyst missile was developed by Chelomey; eventually it was carried by 'Charlie II' class submarines.

Development and construction were protracted partly because the new submarine was intended not to use existing submarine systems, and partly because of delays in supplying hull material. The concept design and TTZ were approved early in 1960. All versions developed at that time had two-reactor twin-screw propulsion, titanium hulls operable at 400m depth, missile and torpedo armament, a new high-pressure air system (400kg/cm^2), a new high-pressure hydraulic system (150kg/cm^2), a silver-zinc battery, and new navigation (Sigma-61) and sonar (Rubin) systems.

After model tests, the stern was divided (into a 'double-Hoggner' form) to reduce self-noise. The pressure hull was cylindrical (29ft 6in [*9m*] diameter amidships), but at the bow it had a figure-8 section (two 18ft 1in [*5.5m*] cylinders), with the cruise missiles, in individual tubes, alongside, between the pressure and outer hulls. One cylinder contained the torpedo tubes (with a power reloader); the one below contained the big sonar transducer as well as the forward battery.

Machinery was divided into two independent sets, port and starboard, each including a 3000kW turbo-generator. The submarine made 42kts on her official trials, but above 35kts she suffered from severe buffeting ('hydrodynamic noise', according to a Russian design history). After a 12-hour full-speed trial, the conning tower entry door, three hatches, and a buoy fairing were broken. During a series of experimental patrols (January 1970–December 1971), she attained 44.7kts submerged, which is still a record.

The Russian design history states that no further ships of this class were ordered due to the excessive noise level, the tactical shortcomings (unspecified) of the missile system, and because of the inadequate service lifetimes of the primary equipment. Building time was too lengthy, which probably meant that 'Papa' was far too costly in terms of resources.

'Papa' class USN

50 Let SSR

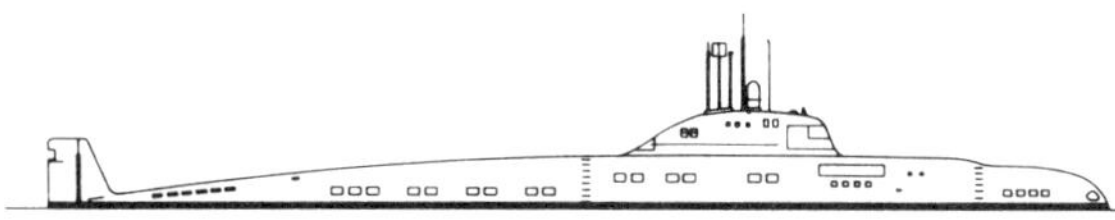

'VICTOR I' class *attack submarines* (Project 671, 671V, 671K Ersh)

Displacement:	4300t surfaced; 5100t submerged
Dimensions:	311ft 8in × 32ft 10in × 23ft *95.0m × 10.0m × 7.0m*
Machinery:	1-shaft nuclear: 1 reactor, 30,000shp = 30kts; 2 small propellers for low speeds
Armament:	6–533mm TT (bow, including SS-N-15 missiles, 18 torpedoes)
Sensors:	Radar Snoop Tray; sonar Shark Teeth L/F
Complement:	94 (32 officers)

Class:

K-38 (renamed *50 Let SSSR* in 1970; begun Jan 1965, launched Oct 1965, delivered 5.11.67), *K-69* (renamed *B-369*), *K-147, K-53, K-306, K-323, K-370, K-438, K-367, K-314, K-398, K-454, K-462, K-469, K-481*. (Total 15 units)

Project 671V: *K-314, K-454, K-469*

Nuclear attack submarines originally in the *K*-series were renamed in the *B*-series on 29 August 1991. All built at Admiralty, Leningrad; two per year, 1968–75. Most served with the Pacific Fleet. The first unit was once named *Yorsh*; other names, possibly unofficial, were *Del'fin, Sazan, Treska, Ugor. K-371* (Pacific Fleet) was discarded after severe damage while being recored, 10 August 1985. Most were discarded rather than being recored, from 1991 on. Only one or two were left in 1995.

Project 671 emerged from a larger programme of follow-ons to the 'November' class. The Malakhit KB (SKB-143), which had designed 'November', conducted numerous studies of possible future nuclear submarines during 1956–59. In May 1958 the new Bureau Chief, VI Dubovichenko, decided to offer one of the sketch designs to the State Committee on Shipbuilding. It was a US-style single-hull submarine (with a cylindrical, rather than a curved, pressure hull) with a single reactor, a single turbine, and a single propeller, ballast tanks only at the ends, and a boxlike superstructure. Unlike US nuclear submarines, the design carried some ballast between internal transverse bulkheads. The double bulkheads also accommodated escape hatches leading into escape capsules in the superstructure. Unlike earlier Soviet submarines, this one would not be designed to the usual three-compartment standard, *ie*, to survive surface collisions. The designers argued that surface considerations were irrelevant to a submarine designed to spend most of its time submerged.

The State Committee showed no interest in the proposal, but it did announce a competition for four next-generation submarines: Project 667, an SSBN; Project 669, a large torpedo submarine (a direct successor to 'November'); Project 670, a small torpedo submarine for mass production; and Project 671, an ASW submarine. Three design bureaus competed: SKB-143, TsKB-18 (Rubin), and TsKB-112 (Malakhit). SKB-143 would be assigned Project 671; Malakhit would get Project 670, and Rubin would get Project 667. Project 669 was amalgamated with Project 671.

Project 671 seems to have been inspired by the contemporary US *Tullibee* class. The competition requirement called for a powerful new sonar, four torpedo tubes (eight torpedoes), a speed of about 30kts, and a test depth of 300m, all within a normal displacement of about 2000t.

By this time SKB-143 had designed three nuclear submarines: Project 627 ('November'), Project 645 (the liquid-metal version of 'November'), and the abortive Project 639 (an SSBN). The latter had introduced AC rather than DC electric power, and a larger-diameter pressure hull. Project 645 introduced self-contained turbogenerators (in Project 627 and its contemporaries, they were driven directly by the main turbines).

SKB-143 proposed a derivative of its proposed SSN, with a single shaft. At about the same time, the main naval staff developed standards for new SSNs. It assumed that the usual surface survivability standard would be maintained, and it asked for two shafts and two reactors, for reliability. SKB-143 argued instead that underwater performance should be paramount. Presumably its designers had reached much the same conclusions as their US counterparts (after a series of model tests, culminating in the construction of the *Albacore*). They pressed for a single-screw design with a body-of-revolution hull, for minimum ballast tankage (to minimise wetted surface) and for choosing the number of reactors only on the basis of required power. The State Committee agreed, but the naval staff and its own technical experts resisted.

Meanwhile Project 669 was dropped; clearly Project 671 would have to be a dual-purpose submarine. To provide it with sufficient speed and armament, the TTZ was revised: maximum displacement could rise to the limit set by the capacity of the White Sea–Baltic Canal. The SKB won its argument for a single shaft, on the basis that it would minimise noise and overall displacement while providing the greatest possible speed on the available power. Admiral Gorshkov personally approved the single shaft as a one-time exception, but in fact the same principle was applied to the parallel Project 670. In retrospect, the designers were particularly proud that they had been able to choose dimensions based on hydrodynamics, rather than adopting the usual practice of adding up the lengths of compartments. For example, to minimise length they placed the reactors side by side, a practice repeated in later Soviet SSNs. As in contemporary US SSNs, the new hull form drastically reduced wetted surface

'Victor I' class Apr 1975 USN

compared to displacement: Project 671 displaced about 30 per cent more than 'November', but had about the same wetted surface. Test depth is 270m (collapse depth is 350m).

The bow presented a particular problem. Standard practice, as in other navies, was to run the torpedo tubes in two vertical rows. That would have left little space for the big sonar array required. The solution adopted for Project 671 was to turn the two rows horizontally (two above four), with the sonar in the chin position. Coincidentally, this was much the solution adopted in the US *Skipjack* class a few years earlier. The Russians went further, adopting power loading from a moving tray (as in later US designs). Soviet practice apparently differed from US practice in that four of the torpedo tubes were accommodated in an inner hull tube extending towards the bow; two more (the lower outer pair) were apparently non-reloadable tubes outside the extended inner hull. In later versions of the design, the reloadable quartet of tubes was increased in diameter and length to accommodate 65cm weapons.

As for power, all of the second-generation submarines had a new reactor which had nearly the same output as the two reactors of the 'November' class. The SKB-143 designers estimated that one such reactor would give good speed, but they chose two in order to provide a margin for future growth. Slightly earlier the US submarine designers had made the opposite choice in the *Skipjack* and *Thresher* classes. They would regret it in the 1960s, as the submarines had to grow to accommodate new equipment, and lost too much speed in the process. The main propeller of a 'Victor I' has five blades, and there are two auxiliary propellers for maneuvering.

The technical design was approved on 3 November 1959.

Apparently this submarine, already in production, was hurriedly adapted to take the SS-N-15 ASW missile (equivalent to the US SUBROC, the design of which had been compromised); the modified design was Project 671R (R for rocket). *K-314, K-454*, and *K-469* were Project 671V. There was also a modified Project 671K version.

'Victor I' is externally distinguishable from 'Victor II' by the boxy hump in the casing of the latter, between bow and sail. Given the very careful attention to hydrodynamics in the Project 671 design, it seems reasonable to guess that the hump was needed to accommodate additional equipment. 'Victor II' has also been lengthened, probably to accommodate 65cm weapons.

An unofficial Russian drawing which appears to be of a 'Victor I' shows a short flank array in addition to the big bow array (which occupies the lower two-thirds of the bow), plus sound-transparent windows on the forward and after parts of the sail.

'Victor II' class

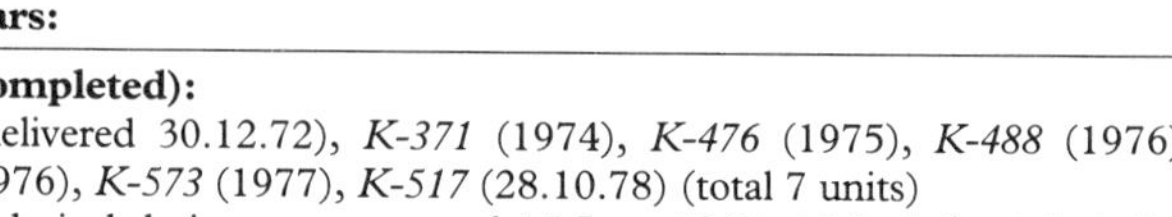

'VICTOR II' class *attack submarines* (Project 671RT)

Displacement:	4500t surfaced; 5700t submerged
Dimensions:	334ft 6in × 32ft 10in × 23ft *102m × 10.0m × 7.0m*
Complement:	100 (20 officers)
Other particulars:	As 'Victor I' class

Class (completed):

K-387 (delivered 30.12.72), *K-371* (1974), *K-476* (1975), *K-488* (1976), *K-495* (1976), *K-573* (1977), *K-517* (28.10.78) (total 7 units)

The technical design was approved 15 June 1967. All built by Admiralty, Leningrad; launched 1972, 1974, 1975, 1976 (two), 1977, in 1978. By 1994 at least four had been stricken because there were no resources to recore and refit them.

This is a 'Victor' redesigned to take the follow-on submarine-launched ASW weapon, the 65cm SS-N-16. This weapon could deliver a homing torpedo; hence the RT suffix to the project number. As in 'Victor I', only four tubes are reloadable. Reportedly two are 53cm diameter, presumably because when this type was developed no 65cm torpedoes were as yet in service (the first such weapon was 65-76). Presumably the extra length was needed to accommodate the longer weapon; it cost about 2kts in maximum speed.

Reportedly the usual load-out is two 65cm weapons per tube; in wartime a third weapon would be carried in the tube. The 53cm tubes can carry homing torpedoes or Shkvals or full-diameter decoys (MG-74 Korun). The sonars in this version are MT-70 and Skat.

'VICTOR III' class *attack submarines* (Project 671RTM Shchuka)

Displacement: 4900t surfaced; 6000t submerged
Dimensions: 341ft 2in × 32ft 10in × 23ft
104m × 10.0m × 7.0m
Other particulars: As 'Victor I' class

Class:
K-138, K-218, K-242 (named 50 *Let Komsomolska-na-Amur* May 1982), *K-244, K-247, K-251, K-254, K-255, K-264, K-292, K-298, K-299, K-305, K-324, K-355, K-358* (named *Murmaskiy Komsomolets* 1990–91), *K-360, K-388, K-412, K-414, K-448, K-492, K-502, K-507, K-524* (named *60 Let Shefstva VLKSM* 1982–91), *K-527* (26 units total)

Built by Admiralty, Leningrad and at Komsomolsk; launchings: one 1977, two 1978, three 1979, four 1980, three 1981, three 1982, two 1983, two 1984, one July 1985, one each 1987–91. Note that production continued after the introduction of the new 'Sierra' and 'Akula' classes. The technical design of this vessel was approved in 1975.

The 'M' suffix means modernised: with a new sonar suite (including a bow sphere, flank and towed arrays), four 65cm torpedo tubes, and silenced (reportedly partly rafted) machinery. There is also a wake sensor. Some have non-acoustic sensors on the starboard side of the leading edge of the sail. According to an unofficial Russian account, the sonar is Skat-KS plus MT-70, the latter probably for communication. This class began the current Russian trend towards much quieter submarines. The longer hull presumably accommodates new fire control and command systems. The timing of construction suggests that this was the first class of submarines (other than 'Alfa') to have a 'Second Captain' combat direction system. Units were further modified (as Project 671RTMK) with a new Viking command system reportedly based on stolen details of the Norwegian system in the *Ula* class. These submarines appeared at the same time as a family of 65cm weapons, including both wake-following torpedoes and a torpedo tube version of the SS-N-22 (P-100) missile. Reportedly different units are specialised either as ASW or as anti-ship platforms, with no external differences.

One unit, *K-292* (Project 6717), with a 32ft (*10m*) fairing on deck forward of the sail, test-fired the SS-N-21 missile. It is not clear whether the fairing was possibly a means of quickly equipping numerous attack submarines with this weapon. These submarines have a towed VLF buoy in the casing abaft the sail.

Early units had seven-bladed propellers, but most (and some 'Victor IIs') have an unusual tandem propeller consisting of two four-blade units oriented 22.5 degrees apart and turning together on the same shaft. Similar propellers were used in fast torpedo boats built early in this century to avoid cavitation. Diving depth: 350m (400m maximum).

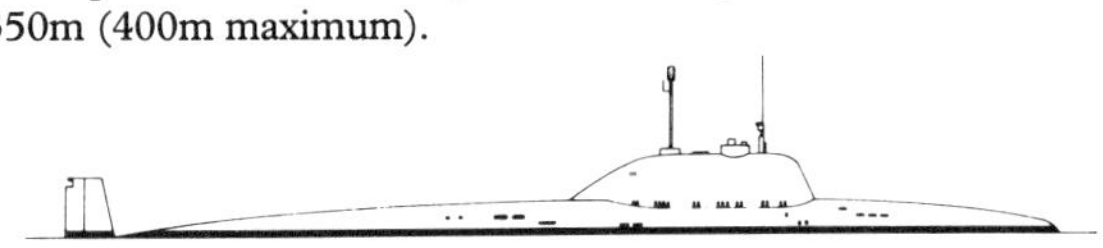
'Alfa' class

'ALFA' class *attack submarines* (Project 705 Lira)

Displacement: 2800t surfaced; 3680t submerged
Dimensions: 267ft 1in × 31ft 2in × 23ft
81.4m × 9.5m × 7.0m
Machinery: 1-shaft nuclear, 2 reactors, 45,000shp = 43–45kts submerged
Armament: 6–533mm TT (18 torpedoes or 36 mines) or SS-N-15 missiles
Sensors: Radar Snoop Tray; sonar L/F bow
Complement: 31 (all officers)

Class:
K-377, K-316, K-373, K-123, K-432*, K-463, K-493**. Starred units are Project 705K built at Severodvinsk.

The project was approved by the Council of Ministers on 27 May 1961; TTZ approved 11 December 1961. The first unit, *K-377*, was built at Admiralty/Sudomekh, Leningrad (Yard No. 900). The first Severodvinsk unit was *K-123* (Yard No. 105, delivered 26 December 1977). The Western count showed seven ships, but probably the last was actually the sole 'Mike'. Reported launch dates: 1967 (in service 1972), 1974 (in service 1979), 1976, two 1978, 1980, 1981 (in service 1983). The first unit became known to the West about 1971. It was a failure (reportedly due to a freeze-up of the reactor coolant metal) and was broken up in 1974. *K-123* had her primary reactor loop destroyed 8 April 1982 and was in refit 9 years; she was probably the unit (705ZhMT) which had a new pressurised-water reactor installed. Reactors: OK-550 on units built at Leningrad, BM-40A on units built at Severodvinsk. Power rating is 155mW.

'Alfa' was part of the programme of radical new types of ships authorised in 1956. A high-level committee developed a TTZ specifically to exceed the likely performance of Western submarines which might appear during the next 10 to

'Alfa' class 1982 Courtesy Ian Sturton

15 years. Russian accounts emphasise the degree of automation achieved rather than the high speed (due to liquid-metal reactors) or deep diving depth (due to adoption of a titanium hull). The crew size is variously given as 45 or even 29, and apparently all are officers. Reportedly this class is not designed for underway maintenance. For example, submarine size may have been held down by limiting reactor shielding to a layer between the reactor compartment and the living spaces, accepting that the reactor space would always be too 'hot' to enter while the submarine was at sea.

Two liquid-metal (lead-bismuth) reactors drive a pair of turbo-generators, which in turn drive a motor. For compactness, the Soviets chose AC electric power. The US Navy used DC, which entailed a far more massive powerplant, on the grounds that an AC plant could not quickly back down in an emergency. Beside the main propeller, there are two auxiliary propellers near the tips of the stern planes. Power output has also been reported as 47,000shp, for a rated speed of 14kts/42kts. A recent Russian book quotes 40,000shp = 14kts/41kts.

The design may have been influenced by contemporary US submarine thinking, which in 1956 favoured automation, increased speed, and turbo-electrical drive for silencing. The cylindrical 'Kerch' sonar introduced in this class (and used in many later designs) may also reflect US thinking of this period. It is interesting that the Soviets did *not* quickly adopt the next series of US ideas (which emerged about 1957–58): the spherical bow array and rafted propulsion.

The sonar system reportedly consists of a big cylinder (Kerch) in the lower part of the bow and a smaller active cylinder in the forward part of the sail. However, a recent unofficial Russian book credits this class with a unique sonar suite consisting of an Okean sonar and Yenisei passive sonar.

Apparently for a time the Soviets thought that all next-generation submarines would have titanium hulls, since they adopted this material for the slightly later missile submarine, 'Papa' (see above). Development of methods to work the new material turned out to be difficult, probably much delaying this programme (and distorting its apparent position in the Soviet design sequence). However, titanium did make 'Alfa' the deepest-diving of all operational submarines, with a reported rated diving depth of 700m (about 2300ft); recently reported as 700m operating depth, 750m crush depth. The appearance of such a submarine was extremely troubling to Western navies. It inspired US and British interest in deeper-diving and faster torpedoes, such as Mk 48 ADCAP, Mk 50, and Spearfish.

The prototype was Project 705. The other six units were built to a 'corrected' design, Project 705K. It appears that three had titanium hulls; the other three had steel hulls (they may have been designated 705Zh). Apparently the highly-automated 'Alfa' proved unsuited to lengthy patrols, and trouble with the powerplant persisted. One was reported broken up in 1988. Another was recommissioned in late 1989 after a five-year refit, which may have involved installation of a conventional pressurised-water reactor. The new project number, 705ZhMT, suggests an improved version of 705Zh. Here T probably does *not* mean torpedo.

'OSCAR' class *cruise missile submarines* (Project 949 Antey)

Displacement: 12,500t surfaced; 14,600t submerged ('Oscar II': 13,900t surfaced; 16,000t submerged)
Dimensions: 469ft 2in ('Oscar II': 505ft 2in) × 59ft 9in × 29ft 6in
143m (154m) × 18.2m × 9m
Machinery: 2 reactors, 2 turbines (7-blade propellers), 90,000shp = 30kts ('Oscar II': 28kts). Speed has also been reported as 33kts
Armament: 24 SS-N-19, 2–533mm and 4–650mm torpedo tubes (24 tube-launched weapons); also reported as 4 of each calibre
Complement: 130

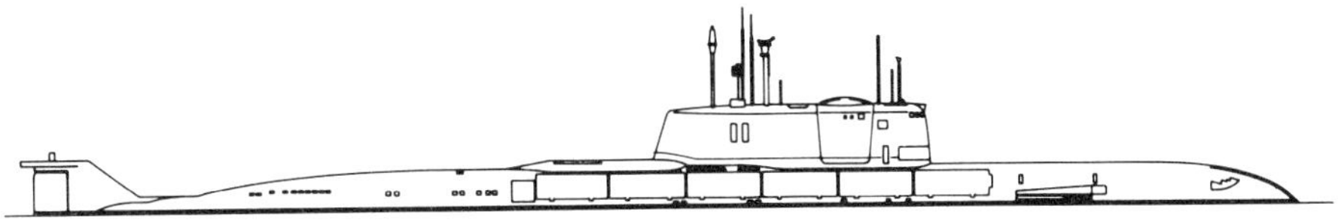
'Oscar II' class

Class: (all built at Severodvinsk)
'Oscar I' (Project 949 Antey): *K-208*, *K-523* (was named *Minskiy Komsomolets*). Launchings: one spring 1980 (laid down 1978, trials late 1980), second 1982.
'Oscar II' (Project 949A Antey-II): *K-148* (launched Aug 1986, comp. 1987), *K-380* (launched 1986, comp. 1987), *K-512* (was named *70 Let VLKSM*, launched Dec 1987, comp. 1988), *K-530* (comp. Dec 1989), *K-119* (comp. 1991), *Orel* (comp. 28.1.93), *Kasatka* (comp. 12.9.93), one other (launched May 1993, comp. 1993/4). The codename 'Granit', suggested for this class, probably refers to their SS-N-19 missile.

These are, in effect, successors to 'Echo', capable of launching the next-generation SS-N-19 missile while submerged. They were originally to have worked on the basis of radar satellite data (transmitted through the Punch Bowl antenna in their sails), but the satellite systems are gradually collapsing. They can also obtain targeting information via a towed VLF buoy housed abaft the sail. The missiles are in tubes canted 40 degrees to the vertical, one door covering each pair. As in 'Papa' and 'Charlie', the missiles occupy the space between the inner and outer hulls; inner hull volume must be somewhat limited (standoff is about 11.5ft [*3.5m*]; reported pressure hull diameter is 27.9ft [*8.5m*]). The torpedo room forward may therefore not have much greater capacity than that of a 'Victor III'. The sonar suite is probably that of a 'Victor III', including an array streamed through a pipe atop the vertical stabiliser. The technical project for this class was approved in 1969.

The reactors are reportedly the same *Artika* type as in 'Typhoon', with 190mW output. Note that in 1975 a high-level decision was made specifically to use the nuclear icebreakers (*eg*, 'Arktika' class) to develop new submarine reactors, and new cores for existing reactors. The new type of reactor was first used in this class and in 'Typhoon'.

As of 1991, the reported planned programme called for twelve 'Oscar' class submarines (*ie*, ten 'Oscar II'). However, in January 1992 the Russian Ministry of Defence reported that six units were being broken up incomplete. For some years a companion short-range missile-launching submarine was expected, but it has never appeared; presumably conventional attack submarines armed with the torpedo-tube version of SS-N-22 (P-100) perform this role.

'TYPHOON' class *strategic submarines* (Project 941 Akula)

Displacement:	18,500t surfaced; 25,000t submerged
Dimensions:	541ft 4in wl, 562ft 9in oa × 75ft × 40ft *165m, 171.5m × 22.8m × 12.2m*
Machinery:	2 OK-650 reactors, 2 shafts (shrouded 7-blade propellers), 90,000shp = 25kts
Armament:	20 SS-N-20, 2–533mm and 4–650mm torpedo tubes (36 torpedoes and SS-N-15/16)
Complement:	150 (50 officers)

Name	Builder	Laid down	Launched	Comp	Fate
TK-208	Severodvinsk	3.3.77	23.9.80	12.12.81	Extant 1995
TK-202	Severodvinsk	1.10.80	26.4.82	28.12.83	Extant 1995
TK-12	Severodvinsk	27.4.82	17.12.83	27.12.84	Extant 1995
TK-13	Severodvinsk	5.1.84	21.2.85	29.12.85	Extant 1995
TK-17	Severodvinsk	24.2.85	Aug 1986	6.11.87	Extant 1995
TK-20	Severodvinsk	6.1.87	Jul 1988	Sep 1989	Extant 1995

TK-208 was operational by 1983; she launched two missiles within 15 seconds in October 1982. Reportedly seven more were cancelled, possible names being *TK-111*, *TK-11*, and *TK-213*. The technical design for this submarine was approved in 1972.

This submarine, designed by Firma Rubin (ex-TsKB-18), was intended for the final phase of a sustained nuclear war. It would go out to sea, dive under the ice, and remain on the bottom for a year, surviving during a nuclear exchange. It would then be available for a reprisal attack on a recovering enemy. Hence the enormous emphasis on habitability; the submarine is extraordinary roomy, with a sauna and even an aviary. However, according to an article in the Russian press, the effective limit on crew endurance is 100 days. Construction of 'Delta IV' class strategic submarines continued in parallel with this project, since they had a very different role, to fight early in the war. 'Typhoon', which became the NATO nickname of the class, is actually the Russian code-name (Tayfun) for the combination of submarine and missile; the submarine itself is Akula. The West first learned of this project in 1974, when Leonid Brezhnev, meeting President Gerald R Ford at Vladivostok, cautioned that 'Typhoon' would go ahead if the United States persisted with Trident. For some years there was speculation as to whether Typhoon was a new missile or a new submarine, or both. Components were so large that when they were being assembled at Severodvinsk some analysts thought they were intended for an aircraft carrier.

The unique configuration consists of five cylinders, mostly wrapped in the outer hull: two main units running the length of the submarine; one forward for the six TT, the torpedoes (and tactical missiles) and the bow sonar sphere; one for command/control (under the sail); and one aft for steering. Each of the main hulls contains a 190mW reactor (of the type used in 'Arktika' class icebreakers), a 45,000shp turbine, and an 800kW diesel generator. Apparently it was impossible to obtain sufficient volume in a single-cylinder of practicable dimensions, and the missiles require such large hull penetrations that it would have been impossible to place two rows side-by-side in a conventional hull. Flat decking abeam the sail probably contains crew escape modules. The first unit built has a pair of pyramidal objects near the paired communication-buoy hatches abaft her sail. They are probably intended to create turbulence forward of the propeller, for silencing.

'Typhoon' class MoD

Placing the missiles forward of the sail, control centre, and machinery has an important advantage for the sort of lengthy programme represented by the earlier 'Yankee/Delta' classes, in which production switches periodically to a new missile. Missiles placed between control centre and machinery require that the submarine be extensively re-engineered when the missile section changes. Much less need be done if the missiles are carried forward. The fin-aft configuration is not altogether unique; it was proposed in the early 1960s for a NATO diesel-electric ballistic missile submarine (to avoid piping exhaust gas through the missile compartment).

The sonar suite (Skat) consists of a large active-passive array in the bow below the torpedo tubes and an active sonar behind a large window in the sail. Photos do not show a towed array. Since the ship is fitted with the 'Rim Hat' ESM/radar system, she is probably fitted to fire SS-N-22 (P-100) missiles as well as defensive ASW weapons.

TK-208 began a modernisation refit in 1991; she is to be fitted with a follow-on missile with about one-fifth the CEP of the SS-N-20. Another unit was reportedly damaged when a missile launch failed in September 1991. Yet another may have suffered fire damage when a torpedo being loaded was dropped on her deck.

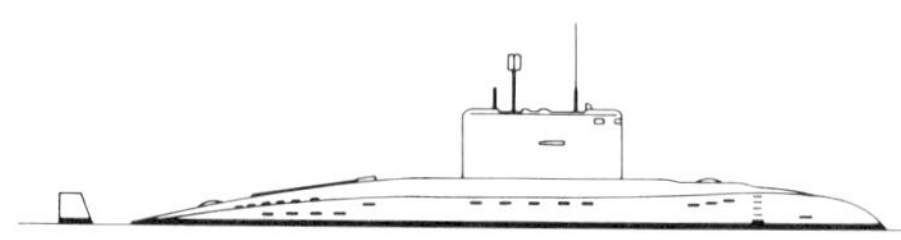
'Kilo' class

'KILO' class *attack submarines* (Project 877)

Displacement:	2325t surfaced; 3076t submerged
Dimensions:	238ft 2in × 32ft 6in × 29ft 4in *72.6m × 9.9m × 6.2m*
Machinery:	1 shaft, 6-bladed propellor: 2 D-42 diesels, 3650bhp, 1 main electric motor, 5790hp (plus 1 130hp creep motor and 2 102hp emergency motors), – 10kts/17kts. Range 6000nm at 7kts surfaced, 400nm at 3kts submerged
Armament:	1 SA-N-5/8 (8 missiles), 6–533mm TT (bow, 18 torpedoes or 24 mines; 2 tubes fitted for wire-guided torpedoes)
Sensors:	Sonar MGK-400 (NATO Shark Teeth)
Complement:	53 (12 officers)

NB Data apply to 877EM version.

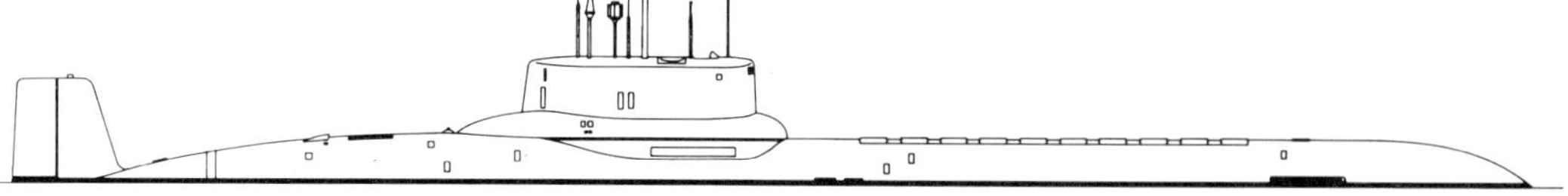
'Typhoon' class

'Kilo' class *431* 20.5.94 L & L van Ginderen

Class:
B-177 (*Del'fin*), *B-187*, *B-190*, *B-219*, *B-224*, *B-227*, *B-229*, *B-248*, *B-260*, *B-345*, *B-354*, *B-394* (named *Komsomolets Tadjikistana* 26.8.80–1991), *B-401*, *B-402*, *B-404* (named *Tyumenskiy Komsomolets* until 1991), *B-405*, *B-425*, *B-437* (named *Magnitogorskiy Komsomolets* 9.6.80–1991), *B-439*, *B-445*, *B-459*, *B-464*, *B-468*, *B-470*, *B-471*, *B-494*, *B-800*, *B-871* (total 28 units, plus exports)

Exports: Algeria (two, in 1987 and in Jan 1988), China (two; see below), India (eight in 1986-89), Iran (two, including *B-219* in Oct 1992), Poland (one, 21.6.86), Romania (one). A boat initially ordered by Romania was sold to Iran, but not delivered because payment could not be made. She was delivered to China in June 1995. China is to receive a second unit late in 1995. At this writing (June 1995), another ten are to be built in Russia for China, to be followed by at least ten built under license in China. The programme is likely to be quite large, since China will probably be replacing large numbers of elderly submarines. The new-construction units may be Project 636 rather than the existing Project 877. Early in 1994 China reportedly ordered two of the follow-on Project 636. India is reportedly interested in building 'Kilo' under license. The two Iranian boats were apparently originally intended for Soviet service. Another two were ordered, but failing Iranian finances killed the deal, and apparently these were the boats sold to China. In 1994 Firma Rubin quoted a price of $250 million for a Project 877EKM submarine.

Designed by Firma Rubin (ex-TsKB-18) as two parallel projects, 'Granay' for the Soviet Navy and 'Warshavyanka' for export, particularly to Warsaw Pact allies. The first boat, built at Komsomolsk, was launched 12 September 1980 (in service April 1982). This class was also built a Gorkiy (Nizhny Novgorod) and at Leningrad (St Petersburg). Launchings: one is Sep 1980 (in service Apr 1982), one in Aug 1981, two in 1983, three in 1984 (one export), four in 1985 (two export), four in 1986 (three export), four in 1987 (two export), four in 1988 (two export), four in 1989 (one export), four in 1990 (one export), three in 1991 (one export), two in 1992 (one export), one 1993 (Komsomolsk, 6 October 1993). Submarine production at Nizhny has now shut down, and the unit at Komsomolsk was said to be the last 'Kilo' to be built there; further export units will presumably be built at a slow rate at Admiralty, St Petersburg. *Del'fin*, the last Russian Navy unit from Nizhny Novgorod, left the Black Sea for the Pacific Fleet in March 1992. Total for the Russian Navy: twenty-four. There may be a follow-on programme of the successor Project 636 class. During 1990 it was reported that at least some of these submarines had been cancelled, and that complete 'Kilo' power trains were being offered for sale within the Soviet Union, for civilian purposes.

This submarine was designed specifically to exploit the new 'Second Captain' automation technology, in which a single central computer would perform both fire control and ship control functions. For example, fire control data were entered automatically, rather than manually as in the 'Leningrad' system of the preceding 'Foxtrot' class. Diving and machinery control were also automated. Most functions are controlled from a single central panel. Manning could therefore be reduced drastically, to a total of 52 on a 3-watch basis.

Silencing (including magnetic and presumably UEP) was emphasised. The diesel generators are suspended (rafted). The hull has an anti-sonar coating. Flooding ports were eliminated from the forebody of the submarine and the fore planes were moved aft to reduce noise, particularly noise interfering with the bow sonar.

The design emphasised underwater performance; it was not a submersible like a 'Foxtrot', which would spend much of its time surfaced. Therefore Rubin adopted an *Albacore* hull and accepted limited surface speed. This hull form was probably also attractive because it could meet the requirements of both types of Soviet export customer: shallow-water navies requiring short hulls, and deep-water navies requiring considerable internal volume. Previous diesel submarines had been single-decked, and therefore could add usable space only by adding length. As in other Russian designs, this one is compartmented for survivability; Rubin claims that it can remain afloat (on the surface) with one compartment and two adjacent main ballast tanks flooded. Operating depth is 240m (about 790ft), maximum depth is 300m; periscope depth is 17.5m.

The Project 877EM combat system is built around a general-purpose computer (MVU-110EM) which receives sonar data and transmits torpedo settings. Torpedoes can also be reloaded from the central control station (using a Murena panel to control an automatic rapid loader); they can be launched at any depth, presumably by water slug. The computer can automatically track and engage two targets using wire-guided torpedoes (only two tubes are fitted to fire such weapons); three more targets can be tracked manually. The central computer can also hold the submarine on an ordered course, and can solve navigational problems (*eg*, presumably, closest point of approach).

Reportedly the original Project 877 for the Soviet Navy had four internal reloadable torpedo tubes and two external tubes (two tubes over four) like contemporary Soviet SSNs. None of the tubes could fire wire-guided weapons. Later versions have six internal tubes (also two tubes over four). Export units have a position for a shoulder-fired SA-N-5/8 missile in the sail. Some boats may have automated four-round launchers, which can be used when the boat is surfaced, for a total of twelve missiles.

'Kilo' was the first Soviet submarine to use true diesel-electric propulsion. It was probably adopted because it simplified silencing and because it allowed two compact diesels to drive the single screw that was hydrodynamically preferable. The batteries were a new high-capacity type (9700kW each). The propeller shaft can be turned by a creep motor. There are also two water channels let into the hull aft. Small low-speed propellers in them can be used in an emergency or for manoeuvering.

The TTZ for this design was issued in 1974. Reportedly the first boats (Project 877) had three diesel generators (probably 1350bhp each) and a 5000hp motor. The corresponding export version, 877E, had six internal tubes, but (as in 877) none was fitted for wire-guided torpedoes. This version was supplied to Algeria, India (first three), Poland, and Romania. An improved Project 877M for the Soviet Navy still had four internal tubes, but all were fitted to fire wire-guided torpedoes (presumably the associated combat system computer could track and engage four targets simultaneously). The corresponding export Project 877EM (five to India), the type described here, has six internal tubes, two of which can fire wire-guided torpedoes. Project 877MK is an improved Soviet Navy version with a new command system based on pirated details of the Norwegian MSI-90U of the *Ula* class. Finally, Project 877EMK is the corresponding export version. Reportedly the first seven Russian boats were 877s, the next seven were 877M, the next seven were 877K, and the last two were 877EMK; the other five boats of this last series were, reportedly, offered to Iran (however, note that the list of completions shows twenty-four Russian boats).

Project 636 is a proposed follow-on offered for export. It is slightly longer (242ft 2in/*73.8m*) and will displace 50t more than a Project 877. There would be a more powerful air-conditioning plant, for Mideast customers. The 5500hp motor will drive the boat at 19kts, and there will also be a 125hp (95kW) creep motor and a pair of 100hp (75kW) manoeuvering motors. Range: 7500nm at 7kts (snorkelling), 400nm at 3kts submerged. One version of this design may have a vertical cylinder abaft the sail to accommodate four vertically-launched cruise missiles. The torpedo armament offered matches that of a 'Kilo'. The tail planes are in an inverted 'Y' configuration. China reportedly has ordered two boats.

'Amur' is a down-sized Project 877 also offered by the Rubin bureau. Rubin claims that it is built around a new kind of compact powerplant, presumably a diesel driving an AC generator, which can charge the battery through a solid-state inverter. Further automation will reduce the crew to 45 (3-shifts, 30 for two shifts). Displacement is cut to 1400–1900t, the boat can town an array, and weapons capacity is sixteen (six tubes).

'Sierra' class

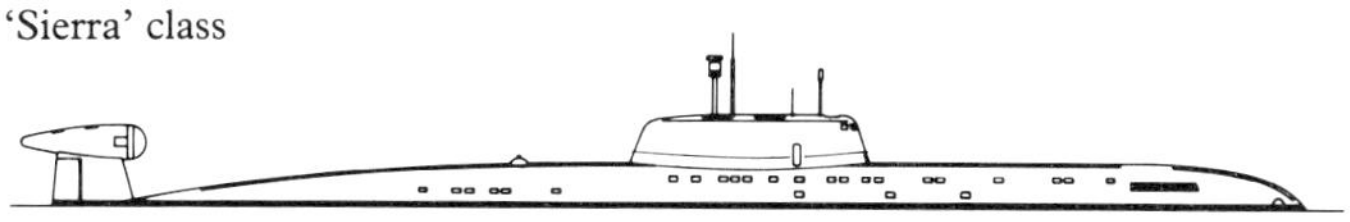

'SIERRA' class *attack submarines* (Project 945A, 945B Barrakuda)

Displacement: 7900t submerged
Dimensions: 351ft x 39ft 5in x 28ft 11in
107m x 12m x 8.8m
Machinery: 1 shaft, 1 reactor (7-blade propeller), 43,000shp = 18kts/35kts
Armament: 2–533mm and 4–650mm TT
Complement: 60 (31 officers)

Name	Builder	Laid down	Launched	Comp
Project 945 ('Sierra I')				
BARRAKUDA	Gorkiy	May 1982	Jun 1983	Jun 1984
KONDOR	Gorkiy	Aug 1983	Apr 1985	1987
Project 945B ('Sierra II')				
K-239	Gorkiy	–	May 1989	1990
K-276	Gorkiy	–	1992	1993

Barrakuda was launched at Gorkiy (now Nizhniy Novgorod) but transferred by river and canal for completion at Severodvinsk. The TTZ for this class was approved in March 1972.

This titanium-hulled submarine (test depth 1km) was the follow-on to 'Victor III'. Like 'Akula', it combines the 'Victor III' sonar suite with a very quiet (fully rafted) powerplant incorporating the *Arktika* reactor. 'Environmental' (non-acoustic) sensors are in the sail. A crew escape pod occupies the centre of the rather wide sail. Much to the disgust of the design organisation, Lazurit, construction of this type was cancelled in favour of 'Akula', which was apparently much less expensive.

This is essentially a 'Victor III' follow-on. Reportedly the bow tubes are splayed out around the bow sonar sphere, which may be larger than that in a 'Victor III'. The hull is titanium, and reported designed test depth was 1000m (3280ft). Another report gave diving depth as 700m (crush depth 800m).

The second unit has a low extension to the after end of the sail. The third unit ('Sierra II') was lengthened 16ft 5in (*5m*) to 367ft 5in (*112m*; displacement submerged increased to 8200t); her sail is 19ft 8in (*6m*) longer and is noticeably wider and blunter. Masts are mounted on the starboard side, leaving space on the port side for two buoyant crew escape chambers.

In January 1992 the Russian Ministry of Defence announced that three submarines of this type were being scrapped before completion. This figure is plausible, on the theory that the first series ordered was a full brigade of seven.

'Akula' class

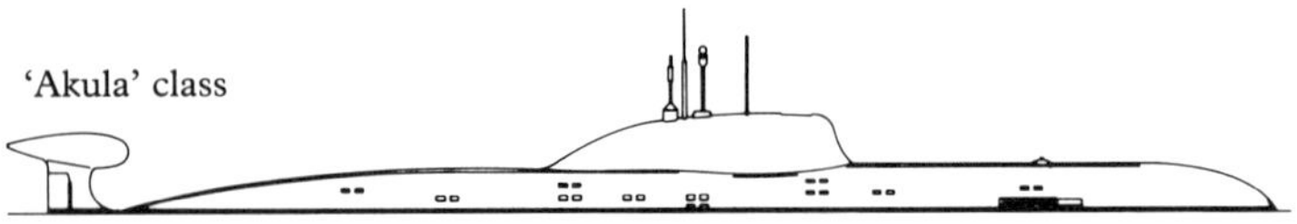

'AKULA' class *attack submarines* (Project 971 Bars)

Displacement: 7500t surfaced; 9100t submerged
Dimensions: 377ft 4in × 45ft 11in × 34ft 1in
115m × 14m × 10.4m
Machinery: 1 shaft, 1 OK-650B reactor, 7-blade propeller, 43,000shp = 20kts/35kts
Armament: 2–533mm and 4–650mm TT

Class (by builder, completed):
Severodvinsk: *K-284* (launched 1.7.84, completed 30.12.84), *K-263* (1985), *K-322* (1986), *K-391* (1987), *K-331* (1988), *K-419* (1989), *K-267 Drakon* (launched 4.8.94, to complete 1995); two others under construction as of 1995 for completion 1996 and 1997.

Komsomolsk: *K-480 Bars* (launched 1983, completed 1984), *K-317 Pantera* (launched May 1990, in service 1990), *K-461 Volk'* (launched 11.6.91, completed 1991), *K-328 Leopard* (launched 6.10.92, completed 28.2.93), *K-157* Tigr (launched 10.7.93, completed 1994), *Rys'* (launched 1992, completed 1994), *Vepr'* (laid down 1992, probably launched 1993)

The name *Yaguar* has also been associated with this class, perhaps for one of the Severodvinsk units.

Despite its apparent attack submarine appearance, this submarine was intended primarily as an SS-N-21 platform. It therefore has only two 65cm TT (the other six tubes are 53cm, to fire SS-N-21s). Recent units have six horizontal bow external tubes above the pressure hull, providing additional missile or torpedo capacity. 'Sierra' was the corresponding attack submarine. However, the steel 'Akula' is far less expensive than the titanium 'Sierra', so it is being produced as a general-purpose submarine. The sonar suite matches that of a 'Victor III' (but with an improved flank array, Akula, extending a third the length of the hull), but the powerplant is new (it is probably the *Arktika* reactor plus fully-rafted machinery). The strategic mission required a communications buoy, which is carried in the streamlined after extension of the sail.

The single reactor is the same type as in 'Sierra', 'Oscar', and 'Typhoon'. According to Russian sources, designed core life was 5 years, but 'Akulas' apparently steamed more high-speed hours than expected (they were used to trail US carrier battle groups) and actual lifetime was often as little as two and a half years. Because core production was planned for a five-year lifetime, cores had to be diverted from other classes, and ships laid up *eg*, four 'Oscars' on the Kola Peninsula in 1994. US cores are said to last at least 10–15 years.

This design was developed by Firma Rubin (ex-TsKB-18). Operating depth is reportedly 900m.

The first two units tested SS-N-21 during the winter of 1987–88. In January 1992 the Russian Ministry of Defence announced that six submarines of this type were being broken up before completion.

Komsomolets 1984

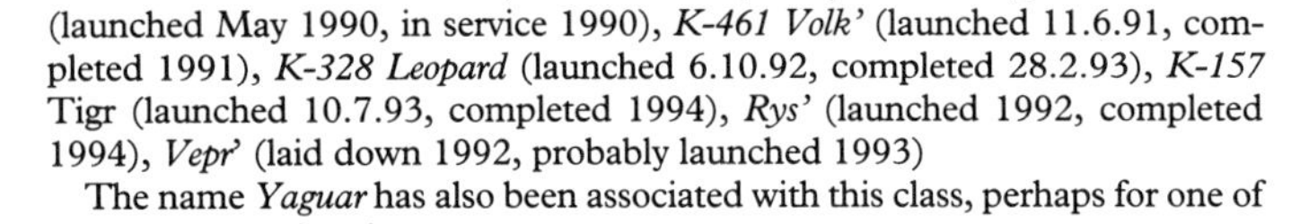

'MIKE' class *attack submarine* (Project 685 Plavnik)

Displacement: 5880t surfaced; 8500t submerged
Dimensions: 385ft 6in × 35ft 1in × 26ft 3in
120m × 11m × 9m
Machinery: 1 shaft, 1 OK-650B-3 reactor, 43,000shp = 14kts/30.6kts
Armament: 6–533mm TT (12 weapons)
Complement: 57 (29 officers)

Class:
K-278 (named *Komsomolets*: laid down 22.4.78, launched 9.5.83, completed 20.10.83, lost 7.4.89)

This submarine was designed by Firma Rubin (TsKB-18). The TTZ was approved in 1966, and the technical design on 16 December 1974.

This was described officially as a deep-diving experimental submarine with a titanium hull, to test several innovative technologies. One was probably powerplant automation. The absence of a stern pod for a towed array suggests that she was a deeper-diving equivalent to 'Victor I' or 'Victor II'. She apparently had the new reactor used in third-generation submarines (early reports credited her with a liquid-metal type, then with a relatively small pressurised-water reactor as in a 'Charlie'). Reported design depth was 1000m (1250m crush depth). She was equipped with an escape sphere which broke loose when she hit the bottom at about 4900ft; reported design test depth was 1km (3280ft). She apparently did not suffer significant structural damage until she hit bottom. Given her two-over-four torpedo tube arrangement (as in a 'Victor'), she probably had four reloadable 53cm tubes and two external decoy tubes. The Soviets announced that at the time of loss she was carrying ten conventional and two nuclear weapons (probably SS-N-15s), as well as seven 'special devices', most likely decoys (such as BGTs) for use from the external tubes and from signal ejectors. However, according to a recent Russian publication, a typical load-out comprised two RK-55 (SS-N-21) missiles, two Shkval, two ASW homing torpedoes (SAET-60M), and six anti-ship missiles and ten anti-ship homing torpedoes.

Project numbers 705M and 943 have been associated with this submarine. The former was presumably a modified 'Alfa' with a pressurised-water reactor, an intermediate step before this submarine design; the latter may have been a planned production version.

Project 885 *attack submarine*

Displacement: 5800t surfaced; 8200t submerged
Dimensions: 364ft 2in × 39ft 4in (49ft 4in over stabilisers) × 27ft 7in
111.0m × 12.0m (15.0m) × 8.4m
Machinery: 1 reactor, 1 shaft, 43,000shp = 19kts/31kts
Armament: 12 missile tubes, 4–650mm TT (bow), 2–533mm TT (total of 24–65cm weapons carried including Oniks SSM and SS-N-16)
Sensors: See notes
Complement: 50 (22 officers)

Class:
Severodvinsk (laid down Dec 1993 at Severodvinsk for planned launch 1996–97)

This is a new multi-purpose submarine, designed by Firma Rubin (TsKB-18). It has an 'Akula' powerplant (with a pumpjet propulsor) and a new combat system. Like the US *Seawolf*, it has separate part-spherical transmitting and receiving arrays. The combination is more massive than that in an 'Akula', so the submarine's bow tubes (four 65cm, two 53cm) are angled around it (as in a 'Sierra'). There are also twelve vertical tubes abaft the sail. In some ways, then, this is a Russian equivalent of the US Improved *Los Angeles*, with its combination of conventional torpedo tubes and vertical tubes for cruise missiles. The Russians have advertised submarine vertical tubes for their new generation of medium-range anti-ship missiles (Chelomey 'Yakhont' and 'Alfa').

MIDGET *submarines* MS 520 and 521 (Project 865, Losos)

Two midget submarines were completed at the Admiralty yard, Leningrad, in 1986 and turned over to the Soviet Navy in 1988. They were assigned to the Baltic bases of Liepaja and Paldiski. Their existence was revealed early in 1991, in the context of waste, fraud, and abuse; the whistle-blower was Cdr 3rd rank AS Shahov. Each 219t submarine has a range of 1000km (about 540nm) and an endurance of ten days; speed is about 6kts. Maximum diving depth (collapse depth?) is 200m (660ft). The submarine is designed both to carry torpedoes (2×53cm tubes) and to transport three divers. Dimensions: 91ft 10in × 13ft 1in, *28.0m × 4.0m*.

Project 865 was established in 1973. Design by TsKB-16 (Malakhit). Shahov stated that he had been invited to be a chief assistant in the project in October 1982, his training beginning in January 1983. Completion was scheduled for early 1984, but tests did not begin until 1986. Apparently they were not successful, and Shahov protested to the Party control group in November 1987. All tests were suspended early in 1988, but 'certain people in the military hierarchy and industry' demanded that the mini-submarines be turned over to the navy by 31 December 1988. Two commissions investigated problems in the mini-submarines in 1989 and 1990, and on 5 September 1990 the Chief of Staff and Chief of Combat Training, Vice-Admiral A Kuzmin, decided that the mini-submarines would not be allowed to go to sea. For example, the craft could not maintain depth while divers were released through the lock-out chamber. Kuzmin wanted the crew increased to five, and the commissions found the electric and air pressure systems too weak. Total expenditure to early 1991 was 40m rubles.

Project 865 was clearly the sort of submersible so often suspected of violating Swedish territorial waters. Apparently it was never operational, and the Soviets used the occasion of its revelation to state that the Baltic Fleet had never included any operational minisub or tracked sub system (nor was Project 865 ever operational). Such details would not necessarily have covered Spetznaz or KGB operations, however.

The Soviets certainly captured both German and Italian midget submarines, and during the 1950s there were periodic rumours that such craft had been put into mass production. There may have been some confusion between true midgets and the small M class, built both before and after the war. Midget submarines based on German designs were unofficially reported as having taken part in April 1957 fleet manoeuvres, but they did not figure in official intelligence publications.

A recent Russian list of enemy units taken over during and after the Second World War does include four Italian CB-class midget submarines, but it also indicates that they never entered service postwar. No German midgets are listed.

Other Russian submersible projects were:

Project 907 Triton-1 and Project 908 Triton-2: *B-483–490*, *B-520–543*. Triton-1: 1.6t, 16ft 5in × 3ft 11in, *5m × 1.2m*; Triton-2: 5.7t, 31ft 2in × 6ft 4in, *9.5m × 1.9m*. Diving depth 40m. Crew 2 in Triton-1, 6 in Triton-2; endurance 30 (60)nm. Built at Nizhni Novgorod 1974–1980.

Project 1806 Poisk: *AGS-6*, *AGS-17*, *AGS-37*, *AGS-38*. Diving depth 4000m. On *Kommuna*, *El'Brus*, *Alagezhe* salvage ships.

Project 1832 Sever: *LS-17*, *LS-8*, *LS-28*, *LS-34*: 28t surfaced; 40t submerged, 41ft × 8ft 10in × 12ft 6in, *12.5m × 2.7m × 3.8m*; dive to 2000m, capable of 4kts, 5 crew.

Project 1837 and 1837K: *APS-5*, *APS-11*, *APS-18–27*: 35t, length 39ft 8in, *12.1m*, diving depth 2000m; accommodation for 11 on board. For 'India' class as well as submarine salvage ships of *El'brus* class and *Kommuna*.

Project 1839 and 1839.1 (NATO 'El'brus'): *ARS-1–4*, *ARS-7*, *ARS-9*, *ARS-10*, *ARS-12–16*, *ARS-29–33*, *ARS-35*, *ARS-36*: 45t, 44ft 11in, *13.7m* long, diving depth 2200m. *ARS-14* on rescue ship *G Koz'min*, others on *El'brus* class.

Project 1855 (Mir): *Mir-1–Mir-5*: 18.7t, 25ft 7in × 24ft 2in × 10ft 6in, *7.8m × 2.9m × 3.2m*. Deep divers (maximum 6100m) built from 1987 on in Finland.

'Bravo' class

AUXILIARY SUBMARINES

The Soviets built a variety of experimental and auxiliary submarines during this period. 'Papa' probably does not come into this category (she was probably conceived as an operational type, although she did test a variety of new systems). 'Mike' is a borderline case, in that she was apparently conceived as an experimental submarine but was provided with a real weapon system.

Current non-operational classes (which may have had important wartime operational roles) are:

'Bravo' (Project 690 Kefal'): *SS-368*, *SS-256*, *SS-310* (completed 31.10.70), *SS-356*. Lazurit design for a padded ASW target, built at Komsomolsk (one completed 1967, two 1968, one 1970) and distributed among the four fleets. Displacement 2750t/3300t, dimensions 239ft 6in × 32ft 2in × 23ft 11in, *73m × 9.8m × 7.3m*, powerplant 1 shaft, 3500/3000hp = 15kts. Armament: 1–533mm, 1–406mm TT. Complement 60. The unusual hull form, with maximum sectional area abaft the sail, is probably intended to protect the machinery.

'India' (Project 940: *BS-203*, *BS-486* [until 1992 named *Komsomolets Uzhbekistan*]): a diesel-electric rescue submarine carrying two submersibles. Both units were built at Komsomolsk (one completed 1979, one 1980); one was assigned to the Northern and one to the Pacific Fleet. Fin-mounted diving planes, which are unusual in Soviet practice, would be particularly helpful in maintaining trim while launching the submersibles. They displace about 3900t surfaced and 4800t submerged (347ft 8in × 32ft 10in; *106.0m × 10.0m*); submerged speed is 15kts. There are two types of submersible. 37ft and 40ft long.

'Lima' (Project 1840: *BS-555*, built at Sudomekh; completed 1979) was a small experimental diesel submarine with her sail set well back from the bow of her unusually bulky hull. A forward extension of the sail housed an active sonar transducer, and there was also a bow-mounted sonar transducer. The radar mast was fixed. Displacement 1800t/2300t, dimensions 255ft 11in × 26ft 3in × 19ft 8in, *78.0m × 8.0m × 6.0m*, diesel-electric drive (2 generator sets, 2000bhp each, 1 shaft), 2500hp = 17/14kts. Complement 42. Served in the Black Sea 1978–89, then returned to St Petersburg; may have been abandoned 1993.

'Uniform' (Project 1910, *AS 15* and *AS-16*: both built by Sudomekh; one launched November 1982 and completed June 1983, one launched April 1988 and completed November 1989) is a small nuclear submarine built at Sudomekh. She displaces 1900t/2500t submerged (239ft 6in × 22ft 11in × 16ft 6in, *73.0m × 7.0m × 5.0m*). Suggested roles are specialised research (to replace the converted conventional submarines) and Spetznaz support. This was the first Soviet single-hulled submarine. In 1995 it was revealed that the Soviets had built their small deep-diving submarines specifically to gain access to Western undersea cables, such as those which connect SOSUS arrays to shore stations.

'X-Ray' (Project 1851): *AS-11*. 550t surfaced; 1000t submerged, 131ft 3in × 17ft 5in × 16ft 5in, *40.0m × 5.3m × 5.0m*, 1 reactor, 10MW, built by Sudomekh 1982. A small deep-diving nuclear research submarine, completed at Sudomekh in 1983. She is comparable to the US NR-1.

There was also an earlier submersible, *AS-4*, which worked with salvage ships.

Project 10831 (no NATO name): *AS-12*: 600t surfaced; 1100t submerged, 131ft 3in × 19ft 8in × 16ft 9in, *40.0m × 6.0m × 5.1m*, 1 reactor, 10,000hp = 30kts; 20 crew. Diving depth 1000m. Built at Severodvinsk. No date of construction given.

'Paltus' (Project number unknown) is a small nuclear submarine (1520t submerged, 154ft 2in × 13ft 1in × 13ft 1in, *47.0m × 4.0m × 4.0m*) built at Sudomekh, launched April 1991, and probably intended for sea-floor work and/or salvage.

'Beluga' (Project 1710 Makrel', *SS-533*) is an experimental submarine built at Sudomekh (Leningrad) in 1987 and operated in the Black Sea: 1900t/2400t, 203ft 5in × 28ft 7in × 19ft 8in, *62.0m × 8.7m × 6.0m*, 200/2500shp = 10/28kts, complement 40. The hull form resembles that of an 'Alfa' but she is about half the size. The reported power seems insufficient for the reported speed. Officially described as intended for 'advanced hull form, propulsor, and boundary-layer trials'. The latter phrase suggests some form of drag reduction (*eg*, polymers). Like the old US *Albacore*, she is most likely a control test vehicle, to develop means of controlling very fast submarines. Speculation that she tests air-independent propulsion plants is unlikely to be correct, since such plants generally provide a submarine with long endurance at very low speed (in which case there would have been no point in the sort of elaborate streamlining this submarine shows).

AMPHIBIOUS WARFARE VESSELS

'MP 2' class *medium landing ships* (Project 450)

Displacement: 600t standard; 750t full load
Dimensions: 182ft 9in × 23ft × 8ft 6in
55.7m × 7.0m × 2.6m
Machinery: 1-shaft diesel, 1200bhp = 15kts. Range 6000nm at 10kts
Armament: 6–25mm (3×2)
Complement: 50 (plus 200 troops)

Class:
MDK-16, *MDK-17*, *MDK-23*, *MDK-40*, *MDK-41*, *MDK-46*, *MDK-47*, *MDK-48*, *MDK-49*, *MDK-50*, *MDK-52*, *MDK-53*, *MDK-114–116*, *MDK-124*, *MDK-142*, *MDK-144*, *MDK-145*, *MDK-147–150*, *MDK-162* (23 units)

First specially-built Soviet amphibious ships; twenty-three built, first laid down 1956, completed 1958. Could carry four tanks or 200t. MP was a NATO codename, not a Soviet designation, as in the case of the P-series of MTBs. Now stricken.

'MP 4' class *medium landing ships* (Project 450bis)

Displacement: 780t full load
Dimensions: 180ft 9in × 29ft 2in × 11ft 2in
55.1m × 8.9m × 3.4m
Machinery: 1-shaft diesel, 1100bhp = 12kts. Range 5500nm at 12kts
Armament: 4–25mm (2×2)
Complement: 50

Class:
MDK-128, *MDK-130–136*, *MDK-143*, *MDK-146* (10 units)

Converted coastal motor transports: ten built, laid down from 1956 and first completed 1957–58. One of this type was built in Egypt. Can carry eight APCs or 550t of cargo. Originally given NATO codename 'Uzka', also referred to as 'Kumos' class; five reported in reserve 1982. Since stricken.

'MP 6' class *landing ships* (Project 512)

Displacement: 2000t
Dimensions: 246ft 1in × 36ft 1in × 13ft 1in
75.0m × 11.0m × 4.0m
Machinery: Diesels = 10kts. Range 3300nm at 9kts
Armament: 4–45mm (1×4), 6–10 tanks
Complement: ?

Class:
SDK-2, *SDK-4*, *Bureya*, *Khoner*, *Bira*, *Irtiz*, *Vologda*, *SDK-38* (8 units). *Bureya* and *Khoner* were adapted to transport rockets as *VTR-296* and *-297*.

Converted 'Bira' class freighters: eight ships rebuilt 1958–61. Classed as VTR (*Voyennyy Transport* or Military Transport). They originally had bow doors, but later these were welded shut, and they served as coastal freighters. Original gun armament optional, now credited with 6–37mm (3×2). There are also several other classes of Soviet freighters classified as VTR; all are unarmed. One was transferred to Indonesia in 1962.

'MP 8' class *medium landing ships* (Project 188)

Displacement: 800t standard; 1200t full load
Displacement: 236ft 3in × 34ft 1in × 8ft 10in
72.0m × 10.4m × 2.7m
Machinery: 2-shaft diesel, 2500bhp = 14kts. Range 8000nm at 10kts
Armament: 4–57mm (2×2)
Complement: ?

Class:
SDK-7, *SDK-8*, *SDK-9*, *SDK-12*, *SDK-24–35*, *SDK-75*, *SDK-76* (19 units). All completed 1958–63. Designed by TsKB-50.

Enlarged 'MP 2'. Could carry up to twelve APCs or 400T of cargo. Now stricken.

'Vydra' class

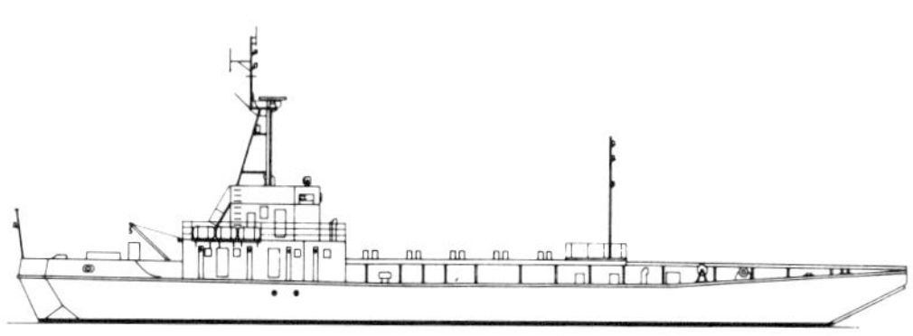

'MP 10/SMB-1' class *landing craft* (Project 106)

Displacement: 420 tons full load
Dimensions: 157ft 6in × 19ft 9in × 6ft 6in
47.7m × 6.4m × 2.0m
Machinery: 2-shaft diesels, 400bhp = 11kts. Range 500nm at 8kts

These were minor utility landing craft comparable to the US LCU, but derived from the wartime German MFP-D. MP-10 differed in having a small cargo hold forward of the bridge, served by a crane; the reduced cargo deck held only one tank or two trucks. The bow ramp sealed the cargo deck; there was no separate cargo door. Total production is not known, but in the late 1970s naval SMB-1 and MP-10 totalled about forty-six. All MP-10 were gone by 1982.

'SMB-1' was a slightly larger open-deck bow ramp commercial beaching craft (1m longer cargo space) built 1960–65; some were operated by the Soviet Navy: 180t standard/300t full load, 159ft 1in × 21ft 4in × 6ft 7in, *48.5m × 6.5m × 6m*, 600hp = 10kts, 16 crew. Capacity: 150t or 3 tanks. Five were transferred to Egypt. Fifteen remained in service in 1990.

The successor Project 106K ('Vydra') *MPK-155–166*, *MPK-168–170*, *MPK-172–175*, *MPK-172–217* built 1967–69: 308t standard/442t full load, 178ft 10in × 25ft 3in × 7ft 10in, *54.5m × 7.7m × 2.4m*, 1000bhp = 10.5kts, range 2500nm at 10kts. Capacity 100 troops or 200t cargo. Like a US LCU, it had an open tank deck. Of fifty-six built in East Germany, nine were left in Russian service in 1992. They were succeeded by the big air-cushion vehicles.

Transfers: ten each to Bulgaria and Egypt; four SMB-1 to Egypt.

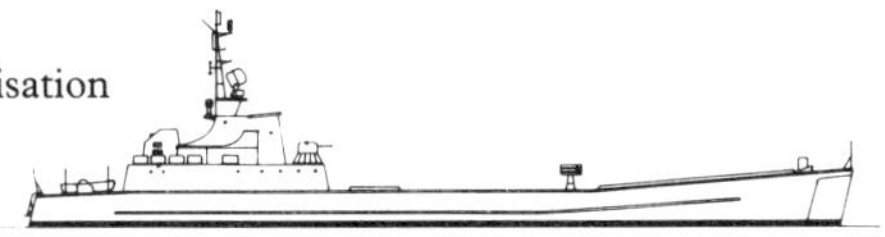

'Polnocny A' after modernisation

'POLNOCNY' class *medium landing ships*

Displacement: 770t full load (800t Type B, 1150t Type C)
Dimensions: 239ft 6in (242ft 9in Type B, 266ft 9in Type C) × 28ft 3in (33ft 1in Type C) × 6ft 7in (6ft 11in Type C)
73.0m, 74.0m, 81.3m × 8.6m, 10.1m × 2.0m, 2.1m
Machinery: 2 shafts, 2 diesels, 4000bhp = 19kts. Range 1500nm at 14kts, 900nm at 18kts (5000bhp = 18kts. Range 900nm at 17kts Type C)
Armament: 2 SA-N-5 (2×2, 16 missiles) in many; 4 SA-N-5 (4×2, 32 missiles) in one; 2–14.5mm MG (1×2) in some, 4–30mm (2×2) in some; 2–140mm RL (2×18). Some unarmed. Types B and C: 4 SA-N-5 (4×4, 32 missiles) in most, 1 or 2–30mm (2×2), 2–140mm RL (2×18)
Sensors: Radar Spin Trough or Drum Tilt
Complement: 35 (40 in B and C plus 100 or 180 troops)

Class:
'Polnocny A' (14 units): *MDK-36, MDK-37, MDK-68, MDK-78, MDK-79, MDK-81, MDK-85, MDK-86, MDK-87, MDK-88, MDK-92, MDK-93, MDK-94, MDK-107*

'Polnocny B' (29 units): *SDK-19, SDK-21, SDK-39, SDK-44, SDK-45, SDK-70–74*, SDK-83, *SDK-84, SDK-96, SDK-99, SDK-102, SDK-111, SDK-112, SDK-119–121, SDK-137, SDK-171, SDK-172*

'Polnocny C' (8 units): *SDK-20, SDK-22, SDK-61, SDK-72, SDK-82, SDK-88, SDK-154, SDK-164*

All built at Polnocny shipyard, Gdansk, Poland, completed 1961–73. Beside transfers, twenty-three were built for the Polish Navy. Transfers: 'Polnocny A': Algeria (one, 1976), Egypt (three, 1974), India (two, 1966), Somalia (one, 1976). 'Polnocny B': Angola (three, 1977–79), Cuba (two, 1982), Ethiopia (one, 1981, one, 1983), Syria (three, 1984–85), Vietnam (three, 1979–80), South Yemen (two, 1973, one, 1977, one 1979). 'Polnocny C': Iraq (four, 1977–79), Libya (one, 1977, two, 1979). Disposals: seven discarded 1990–91. Polish designation: Project 770, later modified to 771 and 773; 'Polnocny C' amphibious command ship is Project 776. Soviet designation: SDK, or *Srednyy Desantnyy Korabl'*, Medium Landing Ship. Cargo: 180t in A and B, 250t in C. Early ships carry up to eight APCs, Type C eight. The decision to provide rockets aboard the landing ship for close-in support would seem to be a major shift in Soviet doctrine. It is possible that the earlier craft were intended for unopposed landings, merely moving tanks more expeditiously than would be possible by land. Note, too, that the 'Polnocny' was the first Soviet amphibious ship apparently designed for troop carrying. Some of the exported ships have helicopter pads immediately forward of the superstructure. SA-N-5s are recent installations.

These craft were designed in Poland, and had Polish project numbers: 770 for Polnocny-A, 771 for Polnocny-B, 774 for Polnocny-C, 776 for Polnocny-D. They are classed as SDKs, medium landing (D, *densantyi*) ships, numbered in an SDK series.

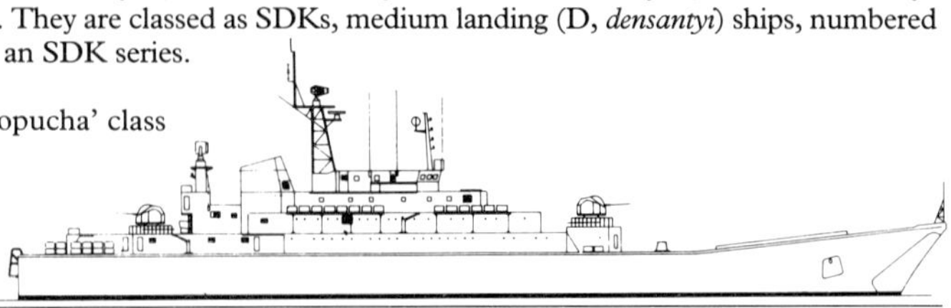

'Ropucha' class

'ROPUCHA' class *large landing ships* (Project 775)

Displacement: 3200t full load
Dimensions: 370ft 9in × 45ft 11in × 9ft 6in
113.0m × 14.0m × 2.9m
Machinery: 2 shafts, 2 diesels, 10,000bhp = 18kts. Range 3500nm at 16kts
Armament: 4 SA-N-5 (4×4, 32 missiles) on 2 ships, 4–57mm (2×2)
Sensors: Radar Strut Curve, Muff Cob, Don 2
Complement: 70 (plus 230 troops)

'Ropucha' class 6.9.91 — L & L van Ginderen

Class (completed):
*BDK-11** (Apr 1990), *BDK-14* (31.8.81), *BDK-32, BDK-43, BDK-45, BDK-46, BDK-57, BDK-48* (30.6.75), *BDK-54*, BDK-55, BDK-56* (named *Konstantin Ol'shanskii*), *BDK-58, BDK-60, BDK-63* (30.7.75), *BDK-64* (named *Tsezar Kunikov*), *BDK-67, BDK-90* (30.11.75), *BDK-91, BDK-99, BDK-105, BDK-119, BDK-121, BDK-122*, BDK-181, BDK-182, BDK-197* (17.8.78), *BDK-200* (28 units)
Starred units are 'Ropucha II' (Project 775-III). Some unidentified units are Project 775-II.

Built by Polnocny Shipyard, Gdansk, Poland at about two a year. Can carry 450t of cargo as well as vehicles and equipment for the 230 troops, 75 per cent of the much larger USN *Newport* class LST capacity. Soviet designation: BDK, *Bolshoi Desantny Korabl*, large landing ship. Positions for two bombardment RLs forward, although no unit has been fitted with them. A combination of bow and stern doors permits roll-on/roll-off operation. Frequently deployed in the Mediterranean, Indian Ocean and off W Africa. One transferred to South Yemen 1980. Production resumed in 1982, and as of 1991 there were about twenty-five in Soviet naval service. Several were fitted with SA-5/8 launchers.

Like the 'Polnocnys', these are Polish designs, so the project numbers are also probably Polish rather than Russian.

Later units, 'Ropucha II' (Project 775-III: three built), have 1–76mm gun (AK-176) forward in place of two twin 57mm, 2 AK-630 (30mm Gatlings), a Cross Round radar, and 2 UMS-73 Grad rocket launchers.

These ships are classified as large (B) landing ships, BDK. Twenty-four ('Ropucha I') were in Russian service in 1992: known names were *Tsezar Kunkov* (ex-*BDK-64*), *Aleksandr Shafalin* (ex-*BDK-100*), *Konstantin Ol'shanskiy*, *Bobusk*, *BDK-11*, *BDK-23*, *BDK-46*, *BDK-67*, *BDK-108*, *BDK-122*, fourteen others.

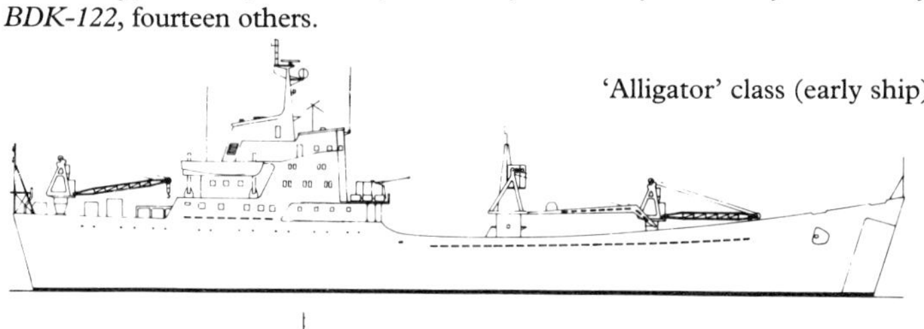

'Alligator' class (early ship)

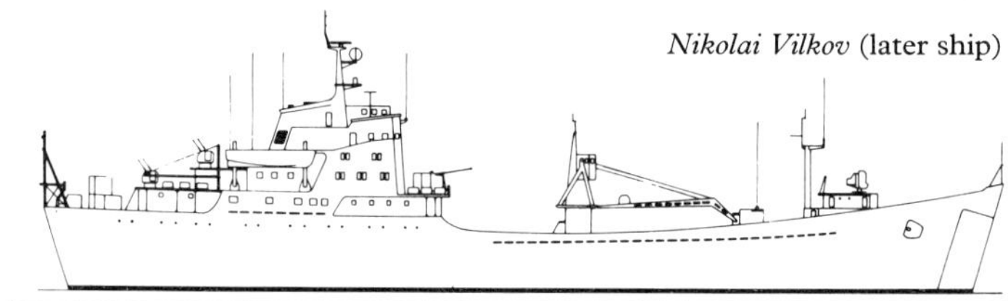

Nikolai Vilkov (later ship)

'ALLIGATOR' class *large landing ships* (Project 1171, Nosorog)

Displacement: 3400t standard; 4700t full load
Dimensions: 370ft 1in × 50ft 2in × 14ft 5in
112.8m × 15.3m × 4.4m
Machinery: 2 diesels, 8000bhp = 18kts. Range 14,000nm/10kts, 9000nm/16kts
Armament: 3 SA-N-5 (3×2, 24 missiles), in some; 2 SA-N-5 (2×2, 16 missiles) in *Petr Il'ichyev*; 2–37mm (1×2, 12 ships), 4–25mm (2×2, last 2 ships), 1–122mm RL (1×40, some ships)
Sensors: Radar Spin Trough or Don 2, Muff Cob
Complement: 75 (plus 300 troops)

Class:
Type 1: *BDK-10 Voronezhskiy Komsomolets* (comp 18.6.66), *Krysmkiy Komsomolets, BDK-13 Tomskiy Komsomolets, Komsomolets Karelii*
Type 2: *BDK-66, Servei Lazo, BDK-69*
Type 3: *BDK-77, Ilya Azarov, Donetskiy Shakhter, Krasnaya Presnya, 50 Let Shefstva VLKSM (Komsomol), Aleksandr Tortsev, Petr Il'yichev*
Type 4: *Nikolai Vilvov, Nikolai Fil'chenkov*

All designed by TsKB-50 and built at Kaliningrad 1964–67.

Soviet designation, BDK, large landing ship. These ships have bow and stern ramps for unloading vehicles, and are essentially of freighter configuration. Can carry 25–30 APCs or 1500t of cargo. Early units have three cranes (15t or two 5t); later ships have only one crane. Later units have an enclosed bridge and a RL forward (one of these bombarded Eritrean positions around Massawa, Ethiopia, in January 1978), and the last two have 25mm guns.

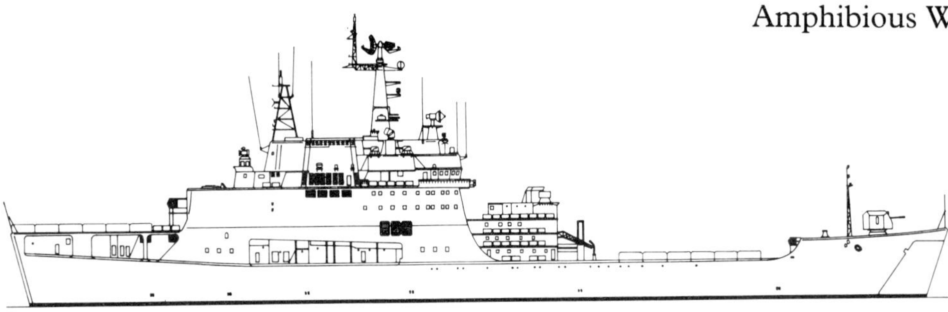

Ivan Rogov as completed

IVAN ROGOV class *assault ships* (Project 1174, Yedinorog)

Displacement:	11,000t standard; 13,000t full load
Dimensions:	518ft 4in × 78ft 9in × 26ft 11in *158.0m × 24.0m × 8.2m*
Machinery:	2-shaft geared gas turbines, 20,000shp = 23kts. Range 12,000nm at 14kts, 8000nm at 20kts
Armament:	1 SA-N-4 (1×2, 20 missiles), 2–76mm (1×12), 4–30mm Gatling, 2–122mm RL (1×2, 40 tube), 4 helicopters
Sensors:	Radar Head Net C, Owl Screech, Pop Group, 2 Bass Tilt; ECM 4 Bell Series
Complement:	200 (plus 550 troops)

Class (completed):
Ivan Rogov (15.6.78), *Aleksandr Nikolaev* (30.12.82), *Mitrofan Moskalenko* (1989)

When she appeared in 1978, *Ivan Rogov* (launched at Kaliningrad 1976 and three times the size of an 'Alligator') appeared to presage a Soviet shift towards amphibious intervention overseas. She can carry a full naval infantry battalion including ten tanks and thirty APCs and has a unique pair of 'Lebed' class hovercraft to put them ashore as well as four Kamov Ka-25 Hormone helicopters. However, construction of the class was relatively slow, the second ship not emerging until 1983, although there were reports of a third unit at that time. She has an integral shore bombardment capability in her 122mm automatic RL. Transferred with *Minsk* from the Black Sea to the Pacific Fleet in the first half of 1979. Sometimes she is compared with the US LPH, although the small hangar in the superstructure is in no way comparable with the standard US hangar deck. Below that deck there is a tank deck forward, and a well deck (with a stern gate) aft. In this sense the concept matches that of the LHA, albeit on a much smaller scale. Unlike the US ship, however, this one beaches to discharge her vehicles through a bow door. Soviet designation: BDK.

Ivan Rogov Mar 1979 RN

'Aist' class
NB 1/750 scale

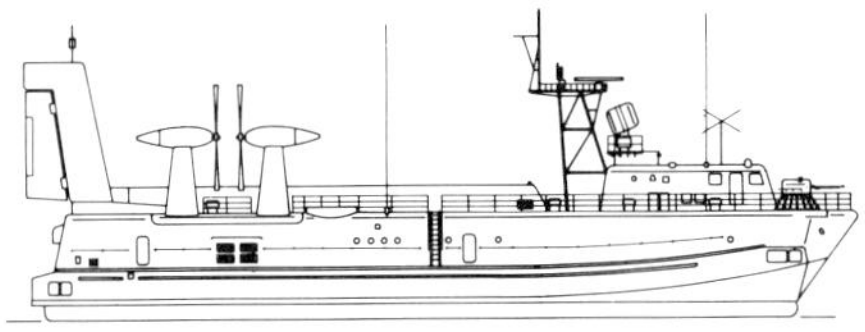

'Pomornik' class
NB 1/750 scale

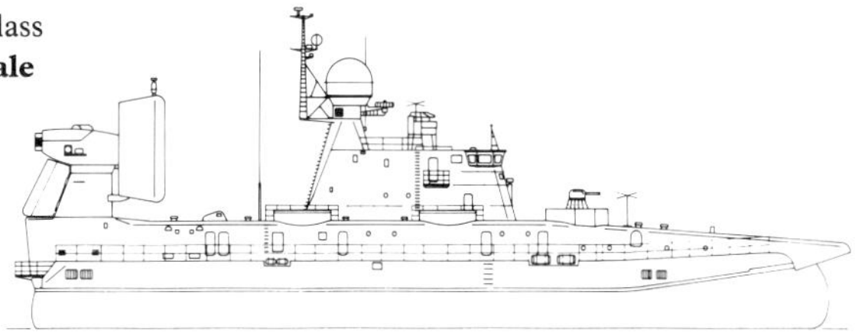

'Gus' class
NB 1/750 scale

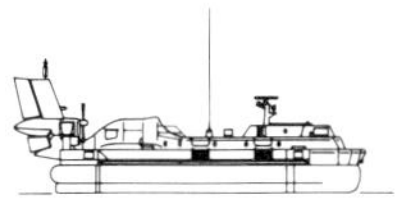

AIR CUSHION LANDING CRAFT (Project 1232 Series)
These very large craft are the functional successors to the MP-10 and 'Vydra' series. Types were:

'Aist' (Project 1232.1 Dzheyran): 19 units: *MDK-18, MDK-80, MDK-82, MDK-89, MDK-100, MDK-102, MDK-110, MDK-111, MDK-122, MDK-125, MDK-126, MDK-127, MDK-136, MDK-138, MDK-141, MDK-172, MDK-173, MDK-516, MDK-518.* 250t standard/310t full load, 156ft 10in × 57ft 5in × 5ft 3in, *47.8m × 17.7m × 1.6m*, 2 MT-70 gas turbines for lift and propulsion (4 propellors), 32,000hp = 70kts (50kts cruise), range 350nm at 60kts. Armament: 2 twin AK-630 (30mm Gatlings), 2 quadruple SA-N-5/8 in later units. Complement 10. Capacity: 4 PT-76 light tanks and 50 men or 2 main battle tanks and 200 men. Built in Leningrad: 1970 (one), 1973 (one), one per year to 1979, then three per year. Similar to British SR.N4, with bow and stern doors. Spin Trough and Bass Tilt radars. Largest such craft in the world when completed. One discarded 1990–91.

'Pomornik' (Project 1232.2 Zubr): First unit was accepted in 1988, 3 years late. Ten completed by 1992. Programme: Eight Baltic (two series of four), three Black Sea. Class (13 units): *MDK-95, MDK-97, MDK-103, MDK-104, MDK-106, MDK-108, MDK-109, MDK-116, MDK-117, MDK-118, MDK-123, Donetsk* (completed 26.6.93), *Ivan Bogun*. Planned service life, 16 years. 400t standard/550t full load, 188ft × 84ft × 6ft 7in, *57.3m × 25.6m × 2m*, 3 MT-70 gas turbines for propulsion (3 propellors in shrouds), 40,500hp; 2 gas turbines for lift, 27,000hp; maximum 63kts (55kts cruise). Stores for 5 days (1 day if carrying troops); range 300nm. Complement 31. Capacity: 3 tanks or 10 BTR (APCs) or 360 men and 25t of cargo. Armament: 2 AK-630 (30mm Gatlings), 4 SA-N-5/8, 2–22-tube barrage rocket launchers. Positive-E radar in first unit, Bass Tilt for fire control. Largest Soviet air cushion vehicle.

The smaller air cushion vehicles were, in effect, successors to small ramped landing craft broadly equivalent to the US LCM (see below).

MINOR LANDING CRAFT
From 1954 to 1974 the Soviets built large numbers of 'T-4' (Project 306) landing craft roughly equivalent to the US LCM, but larger (62t) and more seaworthy. They can be carried as deck cargo on board conventional freighters. Transfers: Angola (five 1976), Cuba (seven 1967–74), Ethiopia (four 1977–78, two 1984), Guinea (four 1974), Guinea-Bissau (two 1975, two 1978), Kampuchea (two 1984–85), Somalia (two 1968, two 1979), Vietnam (fifteen 1967–69), North Yemen (two 1970), South Yemen (three 1970, two 1982). All surviving Soviet units discarded by 1990.

Later units are Project 1784: (85 units): *D-501–515, D-519–581, D-601–608.* 70t, 62ft 4in × 14ft 1in × 3ft 3in, *19.0m × 4.3m × 1.0m*, 150hp, 2 crew. Cargo capacity 30t. Ten were in service in 1992.

From 1979 onwards, 'T-4' was succeeded by a new 90t 'Ondatra' class (Project 1176 Akula): 115t, 80ft 5in × 17ft 1in × 5ft 1in, *24.5m × 5.2m × 1.55m*, 2 engines, 300hp = 10kts, range 500nm at 5kts (stores for 2 days). Cargo capacity: 50t or 1 tank; well deck is 44ft 11in × 12ft 10in (*13.7m × 3.9m*). Forty were in service in 1992. *Ivan Rogov* carries one as a tug for her 'Lebed' class hovercraft. In all, forty-nine were built.

The true successors to these craft are small hovercraft which can be carried in the well decks of *Ivan Rogov* class ships: Projects 1205, 1206, and 1209. They are classed as landing (D) cutters (KA).

'Gus' (Project 1205 Skat): (*D-418–447*) 27t, 70ft 3in × 23ft 11in × 1ft 8in, *21.4m × 7.3m × 0.5m*, 2 TVD-10 gas turbines for propulsion, 2 propellors, 1800hp = 60kts; 1 TVD-10 for lift, 780hp; range 230nm at 43kts, 185nm at 50kts. Capacity 25 men. Complement 4. Built 1969–74; thirty in service 1992. Military version of civilian 'Skate' air-cushion vehicle. There is a training version. Unlike the later air-cushion vehicles, these carry only troops. Disposals: two in 1988–89, twelve in 1990–91.

'Lebed' (Project 1206 Kalmar): Built 1975–81 at Fedodosiya (fifteen) and Leningrad (five). (*D-52, D-277, D-347, D-348, D-435*, 13 others) 87t, 82ft × 36ft 1in × 4ft 3in, *25m × 11m × 1.3m*, 2 AI-20K gas turbines for lift and cruise, 2 shrouded propellors, 20,000hp = 70kts, range 100nm at 60kts (stores for 1 day), 2–12.7mm (1×2), complement 6. Capacity 2 light tanks or 40t and 120 men. This is the type carried by *Ivan Rogov*. One discarded 1990–91.

'Pelikan' (Project 1206T Kasatka) is a mine countermeasures variant. Three were in service in 1992. 91t, 80ft 9in × 37ft 1in × 3ft 11in, *24.6m × 11.3m × 1.2m*, 2 gas turbines, 16,400hp, 1 AK-630M (30mm Gatling), complement 5–6.

'Tsaplya' (Project 1206.1 Murena) is a lengthened version for riverine work, eight of which were built at Feodosiya and Khabarovsk 1984–92 (*D-453*, (completed 1985), *D-458* (completed 1986), *D-259* (completed 1987), *D-285* (completed 1988), *D-447* (completed 1989), *D-323* (completed 1990), *D-142* (completed 1991), *D-143* (1992)): 170.5t, 118ft 1in × 42ft 8in × 5ft 3in, *36.0m × 13.0m × 1.6m*, 2 gas turbines, 18,000hp, 2 AK-630 (30mm Gatlings), 2–40mm grenade launchers, crew 6. Capacity 50t (1 tank and 80 men).

'Utenok' (Project 1209 Omar): Two (*D-346*, *D-349*) built at Feodosiya 1981–82: 25/70t, 86ft 3in × 38ft 5in × 1ft 3in, *26.3m × 11.7m × 1m*, 1 MT-70 gas turbine for lift and cruise, 2 aircraft propellors, 12,000hp = 65kts, 1 twin 12.7mm MG and 1–40mm grenade launcher (BP 30), complement 7. Capacity: 60 men. One (*DKA-519*) and probably both have been discarded.

WING-IN-GROUND EFFECT CRAFT

'Lun' class (Project 902): *MD-160* (completed 1987), *S-31* (completed 1993), *Spasatel'* (completed 1995). This is the 'Caspian Sea Monster', powered by 4 or 8 turbojets (equivalent to 30,000bp or 60,000hp) on stub wings just abaft the cockpit. 400t, 241ft 2in × 144ft 4in (wingspan) × 8ft 10in, *73.5m × 44.0m × 2.7m*, capable of 300kts; range 3000km. *MD-160* is the troop-carrier version, with four engines forward. *S-31* is the missile-launcher, with three twin Moskit launchers (which can be fired at 250kts) and eight jet engines forward, and with targeting radar in the tail.

'Orlan' class (Project 904): *S-20* (*Dubl'*, completed 1972), *S-23* (completed 1975), *S-21* (completed 1978), *S-25* (completed 1979), *S-26* (completed 1980); 125t (140t full load); 190ft 3in × 103ft 4in × 3ft 3in; *58.0m × 31.5m × 1.0m*; 190kts; range 1080nm at 146kts. Can carry troops and light vehicles (payload 28t in a 82ft x 10ft 10in × 9ft 10in; *25.0m × 3.3m × 3.0m* cargo bay, with a hinged nose). Powerplants: two NK-8 turbojets (23,150lbs thrust each) and one NK-12M turboprop (14,795ehp). Armament: one twin 23mm gun mount. Lift turbojet exhausts swivel down to direct air beneath the wings at takeoff. Normal operating altitude is 5–15m, but the WIG can rise out of ground effect. The program never reached series production because of serious corrosion problems and very limited payload and range. In 1991 a commercial version, A-90-150, was offered with a normal takeoff weight of 110t (125t maximum).

SMALL SURFACE COMBATANTS

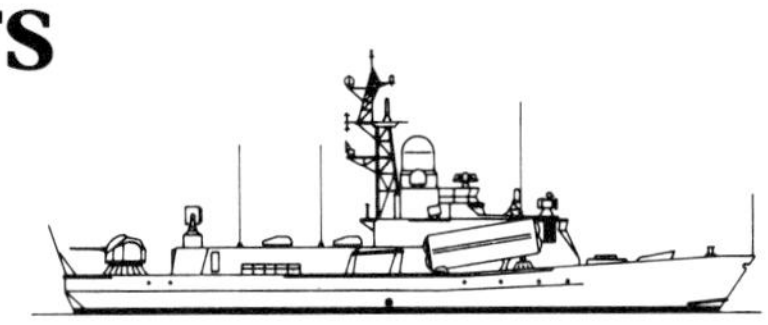

'Nanuchka I' class

'NANUCHKA' class *missile corvettes* (Project 1234 Ovod)

Displacement:	560t standard; 660t full load
Dimensions:	194ft 7in × 41ft 4in × 7ft 11in *59.3m × 12.6m × 2.4m*
Machinery:	3 paired M504 diesels, 30,000bhp = 32kts. Range 2500nm/12kts, 900nm/30.5kts
Armament:	6 SS-N-9 (2×3), 1 SA-N-4 (1×2, 20 missiles), 2–57mm (1×2), 1–76mm and 1–30mm Gatling in 'Nanuchka III')
Sensors:	Radar Band Stand, Peel Pair, Pop Group, Muff Cob (Bass Tilt in 'Nanuchka III'); ECM 2 passive arrays, 2 chaff RL (2×16)
Complement:	60

Class:

'Nanuchka I' (Project 1234 Ovod): *Briz, Buran, Burun, Burye, Vikhr, Volnya, Grad, Grom, Groza, Zarnitsa, Metel, Molniya, Musson* (lost in Pacific 1987), *Taifun, Tsiklon, Shkval, Shtorm*

'Nanuchka II' (Project 1234E) export

'Nanuchka III' (Project 1234.1 Ovod I): *Aisberg, Bora, Zarya, Zeb, Priliv, Purga, Raduga, Priboi, Meteor, Mirazh, Smerch, Tucha, Uragan, Rassvet, Veter, Otliv, Liven, Passat*

'Nanuchka IV' (Project 1234.2): *Nakat*

Zeb was renamed *Komsomolets Moldovii* on 28 April 1982. *Burye* and *Briz* were named *MRK 3* and *MRK 7* until 25 April 1970, respectively.

Builders: construction began 1967; the first 'Nanuchka I' was completed 1969. Petrovskiy (Leningrad): fourteen 'Nanuchka I' followed by three 'Nanuchka II' for India (assigned Russian names *Uragan-I* [comp 25.3.75], *Priboy-I* [comp 30.3.76], *Priliv-I* [comp 1.7.76] and presumably intended as 'Nanuchka I''; these names were later used for for 'Nanuchka IIIs') then seventeen 'Nanuchka III' (including the sole 'Nanuchka IV') for the Soviet Navy and then seven 'Nanuchka II' for Algeria and Libya.

'Nanuchka I' class 1975 MoD

Ulis (Vladivostok): Three 'Nanuchka I', four 'Nanuchka III'. Presumably one 'Nanuchka III' intended for Soviet use was completed instead as a 'Nanuchka II' for Algeria. One planned 'Nanuchka III', *Koreets*, was broken up incomplete in 1991. Known completion dates: Leningrad first-series units: *Vikhr* (31.8.77), *Zarnitsa* (5.2.73), *Zarya* (12.3.74), *Metel'* (24.4.74), *Shtorm* (6.9.75), *Raduga* (19.3.75). Second series: *Buron* (30.12.77), *Veter* (7.2.77), *Zyb* (16.2.79), *Aysberg* (24.10.80), *Tucha* (24.10.80), *Uragan (II) (15.12.83), Mirazh* (9.1.84). Vladivostok: first series: *Tsiklon* (31.12.72), *Taifon* (30.12.79), *Musson* (30.12.81). *Musson* sank after having been hit accidentally by a cruise missile during a practice firing in March 1987; 27 men were lost. *Groza*, completed before 1973, went into reserve 1980. *Burye* went into reserve 1990 and was broken up in September 1991. One other unit may have been discarded in 1989.

The largest craft the Soviets described as 'rocket-cutters' (RK), these vessels were often seen far from coastal waters, as in the Mediterranean. They carried a new anti-ship missile, SS-N-9 (probably initially P-50 and then P-120) and a significant defensive weapon, SA-N-4 on the foredeck. The ship tracks the outgoing missile using its Band Stand radar and can transmit course corrections via Fish Bowl data link. Later units have a Light Bulb data link in place of Fish Bowl. Because SS-N-9 could not be exported, the export version ('Nanuchka II') is armed with SS-N-2C and carries the required Square Tie targeting radar in its Band Stand radome; in some cases the radome also covers ESM for targeting. 'Nanuchka III' is a modified version for the Soviet Navy, carrying a 76mm gun in place of the 57mm weapon of the 'Nanuchka I'; the fire control radar is Bass Tilt rather than Muff Cob. This version also has a 30mm Gatling gun aft. The single 'Nanuchka IV' has canvas-shrouded missile launching rails rather than the usual triple launcher. Its role is unclear, and it may have been converted as a film prop. Reported armament: two sextuple SS-N-25 Uran or Nakate (Kh-59 naval version missiles.

'Nanuchka' is reportedly a poor sea boat. Its engines, M504 diesels paired (as M507s) on three shafts, are considered unreliable. Early units had paired M503s, for a total of 24,000bhp.

'Tarantul I' class
NB 1/750 scale

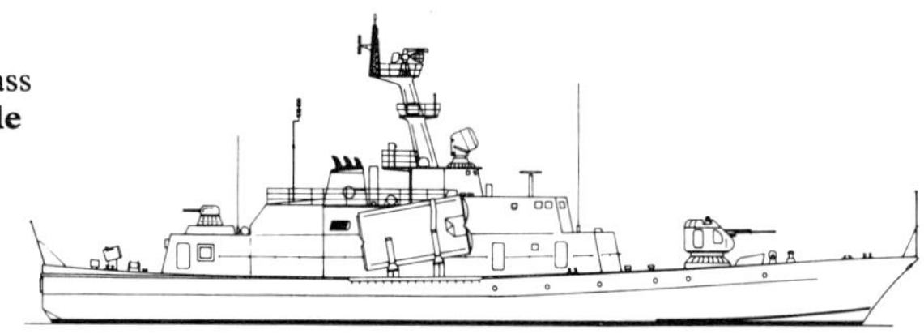

'Tarantul II' class
NB 1/750 scale

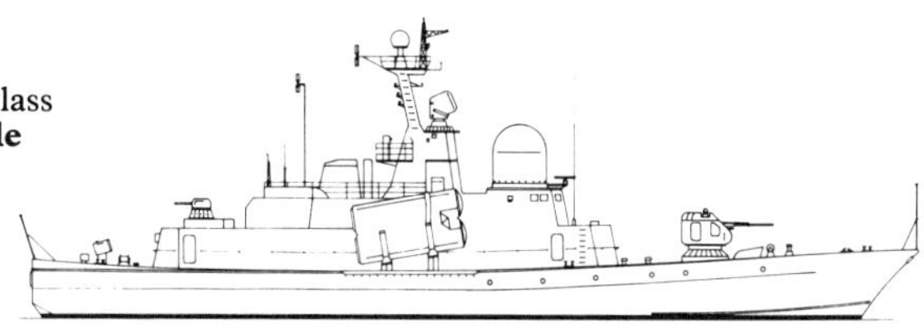

'TARANTUL' class *missile corvettes* (Project 1241.1, Molniya-1)

Displacement:	480t standard; 540t full load
Dimensions:	170ft 7in wl, 183ft 9in oa × 34ft 5in × 8ft 2in *52.0m, 56.0m × 10.5m × 2.5m*
Machinery:	see notes
Armament:	4 SS-N-2C (2×2), 1 SA-N-5 (1×4, 8 missiles), 1–76mm, 2–30mm Gatling (2×1)
Sensors:	Radar Spin Trough, Bass Tilt, 1 targeting radar; ECM 4 arrays, 2 chaff RL (2×16)
Complement:	50

Class:

R-5 (*Kaliningradskiy Komsomolets* 9.6.80–1992), *R-11, R-14, R-16, R-18, R-19, R-20, R-24, R-26, R-28, R-33, R-42, R-45, R-46, R-47, R-48, R-52, R-54* (*Krasnodarskiy Komsomolets*), *R-56, R-58, R-60, R-63* (*Kuibyshevskiy Komsomolets*), *R-66, R-69, R-71, R-74, R-75, R-76, R-79, R-83, R-85, R-86, R-88, R-98, R-99, R-100, R-103, R-104, R-109, R-113, R-115, R-141* (*Kronshdatskiy Komsomolets*), *R-142, R-143, R-144, R-145, R-158, R-160* (*Poltavskiy Komsomolets*), *R-162, R-164, R-176, R-177, R-179, R-229, R-230, R-239, R-240, R-255* (*Kirovskiy Komsomolets*), *R-261, R-271, R-297, R-298, R-334, R-442*

R stands for RKA, 'rocket cutter'.

Production: twenty-three 'Tarantul II' built for Soviet Navy 1981–86, sixteen-plus completed from 1987 on. Builders: Volodarskiy (Rybinsk: 'Tarantul I' only), Petrovskiy (Leningrad: prototypes and production boats), Srednyy Neva (Kolpino), Ulis (Vladivostock).

Transfers: India (seven), East Germany (five), Poland (seven), Yemen (two). This class is also being built under licence in India (twenty-three built or planned).

The Almaz KB, which had developed the 'Osa' and 'Komar' classes, began work on a new missile boat in 1970. Experience with the earlier boats showed a need for a more seaworthy hull, better gun and anti-aircraft armament, and a longer-ranged and more sophisticated anti-ship missile. Apparently early studies envisaged a modified 'Osa', but in 1973 a new design, to use a missile (presumably the P-80 version of SS-N-22, then being fitted to *Sovremennyy* class destroyers), was authorised. The new missile required a more advanced target designation system (*ie*, the sort of talk-back system already employed by

'Nanuchkas'). The prototype Project 1241 boat (later 1241.0) was completed in Leningrad in 1978 for 1978–80 trials. Trials soon demonstrated that the new long-range missile could not be accommodated. The later 3M80 (also designated SS-N-22 by NATO) may have been developed specifically for this class.

In the interim, boats armed with SS-N-2 missiles were built both for the Soviet Navy and for export. The export version was 'Tarantul I' (Project 1241.1, also known as 1241PE). Soviet boats ('Tarantul II': 1241.1M) had the Band Stand and Light Bulb associated with talk-back missile systems; apparently they are armed with modified Styx, P-27. The Soviets retained two 'Tarantul I' for training. 'Tarantul III' (1241.1MR, also called 1241RE, Molniya M) is armed with four SS-N-22 (3M80).

These boats are driven by a two-shaft COGAG powerplant using two 12,000shp NK-12 aircraft gas turbines, with two 4000shp cruise turbines. Main engine exhaust ports in the transom presumably add some thrust. The two shafts are physically connected by a cross-shaft, so the boat can loiter on one engine, to save fuel. Endurance is, however, limited because the hull is optimised for very high speed (it is rated at 43kts, and has exceeded that figure on trials). The hull form also limits manoeuverability. Rated cruising range is 1600nm at 14kts (10 days' stores endurance, so the boat can loiter). Two boats were experimentally equipped with an alternative cruise powerplant, a pair of 5000hp M-510 engines (as on missile boats). On both diesels and gas turbines the boats could achieve 41kts, but the loss of speed was more than made up for by an increase in cruising range to 2400nm. This powerplant was adopted for 'Tarantul III'.

As of 1994, two further versions, 12417 (1241.7) and 12418 (1241.8) were being offered for export. In 12417 ('Tarantul 4'), the AK-630 Gatlings are replaced by the new 'Kashtan' (CADS-N-1) defensive system. The ship's main radar is replaced by the associated Cross Round target acquisition/search set. A production 'Tarantul' (R-71) was refitted to this standard. In the alternative 12418, the missiles are replaced by four quadruple SS-N-25 (Kh-35) canister launchers. This version is essentially a modified 'Tarantul I' with a Cross Round main radar. The missile offers considerably greater range than the earlier Styx.

Given Almaz' experience with earlier multi-purpose small combatant designs (Project 200 and Projects 205/206), this hull was probably designed from the outset for possible use for ASW, as 'Pauk' below.

'Pauk' class
NB 1/750 scale

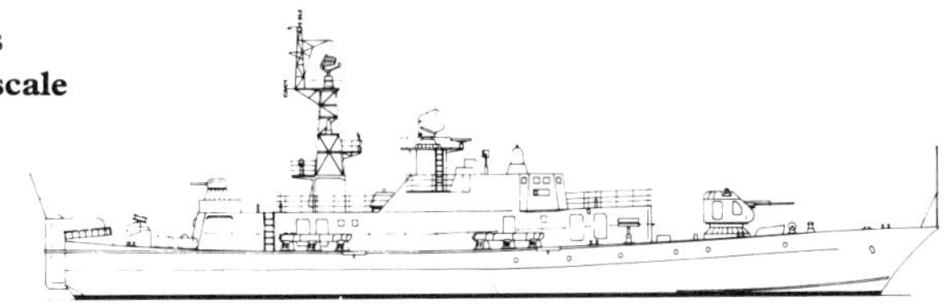

'PAUK' class *corvettes* (Project 1241.2, Molniya-2)

Displacement:	500t standard; 580t full load
Dimensions:	190ft 3in × 30ft 10in × 8ft 2in *58.0m × 9.4m × 2.5m*
Machinery:	2-shaft diesel, 20,000bhp = 28–34kts
Armament:	1 SA-N-5 (1×4, 8 missiles), 1–76mm, 1–30mm Gatling, 2 RBU-1200 ASW RL (2×25), 2 DC racks (12 DC), 4–406mm TT (4×1)
Sensors:	Radar Spin Trough, Bass Tilt, new air-surface search; sonar M/F hull and dipping; ECM passive arrays, 2 chaff RL (2×16)
Complement:	40

Class:
45 units: *G Gnatenko*, *G Kuropyatnikov*, *MPK-11*, *MPK-12*, *MPK-15*, *MPK-17*, *MPK-19*, *MPK-50*, *MPK-60* (*Komsomolets Bashkirii* until 20.4.92), *MPK-61*, *MPK-62*, *MPK-68*, *MPK-72*, *MPK-75*, *MPK-76*, *MPK-88*, *MPK-90*, *MPK-93*, *MPK-95K*, *MPK-100*, *MPK-106*, *MPK-116*, *MPK-124*, *MPK-144*, *MPK-146*, *MPK-146*, *MPK-157*, *MPK-159*, *MPK-222*, *MPK-291*, *MPK-345*, *PSKR-801–815*. PSK units are for KGB Maritime Border Guards; MPKs are small ASW ships. All built at Yaraslavl and Ulis.

This ASW patrol boat is built on a 'Tarantul' hull built 1977–82. The MF sonar suite (hull and stern dipper) is designated Bronza. A helicopter dipper (HF) was later added in a cabinet on the starboard side near the stern; it might be more useful in shallow water. A variant seen in 1982 in the Baltic had its pilot house a half-deck higher. Machinery: 2 paired M504 (M517) diesels.

Transfers: Libya, Cuba (one in 1990), Bulgaria (MPK-146 in 1989, MPK-124 in 1990), India (four). The export version (1124PE for Cuba and India) has Cross Round radar atop the mast and two 53cm rather than four 40cm TT. One unit, possibly intended for Cuba, was retained by Russia for training. Reportedly India planned license production, but as of 1995 none had been seen.

'SVETLYAK' class *ASW corvettes* (Project 1140.0)

Displacement:	500t
Dimensions:	164ft 1in × 30ft 2in × 8ft 2in *50m × 9.2m × 2.5m*
Machinery:	3 M-504 diesels, 15,000hp = 33kts
Armament:	1–76mm AK-176, 1–30mm AK-630 Gatling, 2–406mm TT (2×1)
Sensors:	?
Complement:	38

Class:
AK-232, *AK-233*, *AK-235–267*

Thirty-five (Project 10410) built, beginning 1989 at Leningrad and 1990 at Vladivostok (Ulis). These were KGB Maritime Border Guards craft to replace the existing Project 205Ps. For the Black Sea and Pacific Fleets and the Caspian Flotilla. Two derivatives have been offered for export: Project 10411 (missile boat) and Project 10412 (gunboat). They are, in effect, successors to the various versions of Project 205, particularly 'Stenka' (Project 205P). Project 10412 is armed with an AK-176M forward and with an AK-630M Gatling gun aft. Project 10411 is armed with eight Kh-35 (SS-N-25) anti-ship missiles, in two quadruple canister launchers, in addition to the guns; the targeting radar is Plank Shave (Garpun-BAL).

'Poti' class

'POTI' class *corvettes* (Project 204)

Displacement:	500t standard; 580t full load
Dimensions:	194ft 11in × 25ft 11in × 6ft 7in *59.4m × 7.9m × 2.0m*
Machinery:	2-shaft CODAG: 2 gas turbines, 30,000shp plus 2 M503A diesels, 8000bhp = 38kts. Range 4500nm at 10kts, 520nm at 37kts
Armament:	2–57mm (1×2), 2 RBU-6000 ASW RL, 4–406mm TT (4×1)
Sensors:	Radar Strut Curve, Muff Cob, Spin Trough; sonar H/F Herkules, dipping; ECM 2 Watch Dog
Complement:	80

Class:
MPK-1, *MPK-10–12*, *MPK-14–20*, *MPK-38*, *MPK-44*, *MPK-50*, *MPK-58–62*, *MPK-68–77*, *MPK-85*, *MPK-88–100*, *MPK-103*, *MPK-106*, *MPK-107*, *MPK-109*, *MPK-111*, *MPK-112*, *MPK-114*, *MPK-124*, *MPK-125*, *MPK-128*, *MPK-134*, *MPK-136*, *MPK-148*, *MPK-156*, *MPK-157*, *MPK-158*, *MPK-159*, *MPK-160*, *MPK-215*, *MPK-375*, *MPK-431*, *MPK-435*, *MPK-440*, *MPK-446*, *MPK-449*, *MPK-481*. *MPK-20* was later renamed *SM-442*, and *MPK-107* became *SM-450*. Total of sixty-nine built 1961–68 at Kerch (twenty-four), Zelenodolsk (thirty-two), and Khabarovsk (eight in 1964–68). Transfers: Bulgaria (three, 1973), Romania (three, 1967).

These were the first large Soviet gas turbine warships, using a new M-2 engine to achieve 38kts; they were also the fastest ASW ships the Soviets ever built. The two propellors are mounted in thrust tubes the length of the poop, which contains the two turbines; the latter exhaust through ports above them in the transom, and also drive air compressors whose output is exhausted into the propellor tubes, to add thrust. This is similar to the 'Mirka' plant.

The first units carried the earlier open twin 57mm mount and RBU-2500s; most units had the enclosed twin mount and RBU-6000s, plus two or four single 40cm TT. The ships transferred to Romania had the original battery, with triple 53cm TT rather than 40cm.

'DERGACH' class *missile corvette* (Project 1239 Sivuch)

Displacement:	1000t standard; 1260t full load
Dimensions:	261ft 10in × 41ft 8in (also given as 211ft 7in × 55ft 9in) × 5ft 7in *79.8m × 12.7m (64.5m × 17m) × 1.7m*
Machinery:	2 20,000shp gas turbines plus 6 5000hp diesels, 70,000shp = 50kts (27–30kts on diesels)
Armament:	8 SS-N-22 (2×4), 2 SA-N-4 (1×2), 1–76mm AK-176, 2–30mm AK-630 Gatling
Sensors:	Radar Cross Round, Band Stand/Light Bulb, Bass Tilt, Pop Group
Complement:	35

Class:
Bora (ex-*MPK-27*, 18.3.92), *MPK-17*, completed Zelenodolsk 1987 and 1991. One other planned but not built.

Reportedly this class was conceived as a coastal forces command type, to lead groups of 'Tarantuls'. The prototype reportedly suffered from heavy vibration. Furthermore, she had to stop in order to switch from diesels to gas turbines. At least five test platforms (NATO Kerch-1, -2, etc) preceded this type. No details are available.

'P 2' class *fast attack craft* (Project 200) and 'MO-V' class *sub-chasers* (Project OD-200)

Displacement: 48.2t
Dimensions: 77ft 9in × 15ft 5in × 8ft 2in
23.7m × 4.7m × 2.5m
Machinery: 2 gasoline engines, 2500hp = 38kts. Range 600nm at 25kts, 100nm at 8kts
Armament: 2–533mm TT (2×1), 4–14.5mm MG (2×2)
Complement: ?

Class:
Nineteen TM-200 and TM-200bis (steel, with gasoline engines) laid down 1943 at Rybinsk, in service 1946: *TK-455–459, TK-708–711, TK-741–750*

167 Project 200bis (TD-200/TD-200bis) in service 1947–52, built at Leningrad: *TK-16–20, TK-31, TK-32, TK-36, TK-37, TK-40, TK-42, TK-47, TK-49–54, TK-74–78, TK-86, TK-91, TK-92, TK-96, TK-121, TK-126, TK-134, TK-149, TK-169, TK-170, TK-180–187, TK-202–220, TK-255–266, TK-421–449, TK-701–720, TK-839–845, TK-890–939, TK-1090–1106*

Sixty-three OD-200 (NATO 'MO-V'): *MO-269–300, MO-310, MO-471–474, MO-641–663, MO-691, MO-692, MO-694*

This design was conceived by a new wartime KB formed at Sosnovka under NN Isanin, as a common hull and powerplant combination which could be used either as a small torpedo-boat (MTKA) or as a sub-chaser. TD-200 was a three-shaft diesel version (1000 HPM-50s) with a wooden hull, designed in Leningrad by a subsidiary of TsKB-32. From 1950, forty-six TD-200 (probably TD-200B) received radar (Zarnitsa), the first for a Soviet MTB, and IFF. The sub-chaser version, OD-200, was derived from TD-200. The hull form was similar to that of a US Higgins boat. Large numbers were reportedly exported to China. This class was originally designated the '603 Class' by the US Navy and then 'P-2'.

'P 4' class *fast attack craft (torpedo)* (Project 123bis, M-123bis, 123K Komsomolets)

Displacement: 22t standard; 25t full load
Dimensions: 72ft 2in × 15ft 5in × 4ft 11in
22.0m × 4.7m × 1.5m
Machinery: 2 M503 diesels, 2400bhp = 42kts. Range 400nm at 13kts
Armament: 2–457mm TT, 2–12.7mm or 14.5mm MG (1×2), 4–8 DC
Sensors: Radar Skin Head
Complement: 12

Class:
TK-1–6, TK-8–12, TK-21–30, TK-46, TK-95, TK-188–191, TK-271–301, TK-317–320, TK-327–332, TK-337–350, TK-352–380, TK-386–420, TK-460–471, TK-482–512, TK-551–558, TK-574, TK-575, TK-597, TK-599, TK-601–609, TK-637–660, TK-677–700*, TK-721–740, TK-751–771*, TK-809–838, TK-940–957, TK-972, TK-980–995* (349 units)

TK-607 was named *Komsomolets Kazakhstana*; *TK-608*, was *Pioneer Kazakhstana*. Asterisks indicate M-123bis version.

Eighty-eight Project 123B built at Tyumen 1946–48, fifty built at Feodsosiya 1949–51, then another 211 Project 123K at Feodosiya.

This small MTB (MTKA) was designed by a new KB formed during the war in Tyumen. After developing plans for late-production G-5 class MTBs (Project 116 series XII and XIII, with US-supplied Packard engines and ahead-firing torpedoes rather than the earlier aft-firing troughs), it designed a new boat. Project 123K had radar. Eight of this version were experimentally fitted with shallow hydroplanes. The last Soviet stepped-hull planing craft, these aluminium boats are direct descendants of prewar Soviet MTBs initially based on seaplane practice. The most striking visual feature is the long foredeck, the cockpit being almost amidships.

This design typified the small MTB (MTKA), one of the two types selected for production under the 1945–55 plan. The other was a large MTB, BTKA. The MTKA was expected to be small enough to be transported on flatcars, operating near a base (range 500nm, stores endurance 1–1.5 days, two 45cm TT). The large MTB would need better seakeeping qualities and greater internal hull volume (hence it had to have a conventional rather than a planing hull); it nedded a range of 1000nm and an endurance of 5–6 days. It would be armed with a pair of 53cm TT. Both types were to have two or three anti-aircraft mounts (twin turrets with 12.7mm or 14.5mm machine guns). For ASW, some were to be fitted with two depth charge racks, and some could carry mines as alternatives to torpedoes. Project 123 turned out to be the last of the small MTB series.

Transfers: Albania (six in 1956, plus six from China 1965), Bangladesh (four from China 1983), Benin (two in 1979 from North Korea), Bulgaria (eight in 1956), China (see below), Cuba (twelve in 1962–64), Cyprus (six in 1965–65), Egypt (four from Syria, 1970), North Korea (twelve in 1952–53), North Vietnam, Romania (twelve), Somalia (four in 1972), Syria (seventeen in 1958–60), Tanzania (four in 1972–73), North Yemen (one in 1960, three in 1970–71, South Yemen (four in 1978). It is not certain to what extent the seventy in Chinese service were home-built. At least fifteen were built in North Korea ('Iwon' class). Reportedly all were stricken from Soviet service by about 1975, although they were still carried in Western naval handbooks after that date.

'Komar' class
NB 1/750 scale

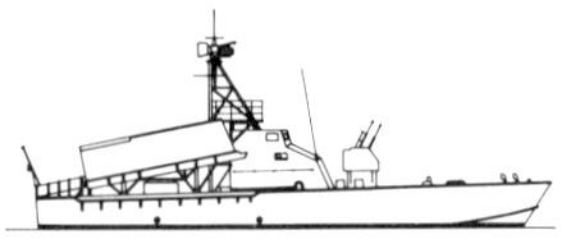

'P/6/8/10' class *fast attack craft (torpedo)* (Project 183): MO-VI class *sub-chasers* Project 199): 'KOMAR' class *fast attack craft (missile)* (Project 183R)

Displacement: 61.5t normal; 66.5t full load
Dimensions: 83ft 4in × 20ft 6in × 4ft 1in
25.4m × 6.24m × 1.24m
Machinery: 4 shaft M-50F 4-stroke diesels (M-50F-1 and M-50FTK), 4800bhp = 44kts. Range 600nm at 32kts, 1000nm at 14kts
Armament: 2–533mm TT (2×1), 4–25mm (2×2, 2M-3 and later 2M-3M mounts)
Complement: ?
NB Data are for a Project 183 MTB

Class:
'P-6/8/10' class (622 units): *T-61–67, T-69, T-72, T-112–114, T-151–161, T-172–175, T-192–199, T-221–254, T-266–271, T-701–749, T-851–889, T-901–939, T-962–981, T-1000–1316, T-1330–1450.* All initially designated TK, with the same numbers. Transfers: China (twelve, plus eighty built under license), Cuba (1962: *TK-61, TK-1331, TK-1341, TK-1346, TK-1347, TK-1349, TK-1350, TK-1384, TK-1385, TK-1432, TK-1433*), Egypt (twenty-four in 1956 and 1960), Guinea (four), East Germany (thirteen on 5.11.57, eighteen in 1958–59), Indonesia (eight in 1961–62, including *TK-1409, TK-1416, TK-1430, TK-1422, TK-1429, TK-1431*), Iraq (two in 1959, four in Nov 1960 and six in Jan 1961), North Korea (ten), Poland (twenty in 1957–58), Somalia (four), Yemen (two). This total includes Project 183T and Project 183Ts. As targets, the sixty Project 183Ts units were: *KTs-4–9, KTs-23–25, KTs-29–31, KTs-37–39, KTs-44–48, KTs-50–52, KTs-54–61, KTs-74–103.* One of the two boats converted in 1957 to test the P-15 missile (Styx) was *T-69*.

Forty units were converted into naval yachts about 1964, including *Albatros, Burevestnik, Sokol, Taifun, Tsiklon, Chaika* (68t, 39kts)

'MO-VI' class (60 units): *P-301–351, P-544–556* (with Tamir-10 sonar)

'Komar' class: *R-63, R-64, R-66, R-69, R-70, R-71, R-72, R-81, R-88, R-90, R-94–96, R-98, R-101–105, R-115–120, R-124–140, R-148–159, R-165–168, R-170, R-171, R-174, R-187–191, R-197, R-198, R-200, R-202, R-203, R-205, R-207–209, R-211, R-213–217, R-231–238, R-261–282*

This was the large MTB (BTKA) designed for construction under the 1946–55 Plan. It was designed by a special NKVD bureau, SKB-5, which later became TsKB-5 and then the current Almaz. The lead boat was completed in 1949, and a total of 674 (including modified versions) was built in 1952–60. Project 183 was the first Soviet MTB designed with radar (Zarnitsa). It had a large stepless semi-planning wooden hull.

Variants:
Project 183T: first Soviet gas-turbine boat, tried 1953 with 400hp M-1 gas turbine plus diesels; achieved 50kts. This was probably the NATO 'P-8'.

Project 183TK: follow-on to 183T: twenty-five boats built at Leningrad with 4000hp M-1 gas turbines to boost their diesel powerplants. At 83t, they could achieve 52kts. However, struts for their five propellers created considerable drag. Two units of this type had a shallow bow plane, but these stainless steel planes proved not sturdy enough. This was probably 'P-10'. The hydrofoil version was tentatively designated 'P-12'.

Project 183Ts: About sixty radio-controlled targets.

Project 183A: one boat covered in 'arctilite', a plastic preservative like that used on US boats supplied under Lend-Lease. The project was abandoned when the material was taken out of production.

Project 183U: enlarged design (1958), an attempt to modernise this type of boat with two new fast-running diesels. It failed, and was not recommended for production. Presumably it was an alternative to the new round-chine Project 205/206.

Project 183R: Komar missile boat (see below)

Project 199: A small interceptor for the KGB Maritime Border Troops, with a Tamir sonar and an ASW suite (it was to work in conjunction with shore sonar sites, such as harbour defence sonars). Although generally identified as the 'Zhuk' patrol boat, this was almost certainly the boat NATO called 'MO-VI'. A total of fifty-two were built.

There was also a Project 183E, details of which are not available.

While Project 183 was in production, TsKB-19 offered two other designs, Project 184 and 125. Two of each were built at Feodosiya (see below for Project 125 as used by the KBG Maritime Border Guards). Project 184 (1956) was a lightweight MTB with a stepless aluminium hull: 34t, 70ft 10in × 16ft 5in × 6ft 11in (*21.6m × 5.0m × 2.1m*), max speed about 46kts, cruise range about 300nm at 45kts, 500nm at 33kts, stores endurance 2 days. Armament: 2–533mm TT and 4–14.5mm MG (2×2). It did not enter production.

Transfers: Algeria (ten in 1963–68), China (few), Cuba (twelve in 1962), East Germany (twenty-seven in 1957–60), Egypt (twenty in 1956–58), Guinea (four in 1965–67), Guinea-Bissau (four in 1975–76), Indonesia (fourteen in 1961–62), Iraq (ten in 1960–62), North Korea (forty-five in 1960s), North Vietnam (three in 1967), Poland (twenty in 1957–58), Somalia (four in 1968), South Yemen (two in 1973), Tanzania (three from East Germany 1974–75). About eighty were built in China (1956–66), and at least another twenty in North Korea ('Sinpo' class). 'MO-VI' transfers: Bulgaria (ten), Guinea (four reported 1972–75), Nigeria (three in 1967), North Korea (twenty in 1957–60).

'Komar' (Project 183R), the world's first missile attack boat, seens to have begun as the test platform for the new P-15 (Styx) missile. Missile development was completed in 1956. Work on a parallel missile boat, Project 205 'Osa', was already underway. It appears that in 1956 or 1957 it was decided that the test platforms for the new SS-N-1 (KSShch) and SS-N-2 (P-15) missiles would be developed into interim tactical platforms. The design project for the adapted Project 183 was formally approved in August 1957. By that time work was already proceeding on a boat to test SS-N-2. Because no one really knew what the effect of the missile's blast (both heat and pressure pulse) would be, the first trial shots in the Black Sea were fired remotely from another boat.

The two single fixed torpedo tubes were replaced by missile launchers, a new surface search radar (Square Tie) was installed, and the boat had only one of the usual two twin 25mm 2M-3M guns. Displacement increased slightly (to 66.5t/77.5t); speed was 38kts (26kts cruise). Missiles could be fired in Sea State 4. In all, 112 were built (not converted from existing hulls) between December 1959 and the end of 1965.

Transfers: Algeria (six in 1967), China (eight in 1960–61), Cuba (eighteen in 1962–66), Egypt (seven in 1962–67), Indonesia (twelve in 1961–65), Iraq (three in 1972), North Korea (ten), Syria (nine in 1963–66), Vietnam (four in 1972). About forty were built in China, plus about 80–85 'Hoku' (Project EM1A) class, a steel-hulled improved version (which is also offered for export carrying the C801 missile; six were transferred to Egypt (1984) and four to Pakistan (two in 1981, two in 1982). North Korea built eight of her own steel-hulled version ('So Hung' class, 1980–81). Egypt built six derivatives of the 'Komar' as her *6 October* class, equipped with Western weapons, 1980–81.

PROJECT 184 *torpedo boats*

Displacement:	31.5t standard; 34.1t full load
Dimensions:	70ft 10in × 16ft 5in × 6ft 11in *21.6m × 5.0m × 2.1m*
Machinery:	3 shafts, 3 M-50F diesels, 3600bhp = 46kts. Range 500nm at 33kts, 300nm at 45kts
Armament:	2–533mm TT, 2–14.7mm (1×2)
Complement:	10 (1 officer)

Class (completed):
TK-327 (27.8.56), *TK-328*

No NATO code name assigned.

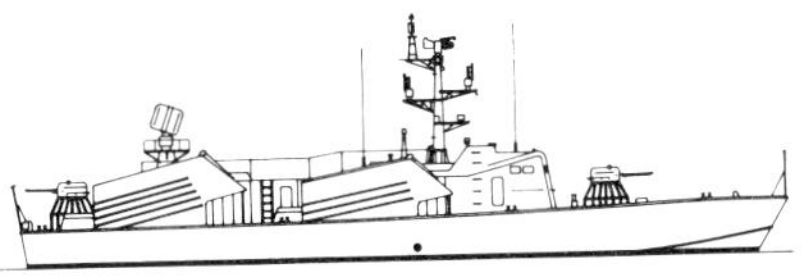
'Osa I' class
NB 1/750 scale

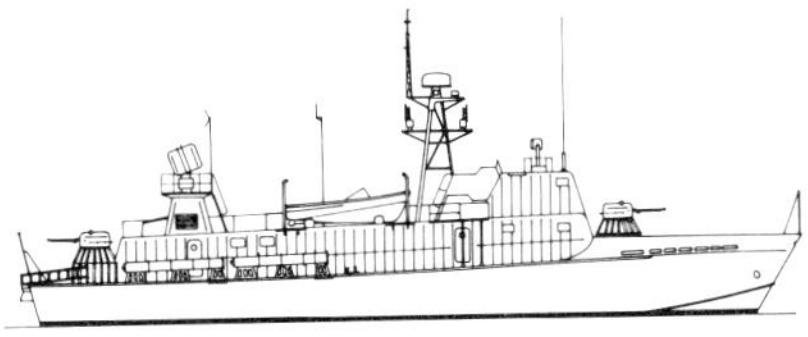
'Osa II' class
NB 1/750 scale

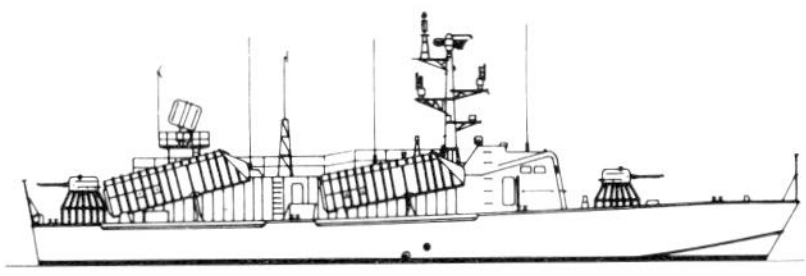
'Stenka' class
NB 1/750 scale

'OSA' class *fast attack craft (missile)* (Project 205, 205M): 'STENKA' class *ASW craft* (Project 205M, 205P Tarantul)

Displacement:	172t standard; 209t full load (184t/229t with M-504B diesel)
Dimensions:	123ft × 24ft 11in × 12ft 6in *37.5m × 7.64m × 3.8m*
Machinery:	3 shaft M-503A diesels, 12,500bhp = 38.5kts (40kts with M-504B diesel)
Armament:	4 SS-N-2 SSM (4×1), 4–30mm (2×2)
Sensors	?
Complement:	26 (28 in M-504B version

Class:
'Osa I/II' class (R for RKA, 'rocket cutters'): *R-1–5*, *R-6* (*Brestskiy Komsomolets*), *R-7–10*, *R-12–14*, *R-17–23*, *R-24* (*Tambovskiy Komsomolets*), *R-25–32*, *R-34–41*, *R-43*, *R-49–51*, *R-53*, *R-55* (*Komsomolets Tatariy*), *R-57*, *R-59*, *R-65*, *R-67*, *R-68*, *R-70–73*, *R-77*, *R-78*, *R-79* (*Kirovskiy Komsomolets*), *R-80–84*, *R-87* (*Amurskiy Komsomolets*), *R-89*, *R-91–93*, *R-97*, *R-104–114*, *R-121–123*, *R-141* (*Kronshtadskiy Komsomolets*), *R-146*, *R-147*, *R-161*, *R-163*, *R-166*, *R-167*, *R-169*, *R-172*, *R-173*, *R-175* (*Michurinskiy Komsomolets*), *R-176*, *R-180*, *R-183*, *R-189*, *R-193*, *R-194*, *R 196* (completed 24.10.67), *R-199* (completed 26.12.67), *R-201*, *R-204*, *R-206*, *R-210* (completed 19.11.68), *R-212*, *R-218*, *R-220–230*, *R-288–333*, *R-460–474*, *R-494–537*, *R-540–552*

'Stenka' class: *AK-234*, *AK-301–350*, *AK-355–370*, *AK-374–376*, *AK-378*, *AK-379*, *AK-381*, *Ivan Folubets*, *Nikolai Kaplunov*, *P-616*, *P-623*, *P-629*, *P-630*, *P-631*, *P-635–638*, *P-641–645*, *P-648–652*, *P-657*, *P-659*, *P-665*, *P-695*, *P-702*, *P-705*, *P-709*, *P-720*, *P-722* (138 units)

'Komar' class 1962 CMP

'Stenka' class Apr 1993 L & L van Ginderen

Transfers: *AK-234* and *AK-374* to Azerbaijan, *P-623*, *P-629*, *P-630*, *P-635–637*, *P-642*, *P-722* to Ukraine after the breakup of the Soviet Union. Built at Leningrad and Vladivostok.

Project 205 Tsunami was the fast missile boat initially planned to fire the P-15 missile. Early in the project stage TsKB-5 (now Almaz), which was also responsible for a new torpedo boat (Project 206), decided to use the same round-chine hull for both. Work on an extended-range version of P-15 (P-15M, range 80km as against 40km) was conducted in parallel with the Project 205 design, although it appears that early boats had the short-range missile. These craft were designed to operate under nuclear/chemical conditions, with fully-enclosed guns. 'Osa I' (Project 205) had hangar-type launchers, but in 'Osa II' (Project 205M) they were replaced by containers.

The new hull (round-bilge with hard-chine afterbody) was intended to operate in Sea State 4 without any speed restriction, and in Sea State 5 at 30kts. New M-503 (4000bhp) and M-504 (5000bhp) high-speed diesels were adopted (they were later replaced by improved M-503A and M-504B versions). M-503A could operate at 4000bhp (2200rpm) for 10 per cent of engine time between overhaul (TBO); at 3600bhp (1700rpm) for 15 per cent of TBO; at a continuous rating of 3300bhp (1500rpm) for an unlimited time. TBO is normally 600hr, but can be extended for another 150hr if necessary. These engines are described as light but also noisy and hot. They required an hour of warm-up before running, and breakdowns (*eg*, in rubber joints of the fuel system) were relatively common above 1500rpm.

Over 400 boats were built (excluding 'Stenka'); the lead boat was commissioned in 1960. Modified versions included:

205U: Project 205 with containerised P-15U missile (with folding wings and improved guidance: NATO SS-N-2B): 'Osa II'

205M: 'Osa II' with new diesel (M-504) and P-15M missile (NATO SS-N-2C)

205MA: 'Slepen' trials boat; see below

205MR: 'Osa II' with new missile, probably modifed P-15M

205K: Project 205 rebuilt to 205M standard, possibly also designated 205A

205P: 'Stenka', see below

205Ch: 400Hz AC power rather than the 220v DC standard in the rest of the class; this boat has P-15M missiles

205.4: Modified export version with twelve SS-N-25 (Kh-35)

About 175 'Osa I' (1959–66) and 114 'Osa II' (1966–70) were built in the Soviet Union, together with about 125 'Osa I' in China (*Huangfen* class: four to North Korea 1982 and four to Pakistan 1984) and at least four in North Korea (*Soju* class, 1981–85).

Transfers: 'Osa I' Algeria (three in 1967), Benin (two in 1979), Bulgaria (three in 1970–71), China (four 1960s), Egypt (ten in 1966), East Germany (fifteen in 1966), India (eight in 1971), Iraq (two in 1971–2, two in 1973, two in 1974), North Korea (twelve in 1968–73), Poland (thirteen in 1966–67), Romania (six in 1964), Syria (eight in 1966), Yugoslavia (ten in 1965–69).

'Osa II': Algeria (eight in 1976–78, one in 1980), Angola (six in 1982–83),

Bulgaria (six in 1978, 1982, 1984), Cuba (two in 1977, three in 1978, two in 1979, two in 1981, four in 1982), Ethiopia (one in 1978, one in 1980, one in 1981, one in 1982), Finland (four in 1975), India (eight in 1976), Iraq (two in 1974, two in 1975, two in 1976, two in 1977, two in 1984), Libya (one in 1976, four in 1977, one in 1978, three in 1979, three in 1980), Somalia (two in 1975), Syria (two in 1978, four in 1979, two in 1982, two in 1984, two in 1985), Vietnam (two in 1979, four in 1980, two in 1981), South Yemen (eight in 1979–83), North Yemen (two in 1982).

As of 1982, *c* seventy 'Osa I' and *c* fifty 'Osa II' remained in Soviet hands. These figures had fallen to twenty-two 'Osa I' and twenty-four 'Osa II' by 1990. Recent disposals: nine Osa in 1987–88, twenty-five in 1990–91.

Project 205P Tarantul (NATO 'Stenka') was a successor to Project 199 (NATO 'MO-VI'). Boats are now designated gunboats (AK), but are primarily ASW platforms. They were originally designated PSKR, Pogranichniy Storozhevoi Korabl. A total of 114 were built, beginning with five in 1967, at Leningrad (Petrovskiy) and Vladivostok. Known current designations: *AK-312, AK-314, AK-318, AK-330, AK-332, AK-333, AK-342.* These craft are powered by both M-503A and M-504 diesels. The dipping sonar is that used on a 'Hormone-A' helicopter, and the armament is 4 single 40cm TT, 2 DC racks (12 charges), and 4–30mm guns (2 AK-230 mounts). Transfers: four to Cuba (1985), three to Kampuchea (1 October 1985, others by 1987). Some boats lack torpedo tubes (and, probably, sonar) and carry a motor launch instead; some have a Peel Cone radar in place of the usual Pot Drum. Transferred units lack torpedo tubes or sonar; they are coastal guardships.

Project 205PE (NATO 'Slepen'), *AK-225*, was a single hydrofoil trials craft (powered by M504 diesels) built in 1970 by Petrovskiy (Leningrad), it was armed with a twin 57mm gun forward (for 'Grisha III' trials). A second unit (probably Project 205MA, named *Ivan Evteyev*) was built at Vladivostok in 1975, armed with a 76mm gun for tests for the 'Matka' and 'Nanuckha' classes. Each boat retained her AK-230 aft.

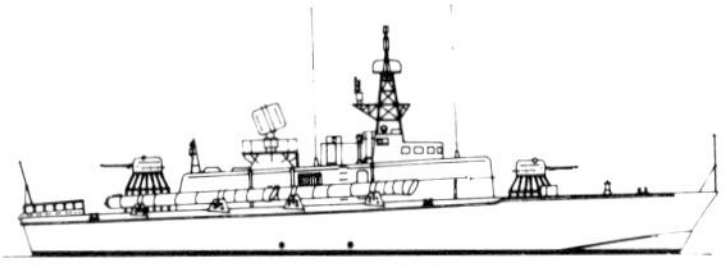

'Shershen' class
NB 1/750 scale

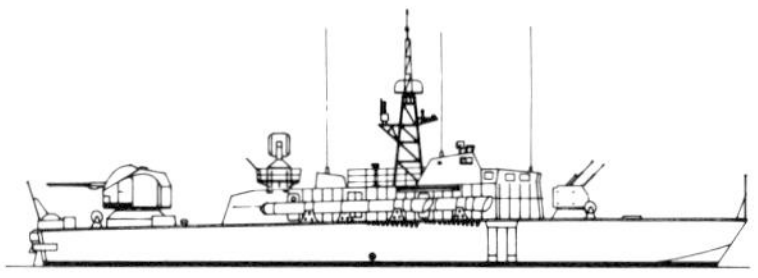

'Turya' class
NB 1/750 scale

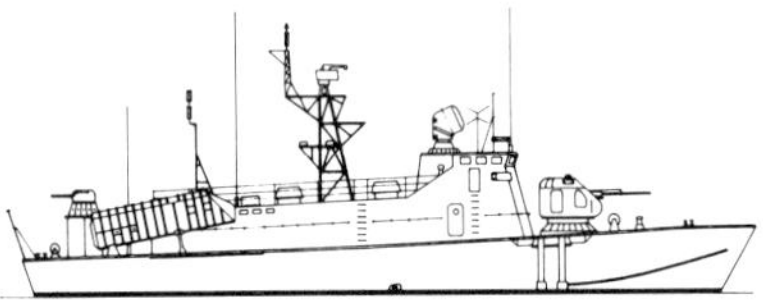

'Matka' class
NB 1/750 scale

'SHERSEN' class *fast attack craft (torpedo)* (Project 206): 'MOL' class *fast attack craft (torpedo)* (Project 206E): 'BOGOMOL' class *gunboats* (Project 206EP): 'TURYA' class *torpedo hydrofoils* (Project 206M Shtorm): 'MATKA' class *missile hydrofoils* (Project 206MR)

Displacement: 220t normal; 250t full load
Dimensions: 129ft 9in × 24ft 9in (41ft over foils) × 6ft 7in (13ft 1in over foils)
39.6m × 7.6m (12.5m) × 1.8m (4m)
Machinery: 3 shaft M503 diesels (later M503A or M-504), 15,000bhp = 40+kts. Range 600nm at 37kts, 1450nm at 14kts
Armament: 2–533mm TT (2×1), 2–57mm (1×2, AK-275 mount), 2–25mm (1×2, 2M-3M mount)
Complement: 30
NB Data are for Project 206M

Class:
Project 206 ('Shershen'): *T-3, T-7, T-12–14, T-17–19, T-21, T-22, T-25–31, T-33, T-34, T-38, T-39, T-57, T-83–86, T-97–99, T-105, T-113–115, T-118, T-125, T-129, T-131, T-132, T-162, T-165, T-168, T-169, T-171, T-172, T-176, T-179, T-181, T-182, T-184, T-185, T-188–191, T-200, T-213, T-217, T-219, T-220, T-295, T-296, T-337, T-359, T-360, T-363, T-373, T-526–542* (87 units). Built at Zelenodolsk, Yaroslavl (15 units, 1960–70). Transfers: Angola (four in 1977–78), Bulgaria (six in 1970–71), Cape Verde Islands (two in 1979), Congo (three, also reported as one in 1979), Egypt (six in 1967–68), East Germany (eighteen in 1968–76), Guinea (two, also reported as three in 1978–79), Guinea-Bissau (two, also reported as one in 1978), North Korea (three, also reported as four in 1970–71), Vietnam (four in 1979, four in 1980, two in 1981, four in 1983), Yugoslavia (two plus thirteen transferred from Soviet service; also reported as four in 1965 plus ten built under license).

Project 206M Shtorm ('Turya'): *T-68, T-70, T-72, T-73, T-75, T-88* (completed 1974), *T-96* (completed Dec 1972), *T-100* (completed 1972), *T-116, T-117, T-119, T-126* (completed Aug 1971), *T-140* (completed 1971), *T-150* (completed 1973), *T-163, T-164, T-166, T-177, T-178, T-183, T-201, T-252* (completed 1975), *T-253* (completed 1975), *T-272* (completed 1976), *T-273* (completed 1976), four others (29 units). Transfers: Cuba (two each in Feb 1979, Feb 1980, Feb 1981, Jan 1983, one in Nov 1983), Ethiopia (one each in 1985 and in Mar 1986), Kampuchea (one each in 1984 and 1985), Seychelles (one in Aug 1986), Vietnam (three in 1984, two in 1986). *T-72* and *T-117* to Latvia 1993.

Project 206MR Vikhr' ('Matka'): *R-15, R-25, R-27, R-30, R-44, R-50, R-85, R-161, R-166, R-172, R-174, R-180, R-185, R-221, R-240, R-251, R-254, R-260, R-262,* (*Tula*), *R-265* (*Komsomolets Tatariy*)

TsKB 5 began Project 206 in 1956 as a round-bilge successor to Project 183. This boat would have been somewhat smaller than an 'Osa' (Project 205), about 160t, 110ft 3in × 22ft 1in × 12ft 4in, *33.6m × 6.74m × 3.75m*, to make 45kts (cruise at 35kts, range 1000nm at this speed). Because this was slightly smaller than (but comparable to) an 'Osa', the decision was taken to use the hull of the larger boat. The 4–533mm TT were radar-controlled by a torpedo fire control system, OTA-53-206. The guns were remote-controlled by system, 2 twin AK-230 (remote controlled by MP-104 FCS). The lead boat was built at Yaroslavl, with others at Leningrad and Vladivostok, for a total of 123.

Project 206E 'Mol' was an export version (presumably simplified; powered by M503A or M504 diesels, rated at 41kts). It was armed with 4–53cm TT and 2 AK-230 guns. The Soviet Navy retained one (*TK-127*) in the Black Sea, probably to train foreign crews. Transfers: Ethiopia (two in 1978), Iraq (two modified units possibly in 1990), Somalia (four in 1976), Sri Lanka (one in 1975). Boats previously reported transferred to South Yemen were those sent to Ethiopia. The previously reported designation, Project 205E, seems less plausible than the one listed here. Project 206E may later have been redesignated 206ET to distinguish it from 206EP, 'Bogomol'.

Project 206M had a bow hydroplane added; series production began in 1970 (completions 1972–76). She was able to operate freely at 40kts in Sea State 4, and at 35kts in Sea State 5. She reverted to the earlier over-and-under 25mm mount forward, and had a twin 57mm aft; there were four 53cm TT. Thirty were completed for the Soviet Navy by Srednyy Neva (Kolpino) and Ulis (Vladivostok). These boats have a helicopter-type (Rat Tail) dipping sonar in a sponson over the starboard quarter. Since 1990 an ESM set has been added. By 1993 at least two Baltic Fleet units had had their hydrofoils removed. Note that the TT apparently fire both ASW and ASuW torpedoes.

Project 206ME is the export version, without dipping sonar: for Cuba (two in 1979, two in 1980, two in 1981, three in 1983), Ethiopia (one in 1985), Kampuchea (one in 1984, two in 1985), Seychelles (one in 1986), Vietnam (five in 1984–86).

Project 206MR Vikhr' is a missile hydrofoil version ('Matka') armed with two P-15M missiles, sixteen of which were built 1977–83 at Leningrad and Kolpino. She has 1–76mm gun forward (AK-176 mount) and a 30mm Gatling aft (AK-630).

The Project 206EP gunboat is in effect a missile-less 'Matka' without hydrofoils. Transfers: four to Guinea-Bissau (two February 1988, two by February 1989). Built by Ulis (Vladivostok).

The current export version on offer is Project 02065 (*ie*, 206.5), a version without hydrofoils armed with four SS-N-25. There is also a proposed 264kt patrol boat version (crew 25, range 2500nm at 11–12 kts).

PROJECT 125 *fast attack craft (torpedo)*: 'PCHELA' class *fast patrol boats* (Project 125P)

Displacement: *c*60t
Dimensions: 82ft × 16ft 1in × 8ft 6in
25m × 4.9m × 2.6m
Machinery: 2 shaft M-503 diesels 8000bhp, plus gas turbines with afterburner for 400kg thrust = 68kts
Armament: 2–533mm TT (2×1), 4–14.5mm MG (2×2)

Class:
T-330 (Project 125), *P-575–580, P-589–596*

This was a hydrofoil derivative of the failed Project 184 with a gas turbine whose TRD-20P afterburner added thrust directly. On trials, the boat showed that it could be used at up to 45kts in Sea State 3. Two prototype torpedo boats were built for the Soviet Navy, then sixteen for the KGB Maritime Border Guards (as Project 125P); they were, in effect, successors to Project 199 ('MO-VI'). A few had helicopter-type dipping sonars.

The prototype was completed in 1960, and the KGB boats reportedly in 1964–65. They employ a half-submerged forward foil and a fully-submerged after foil.

'Mukha' class 1994 L & L van Ginderen

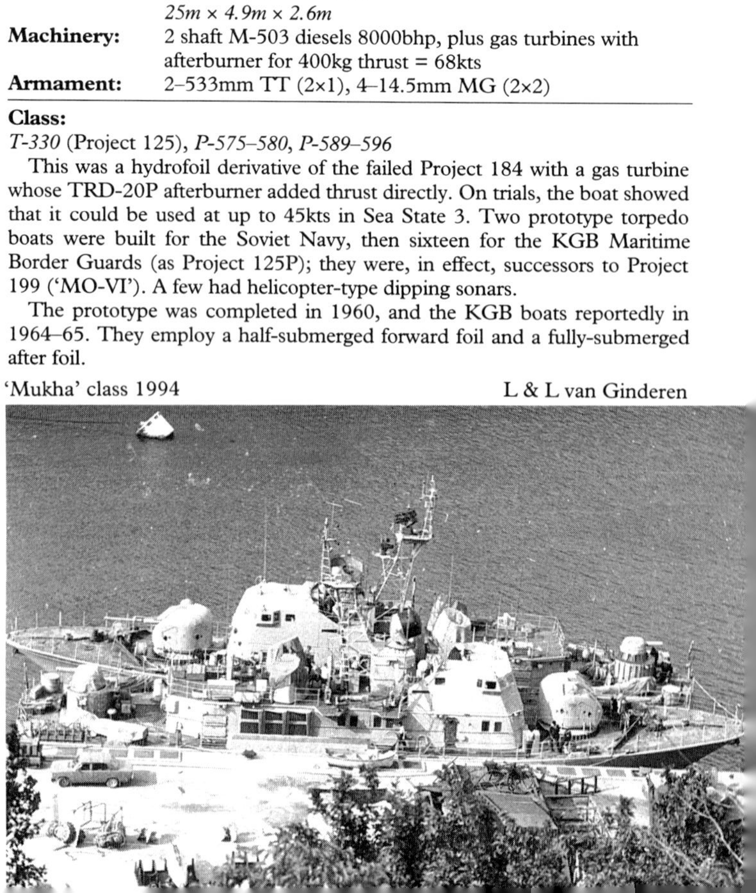

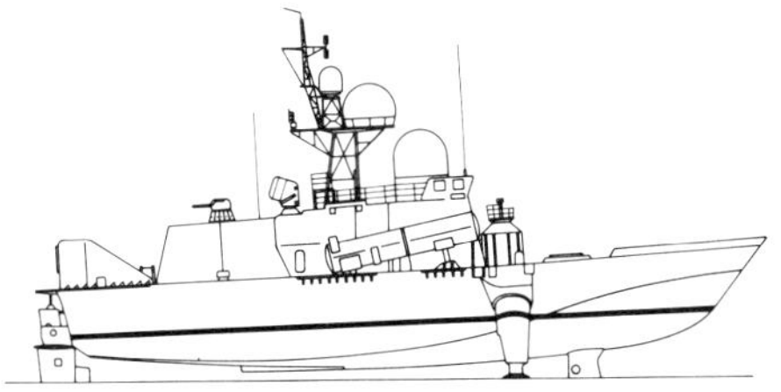

'Sarancha' class
NB 1/750 scale

'SARANCHA' class *missile hydrofoil* (Project 1240, Uragan)

Displacement: 280t standard; 320t full load
Dimensions: 175ft 11in (166ft 1in foils extended) × 37ft (74.5ft foils extended) × 8ft 6in (23ft 11in foils extended)
53.6m, 50.6m × 31.3m, 23.5m × 2.6m, 7.3m
Machinery: 2 gas turbines, 4 shafts, 30,000shp = 58kts
Armament: 2 SS-N-9 (2×2), 1 SA-N-4 (1×2, 20 missiles), 1–30mm Gatling
Sensors: Radar Band Stand, Pop Group, Bass Tilt
Complement: 35

Class:
MRK-5

Completed by Petrovskiy, Leningrad, 1977. Unlike earlier Soviet hydrofoils, this one had fully-submerged foils. They required advanced control systems, but should have confered very superior rough-water performance. In principle, 'Sarancha' would be comparable to the US *Pegasus* (PHM), although actual performance may have been very different. Propulsion was by propellers in pods on the after foils. Although displacement was less than half that of a 'Nanuchka', 'Sarancha' had 4 SS-N-9 missiles and an SA-N-4 for self-defence, as well as a 'Gatling' gun aft. 'Sarancha' was apparently too large and complex for production; 'Tarantul III' was bought instead. She was discarded in 1990.

'BABOCHKA' (Project 1141) and 'MUKHA' (Project 1141.1) class *ASW hydrofoils*

Displacement: 320t standard; 400t full load
Dimensions: 164ft × 27ft 1in (53ft 2in over foils) ×13ft 1in (19ft 4in foilborne)
50m × 8.5m (16.2m) × 4m (5.9m)
Machinery: 2 gas turbines, 22,600shp = 40kts.
Armament: 1–76mm AK-176, 1–30mm AK-630 Gatling
Sensors: ?
Complement: 23

Class:
'Babockha' is *Aleksandr Kunakhovich*; the two 'Mukhas' are *MPK-215* and *MPK-220* (one was named *Sokol*). 'Babochka' was built 1977 at Feodosiya; the two 'Mukhas' were built 1986–87.

'Babochka' was the largest Soviet hydrofoil to date, and like 'Sarancha' she had submerged foils. The two nests of TT (two above two) were located on either side, forward of the superstructure in 'Babochka', well aft in 'Mukha' (whose deckhouse and general arrangement resemble that of a 'Matka' class missile boat). 'Babochka' had only a pair of Gatlings, whereas 'Mukha', the production version, had the 76mm gun forward.

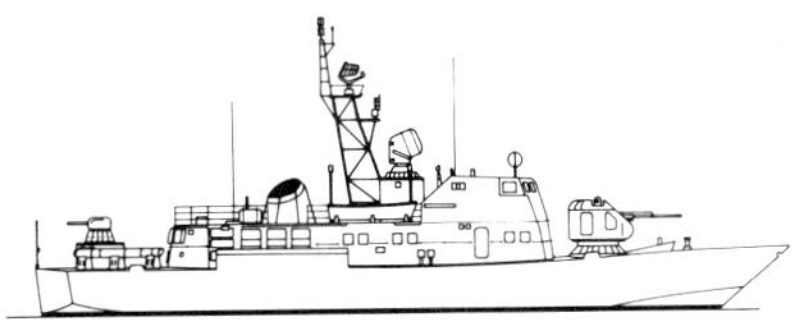

'Muravey' class
NB 1/750 scale

'MURAVEY' class *hydrofoil patrol boat* (Project 133, Antares)

Displacement: 180t standard; 230t full load
Dimensions: 126ft 8in × 24ft 11in (29ft 6in over foils) × 6ft 3in (13ft 1in foilborne
40.0m × 7.6m (8.99m) × 1.9m (3.99m)
Machinery: 2 gas turbines, 22,600shp = 40kts
Armament: 1–76mm AK-176, 1–30mm AK-630 Gatling, 2–406mm TT (2×1, aft)
Sensors: ?
Complement: 34

Class:
Del'fin, P-102, P-103, P-105, P-107–110, P-115–121 (16 units)

All built at Feodosiya 1983–89 as KGB guardships (PSK). They were presumably intended to replace 'Stenkas' in the Black Sea.

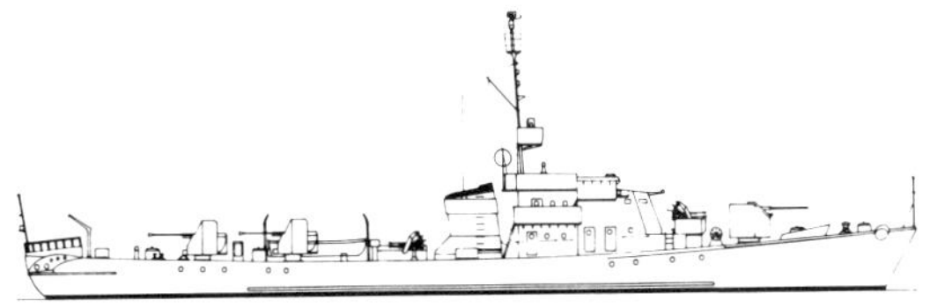

'Kronstadt' class
NB 1/750 scale

'KRONSTADT' class *large patrol craft* (Project 122 bis)

Displacement: 300t normal; 330t full load
Dimensions: 170ft 7in × 21ft 4in × 7ft 3in
52.1m × 6.5m × 2.2m
Machinery: 3 diesels, 3300bhp = 18kts. Range 3500nm at 14kts
Armament: 1–85mm/52, 4–37mm (2×2), 2 MG, 2 RBU-900 ASW RL (RBU-1800 after *c*1960), 2 DC racks
Sensors: Sonar H/F hull (Tamir)
Complement: 35–40

Class:
BO-143–179, BO-181–202, BO-245–255, BO-266–297, BO-348, BO-357, BO364, BO-424–471*, BO-531–539* (157 units)

Starred units had a pair of 5-barrel RBU-1200.

Designed by TsKB 51 in Gorkiy, 1948. Built at Zelenodolsk and Severodvinsk through 1955. Transfers: Albania (four 1961), Bulgaria (two 1957), China (ten 1955; she built fourteen more), Cuba (six 1962), Indonesia (fourteen 1958), Poland (four 1950, four 1956), Romania (three 1960). In 1955–56 twenty became 'Libau' class AGL (administrative and communications auxiliaries); many served with the KGB.

Postwar successors to the unsuccessful wartime 'Artillerist' class large submarine-chasers, these ships were originally numbered in a BO (*Bolshoi Okhotnik*, or large hunter) series, later designated MPK (*Maly Protivo-Lodochny Korabl'*, or small ASW ship). The prototype appeared in 1948, production ending in 1956 at Zelenodolsk, the Soviet lead yard for small combatants; their production (and functional) successor was the 'SO I'. There were two series, the first with combined DC and mine armament, the second solely with DC. Some had the RBU-900 'Mousetrap' ASW launcher; after 1960 some had two RBU-2500s each.

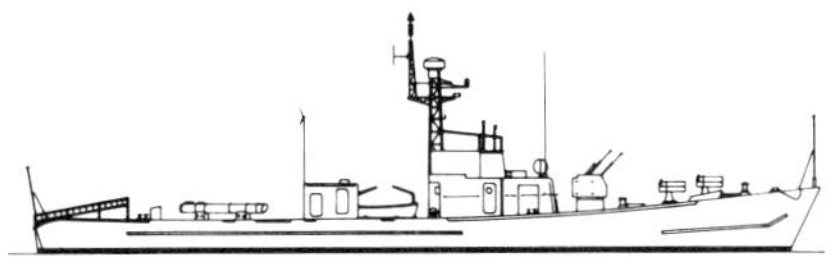

'SO I' class
NB 1/750 scale

'SO I' class *large patrol craft* (Project 201, 201M, 201P)

Displacement: 170t standard; 215t full load
Dimensions: 137ft 10in × 20ft × 6ft 3in
42.0m × 6.1m × 1.9m
Machinery: 3-shaft diesels, 7500bhp = 28kts. Range 1920nm/7kts, 1100nm/13kts, 3400nm/28kts
Armament: 4–25mm (2×2), 4 RBU-1200 ASW RL, 2 DC racks, 2–406mm TT (2×1) in some ships replacing twin 25mm mount aft, 10 mines max
Sensors: Radar Pot Head; sonar H/F hull (Tamir)
Complement: 31

Class:
P-175–190, P-255–259, P-260, P-261–270, P-325–328, P-450–490, P-500–543, P-550–571.* Total 124 units. The asterisk indicates the sole Project 201; all others were 201Ms. Project 201T (units not known) had two 406mm torpedo tubes; all others had RBUs and depth charges only. Sonar was Tamir-2 (export units: Tamir-11). Built 1957–68 at Zelenodolsk, Kerch. Some units had a single 57mm gun in place of the 25mm guns forward; others had the two torpedo tubes aft in place of the after 25mm guns. Transfers: Albania (four), Algeria (six 1965–67), Bulgaria (six 1963–64), China (two 1960), Cuba (eighteen 1964–67), East Germany (sixteen 1960–61), Egypt (twelve 1962–67), Iraq (three 1962), North Korea (six 1957–61, and twelve more built there), Vietnam (eight plus 1980–83), South Yemen (two 1972).

This was the production successor to the 'Kronstadt'. They are generally described as poor seaboats, and are rather smaller than their predecessors, at (according to one source) about 170t rather than 310t. Gun armament was sharply reduced to two twin 25mm AA, as in contemporary coastal craft, but there were 4RLS (RBU-1200, former MBU-1800). The first ships, also known as the 'Misha' class, had a 45mm gun and DC aft, but no RL. Compared to the 'Kronstadts', they were considerably faster, with over twice the power, but needed 20 per cent fewer men, in a smaller hull. Late units receiving two single 406mm ASW TT in place of the after twin 25mm gun, with shorter DC racks. Thirty-five were extant with KGB in 1982. All since stricken.

'POLUCHAT I' class *coastal patrol craft* (Project 368)

Displacement: 90t full load
Dimensions: 97ft 1in × 19ft × 4ft 11in
29.6m × 5.8m × 1.5m
Machinery: 2-shaft M50 diesels, 2400bhp = 18kts. Range 450nm at 17kts, 900nm at 10kts
Armament: 2–14.5mm MG (1×2)
Complement: 20

KGB version of the standard Soviet torpedo recovery boat, thirty-four completed 1953–56. Transfers: Albania (four), Algeria (one 1979), Angola (two 1979), Bangladesh (one 1973), Congo (two 1978–79), Cuba (one or more), Egypt (two), Ethiopia (one 1982), Ghana (four 1967), Guinea (two 1972–73), Guinea-Bissau (two 1978), India (five 1967–69), Iraq (two 1956), Libya (one 1985), Mozambique (one 1977), Somalia (five 1968–69), Tanzania (one 1970s), Vietnam (two 1963), North Yemen (four 1970), South Yemen (one 1970s).

'ZHUK' class *coastal patrol craft* (Project 1400, 1400M Grif)

Displacement: 50t full load
Dimensions: 75ft 2in × 16ft 1in × 4ft 11in
22.9m × 4.9m × 1.5m
Machinery: 2-shaft M50 diesels, 2400bhp = 30kts
Armament: 2–14.5mm MG (2×1, 2×2 in some)
Sensors: Radar Spin Trough
Complement: Approx 28

Class:
AK-10–55, P-60–101, P-125, P-141, P-508, P-512, P-517, P-519, P-526–534, P-547, P-550, P-555, P-558, P-562, P-574, P-579 (85 units). Built 1971–1986. *AK-55* to Azerbaijan after the breakup of the Soviet Union.

In production since 1970. Sometimes erroneously associated with Project 199, the ASW version of the Project 183 torpedo boat built before 1960. In 1995, thirty were in Russian service; over 110 had been exported. Most are armed with twin 12.7mm MG (UTES-M mounting); earlier non-export units had one (sometimes two) over-and-under 14.5mm mounts (2M-7); Maritime Border Guards units often have a twin 12.7mm forward and a searchlight aft. The hull is aluminium alloy. Project 1400ME is a lengthened version (54.5t, 93ft 10in × 19ft 4in × 3ft 3in, *28.6m × 5.9m × 1.0m*, armed with 1 AK-630M Gatling gun) offered for export. Project 1400.6 is another proposed export version, a hydrofoil powered by a 2000shp COTA 2500 gas turbine (with a cruise diesel). Early units had two M-50F4 diesels (1200bhf each); later ones have M-240s (1500bhp each).

Transfers: Algeria (one 1981), Angola (one 1977), Benin (four 1977–80), Bulgaria (five 1977), Cape Verde (one 1980), Cuba (forty 1971–88), Ethiopia (two Oct 82, two Jun 90), Guinea (two Jul 87), Iraq (five 1974–5), Kampuchea (three 1985–87), Mauritius (two Jan 90), Mozambique (five 1978–80), Nicaragua (eight 1982–86), Seychelles (two 1981–82, Somalia (one 1974), Syria (six 1981–84), Vietnam (nine 1978–88, five 1990), North Yemen (five 1978–87), South Yemen (five 1975).

PROJECT 194 *sub-chasers*

Displacement: 110t standard; 161t full load
Dimensions: 88ft 7in × 20ft 4in × 5ft 7in
27.0m × 6.2m × 1.7m
Machinery: 3 shafts, 2 M-50F diesels, 2400bhp, plus 1 M-50E diesel, 600bhp = 24kts. Range 1500nm at 10.5kts
Armament: 1–45mm, 1–37mm, 2–12.7mm MG (2×1), 4 DCT, 2 DC tracks (18 DC)
Sensors: Sonar Tamir
Complement: 27

Class:
MO-105, MP-106

No NATO code name. Both built at Zelenodolsk in 1955, to replace the 'MO-4' class.

PROJECT 191 *river gunboats* (no NATO designation)

Displacement: 55t normal; 57t full load
Dimensions: 85ft × 14ft 1in × 2ft 4in (2ft 2in in Project 191M)
25.9m × 4.3m × 0.7m (0.65)
Machinery: 2000bhp = 24kts. Range 275nm at 15kts. Stores endurance 5 days
Armament: 1–85mm (MK-85 turret), 2–14.5mm MG
Complement: ?

Class:
119 built at Perm' and Leningrad 1947–52: *BK-166–174, BK-275–300, BK-491–499, BK-544–549, BK-571–580, BK-590–598, BK-603–615, BK-617–628, BK-632, BK-633, BK-634, BK-925–960.* Boats were redesignated AK, artillery (rather than armoured) cutters, with the same numbers, in 1965. At least some survived into the 1980s. Production included eighteen Project 191M.

In 1946 TsKB-19 was assigned to develop new large and small armoured boats to replace existing Types 1124/1125, using 85mm tank turrets (two on the large boat, one one the small). Each also had two coaxial armoured turrets for 12.7mm heavy machine guns, and the standard powerplant was the 600hp V-12 diesel which powered the T-34 tank (three in the large, two in the small armoured boats, for 16kts). Estimated displacements were 86t (draft 0.95m) and 53t (0.65m). Then the large boat was dropped, leaving only the Project 191, the prototype of which was built at Perm in 1946.

Further boats were built under Project 191M. The main changes were installation of M-50 (1000bhp) diesels and 14.5mm machine guns instead of 12.7mm. Weight was reduced to keep from increasing draft as armour was increased. To hold down armour weight, the 7mm side armour was angled at 10 degrees to the vertical. Deck armour increased from the 4mm of 1124/1125 to 10mm.

Project 192 was an improved version, with thicker armour (angled at 15 degrees); 98ft 5in × 16ft 5in × 3ft 1in, *30.0m × 5.0m × 0.95m*, 3 M-50 diesels, 3000bhp = 23kts, range 200nm at 18kts. Class: *BK-922, BK-924*, both built at Kolpino 1951. Designed by TsKB-19. After several had been built at Leningrad in 1951, construction was stopped. Work on further armoured boats ended, and by 1958 the river flotillas had been disbanded.

'SHMEL' class *river monitor* (Project 1204)

Displacement: 77.4t full load
Dimensions: 90ft 11in × 14ft 2in × 6ft 7in
27.7m × 4.32m × 2m
Machinery: 2 M50-F diesels = 24kts. Range 300nm at economical speed. Stores endurance 7 days
Armament: 1–76mm (PT76 turret), 1–14.5mm MG, 1–140mm BM-14/17 RL, 4–30mm Plamya grenade launchers (4×1)
Complement: 14

Class:
119 built 1967–74 at Kerch and Khabarovsk: *AK-197–199, AK-201–203, AK-205–209, AK-211, AK-222–225, AK-234, AK-241–250, AK-255–273, AK-354, AK-374, AK-385, AK-387, AK-397–399, AK-404–409, AK-483–485, AK-489, AK-500–509, AK-527, AK-563, AK-564, AK-581–585, AK-587–589, AK-599–602, PSKR-336–377*

Work on river craft reviewed about 1965, presumably in response to the growing tension between the Soviet Union and China, *eg* on the Amur River. TsKB-5 (now Almaz) developed a new Project 1204 boat (NATO Shmel'). The lead boat was completed in Kerch in 1967; about 100 were built there, plus about twenty at Nikolaev. Protection: 10mm vertical armour over wheelhouse and barbettes, 8mm hull and bulkhead, 5mm deck and wheelhouse top. Three 'Shmels' could embark a battalion of naval infantry, although they would have to be billeted aboard another ship. Used on the Danube, Amur, and Ussuri Rivers as well as in the Caspian Flotilla (part of the Black Sea Fleet). Transfers: four to Kampuchea 1984–85.

'YAZ' class *river monitors* (Project 1208 Slepen')

Displacement: 450t full load
Dimensions: 183ft 9in × 29ft 6in × 4ft 9in
56m × 9m × 1.45m
Machinery: 4600bhp = *c*23kts. Range 1000nm at 10kts
Armament: 2–100mm or 115mm (tank turrets), 2–30mm AK-360, 2–12.7mm MG, 2 30mm grenade launchers, 1–140mm ZiF-121M RL
Complement: 32

Class (completed):
Twenty-one built 1978–90. *MAK-1* (1975), *MAK-2* (1975), *MAK-3* (1982), *60 Let VChK* (1978), *MAK-5* (1979), *MAK-6* (1976), *MAK-7* (1978), *MAK-8* (1984; *Khabarovskiy Komsomolets* until Apr 1992), *Imenii 60-Letiya Pogranivoiysk, MAK-10* (1981), *MAK-11* (1983), 10 others.

Almaz (TsKB-5) began work on this large gunboat in 1967, but the prototype (built at Kerch) did not appear until 1975 because there was no design background similar to that provided by Projects 191/192 for Project 1204 ('Shmel'). Protection was 35mm.

The Soviet Navy then abandoned building specialist riverine craft in favour of air-cushion craft which could transport the army's self-propelled artillery and missile launchers.

'PIYAVKA' class *river patrol boats* (Project 1212)

Displacement: 150t
Dimensions: 136ft 5in × 23ft × 3ft 3in
41.6m × 7m × 1m
Machinery: 3 M-401 diesels, 4200bhp = 18kts
Armament: 1–30mm AK-630M, 1–30mm BP-30 grenade launcher
Complement: ?

Class (completed):
Eight built 1980–85 at Vladivostok on 'Vosh' hulls. Names: *PSKR-50* (1979), *PSKR-51* (1980), *PSKR-52* (1981), *PSKR-53* (1982), *PSKR-54* (1982), *PSKR-55* (1983), *PSKR-56* (1983), *PSKR-57* (1984). For the Amur flotilla.

'Vosh' class 1994 L & L van Ginderen

'VOSH' class *river monitors* (Project 1248 Moskit)

Displacement: 150t
Dimensions: 137ft 10in × 23ft × 3ft 11in
42m × 7m × 1.2m
Machinery: 3 M-401 diesels 4200bhp = 18kts
Armament: 1–76mm or 115mm (tank turret), 2–12.7mm MG (1×2, UTES-M), 1–30mm grenade launcher, 1 ZiF-21 RL

Class:
Seven built 1980–84 at Vladivostok and Kerch. Names: *BAK-313, BAK-320, BAK-440, BAK-459, BAK-467, BAK-469,* 1 other. Four others. For the Amur flotilla.

'SAYGAK' class *river launches* (Project 1408.1)

Displacement: 13t
Dimensions: 45ft 11in × 11ft 6in × 1ft 10in
14m × 3.5m × 0.55m
Machinery: 1 M401B diesel, 1 waterjet, 1000hp = 35kts. Range 200nm at 35kts
Complement: 2 (plus 4–8 police)

Total of twenty built at Perm' 1986. Aluminium hull, rail transportable. Offered for export 1993.

MINE WARFARE VESSELS

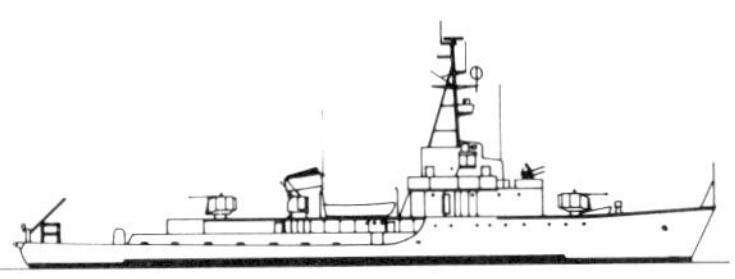

'T43' class

'T 43' class *ocean minesweepers* (Project 254, 254A)

Displacement: 500t standard; 535t normal; 569t full load; 602t max
Dimensions: 177ft 2in wl, 190ft 3in oa × 27ft 11in × 7ft 1in
54m, 58m × 8.5m × 2.15m
Machinery: 4-stroke turbo-charged Type 9D diesels, 2200bhp = 14kts (8.3kts with sweep in deep water). Range 3800nm at 10kts
Armament: 4–37mm (2×2), 2–12.7mm MG (1×2), 2 BMB-1 DC mortars (10 DC), or 10 KB-3 or up to 16 M08/39 mines
Sensors: See notes
Complement: ?

Class:
MT-32, MT-43–49, MT-51, MT-52, MT-54–57, MT-59–69, MT-74–76, MT-78–86, MT-88, MT-89, MT-91–99, MT-102–107, MT-109, MT-131–141, MT-415–433, MT-480–500, MT-502, MT-503, MT-505–529, MT-770–752, MT-755, MT-801–837, Vsevolod Vishnevskii, Ivan Fioletov, Ivan Rogov, Ivan Maslov, Professor Papkovich (178 units). Names applied to numbered units after completion: *Dmitri L'sov* (*MT-46*), *Komsomolets Kalmikii* (*MT-69*, renamed 14.4.76), *Evgenii Nikonov* (*MT-94*), *Kontr-Admiral Khoroshkhin* (*MT-109*), *Komsomolets Estonii* (*MT-132*), *Nikolai Markin* (*MT-496*), *Komsomolets Belorussii* (*MT-502*, renamed 12.10.62), *Komsomolets Latvii* (*MT-529*), *Sakhalinskii Komsomolets* (*MT-751*), *Stepan Shaumyan* (*MT-805*), *Meshadi Azizbekov* (*MT-818*), *Kaliningradskii Komsomolets* (*MT-833*), *Kontr-Admiral Yurkovskii* (*MT-834*), *A Nikolaev* (*MT-837*). Built 1949–57 at Kamysh Burun (Kerch: sixty-one units) and Izhora, Leningrad. In addition, twelve were built in Poland 1955–60 and about twenty in China from 1956 on. Transfers: Algeria (two in 1968), Albania (two in 1960), Bulgaria (three in 1953), China (four in 1955), Cuba (*MT-416, MT-432, MT-433*), Egypt (seven in 1962), Indonesia (six in 1962), Iraq (two in 1969), Syria (two).

Conversions to patrol ships 1963–72: *MT-98, MT-419, MT-421, MT-751, MT-801, MT-802, MT-832* became *P-499 Kirov, P-502 Dzerzhinskii, Brilliant, Korund, Sapfir, Izumrud, Rubin.*

Conversions to Project 258 1956–65: *KVN-1* (*MT-482*), *KVN-2* (*MT-483*), *KVN-3* (*MT-496*), *KVN-9* (*MT-807*), *KVN-12* (*MT-803*), *KVN-13* (*MT-804*), *KVN-14* (*MT-814*), *KVN-15* (*MT-510*), *KVN-16* (*MT-511*), *KVN-17* (*MT-513*), *KVN-18* (*MT-517*), *KVN-19* (*MT-422*), *KVN-22* (*MT-821*), *KVN-24* (*MT-474*), *KVN-25* (*MT-82*), *KVN-27* (*MT-820*), *KVN-50.*

SS conversions: *SS-1, SS-2* (*Ivan Rogov*), *SS-3* (*MT-47*), *SS-13* (*Chernomorets*).

Project 254 was a coastal (later ocean) minesweeper assigned to TsKB-363 in Leningrad (it later merged with TsKB-50 to become the Western Planning and Design Office). This organisation had alrady developed some mine countermeasures ships. Work began in 1946 and the design was approved the same year; it was specially adapted for mass production. The first unit was completed in 1948. Ships were built in Leningrad and Kerch.

The design was specially degaussed (to limit signature to within 20 milliorsteds at 6m depth), but no attempt was made to limit acoustic or pressure signatures. Sweeps: MT-1 ocean mechanical sweep; the ship could also tow the MTSh modernised sweep. Overall sweep capacity was three times that of Project 53-U (*Fugas* class).

Modified versions: Project 254-K: lead ship completed 1952, with new sweeps: MT-2 mechanical, Lin', Fakel-1, Fakel-M3, Ryf-K radars; Tamir-II sonar; Koordinator system receivers; increased electric power.

Project 254-M: lead ship completed 1955. They had a deep mechanical sweep, magnetic (coil) sweep, acoustic sweep, and more modern ASW mortars; and 25mm guns in place of 12.7mm MG. They could tow magnetic and acoustic sweeps at 12kts rather than 7kts. The coil sweep controls could vary the sweep field with the acoustic field. Variable-pitch propellors were fitted. Full load displacement increased to 600t.

Project 254-A: modified for atomic protection; lead ship delivered 1957. The hull was additionally reinforced and the internal space partly pressurised to keep fallout out. Battle stations had closed-cycle ventilation, and a washdown system was fitted. The 37mm guns were replaced by single 45mm guns.

The SKR version for the KGB was Project 513.

A radar picket version (Project 254PGR, at least some converted from Project 254M) appeared in 1956, armed with 4–37mm (2×2, and a twin 25mm in most). The initial radar installation was Knife Rest (fore and aft), but later some had the dorward radar replaced by a Big Net, and some at least had only Big Net, on a massive lattice mast aft. Twenty were converted.

Other units were converted to noise-measurement ships (Project 513/513M, with GKS designators), diving tenders, and trials ships.

T 58/PGR class

'T 58' class *ocean minesweepers* Project 264)

Displacement: 790t standard; 860t full load
Dimensions: 230ft × 29ft 6in × 7ft 11in
70.1m × 9m × 2.4m
Machinery: 2 37DR diesels, 4000bhp = 18kts
Armament: 4–57mm (2×2), 4–25mm (2×2), 1 MBU spigot mortars, DC mortars
Complement: 82

Class:
T-1–12, T-14, T-15, T-22, T-50, T-71, T-108, T-110, T-112–130, Pavel Khokhryakov, Vasilii Gromov, Timofei Ul'yantsev (completed 16.10.60), *F Mitrofanov, Pavlin Vinogradov* (completed 16.1.61), *St. L-T Vladimirov, St. L-T Pekarev* (45 units).

T-6 (completed 14.4.60) was named *Vladimir Polukhin. T-9* (laid down 24.9.57, launched 5.9.58, completed 6.8.60) was named *Primskii Komsomolets* in 1975. Some were converted to guard ships (SKR) in 1975: *T-50* (*KVN-21*), *T-112* (*KVN-23*), *T-4* (*SKR-132*); two others were renumbered *SKR-13* and *SKR-137.*

Project 513 acoustic measurement ships (19 units): *GKS-11* (launched 20.6.56), *GKS-12–26, GKS–45, GKS-46.*

Project 532 and 532A submarine rescue ships (14 units): *Gidrolog, Khibin', Valdai, Zangezur, Kazbek, SS-30, SS-35, SS-40, SS-47, SS-48, SS-50* (*Shakhter*), *SS-51, SS-53.*

Names of KGB units: *Dzerzhinskiy, Menzhinshiy, Kirov, Korund, Pavlik, Vinogrado, Primorsky Komsomolets, Starschiy Leitenant Vladimirov*; others numbered

Two series, begun 1957 and 1959, completed 1960 and 1962. A total of forty-five were built, probably no more than five in the initial series; fourteen, probably in this series, were converted to 'Valdai' class submarine rescue ships.

The Project 264 TTZ was approved in 1949; the lead ship was delivered in 1957. The objective was increased sweep capacity, using improved influence sweeps and a better mechanical sweep deployable at a higher speed. The new design would also have betterself-defence weapons and electronics, including sonar. The basis arrangement of Project 254 was unchanged, but power-to-weight ratio had to be doubled. Unlike the powerplant of a Project 254 ship, in this case the powerplant was on two levels, to minimise its length. As in the earlier design, there was no special provision to minimise acoustic or pressure signature.

Sweeps: MKT-1 ocean mechanical sweep, TEM-52 magnetic (coil) sweep, BGAT deep high-speed acoustic sweep.

Production units were Project 264A, with better protection against nuclear attack, as well as new sweeps. Harbour diesel output was quadrupled, degaussing gear was modernised, the hull near the engine rooms was covered with rubber for silencing, the engines and generators were sound-mounted, and the propellors were provided with bubble silencers. RBU-2500 replaced the Hedgehog, the 25mm guns were replaced by 30mm weapons, and a new sonar was installed. Displacement increased to 868t. Construction was approved in 1956, and the first unit was delivered in 1958.

These ships were reclassified as patrol ships in 1975.

Three units were converted into radar pickets. They were probably intended as range security ships rather than as radar pickets. They are armed with two SS-N-5 (2×4, 16 missiles) and 2–57mm (1×2), 4–30mm (2×2), and designated KVN, Korabi Vozdushnogo Nabyudeniya, Radar Surveillance Ships. The first appeared in 1974, converted at Izhora, Leningrad.

Nineteen units were redesignated corvettes (Project 264PGR), for the navy and for KGB Maritime Border Guards (one was transferred to India in 1971). Five remained in 1992: *St Leitenant Vladimirov*, *M Gromov*, *St Leitenant Lekarev*, *KVN-36*, one other. They were armed with two SA-N-5, 4–57mm (2 twin ZiF-31B), 4–25mm (2M-3M), and 2 RBU-1200 rocket launchers. Equipment included a Tamir-2 sonar.

At least seven were completed as submarine rescue vessels (Project 532A, *Valday* class). Six later served as target service vessels (NATO 'Potok' class, with OS designators).

Several served as intelligence collectors (AGIs).

Transfers as minesweepers: one each to Guinea and Yemen. One submarine rescue vessel was transferred to India.

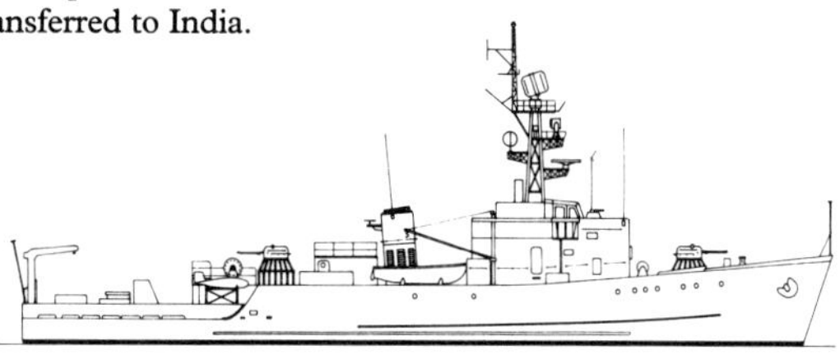

'Yurka' class
NB 1/750 scale

'YURKA' (Project 266 Rubin) and 'NATYA' (Project 266M Avkamarin) class *ocean minesweepers*

Displacement: 520t standard; 560t full load
Dimensions: 170ft 5in × 30ft 10in × 8ft 6in
52m × 9.4m × 2.6m
Machinery: 2 M-503 diesels, 5000bhp = 16kts (14kts with one sweep). Range 1500nm at 12kts. Stores endurance 7 days
Armament: 4–30mm (2×2), 2 RBU-2500 ASW RL
Sensors: ?
Complement: ?
NB Data are for Project 266

Class (completed):
Project 266 and 266D (41 units): *MT-6*, *MT-18*, *MT-27*, *MT-47*, *MT-58*, *MT-62*, *MT-63*, *MT-72*, *MT-73*, *MT-80*, *MT-82*, *MT-86*, *MT-94* (named *Evgenii Nikonov* 1965), *MT-163*, *MT-179*, *MT-193*, *MT-200*, *MT-205*, *MT-208*, *MT-209*, *MT-214*, *MT-219*, *MT-221* (31.8.70), *MT-223* (13.4.68 named *Grigorii Vakulenchik*), *MT-226* (named *Petr Mal'kov* 1968), *MT-238* (13.10.67), *MT-242* (11.6.68), *MT-253*, *MT-257* (26.12.69; named *Vitse-Admiral Sabaneev* 1984), *MT-262–264*, *MT-284* (12.9.68), *Aleksandr Kazarskii*, *Arsenii Raskin*, *A Sokolov*, *Afanasii Matioshenko*, *Boris Safonov*, *Ivan Maslov* (1966), *Ivan Sivko*, *Michman Pavlov* (31.12.70)

Project 266M (45 units): *MT-264*, *MT-265*, *MT-303**, *MT-417**, *MT-418**, *MT-461**, *MT-463** (renamed *MT-749* in 1990), *MT-712**, *Artillerist*, *Vsevelod Vishnevskii*, *Valentin Pikul'*, *Vitse-Admiral Zhukov* (ex-*Elektrik*), *Vitse-Admiral Zakharin*, *Dmitrii Li'sov*, *Dizelist*, *Desantnik*, *Dal'nomershik*, *Zapad* (30.11.75), *Zapyad* (30.12.76), *Zentchik*, *Komendor*, *Kontr-Admiral Vlasov*, *Kontr-Admiral Khoroshkhin* (1972), *Kontr-Admiral Pershin* (1970), *Mashinist*, *Mina*, *Miner*, *Marsovii*, *Motorist*, *Navodchik* (named *Kurskii Komsomolets* Jun 1982 – 18.3.92), *Pudemetchik*, *Paravan* (31.10.73), *Raketchik*, *Rulevoi*, (*Khar'kovskii Komsomolets* until 18.3.92), *Razvedchik*, *Svyazist*, *Semen Roshal* (launched 27.10.70), *Signal'shik*, *Snaiper*, *Strelok***, *Tral* (30.10.72), *OS-99* (ex-*Torpedist*), *Turbinist* (named *Komsomolets Estonii* 1975), *Yakor* (30.9.74). Starred units are 266ME; the double-starred units is 266DM.

'Natya' class 1990 L & L van Ginderen

The preliminary design for the next ocean minesweeper, Project 266, was completed in 1957; the engineering design was approved in 1959. The lead ship was delivered in 1963. The main improvement, compared to previous designs, was internal protection against mine explosions. Signatures were also drastically reduced: compared to those of Project 264A, magnetic by a factor of 40, electric (UEP) by a factor of 10–15, acoustic by a factor 3–4, and hydrodynamic (pressure) by 20 to 30 per cent. The hull was built of low-magnetic steel and an automatic control adjusted the degaussing field for, *eg*, eddy currents generated in the hull as the ship rolled. The design used slow-turning quite propellors as well as rubber damping materials on machinery foundations. Project 266 had special electro-optical sights to detect floating mines.

Project 266 was designed to employ new sweeps, to deal with influence mines at depths from 25m to 150m. The ship also had a capacity to take extra sweeps on board (*eg*, net sweeps, bottom sweeps). A recent design history mentions automated operation and remote control of sweeps, guns, engines, and other equipment.

The main gap in the design was the absence of a sonar to detect bottom mines.

There is also a 266D variant.

In all, fifty-two were built 1962–69. Transfers: Egypt (four in 1969), Vietnam (two in Dec 1979). They are similar to 'Natya' (266M) but without the stern ramp. The broad funnel shows that the engines are mounted side by side; 'Natya' has a more conventionally-shaped funnel. One was lost in an explosion in the Black Sea, 19 August 1989. Five were discarded 1991, and another five in 1992; by 1995 only seventeen were left.

Project 266M was an improved version in which the coil sweep was replaced by a deep sweep with new control gear; a sweep was provided to deal with active (rising) mines; a 3-channel broadband towed television mine detection/sweep system was added; and a sonar was fitted to detect bottom mines. The new ramp stern made it easier to set and recover sweeps. The ship was air conditioned. The main engines (2 M-504) were mounted on sound-damping beams and the quiet propellors were shrouded to further reduce sound radiation (a special cavitation monitor was installed). Adjustable protectors compensated for the electric field. Two twin 25mm guns (2-M3-M) were added. The hull had to be lengthened by 29ft 6in (*9m*), widened by 2ft (*0.6m*), and displacement increased to 790t fully loaded. The Soviets estimated that a Project 266M minesweeper was nearly 50 per cent more effective than its predecessor Project 266.

The first unit, *Semen Roshal'*, was delivered in the autumn of 1970. A total of thirty-five ships were built at Kolpino (thirty-two) and Khabarovsk (three); production ended 1982.

The single 266DM (NATO 'Natya II') was built at Kolpino in 1981 with an aluminium hull. She has an elongated deckhouse and no RBUs or 25mm guns. Presumably she is a test ship for a mine hunting system deployed from a deckhouse on her fantail, which replaces the standard minesweeping winch of this class.

Three ships were further modified (NATO 'Natya III') with AK-630M Gatling guns instead of AK-230s.

Transfers: India (two in 1978, two in 1979, two in 1980, one in 1986, three in 1987, two in 1988), Libya (two in 1981, three in 1983, one in 1984, one in 1985, one in 1986), Syria (one in 1985, as a training ship), Yemen (one). As of 1995, disposals included *Artillerist*, *Ivan Maslov*, *Miner*, *MT-159*; *Rulevoy* and *Signalshchik* were stricken 1994. *Torpedist* became an experimental ship (*OS-99*) 1990.

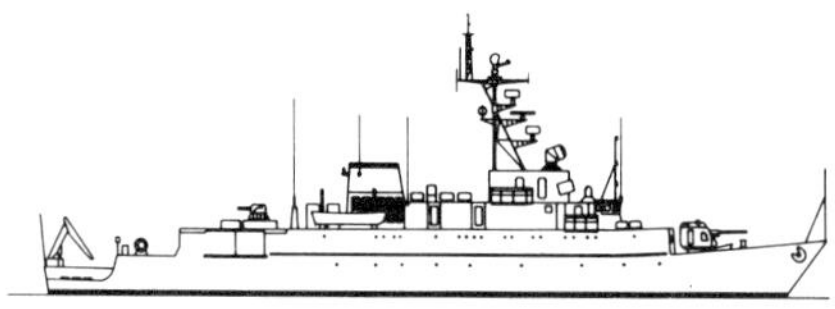

'Gorya' class

'GORYA' class *minehunter* (Project 1260)

Displacement: 900t standard; 1100t full load
Dimensions: 229ft 6in × 39ft 4in × 13ft 1in
70.0m × 12.0m × 4.0m
Machinery: 2-shaft diesels, 5000bhp = 16kts; see notes
Armament: 1–76mm, 1–30mm Gatling, 2 SA-N-14 (2×4), 2×2 special torpedo apparatus
Sensors: Radar Palm Frond, Nyada, Bass Tilt; assumed HF minehunting sonar
Complement: 60

Class (no):
A Zheleznyakov (MT-901), *A Gumanenko*, 1 other

The prototype, *Zhelezhnyakov*, was built at Svedniy Neva (Kolpino); launched August 1987 and completed 1988. A class of twenty units (to be built over five years) was planned, but the programme was deferred (more likely abandoned) as too expensive. Total cost of the lead ship was 41m Rubles, of which only 5m went into the hull. The ship was intended partly to sweep US deep-ocean captor mines (in which sense it was analogous to the current generation of new US deep-water mine hunters, designed to deal with Soviet rising mines). She was intended to carry two submersibles (CTICH, meaning 'self-propelled remote-controlled seeker-destructor' in Russian), each carrying two 100kg demolition charges. A total of 20 such charges would have been embarked. Presumably the submersibles would have been launched by cranes extending out through the two sliding doors in the deckhouse. The planned outfit also included two special deep-water sweeps, Anaconda and Cobra. Each would have employed a pair of old SET-40 torpedoes towing a sweep wire between them. Anaconda would have cut through mine cables; Cobra would have used cable-cutting charges.

Power also reported as 16,000bhp (presumably four diesels) for 24kts. Such speed would be useful for a valuable ship transiting between mine hunting assignments.

'T 301' class *inshore minesweepers* (Project 253L)

Particulars: See 1922–1946 volume

Production of this wartime class continued at Rybinsk through 1956, very large numbers having been built. In May 1955, for example, the ONI listed a total of 263. These craft were designed specifically for rapid production, without compound curves in their hulls. Production began in 1944, and about fifty had been completed by the end of the Second World War, the final total being over 200. Reportedly machinery for the early units was provided by the US under Lend-Lease, later units with Soviet diesels being about 3m longer. A total of eighty was reported still in service in 1968. Only about five were left by 1977.

Transfers: Albania (seven 1957–60), Bulgaria (four 1955), Egypt (two 1962–63), Indonesia (one about 1962), Poland (eight 1957), Romania (fourteen 1956–59).

Project 255/255K *coastal minesweepers*

Displacement:	148t standard; 160t full load (255K: 150t standard; 165t full load)
Dimensions:	118ft 1in wl × 18ft 8in × 5ft 3in *36.0m × 5.7m × 1.6m*
Machinery:	3 shafts, 3 Superior diesels, 690bhp = 12kts (255K: 12.5kts), 7kts when sweeping
Armament:	2–45mm (2×1), 4–12.7mm MG (2×2), could carry but not lay mines
Complement:	35

This was a modified version of the Project 253 (NATO 'T-301') '100-tonner' designed immediately after the Second World War. Production began in 1946 at Rybinsk. These boats had Tamir-10 searchlight sonars to detect moored mines. Sweeps: OPT single bottom sweep, PEMT-4 loop coil magnetic sweep, two BAT-2 towed acoustic sweeps. The steel boat's degaussing coil reduced its magnetic signature by up to 20mErg at 4m below the designed waterline. Note that the diesels were an imported US type. They proved inadequate; the boat was redesigned in 1950 (as Project 255K) with Russian-made 2D-12 (300 bhp) engines and a separate diesel generator. The first unit was completed in 1951 at Rybinsk.

'SASHA' class *coastal minesweepers* (Project 265)

Displacement:	245t standard; 269t full load
Dimensions:	147ft 8in × 20ft 4in × 5ft 11in *45.1m × 6.2m × 1.8m*
Machinery:	2 diesels, vertical cycloidal propellers, 2200bhp = 19kts. Range 2100nm at 12kts, 1300nm at 18.5kts
Armament:	1–45mm or 57mm, 4–25mm (2×2), 12 mines max
Sensors:	Radar Ball End
Complement:	25

Class:
T-17, *T-19*, *T-20*, *T-23*, *T-156–162*, *T-164–176*, *T-178*, *T-180–192* (37 units)

These steel coastal sweepers (BT, *Beydovoy Tralchik*) were built 1954–63 to replace the wartime 'T-301' and the interim Project 255; they were also used as patrol craft. Only two were left by 1992. The TTZ was issued in 1946; it called for mine countermeasures up to 60nm from a base and also for minelaying. The lead ship was completed at Rybinsk in 1953. The main improvements, compared to Project 255, were a better sonar (Tamir-11) and better degaussing. The armament comprised one new twin 45mm mount (SM-17) instead of two single 21-KM, and two new twin 14.5mm machine guns (2M-7P type). The engines were two much more powerful (900bhp) 33-D diesels. On sea trials, the lead unit proved quite wet. In the improved Project 265K, the shape of the bow was changed, bilge keels were added, and one 25kW diesel sweep generator was replaced by a 50kW unit. The 14.5mm machine guns were replaced by twin 25mm weapons, a new contact sweep was introduced, and also a new SEMT solenoid magnetic sweep. The next step (265A) was to provide nuclear protection and also to increase engine power. A new acoustic sweep (AT-2) was introduced, and the sonar was replaced by a new MG-11-265. The last ships of the series were armed with 57mm guns. Tonnage now increased to 319t. The first unit was completed in 1960. The last three ships of the group were built as minehunters (Project I-265), with prototype IT-3 towed three-channel televisions.

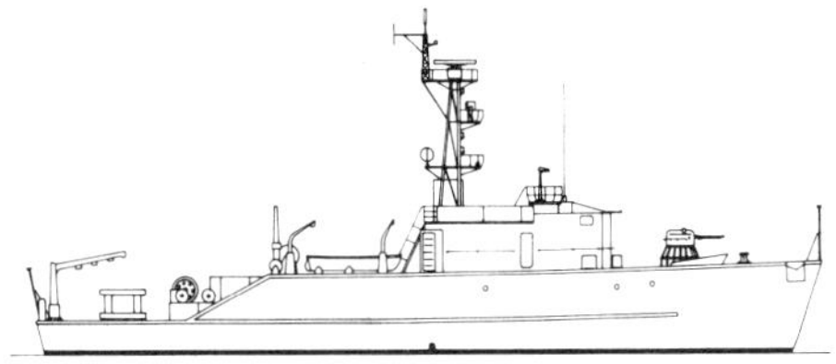

'Vanya' class
NB 1/750 scale

'VANYA' class *coastal minesweepers* (Project 257D, 257DM, 257DT and 699B)

Displacement:	200t standard; 260t full load ('Vanya II' 255t)
Dimensions:	131ft 7in × 25ft 11in × 5ft 7in *40.2m × 7.9m × 1.7m*
Machinery:	2-shaft diesels, 2200bhp = 16kts. Range 2400nm at 10kts, 1400nm at 14kts
Armament:	2–30mm (1×2), 8 mines max ('Vanya II' 12)
Sensors:	Radar Don-2
Complement:	30

Class (completed):
BT-33–37, *BT-42*, *BT-44*, *BT-45* (*Komsomolets Kirgizii*), *BT-70*, *BT-77*, *BT-87*, *BT-101*, *BT-111*, *BT-116*, *BT-142–149*, *BT-168*, *BT-194–199*, *BT-203*, *BT-206*, *BT-216*, *BT-218*, *BT-220*, *BT-222*, *BT-224*, *BT-227–231*, *BT-233–237*, *BT-239*, *BT-240*, *BT-293*, *BT-314** (20.6.70), *BT-316** (30.9.70), *BT-415*, *BT-416* (47 units). Starred units are Project 257DM.

Built 1960–73 at Kolpino and Vladivostok. The lead unit was delivered at the end of 1961. In 1995, about thirty were left (Projects 257D and 257DM). This new coastal unit was conceived as a minehunter with substantially reduced signatures (magnetic, electric [UEP], acoustic, and hydrodynamic [pressure]). Project 257D was the wooden version; Project 257M was the version built of low-magnetic signature steel. The wooden version was adopted. The designers claimed that, in comparison with Project 265, they had reduced the magnetic signature by a factor of nearly 50, the acoustic signature by a factor of 8-10, and the UEP signature by a factor of 5. They also claimed a significant reduction in the hydrodynamic (pressure) signature. The degaussing system included automatic coil control; the UEP signature was reduced by using dielectric coatings and insulation, and special sound damping machinery foundations were used. The low-noise propellers used air bubble shields (like the contemporary US Masker system). Sweep equipment was handled, for the first time, using hydraulics. The single gun was a special non-magnetic version of the AK-230.

Sweep equipment was drastically simplified: the ships had only the television hunting tow, an electromagnetic sweep (described as a hunter/killer for bottom mines), and a contact sweep, only one of which could be used at a time.

It turned out that the ship and powerplant could accommodate more elaborate sweeps, so a new Project 257DM was begun in the spring of 1961. It could now deal with mines with combination magnetic-acoustic fuses (using one of three combination sweeps: loop coil/acoustic, solenoid/acoustic, or contact/acoustic), and also active mines (*ie*, rising mines, which the Soviets were then developing), using a contact sweep combined with acoustic sweeps. The changes made it possible for Project 257DM units to convoy ships in coastal waters. The export version was Project 257DME.

Project 699 was an improved version, built in small numbers at Petrozavodsk, powered by two 2500bhp diesels rather than by the M-870-FTK 1200bhp lightweight high-speed diesels of the earlier Project 257. They made it possible for the ship to tow heavier sweeps such as barge-sweeps (as against pressure mines) and twin bottom contact sweeps. Three modernised versions (Project 699B 'Vanya II') were built at Kolpino in 1976. They are 3ft 3in (*1m*) longer, with larger diesel exhaust amidships, and are armed with a twin 25mm gun (2M-3M) in place of the earlier AK-230. Endurance is slightly greater (1450nm at 14kts rather than 1400nm at 14kts) and tonnage is slightly smaller (195t/255t rather than 200t/260t). One unit was modified 1974, perhaps as a minehunter; sweep gear was removed and a small mast raised amidships. She was armed with 2–25mm (1×2), which might be more accurate than 30mm for mine disposal. Transfers: Bulgaria (one in 1971, one in 1972, five in 1980s), Syria (two in Dec 1972). *BT-251* sank 19 August 1989 after an explosion.

'ZHENYA' class *coastal minesweepers* (Project 1252 Izmrud)

Displacement:	220t standard; 300t full load
Dimensions:	139ft 1in × 25ft 11in × 5ft 11in *42.4m × 7.9m × 1.8m*
Machinery:	2-shaft diesels, 2400bhp = 16kts. Range 2400nm at 10kts, 1400nm at 14.5kts
Armament:	2–30mm (1×2)
Sensors:	Radar Spin Trough
Complement:	40

Class (completed):
Komsomolets Turkmenii (1966), *BT-215* (1968; *Komsomolets Buryatii* until 18.3.92), *BT-177* (1969). *Komsomolets Turkmenii* blew up 19 August 1989.

This was a further attempt to reduce signature through adoption of glass-reinforced plastic (GRP). The contract design was approved for production in 1961, and the first unit was completed in 1966.

Project 1253 ALMAZ *unmanned minesweeper*

Displacement: 112t standard; 116t full load
Dimensions: 82ft × 17ft 9in × 5ft 4in
25.0m × 5.4m × 1.63m
Machinery: 1 diesel, 600bhp, with waterjet; 2 500kW sweep generators
Armament: 4–25mm (2×2)

This unmanned sweeper was conceived as a counter to new kinds of mine fuzing mechanisms which were entering service. It was designed to lay a line charge (cord charge), which would, in theory, destroy any bottom mine. The specification was completed in the summer of 1959, the preliminary design in 1960, and the contract design in 1962. Special emphasis was laid on reducing the ship's signature, so that it could pass through an influence field while laying its charge. An attempt to develop a GRP hull failed, so the boat was built of wood.

The lead unit was laid down in the fall of 1962 at Petrozavodsk and commissioned in the fall of 1966; the first production unit followed in 1969. In appearance, this boat resembled a slightly enlarged 'Olya'.

'SONYA' class *coastal minesweepers* (Project 1260 Yakhut)

Displacement: 450t full load
Dimensions: 160ft 1in × 28ft 10in × 6ft 11in
48.8m × 8.8m × 2.1m
Machinery: 2-shaft diesels, 2400bhp = 15kts. Range 3000nm at 10kts, 1600nm at 14kts
Armament: 2–30mm (1×2), 2–25mm (1×2)
Sensors: Radar Spin Trough
Complement: 43

Class:
BT-16 (*Astrakhanskii Komsomolets* until Apr 1992), *BT-21*, *BT-23* (ex-*Sakhalinskii Komsomolets*), *BT-51*, *BT-56*, *BT-78*, *BT-96*, *BT-100*, *BT-114*, *BT-121*, *BT-132*, *BT-215*, *BT-232*, *BT-243–249*, *BT-251*, *BT-252*, *BT-255–261*, *BT-263*, *BT-268–270*, *BT-285* (*Kolomenskii Komsomolets*), *BT-79* (named *Khersonskii Komsomolets* 24.5.76–18.3.92), *BT-202* (named *Sevastopolskii Komsomolets* until 18.3.92), *BT-126* (named *Orenburgskii Komsomolets* 24.6.76–18.3.92), *BT-155*, *BT-241**, *BT-260*, *BT-267*, *BT-325* (named *Komsomolets Buryatii* 30.12.85), *Vladimir Polukhin**, *Aleksei Lebedev**, *BT-294–296*, *BT-298–302*, *BT-315*, *BT-317–325*, *BT-327–330*, *BT-340–343*, *BT-350*, *BT-438*, *BT-470*, *BT-734* (72 units)

Starred units are Project 12655 (1265.5). Built 1971–1991 at Vladivostok (Ulis: twenty-two units) and at Petropavlovsk (Avangard). Export units were Project 1265E. Transfers: Azerbaijan (three in 1992), Bulgaria (four in 1983–86), Cuba (two in 1980, two in 1985), Syria (one in 1986), and Vietnam (one in Feb 1987, one in Feb 1988).

These ships were designed around a new mine countermeasures weapon. They had a mine detection sonar and an underwater communications link. Machinery is remote-controlled, presumably so that all personnel can be concentrated in a safe control station.

'Sonya' class 1994 — L & L van Ginderen

'TR 40' class *inshore minesweepers* (Project 151)

Displacement: 46.3t standard; 49.2t full load
Dimensions: 89ft 11in × 13ft 1in × 1ft 8in
27.4m × 4.0m × 0.5m
Machinery: 2-shaft diesels, 600bhp = 15kts. Range 500nm at 10kts
Armament: 2–25mm (1×2), 2–12.7mm MG (1×2), 20 mines max
Complement: 16

Class:
RT-20–28, *RT-33–35*, *RT-145–154*, *RT-157–170*, *RT-175–196*

About fifty-five were built in Poland in the late 1950s for river and coastal operations, as a patrol boat, sweeper, and minelayer; seven remained in Polish service, and eight went to Rumania. TR is a Polish designation meaning *Tralowiec Redowy*, or roadstead sweeper, the equivalent of the Soviet term.

'K 8' class *inshore minesweepers* (Project 361T)

Displacement: 19.4t standard; 26t full load
Dimensions: 55ft 5in × 10ft 6in × 2ft 8in
16.9m × 3.2m × 0.8m
Machinery: 2-shafts, 2 3D6 diesels, 3000bhp = 12kts. Range 300nm at 9kts
Armament: 2–14.5mm MG (1×2)
Complement: 6

A total of about 130 was built by Polnocny (Gdansk, Poland), 1954–59; replaced by the 'Yevgenya' class. Wooden hulls. The K designation is Polish, standing for Kuter (cutter). With minor modifications these craft became motor gunboats and minesweepers. Poland operated about thirty in her navy (as sweepers) as well as a like number in her coast guard. Seventy went to the Soviet Union. Transfers: Cuba (one 1978), Nicaragua (four 1984), Vietnam (five 1980). About twenty-five remained in Soviet service in 1995.

NOTE: There was also a small coastal motor boat, 'PO 2', used by the Soviets for sweeping; no details are available. Transfers: twenty-four to Bulgaria and two to Egypt. It dates from the early 1950s.

'YEVGENYA' class *inshore minesweepers* (Project 1258 Korund)

Displacement: 88.5t standard; 90t normal; 91.5t full load; 94t maximum
Dimensions: 85ft 9in oa, 75ft 5in bp, 79ft 3in wl × 19ft 4in ext, 17ft 8in wl × 4ft 5in norm, 4ft 6in full
26.13m, 23m, 24.16m × 5.9m, 5.4m × 1.35m, 1.38m
Machinery: 1 diesel, 850bhp = 12kts at 90 tons. Range 300nm at 10kts
Armament: 1 twin 14.5mm MG (2M7) (2000 rounds). Sweeps: MT-34, AT-2, SEMT-1, Neva, GKT-3
Complement: 10

Class (completed):
RT-29, *RT-32*, *RT-41* (1980), *RT-52* (1980), *RT-136*, *RT-155*, *RT-172*, *RT-297*, *RT-349*, *RT-361*, *RT-402* (1978), *RT-403* (1978), *RT-420*, *RT-471* (1979), *RT-435*, *RT-473*, *RT-583*, *RT-588*, *RT-602*, *RT-823*, *RT-1202–1215*, *RT-1244*, *RT-1256*, *RT-1258* (45 units). All built 1976–80.

In 1995, forty-five remained in the Russian Navy; all were built 1967–76. Boats built since 1976 were all for export. The only sonar is MG-7, a high-frequency type used as an anti-swimmer device on board larger ships such as frigates and destroyers. These plastic-hulled minehunters tow an underwater television.

Transfers: Azerbaijan (two in 1992): Cuba (two in 1978, two in 1979, two in 1980, one in 1982, four in 1984), India (three in 1983, three in 1984), Iraq (three in 1975), Mozambique (two in 1985), Nicaragua (four, of which two or more from Cuba), Syria (one in 1978, one in 1981, two in 1985, one in 1986), Vietnam (two in Nov 1984), North Yemen (two in 1982). Two transferred to Estonia were not placed in service. The Caspian Flotilla units (*RT-136*, *RT-173*) were to be transferred to Azerbaijan.

'OLYA' class *inshore minesweepers* (Project 1259 and 12592 Malakhit)

Displacement: 50t full load
Dimensions: 80ft 9in × 13ft 9in × 3ft 3in
24.6m × 4.2m × 1.0m
Machinery: 2 shaft diesels, 400bhp = 18kts. Range 500nm at 10kts
Armament: 2–25mm (1×2)
Sensors: Radar Spin Trough
Complement: 15

First appeared 1975. Classed as RT, roadstead sweeper. Four completed 1976. All originally served at Kronstadt; in 1995 the survivors all served at Batysk. This type was also built under licence in Bulgaria (four built 1988–92).

'LIDA' class *inshore minesweepers* (Project 10750 Sapfir)

Displacement: 130t
Dimensions: 101ft 8in × 21ft 3in × 5ft
31.0m × 6.5m × 1.5m
Machinery: 2-shaft diesels, 900bhp = 11kts
Armament: 1–30mm Gatling
Complement: 12–14

Class:
BT-756–775. (20 units)

'Yevgenya' class replacement, which first appeared in the Baltic in 1991. By 1992 ten were in service, all built at Kolpino. They are probably built of GRP and they may carry SA-N-5 or -8 hand-held missiles.

'ANDRYUSHA' class *special minesweepers* (Project 1256 Topaz)

Displacement: 373t standard; 400t full load
Dimensions: 154ft 3in × 28ft 6in × 7ft 9in
47.0m × 8.7m × 2.36m
Machinery: 2-shaft CODOG: 1 20,000shp gas turbine, 2×2475bhp diesels; maximum speed 38kts. Range (presumably on diesels) 1000nm at 15kts
Armament: 1–30mm Gatling
Sensors: Radar Spin Trough
Complement: 16

Class:
BT-10 (named *Altaiskii Komsomolets* 13.3.75–1992), *BT-150* (named *Komsomolets Estonii* 11.6.75–Apr 1991)

These ships were conceived to escort larger ships through mined waters, dealing with mines with combination (pressure, acoustic, magnetic) fuzes. The unusually high power was intended to allow them to move at high speed to create a turbulent wake which would trigger pressure mines. In theory, the sweeper would first sweep for magnetic and acoustic mines, then escort ships in the hydrodynamic field of its own wake, manoeuvering to distort the pressure fields of the ships so as not to trigger the mines. Alternatively, the sweeper could create very strong fields of its own, to trigger mines beyond ranges at which they would be lethal. Hence the solenoid coils which are partly visible on the ship's side. Machinery noise could be amplified to trigger acoustic mines. Because there was always a change that a triggered mine would destroy the sweeper, remote-control gear was installed.

They have wooden or plastic hulls, and large cable ducts running down both sides. A prominent stack serves a gas turbine generator, the propulsion diesels exhausting through the hull sides. The cables and generator presumably indicate magnetic sweeping.

'ILYUSHA' class *special minesweepers* (Project I-376, Yaroslavets)

Displacement: 50t standard; 70t full load
Dimensions: 80ft 1in × 16ft 1in × 4ft 7in
24.4m × 4.9m × 1.4m
Machinery: 1-shaft diesel, 450bhp = 12kts
Armament: None
Sensors: Radar: Spin Trough
Complement: 10

These 10 drone minesweepers entered service 1966–69. They are radio-controlled while sweeping, but are manned for transit.

Known designations: *BT-168*, *BT-293*, 8 others. This may be Project 14.376.

There are also towed minesweeping drones (about 25t, 39ft 3in × 13ft 1in [*12m × 4m × 4.9m*]), which deploy magnetic sweeps or line charges to clear lanes.

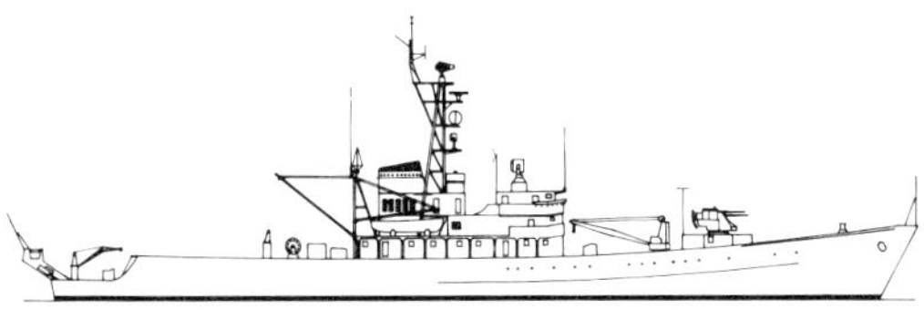
'Alesha' class

'ALESHA' class *minelayers* (Project 317)

Displacement: 2900t standard; 3500t full load
Dimensions: 318ft 3in × 45ft 11in × 17ft 9in
97.0m × 14.0m × 5.4m
Machinery: 4 diesels, 8000bhp = 17kts. Range 4000nm/16kts, 8500nm/8kts
Armament: 4–57mm (1×14), 300 mines max
Sensors: Radar Don-2, Strut Curve, Muff Cob
Complement: 190

Class:
Pripyut, *Sukhona*, *V'chegda*

Three completed 1967–69, probably on the Black Sea. Soviet designation ZM Zagraditel Minnyy, or minelayer. These ships are used to lay both defensive minefields and ASW harbour defence nets.

'RYBAK' class *inshore minesweepers* (Project 1338 Nel'ma)

Displacement: 60t
Dimensions: 72ft 2in × 19ft 8in × 3ft 3in
22m × 6 × 1m
Machinery: 1 36-D diesel, 150hp = 9kts. Stores endurance 7 days
Armament: 2–12.7mm MG (1×2)
Complement: 7

Five boats in service 1992, including *PT-223*. This is apparently a fishing boat modified either as a mobilisation prototype or as a COOP boat for Q-route work.

'BALTIKA' class *inshore minesweeper* (Project 1380)

Displacement: 150t standard; 210t full load
Dimensions: 83ft 4in × 22ft 4in × 7ft 10in
25.4m × 6.8m × 2.4m
Machinery: 1 NDB-26 diesel, 300hp = 10kts. Range 1350nm at 10kts,
Armament: 2–12.7mm MG (2×1)
Complement: 10

Class:
BT-225

Prototype base minesweeper completed at Kiev in 1983.

SChS-150 class *minesweeper* (Project 572)

Displacement: 80t
Dimensions: 82ft 8in × 18ft 4in × 3ft 11in
25.2m × 5.6m × 1.2m
Machinery: 1 diesel, 110hp = 9kts. Stores endurance 8 days
Complement: ?

No armament listed. For Caspian Flotilla.

Project 696/696DB *inshore minesweepers*

Displacement: 293t standard; 315t full load
Dimensions: 147ft 8in × 24ft 7in × 4ft 11in
45.0m × 7.5m × 1.5m
Machinery: 2 diesels, 2200bhp = 16.5kts. Range 1000nm at 12kts
Armament: 1–30mm AK-630, 20 mines

Four units built 1976. GRP hull, stabilised.

Successor States to the former Soviet Union

All naval forces of the successor states are comprised of former Soviet vessels or, in the case of the Baltic states, ships transferred from Scandinavian navies or the former GDR. For particulars and illustrations of the various vessels below, see the entries for the originating state.

AZERBAIJAN

A naval force was set up in July 1991, based at Baku on the Caspian Sea. It had been decided that half of the former Soviet Caspian Sea flotilla would be divided between Russia and Azerbaijan, with the other half to be shared later between Turkmenistan and Kazakhstan. Other units came from the KGB Border Guards. The following vessels were reported extant in 1995:

One 'Petya II' class frigate, *404* (ex-*SKR-16*), transferred 1992.

One 'Polnocny B' class medium landing ship, transferred August 1992.

Three 'Polnocny A' class medium landing ships, transferred August 1992 from the Caspian flotilla. Believed to be in poor condition.

One 'T 4' class landing craft transferred July 1992.

Two 'Osa II' class fast attack craft (missile), transferred August 1992. One believed to have been a training vessel prior to transfer.

One 'Zhuk' class coastal patrol craft.

Two 'Stenka' class fast attack craft (torpedo). Transferred July 1992 from KGB Border Guards. Likely that sonar and ASW TT removed.

Three 'Sonya' class coastal minesweepers transferred August 1992.

Two 'Yevgenya' inshore minesweepers transferred 1992.

ESTONIA

A small coastal patrol force was established in April 1992, with ex-Swedish, ex-Finnish, Soviet and East German vessels, based at Tallinn. Extant in 1995 were:

One ex-Danish *Maagen* class fishery protection vessel. Transferred 29 March 1994. Renamed *Ahti.*

Six ex-GDR 'Osa II' class fast attack craft (missile). To be transferred from Germany disarmed in 1995. Planned to be re-armed with light weapons only for use as patrol craft.

Two ex-GDR 'Kondor I' class intelligence collection ships. Offered to Estonia 1993. Transferred disarmed.

Two ex-Russian 'Zhuk' class coastal patrol craft. Originally based at Tallinn as part of the Russian Baltic Fleet and transferred to Estonia 1992. Only one inoperable unit reported extant 1 995.

One ex-Norwegian *Storm* class FAC(M): *Torm* (ex-*Arg*), transferred disarmed December 1994.

Three Swedish Kbv 236 class patrol boats. First unit transferred April 1992. Second later in 1992 and the third in October 1993. First Estonian naval units.

Three ex-Finnish patrol craft transferred November 1992. Operated by the Coast Guard.

GEORGIA

Although the intention to establish a naval force, based at the port of Poti on the Caspian Sea, was announced in 1992 following the break-up of the Soviet Union, the civil strife that has beset the country in recent years has apparently prevented this. No transfers have been announced by the Russian Navy as of 1995.

KAZAKHSTAN

Kazakhstan was due to receive up to one-quarter of the units of the former Soviet Caspian Sea flotilla (see under Azerbaijan above) and naval bases were under construction at Fort Chevchenko on the Caspian Sea and Aralsk on the Aral Sea, but no transfers have been reported as of 1995.

LATVIA

A Coast Guard was established in 1992, operating three ex-GDR 'Osa I' FAC(M)s transferred unarmed in 1993, two ex-GDR 'Kondor I' class minesweepers transferred in August 1993 and one ex-Norwegian *Storm* class FAC(M) transferred disarmed in December 1994, as well as a number of converted trawlers and smaller patrol craft.

LITHUANIA

Re-established on 17 November 1992, the Lithuanian Navy has a total of 200 personnel. Although Germany offered former GDR ships in 1993, including two 'Koni' class frigates and other patrol craft, this was rejected due to lack of resources to operate them. Naval units extant 1995 were:

Two 'Grisha III' class frigates, *Zemaitis* (F 11, ex-*MPK-44*) and Aukstaitis (F 12, ex-*MPK-108*). Transferred from the Russian Baltic fleet in November 1992. Probably in use as stationary training vessels.

One ex-Norwegian *Storm* class FAC(M): *Dzukas* (ex-*Glimt*), transferred unarmed December 1994.

A number of smaller patrol craft and auxiliaries are also operated.

TURKMENISTAN

Turkmensistan was also to receive units of the former Caspian Sea flotilla, but no transfers have been reported as of 1995.

UKRAINE

After achieving independence in 1991, Ukraine initially claimed ownership of the entire former Soviet Black Sea Fleet, which was based almost entirely in what had become Ukrainian ports. This was opposed by Russia, which proclaimed the fleet to be a strategic asset of the Commonwealth of Independent States. Since then, there have been continuing negotiations, interspersed with defections of vessels and in some cases outbreaks of fighting between Ukrainian personnel and Russian sailors. On 3 August 1992, an agreement was signed allowing for joint control of the Black Sea Fleet until 1995, when division of forces would be agreed, Ukraine expecting to receive the majority. However, the continuing disagreements between the states have lead to renegotiation of the deal. In the latest recorded agreement, dated 15 April 1994, Russia was to retain control of half of the fleet outright, leasing a further third from Ukraine whilst the remainder went to the new state. Disputes have continued, however, about which state gets which vessels, and about where the forces of each state are to be based.

Ukraine has considerable potential naval capability, controlling as she does both the bases in the Black Sea, including Sevastopol, and the main shipyard for building major surface vessels at Nikolayev, which seriously affected the Russian Navy, leading to the deletion of the *Kiev* class air-capable ships, as the facilities for their repair were no longer in Russian hands. Ships building or refitting at Nikolayev have been taken over by Ukraine for completion for her navy (see below). Shortage of resources, however, may limit how far Ukraine is able to exploit its potential strength.

The major fleet units are based at Sevastopol, with Border Guard and other minor forces at Odessa and smaller ports. Approximate personnel strength was 3500 in 1993. The Border Guard also operates a mixed force of former KGB patrol craft (see below). Units known to be in Ukrainian service in 1995 were as follows:

One ex-Russian *Slava* class cruiser: *Vilna Ukraina* (ex-*Admiral Lobov*), taken over in March 1993 while building at Nikolayev. Completing for the Ukrainian Navy 1995. To be the fleet flagship. The cruiser *Slava,* refitting at Nikolayev since December 1990, was taken over incomplete, but work has reportedly not restarted.

Two ex-Russian 'Kara' class cruisers, Russian *Nikolayev* and *Tashkent,* at Nikolayev since 1987, were stricken from the Russian Navy in 1992. There were reports that one was to be recommissioned into Ukrainian service, but both were sold to India for BU in August 1994.

Two ex-Russian 'Krivak III' class frigates: *Hetman Petro Sagaidachny* (ex-*Latsis*), *Hetman Badya Vyshnevetskyy* (ex-*Krasny Vympel*). Both taken over incomplete 1992. *Sagaidachny* entered Ukrainian service in March 1993, and deployed to the Mediterranean in 1994 and to the Indian Ocean in early 1995. Largest operational unit of the Ukrainian Navy. *Vyshnevetskyy* due to complete 1995.

Four 'Grisha V' class frigates: *Lutsk, Ternopol, Zaporozkaya Sech, Lvov. Lustk* taken over whilst under construction 1993 and never part of Russian fleet. Extant 1995. Remainder building at Kiev.

Two ex-Russian 'Pomornik' class air-cushion vehicles: *Donets, Ivan Bohun.*

MARITIME BORDER GUARD

Forty-two former KGB patrol craft handed over to Ukrainian control February 1992. Known to include twenty 'Stenka' class patrol craft, plus 'Shmel' class river gunboats and possibly several 'Svetlyak' class corvettes from the Black Sea Fleet.

Spain

The year 1947 saw Spain's fortunes at a low ebb. The Civil War had cost more than half a million lives, with hundreds of thousands more forced into exile. Recovery was hampered by the great material destruction, particularly to transport, and by the almost immediate onset of the Second World War. Franco was able to keep the devastated and impoverished country neutral between 1939 and 1945, although his sympathies were with the Axis Powers, which had contributed so largely to his victory in 1938–39, and his Blue Division was formed to fight on the Eastern Front (1941–43). When in 1943 Axis defeat became a certainty, he shifted to a policy of strict neutrality which earned him little credit, and no reward, from the victors. Spain was excluded from the United Nations in 1945 and ostracised by the Western Democracies, while the Communist Bloc, which recognised the defunct Republic, remained implacable foes. Although VE and VJ Days were not followed by a blockade of Spain (or even the invasion which many on the Left would have liked), her economic and political isolation brought grim years of near famine, rural misery and guerilla activity.

Franco, however, was not destined to be overthrown by his enemies without or within; the start of the Cold War was Madrid's opportunity and led gradually to the establishment of normal relations with the West. Internally, the regime relaxed and slightly released its grip, with the Falange virtually reduced to impotence, control of education returned to the Church, a fundamental law on personal freedoms and the promise of a return to constitutional monarchy at a future date. The state controls and stagnant economic nationalism of the corporate state were replaced by a normal capitalist system, and rapidly growing foreign investment. The 1953 Military Aid agreement with the United States gave Franco extensive military and financial aid in return for four US air and naval bases; with this accolade, Spain returned to international respectability.

The remaining Franco years (1953–75) were a time of political stagnation, as the many forces moving towards change marked time and waited for the old Caudillo's death. His mastery of the army and the rubber-stamp parliament, nominated without political parties or normal elections, guaranteed that there would be no change in the leadership or the regime in his lifetime. The economic growth and prosperity of the 1950s ensured that the bulk of the population had no wish for change, although the high growth rate brought the usual associated stresses: a wage-price spiral, inflation and labour problems. Basque separatism, harshly crushed in 1938–39, revived and developed a ruthless terrorist element.

Franco's death in 1975 was the signal for rapid political changes. The monarchy, considered an important element of continuity and stability, was restored and the formation of political parties allowed, while a new constitution, free elections and a proper parliament brought Spain into line with Western Europe. The changes were achieved smoothly and peacefully, diehard generals in the army preferring to protest and acquiesce. The military command structure was modernised from 1983 and by the end of 1986 the last senior officers with civil war experience retired, minimising the likelihood of a military coup. The Socialist government of Felipe Gonzales, installed in October 1982, remained in power in 1995; his government continued the political, economic and military integration of post-Franco Spain into mainline Western Europe.

The very rapid economic growth of the 1950s and 1960s ended with the rise in oil prices and world recession. The economy remains generally buoyant, with a good growth rate, but weak; the problems are manifested in a higher than average inflation rate, high unemployment (22 per cent in 1985), and a current-account balance of payments deficit. The country nevertheless successfully joined the EEC (1 January 1986) and the European Monetary Union, the so-called currency 'snake' (1989).

In external relations, Spain's return to the Western camp from 1948 and subsequent close relationship with the United States as a staunch anti-Communist ally were maintained, but membership of NATO did not follow until 30 May 1982. While in opposition, Felipe Gonzales had campaigned against NATO membership and the presence of US bases in Spain; in power, he promised a referendum on continuing NATO membership, suspending Spanish participation in the meantime. In 1984, considerations of Western European solidarity and the military modernisation of Spain led him to reverse his opposition to NATO, and to introduce certain conditions for approving membership: these included remaining outside the fully integrated NATO command structure, reducing the US military presence in Spain, and a ban on the 'introduction, maintenance or installation' of nuclear weapons in the country.

The referendum (12 March 1986) gave a 52–40 per cent majority in favour of membership. The USA withdrew from the Torrejon air base, but facilities at other US bases and communications centres, including the naval base at Rota, were unaffected (in the mid–1980s, the US Navy had stopped using Rota as its major overseas ballistic missile submarine base). In 1988 bilateral defence agreements were signed with Portugal and Britain: a Portuguese admiral could normally command Spanish forces in the Eastern Atlantic, while, pending settlement of the Gibraltar question, Britain and Spain would both deal with NATO Southern Command in Naples in preference to each other. In November 1988, Spain and Portugal entered the Western European Union (WEU), signifying Spanish acceptance in theory of NATO's doctrine of nuclear deterrence. In practice, to placate unilateralists, the storage of nuclear weapons within the country or overflights by planes carrying them would not be allowed, but, continuing existing practice, ships which might be carrying such weapons would be admitted without question into Spanish ports. In 1995 Spain is still outside the NATO military command structure, but otherwise is fully involved in almost every NATO activity, for example participating in naval exercises from 1987 onwards.

Unlike neighbouring Portugal, Spain did not cling to her colonies, those remaining after the 1898 debacle being quietly relinquished during the later Franco years. The Spanish Protectorate over part of Morocco ended in 1956 (and the Ifni enclave was ceded in 1969), Equatorial Guinea was turned over to African rule in 1968, and the Spanish Sahara was surrendered to Moroccan pressure in late 1975. In giving up her Moroccan protectorate, however, Spain insisted on keeping the coastal cities of Ceuta and Melilla, Spanish since the sixteenth century, despite periodic Moroccan demands for the cession of both them and the Canary Islands, a short distance off the Moroccan coast. Meanwhile, Spanish claims to Gibraltar, in British hands since 1704, have been pressed with equal insistance and constancy. The land frontier with Spain was closed completely between 1969 and 1985, and a satisfactory settlement still seems remote.

NAVAL DEVELOPMENTS 1947–89

The progress of the Spanish Navy in this period must be related to the politico-economic background. The navy of 1947 was very much the navy of 1939, the World War having made little difference to the fleet except to propel it almost a decade further into obsolescence. The light cruisers of the *Alfonso* class had undergone minor face-lifts, and the older *Mendez Nunez* a more substantial reconstruction as an anti-aircraft cruiser, but in general the major units had outdated AA and fire-control systems and were without radar and sonar, besides having none of the wartime 'A's and 'A's characteristic of units involved in hostilities. Ships authorised in the 1920s and 1930s but not begun until just before the Civil War were still on the stocks, construction having been suspended between 1936 and 1939, and again between 1940 and 1944. A modest start to construction was allowed in 1944, when the completion of two destroyers and three submarines and the building of eight AA escorts was authorised.

A much more ambitious programme was announced in 1945: nine destroyers, nine large torpedo-boats, six corvettes, six submarines, six motor torpedo-boats and twelve patrol vessels were to be put in hand to replace older units and tonnage worn out in the Civil War. This programme, however, was delayed for many years because of limited funds and fluctuations in the availability of what could be allocated. Six destroyers, six submarines and ten patrol vessels were cancelled; the designs were distinctly old-fashioned (essentially pre-Second World War) and most of the ships which survived the financial stringencies were still fitting out when the 10-year 1953 Agreement with the United States marked the beginning of US assistance under the Mutual Defense Assistance Program, and a new era for the Spanish Navy.

The effect of this agreement and its extensions were to give Spain access to recent developments in ship systems design and weapon technology. Naval Estimates more than doubled between 1954 and 1961, and personnel increased from 36,000 to 42,000 (the figure for 1945 was only 26,000). The acquisition of modern or renovated US warships started almost at once; twelve new coastal minesweepers were tranferred between 1954 and 1959, and five *Fletcher* class destroyers between 1957 and 1960. About thirty Spanish ships were modernised betwen 1959 and 1963, with new anti-aircraft guns, anti-submarine weapons, radar and other electronic equipment (some while still fitting out to their original designs): two *Alava* class destroyers, nine *Audaz* class fast frigates, two *Pizarro* class frigates, two *Jupiter* class frigate-minelayers, five *Atrevida* class corvettes and seven *Almanzora* class minesweepers, plus two submarines streamlined.

Similar plans announced for four old *Churruca* class destroyers were not implemented because their useful lives after modernisation would have been too short and, instead, a new construction programme of four destroyers, four escorts and twelve patrol vessels was put forward. Orders were not placed immediately, but negotiations were started with Britain for the construction of four *Leander* class frigates and possibly *Oberon* class submarines (Spain was anxious to obtain experience of non-American techniques and practices, and Britain had supplied radar equipment for the new destroyer *Oquendo*. Unfavourable UK Press and Parliamentary comments on these proposals (and on the Spanish regime) killed the negotiations (1964), and the orders went to the United States and France. Agreements were signed in 1966 for the construction in Spain of five modified *Knox* class frigates and two *Daphne* class submarines, with three more proposed. The strength envisaged at this time for the Spanish Navy in 1975 was two aircraft carriers, two cruisers, eight destroyers, twelve frigates, eight submarines, sixty minesweepers and 100 auxiliaries with twenty-seven anti-submarine patrol aircraft and forty-eight helicopters.

A major reinforcement for the fleet arrived in 1967, when the light aircraft carrier *Dedalo*, ex-USS *Cabot*, was transferred. The first Spanish carrier (her namesake of the 1920s, an ex-merchantman fitted for transporting balloons and seaplanes, was not a front-line ship), she initially operated anti-submarine helicopters only, but an anti-ship capability was developed from 1973 with the order of Harrier AV-8 VSTOL aircraft (Matadors) from the US Marines; two of the first five were delivered in 1976 and the order of ten was complete by 1981. Smaller ships were also acquired. The United States agreed in August 1970 to supply two submarines, five destroyers, four ocean minesweepers, three tank landing and one dock landing ships (replacing an ammunition ship, while an oiler was not transferred), and the last of these were transferred in October 1973. Under a further Agreement in 1976, the United States was to provide four more ocean minesweepers and a repair ship, but these units were not handed over.

Meanwhile, in January 1973, a Spanish new construction programme was announced, consisting of one light cruiser, three guided missile destroyers, ten corvettes, two *Agosta* class submarines, some PF type craft, twelve patrol craft for fishery protection, seven auxiliaries and several landing craft. For financial reasons, part of this programme, including the cruiser, the guided missile destroyers, two corvettes and the PF craft, was deferred until 1977–78 and was then ordered in modified form; the cruiser as a light aircraft carrier of the US Sea Control Ship type, and the destroyers as US 'FFG 7' class frigates.

The ultimate target strength of the fleet is two task groups, each consisting of one light carrier, one cruiser and four missile frigates, and four escort groups, each of two missile and two *Descubierta* class frigates. To achieve this, the next (1983–90) phase of the construction programme, as first announced, proposed the building of a second sea control ship, two cruisers, eight more missile frigates (either FFG or enlarged *Descubierta* type), two Tripartite minehunters, besides auxiliaries, ten more Matadors, six maritime patrol and four anti-submarine aircraft. Once again, however, financial considerations intervened, and the phase was modified to provide two destroyers, seven frigates, the minehunters and a support ship, together with auxiliaries and additional aircraft.

In the following year, the phase was further modified and extended (1984–90/95), becoming three submarines, one or two (later definitely two) destroyers of the US *Spruance* (DD-963) class, five frigates (later four, one FFG-7 and three Type 552/A, a modified *Descubierta*), six FAC (missile), twelve (later sixteen) tripartite minehunters and sweepers, the support ship and LPD, LSTs, plus OPVs, smaller ships and auxiliaries.

Since joining NATO, defence spending has averaged around 2 per cent of GNP, at the lower end of the NATO range. In 1986, naval spending was increased from 20 per cent to 30 per cent of the defence budget to replace obsolete tonnage, the money being taken from the army. In spite of this increase, actual orders under the 1984–90/95 phase were minimal before it was superseded by the 1988–96 phase, which postponed the projected submarines, and substituted five NATO NFR 90 AAW frigates for the destroyers and frigates proposed earlier; it was tentatively proposed that the prototype NFR 90 would be built by Bazan, for delivery in 1998. New patrol vessels would be substituted for the *Atrevida* quartet.

THE ALTA MAR PLAN, 1989–2002

The most recent new construction programme, the 1989–2002 *Alta Mar* ('High Seas') Plan, recast in 1990, included the following: (a) to replace the obsolete *Gearing* class, four F-100 type ASW-optimised frigates would be built between 1994 and 2001, the first entering service in 1998; (b) to replace the modernised *Baleares* class, five F-110 type AAW-optimised frigates would be built between 1998 and about 2007, the first entering service in 2003 (Spain was one of the last countries to abandon the NFR 90 frigate project, 2 January 1990); (c) the postponed submarines to replace the *Daphne* class would be delivered from 2002, the *Agosta* class being modernised in the interim; (d) the first of twelve GRP MCMV (eight hunters, four sweepers) to replace the existing elderly ex-US craft would be delivered in 1994; (e) a fast replenishment ship for the carrier task force, designed by Bazan and the Dutch Nevesbu consortium and based on the Dutch *Zuiderkruis* was to be ready in 1994 (an interim fleet oiler was fitting out); and (f) modernisation of the amphibious warfare force would begin with the construction of a LPD in 1998.

No mention was made of a second carrier (although the ultimate target strength remains two task groups), nor was the modified *Descubierta* class included. The *Alta Mar* Plan was formulated before the end of the Cold War and the onset of the world economic recession, and was rapidly in trouble. 1990 budget cuts caused a one-year delay in the start of the MCM vessels, and a slowing of the fifth and sixth FFG-7 frigates, and the order dates of the new types have all slipped.

Present (September 1995) thinking is that the first F-100 class frigate will not enter service until at least 2001, while the follow-on F-110 class may not materialise; the MCM programme has been curtailed drastically, but the new replenishment ship will now be completed in mid-1995 and construction of the first LPD is planned to begin in 1995.

TECHNICAL MATTERS

In technical matters, the Spanish Navy attaches great importance to attaining self-sufficiency in all aspects of ship design and construction, but the limited development of the defence industries has meant the building of factories and plant for manufacturing foreign equipment under licence. Taking the *Descubierta* class of light frigates as examples, the basic ship characteristics were obtained by 'stretching' a German design for Portugal, the machinery is German in origin, the gun armament Italian and Swedish, the missile equipment American and the radar and fire control systems Dutch. The highly successful end-product is an excellent example of cooperation by the international armaments industry, but hardly a monument to Spanish self-sufficiency. More recently, however, Spain has itself originated an advanced weapon system, the Meroka CIWS, consisting of two non-rotating six-barrelled Oerlikon 20mm R TG guns, with Lockheed Electronics fire control incorporating Doppler radar and stabilised optical sights. In the present economic climate, self-sufficiency is becoming impracticable, and various multinational cooperation projects and feasibility studies are in hand, with the USA, Netherlands, Germany, France and Britain as the main partners. Since being set up in 1947 the nationalised Empresa Nacional Bazán has built all of Spain's locally constructed warships – frigates at the Ferrol yard, surface ships and submarines at Cartagena, and smaller craft at La Carraca, Cadiz.

STRATEGY AND OPERATIONS

The functions of the navy include protecting national territory and interests – remembering Spain's highly strategic location – and projecting the power and prestige of the state. Between 1947 and 1982 warships participated in colonial defence on several occasions. The Ifni enclave, attacked sporadically by Moroccan irregulars, was the scene of cooperation with the army and air force, including shore bombardments, in 1957, 1958, 1962, 1964 and 1965. In an ostentatious display of force during one of these clashes, the cruisers *Canarias*, *Miguel de Cervantes* and *Mendez Nunez*, with destroyers and other warships, entered Moroccan territorial waters off Agadir and anchored 100 yards from the harbour entrance. In 1961 *Canarias* was sent further south, to support the administration in troubled Fernando Po and Rio Muni. When these territories became independent, *Canarias* returned to cover the evacuation of the remaining Spanish settlers. In 1969 and 1970 Spanish claims to Gibraltar were pointedly expressed by sending the carrier *Dedalo* and escorts to anchor in the Bay of Algeciras near the Rock. The British responses were predictable and traditional: in 1969, for example, the carriers *Eagle* and *Hermes* were ordered to Gibraltar, arriving just before the Spanish ships. In the early 1980s, the main units of the fleet were based at Ferrol, except for the *Descubierta* class frigates at Cartagena. The older FRAM I destroyers and the submarines operated in the Mediterranean.

With Spanish membership of NATO, the perceived threats changed. Two types of conflict were envisaged, the local and the large-scale. The former, a possible consequence of the Islamic revival in the Magreb, might include (a) an attack on Ceuta and Melilla, (b) less probably, an assault on the Canary Islands, or (c) serious attacks on the fishing fleet; such situations would be dealt with by Spain alone. In the event of an East-West conflict, Spain would act in close coordination with her NATO partners; in naval terms, the East Atlantic was regarded as more important than the Mediterranean, and the Gibraltar Straits was a particularly vital focus of major trade routes.

In line with the usual NATO interpretation of the likely course of a major conflict in the 1980s, Spanish ports might have had to replace more vulnerable ports further east for the unloading of reinforcements and supplies from the USA.

The disposition of Spanish ships was altered to reflect the changed strategy. At present the main units of the fleet, the carrier and FFG-7 frigates, are based at Rota, Cadiz, while the *Baleares* class frigates and (until they were withdrawn) the FRAM I destroyers are at Ferrol, with the smaller *Descubierta* class frigates, submarines and MCMV at Cartagena for the Mediterranean. The fleet command centre was moved from Ferrol to Rota in 1987. The amphibious force is based at Puntales, Cadiz. There are also bases at Tarifa, Cadiz (for small craft), Port Mahon on Minorca and Las Palmas in the Canary Islands.

Spanish warships have not been used operationally in recent years, apart from EEZ and anti-terrorist patrols and in fishery protection. Patrol craft also intercept would-be drug smugglers and illegal immigrants. Following Iraq's occupation of Kuwait (2 August 1990), the frigates *Santa Maria*, *Descubierta* and *Cazadora* were sent to the Red Sea and Gulf as part of the International Force; they were relieved in November by the *Numancia*, *Diana* and *Infanta Cristina*. Spanish warships did not participate in 'Desert Storm' and subsequent active operations. More recently, Madrid has sent warships to assist in the blockade of former Yugoslavia.

NAVAL AIRCRAFT

During the past decade, considerable effort has been devoted to the Naval Air Service. Eleven EAV–8B (Harrier II) STOVL aircraft for *Principe de Asturias* were delivered from late 1987; these aircraft are to be uprated to Harrier II Plus standard with APG–65 radar, and eight new Harrier II Plus are on order for delivery from 1996; the first AEW version of the Sea King began trials in September 1987. Six Seahawk LAMPS III helicopters for the FFG–7 frigates were delivered in 1988–89 and another four are on order. P–3 maritime patrol aircraft are operated by the air force under the navy's operational control; seventy-two EF–18 A/B provide land-based cover for warships in the western Mediterranean or off Morocco's Atlantic coast.

CONCLUSIONS

In the past, Spain's Navy often had a poor reputation for efficiency, upkeep and fighting spirit. Sheridan's eighteenth-century comment, 'the Spanish Fleet you cannot see, because it is not yet in sight,' reflected foreign opinion, and its miserable performance in the Spanish-American War did nothing to change this view. Rapid improvements have, however, taken place in the present century, particularly during the past twenty years, and the Spanish Navy of 1995 is a balanced, well-trained, carefully maintained and up-to-date force.

FLEET STRENGTH 1947

CRUISERS

No	Name	Launched	Disp (full load)	Fate
Eugenic class				
–	NAVARRA (ex-*Republica*, ex-*Reina Victoria Eugenia*)	21.4.20	6500t	Stricken 1956
Nunez class				
C 01	MENDEZ NUNEZ	27.7.22	6045t	Stricken Dec 1963
Alfonso class				
C 11	GALICIA (ex-*Libertad*, ex-*Principe Alfonso*)	23.1.25	9900t	Stricken 2.2.70
C 12	ALMIRANTE CERVERA	16.10.25	9660t	Stricken 1966
C 13	MIGUEL DE CERVANTES	19.5.28	9900t	Hulked 1963
Canarias class				
C 21	CANARIAS	28.5.31	13,500t	Stricken 17.12.75

Nunez class
During the 1950s the number of 20mm AA guns was altered to 16 (4×4) and later to 23 (4×4, 7×1). In the mid-1950s she received 2 DCT and 1 DC rack. Allocated pennant number C 01 in 1961. In the 1960s she served as an accommodation hulk for minesweeper crews at Las Palmas, and the forward 120mm guns were probably removed. Finally used as an accommodation hulk at Cartagena.

Alfonso class
Reconstructed from 1940 onwards: *Galicia* rearmed July 44, *Cervantes* July 45. Both reached 31.5kts in 1946. *Galicia* and *Cervantes* 8–90mm (4×2) replaced c1952 by 8–37mm AA (4×2). In 1957 (?1960) 12 TT (*Cervera*) or 6 TT (*Galicia* and *Cervantes*) were removed. In late 1950s *Galicia* received a radar scanner just abaft the main director on the bridge on a small lattice tower. In *Cervera* 4–105mm AA were replaced in late 1950s by additional 8–37mm AA (4×2) in lieu of planned US 76mm guns. In 1961 they were allocated the new pennant numbers in the table above.

Canarias class
Reconstructed at Ferrol between October 1952 and February 1953, when the broad trunked funnel was replaced by two vertical funnels, a return to the original plan which had never been carried out. 1960s plans for a major reconstruction as a guided missile cruiser were not implemented, as by then she was considered too old. Details seem not to have been finalised, but the result would probably have been on the lines of the Dutch *De Zeven Provincien* or Italian *Giuseppe Garibaldi*, with a Terrier installation aft, modern AA guns either side amidships and perhaps the two twin 8in turrets forward. Instead, a limited modernisation in 1969 provided a full complement of radars and a CIC; final characteristics were:

Displacement: 13,750t full load
Armament: 8–8in/50 (4×2), 4–4.7in/45 (4×2), 4–40mm/70 (4×1), 4–37mm/70 (4×2), 1 or 2–20mm, 2 DCT, 1 DC rack
Sensors: Radar SG-6B, MLA-1B, Marconi surface search, Decca 12

The pennant number C21 was allocated in 1961. *Canarias* was sold on 14 September 1977 and broken up in 1978.

DESTROYERS

No	Name	Launched	Disp (full load)	Fate
Alsedo class				
–	ALSEDO	26.10.22	1145t	Stricken 1957
–	JUAN LAZAGA	Mar 1924	1145t	Stricken 1961
–	VELASCO	June 1923	1145t	Stricken 1957
Churruca class – 1st Group				
D 11 (11)	SANCHEZ BARCAIZTEGUI	24.7.26	2087t	Stricken 1965
D 12 (12)	JOSE LUIS DIEZ	25.8.28	2087t	Stricken 1966
– (32)	LEPANTO	7.11.28	2087t	Stricken Aug 1957
– (34)	ALCALA GALIANO	12.4.30	2087t	Stricken 1957
– (42)	ALMIRANTE VALDES	8.9.31	2087t	Stricken 1957
D 13 (33)	CHURRUCA	Jun 1929	2087t	Stricken 1964
Churruca class – 2nd Group				
D 14 (13)	ALMIRANTE ANTEQUERA	29.12.30	2175t	Stricken 1969
D 15 (31)	ALMIRANTE MIRANDA	20.6.31	2175t	Stricken 2.3.70
D 16(22)	GRAVINA	24.12.31	2175t	Stricken 1964
D 17(24)	ESCANO	28.6.32	2175t	Stricken 1964
D 18 (43)	ULLOA	24.7.33	2175t	Stricken 1964
– (14)	JORGE JUAN	28.3.33	2175t	Stricken 1959
– (41)	CISCAR	26.10.33	2175t	Lost 17.10.57
Alava class				
D 23 (23)	ALAVA	19.6.47	2170t	Stricken 1978
D 21 (21)	LINIERS	1.5.46	2170t	Stricken 1982

Alava after modernisation

Navarret Collection

No	Name	Launched	Disp (full load)	Fate
Ex-Italian Poerio class				
–	HUESCA (ex-*Alessandro Poerio*)	4.8.14	1012t	Stricken 1953
–	TERUEL (ex-*Guglielmo Pepe*)	17.9.14	1012t	Stricken 1947
Ex-Italian Aquila class				
–	CEUTA (ex-*Falco*)	16.8.19	1730t	Stricken 1949
–	MELILLA (ex-*Aquila*)	27.6.19	1705t	Stricken 1950

Churruca class (1st and 2nd groups)
Armament of all surviving ships modified in the early 1940s to increase AA fire: 4–120mm/45 (4×1), 1–76mm/40 AA (only on 2nd group), 4–37mm AA (2×2), 4–20mm AA, 6–533mm TT and 4 DCT. Later some of 37mm and 20mm guns were removed. Originally had initials painted on bows, but allocated pennant numbers *c*1953. In 1961 the new 'D' pennant numbers quoted above were introduced. *Ciscar* ran aground in fog and broke her back off Ferrol and was discarded in 1958.

Modernisation of four *Churruca* class destroyers with new AA guns, ASW armament and new electronics, which was to be carried out under MDAP and planned for late 1950s and early 1960s, was later cancelled.

Alava class
Commissioned 21 December 1950 and 27 January 1951 respectively. Allocated pennant numbers 23 and 21 in *c* 1952–53 respectively, and D 52 and D 51 after modernisation. As modernised 1960–62: 3–76mm/50 AA Mk 22 on Mk 34 mounts (3×1), 3–40mm/70 AA SP48 (3×1), 2 side-launching racks for 6 ASW torpedoes, 2 Hedgehogs, 8 DCT, 2 DC racks. Refitted with lattice foremast and caps on after funnels. Forecastle lengthened aft *c*8m. Classified as ASW or fast frigates, with a displacement of 1842t standard and 2287t full load (31,500shp = 29kts max and oil 370/?540t). Completed modernisation 17 January 1962 *(Alava)* and 18 September 1962. Air search MLA-1B radar, one SG-6B surface search set, Decca TM-626 navigational and 2 Mk 63 fire control. One hull-mounted sonar SQS-30A. *Liniers* was used for training midshipmen at the Naval Academy until 1982.

Ex-Italian (acquired during 1938–39)
Probably never had initials on bows. *Huesca* used as TS in 1950s.

GUNBOATS AND MINELAYERS

No	Name	Launched	Disp (full load)	Fate
Castillo class gunboats				
–	CANOVAS DEL CASTILLO	21.1.22	1314t	Stricken 1959
–	CANALEJAS	1.12.22	1314t	Stricken 1953
–	DATO	1923	1314t	Stricken 1954
Durango type gunboat				
–	CALVO SOTELO (ex-Mexican *Zacatecas*)	27.8.34	2000t	Stricken 1957
Jupiter class gunboat-minelayers				
F 11 (16)	JUPITER	14.9.35	2600t	Stricken 1974
F 01 (39)	MARTE	19.6.36	2600t	Stricken 1972
F 02 (17)	NEPTUNO	19.12.37	2600t	Stricken 1970
F 12 (38)	VULCANO	12.10.35	2600t	Stricken 1977
Eolo class gunboat-minelayers				
F21 (28)	EOLO	30.8.39	1900t	Stricken 1972
F 22 (29)	TRITON	24.2.40	1900t	Stricken 1972
Pizarro class gunboats				
F 32 (18)	HERNAN CORTES	3.8.44	2246t	Stricken 2.12.71
F 34 (37)	MARTIN ALONSO PINZON	3.8.44	2246t	Stricken 1968
F 31	PIZARRO	3.8.44	2246t	Stricken 1968
F33	VASCO NUNEZ DE BALBOA	3.8.44	2246t	Stricken 1972
F 42 (27)	LEGASPI	8.8.44	2246t	Stricken 1978
F 35 (26)	MAGALLANES	8.8.44	2246t	Stricken 1972
F 36 (36)	SARMIENTO DE GAMBOA	8.8.44	2246t	Stricken 1973
F 41 (19)	VINCENTE YANEZ PINZON	8.8.44	2246t	Stricken 1982

Castillo class
Dato sunk in 1936 by *Jaime I* but refloated and reconstructed at Cadiz. *Canovas del Castillo* was rated as *cañonero* until 1958, when she was re-rated as *fragata*.

Durango type
Two of four 102mm guns removed ('B' and 'X' positions) and after superstructure reduced in late 1940s. Disarmed from 1952.

Jupiter class
All four allocated 'F' pennant numbers in 1961. *Jupiter* and *Vulcano* were modernised 1959–61 and in the latter year all were reclassified as frigates. The modernisation of *Jupiter* (with lattice foremast and SPS-6 search and tactical radar) was completed on 28 October 1960, and of *Vulcano* on 28 February 1961. Both ships: 4–76mm/50 AA Mk 26 (4×1), 4–40mm/70 AA (4×1), 2 Hedgehogs, 8 DCT, 2 DC racks, 254 (first) or 238 (later ships) mines. From the late 1950s *Marte (Neptuno* in brackets) were armed with: 4–120mm, 2(0)–76mm AA, 3(4)–37mm AA, 2(3)–20mm AA, 2 DCT and 264 mines. Modernised ships had air search MLA-1B and tactical SG-6B radars. *Neptuno* was employed as a training ship for midshipmen (51 cadets).

Eolo class
Dual-purpose frigates or gunboats and minelayers. Allocated 'F' pennant number in 1961. Armed with 4–105mm (not 102mm) AA Krupp Model 1933 (4×1), 4–37mm AA (2×2; not 4×1), 4–13mm AA (probably 2×2), 170 (*Triton* 180) mines, 2 DCT. In 1960s probably fitted with 2 fixed TT for ASW torpedoes. Removed from the effective list on 2 February 1972 and 2 March 1972 respectively and were scrapped.

Pizarro class
Some (for example *Pizarro*, *Vasco Nuñez de Balbao*, *Vicente Yañez Pinzon*) were armed originally with 2–105mm (2×1) in 'A' and 'X' positions instead of 6–120mm AA. Later all eight armed with 6–120mm AA (3×2), 8–37mm AA (4×2), 6–20mm AA (2×2, 2×1), 4 DCT and 30 mines. In 1958 were officially re-rated as frigates, and allocated 'F' pennant number in 1961. *Legaspi* and *Vicente Yañez Pinzon* as modernised: 2–127mm/38DP (2×1), 4–40mm/70 AA (4×1), 2 side-launching ASW torpedo racks (6 torpedoes), 2 Hedgehogs, 8 DCT, 2DC racks. Completed modernisation 14 January 1960 and 25 March 1960 respectively. Received lattice foremast, SPS-5B surface search radar, MLA-1B air search radar and Decca TM-626 navigational radar.

Hernan Cortes and *Sarmiento de Gamboa* (for disposal 1971) were not modernised. *Vicente Yañez Pinzon* was re-classified as *patrullero de altura* with new pennant number PA 41 *c*1981.

SUBMARINES

Name	Launched	Disp (surface)	Fate
'C' class			
C 2	4.5.28	916t	Discarded 1952
'General' class (ex-Italian)			
GENERAL MOLA (ex-*Evangelista Torricelli*)	27.5.34	880t	Stricken 1959
GENERAL SANJURJO (ex-*Archimede*)	1.10.33	880t	Stricken 1959
Type VIIC (ex-German)			
G 7 (ex-*U 573*)	17.4.41	757t	Stricken 2.5.70
'D' class			
D 1	11.5.44	1065t	Stricken 1965
D 2	21.12.44	1065t	Stricken 1971
D 3	20.2.52	1065t	Stricken 1971

Proposals announced in 1945 to build submarines *G 1–G 6*, generally similar to *G 7*, were abandoned *c*1960.

'General' class (ex-Italian)
Wore the pennant numbers C 5 and C 3 respectively from the early 1950s; the after 100mm gun was removed at about the same time.

Type VIIC (ex-German)
Allocated the pennant number S 01 in 1961; 20mm gun later removed.

'D' class
The three 'D' class boats were completed on 18 March 1947, 2 April 1951 and 20 February 1954 respectively. They were given the pennant numbers S 11, S 21 and S 22 in 1961.

D 1 was not modernised but the gun was removed in the early 1960s. The other two completed 2-year reconstructions on 10 December 1963 and 14 March 1963 respectively, in which the hull and conning tower were streamlined, and the deck gun and after TT removed.

MINESWEEPERS

No	Name	Launched	Disp (full load)	Fate
Bidasoa class				
M 01	BIDASOA	15.9.43	775t	Stricken 1972
–	GUADALETE	18.10.44	775t	Foundered 25.3.54
M 03	LEREZ	21.12.44	775t	Stricken 1971
M 02	NERVION	15.4.44	775t	Stricken 1972
M 05	SEGURA	15.5.45	775t	Stricken 1972
M 04	TAMBRE	18.10.44	775t	Stricken 1972
M 06	TER	15.4.45	775t	Stricken 1972

Guadalete was lost in a storm off Gibraltar while employed as a coast guard vessel. In 1961 their original pennant numbers (DM 1–DM 7 respectively) were changed to those noted above.

OTHER VESSELS
LT 11, *12* and *14*, the three surviving ex-German S-boats of the *S 2* type sold to Spain during the Civil War, were discarded in the 1940s. The motor torpedo-boat *LT 17* (ex-Italian *Napoli*, MAS 100), transferred in 1937, was stricken in 1951. *LT 21–LT 26* (the ex-German S-boats *S 73*, *S 78*, *S 124–S 126* and *S 128*, sold to Spain in 1943) were stricken between 1955 and 1958. The First World War-built 177t torpedo-boats Nos *14* and *17* were BU in 1952 after service as minelayers and training craft. There was also a large number of small patrol boats and motor launches, for fishery protection, customs and coast-guard work.

MAJOR SURFACE SHIPS

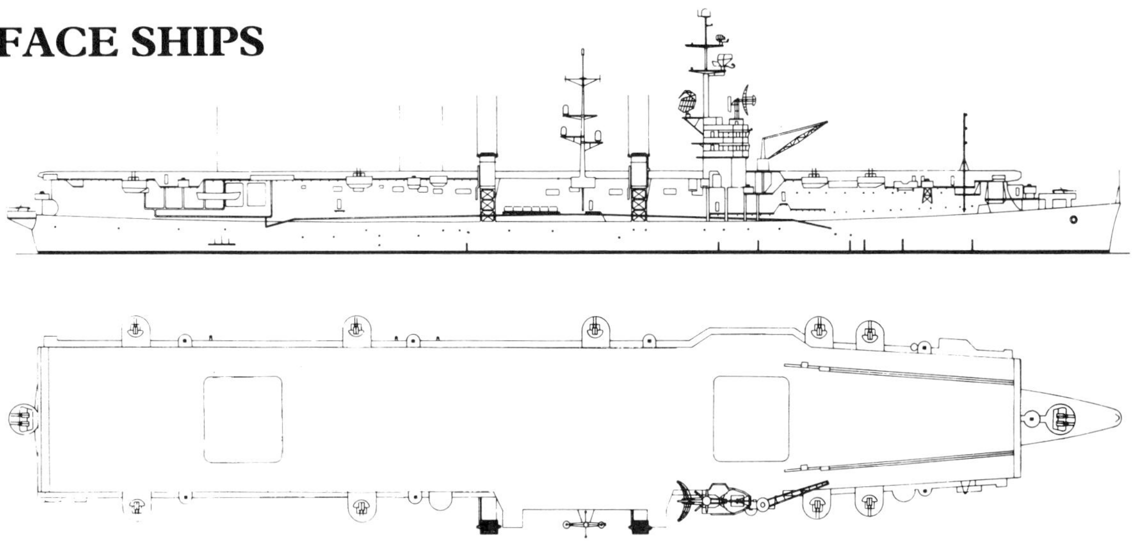

Dedalo 1974

DEDALO *light aircraft carrier*

Displacement: 13,000t standard; 16,416t full load
Dimensions: 600ft wl, 623ft oa × 71ft 6in wl, 109ft 2in oa × 26ft
182.9m, 189.9m × 21.8m, 33.3m × 7.9m
Machinery: 4-shaft General Electric geared turbines, 4 Babcock & Wilcox boilers, 100,000shp = 31kts. Oil 1800t. Range 7500nm at 15kts
Armour: Belt 5in on 0.63in STS, bulkheads 5in; amour deck over belt 2in
Armament: 26–40mm/60 AA (2×4, 9×2), 7 VSTOL aircraft, 20 helicopters
Sensors: Radar SPS-6, SPS-8, SPS-10, SPS-40, Mk 28 and Mk 29, Tacan
Complement: 1112 (excluding aircrew)

No	Name	Builder	Laid down	Launched	Comp	Fate
R 01	DEDALO (ex-*Cabot*)	New York SB	13.3.42	4.4.43	24.7.43	Preserved 1989

US *Independence* class light aircraft carrier, transferred to Spain under MDAP on 30 August 1967 and purchased outright in 1973. The transfer of *Thetis Bay* (LPH 6) was originally contemplated. Extensively refitted and modernised before transfer, with strenthened decks and lifts, new electronic equipment, and funnels 1 and 3 removed. Hangar capacity 20 aircraft. In later years, best speed reported as 24 kts. Suggestions that this ship would be fitted with Meroka CIWS and kept in service after completion of *Principe de Asturias* were incorrect. Returned to US in 1989 for preservation at New Orleans as a museum.

*Dedalo c*1978 Spanish Navy

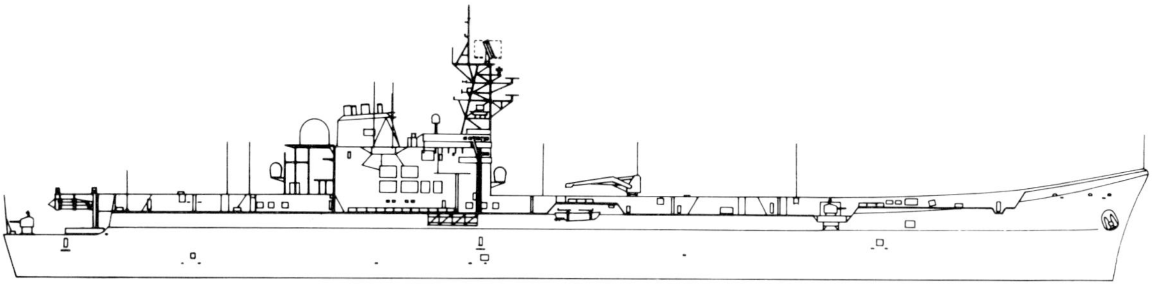

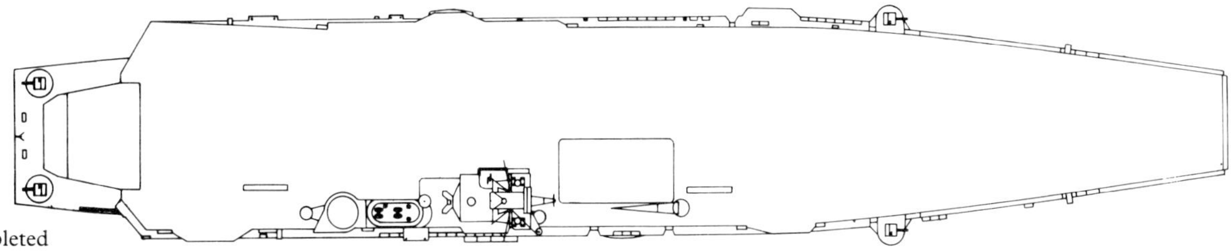

Principe de Asturias as completed

Designed by Gibbs and Cox, New York, as an updated Sea Control Ship, a type proposed for the US Navy in FY75 but not funded by Congress; the 12 degree ski-jump allows more effective use of the EAV–8B Matador II (Bravo) aircraft, delivered in 1987–88. The flight deck is 575ft 1in × 95ft 1in (*175.3m* × *29.0m*); the two lifts are right aft and forward of the island, the latter offset to starboard. The nominal aircraft complement is listed in the table; the maximum capacity is thirty-seven VSTOL aircraft and helicopters, but only twenty-four of these can be operated at once and a normal operational total is twenty; flying hours in 1989 exercise reported as limited to 18 out of the 24 by aircrew accommodation; in 1990 refit, flying control position was enlarged and extended, and accommodation increased (for additional fifty-six persons). As in the FFG–7 class, two small electric motors with retractable propellor pods provide emergency 'get home' facility. Fitting-out was delayed by the installation of a flag (lower) bridge and consequent modifications to command and control systems. Fleet flagship.

PRINCIPE DE ASTURIAS *light aircraft carrier*

Displacement: 16,700t full load
Dimensions: 615ft 2in pp, 642ft 8in oa × 79ft 8in × 30ft 10in
187.5m, 195.9m × 24.3m × 9.4m
Machinery: 1 shaft, 2 LM 2500 gas turbines, 46,400shp = 26kts. Range 6500nm at 20kts
Armament: 4–20mm Meroka CIWS, 2 saluting guns, 17 VSTOL aircraft and helicopters (see notes)
Sensors: Radar SPN-35A, SPS-52C, SPS-55, Tacan
Complement: 763

No	Name	Builder	Laid down	Launched	Comp	Fate
R 11	PRINCIPE DE ASTURIAS	Ferrol	8.10.79	22.5.82	30.5.88	Extant 1995

Principe de Asturias as completed

Rayo as reconstructed — Navarret Collection

Design based on the pre-war French *Le Fier* class of large torpedo-boats or light destroyers, as modified by the Germans. *Audaz*, *Meteoro*, *Osado* and *Rayo* completed with the quoted armament, although the 21in TT were apparently omitted, and were reconstructed as fast ASW frigates between 1959 and 1963, when particulars became: 1227t standard, 1548t full load, 31.6kts, 2–3in/50 AA (2×1), 2–40mm/70 AA, 2 Hedgehogs, 8 DC mortars, 2 DC racks, 2 side-launching racks for ASW torpedoes; the other five units were delivered with these data. Reported as unstable on first trials. *Ariete* grounded on the Galician coast and became a total loss.

AUDAZ class *destroyers*

Displacement: 1106t standard; 1474t full load
Dimensions: 295ft 3in pp, 308ft 2in oa × 30ft 9in × 10ft
90.0m, 93.9m × 9.4m × 3.0m
Machinery: 2-shaft Rateau-Bretagne geared turbines, 3 3-drum boilers, 30,800shp = 33kts. Oil 290t. Range 3800nm at 14kts
Armament: 3–4.1in/45 DP (3×1), 4–37mm AA, 8–20mm AA, 6–21in TT(3×2)
Sensors: Radar MLA-18, SPS-5B,SPG-34; sonar QHBa
Complement: 145

No	Name	Builder	Laid down	Launched	Comp	Fate
D 36	ARIETE	Ferrol	3.8.45	24.2.55	7.2.61	Wrecked 25.2.66
D 31	AUDAZ	Ferrol	26.9.45	24.1.51	30.6.53	Stricken 1974
D 34	FUROR	Ferrol	3.8.45	24.2.55	9.9.60	Stricken 1974
D 38	INTREPIDO	Ferrol	14.7.45	15.2.61	25.3.65	Stricken 1982
D 33	METEORO (ex-*Atrevido*)	Ferrol	3.8.45	4.9.51	30.11.55	Stricken 1975
D 32	OSADO	Ferrol	3.8.45	4.9.51	26.1.55	Stricken 1972
D 35	RAYO	Ferrol	3.8.45	4.9.51	26.1.56	Stricken 1974
D 39	RELAMPAGIO	Ferrol	14.7.45	26.9.61	7.7.65	Stricken 1975
D 37	TEMERARIO	Ferrol	14.7.45	29.3.60	16.3.64	Stricken 1975

Nine units of this class, based on the pre-war French *Le Hardi* class, were ordered in 1947–48, but six, *Blas de Lezo*, *Blasco de Garay*, *Bonifaz*, *Gelmirez*, *Langara* and *Recalde*, were cancelled in 1953. The 8–4.1in guns (4×2), 7–21in torpedo tubes (1×3, 2×2) and 2 DC throwers originally planned were replaced by the 4.7in guns and ASW weapons in the Table while the ships were building. Trials of *Oquendo* showed a dangerous lack of stability, so she was taken in hand for ballasting and reduction of top weight. This work was completed on 22 April 1963, and revised particulars were: 2582t standard, 3005t full load, 12ft 8in (*3.8m*) draught, 32.4kts, with armament reduced by the loss of 'X' turret (2–4.7in) and the 20mm AA. British radars and fire-control equipment were installed. *Roger de Lauria* and *Enseneda* were dismantled and completed to a much altered and enlarged design (see below).

OQUENDO class *destroyers*

Displacement: 2050t standard; 2765t full load
Dimensions: 363ft 6in pp, 382ft 1in oa × 36ft 5in × 11ft 10in mean
110.8m, 116.5m × 11.1m × 3.6m
Machinery: 2-shaft Rateau-Bretagne geared turbines, 3 3-drum boilers, 60,000shp = 39kts. Oil 659t. Range 5000nm at 15kts
Armament: 6–4.7in/50 DP (3×2), 6–40mm/70 AA (6×1), 4–20mm AA, 2 Hedgehogs, 2 side-launching racks for ASW torpedoes
Sensors: Radar Types 262, 275, 293, SNW-10; sonar QHBa
Complement: 267

No	Name	Builder	Laid down	Launched	Comp	Fate
D 41	OQUENDO	Ferrol	15.6.51	5.9.56	13.9.60	Stricken 1978
D 42	ROGER DE LAURIA	Ferrol	4.9.51	12.11.58	(30.5.69)	See notes
D 43	MARQUES DE LA ENSENEDA	Ferrol	4.9.51	15.7.59	(10.9.70)	See notes

Roger de Lauria as completed — Navarret Collection

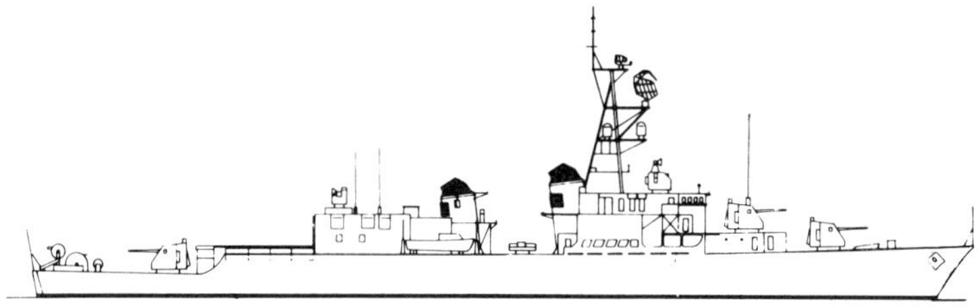

Roger de Lauria 1979

ROGER DE LAURIA class *destroyers*

Launched as units two and three of the *Oquendo* class (above) but completed at Cartegena to an enlarged and modernised design; dates of relaunching after hulls were lengthened and widened are given in the Table. Full US weapon and electronic fit, and generally to USN *Gearing* class FRAM II standards.

D 43 repaired after terrorist bomb attack on 2 October 1981, parts being taken from *Roger de Lauria*, but was never fitted with the proposed SAM or Meroka CIWS. Speed reduced by recurrent boiler problems.

Displacement: 3012t standard; 3785t full load
Dimensions: 391ft 6in oa × 42ft 8in × 18ft 4in
119.3m × 13m × 5.6m
Machinery: 2-shaft Rateau-Bretagne geared turbines, 3 3-drum boilers, 60,000shp = 31kts. Oil 700t. Range 4500nm at 15kts
Armament: 6–5in/38 DP (3×2), 6–12.75in TT (Mk 44 ASW torpedoes; 2×3), 2–21in TT (Mk 37 ASW torpedoes),
1 helicopter
Sensors: Radar SPS-10, SPS-40, RM-426, Mk 25 and Mk 35; sonar SPS-32C, SQS-10
Complement: 318

No	Name	Builder	Laid down	Launched	Comp	Fate
D 42	ROGER DE LAURIA	Ferrol/Cartagena	(4.9.51)	29.8.67	30.5.69	Stricken 1982
D 43	MARQUES DE LA ENSENEDA	Ferrol/Cartagena	(4.9.51)	2.3.68	10.9.70	Stricken Nov 1988

Projected guided missile destroyers

Three guided missile destroyers (projected under the January 1973 Programme) were in an advanced state of planning during the 1970s, but never laid down (commencement date probably 1974). Likely details are: SSM and SAM systems, 127mm DP OTO Melara guns, ASW helicopters, COGOG propulsion giving a speed of 34kts.

Particulars: Never published

Ex-US FLETCHER class *destroyers*

Built 1941–44 and transferred under MDAP and purchased outright in 1972. *Galiano* and *Jorge Juan* are of later *Fletcher* class.

Armament: 4–5in/38 (4×1) (*Lepanto, Ferrandiz* 5–5in/38), 6–3in/50 Mk 33 (2×3) (*Lepanto, Ferrandiz* 6–40mm/60 AA, 6–20mm/70 AA), 3–21in TT (1×3, not in *Lepanto, Ferrandiz*), 6–12.75in Mk 32 ASW TT (2×3), 2 Hedgehogs, 1 DC rack (*Lepanto* 2 DC racks)
Sensors: Radar SPG-34, SPS-6C, SPS-10 Mk 35; sonar SQS-4 or SQS-29
Complement: 290
Other particulars: As US *Fletcher* class

No	Name	Builder	Acquired	Fate
D 24	ALCALA GALIANO (ex-*Jarvis*)	Seattle-Tacoma	3.11.60	Stricken Oct 1989
D 22	ALMIRANTE FERRANDIZ (ex-*David W Taylor*)	Gulf SB	15.5.57	Stricken 17.11.87
D 23	ALMIRANTE VALDES (ex-*Converse*)	Bath Iron Wks	1.7.59	Stricken Oct 1986
D 25	JORGE JUAN (ex-*McGowan*)	Federal, Kearny	1.12.60	Stricken Sep 1988
D 21	LEPANTO (ex-*Capps*)	Gulf SB	15.5.57	Stricken Dec 1985

Ex-US GEARING class *destroyers*

Built in 1944–45 and transferred under MDAP and purchased outright on 17 May 1978; all FRAM I modernisations. *Blas de Lezo*, with both gun turrets forward and no ASROC, was reported the only unit to carry the Hughes 500M helicopter; the others had turrets in 'A' and 'Y' positions. No further modernisation, for example installation of SSM or CIWS, was attempted.

Armament: 4–5in/38 (2×2), 1 ASROC ASW system, 6–12.75in ASW TT (2×3), 1 helicopter (see notes)
Sensors: Radar SPS-10, SPS-37 or SPS-40, Mk 25 or 28; sonar SQS-23
Complement: 274
Other particulars: As US *Gearing* class (FRAM I type)

No	Name	Builder	Acquired	Fate
D 65	BLAS DE LEZO (ex-*Noa*)	Bath Iron Wks	31.10.73	Stricken 1991
D 61	CHURRUCA (ex-*Eugene A Green*)	Federal, Kearny	31.8.72	Stricken 15.9.89
D 62	GRAVINA (ex-*Furse*)	Consolidated, Orange	31.8.72	Stricken 1991
D 64	LANGARA (ex-*Leary*)	Consolidated, Orange	31.10.73	Stricken 1992
D 63	MENDEZ NUNEZ (ex-*O'Hare*)	Consolidated, Orange	31.10.73	Stricken 1992

Baleares as completed — Spanish Navy

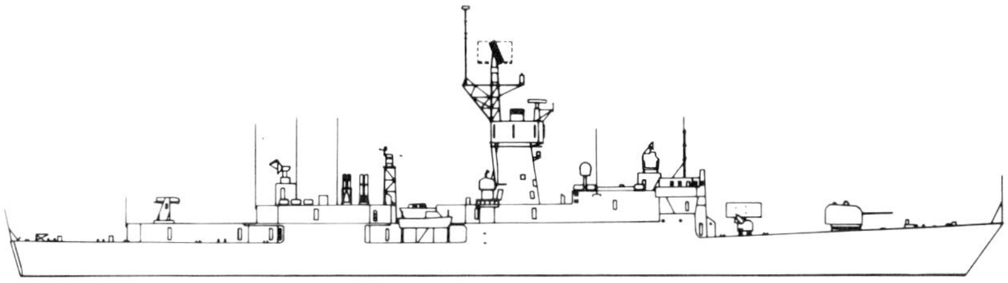

Baleares 1992

Generally very similar to US *Knox* (FF-1052) class, but with medium-range Tartar/Standard system replacing helicopter facilities and BPDMS. Hulls and machinery were fabricated in Spain, but the USN supplied weapons and sensors. All torpedo tubes fitted internally and fixed, the 12.75in being inclined at 45°, while the 21in are in the stern. All underwent mid-life modernisation between 1985 and 1991, the main visible changes being the addition of Harpoon SSM (only four of the possible eight missiles usually fitted) and Meroka 20mm CIWS with associated RAN 12L fire control radar. The SAM fire control, data collection and data link facilities were enhanced, and SPS 56 sonar replaced SPS 23.

BALEARES class *frigates*

Displacement: 3015t standard; 4177t full load
Dimensions: 415ft pp, 438ft oa × 46ft 9in × 24ft 7in (max over sonar
126.5m, 133.5m × 14.25m × 7.5m
Machinery: 1-shaft Westinghouse geared turbines, 2 V2M boilers, 35,000shp = 28kts. Range 4500nm at 20kts.
Armament: 1–5in/54 Mk 42, 1 Standard SAM (16 SM1-MR missiles), 8 Harpoon SSM, 2–20mm Meroka CIWS, 1 ASROC ASW system, 4–12.75in TT (Mk 44 ASW torpedoes), 2–21in TT (Mk 37 ASW torpedoes)
Sensors: Radar SPS-10, SPS-52A, SPG-51C, SPG-53B; sonar SQS-56, SQS-35V VDS
Complement: 256

No	Name	Builder	Laid down	Launched	Comp	Fate
F 72	ANDALUCIA	Ferrol	2.7.69	30.3.71	23.5.74	Extant 1995
F 74	ASTURIAS	Ferrol	30.3.71	13.5.72	2.12.75	Extant 1995
F 71	BALEARES	Ferrol	31.10.68	20.8.70	24.9.73	Extant 1995
F 73	CATALUNA	Ferrol	20.8.70	3.11.71	16.1.75	Extant 1995
F 75	EXTREMADURA	Ferrol	3.11.71	21.11.72	10.11.76	Extant 1995

The proposed lease of two ships, agreed in December 1993 subject to approval by the US Congress, was cancelled in 1994.

Ex-US KNOX class *frigates*

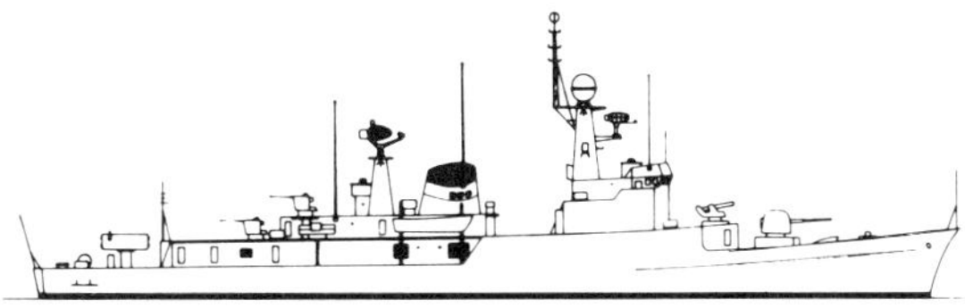

Descubierta as completed

Developed by Blohm & Voss from the Portuguese *Joao Coutinho* design, and subsequently modified by Bazan. Four ships were ordered in December 1973, and four more in May 1976. Two of the class, *Centinela* (F 37) and *Serviola* (F 38), were sold to Egypt before completion; another was built for Morocco. Several frigates of an improved *Descubierta* design prepared in the early 1980s (1879t full load, length *c*100m oa, weapons as above plus helicopter and hangar, gas turbine propulsion, 30+ kts) were proposed, but orders placed were for the much larger FFG-7 type.

DESCUBIERTA class *frigates*

Displacement: 1233t standard; 1482t full load
Dimensions: 278ft 10in pp, 291ft 3in oa × 34ft 1in × 12ft 6in
85m, 88.8m × 10.4m × 3.8m
Machinery: 2 shafts, 4 MTU-Bazán 16V956 TB 91 diesels, 16,000bhp = 25kts. Range 4000nm at 18kts
Armament: 1–3in/62 OTO Melara, 2–40mm/70 Bofors (2×1), 8 Harpoon SSM, 1 Sea Sparrow SAM (1×8), 1–375mm Bofors ASW mortar (1×2), 6–12.75in ASW TT(2×3)
Sensors: Radar DA-05/2, ZW-06, WM-22/41 or WM25; sonar DE-1160B
Complement: 118

No	Name	Builder	Laid down	Launched	Comp	Fate
F 35	CAZADORA	Ferrol	14.12.77	17.10.78	2.7.81	Extant 1995
F 31	DESCUBIERTA	Cartagena	16.11.74	8.7.75	18.11.78	Extant 1995
F 32	DIANA	Cartagena	8.7.75	26.1.76	30.6.79	Extant 1995
F 34	INFANTA CRISTINA	Cartagena	14.9.76	25.4.77	24.11.80	Extant 1995
F 33	INFANTA ELENA	Cartagena	26.1.76	14.9.76	12.4.80	Extant 1995
F 36	VENCEDORA	Ferrol	1.5.78	24.4.79	27.3.82	Extant 1995

Infanta Elena 19.10.85
L & L van Ginderen

SPAIN

First three ordered 29 June 1977, but construction was deferred to give priority to the new carrier; the original names listed above for F 81, F 82 and F 83, and a suggested allocation of *Nina* and *Pinta* to F 85 and F 86, are not officially confirmed. *Reina Sofia*, approved in 1985, was ordered 19 June 1986, and the last pair on 26 December 1989. Long-hulled FFG-7 type, with extra beam to allow more top-weight. Escorts for *Principe de Asturias*.

US OLIVER HAZARD PERRY class *frigates*

Displacement: 4017t full load
Dimensions: 451ft 2in oa × 46ft 11in × 24ft 7in
137.7m × 14.3m × 7.5m
Machinery: 1 shaft, 2 LM 2500 gas turbines, 40,000shp = 29kts. Range 4500nm at 20kts
Armament: 1–3in/62 OTO Melara, 1 Standard SAM (32 SM1-MR SAM and 8 Harpoon SSM), 1–20mm Meroka CIWS, 6–12.75in Mk 32 ASW TT (2×3), 2 LAMPS helicopters
Sensors: Radar SPS-49, Mk 92 fire control system; sonar SQS-56, SQR-19
Complement: 223

No	Name	Builder	Laid down	Launched	Comp	Fate
F 83	NUMANCIA (ex-*Nina*, ex-*Murcia*)	Ferrol	8.1.86	30.1.87	8.11.88	Extant 1995
F 84	REINA SOFIA (ex-*America*)	Ferrol	12.12.87	19.7.89	18.10.90	Extant 1995
F 81	SANTA MARIA (ex-*Navarra*)	Ferrol	23.5.82	24.11.84	12.10.86	Extant 1995
F 82	VICTORIA (ex-*Pinta*, ex-*Leon*)	Ferrol	16.8.83	23.7.86	11.11.87	Extant 1995
F 85	NAVARRA	Ferrol	15.4.91	23.10.92	Jun 1994	Extant 1995
F 86	CANARIAS	Ferrol	15.4.92	21.6.93	Dec 1994	Extant 1995

Reina Sofia 17.7.94
L & L van Ginderen

Designed in collaboration with German and Dutch shipyards; type of air and surface search radars, and hull-mounted medium frequency active sonars, not yet announced. The first of four units is planned for completion in 2001.

F 100 type *frigates*

Displacement: 4500t deep load
Dimensions: 419ft 3in × 50ft 11in × 15ft 5in
127.8m × 15.5m × 4.7m
Machinery: 2-shaft CODOG, 28 kts
Armament: 8 Harpoon SSM (2×4), SAM, 1–3in/62 OTO Melara, 1–20mm Meroka CIWS, 1 LAMPS III helicopter
Sensors: See notes
Complement: 200

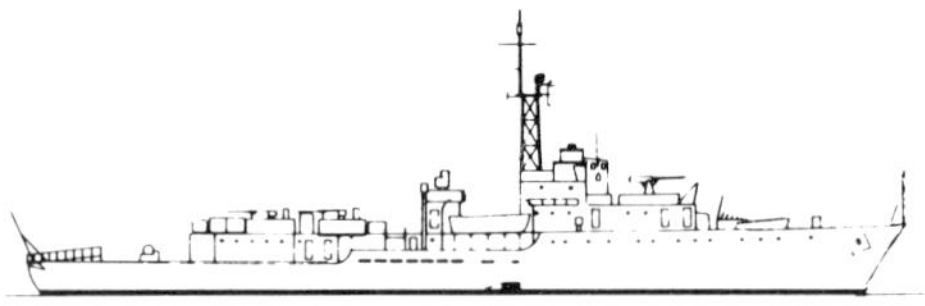

Atrevida 1976

Atrevida and *Descubierta* were completed with the above armament, but the former was modernised for ASW duties in 1959–60, when particulars became: 997t standard, 1135t full load, 9ft (*2.7m*) draught, 1–3in/50 DP, 3–40mm/70 AA (3×1), 2 Hedgehogs, 8 DC mortars, 2 DC racks; the four later units were delivered with these data. Rated as corvettes although given 'F' pennant numbers. The four survivors were refitted in 1979 for offshore patrol and fishery protection duties, with ASW weaponry removed and revised particulars: 1031t standard, 1118t full load, 9ft 10in (*3.0m*) draught, 1–3in/50 DP, 3–40mm/70 AA (3×1), 2–20mm AA (1×2).

ATREVIDA class *corvettes*

Displacement: 912t standard; 1022t full load
Dimensions: 223ft 1in pp, 247ft 8in oa × 33ft 6in × 8ft 8in
68m, 75.5m × 10.2m × 2.6m
Machinery: 2 shafts, Sulzer diesels, 3000bhp = 18.5kts. Oil 100t. Range 8000nm at 10kts
Armament: 1–4.1in/45 DP, 2–37mm AA (1×2), 12–20mm AA (3×4), 4 DCT, 20 mines
Sensors: Radar SPS-5B; sonar QHBa
Complement: 132

No	Name	Builder	Laid down	Launched	Comp	Fate
F 61	ATREVIDO	Cartagena	26.6.50	2.12.52	19.8.54	Stricken 1992
F 51	DESCUBIERTA	Cartagena	26.6.50	9.6.52	1.2.55	Stricken 1978
F 63	DIANA	Cartagena	27.7.53	29.4.55	13.5.60	Stricken 1973
F 64	NAUTILUS	Cadiz	27.7.53	23.8.56	15.12.59	Stricken 1991
F 62	PRINCESA	Cartagena	18.3.53	31.3.56	3.10.59	Stricken 1991
F 65	VILLA DE BILBAO (ex-*Favorita*)	Cadiz	18.3.53	19.2.58	2.7.60	Stricken 1991

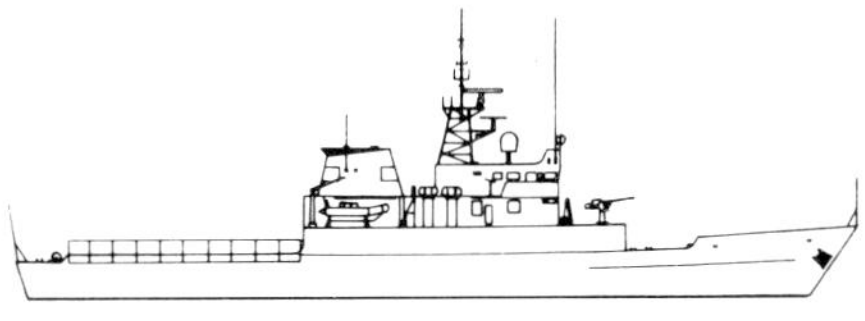

Serviola as completed

These ships, of modified *Halcon* type, were originally formulated in late 1988 as the *Milano* class (Type B–215), slightly heavier, faster and more lightly armed than the design finally adopted (1360t full load, 214ft 3in × 33ft × 10ft 6in (*65.3m × 10.1m × 3.2m*), 20kts, 1–40mm/70 AA, 2MG, 1 Hughes 500M helicopter). Replaced *Atrevida* class.

SERVIOLA class *corvettes*

Displacement: 836t standard; 1106t full load
Dimensions: 206ft 8in pp, 225ft 5in oa × 34ft × 11ft
63.0m, 68.7m × 10.4m × 3.4m
Machinery: 2 shafts, MTU/Bazán 16V 956 TB91 diesels, 7500bhp = 19kts. Range 8000nm at 12kts
Armament: 1–3in Mk 26, 2 MG, 1 helicopter
Sensors: Decca 2459
Complement: 42

No	Name	Builder	Laid down	Launched	Comp	Fate
P 49	ATALAYA	Ferrol	14.12.90	22.11.91	29.6.92	Extant 1995
P 72	CENTINELA	Ferrol	12.12.89	30.3.90	24.9.91	Extant 1995
P 71	SERVIOLA	Ferrol	17.10.89	10.5.90	22.3.91	Extant 1995
P 73	VIGIA	Ferrol	30.10.90	12.4.91	24.3.92	Extant 1995

SUBMARINES

The construction of four new submarines *G 1–G 4* (*G 5* and *G 6* were also projected) ordered in 1945 under the 1943 Programme for construction at Cartagena. They were to have been generally similar to *G 7* (ex-German *U 573*), but the plan was abandoned, although some may have been laid down. Ex-US submarines were expected to be acquired under MDAP in their place.

'G' class *submarines*

Displacement: 766t normal; 930t submerged
Dimensions: 220ft 2in × 20ft 7in × 14ft 2in–15ft 9in
67.1m × 6.3m × 4.3–4.8m
Machinery: 2-shaft diesels plus electric motors, 3000bhp/750shp = 17.5kts/8kts. Oil 120t. Range 9000nm/6500nm at 10kts/12kts
Armament: 5–533mm TT (4 bow, 1 stern), 1–88mm AA, 1–20mm AA
Complement: 44

Class (fate):
SA 41 (stricken 1971), *SA 42* (stricken 1971)

Launched in 1957 and originally rated as experimental submarines, known as *Foca I* (F I) and *Foca II* (F II) respectively; re-rated as assault submarines and numbered as above in 1963. Built at Cartagena and based on the German 'Seehund' type.

FOCA class *midget submarines*

Displacement: 16t surfaced; 20t submerged
Dimensions: 45ft 5in oa × 6ft × 5ft
13.9m × 1.8m × 1.5m
Machinery: 1-shaft Pegaso diesel plus Siemens electric motor, 160bhp/110shp = 9.2kts/12kts. Range 700nm at 7kts
Armament: 2–21in TT
Complement: 3

Class (fate):
SA 51 (stricken 1977), *SA 52* (stricken 1977)

Launched in 1958 and originally rated as experimental submarines, known as *Tiburon I* and *Tiburon II* respectively; re-rated as assault submarines and numbered as above in 1963. Built at Cartagena.

TIBURON class *midget submarines*

Displacement: 78t surfaced; 81t submerged
Dimensions: 70ft 6in oa × 9ft × 9ft
21.5m × 2.7m × 2.7m
Machinery: 1-shaft Pegaso diesels plus electric motors, 4000bhp/400shp = 10kts/14.5kts
Armament: 2–21in TT
Complement: 5

Originally built 1943–44 and transferred in 1959 after modernisation. Further modernised in 1964–65, when armament became 6–21in TT and 4 TT for ASW torpedoes; can carry 22 torpedoes. Although due for deletion in 1975, this boat was retained in service until 1982, because of machinery defects in *Monturiol* (below). Name abbreviated to *A G Reyes*.

Ex-US BALAO class *submarine*

Particulars: As US *Balao* class (Fleet Snorkel type)

No	Name	Builder	Acquired	Fate
S 31	ALMIRANTE GARCIA DE LOS REYES (ex-*E 1*, ex-*Kraken*)	Manitowoc	24.10.59	Stricken 1982

Ex-US BALAO class *submarines*

Built as *Balao* class submarines in 1943–44, but extensively modernised in 1953–54. Transferred in 1971 (S 32), 1972 (S 33, 34) and 1974 (S 35). S 33 laid up with mechanical defects in 1975; when deleted in 1977 her name was transferred to S 35.

Particulars: As US *Balao* class (GUPPY IIA type)

No	Name	Builder	Acquired	Fate
S 34	COSME GARCIA (ex-*Bang*)	Portsmouth N Yd	1.10.72	Stricken 2.6.84
S 32	ISAAC PERAL (ex-*Ronquil*)	Portsmouth N Yd	1.7.71	Stricken 1982
S 33	NARCISO MONTURIOL (ex-*Picuda*, ex-*Obispo*)	Portsmouth N Yd	1.10.72	Stricken 1977
S 35	NARCISO MONTURIOL (ex-*Jallao*)	Manitowoc	26.6.74	Stricken 31.12.84

Narval (outboard) alongside *Narcisco Monturiol* 1980 — M G Photographic

French DAPHNE type *submarines*

Two ordered in December 1966, and a second pair in March 1970, and a fifth was originally envisaged. Built with extensive French assistance, under Agreement of July 1966. All modernised between 1983 and 1988, with DUUA 2A sonar replacing DUUA 1, and updated fire controls. New, larger sonar dome on forward casing repositioned slightly further aft.

Displacement: 869t surfaced; 1069t submerged
Dimensions: 190ft oa × 22ft × 15ft
57.8m × 6.8m × 4.6m
Machinery: 2 shafts, 2 SEMT-Pietstick diesels plus 1 electric motor, 1300bhp/1600shp = 13.2kts/15.5kts. Range 4300nm at 7.5kts
Armament: 12–550mm TT (8 bow, 4 stern)
Sensors: Sonar DUUA 2A, DSUV 22
Complement: 47

No	Name	Builder	Laid down	Launched	Comp	Fate
S 61	DELFIN	Cartagena	13.8.68	25.3.72	3.5.83	Extant 1995
S 63	MARSOPA	Cartagena	19.3.71	15.3.74	12.4.75	Extant 1995
S 64	NARVAL	Cartagena	24.4.72	14.12.74	22.11.75	Extant 1995
S 62	TONINA	Cartagena	2.3.70	3.10.72	10.7.73	Extant 1995

French AGOSTA type *submarines*

Two ordered in May 1975, and a second pair in June 1977. Built with some French assistance, and modernised in the mid-1990s with towed array sonars, improved fire control and ECM, and new batteries and periscopes. Plans to fit SM 39 SSM may be dropped.

Displacement: 1490t surfaced; 1740t submerged
Dimensions: 222ft oa × 22ft × 18ft
67.6m × 6.8m × 5.5m
Machinery: 1 shaft, 2 SEMT-Pielstick 16 PA diesels plus electric motors, 3600bhp/4600shp = 12kts/20kts max
Armament: 4–533mm TT (bow)
Sensors: Sonar DUUA 2A/2B, DSUV 22, DUUX 2A or DUUX 5
Complement: 54

No	Name	Builder	Laid down	Launched	Comp	Fate
S 71	GALERNA	Cartagena	5.9.77	5.12.81	22.1.83	Extant 1995
S 73	MISTRAL	Cartagena	30.5.80	14.11.83	5.6.85	Extant 1995
S 72	SIROCO	Cartagena	27.11.78	13.11.82	5.12.83	Extant 1995
S 74	TRAMONTANA	Cartagena	10.12.81	30.11.84	27.1.86	Extant 1995

S 80 type *submarines*

Intended to replace *Daphne* class; project definition completed jointly by Bazan and DCN/Dubigeon, France, 1986, but orders were postponed, and the *Agosta* class modernised instead. Current proposals are for a recast S 80, designated *Scorpene*, also designed by Bazan and DCN, and displacing about 2000t; construction of the first unit is planned to start in 1999.

Displacement: 2410t submerged
Dimensions: 249ft 4in oa × 25ft 7in 00ft
76.0m × 7.8m × 00m
Machinery: 1 shaft, 2 diesels plus 1 electric motor = 12kts/25kts. Range 10,500nm
Armament: 4/6–533mm TT
Complement: 35

No	Name	Builder	Laid down	Launched	Comp	Fate
S 81	No I	Cartagena	-	-	(1992)	-
S 82–4	Nos II, III,IV	Cartagena	-	-	-	

AMPHIBIOUS WARFARE VESSELS

Ex-US CABILDO class *dock landing ship*

Particulars: As US *Cabildo* class

No	Name	Builder	Acquired	Fate
L 31	GALICIA (ex-*San Marcos*)	Philadelphia N Yd	1.7.71	Stricken 12.1.88

Originally built in 1944–45, transferred 1971 and purchased outright in 1974.

Ex-US TERREBONNE PARISH class *tank landing ships*

Particulars: As US *Terrebonne Parish* class

No	Name	Builder	Acquired	Fate
L 11	VELASCO (ex-*Terrebonne Parish*)	Bath Iron Wks	29.10.71	Stricken 1994
L 12	MARTIN ALVAREZ (ex-*Wexford County*)	Christy Corp	29.10.71	Extant 1995
L 13	CONDE DE VENADITO (ex-*Tom Green County*)	Bath Iron Wks	5.1.72	Stricken 1990

Originally built 1952–53; all transferred in 1971–72 and purchased outright on 5 August 1976. L 11 and L 12 retained in service until replaced by two ex-US *Newport* class LSTs in 1994–95 (see below).

Ex-US HASKELL class *attack transport*

Particulars: As US *Haskell* class

No	Name	Builder	Acquired	Fate
TA 11	ARAGON (ex-*Noble*)	USA	19.12.64	Stricken 1987

Built in 1944–45; transferred in 1964. Name transferred to new transport in 1980 and ship laid up.

Ex-US ANDROMEDA class *attack cargo ship*

Particulars: As US *Andromeda* class

No	Name	Builder	Acquired	Fate
TA 21	CASTILLA (ex-*Achernar*)	USA	2.2.65	Stricken 1987

Launched 1943; transferred in 1965. Name transferred to new transport in 1980 and ship laid up.

Ex-US PAUL REVERE class *amphibious transports*

Particulars: As US *Paul Revere* class

No	Name	Builder	Acquired	Fate
L 21	CASTILLA (ex-*Paul Revere*, ex-*Diamond Mariner*)	New York SB	17.1.80	Extant 1995
L 22	ARAGON (ex-*Francis Marion*, ex-*Prairie Mariner*)	New York SB	11.7.80	Extant 1995

Originally C4–S–1 cargo vessels, launched in 1953–54; for conversion notes, see under United States. Both purchased by Spain in 1980.

Ex-US NEWPORT class *amphibious transports*

Particulars: As US *Newport* class

No	Name	Builder	Acquired	Fate
L 41	HERNAN CORTES (ex-*Barnstable County*)	San Diego	26.8.94	Extant 1995
L 42	PIZARRO (ex-*Harlan County*)	San Diego	May 1995	Extant 1995

These ships replaced *Velasco* and *Martin Alvarez* (above). Meroka CIWS may replace Phalanx.

Dutch ROTTERDAM class *assault ships*

The construction of the first of two 12,000t (full load) amphibious assault ships, designed jointly by Bazan and the Dutch Nevesbu consortium, is scheduled to begin in mid-1995, for completion in 1997. Particulars will be as for the Dutch *Rotterdam*, but with two 20mm Meroka CIWS replacing Goalkeeper. No dates for the second ship are available.

OTHER AMPHIBIOUS WARFARE CRAFT

Three US medium landing ships were transferred in 1960 and given Spanish numbers (*LSM 1*, ex-*LSM 329*, *LSM 2*, ex-*LSM 331*, *LSM 3*, ex-*LSM 343*). *LSM 3* was stricken in 1974 and the other two were hulked as accommodation ships in 1976.

The three Spanish-built tank landing craft (*A 6–8*, ex-*LCT 6–8*, ex-*BDK 6–8*) of the French EDIC type, built 1965–66, were extant in 1995 (665t, 1–20mm, 2–12.7mm MG, 9.5kts) The earlier Spanish-built tank landing craft *LCT 3–5* (ex-*BDK 3–5*, ex-*K 3–5*), of the British LCT (4) type, were stricken in 1984 (*LCT 3*) and 1985 (602t, 2–20mm, 8.5kts). The ex-British *LCT 2* (ex-*BDK 2*, ex-*Marsa*, *K2*) of the LCT(4) type was stricken in 1983.

In service in 1983 were a total of two LCUs, twenty-two LCM(6)s, six LCM(8)s, twenty-three LCPs and seventy LCVPs; the 1995 figures, including new construction, were two LCUs, fourteen LCM(6)s, eight LCM(8)s, nine LCPs, twenty LCVPs and one VCA (cushioncraft).

LCT 6 Aug 85 L & L van Ginderen

SMALL SURFACE COMBATANTS

German S-BOAT type *fast attack craft*

Displacement: 100t standard; 116t full load
Dimensions: 114ft 2in oa × 16ft 9in × 4ft 7in
34.8m × 5.1m × 1.4m
Machinery: 3 shafts, Mercedes-Benz diesels, 7500bhp = 40kts. Oil 20t. Range 650–800nm at 30kts
Armament: 2–20mm AA, 4 MG, 2–21in TT
Sensors: Radar RM-914 (in some boats)
Complement: 22

Class (fate):
LT 27 (stricken 1963), *LT 28* (stricken 1963), *LT 29* (stricken 1963), *LT 30* (stricken 1977), *LT 31* (stricken 1977), *LT 32* (stricken *c*1978)

All built by La Carraca, Cadiz, in 1952–56, to the Lürssen S-boat design of 1944. *LT 27–29* completed 1953, *LT 30* 1954, *LT 31* and *32* 1957.

LAZAGA class *fast attack craft*

Displacement: 275t standard; 399t full load
Dimensions: 178ft 6in wl, 190ft 7in oa × 24ft 11in × 9ft 2in
54.4m, 58.1m × 7.6m × 2.8m
Machinery: 2 shafts, MTU-Bazán MA15 TB 91 diesels, 8000bhp = 30kts. Range 2260nm/4200nm at 27kts/17kts
Armament: 1–76mm/62 OTO Melara compact, 1–40mm/70 Breda Bofors, 2–20mm Oerlikon GK 204 (2×1): 6–12.75in Mk 32 ASW TT, DC racks
Sensors: Radar WM-22; sonar ELAC
Complement: 34

Class:
Lazaga (PC01), *Alsedo* (PC 02), *Cadarso* (PC 03), *Villamil* (PC 04), *Bonifaz* (PC 05), *Recalde* (PC 06)

Ordered in 1972, to the basic Lürssen T 57 design, the first unit being built by Lürssen, Vegesack and the remainder by Bazán, La Carraca, between 1974 and 1977. For fishery protection in peacetime, but being relieved in these duties by the new *Anaga* class. Planned Harpoon SSM never fitted; ASW weapons not installed in all units. All were stricken in 1993.

CORMORAN *fast attack craft*

Displacement: 374t full load
Dimensions: 185ft 8in oa × 24ft 8in × 6ft 6in
56.6m × 7.5m × 2.8m
Machinery: 3 shafts, MTU-Bazán MA-16V 956 TB91 diesels, 13,500bhp = 34kts
Armament: 1–40/70 Bofors, 1–20mm Oerlikon
Sensors: Surface search radar only
Complement: ?

Class:
Cormoran (P 41)

Launched by Bazan, Ferrol in 1985 as a speculation. Taken over by Spanish Navy in 1989 as testbed and patrol vessel; may be returned to Bazan if a purchaser is found. Similar ships built for Morocco and on order for the Philippines.

BARCELO class *fast attack craft*

Displacement: 134ft full load
Dimensions: 118ft 9in oa × 19ft × 8ft 2in
36.2m × 5.8m × 2.5m
Machinery: 2 shafts, MTU-Bazán MD-16V TB 90 diesels, 5760 bhp = 36kts. Range 600nm/1200nm at 33.5kts/16kts
Armament: 1–40mm/70 Breda Bofors, 2–20mm Oerlikon GAM 204, 2–12.7mm MG, 2–21in TT
Sensors: Radar Raytheon 1620/6
Complement: 40

Class:
Barcelo (PC 11), *Laya* (PC 12), *Javier Quiroga* (PC 13), *Ordoñez* (PC 14). *Acevedo* (PC 15), *Candido Perez* (PC 16)

Ordered in 1973, the first unit being built by Lörssen, Vegesack, and the remainder by Bazán, La Carraca, between 1975 and 1977. For fishery protection in peacetime TT are not fitted. All were in service 1995.

Ex-US PC type *patrol vessel*

Armament: 1–3in DP, 1–40mm
Other particulars: As US PC type

No	Name	Builder	Acquired	Fate
–	JAVIER QUIROGA	Luders Marine	1956	Stricken 1971

Built in 1943; transferred to Spain in 1956. Rearmed with 2–37mm AA *c*1965.

ANAGA class *large patrol craft*

Displacement: 296.5t full load
Dimensions: 145ft 7in oa × 21ft 7in × 8ft 2in
44.4 × 6.6m × 2.5m
Machinery: 1 shaft, MTU diesel, 4300bhp = 22kts. Range 4000nm at 15kts
Armament: 1–76mm/62 OTO Melara compact, 2–Mk 100
Sensors: Navigational radar only
Complement: 25

Class:
Anaga (PVZ 21), *Togomago* (PVZ 22), *Marola* (PVZ 23), *Monro* (PVZ 24), *Grosa* (PVZ 25), *Medas* (PVZ 26), *Izaro* (PVZ 27), *Tabarca* (PVZ 28), *Deva* (PVZ 29), *Bergantin* (PVZ 210)

All ordered from Bazán, La Carraca in 1978 and completed between 1980 and 1982. Specially designed for fishery protection and EEZ patrol service, relieving the *Lazaga* class. All were in service in 1995.

Ex-US AMS class *large patrol craft*

The former coastal minesweepers *Naron*, *Turia*, and *Ulla*, transferred to patrol duties and renumbered in 1980, were stricken in 1993. *Mino* and *Sil*, also transferred in 1980, reverted to coastal minesweeper classification in 1984. For details, see Mine Warfare Vessels.

P 50 class *large patrol craft*

Displacement: —t standard; —t full load
Dimensions: 182ft 5in × 36ft 1in × —ft
55.6m × 11m × 0m
Machinery: 2 shafts, Deutz SBV-8M-628 diesels, 4450bhp = 16kts. Range 7500nm at 12kts
Armament: 1–40mm/70 Breda Bofors, 2–20mm Oerlikon, 1 helicopter
Sensors: Surface search and navigational radar
Complement: 34

Class:
P 50–P 55

All scheduled for ordering 1990, but postponed and then cancelled. AB 212-sized helicopter in telescopic hangar. Would have been financed in the civil budget, and operated by the navy for fishery protection.

OTHER LIGHT CRAFT

In addition to the above, the following light craft are included:

The patrol vessels *Pegaso* and *Procyon*, built at Cartagena between 1947 and 1951 and rated as coast guard vessels, were attached to the Naval School and used as training ships *c*1970 (503t full load, 1–3in AA, 2–20mm, 12kts. 3in gun removed in late 1960s). These were the only units built of the twelve *Rigel* class coastguard vessels projected in 1945.

The patrol vessels *Centinela* and *Serviola*, completed at Ferrol in 1953 and rated as fishery protection vessels, were transferred to Mauritania in 1977 (28t full load, 2–37mm, 12kts).

The trawler-type patrol vessels *Cies* and *Salvora*, purchased in 1952 and rated as fishery protection vessels, were stricken in 1973 and 1990 respectively (270t full load, 12kts, 1–20mm).

The trawler-type large patrol craft *Chilreu*, launched in 1988 and purchased as a fishery protection vessel, was in service in 1995 (1157t full load, 1 MG, 13kts).

The large patrol craft *Candido Perez* (*W 11*, ex-*SC 679*), a former US submarine-chaser of the Second World War. This '110ft' wooden-hulled type transferred in 1956, was stricken in 1974 (138t full load, 1–40mm, 3–20mm, 2 DCT, 2 Mousetraps, 15.6kts).

Some ten coastal and inshore patrol craft of the motor launch type were also in service in 1995. The largest were the four boats of the *Conejara* class (*P 31–34*, 1981–82, 85t, 2–20mm, 25kts).

SURFACE EFFECT SHIP

The order for a 360t missile-armed surface-effect ship, postponed for budgetary reasons, was cancelled in 1993.

MINE WARFARE VESSELS

GUADIARO class *minesweepers*

Displacement: 671t standard; 770t full load
Dimensions: 243ft 10in oa × 33ft 6in × 12ft 3in max
74.3m × 10.2m × 3.7m
Machinery: 2-shaft triple expansion plus auxiliary turbine, 2 Yarrow boilers, 2400ihp = 16kts. Range 1000nm at 16kts.
Armament: 1–3.5in, 1–37mm AA, 2–20mm AA
Sensors: Radar TM-626 or RM-914; sonar UQS-1
Complement: 79

No	Name	Builder	Launched	Fate
M 14	ALMANZORA	Cartagena	22.7.53	Stricken 1977
M 17	EO	Cadiz	22.9.53	Stricken 1978
M 13	EUME	Cartagena	22.7.53	Stricken 1977
M 16	GUADALHORCE	Cartagena	18.253	Stricken 1978
M 11	GUADIARO	Cartagena	26.6.50	Stricken 1977
M 15	NAVIA	Cadiz	28.7.53	Stricken 1979
M 12	TINTO	Cartagena	26.6.50	Stricken 1976

Based on the German 1940 type minesweepers, and very similar to the earlier *Bidasoa* class. The auxiliary turbine utilised the exhaust steam from the reciprocating engines. Modernised 1959–61 with removal of the 3.5in and 37mm guns. Used as patrol craft in their final years.

Ex-US AMS type *coastal minesweepers*

Particulars: As US AMS type

Class (no, former Spanish no, ex-US name/no):
Duero (M 23, M 28, ex-*Spoonbill*, MSC 202), *Ebro* (M 22, M 26, ex-*MSC 269*), *Genil* (M 25, M 31, ex-*MSC 279*), *Jucar* (M 21, M 23, ex-*AMS 220*), *Llobregat* (–, M 22, ex-*AMS 123*), *Mino* (PVZ 53, M 25, ex-*AMS 266*), *Nalon* (PVZ 51, M21, ex-*AMS 139*), *Odiel* (M 26, M 32, ex-*MSC 288*), Sil (PVZ 55, M 29, ex-*Redwing*, MSC 200), *Tajo* (M 24, M 30, ex-*MSC 287*), *Turia* (PVZ 54, M 27, ex-*AMS 130*), *Ulla* (PVZ 52, M 24, ex-*AMS 265*)

Wooden-hulled minesweepers of the *Adjutant*, *Redwing* and *MSC 268* classes, launched 1953–58 and transferred in 1954 (M 21, M 22), 1955 (M 27), 1956 (M 23–25), 1958 (M26) and 1959 (other five). *Llobregat* was discarded in 1980 after a fire, five were transferred to patrol duties in 1980 and given PVZ numbers as above (these numbers were later given the P prefix). Two, *Mino* and *Sil*, reverted to minesweeping duties in 1984, with new numbers M 28 and M 27 respectively; together with the six minesweepers not transfer-red to patrol duties, they were in service in 1995. They have Decca TM-626 or RM-914 tactical radar and UQS-1 sonar.

Duero 1960 L & L van Ginderen

Ex-US AGGRESSIVE class *ocean minesweepers*

Particulars: As US *Aggressive* class

No	Name	Builder	Acquired	Fate
M 41	GUADALETE (ex-*Dynamic*)	Colbert, Stockton	1.7.71	Extant 1995
M 42	GUADALMEDINA (ex-*Pivot*)	Wilmington BW	1.7.71	Extant 1995
M 43	GUADALQUIVIR (ex-*Persistant*)	Tacoma	1.7.71	Extant 1995
M 44	GUADIANA (ex-*Vigor*)	Burgess, Manitowoc	4.4.72	Extant 1995

Built in 1952–55; transferred under MDAP in 1971 and 1972. All purchased 1974. Have SQQ-14 VDS and therefore mine classification capability, but no PAP or diving facilities. *Guadalete* transferred to patrol duties in 1980, and was renumbered PVZ 41, but reverted to minesweeping duties and number M 41 in 1981. All were in service in 1995.

CME type *minehunters*

Displacement: 530t full load
Dimensions: 177ft 2in oa × 35ft 1in × 7ft 3in
54m × 10.7m × 2.2m
Machinery: 2 shafts, 2 MTU-Bazán TB83 diesels, 1523bhp = 14kts (plus 2 motors for hunting, 7kts, and 2 side thrusters)
Armament: 1–20mm GAM-B01
Sensors: Navigation radar; sonar 2093 or SQQ 32
Complement: 40

The existing twelve elderly ex-US minesweepers are to be replaced by new construction. After several years' consideration of various proposals (Tripartite type, Italian *Lerici* class, US MHC 51 design, indigenous 'Cazador' design), a contract was signed (July 1989) between Vosper Thornycroft and Bazán for transfer of *Sandown* SRMH GRP technology transfer. It was envisaged that an eventual eight hunters and four sweepers of modified UK *Sandown* class would be built, but the proposals were changed; four hunters of revised type with the above data were ordered in 1993, and will incorporate the results of further GRP technical assistance, from DCN. The first unit is due to be built between 1995 and 1998. A further four are proposed.

Sri Lanka

The Royal Ceylon Navy was established on 9 December 1950 with its principal base at the historic port of Trincomalee. No part of this island paradise, less than a third the size of Britain, is more than 80 miles from the sea and India lies 35 miles across the Palk Strait. Independence came on 4 February 1948 and on 22 May 1972 Ceylon became a republic, being renamed Sri Lanka ('Resplendent Island'), the navy likewise.

Until 1971 Ceylon relied on the British presence East of Suez for external security under the 1947 Defence Agreement (still valid today). A very modest military expansion began after an internal revolt that year and the navy began adding some Chinese patrol craft to its all-British fleet. Manpower remained relatively static until the late 1980s, when the Tamil insurgency led to an expansion from under 3000 in 1982 to 7000 in 1991. The major recent acquisitions are four 'command ships' to co-ordinate light patrol craft in the counter-insurgency campaign.

Ex-British 'RIVER' class *frigates*

Name	Builder	Acquired	Fate
GAJABAHU (ex-*Misnak*, ex-*Hallowell*)	Canadian Vickers	1959	For disposal 1978
MAHASENA (ex-*Mivtakh*, ex-*Violetta*, ex-*Orkney*)	Canadian Yarrow	1959	BU 1964

Two elderly ex-Canadian 'River' class frigates were bought from Israel in 1959. The former armament of three 4.7in was replaced by a single RN-pattern 4in/45 Mk 5 HA in 'B' position. *Mahasena* sold to Hong Kong breakers for £20,000 in June 1964. About 1972 *Gajabahu* was reported to be non operational.

Ex-British 'FORD' class *seaward defence boat*

Class (former name):
Kotiya (ex-*Desford*)

Transferred from RN 1955; wrecked off Colombo 22 December 1964.

JAYESAGARA class *patrol craft*

No	Name	Builder	Launched	Fate
P 601	JAYESAGARA	Colombo DY	26.5.83	Extant 1995
P 602	SAGAWARDENE	Colombo DY	20.11.83	Sunk 19.9.94

Two offshore patrol vessels ordered 31 December 1981 and first laid down 12 October 1982. 330t, 130ft 6in × 23ft × 7ft, (*39.8m × 7m × 2.1m*), 2 shafts, MAN diesels, 2100bhp = 16kts. Range 3000nm at 11kts; armament: 2–25mm (1×2), 2–14.5mm (1×2). *Sagawardene* sunk by Tamil separist suicide attack 19 September 1994, *Jayesagara* extant 1995.

Ex-Chinese 'SHANGHAI II' class *large patrol craft*

Class:
Balawatha, Daksaya, Ramakami, Sooraya, Weeraya, Jagatha, Rakshata, Ranasuou, Ranawini, Ranasim

Transferred February 1972 and 1975 (first five). Last two in 1980 commissioning 30 November 1980. *Daksaya* and *Balatha* stricken; remainder extant 1995.

Ex-Soviet 'MOL' class *fast attack craft*

Samudra Devi handed over December 1975 without TT, radar or IFF. Deleted 1983.

Vosper Type 101 *coastal patrol craft*

Displacement: 15t full load
Dimensions: 45ft 3in × 11ft 6in × 2ft 11in
13.8m × 3.6m × 0.9m
Machinery: 2 shafts, 2 GM 6–71 diesels, 500bhp = 25kts
Armament: 1 MG
Complement: 6

Class:
P 102–110, P 201–211

Two ordered 1965, seven in 1966; hard chine construction with double teak planking, twenty-one built by Vosper Thornycroft of Singapore and assembled in Sri Lanka by September 1968. Radar, radio and searchlight equipped. *P 102* lost 1979, *P 106, 109, 201, 204, 205, 207, 208, 210* stricken 1978–79, eleven extant 1982, in two squadrons at Kalpitiya and Kavainagar, *103, 104, 108* and *109* only on harbour duties. Six stricken 1983; only one operational 1992 and stricken subsequently.

BELIKAWA class *coastal patrol craft*

Displacement: 22t full load
Dimensions: 55ft 11in × 14ft 9in × 3ft 11in
17m × 4.5m × 1.2m
Machinery: 2 shafts, 2 GM 8V-71 T1 diesels, 800bhp = 23.6kts. Range 1000nm/790nm at 18kts/12.2kts
Armament: 3–7.62mm MG
Sensors: Radar Decca
Complement: 7

Class (no, in service):
Belikawa (P 421, Apr 1977), *Diyakawa* (P 422, Jun 1977), *Korawakka* (P 423, Jul 1977), *Seruwa* (P 424, Sep 1977), *Tarawa* (P 425, Oct 1977)

Five built by Cheverton Workboats of Cowes (UK) for customs work but transferred to general naval patrol duties. Reverted to numbers in 1980s; extant 1995.

PRADEEPA class *coastal patrol craft*

Displacement: 40t standard; 44t full load
Dimensions: 66ft × 10ft × 7ft
20.1m × 3.05m × 2.1m
Machinery: 2 shafts, 2 GM 8V-71 TI diesels, 800bhp = 19kts. Range 1200nm at 14kts
Armament: 2–20mm (2×1)
Complement: 10

Class (no, in service):
Pradeepa (P 431, 1980), *P 432* (1980), *P 433* (30.9.80), *P 434* (9.10.80), *P 435* (11.2.81), *P 436* (11.4.81), *P 445* (30.9.82), *P 446* (17.9.82), *P 447*, *P 448*, *P 449* (renumbered *P 231–236*, *P 241–245*)

First pair ordered June 1976 from Colombo DYd, second in November. Very slow building programme. Five more 22kt, 1240bhp, beamier units ordered in 1979. *P 236* stricken 1990, remainder extant 1995.

Israeli 'DVORA' and 'SUPER DVORA' type *coastal patrol craft*

Class (no, in service):
P 453 (1984), *P 454* (1984), *P 455* (Oct 1986), *P 456* (Oct 1986), *P 457* (Oct 1986), *P 458* (Oct 1986), *P 463* (Nov 1987), *P 464* (Nov 1987), *P 465* (1988), *P 466* (1988), *P 467* (1988), *P 468* (1988)

The second batch, ordered in October 1988 from Israeli Aircraft Industries, have slightly more powerful MTU diesels. Extant 1995.

South Korean 'KILLER' type *coastal patrol craft*

Class (no, in service):
P 473 (Feb 1988), *P 474* (Feb 1988), *P 475* (Feb 1988)

Bought from Korea Shipbuilding & Engineering, of Busan. Extant 1995.

Ex-British ALGERINE class *minesweepers*

Class (former name, acquired, fate):
Parakrama (ex-*Pickle*, Jul 1958, BU 1964), *Vijaya* (ex-*Flying Fish*, ex-*Tillsonburg*, Sep 1949, returned to RN 1964)

Parakrama was sold for scrapping in Hong Kong but *Vijaya* was returned to RN for scrapping.

MISCELLANEOUS

Two unarmed *Hansaya* class coastal patrol craft (36t, 66ft, 16kts) built by Korody Marine Corporation of Venice in 1956 were stricken *c*1980.

Ex-merchantman *Kota Rukun* acquired 1986 and converted to command tender *A 526*. Three roll-on/roll-off ships, *Carinia*, *Francesca* and *Delicia* acquired August 1984 and converted to surveillance command ships *Abheetha* (P 714), *Edithara* (P 715) and *Wickrama* (P 716). All four extant 1995.

Sudan

The Sudan, by area the largest African state, became independent in 1956, when the Anglo-Egyptian condominium imposed after the fall of the Mahdist rule was terminated. Her subsequent history has been strongly influenced by her geography and demography, as she straddles the major natural link between the Arab world and tropical Africa, and has a population of over 26 million divided between Arab Muslims (about 70 per cent) and various black minorities. Resentment of Arab rule by the partly Christian negroes of the south led to open revolt before independence, and trouble flared again repeatedly before the southern separatists were mollified in 1970 by the promise of autonomy. The 14-year civil war ended in 1972 after 500,000 deaths and 250,000 refugees, but has since restarted and is currently affecting much of southern Sudan. By 1994 the warfare had left an estimated one million dead, including 300,000 who died from the war-induced famine of 1988. Some three million refugees have fled the fighting, mainly to neighbouring states.

Two periods of civilian, parliamentary government were conspicuously unsuccessful, and the Army has been almost continuously in power but presidential and legislative elections were scheduled for 1995. Foreign aid has come from the Eastern bloc (including the first major Soviets arms exported south of the Sahara) and, after the reversal of policy following the abortive Communist coup in 1971, from the West and the oil-rich Arab states. Even so the economy shrank by three per cent in 1980. Sudan's foreign policy attempts to conciliate all her seven neighbours, but Gaddafi's Libya was suspected of plotting to topple the Sudanese leader.

Militarily, almost all expenditure has been on the 65,000-strong army; assistance came from Egypt, with which there is a 1977 defence agreement. The navy (600 volunteers since 1980) was set up in 1962 to protect the Red Sea coast (Port Sudan), and also operates on the Nile (Khartoum). The original Yugoslav training mission left in 1972. Attempts to purchase fast attack craft (missile) from Spain in 1984, and 110ft patrol boats from the USA in 1987 came to nothing and by 1995 the navy had all but disappeared.

EL GIHAD class *large patrol craft*

Displacement: 86t standard; 100t full load
Dimensions: 103ft pp, 115ft oa × 16ft 1in × 5ft 2in
31.4m, 35.0m × 4.9m × 1.7m
Machinery: 2 shafts, 2 MTU 12V 493 diesels, 1820bhp = 20kts. Range 1400nm at 12kts
Armament: 1–40mm, 1–20mm, 2–7.62mm MG
Complement: 20

Class (no):
El Gihad (PB 1, 1961), *El Horriya* (PB 2, 1961), *El Istiqlal* (PB 3, 1962), *El Sahab* (PB 4, 1962)

Built by Mosor, Yugoslavia, as the Sudanese Navy's first ships. All out of service by 1987.

Ex-Yugoslav 'KRALJEVICA' class *large patrol craft*

Class (no):
El Fasher (522), *El Khartoum* (523)

Transferred in 1969. Hulked in 1981.

MISCELLANEOUS

Ex-Iranian Coast Guard patrol craft *Sheikan* (128), *Kadir* (129), *Karari* (130), acquired 1975. Only *Kadir* and *Karari* extant 1995.

Six ex-Yugoslav Type 101 MTBs (gun version with 2–40mm, 2–20mm) acquired 1970 and discarded 1981.

Two ex-Yugoslav 'DTM 221' class LCTs *Sobat* and *Dinder* transferred 1969. Both discarded by 1986.

Four ex-Iranian 40ft patrol craft *Fijab*, *Helote*, *Maroub* and *Satak*, transferred 1974, all reported stricken *c*1990.

Surinam

Surinam (formerly Dutch Guiana) lies on South America's northern coast covering an area of 63,037 square miles. Of her three neighbours, Brazil, French Guiana and Guyana, Surinam has had border disagreements with the two latter named. The cause of disagreement in both cases has been over mineral deposits lying within Surinam's territory. Granted independence in 1975, Surinam still relies heavily for defence on the Netherlands. The civilian government was overthrown by a 1980 military coup, and a mainly civilian government was appointed in 1982. The president was overthrown by a military coup in December 1990, and a civilian supporter of the military was installed as president. Elections of May 1991 failed to secure the necessary majority to elect a new president, but a special sitting of the assembly elected a new president in September. The constitution was amended in 1992 to limit the power of the military, and later the same year the government signed a peace agreement with the two largest guerrilla movements. Under this agreement guerrilla group members were integrated into the police force. Surinam operates a small navy, employing 240 officers and men and based at Paramaribo. The majority of the navy's craft were acquired from the Netherlands.

Fleet strength:

Three 32m, 127t patrol craft built in 1976–77 by De Vries, Aalsmeer: *S 401 – S 403*, armed with 2–40mm (2×1). New pennant nos. *P 401–P 403*.

Three 22m, 70t patrol craft built by Schottel in 1976: *C 301 – C 303*, armed with 3 MGs. *C 302* stricken.

Surinam also operates three 12.6m river patrol craft; *Bahadoer* (RP 201), *Fajablow* (RP 202), *Korangon* (RP 203). Built by Schottel, Netherlands and commissioned December 1975 (RP 201, RP 202) and February 1976; 15t, 14kts, 1 Dorman 8JT diesel, 280hp, 1–12.7mm, 4 crew.

One 10m river patrol craft, ordered December 1974 and entered service August 1975. Built by Schottel, Netherlands, 14kts, 1 Dorman 8JT diesel, 280hp.

Sweden

Sweden's Fleet dominated the Baltic in 1947 and for some years after while her war-ravaged neighbours recovered. New ships and vessels, from heavily armed, very fast cruisers to small high speed MTBs were built, delivered and commissioned. New classes of destroyers, submarines, torpedo-boats and minesweepers joined the Royal Swedish Navy. Uniquely, it acquired nuclear weapon proof bunkers carved out of cliffsides to shelter and hide submarines and ships up to destroyer size.

From 1958 the five year defence plans changed to the disadvantage of the navy – and later also of the army and the air force. The idea of a 'light navy', no longer including coastal battleships or cruisers as backbones of the three task forces, was followed by a theory of light task forces of only one or two destroyers with one or two squadrons of large torpedo boats and, of course, submarines, coast defence flotillas of small MTBs, minesweepers, etc. However, this 'lighter' navy could not be maintained. The last cruiser , *Göta Lejon*, was sold to Chile in 1971. Two missile destroyers and one ocean minelayer were cancelled. Destroyers were reclassified as frigates after or without being equipped with modern ASW weapons. The submarine force fell by half between 1975 and 1982.

During the Cold War, the Swedish Navy had to deal with almost constant intrusions into Swedish territorial waters, there being forty such incidents in 1982 alone. Despite heavy depth-charging and detonation of ground mines, all these intruders escaped. Combating this problem was an important task for the navy, and a mini-sub was built in the late 1980s as an ASW target, simulating the type of vessel believed to be involved.

By 1985 the two remaining destroyers had been disposed of, but numbers of light craft had increased, with torpedo boats being fitted with a new Saab SSM and new classes of submarine being commissioned. Uniquely, the Swedes persisted with work on air-dependent conventional propulsion systems for submarines, when the major navies had abandoned them for nuclear boats, and in 1996 the first operational class of submarines with such propulsion systems will enter service (the *Götland* class). The Swedish submarine industry also scored an export success in 1987 with the sale of the *Collins* class boats to the Royal Australian Navy, claimed to be the most sophisticated conventional submarines in the world. Future plans for the 'Submarine 2000' class include air-independent main engines and radical new hull forms.

In 1995 naval personnel totalled 9150, plus 6000 reserves. There is also the Coastal Artillery Service with 3900 active personnel organised into five regiments and 3000 reservists, which operates both static and mobile coast-defence gun and SSM batteries, as well as a number of mine warfare and patrol vessels. Principal naval bases are at Karlskrona and Muskö. Although a 'lighter' fleet than in 1947, the Swedish Navy has remained a capable and powerful force centred around submarines, missile combatants and mine warfare vessels.

FLEET STRENGTH 1947

COASTAL BATTLESHIPS

Name	Launched	Disp (std)	Fate
Aran class			
ARAN	14.8.01	3800t	Stricken 13.6.47, BU 1955
TAPPERHETEN	7.11.01	3800t	Stricken 13.6.47, BU 1952
MANLIGHETEN	1.12.03	3850t	Stricken 24.2.50, pontoon pier at Gullmarsbasen naval base
Oscar II class			
OSCAR II	6.6.05	4320t	Stricken 24.2.50, school hulk, BU 1974
Sverige class			
SVERIGE	3.5.15	7850t	Stricken 30.1.53, BU 1958
DROTTNING VICTORIA	15.9.17	7920t	Stricken 1.4.57, BU 1959
GUSTAF V	31.1.18	7980t	Stricken 1.4.57, school hulk, BU 1970

CRUISERS

Name	Launched	Disp (std)	Fate
FYLGIA	20.12.05	4400t	Stricken 30.1.53, BU 1957
GOTLAND	14.9.33	4775t	Stricken 1.7.60, BU 1962
Tre Kronor class			
TRE KRONOR	16.12.44	8200t	Stricken 1.1.64, BU 1968
GOTA LEJON	17.11.45	8200t	Sold to Chile 18.9.71

DESTROYERS

Name	Launched	Disp (std)	Fate
Sigurd class (stricken 13.6.47)			
SIGURD	19.9.08	460t	Sold 1951, BU
RAGNAR	30.5.08	460t	Sold 1951, BU
VIDAR	9.6.09	460t	Sunk as target 1961
Hugin class (stricken 13.6.47)			
HUGIN	10.12.10	460t	BU 1949
Wrangel class(stricken 13.6.47)			
WRANGEL	25.9.17	587t	Research and target ship, BU 1960
WACHTMEISTER	19.12.17	587t	BU 1950
Ehrensköld class (rebuilt as frigates 1951, stricken 1.4.63)			
EHRENSKOLD	25.9.26	1050t	BU 1964
NORDENSKJOLD	19.6.26	1050t	BU 1964
Psilander class (stricken 13.6.47)			
PSILANDER (ex-Italian *Giovanni Nicotera*)	24.6.26	1250t	BU 1949
PUKE (ex-Italian *Bettino Ricasoli*)	29.1.26	1250t	BU 1949
Klas Horn class			
KLAS HORN	13.6.31	1020t	Stricken 15.8.58, school hulk, BU 1968
Romulus class (rebuilt as frigates 1953, stricken 15.8.58)			
ROMULUS (ex-Italian *Spica*)	11.3.34	870t	BU 1959
REMUS (ex-Italian *Astore*)	22.4.34	870t	BU 1959
Göteborg class (rebuilt* 1961–63, or re-rated† as frigates 1961)			
GÖTEBORG	14.10.35	1040t	Stricken 5.8.58, sunk as target 14.8.62
STOCKHOLM†	24.3.36	1040t	Stricken 1.1.64, BU 1965
MALMO*	22.9.38	1200t	Stricken 1.1.67, BU 1970
KARLSKRONA*	16.6.39	1200t	Stricken 1.7.74, BU 1979
NORRKOPING	5.9.40	1200t	Stricken 1.2.65, BU
GAVLE†	25.9.40	1200t	Stricken 6.12.68, from 1969 generator at Simpevarp nuclear power stn
Mode class (rebuilt as frigates 1954–55)			
MODE	11.4.42	750t	Stricken 1.7.70, school hulk, BU 1978
MAGNE	25.4.42	750t	Stricken 1.1.66, BU 1973
MJOLNER	9.4.42	750t	Stricken 1.4.66, sold 1969, civil hulk
MUNIN	27.5.42	750t	Stricken 6.12.68, BU 1970
Visby class			
VISBY	16.10.42	1150t	Rebuilt to frigate 1966, stricken 1.7.82
SUNDSVALL	20.10.42	1150t	Rebuilt to frigate 1968, stricken 1.7.82
HALSINGBORG	23.3.43	1150t	Rated frigate 1.1.65, stricken 1.7.78, BU 1979
KALMAR	20.7.43	1150t	Rated frigate 1.1.65, stricken 1.7.78, air force target
Oland class (rated frigates 1975)			
OLAND	15.12.45	2000t	Stricken 1.7.78
UPPLAND	15.11.46	2000t	Stricken 1.7.78, BU 1983

SUBMARINES

Name	Launched	Disp (std)	Fate
Draken class			
DRAKEN	20.10.26	667t	Stricken 28.10.48, BU 1951
GRIPEN	21.8.28	667t	Stricken 13.6.47, BU 1957
Delfinen class (stricken 24.2.53)			
DELFINEN	20.12.34	540t	BU 1957
NORDKAPAREN	9.2.35	540t	BU 1958
SPRINGAREN	27.4.35	540t	BU 1956
Sjolejonet class			
SJOLEJONET	25.7.36	580t	Stricken 15.5.59, BU 1962
SJOBJORNEN	15.1.38	580t	Stricken 1.11.64, BU 1967
SJOHUNDEN	26.11.38	580t	Stricken 1.7.60, BU 1962
SVARDFISKEN	18.5.40	580t	Stricken 15.5.59, BU 1961
TUMLAREN	8.9.40	580t	Stricken 1.1.64, BU 1967
DYKAREN	17.12.40	580t	Stricken 1.12.59, BU 1962
SJOHASTEN	19.10.40	580t	Stricken 1.4.63, BU 1963
SJOORMEN	5.4.41	580t	Stricken 1.1.64, BU 1965
SJOBORREN	14.6.41	580t	Stricken 15.6.59, BU 1961
Neptun class (stricken 1.4.66)			
NEPTUN	17.11.42	550t	BU 1971
NAJAD	26.9.42	550t	BU 1971
NACKEN	26.9.42	550t	BU 1971
U 1 class			
U 1	14.6.41	367t	Stricken 1.7.60, experiment boat, BU 1964
U 2	16.5.42	367t	Stricken 1.12.61, BU 1965
U 3	11.6.42	367t	Stricken 1.11.64, from 1965 museum boat at Malmö
U 4	5.6.43	367t	Stricken 1.7.70, BU
U 5	8.7.43	367t	Stricken 1.7.76, BU 1977
U 6	18.8.43	367t	Stricken 1.7.74, BU 1976
U 7	25.11.43	367t	Stricken 1.7.73, BU 1975
U 8	25.4.44	367t	Stricken 1.7.76, BU 1977
U 9	23.5.44	367t	Stricken 1.7.76, BU 1977

U 1 class
U 4–U 9 rebuilt 1963–64 as streamlined hunter-killer boats of 388t with single hull and snorkel. Renamed *Forellen, Abborren, Siken, Gäddan, Laxen* and *Makrillen* respectively.

AMPHIBIOUS WARFARE VESSELS

Ane class landing craft (135t; launched 1943–45)
Ane (J 324), *Balder* (J 325), *Loke* (J 326), and *Ring* (J 327). All deleted 1992.

Alvsnabben 1954 — L & L van Ginderen

MINELAYERS

Name	Launched	Disp (std)	Fate
CLAS FLEMING	14.12.12	1735t	Stricken 1.1.59, BU 1961
ALVSNABBEN	19.1.43	4250t	Stricken 1.7.82, BU

Coastal minelayers: *Mul 3*, 99t, stricken 1956; *Mul 5*, 108t, stricken 1952; *Mul 6*, 108t, stricken 1951; *Mul 7*, 109t, stricken 1953; *Mul 8*, 109t, stricken 1953; *Mul 9*, 120t, stricken 1957; *Mul 10*, 166t, stricken 1970; *Mul 11*, 200t, launched 1946, renamed *Kalusund* and extant 1995.

MOTOR TORPEDO-BOATS

T 11–14 (ex-Italian *MAS 506, 508, 511, 524*), launched 1937–38, purchased 1940, 17t. *T 11* stricken 28.6.46, target, sunk, salvaged, private motor yacht; *T 12* stricken 1.4.49, sold; *T 13* stricken 28.6.46, sold; *T 14* stricken 28.6.46, experiment boat, burned and sunk 1948.
T 3 and *T 4* Vosper boats, launched 1939, stricken 1.4.49, sunk as targets 1963; *T 4* salvaged, private motor yacht.
T 15–18, built at Kockums, launched 1941, 22.5t; all stricken 1.10.56.
T 21–31, built at Kockums, launched 1942, 27t; *T 21–24* stricken 1.1.59. *T 25, 26, 27* and *30* stricken 10.10.57, *T 28, 29* and *31* stricken 15.5.59.

MINESWEEPERS

Styrbjörn, Starkodder (ex-Norwegian whalers *Klo, Graham*) launched 1923 and 1925, 350t and 357t, stricken 1.12.59; *Styrbjörn* sold to tug company; *Starkodder* BU 1961.
Arholma, Landsort, launched 1937, 367t. *Arholma* stricken 1.4.60, target, sunk; *Landsort* stricken 1.1.64, BU 1965.
Bremön class: *Bredskär, Bremön, Grönskär, Holmön, Koster, Kullen, Ramskär, Sandön, Ulvön, Ven, Vinga, Örskär*, launched 1940–41, 460t, stricken 1964–68. *Bredskär, Holmön, Sandön, Ven, Vinga* and *Örskär* targets, all scrapped except *Bremön*, school hulk and museum ship in Karlskrona.
M 1, M 2, launched 1937, 61t, stricken 1952 and 1953 respectively, both to the navigation administration, later BU.
M 3–M 14, launched 1940–41, 60t, stricken 1955–60. *M 3, 4, 5, 6, 9* and *10* surveying vessels, later BU, the last one in 1971. *M 7* and *M 8* training vessels, sold 1968 to private owners. *M 11, 12* and *14* transferred to Central Customs Authority. *M 13* transferred to Royal Science Academy as research vessel *Sagitta*.
M 15–26, launched 1941, 70t. *M 15, 16* and *21–26* from 1976 clearance diving tenders. *M 17* and *18* from 1961 tenders *Lommen* and *Spoven*, both stricken 1984. *M 19* stricken 1965 and transferred as tender to Central Torpedo Workshop. *M 20* rebuilt 1969 as mine research vessel *Skuld*, extant 1995. *M 15* and *M 16* stricken 1984, *M 23* and *M 26* stricken 1989; remainder extant 1995.

PATROL BOATS

V 27, 28, 30 and *35*, ex-first class torpedo-boats, launched 1898–1903, 102–116t, all stricken 13.6.47. BU.
V 39 Iris, V 40 Thetis, V 41 Spica, V 42 Astrea, V 43 Antares, V 45 Altair, V 47 Polaris and *V 48 Perseus*, ex-first class torpedo-boats, *Plejad* type, launched 1908–10, 115–124t, all stricken 13.6.47, *Astrea* sunk as target, remaining BU.
V 5, 8, 11, 12, 14 and *15*, ex-second class torpedo-boats, launched 1906–09, 60t, stricken 1953, 1954, 1954, 1947, 1954 and 1947 respectively, successively BU.
V 19, launched 1914, 67t, stricken 1947, BU.
V 47 Sökaren (ex-*V 16*), *V 48 Sveparen* (ex-*V 17*, later *A 253*), *V 49 Spängaren* (ex-*V18*, later *A 254*), launched 1917–18, 226t. *Sökaren* sunk 1953. *Sveparen* stricken 1964, sold 1967. *Sprängaren* stricken 1961, civil harbour tug.
J 71 Kanon, J 72 Krut, ex-Norwegian whalers, launched 1924 and 1923 respectively, 300t, stricken 1958, BU.
J 21 Granat, J 22 Harpun, ex-Norwegian whalers, launched 1925, 440t, stricken 1963 and 1965 respectively, sold.
V 71 Asköfjärd, V 72 Baggensfjärd, V 73 Nämdöfjärd, launched 1931–32, 25t, stricken 1954.
V 74 Kanholmsfjärd, V 75 Lidöfjärd, V 76 Edöfjärd, launched 1933, 28t, stricken 1954.
V 41 (ex-*V 21*) *Jägaren*, *V 42* (ex-*V 22*) *Kaparen*, *V 43* (ex-*V23*) *Snapphanen*, *V 44* (ex-*V 24*) *Väktaren*, launched 1932–34, 310t. *Snapphanen* sold 1959, rebuilt, sold to Guatemala as *José Francisco Barrundia*, remaining three stricken 1.1.59, later BU.
SVK 1–5, launched 1944, 19t, *SVK 2, 4* and *5* stricken 1980.
V 51–56, launched 1944–45, 145t, *V 51, 52, 53* and *56* stricken 16.11.66, *V 54* and *55* stricken 17.6.65.

ICEBREAKERS

Atle (ex-*Statsisbrytaren*), launched 1925, 1750t, stricken 23.9.66, sold 1967.
Ymer, launched 1932, 3465t, stricken 15.2.76.

MAJOR SURFACE SHIPS

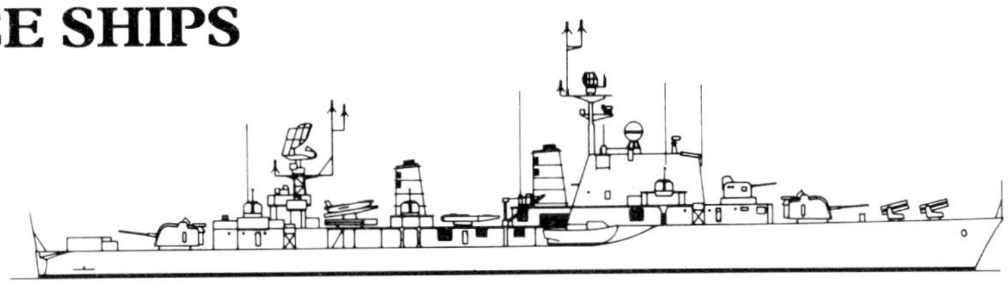

Halland 1975

Halland and *Småland* voted 19 November 1948. First non-Soviet destroyers with anti-ship missiles (1967), first Rb 315, then Rb 08 and later Rb 20. *Halland* modernised 1962 and 1969, *Småland* 1964 and 1967. Bofors 120mm guns fully automatic (40rpm) and also AA. *Lappland* and *Värmland* voted 1955. PEAB 9LV 200 Mk 2 fire control radar replaced original LW-03 air search and M 45 fire control sets. Missile magazine below aft superstructure with launcher above the aft triple TT. Two sisters built for Colombia. *Halland* served as training ship from 1980 until decommissioned in 1982.

HALLAND class *destroyers*

Displacement: 2630t standard; 3400t full load
Dimensions: 381ft wl, 397ft 2in oa × 41ft 4in × 18ft
116.0m , 121.0m × 12.6m × 5.5m
Machinery: 2-shaft de Laval geared turbines, 2 Penhoët 40 atö boilers, 58,000shp = 35kts. Oil 524t. Range 3000nm/445nm at 20kts/35kts
Armament: 1 Saab Rb08A SSM (Mk 20 launcher), 4–120mm/50 (2 × 2), 2–57mm/60 AA (1 × 2), 6–40mm70 AA (6 × 1), 8–533mm TT (1 × 5 + 1 × 3), 2–375mm ASW RL (2 × 4), mines, 6–103mm RFL on 120mm turret
Sensors: Radar Scanter 009, Thomson-CSF Saturn, HSA M22; sonars search and attack sets
Complement: 290

No	Name	Builder	Laid down	Launched	Comp	Fate
J 18	HALLAND	Götaverken, Göteborg	1951	16.7.52	8.6.55	Decomm 29.9.82, BU 1985
J 19	SMALAND	Eriksberg, Göteborg	1951	23.10.52	12.1.56	Deleted 1985
–	LAPPLAND	–	–	–	–	Cancelled 1958
–	VARMLAND	–	–	–	–	Cancelled 1958

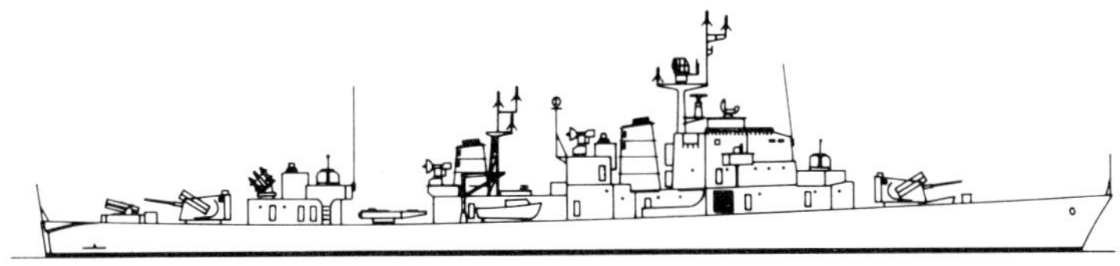

Halsingland 1975

Voted 1953. Modernised: *Ostergötland* 1963 and 1971, *Södermanland* 1962 and 1967, *Gästrikland* 1965, *Hälsingland* 1968. From 1963 3–40mm were replaced by one Seacat SAM (1 × 4), known in Swedish service as Rb 07. Pennant numbers (J for *Jagare*, destroyer) painted on bows from 1966.

OSTERGOTLAND class *destroyers*

Displacement: 2150t standard; 2600t full load
Dimensions: 351ft wl, 367ft 6in oa × 36ft 10in × 12ft
107.0m, 112.0m × 11.2m × 3.7m
Machinery: 2-shaft de Laval geared turbines, 2 Babcock & Wilcox 40 atö boilers, 47,000shp = 35kts. Oil 330t. Range 3000nm at 20kts
Armament: 4–120mm/50 (2 × 2), 7–40mm (7 × 1), 6–533mm TT (1 × 6), 1 Squid ASW Mk 3 mortar, 60 mines
Sensors: Radar Thomson-CSF Saturn, HSA M44 and M45; sonars search and attack sets
Complement: 244

No	Name	Builder	Laid down	Launched	Comp	Fate
J 20	OSTERGOTLAND	Götaverken, Göteborg	1.9.55	8.5.56	3.3.58	Stricken 1.7.82
J 21	SODERMANLAND	Eriksberg, Göteborg	1.6.55	28.5.56	27.7.58	Stricken 1.7.82
J 22	GASTRIKLAND	Götaverken, Göteborg	1.10.55	6.6.56	14.1.59	Stricken 1.7.82
J 23	HALSINGLAND	Eriksberg and Kockums, Malmö	1.10.55	14.1.57	17.6.59	Stricken 1.7.83

Smaland as completed Bofors

Ostergötland 1962 L & L van Ginderen

SUBMARINES

Ex-RN midget submarine sold to Sweden, used as training platform for attack divers. Stricken 27 January 1970, used as experimental boat. Presented to Imperial War Museum, London, 1975.

Ex-British STICKLEBACK class *midget submarine*

Name	Builder	Acquired	Fate
SPIGGEN (ex-X 51 *Stickleback*)	Vickers-Armstrong, Barrow	15.7.58	Museum boat 1975

First three voted 1949; last three voted 1952. Swedish design derived from German *U 3503*, a type XXI submarine sunk by own crew outside Göteborg 8 May 1945, and salvaged by the Swedish Navy. Three-letter markings (Haj, Säl, Val, Bäv, Iln and Utn) on sail instead of pennant numbers.

HAJEN class *submarines*

Displacement: 720t standard; 785t max; *c*900t submerged
Dimensions: 211ft 7in wl, 216ft 6in oa × 16ft 8in × 16ft 5in
64.5m, 66.0m × 5.1m × 5.0m
Machinery: 2 shafts, 2 Pielstick diesels, 2 ASEA electric motors, 1660bhp/1700kW = 16/20kts
Armament: 4–533mm TT (bow, 8 torpedoes), mines
Complement 44

Name	Builder	Laid down	Launched	Comp	Fate
HAJEN	Kockums	1953	11.12.54	28.2.57	Stricken 1.7.80
SALEN	Kockums	1953	3.10.55	8.4.57	Stricken 1.7.80
VALEN	Karlskrona NYd	1953	21.4.55	4.3.57	Stricken 1.7.80
BAVERN	Kockums	1955	3.2.58	29.5.59	Stricken 1.7.80
ILLERN	Karlskrona NYd and Kockums	1955	14.11.57	31.8.59	Stricken 1.7.80
UTTERN	Kockums	1955	14.11.58	15.3.60	Stricken 1.7.80

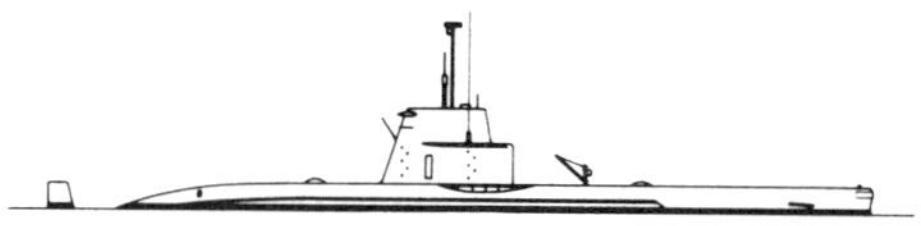

Draken as completed

First three voted 1956, last three voted 1957. *Draken* and *Gripen* modernised 1971 but not rebuilt later. *Vargen* modernised 1971, last three 1970. These four extensively modernised from 1981 and remained in service until the *Västergotland* submarines were commissioned. The class were an improved *Hajen* type with only one, larger 5-blade propeller, giving reduced noise and higher speed. Pennant letters Dra, Gri Vgn, Del, Nor, Spr.

DRAKEN class *submarines*

Displacement: 770t standard; 835t max; *c*950t submerged
Dimensions: 221ft 6in wl, 226ft 6in oa × 161ft 9in × 17ft 5in
67.5m, 69.3m × 5.1m × 5.3m
Machinery: 1 shaft, 2 Pielstick type diesels, 2 ASEA electric motors, 1600bhp/1680kW = 17/22kts
Armament: 4–533m TT (12 torpedoes), mines
Complement: 36

Name	Builder	Laid down	Launched	Comp	Fate
DRAKEN	Kockums	1957	1.4.60	4.4.62	Stricken 1.7.81
GRIPEN	Karlskrona NYd	1958	31.5.60	28.4.62	Stricken 1.7.81
VARGEN	Kockums	1958	20.5.60	15.11.61	In reserve 1988, stricken
DELFINEN	Karlskrona NYd	1958	7.3.61	7.6.62	In reserve 1988, stricken
NORDKAPAREN	Kockums	1959	8.3.61	4.4.62	In reserve 1988, stricken
SPRINGAREN	Kockums	1960	31.8.61	7.11.62	In reserve 1988, stricken

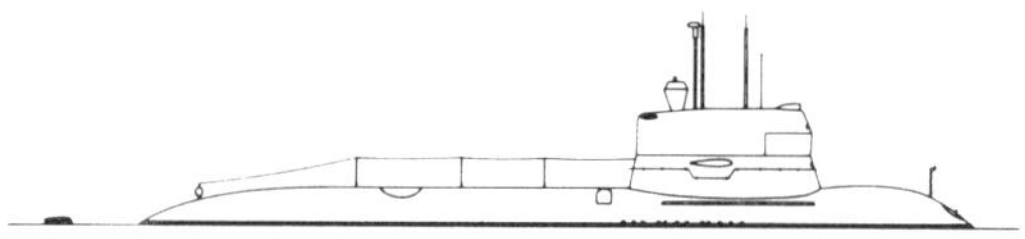

Sjöormen as completed
NB 1/750 scale

Voted 1961; two decks, bow planes on the sail and X-configured stern planes. Diving depth *c*150m. 'Teardrop' hull shape, 5-blade propeller. Attack submarines, A 11 type. Pennant letters Sor, Sle, Sbj, Shu, and Shä. Received new fire control system (Ericsson IBS-A17) and CSU-83 sonar suite 1984–85. *Sjöhunden* and *Sjölejonet* modernised 1992–94, receiving improved electronics and towed-array sonar. Remaining three were to have remained in service until replaced by *Gotland* class, but laid up 1993 due to lack of operating funds.

SJOORMEN class *submarines*

Displacement: 1125t standard; 1400t submerged
Dimensions: 164ft wl, 165ft 7in oa × 20ft × 19ft
50.0m, 50.5m × 6.1m × 5.1m
Machinery: 1 shaft, 2 Pielstick 12 PA-4 diesels, 1 ASEA electric motor, 220bhp/1270kW = 15/20kts. Endurance 21 days
Armament: 4–533mm TT (bow), 2–400mm TT (for Type 42 ASW torpedoes or mines)
Complement: 23

Name	Builder	Laid down	Launched	Comp	Fate
SJOORMEN	Kockums	1965	25.1.67	31.7.68	Laid up 1993
SJOLEJONET	Kockums	1965	29.6.67	16.12.68	Extant 1995
SJOBJORNEN	Karlskrona NYd	1966	9.1.68	28.2.69	Laid up 1993
SJOHUNDEN	Kockums	1966	21.3.68	25.6.69	Extant 1995
SJOHASTEN	Karlskrona NYd	1966	6.8.68	15.9.69	Laid up 1993

Sjolejonet 1981
L & L van Ginderen

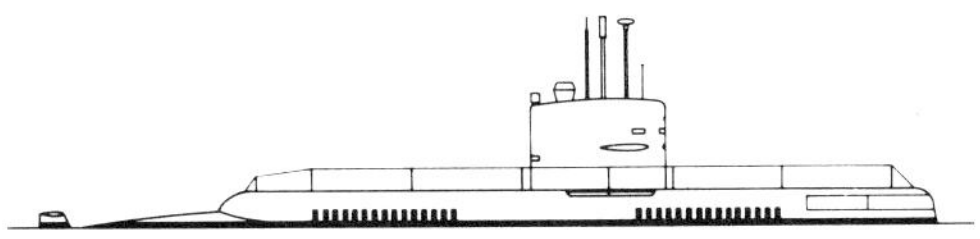

Näcken as completed
NB 1/750 scale

Voted 1972, contracted 22 March 1973. They have Kollmorgen periscopes, two decks, a diving depth *c*150m and 'tear-drop' hull shape. Data Saab NEDPS system uses two Censor 932 computers to give engine and tactical information. Attack submarines, A14 type, with pennant letters Näk, Nep and Naj.

In 1987–88 *Näcken* fitted with two United Stirling Type V4–275 closed-cycle engines, overall length increasing by 26ft 3in (*8m*). With this air-independent powerplant, endurance submerged increased to 14 days. All units of this class are to be modernised to same standard as the *Västergötland* class.

NACKEN class *submarines*

Displacement: 980t standard; 1150t submerged
Dimensions: 144ft 4in wl, 162ft 5in oa × 18ft 8in × 18ft
44.0m, 49.5m × 5.7m × 5.5m
Machinery: 1 shaft, MTU 16V 652 diesel, Jeumont-Schneider electric motor, 2100bhp/1150kW = 20/25kts
Armament: 6–533mm TT, 2–400mm ASW TT (all bow, 12 torpedoes)
Sensors: Large bow sonar
Complement: 19 (5 officers)

Name	Builder	Laid down	Launched	Comp	Fate
NACKEN	Kockums and Karlskrona NYd	1976	17.4.78	25.4.80	Extant 1995
NEPTUN	Kockums and Karlskrona NYd	1976	6.12.78	5.12.80	Extant 1995
NAJAD	Kockums and Karlskrona NYd	1976	13.8.79	26.6.81	Extant 1995

Type A 17 attack submarines. Replacements for *Draken* class. Designed by Kockums, Malmö, and ordered in 1981, with Kockums building the mid-section of the hulls and Karlskronavarvet building the bows and sterns. All TT are bow-mounted, with the six 533mm tubes above the three 400mm tubes. Can carry up to 22 mines in externally-mounted containers. Plans to equip these boats with four SSM vertical launch tubes in the sail have been abandoned. Anechoic hull coating. These boats may be retro-fitted with the same air-independent propulsion system as *Näcken. Västergötland* has a prototype TV camera on a telescoping mast replacing one periscope. *Östergötland* damaged in a collision while on trials 7 November 1989. Pennant letters Vgd, Hgd, Söd and Ögd.

VASTERGOTLAND class *submarines*

Displacement: 1070t surfaced; 1143t submerged
Dimensions: 159t 1in × 19ft 11in × 18ft 4in
48.5m × 6.06m × 5.6m
Machinery: 1 shaft, 2 Hedemora V12A/15-Ub diesels, 2160bhp, 1 electric motor, 1800shp = 11kts/20kts. Endurance 45 days
Armament: 6–533mm TT (bow, 12 torpedoes), 4–400mm ASW TT (bow, 6 torpedoes), 22 mines
Sensors: Radar Terma; sonar Atlas CSU83 hull-mounted, flank array; ECM Argo AR-700-S5
Complement: 20 (5 officers)

Name	Builder	Laid down	Launched	Comp	Fate
VASTERGOTLAND	See notes	10.1.83	17.9.86	27.11.87	Extant 1995
HALSINGLAND	See notes	1.1.84	31.8.87	20.10.88	Extant 1995
SODERMANLAND	See notes	1985	12.4.88	21.4.89	Extant 1995
OSTERGOTLAND	See notes	1986	9.12.88	10.1.90	Extant 1995

Two *Västergötland* class submarines as completed David Millar

Type A 19 attack submarines. Design is updated version of *Västergötland* class with air-independent propulsion system and improved electronics. Originally ordered 28 March 1990, but specification changed 5 September 1991 to incorporate Stirling air-independent propulsion system, resulting in a 200t increase in displacement and overall length increasing by 24ft 7in (*7.5m*). Dimensions given in table are for enlarged design as completed. Three boats under construction, to enter service 1996–98. Plans for a further two boats were cancelled. Pennant letters Gld, Und, Hnd.

GOTLAND class *submarines*

Displacement: 1240t surfaced; 1490t submerged
Dimensions: 196ft 10in × 19ft 11in × 18ft 4in
60.0m × 6.06m × 5.6m
Machinery: 1 shaft, 2 Hedemora V12A/15-Ub diesels, 3600bhp, 1 electric motor, 1800shp, 2 Stirling V4-275R Mk II air-independent generator sets = 11kts/20kts
Armament: 4–533mm TT (bow, 12 torpedoes), 2–40mm ASW TT (bow, 4 torpedoes), 22 mines
Sensors: Radar Terma; sonar Atlas CSU-90 suite; ECM Manta intercept
Complement: 20

Name	Builder	Laid down	Launched	Comp	Fate
GOTLAND	Kockums	27.11.92	2.2.95	Jul 1996	Completing
UPPLAND	Kockums	14.1.94	Jan 1996	–	Building
HALLAND	Kockums	21.10.94	Sep 1996	–	Building

URF *rescue submarine*

Displacement: 52t standard
Dimensions: 44ft 3in × 14ft 9in × 9ft 6in
13.5m × 4.5m × 2.9m
Armament: None
Complement: 5

URF = 'Ubåtsräddningsfarkost' (submarine rescue vessel). Launched 24 August 1978 at Kockums, commissioned 1978. Can be transported on trailer to launching place or be towed at 10kts. Television cameras and other equipment for finding submarines. Can take twenty-six men on board on each dive. Her HY 130 steel pressure hull can dive to 460m and has a collapse depth of 900m.

SPIGGEN II *midget submarine*

Displacement: 12t surfaced; 14t submerged
Dimensions: 36ft 1in × 5ft 7in × 8ft 10in
11.0m × 1.7m × 2.7m
Machinery: 1 shaft diesel-electric = 10kts/5kts. Endurance 24 hrs
Armament: None
Complement: 2

Launched 19 June 1990 by K A Johanssons AB. Designed to simulate the type of submersible believed to violate Swedish territorial waters. Used as target. Extant 1995.

AMPHIBIOUS WARFARE VESSELS

GRIM class *landing craft*

Displacement: 327t standard
Dimensions: 124ft × 28ft 2in × 8ft 6in
37.8m × 8.6m × 2.6m
Machinery: 2 shafts, 2 diesels, 800bhp = 12kts
Armament: 2–20mm (2 × 1)

Name	Builder	Launched	Fate
BORE	Asiverken	1966	Extant 1995
GRIM	Kalmar Varv	1961	Extant 1995
HEIMDAL	Asiverken	1966	Extant 1995
SKAGUL	Hammarbyverken	1960	Extant 1995
SLEIPNER	Hammarbyverken	1959	Stricken 1992

LCM type ships, all of car ferry design. *Bore* and *Heimdal* completed 1967, *Grim* 1961. *Skagul* and *Sleipner* completed 1960 (275t standard, 335t full load) slightly shorter and narrower with 640bhp diesels giving 10kts. Operated by the Coastal Artillery.

MINOR LANDING CRAFT

LCU type
The Coastal Artillery had eighty-one of this 31t type (nos *201–276* and *208–284*, launched 1956–77); seventy-two remained in service 1995, and twelve of a 33t type (nos *401–412*). Five older 30t craft (*L 51–L 55*) built in 1968 were stricken *c*1985.

LCA type
There were fifty-four craft (built 1956–73, nos *301–354*) of this 4t type, operated by the Coastal Artillery. All now stricken. Fifteen older 10t craft (*L 1–L 15*), launched 1949–50 were disposed of 1959–60.

Fourteen 13.2t 'Stridsbåt-90H' class fast landing craft (*801–814*) in service 1992 and extant 1995. Armed with up to 3–12.7mm MG and provision also for 81mm mortar. Carry up to 20 troops. Two modified 1993 to act as command boats, with communications equipment replacing troop accommodation and armament. Second series of sixty-four boats (*815–878*) ordered 23 January 1992. Six of these are to be command variant. Eighteen extant 1995.

822 Jul 1995 L & L van Ginderen

SMALL SURFACE COMBATANTS

Ex-US 'HIGGINS' type *MTB*

Experimental boat *T 1*, launched 1945 at Annapolis Yacht Yard, USA, 50t standard, 72ft × 21ft 4in × 4ft (*22.0m* × *6.5* × *1.2m*), wooden hull, 3 Packard petrol motors, 405 bhp = 40kts. 'Higgins' type bought 1950, burned 1957, sunk, salvaged, sold 1958.

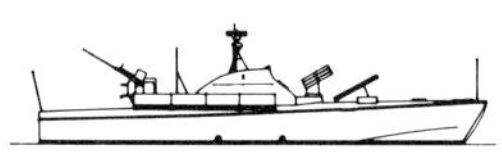

T 32 1967
NB 1/750 scale

T 32 class *MTBs*

Displacement: 38.5t standard
Dimensions: 75ft 6in × 18ft 5in × 4ft 6in
23.0m × 5.6m × 1.4m
Machinery: 3 shafts, 3 Isotta-Fraschini 184C petrol motors, 4500bhp = 45kts
Armament: 2–533mm TT, 1–40mm
Complement: 16

Class:
T 32–41

Built 1951–52 with all-welded steel hulls at Kockums and Karlskrona NYd, commissioned 1951–52. *T 32–37* stricken 1 July 1972, *T 38–40* stricken 1 January 1975. Hull of *T 41* used for building of patrol boat *Viken* (V 04) at Karlskrona 1966–67, stricken 1991.

T 201 *experimental MTB*

Built at Fisksätra Varv, Saltsjöbaden, 1952; commissioned 1953; wooden hull, 17t. *16.3m* × *3.8m* × *1.5m*, 3 Isotta-Fraschini 184C petrol motors, 3 shafts 4500bhp = 44kts. Armament: 2–457mm TT, later changed to 2–533mm TT. Stricken 13.10.64. Target boat at missile experimental station at Karlsborg, Lake Vättern; sold 1980.

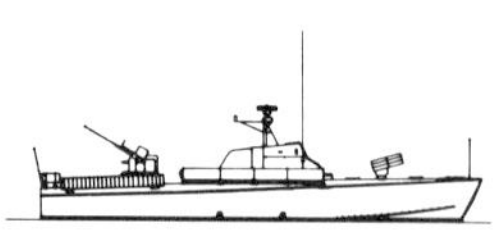

T 46 as completed
NB 1/750 scale

T 42 class *MTBs*

Displacement: 40t standard; 44.5t full load
Dimensions: 75ft 6in × 19ft 5in × 4ft
23.0m × 5.9m × 1.4m
Machinery: 3 shafts, 3 Isotta-Fraschini 184C petrol motors, 4500bhp = 45kts
Armament: 2–533mm TT, 1–40mm
Complement: 16

Class:
T 42–T 56

Built 1955–57 at Kockums, hulls of *T 53*, *54* and *55* at Karlskrona NYd. Steel hulls of *T 42–44* used for building of patrol boats *Skanör* (V 01), *Smyge* (V 02), *Arild* (V 03) at Karlskrona 1976–77. *Skänor* and *Smyge* stricken 1989, *Arild* stricken 1991. Hulls of *T 47*, *48*, *50* and *51* used for building of patrol boats V 05–V 08 (*Oregrund*, *Slite*, *Sandhamn* and *Lysekil*), launched 1983 at Karlskronavarvet. All four stricken 1993–94. Plans to convert four of remaining six hulls to patrol boats abandoned 1983 in favour of new construction (see *Dalerö* class below) and all stricken.

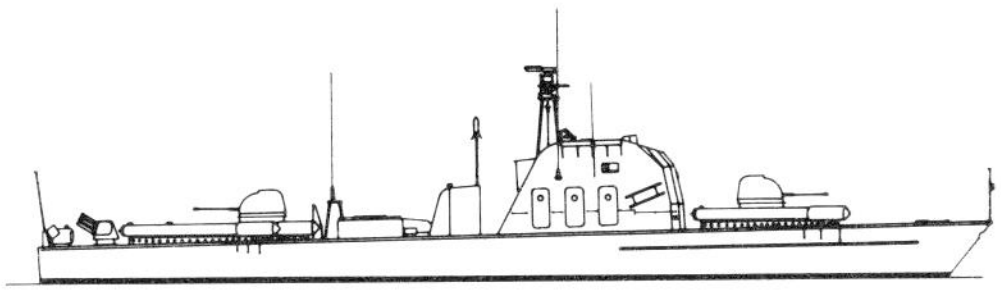

Astrea 1968
NB 1/750 scale

PLEJAD class *fast attack craft (torpedo)*

Displacement: 155t standard; 170t full load (*Perseus* 145t standard)
Dimensions: 148ft 3in wl, 157ft 6in oa × 19ft × 5ft 3in
45.2m, 48.0m × 5.8m × 1.6m
Machinery: 3 shafts, MTU 20V 672 diesels (2 in *Perseus* and 1 Proteus gas turbine, 4500bhp + 6000shp), 9000bhp = 37.5kts. Range 600nm at 30kts
Armament: 2–40mm (2 × 1), 6–533mm TT (4 in *Perseus*), 4–103mm (4 × 1) and 12–57mm (1 × 12) RFL
Complement: 33

No	Name	Builder	Launched	Fate
T 101	PERSEUS	Karlskrona NYd	23.3.50	Stricken 1.1.67
T 102	PLEJAD	Lürssen-Karlskrona NYd	21.11.53	Stricken 1.7.77
T 103	POLARIS	Lürssen-Karlskrona NYd	18.1.54	Stricken 1.7.77
T 104	POLLUX	Lürssen-Karlskrona NYd	3.3.54	Stricken 1.7.77
T 105	REGULUS	Lürssen-Karlskrona NYd	15.4.54	Stricken 1.7.77
T 106	RIGEL	Lürssen-Karlskrona NYd	16.6.54	Stricken 1.7.77, sold 1978
T 107	ALDEBARAN	Lürssen-Karlskrona NYd	16.8.54	Stricken 1.7.81
T 108	ALTAIR	Lürssen-Karlskrona NYd	24.5.56	Stricken 1.7.77
T 109	ANTARES	Lürssen-Karlskrona NYd	14.8.56	Stricken 1.7.77
T110	ARCTURUS	Lürssen-Karlskrona NYd	5.11.56	Stricken 1.7.81
T 111	ARGO	Lürssen-Karlskrona NYd	18.1.57	Stricken 1.7.77
T 112	ASTREA	Lürssen-Karlskrona NYd	15.3.57	Stricken 1.7.81

Perseus was a much altered prototype commissioned 1951 (145t standard). Frequently rebuilt for MTB experiment. First group of six commissioned 1954–55, rebuilt 1972, 1970, 1971, 1972, 1971 and 1965 respectively. Second group of five commissioned 1956–58, rebuilt 1965, 1964, 1964, 1964 and 1964 respectively. Could be fitted as minelayers if TT removed.

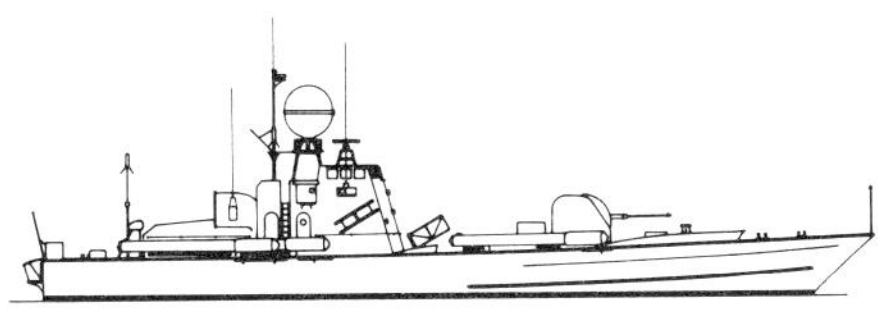

Capella 1970
NB 1/750 scale

SPICA class *fast attack craft (torpedo)*

Displacement: 200t standard; 235t full load
Dimensions: 134ft 6in wl, 140ft 1in oa × 23ft 4in × 5ft 6in
41.0m, 42.5m × 7.1m × 1.6m
Machinery: 3-shaft Bristol Proteus 1274 gas turbines (3 CP propellers), 12,750shp = 40kts
Armament: 1–57mm/70, 6–533mm TT, 4–103mm (4 ×1) and 5–67mm (1 × 6) RFL
Sensors: Radar Scanter 009, HSA M22
Complement: 30

No	Name	Builder	Launched	Fate
T 121	SPICA	Götaverken	26.4.66	Stricken 1989
T 122	SIRIUS	Götaverken	26.4.66	Stricken 1.7.85
T 123	CAPELLA	Götaverken	26.4.66	Stricken 1989
T 124	CASTOR	Karlskrona NYd	25.10.65	Stricken 1.7.85
T 125	VEGA	Karlskrona NYd	12.3.66	Stricken 1989
T 126	VIRGO	Karlskrona NYd	10.9.66	Stricken 1989

Voted 1961, laid down 1964–65, commissioned 1966–68. Could be minelayers if TT Removed. Plans to equip class with SSM abandoned and all stricken by 1989. Four reported for sale 1993.

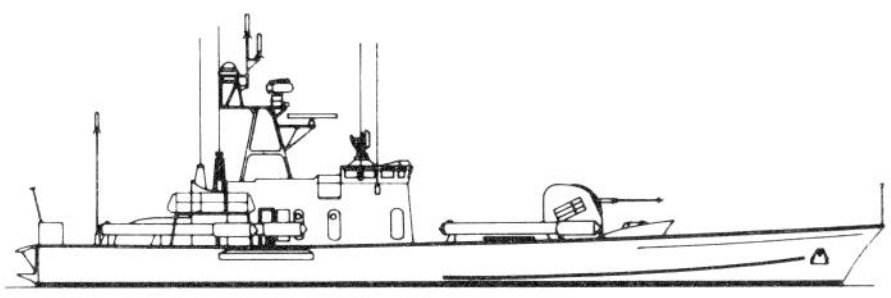

Norrköping 1973
NB 1/750 scale

NORRKOPING ('SPICA II') class *fast attack craft (torpedo/missile)*

Displacement: 220t standard; 255t full load
Dimensions: 137ft 9in wl, 143ft oa × 23ft 4in × 5ft 3in
42.0m, 43.6m × 7.1m × 1.6m
Machinery: 3-shaft Rolls-Royce Marine Proteus gas turbine (3 CP propellers), 12,750shp = 41kts
Armament: 1–57mm/70, 6–533mm TT
Sensors: Radar Scanter 009, PEAB 9LV 200 Mk 1
Complement: 28

No	Name	Builder	Launched	Fate
T 131	NORRKOPING	Karlskrona	16.11.72	Discarded 1993
T 132	NYNASHAMN	Karlskrona	24.4.73	Discarded 1993
T 133	NORRTALJE	Karlskrona	18.9.73	Extant 1995
T 134	VARBERG	Karlskrona	2.2.74	Extant 1995
T 135	VASTERAS	Karlskrona	15.5.74	Extant 1995
T 136	VASTERVIK	Karlskrona	2.9.74	Extant 1995
T 137	UMEA	Karlskrona	15.1.75	Extant 1995
T 138	PITEA	Karlskrona	12.5.75	Extant 1995
T 139	LULEA	Karlskrona	19.8.75	Extant 1995
T 140	HALMSTAD	Karlskrona	17.10.75	Extatn 1995
T 141	STROMSTAD	Karlskrona	26.4.76	Extant 1995
T 142	YSTAD	Karlskrona	3.9.76	Extant 1995

Laid down 1971–75. In service dates T 131 (11.5.73), T 132 (28.9.73), T 133 (1.8.74), T 134 (13.6.74), T 135 (25.10.74), T 136 (15.1.75), T 137 (15.5.75), T 138 (12.9.75), T 139 (28.11.75), T 140 (9.4.76), T 141 (24.9.76), T 142 (10.12.76). All fitted with Saab RBS-15 SSM 1982–84 (maximum of eight missiles, in place of four TT; only two SSM normally carried). Also equipped with new electronics including Sea Giraffe 50HC radar and MARIS 880 weapons control system allowing OTH targetting of SSM with data transmitted from a helicopter. Mines can be carried in place of TT and SSM. All originally to have been stricken 1993, but later announced that six (T 131, T 132, T 138, T 139, T 140, T 142) would be modernised to keep them in service until 2010. First unit scheduled to complete December 1995.

Umea 15.8.85 L & L van Ginderen

HUGIN class *fast attack craft (missile)*

Displacement: 120t standard (*Jägaren* 115t); 150t full load (*Jägaren* 140t)
Dimensions: 118ft wl, 120ft oa × 20ft 8in × 5ft 7in (*Jagaren* 4ft 11in)
36.0m, 36.4m × 6.2m × 2.0m (1.5m)
Machinery: 2 shafts, 2 MTU MB20V 672 TY90 diesels, 7200bhp = 35kts. Range 550nm at 35kts
Armament: 6 Penguin Mk 2 SSM (6 × 1), 1–57mm/70, 103mm RFL, 24 mines or 2 DC racks instead of missiles
Sensors: Radar Scanter 009, PEAB 9LV200 Mk 2; sonar Simrad SQ 3D/SF; ECM Saab-Scania EWS-905
Complement: 18

Munin May 1983 L & L van Ginderen

No	Name	Builder	Launched	Fate
P 150	JAGAREN	Bergens Mekaniske Verksteder	8.6.72	Patrol boat 1988, extant 1995
P 151	HUGIN	Bergens Mekaniske Verksteder	3.6.77	Discarded 1993
P 152	MUNIN	Bergens Mekaniske Verksteder	5.10.77	Discarded 1993
P 153	MAGNE	Bergens Mekaniske Verksteder	9.1.78	Discarded 1993
P 154	MODE	Westermoen Hydrofoil, Mandal	8.8.78	Discarded 1993
P 155	VALE	Westermoen Hydrofoil, Mandal	30.10.78	In reserve Jul 1994
P 156	VIDAR	Westamarin, Mandal	6.3.79	Extant 1995
P 157	MJOLNER	Westamarin, Mandal	12.6.79	In reserve Jul 1994
P 158	MYSING	Westamarin, Mandal	18.9.79	In reserve Jul 1994
P 159	KAPAREN	Bergens Mekaniske Verksteder	8.8.79	Extant 1995
P 160	VAKTAREN	Bergens Mekaniske Verksteder	12.11.79	Extant 1995
P 161	SNAPPHANEN	Bergens Mekaniske Verksteder	18.3.80	Extant 1995
P 162	SPEJAREN	Bergens Mekaniske Verksteder	13.5.80	Extant 1995
P 163	STYRBJORN	Bergens Mekaniske Verksteder	15.8.80	Extant 1995
P 164	STARKODDER	Bergens Mekaniske Verksteder	11.11.80	Extant 1995
P 165	TORDON	Bergens Mekaniske Verksteder	3.2.81	Extant 1995
P 166	TIRFING	Bergens Mekaniske Verksteder	14.9.81	Extant 1995

Built in Norwegian yards (order placed for P 151–P 160 on 15.5.75) for trade compensation reasons and similar to Norway's *Snögg* class FAC. *Jägaren* originally had the pennant number P 151 (to 1977) as prototype boat undergoing 3 years trials in Swedish waters. In 1988 armament was changed to 1–40mm/70 Bofors plus mines and she was reclassified as a patrol craft. The missiles are a Swedish-Norwegian development of the original Penguin. The original diesels fitted were rebuilt and modernised diesels from Swedish *Plejad* class torpedo boats. The *Hugin* class boats are successors to the 'T 42' class MTBs as part of the coastal defence system. In service dates P 150 (24.11.72), P 151 and P 152 (3.7.78), P 153 (12.10.78), P 154 (12.1.79), P 155 (26.4.79), P 156 (10.8.79), P 157 (24.10.79), P 158 (14.2.80), P 159 (7.8.80), P 160 (19.9.80), P 161 (14.1.80), P 162 (2.3.80), P 163 (15.6.80), P 164 (24.8.81), P 165 (26.10.81), P 166 (2.1.82).

In 1991–94 eight of the class (P 159–P 166) were modernised with new engines and improved sonar including Toadfish VDS, and are all extant 1995.

Malmö Jan 1995 L & L van Ginderen

STOCKHOLM ('SPICA III') class *corvettes*

Displacement: 310t standard; 335t full load
Dimensions: 152t 11in wl, 165ft 8in oa × 24ft 7in × 6ft 7in
46.6m, 50.5m × 7.5m × 2.0m
Machinery: 3-shaft CODAG: 1 Allison 570KF gas turbine, 7170shp = 32kts, 2 MTU 16V396 TB 93 diesels, 4190bhp = 20kts.
Armament: 8 RBS-15 SSM (4 × 2), 1–57mm/70 Bofors, 1–40mm/70 Bofors, 2–533mm TT (2 × 1), 4 ASW-600 ASW RL, up to 40 mines. See notes
Sensors: Radar Terma PN-612 nav, Sea Giraffe 50HC search, Celsius Tech 9LV 200 Mk 2 fc; sonar SS 304 Spira hull-mounted, TSM 2642 Salmon VDS; ECM EWS-905 intercept, 2 decoy RL
Complement: 30 (7 officers)

No	Name	Builder	Launched	Fate
K 11	STOCKHOLM	Karlskronavarvet	24.8.84	Extant 1995
K 12	MALMO	Karlskronavarvet	21.3.85	Extant 1995

Development of *Norrköping* class, ordered from Karlskronavarvet in September 1981, entering service in 1985. Armament fit is variable to suit mission; mine rails replace two twin SSM canisters, and 4–400mm TT can replace all SSM. Classed as corvettes. In 1989 *Stockholm* tested the Bofors Trinity 40mm CIWS, in place of the after 40mm Bofors. Basis for later *Göteborg* class. Pennant letters Sto and Mmö.

GOTEBORG class *corvettes*

Displacement: 380t standard; 425t full load
Dimensions: 164ft 1in wl, 187ft oa × 26ft 3in × 6ft 4in
50.0m, 57.0m × 8.0m × 1.93m
Machinery: 3 shafts: 3 MTU 16V 396 TB94 diesels, 3 KaMeWa 80-S62/6 waterjets, 8640bhp = 32kts
Armament: 8 RBS-15 SSM (4 × 2), 1–57mm/70 Bofors, 1–40mm/70 Bofors, 4–400mm ASW TT (2 × 2), 4 ASW 600 ASW RL, mines
Sensors: Radar Terma PN-612 nav, Celsius Tech Pilot search, 2 Celsius Tech 9GR-400 fc; sonar SS 304 Spira hull-mounted, TSM 2642 Salmon VDS; ECM EWS-905 intercept, 2 decoy RL
Complement: 30 (7 officers)

No	Name	Builder	Launched	Fate
K 21	GOTEBORG	Karlskronavarvet	12.4.89	Extant 1995
K 22	GALVE	Karlskronavarvet	12.8.89	Extant 1995
K 23	KALMAR	Karlskronavarvet	1.11.90	Extant 1995
K 24	SUNDSVALL	Karlskronavarvet	29.11.91	Extant 1995

Four units ordered 1 December 1985. Proposed further pair not proceeded with. A development of the *Stockholm* class, to replace the remaining *Spica* class FACs. Include measures to reduce radar, IR and acoustic signatures. ASW TT are mounted to starboard, one pair firing aft and one forward. *Göteborg* was fitted with the 40mm Bofors Trinity CIWS in 1992. Pennant letters Gbg, Gle, Kmr and Svl.

Göteborg 25.4.90 L & L van Ginderen

Smyge Jun 1993 L & L van Ginderen

SMYGE (YS-2000 class) *surface effect ship*

Displacement: 140t full load
Dimensions: 88ft 7in wl, 99ft 9in oa × 37ft 5in × 6ft 3in (2ft 4in on cushion)
27.0m, 30.4m × 11.4m × 1.9m (0.7m)
Machinery: 2 waterjets, 2 MTU 16V396 TB94 diesels, 5550bhp, 2 lift fans, 2 Saab DSI-14 diesels, 1250bhp = 50kts
Armament: 2 RBS-15 SSM (2 × 1), 1–40mm/70, 1–40mm ASW TT
Sensors: Radar navigational; sonar TSM-2643 VDS, minehunting system; ECM intercept array on retractable mast (planned)
Complement: 14 (6 officers)

Class (launched):
Smyge (Mar 1991)

Test surface-effect craft built by Karlskronavarvet for signature reduction trials. Her name is Swedish for 'Stealth'. Hull form has low radar profile and is coated with radar-absorbent materials. Has tested the RBS-15 SSM and a VDS and minehunting sonar system. It is planned to develop two classes of ships from the data obtained from *Smyge*: the 180ft (*55m*) 'YS-2000' class FAC(M), of which a total of sixteen are planned to replace the older FAC classes, and the 'YSB' class minesweepers/minehunters (see under Mine Warfare Vessels, below). The first four 'YS-2000' class were ordered in 1994, with the last to be delivered in 2007. Originally planned as a twin-hull design, increases in displacement compelled a change to a conventional monohull configuration, incorporating 'stealth' features tested on *Smyge*. They are to be armed with SSM, plus a point-defence SAM system.

V 57 *patrol boat*

Displacement: 115t standard; 135t full load
Dimensions: 98ft wl, 106ft 7in oa × 17ft 4in × 6ft 10in
29.9m, 32.5m × 5.3m × 2.3m
Machinery: 1 shaft, 1 Nohab diesel MG-8, 500bhp = 13.5kts
Armament: 1–20mm
Complement: 17

Built 1953, at Stockholm N Yd, commissioned 1954. Fitted for minelaying and attached to Coastal Artillery. Deleted 1985.

BEVAKNINGSBAT 61 class *patrol boats*

Displacement: 28t full load
Dimensions: 69ft 2in × 15ft × 4ft 3in
21.1m × 4.6m × 1.3m
Machinery: 3 shafts, 3 Volvo-Penta DSIR82 or Scania Vabis TIMD96B diesels, 535-705bhp = 18kts
Armament: 1–20mm
Sensors: Radar navigation set
Complement: 8

Class:
Bevakningsbåt 61–77

Built at Bröderna Larsson Varv & Verkstad, Kristinehamn (No *61–70*) in 1960–61 and Coastal Artillery Yard at Fårösund, Gotland (No *71–77*) in 1966–67. The majority of this class have been refitted with a new pilot house and a taller tripod mast. Extant 1995 with Coastal Artillery.

DALERO class *patrol boats*

Displacement: 50t full load
Dimensions: 76ft 9in × 16ft 9in × 3ft 5in
23.4m × 5.1m × 1.05m
Machinery: 3 shafts: 2 MTU 8V396 TB83 diesels, 2100bhp = 30kts
Armament: 1–40mm/70 Bofors, 2–7.62mm MG (2 × 1), mines
Sensors: Radar navigational
Complement: 7

Class (no, in service):
Dalerö (V 09, 21.9.84), *Sandhamn* (V 10, 5.12.84), *Osthammar* (V 11, 1.3.85)

Ordered 28 February 1982 from Djupviks Varvet, Rönnäg, in place of further conversions of torpedo boats to patrol vessels (see *T 42* class above). All three extant 1995.

TAPPER class *patrol boats*

Displacement: 54t standard; 60t full load
Dimensions: 64ft 9in pp, 71ft 8in oa × 17ft 9in × 3ft 5in
19.73m, 21.85m × 5.4m × 1.05m
Machinery: 2 waterjets, 2 MWM TBD 234 V16 diesels, 2440bhp = 28kts
Armament: 2–12.7mm MG (2 × 1), 4 RBS-12 ASW RL, 8 DC, mines
Sensors: Radar navigational; sonar Simrad hull-mounted
Complement: 8

Class (no, in service):
Tapper (81, 4.2.93), *Djärv* (82, 1993), *Dristig* (83, 1994), *Händig* (84, 1994), *Trygg* (85, 1995), *Modig* (86, 1995), *Hurtig* (87, 1995)

Ordered 1991 to replace *Bebakningsbåt 61* class in service with the Coastal Artillery. All extant 1995.

MINE WARFARE VESSELS

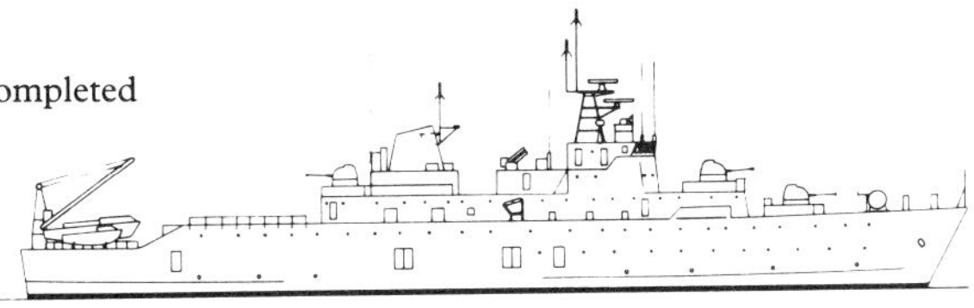

Alvsborg as completed

ALVSBORG class *minelayers*

Displacement: 2500t standard; 2660t full load (2400t; 2540t *Visborg*)
Dimensions: 301ft 10in wl, 303ft 1in oa × 48ft × 13ft 2in
92.0m, 92.4m × 14.7m × 4.0m
Machinery: 1 shaft (CP propeller), 2 Nohab-Polar 12cyl diesels, 4200bhp = 17kts
Armament: 3–40mm (3 × 1), 300 mines, 1 helicopter
Sensors: Radar Scanter 009, Raytheon, HSA M22; ECM 2 decoy RL
Complement: 95

No	Name	Builder	Launched	Fate
M 02	ALVSBORG	Karlskrona N Yd	10.11.69	Extant 1995
M 03	VISBORG	Karlskrona N Yd	22.1.74	Extant 1995

Alvsborg ordered 1968 (after a predecessor of the same name was voted 1955 and cancelled 1958), laid down Nov 1968, commissioned 6 April 1971. Also used as a submarine tender with accommodation for five crews (205 men). *Visborg* ordered 1972, laid down 16 October 1973, commissioned 6 February 1976. Also used as a command ship for C-in-C Baltic Fleet with accommodation for 155 staff members.

CARLSKRONA *minelayer*

Displacement: 3130t standard; 3300t full load
Dimensions: 321ft 6in wl, 346ft 8in oa × 49ft 10in × 13ft 1in
98.0m, 105.7m × 15.2m × 4.0m
Machinery: 2 shafts (CP propellers), 4 Nohab-Polar F212-D825 12cyl diesels, 10,560bhp = 20kts
Armament: 2–57mm (2 × 1), 2–40mm (2 × 1), 105 mines, 1 helicopter
Sensors: Radar Scanter 009, Raytheon, Thomson-CSF Saturn, 2 PEAB 9LV200 Mk 2; sonar Simrad SQ3D/SF
Complement: 186 (as training ship)

No	Name	Builder	Launched	Fate
M 04	CARLSKRONA	Karlskrona N Yd	28.5.80	Extant 1995

Ordered 25 November 1977, laid down 8 February 1979, commissioned 19 March 1982. Reinforced hull and 14 watertight compartments. Also cadet training ship for the annual cruise when complement includes 136 cadets and conscripts. Also has a bow thruster. Two combat information centres on board for *Hugin* and 'Spica II' class FAC. Ericsson Sea Giraffe HC50 air/surface search radar now fitted in place of Thomson-CSF Saturn. A helicopter deck but no hangar.

Carlskrona as completed RAN

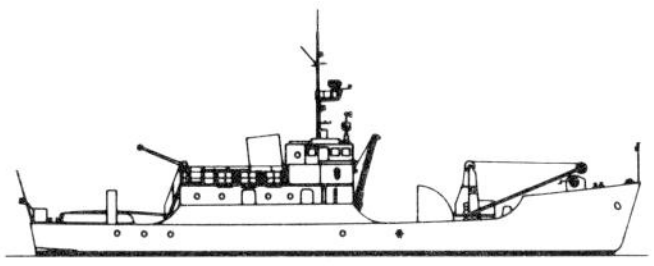

Mul 19 1979
NB 1/750 scale

COASTAL MINELAYERS
Coastal minelayers (*Mul* = *minutläggare*) all operated by the Coastal Artillery: *Mul 12–19*, launched 1952–56, 245t standard, 102ft 4in oa × 24ft 4in × 10ft 2in, *31.2m* × *7.4m* × *3.1m*, 2 diesels, 460bhp = 10.5kts, 1–40mm, mines, 18 crew. Given names in 1985; respectively *Arkosund*, *Kalmarsund*, *Alnösund*, *Grundsund*, *Färösund*, *Skramsösund*, *Öresund* and *Bårösund*. *Färösund* (Mul 16) stricken 1992, remainder extant 1995. *Furusund* (Mul 20) laid down 1982, 245t full load, 106ft 4in × 27ft 7in × 5ft 11in, *32.4m* × *8.4m* × *1.8m*, 2 diesels, 670bhp = 11kts, 1–20mm/70 Oerlikon, 2–7.62mm MG (2 × 1), 24t of mines, crew 10. Proposed class of ten (Mul 21–30) not built due to bankruptcy of builder. *Furusund* extant 1995.

The Coastal Artillery also operates forty-two 15t minelaying launches (*M 501–M 542*), built 1969–71, and six 2.5t small minelaying launches (*1879–1884*), in service 1983–84.

ORUST class *inshore minesweepers*

Displacement: 95t standard; 110t full load
Dimensions: 62ft 4in × 19ft 8in × 4ft 6in
19.0m × *6.0m* × *1.5m*
Machinery: 1 shaft, 1 diesel, 210bhp = 9kts
Armament: 2–25mm
Complement: 10

Class:
Orust (M 41), *Tjorn* (M 42)

Launched 1948, wooden hulls. Stricken 1 July 1977.

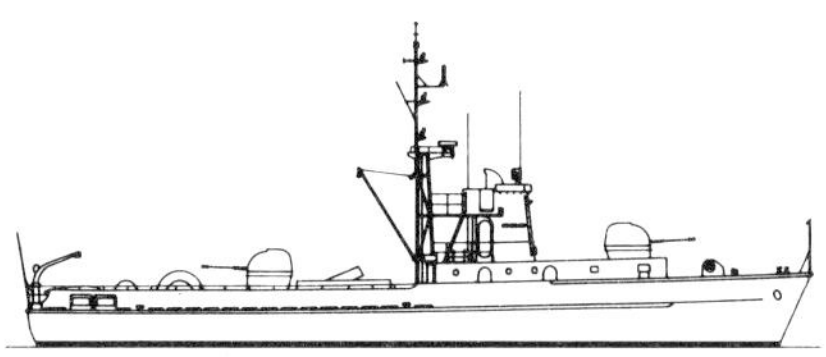

Tjurkö 1969
NB 1/750 scale

HANO class *coastal minesweepers*

Displacement: 280t standard; 295t full load
Dimensions: 131ft 2in × 23ft × 8ft
40.0m × *7.0m* × *2.4m*
Machinery: 2 shafts, 2 Nohab-Polar MG-7 diesels, 910bhp = 14.5kts
Armament: 2–40mm (2 × 1), (*Utö* 1–40mm), 6–57mm RFL
Complement: 25

Class:
Hanö (M 51), *Tärnö* (M 52), *Tjurkö* (M 53), *Sturkö* (M 54), *Ornö* (M 55), *Utö* (M 56)

Launched 1952–53, commissioned 1953. Built at Karlskrona N Yd; *Hanö*, *Tärnö* and *Tjurkö* completed at Kalmar Varv. Steel hulls. All reclassified 1 July 1979 as patrol boats ('vedettbåtar') with same numbers but M changed to V. *Hanö* and *Uto* stricken 1 July 1982, *Tärnö* and *Tjurkö* in 1989 and *Ornö* in April 1992. *Sturkö* was decommissioned in 1987 and was extant 1995 as a youth training ship.

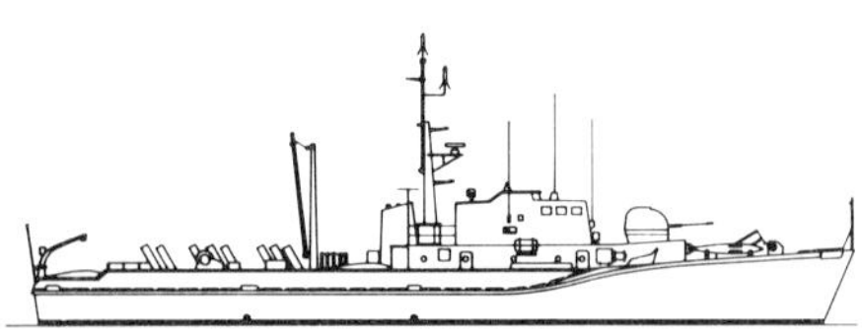

Skaftö 1966
NB 1/750 scale

ARKO class *coastal minesweepers*

Displacement: 272t standard; 300t full load
Dimensions: 131ft pp, 144ft 6in oa × 23ft × 8ft
39.9m, *44.0m* × *7.0m* × *2.4m*
Machinery: 2 shafts, 2 Mercedes-Benz diesels, 1600bhp = 14.5kts
Armament: 1–40mm, 6–57mm RFL
Complement: 25

Class (no, in service):
Arkö (M 57, 1958), *Spårö* (M 58, 1958), *Karlsö* (M 59, 1958), *Iggö* (M 60, 1961), *Styrsö* (M 61, 1962), *Skaftö* (M 62, 1962), *Aspö* (M 63, 1962), *Hasslö* (M 64 1962), *Vinö* (M 65, 1963), *Vållö* (M 66, 1963), *Nåmdö* (M 67, 1964), *Blidö* (M 68, 1964)

M 57–M 59 were built at Oskarshamns Varv, M60–M62 at Kalmar Varv, M 63–M 65 at Karlskrona NYd and M 66–68 at Helsingborgs Varv. M 57 to M 60 have one strake on the wooden hull, later units two. M 66 stricken 1 July 1980, M 65 1 July 1981. Remainder all stricken 1990–94.

Gassten 6.5.83 L & L van Ginderen

HISINGEN class *inshore minesweepers*

Displacement: 130t standard; 140t full load
Dimensions: 62ft 4in wl, 72ft oa × 21ft 4in × 4ft 7in
19.0m, *22.0m* × *6.4m* × *1.4m*
Machinery: 1 shaft (CP propeller), 1 Skandia F35T-4MA diesel, 400bhp = 9kts (*Hisingen* 2 Scania Vabis DS1 10R80 diesels, 37bhp)
Armament: 1–40mm
Complement: 10

Class:
Hisingen (M 43), *Blackan* (M 44), *Dämman* (M 45), *Galten* (M 46), *Gillöga* (M 47), *Rödlöga* (M 48), *Svartlöga* (M 49)

First four launched at different small yards 1960–61, commissioned 1960–61. Last three launched with higher bridges and bluffer bows at Karlskrona N Yd 1964, commissioned 1964. Wooden hulls. This class is a trawler type; used to train conscripts in the mobilisation of similar civil boats. Rearmed with 1–20mm Oerlikon in place of 40mm. All fitted with a new deckhouse amidships in 1987. All seven extant 1995.

GASSTEN class *inshore minesweepers*

Displacement: 120t standard; 135t full load
Dimensions: 79ft × 21ft 4in × 12ft 1in (*Viksten* 83ft, *25.3m* long)
24.0m × *6.5m* × *3.7m*
Machinery: 1 shaft (CP propeller), 2 Scania Vabis DSI 11R92 diesels, 460bhp = 9kts
Armament: 1–40mm
Complement: 10

Class:
Gåssten (M 31), *Norsten* (M 32), *Viksten* (M 33)

Wooden hull class similar to *Hisingen* boats. M 31 launched Knippla SY 17 November 1972 and commissioned 16 November 1973. M 32 launched Häl-levikstrands Varv 24 April 1973 and commissioned 12 October 1973. M 33 launched Karlskrona 18 April 1974 and commissioned 20 June 1974 as GRP forerunner for new minehunters which were in the event not built. All extant 1995.

LANDSORT class *inshore minesweepers/minehunters*

Displacement: 310t standard; 360t full load
Dimensions: 147ft 8in pp, 155ft 10in oa × 31ft 5in × 7ft 7in
45m, *47.5m* × *9.6m* × *2.3m*
Machinery: 2 shafts: 4 Saab-Scania DSI-14 diesels, 1456bhp = 15 kts. Range 2000nm at 12kts
Armament: 1–40mm/70 Bofors, 4 ASW 600 ASW RL (4 × 9), mines. See notes
Sensors: Radar Terma Scanter; sonar TSN 2022; ECM Matilde intercept, 2 decoy RL
Complement: 26 (12 officers)

Kullen 14.8.87 L & L van Ginderen

Class (no, in service):
Landsort (M 71, 19.3.84), *Arholma* (M 72, 23.11.84), *Koster* (M 73, 30.5.86), *Kullen* (M 74, 3.7.87), *Vinga* (M 75, 22.11.87), *Ven* (M 76, 12.12.88), *Ulvön* (M 77, 9.10.92)

Ordered in three groups from Karlskronavarvet: first two on 25 February 1981, four more on 31 January 1984 and the last one in October 1988. A proposed eighth unit was cancelled. GRP hulls. As well as armament listed above, have provision for a shoulder-launched RBS-70 SAM. Equipped with two 'Sea Owl' remote-controlled mine disposal vehicles and can also control up to three SAM class drone MCM craft (see below). Pennant letters, respectively, Ldo, Arh, Ksr, Kln, Vin, Ven, Uln. All extant 1995.

'SAM' class *remote-controlled MCM craft*

Displacement: 20t full load
Dimensions: 59ft 1in × 19ft 8in × 2ft 4in
18.0m × 6.0m × 0.7m
Machinery: 1 shaft, 1 diesel, 210bhp = 10kts. Range 330nm at 7kts

Class (in service):
SAM 01–SAM 05 (1983)

Remote-controlled craft, controlled from *Landsort* class minesweepers. Original *SAM 03* and *SAM 05* sold to USN in 1991 for use in the Persian Gulf. Replacements in service 1992. A total of twenty planned. Five extant 1995.

MISCELLANEOUS

THULE *icebreaker*

Displacement: 2280t standard; 2820t full load
Dimensions: 187ft wl, 204ft 4in oa × 52ft 10in × 20ft 4in
57.0m, 62.2m × 16.1m × 6.2m
Machinery: 3 shafts (1 fwd), 3 Nohab K 58 M diesels, 5040bhp = 14kts
Armament: (2–40mm), 1 helicopter
Complement: 43

Specially designed for the Sound and other narrow waters. Launched 12 October 1951, commissioned 1953. Built at Karlskrona N Yd. Also used as a depot ship for minesweepers. Armament wartime only. Stricken *c*1988.

ODEN *icebreaker*

Displacement: 4950t standard; 5220t full load
Dimensions: 272ft 10in × 63ft 8in × 22ft 8in
83.2m × 19.4m × 6.9m
Machinery: 4 shafts (2 fwd), 6 Nohab SN 19 diesels, 10,200bhp = 16kts. Oil 740t
Armament: (4–40mm), 1 helicopter
Complement: 75

Launched 16 October 1956 by Wärtsilä, Helsinki, and similar to Finnish *Tarmo* class. In service since 19 November 1957. Stricken 1989.

TOR class *icebreakers*

Displacement: 4980t standard; 5290t full load
Dimensions: 277ft 2in × 69ft 6in × 22ft 8in
84.5m × 21.2m × 6.9m
Machinery: 4 shafts (2 fwd), 4 Wärtsilä-Sulzer 8H51 (*Njord* 9MH51) diesels, 13,820shp = 18kts
Armament: (4–40mm (2 × 1, 1 × 2)), 1 helicopter
Complement: 75

Name	Builder	Launched	Comp	Fate
TOR	Wärtsilä, Turku	25.5.63	31.1.64	Extant 1995
NJORD	Wärtsilä, Helsinki	3.10.68	8.10.69	Extant 1995

Similar to *Oden* and the Finnish *Tarmo* class.

ALE *icebreaker*

Displacement: 1320t standard; 1488t full load
Dimensions: 154t 2in × 41ft 8in × 17ft 4in
47.0m × 12.7m × 5.3m
Machinery: 2 shafts, 4 Wärtsilä-Sulzer 824TS diesels, 5360bhp = 14kts
Armament: (1–40mm)
Complement: 21

Specially designed for use on Lake Vänern in Central Sweden. Launched 1 June 1973, commissioned 19 December 1973. Built at Wärtsilä, Turku, completed in Helsinki. Also used as a surveying vessel. Extant 1995.

ATLE class *icebreakers*

Displacement: 7300t standard; 9000t full load
Dimensions: 324ft 9in wl, 343ft 1in oa × 78ft 1in × 23ft 11in
99.0m, 104.6m × 23.8m × 8.3m
Machinery: 4 shafts (2 fwd), 5 Wärtsilä-Sulzer 12 PC-2-2V diesels, 25,000bhp = 18kts
Armament: (4–40mm), 2 helicopters
Complement: 54

Name	Builder	Launched	Fate
ATLE	Wärtsilä, Helsinki	27.11.73	Extant 1995
FREJ	Wärtsilä, Helsinki	10.12.74	Extant 1995
YMER	Wärtsilä, Helsinki	3.9.76	Extant 1995

Atle laid down 10 May 1973 and commissioned 21 October 1974; *Frej* commissioned 30 September 1975. *Ymer* ordered 24 March 1975 and laid down 12 February 1976, commissioned 25 October 1977. Sister ships of Finnish *Urho* class. Crew live and work above main deck.

ODEN *icebreaker*

Displacement: 10,300t standard; 12,900t full load
Dimensions: 305ft 9in pp, 353ft 8in oa × 101ft 8in × 23ft
93.2m, 107.8m × 31.0m × 7.0m
Machinery: 2 shafts, 4 Cegielski-Sulzer 8ZAL40S 8cyl diesels, 24,480bhp = 17kts. Range 30,000nm at 13kts
Armament: (4–40mm/70 Bofors [4 × 1], mines)
Sensors: Radar navigational
Complement: 32

Laid down Gotaverken Arendal 19 October 1987, launched 5 August 1988 and entered service on 29 January 1989. Rectangular hull form. Helicopter deck aft and can be equipped as minelayer. Planned second unit of class, to be named *Thule*, not yet funded. *Oden* extant 1995.

Switzerland

From time to time, and particularly in the fifteenth and sixteenth centuries, the Swiss Confederation has maintained quite large fleets on the lakes. These all have the same purpose, to discourage unwelcome people and especially armies from crossing into Switzerland. The modern Swiss Navy has exactly the same purpose. Ten 8-ton 12-metre, 23kt diesel patrol craft (8 men and 2 MG) were built in 1942 to discourage the Germans and anyone else from crossing the lakes or Upper Rhine without permission, and they were used since the Second World War against smugglers and illegal immigrants. Between 1978 and 1981, these were gradually replaced by new boats of the 'Patrouillenboat 80' class (6t full load, 35ft, *10.7m*, 35kts, 2–12.7mm MG (2 × 1), crew 8): *Antares*, *Aquarius*, *Castor*, *Mars*, *Orion*, *Perseus*, *Pollux*, *Saturn*, *Sirius*, *Uranus*, *Venus*. All eleven extant 1995, and operate on Lakes Constance, Geneva and Maggiore. A further six 11m patrol craft are operated by the Swiss police.

Syria

Syria has long maintained a small naval force equipped mainly with Soviet-supplied ships and equipment; in 1980 the USSR signed a treaty of friendship and co-operation formalising links with Syria going back to 1954. Defence programmes concentrated on the army and air force, partly perhaps because of the navy's poor showing in the October 1973 Arab–Israeli War and its subsequent inability to interfere with Israeli naval operations against the Lebanese coast; the Soviet Navy's massive and long standing presence at Syrian ports after Egypt's *volte-face* in 1976 was also a factor. Consequently naval equipment did not feature prominently among acquisitions of modern defence material, which has left the navy with a rather modest force of negligible fighting value. Bases are at Latakia, Baniyas, Tartus (once heavily used by the Soviet Mediterranean Fleet) and Al-Mina-al-Bayda.

In the 1990s the Syrian Navy remains of secondary importance in the country's military strategy. Nevertheless, in the 1980s she received several warships from the Soviet Union, including three submarines of the 'Romeo' type. There are persistent rumours that the acquisition of three modern 'Kilo' type submarines is planned but up to 1992 these have not materialised, and the current situation in the former Soviet Union makes the subsidised transfer of such boats extremely unlikely. The naval air arm has grown slightly over the past ten years and now includes twelve to twenty Mil Mi-14 and six to ten Ka 25 'Hormone' helicopters. Coast defence batteries are equipped with vehicle-mounted Soviet anti-ship missiles of SS-C-1b Shaddock (land-based version of SS-N-3 missile) and SS-C-3 Styx (land-based version of the widely used SS-N-2 missile).

MAJOR SURFACE SHIPS

Ex-Soviet 'PETYA I' class *frigates* (Project 159)

Two of these light frigates, *Al Hirasa* (14) and *12*, were transferred in 1975–76. Built 1963. Extant 1995.

Ex-Soviet 'NATYA' class *corvette* (Project 266M)

One 'Natya' class minesweeper converted to a corvette with all sweeping gear removed was acquired from the Soviet Navy on 26 January 1985. She is armed only with two 30mm AK 230 automatic guns and two old twin 25mm mounts. Extant 1995.

SUBMARINES

Ex-Soviet 'ROMEO' class *submarines* (Project 633)

Three 'Romeo' class submarines were transferred in the 1980s; the first two in November 1985 and the third boat in December 1986. These boats are officially classed in the Soviet Navy as 'Project 633' type. An older submarine of the 'Whiskey' class (Project 613 type) was transferred in November 1986 to serve as battery-charging hulk to support these boats. Extant 1995.

AMPHIBIOUS WARFARE VESSELS

Ex-Soviet 'POLNOCNY B' class *medium landing ships* (Project 771)

The first 'Polnocny B' class medium landing ship was transferred from USSR on 15 January 1984 and two more in February 1985. Built in the late 1960s by Polish shipyard Stocznia Polnocna, Gdansk, as were all 'Polnocny' class ships for the Soviet Navy. Extant 1995.

SMALL SURFACE COMBATANTS

Ex-Soviet 'KOMAR' class *fast attack craft (missile)* (Project 183R)

Nine transferred from the USSR in 1963–66. Pennant nos 41–49. Three sunk during the October 1973 War. All laid up 1985–86 and five returned to service in the early 1990s.

Ex-Soviet 'OSA I' class *fast attack craft (missile)* (Project 205)

Soviet designation Project 205. Eight transferred in the late 1960s. Two sunk during October 1973 War and later replaced by USSR. Pennant nos 21–26. Four (23–26) extant 1995.

Ex-Soviet 'OSA II' class *fast attack craft (missile)* (Project 205M)

Six launched in early 1970s were transferred, two in 1978; two in October 1979 and two in November 1979; two in 1982, two in May 1984 and two in 1985. Four stricken by 1987. Others extant 1995.

Ex-Soviet 'P 4' class *fast attack craft (torpedo)* (Project 123)

USSR transferred seventeen during 1957–60: three sunk in October 1973 War; four transferred to Egypt in 1970. All stricken 1985–86.

Ex-Soviet 'P 6' class *fast attack craft (torpedo)*

One transferred from USSR. Stricken.

Ex-French CH 5 type *submarine chasers*

Class (former name):
Akba Ben Naseh (ex-*CH 10*), *Al Harissi* (ex-*CH 19*, ex-*RA 1*, ex-*CH 19*), *Tarek Ben* (ex-*CH 13*)

Three built in 1938–41 by AC de France and AC Seine Maritime, were transferred in 1949 (1952?) to form the nucleus of the Syrian Navy. Rebuilt by FC de la Méditerranée in 1955–56 when funnels removed. *CH 15* was captured by Germans and retroceded incomplete in 1945. All stricken 1984.

Ex-Soviet 'ZHUK' class *patrol craft* (Project 199)

Three were delivered by the USSR in December 1983, three in December 1984 and the last three in January 1985. Extant 1995.

MINE WARFARE VESSELS

Ex-Soviet 'T 43' class *ocean minesweepers* (Project 254)

Class:
Yarmouk, Hittine

Transferred in 1962. *Hittine* sunk in October 1973 War off Latakia in the first naval action.

Ex-Soviet 'VANYA' class *coastal minesweepers* (Project 257D)

Two transferred from USSR in December 1972. Names *Kadisia* (775) and *Yarmuk* (776). Extant 1995. There is also an ex-Soviet *Nyryat* type ship used as a divers' base ship.

Ex-Soviet 'SONYA' class *coastal minesweeper* (Project 126S)

One standard Soviet wooden-hulled (GRP-sheated) 'Sonya' class coastal minesweeper was transferred to Syria in January 1986. Extant 1995.

Ex-Soviet 'YEVGENYA' class *inshore minesweepers*

First of these GRP construction inshore minesweepers was transferred from the USSR in 1978, the second in 1981, third and fourth in 1985 and fifth in 1986. Fitted with television minehunting system good for depths up to 30m. Extant 1995.

Taiwan

After the failure to stop the Communist forces crossing the Yangtze in April 1949, the Kuomintang commanders met on 16 July and decided to evacuate their remaining troops and ships 90 miles from the mainland to Taiwan and the other offshore islands. This was completed by 8 December to the tune of two million people, and, although Hainan Island fell to Lin Piao's Communists in 1950, Quemoy and Matsu remained under Kuomintang (Nationalist) control with the help of the US Seventh Fleet. Since then Taiwan, from her capital Taipei, has never ceased to claim to be the lawful ruler of all China (similarly the mainland claims sovereignty over Taiwan). With American help, Taiwan retained China's seat in the United Nations and it was not until the American *rapprochement* with Peking that Taiwan was expelled from the UN and the People's Republic took her place on 26 October 1971.

Although about half the Chinese Navy either defected to the Communists or was captured by them, Taiwan retained all the larger ex-USN warships, and these formed the basis of her fleet for the early 1950s. Unlike the Communists, who rearmed all their ex-Japanese vessels, Taiwan only rearmed those in the best condition, and the others were left to rot until they were finally scrapped. A few ex-British and Canadian vessels that had been demilitarised and sold to Chinese owners were also rearmed at this time.

After the end of the Korean War the Americans began building up the armed forces of their client states in the Far East, and Taiwan received several ex-Second World War US destroyers, destroyer escorts and smaller craft. Her amphibious forces were also built up at this time, so that they would be able to invade the mainland if a suitable opportunity offered. There were a number of small clashes with the Communist Navy in which the Nationalists usually came off badly, but the US Seventh Fleet provided protection against a major offensive. This was demonstrated in 1958, when the Communists made a second major attempt to take over the offshore islands of Quemoy (which dominates the mainland port of Amoy) and Matsu. The US involvement in the Vietnam War, and the logistic support offered by Taiwan, led to an increase in the numbers of warships transferred from the USN.

In the aftermath of the Vietnam War US policy shifted from supporting Taiwan to developing relations with China, which lead to the loss of Taiwan's seat on the UN Security Council. However, to counterbalance this, military aid was increased and large numbers of Second World War vintage ships were transferred, from destroyers downwards. Chiang Kai-shek died in 1975 and his son took over the presidency. At this time, as a result of the US *rapprochement* with China, it was decided to seek arms supplies from other countries, to reduce the former total dependency on what now seemed to be a less than unswerving ally. However, these attempts were hindered by other countries' desire not to offend China. Co-operation with Israel lead to the development of the Hsiung Feng I SSM (a copy of the Gabriel II) and a class of fast attack craft, but Chinese pressure has continued to frustrate Taiwan's attempts to acquire modern submarines. For example, in 1984 the Netherlands refused to sell more than the initial two of a proposed four *Zwaardvis* class boats and in March 1993 the French government withdrew an offer of ten *Agosta* class submarines as a direct result of Chinese opposition. Economic problems and a lack of expertise also hindered attempts to build warships in Taiwan. Relations with the US reached their lowest point in 1978 when the Carter administration withdrew diplomatic recognition and abrogated the 1955 Mutual Defense Treaty, the same year in which it signed a technological aid agreement with mainland China, but under Reagan the situation greatly improved and by the 1990s new surface ships were being transferred and Taiwan was building its own frigates, with US assistance.

The navy's role has always been conditioned by the strengths and weaknesses of the Chinese Navy. In the early part of this period, the main strength of the Chinese fleet lay in its submarine force, and so the Taiwanese concentrated on acquiring ASW capability. The later growth of the Chinese surface fleet lead to the arming of Taiwanese destroyers with SSM and their own atttempts to create a force of submarines. The need to support the offshore islands (over 100 miles from Taiwan itself) and the expressed intention to re-invade the mainland has meant that a large amphibious force has been maintained since 1949. The Taiwanese marine corps is the second largest in the world at 35,000 men (plus an equal number of reservists) in three divisions. In 1995 naval personnel numbered 37,000 active (plus 32,000 reservists). The fleet is organised into three naval districts, with their bases at Tsoying, Makung in the Pescadores Islands and Keelung.

Chinese pressure prevented Taiwan fully diversifying its arms procurement in the 1970s and continues to prevent countries other than the USA being keen to provide arms. A recent example has been the construction of frigates in France, which has been seriously affected by a French desire not to offend China. Although Taiwan now builds its own missiles (the later Hsiung Feng II SSM, equivalent to the US Harpoon, and a projected SAM), the dependence on the USA for both hulls and equipment remains unbroken. With this US help, the Taiwanese Navy is set to continue to expand, in parallel to the expansion of the Chinese Navy and its seemingly more aggressive stance (*eg* over the Spratley Islands).

MAJOR SURFACE SHIPS

Ex-Japanese KAGERO class *destroyer*

Armament:	(1952) 6–127mm/50 (3 × 2), 8–25mm (4 × 2) (1959) 3–127mm/38 (3 × 1), 2–76mm (2 × 1), 17–40mm
Sensors:	Radar, Sonar
Complement:	290
Other particulars:	As Japanese *Kagero* class

Name	Builder	Acquired	Fate
TAN YANG (ex-*Yukikaze*)	Sasebo DY	1947	BU 1971

After being disarmed from 1945–51, *Tan Yang* was given an extensive refit during 1951–52, being rearmed with three twin 127mm mounts (one forward and two aft), her main gun armament when she was the IJN *Yukikaze*. Her machinery was refurbished, but the maximum speed that she achieved on trials in February 1953 was 27.5kts. She was refitted with US weapons and sensors in 1959. At the same time the forecastle was extended slightly aft. For many years she was Taiwan's largest operational warship. She ran aground during a typhoon in May 1970, and because of her age and the damage it was decided to scrap her.

Han Yang 22.6.54 L & L van Ginderen

Ex-US BENSON/GLEAVES class *destroyers*

Armament:	4–127mm/38 (4 × 1) (*Hsien Yang* 3–127mm/38 (3 × 1), 4–40mm, 4–6–20mm, 5–533mm TT (1 × 5) (*Nan Yang* only)
Sensors:	Radar SC, SG, Mk 4; sonar
Complement:	250
Other particulars:	As US *Benson/Gleaves* class

Name	Builder	Acquired	Fate
LO YANG (ex-*Benson*)	Bethlehem, Quincy	26.2.54	Deleted 1975
HAN YANG (ex-*Hilary P Jones*)	Charleston NY	26.2.54	Deleted 1975
NAN YANG (ex-*Plunkett*)	Federal, Kearney	16.2.59	Deleted 1975
HSIEN YANG (ex-*Rodman*)	Federal, Kearny	28.7.55	Expended 1976

Lo Yang and *Han Yang* were presented to Taiwan. *Hsien Yang* and *Nan Yang* were transferred on loan from the USN. In addition the ex-Japanese destroyers *Hatakaze* and *Asakaze* were transferred to Taiwan in August 1970 to provide spares for the first four. Neither saw service, being immediately cannibalised. *Hsien Yang* and ex-*Hatakaze* were both rebuilt as destroyer minesweepers in 1944, losing a 127mm gun and the TT to provide space for the minesweeping gear. They were reclassified as destroyers in 1954. The others were all standard units of their class. *Nan Yang* was the only one to retain TT, otherwise they were unmodernised. They were replaced by larger modernised ex-US destroyers. *Hsien Yang* was sunk for film purposes.

Ex-US FLETCHER class *destroyers*

Armament:	5–127mm/38 (5 × 1), 6–40mm (3 × 2), 6–20mm (3 × 2), 5–533mm TT (1 × 5), 2 Hedgehog Mk 2. See notes
Sensors:	Radar SPS-6, SPS-10, Mk 25, Mk 37. See notes. Sonar SQS-4 or SQS-29; ECM BLR-1, 4 chaff RL
Complement:	250–275

No	Name	Builder	Acquired	Fate
997	AN YANG (ex-*Kimberly*)	Bethlehem, Staten Island	2.6.67	Stricken 1994
947	CHIANG YANG (ex-*Mullany*)	Bethlehem, San Francisco	6.10.71	Stricken 1994
934	KUN YANG (ex-*Yarnell*)	Bethlehem, San Francisco	10.6.68	Stricken 1994
956	KWEI YANG (ex-*Twining*)	Bethlehem, San Francisco	16.8.71	Stricken 1994

An Yang and *Kun Yang* were transferred on loan. Both were purchased in January 1970. *Chiang Yang* and *Kwei Yang* were bought direct. A fifth unit, ex-USS *Sproston*, was transferred in 1968 for cannibalisation for spares. The armament given in the table refers to *Kun Yang* as transferred. *Chiang Yang* was transferred with modified armament and sensors. She had 4–127mm/38 (4 × 1) in 'A', 'B', 'X' and 'Y' positions, 6–76mm (3 × 2) in 'Q' position and between the funnels, two Hedgehog and 6–324mm ASW TT (2 × 3). She was fitted with SPS-10, SPS-12, Mk 25, Mk 37, Mk 56 and Mk 63 radars. All ships of this class were heavily modernised during their careers. *Kun Yang* received a quadruple Sea Chaparral SAM launcher (sixteen missiles) in place of a twin 40mm mount between 'Q' and 'X' turrets. *An Yang* and *Kwei Yang* received a similar upgrade, but also had their 533mm TT replaced with 324mm ASW TT. In 1988 the class was equipped with three Hsiung Feng I SSM (1 × 3), replacing 'X' 127mm turret, and at this time *Chiang Yang* was brought up to the same standard as the rest of the class. Finally *Kwai Yang* and *Chiang Yang* had 'B' turret replaced with a 76mm OTO Melara. Fire control electronics were also updated. By the 1990s these ships were rarely seen at sea as they were approaching the end of their useful lives and were stricken in 1994 as new ships came into service. On decommissioning their pennant numbers were 918, 909, 919 and 908 respectively.

Ex-US ALLEN M SUMNER class *destroyers*

Armament: 6–127mm × 38 (3 × 2), 6–76mm (2 × 2, 2 × 1), 2 Hedgehog Mk 2, 6–324mm Mk 32 ASW TT (2 × 3), DC in some ships. See notes
Sensors: Radar SPS-10, SPS-6C, Mk 25. See notes. Sonar SQS-29. See notes; ECM WLR-1, 4 chaff RL
Complement: *c*275

No	Name	Builder	Acquired	Fate
876	HENG YANG (ex-*Samuel N Moore*)	Bethlehem Staten Island	Feb 1970	Stricken 1993
986	HSIANG YANG (ex-*Brush*)	Bethlehem, Staten Island	9.12.69	Stricken 1984
988	HUA YANG (ex-*Bristol*)	Bethlehem, San Pedro	9.12.69	Stricken 1993
972	HUEI YANG (ex-*English*)	Federal, Kearny	11.8.70	Extant 1995
949	LO YANG (ex-*Taussig*)	Bethlehem, Staten Island	6.5.74	Stricken 1994
954	NAN YANG (ex-*John W Thomason*)	Bethlehem, San Francisco	6.5.74	Extant 1995
928	PO YANG (ex-*Maddox*)	Bath Iron Wks	6.7.72	Stricken 1985
944	YUEN YANG (ex-*Haynsworth*)	Federal, Kearney	12.5.70	Stricken 1993

These ex-USN units had been modified in the 1950s with 76mm guns, ASW TT and improved sensors. *Lo Yang* and *Nan Yang* were converted to FRAM II configuration in September 1962 and January 1961 respectively, giving improved ASW capability. Since their transfer with armament as in the table above (except without 76mm guns), they have been equipped with Hughes 500 ASW helicopters, which are small enough to use the existing DASH ASW drone helicopter landing pad and hangar. In 1985 they were further modified with 1–76mm OTO Melara replacing 'B' 127mm turret, five Hsiung Feng I SSM (1 × 3 replacing 'X' 127mm turret, 2 × 1 on after superstructure), and a quadruple Sea Chaparral SAM launcher right aft. Sonar in 1995 was a Krupp-Atlas DSQS-21CZ hull-mounted and radars were SPS-58A and SPS-29. Both were extant 1995 with pennant numbers 914 and 917 respectively.

The other six ships were transferred with the weapons and equipment shown in the table. *Huei Yang* (1995 pennant number 906) had 'B' 127mm turret replaced with a 76mm OTO Melara and was fitted with 5 Hsiung Feng I SSM and Sea Chaparral SAM, but retained her after turret, the SAM launcher being mounted atop the after superstructure. She was extant in 1995.

The remaining five ships all received the 'Tien Shi' modernisation programme, in which six Gabriel (later Hsiung Feng I) SSM (2 × 3) and Sea Chaparral were mounted aft, replacing the 76mm AA and X 127mm turret, while retaining 'A' and 'B' 127mm turrets. *Hsiang Yang* (901 on deletion) was stricken in 1984 and *Po Yang* (910) in 1985, the remainder being stricken in 1993. On deletion their pennant numbers were as follows: *Hen Yang* 902, *Hua Yang* 903 and *Yuen Yang* 905.

Ex-US GEARING class *destroyers*

Armament: *Fu Yang*: 6–127mm/38 (3 × 2), 6–324mm ASW TT, 2 Hedgehog *Dang Yang:* 4–127mm/38 (2 × 2), 6–324mm ASW TT, 1 Hedgehog
Others: 4–127mm/38 (2 × 2), 1 ASRO C ASW system (1 × 8) (NB not on *Han Yang*, *Kai Yang*, and *Lai Yang*), 6–324mm ASW TT (3 × 2). See notes.
Sensors: Radar SPS-10, SPS-37, Mk 37; sonar SQS-23; ECM WLR-1 or WLR-3, 4 chaff RL. See notes.

Suei Yang 4.6.89 L & L van Ginderen

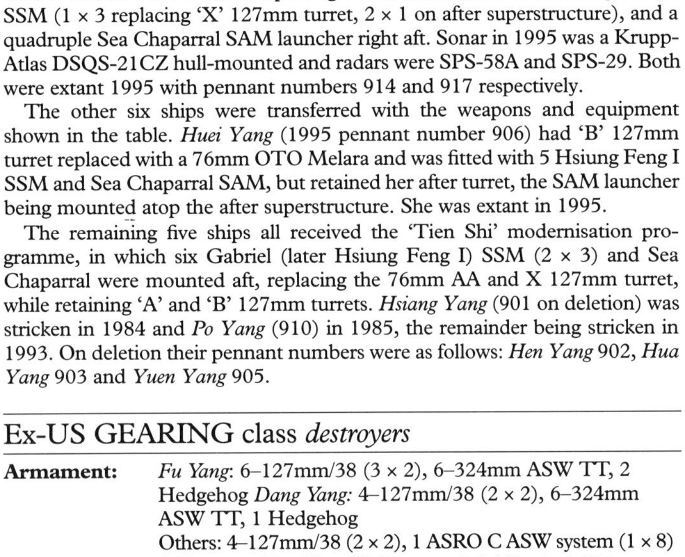

Te Yang 11.6.92 L & L van Ginderen

No	Name	Builder	Acquired	Fate
963	FU YANG (ex-*Ernest G Small*)	Bath Iron Wks	19.2.71	Extant 1995
978	HAN YANG (ex-*Herbert J Thomas*)	Bath Iron Wks	19.2.71	Extant 1995
966	DANG YANG (ex-*Lloyd Thomas*)	Bethlehem, San Francisco	12.10.72	Extant 1995
921	CHIEN YANG (ex-*James E Kyes*)	Todd, Seattle	18.4.73	Extant 1995
–	LAO YANG (ex-*Shelton*)	Todd, Seattle	18.4.73	Extant 1995
938	LIAO YANG (ex-*Hanson*)	Bath Iron Wks	18.4.73	Extant 1995
915	KAI YANG (ex-*Richard B Anderson*)	Todd, Seattle	10.6.77	Extant 1995
–	CHAO YANG (ex-*Rowan*)	Todd, Seattle	10.6.77	Cannibalised 1977
932	SHEN YANG (ex-*Power*)	Bath Iron Wks	1.10.77	Extant 1995
925	TE YANG (ex-*Sarsfield*)	Bath Iron Wks	1.10.77	Extant 1995
981	LAI YANG (ex-*Leonard F Mason*)	Bethlehem, San Francisco	10.3.78	Extant 1995
927	YUN HANG (ex-*Hamner*)	Federal, Newark	27.2.81	Extant 1995
928	CHEN YANG (ex-*Johnston*)	Consolidated, Orange	27.2.81	Extant 1995
929	SHAO YANG (ex-*Hollister*)	Todd, Seattle	17.4.83	Extant 1995
930	TSU YANG (ex-*Hawkins*)	Consolidated, Orange	17.4.83	Extant 1995

Fu Yang had been converted into a radar picket destroyer in 1952 and was given a FRAM II conversion in August 1961. Her SPS-30 radar was removed prior to transfer and on arrival in Taiwan she was fitted with 8–40mm (4 × 2) and 4–12.7mm MG (1 × 4). She later received a triple Gabriel SSM launcher. Further modernisation included the replacement of 'B' 127mm turret with 1–76mm OTO Melara, five Hsiung Feng SSM (1 × 3, 2 × 1) in place of the Gabriel and Sea Chaparral SAM replacing the lighter AA guns, reduced to 2–40mm (2 × 1). The SPS-10 radar was replaced with SPS-58 and updated EW equipment fitted. She was extant in this configuration in 1995, serving as fleet flagship with pennant number 907.

Dang Yang was completed as a special ASW destroyer and underwent FRAM II modernisation in 1961. Transferred to Taiwan in 1972, she was fitted with 4–40mm (2 × 2) and 4–12.7mm (1 × 4), and equipped with a Hughes 500 ASW helicopter (using the DASH drone pad and hangar). By 1995 she had been extensively modernised, with Israeli assistance. A single 76mm OTO Melara had replaced the forward 127mm turret, five Hsiung Feng SSM were fitted (1 × 3 in 'B' turret position, 2 × 1 atop the hangar) and her lighter guns had been reduced to two single 40mm mounts. Radars and fire control equipment had also been modernised. In 1995 she carried pennant number 911.

The remaining twelve ships had all received FRAM I modernisations. *Han Yang* had been converted as a NBCW test ship in 1963–64 and retained the additional superstructure on transfer. She also lacked ASROC, as did *Kai Yang* and *Lai Yang*, its place being taken by 4–40mm (2 × 2). The latter two ships also differed in having two twin 127mm turrets forward in 'A' and 'B' positions, rather than one each fore and aft. A further unit, *Chao Yang*, ran aground whilst under tow to Taiwan on 22 August 1977 and was so badly damaged that she was cannibalised to provide spares for the rest of the class. Another ship (ex-*Warrington*) had been acquired for the same purpose in 1972. Some of these ships were in poor condition when they were transferred. *Chien Yang*, *Lao Yang* and *Liao Yang* had been offered to the Spanish Navy in 1973, but had been rejected on grounds of their bad condition. All these ships (except *Tsu Yang* which for unknown reasons remains as transferred) have been extensively modernised in Taiwanese service. All operate a Hughes 500 ASW helicopter and have Raytheon DE-1191 sonar. Four ships (*Han Yang*, *Lai Yang*, *Kai Yang* and *Sui Yang*) have been equipped with 1–76mm OTO Melara in 'B' position above 2–127mm (1 ×2), five Hsiung Feng SSM (1 × 3 in place of after 127mm turret, 2 × 1 atop hangar), one quadruple Sea Chaparral SAM launcher (16 missiles) on the fantail and 2–40mm Bofors (2 × 1). All retain the ASW TT and *Sui Yang* has ASROC. They received the Israeli-built Elbit NTCCS (Naval Tactical Command and Control System) as part of the same refit. The remaining seven ships have had all 127mm guns deleted and a single 76mm OTO Melara mounted in 'A' position. They are armed with ten US Standard SM-1 SAM (which can also engage surface targets). Two twin box launchers are mounted forward of the bridge and two triple launchers aft, with a 10mm Phalanx CIWS between them. Sensors in 1995 included: radar SPS-58A, HSA DA-08/2, HSA STIR-18 missile control and HW-160 gun fire control; sonar as above. Pennant numbers in 1995 were as follows: *Chien Yang* 912, *Han Yang* 915, *Lai Yang* 920, *Liao Yang* 921, *Shen Yang* 923, *Kai Yang* 924, *Te Yang* 925, *Sui Yang* 926, *Yun Yang* 927, *Chen Yang* 928, *Shao Yang* 929, *Tsu Yang* 930.

Ex-British 'CASTLE' class *frigates*

Name	Builder	Acquired	Fate
TE AN (ex-*Shi Lin*, ex-*Hsi Ling*, ex-*Ta Tung*, ex-*Orangeville*, ex-*Hedingham Castle*)	Robb	1.6.51	Deleted c1970
KAO AN (ex-*Chiu Chin*, ex-*Ta Ting*, ex-*Tillsonburg*, ex-*Pembroke Castle*)	Ferguson	1.6.51	Deleted 1963

RN corvettes built in 1944 that had served in the RCN and were sold as merchant ships to China in 1947. Two were rearmed with 1–120mm, 1–76mm, 4–40mm (2 × 2), 4–20mm (2 × 2) by Taiwan on 1 June 1951 at much the same time that Communist China also rearmed two.

Ex-US BUCKLEY and RUDDEROW (TE and TEV type) class *destroyer escorts (frigates)*

Armament: 1–127mm/38, 6–40mm (3 × 2), 4-LCVP. See notes (*Tai Yuan* 2–127mm/38 (2 × 1), 4–40mm (2 × 2), 4–20mm (4 × 1), 1 Hedgehog Mk 2, 6–324mm Mk 32 ASW TT (2 × 3), 2 Mk 9 DC racks
Sensors: Radar SPS-5 (*Tai Yuan* SPS-5, SPS-6, Mk 26); sonar
Complement: 200 + 160 troops (except *Tai Yuan*)

No	Name	Builder	Acquired	Fate
845	CHUNG SHAN (ex-*Blessman*)	Bethlehem, Hingham	Aug 1967	Extant 1995
838	FU SHAN (ex-*Truxtun*)	Charleston NY	Mar 1966	Extant 1995
322	HENG SHAN (ex-*Raymond W Herndon*)	Bethlehem, Quincy	11.7.66	Discarded 1976
854	HUA SHAN (ex-*Donald W Wolf*)	Defoe, Bay City	May 1965	Stricken 1993
323	KANG SHAN (ex-*George W Hingham*)	Bethlehem, Hingham	19.5.67	Discarded 1978
821	LU SHAN (ex-*Bull*)	Defoe, Bay City	Aug 1966	Extant 1995
325	LUNG SHAN (ex-*Schmitt*)	Bethlehem, Quincy		Discarded 1976
893	SHOU SHAN (ex-*Kline*)	Bethlehem, Quincy	Mar 1966	Extant 1995
878	TAI SHAN (ex-*Register*)	Charleston NY	Oct 1966	Stricken 1991
959	TAI YUAN (ex-*Riley*)	Bethlehem, Hingham	10.7.69	Extant 1995
615	TIEN SHAN (ex-*Kleinsmith*)	Defoe, Bay City	Jun 1967	Extant 1995
834	WEN SHAN (ex-*Gantner*)	Bethlehem, Hingham	May 1966	Stricken 1991
826	YU SHAN (ex-*Kinzer*)	Charleston NY	Apr 1962	Extant 1995

Chung Shan (ex-PF 43), *Kang Shan* (ex-PF 43), *Lu Shan* (ex-PF 36) and *Wen Shan* (ex-*PF 34*) are ex-USN 'TE' class DEs. They have higher bridges than the others which are ex-USN 'TEV' class DEs. Apart from that, they are almost identical. All except *Tai Yuan* (ex-PF 27) were converted into high speed transports (APD) either before or just after completion. *Tai Yuan* was never converted, and differs from the others by having a tripod mast, no davits for LCVP or troop quarters, and more sophisticated sensors. She was finally bought in March 1974. She was later fitted for minelaying.

In 1991 all armament except for 2–40mm (1 × 2) removed and they were extant 1995 as offshore patrol vessels.

Wen Shan collided with ex-*Walter B Cobb* whilst under tow to Taiwan on 17 April 1966, and the latter foundered four days later. She was replaced by *Lu Shan*. *Tien Shan* (meaning 'Heavenly Mountain') was not bought by Taiwan until 1974. Whereas the other APDs were fitted with a second 127mm/38 gun aft instead of from 1970, *Tien Shen* was not given it until she was actually bought. *Hua Shan* (ex-PF 33) had 8–40mm (4 × 2) and 8–20mm (4 × 2). Some of the others may have been fitted similarly. A few have Decca 707 radar, but they are very sparsely equipped with sensors, having only optical directors for their guns. Because of the extra topweight caused by the second 127mm/38 gun, they only carry 2-LCVP. Taiwan rates them as frigates rather than high speed transports and uses them accordingly. Pennant numbers changed in 1976. *Yu Shan* was PF 32, *Tai Shan* PF 38 and *Shou Shan* PF 37.

Ex-US KNOX class *frigates*

Particulars: See US *Knox* class frigates

No	Name	Builder	Acquired	Fate
932	CHI YANG (ex-*Robert E Peary*)	Lockheed, Seattle	23.7.92	Extant 1995
933	FENG YANG (ex-*Brewton*)	Avondale	7.8.92	Extant 1995
934	FEN YANG (ex-*Kirk*)	Avondale	6.8.93	Extant 1995
935	– (ex-*Joseph Hewes*)	Avondale	30.6.94	Extant 1995
936	– (ex-*Cook*)	Avondale	1994	Extant 1995
937	– (ex-*Barbey*)	Avondale	1994	Extant 1995
–	– (ex-*Whipple*)	Todd, Seattle	–	For transfer FY1995
–	– (ex-*Stein*)	Lockheed, Seattle	–	For transfer FY1995
–	– (ex-*Bagley*)	Todd, Seattle	–	For transfer FY1995

Nine ships transferred on five-year leases. All are to be purchased when their lease periods end. A further three units are scheduled for transfer in FY1996. Identical to ships transferred to Greece and Turkey. In 1993 the USA announced the sale of Harpoon SSM to Taiwan to equip this class (four missiles carried, fired from ASROC launcher). The Taiwanese Navy is to receive the largest number of decommissioned *Knox* class frigates.

(For illustration see under US *Oliver Hazard Perry* class)

KWANG HUA I/PFG-2 BATCH I class *frigates*

Displacement: 3100t standard; 4200t full load
Dimensions: 455ft 5in oa, 413ft 1in wl × 45ft × 18ft 8in
138.8m, 125.9m × 13.72m × 5.7m
Machinery: 1 shaft (CP propeller); 2 GM LM-2500 gas turbines, 41,000shp = 29kts. 2 auxiliary electric propulsors, 720shp = c6kts. Range 5000nm at 18kts
Armament: 8 Hsiung Feng II SSM (2 × 4), 1 Standard SAM (40 Standard SM-1 MR SAM), 1–76mm OTO Melara, 2–40mm Bofors (2 × 1), 1–20mm Phalanx CIWS, 3–20mm (3 × 1), 6–324mm ASW TT (2 × 3), 1 or 2 helicoptors
Sensors: Radar SPS-55, SPS-49(V)5, Mk 92, Mod 6, HSA STIR-24; sonar DE-1160B hull-mounted (1105 on, see notes); ECM SLQ-32(V)5, Mk 36 SRBOC decoy system, SLQ-25 Nixie torpedo decoy
Complement: 206 (13 officers)

No	Name	Builder	Launched	Fate
1101	CHEUNG KUNG	China SB Co	27.10.91	Extant 1995
1103	CHENG HO	China SB Co	15.10.92	Extant 1995
1105	CHI KUANG	China SB Co	27.9.93	Extant 1995
1106	YUEH FENG	China SB Co	31.7.94	Completing
1107	TZU I	China SB Co	Jun 1995	Completing
1108	PAN CHAO	China SB Co	1996	Building
1109	CHANG CHIEN	China SB Co	1997	On order

Slightly modified version of the 'long-hulled' US *Oliver Hazard Perry* class, ordered in May 1989 from the China Shipbuilding Corporation, Kaohsiung. The first two units were constructed from prefabricated kits supplied by Bath Iron Works, which is providing help with the remainder of the class. It was originally planned to build only four of this basic version, but development problems with the electronics and SAM system for the advanced variant (see below) has meant that the programme has been extended. The Hsiung Feng II launchers are mounted atop the superstructure, just abaft the bridge, firing across the ship. From 1105 these ships were to be equipped with a towed linear hydrophone array, either the Edo SQR-18A(V)2 or the Plessey-Thomson ATAS/Lamproie. There have also been delays in acquiring helicopters for this class.

KWANG HUA I/PFG-2 Batch II class *frigates*

Displacement: c4300t full load
Dimensions: 427ft 5in oa, 433ft 1in wl × 45ft × 18ft 8in
144.0m, 132.0m × 13.72m × 5.7m
Machinery: 2 shafts (CP props); 2 GE LM-2500 gas turbines, c55,000shp = 28kts. Range 4200nm at 20kts
Armament: 8 Hsiung Feng II SSM (2 × 4), 1 Mk 41 VLS (48 Tien Kung I or Standard SM-1 SAM), 16 Sea Sparrow SAM (1 × 16, see notes), 1–76mm OTO Melara, 2–40mm Bofors (2 × 1), 2–20mm Phalanx CIWS (2 × 1), 6–324mm ASW TT (2 × 3), 1 helicopter
Sensors: Radar SPS-55, Martin-Marietta ADAR-2N 'Mini-Aegis', 2 HSA STIR-24; sonar as Batch I (see above); ECM Chang Feng III, 4 King Feng decoy RL, SLQ-25 Nixie towed torpedo decoy
Complement: ?

No	Name	Builder	Launched	Fate
110	TIEN TAN	China SB Corp	1998	On order

An updated version of the Batch I design, lengthened by a 5m 'plug' to accommodate a VLS SAM system, first ordered in 1989. Details given above are provisional. The 76mm gun will be mounted on the bow, forward of the Mk 41 VLS, and a second Phalanx CIWS forward of the bridge. A 16-cell VLS for Sea Sparrow will be mounted aft, replacing one of the helicopter hangars. The SSM launchers will be carried amidships. The 'Mini-Aegis' system will be able to track 300 targets simultaneously and engage sixteen of them at once. The two-shaft arrangement corrects one of the perceived weaknesses of the original *Oliver Hazard Perry* class. This programme has been delayed by problems with the radars and the Tien Kung I SAM and the first unit is not scheduled to complete until 1999. It is proposed to build another four ships.

(For illustration see under French *La Fayette* class)

KWANG HWA II (LA FAYETTE) class *frigates*

Armament: 8 Hsiung Feng II SSM (2 × 4), 1–76mm, 6–324mm ASW TT (3 × 2), 1 helicopter
Other particulars: See French *La Fayette* class

Class (no):
Hsi Ning (1202), *Kang Ting* (1203), *Wuchang* (1205), *Ti Hua* (1206), *Kunming* (1207), 1 other

Initial order for six units placed with DCAN, Lorient in September 1991, with an option for a further fourteen. Design is called 'MOP-1' in France and is based on the *La Fayette* class frigates. The plan called for six hulls to be supplied from France for fitting out in Taiwan, but in 1993 it was decided that the ships would be completed in France to save money. In October 1993 the French government approved the sale of the complete *La Fayette* weapons fit to Taiwan for these ships, but the response of the Taiwanese is not known. In January 1994, it was announced by both France and the People's Republic of China that no more orders for this class would be accepted, but this was later clarified to mean that the ships under option, but no more, would be completed. The first unit is scheduled for delivery in early 1996 and the last in 1998.

SUBMARINES

Ex-US 'GUPPY II' type *attack submarines*

No	Name	Builder	Acquired	Fate
736	HAI SHIH (ex-*Cutlass*)	Portsmouth NY	12.4.73	Extant 1995
794	HAI PAO (ex-*Tusk*)	Cramp, Philadelphia	18.10.73	Extant 1995

Formerly USN *Tench* class submarines, launched 1944–45. Both converted to GUPPY II configuration in 1948 with increased battery capacity, a streamlined fin and bow and improved sonar. The US intended that they should only be capable of being used for ASW training, so the TT were sealed up before they were transferred. They were returned to operational status, however, with modern Italian torpedoes supplied in 1976. They were scheduled for replacement by Dutch-built boats, but the problems with that programme (see below) has meant that they have remained in service, despite being obsolete.

HAI LUNG class *attack submarines*

Displacement: 2376t surfaced; 2660 submerged
Dimensions: 219ft 7in × 27ft 7in × 22ft
66.92m × 8.4m × 6.7m
Machinery: 1 shaft diesel-electric: 3 Brons/Stork-Werkspoor 12 ORUB 215 diesels, 1 Holec 3800kW motor, 4050hp = 11kts/20kts. Range 10,000nm at 9kts surfaced
Armament: 6–533mm TT (bow)
Sensors: Radar HSA ZW-06; sonar HSA SIASS integrated system
Complement: 67 (8 officers)

No	Name	Builder	Launched	Fate
793	HAI LUNG	Wilton Fijenoord	4.10.86	Extant 1995
795	HAI HU	Wilton Fijenoord	20.12.86	Extant 1995

Modified versions of Netherlands *Zwaardvis* class ordered in 1980 despite protests from mainland China. These were the first new major warships built for Taiwan since 1949. Construction was delayed by financial difficulties at the builder. Boats delivered as deck cargo in 1987 and 1988 respectively. A highly automated design. A later order for two more (with option for a further pair) was turned down by the Netherlands government in 1984 due to Chinese pressure. (See Introduction to this section for further details of the problems faced by the Taiwanese submarine programme).

Italian SX 404 class *midget submarines*

Three were used to land agents on the Chinese mainland. All discarded 1974–75. See Pakistan for data.

AMPHIBIOUS WARFARE VESSELS

Ex-US *dock landing ships*

No	Name	Builder	Acquired	Fate
639	CHUNG CHENG (ex-*White Marsh*	Moore, Oakland	17.11.60	Deleted c1988
618	CHEN HAI (ex-*Fort Marion*)	Gulf SB	15.4.77	Extant 1995

Chung Cheng (ex-191) formerly *Tung Hai*, finally bought in May 1976. *Chen Hai* stricken from USN October 1974 having received FRAM II modernisation December 1959–April 1960. Has a helicopter platform over docking well and SPS-5 radar. Now fitted with Sea Chaparral SAM (1 × 4).

Ex-US LST class *tank landing ships*

Class (former name/no, new nos, transfer):
Chang Hai (ex-*LST 755*, 697/201, Apr 1946), *Chung Ting* (ex-*LST 537*, 673/203, Mar 1946), *Chung Hsing* (ex-*LST 557*, 684/204, Mar 1946), *Chung Chien* (ex-*LST 716*, 679/205, Jun 1946), *Chung Chi* (ex-*LST 1017*, 626/206, Dec 1946), *Chung Shun* (ex-*LST 732*, 624/208, Mar 1946), *Chung Lien* (ex-*LST 1050*, 691/209, Jan 1947), *Chung Yung* (ex-*LST 574*, 657/210, Mar 1959), *Chung Kuang* (ex-*LST 503*, 646/216, Jun 1960), *Chung Suo* (ex-*Bradley County*/*LST 400*, 667/217, Sep 1958), *Chung Chih* (ex-*LST 279*, 218, Jun 1960), *Chung Chuan* (ex-*LST 1030*, 651/221, Feb 1948), *Chung Sheng* (ex-*LST 1033*, 686/222, Dec 1947), *Chung Fu* (ex-*Iron County*/LST 840, 619/223, Jul 1958), *Chung Cheng* (ex-*Lafayette County*/LST 859, 676/224, Aug 1958), *Chung Chiang* (ex-*San Bernadino County*/LST 1110, 635/225, Aug 1958), *Chung Chih* (ex-*Sagadahoh County*/LST 1091, 655/226, Oct 1958), *Chung Ming* (ex-*Sweetwater County*/LST 1152, Oct 1958), *Chung Shu* (ex-*LST 520, 228*, Sep 1958), *Chung Wan* (ex-*LST 535*, 654/229, Sep 1958), *Chung Pang* (ex-*LST 578*, 629/230, Sep 1958), *Chung Yeh* (ex-*Sublette County*/LST 1144, 699/231, Sep 1961).

Fifteen were transferred from the USN during 1957–61 to complement those surviving from before 1949. All were built during 1942–45. A number were scrapped in the 1950s and 1960s and one was sunk by Communist MTBs off Quemoy on 25 August 1958. Four others were transferred from the mercantile service. One was used as a workshop by the CIA-run Air America in Taiwan during the early 1950s. The survivors were extensively rebuilt during the late 1960s and are now the equivalent of new ships. They have davits for 2–6 LCVP. Most are armed with up to ten 40mm and six 20mm, but others have been rearmed with 2–76mm (2 × 1) and 6–40mm (3 × 2). *Chung Chih* (218) was deleted in 1978 and *Cheung Cheng* (224) in 1989. Between 1984 and 1991, six ships (*Chung Ting* 203, *Chung Chi* 206, *Chung Lien* 209, *Chung Chiang* 225, *Chung Shu* 228 and *Chung Wan* 229) were placed in reserve, all being deleted in 1993. Thirteen ships were extant 1995.

Kao Hsiung (ex-*Chung Hai*, ex-USS *Dukes County*, ex-*LST 735*), which was transferred as an LST in May 1957, was converted into an amphibious flagship with additional radar and communications equipment in 1964 and purchased in November 1974. She was extant 1995.

Ex-US LSM 1 class *landing ships*

Class (no, former name, transfer):
Mei Chin (241, ex-*LSM 155*), *Me Heng* (245, ex-*LSM 456*), *Mei Ho* (248, ex-*LSM 13*), *Mei Hung* (246, ex-*LSM 442*), *Mei I* (243, ex-*LSM 76*), *Mei Chien* (249, ex-*LSM 76*), *Mei Hwa* (250, ex-*LSM 256*), *Mei Chen* (251, *LSM 422*), *Mei Kun* (252, ex-*LSM 478*), *Mei Ping* (253, ex-*LSM 471*), *Mei Wen* (254, ex-*LSM 472*), *Mei Han* (255, ex-*LSM 474*), *Mei Lo* (637/256, ex-*LSM 362*, May 1962), *Mei Ping* (659/253, ex-*LSM 471*, Nov 1956), *Mei Sung* (694/247, ex-*LSM 457*, 1946), *Mei Tseng* (649/244, ex-*LSM 431*, 1946)

Mei Tseng also formerly numbered 341. Rebuilt in Taiwan late 1960s and have SO-8 radar. *Mei Lo* replacement for namesake (ex-*LSM 157*) damaged by

Communist artillery and beached on Quemoy Island 8 September 1958. Only last four extant 1995, rest deleted 1970–75.

Ex-US LSIL and LSSL type *landing craft*

Class (no, former name):
Mei Chien (249, ex-*LSM 76*), *Mei Hwa* (250, ex-*LSM 256*), *Lien Chu* (261, ex-*LSIL 233*), *Lien Li* (262, ex-*LSIL 417*), *Lien Sheng* (263, ex-*LSIL 418*), *Lien Chang* (466, ex-*LSIL 1017*)

First five transferred in 19 February 1954 after being on loan to the French in Indo-China. Another five transferred to provide spare parts. All deleted in early 1970s. *Lien Chang* transferred in 1958 and served as survey ship with 1–40mm and 4–20mm (4 × 1) until she was stricken *c*1994.

Ex-US LCT (6) class *tank landing craft*

Class (no, former name):
Ho Chung (485, ex-*LCU 512*), *Ho Chao* (406, *LCU 1429*), *Ho Cheng* (486, ex-*LCU 1145*), *Ho Chi* (401, ex-*LCU 1212*), *Ho Chie* (SB 1, ex-*LCU 700*), *Ho Chien* (496, ex-*LCU 1278*), *Ho Chun* (494, ex-*LCU 1225*), *Ho Chung* (484, ex-*LCU 849*), *Ho Deng* (404, ex-*LCU 1367*), *Ho Feng* (405, ex-*LCU 1397*), *Ho Hoei* (402, ex-*LCU 1218*), *Ho Shun* (494, ex-*LCU 1225*), *Ho Teng* (407, ex-*LCU 1452*), *Ho Tsung* (482, ex-*LCU 1213*), *Ho Yao* (403, ex-*LCU 1244*), *Ho Yung* (495, ex-*LCU 1271*)

Ex-USN units later reclassified as LCU 501 class. Three transferred in January–February 1958 and six more in November–December 1959 to add to the six left in Nationalist hands after 1949. Ex-LCUs *1212, 1218, 1367, 1397* and *1452* transferred under MAP in 1964. All rearmed with 2–12.7mm in addition to existing 2–20mm, all still in service in 1995. Built during 1942–43, and finally bought on 3 April 1978.

US LCU 1466 class *utility landing craft*

Displacement: 130t light; 268t full load
Dimensions: 115t 2in oa × 35ft × 4ft 2in
35.1m × 10.7m × 1.2m
Machinery: 3-shafts, Gray Marine 64/65 YTL diesels; 675bhp = 10kts. Oil 11t. Range 1200nm at 6kts
Armament: 3–20mm (3 × 1) (some also have 2–12.7mm)
Complement: 15

Class (no, former name):
Ho shan (488, ex-*LCU 1596*), *Ho Chuan* (489, ex-*LCU 1597*), *Ho Seng* 490, ex-*LCU 1598*), *Ho Meng* (491, ex-*LCU 1599*), *Ho Mou* (492, ex-*LCU 1600*), *Ho Shou* (493, ex-*LCU 1601*)

All six built by Ishikawajima-Harima, Tokyo, under US Offshore Procurement Program for transfer to Taiwan. In service March 1955 and all extant 1995. Can carry 167t of cargo. All originally numbered in 200 series *eg Hou Mou* (292)

MINOR LANDING CRAFT
About 250 US LCM (6) class built in Taiwan.
Eight ex-US LCM (3) class.
About 100–120 US LCVP class most attached to the LSTs.
Some 25–30 Taiwan Type 272 (5t full load, some 2 MG, 3 crew, built 1970s).

SMALL SURFACE COMBATANTS

Ex-US RAVEN and AUK class *fleet minesweepers (now corvettes)*

No	Name	Builder	Acquired	Fate
896	CHU YUNG (ex-*Waxwing*)	American SB, Cleveland	18.11.65	Stricken 1992
867	PING JIN (ex-*Steady*)	American SB, Cleveland	Mar 1968	Deleted 1993
884	WU SHENG (ex-*Redstart*)	Savannah Machine & Foundry, Georgia	22.7.65	Stricken 1992
	CHIEN MEN (ex-*Toucan*)	American SB, Cleveland	22.12.64	Sunk 6.8.65

After transfer to Taiwan all sweeping gear removed and second 76mm gun mounted aft. *Chien Men* sunk by Communist warships south of Quemoy and replaced by *Ping Jin*. *Chu Yung* fitted for minelaying (80 mine capacity?) 1975. Fitted with SPS-5 radar and SQS-17 sonar. ASW Hedgehog and 2 Mk 9 DC racks supplemented by triple Mk 32 324mm TT.

New missile corvettes

In 1994 a class of sixteen missile corvettes of *c*1250t were to be ordered. A design based on the USCG *Bear* class patrol vessel had been selected, but at the time of writing no contract had been awarded. These ships will augment Taiwan's coastal patrol and ASW capability.

FUH CHOW class *fast attack craft (torpedo)*

Displacement: 33t light; 40t full load
Dimensions: 69ft oa × 19ft 10in × ?
22.6m × 6.5m × ?m
Machinery: 3 shafts, 3 petrol engines; ? hp = 40kts, 27kts cruising
Armament: 1–40mm, 2–20mm (2 × 1), 2–457mm TT (2 × 1)
Complement: 12

Class (no, transferred):
Fuh Chow (PT 1/PT 511, 1.6.57), *Hsueh Chih* (PT 2/PT 512, 6.11.57)

Both built by Mitsubishi SB Co, Japan. They are conventional torpedo-armed FAC with Italian Type 184 weapons. Both stricken 1982.

Ex-US 'Vosper' type 70ft *fast attack craft (torpedo)*

Class (no, former name, transfered):
Fan Kong (PT 513, ex-*PTC 32*, 19.8.57), *Sao Tong* (PT 514, ex-*PTC 33*, 1.11.57)

Both PT boats built by Annapolis Yacht Yd, Annapolis, 1943. Wooden hulls, stricken 1982. Armament 1–20mm, 4–12.7mm (2 × 2), 2 TT.

Ex-US 'Elco' type 80ft *fast attack craft (torpedo)*

Class (former name, transferred):
Fu Kuo (ex-*PTC 34*, 1.9.57), *Tian Kuo* (PT 516, 1.9.57)

Both Second World War PT boats built by Hutchins Yacht Co, Jacksonville, Florida.

LUNG CHIANG class *fast attack craft (missile)*

Displacement: 218t standard; 250t full load
Dimensions: 164ft 6in oa 154ft pp × 23ft 9in × 7ft 5in
50.1m × 46.9m × 7.2m × 2.3m
Machinery: 3 shafts CODOG: (3 CP propellers), 3 AVCO-Lycoming TF-40A gas turbines, 15,000shp = 40kts plus 3 GM 12V149 TI diesels, 3600bhp = 20kts. Range 700nm at 40kts (gas), 2700nm at 12kts (diesel), 1900nm at 20kts (3 diesels)
Armament: 4 Hsiung Feng I SSM (4 × 1), 1–76mm OTO Melara Compact, 2–30mm Emerlec (1 × 2), 2–12.7mm (2 × 1)
Sensors: Radar SPS-58A, Selenia RAN 11 LX, IPN 10
Complement: 35

Class (no, in service):
Lung Chiang (PGG 581, 15.5.78), *Sui Chiang* (PGG 582, 1979)

The first of this class was built to Tacoma's design at their yard in Washington State, USA. The PSMM-5 design is developed from the US *Asheville* class FAC and is the same as the Korean *Paek Ku* class. There were thirteen planned from Taiwanese yards but they have been cancelled because of the unavailability of US or Italian missiles. The two ships built have been armed with the shorter-ranged Israeli Gabriel SSM, which Taiwan calls Hsiung Feng I. Both extant 1995.

HAI OU class *fast attack craft (missile)*

Displacement: 47t full load
Dimensions: 70ft 11in × 18ft × 3ft 1in
21.62m × 5.49m × 0.94m
Machinery: 2 shafts: 2 MTU 12V331 TC81 diesels, 2720bhp = 36kts. Range 700nm at 32kts
Armament: 2 Hsiung Feng I SSM (2 × 1), 2–12.7mm MG (2 × 1)
Sensors: Radar LN-66, CS/SPG-24
Complement: 8–10

Class:
FABG 1–FABG 50

Design based closely on Israeli *Dvora* class. Fifty units built 1980–84 by China SB, Kaohsiung. Aluminium hull. Early units have pylon mast and SSM launchers mounted astern, later units have lattice mast and SSM amidships, with mount for 20mm gun aft. All extant 1995.

Ex-US PC (173ft) class *submarine chasers (large patrol craft)*

Class (former name):
Pei Kiang† (ex-*Hanford*, *PC 1142*), *Po Kiang** (ex-*PC 1254*), *To Kiang* (ex-*Milledgeville*, PC 1263), *Tung Kiang*† (ex-*Placerville*, PC 1087), *Yuan Kiang** (ex-*PC 1182*), *Chang Kiang** (ex-*PC 1232*), *Chih Kiang** (ex-*PC 1078*), *Ching Kiang** (ex-*PC 1168*), *Chu Kiang** (ex-*PGM 31*, ex-*PC 1567*), *Chung Kiang** (ex-*PC 1262*), *Hang Kiang*† (ex-*Vandalia*, PC 1175), *Hsi Kiang*† (ex-*Susanville*, PC 1149), *Hsiang Kiang** (ex-*PC 786*), *Li Kiang** (ex-*PC 1208*), *Lin Kiang*† (ex-*Escondido*, PC 1169)

These fifteen ex-USN units (launched 1942–44), named after rivers, were added to those left in Nationalist service from before 1949. * = 1954 transfers (nine). † = 15 July 1957 transfers (five). *To Kiang* was transferred in July 1959. *Chang Kiang* was sunk by Communist warships south of Quemoy on 6 August 1965. Most other units were discarded in the early 1970s, except *Yuan Kiang* (deleted 1966), but *Hsi Kiang, Pei Kiang* and *Tung Kiang* were transferred to the Customs *c*1973. They were sold in 1976.

COASTAL PATROL CRAFT

Two ex-US 'AVR' type (32.3t full load, 63ft, 22kts, 4 MG, 8 crew, transferred 1.6.69), *Chie Kuo* (PTC 35) and PTC 36.

Fifteen Type 42 (10.5t standard, 40kts, 1–40mm) *PB 60–74* built in Taiwan late 1960s and early 1970s. Extant 1995.

PTC 38 (30t, 21kts, 1–40mm, 1–20mm) unsuccessful design built at Chien Hwa SY 1962.

Ten 15m new patrol craft (12t, 25kts, 1–40mm, 2 diesels and water jet) built at China SB Kahosung (Taiwan) since 1971. 40mm later replaced by 2–12.7mm MG. Extant 1995.

Twenty-two 32m patrol boats (100t, 40kts, 1–40mm, 2–12.7mm MG, 2 DC racks) *PCL 1–PCL 22.* Designed by Vosper-QAF, Singapore and built by China SB, Kaohsiung. All extant 1995.

Seven PBC series patrol craft (28t, 30kts, 2–12.7mm MG) *PBC 5501–PBC 5507*, built 1989–91 by China SB, Kaohsiung. All extant 1995.

Twenty-six PBC 3501 class (28t, 40kts, 2–12.7mm MG) PBC 3501–PBC 3526. PBC 3501 (prototype) built by Vosper-QAF, Singapore, remainder by China SB, Kaohsiung, since 1990.

MINE WARFARE VESSELS

US ADJUTANT class *coastal minesweepers*

Class (former names, nos, in service):
*Yung An** (ex-*MSC 123*, 449, 4.6.55)
Yung Chen (ex-*Maaseick*, ex-*MSC 78*, 441, Jul 1953)
Yung Chi (ex-*Charleroi*, ex-*MSC 152*, 497/166, Feb 1954)
Yung Ching (ex-*Eakloo*, ex-*MSC 101*, May 1953)
Yung Chou (ex-*MSC 278*, 423/158, 10.6.59)
*Yung Fu** (ex-*Diest*, ex-*Macaw* MSC 77, May 1953)
Yung Hsin† (ex-*MSC 302*, 488, 5.3.65)
Yung Ju† (ex-*MSC 300*, 457, 15.4.65)
Yung Jen (ex-*St Nicholas*, ex-*MSC 64*, 485, Feb 1954)
Yung Lo† (ex-*MSC 306*, 469/161, 18.3.66)
*Yung Nien** (ex-*MSC 277*, 479, 10.6.59)
*Yung Ping** (ex-*MSC 140*, 4.6.55)
Yung Shan (ex-*Lier*, ex-*MSC 63*, 476 Jul 1953)
Yung Sui (ex-*Diksmunde*, ex-*MSC 65*, Feb 1954)

US-built wooden minesweepers for transfer to friendly navies. * = MSC 268 with less powerful machinery. † = MSC 289 with lower bridges and taller funnels. The other six were built for Belgium during 1953–56 as part of the *Adjutant* class. They were transferred to Taiwan in November 1969 and can be distinguished from their sisters by the pole mast aft. Original armament of 1–40mm was replaced by 2–20mm (2 × 1) on transfer. Decca 45 or 707 radar is carried and UQS-1D sonar. A further unit, *De Panne*, was transferred for cannibalisation and has been stripped for parts. *Yung Ping* stricken 1982, *Yung Chi, Yung Ching* and *Yung Ju* stricken by 1991 and *Yung Fu* in 1992. Nine extant 1995.

Ex-US *minesweeping boat*

MSB 12 (ex-*MSB 4*) transferred in December 1961 and extant 1995.

Ex-US *minesweeping launches*

MSML 1–12 transferred in March 1961 (built 1943–45). Eight extant 1995.

MWW 50 class *minehunters*

Displacement:	500t full load
Dimensions:	160ft 9in × 35ft 5in × 9ft 2in *49.0m × 10.8m × 2.8m*
Machinery:	2 shafts: 2 MTU 8V396 TB 93 diesels, 218 bhp = 15kts. Range 3500nm at 15kts
Armament:	1–40mm, 2–20mm (2 × 1)
Sensors:	Sonar SA950 hull-mounted
Complement:	45

Class (no, acquired):
Yung Fung (MHC 1301, 1990), *Yung Chia* (MHC 1302, 1990), *Yung Nien* (MHC 1303, May 1991), *Yung Shueng* (MHC 1305, May 1991)

Ordering from Abeking & Rasmussen, Germany in 1989 to replace aging ex-US minesweepers. First four units transferred unarmed and without MCM gear as 'Multi-purpose offshore vessels' for the China Petroleum Company. It was planned to order eight more, but a scandal concerning the previous order has delayed this. By 1991 the four extant ships had received mechanical sweeping gear and German-built remote-controlled submersibles, but not armament. All four extant 1995.

Ex-US AGGRESSIVE class *ocean minesweepers*

It has been reported that four ex-USN *Aggressive* class minesweepers are to be transferred to Taiwan during 1995. Units selected were *Implicit* (MSO 455), *Conquest* (MSO 488), *Excel* (MSO 488) and *Pledge* (MSO 492). There was said to be some resistance in the Taiwanese Navy to the acquisition of such elderly ships.

Tanzania (inc Zanzibar)

Tanzania was created in March 1964 by the union of mainland Tanganyika, which had gained independence in 1961, and the small island state of Zanzibar, where a left-wing African uprising had unseated the Arab Sultan soon after independence in December 1963. The ruler of Tanzania since her formation was Julius Nyerere, a strong and uncompromising proponent of African socialism, to which he gave the name 'Ujamaa'.

Internationally, Tanzania followed an erratic but generally left-wing course, quarrelling frequently with western countries, and less often with African and Communist states, for a variety of reasons. Chinese influence increased considerably after 1969, while the Tanzam railway from Dar es Salaam to the Zambian copperbelt was being built, but Nyerere resolutely refused to be the puppet of any nation or faction. Zanzibar has internal autonomy and retains a small degree of freedom in international relations. Its Marxist government mellowed greatly in recent years. In 1992 a presidential bill endorsing multiparty politics was legalised, with all parties active in both the mainland and in Zanzibar.

The armed forces are mainly for internal security duties, although the Chinese-trained army performed well in its 1979 conquest of Uganda. The navy is small, with volunteer personnel of about 1050 in 1995, and operates from a main base built in 1971 at Dar es Salaam under Chinese supervision. Other coastal ports are Mtwara, Lindi and Tanga. Lake Victoria is patrolled from Mwanza, Bukota and Musoma while Lake Tanganyika is watched from Kigoma. Tanzanian warships, of the fast attack craft and patrol boat types, are from the Soviet Union, China and East Germany (1966–75), with two small craft from West Germany which with Indonesia and the Netherlands has also given training; no substantial changes are probable in the near future.

Six ex-Chinese 'Shanghai II' class fast attack craft (gun) transferred 1970–71. Numbered JW 9861–JW 9866. Two more were transferred in June 1992. Only one extant 1995.

Four ex-Chinese 'Huchwan' class FAC (torpedo) transferred 1975 without hydrofoils (JW 9841–9844). All extant 1995.

Three ex-Soviet 'P 6' class FAC (gun) transferred from E Germany 1974–75 with TT removed. All discarded in 1980s.

Four ex-Soviet 'P 4' class FAC (torpedo) transferred from USSR and E Germany 1970–73. Some minus TT. All discarded 1980s.

Ex-Soviet 'Poluchat' class large patrol craft. Discarded by 1986.

Two ex-E German *Schwalbe* class inshore minesweepers transferred January 1966 and January 1967 minus minesweeping gear and radar. Named *Araka* and *Salaam*. Out of service by 1995.

Two ex-W German coastal patrol craft bought 1967 via Portugal. Named *Rafiki* and *Uhuru*. Since stricken.

Four ex-Chinese 'Yulin' class coastal patrol craft transferred November 1966 and operated on Lake Victoria. Only one in service 1995.

Two ex-Chinese LCMs. Probably no longer operational.

Japanese LST *Maendelo* launched 13.2.80.

Four coastal patrol boats loaned by West Germany in 1963 were discarded in 1967.

Five ex-North Korean *Kinjan* class patrol craft, transferred 1987–88. Only one remains in service in 1995.

Ex-Yugoslav 'Type 16' patrol craft. Extant 1995.

Zanzibar has four Vosper Thornycroft 75ft type GRP coastal patrol craft (*Unguja* and *Pemba* delivered 6.7.73, *Mnemba* in March 1974 and *Miwi* in April). One of the first orders for this Keith Nelson design, they displace 70t, can manage 24.5kts and have 2–20mm (2 × 1).

Thailand

The Kingdom of Thailand has many difficult internal and external defence problems to resolve with limited means. She has nearly 700 miles of navigable waterways, increased by half in the monsoon. Up to 1939 Thailand was known in the English-speaking world as Siam. The existence of an anti-Japanese underground movement won Thailand moderate treatment at the end of the war, though she was forced to return the territories acquired in 1941 and 1943 to French and British control. The Royal Thai Navy was left unmolested and the

Meklong 7.3.82 RAN

ships served the new Thai state for many years. One coast defence ship, *Sri Ayuthia*, was lost during the 1951 rebellion. Manpower rose from 10,000 to 18,000 in the 1950s.

Thailand was one of the original signatories of the 1954 SEATO Manila Pact agreement, having already lost a frigate in the Korean War, and later provided troops, air bases and equipment to assist American forces in Vietnam. Currently Thailand, one of the instigators of the ASEAN agreement, has formed close ties with both Malaysia and the Philippines.

During the 1980s the navy received numerous fighting ships including US-built *Ratanakosin* class fast corvettes, German-built minesweepers/hunters and Thai-built patrol boats of the *Sattahip* class. Four frigates of the 'Jianghu' type were completed in China in the 1990s along with *Kamronsin* class corvettes built in Thai shipyards to a British design. The Chinese ships cost only half as much as Western-built equivalents, but their combat effectiveness is lower because of their obsolescent armament and equipment. The Italthai shipyard, on the other hand, is delivering big modern and capable landing ships of the *Sichang* class. Thailand, like other countries in the region, is trying to build up its shipbuilding industry. The importance of the navy in Thai military thinking was emphasised by the order in 1991 of a 7800t helicopter carrier to be built by Blohm & Voss, Germany at a cost US $225m. This order was later cancelled.

However, in March 1992 a new light aircraft carrier based on the larger Spanish *Principe de Asturias* design was ordered from Bazan, Ferrol. With this vessel, plus the new frigates and 20,000t replenishment tanker building in China and the planned acquisition of ex-US *Knox* class frigates, the Thai Navy will acquire a true blue water capability to project power in the region.

FLEET STRENGTH 1947

DESTROYER

Name	Launched	Disp (std)	Fate
PHRA RUANG (ex-*Radiant*)	25.11.16	1035t	Stricken 1959

FRIGATE

Name	Launched	Disp (std)	Fate
MEKLONG	27.11.36	1400t	TS 1976; Extant 1995

COAST DEFENCE SHIPS

Name	Launched	Disp (std)	Fate
SRI AYUTHIA	21.7.37	2265t	Sunk 3.7.51
RATANAKOSINDRA	21.4.25	886t	Stricken 1968
SUKHOTHAI	19.11.29	886t	Stricken *c*1972

TORPEDO-BOATS

Name	Launched	Disp (std)	Fate
CHANDRABURI	18.1.37	318t	Stricken 1978
CHUNPHORN	18.1.37	318t	Stricken *c*1976
PATANI	16.10.36	318t	Stricken 1978
PUKET	28.9.35	318t	Stricken *c*1976
RAYONG	11.1.37	318t	Stricken 1978
SURASDRA	28.11.36	318t	Stricken 1978
TRAD	26.10.35	318t	Stricken *c*1976
KANTAN	26.3.37	110t	Stricken 1976
KYLONGYAI	26.3.37	110t	Stricken 1976
TAKBAI	26.3.37	110t	Stricken *c*1973

SUBMARINES

Name	Launched	Disp (std)	Fate
BALAJUNBOL	14.5.36	370/430t	Decommissioned 1955
MACHANU	24.12.36	370/430t	Decommissioned 1955
SINSAMUDAR	14.5.36	370/430t	Decommissioned 1955
VILUN	24.12.36	370/430t	Decommissioned 1955

MINELAYERS

Name	Launched	Disp (std)	Fate
BANGRACHAN	1936	368t	Stricken 1980
NHONGARHAI	22.7.36	368t	Stricken 1980

MINESWEEPER

Name	Launched	Disp (std)	Fate
CHOW PHRAYA (ex-*Havant*)	24.3.19	840t	Stricken 24.8.71

COASTAL CRAFT

Name	Launched	Disp (std)	Fate
CMB 6–17	1935–37	16t	Stricken 1950s
SARASINDHU	1936	50t	Stricken 1960s
THIEW UTHOCK	1936	50t	Stricken 1960s
TRAVANE VARI	1936	50t	Stricken 1960s
SAMET (ex-*Pi Sua Nan*)	1906	150t	Stricken 1960s

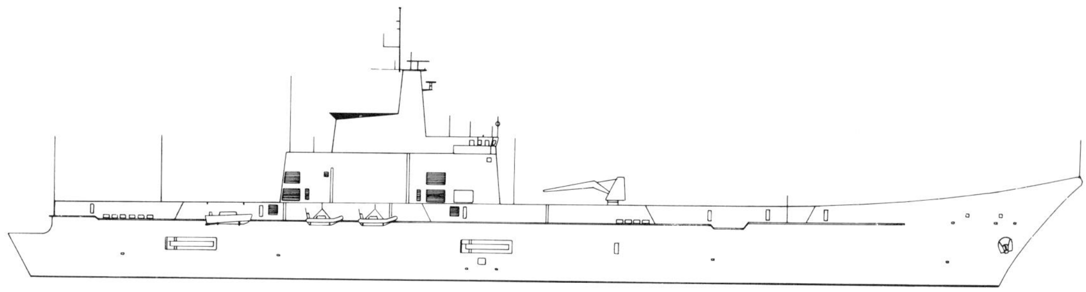

Chakri Naruebet design

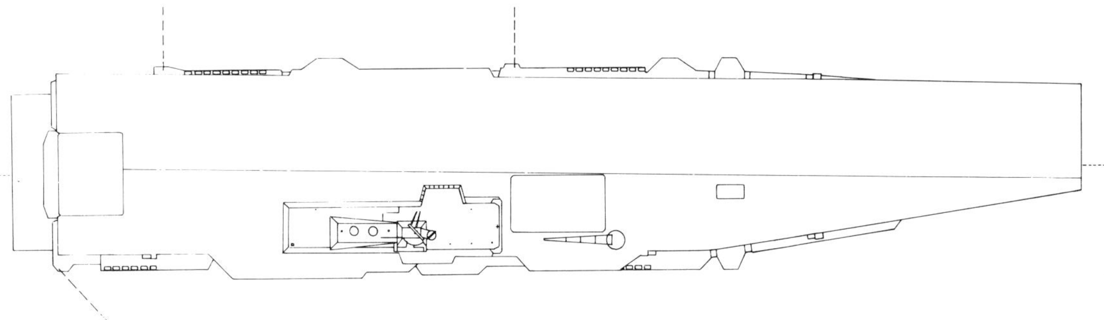

MAJOR SURFACE SHIPS

CHAKRI NARUEBET *light aircraft carrier*

Displacement: 11,485t full load
Dimensions: 538ft 5in wl, 599ft 1in oa × 100ft 1in max, 73ft 10in wl × 20ft 4in
164.1m, 182.6m × 30.5m, 22.5m × 6.2m
Machinery: 2 shafts (CP propellers) CODOG, 2 GE LM 2500 gas turbines, 44,250shp = 27kts, 2 MTU 16V 1163 TB83 diesels, 11,780bhp = 16kts. Range 10,000nm at 12kts
Armament: 1 Mk 41 VLS for Sea Sparrow SAM, 2–30mm (2×1), 4–20mm Vulcan Phalanx CIWS, up to 12 VSTOL aircraft or 14 helicopters
Sensors: Radar SPS 52C or SPS 48 air search, SPS 64 surface search, fire control, aircraft control; sonar hull-mounted MF
Complement: 455 (62 officers) plus 146 air crew

No	Name	Builder	Laid down	Launched	Comp	Fate
911	CHAKRI NARUEBET	Bazan	Sep 1994	–	–	Building

The Thai Navy had already been looking for a large air-capable ship for some years when a 7800t ship initially ordered from Bremer Vulkan was cancelled as unsatisfactory on 20 July 1991. This was replaced by an order for a larger ship placed with the Spanish shipbuilders Bazan on 27 March 1992. First steel was cut in October 1993. The ship is a smaller version of the *Principe de Asturias* design (see Spain) which was itself derived from the abortive US Sea Control Ship project of the 1970s. The Thai vessel differs from the Spanish carrier in the arrangement of its propulsion machinery, having two shafts instead of one.

This very interesting design has a 12° ski jump and two 20t capacity lifts, positioned just forward of the island superstructure and on the after part of the flight deck. Armament may be fitted in Thailand. Hangar capacity is up to ten AV-8B Harrier II VSTOL aircraft or Sea King helicopters. Missions include SAR coordination and EEZ patrol. The ship is also to have special accommodations fitted for the Thai royal family. This ship, once commissioned (planned for 1997), will greatly enhance the power-projection capability of the Thai Navy, being the only ship of its kind in the region apart from the two Indian carriers.

Ex-British Modified 'FLOWER' class *frigates*

No	Name	Builder	Acquired	Fate
PF 4	BANGPAKONG (ex-*Gondwana*, ex-*Burnet*)	Ferguson	15.5.47	Stricken 1984
	PRASAE(i) (ex-*Sind*, ex-*Betony*)	Hall	15.5.47	Sunk 13.1.51

Both 1943-built units. Served in Royal Indian Navy from 1945 before transfer to the Royal Thai Navy on 15 May 1947. *Prasae* lost in Korean War on 13 January 1951. On *Bangpakong* 76mm replaced the old 102mm gun and 40mm guns replaced 20mm. *Bangpakong* served as TS.

Ex-US TACOMA class *patrol frigates*

No	Name	Builder	Acquired	Fate
1	TAHCHIN (ex-*Glendale*)	Consolidated Steel, San Pedro	29.10.51	Extant 1995
2	PRASAE (ii) (ex-*Gallup*)	Consolidated Steel, San Pedro	29.10.51	Extant 1995

Similar to the British-designed 'River' class frigates. Built in 1943–44. Refitted early 1970s. *Prasae* has never been fully repaired since she sustained collision damage in January 1972.

Ex-US CANNON ('DET') class *destroyer escort*

No	Name	Builder	Acquired	Fate
3	PIN KLAO (ex-*Hemminger*)	Western Pipe & Steel, San Pedro	Jul 1959	Extant 1995

Built in 1943–44 and transferred at New York Navy Yd July 1959, under MDAP and finally purchased on 6 June 1975. The 3–533mm TT were removed and the 4–20mm were replaced by 4–40mm. The 6–324mm Mk 32 ASW TT (2×3) were fitted in 1966.

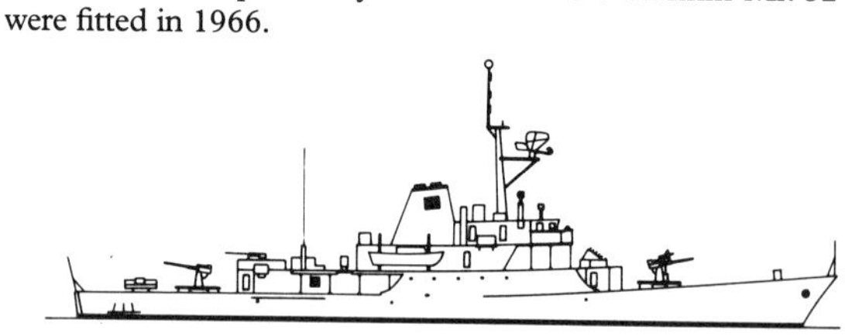

Tapi 1975

TAPI class (US PF 103 class) *frigates*

Armament: 2–76mm/50 Mk 34 (2×1), 2–40mm/60 (1×2), 6–324mm Mk 32 ASW TT (2×3), 1 Hedgehog Mk 2, 1 DC rack Mk 9
Complement: 150
Other particulars: As Iranian *Bayandor* class

No	Name	Builder	Launched	Fate
5	TAPI (ex-*PF 107*)	American SB Co, Toledo	17.10.70	Extant 1995
6	KHIRIRAT (ex-*PF 108*)	Norfolk SB	2.6.73	Extant 1995

Sisters to the Iranian *Bayandor* class frigates. *Tapi* ordered 27 June 1969 (laid down 1 April 1970) and *Khirirat* on 25 June 1971 (laid down 18 February 1972). Completed 1 November 1971 and 10 August 1974 respectively.

This type was originally based on the Italian-built *Pattimura* class ordered by Indonesia. *Tapi* was modernised in 1983 and *Khirirat* in 1985–87. They were partly rearmed with an Italian 76mm/62 OTO Melara gun in the bows replacing the old US 76.2mm mount and the 40mm Bofors gun replacing the aft 76.2mm gun. Two single 20mm mounts replaced their original twin 40mm gun mountings. The modern fire control radar of HSA, WM–25 type is now mounted on the bridge structure. Hedgehog ASW mortar was removed. Reportedly funnel was heightened. Main surveillance radar is SPS-6 and sonar SQS–17A.

Khiriat May 1990 Courtesy Ross Gillett

Makut Rajakumam as completed Yarrow

Makut Rajakumarn 1975

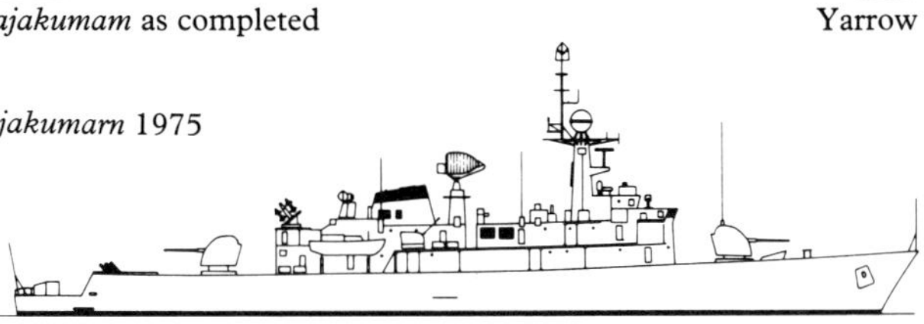

MAKUT RAJAKUMARN (Yarrow type) *frigate*

Displacement: 1650t standard; 1900t full load
Dimensions: 320ft oa, 305ft 1in pp × 36ft × 18ft 1in
97.6m, 93.0m × 10.97m × 5.5m
Machinery: 2-shaft CODOG (CP propellers): 1 Rolls-Royce Olympus TM 3B gas turbine, 23,000shp = 26kts plus 1 Crossley-Pielstick 12 PC2V diesel 6000bhp = 18kts. Range 5000nm/1200nm at 18/26kts
Armament: 2–4.5in/55 Mk 8 (2×1), 2–40mm/60 Bofors (2×1), 4 Seacat SAM (1×4), 1 Limbo, 1 DC rack, 2 DCT
Sensors: Radar HSA LW-04, HSA WM-22, Seacat control M 44, Decca 626; sonar type 170B, 162 and Plessey MS 27
Complement: 140

No	Name	Builder	Launched	Fate
7	MAKUT RAJAKUMARN	Yarrow	18.11.71	Extant 1995

A general purpose frigate was ordered from Yarrow on 21 August 1969 similar to the earlier, shorter and more lightly armed Malaysian *Rahmat*. Laid down 11 January 1971 and completed 7 May 1973. The ship is highly automated and proved successful in service. Fitted as flagship.

The ship was modernised in 1985–88; fitted with new sonar Krupp-Atlas DSQS-21B, new air search radar HSA DA-05, while Sea Cat missiles and their director were removed. Limbo ASW mortar was replaced by six 324mm Mk 32 ASW TT with British Stingray torpedoes. In 1992 it was planned to change armament further: aft 4.5in gun was to be replaced by Harpoon SSM and Sea Sparrow SAM together with Vulcan-Phalanx Mk 15 type CIWS. These plans have been scrapped, and she now serves as a training ship. A severe fire in the engine room in 1984 resulted in additional work, including replacement of the Olympus gas turbine, repair to engine room control room and central electric switchboard.

Chinese 'JIANGHU III' (053HT, 053HT(H)) type *frigates*

Displacement: 1676t standard; 1924t full load
Dimensions: 338ft 7in × 37ft 1in × 10ft 2in
103.2m × 11.3m × 3.1m
Machinery: 2 shafts, 2 CP propellers, 4 MTU 20V1163 TB83 diesels, 29,440bhp = 30kts. Range 3500nm at 18kts
Armament: 8–Ying Ji SSM (4×2), 4–100mm/56 (2×2; only 1×2 in 457–458), 8–37mm (4×2), 2–RBU 1200 ASW mortars (2×5), 4 BMB–2 DC mortars, 2 DC racks, mines, 1 helicopter (in 457–458)
Sensors: Radar Type 354 Eye Shield, Type 352C Square Tie, Type 434 Sun Visor, Type 341 Rice Lamp, Racal Decca 1290; sonar SJD–5A
Complement: 168 (22 officers)

No	Name	Builder	Launched	Fate
455	CHAO PHRAYA	Hudong SY, Shanghai	24.6.90	Extant 1995
456	BANGPAKONG	Hudong SY, Shanghai	25.7.90	Extant 1995
457	KARABURI	Hudong SY, Shanghai	28.12.90	Extant 1995
458	SAIBURI	Hudong SY, Shanghai	27.8.91	Extant 1995

First two ordered on 18 July 1988 for delivery of first to be within 30 months. First and second ship laid down in April 1989, with *Karaburi* and *Saiburi* in 1990. Essentially identical to the original 'Jianghu III' type except for the more powerful machinery of four MTU engines. First two (Type 053HT) are to be employed as training ships. The second pair (Type 053HT H) allow much more flexible use due to their helicopter deck (hangar is not provided). These are sturdy ships but the Chinese armament is completely obsolete; the gun armament is of a technical sophistication typical of Allied ships of the late Second World War. Perhaps the SSM of the Ying Ji (Strike Eagle) type will prove useful. The chief advantage of this order for the Thais is the very low price quoted by the Chinese. Originally the Thai Navy sought the ships without armament, but the Chinese insisted on delivery of fully armed vessels. There are plans for radical post-delivery modernisation, but there is no indication of when this could be implemented.

Naresuan as completed

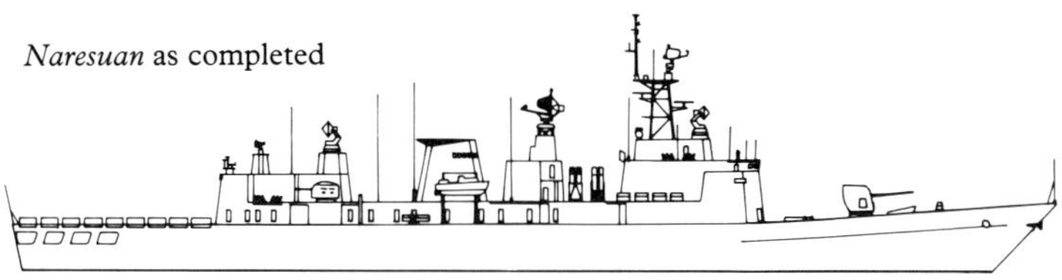

NAREUSAN (25T type) *frigates*

Displacement: 2500t standard; 2980t full load
Dimensions: 393ft 8in × 42ft 8in × 12ft 6in
120.0m × 13.0m × 3.8m
Machinery: 2 shafts, CP propellers, CODOG, 2 LM 2500 gas turbines, 2 MTU 20V1163 TB83 diesels, 44,000bhp=32kts, 14,720bhp=20kts. Range 4000nm at 18kts
Armament: 8 Harpoon SSM (2×4), 1-Mk 41 VLS launcher (1×8) Sea Sparrow SAM, 1–127mm/54 FMC Mk 45 Mod2, 4–37mm/63 (2×2), 6–TT 324mm Mk 32 ASW (2×3), 1 helicopter Kaman SH-2F Seasprite
Sensors: Radar HSA LW-08, Type 360 (Chinese), Raytheon SPS 64(V)5, 2–HSA STIR, type 374G (Chinese); sonar SJD-7 (Chinese)
Complement: 150

No	Name	Builder	Launched	Fate
621	NAREUSAN	Zhonghua SY, Shanghai	27.4.93	Extant 1995
622	TAKSIN	Zhonghua SY, Shanghai	May 1994	Completing

These ships were ordered on 21 September 1989 from China State Shipbuilding Corporation (CSSC). They were designed by the Thai Navy and CSSC. The design incorporates much Western machinery, equipment and armaments or electronics and are much more powerful than the smaller frigates of the 'Jianghu' type. Laid down in 1991, they were to be completed in 1994. These ships are fitted with fin stabilisers. The hangar and flight deck accommodates such helicopters as British Lynx or the US Kaman Sea Sprite (SH-2F). European and US weapon systems are to be fitted in Thailand after delivery of the frigates. General layout of the *Nareusan* class frigates resembles typical western European ships of this size.

Ex-US KNOX class *frigates*

No	Name	Builder	Launched	Acquired	Fate
461	PHUTTHAYOTFA CHULAOK (ex-*Truett*)	Avondale	3.2.73	30.7.94	Extant 1995
462	PHUTTHALOETLA NAPHALAI (ex-*Ouellet*)	Avondale	17.1.70	1996	–

Two surplus *Knox* class frigates for transfer on lease to the Thai Navy, first on 30 July 1994, second planned for 1996. There are plans to acquire a further pair of these ships after initial familiarisation of Thai crews with their high-pressure steam powerplants.

AMPHIBIOUS WARFARE VESSELS

Ex-US LST type *tank landing ships*

Class (former name, transferred):
Angthong (ex-*LST 294,* 1950s), *Chang* (ex-*Lincoln County,* LST 898, Aug 1962), *Pangan* (ex-*Stark County,* LST 1134, May 1966), *Lanta* (ex-*Stone County,* LST 1141, 12.3.70), *Prathong* (ex-*Dodge County,* LST 722, 17.12.75)

Five USN units built in 1943–45. Numbered LST 1–5 in Thai service. *Angthong* employed as TS since 1978. *Chang* has reinforced bow and hull for previous Arctic service. *Angthong* stricken 1980s, others extant 1995.

Ex-US LSM 1 type *medium landing ships*

Class (former name, transferred):
Kut (ex-*LSM 338,* Oct 1946), *Phai* (ex-*LSM 333,* Oct 1946), *Kram* (ex-*LSM 469,* 25.5.62).

Built in 1944–45. *Kram* transferred under MAP. Have SO-8 or SPS-5 radar. *Phai* stricken 1990, remainder extant 1995.

Ex-US LCI *infantry landing craft*

Class:
Prab (ex-*LCI 670*), Satakut (ex-*LCI 739*)

Built in 1944 and transferred in 1950. SO-8 radar. Extant 1995.

Ex-US LCT(6) type *tank landing craft*

Class:
Ardang (ex-*LCU 10*), *Kolum* (ex-*LCU 12*), *Mataphon* (ex-*LCU 8*), *Phetra* (ex-*LCU II*), Rawi (ex-*LCU 9*), Talibong (ex-*LCU 13*)

All built 1944. Transferred October 1946–November 1947. Employed as transport ferries on the Chao Phraya river, *Kolum* stricken 1984; others extant 1995.

Ex-US LSSL type *support landing craft*

Nakka (LSSL 3, ex-*LSSL 102*) acquired in October 1966 when Japan returned her as *Himawari* to USA and classed as support gunboat. Modified LCI with following armament: 1–76mm, 4–40mm, 2–20mm, 4–81mm mortars (4×1). Raytheon 1500B Pathfinder radar. Used mainly as small craft tender. Extant 1995.

SICHANG class *tank landing ships*

Displacement:	3540t normal; 4235t full load
Dimensions:	337ft 11in oa, 300ft 9in pp × 51ft 4in × 11ft 6in *103.0m, 91.65m × 15.65m × 3.52m*
Machinery:	2 shafts, 2 MTU 20V1163 TB 82 diesels, 11,000bhp = 16kts. Range 4000nm/14kts, 7000nm/12kts.
Armament:	1–40mm, 2–20 (2×1), 2–12.7mm, 1–81mm mortar
Sensors:	Radar Decca 1226
Complement:	52 plus 339 troops

No	Name	Builder	Launched	Fate
6	SICHANG	Italthai SY	14.4.87	Extant 1995
7	SURIN	Bangkok Dock	1988	Extant 1995

These ships are built under license to the modified French PS 700 design. Similar landing ships are in the Libyan Navy. First ordered in 1984 and the second in 1987. A total of six is planned but the programme is well behind schedule. Payload is 850t (fourteen tanks). Fitted with helicopter platform aft, and a 17m long bow ramp. Italthai SY in Sampatprakarn is assisted by Tacoma Boatbuilding. *Surin* reportedly is longer (possible 31ft–9.5m) and fitted with MWM engines. Biggest naval ships so far built in Thailand.

THONG KAEO class *utility landing craft*

Displacement:	193t normal; 396t full load
Dimensions:	134ft 6in oa × 29ft 6in × 6ft 11in *41.0m × 9.0m × 2.1m*
Machinery:	2 shafts, 2 GM 16V71M diesels, 1400bhp = 10kts Range 1200nm/10kts
Armament:	2–20mm (2×1), 2–7.62mm (2×1)
Complement:	31–32

Class (no, in service):
Thong Kaeo (7, 23.12.82), *Thong Lang* (8, 19.4.83), *Wang Nok* (9, 16.9.83), *Wang Nai* (10, 11.11.83)

A class of Thai-designed utility landing craft modelled on the US 'LCU 1626' type. Four ordered in 1980. Fitted with navigation radar. All extant 1995.

MINOR LANDING CRAFT

There are also twenty-six ex-US LCM 6 type landing craft (*14–16, 61–68, 71–78, 81–82, 85–87)* of which twenty-one were delivered in February-April 1969 and eight LCVP transferred in March 1963 and carried by the four LSTs. Also one Thai-built LCA with glass fibre hull and a bow ramp (10t, 39ft 5in × 9ft 9in (*12.0m × 3.0m*), 2 Chrysler diesels, 25kts, 35 troops).

SMALL SURFACE COMBATANTS

RATANAKOSIN class *corvettes*

Displacement:	840t normal; 960t full load
Dimensions:	252ft oa × 31ft 4in × 8ft *76.82m × 9.55m × 2.44m*
Machinery:	2 shafts, 2 MTU 20V1163 TB93 diesels, 16,000bhp = 26kts. Range 3000nm at 16kts
Armament:	8 Harpoon SSN (2×4), 1 Albatros SAM system (1×8) with Aspide missiles, 1–76mm/62 OTO Melara, 2-40mm Breda (1×2), 2–20mm (2×1), 6–324mm Mk 32 TT ASW
Sensors:	Radar Decca 1126, HSA WM–25, HSA ZW–06, HSA DA-05; sonar Krupp-Atlas DSQS-21B
Complement:	87

No	Name	Builder	Launched	Fate
1	RATANAKOSIN	Tacoma BB, USA	11.3.86	Extant 1995
2	SUKHOTHAI	Tacoma BB, USA	20.7.86	Extant 1995

Ordered on 9 May 1983 from Tacoma Boatbuilding. PFMM Mk 16 design. Plans to build a third ship in Thailand cancelled. *Ratanakosin* laid down on 6 February 1984 and completed on 26 September 1986, and *Sukhothai* laid down on 26 March 1984 and completed on 19 February 1987. These heavily armed ships can be considered as an enlarged version of the Saudi 'Badr' design. Fitted with HSA Mini-SADOC weapons control system and LIROD optronic director. May be used as flagships for squadrons of fast attack craft.

Sukhothai Mar 1995 — L & L van Ginderen

KAMRONSIN class *small corvettes*

Displacement:	337t normal; 475t full load
Dimensions:	186ft pp, 203ft 5in oa × 27ft 1in × 7ft 4in *56.7m, 62.0m × 8.26m × 2.25m*
Machinery:	2 shafts (CP propellers), 2 MTU 12V1163 TB92 diesels, 7500bhp = 25kts. Range 2500nm at 15kts
Armament:	1–76mm/62 OTO Melara, 2–30mm Breda (1×2), 2–12.7mm (2×1), 6 324 mm ASW TT Plessey PMW-49A (Stingray torpedoes), 1 DC rack, mines
Sensors:	Radar Decca 1226, Plessey AWS-4; sonar Krupp-Atlas DSQS-21C
Complement:	57

No	Name	Builder	Launched	Fate
1	KAMRONSIN	Italthai SY	15.8.89	Extant 1995
2	THAYANCHON	Italthai SY	7.12.89	Extant 1995
3	LONGLOM	Royal Thai NDY, Bangkok	8.8.89	Extant 1995

These corvettes were ordered on 29 September 1987. Design produced by Vosper Thornycroft on the basis of experience with Omani *Dhofar* class fast attack craft. To be fitted with Sea Archer 1A Mod 2 optronic fire control system and Plessey NAUTIS–P combat data system. Completed for trials in 1991. Plans for further units have been dropped.

Lürssen 'TNC 45' type *fast attack craft (missile)*

Particulars: As Singapore

Class (no, launched, in service):
Prabparapak (1, 29.7.75, 28.7.76), *Hanhak Sattru* (2, 28.10.75, 6.11.76), *Suphairin* (3, 20.2.76, 1.2.77)

Ordered June 1973. Of basic Lürssen TNC 45 type but very similar to Singapore's *Sea Wolf* FAC (also built by Lürssen's Hong Leong, Singapore subsidiary). Fitted with HSA WM 28 fire control and one navigation radar. All three extant 1995.

Ratcharit as completed

NB 1/750 scale

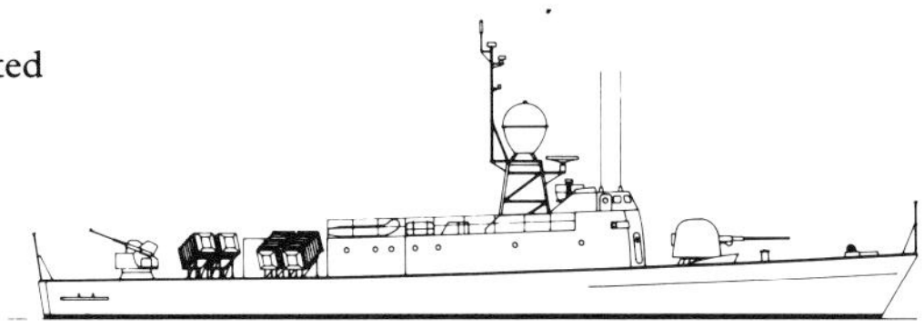

RATCHARIT class *fast attack craft (missile)*

Displacement:	235t standard; 270t full load
Dimensions:	163ft 5in × 24ft 6in × 7ft 6in *49.8m × 7.5m × 2.3m*
Machinery:	3 shafts (CP propellers), 3 MTU MD 20V538 TB91 diesels, 13,500bhp = 37kts. Range 2000nm at 15kts, 650nm at 36kts
Armament:	4 MM38 Exocet SSM (4×1), 1–76mm/62 OTO Melara, 1–40mm/70 Breda-Bofors
Sensors:	Radar WM-20, navigation
Complement:	45

Class (no, launched, in service):
Ratcharit (4, 30.7.78, 10.8.79), *Witthayakhom* (5, 2.9.78, 12.11.79), *Udomet* (6, 28.9.78, 21.2.80)

Ordered 23 July 1976. Built by CN Breda of Venice. Belong to Breda's MV 250 type. Can make 20kts on one engine. All three extant 1995.

Chonburi as completed

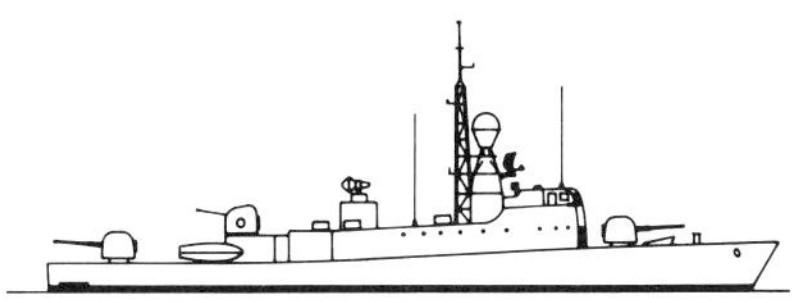

CHONBURI class *fast attack craft (gun)*

Displacement:	400t standard; 450t full load
Dimensions:	188ft 7in pp, 198ft oa × 29ft × 15ft *57.5m, 60.4m × 8.8m × 4.5m*
Machinery:	3 shafts (CP propellers), 3 MTU 20V 538 TB91 diesels, 12,600bhp = 29kts. Range 2500nm at 18kts, 900nm at 29kts
Armament:	2–76mm/62 OTO Melara (2×1), 2–40mm/70 Breda-Bofors (1×2), 4 Hycor Mk 135 chaff RL
Sensors:	Radar 3RM-series WM-22/61, Elsag-Selenia Dando; ECM passive intercept
Complement:	45

Class (no, launched):
Chonburi (1, 7.6.82), *Songkhla* (2, 6.9.82), *Phuket* (2, 3.2.83).

First pair ordered from CN Breda of Venice in November 1979, third in 1981. Italian MV 400 TH type FAC (gun) design. Able to have SSM. Steel hull and aluminium superstructure. Extant 1995.

Ex-US PC 461 (173ft) type *submarine chasers (large patrol craft)*

Class:
Khamronsin (ex-*PC 609*), *Liulom* (7, ex-*PC 1253*), *Longlom* (8, ex-*PC 570*), *Phali* (4, ex-*PC 1185*), *Sarasin* (1, ex-*PC 495*), *Sukrip* (5, ex-*PC 1218*), *Thayanchon* (2, ex-*PC 575*), *Tongpliu* (6, ex-*PC 616*)

Eight transferred to Thai Navy March 1947 to December 1952. SPS-25 radar in some. All except *Sarasin* have 2–324mm Mk 32 ASW TT (2×1). Larger crews (62–71) than original USN complement. *Khamronsin* stricken 1956, *Thayanchon* 1982, *Longlom* 1984, *Sarasin* 1992, *Pheli* 1992 and *Tongpilu* 1994. Others extant 1995.

Ex-US SC (110ft) type *submarine chasers (large patrol craft)*

Class:
SC 31 (ex-*SC 1632*), *SC 32* (ex-*SC 1633*), *SC 33* (ex-*SC 1634*)

Three built in 1945 and transferred in 1950s. Renamed *SC 7* and *SC 8* (ex-*SC 31, 32*). *SC 33* BU 1962 and *SC 7* (ex-*SC 31*) in 1973.

US 'CAPE' class *coast guard cutters (large patrol craft)*

Class:
T 81–84 (ex-*CG 13–16*)

New Coast Guard cutters transferred from USA in 1954. Built in 1953. Cost $475,000 each. Similar to USCG 95ft type. Built by US Coast Guard Yard, Curtis Bay. Renamed *CGC 3–6* and later *T 81–84. T 83* stricken 1982 and all others since.

US PGM 71 type *large patrol craft*

Class (former no, launched, in service):
T 11 (ex-*PGM 71,* 22.5.65–1.2.66), *T 12* (ex-*PGM 79,* 18.12.65–1967), *T 13* (ex-*PGM 107,* 13.4.–28.8.67), *T 14* (ex-*PGM 116,* 13.6–18.8.69), *T 15* (ex-*PGM 117,* 24.6–18.8.69), *T 16* (ex-*PGM 115,* 24.4.69–12.2.70), *T 17* (ex-*PGM 113,* 3.6.69–12.2.70), *T 18* (ex-*PGM 114,* 24.6.69–12.2.70), *T 19* (ex-*PGM 123,* 14.5–25.12.70), *T 20* (ex-*PGM 124,* 22.6–Oct 1970)

Ten US PGM type launched May 1965–June 1970 by Peterson, Sturgeon Bay, USA and on completion transferred to the Thai Navy. *T 11, 12* and *13* since stricken. Seven extant 1995.

T 93 class *large patrol boats*

Displacement:	117t standard; 125t full load
Dimensions:	141ft 1in oa, 105ft wl × 18ft 8in × 4ft 7in *43.0m, 32.0m × 5.7m × 1.4m*
Machinery:	2 shafts, 2 MTU 12V538 TB91 diesels, 3300bhp = 25kts
Armament:	2–40mm/60 (2×1), 2–12.7mm (2×1)
Complement:	16

Class (in service):
T 93 (launched 1973), *T 94* (16.9.81), *T 95* (1981), *T 96* (1982), *T 97* (16.9.83), *T 98* (1984), *T 99* (May 1987)

Enlarged version of the earlier *T 91* design. Built by Royal Thai NDY at Bangkok. *T 99* has 20mm gun aft instead of 40mm. All extant 1995.

SATTAHIP class *large patrol craft*

Displacement:	270t standard; 300t full load
Dimensions:	164ft 4in oa, 154ft 10in × 24ft × 5ft 2in (5ft 11in props) *50.1m, 47.2m × 7.3m × 1.58m, 1.8m*
Machinery:	2 shafts, 2 MTU 16V538 TB91 diesels, 6800bhp = 22kts. Range 2500nm/15kts
Armament:	1–76mm US Mk 26 (three received Italian 1–76mm/62 OTO Melara), 1–40mm, 2–20mm (2×1), 2–12.7mm (2×1)
Complement:	56

Class:(no, in service):
Sattahip (4, 16.9.83), *Klongyai* (5, May 1984), *Takbai* (6, Jul 1984), *Katang* (7, 14.10.85), *Thepa* (8, 17.4.86), *Thai Muang* (9, 17.4.86)

First four units ordered on 9 September 1981 and others on 27 December 1983 and 31 August 1984 at Italthai SY at Samutparakarn, Bangkok, according to the radically modified US 'PSMM Mk 5' design. Fitted with Decca 1226 radar and optronic NA 18 gun fire control system. All launched in 1983–85. Extant 1995.

SPINAKARIN class *large patrol craft*

Displacement:	630t full load
Dimensions:	203ft 5in × 26ft 11in × 8ft 2in *62.0 × 8.2m × 2.5m*
Machinery:	2 shafts (KaMeWa CP props), 2 Deutz MWM BV 16M628 diesels, 9524bhp = 25kts. Range 2500nm at 15kts
Armament:	1–30mm/82 Breda, 2–20mm Oerlikon GAM BO1 (2×1)
Complement:	45

Class (no):
Srinakarin (1804)

Ordered September 1989 from Italthai Marine, Bangkok and delivered April 1992 for use by the marine police force. Design derived from *Khamronsin* class corvettes built for Thai Navy, with different machinery, armament and fittings. Planned second ship of class not built because of shortage of funds.

COASTAL PATROL CRAFT

Two ex-US YP type: *CGC 11, CGC 12.* Former US Coast Guard cutters of YP type (44.5t, 20.5kts, 1–20mm). Renamed *CGC 1* and *CGC 2. CGC 12* stricken in 1978 and *CGC 11* in 1973.

Two Thai-built T 91 type built by Naval Dockyard, Bangkok in 1965–73 in two batches. Characteristics: 88t, 104ft 4in × 17ft 6in × 5ft 6in (*31.8m × 6.4m × 1.9m*), MTU diesels, 25kts, 1–40mm, 4–20mm (4×1), 2 MG (21 men). *T 91* and *T 92* extant 1995.

Six MPB built in Bangkok in 1940s (16t, 12.5kts, 1–20mm): *MPB 1, 3* stricken in 1956 and the rest in 1961.

Five ex-US 36ft RPC type transferred in March 1967; (10.4t, 14kts, 2 twin MG). Used on Mekong river. Three extant 1995.

Thirty-seven ex-US PBR Mk 2 type, twenty transferred 1966–67, ten in 1972. All extant 1995 and used on Mekong.

Twelve ex-US 'Swift' class: *T 21–29, T 210–212*, transferred 1967–75.

Sixteen Thai-built 64ft units (34t full load, 1–20mm, 1–81mm mortar, 2

MTU diesels, 1300bhp = 22kts, 8 men); built by Ital-Thai Development Co, Bangkok 1980–81. *T 213–228*, first three in service 29 August 1980, last four in 1984. Fishery protection role. Fourteen extant 1995.

One Hysucat 18 experimental hydrofoil catamaran built of GRP in Thailand in 1986, designed with help from Fr. Lürssen, unsuccessful, planned series construction cancelled; 39t (*18.3m*), 2 MWM diesels, 1640bhp = 36kts, 1–20mm Oerlikon (originally 1–20mm Sea Vulcan). Extant 1995.

Two Hameln type large patrol craft *Damrong Rachanuphat* (1802) and *Lopburi Ramas* (1803) were delivered in 1969 and 1972 by German yard Schiffswerf Hameln and are used by maritime police; 430t, 56.7m, 2 MTU diesels, 4400bhp = 23kts, 1–76mm/50, 2–20mm Oerlikon. Extant 1995.

Two Sumidagawa type coastal patrol boats Chasanyabadee (1101) and Promothee (1103) were delivered in 1972–73 by Sumidagawa, Japan, used by maritime police; 130t full load, 34.0m, 3 diesels, 4050bhp = 32kts, 2–12.7mm. Extant 1995.

Other patrol boats operated by maritime police include *Chawengsak Songkram* (1102, 1973, 190t, 32kts, 2–20mm Oerlikon), *Sriyanont* (901, 1986, 52t, 23kts, 1–20mm Oerlikon, 2–7.62mm MG); three Halter type *Phrongkamrop* (807), *Picharnpholakit* (808) and *Raminthra* (809, 34t, 25kts, 1–20mm Oerlikon, 2–7.62mm MG); three Technautic type *No 810–812* (50t, 27kts, 1–20mm Oerlikon, 2–7.62mm MG) and over fifty smaller patrol craft and boats which can easily be used by the navy in case of hostilities.

MINE WARFARE VESSELS

Ex-British ALGERINE class *ocean minesweeper*

Phosamton (ex-*Minstrel*) transferred in April 1947. The 20mm guns were increased from three to six and DCTs from two to four in 1966. Extant 1995 as TS.

Ex-US YMS type *motor coastal minesweepers*

Class:
Bangkeo (ex-*YMS 334*), *Ladya* (ex-*YMS 138*), *Tadindeng* (ex-*YMS 21*)

Transferred to Thai Navy in 1947–51. Wooden construction. All three stricken 1964–65.

US BLUEBIRD class *coastal minesweepers*

Class (no, former no, in service):
Bangkeo (6, ex-*MSC 303*, 9.7.65), *Donchedi* (8, ex-*MSC 313*, 17.9.65), *Ladya* (5, ex-*MSC 297*, 14.12.63), *Tradindeng* (7, ex-*MSC 301*, 26.8.65)

Built by Peterson Builders Inc and Tacoma Boatbuilding Co in 1962–65. Similar to *Bluebird* class.

THALANG *mine countermeasures support ship*

Displacement: 1000t full load
Dimensions: 182ft 8in × 32ft 10in × 10ft
55.7m × 10m × 3.1m
Machinery: 2 shafts, 2 MTU diesels, 1310bhp = 12kts
Armament: 1–40mm, 2–20mm (2×1), 2–12.7mm (2×1), mines
Sensors: Radar Decca TM 1226
Complement: 77

No	Name	Builder	Launched	Fate
1	THALANG	Bangkok Naval DY	1979	Extant 1995

Designed by Ferrostaal of Essen, W Germany. In service 25 June 1980. She carries two 3t cranes and four spare MCM kits for minesweepers.

BANGRACHAN class *mine hunter/sweepers*

Displacement: 414t light; 443t full load
Dimensions: 157ft 6in oa, 149ft 11in pp × 30ft 6in × 9ft
48.0m, 45.7m × 9.3m × 2.75m
Machinery: 2 shafts (CP propellers), 2 MTU 16V396 TB83 diesels, 3223bhp = 18kts, auxiliary electric propulsion = 7kts
Armament: 3–20mm Oerlikon GAM-BO1 (3×1), mines
Sensors: Radar Decca 1229; sonar Krupp-Atlas DSQS-11H
Complement: 40

Class (no, in service):
Bangrachan (2, Oct 1987), *Nongsarai* (3, Jan 1988)

These ships of the Lürssen Vegesack's 'M 48' design were ordered 31 August 1984 and 5 August 1985. There was an option for two more. They have composite hull construction with non-magnetic metal framing and wooden skin. Fitted with Krupp-Atlas MWS-80R mine countermeasure system. Carry two Gaymarine Pluto remote controlled submersibles. Also equipped with conventional mechanical, magnetic, acoustic sweep gear. Carry thirty disposal charges for the Pluto submersible. For better control fitted with Becker flap rudders and Motorola Mini Ranger navigational positioning system. There are persistent rumours that the minehunting system in general and particularly the DSQS-11H sonar did not prove effective in practice. Extant 1995.

MINESWEEPING BOATS
Five *MSML 1–5* (25t, 2–20mm) built in Thailand and stricken 1980. Five ex-US 50ft MSB (21t full load, 8kts, MGs, 6 crew) transferred 1963–66 and extant 1995.

Togo

The West African republic of Togo, independent from France in 1960 was under a moderate military regime from 1967. In recent years, particularly since 1990, there has been much civil unrest amid popular support for a multiparty political system which eventually came into being. The country has a short coastline of only 35 miles. The 200 volunteer naval personnel under a French Navy lieutenant operate two wooden coastal patrol craft, *Mono* and *Kara* (105ft, 30kts, 1–40mm, 1–20mm), built in France and delivered to Lomé on 5 October 1976 being named after the country's rivers.

Tonga

The Kingdom of Tonga or Friendly Islands, a collection of seven main groups of islands spread over nearly 300 square miles of the South Pacific, 450 miles ESE of Fiji, was annexed by Britain in 1899. It became independent in June 1970. King Taufa'ahau Tupou IV commissioned the 15t coastal patrol craft *Ngahau Koula* (P 101) into the Maritime Division of the Royal Tongan Defence Force on 10 March 1973. She was joined by her sister *Ngahau Siliva* (P 102) on 10 May 1974. They are both Brooke Marine-built 45ft vessels capable of 21kts (800–1000nm range) with 2 12.7mm Browning MGs, Decca 101 radar and seven volunteer crew. They have been joined by three ASI 315 'Pacific Forum' patrol boats – *Voea Neiafu* (P201), *Voea Pangai* (P202) and *Voea Savea* (P203) – originally ordered for Kiribati but completed in Australia for Tonga. The LCM 8 class *Late* was transferred from Australia in 1992, and the *Mannheim* class LCU was transferred from Germany in 1991. All are extant.

Voea Savea Jul 1994 L & L van Ginderen

Trinity as completed — Vosper Thornycroft

Trinidad and Tobago

The most southerly of the West Indies, Trinidad and Tobago became independent on 31 August 1962 and a republic in 1976. The regime in the islands is a stable one although internal tensions brought about by economic problems caused rioting to take place in the early 1970s. Although independent, Trinidad and Tobago still relies on Britain in matters of defence. The army has about 800 personnel and the Coast Guard about 650; the main base is at Staubles Bay.

Swedish CG 40 type *large patrol craft*

Displacement: 200t
Dimensions: 133ft 3in × 21ft 11in × 5ft 3in
40.6m × 6.7m × 1.6m
Machinery: 2 shafts, 2 Paxman-Valenta 16RP200 diesels, 8000hp = 27kts. Range 3000nm at 15kts
Armament: 1–40mm Bofors, 1–20mm
Sensors: Radar Decca TM 1226
Complement: 22

No	Name	Builder	Comp	Fate
CG 5	BARRACUDA	Karlskronavarvet	15.6.80	Extant 1995
CG 6	CASCADURA	Karlskronavarvet	15.6.80	Extant 1995

Ordered August 1978. Fitted with optronic fire control for the 40mm. The first export success for this Swedish design.

Vosper 103ft type *large patrol craft*

Displacement: 96–100t
Dimensions: (CG 1 and CG 2) 102ft 7in × 19ft 8in × 5ft 6in
31.4m × 5.9m × 1.7m
(CG 3 and CG 4) 103ft × 19ft 10in × 5ft 10in
31.5m × 5.9m × 1.8m
Machinery: 2 shafts, 2 Paxman 12 YJCM Ventura diesels, 2900hp = 24kts. Range 1800nm at 13.5kts. Oil 18t
Armament: 1–40mm (CG 1 and CG 2), 1–20mm (CG 3 and CG 4)
Complement: 17 (19 in CG 3 and CG 4)

No	Name	Builder	Comp	Fate
CG 1	TRINITY	Vosper	20.2.65	Deleted 1986
CG 2	COURLAND BAY	Vosper	20.2.65	Deleted 1986
CG 3	CHAGUARAMAS	Vosper	18.3.72	Deleted 1988
CG 4	BUCCO REEF	Vosper	18.3.72	Deleted 1992

All units fitted with roll damping fins, air conditioned.

Trinidad and Tobago also operates: two coastal patrol craft, *Naparima* (CG 01) and *El Tucuche* (CG 25) commissioned 13 August 1976 and 1976 respectively; CG 25 slightly larger at 55ft compared to CG 01, 50ft; 2 GM 8V71 diesels (TL 71, CG 25). Built by Tugs and Lighters Ltd, Port of Spain.

Five Souter Wasp 17m coastal patrol craft *Plymouth* (CG 27), *Caroni* (CG 28), *Galeota* (CG 29), *Moruga* (CG 30), *Cedros* (CG 35), four delivered by Souter in 1982 and CG 35 in 1984 (19.3t, 1300bhp = 32kts, 1 MG). Extant 1995.

Two Souter Wasp 20m coastal patrol craft, *Kairi* (CG 31) and *Moriah* (CG 32) delivered by Souter in 1982 (32t, 2400bhp = 30kts, 2 MG). Extant 1995.

One coastal patrol craft *Carenage* (CG 37) delivered in 1980 (14.9t, 23.5kts, 2 MG). Extant 1995.

There is also a 23ft fibreglass coastal patrol craft *Fort Chacon* (CG 9) deleted 1991, built in Trinidad, 1 Caterpillar diesel, 27kts; two coastal tugs, *Snapper* and *Bonito*, launched 16 January 1980 and 1981 respectively, built by Bodewes, Millengen and a sail training vessel, *Hummingbird II*, a 40ft ketch built in Trinidad 1966.

The first craft operated by the service, the 60ft *Sea Hawk* and the 45ft *Sea Scout* (both built by J Taylor, Shoreham, UK) were stricken in 1975–76.

Tunisia

Tunisia became independent from French protection in 1956, and has managed to keep her reputation of being the most peaceful and civilised of North African countries. Economic development has been generally successful, but tension caused by the widening disparity between rich and poor led to serious riots in 1979, and further strains have resulted from the growth in Islamic fundamentalism.

External policies have been generally neutralist but pro-Western, although relations with France in the early years of independence were sometimes very bad. In 1961 French paratroops caused 1000 casualties among Tunisian demonstrators protesting at France's retention of the Bizerta naval base, and matters again came to a head in 1964 on Tunisian nationalisation of foreign landholdings. More recently, it is relations with Libya that have caused concern, particularly after the 1980 *Gafsa* incident, and Tunisia now relies on France for military protection against her ambitious neighbour.

The military budget, expressed both absolutely and as a percentage of GNP, is much smaller than those of Tunisia's neighbours, so her armed forces are correspondingly weaker and more poorly equipped. The navy, subordinate to the Tunisian National Army, comprises a number of patrol craft headed by an obsolete ex-US destroyer escort, and is no more than a coast defence force. This does not matter while France is prepared to send warships into Tunisian territorial waters whenever needed.

Tunisia has nine naval bases and twenty-two secondary ports along her 700-mile shore. The Soviet Mediterranean Fleet made heavy use of Menzel-Bourguiba for repair work from 1977 and the expulsion from Egypt but gained no permanent facilities. There seems little chance of significant Tunisian naval expansion in the near future, although some modernisation may be possible, perhaps with US assistance. The personnel, totalling 4600 one-year conscripts in 1995 (2000 in 1974) with 3000 reservists, is considered fairly well trained and educated, and loyal to the regime, but likely to be unaggressive in action.

Ex-US SAVAGE class *radar-picket (frigate)*

No	Name	Builder	Acquired	Fate
E 7	INDAKH (ex-*President Bourguiba*, ex-*Thomas J Gary*)	Consolidated, Orange	27.10.73	See notes

Ex-US *Edsall* class DE (launched 21.8.43) converted to a *Savage* class radar-picket (SPS-10 and SPS-29 radars) in 1958 (DE 326, becoming DER 326). Transferred to Tunisia as the navy's flagship in 1973. Hedgehog was replaced by 6–12.75in Mk 32 ASW TT (2 × 3). Suffered a major fire on 16 April 1992 and is not operational as of 1995.

President Bourguiba 1983 — L & L van Ginderen

Ex-French CHAMOIS class *corvette*

No	Name	Builder	Acquired	Fate
E 71	DESTOUR (ex-*Chevreuil*)	Arsenal de Lorient	13.10.59	Hulked 1972

Ex-Chinese 'SHANGHAI II' class *fast attack craft (gun)*

Class (no):
Gafsa (P 305), *Amilcar* (P 306)

Transferred late 1977. Re-engined in Bizerta 1984. In service 1995.

French P 48 class *large patrol craft (missile)*

No	Name	Builder	Launched	Fate
P 301	BIZERTA	Ch Franco-Belges	20.11.69	Extant 1995
P 302	EL HORRIA (ex-*Liberté*)	SFCN	12.2.70	Extant 1995
P 304	MONASTIR	SFCN	25.6.74	Extant 1995

Laid down 5 June 1969, 20 November 1969 and 1 February 1974, completed 10 July 1970, October 1970 and 25 March 1975.

French LA COMBATTANTE III type *fast attack craft (missile)*

Displacement:	395t standard; 425t full load
Dimensions:	186ft 4in × 26ft 11in × 8ft 2in *56.8m × 8.2m × 2.5m*
Machinery:	4 shafts, 4 MTU diesels, 19,300bhp = 38.5kts
Armament:	8 MM40 Exocet SSM (2 × 4), 1–76mm OTO Melara, 2–40mm Breda (1 × 2), 4–30mm Oerlikon (2 × 2)
Sensors:	Radar Thomson Triton-S, Laster
Complement:	35

Class (no, in service):
La Galite (P 501, 27.2.85), *Tunis* (P 502, 28.3.85), *Carthage* (P 503, 29.4.85)
Ordered 1981 from CMN, Cherbourg. Similar craft in service with Israel, Greece and South Africa. All extant 1995.

Tunis 1992 L & L van Ginderen

French LE FOUGEUX *submarine chaser*

Built as *P 9* by A C Dubigeon, Nantes, under US offshore order. Served as the West German ASW trials vessel *UW 12* 12 March 1967 to 15 September 1967. Bought by Tunisia 16 June 1970 after transfer in December 1969. Hedgehog and all four DCT removed mid-1970s. Renamed *Sakiet Sidi Youssef*. Discarded 1988.

TAZARKA class *large patrol craft*

Displacement:	100t standard; 120t full load
Dimensions:	103ft oa × 19ft 6in × 5ft 6in *31.4m × 5.9m × 1.7m*
Machinery:	2 shafts, 2 MTU diesels, 4000bhp =kts. Range 1500nm at 14kts
Armament:	2–20mm
Sensor:	Radar Decca 916
Complement:	24

Class (no):
Tazarka (P 205), *Menzel Bourguiba* (P 206)
Ordered from Vosper Thornycroft 9 September 1975 and launched 19 July 1976 and in service since 27 October 1977. Original armament planned was 1–40mm. Extant 1995.

Tazarka as completed Vosper Thornycroft

US ADJUTANT class *coastal minesweepers*

Class (former name and no):
Hannibal (ex-*Coquelicot*, M 673), *Sousse* (ex-*Marjolaine*, M 640, ex-*Aconit*)
Built for France in US shipyards under MDAP; *Hannibal* transferred in 1973, *Sousse* in July 1977. Both used as patrol craft and for fishery protection, minesweeping gear having been removed. Both discarded 1988.

MISCELLANEOUS
Four French 32-metre coastal patrol craft (60t, 28kts, 2–20mm, 17 men): *Istikal* (P 201, ex-French *VC 11* transferred Mar 1959); *Joumbouria* (P 202, ex-French *P 761* transferred Jan 1961), *Al Jala* (P 203, in service Nov 1963) and *Remada* (P 204, in service Jul 1967).
Eight French 25-metre coastal patrol craft (38t, 23kts, 1–20mm, 10 men): *V 101* to *V 106* in service 1961–63, *V 107* and *V 108* disarmed and transferred to Fisheries Administration in 1971. *V 108* lost 7 July 1989.

Turkey

In 1946 the USSR tried to make Turkey revise their mutual border, and also attempted to change the Straits agreements to enable Soviet ships to leave the Black Sea more easily. It was not surprising, therefore, that Turkey signed an aid treaty with the United States in 1947, and that she later joined NATO (with Greece) in 1952. Turkey's main problems during the last three decades, however, have not been with the USSR but with Greece. She has also suffered considerable internal unrest including a number of military *coups*, some of which have been successful.

PROBLEMS WITH GREECE
There have been problems with Greece both over the Greek islands in the Aegean and in Cyprus, where the Turks are in a minority of one third. Over Cyprus there has been continuous Greek pressure for *Enosis* (union with Greece), and after Cyprus became independent in 1960 the troubles escalated. In 1964 Turkey and Greece intervened militarily in Cyprus and had to be separated by the United Nations' peace-keeping force, and in 1975 Turkey mounted an amphibious invasion of Cyprus. The only casualty the Turkish Navy suffered was the destroyer *Kocatepe*, sunk in mistake for a Greek ship by the Turkish Air Force. An unlooked for side-effect of this invasion for the Turkish armed forces was the imposition of an armed embargo by the US Congress in 1975, which was lifted in 1980.

The development of the Turkish Navy in this period has largely mirrored that of its main potential adversary, Greece. As with the Greek Navy, considerable numbers of ex-RN and USN vessels were transferred in the post war period, forming the basis of the surface fleet for the next forty years. In the 1970s Turkey acquired fast attack craft and submarines, suited to the restricted waters of the Aegean, and maintains a sizeable amphibious force, put to use in 1975 in Cyprus, and in the later 1980s sought to acquire more modern surface vessels and the capacity to build them in local shipyards, just as Greece did. Turkey appears to have been rather more successful in establishing a local warship building industry, albeit with considerable foreign financial and technical support. Modern submarines, MEKO 200 type frigates, three classes of LST and numerous LCUs and other landing craft have all been built in Turkish naval yards. Although the navy has benefitted from the post-cold war shedding of vessels by the major NATO navies, acquiring several German ships plus ex-US *Knox* class frigates, the increasing sophistication of Turkey's shipbuilding industry has meant that the navy is not wholly dependent on this source for more modern ships. In 1991, therefore, Turkey refused the offer of four ex-USN *Charles F Adams* class destroyers of the type transferred to Greece, citing their poor condition and general age. There are reports that the acquisition of the *Knox* class frigates was also opposed within the navy for similar reasons. In 1995 the Turkish Navy had 59,800 active personnel (plus 70,000 reservists) and a 4000-man marine corps.

FLEET STRENGTH 1947

BATTLECRUISER

Name	Launched	Disp	Fate
YAVUZ SULTAN SELIM (ex-*Goeben*)	23.3.11	22,980t	BU 1971

Yavuz was immobile at Izmir from 1948. She was decommissioned in 1960 and several attempts were made to preserve her.

DESTROYERS

Name	Launched	Disp	Fate
Kocatepe class			
KOCATEPE	19.3.31	1630t	Discarded *c*1950
APATEPE	7.2.31	1650t	Discarded *c*1950
Tinaztepe class			
TINAZTEPE	27.7.31	1610t	Discarded 1957
ZAFER	27.7.31	1610t	Discarded 1957
Gayret class			
DEMIRHISAR	1941	1880t	Discarded 1960
MUAVENET	24.2.41	1880t	Discarded 1960
SULTANHISAR	1941	1880t	Discarded 1960
Ex-British 'O' class			
GAYRET (ex-*Oribi*)	14.1.41	2430t	Discarded 1965

The *Kocatepe* and *Tinaztepe* classes were built in Italy, based on contemporary Italian destroyer designs.

The *Gayret* class was based on British 'H' class destroyers. *Muavenet* served in the RN as *Inconstant* before being delivered to Turkey in 1945. Her sister, *Gayret*, was never delivered to Turkey, having been lost as *Ithuriel* on 28 November 1942. She was replaced by the British 'O' class destroyer *Oribi*, which was transferred to Turkey on 18 June 1946 and renamed *Gayret*.

SUBMARINES

Name	Launched	Disp	Fate
Birindci Inönü class			
BIRINDCI INONU	1.2.27	620t	Discarded *c*1950
IKINDCI INONU	12.3.27	620t	Discarded *c*1927
Dumlupinar class			
DUMLUPINAR	4.3.31	1150t	Discarded *c*1950
Sakarya class			
SAKARYA	2.2.31	940t	Discarded *c*1950
Gur class			
GUR	1932	960t	Discarded *c*1950
'Ay' class			
SALDIRAY	23.7.38	1210t	Discarded 1957
YILDIRAY	26.8.39		Discarded 1957
Oruc Reis class			
BURAC REIS	19.10.40	861t	Discarded 1957
MURAT REIS	20.7.40	861t	Discarded 1957
ORUC REIS	19.7.40	861t	Discarded 1957

The *Birindci Inönü* class were German-designed, Dutch-built coastal submarines. *Dumlupinar* and *Sakarya* were built in Italy and resembled contemporary Italian submarines. *Gur* was built to German designs in Spain. The 'Ay' class were modified versions of the German *IXA* design. Sister *Atilay* was lost by accident on 14 July 1942.

The *Oruc Reis* class were reduced versions of the British 'S' class. They were taken over by the RN and became *P 614, P 612* and *P 611* respectively. *Murat* and *Oruc* were delivered to Turkey in 1942 and *Burac* in 1945. Sister *Ulac Ali Reis* was sunk as *P 615* on 18 April 1943.

MINELAYERS

Name	Launched	Disp	Fate
ATAK	1938	500t	Discarded early 1960s
Sivrihisar class			
SIVRIHISAR	1940	350t	Discarded 1960
TORGUD REIS	1940	350t	Discarded 1960

Atak and the *Sivrihisar* class carried 40 mines.

COASTAL MINESWEEPERS

Name	Launched	Disp	Fate
Kavak class			
KAVAK	1937	32t	Discarded 1950s
CANAK	1937	32t	Discarded 1950

COASTAL FORCES

Name	Launched	Disp	Fate
MTB class (launched 1942, 70t)			
MTB 1–10			Discarded 1950s
Fairmile 'B' type (launched 1940–42, 86t)			
AB 1–8			Discarded 1970–73
HDML type (launched 1940–44, 54t)			
LSB 1–8			Discarded 1960s

MTB 1–10 were armed with 2–21in (533mm) TT (2 × 1).

Fairmile 'B' type motor launches were transferred from Britain in 1944–45. *AB 1–8* were RN *ML 386, ML 584, ML 836, ML 837, ML 838, ML 842, ML 862* and *ML 863* respectively. HDML type harbour defence motor launches were also transferred from Britain in 1944–45.

MAJOR SURFACE SHIPS

(For illustration see under countries of origin in this and 1922–46 volume)

Ex-US GLEAVES class *destroyers*

Particulars: As US *Gleaves* class

No	Name	Builder	Acquired	Fate
D 344	GAZIANTEP (ex-*Lansdowne*)	Federal, Kearney	1950	Discarded 1973
D 346	GELIBOLU (ex-*Buchanan*)	Federal, Kearney	29.4.49	Discarded 1976
D 347	GEMLIK (ex-*Lardner*)	Federal, Kearney	1950	Stricken 1981
D 345	GIRESUN (ex-*McCalla*)	Federal, Kearney	29.4.49	Discarded 1973

All were ex-USN *Gleaves* class destroyers. They were modernised in the USA 1957–58, and were fitted with a tripod foremast and an enclosed bridge. *Gelibolu* and *Giresun* had their 'X' position 5in (127mm) and their AA guns replaced by 4–3in (76mm). Although *Gemlik* was discarded in 1974, her hulk remained in use as a tender to the Turkish Navy's Engineering School until 1981.

Ex-British 'M' class *destroyers*

Particulars: As British 'M' cass

No	Name	Builder	Acquired	Fate
D 348	ALP ARSLAN (ex-*Marne*)	Scotts	29.6.59	Discarded early 1970s
D 350	KILIC ALI PASHA (ex-*Matchless*)	Stephen	29.6.59	Discarded early 1970s
D 349	MARESAL FEVZI CAKMAK (ex-*Meteor*)	Vickers-Armstrong, Tyne	29.6.59	Discarded early 1970s
D 351	PIYALE PASA (ex-*Milne*)	Stephen	29.6.59	Discarded early 1970s

Ex-British 'M' class destroyers transferred to Turkey after refitting in 1958–59. During the refit the aft 4–21in (533mm) TT (1 × 4) and the AA guns were removed, and a new deckhouse and AA and ASW armament were installed. In Turkish service they were armed with 6–4.7in (120mm) (3 × 2), 6–40mm (1 × 2, 4 × 1), 1 Squid ASW mortar (1 × 3) and 4–21in (533mm) TT (1 × 4). *Piyale* was fitted as flotilla leader.

Ex-US FLETCHER class *destroyers*

Particulars: As US *Fletcher* class

No	Name	Builder	Acquired	Fate
D 344	ICEL (ex-*Preston*)	Bethlehem, San Pedro	15.11.69	Stricken 1981
D 343	ISKENDERUN (ex-*Boyd*)	Bethlehem, San Pedro	1.10.69	Stricken 1981
D 340	ISTANBUL (ex-*Clarence K Bronson*)	Federal, Kearney	14.1.67	Stricken 1987
D 341	IZMIR (ex-*Van Valkenburgh*)	Gulf SB	28.2.67	Stricken 1987
D 342	IZMIT (ex-*Cogswell*)	Bath Iron Wks	1.10.69	Stricken 1980

All were standard ex-US *Fletcher* class destroyers except *Icel*, which had a modified bridge; she served as a harbour training ship at Gölcük.

Muavenet was completed as a *Robert H Smith* class (modified *Allen M Sumner* type) minelayer. She was modernised before her transfer to Turkey but had no ASW equipment. She received new radar and armament in 1982–83, with SPS-40 replacing SPS-6 and 2–76mm (1 × 2) replacing the after quad 40mm mount. She retained mine rails for up to 80 mines. On 2 October 1992, while on NATO exercises, she was severely damaged by a Sea Sparrow missile accidentally fired from the carrier USS *Saratoga*. Repairing her was considered too expensive and she was stricken in 1993. *Zafer* was given a FRAM II modernisation by the USN and used as a sonar trials ship prior to being transferred to Turkey on 22 October 1971. She was fitted with 4–40mm (2 × 2) between the funnels in 1977 and a twin 35mm Oerlikon mount replaced the single 40mm on the helicopter flight deck in 1979. She was stricken in 1993.

Ex-US ALLEN M SUMNER class *destroyers*

Particulars: As US *Allen M Sumner* class (FRAM II type)

No	Name	Builder	Acquired	Fate
DM 357	MUAVENET (ex-*Gwin*)	Bethlehem, San Pedro	22.10.71	Damaged 2.10.92, stricken 1993
D 356	ZAFER (ex-*Hugh Purvis*)	Federal, Kearny	15.2.72	Stricken 1993, BU 1994

These were all ex-US *Gearing* class destroyers, originally launched 1944–45. *Adatepe* was purchased on 15 February 1973, two years after first being transferred. The first *Kocatepe* was sunk in error by Turkish Air Force jets on 22 July 1974 during the landing on Cyprus, having been mistaken for a Greek ship. The ex-*Norris*, which had been bought for spares, was hurriedly refitted and renamed *Kocatepe*, commissioning on 20 July 1975. Another ship originally purchased for cannibalisation, the ex-*Meredith*, was also refitted and commissioned as *Savastepe* on 20 July 1981. The ex-*Robert H McCord* and *Fiske* were leased to Turkey in 1980 and were purchased on 6 August 1987, along with ex-*Orleck* acquired in 1982. An eleventh *Gearing*, the ex-*McKean*, previously damaged in a collision, was transferred for cannibalisation in 1982.

Kocatepe (ex-*Norris*) and *Tinaztepe* (FRAM II conversions) were fitted with 2–40mm (2 × 1) on the helicopter deck aft in 1974 (later replaced by 2–35mm Oerlikon (1 × 2) on *Kocatepe* in 1980) and 4–40mm (2 × 2) between the funnels in 1979. ASW armament was one trainable Hedgehog forward, 6–324mm ASW TT (2 × 3) plus one DC rack. *Tinaztepe* was badly damaged in a collision in 1984 and never returned to service. *Kocatepe* was stricken in 1993.

The remainder were all FRAM I conversions. *Cakmak*, *Gayret* and *Adatepe* all received 2–40mm (1 × 2) just forward of the bridge and 2–35mm Oerlikon (1 × 2) on the helicopter deck aft in the mid-1970s. The twin 40mm mount was later replaced by a second 2–35mm Oerlikon (1 × 2). *Savastepe* when transferred had both her 127mm turrets forward and had a twin 35mm mount fitted on the fantail, retaining her helicopter deck (as does *Yucetepe*). ASW armament comprised an ASROC 8-cell launcher (not on *Savastepe*), 6–324mm ASW TT (2 × 3) and a DC rack. Due to their being on lease, *Kilic Ali Pasa*, *Piyale Pasa* and *Yucetepe* were little modified on transfer, though they have received 2–35mm Oerlikon (1 × 2) aft plus a DC rack. *Cakmak* and *Gayret* were fitted with eight Harpoon SSM (2 × 4) on the helicopter deck, forward of the 35mm guns in 1986 and other units of the class may have been similarly equipped. A plan to fit four of these ships with a Sea Sparrow SAM VLS plus associated radars has been abandoned. Six were extant 1995.

GEARING class *destroyers*

Particulars: As US *Gearing* class (FRAM I and II types)

No	Name	Builder	Acquired	Fate
D 353	ADATEPE (ex-*Forrest Royal*)	Bethlehem, Staten Is	27.3.71	Stricken 1993, BU 1994
D 351	MARESAL FEVZI CAKMAK (ex-*Charles H Roan*)	Bethlehem, Quincy	29.9.73	BU 1995
D 352	GAYRET (ex-*Eversole*)	Todd-Pacific, Seattle	11.7.73	Extant 1995
D 354	KOCATEPE (i, ex-*Harwood*)	Bethlehem, San Pedro	17.12.71	Sunk 22.7.74
D 354	KOCATEPE (ii, ex-*Norris*)	Bethlehem, San Pedro	7.7.74	Stricken 1993
D 355	TINAZTEPE (ex-*Keppler*)	Bethlehem, San Francisco	30.6.72	Laid up 1984
D 347	SAVASTEPE (ex-*Meredith*)	Consolidated, Orange	7.12.79	Extant 1995
D 349	KILIC ALI PASA (ex-*Robert H McCard*)	Consolidated, Orange	5.6.80	Extant 1995
D 350	PIYALE PASA (ex-*Fiske*)	Bath Iron Wks	2.6.80	Extant 1995
D 345	YUCETEPE (ex-*Orleck*)	Consolidated, Orange	1.10.82	Extant 1995

Savastepe 4.11.85 L & L van Ginderen

The *Carpenter* class was a specialised ASW version of the *Gearing* class and launched in 1949. Received FRAM I modernisation in 1962. Armament on transfer was 2–127mm (1 × 2), ASROC (1 × 8) and 6–324mm ASW TT (2 × 3). They later received 2–35mm Oerlikon (1 × 2) forward of the bridge and 2–76mm 91 × 2) on the fantail. Both extant 1995.

Ex-US CARPENTER class *destroyers*

Particulars: See US *Carpenter* class

No	Name	Builder	Acquired	Fate
D 347	ANITEPE (ex-*Carpenter*)	Consolidated, Orange	20.2.81	Extant 1995
D 346	ALCITEPE (ex-*Robert A Owens*)	Bath Iron Wks	16.2.82	Extant 1995

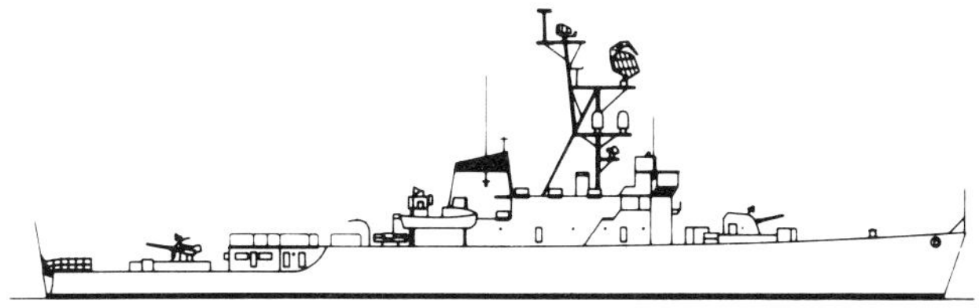

Berk 1976

These were the first major warships built in Turkey since before the First World War, and are based on the USN *Claud Jones* class frigate design. They have a slightly different hull, however, a more powerful armament, different machinery and one large funnel. There is a helicopter pad at the break of the forecastle aft, but no hangar.

BERK class *frigates*

Displacement: 1450t standard; 1950t full load
Dimensions: 312ft 4in oa × 39ft 10in × 18ft 1in
95.2m × 12.1m × 5.7m
Machinery: 1 shaft, 4 Fiat diesels, 24,000bhp = 25kts. Range 10,000nm at 9kts
Armament: 4–3in/50 (2 × 2), 6–324mm TT (2 × 3), 2 Hedgehog
Sensors: Radar SPS-40, SPS-10, SPG-4; sonar SQS-11
Complement: 175

No	Name	Builder	Laid down	Launched	Comp	Fate
D 358	BERK	Gölcük N Yd	9.3.67	25.6.71	12.7.72	Extant 1995
D 359	PEYK	Gölcük N Yd	18.1.68	7.6.72	24.7.75	Extant 1995

Peyk 14.9.83 L & L van Ginderen

Two units transferred 1983. Sister ship *Lübeck* was transferred 1 December 1988 and was to be commissioned into service, but was cannibalised for spares instead. Fourth unit *Braunschweig* delivered to Turkey 4 July 1989, also for cannibalisation.

Ex-German KOLN class *frigates*

Particulars: See German *Köln* class

No	Name	Builder	Acquired	Fate
D 360	GELIBOLU (ex-*Karlsruhe*)	H C Stülcken, Hamburg	28.3.82	Extant 1995
D 361	GEMLIK (ex-*Emden*)	H C Stülcken, Hamburg	23.9.83	BU 1994

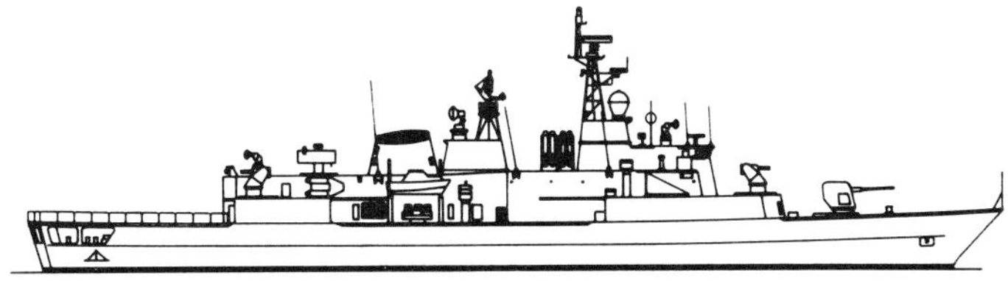

Yavuz 1987

Four units of the MEKO 200 design ordered in April 1983. Two were built in Germany, whilst Blohm & Voss provided technical assistance for the second pair to be built at the Gölcük Naval Shipyard. Similar vessels have been built for Greece and Portugal. The 25mm Sea Zenith CIWS is in use only with the Turkish Navy and replaces the single 20mm Vulcan Phalanx projected for this class. They each carry a single AB-212 helicopter equipped with Sea Skua air-to-surface missiles. These were the first new major surface units built for Turkey since the *Berk* class and are modern, capable ships.

YAVUZ class *frigates*

Displacement: 2700t standard; 2994t full load
Dimensions: 362ft 6in oa, 335ft 4in pp × 43ft 6in × 12ft 11in
110.5m, 102.2m × 13.25m × 3.94m
Machinery: 2 shafts (CP propellers); 2 MTU 20V1163 TB93 diesels, 40,000bhp = 27kts. Range 4000nm at 20kts
Armament: 8 Harpoon SSM (2 × 4), 1 Sea Sparrow SAM (1 × 8, 16 missiles), 1–127mm/54, 12–25mm Oerlikon-Contraves Sea Zenith GM 25 CIWS (3 × 4), 6–324mm ASW TT (2 × 3), 1 helicopter
Sensors: Radar TM 1226, AWS-6, HSA DA-08, HSA WM-25, HSA STIR-24; sonar SQS-56; ECM HSA Rapid/Ramses suite, Mk 36 SRBOC decoy RL, SLQ-25 Nixie towed torpedo decoy
Complement: 180 (26 officers)

No	Name	Builder	Laid down	Launched	Comp	Fate
F 240	YAVUZ	Blohm & Voss, Hamburg	31.5.85	7.11.85	17.7.87	Extant 1995
F 241	TURGUT REIS	Howaldtswerke, Kiel	20.9.85	30.5.86	4.2.88	Extant 1995
F 242	FATIH	Gölcük N Yd	1.1.86	24.4.87	22.7.88	Extant 1995
F 243	YILDERIM	Gölcük N Yd	24.4.87	22.7.88	21.7.89	Extant 1995

Barbados 28.2.95
L & L van Ginderen

Lenghtened version of MEKO 200 class frigate. First two ordered 19 January 1990, second pair on 14 December 1992. They have a CODOG propulsion plant, improved electronics and better NBC protection as compared to the previous four ships. Currently armament is as the earlier ships detailed above, but both F 244 and F 245 were built to receive a 16-cell Sea Sparrow VLS SAM system in place of the octuple Mk 29 launcher and the second pair will have this system fitted from the start.

BARBAROS class *frigates*

Displacement: 3100t standard; 3350t full load
Dimensions: 351ft 8in pp, 382ft 11in oa × 48ft 7in × 13ft 11in
107.2m, 116.72m × 14.8m × 4.25m
Machinery: 2 shafts (CP propellers) CODOG: 2 GE LM-2500-30 gas turbines, 60,000shp = 32kts, 2 MTU 12V1163 TB83 diesels, 13,060bhp = 22kts. Range 4100nm at 18kts (diesel)
Armament: As *Yavuz* class above
Sensors: Radar Decca 2040 BT ARPA, AWS-6, AWS-9, HSA STIR-24, HSA STIR-18, 2 Contraves TMX; sonar SQS-56 hull-mounted; ECM Racal Cutlass B1, Racal Scorpion B, Mk 36 SRBOC decoy RL, SLQ-25 Nixie towed torpedo decoy
Complement: 180 (24 officers)

No	Name	Builder	Laid down	Launched	Comp	Fate
F 244	BARBAROS	Blohm & Voss, Hamburg	Apr 1992	26.10.94	Mar 1995	Extant 1995
F 245	ORUC REIS	Gölcük N Yd	23.7.92	28.7.94	1996	Completing
F 246	–	Blohm & Voss, Hamburg	1994	1996	1997	Building
F 247	–	Gölcük N Yd	1994	–	1998	Building

Eight units of the *Knox* class transferred initially on lease. Ex-*Fanning* was transferred to replace the damaged destroyer *Muavenet* (see above). Identical vessels have been transferred to Greece, Taiwan and Egypt. In July 1993 Harpoon missiles were sold to Turkey to equip this class (to be fired from the ASROC launcher: four missiles carried in ASROC reload magazine). Towed array sonars for these ships were not transferred to Turkey. *Muavenet* does not have the spray strakes and bow bulwarks fitted to other units of this class to improve sea-keeping qualities. There is reported to have been resistance within the Turkish Navy to the acquisition of these ships, which were considered lightly armed and in poor condition.

Ex-US KNOX class *frigates*

Particulars: See US *Knox* class

No	Name	Builder	Acquired	Fate
F 250	MUAVENET (ex-*Capodanno*)	Avondale	30.7.93	Extant 1995
F 251	ADATEPE (ex-*Fanning*)	Todd, San Pedro	31.7.93	Extant 1995
F 252	KOCATEPE (ex-*Reasoner*)	Lockheed, Seattle	30.8.93	Extant 1995
F 253	ZAFER (ex-*Thomas C Hart*)	Avondale	30.8.93	Extant 1995
F 256	EGE (ex-*Ainsworth*)	Avondale	20.5.94	Extant 1995
F 255	KARADENIZ (ex-*Bowen*)	Avondale	27.5.94	Extant 1995
F 257	AKDENIZ (ex-*Donald B Beary*)	Avondale	3.6.94	Extant 1995
F 254	TRAKYA (ex-*McCandless*)	Avondale	5.6.94	Extant 1995

SUBMARINES

They were all US *Balao* class submarines, built in 1943–44. The ones transferred in 1948 were more or less unmodified, but *Canakkale* and subsequent boats were semi- or fully steamlined, and most of them were brought up to the fully streamlined 'Fleet Snorkel' standard eventually. *Dumlupinar* was lost accidently in the Dardanelles on 4 April 1953. *Turget* was purchased on 15 February 1973 and *Hizir* in August 1973, and they were cannibalised for spare parts after being discarded.

Ex-US BALAO class *submarines*

Particulars: As US *Balao* class (unmodified and Fleet Snorkel types)

No	Name	Builder	Acquired	Fate
S 330	BIRINCI INONU (ex-*Brill*)	Electric Boat	1948	Discarded 1973
S 333	CANAKKALE (ex-*Bumper*)	Electric Boat	1950	Discarded 1973
S 341	CERBE (ex-*Hammerhead*)	Manitowoc	7.8.54	Discarded 1973
S 335	DUMLUPINAR (ex-*Blower*)	Electric Boat	1950	Lost 4.4.53
S 334	GUR (ex-*Chub*)	Electric Boat	1948	Discarded 1974
S 344	HIZAR REIS (ex-*Mero*)	Electric Boat	20.4.60	Discarded 1977
S 331	IKINCI INONU (ex-*Blueback*)	Electric Boat	1948	Discarded 1973
S 343	PIRI REIS (ex-*Mapiro*)	Manitowoc	20.4.60	Discarded 1973
S 340	PREVEZE (ex-*Guitarro*)	Manitowoc	7.8.54	Discarded *c*1970
S 332	SAKARYA (ex-*Boarfish*)	Electric Boat	1948	Discarded 1974
S 342	TURGET REIS (ex-*Bergall*)	Electric Boat	17.10.58	Discarded 1977

She was converted by the USN to GUPPY IA standard in the early 1950s, and was transferred to Turkey on 24 August 1972. *Dumlupinar* was badly damaged in a collision on 1 September 1976, and had to be beached. She was eventually repaired and put back into service until she was stricken in 1987.

Ex-US BALAO class *submarine*

Particulars: As US *Balao* class (GUPPY IA type)

No	Name	Builder	Acquired	Fate
S 339	DUMLUPINAR (ex-*Caiman*)	Electric Boat	24.8.72	Stricken 1987

Murat and *Uluc Ali* had been used by the USN as ASW targets. All these *Balao* class submarines had been converted by the USN to GUPPY IIA standard. *Cerbe* is the only operational ex-USN GUPPY III to retain the original stepped sail.

Ex-US BALAO class *submarines*

Particulars: As US *Balao* class (GUPPY IIA type)

No	Name	Builder	Acquired	Fate
S 346	BIRINCI INONU (ex-*Thredfin*)	Portsmouth N Yd	15.8.73	Extant 1995
S 335	BURAK REIS (ex-*Sea Fox*)	Portsmouth N Yd	14.12.70	Extant 1995
S 340	CERBE (ex-*Trutta*)	Portsmouth N Yd	24.8.72	Extant 1995
S 336	MURAT REIS (ex-*Razorback*)	Portsmouth N Yd	17.11.70	Extant 1995
S 337	ORUC REIS (ex-*Pomfret*)	Portsmouth N Yd	3.5.72	Stricken 1987
S 345	PREVEZE (ex-*Entemedor*)	Electric Boat	24.8.73	Stricken 1987
S 338	ULUC ALI REIS (ex-*Thornback*)	Portsmouth N Yd	24.8.73	Extant 1995

The USN modernised these two war-built *Balao* class submarines to GUPPY III standards in 1962. Two more GUPPY IIIs – *Clamagore* and *Tiru* – were to have been transferred to Turkey, but this was prevented by the US Congress imposing an arms embargo on Turkey in 1975.

Ex-US BALAO class *submarines*

Particulars: As US *Balao* class (GUPPY III type)

No	Name	Builder	Acquired	Fate
S 341	CANAKKALE (ex-*Cobbler*)	Electric Boat	21.11.73	Extant 1995
S 333	IKINCI INONU (ex-*Corporal*)	Electric Boat	21.11.73	Extant 1995

Tang was to have been sold to Iran but following the revolution this was cancelled and she was leased to Turkey for an initial period of five years. A second boat was leased in 1983 and both were purchased by Turkey on 6 August 1987.

Ex-US TANG class *submarines*

Particulars: See US *Tang* class

No	Name	Builder	Acquired	Fate
S 343	PIRI REIS (ex-*Tang*)	Portsmouth N Yd	29.2.80	Extant 1995
S 342	HIZIR REIS (ex-*Gudgeon*)	Portsmouth N Yd	30.9.83	Extant 1995

Similar to the Type 209 German-designed submarines serving in the Greek and several South American navies. A single hull, medium-sized design capable of diving to more than 650ft (*200m*). A total of twelve was originally planned, with the first three built in Germany and later units in Turkey with German technical assistance. In the event, three were built to this design at the Gölcük Naval Shipyard and a larger design was then selected for further construction (see below). *Yildiray* was the first submarine to be built in Turkey. The first two have Hollandse Signaal M8 torpedo fire control, the other four having HSA SINBADS.

ATILAY class *submarines*

Length: 183ft 9in oa *56m*
Other particulars: As Greek *Glavkos* class

No	Name	Builder	Laid down	Launched	Comp	Fate
S 347	ATILAY	Howaldstwerke, Kiel	2.8.72	23.10.74	29.7.75	Extant 1995
S 348	SALDIRAY	Howaldtswerke, Kiel	2.1.73	14.2.75	21.10.75	Extant 1995
S 349	BATIRAY	Howaldtswerke, Kiel	11.6.75	26.10.77	20.7.78	Extant 1995
S 350	YILDIRAY	Gölcük N Yd	1.5.76	20.7.77	20.7.81	Extant 1995
S 351	DOGANAY	Gölcük N Yd	21.3.80	16.11.83	16.11.85	Extant 1995
S 352	DOLUNAY	Gölcük N Yd	16.11.83	21.7.88	21.7.89	Extant 1995

An enlarged version of the standard Type 209/1400 design, first two ordered 1987, second pair in 1994, replacing later units of the *Atilay* class. Diving depth of 1050ft (*320m*). Torpedo tubes have capability to launch Sub-Harpoon and to lay mines.

PREVEZE class *submarines*

Displacement: 1464t surfaced; 1586t submerged
Dimensions: 203ft 5in × 20ft 4in × 18ft 1in
62.0m × 6.2m × 5.5m
Machinery: 1 shaft diesel-electric: 4 MTU 12V493 A280 AG diesels, 1 electric motor, 8200hp = 11kts/21.5kts
Range 10,000nm at 8kts snorkelling
Armament: 8–533mm TT (bow; 14 torpedoes or Sub-Harpoon SSM or mines)
Sensors: Sonar Krupp-Atlas CSU-85/1 (flank and towed passive arrays)
Complement: 35 (8 officers)

No	Name	Builder	Laid down	Launched	Comp	Fate
S 353	PREVEZE	Gölcük N Yd	27.7.89	27.11.93	27.7.94	Extant 1995
S 354	SAKARYA	Gölcük N Yd	4.4.90	28.7.94	1995	Extant 1995
S 355	MART	Gölcük N Yd	28.7.94	–	–	Building
S 356	ANAFARTALAR	Gölcük N Yd	28.7.94	–	–	Building

AMPHIBIOUS WARFARE VESSELS

Ex-US LST 511 class *tank landing ships*

Particulars: As US LST type (*LST 511–1152* group)

Class:
Bayraktar (NL 120), *Sancaktar* (NL 121)

Bayraktar (ex-*Bottrop*, ex-*Saline County*, ex-*LST 1101*) and *Sancaktar* (ex-*Bochum*, ex-*Rice County*, ex-*LST 1089*) were originally transferred from the USN to West Germany in 1961. There they were converted to minelayers. They were then transferred to Turkey on 13 December 1972 and 12 December 1972 respectively and converted back to LSTs in 1974–75. They were replaced in the mine warfare category once more in 1980. Both extant 1995.

Ex-US TERREBONNE PARISH class *tank landing ships*

Particulars: As US *Terrebonne Parish* class

Class:
Ertugrul (L 401), *Serdar* (L 402)

Two ex-USN *Terrebonne Parish* class LSTs were transferred to Turkey, *Ertugrul* (ex-*Windham County*, ex-*LST 1170*) was transferred on 4 June 1973 and *Serdar* (ex-*Westchester County*, ex-*LST 1167*) was transferred on 27 August 1974. They are armed with 6–76mm/50 (3 × 2). Both extant 1995.

CAKABEY class *tank landing ship*

Displacement: 1600t
Dimensions: 253ft 6in oa × 39ft 5in × 7ft 6in
77.3m × 12.0m × 2.3m
Machinery: 3 shafts, 3 diesels, 4320 bhp = 14kts
Armament: 4–40mm (2 × 1), 4–20mm (2 × 2)
Sensors: Radar navigational
Complement: ?

Class: *Cakabey* (NL 122)

Cakabey was built at Taskizak N Yd, in Istanbul and launched on 30 June 1977. Her design is based on that of the USN LSTs but she has a higher freeboard and a helicopter landing deck amidships and has a minelaying capacity (150 mines). She can transport 400 troops, nine tanks and ten jeeps; two LCVPs are carried under davits. Extant 1995.

SARUCABEY class *tank landing ships*

Displacement: 2600t full load
Dimensions: 301ft 10in × 45ft 11in × 7ft 7in
92.0m × 14.0m × 2.3m
Machinery: 3 shafts, 3 diesels, 4320 bhp = 14kts
Armament: 3–40mm/70 Bofors (3 × 1), 4–20mm/70 Oerlikon (2 × 2), 150 mines
Sensors: Radar navigational
Complement: ?

Class (no, launched):
Sarucabey (NL 123, 30.7.81), *Karamurselbey* (NL 124, 26.7.84)

Enlarged version of the *Cakabey* class, both built by the Taskizak N Yd in Istanbul. They entered service on 26 July 1984 and 26 July 1985 respectively. Military load of eleven tanks, twelve trucks and 600 troops. Two LCVPs carried amidships, with a large crane to handle them. Two stern minelaying ports.

OSMAN GAZI class *landing ships*

Displacement: 3773t full load
Dimensions: 344ft 6in × 52ft 10in × 15ft 9in
105.0m × 16.1m × 4.8m
Machinery: 2 shafts, 4 MTU 8V396 diesels, 10,000bhp = 17kts. Range 4000nm at 15kts
Armament: 3–40mm/70 Bofors (3 × 1), 2–35mm/90 Oerlikon (1 × 2), mines
Sensors: Radar navigational
Complement: ?

Class (no, in service):
Osman Gazi (NL 125, 27.7.94), *Orhan Gazi* (NL 126, Oct 1994)

Built by Taskizak N Yd, Istanbul. Designed to have 50 per cent greater military lift capacity than the *Sarucabey* class (fifteen tanks, 900 troops). Equipped as amphibious command ships and have a helicopter deck aft and minelaying capability. Both extant 1995, with a third ship planned.

Osman Gazi Oct 1994 — L & L van Ginderen

LANDING CRAFT

Name	Launched	Disp	Fate
Ex-British LCT (4) type			
C 101–106	1942	700t	*C 101, C 102, C 104–C 106* BU 1981–82; *C 103* stricken *c*1985
EDIC type			
C 107–138	1966–80	580–600t	See notes
Ex-US LCU 501 class			
C 201–204	1943–44	320t	All stricken by 1986
Turkish LCU type			
C 205–216	1965–66	405t	Stricken 1991–93
US LCM (8) type			
C 301–331	1965	113t	22 extant 1995
Turkish C 139 LCUs			
C 139–C 155	1984–93	600t	Extant 1995

C 101–C 106 were transferred on 25 September 1967. They were modernised on arrival in Turkey. All BU by 1985.

C 107–C 138 were based on the French EDIC type. *C 119–C 137* were slightly larger than previous vessels of the class and *C 121–C 138* had an MG position above the bridge. *G 134* transferred to Libya 1979, *C 129* and *C 130* in 1983. *C 136* lost in 1985. *C 107, C 108, C 111–C 116* and *C 121* deleted by 1991, *C 107* and *C 118* stricken 1992. Remainder extant 1995.

C 201–C 204 transferred to Turkey in June 1967. Turkish-built *C 205–C 216* differed from US LCUs in having the conning tower forward.

C 301–C 331 built in Turkey to the US LSM (8) design. Eight stricken 1991 and one in 1992. Twenty-two extant 1995.

C 139–C 155 were a development of the *C 107* type design, larger and with taller superstructure. Military load of six tanks and 100 troops. Seventeen extant 1995.

SMALL SURFACE COMBATANTS

Ex-US AUK class *patrol vessels*

Particulars: As US *Auk* class

No	Name	Builder	Acquired	Fate
	CANDARLI (ex-*Frolic*)	Alameda	Mar 1947	Stricken 1986
M 507	CARDAK (ex-*Tourmaline*)	Gulf SB	Mar 1947	Discarded 1974
	CARSAMBA (ex-*Tattoo*)	Associated	Mar 1947	Stricken *c*1985
M 505	CESME (ex-*Elfreda*)	Associated	Mar 1947	Discarded 1974
M 509	EDINCIK (ex-*Grecian*)	Associated	Mar 1947	Discarded 1974
M 512	EDREMIT (ex-*Chance*)	Associated	Mar 1947	Discarded 1973
	ERDEMLI (ex-*Catherine*)	Associated	Mar 1947	Discarded 1963
	EREGLI (ex-*Pique*)	Gulf SB	Mar 1947	Discarded 1973

These eight ships were built in the United States as *Auk* class minesweeping escorts, and were transferred to the RN on completion under lease-lend as the *Catherine* class. They were re-transferred to Turkey in March 1947 and were employed mainly as escorts. By the late 1960s they were being used in auxiliary roles. *Candarli* (A 593) and *Carsamba* (A 594) were converted to survey ships, with a deckhouse between the funnels and only 2–40mm (2 × 1). The others were used either as command ships for light craft (*Cesme, Cardak* and *Edremit*), or training (*Edincik* and *Erdemli*) or support (*Eregli*) ships.

Ex-British BATHURST class *patrol vessels*

Particulars: As British *Bathurst* class

No	Name	Builder	Acquired	Fate
M 501	ALANYA (ex-*Broome*)	Sydney	Aug 1946	Discarded 1975
M 502	AMASRA (ex-*Pirie*)	Broken Hill	Aug 1946	Discarded 1975
	AYVALIK (ex-*Gawler*)	Broken Hill	Aug 1946	Discarded 1963
M 500	AYVALIK (ex-*Antalya*, ex-*Geraldton*)	Brisbane	Aug 1946	Discarded 1975
M 503	HAMIT NACI (ex-*Ayancik*, ex-*Launceston*)	Sydney	Aug 1946	Discarded 1965

These Australian-built *Bathurst* class fleet minesweepers were transferred to Turkey in 1946. They were mainly used as escorts. *Antalya* took the original *Ayvalik*'s name when the latter was discarded in 1963. By the late 1960s the survivors were used as logistic support ships.

Ex-British BANGOR class *patrol vessels*

Particulars: As British *Bangor* class

No	Name	Builder	Acquired	Fate
P 121	BAFRA (ex-*Nipigon*)	Dufferin	29.11.57	Discarded early 1970s
P 129	BANDIRMA (ex-*Kenora*)	Port Arthur	29.11.57	Discarded early 1970s
P 130	BARTIN (ex-*Kentville*)	Port Arthur	29.11.57	Discarded early 1970s
P 122	BEYKOZ (ex-*Blairmore*)	Port Arthur	29.3.58	Discarded early 1970s
P 123	BEYLERBEYI (ex-*Mahone*)	North Vancouver	29.3.58	Discarded early 1970s
P 124	BIGA (ex-*Medicine Hat*)	Canadian Vickers	29.11.57	Discarded 1963
P 125	BODRUM (ex-*Fort William*)	Port Arthur	29.11.57	Discarded early 1970s
P 126	BORNOVA (ex-*Westmount*)	Dufferin	29.3.58	Discarded early 1970s
P 127)	BOZCAADA (ex-*Swift Current*)	Canadian Vickers	29.3.58	Discarded early 1970s
P 128	BUYUKDERE (ex-*Sarnia*)	Dufferin	29.3.58	Discarded early 1970s

All built in Canada as *Bangor* class fleet minesweepers, but re-rated as coastal escorts in 1953. They were transferred to Turkey in 1957 and early 1958, and sailed to Turkey on 19 May 1958.

Norwegian NASTY type *fast attack craft*

Particulars: As Norwegian *Tjeld* class

Class:
Dogan (P 327), *Marti* (P 328)

Two ex-West German *Nasty* type (*Tjeld* class) MTBs, built by Boat Batservice, Norway, in 1959–60, were transferred to Turkey as part of the German-Turkish war reparations. *Dogan* (ex-*Hugin*) and *Marti* (ex-*Munin*) were discarded in 1973.

Ex-West German JAGUAR class *fast attack craft*

Particulars: As West German *Jaguar* class

Class:
Firtina (P 330), *Tufan* (P 331), *Kilic* (P 332), *Mizrak* (P 333), *Yildiz* (P 334), *Kalkan* (P 335), *Karayel* (P 336)

Seven ex-West German *Jaguar* class, which can be distinguished from the later *Kartal*s by their shorter, stepped deckhouse, were transferred to Turkey between late 1975 and early 1976. These were *Firtina* (ex-*Pelikan*), *Kalkan* (ex-*Wolf*), *Karayel* (ex-*Pinguin*), *Kilic* (ex-*Löwe*), *Mizrak (ex-Hähner)*, *Tufan* (ex-*Storch*) and *Yildiz* (ex-*Tiger*). In addition, ex-*Alk*, ex-*Fuchs* and ex-*Reiher* were transferred at the same time to provide spares. *Firtina, Kilic* and *Yildiz* stricken in 1982, *Turfan* and *Karayel* by 1992 and the last pair in 1993.

KARTAL class *fast attack craft*

Dimensions: 140ft 6in oa × 23ft 6in × 7ft 2in
42.8m × 7.2m × 2.2m
Armament: 4 Penguin 2 SSM (4 × 1), 2–40mm/70 (2 × 1), 2–21in TT. See notes.
Other particulars: As West German *Jaguar* class

Class:
Denizkusu (P 321), *Atmaca* (P 322), *Sahin* (P 323), *Kartal* (P 324), *Melten* (P 325), *Pelikan* (P 326), *Albatros* (P 327), *Simsek* (P 328), *Kasirga* (P 329)

Nine *Kartal* class convertible patrol boats were built by Lürssen for Turkey during 1967–71. *Albatros, Atmaca, Denizkusu, Kartal, Kasirga, Melten, Pelikan, Sahin* and *Simsek* were almost identical with the West German *Jaguar* class when first built, but *Albatros, Melten, Pelikan* and *Simsek* were fitted with SSM in place of the aft pair of 21in (533mm) TT in 1975 and the other five have been similarly fitted since. *Melten* was sunk in a collision with a Soviet ship on 25 September 1985. Remainder extant 1995.

Dogan 1977 Lürssen

DOGAN class *fast attack craft*

Displacement: 353t normal; 434t full load
Dimensions: 190ft 6in × 25ft × 8ft 8in
58.1m × 7.6m × 2.7m
Machinery: 4 shafts, 4 MTU 16V956 TB91 diesels, 18,000bhp = 36.5kts. Range 700nm/3000nm at 35kts/16kts
Armament: 8 Harpoon SSM (2 × 4), 1–76mm OTO Melara, 2–35mm Oerlikon (1 × 2), 2–7.62mm MG (2 × 1)
Sensors: Series I: Radar 1226, WM 28–41; ECM MEL SUSIE-1, 2 decoy RL
Series II: Radar AWS-6, TMX-CW; ECM Racal Cutlass B1, Mk 36 SRBOC decoy system
Complement: 38

Class (no, in service):
Series I: *Dogan* (P 340, 15.6.77), *Marti* (P 341, 28.7.78), *Tayfun* (P 342, 19.7.79), *Volkam* (P 343, 25.7.80), *Ruzgar* (P 344, 17.12.84), *Poyraz* (P 345, 7.2.86), *Gayret* (P 346, 22.7.88), *Firtina* (P 347, 23.10.88)
Series II: *Yildiz* (P 348, 27.7.94), *Karayel* (P 349, 1995), plus three more

Design based on German FPB 57 class, initially ordered in August 1973. Prototype *Dogan* built by Lürssen, Vegesack and the remainder at the Taskizak N Yd, Istanbul, at an initial rate of one per year. Series II, ordered June 1991 with improved electronics. Normally only four Harpoon SSM are carried. Similar to the Israeli *Reshef* class FAC(M)s. All extant 1995.

US PC type *patrol vessels*

Particulars: As US PC type

Class:
Sultanhisar (P 111), *Demirhisar* (P 112), *Yarhisar* (P 113), *Akhisar* (P 114), *Sivrihisar* (P 115), *Kochisar* (P 116)

Six PC boats were built for Turkey in 1963–65 to the same design as the wartime PC Boats. All were built by Gunderson in the United States, except *Kochisar* which was built by Gölcük N Yd in Turkey. They are armed with 1–40mm, 4–20mm (2 × 2), one Hedgehog ASW mortar forward and four DCT. All were in service in 1995.

Ex-US PGM 71 class *patrol boats*

Particulars: As US *PGM 71* class

Class:
AB 21 (P 1221), *AB 22* (P 1222), *AB 23* (P 1223), *AB 24* (P 1224)

AB 21–24 (ex-PGM 104–108) are four USN *PGM 71* type patrol boats built in the United States and taken over by Turkey on completion in 1967–68. Pennant numbers were originally P 117–P 120. They are armed with 1–40mm, 4–20mm (2 × 2), 4–12.7mm MG (2 × 2), 2 Mk 22 Mousetrap ASW, and two DC racks. All were in service in 1995, with pennant numbers P 122–P 124.

Ex-US ASHEVILLE class *patrol vessels*

Particulars: As US *Asheville* class

Class:
Yildrim (P 338), *Bora* (P 339)

Two ex-US *Asheville* class patrol vessels were transferred to Turkey, *Bora* (ex-*Surprise*) on 28 February 1973 and *Yildrim* (ex-*Defiance*) on 11 June 1973. *Yildrim* blew up and sank off Lesbos on 11 April 1985. *Bora* extant 1995.

GIRNE *patrol boat*

Particulars: As Spanish *Lazaga* class

Class:
Girne (P 140)

Girne was built at Taskizak, Istanbul in 1975–76. The PB 57 design by Lürssen, was based on the Spanish *Lazaga* class but with lighter armament. She was originally to have been armed with 1–40mm, 2–35mm (1 × 2), 1 Mousetrap ASW mortar and 4–21in (533mm) TT (4 × 1), but is now reported to carry 2–40mm/70 Bofors (2 × 1) and 2–12.7mm MG (2 × 1). ASW weapons comprise 4 Mk 20 Mousetrap launchers, 2 DCT and 2 DC racks. She was to have been the first of a class, but she has not been successful in service, and no more have been built. Extant 1995.

USCG 83ft Cutter type *patrol boat*

Particulars: As US 83ft Coast Guard Cutter type

Class:
LS 9–LS 19

LS 9–12 are four US 83ft Coast Guard cutters built in the United States for Turkey in 1953. Pennant numbers were changed from P 339, P 308–P 310 to P 1209–P 1212. All were in service in 1995, with pennant numbers P 109–P112.

AB 25 class *patrol boats*

Displacement:	150t normal; 170t full load
Dimensions:	132ft × 21ft × 5ft 6in *40.2m × 6.4m × 1.7m*
Machinery:	2-shaft diesels, 4800bhp = 22kts.
Armament:	2–40mm (2 × 1) or 1–40mm, 1–20mm
Sensors:	?
Complement:	?

Class:
AB 25–AB 36 (P 1225–P 1236)

AB 25–36 were built at Taskizak during 1967–70 with French assistance. Some have Mousetrap ASW launchers and DC racks. Pennant numbers were changed in 1991 to P 125–P 136. Fourteen more units of this class have been built for the Turkish Gendarmerie. All extant 1995.

MINE WARFARE VESSELS

NOTE
Turkey operates a number of LSTs in the dual minelaying/landing ship role (see under Amphibious Warfare Vessels). The destroyer *Muavenet* also retained minelaying capability (see under Major Surface Ships), as do the ex-German *Köln* class frigates and *Rhein* class support ships. Minesweepers transferred from the British, Canadian and US navies were used principally as escorts and patrol vessels (see under Small Surface Combatants). There is also the 540t controlled mineplanter *Mehmetcik* (N 115), built by Higgins, New Orleans in 1958. She is normally unarmed.

Danish FALSTER class *minelayer*

Particulars: As Danish *Falster* class

No	Name	Builder	Acquired	Fate
N 110	NUSRET	Frederikshavn N Yd	1964	Extant 1995

Laid down in 1962, *Nusret* was built in Denmark for Turkey under MDAP. She is almost identical with the Danish *Falster* class minelayers, but has no Sea Sparrow SAMs. Pennant number was N 108.

Ex-US LSM 1 class *coastal minelayers*

Particulars: See Danish *Beskytteren* class

Class:
Mordogan (N 101), *Meric* (N 102), *Marmaris* (N 103), *Mersin* (N 104), *Mürefte* (N 104)

Marmaris (ex-*LSM 481*), *Meric* (ex-*LSM 490*) and *Mordogan* (ex-*LSM 484*) were built by the USN as LSM in 1945. They were rebuilt as coastal minelayers and transferred to Turkey in 1952. Two more were supplied to Norway as *Vidar* (ex-*LSM 494*) and *Vale* (ex-*LSM 492*). They were retransferred to Turkey on 1 November 1960 and renamed *Mersin* and *Mürefte* respectively. Armament is now 6–40mm (3 × 2) , 5–20mm (5 × 1), 400 mines. *Meric* and *Marmaris* stricken 1988 and *Mordogan* and *Murefte* in 1993. *Mersin* extant 1995.

US REDWING/ADJUTANT type *coastal minesweepers*

Particulars: As US *Redwing* class and *Adjutant* class

Class:
Redwing early type – *Samsun* (M 510), *Sinop* (M 511), *Surmene* (M 512), *Seddul Bahr* (M 513)
Redwing later type – *Silifke* (M 514), *Saros* (M 515), *Sigacik* (M 516), *Sapanca* (M 517), *Sariyer* (M 518)
Adjutant type – *Seymen* (M 507), *Selcuk* (M 508), *Seyhan* (M 509)

Four US-built *Redwing* class coastal minesweepers were transferred to Turkey as they were completed in 1958–59 under MDAP: *Samsun* on 30 September 1959, *Sinop* in February 1959, *Surmene* on 27 March 1959 and *Seddul Bahr* in May 1959. Five more were built for Turkey during 1964–67: *Sigacik*, transferred in June 1965, *Sapanca* on 26 July 1965, *Silifke* in September 1965, *Saros* in February 1966 and *Sariyer* on 8 September 1967. This later five were of the MSC 294 type, with a slightly larger hull, lower superstructure, taller funnels and different machinery. Three of the earlier *Adjutant* type were acquired in 1970. *Seyhan* (ex-*Renoncule*) and *Selcuk* (ex-*Pavot*) on 24 March 1970), and *Seymen* (ex-Belgian *De Panne*) on 19 November 1970. These were transferred from France and Belgium to Turkey via the United States. All twelve were in service in 1995.

Ex-Canadian 'BAY' class *coastal minesweepers*

Particulars: As Canadian 'Bay' class

Class:
Trabzon (M 530), *Terme* (M 531), *Tirebolu* (M 532), *Tekirdag* (M 533)

Four Canadian 'Bay' type coastal minesweepers, all built in 1951–53, were transferred from Canada to Turkey under MDAP on 19 May 1958. They were *Tekirdag* (ex-*Ungava*), *Terme* (ex-*Trinity*), *Tirebolu* (ex-*Comax*), and *Trabzon* (ex-*Gaspé*). In 1991, *Trabzon, Terme* and *Tirebolu* were redesignated as patrol craft in 1991 and *Tekirdag* as an intelligence collecting ship and their MCM gear was removed. All were extant in these roles in 1995.

French MERCURE class *coastal minesweepers*

Particulars: As French *Mercure* class

Class:
Karamürsel (M 520), *Kerempe* (M 521), *Kilimli* (M 522), *Kozlu* (M 523), *Kusadasi* (M 524), *Kemer* (M 525)

Six ex-West German *Vegesack* class coastal minesweepers, built to the French *Mercure* class design during 1959–60, were transferred to Turkey; *Karamürsel* (ex-*Worms*), *Kerempe* (ex-*Detmold*), *Kilimli* (ex-*Siegen*), *Kozlu* (ex-*Hameln*) and *Kusadasi* (ex-*Vegesack*) between late 1975 and early 1976, and *Kemer* (ex-*Passau*) in 1979. Fitted with sonar from 1988. All extant 1995.

Ex-US YMS type *inshore minesweeper*

Particulars: As USN *YMS* type

Eight ex-*YMS* type motor minesweepers were acquired by Turkey from the United States in 1947. *Kas* (ex-*YMS 79*) and *Kilimi* (ex-*YMS 289*) were discarded in 1963, *Kozlu* (ex-*YMS 348*) and *Kusadasi* (ex-*YMS 468*) in 1965 and *Kemer* (ex-*YMS 228*), *Kerempe* (ex-*YMS 239*), *Kirte* (ex-*YMS 307*) and *Karamürsel* (ex-*Küllük*, ex-*YMS 375*) in 1966.

US CAPE class *inshore minesweepers*

Particulars: As USN 'Cape' class

Class:
Foca (M 500), *Fethiye* (M 501), *Fatsa* (M 502), *Finike* (M 503)

Four US 'Cape' class inshore minesweepers (*MSI 15–MSI 18*) were constructed under MDAP and were transferred to Turkey on completion between August and December 1967. All four were in service in 1995.

MISCELLANEOUS

Ex-German RHEIN (Type 401) and ISAR (Type 402) class *support ships*

Particulars: See German *Rhein* class

Class (no, former name, transferred):
Cezayirli Gazi Hasan Pasa (i) (A 579, ex-*Ruhr*, 9.11.76), *Sokulla Mehmet Pasa* (A 577, ex-*Isar*, 30.9.82), *Cezayirli Gazi Hasan Pasa* (ii) (A 579, ex-*Elbe*, 1993)

Ex-German FAC and minesweeper support ships transferred to Turkey as officer training vessels. The first A 579 was deleted in 1993 (broken up 1994) and replaced by the ex-*Elbe* in the same year. They can also serve as escorts or minelayers in wartime. *Sokulla Mehmet Pasa* and *Cezayirli Gazi Hasan Pasa* (ii) extant 1995.

Cezayirli Gazi Hasan Pasa 19.7.93 L & L van Ginderen

Tuvalu

Formerly the Ellice Islands, administered as a British colony with the Gilbert Islands (now Kiribati), Tuvalu became independent in 1978. A small patrol boat force is maintained, comprising one Australian-built ASI 315 class patrol boat *Te Mataili* (165t full load, 103ft 4in × 26ft 7in × 6ft 11in, *31.5m × 8.1m × 2.12m*, 21kts, crew 17), in service 8 October 1994, extant 1995, and one small patrol craft *Nivanga* (no details known), in service 1991 and extant 1995.

United Arab Emirates

The United Arab Emirates (formerly the Trucial States) is composed of seven emirates – Abu Dhabi, Ajman, Dubai, Fujairah, Ras al Khaimah, Sharjah and Umm al Qaiwain. Six of these came together as an independent state on 2 December 1971 when they ended their individual special treaty relationships with Britain. Ras al Khaimah joined the other six on 10 February 1972.

Six of the emirates, which have a total area of 32,278 square miles, lie on the shore of the Gulf between the Musandam peninsula in the east and the Qatar peninsula in the west, while Fujairah lies on the Gulf of Oman. The 1992 UN population estimate was 2,173,000.

The armed forces all came under the control of the Supreme Defence Council when the emirates merged. The Ministry of Defence is located in Dubai and general headquarters is in Abu Dhabi. The naval force of Abu Dhabi had been formed in 1968 and the UAE naval force was created on 1 February 1978. Its function is to patrol territorial waters and oil installations in UAE marine areas. The main naval base is at Taweela, between Abu Dhabi and Dubai.

In the early 1980s the total personnel comprised 1200 volunteer officers and ratings (commanded by a brigadier). By 1989 this had increased to 100 officers and 1400 ratings, and in 1994, 150 and 1850.

Main ports are Abu Dhabi, Dubai, Ras al Khaimah and Sharjah.

During the past fifteen years fighting capability of the navy increased significantly. Six Lürssen 'TNC 45' fast attack craft were commissioned in 1980–81, two similar Lürssen craft of the *Mubarraz* class entered service in 1990. The most important acquisitions were two small fast missile-armed helicopter-capable corvettes of the Lürssen 'FPB 62' type commissioned in 1990–91. Two very similar corvettes entered service in the navy of Bahrain in 1987–88. It is noteworthy that most of the attack craft delivered in recent years to the Gulf states were built by Fr Lürssen, Vegesack, Germany.

Current plans are for the acquisition of more missile-armed fast attack craft as well as some hovercraft. Interest has been expressed in the acquisition of new or second-hand frigates of *c*2500t displacement, and it is planned to lease two ex-US *Oliver Hazard Perry* class frigates in 1996. There is also a requirement for at least three minehunters.

Lürssen 'FPB 62' type *missile corvettes*

Displacement: 560t standard; 630t full load
Dimensions: 214ft 3in oa × 30ft 6in × 9ft 6in
65.3m × 9.3m × 2.9m
Machinery: 4 shafts, 4 MTU 16V 538 TB92 diesels, 13,640bhp = 32kts. Range 4000nm/16kts
Armament: 8 MM40 Exocet SSM (2×2), 8 Crotale Navale SAM (1×8), 1–76mm/62 OTO Melara Super Rapido, 1 Goalkeeper 30mm 7-barrelled CIWS, Dagaie chaff RL, 1 Alouette SA 316 helicopter
Sensors: Radar BEAB (Bofors Electronic) Sea Giraffe, Racal Decca 1226, fire control BEAB 9LV331, Thomson-CSF DRBV 51C; ESM/ECM Racal Cutlass/Cygnus
Complement: 43

Muray Jip Jun 1990 L & L van Ginderen

No	Name	Builder	Commissioned	Fate
P 6501	MURAY JIP	Lürssen, Vegesack	Nov 1990	Extant 1995
P 6502	DAS	Lürssen, Vegesack	Jan 1991	Extant 1995

Small fast missile corvettes ordered by the end of 1986 to the modified FPB 62 design. Launched in 1989–90. Similar ships were also built for Bahrain, but fitted with shorter tower mast. These corvettes are specially adapted for conditions of the Gulf, being well armed and air-capable. Their armament includes a SAM battery of Crotale Navale type together with a fully-fledged gunnery CIWS system which is rare for ships of their size. The helicopter can itself be armed with light anti-ship missiles.

MUBARRAZ class *fast attack craft (missile)*

Displacement: 225t standard; 260t full load
Dimensions: 147ft 4in oa × 23ft × 7ft 11in (prop)
44.9m × 7.0m × 2.4m
Machinery: 2 shafts, 2 MTU 20V538 TB93 diesels, 9400bhp = 40kts. Range 500nm/38kts
Armament: 4 Aerospatiale MM40 Exocet SSM (2×2), 1 Matra Sadral (Mistral Sam) (1×6), 1–76mm/62 OTO Melara Super Rapido, Dagaie chaff RL
Sensors: Radar BEAB Sea Giraffe, Racal Decca 1226, fire control BEAB 9LV331; ECM/ESM Racal Cutlass/Cygnus
Complement: 40 (5 officers)

Class (no, in service):
Mubarraz (P 4401, Aug 1990), *Makasib* (P 4402, Aug 1990)

Heavily modified Lürssen 'TNC 45' type fast attack craft ordered late 1986 in Germany (Lürssen) at the same time as the two 'FPB 62' type corvettes. Launched in 1989–90. Fitted with the first export installation of new French Mistral SAM complete with CSEE Najir optronic director. Air/surface search and fire control radars delivered by new Swedish company BEAB now owned by Bofors (Philips-PEAB was sold out to Bofors, hence the new name). SSM placed in a short cut in the long superstructure. According to some reports main engines develop 14,000bhp for designed 40kts speed. Both extant 1995.

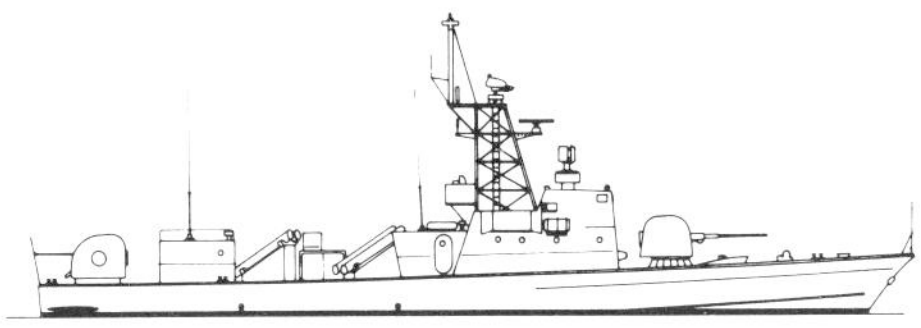

Rodqum as completed
NB 1/750 scale

Lürssen 'TNC 45' type *fast attack craft (missile)*

Displacement: 231t standard; 259t full load
Dimensions: 147ft 4in oa, 138ft 9in pp × 23ft × 8ft 2in
44.9m, 42.3m × 7.0m × 2.5m
Machinery: 4 shafts, 4 MTU 16V538 TB92 diesels, 13,640bhp = 40kts. Range 500nm/38kts, 1500nm/16kts
Armament: 4 MM40 Exocet SSM (2×2), 1–76mm/62 OTO Melara, 2–40mm/70 Breda (1×2), 2–7.6mm MG, 1 CSEE Dagaie chaff RL
Sensors: Radar Racal Decca TM 1226, PEAB 9LV200 Mk 2 (2 antennas: search and tracking); ECM/ESM Racal Cutlass/Cygnus
Complement: 32 (5 officers)

Class (no, in service):
Banyas (P 4501, Nov 1980), *Marban* (P 4502, Nov 1980), *Rodqm* (P 4503, Jul 1981), *Shaheen* (P 4504, Jul 1981), *Saqar* (P 4505, Sep 1981), *Tarif* (P 4506, Sep 1981)

Lürssen 'TNC 45' type fast attack craft ordered in late 1977, launched 1980–81 and shipped in pairs on merchant ships. First regular warships armed with new French MM40 Exocet missiles. For 40mm guns the CSEE Panda director is provided. Carry 350 rounds 76mm, 1800 rounds 40mm, and 6000 rounds MG ammunition. All extant 1995. Similar (modified) boats were in service with Kuwait Navy.

Vosper Thornycroft 110ft type *large patrol craft*

Displacement: 110t standard; 175t full load
Dimensions: 103ft 4in pp, 109ft 11in oa × 21ft × 5ft 3in–6ft 7in
31.5m, 33.5m × 6.4m × 1.6–2.0m
Machinery: 2 shafts, 2 Rushton-Paxman Valenta RP200M diesels, 5400hp = 29kts. Range 1800nm at 14kts
Armament: 2–30mm/75 BMARC Oerlikon A32 (1×2), 1–20mm Oerlikon A41A, 2–51mm RFL
Sensors: Radar Decca RM-916
Complement: 26

Class (no, launched, in service):
Ardhana (P 1101, 7.3–24.6.75), *Zurara* (P 1102, 13.6–14.8.75), *Murban* (P 1103, 16.9.75), *Al Ghulian* (P 1104, 16.9.75), *Radoom* (P 1105, 15.12.75, 1.7.76), *Ghanadhah* (P 1106, 1.3–1.7.76)

Round bilge steel hull construction. All built by Vosper Thornycroft, Portsmouth, for Abu Dhabi. P 1103 and P 1104 sailed out, rest transported by freighter. All extant 1995.

UNITED ARAB EMIRATES

US-built *customs patrol craft*

Displacement: 70t full load
Dimensions: 76ft 9in × 18ft 1in × 4ft 11in
23.4m × 5.5m × 1.5m
Machinery: 2 shafts, 2 General Motors 12V 71T diesels, 1400hp = 25kts. Range 750nm at 25kts
Armament: 2–20mm (2×1)
Complement: ?

Class:
No 21–No 25

Built by Camcraft, New Orleans. In service since September 1975 with the Coast Guard. US P-77A design. New numbers 753–757. All extant 1995.

KAWKAB class (Keith Nelson type) *coastal patrol craft*

Displacement: 25t standard; 32t full load
Dimensions: 52ft pp, 57ft 6in oa × 15ft 6in × 4ft 6in
15.8m, 17.5m × 4.7m × 1.4m
Machinery: 2 shafts, 2 Caterpillar diesels, 750bhp = 19kts. Range 300nm at 10kts. Endurance 7 days
Armament: 2–20mm (2×1)
Sensors: Radar Decca RM 916
Complement: 11

Class (no, in service):
Kawkab (P 561, 7.3.69), *Thoaban* (P 562, 7.3.79), *Baniyas* (P 563, 27.12.69)

Glass fibre hull construction, designed and built by Keith Nelson, Bembridge, a division of Vosper. Originally operated by Abu Dhabi. Some sources give displacement 32t standard and 38t full load and range 445nm at 15kts. Used for coastal patrol, hydrographic survey, and oilfield surveillance. Fresh water evaporators provides 900 litres daily. All extant 1995.

Cheverton type 50ft *customs patrol craft*

Displacement: 20t
Dimensions: 50ft 2in × 14ft 1in × 4ft 7in
15.3m × 4.3m × 1.4m
Machinery: 2 shafts, 2 GM diesels, 850bhp = 23kts. Range 1000nm at 20kts
Armament: 1–7.62mm MG
Complement: 8

Class:
Al Shaheen, Al Aqab

GRP hull. Built by Cheverton Ltd, Isle of Wight. Commissioned February 1975. Originally delivered to the Sharjah Marine Police. Extant 1995.

DHAFEER class *coastal patrol craft*

Displacement: 10t
Dimensions: 40ft 4in × 11ft 3in × 3ft 7in
12.3m × 3.4m × 1.1m
Machinery: 2 shafts, 2 Cummins diesels, 3700bhp = 19kts. Range 350nm at 15kts
Armament: 2–7.62mm MG or 1–12.7mm MG
Complement: 6

Class (no, launched, in service):
Dhafeer (P 401, Feb–1.7.68), *Ghadunfar* (P 402, May–1.7.68), *Hazza* (P 403, May–1.7.68), *Durgham* (P 404, Sep 1968, 7.6.69), *Timsah* (P 405, Sep 1968, 7.6.69), *Murayjib* (P 406, Feb–7.7.70)

Glass fibre hull construction, built by Keith Nelson, Bembridge. Old sources give range 150nm/12kts. Transferred to Ministry of Interior (Marine Police), July 1977. All deleted 1991.

COASTAL PATROL BOATS (COAST GUARD)

Six Italian Baglietto 'GC23' type fast patrol craft, *No 758–763*, delivered in pairs in 1986, 1987, 1988; Coast Guard, aluminium hull and superstructure (50.7t full load, 79ft, 3260bhp = 42kts, 1–20mm Oerlikon, 2–7.62mm MG).

Two Baglietto '13FC' patrol boats delivered in 1988; Coast Guard, 11t, 43ft, 760bhp = 43kts, 2–7.62mm MG). Extant 1995.

Ten Watercraft Mk II patrol boats delivered in 1983; Coast Guard, GRP hull, (25t full load, 45ft oa, 1300bhp = 26kts, 2–7.62mm MG). Extant 1995.

Seventeen Watercraft 'P 1220' patrol boats, first ten delivered in 1979–80 and seven ordered in 1980, GRP hull (10t, 39ft, 800bhp = 29kts, 2–7.62mm MG). All deleted by 1993.

Sixteen Camcraft 65ft patrol boats delivered from USA in 1979–83; Coast Guard (50t full load, 65ft oa, 1630bhp = 25kts or 700bhp in two units, 1–20mm Oerlikon).

One 'Posilipo' patrol boat, delivered in 1984; Coast Guard, GRP hull, (35.5t full load, 65ft, 1630bhp = 24kts, 1–20mm Oerlikon).

Three Baglietto 59ft type patrol craft Nos 501–503 (22t full load, 59ft (*18.1M*) oa, 2 MTU diesels = 40kts, 2–7.62mm MG), delivered 1993. All extant 1995.

LANDING CRAFT

Two Vosper QAF type landing craft, *Jananah* and *Dayyinah*, completed 1988 (350dwt, 177ft oa, 2 MTU diesels, load: four medium tanks), two more ordered 1990. Built by Argos Eng, Singapore.

Landing craft *Al Feyi*, built Singapore 1987 (650t full load, 164ft oa, 11kts).

Landing craft *Ghagha II*, built 1987 (100t full load, 131ft oa, 9kts).

MISCELLANEOUS

One diving tender *D 1051*, built by Crestitalia 1987, used primarily for mine clearance; lengthened version of Italian *Alcide Pedretti* (100t full load, 103ft oa, 26kts).

United Kingdom

Although for the past forty-eight years Great Britain has not been at war with her neighbours, the years 1947–95 were traumatic for the Royal Navy, and as full of change as any in its previous history. The rapid rundown after the end of the Second World War hit the RN harder than any of the other victors: not only was the British economy in a parlous state, but the colossal subsidies of the Lease-Lend programme were suddenly withdrawn, and the urgent need to release manpower from the armed forces meant that large numbers of ships had to be laid up at short notice.

In 1948 a large scale pruning of the Reserve Fleet removed all but five of the surviving battleships, all cruisers earlier than the *Southampton* class and all prewar destroyers. Almost every one of these ships was scrapped, but a handful found their way into other navies.

With more vessels than complements to man them, and a desperate shortage of merchant ships, there was no priority for new construction, but work on battleship design continued until as late as 1948. This was largely because of rumours about Russian capital ships, but there was a tacit admission that battleships were beyond the RN's means and were in any case of limited value. With less validity the same reasoning was applied to cruiser designs, for at this time Western strategic thought was dominated by fears of a nuclear holocaust. Many military writers and theorists believed that warships had no role to play since a future war would be decided by long-range bombers armed with nuclear weapons.

It was recognised by the more perceptive planners that the aircraft carrier could still play a major role, and so permission was given to continue work on two large and four intermediate hulls from the wartime programme. The very big *Malta* class had been cancelled before any work had been done on them, but the hulls chosen, the smaller 36,800t *Africa* class, were a better-balanced design, probably had more 'stretch' and were therefore more cost-effective, and had enclosed hangar space, a necessity for nuclear war.

The modernisation of the six armoured carriers was looked into, but it was found that the restricted hangar height ruled out the newest, the two *Implacable*s and *Indomitable*. Of the other three, *Illustrious* and *Formidable* had suffered heavy war damage, leaving only *Victorious*. It is interesting to note that before the survey, made in 1949, which settled her fate, modernisation plans had been drawn up for *Formidable*: she would have had a fully angled flight deck, some two years before the RAE Bedford Conference at which the angled deck is claimed to have been 'invented'. In fact, it was the intention to modernise all six carriers; *Victorious* was uniquely the prototype.

REFITS AND MODERNISATIONS

The Admiralty was well aware that the fire control and radars of the current ships were obsolescent, and this, coupled with their strenuous war service, poor anti-corrosion protection and general lack of amenities, made major overhauls imperative. There was also the question of replacing out-of-date weaponry, and to correct the situation a series of major refits was put in hand in the early 1950s. The battleships were considered for conversion to missile ships, but, as in the United States, the cost was seen to be prohibitive and out of all proportion to any improvement in fighting value, so work on the battleships was confined to routine maintenance. Finally, the modernisation programme was cut back to one carrier and as many cruisers and destroyers as possible.

The *Southampton* and 'Colony' classes and *Belfast* were all given partial modernisations between 1950 and 1960, but mounting costs and the growing age of the ships caused the programme to be cut back at *Swiftsure*. It has been suggested that *Swiftsure* turned out to have been inadequately surveyed after sustaining severe damage in collision with the destroyer *Diamond* during the big NATO exercise 'Mainbrace' in 1953, and that by the time the extent of the damage became evident in 1960 the ship's modernisation had been almost

The British carrier force of the 1950s: *Eagle*, *Bulwark* and *Albion* 20.10.56 CMP

completed; nothing could be done to remedy the situation and she was towed away to the scrapyard.

Apart from the 'Battle' class and later destroyers, virtually all the wartime 'Emergency' destroyers were unsuitable for modern warfare because of inadequate air defence and fire control. Current needs for fleet escorts were met by the big destroyers, but there were no fast escorts to cope with the new Soviet 'Whiskey' class submarines. The 'Emergency' destroyers were ideal for conversion and so the Type 15 frigate design by N G Holt, was drawn up as a stopgap until the new frigate designs were ready. Despite the end of the Korean War, East–West tension remained high throughout the 1950s, and to make up numbers other simpler and cheaper conversions were also put in hand.

It is not often realised that corrosion was a major problem with war-built ships. Before the war, mill scale was removed by pickling in dilute acid; thin plating, particularly in destroyers, was galvanised. To save time, war-built ships did not receive either treatment, and by the end of the war were usually in a worse condition than older vessels.

NEW CONSTRUCTION

The major surface threat in the 1950s was seen as the Soviet *Sverdlov* class cruisers operating against Western trade. So seriously was this possibility taken that two new designs were drawn up to deal with it, a cruiser and a super-destroyer. The cruiser was intended to provide air-cover with guided missiles and relied on rapid-firing 6in guns as its main armament, while a new rapid-firing single 5in gun was designed for the destroyers.

The cruiser would have displaced 18,000t (standard) and would have carried Seaslug missiles aft and automatic 6in guns forward. The destroyer, on the other hand, relied on three 5in guns for air defence, and when the project was finally dropped it was because it had no missile defence. The gun derived from the British Army's land-mobile 5in, reached an advanced design stage, but was stopped when the Ordnance Board recommended an end to 'gunnery solutions to AA defence' in 1958.

All that remained of this interesting design was the advanced steam machinery which was developed for the fast escort, stopped, and then cut down and married with the gas turbines for the 'County' class, the larger fleet escort for which the Board opted, after much thought. The first two were ordered in 1955, followed by two further units in 1957. The more urgent need of frigates had meant that the first six Type 12 ships, intended as prototypes, had been ordered in 1951, followed by another twelve of slightly modified design in 1954. These frigates were intended to be cheap, but this proved to be a vain hope. They were, however, a great success and are still widely regarded as one of the finest escort designs of the period in any navy. In 1952–55 a dozen cheaper, 'utility' Type 14 ships were laid down to supplement the Type 12s, and a simpler destroyer conversion, the Type 16, was put in hand.

Two other new classes of frigates were ordered in 1951, a standard diesel-engined hull in two variants, one for anti-aircraft defence and the other for aircraft-direction. It had been intended to make up numbers in both categories with more conversions of destroyers, but these plans were shelved when a change of heart by the Staff indicated that it would be possible to go back to building general-purpose warships. This change of policy was influenced by new trends in technology, with more compact machinery and electronics on the way. Despite cutbacks and delays, however, the Korean War building programme had created an impressive escort force. By the end of 1958, twenty-four new frigates had been completed and thirty-four destroyers had been converted; in addition, twenty-one 'C' class destroyers had received varying degrees of modernisation to improve their escort capability, and many of the slower 'Loch' class had been rearmed.

The other area in which the Royal Navy invested heavily was mine warfare. Having suffered considerable casualties from the mining of harbours and estuaries, the British were understandably more sensitive about this than the Americans (who in any case had much deeper coastal waters) and when it was learned during the Korean War that Russian magnetic mines were proving immune to degaussing measures it was realised that a large number of new minesweepers were needed to replace the steel-hulled *Algerine*s. Between 1950 and 1960 some one hundred and twenty coastal minesweepers were built, concurrently with one hundred inshore types. The CMS design proved an outstanding success, robust and sufficiently seaworthy to be used on offshore patrol, but the IMS design proved too small to accommodate the generating power needed for the latest anti-acoustic and anti-magnetic sweeping devices. A further improvement was the introduction of 'minehunting', using a special high-frequency sonar to detect and classify mines on the seabed so that they could be destroyed by a demolition charge. When this equipment had been perfected a number of coastal minesweepers were converted to minehunters and over the years the new technique has largely replaced traditional sweeping as the major method of dealing with mines in coastal waters.

SUBMARINES

By 1945 the Royal Navy had built up a formidable submarine force and there was no intention of frittering away such hard-won expertise. Apart from the possibility of offensive use in wartime, an active force of submarines was ruled to be essential in training anti-submarine forces, and great attention was paid to German developments. After prolonged tests with some Type XXI U-boats and the Type XVIIB *Meteorite*, it was decided to build two experimental boats with Walter hydrogen peroxide turbine propulsion, and to modernise the 'A' class and the welded boats of the 'T' class along the lines of the Type XXI, with enlarged batteries and streamlined hulls.

The two peroxide boats proved expensive, but the lessons learned from them were incorporated in a new design of diesel-electric boats, the *Porpoise* class. This hull form was very successful and was repeated in the improved *Oberon* class, one of the most outstanding 'conventional' designs of the period. What put paid to peroxide propulsion was the successful adaptation of nuclear power, for this gave virtually unlimited thermal power without oxygen from any source. When USS *Nautilus* proved such a triumph in 1955, the Royal Navy immediately decided to switch to nuclear propulsion.

Although the intention to 'go nuclear' was there even before the two peroxide boats were built, work on a prototype reactor, at Dounreay in Scotland, did not start until 1957, and it fell so far behind schedule that the Admiralty decided to buy a reactor from the United States for the first British boat, and to put the Dounreay type into two later classes. Since then a fourth and a fifth class has been completed, making the RN one of only five navies operating nuclear-propelled ships.

For some years it was believed that no more diesel-electric 'submersibles' would be built, but the cost not only of building nuclear boats but also of manning them forced the RN to return to a conventional design, the 2400t SSK (submarine, hunter-killer) announced in 1979. These carried a similar armament to the nuclear boats and entered service in the early 1990s.

In 1963 the momentous decision was taken to transfer responsibility for the British nuclear deterrent from the RAF to the RN, in the form of Polaris underwater-launched ballistic missiles. Five submarines were ordered in 1963–64, but the incoming Labour Government cancelled the fifth boat in February 1965 as a gesture of appeasement to its left wing. The entire project, including training the crews, building four submarines and providing base and support facilities, was carried through with great rapidity, making Polaris one of the few major British defence projects to keep within its financial limits and become operational on time.

In July 1980 it was announced that four new submarines would be built, armed with Trident II missiles. These would have a range of 6000 miles, well outside the range of Soviet anti-submarine forces, and would have eight independently targeted warheads instead of three. The major components of the missile system were to be bought from the United States, but many sub-systems would be provided by British subcontractors and the missiles themselves would be leased.

A NEW GENERATION OF WARSHIPS

Although the 'electronic' component of ships had been growing rapidly since 1940, during the 1950s there were signs that the trend was beginning to slow down to some extent because of the reduction in size of individual items of electronic equipment. This meant, for example, that the previous policy of building single-function escorts could be discarded. The unsophisticated 'Tribal' class showed that it was now possible to build a general-purpose frigate once more, and so in 1960 the decision was made to build the last three *Rothesay* class frigates to a totally recast design. The hull was identical, but it was now possible to include a light helicopter and a long-range surveillance radar, combining a high-level anti-submarine capability with a good measure of air warning and aircraft-direction.

It was now recognised that no small escort could do more than defend herself against air attack (point defence), and a new class of larger air-defence destroyers (DLGs) was planned to provide area defence for the new aircraft carriers that were scheduled. Known as the Type 82, four were planned as escorts for two carriers, armed with a new surface-to-air missile, but propelled by a combined steam and gas turbine plant as in the 'County' class. When the carrier programme was cancelled in 1966 the class was cut to one ship (to be used as a testbed for new weapons and machinery), but she was followed by a new class of smaller destroyers (DDGs) the function of which would be AAW and not general-purpose.

The cancellation of the aircraft carriers in 1966 was a traumatic experience for the Royal Navy, involving the resignation of the First Sea Lord and the loss of its chief striking force. The carrier *Victorious* went quickly (after fire damage), followed later by *Eagle*, and the responsibility for naval air strike was handed over to the RAF. All the other inter-service rivalry flared up, for it was felt by many supporters of naval air power that the RAF had unfairly (and inaccurately) convinced the Secretary of State for Defence that the carriers were vulnerable to strikes from land-based inter-continental ballistic missiles. The Navy kept its helicopters, but was to hand over its Phantom and Buccaneer aircraft to the Air Force, with a contingent of RAF aircrew mixing with the surviving carrier's air group; thereafter shore-based RAF squadrons would take over the mission of protecting ships at sea.

As had been predicted by many, the scheme proved unworkable. The RAF was already overstretched in providing for the air defence of the United Kingdom, and even if the aircraft earmarked for naval support were serviceable, a few minutes' delay in takeoff meant that they arrived on the scene too late. But even after both services realised that the situation could not continue, the politicians clung to their ruling that there could never be any more aircraft carriers. What was certain, however, was that the RN could not afford to build or operate a large fixed-wing aircraft carrier comparable to those in the US Navy. The original decision had been the right one, for the cost of the aircraft alone would have been crippling.

The answer would have to be found elsewhere. In 1967 serious consideration was given to a 'command cruiser' carrying six large ASW helicopters, similar to the French *Jeanne d'Arc*. Further studies showed that greater efficiency would be obtained with nine helicopters, but in an internal hangar; logically this led to a starboard island, lifts and a clear flight deck. But to avoid political disputes with the RAF and the politicians, the euphemism 'through-deck cruiser' was coined, lest it be said that an aircraft carrier was under consideration. A further complication was the existence of the Harrier V/STOL aircraft, a naval version of which had been under development, and it was decided to make sure that the design of the new ship could accommodate the Sea Harrier if required. Thereafter the battle became purely political, and the Sea Harrier was not ordered until May 1975, more than two years after the ship had been laid down. In 1980 the first two squadrons were formed, giving the RN control over its own air power once more, and the debate now revolves around the aircraft to follow the Sea Harrier.

By the mid-1970s the growing strength of the Soviet Navy in nuclear submarines was focusing attention on the so-called GIUK (Greenland–Iceland–UK) Gap. This 'choke point', through which submarines from the Northern Fleet must pass to attack targets in the North Atlantic, had already been 'bugged' with networks of passive acoustic sensors laid on the seabed to track the passage of hostile submarines. Known as SOSUS (Sound Surveillance System), the system needed big, weatherly anti-submarine ships to patrol the area and pounce on targets in areas defined by the SOSUS arrays. While the US Navy built the big *Spruance* class destroyers, the RN opted for 4000t frigates armed with two helicopters, surface missiles and a new point-defence missile capable of defending against other missiles, and fitted with elaborate command, control and communications (C^3I). The Sea Wolf missile had originally been conceived as an anti-aircraft point-defence weapon, but during development it had been tailored to meet the threat from underwater-launched missiles and sea-skimmers, and for some years was the only anti-missile missile system in service.

Despite the return to a balanced navy, time was running out for the RN. The cost of warships was rising beyond control, and in particular the cost of protecting them against anti-ship missiles – for example, computer studies showed that a double-ended Sea Dart destroyer would displace over 8000 tons and cost hundreds of millions of pounds. To make matters worse the overheads of the Royal Dockyards made the cost of half-life refits exorbitant: fitting Sea Wolf and Type 2016 sonar to a *Leander* class frigate cost more than £70m, and even a modest refit for a 'County' class DLG in 1981 cost £64m.

In June 1981 the axe fell, when the new Secretary of State for Defence, John Nott, announced that the surface fleet would be cut back. One of the three support carriers would be laid up or sold, the escort force would be cut to fifty ships, and no more half-life modernisations would be approved. The Type 22 programme would be cut short and replaced by a new utility escort known as Type 23, but to compensate, the nuclear hunter-killer programme would be accelerated and the Type 2400 diesel-electric submarines would proceed. As things stood at the beginning of 1982, the old commando carrier *Bulwark* had been paid off, *Hermes* was to pay off shortly, *Invincible* was 'on offer' to the Australian Navy, and two DLGs and two *Leander* class frigates had been sold. All the older frigates and several of the *Leanders* were also to be sold. Such was the situation when the Falklands crisis intervened.

THE FALKLANDS CAMPAIGN

When Argentine forces occupied the Falkland Islands and South Georgia on 2 April 1982, the British Government immediately dispatched a task force to recapture the islands. Although the assembling and equipping of such a force around the carriers *Hermes* (flagship) and *Invincible* was a masterpiece of speed and efficiency (the carriers and several warships sailed and 5000 tons of military stores were loaded, all within seventy-two hours of the order to embark), the fighting around the Falklands showed that the Royal Navy was no longer adequately equipped for such operations. The fact that the Falklands are 8000 miles from Britain and that the nearest base was Ascension caused particular strain, but air attacks by Argentine shore-based aircraft showed that a small air group of fewer than thirty Sea Harriers was no substitute for a large carrier with high-performance aircraft.

The first blow came on 4 May, when the DDG *Sheffield* was hit by an air-launched Exocet missile from a Super Etendard strike aircraft. The missile failed to detonate, but it set the fuel tanks on fire and the ship had to be abandoned some hours later. She was taken in tow because the navy hoped to have the damage examined by experts at South Georgia, but when the weather deteriorated on 10 May the burned-out hulk was scuttled. The strike against the Task Force was undoubtedly provoked by the loss of the old cruiser *General Belgrano*, which was torpedoed by the nuclear submarine *Conqueror* on 2 May. According to later claims by the British Government, *Belgrano* and her Exocet-armed destroyers were the southernmost element of a three-pronged attack on the Task Force, and the submarine's CO was given permission to attack any hostile targets. Surprisingly the elderly Mk 8 torpedo was used, at a range of about 3000yds, rather than the long range Mk 24.

Once the amphibious assault had got under way at San Carlos on 21 May, fierce air battles became an almost daily occurrence, and bombing or rocket attacks accounted for the DDG *Coventry* and the frigates *Ardent* and *Antelope*. The Navy found that the lack of airborne early warning aircraft made the Sea Harriers' task much harder, and the older radars, 965 and 992Q, had great difficulty in tracking fast, low-flying air targets. The need for stronger close-range defence led to the use of troops and their GPMGs to provide AA fire, and ships still in the UK were hurriedly rearmed with twin 30mm and single 20mm guns, as well as additional chaff-launchers to give protection against guided missiles. The Vulcan Phalanx 20mm 'Gatling' was acquired from the United States and two were put on the flight deck of the new carrier *Illustrious*.

A large number of merchant ships were chartered or requisitioned, including the luxury liners *Canberra* and *Queen Elizabeth 2*, with helicopter platforms, naval-standard communications and, in some cases, 20mm guns, added. The most elaborate conversions were the ro-ro container ships *Atlantic Conveyor* and *Atlantic Causeway*. The former took RAF Harriers to the South Atlantic and then acted as a 'spare deck' for aircraft and helicopters in San Carlos Water, while the latter was given a hangar and four anti-submarine Sea King helicopters. *Atlantic Conveyor*'s loss to an Exocet attack was a major setback to the timetable of Operation 'Corporate' for she carried a portable air-strip and refuelling equipment intended to permit the

RFA landing ships alongside *Fearless* in San Carlos Water, May 1982 CMP

ground-support aircraft to operate away from the carriers. Although the land forces got safely ashore, the Navy's presence remained vital throughout, to move supplies and to provide cover against Argentine air strikes. A rash attempt by 5 Brigade to move without Sea Harrier cover led to the crippling of the LSLs *Sir Galahad* and *Sir Tristram* at Fitzroy, arguably the worst moment of an otherwise brilliant campaign in which numerically inferior forces had repeatedly out-thought and out-fought the opposition.

Two days before the surrender on 24 June, the DLG *Glamorgan* was hit by a land-launched Exocet missile, which wrecked her hangar and killed fifteen men, but by a near-miracle she escaped serious damage. The missile was detected on radar and correctly identified, giving her a vital seventy seconds to turn away and try to get out of range. As the Exocet was nearly out of fuel it did not strike with full force, and it seems that the warhead may have detonated only partially.

POST–1982

Although the Falklands campaign halted the rundown temporarily, the 1981 cuts were merely postponed. The collapse of the Warsaw Pact in the late 1980s provided the Treasury with a strong case for reducing defence expenditure even faster. The 'Options for Change' policy review in 1990–91 reaffirmed the need for amphibious ships but the commitment made in the mid-1980s to maintain 'about fifty' escorts soon shrank to 'about forty', and older frigates were retired or sold off rapidly. A major casualty was the planned new nuclear attack submarine design, but the Trident programme remained sacrosanct.

The Navy confronted the crisis by offering a major cut in personnel to secure the Type 23 frigate programme and the planned replacements for the Type 42 destroyers. In 1989 the UK was the first to withdraw from the NFR–90 collaborative frigate programme, effectively wrecking it. In late 1991 a new goal was announced, an Anglo-French-Italian air defence frigate, to fill the gap.

Two defence reviews in the early 1990s, 'Options for Change' and 'Front Line First' imposed severe financial cuts on the RN. Although a reduction in numbers was inevitable, the Navy evolved a credible alternative to the cold war/northwest European strategy which had been in place for 20 years. Profiting by experience in the Falklands and in the Gulf in 1990–91, the Navy offered a limited out-of-area capability to deal with sudden crises, and to support United Nations' peace-keeping operations. The result is renewed emphasis on amphibious warfare, and a reduction of the anti-submarine role in NATO's Eastern Atlantic theatre.

MINE WARFARE

The Soviet Navy's interest and expertise in mine warfare showed no sign of decreasing in the 1970s, and the replacements designed for the RN's ageing coastal minesweepers proved to be, ton for ton, among the most complex and expensive warships ever built. To reduce their magnetic signature virtually to nil, it was necessary to design non-magnetic diesels and build the hull of glass-reinforced plastic (GRP). GRP is expensive and difficult to fabricate on such a large scale, and the *Brecon* class are currently the world's largest ships built of this material. They are known as Mine Counter Measures Vessels (MCMVs) because they can accommodate both hunting and sweeping functions in one hull.

Since 1982 MCM forces have been in action four times, including the clearance of Port Stanley, hunting for Libyan mines in the Gulf of Suez, dealing with Iranian mines in 1987, and finally, supporting the Coalition forces in the 1991 Gulf conflict. Five 'Hunt' class performed outstanding service in dealing with Iraqi mines laid off Kuwait City.

As with larger warships, the government commitment to maintain a force of fifty modern MCM craft has not been honoured. The thirteen 'Hunts' had been supplemented by twelve 'River' class MSFs and five *Sandown* class SRMHs, but by 1995 the force of 'Tons' had disappeared and there was no further role for the 'Rivers'.

COMMONWEALTH NAVIES

The disappearance of the Empire weakened the links which bound the navies of Canada and Australia to that of Great Britain, and there was a steady drift away from a reliance on British equipment. The RCN was the first, purchasing USN weapons and electronics to make it easier to get spares in an emergency, and going as far as to design its own escort destroyers. The RAN, on the other hand, continued to build RN designs in its own shipyards for some years.

An attempt to follow the Canadian example in designing a special class of missile destroyer to meet Australian specifications came to nothing, and since 1962 standard US Navy or other designs have been bought. Co-operation with the RN continues, however, and the Ikara anti-submarine missile system and other equipment have been jointly developed. While South Africa remained in the Commonwealth the SAN relied totally on British *matériel*, but after 1961 political pressure to prevent them from buying military equipment forced the South Africans to turn elsewhere.

The 'New Commonwealth' navies, principally India and Pakistan, fitted the traditional mould for a longer period because they lacked indigenous resources, but Pakistan turned increasingly to China for help and India to the Soviet Union. India also pursued a sensible policy of building up local resources to a point at which British designs could be expanded and even equipped with Soviet weaponry.

WEAPONRY

The Royal Navy continued its policy of demanding high standards in its weapons and equipment, but the underlying economic weakness which had become particularly evident after 1948 largely frustrated this. Lack of money to buy equipment led to endless delays, some imposed by the navy's bureaucratic machinery and others by the Treasury. In the end the result was always the same: equipment came into service far too late, making it obsolescent almost from the day it became operational. Other problems were caused by the temptation to allow too many committees to make alterations to specifications. Thus the Seaslug missile was made extraordinarily bulky on the advice of the Royal Aircraft Establishment, Farnborough, which did not believe reports that the American Terrier missile had its booster motor mounted in tandem. For political reasons it is almost impossible to purchase foreign equipment, but when it is bought the immediate temptation has been to 'Anglicise' it; this happened with the American Mk 56 fire control system, which after many years of painful development re-emerged as the MRS3.

The beam-riding Seaslug gave way eventually to the much superior semi-active homer Sea Dart, which followed American ideas more closely, and this weapon is currently being upgraded. The experience with lighter missiles has been happier, for the short-range Seacat was designed to replace the 40mm Bofors gun, and by being tailored specifically to one role it emerged as a relatively cheap and cost effective close-range defence system for ships down to frigates. Its replacement was to have been Sea Wolf, but the lengthy and costly development enabled the designers to give it a capability against sea-skimming missiles. The system was further developed into a vertical launch (VL) system for the Type 23 frigates and AORs. A lighter Type 911 tracker was produced, and the missile's range has been increased from 10,000m to 14,000m.

The problem for the RN is that it has shrunk in size without losing its front-line commitments, and as a result its production runs are short. This pushes up cost and explains why so many good ideas have had to depend on export orders before they can go ahead. Originally it was hoped to persuade the Netherlands to buy Sea Dart in exchange for a Signaal 3-D radar, and the Type 22 frigate was also begun as a joint venture with the Dutch.

Since 1947 British torpedoes have earned a poor reputation, most of it undeserved. Although the Mk 8 submarine torpedo was by far the most efficient and cost-effective torpedo used in the Second World War, it was hoped to replace it with faster and more accurate weapons, using British wartime developments as well as ideas found in Germany in 1945. The most promising development was High-Test Peroxide (HTP), and this was used to drive the high-speed 'Fancy' or 21in Mk 12. This weapon had to be withdrawn after a propellant explosion sank the submarine *Sidon* in 1955, but was subsequently redeveloped by the Royal Swedish Navy into the Tp61, a very safe and effective torpedo.

Several torpedoes were designed and then axed for financial reasons, notably the Pentane and the Mk 30 Mod 1. The desperate need of a lightweight homing torpedo for aircraft, helicopters and frigates led to the purchase of American Mk 43, Mk 44 and Mk 46 torpedoes, but the Mk 43 became obsolete many years ago and the other two have been replaced by the Stingray. Despite many problems, the Mk 24 Tigerfish came into service in the 1970s and is being followed in the 1990s by the new heavyweight Spearfish. The cancellation of the Undersea Guided Weapon (USGW) forced the navy to buy the USN's Sub-Harpoon to give its nuclear submarines a long-range anti-ship weapon, and this started trials in October 1981.

The purchase of Sub-Harpoon was followed by the selection of the latest surface version, RGM–84C, as the successor to Exocet. The Netherlands 30mm Goalkeeper anti-missile 'Gatling' gun (CIWS) was bought in exchange for the sale of Marine Spey gas turbines to the RNIN. The experience of 1982 led to increased expenditure on electronic warfare, and in 1991 the RN's performance in the Gulf conflict proved that the money had been well spent.

BRITISH NAVAL GUNS 1947–92*

Type	Remarks
20mm Oerlikon Mk 7A single	Obsolete in 1945, but brought back as close-range surface weapon in 1964 and 1982
20mm Oerlikon twin	Obsolescent in 1945, but used in CMSs
30mm single	Naval Rarden, on trials 1980
30mm Mk 1	Defence Systems (ex-Laurence Scott) mounting with Oerlikon/BMARC KDB 30mm gun; replacing all 40mm from 1987 onwards
40mm/60 Mk 3 single	Obsolete in 1945, but brought back as armament of OPVs 1977
40mm/60 Mk 4 twin	Hazemeyer triaxially stabilised; obsolescent in 1945 and replaced by STAAG Mk 2 *c*1950
40mm/60 STAAG Mk 2 twin	Radar-controlled automatic AA mounting, replaced Mk 4
40mm/60Mk 5 twin	Utility mounting which proved more reliable than the STAAG; still in service 1980
40mm/60 Mk 6 6-barrelled	Heavy-duty multiple with auto-feed for capital ships and carriers; obsolete by mid-1960s
40mm/60 Mk 7 single	Power-mounted but hand loaded; replaced by Mk 9 post-1950
40mm/60 Mk 8 single	Battery-driven version of Mk 7; not operational, re-engineered Mk 7
40mm/60 Mk 9 single	Replacement for Mk 7; still in service
40mm/70 Mks10, 11, 12	Single, twin and sextuple mountings with L/70 Bofors and new ammunition; order for 350 barrels cancelled *c*1957
6pdr (57mm) 7 cwt Mk 2	Coastal Forces power-mounting; obsolescent in 1945, but not withdrawn until 1950s
3in/50 Mk 34	USN mounting provided under MDAP for 'Loch' class modernisation; reallocated to new fleet minesweepers 1954 but order cancelled
3in/50 Mk 22 twin	USN mounting provided with shields and modified Mk 63 radar fire control; *Victorious* only. Originally intended for Type 62 frigate conversions
3in/70 Mk 6 twin	British production based on USN experimental mounting; intended for frigates, but used only in *Tiger* class cruisers and by RCN
3.3in CF Mk 2	Unsuccessful adaptation of Centurion tank's 20pdr, intended for *Brave* class
4in/45 Mk 5 HA single	Obsolete destroyer gun on high-angle mounting; obsolete in 1945, but still in service 1950s
4in/40 Mk 12 single	Submarine gun; in service to 1950s
4in/40 Mk 20 single	Submarine gun; in service to 1950s
4in/45 Mk 16 twin	Obsolescent, but continued in service to 1960s; some fitted with US Mk 63 radar on shield
4in/45 single	Vickers automatic type, developed from Army gun for East Coast Gunboat *c*1950; cancelled, but sold to Chile
4.5in/45 Mk 5 single	Mk 4 gun, replacement for 4.7in Mk 9; adapted to RPC and in service until 1980s
4.5in/45 Mk 6 twin	Standard frigate and destroyer mounting of 1950s and 1960s, continued in place of 3in Mk 6 when that gun did not appear on time. Withdrawn with last of *Leanders*
4.5in/25 8 cwt CF Mk 1	Adaptation of 95mm demolition howitzer toreplace 57mm (6pdr) in Coastal Forces; obsolete by mid-1950s
4.5in/55 Mk 8	New surface bombardment gun for DDGs, DLGs and frigates, with modest performance but higher reliability; introduced in mid-1960s and still in service
5in/56 single Mk 2	Naval version of Army's 'Green Mace' intended as new DP gun for destroyers in early 1950s. Cancelled *c*1957
5in/56 twin Mk 1	Twin version of same gun, intended for cruisers; design stage only
5.25in/50 QF Mk 1 twin	DP mounting in capital ships and AA cruisers; obsolescent by 1950s, but retained until 1968
6in/50 QF Mk 23 triple	Standard cruiser turret-mounting; obsolete by 1950s
6in/50 QF Mk 25 triple	Designed for *Neptune* class, but work ceased after 1946 in favour of Mk 26
6in/50 Mk 26 twin	Rapid-fire twin 80 degrees elevation, water-cooled; *Tiger* class
8in/50 Mk 8 twin	Obsolescent in 1945, but remained in service in old heavy cruisers until 1954
14in/45 Mk 7 quad and twin	*King George V* class
15in/42 Mk 1 (N) twin	Original 1912 model with elevation increased to 30 degrees, improved shell and supercharges; used in *Vanguard* with RPC
16in/45 Mk 1 triple	Obsolescent by 1945; *Nelson* only
16in/45 Mk 4 triple	Projected for 'Improved *Lion*' class but never got beyond design stage

*Roman numerals replaced by Arabic Mark numbers about 1948; new form used throughout for clarity

BRITISH TORPEDOES 1947–92

Torpedo	Weight (lb)	Explosive (lb)	Range (yds)/ at (kts)	Remarks
21in Mk 8	3353	750	5000/44 or 7000/40	Introduced 1927, but in service in submarines until 1980s
21in Mk 9	3731	750	10,000/40 or 14,000/35	Surface version of Mk 8; in service until 1960s
21in Mk 11	?	750	?	Copy of German G7e; replaced 1950s by Mk 12, etc
21in Mk 12	?	750	?	Code-named 'Fancy SR'. HTP version of Mk 8; withdrawn as unsafe 1959
21in Mk 20 (E)	1800	200	12,000/20	'Bidder', ASW single-speed for frigates; withdrawn 1950s
21in Mk 20(S)	1800	200	12,000/20 8000/12	Two-speed submarine homer variant of Bidder; still in service in improved form
21in Mk21	?	300	12,000/30	'Pentane'; dropped during development
21in Mk 22	1800	200	12,000/20	Cable-set version of Mk 20; cancelled
21in Mk 23	2000	?	12,000/78	Wire-guided submarine type; replaced Mk 20 in 1970
21in Mk 24	3240	750	12,000/35 or 23,000/24	Tigerfish heavyweight for submarines, also sold to Brazil and Turkey. Mod 0 superseded by Mod 1 (1981) with anti-surface capability and Mod 2 (1986) with under-ice capability
21in Spearfish	4077	661	*c*25,000/55	Successor to Mk 24 Mod 2, ordered late 1981 (100 pre-production models), but main production order (estimated 300) not placed until end of 1994. Designed to run at 3000ft at full speed, using HAP-Otto thermal fuel and Sundstrand turbine
18in Mk 30	630	?	2500/19 or 8000/12	Ex-'Dealer B', air-dropped ASW homer withdrawn 1971; Mod 1 cancelled 1955
12.75in Mk 31	?	?	?	Cancelled 1971
10in Mk 43	?	50	?	Mod 3, US air-dropped weapon; bought as interim replacement for Mk 30
12.75in UK Mk 44	512	75	5500/30	Electric ASW homer built under US licence for ASW aircraft and ships; brought into service late 1960s
12.75in Mk 46	510	90	12,000/40	Otto-fuel version of Mk 44; bought 1971 to replace Mk 30 Mod 0 and Mk 31
12.75in Stingray	288	77	12,000/45	Replacement for Mk 44/46; entered service 1980 and currently undergoing major upgrade

UNITED KINGDOM

BRITISH RADARS AND EW SYSTEMS 1947–92

Type	Band	Remarks
167	10cm	Submarine. Seaguard 'cheese', also fitted to periscope
242	IFF A–Band	
253	IFF A–Band	
262	3cm	Tracker for STAAG, CRBF, etc, with dish antenna; obsolescent by 1982
267	P/X–Band	Surface search for submarines; also in 267MW and 267PW versions
274	10cm	Surface warning for capital ships and cruisers; obsolete by 1964
275	10cm	Tracker fitted to Mks 37, 6 and 6M fire control directors
277P	10cm	Height-finder introduced 1943; replaced mid-1950s by Type 277Q/278
277Q	10cm	Replacement for Type 277 in major surface ships
278	10cm	Solid-state replacement for 277, using same antenna ('County' class DLGs)
79/279	metric	Obsolete air warning set, but remained in active ships until 1948
79B/281	metric	Still in use until 1951
291	10cm	Air warning, dipole antenna for submarines
293P/Q	10cm	Surface/low-angle air search and target-indication
667		'Cooky' spot noise jammer introduced in early 1960s
668/669		Jammer elements of UA-8/9 EW system, possibly improved versions of Type 667
670		Racal's 'Heather', fitted in Sea Wolf *Leanders* and Type 42 DDGs and Type 22 frigates
675		Thorn-EMI replacement for 670 and the cancelled Millpost system. The latest version is 675(2), marketed commercially as Guardian
695		ECM (jammer) element of the new system intended for the Type 23 frigates
801	?	Projected surveillance radar; cancelled mid-1950s
901	5–7.5cm	Seaslug tracker
903/904	1.5–3.75cm	Tracker for MRS3, with dish antenna
909	1.5–3.75cm	Sea Dart SAM tracker; cassegrain antenna
909I		Updated version of 909 tracker, with new dish antenna; retrofits started in 1989
911		Fire control radar for lightweight version of Sea Wolf SAM system, marketed commercially as Marconi 805SW. In later Type 22 frigates
912	3–3.75cm	RN designation for Italian RTN-10X tracker
931	1.5cm	Auxiliary surface gunnery set to improve Type 274; not operational
932	K–Band	Canadian splash-spotting set
940	G–Band	Late Second World War IFF
960	Metric	Long-range air warning, introduced *c*1950 replaced by 1965
961/ 962X/963	?	Early forms of Carrier Controlled Approach (CCA) radars
965	1.5/200MHz	Anglicised version of SPS-6B, in service 1961 onwards; in single 'bedstead' AKE-1 or double AKE-2 aerial. Being replaced 1022 (966 is designation sometimes used for 'cleaned up' version)
967/968	7.5–30cm	Sea Wolf surveillance, mounted back-to-back; in service 1978
973	?	Periscope-mounted ranging set for submarines
974	3cm	Decca 12 navigation set
975	1.5–3.75cm	Surface search set for small ships; Type 975ZW used for position-fixing in MCMVs
977	?	Shore-based radar, to detect aerial minelaying
978	3cm	Navigation radar for frigates
980/981 982/983	7.5–15cm	Formerly combined as WCB-1 for carriers, and used in *Eagle*; subsequently separated into height-finder and 'Hayrake' for carrier air warning
984	?	Comprehensive Display System (CDS), 3-D using massive 'searchlight' antenna; obsolete since 1966
985	?	Experimental fixed-array 3-D surveillance set under development *c*1959, but later dropped
986/987	?	Projected successors to 982/983
988	?	Projected 'Broomstick' Anglo-Dutch 3-D set intended for *Bristol* class DLGs and CVA-01
992/992Q and R	10cm	Surface/low-level air search and target-indication, replacing 293 in large ships
993	10cm	Surface/low-level air search and target-indication, replacing 293 in small ships
994	10cm	Solid-state version of 993, using Plessey AWS-4 processing equipment
996	25cm?	3-D S-band surveillance and target-indicating radar (STIR) which replaces Type 992 series in Type 42 DDGs and the carriers. Marketed by Siemens-Plessey ASW-9, and came into service in Type 23 frigates in 1990
1000/1001	?	Submarine search set in 1950s and1960s; replaced by 167
1002	X–band	Submarine navigation set in *Porpoise* class
1006	I–Band	Navigation set, replacing 974; also exists in submarine version
1007	X–Band	Replacement for Type 1006 navigation set
1008	E/F–Band	New Racal navigation set
1010		Alternative designation for Cossor Mk 10 IFF
1011		IFF transponder
1013		Lightweight non-magnetic IFF transponder for MCM craft
1022	15–30cm	Interim Surveillance and Target-Indication Radar (STIR) using Signaal LW-08 processing with antenna for 1030; replacement for 965; at sea 1979
1030	25cm	STIR using back-to-back antennas; at sea 1980, but cancelled 1981
AN/SPG-34		USN fire-control set used with 4in AA guns in *Fiji* class cruisers and RCN 'Tribals'
AN/SPS-6		USN air warning set fitted in *Fleetwood* and *Warrior*
AN/SPN-35		USN Carrier Controlled Approach (CCA) radar in carriers *Ark Royal* and *Melbourne*
UA-1		Post-1945 direction-finder working in S-Band
UA-4		Submarine version of UA-1
UA-8/9/10		Passive intercept system code-named 'Porker', in service from early 1970s
UA-11/12		ESM systems for SSBNs
UA-13		Passive intercept system mounted in new *Leander* class frigates
UA-15		Additional element in ESM systems to cover 12–18GHz band
UAA(1)/ UAA(2)		MEL 'Abbey Hill' ESM system fitted in the late 1970s; UAA(2) provides a higher degree of automated threat-identification
UAB		Racal ESM system in SSBNs and SSNs; being replaced by UAP
UAC		ESM set for later SSNs and *Upholder* class SSKs; UAC(3) will be fitted in the *Vanguard* class SSBNs
UAF(1)		Racal ESM suite for the early Type 23 frigates; superseded by UAT
UAG(1)		Integrated ESM/passive countermeasures system for AORs and other auxiliaries
UAH		ESM set for modernised *Oberon* class SSKs (marketed commercially by MEL/Thorn EMI as Manta)
UAJ		Racal ESM suite, the equivalent of the commercial Porpoise set
UAL		ESM set in *Churchill* class SSNs
UAP		ESM system for submarines, marketed commercially as Sabre; UAP(1) is for new SSNs and SSKs, UAP(2) is an update of UAC
UAR		ESM system for RFAs
UAS(1)		Surveillance receiver
UAT		MEL (later Thorn EMI and now Racal) ESM system selected to replace UAF(1) in the Type 23 frigates
Matilda		MEL threat-warning receiver added to MCMVs for Gulf duty in 1987

BRITISH SONARS* 1947–92

Type	Remarks
129	10kc 'chinstrap' submarine search set
138	15kc 'dustbin' submarine search set
144	Standard escort set; introduced in 1943, obsolete by 1950s
144Q	Depth-finding set, used to provide settings for Squid; replaced by 147
147	Attack set, linked direct to 144; replaced 144Q
162	Bottom search set; still in use 1995
165	Original experimental towed array; became 2038
166	Proposed for 1954 fleet minesweepers but dropped
170	Attack set used to control Squid and Limbo; replaced 147 in 1950s
174	Medium-range search set used in frigates from 1950; replaced by 177
177	Replacement for 174; obsolescent

Type	Remarks
182	Torpedo decoy
184	Medium-range search set, introduced in *Leander* class
184M	Solid-state version of 184
185	Underwater telephone
186	Non-steerable VLF array, fixed around submarine ballast tanks
187	Submarine attack set; bow array, 2.5/10kHz
193	Minehunting set; introduced in 1960s
193M	Solid-state version of 193
195	Helicopter 'dunking' set
196	Escape underwater telephone
197	Submarine passive ranging set, for detection of sonar
199	Variable-depth set; RN designation for Canadian SQS-505
719	H/F fast turning array for submarines
2001	Active/passive DIMUS (digital multi-beam steered) bow set for SSNs, developed by AUWE with USN assistance in late 1950s and early 1960s. Upgraded in late 1980s; succeeded by 2020
2007	Long-range passive search set (upgraded Type 186) using steerable beams, working in 1–3kHz; fitted in *Oberon* class SSKs and later in *Trafalgar* class SSNs
2016	Hull-mounted active/passive search set for surface ships; developed from prototype tested in *Matapan* in 1979–83; replaced Type 184 and has since been succeeded by 2050
2020	Bow mounted active/passive 360deg set working in 2–16kHz (passive) and 3–13kHz (active), for SSNs. To be replaced by 2076
2023	US BQR-15 towed array for *Resolution* class SSBNs, with Plessey analyser
2024	Copy of BQR-15 with British processor; replaced by 2026
2026	Passive towed array developed by ARE Portland ans GEC Sonics and Plessey Marine for SSNs
2030/2035	RN designation for two proposed variants of BQR-23; not bought from US
2031	Passive towed array for surface ships; 2031I prototype developed by GEC Sonics and installed in four *Leander* class frigates; 2031Z used improved 'Curtis architecture' and developed by ARE Portland and Dowty Maritime and Waverley Electronics; installed in Type 22 and Type 23 frigates
2032	Processor designed to enhance 2020 in the *Swiftsure* and *Trafalgar* class SSNs
2034	?
2040	Triple-tier version of French Eledone/Argonaute and Netherlands Octopus submarine attack set; installed in *Upholder* SSKs; was to be replaced by 2075
2043	Processor enhancement to 2020
2046	Passive towed array for submarines; replaces 2026 in SSNs and fitted in *Upholder* class SSKs; twenty-seven ordered
2050	Bow- and hull-mounted long-range active/passive search set for carriers, destroyers and frigates; retains 2016 transducer with new modular processing and new displays; thirty-four ordered
2051	Narrow-band active/passive search set for submarines; part of Triton update for RN *Oberon* class SSKs and also supplied to Canada for SOUP update
2052	Passive broad-band towed array retrofitted to early SSNs
2054	Passive/active search set for *Vanguard* class SSBNs; optimised for evasive tactics; first set delivered 1989, operational from 1992 onwards
2057	New passive towed array intended for submarines and surface ships; under development in early 1990s
2059	Osprey Electronics' 'pinger' for tracking DSRV rescue vehicles
2074	Passive/active search set, update of 2020 in *Swiftsure* and *Trafalgar* classes in mid-1990s; ordered 1991
2075	Ferranti-Thomson passive/active bow set intended as replacement for 2040 in second batch of *Upholder* class SSKs; ordered 1989 but cancelled 1991
2076	Integrated sonar system for Batch 2 *Trafalgar*s and for retrofitting to older SSNs
2077	HF sonar project under development in 1990
2078	Integrated passive/active search set for SSN-20 or its successor in mid-1990s
2087	Combined LF active/passive towed array for Type 23 frigates; under development from 1995
2093	Variable-depth minehunting set; in *Sandown* class SRMH
AN/LIQC	USN-pattern underwater telephone, replaced by 185

*Known as Asdics until 1948, when USN term became standard

CODENAMES AND ACRONYMS

AIO	Action Information Organisation
Ben	Early codename for 'Longshot'; later renamed 'Popsy'
Bidder	1945 torpedo project; 'Bidder–A' was an 18in homer which became 'Dealer–B' and ultimately the Mk 30
CAAIS	Computer-Aided Action Information System, developed by Ferranti for frigates in 1970s
CACS	Computer-Aided Command system, developed by Ferranti for later *Broadsword* class frigates. CACS-1 trials in *Boxer* and *Beaver* 1984–85; CACS-2/3 intended for modernised Type 42 DDGs and Types 43/44 (cancelled 1981); CACS-4 intended for Type 23 frigates and AORs (cancelled 1987); CACS-5 installed in *Cornwall* class frigates
CDS	Comprehensive Display System for Type 984 radar
CF299	Project-number for Sea Dart GWS30 missile system
Chevaline	1970s programme update warheads for British A–3 Polaris missiles
CLOS	Command to Line of Sight
CRBF	Close Range Blind Fire, designation for radar-controlled light fire control system developed in the 1950s
DACR	Multiple 6-barrelled 40mm Bofors gun-mounting projected in early 1950s; possibly the L/70 Mk 12
DCA	AIO system for submarines
DCB	Submarine fire control system for Mk 24 torpedoes
DCC	Fire control system for Type 2400 submarines
DCG	Processor added to DCB
DCH	Replaced TCSS-10/11 in upgraded *Oberon* class (Ferranti COMKAFS-D)
DNA	RN designation of BAe-SEMA SSCS command system in Type 23 frigates
Dealer-B	Early name for 18in Mk 30 torpedo, under development 1945–59
Flyplane	Computer processing equipment associated with Mk 6M fire control system
GMS1	Guided Missile System No 1; original designation of Seaslug GWS1 missile system
Green Light	Project name for Seacat GWS20 system
Grog	Formerly 'Mackle', became the 21in Mk 23 torpedo in 1971
GWS20/21 22/23/24	Variants of Seacat short-range missile system
GWS25	Sea Wolf point-defence missile system
GWS30	Sea Dart area-defence missile
GWS31	Sea Dart Mk 2 (cancelled 1981)
GWS50	MM-38 Exocet surface-to-surface missile (1974)
GWS60	RGM-84C Harpoon surface-to-surface missile (1985)
Hot Shot	Later codename for 'Long Shot', ex-'Popsy', ex-'Ben'
Limbo	Project name for Mk 10 anti-submarine mortar
Long Shot	Early 1950s project to adapt air-to-air missile (probably Fireflash) for surface-to-air defence; ex-'Popsy', ex-'Ben'
Lop/Gap	1944 beam-riding missile, later redesignated RTV and then Project 502, leading to Seaslug
LRS1	Long-range System No 1, a 1944 projected development of Type 901 radar for guns up to 5.25in; dropped 1949
Mackle	Original name for Mk 20 torpedo, started in early 1950s; renamed 'Grog' in 1956
MRS1	Medium-Range System No 1, a 7000yd AA fire control system based on Type 262 radar, first in STAAG mountings and then in CRBF systems
MRS3	Medium-Range System No 3, adaptation of US Navy's Mk 56 director to replace MRS1
MRS8	Medium-Range System No 8, a CRBF director with predictor units replaced by a computer, for controlling 4in and 40mm guns
NSR	Naval Staff Requirement, the specification which precedes construction of a warship for the RN
NST	Naval Staff Target, the outline specification which leads to a Staff Requirement
Ongar	Original codename for the Mk 24 Mk 1 Tigerfish torpedo
Orange/Yellow Crop	Helicopter-borne electronic warfare systems
Popsy	Ex-'Ben', became 'Long Shot'
Project 502	Early designation for Seaslug missile system
STD	Simple Tachymetric Director, optical sight for 40mm guns
STIR	Surveillance and Target Indication Radar, developed into Type 1022 and Type 1030 radars
STWS	Ship's Torpedo Weapon System, British development of US Navy's Mk 32 torpedo-launcher
TCSS–3	Torpedo Control System for Submarines; fitted in *Porpoise* class and older submarines
TCSS–5	Hybrid TCSS–3 system adapted to allow RCN *Oberon* class to fire Mk 37 torpedoes
TCSS–7	First umbilical-link set FCS in RN; fitted in *Oberon* class; replaced by TCSS–9
TCSS–9	Action information-handling system in *Oberon* class
TCSS–10	Update of TCSS–9 to allow *Oberon* class to handle Mk 24 Mod 1 torpedoes
Tom	Tachymetric One-Man Director
Type B	Improved 'Squid' anti-submarine mortar

Type D	Triple trainable anti-submarine mortar, later Mk 10 'Limbo'

FLEET STRENGTH 1947

BATTLESHIPS

Name	Launched	Disp (deep load)	Fate
Nelson class			
NELSON	3.9.25	44,054t	BU 1948
King George V class			
KING GEORGE V	21.2.39	44,460t	BU 1958
DUKE OF YORK	28.2.40	44,790t	BU 1958
ANSON	24.2.40	45,360t	BU 1958
HOWE	9.4.40	44,510t	BU 1958
Vanguard class			
VANGUARD	30.11.44	51,420t	BU 1960

Nelson class

Because *Nelson* had been refitted in the United States at the end of 1944 she was in sufficiently good shape to serve in the postwar fleet, unlike her sister *Rodney*, which was in such poor repair by 1945 that she was laid up immediately. Apart from removal of the 20mm Oerlikons *Nelson* remained unaltered from the last days of the war.

According to Winston Churchill's memoirs, a major modernisation was discussed to enable *Nelson* to serve for several years in the postwar fleet, but no other details have survived. In any case she was too slow and there was no front-line role for battleships any more. After a brief spell as flagship of the Home Fleet at the end of 1945, she joined the Training Squadron at Portland in August 1946 as Flagship, RA Training Battleships for a year. She was sold in February 1948 and used as a bombing target in the Firth of Forth before scrapping commenced in March 1949.

King George V class

The four survivors were immediately stripped of their smaller light AA guns, but otherwise remained unaltered until scrapped. *King George V* and *Howe* had 64–2pdr pompoms (8×8), but *Duke of York* and *Anson* had an extra 4-barrelled pompom on each side of the bridge and on the forecastle deck.

King George V served as flagship of the Home Fleet in 1946 and then as a private ship until paid off into reserve in 1950. After a spell with the Training Squadron at Portland, *Anson* paid off in 1949, while *Howe* served as a harbour TS at Portsmouth from 1950 to 1951, after a year as Flagship, RA Training Battleships (1947). *Duke of York* served with the Home Fleet as flagship until 1949 and remained in commission as flagship of the Reserve Fleet until 1951. All four were eventually laid up in the Gareloch and schemes were discussed in the mid-1950s for converting them to missile ships, but the cost would have been prohibitive and in 1957 they were sold for scrapping.

Vanguard

Only two minor changes were made to the ship during her career: the removal of the 11–40mm single Bofors guns and the temporary replacement of the twin STAAG Bofors mounting on 'B' turret by an observation and saluting platform for the Royal Tour to South Africa early in 1947. She refitted at Devonport in 1947–48 and went to the Mediterranean in January 1949 for six months. She became a training ship at Portland, acting as temporary Home Fleet flagship during exercises. After a refit at Devenport in 1954 she paid off into reserve at Portsmouth, becoming the flagship of the Reserve Fleet and NATO HQ ship. She was sold for scrapping in 1960.

MONITORS

Name	Launched	Disp (deep load)	Fate
ROBERTS	1.2.41	9500t	BU 1965
ABERCROMBIE	31.3.42	9900t	BU 1954

Roberts was initially used as a turret drill ship at Devonport and later as an accommodation ship for the Reserve Fleet. Although removed from the Navy List in 1955 she lasted another ten years until sold for scrapping in 1965. Neither she nor *Abercrombie* underwent any further change apart from items stripped over the years. Although nominally a sister to *Roberts, Abercrombie* differed in several respects. In July 1946 she replaced the old *Erebus* as turret drill ship at Chatham, and subsequently served as an accommodation ship for the Nore Reserve Fleet. She was towed to Portsmouth in 1953 and sold for scrapping the following year.

AIRCRAFT CARRIERS

Name	Launched	Disp (deep load)	Fate
Illustrious class			
ILLUSTRIOUS (R 87)	5.4.39	31,630t	BU 1956
VICTORIOUS (R 38)	14.9.39	31,630t	Reconstructed (see notes)
FORMIDABLE (R 67)	17.8.39	31,630t	BU 1953
Modified Illustrious class			
INDOMITABLE (R 92)	26.3.40	29,730t	BU 1955
Implacable class			
IMPLACABLE (R 86)	10.12.42	32,110t	BU 1955
INDEFATIGABLE (R 10)	8.12.42	32,800t	BU 1956
Unicorn			
UNICORN (R 72)	20.11.41	20,300t	BU 1959
Colossus class			
GLORY (R 62)	27.11.43	18,040t	BU 1961
OCEAN (R 68)	8.7.44	18,040t	BU 1962
VENERABLE (R 81)	30.12.43	18,040t	To Netherlands 1948
VENGEANCE (R 71)	23.2.44	18,040t	To Brazil 1956
THESEUS (R 64)	6.7.44	18,300t	BU 1962
TRIUMPH (R 16)	2.10.44	18,300t	Repair ship 1964, BU 1982
WARRIOR (ex-*Brave*) (R 31)	20.5.44	18,300t	To Argentina 1958
Perseus class			
PERSEUS (ex-*Edgar*) (R 51)	26.3.44	16,475t	BU 1958
PIONEER (ex-*Mars*) (R 76)	20.5.44	16,475t	BU 1954
Majestic class			
MAGNIFICENT (R 36)	16.11.44	19,550t	To Canada 1948
TERRIBLE (R 93)	30.9.44	19,550t	To Australia 1948
POWERFUL (R 95)	27.2.45	20,000t	To Canada 1952
MAJESTIC (R 77)	28.2.45	20,000t	To Australia 1949
HERCULES (R 49)	22.9.45	19,500t	To India 1957
LEVIATHAN (R 97)	7.6.45	17,780t	BU incomplete 1968

Illustrious class

All three ships had seen strenuous war duties and only *Victorious* was still serviceable. *Formidable* had suffered severe damage off Okinawa and was laid up on her return from trooping in 1947, while *Illustrious* had been in hand for repairs and a major overhaul since June 1945. When she recommissioned in mid-1946 she had a more powerful catapult and 30 per cent more aviation fuel, and the forward and after ends of her flight deck had been remodelled to

Indomitable 1953 CMP

extend its area. She then served for eight years as the trials and training carrier for the Home Fleet, occasionally embarking front-line squadrons for exercises. During this time she launched and landed the first jet aircraft, a modified Vampire. The armament of *Illustrious* was altered in 1946, the single 20mm guns being replaced by 21–40mm single Bofors, although the six 8-barrelled pompoms were retained.

In 1949 plans were drawn up for a full modernisation of *Formidable*, including an angled deck, but a survey revealed that her hull had suffered extensive strain and in 1950 she was towed to the Motherbank and lay there derelict until sold in 1953.

Victorious returned to the United Kingdom from Australia in October 1945, but two months later, after a short refit, she returned to the Far East to begin repatriation of British troops. When this was over in January 1947 she was paid off into reserve because of the rapid demobilisation of 'hostilities only' ratings, but recommissioned in October the same year. Until March 1950 she served as the Home Fleet training carrier, based at Portland, and was then paid off for modernisation. For details of this see under Major Surface Ships.

Modified Illustrious class
Indomitable returned to the United Kingdom in November 1945, having landed her aircraft to make room for repatriated personnel, and made two more trooping runs in 1946. From 1947 to 1950 she was refitted to the same standard as *Illustrious* and joined the Home Fleet for a three-year commission. During this period her close-range armament consisted of 40–2pdr pompoms (5×8), 24–40mm Bofors (24×1) and 4–3pdr saluting guns. Aircraft embarked included Fireflies, Sea Furies, Sea Hornets, Firebrands and the first ASR helicopters. In October 1953 she was reduced to reserve and laid up in the Clyde until October 1955 before being sold for scrapping.

Implacable class
Implacable served with the British Pacific Fleet (BPF) after the war's end and did not leave for the UK until April 1946. On her arrival in June she immediately recommissioned in the Home Fleet for deck-landing trials. Late in 1947 her air group was allocated, but teething troubles meant that she did not finally embark her thirteen Sea Hornets and twelve Firebrands until March 1949. In the autumn of that year she embarked four Sea Vampires of the Jet Fighter Evaluation Unit and the following summer she operated twelve Barracudas in the ASW role, but only three months later, in September 1950, she was paid off into reserve. After service from 1952 to 1954 as a training carrier she was sold for scrapping in 1955, having given only eight years of active service.

Indefatigable repatriated Allied ex-prisoners from Japan and then re-embarked her air group for a cruise to New Zealand. She left Sydney at the end of January 1946 for the UK, but made a trip to the Far East to repatriate troops before paying off into reserve. From 1950 she served as a training carrier, with the hangars converted to classrooms and messdecks, and with *Implacable* paid off for the last time in August 1954, being sold for scrapping in 1956. Although completed late in the war, neither she nor her sister were suitable for operating the new generation of large aircraft because of their lack of hangar height. As with *Victorious*, they could have been stripped to the hangar deck and reconstructed, but cost and time stifled the programme. As with earlier carriers the close-range AA armament was reduced in 1946, and as training carriers all 8-barrelled pompoms were removed as well. *Indefatigable* had a conspicuous box structure built at the after end of the flight deck and had the large D/F top removed from her foremast.

Unicorn
On her return from the Far East in January 1946, *Unicorn* was immediately put into reserve, but she recommissioned in the summer of 1949 to carry aircraft and spares for *Triumph* in the Far East. After the outbreak of war in Korea in June 1950 her commission was extended, and she did not return to Devonport until November 1953. Shortly before her departure she bombarded land targets in North Korea with her 4in guns, a rare event for a carrier. Despite her short service life of only seven years she did not recommission again and was sold for scrapping in 1959. In June 1953 she had been re-rated as a ferry carrier.

Colossus class
Apart from the name-ship, which had been transferred to France as *Arromanches* in 1946, *Venerable* was sold to the Royal Netherlands Navy on 1 April 1948, replacing the escort carrier *Nairana* and taking the same name, *Karel Doorman*; *Warrior* had been lent to the Royal Canadian Navy on completion in 1946, but she was returned in 1948 when *Magnificent* was ready, and was recommissioned in the RN; and in 1952 *Vengeance* was lent to the Royal Australian Navy, pending completion of *Melbourne*.

Glory, Ocean, Theseus and *Triumph* served with great distinction in the Korean War in 1950–53, proving that they were among the most successful utility carrier designs ever built, but they were rapidly made obsolescent by the new generation of jet aircraft coming into service at the end of the 1950s. *Glory, Ocean* and *Theseus* were all scrapped 1961–62, *Vengeance* having already been sold to Brazil in 1956 and renamed *Minas Gerais*.

On her return from the RCN in 1948, *Warrior* was used for deck-landing trials and in 1948–49 she was fitted with a flexible landing deck to allow jet fighters fitted with a skid undercarriage to make 'soft' landings. This flexible deck consisted of a rubber sheet supported on air bags and extended from abaft the island to within 150ft of the stern. Abaft this 'carpet' the flight deck was built up to the same level with a light steel ramp rising to 2ft 6in high, with a single arrester wire. In 1952–53 the ship was modernised with a new enlarged bridge and a lattice foremast, and two years after that she was again refitted with a partial angle deck and stronger arrester gear to allow her to continue trials with the latest aircraft. Her ten arrester wires could stop a 15,500lb aircraft landing at 60kts and the single catapult could launch a 20,000lb aircraft at the same speed (she had originally been designed to take 14,000lb aircraft). She acted as HQ ship for the nuclear test programme at Christmas Island in the Pacific from February to October 1957 and on her return was offered for sale to Argentina. The deal was concluded in July 1958 and she sailed for Argentina three months later, to be renamed *Independencia* at Puerto Belgrano. In RN service she had been armed with four twin 40mm Bofors and twenty singles, but these were reduced and when she entered Argentine service she had only eight single 40mm. With all the modifications her full load tonnage had risen to 19,600 and speed had dropped to 23kts.

After her return from Korea in 1953, *Triumph* was employed as a training ship for officer cadets, but under the 1956–57 Estimates funds were allocated to convert her to a heavy repair ship. This involved filling the hangar with machinery and workshops to enable her to carry out major maintenance work for escorts and helicopters on foreign stations, without the assistance of a naval dockyard. It was not a priority job, and with numerous delays the work finally cost £10.2m and lasted seven years. She was laid up at Chatham in a high state of preservation, ready to provide emergency repair facilities if needed, but was stripped for disposal in 1981, and scrapped the following year.

Perseus class
Laid down and launched as light fleet carriers of the *Colossus* class, these two ships were completed as aircraft maintenance ships in the light of Pacific Fleet requirements. This involved the provision of large-scale workshops and repair facilities for other carriers' aircraft. Unlike *Unicorn*, however, they could not operate aircraft as they had two heavy cranes and deckhouses on the flight deck. Armament consisted of 16–2pdr pompoms (4×4) and 2–20mm Oerlikons (2×1), but the latter were replaced by 40mm Bofors in 1946–47. Standard displacement was 26,265t and dimensions 633ft pp, 694ft 6in oa × 80ft 4in wl × 23ft max.

In 1950–51 *Perseus* was modified with a steel structure at the forward end of the flight deck carrying the prototype Mitchell steam catapult. These trials proved that the heaviest aircraft could now be launched, even when the carrier was not under way, by using steam from the ship's own boilers. Both ships were re-rated as Ferry Carriers in June 1953 together with *Unicorn*, but *Pioneer* was sold for scrapping the following year. *Perseus* remained in reserve for another four years before she too was scrapped.

Majestic classs
Work on all six of this class was stopped after the end of hostilities, but when the Australian and Canadian navies expressed interest in acquiring two, work restarted on *Terrible* and *Magnificent*. These were handed over in 1948 without modification of the original design, and the following year work also started on *Majestic* for the Australians, but to a greatly modified design. Then in 1952 the Canadians bought *Powerful* and she too underwent major reconstruction. Finally in 1957 the Indian Navy bought *Hercules*, leaving only *Leviathan* which was progressively cannibalised to provide spares for her sisters (for example, her boilers were removed and sold to the Netherlands to refit *Karel Doorman*).

For further details of transferred *Colossus* and *Majestic* class carriers, see under the recipient navy.

Campania
Sole survivor of three largely similar fast refrigerated cargo ships converted to escort carriers in 1942–44, she was the first British CVE to have an Action Information Organisation (AIO) and Type 277 height-finding radar. After a period in reserve she recommissioned as a ferry carrier and special transport, but was lent out for the Festival of Britain in 1951 and refitted by Cammell Laird. As a special exhibition ship she visited ports in the UK and Europe, her 515ft × 71ft hangar filled with exhibits and displays. She was returned to the Admiralty and after another short period in reserve recommissioned as a transport to take scientists and observers out to the Monte Bello Islands for the first British nuclear tests in 1952. On her return in December 1952 she was laid up and sold for scrapping in 1955.

As a ferry carrier she could carry ninety aircraft. Her original armament was replaced by two twin Mk 5 Bofors mountings forward and two single Bofors aft, sponsoned to port and starboard.

CRUISERS

Name	Launched	Disp (deep load)	Fate
Kent class			
CUMBERLAND (C 57)	16.3.26	13,540t	BU 1959
London class			
LONDON (C 69)	14.9.27	14,580t	BU 1950
DEVONSHIRE (C 39)	22.10.27	14,300t	BU 1954
Norfolk class			
NORFOLK (C 78)	12.12.28	14,600t	BU 1950
Leander class			
ACHILLES (C 70)	1.9.32	9740t	To India 1948
Arethusa class			
AURORA (C 12)	20.8.36	7400t	To China 1948
Southampton class			
BIRMINGHAM (C 19)	1.9.36	12,190t	BU 1960
GLASGOW (C 21)	20.6.36	12,190t	BU 1958

Cumberland as trials ship 1958 CMP

Name	Launched	Disp (deep load)	Fate
NEWCASTLE (C 76)	23.1.36	12,190t	BU 1959
SHEFFIELD (C 24)	23.7.36	12,190t	BU 1967
Gloucester class			
LIVERPOOL (C 11)	24.3.37	12,330t	BU 1958
Edinburgh class			
BELFAST (C 35)	17.3.38	14,900t	Preserved 1971
Dido class			
ARGONAUT (C 61)	6.9.41	7515t	BU 1955
CLEOPATRA (C 33)	27.3.40	7515t	BU 1958
DIDO (C 37)	18.7.39	7515t	BU 1958
EURYALUS (C 42)	6.6.39	7515t	BU 1959
PHOEBE (C 43)	25.3.39	7515t	BU 1956
SIRIUS (C 82)	18.9.40	7515t	BU 1956
Fiji class			
BERMUDA (C 52)	11.9.41	11,090t	BU 1965
CEYLON (C 30)	30.7.42	11,090t	To Peru 1959
GAMBIA (C 48)	30.11.40	11,090t	BU 1968
JAMAICA (C 44)	16.11.40	11,090t	BU 1960
KENYA (C 14)	18.8.39	11,090t	BU 1962
MAURITIUS (C 80)	19.7.39	11,090t	BU 1965
NEWFOUNDLAND (C 59)	19.12.41	11,125t	To Peru 1959
NIGERIA (C 60)	18.7.39	11,040t	To India 1957
Bellona class			
DIADEM (C 84)	26.8.42	7560t	To Pakistan 1956
ROYALIST (C 89)	30.5.42	7360t	To New Zealand 1954
Swiftsure class			
SWIFTSURE (C 08)	4.2.43	11,240t	BU 1962
Superb class			
SUPERB (C 25)	31.8.43	11,546t	BU 1960

Kent class
Cumberland was converted to a trials ship in 1949–51, with lattice masts, new directors and added accommodation. The original armament was entirely removed, but individual light weapons were added from time to time for trials. At first she carried one, then two single 4.5in Mk 4 guns and two STAAG 40mm Bofors mountings on the starboard side with a 'Battle' class Mk 37 director on the starboard bridge wing, but these were later removed. From 1953 she carried first the twin 3in/70 Mk 6 aft with an MRS3 director on the bridge and then the 6in Mk 26 twin automatic turret on 'B' barbette. At the end of the 1958 trials programme, which tested a new stabiliser design and the first pre-welling NR propellers, she was finally reduced to reserve and was sold the following year.

London class
London served in the Far East in 1947–49 and was damaged by Chinese shore batteries while trying to rescue the sloop *Amethyst* in 1949. The 20mm guns had been removed in 1945, but little else was done to alter her. *Devonshire* was rebuilt as a cadet TS in 1946–47, with only 'A' turret and the secondary battery retained. This was altered to two twin 4in AA mountings (port and starboard amidships), a quadruple 2pdr pompom, a single 40mm Bofors, three 20mm and four saluting guns. The original machinery remained unchanged, but only four boilers were used, giving a speed of 21kts max. In April 1947 she recommissioned as a sea-going training cruiser for officer cadets and finally paid off at the end of 1953.

Norfolk class
Norfolk became flagship of the East Indies Squadron postwar, but returned home to pay off in 1949.

Leander class
Achilles was bought by the newly independent Indian Navy and renamed *Delhi*. She was handed over on 5 July 1948, her sisters *Ajax* and *Leander* having been offered as well, but rejected on grounds of cost and lack of manpower.

Arethusa class
Aurora was transferred to China in 1948 and was renamed *Chungking*, but a year later she defected to the Communists. Renamed *Huang Ho* and finally (allegedly) *Pei Ching*, she was reported to have been sunk by bombing in Taku by Nationalist aircraft, but was then salved and recommissioned by late 1951. By the mid-1950s she was reported to be laid up as a hulk, and it is most unlikely that she ever again became operational.

Southampton class
In 1950–52 *Birmingham* and *Newcastle* were modernised, with new bridgework, a lattice foremast and air conditioning to suit them for service in the Far East. To control the 4in AA they were given two Mk6 high-angle directors at the corners of the former aircraft hangar. *Sheffield* was similarly modernised in 1956–57, but *Glasgow* remained very much as she had finished the war. All four ships were actively employed in the postwar fleet, but were considered to be too old to justify any major modernisation.

Gloucester class
Liverpool served for a few years in the Mediterranean, but never received any modernisation.

Edinburgh class
Belfast recommissioned for the Korean War and was then scheduled to receive full modernisation in 1955–56. This would have involved upright funnels, a new large bridge and a new secondary armament of Mk 11 40mm/77 Bofors guns, but it had to be shelved because of a lack of manpower at Portsmouth Dockyard. However, the ship was given a partial modernisation in 1956–59 during which she received a new *Tiger*-style bridge. Her secondary armament was unaltered, but the close-range AA was changed to six twin 40mm Mk 5 Bofors mountings, with directors (MRS8, externally similar to CRBF) controlling them in four groups. Although the original raked funnels remained, two upright lattice masts replaced the tripods, giving her an ungainly profile.

In 1963 the forward pair of 40mm mountings was removed, but very few other changes had been made by the time the ship was paid off. In that year she became flagship of the Reserve Fleet at Portsmouth and in 1971 was acquired by a charitable trust for conversion to a museum. She now lies in the River Thames below London Bridge.

Dido class
Although modern, these ships were regarded as too cramped and insufficiently stable to receive new equipment, and therefore the majority were used for a few years and then scrapped. *Dido* and *Sirius* retained the fifth 5.25in turret in 'Q' position, and *Phoebe* and *Cleopatra* had US-pattern quadruple 40mm Bofors mountings in 'Q' position and in the waist. *Phoebe* was the first to be scrapped, partly as a result of damage suffered in a collision in the Mediterranean and partly due to her very poor condition.

Fiji class
The main complaints against these ships were that their accommodation was very cramped (hence there was no space for additional equipment) and that their side armour was thin, but in spite of this several served on foreign stations as flagships. *Nigeria* was the only one to retain 'X' triple 6in turret after the war, but it was removed when she was bought by India in 1954. The survivors were all variously modified, some with covered bridges and others with new CRBF directors for their 4in guns, but only *Ceylon* and *Newfoundland* received a modernisation on the lines of that given to the *Southampton*s and to *Mysore* (the renamed *Nigeria*). *Newfoundland* was modernised at Devonport in 1955–56 with two lattice masts and Mk 6 high-angle directors, but *Ceylon* had only a lattice foremast and CRBF directors. Both were sold to Peru, *Newfoundland* becoming *Almirante Grau* on 30 December 1959 and her sister became *Coronel Bolognesi* on 9 February 1960.

Bellona class

In 1954 *Royalist* was taken in hand for modernisation, and when at the end of 1955 *Bellona* returned to the UK her crew transferred to *Royalist* to take her back to New Zealand. The only other member of the class to be modernised was *Diadem*, which had a fairly extensive refit (new bridge, lattice masts) at Portsmouth in 1956–57 before transferring to Pakistan as *Babur*.

Swiftsure

Shortly after the war the Admiralty gave serious consideration to the problem of what to do with their latest cruisers. *Swiftsure, Ceylon* and *Newfoundland*, it was hoped, could have their triple 6in Mk 23 turrets replaced by two twin automatic 6in Mk 26 mountings, and their 4in Mk 19 mountings replaced by three twin 3in Mk 6. Lack of money and the slow production of new radars and gun mountings made the scheme a poor starter, but at the end of 1953 it was still hoped to begin *Swiftsure*'s modernisation in January 1956. Her refit started in February 1957 at Chatham, but in August 1959, with her new bridge structure and lattice masts in position, all work was halted; it was costing far too much to produce an obsolete ship.

Superb

Superb was under consideration for a similar modernisation to *Swiftsure*, bringing her up to *Tiger* standard. Work was to have begun in April 1957, but this was never done, leaving her erstwhile sisters to be completed to a radically different design (see Major Surface Ships).

CRUISER-MINELAYERS

Name	Launched	Disp (deep load)	Fate
Abdiel class			
MANXMAN (N 70)	5.9.40	4000t	BU 1971
Apollo class			
APOLLO (N 01)	5.4.43	4000t	BU 1962
ARIADNE (N 65)	16.2.43	4000t	BU 1964

Abdiel class

Manxman served in the postwar fleet, but principally as a flagship in the late 1940s and early 1950s. In 1955 she had the after twin 4in replaced by an extended deckhouse. In 1962–63 she was converted at Chatham to a parent ship for coastal minesweepers. To enable her to maintain and repair up to sixteen minesweepers the capacious mine deck was converted to storerooms and offices. Half the boilers were removed, allowing the forward boiler room to be converted to accommodate extra generators and evaporators. The forward funnel was retained to carry ventilator trunking and the diesel exhausts. Her armament was reduced to 6–40mm Bofors, one twin Mk 5 on the centreline abaft the mainmast and four single Mk 7s sided amidships. On her return from her last commission in the Far East in 1968 she was used for a short while for sea training of engineer officers.

Apollo class

Like *Manxman, Apollo* was used for several years as a flagship, but in 1954 the after deckhouse was raised by one deck and so she retained the after twin 4in gun. *Ariadne* was laid up at Sheerness in 1946 and never recommissioned.

DESTROYERS

Name	Fate	Name	Fate
Napier class: launched 1940–41, 2540–2555t deep load			
NAPIER (D 297)	BU 1956	NOBLE (ex-*Piorun*, ex-*Nerissa*) (D 165)	BU 1955
NEPAL (D 125)	BU 1956	NORMAN (D 149)	BU 1958
NIZAM (D 38)	BU 1955		
Milne class: launched 1940–41, 2810–2840t deep load			
MARNE (D 135)	To Turkey 1958	MILNE (D 58)	To Turkey 1959
MATCHLESS (D 252)	To Turkey 1959	MUSKETEER (D 186)	BU 1955
METEOR (D 273)	To Turkey 1959		
Onslow and Pakenham classes: launched 1941–42, 2365–2430t deep load			
OBDURATE (D 139)	BU 1964	OPPORTUNE (D 180)	BU 1955
OBEDIENT (D 248)	BU 1962	ORWELL	Type 16 (see notes)
OFFA (D 29)	To Pakistan 1959	PALADIN	Type 16
ONSLAUGHT (D 04)	To Pakistan 1951	PETARD	Type 16
ONSLOW (D 17)	To Pakistan 1949		
Quillam and Rotherham classes: launched 1941–42, 2425–2480t deep load			
RAIDER (D 15)	To India 1949	RELENTLESS	Type 15
REDOUBT (D 141)	To India 1949	ROCKET	Type 15
ROTHERHAM (D 209)	To India 1949	ROEBUCK	Type 15
RAPID	Type 15 (see notes)		
Saumarez, Troubridge, Grenville, Valentine and Kempenfelt classes: launched 1942–43, 2505–2545t deep load			
SAUMAREZ (D 12)	BU 1950	URCHIN	Type 15
SAVAGE (D 27)	BU 1962	URSA	Type 15
TROUBRIDGE	Type 15	VENUS	Type 15
TEAZER	Type 16	VERULAM	Type 15
TENACIOUS	Type 16	VIGILANT	Type 15
TERMAGANT	Type 16	VIRAGO	Type 15
TERPSICHORE	Type 16	VOLAGE	Type 15
TUMULT	Type 16	KEMPENFELT (ex-*Valentine*) (D 103)	To Yugoslavia 1956
TUSCAN	Type 16	WAGER (D 298)	To Yugoslavia 1956
TYRIAN	Type 16	WAKEFUL (ex-*Zebra*)	Type 15
GRENVILLE	Type 15	WESSEX (ex-*Zenith*)	To South Africa 1950
ULSTER	Type 15	WHELP (D 237)	To South Africa 1953
ULYSSES	Type 15	WHIRLWIND	Type 15
UNDAUNTED	Type 15	WIZARD	Type 15
UNDINE	Type 15	WRANGLER	Type 15
URANIA	Type 15		
Myngs and Cavendish classes: launched 1943–44, 2510–2575t deep load			
MYNGS (D 06)	To Egypt 1955	CAVENDISH (ex-*Sibyl*)	Modernised (see notes)
ZAMBESI (D 66)	BU 1959	CAESAR (ex-*Ranger*)	Modernised
ZEALOUS (D 39)	To Israel 1955	CAMBRIAN (ex-*Spitfire*)	Modernised
ZEBRA (ex-*Wakeful*) (D 81)	BU 1959	CAPRICE (ex-*Swallow*)	Modernised
ZENITH (ex-*Wessex*) (D 95)	To Egypt 1955	CARRON (ex-*Strenuous*)	Modernised
ZEPHYR (D 19)	BU 1958	CARYSFORT (ex-*Pique*)	Modernised
ZEST	Type 15	CASSANDRA (ex-*Tourmaline*)	Modernised
ZODIAC (D 54)	To Israel 1955	CAVALIER (ex-*Pellew*)	Modernised
Childers, Cossack and Cromwell classes: launched 1944–45, 2510–2535t deep load			
CHAPLET (D 52)	BU 1965	COMET (D 26)	BU 1962
CHARITY (D 29)	To Pakistan 1958	COMUS (D 20)	BU 1958
CHEQUERS (ex-*Champion*) (D 61)	BU 1966	CONCORD (ex-*Corso*) (D 03)	BU 1962
CHEVIOT (D 90)	BU 1962	CONSORT (D 76)	BU 1961
CHEVRON (D 51)	BU 1969	CONSTANCE (D 71)	BU 1956
CHIEFTAIN (D 36)	BU 1961	CONTEST (D 48)	BU 1960
CHILDERS (D 91)	BU 1963	COSSACK (D 57)	BU 1961
CHIVALROUS (D 21)	To Pakistan 1954	CREOLE (D 82)	To Pakistan 1958
COCKADE (D 34)	BU 1964	CRISPIN (ex-*Craccher*) (D 168)	To Pakistan 1958
'Battle' class (1st group): launched 1943–45, 3290–3300t deep load			
ARMADA (D 14)	BU 1965	LAGOS (D 44)	BU 1967
BARFLEUR (D 80)	BU 1966	ST JAMES (D 65)	BU 1961
CADIZ (D 79)	To Pakistan 1957	ST KITTS (D 18)	BU 1962
CAMPERDOWN (D 32)	BU 1970	SAINTES (D 84)	BU 1972
FINISTERRE (D 55)	BU 1967	SLUYS (D 60)	To Iran 1966
GABBARD (D 49)	To Pakistan 1957	SOLEBAY (D 70)	BU 1967
GRAVELINES (D 24)	BU 1961	TRAFALGAR (D 77)	BU 1970
HOGUE (D 74)	BU 1962	VIGO (D 231)	BU 1964
'Battle' class (2nd group): launched 1945–46, 3400–3420t deep load			
AGINCOURT	Radar picket (see notes)	CORUNNA	Radar picket
AISNE	Radar picket	DUNKIRK (D 09)	BU 1965
ALAMEIN (D 17)	BU 1964	JUTLAND (ex-*Malplaquet*) (D 62)	BU 1965
BARROSA	Radar picket	MATAPAN (D 43)	BU 1979
'Weapon' class: launched 1945–46, 2825t deep load			
BATTLEAXE	Radar picket (see notes)	CROSSBOW	Radar picket
BROADSWORD	Radar picket	SCORPION (ex-*Tomahawk*, ex-*Centaur*)	Radar picket

Napier class
All except *Nepal* had a lattice foremast by 1945 and were laid up in 1945–46. *Nepal* carried out sweeping trials from 1949 and was fitted with twin davits and a big sweep winch on the quarterdeck. All five vessels were considered for Type 16 or 18 frigate conversions (see below), but they were not worth the cost as they were more than half way through their hull-lives.

Milne class
All were laid up in 1946, but in 1951 *Musketeer* had a short refit to become an accommodation ship for coastal forces at Harwich. The remaining four were modernised and sold to Turkey; *Marne, Matchless, Meteor* and *Milne* became *Maresal Fevzi Cakmak, Kilic Ali Pasa, Piyale Pasa* and *Alp Arslan*. They had been under consideration for conversion to Type 62 aircraft direction frigates.

Onslow and Pakenham classes
Obedient and *Obdurate* retained their minelaying capability postwar. *Orwell, Paladin* and *Petard* were converted to Type 16 frigates in 1951–56 (see under Major Surface Ships). *Offa, Onslow* and *Onslaught* became the Pakistani *Tariq, Tippu Sultan* and *Tughril.*

Rotherham class
Apart from three sold to India, the remainder of the class were converted to frigates in 1949–53 (see under Major Surface Ships). *Raider, Rotherham* and *Redoubt* became the Indian *Rana, Rajput* and *Ranjit.*

Saumarez, Troubridge, Grenville, Valentine and Kempenfelt classes
Saumarez and *Volage* were mined by Albanian pirates in the Corfu channel in October 1946; the former was so badly damaged that she was laid up and scrapped four years later, but the latter had her bow rebuilt and rejoined the Mediterranean Fleet. There was a proposal to weld the wrecks together as HMS *Sausage*! Postwar, *Savage* was used for trials of propellers and shafts; these trials and those conducted later with *Teazer*, together with the associated R&D, gave the RN a 10-year lead in the design of quiet propellers. *Savage* was laid up in 1952 and although considered for conversion to a frigate she was put on the disposal list in 1958 and sold in 1962. *Wessex* and *Whelp* were sold to South Africa in 1950–53 and *Kempenfelt* and *Wager* to Yugoslavia in 1956; the remainder were converted to frigates between 1950 and 1956 (see under Major Surface Ships).

Myngs and Cavendish classes
Both classes were laid up shortly after the war, but from 1957 the *Cavendish* class received a major modernisation (see under Major Surface Ships). In 1955 four of the *Myngs* class were sold to Egypt and Israel; one of the survivors became a frigate and the others were scrapped. *Myngs* and *Zenith* became the Egyptian *El Fateh* and *El Qaher*, while *Zealous* and *Zodiac* became the Israeli *Elath* and *Yaffo*.

Childers, Cossack and Cromwell classes
Creole and *Crispin* had a prominent W/T cabin in place of 'B' gun, but the majority were refitted for ASW with 'Y' gun replaced by two Squid mortars and a handling room at the after end of the deckhouse. Light AA was modernised in all of them, a typical fit being a twin 40mm Mk 5 amidships, two single Mk 7s in the bridge wings and two more abaft the funnel. They retained the Mk 6 fire control and did not receive the same degree of modernisation as the *Cavendish* group. In 1954 *Charity* and *Chivalrous* were lent to Pakistan, followed by *Creole* and *Crispin* four years later. The rest of the class saw considerable service in home waters and on foreign stations in the late 1940s and 1950s. *Charity, Chivalrous, Creole* and *Crispin* became the Pakistani *Shah Jehan, Taimur, Alamgir* and *Jahangir*.

'Battle' class (1st group)
All ships fitted with the 4in starshell gun in 'Q' position had it replaced by two single 40mm AA guns, and the 40mm Mk 4 Hazemeyer twins were replaced by two STAAG Mk 2 mountings aft. In 1947 *Saintes* had her 4.5in Mk 4 BD mounting replaced temporarily by the Mk 6 twin, the prototype for the *Daring* class. In all ships the single 40mm on the quarterdeck was replaced by a Squid mortar on the starboard side. In the early 1960s the STAAGs were replaced by twin Mk 5s in *Camperdown* and *Saintes*. All the class were actively employed in the post-war fleet. *Finisterre* (and later *Vigo*) was used as a gunnery firing and training ship and had a square deckhouse added on the platform between the torpedo-tubes. In 1956 *Cadiz* and *Gabbard* were sold to Pakistan, becoming *Khaibar* and *Badr*, and in 1966 *Sluys* was sold to Iran, being renamed *Artemiz*. In 1959 *Hogue* was badly damaged in collision with the Indian cruiser *Mysore* off Ceylon. She was laid up disarmed at Singapore until 1961. The remainder were scrapped without receiving any further modernisation.

Seraph 1948 CMP

'Battle' class (2nd group)
Four were selected for conversion to radar pickets in 1960–62 (see under Major Surface Ships) and *Alamein, Dunkirk* and *Jutland* were scrapped. *Matapan* was laid up immediately after completing trials in September 1947 and did not commission until 1971, when she was taken in hand for a two-year conversion to a sonar trials ship.

'Weapon' class
All four ships served with the Fleet from 1948. In the mid-1950s *Scorpion* had her Squid mortars replaced by a single Limbo mortar of greater range and wider training arcs. She also had an experimental shield fitted to 'A' gun, but this was subsequently removed. In 1958–59 all four were converted to radar pickets (see Major Surface Ships).

Ex-GERMAN DESTROYERS
The *Leberecht Maass* class ships *H 97* (ex-*Richard Beitzen*) and *R 38* (ex-*Hans Lody*), and the *Z 23* class *Z 30* were all evaluated and cannibalised for technical information. *Z 30* was in very poor condition and was sent to Rosyth for structural experiments in February 1946; she arrived at Troon for scrapping in September 1948. *R 38* was renumbered *H 40* to release the pennant number for HMS *Crystal* in February 1946. She was used as an accommodation ship at Woolston from October 1946 until January 1949. *H 97* had to be beached after her bottom plating corroded, and she was put on the disposal list in January 1947. *Z 38*, later *Nonsuch*, was used for trials on the Clyde in 1945, then laid up at Portchester until renamed early in 1947. She recommissioned after repairs in September 1948 and was used for trials until November 1949.

SUBMARINES

Name	Fate	Name	Fate
'T' class (2nd group): launched 1942–44, 1327t/1575t			
TRESPASSER (S 12)	BU 1961	TAPIR (S 35)	To Netherlands 1948, streamlined (see notes)
TACTICIAN (S 14)	BU 1963	TAURUS (S 39)	To Netherlands 1948
'T' class (3rd group): launched 1942, 1327t/1575t			
TEMPLAR	BU 1959	TALLY HO (S 87)	BU 1967
TRUCULENT	Sunk in collision 12.1.50		
'T' class (4th group): launched 1942–43, 1319t/1571t			
THULE (S 25)	BU 1962	TOKEN	Streamlined
TUDOR (S 126)	BU 1963	TRADEWIND (S 29)	BU 1955
TIRELESS	Streamlined	TRENCHANT (S 31)	BU 1963
'T' class (5th group): launched 1943, 1319t/1571t			
TELEMACHUS (S 21)	BU 1961	THOROUGH (S 24)	BU 1961
'T' class (6th group): launched 1944–45, 1319t/1571t			
TIPTOE	Converted (see notes)	TALENT (ex-*Tasman*)	Streamlined
TRUMP	Converted	TEREDO	Streamlined
TACITURN	Converted	TABARD	Converted
'T' class(8th group): launched 1943–45, 1319t/1571t			
TOTEM	Converted	TURPIN	Converted
TRUNCHEON	Converted	THERMOPYLAE	Converted
'S' class (3rd group): launched 1941–45, 814–842t/990t			
SATYR (S 84)	To France 1952	SEA SCOUT (S 153)	BU 1965
SERAPH (S 89)	BU 1965	SELENE (S 154)	BU 1960
SIRDAR (S 76)	BU 1965	SENESCHAL (S 75)	BU 1960
SPITEFUL	To France 1952	SENTINEL (S 56)	BU 1962
SPORTSMAN	To France 1951	SAGA	To Portugal 1948
STOIC	BU 1950	SCORCHER (S 58)	BU 1962
SCYTHIAN (S 137)	BU 1960	SIDON (S 59)	Sunk 16.6.59
SCOTSMAN (S 143)	BU 1964	SLEUTH (S 61)	BU 1958
SEA DEVIL (S 44)	BU 1965	SOLENT (S 62)	BU 1962
STATESMAN (S 46)	To France 1952	SPEARHEAD	To Portugal 1948
STURDY (S 48)	BU 1958	SPRINGER (S 64)	To Israel 1958

Name	Fate	Name	Fate
SUBTLE (S 51)	BU 1959	SPUR	To Portugal 1948
SANGUINE (S 66)	To Israel 1958		
'U' class (2nd group): launched 1942–43, 646t/732t			
UNTIRING (ex-*Xifias*, ex-*P 59*)	Sunk as target 1957	UPSTART (ex-*Amfitriti*, ex-*Upstart*, ex-*P 65*)	Sunk as target 1957
Acheron class: launched 1944–47, 1385t/1620t			
ACHERON	Modernised (see notes)	ANDREW	Modernised
ALCIDE	Modernised	AUROCHS (S 26)	BU 1967
ALDERNEY	Modernised	AENEAS	Modernised
ALLIANCE	Modernised	AMPHION (ex-*Anchorite*)	Modernised
AMBUSH	Modernised	ALARIC	Modernised
AURIGA	Modernised	ASTUTE	Modernised
AFFRAY (S 20)	Sunk 17.4.51	ARTEMIS	Modernised
ANCHORITE (ex-*Amphion*)	Modernised	ARTFUL	Modernised
XE 1 class: launched 1942–45, 30.3t/33.6t			
XE 7	BU 1952	XE 9	BU 1952
XE 8	BU 1952	XE 12	BU 1952

'T' class
From 1951 a series of improvements was put in hand. Those with riveted hulls were limited in diving depth, and got the 'T' class 'Streamlined' conversion; this involved fitting new batteries, streamlining the conning tower into a new 'fin', and reshaping the bow casing to give higher speed. Both these boats and the unmodernised ones had the 'snort' (snorkel) mounted on the port side, hinged at deck level. The 20mm gun was removed from all, but several retained their 4in guns.

Tactician, Telemachus and *Thorough* were lent to the Royal Australian Navy in 1951 for ASW training, but remained under RN control. *Tapir* and *Taurus* became the Netherlands *Zeehund* and *Dolfijn* from June 1948 to June 1953 and then reverted to the RN; they were scrapped in 1966 and 1960 respectively. *Truculent* was run down by the Swedish tanker *Divina* in the Thames Estuary on 12 January 1950 and sank with heavy loss of life; she was salved two months later and scrapped.

The class differed widely in appearance, some retaining the deck gun and others having a streamlined fin and modified casing. The 6th group had welded hulls and so most underwent a full modernisation from 1951 (see under Submarines).

'S' class
This class was not scheduled for any major modernisation, but several boats underwent modification for trials. *Seraph* had light padding over her saddle tanks to protect her from practice torpedoes, while *Scotsman* was given a series of modifications, including the removal of her CT followed by a streamlined fin and casing to give her higher speed while carrying out sonar trials. Most had the deck gun removed, but the three sold to Portugal retained a 4in gun in a revolving shield, as in the early 'T' class.

The three Portuguese boats were transferred in 1948–49 and in 1951–52 four more were lent to France; Israel bought two in 1958. *Sidon* was lying alongside the depot ship *Maidstone* in Portland on 16 June 1959 when a torpedo exploded and sank her. The cause was officially claimed to be a compressed air bottle, but many years later it was admitted that the real cause was high-test peroxide (HTP) in a Mk 12 'Fancy' torpedo. *Sidon* was raised and sunk as an Asdic target off Portland in June 1957.

Spur, Saga and *Spearhead* became the Portuguese *Narval, Nautilo* and *Neptune*; *Satyr, Spiteful, Sportsman* and *Statesman* became the French *Saphir, Sirène, Sibylle* and *Sultane* and *Sanguine* and *Springer* became the Israeli *Rahav* and *Tanin*.

'U' class
The rest of the class had all been disposed of, but in 1952 *Amfitriti* and *Xifias* were returned by the Royal Hellenic Navy and were reinstated in the RN under their original names. Both were used as bottom targets, off the Isle of Wight and Start Point respectively.

Acheron class
As with the 'T' class all 20mm guns were removed shortly after the war and those boats with low bow casings had bow buoyancy tanks fitted. Subsequently the external bow tubes were blanked off. All except *Aurochs* underwent conversion from 1955 (see under Submarines).

XE class
No alteration was made to these midget submarines, *XE 9* was the last to be operational, being retained for training until the replacement came into service.

EX-GERMAN SUBMARINES

These comprised the Type VIIC: *N 86* (ex-*U 249*), *N 83* (ex-*U 1023*), *N 65* (ex-*U 776*) and *U 953*; the Type VIIC/41–42: *N 16* (ex-*U 1105*), *N 19* (ex-*U 1171*) and *U 1108*; the Type XVIIB: *Meteorite* (ex-*N 25*, ex-*U 1407*); the Type XXI: *N 41* (ex-*U 3017*) and *U 2518*; and the Type XXIII: *U2348*. Most were 'cannibalised' to evaluate components, but the Type XXI and Type XVIIB were run to evaluate their impact on ASW tactics and to gain experience for new designs. The fates of most are obscure, but it must be assumed that the majority were finally scuttled as bottom targets and then raised and scrapped *c*1948–49 (they disappear from the Navy List in 1948). *U 2518* was given to France in 1947 and became the *Roland Morillot*. *N 41* was broken up in 1949.

Although the Type XXI was the most potent U-boat of all and the most influential on future designs, Allied scientists were fascinated by the possibilities of the Walter high-test peroxide turbine, in which an enriched oxidant called perhydrol was decomposed by a catalyst to produce steam and oxygen at 1765° F. This mixture was then burned with oil fuel at a lower temperature (986° F) to produce a gas which drove a conventional geared turbine. There were two drawbacks: the unpleasant nature of perhydrol, which is corrosive and unstable, and the limited range of only 114 miles.

Despite these drawbacks, the RN worked hard to get *Meteorite*, their Type XVIIB, going but she was never regarded as more than '75 per cent safe'. She was commissioned as early as 25 September 1945, but did not receive her name until 1947, presumably once she was running. She provided valuable experience for the *Explorer* class. She was broken up in 1949.

FRIGATES

Name	Fate	Name	Fate
'Hunt' class (Type 1): launched 1939–40, 1420–1450t deep load			
ATHERSTONE (F 05)	BU 1957	GARTH (F 120)	BU 1958
BLENCATHRA (F 24)	BU 1957	HAMBLEDON (F 137)	BU 1957
BROCKLESBY (F 142)	BU 1968	HOLDERNESS (F 148)	BU 1956
CATTISTOCK (F 135)	BU 1957	MENDIP (F 160)	To China 1948, to Egypt 1950
CLEVELAND (F 146)	Wrecked 28.6.57	MEYNELL (F 182)	To Ecuador 1954
COTSWOLD (F 154)	BU 1958	PYTCHLEY (F 192)	BU 1956
COTTESMORE (F 178)	To Egypt 1950	QUANTOCK (F 158)	To Ecuador 1954
EGLINTON (F 87)	BU 1956	SOUTHDOWN (F 25)	BU 1956
FERNIE (F 111)	BU 1956	WHADDON (F 145)	BU 1959
'Hunt' class (Type 2): launched 1940–42, 1580–1625t deep load			
AVON VALE (F 206)	BU 1958	FARNDALE (F 70)	BU 1962
BEAUFORT (F 14)	To Norway 1952	LAMERTON (F 88)	To India 1953
BEDALE (ex-*Slazak*, ex-*Bedale*) (F 126)	To India 1953	LEDBURY (F 190)	BU 1958
BICESTER (F 134)	BU 1956	MIDDLETON (F 174)	BU 1957
BLACKMORE (F 143)	To Denmark 1952	OAKLEY (ex-*Tickham*) (F 198)	To Germany 1958
BLANKNEY (F 130)	BU 1959	SILVERTON (ex-*Krakowiak*, ex-*Silverton*) (F 55)	BU 1959
CALPE (F 171)	To Denmark 1952	TETCOTT (F 199)	BU 1956
CHIDDINGFOLD (F 131)	To India 1953	WHEATLAND (F 22)	BU 1959
COWDRAY (F 152)	BU 1959	WILTON (F 128)	BU 1959
CROOME (F 162)	BU 1957	ZETLAND (F 59)	To Norway 1954
EXMOOR (ex-*Burton*) (F 08)	To Denmark 1953		
'Hunt' class (Type 3): launched 1941–43, 1545–1590t deep load			
ALBRIGHTON (F 112)	To Germany 1958	HAYDON (F 75)	BU 1958
BELVOIR (F 132)	BU 1957	MELBREAK (F 173)	BU 1956
BLEASDALE (F 150)	BU 1956	STEVENSTONE (F 16)	BU 1959
EASTON (F 109)	BU 1953	TALYBONT (F 118)	BU 1961
EGGESFORD (F 15)	To Germany 1958		
'Hunt' class (Type 4): launched 1942, 1700–1750t deep load			
BRECON (F 176)	BU 1962	BRISSENDEN (F 79)	BU 1965
Grimsby class: launched 1936, 1480–1510t deep load			
FLEETWOOD (F 47)	BU 1959		
Bittern class: launched 1936, 1790t deep load			
STORK (F 81)	BU 1958		
Egret class: launched 1938, 1790t deep load			
PELICAN (F 86)	BU 1958		
Black Swan class: launched 1939–42, 1770–1950t deep load			
BLACK SWAN (F 57)	BU 1956	WILD GOOSE (F 45)	BU 1956
ERNE (F 03)	BU 1965	WOODCOCK (F 90)	BU 1955
FLAMINGO (F 18)	To Germany 1959	WREN (F 28)	BU 1956
WHIMBREL	To Egypt 1949		
Modified Black Swan class: launched 1942–46, 1880–1950t deep load			
ACTAEON (F 07)	To Germany 1958	MODESTE (F 42)	BU 1961
ALACRITY (F 60)	BU 1956	NEREIDE (F 64)	BU 1958
AMETHYST (F 116)	BU 1957	OPOSSUM (F 33)	BU 1960
CRANE (F 123)	BU 1965	PEACOCK (F 96)	BU 1958

Name	Fate	Name	Fate
CYGNET (F 38)	BU 1956	PHEASANT (F 49)	BU 1963
HART (F 58)	To Germany 1959	REDPOLE (F 69)	BU 1960
HIND (F 39)	BU 1958	SNIPE (F 20)	BU 1960
MAGPIE (F 82)	BU 1959	SPARROW (F 71)	BU 1958
MERMAID (F 30)	To Germany 1959	STARLING (F 66)	BU 1965
'Captain' class: launched 1943, 1823t deep load			
HOTHAM (F 583)	BU 1956		
'River' class: launched 1941–43, 1920–2180t deep load			
AVON (F 97)	To Portugal 1949	NITH (F 215)	To Egypt 1948
AWE (F 526)	To Portugal 1949	ODZANI (F 356)	BU 1957
BALLINDERRY (F 155)	BU 1961	PLYM (F 271)	Expended 3.10.52
CHELMER (F 221)	BU 1957	RIBBLE (ex-*Duddon*) (F 525)	BU 1957
DART (F 21)	BU 1957	ROTHER (F 224)	BU 1955
DERG (F 257)	BU 1960	SPEY (F 246)	To Egypt 1948
DOVEY (ex-*Lambourne*) (F 523)	BU 1955	SWALE (F 217)	BU 1955
ETTRICK (F 254)	BU 1953	TAFF (F 367)	BU 1957
EXE (F 92)	BU 1956	TAVY (F 272)	BU 1955
FAL (F 266)	To Burma 1948	TAY (F 232)	BU 1956
HALLADALE (F 417)	Sold 1949	TEES (F 293)	BU 1955
HELFORD (F 252)	BU 1956	TEST (ex-*RIN Neza*, ex-*Test*) (F 56)	BU 1955
HELMSDALE (F 253)	BU 1957	TEVIOT (F 222)	BU 1955
JED (F 235)	BU 1957	TOWY (F 294)	BU 1956
KALE (F 241)	BU 1956	USK (F 295)	To Egypt 1948
LOCHY (F 365)	BU 1956	WAVENEY (F 248)	BU 1957
MEON (F 369)	BU 1966	WEAR (F 230)	BU 1957
NENE (F 270)	BU 1955	WYE (F 371)	BU 1955
NESS (F 219)	BU 1956		
'Loch' class: launched 1943–45, 2260t deep load			
LOCH ACHANALT (F 424)	To New Zealand 1948	LOCH KILLISPORT (F 628)	BU 1970
LOCH ACHRAY (F 426)	To New Zealand 1948	LOCH LOMOND (F 437)	BU 1968
LOCH ALVIE (F 428)	BU 1965	LOCH MORE (F 639)	BU 1963
LOCH ARKAIG (F 603)	BU 1960	LOCH MORLICH (F 517)	To New Zealand 1949
LOCH CRAGGIE (F 609)	BU 1963	LOCH QUOICH (F 434)	BU 1957
LOCH DUNVEGAN (F 425)	BU 1960	LOCH RUTHVEN (F 645)	BU 1966
LOCH ECK (F 422)	To New Zealand 1948	LOCH SCAVAIG (F 648)	BU 1959
LOCH FADA (F 390)	Sold 1968, but used for trials until 1976	LOCH SHIN (F 421)	To New Zealand 1948
LOCH FYNE (F 429)	BU 1970	LOCH TARBERT (F 431)	BU 1959
LOCH GLENDHU (F 619)	BU 1957	LOCH TRALAIG (F 655)	BU 1963
LOCH GORM (F 620)	Sold 1961	LOCH VEYATIE (F658)	BU 1965
LOCH INSH (F 433)	To Malaysia 1964	DERBY HAVEN (ex-*Loch Assynt*)	To Iran 1949
LOCH KATRINE (F 625)	To New Zealand 1949	WOODBRIDGE HAVEN (ex-*Loch Torridon*) (F 654)	BU 1965
LOCH KILLIN (F 391)	BU 1960		
'Bay' class: launched 1944–45, 2420t deep load			
BIGBURY BAY (ex-*Loch Carloway*) (F 06)	To Portugal 1959	ST AUSTELL BAY (ex-*Loch Lyddoch*) (F 634)	BU 1959
BURGHEAD BAY (ex-*Loch Harport*) (F 622)	To Portugal 1959	ST BRIDES BAY (ex-*Loch Achilty*) (F 600)	BU 1962
CARDIGAN BAY (ex-*Loch Laxford*) (F 630)	BU 1962	START BAY (ex-*Loch Arklet*) (F 604)	BU 1958
CARNARVON BAY (ex-*Loch Maddy*) (F 636)	BU 1959	TREMADOC BAY (ex-*Loch Arnish*) (F 605)	BU 1959

Name	Fate	Name	Fate
CAWSAND BAY (ex-*Loch Roan*) (F 644)	BU 1959	VERYAN BAY (ex-*Loch Swannay*) (F 651)	BU 1959
ENARD BAY (ex-*Loch Bracadale*) (F 35)	BU 1957	WHITESAND BAY (ex-*Loch Lubnaig*) (F 633)	BU 1956
LARGO BAY (ex-*Loch Fionn*) (F 423)	BU 1958	WIDEMOUTH BAY (ex-*Loch Frisa*) (F 615)	BU 1957
MORECAMBE BAY (ex-*Loch Heilen*) (F 624)	To Portugal 1961	WIGTOWN BAY (ex-*Loch Garasdale*) (F 616)	BU 1959
MOUNTS BAY (ex-*Loch Kilbirnie*) (F 627)	To Portugal 1961	ALERT (ex-*Dundrum Bay*, ex-*Loch Scamadale*) (F 647)	BU 1981
PADSTOW BAY (ex-*Loch Coulside*) (F 608)	BU 1959	SURPRISE (ex-*Gerrans Bay*, ex-*Loch Carron*) (F 436)	BU 1965
PORLOCK BAY (ex-*Loch Seaforth*, ex-*Loch Muick*) (F 650)	To Finland 1962		
Modified 'Flower' class: launched 1943–44, 1350–1370t deep load			
ARABIS (F 85)	BU 1951	ARBUTUS (F 103)	BU 1951
'Castle' class: launched 1943–44, 1590–1630t deep load			
ALLINGTON CASTLE (ex-*Amaryllis*) (F 89)	BU 1959	KENILWORTH CASTLE (F 420)	BU 1959
ALNWICK CASTLE (F 105)	BU 1958	KNARESBOROUGH CASTLE (F 389)	BU 1956
AMBERLEY CASTLE (F 286)	Weather ship 1960	LANCASTER CASTLE (F 691)	BU 1960
BAMBOROUGH CASTLE (F 12)	BU 1959	LAUNCESTON CASTLE (F 397)	BU 1959
BERKELEY CASTLE (F 387)	BU 1955	LEEDS CASTLE (F 384)	BU 1958
CAISTER CASTLE (F 690)	BU 1956	MORPETH CASTLE (F 693)	BU 1960
CARISBROOKE CASTLE (F 379)	BU 1958	OAKHAM CASTLE (F 530)	Weather ship 1958
DUMBARTON CASTLE (F 388)	BU 1961	OXFORD CASTLE (F 692)	BU 1960
FARNHAM CASTLE (F 413)	BU 1958	PEVENSEY CASTLE (F 449)	Weather ship 1960
FLINT CASTLE (F 383)	BU 1958	PORTCHESTER CASTLE (F 362)	BU 1958
HADLEIGH CASTLE (F 355)	BU 1959	RUSHEN CASTLE (F 372)	BU 1960
HEDINGHAM CASTLE (ex-*Gorey Castle*) (F 386)	BU 1958	TINTAGEL CASTLE (F 399)	BU 1958

'Hunt' classes

These former escort destroyers were re-rated as anti-aircraft frigates in 1947, but very few saw postwar service in the RN, apart from training and occasional recommissioning (*eg Eglinton* in Exercise 'Sleeping Beauty'). *Brocklesby* was disarmed for service with AUWE at Portland, and from 1953 carried out trials with a large variable-depth sonar on the quarterdeck. *Mendip* served as the Chinese *Lin Fu* in 1948–49 and on her return was sold to Egypt as *Mohamed Ali el Kebir*, with *Ibrahim el Awal* (ex-*Cottesmore*); Denmark bought *Blackmore, Calpe* and *Exmoor* in 1952–53 and renamed them *Esbern Snare, Rolf Krake* and *Valdemar Sejr*; India was lent *Chiddingfold, Bedale* and *Lamerton* and renamed them *Ganga, Godavari* and *Gomati*; Norway received *Badsworth, Beaufort* and *Zetland*, renaming them *Arendal, Haugesund* and *Tromsö*; Ecuador received *Quantock* and *Meynell* as *Presidente Alfare* and *Presidente Velasco Ibarra*; and *Albrighton, Eggesford* and *Oakley* became *Raule, Brommy* and *Gneisenau* in the Federal German Navy in 1958.

Grimsby class

The former sloop *Fleetwood* was disarmed in 1947 and fitted out as trials ship for the Admiralty Signal and Radar Establishment with a lattice foremast to carry extra arrays and deckhouses added forward and aft. In 1950 she was refitted and later had a lattice tower added aft.

Bittern and Egret classes

The ex-sloop *Stork* was laid up from 1946; *Pelican* served for a short while postwar but was then laid up.

Black Swan class

These former sloops served postwar but were paid off in the early 1950s, *Erne* becoming the RNVR drillship *Wessex* in 1952. *Whimbrel* became the Egyptian *El Malek Farouq*, and *Flamingo* became the Federal German *Graf Spee*.

Modified Black Swan class

Most of this class saw considerable service on foreign stations from 1940s to the early 1960s. *Amethyst* was severely damaged by Chinese field artillery on the Yangtze in 1949, and like her sisters had the midships Mk 4 Hazemeyer Bofors guns replaced by Mk 5 mountings. Only *Pheasant* and *Redpole* retained the tripod foremast. In 1948–49 *Redpole* and *Starling* were disarmed and converted to tenders for navigational training. *Redpole* had a tripod foremast and a radar scanner on a short lattice mainmast while *Starling* had two tall lattice masts. Foreign sales of these ex-sloops were limited to *Actaeon, Hart* and *Mermaid*, which became the Federal German *Hipper, Scheer* and *Scharnhorst* in 1958–59.

'Captain' class

In theory all the Lease-Lend warships had to be returned to the United States immediately after the war, but inevitably some slipped through the net. The ex-*Buckley* class DE *Hotham* was a particular exception for she remained on the Navy List until 1956. Early in 1945 she had been stripped of fittings to allow her to be used as an emergency generating station for Singapore. When it turned out that she was not needed she was sent to Hong Kong as a guardship, although subsequently she returned to Singapore. In 1951 she returned to the UK, and was then earmarked as trials ship for gas turbines. The machinery – twin-shaft EL60A turbines developing 6000shp each – was built by English Electric, but was never installed because more modern turbines became available. Finally the ship was nominally returned to the USN in 1956, but scrapped in the UK.

'River' class

Virtually all these frigates paid off in the postwar years and never recommissioned, as there were more than sufficient 'Lochs', 'Bays' and 'Castles' to meet peacetime needs. Those that were refitted had 40mm singles in place of the 20mm guns, and a Type 277 height-finder radar in place of the original Type 271 above the bridge. *Helmsdale* was disarmed for trials, and had two Squid DC mortars in 'B' position. *Meon* and *Waveney* were used for propeller silencing trials and as Landing Ships Headquarters, Small (LSH (S)), with armament reduced to three 40mm and four 20mm guns; the after deckhouse was enlarged to provide offices for Army and Royal Marines personnel. *Exe* was earmarked for a similar conversion, but this does not appear to have been undertaken. Scrapping of the survivors began in the mid-1950s, but the last did not go to the breakers until 1966.

'Loch' class

This class proved ideal for postwar duties, with modern sensors and weapons, and they survived to the mid-1960s. For service in the Persian Gulf nine were modernised in the early 1950s: *Loch Alvie, Loch Fada, Loch Fyne, Loch Insh, Lock Killisport, Loch Lomond, Loch Ruthven, Loch Tralaig* and *Loch Veyatie*. This involved replacing the single 4in Mk 5 HA gun with a twin 4in Mk 19 and updating the secondary armament and radars. The secondary armament now consisted of four single 40mm Mk 9 (in bridge wings and at the forecastle break) and a twin 40mm Mk 5 aft. The Type 277 height-finder aerial at the masthead was replaced by a Type 277Q. *Loch Insh* became the Malaysian *Hang Tuah* in 1964. *Loch Fada* was used to test Vertically Launched Sea Wolf in 1976.

Two ex-'Loch' class frigates were completed in 1945 as depot ships (2160t full load; 2–40mm Mk 5) for small minesweepers and coastal forces. *Derby Haven* became the Iranian *Babr* in 1949 but *Woodbridge Haven* was actively employed for most of her life.

'Bay' class

Although also used widely on peacetime duties this class did not prove quite as successful as the 'Lochs' because of their weak anti-submarine armament, and they were never given any form of major modernisation. *Porlock Bay* became the Finnish *Matti Kurki* in 1962; and *Burghead Bay, Morecambe Bay, Bigbury Bay* and *Mounts Bay* became the Portuguese *Alvares Cabral, Don Francesco de Almeida, Pacheco de Pereira* and *Vasco da Gama* in 1959–61.

Two ex-'Bay' class frigates were completed in 1946 as dispatch vessels and C in C's yachts on foreign stations. *Alert* and *Surprise* had a full load displacement of 2440t and were armed with 2–4in/45 Mk 16 (2×2), 2–40mm Mk 7 (2×1), 4–3pdr saluting. For the Coronation Naval Review on 15 June 1953 *Surprise* was converted to a temporary Royal Yacht, with the twin 4in gun replaced by a glassed-in saluting platform for HM The Queen. Greatly increased accommodation was provided aft, with a broad shelter deck extending from the break of the forecastle to the stern.

Modified 'Flower' class

These two ex-corvettes were returned by the Royal New Zealand Navy in 1948, and were retained on the Navy List for three years. They were thus the only survivors of the large number of corvettes operated by the RN from 1940 onwards. The remainder were disposed of in large numbers from 1946, some to be converted to whalecatchers and small coasters, but the majority to be scrapped.

'Castle' class

Like other escorts these vessels were absorbed into the frigate category in 1947. Many served with the Training Squadron at Portland, but the hulls were too small for any sort of modernisation, and they went for scrap from the mid-1950s. *Berkeley Castle* capsized in dry dock at Sheerness during the disastrous floods in February 1953 and was never repaired. The weather ships were lent to the Air Ministry with the 40mm Bofors guns and the Squid DC mortar cocooned, but never returned to naval service.

TRAWLERS

'Isles' class

Annet, Bardsey, Bern, Caldy, Coll, Fetlar, Flatholm, Foulness, Gorregan, Graemsay, Erraid, Gateshead, Lindisfarne, Lundy, Neave, Orsay, Ronay, Sandray, Shillay, Skomer, Skye, Steepholm, Sursay, Switha, Tahay, Tiree, Tocogay, Trodday, Trondra, Vacesay, Vallay, Wiay

Survivors of a large class of Admiralty trawlers. Many served postwar as danlayers, wreck dispersal vessels, experimental anti-submarine vessels, etc. They were scrapped 1957–67, apart from a few converted to tank-cleaning vessels.

MINESWEEPERS

Halycon class: launched 1936–38, 1330t deep load

Franklin, Scott, Seagull and *Sharpshooter* reverted to surveying duties postwar, *Sharpshooter* being renamed *Shackleton* in 1953. They were scrapped between 1956 and 1965.

Algerine class: launched 1941–45, 1225–1325t deep load

Eighty-three of these successful sweepers were retained; many were transferred to other navies, and served as training ships. They were surplus to requirements after it was realised during the Korean War that steel-hulled sweepers could no longer rely on degaussing for protection against magnetic mines.

Cadmus, Fancy, Liberty, Ready, Rosario and *Spanker* became the Belgian *Georges Lecointe, A F Dufour, Adrien de Gerlache, Jan van Haverbeke, De Moor* and *De Brouwer* (1949–53); *Hare* became the Nigerian *Nigeria* (1959); *Flying Fish* and *Pickle* became the Ceylonese *Vijaya* and *Parakrama* (1949–58); *Fly* became the Iranian *Palang* (1949); *Pelorus* and *Rosamund* became the South African *Pietermaritzburg* and *Bloemfontein* (1949); *Arcturus, Aries, Gozo, Lightfoot* and *Postillion* became the Royal Hellenic Navy's *Pyrpolitis, Armatolos, Polemistis, Navamachos* and *Machitis* (1947); *Mariner* became the Burmese *Yan Myo* (1958); and *Minstrel* became the Royal Thai Navy's *Phosampton* (1947).

COASTAL FORCES

Vosper 73ft type: 47t deep load

MTB 386, 392, 502, 523–527, 529, 530, 532, 533

These were renumbered as fast patrol boats in 1950, *MTB 386* and *392* becoming *FPB 1001* and *1002*, and *523–527, 529, 530, 532* and *533* becoming *FPB 1023–1027, 1029, 1030, 1032* and *1033*. *FPB 1030* sank after a collision with her sister *FPB 1032* off the Netherlands coast on 28 March 1950, while *FPB 1023* caught fire and blew up in Aarhus harbour, Denmark, on 17 May 1953. Only *FPB 1002* survived (as a hulk) by 1956. The armament was mixed, some having a 6pdr forward, others twin 20mm. *FPB 1032* received an experimental 4.5in/25.

British Power Boat 71¾ft type: 53t deep load

MTB 470, 480, 496, 498, 505–509, 519, 521, 522, 596, 598

Nine were renamed in 1947: *Proud Fusilier* (ex-*MTB 505*), *Proud Grenadier* (ex-*MTB 506*), *Proud Guardsman (*ex-*MTB 507*), *Proud Highlander* (ex-*MTB 508*), *Proud Knight (*ex-*MTB 509*), *Proud Lancer* (ex-*MTB 519*), *Proud Legionary* (ex-*MTB 522*), *Proud Patriot* (ex-*MTB 596*) and *Proud Patroller* (ex-*MTB 598*), with 1000 added to their numbers. The others were all stricken by 1947, except *MTB 470, 480, 496* and *498*, which were renumbered as fast patrol boats *FPB 1570, 1580, 1596* and *1598*.

Camper & Nicholson type: 102.5t deep load

MGB 2009

Ordered in 1941 and laid down in May 1942, was completed in 1945 for experimental purposes. She displaced 86.5t standard, was 117ft 3in oa × 19ft 1in × 4ft 4in max (*35.7m × 5.8m × 1.3m*) and was fitted with 3-shaft gas turbine/gasoline machinery (one Gatric gas turbine plus two Packard engines, 5200shp = 34kts); she was unarmed and had a complement of thirty-one. Renumbered *P 5559*, she ran a series of trials in 1947 before being sold *c*1948.

Fairmile 'D' type: 118t deep load

MTB 731, 750, 758, 759, 780, 790, 793, 794, 795

All were renumbered, *MTB 731, 750, 758, 779, 780, 785, 790* and *793–795*, becoming *FPB 3001, 3002, 5001, 5003, 5010, 5013, 5031, 5032, 5033* and *5035–5037; 3001, 3002, 3050* and *3053* served as motor anti-submarine boats, (MA/SBs) the others as convertible MTBs/MGBs with four TT or two guns, 6pdrs in most, but 4.5in/25 in *5036*. Only *FPB 5001, 5035* and *5036* were still in service by 1958 and they were put on the disposal list shortly afterwards.

Modified Fairmile 'D' type: 118t deep load

MTB 5002, 5005, 5007, 5008, 5009, 5015, 5020

These craft had numbers in the same series as the ordinary 'D' type, and so were not renumbered, merely reclassified as FPBs. They were armed as convertible MTBs/MGBs, with 6pdr guns and 21in TT, although *5009* mounted two 4.5in/25 by 1958 only *FPB 5015* and *5020* survived, but they were soon put on the disposal list.

Fairmile 'F' type

MGB 2001

Prototype 'F' with four Bristol Hercules engines.

Fairmile 'B' Type Motor Launches: 85.6t deep load

ML 2154–2156, 2217, 2220, 2221, 2223, 2237, 2248, 2250, 2337, 2338, 2342, 2357, 2360, 2461, 2462, 2489, 2491, 2493, 2564, 2567, 2568, 2571, 2575–2577, 2581–2583, 2585, 2586, 2592, 2593, 2595, 2866, 2869, 2882, 2886, 2901, 2906, 2912, 2919, 2921; RML 495, 496, 498, 512, 515, 529

Most were scrapped or sold in the 1950s and only five were in service by 1958.

Harbour Defence Motor Launches (HDML): 54t deep load
SDML 1001, 1053, 1081, 1085, 1091, 1301, 1323, 1326, 1329, 1393, 1411, 3328, 3501–3503, 3505–3508, 3510–3516, Meda (ex-*SML 352*, ex-*ML 1301*)
Most were scrapped in the 1950s, only nine remaining by 1958. Several served in the Far East as gunboats, with a power-operated 6pdr forward and a 20mm aft. Latterly they were re-rated as Seaward Defence MLs.

126ft Motor Minesweepers: 430t deep load
MMS 1002–1004, 1011, 1017, 1030, 1034, 1038, 1042, 1044, 1048, 1060, 1061, 1075, 1077, 1089, 1090

105ft Motor Minesweepers: 295t deep load
MMS 1501, 1505, 1532, 1534–1536, 1546, 1548, 1550, 1553, 1556, 1558, 1569, 1579, 1583, 1584, 1586, 1599, 1600, 1602–1606, 1609, 1610, 1635, 1644, 1667, 1672, 1681, 1685, 1717, 1724, 1728, 1733, 1736, 1761, 1763, 1771, 1772, 1775, 1783, 1785, 1786, 1788–1791, 1794
Both the 126ft and 105ft types were replaced by the 'Ton', 'Ham' and 'Ley' classes in the 1950s, although four of the large type served for longer as degaussing vessels *DGV 401–404*.

EX-GERMAN MOTOR TORPEDO-BOATS

At the end of the Second World War about 34 ex-*Schnellboote* or 'E-Boats' were handed over. Most were scrapped, but three were retained for experimental purposes, *S 130, S 208* and *S 212*. These were renumbered *P 5230, P 5208* and *P 5212*. Displacing 105t normal load and 114ft 6in oa × 16ft 9in × 6ft mean (*34.9m × 5.1m × 1.8m*), they were fitted with 3-shaft diesels (*P 5230* Mercedes-Benz, 7500bhp; *P 5212* 1 Mercedes-Benz + 2 Napier Deltic, 9000bhp; *P 5208* 3 Napier Deltic, 9000bhp) giving 39kts. Oil was 20t and complement about fifteen.

P 5212 operated in British waters until paid off for scrapping in 1956. The other two were employed on what were officially described as fishery protection duties under the British Flag Officer, Germany. What this actually involved was surveillance of the Soviet Navy in the eastern Baltic. Both craft, unarmed and fitted with equipment for monitoring signals, observed manoeuvres, trusting to their big margin of speed to get away if detected. In 1957 these marine equivalents of the U-2 spy-plane were handed over to the newly-formed *Bundesmarine* in 1957 and renumbered *UW 10* and *UW 11*. They were rearmed for training duties with two TT and were finally scrapped in 1964–65.

STEAM GUNBOATS

Grey Seal class: launched 1941–42, 255t deep load
Grey Seal, Grey Fox, Grey Wolf, Grey Goose
Three of the survivors were sold and scrapped in 1948–49, but *Grey Goose* was selected to replace the frigate *Hotham* as the trials ship for the prototype Rolls-Royce PM60 gas turbine. She was taken in hand in 1952 and given two funnels with sharply raked tops. In 1954–55 she ran a series of intensive trials, paving the way for the successful 'marinising' of the G2 and Proteus for small craft and the G6 for large ships. When the trials were over in 1955 she was stripped and laid up. She was sold in December 1957.

AMPHIBIOUS WARFARE SHIPS

LST (3): launched 1944–45, 4980t deep load
*Frederick Clover (LST 3001), Anzio (LST 3003), Tromsö (LST 3006), LST 3008, Reginald Kerr (LST 3009), Avenger (LST 3001), Ben Nevis (LST (Q) 1, ex-*LST 3012), Ben Lomond (LST (Q) 2, *ex-LST 3013), LST 3014, Battler (LST 3015), Dieppe (LST 3016), LST 3017, Vaagsö (LST 3019), Charles McLeod (LST 3021), LST 3022, Maxwell Brander (LST 3024), Bruizer (LST 3025), Charger (LST 3026), Lofoten (LST 3027), Snowden Smith (LST 3028), Chaser (LST 3029), LST 3031, LST 3033, LST 3035, Puncher (LST 3036), Evan Gibb (LST 3037), Fighter (LST 3038), LST 3040, LST 3041, Hunter (LST 3042), Messina (LST 3043), Narvik (LST 3044), LST 3501, Pursuer (LST 3504), Ravager (LST 3505), LST 3507, Scorcher (LST 3508), Humphrey Gale (LST 3509), Slinger (LST 3510), Reggio (LST 3511), LST 3512, Salerno (LST 3513), Smiter (LST 3514), Stalker (LST 3515), Striker (LST 3516), St Nazaire (LST 3517), Suvla (LST 3518), LST 3519, Thruster (LST 3520), Tracker (LST 3522), Trouncer (LST 3523), Trumpeter (LST 3524), Walcheren (LST 3525), Zeebrugge (LST 3532), LST 3524*
Numbered craft were all sold by 1950 or transferred – *3017, 3035, 3501* became the Australian *Tarakan, Labuan* and *Lae* in 1951 – and all vessels with generals' names (*Frederick Clover*, etc) had been on charter to the War Department from 1946. *Avenger* became Indian *Magar* in 1951 and *Smiter* was wrecked off Lagos, Portugal, on 25 April 1949. In 1956 *Tromsö, Attacker, Battler, Charger, 3033, Fighter, 3041, Hunter, Pursuer, 3507, Slinger, 3512, St Nazaire, 3519, Thruster, Trouncer, Trumpeter, Walcheren* and *3524* were chartered commercially and renamed *Empire Gannet, Empire Cymric, Empire Puffin, Empire Nordic, Empire Shearwater, Empire Grebe, Empire Doric, Empire Curlew, Empire Tern, Empire Gaelic, Empire Kittiwake, Empire Celtic, Empire Skua, Empire Baltic, Empire Petrel, Empire Gull, Empire Fulmar, Empire Guillemot* and *Empire Cedric*. The rest were broken up from 1970 onwards. In 1964 *Lofoten* was converted to a helicopter support ship, with a flight deck to enable her to carry six Wessex ASW helicopters out to sea for deep-water training. Last in service was the RFA *Empire Gull*, scrapped in 1980. *Lofoten*, latterly in service as a dockyard hulk, was broken up in 1994.

LCM (7): launched 1943–44, 63t
About 50, from *LCM 7007* to *7216*, survived in 1953. By 1967 only five were left, three serving as naval servicing boats (NSB) and stores carriers. Several had been re-engined.

LCT (8): launched 1945–47, 810t loaded
LCT 4001, 4002, 4025, 4037–4045, 4049, 4050, 4061–4064, 4073, 4074, 4085, 4086, 4097–4099, 4128, 4148, 4156, 4164, 4165
LCT 4001, 4002, 4037–4041, 4043, 4044, 4049, 4061, 4062, 4064, 4073–4074, 4085, 4086, 4097, 4099, 4128 and *4164* were renamed *Redoubt, Agheila, Rampart* (later *Akyab*), *Citadel, Parapet, Bastion, Abbeville, Counterguard, Portcullis, Audemer, Aachen, Sallyport, Ardennes, Antwerp, Agedabia, Arromanches, Andalnes, Buttress, Arezzo* and *Arakan* in 1956. Numbered craft were put on the disposal list in 1958–60, *Counterguard* became the Malaysian *Sri Langkawi* in 1965, *Buttress* was sold to France as *L 9061* in 1965 and *Bastion* was transferred to Zambia in 1966. The remainder were disposed of from 1965 onwards but in 1980 *Abbeville, Agheila* and *Audemer* were still being operated by the Royal Corps of Transport.

LCT (3): launched 1941–44, 640t loaded
LCT (E) 341, LCT (E) 413
341 was lent to the Royal Malaysian Navy as *Malaya* (*MRC 1401*); *MRC 1413* (ex-*LCT(E) 413*) was scrapped.

LCT (4): launched 1942–45, 640t loaded
LCT 403 (ex-*LCT 1220*), *404*, (ex-*1231*), *405* (ex-*523*), *406* (ex-*941*), *407* (ex-*1186*), *408* (ex-*1202*); *MCR 1013, 1015, 1023, 1097, 1098, 1100, 1109, 1110, 1119, 1120, 1122*
All were broken up in the 1960s.

LCA: built 1940–44, 13t
A total of 286, ranging from *LCA 1010* to *1992*, were on strength in 1947; all were scrapped in the 1960s.

LCH: built 1943, 5t
LCH 243 was broken up *c*1955.

LCP(L): built 1942, 10¾t
The ten survivors, ranging from *LCP(L) 519* to *561* and *501*, were scrapped in the 1960s.

MISCELLANEOUS VESSELS

The river gunboat *Locust* became an RNVR drill ship in 1951, and was subsequently broken up. The netlayer *Guardian* was scrapped in 1962, while *Protector* was employed as an Antarctic patrol ship 1955–68, being broken up in 1970.

The following minelayers were in service in 1947: *Plover* (BU 1968), *Blackbird* (ex-*Sheppey*) (sold 1949), *Dabchick* (ex-*Thorney*) (to Malaya 1954), *Stonechat* (BU 1967), *Redshank* (ex-*Turbot*) (BU 1957), *Ringdove* (sold 1950), *Linnet* (BU 1964), *Miner I* (ex-*M 1*) (*Minstrel* 1962, sold 1967), *Miner II* (ex-*M 2*) (*Gossamer* 1949, sunk as target 1970), *Miner III* (ex-*M 3*) (BU 1977), *Miner IV* (ex-*M 4*) (BU 1964), *Miner V* (ex-*M 5*) (PAS *Britannic* 1960, RAF target 1979), *Miner VI* (ex-*M 6*) (sold 1966), *Miner VII* (ex-*M 7*) (PAS *Steady* 1959, BU 1980) and *Miner VIII* (*Mindful* 1963, sold 1965), and the depot ships *Resource* (BU 1954), *Woolwich* (BU 1962), *Maidstone* (BU 1978), *Forth* (BU 1985), *Tyne* (BU 1973), *Adamant* (BU 1970) and the aircraft transports *Blackburn* (RNVR drill ship 1950, sold 1968), *Ripon* (BU 1959), *Roc* (sold 1959), *Seafox* (sold 1958) and *Walrus* (renamed *Skua* 1953, sold 1962).

NATO PENNANT NUMBERS
Carried from 1950 by ships in commission, the most important are noted after the relevant vessel in the lists above. However, the following landing ships and minesweepers were also allocated numbers:

Amphibious warfare ships

L 101	ANZIO	L 119	REGGIO
L 102	ATTACKER	L 120	ZEEBRUGGE
L 104	BEN NEVIS	L 121	SALERNO
L 105	BEN LOMOND	L 123	SLINGER
L 106	BUCHAN NESS	L 125	ST NAZAIRE
L 107	CHARGER	L 126	STALKER
L 108	DIEPPE	L 127	BRUISER
L 109	FIGHTER	L 128	STRIKER
L 110	KEREN	L 129	SUVLA
L 111	LOFOTEN	L 130	TRACKER
L 112	MESSINA	L 132	CHASER
L 114	NARVIK	L 133	TROUNCER
L 115	PUNCHER	L 134	TRUMPETER
L 116	PURSUER	L 135	WALCHEREN
L 117	RAVAGER	L 180	HUNTER
L 118	BATTLER		

Minesweepers

M 01	ALBACORE	M 46	PLUTO
M 11	BRAMBLE	M 49	COURIER
M 17	SKIPJACK	M 53	FIERCE
M 29	COCKATRICE	M 81	TOCOGAY

M 88 CHEERFUL
M 163 SKYE
M 216 ESPIEGLE
M 221 ONYX
M 225 RINALDO
M 276 LENNOX
M 277 ORESTES
M 289 MELITA
M 293 PICKLE
M 294 PINCHER
M 295 PLUCKY
M 298 RECRUIT
M 299 RIFLEMAN
M 302 THISBE
M 303 TRUELOVE
M 304 WATERWITCH
M 307 HOUND
M 325 PROVIDENCE
M 329 MOON
M 333 SEABEAR
M 341 WIAY
M 350 COQUETTE
M 354 SERENE
M 356 WELFARE
M 360 MARY ROSE
M 367 STORMCLOUD
M 376 GOLDEN FLEECE
M 379 CORNFLOWER
M 380 MARINER
M 381 MARMION
M 382 SYLVIA
M 383 TANGANYIKA
M 384 ROWENA
M 385 WAVE
M 386 WELCOME
M 377 LIONESS
M 387 CHAMELEON
M 389 HARE
M 390 JEWEL
M 422 IMERSAY
M 424 SANDRAY
M 426 SHILLAY
M 427 SURSAY
M 428 JASEUR
M 429 RONA
M 431 TRODDAY
M 432 VACEASAY
M 433 LAERTES
M 434 VALLAY
M 435 MÆNAD
M 436 MAGICIENNE
M 438 MANDATE
M 443 MARVEL
M 444 MICHAEL
M 448 PYRRHUS
M 449 ROMOLA
M 450 ORSAY
M 452 TAHAY
M 454 MYRMIDON
M 455 MYSTIC
M 456 NERISSA
M 462 ORCADIA
M 463 OSSORY

MAJOR SURFACE SHIPS

Modified LION class *battleship*

Although officially cancelled in 1944 (all serious work was stopped following the sinking of *Tirpitz*), the *Lion* class were kept on the drawing-board as the Admiralty hoped to be able to restart two of them postwar. The design was completely modified in the light of wartime experience, the principal changes being to broaden the beam, to deepen the torpedo protection and replace the original 16in/45 Mk II gun with a projected Mk IV/45 gun. Little else is known about the design except that by the time the work stopped (*c*1950) the thickness of deck armour had risen to 12in. Displacement was 49,000t (standard), 56,500t (full load) and dimensions 810ft × 115ft × 34ft 3in mean (264.9m × 35.1m × 10.4m).

Albion and *Bulwark* (foreground) as commando carriers, 1962 CMP

Four completed units of a projected class of eight 'intermediate fleet carriers' were retained for the postwar fleet. One was so delayed that she eventually formed a separate type, but the other three, redesigned to incorporate as many wartime lessons as possible, were launched in 1947–48 and completed in somewhat leisurely fashion six years later. Almost as soon as the vessels were completed they were taken in hand for modification. An 'interim' or 5 degrees angled deck was added and this necessitated the removal of three twin Bofors guns from the port side. After the Suez operations in 1956 the ships were given steam catapults and stronger arrester wires, and the six-barrelled Mk 6 Bofors mounting abaft the island was removed to provide deck space.

They proved too small to operate the new generation of aircraft coming into service in the late 1950s, and in 1959–60 *Bulwark* was converted to a commando carrier. This involved the removal of her arrester gear and catapults, the provision of four assault landing craft (LCAs) in davits aft and the modification of her accommodation to allow 733 Royal Marines to embark. With eight helicopters in place of her former air group, she could land troops and vehicles ashore rapidly, and proved such a success that her sister *Albion* was similarly converted in 1961–62. *Albion*'s conversion was slightly more elaborate, and allowed her to embark 900 commandos. Plans to convert *Centaur*, however, were shelved because of rising costs, and she was paid off into reserve in 1966, to be scrapped six years later.

The air groups varied during their careers. At Suez, for example, *Albion* was flying Sea Venom and Sea Hawk fighters, Skyrider AEW aircraft and Sycamore helicopters. At the end of her career *Centaur* carried four Gannet AEW aircraft and eight Whirlwind helicopters in addition to her Sea Vixens and Scimitars.

In April 1976 *Bulwark* paid off into reserve, but as a contingency against late delivery of the *Invincible* class she was maintained in the best condition possible. A rumoured sale to Peru did not materialise and in 1980 she recommissioned to release *Hermes* from the amphibious role when the latter embarked her Sea Harrier squadron. Under the 1981 defence cuts, however, she was paid off and put on the disposal list.

CENTAUR class *aircraft carriers*

Displacement: 22,000t standard; 27,000t full load
Dimensions: 737ft 9in oa × 123ft max × 27ft max
224.8m × 37.5m × 8.2m
Machinery: 2-shaft geared steam turbines, 4 3-drum boilers, 78,000shp = 28kts
Armour: Flight deck, hangar deck (amidships) and main deck (ends), 1in–2in, uptakes 1in
Armament: 32–40mm (2×6, 8×2, 4×1), 26 aircraft
Sensors: Radar 982, 983, 275, 974
Complement: 1028/1102 + about 300 air group

No	Name	Builder	Laid down	Launched	Comp	Fate
R 07	ALBION	Swan Hunter	23.3.44	6.5.47	26.5.54	BU 1972
R 08	BULWARK	Harland & Wolff	10.5.45	22.6.48	4.11.54	BU 1984
R 06	CENTAUR	Harland & Wolff	30.5.44	22.4.47	1.9.53	BU 1971

Centaur 1964 CMP

Victorious was virtually rebuilt from 1950 to the end of 1957 to allow her to operate modern jet aircraft. During the process the hull was widened, deepened and lengthened and the machinery and boilers replaced, although it was only decided in 1955 to replace the boilers (!). Two steam catapults, an angled deck and mirror landing sights were provided, as well as a 3-D radar and faster lifts.

The major item of the reconstruction was the provision of a massive sponson on the port side for the 8¾ degrees angled deck. This extended the flight deck 35ft 6in outboard and was the first fully angled deck fitted in a British carrier. The new island was surmounted by a huge 'searchlight' array for the Type 984 3-D radar, the first to go to sea. Two parallel-track 145ft steam catapults were provided and because of the angled deck there were only four arrester wires, each with an average span of 80ft. A major internal improvement was to nearly double the generating capacity to 4200kW, but even this had to be increased to 5000kW in 1962–63. The former armament was completely replaced by six twin 3in/50 Mk 33 automatic gun mountings purchased from the US Navy, mounted one on each side forward and two on each side aft. In addition one six-barrelled 40mm Mk 6 automatic mounting was positioned on the starboard side between the boat crane and the after 3in guns. Each gun mounting was controlled by its own CRBF (Close Range Blind Fire) director positioned alongside. During a May 1962–August 1963 refit two 3in mountings and the Mk 6 Bofors were removed to compensate for the weight added.

During the 1962–63 refit the flight deck was strengthened and an access gangway was provided outboard of the island. The 'flyco' on the inboard side of the island was extended and enlarged and the mirror sights were replaced by projector sights.

When she recommissioned on 14 January 1958, *Victorious* was the most modern British carrier, but although it had originally been hoped that she would operate fifty-four aircraft, the rapid growth in size of the latter meant that she never carried more than twenty-eight (plus eight helicopters) and finally carried only twenty-three (eight Buccaneers, eight Sea Vixens, two Gannets and five Wessex). In 1968 she suffered a minor fire while refitting, and as the decision had already been taken to run down the carrier force she was summarily put on the disposal list and sold for scrapping the following year.

VICTORIOUS *aircraft carrier*

Displacement: 30,530t standard; 35,500t full load
Dimensions: 740ft pp, 781ft oa × 103ft 6in wl, 157ft fd × 31ft
225.5m, 238.0m × 31.5m, 47.8m × 9.5m
Machinery: 3-shaft Parsons geared steam turbines, 6 Foster Wheeler boilers, 110,000shp = 31kts
Armour: Belt and hangar side 4in, flight deck 3in, hangar deck 2in
Armament: 12–3in/50 Mk 33 (6×2), 6–40mm Mk 6 (1×6), 36 aircraft. See notes
Sensors: Radar 984, 974, 293Q
Complement: 2400

No	Name	Yard	Reconstructed	Fate
R 38	VICTORIOUS	Portsmouth DYd	Oct 1950–1957	BU 1969

Victorious 1.6.67 CMP

EAGLE *aircraft carrier*

Displacement: 43,060t (finally nearly 45,000t) standard; 53,390t full load
Dimensions: 720ft pp, 803ft 9in (finally 811ft 9in) oa × 171ft fd, 112ft 8in wl × 36ft
219.5m, 245.0m (247.4m) × 52.1m, 34.3m × 11.0m
Machinery: 4-shaft geared steam turbines, 8 Admiralty 3-drum boilers, 152,000shp = 31kts. Oil 5500t
Armour: Waterline belt 4in, flight deck 4in–1in, hangar side 1in, hangar deck 2in–1in. See notes
Armament: (1952) 16–4.5in/45 Mk 6 DP (8×2), 48–40mm/60 Mk 6 AA (8×6), 4–40mm/60 Mk 5 AA (2×2), 9–40mm/60 Mk 7 AA (9×1), 60 aircraft; (1964) 8–4.5in (4×2), 6 Seacat GWS22 SAM (6×4), 45 aircraft
Sensors: Radar 960, 982, 983, 275, 262, 974 (later 984, 963, 965 (AKE-2))
Complement: 2750 (inc air group)

No	Name	Builder	Laid down	Launched	Comp	Fate
R 05	EAGLE (ex-*Audacious*)	Harland & Wolff	24.10.42	19.3.46	1.10.51	BU 1978

Originally intended to form one of a class of four 32,500t improved *Implacables*, *Audacious* was quite far advanced at the end of the war and was therefore earmarked for completion more or less as designed. Her name was changed to *Eagle* in January 1946 because the original *Eagle* had just been cancelled. Her surviving sister *Ark Royal*, however, was held up to allow numerous improvements to be incorporated, and so the two ships differed in many details. A new legend was drawn up in 1946 for both ships, raising displacement to 36,970t (standard), 45,720t (deep).

Eagle's design followed closely the pattern of the wartime armoured carriers, with eight twin 4.5in DP gun mountings in quadrantal sponsons, but the close-range armament was the most modern available. Eight six-barrelled 40mm Bofors Mk 6, each with its own CRBF fire control director, were mounted, four on the port side, two on the starboard side and one at each end of the island. In addition there were nine single 40mm Mk 7 on the island and on the port side, and before completion two twin 40mm Mk 5 were added in a sponson under the overhang of the flight deck at the stern. This gave the phenomenal total of twelve radar directors controlling 16–4.5in and 48–40mm barrels, with a further 13–40mm barrels under the local control.

As the only large modern carrier, *Eagle* was a major addition to the RN's strength when she started her builder's trials in October 1951. On trials she made 30.53kts at 44,250t with 156,630shp and 29.6kts at deep load (49,950t). Endurance was calculated at 4500nm at 24kts. She took part in her first major operation, the NATO exercise 'Mainbrace', at the end of 1952, and then went to the Mediterranean. From June 1954 to February 1955 she was under refit, emerging with a 5° angled deck and the new mirror landing sight. Apart from realignment of the arrester wires and other flight deck fittings, the change necessitated the removal of one Mk 6 Bofors mounting from the port side; the three single Mk 7s were also removed from the port side of the island.

On 1 November 1956 *Eagle*'s Sea Hawks, Wyverns and Sea Venoms began six days of attacks on Egyptian airfields and positions during the Anglo-French attempt to seize the Suez Canal. The air group flew 621 sorties despite having a defective catapult. On her return to Devonport she had a short refit and then returned to the Mediterranean, where she spent most of the remainder of the commission.

From mid-1959 to May 1964 she underwent complete rebuilding at Devonport to bring her up to the standard of *Victorious*: a fully (8°) angled deck, steam catapults and the long-range 3-D air warning Type 984 rader were fitted. The opportunity was taken to renew her boiler firebricks, for these had given constant trouble since completion. In addition she was given a short lattice foremast carrying a Type 965 (AKE-2) 'double bedstead' surveillance radar, and the forward 4.5in sponsons were removed to increase internal accommodation. The close-range armament was entirely replaced by six quadruple Seacat GWS22 SAM systems, four sponsoned port and starboard forward and aft, one right aft under the flight deck overhang and one on the starboard side between the island and the after sponson. A less obvious but important improvement was to boost the generating capacity to 8250kW, nearly double the original power, while flight deck armour was reduced from 4in to 1in.

With the growing size and complexity of aircraft, and as the after half of the lower hangar was now given over to workshops, the complement dropped to thirty-five fixed-wing aircraft and ten helicopters, half the complement envisaged in 1942. In 1954 she had operated fifty-nine aircraft, including Sea Hawks, Avengers, AEW Skyraiders and a Dragon-fly helicopter; ten years later she had Sea Vixens, Scimitars and Gannets.

In 1964 the ship sailed for the Far East to help contain the Indonesian Confrontation by maintaining standing air patrols over the Malacca Strait. No sooner was this crisis over than she sailed for Aden to provide reinforcements and then had to provide air defence for Zambia against possible invasion from Rhodesia. The 'Beira Patrol', a blockade to prevent oil from reaching Rhodesia, involved the identification of more than seven hundred ships and the covering of 200,000 square miles of sea each day, and it kept *Eagle* at sea for seventy-one days.

Although the most modern carrier, she fell victim to the 1966 decision to axe the RN's carriers, on the flimsy pretext that she would cost too much to convert to operate Phantoms; in fact she had already operated Phantoms and needed only minor modifications to her flight deck equipment. She was paid off at Portsmouth in January 1972 and was laid up at Devonport later the same year. Although *Eagle* was officially in fully maintained reserve, the difficulty of keeping *Ark Royal* running meant that she was progressively 'cannibalised', and she was finally towed away to the breakers in 1978.

Eagle 1.11.57 CMP

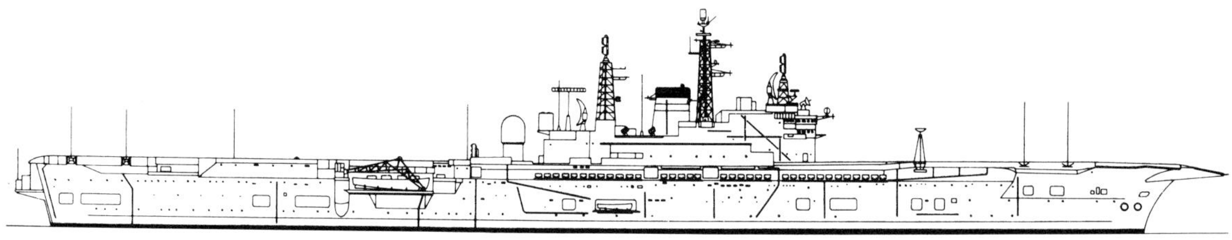

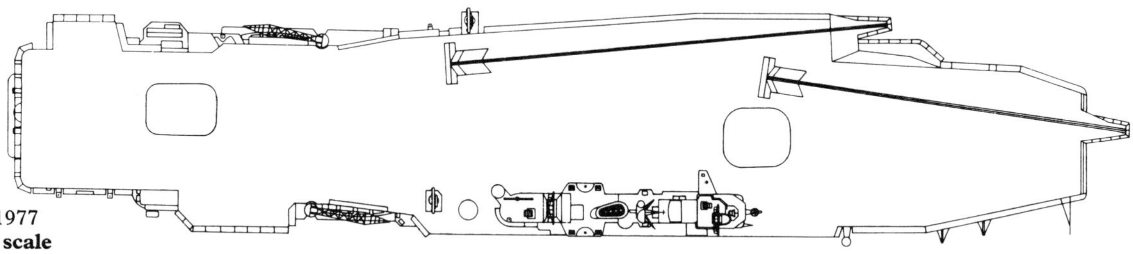

Ark Royal 1977
NB 1/1750 scale

ARK ROYAL *aircraft carrier*

Displacement:	43,340t standard; 53,060t full load
Dimensions:	As *Eagle* except 164ft 6in fd *50.1m*
Machinery:	As *Eagle*
Armour:	As *Eagle*
Armament:	(1955) 12–4.5in Mk 6 DP (6×2), 36–40mm Mk 6 AA (6×6), 4–40mm Mk 5 AA (2×2), 6–40mm Mk 7 AA (6×1), 50 aircraft; (1959) 8–4.5in (4×2), 28–40mm AA (4×6, 2×2), 48 aircraft; (1969) no guns, fitted for, but not with, 4 Seacat GWS22 SAM (4×4), 36 aircraft
Sensors:	Radar 960, 982, 983, 275, 262, 974; later 965 (AKE-2), SPN-35
Complement:	2637 (inc air group)

No	Name	Builder	Laid down	Launched	Comp	Fate
R 09	ARK ROYAL	Cammell Laird	3.5.43	3.5.50	25.2.55	BU 1980

Although often claimed to have been projected as HMS *Irresistible*, there is no official record of this name. The ship was in fact laid down 18 months after the previous *Ark Royal* was sunk, and there was every reason to reserve that name for a new carrier. She was not very far advanced at the end of the war and so the opportunity was taken to change the design considerably, but this made her completion more and more protracted, until she became the butt of comedians' jokes. She finally started trials in June 1954 and entered service in February 1955.

The most important alteration was to provide her with a deck-edge lift, as in US carriers. It was never a great success because it served the upper hangar only, and in any case the device was not really suited to the British type of closed hangar. The port forward 4.5in guns were never installed because the ship went to sea with a 5° angled deck, and this also meant a smaller number of 40mm guns on the port side, as in *Eagle*. She had a tall lattice foremast, and a different island.

As with *Eagle* it proved difficult to accommodate two 210ft steam catapults on the foredeck, and so when these were installed in 1967–70 an unusual deposition was adopted, with one on the port side forward and the other in the waist.

Ark Royal underwent many more changes than *Eagle*. In 1956 the starboard 4.5in turrets were also removed, followed by the forward pairs (port and starboard) of the after group in 1964 and the rest in 1969. During the 1959 refit the side lift was removed to allow the hangar to be enlarged, and in 1961 a deck-landing projector sight was installed, with the 'Hilo' long-range guidance system and stronger steam catapults. Between March 1967 and February 1970 she underwent a major refit to allow the operation of Phantom F-4K interceptors and Buccaneer S2 strike aircraft. This involved removing the last of the 40mm guns; it was planned to replace them with four Seacat missile systems, but these were never installed.

The radar outfit always differed from *Eagle*'s. As built, *Ark Royal* had two Type 982 'hayrake' scanners and two Type 983 height-finders, with Type 262 fitted to the directors controlling the 40mm guns, and Type 275 controlling the 4.5in. In 1959 a 'double bedstead' Type 965M aerial was stepped on a short lattice mast forward of the mainmast; in 1967–70 a second 965M was added on a much taller lattice mainmast, and a large dome abaft the island, housing a Carrier Controlled Approach (CCA) radar of US origin, was fitted.

Although completed later than *Eagle*, the 'Ark' was never as mechanically reliable, and had the 1966 Defence Review not cancelled her replacement (CVA-02) she would have been scrapped first. In 1966 it was felt that she might last until 1972 without major repairs to her machinery, and the fact that she staggered on for another six years is testimony to the heroic efforts of her engineers. Although hard-worked she missed all the important actions except the Indonesian Confrontation, and had it not been for a thirteen-week TV serial called 'Sailor' she might never have endeared herself to the public in the way that she did. Despite efforts to raise money to preserve her the MoD steadfastly refused to sanction any unworthy scheme; in any case the sums talked of were wildly unrealistic, and she was towed away in the autumn of 1980. After she had been paid off at Devonport in December 1978 her name was given to the third of the *Invincible* class.

Ark Royal 1961 CMP

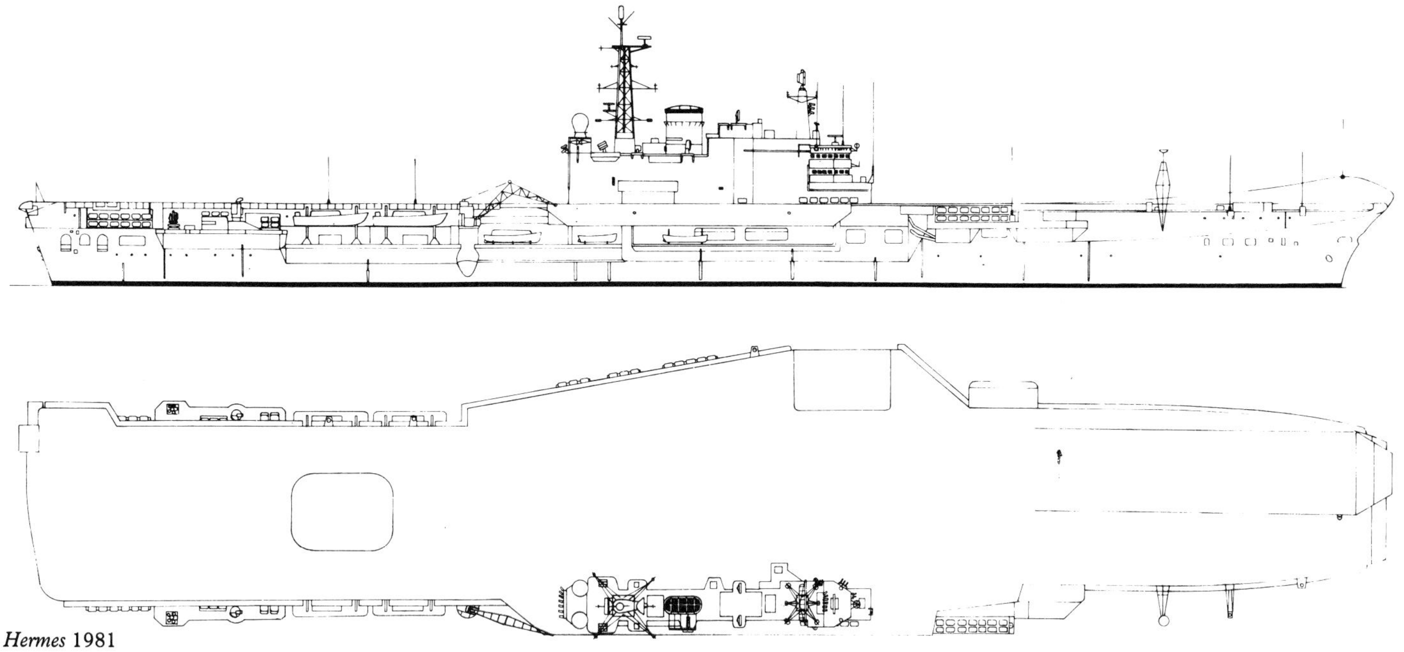

Hermes 1981

When the original order for *Hermes*, sixth of the *Centaur* class, was cancelled in October 1945, the name was given to *Elephant* of the same class; very little work was done on the hull, however, giving time for the specification to be revised. Thus, while the three *Centaur* class went ahead with their 'interim' standard of equipment, *Hermes* was radically redesigned to include a 6° angled deck, a side lift and a Type 984 3-D long-range radar with Comprehensive Display System (CDS).

HERMES *aircraft carrier*

Displacement: 23,900t standard; 28,700t full load
Dimensions: 650ft pp, 744ft 4in oa × 160ft (increased from 144ft 6in in 1964–66) × 29ft mean
198.1m, 226.9m × 48.8m (44.1m) × 8.8m
Machinery: 2-shaft geared steam turbines, 4 Admiralty 3-drum boilers, 76,000shp = 28kts. Oil 4200t
Armour: Mantlets over magazines 40mm, reinforcing to flight deck 19mm
Armament: 10–40mm Mk 5 (5×2) (replaced 1964–66 by 2 Seacat GWS22 SAM (2×3)), 28 aircraft (as completed)
Sensors: Radar 984, 982, 262, 974
Complement: 1830 + 270 air group (original)

No	Name	Builder	Laid down	Launched	Comp	Fate
R 12	HERMES (ex-*Elephant*)	Vickers-Armstrong, Barrow	21.6.44	16.2.53	18.11.59	To India 1987

Although the deck-edge lift was unusual in British carriers (the first was still being installed in *Ark Royal*) it was dictated by the need to find space for the steam catapults at the forward end of the flight deck. Once the forward centreline lift was suppressed the side lift was needed to work aircraft, and it also provided a convenient extension to the angled deck area.

The appearance of the ship was very different from that of her former sisters, with a massive 'dustbin' scanner for the Type 984 radar over the island. The superstructure was considerably enlarged, but the armament was reduced to five twin 40mm Bofors (two port, three starboard) controlled by MRS8 directors. In 1964–66 she was refitted and the Bofors were replaced by two quadruple Seacat close-range SAM defence systems, in the after port and starboard sponsons. Another improvement was the addition of the 'Alaska highway' outboard of the island. In 1971–73 the Type 984 scanner was replaced by a Type 965 'bedstead' and a massive array of deck-landing lights was provided on the foremast and bridge.

The main difficulty with *Hermes* was her small size, for when she commissioned at the end of 1959 she could only operate twenty Sea Vixen, Scimitar and Buccaneer and eight Gannet ASW aircraft. After serving in the Atlantic, Mediterranean, Indian Ocean and Far East she had to be withdrawn from front-line service because she could not operate the new F-4K Phantom, which had replaced the Sea Vixen. In 1971 she paid off for conversion to a commando carrier, losing her arrester wires, catapults and 3-D radar. In 1977 she was rerated as an ASW carrier, with Sea King helicopters in place of the Wessex Mk Is, but she has now undergone conversion to operate a squadron of Sea Harrier STOVL aircraft. This conversion also includes the provision of a 7° ski-jump ramp, extended the full width of the flight deck.

In her new role *Hermes* bridged the gap until completion of *Ark Royal* and served until 1984. Her last air group as an ASW carrier consisted of nine Sea Kings and four Wessex Mk 5s, but at the end of 1980 she embarked five Sea Harriers and nine Sea King Mk 2s. Although threatened with premature scrapping under the 1981 defence cuts, she acted as flagship of the Falklands Task Force April–June 1982, during which time she embarked twenty Sea Harriers and ten Sea Kings. Paid off in 1984, but in 1986 she was refitted at Devonport and sold to India; transferred at Devonport on 12 May 1987 and renamed *Viraat*.

Hermes as completed CMP

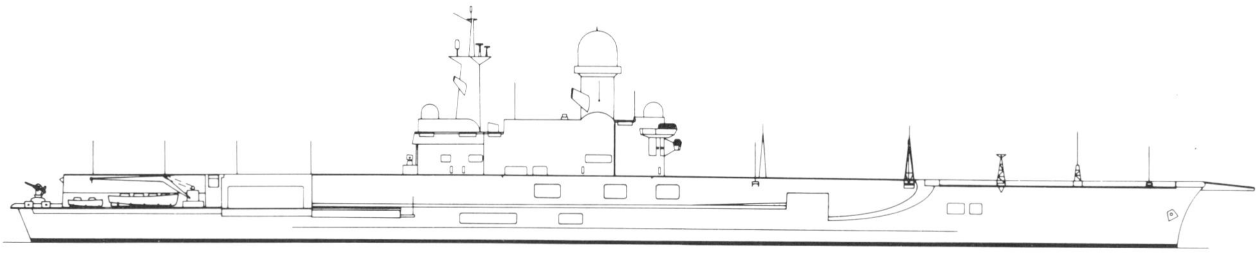

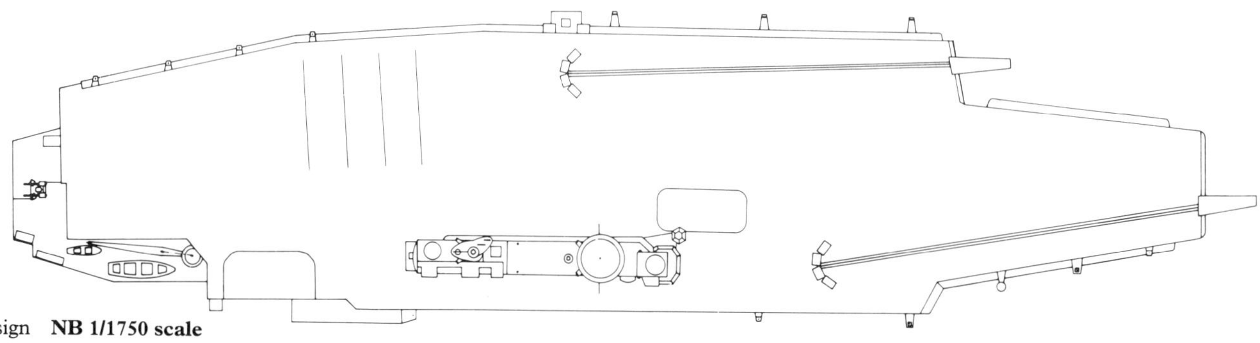

CVA-01 final design **NB 1/1750 scale**

The story of the last class of fleet carriers designed for the RN is a harrowing one – although it must be said that the cancellation of the programme in 1966 was absolutely necessary, if only to save the Navy from getting the wrong carrier – and it will remain a classic example of how a design can be ruined by constraints.

The MoD needed two new fleet carriers to replace *Victorious* and *Eagle* in the early 1970s and their role was seen as providing strike power against airfields and hostile ships, and air defence for the fleet. It was hoped that the project would be backed by the RAF as well, using the planned joint RAF/FAA swing-wing 'strike fighter' as the principal unit of the air group. When that aircraft died stillborn the new carrier design (designated CVA-01 or Attack Carrier, as in the USN) hit its first problem. The second was purely political, for both Lord Mountbatten and his VCNS were certain that the Treasury and the 'bomber lobby' in the RAF would oppose a large carrier. This proved to be the case, and when it was pointed out that a carrier capable of accommodating the latest aircraft and catapults of the necessary length would displace over 60,000t, the politicians asked for the length of catapult to be reduced (!).

To cope with such ill-considered requests the designers cut the catapults from four to two and eliminated the deck park, reducing displacement to 55,000t, but there was still pressure to cut the size to 30,000t, half as big as the original target. Even though the design was changed to a 'furniture van' concept, with very light structure, no armour and only a novel form of underwater protection against torpedoes, it could not be reduced below 50,000t. Speed was restricted to 27–28kts, but after doubts were raised about putting so much power out through two shafts, the old 3-shaft system of the *Illustrious* class was re-adopted, so that one engine could be closed down for maintenance and still leave enough power for Fleet speed.

The flight deck layout was unusual, offset to port, but angled at only 3½°. The passageway (known as the 'Alaska highway') fitted in *Hermes* and other carriers was made even wider, allowing aircraft as well as vehicles to be moved forward and aft without disturbing the deck park. There were other novel features, such as doors at the after end of the hangar to allow the engines to be run inside, and a new 'scissors' type of lift. It had been hoped to have two Anglo-Dutch 'Broomstick' radars housed in radomes forward and aft, but it was soon discovered that two radars of this type would have to be kept very far apart – miles in fact! – to avoid mutual interference, and so one was dropped. Two BS6 catapults were planned, with water-spray type arrester gear.

Displacement had been kept to 50,000t but when 2000t of armour were added, the total continued to be referred to as 50,000t for political reasons. Things got worse. To fit all the requirements in such a small hull, many ingenious ideas were adopted. This was a recipe for disaster, as experience shows that only some 20–25 per cent novelty can be accepted in a design; more than this raises insoluble problems. With hindsight the biggest blunder was the political one of confusing size with cost, but given the Treasury's insistence that CVA-01 should not exceed the full load tonnage of *Eagle* there was very little that either Staff or Ship Department could do. Not for nothing did the project leader refer to the cancellation as the happiest day of his life. The design work was authorised in July 1963 and ended in February 1966 without an order being placed.

CVA-01 class *fleet aircraft carriers*

Displacement: 53,000t standard; *c*63,000t full load
Dimensions: 925ft oa × 184ft fd, 122ft wl
281.9m × 56.1m, 37.2m
Machinery: 3-shaft geared steam turbines, 6 Foster Wheeler boilers, 135,000shp = 28kts
Armour: Some deck and side armour and underwater protection
Armament: Probably 1 Sea Dart GWS30 SAM (1×2), 4 Seacat GWS20/22 SAM (4×4), ?40 aircraft
Sensors: New CDS, 903, 974, etc.
Complement: 2720

No	Name	Builder	Laid down	Launched	Comp	Fate
CVA 01	QUEEN ELIZABETH	–	–	–	–	Cancelled 1966
CVA 02	DUKE OF EDINBURGH	–	–	–	–	Cancelled 1966

Illustrious Jun 1994

C & S Taylor

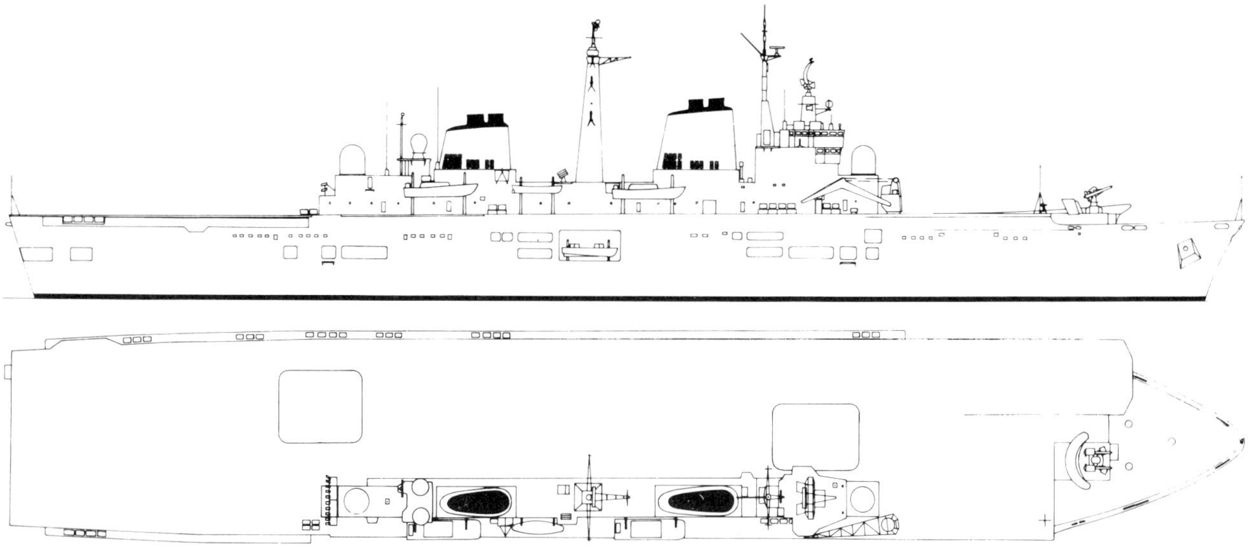

Invincible
as completed

INVINCIBLE class *support carriers*

Displacement: 16,000t standard; 19,500t full load*
Dimensions: 632ft pp, 677ft oa x 115ft fd, 90ft wl x 24ft full load
192.6m, 206.3m x 35.0m, 27.5m x 7.3m
Machinery: 2-shaft COGOG: 4 Olympus gas turbines, 112,000shp = 28kts. Range 5000nm at 18kts. Fuel unknown
Armour: None
Armament: 1 Sea Dart GWS30 SAM (1x2), 2 Phalanx CIWS (added 1982), 14 aircraft. See notes
Sensors: Radar 1022, 909, 1006, etc; sonar 184M
Complement: 557 + 318 air group

No	Name	Builder	Laid down	Launched	Comp	Fate
R 05	INVINCIBLE	Vickers-Armstrong, Barrow	20.7.73	3.5.77	11.7.80	Extant 1995
R 06	ILLUSTRIOUS	Swan Hunter	7.10.76	1.12.78	20.6.82	Extant 1995
R 09	ARK ROYAL (ex-*Indomitable*)	Swan Hunter	14.12.78	2.6.81	1.11.85	Extant 1995

* It has been suggested that these figures are intentionally misleading. If the standard displacement is 19,500t as suspected, her full load tonnage will be about 23,000t.

The death of the fleet carrier replacement programme in February 1966 did not mean the end of plans for air-capable ships. By 1967 an outline Staff Requirement had produced studies for a 12,500t command cruiser carrying six Sea King helicopters. The ship was designated CCH, or helicopter-carrying cruiser, and that is what she was: a derivative of the *Tiger* class, with a missile/gun armament forward and a large external hangar aft, as in the French *Jeanne d'Arc*.

Simple logic dictated that 'cleaning up' the superstructure and moving it to the starboard side would provide more deck space, although any attempt to do this was likely to be regarded by the politicians as a resurrection of the fleet carrier. But studies continued to show that increasing the air group to more than nine helicopters must lead to a 'through deck' or clear flight deck and island superstructure, if only to provide sufficient internal space for the hangar and workshops. Thus, despite the difficult climate, the design staff prepared designs for what were to be called 'through-deck cruisers' rising from 17,500t to 19,500t and with the growth in size it became possible to consider the possibility of using short take-off/vertical landing (STOVL) strike aircraft. The RN had refused to accept the original P1154 STOVL fighter on the grounds that its lack of navigational radar and short endurance would give it a very low capability at sea. But in the interim the designers of the successful subsonic Harrier had pressed ahead with a Sea Harrier equipped with a more powerful engine, more fuel and new avionics which would provide the pilot with computer-aided navigation and attack data. In effect, advances in electronics have made it possible to do away with the observer while still reducing the pilot's work-load.

In May 1975 it was officially announced that the new TDC would carry Sea Harriers, although the capability had long since been designed into the ship to avoid subsequent delays; in service nine Sea Kings and five Sea Harriers are embarked. The Sketch Design was ready at the end of 1970, but the order was not placed until April 1973. For a long time there was no mention of any sisters, but in May 1976 No 2 (*Illustrious*) was ordered, followed by No 3 in December 1978. The third ship was planned to be called *Indomitable*, but immediately after the old *Ark Royal* paid off it was announced that her name would go to the new ship.

There are many novel features in the design. First, it is the largest gas turbine-powered ship in the world, with four Rolls-Royce Olympus units coupled to two shafts. Second, the emphasis is on large internal volume to allow the maintenance of virtually all equipment by exchange; in other words, everything down to the gas turbines and operating consoles has to be removable. This had already been achieved in other gas turbine ships, but the *Invincible* design carries it much farther. Because of the lack of armour and the use of much lighter alloy, the volume of the hull is big for its tonnage.

Invincible is the first ship equipped with the new Type 1022 long-range air warning radar and she also has a double-headed Sea Dart guided weapon-system, with Type 910 trackers at each end of the island and a twin-arm launcher on the centreline forward. The provision of Sea Dart had been criticised for an alleged reduction of the flying capability, but it must be remembered that the Sea Harrier is a subsonic strike aircraft, not a supersonic interceptor, and therefore the ship (and other vessels in company) needs a long-range, rapid-reaction defence against air attack.

While *Invincible* was under construction a very simple improvement was perfected which revolutionised flying at sea. This is the 'ski-jump', a ramp which gives the Sea Harrier sufficient lift to enable it to take off using less fuel or carrying more payload. In *Invincible* and *Illustrious* it takes the form of a 47t light steel structure welded on to the port side of the flight deck, rising at an angle of 7° over the bow, but it has the effect of adding 1500lb of payload to the Sea Harrier; *Ark Royal* will have a steeper ramp to improve performance further.

The euphemism 'through-deck cruiser' (CAH) was subsequently changed to 'command cruiser' which was a reasonable title: much of the cost of *Invincible* is accounted for by 'Cu23' (Communications, Command and Control), and above all she is a flagship, having effective control of submarines and surface ships. By the time that *Invincible* came into service, however, the term 'carrier' was back in favour, and she is now rated as a support carrier (CVS).

It remains to be seen what developments in STOVL aircraft will take place in the next decade or so, but the *Invincible* class promise to be very effective anti-submarine ships, taking the most potent ASW helicopters far out to sea, but able to protect them from attack by enemy bombers. Their principal role is clearly to patrol the Greenland-Iceland-UK Gap in support of ASW forces operating behind the SOSUS barrier. As the first Western ships designed to exploit fixed-wing STOVL aircraft they are having a marked influence on other navies.

Following the 1981 Defence Review *Invincible* was offered to the Royal Australian Navy as a replacement for *Melbourne*. The offer was accepted, subject to some modifications to fuel stowage, communications, etc, when the Falklands Crisis intervened. She sailed with *Hermes* on 5 April, but did not return until relieved by *Illustrious* in late August. During the campaign she embarked nine Sea Harriers and ten helicopters. *Illustrious* was rushed to completion with the addition of two 20mm Phalanx 'Gatling' guns (one forward to starboard of Sea Dart, one right aft on starboard side of the flight deck). *Invincible* was similarly fitted but *Ark Royal* was fitted with three Goalkeeper CIWS (one in bows on centreline, one sponsoned on port side at after end of the flight deck, and one sponsoned to starboard of the island).

Invincible refitted at Devonport during April 1986 – January 1989 to similar standard as *Ark Royal*, with a longer ski-jump and 12 degrees exit angle, enlarged accommodation etc. *Illustrious* was to follow in 1989 but she was laid up in reserve for two years because of financial problems. Preliminary work started on this £100m refit in August 1991, and full reconstruction in mid-November. She and her sisters will also receive enhanced EW and C³I systems.

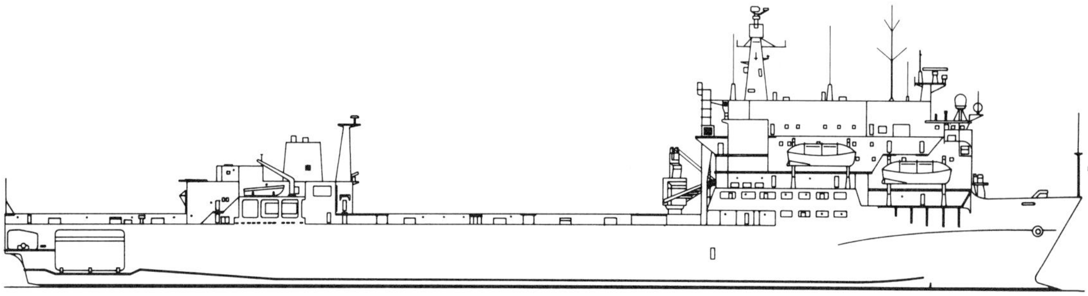

Argus 1989

Although not strictly an aircraft carrier, being a replacement for RFA *Engadine*, this former roll-on/roll-off and containership functions as an auxiliary carrier, as the choice of an old carrier name signifies. She has a wartime role as a ferry carrier and has taken part in exercises as a 'spare deck'. In the Gulf conflict in 1991 she was hurriedly converted to a hospital ship but also operated HC4 Commando troop-carrying helicopters. Her peacetime role is to take ASW helicopters out to their deepwater exercise areas (saving fuel and flying time), and in wartime she can ferry up to twelve Sea Harrier STOVL aircraft. She has a naval-standard ops room and AIO, and full communications equipment to enable her to operate with the Fleet. She is civilian-manned by the RFA, with naval air group and command team embarked.

She was taken up from trade in 1982 and returned to her owners, but purchased outright in 1984. Converted to ATS by Harland & Wolff during April 1984–March 1988, she was dedicated on 1 June 1988, relieving *Engadine* the following year. During conversion three watertight bulkheads were added and to compensate for weights and a 5ft layer of concrete was also added above the hangar (serving incidentally as splinterproof protection).

ARGUS *aviation training ship*

No	Name	Comp	Fate
A 135	ARGUS (ex-*Contender Bezant*)	1989	Extant 1995

Argus 1993 Harland & Wolff

The *Neptune* design of 1944 was dropped at the end of the war, and in its place a new design was developed, known as the *Minotaur* class. Instead of the triple 6in Mk 25, the ships would have had five twin Mk 26 mountings, capable of rapid fire (12–15rpm) at 80° elevation. The layout would have been virtually an enormously expanded *Dido* with three turrets forward and two aft, but there would also have been a heavy secondary battery of 3in guns and TT. They would have been much inferior to the US *Worcester*s owing to their heavy electrical equipment.

The most important innovations were a flush weather deck and a 'staggered' machinery layout. The design was clearly influenced not only by war experience, but by the magnificent designs produced by the US Navy for the Pacific. Three main design stages were reached, following a Controller's meeting early in 1946. Each time the design was slightly reduced in size, but finally lack of money forced the Navy to abandon any idea of building a new cruiser class for the time being.

Although four ships, *Minotaur*, *Centurion*, *Edgar* and *Mars* had been planned in 1945, no contracts were placed.

A decision was deferred on the *Minotaur* class, allowing work to continue on refining the design. The last revision found is dated 1951, by which time the need for other types of warship was taking priority. It must be assumed that the project had died by 1952–53.

MINOTAUR class (1947 design) *cruisers*

Displacement: 14,300t standard; 17,260t deep load
Dimensions: 616ft wl × 74ft × 24ft mean
187.7m × 22.5m × 7.3m
Machinery: 4-shaft geared steam turbines, 4 Admiralty 3-drum boilers, 120,000shp = 31kts. Oil 2640t
Armour: Side 3in, deck 1in, bulkheads 3in–2in, turrets 4in
Armament: 10–6in/50 QF Mk 26 (5×2), 16–3in/70 Mk 6 (8×2), 16–21in TT (4×4)
Complement: about 1300

MINOTAUR class (1951 design) *cruisers*

Displacement: 15,280t standard; 17,940t deep load
Dimensions: 645ft wl × 75ft × ?26ft
196.6m × 22.9m × 7.9m
Machinery: 4-shaft geared steam turbines, 4 boilers, 100,000shp = 31kts. Oil 2300t
Armour: Side 3in, deck over machinery 1in (1in elsewhere), magazines 1in
Armament: 10–6in/50 QF Mk 26 (5×2), 8–3in/70 Mk 6 (4×2), 15 DC, 4–3pdr saluting (4×1)
Complement: about 1300

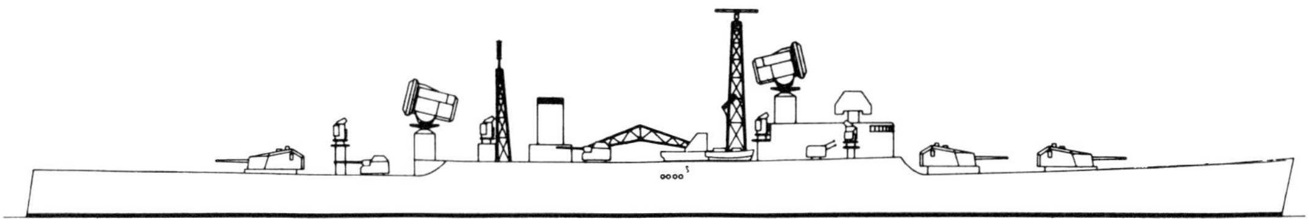

The 1960 medium cruiser design

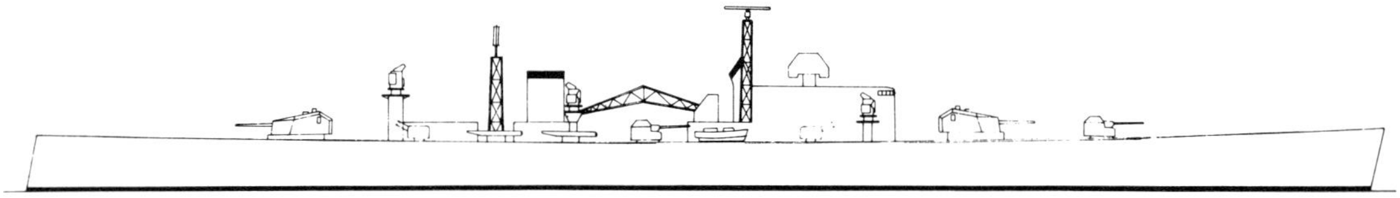

The 1960 small cruiser design

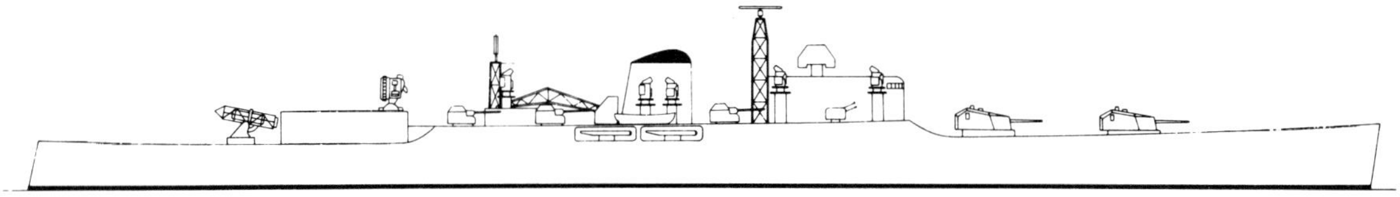

The 1960 missile-armed large cruiser design

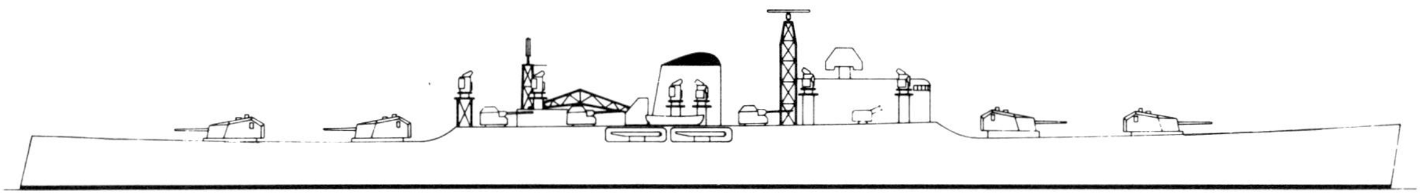

The 1960 gun-armed large cruiser design

In the years immediately after the war it was felt that a traditional cruiser with the ability to operate independently for long periods was still necessary to carry out the duties of trade protection and area anti-aircraft defence of fleets or convoys. However, it was also believed that such a ship had an offensive function as well, and these joint requirements led to a series of design studies around 1948–49 for ships which could be in service in about ten years' time, and were therefore referred to as '1960 cruisers'. They were to have some or all of the following features: surface gun armament; long and short range AA armament (with missiles if possible); speed, endurance and seakeeping; at least splinter protection to vitals; torpedo, and ASW armament; aircraft direction capability; command and communications facilities. Radical solutions, such as nuclear propulsion, were deliberately excluded in order to meet the proposed schedule, which in turn was determined by the delivery of major new weapon systems – the twin 6in Mk 26 and 3in/70 in 1953, 5in/70 and a new close-range weapons (DACR) in 1957 and the Seaslug SAM in 1958. Conventional steam machinery was adopted (improved *Daring* type), and the armament was arranged to obviate all physical fouling between guns (necessary because of high training speeds and automatic control).

Details of four variants are included here, although there is no evidence that the construction of any type was ever seriously planned. A light cruiser design was under consideration in mid-1949, but very little information has survived.

1949 Light Cruiser

Displacement: 4500t deep load
Dimensions: ?
Armament: ? × 5in/80 (?×1), TT
Sensors: Radar 2 Type 984
Complement: ?

1960 Medium Cruiser

Displacement: 12,250t standard; 14,750t deep load
Dimensions: 600ft wl × 70ft × 23ft
182.9m × 21.3m × 7.0m
Machinery: 4-shaft geared steam turbines, 90,000shp = 31kts
Armament: 6–6in/50 Mk 26 (3×2), 8–3in/70 Mk 6 (4×2), 2 DACR mountings (2×6?), 8 TT (16 torpedoes)
Sensors: Radar 984, 960, 992, gunnery sets; sonar not known
Complement: ?

1960 Small Cruiser

Displacement: 10,500t standard; 13,000t deep load
Dimensions: 585ft wl × 68ft × 22ft
178.3m × 20.7m × 6.7m
Machinery: 4-shaft geared steam turbines, 90,000shp = 31kts deep clean
Armament: 4–5in/56 DP (2×2), 6–3in/70 Mk 6 (3×2), 4 DACR mountings (4×6?), 16–21in QR TT (4×4)
Sensors: Radar 960, 992, gunnery sets; sonar not known
Complement: ?

1960 Large Cruiser (Missile-Armed)

Displacement: 14,000t standard; 17,000t deep load
Dimensions: 630ft wl × 73ft × 22ft
192.0m × 22.3m × 6.7m
Machinery: 4-shaft geared steam turbines, 95,000shp = 31kts
Armament: 4–5in/56 DP (2×2), 12–3in/70 Mk 6 (6×2), 2 DACR mountings (2×6?), 1 Seaslug SAM (1×2; 48 missiles)
Sensors: Radar 960, 992, 901, gunnery sets; sonar not known
Complement: about 900

1960 Large Cruiser (Gun-Armed)

Displacement: 14,500t standard; 17,500t deep load
Dimensions: 645ft wl × 74ft × 23ft
195.6m × 22.6m × 7.0m
Machinery: 4-shaft geared steam turbines, 90,000shp = 31kts deep clean
Armament: 8–5in/56 (4×2), 12–3in/70 Mk 6 (6×2), 2 DACR mountings (2×6?), 16–21in QR TT (4×4)
Sensors: Radar 960, 992, fire control sets; sonar not known
Complement: 900–950

Missile Cruiser conversions

Numerous proposals were looked at in the post–1945 period for converting existing hulls into missile cruisers. Space allows no more than a summary, but they ranged from the four *King George V* class battleships and *Vanguard* to the 'Town' class cruisers and even carriers. Apart from the very high cost, most proposals suffered from the comparatively small number of missiles which could be installed. Existing cruisers were either too old (the 'Town' class) or too cramped (the 'Colony' group) to accommodate the large number of Sea Slug missiles required.

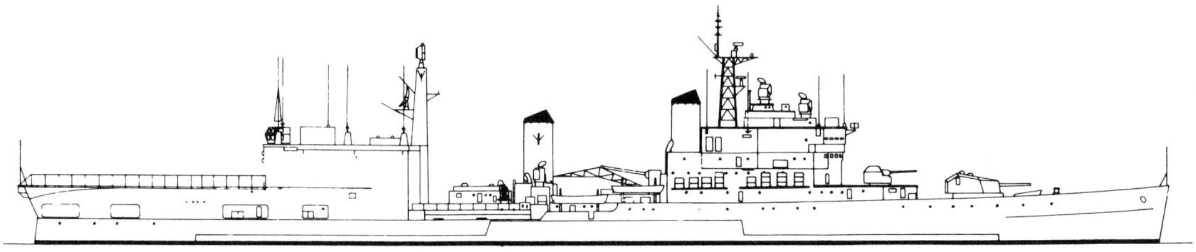

Blake 1969

TIGER class *cruisers*

Displacement: 9550t light load; 11,700t deep load. See notes
Dimensions: 538ft pp, 555ft 6in oa × 64ft × 23ft
163.8m, 169.3m × 19.5m × 7.0m
Machinery: 4-shaft geared steam turbines, 4 Admiralty 3-drum boilers, 80,000shp = 31.5kts. Range 2100nm/6500nm at 31.5kts/13kts
Armour: Belt 3in–3in, bulkheads 2in–1in, turrets 2in–1in, crowns to engine rooms and magazines 2in
Armament: 4–6in/50 Mk 26 (2×2), 6–3in/70 Mk 6 (3×2). See notes
Sensors: Radar 960, 992Q, 277Q, 903. See notes
Complement: 880

No	Name	Builder	Laid down	Launched	Comp	Fate
C 20	TIGER (ex-*Bellerophon*)	John Brown	1.10.41	25.10.45	18.3.59	BU 1986
C 34	LION (ex-*Defence*)	Scotts	24.6.42	2.9.44	20.7.60	BU 1975
C 99	BLAKE (ex-*Tiger*, ex-*Blake*)	Fairfield	17.8.42	20.12.45	8.3.61	BU 1982

Although the decision to complete the last three of the *Tiger* class to a new design was made by the Board of Admiralty in 1951 (a move strongly opposed by many), the dismantling of the old structure did not start until 1954. The new design had two twin 6in Mk 26 DP automatic mountings, one forward and one aft, with three twin 3in Mk 6 AA mountings in 'B' position and abreast the second funnel.

The internal fittings reflected the needs of a new generation of naval warfare; air conditioning throughout, a 200-line telephone exchange and a computer-aided Action Information Organisation (AIO). Each mounting had its own MRS3 fire-control set (basically an Anglicised version of the US Navy's Mk 56 with a Type 903 dish scanner): there were two above the bridge, one aft and one on each side amidships. Radars included a Type 960 masthead dipole array for air warning (range 170 miles), Type 992 for surface search at the masthead (range 30 miles), and a Type 277Q height-finder on the mainmast (range 120 miles).

The armament was impressive, each water-cooled 6in barrel capable of delivering 20rpm. The 3in could theoretically deliver up to 120rpm; in practice 90rpm was the best achieved.

Like most attempts to produce perfection, the *Tigers* were too late, arriving on the scene just when the gun solution to AA defence was going out of fashion. Although fine, modern ships, there was little they could do other than engage smaller warships with gunfire or bombard shore positions. Their excellent command and communications facilities, however, still made them useful as task group flagships, and it was decided to give them more flexibility by replacing one 6in turret with a hangar and flight deck for four helicopters. This was approved, and in 1965–69 *Blake* was refitted at Portsmouth, followed by *Tiger* at Devonport in 1968–72. Lack of money prevented *Lion* from being taken in hand and she was put on the disposal list in 1975, but not before so much equipment had been 'cannibalised' that her sister was nicknamed 'Liger'. As reconstructed, displacement rose to 9975t standard, 12,080t deep load, and armament was now 2–6in Mk 26 (1 × 2), 2–3in Mk 6 (1 × 2), 2 Seacat GWS22 SAM systems (2 × 4) and four Wessex (later Sea King) helicopters. New sensors consisted of radar Types 965, 992Q, 278 and 903.

Lion as completed CMP

As refitted *Blake* had a particularly ungainly profile, with a huge box hangar from mainmast to stern. The midships 3in turrets had to go, leaving her with the somewhat anomalous feature of two different calibres of gun in 'A' and 'B' positions. When *Tiger* emerged in 1972 her funnels were considerably taller, with squared-off caps, and after another commission this improvement was given to *Blake* as well. Both ships proved useful in their new role, but their big crews made them expensive to maintain, and when the manpower shortage hit the Royal Navy in the late 1970s they were prime candidates for the disposal list. In any case the recommissioning of *Hermes* and *Bulwark* had remedied the previous shortage of ASW helicopter ships, and so in 1979 *Tiger* was put on the disposal list, while *Blake* was laid up at Chatham in January 1980. *Blake* was not sold until 1982 while *Tiger* lasted until 1986.

Daring 1954 MoD

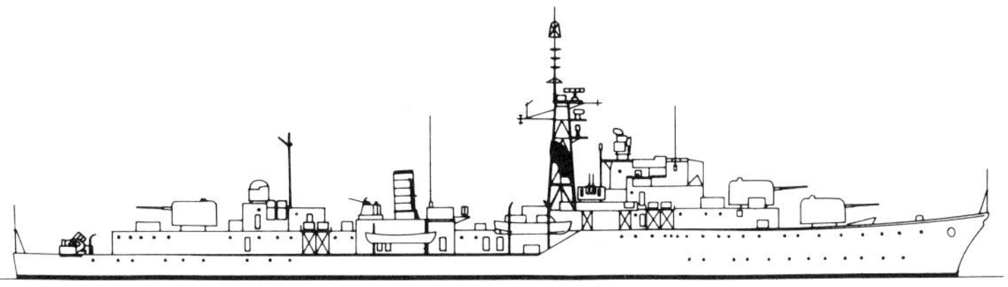

Defender 1969

DARING class *destroyers*

Displacement: 2830t standard; 3580t deep load
Dimensions: 366ft pp, 390ft oa × 43ft × 13ft 7in mean deep load
111.6m, 118.8m × 13.1m × 4.1m
Machinery: 2-shaft double-reduction geared turbines, 2 Foster Wheeler (*Daring, Delight, Decoy, Diana* Babcock & Wilcox) boilers, 54,000shp = 34.75kts. Oil 590t
Armament: 6–4.5in/45 Mk 6 (3×2), 2 to 6–40mm/60 Mk 5 (1–3×2) (*Decoy, Diamond, Diana, Duchess* 2–40mm/60 Mk 9), 1 Squid
Sensors: Radar 903, 262, 293; sonar 170, 174/177
Complement: 297–330

No	Name	Builder	Laid down	Launched	Comp	Fate
D 108	DAINTY	White	17.12.45	16.8.50	26.2.53	BU 1972
D 05	DARING	Swan Hunter	29.9.45	10.8.49	8.3.52	BU 1971
D 106	DECOY (ex-*Dragon*)	Yarrow	22.9.46	19.3.49	28.4.53	To Peru 1970
D 114	DEFENDER (ex-*Dogstar*)	Stephen	22.3.49	27.7.50	5.12.52	BU 1972
D 119	DELIGHT (ex-*Disdain*, ex-*Ypres*)	Fairfield	5.9.46	21.12.50	9.10.53	BU 1971
D 35	DIAMOND	John Brown	15.3.49	14.6.50	21.2.52	Harbour TS 1970, BU 1981
D 126	DIANA (ex-*Druid*)	Yarrow	3.4.47	8.5.52	29.3.54	To Peru 1970
D 154	DUCHESS	Thornycroft	2.7.48	9.4.51	23.10.52	To Australia 1964

Although similar big destroyers were built abroad, in 1953 the Admiralty decided that they should be rated as '*Daring* type ships' and manned as light cruisers, under a captain, with bugle-calls instead of pipes, etc. These ships were divided into two groups: *Dainty, Daring, Defender* and *Delight* had a 220v DC electrical installation whereas the others had the new 440v AC system which was to become standard. Starting in 1963 the class was modernised, with all TT removed (the after bank had been removed in 1958–59), the STAAG Bofors mounting in the bridge wings removed and the Mk 6M director replaced by an MRS3 type. The DC group received two more twin Mk 5 mountings in place of the STAAG, but the others had Mk 7 singles in the bridge wings.

In 1963 *Decoy* was fitted with a quadruple Seacat short-range missile launcher abaft the second (visible) funnel, but although firing trials were successful it was decided not to refit the rest of the class, and the launcher was removed. The system was MRS8, a reworked CRBF director with new below-decks equipment.

Diana and *Daring* were temporarily fitted with a streamlined after funnel casing a short while after completion, at the personal wish of Lord Mountbatten; it improved their looks, but restricted the arcs of fire of the aft twin Bofors and was removed after a few years. The ships served on all stations and gave good service, but as the design had already been stretched there was insufficient room for new sensors and weaponry. *Duchess* was transferred to the Royal Australian Navy in 1964 to replace *Voyager* and retained her name, and in 1970 *Decoy* and *Diana* were sold to Peru, becoming *Ferré* and *Palacios*.

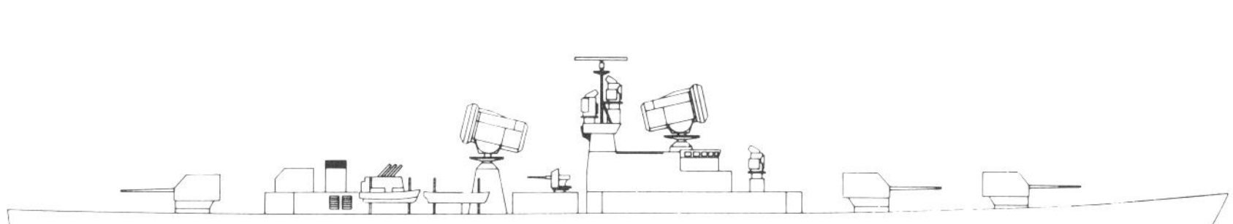

The 1953 destroyer design

1953 Destroyer

Displacement: about 3000t standard; about 3800t deep load
Dimensions: 450ft pp × 47ft 3in × 12ft 8in
137.2m × 14.4m × 3.9m
Machinery: 2-shaft geared steam turbines, 60,000shp = 38kts
Armament: 3–5in/56 Mk 2 DP (3×1), 2–3in/50AA (2×1), 1 Ruler, 1 Limbo Mk 10, 8–21in TT
Sensors: Radar 984 (2), 992, 903; sonar not known
Complement: ?

To follow the *Daring* class it was proposed to build another class of large destroyers, but the Board of Admiralty's obsession with the threat from Soviet *Sverdlov* class cruisers (abetted by the usual exaggerated intelligence assessments) led to pressure to provide gunpower. After numerous studies an attempt was made to find commonality with the Army's Green Mace gun design. This was a massive mobile 5in/56 AA gun intended to use folding-fin, discarding-sabot (FFDS) ammunition and fire 96rpm. A constructor was lent to RARDE at Fort Halstead to study problems of adapting the installation to naval use.

Although the Naval Staff originally wanted a twin mounting for this class, a single mounting proved adequate. Even when scaled down from 96 to 80rpm the rate of fire was calculated to be sufficient to overwhelm the *Sverdlov*'s dozen 6in guns. To enable the ship to close the range rapidly and make herself a difficult target she was intended to make 38kts.

The result, after many changes of design (sixty-four different design stages are recorded), was a large and remarkably modern-looking warship, more reminiscent of the later USN CGs than their equivalents in the RN, the 'Counties'. They would have had two pyramidal 'macks', fixed TT in the superstructure and a 'Ruler' anti-torpedo system aft. But a growing belief by Lord Mountbatten that guns could not cope with the air threat led to its being dropped in 1955 in favour of a DLG design.

The 1951 design shown is an intermediate stage, with two Type 984 3-D comprehensive display radars, a single 'mack' and three MRS3 directors. The 'Ruler' shown aft is an anti-torpedo system similar to Hedgehog, although with proximity rather than contact fuzing, firing multiple charges to detonate in the path of a torpedo. What is also unusual is the pair of single 3in/50 US-pattern guns amidships.

CAVENDISH class *destroyers*

Displacement: 1964–2053t light; 2675t deep load
Dimensions: 339ft 6in pp, 362ft 9in oa × 35ft 8in × 14ft 3in–14ft 6in mean deep load
103.5m, 110.6m × 10.9m × 4.3m–4.4m
Machinery: 2-shaft Parsons geared turbines, 2 Admiralty 3-drum boilers, 40,000shp = 36.75kts. Range 2800nm at 20kts
Armament: 3–4.5in/45 Mk 4 on CPS mountings (3×1), 2–40mm Mk 5 AA (1×2), 2–40mm/60 Mk 9 (2×1), 2 Squid, 4–21in Mk 8 TT (1×4)
Sensors: Radar 275, 293P or Q, 978/974, 944
Complement: 192 peace; 264 max

No	Name	Builder	Laid down	Launched	Comp	Fate
D 07	CAESAR	John Brown	3.4.43	14.2.44	5.10.44	BU 1967
D 15	CAVENDISH	John Brown	19.5.43	12.4.44	13.12.44	BU 1967
D 85	CAMBRIAN	Scotts	14.8.42	10.12.43	17.7.44	BU 1970
D 30	CARRON	Scotts	26.11.42	28.3.44	6.11.44	BU 1967
D 01	CAPRICE	Yarrow	28.9.42	16.9.43	5.4.44	BU 1979
D 10	CASSANDRA	Yarrow	30.1.43	29.11.43	28.7.44	BU 1967
D 25	CARYSFORT	White	12.5.43	25.7.44	20.2.45	BU 1970
D 73	CAVALIER	White	28.2.43	7.4.44	22.11.44	Preserved 1977

In 1953 the first of the class (*Carron*) was taken in hand for modernisation and by 1961 all eight had been recommissioned. The conversion enhanced the ships' capabilities as general-purpose escorts, with the latest Mk 6M fire control, remote power control (RPC) fitted to the guns, and the after TT replaced by a deckhouse with two Squid ASW mortars and a twin Mk 5 Bofors AA gun. *Caesar*, *Cambrian*, *Caprice* and *Cassandra* had an enclosed frigate-style bridge, which was left open on the others of the class. In 1966 *Caprice* and *Cavalier* received a quadruple Seacat GWS20 missile system in place of the Mk 5 Bofors, and this necessitated a big handing room on top of the original deckhouse. Shortly after recommissioning *Carron* was attached to the Dartmouth Training Squadron, with a large navigating bridge in place of 'B' gun. In 1960 she became a navigation tender attached to *Dryad* and had the rest of her armament removed.

Cavalier paid off in 1972, but avoided the breakers. Being the last RN 'conventional' destroyer, strenuous efforts were made to preserve her as a memorial. Finally she was acquired by the HMS *Cavalier* Trust in 1977 and is now open to the public on the Tyne.

Cavalier 25.7.79 — MG Photographic

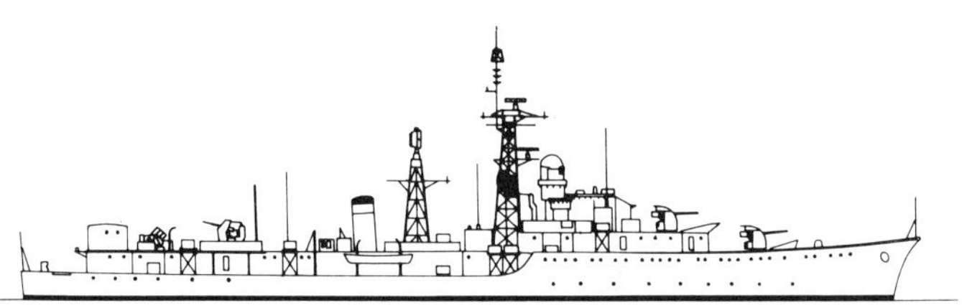

Crossbow 1964

'WEAPON' class *radar pickets*

Displacement: 2280t normal; 2935t full load
Dimensions: 341ft 6in pp, 365ft oa × 38ft × 14ft 8in mean deep load
104.1m, 111.2m × 11.6m × 4.5m
Machinery: 2-shaft Parsons geared turbines, 2 Foster Wheeler boilers, 40,000shp = 34kts. Range 5000nm at 15kts
Armament: 4–4in Mk 16 AA (2×2), 4–40mm/60 STAAG Mk 2 AA (2×2), 2–40mm/60 Mk 9 AA (2×1), 2 Squid (*Scorpion* 1 Limbo Mk 10)
Sensors: Radar 965 (AKE-1), 293, 262; sonar 170, 174/177
Complement: 234

No	Name	Builder	Laid down	Launched	Comp	Fate
D 118	BATTLEAXE	Yarrow	22.4.44	12.6.45	23.10.47	BU 1964
D 31	BROADSWORD	Yarrow	20.7.44	5.2.46	4.10.48	BU 1968
D 96	CROSSBOW	Thornycroft	26.8.44	20.12.45	4.3.48	BU 1972
D 64	SCORPION (ex-*Tomahawk*, ex-*Centaur*)	White	16.12.44	15.8.46	17.12.47	BU 1971

In 1958–59 all four vessels were converted to radar-pickets, with a tall lattice mast between the (visible) funnel and the foremast, carrying a Type 965 long-range radar array. The TT were removed and a continuous deckhouse ran aft as far as the original superstructure. The opportunity was taken to move *Battleaxe*'s and *Broadsword*'s twin Squid mortars and handing room from 'B' position back aft, and to resite the twin 4in gun forward; they were now identical with the other two ships in layout. The old Mk 6 director over the bridge was replaced by an MRS8 director.

On 1 August 1962 *Battleaxe* was badly damaged in a collision with the frigate *Ursa* in the Clyde and was laid up. The class were superseded by the 'Battle' conversions, with their more elaborate air-plotting facilities, and were put on the disposal list 1964–66.

Broadsword 1962 CMP

In 1960 four of the second group were taken in hand for conversion to radar-pickets. This involved stepping a larger lattice foremast to carry the Type 965 radar array, with its 'double bedstead' AKE-2 antenna. The original AA and TT armament was removed and a new after superstructure built, with a short lattice mast carrying a Type 978 height-finder array. Abaft this was an MRS8 director and a quadruple Seacat missile launcher. The original Mk 37 fire control director was left and so was the Squid ASW mortar on the quarter-deck. Internally the ships were totally rebuilt to improve habitability and to provide a new AIO.

All four conversions were completed by the spring of 1962, but the ships were all on the disposal list in 1970. They were originally DC (direct current) ships and had to converted to AC (alternating current) to provide power for the new radars and the Seacat missile system.

'BATTLE' class *radar pickets*

Displacement: 2780t standard; 3430t full load
Dimensions: 355ft pp, 379ft oa × 40ft 6in × 15ft 4in mean deep load
108.2m, 115.5m × 12.3m × 4.7m
Machinery: 2-shaft Parsons geared turbines, 2 Admiralty 3-drum boilers, 50,000shp = 35.75kts. Range 4400nm at 15kts
Armament: 4–4.5in Mk 5 DP (2×2), 2–20mm (2×1), 1 Seacat GWS22 SAM (1×4), 1 Squid
Sensors: Radar 965 (AKE-2), 293, 262 (Seacat), 278; sonar 174/177, 170
Complement: 268

No	Name	Builder	Laid down	Launched	Comp	Fate
D 86	AGINCOURT	Hawthorn Leslie	12.12.43	29.1.45	25.6.47	BU 1974
D 22	AISNE	Vickers-Armstrong, Tyne	26.8.43	12.5.45	20.3.47	BU 1970
D 68	BARROSA	John Brown	28.12.43	17.1.45	14.2.47	BU 1978
D 97	CORUNNA	Swan Hunter	12.4.44	29.5.45	6.6.47	BU 1975

Agincourt Sep 1964 MG Photographic

Recent research by the World Ship Society has unearthed a hitherto unsuspected project, the 1946 Fleet Aircraft Direction Escort (FADE). It was to be longer than the *Daring*s, with a maximum speed of 31.5kts, and would carry two large Type 980 radars (later replaced by Type 960). Priority was given to the Type 41 and Type 61 frigates, so design work was slow, and the project appears to have been dropped in 1949. At least two ships were planned, and in 1947 a FADE conversion of an *Abdiel* class minelayer was considered. The FADE mission would have been to extend early warning of hostile aircraft, to extend the range at which friendly aircraft could be controlled and to act as a rallying point for returning aircraft.

1946 Fleet Aircraft Direction Escort Project

Displacement: *c*2500t
Dimensions: 391ft × ? ×
119.17m × ? × ?
Machinery: 2-shaft geared steam turbines, 2 boilers, 54,000shp = 31.5kts. Range 4400nm at 20kts
Armament: 4–4.5in/45 Mk 6 (2×2) or 4–3in/70 (2×2), 4–40mm/60 STAAG (2×2), 2–40mm/60 Mk 7 (2×1), 1 Type D mortar
Sensors: (i) Radar 2 Type 980, 2 Type 981, 275, 262(5), and 293
(ii) Radar 2 Type 960, 982, 983
(iii) Radar 2 Type 984
Complement: 344 rising to 397+

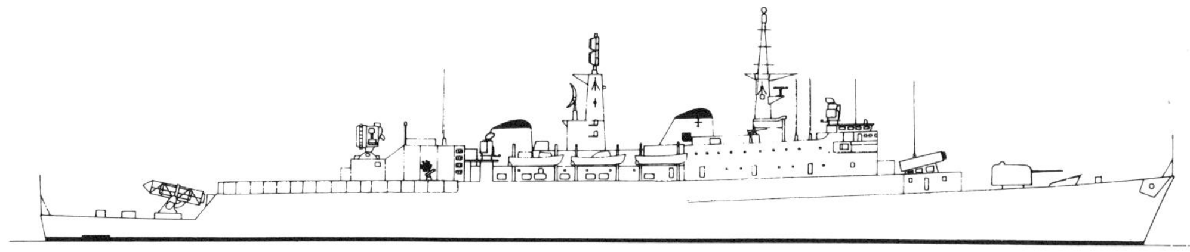

Glamorgan 1976

'COUNTY' class *guided missile destroyers*

Displacement: 6200t normal; 6800t full load
Dimensions: 505ft pp, 521ft 6in oa × 54ft × 20ft 6in max
153.9m, 158.9m × 16.4m × 6.2m
Machinery: 2-shaft COSAG: geared steam turbines, 2 Babcock & Wilcox boilers, plus 4 G6 gas turbines, 30,000shp + 30,000shp = 30kts. Range 3500nm at 28kts
Armament: 4–4.5in/45 Mk 6 (2×2), 1 Seaslug SAM (1×2), 2 Seacat GWS21 (later ships GWS22) SAM (2×4), 2–20mm (2×1), 1 helicopter. See notes
Sensors: Radar 965 (AKE-1 or -2), 277/278, 992Q/993, 901, 903/904; sonar 170, 174/177 (replaced by 184)
Complement: 440–471

No	Name	Builder	Laid down	Launched	Comp	Fate
D 02	DEVONSHIRE	Cammell Laird	9.3.59	10.6.60	15.11.62	Sunk as target 17.7.84
D 06	HAMPSHIRE	John Brown	26.3.59	16.3.61	15.3.63	BU 1979
D 16	LONDON	Swan Hunter	26.2.60	7.12.61	14.11.63	To Pakistan 1982
D 12	KENT	Harland & Wolff	1.3.60	27.9.61	15.8.63	HS 1980
D 20	FIFE	Fairfield	31.5.62	9.7.64	21.6.66	To Chile 1987
D 19	GLAMORGAN	Vickers-Armstrong, Tyne	13.9.62	9.7.64	13.10.66	To Chile 1986
D 18	ANTRIM	Fairfield	20.1.66	10.10.67	14.7.70	To Chile 1984
D 21	NORFOLK	Swan Hunter	15.3.66	16.11.67	7.3.70	To Chile 1982

Devonshire 1977 C & S Taylor

The term 'destroyer' was applied to these ships to obtain Treasury approval, but they were later rated as DLGs, from the American term 'Destroyer Leader, Guided Missile', applied to the new generation of big fleet escorts. As in the USN ships, however, destroyer principles governed their design: there was no longitudinal subdivision and the ships had relatively light scantlings to keep down size and weight.

Like the *Whitbys* the design was volume-critical, and in particular it centred on the Seaslug missile system. Seaslug was a first-generation beam-rider, with its four booster-motors wrapped around the body. British guided weapons of the period relied on wrap-around boosters because it was feared (incorrectly) that end-on mounting of the booster would produce instability in the missile after launch.

Thus although the new DLG had ample vertical space between decks for upright stowage and loading through deck-hatches, she had to be given a horizontal 'tunnel' through the superstructure. The missiles are brought up from the forward magazine on a hoist and then travel through the tunnel, undergoing checkouts, fitting of wings, etc, before being run out on to the launcher. This was a somewhat over-engineered, massive lattice-work affair weighing as much as a gun turret. Although the figure of thirty-six rounds is quoted, it is believed to be less in view of their bulk.

Four ships were voted under the 1955–56 and 1956–57 Estimates, followed by another four under the 1961–62 and 1964–65 Estimates. The second group had the slightly improved Seaslug Mk 2, externally identical, but having some capability against surface targets. They were also fitted with the 'double bedstead' AKE-2 version of the Type 965 long-range radar and had GWS22 directors controlling the Seacat missiles in place of GWS21. For the first time in a British warship, the design included provision for a medium helicopter, but arrangements for housing the Wessex – a hangar with a folding vertical door on the port side of the superstructure – were rather cumbersome.

The 'Counties' were, together with the 'Tribals', the first ships to have Combined Steam and Gas turbine (COSAG) machinery; main drive is supplied by twin HP and LP steam turbines and twin gas turbines are geared to the same shafts for boost. The system has the advantage of permitting the ship to get under way at short notice, as well as providing more power at high speed without excessive weight.

In 1972 *Norfolk* began a conversion to operate the new French Exocet surface-to-surface missile: four MM38 canisters were mounted in 'B' position, with special deflectors to avoid damage to the bridgework from the booster. She carried out firing trials in 1974, making her the first RN operational warship to fire an SSM. She was also refitted with a lighter tubular foremast carrying a Type 992Q radar scanner, a modification subsequently extended to *Antrim*, *Fife* and *Glamorgan* which also received Exocet. The older ships were refitted with GWS22 Seacat directors.

The limitations of the Seaslug missile system, inherent in the fact that it was a beam-rider, mean that these handsome ships were of diminishing value; furthermore, they were expensive to man and maintain, and so *Hampshire* was paid off in 1976 to be cannibalised. *Devonshire* was under consideration for sale to Egypt after conversion to carry six Lynx helicopters in the former missile loading space in the main deck, but this fell through in 1978 and she too went on the disposal list. *London* had been earmarked for conversion to a minelayer – again the Seaslug 'tunnel' would prove ideal – to meet the need for operations behind the SOSUS barriers in the Greenland-Iceland-UK (GIUK) Gap. When this was cancelled she was sold to Pakistan in 1982. *Kent* was deleted in April 1980, but remained in service for harbour training. The later ships had good command facilities and were intended to last until the late 1980s, until all three *Invincible* class carriers were in service, but under the 1981 defence cuts they went earlier. *Antrim* and *Glamorgan* were damaged, one by bombs, the other by bombs and an Exocet, in the Falklands.

All four of the later ships were sold to Chile. *Norfolk* was transferred in April 1982; *Antrim* in June 1984, *Glamorgan* in October 1986, and *Fife* in August 1987.

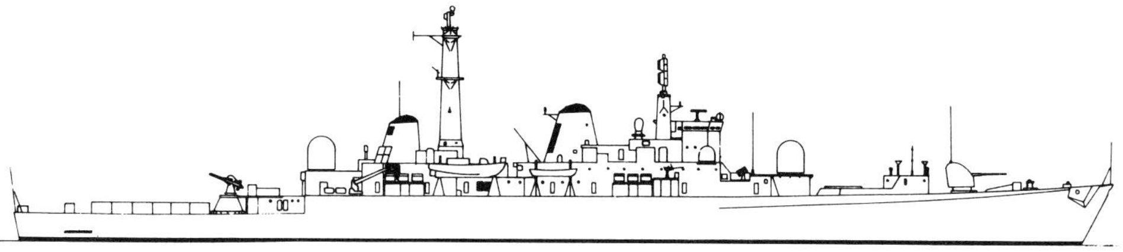

Bristol 1979

BRISTOL class *guided missile destroyers*

Displacement: 6700t normal; 7700t full load
Dimensions: 490ft pp, 507ft oa × 55ft × 22ft 6in max
149.3m, 154.3m × 16.8m × 6.8m
Machinery: 2-shaft COSAG: geared steam turbines, 2 Babcock & Wilcox boilers, plus 2 Olympus TMIA gas turbines, 30,000shp + 44,000shp = 30kts. Range 5000nm at 18kts
Armament: 1–4.5in/55 Mk 8, 1 Sea Dart GWS30 SAM system (1×2, 40 missiles), 1 Ikara ASW system (32 missiles), 1 Limbo Mk 10
Sensors: Radar 909(2), 965 (AKE-2), 978, 1006, 992/993; sonar 184, 162
Complement: 407

No	Name	Builder	Laid down	Launched	Comp	Fate
D 23	BRISTOL	Swan Hunter	15.11.67	30.6.69	31.3.73	HS 1991
	–					Cancelled 1967
	–					Cancelled 1967
	–					Cancelled 1967

The new carriers planned for the 1970s (see *CVA-01*) would need escorts to provide them with anti-aircraft and anti-submarine defence and so four new DLGs were planned to follow the 'County' class. Although rated as DLGs, for some reason they were given Type No 82 in the general-purpose frigate series (originally known as CF299 frigates). With the cancellation of *CVA-01* in 1966 it would have been logical to cancel them as well, and indeed the Board of Admiralty was willing to accept seven extra *Leander* class frigates in their place, but a trials ship had been promised for the Anglo-Netherlands 'Broomstick' radar project in the hope that the Netherlands would order the Olympus gas turbine. Finally, the class was cut back to one unit, the reason being that she could act as a trials ship for new weapons.

Although superficially resembling a 'cleaned-up' version of the 'County' class, the design incorporated another generation of equipment and ideas. The Sea Dart missile system used vertical stowage and a twin-arm zero-length launcher like USN systems, and is a semi-active homing missile with a much greater capacity to handle multiple targets. Although the same type of COSAG machinery was adopted, much more attention was paid to providing good access to the gas turbines. The result was that the design changed from a single-funnel to a two-funnel configuration and finally to three funnels, the after uptakes being split into two separate funnels to allow access to the Olympus gas turbines. *Bristol* was thus the first RN ship to be designed around the new concept of upkeep by replacement, which reduces the time spent on machinery overhauls. Although nearly 1000t bigger than the 'Counties', she needed seventy men fewer to run her.

She was also the first RN ship to go to sea with the Anglo-Australian Ikara ASW missile system, mounted in a 'zareba' forward of the bridge. The new 4.5in Mk 8 gun in a Vickers mounting replaced the old Mk 6 twin, but only one was provided, in 'A' position. Work had been proceeding on a joint Anglo-Netherlands project to produce a Comprehensive Display System (CDS) radar for air warning, surveillance and aircraft direction, and it was to be mounted over the bridge, but the rising cost of research and development, combined with the belief that there would be no need for such a system because there would be no carrier-aircraft to direct, led to Britian withdrawing from the scheme. At the same time, the Netherlands was disenchanted with the spiralling cost (and volume) of the Sea Dart missile system; by mutual consent the CDS was dropped, and *Bristol* went to sea with the less capable Type 965 (AKE-2).

Although sometimes criticised as being a 'white elephant' she was an extremely versatile ship, with sophisticated command and communications to back up her weapon systems. The only serious shortcoming was the lack of a helicopter hangar, omitted because she was originally intended to operate with carriers. After completion she was used for weapon and system trials, and as a result never received the minor improvements given to other large ships.

The heart of *Bristol*'s weapon systems was the Action Data Automation Weapon System Mk 2 (ADAWS-2), which co-ordinated data from the radar and sonar plots, the inertial navigation system, and any other source, and processed it via a suite of digital computers.

In 1973 the ship suffered serious, almost fatal, damage when her steam plant was virtually destroyed by fire. For a time it was feared that she would be scrapped, but she was able to complete the Ikara and Sea Dart trials programme running only on the two gas turbines. This helped to convince doubters about the wisdom of the change to all-gas turbine propulsion, and since then the steam plant has been repaired. She was intended to receive a big half-life modernisation in 1981 but this was rescinded. However Type 1022 radar replaced Type 965 and other improvements were made. She served for a while as a training ship.

Bristol 1974 C & S Taylor

Sheffield 1975

SHEFFIELD class *guided missile destroyers*

Displacement: 3850t normal; 4350t full load
Dimensions: 293ft wl, 410ft oa × 46ft × 19ft max
119.5m, 125.0m × 14.0m × 5.8m
Machinery: 2-shaft COGOG: 2 Olympus TM3B gas turbines, 50,000shp = 30kts; 2 Tyne RMIA gas turbines, 8000shp = 18kts
Armament: 1–4.5in/55 Mk 8, 2–20mm (2×1), 1 Sea Dart GWS30 SAM system (1×2, 22 missiles), 1 helicopter, 6–12.75in TT (2×3). See notes
Sensors: Radar 909 (2), 965 (AKE-2) or 1022, 992Q/993, 1006; sonar 184M, 162
Complement: 299–312

No	Name	Builder	Laid down	Launched	Comp	Fate
D 80	SHEFFIELD	Vickers-Armstrong, Barrow	15.1.70	10.6.71	16.2.75	Sunk 10.5.82
D 86	BIRMINGHAM	Cammell Laird	28.3.72	30.7.73	3.12.76	Extant 1995
D 108	CARDIFF	Vickers-Armstrong, Barrow	6.11.72	22.2.74	24.9.79	Extant 1995
D 118	COVENTRY	Cammell Laird	29.1.73	21.6.74	20.10.78	Sunk 26.5.82
D 87	NEWCASTLE	Swan Hunter	21.2.73	24.4.75	23.3.78	Extant 1995
D 88	GLASGOW	Swan Hunter	16.4.74	14.4.76	24.5.79	Extant 1995
D 89	EXETER	Swan Hunter	22.7.76	25.4.78	19.9.80	Extant 1995
D 90	SOUTHAMPTON	Vosper Thornycroft	21.10.76	29.1.79	31.10.81	Extant 1995
D 91	NOTTINGHAM	Vosper Thornycroft	6.2.78	18.2.80	14.4.83	Extant 1995
D 92	LIVERPOOL	Cammell Laird	5.7.78	25.9.80	9.7.82	Extant 1995

The cancellation of *CVA-01* led to a new Staff Requirement for a smaller fleet escort capable of providing area defence. The result was the Type 42 guided missile destroyer (DDG), which achieved significant savings on cost and displacement by dropping the Ikara long-range ASW missile and Limbo mortar and adopting an all-gas turbine (COGOG) propulsion system, using Rolls-Royce Olympus turbines for main drive and Tynes for cruising.

Although lacking Ikara, the ASW capability was greatly improved over previous ships by providing a hangared Lynx light helicopter (armed with torpedoes and missiles). Unlike *Bristol* the forecastle deck extends right aft to form the helicopter flight deck, leaving a small covered quarterdeck below for handling mooring wires. Close-range ASW defence is provided by triple TT similar to the USN Mk 32, firing Mk 44 or Mk 46 weapons, but eventually to fire the new Stingray lightweight torpedo. Apart from having the Sea Dart twin-arm launcher forward (a lighter hand-operated version of the type in *Bristol*), the configuration is similar to *Bristol*: Type 909 Sea Dart trackers forward and aft and a Type 965 (AKE-2) air warning radar above the forward superstructure.

The appearance is quite different from previous DLGs, with a single wide funnel. *Sheffield* appeared with prominent 'Loxton Bends' on each side, an attempt to keep her heat signature down, but this was not repeated in following ships. *Exeter* commissioned in 1980 with the new Type 1022 interim STIR radar in place of the double 965, and this set was retro-fitted to the earlier ships. *Sheffield* did not receive her STWS TT, but the others carried them on small platforms abreast the mainmast.

The main weakness of the *Sheffield* class lay in the constraints placed on dimensions during the design phase. As a result of Treasury pressure the length and beam were kept down, resulting in cramped accommodation and reduced endurance. The problem of endurance was shrugged aside on the grounds that the ships had no 'east of Suez' role and would always be within reach of a replenishment group around the coasts of north-western Europe, but the short forecastle affects their seaworthiness by making them very wet forward. This fault has been rectified in later vessels (see below), but it places severe limits on what can be added to the surviving ships.

With ten ships completed between 1975 and 1981 they were the largest class of major ships built for the RN since 1967. During the Falklands Campaign the survivors were fitted with 4–30mm Oerlikons (2×2) and two single 20mm amidships, had their EW enhanced and US chaff systems added. In addition, the Lynx helicopter received the Sea Skua air-to-surface missile, rushed into service. Later the 30mm guns were replaced by Mk 15 Phalanx CIWS.

The high degree of automation has resulted in a reduction of one hundred men over *Bristol*'s complement. The layout of the machinery compartments allows easy removal routes, and a complete change of gas turbine can be carried out by the ship herself: all that is needed is a sheltered anchorage and a crane capable of lifting the turbine in its module.

Cardiff was seriously delayed by a shortage of skilled manpower at Barrow-in-Furness and had to be towed to the Tyne in February 1976 for completion by Swan Hunter. *Liverpool* was the first to be built by the 'extrusion' method: large sections of hull were built separately and then moved on to the slipway. This enabled Cammell Laird to cut delivery time by a year.

Sheffield was disabled and set on fire by an Argentine AM39 Exocet missile on 4 May 1982; the wreck was scuttled 10 May. *Coventry* was sunk by bombs on 25 May.

All surviving ships have now been fitted with Type 996 radar in place of Type 992Q/993, 1022 in place of 965 and Phalanx CIWS P&S amidships. *Southampton* was badly damaged on 3 September 1988 in collision with a roll-on/roll-off ship in the Gulf: she was rebuilt on the Tyne during 1989–91.

Nottingham Jul 1992 — C & S Taylor

Gloucester 1989

The shortcomings of the *Sheffield* class could easily be remedied by restoring the length cut off the original design and slightly increasing the beam, and in November 1978 an order was placed for the first of four 'stretched Type 42s'. In configuration they are virtually identical with the later *Sheffield* class, with Type 1022 air warning radar. They also have the later version of the STWS defence against submarines using the Stingray torpedo and have four 30mm Oerlikons or Mk 15 Phalanx (depending of type of deployment) added. Because of the longer hull they are marginally faster than the *Sheffield* class.

During the Gulf conflict of 1991 *Gloucester* engaged and destroyed a 'Silkworm' anti-ship missile. *Edinburgh* was to be the first DDG armed with lightweight Seawolf, and in 1990 had a spray-deflecting bulwark and a single Phalanx fitted forward. In 1995 she reverted to the same configuration as her sisters.

MANCHESTER class *guided missile destroyers*

Displacement: 4750t normal; 5350t full load
Dimensions: 434ft wl, 463ft oa × 49ft × 19ft max
132.3m, 141.1m × 14.9m × 5.8m
Machinery: As *Sheffield* class
Armament: As *Sheffield* class. See notes
Sensors: Radar as *Sheffield* class, but 1022, in place of 965
Complement: 312

No	Name	Builder	Laid down	Launched	Comp	Fate
D 95	MANCHESTER	Vickers-Armstrong, Barrow	19.5.78	27.11.80	16.12.82	Extant 1995
D 96	GLOUCESTER	Vosper Thornycroft	25.10.79	2.11.82	11.9.85	Extant 1995
D 98	YORK	Swan Hunter	18.1.80	21.6.82	9.8.85	Extant 1995
D 97	EDINBURGH	Cammell Laird	8.9.80	14.4.83	17.12.85	Extant 1995

Manchester Jun 1984 C & S Taylor

TYPE 43 *guided missile destroyers*

Displacement: 6000t full load
Dimensions: Not known
Machinery: 2-shaft COGOG, 4 Olympus TM3B gas turbines, 94,000shp = 30kts
Armament: 2 Sea Dart GWS30 SAM, 4 Sea Wolf GWS25 SAM, 4 MM38 Exocet SSM, 1–4.5in Mk 8 gun, 2–20mm (2×1), 6–324mm TT (2×3), helicopter
Sensors: Radar 1022, 1006, 909, 910; sonar 184, 162M, 185
Complement: Not known

The successor to the Type 42 DDG was intended to cure all the ills of that class, a double-ended Sea Dart system, heavy point defence, and a good offensive capability. However at an estimated cost of at least £200 million per ship it fell victim to the 1974–75 Defence Review. An attempt to replace it with an austere derivative (only one Sea Dart system), the Type 44, was also vetoed by the Treasury.

It was an impressive and unorthodox design, with a flush deck, hangar between the funnels, and a COGAG propulsion system similar to that in the *Invincible* class carriers, but using Spey gas turbines.

The 1945 Frigate projects

The postwar frigate designs derived from studies carried out in 1944–45 by Herbert Pengelly and Ken Purvis. Wartime experience showed that escorts needed to be robust, with outer bottom plating at least 12mm (0.3in) thick and a strong forecastle deck. They should be easy to build and capable of being altered from the anti-submarine (ASW) to the anti-aircraft (AA) role at a late stage in building. Typical of a large number of studies was TSD 2001/45 for a '25kt sloop'.

Displacement: 1400t standard
Endurance: 4500nm at 18kts (six months out of dock)
Armament: ASW – one twin 4.5in Mk 6, one twin 20mm, six 'Bidder' Mk 20 (E) torpedoes, two DC throwers; AA – two twin 4.5in or three twin 4in, four twin 20mm, 'Ben', four DC throwers.

More detailed calculations for the ASW type suggested a ship 335ft × 43ft (102.1m × 13.1m) on 1650t standard, 2300t deep load.

By February 1946 an aircraft-direction (AD) frigate had evolved, equipped with Type 960, 277Q and 292Q radars and armed with a close-range AA missile known as 'Long shot' (formerly 'Ben', then 'Popsy') or a twin 4.5in, four single Bofors and a Squid mortar. By the following August the requirements for 'Longshot' (later known as 'Hot Shot') had been dropped and there was even talk of dropping the 40mm gun as the Bofors was considered to be obsolete. Types 980 and 981 radars were specified for the AD version, and a Type 'D' mortar (Limbo) had replaced the Improved or 'B' Type Squid.

In March 1947 it was decided that the new frigates could not be given steam machinery as the British heavy engineering industry no longer had sufficient capacity and because it was doubtful whether sufficient skilled personnel could be recruited to run advanced turbine machinery. Although the ASW design had to be steam-powered to get the required speed of 27kts, the ASW and AD versions were slower and could be given diesels.

A simple code was drawn up to distinguish the three streams of design: type-numbers 11–39 were allocated to ASW frigates, 41–59 to AA and 61–79 to AD. Although much modified (there have been at least two Types 18 and two Types 42 already), the basic system was still in use in 1995.

UNITED KINGDOM

TYPE 11 class *frigates*

The original 1945 Frigate Concept called for a standard steam-engined hull capable of 25kts sea speed ('25kts sloop'), with two variants for ASW and AA. An AD variant appeared in February 1946. As the AA and AD versions appeared there must at some stage have been a Type 11 ASW, but no trace has survived. As mention of it stops so early in the post-1945 frigate papers, we must conclude that it was dropped as soon as news of the Soviet 'Whiskey' class became known, before any detailed design work was done.

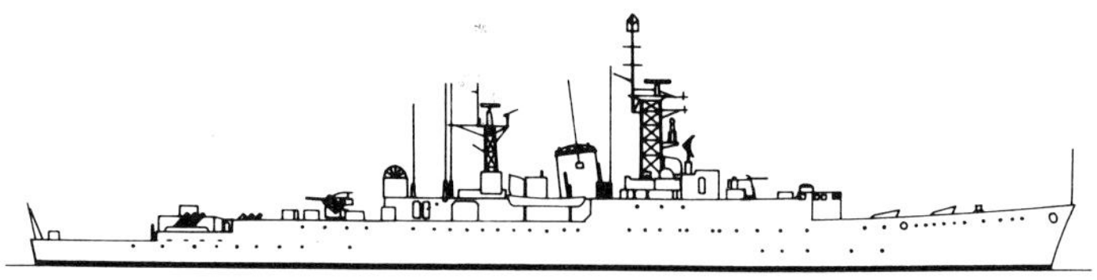

Relentless 1967

RAPID class *frigates*

Displacement: 2240t normal; 2850t full load
Dimensions: 339ft 6in pp, 358ft 3in oa × 35ft 8in × 13ft 10in–14ft 1in
103.5m, 109.2m × 10.9m × 4.2m–4.3m
Machinery: 2-shaft Parsons geared turbines, 2 Admiralty 3-drum boilers, 40,000shp = 36.75kts. Range 3000nm at 20kts
Armament: 2–4in/45 QF Mk 16 AA (1×2), 2–40mm Mk 5 AA (1×2), 2–21in TT (2×1), 2 Squid Mk 4 (later ships Limbo Mk 10)
Sensors: Radar 278, 262, 293; sonar 170, 174
Complement: 195

No	Name	Builder	Laid down	Launched	Comp	Fate
F 197	GRENVILLE	Swan Hunter	1.11.41	12.10.42	27.5.43	BU 1983
F 138	RAPID	Cammell Laird	16.6.41	16.7.42	20.2.43	Sunk as target 3.9.81
F 185	RELENTLESS	John Brown	20.6.41	15.7.42	30.11.42	BU 1971
F 191	ROCKET	Scotts	14.3.41	28.10.42	4.8.43	BU 1967
F 195	ROEBUCK	Scotts	19.6.41	10.12.42	10.6.43	BU 1968
F 09	TROUBRIDGE	John Brown	10.11.41	23.9.42	8.3.43	BU 1970
F 83	ULSTER	Swan Hunter	12.11.41	9.11.42	30.6.43	BU 1980
F 17	ULYSSES	Cammell Laird	14.3.42	22.4.43	23.12.43	BU 1970
F 53	UNDAUNTED	Cammell Laird	8.9.42	19.7.43	3.3.44	Sunk as target Nov 1978
F 141	UNDINE	Thornycroft	18.3.42	1.6.43	23.12.43	BU 1965
F 08	URANIA	Vickers Armstrong, Barrow	18.6.42	19.5.43	18.1.44	BU 1971
F 196	URCHIN	Vickers Armstrong, Barrow	26.3.42	8.3.43	24.9.43	BU 1967
F 200	URSA	Thornycroft	2.5.42	22.7.43	1.3.44	BU 1967
F 50	VENUS	Fairfield	12.1.42	23.2.43	28.8.43	BU 1972
F 29	VERULAM	Fairfield	26.1.42	22.4.43	10.12.43	BU 1972
F 93	VIGILANT	Swan Hunter	31.1.42	22.12.42	10.9.43	BU 1965
F 76	VIRAGO	Swan Hunter	16.2.42	4.2.43	5.11.43	BU 1965
F 41	VOLAGE	White	13.12.42	15.12.43	26.5.44	BU 1976
F 159	WAKEFUL (ex-*Zebra*)	Fairfield	3.6.42	10.6.43	17.2.44	BU 1971
F 187	WHIRLWIND	Hawthorn Leslie	31.7.42	20.8.43	20.7.44	Sunk as target 1974
F 42	WIZARD	Vickers-Armstrong, Barrow	14.9.42	29.9.43	30.3.44	BU 1967
F 157	WRANGLER	Vickers-Armstrong, Barrow	23.9.42	30.12.43	14.7.44	To South Africa 1956
F 102	ZEST	Thornycroft	21.7.42	14.10.43	20.7.44	BU 1970

The urgent need of fast frigates to counter the new generation of fast Soviet 'Whiskey' class submarines could most readily be met by adapting the numerous 'Emergency' destroyer hulls. By the late 1940s these were totally unsuited to modern warfare, the majority having low-angle guns and primitive fire control. Accordingly in 1949 plans were drawn up to convert two prototypes, *Rocket* and *Relentless*.

The change was dramatic, with a new forecastle deck extending almost to the stern and a superstructure of riveted aluminium. A new type of enclosed bridge was provided at the forward end of the superstructure, directly ahead of the Operations Room, mainly in order to be close to the latter, but also to provide protection from the weather. The captain was provided with a periscope to permit him to con the ship from the Ops Room and even the lookouts were given transparent plastic domes in the bridge wings. The armament was light – a twin 4in Mk 16 with its own CRBF director aft and a twin Bofors above the bridge – but the ASW armament was nearly on a par with the as yet unbuilt Type 12, consisting of search and attack sonars of the latest type and two Squid mortars aft (replaced by an ASW mortar Mk 10 in later ships); *Rocket* and *Relentless* had the Mk 10 prototype. It was hoped to supplement this with eight 21in ASW homing torpedoes, but the failure of the 'Bidder' Mk 20 (E) torpedo caused the tubes to be removed from the few ships so fitted.

This Type 15 conversion ranks as

Ulster 1960 CMP

one of the outstanding successes of the postwar period, wringing as much as an extra twenty-five years out of obsolescent hulls. They were regarded as expensive in their day, but the cut-price Type 16s which followed them were not nearly so effective. They proved weatherly and robust, and made up numbers of escorts at a time when new construction was coming into service very slowly. Endurance was reduced to 4750nm at 15kts (476t fuel).

Rocket and *Relentless* completed their conversion in 1952 and were followed by the others in 1952–57. Yards known to have been involved are: Portsmouth DYd – *Relentless* 1950–52, *Verulam* 1952, *Troubridge* (completed by White, 1955–57); Devonport DYd – *Rocket* 1950–52, *Roebuck* 1952–53, *Ulysses* 1952–53, *Venus* 1953, *Wizard* 1954; Chatham DYd – *Ulster* 1952–53, *Zest* 1954–56; *Grenville* 1953–54, *Virago* 1952; Scotts – *Wakeful* 1952–53; Palmers – *Whirlwind* 1952–53, *Ursa* 1952–54; Harland & Wolff, Belfast – *Wrangler* 1951–52; Harland & Wolff, Liverpool – *Urania* 1953–54; White – *Volage* 1952, *Undaunted* 1952–54; Barclay Curle – *Urchin* 1952–54; Stephens – *Rapid* 1952–53; Thornycroft – *Vigilant* 1952; Yarrow – *Undine* 1954. The main modification was to provide a new enclosed bridge above the ops room and in three of the ships (*Troubridge, Ulster, Zest*) the 40mm mounting was resited in an embrasure at the forward end of 01 deck. *Wrangler* became the South African *Vrystaat* on transfer to that navy.

In their later years the survivors were used for training and some were disarmed. *Grenville* and *Undaunted* were fitted with helicopter platforms to test the new concept of using light helicopters to drop homing torpedoes against submarines.

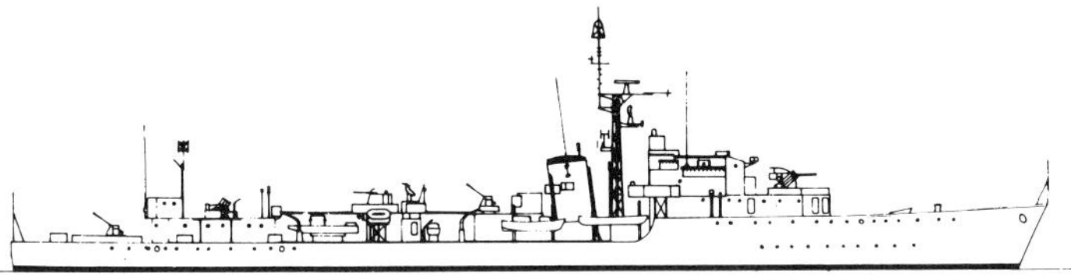

Termagent as converted

The cost of the *Rapid* or Type 15 'full' conversion forced the Admiralty to order a Type 16 'limited' conversion. It had the virtue that it could be accomplished quickly because the original destroyer hull was retained, but capability was much lower. The twin 4in gun on 'B' position had to rely on a Simple Tachymetric Director (STD) above the bridge, and the main sensor was a single Type 293 surface/air search and target-indication radar at the masthead. To provide a surface capability (of dubious value), one quadruple set of TT was retained, and *Orwell* and *Paladin* retained their minelaying rails. Twin Squids and a handing room, as in the 'C' class destroyers, were sited on the after superstructure. *Tenacious* completed her prototype conversion in 1952 and the rest were completed by 1956. Yards known to be involved were: Rosyth DYd – *Orwell* and *Tenacious* 1952, *Paladin* 1956; Harland & Wolff, Belfast – *Petard* 1956; Thornycroft – *Terpsichore* 1952–53; Mountstuart – *Teazer* 1952, *Tuscan* 1953; Grayson Rollo – *Termagant* 1953, *Tumult* 1954; Harland & Wolff, Liverpool – *Tyrian* 1953. Maximum draught was 16ft 1 in and endurance 5225nm at 15kts (588t oil).

In the first ships the original destroyer bridge was retained, but *Teazer, Terpischore* and *Tumult* were given a higher and narrow bridge, as in the new Type 12 frigates.

TENACIOUS class *frigates*

Displacement: 1793t normal; 2593t full load
Dimensions: As *Rapid* class
Machinery: As *Rapid* class; range 1700nm at 20kts (*Orwell, Paladin* and *Petard*)
Armament: 2–4in/45 Mk 19 AA (1×2), 2–40mm/60 Mk 5 AA (1×2), 3–40mm/60 Mk 9 AA (3×1), 4–21in TT (1×4), 2 Squid Mk 3
Sensors: Radar 293, 974; sonar 170, 174P
Complement: 170–180

No	Name	Builder	Laid down	Launched	Comp	Fate
F 98	ORWELL	Thornycroft	16.5.40	2.4.42	7.10.43	BU 1965
F 169	PALADIN	John Brown	22.7.40	11.6.41	12.12.41	BU 1962
F 26	PETARD (ex-*Persistent*)	Vickers-Armstrong, Tyne	26.12.39	27.3.41	14.6.42	BU 1967
F 23	TEAZER	Cammell Laird	20.10.41	7.1.43	13.9.43	BU 1965
F 44	TENACIOUS	Cammell Laird	3.12.41	24.3.43	30.10.43	BU 1965
F 189	TERMAGANT	Denny	25.11.41	22.3.43	18.10.43	BU 1965
F 19	TERPSICHORE	Denny	25.11.41	17.6.43	20.1.44	BU 1966
F 121	TUMULT	John Brown	16.11.41	9.11.42	2.4.43	BU 1965
F 156	TUCAN	Swan Hunter	6.9.41	28.5.42	11.3.43	BU 1966
F 67	TYRIAN	Swan Hunter	15.10.41	27.7.42	8.4.43	BU 1965

Teazer 1957 CMP

The frigate designs of the 1950s and 1960s

Official drawing of Type 18 ASW conversion NMM

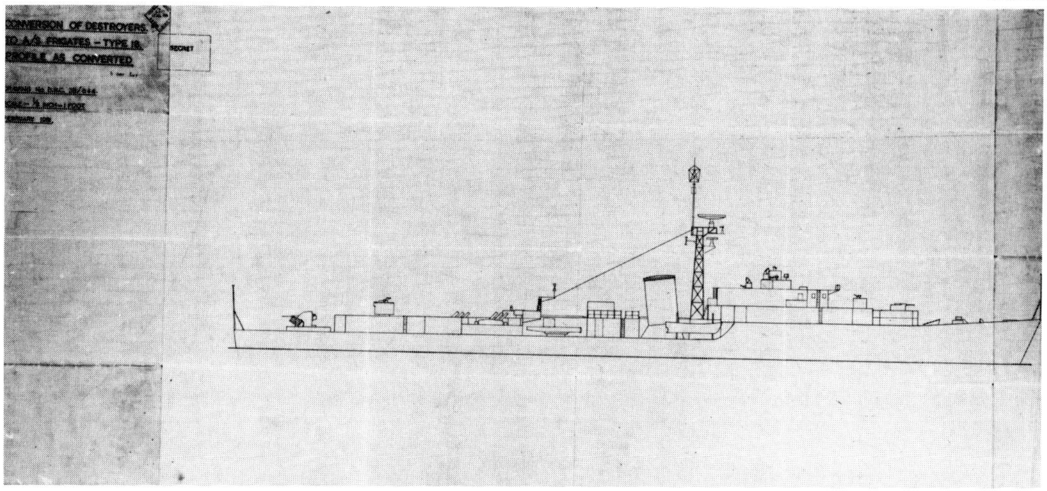

Type 17. Three separate designs received this designation. The first, *c*1950, was a 3rd Rate ASW frigate, intended as a less capable edition of the *Blackwood* (Type 14) but bigger to improve habitability. Armament would have been one Limbo Mk 10 mortar and one 40mm Bofors, on a length of 300ft (91.4m) and 10,000shp single-shaft steam machinery.

The second Type 17 was a 3000t ASW frigate design of the mid-1960s, and in the late 1960s a further design emerged. This was a large ASW ship displacing 6000t. No information has been released about these designs, but they were produced in answer to specific Staff Requirements.

From the Type 17 evolved the Common Hull Frigate, intended for mass production with optional ASW/AA armament. The hull was a derivative of the *Black Swan*, but with Y111 twin-shaft steam turbines. Armament would have been two twin 4in Mk 16, two twin 40mm Bofors Mk 5 and Squid, and the ASW version would have had Limbo Mk 10 plus two twin Bofors.

What was a very sound idea had to be dropped in 1954 because the concept of mass production in wartime was not feasible: war would last less than the four years needed to set up a production line.

Type 18. Again the designation covers more than one design. The first appears to have been a modified version of the *Rapid* (Type 15) destroyer conversion, with Limbo Mk 10 mortars positioned where the 21in TT were in *Rapid.* A second Type 18 was a 'T' class Type 16 with a modified bridge.

Type 19. The confrontation with Indonesia in 1964 led to a very unusual Staff Requirement: an escort capable of 40kts and having a propeller life long enough to get to Singapore. The fighting qualities were to equal those of the *Whitby* frigate design. With hindsight one can only wonder what the utility of such a ship might be, for even if such contradictory requirements could be reconciled the penalties paid in vibration and unreliability could have been enormous. It is said that the design team argued that the propellers' life expectation would be only twenty hours; the Admiralty Experiment Works at Haslar were less generous, allowing twenty minutes! Four Olympus gas turbines would have produced about 112,000shp on 4000t. Fortunately this product of diseased Staff thinking died under the 1966 Defence Review.

Type 20. Nothing is known about this project, but from its date (mid-1960s) it was possibly an early study for a *Leander* replacement.

Type 42. Better known as the East Coast Gunboat, this was a 1950 project for an AA frigate to defend coastal convoys. She would have carried three single 4in of a new Vickers automatic type controlled by several MRS3 fire-control systems, on a displacement of 2000t. The 300ft hull would have been diesel-propelled at 25kts, and the gun was developed from an Army AA system (it was eventually installed in two Chilean destroyers).

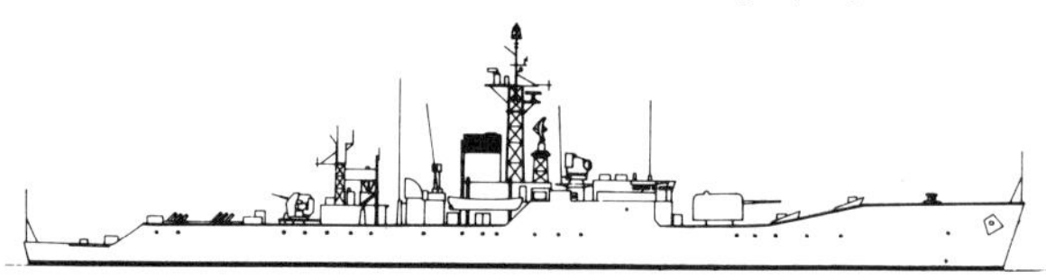

Torquay 1957

WHITBY class *frigates*

Displacement: 2150t normal; 2560t deep load
Dimensions: 360ft pp, 370ft oa × 41ft × 17ft max
109.7m, 112.7m × 12.5m × 3.9m
Machinery: 2-shaft geared steam turbines, 2 Babcock & Wilcox boilers, 30,000shp = 29kts. Range 4500nm at 12kts
Armament: 2–4.5in/45 Mk 6 DP (1×2), 2–40mm STAAG Mk 2 (1×2) (later 1–40mm Mk 7), 12–21in TT (2×2, 8×1), 2 Limbo Mk 10. See notes
Sensors: Radar 275, 262, 293, 277, 974/978; sonar 162, 170, 174/177
Complement: 152–231

No	Name	Builder	Laid down	Launched	Comp	Fate
F 77	BLACKPOOL	Harland & Wolff	20.12.54	14.2.57	14.8.58	BU 1978
F 73	EASTBOURNE	Vickers-Armstrong, Tyne	13.1.54	29.12.55	9.1.58	BU 1986
F 63	SCARBOROUGH	Vickers-Armstrong, Tyne	11.9.53	4.4.55	10.5.57	BU 1977
F 65	TENBY	Cammell Laird	23.6.53	4.10.55	18.12.57	BU 1977
F 43	TORQUAY	Harland & Wolff	11.3.53	1.7.54	10.5.56	BU 1987
F 63	WHITBY	Cammell Laird	30.9.52	2.7.54	10.7.56	BU 1979

The so-called 1945 Sloop studies had envisaged a 25kt ASW version of the standard hull, but the need for a Type 11 frigate very quickly disappeared with the postwar appearance of the fast submarine. If such a design existed it had a short life, for the Staff Requirements raised the sea speed to 27kts, capable of being maintained in a seaway. Out of this emerged the Type 12 design, and the AA and AD, the downgraded, diesel-engined versions of the standard design.

The Type 12 hull-form was, and remains, one of the outstanding small warship designs of the century. A sharp 'V' form forward permitted the hull to be driven into a head sea without the 'panting' and 'slamming' which went with traditional destroyers, but amidships the hull widened to a deep square form. This acted against high smooth-water speed, but made for good sustained speed, ideal for pursuit of a submarine in rough weather. To reinforce this the big (12ft diameter) propellers were slow-running, making them quieter and more efficient, and a raised topgallant forecastle kept spray and green water off the bridge. Range was to be 4500nm at 12kts.

The armament was intended to be the twin 3in/70 Mk 6 AA gun, a 40mm STAAG, two ASW Mk 10 triple mortars and no fewer than twelve TT for launching homing torpedoes, eight fixed and two twin training tubes. The *Whitbys* were prototypes and so went to sea with the existing,

Scarborough as completed CMP

somewhat cumbersome 4.5in, although production vessels were scheduled to get 3in gun. The 21in Mk 20 (E) torpedo (formerly 'Bidder') never became operational.

Because of the fine lines the armament and bridgework had to be far aft (this made for greater crew efficiency, but it did mean a crowded upper deck). In the first three ships (*Whitby, Torquay* and *Scarborough*) the funnel was a thin 'stovepipe' close to the lattice foremast (to resist atomic blast), and as the radar arrays suffered from smoke-corrosion (and to improve the ship's appearance) *Eastbourne* was given a thicker, raked funnel with a special domed cap. It was liked, and was subsequently retro-fitted to all three.

The need to keep down weight meant that plating was very thin, with the result that all six ships had to be partially replated soon after completion. Only four vessels were fitted with TT and these were removed by the early 1960s. The STAAG was also removed and replaced by a single 40mm Mk 9. *Torquay, Eastbourne, Scarborough* and *Tenby* were altered for service in the Dartmouth Training Squadron, with a large deckhouse aft for classrooms and offices. None of the class received any further modernisation: *Scarborough* and *Tenby* were temporarily transferred to Pakistan, but when the refit cost rose to £10m each the deal was cancelled. *Blackpool* was on loan to New Zealand from 1966 to 1971. *Torquay* was used for navigation and direction training, and carried out trials of the first Computer Assisted Action Information System (CAAIS).

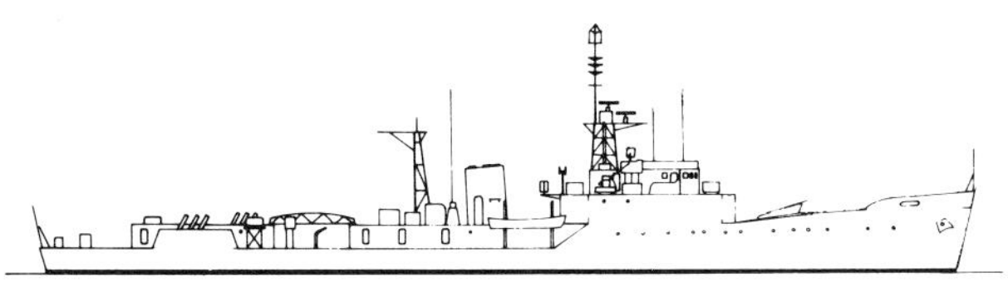

Hardy 1970

BLACKWOOD class *frigates*

Displacement: 1180t normal; 1535t deep load
Dimensions: 300ft pp, 310ft oa × 35ft × 15ft 6in max
91.4m, 94.5m × 10.7m × 4.7m
Machinery: 1-shaft geared steam turbines, 2 Babcock & Wilcox boilers, 15,000shp = 27kts. Range 4500nm at 12kts
Armament: 3–40mm/60 Mk 9 (3×1), 2 Limbo Mk 10, 4–21in TT (2×2)
Sensors: Radar 974, 291; sonar 174, 170, 162
Complement: 140

No	Name	Builder	Laid down	Launched	Comp	Fate
F 78	BLACKWOOD	Thornycroft	14.9.53	4.10.55	22.8.57	BU 1976
F 80	DUNCAN	Thornycroft	17.12.53	30.5.57	21.10.58	BU 1985
F 48	DUNDAS	White	17.10.52	25.9.53	9.3.56	BU 1983
F 84	EXMOUTH	White	24.3.54	16.11.55	20.12.57	BU 1979
F 51	GRAFTON	White	25.2.53	13.9.54	11.1.57	BU 1971
F 54	HARDY	Yarrow	4.2.53	25.11.53	15.12.55	HS 1971, sunk as target 1983
F 85	KEPPEL	Yarrow	27.3.53	31.8.54	6.7.56	BU 1979
F 88	MALCOLM	Yarrow	1.2.54	18.10.55	12.12.57	BU 1978
F 91	MURRAY	Stephen	30.11.57	22.2.55	5.6.56	BU 1970
F 94	PALLISER	Stephen	15.3.55	10.5.56	13.12.57	BU 1983
F 62	PELLEW	Swan Hunter	5.11.53	29.9.54	26.7.56	BU 1971
F 97	RUSSELL	Swan Hunter	11.11.53	10.12.54	7.2.57	BU 1985

The high unit cost of the *Whitby* design – £3.5m – led the Admiralty to consider cheaper vessels which could also be built more quickly: the result was the Type 14 (there was no Type 13), with the same ASW capacity as the Type 12, but with only half the power and virtually every other capability eliminated. All were ordered in March 1951.

The result illustrates vividly the problems of building a 2nd rate frigate, for although the *Blackwoods* achieved exactly what the Staff Requirement demanded – good ASW qualities at half the cost of the *Whitbys* – they were not popular in the Navy. Described as a utility design (as against the 'quality' Type 12) they were mocked as 'futility frigates', for their single-mission design made them inflexible for peacetime use. What was ignored was the fact that if hostilities had broken out with the Soviet Union in the late 1950s (as many Western defence planners envisaged) there would have been an extra half dozen Type 14s and ten Type 16s. In ASW exercises the *Blackwoods* often proved superior to more sophisticated frigates, mainly because their inflexibility permitted them to be left to continue their ASW training to a high level, whereas other frigates frequently had their training programmes interrupted.

With only half the machinery of the *Whitby* class the designers had to go for finer lines and less weight to reach 25kts, and so the short forecastle reappeared. Although later strengthened after experiencing exceptionally heavy weather off Iceland, they proved very seaworthy, and many gave good service in the Fishery Protection Squadron.

Exmouth as converted 1969 CMP

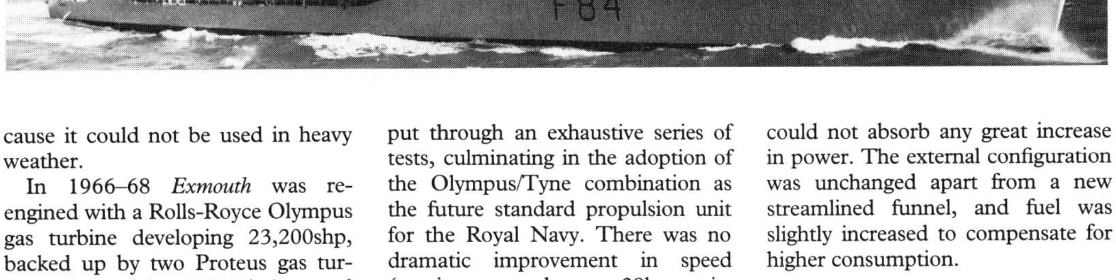

Blackwood, Exmouth, Malcolm and *Palliser* were completed with two twin 21in TT amidships to supplement the Mk 10 mortars, but the tubes were subsequently removed. After a few years in service the after 40mm gun was removed as well, mainly because it could not be used in heavy weather.

In 1966–68 *Exmouth* was re-engined with a Rolls-Royce Olympus gas turbine developing 23,200shp, backed up by two Proteus gas turbines for cruising (8500shp) geared to the same shaft. The ship was then put through an exhaustive series of tests, culminating in the adoption of the Olympus/Tyne combination as the future standard propulsion unit for the Royal Navy. There was no dramatic improvement in speed (maximum speed rose to 28kts, cruising speed of 22kts) because the hull could not absorb any great increase in power. The external configuration was unchanged apart from a new streamlined funnel, and fuel was slightly increased to compensate for higher consumption.

TYPE 62 *air defence frigates*

Displacement: ?
Dimensions: As 'M' and 'T–Z' classes
Machinery: As 'M' and 'T–Z' classes
Armament: 2–3in/50 (2×3), 2–40mm Mk 5 (1×2), 1 Squid mortar
Sensors: Radar 982, 983 etc; sonar 147F, 166
Complement: ?

Work began on the design of an air defence frigate in 1949, and by 1952 an outline staff requirement was ready. Initial proposals were to convert the five surviving 'M' class destroyers *Marne*, *Matchless*, *Meteor*, *Milne* and *Musketeer*, and seven later 'Emergency' type, *Myngs*, *Kempenfelt*, *Troubridge*, *Wager*, *Whelp*, *Savage* and *Grenville*. Armament was to be one twin 4in Mk 19 with CRBF director, a twin 40mm Bofors with STD and a single Squid. Asdic Type 162 was dropped.

It was hoped to have two or even three intercept radars, including the new Types 982 and 983 air intercept sets, but the 'Emergencies' lacked sufficient beam, and they were to be restricted to a pair of Type 277Q height-finders. Even the 'M' design would be seriously overloaded, with the 982 and 983 sets sited abaft the mainmast, or a 277Q height-finder above the bridge and the two big radars on the centreline aft.

Ursa replaced *Grenville*, and in March 1952 all the 'Emergencies' (only four remained) were removed from the programme, and the total of 'Ms' was down to four. In December 1953 it was decided to replace the twin 4in with twin USN-pattern 3in, supplied under the FY 1953 Mutual Defense Assistance Program (MDAP). Serious doubts were raised about shock-resistance of these pre-war destroyers, and the programme was finally cancelled in April 1954.

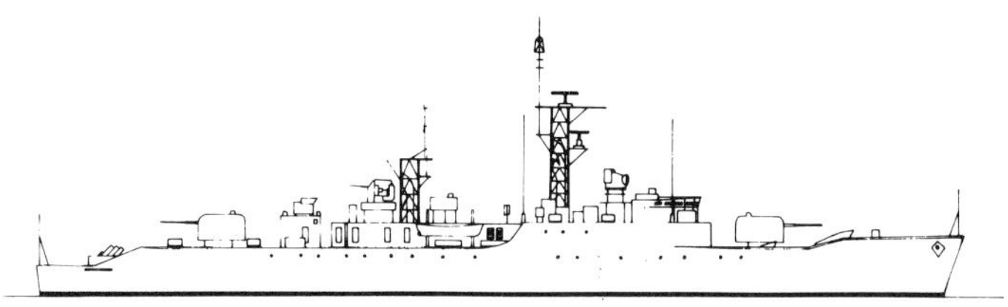

Puma as completed

LEOPARD class *frigates*

Displacement: 2300t normal; 2520t deep load
Dimensions: 330ft pp, 340ft oa × 40ft × 11ft 10in
100.6m, 103.6m × 12.2m × 3.6m
Machinery: 2-shaft diesel: 8 ASR1 diesels, 14,400bhp = 25kts. Range 7500nm at 16kts
Armament: 4–4.5in/45 Mk 6 DP (2×2) 2–40mm/60 STAAG Mk 2 AA (1×2), 1 Squid
Sensors: Radar 275, 960, 262, 974, 992Q (later 965 (AKE-1)), 978, 993; sonar 170, 174, 162
Complement: 205

No	Name	Builder	Laid down	Launched	Comp	Fate
F 14	LEOPARD	Portsmouth DYd	25.3.53	23.5.55	30.9.58	BU 1977
F 27	LYNX	John Brown	13.8.53	12.1.55	14.3.57	To Bangladesh 1982
F 34	PUMA	Scotts	16.11.53	30.6.54	27.4.57	BU 1976
F 37	JAGUAR	Denny	2.11.53	20.7.57	12.12.59	To Bangladesh 1978
–	PANTHER (i)	John Brown	–	–	–	To India 1953
–	PANTHER (ii)	–	–	–	–	Cancelled 1957

For the remainder of the 1945 frigates it was decided to adopt diesel propulsion because it was felt that so many steam plants would overstretch the capacity of British industry. The Type 41 AA frigate was intended to support the Type 12, along with its half-sister, the Type 61 AD version, and the first orders were placed in August 1951. The 41/61 in steam form was the basic design and the 12 was developed from it. The raised forecastle of the Type 12 was adopted in order to enable speed to be kept up in a head sea, but was somewhat shorter and therefore lacking in elegance. Another unattractive feature was the absence of a funnel, for the diesel exhaust-trunking was brought up inside both lattice masts. This proved unsuccessful as well as unsightly, and to prevent the corrosion of masthead equipment *Puma* had her mainmast plated. The improvement was retro-fitted to the other ships.

By the standards of the day a most elaborate anti-aircraft defence was provided, with a Type 960 long-range air warning radar on the mainmast, a Mk 6M fire control system controlling two twin 4.5in automatic mountings backed up by a CRBF director aft, and a STAAG autonomous close-range 40mm mounting. The STAAG was not reliable, how-ever; *Jaguar* entered service with a single 40mm gun in its place and the rest followed suit. *Jaguar* was the first to have the more effective Type 965 (AKE-1) long-range radar, but in time all four received it.

A fifth unit, *Panther*, was sold (before being laid down) to India as *Brahmaputra*; a replacement was ordered, but subsequently cancelled.

In 1978 *Jaguar* was transferred to Bangladesh as *Ali Haider*, followed by *Lynx* in 1982 (renamed *Abu Bakr*). Although effective ships when built, the Admiralty Standard Range 1 diesel was not a great success.

Lynx 1977 C & S Taylor

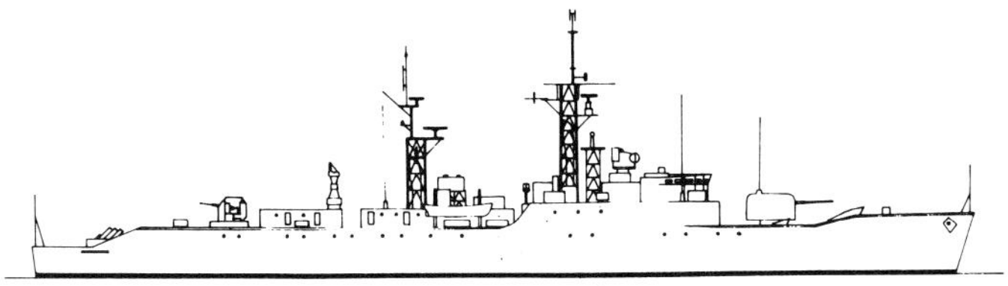

Salisbury as completed

On the same hull as the Type 41, the Type 61 design sacrificed the after 4.5in twin gun in favour of a much more comprehensive air warning radar outfit. The ships were thus a much more valuable addition to the fleet than the *Leopard* class, for with adequate radar coverage carrier aircraft provided much better defence against air attack than guns. This was recognised quite early, and the *Leopard* class was stopped at four units whereas the *Salisbury* class was to have been expanded to eight units. Furthermore, the survivors had received a more comprehensive modernisation.

The original STAAG Mk 2 was later replaced by 1–40mm Mk 9, and the fourth ship, *Lincoln*, was completed later, with a big deckhouse aft to take a GWS20 director and a quadruple Seacat missile launcher (a single 40mm gun was mounted initially). In 1962 *Salisbury* was refitted with a plated mainmast carrying a 'single bedstead' Type 965 antenna. In 1964 *Chichester* appeared with both masts plated, the mainmast carrying a 'double bedstead' Type 965 array and Seacat GWS20; *Llandaff* was similarly fitted in 1966, followed by *Lincoln* in 1968 and *Salisbury* in 1970.

Two more of the class were cancelled under the 1957 Defence Review and another (*Coventry*) was suspended. It was hoped to order her in 1961, but by then the more flexible *Leander* design was ready and so she was replaced by *Penelope*. *Llandaff* transferred to Bangladesh as *Umar Farooq* in December 1978, followed by *Lincoln* in 1982. *Salisbury* was sunk as a target on 30 September 1985.

SALISBURY class *frigates*

Displacement: 2170t normal; 2350t deep load
Dimensions: As *Leopard* class
Machinery: As *Leopard* class
Armament: 2–4.5in/45 Mk 6 DP (1×2), 2–40mm/60 STAAG Mk 2 (1×2), 1 Squid
Sensors: Radar 275, 277, 960, 982, 293, 974 (later 965 (AKE-2)), 993, 978, etc; sonar 170, 174, 162
Complement: 207

No	Name	Builder	Laid down	Launched	Comp	Fate
F 32	SALISBURY	Devonport DYd	23.1.52	25.6.53	27.2.57	Sunk as target 1985
F 59	CHICHESTER	Fairfield	26.6.53	21.6.55	16.5.58	BU 1981
F 61	LLANDAFF	Hawthorn Leslie	27.8.53	30.11.55	11.4.58	To Bangladesh 1976
F 99	LINCOLN	Fairfield	1.6.55	6.4.59	7.7.60	To Bangladesh 1982
–	COVENTRY	Vickers-Armstrong, Tyne	–	–	–	Cancelled 1961
–	EXETER	–	–	–	–	Cancelled 1957
–	GLOUCESTER	–	–	–	–	Cancelled 1957

Llandaff 1977 C & S Taylor

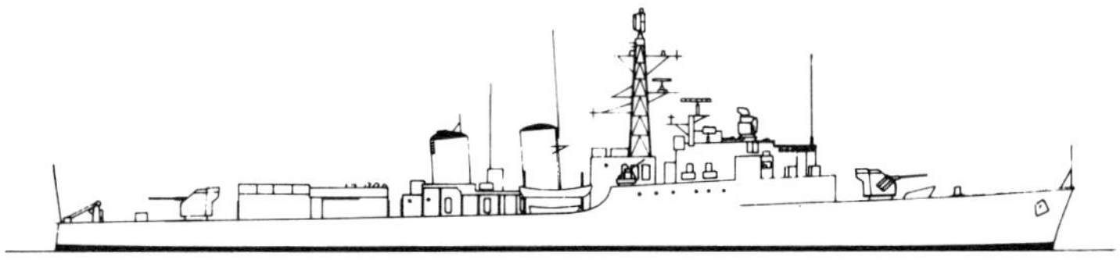
Mohawk as completed

'TRIBAL' class *frigates*

Displacement: 2300t normal; 2700t deep load
Dimensions: 350ft pp, 360ft oa × 42ft 3in × 13ft 3in max
106.7m, 109.7m × 12.9m × 4.0m
Machinery: 1-shaft COSAG: geared steam turbine, 1 Babcock & Wilcox boiler, plus 1 AEI G6 gas turbine, 12,500shp + 7500shp = 27kts. Range 4500nm at 12kts
Armament: 2–4.5in/45 Mk 5 (2×1), 2–40mm/60 Mk 7 or 2–20mm (2×1), 1 Limbo Mk 10, 1 helicopter
Sensors: Radar 965 (AKE-1), 993, 903, 974/978; sonar 170, 177, 162 (*Ashanti* and *Gurkha* also 199)
Complement: 253

No	Name	Builder	Laid down	Launched	Comp	Fate
F 117	ASHANTI	Yarrow	15.1.58	9.3.59	23.11.61	HS 1980, sunk 1988
F 119	ESKIMO	White	22.10.58	20.3.61	21.2.63	BU 1992
F 122	GURKHA	Thornycroft	3.11.58	11.7.60	13.2.63	To Indonesia 1984
F 125	MOHAWK	Vickers-Armstrong, Barrow	23.12.60	5.4.62	29.11.63	BU 1983
F 131	NUBIAN	Portsmouth DYd	7.9.59	6.9.60	9.10.62	Sunk as target 1987
F 133	TARTAR	Devonport DYd	22.10.59	19.9.60	26.2.62	To Indonesia 1985
F 124	ZULU	Stephen	13.12.60	3.7.62	17.4.64	To Indonesia 1985

Inspired by the work already done on the Common Hull Frigate (see Type 42 frigates), the Naval Staff drew up a project known as a Common Purpose Frigate (later called a Sloop) having a second-rate AA, ASW and AD capability; in effect this meant a reversion to general-purpose ships, an admission that the breakdown into single-purpose ships was wrong. The armament was to be two twin 4in Mk 16, a twin 40mm Mk 5, a Limbo Mk 10 mortar and eight ASW TT. The standard sonar and radars were to be provided, but a measure of aircraft-direction was to be provided by a USN SPS-6C long-range radar.

Several changes of armament were proposed: a twin 3in/70 Mk 6 and two Vickers single automatic 4in in turn, but these were rejected to avoid pushing up dimensions. Then it was decided to Anglicise the SPS-6C radar, and until the weight and dimensions of the new Type 965 radar aerial were known the lattice mast could not be designed; allowance was made for an aerial weighing 2 times the manufacturers' 1956 estimate, and this turned out to be sufficient! The result was a very light mast which actually weighed less than the array it carried. Less difficulty was encountered with the fire control, for the Type 802 system was cancelled and replaced by the light MRS3, again Anglicised USN equipment, this time based on their Mk 56 director.

The designation 'sloop' was chosen deliberately because 'frigate' denoted a single-mission ship, and from the unused general purpose 80 series came the later designation Type 81. Subsequently there was political pressure to show that the RN was capable of increasing its frigate force, and to bow to this whim the worthy designation of sloop was changed to frigate. There was a need to replace the nine 'Loch' class vessels in the Persian Gulf, and this became the prime mission for the new design. Several more attempts were made to upgrade the design, mainly to increase speed, but these were ruled out.

The most important innovation was the adoption of COSAG machinery, half that of the *Kent* class, whose design was proceeding at the same time. *Ashanti* was thus in effect the test-bed for the new machinery and the weight saved could be used for a more effective armament. Air warning radar could be accommodated, and as the trials with light helicopters in Type 15 frigates had proved successful it was decided to try to incorporate a hangar for a Fairey Ultra Light helicopter. This hangar was ingenious, the lift forming both the roof of the hangar and the flight deck. When the helicopter (in fact a Wasp) is stowed the hangar is closed by portable roof-sections.

The adoption of COSAG machinery necessitated two funnels, the foremost serving the boiler room and the aftermost the gas turbine room. The ships were also unusual in having a tapering flush deck, giving good freeboard forward on a small displacement without sacrificing strength or covered access. On the displacement and cost-limits it was possible to provide only two single hand-loaded semi-automatic 4.5 Mk 5 guns (for shore bombardment) but provision was made for two Seacat (2 × 4) launchers, which replaced the 40mm weapons. As completed, only *Zulu* carried Seacat, with the prominent GWS21 version of the CRBF directors on platforms abreast the second funnel; by 1977, however, the remainder had been fitted as well. Some had VDS.

Gurka 1977 — C & S Taylor

Although successful ships, the 'Tribals' fell victim to the RN's 1979 manpower crisis, and all were paid off from the end of that year, *Gurkha* being the first to be put on the disposal list; she recommissioned in 1982, however, in the wake of the Falkands Crisis, and some of her sisters were brought back into service for a short while. Three sold to Indonesia: *Zulu* in May 1985, *Gurkha* in October 1985, and *Tartar* in April 1986.

Plymouth 1976 — C & S Taylor

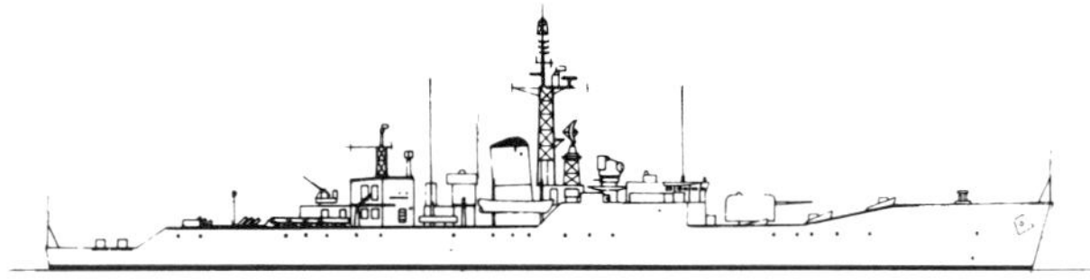

Londonderry 1960

ROTHESAY class *frigates*

Displacement: 2150t (later 2380t) normal; 2560t (later 2800t) deep load
Dimensions: As *Whitby* class
Machinery: As *Whitby* class
Armament: As *Whitby* class. See notes
Sensors: Radar 975, 977, 293, 974, 262 (later 903), 994, 978; sonar 170, 174/177, 162
Complement: 200–235

No	Name	Builder	Laid down	Launched	Comp	Fate
F 107	ROTHESAY	Yarrow	6.11.56	9.12.57	23.4.60	BU 1988
F 108	LONDONDERRY	White	15.11.56	20.5.58	18.10.61	Sunk as target 1989
F 106	BRIGHTON	Yarrow	23.7.57	30.10.59	28.9.61	BU 1985
F 113	FALMOUTH	Swan Hunter	23.11.57	15.12.59	25.7.61	BU 1989
F 101	YARMOUTH	John Brown	29.11.57	23.3.59	26.3.60	Sunk 1987
F 129	RHYL	Portsmouth DYd	29.1.58	23.4.59	31.3.60	Sunk 1985
F 103	LOWESTOFT	Stephen	9.6.58	23.6.60	26.9.61	Sunk as target 1986
F 115	BERWICK	Harland & Wolff	16.6.58	15.12.59	1.6.61	Sunk as target 1986
F126	PLYMOUTH	Devonport DYd	1.7.58	20.7.59	11.5.61	Preserved 1989
–	WEYMOUTH	Harland & Wolff	10.4.59	–	–	Cancelled 1960
–	FOWEY	Cammell Laird	19.10.59	–	–	Cancelled 1960
–	HASTINGS (i)	Thornycroft	–	–	–	To New Zealand
–	HASTINGS (ii)	Yarrow	2.12.59	–	–	Cancelled 1960

These were repeats of the *Whitby* design, ordered under the 1954–55 Programme, but 'tidied up' internally. The raked funnel fitted in *Torquay* was incorporated during construction, but there were few other external differences. The twelve TT were removed and only the early ships carried the STAAG 40mm mounting. In its place was put a large deckhouse to carry a Seacat GWS20 missile system, but for some years all ships carried a single 40mm Bofors gun.

The last three hulls were cancelled and completed to the *Leander* design, the success of which prompted a major modernisation of the *Rothesays*. The major improvement was the incorporation of a Wasp light helicopter (Mk 44/46 homing torpedoes), while the opportunity was taken to update the fire control by replacing the Mk 6M with the MRS3 system. The appearance was slightly changed by providing a new plated foremast and raising the funnel, but in other respects the layout remained the same. Armament was now 2–4.5in, one Seacat GWS20 SAM (1 × 4), 2–20mm, one Limbo Mk 10.

In 1978 *Falmouth* carried out trials of the winch gear for the new towed array sonar on the quarterdeck. From November 1975 to October 1979 *Londonderry* was reconstructed to serve as a trials ship for the Admiralty Surface Weapons Establishment (ASWE). This involved the removal of the 4.5in guns, altering the propulsion system to waterjets for quietness, and stepping a large, plated mizzenmast to carry the new Type 1030 STIR radar. In the autumn of 1980 she also took the 30mm Naval Rarden gun to sea.

The first *Hastings* (ordered February 1956) became the Royal New Zealand Navy's *Otago* in February 1957; another ship was ordered in her place, but completed as a *Leander*. The 1981 Defence Review spelled the end of the class, although some lasted longer as a result of the Falklands. *Lowestoft* was fitted with the first Type 2031I towed sonar array in 1981–82. *Plymouth* was handed over to the Warship Preservation Trust in April 1989 and was towed to the Clyde in June the following year.

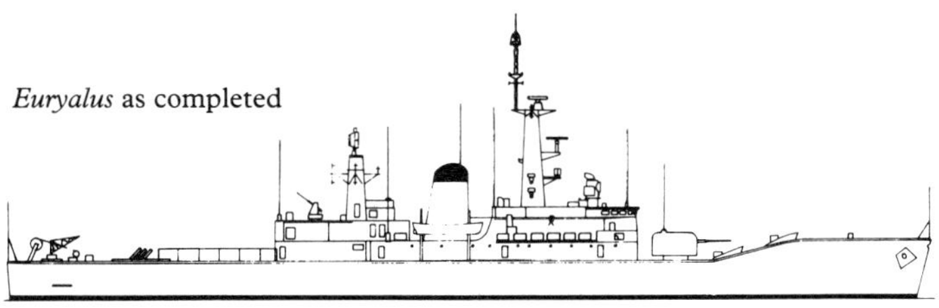

Euryalus as completed

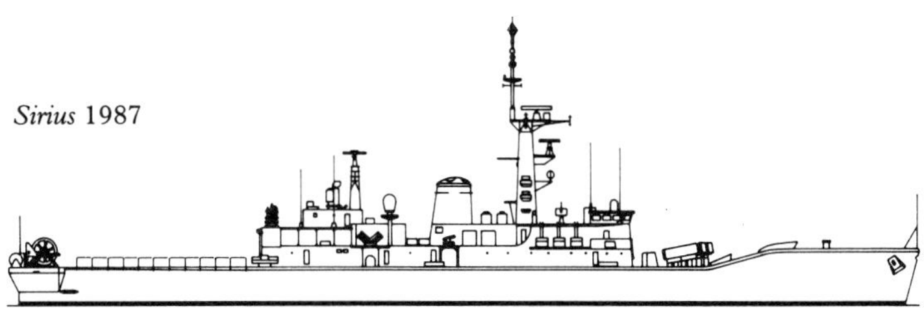

Sirius 1987

LEANDER class *frigates*

Displacement: 2350t normal; 2860t deep load
Dimensions: 360ft pp, 372ft oa × 41ft × 18ft max
109.7m, 113.4m × 12.5m × 5.5m
Machinery: 2-shaft geared steam turbines, 2 Babcock & Wilcox boilers, 30,000shp=28kts. Range 4000nm at 15kts
Armament: 2–4.5in/45 Mk 6 DP (1×2), 2–40mm/60 Mk 9 AA (2×1) (later 1 Seacat GWS22 SAM (1×4), 2–20mm (2×1) added), 1 Limbo, 1 helicopter
Sensors: Radar 965 (AKE-1), 992Q, 903, 974/978; sonar 162, 184, 199
Complement: 251–263

No	Name	Builder	Laid down	Launched	Comp	Fate
F 109	LEANDER (ex-*Weymouth*)	Harland & Wolff	10.4.59	28.6.61	27.3.63	Sunk as target 1989
F 114	AJAX (ex-*Fowey*)	Cammell Laird	19.10.59	16.8.62	11.12.63	BU 1988
F 104	DIDO (ex-*Hastings (ii)*)	Yarrow	2.12.59	22.12.61	18.9.63	To New Zealand 1983
F 127	PENELOPE	Vickers-Armstrong, Tyne	14.3.61	17.8.62	1.11.63	To Ecuador 1991
F 10	AURORA	John Brown	1.6.61	28.11.62	9.4.64	BU 1990
F 15	EURYALUS	Scotts	2.11.61	6.6.63	16.9.64	BU 1993
F 18	GALATEA	Swan Hunter	29.12.61	23.5.63	25.4.64	Sunk as target 1988
F 38	ARETHUSA	White	17.9.62	5.11.63	29.9.65	Sunk 1991
F 39	NAIAD	Yarrow	30.10.62	4.11.63	15.3.65	Sunk 1990
F 42	PHOEBE	Stephen	3.6.63	8.7.64	15.4.66	Sold 1992
F 28	CLEOPATRA	Devonport DYd	19.6.63	25.3.64	1.3.66	BU 1994
F 45	MINERVA	Vickers-Armstrong, Tyne	27.7.63	19.12.64	13.5.66	BU 1994

The advances made in the 'Tribal' class pointed the way to a return to the concept of a general-purpose escort instead of the prevailing idea of specialised escorts operating in a group. Work done on a revision of the Type 12 design for New Zealand, incorporating two helicopters and full air-conditioning, provided further impetus to studies by M K Purvis. The design which evolved managed to incorporate the long-range air warning radar and the helicopter of the 'Tribal' class, but with the armament, fire control and rough-weather capability of the *Whitby*/*Rothesay* design.

The secret was to replace the separate ballast tanks of the Type 12 with self-compensating fuel tanks, in which the fuel was displaced by sea water as it was consumed. As many heavy items as possible were moved lower down in the ship, increasing not only the margin of stability, but also the amount of space for the operations room, extra generating capacity and auxiliary machinery. The forecastle deck was extended right aft and a new superstructure deck was built above that, extending from the bridge back to the helicopter hangar. This transformed the appearance and as it was complemented by a well-balanced funnel the *Leander* set a new standard of good looks sadly lacking in some previous British designs. Building of the class was expedited by stopping work on the last three *Rothesays* at an early stage, and a further thirteen ships were laid down under the 1960–64 Programmes.

The class are generally considered to be the finest escorts ever built for the Royal Navy and some of the best in the world. To increase the margin for future growth the next group of *Leanders* ordered was given 2ft more beam (see below).

The machinery of the original design was improved, providing a greater degree of automatic control: the first ten had the Y-100 type and the rest Y-111. One of the benefits conferred by having advanced machinery control was the remarkable manoeuvrability and acceleration which was such a feature of the *Leanders*. During the 'Cod War' the *Leanders* proved that they could outmanoeuvre the Icelandic gunboats by changing speed rapidly.

All except *Diomede* received the Type 199 variable depth sonar (VDS) in a stern well. In some ships the well remained, but in others it was plated in to provide extra accommodation.

While serving as a training ship *Diomede* had her 4.5in guns cocooned and the Type 965 radar removed. In 1966 *Penelope* was disarmed for trials in the Mediterranean. This involved removing her propellers and towing her at the end of a mile-long nylon hawser to measure flow noise and hull drag. She was later fitted with the prototype Sea Wolf GWS25 missile system, with the six-cell launcher on the quarterdeck, Types 967/968 radar on the mainmast and a Type 910 tracker on a deckhouse aft. The equipment was removed in 1979 and she went into dockyard hands for the Exocet conversion.

Although no longer bearing any relation to the *Whitby* design (apart from the hull-form), they are incorrectly referred to as Type 12 frigates.

No	Name	Builder	Laid down	Launched	Comp	Fate
F 40	SIRIUS	Portsmouth DYd	9.8.63	22.9.64	15.6.66	For disposal 1992
F 52	JUNO	Thornycroft	16.7.64	24.11.65	18.7.67	TS 1985, BU 1995
F 56	ARGONAUT	Hawthorn Leslie	27.11.64	8.2.66	17.8.67	For disposal 1993
F 47	DANAE	Devonport DYd	16.1.65	21.10.65	10.10.67	To Ecuador 1991

Euryalus (Ikara conversion) 1977 C & S Taylor

Ikara conversions
Eight ships received the Batch 1 conversion, which involved replacing the 4.5in gun mounting with a 'zareba' containing an Ikara GWS40 ASW missile system and its magazine and loading gear below. The air warning radar was removed from the mainmast because the ships were intended to operate in mixed escort groups. The Seacat arrangement aft was improved, with the GWS22 director moved to the centreline, leaving room for two quadruple launchers to port and starboard.

Leander was recommissioned in December 1972 and was followed by *Ajax* (September 1973), *Galatea* (September 1974), *Naiad* (July 1975), *Aurora* and *Euryalus* (March 1976), *Arethusa* (April 1977) and *Dido* (October 1978). Displacement was now 2450t normal, 2960t deep load, and complement 257; sensors consisted of radar Types 994, 903 amd 1006 and sonar Types 184, 162 and 170; 2–40mm/60 Mk 9 (2 × 1) and 6–12.75in TT (2 × 3) were fitted and one Wasp helicopter was carried.

Exocet conversions
Although twelve Ikara conversions were planned the programme was cut short at eight units, for the RN needed to improve its surface-strike capability. The eight Batch 2 ships were given four MM38 Exocet missiles in place of the 4.5in guns, and at the same time the helicopter hangar was enlarged to take the new Lynx, and the new STWS-1 torpedo defence was added; 2–40mm/60 Mk 9 (2 × 1) and 6–12.75in TT (2 × 3) are also carried. Unlike the Ikara conversions the Type 965 air warning radar (AKE-1) was retained; other sensors consisted of radar Types 903, 1006 and 994 and sonar Types 184/184M and 162. *Cleopatra* was selected as the trials ship for the Type 2031 towed array sonar, and conversion began in 1980.

The provision of self-compensating fuel tanks did not prove successful, and as the Exocet missiles made relatively low demands on topweight it was possible to revert to separate ballast tanks. This eliminated problems with marine growths in fuel tanks and incidentally increased the usable fuel by nearly 50 per cent, providing a much-needed improvement in endurance: the original design called for 4000nm at 15kts, but the effective endurance was nowhere near this.

Apart from the four Exocets forward, the appearance was close to the original profile than the Batch 1 ships. The Seacat arrangements on the after superstructure were altered to a centreline director and two launchers port and starboard, and a third launcher was positioned on the centreline ahead of the Exocets. Displacement became 2790t normal and 3200t full load, and complement fell to 223.

Recommissioning dates were: *Cleopatra* November 1975, *Sirius* April 1977, *Phoebe* October 1977, *Minerva* March 1979, *Argonaut* March 1980, *Danae* September 1980 and *Penelope* 22 January 1982. *Juno*'s Batch 2 conversion was cancelled and she was disarmed as a training ship at Rosyth. *Phoebe* was again refitted in 1981–82 when the Type 965 radar was removed, the Exocet ramps lowered and Type 2031 towed array sonar fitted. *Argonaut* was damaged by bombs during the Falklands fighting.

By the end of 1991 only six of the class remained. *Juno* paid off in 1980; disarmed as training ship and replaced *Torquay* in 1985. *Dido* became HMNZS *Southland* in 1983. *Naiad* was used for 'Hulvul' hull vulnerability trials in 1990 and then sunk as target. *Penelope* was sold to Ecuador in 1991 and renamed *Presidente Eloy Elfaro*. Many were sunk as targets for a new generation of weapons.

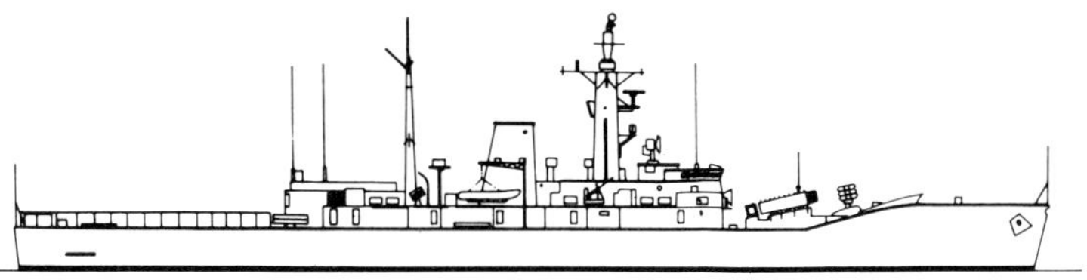

Andromeda (Sea Wolf conversion) 1981

To provide more fuel stowage and a better margin for future growth, the last ten of the *Leander* class were 'kippered' by adding 2ft to the beam. This reduced speed slightly, but improved stability and endurance, both items of greater fighting value. Although referred to in the RN as Type 12s or 'broad-beam *Leanders*' they formed a separate class, with improved machinery, a difference enhanced by the radical reconstruction from 1977. The steady improvement in machinery stated in the original class continued, and these ships had the Y-136 plant. Speed dropped by half a knot because of increased displacement, but the improved degree of automation made for more efficient use of the machinery.

As built the ten ships were identical with the *Leander* class in equipment, but the extra internal volume meant that they could be reconstructed to carry two important new systems: the long-range Type 2016 sonar and the Seawolf GWS25 Mod 0 missile system, capable of destroying sea-skimming missiles as well as supersonic aircraft. The two systems were inter-dependent because the Type 2016 was intended for detecting fast submarines in the GIUK Gap, far from the support of shore-based aircraft. The most likely opposition to be encountered is surface attack, against which Exocet missiles provided a long-range defence, and missile attack, either from bombers or from submarines. The Sea Wolf system was intended particularly to counter the 'pop-up' missile fired from Soviet 'Charlie' and 'Papa' class submarines. Other sensors included radar Types 967, 968, 910, 294 and 1066 and sonar Type 162M. A Lynx helicopter armed with Sea Skua ASMs or ASW torpedoes is carried.

'Broad beam LEANDER' class *frigates*

Displacement: 2500t (later 2790t) normal; 2962t (later 3300t) deep load
Dimensions: 360ft pp, 372ft oa × 43ft × 19ft max
109.7m, 113.4m × 13.1m × 5.5m
Other particulars: As *Leander* class

No	Name	Builder	Laid down	Launched	Comp	Fate
F 58	HERMIONE	Stephen	10.12.65	26.4.67	11.7.69	For disposal 1992
F 57	ANDROMEDA	Portsmouth DYd	25.5.66	24.5.67	4.7.68	To India 1995
F 60	JUPITER	Yarrow	3.10.66	4.9.67	13.8.69	For disposal 1991
F 69	BACCHANTE	Vickers-Armstrong, Tyne	25.10.66	29.2.68	5.12.69	To New Zealand 1982
F 75	CHARYBDIS	Harland & Wolff	27.1.67	28.2.68	6.6.69	Sunk as target Jun 1993
F 71	SCYLLA	Devonport DYd	17.5.67	8.8.68	14.2.70	For disposal 1994
F 12	ACHILLES	Yarrow	27.10.67	21.11.68	9.6.70	To Chile 1990
F 16	DIOMEDE	Yarrow	30.1.68	15.4.69	2.4.71	To Pakistan 1988
F 70	APOLLO	Yarrow	21.11.69	15.10.70	28.5.72	To Pakistan 1988
F 72	ARIADNE	Yarrow	1.5.70	10.9.71	10.2.73	To Chile 1992

The conversion was a tight fit, and it involved virtually gutting the ship internally and resiting bulkheads. Every spare ton of topweight had to be reduced, by such expedients as cutting off the funnel-top and putting the 6–12.75in TT (2 × 3) a deck lower. The 6-cell Sea Wolf launcher was mounted forward of the bridge, with a heavy movable blast-shield to protect the four Exocet MM38 SSMs. A single director was rather lighter than the original in the *Broadsword* class, is mounted over the bridge, with the back-to-back Type 967 and Type 968 surveillance radars at the masthead. As in the Exocet conversions the Limbo mortar was removed to allow for a larger flight deck and hangar. The generating capacity had to be increased over the 200kW in the Batch 2 *Leanders*, for five computers are required to handle the data.

Andromeda was the lead-ship, having gone into Devonport in late 1977; she completed in December 1980, and the rest followed in 1982–84. Following the 1981 defence cuts, the programme was reduced, leaving the remaining ships unconverted. *Bacchante* transferred to New Zealand in 1982; *Apollo* and *Diomede* went to Pakistan in 1988; *Achilles* went to Chile in 1990 and was followed by *Ariadne* in 1992; *Danae* went to Ecuador in 1991; was the first of the modernised ships to go for disposal in October 1991, and the others went in 1992–93.

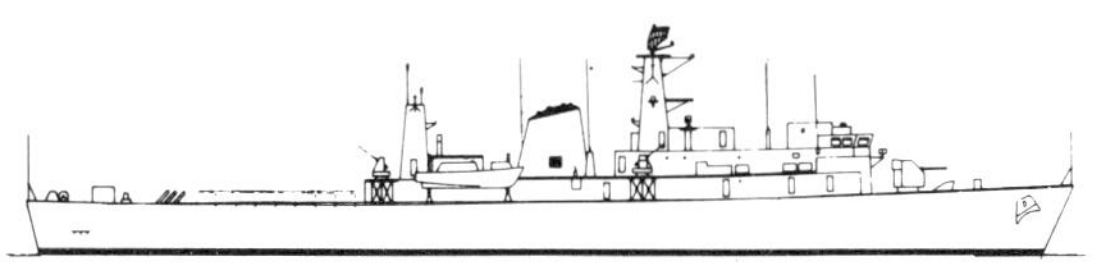

Mermaid as completed

Ordered by Ghana to serve as flagship and presidential yacht for President Nkrumah, but cancelled when a *coup* toppled him from power. She was launched without being named and completed fitting out in 1968, but without armament. Negotiations to sell her abroad failed and after being laid up at Scotstoun she was bought by the RN in April 1972 for conversion to a training ship; refitting started at Chatham in October that year.

The design was based on the Type 41 or *Leopard* class, but without the raised forecastle and with a flush forecastle deck and a funnel. Although successful in RN service, she was an odd number in the inventory and was sold to Malaysia in May 1977 as *Hang Tuah*, replacing the previous ship of that name.

MERMAID *frigate*

Displacement: 2300t normal; 2520t deep load
Dimensions: 320ft pp, 339ft 3in oa × 40ft × 12ft
97.5m, 103.4m × 12.2m × 3.7m
Machinery: 2 shafts, 8 Admiralty Standard Range 1 diesels, 14,400bhp = 24kts. Range 4800nm at 18kts
Armament: 2–4in/45 Mk 16 AA (1×2), 4–40mm/60 Mk 9 (4×1), 1 Limbo Mk 10
Sensors: Radar ASW-1, 978; sonar 174, 170B
Complement: about 235

No	Name	Builder	Laid down	Launched	Comp	Fate
F 76	MERMAID	Yarrow	1965	29.12.66	16.5.73	To Malaysia 1977

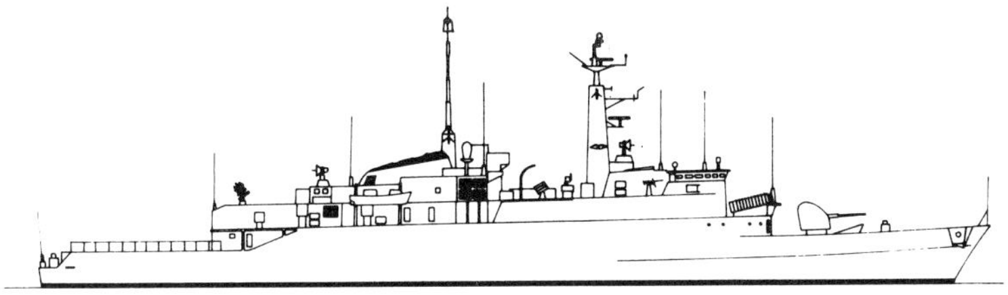

Alacrity 1980

AMAZON class *frigates*

Displacement: 3100t normal; 3600t deep load
Dimensions: 360ft pp, 384ft oa × 41ft 9in × 19ft max
109.7m, 117.0m × 12.7m × 5.8m
Machinery: 2-shaft COGOG: 2 Rolls-Royce Olympus TM3B gas turbines, 56,000shp = 30kts, 2 RM1A Tyne gas turbines, 8500shp = 17kts. Range 4000nm at 17kts
Armament: 4 MM38 Exocet SSM (except *Amazon* and *Antelope*), 1–4.5in/55 Mk 8, 2–20mm (2×1), 1 Seacat GWS24 SAM (1×4), 6–12.75in TT (2×3), 1 helicopter
Sensors: Radar 992Q, 912, 978, sonar 184M, 162M
Complement: 175–192

No	Name	Builder	Laid down	Launched	Comp	Fate
F 169	AMAZON	Vosper Thornycroft	6.11.69	26.4.71	11.5.74	To Pakistan 30.9.93
F 170	ANTELOPE	Vosper Thornycroft	23.3.71	16.3.72	19.7.75	Sunk 24.5.82
F 171	ACTIVE	Vosper Thornycroft	23.7.71	23.11.72	17.6.77	To Pakistan 23.9.94
F 172	AMBUSCADE	Yarrow	1.9.71	18.1.73	5.9.75	To Pakistan 28.7.93
F 173	ARROW	Yarrow	28.9.72	5.2.74	29.7.76	To Pakistan 1.3.94
F 174	ALACRITY	Yarrow	5.3.73	18.9.74	2.7.77	To Pakistan 1.3.94
F 184	ARDENT	Yarrow	26.2.74	9.5.75	13.10.77	Sunk 22.5.82
F 185	AVENGER	Yarrow	30.10.74	20.11.75	15.4.78	To Pakistan 23.9.94

The Type 21 design had its origin in the need to replace the diesel-engined frigates of the *Leopard* and *Salisbury* classes. At the same time private shipbuilders were lobbying for a chance to build a frigate for the RN without what they claimed to be excessively high standards imposed by the MoD's Ship Department at Bath. It was claimed that such a frigate would not only be cheaper (£3.5m as against £5m for a *Leander* at contemporary prices), but also a good export item. The Controller of the Navy backed this claim wholeheartedly, even to the point of trying to cut the Ship Department out of the design process.

The result was a wasted opportunity to bring private and official designers together. Details of the design did not meet certain naval requirements, the cost was £14.4m for the first ship, and the vessels lacked sufficient 'stretch' to take the next generation of weapons (specifically the Type 2016 sonar, a long-range radar and the Sea Wolf missile). They were popular with the RN, however, particularly because of their comfortable officers' accommodation (in contrast, the ratings' accommodation was below the standard in contemporary RN ships) and because they handled easily.

From *Active* onwards, four MM38 Exocet SSMs were fitted in 'B' position, toed inwards to allow the blast to vent outboard to port and starboard. The triple STWS-1 close-range torpedo defence was also fitted after completion. The fire control system was a new departure: as no suitable British light system existed, the RN bought the Italian Selenia RTN-10X radar tracker and combined it with the British Ferranti WSA-4 below-decks equipment, providing control for both the 4.5in gun and the Seacat GWS24. Although there was talk of the design allowing the replacement of Seacat by Sea Wolf this was never possible because the ships were delivered overweight and did not have the margin of stability necessary for such a big system. A Lynx helicopter armed with lightweight torpedoes was carried.

The Royal Australian Navy contributed a large sum (believed to be about 35 per cent) to the design costs, with the intention of building five ships in Australian yards; this project never came to fruition, however, and although Argentina showed some interest, no export orders were ever won for this design. In 1980 it was confirmed that the class would not receive a modernisation, but all six survivors were stiffened post-1982.

In 1977 in the Far East *Amazon* suffered a bad fire, and this drew attention to the risk of building warships with all-aluminium superstructures, a risk underlined by the Falklands conflict. The designers maintained that they could not have met the requirement to raise the weapon load without saving as much topweight as possible. To this their critics replied that the flimsiness of aluminium accentuates the vulnerability of modern warships to relatively minor damage. Whatever the merits of the argument, after the construction of the *Amazon* class the British Naval Staff changed their minds about aluminium, and stipulated that future ships use steel only. During the Falklands conflict *Ardent* was badly damaged by air-launched unguided rockets on 21 May 1982 and sank shortly afterwards; *Antelope* was hit by bombs on 23 May, and blew up on the following day while an explosives expert was trying to defuze an unexploded bomb. The six remaining were sold to Pakistan in 1993–94.

Active Mar 1981

C & S Taylor

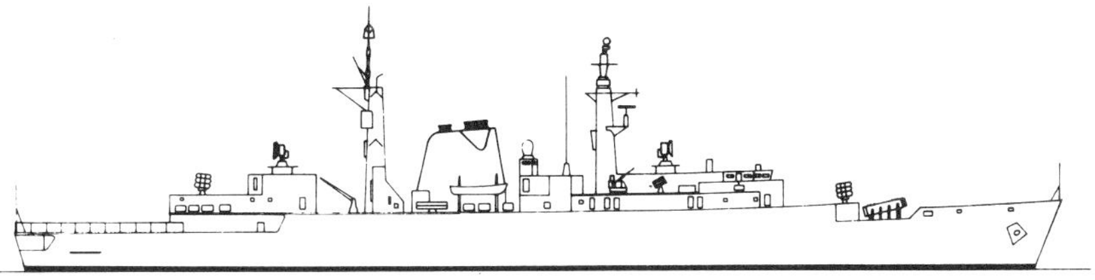

Battleaxe as completed

BROADSWORD class *frigates*

Displacement: 4000t normal; 4400t deep load
Dimensions: 410ft wl, 430ft oa × 48ft 6in × 19ft 10in max
125.7m, 131.2m × 14.8m × 6.1m
Machinery: 2-shaft COGOG: 2 Rolls-Royce Olympus TM3B gas turbines, 54,600shp = 30kts, plus 2 Rolls-Royce Tyne RMIC gas turbines, 9700shp = 18kts. Range 4500nm at 18kts
Armament: 4 MM38 Exocet SSM, 2 Sea Wolf GWS25 Mod 0 SAM (2 × 6), 2–40mm/60 Mk 9 (2×1), or 2–30mm Mk 1 (2×1), 6–12.75in (324mm) TT (2×3) for Mk 46 or Stingray torpedoes, 2 WG 13 Lynx helicopters
Sensors: Radar 967, 968, 910, 1006; sonar 2016/2050, 162, 165
Complement: 222 (17 officers, 205 ratings)

No	Name	Builder	Laid down	Launched	Comp	Fate
F 88	BROADSWORD	Yarrow	7.2.75	12.5.76	3.5.79	To Brazil 30.6.95
F 89	BATTLEAXE	Yarrow	4.2.76	12.5.77	28.3.80	To Brazil 1997
F 90	BRILLIANT	Yarrow	25.3.77	15.12.78	15.5.81	To Brazil 1996
F 91	BRAZEN	Yarrow	18.8.78	4.3.80	2.7.82	To Brazil 1996

Originally conceived as a follow-on to the *Leander* class and intended to be built for the Royal Netherlands Navy as well, the Type 22 design was forced to grow to meet new threats and to accommodate new technology. The increasing inability of medium-range active sonars to cope with fast nuclear submarines led to the development of a new 'fleet' sonar, capable of long-range passive detection and use of convergence zones and 'bottom-bounce' techniques. The threat from SS-N-7 'pop-up' missiles dictated the development of Sea Wolf anti-missile system, and finally the creation of the SOSUS barriers of underwater listening arrays in the GIUK Gap required a ship with a high degree of autonomy and good seakeeping.

As a result the *Broadsword* hull had much more freeboard than the *Sheffield* class DDG, a double hangar to accommodate a second Lynx (a late addition), a double-headed Sea Wolf system, and a main armament restricted to four MM38 Exocet SSM. The hull was in fact an expansion of the original Type 12 design, intended to be driven in rough weather with little reduction in speed. The processing equipment needed for all this required seven computers, but there was another reason for the spaciousness: experience has shown that maintenance is the biggest headache in a warship, and that the 'payload' has to be updated once or even twice in the lifetime of the 'platform' or hull. Thus the *Broadsword* design left additional space for future growth, in much the same way as the American *Spruance* design.

Although rated as frigates they were bigger than the DDGs, and proved magnificent ASW ships. The first two ships were not originally fitted with TTs. They were to be fitted with the Type 2031Z towed array, but the weight of the winch would have strained the hull girder. The enormous number of inputs from towed arrays also showed that the original Type 22 AIO was not big enough for the task. Large as these ships were, a second 'stretched' batch was ordered. For the first time the lead yard received all subsequent orders, as it is now recognised that this saves a considerable amount of time and money. *Broadsword* was damaged during the Falklands conflict, but both she and her sister *Brilliant* performed very well, the latter being the first to use Sea Wolf operationally when she shot down two Skyhawks on 12 May 1982, she was also the first ship to use Sea Skua in anger.

In 1988–89 both *Broadsword* and *Battleaxe* were refitted by DML, with new funnels similar to their later sisters and additional accommodation for training. In 1991 it was suggested that these fine ships might be offered for sale, and in November 1994 all four were bought by Brazil.

Battleaxe May 1985 — C & S Taylor

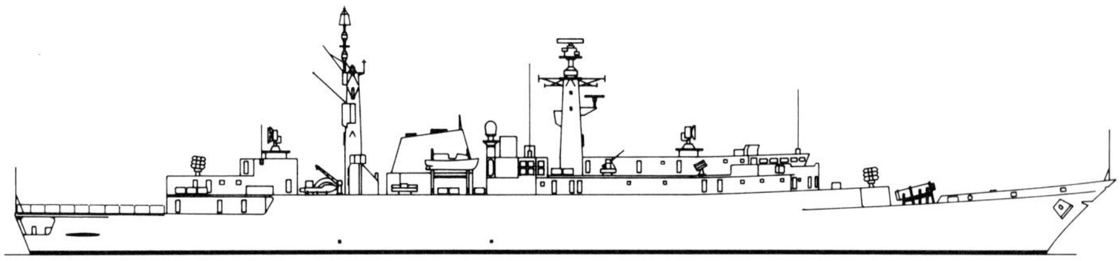

Coventry 1989

Although an initial order of twelve *Broadswords* was talked of, after four units had been ordered the decision was made to lengthen the subsequent ships in order to accommodate the Type 2031Z towed array sonar. Orders for the first two of these improved *Broadswords* were placed in April 1979, and it was planned to build up to twenty-two units, but once again the design was revised in the light of Falklands experience, and the class only ran to ten units. The major design changes in these Batch 2 ships included an expanded Operations Room and AIO to handle Type 2031Z data, the CACS-1 combat system in place of CAAIS, and moving the Type 2016/2050 sonar from its hull position to the bow (hence the marked overhang and the stem anchor). From *Brave* onwards the flight deck was enlarged to accommodate a single Sea King or the new EH.101 Merlin helicopter, and the stern was redesigned with a square transom. *Boxer* and *Beaver* were trials ships for the CACS-1 command system and *Brave* had Marine Spey SM1A gas turbines in place of Olympus TM3Bs. In 1990–91 she was given the uprated SM1C version, developing more power. *Bloodhound* was renamed *London* at the request of the Lord Mayor of London, and the last pair were named in honour of the destroyers lost in 1982.

BOXER class *frigates*

Displacement: 4100t normal; 4800t deep load
Dimensions: *Boxer* and *Beaver*
485ft 10in oa, 451ft 1in pp × 48ft 6in × 21ft max
148.1m, 137.5m × 14.8m × 6.4m
Brave onwards
490ft 6in oa
149.5m
Machinery: 2-shaft COGOG as in Batch 1, except *Brave*: 2 Rolls-Royce SM1A (later SM1C) Spey gas turbines, 37,540 (52,300 shp) gas turbines = 28.5kts; plus 2 Tyne RM1C gas turbines, 9700 shp = 18kts. Range 8000nm at 18kts (later ships)
Armament: As Batch 1, GWS25 Mod 3, 2 Lynx helicopters or 1 Sea King
Sensors: Radar 967 or 967M, 968, 911, 1006; sonar 2050, 2031Z, 162, 165
Complement: 273 (30 officers, 243 ratings)

No	Name	Builder	Laid down	Launched	Comp	Fate
F 92	BOXER	Yarrow	1.11.79	17.6.81	14.1.84	Extant 1995
F 93	BEAVER	Yarrow	20.6.80	8.5.82	18.12.84	Extant 1995
F 94	BRAVE	Yarrow	14.5.82	19.11.83	4.7.86	Extant 1995
F 95	LONDON (ex-*Bloodhound*)	Yarrow	7.2.83	27.10.84	5.6.87	Extant 1995
F 96	SHEFFIELD	Swan Hunter	29.3.84	26.3.86	26.7.88	Extant 1995
F 98	COVENTRY	Swan Hunter	29.3.84	8.4.86	14.10.88	Extant 1995

Brave Jun 1986 C & S Taylor

Cumberland Jun 1991 C & S Taylor

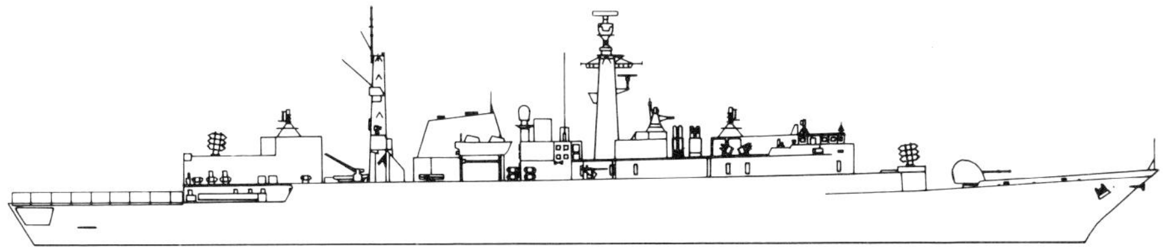

Cornwall as completed

Four more Type 22s were announced at the end of 1982, to replace Falklands losses, but they were redesigned to form Batch 3, with improved command facilities (including flag accommodation), new SSMs, a 4.5in gun and a close-in weapon system (CIWS). The hull of the Batch 2 design was retained, but armament was rearranged to provide space for the medium-calibre gun. The eight RGM-84C Harpoon missiles were positioned athwartships behind the wheelhouse, with the 30mm Goalkeeper CIWS one deck up. They were also the first frigates to be completed with the new single 30mm Mk 1 in place of the elderly 40mm Bofors.

To mark the numerous differences and enhanced capability the 'B' series names gave way to a 'C' series. They have the CACS-5 command system (a version of CACS-1) and the gun is controlled by two GSA 8A Sea Archer electro-optical directors (port and starboard above the wheelhouse).

CORNWALL class *frigates*

Displacement: 4200t normal; 4900t deep load
Dimensions: As Batch 2
Machinery: As Batch 2
Armament: 8 RGM-84C Harpoon SSM (2×4), 1 4.5in/55 Mk 8 gun, 1–30mm Goalkeeper CIWS (1×7), 2/4–30mm Mk 1 (2/4×1), 6–12.75in (324mm) TT for Stingray torpedoes (2×3), 1 Sea King helicopter
Sensors: As Batch 2
Complement: 250 (31 officers, 219 ratings)

No	Name	Builder	Laid down	Launched	Comp	Fate
F 99	CORNWALL	Yarrow	14.12.83	14.10.85	23.4.88	Extant 1995
F 85	CUMBERLAND	Yarrow	12.10.84	21.6.86	10.6.89	Extant 1995
F 86	CAMPBELTOWN	Cammell Laird	4.12.85	7.10.87	27.5.89	Extant 1995
F 87	CHATHAM	Swan Hunter	12.5.86	20.1.88	4.5.90	Extant 1995

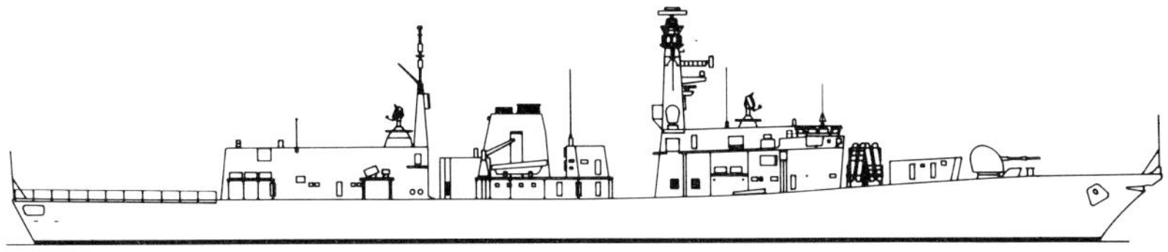

Norfolk 1989

Although the Type 22 is recognised as a highly efficient ASW platform, alarm at its high cost led to a call for a cheaper frigate, relying on the Type 2031Z towed array sonar rather than the big Type 2050 and Lynx helicopters. Out of this has emerged the Type 23, expanded to include a bow-mounted sonar, a large ASW helicopter and more general-purpose qualities. Although Falklands experience resulted in improved fire precautions and damage control, the decision to incorporate a double-headed Sea Wolf SAM system predated that conflict. By the time the design was finalised BAe's vertically launched version was nearly ready for trials, and it was selected in place of the earlier lightweight version of GWS25.

The design is optimised for towed array operations, so silencing of machinery was a top priority. Finally a unique combined diesel electric and gas turbine (CODLAG) was chosen, using Spey gas turbines for main drive, and switching to a diesel-electric drive for slow but silent running when using the towed array. The advantage of CODLAG is the elimination of gearbox noise, and the generators are positioned above the waterline to reduce radiated noise.

The opportunity was taken to incorporate 'stealth' features to enhance survivability in surface warfare. This takes the form of a flared hull and the elimination of right angles and radar-reflecting corners in the superstructure, etc. It is claimed that the Type 23 is 'invisible' to the seekers of the current range of anti-ship missiles, and it also claimed to be the quietest ASW ship in the world.

In all sixteen ships are planned. From the *Richmond* onwards SM1C Spey gas turbines replaced the SM1A. The first ships went to sea without their planned CACS-4 command system but its successor, the DNA (1) system, is being retrofitted from 1995 onwards The first six ships have Racal's UAF-1 ESM system, but later ships will have the Thorn EMI (now Racal) UAT equipment (earlier ships will have this system retrofitted; all will have 675 (2) ECM).

'DUKE' class *frigates*

Displacement: 3500t normal; 4200t deep load
Dimensions: 436ft 4in oa, 403ft 6in wl × 52ft 10in × 24ft max
133m, 123m × 16.1m × 7.3m
Machinery: *Norfolk* and early ships 2-shaft CODLAG: 2 Rolls-Royce SM1A/C Spey gas turbines, 37,540shp = 28kts; plus 2 GEC 1.5MW electric motors, 4 Paxman Valenta 12RPA 200CZ generators, 7000bhp = 15kts. Range 8500nm on diesel-electric
Armament: 8 RGM-84C Harpoon SSM, 32 GWS26 Mod 1 VL Seawolf SAM, 1–4.5in/55 Mk 8, 2–30mm Mk 1 (2×1), 4–12.75in (324mm) TT (2×2), Stingray torpedoes, 1 EH 101 Merlin helicopter
Sensors: Radar 996(1), 911, 1007; sonar 2050, 2031Z
Complement: 169 (12 officers, 157 ratings)

No	Name	Builder	Laid down	Launched	Comp	Fate
F 230	NORFOLK	Yarrow	19.12.85	10.7.87	1.6.90	Extant 1995
F 231	ARGYLL	Yarrow	20.3.87	8.4.89	31.5.91	Extant 1995
F 229	LANCASTER	Yarrow	18.12.87	24.5.90	9.11.91	Extant 1995
F 233	MARLBOROUGH	Swan Hunter	22.10.87	21.1.89	14.6.91	Extant 1995
F 234	IRON DUKE	Yarrow	12.12.88	2.3.91	20.5.93	Extant 1995
F 235	MONMOUTH	Yarrow	1.6.89	23.11.91	24.9.93	Extant 1995
F 236	MONTROSE	Yarrow	1.11.89	31.7.92	May 1994	Extant 1995
F 237	WESTMINSTER	Swan Hunter	18.1.91	4.4.92	13.5.94	Extant 1995
F 238	NORTH-UMBERLAND	Swan Hunter	4.4.91	4.4.92	May 1994	Extant 1995
F 239	RICHMOND	Swan Hunter	18.2.92	6.4.93	8.11.94	Extant 1995
F 280	SOMERSET	Yarrow	12.10.92	25.6.94	Jul 1995	Extant 1995
F 282	GRAFTON	Yarrow	13.5.93	3.12.94	–	Fitting out 1995
F 281	SUTHERLAND	Yarrow	14.10.93	–	–	Building 1995
–	–	–	–	–	–	Projected
–	–	–	–	–	–	Projected
–	–	–	–	–	–	Projected

Argyll Jun 1994 C & S Taylor

UNITED KINGDOM

NFR-90 *frigates*

The Royal Navy participated in the 1980s in the project to build a common hull frigate to meet the NATO Frigate Requirement for the 1990s (NFR-90). The RN requirement was for twelve ships to replace the Type 42 destroyers after the turn of the century, but the project was beset by difficulties. The different timescales of eight navies proved impossible to reconcile, and little or no progress had been made on defining the crucial combat system and main weapon system. The RN withdrew from the project in September 1989, followed by France and Italy, and the remaining partners agreed to wind up the project after US $85 million had been spent on the design studies.

'Horizon' Frigate Project

In 1991 work was well advanced on a new air defence frigate to meet the RN's needs. The design was to be tailored to meet French needs as well, but a strong lobby in the RN wanted a 6000-tonne double-ended design. The cost problems which wrecked the Type 43 project (see above) no longer apply. Software now accounts for 85 per cent of electronic system development, but replication costs virtually nothing, so the cost of giving a ship two VLS installations involves little more than the cost of an extra tracker and the second VL silo. However, agreement was reached with the French Navy to proceed with a joint design, known as the Common New Generation Frigate (CNGF) or Project 'Horizon'. Italy subsequently joined the programme.

Artist's impression of 'Horizon' frigate

SUBMARINES

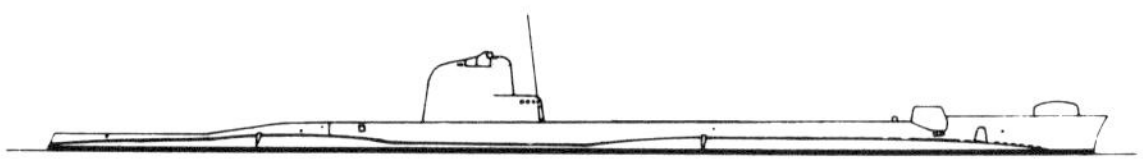

Thermopylae 1953

'T Conversion' type *patrol submarines*

Displacement: 1261–1460t normal; 1554–1588t/1734t
Dimensions: 285ft pp, 290ft 6in (*Tabard* group 293ft) oa × 26ft 7in × 16ft max
86.9m, 88.5m (89.5m) × 8.1m × 4.9m
Machinery: 2-shaft diesel-electric: 4 supercharged ASR diesels, plus 4 electric motors, 2800bhp/6000shp = 14kts/15.4kts. Oil 200t
Armament: 6–21in TT (20 torpedoes) or 12 Mk 2 mines
Sensors: Search and navigation radar; sonar 186, 187, 197
Complement: 68

No	Name	Builder	Laid down	Launched	Comp	Fate
Taciturn group						
S 34	TACITURN	Vickers-Armstrong Barrow	9.3.43	7.6.44	7.10.44	BU 1971
S 55	THERMOPYLAE	Chatham DYd	26.10.43	27.6.45	5.12.45	BU 1970
S 54	TURPIN	Chatham DYd	24.5.43	5.8.44	18.12.44	To Israel 1965
S 52	TOTEM	Devonport DYd	22.10.42	28.9.43	9.1.45	To Israel 1964
Tabard group						
S 42	TABARD	Scotts	6.9.44	21.11.45	25.6.46	HS 1969, BU 1974
S 32	TIPTOE	Vickers-Armstrong, Barrow	10.11.42	25.2.44	13.6.44	BU 1975
S 33	TRUMP	Vickers-Armstrong, Barrow	31.12.42	25.3.44	9.7.44	BU 1971
S 53	TRUNCHEON	Devonport DYd	5.11.42	22.2.44	25.5.45	To Israel 1968

Eight boats of the *Taciturn* and *Tabard* groups were given conversions in 1951–56, following the 'S' class. The pressure hull was cut in two and lengthened by a 20ft (14ft in *Taciturn*, 12ft in *Thermopylae*, *Turpin* and *Totem*) section containing a second pair of electric motors, with clutches to convert them from direct drive to diesel electric, and fourth battery section containing 6560 type cells. This was to provide an underwater speed of 15kts, more than twice the previous normal speed. All guns, external TT and fittings were removed, the periscopes, snorkel radar and R/T masts were enclosed in a streamlined 'fin', and the bow and stern were reshaped. In *Tabard* and *Trump* the fin was fully enclosed, but the others had a small low-set bridge.

The torpedo armament, reduced by the removal of the five external tubes, was now four bow and two stern 21in. They were intended to receive the Mk 20(S) ('Bidder') electric passive homing torpedo, but received the wire-guided version, Mk 23 ('Grog') instead, and after the withdrawal of the Mk 12 ('Fancy'), the elderly Mk 8 was reintroduced to provide some anti-surface capability, because the Mk 23 did not function particularly well against surface targets.

Turpin, Totem and *Truncheon* became the Israeli *Leviathan, Dakar* and *Dolphin.*

Explorer 19.3.57 CMP

Explorer as completed

EXPLORER class *experimental submarines*

Displacement: 776t standard; about 980t/1076t
Dimensions: 209ft 5in pp, 225ft oa × 15ft 8in × 18ft 2in
63.8m, 68.7m × 4.8m × 5.5m
Machinery: 2-shaft Vickers hydrogen peroxide turbine, plus electric motor, 15,000shp/400bhp = 27kts/18kts. Fuel 39t perhydrol
Armament: None
Complement: 41–49

No	Name	Builder	Laid down	Launched	Comp	Fate
S 30	EXPLORER	Vickers-Armstrong, Barrow	20.7.51	5.3.54	28.11.56	BU 1965
S 40	EXCALIBUR	Vickers-Armstrong, Barrow	13.2.52	25.2.55	22.3.58	BU 1970

Although the RN was well aware of the importance of nuclear propulsion, there was an acute need of high-speed underwater targets to evaluate new ASW tactics, and as it was believed that the Soviet Navy was developing the Walter turbine, an order was placed in August 1947 for two improved versions of *Meteorite*. They were unarmed and purely experimental, and although much safer than the ex-U-boat they were unofficially known as 'Exploder' and 'Excruciator' because of their frequent minor mishaps. Diving depth was 550ft and surface endurance 500nm.

The nature of perhydrol meant that it had to be stored in plastic-lined tanks, open-topped to prevent the build-up of gas and to permit drenching with water in the event of spontaneous combustion. The slightest fragment of dust, rust or any other impurity was sufficient to start the release of oxygen, and a drop spilt would corrode metal, cloth or flesh. In view of the fact that both the Americans and the Russians gave up their attempts to get HTP propulsion to work, the achievement of the RN was considerable, but there can have been few tears at the boats' paying-off.

STICKLEBACK class *midget submarines*

Displacement: 32t standard; 35.32t/39.27t
Dimensions: 50ft 8in pp, 53ft 10in oa × 6ft × 7ft 6in
15.4m, 16.4m × 1.8m × 2.3m
Machinery: 1-shaft diesel-electric: 1 Perkins P6 6cyl diesel, plus 1 electric motor, 50bhp/44shp = 6.5kts/6kts. Oil 1.19t
Armament: 2 detachable 2t side-charges
Complement: 5

No	Name	Builder	Laid down	Launched	Comp	Fate
X 51	STICKLEBACK (ex-*X 51*)	Vickers-Armstrong, Barrow	?	1.10.54	?	Preserved 1977
X 52	SHRIMP (ex-*X 52*)	Vickers-Armstrong, Barrow	?	30.12.54	?	BU 1965
X 53	SPRAT (ex-*X 53*)	Vickers-Armstrong, Barrow	?	1.3.55	?	BU 1966
X 54	MINNOW (ex-*X 54*)	Vickers-Armstrong, Barrow	?	5.5.55	?	BU 1966

It had previously been assumed that these craft were ordered merely as replacements for worn-out XE-craft, but research into declassified papers in 1994 showed a much more complex history. In 1954 the Royal Navy conceived the idea that a nuclear sea mine could be laid by 'sneak' craft in the approaches to Soviet naval bases such as Kronstadt. At first it was assumed that the payload would be the centre section of the RAF's Mk 1 'Blue Danube' 10,000lb 20 kiloton bomb, but the following year the physics package of a modified 'Red Beard' 2000lb bomb was substituted.

Under the revised Staff Requirement the so-called 'Cudgel' sea mine was to be carried by one of the *X-51* class (given names in December 1954), and laid at depths down to 300ft (*c*90m). Cudgel's timer allowed settings at half-hour intervals from 30 minutes to 12 hours after laying, or at 12-hour intervals from 12 hours to seven days after laying.

The project proved unaffordable because of the shortage of fissile material, and 'Cudgel' was cancelled sometime between late 1955 and mid-1956. *Stickleback* was sold to the Royal Swedish Navy in 1958 and renamed *Spiggen*, and *Sprat* was lent to the US Navy at about the same time, to test harbour defences. Although stricken in 1970 the *Spiggen* was returned to the UK in 1977 for preservation at the Submarine Museum in Gosport.

Sprat 25.6.57 CMP

Five riveted boats were given a partial modernisation on the lines of the 'T Conversions', with a streamlined casing and fin, guns removed, but no change in propulsion; they were a logical development of the 'S' class streamline. Although not so capable as the full conversions they were useful for ASW training because they were quieter and faster than before: in October 1952, trials between the prototype *Tireless* and the unconverted *Tudor* showed an increase of 1.4kts in submerged speed. Higher capacity 6560amp/hr batteries replaced the former 5350amp/hr type. The sensor suite is unknown, but photographs indicate earlier marks than fitted in the 'T Conversions'. The original Mk 8 torpedoes were replaced by Mks 20 and 23.

'T Streamline' type *patrol submarines*

Displacement: 1090t standard; 1424t/1571t
Dimensions: 265ft pp, 273ft 6in oa × 26ft 7in × 15ft 10in max
80.8m, 83.4m × 8.1m × 4.8m
Machinery: 2-shaft diesel-electric: 2 ASR diesels, plus 2 electric motors, 2500bhp/1450shp = 15.25kts/9.5kts. Oil 209t
Armament: 6–21 TT (11 torpedoes)
Sensors: See notes
Complement: 50

No	Name	Builder	Laid down	Launched	Comp	Fate
S 37	TALENT	Vickers-Armstrong, Barrow	21.3.44	13.2.45	26.7.45	BU 1970
S 28	TOKEN	Portsmouth DYd	6.11.41	19.3.43	15.12.45	BU 1970
S 77	TIRELESS	Portsmouth DYd	30.10.41	19.3.43	18.4.45	BU 1968
S 35	TAPIR	Vickers-Armstrong, Barrow	29.3.43	21.8.44	30.12.44	BU 1966
S 38	TEREDO	Vickers-Armstrong, Barrow	17.4.44	27.4.45	5.4.46	BU 1965

Tireless as streamlined MoD

Artemis 1967

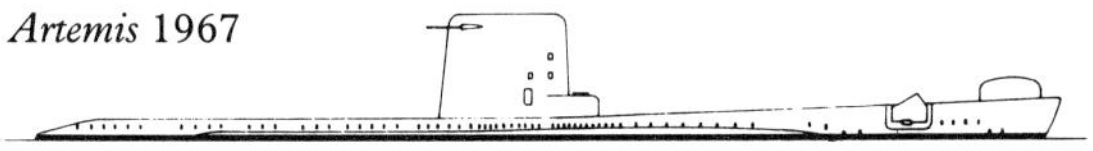

Artful underwent modernisation in 1955–56, along the lines of the 'T Conversions', with enlarged batteries and rebuilt bow and stern and streamlined fin, etc but retaining the original motors. All except *Aurochs* had undergone a similar conversion by 1960. The Mk 8 torpedoes were subsequently replaced by Mk 23, and some boats could carry twenty-six Mk 5 mines. *Alliance* was put on concrete trestles at HMS *Dolphin*, Gosport, in 1979 as a permanent submarine memorial.

Before being scrapped *Aeneas* was lent to Vickers for trials with the SLAM (Submarine Launched Anti-aircraft Missile). This is a TV-guided version of the Blowpipe missile, mounted in the fin and controlled from below, to provide a defence against helicopters. The system was not adopted by the RN and after the trials the boat was scrapped.

'A' class (modernised) *patrol submarines*

Displacement: 1120t standard; 1370t light; 1443t/1620t
Dimensions: 221ft pp, 282ft oa × 22ft 3in × 18ft 1in max
67.4m, 85.9m × 6.8m × 5.5m
Machinery: 2-shaft diesel-electric: 4 supercharged Vickers-Admiralty diesels, plus 2 electric motors, 4300bhp/1250shp = 19kts/15kts. Oil 213t
Armament: 6–21in TT (18 torpedoes) or 18 Mk 2 mines
Sensors: As 'T Conversions'
Complement: 63

No	Name	Builder	Laid down	Launched	Comp	Fate
S 61	ACHERON	Chatham DYd	26.8.44	25.3.47	17.4.48	BU 1972
S 72	AENEAS	Cammell Laird	10.10.44	25.10.45	31.7.46	BU 1974
S 41	ALARIC	Cammell Laird	31.5.44	18.2.46	11.12.46	BU 1971
S 65	ALCIDE	Vickers-Armstrong, Barrow	1.2.45	12.4.45	18.10.46	BU 1974
S 66	ALDERNEY	Vickers-Armstrong, Barrow	6.2.45	25.6.45	10.12.46	BU 1972
S 67	ALLIANCE	Vickers-Armstrong, Barrow	13.3.45	26.7.45	14.5.45	Preserved 1979
S 68	AMBUSH	Vickers-Armstrong, Barrow	17.5.45	24.9.45	22.7.47	BU 1971
S 43	AMPHION (ex-*Anchorite*)	Vickers-Armstrong, Barrow	14.11.43	31.8.44	27.3.45	BU 1971
S 64	ANCHORITE (ex-*Amphion*)	Vickers-Armstrong, Barrow	19.7.45	22.1.46	18.11.47	BU 1970
S 63	ANDREW	Vickers-Armstrong, Barrow	13.8.45	6.4.46	16.3.48	BU 1977
S 49	ARTEMIS	Scotts	28.2.44	26.8.46	15.8.47	BU 1972
S 96	ARTFUL	Scotts	8.6.44	22.5.47	2.2.48	BU 1972
S 47	ASTUTE	Vickers-Armstrong, Barrow	4.4.44	30.1.45	30.6.45	BU 1970
S 69	AURIGA	Vickers-Armstrong, Barrow	7.6.44	29.3.45	12.1.46	BU 1974

Andrew with deck gun, Mar 1974 C & S Taylor

Porpoise Oct 1966 CMP

Apart from the *Explorer* class these were the first British postwar submarines and they incorporated experience gained from wartime operations, trials with surrendered U-boats and the 'T Conversions'; the first batch was ordered in April 1951. They proved a great success (mainly because they were very quiet) and only started to go out of service because the 1975 Defence Review cut back the operational strength of the RN and because the new nuclear boats were making demands on manpower.

The two sets of batteries were linked in series to give 880v for short bursts of high speed. They had a range of 9000nm on the surface, sufficient for world-wide deployment. Mk 8 and Mk 20 torpedoes were carried. *Sealion* and *Walrus* were ordered to the *Oberon* design on 11 February 1954, and the rest on 19 April 1951. *Walrus* was bought for refit and resale; renamed *Seaforth B* and still extant late 1991 (laid-up at Immingham).

PORPOISE class *patrol submarines*

Displacement: 1565t standard; 1975t/2303t
Dimensions: 241ft pp, 290ft 3in oa × 26ft 6in × 18ft 3in
73.4m, 88.5m × 8.1m × 5.6m
Machinery: 2-shaft diesel-electric: 2 ASR 1 16cyl diesels, plus 2 English Electric motors, 3680bhp/6000shp = 12kts/17kts. Oil 258t
Armament: 8–21in TT (6 fwd, 2 aft; 30 torpedoes)
Sensors: Search and navigation radar; sonar 187, 2007, 2019
Complement: 71

No	Name	Builder	Laid down	Launched	Comp	Fate
S 01	PORPOISE	Vickers-Armstrong, Barrow	15.6.56	25.4.56	16.4.58	Sunk as target 1985
S 02	RORQUAL	Vickers-Armstrong, Barrow	15.1.55	5.12.56	24.10.58	BU 1977
S 03	NARWHAL	Vickers-Armstrong, Barrow	15.3.56	25.10.57	4.5.59	Sunk as target 3.8.83
S 04	GRAMPUS	Cammell Laird	16.4.55	30.5.57	19.12.58	BU 1980
S 05	FINWHALE	Cammell Laird	18.9.56	21.7.59	19.8.60	BU 1988
S 06	CACHALOT	Scotts	1.8.55	11.12.57	1.9.59	BU 1980
S 07	SEALION	Cammell Laird	5.6.58	31.12.59	25.7.61	BU 1990
S 08	WALRUS (ex-*Whale*)	Scotts	12.2.58	22.9.59	8.2.61	Sold 1987

Although plans for a nuclear submarine had been in embryo as early as 1943, a Government directive to the Admiralty to wait until the land-based nuclear power stations were ready meant that nothing could be done until the mid-1950s. Delays with the British Dounreay reactor threatened to delay the programme even further, and so in 1958 the US Navy was asked to supply a Westinghouse S5W reactor of the same type as in the *Skipjack* class. This meant that the after end of *Dreadnought* is virtually an adaptation of that design to British industrial practice grafted on to British equipment and weapon systems, although a considerable amount of information about advanced British design procedures was supplied to the USN in exchange.

The reactor was built by Rolls-Royce and the turbines by English Electric. She was fitted with Type 2001 sonar in the 'chin' position, Type 2007 for long-range passive detection and Type 2019 for ranging and interception. Because *Dreadnought* has one shaft and rudder, all the TT are forward. Early in 1982 it was announced that she had reached the end of her hull life and would be laid up after her reactor had been dismantled and the core removed.

Dreadnought paid off at Chatham and was towed to Rosyth in 1983, where she remains.

DREADNOUGHT *nuclear attack submarine*

Displacement: 3000t standard; 3500t/4000t
Dimensions: 265ft 9in pp × 32ft 3in × 26ft
81.0m × 9.8m × 7.9m
Machinery: 1-shaft nuclear: 1 S5W reactor plus geared steam turbine, 15,000shp, plus diesel-electric auxiliary = 15kts/28kts
Armament: 6–21in TT (24 torpedoes)
Sensors: Search and navigation radar; sonar 2001, 2007, 2019
Complement: 88

No	Name	Builder	Laid down	Launched	Comp	Fate
S 101	DREADNOUGHT	Vickers-Armstrong, Barrow	12.6.59	21.10.60	17.4.63	For disposal 1982

Dreadnought on trials, Dec 1962 CMP

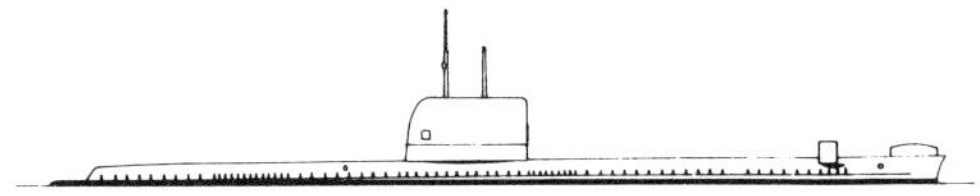

Otus 1969

OBERON class *patrol submarines*

Displacement: 1610t normal; 2030t/2410t
Dimensions: As *Porpoise* class
Machinery: As *Porpoise* class
Armament: Originally as *Porpoise* class, but are being altered to fire Mk 24 Mod 1 torpedoes
Sensors: Search and navigation radar; sonar 187, 2007, 2019
Complement: 64

No	Name	Builder	Laid down	Launched	Comp	Fate
S 09	OBERON	Chatham DYd	28.11.57	18.7.59	24.2.61	Sold 1987
S 10	ODIN	Cammell Laird	27.4.59	4.11.60	3.5.62	BU 1991
S 11	ORPHEUS	Vickers-Armstrong, Barrow	16.4.59	17.11.59	25.11.60	HS 1987; BU 1994
S 12	OLYMPUS	Vickers-Armstrong, Barrow	4.3.60	14.6.61	7.7.62	To Canada 1989
S 13	OSIRIS	Vickers-Armstrong, Barrow	26.1.62	29.11.62	11.1.64	BU 1993
S 14	ONSLAUGHT	Chatham DYd	8.4.59	24.9.60	14.8.62	BU 1991
S 15	OTTER	Scotts	14.1.60	15.5.61	20.8.62	Sold 1992
S 16	ORACLE	Cammell Laird	26.4.60	26.9.61	14.2.63	Sold 1992
S 17	OCELOT	Chatham DYd	17.11.60	5.5.62	31.1.64	Preserved 1992
S 18	OTUS	Scotts	31.5.61	17.10.62	5.10.63	Sold 1992
S 19	OPOSSUM	Cammell Laird	21.12.61	23.5.63	5.6.64	For disposal 1994
–	ONYX (i)	Chatham DYd	27.9.62	–	–	To Canada 1963
S 20	OPPORTUNE	Scotts	26.10.62	14.2.64	28.12.64	Sold 1992
S 21	ONYX (ii)	Cammell Laird	16.11.64	18.8.66	25.11.67	BU 1992

Although outwardly identical with the *Porpoise* class, these boats bear the same relationship to them as the *Rothesay* class frigates did to the *Whitby* class: a number of detailed improvements were incorporated to make a considerable overall improvement in efficiency. *Orpheus* was given an aluminium casing, but later boats adopted glass-fibre laminate. The particular feature of this class is its quietness, with all equipment soundproofed, but in addition a higher grade of steel permits a greater diving depth than the *Porpoise* class (believed to exceed 1000ft). *Onyx* was bought on the stocks by the Royal Canadian Navy and launched as ashmcsxs *Ojibwa* in 1964; a replacement was added to the programme. Ten were to be retained until replaced by the *Upholder* class, but when the SSK force was cut in 1991 they went out of service early. *Osiris* was cannibalised in 1992 to provide spares for Canadian *Oberons*.

The class was planned to be replaced by ten *Upholder* class SSKs in the late 1980s but the rapid rundown of submarine strength meant that most were retired without being replaced. In the 1980s most were modernised, with Type 2051 sonar, DCH combat system and UAL electronic warfare system, as well as the Type 2046 clip-on towed array. In 1986 *Oberon* was bought for refit and resale; renamed *Seaforth A* and was laid up at Immingham. *Otus* carried out trials with Sub-Harpoon in 1989. *Olympus* was sold to Canada in August 1989 for harbour training at Halifax. *Onyx* was acquired by the Warship Preservation Trust 1991, to be on display at Birkenhead, and *Ocelot* was transferred to the Chatham Historic Dockyard Trust in 1992. *Otter* and *Otus* were sold in 1992 to the Pounds organisation at Portsmouth.

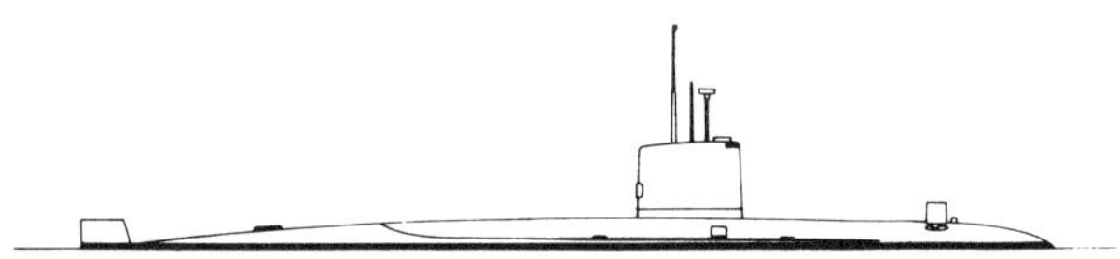

Valiant 1977

VALIANT class *nuclear attack submarines*

Displacement: 4000t normal; 4400t/4900t
Dimensions: 285ft pp × 33ft 3in × 27ft
86.9m × 10.1m × 8.2m
Machinery: 1-shaft nuclear: 1 pressurised water-cooled reactor (PWR 1) plus geared steam turbine, 15,000shp, plus diesel-electric auxiliary = 20kts/28kts
Armament: 6–21in TT (26 torpedoes)
Sensors: Search and navigation radar; sonar 2001, 2007, 183, 197, 2019
Complement: 103

No	Name	Builder	Laid down	Launched	Comp	Fate
S 102	VALIANT (ex-*Inflexible*)	Vickers-Armstrong, Barrow	22.1.62	3.12.63	18.7.66	For disposal 1992
S 103	WARSPITE	Vickers-Armstrong, Barrow	10.12.63	25.9.65	18.4.67	For disposal 1990
S 46	CHURCHILL	Vickers-Armstrong, Barrow	30.6.67	20.12.68	15.7.70	For disposal 1991
S 48	CONQUEROR	Cammell Laird	5.12.67	28.8.69	9.11.71	For disposal 1990
S 50	COURAGEOUS (ex-*Superb*)	Vickers-Armstrong, Barrow	15.6.68	7.3.70	16.10.71	For disposal 1992

This class was a slightly enlarged edition of the *Dreadnought* to accommodate the bulkier British reactor developed at the Admiralty Reactor Test Establishment at Dounreay. The completion of *Valiant* was delayed by a year because of the priority given to the Polaris programme. As in the *Dreadnought*, conventional Laurence-Scott electric motors and a 112-cell battery provided emergency power, with a Paxman diesel generator. All TT are forward, and Mk 24 torpedoes have replaced the earlier Mks 8 and 23.

Courageous was the first of the class to be refitted with the new Type 2020 long-range passive sonar, but the whole class received Sub-Harpoon missiles and *Courageous* tested this weapon in 1981, and *Churchill* also evaluated the Mk 48 Mod 3 torpedo, with *Otus*. *Courageous* was the first to deploy Sub-Harpoon operationally, in 1981.

During the Falklands conflict *Conqueror* sank the Argentine cruiser *General Belgrano* on 2 May 1982 using two Mk 8 torpedoes. The age of the reactors started to give concern in 1988, and the cutback in submarine strength meant that none received any further planned modernisation.

Conqueror 1978 L & L van Ginderen

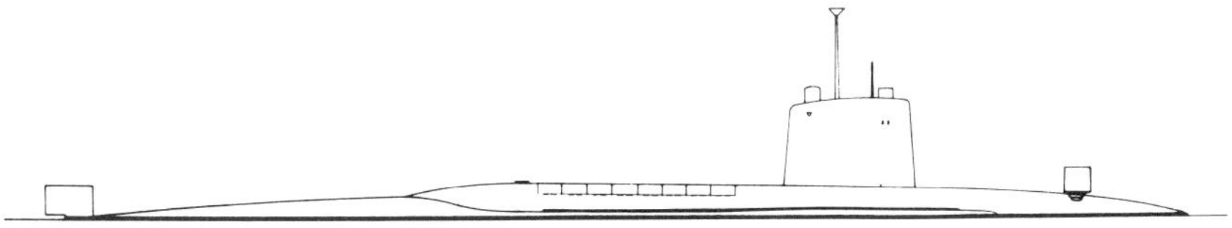

Renown 1969

RESOLUTION class *ballistic missile submarines*

Displacement: 7500t/8500t
Dimensions: 360ft pp, 425ft oa × 33ft × 30ft
109.7m, 129.5m × 10.1m × 9.1m
Machinery: 1-shaft nuclear: pressurised water-cooled PWR 1 reactor, plus 2 geared steam turbine, 15,000shp, plus Paxman auxiliary diesel, 4000bhp = 20kts/25kts
Armament: 16 Polaris A-3 UGM-27C SLBM, 6–21in TT
Sensors: Radar 1006; sonar 2001, 2007, 183, 197, 2019, 2024 (replaced by 2046)
Complement: 143 (13 officers, 130 ratings)

No	Name	Builder	Laid down	Launched	Comp	Fate
S 22	RESOLUTION	Vickers-Armstrong, Barrow	26.2.64	15.9.66	2.10.67	For disposal 1994
S 23	REPULSE	Vickers-Armstrong, Barrow	12.3.65	4.11.67	28.9.68	Extant 1995
S 26	RENOWN	Cammell Laird	25.6.64	25.2.67	15.11.68	Extant 1995
S 27	REVENGE	Cammell Laird	19.5.65	15.3.68	4.12.69	For disposal 1992
–	RAMILLIES	–	–	–	–	Cancelled 1964

At the 1962 Nassau Conference, the momentous decision was taken to transfer responsibility for the United Kingdom's nuclear deterrent from the RAF to the RN, following the cancellation of the airborne Skybolt system. Instead the RN was to build five nuclear ballistic missile submarines (SSBNs) based broadly on the missile compartment of the US Navy's *Lafayette* class, with British equipment and machinery, but armed with missiles, tubes and fire control systems supplied from the United States.

A special Polaris Executive was set up to supervise the building of the boats and the creation of training and support facilities, and so successful was the collaboration between the administrative and technical sides that the first boat, *Resolution*, went on patrol as planned in 1968. With *Dreadnought* it was the only British post-1945 defence programme to remain within its financial budget. Only one problem was encountered: the figure of five boats was arrived at after a careful study of refit schedules, and when the new Labour Defence Minister, Denis Healey, cancelled the order for the fifth boat as a gesture of appeasement, it became difficult to guarantee that one Polaris 'bomber' would remain on patrol at all times. Up to 1992, however, there had officially been no gap in the patrol.

Apart from the use of British equipment and design features from the *Valiant* class, the main departure from the *Lafayette* class was the rejection of diving planes on the fin. Although these are standard in USN submarines, British designers maintain that their advantages of better slow-speed control do not outweigh their disadvantages. These submarines will be replaced after 1992 by four new SSBNs armed with Trident missiles, but to improve the ability of the existing Polaris missiles to penetrate defences a new warhead was developed: code-named 'Chevaline' (A3TK), it consists of three independently targeted re-entry warheads capable of hitting points 40km apart. The completion of this programme was announced in 1980, shortly before confirmation of the Trident purchase. Fitted with DCB combat system and UAB/UAC electronic warfare systems.

Resolution as completed MoD

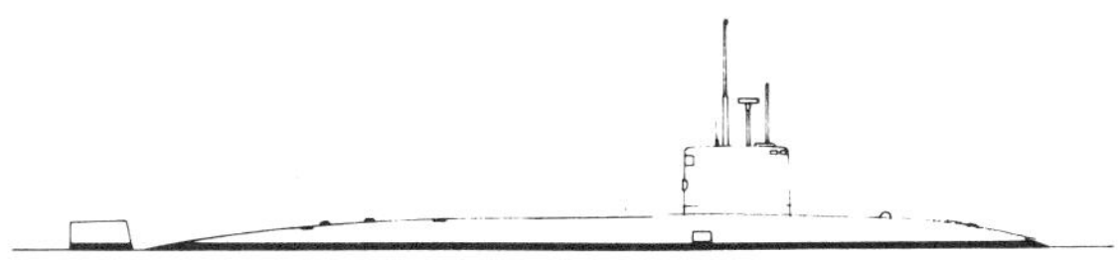

Superb 1977

SWIFTSURE class *nuclear attack submarines*

Displacement: 4400t normal; 4400t/4900t
Dimensions: 272ft pp × 32ft 4 in × 27ft
82.9m × 9.8m × 8.2m
Machinery: 1-shaft nuclear: 1 PWR 1 reactor, plus 2 GEC geared steam turbines, 15,000shp, plus Paxman diesel, 4000bhp = 20kts/30kts
Armament: 5–21in TT (21 Tigerfish Mk 24 Mod 2 torpedoes and 4 Sub-Harpoon missiles)
Sensors: Radar 1007; sonar 2001/2020, 2007, 2019, 2046
Complement: 116 (13 officers, 103 ratings)

No	Name	Builder	Laid down	Launched	Comp	Fate
S 126	SWIFTSURE	VSEL, Barrow	6.6.69	7.9.71	17.4.73	For disposal 1992
S 108	SOVEREIGN	VSEL, Barrow	18.9.70	17.2.73	11.7.74	Extant 1995
S 109	SUPERB	VSEL, Barrow	16.3.73	30.11.74	13.11.76	Extant 1995
S 104	SCEPTRE	VSEL, Barrow	19.2.74	20.11.76	14.2.78	Extant 1995
S 111	SPARTAN	VSEL, Barrow	26.4.76	7.4.78	22.9.79	Extant 1995
S 112	SPLENDID (ex-*Severn*)	VSEL, Barrow	23.11.77	5.10.79	21.3.81	Extant 1995

Shorter and of fuller hull-form than previously, and designed to dive deeper. The fine bow-section means that only five tubes can be fitted, but in other respects the vessels are merely improved *Valiant*s. It is confirmed that reloading of torpedoes takes fifteen seconds and that they are fitted with the Mk 24 Mod 2 and DCB fire control system (DCA in *Swiftsure*). The class is having Type 2001 upgraded to 2020, and will eventually receive the integrated Type 2074 system during the mid-life modernisation planned for the mid-1990s. Sub-Harpoon missiles have replaced some of the torpedoes, and the class will eventually receive the new Spearfish torpedo.

Orders placed in November 1967, May 1969, May 1970, November 1971, February 1973 and May 1976. UAC electronic warfare equipment is being upgraded to UAP.

Superb 1978 MoD

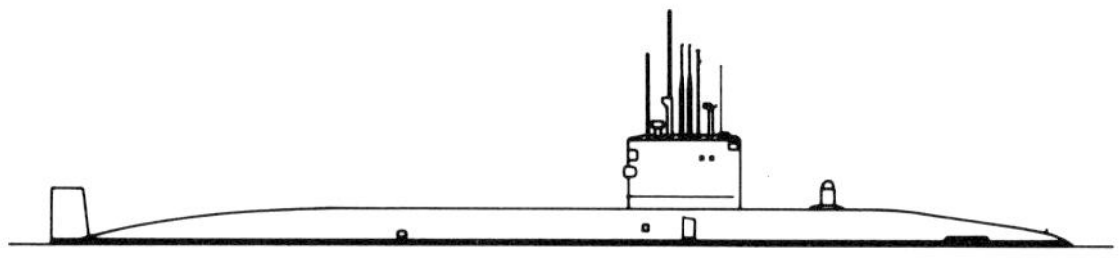

Turbulent 1985

The first of a new class of seven fleet nuclear submarines was ordered in September 1977, followed by a second in July 1978, third in July 1979 and the remainder in June 1981, March 1983, September 1984 and January 1989; two more are planned. They resemble the *Swiftsure* class, but with much more attention paid to silencing: *Turbulent* and later boats have shrouded propulsor instead of a 7-bladed propeller, and all hulls are coated with anechoic tiles.

TRAFALGAR class *nuclear attack submarines*

Displacement: 4700t/5200t
Dimensions: 280ft 2in pp × 32ft 2in × 31ft 3in
85.4m × 9.8m × 9.5m
Machinery: 1-shaft nuclear: 1 PWR 1 reactor, plus 2 GEC geared steam turbines, 15,000shp, plus Paxman diesel and 2 Chloride 112-cell batteries, 4000bhp = 20kts/30kts
Armament: 5–21in TT (21 Tigerfish Mk 24 Mod 2 and 4 Sub-Harpoon missiles)
Sensors: Radar 1007; sonar 2020, 2019, 2007, 2046 etc
Complement: 97 (12 officers, 85 ratings)

No	Name	Builder	Laid down	Launched	Comp	Fate
S 107	TRAFALGAR	VSEL, Barrow	25.4.79	1.7.81	27.5.83	Extant 1995
S 87	TURBULENT	VSEL, Barrow	8.5.80	1.12.82	28.4.84	Extant 1995
S 88	TIRELESS	VSEL, Barrow	1981	17.3.84	5.10.85	Extant 1995
S 90	TORBAY	VSEL, Barrow	1982	8.3.85	7.2.87	Extant 1995
S 91	TRENCHANT	VSEL, Barrow	1984	3.11.86	14.1.89	Extant 1995
S 92	TALENT	VSEL, Barrow	1986	15.4.88	12.5.90	Extant 1995
S 93	TRIUMPH	VSEL, Barrow	1987	16.2.91	12.10.91	Extant 1995

Trafalgar 31.8.83
L & L van Ginderen

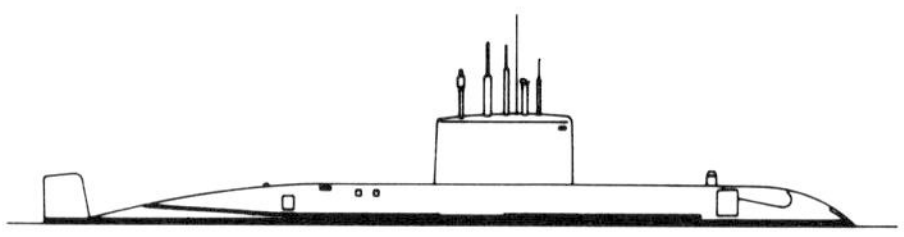

Upholder 1989

A new type of 'conventional' submarine was announced in September 1979, to replace the *Oberons* and to make up numbers for training. Although at one time twelve boats were planned this dropped to ten and then nine, but under the 1990–91 defence cuts the class was stopped at four units.

The first four were fitted with Type 2040 sonar, derived from the French Eledone and Netherlands Octopus, but plans to replace this with Type 2075 were dropped in 1991. This class was declared surplus to requirements in 1994.

UPHOLDER class *patrol submarines*

Displacement: 2220t/2455t
Dimensions: 230ft 7in oa × 25ft × 17ft 8in
70.3m × 7.6m × 5.5m
Machinery: 1-shaft diesel-electric: Paxman Valenta 1600 RPA SZ diesels, 1 GEC electric motor, 2 120-cell Chloride batteries = 12kts/20kts (12kts snorting). Range 8000nm at 8kts.
Armament: 6–21inch TT (14 Tigerfish Mk 24 Mod 2 torpedoes, 4 Sub-Harpoon missiles)
Sensors: Radar 1007; sonar 2040, 2019, 2007, 2046
Complement: 47 (7 officers, 40 ratings)

No	Name	Builder	Laid down	Launched	Comp	Fate
S 40	UPHOLDER	VSEL, Barrow	Nov 83	2.12.86	9.6.90	For disposal 1995
S 41	UNSEEN	Cammell Laird	12.7.87	14.11.89	19.7.91	For disposal 1995
S 42	URSULA	Cammell Laird	10.1.89	28.2.91	8.5.92	For disposal 1995
S 43	UNICORN	Cammell Laird	13.3.90	16.4.92	25.6.93	For disposal 1995

Unicorn Jun 1993 C & S Taylor

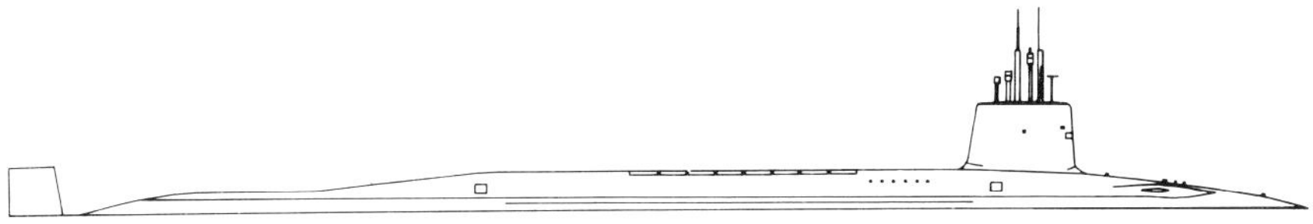

Vanguard class design

In 1980 the MoD announced that it would build four submarines armed with the Trident UGM-93A, replacing the four Polaris-armed *Resolutions* towards the end of the decade. In general they resemble the *Resolution* class but are much bigger. The first, SSBN 05, was ordered in April 1986, followed by SSBN 06 in October 1987, and SSBN 07 in November 1990. The still-unnamed SSBN 08 was partially funded in advance of the order in 1991.

The major improvements are the more powerful PWR 2 reactor and the integrated Type 2054 sonar system. The emphasis on both weapons and platform is higher reliability, to reduce the time spent in refit. The interval between recoring is estimated to be eight to nine years. It was hoped to buy the C4 Trident I missile system but this is no longer in production for the USN, so the D5 Trident II had to be bought instead. Maintenance of the missiles is to be done by the USN at King's Bay, Georgia, to reduce cost to the RN.

VANGUARD class *ballistic missile submarines*

Displacement: *c*14,000t/16,000t
Dimensions: 491ft 9in pp × 42ft × 39ft 4in
149.9m × 12.8m × 12m

No	Name	Builder	Laid down	Launched	Comp	Fate
S 28	VANGUARD	VSEL, Barrow	3.9.86	4.3.92	14.8.93	Extant 1995
S 29	VICTORIOUS	VSEL, Barrow	3.12.87	29.9.93	7.1.95	Extant 1995
S 30	VIGILANT	VSEL, Barrow	16.2.91	–	–	Building
S 31	VENGEANCE	VSEL, Barrow	1.2.93	–	–	Building

Victorious as completed

MoD

SSN-20 class *nuclear attack submarines*

Design studies for a successor to the *Trafalgar* class began in 1987, known as *SSN-20* but unofficially designated the 'W' class Up to seven were planned, using a modified version of the PWR 2 reactor plant. Project-definition started late in 1989, but by 1990 it was clear that costs were spiralling (in excess of £400 million per hull, excluding R & D). The programme fell foul of Treasury expenditure cuts in 1991, and a cheaper, cost-effective design is to replace it in the mid-1990s.

In November 1991 VSEL was awarded a £6 million contract for a twelve-month design study for the Batch 2 *Trafalgar* project. Cost reductions will be achieved by retaining proven features from the upgraded 'T' design, including SMCS, Type 2076 sonar and the Spearfish torpedo. However the hull-shape will probably be improved quite considerably, and internal layout will take advantage of developments in non-penetrating masts (*eg* repositioning the control room). It is hoped to have the first of class in service by 2003.

AMPHIBIOUS WARFARE VESSELS

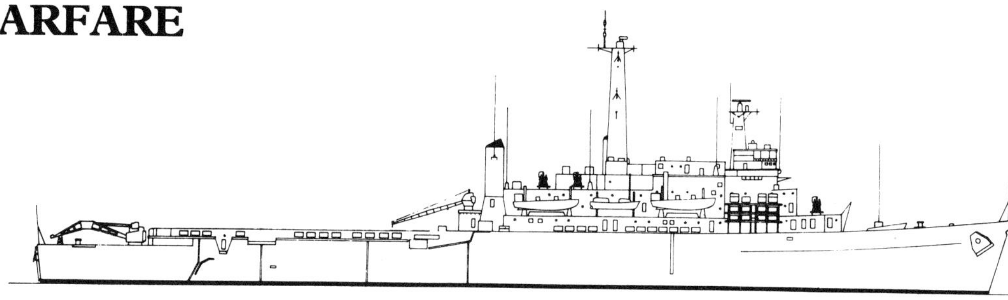

Intrepid 1981

FEARLESS class *assault ships*

Displacement: 11,060t normal; 12,120t deep load; 16,950t ballasted
Dimensions: 500ft wl, 520ft oa × 80ft × 20ft 6in normal
152.4m, 158.5m × 24.4m × 6.3m
Machinery: 2-shaft geared steam turbines, 2 boilers, 22,000shp = 21kts. Range 5000nm at 20kts
Armament: 2–40mm/60 Mk 9 Bofors (2×1), 4 Seacat GWS22 SAM (4×4)
Sensors: Radar 978, 993
Complement: 580 + 400–700 troops

No	Name	Builder	Launched	Fate
L 10	FEARLESS	Harland & Wolff, Belfast	19.12.63	Extant 1995
L 11	INTREPID	John Brown, Clydebank	25.6.64	Extant 1995

To provide the Royal Marines with greater capability, two updated versions of the wartime dock landing ship were ordered in 1962. Although launched in surroundings of theatrical secrecy they were simply smaller versions of the contemporary *Raleigh* class building by the USN. The principal advantage over the existing LSTs was the seaworthiness, range and capacity, for they could accommodate 400 troops on board (700 for short periods) as well as, typically, fifteen tanks and twenty-seven vehicles. *Fearless* was laid down on 25 July 1962 and completed on 25 November 1965; *Intrepid* was laid down on 19 December 1962 and completed on 11 March 1967.

Despite being large and valuable ships the armament is light. Each ship carries four LCM(9) landing craft in davits amidships.

In 1976 a defence cutback caused *Intrepid* to be paid off into reserve and *Fearless* to be downgraded to the Dartmouth Training Squadron, but every effort was made to keep them available for an emergency. In 1979 *Intrepid* was brought forward from reserve to allow *Fearless* to be refitted. *Intrepid* had de-stored in late 1981 and was an offer to Argentina, but in April 1982 she was hurriedly recommissioned to join *Fearless* in the Falklands Task Force.

Intrepid was refitted in 1984–85, and *Fearless* was given a major modernisation in 1980–90. *Fearless* now has two Mk 15 Phalanx CIWS and the Nautis-L command and control system but the poor state of *Intrepid* makes it unlikely that she will be given a similar modernisation. A design study contract was awarded in late 1991 for the replacement ships, designated LPD(R)s, and it is hoped to order them in 1995.

OCEAN *assault helicopter carrier*

Displacement: *c*20,000t deep load
Dimensions: 668ft oa × 107ft max × 21ft 4in
203.6m × 32.6m × 6.5m
Machinery: 2 shafts, 2 Crossley-Pielstick diesels = 17kts
Armament: 3–20mm Mk 15 Phalanx CIWS (3×1), 12 HC 4 Commando helicopters
Sensors: 250 + 700 troops

No	Name	Builder	Launched	Fate
L 12	OCEAN	Kvaerner Govan/ VSEL	Oct 1995	Completing

In 1989 three yards submitted tenders for an Aviation Support Ship (ASS) to provide the Royal Marines with a helicopter carrier. Proposals ranged from a conversion of a roll-on/roll-off container ship similar to the Aviation Training Ship RFA *Argus*, to a mercantile hull built from-the-keel-up, and it was even suggested that the bare hull of a *Wasp* class LHD could be bought from the US Navy and fitted out to RN/RM requirements. It was hoped to order the ASS in mid-1990 but this was deferred to 1992. The ADAWS-2000 command and control system will be fitted, but armament will be light.

Tenders were invited from Swan Hunter and VSEL in February 1992 and an order was placed with VSEL a year later for about £170m. The hull was subcontracted to Kvaerner Govan Ltd on the Clyde. VSEL will install electronics and weapons and complete final fitting out. DLJ (2) chaff/infrared decoy launchers will be fitted. Delivery is planned for 1997.

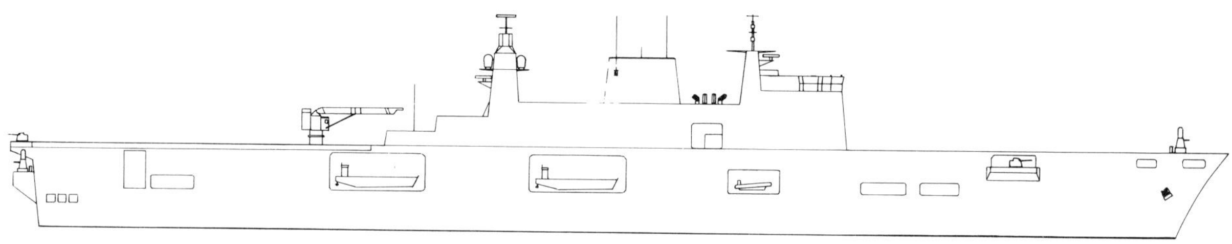

Ocean design

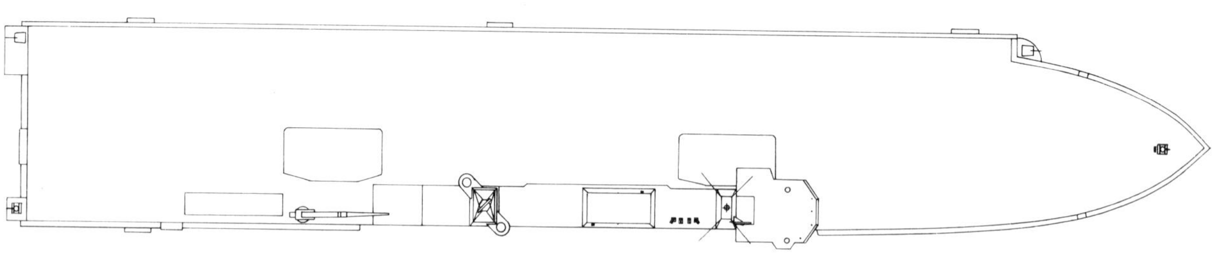

Fearless May 1994 C & S Taylor

Sir Percivale Jun 1994 C & S Taylor

SIR LANCELOT class *logistic landing ships*

Displacement: 3270t light; 5674t full load
Dimensions: 380ft wl, 412ft 1in oa × 59ft 8in × 13ft
115.8m, 125.1m × 19.6m × 4.3m
Machinery: 2 shafts, 2 Mirrlees 10cyl diesels, 9400bhp (*Sir Lancelot* Denny-Sulzer diesels, 9520bhp) = 17kts. Range 8000nm at 15kts
Armament: Fitted for 2–40mm (2×1), but not normally carried
Complement: 68 + 340–534 troops

No	Name	Builder	Launched	Fate
L 3029	SIR LANCELOT	Fairfield	25.6.63	Sold 1989
L 3004	SIR BEDIVERE	Hawthorn Leslie	20.7.66	Extant 1995
L 3005	SIR GALAHAD	Stephen	19.4.66	Bombed 8.6.82, scuttled
L 3505	SIR TRISTRAM	Hawthorn Leslie	12.12.66	Extant 1995
L 3027	SIR GERAINT	Stephen	26.1.67	Extant 1995
L 3036	SIR PERCIVALE	Hawthorn Leslie	4.10.67	Extant 1995

Sir Lancelot was the prototype of a class of six landing ships ordered by the Ministry of Transport for use by the Army. They were initially chartered to and operated by shipping companies, but were taken over by the Royal Fleet Auxiliary in 1980. They can carry, and operate, helicopters, sixteen main battle tanks, thirty-four mixed vehicles and up to 534 troops; they can also lay out pontoons. They were designed to mercantile specifications, to operate virtually as roll-on/roll-off ferries in support of the Army, and were never intended to take part in opposed landings – hence the generally lightweight construction and aluminium superstructure. *Sir Galahad* and *Sir Tristram* were bombed at Fitzroy during the Falklands conflict; the former was so badly damaged that she was later towed out to sea and torpedoed by a submarine, but *Sir Tristram* survived. After some time in the Falklands as an accommodation hulk she was brought home in 1983. She was lengthened by 29ft and given a bigger flight deck and a new steel bridge. *Sir Bedivere* started a similar but more elaborate conversion in 1994. Re-engined with Stork-Wärtsilä diesels. It is hoped to bring the remaining two LSLs up to the same standard in the late 1990s. *Sir Lancelot* was sold commercially in 1989 and then to the Singapore navy.

Chartered *logistic ships*

No	Name	Launched	Fate
L 3522	SIR CARADOC (ex-*Grey Master*, 1899grt)	11.8.72	Returned 1987
L 3532	SIR LAMORAK (ex-*Lakespan Ontario*, 1566grt)	1972	Returned 1986

These were roll-on/roll-off ships chartered from Norwegian and Canadian owners early in 1983 as temporary replacements for LSLs damaged in the Falklands. They were returned to their owners during 1986–87.

Sir Galahad 1989

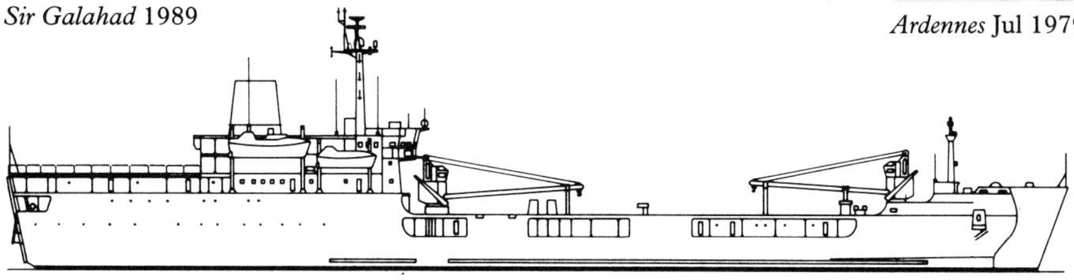

SIR GALAHAD *logistic landing ship*

Displacement: 6200t light; 8585 full load
Dimensions: 461ft oa × 64ft × 14ft 2in
140.5m × 19.5m × 4.3m
Machinery: 2-shaft diesel, 2 Mirrlees Blackstone, 13,320bhp = 18kts. Range 13,000nm/15kts
Armament: 2 Bofors 40mm/60 (2×1), 3 Oerlikon 20mm GAM B03, 1 Sea King HC 4 helicopter
Complement: 49 + 343/537 troops

No	Name	Launched	Fate
L 3005	SIR GALAHAD	13.12.86	Extant 1995

Ordered in September 1984 from Swan Hunter as a replacement for the LSL damaged in 1982 and subsequently sunk as a war grave off Port Stanley. There are many improvements over the original design, including enlarged helicopter deck, better firefighting and steel bridge, as well as greater capacity.

LANDING CRAFT

Name	Fate	Name	Fate
Ardennes class: launched 1976–77, 1413t full load			
ARDENNES (L 4001)	Extant 1995	ARAKAN (L 4007)	Extant 1995
Avon class: launched 1961–67, 100t full load			
AVON (RPL 01)	For disposal 1984	HAMBLE (RPL 08)	Stricken *c*1980
BUDE (RPL 20)	For disposal 1983?	ITCHEN (RPL 09)	For disposal 1987?
CLYDE (RPL 03)	For disposal 1986?	KENNET (RPL 10)	For disposal 1984
DART (RPL 04)	For disposal 1987?	LODDEN (RPL 11)	For disposal 1984
EDEN (RPL 05)	For disposal 1994	MEDWAY (RPL 12)	For disposal 1994
FORTH (RPL 06)	For disposal 1994	GLEN (RPL 07)	Stricken *c*1980

Ardennes Jul 1979 C & S Taylor

Name	Fate	Name	Fate
Arromanches class: launched 1981–87, 100t full load			
ARROMANCHES (L 105)	Extant 1995	AKYAB (L 109)	Extant 1995
		AACHEN (L 110)	Extant 1995
ANTWERP (L 106)	For disposal 1994	AREZZO (L 111)	Extant 1995
ANDALSNES (L 107)	Extant 1995	AGHEILA (L 112)	For disposal 1994
ABBEVILLE (L 108)	Extant 1995	AUDEMAR (L 113)	Extant 1995
LCM(9) type: launched 1963–66, 176t full load			
L 700–L 715	Extant 1995	L 3508	Extant 1995
LCVP Mk 4 type: built 1981–86, 16t loaded			
LCVP Mk 5 type: ordered 1995, 20t loaded			

Ardennes class
Operated by the Army's Royal Corps of Transport, more as ferries than as warships, although they are designed to replace the wartime LCT(8) type. They are unarmed, but can carry five 70t tanks or up to twenty-four standard 20ft containers, plus thirty-four troops. They were built by Brooke Marine.

Avon class
Also operated by the Royal Corps of Transport, they were officially Ramp Powered Lighters, hence the RPL pennant numbers. Built by White and Saunders-Roe, they have largely been replaced by the *Arromanches* class.

Arromanches class
The first two were launched by Brooke Marine in 1981. Seven more were built by James & Stone, Brightlingsea, in 1983–87, as replacements for the *Avon* class.

LCM(9) type
Vospers built the prototypes *L 3507* and *L 3508* in 1963, and the remainder were built by Dunston, Brooke Marine or J Bolson in 1964–66. They were conventional LCM type craft and could carry two tanks or 100 tons of cargo. Four each of the *L 700* series are carried by *Fearless* and *Intrepid*. *L 3507* was lost during the Falklands conflict. Later reclassified as LCUs.

Two LCM (9)s from *Fearless*, Jul 1982 — C & S Taylor

LCVP Mk 4 type
The first of a new type was ordered on 8 February 1980 from Fairey Allday Marine. Twenty more were ordered from Souters of Cowes from July 1980 onwards, as replacements for the surviving LCVP Mk 3, three of which were extant in 1992. *L 8031*, *L 8401*, *L 8403–8408* and *L 8410–8418* are operated by the Royal Marines, *L 8402*, *L 8409* and *L 8619–8620* by the Army (RCT). They are redesignated LCSs of about 16t full load, and can carry thirty-five troops or two Land Rovers.

LCVP Mk 5 type
First unit ordered from Vosper Thornycroft in January 1995, with options for four more.

Other minor landing craft
In the early 1980s remaining minor landing craft were two ex-LCM(7)s, employed as naval servicing boats and three LCP(L) (3)s, plus a few Raiding Landing Craft (LCRs).

Droxford 1.4.81 — MG Photographic

SMALL SURFACE COMBATANTS

'FORD' class *seaward defence boats*

Displacement: 120t normal; 160t deep load
Dimensions: 110ft pp, 117ft 3in oa × 20ft × 5ft
33.5m, 35.7m × 6.1m × 1.5m
Machinery: 3 shafts, 2 12 cyl Paxman diesels plus 1 Foden, 1100/100bhp = 18kts/8kts
Armament: 1–40mm/60 Mk 7, 2 DCT
Sensors: Radar 978; sonar 144
Complement: 19

Class
Shalford (P 3101), *Aberford* (P 3102), *Axford* (P 3103), *Beckford* (P 3104), *Brayford* (P 3105), *Bryansford* (P 3106), *Camberford* (P 3107), *Desford* (P 3108), *Greatford* (P 3109), *Gifford* (P 3111), *Droxford* (P 3113), *Mayford* (P 3114), *Hinksford* (P 3115), *Ickford* (P 3116), *Dubford* (P 3119), *Glassford* (P 3120), *Kingsford* (P 3121), *Marlingford* (P 3122), *Tilford* (P 3123), *Montford* (P 3124)

Designed by W Holt as inshore ASW craft to replace the wartime HDMLs, with main drive diesels on wing shafts and a cruising diesel on the centre shaft. Appearance was distinctive, with two funnels side by side and a lattice mast. They were designed to carry a single-barrel Squid but this weapon did not materialise and so the lead-craft, *Shalford*, was fitted with a Squid triple DC mortar aft; the remainder had a standard armament of two depth-charge throwers and two racks. *P 3101–3102* were ordered on 6 April 1951, *P 3103–3108* on 3 May 1951, *P 3109, 3111–3113* and *3119* on 11 May 1951, *P 3114–3116* and *P 3121–3122* on 31 May 1951, and *P 3123–3124* on 24 May 1955, and were built by various yards during 1951–57.

Foreign transfers: *Brayford* and *Glassford* (South African *Gelderland* and *Nautilus*, 1954–55); *Desford* (Ceylonese *Kotiya*, 1955); *Aberford* (Kenyan *Nyati*, 1964), and *Axford*, *Hinksford*, *Gifford* and *Dubford* (Nigerian *Kaduna*, *Benin*, *Bonny* and *Sapele*, 1966–68). Most of the class were put on the disposal list in 1966–67, but *Beckford* (renamed *Dee* in 1965) and *Droxford* (renamed *Dee* 1955–65) became tenders to Liverpool and Glasgow University reserve units.

Ex-RAF *patrol craft*

No	Name	Transferred	Fate
P 256	CORMORANT (ex-*Sunderland*)	28.8.85	Sold 1991
P 257	HART (ex-*Stirling*)	28.8.85	Sold 1991
P 259	REDPOLE (ex-*Sea Otter*)	30.10.84	For disposal 1994

These are former *Spitfire* and *Seal* class tenders transferred from the RAF Marine Section in 1984–85.

KINGFISHER class *patrol craft*

Displacement: 187t normal; 194t deep load
Dimensions: 110ft 11in wl, 120ft oa × 21ft 8in × 6ft 6in
33.8m, 36.6m × 6.6m × 2.0m
Machinery: 2 shafts, 2 Paxman 16 YCJM diesels, 4000bhp = 25kts. Range 2000nm at 14kts
Armament: 1–40mm/60 Mk 9, 2–7.62mm MG
Complement: 19

No	Name	Builder	Launched	Fate
P 260	KINGFISHER	Richard Dunston	20.9.74	For disposal 1994
P 261	CYGNET	Richard Dunston	26.10.74	For disposal 1994
P 262	PETEREL	Richard Dunston	14.5.76	BU 1991
P 263	SANDPIPER	Richard Dunston	20.1.77	BU 1992

An unsuccessful attempt to adapt an RAF air-sea rescue design, the *Seal* type, these craft have proved unstable and suffer badly from vibration. Following problems with *Kingfisher*, the later craft were given a lighter mast and had no scuttles. Two were immediately relegated to RNR training tenders and all will be disarmed for service as inshore surveying craft.

JURA *offshore patrol vessel*

Displacement: 778t normal; 1285t deep load
Dimensions: 195ft 3in oa × 35ft × 14ft 4in
59.5m × 10.7m × 4.4m
Machinery: 1 shaft, 2 British Polar SP11 2VS-F diesels, 4200bhp = 17kts
Armament: 1–40mm/60 Bofors Mk 3
Complement: ?

Under considerable political pressure to provide defence for offshore resources in the new 200-mile Exclusive Economic Zone (EEZ), the Royal Navy initiated studies of sea-keeping characteristics. These established criteria of length, beam, freeboard and subdivision, and thereafter commercial designs were sought which approximated to the parameters. The best answer appeared to be a pair of trawler-type fishery protection vessels, *Jura* and *Westra*, operated by the Scottish Department of Fisheries. *Jura* (P 296) was chartered in 1975 and armed with a 40mm gun. She proved successful and the design of the 'Island' class was adapted from it. When HMS *Jersey* became operational in January 1977, *Jura* was returned to her owners.

Orkney 1980
NB 1/750 scale

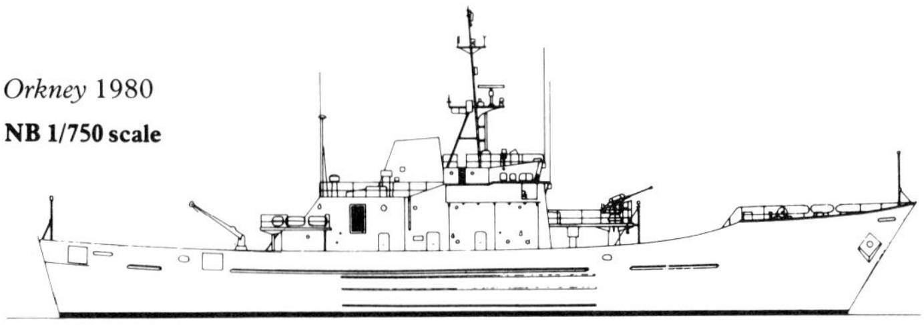

'ISLAND' class *offshore patrol vessels*

Displacement: 1000t normal; 1280t deep load
Dimensions: 176ft wl, 195ft 3in oa × 36ft × 14ft
53.6m, 59.5m × 11.0m × 4.2m
Machinery: 1 shaft, 2 Ruston 12-RK 3 CM diesels, 4380bhp = 16kts. Range 11,000nm at 12kts
Armament: 1–40mm/60 Bofors Mk 3, 2–7.62mm MG
Sensors: Radar 1006; sonar Simrad SU 'sidescan'
Complement: 35

No	Name	Builder	Launched	Fate
P 295	JERSEY	Hall Russell	18.3.76	To Bangladesh 28.1.94
P 297	GUERNSEY	Hall Russell	17.2.77	Extant 1995
P 298	SHETLAND	Hall Russell	22.11.76	Extant 1995
P 299	ORKNEY	Hall Russell	29.6.76	Extant 1995
P 300	LINDISFARNE	Hall Russell	1.6.77	Extant 1995
P 277	ANGLESEY	Hall Russell	18.10.78	Extant 1995
P 278	ALDERNEY	Hall Russell	27.2.79	Extant 1995

To patrol fishing grounds and oilfields in the new 200-mile EEZ, the RN ordered five modified trawler-type vessels in February 1975. They were based on the Scottish Department of Fisheries' *Westra* and *Jura* (see above), giving time for studies for a more capable design. Although much criticised for their low speed and weak armament they proved so successful that two more were ordered in October 1977, equipped with stabilisers but otherwise identical. The philosophy is quite simple: sea-keeping and endurance are the most important attributes, and weaponry is almost irrelevant. They are fitted to replenish at sea from naval oilers and have a sophisticated computer-aided navigation system (CANE). Stabilisers will be fitted to the first five.

SPEEDY *patrol hydrofoil*

Displacement: 117t
Dimensions: (Foils up – hullborne) 101ft oa × 30ft × 6ft
30.8m × 9.1m × 1.8m
(Foils down) 90ft × 30ft × 17ft (hullborne), 7ft 10in
27.4m × 9.1m × 5.2m, 2.4m
Machinery: (Foilborne) twin waterjets, 2 Allison 501–K20A gas turbines, 6600shp = 43kts (50kts dash); 2 shafts, 2 GM 8U92 T1 diesels, 800bhp = 14kts (hullborne). Range 560nm/1500nm at 43kts/14kts
Armament: 2–7.62mm MG
Complement: 18

No	Name	Builder	Launched	Fate
P 296	SPEEDY	Boeing Marine, Seattle	9.7.79	Sold 1986

Speedy 1980 MoD

The Royal Navy's first operational hydrofoil was ordered in June 1978 following an evaluation a year earlier of the commercial jetfoil *Flying Princess*. Boeing had already completed a design-study for the Japanese Maritime Self-Defence Force to adapt the jetfoil for the fishery protection role, and so the RN was able to acquire a hydrofoil at minimum cost (£6.2m). Originally two 7.62mm machine-guns were to have been carried, but *Speedy* commissioned in June 1980 without armament. Her radar (Type 1006) and CANE computer-aided navigation system were similar to the 'Island' and 'Castle' classes, with which vessels she was working. She was also to have been used as a high-speed target, but in 1982 it was announced that she was up for sale. She was leased to her builders for commercial service during 1984–86 and then sold to a Hong Kong owner.

'CASTLE' class *offshore patrol vessels*

Displacement: 1427t
Dimensions: 246ft wl, 265ft 9in oa × 37ft 9in × 11ft 3in mean
75.0m, 81.0m × 11.5m × 3.4m
Machinery: 2 shafts, 2 Paxman diesels, 2820bhp = 20kts. Range 10,000nm at 10kts
Armament: 1–40mm, 2–7.62mm MG
Sensors: Radar 1006
Complement: about 40

No	Name	Builder	Launched	Fate
P 258	LEEDS CASTLE	Hall Russell	29.10.80	Extant 1995
P 265	DUMBARTON CASTLE	Hall Russell	3.6.81	Extant 1995

The need of a properly designed OPV led to Naval Staff Target 7040 being prepared. Experience with the 'Island' class showed that more length would make for greater crew efficiency on extended patrol and enable the ship to have alternative roles. What has emerged is a substantial (but still cheap) ship which can be given a selection of weapons and sensors to meet customers' differing needs. In the RN version it was decided not to install a helicopter hangar, but to use the space instead to provide a bigger flight deck on which a Sea King can land and refuel if necessary. *Leeds Castle* was laid down in August 1979 and *Dumbarton Castle* in 1980, before they were officially ordered. In wartime they could ship an OTO Melara 76mm/62 compact.

Both ships acted as dispatch vessels in the South Atlantic during the Falklands conflict, proof of their sea-keeping abilities. They were given extra fuel stowage and other minor improvements, and both now serve as Falklands guardships. The 40mm/60 Mk 3 Bofors were replaced by 30mm Mk 1.

Leeds Castle Feb 1994 C & S Taylor

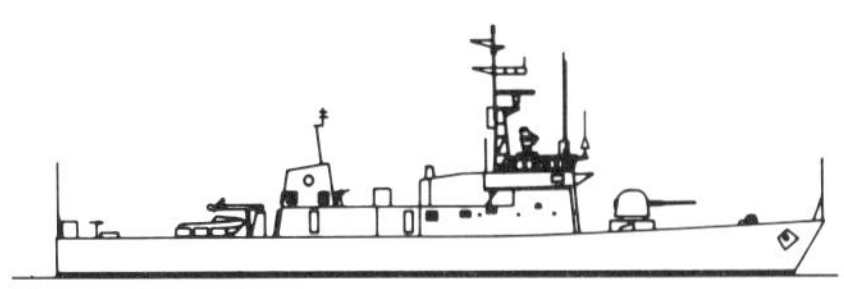

Peacock 1983

PEACOCK class *patrol vessels*

Displacement: 690t
Dimensions: 204ft (oa) × 32ft 10in × 18ft
62.2m × 10m × 5.5m
Machinery: 2 shafts, 2 Crossley Pielstick 18PA6 V280 diesels, 14,000bhp = 25kts
Armament: 1–76mm/62 DP, 4–7.62mm MGs
Complement: 31 (7 officers, 24 ratings)

No	Name	Builder	Launched	Fate
P 239	PEACOCK	Hall Russell	1.12.82	Extant 1995
P 240	PLOVER	Hall Russell	12.4.83	Extant 1995
P 241	STARLING	Hall Russell	11.9.83	Extant 1995
P 242	SWALLOW	Hall Russell	31.3.84	To Ireland 1988
P 243	SWIFT	Hall Russell	11.9.84	To Ireland 1988

Jointly funded by RN (25 per cent) and the Hong Kong Government (75 per cent) to replace five elderly CMSs acting as gunboats. The design was prepared by Hall Russell.

First order placed June 1981 and keel of *Peacock* laid in January 1982. Completed 1984–85. *Swallow* and *Swift* declared surplus to requirements early in 1988 and sold to Irish Naval Service. They were recommissioned as *Ciara* and *Orla* on 21 November 1988. Survivors for disposal end of 1996.

Falkland Islands *patrol vessels*

No	Name	Fate
P 244	PROTECTOR (ex-*Seaforth Saga*)	Sold 1988
P 245	GUARDIAN (ex-*Seaforth Champion*)	BU 1987
P 246	SENTINEL (ex-*Seaforth Warrior*)	Sold 1992

Former oil rig support vessels bought in March 1983 for service in Falklands waters. *Protector* and *Guardian* were commissioned on 21 October 1983, and *Sentinel* on 14 January 1984. They were armed with single 40mm/60 Mk 3 Bofors in bridge wings and 2/3–7.62mm MGs.

MTB 538 *fast attack craft*

Displacement: 36t normal; 45t deep load
Dimensions: 68ft pp, 74ft 2in oa × 20ft 3in × 6ft
20.7m, 22.6m × 6.2m × 1.8m
Machinery: 3-shaft Packard petrol engines, 4050bhp = 40kts max
Armament: 2–20mm (1×2), 4–18in TT, or 1–4.5in/25 8cwt
Complement: 15

MTB 539 *fast attack craft*

Displacement: 43t normal; 50t deep load
Dimensions: 72ft pp, 75ft 3in oa × 19ft 9in × 5ft 6in
21.9m, 22.9m × 3.0m × 1.7m
Machinery: 3-shaft Packard petrol engines, 4050bhp = 42kts
Armament: As *MTB 538*
Complement: 16

These two vessels were planned at the end of the Second World War, but in 1946 the two firms involved, Vosper at Portsmouth and Saunders-Roe at Beaumaris, were given a free hand to develop their own ideas. In *538* (completed August 1948) Vosper developed an advanced, all-glued wood hull, but Saunders-Roe used aluminium alloy in *539*. Both boats were intended to serve as convertible MTBs/MGBs, with four TT or the 4.5in 8cwt short gun as main armament. On entering service they were re-rated as *FPB 1601* and *FPB 1602* respectively. Although they provided experience for the 'Gay' class they did not last long and were put on the sale list in 1956.

'GAY' class *fast attack craft*

Displacement: 50t standard; 65t deep load
Dimensions: 71ft 6in pp, 75ft 2in oa × 20ft 1in × 4ft 1in
21.8m, 22.9m × 6.1m × 1.2m
Machinery: 2-shaft Packard petrol engines, 5000bhp = 40kts
Armament: (As gunboats) 1–4.5in/25 8cwt, 1–40mm/60 Mk 7; (as torpedo-boats) 2–40mm/60 Mk 9 (2×1), 2–21in TT
Complement: 13

Class (no, builder, fate):
Gay Archer (FPB 1041, Vosper, sold 1963), *Gay Bombardier* (FPB 1042, Vosper, sold 1963), *Gay Bowman* (FPB 1043, Vosper, sold 1963), *Gay Bruiser* (FPB 1044, Vosper, sold 1962), *Gay Carabineer* (FPB 1045, Thornycroft, sold 1963), *Gay Cavalier* (FPB 1046, Taylor, sold 1963), *Gay Centurion* (FPB 1047, Thornycroft, sold 1962), *Gay Charger* (FPB 1048, Morgan Giles, sold 1967), *Gay Charioteer* (FPB 1049, Morgan Giles, sold 1972), *Gay Dragoon* (FPB 1050, Taylor, sold 1962), *Gay Fencer* (FPB 1051, McGruer, target 1968), *Gay Forester* (FPB 1052, McGruer, sold 1962)

Short-hulled wooden convertible MTBs/MGBs, these were the last class with petrol engines. The design was developed from Vosper's *MTB 538* and was basically an interim one (after the 'Darks') since development of the Deltic diesel was being delayed. Pennant numbers later changed to 'P', although the number remained the same. Several hulls were used commercially after disposal and others were hulked.

FPB 1041–1044 were ordered on 27 February 1951 and *FPB 1045–1052* on 4 May 1951. All were completed in 1953.

Gay Bombardier as completed CMP

Dark Adventurer as completed
NB 1/750 scale

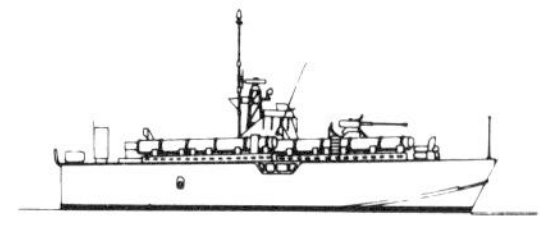

'DARK' class *fast attack craft*

Displacement: 50t standard; 64t deep load
Dimensions: 67ft pp, 71ft 8in oa × 19ft 5in × 6ft 1in max
20.4m, 21.8m × 5.9m × 1.8m
Machinery: 2-shaft Napier Deltic 16cyl diesels, 5000bhp = 40kts. Oil 8t
Armament: (As gunboats) 1–4.5in/25 8cwt, 1–40mm/60 Mk 7; (as torpedo-boats) 1–40mm/60 Mk 7, 4–21in TT
Complement: 15

Class (no, builder, fate):
Dark Adventurer (FPB 1101, Saunders-Roe, sold 1967), *Dark Aggressor* (FPB 1102, Saunders-Roe, sold 1961), *Dark Antagonist* (FPB 1103, Saunders-Roe, sold 1967), *Dark Avenger* (FPB 1105, Saunders-Roe, sold 1966), *Dark Biter* (FPB 1104, Saunders-Roe, sold 1966), *Dark Buccaneer* (FPB 1108, Vosper, sold 1966), *Dark Clipper* (FPB 1109, Vosper, sold 1967), *Dark Fighter* (FPB 1113, Taylor, sold 1967), *Dark Gladiator* (FPB 1114, Taylor, sunk as target 1976), *Dark Hero* (FPB 1115, McGruer, target 1977, for sale 1985), *Dark Highwayman* (FPB 1110, Vosper, sold 1967), *Dark Hunter* (FPB 1117, Miller, sold 1962), *Dark Hussar* (FPB 1112, Thornycroft, sold 1966), *Dark Intruder* (FPB 1118, Morgan Giles, sold 1966), *Dark Invader* (FPB 1114, Morgan Giles, sold 1966), *Dark Killer* (FPB 1115, Thornycroft, sold 1966), *Dark Rover* (FPB 1107, Vosper, sold 1966), *Dark Scout* (FPB 1106, Saunders-Roe, sold 1961).

The first diesel-engined FPBs built for the RN and the first to use the remarkable Deltic with an exceptionally good power-to-weight ratio (4.2hp per lb). They were built of alloy framing and wooden skin, apart from the all-aluminium *Dark Scout*. Conforming to previous practice, they were convertible MTBs/MGBs and could also lay six ground mines. *FPB 1101–1104* were ordered on 27 February 1951, *FPB 1105–1115* on 18 December 1951 (1 cancelled 8 April 1952), *FPB 1114* was re-ordered on 22 January 1952, with *FPB 1118*, and *FPB 1116* ordered on 2 May 1953. Several were renamed: *Dark Aggressor* was ex-*Dark Bowman*, *Dark Biter* ex-*Dark Attacker*, *Dark Buccaneer* ex-*Dark Chaser*, *Dark Clipper* ex-*Dark Chieftain*, *Dark Highwayman* ex-*Dark Crusader*, *Dark Killer* ex-*Dark Defender*, *Dark Hussar* ex-*Dark Explorer*, *Dark Hunter* ex-*Dark Hero*. The first *Dark Horseman* was cancelled on 8 April 1952 and replaced with a second of the same name, which was cancelled on 5 November 1957.

The above-water exhaust caused problems and after a short while all the class were painted black. Flag superior later changed to 'P', although the numbers remained unchanged. Although noisy and lively they were generally regarded as successful, but most were laid up in 1958 when Coastal Forces were disbanded, and all were either hulked or sold commercially by the mid-1960s.

'BOLD' class *fast attack craft*

Displacement: 130t standard; 150t deep load
Dimensions: (*Bold Pathfinder*) 117ft pp, 122ft 8in oa × 20ft 5in × 6ft 7in
35.7m, 37.4m × 6.2m × 2.0m
(*Bold Pioneer*) 116ft 3in pp, 121ft oa × 25ft 6in × 6ft 10in
35.4m, 36.9m × 7.8m × 2.1m
Armament: (As gunboats) 2–4.5in/25 8cwt (2×1), 1–40mm/60 Mk 9; (as torpedo-boat) 1–40mm/60 Mk 9, 4–21in TT
Complement: 20

Class (former no, builder, fate):
Bold Pathfinder (ex-*MTB 5720*, Vosper, sold 1964), *Bold Pioneer* (ex-*MTB 5701*, White, sold 1958)

Experimental prototypes built 1950–53 and served as test-beds for development of the Gatric gas turbine first installed in *MGB 2009*, although they were initially fitted with ex-S-boat Mercedes diesels, which proved very unreliable. They were an earlier design than the 'Dark' and 'Gay' classes, but were bigger and more seaworthy. They were also very impressive, with twin funnels and a lattice foremast. They were both aluminium framed with a wooden shell, but *Bold Pathfinder* had a hull of round bilge form whereas *Bold Pioneer* had a hard chine hull, the object being to compare the two configurations. *Bold Pioneer* also tested the experimental 3.3in gun for the 'Brave' class (see below) but this was removed in 1958 when she was put on the disposal list. *Bold Pathfinder* was laid up in 1962. *P 5701* was ordered on 12 January 1949, *P 5720* was ordered on 2 May 1949; named February 1952.

Bold Pioneer Mar 1957 CMP

'BRAVE' class *fast attack craft*

Displacement: 89t standard; 114t deep load
Dimensions: 90ft pp, 98ft 10in oa × 25ft 6in × 7ft max
27.4m, 30.1m × 7.8m × 1.8m
Machinery: 3 shafts, 3 Rolls-Royce Proteus gas turbines, 10,500shp = 50kts. Fuel 25t
Armament: 1–40mm/60 Mk 9, 4–21in TT
Complement: 20

Class (no, builder, fate):
Brave Borderer (P 1011, Vosper, for sale 1979), *Brave Swordsman* (P 1012, Vosper, preserved 1979)

The last operational fast patrol boats built for the RN, these had welded aluminium frames, a double mahogany skin and fibreglass sheathing below the waterline. They were built 1957–60. A feature of the design was the great beam (to accommodate the three engines), which with the hydraulic flap in the transom helped to maintain the trim.

Although the requirement for FPBs had officially lapsed with the demise of Coastal Forces they played a useful role as targets and in fishery patrol duties. They were originally intended to have the 3.3in CFS-2 (Coastal Forces System Mk 2) stabilised gun system, an adaptation of the Centurion tank's 20pdr. The prototype tested in *Bold Pioneer* had many teething troubles and when Coastal Forces were axed the development was not pursued. Although originally intended to have four TT they were given side-launching cradles to save weight.

Both were laid up in 1970, and in 1979 *Brave Swordsman* was presented to the Haydon-Baillie Collection.

TENACITY *patrol craft*

Displacement: 165t normal; 220t deep load
Dimensions: 144ft 6in oa × 26ft 6in × 7ft 9in
44.0m × 8.1m × 2.4m
Machinery: 3 shafts CODOG: 3 Rolls-Royce Proteus gas turbines, 12,750shp = 39kts; 2 Paxman Ventura cruising diesels, 1200bhp = 16kts
Armament: 2–7.62mm MG. See notes
Sensors: Navigation radar only
Complement: 32

An experimental private venture fast patrol boat built at Vosper Thornycroft's Camber yard in Portsmouth 1967–69. Designated P 276, she was not bought by the RN until January 1972, but was used for demonstrations and was twice chartered by the RN in 1971. Although designed for a heavy armament this was never fitted (the twin 35mm mounting and quad SSMs credited to her in *Jane's* being dummies), and when taken in to the RN she was given only two light machine-guns for fishery protection duties. The main alteration was the fitting of two davits for handling inflatable dinghies on the forecastle, but in 1978 she received a 40mm Bofors gun on the forecastle. She was reported to be for sale, but did not arrive at Southampton from Rosyth until 1985.

Tenacity Apr 1974 C & S Taylor

SCIMITAR class *fast patrol boats*

Displacement: 80t normal; 102t deep load
Dimensions: 100ft pp, *c*106ft oa × 26ft 8in × 6ft 5in
30.5m, 32.3m × 8.1m × 1.9m
Machinery: 2-shaft CODAG: 2 Rolls-Royce Proteus plus 2 Foden 6cyl diesels, 8500shp + 240bhp = 40kts
Armament: None
Complement: 12

Class (no, builder, fate):
Cutlass (P 274, Vosper, for disposal 1982), *Sabre* (P 275, Vosper, for disposal 1982), *Scimitar* (P 271, Vosper, sold commercially 1983)

Slightly smaller editions of the 'Brave' class, ordered in 1968 to serve as targets. The design permits installation of a third Proteus turbine and a light armament of missiles or guns if needed for the operational role, but they proved extremely useful simulating attacks by fast missile craft. In 1979 *Scimitar* was sent to Hong Kong, and in July 1980 *Sabre* lost her bow in a collision with a breakwater at Alderney. They were placed in storage ashore in 1981.

MINE WARFARE VESSELS

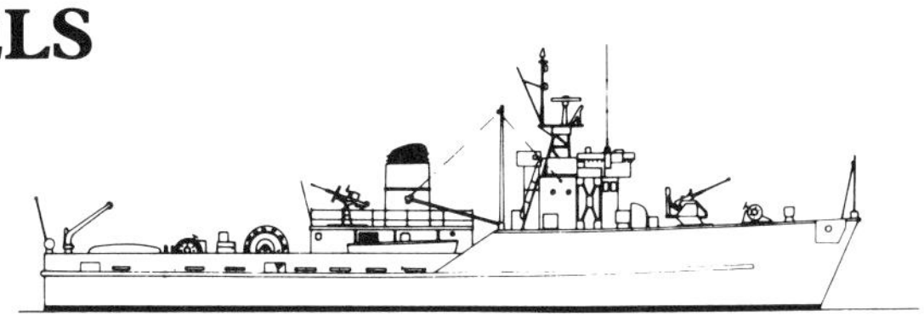

Bossington 1957
NB 1/750 scale

'THORPE' class *minesweepers*

Three ships designated AMS 01–03 were ordered in the 1950 programme, but cancelled in 1953. No details have come to light about their function, equipment or particulars, but the insect names originally selected suggest that they had 'Ton' hulls but a different function such as control ships. *Black Ant* was renamed *Cutthorpe*, *Grey Ant* was renamed *Ainthorpe* and *Silver Ant* was renamed *Bilsthorpe.*

'TON' class *coastal minesweepers*

Displacement: 360t normal; 425t deep load
Dimensions: 140ft pp, 152ft oa × 28ft 9in × 8ft 3in max
42.7m, 46.3m × 8.8m × 2.5m
Machinery: 2 shafts, 2 12cyl Mirrlees or 18cyl Napier Deltic diesels, 2500bhp or 3000bhp = 15kts. Range 3000nm at 8kts
Armament: 1–40mm/60 Mk 7, 2–20mm (1×2)
Sensors: Radar 978 (later 1006); sonar 193 (minehunters only)
Complement: 29 (minehunters 38)

No	Name	Builder	Launched	Fate
M 1101	CONISTON	Thornycroft	9.7.52	BU 1970
M 1102	ALCASTON	Thornycroft	5.1.53	To Australia 1961
M 1103	ALFRISTON	Thornycroft	29.4.53	BU 1988
M 1104	ALVERTON	Camper & Nicholson	18.11.52	To Irish Naval Service 1971
M 1105	AMERTON	Camper & Nicholson	16.3.53	BU 1971
M 1106	APPLETON	Goole SB	4.9.52	BU 1972
M 1107	BEACHAMPTON	Goole SB	29.6.53	Sold Hong Kong 1985
M 1108	BEVINGTON	White	17.3.53	To Argentina 1967
M 1109	BICKINGTON	White	14.5.52	BU 1988
M 1110	BILDESTON	Doig	9.6.52	BU 1988
M 1111	EDDERTON	Doig	1.11.52	Survey ship *Myrmidon*
M 1112	BOURSTON	Richards IW	6.10.52	BU 1975
M 1113	BRERETON	Richards IW	14.3.55	BU 1992
M 1114	BRINTON	Cook, Welton & Gemmell	8.8.52	For disposal 1993
M 1115	BRONINGTON	Cook, Welton & Gemmell	19.3.53	Museum ship 1989
M 1116	BURNASTON	Fleetlands	18.12.52	BU 1971
M 1117	BUTTINGTON	Fleetlands	11.6.53	BU 1970
M 1118	CALTON	Wivenhoe	24.10.53	BU 1968
M 1119	CARHAMPTON	Wivenhoe	21.7.55	BU 1970

Cutlass Jun 1979 C & S Taylor

No	Name	Builder	Launched	Fate
M 1120	CAUNTON	Montrose	20.2.53	BU 1970
M 1121	CHEDISTON	Montrose	6.10.53	To Australia 1968
M 1122	CHILCOMPTON	Herd & McKenzie	23.10.53	Sold 1971
M 1123	CLARBESTON	Richards IW	18.2.54	BU 1970
M 1124	CRICHTON	Doig	17.3.53	BU 1987
M 1125	CUXTON	Camper & Nicholson	9.11.53	BU 1992
M 1126	DALSWINTON	White	24.9.53	BU 1973
M 1127	DARLASTON	Cook, Welton & Gemmell	25.9.53	To Malaysia 1960
M 1128	DERRITON	Thornycroft	22.12.53	BU 1970
M 1129	OULSTON	Thornycroft	20.7.54	To Irish Naval Service 1971
M 1130	HIGHBURTON	Thornycroft	2.6.54	BU 1978
M 1130	BLAXTON	Thornycroft	26.1.55	To Argentina 1967
M 1132	HICKLETON	Thornycroft	26.1.55	To Irish Naval Service 1971
M 1133	BOSSINGTON (ex-*Embleton*)	Thornycroft	2.12.55	BU 1988
M 1134	ESSINGTON	Camper & Nicholson	Sep 1954	To Malaysia 1964
M 1135	FENTON	Camper & Nicholson	10.3.55	BU 1968
M 1136	FITTLETON	White	3.2.54	Sunk in collision 1977, salved and BU
M 1137	FLOCKTON	White	3.6.54	BU 1969
M 1138	FLORISTON	Richards IW	26.1.55	Sold 1968
M 1139	SOMERLEYTON (ex-*Gamston*)	Richards IW	1.7.54	To Australia 1961
M 1140	GAVINTON	Doig	27.7.53	BU 1991
M 1141	GLASSERTON	Doig	3.12.53	BU 1988
M 1142	HAZLETON	Cook, Welton & Gemmell	6.2.54	To South Africa 1955
M 1143	HEXTON	Cook, Welton & Gemmell	1954	To Malaysia 1963
M 1144	DUNKERTON	Goole SB	8.3.54	To South Africa 1955
M 1145	DUFTON	Goole SB	13.11.54	BU 1977
M 1146	HODGESTON	Fleetlands	6.4.54	BU 1988
M 1147	HUBBERSTON	Camper & Nicholson	14.9.54	BU 1992
M 1148	ILMINGTON	Camper & Nicholson	8.3.54	To Argentina 1967
M 1149	BADMINTON (ex-*Ilston*)	Camper & Nicholson	14.10.54	BU 1970
M 1150	INVER-MORISTON	Dorset Yacht	2.6.54	BU 1971
M 1151	IVESTON	Philip	1.6.54	TS 1993
M 1152	JACKTON	Philip	28.2.55	To Australia 1961
M 1153	KEDLESTON	Pickersgill	21.12.53	BU 1992
M 1154	KELLINGTON	Pickersgill	12.10.54	TS 1993
M 1155	MONKTON (ex-*Kelton*)	Herd & McKenzie	30.11.55	Sold Hong Kong 1985
M 1156	KEMERTON	Harland & Wolff	27.11.53	BU 1971
M 1157	KIRKLISTON	Harland & Wolff	18.2.54	BU 1991
M 1158	LALESTON	Harland & Wolff	18.5.54	BU 1985
M 1159	LANTON	Harland & Wolff	30.7.54	BU 1970
M 1160	LETTERSTON	Harland & Wolff	26.10.54	BU 1971
M 1161	LEVERTON	Harland & Wolff	2.3.55	BU 1972
M 1162	KILDARTON (ex-*Liston*)	Harland & Wolff	23.5.55	BU 1969
M 1163	LULLINGTON	Harland & Wolff	31.8.55	To Malaysia 1966
M 1164	MADDISTON	Harland & Wolff	27.1.56	BU 1975
M 1165	MAXTON	Harland & Wolff	24.5.56	BU 1989
M 1166	NURTON	Harland & Wolff	22.10.56	Sold 1995
M 1167	REPTON (ex-*Ossington*)	Harland & Wolff	1.5.57	Sold 1982
M 1168	DILSTON (ex-*Pilston*)	Cook, Welton & Gemmell	15.11.54	To Malaysia 1964

No	Name	Builder	Launched	Fate
M 1169	PENSTON	Cook, Welton & Gemmell	9.5.55	BU 1970
M 1170	PICTON	Cook, Welton & Gemmell	20.10.55	BU 1969
M 1171	ALDINGTON (ex-*Pittington*)	Camper & Nicholson	15.9.55	To Ghana 1964
M 1172	THANKERTON	Camper & Nicholson	4.9.56	To Malaysia 1966
M 1173	POLLINGTON	Camper & Nicholson	10.10.57	BU 1987
M 1174	PUNCHESTON	Richards IW	20.11.56	BU 1977
M 1175	QUAINTON	Richards IW	10.10.57	BU 1979
M 1176	RENNINGTON	Richards IW	27.11.58	To Argentina 1967
M 1177	RODDINGTON	Fleetlands	24.2.55	BU 1972
M 1178	SANTON	Fleetlands	18.8.55	To Argentina 1967
M 1179	SEFTON	Fleetlands	15.9.54	BU 1968
M 1180	SHAVINGTON	White	25.4.55	BU 1987
M 1181	SHERATON	White	20.7.55	For disposal 1993
M 1182	SHOULTON	Montrose	10.9.54	BU 1981
M 1183	SINGLETON	Montrose	18.11.55	To Australia 1961
M 1184	SULLINGTON	Doig	7.4.54	Survey ship *Mermaid* 1964, BU 1970
M 1185	SWANSTON	Doig	1.7.54	To Australia 1954
M 1186	TARLTON	Doig	10.11.54	To Argentina 1967
M 1187	UPTON	Thornycroft	15.3.56	BU 1991
M 1188	WALKERTON	Thornycroft	21.11.56	BU 1990
M 1189	WASPERTON	White	28.2.56	Sold 1986
M 1190	WENNINGTON	Doig	6.4.55	To India 1956
M 1191	WHITTON	Fleetlands	30.1.56	To India 1956
M 1192	WILKIESTON	Cook, Welton & Gemmell	26.6.56	BU 1976
M 1193	WOLVERTON	Montrose	22.10.56	Sold Hong Kong 1986
M 1194	WOOLASTON	Herd & McKenzie	6.3.58	BU 1981
M 1195	WOTTON	Philip	24.4.56	BU 1992
M1196	YARNTON	Pickersgill	26.3.56	Sold 1986
M 1197	OVERTON	Camper & Nicholson	28.1.56	To India 1956
M 1198	ASHTON (ex-*Cheriton*)	White	5.9.56	BU 1977
M 1199	BELTON	Doig	3.10.55	BU 1974
M 1200	SOBERTON	Fleetlands	20.11.56	For disposal 1991
M 1201	DURWESTON	Dorset Yacht	18.8.55	To India 1956
M 1203	MARYTON	Montrose	3.4.58	BU 1969
M 1204	STUBBINGTON	Camper & Nicholson	8.8.56	BU 1989
M 1205	WISTON	Wivenhoe	3.6.58	BU 1982
M 1206	FISKERTON	Doig	12.4.57	BU 1977
M 1207	CASTLETON	White	26.8.58	To South Africa 1959
M 1208	LEWISTON	Herd & McKenzie	3.11.59	BU 1986
M 1209	CHAWTON	Fleetlands	24.9.57	BU 1977
M 1210	STRATTON	Dorset Yacht	29.7.57	To South Africa 1959
M 1211	HOUGHTON	Camper & Nicholson	22.11.57	BU 1971
M 1212	DUMBLETON	Harland & Wolff	8.11.57	To South Africa 1958
M 1213	OAKINGTON	Harland & Wolff	10.12.58	To South Africa 1959
M 1214	PACKINGTON	Harland & Wolff	3.7.58	To South Africa 1959
M 1215	CHILTON	Cook, Welton & Gemmell	15.7.57	To South Africa 1958
M 1216	CROFTON	Thornycroft	7.3.58	BU 1987

During the Korean War it was learned that Russian magnetic influence mines could defeat degaussing, making the steel-hulled *Algerine* class 'fleet' sweepers obsolete and raising fears of a mining offensive against European ports. To counter this threat, in 1951 a massive programme of 118 coastal minesweepers was authorised, backed up by inshore sweepers for working in estuaries. The ships were launched during 1952–59. The original plan was to name all coastal minesweepers after insects, with distinguishing colours for differing equipment fits, but eventually names of villages ending in 'ton' were adopted.

M 1101–1115 were ordered on 9 September 1950 and were to be *Red Ant, Blue*

Ant, Green Ant, Golden Ant, Red Aphis, Blue Aphis, Green Aphis, Golden Aphis, Red Bee, Blue Bee, Green Bee, Golden Bee, Red Beetle, Blue Beetle and *Green Beetle*. *M 1116–1117* were ordered on 20 October 1950, and were to be *Golden Beetle* and *Red Butterfly*. *M 1118–1127*, ordered on 4 April 1951, were to be *Blue Butterfly, Green Butterfly, Golden Butterfly, Red Centipede, Blue Centipede, Green Centipede, Golden Centipede, Red Cicala, Blue Cicala* and *Green Cicala*. *M 1128–1144*, ordered on 28 September 1951, were to be *Golden Cicala, Red Cockchafer, Blue Cockchafer, Green Cockchafer, Golden Cockchafer, Red Cricket, Blue Cricket, Green Cricket, Golden Cricket, Red Dragonfly, Blue Dragonfly, Green Dragonfly, Red Firefly* and *Blue Firefly*. Village names were allocated from March 1952; *M 1145–1149* were ordered on 14 February 1952; *M 1150–51*, on 19 March 1952, *M 1152–1154* on 21 April 1952, *M 1155–67* on 22 March 1952 and *M 1168–1186* on 17 June 1952.

The design called for an aluminium-framed, wooden planked hull with non-magnetic fittings, capable of undertaking ocean passages, and the result was a sturdy craft which is still in service all over the world. Early vessels had Mirrlees diesels, but from *Highburton* onwards the more powerful Napier Deltic was fitted, and eventually the original 'Tons' were re-engined. Early members of the class had an open bridge and the lead-ship *Coniston* had no top to her funnel and a short lattice mast, but subsequent vessels had a double-finned funnel top and covered bridges were progressively introduced. The last major external change was to revert to a tripod mast, for a new radar.

In 1964 *Kirkliston* was converted to a 'minehunter', with LL sweep gear removed and a minehunting Type 193 sonar installed in the hull beneath the bridge. Active rudders were fitted to allow her to position herself over the mine, and four divers and two inflatable boats were carried to permit the mine to be blown up by hand. In *Shoulton* an auxiliary diesel-hydraulic pump-jet system was installed in 1965–66 to provide quieter propulsion. Another fifteen were modified to minehunters, some with the Sperry Towed Acoustic Generator (TAG). Six were transferred to the RAN in 1961, four to the Indian Navy, one to Ghana, six to South Africa, six to Argentina in 1968 and three to the Republic of Ireland in 1971. Two, *Edderton* and *Sullington*, became the surveying vessels *Myrmidon* and *Mermaid*, while *Invermoriston* became an air-sea rescue vessel and *Laleston* was converted to a diving tender. The French, Canadians and Dutch were given the drawings to copy.

Many have served as RNVR/RNR tenders, reverting to their original names when replaced by others: *Thames* (*Alverton, Buttington, Woolaston*), *Venturer* (*Hodgeston, Buttington*), *Burnicia* (*Kedleston*), *Clyde* (*Amerton, Crichton*), *Humber* (*Bronington*), *Mersey* (*Amerton, Pollington*), *Solent* (*Warsash*, ex-*Crofton*), *Curzon* (*Bickington, Fittleton*), *Killiecrankie* (*Bickington, Derriton*), *St David* (*Crichton*), *Kilmorey* (*Alfriston, Kirliston*), *Warsash* (*Alfriston, Boulston, Crofton*), *Montrose* (*Dalswinton, Nurton, Chediston*) and *Northumbria* (*Quainton, Hodgeston*).

Swanston, Somerleyton, Singleton, Alcaston and *Jackton* became the Australian *Gull, Hawk, Ibis, Snipe* and *Teal*; *Aldington* became the Ghanaian *Ejura*; *Whitton, Wennington, Durweston* and *Overton* became the Indian *Cannanore, Cuddalore, Kakinada* and *Karwar*; *Thankerton, Dilston, Essington, Hexton, Darlaston* and *Lullington* became the Malaysian *Brimchang, Jerai, Kinabulu, Ledang, Mahamiru* and *Tahan*; *Rennington, Santon, Ilmington, Hickleton, Tarlton* and *Bevington* became the Argentine *Chaco, Chubut, Formosa, Neuquem, Rio Negro* and *Tierra del Fuego*; and *Alverton, Blaxton* and *Oulston* became the Irish *Banba, Fola* and *Grainne*.

Late in 1971 five were converted to coastal patrol craft with 'P' numbers – *Beachampton* (P 1007), *Monkton* (P 1055), *Wasperton* (P 1089), *Wolverton* (P 1093) and *Yarnton* (P 1096), and a second 40mm/60 Mk 7 fitted abaft the funnel; all five were sent to Hong Kong. Minehunter conversions consisted of *Kirkliston, Shoulton, Bossington, Brereton, Bronington, Derriton, Glasserton, Highburton, Hubberston, Iveston, Kellington, Sheraton, Bilderston, Brinton, Gavinton, Kedleston, Maxton* and *Nurton*. *Iveston* was the last CMS with Type 193 Mod 0 sonar.

1954 *fleet minesweepers*

Displacement:	1522t deep load
Dimensions:	237ft oa × 38ft 6in × 11ft 5in *72.2m × 11.7m × 3.5m*
Machinery:	2-shaft geared steam turbines, 5000shp = 17.5kts (11.5kn while sweeping)
Armament:	1–3in/50, 2–40mm/60 Bofors (2×1), 1 Squid mortar
Sensors:	Radar 974; sonar 147, 164
Complement:	?

In mid-1952 a Staff Requirement for an *Algerine* replacement was drawn up. Although the legend was approved in January 1953 and the first five were to be ordered under the 1953–54 Estimates the design was cancelled in Jan 1955.

Although priority was given to mine countermeasures, with provision for silenced propellers, an improved degaussing system and automatic course correction, they were clearly intended to serve as coastal escorts as well. The 3in guns were to be supplied by the United States under MDAP to provide AA defence (with Mk 63 directors).

'HAM' class *inshore minesweepers*

Displacement:	120t normal; 159t deep load
Dimensions:	100ft pp, 106ft 6in–107ft 6in oa × 21ft–22ft × 5ft 6in–5ft 9in *30.5m, 32.5m–32.75m × 6.4m–6.7m × 1.67m–1.75mm*
Machinery:	2 shafts, 2 Davey Paxman 12cyl diesels, 1100bhp = 14kts. Range 2350nm at 9kts
Armament:	1–40mm/60 Mk 7 or 1–20mm
Sensors:	Radar 978
Complement:	15

Class:
Inglesham (M 2601), *Altham* (M 2602), *Arlingham* (M 2603), *Asheldham* (M 2604), *Bassingham* (M 2605), *Bedham* (M 2606), *Bisham* (M 2607), *Blunham* (M 2608), *Bodenham* (M 2609), *Boreham* (M 2610), *Bottisham* (M 2611), *Brantingham* (M 2612), *Brigham* (M 2613), *Bucklesham* (M 2614), *Cardingham* (M 2615), *Chelsham* (M 2616), *Chillingham* (M 2617), *Cobham* (M 2618), *Darsham* (M 2619), *Davenham* (M 2620), *Dittisham* (M 2621), *Downham* (M 2622), *Edlingham* (M 2623), *Elsenham* (M 2624), *Etchingham* (M 2625), *Everingham* (M 2626), *Felmersham* (M 2627), *Flintham* (M 2628), *Damersham* (M 2629), *Fritham* (M 2630), *Glentham* (M 2631), *Greetham* (M 2632), *Halsham* (M 2633), *Harpham* (M 2634), *Haversham* (M 2635), *Lasham* (M 2636), *Hovingham* (M 2637), *Cranham* (M 2701), *Frettenham* (M 2702), *Isham* (M 2703), *Kingham* (M 2704), *Hildesham* (M 2705), *Ledsham* (M 2706), *Littlesham* (M 2707), *Ludham* (M 2708), *Mersham* (M 2709), *Mickleham* (M 2710), *Mileham* (M 2711), *Neasham* (M 2712), *Nettleham* (M 2713), *Ockham* (M 2714), *Ottringham* (M 2715), *Pagham* (M 2716), *Fordham* (M 2717, ex-*Pavenham*), *Petersham* (M 2718), *Pincham* (M 2719), *Powderham* (M 2720), *Pulham* (M 2721), *Rackham* (M 2722), *Reedham* (M 2723), *Rendlesham* (M 2724), *Riplingham* (M 2725), *Shipham* (M 2726), *Saxlingham* (M 2727), *Shrivenham* (M 2728), *Sidlesham* (M 2729), *Stedham* (M 2730), *Sparham* (M 2731), *Sulham* (M 2732), *Thakeham* (M 2733), *Tibenham* (M 2734), *Tongham* (M 2735), *Tresham* (M 2736), *Warmingham* (M 2737), *Wexham* (M 2738), *Whippingham* (M 2739), *Wintringham* (M 2777), *Woldingham* (M 2778), *Wrentham* (M 2779), *Yaxham* (M 2780), *Portisham* (M 2781), *Popham* (M 2782), *Odiham* (M 2783), *Puttenham* (M 2784), *Birdham* (M 2785), *Rampisham* (M 2786), *Abbotsham* (M 2787), *Georgeham* (M 2788), *Malcham* (M 2789), *Thatcham* (M 2790), *Sandringham* (M 2791), *Polsham* (M 2792), *Thornham* (M 2793).

The class was originally allocated bird names with distinguishing colours. *M 2601–2618*, ordered on 29 September 1950, were to be *Red Bantam, Blue Bantam, Green Bantam, Golden Bantam, Red Bullfinch, Blue Bullfinch, Red Chaffinch, Blue Chaffinch, Green Chaffinch, Golden Chaffinch* and *Red Crow* (*M 2612–2613* do not appear to have had bird names), *Blue Crow, Green Crow, Golden Crow, Red Cuckoo* and *Blue Cuckoo*. *M 2701–2706* were to be *Green Cuckoo, Golden Cuckoo, Red Dove, Blue Dove, Green Dove* and *Golden Dove* (ordered 23 April 1951). *M 2625–2647* were ordered 17 October 1951; *M 2648* was ordered 20 November 1951; *M 2649–2676* were ordered 14 January 1953. *M 2677–2684* were ordered 4 March 1954; *M 2685–2691* were ordered 19 February 1952 and *M 2692–2693* were ordered 5 May 1955.

In parallel with the coastal minesweeper programme, thirty-seven inshore sweepers were laid down, to operate in estuarine waters. J Samuel White acted as the lead-yard and as in the CMSs wooden and aluminium construction was adopted. A second series (mine disposal craft) followed, M 2701–M 2739, with 9in more beam and slightly deeper draught, while the third, all-wood series, M 2777–M 2793, were 1ft longer and 1ft beamier. *Reedham* was fitted with fin stabilisers and these were so successful that the rest of the class were similarly retrofitted. Series 2 and Series 3 vessels had a prominent rubbing strake. All were ordered between September 1950 and November 1951.

Although sturdy and seaworthy craft they were found to be too small to cope with the latest electrical and acoustic sweep gear, as they lacked generating capacity. They were accordingly downgraded to subsidiary duties within a few years, serving as diving tenders, torpedo recovery vessels, degaussing vessels, etc. A number are still in service as auxiliary craft. *Bisham* and *Edlingham* were badly damaged by fire while laid up in reserve in 1955 and were scrapped in 1957.

Foreign transfers consisted of *Boreham, Felmersham, Bedham, Asheldham, Brantingham* and *Altham* (Malaysian *Jerong, Todak, Langa, Suka, Sri Perlis, Temasek* and *Sri Johore*, 1958–59); *Greetham* and *Harpham* (Libyan *Zuara* and *Brak*, 1962); and *Bottisham, Chelsham* and *Halsham* (transferred to the RAN in 1965–66 and renumbered *5000–5002*). Under MDAP *Frettenham, Wexham, Mersham, Isham, Kingham, Stedham, Rendlesham, Petersham, Mileham, Tibenham, Sparham, Whippingham, Sulham, Riplingham* and *Pincham* became the French *M 771–776* and *M 781–789* in 1954–55. *Littleham* and *Hildersham* became the Indian *Bassein* and *Bimlipatam* in 1955, *Ottringham* became the Ghanaian *Afadzato* in 1959.

Pagham Jun 1977 MG Photographic

'LEY' class *inshore minesweepers*

Displacement: 123t normal; 164t deep load
Dimensions: 100ft pp, 106ft 9in max, 21ft 9in × 5ft 6in
30.5m, 32.5m × 6.6m × 1.67m
Machinery: 2 shafts, 2 8cyl Davey Paxman diesels, 7000bhp = 13kts. Oil 15t
Armament: 1–40mm/60 Mk 7 or 1–20mm
Sensors: Radar 978
Complement: 15

Class:
Dingley (M 2001), *Aveley* (M 2002), *Brearley* (M 2003), *Brenchley* (M 2004), *Brinkley* (M 2005), *Broadley* (M 2006), *Broomley* (M 2007), *Burley* (M 2008), *Chailey* (M 2009), *Cradley* (M 2010), *Edgley* (M 2011, ex-*Wrentham*).

M 2001–2003 were ordered on 9 September 1950, and were to be *Black Bantam, Grey Bantam* and *Silver Bantam*. *M 2004–2008*, ordered 23 April 1951, were to be *White Bantam, Black Bullfinch, Grey Bullfinch, Silver Bullfinch* and *White Bullfinch*. *M 2009–2013* were ordered 17 October 1951; *M 2014–2016* on 14 July 1953. Ex-*M 2011–2013* of 1951 programme and ex-*M 2014–2016* of 1952 programme were re-ordered as 'Ham' class.

Very similar to the 'Ham' class, composite-built with aluminium alloy framing and wooden planked hulls. The programme was cut short with the cancellation of *Edgley*. They survived longer than the 'Ham' type because they were re-equipped for minehunting, minus sweep gear; all were either reduced to training or had been sold by the mid 1970s. *Broadley* was sold in 1957 following severe damage on 29 September 1956 while laid up in reserve.

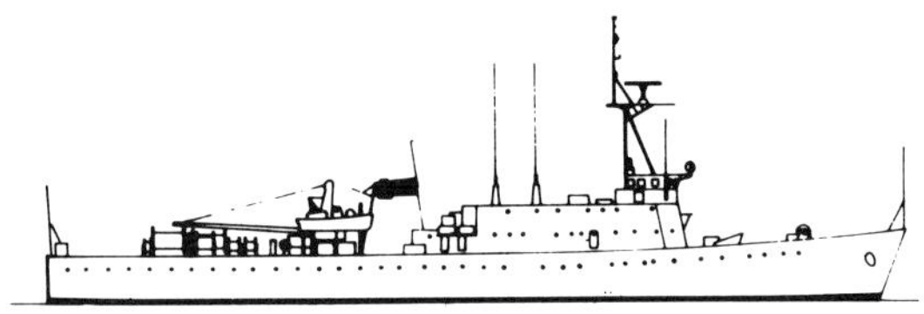

Abdiel 1970

ABDIEL *exercise minelayer*

Displacement: 1375t standard; 1460t full load
Dimensions: 265ft oa × 38ft 6in × 10ft
80.8m × 11.7m × 3.0m
Machinery: 2 shafts, 2 16cyl Paxman Ventura diesels, 2690bhp = 16kts
Armament: 1–40mm/60 Mk 7, 44 mines
Sensors: Radar 978
Complement: 98

No	Name	Builder	Launched	Fate
N 21	ABDIEL	Thornycroft	22.1.67	Sold 1988

Designed to lay exercise minefields and to support mine countermeasures forces, *Abdiel* was ordered in Jun 1965 and commissioned on 17 Oct 1967. She received a major refit in 1978.

WILTON *minesweeper/hunter*

Particulars: As 'Ton' class

Wilton (M 1116) was the prototype Glass Reinforced Plastic (GRP) minesweeper/hunter, and the first of her kind in the world. To test the new non-magnetic material against wooden construction a copy of the 'Ton' class was built in GRP and equipped with machinery and gear stripped from *Derriton* before the latter was broken up. *Wilton* was laid down by Vosper Thornycroft on 18 January 1972. She is a great success, having weathered two collisions without serious damage. In 1991 she was transferred to the Dartmouth Training Squadron. For disposal 1994.

Wilton 1977 C & S Taylor

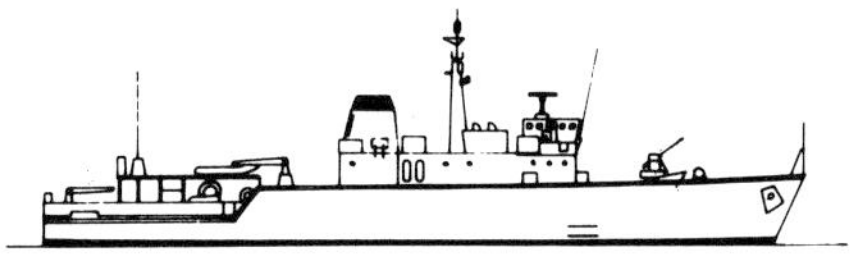

Brecon as completed

'HUNT' class *mine countermeasures vessels*

Displacement: 750t deep load
Dimensions: 187ft wl, 197ft oa × 32ft 3in × 7ft 3in
57.0m, 60.0m × 9.8m × 2.2m
Machinery: 2-shaft diesel: 2 Paxman 9–59K Deltic diesels, 3540bhp = 17kts (plus hydraulic drive, 18kts)
Armament: 1–40mm/60 Mk 9 (now 30mm Mk 1)
Sensors: Radar 1006; sonar 193M
Complement: 45 (6 officers, 39 ratings)

No	Name	Builder	Launched	Fate
M 29	BRECON	Vosper Thornycroft	21.6.78	Extant 1995
M 30	LEDBURY	Vosper Thornycroft	5.12.79	Extant 1995
M 31	CATTISTOCK	Vosper Thornycroft	22.1.81	Extant 1995
M 32	COTTESMORE	Yarrow	9.2.82	Extant 1995
M 33	BROCKLESBY	Vosper Thornycroft	12.1.82	Extant 1995
M 34	MIDDLETON	Yarrow	27.4.83	Extant 1995
M 35	DULVERTON	Vosper Thornycroft	3.11.82	Extant 1995
M 36	BICESTER	Vosper Thornycroft	4.6.85	Extant 1995
M 37	CHIDDINGFOLD	Vosper Thornycroft	6.10.83	Extant 1995
M 38	ATHERSTONE	Vosper Thornycroft	1.3.86	Extant 1995
M 39	HURWORTH	Vosper Thornycroft	25.9.84	Extant 1995
M 40	BERKELEY	Vosper Thornycroft	3.12.86	Extant 1995
M 41	QUORN	Vosper Thornycroft	23.1.88	Extant 1995

Concurrent with the building of *Wilton*, work began on a new generation of MCMVs or mine countermeasures vessels, a new term signifying that the vessels were capable of both sweeping and hunting. They were the largest GRP ships in the world, with Plessey Type 193M sonar, elaborate position-fixing gear, hydraulic drive and a bow-thruster to permit manoeuvring at low speed. Every effort was made to eliminate the magnetic 'signature' and to cut out radiated noise (including a near non-magnetic version of the Deltic diesel) but at the same time providing a ship capable of deep-sea passages. On a cost per ton basis, they were some of the most expensive warships ever built, £24m for *Brecon* and £15m for the rest.

Two special GRP complexes were built, one at Woolston and the other at Scotstoun, permitting two MCMVs to be built in each. M 30 was ordered on 31 March 1977; M 31–32 and M 34 on 29 January 1979; M 33, M 35, M 37, M 39 on 19 June 1980 and M 36, M 38 on 14 December 1982.

Brecon and *Ledbury* swept two minefields at Port Stanley after the recapture of the Falklands. and others operated in clearing the Suez Canal and the Persian Gulf in 1985–91. All have had their sonars updated to Mod 1 standard, and a mid-life modernisation is planned for the mid-1990s. Numerous minor changes made in 1990 for Gulf operations, including extra 20mm guns and ESM equipment.

Brocklesby Jul 1994 C & S Taylor

VENTURER class *trawler-minesweepers*

Tonnage: 382t gross; 134t net
Dimensions: 120ft 9in wl × 29ft 3in × 12ft 10in
36.8m × 8.9m × 3.9m
Machinery: 2 shafts, 2 Mirrlees-Blackstone EWSL8M(A) diesels, 2000bhp = 11kts. Oil 21,000gal
Armament: Nil
Sensors: Commercial navigational radar only
Complement: 24

Class (no, former name, builder, fate):
Venturer (M 08, ex-*Suffolk Harvester*, Cubow, Woolwich, returned 1984), *St David* (M 07, ex-*Suffolk Monarch*, Cubow, Woolwich, returned 1984).

To combat the threat from deep-laid mines (mainly to submarines) the RN chartered two commercial stern trawlers in 1977 and four more in 1978, using them with their original crews, but with naval mine warfare specialists on board. These 'Highland Fling' exercises showed that sweep wires could be operated at far greater depths than had been thought, and so a new concept of Extra-Deep Armed Team Sweeping (EDATS) came into being. Special depth-keeping gear allows the sweep wire to follow the contour of the sea-bed, and the depth is believed to be well over 1000ft. Explosive cutters are attached at 50-fathom (300ft) intervals.

In 1978 two trawlers (built in 1973) were chartered from Small & Co of Lowestoft for manning by RNR crews from the South Wales and Severn Divisions. They were returned to their owners late in 1984. They had been successful as sweepers but power-generating capacity fell far short of naval requirements.

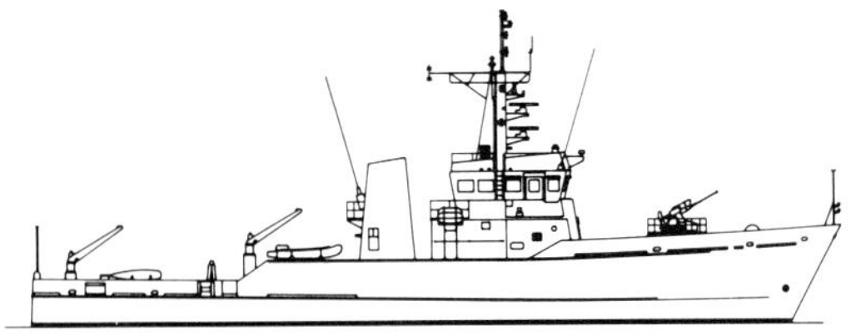

Ribble 1989

'RIVER' class *fleet minesweepers*

Displacement: 890t deep load
Dimensions: 156ft oa × 34ft 6in × 9ft 6in
47.5m × 10.5m × 2.9m
Machinery: 2 shaft diesel, 2 Ruston 6RKCM diesels, 3040bhp = 14kts. Range 4500nm/10kts
Armament: 1–40mm/60 Bofors Mk 3
Sensors: Radar: 2 Racal-Decca TM1226C
Complement: 30 (7 officers, 23 ratings)

No	Name	Builder	Launched	Fate
M 2003	WAVENEY	Richards (L)	8.9.83	To Bangladesh 3.10.94
M 2004	CARRON	Richards (GY)	23.9.82	To Bangladesh 3.10.94
M 2005	DOVEY	Richards (GY)	7.12.83	To Bangladesh 3.10.94
M 2006	HELFORD	Richards (GY)	16.5.84	To Bangladesh 3.10.94
M 2007	HUMBER	Richards (L)	17.5.84	To Brazil 1995
M 2008	BLACKWATER	Richards (L)	29.8.84	Extant 1995
M 2009	ITCHEN	Richards (L)	30.6.84	Extant 1995
M 2010	HELMSDALE	Richards (L)	11.1.85	To Brazil 1995
M 2011	ORWELL	Richards (GY)	7.2.85	Extant 1995
M 2012	RIBBLE	Richards (GY)	7.5.85	To Brazil 1995
M 2013	SPEY	Richards (L)	22.5.85	Extant 1995
M 2014	ARUN	Richards (L)	20.8.85	Extant 1995

(GY) = Great Yarmouth, (L) = Lowestoft

In 1980 it was hoped to order twelve EDATS minesweepers at a cost of £2.4m each. The options were a Navy-designed hull or an adapted commercial design such as a stern trawler. Hall Russell, for example, submitted a derivative of their torpedo recovery vessel, but finally Richards' adapted oil-rig support vessel proposal was accepted. Four were ordered on 23 September 1982, followed by six in May 1983 and the last pair on 25 February 1984.

Plans to order three more for the Fishery Protection Squadron were deferred indefinitely in 1991. Originally to be the *Amethyst* class, with names of precious stones. All except *Blackwater* were allocated to RNR Divisions, but belonged to 10th MCM Squadron. The Bofors gun was replaced in some by 30mm Mk 1.

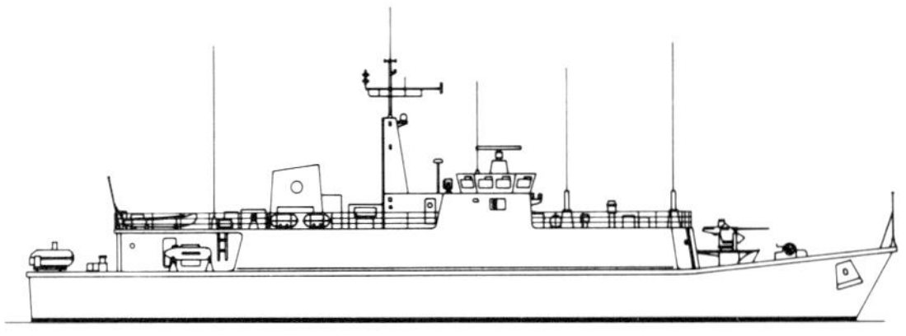

Sandown 1989

SANDOWN class *minehunters*

Displacement: 450t normal; 480t deep load
Dimensions: 172ft 10in oa × 34ft 5in × 6ft 10in
52.7m × 10.5m × 2.1m
Machinery: 2 Paxman Valenta 6-RP200E diesels driving Voith Schneider cycloidal propellers, 1500bhp = 13kts. Range 3000m/12kts
Armament: 1–30mm Mk 1
Sensors: Radar 1007; sonar 2093
Complement: 34 (7 officers, 27 ratings)

No	Name	Builder	Launched	Fate
M 101	SANDOWN	Vosper Thornycroft	18.4.88	Extant 1995
M 102	INVERNESS	Vosper Thornycroft	27.2.90	Extant 1995
M 103	CROMER	Vosper Thornycroft	6.10.90	Extant 1995
M 104	WALNEY	Vosper Thornycroft	25.11.91	Extant 1995
M 105	BRIDPORT	Vosper Thornycroft	31.7.92	Extant 1995
M 106	PENZANCE	Vosper Thornycroft	–	Ordered 1995
M 107	PEMBROKE	Vosper Thornycroft	–	Ordered 1995
M 108	GRIMSBY	Vosper Thornycroft	–	Ordered 1995
M 109	BANGOR	Vosper Thornycroft	–	Ordered 1995
M 110	RAMSEY	Vosper Thornycroft	–	Ordered 1995
M 111	BLYTH	Vosper Thornycroft	–	Ordered 1995
M 112	SHOREHAM	Vosper Thornycroft	–	Ordered 1995

Designed as a cheaper alternative to the 'Hunt' class MCMVs by reducing the number of roles, hence the designation 'Single Role Minehunter' (SRMH). Despite this the unit cost was some £40m in 1990, allowing only three to be built for the price of two 'Hunts'. The modular design permits them to be equipped as minesweepers if required.

Problems with the Type 2093 variable-depth sonar delayed full acceptance of *Sandown* until late 1991 or early 1992, but the design is considered very successful, and six have been ordered for Saudi Arabia. Spain has also negotiated a technology-transfer deal to permit a variant of the design to be built locally.

First of class was ordered in August 1985, followed by four on 23 July 1987. Invitations to tender for up to seven more were withdrawn by the UK MoD in 1991 as an economy measure, but construction resumed in 1995. They were the first MCM vessels equipped with the new PAP 104 Mk 5 remote mine disposal vehicle. *Penzance* laid down 25 September 1995.

United States of America

In 1945 the mission of the US Navy changed dramatically. Before the Second World War, a campaign was envisaged in which the USN would gain control of the Pacific, and then use that control to defeat a major maritime power, Japan. With the defeat of Japan, there was no longer any opposing sea power. This did not end the threat of war; certainly by early 1946 it was clear that the Soviet Union had resumed her prewar hostility to the West. However, unlike the prewar Japanese and German threats, the Soviet threat was two-pronged. The USSR maintained powerful military forces, but she also hoped that her communist ideology would be attractive to Western populations. Its argument was that the Western political-economic system was bankrupt.

The problem facing the West was twofold. Postwar economic recovery, which was essential to ward off the ideological attack, clearly would require heavy investments, funds for which could be freed by demobilisation. Any rearmament, to fend off the purely military threat, had to detract from essential economic recovery and growth. The military problem was further complicated in that considerable standing forces were needed to deal with the instability resulting from the Second World War. More money was needed to develop the radically new technologies, such as jet engines and nuclear weapons, which had been created in wartime.

At first the cost of new weapons development was relatively low. The United States was able to maintain the large standing naval forces required to provide substantial presence in unstable areas of importance, such as the Mediterranean and the Far East. However, as the new weapons, particularly new jet aircraft, entered production they began to compete with the standing forces. By 1948 the United States economy was in recession and the Truman Administration decided to cap defence spending. At this time it was also stockpiling strategic material, partly in fear that the ongoing Soviet advance would block Western access to some of them. Standing forces had to be cut drastically; the US Government shifted to heavy reliance on nuclear deterrence, largely exercised by the US Air Force. Many warships, including most of the carriers, were laid up. The new carrier *United States*, which would have supported a naval nuclear strike role, was cancelled.

The shift towards reliance on nuclear threats followed a new national policy. In 1947 the US State Department proposed a way to solve these problems: containment, the fundamental US policy of the next four decades. The fundamental assumption, which ultimately proved correct, was that the Soviet system could not survive the strain of economic competition if it could be denied control of the two vast industrial concentrations still beyond its grasp, in western Europe and in Japan. Its internal contradictions would be exacerbated by the denial of economic ties with the West. Clearly the USSR would not happily accept such isolation, and clearly she might try to attack. Containment envisaged a minimum investment in deterrent forces. Cold war, however, would certainly last a very long time; the West could not sustain forces comparable to those of the Soviet Union, at least not if it tried to maintain a free social system.

At first it seemed unlikely that the USSR, terribly damaged by the Second World War, would present an immediate military threat. After all, the United States had a monopoly on the atomic bomb. However, Stalin soon showed that the bomb's deterrent value was limited. Soviet-supported communists overthrew the Czech government in 1948: serious contingency war planning began, in conjunction with the United Kingdom. The NATO treaty was signed the following year, against the background of Stalin's blockade of Berlin. Initially the alliance was intended as a long-term means of reassuring the western Europeans. However, the Soviet Union broke the atomic monopoly in 1949, and it no longer seemed quite so obvious that Stalin could be deterred. Given budget pressures, however, the US Government continued to hope that its expanding nuclear strike forces would deter Stalin from any major advance.

By this time it was no longer clear that there was any long-term future for the US Navy. In the short term, the air force would have to prosecute its nuclear war from overseas bases, and at the very least they would have to be supplied (the Navy argued that a pure nuclear strategy would tend to discourage allies, whose countries would be destroyed in its execution). Ultimately, however, the air force expected to fly longer-range bombers (B-36s) from bases in the United States, in which case there would no longer be much point in securing sea communications abroad. In June 1950, Stalin demonstrated that nuclear weapons could not, in themselves, deter him: his ally, North Korea, invaded South Korea. Many in the West believed that, once the West was tied down in Asia, Stalin would move in Europe, and that World War III was scheduled for 1953 or 1954 (Stalin died before this idea could be tested). NATO was soon transformed into the military alliance which saw through the Cold War.

Policy reversed to favour major rearmament, including revived conventional forces. The United States never again relied completely on a nuclear threat. A much-enlarged defence budget accommodated a revived navy, including numerous new and rebuilt carriers. On the other hand, the budget never increased sufficiently to match Soviet conventional forces in Europe. Nor were the NATO allies willing to build up to the required force levels. The solution in Europe seemed to be a limited conventional defence backed by tactical nuclear weapons, the only means of stopping a massive Soviet army assault. On the Eurasian periphery, the United States adopted a policy of resisting Soviet expansionism, by force if necessary, as in Korea. Although US nuclear weapons backed the new policy, it had a large non-nuclear component which justified a large operating fleet.

Nuclear deterrence itself changed over time. By the mid-1950s, the USSR had a considerable nuclear arsenal of their own, and it was less and less clear that the United States could actually engage in nuclear war. Similarly, the Soviet Union could not lightly attack the United States. Khrushchev suggested a way out of this impasse: he could support local revolutionary movements in hopes of weakening the West, for example by drying up sources of raw materials. By the late 1950s it was clear that nuclear weapons could not be anything like an appropriate response (though for a time it still seemed that they could be used to defend Europe). The Kennedy Administration in particular began to emphasise conventional forces specifically for operations in the Third World. It intervened in Vietnam, largely as a way of discouraging Khrushchev's 'wars of national liberation'. Ultimately this intervention failed, but proponents could argue that the opposition had suffered so badly that it also was forced to give up on its strategy. The US investment in Vietnam probably also encouraged the anti-communist coup in Indonesia. About five years after the collapse of South Vietnam, the Reagan Administration accepted much the same logic (seeking to escape from the nuclear impasse) to support rebels fighting Soviet-backed forces in, among other places, Afghanistan and Angola. This time the logic worked: in combination with the sort of internal contradictions foreseen by the 1947 policy of containment (which were presumably intensified by attempts to match accelerated US defence spending), the Soviet disaster in Afghanistan helped to pull down the Soviet system. By the end of 1991 it was quite clear that the Cold War had ended and that the United States and the West had won.

The US Navy developed after 1945 had two quite different missions. One was direct confrontation with the USSR in the event deterrence failed. The other was presence/combat in the Third World (Korea demonstrated that deterrence was unlikely to have much effect there). The central problem was how to reconcile the two missions, given limited resources.

The Soviet problem was twofold. On the one hand, the fleet could contribute directly to the nuclear threat (or, if necessary, to a nuclear strike). Although carrier- and submarine-based weapons might not amount to a large fraction of total US striking power, their mobility could be quite valuable. It could force the Soviet defences to cover a much wider arc of their country than would have been necessary had they faced only bombers based in, say, North America and England. Similarly, the presence of submarines in the Mediterranean forced the USSR to extend anti-missile defences all the way around targets such as Moscow, whereas only some arcs would have been required against missiles from North America. In each case, given the limited Soviet resources, such additional threats would tend to thin out defences throughout the country, or at the least to raise the cost of Soviet defence, and thus the strain it placed on the Soviet economy. Resources spent on strategic defence could not be applied to forces facing Europe.

The second contribution was to fighting power in war. Although NATO proclaimed that it would answer a massive Soviet invasion with nuclear weapons, it was clear that a war might well never go to that level. In that case the success of reinforcement across the Atlantic might well determine whether the alliance survived. From the beginning of NATO, then, a major naval question was how to counter the very large Soviet anti-shipping force, consisting of both submarines and anti-ship bombers. Given the relative ineffectiveness of existing ASW measures, the initial response emphasized carrier strikes 'at source' on both the submarines and the bombers. One effect of an early naval offensive would be to demoralise the Soviet Navy. Another might well be to attract Soviet resources to attack the carrier groups rather than NATO resupply shipping; presumably the carrier groups would be far better equipped to destroy submarines and bombers attacking them.

In current terms, this strategy used power projection to gain sea control: strikes would be used both to destroy the enemy anti-ship forces in their bases, and to draw the surviving elements of that force into combat on terms most favourable to NATO. One advantage of this policy was that it emphasised exactly those strike forces which could also be used in the Third World, thus solving the difficult economic problem posed by the dual naval mission.

Until about 1970, then, US and NATO naval policy called for early carrier strikes against the base areas of the Soviet fleet, in the north and on the Black Sea. There was also a pure ASW component, consisting of submarine barriers across choke points (and off the US coast), maritime patrol aircraft supported by underwater surveillance (SOSUS), and convoy escorts. The submarine barriers and SOSUS both depended heavily on passive sonar, that is, on the relative noisiness of Soviet submarines. After about 1960 convoy operations became less and less attractive because the large number of surviving Second World War escorts was no longer a factor (these ships could not accommodate the massive new active sonars). Replacement escorts could not be built in sufficient numbers.

Then policy changed. Admiral Elmo Zumwalt took office as CNO. He was much impressed by the new Soviet surface presence in the Mediterranean, which threatened the strike carriers there. He was also painfully aware that, after the end of the war in Vietnam, defence funds would be cut, and inter-service rivalry would revive. He chose to define sea control and power projection as separate, perhaps even opposed, naval missions. Sea control was largely defined as convoy warfare (Zumwalt suspected that submarine barriers and SOSUS would both become ineffective as Soviet submarines became quieter). In Zumwalt's view, the big carriers would make their main contribution to success in general war by helping to defend convoys against air attacks; they would operate singly or perhaps in pairs, rather than in the concentrations (three or four together) previously envisaged. Carriers might well also be grouped with amphibious forces in the separate Third World power projection role, though this role seemed less and less important as the United States turned away from intervention in the Third World after the disaster in Vietnam.

United States interest increasingly focused on Europe. The navy's role began to blur. Given increased Soviet bomber and submarine strength, some questioned the viability of seaborne reinforcement. United States *matériel* for several divisions was stockpiled in Germany to await the arrival (by air) of troops. Clearly seaborne reinforcement would be necessary if a war dragged on, but in the late 1970s many argued that any conventional defence of western Europe would soon end with either a nuclear threat or with resort to nuclear weapons, in which case sea lane defence would be irrelevant.

By the late 1970s, however, the view was gaining ground that a war, once it began in Europe, might well never escalate to nuclear attacks. Studies suggested that NATO would not win simply by threatening or executing nuclear attacks. Instead, whichever side was losing in the conventional phase would probably keep losing (only at a much greater cost) if the war became nuclear. This apparently abstract argument had enormous consequences. By this time NATO had relatively short supplies of munitions, enough to fight for only a few weeks. This level of supply had been justified by the assumption that no war could last very long and that the nuclear threat or reality would soon bring both sides to their senses. Short war meant that reinforcement was not vitally important, and thus that naval forces could be neglected. Money would be spent instead on ground and tactical air forces which would, it was hoped, achieve enough at the outset of a war to convince the Soviet forces to halt any aggression.

Intense Soviet efforts to modernise conventional forces in Europe convinced some in the US defence establishment that the short-war escalation strategy would fail in wartime. The war in Central Europe would have to be fought to a non-nuclear finish, involving protracted reinforcements. In this case security of the sea communications between Europe and North America would be extremely important. Many argued for large purchases of convoy escorts, mainly frigates, and also for nuclear submarines to form barriers against Soviet submarines in North Atlantic.

The navy argued differently. By the late 1970s it was rejecting the division between sea control (now understood mainly as ASW) and power projection (carriers and amphibious ships); the time was ripe for a return to the classical ideas of the 1950s and 1960s. The navy could do much more than merely protect the shipping reinforcing Europe.

The central idea was that, although naval forces could not match the Soviet Navy one for one, their ability to threaten widely separated areas of the Soviet empire might well tie down substantial forces, including ground and tactical air forces which otherwise might fight in western Europe. For example, in the late 1970s there was a serious proposal that, in the event of war in Europe, the bulk of the US Seventh Fleet would transfer into the Atlantic. The counter-argument, which won out, was that pressure on the Soviet Far East would tie down enormous forces. US carrier strikes in the area would, among other things, reinforce Soviet fears of ultimate Chinese intervention, tying down very large numbers of troops on the Sino-Soviet border.

More generally, the Soviet strategists were already quite sensitive about the prospect of amphibious assault, even though it was always clear that the assault forces would be relatively limited. Any encouragement of such fears would tie down defensive forces which might otherwise act offensively on the Central Front. Similarly, US naval strategists began to argue that the security of the flanks of the Central Front, both in the north (Norway) and in the south (the Mediterranean) was essential to any success in protracted war. Forces could not simply be concentrated in the centre. Thus it was argued that unless NATO won the battle of the Norwegian Sea at the outbreak of war, forces on the Central Front could not be resupplied (air bases captured by Soviet forces in Norway would threaten the resupply ports). A NATO ability to counter Soviet attacks on Norway would also force the Soviet Union to concentrate more of her own finite forces in the north, away from the main ground theatre in the centre.

Such arguments formed the beginning of the Maritime Strategy, which was essentially a revival of the classical naval ideas of the pre-Zumwalt era. Once more the navy argued that sea control and power projection were (or could be) two sides of the same coin; that the ability to project power (including amphibious power) around the Eurasian land mass would in itself help tie down and weaken the Soviet Union.

Moreover, the USSR now had a naval asset, the strategic submarine fleet, worth protecting. Soviet strategists expected US forces to attack it, and the Maritime Strategy tended to reinforce that fear. It therefore might help to convince the Soviet planners to keep their submarines out of the sea lanes to Europe, at least early in a war. Eventually the Soviets began to discuss anti-shipping operations, not least as a way of levering the US forces away from their SSBN holding areas. However, the designs of their submarines were not really well suited to shipping warfare (they carry too few torpedoes, and the torpedoes are not dual-purpose).

Thus the object of early offensive naval operations was largely to pin down Soviet anti-shipping units in home waters. Cuba was particularly significant, since it was the one place where Soviet naval forces were forward-deployed. Ideally, Cuba would be attacked early in a general war, but it was quite possible that she would choose to remain neutral, and thus tie down substantial US forces. For example, the Second Fleet, normally used to work up ships before they deployed abroad (to the Sixth Fleet in the Mediterranean), would provide much of the firepower of the Strike Fleet assigned to make early attacks on the Soviet Northern Fleet area. It was also the main US force near Cuba; if it could be tied down, the Atlantic portion of the Maritime Strategy could not be executed.

Cuba and the new semi-satellite state of Nicaragua might also be particularly significant because of their positions near both the Panama Canal and the mouth of the Mississippi. Much of the US *matériel* needed to support a NATO war in Europe would have to come through the Canal from the West Coast; much of the remainder would come down the Mississippi. The Gulf of Mexico is shallow; it is a difficult area for ASW. Thus, although the Maritime Strategy was concerned mainly with heavy carrier battle groups, it also had a narrow-seas element.

The new maritime strategy became official policy under President Reagan; Secretary of the Navy John Lehman used it to justify his new shipbuilding programme. The strategy was not universally popular. The navy had never publicly articulated its strategy in the pre-

Zumwalt years, so many critics did not acknowledge that the Maritime Strategy was a return to classical ideas, not an aggressive new departure. Many in the NATO alliance considered the strategy provocative, despite the fact that it could be seen merely to express ideas which had been standard NATO currency since 1949.

The Cold War equivalent of the Maritime Strategy was to support anti-Soviet guerrilla movements around the periphery of the Soviet Union. As in the wartime strategy, none of the operations could in itself defeat the main enemy. All of them, however, would draw off resources. The crucial perception, central to containment itself, was that those resources were finite. In this sense the main theatre of engagement throughout the early 1980s was the technological arms competition, particularly in strategic offence and defence. The USSR found herself badly stretched in this central arena; the naval component of strategic forces was far too costly. Failure to match the West here was aggravated by public failure on the periphery in Afghanistan, and Mikhail Gorbachev was forced to retreat politically. As the designers of containment had supposed, retreat was fatal. The Soviet system was too rigid to adapt.

The size of the Reagan Administration's building programme, to achieve a six hundred-ship fleet, was predicated on the perception that Congress could be convinced to buy twenty ships per year, and that average ship life was thirty years. However, the content of the programme came out of the Maritime Strategy. Exercises tested it. New developments were justified by their place in the overall forward strategy. The navy had not had an explicit strategy since the Orange war plans of the pre-1941 era, and the Maritime Strategy was intended to perform a similar function, unifying the force and giving it direction.

With the collapse of the Soviet Union in 1990–92, the direct military threats to Europe and Japan, which had motivated postwar US strategy in the first place, were essentially gone. The Maritime Strategy was an application of basic naval concepts to a particular strategic problem; now that problem seems to be over. For the near-term future, the United States faces a variety of disturbances in the Third World, not unlike those which followed the end of the Second World War: unpredictable, sudden, and unrelated. It seems likely that the navy built to fight the USSR will have to deal with most of them. After all, only ships can provide long-term presence off a foreign coast, then apply military pressure if and when it is needed. The alternatives, land-based aircraft and ground troops, require the consent of a country's neighbours, and their presence is intermittent at best. Aircraft in particular cannot provide the sort of quiet pressure which warships so easily impose.

With the end of the Cold War, what amounted to postwar demobilisation has set in. It seems likely that forces specially designed to deal with the Soviet threat will be cut much more drastically than others: this applies to ASW convoy escorts, sonar surveillance ships, many maritime patrol aircraft and, probably, to many submarines. It should apply far less to carriers and to large amphibious ships. There has also been pressure to cut back those sophisticated systems which presumably are not as important in the face of Third World forces, such as F-14s.

The Gulf conflict which began in August 1990 was the first post-Cold War test of the US fleet. Six carriers and very large numbers of other ships were deployed. Tomahawk land-attack missiles were used for the first time; some were fired by submarines. Although no amphibious assault was made, the palpable threat of such an attack did tie down substantial numbers of Iraqi troops. The one apparently unexpected feature of the war was the large Iraqi mining campaign. There is no question but that US mine countermeasures technology generally matched that of other navies. Even so, it seems that mines just cannot be neutralised quickly enough for an amphibious assault. Given that such assaults are probable in the future, a new type of mine countermeasures had become a current priority.

Theodore Roosevelt in the Gulf, March 1991 L & L van Ginderen

As reflected in these pages, the cuts will leave a US fleet oriented mainly towards carrier battle groups and major amphibious ships, with many fewer ASW ships. It is not yet clear to what extent the submarine force will be considered primarily an ASW asset, and thus deserving major cuts. The submariners have argued strongly that they offer far broader capabilities, some of them particularly valuable in an era of Third World conflict. Examples are covert intelligence-gathering (including minefield reconnaissance using underwater vehicles) and the covert insertion of special forces.

Although the carriers themselves have not been cut too badly, their attack bomber forces have suffered. During the 1980s, the navy had to choose between buying a new-generation 'stealth' bomber, buying a two-seat bomber version of the F/A-18, and upgrading the existing A-6. The choice was to release money for the next-generation bomber by limiting near-term purchases to the upgraded A-6F. One argument was that no F/A-18 variant could carry anything like enough bombs to a sufficient range. Moreover, the new technology promised that even a few carrier aircraft could penetrate most Third-World national air defences.

Then the new-generation programme, the A-12, collapsed. It had proven too ambitious; it is also possible that the intense secrecy surrounding all 'stealth' programmes precluded the normal review process while pushing cost up too sharply. An attempt to build a more modest yet stealthy long-range attack bomber, F/A-X, also failed. That left only the drastically modified F/A-18, the E/F model (the A-6 upgrade was rejected). Some F-14s are being modified as bombers, but they, too, cannot be considered full replecements for the A-6 force.

The main effect of these developments is that the carriers are quickly losing their long-range strike capability. To some extent Tomahawk missiles fired (eg by submarines) close offshore maintain a deep strike capability against land targets, but they cannot deliver anything remotely like the firepower of the bombers. It is possible that the F/A-18 will suffice in a post-Cold War era, but it also seems possible that carriers trying to keep far enough offshore to avoid enemy tactical aircraft and other nuisances will find themselves unable to strike well inland. That limitation may begin to erode the rationale for the carriers.

An attempt to rely mainly on missiles, such as Tomahawk, for deep strikes is unlikely to succeed, mainly because each missile carries only a limited payload. Now that nuclear weapons have become virtually unusable (in the wake of the Cold War), it is likely to become more obvious that such small payloads buy relatively little, no matter how precisely they are delivered.

THE US CARRIER FORCE

The value of resupply in a major war depends upon the expected length of that war. Studies carried out in about 1949 envisaged a lengthy conflict. Although nuclear weapons might be used, they would not be available in sufficient quantity to end the war, and therefore the war would probably resemble the Second World War, but fought with more modern weapons. Convoy escort, therefore, was a major consideration, and new convoy escorts such as the *Dealey* class were designed with mass production in mind. In view of inter-service rivalry with the newly independent air force, the navy justified its desire to buy larger carriers on the basis of the submarine problem: their heavy attack planes would be able to destroy Soviet submarines at base. American war planning even included the mass production of ASW submarines (SSK), which would operate in the approaches to Soviet bases.

From 1952 onwards, the development of H-bombs seemed to indicate a rather different type of war, which might be concluded very rapidly with a series of massive nuclear strikes. By this time the infusion of funds from the Korean War mobilisation had largely healed the Navy–Air Force controversy over strategic missions, and the carrier force was assigned a strategic attack role in the general war plan. Heavy attack aircraft such as the North American AJ Savage and then the Douglas A3D Skywarrior had much the same role as the air force's B-47, and the navy's attention turned largely to the development of fast attack carrier task forces. Such formations still had a sufficient margin of speed as to be almost immune to submarine attack. Although conventional ASW continued to be of great importance, the new mobilisation-design escorts were not built in very large numbers.

During the late 1940s and early 1950s the greatest threat to the fast carriers was air attack, and much of the large expenditure which could be assigned to the carrier task forces went on countermeasures. In 1945 the USSR captured German air-to-surface missile technology, which had succeeded during the war in sinking several Allied warships, as well as the Italian battleship *Roma*. By 1946 there was a feeling that the prime threat of the future was the missile-equipped bomber, attacking by night or in bad weather, using radar. Ultimately there were two interconnected approaches to task force air defence. One was the carrier fighter. The carrier and her aircraft could be equipped for all-weather operation, and the efficacy of fighter control improved by equipping destroyers as radar pickets. For combat against high-performance bombers, which might be carrying nuclear weapons, the fighters were equipped first with unguided air-to-air rockets and then with missiles – Sidewinder, Sparrow and ultimately, the very long range Phoenix. The evolution of the naval fighter from a day fighter-bomber to an all-weather Fleet Air Defence interceptor was a very expensive one, with important financial consequences for the balance between aircraft and ship procurement budgets.

The other major element of fleet air defence was the missile-armed surface ship. The US Naval SAM programme started with Project 'Bumblebee' in December 1944, and ultimately produced a family of three weapons (Terrier, Talos and Tartar, in order of appearance) which could arm anything, from a destroyer escort to a cruiser. Generally the radar and computer system associated with anti-aircraft missiles also suffices for fighter control, so that with the general availability of missile escort for fast carriers the specialised picket destroyers were phased out in the early 1960s. The SAM programme was huge, and it was considered extremely urgent. Hulls were relatively expensive, and the first US missile ships were all conversions of existing cruiser hulls, with the usual attendant inefficiencies. More compact missile systems went aboard destroyers and the enlarged destroyers (or fast task force escorts) which were then designated frigates, from the *Dewey* class onwards.

Of the several new technologies the Navy developed after 1945, shipboard anti-aircraft missiles appear in retrospect to have been the most expensive. The programme produced no really effective systems until about 1965, although by that time eleven existing ships had been converted, one nuclear cruiser built, and twenty-one missile frigates and twenty-three missile destroyers completed. Indeed, there was no question but that such a programme was needed. The problem was that even as the problems of the first generation were being solved, entirely new systems (Typhon, which evolved into the present Aegis) were in hand. Cost escalation in new construction, particularly in the guided missile ships, led to a considerable curtailment of the shipbuilding programme in the late 1950s; for example, three planned missile cruiser conversions could not be carried out. The cost squeeze became particularly acute with the advent of Polaris, since funds for the new crash programme had to come from a budget previously designed to cope only with ASW and with the carrier force.

The other major element of the carrier attack force was underway replenishment, or UNREP. Although a carrier might well be able to remain at sea for several months, her aviation fuel and ordnance would not support air operations for more than a few days at a time. Strike tactics therefore called for a run into a target area, followed by intense operations, then withdrawal to a safe area well out to sea and replenishment in preparation for another series of strikes. Such tactics were first used in 1945, and they continue to be important except that the distance to which the task force must withdraw for safety is now much greater, given the reach of bombers, coast defence vessels and missiles. UNREP modernisation was then an important element of the fast carrier programme from 1945 onwards, with the emphasis on reducing the period of vulnerability during replenishment. Ultimately the navy adopted the concept of 'one-stop' replenishment, in which a single large auxiliary provided both fuel and ammunition simultaneously. Between 1964 and 1970, four large fast combat support ships (AOE), each rated at 26kts, were built. In principle they in turn would be supported by slower (20kts or 21kts) bulk stores carriers, combat stores ships (AFS), ammunition ships (AE) and fleet oilers (AO), the AOEs shuttling between such ships and the fast carriers. A new generation of AOEs is now appearing.

Although the carriers have been the core of the surface navy since 1945, their role and even their existence has been subject to almost continuous debate. The programme of heavy strategic carriers was cancelled with the *United States* in 1949, largely, it appears, because of air force objections. Even though the carrier programme was restored after the outbreak of the Korean War, the air force continued to argue throughout the 1950s that carriers were inherently inefficient, given the large investment in ships and aircraft required merely to protect them. After the authorisation of the large carrier *Enterprise*, the rising cost of new carrier construction led to the virtual suspension of the previous programme of annual authorisations, and to a debate between navy advocates of nuclear power and a Secretary of Defense determined to economise by returning to oil fuel. Secretary of Defense McNamara also questioned the need for so large a carrier force, given the transfer of strategic attack functions to missiles. However, the demonstration of carrier utility in Vietnam convinced him, and by the end of his term he advocated a new nuclear carrier programme – which, however, was not to be completed for many years. Thus no carrier was authorised between 1973 (CVN 70, *Carl Vinson*) and 1979, when CVN 71 (*Theodore Roosevelt*) was approved despite Administration efforts to substitute a more limited ship, the abortive CVV. In 1982 the big carrier's future remained in doubt, but nevertheless Congress authorised two more large-deck carriers, CVN 72 and 73.

The rate of procurement has fallen sharply since the 1950s, in part because of escalating costs. With *Forrestal* a 'one carrier per year' policy was laid down, culminating in *Enterprise* in FY58. Attempts to design a much less expensive nuclear carrier failed, and no new carrier was authorised until FY61 (CVA 66, *America*). A policy of buying a carrier every other year was then formulated; CVA 67, *John F Kennedy*, was authorised in FY63. However, the procurement of the next class of nuclear carriers (*Nimitz* class) was stretched out: CVN 68 in FY67, CVN 69 in FY71 and CVN 70 in FY74. With the construction of CVN 71, were thirteen modern large-deck carriers available.

In fact the planned size of the carrier force has varied considerably since the Korean War. For example, as of late 1950 the target was a total of twelve modern attack carriers (*Midway*s and rebuilt *Essex*s) by June 1952. All would be able to launch nuclear-armed long-range attack aircraft, but fifteen rather than nine *Essex* class carriers were rebuilt, so that with the introduction of the first two *Forrestal*s the attack carrier force, including ships under reconstruction, was eighteen to twenty strong for a short time in the late 1950s. After that, *Essex* class attack carriers were transferred to ASW duties as the new carriers joined, and the force objective settled to fifteen, a figure constant until the early 1970s, when the retirement of *Oriskany* and *Franklin D Roosevelt* cut it to thirteen. Generally three carriers are required to keep one on a forward station: one on station, one working up after overhaul, and one in overhaul. Thus fifteen permitted the usual deployment of two in the Mediterranean and three in the western Pacific, with virtually no leeway for contingencies, or for that matter, for the oft-discussed Indian Ocean Fleet. During the 1960s and 1970s there was some slack, to be sure, in the existence of nine *Essex* class ASW carriers capable of limited attack duty; one actually served in Vietnam in this role. However, the last of these ships was retired in 1974. An active twelve-carrier force in practice means only two carriers forward-deployed in the two principal operating areas, unless time on station is drastically increased (to the detriment of morale and, importantly, re-enlistment) or overhauls are curtailed.

Under the Reagan Administration, the force goal of fifteen deployable carriers (which would not include those under long-term overhaul) was revived. Two nuclear carriers (CVN 72 and 73) were authorised under the FY83 programme, and *Midway* was retained in service well beyond her expected retirement date of about 1985. Another pair of carriers (CVN 74 and 75) were authorised under the FY88 programme. The goal was cut back to twelve (one of which would be manned by reservists) after the end of the Cold War. At present the intention is to replace all the fossil-fuel carriers with nuclear-powered ships; a total of ten nuclear carriers are currently either in service or under construction. CVN 76 was authorised in FY95, and the current plan is to buy CVN 77 under the FY02 (2002) programme. She may be built to a new design; a CVN(X) study is currently underway.

One major effect of the post Cold War cuts has been a drastic cut in carrier aircraft procurement. With the retirement of the A-6 attack aircraft (without any immediate successor), the fighter-attack mix on carriers is being cut to fifty aircraft, fourteen F-14 Tomcats and thirty-six F/A-18. In effect, this is the old carrier air wing minus the heavy attack (A-6E) squadron. Some F-14s are being given attack capability, but the long all-weather reach of the A-6E is being lost. The proposed A-6 successor, the stealthy A-12, died of excessive cost, and a somewhat more austere stealthy bomber, A-X (later F/A-X), also died. As of 1995, the only attack aircraft likely to appear in the near term is the stretched F/A-18E/F version of the existing short-range F/A-18.

The first two carrier air wings and one reserve wing shifted to the new mix in FY94. By the end of FY95, too, the helicopter mix on the carriers will be cut to three SH-60B and two HH-60 (for search/ rescue). The other aircraft on board are E-2Cs (early warning), S-3s (ASW) and EA-6Bs (electronic warfare).

The size of the US carrier force in turn determines, to a considerable extent, the size of the surface navy. High-quality missile escorts are bought to accompany fast carrier battle groups, and the slower ASW escort force is sized to protect the UNREP groups (as well as the amphibious assault force). To some extent even the attack submarine force is sized for fast carrier warfare, as current practice is to provide a nuclear submarine as part of the battle group, for distant ASW escort. In fact, the ideal ratio between escorts and carriers varies considerably, for example with developments in shipboard weapons and with the character of the threat: the balance between anti-air and anti-submarine within the battle group has changed considerably with the advent of a large Soviet nuclear attack submarine force, part of it now armed with short-range anti-ship missiles.

Despite a series of violent changes in the rationale for the carrier force, its development has been remarkably consistent since the Second World War. Perhaps the principal reason is the great flexibility of the carrier, given the variety of air groups it can support. *United States* was designed in the late 1940s specifically for nuclear attack, but the great size required merely to support a heavy bomber made for great capacity for conventional ordnance and aviation fuel. Thus when the Korean War demonstrated the value of carriers in conventional ground support, the navy could argue that a new large carrier, which became *Forrestal*, would be extremely effective in just that mission. Since the Second World War the carrier has been the only means by which the United States has been able to intervene abroad in the absence of friendly bases and base agreements. With the loss of many bases from the mid-1950s onwards, this has been an increasingly valuable capability, and it has been demonstrated many times.

Although the large carriers were originally designed for long-range nuclear attack, there was always considerable navy sentiment against such a strategy. By the mid-1950s navy long-range planners considered general nuclear war unlikely, in view of the probable development of the Soviet nuclear arsenal. They therefore adopted as a planning assumption the expectation that war in the mid and late 1960s – for which the navy was buying prototype equipment and ships in the mid-1950s – would be limited, quite possibly involving Soviet proxies rather than the Soviet Union itself. The fleet would have to fight around the Eurasian periphery, and its role would be a combination of tactical support of marine amphibious operations and strikes at moderate range. Strategic attack, although a vital function, might well be shifted from carrier aircraft to missiles, particularly as suface-to-air missiles were beginning to limit the chances of survival of even advanced long-range bombers.

Such conclusions were particularly important because of their implications for carrier and aircraft design; in 1955 carriers were very large principally because they had to accommodate the 70,000lb supersonic A3D Skywarrior, and the Bureau of Aeronautics looked towards a 100,000lb supersonic bomber in ten years' time. However, if the fleet's long-range attack mission were shifted to a missile such as Regulus, and much more modest demands placed on the attack aircraft, the carrier of the future might be much smaller and less expensive. In about 1956 this concept led to the design of the A2F Intruder, which became the A-6, a bomber which would survive by stealth rather than by high-speed, high altitude penetration of enemy defences. It also appeared that fleet air defence would most efficiently be conducted by a low-performance fighter with very long range missiles, the F6D Missileer. In fact the change in carrier design never occurred, primarily because it transpired that large carriers were needed to support intensive ground attacks, even by relatively undemanding aircraft. In 1956 it still appeared that tactical nuclear weapons might be used quite freely, and that the total ordnance load of a carrier therefore would be small. Ten years later that was inconceivable, and a thousand tons of carrier ordnance lasted only a few days in Vietnam. Moreover, Missileer was replaced by a manned fighter of much higher performance, still carrying the very long range missile – the present F-14 Tomcat.

Unlike Missileer, the high-performance F-14 could escort strikes and dogfight. The latter proved important in Third World operations, as when F-14s shot down Libyan fighters on two occasions. The next step was clearly a fighter suited both to air-to-air and air-to-ground work, the F/A-18 (a programmable radar is the key to true dual capability). Unfortunately, the F/A-18 was conceived as a supplement to the F-14 and as a replacement for carrier-based light bombers (A-7s), rather than as a replacement for both the F-14 and the long range A-6. As a consequence, it was designed as a relatively inexpensive lightweight fighter, with only limited range and weapons capacity. Thus it cannot provide the sort of very long-range air-to-air firepower offered by the ageing F-14, nor can it reach A-6 ranges with A-6 bomb loads.

As this is written, it seems unlikely that a true F-14 *or* A-6 successor will soon emerge; the joint navy/air force programme for a new stealthy attack aircraft died of excessive cost. There is considerable interest in a future low-cost fighter, which may be capable of STOVL performance. Its emergence may push down the unit cost of future carriers, particularly if some of the deep strike mission is transferred to missile-firing surface ships and submarines. However, no missile can even approach the economy offered by a reusable aircraft; the only exception is the now virtually unusable nuclear weapon, which offers so much firepower in so small a package. It is not clear how this dilemma will be resolved early in the next century.

ANTI-SUBMARINE WARFARE

Throughout the 1950s the character of ASW began to shift, both as to tactics and as to rationale. There had always been a tension in ASW thinking between offensive (sub-hunting) and defensive (convoy) operations, with the use of submarine 'ambushers' in an intermediate position. During the Second World War offensive ASW was largely based on the success of codebreaking and on the means of detecting submarines at long range, either by high-frequency direction-finding (HF/DF) or by radar (against surfaced submarines and, at much shorter ranges, snorkels or even periscopes). In each case aircraft were extremely valuable because of their ability rapidly to close a fleeting contact, attacking before the submarine might escape. The German development of fast submarines capable of extended underwater operation (Type XXI) materially reduced the effective detection range of existing airborne radars. A burst transmitter, 'Kurier', made shipborne HF/DF difficult if not impossible. Both developments fell into Soviet hands in 1945, and there could be no guarantee of success against Soviet codes in any future war. Although the Soviet submarine fleet actually in service in 1945–50 was composed of conventional submarines technically inferior to those the Germans had had in service in mid-war, it was clear that the USSR had or would soon have the potential to produce a much more ominous fleet.

Thus during the first five years after victory the US Navy concentrated on efforts to overcome the new generation of fast submarines. Larger airborne radars were adopted to balance the much reduced radar cross-section presented by a snorkel, and sonobuoys were developed to permit a hunting aircraft some indication of the presence of a target. Indeed, forced back to convoy tactics, the navy considered using aircraft to plant a sonobuoy barrier around a moving convoy, or helicopters to 'dunk' sonars ahead of it. Major efforts went into the development of larger ship-mounted sonars, culminating in the present SQS-26/53, which has a considerable impact on the design of ships carrying it. In the absence of any reliable means of detecting submarines at long range, the only effective tactic for the protection of shipping was convoy, a tactic extremely ineffecient in terms of submarines killed per unit effort. Given the sophistication of the new submarines, the ASW ships would become more and more expensive, so that it might be expected that ultimately their procurement and maintenance would strangle the naval budget.

This issue was resolved by the advent of fixed underwater arrays. Fixed to the ocean bottom, SOSUS (Sound Surveillance System) hydrophones could detect submarines at a considerable distance, their effective range varying with the noisiness of the individual target, signal processing techniques and ocean sound conditions. SOSUS was first tested in 1954, and during the next decade the system was extended and its sensitivity improved. Offensive ASW tactics now became extremely attractive, with aircraft far more effective than surface ships in many cases, given their ability to reach a contact position more quickly. Generally the prosecution of a SOSUS contact was a somewhat extended process, as the original contact would be relatively imprecise and the attacker would therefore have to re-acquire the target. Thus very long range patrol aircraft such as the Lockheed Neptune (P-2) and Orion (P-3) became particularly important elements of the ASW team.

The specialist ASW carriers, originally intended to support convoy tactics, now were used to close mid-ocean gaps in patrol aircraft coverage, and the endurance of hunter-killer (HUK) task groups built around them had to be increased. By the late 1950s *Essex* class fleet carriers were operating in the ASW role as support carriers (CVS), screened by converted destroyers. In analogy to the development of

An ASW support group built around the *Yorktown* 1967 USN

the fast carrier UNREP force, smaller one-stop UNREP ships (AOR) were designed to support the HUK forces. However, as P-3 Orions replaced the P-2s, their increased range eliminated the mid-Atlantic gap in air cover, and the CVSs were reassigned to protect the fast carrier forces from an increasing Soviet submarine threat.

Offensive ASW was more than a matter of economics. From the late 1950s onwards the USSR was known to possess strategic attack submarines, and navy tasks expanded to include peacetime trailing and wartime destruction of these craft. It should perhaps be emphasised that this mission was by no means the oft-publicised one of instant destruction of all enemy strategic submarines upon the outbreak of war. Rather, given the fleeting and statistical character of SOSUS contacts, it was to eliminate these craft within a reasonable period after outbreak, assuming that the war would not begin with a surprise missile bombardment. This mission has received considerably less publicity in recent years, as advocates of theories of strategic stability have claimed that the very invulnerability of strategic submarines in some way guarantees that they will not be used. But the threat of US offensive ASW appears to be among the chief motivations of Soviet fleet development.

Submarines were another vital element of ASW strategy. They alone could occupy forward positions off Soviet submarine base areas, detecting and sinking enemy submarines as they emerged, and so reducing losses to what seemed, initially, to be almost undetectable craft. Such operations, however, would require hundreds of specialised killer submarines, figures as high as 250 frequently being cited in the late 1940s. To some extent the number could be made up by emergency conversions of existing fleet submarines, but such tactics were fundamentally unattractive, given the technology of the first postwar years.

Later, however, advances in submarine sonar technology made it possible to move the submarines back from Soviet coastal waters to such 'choke points' as the Greenland-Iceland-UK Gap, where more sophisticated craft in smaller numbers (but equipped with long-range weapons to match their sensors) could inflict severe losses as Soviet submarines made for the North Atlantic. From the *Thresher* class onwards the special SSK submarine ASW designation was dropped, and all new nuclear submarines were optimised for the ASW role, with very long range sonar and such weapons as the SUBROC ASW missile. The transformation was a fortunate one for the submarine force which, after a great triumph over Japan, faced an enemy with little essential seaborne traffic and no great pretensions towards conventional surface fleet development.

In principle, in a war lasting weeks or months, most Soviet submarines would be lost to a combination of the submarine barriers and the aircraft prosecuting SOSUS contacts. Surface ASW escorts, the most numerous and also the most expensive element of the net ASW system, would act primarily as a back-up, preserving shipping until the attrition enforced by the two primary systems had taken its toll. The consequences of this conceptual development first became apparent in the late 1960s, when the number of escorts on hand in reserve fell well below the level required for a combination of convoy duty and the protection of fleet units. It would appear that this decline represented a conscious decision, apart from the severe economic forces acting against surface ship procurement at the time.

Convoy protection was not as important as the protection of naval formations because, it could be argued, there were so many merchant ships in Western service that even substantial losses early in a war could be sustained. High-value naval units such as attack carriers and amphibious ships were not nearly so expendable, and so required close escort, even if the latter tactic was, on the whole, inefficient. Thus from the 1950s onwards the specialised ASW escorts began to have less the aspect of austere mobilisation ships and more that of high-quality units which might well on occasion replace destroyers. For example, their speed rose to permit them to combine tactically with CVS task groups, or with fast (and therefore valuable) merchant ships. The fast carrier escorts (frigates in then-current US naval terminology), which had been optimised for the anti-air defence of the carriers, received more powerful and more sophisticated ASW batteries. By the late 1960s the Major Fleet Escort study would show that three missile anti-air ships per carrier would suffice. This was the genesis of the *Spruance*, a pure ASW ship which was fast enough to operate in the carrier screen. As early as 1960, new destroyer escorts were sometimes being described as destroyer replacements, capable of about 30kts at full power.

STRATEGIC MISSILE SUBMARINES

In 1959 a third major element was added to the fleet – the Polaris submarine. Naval interest in missiles for strategic attack dated from immediately after the Second World War, when copies of German V1s were fired from surfaced submarines and experiments were conducted to develop a capsule in which a submarine might tow a V2 ballistic missile. A strategic attack cruise missile, Regulus, was designed and built, and several submarines were converted to carry 'Trounce' guidance equipment for it. However, the Regulus programme did not have nearly the impact on the fleet as a whole that Polaris had. To some extent, indeed, Regulus appealed merely to offer the possibility that the fleet would be able to conduct strategic attack aircraft, a possibility recognised in the 1955 long-range study. Polaris offered something new, a survivable missile base which could reliably endure an enemy first strike. Moreover, Polaris was a separate system, linked to the remainder of the fleet only in the most tenuous way.

Polaris had two major impacts on the fleet. First, it permitted the release of the attack carriers from the strategic attack mission: the new long-range bomber, the A-5 Vigilante, was converted for reconnaissance. However, a navy already reeling from cost escalation in the surface missile ships found it extremely difficult to pay simultaneously for Polaris (including a large nuclear submarine programme), the mass of new and sophisticated ASW ships, the reconstruction of the numerous wartime destroyers still in service, and large numbers of advanced naval aircraft: too many programmes coincided. Thus Polaris had a severe if indirect effect on the modernisation of the fleet, an effect which could only be reduced by the continued employment of obsolete warships.

AMPHIBIOUS WARFARE

Carriers were only one aspect of the new land-attack emphasis developed during and after the war. The other was amphibious assault, which before 1942 had been limited to a Marine Corps battalion on each coast, for what would now be called counter-insurgency. In 1945 it appeared that the day of large landings was over, given the vulnerability of concentrated shipping to nuclear attack. But the marines and the navy persisted, and from 1948 onwards set a goal of replacing the existing 10kt troop lift with an integrated 20kt amphibious force. Ideally many of the vast number of extemporised wartime types would be eliminated in favour of a small number of carefully designed specialised types. This ideal was met only in the late 1960s and even then only a far smaller scale than had originally been envisaged.

In any case, one of the surprises of the Korean War was the continued utility of large-scale amphibious operations, demonstrated in the famous 'end-run' of Inchon. New amphibious ships were built during the 1950s, and the marines began to develop a new technique of helicopter assault ('vertical envelopment'). It was intended specifically as a means of achieving tactical concentration at the objective without requiring a concentration of amphibious ships vulnerable to nuclear attack, and it led to the conversion and construction of specialised helicopter carriers. To some extent vertical envelopment was inspired, too, by the success of the extemporised British helicopter operation at Suez in 1956.

Vertical envelopment did not make earlier classes of amphibious ships obsolete, because the helicopters still could not move heavy equipment and supplies over a beach, and so the amphibious modernisation programme which began in the late 1950s included new dock landing ships and then a new class of fast LSTs, the latter finally realising the postwar goal of a 20kt amphibious force. The latest US amphibious ships, the *Tarawa* class LHAs, in effect combine the helicopter carrier with a dock landing ship capable of launching smaller landing craft which can actually move tanks and other vehicles over a beach.

Amphibious operations grew considerably more important in US naval planning as the United States and the Soviet Union approached strategic parity, at least to the extent that massive nuclear warfare grew increasingly unlikely. By the early 1960s the navy's vision of future warfare, limited to proxy wars on the Eurasian periphery, was widely accepted, and the Kennedy Administration expected the most likely war to be a counter-insurgency. Both Vietnam and the short operation in the Dominican Republic fitted this pattern, and both came at a time of increasing investment in landing ships and craft. That investment seems, in retrospect, somewhat one-sided. By the mid-1960s the US amphibious force consisted not only of landing ships, but also of fire support vessels and mine warfare craft which could assure the security of the beach area. However, the fire support ships were all remnants of the large wartime building programmes, increasingly expensive to operate and with only a very limited remaining life. Mine craft were also proving expensive to operate, and their numbers were beginning to decline. To some limited extent the development of helicopter mine countermeasures appeared to solve the mine problem at the beach, but attempts to develop a new specialist fire support ship (LFS) failed, and a later decision to mount a single lightweight 8in gun aboard the new destroyers also proved abortive.

'LOW END' AND 'HIGH END' CAPABILITIES

These themes, carrier attack, ASW, the strategic submarine, and amphibious lift, largely describe the US Navy as it emerged from the postwar developmental period, and as it existed in 1982. By 1960 the new technology and the new ideas of the immediate postwar period had largely been assimilated and, in some cases, surpassed. Nuclear power was a reality for both submarines and surface ships, and long-range sonar was well along. Fixed systems and long-range aircraft had transformed the ASW problem and indeed had made it possible for the United States to wage an ASW war without the long period of mobilisation required in the past, given the balance between forces maintained in peacetime and wartime requirements. The new defensive missiles, although not yet satisfactory, did exist in considerable numbers, and, at least for the present, the fleet had a large number of missile escorts either in service or on order. The US Navy of 1960 was powerful and could look forward to increasing power and capability.

There was one major problem, and this problem has dominated US naval development since 1960: bloc obsolescence. By 1960 the fleet was ageing very rapidly: the great majority of its hulls were still of wartime construction, at least fifteen years old. The vast force of more than two hundred destroyers was already so badly worn that emergency measures were required, and many were rebuilt under the FRAM (Fleet Rehabilitation and Modernisation) programme which also applied to submarines and ASW carriers, and to some amphibious ships. It was intended to prolong service life by up to eight years, and thus to forestall the immense replacement programme which would ultimately be required. Even as large-scale replacement was being planned, however, the United States entered the Vietnam War. New naval construction was curtailed as limited funds went to operation and ammunition, while the war itself wore out the large remaining force of wartime destroyers and cruisers. The replacement programme planned in the late 1960s was ultimately reduced to thirty *Spruance* class destroyers, and a second and more austere project resulted in the *Perry* class frigates. In 1981 President Reagan's Secretary of the Navy, John Lehmann, called for a '600 ship' navy, a figure which seemed practicable since Congress would probably provide an average of twenty ships a year (and ship lifetime would average 30 years). Even this ambitious programme could not bring the fleet back to the numerical levels of pre-Vietnam days, when as many as nine hundred ships were active, including (1964) sixteen cruisers, twenty-five frigates (now rated as cruisers and missile destroyers), twenty-one missile destroyers, 179 all-gun destroyers and forty ocean escorts.

One useful description of the procurement problems of the postwar navy is to describe the fleet, in Admiral Zumwalt's terms, as a 'high-low' mix – a mixture of more and less capable ships. During the Second World War the 'high' end of the mix consisted of the new ships of the fast carrier task forces, the ships designed in about 1940 and built in such vast numbers in wartime. The 'low' end included both the ASW forces (destroyer escorts and old destroyers) and the amphibious and amphibious-support forces, including elderly battleships for shore bombardment. If the *Essex* class carriers represented the 'high' end, the escort carriers were the 'low' end. One important consequence of the strategic and technological shifts of 1945 was that the 'high' end of the wartime mix was swiftly shunted downwards to become the low-capability end of the postwar mix; the 'low' end of the wartime mix had so little capability that it was soon either discarded or laid up. Only the amphibious force maintained its place in the previous spectrum.

In effect the postwar programmes were devoted almost entirely to the 'high' end of the naval spectrum; quite naturally, they could not replace the vast number of 'low-end' ships remaining after the end of the war. However, it has proven extremely difficult for the navy deliberately to design and build austere 'low-end' ships, given the 'high-end' character of almost the entire postwar programme. Moreover, the 'high-end' origins of the ships remaining in service postwar served to conceal from most observers their effective 'low-end' status. Thus when Admiral Zumwalt tried to build a new 'low-end' in the *Perry* class, he was strongly opposed. *Perry* class frigates were bought in some considerable numbers, but criticism of the austerity measures inherent in their design has increased their cost.

The disappearance of the 'low end' had particularly severe consequences for the amphibious forces, which had long counted on 'low-end' ships such as surviving all-gun cruisers for fire support during opposed landings. The postwar navy was above all a missile navy, designing some ships entirely without guns even as early as 1956. It certainly had no place for a new all-gun cruiser, especially as there was no important naval surface opponent in sight, at least until the 1970s. Moreover, there was considerable ambiguity concerning the amphibious forces and their need of gunfire support. Thus the LFS, first proposed about 1965 as a replacement for the rapidly ageing cruisers, was never really able to compete with the carrier escorts and the new ASW ships; all that the marines could secure was a promise (since broken) that some of the new destroyers would mount a new lightweight 8in gun specifically for shore bombardment. Even this hope must have seemed dim, given the paucity of highly capable ASW escorts and the hazards of shore bombardment.

Thus the issue of the mass replacement of the 'low-end' Second World War fleet was resolved: there will be no mass replacements, unless the United States makes a sudden decision in favour of massive increases in defence spending. To be sure, there are always schemes for the emergency use of commercial shipping (eg 'Arapaho' for placing V/STOL fighters aboard large container ships), but there appear to be no serious plans for the expansion and simplification of any current building programmes: the *Perry* is not a mobilisation prototype like the *Dealey* of a quarter of a century earlier. On the other hand, the navy does enjoy the support of an increasing variety of shore-based systems such as SOSUS, which may take the lack of numbers of small warships far more tolerable, particularly in an extended war of attrition at sea. Moreover, it can be argued that a modern war would quite likely end before the US production machine could be mobilised to produce a flood of escorts. In that case the navy will have been proven correct in emphasising high unit capability.

The Reagan Administration navy was generally a 'high-end' fleet, but its new surface combatants were expected to operate alongside large numbers of existing modern 'low-end' frigates. The question in the late 1980s was whether the latter should be replaced by new frigates as they wore out. The United States was supporting a NATO frigate programme, which included a new frigate AAW weapon system, NAAWS. About 1987 a pair of studies, SOCS and SCIFRS, was conducted. SOCS sketched the characteristics of the surface combatant of the future; SCIFRS sketched the fleet of the future. Both were part of a larger programme, the 'Revolution at Sea', that envisaged Aegis and Tomahawk as the beginning of a new type of surface warship, much as the gunnery revolution at the turn of the century had led to the Dreadnought.

SCIFRS concluded that there was no point in any follow-on frigate. Instead, at mid-life surface combatants would be reassigned to the 'protection of shipping' (ie frigate) role. The low end of the future US fleet would be the older ships built initially as the high end. This approach had several advantages. It promised much more standardisation and much longer production runs; the object was to keep surface combatants continuously in production. Their designs would evolve over time. The alternative, sudden spurts of construction and (much later) replacement, was basically uneconomical and also sacrificed corporate memory of any design lessons learned with any one class.

The SCIFRS idea could also be justified historically. The United States did not build 'low end' ocean surface combatants until the Second World War. In 1941, the escort force consisted mainly of modified First World War-built destroyers. Much the same could be said of the bulk of the fast escorts of the 1960s, the FRAM destroyers. They were no longer suited to battle group missions, so they were modernised as ASW (and, it turned out, gunfire support) ships. The new frigates of the 1960s and 1970s did not replace the FRAMs on a one-for-one basis; the fleet merely shrank. It was surely better to go for a steady replacement rate. This philosophy of steady production was applied to the *Arleigh Burke* class. It remains to be seen whether pressures for lower costs ultimately lead the navy back into frigate production. It may be significant that the Royal Navy, relatively close to the US Navy in philosophy, has steadily increased the size of its own frigates in recent years, in effect admitting that small 'low end' ships are not worthwhile.

AIRCRAFT CARRIER DESIGN

Since early in the Second World War, the carrier has been the principal instrument of US sea power, both for power projection against shore targets and for sea control (ASW and, during the war, anti-ship warfare). In 1945, with no prospective enemy fleet in sight, interest within the navy shifted to carrier-based nuclear strikes, and a very large ship, the *United States*, was designed specifically to accommodate a new generation of long-range carrier bombers. At the same time existing ships were to be rebuilt to operate the new jet-propelled tactical aircraft: the *Essex* class carrier *Oriskany*, construction of which was suspended immediately after the war, became the prototype for a series of SCB-27A and SCB-27C carrier reconstructions, and the larger *Midway*s were later rebuilt as well. Meanwhile work proceeded on carrier ASW operations, using war-built escort and light fleet carriers at first. However, an unconverted *Essex* was tested in the ASW role in 1952, and as rebuilt *Essex*s joined the fleet their unconverted sisters were redesignated as ASW support carriers, or CVSs. Later the converted *Essexs* succeeded them, as new large attack carriers replaced them in the power projection role.

As for the *United States*, she fell victim to inter-service rivalry, because the Secretary of Defense, Louis Johnson, decided in favour of land-based (air force) strategic bombers. With the outbreak of the Korean War, however, it became more evident that carriers were valuable, and a new large-carrier design, which became the *Forrestal*, was begun in the autumn of 1950. She benefited greatly from two British inventions, the angled deck and the steam catapult, and she was the prototype for eight conventionally-powered attack carriers. For a time, it appeared that one would be authorised each year, this pace continuing from FY52 to FY58 (*Enterprise*), but the FY58 ship was so expensive that no carrier was approved in FY59 or in FY60; one problem was the issue of whether the next carrier should or should not be nuclear-powered. Conventional carriers were authorised in the FY61 and FY63 budgets, in the latter case very much against the professional advice of the Chief of Naval Operations. In addition, rising carrier cost and the increased firepower of Soviet anti-carrier forces both made carrier construction far more controversial in the late 1960s and the 1970s than in the 1950s.

Thus no carrier was authorised between FY63 and FY67, when the first of a new class of nuclear carriers, *Nimitz*, was approved. At the time a class of three was to have been built on a single-contract basis; authorisations stretched out to FY70 for the second unit and FY73 for the third. After that the Carter Administration proclaimed the end of large carrier construction, seeking instead to build a 60,000t CVV or CTOL-V/STOL carrier. When Congress disagreed, President Carter went so far as to veto a Defense Bill to avoid the authorisation of an additional nuclear carrier. However, that ship was finally voted in the FY80 budget as CVN 71, and in 1981 Congress approved long-lead items for CVN 72, and CVN 73 planned by the Reagan Administration. Buying two carriers together entailed a considerable saving on total cost, and two carriers were again authorised under the FY88 programme (CVN 74 and 75).

The advent of V/STOL aircraft has made much smaller carriers attractive. As CNO, Admiral Zumwalt advocated the construction of a very austere V/STOL-helicopter ASW escort carrier, which he called the Sea Control Ship (SCS), and which was never approved by Congress. One of its political problems was that the carrier navy saw it as an attempt to abandon the large-deck carrier with its high capability conventional take-off air wing. Ultimately Spain purchased plans for the SCS and built one, as *Principe de Asturias*. *Guam* served as the trials ship for the SCS concept, and Congressional interest in small carriers survived into the 1980s. To some extent the decision to equip the LHD to operate an alternative sea control air wing was a response to this pressure. However, the Marines were also more than glad to have organic shipboard combat aircraft. *Nassau* operated a squadron of Marine Harriers to augment a carrier group in the Mediterranean for a time in 1981. A decade later she used similar aircraft to attack ground targets in Kuwait, demonstrating the value of locating Marine strike fighters aboard the largest Marine amphibious ship.

With the drawdown of the carrier force, and with increasing demands on a smaller US fleet for presence abroad, in the early 1990s the navy became interested in using the large amphibious carriers for presence missions, almost as austere carriers. They lack such vital carrier capabilities as airborne early warning, however. Unfortunately, in the inter-service warfare incident on drastic cuts in the US defence budget, the air force chose to propose that the approved force of twelve deployable carriers include not only large decks but also the LHAs and LHDs. As of early 1995 this assault had been beaten back.

Given the long lead time inherent in building a large carrier, and the need of sea-based air power in a sudden crisis, several plans for emergency carrier construction or operation have been proposed. One is 'Arapaho', in which a container ship is provided with a flat deck and containerised maintenance facilities for V/STOL fighters. The Reagan Administration also proposed the reactivation of two *Essex* class carriers, *Oriskany* and *Bon Homme Richard*, although Congress later rejected the plan. One problem was that these ships could not operate anything like the full range of modern naval aircraft; for example, the air group proposed for *Oriskany* consisted entirely of A-4M Skyhawks. These ships were later stricken.

FLEET ESCORT CLASSES

Since the early 1950s the US Navy has used type designations such as 'cruiser' and 'destroyer', but in practice the distinction between fleet, *ie* fast task force, escorts and convoy or ASW or 'ocean' escorts has been far more important: there are no longer traditional cruisers really suitable for independent operations, and the failure of the Strike Cruiser (CSGN) project suggests that such independent ships are not likely to be built in the near future either. The *Mitscher* class 'frigates' (or large destroyers) were the first specialised fast task force escorts, much of the addition in size compared with Second World War destroyers going into higher speed and better sea-keeping. In 1954 the Schindler Committee on the Long Range Shipbuilding and Conversion Program went so far as to recommend the elimination of the traditional destroyers in favour of a mixture of fast task force escorts (frigates), ASW ships (for ASW task force operations: DDC or 'corvettes' converted from existing destroyers), and extremely austere 'ocean escorts' which would fill in the numbers in convoy screens and perhaps in the ASW Task Force. Although new classes of cruisers were proposed from time to time, the only ship of that category to be built after 1945 was the nuclear *Long Beach*, designed originally as a nuclear missile frigate.

At the upper end of the size spectrum, eleven Second World War cruisers were converted to missile ships specifically to screen fast carrier forces. Their great size was dictated, first, by the availability of hulls of particular sizes exceeding the minimum required for conversion and, second, by the considerable size requirements of missiles such as Talos. By the mid-1960s however, it was clear that much smaller (large destroyer) hulls could accommodate much the same missile firepower, although not the same flag facilities or the same magazine stowage, at a much lower cost in manpower. Cruiser conversion ceased, and ships were withdrawn from service as their much smaller replacements were completed. From a functional point of view, then, there is little to choose between a *Cleveland* class Terrier cruiser such as the *Topeka* and the more destroyer-like *Leahy* or *Belknap*. This circumstance was explicitly recognised when most of the former frigates (DL) were reclassified as cruisers (CG) in 1975.

Even the distinction between those ships still classified as destroyers and the cruisers is at best a tenuous one. For example, in the mid-1960s there were projects for a new missile destroyer (DDG) to replace the *Charles F Adams* class; necessary additions for such systems as NTDS drove displacement up to almost 8000t. Only nuclear-powered versions of this ship were built, as the *California* class – and they are now designated as cruisers. The *Spruance* class are clearly cruiser-size specialist ASW ships, and their missile version (*Kidd* class) is difficult to distinguish from the *Virginia* class nuclear cruiser, at least in AAW capability. The *Spruance*s are unique among fleet escorts built since the *Mitscher*s in that they have no area AAW weapon. They reflect the growth of the Soviet fast submarine force to the point at which it achieved a substantial capability even against fast-moving carrier forces. Given the relatively short range of ASW sensors, it is necessary almost to surround a moving force with ASW ships, whereas relatively fewer AAW launchers are needed. It was, then, decided in 1968 to build a mix of AAW-ASW fleet escorts and pure ASW escorts, the latter becoming the *Spruance*s.

Thus the requirements for a ship to be considered as a fleet escort amount to high speed (at least 30kts in rough weather), long range, and good sea-keeping; the weapons and sensors do not always differ greatly from those of other categories. Even the speed limit may be relaxed in some cases to achieve affordable ships in sufficient numbers. Moreover, given the limited numbers of satisfactory escorts currently available, ocean escorts which marginally meet the speed requirement (such as *Garcia* and *Perry* class frigates) sometimes operate with fast carrier formations.

To an increasing extent fast attack submarines, particularly those of the *Los Angeles* class, operate as distant escorts for naval formations, in the 'direct support' role. They are excellent sonar platforms, and they can keep up with the force, generally operating some considerable distance ahead in the area from which an enemy submarine would have to attack. For example, in some formulations the typical carrier battle group with a nuclear carrier includes two nuclear cruisers and a nuclear submarine.

The gun-armed cruisers left over from the Second World War programme clearly do not quite fit this account. They functioned as AAW/anti-surface escorts within carrier groups postwar, and in Korea and Vietnam they had an important shore bombardment role. Some were also refitted as flagships, losing some of their secondary guns in favour of electronics and accommodation. As a category, however, they died out as the navy shifted more and more from guns to missiles, and as the threat to US naval forces came more and more to be an airborne or submarine-borne one: even the 1980s Soviet surface fleet was likely to attack by missile rather than by surface gun. Thus, by 1982 the US cruiser force had been reduced to two large 8in ships which had been in reserve since 1960–61.

The missile cruisers were as important as air control platforms as they were as platforms for defensive weapons. In the Vietnam War, an NTDS ship operated on station in the Tonkin Gulf, controlling US aircraft operating over North Vietnam. The radar buoy in the Gulf defined a Positive Identification Radar Advisory Zone (PIRAZ), and the ship on PIRAZ duty was often responsible for successful interceptions by air force fighters. For example, one enlisted air controller aboard the cruiser *Chicago* was decorated for arranging the destruction of 12 MiGs. The track capacity of the NTDS system made it possible for US aircraft to operate successfully in an environment of neutral airliners as well as friendly and North Vietnamese fighters. Conventional radar could not have succeeded, since the neutral airliners would not have responded to IFF challenges.

The missiles deserve some mention. The early Terrier had a range of about 10nm, compared to about 50nm for Talos, but in the late 1950s Terrier and Talos ranges were both doubled, and the last production Terriers had a new engine which again doubled their range, to about 40nm. The much smaller Tartar underwent a similar evolution, from 7.5nm to 17.5nm in the Improved Tartar and then to about 25nm in TRIP (Tartar Reliability Improvements Program) and in the Standard Missile (SM-1). These Tartar and Terrier ranges are line-of-sight figures; in the new Aegis system command guidance permits the use of more energy-efficient trajectories, for range gains of, respectively, about 70 and over 100 per cent, for maximum ranges of about 42nm and 80nm: a Tartar-sized missile with Aegis guidance has Terrier range. Moreover, missile effectiveness depends upon fire control system capability and reliability: although some SPG-51D (Tartar-D) ships have fewer fire control channels than have earlier ships, these channels are available for a much greater proportion of the time.

The major recent development in missile systems is the integration of all of a ship's sensors to improve the probability of detection of an incoming weapon. This is possible because the new generation of naval radars, such as SPS-49 and updated versions of SPS-40, -48 and -52, automatically detect targets, which data they can automatically feed into a central computer for comparison with other radar data, and for correlation into track files. Given this data, the central computer can evaluate the degree of threat presented by each radar target, and can decide which target to engage. Systems of this type were developed from the late 1960s on in response to the threat of Soviet anti-ship missiles, under the New Threat Upgrade (NTU) programme; the central combat system integration computers are SYS-1 and -2. New Threat Upgrade soon incorporated the new SM-2 missile being designed for Aegis ships, with its command-programmable autopilot. Like an Aegis ship, an NTU ship can, in effect, time-share its tracker-illuminators, keeping many missiles airborne simultaneously. Unlike an Aegis ship, it cannot use main search radars to track targets so precisely that the illuminators can be slaved. Instead, each must acquire its target before the missile can home, so each requires more time per target than the slaved illuminator in an Aegis ship.

NTU was applied to both the single-stage SM-2(MR) and to the two-stage SM-2(ER) missile. During the Cold War, the primary threat was massive Soviet missile attacks mounted from long range. Increasing overall fleet firepower was vital. Short-range effectiveness was less important. After the Cold War, the main threat was small numbers of weapons fired from relatively short range. Reaction time became more important. Unfortunately, the firing rate of the two-stage missile was limited because the boosters had to be finned manually. Recently-converted NTU ships (*Leahy* and *Belknap* class cruisers) were, therefore, retired and discarded (they were also disposed of to eliminate manpower-intensive high-pressure steam plants).

OCEAN ESCORT TYPES

After the Second World War the former destroyer escort category was renamed 'ocean escorts'; functionally, this group also includes the long series of specialised ASW destroyer conversions, including the FRAM conversions of *Summer* and *Gearing* class general-purpose destroyers. Originally it was expected that a new class of fast ASW ships would be complemented by a replacement for the coastal escorts (PC and PCE) of the Second World War; indeed, the *Dealey*s were conceived at first as a PC replacement, the DE replacement being capable of 30kts with greater endurance. The problem was that the new fast submarines (such as the German Type XXI, the Soviet 'Whiskey' and the US GUPPY and *Tang*) could outrun destroyer escorts and lesser craft in bad weather, for at least long enough to escape from sonar range. Thus a 5–6kt speed improvement over war-built types was desired (27kts for the coastal escort, 30 for the ocean-going type). With an enormous backlog of existing ASW craft, however, Congress was clearly unwilling to vote funds for new construction, particularly in the numbers needed by the Navy.

The first step, then, was a conversion programme: *Fletcher* class destroyers were to be converted to a new destroyer escorts (DDE). Indeed, it was hoped that all surviving *Fletcher*s would be so modified, and most of them retained in reserve against an emergency. In practice, the reconstruction programme was halted after a short time in favour of increased expenditure on research as well as modest ASW improvement to the active general purpose destroyer force.

At the same time, *Gearing* class destroyers were modified or completed to ASW designs, essentially for experimental purposes, as part of a programme of carrier-destroyer hunter-killer groups. These latter were originally intended primarily to support convoys, but later evolved into specialised ASW task forces which prosecuted SOSUS contacts. Task force operation with converted *Essex* class carriers imposed a minimum speed limit of about 27kts, in itself a major influence on ASW escort development. The FRAM destroyers of the 1960s were in effect a mass-produced version of the DDE of a decade earlier, in that they represented an ASW modification of a general-purpose destroyer, all AAW capability (in effect) having been sacrificed. Although they were useful shore bombardment assets, and although they were considerably faster than the frigates which replaced them, they were closer to frigates than to fleet escorts, and so are listed as ocean escorts.

Until the 1950s, the problem of the ocean escorts had been that they were not sufficiently capable of operating with carrier strike forces in peacetime, and indeed were rather expensive to maintain, in terms of the capability they possessed. However, very large numbers would be required upon mobilisation. The only possible solutions were specialised mass production design (in the case of the *Dealey* and *Claud Jones* classes) and prototypes for the mass conversion of the remaining mothballed Second World War ships (as in the *Fletcher* DDE and in some ASW conversions of wartime destroyer escorts). In fact the solution chosen was to abandon most forms of convoy operation in favour of massive use of aircraft cued by long-range sensors such as SOSUS, with surface escorts reserved only for the most essential missions. The surface escorts could, in turn, be made far more sophisticated, to counter more and more sophisticated submarines.

The evolution of the ocean escorts is symbolised by the evolution of their weapons. Until the mid-1950s average sonar range was about 1500yds; typical weapons were Hedgehog (about 250–300yds) and Weapon Alfa (800yds), plus homing torpedoes. With the advent of the SOS-4 sonar, ranges of 5000–10,000yds became practicable, and long-range ASW weapons were developed: the RAT rocket-thrown torpedo (5000yds) and the DASH drone helicopter (up to about 15,000yds). FRAM II conversions generally provided a variable-length sonar in addition to SQS-4, for reliable operation even below the thermal layer; the primary stand-off weapons were DASH and a long-range ASW torpedo, Mk 37. FRAM I ships, as well as many new ships, had the larger SQS-23 sonar. With a reliable range generally given as 10,000yds; the associated weapons were ASROC (rocket-thrown torpedo or nuclear depth-charge) and the DASH drone. By the late 1950s there was another new sonar, SQS-26 (in transistorised form, SQS-53) with a reliable direct-path range probably exceeding 20,000yds, and theoretically capable of reaching the first convergence zone, at about 70,000yds. There was no associated weapon at first, although a longer-range ASROC (about 20,000yds) was proposed; but the solution to convergence-zone attack turned out to be manned LAMPS I helicopter. Finally, in the towed arrays, a beyond first convergence zone capability (passive, to be sure) is claimed. Prosecution requires a long-range helicopter, LAMPS III. One might suggest that confidence in the towed array is so great that its primary carrier, the *Perry* class frigate, has only a very short-range active sonar. With each advance in weaponry and sonar the minimum size of the platform to carry these devices has grown. For example, it was possible to fit SQS-4 into a Second World War destroyer escort, but not SQS-23. SQS-26 required very nearly a cruiser-size hull for high speed, the *Bronstein* class frigate representing a bare minimum.

Ocean escorts had a variety of designations. In the late 1940s a distinction was drawn between submarine-killers (DDK) and escorts (DDE) with somewhat less developed ASW capabilities; all later became DDE. In 1975, in line with most other navies, the United States redesignated its ocean escorts (DE) as frigates (FF), the missile-armed units becoming FFG (formerly DEG). The *Perry* class were briefly designated patrol frigates (PF), presumably as a way of distinguishing them from the unpopular but similar sized *Knox* class.

ATTACK SUBMARINE CLASSES

At the end of the Second World War the United States had a very large but somewhat confused submarine force, since its next potential adversary presented few significant ship targets. During the next decade two divergent lines of development were pursued: submarine ASW and submarine strategic warfare. At first separate series of attack submarines loosely modelled on the new German Type XXI (*Tang* and her successors) and specialised ASW submarines (*Barracuda* class and fleet submarine conversions) were built; the latter emphasised long-range passive sonars over speed, as a means of detecting enemy high-performance submarines snorkelling en route to operational areas. To some extent, too, the construction of high-performance anti-ship submarines was justified as a means of maintaining the kind of technology needed to understand an adversary with a large modern submarine force; the first GUPPY conversions, for example, were pressed ahead in large part to provide realistic targets for the surface ASW force. They were also considered a mobilisation reserve against the need for large numbers of specialised ASW submarines.

The two major new items of submarine technology of the time were nuclear power and the new *Albacore* hull form, which promised much higher speed and manoeuvrability at fixed power levels. A third line of development, towards closed-cycle conventional power plants for high performance, was suspended, as was the development of advanced steam plants for radar-picket submarines. The latter did, however, lead to the pressure-fired boilers of the *Garcia* class frigates. It appears that nuclear power was initially pressed forward because of its obvious advantages for attack submarine operation, without much reference to US tactical requirements: the first nuclear submarines were anti-ship types. The *Tullibee* was an attempt to combine nuclear power with a long-range sonar; she had the relatively low speed of her conventional forebears, the contemporary *Skipjack* class adopting the high-speed *Albacore* hull form.

By about 1957, it had become clear that Soviet submarines would be the primary submarine targets. The next submarine class combined deeper diving depth and a very long-range sonar with a new long-range weapon, SUBROC, a ballistic rocket with a nuclear warhead; it also had the new high-speed hull form. It would appear that the BQQ-2 active-passive sonar of these *Thresher* class submarines could reach out to at least the first convergence zone, since SUBROC range is usually quoted as about 30–40nm. Since the new ships had the same power plant as the much smaller *Skipjack*s, they registered some loss of speed. With the appearance of the *Thresher*s with their great ASW capability, all submarines were reassigned the ASW mission, and the specialised ASW submarine designation (SSK) was cancelled. Submarine tactics continued along the lines familiar more than two decades later, with a technical emphasis on quietness and passive operation; most submarines would form ambush barriers across choke points such as the Greenland–Iceland–UK Gap.

Major attention was devoted to submarine noise reduction. The *Thresher*s were already very quiet, but during the 1960s two submarines were built to explore alternative means of silencing two major sources of noise: turbine reduction gearing (*Glenard P Lipscomb* had turbo-electric propulsion) and reactor pumps (*Narwhal* had a natural circulation reactor). Because these submarines were not completed until, respectively, 1969 and 1974, their operation could not contribute materially to the design of the next submarine class, the *Los Angeles*; essentially the *Los Angeles* was an attempt to recapture the speed lost in the transition from *Skipjack* to *Thresher*, at the cost of twice the installed power and a considerable increase in size. Almost certainly there have been advances in silencing as well, but nothing approaching what a major change in power plant (as in the experimental submarines) would accomplish. The chief virtue of high speed is that the submarine can operate in direct support (D/S) of fast carrier battle groups, patrolling ahead of them to destroy submarines in their path, or to vector carrier-based aircraft to attack such enemy submarines.

In the early 1970s an abortive successor to the *Los Angeles* class was designed. At that time a submarine-launched tactical missile (STAM) was in the early stages of development; it would have been a combination SUBROC and anti-ship missile with a torpedo warhead. Given the limited number of torpedo reloads carried by even a large attack submarine, it was to have had additional vertical launch tubes for its missiles. The length of the launch tube compartment contributed to the size of the submarine; in addition, it had further silencing and a higher maximum speed, all of which made for a submerged displacement of about 14,000t. As the *Los Angeles* class, of only about 6900t, already strained navy budgets, Admiral Zumwalt had the new type cancelled. Indeed, during the Carter Administration a new austere attack submarine, tentatively described as the F/A (Fleet Attack, but derisively named 'Fat Albert' by some) was proposed. It was never completely defined, and *Los Angeles* class construction continued. However, the issue of limited weapon capacity was addressed: late units have twelve vertical launch tubes for Tomahawk missiles, mounted between the pressure hull and the bow spherical sonar array.

By the early 1980s, there was sufficient money to start a new submarine class which would address the defects of the *Los Angeles*. The new design incorporates a new reactor; it is the *Seawolf*. It proved extremely expensive (particularly after Trident construction ended, since then each attack submarine had to absorb more industrial overhead).

The programme to develop a new less expensive attack submarine has had several names: Centurion, NAS (new attack submarine) and NSSN (new SSN). As of 1994, it appeared that the first such submarine might be ordered under the FY98 programme, after the three *Seawolf*s. However, it is by no means clear that design economies will make such craft significantly cheaper than *Seawolf*, and it seems quite possible that *Seawolf* construction will continue well into the next century.

STRATEGIC SUBMARINE CLASSES

Strategic attack was a major potential submarine function as early as 1945, when the United States captured German plans for a submarine-towed submersible barge containing a V2 rocket. At that time there was already a programme for a US version of the German V-1, which the navy called Loon, and one was fired from the submarine *Cusk* on 18 February 1947. By this time a programme for air-breathing long-range weapons suited to submarine launch was well underway, resulting in the Regulus cruise missile, ultimately deployed aboard five specialised submarines (two conversions of fleet boats and three specially built submarines, one of them, *Halibut*, nuclear-powered). All operated in the deterrent role in the Pacific 1957–64, ultimately being replaced by the ballistic Polaris system; a supersonic follow-on cruise missile, Regulus II, was cancelled.

Thus most US strategic submarines have operated with ballistic missiles: Polaris, Poseidon, and the new Trident. Polaris itself carried out as a crash programme, *George Washington* class submarines being conversions of *Skipjack* class hulls already under construction, with 130ft added abaft the fin. They were followed by the improved *Ethan Allen* class (which carried Polaris A-2 from the first) and then by the *Lafayette* class; the entire Fleet Ballistic Missile Program of forty-one submarines was completed between 1959 and 1966. There followed a long hiatus, while development of new missiles (Poseidon) was not accompanied by any new submarine project. In 1967 there was a new strategic study (STRAT-X) advocating a new very long-range missile, initially envisaged as externally (and perhaps horizontally) carried, floating to the surface for firing. The new submarine would be very large but relatively inexpensive, capable of only very low speed and with a minimal crew. However, this ULMS (Undersea Long-range Missile System) concept gradually evolved into a Polaris-like submarine, with more missiles per submarine to obtain greater cost-effectiveness: the 24-tube *Ohio* is the result. Trident I missiles were retro-fitted to some existing Poseidon-capable submarines.

With the end of the Cold War, it was decided the *Ohio* production would end with the eighteenth unit; some early units may be modified for alternative roles. All earlier strategic submarines were discarded (two were converted to support SEAL operations).

RADAR-PICKETS

One of the major lessons of the Second World War was the immense value of early warning against air attack, whether against cities or against moving naval forces. Picket forces were improvised, for example to protect the fleet off Okinawa and to protect Britain from V-1 attack in 1944–45; to some extent they were supplemented postwar by specialised radar-picket (airborne early warning) aircraft, the ancestors of the current E-2C Hawkeye and AWACS. Three main categories of seaborne radar-pickets developed postwar: tactical (surface) pickets, strategic pickets and picket submarines. The tactical units were direct descendants of the twenty-four radar-picket destroyers of 1945. Twelve more *Gearing* class destroyers were converted under the FY52 programme, and all Fleet Escorts were provided with height-finding radars and CICs suited to air (fighter) control. The reversion of all but four of the converted *Gearing* class destroyers to ASW roles under FRAM can be read as an indication of the increasing reliability of carrier-based radar-picket aircraft from the early 1960s onwards.

The next wave of radar-pickets was converted to serve the US Continental Air Defense System as a seaward barrier. Conversions began under the FY49 programme, their priority so high that the first had two each of the scarce SPS-6 radars. In the interests of endurance, nearly all were diesel-powered; the incomplete steam-powered *Vandiver* and *Wagner* appear to have been completed as pickets as mobilisation prototypes. The destroyer escort conversions were both expensive and relatively uncomfortable for lengthy stays on the picket line; in the mid-1950s sixteen 'Liberty' Ship (merchant) hulls were converted, and provided with long-range radars and a CIC (YAGR1–16, FY55–58). There were even proposals for a follow-on armed picket force of radar/sonar/Talos or Terrier ships, perhaps based on 'Victory' hulls, and provisionally designated the PBG. However, with the fading of the bomber threat in the 1960s the DERs were transferred to surface patrol duties (largely Operation 'Market Time' blockade, off Vietnam) and the 'Libertys' decommissioned and disposed of (April–September 1965).

Finally, there were submarine pickets. They were originally proposed as a result of the carnage on the picket line at Okinawa: a submarine could not match the radar suit of a destroyer, but it could submerge to avoid attack. Two submarines were converted (albeit without special radars) before VJ-Day, with four more under conversion; in 1946 the first two specialised conversions, *Spinax* and *Requin*, were carried out, under Project 'Migraine'. Ultimately twelve fleet submarines were converted, and three specialised submarine radar pickets built. All the diesel submarines had the basic defect of low speed and thus could not really accompany the fast task force; but the nuclear picket, *Triton*, for many years the largest submarine in the would, could.

Interest in submarine picket operations declined sharply towards the end of the 1950s, the converted submarines losing their radars and their special status in 1959. The two specially built diesel submarines reverted to attack status in 1961 and underwent FRAM modernisation in 1964. The immense *Triton* was also reclassified. A project to use her as a National Emergency Command Post Afloat (NECPA), natural in view of her large CIC, was dropped.

AMPHIBIOUS SHIP TYPES

In about 1948 a goal of a 20kt amphibious force was set, although there was no substantial new amphibious construction for the time being, given the mass of existing tonnage and considerable scepticism as to the viability of amphibious forces in a nuclear environment. The success at Inchon in 1950 (and, indeed, the entire Korean experience) showed that amphibious operations remained important, and the amphibious force was gradually replaced during the 1950s and 1960s. Meanwhile the maze of wartime types was rationalised, and the short-range non-transportable craft (LCI, LCS, LSM) discarded. In addition, the marines added the 'vertical envelopment' concept and, with it, a series of totally new types and tactics.

Long-range amphibious operations during the Second World War required a combination of heavy-lift non-beaching ships and beaching craft to bring their equipment over the shore. The ships (except for dock landing ships, LSDs) could carry personnel-vehicle (LCVP) or small mechanized equipment (LCM) beaching craft; large tanks and vehicles required a larger beaching craft, the LCT, which postwar became the LSU and then the LCU (landing craft, utility). It could not be launched from a davit, and had therefore to be carried in a floodable well deck, as in an LSD. The non-beaching ships were attack (personnel) transports (APA, later LPA) and attack cargo ships (AKA, later LKA). There were in addition large beaching ships, LSTs, which could carry relatively less on their displacement than the AKAs, but which could carry very heavy equipment and could disgorge vehicles far more quickly, and which therefore remained valuable.

In the early 1950s the Marine Corps concluded that amphibious operations under the threat of atomic attack required the dispersal of the amphibious fleet, which was impossible unless some means were found of moving troops and their equipment far more rapidly than had been possible with slow beaching craft. Their solution was the helicopter: although a helicopter could not carry very heavy equipment, it could concentrate troops very rapidly. In ship terms, carriers could be converted to carry troops and their helicopters beginning with the escort carrier *Thetis Bay*, which was followed by a series of early *Essex* class carriers (formerly CVSs) and then by the specialised *Iwo Jima* class. Their major defect was that they could not operate in bad weather, since they carried no landing craft. The solution was a combination of LPH and LSD with extra cargo space, the LHA, or amphibious assault ship, with a substantial well deck in addition to a large flight deck and hangar.

Perhaps the major technical problem faced by the 20kt programme was the LST, traditionally very slow because of its need to provide bow doors. The first attempt to design a fast LST called for a bow propeller and a stern ramp, but it was rejected in favour of a compromise, a 15–17.5kt conventional LST, twenty-two of which were built in the 1950s. However, twenty new 20kt *Newport* class LSTs were built under the FY65–67 programmes, with seven more planned for FY71 but deferred. They achieve a good hull form by using a ramp extended *over* their bows.

The other major classes developed less spectacularly. Shortly after the end of the war the Maritime Commission developed a new fast cargo ship, the 'Mariner', which, like its prewar predecessors, was well adapted to naval conversion: two became LPAs, and another became an LKA. Although several classes were developed to carry personnel, the cargo function was unique, and five more attack cargo ships (*Charleston* class) were built under the FY65 and FY66 programmes. Postwar LSDs were generally adaptations of the wartime LSD concept to higher speeds, with eight built in the 1950s (*Thomaston* class) and five more late in the 1960s (*Anchorage* class); in the 1980s a new LSD 41 class was begun. Between the two LSD classes there was an attempt to combine LSD and attack transport roles in the LPD, a total of fifteen being built. LPD 16 (FY66 programme) was deferred in favour of the LHA programme and then cancelled February 1969. However, the LPD concept remained valid. The new amphibious ship, initially called L(X), will be LPD 17. The mix of types (LHD, LSD, LPD) is determined by the mix of landing craft and helicopter 'spots' and troop, equipment, and supplies needed to support an operation.

Another major line of development was higher speed for the beaching craft. During the 1950s hydrofoil LCVPs were tested, but they were never adopted. There were also tests of very large amphibious tractors, similar in concept to the LVT; again, they were not adopted.

By the early 1980s the Marines had evolved a new 'over the horizon' technique of amphibious assault, based on the high speed of their new LCAC (landing craft air cushion). In theory, LCACs could be launched as much as 50 miles from a beach, and they could use beaches impassable to earlier beaching craft. These considerations promised to return an element of surprise which, until then, had been lost. The new generation of well-deck ships (LHDs and LSD 41 and 49 classes) all were designed to accommodate LCACs.

However, there were still problems. The LCAC is entirely unarmoured. It cannot assault enemy troops directly. The Marines use amphibious tractors (LVTs, later reclassified AAVs) for that purpose. Unfortunately the AAV is very slow. It must therefore be launched close to a beach, for example either by an LCU or an LST. Either way the speed and surprise inherent in an LCAC assault will be lost. As of 1994, the Marines were considering several alternative designs for AAVs capable of running through water at high speed, then converting near the beach to run overland. The Marines are most reluctant to abandon the one vehicle they can use to move well inland from a beach in the face of enemy troops, so they consider the new Advanced AAV a key programme, if they are to continue to be capable of assaulting defended beaches. The LCAC is also quite expensive, and it is not clear that enough have been or can be bought. The original quantities were predicated on requirements to bring cargo to a beach, and did not take combat losses into account. Clearly the hope was that, given a very wide variety of available beaches, the Marines could find some which were undefended. This would be reasonable on a long enemy coast, but not in a place such as Kuwait, where the beach was relatively short.

PATROL CRAFT

Small fighting craft have had a very mixed career in the US Navy, generally being improvised in wartime and discarded as soon as the immediate need of them ended. Very generally, small craft cost more to operate per ton than do larger ones, but they are much easier to build in numbers. Often they deteriorate rapidly because of their relatively flimsy construction. Moreover, in peacetime the navy generally is far more concerned with its mid-ocean mission than with operations near its own or an enemy coast. Thus the large sub-chaser fleet of the First World War was almost completely discarded within a few years of the end of the war. After 1945 all but four of the hundreds of PT-boats were scrapped, although four postwar prototypes were built. As for the inshore ASW craft, most of them were also soon discarded, and repeated proposals for new PCE, PC, PCS, and SC designs produced no craft. On the other hand, the United States did build a variety of smaller combatants for its allies, including a series of motor gunboats (PGM).

Reasons for limited interest in coastal craft included very limited naval spending, most of it consumed by the rising cost of the new naval technology of the 1950s. There was no enemy fleet for a new generation of PTs to attack, and it appeared that small craft were grossly inadequate for any type of ASW, even off US ports. Thus interest in small combatants did not really begin to rise until the Kennedy Administration, which combined official interest in them with a pair of operational considerations: Cuba and southeast Asia. The threat of Soviet-built small attack craft based in Cuba had a great effect on the US fleet of the time, leading to studies of the anti-ship efficacy of anti-aircraft missiles – and to the construction of the *Asheville* class motor gunboats. Once the United States found itself engaged in Vietnam, the coastal and river operations led to the construction of a great variety of small craft, some of which have been retained – albeit on an experimental basis.

The only other major postwar small combatants were the hydrofoils, built at first as experiments in a new technology. Ultimately these experiments led to a proposal for a NATO-wide class of missile-armed hydrofoils (PHM), whose primary purpose would have been to deny access to the narrow straits around Europe. Costs rose very steeply, however, and only the American programme of six craft survived at all. As in the 1950s, questions were soon raised as to their function in a fleet the main threats to which were Soviet air and submarine forces. At one point the hydrofoil advocates went so far as to propose lightweight fleet air defence systems (such as Sea Phoenix) for hydrofoil operation, but ultimately they reverted to single-system operation (Harpoon). During the mid-1980s, however, there was increased US interest in possible Caribbean operations using hydrofoils.

One other category deserves mention here. From the Vietnam War on, special craft were built or adapted to support SEAL (naval special forces) operations. An air-cushion SEAL support craft, unofficially called Sea Viking, was designed and begun, but it suffered from overweight and its contractor, Rohr, went into liquidation before it could be completed. Eventually a series of conventional craft, modelled on the British-built 'Province' class missile boats, was bought. They are too large and too slow to support truly covert operations. To some extent the new air-transportable PB Mk V fills this requirement.

MINE WARFARE VESSELS

The postwar period saw, first, the decline of a massive war-built mine force, then its resuscitation under Cold War conditions and under the impetus of mine warfare in Korea, after which a second slow decline set in. One might well speculate that the current low estate of mine warfare in the US Navy is intimately connected with the limited prestige of offensive mining in a navy which has been quite successful at just that tactic in the past. One reason for the latter, some wags would say, is that it is not possible to win command of a mine; with the end of the specialised surface minelayers (which have been replaced by submarines and aircraft, neither of them dedicated to the mine mission) there are no major mine warfare commands afloat. The high cost of an active minesweeper force was also a major factor: the wooden hulls of non-magnetic sweepers were at best difficult to maintain, and reportedly the non-magnetic engines were a constant source of trouble. A Navy hard-pressed to maintain its most vital forces (for ocean ASW and for carrier attack) could not operate a large force of labour-intensive small craft. In some ways, then, the same forces which doomed the wartime PT force also doomed the mine force. In 1982, for example, there was a total of only twenty-five ocean sweepers (MSO), of which twenty-two were operated by the Naval Reserves, and seven minesweeping boats. By way of comparison, at the end of the post-1945 rundown, there were ten destroyer minelayers (four active), twenty destroyer minesweepers (twelve active), one large minelayer (in reserve), 123 ocean sweepers (ten active), and thirty-seven coastal sweepers (twenty-seven active, sixteen training reservists). In 1958, after a massive new construction programme, the former destroyers were gone and the minelayer remained in reserve; in addition there were sixty-five new ocean and twenty-two new coastal sweepers, all active, and 116 old steel ocean and thirty-seven old coastal sweepers (one ocean type active). There were also thirty-three minehunters (five active), fifty-two minesweeping boats, and five mine support ships (all inactive).

For the US Navy, the mine threat breaks down into a relatively weak threat against US ports (mainly from submarine minelayers), a potential deep-water threat against stategic submarines, and a very considerable threat to amphibious forces assaulting defended areas. The threat at home was taken fairly seriously throughout the 1950s, particularly before SOSUS made it difficult for Soviet submarines to approach the US coast undetected. For example, a minesweeper conversion of the standard navy 50ft motor launch was developed and tested. About 1959 the navy seems to have accepted that if major war came, it would be short and nuclear; the mine threat to domestic ports could be discounted. To some extent Cuba outflanked SOSUS, in that it lay on the Continental Shelf and the system han only limited shallow-water capability. Therefore Fidel Castro's victory in Cuba did in theory revive the coastal mine threat, but the more general stategic conclusion (short and nuclear) remained.

Only in the early 1980s did it become apparent that war might be both global and non-nuclear (ie sustained). That revived interest in port-protection. Early estimates of the sheer size of the force needed to insure that US ports could be kept open seemed prohibitive. Eventually the planned force was scaled down considerably, partly by using converted civilian craft (COOPs, craft-of-opportunity). Part of this strategy was to cut costs by dividing the mine countermeasures force into coastal and ocean-going units.

By this time there was also real interest in ocean mine counter-measures. In about 1975 the US Navy became aware of two new Soviet ASW mines, a rising weapon and an influence (UEP, under-water electronic potential) device. Both seemed to threaten US strategic submarines proceeding to sea or through strategic straits. Rising mines might also attack surface ships passing through straits. The new deep ocean sweeper/hunter would be equipped with a deep-running remote-controlled submersible, which ultimately became the current SLQ-48. Initially this unit would have been about the size of a small frigate, but it was considered prohibitively expensive. Also, it would probably have had signatures too large to permit its use in coastal waters.

By 1959 the navy had concluded that future conventional warfare would focus on foreign coastal waters, most likely around Eurasia. Amphibious assault would continue to be extremely important. As in the past, it would be opposed by mines, and therefore minesweepers were included in the amphibious assault force. Unfortunately, they were inherently rather slow, a problem given the postwar drive towards a 20kt amphibious force.

The Soviet mine threat to West European ports was a major driving force in postwar US minesweeper development, as the United States sought to provide her European allies with the means to keep their ports open. She therefore designed and built (both at home and, under an assistance programme, abroad) a coastal sweeper, a lineal successor to the wartime YMS. In effect this was a way to avoid the US manpower problem while still providing the initial capital outlay represented by the sweeper. Relatively few of these AMS or MSC were retained by the US Navy, whereas many of the larger AM or MSO (ocean sweepers), direct successors to the wartime 180-footer, were kept.

Mine development drove both programmes. Sweepers in service in 1945 could be defeated by the sensitive magnetic mines then entering service; non-magnetic construction was a major issue in postwar construction, with specified levels of magnetic signature from the earliest design phase. Where wooden construction had previously been adopted (in the wartime YMS) to make mass production easier, in postwar craft it was part of a careful programme of signature reduction. In both cases, large generators were required to power magnetic sweep gear; in some cases postwar that meant gas turbines. Pressure mines were a major problem. The US Navy developed a large towed countermeasure, XMAP, which was to simulate the pressure signature of a merchant ship, and postwar sweepers carried among their staff requirements the ability to tow it. Ultimately it proved insufficiently blast-resistant, and was abandoned, but the interest in unusual countermeasures continued, with experiments with blast-resistant 'Liberty' ship hulls which could be operated (and propelled) entirely from above decks.

The problems associated with more sophisticated mines led to the development of minehunters, which might identify such devices one by one on the seabed for individual destruction. That is, the classical minesweeping technique of simulating a ship signature in order to neutralise mines in quantity (without destroying the simulator) was at least tentatively abandoned in favour of high-definition sonar and special mine-destruction devices. This programme began in 1945 with landing craft conversions: AMCU 1–6 were former tank landing craft, or LCT(6), and AMCU 7–11 were former infantry landing craft, LCIL; all were wartime conversions. AMCU 12–14, 44–50 (*Harkness* class) were YMS/PCS conversions. AMCU 15–42 (*Accentor* class) were LSIL conversions under the FY52 programme. In fact the conversions of AMCU 15, 23, 27, 31, 32, 36 and 40 were cancelled. Finally, a new-construction minehunter, the *Bittern*, was built as AMCU (later MHC) 43. All the existing AMCU were redesignated MHC (minehunter, coastal) on 7 February 1955. Two sister ships, which would have become MHC 44 and 45, were not built. Finally, in January and February 1954, five auxiliary motor minesweepers (AMC, former YMS) were redesignated as minehunters (AMCU 46–50), although it does not appear that any modifications were made to them. Early in 1955, when it became apparent that MHC 44 and 45 would not be built, these hull numbers were applied to two more motor minesweepers, AMS 2 and 3.

The primary problem of the minesweepers was their identification with the amphibious force, which meant an ultimate goal of a 20kt mine force. That in turn suggested the development of small sweepers which could be carried to the objective area aboard large specialised or amphibious craft. Two approaches were tried. In one, small minesweeping boats (MSB) could be carried in the well decks of an LSD or on the deck of a cargo ship with heavy-lift booms. Alternatively, a 36ft MSL could be carried in the davits of standard amphibious ships. In 1955 a specialised mother ship for such craft was planned, as a conversion of the wartime large minelayer/motorised transport (CM/LSV) hull. Two ships were converted in 1963–66; they could carry twenty MSL, two minesweeping helicopters, and spare gear (including that for MSLs), and could control them in an assault sweep. One LSD and one LST were also converted, but with lesser boat capacity and no helicopters.

The helicopters represented the next attempt at a fully mobile assault sweeping force. During the early 1960s a special helicopter-drawn sled, accommodating magnetic and other countermeasures, was developed; it could be towed by the standard CH-53 Marine helicopter, which was operated by a variety of fleet amphibious units, most of them capable of a sustained passage at 20kts. Later even a minehunting sonar, AQS-14, was developed for helicopter towing. The Mk 105 sled was used in mine clearance operations off Hanoi in 1973.

For a time it seemed that the helicopters might replace surface sweepers altogether. However, this did not prove practical; instead, the two elements worked together. Helicopters were used as precursor sweepers, to deal with mines deliberately set to destroy the surface sweepers (since the mines could not destroy the helicopters, they could be swept). This made it relatively safe for the sweepers and hunters.

When mine countermeasures requirements were reviewed in the early 1980s, it seemed that very large numbers would be required merely to assure free use of US ports, not to mention strategic straits, in wartime. They were unaffordable, even after the beginning of the Reagan Administration build-up. The solution was a threefold high-low mix. The highest end was the MCM, which could deploy abroad. It became the *Avenger* class. The navy would also build a class of coastal minehunters (MSH) operating out of US ports. Because not nearly enough could be built, a third tier was conceived; the COOP (craft-of-opportunity), a converted fishing boat (or, it turned out, a training boat [YP]). In peacetime, the COOPs assigned to various ports would conduct bottom surveys of chosen Q-routes. Such surveys are essential because mines are not necessarily easily distinguished from other bottom junk in an industrial area. However, if the bottom has been surveyed, changes can certainly be detected. In wartime, the COOPs would have towed simple sweeps.

The MSH was also an opportunity to introduce new technology, in the form of a surface effect (air cushion) hull and GRP material. The surface effect hull was attractive because it should have been inherently safer against mine explosions. It also promised high transit speed between hunts, so that a smaller number of hunters could deal with a large coastal area (Norway later adopted just this logic). Unfortunately the test hull section dismally failed its shock test, and the design was badly overweight, a fatal defect in an air-cushion craft.

In the end, the MSH was replaced by an MHC (coastal minehunter), a modified and considerably enlarged version of the Italian *Lerici*, not far short of what would be needed for overseas deployment (later ones will be even larger); the gap between MCM and MHC is no longer so clear. This is fortunate, since the odds on mining US ports lengthened dramatically just as the MHC programme hit production. At the same time the logic of the COOP program no longer held particularly strongly, and it was virtually abandoned.

The run-down of new mine craft construction during and immediately after the Korean War was extremely rapid. Thus the FY50 programme included only one XMAP anti-pressure-mine device, with another (and an ocean sweeper) authorised but deferred. In addition, the army coastal controlled mineplanting function was turned over to the navy, the latter receiving six coastal mineplanters, two local mineplanters (YMP), and four minesweeping boats (MCM, later MSB 1–4). Funds were released with the outbreak of the Korean War, however, and under the FY51 and three supplemental programmes a total of nineteen ocean sweepers and eighteen MSB were authorised; another thirteen sweepers and two coastal sweepers were authorised but deferred, and the FY50 sweeper awarded. All of the deferred ocean and coastal sweepers were purchased under the FY52 programme, which included seventeen more coastal sweepers, thirty MSHs, and the conversion of thirty-one craft to minehunters (AMCU, later MHC). FY53 included nine ocean and twenty coastal sweepers (plus one ocean sweeper deferred), and FY54 included four ocean sweepers and the specialized minehunter *Bittern*. Small boat funds bought four MSL, coastal sweepers somewhat smaller than the MSB. By this time funds were beginning to tighten, as the carrier and guided missile programmes began very badly to overrun original estimates. Thus FY55 included only three ocean sweepers, and two MSI (replacements for the original postwar coastal sweeper) had to be deferred to FY56 before programme submission. The latter programme included no ocean sweepers at all, although twenty-six MSL were included among small boats authorised. No minecraft appeared in the FY57 programme; in FY58 two coastal craft were built to replace two transferred.

The threat of North Vietnamese riverine mines in Vietnam inspired the development of a variety of small craft, only the designations of which remained a few years later: drone minesweepers (MSD), river minesweepers (MSM), and patrol minesweepers (MSR). However, the minesweeping launches (MSL) remained, and there were headquarters ship conversions (MCS) under the FY63 and FY64 programmes. However, the reader might note that this latter programme was originally far larger, all of the *Terror* class hulls (except *Terror* herself) being redesignated in the MCS series as early as 1955.

All were stricken 1 September 1961, but *Catskill* (MCS 1) and *Ozark* (MCS 2) were reinstated (and actually converted) 1 June 1964 and 1 October 1963, respectively. The LSD *Epping Forrest*, used as a mine escort tender, became MCS 7 on 30 November 1962. Similarly, the tank landing ship *Orleans Parish* was reclassified MCS 6 on 19 January 1959, reverting to LST status on 1 June 1965. It might be argued that these redesignations themselves reflected the new mine threat encountered in Vietnam: a coastal and riverine threat requiring countermeasures by small craft.

COAST GUARD CUTTERS

Although the US Coast Guard functions as a civilian organisation in peacetime, it is amalgamated with the navy in wartime. Its cutters are therefore designed with secondary naval (generally escort) missions in mind, and in fact some of them operated in a naval role in Vietnam, the 'Cape' class coastal craft forming part of the riverine force. In addition, the Coast Guard did operate twelve ex-navy destroyer escorts (as ocean weather ships) during the Korean War, and for much of the postwar period included an ASW weapon, Hedgehog, among the standard equipment of its cutters.

Primarily because of the size of its force in 1945, the Coast Guard received no major new units until the mid-1960s, when twelve long-endurance cutters (WHEC, formerly WPG) were laid down. They were to replace an existing force of *Owasco* and *Campbell* class cutters, the latest of which had been completed at the end of the Second World War; the Coast Guard of the postwar period had also operated a force of eighteen former small seaplane tenders transferred to it in 1947–48. Of these, seven were transferred to South Vietnam beginning in 1970, the others being decommissioned from 1968 onwards. The other major postwar re-equipment programmes were for smaller ships: the 210ft *Reliance* class medium endurance cutter (WMEC, ex-WPC) and the newer 270ft *Bear* class medium endurance cutters (WMEC). From a naval point of view, the other major postwar Coast Guard programme was the patrol boat (WPB), built in both the *Cape* (95ft) and *Point* (82ft) classes, the latter directly replacing a series of 83-footers which served prominently during the Second World War. The 95ft WPB formed the basis for a US motor gunboat (PGM) design which was exported to many South American and Asian navies.

The extent to which Coast Guard designs are compatible with naval missions is shown by the series of studies of *Hamilton* (WHEC) class hulls as the basis for what became the *Perry* class FFG. In addition, the *Bear* class was proposed as a basis for an ASW-oriented corvette which the Carter Administration proposed as the FFX; all were to have been operated in peacetime by the Naval Reserve, a concept reminiscent of the origins of the Second World War PC. Among US warships of their size, the medium endurance cutters are unique in their very considerable helicopter capabilities, necessary for the Coast Guard sea rescue mission. Their armament is justified in peacetime by the need to maintain security in US coastal waters. They do not, therefore, have very powerful sonars, although the SQS-36/38 sonar of the *Hamilton*s matches the SQS-56 of the FFGs. With the advent of the towed array system, however, there was the possibility of greatly enhancing their wartime ASW capability at very low cost in peacetime capability. In that case the ability to operate large helicopters, particularly in rough weather, would have been a major asset. With the end of the Cold War, all ASW equipment was removed from Coast Guard cutters.

For most of its existence the Coast Guard has been an arm of the Department of the Treasury; it is, indeed, descended from the Revenue Service (anti-smuggling). In recent years, however, it has emphasised its sea safety role and it became part of the new Department of Transportation on 1 April 1967.

MAJOR US NAVAL ELECTRONIC EQUIPMENT

US electronic equipment developed since 1945 has been designated in a tri-service series, in which the three prefix letters indicate, respectively, platform, equipment type, and function. They in turn are prefixed by 'AN/', indicating originally, 'Army-Navy', *ie*, multi-service. These letters are omitted for clarity here, so that, for example, AN/SPS-30 is referred to as SPS-30. The principal naval platform prefixes are 'B' (submarine), 'S' (surface ship), 'U' (land and sea 'universal') and 'W' (waterborne, *ie*, submarine and surface ship); 'A' indicates an aircraft system. Major type designators are 'L' (Countermeasures), 'P' (radar), 'Q' (sonar), and 'Y' (computer system); major functions include 'G' (fire control), 'Q' (multi-role or specialised), 'R' (passive detection), 'S' (search), and 'Y' (multi-function electronically-scanned radar). For example, WLR-1 is the first 'waterborne' countermeasures receiver in the series; SPQ-5 is a special or multiple-function ship-borne radar, in this case a Terrier beam-guidance and tracking radar. (Full details can be found in the author's *Naval Institute Guide to World Naval Weapons Systems* (Annapolis: Naval Institute Press, various editions 1989–94.)

Perhaps the most important post-1947 electronic development was a digital combat directional system, NTDS (Naval Tactical Data System), which had an airborne equivalent, ATDS. On a ship, NTDS compiles and maintains a current plot of air, surface, and underwater activity. Earlier manual systems (vertical plots) could be saturated by a few distinct targets; even in its early form, NTDS could handle over a hundred. NTDS ships (and the ATDS airplane, the E-2C) exchange digital target data via a digital link, Link 11, which allows a force to maintain a consistent picture of activity around it. NTDS also reports when a ship has engaged a target, and it can be used to order engagements. NTDS went through at least four generations, using progressively more powerful computers. JPTDS (Junior Participating Tactical Data System) was a version adapted to *Adams* class missile destroyers.

The US Navy encouraged NATO allies to adopt similar systems: Australia and Canada bought NTDS, France bought SENIT, West Germany bought AGIS and SATIR, Italy bought SADOC, and Japan bought NTDS. Britain and the Netherlands bought similar systems (ADAWS and SEWACO), but they used less powerful computers (24- rather than 30-bit) and, for many years, a less powerful link (Link 10). By the late 1980s Link 11 had become a NATO standard, and the ability of Allied ships to share a common tactical picture was extremely important.

NTDS was primarily a means of maintaining a tactical picture, data from which could be passed to a Weapons Direction System (WDS). As computers became more powerful, it became possible to imagine automatic insertion of tactical advice, given on the basis of pre-programmed rules.

Aegis ships were built with two complementary computers, one to maintain the picture and another (the Command and Decision computer, or C&D computer) to provide tactical advice. Ships with more powerful computers (UYK-43s rather than UYK-7s) have intelligence data, such as weapons characteristics, included in their C&D computers. In either case the system is called a Combat Direction System (CDS) rather than a tactical data system (TDS).

NTDS has been superseded by an Advanced Combat Direction System (ACDS), initially mainly NTDS adapted to the more powerful UYK-43 computer. A later version incorporates a separate C&D computer. Another current development is provision for JTIDS (Joint Tactical Information Distribution System), a higher-capacity data link. To accommodate it, and to allow the tactical system to run faster, link processing is removed from the ships' central computers and a separate command and control processor (C2P) added.

A second important current theme is over-the-horizon targeting (OTH-T), which became important with the advent of the Tomahawk anti-ship missile. Effective targeting requires that ships maintain a tactical picture extending well beyond the range of the sensors they carry. US practice has been to supply ships with the fruit of national-level systems (such as satellites and long-range electronic reconnaissance aircraft). These data are called time-late (force sensors are real-time, or near-real-time). Shore and ship terminals form an OTH-T net, using special satellite links (OTCIXS, TADIXS), broadly analogous to the NTDS/Link 11 combination. Initially data were virtually all collated ashore (the shipboard end was a USQ-81 terminal, often part of the Tomahawk system); later ships were provided with sufficiently powerful computers (JOTS) that the OTH-T picture could be built co-operatively. These considerations affect both surface ships and submarines.

Type	Remarks
RADARS (Note surface-search sets are *not* included)	
BPS-2	Radar-picket submarine L-band air search radar, 1953
BPS-3	S-band height-finder for submarine radar-pickets
SPG-49	Talos guidance radar (target tracker) operating in conjunction with SPW-2, with one radar per missile launcher arm
SPG-51	Tartar tracking and illuminating radar
SPG-55	Terrier guidance radar, successor to SPQ-5
SPG-60	Fire control radar of Mark 86 Fire Control System, for gun and missile control
SPG-62	Slaved illuminator of the Aegis system, operating with SPY-1

Type	Remarks
SPQ-5	Terrier fire control radar, essentially a modified SPG-49
SPQ-9	Search and track radar of Mark 86 Fire Control System
SPS-2	Very large long-range air search radar for task force defense. Only aboard *Northampton* and missile cruiser *Little Rock*
SPS-6	First major postwar air search radar; L-band
SPS-8	Nodding height-finder with 'orange-peel' antenna; SPS-8B had a dish and formed a link to the later SPS-30
SPS-10	Standard surface-search radar; newer solid-state version is SPS-67
SPS-12	Postwar L-band air search radar, similar in general characteristics to SPS-6 but considered superior. The Canadian SPS-501 consists of an SPS-12 transmitter with a Dutch antenna
SPS-17	First major postwar P-band air search radar, a return to late-Second World War concepts. SPS-17A had a much-enlarged antenna, for radar-pickets. These ideas were developed through SPS-28 (lightweight version for destroyers), SPS-29, SPS-37, and SPS-43. The latter two had large-antenna versions (SPS-37A, -43A) which were standard on US missile cruisers and carriers. Now being replaced by SPS-49
SPS-30	Dish for height-finding and fighter control, aboard carriers and some missile cruisers
SPS-32/33	Hughes SCANFAR electronically-scanned combination of, respectively, search and three-dimensional radars. Aboard *Long Beach* and *Enterprise* only; replaced by conventional radars 1981, partly because of operational problems
SPS-39	Three-dimensional electronically-scanned (in elevation) radar for missile ships and area defence missile systems. SPS-52 is a modified version, using the same planar-array antenna as the improved SPS-39s. Lighter than, but inferior to, SPS-48. SPS-42 is a version specifically adapted to interface with NTDS
SPS-40	UHF air search radar, in effect a replacement for SPS-6 series
SPS-48	Long range three-dimensional radar, with electronic scanning in elevation. Replaced SPS-30 in many carriers
SPS-49	Current long range two-dimensional air search radar, with stabilised mounting, adapted to current automated combat systems. Replaced the SPS 29/37/43 series
SPS-58/65	Search radar for the Sea Sparrow Point Defense Missile System. Later versions were designated SPS-65, and in many cases used the same antenna as the standard SPS-10 surface search system
Mk 37	Second World War 5in/38 fire control system, the first in US service to use a below-decks computer
Mk 56	Fire control system, usually for 3in/50, sometimes for 5in guns as well
Mk 68	Successor to Mk 37 (Mk 67 was an interim version). Some could also control SM-1 missiles
Mk 86	Gun fire control system, in some cases with missile guidance capability. This is a longer-range alternative to Mk 92
Mk 91	Fire control system for Sea Sparrow. The radar has no separate designator. The *earlier* basic point defence system used a much earlier Mk 115 fire control system
Mk 92	US version of the Signal WM20-series 'egg', on board the hydrofoils and the *Perry* class frigates. In this class the tracker in the egg is supplemented by a Separate Tracking and Illuminating Radar (STIR), carried abaft the mast so that it will survive damage to the egg
Mk 99	Aegis fire control system
TAS Mk 23	Target Acquisition System, an alternative to (and often a successor to) SPS-58/65. It is generally used to indicate targets to the Mk 91
NTU	New Threat Upgrade, a modification to existing Terrier/Tartar fire control systems to engage many more targets nearly simultaneously. It controls SM-2 missiles, which incorporate commandable autopilots. The missiles are commanded to fly into acquisition 'baskets' around the expected target positions, and only then require illumination by fire control radars. This is much the same concept as in Aegis, but it is not quite as efficient. Because conventional search radars cannot locate targets very precisely, the fire control illuminators cannot be slaved to them, and the missile requires more illumination time to find the target. Even so, NTU greatly increases ship firepower at a relatively low cost. It is associated with the Mk 14 Weapons Direction System
SSDS	Ship Self Defense System, if effect replacing Mk 91/NATO Sea Sparrow. This series of systems was inspired by the attack on USS *Stark*, which seemed to show that much quicker reactions and better integration of ECM and active defense were needed. SSDS was also inspired by experiments in which the radars of Phalanx CIWS were used to control other systems, such as RAM missiles, via dedicated local area networks. SSDS grew out of RAIDS (Rapid-Reaction Integrated Defensive System), which provided command advice on how to combine ECM and other defensive measures aboard *Spruance* and *Perry* class ships. The next step, tested on board an LSD, was to combine all available sensors (including the experimental US–Canadian SAR-8 IR search/track device), using the RAM missile. The new LPD-17 design shows an operational form of SSDS, in which the horizon search radar is a modified SPQ-9 (SPQ-9B); the weapons are RAMs and, ultimately, Evolved Sea Sparrows or some other SSDS missile, fired outof a 16-round vertical launcher.
SLQ-17	Carrier self-protection jammer, now being superseded by SLQ-32(V)4
SLQ-32	Standard ESM/ECM system, originally the Design to Price EW System. It triumphed over a rival SLQ-31, essentially a smaller version of SLQ-17. SLQ-32(V)3 incorporates a jammer. There is also a version with a separate jammer, Sidekick, now called SLQ-32(V)5. SLQ-32 is generally associated with SRBOC, the super-rapid blooming offboard chaff (a mortar, not a rocket)
ULQ-6	Standard self-protection jammer of the Vietnam era, typically carried on a latticework stub on either side of the ship. It was often associated with Chaffroc, Zuni chaff rockets launched from a modified aircraft pod. Such pods are recognisable because they are long and trainable
SPY-1	Aegis search and fire control radar
SYS-1, -2	Computer systems which integrate the search and fire control radars of an existing ship, for automatic target detection and tracking, threat assessment, and assignment of defensive weapons. Currently being installed aboard existing US missile-armed major warships
SONARS	
BQG-1, -4	PUFFS, submarine passive fire control sonar systems, consisting of hydrophones distributed along the length of the submarine for triangulation
BQR-2	Listening array for postwar submarines: 6ft circular array
BQR-4	Conformal array based on German GHG for ASW submarines: 10ft wide, 20ft long horseshoe
BQR-15	Towed array for ballistic missile submarine self-defence
BQQ-2	Integrated sonar system for *Thresher* and later submarines, now being upgraded to BQQ-5. It included a bow spherical sonar as well as a conformal array in the hull
BQQ-5	Replacement for BQQ-2, in *Los Angeles* class; includes a towed array
BQQ-6	Trident missile submarine sonar, similar to BQQ-5 but without any active element

Type	Remarks
SQQ-14	Standard minehunting sonar of the 1960s, a modernised version of the earlier UQS-1
SQQ-30	Modernised SQQ-14, interim until SQQ-32 appears
SQQ-32	Minehunting sonar of the current minehunters
SQQ-89	New ASW sonar suit. In destroyers and cruisers it combines the SQS-53 bow sonar, the SQR-19 towed array, and the LAMPS helicopter; it also incorporates a sonar performance estimator (so operators can better guess at which convergence zone their passive sonar is working) and better multi-target tracking in a late version of the Mk 116 underwater fire control system. Frigates have a simpler version omitting the hull sonar altogether
Mk 309	Fire control panel, the simple underwater fire control in frigates and large Coast Guard cutters tied to a short range sonar like SQS-56. It is designed for self defence: given a submarine detection, it fires a Mk 46 torpedo to discourage pursuit and also advises the CO of which evasive course to steer
SQR-14	Very long range towed array sonar, ITASS (Interim Towed Array Surveillance System)
SQR-15	Successor to SQR-14, for special surveillance ships (AGOS)
SQR-18	Interim Tactical Towed Array Surveillance System (TAC-TASS), widely deployed aboard frigates, often towed from the SQS-35 'fish'
SQR-19	Successor to SQR-18
SQS-4	First major low-frequency sonar system, with a nominal range in its later versions of 5–8000yds. In FRAM II ships it was associated with the DASH ASW drone. There were many improved versions, their designations within a series depending on the frequency, between 8 and 12 kc: SQS 29–32 (RTD), SQS 39–42 (Improved reliability and maintainability); SQS 43–46 (VDS, operating off the same 'stack'), and SQS 49–52 (improved reliability and maintainability)
SQS-23	Lower frequency sonar (5 kc) with a nominal range of 10,000 yards, associated with ASROC. AQQ-23 was an active-passive system with a second transducer, usually in a second sonar dome. Many ships with SQS-23 are now being modified to a single-dome SQQ-23 configuration
SQS-26/53	The largest current US surface ship sonar, credited with very long range: about 20,000yds direct-path, as well as bottom-bounce and first convergence zone (about 35nm). SQS-53 is a transistorised version
SQS-35	Independent variable depth sonar (IVDS), a 'fish' operating from its own sonar 'stack'. Aboard many *Knox* class ocean escorts
SQS-36/38	Hull sonar developed from SQS-35; in Coast Guard ships. SQS-38 is a transistorised development
SQS-56	Hull sonar of the *Perry* class. Although often described as a transistorised SQS-23, it operates at a considerably higher frequency (8 kc) and is closer to a modernised SQS-36/38

FLEET STRENGTH 1947

BATTLESHIPS

No	Name	Fate
Iowa class: launched 1942–44, 48,000t standard		
BB 61	IOWA	Active 1951–58, 1983–90
BB 62	NEW JERSEY	Active 1950–57, 1967–69, 1982–91
BB 63	MISSOURI	Active to 1954, 1986–92
BB 64	WISCONSIN	Active 1951–58, 1988–91

Missouri served as training ship 1948–50. All four were recommissioned for Korea, with little change apart from elimination of catapults and aircraft. *New Jersey* was recommissioned for Vietnam; all lighter weapons were removed, and ECM equipment (ULQ-6 and Chaffroc launchers) added. All four were modernised and recommissioned under the Reagan Administration build-up: secondary battery reduced to 12×5in, but 8×4 Tomahawk and 4×4 Harpoon missiles added, as well as four Phalanx. The main air search radar is SPS-49 forward, with LAMPS data link above it, and SLQ-32 in the expanded fire control tower. The original fire control system is retained, but supplemented

New Jersey 30.9.88 L & L van Ginderen

by modern computers. Plans for more ambitious Phase Two reconstruction were abandoned. There was also some interest in a more modest upgrade in which the twin 5in guns would have been replaced by lightweight Mk 45s and Sea Sparrow would have been added. *Wisconsin* and *Missouri* fought in the Gulf conflict of 1991. These ships were laid up for future use, but on 13 January 1995 all four were stricken. They are being offered as museums. As of spring 1995 an attempt was being made in Congress to restore the battleships to the Navy List, but success seemed unlikely.

AIRCRAFT CARRIERS

No	Name	Fate
Essex class: launched 1942–45, 27,100t standard		
CV 9	ESSEX	SCB-27A 1951, CVS 1960, FRAM FY62, decommissioned 1969, stricken 1.6.73, BU
CV 10	YORKTOWN	SCB-27A 1953, CVS 1957, FRAM FY66, decommissioned 1970, memorial
CV 11	INTREPID	SCB-27C 1954, CVS 1962, FRAM FY65, decommissioned 1974, museum 1982
CV 12	HORNET	SCB-27A 1953, CVS 1958, FRAM FY65, decommissioned 1970, reserve 1980
CV 14	TICONDEROGA	SCB-27C 1954, CVS 1969, decommissioned 1973, stricken 16.11.73, BU 1975
CV 15	RANDOLPH	SCB-27A 1953, CVS 1959, FRAM FY61, decommissioned 1969, stricken 1.6.73
CV 16	LEXINGTON	SCB-27C 1955, CVS 1962, decommissioned Feb 1992
CV 18	WASP	SCB-27A 1951, CVS 1956, FRAM FY64, decommissioned 1972, stricken 1.7.72, BU 1973
CV 19	HANCOCK	SCB-27C 1954, decommissioned 1956, stricken 31.1.76, BU
CV 20	BENNINGTON	SCB-27A 1952, CVS 1959, FRAM FY63, decommissioned 1970, reserve 1980
CV 21	BOXER	CVS 1955, LPH 1959, FRAM FY63, decommissioned 1969, stricken 1.12.69, BU
CV 31	BON HOMME RICHARD	SCB-27C 1955, decommissioned 1971, reserve 1980, BU 1992
CV 32	LEYTE	CVS 1953, decommissioned 1959, stricken 1969, BU
CV 33	KEARSARGE	SCB-27A 1952, CVS 1958, decommissioned 1970, stricken 13.2.71, BU
CV 34	ORISKANY	SCB-27A 1950, decommissioned 1976, reserve 1980
CV 36	ANTIETAM	CVS 1953, decommissioned 1963, stricken 1.5.73, BU; angled deck prototype
CV 37	PRINCETON	CVS 1953, LPH 1959, FRAM FY61, decommissioned 1970, stricken 30.1.70, BU
CV 38	SHANGRI LA	SCB-27C 1955, CVS 1959, decommissioned 1971, stricken 15.7.82
CV 39	LAKE CHAMPLAIN	SCB-27A 1952, CVS 1957, decommissioned 1966, stricken 1969, BU 1972
CV 40	TARAWA	CVS 1955, decommissioned 1960, stricken 1967, BU
CV 45	VALLEY FORGE	CVS 1953, LPH 1961, FRAM FY64, decommissioned 1970, stricken 15.1.70, BU
CV 47	PHILIPPINE SEA	CVS 1955, decommissioned 1958, stricken 1969, BU 1971
Midway class: launched 1945–46, 45,000t standard		
CVB 41	MIDWAY	Decommissioned 11.4.92
CVB 42	FRANKLIN D ROOSEVELT	Decommissioned 1977, stricken 1.10.77, BU
CVB 43	CORAL SEA	Decommissioned 1990, BU 1993–95

No	Name	Fate
Independence class: launched 1943, 10,600t standard		
CVL 26	MONTEREY	Active 1950–56, stricken 1.6.70, BU 1971
CVL 28	CABOT	ASW carrier, active 1945–48, stricken 1959, Spanish *Dedalo* 1967
CVL 29	BATAAN	ASW carrier, active 1950–54, stricken 1959, BU
Saipan class: launched 1945, 14,500t standard		
CVL 48	SAIPAN	Decommissioned 1957, recommissioned 1966 as *Arlington* (AGMR 2), decommissioned 1970, stricken 15.8.75, BU
CVL 49	WRIGHT	Active to 1956, and 1963–70, stricken 31.12.77
Bogue class: launched 1942–43, 9400t standard		
CVE 11	CARD	Active 1958–70, stricken 15.9.70, BU 1971
CVE 13	CORE	Active 1958–69, stricken 15.9.70, BU 1971
CVE 23	BRETON	Active 1958–71, stricken 6.8.71, BU 1972
CVE 25	CROATAN	Active 1958–69, stricken 15.9.70, BU 1971
Casablanca class: launched 1943–44, 8200t standard		
CVE 58	CORREGIDOR	Active 1951–58, stricken 1958, BU 1960
CVE 64	TRIPOLI	Active 1952–58; stricken 1959, BU 1960
CVE 86	SITKOH BAY	Active 1950–54, stricken 1960, BU 1961
CVE 88	CAPE ESPERANCE	Active 1950–59, stricken 1959, BU 1961
CVE 90	THETIS BAY	Active 1956–64, stricken 1964, BU 1966
CVE 92	WINDHAM BAY	Active 1951–59, stricken 1959, BU 1961
Commencement Bay class: launched 1944–45, 18,900t standard		
CVE 106	BLOCK ISLAND	Active 1951–54, stricken 1959, BU 1960
CVE 107	GILBERT ISLANDS	Active 1951–55, *Annapolis* 1964–69, stricken 15.10.76
CVE 108	KULA GULF	Active 1951–55, 1965–69, stricken 15.9.70, BU 1971
CVE 110	SALERNO BAY	Active 1951–54, stricken 1961, BU 1962
CVE 112	SIBONEY	Active 1950–56, stricken 1970, BU 1971
CVE 114	RENDOVA	Active 1945–50, 1951–55, stricken 1.4.71, BU 1973
CVE 115	BAIROKO	Active 1945–55, stricken 1960, BU 1961
CVE 116	BADOENG STRAIT	Active 1947–57, stricken 1970, BU 1972
CVE 118	SICILY	Active to 1954; stricken 1960, BU 1961
CVE 119	POINT CRUZ	Active 1951–56, 1965–69, stricken 1970, BU 1971
CVE 120	MINDORO	Active 1945–55, stricken 1959, BU 1960
CVE 122	PALAU	Active 1946–54, stricken 1960, BU 1960

Essex class

All SCB-27s were laid up 1947, reactivated for conversion, except for *Kearsarge*, which was active until conversion. All but CV 39 were fitted with enclosed bows and angled decks. Proposals to reactivate CV 34 and CV 31 failed in Congress in 1981–82. Full details of conversions are given below under Major Surface Ships.

Midway class

All remained active from commissioning onwards, except for SCB-110/110A and later reconstructions. In 1982 *Midway* was forward-deployed in Japan. Full details of conversions are given below under Major Surface Ships.

Saipan class

Saipan became a major communications relay ship and was recommissioned as *Arlington* in 1966 and served in Vietnamese waters. *Wright* became a NECPA command cruiser 1963 (see under Amphibious Warfare Vessels).

Bogue class

All four remaining escort carriers of this class served as aircraft transports until decommissioned.

Casablanca class

Thetis Bay was converted to a helicopter carrier (CVHA 1) at San Francisco N Yd June 1955–July 1956; she was reclassified LPH 6 on 28 May 1959. Details were: 10,866t full load; 8–40mm (4×2), 20 helicopters (max), 1004 troops. The other escort carriers of this class served as aircraft transports for the whole of their postwar active lives.

Commencement Bay class

Gilbert Islands was converted to the major communications relay ship *Annapolis* (AGMR 1) 1962–63 and served until 1969. Details were: 11,473t standard, 22,500t full load; 8–3in/50 (4×2); complement 710. *Kula Gulf* and *Point Cruz* served as MSTS aircraft transports 1965–69.

CRUISERS

No	Name	Fate
Atlanta class: launched 1942–46, 6700t standard		
CL 95	OAKLAND	Decommissioned 1949, stricken 1959, BU 1962
CL 98	TUCSON	Decommissioned 1945, stricken 1966, BU 1971
CL 119	JUNEAU	Decommissioned 1955, stricken 1959, BU 1961
CL 120	SPOKANE	Decommissioned 1950, stricken 15.4.72, BU 1973
CL 121	FRESNO	Decommissioned 1949, stricken 1965, BU 1966
Cleveland class: launched 1943–45, 11,500t standard		
CL 65	PASADENA	Decommissioned 1950, stricken 1.12.70, BU 1972
CL 66	SPRINGFIELD	Decommissioned 1950, reclassified CLG May 1957, stricken 30.9.78, BU
CL 67	TOPEKA	Decommissioned 1949, reclassified CLG May 1957, stricken 1.12.73
CL 82	PROVIDENCE	Decommissioned 1959, reclassified CLG May 1957, stricken 30.9.78, BU
CL 83	MANCHESTER	Decommissioned 1956, stricken 1960, BU 1961
CL 87	DULUTH	Decommissioned 1949, stricken 1960, BU 1961
CL 90	ASTORIA	Decommissioned 1959, stricken 1.11.69, BU 1971
CL 91	OKLAHOMA CITY	Decommissioned 1947, reclassified CLG May 1957, stricken 15.12.79
CL 92	LITTLE ROCK	Decommissioned 1949, reclassified CLG May 1957, stricken 21.11.76, preserved 1977
CL 93	GALVESTON	Never commissioned as CL, reclassified CLG May 1957, stricken 21.12.73
CL 102	PORTSMOUTH	Decommissioned 1949, stricken 1.12.70
CL 104	ATLANTA	Decommissioned 1949, expended in explosive tests 1965–70 as *IX-304*, stricken 1.1.70
CL 105	DAYTON	Decommissioned 1949, stricken 1961, BU 1962
CL 106	FARGO	Decommissioned 1950, stricken 1.3.70
CL 107	HUNTINGTON	Decommissioned 1949, stricken 1961, BU 1962
Worcester class: launched 1947, 14,700t standard		
CL 144	WORCESTER	Decommissioned 1958, stricken 1.12.70, BU 1972
CL 145	ROANOKE	Decommissioned 1958, stricken 1.12.70, BU 1972
Baltimore class: launched 1942–45, 14,000t standard		
CA 68	BALTIMORE	Active to 1956, stricken 15.2.71
CA 69	BOSTON	Decommissioned 1946, reclassified CAG 1 1.11.55, stricken 4.1.74
CA 70	CANBERRA	Decommissioned 1947, reclassified CAG 2 15.6.56, stricken 31.7.78, BU
CA 71	QUINCY	Active 1952–54, stricken 1.10.73, BU 1975
CA 72	PITTSBURG	Active 1951–56, stricken 1.7.73
CA 73	SAINT PAUL	Active to 1971, stricken 31.7.78, BU
CA 74	COLUMBUS	Active to 1959, recommissioned as CG 12 1.12.62, stricken 9.8.76, BU
CA 75	HELENA	Active to 1963, stricken 1.1.74, BU 1975
CA 130	BREMERTON	Decommissioned 1948, active 1951–60, stricken 1.10.73
CA 131	FALL RIVER	Decommissioned 1947, stricken 19.2.71, BU 1972
CA 132	MACON	Decommissioned Apr 1950, recommissioned Oct, active until 1961, stricken 1.11.69
CA 133	TOLEDO	Active to 1960, stricken 1.1.74, BU 1975
CA 135	LOS ANGELES	Active 1951–63, stricken 1.1.74
CA 136	CHICAGO	Decommissioned 1947, recommissioned as CG 11 2.5.64
CA 123	ALBANY	Active to 1958, recommissioned as CG 10 3.11.62, stricken 30.6.85
CA 124	ROCHESTER	Active to 1961, stricken 1.10.73, BU 1975
Des Moines class: launched 1946–47, 17,200t standard		
CA 134	DES MOINES	Active to 1961, stricken 1992
CA 139	SALEM	Active to 1959, stricken 1992
CA 148	NEWPORT NEWS	Decommissioned 1975, stricken 31.7.78, BU 1993

Cleveland class

Full details of the ships converted to missile cruisers (CLG) are given below under Major Surface Ships.

Baltimore class

Helena, *Macon*, *Toledo* and *Los Angeles* were equipped to fire Regulus cruise missiles. Full details of the ships converted to missile cruisers (CG) are given below under Major Surface Ships.

DESTROYERS

No	Name	Fate
Benson-Livermore class: launched 1941–42, 1800t standard		
DMS 19	ELLYSON	Japanese *Asakaze* 1954, to Taiwan 1970
DMS 20	HAMBLETON	Decommissioned 1955, stricken 1.6.71
DMS 21	RODMAN	Taiwanese *Hsieng Yang* 1955
DMS 23	MACOMB	Japanese *Hatakaze* 1954, to Taiwan 1970
DMS 25	FITCH	Decommissioned 1956, stricken 1.7.71
DMS 26	HOBSON	Lost in collision with *Wasp*, 26.4.52
DMS 33	CARMICK	Decommissioned 1954, stricken 1.7.71
DMS 34	DOYLE	Decommissioned 1955, stricken 1.7.71
DMS 35	ENDICOTT	Decommissioned 1955, stricken 1969, BU
DMS 27	JEFFERS	Decommissioned 1955, stricken 1.7.71
DMS 38	THOMPSON	Decommissioned 1955, stricken 1.7.71
DMS 30	GHERARDI	Decommissioned 1955, stricken 1.6.71

Colahan 1948 L & L van Ginderen

Fletcher class: launched 1942–44, 2300t standard

DD 445	FLETCHER	DDE 1949–69, stricken 1.8.67, BU 1972
DD 446	RADFORD	DDE 1949–69, stricken 15.7.69, BU
DD 447	JENKINS	DDE 1949–69, stricken 2.7.69, BU
DD 449	NICHOLAS	DDE 1951–70, stricken 30.1.70, BU
DD 450	O'BANNON	DDE 1951–70, stricken 30.1.70, BU 1972
DD 465	SAUFLEY	DDE 1949–65, stricken 1966, target 1968
DD 466	WALLER	DDE 1950–69, stricken 1969, target 1970
DD 468	TAYLOR	DDE 1951–69, Italian *Lanciere* 1969, spares
DD 470	BACHE	DDE 1951–68, grounded, stricken 1.3.68
DD 471	BEALE	DDE 1951–68, stricken 1968, target 1969
DD 498	PHILIP	DDE 1950–68, stricken 1.10.68, sunk in storm 1972
DD 499	RENSHAW	DDE 1950–70, stricken 14.2.70, BU
DD 507	CONWAY	DDE 1950–69, stricken 1969, target
DD 508	CONY	DDE 1949–69, stricken 1969, target 1970
DD 510	EATON	DDE 1951–69, stricken 1969, target 1970
DD 517	WALKER	DDE 1950–69, Italian *Fante* 1969
DD 519	DALY*	Active 1951–60, stricken 1.12.74, BU 1976
DD 520	ISHERWOOD*	Active 1951–61, Peruvian *Guise* 1961
DD 521	KIMBERLY	Active 1951–54, Taiwanese *An Yang* 1967
DD 527	AMMEN*	Active 1951–60, stricken 1960, BU 1961
DD 528	MULLANY*	Active 1951–71, Taiwanese *Ching Yang* 6.10.71
DD 530	TRATHAN*	Active 1951–65, stricken 1.11.72, target
DD 531	HAZELWOOD	Active 1951–65, stricken 1.12.74, BU 1976
DD 532	HEERMAN*	Active 1951–57, Argentinian *Brown* 1961
DD 534	McCORD	Active 1951–54, stricken 1.10.72
DD 535	MILLER*	Active 1951–54, stricken 1.12.74
DD 536	OWEN	Active 1951–58, stricken 15.4.73, BU 1974
DD 537	THE SULLIVANS*	Activated for Korean War, stricken 1.12.74, preserved 1977
DD 538	STEPHEN POTTER*	Active 1951–58, stricken 1.12.72
DD 539	TINGEY	Active 1951–53, stricken 1965, target 1966
DD 540	TWINING	Active (reserve TS and full service) to 1971, Taiwanese *Kwei Yang* 16.8.71
DD 541	YARNALL	Active 1950–58, Taiwanese *Kuen Yang* 1968
DD 544	BOYD*	Active 1950–69, Turkish *Iskenderun* 1969
DD 545	BRADFORD	Active 1950–61, Greek *Thyella* 1962
DD 546	BROWN	Active 1950–62, Greek *Navarinon* 1962
DD 547	COWELL*	Active 1951–71, Argentinian *Almirante Storni* 17.8.71
DD 556	HAILEY*	Active 1951–60, Brazilian *Pernambuco* 1961
DD 558	LAWS	Active 1951–64, stricken 15.4.73
DD 561	PRITCHETT*	Active 1951–70, Italian *Geniere* 1970
DD 562	ROBINSON	Active 1951–64, stricken 1.12.74
DD 563	ROSS	Active 1951–59, stricken 1.12.74
DD 564	ROWE*	Active 1951–59, stricken 1.12.74
DD 565	SMALLEY	Active 1951–57, stricken 1965, BU 1966
DD 566	STODDARD*	Active 1950–69, stricken 1.6.75
DD 567	WATTS	Active 1951–69. stricken 1.2.74, BU 1975
DD 568	WREN	Active 1951–63, stricken 1.12.74, BU 1975
DD 576	MURRAY	DDE 1951–65, stricken 1965, BU 1966
DD 577	SPROSTON	DDE 1950–65, stricken 1968, BU 1972
DD 596	SHIELDS	Active (reserve TS and full service) to 1972, Brazilian *Maranhao* 1972
DD 629	ABBOT*	Active 1951–65, stricken 1.12.74
DD 630	BRAINE*	Active 1951–71, Argentinian *Domecq Garcia* 18.8.71
DD 631	ERBEN	Active 1951–58, Korean *Chung Mu* 1963
DD 642	HALE*	Active 1951–60, Columbian *Antioquia* 1961
DD 643	SIGOURNEY	Active 1951–1960, stricken 1.12.74
DD 644	STEMBEL*	Active 1951–58, Argentinian *Rosales* 1961
DD 650	CAPERTON*	Active 1951–60, stricken 1.12.74
DD 651	COGSWELL*	Active 1951–69, Turkish *Izmit* 1969
DD 652	INGERSOLL*	Active 1951–70, stricken 20.1.70, target 1974
DD 653	KNAPP	Active 1951–57, stricken 6.3.72, BU 1974
DD 654	BEARSS	Active 1951–74, stricken 1.12.74, BU 1976
DD 655	JOHN HOOD*	Active 1951–74, stricken 1.12.74, BU 1976
DD 656	VAN VALKENBURGH	Active 1951–53, Turkish *Izmir* 1967
DD 657	CHARLES J BADGER	Active 1951–57, to Chile for spares 17.5.74
DD 658	COLAHAN	Active (reserve training and full service) to 1966, stricken, target 1966
DD 659	DASHIELL*	Active 1951–60, stricken 1.12.74, BU 1975

No	Name	Fate
DD 661	KIDD	Active 1951–64, stricken 1.12.74
DD 666	BLACK*	Active 1951–69, stricken 21.9.69
DD 667	CHAUNCEY	Active 1950–54, stricken 1.10.72
DD 668	CLARENCE K BRONSON	Active 1951–60, Turkish *Istanbul* 1967
DD 669	COTTEN*	Active 1951–60, stricken 1.12.74
DD 670	DORTCH*	Active 1951–57, Argentinian *Espora* 1961
DD 671	GATLING	Active 1951–60, stricken 1974, BU 1978
DD 672	HEALEY	Active 1951–58, stricken 1.12.74, BU 1976
DD 673	HICKOX	Active 1951–57, Korean *Pusan* 1968
DD 674	HUNT*	Active 1951–63, stricken 1.12.74, BU 1975
DD 675	LEWIS HANCOCK	Active 1951–57, Brazilian *Piaui* 1967
DD 676	MARSHALL	Active 1951–69, stricken 1969, BU 1970
DD 677	MC DERMUT*	Active 1950–63, stricken 1965, BU
DD 678	MC GOWAN*	Active 1951–60, Spanish *Jorge Juan* 1960
DD 679	MCNAIR*	Active 1951–63, stricken 1.12.74, BU 1975
DD 680	MELVIN	Active 1951–54, stricken 1.12.74, sold 1975
DD 681	HOPEWELL*	Active 1951–70, stricken 1970, target 1972
DD 682	PORTERFIELD	Active 1951–69, stricken 1.3.75
DD 683	STOCKHAM	Active 1951–57, stricken 1.12.74
DD 684	WEDDERBURN	Active (reserve and full service) to 1969, stricken 1969, BU
DD 685	PICKING*	Active 1951–69, stricken 1.3.75
DD 686	HALSEY POWELL	Active 1951–68, Korean *Seoul* 1968
DD 687	UHLMANN*	Active (reserve training and full service) to 1972, last active *Fletcher*, stricken 15.7.72
DD 688	REMEY	Active 1951–53, stricken 1.12.74, BU 1975
DD 689	WADLEIGH*	Active 1951–62, Chilean *Blanco Encalada* 1962
DD 793	CASSIN YOUNG*	Active 1951–60, stricken 1.12.74
DD 794	IRWIN*	Active 1951–58, Brazilian *Santa Catarina* 1968
DD 795	PRESTON*	Active 1951–69, Turkish *Icel* 1969
DD 796	BENHAM*	Active 1951–60, Peruvian *Villar* 1961
DD 797	CUSHING	Active 1951–60, Brazilian *Parana* 1961
DD 798	MONSSEN	Active 1951–57, stricken 1963, BU 1964
DD 799	JARVIS*	Active 1951–60, Spanish *Alcala Galiano* 1960
DD 800	PORTER	Active 1951–53, stricken 1.10.72
DD 802	GREGORY	Active 1951–64, stricken 1966, non-operable training hulk *Indoctrinator* 1966–71 at San Diego, experimental hulk from 1971
DD 804	ROOKS*	Active 1951–62, Chilean *Cochrane* 1962

Allen M Sumner class: launched 1943–44, 2600t standard

DD 692	ALLEN M SUMNER*	Decommissioned 1973, stricken 15.8.73, memorial
DD 693	MOALE*	Decommissioned 1973, stricken 2.7.73, BU 1975
DD 694	INGRAHAM*	Greek *Miaoulis* 15.6.71
DD 696	ENGLISH	Taiwanese *Hui Yang* 1971
DD 697	CHARLES S SPERRY*	Stricken 1973, Chilean *Ministro Zenteno* 1974
DD 698	AULT*	Decommissioned briefly 1950, stricken 16.7.73, BU 1975
DD 699	WALDRON*	Decommissioned briefly 1950, stricken 1973, Colombian *Santander* 1973
DD 700	HAY-NESWORTH	Decommissioned briefly 1950 Taiwanese *Yuen Yang* 1970
DD 701	JOHN W WEEKS	Stricken 12.8.70, BU
DD 702	HANK	Argentinian *Segui* 1972
DD 703	WALLACE L LIND*	Korean *Dae Gu* 1973
DD 704	BORIE*	Argentinian *Bouchard* 1972
DD 705	COMPTON	Brazilian *Matto Grosso* 1972
DD 706	GAINARD	To Iran Mar 1971
DD 707	SOLEY	Stricken 13.2.70, BU
DD 708	HARLAN R DICKSON	Stricken 1.7.72
DD 709	HUGH PURVIS*	Turkish *Zafer* 1972
DD 722	BARTON	Active 1949–68, stricken 1968, target
DD 723	WALKE*	Active 1950–70, stricken 1.2.74
DD 724	LAFFEY*	Active 1951–75, stricken 29.3.75
DD 725	O'BRIEN*	Active 1950–72, stricken 18.2.72, target
DD 727	DE HAVEN*	Active to 1971, Korean *Inchon* 1973
DD 728	MANSFIELD*	Active 1971, to Argentina 4.6.74
DD 729	LYMAN K SWENSON*	Active to 1971, to Taiwan 6.5.74
DD 730	COLLETT*	Active to 1970, to Argentina 4.6.74
DD 731	MADDOX	Active to 1972, Taiwanese *Po Yang* 1972
DD 732	HYMAN	Stricken 1969, BU 1970
DD 734	PURDY	Stricken 2.7.73

No	Name	Fate
DD 736	THOMAS E FRASER	Active to 1955, minelayer DM 24, stricken 1.11.70
DD 737	SHANNON	Active to 1955, minelayer DM 25, stricken 1.11.70
DD 738	HARRY F BAUER	Active to 1956, minelayer DM 26, stricken 15.8.71
DD 744	BLUE*	Active 1949–71, stricken 1.2.74
DD 745	BRUSH	Taiwanese *Hsiang Yang* 1969
DD 746	TAUSSIG*	To Taiwan 6.5.74
DD 747	SAMUEL N MOORE	Taiwanese *Heng Yang* 1969
DD 748	HARRY E HUBBARD	Stricken 1969, BU 1970
DD 750	SHEA	Active to 1958, minelayer DM 30, stricken 1973
DD 752	ALFRED A CUNNINGHAM	Active 1950–71, stricken 1.2.74
DD 753	JOHN R PIERCE	Stricken 2.7.74, BU 1975
DD 754	FRANK E EVANS*	Active 1950–69, sunk in collision with HMAS *Melbourne* 30.6.69
DD 755	JOHN A BOLE*	To Taiwan 6.5.74
DD 756	BEATTY	Venezuelan *Carabobo* 1972
DD 757	PUTNAM*	Laid up 1949–50, stricken 6.8.73
DD 758	STRONG*	Brazilian *Rio Grande de Norte* 31.10.73
DD 759	LOFBERG*	To Taiwan 6.5.74
DD 760	JOHN W THOMASON*	To Taiwan 6.5.74
DD 761	BUCK*	Brazilian *Alacoas* 16.7.73
DD 762	HENLEY	Briefly laid up 1950, stricken 2.7.73
DD 770	LOWRY*	Active 1950–73, Brazilian *Espiritu Santo* 1973
DD 772	GWIN	Decommissioned 1958, minelayer DM 33, Turkish *Muavenet* 1971
DD 775	WILLARD KEITH	Active 1950–72, Columbian *Caldas* 1.7.72
DD 776	JAMES C OWENS*	Briefly laid up 1950, decommissioned 1973, Brazilian *Sergipe* 16.7.73
DD 777	ZELLARS*	Decommissioned 1971, Iranian *Barb* 1971
DD 778	MASSEY*	Decommissioned 1969, stricken 19.9.73, BU 1975
DD 779	DOUGLAS H FOX*	Briefly laid up 1950, Chilean *Ministro Portales* 1974
DD 780	STORMES*	Briefly laid up 1950, decommissioned 1970, Iranian *Palang* 16.2.72
DD 781	ROBERT K HUNTINGTON*	Stricken 1973, Venezuelan *Falcon* 1973
DD 857	BRISTOL	Taiwanese *Hwa Yang* 1969

Gearing class: launched 1944–46, 2600t standard

No	Name	Fate
DD 710	GEARING	FRAM 1, stricken 1.7.73, BU 1975
DD 711	EUGENE A GREENE	DDR 1952, FRAM I, Spanish *Churruca* 1972
DD 712	GYATT	DDG 1956, no FRAM, stricken 1969, target
DD 713	KENNETH D BAILEY	DDR 1953, FRAM II, to Iran 13.1.75 for spares
DD 714	WILLIAM R RUSH	DDR 1953, FRAM I, Korean *Kang Won* 1974
DD 715	WILLIAM M WOOD	DDR 1953, FRAM I, stricken 1.12.76
DD 716	WILTSIE	FRAM I, Pakistani *Tariq* 1977
DD 717	THEODORE E CHANDLER	FRAM I, stricken 1.4.75, sold 1.12.75
DD 718	HAMNER	FRAM I, Taiwanese *Chao Yang* 1981
DD 719	EPPERSON	DDE, FRAM I, Pakistani *Taimur* 1977
DD 742	FRANK KNOX	DDR 1945, FRAM II, Greek *Themistocles* 30.1.71
DD 743	SOUTHERLAND	DDR 1945, FRAM I, for sale 1981
DD 763	WILLIAM C LAWE	FRAM I, stricken 1983
DD 764	LLOYD N THOMAS	DDE 1949, FRAM I, Taiwanese *Dang Yang* 1972
DD 765	KEPPLER	DDE 1949, FRAM II, Turkish *Tinaztepe* 1972
DD 782	ROWAN	FRAM I, Taiwanese *Chao Yang* 1977
DD 783	GURKE	FRAM I, Greek *Tombazis* 1977
DD 784	MCKEAN	FRAM I, for sale 1981
DD 785	HENDERSON	FRAM I, Pakistani *Tughril* 1980
DD 786	RICHARD B ANDERSON	FRAM I, Taiwanese *Kai Yang* 1975
DD 787	JAMES E KYES	FRAM I, Taiwanese *Chien Yang* 1.4.73
DD 788	HOLLISTER	FRAM I, stricken 1.10.79
DD 789	EVERSOLE	FRAM I, Turkish *Gayret* 11.7.73
DD 790	SHELTON	FRAM I, Taiwanese *Lao Yang* 1.4.73
DD 805	CHEVALIER	DDR 1945, FRAM II, Korean *Chung Buk* 1972
DD 806	HIGBEE	DDR 1945, FRAM I, stricken 15.7.79
DD 807	BENNER	DDR 1945, FRAM II, stricken 1974, target

No	Name	Fate
DD 808	DENNIS J BUCKLEY	FRAM I, stricken 2.7.73
DD 817	CORRY	DDR 1954, FRAM I, Greek *Kreizis* 1981
DD 818	NEW	FRAM I, Korean *Taejon* 1977
DD 819	HOLDER	FRAM I, Ecuadorian *Alfaro* 1978
DD 820	RICH	FRAM I, stricken 15.12.77, BU
DD 821	JOHNSTON	FRAM I, Taiwanese 1981. Sold National Metal & Steel Corp as Terminal Stand 13.10.66
DD 822	ROBERT H McCARD	FRAM I, Turkish *Kilic Ali Pasa* 1980
DD 823	SAMUEL B ROBERTS	FRAM I, stricken 2.11.70, BU
DD 824	BASILONE	DDE, FRAM I, stricken 1.11.77, BU
DD 825	CARPENTER	DDE, FRAM I, stricken 1980, Turkish *Anitepe* 1981
DD 826	AGERHOLM	FRAM I, stricken 1.12.78, target
DD 827	ROBERT A OWENS	DDE, FRAM I, Turkish *Aleiptepe* 1982
DD 828	TIMMERMAN	Experimental destroyer (machinery), stricken 1958
DD 829	MYLES C FOX	DDR 1945, FRAM II, Greek *Apostolis* 1980
DD 830	EVERETT F LARSON	DDR 1945, FRAM II, Korean *Jeong Buk* 30.10.72
DD 831	GOODRICH	DDR 1945, FRAM II, stricken 1974, BU 1978
DD 832	HANSON	DDR 1945, FRAM I, Taiwanese *Liao Yang* 1.4.73
DD 833	HERBERT J THOMAS	DDR 1945, FRAM I, Taiwanese *Han Yang* 1974
DD 834	TURNER	DDR 1945, FRAM II, stricken 26.9.69, BU 1970
DD 835	CHARLES P CECIL	DDR 1945, FRAM I, to Greece for spares 1980
DD 836	GEORGE K MACKENZIE	FRAM I, stricken 1.10.76, target
DD 837	SARSFIELD	FRAM I, Taiwanese *Te Yang* 1.10.77
DD 838	ERNEST G SMALL	DDR 1952, FRAM II, Taiwanese *Fu Yang* 1971
DD 839	POWER	FRAM I, Taiwanese *Shen Yang* 1.10.77
DD 840	GLENNON	FRAM I, stricken 1.10.76, BU
DD 841	NOA	FRAM I, Spanish *Blas de Lezo* 1975
DD 842	FISKE	DDR 1952, FRAM I, Turkish *Piyale Pasa* 1980
DD 843	WARRINGTON	FRAM I, to Taiwan 24.4.73
DD 844	PERRY	FRAM I, stricken 1973 (FRAM I prototype)
DD 845	BAUSSELL	FRAM I, stricken 30.5.78
DD 846	OZBOURN	FRAM I, stricken 1.6.75, sold 1.12.71
DD 847	ROBERT L WILSON	FRAM I, stricken 30.9.74
DD 848	WITEK	No FRAM, stricken 17.9.68, target
DD 849	RICHARD E KRAUS	FRAM I, Korean *Kwang Ju* 1972
DD 850	JOSEPH P KENNEDY, JR	FRAM I, stricken 2.7.73, memorial 1973
DD 851	RUPERTUS	FRAM I, Greek *Kontouriotis* 1973
DD 852	LEONARD F MASON	FRAM I, Taiwanese *Lai Yang* 1976
DD 853	CHARLES H ROAN	FRAM I, Turkish *Cakmak* 1973
DD 858	FRED T BERRY	DDE 1949, FRAM II, stricken 15.9.70, BU
DD 859	NORRIS	DDE 1949, FRAM II, Turkish *Kocatepe* 1974
DD 860	McCAFFERY	DDE 1949, FRAM II, stricken 30.9.73
DD 861	HARWOOD	DDE 1949, FRAM II, Turkish *Kocatepe* 1.12.71, sunk in error by Turkish aircraft 7.7.74
DD 862	VOGELGESANG	FRAM I, Mexican *Quietzéalcoatl* 1982
DD 863	STEINAKER	DDR 1953, FRAM I, Mexican *Netzahualcoyotl* 1982
DD 864	HAROLD J ELLISON	FRAM I, Pakistani *Shahjahan* 1983
DD 865	CHARLES R WARE	FRAM I, stricken 1974
DD 866	CONE	FRAM I, Pakistani *Alamgir* 1982
DD 867	STRIBLING	FRAM I, stricken 1.7.76
DD 868	BROWNSON	FRAM I, stricken 30.9.76, BU 1977
DD 869	ARNOLD J ISBELL	FRAM I, Greek *Sachtouris* 1973
DD 870	FECHTELER	DDR 1953, stricken 11.9.70, BU 1972
DD 871	DAMATO	FRAM I, Pakistani *Tippu Sultan* 1980
DD 872	FORREST ROYAL	FRAM I, Turkish *Adatepe* 1973
DD 873	HAWKINS	DDR 1945, FRAM I, stricken 1979, BU 1983
DD 874	DUNCAN	DDR 1945, FRAM II, stricken 1973, target
DD 875	HENRY W TUCKER	DDR 1945, FRAM I, Brazilian *Marcilio Dias* 1973
DD 876	ROGERS	DDR 1945, FRAM I, South Korean *Jeong Ju* 1981

No	Name	Fate
DD 877	PERKINS	DDR 1945, FRAM II, Argentinian *Py* 1973
DD 878	VESOLE	DDR 1945, FRAM I, stricken 1.12.76
DD 879	LEARY	DDR 1945, FRAM I, Spanish *Langara* 1975
DD 880	DYESS	DDR 1945, FRAM I, stricken 1980, Greek *Themistocles* 1981
DD 881	BORDELON	DDR 1945, FRAM I, stricken 1.2.77
DD 882	FURSE	DDR 1945, FRAM I, Spanish *Gravina* 1972
DD 883	NEWMAN K PERRY	DDR 1945, FRAM I, South Korean *Kyong Kai* 1981
DD 884	FLOYD B PARKS	FRAM I, stricken 2.7.73
DD 885	JOHN R CRAIG	FRAM I, stricken 27.7.79, target
DD 886	ORLECK	FRAM I, to Turkey 1982
DD 887	BRINKLEY BASS	FRAM I, Brazilian *Marriz e Barros* 1973
DD 888	STICKELL	FRAM I, DDR 1953, Greek *Kanaris* 1973
DD 889	O'HARE	DDR 1953, FRAM I, Spanish *Mendez Nunez* 1975
DD 890	MEREDITH	FRAM I, Turkish *Savastepe* 1979

Benson-Livermore class
Only the minesweeper conversions listed remained in service.

Fletcher class
Other ships of the class were also transferred abroad, but since they were never active in the US Navy after 1947 they have not been listed. Ships of this class marked* were modified with 4–5in/38 (4×1), 6–3in/50 (3×2), 5–21in TT (1×5).

Allen M Sumner class
Unless otherwise noted, all ships listed were in continuous service until stricken or transferred. Ships of this class marked* received a FRAM II modernisation.

Gearing class
All ships remained active after 1945. The remaining ships in 1982 were serving in the Naval Reserve Force as training units and were sold in 1983–84.

DESTROYER ESCORTS

No	Name	Fate
Buckley class (TE type): launched 1943–44, 1400t standard		
DE 213	WILLIAM T POWELL	Active (training) to 1958 as DER conversion 1945, stricken 1965
DE 217	COOLBAUGH	Active to 1959, stricken 1.7.72, BU 1973
DE 218	DARBY	Active 1950–58, then training, stricken 1968, target 1970
DE 219	J DOUGLAS BLACKWOOD	Active 1951–58, then training, active again during Berlin crisis 1961–62, stricken 30.1.70, target
DE 634	WHITEHURST	Active 1950–62, stricken 1969, BU
DE 644	VAMMEN	Active 1952–69, special ASW conversion
DE 667	WISEMAN	Active 1950–59, 1961–62, stricken 1973
DE 679	GREENWOOD	Active to 1958, then 1961–66, stricken 1967, BU
DE 680	LOESER	Active 1951–68, stricken 1968, BU
DE 696	SPANGLER	Active to 1958, stricken 1.3.72, BU 1973
DE 697	GEORGE	Active to 1958, stricken 1969, BU 1970
DE 698	RABY	Active to 1953, stricken 1968, BU
DE 699	MARSH	Active to 1958, then 1961–69, stricken 1971
DE 700	CURRIER	Active to 1960, stricken 1966, target
DE 704	CRONIN	DEC 1951–53, stricken 1.6.70, BU
DE 705	FRYBARGER	DEC 1950–54, stricken 1.12.72
DE 791	MALOY	DEC 1950–54, BU 1966
Rudderow class (TEV type): launched 1944, 1400t standard		
DE 583	GEORGE A JOHNSON	Active 1950–57, stricken 1965, BU
DE 585	DANIEL A JOY	Active 1950–65, stricken 1966, BU
DE 587	THOMAS F NICKEL	Active 1950–58, stricken 1.12.72
DE 684	DE LONG	Active 1951–69, stricken 1969, target
DE 685	COATES	Active 1951–70, stricken 30.1.70, BU
DE 708	PARLE	Active 1951–62, then training, last active Second World War DE, stricken 1.7.70, target
John C Butler class (WGT type): launched 1943–44, 1400t standard		
DE 255	SELLSTROM	DER 1956–60, stricken 1965, BU
DE 316	HARVESON	DER 1951–60, stricken 1966, target
DE 317	JOYCE	DER 1951–60, stricken 1.12.72, BU 1974
DE 318	KIRKPATRICK	DER 1952–60, stricken 1.8.74, target
DE 322	NEWELL	DER 1957–68, stricken 1968, BU 1972
DE 324	FALGOUT	CG service 1951–54, DER 1955–75, stricken 1.6.75, target
DE 325	LOWE	CG service 1951–54, DER 1955–68, stricken Sep 1968

No	Name	Fate
DE 326	THOMAS J GARY	DER 1957–73, Tunisian *President Bourgiba* 1973
DE 327	BRISTER	DER 1956–68, stricken 23.9.68, BU 1972
DE 328	FINCH	CG service 1951–4, DER 1956–73, stricken 1.2.74, BU 1975
DE 329	KRETCHMER	CG 9151–54, DER 1956–73, stricken 30.9.73
DE 331	KOINER	CG 1951–54, DER 1955–68, stricken 1968, BU
DE 332	PRICE	DER 1956–60, stricken 1.8.74, target
DE 333	STRICKLAND	DER 1952–60, stricken 1.12.72, sold 1975
DE 334	FORSTER	CG 1951–4, DER 1956–71, Vietnamese *Tran Kanh Du* 25.9.71
DE 336	ROY O'HALE	DER 1957–63, stricken 1.8.74, target
DE 339	JOHN C BUTLER	Active 1950–57, stricken 1.6.70, BU
DE 341	RAYMOND	Active 1951–58, stricken 1.7.72
DE 351	MAURICE J MANUEL	Active 1951–57, stricken 1.5.66, target
DE 352	NAIFEH	Active 1951–60, stricken 1966, target
DE 354	KENNETH M WILLETT	Active 1951–59, stricken 1.7.72, target 1974
DE 357	GEORGE E DAVIS	Active 1951–4, stricken 1.12.72
DE 359	WOODSON	Active 1951–52, stricken 1.7.65, BU
DE 360	JOHNNIE HUTCHINS	Active 1950–58, stricken 1.7.72, BU 1974
DE 361	WALTON	Active 1951–68, stricken 1968, target 1969
DE 364	ROMBACH	Active to 1958, stricken 1.3.72
DE 365	McGINTY	Active 1951–58, 1961–62, stricken 1968, BU
DE 366	ALVIN C COCKRELL	Active 1951–68, stricken 1968, target 1969
DE 369	THADDEUS PARKER	Active 1951–67, stricken 1967, BU
DE 406	EDMONDS	Active 1951–72, stricken 15.5.72, BU 1974
DE 414	LE RAY WILSON	Active 1951–59, stricken 15.5.72, BU 1974
DE 416	MELVIN R NAWMAN	Active 1951–60, stricken 1.7.72
DE 418	TABBERER	Active 1951–60, stricken 1.7.72, BU 1974
DE 419	ROBERT F KELLER	Active 1950–59, 1961–62, stricken 1.7.72, BU 1974
DE 421	CHESTER T O'BRIEN	Active 1951–59, stricken 1.7.72
DE 422	DOUGLAS A MUNRO	Active 1951–60, target 1966
DE 424	HAAS	Active 1951–58, stricken 1966, BU
DE 440	McCOY REYNOLDS	Active 1951–57, Portuguese *Corte Real* 1957
DE 441	WILLIAM SEIVERLING	Active 1950–57, stricken 1.12.72
DE 442	ULVERT M MOORE	Active 1951–58, stricken 1965, sunk as target Jul 1966
DE 444	GOSS	Active 1951–58, stricken 1.3.72, BU 1973
DE 445	GRADY	Active 1951–57, stricken 1968, BU
DE 446	CHARLES E BRANNON	Active 1950–60, stricken 1968, BU
DE 447	ALBERT T HARRIS	Active 1951–68, stricken 1968, BU
DE 448	CROSS	Active 1951–58, stricken 1966, BU
DE 449	HANNA	Active 1950–59, stricken 1.12.72
DE 508	GILLIGAN	Active 1950–59, stricken 1.3.72, BU 1973
DE 509	FORMOE	Active 1951–57, Portuguese *Diogo Cao* 1957
DE 510	HEYLIGER	Active 1951–58, stricken 1966, BU
DE 531	EDWARD H ALLEN	Active 1951–58, stricken 1.7.72
DE 532	TWEEDY	Active 1952–69, ASW conversion, stricken Jul 1969, target 1970

Tweedy (ASW conversion) 1958 USN

No	Name	Fate
DE 534	SILVERSTEIN	Active 1951–59, stricken 1.12.72
DE 535	LEWIS	Active 1952–60, ASW conversion, stricken 1.1.66, target
DE 537	RIZZI	Active 1951–58, stricken 1.8.72
DE 538	OSBERG	Active 1951–57, stricken 30.6.74, sold 1974
DE 539	WAGNER	Completed as DER 1955, decommissioned 1960, stricken 1.6.75, BU 1978
DE 540	VANDIVER	Completed as DER 1955, decommissioned 1960, stricken 1.11.74
Cannon class (DET type): launched 1943–44, 1200t standard		
DE 745	SNYDER	Active 1950–60, stricken 1.8.72, BU 1973
Edsall class (FMR type): launched 1942–43, 1250t standard		
DE 746	HEMMINGER	Active 1950–58, Thai *Pin Klao* 1959
DE 748	TILLS	Active 1950–68, stricken 1968, target
DE 749	ROBERTS	Active 1950–64, stricken 1968, BU
DE 750	McCLELLAND	Active 1950–60, stricken 1.8.72, BU 1974
DE 765	EARL K OLSEN	Active 1950–58, stricken 1.8.72
Edsall class (FMR type): launched 1942–43, 1250t standard		
DE 133	PILLSBURY	DER 1955–60, stricken 1965, BU
DE 142	FESSENDEN	DER 1952–60, stricken 1966, target 1967
DE 145	HUSE	Active 1951–65, stricken 1.8.73
DE 147	BLAIR	Active 1951–56, DER 1957–72, stricken 1.12.72, sold 1975
DE 239	STURTEVANT	Active 1951–56, DER 1957–60, stricken 1972, BU 1974
DE 244	OTTERSTETTER	DER 1952–60, stricken 1.8.74, target
DE 246	SNOWDEN	Active 1951–60, 1961–62, stricken 1968, BU
DE 251	CAMP	DER 1956–71, Vietnamese *Tran Hung Dao* 1970, escaped to Philippines 1975, became *Rajah Lakandula*
DE 252	HOWARD D CROW	Active 1951–58, stricken 1965, BU 1970
DE 382	RAMSDEN	CG 1951–54, DER 1957–60, stricken 1.8.74, BU
DE 383	MILLS	DER 1957–70, stricken 1.8.74
DE 384	RHODES	DER 1955–63, stricken 1.8.74, target
DE 386	SAVAGE	DER 1955–69, stricken 1.6.75, target
DE 387	VANCE	CG 1951–54, DER 1956–69, stricken 1.6.75, target
DE 388	LANSING	CG 1951–54, DER 1956–65, stricken 1.2.74, BU 1975
DE 389	DURANT	CG 1951–54, DER 1955–74, stricken 1.4.74, BU 1975
DE 390	CALCATERRA	DER 1955–73, stricken 2.7.73
DE 391	CHAMBERS	CG 1951–54, DER 1955–60, stricken 1.3.75, sold 1975
DE 393	HAVERFIELD	DER 1955–69, stricken 1969, BU 1972
DE 397	WILHOITE	DER 1955–69, stricken 1969, BU
DE 400	HISSEM	DER 1956–70, stricken 1.6.75, target
DE 797	WEEDEN	Active 1950–58, stricken 1968, BU
DE 800	JACK W WILKE	Active to 1960, stricken 1.8.72, sold 1974

Active periods do not always include period of Naval Reserve Training. Ships were reactivated in 1961 for the Berlin Crisis.

Lansing (DER conversion) 16.11.63 USN

FRIGATES

No	Name	Fate
Tacoma class: launched 1943, 1500t standard		
PF 3	TACOMA	Activated 1950, Korean *Taedong* 1951
PF 4	SAUSALITO	Activated 1950, Korean *Imchin* 1953
PF 5	HOQUIAM	Activated 1950, Korean *Naktong* 1951
PF 7	ALBUQUERQUE	Activated 1950, Japanese *Tochi* 1953
PF 8	EVERETT	Activated 1950, Japanese *Kiri* 1953
PF 21	BAYONNE	Activated 1950, Japanese *Buna* 1953
PF 22	GLOUCESTER	Activated 1950, Japanese *Tsuge* 1953
PF 27	NEWPORT	Activated 1950, Japanese *Kaede* 1953
PF 36	GLENDALE	Activated 1950, Thai *Tachin* 1951
PF 46	BISBEE	Activated 1950, Colombian *Capitan Tono* 1952
PF 47	GALLUP	Activated 1950, Thai *Prasae* 1951
PF 51	BURLINGTON	Activated 1950, decommissioned at Yokosuka 1952, Colombian *Almirante Bron* 1953
PF 70	EVANSVILLE	Activated 1950, Japanese *Keyaki* 1953

Second World Patrol Frigates (PF) with reciprocating engines. All were stricken after the war, but some, returned by the Soviet Union, were in Japan at the outbreak of the Korean War and were re-acquired by the United States for local escort duties. Each ship was active for a year or more under the US flag, some operating during the Wonsan landings. Two frigates, PF 48 and PF 49, were lent to South Korea without US use in 1950.

SUBMARINES

No	Name	Fate
Gato class: launched 1941–43, 1500t standard		
SS 214	GROUPER	Active to 1968, SSK conversion 1951, stricken 1968, BU 1970
SS 220	BARB	Active 1951–54, converted to GUPPY 1B and transferred to Italy as *Enrico Tazzoli* 1955
SS 222	BLUEFISH	Active 1952–53, stricken 1959
SS 229	FLYING FISH	Active to 1954, AGSS 1950, stricken 1958, BU 1960
SS 230	FINBACK	Decommissioned 1950, stricken 1958, BU 1959
SS 240	ANGLER	Active 1951–68, SSK 1953, stricken 15.12.71
SS 241	BASHAW	Active 1949, then 1951–69, SSK 1953, stricken 1969
SS 242	BLUEGILL	Active 1951–69, SSK 1953, stricken 1969, scuttled 1970
SS 243	BREAM	Active 1951–69, SSK 1953, stricken 1969, target
SS 244	CAVALLA	Active 1951–68, SSK 1953, stricken 1969, memorial
SS 246	CROAKER	Active 1951–68, SSK 1953, stricken 20.12.71, memorial 1976
SS 247	DACE	Active 1951–54, converted to GUPPY 1B for transfer to Italy as *Leonardo da Vinci* 1955
SS 267	POMPON	SSR 1951–60, stricken 1960
SS 269	RASHER	SSR 1951–67, stricken 20.12.71, BU 1975
SS 270	RATON	SSR 1952–69, stricken 1969, expended 1969
SS 271	RAY	SSR 1951–58, stricken 1960
SS 272	REDFIN	SSR 1953–67, stricken 1970, BU 1971
SS 274	ROCK	SSR 1953–69, stricken 1969
SS 282	TUNNY	SSG 1953–65, APSS 1966–69, stricken 1969, expended
Balao class: launched 1942–45, 1525t standard		
SS 285	BALAO	Active 1952–53, stricken 1.8.63
SS 287	BOWFIN	Active 1951–54, then training, stricken 1.12.71 memorial 1979
SS 291	CREVALLE	Active 1951–55, 1957–62, stricken 1968, BU 1971
SS 298	LIONFISH	Active 1951–53, then training, stricken 20.12.71, memorial 1972
SS 299	MANTA	Active 1949–55, then training, stricken 1967, target 1969

Redfin Nov 1957 L & L van Ginderen

No	Name	Fate
SS 302	SABALO	Active 1951–71, Fleet Snorkel 1952, stricken 1971, preserved 1973
SS 303	SABLEFISH	Active to 1969, Fleet Snorkel 1951, stricken 1969, BU 1971
SS 307	TILEFISH	Active to 1959, Fleet Snorkel 1960 for Venezuela as *Carite*
SS 309	ASPRO	Active 1951–54, stricken 1962, target
SS 310	BATFISH	Active 1952–60, then training, stricken 1969, memorial 1972
SS 311	ARCHERFISH	Active 1952–55, 1957–68, stricken 1969, target
SS 312	BURRFISH	SSR 1948–56, to Canada as *Grilse* 1961
SS 313	PERCH	SSP 1948–67, stricken 1.12.71
SS 315	SEALION	SSP 1948–60, then training, stricken 15.3.77, test hulk
SS 317	BARBERO	SSA 1948–50, SSG 1955–64, stricken 1.7.64 target
SS 318	BAYA	Active 1948–72 as sonar test ship, stricken 30.10.72, BU 1974
SS 319	BECUNA	Active to 1969, GUPPY 1A 1951, stricken 1973, memorial 1976
SS 320	BERGALL	Active to 1959, Fleet Snorkel 1952, to Turkey as *Turgut Reis* 1958
SS 322	BLACKFIN	Active 1951–72 as GUPPY 1A, stricken 30.6.72, target to Greece 1972
SS 323	CAIMAN	Active to 1972, GUPPY 1A 1951, to Turkey as *Dumlupinar* 1972
SS 324	BLENNY	Active to 1969, GUPPY 1A 1951, stricken 1973, target
SS 325	BLOWER	Active to 1950, Fleet Snorkel for Turkey as *Dumlupinar*
SS 326	BLUEBACK	Active to 1948, to Turkey as *Ikinci Inonu* 1948
SS 327	BOARFISH	Active to 1948, to Turkey as *Sakarya* 1948
SS 329	CHUB	Active to 1948, to Turkey as *Gur* 1948
SS 330	BRILL	Active to 1948, to Turkey as *Birinci Inonu* 1948
SS 331	BUGARA	Active to 1970, Fleet Snorkel 1951, stricken 1.10.70, foundered 1971
SS 333	BUMPER	Active to 1950, Fleet Snorkel for Turkey as *Canakkale* 1950
SS 336	CAPITAINE	Active to 1950, then 1957–66, Fleet Snorkel for Italy as *Alfredo Cappellini* 1966
SS 337	CARBONERO	Active to 1970, SSG 1949, Fleet Snorkel 1952, stricken 1.12.70, target
SS 339	CATFISH	Active to 1971, GUPPY II 1949, Argentine *Santa Fe* 1971
SS 338	CARP	Active to 1968, Fleet Snorkel 1952, stricken 20.12.71
SS 340	ENTEMEDOR	Active to 1948, then 1950–72, GUPPY IIA 1952, Turkish *Preveze* 1973
SS 341	CHIVO	Active to 1971, GUPPY 1A 1951, Argentine *Santiago del Estero* 1.7.71
SS 343	CLAMAGORE	Active to 1975, GUPPY II 1948, GUPPY III 1962, stricken 27.6.75, memorial 1979
SS 344	COBBLER	Active to 1973, GUPPY II 1949, GUPPY 1962, Turkish *Canakkale* 1973
SS 345	COCHINO	GUPPY II 1949, sunk 26.8.49 off Norway
SS 346	CORPORAL	Active to 1973, GUPPY II 1948, GUPPY III 1962, Turkish *Ikinci Inonu* 1973
SS 347	CUBERA	Active to 1972, GUPPY II 1948, Venezuelan *Tiburon* 1972
SS 348	CUSK	Active to 1969, missile test bed 1948, then Fleet Snorkel 1954, stricken 24.9.69, BU 1972
SS 349	DIODON	Active to 1971, GUPPY II 1984, stricken 15.1.71, BU 1972
SS 350	DOGFISH	Active to 1972, GUPPY II 1948, Brazilian *Guanabara* 1972
SS 351	GREENFISH	Active to 1973, GUPPY II 1948, GUPPY III 1961, Brazilian *Amazonas* 1973
SS 352	HALFBEAK	Active to 1971, GUPPY II 1948, stricken 1.7.71, BU 1972
SS 362	GUAVINA	SSO 1950–59, then training, stricken 1967, target
SS 363	GUITARRO	Active 1952–53, Fleet Snorkel for Turkey as *Preveze* 1954
SS 364	HAMMERHEAD	Active 1952–53, Fleet Snorkel for Turkey as *Cerbe* 1954
SS 365	HARDHEAD	GUPPY IIA 1952–72, Greek *Papanikolis* 1972
SS 368	JALLAO	GUPPY IIA 1954–74, then to Spain as *S 35*
SS 377	MENHADEN	Active 1951–71, GUPPY IIA 1953, stricken 1973, target
SS 382	PICUDA	Active 1953–73, GUPPY IIA 1953, Spanish *Narciso Monturiol* 1972
SS 385	BANG	Active 1951–72, GUPPY IIA 1952, Spanish *Cosme Garcia* 1972
SS 391	POMFRET	Active 1952–71, GUPPY IIA 1953, Turkish *Oruc Reis* 1.7.71
SS 392	STERLET	Active 1950–68, Fleet Snorkel 1952, stricken 1968, target 1969
SS 393	QUEENFISH	Active to 1963, stricken, target 1963
SS 394	RAZORBACK	Active to 1952, GUPPY IIA 1954–70, Turkish *Murat Reis* 1970
SS 395	REDFISH	Active to 1968, stricken 1968, target 1969
SS 396	RONQUIL	Active to 1952, GUPPY IIA 1953–71, Spanish *Isaac Peral* 1.7.71
SS 397	SCABBARDFISH	Active to 1948, Fleet Snorkel for Greece as *Trianna* 1965
SS 398	SEGUNDO	Active to 1970, Fleet Snorkel 1951, stricken 8.8.70, target
SS 399	SEA CAT	Active to 1968, Fleet Snorkel 1952, stricken 2.12.68
SS 400	SEA DEVIL	Active to 1948, 1951–54, 1957–64, stricken 1964, target 1965
SS 402	SEA FOX	GUPPY IIA 1953–70, Turkish *Burak Reis* 1970
SS 403	ATULE	GUPPY IA 1951–73, Peruvian *Pacocho* 1974
SS 404	SPIKEFISH	Active to 1963, stricken 1.5.63, target
SS 405	SEA OWL	Active 1969, Fleet Snorkel 1951, stricken 1969, BU 1971
SS 406	SEA POACHER	Active to 1973, GUPPY IA 1952, Peruvian *La Pedrera* 1974
SS 407	SEA ROBIN	Active to 1970, GUPPY IA 1951, BU 1971
SS 408	SENNET	Active to 1968, Fleet Snorkel 1952, stricken 2.12.68
SS 409	PIPER	Active to 1967, Fleet Snorkel 1951, stricken 1.7.70
SS 410	THREADFIN	Active 1953–72, GUPPY IIA 1953, Turkish *Birinci Inonu* 1973
SS 415	STICKLEBACK	Active 1951–58, GUPPY IIA 1953, sunk by ramming 30.5.58
SS 416	TIRU	Completed as GUPPY II 1948, GUPPY III 1959, stricken 1.7.75
Tench class: launched 1944–46, 1570t standard		
SS 417	TENCH	Active 1950–70, GUPPY IA 1951, stricken 1973, to Peru 1976
SS 418	THORNBACK	Active 1953–71, GUPPY IIA 1953, Turkish *Uluc Ali Reis* 1973
SS 419	TIGRONE	SSR 1948–57, also active 1962–75 as AGSS, stricken 25.10.75, target
SS 420	TIRANTE	GUPPY IIA 1952–73, stricken 1.10.73
SS 421	TRUTTA	Active 1951–72, GUPPY IIA 1953, Turkish *Cerbe* 1972
SS 422	TORO	Active 1947–63, stricken 1.4.63, BU 1965
SS 423	TORSK	Active to 1968, Fleet Snorkel 1952, stricken 15.12.71, memorial 1972
SS 424	QUILLBACK	GUPPY IIA 1953–73, stricken 23.3.73
SS 425	TRUMPETFISH	Active to 1973, GUPPY II 1948, GUPPY II 1962, Brazilian *Goias* 1973
SS 426	TUSK	Active to 1973, GUPPY II 1948, Taiwanese *Hai Pao* 1973
SS 435	CORSAIR	Active 1945–63, stricken and BU 1963
SS 475	ARGONAUT	Active to 1968, Fleet Snorkel 1952, Canadian *Rainbow* 1968
SS 476	RUNNER	Active to 1970, Fleet Snorkel 1952, stricken 15.12.71
SS 477	CONGER	Active to 1963, stricken 1.8.63, BU 1964
SS 478	CUTLASS	Active to 1973, GUPPY II 1948, Taiwanese *Hai Shih* 1973
SS 479	DIABOLO	Active to 1964, Fleet Snorkel for Pakistan as *Ghazi* 1964
SS 480	MEDREGAL	Active to 1970, Fleet Snorkel 1952, stricken 1.8.70, BU 1972
SS 481	REQUIN	SSR 1946–59, active to 1968, stricken 20.12.71, memorial
SS 482	IREX	Prototype Fleet Snorkel conversion 1947, active to 1969, BU 1971
SS 483	SEA LEOPARD	Active to 1973, GUPPY II 1949, to Brazil 1973
SS 484	ODAX	Prototype GUPPY I 1947, GUPPY II 1951, active to 1972, Brazilian *Rio de Janeiro* 1972
SS 485	SIRAGO	Active to 1972, GUPPY II 1949, stricken 1.6.72
SS 486	POMODON	Active to 1970, GUPPY I 1947, GUPPY II 1951, stricken 1.8.70, BU 1971
SS 487	REMORA	Active to 1973, GUPPY II 1947, GUPPY III 1962, Greek *Katsonis* 1973
SS 488	SARDA	Active to 1964, BU 1965
SS 489	SPINAX	Completed as SSR, active to 1969, stricken 11.10.69, BU 1972
SS 490	VOLADOR	Active to 1972, completed as GUPPY II 1958, GUPPY III 1963, to Italy as *Gianfranco Gazzana Priarogia* 1972
SS 522	AMBERJACK	Active to 1973, GUPPY II 1947, Brazilian *Ceara* 1973

No	Name	Fate
SS 523	GRAMPUS	Active to 1972, completed as GUPPY II 1950, Brazilian *Rio Grande Do Sul* 1972
SS 524	PICKEREL	Completed as GUPPY II 1949, active to 1972, GUPPY III 1962, Italian *Primo Longobardo* 1972
SS 525	GRENADIER	Completed as GUPPY II 1951, active to 1973, Venezuelan *Picua* 1973

The SSA was a cargo submarine, the SSP were troop-carriers, and the SSO a submarine oiler. These and other conversions are covered in detail in the main tables below.

SSRs and SSKs reverted to fleet submarines about 1960; dates are *entire* active life of the submarine, *not* merely period of specialised service. Other units were very briefly reactivated before transfer abroad; they are not listed. Note, too, that units for Reserve Training Only are not listed; they were carried as 'In Service In Reserve'.

MINE WARFARE VESSELS

After the postwar rundown, only ten large minesweepers (AM 108, 109, 127, 277, 373, 384, 385, 286, 288 and 390) remained active on 1 Jan 1950. The following were recommissioned for the Korean War: AM 62, 100, 101, 104, 107, 110, 111, 112, 114, 117, 119, 122, 123, 126, 131, 220, 239, 249, 252, 255, 261, 275, 283, 296, 315, 316, 317, 319, 320, 323, 341, 372, 374, 375, 376, 377, 378, 380, 382, 383, 387, 389

Similarly, of fifty-seven motor minesweepers (AMS, formerly YMS) on the Navy List in 1950, only twenty-seven were active. All inactive units were recommissioned for Korean War service.

PATROL CRAFT

Most of the wartime patrol craft were discarded after 1945. By 1950 there were 127 173ft PC (of which thirteen were active), as well as twenty-nine 180ft PCE and sixteen wooden 136ft PCS. Four wartime motor torpedo-boats (PT) were retained for experimentation, awaiting the completion of the four experimental metal-hulled type ordered after 1945. As an indication of the decline of these craft, by 1958 only seven PC were active, together with seven PCE and one PCS; no PT were left.

AMPHIBIOUS CRAFT

Even after the postwar rundown, the Navy still owned 945 of the larger amphibious types in 1950. They included thirty attack cargo ships, fifty-seven 'Victory' type attack transports, seventeen former C-3 type attack transports, ninety-two fast transports (APD), twenty-one landing ships dock (LSD), six LSV (which were never reactivated), 135 LST (of which twenty-six were active, with another sixty-five operating as auxiliaries (and two more of postwar construction), 161 LSM, forty-eight LSMR, 106 support landing ships (LSSL, formerly LCSL), 105 infantry landing ships (LSIL, formerly LCIL), and 129 utility landing ships (LSU, formerly LCT). In addition, the Maritime Commission reserve fleet included thirty-six discarded attack cargo ships (ten C2, twenty-six S4) and eighty attack transports (sixty Victory, fifteen S4, the two specially-built *Doyens*, and three older units). As part of the Korean War programme, all thirty AKA were recommissioned, and six more (AKA 3, 4, 96, 97, 99, and 103) were re-acquired from the Maritime Commission. To the forty-two active APA were added twenty-nine more; twelve APD were recommissioned in addition to nineteen active units. All LSD were active by the end of 1951, as were ninety-six LST, twenty-two LSM, and seventeen LSMR. It does not appear that any LSI or LSSL were activated, and transfers abroad of LSSL began at this time. In addition, thirty-two LSIL were converted to Underwater Object Locators (AMCU), ten earlier units having been discarded (leaving one, AMCU-11).

FLAGSHIPS

Of fifteen ships completed as amphibious command ships, five remained active after the postwar rundown: AGC 7, 8, 11, 15 and 17. Two more, AGC 12 and 16, were reactivated for the Korean War, but another pair (AGC 8 and 15) were laid up during the post-Korea rundown. The inactive units were all stricken in 1959–61, but the five active ones continued throughout the Vietnam War. *Eldorado* (AGC 11) was the last, decommissioned in 1973; her four sisters left service in 1969–71. Service was quite varied. For example, *Adirondack* (AGC 15) was flagship of the Operational Development Force in 1945–49 and as command ship for NATO forces in Southern Europe 1951–53. Others served as command ships in atomic tests and as headquarters ships moored abroad.

THE COAST GUARD

The active Coast Guard Force of the postwar period included thirteen *Owasco* (255ft) cutters, decommissioned in 1973–74 (except for *Sebago*, 1972, and *Iroquois*, 1965), six 'Secretary' class (327ft) cutters of prewar construction (four of which, *Bibb, Duane, Ingham* and *Campbell*, remain active as of this writing), three 250ft 'Lake' class cutters launched in 1938–41 (*Itasca*, active 1946–50, *Tampa*, 1947–54, and *Mocoma*, 1947–50), one 165ft 'A' class cutter (*Tahoma*, decommissioned 1953), eleven 165ft 'B' class cutters (*Ariadne*, decommissioned 1968, *Atalanta*, 1950, *Aurora*, 1968, *Cyane*, 1950, *Daphne*, 1950, *Dione*, 1963, *Nemesis*, 1964, *Nike*, 1964, *Pandora*, 1959, *Perseus*, 1959, *Triton*, 1967) and twenty 125ft cutters. It is not clear which of the 230 83-footers remained active postwar.

Note In the tables which follow, all displacement figures are official; they do not represent the results of inclining experiments. Missile magazine capacities include ready-service rounds except where specified. Endurance figures represent design requirements, *not* service figures. Horsepower and speed of nuclear warships are generally classified, and the figures presented are published estimates unless otherwise stated.

MAJOR SURFACE SHIPS

United States design
NB 1/1750 scale

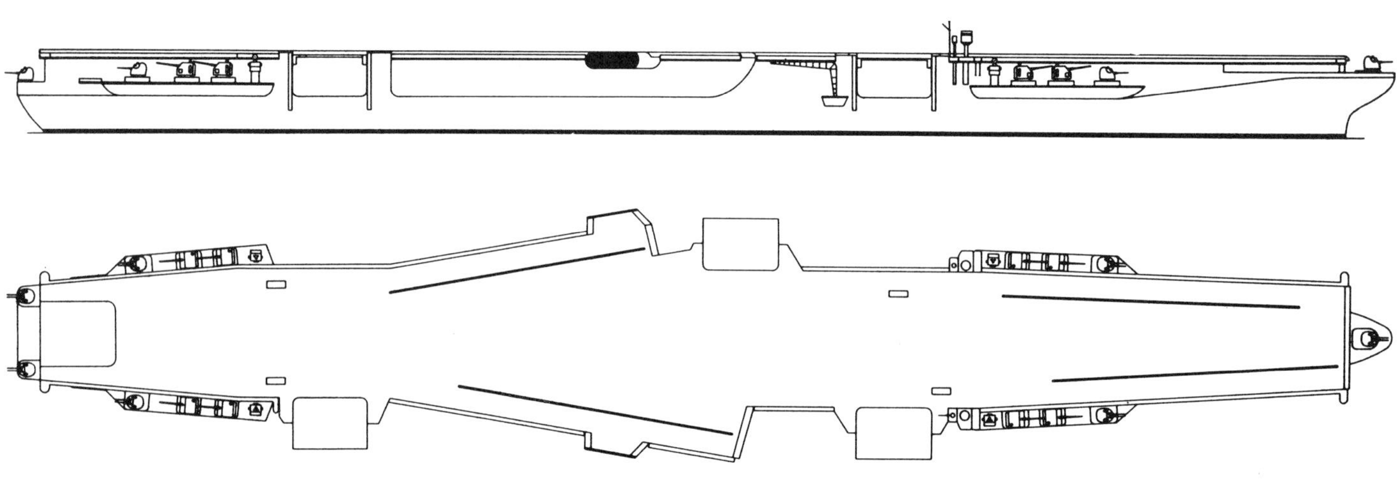

Although she was never completed, *United States* is significant in the course of US carrier development as the progenitor of the *Forrestal*s and their successors. Unlike earlier carriers, she was designed specifically to operate a particular class of aircraft, in her case heavy attack planes intended for strategic nuclear operations. Indeed, the earliest proposals were for a carrier without any hangar at all, because the aircraft would be

UNITED STATES *aircraft carrier*

Displacement: 66,434t standard; 78,946t trials; 83,249t full load
Dimensions: 1030ft wl, 1088ft oa × 125ft wl, 190ft ext × 34ft 6in at trials displacement
314.0m, 331.7m × 38.1m, 57.9m × 10.5m
Machinery: 4-shaft geared turbines, 8 boilers, 280,000shp = 33kts. Range 12,000nm at 20kts
Armament: 8–5in/54, (8×1) 16–3in/70 (8×2), 20–20mm
Sensors: Radar SPS-6, SPS-8
Complement: 4127

No	Name	Builder	Laid down	Fate
CVA 58	UNITED STATES	Newport News	18.4.49	Cancelled 23.4.49

much too large to strike below. Only well into the design were non-strategic roles considered, and indeed *United States* would have had smaller magazines than the later, multi-mission, *Forrestals*. It appears that the hangar was originally included largely to house the fighter escorts for the heavy bombers. Much attention went into flight deck arrangement, the ship being intended to launch two bombers and two fighters simultaneously from a pair of bow catapults and a pair of waist catapults splayed out to each beam. Steam catapults had not yet been invented, and she was to have had a new type of internal-combustion cylindrical catapult, which was also originally specified for the *Forrestals*. A fully flush deck would have simplified flight operations, and her arrangement was to have been superficially similar to that of the later angled deck ships. However, the arrangement of arresting gear would have precluded the simultaneous launch and recovery operations which are the hallmark of the angled deck carrier. The *United States* was designed to carry 500,000 gal (US) of Avgas (aviation fuel) and 2000t of aviation ordnance.

It appears that four ships in all were contemplated, and that later units might have had nuclear power. Certainly smoke disposal from a large flush deck carrier was considered a major and intractable issue. *United States* herself was laid down at Newport News on 19 April 1949, but the Air Force charged that the new carrier would merely duplicate its strategic mission, at a cost of $500m (against a Navy estimate of $189m), and construction was stopped on 23 April, the funds being transferred to the Air Force for bombers. The ensuing fight resulted in one of the Navy's worst peacetime disasters; heavy attack aircraft were restricted to dimensions dictated by the wartime *Midway* class. The major postwar jet heavy attack bomber, the A-3 (A3D) was designed at this time. Only the Korean War brought back the Navy's carrier programme.

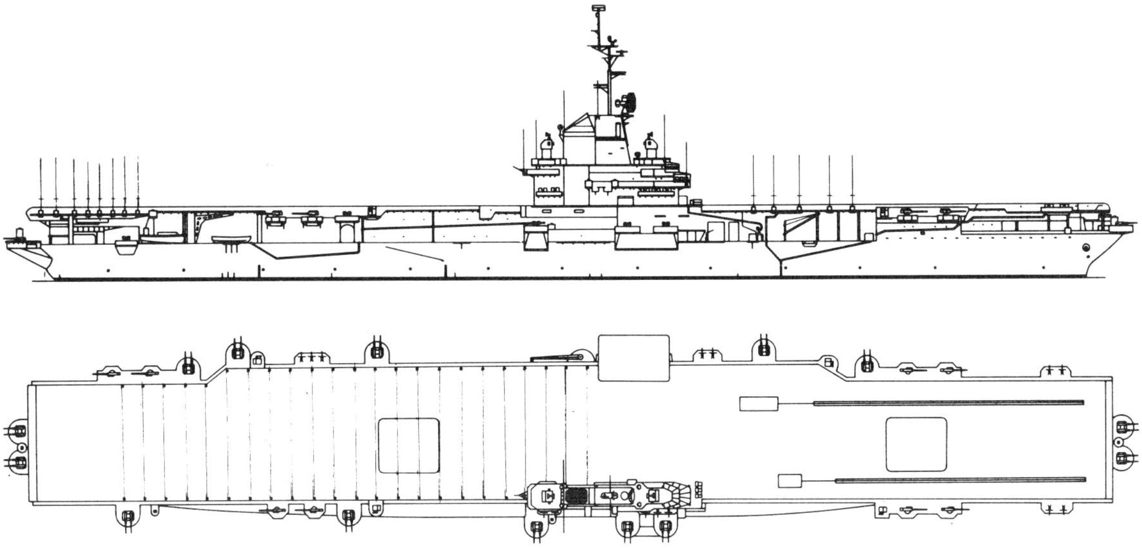

Oriskany 1950
NB 1/1750 scale

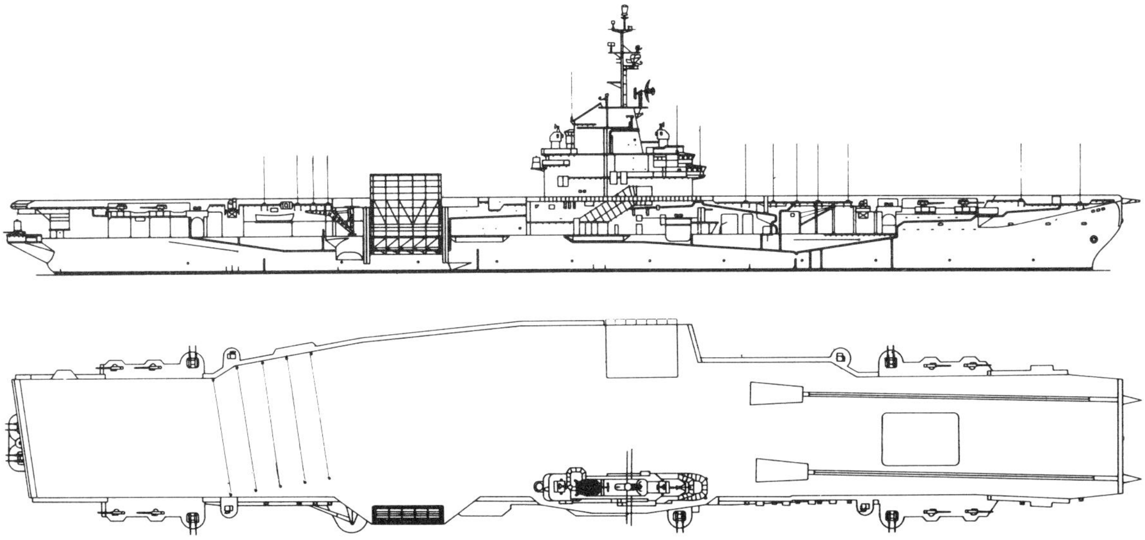

Shangri La 1957
NB 1/1750 scale

ESSEX class (SCB-27A reconstruction) *aircraft carriers*

Displacement: 28,404t light; 40,600t full load
Dimensions: 819ft wl, 898ft 2in oa × 101ft 4in wl, 151ft 11in ext × 29ft 8in
249.7m, 273.8m × 30.9m, 46.3m × 9.1m
Machinery: 4-shaft geared turbines, 8 boilers, 150,000shp = 30kts. Range 15,000nm at 15kts
Armament: 8–5in/38 (8×1), 28–3in/50 (14×2), approx 80 aircraft
Sensors: Radar SPS-6, SC or SR, SPS-8 (SX at first) (SPS-6 was later replaced by SPS-12 and then by SPS-29; SPS-8 was later replaced by SPS-30); sonar SQS-23 (CVSs only)
Complement: 2905

ESSEX class (SCB-27C reconstruction) *aircraft carriers*

Displacement: 30,580t light; 43,060t full load
Dimensions: 820ft wl, 894ft 6in oa × 103ft wl, 166ft 10in ext × 30ft 4in
250.0m, 272.7m × 31.4m, 50.9m × 9.2m
Machinery: 4-shaft geared turbines, 8 boilers, 150,000shp = 29.1kts. Range 15,000nm at 15kts
Armament: 4–5in/38 (4×1)
Sensors: Radar SPS-12, SPS-8 (ultimately SPS-37A, SPS-30), SR (later SPS-12)
Complement: 2585 (ship), 960 (air group)

For names of reconstructed ships and ultimate fates see under Fleet Strength 1947. Original data, including construction particulars, can be found in the 1922–1946 volume. The SCB-27C particulars given above refer to ships with angled decks near the end of their operational lives. CV 11, 14 and 19 were originally converted with axial decks, 8–5in/38, 24–3in/50 (12×2) and open bows.

At the end of the Second World War the United States had what it considered an obsolescent carrier fleet: the twenty-four *Essex* class carriers then either completed or completing could not operate the coming generation of jet aircraft. A follow-on fleet carrier design, completed in 1946, was not built, and efforts were begun to adapt the existing hulls to the new conditions. The incomplete *Oriskany* was suspended pending completion of a new design, and most of the existing wartime ships were laid up in the postwar run-down. As completed, to a revised SCB-27A design in 1950, *Oriskany* lacked the flight deck gun mounts of her near-sisters. Her flight deck was stronger, and she was fitted with the most powerful hydraulic catapults available, H-8s. Four 5in/38 guns were mounted on the starboard side of her flight deck to supplement the four usually mounted to port. There were also important internal changes. For example, the ready rooms, which in standard *Essex* class carriers were in the gallery deck, were moved below the armoured hangar deck and an escalator fitted outboard of the island in these ships. There were also major revisions to the island structure itself, for better radar antenna coverage. The hull was blistered, but the blister was faired into the hull all the way to the hangar deck level, so that it was not visible to a casual observer. Eight other ships were withdrawn from reserve for SCB-27A modernisation (CV 9, 10, 12, 15, 18, 20, 29 and 33).

A proposed austere SCB-27B conversion programme was dropped, and instead six ships were completed to SCB-27C standard, with a slightly wider blister, two steam catapults (C-11), and the after centreline elevator relocated to the after starboard deck-edge. As jet carriers they also had improved arresting gear. After three such conversions (CV 11, 14 and 19), three more ships (CV 16, 31 and 38) were given modified conversions incorporating angled decks and 'hurricane bows'. The first three SCB-27Cs were later brought up to this SCB-125 standard, as was *Oriskany* (SCB-125A). Of the SCB-27As, all but *Lake Champlain* were ultimately given the angled decks and enclosed bows.

Thus a total of fifteen ships was modernised. The two carriers damaged later in the war, *Franklin* and *Bunker Hill*, were never reactivated, although they were both described during the late 1940s as being an excellent material condition, quite well suited to reconstruction. Of eight carriers active in 1945–48 (CV 21, 32, 33, 36, 37, 40, 45 and 47), only one (CV 33) was rebuilt. *Antietam* became test ship for the angled flight deck and then training carrier at Pensacola. *Boxer, Princeton* and *Valley Forge* became helicopter carriers (LPH); their unconverted sisters were the first *Essex*s to be withdrawn from service post-Korea, at the end of the 1950s. As reserve ships were converted to SCB-27A configuration, active unmodernised ships were transferred to ASW duty under the

Lexington Apr 1956 L & L van Ginderen

CVS designation, replacing the Second World War escort and light fleet carriers previously used. Similarly, as the *Forrestals* and the SCB-27C/125s entered service, SCB-27As were reduced to ASW service. The SCB-27C/125 conversions ultimately followed, although *Oriskany*, *Hancock* and *Bon Homme Richard* were still carried as attack carriers at the end of the Vietnam War, and *Intrepid* operated off Vietnam as a 'light' attack carrier.

As modernised the SCB-27As carried 300,000gals (US) of Avgas (aviation fuel) and 725t of aviation ordnance (including 125t of nuclear weapons).

All of these ships lost their gun armament during their service, as topweight was needed for aviation-related items and also for radars. Typically the twin 3in/50 battery went first. Indeed, it appears that few of the SCB-27As had the full fourteen mounts, and that few of the 27Cs long retained their full complement of eleven. Ships generally ended their careers with only two 5in/38s, although some had as many as four. Aside from the enclosed bows, applied to all of the SCB-27s other than *Lake Champlain*, the most noticeable change during the careers of these ships was in their radars, the typical suit by the mid-1960s including both SPS-30 and the very large SPS-37A or -43A, the latter sponsoned out over the side of the island.

Eight ASW carriers were further modernised during the FRAM programme: CVS 9, 10, 11, 12, 15, 18, 20 and 33. Improvements included an SQS-23 bow sonar, a stem hawsepipe and bow anchor to clear it, and an improved and partially automated ASW-adapted CIC.

In 1981 the Reagan Administration proposed the reactivation and reconstruction of *Oriskany*, perhaps to be followed by *Bon Homme Richard* and *Shangri La*. Congress was unenthusiastic, partly because of severe restrictions on the range of aircraft these ships could operate. *Oriskany* was the last in the class to operate as an attack carrier. However, the fighters and early warning aircraft these ships were capable of operating no longer exist in any numbers, and they would not have been suitable for the new 'lightweight' F-18. It is sometimes argued, indeed, that in future the *Essex*s retained in reserve would be valuable primarily as helicopter carriers, if at all. *Lexington* alone remained in commission, as a training carrier off Pensacola; she was decommissioned in December 1991. All surviving ships were stricken.

Forrestal as completed
NB 1/1750 scale

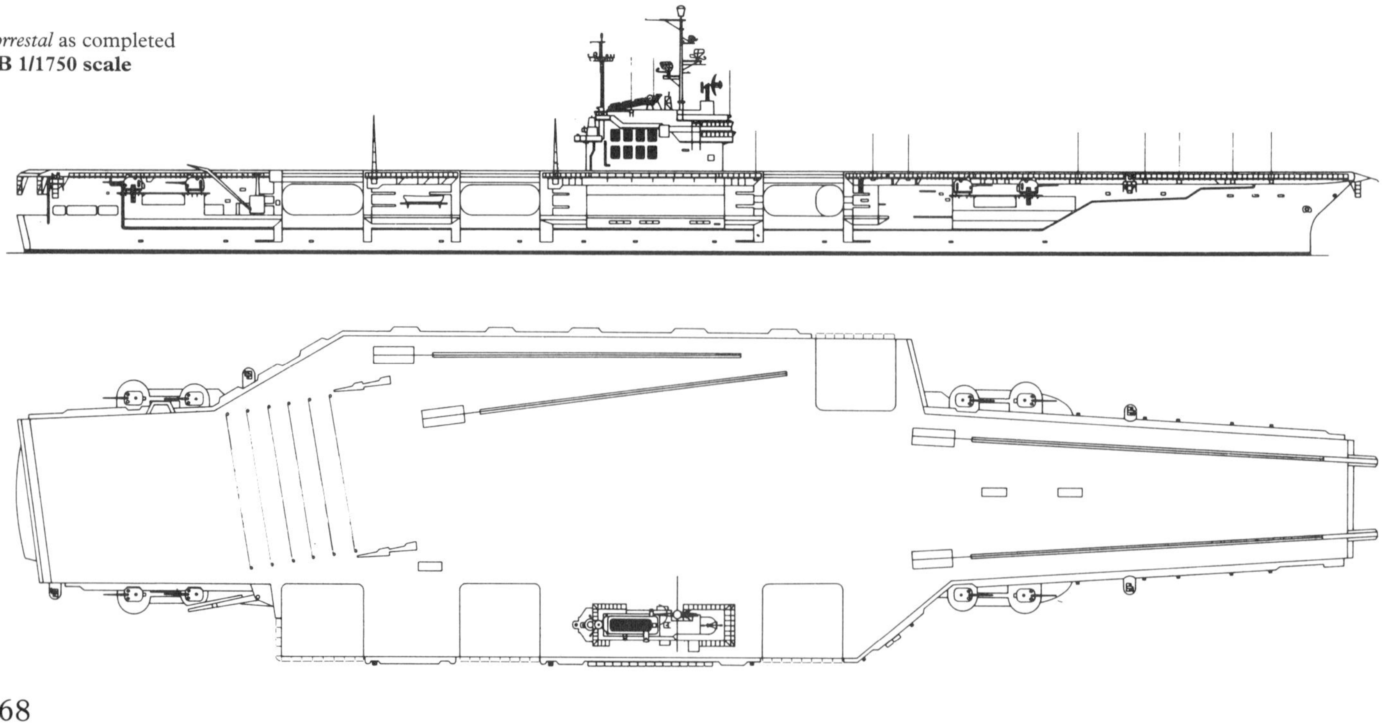

FORRESTAL class *aircraft carriers*

Displacement: 61,163t standard; 78,509t full load
Dimensions: 990ft wl, 1039ft oa × 129ft 4in wl, 250ft ext × 33ft 10in
301.8m, 316.7m × 39.4m, 76.2m × 10.3m
Machinery: 4-shaft Westinghouse geared turbines, 8 Babcock & Wilcox boilers, 280,000shp (260,000shp CVA 59) = 33kts. Oil approx 8570t. Range 12,000nm at 20kts
Armament: 8–5in/54 (8×1), approx 90 aircraft
Sensors: Radar SPS-8, SPS-12 (later SPS-29, then -37A/43A in place of SPS-12, SPS-30 in place of SPS-8; SPS-58 for Sea Sparrow)
Complement: 2764 (ship), 1912 (air group)

No	Name	Builder	Laid down	Launched	Comp	Fate
CVA 59	FORRESTAL	Newport News	14.7.52	11.12.54	1.10.55	Stricken 10.9.93
CVA 60	SARATOGA	New York N Yd	16.12.52	8.10.55	14.4.56	Stricken 20.8.94
CVA 61	RANGER	Newport News	2.8.54	29.9.56	10.8.57	In Reserve 10.7.93
CVA 62	INDEPENDENCE	New York N Yd	1.7.55	6.6.58	10.1.59	Extant 1995

These four ships were a major fruit of the revival of naval shipbuilding during and after the Korean War. *Forrestal* was originally designed as a somewhat smaller version of the cancelled *United States*, Senator Vinson having decreed that he would approve any carrier short of 60,000t. Thus the design called for a flush (island-less) flight deck with four catapults, so that four aircraft could be launched simultaneously, with two in the bow and one angled out from each side in the waist. In order to launch heavy aircraft at high speed, the catapults were to be of a new explosive type, and the stowage of catapult charges was a major design problem. Compared to previous carriers, the sheer capacity of their magazines and aircraft fuel tanks was enormous: for example, *Forrestal* was credited with1650t of aviation ordnance, 750,000 gal Avgas and 789,000gal of JP-5 fuel (compared to 500,000gal for the limited-mission *United States* and 365,000gal for a *Midway*).

As redesigned with an angled flight deck, they have two catapults in the waist, to port, so that aircraft can be spotted simultaneously, although they must be launched in succession. In the first two ships the waist catapults were C-11s, as in the steam-catapult *Essex* class modernisations, whereas the two bow catapults were C-7s. However, all four catapults are C-7s in the last two ships, and in later carriers all four catapults are of the most powerful type. The original defensive battery of eight single 5in/54 rapid-fire guns was reduced, first, by the removal of the forward sponsons, which had reduced ship performance very greatly in heavy weather by generating spray. The four after guns were removed with the installation of the Sea Sparrow SAM (one forward, to starboard, and one aft, to port; sixteen missiles). *Ranger* was the last to have her guns removed, losing two in 1977, and retaining her forward sponsons.

Stacks were raised about 10ft in *Forrestal* and *Independence*.

Major refits, under the Service Life Extension Program (SLEP) at Philadelphia Navy Yard, began with *Saratoga* (1 October 1980–February 1983, but inoperative until 18 November 1983 due to boiler defects), followed by *Forrestal* (21 March 1983–20 May 1985) and *Independence* (18 April 1985–February 1988). As in FRAM, all major systems were overhauled and new electronics (including SPS-48C and SPS-49 radars, improved NTDS, and Tactical Flag Command Center/ Flag Data Display System) fitted. Kevlar armour was added. SLEP was also to have included replacement of C-7 and C-11 catapults by longer C-13s, but that is yet to come. The planned defensive battery is three Sea Sparrow Mk 29 launchers (using Mk 23 TAS and two directors per launcher) and three Phalanx. *Ranger* did not undergo full SLEP but was refitted extensively at Puget Sound (May 1984–June 1985); she received improved evaporators, Halon, acqueous film firefighting systems, Mk 23 TAS, and three Phalanx. She is to be decommissioned during FY93.

Forrestal became training carrier (AVT 59) replacing *Lexington* at Pensacola. However, during a major refit to her new training configuration, she was stricken; she may become a museum. *Independence* replaced *Mid-way* as the carrier forward-deployed at Yokosuka; as of 1994 she was expected to decommission in FY97.

Ranger 1993 L & L van Ginderen

Franklin D Roosevelt (after SCB-110 modernisation) USN

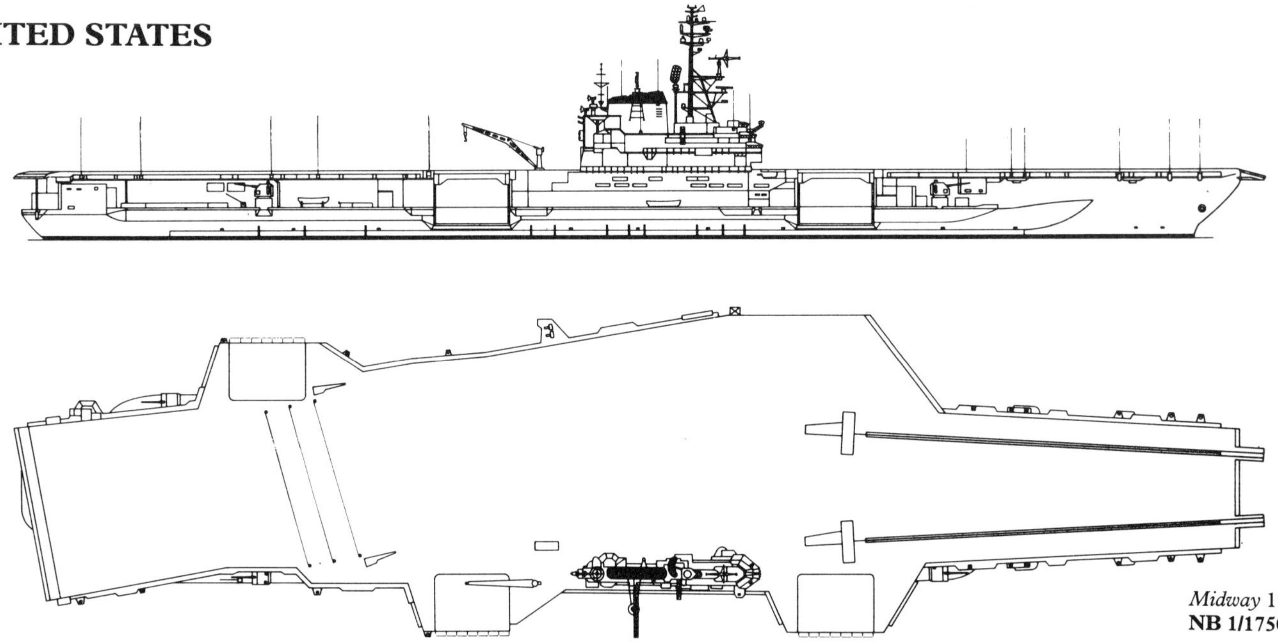

Midway 1971
NB 1/1750 scale

MIDWAY class (SCB-110 reconstruction) *aircraft carriers*

Displacement: 42,710t light; 62,614t full load
Dimensions: 910ft wl, 977ft 2in oa × 121ft wl, 210ft ext × 34ft 6in
277.4m, 297.9m × 36.9m × 64.0m × 10.5m
Machinery: 4-shaft Westinghouse geared turbines, 12 Babcock & Wilcox boilers, 212,000shp = 30.6kts. Range 15,000nm at 15kts
Armament: 10–5in/54 (10×1), 18–3in/50 (9×2)
Sensors: Radar SPS-12, SPS-8 (later SPS-37A/43A, SPS-30; SPS-48 fitted about 1980)
Complement: 4060

For names of reconstructed ships and ultimate fates see under Fleet Strength 1947. Original data, including construction particulars, can be found in the 1922–1946 volume.

Of the US carriers surviving the Second World War, only the *Midways* were capable, unmodified, of operating the first of the new generation of heavy attack aircraft, the AJ Savages. Indeed, *Coral Sea* flew off a Neptune land-based patrol bomber carrying a simulated nuclear bomb, to show that the navy could indeed execute a nuclear mission with existing assets. Even so, by the late 1940s it was clear that some major reconstruction would be necessary, and the SCB-27 project which eventually led to wholesale modification of the smaller *Essex* class carriers was begun with the *Midways* in mind. All three were modernised under the SCB-110 project, but because the modernisations were spread over a considerable period the three ships differed in detail, *Coral Sea* receiving SCB-110A. The first two ships received the two C-11 steam catapults, 'hurricane bow', and angled deck of SCB-27C, and their 5in and 3in batteries were considerably reduced. *Franklin D Roosevelt* was given a new tapered mast, *Midway* a lattice. *Coral Sea* received an additional C-11 in her waist. In each case, too, the after centreline elevator was replaced by an after starboard deck-edge unit and, as in the smaller carriers, elevators and flight decks were considerably strengthened.

By the mid-1960s it was clear that another round of reconstructions would be needed. Unlike the ageing *Essexs*, the *Midways* seemed well-suited to current aircraft and were expected to serve well into the 1980s. *Midway* was the first, and the immense cost of her SCB-101.66 refit (February 1966 to January 1970 at Mare Island, for a total of $202m) deterred the navy from any similar modernisation of *Franklin D Roosevelt*. As the most modern, *Coral Sea* needed rather less in any case. *Midway* received C-13 catapults, an enlarged flight deck, and had her forward centreline elevator removed in favour of a second starboard deck-edge unit, her port deck-edge one replaced by a port deck-edge elevator further aft. NTDS was also fitted. *Franklin D Roosevelt* was given a far more austere refit ($46m), her forward centreline elevator being replaced by a starboard deck-edge unit forward of her island. As she was in the worst material condition of the three, and the least suited for further service, she was stricken in 1977. *Coral Sea* had emerged from SCB-110A with both centreline elevators eliminated in favour of starboard deck-edge units, and she was not heavily rebuilt; in 1982 she remained in active reserve, without a regularly assigned air group, partly for training activities on the West Coast. *Midway* was still an active unit of the Pacific Fleet in 1982.

The original very heavy gun batteries of these ships were rapidly drawn down in service, as both the value of the gun and its cost in space and weight were re-evaluated. It appears that some of their Mk 39 5in/54 mounts were ultimately mounted aboard Japanese destroyers built postwar with US aid. During the late 1940s the 5in batteries of *Midway* and *Franklin D Roosevelt* were reduced from eighteen to fourteen guns, and *Coral Sea* was completed with only fourteen. After SCB-110 *FDR* had only ten 5in/54, and by 1963 that number had been reduced to four; *Midway* was similar. *Coral Sea* had only six after SCB-110A, and only three after 1962. *Midway* retained three guns after her last reconstruction; *FDR* had four. All guns were later removed from *Midway* and *Coral Sea*; in 1982 the former had two Sea Sparrow SAM (2 × 8) and both had three Phalanx CIWS. The original provision for aviation stores was 1376t of ordnance, 356,000gal Avgas and 600,000gal of JP-5 fuel; ordnance was reduced to 1210t after SCB-101.66, but JP-5 was increased to 1,186,000gal JP-5 (no Avgas).

Coral Sea was modernised at Norfolk to keep her in service through 1992 (re-entered service 29 January 1985).

Midway was blistered at Yokosuka during a major refit (1 April–28 November 1986; the blisters were 183m long and 3m wide). Unfortunately this so increased her GM that her roll period was reduced to 9 sec; her motion was considered unacceptable. The projected cure was a waterline slot cut into the blisters (to reduce waterline beam to its original figure, while retaining most of the added buoyancy), but she was retired unmodified.

KITTY HAWK class *aircraft carriers*

Displacement: 60,005t light; 80,945t full load
Dimensions: 990ft wl, 1047ft 6in oa × 129ft 4in wl, 251ft 8in ext × 36ft
301.8m, 319.4m × 39.4m, 76.7m × 11.4m
Machinery: 4-shaft Westinghouse geared turbines, 8 Foster Wheeler boilers, 280,000shp = 33.6kts. Range 12,000nm at 20kts
Armament: 2 Terrier SAM (2×2, 40 missiles per launcher), 3 Sea Sparrow SAM (CVA 67 only, 3×8, no reloads), approx 90 aircraft
Sensors: Radar SPS-37A, SPS-39, SPS-8B (CVA 63 and 64 only; CVA 66 SPS-30; no fighter control in CVA 67, which had SPS-48 in place of -39), SPS-43A (CVA 66 and CVA 67 as completed), SPG-55 (not CVA 67); sonar SQS-23 (CVA 66 only)
Complement: 3306 (ship), 1379 (air group) (CVA 66)

No	Name	Builder	Laid down	Launched	Comp	Fate
CVA 63	KITTY HAWK	New York SB	27.12.56	21.5.60	29.4.61	Extant 1995
CVA 64	CONSTELLATION	New York N Yd	14.9.59	8.10.60	27.10.61	Extant 1995
CVA 66	AMERICA	Newport News	9.1.61	1.2.64	23.1.65	Extant 1995
CVA 67	JOHN F KENNEDY	Newport News	22.10.64	27.5.67	7.9.68	Extant 1995

These ships actually form three subclasses; all are improved *Forrestals*. An entirely new carrier design for CVA 63 was foregone because of the imminence of the introduction of nuclear power, which, it was believed, would limit any new class to only two units. *Kitty Hawk* and *Constellation* differed from the *Forrestals* in having their islands moved aft, so that two instead of one elevator are forward of them, for improved flight deck operations. They also have a lattice radar mast aft of the island structure, and were armed with Terrier missiles (two twin Mk 10 launching systems on the port and starboard quarters, with paired SPG-55 guidance radars on the island structure). *America*

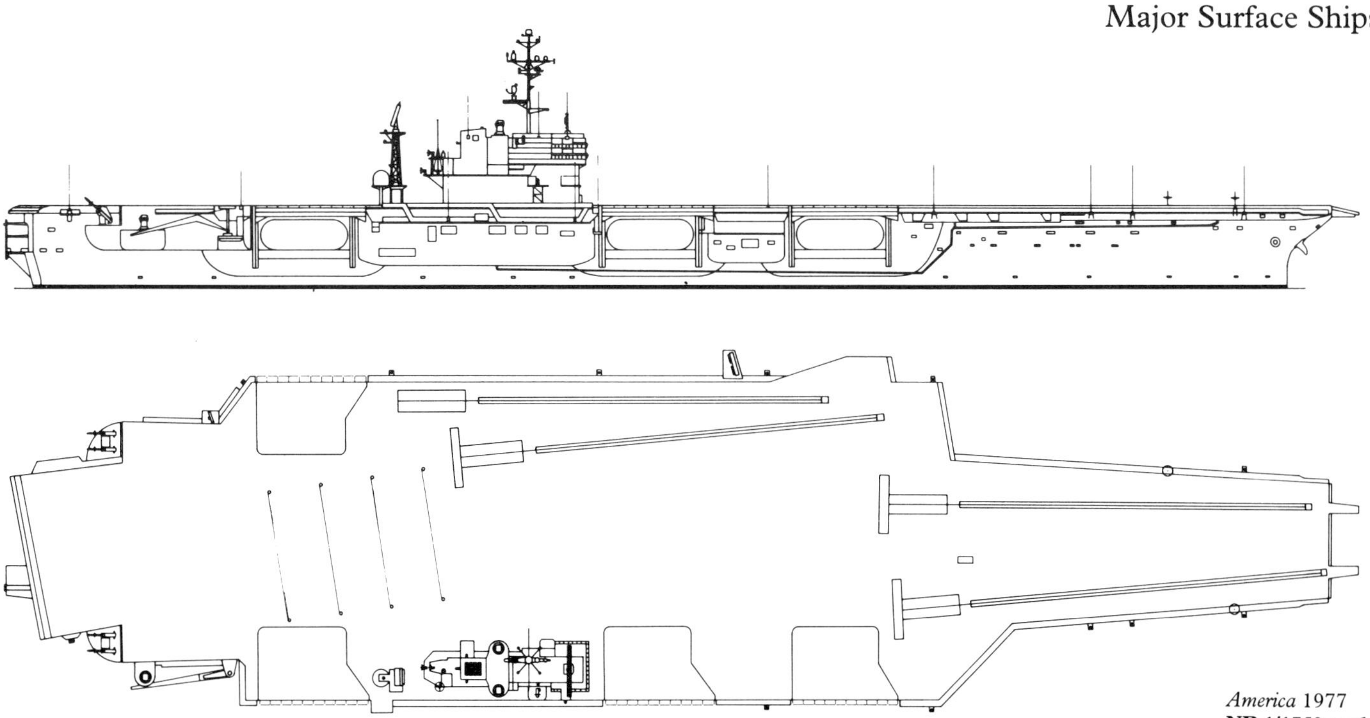

America 1977
NB 1/1750 scale

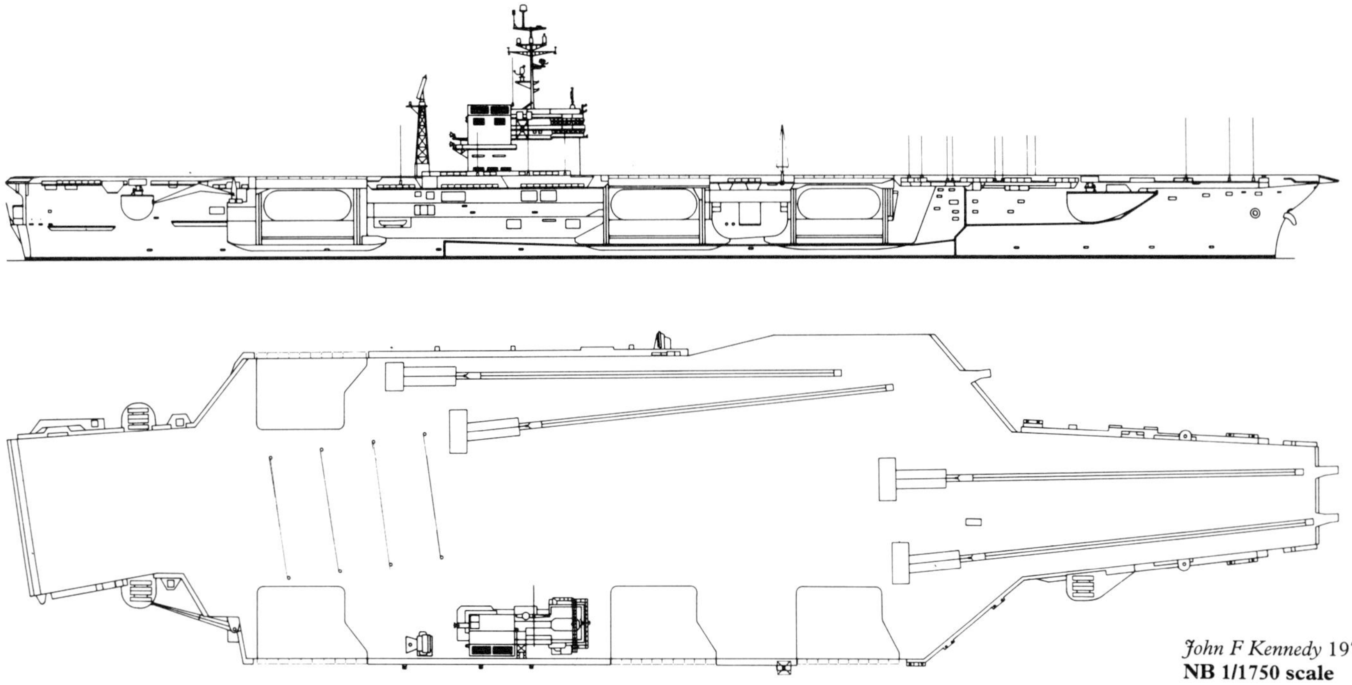

John F Kennedy 1972
NB 1/1750 scale

Kitty Hawk 1978 Stefan Terzibaschitsch

(CVA 66) is very similar, having been built in preference to an austere nuclear carrier in FY61; she received updated Terrier, compatible with the Standard Missile. *John F Kennedy* was built to a revised design, incorporating a new underwater protection system originally developed for the nuclear carriers, and, at first, without defensive weapons. Early in 1969 she was fitted with three Sea Sparrow launchers, and her three half-sisters were similarly refitted. *Kennedy* has her funnel angled to starboard.

America was the first attack carrier with a special integrated CIC/airborne ASW control centre (ASCAC) and, in 1981, retained SM-1 missiles and SPG-55B radar directors, as well as SPS-52 3-D radar instead of the more effective SPS-48 of her two sister-ships.

Kitty Hawk had her Terrier system replaced by two Sea Sparrow (with Mk 91 directors) about 1977. *Constellation* had her Mk 10 launchers and SPG-55B guidance radars removed in a refit at Puget Sound (December 1982–February 1984). *America* had her SM-1 system removed about 1982. Her SPS-52 was replaced by SPS-48 at this time.

By 1990 all were armed with three Mk 29 Sea Sparrow launchers (each with three directors) using TAS Mk 23 (in place of the earlier SPS-58) target indication radar. SPS-43A search radars were replaced by SPS-49 in 1979–80.

America and *John F Kennedy* were fitted for SQS-23 sonar in bow domes, on the theory that in a spread-out fleet formation it would be impossible to maintain a tight sonar screen, so that the carrier might have to detect submarines leaking through; SQS-23 was also fitted to some ASW carriers. Similar reasoning led to the installation of this sonar in *Cleveland* class missile cruisers. SQS-23 was not installed in CVA 67 to reduce cost; presumably the measure lost much of its value as submarine weapon ranges increased beyond torpedo range.

Aviation ordnance and fuel capacity is slightly greater than that of the *Forrestal* class, at 2150t and 1,950,000gal in the *Kennedy*. In 1979 a repeat *Kennedy* was proposed as an alternative to a non-nuclear CVV. Unlike the *Forrestal*s, these ships have four C-13 catapults (one is a longer C-13-1 in CV 66 and 67). Their rearranged flight decks make it far easier and safer to launch and recover aircraft simultaneously.

SLEP refits: *Kitty Hawk* (January 1988–February 1991), *Constellation* (1 July 1990–early 1993). Modifications include new catapult rotary engines, new arresting gear engines, SPS-48E, SPS-49(V)5, Mk 23 TAS, SPN-46 landing aid radar, updated NTDS, torpedo countermeasures, and improved ECM. As of 1990, *America* was scheduled for SLEP in 1996–99, but she was fitted with TAS Mk 23 during a regular overhaul in the late 1980s. As of 1992 it was expected that she would be decommissioned in 1996 instead of being modernised. *Kennedy* received two more Phalanx and Mk 23 TAS during a refit in October 1984–October 1985, and was scheduled to receive SPS-48E radar under the FY92 programme. Instead of a planned 1993–95 SLEP, she had a 14-month 'complete overhaul' (FY93 programme).

Enterprise 1968
NB 1/1750 scale

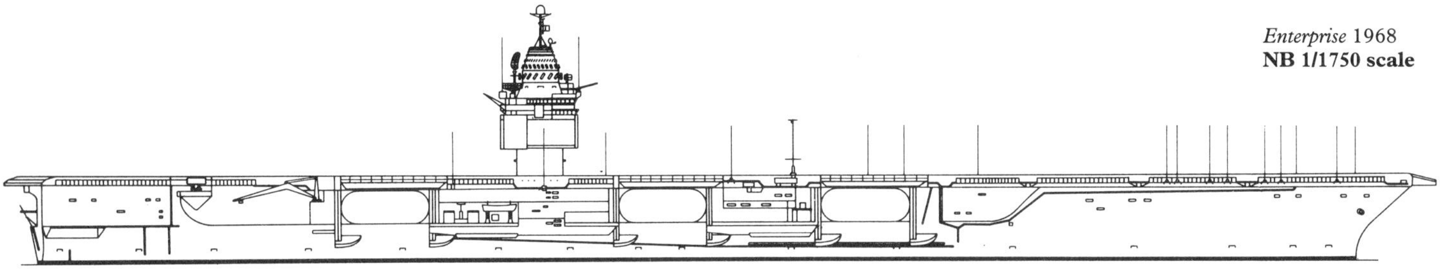

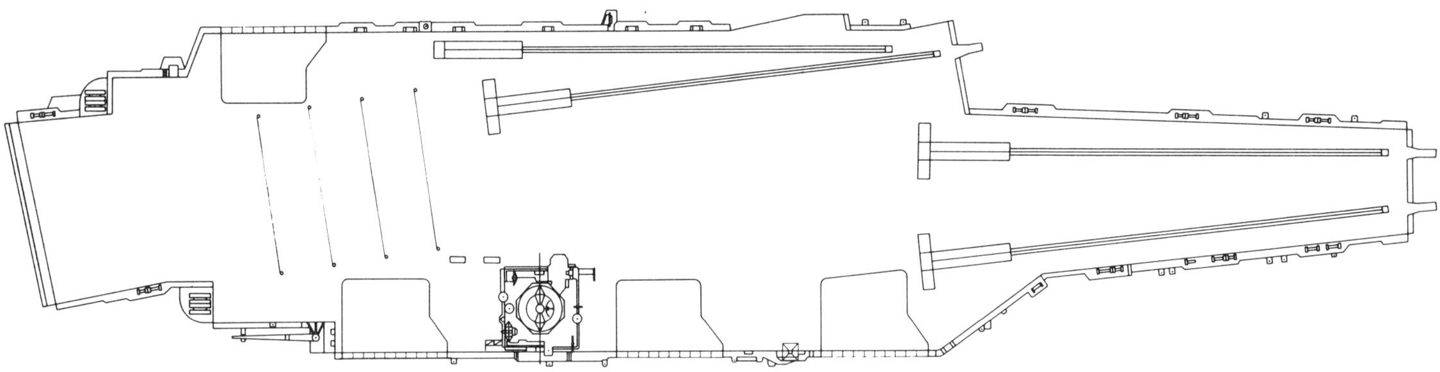

Probably the most spectacular warship of her time, *Enterprise* was the first US nuclear carrier. Her size was dictated by her power plant: although it was not much heavier than equivalent weights for a conventional ship, she still had to carry a large liquid load for underwater protection, and that load greatly increased her displacement. Since it could be used for aircraft fuel, it allowed her to operate an unusually large air group. Reportedly she can sustain twelve days of intense air operations without replenishment. Again, since she was already an outstandingly capable ship, she received other innovative equipment, such as electronically scanned main radars (SPS-32 and SPS-33), with a conventional radar (SPS-12 added in 1968) as back-up. Although space and weight for Terrier (as in CVA 63, 64 and 66) was provided, the missile was not fitted, in an effort to hold costs down; Sea Sparrow was fitted in 1967, with a modified aircraft (F-4) fire control radar. *Enterprise* was refitted in 1980, and her massive fixed radars removed.

ENTERPRISE *aircraft carrier*

Displacement: 71,277t light; 89,084t full load (as of 1968)
Dimensions: 1040ft wl, 1123ft 2in oa × 133ft wl, 255ft ext × 37ft 1in
317.1m, 342.4m × 40.5m, 77.7m × 11.3m
Machinery: 4-shaft nuclear: 8 A2W reactors, 4 Westinghouse geared turbines, approx 280,000shp = 32kts
Armament: None as built (3 Sea Sparrow SAM (3×8) fitted 1967), approx 90 aircraft
Sensors: Radar SPS-32, SPS-33 (SPS-12 added in 1968)
Complement: 3325 (ship), 1891 (air group), 71 Marines

No	Name	Builder	Laid down	Launched	Comp	Fate
CVAN 65	ENTERPRISE	Newport News	4.2.58	24.9.60	25.11.61	Extant 1995

Her aviation fuel stowage is 2,720,000 gal, reflecting the requirements of underwater protection rather than air group operations; in earlier ships aviation fuel stowage was balanced with ship endurance fuel stowage in side protection spaces, and from time to time, it was proposed that ship boilers be modified to burn aviation fuel (JP-5). Aviation ordnance capacity is 2520t, compared to about 2000t in non-nuclear carriers. *Enterprise* has four C-13 steam catapults.

Like *Nimitz*, *Enterprise* has an integrated CIC/ASW command centre (ASCAC). Since 1976 she has carried two British SCOT satellite communications antennae. As refitted, she has the standard carrier radar suit of SPS-48C and SPS-49, with SPS-65 (antenna common with SPS-10) for Sea Sparrow warning. She received three Phalanx mounts.

Enterprise entered Newport News in 1991 for refuelling and refit (but not full SLEP); she was to receive TAS Mk 23 and a third Sea Sparrow launcher, plus an upgraded NTDS, TFCC, and ASCAC. Refuelling is very expensive because *Enterprise* shares reactor components only with the cruiser *Long Beach*. Since the carrier force was being reduced anyway, there was serious talk of laying the ship up instead of refuelling her (OSD considered deleting CVN 76 from the projected programme unless *Enterprise* was laid up). However, the nuclear components had already been bought, so the work began in 1991. One reason was that without this work the naval nuclear industry itself might begin to collapse, given the declining submarine programme.

Enterprise Sep 1977

L & L van Ginderen

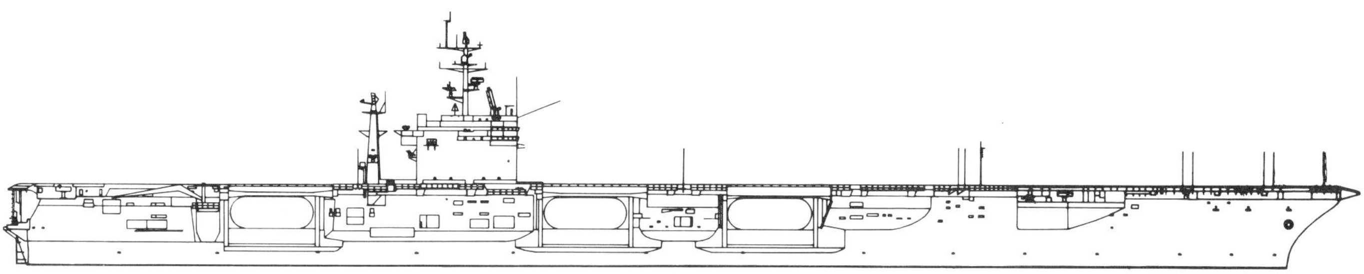

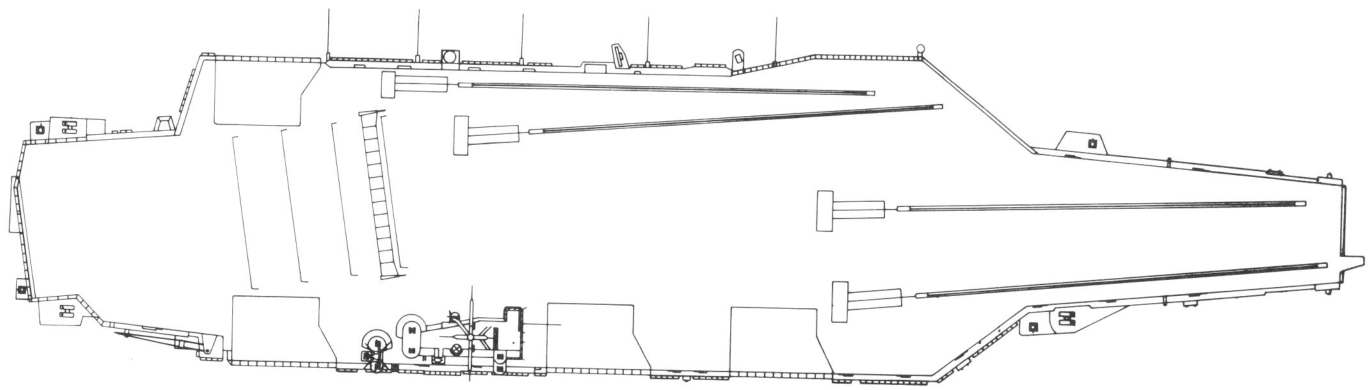

George Washington as completed
NB 1/1750 scale

With these ships the US Navy returned to building nuclear carriers of *Enterprise* size. They incorporate a new two-reactor power plant, developed from a project for a single reactor destroyer/frigate plant, and so have even more internal space than the huge *Enterprise*; in addition they incorporate the more economical (in terms of space) torpedo protection system of the *Kennedy*. Their general arrangement and electronic suit also duplicate those of *Kennedy*, and they are armed with three Sea Sparrow launchers for self-defence. Approximate capacity is 2570t of aviation ordnance and 2,800,000gal of aviation fuel, in each case more than the *Enterprise*, of similar size, can accommodate.

The first three ships (CVN 68–70)

NIMITZ class *aircraft carriers*

Displacement: 73,973t light; 91,440t full load
Dimensions: 1040ft wl, 1088ft oa × 134ft wl, 257ft 6in ext × 36ft 8in
317.1m, 331.7m × 40.8m, 78.5m × 11.2m
Machinery: 4-shaft nuclear: 2 A4W reactors, 4 geared turbines, approx 260,000shp = 30+kts
Armament: 3 Sea Sparrow SAM (3×8, no reloads), approx 90 aircraft
Sensors: Radar SPS-43A, SPS-48 (later ships have SPS-49 in place of SPS-43A)
Complement: 5621 (ship plus air group)

No	Name	Builder	Laid down	Launched	Comp	Fate
CVAN 68	NIMITZ	Newport News	22.6.68	13.5.72	3.5.75	Extant 1995
CVAN 69	DWIGHT D EISENHOWER	Newport News	14.8.70	11.10.75	18.10.77	Extant 1995
CVN 70	CARL VINSON	Newport News	11.10.75	15.3.80	13.3.82	Extant 1995
CVN 71	THEODORE ROOSEVELT	Newport News	31.10.81	27.10.84	25.10.86	Extant 1995
CVN 72	ABRAHAM LINCOLN	Newport News	3.11.84	13.2.88	11.11.89	Extant 1995
CVN 73	GEORGE WASHINGTON	Newport News	25.8.86	21.7.90	4.7.92	Extant 1995
CVN 74	JOHN C STENNIS	Newport News	1.12.90	Mar 1994	Dec 1995	Completing
CVN 75	HARRY S TRUMAN	Newport News	Aug 1992	Dec 1995	Sep 1998	Building
CVN 76	RONALD REAGAN	Authorised in FY95 programme				

are a separate sub-class. They can be distinguished further by the presence of only a single bridle-catching boom in CVN 69 and 70. Current carrier aircraft employ a catapult attachment built into their landing gear, and so do not use the wire bridles of the past. *Carl Vinson* was completed with four Phalanx CIWS; the others had three each.

The Navy claims that these ships can take three times the damage which *Essex* class carriers survived in 1944–45. The first three were later fitted with Kevlar splinter armour (in order of completion, during June 1983–September 1984, October 1985– April 1987, and 1989 refits). SPS-49 replaced SPS-43A in CVN 68 and 70 in 1983, and in CVN 69 in 1986. Mk 23 TAS was added, and all were scheduled to receive SPS-48E radar under the FY91–92 programmes. *Carl Vinson* was completed with three Sea Sparrow launchers. Her two sisters received their third launchers during refits.

They are described as carrying 90 per cent more aviation fuel and 50 per cent more aviation ordnance than a *Forrestal.* All were completed with ASCAC (ASW Classification and Analysis Center), permitting data to be shared between the carrier, ASW aircraft, and escorting ships.

The other ships (*Theodore Roosevelt* class) are considerably improved, particularly in self-defence (Kevlar over vitals, and better hull protection). A reduction in aviation ordnance (to 1954 tons) presumably reflects increased volume devoted to protection. Otherwise these ships are similar to the CVN 68 class, with much the same hull envelope. CVN 72 and 73 have new lower-pressure catapults and SPN-46 landing aids. CVN 74 is to be completed with SPS-48E instead of the -48C of her sisters. Each is armed with three Sea Sparrow launchers (with TAS Mk 23) and four Phalanx. *Harry S Truman* was originally to have been named *United States*.

Dwight D Eisenhower on trials 23.8.77 Stefan Terzibaschitsch

Medium Carrier Project (CVV) *aircraft carrier*

Displacement: 45,192t light; 59,794t full load
Dimensions: 860ft wl, 912ft oa × 126ft wl, 256ft 6in ext × 34ft 7in
262.2m, 278.0m × 38.4m, 78.2m × 10.5m
Machinery: 2-shaft geared turbines, 6 boilers, 140,000shp = 27.8kts
Armament: 3 Phalanx CIWS, 50 aircraft (1191t avord, 2700t JP-5)
Sensors: Radar SPS-48, SPS-49
Complement: 4024

The CVV was proposed by the Carter and Ford Administrations as a means of reducing the unit size and cost (and, incidentally, capability) of US attack carriers. It was ultimately defeated by the Navy and its Congressional allies. The concept originated in the Ford Administration as the Tentative Conceptual Baseline (T-CBL) for a new carrier to replace the ageing *Franklin D Roosevelt.* The Chief of Naval Operations, Admiral Elmo Zumwalt, sought a reversal in the steady growth of warship costs. He was also responsible for an attempt to devise an inexpensive Aegis platform (DG/Aegis, which was abortive), for the Sea Control Ship, and for the limited-cost *Perry* class escort. Admiral Zumwalt's Project SIXTY advisory group, formed at the beginning of his term as CNO, initially examined a new 40,000t carrier, but ultimately chose to propose one of 50,000t to 60,000t. Secretary of Defense Laird issued a Program Decision Memorandum for a $550m (FY73 dollars) carrier on 21 September 1972, studies beginning in December 1972. Although the product of these studies, the T-CBL, was not ordered, interest in such a ship continued, Secretary of Defense Schlesinger (who succeeded Laird) ordering a new round of studies in July 1975. They failed when the ship was redefined as nuclear at Navy request, and President Ford ultimately called for a repeat *Nimitz* rather than for a new smaller ship. However, he did call for the construction of two smaller oil-fired carriers in FY79 and FY81. They were to be adapted for a new generation of VSTOL aircraft, and as such received a new designation, CVV. The incoming Carter Administration found the CVV idea attractive, at least in part as a means

Artist's impression of CVV design USN

of reducing the cost of the shipbuilding programme.

The CVV employed half the power plant of a conventional steam carrier such as the *John F Kennedy*, and was criticised for its relatively low speed. For example, at full power (140,000 shp) the T-CBL would make only 27.8 kts, so that it would not be compatible with fast carriers. With her catapults in use, speed would fall much further, as a large percentage of available steam would have to be fed into the catapults. These problems would have been alleviated in the full CVV, which had two more boilers.

Cost-saving measures included a restriction to two shafts, two catapults, and only two elevators, all of which appeared to limit the survivability of the medium carrier. There were even proposals to forego a large fraction of the electronic installation of the CVV to save initial cost. For example, there would be no SPS-49 long-range air search radar; the SPS-48C three-dimensional radar would have to suffice, at least at first. Similarly, one of the two planned 2500kW ship service turbogenerators would have been deferred, as well as the liquid oxygen plant. However, the CVV design incorporated a new and extremely robust form of magazine protection against cruise missile attack, as well as a new type of under-bottom protection. In this sense the proposed carrier was extremely non-austere. Presumably some of these ideas will eventually be incorporated in the full-deck carriers which have been authorised since the death of the CVV.

Sea Control Ship Project (SCS) *aircraft carrier*

Displacement:	9773t light; 13,736t full load
Dimensions:	585ft wl, 610ft oa × 80ft × 21ft 7in *178.4m, 186.0m × 24.4m × 6.6m*
Machinery:	1 shaft, 2 LM 2500 gas turbines, 45,000shp = 26kts
Armament:	2 CIWS
Sensors:	Radar SPS-52
Complement:	700

The Sea Control Ship was a small helicopter carrier, part of Admiral Zumwalt's 'low-end' emphasis, and as such bitterly fought by much of the naval aviation community, as well as by advocates of nuclear power, led by Admiral H G Rickover. The concept appears to have originated in 1969 with a Long Range Objectives Group proposal for a new class of helicopter-carrying escorts (or DHK) to provide active-sonar screens for convoys and other naval formations. At the time, as now, US wartime ASW strategy was predicated on the capability of passive systems, most notably SOSUS, which were expected to guide P-3 aircraft to attack on Soviet submarines in an area-attrition campaign. It was feared that improvements in Soviet submarine silencing might neutralise SOSUS, and thus force the United States back to more conventional convoy tactics, which in turn would require much larger numbers of surface escorts. Studies as early as 1959 strongly suggested that, with the loss of the mass of the Second World War ships, these numbers would not be attainable in a crisis. Helicopters with active sonars might be an equalizer; the LRO study envisaged both escort carriers (initially converted *Commencement Bays*) and 30kt DHKs of 12,000t to 14,000t, each carrying twelve advanced ASW helicopters, as well as conventional destroyer weapons. There were to be a total of fourteen DHK and fifteen escort carriers. Within a few years passive tactical systems (such as the towed array and many of the sonobuoys) were even more important than in 1969, but interest in small helicopter carriers survived.

As LAMPS became an important fixture of the surface escort ASW system, the small carrier became attractive as a maintenance base and hangar for helicopters operated by the escorts. That is, maintenance could be considered a major barrier to effective sustained helicopter operation. Admiral Zumwalt's programme called for a total of eight sea control ships, to support a total of eight convoys or underway replenishment groups (with their escorts). The SCS designation was adopted in May 1971 and a preliminary design was ready by January 1972. It showed a capability for intermediate-level maintenance, and was expected to cost $175m in FY75 dollars (prototype: follow-on ships would cost $117m each). At this time the prototype was to have been ordered in the FY75 programme, for completion in FY78, with three more in FY76, and two each in FY77 and FY78.

In theory, the helicopters of the SCS were to have localised and attacked submarines detected by the towed arrays streamed by the frigates, which (in 1971) were expected to have convergence zone range. In addition, VSTOL fighter aboard the SCS would be able to protect the convoy from aircraft at ranges of up to 100nm, backed up by the Standard missiles of the frigates. The SCS air group was set at eleven ASW and three early warning helicopters, plus three Harrier fighters: the carrier would maintain two ASW and one AEW helicopter continuously on station, with a ready fighter on deck alert. The concept was tested by the helicopter carrier *Guam*, which was modified between October 1971 and January 1972; major modifications included the provision of an ASW analysis centre and improved aircraft control, direction, and maintenance facilities. Tests, which continued until April 1974, were generally considered successful despite problems with the SH-3G helicopter. However, the evaluators questioned the viability of the SCS, given the absence of both the advanced helicopter and the advanced VSTOL which she was intended to operate. Moreover, she had failed to maintain the two helicopters permanently on station – due, supporters of the SCS suggested, more to the deficiencies of the helicopters (and of the spares policy supporting them) than to basic problems. However, the SCS concept was dropped by Admiral Zumwalt's successor, Admiral Holloway, in favour of an alternative VSTOL support ship, or VSS.

Throughout the SCS design was optimised for minimum cost, the goal being to build eight ships for the cost of one nuclear carrier. Thus Admiral Zumwalt settled for a single shaft powerplant, and limited defensive weapons to two Phalanx close-in defensive guns. With the cancellation of the SCS project, the plans were sold to the Spanish Navy, and a modified version of the basic design was built as *Principe de Asturias* (see under Spain).

VSTOL Support Ship Project (VSS) *aircraft carrier*

Displacement:	20,116t light; 29,130t full load
Dimensions:	690ft wl, 717ft oa × 178ft oa × 24ft 4in *210.4m, 218.6m × 54.3m × 7.7m*
Machinery:	2 shafts, 4 LM 2500 gas turbines, 90,000shp = 30kts
Armament:	2 Phalanx CIWS, 8 Harpoon SSM (2×4)
Sensors:	Radar SPS-52
Complement:	959 ship, 628 aviation

In 1975 the new CNO, Admiral Holloway, called for a new VSTOL carrier, which he hoped would assist in the planned transition to an all-VSTOL navy by early in the next century. Although theoretically not a direct replacement for the conventional carriers, the VSS was required to achieve speeds not far from those of such ships, and as such had to have two shafts. About fifty alternatives were examined in 1974–75, including large ships with catapults and arresting gear. However, the VSS design ultimately adopted was an enlarged SCS, which could operate sixteen SH-53, six LAMPS, and four Harriers. Four (compared to two in the SCS) LM 2500 gas turbines would drive her at about 29kts. The designwas reworked twice: first, to accommodate more advanced VSTOL aircraft, and then to achieve a measure of magazine protection; this last version, VSS III, is the one described in the table. In 1983 it is still sometimes used for comparison with more conventional carrier alternatives, although design work effectively ended in 1978.

Congressional interest in such ships has continued. For example, in FY79 Congress almost funded the conversion of a helicopter carrier to an SCS; the Senate Armed Services Committee added $40 million for conversion, as well as $70 million for long-lead items for a sixth LHA, which would more than replace the LPH withdrawn for the SCS conversion. It can be argued that the LHD (see under Amphibious Warfare Vessels) represents a reaction to this sentiment, in that it is to have a secondary sea control capability.

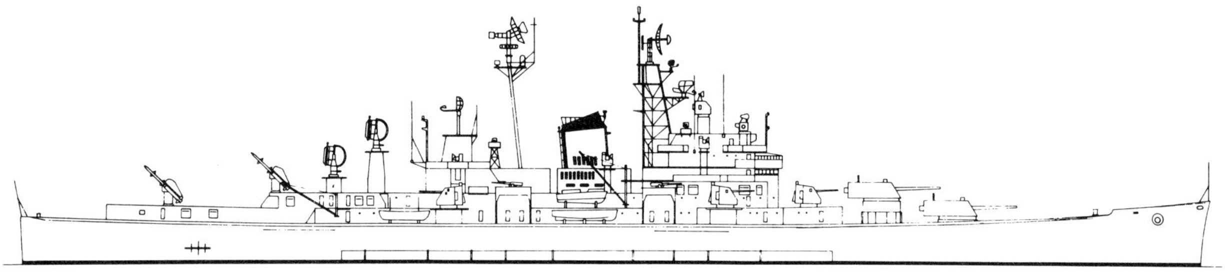

Canberra 1958

BOSTON class *fleet escorts (missile cruiser conversions)*

Displacement: 13,589t light; 17,947t full load
Dimensions: 664ft wl, 673ft 5in oa × 69ft 8in × 24ft 11in
202.4m, 205.3m × 21.25m × 7.6m
Machinery: 4-shaft geared turbines, 4 Babcock & Wilcox boilers, 120,000shp=33kts. Range 7300nm at 20kts. Oil 1448t
Armament: 2 Terrier SAM (2×2, 72 missiles per launcher), 6–8in (2×3), 10–5in (5×2), 8–3in/50 (4×2)
Sensors: Radar SPS-6, SPS-8, SPS-12, CXRX, two SPQ-5 (Mk 25 mod 7 in CAG 1)
Complement: 1544

No	Name	Yard	Recomm	Fate
CAG 1	BOSTON	New York SB	1.11.55	Stricken 1.11.73
CAG 2	CANBERRA	New York SB	15.6.56	Stricken 31.7.78, BU

These two heavy cruisers were the first US anti-aircraft missile ships, rushed to completion to meet a very severe perceived threat. The first ship seriously considered for conversion was the older cruiser *Wichita*, which might have had three launchers, replacing all her heavy gun turrets. The Terrier missile itself was rushed into production; it began as a test vehicle in the Talos programme, and the conversion of *Boston* and *Canberra* was conceived as a first phase, an austere effort which might be followed by the removal of the two forward turrets if it proved successful. The conversions were in fact successful, but instead of giving the ships further refits, extra ships were converted. By 1966 the radar fit was CXRX, SPS-37A, SPS-30, two SPQ-5 (CAG 2). By the 1960s both were obsolete, and a project to modernise them was rejected as too expensive. They were reclassified as cruisers (CA) in May 1968 and relegated to shore bombardment duties; before de-activation, the missile launchers were removed, leaving them with two 8in triple turrets forward.

They were decommissioned in 1970.

Canberra 1964
L & L van Ginderen

Providence 1966 USN

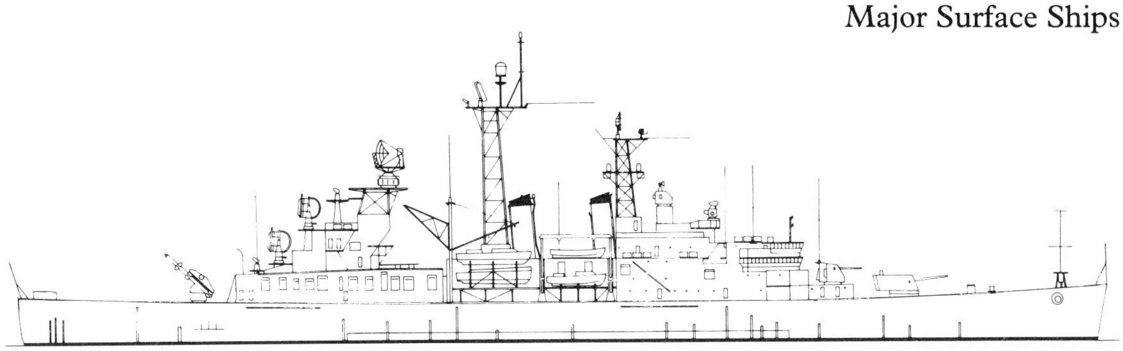

Little Rock 1960

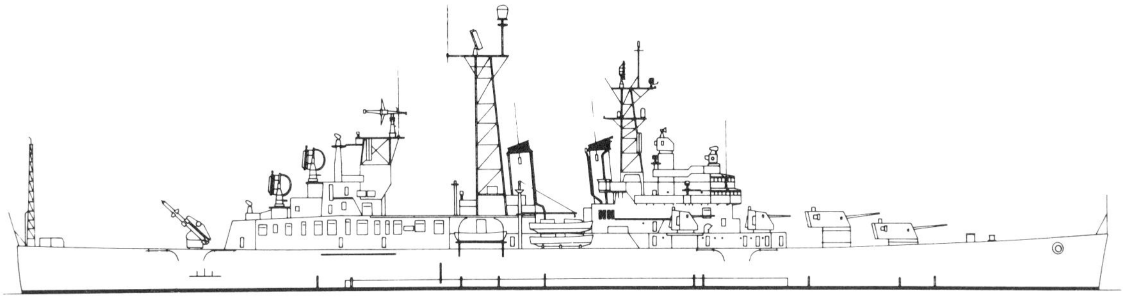

Galveston 1959

Six *Cleveland* class light cruiser hulls were withdrawn from reserve and converted to austere missile ships, the entire missile installation being above the weather deck aft, for either Talos (CLG 3–CLG 8) or Terrier (CLG 6–CLG 8), all being reclassified in May 1957. They were a solution to the problem of missile ship cost, as a series of specially built cruisers would have been far too expensive. Moreover, they were not double-enders as had originally been planned. In CLG 3 and CLG 8 the fore part of the ship, including two 6in turrets and three 5in/38 gunhouses, remained unaltered, except for the provision of a heavy new foremast to carry a long-range air search radar. In the other four, a new forward superstructure carried flag spaces, and only one 6in and one 5in mounting were retained. All carried a lattice mast amidships for a three-dimensional electronically scanned radar, and a shorter one aft for a fighter-control height-finder; in some cases the two were interchanged. In each, too, a pair of guidance radars aft (SPG-49 for Talos, SPQ-5 for Terrier) controlled missiles fired by one twin-arm launcher. Flagships carried helicopters, with a landing pad on the quarterdeck, but had no hangars to stow them. By 1963 SPS-30 had replaced SPS-8B/SPS-2, and all ultimately had SPS-37A/43A rather than SPS-29.

CLEVELAND class *fleet escorts (missile cruiser conversions)*

Displacement: 11,066t light; 15,152t full load (CLG 5)
Dimensions: 600ft wl, 610ft oa × 65ft 8in × 25ft 8in
182.9m, 186.0m × 20.0m × 7.8m
Machinery: 4-shaft geared turbines, 4 Babcock & Wilcox boilers, 100,000shp=32kts. Range 8000nm at 15kts. Oil 2661t
Armament: 3–6in/47 (1×3), 2–5in/38 (1×2) (6–6in/47 (2×3), 6–5in/38 (3×2) in CLG 3 and CLG 8), 1 Talos SAM (1×2, 46 missiles, in CLG 3–CLG 5), 1 Terrier SAM (1×2, 120 missiles, in CLG 6–CLG 8)
Sensors: Radar SPS-9 (SPS-17 in CLG 4), SPS-39, SPS-8B (CLG 3, CLG 5–CLG 8; SPS-2 in CLG 4), 2 SPQ-5 (CLG 6–CLG 8), 2 SPG-49/SPW-2 (CLG 3–CLG 5); sonar SQS-23. See notes
Complement: 1382

No	Name	Yard	Recomm	Fate
CLG 3	GALVESTON	Philadelphia N Yd	28.5.58	Decommissioned 1970, stricken 21.12.73
CLG 4	LITTLE ROCK	New York SB	3.6.60	Decommissioned 1976, memorial
CLG 5	OKLAHOMA CITY	Bethlehem	7.9.60	Decommissioned 1979, stricken 15.12.79
CLG 6	PROVIDENCE	Boston N Yd	17.9.59	Decommissioned 1973, BU 1978
CLG 7	SPRINGFIELD	Boston N Yd	2.7.60	Decommissioned 1973, BU 1978
CLG 8	TOPEKA	New York SB	26.3.60	Decommissioned 1969, stricken 1.12.79

SPS-39 was removed from CLG 4 and 5 by 1967.

All these ships were refitted with SQS-23 hull sonars and with DASH drone control facilities in the early 1960s, as part of a programme to permit more open fleet formations (against nuclear attack): in such formations submarines would probably be able to penetrate the outer ASW screen, and each long-range sonar within that screen would be valuable. Similar reasoning applied to the fleet carrier *America*. As first completed, the former *Clevelands* were considered top-heavy and had to be specially ballasted.

These ships were never entirely satisfactory, given their limited size. The two pure missile cruisers went first, decommissioned in 1969–70 and stricken in 1973. Fleet flagships were considered far more important, but the Terrier ships never received modernised fire control systems (with SPG-55 radars) because of the cost of conversion and the limited remaining life in their hulls; they were both withdrawn in 1973 and later stricken. The two Talos ships were the last to go, *Oklahoma City* ending her service with the Seventh Fleet in 1979.

Oklahoma City (as well as *Chicago* and *Long Beach*) used Talos anti-radar missiles (RGM-8H) to engage North Vietnamese radar sites during the Vietnam War. The number of such attacks has not been released, but it appears to have been small.

Columbus 18.11.63 L & L van Ginderen

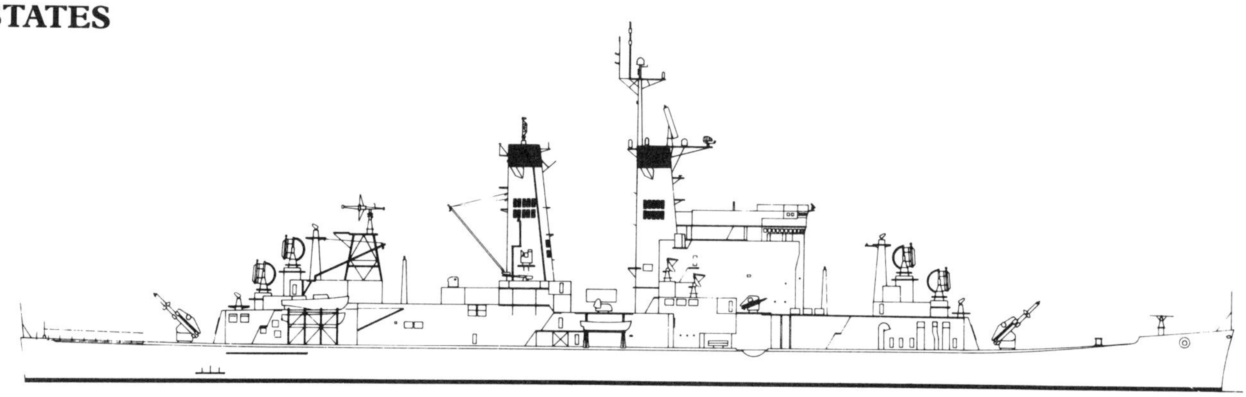

Albany 1970

ALBANY class *fleet escorts (missile cruiser conversions)*

Displacement: 14,394t light; 18.777t full load
Dimensions: 664ft wl, 674ft 11in oa × 69ft 9in × 25ft 9in
202.4m, 205.8m × 21.3m × 7.9m
Machinery: 4-shaft GE geared turbines; 4 Babcock & Wilcox boilers, 120,000shp = 32kts. Range 7000nm at 15kts
Armament: 2 Talos SAM (2×2, 52 missiles each), 2 Tartar SAM (2×2, 42 missiles each), 2–5in/38 (2×1), 1 ASROC ASW system (1×8), 6–12.75in Mk 32 ASW TT (2×3)
Sensors: Radar SPS-43A (in place of original SPS-29), 2 SPS-30, 4 SPG-49/SPW-2, 4 SPG-51, sonar SQS-23. See notes
Complement: 1266

No	Name	Yard	Recomm	Fate
CG 10	ALBANY	Boston N Yd	3.11.62	Stricken 30.6.85
CG 11	CHICAGO	San Francisco N Yd	2.5.64	Decommissioned 1.3.80
CG 12	COLUMBUS	Puget Sound N Yd	1.12.62	Stricken 9.8.76, BU

These three ships were the most elaborate of the postwar US missile cruiser conversions, as well as the most futuristic in appearance. They alone received the kind of Talos launching system originally envisaged, extending its magazine deep into their hulls fore and aft (with fifty-two missiles at each end). Planned entirely without guns, they had a secondary battery of two twin Tartar launchers on their beams (eighty-four missiles), as well as ASROC and the associated SQS-23 sonar system; in effect they were massive missile frigates. The high bridge was necessary to clear the paired SPG-49 missile control radars forward of it, and the unusual 'Mack' arrangement was chosen to reduce radar interference from the funnels. In addition, early planning called for the Regulus II surface-to-surface missile, but it was deleted at the design stage. However, plans for further conversions (which would have been CG 13–CG 15) did call for eight Polaris amidships. CG 10 and CG 11 were modernised with SPS-48 in place of SPS-39, and one SPS-30 was removed.

A pair of open 5in/38 guns was installed amidships at the insistence of the President, John F Kennedy, apparently after he had witnessed the failure of a demonstration of Terrier against a drone; the President may also have been impressed by the absence of any weapon suitable against a small torpedo-boat. Neither at which time she was also fitted with SPS-48 radar in place of her earlier SPS-39, one of her two SPS-30 fighter control radars being removed ASROC nor the Mk 32 TT had reloads.

Three further ships were to have been converted under the FY60 programme as CG 13–CG 15, with a new SPG-56 missile control radar (SCB-173A rather than SCB-173); they were cancelled in favour of the new Typhon missile system, itself abandoned a few years later because of excessive cost. Only the last ship, *Chicago*, was completed with (a primitive) NTDS; *Albany* received a full NTDS system as part of an AAW modernisation (SCB-002) under the FY68 programme (Boston Naval Shipyard, February 1967–June 1969) as weight compensation. Her Talos missile system was modernised with digital fire control. She was refitted in 1974–75 to serve as flagship of the Second Fleet and her remaining SPS-30 replaced by a satellite communications antenna. *Chicago* was modernised in August 1972–August 1973, albeit not on the same scale as *Albany*. Her NTDS was brought up to date, and she was fitted with an SLQ-26 Threat Reactive Anti-Ship Missile Defence system. *Albany* had already had the SLQ-19A ASMD system fitted in 1970, after her SCB-002 modernisation. *Columbus* alone was not modernised, and she was, therefore, the first of the class to be laid up, in May 1975. She was stricken in 1976. Her two sisters were decommissioned in 1980, their Talos missile system having been withdrawn from service to save money, even though initially they had been planned for retention through 1985. Normal practice would have been to strike them at this time, but in view of the international situation the Reagan Administration chose to retain them in reserve. Note, however, that with Talos gone they are reduced to a Tartar missile battery.

Chicago served as primary air protection ship during the mining of Haiphong Harbour in 1972; she and other missile ships were assigned primary responsibility, with carrier fighters acting only as back-ups. When MiGs were detected heading for the low-flying mining aircraft, *Chicago* shot one down with Talos at a range of 48nm, and the others fled.

Long Beach as completed

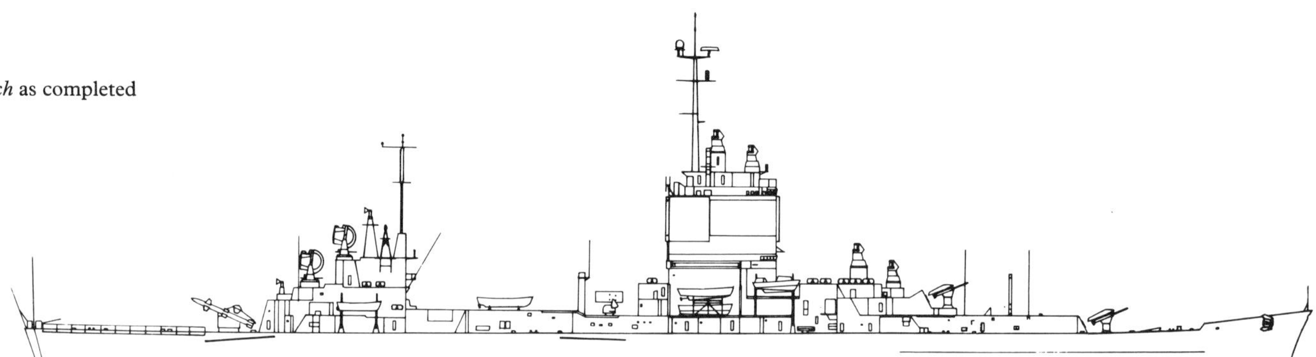

LONG BEACH *fleet escort (missile cruiser)*

Displacement: 15,111t light; 16,602t full load
Dimensions: 690ft wl, 721ft 3in oa × 73ft 4in × 23ft 9in
210.4m, 219.9m × 22.3m × 7.3m
Machinery: 2-shaft nuclear: 2 C1W reactors, 2 GE geared turbines, about 80,000shp = 30kts +
Armament: 1 Talos (1×2, 52 missiles), 2 Terrier SAM (2×2, total 120 missiles), 2–5in/38 (2×1), 1 ASROC ASW system (1×8, 20 missiles), 6–12.75in Mk 32 TT (2×3). See notes
Sensors: Radar SPS-32, SPS-33, 2 SPG-49/SPW-2, 4 SPG-55; sonar SQS-23. See notes
Complement: 1107

No	Name	Builder	Laid down	Launched	Comp	Fate
CGN 9	LONG BEACH	Bethlehem, Quincy	2.12.57	14.7.59	9.9.61	Deactivated 2.7.94

Intended at first as a nuclear frigate, this ship grew in the design stage to large cruiser size; indeed, she was the only cruiser (of traditional size) built as such by the US Navy after the war. She shared with *Enterprise* the futuristic fixed-array SPS-32/33 radar system, and has an SQS-23 long-range sonar in her bow, with the associated ASROC launcher amidships, as well as a pair of single 5in/38 guns, which were not part of the original design. She combined a pair of Terrier launchers forward (one with forty, one with eighty missiles) with Talos aft (fifty-two missiles, as in the *Albany* class), and had a two-reactor power plant. Although *Long Beach* is a substantial ship at over 14,000t, she stands in contrast to much larger nuclear cruisers proposed in the mid-1950s, with four-shaft plants more closely approaching traditional cruiser power. Polaris or Regulus II was included in the original design, but neither was installed. The C1W reactors installed are reportedly very similar to the A2Ws of the *Enterprise*, presumably somewhat uprated, as total *Enterprise* power is reported as 280,000shp.

The SPS-32/33 radar system was combined with one of the earliest NTDS installations. In 1968 a conventional SPS-12 air search radar was added, partly because of maintenance difficulties with the fixed radars, and partly because the fixed sets did not have any integral IFF. A 1970 modernisation provided integral IFF and digital Talos fire control, but the SPS-12 remained.

A proposal for Aegis conversion (*ie* for the conversion of *Long Beach* to a prototype strike cruiser with an 8in gun forward) was rejected in 1977, reportedly out of concern that it would have jeopardised the Strike Cruiser programme itself.

Long Beach Jan 1980 L & L van Ginderen

While operating in the Tonkin Gulf, *Long Beach* shot down two MiGs over North Vietnam at a range of about 65nm, in May and June 1968. At the time these were described as the first recorded combat uses of a naval SAM.

Long Beach was modernised at Puget Sound (6 October 1980 to 26 March 1983): the SPS-32/33 'billboard' radars were removed (although their flat-sided blockhouse superstructure remains, with 1.75in armour replacing the weight of the radars) and SPS-48 and -49 were installed. Two Phalanx replaced the old SPG-49 Talos guidance radars and Harpoon canisters were arranged to fire athwartships (they replace the Talos launcher aft). The old Talos magazine aft was not used. The SQS-23 sonar was modified to SQQ-23 configuration, but without adding a second sonar dome.

A further NTU modernisation, scheduled for FY93, was not carried out; the ship decommisioned instead (1994). The FY93 modernisation would have included installation of TFCC and also replacement of the Harpoons aft by Tomahawks in armoured box launchers (Harpoons would have been relocated to the top of the superstructure).

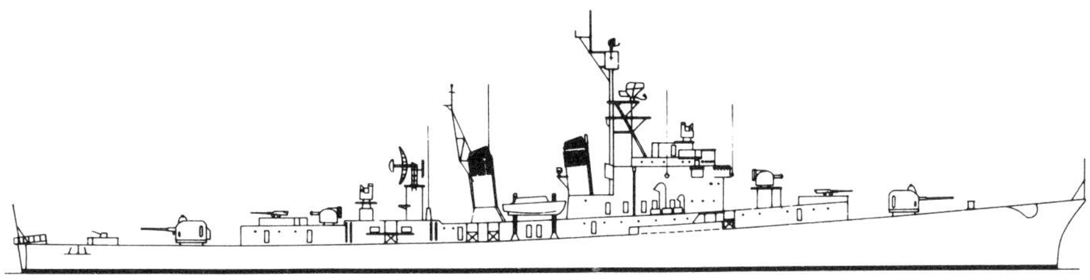

Mitscher 1954

MITSCHER class *fleet escorts (destroyers)*

Displacement: 3331t light; 3642t standard; 4855t full load
Dimensions: 476 ft wl, 490ft oa × 47ft 6in × 14ft 8in
145.1m, 149.4m × 14.5m × 4.5m
Machinery: 2-shaft geared turbines, 4 Foster Wheeler boilers, 80,000shp = 36.5kts. Range 4500nm at 20kts. Oil 740t
Armament: 2–5in/54 (2×1), 4–3in/50 (2×2), 8–20mm (4×2), 4–21in TT, 2 Weapon Alfa ASW, 1 DC track. See notes
Sensors: Radar SPS-6, SPS-8; sonar QHB, SQG-1 (sonar replaced by SQS-4 mid-1950s, SQS-23 in DLG conversions, with SPS-37 and SPS-48 plus two SPG-51; SQS-26 in the two remaining DLs). See notes

No	Name	Builder	Laid down	Launched	Comp	Fate
DL 2	MITSCHER	Bath Iron Wks	3.10.49	26.1.52	15.5.53	DDG 35 1968, stricken 1.6.78
DL 3	JOHN S McCAIN	Bath Iron Wks	24.10.49	12.7.52	12.10.53	DDG 36 1969, stricken 29.4.78
DL 4	WILLIS A LEE	Bethlehem, Quincy	1.11.49	26.1.52	28.9.54	Decommissioned 1969, stricken 15.5.72
DL 5	WILKINSON	Bethlehem, Quincy	1.2.50	23.4.52	29.7.54	Decommissioned 1969, stricken 1.5.74

These four ships, laid down as DD 927–930, were originally designated 'destroyers', then 'frigates', in a new category which persisted until 1975. They owed their large size to the requirement for high speed, deeply loaded, in a seaway, and were originally designed in 1944–46 with conventional destroyer batteries, the only major innovation being the 5in/54 gun in a twin semi-automatic version. However, as the design developed it became clear that fast carrier escorts would require maximum anti-aircraft capability, including not only guns but also fighter-direction equipment (symbolised visually by the heavy stabilised SPS-8height-finder). For a time, it appeared that the gun battery would consist entirely of the new 3in/70 twin automatic anti-aircraft gun, which was considered more effective than the new rapid-fire 5in/54 (itself firing faster than the earlier twin type), but the latter was retained at the personal insistence of Admiral Nimitz, at least partly in view of the inability of the lighter weapon to deal with submarines forced to the surface. In fact the large CIC and the heavy radar were even more important; as early as 1945 it was clear that carrier fighters, controlled by the carrier and by radar-pickets such as the *Mitschers*, would be the primary air defence of the fast carrier task force. Anti-submarine capability was secondary, as the high speed of the task force would in itself guarantee almost complete immunity. For example, in the design of the propellers and of the hull itself, high speed (to keep up with the task force) took priority over silencing. The primary ASW weapon was to be a guided torpedo, fired from a fixed tube, with a pair of rocket launchers (Weapon Alfa) as back-up.

High speed in a small ship meant a very compact machinery installation, and these four ships introduced the 1200lb plant to US service. All four were considered experimental, and gave considerable trouble in service, so that *Mitscher* and *John S McCain* had to be reboilered. Indeed, engine trouble was sometimes given as the reason for the early retirement of the class. On trials, *Mitscher* made 34.84 kts on 75,862shp at 4550t.

All were much modified in service. They were completed with open 3in/50s, as the enclosed 3in/70 was not yet ready. The latter were installed in 1957, and the original depth-charges as well as the after Weapon Alfa were surrendered as weight compensation, because the new mounting was much heavier than its original designed figure. In 1960 all four ships were fitted to operate the DASH drone helicopter, from a helipad which replaced the after 3in/70 mounting; *Wilkinson* landed the first DASH aboard at sea in August 1960. The DASH hangar was built above the torpedo room abaft No 2 funnel and the SPS-8 radar was re-sited to its top. The first two units were fitted with SQS-23 sonars, the last two with prototypes of the new SQS-26, in large bow domes which initially caused some problems. Of the forward 3in/70 and Weapon Alfa, the latter was soon landed in the two SQS-26 units. *Mitscher* abandoned her forward 3in/70 before her Weapon Alfa, but it appears that *John S McCain* retained both until her conversion as a DDG. In each case it was expected that ASROC would ultimately replace at least the forward Weapon Alfa.

As early as 1959 missile conversions were proposed, both Tartar and Terrier appearing feasible. The first two units were converted to Tartar-firing missile *destroyers* under the FY64 programme, but further conversions were abandoned in view of the ineffectiveness of the Tartar SAM system. Conversion entailed the replacement of the surviving superfiring weapons forward by ASROC (1×8, 16 missiles) with 6–12.75in TT and a single-arm Tartar launcher aft (40 missiles). The result was criticised as top heavy, and both ships were stricken after little more than a decade of service. The redesignation of the ships as DDG 35 and 36 rather than as DLG was indicative of the rapid growth in all classes of US warships. The two unconverted units were decommissioned in 1969.

All four were used to some extent as experimental ships. For example, *Mitscher* operated the first helicopter from a US destroyer in 1957, leading to the development of DASH.

John S McCain 1977 USN

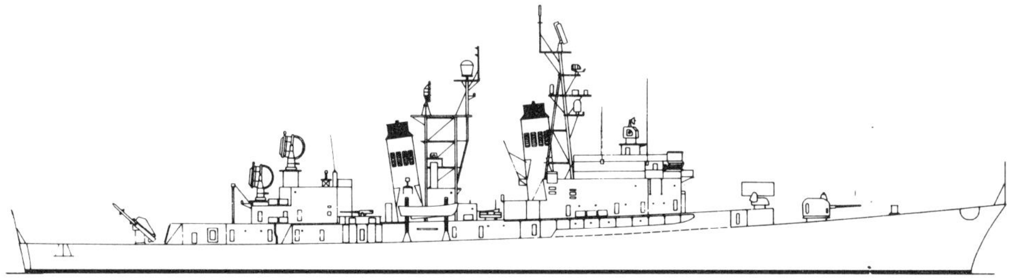

Dewey 1961

FARRAGUT class *fleet escorts (destroyers)*

Displacement: 4167t light; 5648t full load
Dimensions: 490ft wl, 512ft 6in oa × 52ft 4in × 17ft 9in
149.4m, 156.3m × 15.9m × 5.3m
Machinery: 2-shaft geared turbines, 4 Foster Wheeler boilers (Babcock & Wilcox in DLG 9–DLG 15) 85,000shp = 32kts. Range 5000nm at 20kts
Armament: 1–5in/54 RF, 1 Terrier SAM (1×2, 40 missiles), 1 ASROC ASW system (1×8), 6–12.75in Mk 32 ASW TT (2×3, 12 torpedoes)
Sensors: Radar SPS-29, SPS-37, 2 SPQ-5 (SPG-55 in DLG 7–DLG 13); sonar SQS-23
Complement: 360

No	Name	Builder	Laid down	Launched	Comp	Fate
DLG 6	FARRAGUT	Bethlehem, Quincy	3.6.57	18.7.58	10.12.60	Decomm 31.10.89
DLG 7	LUCE	Bethlehem, Quincy	1.10.57	11.12.58	20.5.61	Decomm 1.4.91
DLG 8	MACDONOUGH	Bethlehem, Quincy	15.4.58	9.7.59	4.11.61	Decomm 23.10.92
DLG 9	COONTZ	Puget Sound N Yd	1.3.57	6.12.58	15.7.60	Decomm 4.10.89
DLG 10	KING	Puget Sound N Yd	1.3.57	6.12.58	17.11.60	Decomm 28.3.91
DLG 11	MAHAN	Mare Island N Yd	31.7.57	7.10.59	25.8.60	Decomm 15.6.93
DLG 12	DAHLGREN	Philadelphia N Yd	1.3.58	16.3.60	8.4.61	Damaged 31.7.92
DLG 13	WILLIAM V PRATT	Philadelphia N Yd	1.3.58	16.3.60	4.11.61	Decomm 30.9.91
DLG 14	DEWEY	Bath Iron Wks	10.8.57	30.11.58	7.12.59	Decomm 31.8.90
DLG 15	PREBLE	Bath Iron Wks	16.12.57	23.5.59	9.5.60	Decomm 15.11.91

These were the first missile-armed fleet escorts designed and built as such, although the first three were originally to have been armed only with guns. They were designed on the recommendation of the Schindler Committee on the Long Range Shipbuilding Plan (1954), which argued for the construction of specialised Fast Task Force Escorts primarily for anti-aircraft screening, with submarine detection (for warning) but not destruction as an important secondary task. The resulting design called for four single rapid-fire 5in/54s as well as two twin 3in/50s; ASW weapons were limited to a single depth-charge track aft, a pair of fixed Hedgehogs (as in destroyers), and a quintuple torpedo tube intended primarily 'to embarrass enemy heavy units attempting to attack the carrier task force', as at Samar 1944.

Throughout the design, the installation of Terrier in place of the two after 5in/54s was carried as an alternative, and in 1955 the Chief of Naval Operations, Admiral Carney, personally decided that half of the FY56 ships were to be built as missile frigates. His successor, Admiral Burke, decided to accelerate the introduction of missiles into the fleet by ordering all six FY56 ships as missile ships. Meanwhile development of new anti-surface ship torpedoes was abandoned, and the planned quintuple trainable tubes, which were not compatible with existing long-range anti-submarine torpedoes, were dropped from the project. At about the same time, however, the advent of Soviet nuclear submarines, which were capable of attacking even fast task forces, made ASW a more important function of fast fleet escorts. Thus these ships were redesigned, first for the rocket-assisted torpedo (RAT) and then for ASROC, which replaced one of the two forward 5in/54s which had survived the transformation from all-gun to gun-and-missile frigates. At the same time their sonar was upgraded from the SQS-4 to the much larger SQS-23.

King and *Mahan* tested the prototype NTDS installations in 1961–62. Subsequently all ten ships were subject to extensive AAW modernisation. NTDS was installed and a new guidance radar for Terrier fitted (SPG-55B in lieu of SPQ-5A and SPG-55A in ships as built). Larger ship service turbo-generators were fitted to carry the increased electrical load. *Farragut* alone received reload stowage for her ASROC. Weight compensation included the removal of the two twin 3in/50s formerly carried. These refits were completed between 1969 and 1977, under the FY66–74 programmes (no awards in FY67 and FY73).

Harpoon quadruple launchers replaced the former twin 3in/50 mounts. *Coontz* tested the Vulcan Phalanx CIWS in 1973–74, before modernisation, and in 1975 *King* carried two Vulcans. They now have

Luce 1979 C & S Taylor

two-dome SQQ-23A sonars, and are scheduled to receive SPS-49 radars in place of the former SPS-29C and SPS-37.

Mahan (DDG 42) was test ship for SM-2(ER) in 1979; this system allows a Terrier ship to engage several targets simultaneously, using a new Weapons Direction System Mk 14 and a new missile down-link. In addition, because it incorporates command guidance, the missile can fly a much longer-range energy-efficient trajectory; in effect the SM-2(ER) replaces the earlier Talos. In 1982 *Mahan* tested the New Threat Upgrade, which sharply reduces ship reaction time through the integration of shipboard radars (SPS-48E and SPS-49(V)5), NTDS, SYS-2, and the SM-2(ER) Block II missile. As of early 1983, NTU/SM-2 Block II is scheduled for ten Tartar and twenty-one Terrier ships: *Long Beach*, the *Leahy*, *Belknap*, *California* and *Virginia* class cruisers, and the *Kidd* class.

All were redesignated missile *destroyers* (DDG 37–46 in 1975, and are the only DDGs which fire the two-stage Terrier or Standard (Extended Range) missile.

LEAHY class *fleet escorts (cruisers)*

Displacement: 5146t light; 7590t full load
Dimensions: 510ft wl, 533ft oa × 53ft 4in × 19ft
155.5m, 162.5m × 16.3m × 5.8m
Machinery: 2-shaft geared turbines, 4 Babcock & Wilcox boilers (Foster Wheeler in DLG 19–DLG 24), 85,000shp = 32kts. Range 8000nm at 20kts
Armament: 2 Terrier SAM (2×2, 40 missiles each launcher), 1 ASROC ASW system (1×8, no reloads), 4–3in/50 (2×2), 6–12.75in Mk 32 ASW TT (2×3, 12 torpedoes)
Sensors: Radar SPS-37, SPS-39, 4 SPG-55; sonar SQS-23
Complement: 377

No	Name	Builder	Laid down	Launched	Comp	Fate
DLG 16	LEAHY	Bath Iron Wks	3.12.59	1.7.61	4.8.62	Stricken 1.10.93
DLG 17	HARRY E YARNELL	Bath Iron Wks	31.5.60	9.12.61	2.2.63	Stricken 29.10.93
DLG 18	WORDEN	Bath Iron Wks	19.9.60	2.6.62	3.8.63	Stricken 1.10.93
DLG 19	DALE	New York SB	6.9.60	28.7.62	23.11.63	Stricken 22.9.94
DLG 20	RICHMOND K TURNER	New York SB	9.1.61	6.4.63	13.6.64	Stricken 23.9.94
DLG 21	GRIDLEY	Puget Sound N Yd	15.7.60	31.7.61	25.5.63	Stricken 21.1.94
DLG 22	ENGLAND	Todd, Los Angeles	4.10.60	6.3.62	7.12.63	Stricken 21.1.94
DLG 23	HALSEY	Mare Island N Yd	26.8.60	15.1.62	20.7.63	Stricken 28.1.94
DLG 24	REEVES	Puget Sound N Yd	1.7.60	12.5.62	15.5.64	Stricken 12.11.93

In these ships the guided missile was finally made paramount, with guns little more than an afterthought. The need for large numbers of missiles to protect the Fleet against a mounting air threat and the relatively slow rate of new construction led to a conscious decision that new missile ships be double-ended. It had previously been objected that missile mounts in the bow would be so subject to sea damage as to be useless; in the *Leahy* class special efforts were to protect the forward launcher, including the knuckling of the hull, which was uncharacteristic of previous US practice. The decision to adopt a forecastle type of hull reflects the explosive growth of space (rather than weight) requirements, a characteristic of modern warship development. As in the *Farragut*s, ASW was a distinctly secondary consideration, but the *Leahy*s were designed to incorporate ASROC (without reloads) and had a bow dome for their SQS-23 sonars.

As in the case of earlier Terrier ships, the *Leahy* class suffered from limited effectiveness as built, and was subject to extensive modernisation. NTDS was fitted, with the improved SPG-55B guidance radar and a modernised fire control system capable of controlling the Standard Missile (SM-1). Refits were funded under the FY66 (DLG 16), FY67 (DLG 17, 18, 21, 22 and 24) and FY71 (DLG 19, originally under FY68–DLG 20 and DLG 23) budgets. Later all had their pair of twin 3in/50 removed for replacement by quadruple Harpoon launchers.

All had received SM-2ER capability by 1985. CG 19 was the first to be fitted with SPS-49 in place of the earlier SPS-43 main air search radar (all were later fitted). All also received the SLQ-32(V)3 ECM system. All now have two Phalanx as well as four 0.50 calibre machine guns.

The major more recent improvement is NTU. CG 16, 17, and 19 received this modernisation in FY87–88, followed by CG 18 in FY88–89, CG 22 during refit October 1986–October 1987, CG 20 during refit June 1988–July 1989. CG 21 began NTU refit 9 January 1989, and CG 23 on 1 May 1989.

All were redesignated CG (cruisers) rather than DLG in 1975.

Richmond K Turner 27.9.81
L & L van Ginderen

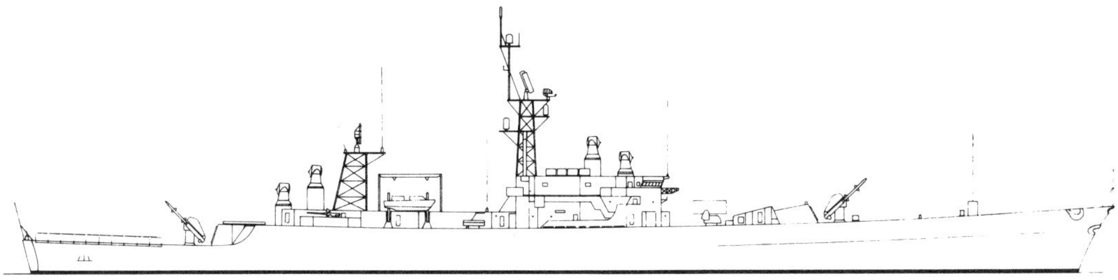

Bainbridge 1962

Bainbridge (DLG 25, CGN 25 from 1975) was essentially a nuclear-powered equivalent of the *Leahy* class, built as the prototype nuclear fast task force escort. At the time she represented the minimum platform for a two-reactor power plant which could be expected to attain destroyer speed; a fast submarine plant, such as that in the large *Triton*, was rejected because it weighed too much for the power it produced to propel a destroyer with a worthwhile payload at an acceptable speed. Destroyer nuclear propulsion was given high priority from about 1956 onwards because destroyer endurance, not carrier endurance, was a limiting factor in carrier task force operations.

AAW modernisation at Puget Sound (30 June 1974–24 September 1976, completed at San Diego April 1977): large deckhouse aft for NTDS, 3in/50 replaced by 20mm guns and then by quadruple Harpoon canisters (1979). Two Phalanx installed during refit October 1983–April 1985, SPS-37 replaced by SPS-49, SLQ-32(V)3 and RBOC chaff launchers installed. She now has the Mk 14 weapons direction system (NTU).

BAINBRIDGE *fleet escort (cruiser)*

Displacement: 7250t light; 7982t full load
Dimensions: 540ft wl, 565ft oa × 56ft × 19ft 5in
164.6m, 172.3m × 17.0m × 5.9m
Machinery: 2-shaft nuclear: 2 D2G reactors, 2 geared turbines, approx 60,000shp = *c*30kts
Armament: 2 Terrier SAM (2x2, 40 missiles each), 1 ASROC ASW system (1×8), 6–12.75in Mk 32 ASW TT(2×3), 4–3in/50 (2×2)
Sensors: Radar SPS-37, SPS-39, 4 SPG-55; sonar SQS-23
Complement: 459

No	Name	Builder	Laid down	Launched	Comp	Fate
DLGN 25	BAINBRIDGE	Bethlehem, Quincy	15.5.59	15.4.61	6.10.62	Decomm FY95

Bainbridge 20.1.83 RAN

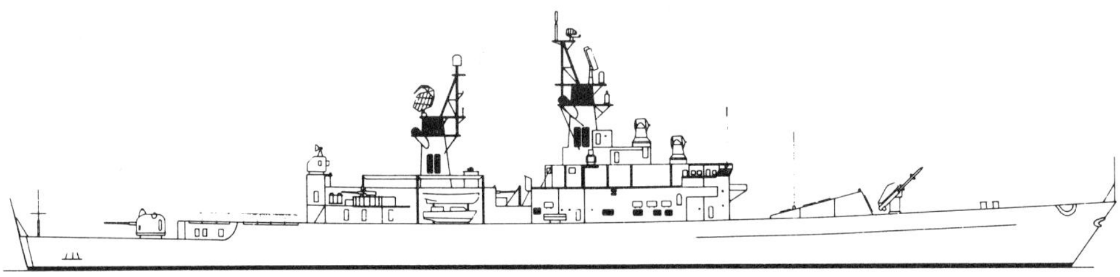

Sterett 1969

These ships suffered a somewhat tortured design history, originating as an attempt to produce a missile destroyer both less expensive than the *Adams* class and more effective in ASW, with the new SQS-26 sonar and endurance increased from under 4500nm to 6000nm; the result, even with a new austere missile system (twelve rather than forty Tartars were planned) was far more expensive than the original destroyer. Quite soon Terrier was substituted, as befitting so expensive a ship, and the existing *Leahy* class hull was adapted to the new design. However, its destroyer origin survived in the requirement for a single 5in/54 rapid-fire gun. Considerable economy in design was achieved by combining the Terrier and ASROC launchers; the system mounted in these ships can accommodate up to sixty Terriers, or

BELKNAP class *fleet escorts (cruisers)*

Displacement: 5409t light; 7890t full load
Dimensions: 524ft wl, 547ft oa × 54ft 9in × 18ft 2in
159.8m, 166.8m × 16.7m × 5.5m
Machinery: 2-shaft geared turbines, 4 Babcock & Wilcox boilers (Combustion Engineering in DLG 28–DLG 31, DLG 33), 85,000shp = 32kts. Range 7100nm at 20kts
Armament: 1 Terrier/ASROC SAM/ASW system (1×2, up to 20 ASROC of a total of 60 missiles), 1–5in/54 RF, 2–3in/50 (2×1), 6–12.75in Mk 32 ASW TT (2×3), 2–21in ASW TT (not fitted in some ships)
Sensors: Radar SPS-43, SPS-48, 2 SPG-55; sonar SQS-26; NTDS in DLG 28–DLG 34
Complement: 388

No	Name	Builder	Laid down	Launched	Comp	Fate
DLG 26	BELKNAP	Bath Iron Wks	5.2.62	20.7.63	7.11.64	Decomm FY95
DLG 27	JOSEPHUS DANIELS	Bath Iron Wks	23.4.62	2.12.63	8.5.65	Stricken 22.1.94
DLG 28	WAINWRIGHT	Bath Iron Wks	2.7.62	25.4.64	8.1.66	Stricken 10.11.93
DLG 29	JOUETT	Puget Sound N Yd	25.9.62	30.6.64	3.12.66	Stricken 28.1.94
DLG 30	HORNE	More Island N Yd	12.12.62	30.10.64	15.4.67	Stricken 4.2.94
DLG 31	STERETT	Puget Sound N Yd	25.9.62	30.6.65	8.4.67	Stricken 24.3.94
DLG 32	WILLIAM H STANDLEY	Bath Iron Wks	29.7.63	19.12.64	9.7.66	Stricken 24.3.94
DLG 33	FOX	Todd, San Pedro	15.1.63	21.11.64	28.5.66	Stricken 15.4.94
DLG 34	BIDDLE	Bath Iron Wks	9.12.63	2.7.65	21.1.67	Stricken 30.11.94

it can accommodate forty Terriers and up to twenty ASROCs. Increasing interest in ASW was also manifested in provision for the DASH ASW drone helicopter, and for Mk 48 torpedo tubes in the transom stern. They were redesignated 'cruisers' in 1975.

Wainwright (DLG 28) was the first to be completed with the Naval Tactical Data System integrated into her weapon control system; the two earlier ships later also received NTDS. *Belknap*, the name ship, was severely damaged by collision with the carrier *John F Kennedy* in the Mediterranean in November 1975, and was rebuilt at Philadelphia (1975–80); however, proposals to fit the new Aegis system were not accepted and she emerged only slightly modified. As of 1981 the modernisation programme called for eight Harpoon launchers (four to port firing forward, four to starboard firing aft) in place of the former 3in/50 secondary guns, and all are to receive a pair of Phalanx Close-In Weapon Systems. The Mk 25 torpedo tubes for long-range ASW torpedoes were removed from all ships of this class. *Fox* was trials ship for Tomahawk box launcher (1977). In addition, *Wainwright* has served as trials ships for SM-2(ER).

Belknap is Sixth Fleet flagship, with TFCC on board. As refitted 1976–80 she had her 3in guns replaced by Harpoon canister launchers, received SPS-48C and SPS-49 radars, improved NTDS, and SQS-53A sonar with digital (Mk 116) FCS. She was again refitted (as a flagship) at Philadelphia (June 1985–March 1986), with a new two-level deckhouse forward of the bridge and her hangar converted to accommodation space.

Fox 10.9.87 — RAN

CG 26, 28, 30, and 31 were all fitted with TFCC in 1983–85.

As of 1992, all are having SPS-48 updated to -48E as part of the NTU modernisation (SYS-2 added). CG 34 was completed July 1987, CG 30 begun September 1988, CG 29 completed 1989, CG 33 September 1989–September 1990; all the others were scheduled to complete by September 1993 (CG 26 last). At least CG 26, 29, and 30 had SM-2ER capability before they underwent NTU conversion.

A plan to carry Tomahawk in armoured box launchers was abandoned.

Both *Sterett* and *Biddle* engaged and shot down North Vietnamese MiGs over the Gulf of Tonkin during 1972. The former was the first US ship to be attacked by the MiGs, on 19 April 1972, while covering *Oklahoma City* and *Higbee* in a shore bombardment. She shot down two MiGs and what was believed to have been a Styx missile, the first anti-ship missile to have been destroyed by another missile in combat. *Biddle* was attacked by five MiGs when on PIRAZ (Positive Identification and Radar Advisory Zone, or air control) duty in the Gulf of Tonkin on the night of 19 July 1972, shooting down two.

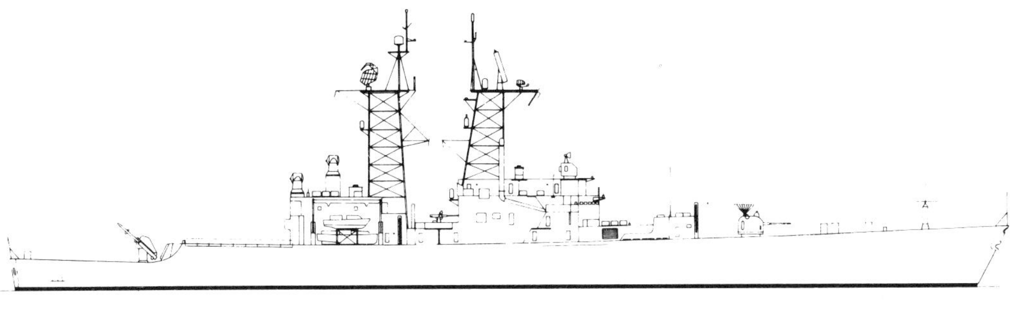
Truxtun 1980

Truxtun (DLGN 35) was essentially a nuclear version of the *Belknap* class, somewhat enlarged and rearranged to accommodate her power plant. The Nuclear Propulsion organisation had originally hoped to duplicate the earlier *Bainbridge*, but this was rejected in view of the need for the improved SQS-26 sonar.

Her 3in/50 were replaced by Harpoon in 1980. Refitted 4 October 1982–April 1984: two Phalanx and new EW suit added (SLQ-32 and SRBOC). She has a Mk 14 weapons direction system (NTU). *Truxtun* is to decommission in 1993.

This ship was designated 'cruiser' (CGN) rather than frigate (DLGN) in 1975.

TRUXTUN *fleet escort (cruiser)*

Displacement: 8149t light; 8927t full load
Dimensions: 540ft wl, 564ft oa × 57ft 10in × 19ft 10in
164.6m, 172.0m × 17.6m × 6.0m
Machinery: 2-shaft nuclear: 2 D2G reactors, 2 geared turbines, approx 60,000shp = *c*30kts
Armament: 1 Terrier/ASROC SAM/ASW system (1×2, total 60 missiles), 1–5in/54 RF, 2–3in/50 (2×1), 6–12.75in Mk 32 ASW TT (2×3)
Sensors: Radar SPS-40, SPS-48, 2 SPG-55; sonar SQS-26
Complement: 490

No	Name	Builder	Laid down	Launched	Comp	Fate
DLGN 35	TRUXTUN	New York SB	17.6.63	19.12.64	27.5.67	Decomm FY95

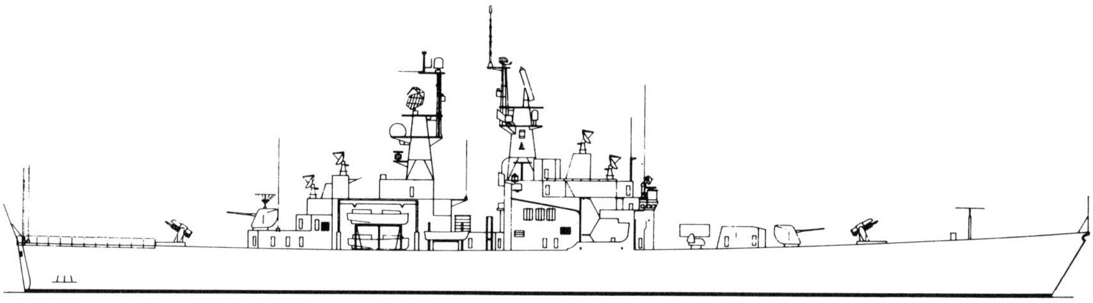

California 1977

These two ships were nuclear versions of the abortive conventionally-powered FY66 guided missile destroyer, built under the FY67 and FY68 programme, the first long lead time items having been bought under the FY66 programme. A third ship, partially funded in FY68, was built instead of a new design. Their construction was quite expensive, and by the late 1960s Secretary of Defense McNamara was unwilling to release funds for them; they were built as a result of Congressional pressure.

The design originally called for a pair of heavyweight Mk 42 5in/54 guns and two torpedo tubes for Mk 48 torpedoes, in the transom; the tubes were discarded and lightweight Mk 45 guns substituted. The missile battery was controlled by the new digital Tartar-D (SPG-51D) fire control system, and, compared to the *Bainbridge* and *Truxtun*, they had improved D2G reactors with about three times the core lifetime. A helicopter landing area was provided aft, but they had no hangar, as they were designed during the gap between DASH and LAMPS. The *Californias* have been described as the first US nuclear surface ships intended for series production. They were redesignated 'cruisers' (CGN) in 1975.

Kevlar armour was added over vitals. NTU: added during FY89 and FY90, SPS-49 replacing the earlier SPS-40 radar.

ASROC was removed in 1993. Both were retained (to decommission 1998–99) because they have been refuelled 1992–93.

CALIFORNIA class *fleet escorts (cruisers)*

Displacement: 10,150t full load
Dimensions: 570ft wl, 596ft oa × 61ft × 20ft 6in
173.8m, 181.7m × 18.6m × 6.3m
Machinery: 2-shaft nuclear: 2 D2G reactors, 2 geared turbines, approx 60,000shp = 30+kts
Armament: 2 Standard SAM (1×2, 40 missiles each), 1 ASROC ASW system (1×8, 24 missiles), 4–12.75in Mk 32 ASW TT (4×1 fixed, 16 torpedoes), 2–5in/54 (2×1)
Sensors: Radar SPS-40, SPS-48, 2 SPG-51, SPG-60, SPQ-9; sonar SQS-26
Complement: 533

No	Name	Builder	Laid down	Launched	Comp	Fate
DLGN 36	CALIFORNIA	Newport News	23.1.70	22.9.71	16.2.74	Extant 1995
DLGN 37	SOUTH CAROLINA	Newport News	1.12.70	1.7.72	25.1.75	Extant 1995

South Carolina 1980 L & L van Ginderen

Redesignated cruisers in 1975, these ships were initially intended as nuclear counterparts to the *Spruance* DDG design (DXGN), although they were completed as slightly improved equivalents of the *Californias*. Four were built under the FY70–75 budgets (with none in FY73 and FY74), but a fifth, longlead items for which had been approved under the FY75 programme, was stopped in FY76 because, without Aegis, she would be obsolete.

The original design called for a separate ASROC launcher and one Mk 13 missile launcher aft; however, it was argued that a nuclear frigate might well have to operate independently and therefore would need two missile launchers, to overcome limited magazine capacity as well as to assure the availability of at least one launcher. The adoption of the Mk 26 missile launcher eliminated the ASROC requirement, but these ships still have most of their battery aft (Mod 1, forty-four weapons), together with both SPG-51D guidance radars; forward they have a single gun fire control system with one SPG-60 tracker-illuminator, providing a possible third missile-control channel. Typically they carry fifty SAM, sixteen ASROC and two test rounds. A hangar below the quarterdeck can accommodate LAMPS helicopters, but its elevator reportedly leaks badly in wet weather, and the helicopter is not usually embarked; this arrangement is often criticised.

Hopes that at least some ships of this class would be fitted with Aegis (then called ASMS) did not materialise, although a modified *Virginia* hull, CGN 42, was for a time a candidate successor to the Strike Cruiser.

These ships carry their Harpoon missiles in canisters firing athwartships, forward of the bridge. All were fitted with eight Tomahawks in armoured box launchers aft, in place of the earlier helicopter facility, with its leaky below-decks hangar.

All completed SM-2 upgrade (beginning 1983); as of 1990 full NTU upgrade was to have been completed by FY95. CGN 40 was the first to have her SPS-40 replaced by SPS-49;

VIRGINIA class *fleet escorts (cruisers)*

Displacement: 11,000t full load
Dimensions: 560ft wl, 585ft oa × 63ft × 21ft
170.7m, 178.3m × 19.2m × 6.4m
Machinery: 2-shaft nuclear: 2 D2G reactors, 2 geared turbines, approx 60,000shp = 30+kts
Armament: 2 Standard SAM/ASW systems (2×2, 24 missiles forward, 44 in ASROC aft), 2–5in/54 (2×1), 6–12.75in Mk 32 ASW TT (2×3, 14 torpedoes), 1 helicopter
Sensors: Radar SPS-40, SPS-55, 2 SPG-51, SPG-60, SPQ-9; sonar SQS-53
Complement: 519

No	Name	Builder	Laid down	Launched	Comp	Fate
DLGN 38	VIRGINIA	Newport News	19.8.72	14.12.74	11.9.76	Deactivated 28.2.94
DLGN 39	TEXAS	Newport News	18.8.73	9.8.75	10.9.77	Deactivated 16.7.93, stricken
CGN 40	MISSISSIPPI	Newport News	22.2.75	31.7.76	5.8.78	Extant 1995
CGN 41	ARKANSAS	Newport News	17.1.77	21.10.78	18.10.80	Extant 1995

she was followed by CGN 41 and then CGN 39. CGN 40 and 41 have SPS-48C instead of -48A. Two Phalanx are fitted at the after end of the forward superstructure.

Kevlar armour was installed during FY82–86 overhauls.

Note that although these ships carry modern ASW sensors, they no longer carry any helicopter; they are primarily AAW ships. That is partly because they are inherently noisy, given the sound generated by the circulation pumps of their nuclear plants. Analogous sources of noise on nuclear submarines are sound-isolated. Note, however, that these were among the first ships to carry the new SQS-53A sonar and its associated digital fire control system, Mk 116.

Each nuclear fuelling is intended to last ten years.

ASROC missiles were removed when the NTU refit was carried out; the update included SPS-48E replacing -48 or -48C. *Virginia* still had her SPS-40B when retired, so presumably she at least never had the full upgrade. The first two ships were deactivated and stricken rather than being refuelled; the other two are to decommission in 1996–97.

Mississippi 4.6.84 L & L van Ginderen

Arkansas 18.7.84 RAN

Barry (after ASW conversion) 1971 USN

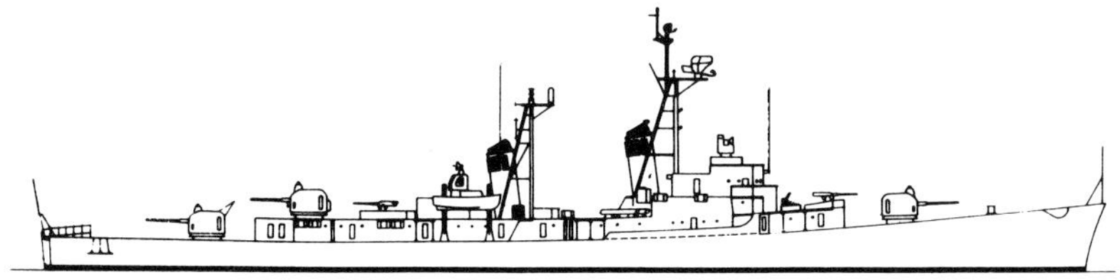

John Paul Jones as completed

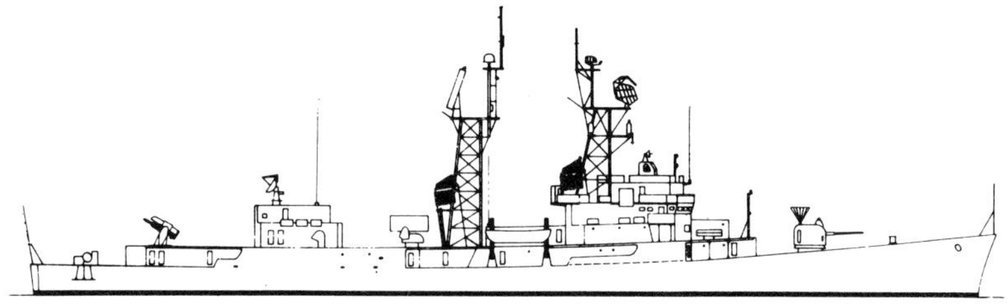

Somers 1968

FORREST SHERMAN class *fleet escorts (destroyers)*

Displacement: 2734t light; 4916t full load
Dimensions: 407ft wl, 418ft 6in oa × 44ft 11in × 15ft
124.1m, 127.6m × 13.7m × 4.6m
Machinery: 2-shaft geared turbines, 4 boilers, 70,000shp = 33kts. Range 4500nm at 20kts
Armament: 3–5in/54 (3×1), 4–3in/50 (2×2), 2 Hedgehog ASW mortars, 4–21in TT (4×1), 6–12.75in Mk 32 ASW TT (2×3)
Sensors: Radar SPS-6; sonar SQS-4
Complement: 324

No	Name	Builder	Laid down	Launched	Comp	Fate
DD 931	FORREST SHERMAN	Bath Iron Wks	27.10.53	5.2.56	9.11.55	Decomm 5.11.82, BU 1994
DD 932	JOHN PAUL JONES	Bath Iron Wks	18.1.54	7.5.55	5.4.56	DDG 32 Sep 1967, decomm 15.12.82, name cancelled 24.3.86
DD 933	BARRY	Bath Iron Wks	15.3.54	1.10.55	31.8.56	Decomm 5.11.82, museum ship Washington 1984
DD 936	DECATUR	Bethlehem, Quincy	13.9.54	15.12.55	7.12.56	DDG 31 Apr 1967, decomm 30.6.83, stricken 16.3.88
DD 937	DAVIS	Bethlehem, Quincy	1.2.55	28.3.56	28.2.57	Decomm 20.12.82, BU 1993
DD 938	JONAS INGRAM	Bethlehem, Quincy	15.6.55	8.7.56	19.7.57	Sunk as target 23.7.88
DD 940	MANLEY	Bath Iron Wks	10.2.55	12.4.56	1.2.57	Decomm 11.3.83, BU 1993
DD 941	DU PONT	Bath Iron Wks	11.5.55	8.9.56	1.7.57	Decomm 11.3.83, BU 1993
DD 942	BIGELOW	Bath Iron Wks	6.7.55	2.2.57	8.11.57	Decomm 5.11.82, BU 1993
DD 943	BLANDY	Bethlehem, Quincy	29.12.55	19.12.56	8.11.57	Decomm 5.11.82, BU 1993
DD 944	MULLINIX	Bethlehem, Quincy	5.4.56	18.3.57	7.3.58	Decomm 11.8.83
DD 945	HULL	Bath Iron Wks	12.9.56	10.8.57	3.7.58	Stricken 1983 as target
DD 946	EDSON	Bath Iron Wks	3.12.56	1.1.58	7.11.58	Decomm 15.12.88, museum ship New York
DD 947	SOMERS	Bath Iron Wks	4.3.57	30.5.58	3.4.59	Decomm 19.11.82
DD 948	MORTON	Ingalls	4.3.57	23.5.58	26.5.59	Decomm 22.11.82, BU 1993
DD 949	PARSONS	Ingalls	17.6.57	19.8.58	29.10.59	DDG 33 Nov 1967, decomm 19.11.82, stricken 15.5.84
DD 950	RICHARD S EDWARDS	Puget Sound N Yd	20.12.56	24.9.57	5.2.59	Decomm 15.12.82
DD 951	TURNER JOY	Puget Sound N Yd	30.9.57	5.5.58	3.8.59	Decomm 22.11.82

These were the only postwar general-purpose gun-armed destroyers, intended originally as a mobilisation version of the expensive *Mitscher* class (1950). However, when a Gibbs and Cox study showed that a simplification of the sophisticated construction of the earlier design would result in a 5000t ship, a new design was begun, concentrating more on anti-aircraft firepower than on ASW. In the end, even a trainable Hedgehog was given up, as it would have required extra length. It appears that the unusual gun arrangement, two aft and one forward, was adopted for dryness; in addition it was well suited for anti-aircraft defence. Given the high rate of fire of the automatic Mk 42 gun mounting in these ships, their battery was more than equivalent to that of the earlier *Gearing* class, although this was not evident at the time.

From DD 936 onwards the sheer line was raised 3ft at the bow for additional dryness; the first ships (FY53 group) were too far along to be modified. ASW weapons were fixed Hedgehogs and four long torpedo tubes, as well as a dropping system (succeeded by the familiar Mk 32 triple tubes) for lightweight torpedoes. However, the Bureau of Ordnance abandoned long torpedoes for surface ships in about 1956, and only the first two units were completed as planned.

By 1954 the state of destroyer and escort development was sufficiently confused to require special analysis, and a Committee of the Long Range Shipbuilding and Conversion Plan proposed the abandonment of general-purpose destroyer construction in favour of a combination of fast task force escorts (which became the *Farraguts*), fast ASW escorts (a proposed 'corvette' conversion of the *Bensons* and austere ocean escorts (destroyer escorts such as the *Dealeys*). In fact *Forrest Sherman* class construction was soon terminated in favour of a missile-armed equivalent, the *Charles F Adams* class, with eighteen general-purpose escorts completed.

By 1959 all were scheduled for conversion to Tartar-armed missile destroyers, a single missile launcher aft replacing both 5in/54 guns there. However, Secretary of Defense Robert S McNamara sharply curtailed the conversion programme in view of high unit cost and the low effectiveness of the Tartar missile; only four were converted, as DDG 31–DDG 34. Of the others, all were to have had ASW modernisation (with one 5in/54 removed, SQS-23 and SQS-35 sonars, and ASROC), but six units (DD 931, 932, 944–946, 951) were dropped because of cost escalation; all received the large SQS-23, however. *Barry* was the test ship for the bow-mounted SQS-23, and *Hull* tested the Mk 71 8in/55 Major Caliber Lightweight Gun. *Bigelow* was a test ship for Vulcan Phalanx. The DDG conversion carried forty Tartar and 16 ASROC rounds, and the ASW conversions carried twenty ASROC; both had 6–12.75in Mk 32 TT (without reloads).

Edson was assigned to Naval Reserve Training on 1 April 1977. During 1982 it was announced that the remainder of the class would be retired to offset the considerable cost of new construction. Reportedly the 1200psi plant was becoming a maintenance problem. *Jonas Ingram* decommissioned in March 1983 to become an engineering test hulk and was later sunk as a target. *Hull* decommissioned in July 1983 for use as a target. In 1992 *Decatur* was partly reactivated for trials of the Ship Self-Defense System (SSDS) at Puget Sound Naval Shipyard: she was equipped with a TAS Mk 23 radar, SLQ-32(V)3 ECM, launchers for RAM and Sea Sparrow (RIM-7P) missiles, and Phalanx (Block 1). She is to be used for live firing trials at sea.

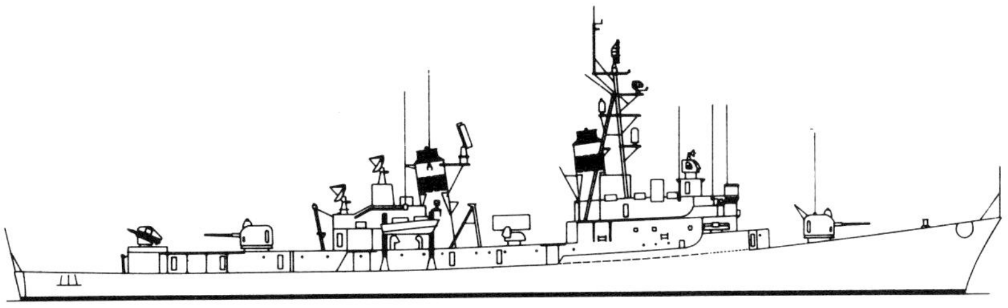

Charles F Adams 1962

CHARLES F ADAMS class *fleet escorts (destroyers)*

Displacement: 3277t light; 4526t full load
Dimensions: 420ft wl, 437ft oa × 47ft × 15ft
128m, 133.2m × 14.3m × 4.6m
Machinery: 2-shaft geared turbines, 4 boilers, 70,000shp = 33kts. Range 4500nm at 20kts
Armament: 2–5in/54 RF, 1 Tartar SAM (1×2, 42 missiles) (1×1, 40 missiles in DDG 15–DDG 24), 1 ASROC ASW system (1×8), 6–12.75in Mk 32 ASW TT (2×3)
Sensors: Radar SPS-29 (SPS-40 in later units), SPS-39, 2 SPG-51; sonar SQQ-23A or SQS-23A (bow mounted in DDG 20–DDH 24)
Complement: 333–350

No	Name	Builder	Laid down	Launched	Comp	Fate
DDG 2	CHARLES F ADAMS	Bath Iron Wks	16.6.58	8.9.59	10.9.60	Decomm 1.8.90
DDG 3	JOHN KING	Bath Iron Wks	25.5.58	30.1.60	4.2.61	Decomm 30.3.90
DDG 4	LAWRENCE	New York SB	27.10.58	27.2.60	6.1.62	Decomm 30.3.90
DDG 5	CLAUDE V RICKETTS (ex-*Biddle*)	New York SB	18.5.59	4.6.60	5.5.62	Decomm 31.10.89
DDG 6	BARNEY	New York SB	18.8.59	10.12.60	11.8.62	Decomm 17.12.90
DDG 7	HENRY B WILSON	Defoe, Bay City	28.2.58	23.4.59	17.12.60	Decomm 2.10.89
DDG 8	LYNDE McCORMICK	Defoe, Bay City	4.4.58	9.9.60	3.6.61	Decomm 1.10.91
DDG 9	TOWERS	Todd-Pacific, Seattle	1.4.58	23.4.59	6.6.61	Decomm 1.10.90
DDG 10	SAMPSON	Bath Iron Wks	2.3.59	9.9.60	24.6.61	Decomm 24.6.91
DDG 11	SELLERS	Bath Iron Wks	3.8.59	9.9.60	28.10.61	Decomm 31.10.89
DDG 12	ROBINSON	Defoe, Bay City	23.4.59	27.4.60	9.12.61	Decomm 1.10.91
DDG 13	HOEL	Defoe, Bay City	1.6.59	4.8.60	16.6.62	Decomm 1.10.90
DDG 14	BUCHANAN	Todd-Pacific, Seattle	23.4.59	11.5.60	7.2.62	Decomm 1.10.91
DDG 15	BERKELEY	New York SB	1.6.60	29.7.61	15.12.62	Decomm 1.10.92, to Greece
DDG 16	JOSEPH STRAUSS	New York SB	27.12.60	9.12.61	20.4.63	Decomm 1.2.90, stricken 1.10.91, to Greece 1992
DDG 17	CONYNGHAM	New York SB	1.5.61	19.5.62	13.7.63	Decomm 26.10.90
DDG 18	SEMMES	Avondale	18.8.60	20.5.61	10.12.62	Decomm 12.4.91, stricken 1991, to Greece 1992
DDG 19	TATTNALL	Avondale	14.11.60	28.8.61	13.4.63	Decomm 18.1.91
DDG 20	GOLDSBOROUGH	Puget Sound N Yd	3.1.61	15.12.61	9.11.63	Decomm Oct 1992, to Australia for spare parts
DDG 21	COCHRANE	Puget Sound N Yd	31.7.61	18.7.61	21.3.64	Decomm 1.10.90
DDG 22	BENJAMIN STODDERT	Puget Sound N Yd	11.6.62	8.1.63	12.9.84	Decomm 20.12.91
DDG 23	RICHARD E BYRD	Todd-Pacific, Seattle	12.4.61	6.2.62	7.3.64	Decomm 1.5.92
DDG 24	WADDELL	Todd-Pacific, Seattle	6.2.62	26.2.63	28.8.64	Decomm 1.10.92, to Greece

These Tartar-armed missile destroyers were originally designed as modified versions of the *Forrest Sherman*: a design requirement of the Tartar missile system was that it could replace a single 5in mounting on a one-for-one basis. The twin 3in/50 nearest the missile launcher was to be replaced by a replenishment-at-sea area for missile replacement, the other by a twin ASW weapon launcher (for the RAT system then in development), and there was to have been a quintuple nest of torpedo tubes between the stacks. However, RAT was abandoned in favour of ASROC, whose eight-round 'pepperbox' replaced the originally contemplated long torpedo tubes; these ships also received the usual launching system for lightweight ASW torpedoes. Quite soon it became evident that the new destroyer would have to be lengthened considerably to accommodate the new SQS-23 sonar and an enlarged CIC (for missile control), and to balance off increased displacement without a loss of speed.

There were several modifications during the construction of the class. FY59 and later ships (DDG 15–DDG 24) had single- rather than twin-arm Tartar launchers for greater reliability; FY60 and FY61 units (DDG 20–DDG 24) had bow-mounted sonars, a change evident from the bow location of their anchors. Australia and West Germany both purchased near-sisters: the Australian DDG 25–DDG 27 had Ikara instead of ASROC amidships, and the West German DDG 28–DDG 30 had 'Macks'.

DDG production terminated with the FY61 programme largely because of the promise of the Typhon programme then in prospect; DDGs contemplated for the FY66–67 programmes were not built.

Some ships were refitted with 4-round ASROC reload magazines to starboard alongside the forward funnel.

These ships were completed without computer combat direction systems. However, by 1971 a new smaller central computer (UYK-7) made possible a compact version of NTDS. It was to have been installed in the entire class (and also in the DDG 31s), but funds were short, and only DDG 9, 12, 15, and 21 were fitted. All were completed by 1973. The system was called Junior Participating Tactical Data System (JPTDS); it was roughly equivalent to the French SENIT and to the German SATIR. The associated Weapons Direction System is Mk 13. Programmes for modernisation were complicated by the very tight design of these ships. Thus a programme which was to have begun with DDG 3 in FY80 was abandoned under Congressional pressure when projected cost per ship rose to $221m; it had already been cut to ten ships. Instead, DDG 15 to DDG 24 will be modernised as part of their regular refits, with work spread over two overhauls, for a total of two years per ship. The first was *Tattnall* (Philadelphia N Yd, begun August 1981); *Goldsborough* and *Benjamin Stoddert* have also been funded, as of early 1983. In the case of the *Tattnall*, the combat systems upgrade was carried out in the first overhaul period, with a hull and machinery (including sonar) upgrade scheduled for a second overhaul about two years later.

The major external changes evident are the installation of the Mk 86 gun/missile fire control system, which adds a third full missile control channel forward (the earlier SPG-53 had a secondary missile control capability); the installation of SLQ-32 and SLQ-20 countermeasures in place of the earlier ULQ-6B; and the installation of new radars (SPS-40C/D in place of the less reliable SPS-40B, SPS-52C in place of SPS-39, and LN-66 for navigation). More significantly, the ships are fitted with a three-computer NTDS and with an integrated combat system (SYS-1). Countermeasures capability is also increased by the installation of Super-Rapid Blooming Offboard Chaff (SRBOC). To service this increased electronic equipment, air conditioning compressor capacity increases from 140 to 225 tons, and ship service turbo-generator capacity is increased by 750kW. Modernised ships have their SQS-23 systems upgraded to two-dome SQQ-23 configuration.

All ships had their Mk 11 or Mk 13 missile launchers modified to accommodate Harpoon (typically four in a Mk 11, six in a Mk 13). In modernised ships, SPS-40 generally replaced SPS-37 (although DDG 5 still had her SPS-37 as late as October 1988), and SLQ-32/SRBOC was added. SPS-39A radars had planar array antennas fitted (it is not clear whether SPS-52 replaced SPS-39).

Waddel 29.7.86 L & L van Ginderen

Leftwich Jul 1986
L & L van Ginderen

Hayler 10.7.94
L & L van Ginderen

A contract for the development and production of thirty ships was awarded on 23 June 1970 to a new yard established by Ingalls Shipbuilding Division of Litton Industries, Pascagoula, Mississippi. The entire class was contracted to a single shipyard to facilitate design and mass-production, but construction was delayed by labour and technical problems. These large destroyers were surely among the more controversial of recent US warships: they are usually described as too large and too poorly armed. In fact they were the direct consequence of an attempt to replace the mass of Second World War destroyers, which were nearing the end of their lives in the mid-1960s.

Initially, it was hoped that most of the ASW destroyers would be replaced by *Knox* class ocean escorts, and that the Navy would build a new class of missile destroyers to make up for the gap in such construction since the early 1960s, a gap caused largely by the failure of the Typhon system planned for FY63 and later units. Analysis suggested, first, that it would be wise to seek maximum commonality between the ASW and missile units, and second, that although each carrier would require six escorts, only three would need anti-aircraft missile capability. In addition, Secretary McNamara strongly espoused a project system in which preliminary design would be done by a contractor.

One of the benefits of a common hull was that, if the air threat were to increase in the future, ASW ships could easily be refitted to meet it. Ultimately, relatively few ships could be ordered, none of them for air defence. However, the *Spruance* remained an anti-aircraft design with its missiles, radars and some of its computers never installed. To some extent, too, its large size was mandated by the requirement for 30kts in rough weather, a requirement of carrier operations.

Plans to install single lightweight 8in/55 guns for amphibious fire support were cancelled in 1978, but the potential for major anti-aircraft upgrade remains. Indeed, Iran ordered six (later reduced to four) anti-aircraft versions of the *Spruance* class in 1973–74. With the Iranian Revolution, they were offered for sale, and all four were purchased as the *Kidd* class in July 1979, receiving DDG hull numbers in the *Spruance* series. As in the case of *Virginia* class cruisers, they have magazine stowage for sixty-eight missiles, with two SPG-51D guidance radars plus an SPG-60. The greatly increased displacement of the *Kidds* is due in part to the addition of armour. A thirty-first *Spruance*, DD 997, was authorised in FY78 with the proviso that it have increased helicopter facilities; however, she was identical to the thirty earlier ships already ordered, except for the provision of Kevlar armour and an SPS-49 air search radar. The Reagan Administration FY83–FY87 programme included three more ships of this type in FY86–FY87, out of a planned total of six. None was built.

On trials *Spruance* made about 32kts; the 1300t increase in the *Kidds* cost about one knot.

Spruance as built

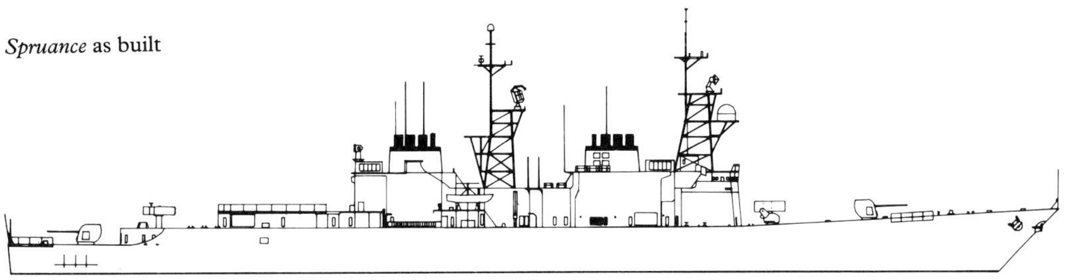

SPRUANCE and KIDD classes *fleet escorts (destroyers)*

Displacement: 5826t light; 7800t full load (*Kidd* class 9200t full load)
Dimensions: 529ft wl, 563ft 4in oa × 55ft × 20ft 6in
161.3m, 171.7m × 16.8m × 6.3m
Machinery: 2-shafts, 4 LM 2500 gas turbines, 80,000shp = 30kts. Range 6000nm at 20kts
Armament: 2–5in/54 (2×1), 1 Sea Sparrow SAM (1×8, 24 missiles), 1 ASROC ASW system (1×8, 24 missiles), 6–12.72in Mk 32 (2×3, 14 torpedoes), 1 helicopter
Kidd class: as above except 2 Mk 26 launchers (2×2, typically 50 Standard SM-2 SAM, 16 ASROC and 2 test missiles) and 2–20mm Phalanx CIWS instead of Sea Sparrow and ASROC launchers
Sensors: Radar SPS-40 (*Kidd* class SPS-48, 2 SPG-51), SPQ-9, SPG-60; sonar SQS-33
Complement: 296

No	Name	Builder	Laid down	Launched	Comp	Fate
Spruance class						
DD 963	SPRUANCE	Litton	17.11.72	10.11.73	20.9.75	Extant 1995
DD 964	PAUL F FOSTER	Litton	6.2.73	23.2.74	21.2.76	Extant 1995
DD 965	KINKAID	Litton	19.4.73	25.5.74	10.7.76	Extant 1995
DD 966	HEWITT	Litton	23.7.73	24.8.74	25.9.76	Extant 1995
DD 967	ELLIOTT	Litton	15.10.73	19.12.74	22.1.76	Extant 1995
DD 968	ARTHUR W RADFORD	Litton	14.1.74	1.3.75	16.4.77	Extant 1995
DD 969	PETERSON	Litton	29.4.74	21.6.75	9.7.77	Extant 1995
DD 970	CARON	Litton	1.7.74	24.6.75	1.10.77	Extant 1995
DD 971	DAVID R RAY	Litton	23.9.74	23.8.75	19.11.77	Extant 1995
DD 972	OLDENDORF	Litton	27.12.74	21.10.75	4.3.78	Extant 1995
DD 973	JOHN YOUNG	Litton	17.2.75	7.2.76	20.5.78	Extant 1995
DD 974	COMTE DE GRASSE	Litton	4.4.75	26.3.76	5.8.78	Extant 1995
DD 975	O'BRIEN	Litton	9.5.75	8.7.76	3.12.77	Extant 1995
DD 976	MERRILL	Litton	16.6.75	1.9.76	11.3.78	Extant 1995
DD 977	BRISCOE	Litton	21.7.75	15.12.76	3.6.78	Extant 1995
DD 978	STUMP	Litton	25.8.75	29.1.77	19.8.78	Extant 1995
DD 979	CONOLLY	Litton	29.9.75	19.2.77	14.10.78	Extant 1995
DD 980	MOOSBRUGGER	Litton	3.11.75	23.7.77	16.12.78	Extant 1995
DD 981	JOHN HANCOCK	Litton	16.1.76	29.10.77	10.3.79	Extant 1995
DD 982	NICHOLSON	Litton	20.2.76	11.11.77	12.5.79	Extant 1995
DD 983	JOHN RODGERS	Litton	12.8.76	25.2.78	14.7.79	Extant 1995
DD 984	LEFTWICH	Litton	12.11.76	8.4.78	25.8.79	Extant 1995
DD 985	CUSHING	Litton	2.2.77	17.6.78	22.9.79	Extant 1995
DD 986	HARRY W HILL	Litton	1.4.77	10.8.78	10.11.79	Extant 1995
DD 987	O'BANNON	Litton	24.6.77	25.9.78	1.12.79	Extant 1995
DD 988	THORN	Litton	29.8.77	14.11.78	12.1.80	Extant 1995
DD 989	DEYO	Litton	14.10.77	27.1.79	22.3.80	Extant 1995
DD 990	INGERSOLL	Litton	5.12.77	10.3.79	12.4.80	Extant 1995
DD 991	FIFE	Litton	6.3.78	1.5.79	31.5.80	Extant 1995
DD 992	FLETCHER	Litton	24.4.78	16.6.79	12.7.80	Extant 1995
Kidd class DDG 993	KIDD (ex-*Kouroush*)	Litton	26.6.78	11.8.79	27.6.81	Extant 1995
DDG 994	CALLAGHAN (ex-*Dargush*)	Litton	23.10.78	1.12.79	29.8.81	Extant 1995
DDG 995	SCOTT (ex-*Nader*)	Litton	12.2.79	1.3.80	24.10.81	Extant 1995
DDG 996	CHANDLER (ex-*Andushirvan*)	Litton	7.5.79	24.5.80	13.3.82	Extant 1995
DD 997	HAYLER	Litton	20.10.80	27.3.82	5.3.83	Extant 1995

Beginning with four ships (FY81), all have been given Kevlar armour around vital spaces (all had been modified by 1986).

Nine ships (including DD 976 in FY83, then DD 974, 979, 983 in FY 84, DD 984, 985, and 989 in FY85) were fitted with paired armoured box launchers for Tomahawk (firing trials by DD 976 January 1981, first operational installation on board DD 974 late 1984). Under the FY86 and later programmes, ships are being fitted with Mk 41 vertical launchers in the spaces originally provided for Mk 26 launchers forward: the sixty-one cells will ultimately accommodate forty-five Tomahawks and sixteen Vertical Launch ASROCs. The first two ships were DD 963 and DD 990. By 1994, vertical launchers had been installed on board DD 963, 964, 966–973, 975, 977, 978, 981, 985, 986, 987, 990– 992 and 997. All but the ships with armoured box launchers will eventually be so modified. The vertical cells can also accommodate Sea Sparrow and SM-2, and in theory the Mk 86 FCS can be modified to control them (as indeed it does on board *Kidd* class destroyers and *Virginia* class cruisers). This modification has not yet been effected on the *Spruance*s.

All will eventually be equipped with TAS Mk 23 and with the SQQ-89 ASW suit, including the SQR-19 towed array and improved versions of the SQS-53 bow sonar.

Ships of this class have tested a variety of new weapons and related devices: DD 971 carries the prototype RAM launcher. DD 976 tested the 30mm EX-83 Gatling gun in 1981. DD 977 had her Mk 86 GFCS modified to test laser-guided shells; she also tested the Sparrow vertical launcher (Mk 48).

The *Kidd* class have all been given NTU modernisation, with SPS-48E and modified SPS-49 radars. Completion dates: DD 995 March 1988, DDG 993 September 1989, DDG 994 July 1990, DDG 996 August 1990. Harpoon was added amidships after commissioning, and also two Phalanx.

TYPHON FRIGATES (DLGN) type *fleet escorts*

Displacement: 10,000t light; 12,000t full load
Dimensions: 650ft wl × 64ft × 21ft
198.2m × 19.5m × 6.4m
Machinery: 2 shafts nuclear: 2 D2G reactors, approx 30kts
Armament: 1 Long Range Typhon SAM launcher (60 missiles, including up to 20 ASROC), 1 Medium Range Typhon SAM launcher (80 missiles), 1–5in/38, 2–21in TT (2×1), 6–12.72in Mk 32 ASW TT (2×3),
Sensors: Radar SPS-49, SPG-59; sonar SQS-26/VDS
Complement: 743

Note data refers to SCB-240.65, the last Typhon DLGN design.

Artist's impression of a Typhon DGLN design, 1961 (USN)

Typhon was the follow-on missile system to the initial '3-T' systems. It was to have replaced both the medium-range Terrier and the long range Talos, with a combination of a Tartar-sized Medium Range weapon and a ramjet-powered Terrier-size Long Range missile. There were several proposals for ships to carry it, but the farthest advanced was the nuclear frigate described here: it is also the closest to the later Aegis ships. The nuclear frigate could be distinguished from alternative cruisers by its lack of flagship facilities; it was also much smaller because it followed frigate rather than cruiser design practice, and thus seemed a much better bargain from the point of view of getting Typhon to sea in significant quantity. Even so, it was extremely expensive, and late in 1963 Secretary of Defense Robert S McNamara cancelled the entire Typhon programme as far too ambitious and expensive. In particular, he was concerned that the existing '3-T' missiles, with much less ambitious planned performance, could not meet their goals, at least in terms of reliability. Money intended for Typhon was re-directed to a 'Get-Well' programme for the existing ships, while a single SPG-59 radar was completed for the test ship *Norton Sound*.

Design problems peculiar to Typhon included the very high power drain of its SPG-59 tracking/search radar, which limited steaming endurance in non-nuclear Typhon ships, and the realisation, in the course of design, that command and control and sensors would far outweigh the missiles proper both in dollar cost and in ship impact. One reason for the provision of a conventional two-dimensional air search radar (SPS-49) was to permit operation without continuously running the big missile control set.

Alternative platforms considered during 1961–63 included Typhon/fleet flagship conversion of the heavy cruiser *Newport News* in which only one turret (No 1) would have been retained, and in which the SPS-43 air search radar would have been wrapped around the SPG-59 tower.

The Typhon DLGN was close to the Aegis cruiser (CSGN/CGN 42) in concept and configuration. The latter, however, showed only a single missile type, corresponding to the Medium Range Typhon: advances in missile design and in guidance considerably expanded that Medium Range. In addition, a high rate of fire was much more important in the Aegis cruiser, which was faced with a variety of short-range ('pop-up') threats. Thus a high rate of fire was much more important in the Aegis design, and the Mark 10 launcher, necessary for a long-range (booster-equipped) missile could not meet it, as the missiles had to be finned manually between the magazine and the launcher.

STRIKE CRUISER (CSGN) type *fleet escorts*

Displacement: 15,902t light; 17,172t full load
Dimensions: 666ft wl, 709ft 7in oa × 76ft 7in × 22ft 3in
203.0m, 216.3m × 23.4m × 6.8m
Machinery: 2-shaft nuclear: 2 D2G reactors, approx 30kts
Armament: 2 Mk 26 Mod 2 missile launchers (2×2; 128 missiles, mix of SM-2 Medium Range and ASROC), 8 Harpoon SSM (2×4), 8 Tomahawk SSM (2×4), 1–8in/55, 6–12.75in Mk 32 ASW TT (2×3), 2 LAMPS III helicopters or VTOL aircraft
Sensors: Radar SPS-49, SPY-1; sonar SQS-53, Towed Array Sonar System
Complement: ?

The strike cruiser was an outgrowth of the US nuclear missile frigates/cruisers, rather than a follow-on to the *Long Beach*. It began in 1973 as a DLGN modified to accommodate the new Aegis missile system – which had been designed in the first place to fit the DLGN 38 class. However, Aegis was so capable a system that its opponents believed that an Aegis ship might be suitable for independent operations, and thus might revive the cruiser role dropped after the Second World War. At first this meant merely that the ship should have a substantial anti-ship capability, in the form of Harpoon and Tomahawk missile cannisters. The initial CSGN design, held to minimum length to reduce displacement, could not make the 30kt speed considered essential for a carrier escort, and during 1975 its hull was lengthened and its D2G frigate power plant upgraded; the final CSGN design was considered fast enough. In this phase, a gun battery (incorporating the 8in lightweight gun) as well as passive protection (armour) unequalled in any postwar US surface combatant were added. It was a measure of the high cost of such protection that it had to be limited to armoured boxes surrounding the above-water magazines and computer spaces. In addition, the hull was large enough to accommodate some vital spaces below the waterline, and it was long enough to permit considerable separation between the two reactors of the power plant. The final Strike Cruiser design showed provision for VSTOL aircraft as well as missiles and guns, and at one point Reuven Leopold, the senior civilian ship designer, proposed a modified version incorporating a flight deck running the length of the ship.

As in the case of the Typhon frigate and cruiser, the sheer size and cost of the Aegis cruiser defeated it. Although two strike cruisers were included in the Ford Administration's five-year programme, that Administration had to abandon the strike cruiser before it left office, substituting a 'cruiser' based on the existing CGN 39 (*California*) design. Even this was never built, although CGN 42 figure in the Reagan Administration Five-Year Shipbuilding Plan. The only other attempt to realise the strike cruiser concept, a proposed reconstruction of *Long Beach*, was also cancelled by the Ford Administration before it left office; in each case the non-nuclear *Ticonderoga* was perceived as a much more economical means of getting Aegis to sea fast enough in sufficient numbers.

The CSGN would have had major fleet flagship facilities, and was valued as such at a time when existing flagships were being stricken; in her absence, the Numbered Fleets are commanded from a variety of auxiliaries, including destroyer tenders and amphibious flagships. The flagship problem was so severe that in the mid- and late-1960s there was some considerable interest in building a new class of cruiser-size flagships with sufficient speed to keep up with major combatants, and with limited self-defensive capability, such as Tartar. This programme was dropped at the time in view of its high cost.

Strike cruiser design, 1975 (USN)

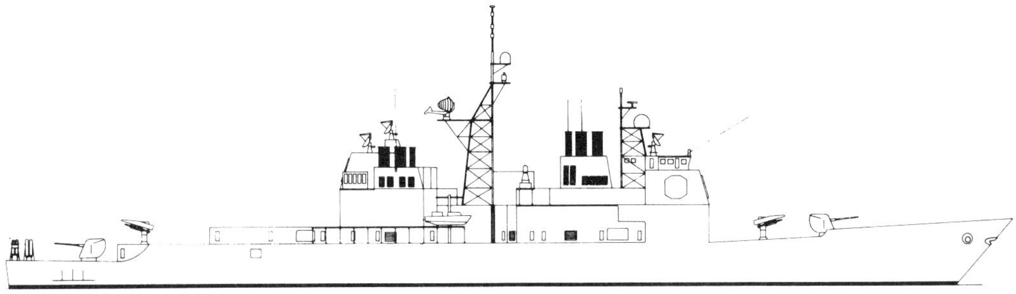

Ticonderoga as built

This series, based on the *Spruance* hull form, is the result of a very lengthy evolution. The Aegis air defence system was originally designed for installation aboard a nuclear frigate (later redesignated cruiser) such as the *California*. However, when Admiral E M Zumwalt Jr became CNO he sought instead a minimum Aegis platform for construction in considerable numbers (DG/Aegis); for a time, it was to have been limited to $100m in FY73 dollars and to 5000t, compared to $45m and 3400t for what became the *Perry* class frigate. On this basis it would have been limited to a single Mk 22 launcher with sixteen missiles, and DG had to be relaxed to 6000t and $125m. There was to have been a single Mk 26 Mod 2 (64-missile) launcher, with cost rising to $200m by early 1973. Advocates of a conventionally-powered Aegis platform rejected this design as making too little use of the power inherent in the missile control system; advocates of nuclear power rejected it as a means of killing off the nuclear ship programme. There was also some scepticism as to the accuracy of the cost estimates for DG/Aegis. Then the Secretary of the Navy asked for studies of a new nuclear version, which eventually became the abortive Strike Cruiser (CSGN) of about 18,000t. Given the high cost of such a ship, the Secretary of Defense demanded a mix of nuclear and conventional Aegis ships, and the conventional study was based on the largest existing destroyer-type hull, the *Spruance*.

The lead ship, *Ticonderoga*, was authorised under the FY78 programme, at the same time that the nuclear strike cruiser died a political death; it was so large and so important that it was redesignated a cruiser in January 1980. Compared to the strike cruiser, it lacks an 8in gun and has only a limited capacity for surface-to-surface missiles, two quadruple canisters. From CG 52 they have an enlarged missile capacity (122 as against 88) with vertical launchers, which carry Tomahawk as well as the SM-2 anti-aircraft missile of the Aegis system.

Authorisations: CG 47 in FY78, CG 48 in FY80, CG 49–50 in FY81, CG 51–53 in FY82, CG 54–56 in FY 83, CG 57–59 in FY84, CG 60–62 in FY85, CG 63–65 in FY86, CG 66–68 in FY87, CG 69–73 in FY88.

TICONDEROGA class *fleet escorts (cruisers)*

Displacement: 6560t light; 8910t full load
Dimensions: 529ft wl, 563ft oa × 55ft × 31ft
161.3m, 171.6m × 16.8m × 9.5m
Machinery: 2 shafts, 4 LM 2500 gas turbines, 80,000shp = 30kts. Range 6000nm at 20kts
Armament: 2 Mk 26 launchers (2×2, 68 Standard SM-1 SAM and 20 ASROC missiles), 8 Harpoon SSM (2×4), 6–12.75in Mk 32 ASW TT (2×3), 2–5in/54 (2×1), 2 LAMPS III helicopters
Sensors: Radar SPY-1A, SPS-49, SPG-62; sonar SQS-53
Complement: 343

No	Name	Builder	Laid down	Launched	Comp	Fate
CG 47	TICONDEROGA	Litton	21.1.80	25.4.81	22.1.83	Extant 1995
CG 48	YORKTOWN	Litton	19.10.81	17.1.83	4.7.84	Extant 1995
CG 49	VINCENNES	Litton	20.10.82	14.1.84	6.7.85	Extant 1995
CG 50	VALLEY FORGE	Litton	14.4.83	23.6.84	11.1.86	Extant 1995
CG 51	THOMAS S GATES	Bath Iron Wks	31.8.84	14.12.85	22.8.87	Extant 1995
CG 52	BUNKER HILL	Litton	11.1.84	11.3.85	20.9.86	Extant 1995
CG 53	MOBILE BAY	Litton	6.6.84	22.8.85	21.2.87	Extant 1995
CG 54	ANTIETAM	Litton	15.11.84	14.2.86	6.6.87	Extant 1995
CG 55	LEYTE GULF	Litton	18.3.85	20.6.86	26.9.87	Extant 1995
CG 56	SAN JACINTO	Litton	22.7.85	14.11.86	23.1.88	Extant 1995
CG 57	LAKE CHAMPLAIN	Litton	3.3.86	3.4.87	12.8.88	Extant 1995
CG 58	PHILIPPINE SEA	Bath Iron Wks	8.5.86	25.4.87	18.3.89	Extant 1995
CG 59	PRINCETON	Litton	15.10.86	25.9.87	11.2.89	Extant 1995
CG 60	NORMANDY	Bath Iron Wks	7.4.87	19.3.88	9.12.89	Extant 1995
CG 61	MONTEREY	Bath Iron Wks	19.8.87	22.10.88	16.6.90	Extant 1995
CG 62	CHANCELLORSVILLE	Litton	24.6.87	15.7.88	4.11.89	Extant 1995
CG 63	COWPENS	Bath Iron Wks	23.12.87	11.3.88	9.3.91	Extant 1995
CG 64	GETTYSBURG	Bath Iron Wks	17.8.88	22.7.89	22.6.91	Extant 1995
CG 65	CHOSIN	Litton	22.7.89	1.9.89	12.1.91	Extant 1995
CG 66	SHILOH	Litton	2.2.89	21.10.90	14.9.91	Extant 1995
CG 67	HUE CITY	Bath Iron Wks	1.8.89	8.9.90	18.7.92	Extant 1995
CG 68	ANZIO	Litton	21.8.89	2.11.90	2.5.92	Extant 1995
CG 69	VICKSBURG	Litton	30.5.90	7.9.91	14.11.92	Extant 1995
CG 70	LAKE ERIE	Bath Iron Wks	14.3.90	16.2.91	24.7.92	Extant 1995
CG 71	CAPE ST GEORGE	Litton	19.11.90	10.1.92	12.6.93	Extant 1995
CG 72	VELLA GULF	Litton	15.4.91	13.6.92	18.9.93	Extant 1995
CG 73	PORT ROYAL	Litton	18.10.91	20.11.92	30.4.94	Extant 1995

Chosin 1993 RAN

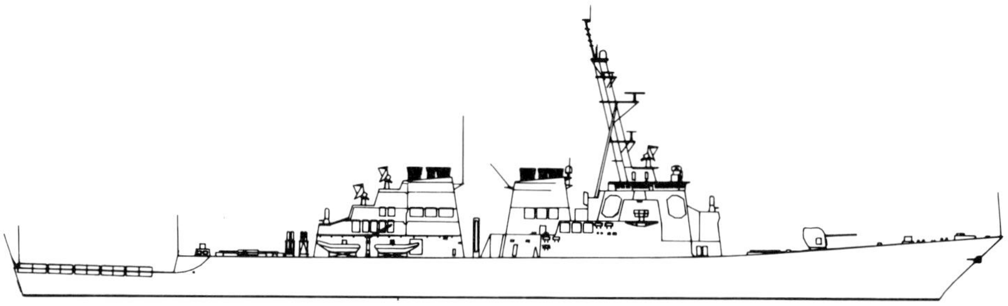

Arleigh Burke as completed

ARLEIGH BURKE class *fleet escorts (destroyers)*

Displacement: 6624t light; 8315t full load (DDG 51; DDG 52 and later are 6682t/8373t)
Dimensions: 504ft 7in × 66ft 11in × 20ft
153.8m × 20.4m × 6.1m
Machinery: 4 GE LM 2500 gas turbines, 100,000shp = over 30kts sustained
Armament: 90 vertical launcher cells, 8 Harpoon missiles, 1–5in/54, 2 Phalanx 20mm guns, 6–12.75in TT
Sensors: Radar SPY-1D, SQQ-89(V)4; sonar SLQ-32(V)2 (V3 from DDG 59 on), ECM
Complement: 341

No	Name	Builder	Laid down	Launched	Comp	Fate
DDG 51	ARLEIGH BURKE	Bath Iron Wks	6.12.88	16.9.89	4.7.91	Extant 1995
DDG 52	BARRY	Litton	26.2.90	10.5.91	12.12.92	Extant 1995
DDG 53	JOHN PAUL JONES	Bath Iron Wks	8.8.90	26.10.91	18.12.93	Extant 1995
DDG 54	CURTIS WILBUR	Bath Iron Wks	12.3.91	16.5.92	19.3.94	Extant 1995
DDG 55	STOUT	Litton	13.9.91	16.10.92	13.8.94	Extant 1995
DDG 56	JOHN MCCAIN	Bath Iron Wks	3.9.91	26.9.92	2.7.94	Extant 1995
DDG 57	MITSCHER	Litton	12.2.92	7.5.93	Dec 1994	Extant 1995
DDG 58	LABOON	Bath Iron Wks	23.3.92	20.2.93	Nov 1994	Extant 1995
DDG 59	RUSSELL	Litton	27.7.92	23.10.93	Apr 1995	Extant 1995
DDG 60	PAUL HAMILTON	Bath Iron Wks	24.8.92	24.7.93	Apr 1995	Extant 1995
DDG 61	RAMAGE	Litton	4.1.93	11.2.94	Jul 1995	Extant 1995
DDG 62	FITZGERALD	Bath Iron Wks	9.2.93	29.1.94	Sep 1995	Completing
DDG 63	STETHAN	Litton	10.5.93	17.6.94	Nov 1995	Completing
DDG 64	CARNEY	Bath Iron Wks	3.8.93	23.7.94	Mar 1996	Completing
DDG 65	BENFOLD	Litton	27.9.93	Nov 1994	Mar 1996	Completing
DDG 66	GONZALEZ	Bath Iron Wks	3.2.94	17.12.94	Jul 1996	Completing
DDG 67	COLE	Litton	28.2.94	Apr 1995	Aug 1996	Completing
DDG 68	THE SULLIVANS	Bath Iron Wks	31.7.94	13.5.95	Dec 1996	Completing
DDG 69	MILIUS	Litton	Aug 1994	Sep 1995	Jan 1997	Building
DDG 70	HOPPER	Bath Iron Wks	Feb 1995	Jan 1996	Jun 1997	Building
DDG 71	ROSS	Litton	Apr 1995	May 1996	Aug 1997	Building
Flight II						
DDG 72	MAHAN	Bath Iron Wks	Jun 1995	–	Nov 1997	Building
DDG 73	DECATUR	Bath Iron Wks	Oct 1995	–	Mar 1998	?
DDG 74	McFAUL	Ingalls	Dec 1995	–	Jun 1998	?
DDG 75	DONALD COOK	Bath Iron Wks	Mar 1996	–	Aug 1998	?
DDG 76	HIGGINS	Bath Iron Wks	Aug 1996	–	Dec 1998	?
DDG 77	–	Bath Iron Wks	–	–	–	?
DDG 78	–	Ingalls	–	–	–	?
Flight IIA	–	Bath Iron Wks	–	–	–	?
DDG 79						

This class was planned to complement the *Ticonderoga* class, and to replace the ageing missile destroyers. Compared to a *Ticonderoga*, they have three illuminators and the entire SPY-1 radar is concentrated in a single deckhouse. Their CIC lacks the unit commander's station of the cruiser. They are also better protected than the cruisers, with steel rather than aluminium upper works, and they carry about 130t of Kevlar armour. Their sides are sloped to reduce their radar cross-section. In 1984 it was hoped that each (after the fifth production ship) would cost about two-thirds as much as a *Ticonderoga*, but would offer three-quarters her anti-aircraft potential.

They lack a helicopter hangar because when they were designed it seemed unlikely that there would ever be enough LAMPS III helicopters to provide one per escort, and also because the ships they replaced had no helicopter facilities at all. They can accept a helicopter on board, and they have LAMPS data link and data processing facilities.

By the late 1980s, the navy planned to build these ships at a steady rate of four or five per year, introducing changes only after several identical ships had been built. At that time a Flight 3 version, which would have had a hangar, was planned. The advantage of steady production would have been steady fleet modernisation, as in the submarine fleet.

When funds tightened dramatically, Flight 3 was abandoned in favour of studies of less expensive versions called DDV (DD variant), for example without the SPY-1 radar. At present Flight II (DDG 72-78) is a slightly modified version of the original Flight I. Flight IIA (DDG 79 and beyond) has been modified to provide a pair of LAMPS III hangars aft, with the helicopter deck beyond. These ships also have a stern wedge, for improved propulsive efficiency. In this version the underway replenishment crane is eliminated from the vertical launcher groups, so that the bow group has sixty-four rather than sixty-one cells, the after group thirty-two rather than twenty-nine, for a total of ninety-six rather than ninety missiles. This version was lengthened by 1.52m by the stern, but the original plan to add a substantial parallel midbody was dropped. Reported full load in 1994 was 8960t, voids being used for fuel; Flight IIA full load was reported as 9217t. Plans to eliminate Phalanx in later ships of the class (in favour of the Evolved Sea Sparrow Missile, quad-packed in standard vertical cells) were dropped because the missile was unlikely to be available in time.

DDG 58 and later ships have an improved fire control system and can fire the Standard IV extended missile. Their combat direction software is Block 5. DDG 68 and later units have the improved SPY-1D radar with a separate Track Initiation Processor (TIP) to filter tracks before they are entered in the ship's combat system. Block II ships have JTIDS (with a C2P processor), the new Combat Direction Finding (SRS-1) system, and add a jammer to their ESM systems (they have SLQ-32(V)3 rather than (V)2).

Authorisation: DDG 51 in FY85, 52–53 FY87 (only two ships bought because Congress bought out the remaining CG 47s), DDG 54 FY88, DDG 55–58 FY89, DDG 59–63 FY90, DDG 64–67 FY91, DDG 68–72 FY92, DDG 73–76 FY93, DDG 77–79 FY94; plans call for DDG 80–82 (two Ingalls, one Bath) FY95, DDG 82–84 FY96, DDG 84–86 FY97, DDG 86–88 FY98, DDG 89–91 FY99.

Arleigh Burke 19.7.91 USN

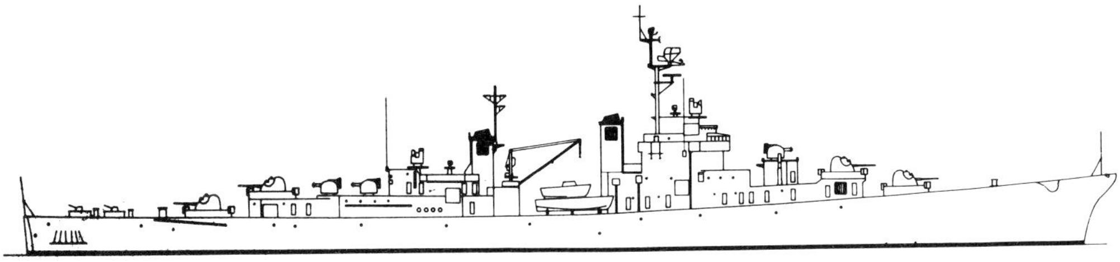

Norfolk 1954

NORFOLK *ocean escort (cruiser)*

Displacement: 4956t light; 5556t standard; 8315t full load
Dimensions: 520ft wl, 540ft oa × 53ft 6in × 19ft
158.5m, 164.6m × 16.3m × 5.8m
Machinery: 2-shaft geared turbines, 4 Babcock & Wilcox boilers, 80,000shp = 33kts. Range 6000nm at 20kts
Oil 1230t
Armament: 8–3in/50 (4×2), 8–20mm (4×2), 8–21in TT (8×1), 4 Weapon Alfa ASW systems
Sensors: Radar SPS-6; sonar QHB, SQG-1 (later SQS-4, then SQS-26)
Complement: 546

No	Name	Builder	Laid down	Launched	Comp	Fate
DL 1	NORFOLK	New York SB	1.9.49	29.12.51	4.3.53	Stricken 1973, BU

Laid down as an ASW cruiser (CLK), *Norfolk* was completed as the first of a new generation of large destroyers or 'destroyer leaders' (DL) or frigates (DL). She was designed specifically as a counter to the new generation of Soviet submarines, which were expected to match the performance of the German Type XXI (and which materialised as the Whiskey). However, given her great size and high cost she could not be duplicated, and a projected sister-ship, CLK 2, was never laid down. Hull design broadly followed that of the *Atlanta* class cruisers, and in details *Norfolk* reflected the experience of the Bikini nuclear tests. For example, special efforts were made to permit fallout to wash from her decks, and her bridgework was enclosed. ASW features included unusually large-diameter, slow-turning propellers for quietness, and her primary ASW weapon was the homing torpedo, fired from fixed tubes in her after superstructure. The visually more prominent Weapon Alfa (automatic rocket launcher) was a closer-range secondary battery. Even her anti-aircraft battery was experimental, four twin 3in/70 having been specified (and four twin 3in/50 fitted at first).

In practice *Norfolk* was unique, serving with the Operational Test and Evaluation Force. She was test ship for ASROC and also for the SPS-26 electronically-scanned radar (the prototype of SPS-39). In 1959 conversion to a Terrier missile ship was proposed, as part of the same programme which produced the *Mitscher* and *Forest Sherman* class missile destroyers (Tartar). *Norfolk* was dropped because of the high cost of conversion to what was perceived at the time as a relatively unreliable weapon, a consideration which curtailed the entire programme.

Norfolk 1965

FLETCHER class DDE conversions *ocean escorts (destroyers)*

Displacement: 2950t full load
Dimensions: 375ft 6in oa × 39ft 8in × 18ft (over sonar dome)
114.8m × 12.1m × 5.5m
Machinery: 2-shaft geared turbines, 60,000shp = 35kts
Armament: 2–5in/38 (2×1), 4–3in/50 (2×2), 8–20mm (4×2), 1 Weapon Alfa ASW System, 1 trainable Hedgehog
ASW mortar, 4–21in TT (4×1). See notes
Sensors: Radar SPS-6; sonar QHB (later fitted with SQS-4)

For names of ships converted see under Fleet Strength 1947. Original particulars, including building dates, can be found in the 1922–1946 volume.

In 1946 it appeared that the USSR would soon mass produce German-type fast submarines, both the 15kt Type XXI and the 25kt Walther U-boat, Type XXVI. To counter the latter, the destroyer escort would have to be upgraded to a type capable of 30kts in rough water, yet there was clearly no inclination to begin a new building programme. Instead, plans were prepared for conversions of existing *Fletcher* class destroyers, which were then being laid up. The programme ended at eighteen units, after an internal Navy study showed that the Soviets had not yet built up their submarine fleet (1950) and that it would be best to conserve funds for research against the day when large numbers of more advanced Soviet submarines appeared. In particular, the principal ahead-launched weapon of the late 1940s, Weapon Alfa, was not yet proved – and seemed not entirely satisfactory in service.

Those ships which were converted received Weapon Alfa and, ultimately, two fixed Hedgehogs, in place of No 2 gunhouse; Nos 3 and 4 5in guns were also removed, and four fixed torpedo tubes fitted. In some units a trainable Hedgehog replaced Weapon Alfa in No 2 position forward. All also received improved (scanning) sonar. Underwater fire control equipment and control spaces were also added. Again, in some units the standard quintuple trainable torpedo tube was retained, at first, as space and weight reservation against the installation of fixed tubes. Three units, DDE 446, DDE 447 and DDE 449, were modernised under the FRAM II programme, with provision for DASH and a variable-depth sonar aft.

Jenkins (after DDE conversion) USN

GEARING class DDE conversions *ocean escorts (destroyers)*

Displacement: 2182t full load
Dimensions: 391ft oa × 40ft 10in × 14ft (full load)
119.2m × 12.5m × 4.3m
Machinery: 2-shaft geared turbines, 60,000shp = 32kts
Armament: 4–3in/70 (2×2), 8–20mm (4×2), 2 Weapon Alfa ASW systems, 4–21in TT (4×1), 1 DC track
Sensors: Radar SPS-6; sonar QHB (later SQS-4)

Data refers to *Carpenter*, the most extreme of the DDE conversions. For names of other ships converted see under Fleet Strength 1947. Original particulars, including building dates, can be found in the 1922–1946 volume

Carpenter 1975 USN

At the end of the Second World War, a total of nine *Gearing* class fleet destroyers were delivered incomplete; four were completed as the prototypes of a new generation of specialised ASW ships. Two of them, *Carpenter* and *Robert A Owens*, were intended as experimental (and temporary) substitutes for the ASW cruiser (CLK) *Norfolk*, with a heavy battery of two Weapon Alfa and a trainable Hedgehog, as well as four fixed torpedo tubes – and a gun battery limited to two twin 3in/50s (later 3in/70s). Two others, *Basilone* and *Epperson*, were completed as prototype escort destroyers, retaining two of their three twin 5in/38 mounts, plus one Weapon Alfa and two Hedgehogs (originally trainable, abreast the forward funnel; later fixed, on the 01 level forward), and four fixed torpedo tubes.

Eleven other *Gearings* received limited ASW conversions as part of the formation of two postwar experimental Task Groups; in each case something comparable to *Basilone* was ultimately planned, but all that was done was the installation of a large trainable Hedgehog in place of No 2 5in mount, as well as the installation of a scanning sonar in place of the former searchlight type: DDE 818–DDE 820, DDE 847 and DDE 871 in 1947, and DDE 764, DDE 765 and DDE 858–DDE 861 under the FY50 programme. Two more ships, *Witek* and *Sarsfield*, had their No 2 gunhouses removed and were used for experiments; *Sarsfield* was a test ship for RAT, for example.

Although there were repeated proposals for DDE completion of the suspended destroyers, that was never done, and the DDE designation was dropped when all surviving *Gearing* class destroyers received what amounted to an ASW conversion under the FRAM programme. It is only fair to note that in 1950 the Navy itself adopted a policy of ASW improvement to all active fleet destroyers, to match the threat of a Soviet undersea fleet rather less developed than had originally been expected. The outward mark of this programme was a pair of Hedgehogs on the 01 level forward; less obvious but just as important was the provision of improved sonar, first a scanning type (QHB), and then SQS-4.

Charles H Roan (after FRAM I conversion)

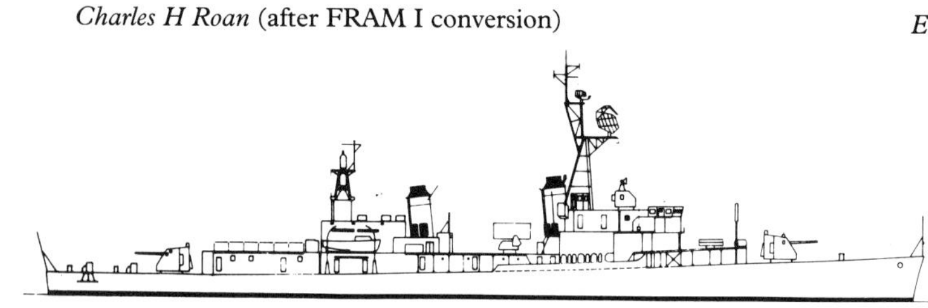

Ernest G Small (after FRAM II conversion, ex-DDR)

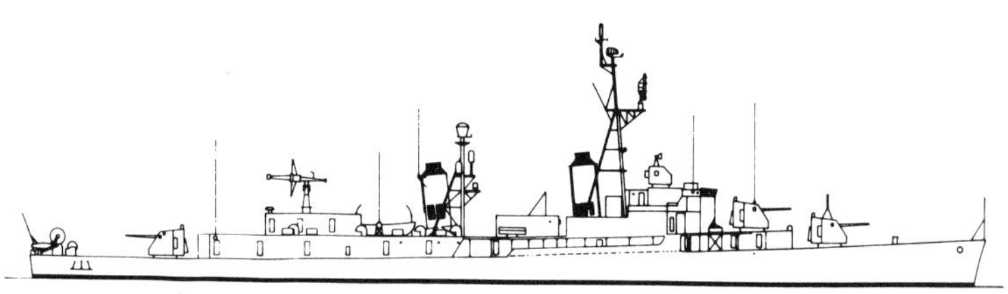

GEARING class FRAM I conversions *ocean escorts (destroyers)*

Displacement: 2406t light; 3493t full load (DD 717)
Dimensions: 383ft wl, 390ft 5in oa × 40ft 11in × 14ft 4in
116.6m, 119.0m × 12.5m × 4.4m
Machinery: 2-shaft geared turbines, 4 Babcock & Wilcox boilers, 60,000shp = 32kts. Range 4000n, at 20kts. Oil 727t
Armament: 4–5in/38 (2×2), 1 ASROC ASW system (1×8, 17 missiles), DASH, 6–12.75in Mk 32 ASW TT (2×3)
Sensors: Radar SPS-29, SPS-37 or SPS-40; sonar SQS-23
Complement: 310

For names of ships converted see under Fleet Strength 1947. Original particulars, including building dates, can be found in the 1922–1946 volume.

By 1958 the Second World War destroyers were beginning to wear out; FRAM was an attempt to extend their lives and so to put off the massive expenditure needed to replace them, in order to maintain the two hundred or more operational destroyers required by the US fleet of that time. In fact, the Vietnam War further delayed replacement, and even cut the one replacement programme undertaken (*Spruance* class), while wearing out the rebuilt ships, so that FRAM *Gearings* remained in (reserve training) service in 1982, more than a decade after their scheduled demise.

FRAM I, the most extensive reconstruction programme, was limited to *Gearings* and was expected to extend their useful lives by eight years. All shipboard components were rehabilitated, and one 5in mount and all lesser guns removed; ASROC, DASH triple ASW torpedo tubes were fitted, as well as the long-range SQS-23 hull sonar. The first eight conversions retained No 2 gunhouse and removed No 3, but it was argued that firepower should be better distributed, and most FRAM Is had 5in guns fore and aft. The only *Gearings* not converted to FRAM I standard were sixteen FRAM II (six radar-pickets, four former radar-pickets and six former DDEs), the experimental *Witek*, and the former missile conversion *Gyatt*.

As converted, these ships retained little modern anti-aircraft capability, particularly in view of increasing aircraft speeds. *Higbee* was bombed and her after 5in mount destroyed during a shore bombardment in the Gulf of Tonkin, 19 April 1972. Efforts were made to provide improved means of self-defence, and Sea Chapparal, which had been rejected as an alternative to Sea Sparrow, was tested aboard the *Floyd B Parks* in May 1972. Later that year, eight other FRAMs were fitted: *Anderson*, *Bausell*, *Gurke*, *Henderson*, *Mackenzie*, *Mason*, *Orleck* and *Tucker*. All missiles were removed as ships returned from South-East Asia the following year. Another special fit for Vietnamese operations was Shrike-on-Board (SOB), in which Shrike anti-radar missiles were fitted on launch rails atop the ASROC box launcher amidships, as a means of countering North Vietnamese coast defence radar. No details of ships fitted are available, and the system was installed on an emergency basis.

FRAM II was originally intended to extend the useful life of a destroyer by five years; it was applied primarily to *Sumner* class destroyers. All three 5in gun mountings were retained, but a new bridge was fitted, together with a new radar and ECM equipment. The existing SQS-4 sonar was retained but improved and relocated forward; it was supplemented by a variable-depth sonar operating from the same signal generator. Short (Mk 32) torpedo tubes were fitted but there was not enough space or weight for ASROC; instead these ships received the DASH drone and a pair of fixed long torpedo tubes (between their funnels) for Mk 37 torpedoes.

As applied to *Lloyd Thomas* (DDE 764, 765 and DDE 858–861) class escort destroyers, FRAM II involved the retention of the trainable Mk 15 Hedgehog forward, with DASH aft and a pair of long torpedo tubes amidships; they also received the new SQS-23 sonar. The six ships retained as radar-pickets (DDR) received limited ASW refits since they were primarily AAW ships: they retained their fixed Hedgehogs and received SQS-4 improvements and a variable-depth sonar, plus a new bridge and an enlarged CIC; new radars were also fitted. The former DDRs generally resembled *Sumner* class FRAM IIs, with long torpedo tubes and DASH, as well as all six 5in/38 guns.

The failure of DASH left these ships without a long-range ASW weapon. Ironically, it had been proposed at one time to omit ASROC from FRAM I ships in view of the potential of the drone helicopter, but the rocket had been retained for all-weather capability.

SUMNER class FRAM II conversions *ocean escorts (destroyers)*

Displacement: 2345t light; 3210t full load (DD 760)
Dimensions: 369ft wl, 376ft 6in oa x 40ft 10in x 14ft 1in
112.5m, 114.8m x 12.5m x 4.3m
Machinery: 2 shaft geared turbines, 4 Babcock & Wilcox boilers, 60,000shp = 33.4kts. Range 3300nm at 20kts. Oil 497t
Armament: 6–5in/38 (3x2), 2 Hedgehog ASW mortars, 2–21in TT (2x1), 6–12.75in Mk 32 ASW (2x3), DASH
Sensors: Radar SPS-29, SPS-37 or SPS-40; sonar SQS-4/RDT/VDS
Complement: 302

For names of ships converted see under Fleet Strength 1947. Original particulars, including building dates, can be found in the 1922–1946 volume.

Laffey Jun 1969 — L & L van Ginderen

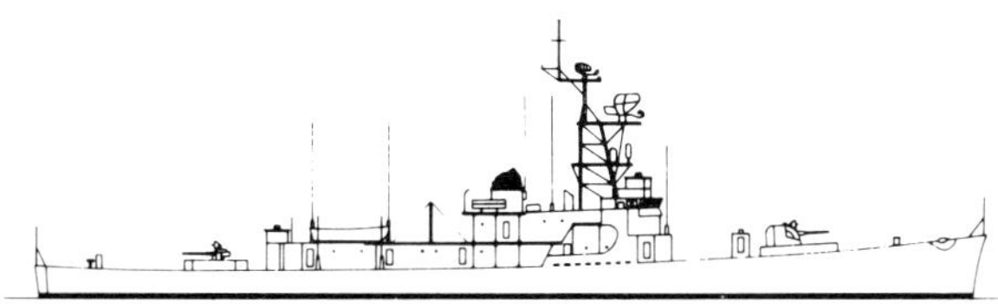

Cromwell 1970

The first postwar destroyer escorts, these ships were initially conceived as upgraded, versions of the wartime 22kt coastal escort (PC) with the additional firepower, sonar and speed required to counter the new generation of submarines. They were also designed with mobilisation in mind, hence the single screw for reduced pressure on the gear-cutting industry. *Dealey* herself was completed with the British Squid ASW mortar, but her sisters all had the US Weapon Alfa (Mk 108 rocket launcher).

They were fast and seaworthy, but were criticised as being underarmed. Under the FY62 and later programme, all but *Dealey*, *Courtney* and *Cromwell* were fitted with SQS-23 sonars, DASH hangars and helicopter pads. Their after twin 3in/50 mounts were removed. Unconverted ships retained all of their guns and had variable-depth sonars complementing their hull SQS-4 series units, as in FRAM II destroyers. In 1967 four Pacific Fleet Units, DE 1023–DE 1026, had single Army-type Bofors guns fitted on their fantails, reportedly on the basis of local initiative. All ended their careers with the usual triple Mark 32 lightweight torpedo tubes, and Weapon Alfa was removed from most towards the end of their careers. No ships of this type served in Vietnam.

Production of the same basic design by foreign navies attested to its basic value: five were built in Norway, and three in Portugal, in each case in somewhat modified form. However, reports circulating in the 1950s to the effect that postwar French frigates (which had US hull numbers because the United States paid for their construction) were modified *Dealeys* were not true.

DEALEY class *ocean escorts (destroyer escorts)*

Displacement: 1314t light; 1877t full load
Dimensions: 308ft wl, 315ft oa x 36ft 8in x 11ft 10in
93.9m, 96.0m x 11.2m x 3.6m
Machinery: 1 shaft geared turbine, 2 Foster Wheeler boilers, 20,000shp = 27kts. Range 6000nm at 12kts
Armament: 4–3in/50 (2x2), Weapon Alfa (2 Squid in DE 1006) ASW system
Sensors: Radar SPS-6; sonar SQS-4
Complement: 173

No	Name	Builder	Laid down	Launched	Comp	Fate
DE 1006	DEALEY	Bath Iron Wks	15.10.52	8.11.53	8.6.54	Uruguayan *18 De Julio* 1972
DE 1014	CROMWELL	Bath Iron Wks	3.8.53	4.6.54	24.11.54	Stricken 1.7.72
DE 1015	HAMMERBERG	Bath Iron Wks	12.11.53	20.8.54	28.2.55	Stricken 31.1.74
DE 1021	COURTNEY	Defoe, Bay City	2.9.54	2.11.55	31.8.56	Stricken 14.12.73
DE 1022	LESTER	Defoe, Bay City	2.9.54	5.1.56	14.6.57	Stricken 31.1.74
DE 1023	EVANS	Puget Sound N Yd	19.9.55	14.9.55	14.6.57	Stricken 3.12.73, BU 1975
DE 1024	BRIDGET	Puget Sound N Yd	19.9.55	24.4.56	24.10.57	Stricken 12.11.73, BU 1975
DE 1025	BAUER	Bethlehem, San Francisco	1.12.56	4.6.57	22.11.57	Stricken 3.12.73, BU 1975
DE 1026	HOOPER	Bethlehem, San Francisco	4.1.56	1.8.57	16.4.58	Stricken 6.7.73
DE 1027	JOHN WILLIS	New York SB	5.7.55	4.2.57	21.2.57	Stricken 14.7.72
DE 1028	VAN VOORHIS	New York SB	29.8.55	28.7.56	15.4.57	Stricken 1.7.72
DE 1029	HARTLEY	New York SB	31.10.55	26.1.57	30.7.57	Colombian *Boyaca* 1972
DE 1030	JOSEPH K TAUSSIG	New York SB	31.1.56	9.3.57	10.9.57	Stricken 1.7.72

Bridget Nov 1963
L & L van Ginderen

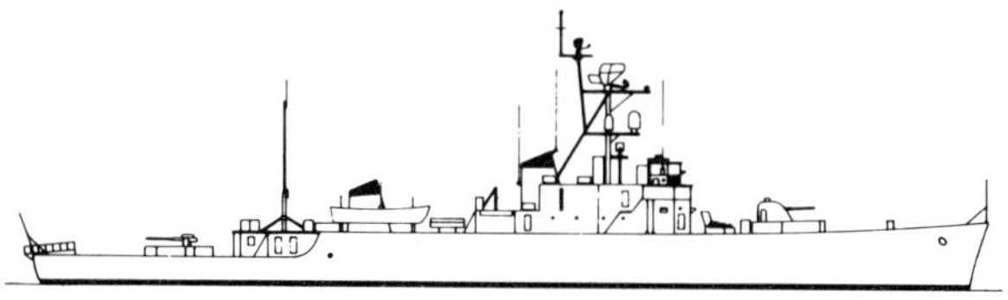

John R Perry 1961

CLAUD JONES class *ocean escorts (destroyer escorts)*

Displacement: 1314t light; 1916t full load
Dimensions: 301ft wl, 312ft oa × 38ft × 12ft 11in
91.8m, 95.1m × 11.6m × 3.9m
Machinery: 2 shafts, 4 diesels, 8700bhp = 21.5kts. Range 7000nm at 12kts
Armament: 2–3in/50 (2×1), 2 Hedgehog ASW mortars
Sensors: Radar SPS-6; sonar SQS-4
Complement: 171

No	Name	Builder	Laid down	Launched	Comp	Fate
DE 1033	CLAUD JONES	Avondale	1.6.57	27.5.58	16.11.58	Indonesian *Mongindisi* 1974
DE 1034	JOHN R PERRY	Avondale	1.10.57	29.7.58	12.1.59	Indonesian *Samadikun* 1973
DE 1035	CHARLES BERRY	Avondale	3.9.57	17.3.59	25.11.60	Indonesian *Martadinata* 1974
DE 1036	McMORRIS	Avondale	1.10.57	26.5.59	4.3.60	Indonesian *Ngurah Rai* 1974

Given the relatively high cost of a *Dealey*, these were an attempt to produce a minimum ocean escort, consonant with ASW requirements and with mobilisation considerations. Diesel power (with two shafts) was therefore chosen, with speed limited to 22kts; weapons were reduced to a pair of fixed Hedgehogs, a lightweight ASW torpedo-dropping system and a single stern depth-charge track. The result was widely disliked, and a call for renewed construction of a more satisfactory escort led back to the *Bronsteins* and their successors. In 1961, the last pair, DE 1035 and DE 1036, received a Norwegian Terne III rocket-thrown depth-charge system as well as light-weight torpedo tubes. However, they were never considered really suitable for ASW, and spent their last years as electronic reconnaissance ships in the Pacific. Two modified versions of this design were built in Turkey (1967–74).

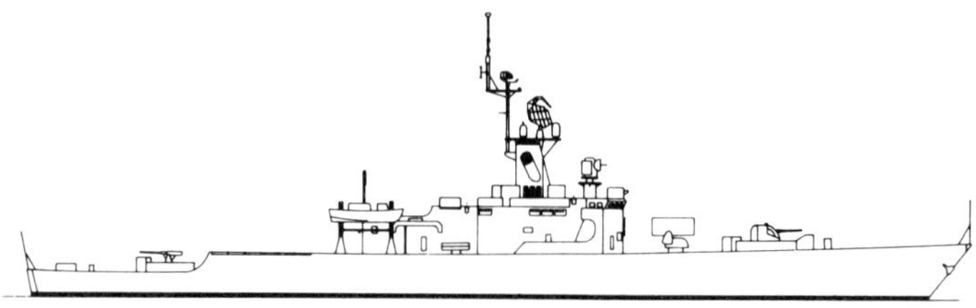

McCloy 1970

BRONSTEIN class *ocean escorts (destroyer escorts)*

Displacement: 1882t light; 2723t full load
Dimensions: 350ft wl, 372ft oa × 41ft × 23ft
106.7m, 113.4m × 12.5m × 7.0m
Machinery: 1 shaft De Laval geared turbine, 2 Foster Wheeler boilers, 20,000shp = 26kts. Range 4000nm at 15kts
Armament: 3–3in/50 (1×2, 1×1), 2–21in TT (2×1), 6–12.75in Mk 32 ASW TT (2×3), 1 ASROC ASW system (1×8), DASH
Sensors: Radar SPS-40; sonar SQS-26
Complement: 191

No	Name	Builder	Laid down	Launched	Comp	Fate
DE 1037	BRONSTEIN	Avondale	16.5.61	31.3.62	16.6.63	Stricken 13.12.90
DE 1038	McCLOY	Avondale	15.9.61	9.6.62	21.10.63	Stricken 17.12.90

These ships, the progenitors of a long line of large US ASW escorts, were conceived as a reaction to the *Claud Jones* class, in effect a *Dealey* with the new SQS-26 sonar, ASROC and DASH, with one of the original four 3in/50 guns traded off for greater ASW effectiveness. One of the penalties was a loss of speed, so that the new escort was slower than the ASW Task Force with which she was to operate; the desire to regain speed (compared to the *Dealey* or even as compared to a 30kt submarine) led to the next series of ships. The after 3in/50 gun was later removed, to provide compensation for a variable-depth sonar; both ships were fitted with a towed array surveillance sonar in the mid-1970s, and *Bronstein* was later fitted with an SQR-15 towed array sonar system. Neither quite fitted with the existing US ASW organisation, given their low speeds.

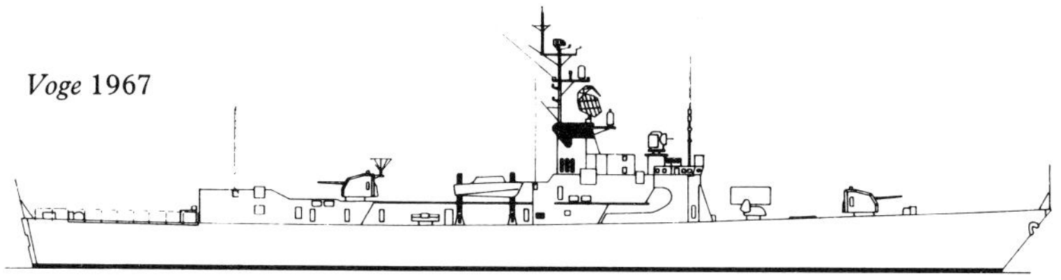

Voge 1967

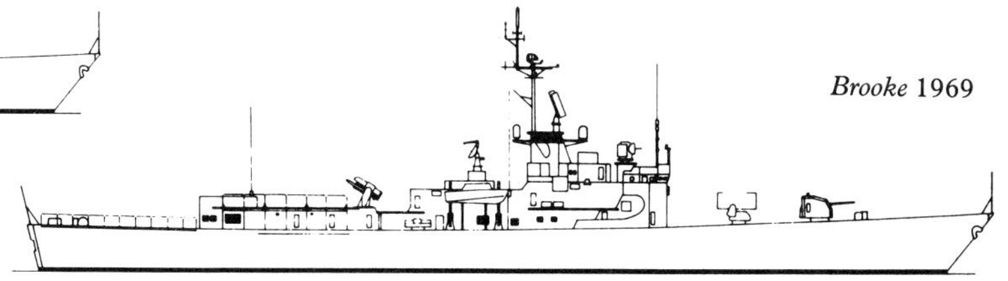

Brooke 1969

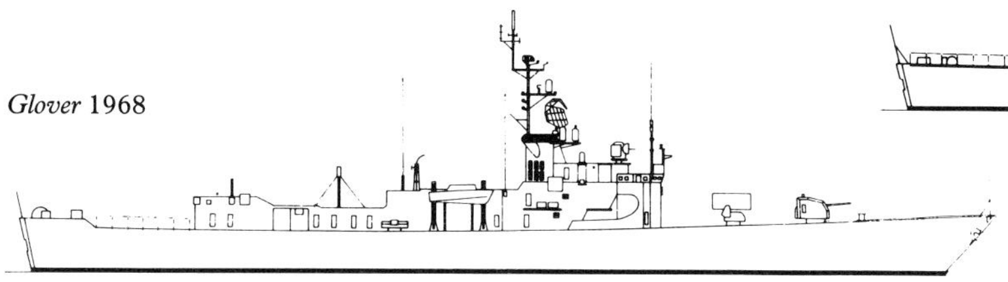

Glover 1968

GARCIA and BROOKE classes *ocean escorts (destroyer escorts)*

Displacement: 2441t light; 3371t full load (*Brooke* 2710t, 3426t)
Dimensions: 390ft wl, 414ft oa × 44ft × 24ft
118.9m, 126.2m × 13.4m × 7.3m
Machinery: 1 shaft Westinghouse geared turbines, 2 Foster Wheeler pressure-fired boilers, 35,000shp = 27kts. Range 4000nm at 20kts
Armament: 2–5in/38 (2×1) (*Brooke* class 1–5in/38, 1 Tartar SAM, 16 missiles), ASROC ASW system (1×8, 16 missiles, except DE 1040–1045, DEG 1–3), 2–21in TT (none in *Brooke* class), 6–12.75in Mk 32 ASW TT (2×3), DASH
Sensors: Radar SPS-40 (*Brooke* SPS-39, SPG-51); sonar SQS-26
Complement: 209 (*Brooke* 228)

No	Name	Builder	Laid down	Launched	Comp	Fate
DE 1040	GARCIA	Bethlehem, San Francisco	16.10.62	31.10.63	21.12.64	Decomm 10.11.88, to Pakistan (*Saif*)
DE 1041	BRADLEY	Bethlehem, San Francisco	17.1.63	26.3.64	15.5.65	Decomm 30.9.88, to Brazil (*Pernambuco*)
DE 1043	EDWARD McDONNELL	Avondale	1.4.63	15.2.64	15.2.65	Decomm 30.9.88, stricken 15.12.92 for BU
DE 1044	BRUMBY	Avondale	1.8.63	6.6.64	5.8.65	Decomm 31.3.89, to Pakistan (*Harbah*)
DE 1045	DAVIDSON	Avondale	20.9.63	2.10.64	7.12.65	Decomm 31.12.88, to Brazil (*Paraiba*)
DE 1047	VOGE	Defoe, Bay City	21.11.63	4.2.65	25.11.66	Decomm 23.9.89, stricken 15.12.92 for BU
DE 1048	SAMPLE	Lockheed, Seattle	19.7.63	28.4.64	23.3.68	Decomm 23.9.88, to Brazil (*Parana*)
DE 1049	KOELSCH	Defoe, Bay City	19.2.64	8.6.65	10.6.67	Decomm 31.5.89, to Pakistan (*Siqqat*)
DE 1050	ALBERT DAVID	Lockheed, Seattle	29.4.64	19.12.64	19.10.68	Decomm 28.9.88, to Brazil (*Para*)
DE 1051	O'CALLAHAN	Defoe, Bay City	19.2.64	20.10.65	3.7.67	Decomm 20.12.88, to Pakistan (*Aslat*)
DEG 1	BROOKE	Lockheed, Seattle	10.12.62	19.7.63	12.3.66	Decomm 16.9.88, to Pakistan (*Khaibar*)
DEG 2	RAMSEY	Lockheed, Seattle	4.2.63	15.10.63	3.6.67	Decomm 1.9.88, for disposal Jan 1993
DEG 3	SCHOFIELD	Lockheed, Seattle	15.4.63	7.12.63	11.5.68	Decomm 8.9.88, for disposal Jan 1993
DEG 4	TALBOT	Bath Iron Wks	4.5.64	6.1.66	2.4.67	Decomm 30.9.88, to Pakistan (*Hunam*)
DEG 5	RICHARD L PAGE	Bath Iron Wks	4.1.65	4.4.66	5.8.67	Decomm 30.9.88, to Pakistan (*Tabuk*)
DEG 6	JULIUS A FURER	Bath Iron Wks	12.7.65	22.7.66	11.11.67	Decomm 10.11.88, to Pakistan (*Badr*)
AGDE 1	GLOVER	Bath Iron Wks	29.7.63	17.4.65	13.11.65	To MSC 15.9.90, deactivated 27.8.92, stricken

By 1959 there was interest in a 30kt successor to the *Bronsteins*; at the same time a new lightweight steam technology, the pressure-fired boiler, was maturing. At first BuShips proposed a conventionally powered 27kt destroyer escort for the FY61 programme, together with an experimental (and faster) pressure-fired ship. This gradually shifted to a requirement that at least one FY61 ship be pressure-fired (27kts), and ultimately to a choice to build an entire series of pressure-fired escorts, even before any one of the new power plants had gone to sea. Other demands at this time were for a replacement for existing Second World War destroyers: given its relatively high speed and good sonar, the pressure-fired escort was a natural candidate; it was therefore given 5in guns in place of the 3in/50s of earlier escorts. There were also calls from the Fleet for a missile battery, and the new ship was designed to accept a minimum Tartar battery in place of its after 5in gun. The result was actually capable of about 30kts on trials, and certainly exceeded mass production limits.

Ten ASW versions (*Garcia* class) and six missile versions (*Brooke* class) were built, as well as a modified research unit, *Glover* (AGDE 1), with a pumpjet propeller, a raised platform above the main deck aft, and no after 5in gun or missile launcher; she was redesignated FF 1098, an operational frigate, in October 1979. The Mk 24 or 25 torpedo tubes originally mounted in the sterns of these ships were removed with the end of the long torpedo programme; DASH facilities were enlarged to accommodate LAMPS. Production of the missile ships ended with the FY63 programme in view of their high cost and limited capability.

Classification of the *Garcia*s was changed to frigate (FF) and the *Brooke*s to FFG on 30 June 1975.

From DE 1047 onwards, these ships were fitted with the sloping bulkhead and ASROC reload magazine developed for the *Knox* class. All but DE 1048 and DE 1050 were refitted to operate the LAMPS helicopter between 1972 and 1975, the two remaining ships being employed to test towed array sonar systems. *Koelsch* and *Voge* were fitted with automated ASW combat systems. *Bradley* served as Sea Sparrow test ship in 1967–68. *Talbot* served as sea-going test ship for the *Perry* class combat system, with an OTO Melara 76mm gun temporarily replacing her 5in/38.

Ships leased to Pakistan (1989) were returned when leases were not renewed due to the dispute over that country's nuclear programme; they were scrapped Singapore. *Ramsey* and *Schofield* were offered to Turkey but rejected. *Glover* was transferred to the Military Sealift Command in 1990 and disarmed, but continued sonar trials, many of them with bistatic systems.

Garcia 1972 USN

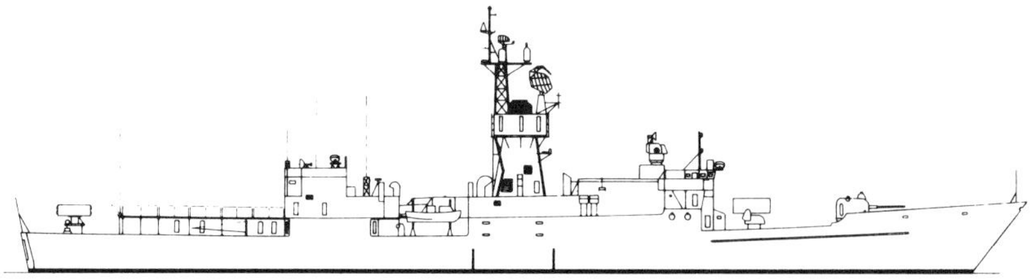

Bagley 1980

These large specialised ASW ships were the subject of considerable controversy, with their single screws and single 5in guns. They were direct descendants of the *Brooke* class, initially modified by 'work study' to reduce their complements and, without increase in hull size, to increase endurance by 500nm. In addition, it was hoped that electronic installations could be consolidated by the use of a series of 'bill-board' antennae spread around the crown of a tall 'Mack'. In practice, the Navy was moved to reject the new pressure-fired boilers, so that hull size had to grow to accommodate a more conventional installation, without loss of speed or endurance. In addition, with the rejection of further missile escort construction, the designers were left with a half-designed escort with only a single gun and with space aft which might be filled by a missile system – or by more DASH drones. In the end, the original single 5in/38 gun was replaced by a 5in/54 (in theory more than equivalent to the two 5in/38s of the *Garcia*) and space and weight was reserved aft for a short-range defensive missile, initially to have been Sea Mauler, but later Sea Sparrow. The generous hangar space originally provided aft made substitution of LAMPS for DASH relatively easy. SQS-35 VDS sonars were fitted to most units (DE, later FF, 1052, 1056, 1063–1071, 1073–1076 and 1078–1097), and their 'fish' later used to tow SQS-18 sonars. They were redesignated FF (frigates) in 1975.

From DE 1078 onwards, all were built by a single yard, Avondale, for economy. The last ten units, DE 1098–DE 1107, authorised in the FY68 programme, were not built. One had been planned as a gas turbine test ship (DE 1101). On the other hand, five missile-armed vari-

KNOX class *ocean escorts (destroyer escorts)*

Displacement: 3020t light; 4066t full load
Dimensions: 415ft wl, 438ft oa × 47ft × 25ft
126.5m, 133.5m × 14.3m × 7.6m
Machinery: 1 shaft Westinghouse geared turbines, 2 CE boilers, 35,000shp = 27kts. Range 4500nm at 20kts
Armament: 1–5in/54, 1 Sea Sparrow SAM (1×8; DE 1052–1069, DE 1071–1083, 1971–75), 1 ASROC ASW system (1×8, 16 missiles), DASH (as completed; later fitted for LAMPS I helicopter), 2–21in TT (2×1, not fitted in most units), 4–12.75in Mk 32 ASW TT (2×2 fixed, 22 torpedoes)
Sensors: Radar SPS-40; sonar SQS-26, SQS-35 (DE 1052, 1056, 1063–1071, 1073–1076, 1078–1097, installation
begun 1972)
Complement: 224

No	Name	Builder	Laid down	Launched	Comp	Fate
DE 1052	KNOX	Todd, ?	5.10.65	19.11.66	12.4.69	Decomm 14.2.92, to Taiwan FY96
DE 1053	ROARK	Todd, Seattle	2.2.66	24.4.67	22.11.69	Decomm 14.12.91
DE 1054	GRAY	Todd, Seattle	9.11.66	3.11.67	4.4.70	Decomm 29.6.91
DE 1055	HEPBURN	Todd, San Pedro	1.6.66	25.3.67	3.7.69	Decomm 20.12.91, to Brazil FY97
DE 1056	CONNOLE	Avondale	23.3.67	20.7.68	30.8.69	Decomm 25.7.92, to Greece
DE 1057	RATHBURNE	Lockheed, Seattle	8.1.68	2.5.69	16.5.70	Decomm 14.2..92
DE 1058	MEYERKORD	Todd, San Pedro	1.9.66	15.7.67	28.11.69	Decomm 14.12.91, to Taiwan FY96
DE 1059	W S SIMS	Avondale	10.4.67	4.1.69	3.1.70	Decomm 6.9.91
DE 1060	LANG	Todd, San Pedro	25.3.67	4.1.69	3.1.70	Decomm 12.12.91, to Taiwan FY96
DE 1061	PATTERSON	Avondale	12.10.67	3.5.69	14.3.70	Decomm 30.9.91
DE 1062	WHIPPLE	Todd, Seattle	24.4.67	12.4.68	22.8.70	Decomm 14.9.92, to Taiwan FY95
DE 1063	REASONER	Lockheed, Seattle	6.1.69	1.8.70	31.7.71	Decomm 30.8.93, to Turkey
DE 1064	LOCKWOOD	Todd, Seattle	3.11.67	5.9.68	5.12.70	To MarAd for disposal 30.9.93
DE 1065	STEIN	Lockheed, Seattle	1.6.70	10.12.70	8.1.72	Decomm 19.3.92, to Taiwan FY95
DE 1066	MARVIN SHIELDS	Todd, Seattle	12.4.68	23.10.69	10.4.71	Decomm 2.7.92
DE 1067	FRANCIS HAMMOND	Todd, San Pedro	15.7.67	11.5.68	25.7.70	Decomm 2.7.92, to Brazil FY94
DE 1068	VREELAND	Avondale	20.3.68	14.6.69	13.6.70	Decomm 30.8.92, to Greece
DE 1069	BAGLEY	Lockheed, Seattle	22.9.70	24.4.71	6.5.72	Decomm 26.9.91, to Taiwan FY95
DE 1070	DOWNES	Todd, Seattle	5.9.68	13.12.69	28.8.71	Decomm 5.6.92
DE 1071	BADGER	Todd, San Pedro	17.2.68	7.12.68	1.12.70	Decomm 20.12.91
DE 1072	BLAKELY	Avondale	3.6.68	23.8.69	18.7.70	Decomm 15.11.91
DE 1073	ROBERT E PEARY	Lockheed, Seattle	20.12.70	23.6.71	23.9.72	Decomm 27.3.92, to Taiwan
DE 1074	HAROLD E HOLT	Todd, Seattle	11.5.68	3.5.69	26.3.71	Decomm 2.7.92

ants of the basic design were built in Spain (with US designations DEG 7–DEG 11). Thirty-one ships (DE 1052–1069 and DE 1070–1083) were fitted with the Sea Sparrow point defence missile system in 1971–75; however, a planned installation of Sea Chapparal in the remaining fourteen, was not carried out. Nor did these ships receive the pair of long torpedo tubes originally specified.

The *Knox* class were always criticised as too wet; in 1980 they began to receive raised bulwarks and spray strakes forward for improved seakeeping.

The whole class, initially less eight training units for reservists, was decommissioned in the aftermath of the cold war. Virtually all these ships were offered to foreign navies. The first transfers (1992) were three each for Greece and Turkey. A bill introduced into Congress late in 1993 approved the following leases: Brazil (four), Egypt (two), Morocco (one), Oman (one), Spain (two), Taiwan (three), Thailand (four), Turkey (four), Venezuela (two), a total of twenty-three ships. However, Brazil, Morocco, Oman, Spain, and Venezuela all rejected the ships earmarked for them. Thailand had to abandon the acquisition of two of the four ships offered due to doubts about her navy's ability to maintain their steam plants. The table shows that Taiwan is to receive more ships (probably a total of twelve); Turkey received an additional ship to compensate for the accidental attack on the old destroyer-minelayer *Muavenet* and will probably receive more.

No	Name	Builder	Laid down	Launched	Comp	Fate
DE 1075	TRIPPE	Avondale	29.7.68	1.11.69	19.9.70	Decomm 30.7.92, to Greece
DE 1076	FANNING	Todd, San Pedro	7.12.68	24.1.70	23.7.71	Decomm 31.7.93, to Greece
DE 1077	OUELLET	Avondale	15.1.69	17.1.70	12.12.70	Decomm 6.8.93, to Thailand FY96
DE 1078	JOSEPH HEWES	Avondale	15.5.69	7.3.70	24.4.71	Decomm 30.6.94, to Taiwan
DE 1079	BOWEN	Avondale	11.7.69	2.5.70	22.5.71	Decomm 30.6.94, to Turkey (*Kavadeniz*)
DE 1080	PAUL	Avondale	12.9.69	20.6.70	14.8.71	Decomm 14.8.92
DE 1081	AYLWIN	Avondale	13.11.69	20.8.70	18.9.71	Decomm 15.5.92
DE 1082	ELMER MONTGOMERY	Avondale	23.1.70	21.11.70	30.10.71	Decomm 30.6.93, to Turkey 15.12.93
DE 1083	COOK	Avondale	20.3.70	23.1.71	18.12.71	Decomm 30.4.92, to Taiwan FY94
DE 1084	McCANDLESS	Avondale	4.6.70	20.3.71	18.3.72	Decomm 5.6.94, to Turkey (*Tvabya*)
DE 1085	DONALD B BEARY	Avondale	24.7.70	22.5.71	22.7.72	Decomm 27.5.94, to Turkey (*Ege*)
DE 1086	BREWTON	Avondale	2.10.70	24.7.71	8.7.72	Decomm 7.8.92, to Taiwan
DE 1087	KIRK	Avondale	4.12.70	25.9.71	9.9.72	Decomm 6.8.93, to Taiwan
DE 1088	BARBEY	Avondale	5.2.71	4.12.71	11.11.72	Decomm 19.3.92, to Taiwan FY94
DE 1089	JESSE L BROWN	Avondale	8.4.71	18.3.72	17.2.73	Decomm 27.6.94, to Egypt (*Damiyet*)
DE 1090	AINSWORTH	Avondale	11.6.71	15.4.72	31.3.73	Decomm 20.5.94, to Turkey (*Akdeniz*)
DE 1091	MILLER	Avondale	6.8.71	3.6.72	30.6.73	Decomm 15.10.91, to Brazil FY94
DE 1092	THOMAS C HART	Avondale	8.10.71	12.8.72	28.7.73	Decomm 30.8.93, to Turkey
DE 1093	CAPODANNO	Avondale	12.10.71	21.10.72	17.11.73	Decomm 30.7.93, to Turkey
DE 1094	PHARRIS	Avondale	11.2.72	16.12.72	26.1.74	Decomm 15.4.92
DE 1095	TRUETT	Avondale	27.4.72	3.2.73	1.6.74	Decomm 31.7.94, to Thailand
DE 1096	VALDEZ	Avondale	30.6.72	24.3.73	27.7.74	Decomm 15.12.91, to Brazil FY95
DE 1097	MOINESTER	Avondale	25.8.72	12.5.73	2.11.74	Decomm 28.6.94, to Egypt (*Rasheed*)

Thomas C Hart 1977
L & L van Ginderen

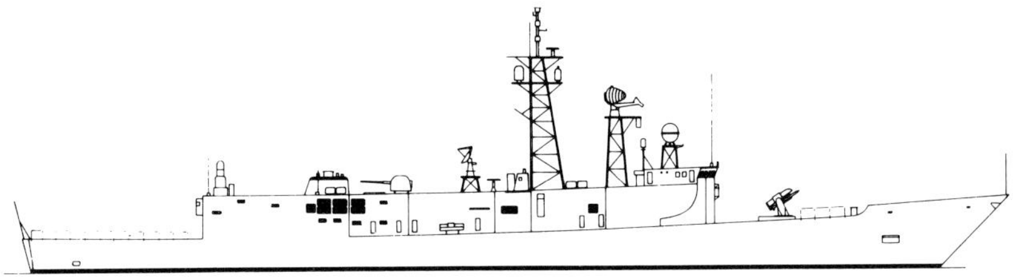

Oliver Hazard Perry as completed

OLIVER HAZARD PERRY class *ocean escorts (frigates)*

Displacement: 2648t light; 3486t full load
Dimensions: 408ft wl, 445ft oa × 47ft 5in × 14ft 5in
124.4m, 135.7m × 14.5m × 4.4m
Machinery: 1 shaft, 2 LM 2500 GE gas turbines, 40,000shp = 28.5kts. Range 4500nm at 20kts
Armament: 1–3in/62, 1 Standard SAM (40 SM-1 SAM or Harpoon SSM missiles), 6–12.75in Mk 32 ASW TT (2×3, 24 torpedoes), 2 LAMPS I or II helicopters
Sensors: Radar SPS-49, Mk 92 fire control system; sonar SQS-56
Complement: 176

No	Name	Builder	Laid down	Launched	Comp	Fate
FFG 7	OLIVER HAZARD PERRY	Bath Iron Wks	6.12.75	25.9.76	30.11.79	Extant 1995
FFG 8	McINERNEY	Bath Iron Wks	16.1.78	4.11.78	15.12.79	Extant 1995
FFG 9	WADSWORTH	Todd, San Pedro	13.7.77	29.7.78	28.2.80	Extant 1995
FFG 10	DUNCAN	Todd, Seattle	29.4.77	1.3.78	24.5.80	Extant 1995
FFG 11	CLARK	Bath Iron Wks	17.7.78	24.3.79	17.5.80	Extant 1995
FFG 12	GEORGE PHILIP	Todd, San Pedro	14.12.77	16.12.78	18.11.80	Extant 1995
FFG 13	SAMUEL ELIOT MORRISON	Bath Iron Wks	4.12.78	14.7.79	11.10.80	Extant 1995
FFG 14	SIDES	Todd, San Pedro	7.8.78	19.5.79	30.5.81	Extant 1995
FFG 15	ESTOCIN	Bath Iron Wks	2.4.79	3.11.79	10.1.81	Extant 1995
FFG 16	CLIFTON SPRAGUE	Bath Iron Wks	30.7.79	16.2.80	21.3.81	Extant 1995
FFG 19	JOHN A MOORE	Todd, San Pedro	19.12.78	20.10.79	14.11.81	Extant 1995
FFG 20	ANTRIM	Bath Iron Wks	21.6.78	27.3.79	26.9.81	Extant 1995
FFG 21	FLATLEY	Bath Iron Wks	13.11.79	15.5.80	20.6.81	Extant 1995
FFG 22	FAHRION	Todd, Seattle	1.12.78	24.8.79	Jan 82	Extant 1995
FFG 23	LEWIS B PULLER	Todd, San Pedro	23.5.79	29.3.80	17.4.82	Extant 1995
FFG 24	JACK WILLIAMS	Bath Iron Wks	25.2.80	30.8.80	19.9.81	Extant 1995
FFG 25	COPELAND	Todd, San Pedro	24.10.79	26.7.80	7.8.82	Extant 1995
FFG 26	GALLERY	Bath Iron Wks	17.5.80	20.12.80	5.12.81	Extant 1995
FFG 27	MAHLON S TISDALE	Tood, San Pedro	19.3.80	7.2.81	13.11.82	Extant 1995
FFG 28	BOONE	Todd, Seattle	27.3.79	16.1.80	15.5.82	Extant 1995
FFG 29	STEPHEN W GROVES	Bath Iron Wks	16.9.80	22.9.80	17.4.82	Extant 1995
FFG 30	REID	Todd, San Pedro	8.10.80	27.6.81	19.2.83	Extant 1995
FFG 31	STARK	Todd, Seattle	24.8.79	30.5.80	23.10.82	Extant 1995
FFG 32	JOHN L HALL	Bath Iron Wks	5.1.81	24.7.81	26.6.82	Extant 1995
FFG 33	JARRETT	Todd, San Pedro	11.2.81	17.10.81	2.7.83	Extant 1995
FFG 34	AUBREY FITCH	Bath Iron Wks	10.4.81	17.10.81	9.10.82	Extant 1995
FFG 36	UNDERWOOD	Bath Iron Wks	30.7.81	6.2.81	29.1.83	Extant 1995
FFG 37	CROMMELIN	Todd, Seattle	30.5.80	1.7.81	18.6.83	Extant 1995
FFG 38	CURTS	Todd, Seattle	1.7.81	6.3.82	8.10.83	Extant 1995
FFG 39	DOYLE	Bath Iron Wks	16.11.81	22.5.85	21.5.83	Extant 1995
FFG 40	HALYBURTON	Todd, Seattle	26.9.80	13.10.81	7.1.84	Extant 1995
FFG 41	McCLUSKY	Todd, San Pedro	21.10.81	18.9.82	10.12.83	Extant 1995
FFG 42	KLAKRING	Bath Iron Wks	19.2.82	18.9.82	20.8.83	Extant 1995
FFG 43	THACH	Todd, San Pedro	6.2.82	18.12.82	17.3.84	Extant 1995
FFG 45	DERWERT	Bath Iron Wks	14.6.82	18.12.82	19.11.83	Extant 1995
FFG 46	RENTZ	Todd, San Pedro	18.9.82	16.7.83	30.6.84	Extant 1995
FFG 47	NICHOLAS	Bath Iron Wks	27.9.82	23.4.83	10.3.84	Extant 1995
FFG 48	VANDERGRIFT	Todd, Seattle	15.10.81	15.10.82	7.9.84	Extant 1995
FFG 49	ROBERT G BRADLEY	Bath Iron Wks	28.12.82	13.8.83	11.8.84	Extant 1995
FFG 50	TAYLOR	Bath Iron Wks	5.5.83	5.11.83	1.12.84	Extant 1995
FFG 51	GARY	Todd, San Pedro	13.12.82	19.11.83	17.11.84	Extant 1995
FFG 52	CARR	Todd, Seattle	26.3.82	26.2.83	27.7.85	Extant 1995
FFG 53	HAWES	Bath Iron Wks	22.8.83	17.2.84	9.2.85	Extant 1995
FFG 54	FORD	Todd, San Pedro	16.7.83	23.6.84	29.6.85	Extant 1995
FFG 55	ELROD	Bath Iron Wks	14.11.83	12.5.84	6.7.85	Extant 1995
FFG 56	SIMPSON	Bath Iron Wks	27.2.84	31.8.84	9.11.85	Extant 1995
FFG 57	REUBEN JAMES	Todd, San Pedro	10.9.83	8.2.85	22.3.86	Extant 1995
FFG 58	SAMUEL B ROBERTS	Bath Iron Wks	21.5.84	8.12.84	12.4.86	Extant 1995
FFG 59	KAUFFMAN	Bath Iron Wks	8.4.85	29.3.86	21.2.87	Extant 1995
FFG 60	RODNEY M DAVIS	Todd, San Pedro	8.2.85	11.1.86	9.5.87	Extant 1995
FFG 61	INGRAHAM	Todd, San Pedro	30.3.87	26.6.88	5.8.89	Extant 1995

These frigates, direct successors to the *Knox* class, suffer from much the same criticisms: a single screw and a single weapon (one Mk 13 missile launcher). However, they benefit from a considerable, if unappreciated, degree of redundancy. For example, the fire control system has two channels (two guidance radars, physically well separated). There is a pair of retractable 325hp motors to bring the ship home at 6kts if she loses her main power plant. Although the hull sonar is a short-range (direct path) SQS-56, the primary ASW is SQR-19 towed array, its contacts prosecuted by the two LAMPS helicopters in a large hangar aft.

The *Perry* class suffers from several ironies. It was first conceived after the truncation of the *Spruance* class failed to replace the FRAM destroyers; Admiral Zumwalt sought alternative weapon suits for AAW, ASW and anti-ship warfare in the hope of reducing unit costs. In fact, the primary anti-ship weapon, Harpoon, could be fired from the primary AAW launcher, the Mk 13. If the ASROC missile (which cannot be fired from the Mk 13) was foregone, helicopter delivery would provide an AAW-oriented frigate with a considerable ASW stand-off capability. Indeed, even though the *Perry* does not share the elaborate silencing of the *Spruance* (for example, it has a diesel ship service generator) or the long-range sonar of previous ocean escorts, it gains a very considerable detection capability in the form of the towed array. Ironically, the ASW frigate was originally rejected on the theory that large numbers of such ships with effective sonars already existed. The advent of the towed array/helicopter combination shows in the decision to double the number of helicopters, which had a considerable effect on the overall design. In the original one-helicopter version, the helipad was placed aft of amidships, in the most favourable position, with the hangar splitting the uptakes. However, two helicopters required a much wider hangar, and both it and the helipad had to be moved aft.

The Mk 92 fire control system is of Dutch origin and is employed, in a somewhat simpler form, in the *Pegasus* class hydrofoils. As installed in the *Perry* class, it is part of a combat system particularly adapted to combat 'pop-up' weapons such as the Soviet SSN-7 underwater-launched missile. The OTO Melara 76mm gun was chosen as superior to a 5in/54 in air defence, and it is complemented by a Vulcan Phalanx close-in defence weapon.

Although all were intended to operate LAMPS III helicopters, FFG 7–35 were completed without the necessary equipment, including an enlarged flight deck (the ship is lengthened 3.16m to accommodate mooring equipment below the flight deck), RAST gear, and fin stabilisers. From the FY79 ships (FFG 36) on, ships were so fitted (RAST was not fitted on completion until FFG 50). RAST (Recovery Assistance, Securing, and Traversing) was tested on board FFG 8 in 1981 (rebuilt as SH-60B trials ship, completed 12 February 1981). FFG 26 conducted stabiliser trials in 1982. Many of the short-hull ships were transferred to the Naval Reserve Force; the others

Lewis B Puller 1989 L & L van Ginderen

are being rebuilt (FFG 7, 15, 28, 29, 32 through FY90).

Mk 92 is being improved in two stages. Near Term Improvement (tested on Board FFG 29, 1984) adds the SM-1 Block 6 missile, and was applied to all ships. Phase Two (Mk 92 CORT) was tested on board FFG 15 in May 1986. FFG 61 was completed with CORT, and six upgrades were funded in FY88 (FFG 50, 51, 53, 55, 56, 61), with three more following in FY90.

All long-hull ships were originally to have received the SQR-19 towed array and the SQQ-89(V)2 sonar suite (including a new LAMPS III system, SQQ-28), but due to delays some ships have SQR-18 and an SQR-17 stand-alone processor.

Ships deployed to the Persian Gulf were fitted with a pair of 25mm 'Bushmaster' guns on the main deck. Some were also equipped with the McDonnell Douglas mast-mounted sight (IRST) atop the pilothouse (display is in CIC).

Displacement is now much increased; FFG 59 was delivered at 4100t, although the design was limited to 3600t with a 39t margin. There was some fear, apparently misplaced, of structural weakness. Complement has also grown. Plans originally called for seventeen officers (too many) and 167 enlisted (too few); FFG 19 and later ships were completed with thirty more enlisted bunks (earlier ships were backfitted).

Protection consists of 19mm aluminium alloy over magazines, 16mm steel over main machinery, and 16mm Kevlar over vital electronics and command spaces.

These ships proved much tougher than expected. *Stark* survived two Exocet hits (17 May 1987; back in service 31 August 1988) and *Samuel B Roberts* survived a contact mine explosion (which very nearly broke her back) on 14 April 1988 (repairs completed October 1989).

Ships of this design were built for Australia, Spain, and Taiwan. In the latter case a much-modified version (PFG 2) was planned, but late in 1994 it was cancelled. It would have had forty-eight vertical launch cells, a phased-array radar (the General Electric ADAR-2N), and a new distributed combat direction system which could be operated, in the event the main CIC was destroyed, from the consoles in the damage control centre. A twin-screw version (each gas turbine is geared to its own shaft) was designed for a proposal to Turkey, and it was also offered to Saudi Arabia (which bought a pair of *La Fayette*s instead, at the end of November 1994).The Turkish offer was still alive as of late 1994, and this design may also have been offered to the UAE. The modified design may have incorporated a Unisys-designed distributed combat direction system.

In the aftermath of the cold war, these ships were initially considered virtually surplus due to their limited capabilities, and for a time in 1993–94 numerous transfers to foreign navies were planned. Reported requests: Oman (one), Turkey (two), UAE (two). Reportedly others were promised to Saudi Arabia. *Duncan* is to transfer to Bahrein in 1996, and two more are to go to Turkey in the same year. However, in the autumn of 1994 the CNO, Admiral Jeremy Boorda, decided to retain the ships for fear that a shrinking fleet could not provide sufficient presence in all the required areas, and further transfers were abandoned, at least for the present.

SUBMARINES

Sea Poacher (GUPPY IA)

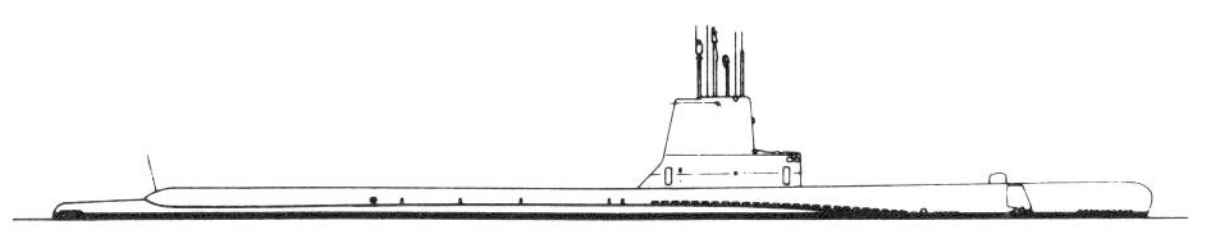

Razorback (GUPPY IIA)

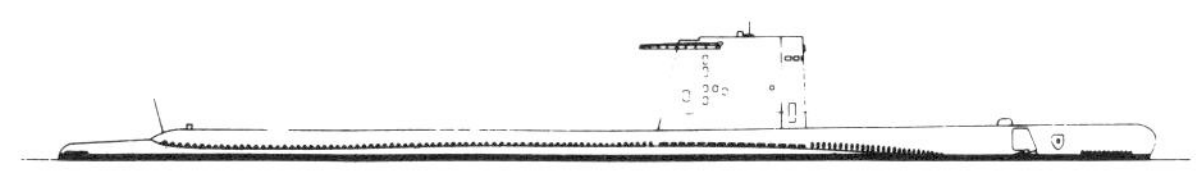

Clamagore (GUPPY III)

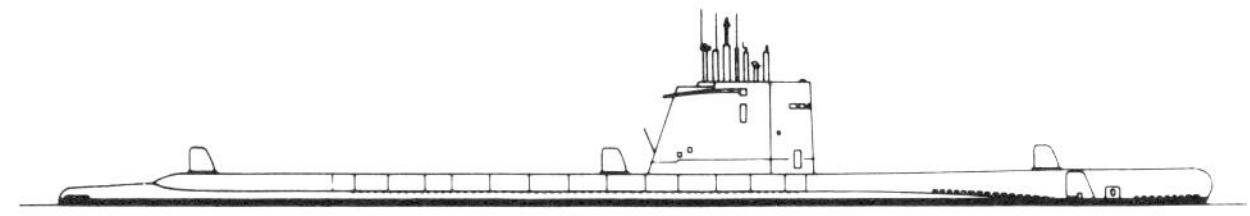

In 1945 the US Navy operated a very large force of conventional 'fleet submarines', which had been extremely successful in the war against Japan, but which had been rendered obsolete by the advent of the German Type XXI. Partly to modernise this force, and partly to provide fast submarines, comparable to Type XXIs, to train the postwar US ASW force, the Navy developed a conversion programme, Project 'GUPPY' (Greater Underwater Propulsive Power): submarines were provided with increased battery capacity (at the cost of four reload torpedoes, some fresh water tanks, and magazine space), and they were streamlined topside, their guns removed

GUPPY fleet submarine conversions *(attack submarines)*

Displacement: GUPPY IA: 1830t surfaced; 2440t submerged
GUPPY II: 1870t surfaced; 2440t submerged
GUPPY IIA: 1848t surfaced; 2440t submerged
GUPPY III: 1975t surfaced; 2450t submerged

Dimensions: 307ft (GUPPY III 326ft 6in) × 27ft × 17ft
93.6m (99.5m) × 8.2m × 5.2m

Machinery: 2-shaft diesels, 4610bhp (GUPPY IIA 3430bhp) = 18kts (surfaced) 9kts (snorkel, on one engine), 16kts (max submerged), 3.5kts (submerged cruise) in GUPPY II; 17.2kts, 6.2kts, 14.5kts, 6.2kts in GUPPY III

Armament: 10–21in TT (6 bow, 4 stern, 24 torpedoes)

Sensors: BQR-2, BQS-2, JT or SQR-3 (BQG-4 PUFFS in many GUPPY III)

Complement: 82 (186 in GUPPY III)

For names of boats converted see under Fleet Strength 1947. Original particulars, including building dates, can be found in the 1922–46 volume. Note SSK were similar to GUPPY IIA, but had BQR-4 passive array in bow. Fleet Snorkel conversions were rated at 10kts underwater (half-hour rate) or 6.5kts snorkelling. In a very few cases, Fleet Snorkel submarines retained their single 5in/25 deck guns as initially converted; they could be distinguished from true GUPPY conversions by their ship-type bows (overall length 312ft) and could make only 3kts sustained underwater.

and their superstructures remodelled. Although snorkels were not fitted to the prototypes *Odax* and *Pomodon* (GUPPY I), all subsequent conversions had them. Trials with the first two units were extremely encouraging, *Pomodon* making 18.2 kts submerged (half-hour battery rate). Streamlining made a very considerable difference, a snorkel GUPPY (Type II) requiring only about 44 per cent of the power of the fleet boat, submerged, at speeds up to 10kts at periscope depth.

Twelve Type II GUPPYs, equivalent to the original Type Is except for their snorkels, were approved in 1947; four of them were submarines which had been suspended incomplete in 1945. Their batteries were extremely advanced and rather expensive, with much shorter working lives than the Sargo II batteries adapted from those earlier US submarines, and in 1949 it was decided that further conversions would all incorporate the more conventional batteries, at a cost in underwater performance (*eg* 15 rather than 16kts at maximum power submerged). The FY51 programme thus included ten GUPPY IA conversions, as well as sixteen more austere 'Fleet Snorkel' conversions, in which the original Fleet Submarine superstructure was replaced by a GUPPY-type fin and a snorkel fitted, but the basic hull form was unchanged; some of these latter units even retained a 5in/25 deck gun. By this time the ASW training role had been eclipsed by a conventional attack submarine function. Two further GUPPY IA conversions were carried out for the Royal Netherlands Navy, *Hawkbill* and *Icefish*.

GUPPY conversions were designed to facilitate later conversions to an SSK. Visually, Fleet Snorkel submarines could be distinguished by their raked bow, where GUPPY bows were rounded; underwater they could not make much more than the 10kts of a fleet submarine, although they could snorkel at 6.5kts (compared to 7.5kts for a GUPPY IA, or 9kts for a GUPPY II). Finally, sixteen more fleet submarines were converted to GUPPY IIA under the FY52 pro-gramme, with improved sonar per-formance, one main engine being removed to allow the relocation of auxiliary machinery away from sonar transducers. Two of the GUPPY IIAs, *Thornback* and *Razorback*, were designed specifically as underwater targets, easily convertible back to standard GUPPY IIA configuration.

Under the FRAM programme, nine GUPPY II submarines were rebuilt to a GUPPY III configuration, lengthened 10ft to allow for a plotting room and a longer conning tower, with new fire control systems to permit them to fire Astor (Mk 45) nuclear ASW torpedoes. They also received a new plastic fin similar to that of the nuclear submarines. During the 1960s, too, US yards converted a number of the surviving fleet boats, as well as several which had already been transferred abroad, to the late 'Fleet Snorkel' configuration, with the new plastic fin. Indeed, Fleet Snorkels remained in service alongside GUPPYs as late as 1973. Many of these ships had the PUFFS (BQG-4) passive sonar system, with its characteristic trio of tall sonar domes.

Finally, the specialised SSK conversions should be mentioned: six fleet boats in FY52 after a prototype conversion in the FY50 programme. All were considered as mobilisation prototypes. In each case two of the four main engines were removed and a snorkel and streamlined fin was fitted; in the prototype, *Grouper*, the passive array was fitted around the forward side of the fin; in Type II (FY52) units, the BQR-4A array was wrapped around the bow, which was clearly blunter than in GUPPYs. There was extensive silencing and sound isolation to permit effective passive operation.

Sabalo (Fleet Snorkel) 1953 USN

Diodon (GUPPY II) 1965 USN

K 1 as completed **NB 1/750 scale**

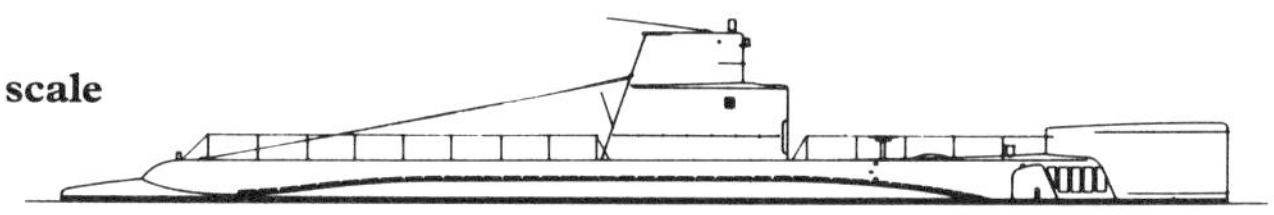

BARRACUDA (ex-'K') class *attack submarines*

Displacement: 765t standard; 1160t submerged
Dimensions: 196ft 1in oa × 24ft 7in × 14ft 5in
59.8m × 7.5m × 4.4m
Machinery: 2 shafts, 2 GM diesels plus GM electric motors, 1050shp = 13kts surfaced, 8.5kts submerged
Armament: 4–21in TT (2 bow, 2 stern)
Sensors: Sonar BQR-4 (in bow dome)
Complement: 37

No	Name	Builder	Laid down	Launched	Comp	Fate
SST 3	BARRACUDA (ex-*K 1*)	Electric Boat	1.7.49	2.3.51	10.11.51	BU 1974
SS 551	BASS (ex-*K 2*)	Mare Island N Yd	23.2.50	2.5.51	11.1.52	BU 1967
SS 552	BONITA (ex-*K 3*)	Mare Island N Yd	17.3.50	21.6.51	11.2.52	BU 1967

These three small diesel submarines were intended as prototypes of a mass produced ASW submarine, which would ambush enemy submarines as they emerged from their bases. The principal requirements were quietness, low cost and an effective passive sonar, the latter being carried in an enlarged bow sonar dome. They were later criticised as too small for habitability or for seakeeping, and *Tullibee* was an attempt to update the ASW submarine (SSK) concept. In view of limited resources available for new construction during the early postwar period, an SSK conversion of fleet submarines was developed as a mobilisation measure, and the GUPPY submarine conversions were all designed to permit later modification to SSK standard.

All three were renamed with 'B' names in December 1955, and in 1959 they were withdrawn from the SSK role, *Barracuda* becoming the training submarine *T 1* (SST 3), her sisters the attack submarine SS 550 and SS 551. They were both stricken in April 1965, *Bonita* having been used as a target (with only superficial damage) in the 1958 nuclear test series. *Barracuda* was stricken on 1 October 1973, having been reclassified formally as an attack submarine on 1 August 1972. All three had their large bow arrays removed in 1959.

Bass Feb 1952 L & L van Ginderen

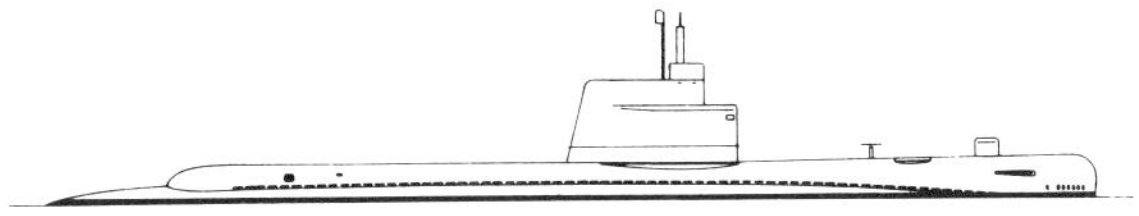

Trigger as completed

TANG class *attack submarines*

Displacement: 1560t standard; 2260t full load
Dimensions: 269ft 2in oa × 27ft 2in (27ft 3in SS 564, 566, 568) × 17ft
82.0m × 8.3m × 5.2m
Machinery: 2 shafts, 3 FM 'pancake' diesels plus 2 electric motors = 15.5kts/18.3kts (designed). See notes
Armament: 8–21in TT (6 bow, 2 stern)
Sensors: Sonar BQS-4 (later BQG-4 in SS 565 and 567)
Complement: 83

No	Name	Builder	Laid down	Launched	Comp	Fate
SS 563	TANG	Portsmouth N Yd	18.4.49	19.6.51	25.10.52	Turkish *Piri Reis* 1980
SS 564	TRIGGER	Electric Boat	24.2.49	3.12.51	31.3.52	Italian *Livio* Piomarta 1973
SS 565	WAHOO	Portsmouth N Yd	24.10.49	16.10.51	30.5.52	Stricken 30.1.80
SS 566	TROUT	Electric Boat	1.12.49	21.8.51	27.6.51	Stricken 19.12.78
SS 567	GUDGEON	Portsmouth N Yd	20.5.50	11.5.52	21.11.52	Turkish *Hizir Reis* 1980
SS 568	HARDER	Electric Boat	20.6.50	14.6.51	19.8.52	Italian *Romeo* Romei 1974

These ships, the US counterpart of the Soviet 'Whiskeys', were the US attempt to assimilate German Type XXI concepts. The first four also had a new compact radial ('pancake') diesel, which itself caused considerable problems in service, the four units (SS 563–SS 566) being lengthened and re-engined with the more conventional power plants of the last pair, 4500bhp diesels and 5600shp electric motors, length 278ft (84.8m); all were later lengthened to 287ft (87.5m) and speed in this configuration was reported as being 16kts surfaced and submerged. Originally conceived almost as an experimental high-speed type (in contrast to the usual fleet boat), they became the basis for postwar development. Compared to wartime fleet submarines, they had much increased battery power for greater submerged speed. Compared to the GUPPY conversions, they benefited from a shorter hull form which in itself reduced resistance underwater. They also had streamlined superstructures, snorkels and improved sonars, although by no means in the SSK category; the new hull form reduced the torpedo tube battery to eight, with only two aft.

There were two competitive designs, SS 563 by Portsmouth Naval Shipyard and SS 564 by Electric Boat, both to the same basic characteristics. The initial Ship Characteristics Board requirements included the potential for re-engining with a closed-cycle or nuclear power plant for an underwater speed of 25kts (17kts with a diesel-electric plant), and an increase in diving depth from the wartime 400ft to 700ft postwar.

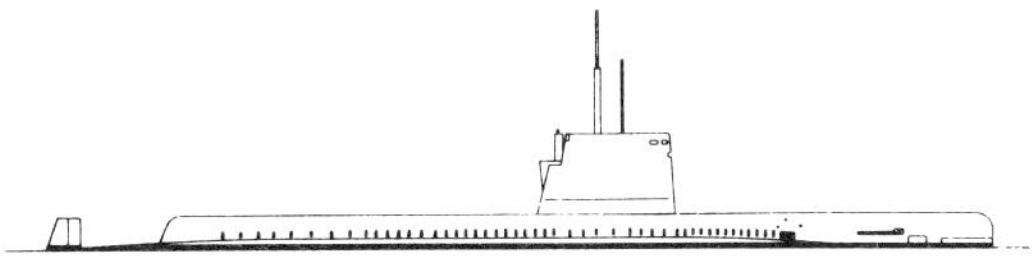

Darter 1967

DARTER *attack submarine*

Displacement: 1872t surfaced; 2372t submerged
Dimensions: 268ft wl, 268ft 7in oa × 27ft 2in × 16ft 9in
81.7m, 81.9m × 8.3m × 5.1m
Machinery: 2 shafts, 3 FM diesels plus 2 Elliott electric motors, 3100bhp/4500shp = 19.5kts/14kts
Armament: 8–21in TT (6 bow, 2 stern)
Sensors: Sonar BQS-4 (later BQG-4)
Complement: 83

No	Name	Builder	Laid down	Launched	Comp	Fate
SS 576	DARTER	Electric Boat	10.11.54	28.5.56	26.10.56	Decommissioned 1982

Darter was an improved version, with console controls; two planned sisters were completed instead as the strategic missile submarines *Grayback* and *Growler*. *Tang*, *Wahoo* and *Trout* were scheduled for transfer to Iran in 1982, 1980 and 1979 respectively, but the Iranian revolution intervened.

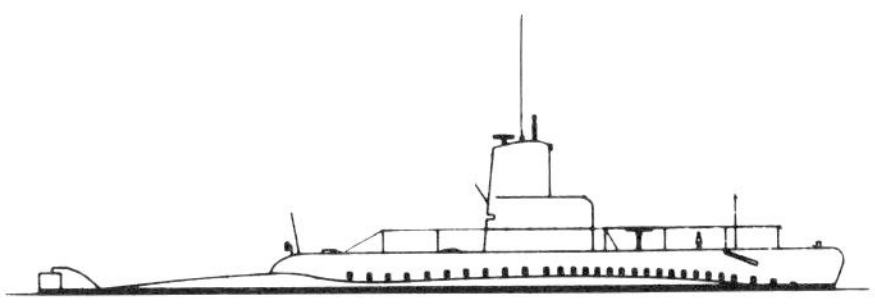

T 2 1955
NB 1/750 scale

MACKEREL (T1) class *target submarines*

Displacement: 303t surfaced; 347t submerged
Dimensions: 131ft 2in oa × 13ft 6in × 12ft 2in
40.0m × 4.1m × 3.7m
Machinery: 1 shaft, 2 GM diesels plus electric motors, 380bhp/1050shp = 8kts/9.5kts. Range 2000nm at 8kts
Armament: 1–21in TT (bow)
Complement: 18

No	Name	Builder	Laid down	Launched	Comp	Fate
SST 1	MACKEREL	Electric Boat	12.5.52	14.10.53	28.11.53	Stricken 31.1.73
SST 2	MARLIN	Portsmouth N Yd	1.4.52	17.7.53	9.10.53	Stricken 31.1.73

SST 1 originally ordered as AGSS 570. Originally numbered simply *T 1* and *T 2*, they were named in 1956.

X 1 *midget submarine*

Displacement: 31t surfaced; 36t submerged
Dimensions: 49ft 2in × 7ft × 7ft
15.0m × 2.1m × 2.1m
Machinery: 1 shaft, diesels plus electric motors, 30bhp = 15kts/12kts. Range over 500nm
Complement: 8 (4 minimum)

No	Name	Builder	Laid down	Launched	Comp	Fate
	X 1	Fairchild, Farmingdale	Jan 1954	7.9.55	6.10.55	Stricken 16.2.73

This project was inspired indirectly by the British X-Craft of the Second World War, one of which visited the USA specifically to provide design references. The original propulsion plant used hydrogen peroxide, this craft being the sole US fruit of the postwar Allied peroxide programme; (see the British *Explorer* class). She was built in 1954–55, but suffered an internal explosion in February 1958, which broke her into three pieces. Rebuilt at Philadelphia, she rejoined the Navy on 14 December 1960 as a special test craft at NSRDC, painted orange.

Seawolf (foreground) and *Nautilus* as completed USN

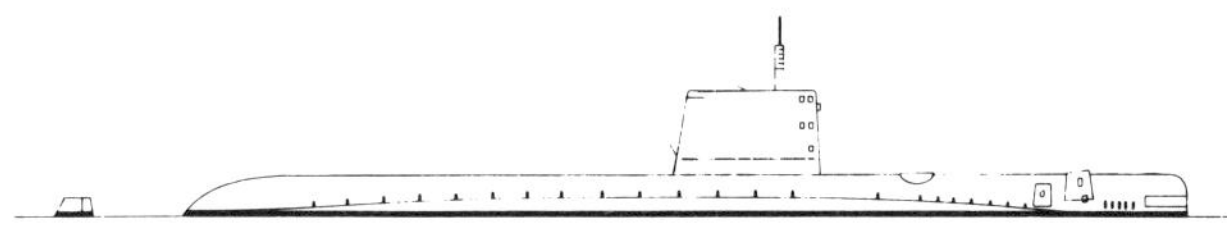

Nautilus 1958

Both built under the FY52 programme, these were the prototype US nuclear submarines, *Nautilus* being by far the more successful of the two. She was the world's first nuclear warship, with an S2W reactor and six torpedo tubes, all of them forward. *Seawolf* was designed around the S2G sodium-cooled reactor, intended as a back-up to the S2W, but abandoned because of operational problems and replaced by an S2Wa in 1959. From 1969 she has been used for research, with four bow thrusters above her pressure hull (two forward and two aft) for precise manoeuvring, and provision for a DSRV research vehicle to be mated to her aft. *Seawolf* also had the last conning tower in a US submarine, within her stepped fin, as later submarines have only a bridge access trunk in their fin.

In 1991 it was reported that *Seawolf* had actually been used for intelligence-gathering, presumably launching a submersible (the thrusters would have been valuable for precise positioning).

NAUTILUS attack submarine

Displacement: 3533t surfaced; 4092t submerged
Dimensions: 319ft 6in wl, 323ft 9in oa × 27ft 8in × 21ft 9in
97.4m, 98.7m × 8.4m × 6.6m
Machinery: 2-shaft nuclear: 1 S2W reactor, approx 15,000shp = approx 23kts submerged
Armament: 6–21in TT (bow)
Sensors: Sonar BQS-4
Complement: 105

No	Name	Builder	Laid down	Launched	Comp	Fate
SSN 571	NAUTILUS	Electric Boat	14.6.52	21.1.54	22.4.55	Museum 1982

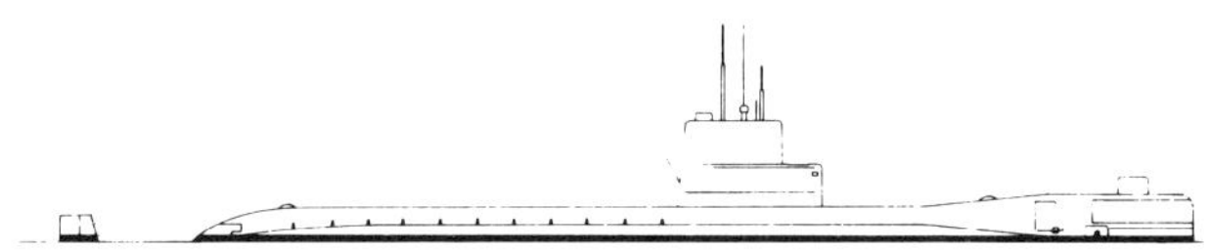

Seawolf 1954

SEAWOLF *attack submarine*

Displacement: 3741t surfaced; 4287t submerged
Dimensions: 337ft 6in × 27ft 8in × 22ft
102.9m × 8.4m × 6.7m
Machinery: 2-shaft nuclear: 1 S2G reactor (later S2Wa), approx 15,000shp = approx 20kts submerged
Armament: 6–21in (bow)
Sensors: Sonar BQR-4, BQS-4
Complement: 105

No	Name	Builder	Laid down	Launched	Comp	Fate
SSN 575	SEAWOLF	Electric Boat	15.9.53	21.7.55	30.3.57	Decomm 31.3.87

Skate as completed

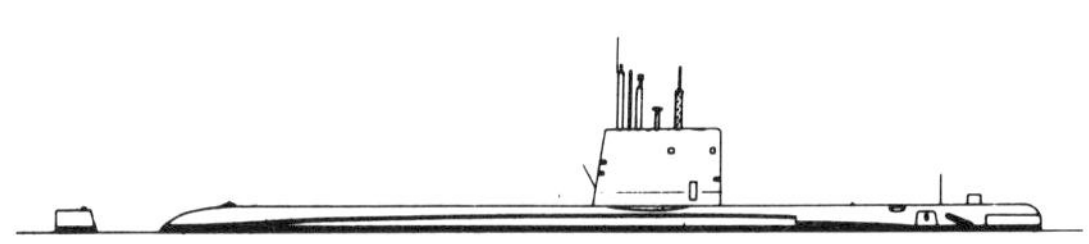

These four attack boats were the first US approach to a 'production' nuclear submarine, considerably smaller than *Nautilus*, with about half her power (S3W in SSN 578 and SSN 583, S4W in SSN 579 and SSN 584); they corresponded approximately to the earlier *Tang* class in dimensions, although they were somewhat heavier. Production was abandoned in favour of the much faster *Skipjack* class, the transition occurring in the FY56 programme.

SKATE class *attack submarines*

Displacement: 2550t surfaced; 2848t submerged
Dimensions: 267ft 8in × 25ft × 20ft 7in
81.6m × 7.6m × 6.3m
Machinery: 2-shaft nuclear: 1 S3W (S4W in SSN 579, 584), 6600shp = approx 20kts submerged
Armament: 8–21in TT (6 bow, 2 stern)
Sensors: BQS-4
Complement: 84

No	Name	Builder	Laid down	Launched	Comp	Fate
SSN 578	SKATE	Electric Boat	21.7.55	16.5.57	23.12.57	Decomm 12.9.86
SSN 579	SWORDFISH	Portsmouth N Yd	25.1.56	27.8.57	15.9.58	Decomm 2.6.89
SSN 583	SARGO	Mare Island N Yd	21.2.56	10.10.57	1.10.58	Decomm 26.2.88
SSN 584	SEADRAGON	Portsmouth N Yd	20.6.56	16.8.58	5.12.59	Decomm 12.6.84

ALBACORE *experimental submarine*

Displacement: 1500t surfaced; 1850t submerged
Dimensions: 210ft 6in oa × 27ft 4in × 18ft 6in
64.2m × 8.3m × 5.6m
Machinery: 1 shaft, 2 GM diesels plus 1 Westinghouse electric motor, 1500bhp/15,000shp = 33kts (submerged in final configuration reportedly)
Complement: 52

No	Name	Builder	Laid down	Launched	Comp	Fate
AGSS 569	ALBACORE	Portsmouth N Yd	15.3.52	1.8.53	5.12.53	Stricken 1.5.80

Although by no means a fighting ship, the experimental submarine *Albacore* deserves inclusion in this book by virtue of her pioneering work in submarine hull form design. She was the first US submarine designed entirely for underwater operation, with a hull form unsuited to surface operation. Conceived initially as a target for surface ASW forces, she was financed partly (as SCB 56 of FY50) by cancelling *Fletcher* class DDE conversions. Her outstanding internal feature was a massive battery capacity which permitted her to operate briefly at extremely high underwater speeds, initially about 27kts and later about 33kts. She was extensively modified during her operational career. Initially she was 203ft 9in long, with conventional stern surfaces. However, Phase III modifications of August 1961 provided her with a new stern with X-form planes, a dorsal rudder, a new bow sonar dome, and dive brakes. Phase IV (December 1962–March 1965) included the installation of a new high-capacity silver-zinc battery and contraprops.

DOLPHIN

The other major experimental submarine in the standard numbering sequence is *Dolphin*, AGSS 555 of the FY61 programme (SCB 207), designed for deep diving, with a 15ft constant-diameter pressure hull. Her experimental torpedo tube was removed in 1970.

Blueback 1967

BARBEL class *attack submarines*

Displacement: 2146t surfaced; 2639t submerged
Dimensions: 218ft 8in wl, 219ft 2in oa × 29ft × 20ft 8in
66.7m, 66.8m × 8.8m × 6.3m
Machinery: 1 shaft, 3 FM diesels plus 1 Westinghouse electric motor, 4800bhp/3150shp = 15kts/21kts
Armament: 6–21in TT (bow)
Sensors: Sonar BQS-4
Complement: 77

No	Name	Builder	Laid down	Launched	Comp	Fate
SS 580	BARBEL	Portsmouth N Yd	18.5.56	19.7.58	17.1.59	Stricken 17.1.90, BU 1994
SS 581	BLUEBACK	Ingalls	15.4.57	16.5.59	15.10.59	Stricken 30.10.90
SS 582	BONEFISH	New York SB	3.6.57	22.11.58	9.7.59	Stricken 28.9.88

These submarines, the last US diesel-electric types built, adopted the streamlined hull form of the experimental *Albacore* to achieve high underwater speed, albeit for only relatively short periods. Stern torpedo tubes had to be eliminated because of the adoption of a single propeller shaft. As built, they had conventional bow diving planes, but they were later modified with planes in their fin structures. They were the first US submarines with all controls centralised in an 'attack center', a practice adopted for all later classes.

Unlike earlier diesel submarines, they were not refitted with PUFFs passive fire control gear, perhaps because the three large sonar domes involved would have ruined streamlining. They differ from *Albacore*-shaped nuclear submarines in that they retain a substantial casing with a noticeable free-flooding slot running most of its length. *Bonefish* was stricken following a serious fire (three fatalities), on 24 March 1988.

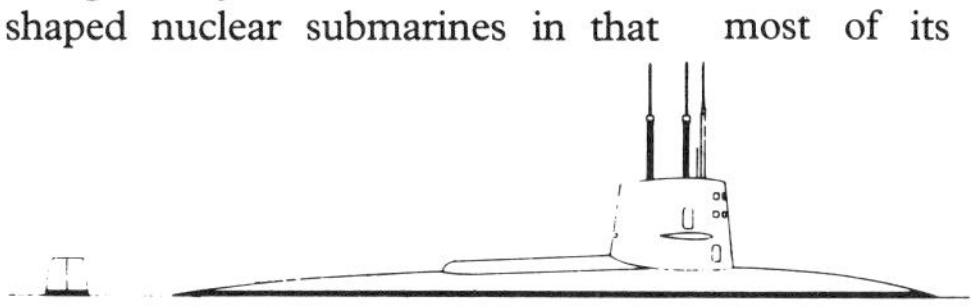

Scamp 1976

SKIPJACK class *attack submarines*

Displacement: 3070t surfaced; 3500t submerged
Dimensions: 246ft wl, 251ft 9in oa × 31ft 8in × 25ft 3in
75.0m, 76.8m × 9.7m × 7.7m
Machinery: 1-shaft nuclear: 1 S5W reactor, 15,000shp = approx 30kts submerged
Armament: 6–21in TT (bow; 24 torpedoes)
Sensors: Sonar BQS-4
Complement: 85

No	Name	Builder	Laid down	Launched	Comp	Fate
SSN 585	SKIPJACK	Electric Boat	29.5.56	26.5.58	15.4.59	Decomm 19.4.90
SSN 588	SCAMP	Mare Island N Yd	23.1.59	8.10.60	5.6.61	Decomm 26.4.88
SSN 589	SCORPION	Electric Boat	20.8.58	19.12.59	27.6.60	Lost May 1968
SSN 590	SCULPIN	Ingalls	3.2.58	31.3.60	1.6.61	Decomm 3.8.90
SSN 591	SHARK	Newport News	24.2.56	16.3.60	9.2.61	Decomm 16.10.86
SSN 592	SNOOK	Ingalls	7.4.58	31.10.60	24.10.61	Decomm 15.5.86

These six ships were essentially nuclear equivalents of the *Barbels*, benefiting from the *Albacore* hull form and also introducing the S5W reactor and single-hull construction to US SSNs. They were the fastest submarines in the US service until the appearance of the *Los Angeles* class, and they formed the basis of the *George Washington* class ballistic missile submarines. A single propeller shaft precluded the installation of stern torpedo tubes, so that they are limited to six bow tubes. Although they do not have the bow sonar or SUBROC of later US submarines, they continue to be highly regarded, and projects for *Skipjack* sonar improvement were considered during the 1960s. *Scorpion* of this class was lost with all hands in May 1968 south-west of the Azores.

Launch of *Albacore* CMP

Sculpin 6.7.65 L & L van Ginderen

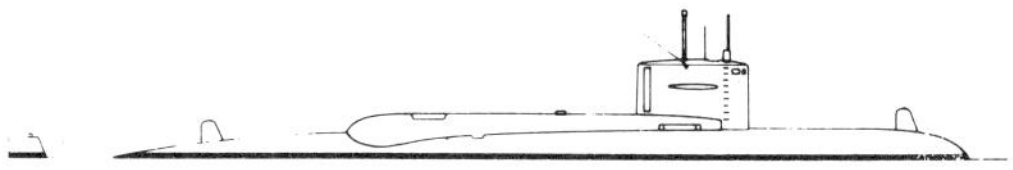

Tullibee 1960

TULLIBEE *attack submarine*

Tullibee was an attempt to build a minimum nuclear submarine specifically for ASW; originally she was to have been even smaller than she turned out to be, and ultimately she was somewhat underpowered, with an S2C reactor reportedly producing only 2500shp, one-third of the *Skate* power plant. She was completed with the BQQ-2 sonar and quartet of hull-mounted torpedo tubes of the *Threshers*, but was not fitted to fire SUBROC. She also had a prototype turbo-electric drive, for quietness. *Tullibee* was later fitted with three large sonar domes for a PUFFS (BQG-4) passive-ranging sonar system.

Displacement: 2316t surfaced; 2607t submerged
Dimensions: 272ft 10in × 23ft 4in × 19ft 4in
83.2m × 7.1m × 5.8m
Machinery: 1-shaft nuclear: 1 S2C reactor, 2500shp = approx 15kts submerged
Armament: 4–21in TT (amidships)
Sensors: Sonar BQQ-2 (later BQG-4 as well)
Complement: 56

No	Name	Builder	Laid down	Launched	Comp	Fate
SSN 597	TULLIBEE	Electric Boat	26.5.58	27.4.60	9.11.60	Decomm 26.6.88

Permit 1970

THRESHER/PERMIT class *attack submarines*

Note SSN 612–615 are usually included in the *Permit* group, but share the dimensions of the *Sturgeon* class, and are therefore listed with it.

This is the largest class of submarines built by the United States since 1945; officially it was subdivided into *Permit* and *Sturgeon* classes; the original name-ship, *Thresher*, was lost on diving trials on 10 April 1963. They combine a stronger hull for deeper diving with an advanced long-range BQQ-2 sonar and its associated SUBROC missile; diving depth is reportedly 1300ft (396.2m). Later units were lengthened to permit the installation of a BQQ-5 sonar in place of BQQ-2; the extra space is free flooding, and underwater displacement rises by about 200t to 350t (SSN 613–615 and most *Sturgeon* class submarines); ultimately all of these units will have BQQ-5. All will also ultimately receive the Mk 117 fire control system in place of their current Mk 113, providing facilities for Harpoon anti-ship missile control. However, since SUBROC is an analogue rather than a digital weapon, and Mk 117 is a fully digital fire con-trol system, the long-range ASW cap-ability (except as provided by the Mk 48 torpedo) is being lost. SSN 613– SSN 615 were, in effect, prototype *Sturgeons* with heavier machinery, taller fins (20ft rather than 13.8ft or 15ft), improved safety features ('Sub-safe' programme), and lengthened hulls (292ft 3in rather than 278ft 6in). SSN 678 and later units are 302ft long. SSN 666, 672 and others were modified to carry a DSRV (submarine rescue vehicle) aft, for launch and recovery while submerged.

As officially described in 1979, *Thresher* class design compromises included a low sail, inadequate numbers of periscopes and antennas, and limited electronics. The reduced sail height was specifically accepted to minimize the speed loss, given the fixed reactor power of the S5W. The SSN 637 class was considerably enlarged for a combination of quieting and additional electronics, as well as improved internal arrangements. Presumably the electronics was a combination of sonar signal processing and torpedo control, one control console being required for each wire-guided torpedo.

These submarines were intended

Displacement: 3705t surfaced; 4311t submerged
Dimensions: 278ft 6in (296ft 9in SSN 605) × 31ft 8in × 25ft 2in
84.9m (90.5m) × 9.7m × 7.7m
Machinery: 1-shaft nuclear: 1 S5W reactor, 15,000shp = approx 27kts submerged
Armament: 4–21in TT (amidships)
Sensors: Sonar BQQ-2
Complement: 94

No	Name	Builder	Laid down	Launched	Comp	Fate
SSN 593	THRESHER	Portsmouth N Yd		9.7.60		Lost 10.4.63
SSN 594	PERMIT	Mare Island N Yd	10.7.59	1.7.61	29.5.62	Stricken 12.6.91
SSN 595	PLUNGER	Mare Island N Yd	20.3.60	9.12.61	21.11.62	Deactivated 10.2.89
SSN 596	BARB (ex-*Pollack*)	Ingalls	9.11.59	12.2.62	24.8.63	Deactivated 20.12.89
SSN 603	POLLACK (ex-*Barb*)	New York SB	14.3.60	17.3.62	26.5.64	Stricken 1.3.89
SSN 604	HADDO	New York SB	9.9.60	18.8.62	16.12.64	Deactivated 1.10.90
SSN 605	JACK	Portsmouth N Yd	16.9.60	24.4.63	31.3.67	Deavtivated 2.10.89
SSN 606	TINOSA	Portsmouth N Yd	24.11.59	9.12.61	17.10.64	Deactivated 15.7.91
SSN 607	DACE	Ingalls	6.6.60	18.8.62	4.4.64	Deactivated 27.2.88
SSN 612	GUARDFISH	New York SB	28.2.61	15.5.65	20.12.66	Deactivated 15.7.91
SSN 621	HADDOCK	Ingalls	24.4.61	21.5.66	22.12.67	Reserve for deactivation 2.4.92

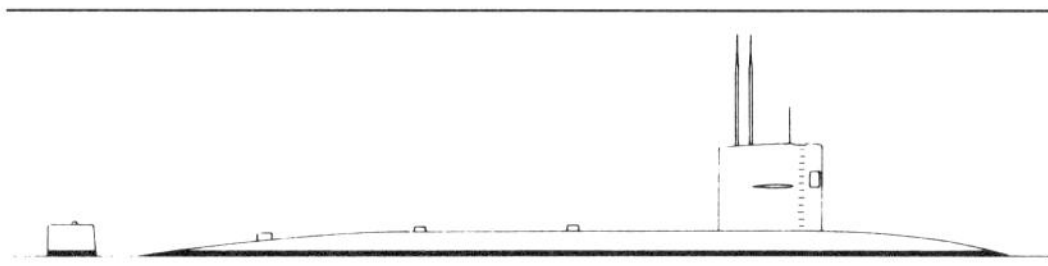

Finback as completed

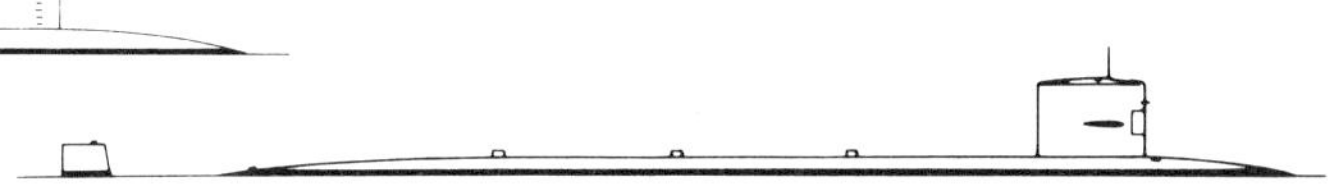

Glenard P Lipscomb as completec

STURGEON class *attack submarines*

Displacement: 4246t surfaced; 4777t submerged
Dimensions: 292ft 3in (302ft from SSN 678) × 31ft 8in × 25ft 6in
89.1m (92.1m) × 9.7m × 7.8m
Machinery: 1-shaft nuclear: 1 S5W reactor, 15,000shp = approx 26kts submerged
Armament: 4–21in TT (amidships)
Sensors: Sonar BQQ-2
Complement: 99

No	Name	Builder	Laid down	Launched	Comp	Fate
SSN 613	FLASHER	Electric Boat	14.6.61	22.6.63	22.7.66	Deactivated 18.7.91
SSN 614	GREENLING	Electric Boat	15.8.61	4.4.64	3.11.67	Reserve for deactivation 31.10.93
SSN 615	GATO	Electric Boat	15.12.61	14.5.64	25.1.68	Extant 1995
SSN 637	STURGEON	Electric Boat	10.8.63	26.2.65	3.3.67	Deactivated 15.4.94
SSN 638	WHALE	General Dynamics, Quincy	27.5.64	14.10.66	12.10.68	Deactivation FY96
SSN 639	TAUTOG	Ingalls	21.1.64	14.4.67	17.8.68	Extant 1995
SSN 646	GRAYLING	Portsmouth N Yd	12.5.64	22.6.67	11.10.69	Extant 1995
SSN 647	POGY	Ingalls	4.5.64	3.6.67	15.5.71	Extant 1995
SSN 648	ASPRO	Ingalls	23.11.64	29.11.67	20.2.69	Deactivated Sep 1994
SSN 649	SUNFISH	General Dynamics, Quincy	15.1.65	14.10.66	15.3.69	Extant 1995
SSN 650	PARGO	Electric Boat	3.6.64	17.9.66	1.5.68	Deactivation FY95
SSN 651	QUEENFISH	Newport News	11.5.64	25.2.66	6.12.66	Deactivated 27.9.90
SSN 652	PUFFER	Ingalls	8.2.65	30.3.68	9.8.69	Extant 1995
SSN 653	RAY	Newport News	1.4.65	21.6.66	12.4.67	Extant 1995
SSN 660	SAND LANCE	Portsmouth N Yd	15.1.65	11.11.69	25.9.71	Extant 1995
SSN 661	LAPON	Newport News	26.7.65	16.12.66	14.12.67	Deactivated 1.10.91

primarily for ASW, with a large spherical sonar array filling their bows, and the torpedo battery, reduced to four tubes, moved aft to abreast the fin, firing diagonally from the hull. One advantage of this arrangement was that a large torpedo room could be provided, permitting a relatively easy choice of weapons for each tube. By the late 1960s, the submarine commander had a choice of long ASW torpedoes, nuclear torpedoes (Astor, or Mk 45, now retired), long anti-ship torpedoes, or SUB-ROC; now there are Harpoon (first tested in *Permit* in 1976) and, very soon, Tomahawk. Unfortunately total weapon capacity is quite small, figures as low as twenty torpedoes in stowage being reported, so that this variety of weapons is very much a mixed bles-sing. Indeed, one advantage of the adoption of the Mk 48 torpedo is that it can replace both long ASW and long anti-ship weapons, and the Mk 45 was discarded partly because it consumed valuable space but was unlikely to be used in action.

Since the power plant, the S5W, duplicated that of the earlier *Skipjack*, and the size of these submarines was much greater (by at least 800t), these ships were considerably slower, their speed reportedly about 26kts. There were three attempts within this class to improve propulsion. *Jack* has contraprops on her single screw, an arrangement never duplicated. *Narwhal* has a natural convection powerplant (S5G rather than S5W); and *Glenard P Lipscomb* has turbo-electric propulsion, both of the last two choices having been made for quietness.

Six were converted to accommodate Dry Deck Shelters for sixteen SEALs and their swimmer delivery vehicles. The test unit was SSN 684 (modified August–December 1982), followed by six more: SSN 678, 679, 684, and then (FY90 programme) 680, 682, and 686. The last will be delivered in 1992. SSN 666 and 672, and others have been modified to carry and launch a DSRV salvage vehicle. Between January 1987 and 1991 *Parche* (SSN 683) was rebuilt at Mare Island for special operations and lengthened by 100ft forward of her sail (total length 393ft, dispacement 6140/7140t).

No	Name	Builder	Laid down	Launched	Comp	Fate
SSN 662	GURNARD	Mare Island N Yd	22.12.64	20.5.67	6.12.68	Deactivated 12.8.94
SSN 663	HAMMERHEAD	Newport News	29.11.65	14.4.67	28.6.67	Deactivated 31.12.93
SSN 664	SEA DEVIL	Newport News	12.4.66	5.10.67	30.1.69	Deactivated 25.2.91, stricken 16.10.91
SSN 665	GUITARRO	Mare Island N Yd	9.12.65	27.7.68	9.9.72	Deactivated 1.10.91
SSN 666	HAWKBILL	Mare Island N Yd	12.9.66	12.4.69	4.2.71	Extant 1995
SSN 667	BERGALL	Electric Boat	16.4.66	17.2.68	13.6.69	Extant 1995
SSN 668	SPADEFISH	Newport News	21.12.66	15.5.68	14.8.69	Extant 1995
SSN 669	SEAHORSE	Electric Boat	13.8.66	15.6.68	19.9.69	Deactivation FY95
SSN 670	FINBACK	Newport News	26.6.67	7.12.68	4.2.70	Extant 1995
SSN 671	NARWHAL	Electric Boat	17.1.66	9.9.66	12.7.69	Extant 1995
SSN 672	PINTADO	Mare Island N Yd	27.10.67	16.8.69	11.9.71	Extant 1995
SSN 673	FLYING FISH	Electric Boat	30.6.67	17.5.69	29.4.70	Extant 1995
SSN 674	TREPANG	Electric Boat	28.10.67	27.9.69	14.8.70	Extant 1995
SSN 675	BLUEFISH	Electric Boat	13.3.68	10.1.70	8.1.71	Extant 1995
SSN 676	BILLFISH	Electric Boat	20.9.68	1.5.70	12.3.71	Extant 1995
SSN 677	DRUM	Mare Island N Yd	20.8.68	23.5.70	15.4.72	Deactivation FY95
SSN 678	ARCHERFISH	Electric Boat	19.6.69	16.1.71	24.12.71	Extant 1995
SSN 679	SILVERSIDES	Electric Boat	13.12.69	4.6.71	5.5.72	Deactivated 14.2.94
SSN 680	WILLIAM H BATES (ex-*Redfish*)	Ingalls	4.8.69	11.12.71	5.5.73	Extant 1995
SSN 681	BATFISH	Electric Boat	9.2.70	9.10.71	1.9.72	Extant 1995
SSN 682	TUNNY	Ingalls	22.5.70	10.6.72	26.1.74	Extant 1995
SSN 683	PARCHE	Ingalls	10.12.70	13.1.73	17.8.74	Extant 1995
SSN 684	CAVALLA	Electric Boat	4.6.70	19.2.72	9.2.73	Extant 1995
SSN 685	GLENARD P LIPSCOMB	Electric Boat	5.6.71	4.8.73	21.12.74	Decomm 11.7.90
SSN 686	L MENDEL RIVERS	Newport News	26.6.71	2.6.73	1.2.75	Extant 1995
SSN 687	RICHARD B RUSSELL	Newport News	19.10.71	12.1.74	16.8.75	Deactivated 1.7.93

Note SSN 671 and 685 are not properly part of this class, but were test vehicles for alternative propulsion: SSN 671 for a natural-circulation (S5G) reactor, SSN 685 for turbo-electric drive (for quietness) – at a considerable cost in size (5800t standard; 6480t submerged, 365ft (111.2m) long) and therefore of speed. SSN 671: 4450t standard; 5350t submerged, 314ft (95.7m) oa.

Tinosa Oct 1979 L & L van Ginderen

Pargo 15.10.83 L & L van Ginderen

Los Angeles as completed

LOS ANGELES class *attack submarines*

Displacement: 6000t surfaced; 6900t submerged
Dimensions: 360ft × 33ft × 32ft 4in
109.8m × 10.1m × 9.8m
Machinery: 1-shaft nuclear: 1 S6G reactor, approx 30,000shp = approx 31kts submerged
Armament: 4–21in TT (amidships; Tomahawk SSMs)
Sensors: Sonar BQQ-5
Complement: 127

No	Name	Builder	Laid down	Launched	Comp	Fate
SSN 688	LOS ANGELES	Newport News	8.1.72	6.4.74	13.11.76	Extant 1995
SSN 689	BATON ROUGE	Newport News	18.11.72	26.4.75	25.6.77	Extant 1995
SSN 690	PHILADELPHIA	Electric Boat	12.8.72	19.10.74	25.6.77	Extant 1995
SSN 691	MEMPHIS	Newport News	23.6.73	3.4.76	17.12.77	Extant 1995
SSN 692	OMAHA	Electric Boat	27.1.73	21.2.76	11.3.78	Extant 1995
SSN 693	CINCINNATI	Newport News	6.4.74	19.2.77	10.6.78	Extant 1995
SSN 694	GROTON	Electric Boat	3.8.73	9.10.76	8.7.78	Extant 1995
SSN 695	BIRMINGHAM	Newport News	26.4.75	29.10.77	20.12.78	Extant 1995
SSN 696	NEW YORK CITY	Electric Boat	15.12.73	18.6.77	10.3.79	Extant 1995
SSN 697	INDIANAPOLIS	Electric Boat	19.10.74	30.7.77	5.1.80	Extant 1995
SSN 698	BREMERTON	Electric Boat	8.5.76	22.7.78	28.3.81	Extant 1995
SSN 699	JACKSONVILLE	Electric Boat	21.2.76	18.11.78	16.5.81	Extant 1995
SSN 700	DALLAS	Electric Boat	9.10.76	28.4.79	18.7.81	Extant 1995
SSN 701	LA JOLLA	Electric Boat	16.10.76	11.8.79	24.10.81	Extant 1995
SSN 702	PHOENIX	Electric Boat	30.7.77	8.12.79	19.12.81	Extant 1995
SSN 703	BOSTON	Electric Boat	11.8.78	19.4.80	30.1.82	Extant 1995
SSN 704	BALTIMORE	Electric Boat	21.5.79	13.12.80	24.7.82	Extant 1995
SSN 705	CITY OF CORPUS CHRISTI	Electric Boat	4.9.79	25.4.81	14.12.82	Extant 1995
SSN 706	ALBUQUERQUE	Electric Boat	27.12.79	13.3.82	21.5.83	Extant 1995
SSN 707	PORTSMOUTH	Electric Boat	8.5.80	18.9.82	1.10.83	Extant 1995
SSN 708	MINNEAPOLIS-ST PAUL	Electric Boat	20.1.81	18.12.82	10.3.84	Extant 1995
SSN 709	HYMAN G RICKOVER	Electric Boat	24.7.81	27.8.83	21.7.84	Extant 1995
SSN 710	AUGUSTA	Electric Boat	1.4.82	21.1.84	19.1.85	Extant 1995
SSN 711	SAN FRANCISCO	Newport News	26.5.77	27.10.79	24.4.81	Extant 1995
SSN 712	ATLANTA	Newport News	17.8.78	16.8.80	6.3.82	Extant 1995
SSN 713	HOUSTON	Newport News	29.1.79	21.3.81	25.9.82	Extant 1995
SSN 714	NORFOLK	Newport News	1.8.79	21.10.81	21.5.83	Extant 1995
SSN 715	BUFFALO	Newport News	25.1.80	8.5.82	5.11.83	Extant 1995
SSN 716	SALT LAKE CITY	Newport News	26.8.80	16.10.82	12.5.84	Extant 1995
SSN 717	OLYMPIA	Newport News	31.3.81	30.4.83	17.11.84	Extant 1995
SSN 718	HONOLULU	Newport News	10.11.81	24.9.83	6.7.85	Extant 1995
SSN 719	PROVIDENCE	Electric Boat	9.10.82	4.8.84	27.7.85	Extant 1995
SSN 720	PITTSBURGH	Electric Boat	15.4.83	8.12.84	23.11.85	Extant 1995
SSN 721	CHICAGO	Newport News	5.1.83	13.10.84	27.9.86	Extant 1995
SSN 722	KEY WEST	Newport News	6.7.83	20.7.85	12.9.87	Extant 1995
SSN 723	OKLAHOMA CITY	Newport News	4.1.84	2.11.85	9.7.88	Extant 1995
SSN 724	LOUISVILLE	Electric Boat	16.9.84	14.12.85	8.11.86	Extant 1995
SSN 725	HELENA	Electric Boat	28.3.85	28.6.86	11.7.87	Extant 1995
SSN 750	NEWPORT NEWS	Newport News	3.3.84	15.3.86	3.6.89	Extant 1995
SSN 751	SAN JUAN	Electric Boat	16.8.85	6.12.86	6.8.88	Extant 1995
SSN 752	PASADENA	Electric Boat	20.5.85	12.9.87	11.2.89	Extant 1995
SSN 753	ALBANY	Newport News	22.4.85	13.6.87	7.4.90	Extant 1995
SSN 754	TOPEKA	Electric Boat	13.5.86	23.1.88	21.10.88	Extant 1995
SSN 755	MIAMI	Electric Boat	24.10.86	12.11.88	30.6.90	Extant 1995
SSN 756	SCRANTON	Newport News	29.6.86	3.7.89	8.2.91	Extant 1995
SSN 757	ALEXANDRIA	Electric Boat	19.6.87	23.6.90	26.6.91	Extant 1995
SSN 758	ASHEVILLE	Newport News	9.1.87	28.10.89	28.9.91	Extant 1995
SSN 759	JEFFERSON CITY	Newport News	21.9.87	17.8.90	28.2.92	Extant 1995
SSN 760	ANNAPOLIS	Electric Boat	15.6.88	19.5.91	11.4.92	Extant 1995
SSN 761	SPRINGFIELD	Electric Boat	29.1.90	4.1.92	9.1.93	Extant 1995
SSN 762	COLUMBUS	Electric Boat	7.1.91	1.8.92	24.7.93	Extant 1995
SSN 763	SANTA FE	Electric Boat	9.7.91	12.12.92	9.1.94	Extant 1995
SSN 764	BOISE	Newport News	25.8.88	23.3.91	7.11.92	Extant 1995
SSN 765	MONTPELIER	Newport News	19.5.89	23.8.91	13.9.93	Extant 1995
SSN 766	CHARLOTTE	Newport News	17.8.90	3.10.92	31.8.94	Extant 1995
SSN 767	HAMPTON	Newport News	2.3.90	3.4.92	6.11.93	Extant 1995
SSN 768	HARTFORD	Electric Boat	27.4.92	4.12.93	Dec 1994	Extant 1995
SSN 769	TOLEDO	Newport News	6.5.91	28.8.93	24.2.95	Extant 1995
SSN 770	TUCSON	Newport News	15.8.91	19.3.94	Aug 1995	Completing
SSN 771	COLUMBIA	?	21.4.93	24.9.94	Sep 1995	Completing
SSN 772	GREENVILLE	Newport News	28.2.92	Sep 1993	Mar 1994	Extant 1995
SSN 773	CHEYENNE	Newport News	6.7.92	1.4.95	Aug 1995	Completing

SSN 689 placed in reserve for scrapping 1.11.93; 692 and 693 to retire during FY95 (ends Oct 1995).

These submarines restore the speed advantage of the *Skipjacks* while improving on the sonar and weapons capability of the *Sturgeons*. The great increase in size (displacement submerged rising from 4650t to 6900t) is due largely to a doubling in reactor power. Even so, some sacrifice had to be made in pressure hull strength, so that these craft cannot dive as deeply as their predecessors. Plans originally called for introducing a new hull steel about 1973, which would have restored the depth in later units, but this programme failed to materialise. Reportedly the *Los Angeles* class gained about 5kts on its predecessor.

This class was designed for an all-digital combat system, but it was not ready in time, so SSN 688–699 had the earlier Mk 113. The Mk 117 was introduced in SSN 700 and eventually installed not only in the earlier ships of this class but also in the *Sturgeons*. Under the FY83 programme Mk 117 was modified to control Subroc missiles (they were retired in 1989). Harpoon missiles were carried from 1978 onwards.

Design work began in 1968 on a new 'mid-1970s' attack submarine, to incorporate both new sensor and propulsion technology and a new stand-off weapon (which would have replaced SUBROC and would have had both anti-ship and anti-submarine roles). In view of the limited internal weapon space available in the torpedo rooms of conventionally designed nuclear submarines, twenty vertical launch tubes would have been provided abaft the fin of the new submarine. Other improvements, such as greatly increased power and improved silencing, added size, so that by December 1972 the new submarine was being described as: 472ft × 40ft (143.8m × 12.2m) (as against 335ft × 33ft (102.1m × 10.1m) for a *Los Angeles*); displacing 13,700t submerged (as against 6630t estimated at that time). It would have cost about 80 per cent as much as a Trident missile submarine, then estimated at $500m in FY73 dollars. Admiral Zumwalt cancelled it and its new missile in 1973, and a submarine-launched version of Harpoon was substituted.

The associated sonar is the new digital BQQ-5, originally comprising a bow sphere and a conformal array similar to that in the *Sturgeon* class, but using digital beam control (beam preforming) instead of analog beamforming. Later a towed array, in an external tube, was added.

Weapon capacity is slightly larger than that of the *Sturgeon* class, thanks largely to a slightly greater hull diameter (reportedly twenty-six rather than twenty-three weapons). In addition, because the *Los Angeles* was designed from the first to fire the Mk 48 torpedo, its torpedo racks are long enough to accommodate both weapon and wire reel. Earlier submarines were designed for either long free-running torpedoes or for short wire-guided ones (Mk 37), and so had to devote a rack to wire canisters when the long wire-guided weapon appeared.

Units from SSN 719 on have twelve vertical launch tubes for Tomahawk missiles, located between the forward end of the pressure hull and the bow sonar sphere. They carry a total of twenty such missiles (earlier units carry eight, displacing eight torpedoes); SSN 701 was the first of the class to be equipped to fire Tomahawk from her torpedo tubes, but SSN 712 was the first with operational Tomahawk capability (30 November 1983).

SSN 751 introduced the new BSY-1 integrated submarine combat system and also had retractable bow planes (rather than sail planes) for under-ice operation. These submarines are described as the '688I' class. SSN 768 introduced improved silencing, additional tail fins (like *Seawolf*), and modified propulsion.

Test units: SSN 691 is being converted to a test submarine. The plans to replace two 21in TT on one side with a single 30in (762mm) tube have been held in abeyance, however. SSN 710 was test ship for the wide aperature array (WAA: BQG-5) from July 1987; it will probably be retrofitted on to some units.

Jacksonville 16.11.82 RAN

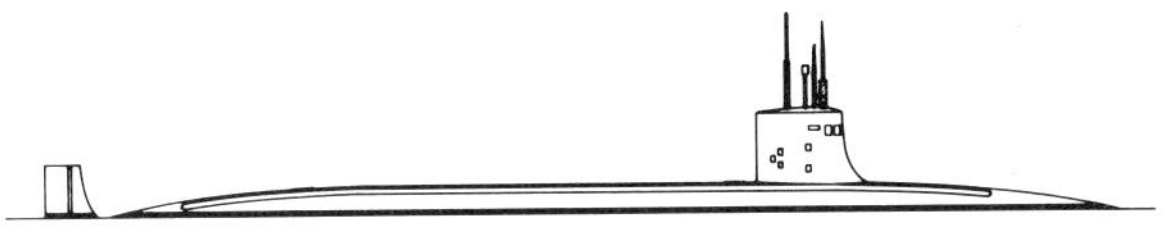

Seawolf design

SEAWOLF class *attack submarines*

This was the first new attack submarine to be designed after the *Los Angeles* class. There had been several proposals during the 1970s, but none was considered sufficiently superior to the gradually evolving *Los Angeles*. The new *Seawolf* combined a new quieter reactor with a new quiet propulsion (a pumpjet), a better hull shape (lower length: beam ratio), and much more firepower, in the form of twice as many tubes and twice as many internally-stowed weapons. *Seawolf* is somewhat faster than *Los Angeles*, but it has a much higher quiet (tactical) speed, reportedly about twice that of the earlier class. The new submarine is equipped with a new distributed combat system, BSY-2, which incorporates a new dual bow sonar (a pair of spheres), the flank array (WAA) installed on late *Los Angeles* class boats, and a pair of towed arrays (a thick TB-16D and a thin TB-29). There is also an active ice- and mine-avoidance sonar, SADS/MIDAS. *Seawolf* was designed for relatively inexpensive maintenance, so that although its unit cost is very high, its total life-cycle cost is intended to match that of earlier submarines. Overhaul (including reactor refuelling) should not be necessary for fifteen years. The first unit was authorised in FY89, for launching in January 1994 and completion in May 1995. These dates probably will not be met, because difficulties were experienced in welding its new HY-100 steel.

Seawolf was also called SSN-21, the nuclear submarine of the twenty-first century, and this new number superseded the original hull number, SSN 774, on numerous official documents. Two more units were authorised in FY91 (one to be built by Newport News, one disputed between the two builders), but in January 1992 Secretary of Defense Dick Cheney announced that the class would be stopped with the first unit, and that the administration would seek authority to reprogramme funds planned for the next two. It had been supposed that procurement of the class might be stretched out, but cancellation came as a surprise; some in Congress are expected to fight it.

Opposition to mass production of so expensive a submarine (ultimately the prototype was expected to cost about $3b) led the navy to work on a less expensive follow-on, initially called 'Centurion'. In early 1992 possible cuts included reductions in overall size, in speed (by using a smaller reactor), and in detail standards (such as silencing or shock-hardening). It is not clear, however, whether production of smaller numbers of less capable submarines will really cost less, given the inevitable industrial overhead load which each submarine would have to carry.

GRAYBACK *strategic submarine*

Displacement: 2287t light; 3638t submerged
Dimensions: 322ft 4in oa × 30ft × 17ft 4in
98.3m × 9.1m × 5.3m
Machinery: 2 shafts, 3 FM diesels, 2 Elliott electric motors, 4500bhp/5600shp = 15kts/12kts (designed; 20kts on surface unofficially reported 1978)
Armament: 4 Regulus I or Regulus II SSM, 8–21in TT (6 bow, 2 stern)
Sensors: Sonar BQS-4 (BQG-4 later installed)
Complement: 84

No	Name	Builder	Laid down	Launched	Comp	Fate
SSG 574	GRAYBACK	Portsmouth N Yd	1.7.54	2.7.54	May 1958	Stricken 16.1.84

These two submarines were laid down as sisters to *Darter*, then cut in two and lengthened by 50ft (15.2m) (*Growler* is 4ft longer) to accommodate two cylindrical hangars for Regulus missiles; both were decommissioned in May 1964 when Regulus was retired. However, *Grayback* was converted to an amphibious transport (LPSS) at Mare Island (November 1967 to May 1969) and lengthened by 11ft 8in (3.5m), and had her former missile hangars converted to 'lock out' swimmers and Swimmer Delivery Vehicles (SDV); her fin structure was raised and three large sonar domes for PUFFS passive sonar (BQG-4)

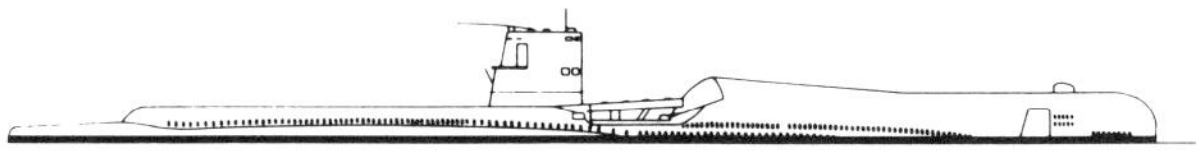

Growler 1959

added. She was reclassified as an attack submarine in 1975, apparently without any further changes. A similar planned conversion of her sister ship was cancelled in view of the high cost of the first conversion. As a commando ship, *Grayback* had a crew of ninety-six and could accommodate ten officers and seventy-five enlisted men.

GROWLER *strategic submarine*

Displacement: 2174t light; 3387t submerged
Dimensions: 317ft oa × 27ft 2in × 17ft
96.6m × 8.3m × 5.2m
Machinery: 2 shafts, 3 FM diesels, 2 Elliott electric motors, 4600bhp/5600shp = 15kts/12kts
Armament: 4 Regulus I or Regulus II SSM, 6–21in TT (4 bow, 2 stern)
Sensors: Sonar BQS-4
Complement: 84

No	Name	Builder	Laid down	Launched	Comp	Fate
SSG 577	GROWLER	Portsmouth N Yd	15.2.55	5.4.57	15.12.58	Stricken 25.5.84, museum ship New York 1990

Grayback (as LPSS) 1975

USN

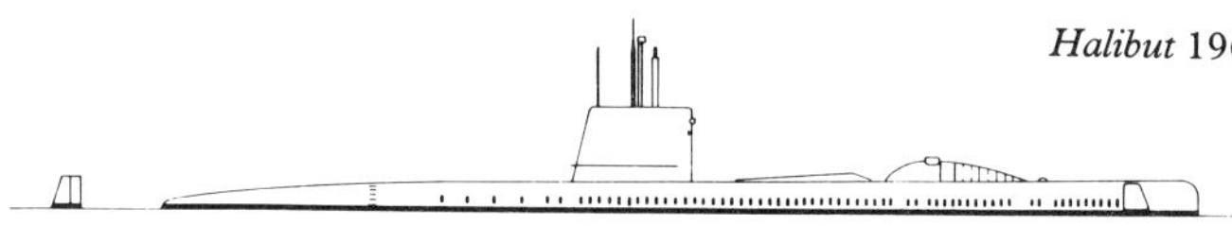

Halibut 1960

Halibut was the sole US nuclear-powered cruise-missile submarine, originally conceived as a near-sister to the *Grayback*s. She was completed with nuclear power and with enlarged missile hangars, for a total of two Regulus II or five Regulus I cruise missiles. The engineering plant was reportedly similar to that of the *Skipjack* class (S3W reactor), with two shafts. The FY58 programme originally included three larger Regulus II submarines, SSGN 594, 595 and 596, with another (SSGN 607) planned for FY59; all completed as *Permit* class attack submarines. This SCB-166A submarine would have carried four Regulus II (or eight Regulus I) in four separate missile hangars forward of the sail. *Halibut*'s hangars were removed after the end of the Regulus programme in 1965 (reclassified to SSN). She was employed in research from then until 1976, with a ducted bow thruster forward and facilities for a deep submergence vehicle (DSRV) aft; all that was left of her missile equipment was a bulged structure on her foredeck. *Halibut* was decommissioned on 30 June 1976 and was in reserve in 1982.

In 1994 it was revealed that *Halibut* had been used for special intelligence-gathering, using a tethered submersible (the DSRV role was a cover); apparently she took the first photographs of the sunken Soviet 'Golf' class missile submarine, parts of which were raised by the *Glomar Explorer*. This disclosure was in the context of discussions of deep sea mining (*eg* for metal nodules), which would require the sort of position-keeping developed for special operations like those *Halibut* had conducted. Other reports suggested that the submersibles had been used to recover Soviet weapons, such as expended missile re-entry vehicles, and that for about two decades the US capacity *covertly* to retrieve objects from very deep water had been highly valued.

HALIBUT *strategic submarine*

Displacement: 3846t surfaced; 4895t submerged
Dimensions: 350ft × 29ft 6in × 20ft 9in
106.7m × 9.0m × 6.3m
Machinery: 2-shaft nuclear: 1 S3W reactor, approx 6600shp = 15kts/15.5kts
Armament: 2 Regulus II or 5 Regulus I SSM, 6–21in TT (4 bow, 2 stern)
Sensors: Sonar BQS-4
Complement: 111

No	Name	Builder	Laid down	Launched	Comp	Fate
SSGN 587	HALIBUT	Mare Island N Yd	11.4.57	9.1.59	4.1.60	Decomm 1976

George Washington 1960

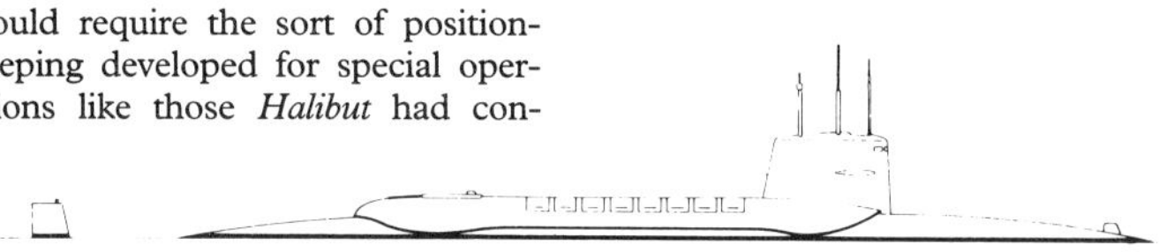

The first US ballistic missile submarines, these vessels were converted from *Skipjack* class attack submarines under construction, by the addition of a 130ft (39.6m) missile section amidships; they have the S5W reactor and the six bow torpedo tubes of the earlier submarines, but reportedly with much reduced torpedo loads. All were converted from Polaris A-1 to A-3 missiles, but could not accommodate Poseidon; they were deactivated in 1980–81 to make way for the *Ohio* class.

Several alternative proposals for future use were made. For example, up to eight cruise missiles can be accommodated in a single Polaris tube. Beginning late in 1980, the missile sections of *Theodore Roosevelt* and

GEORGE WASHINGTON class *strategic submarines*

Displacement: 5959t surfaced; 6709t submerged
Dimensions: 373ft 6in wl, 381ft 8in oa × 33ft × 26ft 8in
113.9m, 116.4m × 10.1m × 8.1m
Machinery: 1-shaft nuclear: 1 S5W reactor, approx 15,000shp = approx 20kts
Armament: 16 Polaris SLBM, 6–21in TT (bow)
Sensors: Sonar BQS-4 (later BQR-19)
Complement: 112

No	Name	Builder	Laid down	Launched	Comp	Fate
SSBN 598	GEORGE WASHINGTON	Electric Boat	1.11.57	9.6.69	30.12.69	SSN Nov 1981, decomm 24.1.85
SSBN 599	PATRICK HENRY	Electric Boat	27.5.58	22.9.59	9.4.60	SSN Oct 1981, decomm 25.5.84
SSBN 600	THEODORE ROOSEVELT	Mare Island N Yd	20.5.58	3.10.59	13.2.61	Decomm 28.2.81
SSBN 601	ROBERT E LEE	Newport News	25.8.58	18.12.59	16.9.60	SSN Mar 1982, decomm 1.12.83
SSBN 602	ABRAHAM LINCOLN	Portsmouth N Yd	1.11.58	14.5.60	11.3.61	Decomm 28.2.81

Abraham Lincoln were physically removed, both units decommissioning on 28 February 1981 at Bremerton, where they remain on the Navy List, in two pieces. Neither had made a missile patrol since 1978, and both required refuelling. The other three units were redesignated attack submarines (SSN). The former missile tubes were filled with cement, and the missile fire control system removed. Note that, compared to *Skipjack* class attack submarines, they were quieter, but considerably slower, and had fewer torpedoes. They lacked the bow sonars of the later attack submarines, but did have towed array passive systems, fitted for self-protection; compared to later SSBNs, they lacked advanced silencing, and could not dive as deep.

Robert E Lee 1966 USN

Cusk 12.9.48 USN

Two submarines, *Cusk* and *Carbonero*, were converted as part of the Loon (copied V1) programme, with launching ramps aft; later they were fitted with cylindrical hangars for two Loon missiles: *Carbonero* was originally only a control craft. Upon termination of the Loon programme in 1953, *Cusk* was converted to a Fleet Snorkel attack submarine, but *Carbonero* was retained for tests with Regulus. Two fleet submarines were converted as operational Regulus I carriers, *Barbero* (a former cargo-carrying submarine) and *Tunny*, a full conversion with a streamlined fin, as in the GUPPYs. She also received a snorkel, and one main engine was removed. By way of contrast, *Barbero*, converted in 1955, never received a streamlined fin; she was considered an interim measure against the availability of specially built submarines such as *Grayback*. Her earlier transport conversion had already cost her two main engines and the stern torpedo tubes; she was fitted with a snorkel and ultimately with a streamlined fin.

Fleet submarine conversions *strategic submarines*

Displacement: 1525t surfaced; 2400t submerged
Dimensions: 312ft oa × 27ft 4in
95.1m × 8.3m
Machinery: 2 shafts, diesel-electric (one main generator engine removed in SSG 282, two removed in SSG 317), 3430bhp (2305bhp in SSG 317)/2740shp (2305shp in SSG 317) = 17 (15) kts/9kts
Armament: 10–21in TT (6 bow, 4 stern; bow TT only in SSG 317)
Complement: 84

For names of boats converted see under Fleet Strength 1947. Original particulars, including building dates, can be found in the 1922–1946 volume.

Carbonero reverted to SS status (from SSG) at the end of the Regulus I test programme in 1961, and the other two reverted in 1965; *Barbero* was stricken, and *Tunny* was converted to a transport submarine (her Regulus hangar proving useful for stowage) at Puget Sound in 1966. She thus joined *Perch* and *Sealion*, converted to submarine transports in 1948, and active in Korea and Vietnam; both had snorkels and Regulus-like LVT/jeep hangars abaft their superstructures, losing all of their torpedoes and two main engines. By way of contrast, *Barbero* was a pure cargo submarine, with cargo spaces in her former torpedo rooms, forward engine room, and some of the after battery compartment. She, too, received a snorkel. The other major submarine auxiliary conversion of the early postwar period was *Guavina*, the oiler, carrying 160,000gal, with all but three torpedo tubes removed (and one forward restricted to discharging cargo) and a snorkel; she had a streamlined fin. She may have been intended to supply forward-deployed SSKs, on the wartime German 'milch cow' principle. In any case, her SSO classification was dropped for a time and then revived briefly in connection with the P6M seaplane programme.

John Marshall (with DDS) 7.5.91
L & L van Ginderen

Ethan Allen 1971

ETHAN ALLEN class *strategic submarines*

Displacement: 6946t surfaced; 7884t submerged
Dimensions: 410ft 5in × 33ft × 27ft 7in
125.1m × 10.1m × 8.4m
Machinery: 1-shaft nuclear: 1 S5W reactor, approx 15,000shp = approx 20kts
Armament: 16 Polaris SLBM, 4–21in TT (bow)
Sensors: Sonar BQS-4, BQR-15, BQR-19
Complement: 110

No	Name	Builder	Laid down	Launched	Comp	Fate
SSBN 608	ETHAN ALLEN	Electric Boat	14.9.59	22.11.60	8.8.61	SSN 1980, decomm 31.3.83
SSBN 609	SAM HOUSTON	Newport News	28.12.59	2.2.61	6.3.62	SSN 1980, decomm 20.9.91
SSBN 610	THOMAS A EDISON	Electric Boat	15.3.60	15.6.61	10.3.62	SSN 1980, deactivated 14.2.92
SSBN 611	JOHN MARSHALL	Newport News	4.4.60	15.7.61	21.5.62	SSN 1981, decomm 23.11.91
SSBN 618	THOMAS JEFFERSON	Newport News	3.2.61	24.2.62	4.1.63	SSN 1981, decomm 24.1.85

The first US ballistic missile submarines designed as such, these ships, like the *George Washingtons*, could not be refitted with the Poseidon missile and were therefore retired as attack submarines. All were originally armed with Polaris A-2, later refitted with A-3; they have towed array passive sonars for self-protection. They were, in effect, ballistic missile submarine versions of the *Threshers*, sharing their deeper-diving hull material and superior silencing. Dates of redesignation to SSN: SSBN 608 (1 September 1980), SSBN 609 (10 November 1980), SSBN 610 (6 October 1980), SSBN 611 (1 May 1981), and SSBN 618 (11 March 1981), in each case at the conclusion of final Polaris cruise. Physical changes began under the FY82 programme (SSBN 608, beginning January 1982): missile tubes were fitted with concrete, and missile fire control system removed. SSBN 609 and 610 were converted under FY83, and the others were done under the FY84 programme. Proposals to install Tomahawk vertical launchers were abandoned. SSBN 609 and 611 were both modified to accommodate a total of sixty-seven SEALs each in double Dry Deck Shelters; they were converted beginning in October 1984.

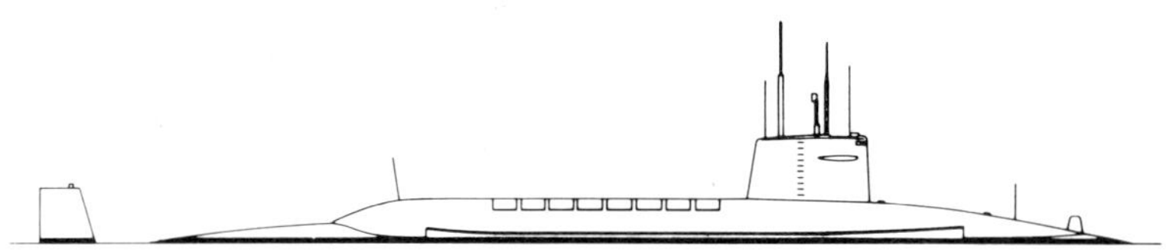

Lafayette as completed

LAFAYETTE class *strategic submarines*

Displacement: 7325t surfaced; 8251t submerged
Dimensions: 425ft × 33ft × 27ft 10in
129.6m × 10.1m × 8.5m
Machinery: 1-shaft nuclear: 1 S5W reactor, 15,000shp = approx 20kts
Armament: 16 SLBM (see notes), 4–21in TT (bow)
Sensors: Sonar BQS-4, BQR-7, BQR-15, BQR-19
Complement: 140

No	Name	Builder	Laid down	Launched	Comp	Fate
SSBN 616	LAFAYETTE	Electric Boat	17.1.61	8.5.62	23.4.63	Deactivated 1.3.91
SSBN 617	ALEXANDER HAMILTON	Electric Boat	26.6.61	18.8.62	27.6.63	Deactivated 1.10.92
SSBN 619	ANDREW JACKSON	Mare Island N Yd	26.4.61	15.9.62	3.7.73	Deactivated 31.8.89 for conversion to ARTB (see below)
SSBN 620	JOHN ADAMS	Portsmouth N Yd	19.5.61	12.1.63	12.5.64	Deactivated 14.9.88
SSBN 622	JAMES MONROE	Newport News	31.7.61	4.8.62	7.12.63	Deactivated 22.9.90
SSBN 623	NATHAN HALE	Electric Boat	2.10.61	12.1.63	23.11.63	Deactivated May 1986
SSBN 624	WOODROW WILSON	Mare Island N Yd	13.9.61	22.2.63	27.12.63	Deactivated 11.1.93
SSBN 625	HENRY CLAY	Newport News	23.10.61	30.11.62	20.2.64	Deactivated 12.3.90
SSBN 626	DANIEL WEBSTER	Electric Boat	23.12.61	27.4.63	9.4.64	Deactivated 27.4.90 (moored trainer)
SSBN 627	JAMES MADISON	Newport News	5.3.62	15.3.63	28.7.64	Deactivated 11.8.91
SSBN 628	TECUMSEH	Electric Boat	1.6.62	22.6.63	29.5.64	Deactivated 15.2.93
SSBN 629	DANIEL BOONE	Mare Island N Yd	6.2.62	22.6.63	23.4.64	Deactivated 1.10.93
SSBN 630	JOHN C CALHOUN	Newport News	4.6.62	22.6.63	15.9.64	Deactivated 1.10.93
SSBN 631	ULYSSES S GRANT	Electric Boat	18.8.62	2.11.63	17.7.64	Deactivated 14.2.92
SSBN 632	VON STEUBEN	Newport News	4.9.62	18.10.63	30.9.64	Deactivated 7.7.93
SSBN 633	CASIMIR PULASKI	Electric Boat	12.1.63	1.2.64	14.8.64	Deactivated 1.10.93
SSBN 634	STONEWALL JACKSON	Mare Island N Yd	4.7.62	30.11.63	26.8.64	Deactivated 10.6.94
SSBN 635	SAM RAYBURN	Newport News	3.12.62	20.12.63	2.12.64	Deactivated 16.9.85
SSBN 636	NATHANIEL GREENE	Portsmouth N Yd	21.5.62	12.5.64	19.12.64	Decomm 12.12.86
SSBN 640	BENJAMIN FRANKLIN	Electric Boat	25.5.63	5.12.64	22.10.65	Deactivated 1.4.93
SSBN 641	SIMON BOLIVAR	Newport News	17.4.63	22.8.64	29.10.65	Deactivated Sep 1994
SSBN 642	KAMEHAMEHA	Mare Island N Yd	2.5.63	16.1.65	10.12.65	SSN 1994
SSBN 643	GEORGE BANCROFT	Electric Boat	24.8.63	5.12.64	22.1.66	Deactivated 1.3.93

These are enlarged versions of the *Ethan Allen* class, all refitted with Poseidon missiles. *Daniel Webster* is unique in having bow, rather than fin, forward diving planes. Patrol endurance is sixty-eight days, with thirty-two day refit periods between patrols, and a sixteen-month yard overhaul every six years; each ship has two crews.

This group is officially divided into three classes: *Lafayette* (SSBN 616), *James Madison* (SSBN 627), and *Benjamin Franklin* (SSBN 640). The SSBN 640s are quieter and have detail improvements. Four more were proposed in the FY65 programme but were not built. Twelve (SSBN 627, 629, 630, 632, 633, 634, 640, 641, 643, 655, 657, 658) were refitted with Trident I (C-5) missiles. SSBN 657 began the first Trident patrol on 20 October 1979. All now have internal countermeasures launchers (eight tubes, four on each side).

Several will be modified with dry deck shelters, to replace the earlier ex-SSBNs, now stricken. SSBN 635 began deactivation in September 1985 for conversion to a non-propelled nuclear training platform (ARTB), but was stricken, unconverted, on 28 August 1989.

The two units reclassified as SSNs were converted to support SEALs; they carry up to sixty-six personnel and their equipment; retain torpedo tubes, carry two dry deck shelters. SSBN 642 was converted at Mare Island (began August 1992); SSBN 645 was completed early in 1994 (at Norfolk). They replaced SSN 609 and 611, which had been converted 1984–86.

No	Name	Builder	Laid down	Launched	Comp	Fate
SSBN 644	LEWIS AND CLARK	Newport News	29.7.63	21.11.64	22.12.65	Deactivated 1.10.91
SSBN 645	JAMES K POLK	Electric Boat	23.11.63	22.5.65	16.4.66	SSN 1994
SSBN 654	GEORGE C MARSHALL	Newport News	2.3.64	21.5.65	29.4.66	Deactivated 13.5.92
SSBN 655	HENRY L STIMSON	Electric Boat	4.4.64	13.11.65	20.8.66	Deactivated 2.11.92
SSBN 656	GEORGE WASHINGTON CARVER	Newport News	24.8.64	14.8.65	15.6.66	Deactivated 2.11.92
SSBN 657	FRANCIS SCOTT KEY	Electric Boat	5.12.64	23.4.66	3.12.66	Deactivated 1.2.93
SSBN 658	MARIANO G VALLEJO	Mare Island N Yd	7.7.64	23.10.65	16.12.66	Deactivation FY95
SSBN 659	WILL RODGERS	Electric Boat	20.3.65	21.7.66	1.4.67	Deactivated 2.11.92

James Monroe 1972 USN

Ohio on trials 1981 USN

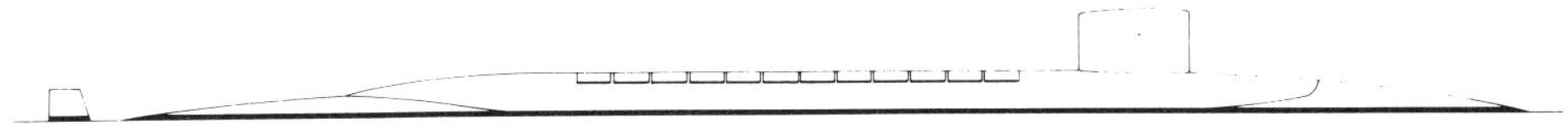

Ohio as completed

OHIO class *strategic submarines*

Displacement: 16,000t surfaced; 18,700t submerged
Dimensions: 560ft × 42ft × 35ft 6in
170.7m × 12.8m × 10.8m
Machinery: 1-shaft nuclear: 1 S8G reactor, 35,000shp = 25kts
Armament: 24 SLBM, 4–21in TT (see notes)
Sensors: Sonar BQQ-6
Complement: 133

No	Name	Builder	Laid down	Launched	Comp	Fate
SSBN 726	OHIO	Electric Boat	10.4.76	7.4.79	11.11.81	Extant 1995
SSBN 727	MICHIGAN	Electric Boat	4.4.77	26.4.80	11.9.82	Extant 1995
SSBN 728	FLORIDA	Electric Boat	9.6.77	14.11.81	18.6.83	Extant 1995
SSBN 729	GEORGIA	Electric Boat	7.4.79	6.11.82	11.2.84	Extant 1995
SSBN 730	HENRY M JACKSON (ex-*Rhode Island*)	Electric Boat	19.1.81	15.10.83	6.10.84	Extant 1995
SSBN 731	ALABAMA	Electric Boat	27.8.81	19.5.84	25.5.85	Extant 1995
SSBN 732	ALASKA	Electric Boat	9.3.83	12.1.85	25.1.86	Extant 1995
SSBN 733	NEVADA	Electric Boat	8.8.83	14.9.85	16.8.86	Extant 1995
SSBN 734	TENNESSEE	Electric Boat	9.6.86	13.2.86	17.12.88	Extant 1995
SSBN 735	PENNSYLVANIA	Electric Boat	2.3.87	23.4.88	9.9.89	Extant 1995
SSBN 736	WEST VIRGINIA	Electric Boat	18.12.87	14.10.89	20.10.90	Extant 1995
SSBN 737	KENTUCKY	Electric Boat		11.8.90	13.7.91	Extant 1995
SSBN 738	MARYLAND	Electric Boat		10.8.91	13.6.92	Extant 1995
SSBN 739	NEBRASKA	Electric Boat		15.8.92	10.7.93	Extant 1995
SSBN 740	RHODE ISLAND	Electric Boat		17.7.93	15.7.94	Extant 1995
SSBN 741	MAINE	Electric Boat		16.7.94	Aug 1995	Completing
SSBN 742	WYOMING	Electric Boat		Jul 1995	Aug 1996	Completing
SSBN 743	LOUISIANA	Electric Boat		Jul 1996	Aug 1997	Building

These are by far the largest US submarines, although they are overshadowed by the Soviet *Typhoons*. They differ from earlier US strategic submarines in having twenty-four rather than sixteen missiles, and incorporate a bow sonar (BQQ-6, similar to BQQ-5 but without active elements) and torpedo tubes set back from the bow, as in the newer attack submarines. The S8G reactor is a natural circulation type, related to the S5G in the *Narwhal*. They are also reported to have turbo-electric drive, which would be another quieting feature. *Ohio* class submarines operate seventy-day patrols, with twenty-five-day refit periods between patrols (shortened by the provision of special access trunks into the hull), with a twelve-month refit every nine years, for an overall force availability of 66 per cent, compared to 55 per cent for Poseidon submarines. The entire programme has suffered from severe delays. *Ohio* did not run trials until June 1981, although delivery was originally scheduled for December 1977. SSBN 726–733 will probably be retro-fitted with the D-5 missile (introduced in SSBN 734). Later units have modified BQQ-5 (without active elements). Reported diving depth is 300m (about 1000ft).

Spinax 27.9.56 USN

Fleet submarine conversions *radar-pickets*

Displacement: Migraine I: 1525t standard; 2410t submerged
Migraine II: 1570t standard; 2410t submerged
Migraine III: 1700t standard; 2308t normal
Dimensions: 312ft (Migraine III: 341ft) oa × 27ft 4in
95.1m (104.1m) × 8.3m
Machinery: 2 shafts, diesels plus electric motors, 4610bhp/1750shp = 17kts/18kts. Range 12,000nm at 10kts
Armament: 1–40mm (not in Migraine III), 4–21in TT (6 in Migraine III; all bow)
Sensors: Migraine I: Radar SS, SV-1, SV-2, BPS-2, YE-2
Migraine II: Radar SS, SV, SV-2, SR-2, YE-3
Migraine III: Radar BPS-2, BPS-3, BPS-4
Complement: Migraine I: 82–90
Migraine II: 82–85
Migraine III: 95–108

For names of boats converted see under Fleet Strength 1947. Original particulars, including building dates, can be found in the 1922–1946 volume.

A total of ten fleet submarines were converted to radar-pickets under three phases of the 'Migraine' programme. In the two prototypes, *Requin* and *Spinax*, the after torpedo room was used as a radar room and CIC, radars being mounted on the hull casing aft and so usable only when the submarine was surfaced. In Migraine I (*Tigrone* and *Burrfish*, FY48–49), the former crew's mess and galley were used as a CIC, and the after torpedo room used for berthing; two of the forward torpedo tubes were also removed, and the battery space reduced by use of GUPPY-type batteries. In Migraine II (a reconstruction of the two original pickets) the main air search radar (as in Migraine I) was raised well above deck, so that it could operate with the submarine awash. A snorkel was fitted. Finally, in Migraine III (FY51–53) six *Gato* class fleet submarines (SS267, 269–272, 274) were cut in half and a new 24ft (7.3m) section added between the forward battery compartment and the control room for a CIC and other electronics; the after torpedo room was converted for berthing space. A streamlined fin with a snorkel was fitted.

All these pickets reverted to conventional submarine status after 1959.

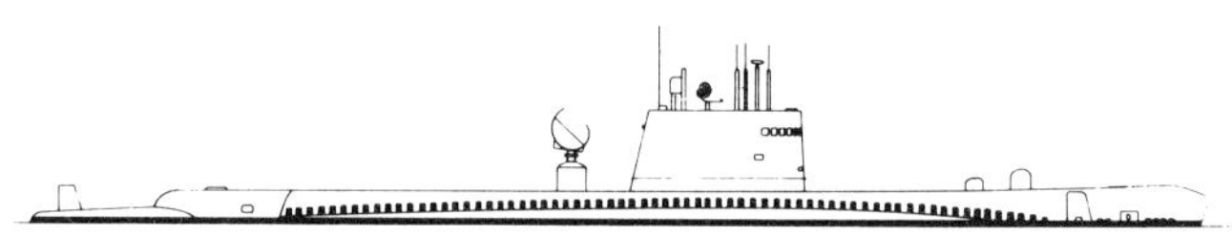

Sailfish 1958

SAILFISH class *radar-pickets*

Displacement: 1900t standard; 318t submerged (surface displacement given as 2625t in 1978)
Dimensions: 350ft 6in oa × 29ft 1in × 16ft 4in
106.9m × 8.9m × 5.0m
Machinery: 2 shafts, 4 FM diesels plus 2 electric motors, 6000bhp/8200shp = 19.5kts/10kts (given as 14kts after removal of radar equipment)
Armament: 6–21in TT (bow)
Sensors: Radar BPS-2, BPS-3; sonar BQG-4 installed after deletion of radar-picket equipment)
Complement: 95

No	Name	Builder	Laid down	Launched	Comp	Fate
SSR 572	SAILFISH	Portsmouth N Yd	8.12.53	7.9.55	May 1956	Stricken 30.9.78
SSR 573	SALMON	Portsmouth N Yd	10.3.54	25.2.56	Dec 1956	Stricken 1.10.77

These large diesel-electric submarines were built to supplement the fleet boat conversions of the 'Migraine' programme, they remained unsatisfactory in that their surface speed did not suffice to keep up with a carrier task force. Throughout the early 1950s the Bureau of Ships continued to work on more efficient powerplants for radar-picket submarines, pressure-fired steam plants showing particular promise.

Both were reclassified as attack submarines on 1 March 1961 and later given FRAM refits. *Salmon* was temporarily reclassified as an auxiliary, as test and evaluation submarine for the DSRV programme (June 1968), but reverted to attack submarine status when that programme was delayed. Both were ultimately fitted with PUFFS passive fire control sonar, and both were stricken under the FY78 programme.

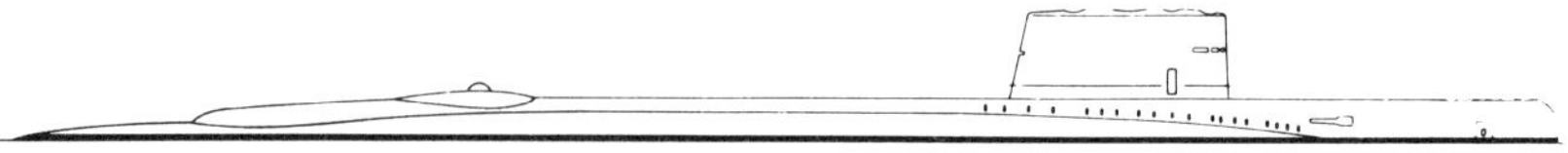

Triton as completed

TRITON *radar-picket*

Displacement: 5963t surfaced; 7773t submerged
Dimensions: 445ft wl, 447ft 6in oa × 36ft 11in × 23ft 6in
135.7m, 136.4m × 11.3m × 7.2m
Machinery: 2-shaft nuclear: 2 S4G reactors, approx 34,000shp = 27kts surfaced, 20kts submerged
Armament: 6–21in TT (4 bow, 2 stern)
Sensors: Radar BPS-2, SPS-26
Complement: 180

No	Name	Builder	Laid down	Launched	Comp	Fate
SSRN 586	TRITON	Electric Boat	29.5.56	17.8.58	10.11.59	Decomm 3.5.69

The construction of this very large nuclear submarine seemed to solve the problem of low surface speed in radar-pickets: with her two-reactor plant, she had a reported speed of 27kts on the surface (but only 20kts submerged). Design features included a very large fin into which air search antennas could retract, and a large CIC for air control. She was also the first US submarine with a three-deck hull. Commissioned in 1959, she was reclassified as an attack submarine on 1 March 1961, and decommissioned 3 May 1969 as too expensive to operate and too unwieldly to be an effective attack submarine. Proposals to use her as an alternative National Emergency Command Post Afloat (as in the cruiser *Northampton*) were abandoned; she was the first nuclear-powered warship to be laid up.

AMPHIBIOUS WARFARE VESSELS

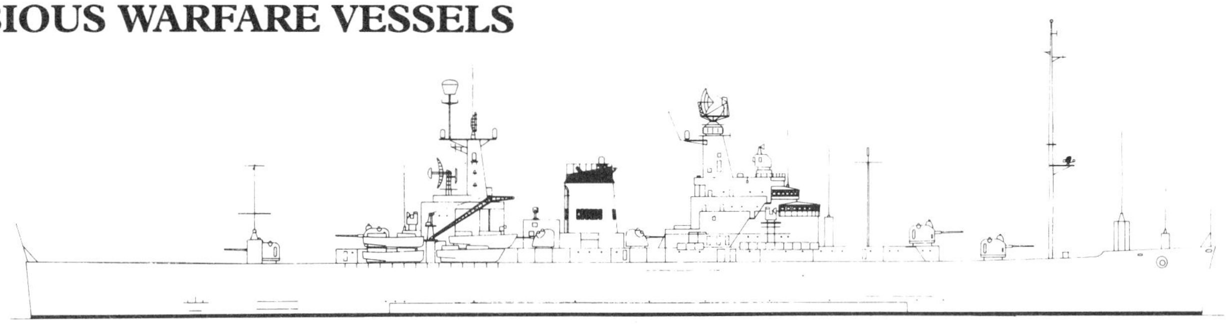

Northampton 1961

NORTHAMPTON *command ship*

Displacement: 12,429t light; 17,049t full load (1964)
Dimensions: 664ft wl, 677ft 2in oa × 69ft 9in × 24ft
202.4m, 206.5m × 21.3m × 7.3m
Machinery: 4-shaft, 4 GE geared turbines, 4 Babcock & Wilcox boilers, 120,000shp = 32.8kts. Range 7000nm at 20kts. Oil 2969t
Armament: 4–5in/54 (4×1), 8–3in/50 (4×2; later 3in/70)
Sensors: Radar SPS-2, SPS-3, SPS-6, SPS-8
Complement: 1635

No	Name	Builder	Launched	Fate
CLC 1	NORTHAMPTON	Bethlehem, Quincy	27.1.51	Stricken 1.12.77

The US Navy first experimented with specialised command and control ships during the Second World War, when it converted a variety of merchant and Coast Guard hulls to Amphibious Command Ships or AGC. In 1946 the incomplete heavy cruiser *Northampton* (laid down on 31 August 1944) was designated as a future AGC; she would have the speed and survivability lacking in the wartime ships. However, as her design progressed her function shifted more and more to that of flagship of a fast carrier force. This transition was only emphasised as the design of the new carriers (*United States* and then *Forrestal*) seemed to show no great capacity for the long-range air search radars then in prospect. That is, the fleet flagship would be valuable because she would be the ship with the long-range radar and therefore the ship controlling Fleet Air Defence. The *Northampton* was, then, designed around the requirements of the largest sea-based radar then in prospect, the massive SPS-2, and a very large Combat Information Centre (CIC) by means of which the Fleet's air defence resources could be concentrated and controlled. Even before she could be completed, proposals were made to complete the suspended battle cruiser *Hawaii* on similar lines, and they were abandoned only in 1954, and then on grounds of cost. Meanwhile the advent of the angled flight deck permitted the construction of conventional islands for the new generation of carriers, and the unique qualities of the radar-equipped fleet command ship lost some of their appeal.

Northampton, completed on 7 March 1953, thus remained unique in the US Fleet. However, the need for specialised fleet flagships with command and control facilities beyond those available in a carrier is reflected in the decision to complete three of the six *Cleveland* class missile cruisers with enlarged command spaces, at the expense of one triple 6in turret and two twin 5in/38. One might also note the longevity of the flagship missile cruisers, as compared to their non-flagship half-sisters, which (from a missile point of view) had identical combat capabilities. Indeed, as the converted flagships neared the end of their service lives, efforts were made to design a new fleet flagship. It was argued that the numbered fleets (*eg* the Sixth in the Mediterranean and the Seventh in the Far East) required more elaborate flag facilities than could be accommodated aboard a carrier. The issue was no longer fleet air defence, which could be handled quite well enough, particularly as the primary radar platforms were now carrier aircraft. Rather, it was strategic planning at the highest levels: it was argued that, after a Soviet nuclear strike, commanders afloat might well be the most senior surviving national authorities, and thus responsible for retaliation. Nuclear strike planning in itself could be justified. At about the same time the increasing age of the amphibious flagships (AGC, late LCC) seemed to demand the construction of new ones, and there were proposals to amalgamate the two tasks. However, the fleet flag role demanded a high enough speed to remain with the fleet at sea, whereas the amphibious force was still designed for 20kts.

In fact only the amphibious flagships were built, as *Blue Ridge* and *Mount Whitney*. *Northampton* decommissioned in 1970.

Northampton 1963 USN

BLUE RIDGE class *command ships*

Displacement: 19,290t full load
Dimensions: 620ft oa × 82ft wl, 108ft ext × 27ft
189.0m × 25.0m, 32.9m × 8.2m
Machinery: 1 shaft General Electric geared turbines, 2 Foster Wheeler boilers, 22,000shp = 20kts
Armament: 2 Sea Sparrow SAM (2×8), 4–3in/50 (2×2)
Sensors: Radar SPS-40, SPS-48 (plus extensive communications arrays)
Complement: 720 (plus 700 flag)

No	Name	Builder	Launched	Fate
AGC 19	BLUE RIDGE	Philadelphia N Yd	4.1.69	Extant 1995
AGC 20	MOUNT WHITNEY	Newport News	8.1.70	Extant 1995

Blue Ridge Oct 1986 RAN

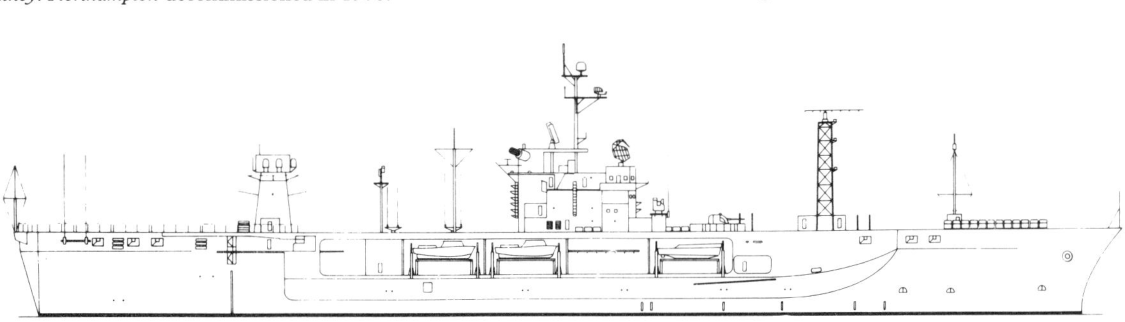

Blue Ridge as completed

These two ships, survivors of a planned class of six, are direct successors to the merchant hull conversions of the Second World War. Their design was determined by a combination of internal volume requirements (for accommodation and for command spaces) and communications requirements (for antenna performance). It is based on, but by no means identical with, the earlier helicopter carrier (LPH) design, with an additional deck level and with an entirely new (full width) superstructure. Special amphibious warfare features included extra control spaces for the amphibious commander. To some extent the decision not to build further LCCs was justified on the ground that the new LHA incorporated considerable command features on its own. In 1982 the two LCCs were being used as fleet flagships, replacing the cruisers formerly employed in that role. *Blue Ridge* replaced the cruiser *Oklahoma City* (Seventh Fleet) in October 1979. *Mount Whitney* is now flagship of Second Fleet, but also carries Commander Amphibious Group Two and Commander Fourth Marine Amphibious Brigade. In contrast to the helicopter carriers, these ships never carried their full designed battery of four twin 3in/50; only two were installed, and in 1974 a pair of octuple Sea Sparrows were added. The 3in guns and Sea Sparrows were all removed in 1992.

WRIGHT and ARLINGTON *command ships/ communications ships*

Displacement: 14,500t standard; 19,600t full load
Dimensions: 664ft wl, 683ft 6in oa × 79ft 9in × 28ft
202.4m, 208.4m × 23.6m × 8.5m
Machinery: 4-shaft geared turbines, 4 boilers, 120,000shp = 33kts
Armament: 8–40mm (3in/50in AGMR 2; 4×2), 6 helicopters (CC 2 only)
Sensors: Radar SPS-6B
Complement: 1317 (ship) plus 522 (flag; CC 2 only)

In 1961 the threat of Soviet ballistic missiles seemed increasingly acute, and it appeared that no fixed US national command centre would be secure in wartime. The Navy proposed the use of surface ships off the coasts. In an emergency, the President could be helicoptered to them, and they could communicate reliably with forces ashore by means of the newly developed tropo-scatter radio, which required a large dish antenna aboard ship. *Northampton* carried out the first tests, and served in the National Emergency Command Post Afloat (NECPA) role until her decommissioning in 1970. Under the FY62 programme the light carrier *Wright* (CC2) was refitted for the same role, on the same 'antenna farm' basis as the amphibious command ships. She did retain her all-weather landing system, presumably to permit the Presidential helicopter to come aboard in any weather. Plans to convert her sister *Saipan* into a third NECPA ship were cancelled, and instead she became a major communications relay ship, *Arlington* (AGMR 2); a fourth NECPA ship was proposed but never reached the appropriation stage. Both NECPA ships were decommissioned in the spring of 1970. Presumably by that time improvements in the Soviet submarine force and, even more, in the number of Soviet submarines operational in the North Atlantic in peacetime, had reduced the value of the NECPA, and attention turned to an airborne command post NECPA. *Arlington* was stricken on 15 August 1975 and *Wright* on 31 December 1977. Proposals to use the very large radar-picket submarine *Triton* as an NECPA did not prove acceptable, perhaps in part because of the difficulty of moving personnel out to her, and the con-siderable limits on her own ability to communicate with the nuclear forces.

Inchon 8.4.93 L & L van Ginderen

IWO JIMA class *helicopter carriers*

Displacement: 10,717t light; 18,004t full load
Dimensions: 556ft wl, 602ft 4in oa × 84ft 2in × 26ft 1in
169.5m, 183.6m × 25.7m × 8.0m
Machinery: 1 shaft Westinghouse geared turbine, 2 Combustion Engineering boilers, 22,000shp = 23.5kts
Armament: 8–3in/50 (4×2). See notes
Sensors: Radar SPS-40
Complement: 667 (plus 2057 troops)

No	Name	Builder	Launched	Fate
LPH 2	IWO JIMA	Puget Sound N Yd	17.9.60	Decomm 31.7.93, stricken
LPH 3	OKINAWA	Philadelphia N Yd	14.8.61	Decomm 19.11.92, stricken for spares
LPH 7	GUADALCANAL	Philadelphia N Yd	16.3.63	Decomm 31.8.94, to be helipad in New York
LPH 9	GUAM	Philadelphia N Yd	22.8.64	Extant 1995
LPH 10	TRIPOLI	Ingalls	31.7.65	Decomm FY95
LPH 11	NEW ORLEANS	Philadelphia N Yd	3.2.68	Extant 1995
LPH 12	INCHON	Ingalls	24.5.69	To become MSC 8 (decomm for conversion 29.9.94)

These ships were the culmination of a long Marine Corps development of vertical envelopment, or helicopter assault. The basic motivation was fear of nuclear bombardment of concentrations of landing craft off relatively restricted beaches; helicopters could approach from greater distances, preserving concentration of the assault troops at the beaches while permitting dispersal of assault shipping. Probably the single greatest defect of the LPH was its in ability to operate landing craft. Thus it could not attack or support troops on the beach in rough weather, and its helicopters could not move heavy equipment over a beach. The *Iwo Jima* class helicopter carriers were designed in 1957 after studies of the conversion of alternative hulls, including some of the wartime fast battleships. In fact three *Essex* class ASW carriers (*Boxer*, *Princeton* and *Valley Forge*) were converted to become austere LPHs (LPH 4, LPH 5 and LPH 8) with cargo stowed on their former hangar decks. They retained their 5in batteries for shore bombardments, a capability not matched by the specially-built LPH (but in fact restored in the later LHA). The basic LPH design incorporated provision for conversion to an ASW helicopter carrier, and in this sense provided a full replacement for the CVS conversions. However, unlike the *Essex* class CVS, the *Iwo Jima*s could not operate fixed-wing aircraft, since the design never incorporated either a catapult or arresting gear. Note that, at the time of design, there was no question of retiring the *Essex* class CVS.

The first special helicopter assault ship was the former escort carrier *Thetis*

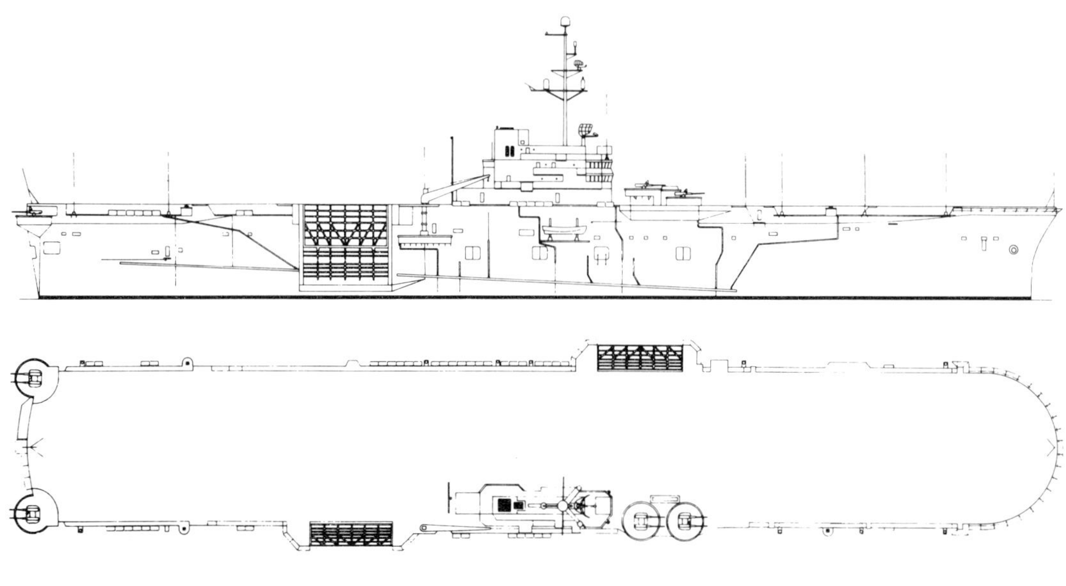

Iwo Jima 1963

Bay (see under Fleet Strength 1947), converted at San Francisco between 1 June 1955 and 1 September 1956, and designated CVHA 1 before being redesignated LPH 2. A second CVE conversion under the FY57 programme, the *Block Island*, was cancelled; she would have been LPH 2. Both the CVE and the CVS conversions were deficient in having insufficient troop spaces and so were not considered true LPHs by the Marines. Even when troops could be accommodated, the subdivision of typical warships made that accommodation less than suitable. For example, the former fleet carrier *Princeton* had Marine spaces ranging in capacity from four to 157 men, and such division would destroy unit cohesion.

The *Iwo Jimas* solved this problem. Their troop spaces, accommodating 2000 troops, were air conditioned, and there were special vehicle stowage areas and special cargo facilities. Flight decks were 'spotted' to accommodate seven CH-46 or four CH-53 helicopters for take-off; the hangar can accommodate, respectively, 19 and 11 of these helicopters. When helicopters are heavily laden, they benefit from take-off runs, so that effective operation limits the number of helicopter 'spots'. *Inchon* (LPH 12) was unique in having davits to accommodate two LCVP landing craft.

In 1971 *Guam* was modified as a prototype Sea Control Ship; she was later refitted as an LPH. *Inchon* operated minesweeping helicopters in Operation Endsweep, the clearance of the ports of North Vietnam in 1973. *Tripoli* was mine countermeasures HQ ship during the Gulf conflict; was mined 18 February 1991. Other units also served in the capacity of countermine mothership. Between 1970 and 1974 all had two of their four twin 3in/50 mounts replaced by Sea Sparrow (BPDMS; 2×8), one on the flight deck forward and one on the port quarter. Each was fitted with two Phalanx CIWS and, later, with a pair of 25mm Bushmasters. The 3in guns were all removed. In 1994 these ships had the last Basic Point Defense Missile Systems in the US Navy, and it seems unlikely that the system will long be retained on board the sole survivor, *New Orleans*.

A modified version of this design was offered to Australia as a replacement for the carrier *Melbourne*; it was considered the most likely candidate, in a competition which included the Spanish (formerly US) Sea Control Ship *Principe de Asturias* and the Italian *Garibaldi*. The original LPH steam turbine plant was to have been replaced by a pair of LM2500 gas turbines of twice the power. This plan was dropped when Britain offered *Invincible*. However, with the change in British naval policy following the Falklands War, the British offer was withdrawn, and as of this writing it remains to be seen whether the Australian competition is reopened.

Inchon (LPH 12) is to be converted into a mine countermeasures support ship (MCS 8); a second conversion planned for FY96 was dropped, as was a plan to acquire a flo-flo merchant ship to carry mine countermeasure craft quickly to combat zones far overseas. As converted, *Inchon* will carry eight MH-53E sweeper/hunter helicopters (hangar capacity is eleven) as well as an EOD (explosive ordnance disposal) team, and will be able to support four *Avenger* or *Osprey* class sweeper/hunters, fuelling them astern while underway or alongside when moored. Perhaps her most important feature will be the ability to process data collected by the helicopters' new AQS-20 minehunting sonars. In theory, the helicopters can quickly survey a potential minefield, leaving decisions as to whether a suspicious object is or is not a mine for the data processors on board the carrier or for individual sweeper/hunters. Other helicopters may use anti-mine torpedoes currently (1994) under development to attack objects believed to be mines. By way of contrast, a surface hunter must stop well short of a suspicious object, examine it, and then neutralise it before moving on. The carrier's new Mine Countermeasures Coordination Center will also be able to process the electronic minehunting charts prepared by the surface units as they work through the field. Conversion entails removal of the ship's BPDMS missile launchers and 3in/50 guns, leaving only Phalanx for self-defence (in theory the ship ought to be able to remain well beyond the horizon of shore defences.

TARAWA class *assault ships*

Displacement: 25,588t light; 38,761t full load
Dimensions: 778ft wl, 820ft oa × 106ft 8in wl, 126ft ext × 25ft 9in
237.2m, 250.0m × 32.5m, 38.4m × 7.8m
Machinery: 2 shaft Westinghouse geared turbines, 2 Combustion Engineering boilers, 70,000shp = 24kts.
Range 10,000nm at 20kts
Armament: 2 Sea Sparrow SAM (2×8), 3–5in/54 (3×1), 6–20mm (6×1). See notes
Sensors: Radar SPS-40, SPS-52, SPG-60, SPQ-9
Complement: 892 (plus 1903 troops)

No	Name	Builder	Launched	Fate
LHA 1	TARAWA	Litton	1.12.73	Extant 1995
LHA 2	SAIPAN	Litton	18.7.74	Extant 1995
LHA 3	BELLEAU WOOD	Litton	11.4.77	Extant 1995
LHA 4	NASSAU	Litton	21.1.78	Extant 1995
LHA 5	PELELIU (ex-*Da Nang*)	Litton	25.11.78	Extant 1995

Nassau Sep 1982 USN

These very large assault ships are the final evolution of the helicopter assault concept, combining the functions of LPH, LPD, and LKA. They have both a very large flight deck (with spots for twelve CH-46 Sea Knights or nine CH-53 Sea Stallions), a full-length hangar deck (for, respectively, thirty and nineteen such craft), space for a reinforced marine battalion, considerable vehicle and

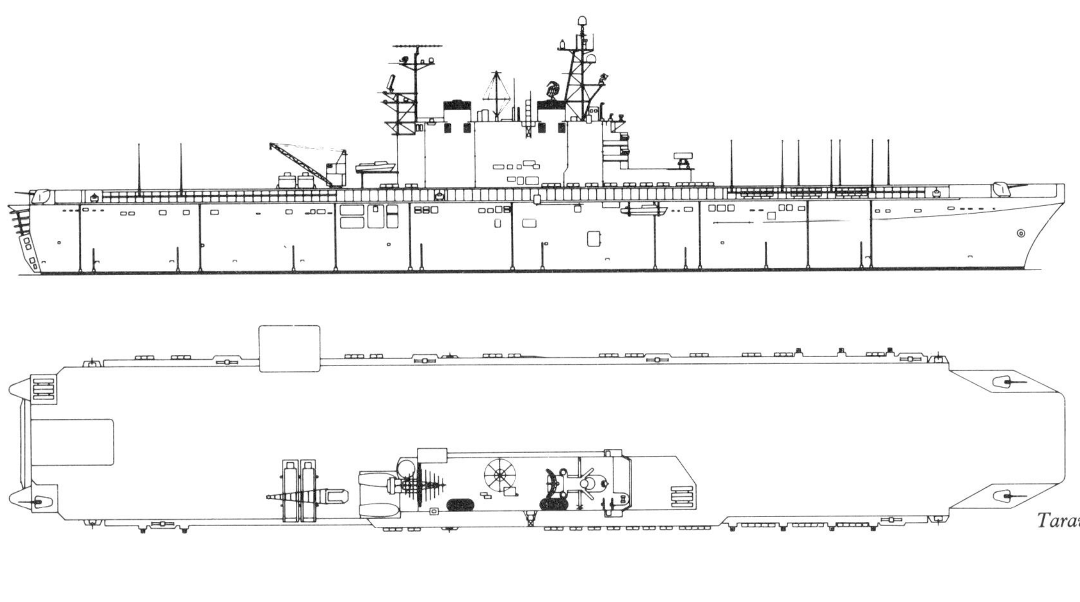

Tarawa as completed **NB 1/1750 scale**

cargo stowage, and a 268ft × 78ft (81.7m × 23.8m) well deck which can accommodate small craft up to LCU size. There is a complex automated cargo-handling system employing conveyors and elevators, and even a 5000sq ft training and acclimatisation room permitting troops to exercise in a controlled environment, so that they can adapt to the climate of their target area en route. Perhaps as importantly, they incorporate substantial command and control facilities, which make up for the shortfall in dedicated amphibious flagships (LCCs). Externally, the emblem of these facilities is the extensive electronic array above the island, including an austere three-dimensional radar (SPS-52) for limited fighter control in the operational area.

Conceived in 1965, the *Tarawa*s were ordered under the Concept Definition Contract Formulation procedure which also produced the *Spruance* class destroyers and the abortive Forward Deployment Ships (FDL). Nine were originally planned, but on 20 January 1971 the Navy announced that LHA 6–LHA 9 would not be built. From time to time the LHA hull, which is about the same size as a Second World War *Essex* class fleet carrier, is proposed as the basis for a limited or VSTOL carrier, and in fact at one point Litton reportedly made such a proposal to the Australian government; later it was scaled down to an LPH derivative. There have also been Congressional attempts to convert an existing LHA to a light carrier, with an additional unit being built to replace it.

Typical small craft capacity is four *LCU 1610* class or two LCU and three LCM-8 or 17 LCM-6, plus forty tracked landing craft (LVT) on the vehicle deck.

LHA 5 was originally to have been named *Da Nang*, but was renamed *Peleliu* on 15 February 1978, presumably in view of popular feelings about the Vietnam War and particularly about its conclusion. Sea Sparrow (BPDMS version) was replaced by Phalanx, and by 1994 the after 5in mount had been removed. All are to have RAM (two launchers each, to port atop the pilothouse and to starboard at the aft end of the flight deck), beginning with LHA 5 (November 1992).

WASP class *assault ships*

Displacement: 28,233t light; 40,532t full load
Dimensions: 844ft × 140ft × 26ft 4in
257.3m × 42.67m × 8.13m
Machinery: 2 sets geared turbines, two boilers, 77,000shp = 24kts. Range 9500nm/20kts
Armament: 2 Sea Sparrow launchers, 3 Phalanx; 30–32 CH-46 helicopters and 6 AV-8B Harriers
Sensors: SPS-49, SPS-52C (SPS-48E in LHD 2 and later), SPN-35, SPN-43
Complement: 1080 (plus 1873 troops)

No	Name	Builder	Launched	Fate
LHD 1	WASP	Ingalls	4.8.87	Extant 1995
LHD 2	ESSEX	Ingalls	4.1.91	Extant 1995
LHD 3	KEARSAGE	Ingalls	26.3.92	Extant 1995
LHD 4	BOXER	Ingalls	13.8.93	Extant 1995
LHD 5	BATAAN	Ingalls	Mar 1996	Building
LHD 6	BONHOMME RICHARD	–	–	–

This is an LHA derivative capable of handling the new air-cushion landing craft. Construction was originally planned for the FY87 programme, but it was moved up to FY84 to fill a perceived gap in Marine assault shipping. LHDs beyond the first five or six were intended to replace the *Iwo Jima*s in the 1990s, so that by 1983 a total of about twelve was envisaged. Currently (1992) the *Iwo Jima* (and *Newport* class LST) replacement is expected to be an entirely new design, LX.

Unlike the original LHA, the LHD was conceived with a secondary sea-control role; they therefore incorporate maintenance facilities for at least twenty Harriers and for four to six ASW helicopters. The flight deck is so long that no ski-jump is required (were one provided, it would interfere with rolling helicopter take-offs). Quite aside from sea control, Marine Harriers might provide very useful support in an amphibious operation. This possibility was demonstrated for the first time in the Gulf conflict, when *Nassau* (an LHA) flew off Harriers to attack Iraqi positions in Kuwait.

LHD 2–4 were ordered as a three-ship option. A similar option was planned for LHD 5–7, and there was some consideration of substituting four LM 2500 (100,000shp) for the existing steam plant. The 1992 budget initially pushed LHD 6 from FY93 out to FY96, but in the end she was restored to the FY93 programme (some funding came from the FY94 budget). At one time eleven were planned. The navy has an option for a seventh ship, and included LHD 7 in its projected FY99 budget. However, advance procurement funds are in the FY95 budget, and the ship may be included in the FY96 programme. The name *Hornet* has been suggested.

Wasp 20.7.93 L & L van Ginderen

'LX' class *assault ship* (LPD 17)

Displacement: 23,000–25,000t full load
Dimensions: 699ft × 103ft
213.05m × 31.4m
Machinery: 4 diesels (2 shafts)
Armament: SSDS system (2 RAM launchers, 2 Phalanx)
Sensors: Radar SPS-48E, SPQ-9B; ECM SLQ-32(V)3, torpedo decoy
Complement: 495 (plus 840–930 troops)

Conceived as a replacement for existing LKA, LPD, LSD 36, and LST classes, L(X) is now tentatively designated LPD 17. The programme was delayed two years in July 1994 and as of late 1994 the lead ship was planned for the FY98 budget; twelve ships are planned (two per year from FY99 on). The well deck will accommodate two LCAC or four LCM(8) or nine LCM(6) or twenty LVT, and there will be two helicopter spots (CH-53 size; alternatively, six CH-46 spots). There will not be any hangar.

L(X) was conceived just as interest in ship self-defence became acute. Hence the elaborate defensive suite. The design also shows space and weight reserved for a 16-cell vertical launcher for the SSDS missile (probably Evolved Sea Sparrow); presumbably there will also be target illuminators for these weapons. The main search radar may also be SPS-49(V)5, but -48E seems likely, since so many such radars have been salvaged from stricken missile cruisers. The hull has been shaped for minimum radar cross section. Reportedly the defensive system will add about $200m to the cost of each ship, so it may well be deleted (estimated cost in FY94 dollars is $850m). Note that this class carries about the same payload as an LPD on a hull twice as large.

LX is currently (1992) planned to replace both the *Iwo Jima*s and the *Newport* class LSTs. The first is scheduled for the FY96 budget. It will displace 13,000t to 15,000t, will be about 600ft long, and will have an LPD-4 size well deck, two helicopter spots, and space for 700 troops.

THOMASTON class *dock landing ships*

Displacement: 6880t light; 11,270t full load (LSD 32–LSD 34, 12,150t)
Dimensions: 510ft oa × 84ft × 19ft
155.5m × 25.6m × 5.8m
Machinery: 2 shaft GE steam turbines, 2 Babcock & Wilcox boilers, 24,000shp = 22.5kts. Range 10,000nm at 20kts
Armament: 16–3in/50 (8×2), 12–20mm (designed). See notes
Sensors: Radar SPS-6
Complement: 404 (plus 341 troops)

Spiegel Grove 1968

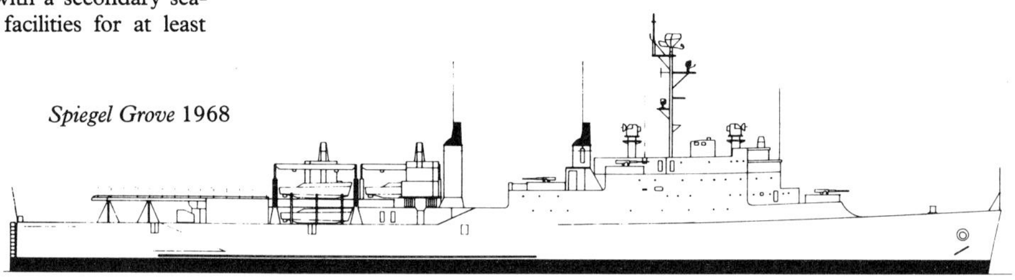

Class (no, launched):
Thomaston (LSD 28, 9.2.54), *Plymouth Rock* (LSD 29, 7.5.54), *Fort Snelling* (LSD 30, 16.7.54), *Point Defiance* (LSD 31, 28.9.54), *Spiegel Grove* (LSD 32, 10.11.55), *Alamo* (LSD 33, 20.1.56), *Hermitage* (LSD 34, 17.6.56), *Monticello* (LSD 35, 10.8.56). All built by Ingalls.

These postwar LSDs were intended to provide a heavy landing craft capacity within the 20kts amphibious force; actually they could make 24kts. Their hull form was more refined, their well decks had a greater depth of water, their machinery (as in an LPD) was located entirely below the well deck rather than in the side walls, and they had mezzanine and super-decks (both portable) for cargo stowage and for helicopter operation. The 391ft × 48ft (119.2m × 14.6m) well deck accommodates three LCU or nine LCM-8 or fifty LVT. The helicopter deck is removable and there are no maintenance facilities; two 50t cranes can reach down to a mezzanine (vehicle stowage) or the well deck. Bows were strengthened for ice navigation, and the last two units were air conditioned. They can carry about 340 troops.

The *Thomaston*s were part of a general revitalisation of the amphibious force in the 1950s, traceable to Korean War experience: four were built under the

Point Defiance 31.5.55 L & L van Ginderen

FY52 programme, two under FY54 and two under FY55. Defensive armament was reduced to six twin 3in/50 during the 1960s and three twin 3in/50 during the 1970s.

Fates: *Thomaston* to MarAd 28.10.91, *Plymouth Rock* to MarAd 8.11.91, *Fort Snelling* to MarAd 7.9.89, *Point Defiance* to MarAd 12.8.91, *Spiegel Grove* to reserve 2.10.89 (stricken Feb 1994), *Alamo* decommissioned and leased to Brazil 2.11.90, *Monticello* to MarAd 2.8.91. Ships in MarAd custody were declared surplus (eg transferable) 24.2.92.

POINT BARROW

One somewhat similar ship requires mention, the T-AKD (cargo ship dock) *Point Barrow*, in effect an unarmed LSD built to a Maritime Administration design (S2-ST-23A) specifically to support US radar stations in the Arctic. She operated with the Military Sea Transport Service (MSTS) until 1965, and then was used until 1970 to carry Saturn rockets and other space equipment from California to Cape Kennedy; she also carried some landing craft to Vietnam. After being laid up in 1971–72, she was returned to general cargo carrying, then again refitted 1974–76 to support the research submersible *Trieste* as AGDS 2, renamed *Point Loma*. The Military Sealift Command (successor to MSTS) also operates a variety of ships which, although unarmed in peacetime, could be used as assault transports in an emergency.

RALEIGH class *dock landing ships*

Displacement:	8051t light; 13,745t full load
Dimensions:	500ft wl, 513ft 1in × 84ft × 20ft 6in *152.4m, 156.4m × 25.6m × 6.3m*
Machinery:	2 shaft De Laval geared turbines, 2 Babcock & Wilcox boilers, 24,000shp = 21kts
Armament:	3in/50 (3×2)
Sensors:	Radar SPS-40
Complement:	501 (plus 932 troops)

Class (no, launched):

Raleigh (LPD 1, 17.3.62), *Vancouver* (LPD 2, 15.9.62), *La Salle* (LPD 3, 3.8.63). All built by New York N Yd, Only *Vancouver* survived as an LPD in 1994. *Raleigh* decommissioned 23 July 1992. *La Salle* (AGF 3) became 6th Fleet flagship 1994.

AUSTIN class *dock landing ships*

Displacement:	10,000t light; 16,900t full load
Dimensions:	570ft oa × 84ft × 23ft *173.8m × 25.6m × 7.0m*
Machinery:	2 shaft De Laval geared turbines, 2 Babcock & Wilcox boilers, 24,000shp = 20kts
Armament:	8–3in/50 (4×2). See notes
Sensors:	Radar SPS-40
Complement:	410–447 (plus 930 troops in LPD 4–6, 840 in others)

Class (no, launched):

New York N Yd – *Austin* (LPD 4, 27.6.64), *Ogden* (LPD 5, 27.6.64), *Duluth* (LPD 6, 14.8.65)

Ingalls – *Cleveland* (LPD 7, 7.5.66), *Dubuque* (LPD 8, 6.8.66)

Lockheed, Seattle – *Denver* (LPD 9, 23.1.65), *Juneau* (LPD 10, 12.2.66), *Coronado* (LPD 11, 30.7.66), *Shreveport* (LPD 12, 25.10.66), *Nashville* (LPD 13, 7.10.67), *Trenton* (LPD 14, 3.8.68), *Ponce* (LPD 15, 20.5.70)

These ships were intended as successors to the Landing Ship Dock with much increased troop (930, 840 in flagships) and cargo (about 2500t) capacity and a smaller well deck (168ft × 50ft (51.2m × 15.2m), accommodating one LCU and three LCM-6 or four LCM-8 or twenty LVT; in addition there is a large crane at the after end of the superstructure which can launch craft carried on the flight deck, typically two LCM-6 or four LCPL). The *Raleigh*s had no helicopter hangar or maintenance facilities, but ships after the *Austin* (LPD 4) have a hangar 58ft to 64ft (17.7m–19.5m) in length and 18ft 6in to 24ft (5.6m–7.3m) in width, extensible to about 80ft. The deck has two helicopter spots, but ships have deployed with up to six CH-46 embarked.

The two classes differ in that the *Austin*s are somewhat enlarged. *La Salle* (LPD 3) was reclassified as a flagship (AGF 3) for the Middle East Force, undergoing conversion during 1972. She received a small helicopter hangar and covered assembly area on her flight deck, as well as a lattice mast for a satellite communications antenna, and admiral's accommodation, plus additional air conditioning. She is painted white to help reduce heating in the Indian Ocean. *La Salle* replaced a succession of small seaplane tenders (AVP) which had operated as Middle East flagships from the late 1940s onwards; *Valcour* (AVP 55) was reclassified as AGF 1 on 15 December 1965 and based at Bahrein, being replaced by *La Salle* in 1972. The LPD *Coronado* served as Middle East Flagship (AGF 11 vice LPD 11) while *La Salle* underwent a major refit from 1980 to 1982.

LPD 7–LPD 13 are fitted as amphibious squadron flagships and have an extra bridge level. A proposed LPD 16, to have been built under the FY66 programme, was deferred in favour of the LHA programme then cancelled in 1969. The other LPDs were built under the FY59 (prototype), FY60 (1), FY61 (1), FY62 (3), FY63 (4), FY64 (3) and FY65 (2) programmes.

Of the original 3in battery, two were removed from the *Austin*s in 1977–78 (port forward, starboard aft), and the others in 1992–93. They were replaced by Phalanx (two each). The sole surviving *Raleigh, Vancouver*, retains 4–3in/50, but also has two Phalanx. Of the *Raleigh* class, LPD 1 was decommissioned 13 December 1991 and turned over to the Maritime Administration 23 July 1992, then reacquired for target tests. LPD 3 became a flagship (AGF 3). LPD 2 survives laid up at Pearl Harbor. All of the *Austin* class are still extant; LPD 11 became a flagship (AGF 11). A plan for SLEP, to extend the life of these ships by 10–15 years, was cancelled by Congress in 1987, but the ships are to be retained beyond the turn of the century.

ANCHORAGE class *dock landing ships*

Displacement:	8600t light; 13,700t full load
Dimensions:	553ft 4in oa × 84ft × 18ft 6in *168.7m × 25.6m × 5.6m*
Machinery:	2 shaft De Laval geared turbines, 2 boilers, 24,000shp = 20kts
Armament:	8–3in/50 (×2). See notes
Sensors:	Radar SPS-40
Complement:	397 (plus 376 troops)

Class (no, launched):

Ingalls – *Anchorage* (LSD 36, 5.5.68)

General Dynamics – *Portland* (LSD 37, 20.12.69), *Pensacola* (LSD 38, 11.7.70), *Mount Vernon* (LSD 39, 17.4.71), *Fort Fisher* (LSD 40, 22.4.72)

The construction of these ships showed, in effect, that the LPD could not quite combine all the capabilities required by the Amphibious Force. Five were built under the FY65 (one), FY66 (three) and FY67 (one) programme. Like their *Thomaston* class predecessors, they have a removable helicopter deck, no helicopter maintenance facilities, and limited troop space (375 men). The

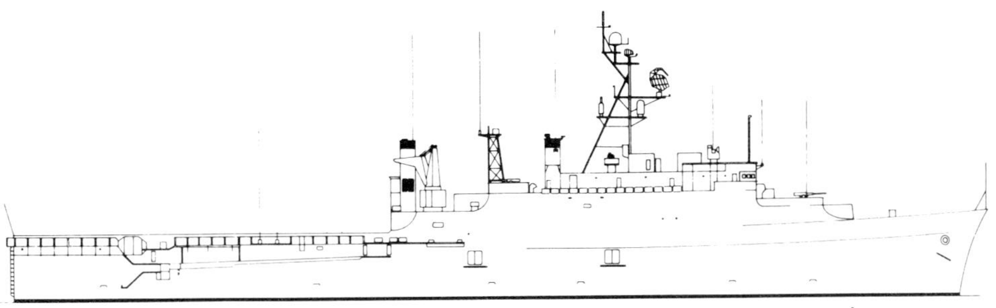

Raleigh as completed

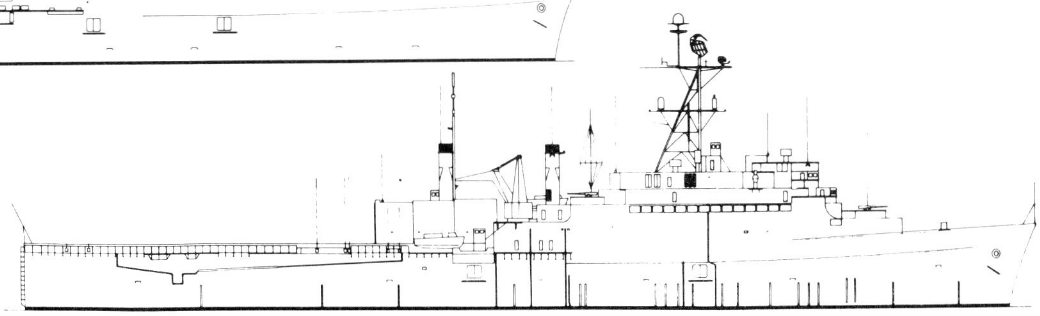

Coronado 1970

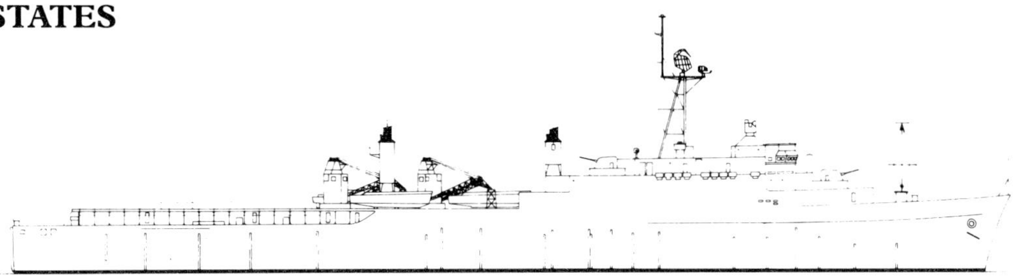

Anchorage 1969

430ft × 50ft well deck accommodates three LCU or nine LCM (8) or about fifty LVT. Two 3in mounts were removed by 1990, and the others in 1993–94; the directors had all been removed in 1977–78. They currently mount a pair of Phalanx and two 25mm Bushmasters. All were extant in 1995.

WHIDBEY ISLAND class *dock landing ships*

Displacement: 11,125t standard; 15,726t full load
Dimensions: 609ft 6in oa, 580ft wl × 84ft × 19ft 8in
185.8m, 176.8m × 25.6m × 6.0m
Machinery: 2 shafts, 4 Colt/Pielstick diesels, 41,600bhp = 22kts
Armament: 2–20mm Phalanx CIWS, 2–20mm Mk 67
Sensors: Radar SPS-67, SPS-40B, LN-66
Complement: 376 (plus 440 troops)

No	Name	Builder	Launched	Fate
LSD 41	WHIDBEY ISLAND	Lockheed, Seattle	10.6.83	Extant 1995
LSD 42	GERMANTOWN	Lockheed, Seattle	29.6.84	Extant 1995
LSD 43	FORT McHENRY	Lockheed, Seattle	1.2.86	Extant 1995
LSD 44	GUNSTON HALL	Lockheed, Seattle	27.6.87	Extant 1995
LSD 45	COMSTOCK	Lockheed, Seattle	16.1.88	Extant 1995
LSD 46	TORTUGA	Lockheed, Seattle	15.9.88	Extant 1995
LSD 47	RUSHMORE	Lockheed, Seattle	6.5.89	Extant 1995
LSD 48	ASHLAND	Lockheed, Seattle	11.11.89	Extant 1995
LSD 49	HARPERS FERRY	Lockheed, Seattle	16.1.93	Completing
LSD 50	CARTER HALL	Lockheed, Seattle	2.10.93	Completing
LSD 51	OAK HILL	Lockheed, Seattle	11.6.94	Completing
LSD 52	PEARL HARBOR	–	–	To complete Jul 1998

A new LSD 41 class was proposed to replace the eight *Thomaston*s, but the Carter Administration deferred construction until Congress voted $41m in long-lead items in FY80; the Administration then announced the construction of six ships in alternate years, beginning in FY81. LSD 41 has an enlarged well deck, 440ft × 50ft (134.1m × 15.2m), adapted to carry four air cushion landing craft (LCAC) or twenty-one conventional LCM (6) landing craft; by way of comparison, the LHD will carry only three LCAC. The first of the class was completed in 1984; she was laid down on 4 August 1981, followed by LSD 42 on 5 August 1982.

LSD 41 was conceived as a near-repeat of *Anchorage* (LSD 36) adapted to LCACs rather than LCMs. That required not only enlargement of the well deck, but also raising the helicopter deck to provide all-round ventilation for LCAC gas turbines. LSD 44–48 are further modified to incorporate collective protection against chemical or biological weapons.

LSD 49 and beyond are a modified *Whidbey Island* design, officially called the LDS 41 Cargo Variant. They were originally requested under the FY88 budget, then withdrawn due to late design changes, and then approved by Congress. The idea originated in an analysis of Marine amphibious assault requirements, which showed that more cargo space was needed, but not in the ratio to LCAC 'spots' implicit in the LSD 41 design. LSD 49 was conceived as a simple substitution of cargo volume for well deck volume (leaving space for two LCAC or ten LCM).

Plans called for a total of ten, beginning with LSD 49 in FY88, LSD 50 in FY90, then one per year after that. The 1992 budget stopped production at LSD 51, but Congress included LSD 52 in the FY92 programme.

Comstock 3.7.90 L & L van Ginderen

TERREBONNE PARISH class *tank landing ships*

Displacement: 5777t full load
Dimensions: 384ft oa × 56ft × 16ft 1in
117.1m × 17.1m × 4.9m
Machinery: 2 shafts, 4 GM diesels, 60,000bhp = 15.5kts
Armament: 6–3in/50 (3×2)
Complement: 205 (plus 391 troops)

Class (no, launched):
Bath Iron Wks – *Terrebonne Parish* (LST 1156, 9.8.52), *Terrell County* (LST 1157, 6.12.52), *Tioga County* (LST 1158, 11.4.53), *Tom Green County* (LST 1159, 2.7.53), *Traverse County* (LST 1160, 3.10.53), Ingalls – *Vernon County* (LST 1161, 25.11.52), *Wahkiakum County* (LST 1162, 23.1.53), *Waldo County* (LST 1163, 17.3.53), *Walworth County* (LST 1164, 18.5.53), *Washoe County* (LST 1165, 14.7.53), Christy – *Washtenaw County* (LST 1166, 22.11.52), *Westchester County* (LST 1167, 18.4.53, *Wexford County* (LST 1168, 28.11.53), *Whitfield County* (LST 1169, 22.8.53), *Windhem County* (LST 1170, 22.5.54).

Probably the greatest disappointment of the early postwar amphibious programme was the failure to develop a fast LST. Quite radical schemes were tried, including a design with bow propellers, which would turn its blunt stern to the beach, and one with a long false bow. Instead, a maximum speed of 15kts was accepted, and these fifteen ships were modified versions of two experimental units (LST 1153 and 1154) laid down at the end of the Second World War, with diesel plants replacing the steam engines of the earlier ships. All had increased troop accommodation, greater tank and cargo capacity and improved discharge arrangements as compared to wartime LSTs. Thus they had a powered vehicle turntable, hydraulic bow doors, heavier ramps, and pontoon causeways. Their hull lines were also finer than those of earlier LSTs, and they had reversible pitch propellers.

The entire LST 1156 class was authorised (as SCB-9) under the FT52 programme, as part of the renaissance of the amphibious force after the demonstration of its continued value during the Korean War. Fates: LST 1166 converted to mine countermeasures craft (MMS 2) 1973, used to test sweeping of Hanoi/Haiphong, then stricken in 1973. LST 1156 to Spain 1971, LST 1159 and LST 1160 to Spain 1972, LST 1168 to Spain 1971; LST 1157 and LST 1169 to Greece 1977; LST 1161 to Venezuela 1973; LST 1167 to Turkey 1974; LST 1170 to Turkey 1973. The remaining ships were stricken and transferred to the Maritime Administration after brief service as MSC cargo carriers.

Of the two very similar steam-driven *Talbot County* (LST 1153) class, launched in 1946–47, *Tallahatchie County* (LST 1154) was converted to an

Terrell County 1969 USN

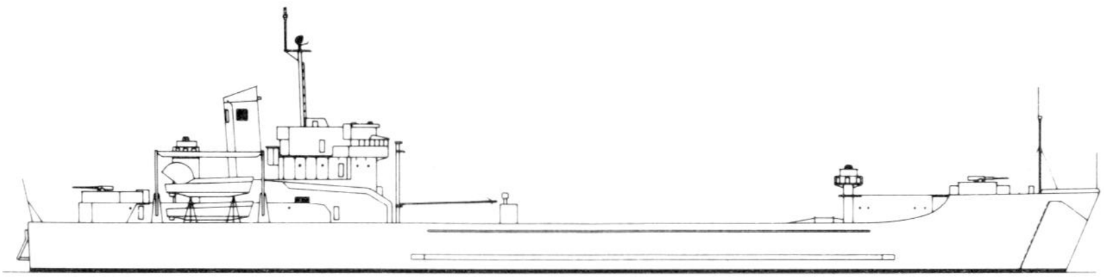

York County 1960

advance aviation base ship (AVB) at the Charleston Navy Yard under the FY60 programme (completed 13 March 1962); she was fitted with a new aluminium superstructure, shops, briefing rooms, and even a portable aircraft control tower. She could support a 270-man aircraft squadron. She was stricken in 1970 and *Talbot County* shortly afterwards.

WAR-BUILT LSTs
Of earlier LSTs still in service postwar, six were converted to small craft tenders (AGP) (LSTs *782, 786, 821, 838* and one other), although three were so designated. *LST 1069* served as a minecraft support ship (MCS 6).

DE SOTO COUNTY class *tank landing ships*

Displacement: 3859t light; 7854t full load
Dimensions: 426ft wl, 442ft oa × 62ft × 13ft 5in
129.9m, 134.8m × 18.9m × 4.1m
Machinery: 2 shafts, 6 FM diesels, 15,000bhp = 17.2kts
Armament: 6–3in/60 (3×2)
Complement: 185 (plus 634 troops)

Class (no, launched):
Avondale – *De Soto County* (LST 1171, 28.2.57), *Grant County* (LST 1174, 12.10.56)
Boston N Yd – *Suffolk County* (LST 1173, 5.9.56)
Newport News – *York County* (LST 1175, 5.3.57), *Graham County* (LST 1176, 19.9.57)
American SB – *Lorain County* (LST 1177, 22.6.57), *Wood County* (LST 1178, 14.12.57).

These SCB-119 LSTs of FY54 (prototype) and FT55 showed little speed improvement over their predecessors (17kts at more than double the power) but were designed for better sea keeping, and in particular for less pounding at high sustained speed. They were fully air conditioned, and two of them could carry as many troops as an APA plus an AKA, as well as more vehicles. They were designed to carry four landing craft in davits, a 175ft (53.4m) pontoon causeway on the side, and two powered pontoon barges and an LCU as deck cargo for side-launching. In addition, they could accept up to five helicopters landing on deck. LST 1172 was cancelled in 1955; all remaining LSTs were assigned 'County' and 'Parish' names that year.

Even at 17kts, these ships could not operate effectively with the 20kt amphibious force, and they were discarded with the advent of the fast *Newport* class LSTs. *Graham County*, scheduled for decommissioning, was instead converted to a tender (AGP-1176) for *Asheville* class gunboats in the Mediterranean at Philadelphia Navy Yard, February–May 1972. Her bow doors were welded shut, the former tank deck becoming machine shops and store rooms. Boat booms were added to her sides, and steam generators fitted to provide steam to gunboats alongside. She was stricken in 1977. *Wood County* was scheduled for a similar conversion to support the *Pegasus* class PHMs (with designator AGHS), but the conversion was deferred in 1977 when the future of the PHM programme came into doubt. The following were transferred: LST 1174 to Brazil 1973, LST 1171 and 1175 to Italy 1972; LST 1173 and 1178 to Greece 1981; LST 1177 to Mexico 1981.

NEWPORT class *tank landing ships*

Displacement: 8342t full load
Dimensions: 522ft 4in oa over hull, 562ft (over derrick arms) × 69ft 6in × 17ft 6in
159.2m, 171.3m × 21.2m × 5.3m
Machinery: 2 shafts, Alco diesels (GM diesels in 1179–1181), 16,500bhp = 20kts. Range over 2500nm (reported cruising range)
Armament: 4–3in/50 (2×2). See notes
Complement: 218 approx (plus 385 troops)

Class (no, launched):
Philadelphia N Yd – *Newport* (LST 1179, 3.2.68), *Manitowoc* (LST 1180, 4.6.69), *Sumter* (LST 1181, 13.12.69), National Steel & Shipbuilding Co – *Fresno* (LST 1182, 28.9.68, *Peoria* (LST 1183, 27.11.68), *Frederick* (LST 1184, 8.3.69), *Schenectady* (LST 1185, 24.5.69), *Cayuga* (LST 1186, 12.7.69), *Tuscaloosa* (LST 1187, 6.9.69), *Saginaw* (LST 1188, 3.2.70), *San Bernardino* (LST 1189, 28.3.70), *Boulder* (LST 1190, 22.5.70), *Racine* (LST 1191, 15.8.70), *Spartanburg County* (LST 1192, 11.11.70), *Fairfax County* (LST 1193, 19.11.70), *La Moure County* (LST 1194, 13.2.71), *Barbour County* (LST 1195, 15.5.71), *Harlan County* (LST 1196, 24.9.71), *Barnstaple County* (LST 1197, 2.10.71), *Bristol County* (LST 1198, 4.12.71).

With these ships the Amphibious Force finally attained its goal of 20kt performance throughout. The many failures to design a fast LCT showed primarily that radical measures were required; in these ships there is a normal ship-shaped bow, but the upper portion splits and a long causeway is lowered through a gallows above; the bow ramp is 112ft (31.4m) long. There is also a stern ramp which permits unloading of LVTs into the water or into a landing craft or on to a pier. Above it is a helicopter deck; there is no hangar. Capacity is 385 troops and 500 tons of cargo; vehicles can be driven down to the lower deck via a ramp, or through the superstructure on to the helicopter deck aft. A 30ft turntable at each end of the tank deck permits vehicles to turn round without having to reverse. There are davits for up to four LCVP. Two Phalanx CIWS are due to be fitted during the 1980s.

The size and speed of the *Newport* class programme suggests the urgency of new amphibious construction in the late 1960s; one analyst has suggested that amphibious craft were one programme that Secretary of Defense McNamara's men strongly supported, as an essential element of the US force projection capability. Thus the prototype was ordered in FY65, followed by eight in FY66, and then by eleven in FY67. They were completed in 1969–72.

With the advent of the LHAs, with their large well decks and vast cargo spaces, the rationale of the LSTs came into question, and as this is written the *Newports* are perhaps the least valuable of the specialised amphibious ships. They may well become obsolescent as amphibious doctrine shifts to distant assault with air-cushion vehicles (LCAC), which can be launched from beyond the horizon, and whose range and speed add considerably to the element of surprise in an amphibious assault.

By 1994 all had had their 3in guns removed (fire controls were removed in 1977–78) and all but LST 1184, 1185, 1190, 1191, and 1195 had one Phalanx. Several (including LST 1187 and 1190) were fitted with 25mm Bushmaster guns. Plans to fit SLQ-32(V)1 ESM system were abandoned, but several were fitted with chaff launchers.

These ships were decommissioned from 1992 on, beginning with *Newport* on 30 September 1992. Transfers (1994): LST 1184 to Chile; LST 1186 to Brazil; LST 1188 and 1193 to Australia; LST 1197 to Spain; LST 1198 to Morocco. Other offers: Argentina (1194), Malaysia (1192), Taiwan (1185, 1190, 1191), and Venezuela (1183, 1187). Transfers were delayed during 1994 due to Congressional resistance to loss of US amphibious capacity, and ships planned for 1995 transfer were retained (LST 1189 to Chile, LST 1196 to Spain). As of August 1994 LST 1183, 1185, 1187, 1190, and two others were to have been retained.

TULARE and PAUL REVERE class *attack transports*

Displacement: 12,000t light; 16,800t full load (*Paul Revere* 10,709/16,838t)
Dimensions: 564ft oa × 76ft × 26ft
172.0m × 23.2m × 7.9m
Machinery: 1 shaft De Laval geared turbines, 2 CE boilers (Foster Wheeler in *Paul Revere* class) 22,000shp = 22kts
Armament: 12/8–3in/50 (6×2; 4×2 in *Paul Revere* class)
Complement: 393 plus 319 troops (307 plus 1657 troops in *Paul Revere* class)

Class (no, launched):
Bethlehem, San Francisco – *Tulare* (AKA, 112, 22.12.53)
New York SB – *Paul Revere* (APA 248, 13.2.54), *Francis Marion* (APA 249, 11.4.53).

These three ships, the only new AKA and APA construction of the 1950s, were conversions of 'Mariner' cargo hulls (C4-S-1A). Of thirty-five 'Mariners' built, the Navy acquired five in all, the other two being the Polaris missile firing

Saginaw 1972

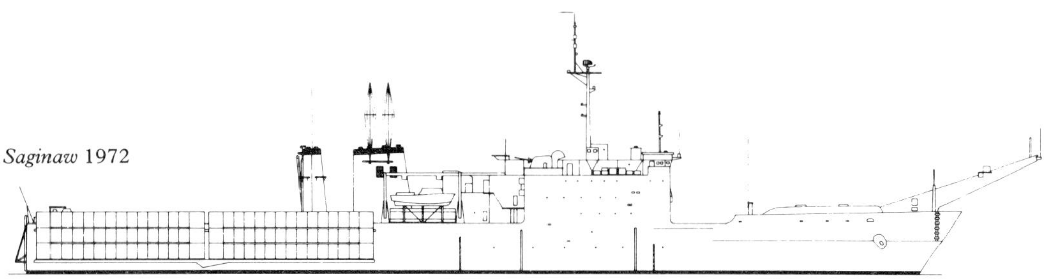

ship *Observation Island* (YAG 57, then EAG 154) and the navigational test ship *Compass Island* (EAG 133), both associated with the Polaris programme. The two attack (personnel) transports, commissioned in 1958 and 1961 respectively, were fitted as amphibious squadron flagships or as troop commander or transport division flagships; they carried seven LCM-6, twelve LCVP, and three LCP(L), and could handle these craft and heavy vehicles with 10t, 30t and 60t cargo booms. In an early example of adaptation to helicopter operation, their helicopter decks were served by cargo elevators. In fact they were the last conventional attack personnel transports, as the transport function was next taken over by a combination of LPH (helicopter carriers) and LPD (landing ship docks with additional troop spaces). They could accommodate 1500 troops each. Both were sold to Spain in 1980.

Tulare was designed primarily for cargo, although she could carry 575 troops (and 300 vehicles), handling up to twenty-seven landing craft with heavy-lift (60t) booms rigged from her quadrupod masts. Like the APAs, she had a helicopter platform. She was stricken on 1 August 1981 and became TS *Bay State* operated by Massachusetts Maritime Academy. The two APAs were both sold to Spain, on 17 January 1980 and 11 July 1980 respectively.

WAR-BUILT ATTACK TRANSPORTS

Although very extensive forces of attack transports and cargo ships were built up during the Second World War, they were largely dispersed postwar, many ships going to the Maritime Administration for lay-up and then disposal. Although some Maritime Administration reserve hulls remain, the shift in commercial shipping from break-bulk (as in an AKA) to containers sharply reduces the degree to which the merchant fleet (both reserve and active) can form a mobilisation reserve. That was not always true: both in FY52 (for Korea) and in FY62 (for Vietnam) six Maritime Reserve fleet AKAs were reacquired and reactivated. Note, however, that in FY65–66 it was necessary to design an entirely new ship as an attack (cargo) transport (*Charleston* class). The Reagan Administration bought and leased numerous modern merchant ships to form a sealift reserve. Many of these were mobilised for the 1990–91 Gulf conflict.

CHARLESTON class *assault cargo ships*

Displacement:	20,700t full load
Dimensions:	575ft 6in oa × 82ft × 25ft 6in *175.5m × 25.0m × 7.8m*
Machinery:	1 shaft Westinghouse geared turbine, 2 CE boilers, 22,000shp = 20kts
Armament:	8–3in/50 (4×2)
Complement:	334 (plus 226 troops)

Class (no, launched):
Charleston (LKA 113, 2.12.67), *Durham* (LKA 114, 29.7.68), *Mobile* (LKA 115, 19.10.68), *St Louis* (LKA 116, 4.1.69), *El Paso* (LKA 117, 17.5.69). All were built by Newport News.

The first assault cargo ships specifically designed for the role, these ships were built under the FY65 (four units) and FY66 programme. They are considerably larger than the 'Mariner' and 'Victory' class conversions they succeed. Designations were changed from AKA to LKA in 1968–69. They carry nine LCM as deck cargo, launching them with two 78t capacity booms; there are also two 40t and eight 15t cranes. There is a helicopter pad aft, but no hangar. They carry 226 troops. One of two twin 3in/50 mounts originally fitted was removed in the late 1970s, with the Mk 56 director, as the 3in was removed from the anti-aircraft role.

All but LKA 16 were transferred to the Naval Reserve Force in 1979–81, but then returned to the regular navy in 1982–83. As of 1994, LKA 112 was in MarAd custody in the James River; the others were in reserve in naval custody in 180-day readiness. As of July 1994 plans called for restoring two ships to five-day readiness. These ships have SLQ-32(V)1 ESM sets and provision for wartime installations of Phalanx; LKA-117 had a towed torpedo decoy (SLQ-25 Nixie).

MINOR LANDING CRAFT

Production of Second World War type landing craft continued after 1945. The LCU (landing craft, utility) are the largest which can be carried by the various dock landing ships. Until the advent of the LCAC, they were the primary means of bringing heavy mobile equipment, such as tanks, over a beach. The first postwar series was *LCU 1466–1609* (completed 1954–57), of which fourteen were completed in Japan for foreign service (*1594–1601* for Taiwan and *1602–1607* for Japan). *LCU 1610–1624* were completed in 1960 and *LCU 1627–1679* in 1967–76 (*LCU 1637* was an all aluminium prototype alternative design; *LCU 1625* was an *LCU 1466* fitted with vertical-axis propellers; and *LCU 1626* was transferred). These later craft have both bow and stern ramps, so that vehicles can drive through. *LCU 1680* and *1681* were ordered in 1985 for Naval Reserves (FY85 programme). Twelve were rehabilitated, beginning with two in FY87. Forty-two (including two transferred from US Army) were in service in 1994, plus three redesignated ASDV (auxiliary swimmer delivery vehicle). Others will probably be built. Although these boats are not normally armed (they can be fitted with two 20mm), they are mentioned here as the lineal descendants of the Second World War LCT, and as the predecessors of the LCAC. They were redesignated LSU from LCT in 1949, and LCU on 15 April 1952.

The step beyond is the fast LCAC, which is intended to move heavy equipment from beyond the horizon over a beach. An LCAC can also cross a wider variety of beaches than an LCU. In the late 1970s this concept was developed (as JEFF B) in competition with a hard-chine planing craft, which would have

Francis Marion 1964 USN

been less expensive, but which could not have carried its cargo past the surf zone. The first two were authorised in FY82, followed by three in FY83, six in FY84, twelve in FY86, sixteen in FY89, twelve in FY90, twenty-four in FY91, twelve in FY92, for a total of ninety-one; the programme initially envisaged a total of 108. Because an LCAC is substantially larger than an LCU, dock landing ships designed to accommodate several of the latter often cannot accommodate more than one LCAC. The new generation of ships (LHD, LSD 41/49, LX) are all sized to take several LCACs.

Because well deck temperatures can be quite high (and noise levels intense), these craft have separate troop compartments (for a total of twenty-five troops). Modular troop compartments, to be acquired 1994–95, will accommodate 180 troops or 150 litters. Several LCACs may be used for mine countermeasures, partly because their well decks offer considerable space to handle gear and partly because the air cushion may reduce the shock effect of mines. Eight minesweeping modules (comparable to equipment on MH-53E helicopters) have been bought (the first three were delivered summer 1993), and eight more may be bought if they are successful.

Compared to an LCU, an LCAC carries less than half as much weight (75t vs 180t) at much higher speed (50kts vs 11kts). It is also far more expensive, relatively delicate, and far more difficult to repair in the event of battle damage or accident. Its rationale is that a force beyond the horizon can mount assaults over so broad an area that it cannot readily be defended against, at least at the surf line: the assault ought therefore to be unresisted (over the horizon assault also limits the exposure of the valuable big amphibious ships to coast defence missiles). Half of the full LCAC force (seventeen of the thirty-four then in service) participated in the Gulf conflict, with 100 per cent availability.

The next smaller type is the LCM, of which two types, LCM(6) and LCM(8) are in service. The former, a Second World War 56ft design, remained in production as late as FY75. LCM(8) was lengthened to 74ft to take an M60 tank or 65t of cargo (compared to 34t in the earlier type). About 300 LCM(8) were built for the Army up to 1967, and 217 were built for the Navy 1949–67. 1967–80 production: ninety-three, including six for Spain. Ten new LCM(8) were authorised in FY82 for use abroad the newT-AKX maritime prepositioning ships. They were cancelled, but eight were included in the FY83 programme, ten FY84, three FY85. Further orders: ten in March 1990, ten more 1991; thirty-four earlier units were rehabilitated under the FY92 programme. The abortive LCM(9) was a hard-chine planing craft.

The shorter LCM(6) is a Second World War design built 1952–80; forty-six were in service in 1994. Production: 626 delivered 1952–67, 220 in 1968–80, including four for Iran, nineteen for Saudi Arabia.

At the low end of the scale were several 36ft craft: the ramped LCVP of Second World War design, built of wood or GRP, and the personnel-carrying LCPL, which had no ramp. The LCVP could carry 4t of cargo and both types could be carried in davits. A total of 1552 LCVP were built 1954–66: 349 with wooden hulls 1954–55, 910 wooden hulls in 1956, twenty-seven plastic hulls (Mk 5) in 1956, nine wooden hulls in 1964, and 257 plastic hulls (Mk 7) in 1967, and thirty-two since 1967. A total of 490 LCVP were converted into gunboats armed with 0.30cal MGs, for Vietnam service 1950–62.

The 10kt LCPL Mk 1 was originally intended to land troops, but with the appearance of the ramped LCVP it was used instead to control boat waves in amphibious assaults. A total of 103 Mk 1 were built in 1956. A total of 233 LCPL Mk 4 (with a cockpit) were built between 1957 and 1967; Mks 2, 3, and 5–10 were all experimental craft, and Mk 11 (194 built 1959–67) was a modified type with a tall windshield, roughly resembling a commercial motorboat. Both types could attain 19kts on a 300bhp diesel. They were followed by Mk 12, the first of which were delivered 1981–84. Another seventy-five were ordered in June 1983, with twenty-three more on option, followed by forty-eight in FY85, fifty in FY86, and sixteen in FY87–88. Eight Mk 13 were included in the FY90 programme. 179 LCPL Mks 11, 12, and 13 were in service in 1994.

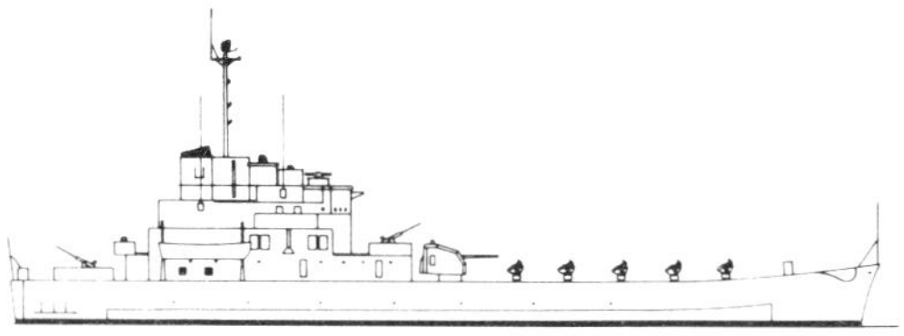

Carronade 1961

CARRONADE *fire support ship*

Displacement: 1040t light; 1500t full load
Dimensions: 245ft × 39ft × 10ft
74.7m × 11.9m × 3.0m
Machinery: 2-shaft diesels, 3100bhp = 15kts
Armament: 1–5in/38, 4–40mm (2×2), 8 rocket launchers
Complement: 139

Class (no, launched):
Carronade (IFS 1, 26.5.53). Built by Puget Sound B & D.

Carronade was a direct successor to the wartime rocket-firing landing ships and craft, built as a result of revived interest in amphibious warfare after Inchon. Design work actually began in 1948, and at one stage a converted DE was considered. Considerations included suitability for mass production and the desire to achieve 20kts (as in other ships of the amphibious convoy then contemplated); the two conflicted, since speeds above about 15kts proved very expensive. In addition, the conflict between a need for large numbers of rockets and the desire to hold down ship size was resolved by placing large numbers of rockets in magazines above the waterline, protected only by 30lb (3/4in) STS. Proposals for a mortar-locating radar, for counter-battery fire, were dropped, the ship being limited to conventional weapons and fire control systems.

Carronade was considered a peacetime luxury, and she was decommissioned in 1960. However, she and three LSMR were reactivated for Vietnam service in 1965, being stricken in 1973. At that time she was classified as an LFR.

Several proposals to build a new-generation amphibious fire support ship (LFS) failed. During the Vietnam War, the LFS was one of the first two naval projects subject to Secretary McNamara's new concept formulation (CONFORM) analysis: all possible alternative ways of accomplishing a mission had to be compared before any one was chosen. In this case, the mission was defined, probably far too simplistically, as destroying an array of targets on a beach. Aircraft seemed far better than gunfire, and the LFS died. It was revived by Congress about 1981, and again died of this sort of analysis (which neglects the important shock and neutralisation effects of shellfire). Also the revival of the battleships provided heavy guns without any need to invest in a new type of ship. With the demise of the battleships, the question of gunfire was revived. This time the availability of precision and bomblet rounds made guns far more attactive; the new fire support programme will probably mean mounting more powerful weapons on board existing destroyers and cruisers.

SMALL SURFACE COMBATANTS

PT 812 type *fast attack craft*

Displacement:
67.9t light; 92.5t full load
Dimensions: 100ft wl, 105ft oa × 18ft 4in × 3ft 11in
30.5m, 32.0m × 5.6m × 1.2m
Machinery: 4 shaft, 4 Packard gasoline engines, 10,000bhp = 38.2kts
Armament: 2–40mm (2×1), 4–20mm (2×2), 4 torpedoes
Complement: 17

Class (no, launched):
Electric Boat – *PT 809* (7.8.50)
Bath Iron Wks – *PT 810* (2.6.50)
Trumpy, Annapolis – *PT 811* (30.11.50)
Philadelphia N Yd – *PT 812* (1.2.51)

These four torpedo-boats were intended to embody wartime design lessons. All were aluminium; *809* was riveted, *810* riveted and welded, and *811* and *812* all-welded; each was designed to mount two single 40mm guns (one in *812*), two twin 20mm, one 81mm mortar, one Mk 107 rocket launcher (*812* only), four torpedo racks, and four depth-charge racks; speed was set at 46kts (43 for the somewhat longer *812*). Although these figures represented a considerable advance on Second World War practice, they still did not make up for the lack of an operational requirement for such coastal craft. *PT 812* was modified as a gas turbine test bed in 1956, with two 4000shp gas turbines on her outboard shafts, and two 800bhp diesels (18kts with controllable-pitch propellers) inboard. *PT 810* and *811* were reactivated in 1963 as *PTF 1* and *PTF 2* in a new Fast Patrol Craft series, officially associated with 'unconventional operations' with SEAL (sea-air-land) teams. Reacquired officially on 21 December 1962, the two ex-PTs were based with six 'Nasty' class acquired from Norway at Subic Bay in the Philippines at the outset of the Vietnam War. *PTF 1* and *PTF 2* were stricken in 1965 and sunk as targets.

PT 812 was transferred to the Army in 1959, to South Korea in 1968. Returning to the USA in 1969 she was later reactivated as *Guardian* and employed until 1974 as an escort for the Presidential Yacht; she became a drone recovery craft (DR-1) in 1975. In that role she is gasoline powered, recovering aerial drones and controlling drone boats; she was named *Retriever* in 1975.

PTF 1 (ex-*PT 810*) 1963 USN

Norwegian NASTY type *fast attack craft*

Displacement: 80.4t
Dimensions: 80ft 4in oa × 24ft 7in × 3ft 10in
24.5m × 7.5m × 1.2m
Machinery: 2 shafts, 2 Napier-Deltic diesels, 6200bhp = 38kts. Fuel 5800gal
Armament: 1–40mm, 2–20mm (2×1), 1–81mm mortar/0.50cal MG
Complement: 17

Class (no, completed):
Batservice, Mandal – *PTF 3–PTF 16* (Dec 1962–Sep 1964)
Trumpy, Annapolis – *PTF 17–PTF 22* (1967–68)

These Norwegian-designed torpedo-boats were acquired by the US Navy for 'unconventional operations'. *PTF 3–PTF 16* were built by Batservice A/B, Mandal, Norway, and the hulls were shipped to the United States for outfitting. During the Vietnam War, these boats were used to carry Vietnamese special forces attacking North Vietnam.

The first two were acquired in December 1962, *PTF 5–PTF 8* on 1 March 1964 and the remainder on 2 September 1964. *PTF 4* was sunk in 1964, *PTF 8, PTF 9, PTF 14, PTF 15* and *PTF 16* in 1966. *PTF 3* and *PTF 5–PTF 7* were leased to South Vietnam on 26 January 1966 and returned (and recommissioned) in 1970. Of this class, one unit (*PTF 13*) was disposed of in 1972; *PTF 3, PTF 6, PTF 7* and *PTF 12* in 1977; *PTF 18* and *PTF 19* in 1980, and the remainder were stricken in 1981.

PTF 17–PTF 22, built by John Trumpy and Sons, Annapolis, duplicated the Norwegian design, entering service in 1968 (*PTF 17–PTF 20*) and 1969 (*PTF 21, PTF 22*).

The Trumpy boats were built under the FY67 programme, the Norwegian units having been acquired under FY64.

Armament varied in all these craft; originally it was reported as two 40mm and two single 20mm, but the forward 40mm was later replaced by the combination 81mm mortar/0.50 cal machine-gun of many Vietnam-era combatants. However, as of 1976 *PTF 18* (and possibly others) retained both 40mm guns.

'OSPREY' type *fast attack craft*

Displacement: 80t light; 105t full load
Dimensions: 94ft 9in oa × 23ft 2in × 7ft
28.8m × 7.0m × 2.1m
Machinery: 2 shafts, Napier-Deltic diesels, 6200bhp = 40kts
Armament: 1–40mm, 2–20mm (2×1), 1–81mm mortar/0.50cal MG
Complement: 19

Class (no, completed):
Sewart Seacraft – *PTF 23–PTF 26* (1967–68)

Under FT67 the Navy bought four 'Ospreys' (*PTF 23–PTF 26*) from Sewart Seacraft of Berwick, Louisiana. They were a commercial, aluminium (rather than wooden) hulled design, sometimes described as improved versions of the 'Nasty'. Although reportedly not completely successful, they were not sold until the 1990s. (*PTF 25* was experimentally fitted with gas turbines in 1978, stricken in 1980.) They could be configured as torpedo-boats, minelayers, or submarine chasers.

CPIC type *patrol craft*

Displacement:
75t full load
Dimensions: 100ft oa × 18ft 6in × 6ft
30.5m × 5.6m × 1.8m
Machinery: 3 shafts, Avco gas turbines, 6750shp (plus Volvo auxiliary diesels on outboard shafts, 500bhp) = over 40kts
Armament: 2–30mm (1×2). See notes
Complement: 11

Developed as a PTF successor, the Coastal Patrol and Interdiction Craft was built by Tacoma Boatbuilding in 1974 and turned over to the South Korean

Navy. She was returned to the United States in 1980 for research. As in the PTFs, there are major weapon positions fore and aft of the bridge structure, as well as pintles for machine-guns along the sides, and guns, missiles, and torpedo tubes can be accommodated. As completed, the CPIC prototype was armed with an Emerson Electric twin 30mm cannon forward.

ASHEVILLE class *fast attack craft*

Displacement: 225t standard; 245t full load
Dimensions: 164ft 6in oa, 23ft 11in × 9ft 6in
50.1m × 7.3m × 2.9m
Machinery: 2 shaft CODOG: 2 Cummins diesels, 1750bhp = 16kts, 1 GE gas turbine, 14,000shp = approx 38kts. Range 1700nm at 16kts
Armament: 1–3in/50, 1–40mm, 4–0.50cal MG. See notes
Complement: 28

Class (no, completed):
Tacoma Boatbuilding – *Asheville* (PG 84, 6.8.66), *Gallup* (PG 85, 22.10.66), *Antelope* (PG 86, 4.11.67), *Ready* (PG 87, 6.1.68), *Crockett* (PG 88, 24.6.67), *Marathon* (PG 89, 11.5.68), *Canon* (PG 90, 26.7.68), *Tacoma* (PG 92, 14.7.69), *Chehalis* (PG 94, 11.8.69), *Benicia* (PG 96, 25.4.70), *Grand Rapids* (PG 98, Aug 1970), *Douglas* (PG 100, Nov 1970)
Petersen – *Defiance* (PG 95, 24.9.69), *Welch* (PG 93, 8.9.69), *Surprise* (PG 97, 17.10.69), *Beacon* (PG 99, 21.11.69), *Green Bay* (PG 101, 5.12.69)

These gas turbine 'motor gunboats' were a product of the Kennedy era, with its Presidential support for small craft and its preoccupation with Cuba. Originally a diesel design was proposed, a natural outgrowth of PGMs built for many US allies and suited for river and coastal operations. However, some within the Bureau of Ships noted that the advent of gas turbines for main propulsion made a much more advanced design possible, their high speed was a dubious advantage, but they did prove extremely useful in riverine warfare, in some cases sustaining considerable damage and in all cases displaying endurance well beyond design goals. They were completed in 1966–71, and their designation changed from PGM to PG in 1967.

The *Asheville* class were also used to test 'special warfare' coastal infiltration techniques, benefiting from their small radar cross-section and in many US and NATO exercises they simulated Soviet small missile craft (Osa and Komar types). In addition, after Vietnam, four units in the Mediterranean were assigned by the then CNO, Admiral Zumwalt, to a 'counter-tattletale' role, following Soviet warships to provide positive warning of any imminent Soviet missile attack on the Sixth Fleet. In this role they were armed with a surface-to-surface variant of the US Standard anti-aircraft missile, with two launchers aft and an additional pair of reloads forward of the launchers. Admiral Zumwalt later used this experience to justify the PHM programme. The ships concerned were PGM 86, 87, 98 and 100, after tests had been carried on aboard PGM 96.

Cavitation generally limited their maximum speed to something less than the design figure always quoted, although at the time it was stated that the provision of new propellers would have allowed them to reach design speed. In service they had a bad reputation for discomfort in a seaway, although they were good sea boats and could ride out very severe storms. Thus proponents of the PHM could argue that the *Asheville*s had demonstrated the utility of small fast surface-attack craft and that hydrofoil design could overcome their bad riding characteristics.

PG 86 and 87 had the Dutch-developed Mk 87 fire control system, a close relative of the Mk 92 of the PHMs and FFGs, employing a computer and automatic aiming; the others all had the Second World War Mk 63 to control their 3in guns. Although up to twenty-two were planned, rising costs and the absence of a clear mission (particularly in the context of the costs of the Vietnam War) cut short the programme, and the absence of a rationale shows in the quick disposal of most of the class. Transfers were as follows: PG 96 to South Korea 1971; PG 95 and 97 to Turkey 1973; PG 85 and 90 (stricken 1976) to Taiwan 1981; PG 92 and 93 to Colombia 1981; PG 86 and 88 to Environmental Protection Agency 1977; PG 100 to Naval Ship Research and Development Center 1979; PG 94 and 98 to Naval Ship Research and Development Center 1975 and 1977 as high-speed tugs *Athena I* and *Athena II*. PG 84 and PG 89 were stricken in 1976 and PG 87 in 1977. The last two survivors were retained to train Saudi personnel who manned somwhat similar craft built in the United States for their navy.

Asheville 11.10.74 L & L van Ginderen

High Point (foreground) and the experimental *Plainview* 1978 USN

HIGH POINT *hydrofoil*

Displacement: 93.9t light; 119.7t full load
Dimensions: 110ft wl, 115ft 9in oa × 31ft 2in × 3ft 11in
33.5m, 35.3m × 9.5m × 1.2m
Machinery: 2 Bristol-Siddeley gas turbines, 6800shp = 45kts; (hull-bourne, 1 diesel 600bhp = 12kts)
Armament: 2–0.50cal MG (1×2), 1 DC rack, 4–12.75in TT (2×2), 4 Harpoon SSM (in 1973–74 tests)
Sensors: Sonar for trials
Complement: 13

Class (no, launched):
High Point (PCH 1, 17.8.62). Built by J M Martinac.

Hydrofoils were the major exception to US Navy lack of interest in fast patrol craft during the 1950s; they were so promising a technology that they could not be ignored. At that time the primary naval problems was ASW, and it was for ASW that the first large US hydrofoil, the *High Point* (PCH 1, SCB 202 of FY60), was designed. It was to function in a 'sprint and drift' mode, listening briefly between high speed dashes. Because it could not really hope to listen while moving at high speed, it would have to operate in pairs, one unit vectoring the other into an attack with homing torpedoes. A large hull sonar would be retracted into the hull at high speed, and extended during low-speed hull-borne operation. There was also an auxiliary sonar, either a variable-depth unit for deep operation, or a towed unit which might provide some limited capability even at speed. There were, in addition, exotic operational concepts, such as buoy-borne MAD for target classification during an attack.

Interest in such ASW hydrofoils dated from about January 1955, although it appears that there was information concerning Soviet experiments the previous year. In February 1956 the Office of Naval Research began operational evaluation of hydrofoil craft systems for ASW. Meanwhile, conventional coastal ASW craft (SC and PCS) were repeatedly included in proposed shipbuilding plans, but were always dropped due to their low priority. In January 1958, however, the CNO asked the Bureau of Ships to study the use of hydrofoil craft for harbour security, and that April the PCH was substituted for the usual combination of PCS and SC prototype in the FY60 budget. Probably largely because of its innovative character, the PCH did survive the FY60 pruning of the budget. Originally several follow-on craft were proposed. For example, an October 1958 proposal for the FY61 budget carried three PCH (as well as a variety of other doomed craft, such as a missile-armed barrier patrol ship, or PBG, five of which were to have been bought).

In fact, only one PCH was built, and it functioned as an experimental boat. Between August 1971 and July 1973 she was converted at Puget Sound as operational test and evaluation platform for the Harpoon missile system, with a quadruple missile cannister on the centreline aft and a new control system, including a steerable forward strut. She was lent to the Coast Guard (WMEH 1) in March 1975 and stricken in 1980.

FLAGSTAFF *hydrofoil*

Displacement: 56.8t
Dimensions: 74ft 5in oa × 21ft 5in × 13ft 6in (foils extended) 4ft 3in (foils retracted)
22.7m × 6.5m × 4.1m, 1.3m
Machinery: 1 shaft, Rolls-Royce Tyne gas turbine, 3620shp = 48kts (hull-borne, 2 GM diesels, 300bhp, water-jet propulsion)
Armament: 1–40mm, 4–0.50cal MG (2×2), 1–81mm mortar
Complement: 13

Class (no, launched):
Flagstaff (PGH 1, 9.1.68)

TUCUMCARI *hydrofoil*

Displacement: 58t
Dimensions: 71ft 10in oa × 19ft 6in × 13ft 11in (foils extended) 4ft 6in (foils retracted)
21.9m × 5.9m × 4.2m, 1.4m
Machinery: Proteus gas turbine, water-jets, 3040shp = 40kts (auxiliary diesel for hull-borne operation, 160bhp)
Armament: 1–40mm, 4–20mm (2×2), 1–81mm mortar (replaced 1971 by another twin 20mm)
Complement: 13

Class (no, launched):
Tucumcari (PGH 2, 15.7.67)

By the early 1960s interest had shifted towards the fast gunboat role, and it appeared that a relatively small hydrofoil could duplicate many of the qualities of the larger *Asheville* class fast gunboat. Two competitive prototypes were built under the FY66 (SCB 252) programme; the Grumman *Flagstaff* and the Boeing *Tucumcari*. At one time a follow-on class of thirty-four hydrofoils was planned, with features borrowed from both. They differed largely in propulsion, Grumman using a geared propeller with the main lifting surface forward; Boeing, water-jets with the main foil surfaces aft. Both were tested in Vietnam, returning home in 1970. Originally armed with 40mm gun forward, *Flagstaff* was tested with an Army 152mm tank-turret gun in 1971. She was loaned to the Coast Guard (WPGH 1) between November and December 1974, sustaining considerable damage in a collision with a whale; she was sold in 1978. *Tucumcari* ran aground on 16 November 1972, was damaged beyond economical repair, stricken, and her hulk used for fire tests by the Naval Ship Research and Development Laboratory.

PEGASUS class *hydrofoils*

Displacement: 231t full load
Dimensions: 131ft 6in (foils extended) × 23ft 2in × 6ft 2in (foils retracted)
40.0m × 8.6m × 1.9m
Machinery: CODOG (water-jet propulsion): 2 Mercedes-Benz diesels (1600bhp = 11kts), 1 GE gas turbine, 18,000shp = over 40kts on foils
Armament: 1–3in/62, 8 Harpoon SSM (2×4)
Sensors: Radar SPS-63, Mk 92 fire control system (Mk 94 in PHM 1)
Complement: 22

Class (no, launched):
Pegasus (PHM 1, 9.11.74), *Hercules* (PHM 2, 13.4.82), *Taurus* (PHM 3, 8.5.81), *Aquila* (PMH 4, 16.9.81), *Aries* (PHM 5, 5.11.81), *Gemini* (PHM 6, 17.2.82). All built by Boeing.

Although neither *Flagstaff* nor *Tucumcari* was reproduced, Boeing did use *Tucumcari* as the basis for the Italian *Sparviero* class missile boats. Israel ordered two 'Super *Flagstaffs*' with Harpoon missiles in 1978. Moreover, Boeing experience with *Tucumcari* presumably shows in the design of the next-generation hydrofoil combatant, the missile hydrofoil *Pegasus*, PHM 1.

The missile hydrofoils were originally conceived by Admiral Zumwalt as successors to the missile-armed *Asheville* class 'tattletales', that is, as a means of countering Soviet missile-armed surface craft, particularly in restricted waters. They had the important advantage of very high speed even in rough weather; it must have appeared that their small size would equate to low unit cost. The PHM programme was, for a time, a NATO-wide one, the Italian and West German governments participating in the design. Germany was to have built twelve. However, as costs rose only the United States, which planned thirty, remained in the programme. Cynics suggested strongly that the PHM was a solution in search of a problem, and (with the programme reduced to six craft) PHM 2–PHM 6 were cancelled on 15 April 1977. However, they were reinstated that August at the request of Congress. PHM 2 was to be unarmed, and by 1981 it appeared unlikely that further units would be built. One of the more curious aspects of the PHM story was the attempt to find alternative missions, a Sea Phoenix variant for fleet air defence coming close to the formal proposal stage within the Department of Defense. These boats were all decommissioned and stricken on 30 July 1993, but were retained pending a decision on their transfer to foreign navies.

PROJECTED HYDROFOILS

The hydrofoil story in the US Navy would be incomplete without mention of some paper projects for much larger types, sometimes designated DE(H) and occasionally Deepwater Escort Hydrofoil. Sketch designs for units as large as 1400 tons have been executed and construction proposed from time to time during the 1970s. Thus the DE(H) competed (apparently unsuccessfully) with the Surface Effect Ship as an Advanced Naval Vehicle; ultimately both died of immature technology and extreme unit cost.

'SWIFT' (PCF) type *patrol craft*

Displacement: Mk I: 42,500lb (19t)
Mk II: 43,035lb
Dimensions: Mk I: 50ft 2in × 13ft 1in × 3ft 6in
15.3m × 4.0m × 1.1m
Mk II: 51ft 4in × 13ft 7in
15.6m × 4.1m
Machinery: 2 shafts, 2 GM diesels, 860bhp = 28kts. Fuel 800 and 828gal, respectively
Armament: 2–0.50cal MG (1×2), 1–81mm mortar/0.50cal MG
Complement: 6

Class:
A total of 193 were built from 1965 onwards: 104 Mk I (six to Philippines); thirty Mk II (eight to Philippines, two Thailand, six Vietnam); thirty-three Mk 3 (three to Philippines, five Thailand, twenty Cambodia). In addition, boats were ordered specifically for transfer: four for Brazil (at least six more built locally), ten Mk II for Thailand, six Mk I or II for Zaire. Only two remained in US service by 1982.

The US Navy, which since 1945 had systematically abandoned coastal warfare, found itself operating a large fleet of very small craft in Vietnamese waters. This fleet was, moreover, extemporised very rapidly. The two principal wartime types were the 'Swift' or PCF and the 'Plastic' or PBR.

The 'Swift' was adapted from a standard commercial boat used to support oil drilling rigs in the Gulf of Mexico. All were built by Sewart Seacraft of Berwick, Louisiana, in 1965–66. The first 104 units were Mark I, the remainder (of about 200 built), Mark II with less sheer, a broken deck line, and the pilot house set somewhat farther back from the bow. A typical armament was a twin 0.50cal machine-gun above the pilot house, with a combination 81mm mortar/0.50cal machine-gun mount aft. PCF denoted Patrol Craft, Fast, a typical rated speed being 28 knots.

PBR type *river patrol boats*

Displacement: Mk I: 6.5t
Mk II: 7.1t
Dimensions: Mk I: 31ft × 10ft 7in × 1ft 11in
9.5m × 3.2m × 0.6m
Mk II: 31ft 11in × 11ft 8in × 2ft 7in
9.7m × 3.6m × 0.8m
Machinery: 2 GM diesels, 432bhp, 2 water-jet pumps = 25kts
Fuel 160gal
Armament: 3–0.50cal MG (2×1, 1×1), 1–40mm grenade launcher, 1–60mm mortar (in some)
Complement: 4

Class:
Production by programme: FY66: 161 Mk I (including prototype): FY67: 81 Mk II (including prototype); FY68: 145 (including thirty-nine army, twenty-seven for Vietnam, one for Brazil); FY69: 37 (including eighteen for Vietnam plus a bare hull); FY70: 23 (including three for the army); FY 71: 30 (ten for Thailand, twenty for USN but originally for Cambodia); FY72: 10 (two for Philippines); FY73: 37 (twenty-two for Cambodia, five for Thailand); FY76: two (for Thailand); FY77: five (for Thailand). In addition, there were six to Burma in 1978, at least twenty-eight to Israel by 1986 (and others built locally as *Yatush* class), a total of thirty-seven for Thailand, and an unknown number of bare hulls to replace ones damaged.

The Patrol Boat, River was smaller than the PCF, with an open-topped conning position, and had a fibreglass hull lined with plastic foam; it was propelled and steered by water-jets, for an extremely shallow draft (3ft compared to 4ft for a Swift). A typical battery consisted of a twin 0.50cal machine-gun forward, a single 0.30cal machine-gun aft, and an infantry-type 40mm grenade launcher. The somewhat larger Mk II typically had three 0.50s (one twin and one single) and a 60mm mortar; it had more powerful engines and enlarged mufflers. A total of 160 Mk I were built by United Boat Builders, Bellingham, Washington between December 1965 and April 1966. The great majority of US PBRs were transferred to South Vietnam. The PBRs were based afloat, LST 846, 821, 838 and 786 being converted to support them. In 1994, twenty-seven PBR Mk II remained. Some had been used during the 1991 Gulf conflict. Some recent versions have GM 6V53T engines for 550bhp (originally 420bhp), to make 30kts (formerly 24kts). One was used to compare two waterjet systems 1993; with new 275bhp diesels she made 31.3kts light.

A PBR 1967 USN

STINGER *river patrol craft*

Displacement: 7.48t full load
Dimensions: 35ft × 9ft 3in × 2ft 2in
10.64m × 2.82m × 0.66m
Machinery: 2 Cummins diesels, 1 water-jet, 600bhp = 38kts (34.6kts sustained)
Armament: See below
Complement: 4 (plus 10 troops)

These aluminium craft replace PBRs for Special Boat Units. They have weapon positions fore and aft for twin or single 0.50cal or 7.62mm MG or single 40mm Mk 19 grenade launcher. Boats can be carried by C-130 aircraft. Programmes: seven FY90, 10 FY92, eight FY93, built by SeaArk Marine of Monticello, Arkansas (first fourteen) and by Swiftships of Morgan City, Louisiana. Boats are organised in 'companies' of seven each, and the first fourteen are based at Camp Lejeune, North Carolina.

RPC type *river patrol boats*

Displacement: 13t
Dimensions: 35ft 9in × 10ft 8in × 3ft 9in
10.9m × 3.25m × 1.14m
Machinery: 2 diesels, total 450bhp = 15kts
Armament: 1–20mm, 1–0.50 cal, 2–0.30 cal, 2–7.62mm
Complement: 6 (plus 12 troops)

The RPC was the first craft built specifically for riverine warfare by the US Navy. It began with a Thai request for modified PT boats; they received six overhauled LCVPs while a better design was evolved. A preliminary design was prepared in mid-1961. Ultimately the RPC was based loosely on a projected (but unbuilt) riverine LCVP called the RCVP. It was not particularly successful; the RPC was criticised as noisy, vulnerable (its hull form tended to attract mines), slow, and underarmed. In all, thirty-four were built under the FY64 programme. All were initially transferred to Vietnam, but six were transferred to Thailand in March 1967.

PB type *patrol boats*

Displacement: Mk I: 26.9t light; 36.3t full load
Mk III: 31.5t light; 41.3t full load
Dimensions: Mk I: 65ft × 16ft × 4ft 10in
19.8m × 4.9m × 1.5m
Mk III: 64ft 11in × 18ft 1in × 5ft 10in
19.8m × *5.5m × 1.8m*
Machinery: Mk I: 2 shafts, 2 GM diesels, 1635bhp = 26kts
Mk III: 3 shafts, 3 GM diesels, 1800bhp = 26kts; range 450nm/26kts, 2000nm/cruising
Armament: Mk I: 6–20mm (1×2, 4×1), or 0.50cal MG
Mk III: 4–0.50cal MG (4×1); 4 Penguin SSM (fitted for evaluation 1981)
Complement: Mk III: 5

Class:
Two Mk I built by Sewart Seacraft 1973; twenty-six Mk III by Marinette Marine and Petersen (four for the Philippines); three Mk IV. No Mk IIs were built. Mk I was a navy version of the enlarged (65ft) Swift built for export from 1966 on: thirteen in FY65 (nine for Korea, one for El Salvador, one for the air force, two for the army), six in FY71 (Philippines), and many others for direct export. Total production of 65-footers, including commercial craft for oil field support, may have been as high as 200. Export customers: Antigua (one), Bahrain (two), Costa Rica (five), Dominica (one), Haiti (three), Honduras (five), Liberia (six), Nigeria (six ordered, four delivered), Panama (two), and St Lucia (one).

The Patrol Boat (PB) was developed as a successor to the 'Swift'. Two Mk I were completed by Sewart Seacraft in 1972 and delivered in 1973. They have one twin and four single gun positions for 20mm or 0.50cal guns. An alternative Mk III design by Peterson Builders of Sturgeon Bay, Wisconsin, won in competition with Mark I and was procured, in 1982, for inshore warfare. Like the Swift, it was adapted from a Gulf of Mexico support boat. The pilot house is offset to starboard to provide maximum deck space, and the nominal armament of four 0.50cal machine-guns can be reinforced by heavier weapons on the main deck. For example, in 1981 the Norwegian Penguin III anti-ship missile was evaluated for them, though it proved too heavy.

Mk III programme: eight in FY73, ten in FY75, eight in FY77; as of September 1993, the fourteen survivors were in land storage at San Diego; one (65PB736) had her engines out. At least one was reactivated for a 1994 exercise in Thailand. Two were transferred to Colombia in 1992. Another, 65PB731, was used for gas turbine trials. By 1994 all had trouble making their design speed.

PB Mk IV is a slightly lengthened (68ft) Mk III, three of which were built by Atlantic Marine of Fort George Island, Florida for Panama Canal Zone defence under the FY84 programme. They had one 20mm cannon forward, replaced by a Mk 38 25mm Bushmaster in 1987. These boats also have 0.50cal MG/81mm mortar combination mounts.

PB Mk V *patrol boats*

This air-transportable SEAL insertion craft in effect replaces the earlier Sea Fox, now discarded; with its trailer, it can be carried aboard a C-5A. Larger than Sea Fox, it should have a far smaller radar cross-section than the big PC (see below). For a time it appeared that the 85ft Israeli 'Shadag' would be bought off the shelf, but then two competitive US prototypes boats were bought: Peterson's 'Sea Stalker' and Trinity-Halter's XFPB (one with three GM diesels, one with two MTU diesels). Late in 1994 the MTU-powered Halter boat was chosen, with an initial order for two and options for up to thirty-eight more (original intention was for fourteen). The first first order was for $11m, including the transporter.

'Sea Stalker' was an 'asymmetric catamaran' based on the British Cougar Marine Cat 2100 'Dark Moon'. It was ordered in August 1993 and delivered that December: 39/75t, 71ft × 18ft × 4ft 6in (21.64m × 5.49m × 1.4m), 2 MTU 16V396 TE94 diesels, 3500bhp = 50+kts. The winning XFPB was a more conventional planing hull: 75t full load, 82ft × 18ft × 4ft 2in (25m × 5.49m × 1.32m), 50+kts. It used three surface-piercing propellers for the GM diesel version, two waterjets for the MTU version.

Assault Support Patrol Boat (ASPB) type *riverine craft*

Displacement: 26.1t full load
Dimensions: 50ft 2in × 15ft 3in over guards × 3ft 9in
15.3m × 4.7m × 1.1m
Machinery: 2 shaft diesels, 850bhp = 14kts
Armament: 1– or 2–20mm (and 2–0.50cal, 1×2, in boats with 1–20mm),
2–0.30cal, 2–40mm grenade launchers, 1–81mm mortar/0.50cal MG (in some)
Complement: 5

Command and Control Boat (CCB) and Monitor type *riverine craft*

Displacement: CCB: 167,000lb (74.5t)
Monitor: 169,000lb
Dimensions: 60ft (oa) × 17ft 6in × 3ft 4in
18.3m × 5.3m × 1.0m
Machinery: 2 diesels, 450bhp emergency, 330bhp continuous = 8.5kts (6kts sustained)
Armament: 1–40mm or 20mm, 1–20mm, 2–0.30cal MG, 1–81mm mortar (Monitor only)
Complement: 11

Riverine warfare required considerable numbers of escort craft to operate directly with the troop carriers. Both shallow-draft 'monitors' (MON) and Assault Support Patrol Boats (ASPB) were developed. The first monitors were converted from LCM(6) (landing craft) hulls, which formed the basis for many of the riverine craft. A typical battery was 1–40mm gun, 1–20mm, 2–0.50cal MGs and 1–81mm mortar (or 2–M10-8 flame throwers). Later units (Mk V) were built as such from the keel up, and had rounded bows without landing craft type bow doors. They used the same hull as the CCB command boats, with a twin 20mm bow turret, an 81mm mortar in a well, and a smaller turret on the superstructure for a 0.30cal MG. Total armament included 4–0.30s. Some had a 105mm howitzer in place of the bow turret and mortar well. Both types were armoured against shaped-charge rockets with characteristic armour slats like Venetian blinds. Numbers built are not available, but all survivors were handed to the South Vietnamese Navy in 1973. They included at least twenty-two converted LCM(6) – transferred 1964–67 – and forty-two purpose-built monitors.

The ASPB (Assault Support Patrol Boat) was the only craft specially designed for the river war, described as the destroyer and minesweeper of that specialized fleet. It was equipped with a chain drag to counter command mines, and armed with a 20mm cannon, a twin 0.50cal MG, two Mk 18

A flotilla of ATCs and a Monitor (right foreground) 1967 USN

grenade launchers, and an 81mm mortar. ASPBs generally led riverine operations, sweeping ahead of the assault formation; they were also used as escorts, patrol craft, and counter-ambush craft. One unusual feature was the underwater exhaust system, adopted for silencing. The ASPB was built in two Marks in 1967–68, with welded steel hulls, and capable of about 14kts (compared to 16kts as designed). The ASPB hull was also used for minesweeping (as the MSR) and for CCB duties. At least eighty-four ASPB were transferred to the South Vietnamese Navy and the remainder discarded.

Troops were generally carried in converted LCM(6)s, ATCs (Armoured Troop Carriers), each carrying a platoon of about 40 men. Some were fitted with helicopter decks as ATC(H), serving as batallion aid stations. In addition, until the arrival of the specially designed ASPB, the ATCs were the minesweepers of the Riverine Force. Each was armed with 1–20mm cannon, 2–0.50cal MGs, and two Mk 18 grenade launchers. Postwar, a new 'mini-ATC' was developed, primarily for Special Forces-type operation, with an aluminium hull and ceramic radar, quiet engines (for 28kts), shallow draft (1ft), and space for up to seven pintle-mounted weapons. By 1981, a total of eighteen had been delivered, one powered by gas turbines.

For command and control, the Riverine Navy had CCBs, converted from LCM(6)s and similar to the monitors except that a command module replaced the 81mm mortar of the latter; it was inserted between the after superstructure and the covered gun position in the bow. The latter could accommodate one 20 or 40mm gun, and there were typically several 0.30 and 0.50cal machine-guns.

Numbers of converted craft are somewhat difficult to fix. The United States initially modified LCMs for Vietnam. The initial series of transfers (planned in 1961) amounted to sixteen LCVP and sixteen LCM (command boats and monitors). They began to arrive in 1963. Ten LCMC command boats were transferred in December 1963, followed by four in 1966. The first seven US monitors were transferred in December 1963, seven more in March and June 1965, and eight more in 1966. There were also LCM troop carriers, denoted LCM-C (LCM-combat). Transfers included at least twelve in December 1963, eight in December 1965, and twenty-four in 1966. The Vietnamese also had similar craft converted by the French or by themselves. When the United States entered the war, new conversions to modified designs were begun. Totals were: twenty-nine monitors (ten Program 4, fourteen Program 5, five Program 6); thirteen command boats (CCB: four Program 4, eight Program 5, one Program 6); 124 armoured troop carriers (ATC: fifty-two Program 4, 64 Program 5, eight Program 6); and eight armoured refuellers (four Program 4, four Program 5, eight Program 6). Program 4, the original US riverine programme for two squadrons, also included two SEAL heavy support craft (HSSC). Ten of their successors, LCMMs, were included in Program 5. Program 4 (two riverine assault squadrons, FY67) had been drastically scaled down and did not include any allowance for losses. Program 5 (FY68) provided another two squadrons plus training and attrition boats. Program 6 was ordered late in January 1972; it included conversions for Cambodia. There were two types of Program 5 monitor, 105mm and flame-thrower.

ASPB production: FY67 (Program 4): thirty-six (plus one test platform), FY68 (Program 5): fifty (fourteen delivered directly to South Vietnam); there were also ten very similar river minesweepers.

Program 5 also included research and development for a next-generation series of riverine craft, which were built only as prototypes.

SEAL *boats and craft*

The SEALs were organised in about 1961. Initially they operated from special swimmer delivery craft, some of them prototypes of the riverine craft described above. Two LCMs were modified as heavily-armoured SEAL support boats (heavy SEAL support craft, or HSSCs) and modified LCPLs became medium SEAL support craft (MSSC). Requirements for replacements (a new MSSC and a light craft, the LSSC) were formally stated in December 1967. Both were modified commercial boats, armed with a variety of small guns on pintle mounts. The LSSC was derived from a commercial 26-footer modified as the STAB (SEAL team assault boat); sixteen Mk I and twenty-four Mk II LSSCs were built. They were later redesignated STABs (strike attack boats). Each had four pintle mounts, normally carrying three M60s and a grenade launcher. In addition, ten 36ft MSSCs were bought in 1968.

All of these boats were interim solutions to the SEAL requirements. At the end of the war two new designs, a LASSC (River Fox) and a mini-ATC, were in the design stage. The LASSC was not built, but the mini-ATC replaced the earlier MSSC. Production: seventeen in FY72, eight in FY76, fourteen in FY78 (ten for the Philippines). These were the last riverine SEAL craft for some time, since after Vietnam the SEAL mission shifted towards attacks mounted from the open sea.

An operational requirement for a new SEAL boat, which became the 36ft Sea Fox, was stated in 1976; the prototype was completed in November 1977. Further production amounted to about forty-five (including seven in FY80, one in FY81, and thirty-six in FY82; a few were transferred to Egypt). Sea Fox has a collapsible superstructure, so that it can be transported by trailer or aboard an aircraft, and it presumably incorporates anti-radar technology. Its main limitation, given the new open-sea role, is its very small size. Sea Fox has been used as a picket boat, for example, off Lebanon. It is typically armed with two 0.50cal machine guns and two M60s. All twenty-four surviving Sea Foxes were retired by January 1993. By that time they were classed as Special Warfare Craft Light (SWCL).

About this time there was also interest in building a new patrol missile boat specifically for Caribbean operations. Initially a study suggested that more PHMs were needed, partly to counter possible operations by the Cuban navy. Then it turned out that a conventional hull was far less expensive than a hydrofoil, and the new programme was tentatively designated PBM. Because it was a relatively shallow-water boat, PBM naturally was also conceived for special warfare roles. This programme was cancelled December 1987.

In 1982 Congress appropriated money for a new SEAL delivery craft, then called PBM. It was conceived as a stealthy mother ship from which small delivery craft could be launched. There were two proposals, a conventional planing hull and a surface effect boat. The latter, unofficially called Sea Viking, won, but it was badly overweight and its builder, Rohr, went bankrupt.

By the late 1980s, a whole new cycle of projects was in train. The delivery boat had to be fairly large, and it seemed wiser to buy something off the shelf than to demand a wholly new design. The Vosper Thornycroft 'Province' class was selected, even though it was far larger (and inherently far less stealthy) than the earlier abortive projects. A total of thirteen, designated PC, was ordered from Bollinger in August 1990. Dimensions: 170ft × 25ft × 7.5ft (300t), powered by four Paxman diesels = 35kts (range 2500nm/12kts). These rather large boats will probably be used mainly for Caribbean anti-drug patrol work (replacing the PHMs) rather than for SEAL insertion, although they do have swimmer facilities aft. Their size and parentage may owe something to an initial desire to revive the earlier PBM concept. Compared to the 'Province' class boats built for Egypt, Kenya, and Oman, they are very lightly armed, with a pair of 25mm Bushmaster (whose mounts are to be replaced by stabilised weapon platforms), six Stinger missiles for air defence, and four weapon stations (one to starboard at the stern and three on the bridge: two to port, one to starboard), each of which can accommodate 0.50cal and 7.62mm machine guns or Mk 19 40mm grenade launchers. The weapons stations may ultimately accommodate Hellfire missiles and folding-fin aircraft rockets. Complement is twenty-eight plus eight SEALS.

The *Cyclone*s were conceived as a class of sixteen to replace the seventeen PB Mk III, replacing earlier SWCM and PCM; they were classed as PBC until June 1991. Programmes: eight in FY90, five in FY91; then the programme was stopped partly because they were too large. Names and completion dates: *Cyclone* (PC 1, 7.8.93), *Tempest* (PC 2, 21.8.93), *Hurricane* (PC 3, 15.10.93), *Monsoon* (PC 4, 22.1.94), *Typhoon* (PC 5, 27.1.94), *Scirocco* (PC6, 11.6.94), *Squall* (PC 7, 4.7.94), *Zephyr* PC 8, Oct 1994), *Chinook* (PC 9, Dec 1994), *Firebolt* (PC 10, Mar 1995), *Whirlwind* (PC 11, Apr 1995), *Thunderbolt* (PC 12, Aug 1995), *Shamal* (PC 13, Nov 1995).

MINE WARFARE VESSELS

AGILE/AGGRESSIVE class *ocean minesweepers*

Displacement:	637t light; 735t full load (MSO 488)
Dimensions:	165ft wl, 171ft 6in oa × 34ft × 13ft 1in *50.3m, 52.3m × 10.4m × 4.0m*
Machinery:	2 shafts, diesels, 1550bhp = 13kts. Range 2400nm at 12kts. See notes
Armament:	1–40mm
Sensors:	Sonar UQS-1
Complement:	70

Class (no):
Agile (MSO 421), *Aggressive* (MSO 422), *Avenge* (MSO 423), *Bold* (MSO 424), *Bulwark* (MSO 425), *Conflict* (MSO 426), *Constant* (MSO 427), *Dash* (MSO 428), *Detector* (MSO 429), *Direct* (MSO 430), *Dominant* (MSO 431), *Dynamic* (MSO 432), *Engage* (MSO 433), *Embattle* (MSO 434), *Endurance* (MSO 435), *Energy* (MSO 436), *Enhance* (MSO 437), *Esteem* (MSO 438), *Excel* (MSO 439), *Exploit* (MSO 440), *Exultant* (MSO 441), *Fearless* (MSO 442), *Fidelity* (MSO 443), *Firm* (MSO 444), *Force* (MSO 445), *Fortify* (MSO 446), *Guide* (MSO 447), *Illusive* (MSO 448), *Impervious* (MSO 449), *Implicit* (MSO 455), *Inflict* (MSO 456), *Loyalty* (MSO 457), *Lucid* (MSO 458), *Nimble* (MSO 459), *Notable* (MSO 460), *Observer* (MSO 461), *Pinnacle* (MSO

Embattle 12.6.63 USN

462), *Pivot* (MSO 463), *Pluck* (MSO 464), *Prestige* (MSO 465), *Prime* (MSO 466), *Reaper* (MSO 467), *Rival* (MSO 468), *Sagacity* (MSO 469), *Salute* (MSO 470), *Skill* (MSO 471), *Valor* (MSO 472), *Vigor* (MSO 473), *Vital* (MSO 474), *Conquest* (MSO 488), *Gallant* (MSO 489), *Leader* (MSO 490), *Persistant* (MSO 491), *Pledge* (MSO 492), *Stalwart* (MSO 493), *Sturdy* (MSO 494), *Swerve* (MSO 495), *Venture* (MSO 496), *Acme* (MSO 508), *Adroit* (MSO 509), *Advance* (MSO 510), *Affray* (MSO 511). They were launched between 12 July 1952 and 18 December 1956. MSO 450–454, 475–487, 498–507, 512–518 were built for other countries under MDAP (eight for France, six for Netherlands, four for Portugal, four for Belgium, two to Norway, and two to Italy).

ABILITY class *ocean minesweepers*

Displacement:	792t light; 934t full load
Dimensions:	182ft wl, 189ft 9in oa x 35ft 1in x 10ft *55.5m, 57.9m x 10.7m x 3.0m*
Machinery:	2 shafts, GM diesels, 2700bhp = 14.5kts
Armament:	1–40mm
Sensors:	Sonar UQS-1
Complement:	83

Class (no):
Ability (MSO 519), *Alacrity* (MSO 520), *Assurance* (MSO 521). Launched 1956–57.

Although minesweeper design continued after 1945, and although the prototype new ocean sweeper was authorised under the FY50 programme, construction on a large scale did not begin until the Korean War, with ten units under the FY51 first supplemental budget, eighteen under the second, and fifteen under the third, another five being planned under the FY52 programme. The new sweeper was somewhat smaller than the 180ft Second World War type, and it was built of plywood for minimum magnetic signature. Two XMAP pressure sweeping caissons could be towed. Given problems in diesel engine procurement under what were then expected to be mobilisation conditions, designs were prepared for either a new Packard light-weight engine or for an existing GM commercial type. Reportedly the rapid development of the Packard (which was also used in the new coastal sweepers) and its aircraft-type design led to many mechanical problems, which were not corrected for as much as eight years after its introduction. Given its unfortunate reputation, the Packard never saw large-scale commercial service, and the Navy suffered again as production ended and spares were largely unavailable. Of ninety-three ships to the SCB-45 A or *Agile* class design, fifty-eight were retained by the US Navy, the others being transferred. They were all reclassified from AM to MSO in 1955.

Some writers split the *Agile* class into sub-classes. *Agile* and thirty-six sisters, transferred abroad had 3040shp Packard diesels. Of fifty-three *Aggressive* class sweepers (MSO 422 series, 2280bhp Packard diesels), twelve were transferred. Four *Dash* class (MSO 431 series, 1520bhp GM diesels) were improved *Agiles* with two rather than four engines.

The *Acme* (MSO 508) and *Ability* (MSO 519) classes were improved and slightly enlarged *Agiles* built under the FY54 and 55 programme, the latter designed as mine division flagships. Ultimately both series were fitted as flagships.

All of the new ocean sweepers were built of wood with bronze and stainless (non-magnetic) steel fittings, with automatic degaussing, to compensate for course changes, as well as electrical insulators in internal piping, lifelines, and stays. All also had a new mine-locating sonar, UQS-1, later replaced by SQQ-14, a variable-depth unit on a rigid retractable rod.

Originally all ocean sweepers in commission in the mid-1960s were to have been modernised; however, only nineteen of the sixty-four ocean sweepers then in service began refits. Modernisation of five (MSO 460, 468, 470, 472 and 519) was cancelled 16 October 1970 and the hulks were scrapped. *Avenge* caught fire during modernisation and had to be scrapped, and another converted ship (MSO 445) burned and sank at sea on 24 April 1973 en route from Subic Bay to Guam. The modernisation programme itself was abandoned with a shift of emphasis to minesweeping helicopters. Modernisation was funded under the FY58 (MSO 433, 441–3, 445, 446, 449, 456) and FY69 (MSO 437, 438, 448, 488, 490) programmes.

Fire was a considerable hazard to the wooden sweepers, and at least six caught fire. In 1981, only twenty-five survived. Units fitted with SQQ-14 sonars have a massive external hoist forward of the bridge, in place of the single 40mm gun formerly mounted there. These units have a single 20mm gun forward; some appear to be entirely unarmed. In most cases, modernisation provided an enlarged superstructure abaft the bridge.

Disposals: MSO 423, 469 stricken 1970: 460, 463 (to Spain), 468, 470, 472, 491 (to Spain) 1971; 426, 435, 436 (to Philippines), 440 (to Philippines), 447, 457, 473 (to Spain) 1972; 432 (to Spain) 1974; 422, 424, 425, 466, 467, 1975; 458, 459, 508, 510 1976; 421, 461, 462, 471, 474, 494, 495, 496, 1966; MSO 445 sunk 24 April 1973. MSO 519 was stricken 1971. MSO 520 and 521 were reclassified as AG 520 and 521 in 1973, their sweeping gear removed, and SQR-15 towed array sonar system fitted for tests. This sonar was operationally tested in the 6th Fleet, 1970–73 by DE 1015, DE 1021, and DE 1022. The two towed array AGs were stricken in 1977. Surviving units were all decommis-sioned for disposal as the new *Avengers* entered service: MSO 439 on 30 Sep-tember 1992, MSO 440 on 16 December 1993, MSO 441 on 2 July 1993, MSO 446 31 August 1992, MSO 455 on 30 September 1994, MSO 488 on 29 June 1994, MSO 489 30 April 1994, MSO 492 on 31 January 1994, MSO 509 on 12 December 1991, and MSO 511 on 31 December 1992. MSO 442, MSO 448, MSO 456, MSO 464 were broken up in 1993. MSO 455, 488, 489, 492 were all transferred to Taiwan; others of this class may also be transferred abroad. An improved MSO design (MSO 523 class) was planned for construction in the late 1960s as SCB-501 (FY66); it was to have combined MSO and MHC capabilities (940t, 184ft x 35ft at waterline). SCB-500.55 was the MSS-1 conversion, and SCB-502 was the MSO 422 class modernisation.

AVENGER class *ocean minesweeper/minehunter*

Displacement:	1312t (full load)
Dimensions:	224ft x 39ft x 12ft 2in *68.3m x 11.9m x 3.7m*
Machinery:	2 shafts: 4 Waukesha (MCM 1–2) or Isotta-Fraschini diesels, 2400bhp = 13.5kts
Sensors:	SQQ-32 sonar (SQQ-30 in MCM 1–9)
Complement:	81

Class (commissioning dates):
Avenger (MCM 1 12.9.87), *Defender* (30.9.89), *Sentry* (MCM 3, 2.9.89), *Champion* (MCM 4, 21.1.91), *Guardian* (MCM 5, 6.12.89), *Devastator* (MCM 6, 6.10.90), *Patriot* (Jul 1991), *Pioneer* (15.12.90), *Scout* (MCM 8, Aug 1992), *Warrior* (3.4.93), *Gladiator* (18.9.93), *Ardent* (18.2.94), *Dextrous* (9.6.94), *Chief* (Nov 1994)

Programmes: MCM 1 in FY82, MCM 2 in FY83, MCM 3–5 in FY84, MCM 6–9 FY85, MCM 10–11 in FY86 (not awarded until January 1989), last three FY90.

These are wooden-hulled craft with glass fibre superstructures. Reportedly they suffered from severe electromagnetic interference problems, and the hull had to be lengthened 7ft due to changes in weight. There were also engine problems (*Avenger* had hers installed wrong way round). There was only a limited stock of US-made Waukesha non-magnetic diesels, but Congress demanded that MCM 10 and later should have US-made units. Congress refused to authorise the three FY88 units because the programme was over a year behind schedule. These ships were reinstated in the FY90 budget. The MCMs are lineal descendants of the earlier MSOs. They carry both a mine neutralisation vehicle (SLQ-48) and standard influence and mechanical sweeps.

From the late 1960s onwards there were several projects for more or less sophisticated new minesweepers, as mine countermeasures technology became more and more elaborate. By the early 1980s it was primarily a matter of locating mines on the seabed, using high-definition sonar (currently SQQ-14), and then destroying mines one by one (minehunting). Success requires, among other things, a very high degree of navigational accuracy. Meanwhile, the function of the minesweepers has been a matter of some controversy. For example, with the appearance of deep-water ASW mines such as CAPTOR, there was a call for deep-water sweepers particularly adapted to keeping the approaches to ballistic submarine bases clear of Soviet mines. On the other hand, the Soviet fleet continued to threaten US ports and coastal waters with more conventional forms of mine warfare; and there is still the threat of mining to stop amphibious assaults – a threat which will abate somewhat as the United States shifts to air-cushion vehicles for such assaults.

A new programme for mine countermeasure (MCM) ship construction began in the 1970s, and the CNO approved a 1640t, 265ft, 18kt ship in Nov 1976. In early 1977 nineteen were planned for construction in FY79–FY82. Its primary purpose wuld have been to counter deep-water ASW mines, and it was abandoned as too expensive. A new design, closely based on that of the existing ocean minesweeper (MSO), was substituted. The first unit, USS *Avenger* (MCM 1), was ordered on 29 June 1982, under the FY82 programme. The next ships were moved from the FY84 to the FY83 programme in view of 'the severity of the mining threat'. In February 1982, the Reagan Administration FY83–87 shipbuilding programme showed four units in each of FY83 and FY84, and five in FY85.

The MCM was envisaged as the high end of a high-low mix of minesweepers, the lower end of which would consist of a combination of coastal minehunters (MSH) and mobilised civilian craft as well as minesweeping helicopters. As in the case of her predecessors, *Avenger* uses a variable-depth sonar (SQQ-30, an improved SQQ-14) for minehunting. Deep mines will be neutralised by remote-control vehicles, operated from an automated CIC, which will plot mine locations with the assistance of a precise integrated navigation system (SSN-2, PINS).

Sentry 1992 L & L van Ginderen

CARDINAL class *coastal minehunters* (MSH)

This air-cushion minehunter was designed by Bell Halter, which received a contract for the prototype, *Cardinal*, in April 1983. At that time it was described as displacing 450–500t fully loaded, and about 150ft long. The big flat deck associated with any air cushion design was considered particularly convenient for a minehunter, which typically requires a large deck area on which to handle its remotely-controlled vehicle. The MSH was intended for five-day missions.

In 1983 the MSH programme consisted of seventeen ships: a prototype actually ordered under the FY84 programme, and four per year for the next five years. An eventual total of thirty-five was planned.

The loser in the MSH competition was Marinette Marine, which offered a modified *Lerici*. When the MSH failed, the *Lerici* variant was picked up as the MHC.

OSPREY class *minehunters*

Displacement: 895t full load
Dimensions: 188ft × 34.8ft × 8.9ft
57.3m × 10.6m × 2.7m
Machinery: 2 Isotta-Fraschini diesels, 3200bhp = 12kts. Range 2500nm at 12kts
Sensors: Sonar SQQ-32
Complement: 51

Class (no, in service):
Osprey (MHC 51, 20.11.93), *Heron* (MHC 52, 6.8.94), *Pelican* (MHC 53, Jun 1995), *Robin* (MHC 54, Dec 1995), *Oriole* (MHC 55, Aug 1995), *Kingfisher* (MHC 56, Aug 1996), *Comorant* (MHC 57, Sep 1997), *Blackhawk* (MHC 58, Feb 1996), *Falcon* (MHC 59, Jul 1996), *Cardinal* (MHC 60, Dec 1996), *Raven* (MHC 61, Aug 1997), *Shrike* (MHC 62, Jun 1998).
Programme: MHC 51 under FY86 programme, MHC 52–53 under FY89, MHC 54–55 under FY90, MHC 56–58 under FY91, MHC 58–60 FY92, MHC 61–62 FY93. In 1991 plans called for the total of twelve (up to MHC 62), to be followed up eight lengthened MHC(V). Then in 1992 the MHC(V) was dropped altogether.

MHC 51 was bought with FY86 funds plus $5m in remaining FY84 funds from the *Cardinal* class failure. Although Marinette ordered the *Lerici* in 1983, these ships are being built by a combination of Intermarine and Hercules Powder. Displacement grew by 110t during final design, and production was well behind schedule.

In contrast to an MCM, which can carry both hunting (SLQ-48) and sweeping gear, an MHC can carry only one or the other. Later units will be modified internally for easier deployment overseas, but they will not be enlarged to accommodate both sweep and hunting gear. Because they were conceived as single-role minehunters, the MHCs lack the generating capacity to operate influence sweeps. They will carry the Modular Influence Sweep System, incorporating its own gas turbine generator.

BLUEBIRD ('ADJUTANT') class *coastal minesweepers*

Displacement: 360t light; 400t full load
Dimensions: 138ft wl, 145ft 5in oa × 26ft 6in × 7ft
42.1m, 44.3m × 8.1m × 2.1m
Machinery: 2 shaft, 2 GM diesels, 1000bhp = 13kts. Range 2500nm at 10kts
Armament: 1–20mm
Sensors: Sonar UQS-1
Complement: 39

Class (no):
Bluebird (MSC 121), *Cormorant* (MSC 222), *Falcon* (MSC 190), *Frigate Bird* (MSC 191), *Hummingbird* (MSC 192), *Jacana* (MSC 193), *Kingbird* (MSC 194), *Limpkin* (MSC 195), *Meadowlark* (MSC 196), *Parrot* (MSC 197), *Peacock* (MSC 198), *Phoebe* (MSC 199), *Redwing* (MSC 200), *Shrike* (MSC 201), *Spoonbill* (MSC 202), *Thrasher* (MSC 203), *Thrust* (MSC 204), *Vireo* (MSC 205), *Warbler* (MSC 206), *Whippoorwill* (MSC 207), *Widgeon* (MSC 208), *Woodpecker* (MSC 209), *Albatross* (MSC 289), *Gannet* (MSC 209).

These ships were essentially versions of the Second World War YMS, initially for transfer abroad under the MDAP programme, and initially designated AMS (later MSC, for coastal). The original design did not show XMAP capability, and 'improved magnetic signature', but because of cost and time considerations, and extreme measures taken in AM design will not be used'. Designs with both the new Packard (total 1200bhp) and a proven GM 8-268A (total 880bhp) diesel were prepared, the latter favoured for foreign customers. They were to be armed with a twin 20mm gun forward, compared to the 3in/50 of the YMS and to the single 40mm of the MSO. Of 159 units of the original *Bluebird* (SCB 69) class launched from 1953 onwards all but twenty were transferred. They are often called the *Adjutant* class. Many more units were built abroad with US assistance, primarily to British designs. Where the US MSCs had wooden frames, the British used aluminium. All had controllable-pitch propellers for agility in minefields. *Albatross* (MSC 289) and *Gannet* (MSC 290) were the last US sweepers authorised until the MCMs of the 1980s (which may not in fact be built) and were slightly enlarged *Bluebirds*. Of 25 units built, only two were retained in US service. They introduced gas turbine generators for magnetic sweeping and modified gear-handling equipment.

Of the US units, only MSC 203–209 had the original GM diesels; MSC 289–290 had updated GM units of about 1000bhp.

Note that US minesweeper (MSC construction for transfer abroad continued after the end of construction for the US Navy. For example, MSC 201–2 were built for Iran, MSC 293–4 for Pakistan, 295, 296, 316, 320, 321 for South Korea, 297, 301, 303, 313 for Thailand; 298, 299, 308–310, 314, 317, 318 for Greece; 300, 302, 306, 307 for Taiwan; 304, 305, 311, 312, 315 for Turkey; and 322–325 for Saudi Arabia; most are rated in a separate MSC 284 class. *Albatross* (MSC 289), and *Gannet* (MSC 290) were intended for transfer to Spain under MDAP, but MSC 200 and 202 were substituted. Disposals: MSC 289 and 290 were stricken 1970, MSC 194 (collision) and MSC 197 in 1972. Six were transferred to Indonesia in 1971: MSC 190–193, 195, 196. MSC 203 and 207 transferred to Singapore 1975. MSC 208 stricken 1973, MSC 122 1974, MSC 121, 198, 199, 201 1975, MSC 204 1977. MSC 205 and 206 to Fiji 1976, MSC 209 to Fiji 1976.

Thrasher 1964 USN

COVE class *inshore minesweepers*

Displacement: 197t light; 232t full load
Dimensions: 105ft wl; 110ft oa × 23ft × 8ft
32.0m, 33.5m × 7.0m × 2.4m
Machinery: 1 shaft, 2 GM diesels, 595bhp = 12.5kts. Range 1000nm at 9kts
Armament: 1 0.50cal MG

Class (no):
Cove (MSI 1), *Cape* (MSI 2). Both were built by Bethlehem, Bellingham, and launched in 1958.

These inshore sweepers (MSI) were intended to replace the MCs. Design work began in 1955, one author describing them as an attempt to fill the gap in shallow-water capability left by the deficiencies of the minesweeping boats (MSB). They were to carry MSB-size magnetic sweep gear and MSC-size moored sweep gear on a hull capable of limited ocean passages, and of overseas passage in an emergency – essentially the requirements levied on the original YMS of the Second World War. the advent of gas turbine generators permitted them to incorporate the basic sweeping capability of the MSC 289 class in a much smaller hull. Of fourteen ships built, two were retained by the US Navy and the rest transferred (SCB 136 type). There were also five units of an improved *Cove* class (MSI 15 type), capable of minehunting as well as of sweeping. All were built in 1965–67 specifically for transfer abroad.

The two US Navy units were relegated to research status, and in the early 1980s were being operated by the Applied Physics Laboratory of Johns Hopkins University (MSI 1) and the Naval Undersea Systems Center, San Diego (MSI 2).

BITTERN *minehunter*

Displacement: 300t standard; 360t full load
Dimensions: 138ft wl; 144ft 6in oa × 28ft × 8ft
42.1m, 44.1m × 8.5m × 2.4m
Machinery: 2-shaft diesels, 1200bhp = 14kts
Armament: 1–40mm
Complement: 40

Class (no, launched):
Bittern (MHC 433, 1957). Built by Consolidated SB.

Bittern, built under the FY54 programme, was a prototype to succeed thirty-one YMS and LSI(L) conversions (then called underwater object locators, or AMCU) under the FY52 (SCB 73) programme. She had no mine destruction capability at all, and was intended rather to locate and plot mine locations, so that small boats or divers could destroy them. To that end she was extremely manoeuvrable, with controllable-pitch propellers and active rudders. Her hull duplicated that of the contemporary *Bluebird* class coastal sweeper.

Proposals for additional minehunter construction (to a somewhat enlarged *Bittern* design, designated SCB 109A rather than 109) never materialised. This

Bittern 1958 USN

was the fate of all the mine craft. For example, an October 1956 sketch list of the FY58 budget showed two ocean and two coastal minesweepers as well as an MSS conversion; but when the budget was cut all had to be deleted. The next April a tentative list for FY59 showed the minehunter, as well as repeat sweepers and an MSS conversion. However, in order of priority the MHC was set below one of three coastal sweepers, and indeed below prototype coastal ASW craft – which were not built. Probably the last budget appearance of the MHC was in early versions of the FY61 budget; as of October 1958 it called for an MSS (then denoted SCB 195), mine support ship (MSC, SCB 123), and a coastal minehunter. However, a FY62–66 shipbuilding programme prepared at about that time showed no new minecraft construction; and all Navy plans of this period were being severely cut by the advent of Polaris.

Bittern herself was decommissioned in September 1965 and leased to a commercial operator in July 1968. She was stricken on 1 February 1972 and broken up.

UNDERWATER LOCATION

Of a total of fifty Underwater Locators, AMCU 1–6 were LCT(6) reclassified on 10 March 1945; AMCU 7–11 were LCI(L) reclassified the same day. AMCU 12–14, MHC 44 and 45, and AMCU 46–50 were all former YMS (coastal minesweeper) hulls, converted under the 1952 programme. AMCU 15–42 were former LCI(L) converted at the same time. Only *Bittern* was a new design. It was intended to build two more, for which the numbers MHC 44 and 45 were allocated. Early in 1954 five YMS were reclassified as AMCU 46–50; when, early in 1955, the two new *Bitterns* were cancelled, AMS (ex-*YMS*) -2 and -3 were the MHC 44 and 45 designations. It is not clear whether these higher numbers actually represent conversions. In addition, the conversion of LSI(L) 1089 to a minehunter under the FY52 programme was cancelled in 1954.

MSB type *minesweeping boats*

Displacement:	42t full load
Dimensions:	57ft 3in oa × 15ft 10in × 4ft 4in *17.5m × 4.8m × 1.3m*
Machinery:	2 diesels, 300bhp = 12kts. Oil 15.8t
Complement:	6

MSB 17 1966 USN

Mks 1 and 2 type *minesweeping launches*

Displacement:	23,100lb (10.3t) for magnetic sweep; 18,900lb for acoustic sweep; 17,600lb for sweep of moored mines (with, respectively, 680, 360 and 340gal of fuel)
Dimensions:	36ft oa × 11ft 7in × 3ft 8in *11.0m × 3.5m × 1.1m*
Machinery:	1 gas turbine, 200hp = 10kts (without sweep gear streamed). Range 100nm (with full power and full load)
Complement:	4

MSL Mk 4 type *minesweeping launches*

Displacement:	22,900lb (10.2t) for magnetic sweep; 20,950lb for acoustic; 18,870lb for moored mines (with in each case, 360gal of fuel)
Dimensions:	36ft × 11ft 7in × 3ft 8in *11.0m × 3.5m × 1.1m*
Machinery:	1 non-magnetic diesel, 160bhp = 10kts. Range 100nm
Complement:	4

Korean War experience showed particularly dramatically how badly the United States needed a shallow-water sweeping capability. In October 1950, 2000 North Korean mines kept a large US fleet out of the otherwise unprotected port of Wonsan for fifteen days, with only six sweepers available. There was already a fair amount of shallow-sweeping experience; during the Second World War 36ft LCVPs had been converted for moored sweeps in water less than 5 fathoms deep, and some were fitted with banks of storage batteries for magnetic sweeping. The army built four shallow-water minesweeping boats (two at Norfolk and two at Mare Island) in 1945; they came into navy hands when the army transferred the mine-planting mission in 1949. The ex-army craft became *MSB 1–4* and formed the design basis for the *MSB* 5 series.

Two new classes of shallow-water craft were designed for assault sweeping; the 57ft MSB and the 36ft MSL, the latter about the size and weight of the LCVPs which were standard for amphibious force Welin davits. The MSB itself grew too heavy for easy handling, partly because plans to use an aluminium hull had to be rejected: eddy currents set up as the boat moved through the earth's magnetic field could themselves generate a magnetic field. One unusual feature of the MSB design was an early use of gas turbines for its magnetic sweep generator. Of fifty ordered (*MSB 5–54*), one (*MSC 24*) was

MSL 32 1968 USN

not built, and *MSB 23* was destroyed by fire on 2 February 1955 and a plastic-hulled replacement was delivered in August 1956. *MSB 29* was an experimental unit built by John Trumpy & Sons, Annapolis, to see how sea-kindly a small sweeper could be; at 82ft it was far too large for overseas transport, but for some years served as flagship of the MSB squadron based at Charleston, South Carolina.

By 1989, only seven MSB remained (the big *MSB 29* plus *MSB 15, 16, 25, 28, 41,* and *51*). Four were sent to the Persian Gulf in 1987, as the Iranians began to lay mines there. *MSB 2* was transferred to Korea and *MSB 4* to Taiwan, both in December 1961. *MSB 25, 28, 29, 41* were all transferred to Panama in March 1993. *MSB 15* and *51* were discarded in 1992. *MSB 17* became a Canal Zone utility boat (57UB753).

The 36ft MSL could indeed fit an AKA davit. Early units were of wood, later ones of plastic. With a very limited size, they could be fitted for moored, magnetic, or acoustic sweeping. They had gas turbines for both propulsion and for magnetic sweeping power, and production continued at least into the 1960s. Surviving MSLs reportedly have diesel engines and a plastic hull; they have been stripped of sweeping gear and are employed as utility launches. Total production was fifty-six during 1946–66: Mk 1 were *MSL 1–4* (1946); Mk 2 were *MSL 5–29* (1948); Mk 3 *MSL 30* (1948); Mk 4 *MSL 31–56* (1966); Mk 5 comprised two Mk 1 and one Mk 2 converted to diesels (1967). The first two Mks were wooden, the rest plastic.

COOP MINE COUNTERMEASURES CRAFT

COOP (Craft of Opportunity Programme) was an attempt to eke out very limited mine countermeasures resources, mainly by conducting frequent bottom surveys of US ports. Special routes in and out of the ports ('Q-routes') were chosen. They had to be surveyed regularly so that any new objects (such as mines) could be identified easily and investigated; to check all objects on the bottom would be prohibitively difficult. The idea was to identify potential wartime COOP craft and to stockpile the necessary imaging (side-scanning) sonars. The boats would be operated by reservists, who would train in peacetime on board specially modified vessels. Several fishing boats were taken up (as well as at least one seized drug-runner), but most of the COOP trainers (designated CT) were converted Naval Academy and Newport training boats (YP series). The first two were converted in 1985. The programme of eighteen boats was to have been completed in 1988, but it had a low priority and it was stretched to 1990. By then the priority had vanished altogether; The navy therefore tried to cancel the programme in 1991, but Congress reinstated it for FY92 and FY93. Two prototype boats (*Frigatebird* and *Albatross*, CT-21 and CT-22, completed 15.10.88 and 5.11.88) were built by Princess Yachts of Tacoma, Washington: 20.4t (light), 58ft 7in × 17ft × 4ft (17.07m × 5.18m × 1.22m), 2 GM diesels, 330bhp = 12kts. By 1993 both had been assigned to Explosive Ordnance Disposal Mobility Units 11 and 17 on the Pacific coast. As of 1994, the other COOPs were twelve converted YPs (CT-1, -2, -4–6, -8–11, -15, -18, and -20) and two ex-fishing boats (CT-2 and -12). The converted YPs have SYQ-12 command systems and tow side-scan sonars. Eleven more conversions were eliminated from the programme, and by 1994 two others (the original CT-1 and CT-19) had been discarded.

Washtenaw County as MSS 2 in 1973 USN

MINE COUNTERMEASURE DEVICES

The threat of pressure mines was so severe that ship-size explosive targets were considered essential. In theory, they would simulate ship signatures, and would survive the subsequent mine explosions. The cylindrical XMAP was described as 'a towed, unmanned minesweeping device for sweeping modern pressure, acoustic and magnetic minefields'; it was to be strong enough to withstand fifty explosions by 2000lb mines in 40 feet of water, which appeared to require the use of 10in steel. Length was about 250ft, beam 29ft, and draft 20ft, with a displacement of 3100 tons, and two could be operated either side by side or in tandem. XMAP design concepts were reflected in the powerful towing engines of early postwar minesweeper designs. However, the prototype was disappointing, and the programme was abandoned in about 1955. On 1 January 1955, XMAP-1 was described as 85.3 per cent complete.

The 'guinea-pig' or 'sperrbrecher' was an alternative concept. As first described in 1948, it was to check-sweep sterilised fields of US pressure-magnetic mines or 'in cases of very great emergency, when no other method of pressure sweeping is available, such vessels may be used to attempt to clear a channel of enemy mines in order that more important vessels may pass'. At a minimum, such ships were to be of 5000 tons gross and 10kts, but larger and faster were considered better. Machinery would be controlled from the bridge, and degaussing coils or an emergency deck would be used to increase her magnetic field so as to explode magnetic mines, with luck beyond their lethal range. Buoyancy material would be fitted to protect the 'guinea-pig' from sinking in the swept channel; timber was an example of material which could be improvised. The 'guinea-pig' was considered very much an emergency device and a last resort, but its development (which continued well into the 1960s) suggests just how difficult minesweeping had become after 1945.

Three Liberty hulls were converted to specialised sweepers under the FY53 programme: two at Yokosuka (YAG 36 and 38), and one (*John L Sullivan*, YAG 37) at Norfolk. The two Japanese conversions may well have been 'guinea-pigs' to determine whether the US mining efforts of the Second World War had actually been neutralised; they were sold for scrap in 1960. YAG 37 was powered by four T34 turbo-props mounted on gunmounts on deck, all other machinery having been removed. After explosion tests in Chesapeake Bay in 1957–58 she was broken up.

The 'guinea-pig' concept continued with the MSS (special devices sweeper) originally proposed in the FY60 programme. It was to sweep 'by its own passage' moored, magnetic, acoustic, and pressure mines in depths safely navigable by ships of this size'. The pressure signature of a 20,000t ship was sought; by sweeping at high speed, it was hoped that the MSS would 'insure that vessels larger than the sweeper will be able safely to traverse the channel at speeds lower than that of the sweeper'. The YAG 37 experiments appeared to show that turbo-prop engines above deck would provide propulsion and steering without the threat of underwater damage. There was some hope that the ship's pressure signature could be increased by special pressure generators, and she was to be able to detonate acoustic and magnetic mines as far away as possible. Conversion of the 'Liberty' Ship *Harry L Glucksman* was authorised under the FY56 programme. Her hull was strengthened and her former superstructure replaced by a shock-mounted pilot house forward; propulsion was by four outboard motors rather than by the turbo-props of YAG 37. Completed on 16 Jun 1969, she was placed in reserve on 15 March 1973; in service, she had a crew of one officer and eight enlisted men. MSS 2 was a much more austere LST conversion carried out to support the clearance of North Vietnamese waters (Operation End Sweep); the major changes were the filling of the tank deck with styrofoam and some shock absorbers. LST 1166 was reclassified as MSS 2 on 9 February 1973, and was used to check-sweep sterilised channels; she was stricken 30 August 1973. Presumably she was a good example of the emergency 'guinea-pig' concept of 1948.

A new generation of sweep devices was bought during and after the Gulf conflict, including two Swedish SAMs (*Gerry* and *Peggy*, acquired January 1991 and sent to the Persian Gulf 5 February 1991). Reportedly the US Navy also investigated the German 'Troika' system.

RIVERINE MINESWEEPING DEVICES

The special requirements of the Vietnam War inspired the development of a considerable variety of small craft, only the designations of which remained a few years later: drone minesweepers (MSD), 70t river minesweepers (MSM), and 38t patrol minesweepers (MSR). The MSM were LCM(6) conversions,

A patrol minesweeper (MSR) with a minesweeping drone (MSD) in the foreground, 1969 USN

Catskill as MCS 1 in 1963 USN

as were many of the other riverine craft. They retained their bow ramps, and were fitted with bar armour against shaped-charge rockets. Typical armament was 2–20mm (not in all), 1–0.50cal MG, and 2–40mm rocket grenade launchers. They were fitted to sweep contact and acoustic mines. The patrol minesweepers, which replaced the MSM, were ASPB conversions (the ASPB is described under Small Surface Combatants above). Some were equipped with projecting pipe frames, to explode contact mines. Others had 4–3.5in bazookas fitted to their twin 0.50cal MG mounts forward. The MSD were special, being part of the force required to keep the approaches to Saigon clear. Eight of them were introduced in February 1969 by Mine Division 113; they drag-swept the Long Tau Shipping Channel linking Saigon to the South China Sea. In practice they were generally towed by larger sweepers to their operational area, then operated by radio control.

Many of these craft were transferred to the Vietnamese, and in 1982 many were in Vietnamese hands. Transfers included eight MSD (transferred 1970), eight MSR, and eight MSM (transferred 1970).

There were also large-ship conversions employed specifically in Vietnam, the Mine Countermeasures Headquarters Ships (MCS). These were planned as early as 1955, at which time all of the *Terror* class hulls (except for *Terror* herself) were redesignated in the MCS series. They were stricken on 1 September 1961, but *Catskill* (MCS 1) and *Ozark* (MCS 2) were reinstated (and actually converted) 1 June 1964 – 1 October 1965, respectively, under the FY64 and FY63 programme. They were armed with 2–5in /38 (2×1), and operated twenty MSCs and two minesweeping helicopters. The tank landing ship *Orleans Parish* was redesignated MCS 6 on 19 January 1959, reverting to LST status 1 June 1965. The LSD *Epping Forrest*, used as a mine escort tender, became MCS 7 on 30 November 1962.

THE COAST GUARD

HAMILTON class *coast guard cutters*

Displacement: 2716t standard; 3050t full load
Dimensions: 350ft wl, 378ft oa × 42ft 9in × 20ft
106.7m, 115.2m × 13.0m × 6.1m
Machinery: 2-shaft CODOG: 2 FM diesels, 7200bhp, 2 P & W gas turbines, 28,000shp = 29kts. See notes
Armament: 1–5in/38, 2–81mm mortars, 2–0.50cal MG, 6–12.75in Mk 32 ASW TT (2×3), 1 helicopter. See notes
Sensors: Radar SPS-29; sonar SQS-36 or SQS-38
Complement: 164

No	Name	Builder	Launched	Fate
WHEC 715	HAMILTON	Avondale	18.12.65	Extant 1995
WHEC 716	DALLAS	Avondale	1.10.66	Extant 1995
WHEC 717	MELLON	Avondale	11.2.67	Extant 1995
WHEC 718	CHASE	Avondale	20.5.67	Extant 1995
WHEC 719	BOUTWELL	Avondale	17.6.67	Extant 1995
WHEC 720	SHERMAN	Avondale	23.9.67	Extant 1995
WHEC 721	GALLATIN	Avondale	18.11.67	Extant 1995
WHEC 722	MORGENTHAU	Avondale	10.2.68	Extant 1995
WHEC 723	RUSH	Avondale	16.11.68	Extant 1995
WHEC 724	MUNROE	Avondale	5.12.70	Extant 1995
WHEC 725	JARVIS	Avondale	24.4.71	Extant 1995
WHEC 726	MIDGETT	Avondale	4.9.71	Extant 1995

These large high-endurance cutters were among the first US warships to be powered by gas turbines, Pratt & Whitney FT 4As in a CODOG (Combined Diesel or Gas turbine) system. Such propulsion was specifically justified by the combination of high endurance (14,000nm at 11kts on diesels) and high maximum speed (29kts on gas turbines, with a range of 2900nm), the latter to run out to the scene of a rescue. There is also a retractable bow manoeuvring unit, described as useful for some oceanographic work. The unusual twin-funnel arrangement can be traced to the desire to place the helicopter pad at the most favourable point in the ship from the point of view of ship motion, about three-quarters of the way aft. The price paid is the narrowness of the hangar, which can accommodate only one HH-52 turbo-shaft helicopter. In the *Perry* class, a very similar configuration was originally selected (albeit with one rather than two shafts), but the uptakes were amalgamated when the design was altered to accommodate two helicopters: the hangar had to be widened and the helicopter pad consequently moved right aft.

'Island' class WPB 1309, Aug 1994 L & L van Ginderen

The design originally called for a Hedgehog superfiring over the single 5in/38 gun forward, but Hedgehogs were removed from those ships which had them; all have the standard pair of triple Mk 32 lightweight torpedo tubes controlled by a Mk 309 Panel. These ships were extensively modernised under the FY85–88 programmes. The 5in gun and its Mk 56 fire control were replaced by an OTO-Melara 76mm gun and a Mk 92 (as in a PHM), the SPS-29D radar was replaced by SPS-40B, and provision was made to carry a LAMPS I helicopter (including installation of an SQR-17 signal processor). Provision was also made for each ship to carry up to eight Harpoons in canisters (WHEC 717 fired the first US Coast Guard Harpoon shots on 16 January 1990). Deleted from original plans were the SLQ-25 towed torpedo countermeasure and modern ECM (SLQ-32 and SRBOC).

With the end of the Cold War, ASW equipment was removed: in July 1992 it was decided to delete the SQS-38 hull sonar, SQR-17 sonobouy analyser, Mk 309 panel, and Mk 32 tubes.

RELIANCE class *coast guard cutters*

Displacement: 950t standard; 1007t full load
Dimensions: 210ft 6in oa × 34ft × 10ft 6in
64.2m × 10.4m × 3.2m
Machinery: 2 shafts, 2 Alco diesels, 5000bhp = 18kts; 5000nm at 13kts (WMEC 620–630, at 14kts). See notes
Armament: 1–3in/50. See notes
Sensors: Sonar (see notes)
Complement: 61

Class (no, launched):
Todd – *Reliance* (WTR 615, 25.5.63), *Diligence* (WMEC 616, 20.7.63), *Vigilant* (WMEC 617, 24.12.63)
Christy, Sturgeon Bay – *Active* (WMEC 618, 31.7.65) Coast Guard Yd – *Confidence* (WMEC 619, 8.5.65), *Venturous* (WMEC 625, 11.12.67), *Durable* (WMEC 628, 29.4.67), *Decisive* (WMEC 629, 14.12.67)

Dallas 3.6.94 L & L van Ginderen

American SB – *Resolute* (WMEC 620, 30.4.66), *Valiant* (WMEC 621, 14.1.67), *Courageous* (WMEC 622, 18.5.67), *Steadfast* (WMEC 623, 24.6.67), *Dauntless* (WMEC 624, 21.10.67), *Dependable* (WMEC 626, 16.3.68), *Vigorous* (WMEC 627, 4.5.68), *Alert* 630, 19.10.68)

These small cutters were designed to operate within about 500 miles of the US coast, with a fifteen day endurance. The first five have CODAG Propulsion (two 1000shp gas turbines, plus two 1500bhp diesels). There is no stack, the diesels exhaust aft. They can tow a 10,000t ship. Although they are too small for a hangar or maintenance facilities, they accommodate a landing pad large enough for the standard Coast Guard HH-52 or HH-3 helicopter. *Alert* was the first US ship fitted with the Canadian 'Beartrap' helicopter system, which provides greatly increased capability in rough water.

From a wartime point of view their armament is extremely limited, only one 3in/50 and two machine-guns; presumably the standard torpedo tubes would be added in wartime. There is no hull sonar in peacetime, but the design makes provision for a wartime installation of SQS-36/38.

Original plans called for the construction of thirty units, but only sixteen were built, all being launched between 1963 and 1968.

BEAR class *coast guard cutters*

Displacement: 1780t full load
Dimensions: 270ft × 38ft × 13ft 6in
82.3m × 11.6m × 4.1m
Machinery: 2-shaft diesels, 7000bhp = 19.5kts
Armament: 1–3in/62
Sensors: See notes
Complement: 95

Class (no, launched):
Bear (WMEC 901, 25.9.80), *Tampa* (WMEC 902, 19.3.81), *Harriet Lane* (WMEC 903, 6.2.82), *Northland* (WMEC 904, 7.5.82), *Spencer* (WMEC 905, 16.6.84), *Seneca* (WMEC 906 16.6.84), *Escanaba* (WMEC 907, 24.8.85), *Tahoma* (WMEC 908, 24.8.85), *Campbell* (WMEC 909, 30.8.86), *Thetis* (WMEC 910, 30.8.86), *Forward* (WMEC 911, 18.8.87), *Legare* (WMEC 912, 18.8.87), *Mohawk* (WMEC 913, 18.5.88).

Programme: two in FY77, two FY78, two FY79, three FY80, one FY81, three FY82.

These 270ft medium endurance cutters were originally designed to replace the 'Secretary' class high endurance units built before the Second World War. As in the case of the *Reliance* class, very large numbers were planned but not built. They lack any on-board ASW sensor or weapon, but their helicopter pad is large enough to accommodate a wartime LAMPS III and they have space and weight for a towed array. They also have space and weight reserved for possible wartime installations of Phalanx and eight Harpoons.

In FY88 Congress paid for an experimental conversion of one ship to emergency ASW platform, with an SQR-18 towed array, LAMPS data link, and SQR-17 helicopter/array processor. WMEC 907 ran trials with a LAMPS III helicopter during 1988.

Up to six ex-Navy undersea surveillance ships (*Stalwart* class) are to be converted into medium-endurance cutters. *Vindicator* (completed 21.11.84) was deactivated in April 1993 for cutter trials, and then commissioned for the Haiti blockade (20.5.94–19.8.94). She and *Persistant* (completed 14.8.85) are to be converted under the FY95 programme. They are currently fitted with SPS-49 air search radars but have no armament. Details: 1600t light (2285t full load), 223ft × 43ft × 15ft (68.28m × 13.1m × 4.57m); 4 diesels, 3200bhp = 11kts, range 300nm at 11kts plus 6480/3 (transit plus cruise on station, as T-AGOS surveillance ships). These ships were intended to conduct 60–90 day patrols, remaining at sea 292 days per year. They will presumably replace *Relience* class cutters.

COAST GUARD PATROL BOATS

During the war in Vietnam, the Coast Guard boats formed part of the Riverine Force, and as such they are included here. There are two series, 82 and 95ft boats. The former were intended to replace a series of Second World War wooden craft. All are steel-hulled, and the 95-footers were fitted with 'Mousetrap' type rocket ASW weapons.

All Coast Guard small craft are numbered in series prefixed by the nominal hull length; for example, CG 95315 is a 95-footer. The 95ft type was built in three series: A, 95300–95311, B, 95312–95320 in 1955–56, and C, 95321–95335, in 1958–59. Two were transferred to Haiti in 1956, two to Ethiopia in 1958, four to Thailand, one to Saudi Arabia, and nine to South Korea; some were new construction transferred under PGM designations, and indeed the 95-footer formed the basis for a PGM 53 was identical to the B type; and PGM 43 series was an austere variant of the B type). All the survivors received 'Cape' names in 1964. Of the three types, the 'C' type, specially adapted for search and rescue, has less armament, electronics, and displacement.

During the 1970s each carried a pair of 0.50cal machine-guns or a combination machine gun/81mm mortar. These weapons were later removed, and 'Cape' class cutters now have a pair of 40mm grenade launchers (Mk 64). Plans to scrap the twenty-five surviving units (several having been scrapped) in favour of thirty new WPBs were cancelled and the survivors were modernised in 1977–81 with new engines.

The last of the 'Cape' class were atricken in 1990–91 (three became navy service craft).

The 82ft cutters were built in four series: A, 82301–82317, built in 1960–61; B, 82318–82331, built in 1961; C, 82332–82370, built in 1962–63 and 1965–67; and D, 82371–82379, built in 1970. In 1965, twenty-six units were deployed with the navy in Vietnam, armed with combination 0.50cal machine-gun/81mm mortars, replacing the original 20mm cannon. As a result seventeen replacement cutters (plus nine already planned) were added to the programme in FY65 (82345–82350) and FY66 (82351–82370). All twenty-six in Vietnam were transferred to the Vietnamese Navy in 1969–70. 'Point' names were assigned in January 1964, later units being named as built, and classified as patrol craft rather than as patrol boats. Their high-speed diesels are controlled from the bridge; the heavier D series makes 22.6 rather than 23.7kts and has reduced endurance (1200/8kts as opposed to 1400nm to 1500nm at that speed in earlier craft). WPB 82314 had two gas turbines and controllable pitch propellers, for up to 26kts, but was later refitted with diesels. In all of these craft the original 20mm gun or combination mount had been removed by 1981, and armament reduced to a pair of 40mm grenade launchers (if that). They are described as well equipped for salvage and towing.

Mass replacement of earlier boats began in 1984, when fifteen Vosper Thornycroft-designed 'Island' class boats were ordered from Bollinger. A sixteenth was ordered the following year, with another sixteen in February 1987. Sixteen more were ordered in 1987 under Congressional Coast Defense Augmentation, five more in 1987 under the Omnibus Drug Act, and the last (WPB 1338–1349) were ordered under the FY90 Navy programme. In 1992 these boats were being rearmed with 25mm Bushmasters (one forward and space for another aft). WPB 1301–1317 displace 117t light (165t full load); WPB 1318–1349 are 107t (155t full load). Dimensions: 110ft × 21ft. All are poweered by two Alco-Paxman Valenta diesels, 5820bhp = 29.7kts. Endurance is 1853nm (24 hours at 26kts transit to patrol area and back, 96 hours at 13.1kts on patrol, a total of five days. Complement is sixteen.

The two prototypes of a new 'Heritage' class, ultimately to number ninety-six, were ordered on 1 July 1989. They have a deep-vee hull form and are expected to last 25 years, compared to 15 for the 'Islands'. Characteristics: 140t (160t full load), 120ft × 23ft, two dicscls, 30kts, rangc 1152nm at 12kts, complement 17. Armament is a 20mm gun plus two 0.50cal machine guns. This programme was cancelled 25 November 1991, although the first unit, *Leopold* (WPB 1400) was nearing launch. As of 1994 a smaller design, to replace the existing 82ft 'Point' class, was being developed; the prototype (of thirty-one) was to be included in the FY95 programme.

FORMER NAVY UNITS

In 1952 the Coast Guard took over twelve former diesel destroyer escorts (see under Fleet Strength 1947 for names and fates), largely for air-sea rescue in the Pacific.

In 1946–48 the Coast Guard took over eighteen former small seaplane tenders (WAVP 370–383 on loan and 384–387 on permanent transfer). They served as cutters into the 1970s: WAVP 370, *Casco*, 371 *Mackinac*, 372 *Humboldt*, 373 *Matagorda*, 374 *Absecon*, 375 *Chincoteague*, 376 *Coos Bay*, 377 *Rock-*

Reliance 16.6.71 US Coast Guard

Bear as completed US Coast Guard

away, 378 *Half Moon*, 379 *Unimak*, 380 *Yakutat*, 381 *Baratraria*, 382 *Bering Strait*, 383 *Castle Rock*, 384 *Cook Inlet*, 385 *Dexter* (ex-*Biscayne*), 386 *McCulloch* (ex-*Wachappreague*), 387 *Gresham* (ex-*Willoughby*). Former Navy designations: AVP 12, 13, 21–25, 29, 26, 31–36, 11, 56, 57. *Dexter* was a former command ship (AGC 18), and *McCulloch* and *Gresham* were former MTB tenders (AGP 8 and AGP 9). In Coast Guard service these ships were armed with a single 5in/38 and two twin 40mm (one quadruple in 374, none in 385). Two twin 20mm were removed in 1957. ASW armament consisted of a Hedgehog (Mk 10) and four DC projectors. Triple 12.75in ASW torpedo tubes were fitted in the mid-1960s. They were used intially as weather ships, with a balloon hangar and launching deck aft. *Dexter* was fitted with four new FM diesels 1957, recommissioned July 1958 as West Coast Training ship. *Rockaway* was adapted as oceanographic ship, 1966. *Unimak* was adapted as East Coast training ship, 1966. She was redesignated as a training cutter (WTR) in 1969, serving as such until her decommissioning in 1975, and was recommissioned as a High Endurance Cutter (WHEC) to enforce the 200m offshore resource zone in 1977, the last of her class. *Gresham*, modified for ocean-station weather work, ultimately had her single 5in/38 removed. All were redesignated High Endurance Cutters (WHEC) in 1966. WHEC 376 returned to Navy 1967 and sunk as target; 371, 373, 385 returned 1968 and sunk as targets; 370 returned 1969 and sunk as target. WHEC 372 and 381 were stricken in 1970. WHEC 380 and 382 were transferred to South Vietnam 1 January 1971, 383 and 394 on 21 December 1971, and 374, 385 and 386 on 21 June 1972. New names: *Pham Ngu Lao* (374), *Ly Thoung Kiet* (375), *Tran Nhat Quat* (380), *Tran Quang Khai* (383), *Tran Binh Trong* (383), *Tran Quoc Toan* (384), and *Ngo Kuyen* (386). In Vietnamese service, all ASW equipment was removed, and 2–81mm mortars replaced the former 40mm battery. Of these ships, the former WHEC 375, 382, 283 and 386 succeeded in fleeing to the Philippines in 1976, upon the fall of South Vietnam. They became the Philippine *Andres Bonifacio*, *Diego Silang*, *Francisco Dagohoy* and *Gregorio de Pillar*.

Uruguay

Economic weakness (and an internal war against Tupamaros guerillas in the 1960s and 1970s) has hampered the growth of the armed forces, and particularly the navy (volunteer personnel 4950 in 1995 including 1000 marines and naval air arm). US aid stopped in 1977 following allegations of human rights violations and Uruguay now renounces all aid, and strongly advocates a policy of non-intervention in other countries. The main naval base is Montevideo.

FLEET STRENGTH 1947

TORPEDO GUNBOAT

Name	Launched	Disp	Fate
URUGUAY	12.4.10	1150t	Discarded 1953

PATROL CRAFT

Name	Launched	Disp	Fate
PAYSANDU	21.7.35	180t	Discarded 1963
RIO NEGRO	22.8.35	180t	Discarded 1969
SALTO	11.8.35	180t	Discarded 1973; survey ship

There was also the ex-US submarine chaser *Maldonado* (ex-*PC 1234*) transferred in 1964 (stricken *c*1969) and the ex-US 63ft ARV *Colonia*, transferred in 1945 and stricken 1986.

MAJOR SURFACE SHIPS

Ex-US DEALEY class *frigate*

No	Name	Builder	Acquired	Fate
3	18 DE JULIO (ex-*Dealey*)	Bath Iron Works	28.7.72	Deleted 1991

Refit 1979–80, Arsenal de Marinha, Rio de Janeiro; first US-built postwar escort constructed and the least modified of her class.

Ex-US CANNON class *frigates*

No	Name	Builder	Acquired	Fate
1	URUGUAY (ex-*Baron*)	Federal SB	May 1952	Deleted 1990
2	ARTIGAS (ex-*Bronstein*)	Federal SB	Mar 1952	Deleted 1988

Modernised in late 1960s and now fitted with SPS-5, SPS-6C (not in *Uruguay*) radar and SQS-4 sonar. Armament in 1982 was 3–3in (3×1), 2–40mm (1×2), 4–20mm (4×1).

Ex-French COMMANDANT RIVIERE class *frigates*

No	Name	Builder	Launched	Acquired	Fate
1	URUGUAY (ex-*Commandant Bourdais*)	Lorient DY	15.4.61	20.8.90	Extant 1995
2	GENERAL ARTIGAS (ex-*Victor Schoelcher*)	Lorient DY	11.10.58	9.1.89	Extant 1995
4	MONTEVIDEO (ex-*Admiral Charner*)	Lorient DY	14.12.62	28.1.91	Extant 1995

Former French frigates of *Commandant Riviere* class; *General Artigas* bought in 1988 and other two in 1990. Exocet missiles and Dagaie chaff launchers removed before transfer, but restored in 1992. Can carry flag officer and staff and embark up to eighty soldiers. *Montevideo* now has pennant no. 3.

AMPHIBIOUS WARFARE VESSELS

Ex-US LCM(6) class *landing ships*

Two units *LD 40* and *LD 41* transferred October 1972.

Uruguay also operates a 12t LCU, *LD 42* and two 31t LCUs *LD 43* and *LD 45*.

SMALL SURFACE COMBATANTS

Ex-British 'CASTLE' class *corvette*

Class (former names):
Montevideo (ex-*Arnrior*, ex-*Rising Castle*).

Transferred 1956; used as a training ship; discarded 1975.

Ex-US AUK class *corvette*

Class (former name):
Commandante Pedro Campbell (ex-*Chickadee*)

Transferred 18 August 1966. Reclassified as corvettes although still has some minesweeping gear. Carried pennant no 4, now 24; extant 1995.

Ex-US AGILE class *corvette*

Class (no, former names):
Maldonado (MS 33, ex-*Bir Hakeim*, ex-*MSO 451*)

A US *Agile* class ocean minesweeper transferred to France in 1964 and then to Uruguay in September 1974. All sweeping gear was removed and she operated as an escort; stricken 1979.

Ex-GDR 'KONDOR II' (Type 89.200) *coastal minesweepers*

Class (no, former name, launched):
Temerario (31, ex-*Riesa*, 2.10.72), *Valiente* (32, ex-*Eilenburg*, 31.8.72), *Fortuna* (33, ex-*Bernau*, 3.8.72), *Audaz* (34, ex-*Eisleben*, 2.1.73).

Former East German minesweepers of 'Kondor II' class, officially called 'Type 89.200'. Taken over by Bundesmarine on Germany's unification and sold to Uruguay in October 1991 together with Polish 'PIAST' class (Type 570) salvage vessel *Otto von Guericke* (now *Vanguardia 26*). Minesweeping gear was retained, but twin 25mm guns removed on transfer to be replaced by 40mm guns. All extant 1995.

Ex-US ADJUTANT class *large patrol craft*

Class (no, former names):
Rio Negro (13, ex-*Marguerite*, ex-*MSC 94*)

Transferred 10 November 1969 from France. Minesweeping gear removed. Deleted 1990.

French VIGILANTE type *large patrol craft*

Displacement: 166t normal; 190t full load
Dimensions: 136ft 2in oa, 124ft 7in pp × 22ft 4in × 6ft 11in
41.5m, 38.0m × 6.8m × 2.1m
Machinery: 2 shafts, 2 MTU diesels, 5400hp = 25kts. Range 2400nm at 15kts
Armament: 1–40mm (provision for 2–20mm)
Sensors: Radar Decca TM 1226C, 1229
Complement: 25

No	Name	Builder	Comp	Fate
5	15 DE NOVIEMBRE	CMN, Cherbourg	16.10.80	Extant 1995
6	26 DE AGOSTO	CMN, Cherbourg	11.2.80	Extant 1995
7	COMODORO COE	CMN, Cherbourg	23.1.81	Extant 1995

Ordered in 1978 and all commissioned 25 March 1981. Have CSEE Panda fire control.

Uruguay also operates an American-built coastal patrol craft, *Paysandu* (60t, commissioned 1968) and a coastal patrol craft, *Carmelo* (70t, commissioned 1957).

Two ex-US 'Cape' class large patrol craft built in 1953 and 1958: *Colonia* (10, ex-*Cape Higgon*), *Rio Negro* (11, ex-*Cape Horn*). Acquired 1990, both extant 1995.

Vanuatu

The former Anglo-French administered New Hebrides in the southwest Pacific became independent on 30 July 1980 as the Republic of Vanuatu. Her Marine Police, based at Port Vila on Efate Island, originally operated a former private yacht *Mala*, which was replaced in 1987 by an Australian-built ASI 315 patrol boat *Tukoro* (for details see under Tuvalu); extant 1995.

Venezuela

The Venezuelan Navy has dramatically increased in size since Second World War. This may be attributed to the development of the nation's vast natural resources, the role played by the navy in the removal of Marcos Perez Jimenez in 1958, and the growing threat of terrorist forces in the Caribbean area. Venezuela also has a long-standing claim on Guyanan territory, currently in abeyance, but supports Belize against Guatemalan claims. The majority of personnel, 14,800 (including 5200 marines and a naval air arm) in 1995, are 18-month conscripts. The main naval bases are Caracas, Puerto Cabella, La Guaira and Puerto de Hierro. Long term development plans include two submarines, two LSMs, four MCM craft, fast attack craft, 800t offshore patrol vessels and small landing ships of LCU type and a replenishment tanker, but financial problems cause delays. Plans to acquire ex-US *Barbel* class submarines came to nothing. Besides the Army and Navy there is also 10,000 strong National Guard (*Fuerzas Armadas de Cooperacion*) with a Marine Wing operating over 100 coastal patrol craft (up to 50t displacement). Coast Guard paramilitary forces include two old frigates *Almirante Clemente* and *General Moran*, one former tug *Miguel Rodriguez* (ex-*Salinan* ATF 161) and twenty-four patrol craft of various sizes.

FLEET STRENGTH 1947

GUNBOATS

Name	Launched	Disp	Fate
GENERAL SOUBLETTE	1925	615t	Discarded 1950
GENERAL URDANETA	1925	615t	Discarded 1950

CORVETTES

Name	Launched	Disp	Fate
CONSTITUCION (ex-*Algoma*)	1946	1060t	Discarded 1962
FEDERACION (ex-*Amherst*)	1946	1060t	Discarded 1956
INDEPENDENCIA (ex-*Dunvegan*)	1946	1060t	Discarded 1953
LIBERTAD (ex-*Battleford*)	1946	1060t	Wrecked 12.4.49
PATRIA (ex-*Oakville*)	1946	1060t	Discarded 1962
VICTORIA (ex-*Westaskawin*)	1946	1060t	Discarded 1962

There were also four ex-US Coast Guard 47t cutters, *Antonio Diaz, Arimendi, Briceno Mendez* and *Brion*, transferred in 1947; *Brion* was renamed *Felipe Santiago Esteves* in 1957.

MAJOR SURFACE SHIPS

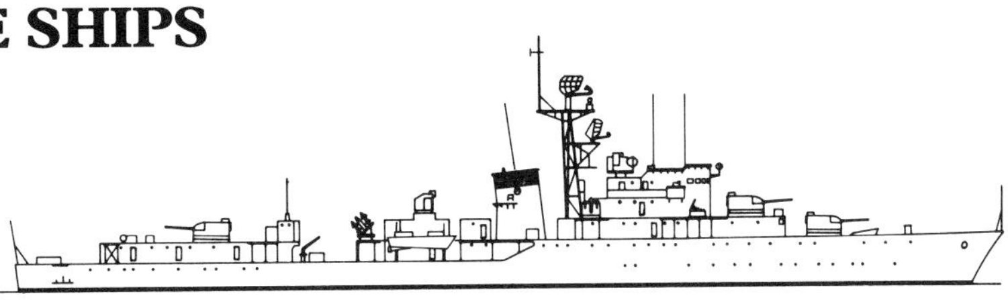

Nueva Esparta 1970

NUEVA ESPARTA class *destroyers*

Displacement: 2600t standard; 3300t full load
Dimensions: 384ft wl, 402ft oa × 43ft × 12ft 9in max
117.1m, 122.5m × 13.1m × 3.9m
Machinery: 2-shaft Parsons geared turbines, 2 Yarrow boilers, 50,000shp = 34.5kts. Range 5000nm at 11kts
Armament: 6–4.5in (3×2), 16–40mm/60 (8×2), 3–21in TT (1×3), 2 DCT, 2 DC racks
Complement: 254

No	Name	Builder	Laid down	Launched	Comp	Fate
D 31	ARAGUA	Vickers-Armstrong, Barrow	29.6.53	27.1.55	14.2.56	Discarded 1975
D 11	NUEVA ESPARTA	Vickers-Armstrong, Barrow	24.7.51	19.11.52	8.12.53	Discarded 1978
D 21	ZULIA	Vickers-Armstrong, Barrow	24.7.51	29.6.53	15.9.54	Discarded 1978

Specifically designed for tropical operations, these units are air conditioned throughout and are quite spacious. The single uptake and stack serviced the two independent engine and boiler rooms. The 4.5in guns were British Mk IV.

Nueva Esparta and *Zulia* were refitted at Palmers, Hebburn and Vickers, Barrow in 1959 when 2 Squid ASW mortars were fitted and TT removed. They went on to New York N Yd in 1960 for improvements to their electronics. *Nueva Esparta* was again refitted in 1968–69, at Cammell Laird, when 2 Seacat SAM (2×4) replaced 12–40mm. Electronics by this time comprised SPS-12, SP6-6 (*Zulia*) or AWS-2 and Type 276, with 2 GWS 20 Seacat directors on *Nueva Esparta*.

Nueva Esparta 1952 USN

Falcon received a FRAM II modernisation while in US service, and now operates a Bell 475 helicopter from the former DASH platform. A single 40mm aft was added to *Carabobo* in 1978.

Ex-US ALLEN M SUMNER class *destroyers*

No	Name	Builder	Acquired	Fate
D 22	FALCON (ex-*Robert K Huntington*)	Todd-Pacific, Seattle	31.10.73	Stricken 1981
D 21	CARABOBO (ex-*Beauty*)	Bethlehem, Staten I	14.7.74	Stricken 1981

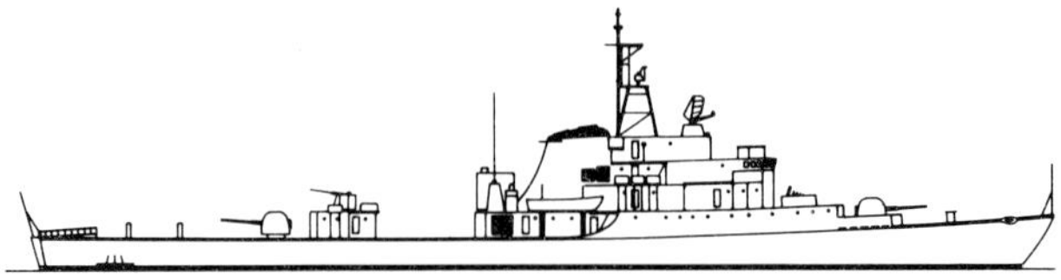

General José Trinidad Moran 1978

This class suffered from many problems, primarily due to their light construction. Ships were air conditioned in living and command spaces. Original 4in main battery was fully automatic and radar controlled. They were rearmed as follows: *Almirante Jose Gacia, Almirante Brion* and *General Jose de Austria* at Ansaldo, Leghorn in 1962 received 1 Lanciabas DC launcher, 8–20mm were removed; *Almirante Clemente, General Jose Trinidad Moran* at Cammell Laird/Plessey from 1968–75 received 2–76mm OTO Melara Compact, in place of 4–4in, and new electronics. The rest of the armament then comprised 2–40mm (1×2), 6–324mm Mk 32 ASW TT (2×3) with Plessey AWS-2, Decca 1226 and Selenia RTN-10X radar and Plessey MS-26 sonar. Best sea speed for all units in later years was 29kts. D 12 and D 22 now re-numbered GC 11 and GC 12 and have been operated by the Coast Guard 1986. Re-engined in Italy in 1984–85 with 2 GMT 16-645 E7CA diesels, 6000bhp = 22kts.

ALMIRANTE CLEMENTE class *frigates*

Displacement: 1300t standard; 1500t full load
Dimensions: 325ft 2in oa × 35ft 5in × 11ft 2in max
99.1m × 10.8m × 3.4m
Machinery: 2-shaft geared turbines, 2 Foster Wheeler boilers, 24,000shp = 34kts. Range 3500nm at 15kts
Armament: 4–4in (2×2), 4–40mm/60 (2×2), 8–20mm/80 (8×1), 2 Hedgehogs, 4 DCT, 2 DC racks, 3–21in TT (1×3)
Complement: 162

No	Name	Builder	Laid down	Launched	Comp	Fate
D 23	ALMIRANTE BRION	Ansaldo Leghorn	12.12.54	4.9.55	1957	Discarded 1978
D 12	ALMIRANTE CLEMENTE	Ansaldo, Leghorn	5.5.54	12.12.54	1956	Extant 1995
D 33	ALMIRANTE JOSE GARCIA	Ansaldo, Leghorn	12.12.54	12.10.56	1957	Discarded 1977
D 32	GENERAL JOSE DE AUSTRIA	Ansaldo, Leghorn	12.12.54	15.7.56	1957	Discarded 1976
D 22	GENERAL JOSE TRINIDAD MORAN	Ansaldo, Leghorn	5.5.54	12.12.54	1956	Extant 1995
D 23	GENERAL JUAN JOSE FLORES	Ansaldo, Leghorn	5.5.54	7.2.55	1956	Discarded 1978

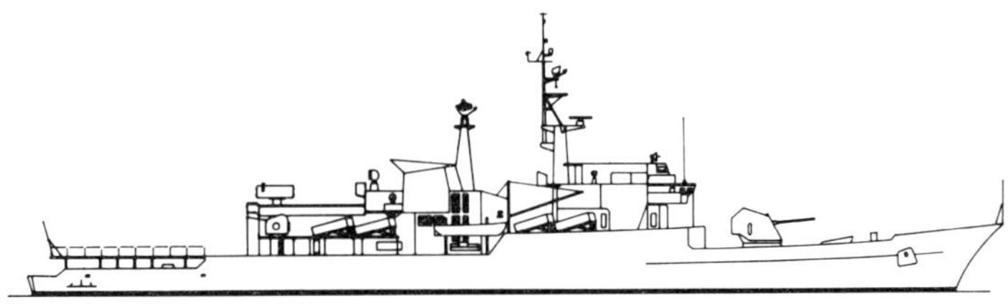

General Soublette as completed

The Venezuelan units are very similar to the standard *Lupos*. However, they have a fixed hangar (for two helicopters) and, therefore, carry no reloads for the Albatros point defence SAM which fires Aspide missiles. In 1981 *Mariscal Sucre* successfully fired an Aspide during a UNITAS exercise. See under Italy for full details. F 26 renamed *Almirante Garcia*.

Italian LUPO class *frigates*

No	Name	Builder	Laid down	Launched	Comp	Fate
F 22	ALMIRANTE BRION	CNR, Riva Trigoso	Jun 1977	22.2.79	7.3.81	Extant 1995
F 25	GENERAL SALOM	CNR, Riva Trigoso	7.11.78	13.1.80	Feb 82	Extant 1995
F 24	GENERAL SOUBLETTE	CNR, Riva Trigoso	26.8.78	4.1.80	5.12.81	Extant 1995
F 23	GENERAL URDANETA	CNR, Riva Trigoso	23.1.78	23.3.79	8.8.81	Extant 1995
F 26	JOSE FELIX RIBAS	CNR, Riva Trigoso	21.8.79	4.12.80	30.7.82	Extant 1995
F 21	MARISCAL SUCRE	CNR, Riva Trigoso	19.11.76	28.9.78	10.5.80	Extant 1995

Almirante José de Garcia Mar 1995
L & L van Ginderen

SUBMARINES

Overhauled at San Francisco Navy Yard 1962; used for training.

Ex-US BALAO class *submarines*

No	Name	Builder	Acquired	Fate
S 11	CARITE (ex-*Tilefish*)	Mare Island N Yd	4.5.60	Discarded 28.1.77

Picuda was overhauled 1979–81 at Puerto Belgrano N Yd in Argentina. S 22 used for harbour training.

Ex-US GUPPY II type *submarines*

No	Name	Builder	Acquired	Fate
S 21	TIBURON (ex-*Cubera*)	Electric Boat	5.1.72	Stricken 1979
S 22	PICUDA (ex-*Grenadier*)	Boston N Yd	15.5.73	Deleted 1990

First two units ordered in 1971 and a second pair on 10 March 1977. They are slightly longer (195ft, *59.5m*) than the standard Type 209, since they incorporate an enlarged bow sonar like the West German Type 206. Second pair planned but not ordered. Refitted at Kiel in 1981 and 1984. Modernised at Kiel 1991–92. Pennant nos. changed to S 31 and S 32 respectively.

West German Type 209 *submarines*

Particulars: As Greek *Glavkos* class

No	Name	Builder	Laid down	Launched	Comp	Fate
S 22	CARIBE	Howaldswerke, Kiel	1.8.73	16.12.75	11.3.77	Extant 1995
S 21	SABALO	Howaldswerke, Kiel	2.5.73	21.8.75	6.8.76	Extant 1995

AMPHIBIOUS WARFARE VESSELS

Ex-US TERREBONNE PARISH class *tank landing ship*

Class (no, former name):
Amazonas (T 51, ex-*Vernon County*).

Transferred on loan 29 June 1973 and purchased 30 December 1977.

Venezuela operates the ex-US repair ship (ex-LST) *Guyana* (T 31, ex-*Quirius*, ex-*LST 1151*); she was used as transport. Deleted 1984. Minor landing craft amount to twelve LCUPs built locally to the standard US design, plus a few ex-US LCVPs and LCPLs.

Four ex-US LSM 1 class landing ships, *Los Monjes* (T 21 ex-T 13; ex-USS *LSM 548*), *Los Roques* (T 22 ex-T 14; ex-USS *LSM 543*), *Los Frailes* (T 23 ex-T 15; ex-USS *LSM 544*), *Los Testigos* (T 24 ex-T 16; ex-USS *LSM 545*) transferred in 1959–60 under MAP; T 21 and T 22 stricken in 1979, T 24 in 1980 and T 23 in 1984.

CAPANA class *tank landing ships*

Displacement: 1800t light; 4070t full load
Dimensions: 344ft × 50ft 6in × 9ft 9in (13ft 9in)
104.8m × 15.4m × 3.0m (4.2m max)
Machinery: 2 shafts, 2 diesels, 7200bhp = 15kts. Range 5600nm at 11kts
Armament: 2–40mm Breda (1×2), 2–20mm Oerlikon
Sensors: Selenia NA 18, radar nav.
Complement: 117 (13 officers)

No	Name	Builder	Launched	Fate
T 61	CAPANA	Korea Tacoma	25.3.83	Extant 1995
T 62	ESEQUIBO	Korea Tacoma	25.3.83	Extant 1995
T 63	GOAJIRA	Korea Tacoma	1984	Extant 1995
T 64	LOS LLANOS	Korea Tacoma	1984	Extant 1995

Modern landing ships designed on lines of the famous WW II LST type ordered in 1982 at Korean shipyard. First pair commissioned on 24 July 1984 and second pair on 20 November 1984. Can carry maximum 1800t, while maximum beaching load is 690t. Can carry 192 fully equipped troops with 10 officers. Each ship has an elevator to the upper deck and tank turntable on the tank deck. All delivered without armament. Have helicopter deck aft. Sister-ships in Indonesia.

Swiftships LCU *landing craft*

Displacement: 390t full load
Dimensions: 130ft × 36ft × 6ft
39.6m × 11.0m × 1.8m
Machinery: 2 shafts, 2 Detroit diesels, 1800bhp = 13kts. Range 1500nm at 10kts
Armament: 3–12.7mm MG
Complement: 26

Class (no, in serv.):
Margarita (T 71, 20 January 1984), *La Orchila* (T 72, 11 May 1984).

Small landing craft of LCU type built by Swiftships Inc., Morgan City. Military load 150t. Carry 100t of fuel. Fitted with 15t crane. Assigned to River Command. Four more were planned.

SMALL SURFACE COMBATANTS

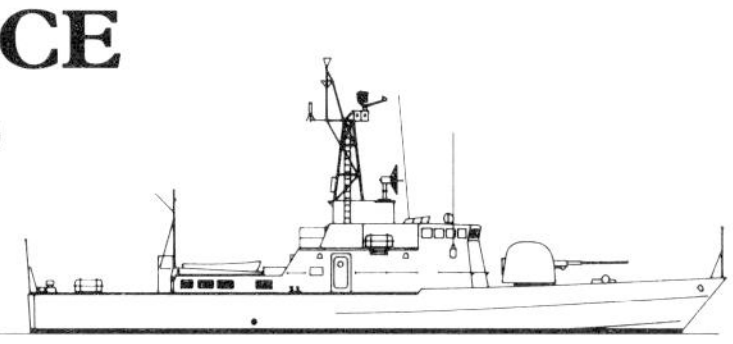

Constitucion as completed
NB 1/750 scale

CONSTITUCION class *fast attack craft*

Displacement: 170t full load
Dimensions: 121ft pp × 23ft 4in × 5ft 11in max
36.90m × 7.10m × 1.80m
Machinery: 2 shafts, 2 MTU diesels, 7200bhp = 31kts. Range 1350nm at 16kts
Armament: 2 Otomat SSM, 1–40mm (*Constitucion, Independencia, Patria*: 1–76mm/62 OTO Melara Compact)
Sensors: Radar SPQ-2D (plus Orion RTN-10X in gun and vessels)
Complement: 17

No	Name	Builder	Launched	Fate
P 11	CONSTITUCION	Vosper Thornycroft	1.6.73	Extant 1995
P 12	FEDERACION	Vosper Thornycroft	26.2.74	Extant 1995
P 13	INDEPENDENCIA	Vosper Thornycroft	24.7.73	Extant 1995
P 14	LIBERTAD	Vosper Thornycroft	5.3.74	Extant 1995
P 15	PATRIA	Vosper Thornycroft	27.9.73	Extant 1995
P 16	VICTORIA	Vosper Thornycroft	3.9.74	Extant 1995

Vosper Thornycroft designed attack craft in gun and missile armed variants. Laid down January 1973 – March 1974 and completed 16 August 1974 – 22 September 1975.

THE NATIONAL GUARD
The National Guard operates a number of small craft, including: 22 *Rio Orinoco* class (65t, 25kts, 1974–78); 12 *Rio Meta* class (45t, 30kts, 1970–1 and 1976–77); 8 *Rio Apure* class (38t, 24kts, 1955), 17 US 'Enforcer' type; 4 new 77ft (*23.5m*) type were ordered in 1980 from R E Direcktor, New York.

Federacion as completed — Vosper Thornycroft

Vietnam

When South Vietnam was conquered by the armed forces of North Vietnam in April 1975 most of her navy's sea-going warships escaped capture by sailing into foreign ports (mainly Philippine). They totalled ten escorts, seven landing ships, three patrol vessels and three repair ships. Other ships were destroyed or were captured by the North Vietnamese in a poor state of repair. For example, the frigate *Tran Khanh Du* (HQ 04) was captured in Saigon during refit; others were scuttled, such as the LSM *Lam Giang* (HQ 402), *Keon Ngua* (HQ 604) and the oiler HQ 474.

It is difficult to give an accurate account of the fates of most former South Vietnamese ships because of the lack of official information covering the last twenty years since the fall of Saigon. However, several larger units have been seen at sea and where more information has come to light about these ships, it has been added to their entries in the South Vietnam section. Large numbers of captured ships have either been stricken or are extant in a very marginal condition due to age and lack of spares.

The Vietnam War (1946–75) is thought of mainly as a ground and air conflict but in both phases, French and American, naval operations, especially amphibious and river ones, were widespread. One whole theatre of war, the Mekong Delta, was a watery maze never dominated by one side or the other. American involvement was gravely increased by the Gulf of Tonkin Incident in August 1964. On 2 August three North Vietnamese ex-Soviet 'P 4' torpedo boats attacked the destroyer USS *Maddox* 30–60 miles off the coast because she had been violating territorial waters (Hanoi claimed 12 miles instead of 3) and more importantly because she had given distant cover to a South Vietnamese 31 July bombardment of the North Vietnamese islands of Hon Me and Hon Nieu, 3–5 miles from the mainland. Communist infiltration of the South was done by sea as well as by the Ho Chih Minh Trail. A second apparent attack (4 August) on *Maddox* and her sister destroyer *Turner Joy* in an atmosphere of considerable electronic confusion precipitated US carrier air strikes that claimed the destruction of twenty-five North Vietnamese warships and fuel bases for the loss of three planes and a pilot.

By this stage South Vietnam's Navy (established 1 January 1955) numbered 15,350 men including a marine corps whereas the North had 2200 sailors (given warships from 1958). Six years later the South's navy had doubled to crew thirty-nine surface combatants, 215 landing vessels, twelve mine launchers and thirty auxiliaries. Their role was increasingly active despite the US Navy's obvious predominance. Both services played a crucial part in blunting the North Vietnamese 1972 offensive into Quang Tri Province with gunfire support and amphibious landings.

The South's naval prowess was demonstrated in January 1974. A Chinese battalion-strong invasion of the Paracel Islands, 225 miles east of Vietnam, cost the Communist superpower two warships sunk and two heavily damaged for one Vietnamese vessel (twenty-three survivors rescued by a Dutch cargo ship). Ten days later the Souths' warships put troops on to the Spratly Islands, 300 miles to the southwest, to prevent their seizure. In April 1976 the Vietnamese Liberation Navy replaced the garrison.

The fall of South Vietnam in early 1975 began in the Central Highlands and her navy, short of fuel and spare parts though not ammunition, could not provide the firepower that had been available in 1972. Evacuation of Hue, Da Nang (6000 marines and 4000 other troops), Qui Nhon and other coastal towns degenerated into chaos and heralded the exodus since of 800,000 'Boat People'.

In 1995 the Vietnamese Navy had some 7000 personnel operating its ageing fleet of ex-Soviet and American ships together with a land-based force of thirteen helicopters and three ASW amphibians. During the brief Sino-Vietnamese war of 1979, the Soviet Pacific Fleet increased its presence at Cam Ranh Bay, and it was reported that nuclear submarines were using the anchorage, but following the withdrawal of all Soviet forces from overseas bases, the loss of Soviet support has hit the navy hard and the operational status of many of its units is questionable.

South Vietnam

MAJOR SURFACE SHIPS

Ex-US EDSALL ('FMR') class *frigates*

Class (no, former name, transferred):
Tran Hung Dao (HQ 01, ex-*Camp,* 6.2.71), *Tran Kanh Du* (HQ 04, ex-*Forster,* 25.9.71).

Former USN destroyer escorts. Built in 1943–44. Extensively converted to radar pickets after 1945. In 1965 large radar antennae was removed and the ships were employed in Indo-China as patrol frigates until transferred to S Vietnam. It has been reported that former-HQ 4, renamed *Dai Ky,* was extant as a training vessel with the Vietnamese Navy in 1994.

Ex-US BARNEGAT class *seaplane tenders (frigates)*

Class (former names):
Tran Quang Khai (HQ 15, ex-*Bering Strait* WHEC 382, ex-*AVP 34*), *Tran Nhat Duat* (HQ 16, ex-*Yakutat* WHEC 380, ex-*AVP 32*), *Tran Binh Trong* (HQ 17, ex-*Castle Rock* WHEC 383, ex-*AVP 35*), *Tran Quoc Toan* (HQ 02, ex-*Cook Inlet* WHEC 384, ex-*AVP 36*), *Tham Ngu Lao* (HQ 03, ex-*Absecon* WHEC 374, ex-*AVP 23*), *Ly Thoung Kiet* (HQ 05, ex-*Chincoteague* WHEC 375, ex-*AVP 24*), *Ngo Kuyen* (HQ 06, ex-*McCulloch* WHEC 386, ex-*Wachapreague* AGP 8, ex-*AVP 56*)

Built 1942–44 as USN seaplane tenders and transferred in 1946–48 to US Coast Guard then to S Vietnam. First two pairs on 1 January 1971 and 21 December 1971. *Tham Ngu Lao* transferred on 15 July 1972 and the last pair on 21 June 1972. These ships were the largest combatants in the S Vietnamese Navy and the only ones to mount a 127mm gun battery. It is believed that all ASW weapons were removed before transfer. In the late 1970s, one unit of this class, former HQ 03, which had been captured in 1975 and renamed *Pham Ngu Lao,* was fitted with 2–SS–N–2 SSM removed from a stricken 'Komar' class FAC(M) and Soviet-built light guns were added. She was reported extant 1995.

AMPHIBIOUS WARFARE VESSELS

Ex-US LST type *tank landing ships*

Class (former names):
Cam Ranh (ex-*Marion County* LST 975), *Da Nang* (ex-*Maricopa County* LST 938), *Thi Nai* (ex-*Cayuga County* LST 529), *Vung Tau* (ex-*Cochino County* LST 603), *Qui Nhon* (ex-*Bullock County* LST 509), *Hna Trang* (ex-*Jerome County* LST 484)

Built in 1943–45, transferred to S Vietnam in 1962 and numbered in order above (HQ 500, HQ 501), 1963 (HQ 502, escaped 1975), 1969 (HQ 503), 1970 (HQ 504, 505). Three reported extant, in poor condition, in 1995.

Ex-US LSM type *medium landing ships*

Class (former nos):
Hat Giang (ex-*LSM 9011*, ex-*LSM 335*), *Han Giang* (ex-*LSM 9012*, ex-*LSM 110*), *Lam Giang* (ex-*LSM 226*), *Ninh Giang* (ex-*LSM 85*), *Huong Giang* (ex-*Oceanside*, LSM 175), *Tien Giang* (ex-*LSM 313*), *Hau Giang* (ex-*LSM 276*)

All seven built in 1944. First two units transferred by USA to France in January 1954 for use in Indo-China and in December 1955 transferred to Vietnam (fitted as hospital ships, LSM-H). Numbered HQ 400–406. *LSM 175* transferred in 1961, *LSM 313* in 1962 and *LSM 276* in 1965. HQ 401 and HQ 404 escaped in 1975. HQ 402 scuttled.

Ex-US LSSL type *landing support ships*

Class (former names):
Nguyen Van Tru (ex-*No Than*, ex-*Framée, LSSL 105*), *Le Trong Dam* (ex-*Linh Kiem*, ex-*Arquebuse*, ex-*LSSL 4*), *Le Van Binh* (ex-*Javeline*, ex-*LSSL 10*), *Doan Ngoc Tang* (ex-*Hallebarde*, ex-*LSSL 9*), *Lulu Phu Tho* (ex-*LSSL 101*), *Nguyen Ngoc Long* (ex-*LSSL 96*), *Nguyen Duc Bong* (ex-*LSSL 129*)

All seven built in 1944–45. Numbered HQ 225–231 by S Vietnam. First four transferred to France in 1951 and two (HQ 225, 226) retransferred to Vietnam in 1955 and 1957. HQ 227–230 transferred in 1953 to Japan and retransferred to Vietnam in 1965. *Nguyen Duc Bong* transferred in 1966. First three officially stricken from the list in 1971. *Le Van Binh* sunk in 1966, *Nguyen Van Tru* sunk in 1970, *Le Trong Dam* sunk in 1970. Three (HQ 228, 229 and 231) escaped in 1975.

Ex-US LSIL type *landing ships*

Class (former nos):
Long Dao (ex-*LSIL 9029*, ex-*LSIL 698*), *Than Tien* (ex-*LSIL 9035*, ex-*LSIL 702*), *Thien Kich* (ex-*LSIL 9038*, ex-*LSIL 872*), *Lio Cong* (ex-*LSIL 9034*, ex-*LSIL 699*), *Tam Set* (ex-*LSIL 9033*, ex-*LSIL 871*)

All five built in 1944, transferred to France in 1951–53 and retransferred to Vietnam in 1956. Numbered HQ 327–331. HQ 329 escaped in 1975.

Ex-US LCU type *utility landing craft*

Class:
HQ 533 (ex-*LCU 1479*), *HQ 534* (ex-*LCU 1480*), *HQ 535* (ex-*LCU 1221*), *HQ 536* (ex-*LCU 1446*), *HQ 537* (ex-*LCU 1501*), *HQ 538* (ex-*LCU 1594*), *HQ 539* (ex-*LCU 1502*), *HQ 540* (ex-*LCU 1475*), *HQ 541* (ex-*LCU 1477*), *HQ 542* (ex-*LCU 1494*), *HQ 543* (ex-*LCU 1493*), *HQ 544* (ex-*LCU 1445*), *HQ 545*, *HQ 144*, *HQ 546* (ex-*YFU 90*, ex-*LCU 1582*), *HQ 547* (ex-*LCU 1481*), *HQ 548* (ex-*LCU 1498*), *HQ 549* (ex-*?*), *HQ 560* (ex-*YLLC 1*, ex-*LCU 1348*), *HQ 561* (ex-*YLLC 5*, ex-*YFU 2*, ex-*LCU 529*), *HQ 562* (ex-*YLLC 3*, ex-*YFU 33*, ex-*LCU 1195*)

Eighteen former USN units 501 series were built during the Second World War with LCT (6) designation; 1466 series were built as LCU (7) during early 1950s. *HQ 533–537* served as French *LCU 9076, 9089, 9086, 9074* and *9887* and were acquired in 1954 from French reparations and other transferred from USA up to 1971. Three units (ex-YLLC) converted while in US service for use as salvage lifting craft fitted with 20t capacity 'A' frame derrick, special anchors, divers' air compressors, welding equipment and salvage pumps. Up to five of the LCU 1466 series are reported extant 1995.

SMALL SURFACE COMBATANTS

Ex-US PCE type *large patrol craft*

Class (no, former name):
Dong Da II (HQ 07, ex-*Crestview*, PCE 895), *Ngoc Hoi* (HQ 12, ex-*Brattleboro*, EPCER 852, ex-*PCER 852*), *Van Kiep II* (HQ 14, ex-*Amherst*, PCER 853)

Former US patrol vessels – escort (PCE), two of which were fitted as rescue ships (PCER) to pick up survivors of convoy sinkings. Built in 1943–44. *Brattleboro* for a time used as an experimental vessel (EPCER). Transferred to Vietnam on 29 November 1961, 11 July 1966 and in June 1970 respectively.

Ex-US ADMIRABLE class *minesweepers (large patrol craft)*

Class (no, former name):
Chi Lang II (HQ 08, ex-*Gayety*, MSF 239, ex-*AM 239*), *Ky Hoa* (HQ 09, ex-*Sentry*, MSF 299, ex-*AM 299*), *Nhut Tao* (HQ 10, ex-*Serene* MSF 300, ex-*AM 300*), *Chi Linh* (HQ 11, ex-*Shelter* MSF 301, ex-*AM 301*), *Ha Hoi* (HQ 13, ex-*Prowess*, IX 305, ex-MSF 280, ex-*AM 280*)

All five built in 1943–44 and later used as patrol vessels. *Prowess* was used as Naval Reserve TS 1962–71 (redesignated IX 305). Transferred to Vietnam in April 1962 (*Chi Lang II*), August 1962 (*Ky Hoa*), January 1964 (*Nhut Tao, Chi Linh*) and June 1970 (*Ha Hoi*). Former *Ha Hoi* believed extant 1995, rearmed with Chinese and Soviet guns.

Ex-US PC (173ft) type *submarine chasers (large patrol craft)*

Class (former names):
Chi Lang (ex-*Mausquet*, ex-*PC 1144*), *Dong Da* (ex-*L'Ardent*, ex-*PC 1167*), *Tay Ket* (ex-*Glaive*, ex-*PC 1143*), *Tuy Dong* (ex-*Trident*, ex-*PC 1146*), *Van Kiep* (ex-*L'Intrépide*, ex-*PC 1130*), *Van Don* (ex-*Anacortes*, ex-*PC 1569*)

Six former USN ships transferred to France after 1945. Subsequently transferred to Vietnam and numbered HQ 01-06; 1955 (*Chi Lang*), 1956 (next four), 1960 (*Van Don*). *Dong Da* and *Chi Lang* deleted in 1961, *Tay Ket* and *Van Kiep* deleted in 1965, *Tuy Dong* deleted in 1971.

Ex-US SC type *submarine chasers*

Class:
HQ 600 (ex-*GC 8*, ex-*SC 1020*), *HQ 601* (ex-*GC 7*, ex-*SC 679*)

Transferred in 1951 to France and retransferred to Vietnam. Both deleted in 1960.

US PGM (100ft) type *large patrol craft*

Class:
Phu Du, Tien Moi, Minh Hoa, Kien Vang, Keo Ngua, Kim Qui, May Rut, Nam Du, Hoa Lu, To Yen, Dinh Hai, Truong Sa, Thai Binh, Thi Tu, Song Tu, Tat Sa, Hoang Sa, Phu Qui, Hon Troc, Tho Chau

Welded steel patrol boats built in the US yards specifically for foreign transfer and assigned PGM numbers (59–70, 72–74, 80–83, 91) for contract purposes. Twenty transferred on completion and numbered HQ 600–629 in 1963–67. Built by J M Martinac SB Corp, Tacoma (HQ 600–605), and Marinette Marine Corp, Washington (the rest). *Phu Du* and *Hon Troc* escaped in 1975.

Ex-US 82ft 'POINT' class *large patrol craft*

Class (former name):
Le Phuoc Dui (ex-*Point Garnet* 82310), *Le Van Nga* (ex-*Point League* 82304), *Huynh Van Cu* (ex-*Point Clear* 82315), *Nguyen Dao* (ex-*Point Gammon* 82328), *Dao Thuc* (ex-*Point Comfort* 82317), *Le Ngoc Thanh* (ex-*Point Ellis* 82330), *Nguyen Ngoc Thach* (ex-*Point Slocum* 82313), *Dang Van Hoanh* (ex-*Point Huason* 82322), *Le Dinh Hung* (ex-*Point White* 82308), *Thuong Tien* (ex-*Point Dume* 82325), *Pham Ngoc Chau* (ex-*Point Arden* 82309), *Dao Van Dang* (ex-*Point Glover* 82307), *Le Dgoc An* (ex-*Point Jefferson* 82306), *Huynh Van Ngan* (ex-*Point Kennedy* 82320), *Tran Lo* (ex-*Point Young* 82303), *Bui Viet Thanh* (ex-*Point Partridge* 82305), *Nguyen An* (ex-*Point Caution* 82301), *Nguyen Han* (ex-*Point Welcome* 82329), *Ngo Van Quyen* (ex-*Point Banks* 82327), *Van Dien* (ex-*Point Lomas* 82321), *Ho Dang La* (ex-*Point Grace* 82323), *Dam Thoai* (ex-*Point Mast* 82316), *Huynh Bo* (ex-*Point Grey* 82324), *Nguyen Kim Hung* (ex-*Point Orient* 82319), *Ho Day* (ex-*Point Cypress* 82326), *Troung Ba* (ex-*Point Marmoc* 82331)

Twenty-six former US Coast Guard 82ft cutters designated WPB (actual length 83ft oa). Transferred to Vietnam in 1969 (HQ 700–707) and 1970 (HQ 708–725). *Huynh Van Cu* (HQ 702) escaped in 1975.

RIVERINE CRAFT

There were over 700 armed craft transferred to South Vietnam by the US Navy from 1965 for use in the riverine and coastal areas of Indo-China. There were also many former French craft which were widely used in the late 1950s and early 1960s. Some of them survived even into the 1970s. The exact number of these small craft serving at any given time is difficult to determine. Below are listed all ex-American craft which were in existence in 1972. There were 107 'Swift' type patrol craft; 293 PBR river patrol boats; twenty-eight RPC river patrol craft; eighty-four ASPB assault support patrol boats; forty-two river monitors; twenty-two LCM 6 landing craft; 100 ATC armoured troop carriers; nine CCB command and patrol boats; four CSB combat salvage boats; and twenty-four minesweeping launches. See US section for data and discussion.

A coastal force of motorised junks was organised with US assistance on 12 April 1960 with 100 junks, twenty-eight groups of junks having been formed by June 1962. Mass production of improved design junks was undertaken to control infiltration of S Vietnamese coastal waters. The latest junks had diesel engines capable of reaching 15kts. They were fitted with armour plates and glass fibre to protect the wooden hull against marine borers and small arms fire. By 1969 about 500 junks were crewed by some 4000 men. Coastal Force (ex-Junk Force) become part of the Vietnamese Navy and no longer a paramilitary organisation in July 1965. Approximately 250 motor-propelled junks were pressed into naval service in January 1972 and these included sixty-two command junks, thirty-one Kien Giang junks, and 153 Yabuta junks. Some of the Yabuta junks were built of ferrous cement. The Yabuta junks usually had two .50 cal MGs; some also had a 60mm mortar. Diesel propulsion permitted them to overtake and search some of the thousands of coastal sailing craft in Indo-China waters.

MINE WARFARE VESSELS

Ex-US BLUEBIRD class *coastal minesweepers*

Class (no, former no):
Ham Tu (HQ 114, ex-*MSC 281*, *Chuong Duong II* (HQ 115, ex-*MSC 282*), *Bach Dang II* (HQ 116, ex-*MSC 283*)

Built in 1958–60 and transferred under MAP to Vietnam in 1959 (first two) and 1960. *Bach Dang II* grounded on 9 October 1970, was stripped and destroyed.

Ex-US YMS type *coastal minesweepers*

Class (no, former name):
Ham Tu (HQ 111, ex-*Aubepine*, ex-*YMS 28*), *Bach Dang* (HQ 113, ex-*Belledone*, ex-*YMS 78*), *Choung Dong* (HQ 112, ex-*Digitale*, ex-*YMS 83*)

Transferred from the French Navy to Vietnam on 11 February 1954. *Ham Tu* deleted in 1958, *Bach Bang* in 1963 and *Choung Dong* in 1964.

Five ex-Soviet 'K 8' class, built in Poland in the late 1950s. Transferred to Vietnam in October 1980. Extant 1995.

North Vietnam

Ex-Soviet 'PETYA' class *frigates*

Class:
HQ 9, HQ 11, HQ 13, HQ 15, HQ 17
First pair were 'Petya III' export type, as transferred to Syria and India, acquired 1978. Remainder were standard 'Petya II' type transferred in April 1981 (second pair) and December 1984. All reported extant 1995.

Ex-Soviet 'POLNOCNY B' class *landing ships*

Class:
HQ 511, HQ 512, HQ 513
Three units transferred May 1979, November 1979 and February 1980 respectively. All reported extant 1995.

OTHER LANDING CRAFT
Twelve ex-Soviet 'T 4' class landing craft transferred 1979.

Ex-Soviet 'P 4' class *fast attack craft (torpedo)*

Twelve transferred to North Vietnam in 1961 and 1964. Some may have been built in China. Three of this class attacked US destroyers in the August 1964 Gulf of Tonkin incident. Several were sunk in action, remainder all now stricken.

'P 6' class *fast attack craft (torpedo)*

Six Chinese-built Soviet 'P 6' class FAC(T) transferred *c*1967. All now stricken.

Ex-Soviet 'KOMAR' class *fast attack craft (missile)*

Three transferred 1972. One lost in action 19 December 1972. Remainder stricken by *c*1982.

Ex-Soviet 'SHERSHEN' class *fast attack craft (torpedo)*

Class:
HQ 301–HQ 314
Two transferred (without TT) in April 1979, followed by eight, in pairs, in September 1979, August 1980, October 1980 and 1981, and a further four in June 1983. All reported extant 1995, but operational status unclear.

Ex-Soviet 'OSA II' class *fast attack craft (missile)*

Eight transferred, in pairs, in October 1979, September 1980, November 1980 and February 1981. All extant 1995.

PATROL BOATS AND CRAFT
Fifteen ex-Soviet 'SO 1' class submarine chasers, used as patrol boats. First group of seven transferred 1960–61 (two), 1964–65 (two) and 1966 (three). One lost in action to US air attack 1 February 1966. All of this group stricken. Second group of eight transferred in pairs March 1980, September 1980, May 1981 and September 1983 respectively. Of this group, four were reported extant in 1995.

Eight ex-Chinese 'Shanghai II' class large patrol boats, transferrred in two groups of four in 1966 and 1968 respectively. None extant 1995.

Four Soviet 'Turya' class patrol boats, transferred May 1984 (one), November 1984 (one) and 1986 (two). All reported extant 1995.

Fifty ex-Chinese 'Swatow' type patrol craft transferred in 1958 (thirty) and 1964 (twenty). Up to twenty-six either lost in action or deleted by 1982, remainder since stricken.

Two ex-Soviet 'Poluchat' type patrol craft transferred 1960. Both since stricken.

Two ex-GDR 'Bremse' class patrol craft transferred *c*1978. Both now stricken.

Fifteen Soviet 'Zhuk' class patrol craft: three transferred 1978, three in 1979, one each in November 1981 and May 1985, three in 1986 and two each in 1989 and 1990. Four have since been stricken. Eleven reported extant 1995, but condition of older units unlikely to be poor.

Two Soviet 'PO 2' class patrol craft, transferred 1980. Both extant 1995.

MINE WARFARE VESSELS
Two ex-Soviet 'Yurka' class ocean minesweepers, both transferred December 1979. Extant 1995, but must be near the end of their useful life.

Four Soviet 'Sonya' class coastal minesweepers. New construction, transferred on 16 February 1987, February 1988, July 1989 and March 1990. All extant 1995.

Two Soviet 'Yevgenya' class inshore minesweepers, delivered November 1984. Operate in Cam Ranh Bay. Both extant 1995.

Yemen

On 22 May 1990 the Yemen Arab Republic (North Yemen) and the People's Democratic Republic of Yemen (South Yemen) united to form the Yemen Arab Republic. Previously the two states had maintained very small naval forces with limited coastal patrol capability and a few hundred personnel. Although North Yemen was by far the richer and more populous of the two, its fleet, based at Hodeida on the Red Sea, was smaller than that of the Peoples' Republic, which benefited from the treaty of friendship signed with the Soviet Union in 1979, which, following the British withdrawal in 1967, used the port of Aden and the island of Socotra until the late 1980s. Clashes between the two states in 1971–73 and 1979 do not seem to have involved their navies, but an attempted *coup* in South Yemen in 1986 lead to the loss of several vessels. After unification, the two states' forces still maintained a fair degree of autonomy, until the outbreak of civil war in May 1994, lasting until July of that year. The war was won by the forces of the North, but losses, defections and damage to base facilities seriously affected the navy's operational capabilities. In 1995 the navy had a reported total of 2000 personnel plus 500 naval infantry. The list below shows vessels in Yemeni service as of 1995.

YEMEN ARAB REPUBLIC 1995

Two 'Tarantul I' class missile corvettes (N)
Three 'Osa II' class FAC(M)s (S)
Three 'Broadsword' class coastal patrol craft (N)
Four 'Zhuk' class coastal patrol craft (N)
One 'Ropucha' class tank landing ship (S)
? Two 'Polnocny B' class tank landing ships (S)
One 'Natya' class ocean minesweeper (N)
Five 'Yevgenya' class inshore minesweepers (3 N, 2 S)

NB (N) = former N Yemen unit; (S) = former S Yemen unit

YEMEN ARAB REPUBLIC (NORTH)

Russian 'TARANTUL I' class *missile corvettes*

Class:
971, 972
Two transferred December 1990 and February 1991 respectively. Standard export version. Both extant 1995.

Ex-Soviet 'P 4' type *fast attack craft (torpedo)*

Four built in the mid-1950s and transferred in late 1960s. All deleted 1983–84.

Ex-Soviet 'Osa II' class *fast attack craft (missile)*

Two transferred May 1982. Both returned to the USSR in 1985.

Ex-Soviet 'POLUCHAT 1' type *large patrol craft*

Four transferred in 1970s to support the 'P 4' type MTBs. Stern ramps for recovering torpedoes. Also used for patrol work. One deleted 1983–84, one in 1985. None reported extant 1995.

'BROADSWORD' class *coastal patrol craft*

Displacement:	91t standard; 110t full load
Dimensions:	105ft × 20ft 7in × 6ft 3in *32.0m × 6.3m × 1.9m*
Machinery:	3 shafts, 3 GM 16V71 T1 diesels, 3200hp = 28kts. Oil 16.3t
Armament:	2–23mm (1×2), 2–14.5mm (1×2), 2–12.7mm MG (2×1)
Sensors:	Radar Decca 914
Complement:	14–20

Class (no):
Sana'a (200), *13th June* (300), *25th September* (400)
Ordered in 1977, built by Halter Marine, New Orleans, USA in 1978. Soviet armament was added after delivery. *13th June* later renamed *Ramadan*, and the class was renumbered 143, 142 and 141 respectively. All extant 1995 but only one reported operational.

Soviet 'ZHUK' type *coastal patrol craft*

Two built in 1978 and transferred in the same year. A further three transferred in January 1987. Unlike other units of this class, the North Yemeni units have their twin AA guns in enclosed gunhouses with hemispherical covers. The four 'Zhuk' class boats reported in Yemeni service in 1995 are believed to be ex-North Yemeni boats.

Russian 'NATYA' class *ocean minesweeper*

Class:
201

Transferred February 1991. Version with two 30mm AK-630M Gatlings (2×1) replacing twin 30mm. Also does not carry 25mm guns. Extant 1995.

Ex-Soviet 'YEVGENYA' class *inshore minesweepers*

Two transferred May 1982 and a third in 1987. All three extant 1995.

Ex-Soviet 'T 4' type *landing craft*

Two transferred 1970 and both deleted 1988.

Ex-Soviet 'ONDATRA' class *landing craft*

Class:
13, 14

Both transferred January 1983. Both extant 1995.

PEOPLE'S DEMOCRATIC REPUBLIC OF YEMEN (SOUTH)

Ex-Soviet 'T 58' type *ocean minesweeper (corvette)*

Used as escort vessel with her minesweeping gear removed. Built in late 1950s and transferred in 1978. Deleted 1984.

Ex-Soviet 'OSA II' type *fast attack craft (missile)*

A total of eight transferred: two in 1979, four in 1980 and two in 1983. Two lost during attempted coup in 1986. Three attempted to defect to Oman in 1994 carrying ex-South Yemeni personnel. Three reported extant 1995 but in poor condition.

Ex-Soviet 'P 6' type *fast attack craft (torpedo)*

Two, pennant nos 111 and 112, transferred in 1971. Replaced by 'MOL' type. Both stricken 1984.

Ex-Soviet 'MOL' type *fast attack craft (torpedo)*

Two transferred in 1978. Based on 'Osa' type FAC. Sisters were in the Somali and Sri Lankan navies. Both stricken 1988.

Ex-Soviet 'SO 1' type *large patrol craft*

Two transferred in April 1972. Both stricken 1983.

Ex-Soviet 'POLUCHAT 1' type *large patrol craft*

One transferred from USSR in early 1970s. Could be used as a torpedo retriever or patrol boat. Stricken 1983.

Ex-Soviet 'ZHUK' type *coastal patrol craft*

Two transferred by the USSR in Feb 1975. Both stricken *c*1990.

British 'TRACKER-2' class *patrol boats*

Displacement:	31t full load
Dimensions:	63ft 2in × 16ft 4in × 2ft 8in *19.25m × 4.98m × 0.8m*
Machinery:	2 shafts, 2 MTU 8V331 TC diesels, 2200hp = 29kts. Range 650nm at 25kts
Armament:	1–20mm
Complement:	11

Five ordered in August 1977 from Fairey Marine, UK for delivery in 1977–78. Used for customs duties. Belonged to Ministry of Interior. Four lost during 1986 attempted coup. One extant 1995.

Ex-British 'HAM' class *inshore minesweepers*

Class (former names):
Jihla (ex-*Al Saqr*, ex-*Bodenham*), *Socotra* (ex-*Al Dairak*, ex-*Blunham*), *Zinqahar* (ex-*Al Ghazala*, ex-*Elsenham*)

Built in 1956 by British private yards. Transferred on independence in 1967. Renamed after local islands in 1975. All three deleted 1984.

Ex-Soviet 'YEVGENYA' class *inshore minesweepers*

Three transferred December 1989. One deleted 1994, remainder extant 1995.

Ex-Soviet 'POLNOCNY B' type *landing ships*

Four medium landing ships of the numerous and successful Polnocny type 'B' version. Built in Poland by Stocznia Pólnocna, Gdańsk. Two delivered in August 1973, one in July 1977 and one in 1979. Two, *Al Wadi'a* (136) and *Siri* (137), reported extant but non-operational 1995.

Ex-Soviet 'ROPUCHA' class *landing ship*

Class:
139

Polish-built after 1975 and transferred 1980. Only 'Ropucha' in service outside the Russian Navy. Reportedly not fitted with SA-N-5 or bombardment rocket launchers as on Russian units. Refitted at Vladivostok 1985. Extant 1995 but reported non-operational.

Ex-Soviet 'T 4' type *landing craft*

Three similar to US LCM(6) type, transferred in November 1970. All now stricken.

Yugoslavia

The only organised political force in Yugoslavia after the expulsion of German occupying troops in 1944–5 was the People's Liberation Army, commanded by Tito, who formally became Prime Minister in March 1945. The monarchy was swiftly abolished and a Federal People's Republic was proclaimed. A new constitution modelled on the Soviet one was adopted in 1946 and 'sovietisation' of the country begun. Industry was nationalised and plans for economic development were introduced, aimed at rapid industrialisation based on Soviet advice and using Soviet credit.

Hostile at first to the West because of the Trieste incident and fears of a royalist counter-revolution, Tito remained closely allied to the Soviet Union. His political ambitions were aimed at the creation of a communist Balkan federation. This was partially attained when Albania became a Yugoslav satellite, while discussions with Bulgaria began and support was given to the communist guerrillas in Greece. Such ambitious policies needed strong armed forces to support them. No wonder the first postwar naval estimate called for authorisation of twelve destroyers, twenty-eight submarines, 120 MTBs, 130 patrol craft and seventy minesweepers. This was regarded as inadequate by the People's Assembly Presidium and a new programme had to be drawn up. It was approved in December 1946 and authorised the construction of four cruisers, twenty destroyers, 140 submarines, 200 MTBs, 100 patrol craft and 100 minesweepers. Such an enormous programme was impossible to complete in an industrially undeveloped country as Yugoslavia then was. Tito's plans failed to be realised when federation with Bulgaria did not materialise because of Soviet opposition. Stalin was unhappy with the foreign policy of Yugoslavia. In June 1948 Yugoslavia was expelled from the Cominform and all communist countries denounced Tito and broke off relations. Stalin's attempts to unseat Tito failed, but the economic boycott had a devastating effect on the Yugoslav economy. This resulted not only in a lowering of living standards but also in the abandonment of the rearmament programme. In those years the Yugoslav Navy was strengthened by the commissioning of four ex-Italian torpedo boats and one recovered ex-Italian submarine, while the necessity of returning the single corvette to Britain was a painful blow.

Faced by economic collapse from one side and the threat of becoming a Soviet satellite from the other, Tito turned to the West for economic aid. Western credits and other innovations revived the Yugoslav economy in the early 1950s, and modernisation of the armed forces could be undertaken. In 1952 the first ten-year naval programme was initiated. Based on the design of the US Higgins PT boats, ninety-six MTBs were built in local yards as well as twenty-four 'Kraljevica' class patrol craft, forty-four landing craft and thirty-two minesweepers. Two 'W' class destroyers were purchased from Britain as well as a *Fougeux* class patrol craft and three *Sirius* class minesweepers from France as the US 'off-shore' procurement, while Yugoslav yards built two further patrol ships and one minesweeper to the above designs. The most impressive warship to join the Yugoslav Navy at that time was the destroyer *Split*, begun in 1939 and completed with US assistance to a revised design. The Yugoslav yards' greatest achievement however was the completion of a pair of locally designed *Sutjeska* class submarines at the closing stage of the first programme.

After Stalin's death, Tito improved relations with the Soviet Union. As a result he was able to balance Yugoslavia between the Soviet Union and the US – playing one off against the other and acquiring economic and military aid from both the great powers at once. In the longer term this policy led Tito to a doctrine of non-alignment while at the same time buying Soviet weapons. These purchases, together with building in Yugoslav yards, enabled the second naval programme to take place during 1962–74. Soviet deliveries comprised ten 'Osa' class missile craft and four 'Shershen' class torpedo craft. Licence production of the latter type began in Yugoslavia and ten boats were finally completed. Four British *Ham* class inshore minesweepers were built in Yugoslavia under the US Military Aid Programme. Yugoslavia's own production consisted of three *Heroj* class submarines, ten patrol boats and seven inshore minesweepers. The emphasis in the navy shifted to submarine and light forces as most of the war-built warships were deleted by the early 1970s.

The third naval programme was started in 1975 and based mainly on locally designed and built warships, fitted with either Soviet or Western armament, sensors and propulsion systems. By the early 1980s the programme reached its half-way point with two 'Koni' class frigates purchased from the Soviet Union and locally built craft as follows: two *Sava* class submarines, eleven Type 601 LCUs, six *Rade Končar* class FAC(M)s and seven *Nestin* class river minesweepers. Increasing growth resulted in the enhancement of the navy's status within the Yugoslav armed forces and in 1980–81 the post of Deputy Chief of the General Staff for Naval Forces was created.

This period was also marked by a growing maturity of the Yugoslav naval shipbuilding industry, which during the 1980s won several foreign contracts for the construction of training frigates (Iraq and Indonesia), midget submarines (Soviet Union, Sweden, North Korea and Libya) FAC(M)s (Libya), patrol craft (Sudan) and river minesweepers (Hungary, Iraq). In 1988 for the first time Yugoslavia exhibited at 'Defendory International', offering submarines and training frigates. In addition various yards repaired and overhauled Soviet and Libyan surface ships and submarines. However further development was restricted by the limited availability of advanced Western weapon systems. Despite dissatisfaction with the performance of the Soviet 'Styx' type missiles, Yugoslavia was not able to obtain either Harpoon, Exocet or Penguin missiles. The frigate built for Indonesia had electronics and modern weapons fitted in the Netherlands.

In the early 1980s a major problem for the Yugoslav Navy was the approaching obsolescence of several of its small combatants, ASW cract, minesweepers, LCTs and the auxiliary vessels, the latter being small and ageing. The construction programmes for the 1980s included *Kotor* class frigates, 'Lora' class submarines, midget submarines of the *Mala* and *Tisa* classes, 'Sliba' class LCTs, a second batch of Type 601 LCUs, 'Kobra' class FAC(M)s, *Biokovo* class patrol craft and *Lubin* class transports/landing craft. Moreover, replacement of outdated 'Samlet' coastal defence missiles with truck-mounted 'Styx' and some naval aviation has been planned. Apparently modernisation of the ASW and MCM forces was planned for the early 1990s, but completion of these programmes was hampered by an economic crisis and domestic unrest since the mid-1980s. The army's budget was reduced from 5.2 per cent to 3.85 per cent of GNP and a reduction of a further 13 per cent was planned. As a result of budgetary cuts the construction of 'Lora' class submarines as well as 'Kobra' class FAC(M)s was cancelled while other elements of the modernisation programme were also affected. Despite these problems, in 1990 the Yugoslav Navy had the largest surface and submarine force among eastern European countries, while the small combatants force was at a similar level to those of Poland and East Germany. However the ASW, MCM and auxiliary forces were inferior and were hampered by maintenance and servicing problems of old and foreign-built ships.

Unlike the other European communist regimes, Yugoslavia commenced the 1990s without significant political upheavals; the comparatively liberal Belgrade regime had provoked no strong popular opposition of the type which had overthrown more oppressive regimes elsewhere in eastern Europe. However, after a series of elections held during 1990 non-communist governments were formed in most of the Yugoslav republics. Freed from ideological ties to the communist federal authority, the republics of Slovenia and Croatia proclaimed their independence in June 1991. This provoked military reaction by the Federal Army, dominated by Serbia, which claimed a large part of Croatian territory. Military operations sparked civil war, which continues.

Before the civil war, the Yugoslav Navy had 12,800 personnel (1500 officers and 11,300 men in 1991), of whom 3600 were afloat, 2300 in coastal defence (three coastal missile batteries armed with three twin launchers of the Soviet SS-C-3 'Styx' type each and a few Swedish RBS-15, and twenty-four artillery batteries of 6in, 130mm and 85mm guns) and 400 in naval aviation; ratings were on one-year conscription. The 2000-strong naval infantry brigade deployed in the Bay of Cattaro was equipped and trained to undertake amphibious assault landings. The fleet headquarters were at Pula, Sibenik and Kumbor, and the main bases at Lora/Split, Pula, Sibenik, Boka Kotorska and Kardljevo. The Danube River Flotilla, with its main base at Novi Sad, as well as several local naval border detachments, were under the operational command of the Army. Naval aviation, with its base at Divulje, consisted of an ASW Helicopter Squadron (five Ka-28 'Helix', fifteen Ka-25 'Hormone') and an air liaison detachment (twenty Mi-8 'Hip' and a few Aerospatiale-Soko SA-341 'Gazelle'). The Naval Air Brigade was part of the Air Force and comprised twenty RJ-1 'Jastreb', eighteen RT-33 surveillance, twenty-four MIG-21 'Fishbed' interceptors, eighteen 'Jastreb/Orao-B' fighter-bombers, two DHC-2 'Beaver' and four CL-215 amphibians.

By 1994 the navy was almost entirely under Serbian control. Based in the Bay of Cattaro (Boka Kotorska), other bases and naval facilities having been lost to Croatia and Montenegro, it had retained some 80 per cent of pre-war warships, with 6000 personnel (including 4500 conscripts). This decline in personnel since 1991 suggests that a large part of the navy may be non-operational. The submarine branch may be having particular difficulties as the majority of sub-mariners were Croatian.

FLEET STRENGTH 1947

SUBMARINES

Name	Launched	Disp	Fate
Hrabri class			
TARA	1927	975t	Stricken 1958
Ex-Italian CB class			
MALISAN	1943	35.4t	Final fate unknown

Hrabri class
Ex-*Nebojsa* renamed after the war; one 4in gun was removed at the end of her career.

Ex-Italian CB class
Midget submarine *CB 20* laid down for the Italian Navy in 1943 and captured by the Germans the same year. Taken over by Yugoslavia in April 1945 and recommissioned.

CORVETTE

Name	Launched	Disp	Fate
Ex-British 'Flower' class			
PARTIZANKA	22.5.40	950t	Returned 1946

Transferred in 1944 and commissioned as *Nada*, then renamed *Partizanka* after the war. Returned to RN and reverted to her old name *Mallow*. Transferred to Egypt 28 October 1948 and commissioned as *Sollum*.

TORPEDO-BOATS

Name	Launched	Disp	Fate
T 1 class			
GOLESNICA	15.12.13	262t	Stricken Oct 1959
T 5 class			
CER	20.3.15	266t	BU 1962

Renamed and rearmed after the war with 2–40mm (2×1), 4–20mm (*Cer* 1–20mm) while the TT were removed.

MTBs

Name	Launched	Disp	Fate
Orjen class			
TC 5	1937	61.7t	Stricken 1963
TC 6	1938	61.7t	Stricken 1963
Ex-US Higgins 78ft type			
MT 1–8	1942	46t	Stricken 1966 (*MT 8* stricken 1955)

Orjen class
Ex-*Durmitor* and *Kajmakcalan* renamed *TC 5* and *TC 6* respectively after the war, then changed to *TC 391* and *TC 392*.

Ex-US Higgins 78ft type
Lend-lease *PT 201, 204, 207–209, 211, 213, 217* transferred from the RN to the Royal Yugoslav Navy at Malta in mid-1944.

MINESWEEPERS

Name	Launched	Disp	Fate
Ex-Italian RD classes			
ML 301–307	1916–18	193–198t	Stricken 1960
Ex-German M-boat 1915 type			
ZELENOGORA	28.3.18	330t	Stricken 1962
Malinska class			
M 31–33	1931	130t	Stricken 1968 (*M 32* in service 1978)

Ex-Italian RD classes
Small minesweepers of various types *RD 6, 16, 21, 25, 27–29* handed over to Yugoslavia in 1947–48 as war reparations.

Ex-German M-boat 1915 type
M 97 purchased in 1921 and commissioned as *Orao*, captured in company of five sisters by the Italians in April 1941 and commissioned by them as *Vergada*. Returned in 1943, renamed *Pionir* after the war, later changed to *Zelenogora*.

Malinska class
Mining tenders *Meljine*, *Malinska* and *Mljet* captured by the Italians in April 1941 and commissioned as *Solta*, *Arbe* and *Meleda*. Returned 1943–44, renamed *M 1* to *M 3* after the war, then *M 31* to *M 33*. Rearmed with one 47mm gun.

LANDING CRAFT

Name	Launched	Disp	Fate
Ex-German MFPs			
D 203, D 204	*c*1942	220t	Stricken *c*1985
Ex-Italian MZ type			
D 206 (ex-*MZ 713*)	1942	239t	Stricken 1979
D 219 (ex-*MZ 717*)	1942	239t	Stricken 1979

There were also some ex-German *Siebel Fähres* in service at that time.

MAJOR SURFACE SHIPS

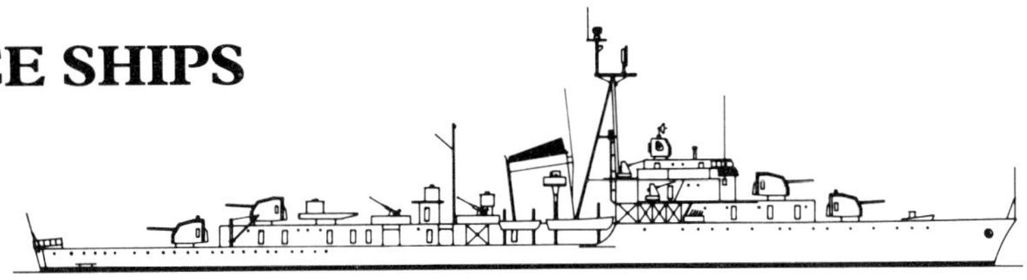

Split as completed

SPLIT *destroyer*

Displacement: 2400t standard; 3000t full load
Dimensions: 376ft 3in pp, 393ft 7in oa × 39ft 4in × 12ft 2in max
114.7m, 120.0m × 12.0m × 3.7m
Machinery: 2-shaft Parsons geared turbines, 2 watertube boilers, 50,000shp = 31.5kts. Oil 590t
Armament: 4–5in/38 (4×1), 12–40mm (1×4, 2×2, 4×1), 5–533mm TT (1×5), 2 Hedgehog, 6 DCT, 2 DC racks, 40 mines
Sensors: Radar SC, SG 1, Mk 12, Mk 22
Complement: 240

No	Name	Builder	Laid down	Launched	Comp	Fate
11	SPLIT (ex-*Spalato*, ex-*Split*)	3 Maj Yd, Rijeka	Jul 1939	Mar 1950	4.7.58	Stricken 1980, to Indonesia 1981

Laid down in the Split Yard to the plans prepared by the French yard Chantiers de la Loire. The original design provided for an armament of 5–140mm/50 (5×1), 10–40mm (5×2), 8–13mm MG, 6–21in TT (2×3) and a speed of 37kts. The incomplete hull was seized by the Italians but they were not able to complete her and she fell into Yugoslav hands again in October 1944. The plans were redrawn to fit the US armament and sensors and the ship was commissioned in 1959. Fire control consisted of Mk 37 set with Mk 12 and Mk 22 radars for the 5in guns while Mk 51 set was provided for the 40mm guns. At the end of her career only one boiler was serviceable which enabled the power plant to produce 25,000shp = 24kts.

Split May 1969
Marek Twardowski Collection

YUGOSLAVIA

Two RN war built destroyers *Kempenfelt* (ex-*Valentine*) and *Wager* were purchased and towed to Yugoslavia in October 1956. They were refitted in a local yard and commissioned as *Kotor* on 10 September 1959 and *Pula* later in that year. Their appearance did not vary greatly from the original configuration.

Ex-British 'W' class *destroyers*

No	Name	Builder	Acquired	Fate
R 21	KOTOR (ex-*Kempenfelt*)	John Brown	Oct 1956	Stricken 1971
R 22	PULA (ex-*Wager*)	John Brown	Oct 1956	Stricken 1971

'Koni' class hulls nos 4 and 8 respectively, with armament of four SS-N-2C 'Styx' SAM (4×1), one SA-N-4 'Gecko' SAM (1×2), 4–76mm (2×2), 4–30mm (2×2), 2 RBU-6000 ASW RL (2×12), mines. VPB = *velici patrolni brod* (large patrol vessel). *Kopar* erroneously reported to be renamed *Zagreb* in 1992. *Split* damaged by the Croatians in November 1991 and since used to provide spares for *Kopar* and the *Kotor* class frigates.

Soviet 'KONI I' class *frigates*

No	Name	Builder	Acquired	Fate
VPB 31	SPLIT	Zelenodolsk	10.3.80	Extant 1995
VPB 32	KOPAR	Zelenodolsk	5.12.82	Extant 1995

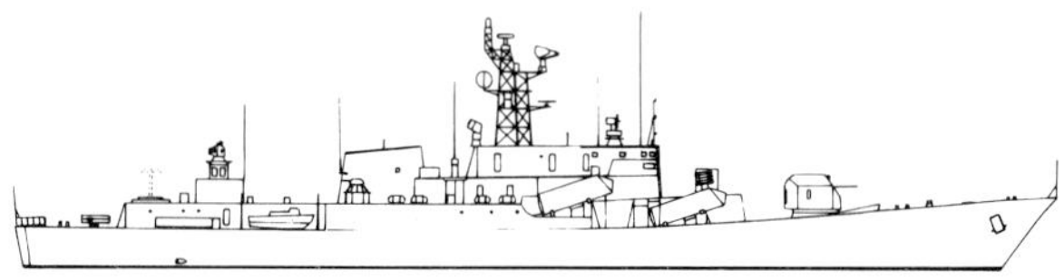

Kotor class

Designed and built in Yugoslavia; the overall design layout is based on the Soviet 'Koni' class (but with considerable differences in structure), power plant of combined Soviet gas turbine and French Pielstick diesels, Soviet missiles, gunnery and sensors, and Whitehead ASW torpedoes.

KOTOR class *frigates*

Displacement: 1700t standard; 1900t full load
Dimensions: 317ft 3in × 42ft × 13ft 9in
96.7m × 12.8m × 4.2m
Machinery: 3 shafts, CODAG 2 SEMT-Pielstick 12PA 6V80 diesels + gas turbine, 9600bhp + 18,000shp = 27kts. Range 1800nm at 14kts
Armament: 4 SS-N-2C SAM (4×1), 1 SA-N-4 SAM (1×2), 2–76mm (1×2), 4–30mm (2×2), 6–324mm Whitehead A 244 ASW TTs (2×3), 2 RBU-6000 ASW RL (2×12), mines, 2 Wallop Barricade chaff RL
Sensors: Radar Strut Curve, Palm Frond, PEAB 9LV200, Drum Tilt, Pop Group, High Pole, Square Head; sonar hull mounted
Complement: 110

No	Name	Builder	Laid down	Launched	Comp	Fate
33	KOTOR	Uljanik, Pula	1981	1984	1988	Extant 1995
34	PULA	Tito, Kraljevica	1982	1986	1988	Extant 1995

SUBMARINES

The former Italian war-built *Nautilo* of the *Flutto* class which sank in Pola when in German hands was raised and commissioned in 1949. Relegated to training duties in 1958 she completed a major refit in 1960 when the 100mm gun was removed and the conning tower streamlined. Deleted 1971.

SAVA *submarine*

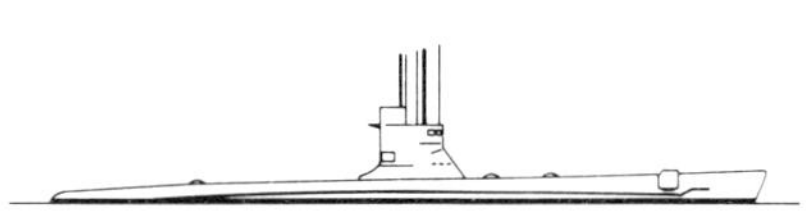

Sutjeska 1964

The first submarines designed and built in Yugoslavia. Modernised later and equipped with Soviet sensors (bow sonar arrays added in early 1970s). Since the early 1980s in training service, and one reported as used for harbour training in 1990. P = *Podnornica* (submarine).

SUTJESKA class *submarines*

Displacement: 700t standard; 820t/945t
Dimensions: 196ft 8in oa × 22ft × 15ft 9in
60.0m × 6.6m × 4.8m
Machinery: 2 (?) shafts, 2 Sulzer diesels, electric motors, 1800bhp = 14kts/9kts. Range 4800nm at 8kts
Armament: 6–533mm TT (4 bow, 2 stern, 8 torpedoes)
Sensors: Radar navigational; sonar hull mounted
Complement: 38

No	Name	Builder	Laid down	Launched	Comp	Fate
P 811	NERETVA	Uljanik Yd, Pula	1957	1959	1962	Stricken 1987
P 812	SUTJESKA	Uljanik Yd, Pula	1957	28.9.58	16.9.60	Stricken 1987

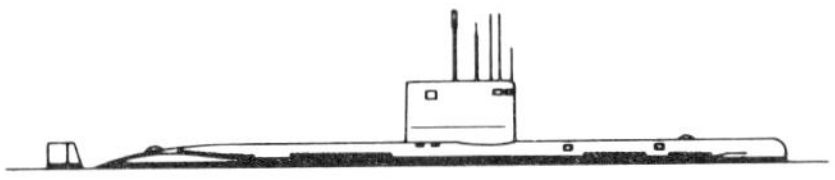

Heroj as completed

HEROJ class *submarines*

Displacement: 1170t/1350t
Dimensions: 210ft × 23ft 6in × 16ft 4in
64.0m × 7.2m × 5.0m
Machinery: 1 shaft, 2 Sulzer diesels, electric motors, 1600hp/1560hp = 10kts/16kts. Range 9700nm at 8kts
Armament: 6–533mm TT (bow, 10 torpedoes or 20 mines)
Sensors: Radar Snoop Group; sonar Krupp Atlas PRS 3 hull mounted; ECM Stop Light
Complement: 55

No	Name	Builder	Laid down	Launched	Comp	Fate
P 821	HEROJ	Uljanik Yd, Pula	1964	18.8.67	1968	Extant 1995
P 822	JUNAK	Split SY	1965	1968	1969	Extant 1995
P 823	USKOK	Uljanik Yd, Pula	1966	Jan 1970	1970	Extant 1995

Incorporated a number of improvements as compared with the earlier *Sutjeska* class, particularly in the shape of the hull and conning tower. Equipped with Soviet senors and armament. *Heroj* was reported as having suffered accidental damage in 1982. Two operational 1995, with the third in reserve and cannibalised for spares.

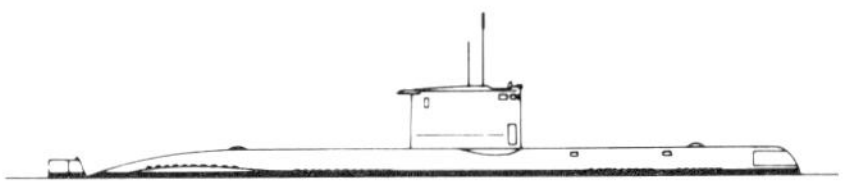

Sava 1979

SAVA class *submarines*

Displacement: 830t/960t
Dimensions: 182ft 7in × 23ft 6in × 18ft 7in
55.7m × 7.2m × 5.1m
Machinery: 1 shaft, 2 Sulzer diesels, electric motors, 1600hp/1560hp = 10kts/16kts. Endurance 28 days
Armament: 6–533mm TT (10 torpedoes or 20 mines)
Sensors: Radar navigational; sonar Krupp Atlas PR5 3, Thomson Sintra hull mounted; ECM Stop Light radar warning
Complement: 27

No	Name	Builder	Laid down	Launched	Comp	Fate
P 831	SAVA	Split SY	1975	1977	1978	Extant 1995
P 832	DRAVA	Split SY	1978	1980	1981	Extant 1995

Improved *Heroj* class, with high degree of automation. Reported fitted with Swedish torpedoes, and radar and ECM of Soviet manufacture but unconfirmed. Diving limit about 1000ft.

'LORA' class *submarines*

Improved 'Sava' class design of 900t patrol submarine capable of Mediterranean deployment. Reported since 1986 under construction at Split to replace the *Sutjeska* class stricken in 1987, but the order was cancelled.

UNA class *midget submarines*

Displacement: 76t/88t
Dimensions: 61ft 7in × 9ft × 8ft 2in
18.8m × 2.7m × 2.5m
Machinery: 1 shaft, 2 electric motors, 68hp = 6kts/8kts. Range 200nm at 4kts
Armament: 4 'R2' SDVs or 6 'R1' chariots or 6 limpet mines
Sensors: Sonar Krupp-Atlas hull mounted
Complement: 6

Class (no):
Una (911), *Tisa* (912), *Zeta* (913), *Soca* (914), *Kupa* (915)

Designed in Yugoslavia and built at Split under designation 'M 100-D' type, the prototype was commissioned in May 1985, and the last in 1989. The programme is possibly still in operation. Fitted with exit/re-entry chambers for divers, these vessels can carry up to eight combat swimmers backed by swimmer delivery vehicles (SDV) or wet chariots. Diving depth 345ft (50m). Designed to defend approaches to small shallow water harbours. *Soca* in Croatian hands since 1991, remainder extant 1995.

MALA class *swimmer delivery vehicles*

Displacement: 1.4t
Dimensions: 16ft 11in × 4ft 7in × 4ft 4in
4.9m × 1.4m × 1.3m
Machinery: 1 shaft, electric motor, 4.5hp = 4.4kts. Range 23nm at 3.7kts
Armament: limpet mines
Sensors: Sonar, navigation echo sounder
Complement: 2

This type, designated 'R2', was designed in Yugoslavia and has been built locally since the early 1980s. Four were delivered to the Yugoslav Navy and several to the Soviet Union, Libya and Sweden. Free flooding submersible craft constructed of aluminium alloy and plexiglass. The pressure hull consists of separate watertight cylinders containing the motor, lead-acid batteries, navigation and electronic equipment. Can be carried by large submarines as well as by 'Una' class midgets. Maximum operating depth 200ft; can carry up to 0.25t of armament (limpet mines). Batteries can be charged only from depot ship or from shore. Two reported captured by the Croatians in 1992, and two extant 1995.

Type R1 *chariots*

Displacement: 0.145t
Dimensions: 12ft 2in × 3ft 5in × 10in
3.72m × 1.05m × 0.26m
Machinery: 1 shaft, electric motor, 1.36hp = 3kts. Range 6nm at 3kts
Armament: mine
Complement: 1

Wet chariots which can be transported in submarine TTs or carried by 'Una' class midgets. Powered by silver-zinc batteries; normal operating depth 200ft.

AMPHIBIOUS WARFARE VESSELS

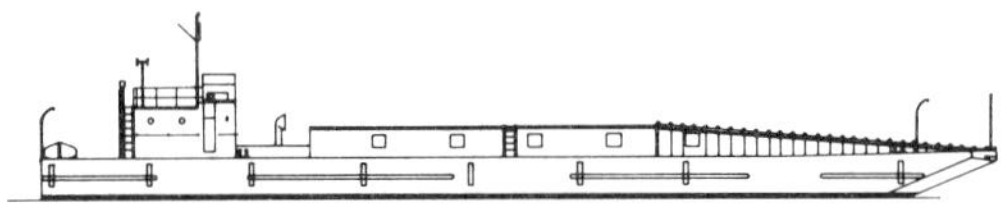

DTM 231 1959
NB 1/750 scale

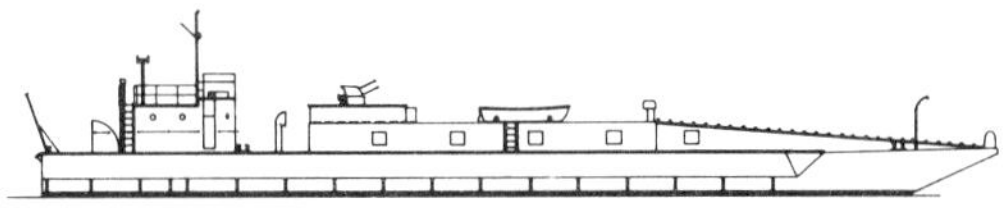

DTM 233 1980
NB 1/750 scale

DTM 221 class *landing craft*

Displacement: 240t standard; 410t full load
Dimensions: 144ft 3in × 19ft 7in × 7ft
47.3m × 6.4m × 2.3m
Machinery: 3 shafts, 3 Gray Marine 64HN9 diesels, 625bhp = 10.3kts. Range 500nm at 9.3kts
Armament: 3–20mm (3×1), 100 mines on 2 rails
Complement: 15 (+200)

Class:
DTM 211-212, 213, 214, 215, 216, 217, 218, 219, 220, 221–223, 224–225, 226, 227, 228–229, 230–231, 232–234, 235–236, 237, DSM 509, 513, 514

Yugoslav built in the 1950s, based on the German Second World War MFP design. *DTM* = LST, *DSM* = LST/minelayer (1m sponsons added for mine rails). Two sold to Sudan in 1969. Thirteen deleted by 1990. *213, 215, 217, 221, 226, 228, 234* and *237* deleted 1990. *233* deleted 1991. *219, 222, 233* and *510* deleted 1992. Croatia took over *217, 501* and two others in 1991, and *229, 232, 513* and *514* remain in service with the Yugoslav Navy.

Catamaran type *landing craft*

Ten 40t boats of the LCU type with a catamaran hull were built in the 1950s. Armed with 4–12.7mm MG (2×2), able to carry one light tank or one field gun. Could carry mines also. Stricken 1978.

Type 21 *landing craft*

Displacement: 32t full load
Dimensions: 69ft 9in × 15ft 7in × 5ft 2in
21.3m × 4.8m × 1.6m
Machinery: 1 shaft, 1 MTU 12V331 TC81 diesel, 1450hp = 23kts. Range 320nm at 22kts.
Armament: 1–20mm
Sensors: Radar Decca 101
Complement: 6

Class:
DJC 601–630

Fibreglass craft built at the Gleben Yard, Korŭla, and delivered between September 1976 and 1979. On the 32sq m cargo space they can carry 6t of cargo or a landing detachment of forty troops. Some of the early vessels have been scrapped, and in 1995 only *DJC 604–606* and *608–610* were in service. Of the others *DJC 629* was deleted in 1991 and *DJC 607* and *611* in 1992. Four of the class were taken over by the Croatians in 1992, together with a further damaged pair which may have been repaired since. DJC = *Desantni Jurisni camac* (LCA).

Type 22 *landing craft*

Displacement: 48t full load
Dimensions: 73ft 2in × 15ft 7in × 3ft 3in
23.3m × 4.8m × 1.0m
Machinery: 2 waterjets, 2 MTU diesels, 1740hp = 35kts. Range 320nm at 22kts
Armament: 2–20mm
Sensors: Radar Decca 101
Complement: 8

Class:
DJC 620–632

Total of twelve of these GRP boats were built locally during the 1980s, with the last completed in 1987. *DJC 622–624* taken over by Croatia 1991, remainder extant 1995.

SILBA class *landing ship*

Displacement: 880t full load
Dimensions: 144ft wl, 163ft 1in oa × 33ft 5in × 8ft 5in
43.8m, 49.7m × 10.2m × 2.6m
Machinery: 2 shafts, 2 Alpha 10V23L diesels, 3100hp = 12kts. Range 1200nm at 12kts
Armament: 4 SA-N-5 Grail SAM (1×4), 4–30mm AK 230 (2×2), 4–20mm M75 (1×4)
Sensors: Radar surface search
Complement: 33

Class (no):
Silba (DBM 241)

A ro-ro design, intended to replace *DTM 211* craft. Built by Brodosplit, Split and commissioned 1990. Can be used as a minelayer, carrying up to ninety-four SAG-1 mines. As a landing ship, can carry six medium tanks or 300 troops. A sister-ship was completed in 1993 for the Croatian Navy. Extant 1995.

SMALL SURFACE COMBATANTS

Ex-Italian CICLONE class *torpedo-boats*

Aliseo and *Indomito* were transferred to Yugoslavia in 1949 as war reparations and commissioned as *Biokovo* and *Triglav* respectively. Fitted with US sensors. Stricken 1971.

Ex-Italian ARIETE class *torpedo-boats*

The prototype of this class, *Ariete*, was transferred in 1949 as war reparations and commissioned as *Durmitor*. Two others of this class *TA 46* (ex-*Balestra*) and *TA 47* (ex-*Fionda*) seized incomplete by the Germans in 1943 were taken over by the Yugoslav Navy in 1948 and renamed *Ucka* and *Velebit* respectively. Only *Ucka* was finally completed in 1949 and unlike *Durmitor* her original armament was changed to 2–100mm/47 (2×1), 3–40mm (3×1), 2 DCT and 28 mines. *Durmitor* was stricken in 1963; *Ucka* in 1968.

Type 108 in early 1950s
NB 1/750 scale

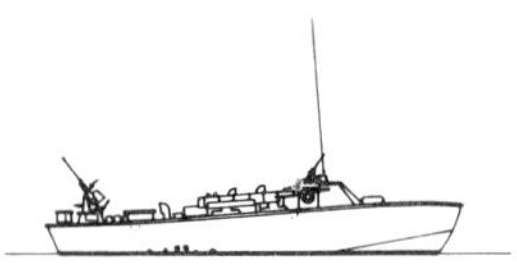

TYPE 108 *fast attack craft (torpedo)*

Displacement: 55t standard; 60t full load
Dimensions: 69ft pp, 78ft oa × 21ft 3in × 7ft 8in
21.0m, 23.8m × 6.5m × 2.4m
Machinery: 3 shafts, 3 Packard petrol engines, 5000bhp = 36kts
Armament: 1–40mm, 2–0.5in MG, 2–18in TT (see notes)
Complement: 14

Class:
P 101 to *P 199*

A total of 96 boats were built in Yugoslavia (1951–60) to the design of the US Higgins 78ft type PT boats. Six sold to Egypt in 1956, two to Ethiopia in 1960, two to Cambodia in 1964 and six to Sudan in 1970. In 1963 about twenty-four boats were converted to MGBs by removal of the TTs and mounting one additional 40mm gun forward instead, being designated Type 158. Their speed did not exceed 24–26kts at that time and they could carry 2–4 mines in addition to the above armament. The unconverted boats were sold to civilian enterprises or subsequently scrapped. The last six converted boats were stricken in 1979.

Soviet 'SHERSHEN' class *fast attack craft (torpedo)*

Class:
Topcider (TC 214), *Ivan* (TC 215), *Jadran* (TC 216)

Soviet built boats, acquired in 1965. TC = *Torpedni Camac* (torpedo boat). *Ivan* extant 1995, others deleted 1993.

TYPE 201 *fast attack craft (torpedo)*

Class:
Biokovac (TC 218), *Crvena Zvezda* (TC 220), *Kornat* (TC 217), *Napredak* (TC 223), *Partizan II* (TC 222), *Partizan III* (TC 221), *Pionir II* (TC 224), *Proleter* (TC 213), *Strjelko* (TC 219), *Partizan* (TC 212), *Pionir* (TC 211)

Licenced version of the Soviet 'Shershen' class, built 1966–71 by the Tito Yard, Kraljevica. Have no DC racks fitted, unlike the Soviet-built boats. *Partizan III* and *Strelko* taken over by Croatia in 1991, though the latter was damaged beyond repair. Five further boats deleted 1992. *Biokovac, Pionir II* and *Proleter* extant 1995.

Soviet 'OSA I' class *fast attack craft (torpedo)*

Class:
Franc Rozman-Stane (RC 309), *Josip Mazar-Sosa* (RC 307), *Karlo Rojc* (RC 308), *Mitar Acev* (RC 301), *Nikola Martinovic* (RC 306), *Petar Drapsin* (RC 303), *Filipovic Steva* (RC 304), *Velimir Skorpik* (RC 310), *Vlado Bagat* (RC 302), *Zikca Jovanovic-Spanac* (RC 305)

Soviet built, transferred during 1965–69. RC = *Raketni Camac* (rocket boat). *Skorpik* and the badly-damaged *Acev* taken over by Croatia. *Bagat*, *Drapsin* and *Rozman-Stane* deleted 1992 and cannibalised. Remainder extant 1995.

Nos 201, 203 *experimental MTBs*

Two experimental boats had been built under the first ten-year programme but further construction was abandoned because of design shortcomings. About 20t displacement and about 30kt speed (1 TT) applied to one of these boats at least.

Rade Koncar as completed
NB 1/750 scale

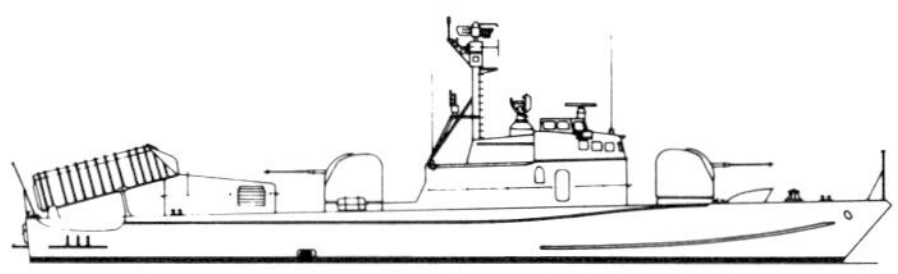

TYPE 240 *fast attack craft (missile)*

Displacement: 240t full load
Dimensions: 147ft 6in × 27ft 6in × 8ft 2in
45.0m × 8.4m × 2.5m
Machinery: 4 shaft CODAG (4 CP propellers): 2 shaft Rolls-Royce gas turbines, plus 2 MTU diesel engines, 7200hp + 8530hp = 39kts. Range 880nm at 23kts
Armament: 2 SS-N-2B Styx SSM (2×1), 2–57mm/70 Bofors (2×1)
Sensors: Radar Decca 1226, Philips TAB, Square Tie
Complement: 30

Class (no, launched):
Rade Koncar (RT 401, 15.10.76), *Vlado Cvetkovic* (RT 402, 28.8.77), *Ramiz Sadiku* (RT 403, 1978), *Hasan Zahirovic Laca* (RT 404, 1979), *Jordan Nikolov-Orce* (RT 405, 1979), *Ante Banina* (RT 406, 1979)

These combined missile-gun combatants were designed by the Naval Shipbuilding Institute of Zagreb and built by the Tito Yd, Kraljevica. Truly international boats close in design to the Swedish *Spica* class and equipped with Bofors guns, Soviet missile system (chosen in preference to Exocet), British turbines, German diesels and Dutch radar systems. RT = *Raketna Topovnjaca*. In service dates RT 401 Apr 1977, RT 402 Mar 1978, RT 403 10.9.78, RT 404 Nov 1979, RT 405 Aug 1979 and RT 406 Nov 1980. *Cvetkovic* taken over by the Croatian Navy in 1991, possibly followed by a second. Of the remainder, one was heavily damaged and only three were operational 1995.

Four of this class were to be ordered by Libya during the late 1980s, but were cancelled.

TYPE 400 'KOBRA' *fast attack craft (missile)*

Displacement: 385t standard; 525t full load
Dimensions: 179ft 10in × 29ft 6in × 8ft 3in
54.8m × 9.0m × 2.5m
Machinery: 4 shafts, 2 Rolls-Royce Proteus gas turbines + 2 MTU diesels, 18,800shp = 40kts. Range 1000/500nm at 20/35kts
Armament: 8 RBS-25 SAM (8×1), 1–3in OTO-Melara, 4–40mm Breda-Dardo (2×2), 8–20mm (2×4)
Sensors: ?
Complement: 30

Designed by the Naval Shipping Institute in Zagreb as a variation of the export FAC(M) ordered by Libya in 1985 at the Tito Yard, Kraljevica. Order for the Yugoslav Navy suspended in 1989 because of economic problems. In 1991 two boats under construction were taken over by the Croatians.

Ex-French FOUGUEUX class *submarine-chaser*

One unit (ex-US *PC 1615*) designated *P 6* (launched 21 December 1954) was delivered by F C Méditerranée, Le Havre, for the Yugoslav Government thanks to US 'offshore' procurement funds. She was completed and commissioned in January 1956 as *Udarnik* (PBR 51). Stricken 1988.

Borac in 1970s
NB 1/750 scale

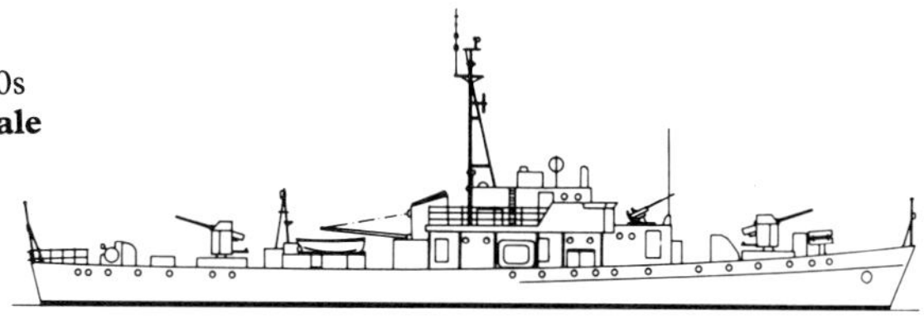

MORNAR class *large patrol craft*

Displacement: 330t standard; 430t full load
Dimensions: 170ft pp, 174ft 8in oa × 23ft × 6ft 6in
51.8m, 53.3m × 7.0m × 2.0m
Machinery: 2 shafts, 4 SEMT-Pielstick diesels, 3240bhp = 24kts. Oil 55t. Range 3000/660nm at 24/12kts
Armament: 2–76mm (2×1), 2–40mm (2×1), 2–20mm (2×1), 4 RBU-1200 ASW RL (4×5), 2 Mk 6 DCT, 2 Mk 9 DC racks
Sensors: Radar Decca 45; sonar Tamir-II
Complement: 60

Class (no, launched):
Mornar (PBR 551, 1958), *Borac* (PBR 552, 1965)

Built by the Tito Yd, Kraljevica, to an improved design of the *Fougueux* class with increased armament. Both modernised 1970–73 by the Naval Repair Yard in Trivat; with Soviet sonar and ASW equipment, the 76mm guns being removed. Deleted 1992.

TYPE 501 'KRALJEVICA' class *large patrol craft*

Displacement: 195t standard; 220t full load
Dimensions: 134ft 5in × 20ft 7in × 5ft 7in
41.0m × 6.3m × 1.7m
Machinery: 2 shafts, 2 diesels, 3000bhp = 17kts. Range 1000nm at 12kts
Armament: 1–3in Mk 22, 1–40mm, 4–20mm (4×1), 2 Mousetrap or 1 Hedgehog, 2 Mk 6 DCT, 2 DC racks
Sensors: Radar Decca 45
Complement: 45

Class:
PBR 501 to *PBR 508*

Built by the Tito Yard, Kraljevica (1953–56), this design was abandoned in favour of the improved Type 509 design. *PBR 502* and *PBR 505* were sold to Bangladesh in 1975. During the later 1970s the US 3in gun and the Hedgehog were removed in favour of 2 RBU-1200 ASW RL (2×5) with Soviet Tamir-11 sonar. *PBR 504* and *PBR 506* reportedly carried names *Napred* and *Junak* respectively. Stricken 1983–86.

Type 509 as completed
NB 1/750 scale

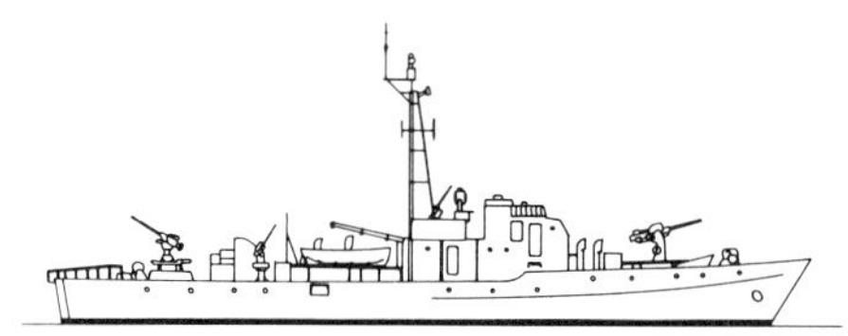

TYPE 509 'KRALJEVICA' class *large patrol craft*

Displacement: 210t standard; 235t full load
Dimensions: 141ft 5in × 20ft 7in × 6ft
43.1m × 6.3m × 1.8m
Machinery: 2 shafts, 2 MAN W8V 30/38 diesels, 3300bhp = 19kts. Range 1000nm at 12kts
Armament: 1–76mm, 1–40mm, 4–20mm (4×1), 2 Mousetrap or 1 Hedgehog, 2 DCT, 2 DC racks
Sensors: Radar Decca 45
Complement: 49

Class:
PBR 509 to *PBR 524*

Built by the Tito Yard, Kraljevica, (1957–59) to the improved and longer design of the Type 501. *PBR 513* to *PBR 518* were sold to Indonesia in 1958–59, *PBR 522* to *PBR 523* to Sudan in 1969 and *PBR 509* to Ethiopia in 1975. The remaining eight including *Streljko* (PBR 520), *Marijan* (PBR 521) and *Proleter* (PBR 524), were stricken 1983–90.

TYPE 131 *coastal patrol craft*

Displacement: 85t standard; 120t full load
Dimensions: 91ft 9in × 14ft 8in × 8ft 3in
28.0m × 4.5m × 2.5m
Machinery: 2 shafts, 2 MTU MB820 diesels, 2200hp = 17kts
Armament: 6–20mm Hispano Suiza HS831 (2×3)
Sensors: Radar Kelvin-Hughes 14/9
Complement: ?

YUGOSLAVIA

Class (no, fate):
Bresice (PC 152, stricken 1983), *Cer* (sold to Malta, Mar 1982), *Durmitor* (PC 139, sold to Malta, Mar 1982), *Granicar* (PC 134, stricken 1992–93), *Kalnik* (PC 132, stricken 1992–93), *Kamenar* (PC 137, stricken 1992–93), *Kozuf* (PC 140, stricken 1992–93), *Lovcen* (stricken 1983), *Romanija* (PC 136, stricken 1992–93), *Rudnik* (PC 135, stricken 1992–93), *Triglav* (PC 135, stricken 1983), *Velebit* (PC 133, stricken 1992–93)

Maritime Border Brigade craft of 85t standard, built 1964–68 by the Trogir Yard.

TYPE 140 'MIRNA' *patrol craft*

Displacement:	120t full load
Dimensions:	105ft × 22ft × 7ft 6in *32.0m × 6.7m × 2.3m*
Machinery:	2 shafts, 2 SEMT-Pielstick 12PA 4200 GDS diesels, 6000bhp = 32kts. Radius 400nm at 20kts
Armament:	1 SA-N-5 SAM (1×4), 1–40mm Bofors, 4–20mm Oerlikon (1×4), 8 DCs, mines, 4 chaff RLs
Sensors:	Radar navigation; sonar hull mounted
Complement:	19

Class (no):
Biokovo (PC 171), *Cer* (PC 180), *Fruska Gora* (PC 177), *Grmec* (PC 175), *Koprivnik* (PC 173), *Kosmaj* (PC 178), *Kozolo* (PC 181), *Muskos* (PC 176), *Pohorje* (PC 172), *Ucka* (PC 174), *Zelenogora* (PC 179)

Built during 1980–85 by Kraljevica Yard as replacement of the 'Type 131' boats in commission with naval border detachments. Fitted with additional outboard electric motor for quiet operation, capable of propelling the vessel at up to 6kts. *Kozolo*, *Biokovac* and *Cer* taken over by Croatia 1991. *Koprivnik* and *Muskos* deleted 1992. Remainder extant 1995. Export version of this class is building under designation 'Type 80'. PC = *Patrolni camac* (patrol craft).

MINE WARFARE VESSELS

GALEB *minelayer/training ship*

Displacement:	5182t standard; 5700t full load (3667 grt)
Dimensions:	384ft 8in × 51ft 2in × 18ft 4in *117.3m × 15.6m × 5.6m*
Machinery:	2 shafts, 2 Burmeister & Wain diesels, 7200bhp = 17kts
Armament:	4–3.5in (4×1), 4–40mm (4×1), 24–20mm (6×4), mines
Complement:	?

Former Italian banana carrier *Ramb III* completed by Ansaldo at Genoa in 1939, sunk as an auxiliary cruiser in 1944. Refloated and reconstructed after the war and commissioned in 1952. The guns were replaced by 6–40mm (6×1) in the late 1960s. Also used presidential yacht and fleet flagship. Extant 1995 as a training ship and minelayer, armed with 4–40mm.

Ex-French SIRIUS class *coastal minesweepers*

Class (no, former name):
Vukov Klanac (M 151, ex-*Hrabri*), *Podgora* (M 152, ex-*Smeli*), *Blitvenica 153, ex-Slobodni*), *Gradac* (M 11, ex-*Snazhi*)

First three delivered under designation *D 25* to *D 27* to Yugoslavia by the A Normand Yard, Le Havre, under the US Offshore Procurement Scheme. Commissioned in September 1957 as *Hrabri*, *Smeli* and *Slobodni* respectively.

The fourth craft was built with French assistance by Mali Losinj Yard and commissioned as *Snazni*. She was experimentally fitted with a gas turbine propulsion. All four were renamed *c*1976. *M 151* and *M 153* were converted to minehunters and fitted with PAP 104 MCMV and hull mounted Plessey Type 193M sonar. *Klanac* heavily damaged November 1991 and taken over by the Croatian Navy. *Gradac* broken up 1993. Others extant 1995.

TYPE 101 *inshore minesweepers*

Displacement:	90t standard; 95t full load
Dimensions:	82ft × 19ft 5in × 6ft 2in *25.0m × 5.9m × 1.9m*
Machinery:	1 shaft, 1 diesel, 135–175bhp = 12kts
Armament:	1–40mm, 1–20mm
Complement:	?

Class:
M 101 to *M 116*, *M 120*, *M 140*

Built 1950–56. All stricken 1966–76.

British 'HAM' class *inshore minesweepers*

Four were built in Yugoslavia (1964–66) under the MS MDAP. Commissioned as *Brsec* (M 142), *Iz* (M 143). *Milset* (M 141), and *Olib* (M 144). *Iz* to Croatia 1991. *Olib* deleted 1992. Remainder extant 1995.

TYPE 117 *inshore minesweepers*

Displacement:	120t standard; 131t full load
Dimensions:	98ft 4in × 18ft × 4ft 9in *30.0m × 5.5m × 1.5m*
Machinery:	2 shafts, 2 GM diesels, 1000bhp = 12kts
Armament:	1–40mm, 3–12.7mm MG (1×2)
Sensors:	Radar Decca 45
Complement:	25

Class:
Grajac, *Ibet*, *Maun* (M 118), *M 120* to *M 123*

Yugoslav built 1964–68. One stricken 1980. *M 117* used for survey, others do coastal patrol as well. Stricken 1987–93.

MISCELLANEOUS

SAVA *river monitor*

Former Yugoslav monitor scuttled in 1944 by the Croatians and raised after the war. Recommissioned in 1952 and rearmed with 2–105mm (2×1), 3–40mm (3×1), 6–20mm. Stricken 1962.

RPC 200 type *river gunboats*

Two 30t craft (15kts and 2–3in guns) launched for the *Donauflotille* in 1945 were completed for the Yugoslav Navy after the war. Stricken late 1970s.

TYPE 16 'BOTICA' *river patrol boats*

Displacement:	23t full load
Dimensions:	55ft 10in × 11ft 8in × 2ft 8in *17.0m × 3.6m × 0.8m*
Machinery:	2 shafts, 2 diesels, 460bhp = 15kts
Armament:	1–20mm Oerlikon, 7–7.62mm MGs
Sensors:	Radar Decca 110
Complement:	7

Class (no):
PC 301–306

Locally built in 1980s, military load of three tonnes. One (reportedly *305*) was stricken by 1990, *303–306* extant 1995.

TYPE 20 *river patrol craft*

Displacement:	55t
Dimensions:	71ft 6in × 17ft 5in × 3ft 9in *21.8m × 5.3m × 1.2m*
Machinery:	2 shafts, 2 diesels, 1200bhp = 16kts. Range 200nm at 15kts
Armament:	2–20mm Oerlikon (2×1), mines
Sensors:	Radar Decca 110
Complement:	10

Class:
PC 211–216

Locally built since 1984, steel hull and GRP superstructure, extant in 1995.

TYPE 15 *river patrol craft*

Displacement:	19.5t full load
Dimensions:	55ft 5in × 12ft 10in × 2ft 4in *16.9m × 3.9m × 0.7m*
Machinery:	2 shafts, 2 diesels, 330bhp = 16kts. Range 160nm at 12kts
Armament:	1–20mm Oerlikon, 2–7.62mm MGs
Sensors:	Radar Decca 110
Complement:	6

Class:
PC 15–1–PC 15–12

Shallow water boats, built since late 1980s. Steel hull and GRP structure. Four delivered to Sudan in 1989. One deleted 1993, remainder extant 1995.

RLM 301 class *river minesweepers*

Class:
M 301–305 (stricken 1980), *M 306–310* (stricken 1986–89), *M 311–313* (stricken 1980), *314, 317* (stricken 1989), *318* (stricken 1990), *319–323* (extant 1995), *324* (stricken 1989)

Built by four Yugoslav yards 1951–53 for service on Danube, 38t, 12kts, 2–20mm.

NESTIN class *river minesweepers*

Displacement: 65t
Dimensions: 88ft 7in × 20ft 8in × 5ft 3in
27.0m × 6.3m × 1.6m
Machinery: 2 diesels, 520bhp = 18kts
Armament: 5–20mm Hispano Suiza (1×3, 2×1)
Sensors: Radar Decca 126
Complement: 17

Class (no, commissioned):
Belegis (M 333, 1976), *Bocut* (M 334, 1976), *Djerdar* (M 336, 1980), *Motajica* (M 332, 18.12.76), *Nestin* (M 331, 20.12.75), *Panonsko More* (M 337, 1980), *Vucedol* (M 335, 1979)

Built 1975–80 by the Brodotehnika Yard, Belgrad. Eight of these 65t standard boats were sold to Hungary, three to Iraq. Extant 1995.

Croatia

In 1990 the Croatians decided to establish their independent state, for the second time in the twentieth century. Despite Serbian opposition (both political and military) independence was declared in July 1991, but was acknowledged by most European states only six months later. The Croatian Navy was established in September 1991 and one of its main tasks was to secure the coastline, as almost all former Yugoslavian coast and naval bases (such as Split, Sibenik, Pola and Ploce) were on Croatian territory and were taken over.

The first ships of the Croatian Navy were two fast attack craft (missile) and a torpedo boat found in Kraljevica. Since then more ex-Yugoslavian ships have been captured and entered Croatian service, including two brand new missile corvettes of the *Kralj* class and a landing ship of the *Silba* class, all three begun for the former Yugoslavian Navy and completed by the Croatians. Also a number of river craft were armed and converted to form a makeshift river flotilla.

In 1995 the Croatian Navy had 1080 personnel and controlled less than 20% of the former Yugoslavian Navy. On the other hand the Croatians took over almost all what was left of Yugoslavia's pre-war merchant shipping. Out of 462 vessels of 3,293,447t gross reported in 1992, the Croatians in 1994 had 224 vessels of 192,682t, while the new Yugoslavia had to be satisfied with just nine ships of 2351t.

Ex-Yugoslavian 'UNA' class *midget submarine*

Soca was taken over by Croatia in 1991.

Modified 'UNA' class *midget submarines*

Displacement: 98t submerged
Dimensions: 73ft 8in × 8ft 9in × 11ft 2in
22.5m × 2.7m × 3.4m
Machinery: Electric motors = 6kts
Armament: 2 *Mala* (R2) class SDVs or mines
Complement: 7

Probably modified *Una* class boats, lengthened to fit auxilliary diesel engine for charging batteries. Two laid down in Brodosplit in 1993. Diving depth 656ft (*200m*). Extant 1995.

Ex-Yugoslavian 'MALA' class *swimmer delivery vehicles*

Two captured by Croatia in 1992.

Ex-Yugoslavian 'Type 201' *fast attack craft (torpedo)*

Class (no):
Vukovar (51)

Partizan III – the 'Shershen' class boat built locally under licence in late 1960s, was taken over by the Croatian Navy in 1991 and renamed. Another of the class – *Strelko* – was heavily damaged when taken over in 1991 and beyond repair.

Ex-Soviet 'OSA I' class *fast attack craft (missile)*

Class (no, former name):
Dubrovnik (41, ex-*Mitar Acev*), *Velimir Skorpik* (42)

Two of the class were taken over in 1991 by Croatians. The former was badly damaged, but repaired at Sibenik.

Ex-Yugoslavian 'KONCAR' (type 240) class *fast attack craft (missile)*

Vlado Cvetkovic was taken over by the Croatian Navy in 1991 and renamed *Sibenic* (no 21). She was followed possibly by another one.

KRALJ (Type 400) class *fast attack craft (missile)*

Displacement: 385t full load
Dimensions: 175ft 9in × 27ft 9in × 7ft 5in
53.6m × 8.5m × 2.3m
Machinery: 3 shafts, 3 M504B–2 diesels, 12500hp = 36kts. Range 1500nm at 20kts
Armament: 4 Saab RBS15 SSM (2×2), 1–57mm Bofors, 1–30mm AK630, 2–20 Oerlikon, 6 mines
Sensors: Radar Racal BT502, BEAB 91V249, Racal 1290A; sonar RIZ PP 10M
Complement: 33

Class (no, commissioned, former name):
Kralj Petar Kresimir IV (11, 1993, ex-*Sergej Masera*), *12* (1994)

Laid down for Yugoslavian Navy in Kraljevica Shipyard as *Kobra* (Type 400) class and taken over in 1991 while on stocks. *Kralj Kresimir* was launched 21 March 1992 and badly damaged in July 1993 when a missile accidentally exploded during loading. The mine rails can be substituted for further two missile launchers.

Ex-Yugoslavian MIRNA class *patrol craft*

Class (no, former name):
Biokovac (64), *Hrvatska Kostajnica* (63, ex-*Cer*), *Solta* (62, ex-*Kozolo*)

Taken over by Croatian Navy in 1991. When captured two of them were damaged (one by missile, another by torpedo), but they were eventually repaired.

Ex-French 'SIRIUS' class *coastal minehunter*

Vukov Klanac, with upper deck heavily damaged in an air raid in autumn 1991, was taken over by the Croatians. Repaired and commissioned in Croatian Navy.

British 'HAM' class *inshore minesweeper*

Iz taken over by Croatia in 1991.

Ex-Yugoslavian 'DTM 221' class *landing craft*

217 was taken over by Croatia in 1990 and renamed *Jastreb*. She was followed by *501* and two others in 1991.

Ex-Yugoslavian 'Type 21' *landing craft*

Four boats were captured in 1991 and commissioned with the Croatian Navy. Two more were damaged, but possibly also repaired and commissioned.

Ex-Yugoslavian 'Type 22' *landing craft*

Former Yugoslavian *DJC 622–624* were taken over by the Croatians in 1991.

Ex-Yugoslavian 'SILBA' class *landing ship/minelayer*

Class (no):
Cetina (81)

Former Yugoslavian *Ran* was launched by Croatian Navy on 10 July 1992 in the Brodosplit Yard and commissioned into Croatian Navy on 19 February 1993.

Slovenia

On 30 January 1993 the Republic of Slovenia announced the formation of her navy (officially entitled the 430th Coast Defence Unit). In 1995 this was reported to consist of the training vessel *Sinji Galeb*, plus a number of small patrol craft. The Customs Service also operates patrol craft. No further details are available.

Zaire

Zaire, formerly Congo-Kinshasa (independent in 1960) and black Africa's largest state, has a short coastline at the mouth of the River Zaire (Congo). Her 1500-volunteer navy (600 marines) operates three flotillas, one coastal (based at Boma and Matadi in the Congo estuary), one river and one lake (mainly Kalemi for Lake Tanganyika). There are seventy river ports on the country's 9320 miles of navigable waterways. Since 1978 Chinese instructors have helped with training (begun by Norway in 1964) which is also obtained from Egypt and France. However, efficiency and maintenance are very low.

Four ex-Chinese 'Shanghai II' class FAC (gun): numbered P 101 to P 104, transferred in pairs September 1976 and September 1978. All out of service 1990.

Three ex-N Korean 'P 4' FAC (torpedo) transferred 1974. All stricken 1986.

Four ex-Chinese 'Huchwan' class hydrofoil FAC (torpedo) transferred in pairs September 1976 and September 1978. All deleted *c*1984.

Two North Korean 'Sin Hung' class torpedo-boats transferred without TT 1987. Extant 1995.

The 70t *Zaire* (ex-*President Mobutu*, ex-*Congo*, ex-*General Olsen*) doubles as a yacht.

Six US Sewart type 65ft coastal patrol craft (33t, 25kts, 6–12.7mm MG, 12 men) bought 1972 and 1974. All deleted *c*1984.

Thirty Arcoa type patrol boats (1.7–7.5t craft, delivered from France since 1975 for river and lake operations).

Zambia

In September 1966 the Zambian Government purchased a tank landing craft of the LCT (8) type, the former HMS *Bastion* (L 4040). She was discarded in February 1969.

Zimbabwe

The Zimbabwean government acquired eight Belgian-built 11,5m patrol craft (unarmed) for patrolling on Lake Kariba. All extant 1995.

Index of Ship Names

Dates are of launching or of acquisition, unless otherwise specified. Anniversary dates used as ships' names are spelt out.